Table I-2

Moduli of Elasticity and Poisson's Ratios

Material	Modulus of Elasticity E GPa	Shear Modulus of Elasticity G GPa	Poisson's Ratio ν
Aluminum alloys	70–79	26–30	0.33
2014-T6	73	28	0.33
6061-T6	70	26	0.33
7075-T6	72	27	0.33
Brass	96–110	36–41	0.34
Bronze	96–120	36–44	0.34
Cast iron	83–170	32–69	0.2–0.3
Concrete (compression)	17–31		0.1–0.2
Copper and copper alloys	110–120	40–47	0.33–0.36
Glass	48–83	19–35	0.17–0.27
Magnesium alloys	41–45	15–17	0.35
Monel (67% Ni, 30% Cu)	170	66	0.32
Nickel	210	80	0.31
Plastics			
Nylon	2.1–3.4		0.4
Polyethylene	0.7–1.4		0.4
Rock (compression)			
Granite, marble, quartz	40–100		0.2–0.3
Limestone, sandstone	20–70		0.2–0.3
Rubber	0.0007–0.004	0.0002–0.001	0.45–0.50
Steel	190–210	75–80	0.27–0.30
Titanium alloys	100–120	39–44	0.33
Tungsten	340–380	140–160	0.2
Wood (bending)			
Douglas fir	11–13		
Oak	11–12		
Southern pine	11–14		

Mechanics of Materials

Ninth Edition, SI

Barry J. Goodno
Georgia Institute of Technology

James M. Gere
Professor Emeritus, Stanford University

Australia • Brazil • Japan • Korea • Mexico • Singapore • Spain • United Kingdom • United States

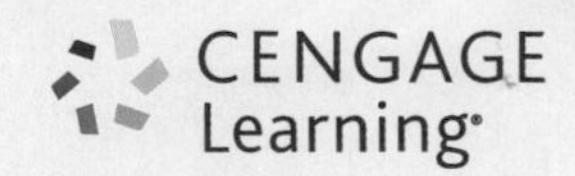

Mechanics of Materials, **Ninth Edition, SI**

Barry J. Goodno and James M. Gere

Product Director, Global Engineering: Timothy L. Anderson

Senior Content Developer: Mona Zeftel

Associate Media Content Developer: Ashley Kaupert

Product Assistant: Teresa Versaggi

Marketing Manager: Kristin Stine

Director, Higher Education Production: Sharon L. Smith

Content Project Manager: D. Jean Buttrom

Production Service: RPK Editorial Services, Inc.

Copyeditor: Patricia Daly

Proofreader: Shelly Gerger-Knechtl

Indexer: Shelly Gerger-Knechtl

Compositor: SPi Global

Senior Art Director: Michelle Kunkler

Cover and Internal Designer: Lou Ann Thesing

Cover and Internal Image: MACIEJ NOSKOWSKI/E+/Getty Images

Intellectual Property
Analyst: Christine Myaskovsky
Project Manager: Sarah Shainwald

Text and Image Permissions Researcher: Kristiina Paul

Manufacturing Planner: Doug Wilke

Library of Congress Control Number: 2016958773

ISBN: 978-1-337-09335-4

Cengage Learning
20 Channel Center Street
Boston, MA 02210
USA

Cengage Learning is a leading provider of customized learning solutions with employees residing in nearly 40 different countries and sales in more than 125 countries around the world. Find your local representative at **www.cengage.com**.

Cengage Learning products are represented in Canada by Nelson Education Ltd.

To learn more about Cengage Learning Solutions, visit **www.cengage.com/engineering**.

Purchase any of our products at your local college store or at our preferred online store **www.cengagebrain.com**.

Printed in the United States of America
Print Number: 02 Print Year: 2017

CONTENTS

*A star attached to a section number indicates a specialized and/or advanced topic.

ABOUT THE AUTHORS

Barry J. Goodno

© Barry Goodno

Barry John Goodno is Professor of Civil and Environmental Engineering at Georgia Institute of Technology. He joined the Georgia Tech faculty in 1974. He was an Evans Scholar and received a B.S. in Civil Engineering from the University of Wisconsin, Madison, Wisconsin, in 1970. He received M.S. and Ph.D. degrees in Structural Engineering from Stanford University, Stanford, California, in 1971 and 1975, respectively. He holds a professional engineering license (PE) in Georgia, is a Distinguished Member of ASCE and an Inaugural Fellow of SEI, and has held numerous leadership positions within ASCE. He is a past president of the ASCE Structural Engineering Institute (SEI) Board of Governors and is also a member of the Engineering Mechanics Institute (EMI) of ASCE. He is past-chair of the ASCE-SEI Technical Activities Division (TAD) Executive Committee, and past-chair of the ASCE-SEI Awards Committee. In 2002, Dr. Goodno received the SEI *Dennis L. Tewksbury Award* for outstanding service to ASCE-SEI. He received the departmental award for *Leadership in Use of Technology* in 2013 for his pioneering use of lecture capture technologies in undergraduate statics and mechanics of materials courses at Georgia Tech. He is a member of the Earthquake Engineering Research Institute (EERI) and has held several leadership positions within the NSF-funded Mid-America Earthquake Center (MAE), directing the MAE Memphis Test Bed Project. Dr. Goodno has carried out research, taught graduate courses and published extensively in the areas of earthquake engineering and structural dynamics during his tenure at Georgia Tech.

Dr. Goodno is an active cyclist, retired soccer coach and referee, and a retired marathon runner. Like co-author and mentor James Gere, he has completed numerous marathons including qualifying for and running the Boston Marathon in 1987.

James M. Gere

Courtesy of James and Janice Gere Family Trust

James M. Gere (1925-2008) earned his undergraduate and master's degree in Civil Engineering from the Rensselaer Polytechnic Institute in 1949 and 1951, respectively. He worked as an instructor and later as a Research Associate for Rensselaer. He was awarded one of the first NSF Fellowships, and chose to study at Stanford. He received his Ph.D. in 1954 and was offered a faculty position in Civil Engineering, beginning a 34-year career of engaging his students in challenging topics in mechanics, and structural and earthquake engineering. He served as Department Chair and Associate Dean of Engineering and in 1974 co-founded the John A. Blume Earthquake Engineering Center at Stanford. In 1980, Jim Gere also became the founding head of the Stanford Committee on Earthquake Preparedness. That same year, he was invited as one of the first foreigners to study the earthquake-devastated city of Tangshan, China. Jim retired from Stanford in 1988 but continued to be an active and most valuable member of the Stanford community.

Jim Gere was known for his outgoing manner, his cheerful personality and wonderful smile, his athleticism, and his skill as an educator in Civil Engineering. He authored nine textbooks on various engineering subjects starting in 1972 with *Mechanics of Materials*, a text that was inspired by his teacher and mentor Stephan P. Timoshenko. His other well-known textbooks, used in engineering courses around the world, include: *Theory of Elastic Stability*, co-authored with S. Timoshenko; *Matrix Analysis of Framed Structures* and *Matrix Algebra for Engineers*, both co-authored with W. Weaver; *Moment Distribution*; *Earthquake Tables: Structural and Construction Design Manual*, co-authored with H. Krawinkler; and *Terra Non Firma: Understanding and Preparing for Earthquakes*, co-authored with H. Shah.

In 1986 he hiked to the base camp of Mount Everest, saving the life of a companion on the trip. James was an active runner and completed the Boston Marathon at age 48, in a time of 3:13. James Gere will be long remembered by all who knew him as a considerate and loving man whose upbeat good humor made aspects of daily life or work easier to bear.

PREFACE

Mechanics of Materials is a basic engineering subject that, along with statics, must be understood by anyone concerned with the strength and physical performance of structures, whether those structures are man-made or natural. At the college level, statics is usually taught during the sophomore or junior year and is a prerequisite for the follow-on course in Mechanics of Materials. Both courses are required for most students majoring in mechanical, structural, civil, biomedical, petroleum, nuclear, aeronautical, and aerospace engineering. In addition, many students from such diverse fields as materials science, industrial engineering, architecture, and agricultural engineering also find it useful to study mechanics of materials.

Mechanics of Materials

In many university engineering programs today, both statics and mechanics of materials are taught in large sections of students from the many engineering disciplines. Instructors for the various parallel sections must cover the same material, and all of the major topics must be presented so that students are well prepared for the more advanced courses required by their specific degree programs. An essential prerequisite for success in a first course in mechanics of materials is a strong foundation in statics, which includes not only understanding fundamental concepts but also proficiency in applying the laws of static equilibrium to solutions of both two- and three-dimensional problems. This ninth edition begins with an updated section on statics in which the laws of equilibrium and an expanded list of boundary (or support) conditions are reviewed, as well as types of applied forces and internal stress resultants, all based upon and derived from a properly drawn free-body diagram. Numerous examples and end-of-chapter problems are included to help students review the analysis of plane and space trusses, shafts in torsion, beams and plane and space frames, and to reinforce basic concepts learned in the prerequisite course.

Many instructors like to present the basic theory of say, beam bending, and then use real world examples to motivate student interest in the subject of beam flexure, beam design, etc. In many cases, structures on campus offer easy access to beams, frames, and bolted connections that can be dissected in lecture or in homework problems, to find reactions at supports, forces and moments in members and stresses in connections. In addition, study of causes of failures in structures and components also offers the opportunity for students to begin the process of learning from actual designs and past engineering mistakes. A number of the new example problems and also the new and revised end-of-chapter problems in this ninth edition are based upon actual components or structures and are accompanied by photographs so that the student can see the real world problem alongside the simplified mechanics model and free-body diagrams used in its analysis.

An increasing number of universities are using rich media lecture (and/or classroom) capture software (such as Panopto and Tegrity) in their large undergraduate courses in mathematics, physics, and engineering. The *many new photos and enhanced graphics* in the ninth edition are designed to support this enhanced lecture mode.

Key Features

The main topics covered in this book are the analysis and design of structural members subjected to tension, compression, torsion, and bending, including the fundamental concepts mentioned above. Other important topics are the transformations of stress and strain, combined loadings and combined stress, deflections of beams, and stability of columns. Some additional specialized topics include the following: stress concentrations, dynamic and impact loadings, non-prismatic members, shear centers, bending of beams of two materials (or composite beams), bending of unsymmetric beams, maximum stresses in beams, energy based approaches for computing deflections of beams, and statically indeterminate beams.

Each chapter begins with a **Chapter Overview** highlighting the major topics covered in that chapter and closes with a **Chapter Summary and Review** in which the key points as well as major mathematical formulas in the chapter are listed for quick review. Each chapter also opens with a photograph of a component or structure that illustrates the key concepts discussed in the chapter.

New Features

Some of the notable features of this ninth edition, which have been added as new or updated material to meet the needs of a modern course in mechanics of materials, are:

- **Problem-Solving Approach**—All examples in the text are presented in a new *Four-Step Problem-Solving Approach* which is patterned after that presented by R. Serway and J. Jewett in *Principles of Physics*, 5e, Cengage Learning, 2013. This new structured format helps students refine their problem-solving skills and improve their understanding of the main concepts illustrated in the example.
- **Statics Review**—The *Statics Review* section has been enhanced in Chapter 1. Section 1.2 includes four new example problems which illustrate calculation of support reactions and internal stress resultants for truss, beam, circular shaft and plane frame structures. Thirty-four end-of-chapter problems on statics provide students with two- and three-dimensional structures to be used as practice, review, and homework assignment problems of varying difficulty.
- **Expanded Chapter Overview and Chapter Summary and Review sections**—The *Chapter Overview* and *Chapter Summary* sections have been expanded to include ***key equations*** and ***figures*** presented in each chapter. These summary sections serve as a convenient review for students of key topics and equations presented in each chapter.
- **Continued emphasis on underlying fundamental concepts** such as equilibrium, constitutive, and strain-displacement/ compatibility equations in problem solutions. Example problem and end-of-chapter problem solutions have been updated to emphasize an orderly process of explicitly writing out the equilibrium, constitutive and strain-displacement/ compatibility equations before attempting a solution.

- **Expanded topic coverage**—The following topics have been updated or have received expanded coverage: stress concentrations in axially loads bars (Sec. 2.10); torsion of noncircular shafts (Sec. 3.10); stress concentrations in bending (Sec. 5.13); transformed section analysis for composite beams (Sec. 6.3); and generalized flexure formula for unsymmetric beams (Sec. 6.5).
- **Many new example and end-of-chapter problems**—More than forty new example problems have been added to the ninth edition. In addition, there are more than 400 new and revised end-of-chapter problems out of the 1440 problems presented in the ninth edition text. The end-of-chapter problems are now grouped as **Introductory** or **Representative** and are arranged in order of increasing difficulty.
- **Centroids and Moments of Inertia review** has moved to Appendix D to free up space for more examples and problems in earlier chapters.

Importance of Example Problems

- Examples are presented throughout the book to illustrate the theoretical concepts and show how those concepts may be used in practical situations. All examples are presented in the **Four-Step Problem-Solving Approach** format so that the basic concepts as well as the key steps in setting up and solving each problem are clearly understood. New photographs have been added showing actual engineering structures or components to reinforce the tie between theory and application. Each example begins with a clear statement of the problem and then presents a simplified analytical model and the associated free-body diagrams to aid students in understanding and applying the relevant theory in engineering analysis of the system. In most cases, the examples are worked out in symbolic terms so as to better illustrate the ideas, and then numeric values of key parameters are substituted in the final part of the analysis step. In selected examples throughout the text, graphical display of results (e.g., stresses in beams) has been added to enhance the student's understanding of the problem results.

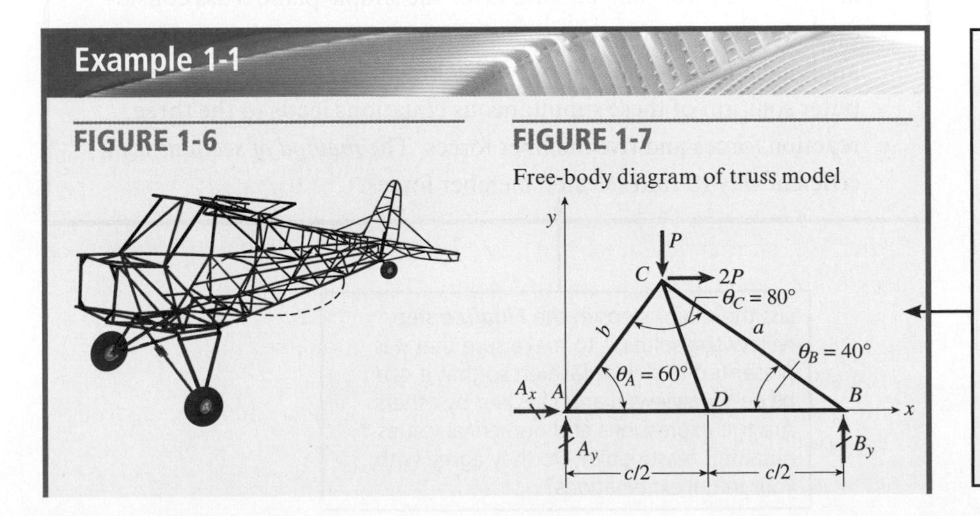

FIGURE 1-6

FIGURE 1-7
Free-body diagram of truss model

In many cases, the problem involves the analysis of a real physical structure, such as this truss structure (Fig. 1-6) representing part of the fuselage of a model air plane. Begin by sketching the portion of the structure of interest showing members, supports, dimensions and loadings. This ***Conceptualization*** step in the analysis often leads to a free-body diagram (Fig. 1-7).

The next step is to simplify the problem, list known data and identify all unknowns, and make necessary assumptions to create a suitable model for analysis. This is the ***Categorize*** step.

Write the governing equations, then use appropriate mathematical and computational techniques to solve the equations and obtain results, either in the form of mathematical formulas or numerical values. The ***Analysis*** step leads to support reaction and member forces in the truss.

List the major steps in your analysis procedure so that it is easy to review or check at a later time.

Solution:

The solution involves the following steps:

1. ***Conceptualize*** [*hypothesize, sketch*]: First sketch a free-body diagram of the entire truss model (Figure 1-7). Only known applied forces at C and unknown reaction forces at A and B are shown and then used in an equilibrium analysis to find the reactions.
2. ***Categorize*** [*simplify, classify*]: Overall equilibrium requires that the force components in x and y directions and the moment about the z axis must sum to zero; this leads to reaction force components A_x, A_y, and B_y. The truss is statically determinate (*unknowns*: $m + r = 5 + 3 = 8$, knowns: $2j = 8$) so all member forces can be obtained using the *method of joints*. . . .
3. ***Analyze*** [*evaluate; select relevant equations, carry out mathematical solution*]: First find the lengths of members AC and BC, which are needed to compute distances to lines of action of forces.

   Law of sines to find member lengths *a* and *b*: Use known angles θ_A, θ_B, and θ_C and $c = 3$ m to find lengths a and b:

$$b = c\frac{\sin(\theta_B)}{\sin(\theta_C)} = (3\text{ m})\frac{\sin(40°)}{\sin(80°)} = 1.958\text{ m},$$

$$a = c\frac{\sin(\theta_A)}{\sin(\theta_C)} = (3\text{ m})\frac{\sin(60°)}{\sin(80°)} = 2.638\text{ m}$$

   Check that computed lengths a and b give length c by using the law of cosines:

$$c = \sqrt{(1.958\text{ m})^2 + (2.638\text{ m})^2 - 2(1.958\text{ m})(2.638\text{ m})\cos(80°)} = 3\text{ m}$$

4. ***Finalize*** [*conclude; examine answer—does it make sense? Are units correct? How does it compare to similar problem solutions?*]: There are $2j = 8$ equilibrium equations for the simple plane truss considered above and, using the *method of joints*, these are obtained by applying $\Sigma F_x = 0$ and $\Sigma F_y = 0$ at each joint in succession. A computer solution of these simultaneous equations leads to the three reaction forces and five member forces. The *method of sections* is an efficient way to find selected member forces.

List the major steps in the ***Finalize*** step, review the solution to make sure that it is presented in a clear fashion so that it can be easily reviewed and checked by others. Are the expressions and numerical values obtained reasonable? Do they agree with your initial expectations?

Problems

In all mechanics courses, solving problems is an important part of the learning process. This textbook offers more than 1440 problems, many with multiple parts, for homework assignments and classroom discussions. The problems are placed at the end of each chapter so that they are easy to find and don't break up the presentation of the main subject matter. Also, problems are generally arranged in order of increasing difficulty, thus alerting students to the time necessary for solution. Answers to all problems are listed near the back of the book.

Considerable effort has been spent in checking and proofreading the text so as to eliminate errors. If you happen to find one, no matter how trivial, please notify me by e-mail (*bgoodno@ce.gatech.edu*). We will correct any errors in the next printing of the book.

Units

The International System of Units (SI) is used in all examples and problems. Tables containing properties of structural-steel shapes in SI units may be found in Appendix F so that solution of beam analysis and design examples and end-of-chapter problems can be carried out in SI units.

Supplements

Instructor Resources

An **Instructor's Solutions Manual** is available online on the Instructor's Resource Center for the book and includes solutions to all problems from this edition with Mathcad solutions available for some problems. The Manual includes rotated stress elements for problems as well as an increased number of free body diagrams. The Solutions Manual is accessible to instructors on http://login.cengage.com. The Instructor Resource Center also contains a full set of **Lecture Note PowerPoints.**

Student Resources

FE Exam Review Problems has been updated and now appears online. This supplement contains 106 FE-type review problems and solutions, which cover all of the major topics presented in the text and are representative of those likely to appear on an FE exam. Each of the problems is presented in the FE Exam format and is intended to serve as a useful guide to the student in preparing for this important examination.

Many students take the *Fundamentals of Engineering Examination* upon graduation, the first step on their path to registration as a Professional Engineer. Most of these problems are in SI units which is the system of units used on the FE Exam itself, and require use of an engineering calculator to carry out the solution. The student must select from four available answers, only one of which is the correct answer. Go to http://www.cengagebrain.com to find the FE Exam Review Problems and the resources below, which are available on the student website for this book:

- *Answers to the FE Exam Review Problems*
- *Detailed Solutions for Each Problem*

S.P. Timoshenko (1878–1972) and J.M. Gere (1925–2008)

Many readers of this book will recognize the name of Stephen P. Timoshenko—probably the most famous name in the field of applied mechanics. A brief biography of Timoshenko appears in the first reference in the *References and Historical Notes* section. Timoshenko is generally recognized as the world's most outstanding pioneer in applied mechanics. He contributed many new ideas and concepts and became famous for both his scholarship and his teaching. Through his numerous textbooks he made a profound change in the teaching of mechanics not only in this country but wherever mechanics is taught. Timoshenko was both teacher and mentor to James Gere and provided the motivation for the first edition of this text, authored by James M. Gere and published in 1972. The second and each subsequent edition of this book were written by James Gere over the course of his long and distinguished tenure as author, educator, and researcher at Stanford University. James Gere started as a doctoral student at Stanford in 1952 and retired from Stanford as a professor in 1988 having authored this and eight other well-known and respected text books on mechanics, and structural and earthquake engineering. He remained active at Stanford as Professor Emeritus until his death in January of 2008.

Acknowledgments

To acknowledge everyone who contributed to this book in some manner is clearly impossible, but I owe a major debt to my former Stanford teachers, especially my mentor and friend, and co-author James M. Gere.

I am grateful to my many colleagues teaching Mechanics of Materials at various institutions throughout the world who have provided feedback and constructive criticism about the text; for all those anonymous reviews, my thanks. With each new edition, their advice has resulted in significant improvements in both content and pedagogy.

My appreciation and thanks also go to the reviewers who provided specific comments for this ninth edition:

Erian Armanios, *University of Texas at Arlington*

Aaron S. Budge, *Minnesota State University, Mankato*

Virginia Ferguson, *University of Colorado, Boulder*

James Giancaspro, *University of Miami*

Paul Heyliger, *Colorado State University*

Eric Kasper, *California Polytechnic State University, San Luis Obispo*

Richard Kunz, *Mercer University*

David Lattanzi, *George Mason University*

Gustavo Molina, *Georgia Southern University*

Suzannah Sandrik, *University of Wisconsin—Madison*

Morteza A.M. Torkamani, *University of Pittsburgh*

I wish to also acknowledge my Structural Engineering and Mechanics colleagues at the Georgia Institute of Technology, many of whom provided valuable advice on various aspects of the revisions and additions leading to the current edition. It is a privilege to work with all of these educators and to learn from them in almost daily interactions and discussions about structural engineering and mechanics in the context of research and higher education. I wish to extend my thanks to my many current and former students who have helped to shape this text in its various editions. Finally, I would like to acknowledge the excellent work of Edwin Lim who suggested new problems and also carefully checked the solutions of many of the new examples and end of chapter problems.

I wish to acknowledge and thank the Global Engineering team at Cengage Learning for their dedication to this new book:

Timothy Anderson, Product Director; Mona Zeftel, Senior Content Developer; D. Jean Buttrom, Content Project Manager; Kristin Stine, Marketing Manager; Elizabeth Brown and Brittany Burden, Learning Solutions Specialists; Ashley Kaupert, Associate Media Content Developer; Teresa Versaggi and Alexander Sham, Product Assistants; and Rose Kernan of RPK Editorial Services, Inc.

They have skillfully guided every aspect of this text's development and production to successful completion.

I am deeply appreciative of the patience and encouragement provided by my family, especially my wife, Lana, throughout this project.

Finally, I am very pleased to continue this endeavor begun so many years ago by my mentor and friend, Jim Gere. This ninth edition text has now reached its 45th year of publication. I am committed to its continued excellence and welcome all comments and suggestions. Please feel free to provide me with your critical input at *bgoodno@ce.gatech.edu*.

Barry J. Goodno
Atlanta, Georgia

PREFACE TO THE SI EDITION

This edition of *Mechanics of Materials* has been adapted to incorporate the International System of Units (*Le Système International d'Unités* or SI) throughout the book.

Le Système International d' Unités

SI units are primarily the units of the MKS (meter–kilogram–second) system. However, CGS (centimeter–gram–second) units are often accepted as SI units, especially in textbooks.

Using SI Units in this Book

In this book, we have used both MKS and CGS units. USCS (U.S. Customary Units) or FPS (foot-pound-second) units used in the US Edition of the book have been converted to SI units throughout the text and problems. However, in case of data sourced from handbooks, government standards, and product manuals, it is not only extremely difficult to convert all values to SI, it also encroaches upon the intellectual property of the source. Some data in figures, tables, and references, therefore, may remain in FPS units.

To solve problems that require the use of sourced data, the sourced values can be converted from FPS units to SI units just before they are to be used in a calculation. To obtain standardized quantities and manufacturers' data in SI units, readers may contact the appropriate government agencies or authorities in their regions.

Instructor Resources

The Instructors' Solution Manual in SI units is available online through the book website at http://login.cengage.com. A digital version of the ISM, Lecture Note PowerPoint slides for the SI text, as well as other resources are available for instructors registering on the book website.

Feedback from users of this SI Edition will be greatly appreciated and will help us improve subsequent editions.

Cengage Learning

MINDTAP ONLINE COURSE

Mechanics of Materials is also available through MindTap, Cengage Learning's digital course platform. The carefully-crafted pedagogy and exercises in this trusted textbook are made even more effective by an interactive, customizable eBook, automatically graded assessments, and a full suite of learning tools.

As an instructor using MindTap, you have at your fingertips the full text and a unique set of tools, all in an interface designed to save you time. MindTap makes it easy for instructors to build and customize their course, so you can focus on the most relevant material while also lowering costs for your students. Stay connected and informed through real-time student tracking that provides the opportunity to adjust your course as needed based on analytics of interactivity and performance. **End-of-chapter quizzes** and **problem sets** test students' knowledge of concepts and numerics. Wrong answers in the algorithmically generated problem sets pop up custom step-by-step solutions to guide students how to solve the problems.

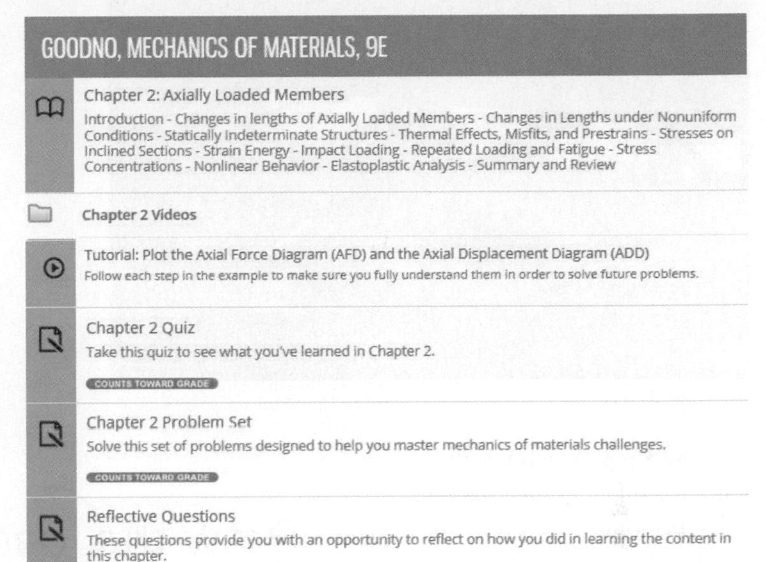

GOODNO, MECHANICS OF MATERIALS, 9E
Chapter 2: Axially Loaded Members
Introduction - Changes in lengths of Axially Loaded Members - Changes in Lengths under Nonuniform Conditions - Statically Indeterminate Structures - Thermal Effects, Misfits, and Prestrains - Stresses on Inclined Sections - Strain Energy - Impact Loading - Repeated Loading and Fatigue - Stress Concentrations - Nonlinear Behavior - Elastoplastic Analysis - Summary and Review
Chapter 2 Videos
Tutorial: Plot the Axial Force Diagram (AFD) and the Axial Displacement Diagram (ADD)
Follow each step in the example to make sure you fully understand them in order to solve future problems.
Chapter 2 Quiz
Take this quiz to see what you've learned in Chapter 2.
COUNTS TOWARD GRADE
Chapter 2 Problem Set
Solve this set of problems designed to help you master mechanics of materials challenges.
COUNTS TOWARD GRADE
Reflective Questions
These questions provide you with an opportunity to reflect on how you did in learning the content in this chapter.
COUNTS TOWARD GRADE

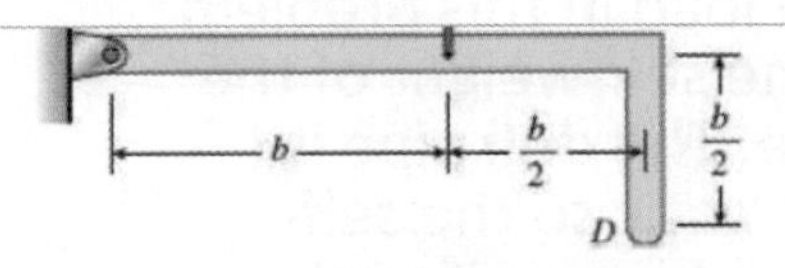

(a) Obtain a formula for the elongation of the spring due to the weight of the arm.

$\delta =$ 3

(b) Repeat pa

$\delta =$ 4

Partially

Try Anoth

Incorrect

(a) Wrong

Solution	
Take first moments about A to find c.g.	$x = \dfrac{\left(\frac{\frac{3}{2}b}{2b}\right)W\left(\frac{3}{4}b\right) + \left(\frac{\frac{b}{2}}{2b}\right)W\left(\frac{3}{2}b\right)}{W}$
	$x = \dfrac{15}{16}b$
Find force in spring due to weight of arm	$\sum M_A = 0$
	$F_k = \dfrac{W\left(\frac{15}{16}b\right)}{b} = \dfrac{15}{16}W$
Find elongation of spring due to weight of arm	$\delta = \dfrac{F_k}{k} = \dfrac{15W}{16k}$

Videos provide views of real world structures discussed in each chapter.

Step-by-Step Tutorials help students master concepts and solve problems explained in examples.

Draw the loads

A prismatic bar is a straight structural member having the same cross section throughout its length.

The load in this problem is the self-weight of the pipe. The drill pipe is prismatic so the self-weight is a uniformly distributed axial load of constant intensity *w* acting along the pipe

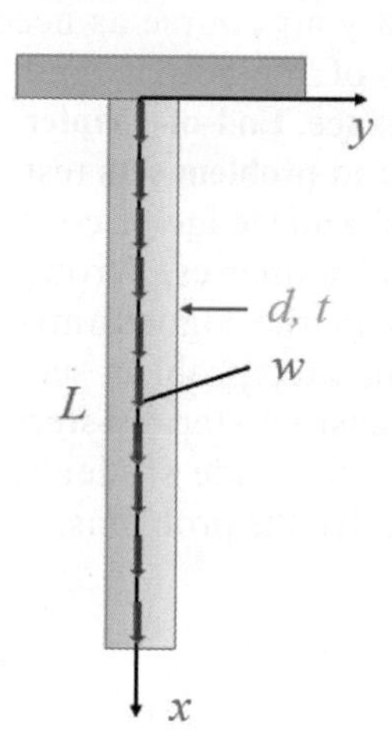

An axial load is a load directed along the axis of the member, resulting in either tension or compression in the bar.

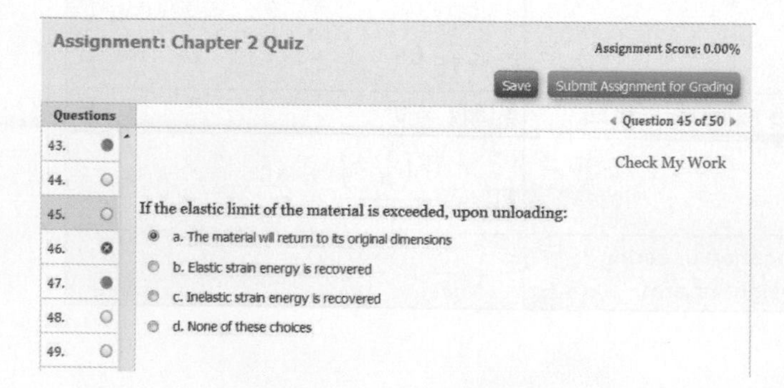

How does MindTap benefit instructors?

- You can build and personalize your course by integrating your own content into the **MindTap Reader** (like lecture notes or problem sets to download) or pull from sources such as RSS feeds, YouTube videos, websites, and more. Control what content students see with a built-in learning path that can be customized to your syllabus.
- MindTap saves you time by providing you and your students with **automatically graded assignments and quizzes**. These problems include immediate, specific feedback, so students know exactly where they need more practice.

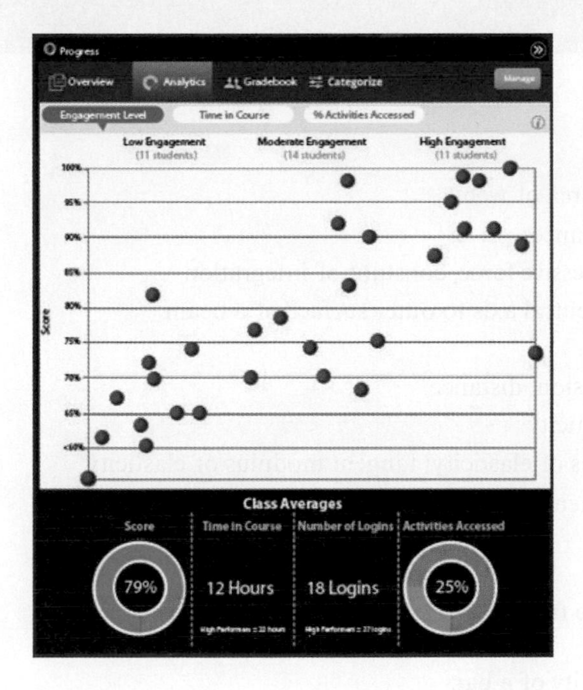

- The **Message Center** helps you to quickly and easily contact students directly from MindTap. Messages are communicated directly to each student via the communication medium (email, social media, or even text message) designated by the student.
- **StudyHub** is a valuable studying tool that allows you to deliver important information and empowers your students to personalize their experience. Instructors can choose to annotate the text with **notes** and **highlights**, share content from the MindTap Reader, and create **Flashcards** to help their students focus and succeed.
- The **Progress App** lets you know exactly how your students are doing (and where they might be struggling) with live analytics. You can see overall class engagement and drill down into individual student performance, enabling you to adjust your course to maximize student success.

How does MindTap benefit your students?

- The **MindTap Reader** adds the abilities to have the content read aloud, to print from the reader, and to take notes and highlights while also capturing them within the linked **StudyHub App**.
- The **MindTap Mobile App** keeps students connected with alerts and notifications while also providing them with on-the-go study tools like Flashcards and quizzing, helping them manage their time efficiently.
- **Flashcards** are pre-populated to provide a jump start on studying, and students and instructors can also create customized cards as they move through the course.
- The **Progress App** allows students to monitor their individual grades, as well as their level compared to the class average. Doing so not only helps them stay on track in the course but also motivates them to do more, and ultimately to do better.
- The unique **StudyHub** is a powerful single-destination studying tool that empowers students to personalize their experience. They can quickly and easily access all notes and highlights marked in the MindTap Reader, locate bookmarked pages, review notes and Flashcards shared by their instructor, and create custom study guides.

SYMBOLS

A	area
A_f, A_w	area of flange; area of web
a, b, c	dimensions, distances
C	centroid, compressive force, constant of integration
c	distance from neutral axis to outer surface of a beam
D	diameter
d	diameter, dimension, distance
E	modulus of elasticity
E_r, E_t	reduced modulus of elasticity; tangent modulus of elasticity
e	eccentricity, dimension, distance, unit volume change (dilatation)
F	force
f	shear flow, shape factor for plastic bending, flexibility, frequency (Hz)
f_T	torsional flexibility of a bar
G	modulus of elasticity in shear
g	acceleration of gravity
H	height, distance, horizontal force or reaction, horsepower
h	height, dimensions
I	moment of inertia (or second moment) of a plane area
I_x, I_y, I_z	moments of inertia with respect to x, y, and z axes
I_{x1}, I_{y1}	moments of inertia with respect to x_1 and y_1 axes (rotated axes)
I_{xy}	product of inertia with respect to xy axes
I_{x1y1}	product of inertia with respect to x_1y_1 axes (rotated axes)
I_P	polar moment of inertia
I_1, I_2	principal moments of inertia
J	torsion constant
K	stress-concentration factor, bulk modulus of elasticity, effective length factor for a column
k	spring constant, stiffness, symbol for $\sqrt{P/EI}$
k_T	torsional stiffness of a bar
L	length, distance
L_E	effective length of a column
ln, log	natural logarithm (base e); common logarithm (base 10)
M	bending moment, couple, mass
M_P, M_Y	plastic moment for a beam; yield moment for a beam
m	moment per unit length, mass per unit length
N	axial force

n	factor of safety, integer, revolutions per minute (rpm)
O	origin of coordinates
O'	center of curvature
P	force, concentrated load, power
P_{allow}	allowable load (or working load)
P_{cr}	critical load for a column
P_P	plastic load for a structure
P_r, P_t	reduced-modulus load for a column; tangent-modulus load for a column
P_Y	yield load for a structure
p	pressure (force per unit area)
Q	force, concentrated load, first moment of a plane area
q	intensity of distributed load (force per unit distance)
R	reaction, radius
r	radius, radius of gyration ($r = \sqrt{I/A}$)
S	section modulus of the cross section of a beam, shear center
s	distance, distance along a curve
T	tensile force, twisting couple or torque, temperature
T_P, T_Y	plastic torque; yield torque
t	thickness, time, intensity of torque (torque per unit distance)
t_{f}, t_{w}	thickness of flange; thickness of web
U	strain energy
u	strain-energy density (strain energy per unit volume)
u_r, u_t	modulus of resistance; modulus of toughness
V	shear force, volume, vertical force or reaction
v	deflection of a beam, velocity
v', v'', etc.	dv/dx, d^2v/dx^2, etc.
W	force, weight, work
w	load per unit of area (force per unit area)
x, y, z	rectangular axes (origin at point O)
x_c, y_c, z_c	rectangular axes (origin at centroid C)
$\overline{x}$, $\overline{y}$, $\overline{z}$	coordinates of centroid
Z	plastic modulus of the cross section of a beam
α	angle, coefficient of thermal expansion, nondimensional ratio
β	angle, nondimensional ratio, spring constant, stiffness
β_R	rotational stiffness of a spring
γ	shear strain, weight density (weight per unit volume)
γ_{xy}, γ_{yz}, γ_{zx}	shear strains in xy, yz, and zx planes
γ_{x1y1}	shear strain with respect to x_1y_1 axes (rotated axes)
γ_θ	shear strain for inclined axes
δ	deflection of a beam, displacement, elongation of a bar or spring

ΔT temperature differential
δ_P, δ_Y plastic displacement; yield displacement
ε normal strain
$\varepsilon_x, \varepsilon_y, \varepsilon_z$ normal strains in x, y, and z directions
$\varepsilon_{x1}, \varepsilon_{y1}$ normal strains in x_1 and y_1 directions (rotated axes)
ε_θ normal strain for inclined axes
$\varepsilon_1, \varepsilon_2, \varepsilon_3$ principal normal strains
ε' lateral strain in uniaxial stress
ε_T thermal strain
ε_Y yield strain
θ angle, angle of rotation of beam axis, rate of twist of a bar in torsion (angle of twist per unit length)
θ_p angle to a principal plane or to a principal axis
θ_s angle to a plane of maximum shear stress
κ curvature ($\kappa = 1/\rho$)
λ distance, curvature shortening
ν Poisson's ratio
ρ radius, radius of curvature ($\rho = 1/\kappa$), radial distance in polar coordinates, mass density (mass per unit volume)
σ normal stress
$\sigma_x, \sigma_y, \sigma_z$ normal stresses on planes perpendicular to x, y, and z axes
σ_{x1}, σ_{y1} normal stresses on planes perpendicular to x_1y_1 axes (rotated axes)
σ_θ normal stress on an inclined plane
$\sigma_1, \sigma_2, \sigma_3$ principal normal stresses
σ_{allow} allowable stress (or working stress)
σ_{cr} critical stress for a column ($\sigma_{\text{cr}} = P_{\text{cr}}/A$)
σ_{pl} proportional-limit stress
σ_r residual stress
σ_T thermal stress
σ_U, σ_Y ultimate stress; yield stress
τ shear stress
$\tau_{xy}, \tau_{yz}, \tau_{zx}$ shear stresses on planes perpendicular to the x, y, and z axes and acting parallel to the y, z, and x axes
τ_{x1y1} shear stress on a plane perpendicular to the x_1 axis and acting parallel to the y_1 axis (rotated axes)
τ_θ shear stress on an inclined plane
τ_{allow} allowable stress (or working stress) in shear
τ_U, τ_Y ultimate stress in shear; yield stress in shear
ϕ angle, angle of twist of a bar in torsion
ψ angle, angle of rotation
ω angular velocity, angular frequency ($\omega = 2\pi f$)

GREEK ALPHABET

Α	α	Alpha	Ν	ν	Nu
Β	β	Beta	Ξ	ξ	Xi
Γ	γ	Gamma	Ο	o	Omicron
Δ	δ	Delta	Π	π	Pi
Ε	ε	Epsilon	Ρ	ρ	Rho
Ζ	ζ	Zeta	Σ	σ	Sigma
Η	η	Eta	Τ	τ	Tau
Θ	θ	Theta	Υ	υ	Upsilon
Ι	ι	Iota	Φ	ϕ	Phi
Κ	κ	Kappa	Χ	χ	Chi
Λ	λ	Lambda	Ψ	ψ	Psi
Μ	μ	Mu	Ω	ω	Omega

CHAPTER 1

Tension, Compression, and Shear

Jan Jirous/Shutterstock.com

This telecommunications tower is an assemblage of many members that act primarily in tension or compression.

Chapter Objectives

- Define *mechanics of materials*, which examines the stresses, strains, and displacements in structures made of various materials acted on by a variety of different loads.
- Study normal stress (σ) and normal strain (ε) in materials used for structural applications.
- Identify key properties of various materials, such as the modulus of elasticity (E) and yield (σ_y) and ultimate (σ_u) stresses, from plots of stress (σ) versus strain (ε).
- Plot shear stress (τ) versus shear strain (γ) and identify the shearing modulus of elasticity (G).
- Study Hooke's Law for normal stress and strain ($\sigma = E\varepsilon$) and also for shear stress and strain ($\tau = G\gamma$).
- Investigate changes in lateral dimensions and volume of a bar, which depend upon Poisson's ratio (ν) for the material of the bar.
- Study normal, shear, and bearing stresses in simple bolted connections between members.
- Use factors of safety to establish allowable values of stresses.
- Introduce basic concepts of design: the iterative process by which the appropriate size of structural members is determined to meet a variety of both strength and stiffness requirements.

Chapter Outline

1.1 Introduction to Mechanics of Materials

Mechanics of materials is a branch of applied mechanics that deals with the behavior of solid bodies subjected to various types of loading. Other names for this field of study are *strength of materials* and *mechanics of deformable bodies*. The solid bodies considered in this book include bars with axial loads, shafts in torsion, beams in bending, and columns in compression.

The principal objective of mechanics of materials is to determine the stresses, strains, and displacements in structures and their components due to the loads acting on them. An understanding of mechanical behavior is essential for the safe design of all types of structures, whether airplanes and antennas, buildings and bridges, machines and motors, or ships and spacecraft. That is why mechanics of materials is a basic subject in so many engineering fields. Most problems in mechanics of materials begin with an examination of the external and internal forces acting on a stable deformable body. First the loads acting on the body are defined, along with its support conditions, then reaction forces at supports and internal forces in its members or elements are determined using the basic laws of static equilibrium (provided that the body is statically determinate).

In mechanics of materials you study the stresses and strains inside real bodies, that is, bodies of finite dimensions that deform under loads. To determine the stresses and strains, use the physical properties of the materials as well as numerous theoretical laws and concepts. Mechanics of materials provides additional essential information, based on the deformations of the body, to solve statically indeterminate problems (not possible using the laws of static equilibrium alone).

Theoretical analyses and experimental results have equally important roles in mechanics of materials. Theories are used to derive formulas and equations for predicting mechanical behavior but these expressions cannot be used in practical design unless the physical properties of the materials are known. Such properties are available only after careful experiments have been carried out in the laboratory. Furthermore, not all practical problems are amenable to theoretical analysis alone, and in such cases physical testing is a necessity.

The historical development of mechanics of materials is a fascinating blend of both theory and experiment—theory has pointed the way to useful results in some instances, and experiment has done so in others. Such famous persons as Leonardo da Vinci (1452–1519) and Galileo Galilei (1564–1642) performed experiments to determine the strength of wires, bars, and beams, although they did not develop adequate theories (by today's standards) to explain their test results. By contrast, the famous mathematician Leonhard Euler (1707–1783) developed the mathematical theory of columns and calculated the critical load of a column in 1744, long before any experimental evidence existed to show the significance of his results. Without appropriate tests to back up his theories, Euler's results remained unused for over a hundred years, although today they are the basis for the design and analysis of most columns (see Refs. 1-1, 1-2, and 1-3).

1.2 Problem-Solving Approach*

The study of mechanics divides naturally into two parts: first, *understanding* the general concepts and principles, and second, *applying* those concepts and principles to physical situations. You can gain an understanding of the general

*The four step problem-solving approach presented here is patterned after that presented by R. Serway and J. Jewett in *Principles of Physics*, *5e*, *Cengage Learning*, *2013.*

concepts by studying the discussions and derivations presented in this book. You can gain skill only by solving problems on your own. Of course, these two aspects of mechanics are closely related, and many experts in mechanics will argue that *you do not really understand the concepts if you cannot apply them*. It is easy to recite the principles, but applying them to real situations requires an in-depth understanding. Problem solving gives meaning to the concepts and also provides an opportunity to gain experience and develop judgment.

A major objective of this text is to assist you in developing a *structured solution process* for problems in statics and mechanics of materials. This process is referred to as a *problem-solving approach* (PSA) and is used in all example problems in the text. The PSA involves the following four steps:

1. ***Conceptualize*** [*hypothesize, sketch*]: List all relevant data and draw a sketch showing all applied forces, support/boundary conditions, and interactions between adjacent bodies. Development and refinement of the *free-body diagram* is an essential part of this step.
2. ***Categorize*** [*simplify, classify*]: Identify the unknowns in the problem and make any necessary assumptions to simplify the problem and streamline the solution process.
3. ***Analyze*** [*evaluate; select relevant equations, carry out mathematical solution*]: Apply appropriate theories, set up the necessary equations for the chosen mathematical model, and then solve for the unknowns.
4. ***Finalize*** [*conclude; examine answer—Does it make sense? Are units correct? How does it compare to similar problem solutions?*]: Study the answers, compare them to those for similar problems you have solved in the past, and test the robustness of the solution by varying key parameters to see how the results change (perhaps even plot the main result as a function of that parameter to investigate the sensitivity of the answer).

You are encouraged to study the *problem-solving approach* presented in the example problems and then apply it to homework and in-class laboratory problems. This structured systematic approach also will be useful during examinations. See Appendix B.2 for further discussion of the Problem Solving Approach summarized above.

All problems appear at the ends of the chapters, with the problem numbers and subheadings identifying the sections to which they belong.

In this book, final numerical results are usually presented with three significant digits when a number begins with the digits 2 through 9, and with four significant digits when a number begins with the digit 1. Intermediate values are often recorded with additional digits to avoid losing numerical accuracy due to rounding of numbers.

1.3 Statics Review

In your prerequisite course on statics, you studied the *equilibrium* of rigid bodies acted upon by a variety of different forces and supported or restrained in such a way that the body was stable and at rest. As a result, a properly restrained body could not undergo rigid-body motion due to the application of static forces. You drew *free-body diagrams* of the entire body, or of key parts of the body, and then applied the *equations of equilibrium* to find external reaction forces and moments or internal forces and moments at critical points. In this section, the basic static equilibrium equations are reviewed and then applied to the solution of example structures (both two and three-dimensional) using both scalar and vector operations (both acceleration and velocity of the body are assumed to be zero). Most problems in mechanics of materials require a static analysis as the first step, so all forces acting on the system and causing its deformation are known. Once all external and internal forces of interest have been found, you can proceed with the evaluation of stresses, strains, and deformations of bars, shafts, beams, and columns as described in subsequent chapters.

Equilibrium Equations

The resultant force R and resultant moment M of *all* forces and moments acting on either a rigid or deformable body in equilibrium are both zero. The sum of the moments may be taken about any arbitrary point. The resulting equilibrium equations can be expressed in *vector form* as:

$$R = \Sigma F = 0 \tag{1-1}$$

$$M = \Sigma M = \Sigma(r \times F) = 0 \tag{1-2}$$

where F is one of a number of vectors of forces acting on the body and r is a position vector from the point at which moments are taken to a point along the line of application of any force F. It is often convenient to write the equilibrium equations in *scalar form* using a rectangular Cartesian coordinate system, either in two dimensions (x, y) or three dimensions (x, y, z) as

$$\Sigma F_x = 0 \quad \Sigma F_y = 0 \quad \Sigma M_z = 0 \tag{1-3}$$

Equation (1-3) can be used for two-dimensional or planar problems, but in three dimensions, three force and three moment equations are required:

$$\Sigma F_x = 0 \quad \Sigma F_y = 0 \quad \Sigma F_z = 0 \tag{1-4}$$

$$\Sigma M_x = 0 \quad \Sigma M_y = 0 \quad \Sigma M_z = 0 \tag{1-5}$$

If the number of unknown forces is equal to the number of independent equilibrium equations, these equations are sufficient to solve for all unknown reaction or internal forces in the body, and the problem is referred to as *statically determinate* (provided that the body is stable). If the body or structure is constrained by additional (or redundant) supports, it is *statically indeterminate*, and a solution is not possible using the laws of static equilibrium alone.

Applied Forces

External loads applied to a body or structure may be either concentrated or distributed forces or moments. For example, force F_B (with units of newtons, N) in Fig. 1-1 is a point or concentrated load and is assumed to act at point B on the body, while moment M_A is a concentrated moment or couple (with units of $N \cdot m$) acting at point A. Distributed forces may act along or normal to a member and may have constant intensity, such as line load q_1 normal to member BC (Fig. 1-1) or line load q_2 acting in the $-y$ direction on inclined member DF; both q_1 and q_2 have units of force intensity (N/m). Distributed loads also may have a linear (or other) variation with some peak intensity q_0 (as on member ED in Fig. 1-1). Surface pressures p (with units of Pa), such as wind acting on a sign (Fig. 1-2), act over a designated

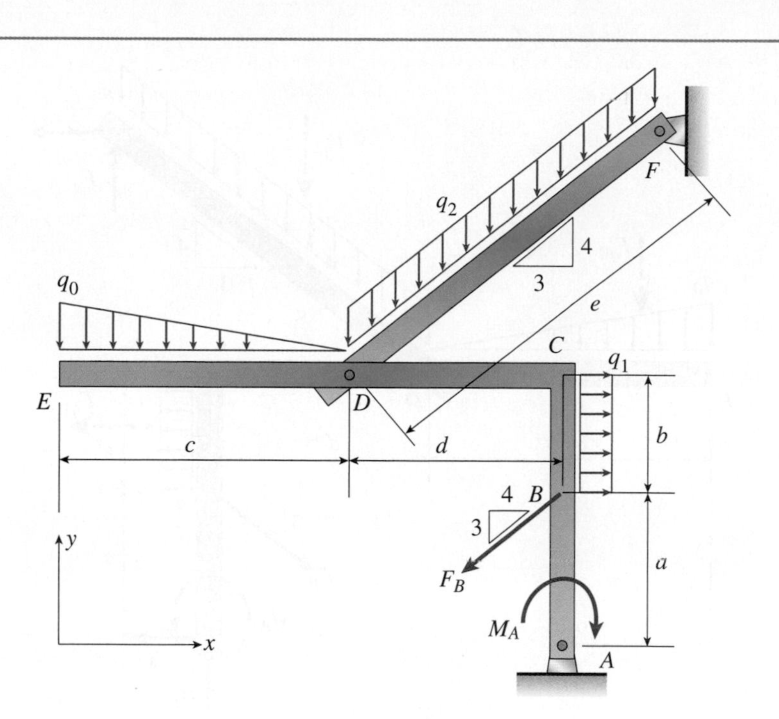

FIGURE 1-1

Plane frame structure

region of a body. Finally, a body force w (with units of force per unit volume, N/m^3), such as the distributed self-weight of the sign or post in Fig. 1-2, acts throughout the volume of the body and can be replaced by the component weight W acting at the center of gravity (c.g.) of the sign (W_s) or post (W_p). In fact, any distributed loading (line, surface, or body force) can be replaced by a statically equivalent force at the center of gravity (or center of pressure for wind) of the distributed loading when overall static equilibrium of the structure is evaluated using Eqs. (1-1) to (1-5).

Free-Body Diagrams

A free-body diagram (FBD) is an essential part of a static analysis of a rigid or deformable body. All forces acting on the body, or component part of the body, must be displayed on the FBD if a correct equilibrium solution is to be obtained. This includes applied forces and moments, reaction forces and moments, and any connection forces between individual components. For example, an *overall* FBD of the plane frame in Fig. 1-1 is shown in Fig. 1-3a; all applied and reaction forces are shown on this FBD and statically equivalent concentrated loads are displayed for all distributed loads. Statically equivalent forces F_{q0}, F_{q1}, and F_{q2}, each acting at the c.g. of the corresponding distributed loading, are used in the equilibrium equation solution to represent distributed loads q_0, q_1, and q_2, respectively.

Next, the plane frame has been disassembled in Fig. 1-3b, so that *separate* FBDs can be drawn for each part of the frame, thereby exposing pin-connection forces at D (D_x, D_y). Both FBDs must show all applied forces as well as reaction forces A_x and A_y at pin-support joint A and F_x and F_y at pin-support joint F. The forces transmitted between frame elements EDC and DF at pin connection D must be determined if the proper interaction of these two elements is to be accounted for in the static analysis.

FIGURE 1-2

Wind on sign

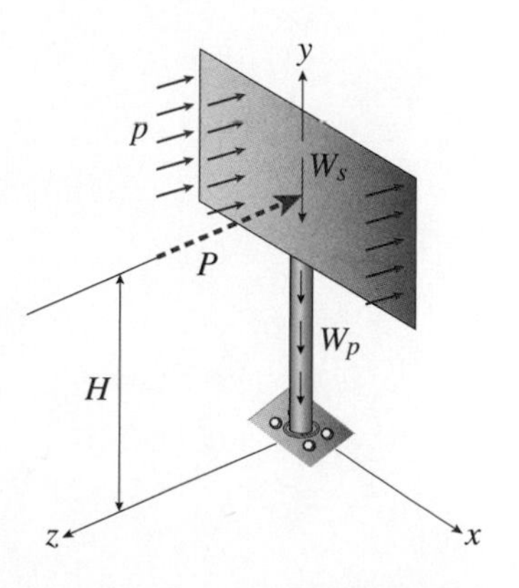

FIGURE 1-3

(a) Overall FBD of plane frame structure from Fig. 1-1, and (b) Separate free-body diagrams of part *ABCDE* and part *DF* of the plane frame structure in Fig. 1-1

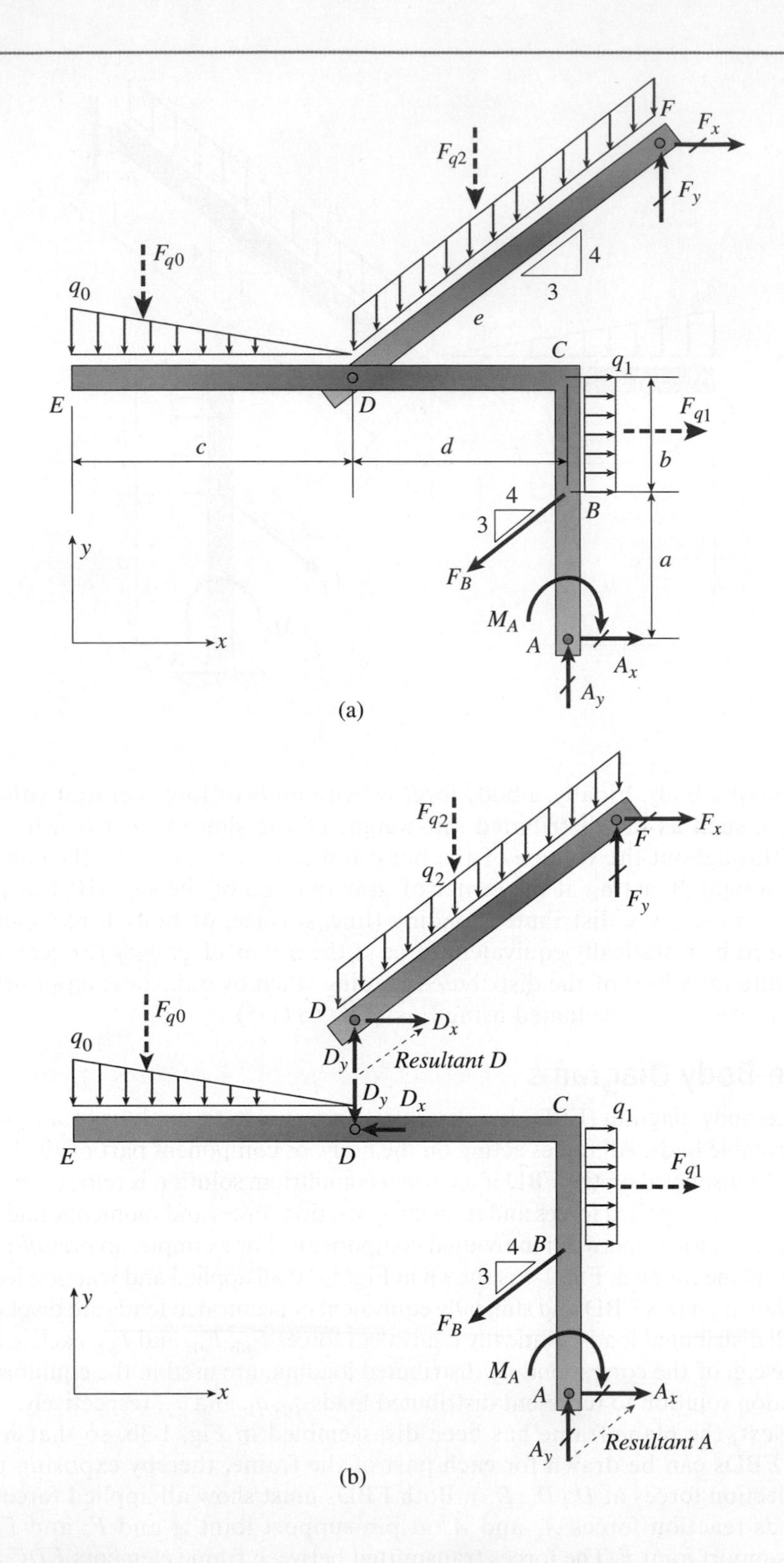

The FBDs presented in Figs. 1-3a and 1-3b are essential parts of this solution process. A *statics sign convention* is usually employed in the solution for support reactions; forces acting in the positive directions of the coordinate axes are assumed positive, and the right-hand rule is used for moment vectors.

Reactive Forces and Support Conditions

Proper restraint of the body or structure is essential if the equilibrium equations are to be satisfied. A sufficient number and arrangement of supports must be present to prevent rigid-body motion under the action of static forces. A reaction force at a support is represented by a single arrow with a slash drawn through it (see Fig. 1-3) while a moment restraint at a support is shown as a double-headed or curved arrow with a slash. Reaction forces and moments usually result from the action of applied forces of the types described above (i.e., concentrated, distributed, surface, and body forces).

A variety of different support conditions may be assumed depending on whether the problem is 2D or 3D. Supports A and F in the 2D plane frame structure shown in Fig. 1-1 and Fig. 1-3 are pin supports, while the base of the 3D sign structure in Fig. 1-2 may be considered to be a fixed or clamped support. Some of the most commonly used idealizations for 2D and 3D supports, as well as interconnections between members or elements of a structure, are illustrated in Table 1-1. The restraining or transmitted forces and moments associated with each type of support or connection are displayed in the third column of the table (these are not FBDs, however). The reactions forces and moments for the 3D sign structure in Fig. 1-2 are shown on the FBD in Fig. 1-4a; only reactions R_y, R_z, and M_x are nonzero because the sign structure and wind loading are symmetric with respect to the y-z plane. If the sign is eccentric to the post (Fig. 1-4b), only reaction R_x is zero for the case of wind loading in the $-z$ direction. (See Problems 1.8-19 and 1.9-17 at the end of Chapter 1 for a more detailed examination of the reaction forces due to wind pressure acting on several sign structures similar to that shown in Fig. 1-2; forces and stresses in the base plate bolts are also computed).

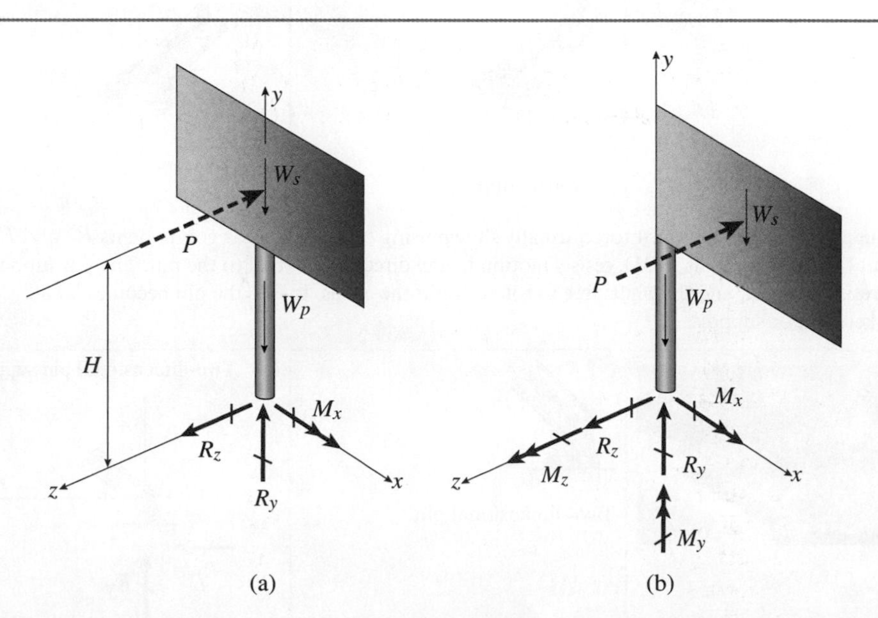

FIGURE 1-4

(a) FBD of symmetric sign structure, and (b) FBD of eccentric sign structure

Table 1-1

Modeling reaction forces and support conditions in 2D or 3D static analysis

Type of support or connection	Simplified sketch of support or connection	Display of restraint forces and moments, or connection forces
1. Roller Support: A single reaction force R is developed and is normal to the rolling surface; force R opposes motion into or away from the rolling surface. The rolling surface may be horizontal, vertical, or inclined at some angle θ. If friction is present, then include a force F opposing the movement of the support and tangential to the rolling surface. In 3D, the roller moves in the *x-z* plane and reaction R_y is normal to that plane.		
 Bridge with roller support (see 1.1, 1.2) 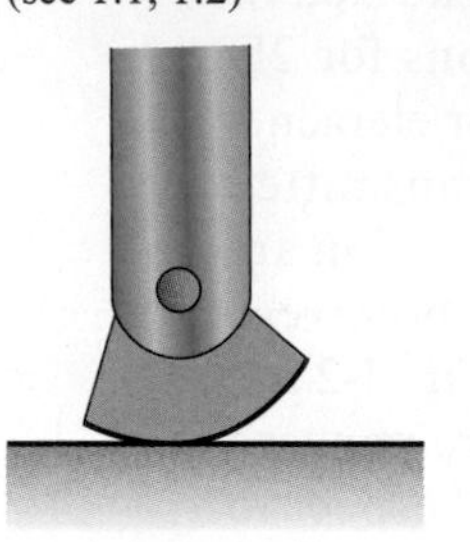 Bridge with *rocker* support (see 1.3)	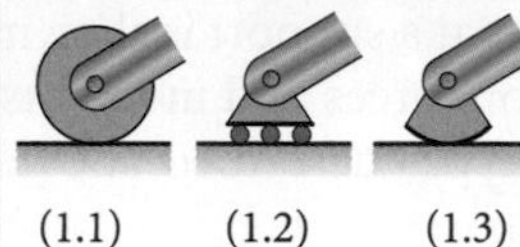 (1.1) (1.2) (1.3) Horizontal ***roller*** support [(1.1), (1.2)]; or alternate representation as ***rocker*** support [(1.3)] Both downward and uplift motions are restrained. 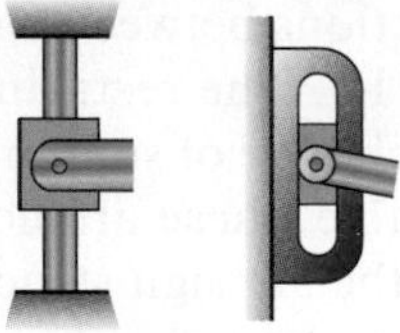 Vertical roller restraints 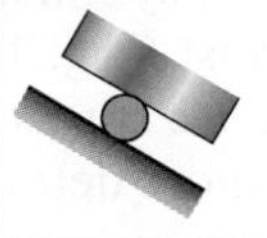 Rotated or inclined roller support 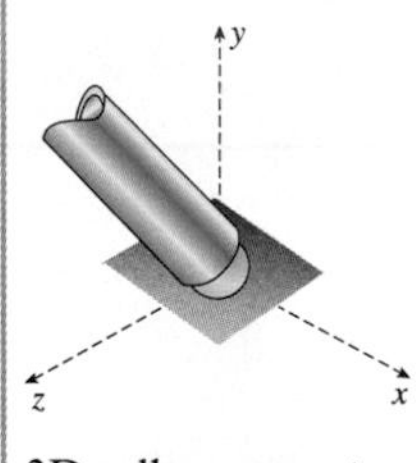 3D roller support	(a) **Two-dimensional roller support** (friction force $\boldsymbol{F} = 0$ for smooth rolling surface) 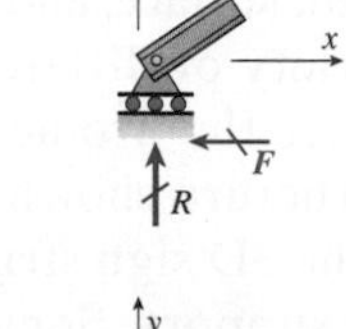 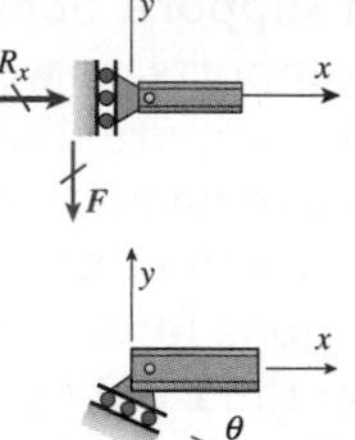 (b) **Three-dimensional roller support** (friction force $\boldsymbol{F} = 0$ for smooth rolling surface; reaction R_y acts normal to plane *x-z* on which roller translates) 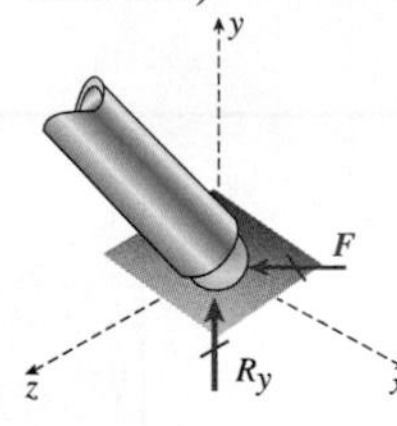
2. Pin Support: A single resultant force, usually shown using two rectangular components R_x and R_y in 2D but three components in 3D, resists motion in any direction normal to the pin. The pin support cannot resist moment, and the pin is free to rotate about the *z* axis. In 3D, the pin becomes a ball-and-socket joint or support.		
	 Two-dimensional pin	(a) **Two-dimensional pin support** 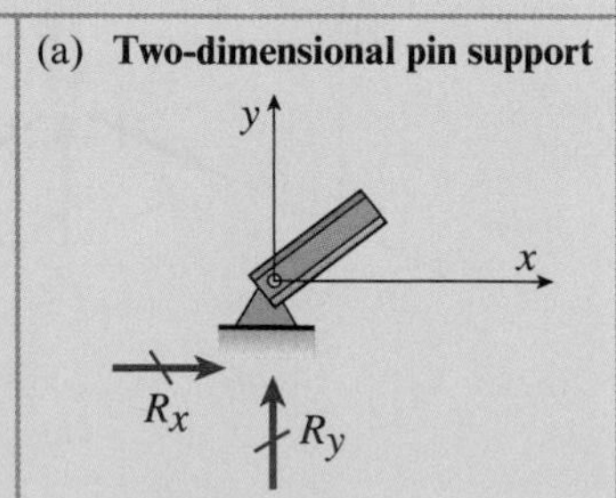

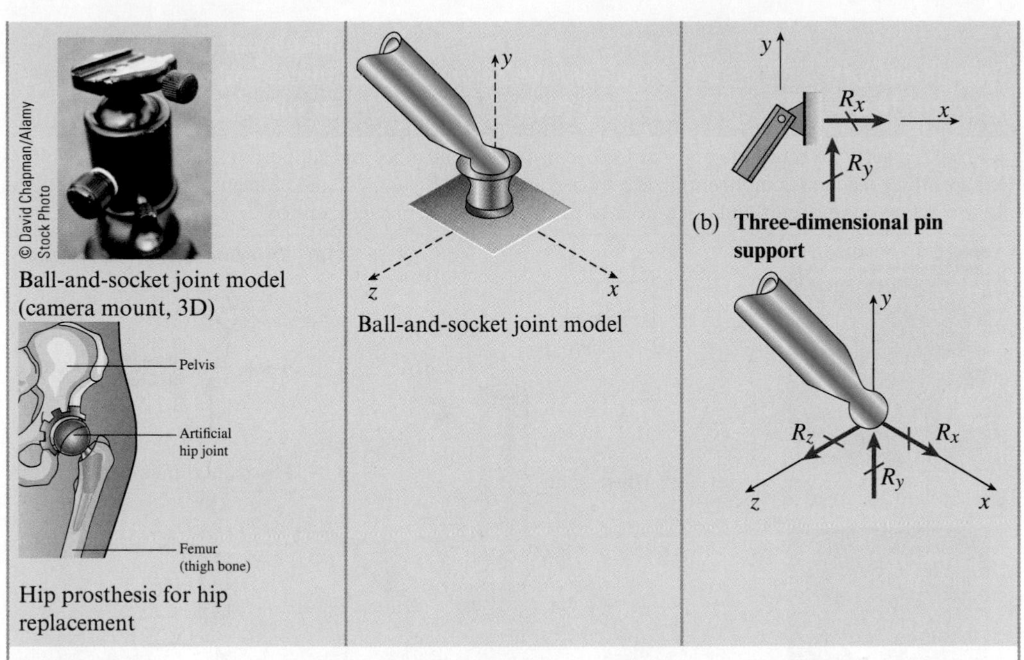

3. Sliding Support: A support that translates without rotation is a sliding support. Examples are a collar sliding along a sleeve or a flange moving within a slot. Reactions in 2D are a force R_x normal to the sleeve and a moment M_z representing resistance to rotation relative to the sleeve. In 3D, the sliding support translates on frictionless plane *y-z* and reaction moment components M_y and M_z prevent rotation relative to that plane.

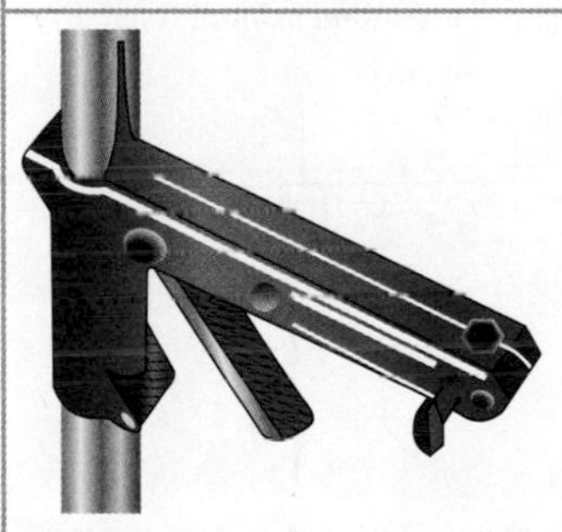

Sliding support for column light stand

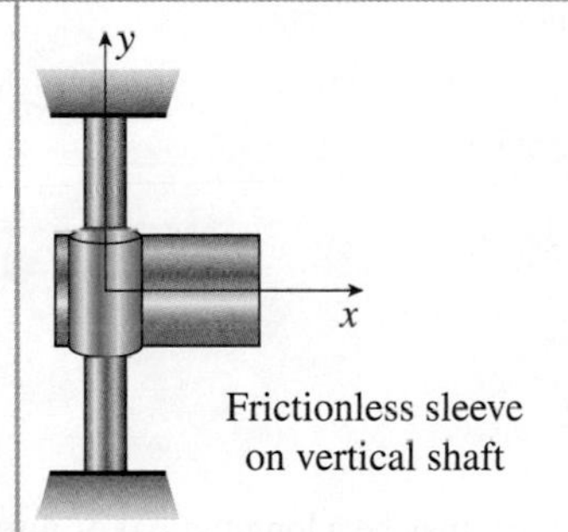

Two-dimensional sliding support

Friction ***F*** opposes motion in $+y$ direction in 2D along sliding surface; ***F*** is zero if smooth surface is assumed.

In 3D, add restraint moment M_x to prevent rotation about *x* axis.

(a) **Two-dimensional sliding support**
(support translates on frictionless path along $+y$ or $-y$ direction)

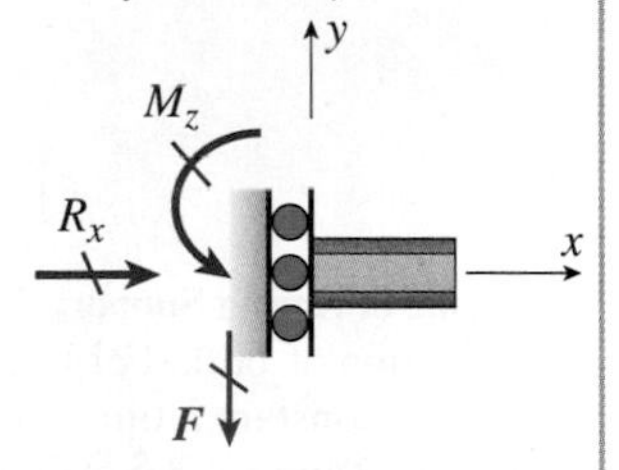

(b) **Three-dimensional sliding support**
(support translates on frictionless *y-z* plane)

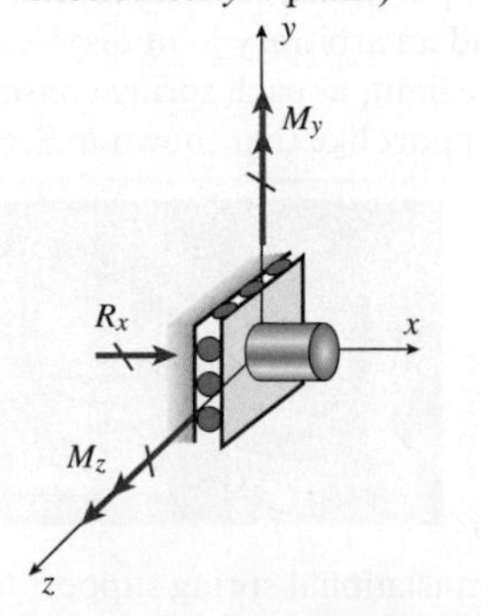

(Continued)

Type of support or connection	Simplified sketch of support or connection	Display of restraint forces and moments, or connection forces
4. Fixed Support: No translation or rotation occurs between member and support in a fixed support. This requires three reaction components in 2D: force components R_x and R_y and moment M_z. In 3D, three force reaction components and three moment reaction components are required.		
 Franco Nadalin/Shutterstock.com Steel bollard anchored in concrete  Radovan1/Shutterstock.com Column bolted to footing	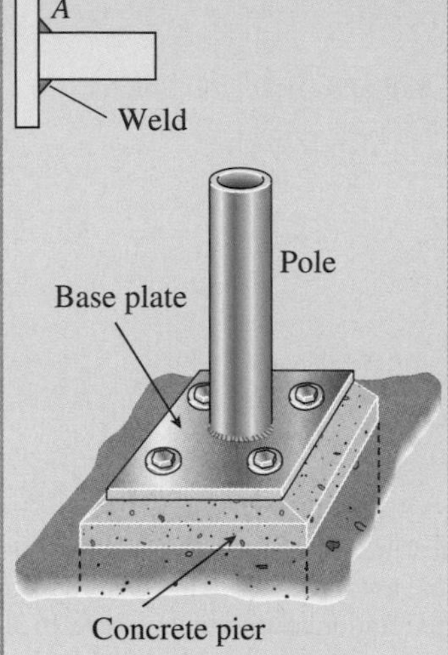 Fixed support at base of sign post	(a) **Two-dimensional fixed support** 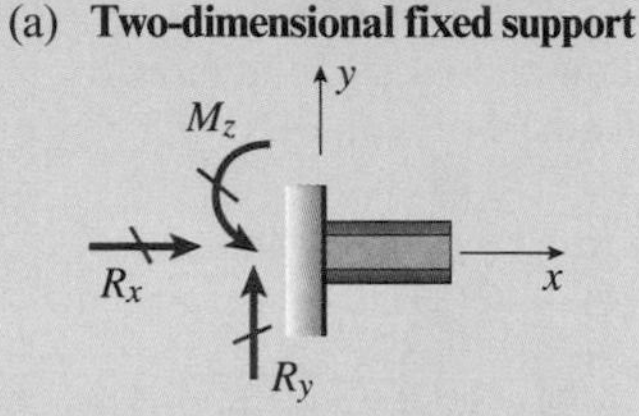 Horizontal member 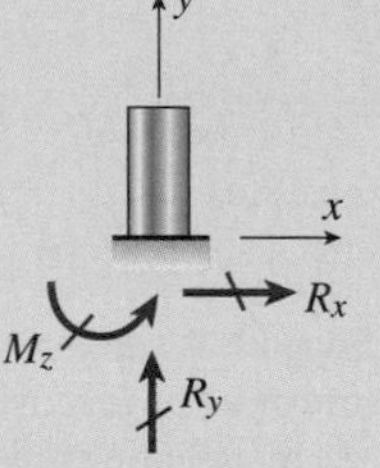 Vertical member (b) **Three-dimensional fixed support** 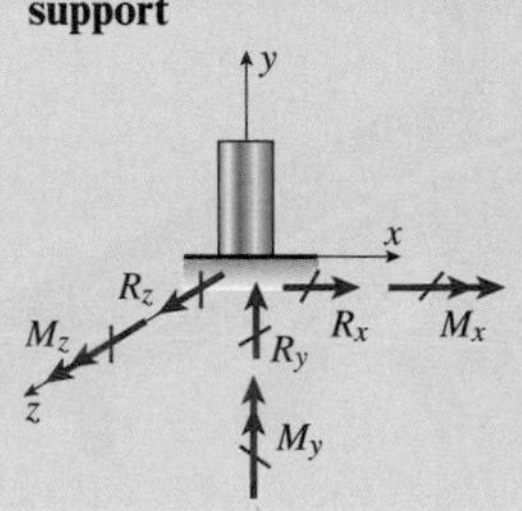
5. Elastic or Spring Support: In 2D, there may be a longitudinal or normal translational spring or a combination of both. For linear springs, the support reaction at the base of the spring is the product of the spring constant k times the displacement δ in the direction of the spring axis. If joint A translates in $+x$ (δ_x) and $+y$ (δ_y) directions, reaction forces R_x and R_y are created in $-x$ and $-y$ directions, respectively, at the supports of linear translational springs. Alternatively, the support may be pinned for translation but have moment spring k_r for rotation. If joint A rotates about the $+z$ axis (θ_z), reaction moment M_z is created in the $-z$ direction at the base of the rotational spring. In 3D, a fully elastic support consists of three translational springs (k_x, k_y, k_z) and three rotational springs (k_{rx}, k_{ry}, k_{rz}), and an arbitrary joint displacement results in three reaction forces and three reaction moments. In the limit, as each spring constant value approaches infinity, the elastic support becomes a fully fixed support like that shown in Section 4b above.		
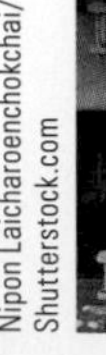 Nipon Laicharoenchokchai/ Shutterstock.com  Translational spring support for heavy equipment	In **3D**, add spring in $+z$ direction k_z with reaction force $R_z = -k_z\delta_z$. In **3D**, add rotational *flexural* spring about $+y$ direction k_{ry} with reaction moment $M_y = -k_{ry}\theta_y$ and add rotational *torsional* spring about $+x$ direction with reaction moment $M_x = -k_{rx}\theta_x$.	(a) **Translational spring (k) in 2D** 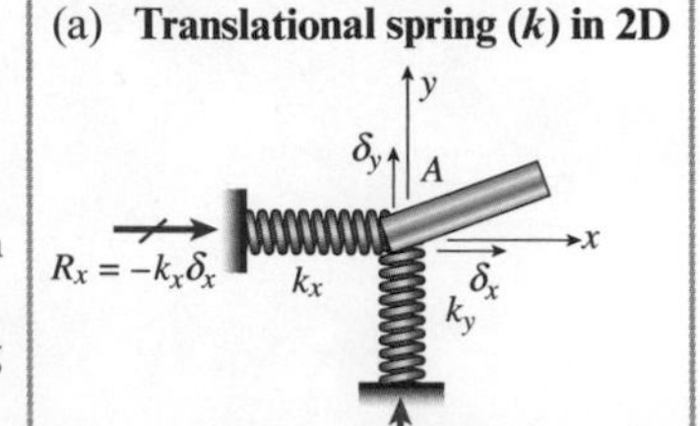

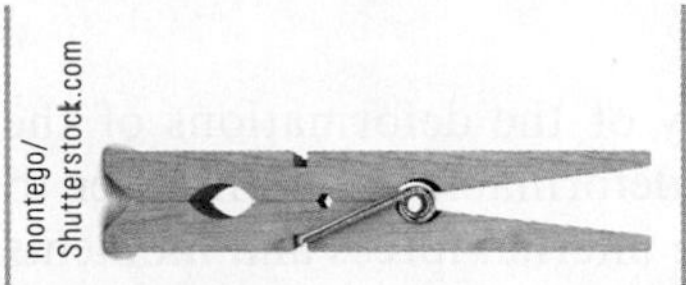

Rotational spring in a clothespin

Torsion springs are found in window shades and as part of the lift mechanism in power garage door-opening systems.

(b) **Rotational spring (k_r) in 2D**

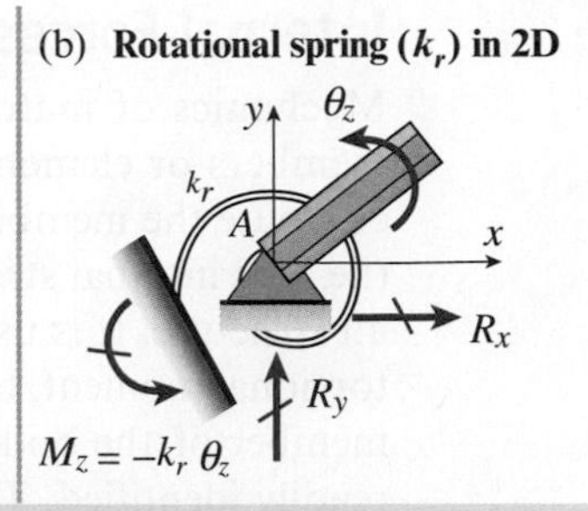

6. Wheel on Rail Support: This support is a particular form of the 3D roller support (see Section 1b above). Now general movement in the *x-z* plane is constrained by normal force R_y and lateral force R_x, both acting normal to the rail or slot on which the wheel travels. If friction is considered, friction force F is added along the rail in the direction opposing the wheel translation.

Wheel rolls on rail or in slot along *z* axis; friction force opposing motion is neglected; R_x is lateral constraint force, R_y is normal force.

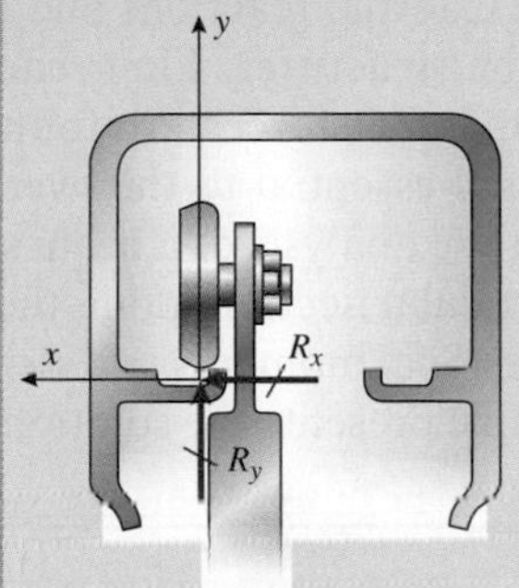

Cross section through guide rail

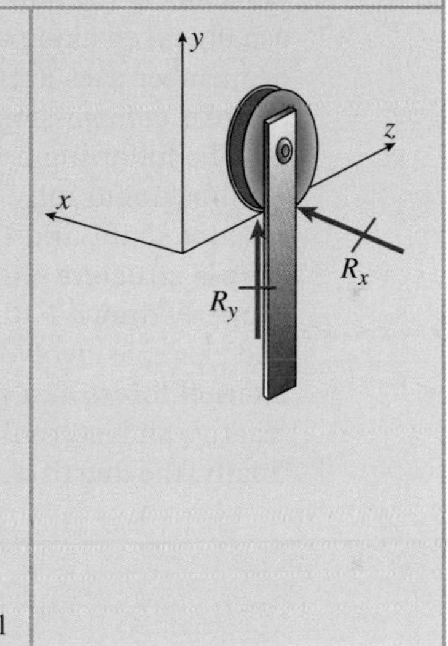

7. Thrust-Bearing Support: A thrust bearing constrains translational motion along the shaft axis while allowing rotary motion to occur about that axis. Support reaction forces and moment components act in all directions except for reaction moment $M_x = 0$ about the thrust axis (in the absence of friction). A special case is the journal bearing for which axial thrust restraint component $R_x = 0$.

Thrust bearing has support reaction force (R_x, R_y, R_z) and reaction moment components (M_y, M_z)—no moment M_x about the thrust or rotation (x) axis.

Pillow block bearing

Journal bearing has no axial thrust reaction force ($R_x = 0$) in addition to ($M_x = 0$).

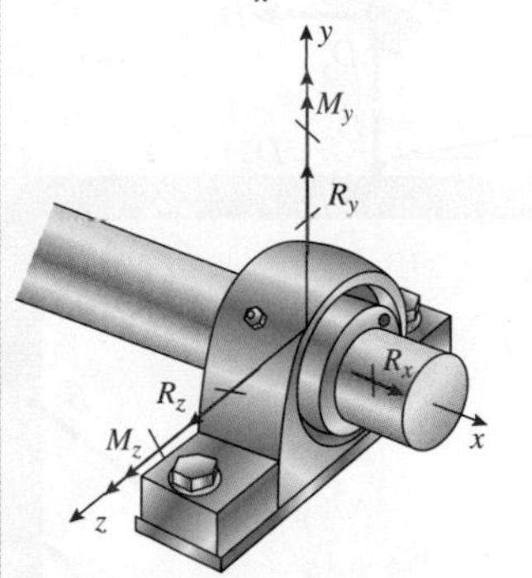

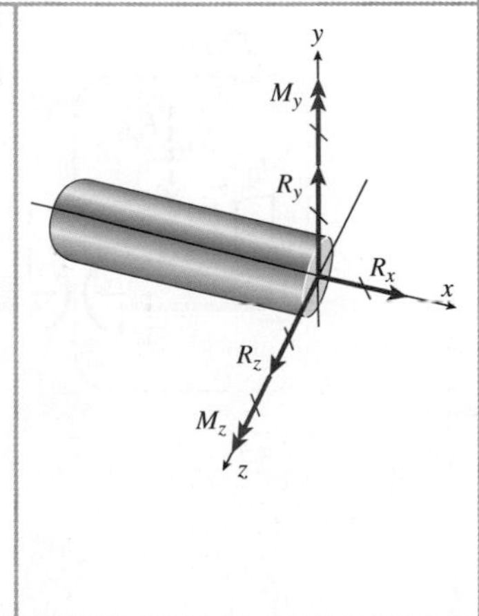

Internal Forces (Stress Resultants)

Mechanics of materials is concerned with study of the deformations of the members or elements that make up the overall deformable body. In order to compute the member deformations, first find the internal forces and moments (i.e., the internal stress resultants) at key points along the members of the overall structure. It is useful to create graphical displays of the internal axial force, torsional moment, transverse shear, and bending moment along the axis of each member of the body so that critical points or regions within the structure are readily identified. The first step is to make a section cut normal to the axis of each member so that a FBD can be drawn that displays the internal forces of interest. For example, Fig. 1-5 shows two cuts made through members *ED* and *DF* in the plane frame; the resulting FBDs now can be used to find N, V, and M in members *ED* and *DF* of the plane frame. Stress resultants N, V, and M are usually taken along and normal to the member under consideration (i.e., local or member axes are used), and a *deformation sign convention* (e.g., tension is positive, compression is negative) is employed in their solution.

The following examples review the application of the equations of static equilibrium to solve for external reactions and internal forces in truss, beam, circular shaft, and frame structures. First reaction forces are computed for a **truss structure** then member forces are found using the *method of joints*. Properly drawn FBDs are essential to the overall solution process. The second example involves static analysis of a **beam structure** to find reactions and internal forces at a particular section along the beam. In the third example, reactive and internal torsional moments in a **stepped shaft** are computed. And, finally, the fourth example presents the solution of a **plane frame structure**.

FIGURE 1-5

FBDs for internal stress resultants in *ED* and *DF*

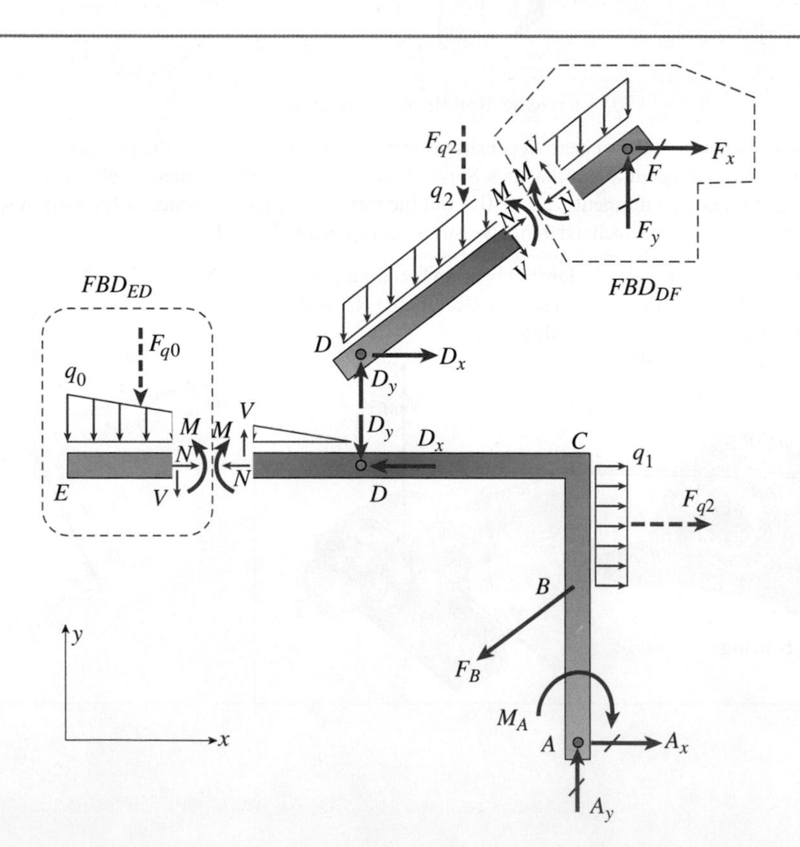

Example 1-1

FIGURE 1-6
Plane truss model

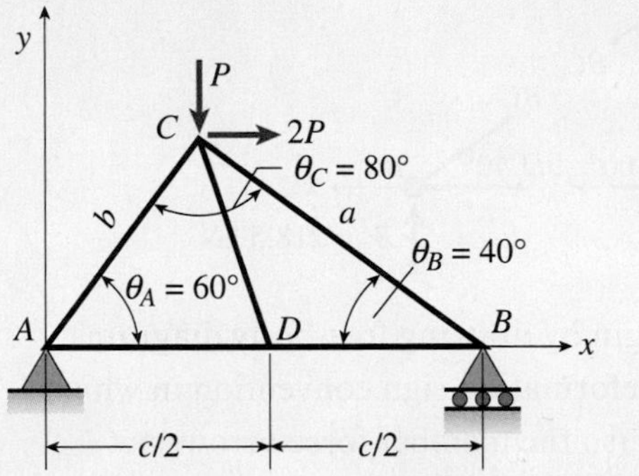

The plane truss shown in Fig. 1-6 has four joints and five members. Find support reactions at A and B and then use the methods of joints and sections to find all member forces. Let $P = 150$ kN and $c = 3$ m.

Solution:

Use the following four-step problem-solving approach.

1. ***Conceptualize*** [*hypothesize, sketch*]: First sketch a free-body diagram of the entire truss model (Fig. 1-7). Only known applied forces at C and unknown reaction forces at A and B are shown and then used in an equilibrium analysis to find the reactions.

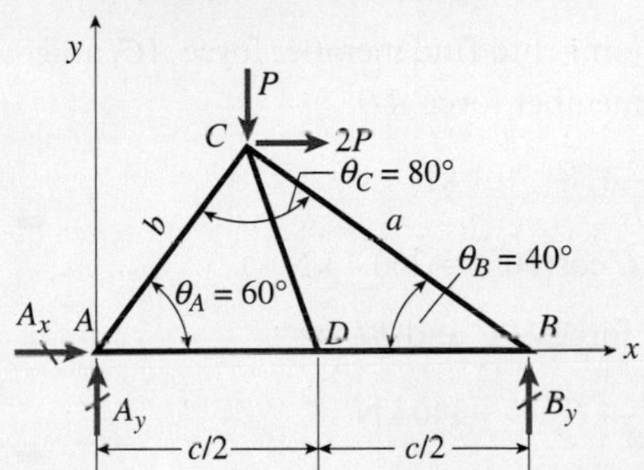

FIGURE 1-7
Free-body diagram of truss model

2. ***Categorize*** [*simplify, classify*]: Overall equilibrium requires that the force components in x and y directions and the moment about the z axis must sum to zero; this leads to reaction force components A_x, A_y, and B_y. The truss is statically determinate (unknowns: $m + r = 5 + 3 = 8$, knowns: $2j = 8$), so all member forces can be obtained using the method of joints. If only a few selected member forces are of interest, the method of sections can be used. Use a statics sign convention when computing external reactions and a deformation sign convention when solving for member forces.

3. ***Analyze*** [*evaluate; select relevant equations, carry out mathematical solution*]: First find the lengths of members AC and BC needed to compute distances to lines of action of forces.

Law of sines to find member lengths *a* and *b*: Use known angles θ_A, θ_B, and θ_C and $c = 3$ m to find lengths a and b:

$$b = c\frac{\sin(\theta_B)}{\sin(\theta_C)} = (3\text{ m})\frac{\sin(40°)}{\sin(80°)} = 1.958\text{ m},\ a = c\frac{\sin(\theta_A)}{\sin(\theta_C)} = (3\text{ m})\frac{\sin(60°)}{\sin(80°)} = 2.638\text{ m}$$

Check that computed lengths a and b give length c by using the law of cosines:

$$c = \sqrt{(1.958\text{ m})^2 + (2.638\text{ m})^2 - 2(1.958\text{ m})(2.638\text{ m})\cos(80°)} = 3\text{ m}$$

Support reactions: Using the truss model free-body diagram in Fig. 1-7, sum forces in x and y directions and moments about joint A:

$$\Sigma M_A = 0 \quad B_y = \frac{1}{c}[P(b\cos(\theta_A)) + 2P(b\sin(\theta_A))] = 218.5\text{ kN}$$

$$\Sigma F_x = 0 \quad A_x = -2P = -300\text{ kN}$$

$$\Sigma F_y = 0 \quad A_y = P - B_y = -68.5\text{ kN}$$

Reaction force components A_x and A_y are both negative, so they act in the negative x and y directions, respectively, based on a statics sign convention.

FIGURE 1-8

Free-body diagram of pin at each truss joint

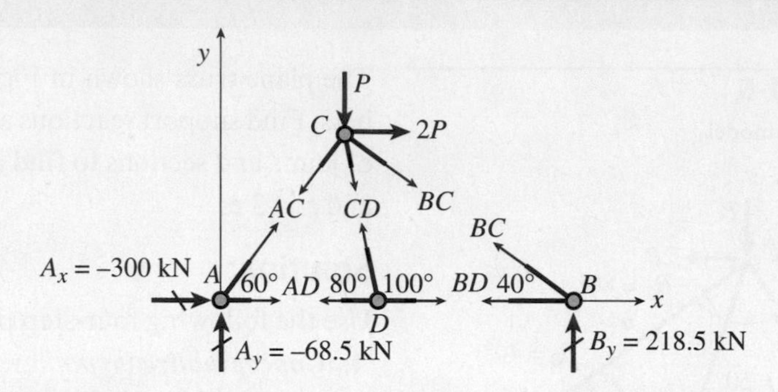

Member forces using method of joints: Begin by drawing free-body diagrams of the pin at each joint (Fig. 1-8). Use a deformation sign convention in which each member is assumed to be in tension (so the member force arrows act away from the two joints to which each member is connected). The forces are concurrent at each joint, so use force equilibrium at each location to find the unknown member forces.

First sum forces in the y direction at joint A to find member force AC, and then sum forces in the x direction to get member force AD:

$$\Sigma F_y = 0 \quad AC = \frac{-1}{\sin(60°)} A_y = 79.1 \text{ kN}$$

$$\Sigma F_x = 0 \quad AD = -A_x - AC\cos(60°) = 260.4 \text{ kN}$$

Summing forces at joint B gives member forces BC and BD as

$$\Sigma F_y = 0 \quad BC = \frac{-1}{\sin(40°)} B_y = -340 \text{ kN}$$

$$\Sigma F_x = 0 \quad BD = -BC\cos(40°) = 260.4 \text{ kN}$$

The minus sign means that member BC is in compression, not in tension as assumed. Finally, observe that CD is a zero-force member because forces in the y direction must sum to zero at joint D.

FIGURE 1-9

Section cut leading to right-hand free-body diagram

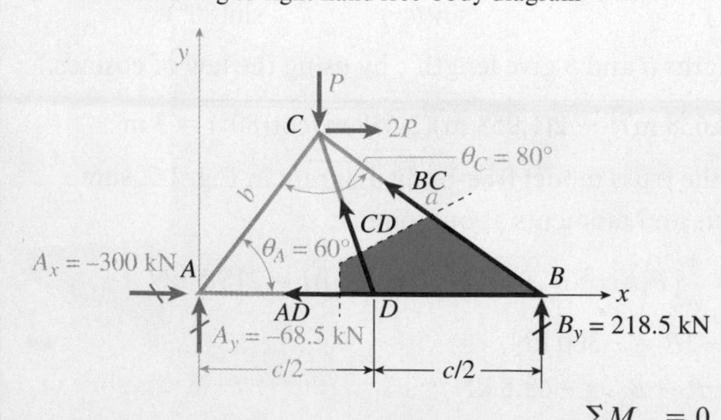

Selected member forces using method of sections: An alternative approach is to make a section cut all the way through the structure to expose member forces of interest, such as AD, CD, and BC in Fig. 1-9. Summing moments about joint B confirms that the force in member CD is zero.

Summing moments about joint C (which is not on the free-body diagram but is a convenient point about which to sum moments because forces CD and BC act through joint C) confirms the solution for force AD as

$$\Sigma M_C = 0 \quad AD = \frac{1}{b\sin(60°)}\left[B_y(a)\cos(40°)\right] = 260.4 \text{ kN}$$

Finally, summing moments about A in Fig. 1-9 confirms member force BC:

$$\Sigma M_A = 0 \quad BC = \frac{1}{c\sin(40°)}[-B_y c] = -340 \text{ kN}$$

4. ***Finalize*** [*conclude; examine answer—Does it make sense? Are units correct? How does it compare to similar problem solutions?*]: There are $2j = 8$ equilibrium equations for the simple plane truss considered, and using the method of joints, these are obtained by applying $\Sigma F_x = 0$ and $\Sigma F_y = 0$ at each joint in succession. A computer solution of these simultaneous equations leads to the three reaction forces and five member forces. The method of sections is an efficient way to find selected member forces. A key step is the choice of an appropriate section cut, which isolates the member of interest and eliminates as many unknowns as possible. This is followed by construction of a free-body diagram for use in the static equilibrium analysis to compute the member force of interest. The methods of sections and joints were used, a common solution approach in plane and space truss analysis.

Example 1-2

A simple beam with an overhang is supported at points A and B (Fig. 1-10). A linearly varying distributed load of peak intensity $q_0 = 160$ N/m acts on span AB. Concentrated moment $M_0 = 380$ N·m is applied at A, and an inclined concentrated load $P = 200$ N acts at C. The length of segment AB is $L = 4$ m, and the length of the overhang BC is 2 m.

Find support reactions at A and B and then calculate the axial force N, shear force V, and bending moment M at midspan of AB.

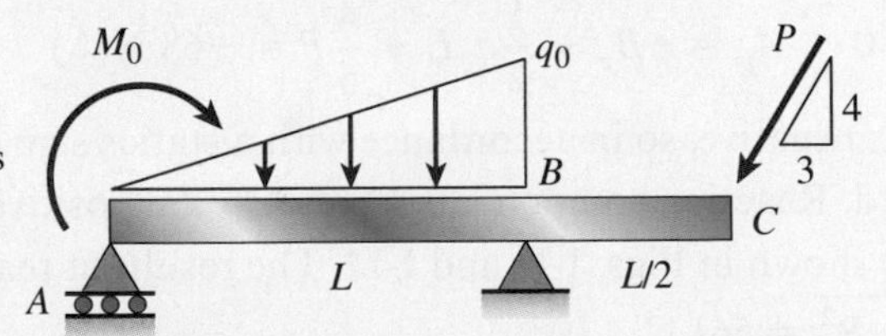

FIGURE 1-10

Beam with an overhang and uniform and concentrated loads

Solution:

Use the following four-step problem-solving approach.

1. ***Conceptualize***: Find the reaction forces A_y, B_x, and B_y using the FBD of the overall structure shown in Fig. 1-11. Internal axial force N, shear force V, and bending moment M at midspan of AB (Fig. 1-12) are obtained by cutting the beam at that location. Either the left-hand or right-hand free-body diagram in Fig. 1-12 may be used to find N, V, and M.

FIGURE 1-11

Free-body diagram of beam

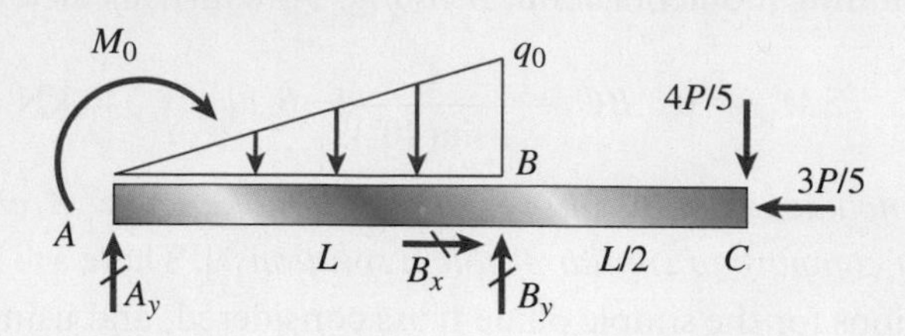

FIGURE 1-12

Left- and right-hand free-body diagrams from section cut at midspan of AB

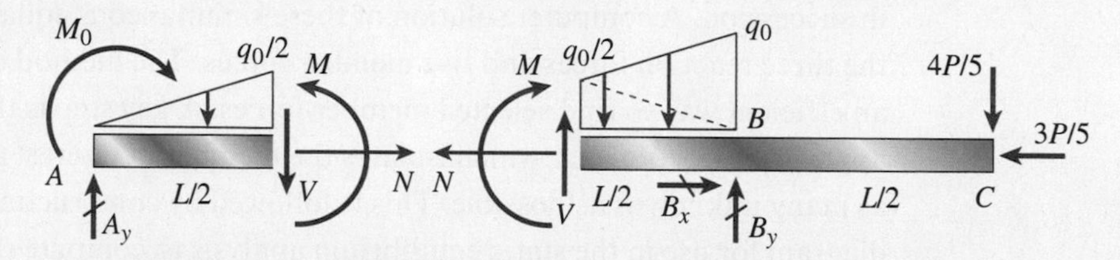

2. ***Categorize***: The free-body diagrams in Fig. 1-12 show internal axial force N, shear force V, and bending moment M in their assumed positive directions based on a deformation sign convention. Start by finding reaction forces A_y, B_x, and B_y then use either the left-hand or right-hand free-body diagram in Fig. 1-12 to find N, V, and M.

3. ***Analyze***:

Solution for external reactions: Sum forces in the x direction to find reaction force component B_x. Next sum moments about A to find reaction component B_y. Finally, sum forces in the y direction to find reaction A_y. Use a statics sign convention in the solution as

$$\Sigma F_x = 0 \quad B_x = \frac{3}{5}P = 120 \text{ N}$$

$$\Sigma M_A = 0 \quad B_y = \frac{1}{L}\left[M_0 + \frac{1}{2}q_0 L\left(\frac{2L}{3}\right) + \frac{4}{5}P\left(L + \frac{L}{2}\right)\right] = 548 \text{ N}$$

$$\Sigma F_y = 0 \quad A_y = -B_y + \frac{1}{2}q_0 L + \frac{4}{5}P = -68 \text{ N}(\downarrow)$$

Reaction A_y is negative, so in accordance with a statics sign convention, it acts downward. Reaction components B_x and B_y are positive, so they act in the directions shown in Figs. 1-11 and 1-12. The resultant reaction force at B is $B_{\text{res}} = \sqrt{B_x^2 + B_y^2} = 561$ N.

Solution for internal axial force N, shear force V, and moment M at midspan of AB: Using the left-hand free-body diagram in Fig. 1-12,

$$\Sigma F_x = 0 \quad N = 0 \quad \Sigma F_y = 0 \quad V = A_y - \frac{1}{2}\frac{q_0}{2}\frac{L}{2} = -148 \text{ N}(\uparrow)$$

$$\Sigma M = 0 \quad M = M_0 + A_y\frac{L}{2} - \frac{1}{2}\frac{q_0}{2}\frac{L}{2}\left[\frac{1}{3}\frac{L}{2}\right] = 190 \text{ N}\cdot\text{m}$$

Alternatively N, V, and M can be obtained if the right-hand free-body diagram is used (Fig. 1-12). Note that the trapezoidal distributed load segment is treated as two triangular loads when moments are summed to find internal moment M:

$$\Sigma F_x = 0 \quad N = B_x - \frac{3}{5}P = 0 \quad \Sigma F_y = 0 \quad V = -B_y + \frac{4}{5}P + \frac{1}{2}\left[\frac{q_0}{2} + q_0\right]\frac{L}{2} = -148\ \text{N}(\downarrow)$$

$$\Sigma M = 0 \quad M = B_y\frac{L}{2} - \frac{4}{5}P\left[\frac{L}{2} + \frac{L}{2}\right] - \frac{1}{2}\frac{q_0}{2}\frac{L}{2}\left[\frac{1}{3}\frac{L}{2}\right] - \frac{1}{2}q_0\frac{L}{2}\left[\frac{2}{3}\frac{L}{2}\right] = 190\ \text{N}\cdot\text{m}$$

The minus sign on internal shear force V shows that it acts opposite to that assumed in Fig. 1-12, as indicated by the arrows in the previous equations.

4. ***Finalize***: The results show that computed internal forces (N and V) and internal moment (M) can be determined using either the left- or right-hand free-body diagram. This applies for any section taken through the beam at any point along its length. Plots or diagrams that show the variation of N, V, and M over the length of the beam are very useful in the design of shafts and beams, because they readily show the critical regions of the beam where N, V, and M have maximum values.

Example 1-3

A stepped circular shaft is fixed at A and has three gears that transmit the torques shown in Fig. 1-13. Find the reaction torque M_{Ax} at A and then find the internal torsional moments in segments AB, BC, and CD. Use properly drawn free-body diagrams in your solution.

FIGURE 1-13

Stepped circular shaft subjected to concentrated torques

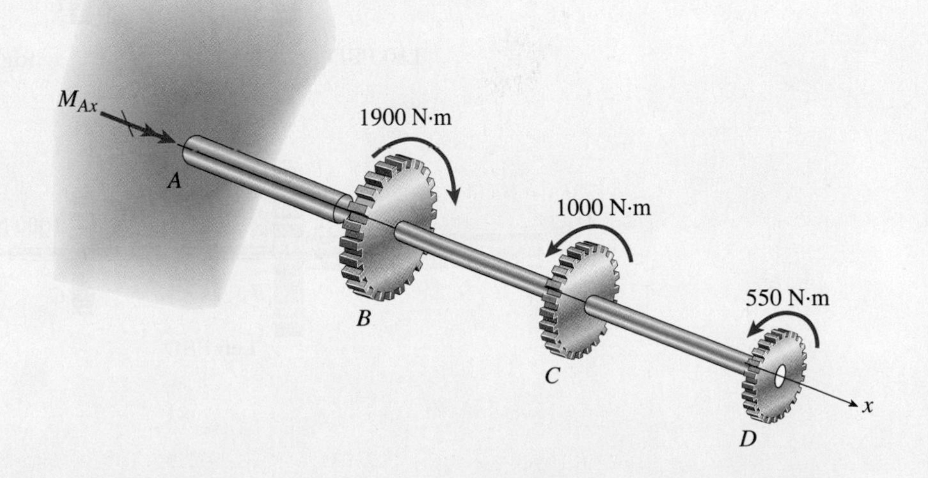

Solution:

Use the following four-step problem-solving approach to find internal torsional moments $T(x)$.

1. ***Conceptualize***: The cantilever shaft structure is stable and statically determinate. The solution for the reaction moment at $A(M_{Ax})$ must begin with a proper drawing of the FBD of the overall structure (Fig. 1-14). The FBD shows all applied and reactive torques. Separate FBDs showing internal torques T in each segment are obtained by cutting the shaft in regions AB, BC, and CD in succession and are given in Fig. 1-15(a–c). Each cut produces a left-hand and a right-hand free-body diagram.

FIGURE 1-14

Free-body diagram of shaft

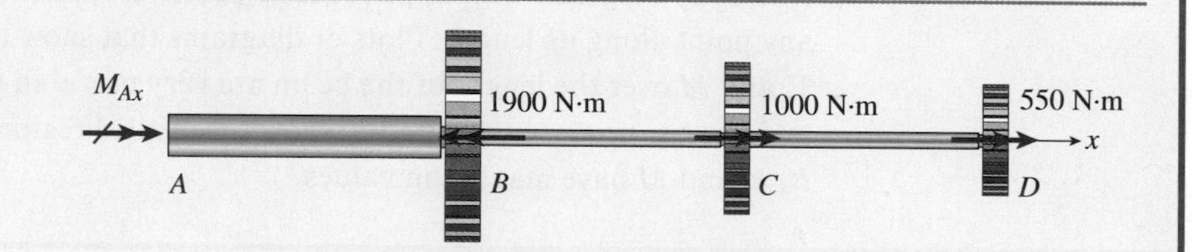

FIGURE 1-15

(a, b, c) Left and right free-body diagrams for each shaft segment

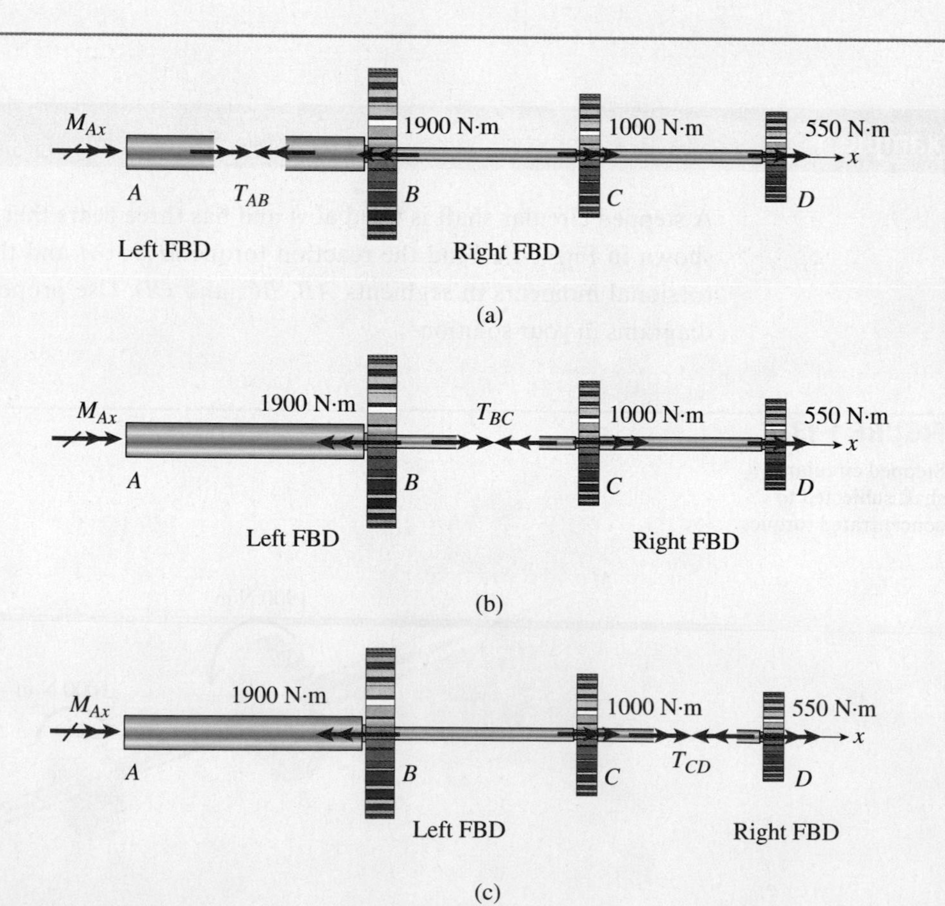

2. ***Categorize***: The shaft is subjected to applied torques that act along the centroidal axis of the shaft, so only internal torsional moment $T(x)$ is present at any section cut along the shaft. There is no distributed torque acting on this shaft, so the internal torsional moment T is constant within each segment.
3. ***Analyze***:

Solution for external reaction moment M_{Ax}:

Sum the moments about the x-axis to find the reaction moment M_{Ax}. This structure is statically determinate because there is one available equation from statics ($\Sigma M_x = 0$) and one reaction unknown (M_{Ax}). A statics sign convention is used [i.e., right-hand rule or counterclockwise (CCW) is positive].

$$M_{Ax} - 1900\ \text{N}\cdot\text{m} + 1000\ \text{N}\cdot\text{m} + 550\ \text{N}\cdot\text{m} = 0$$
$$M_{Ax} = -(-1900\ \text{N}\cdot\text{m} + 1000\ \text{N}\cdot\text{m} + 550\ \text{N}\cdot\text{m})$$
$$= 350\ \text{N}\cdot\text{m}$$

The computed result for M_{Ax} is positive, so the reaction moment vector is in the positive x direction as assumed.

Solution for internal torsional moments T in each shaft segment:

Start with segment AB and use either FBD in Fig. 1-15a to find:

Left FBD:
$$T_{AB} = -M_{Ax} = -350\ \text{N}\cdot\text{m}$$

Right FBD:
$$T_{AB} = -1900\ \text{N}\cdot\text{m} + 1000\ \text{N}\cdot\text{m} + 550\ \text{N}\cdot\text{m} = -350\ \text{N}\cdot\text{m}$$

Next consider segment BC. Summing moments about the x axis in Fig. 1-15b gives

Left FBD:
$$T_{BC} = -M_{Ax} + 1900\ \text{N}\cdot\text{m} = 1550\ \text{N}\cdot\text{m}$$

Right FBD:
$$T_{BC} = 1000\ \text{N}\cdot\text{m} + 550\ \text{N}\cdot\text{m} = 1550\ \text{N}\cdot\text{m}$$

Finally, moment equilibrium about the x axis leads to a solution for the internal torsional moment in segment CD:

Left FBD:
$$T_{CD} = -M_{Ax} + 1900\ \text{N}\cdot\text{m} - 1000\ \text{N}\cdot\text{m} = 550\ \text{N}\cdot\text{m}$$

Right FBD:
$$T_{CD} = 550\ \text{N}\cdot\text{m}$$

In each segment, the internal torsional moments computed using either the left or right FBDs are the same.

4. ***Finalize***: Segment BC has the maximum positive internal torsional moment, and segment AB has the maximum negative torsional moment. This is important information for the designer of the shaft. Properly drawn free-body diagrams are essential to a correct solution. Either the left or right free-body diagram can be used to find the internal torque at any section.

Example 1-4

A plane frame with an overhang is supported at points A and D (Fig. 1-16). (This is the beam of Example 1-2 to which column BD has been added.) A linearly varying distributed load of peak intensity $q_0 = 160$ N/m acts on span AB. Concentrated moment $M_0 = 380$ N · m is applied at A, and an inclined concentrated load $P = 200$ N acts at C. Force P also acts at mid-height of column BD. The lengths of segments AB and BD are $L = 4$ m, and the length of the overhang BC is 2 m.

Find support reactions at A and D and then calculate the axial force N, shear force V, and bending moment M at the top of column BD.

FIGURE 1-16

Plane frame with an overhang and uniform and concentrated loads

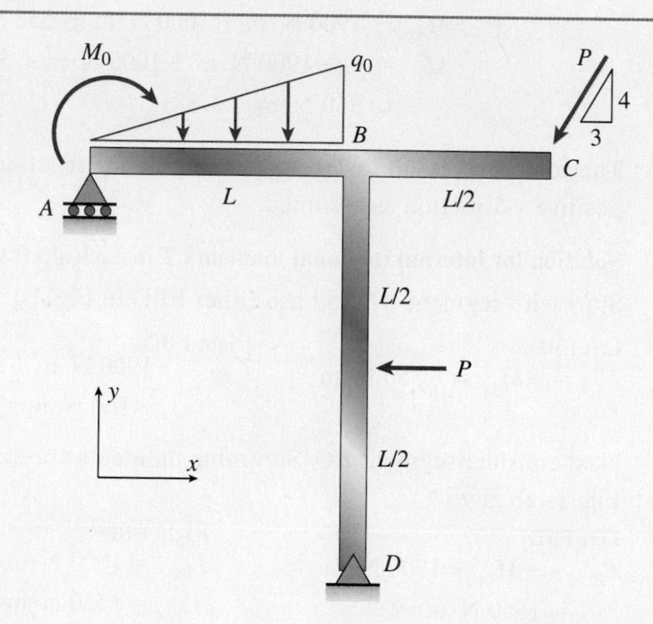

Solution:

The four-step problem-solving approach for this plane frame follows the procedures presented for the beam in Example 1-2.

1. ***Conceptualize***: Find the reaction forces A_y, D_x, and D_y using the FBD of the overall structure shown in Fig. 1-17. Internal axial force N, shear force V, and bending moment M at the top of column BD (Fig. 1-18) are obtained by cutting the column at that location. View column BD with joint D on your left and B on your right to establish the assumed positive directions of N, V, and M on either side of the cut section, as shown in Fig. 1-18.
2. ***Categorize***: First find reaction forces A_y, D_x, and D_y. Then use either the upper or lower free-body diagram in Fig. 1-18 to find N, V, and M. The free-body diagrams in Fig. 1-18 show internal axial force N, shear force V, and bending moment M in their assumed positive directions based on a deformation sign convention.

FIGURE 1-17
Free-body diagram of plane frame

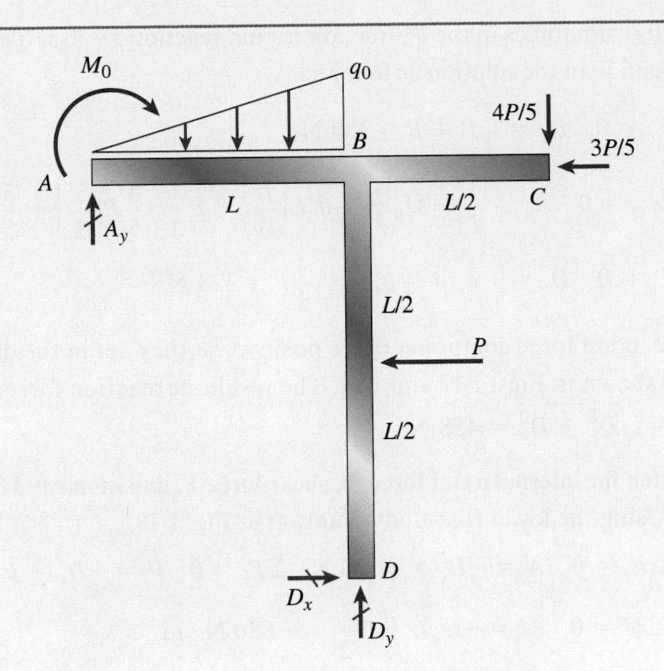

FIGURE 1-18
Upper and lower free body diagrams from section cut at top of *BD*

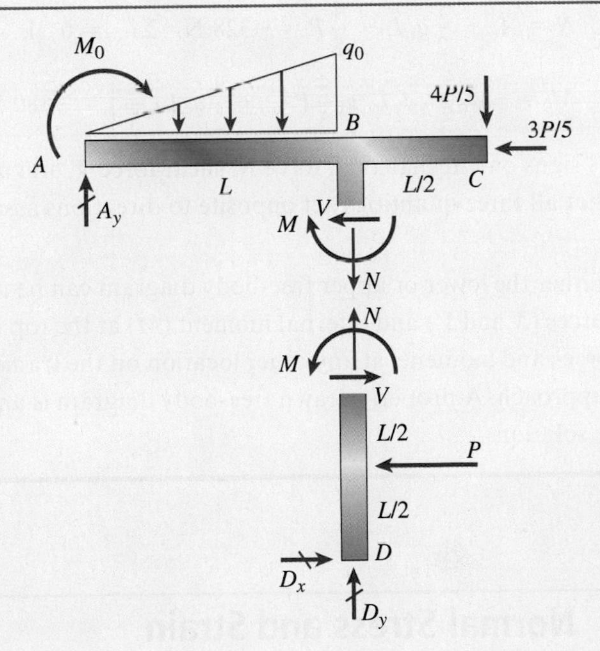

3. ***Analyze***:

Solution for external reactions: Sum forces in the x direction to find reaction force component D_x. Next, sum moments about D to find reaction component A_y.

Finally, sum forces in the y direction to find reaction D_y. Use a statics sign convention in the solution as follows:

$$\Sigma F_x = 0 \quad D_x = \frac{3}{5}P + P = 320 \text{ N}$$

$$\Sigma M_D = 0 \quad A_y = \frac{1}{L}\left[-M_0 + \frac{1}{2}q_0 L\left[\frac{L}{3}\right] + P\frac{1}{2} - \frac{4}{5}P\left[\frac{L}{2}\right] + \frac{3}{5}PL\right] = 152 \text{ N}$$

$$\Sigma F_y = 0 \quad D_y = -A_y + \frac{1}{2}q_0 L + \frac{4}{5}P = 328 \text{ N}$$

All reaction force components are positive, so they act in the directions shown in Figs. 1-17 and 1-18. The resultant reaction force at D is $D_{\text{res}} = \sqrt{D_x^2 + D_y^2} = 458$ N.

Solution for internal axial force N, shear force V, and moment M at top of column BD: Using the lower free-body diagram in Fig. 1-18,

$$\Sigma F_y = 0 \quad N = -D_y = -328 \text{ N} \quad \Sigma F_x = 0 \quad V = -D_x + P = -120 \text{ N}$$

$$\Sigma M = 0 \quad M = -D_x L + P\frac{L}{2} = -880 \text{ N} \cdot \text{m}$$

Alternatively, the upper free-body diagram can be used to compute N, V, and M (Fig. 1-18):

$$\Sigma F_y = 0 \quad N = A_y - \frac{1}{2}q_0 L - \frac{4}{5}P = -328 \text{ N} \quad \Sigma F_x = 0 \quad V = -\frac{3}{5}P = -120 \text{ N}$$

$$\Sigma M = 0 \quad M = -M_0 - A_y L - \frac{4}{5}P\frac{L}{2} + \frac{1}{2}q_0 L\left[\frac{L}{3}\right] = -880 \text{ N}\cdot\text{m}$$

The minus signs on internal axial force N, shear force V, and moment M indicate that all three quantities act opposite to directions assumed in Fig. 1-18.

4. ***Finalize***: Either the lower or upper free-body diagram can be used to find internal forces (N and V) and internal moment (M) at the top of column BD. Section forces and moments at any other location on the frame are found using the same approach. A properly drawn free-body diagram is an important first step in the solution.

1.4 Normal Stress and Strain

Now that reactive and internal forces are known from statics, it is time to examine internal actions more closely for a deformable body of interest. The most fundamental concepts in mechanics of materials are **stress** and **strain**. These concepts can be illustrated in their most elementary form by considering a prismatic bar subjected to axial forces. A **prismatic bar** is a straight structural member having the same cross section throughout its length, and an **axial force** is a load directed along the axis of the member, resulting in either tension or compression

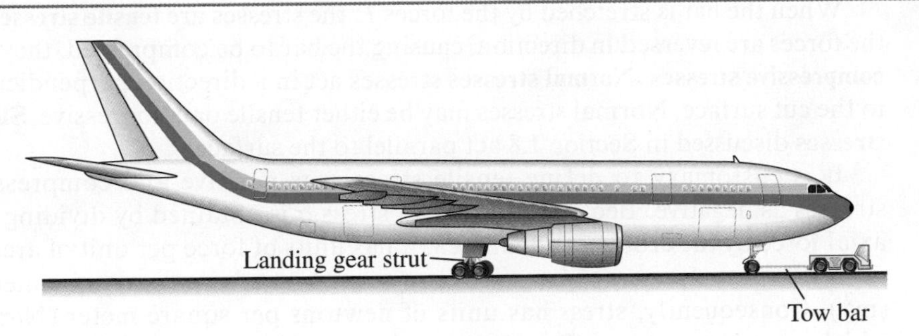

FIGURE 1-19

Structural members subjected to axial loads (the tow bar is in tension and the landing gear strut is in compression)

in the bar. Examples are shown in Fig. 1-19, where the tow bar is a prismatic member in tension and the landing gear strut is a member in compression. Other examples include the members of a bridge truss, connecting rods in automobile engines, spokes of bicycle wheels, columns in buildings, and wing struts in small airplanes.

For discussion purposes, consider the tow bar of Fig. 1-19 and isolate a segment of it as a free body (Fig. 1-20a). When drawing this free-body diagram, disregard the weight of the bar itself and assume that the only active forces are the axial forces P at the ends. Next, consider two views of the bar: the first showing the same bar *before* the loads are applied (Fig. 1-20b) and the second showing it *after* the loads are applied (Fig. 1-20c). The original length of the bar is denoted by the letter L, and the increase in length due to the loads is denoted by the Greek letter δ (delta).

The internal actions in the bar are exposed by making an imaginary cut through the bar at section mn (Fig. 1-20c). Because this section is taken perpendicular to the longitudinal axis of the bar, it is called a **cross section.**

Now isolate the part of the bar to the left of cross section mn as a free body (Fig. 1-20d). At the right-hand end of this free body (section mn) you can see the action of the removed part of the bar (that is, the part to the right of section mn) upon the part that remains. This action consists of continuously distributed *stresses* acting over the entire cross section, and the axial force P acting at the cross section is the *resultant* of those stresses. (The resultant force is shown with a dashed line in Fig. 1-20d.)

FIGURE 1-20

Prismatic bar in tension: (a) free-body diagram of a segment of the bar, (b) segment of the bar before loading, (c) segment of the bar after loading, and (d) normal stresses in the bar

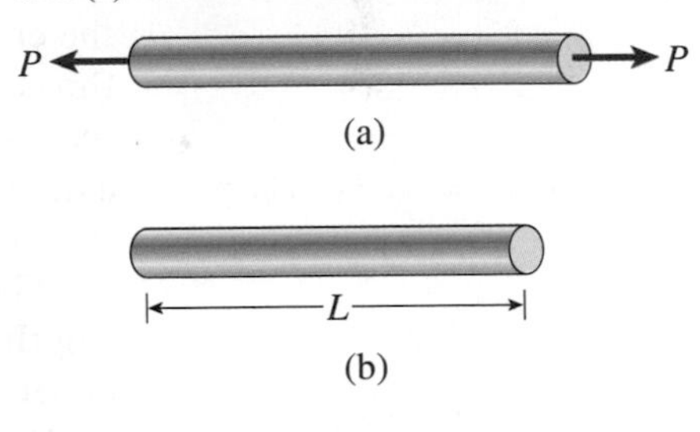

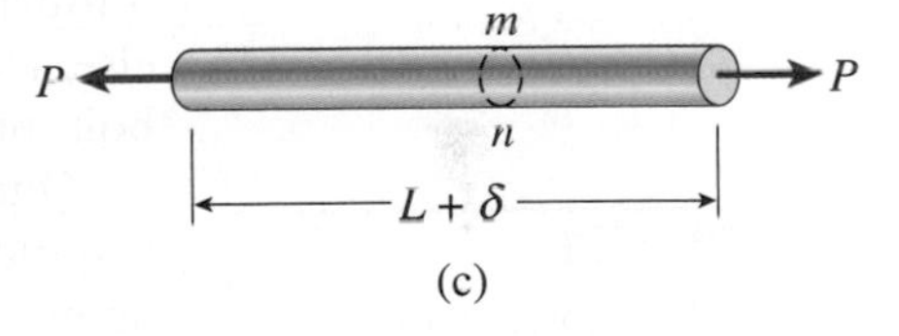

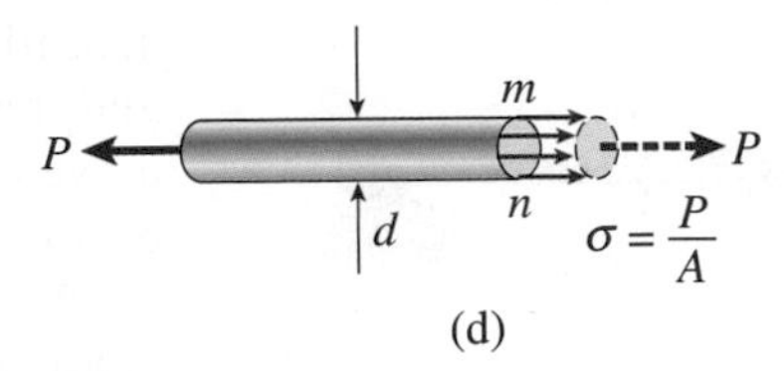

Stress has units of force per unit area and is denoted by the Greek letter σ (sigma). In general, the stresses σ acting on a plane surface may be uniform throughout the area or may vary in intensity from one point to another. Assume that the stresses acting on cross section mn (Fig. 1-20d) are *uniformly distributed* over the area. Then the resultant of those stresses must be equal to the magnitude of the stress times the cross-sectional area A of the bar, that is, $P = \sigma A$. Therefore, you can obtain the following expression for the magnitude of the stresses:

$$\sigma = \frac{P}{A} \tag{1-6}$$

This equation gives the intensity of uniform stress in an axially loaded, prismatic bar of arbitrary cross-sectional shape.

When the bar is stretched by the forces P, the stresses are **tensile stresses**; if the forces are reversed in direction, causing the bar to be compressed, they are **compressive stresses. Normal stresses** stresses act in a direction perpendicular to the cut surface. Normal stresses may be either tensile or compressive. **Shear stresses** discussed in Section 1.8 act parallel to the surface.

It is customary to define tensile stresses as positive and compressive stresses as negative. Because the normal stress σ is obtained by dividing the axial force by the cross-sectional area, it has units of force per unit of area.

In SI units, force is expressed in newtons (N) and area in square meters (m^2). Consequently, stress has units of newtons per square meter (N/m^2), which is equal to a pascal (Pa). However, the pascal is such a small unit of stress that it is necessary to work with large multiples, usually the megapascal (MPa).

Although it is not recommended in SI, you will sometimes find stress given in newtons per square millimeter (N/mm^2), which is a unit equal to the megapascal (MPa).

Limitations

The equation $\sigma = P/A$ is valid only if the stress is uniformly distributed over the cross section of the bar. This condition is realized if the axial force P acts through the centroid of the cross-sectional area, as demonstrated later in this section. However, in this book (as in common practice), it is understood that axial forces are applied at the centroids of the cross sections unless specifically stated otherwise.

The uniform stress condition pictured in Fig. 1-20d exists throughout the length of the bar except near the ends. The stress distribution at the end of a bar depends upon how the load P is transmitted to the bar. If the load is distributed uniformly over the end, the stress pattern at the end is the same as everywhere else. However, it is more likely that the load is transmitted through a pin or a bolt, producing high localized stresses called *stress concentrations*.

FIGURE 1-21

Steel eyebar subjected to tensile loads P

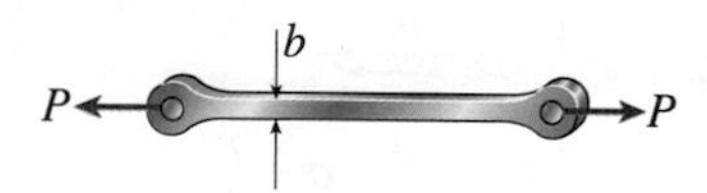

One possibility is illustrated by the eyebar shown in Fig. 1-21. In this instance, the loads P are transmitted to the bar by pins that pass through the holes (or eyes) at the ends of the bar. Thus, the forces shown in the figure are actually the resultants of bearing pressures between the pins and the eyebar, and the stress distribution around the holes is quite complex. However, as you move away from the ends and toward the middle of the bar, the stress distribution gradually approaches the uniform distribution pictured in Fig. 1-20d.

As a practical rule, the formula $\sigma = P/A$ may be used with good accuracy at any point within a prismatic bar that is at least as far away from the stress concentration as the largest lateral dimension of the bar. In other words, the stress distribution in the steel eyebar of Fig. 1-21 is uniform at distances b or greater from the enlarged ends where b is the width of the bar, and the stress distribution in the prismatic bar of Fig. 1-20 is uniform at distances d or greater from the ends where d is the diameter of the bar (Fig. 1-20d). Of course, even when the stress is *not* uniformly distributed, the equation $\sigma = P/A$ may still be useful because it gives the *average* normal stress on the cross section.

Normal Strain

As already observed, a straight bar changes in length when loaded axially, becoming longer when in tension and shorter when in compression. For instance, consider again the prismatic bar of Fig. 1-20. The elongation δ of this bar (Fig. 1-20c) is the cumulative result of the stretching of all elements of the material throughout the volume of the bar. Assume that the material is the same everywhere in the bar. Then half of the bar (length $L/2$) has an elongation equal to $\delta/2$, and one-fourth of the bar has an elongation equal to $\delta/4$.

In general, the elongation of a segment is equal to its length divided by the total length L and multiplied by the total elongation δ. Therefore, a unit length of the bar has an elongation equal to $1/L \times \delta$. This quantity is called the *elongation per unit length*, or **strain**, and is denoted by the Greek letter ε (epsilon). Strain is given by

$$\varepsilon = \frac{\delta}{L} \tag{1-7}$$

If the bar is in tension, the strain is called a **tensile strain**, representing an elongation or stretching of the material. If the bar is in compression, the strain is a **compressive strain** and the bar shortens. Tensile strain is usually taken as positive and compressive strain as negative. The strain ε is called a **normal strain** because it is associated with normal stresses.

Because normal strain is the ratio of two lengths, it is a **dimensionless quantity**, that is, it has no units. Therefore, strain is expressed simply as a number that is independent of any system of units. Numerical values of strain are usually very small, because bars made of structural materials undergo only small changes in length when loaded.

As an example, consider a steel bar having length L equal to 2.0 m. When heavily loaded in tension, this bar might elongate by 1.4 mm, which means that the strain is

$$\varepsilon = \frac{\delta}{L} = \frac{1.4 \text{ mm}}{2.0 \text{ m}} = 0.0007 = 700 \times 10^{-6}$$

In practice, the original units of δ and L are sometimes attached to the strain itself, and then the strain is recorded in forms such as mm/m or μm/m. For instance, the strain ε in the preceding example could be given as 700 μm/m. Strain is sometimes expressed as a percent, especially when the strains are large. (In the preceding example, the strain is 0.07%.)

Uniaxial Stress and Strain

The definitions of normal stress and normal strain are based upon purely static and geometric considerations, which means that Eqs. (1-6) and (1-7) can be used for loads of any magnitude and for any material. The principal requirement is that the deformation of the bar be uniform throughout its volume, which in turn requires that the bar be prismatic, the loads act through the centroids of the cross sections, and the material be **homogeneous** (that is, the same throughout all parts of the bar). The resulting state of stress and strain is called **uniaxial stress and strain** (although lateral strain exists, as discussed later in Section 1.7).

Line of Action of the Axial Forces for a Uniform Stress Distribution

Throughout the preceding discussion of stress and strain in a prismatic bar, the normal stress σ was assumed to be distributed uniformly over the cross section. Note that this condition is met if the line of action of the axial forces is through the centroid of the cross-sectional area.

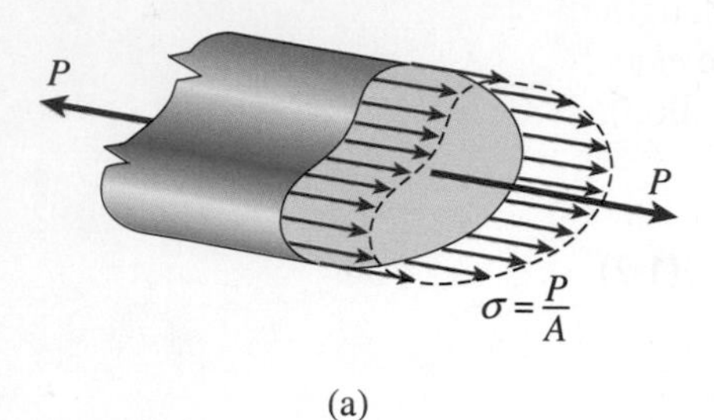

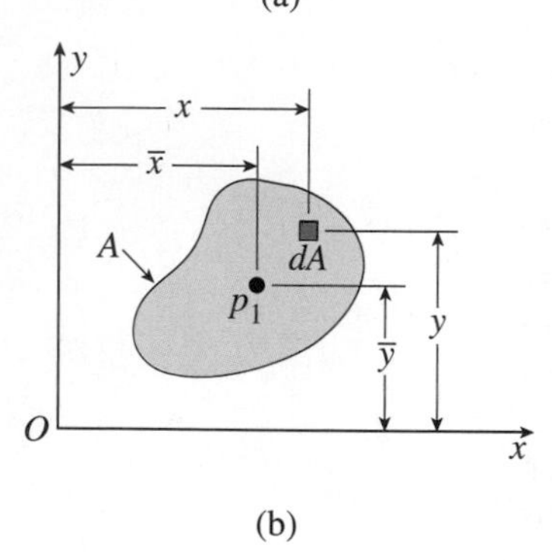

FIGURE 1-22

Uniform stress distribution in a prismatic bar: (a) axial forces P, and (b) cross section of the bar

Consider a prismatic bar of arbitrary cross-sectional shape subjected to axial forces P that produce uniformly distributed stresses σ (Fig. 1-22a). Let p_1 represent the point in the cross section where the line of action of the forces intersects the cross section (Fig. 1-22b). Construct a set of x-y axes in the plane of the cross section and denote the coordinates of point p_1 by $\overline{x}$ and $\overline{y}$. To determine these coordinates, observe that the moments M_x and M_y of the force P about the x and y axes, respectively, must be equal to the corresponding moments of the uniformly distributed stresses.

The moments of the force P are

$$M_x = P\overline{y} \quad M_y = -P\overline{x} \tag{1-8a,b}$$

in which a moment is considered positive when its vector (using the right-hand rule) acts in the positive direction of the corresponding axis.

The moments of the distributed stresses are obtained by integrating over the cross-sectional area A. The differential force acting on an element of area dA (Fig. 1-22b) is equal to σdA. The moments of this elemental force about the x and y axes are $\sigma y dA$ and $-\sigma x dA$, respectively, in which x and y denote the coordinates of the element dA. The total moments are obtained by integrating over the cross-sectional area:

$$M_x = \int \sigma y dA \quad M_y = -\int \sigma x dA \tag{1-8c,d}$$

These expressions give the moments produced by the stresses σ.

Next, equate the moments M_x and M_y obtained from the force P (Eqs. 1-8a and b) to the moments obtained from the distributed stresses (Eqs. 1-8c and d):

$$P\overline{y} = \int \sigma y dA \quad P\overline{x} = -\int \sigma x dA$$

Because the stresses σ are uniformly distributed, they are constant over the cross-sectional area A and can be placed outside the integral signs. Also, σ is equal to P/A. Therefore, you can obtain the following formulas for the coordinates of point p_1:

$$\overline{y} = \frac{\int y dA}{A} \quad \overline{x} = \frac{\int x dA}{A} \tag{1-9a,b}$$

These equations are the same as the equations defining the coordinates of the centroid of an area (see Eqs. D-3a, b in Appendix D). Therefore, the important conclusion here is:

In order to have uniform tension or compression in a prismatic bar, the axial force must act through the centroid of the cross-sectional area.

Always assume that these conditions are met unless it is specifically stated otherwise.

The following examples illustrate the calculation of stresses and strains in prismatic bars. In the first example, disregard the weight of the bar and in the second, include it. (It is customary when solving textbook problems to omit the weight of the structure unless specifically instructed to include it.)

Example 1-5

An elastic spring rests on a base plate that is on top of rigid tube B (see Fig. 1-23). The spring is enclosed by rigid tube A but is longer than tube A by an amount s. Force P is then applied to a cap plate to compress the spring. Both tubes have outer diameter d_O, but the inner diameters are d_A and d_B for tubes A and B, respectively. Assume that spring stiffness $k = 4200$ kN/m, $d_O = 75$ mm, $d_A = 62$ mm, $d_B = 57$ mm, and $s = 3.2$ mm.

(a) If applied load $P = 11$ kN, what are the axial normal stresses in tubes A and B?

(b) Repeat part (a) if $P = 22$ kN.

(c) What is P if the normal stress in tube A is 5.5 MPa? What is the associated stress in tube B?

FIGURE 1-23

Elastic spring inside rigid tubes

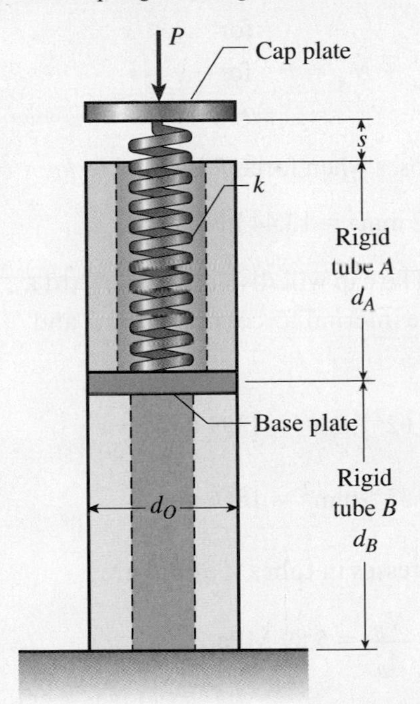

FIGURE 1-24

(a, b) Free-body diagrams ($x < s$, $x = s$)

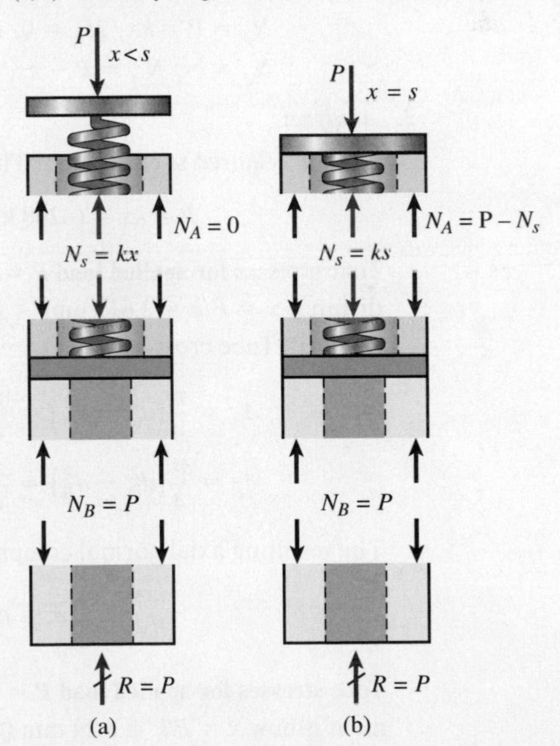

Solution:

Use the following four-step problem-solving approach.

1. ***Conceptualize***: The two possible states of the assemblage are shown in the free-body diagrams in Fig. 1-24. In Fig. 1-24a, an upper-section cut through both the spring and tube A creates an upper free-body diagram that reveals a spring force $(k)(x)$ for the case of downward cap displacement x that is less than gap width s. In Fig. 1-24b, cap displacement x is equal to gap width s, so the spring

force now equals $(k)(s)$. Figure 1-24 also shows lower free-body diagrams for both cases in which a section cut through tube B shows that the internal compressive force in tube B is equal to applied load P. (Internal forces in tubes A and B are shown as two arrows, one at each tube wall, indicating that N_A and N_B are actually uniformly distributed forces acting on the circular cross section of each tube.)

2. ***Categorize***: The force P required to close gap s is $(k)(s)$. This is also the maximum force that can be developed in the spring. If applied force P is too small to close the gap s, force P will be transferred to the base plate and into rigid tube B; tube A will be unaffected by the load. However, if force P is large enough to compress the spring to close the gap s, the spring and tube A will share the load P applied to the cap plate and together will transfer it to tube B through the base plate. In summary, the free-body diagrams in Fig. 1-24 show that, if the spring is compressed by load P an amount x, the compressive internal forces in the spring and the two tubes are

$$N_s = P = kx,\ N_A = 0,\ N_B = P \qquad \text{for} \quad x < s$$
$$N_s = ks,\ N_A = P - ks,\ N_B = N_s + N_A = P \qquad \text{for} \quad x = s$$

3. ***Analyze***:

Force P required to close gap s: The gap closes when force P is equal to ks

$$P = ks = (4200\ \text{kN/m})(3.2\ \text{mm}) = 13.44\ \text{kN}$$

Tube stresses for applied load P = 11 kN: The cap will displace downward a distance $x = P/k = 2.619$ mm ($< s$), so tube internal forces are $N_A = 0$ and $N_B = P$. Tube cross-sectional areas are

$$A_A = \frac{\pi}{4}\left(d_O^2 - d_A^2\right) = \frac{\pi}{4}\left(75^2 - 62^2\right)\text{mm}^2 = 1399\ \text{mm}^2$$

$$A_B = \frac{\pi}{4}\left(d_O^2 - d_B^2\right) = \frac{\pi}{4}\left(75^2 - 57^2\right)\text{mm}^2 = 1866\ \text{mm}^2$$

The resulting axial normal compressive stresses in tubes A and B are

$$\sigma_A = \frac{N_A}{A_A} = 0 \quad \sigma_B = \frac{N_B}{A_B} = 5.89\ \text{MPa}$$

Tube stresses for applied load P = 22 kN: Cap downward displacement is now $x = P/k = 5.24$ mm ($> s$), so tube internal forces are $N_A = P - (k)(s) = (22 - 13.44)\ \text{kN} = 8.56\ \text{kN}$ and $N_B = P$. The normal stresses in tubes A and B are now:

$$\sigma_A = \frac{8.56\ \text{kN}}{1399\ \text{mm}^2} = 6.12\ \text{MPa} \quad \sigma_B = \frac{22\ \text{kN}}{1866\ \text{mm}^2} = 11.79\ \text{MPa}$$

Applied load P if stress in tube A is 5.5 MPa: Force P must exceed $(k)(s) = 13.44$ kN for the gap to close, leading to a force in tube A and a normal stress of $\sigma_A = 5.5$ MPa. The normal compressive force in tube

A is $N_A = (\sigma_A)(A_A) = 7.69$ kN. It follows that applied force P is now $P = N_A + ks = 7.69 \text{ kN} + 13.44 \text{ kN} = 21.1$ kN. Internal force $N_B = P$, so the normal compressive stress in tube B is now $\sigma_B = \dfrac{21.1 \text{ kN}}{1866 \text{ mm}^2} = 11.33$ MPa. ⬅

4. ***Finalize***: If tube A is elastic instead of rigid as assumed here, tube A can be modeled as another spring that is parallel to the spring it encloses. Now a more advanced analysis procedure will be needed to find tube force N_A for the case of $P > (k)(s)$. Force N_A is no longer equal to $P - (k)(s)$, and downward displacement x can be larger than s.

Example 1-6

An antenna and receiver are suspended on a steel wire from a helicopter to measure the effects of wind turbines on a local radar installation (see Fig. 1-25). Obtain a formula for the maximum stress in the wire, taking into account the weight of the wire itself. Calculate the maximum stress in the wire in MPa using the following numerical properties: $L_1 = 6$ m, $L_2 = 5$ m, $d = 9.5$ mm; antenna weight is $W_1 = 380$ N; receiver weight is $W_2 = 700$ N. Note that the weight density γ of steel is 77.0 kN/m^3 (from Table I-1 in Appendix I).

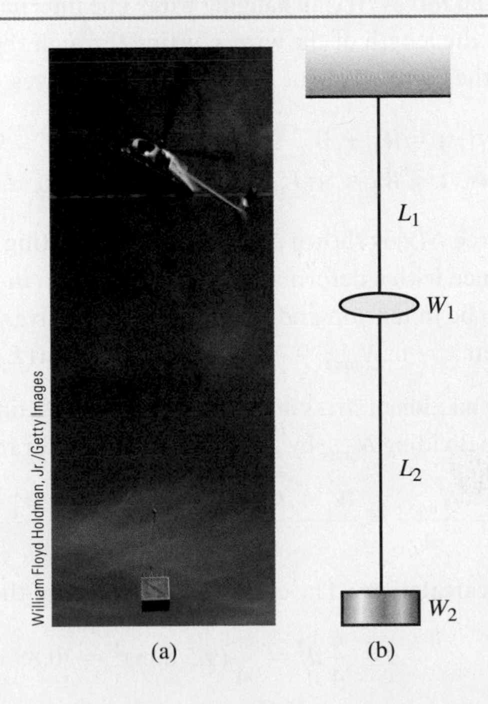

FIGURE 1-25
(a, b) Instruments suspended on wire from helicopter

FIGURE 1-26
Free-body diagram

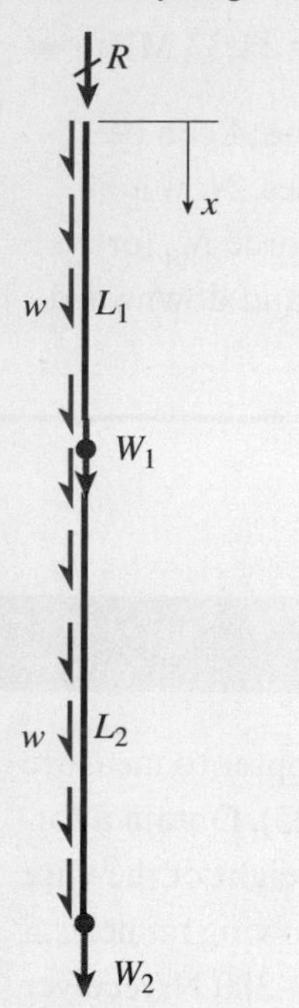

Solution:

Use the following four-step problem-solving approach.

1. ***Conceptualize***: A free-body diagram of the suspended instrument package is shown in Fig. 1-26. The antenna (W_1) and receiver (W_2) weights are concentrated forces at specified locations along the wire; the weight of the wire is a uniformly distributed axial force expressed as $w(x) = \gamma A$, where A is the cross-sectional area of the wire. Cutting the wire at some point x leads to upper and lower free-body diagrams (Fig. 1-27); either can be used to find the internal axial force $N(x)$ at the location of the cut section. The internal axial force in the wire is a maximum at the point at which it is attached to the helicopter ($x = 0$).
2. ***Categorize***: Start by solving for the reaction force R at the top of the wire and then cut the wire a short distance below the support to find N_{max}. The wire is prismatic, so simply divide N_{max} by cross-sectional area A to find the maximum axial normal stress σ_{max}.
3. ***Analyze***:

Reaction force *R*: Use the free-body diagram in Fig. 1-26 to obtain

$$R = -[W_1 + W_2 + w(L_1 + L_2)] = -[W_1 + W_2 + \gamma A(L_1 + L_2)]$$

The minus sign indicates that reaction force R acts in the $(-x)$ direction, or upward in Figs. 1-26 and 1-27.

FIGURE 1-27
Axial force $N(x)$ in wire

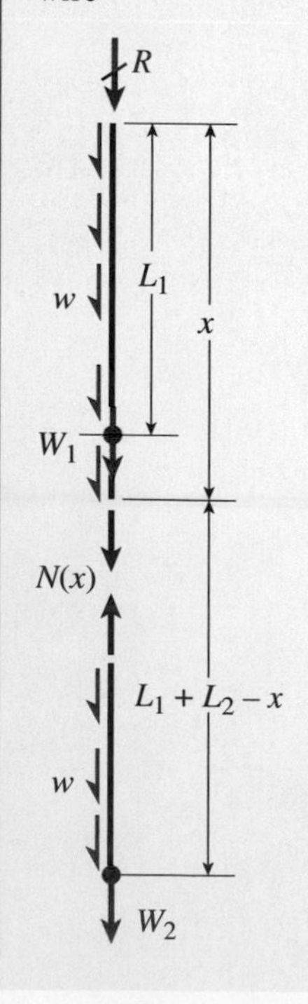

Internal axial forces *N*(*x*) in hanging wire: The internal axial force in the wire varies over the length of the wire. Cutting through the wire in upper and lower segments (the lower segment is cut in Fig. 1-27) gives

$$N(x) = W_1 + W_2 + w(L_1 + L_2 - x) \quad 0 \le x \le L_1$$
$$N(x) = W_2 + w(L_1 + L_2 - x) \quad L_1 \le x \le L_1 + L_2$$

Internal force $N(x)$ is shown as a pair of forces acting away from the cut section in accordance with a deformation sign convention in which the wire is initially assumed to be in tension and that tension is positive. The maximum force in the wire is at $x = 0$: $N_{max} = N(0) = W_1 + W_2 + w(L_1 + L_2)$.

Formula for maximum stress in the wire: The cross-sectional area A of the wire is constant, so dividing N_{max} by A leads to a formula for maximum stress in the wire:

$$\sigma_{max} = \frac{N_{max}}{A} = \frac{W_1 + W_2 + w(L_1 + L_2)}{A} = \frac{W_1 + W_2}{A} + \gamma(L_1 + L_2)$$

Numerical calculations: The cross-sectional area of the wire is

$$A = \frac{\pi}{4}d^2 = \frac{\pi}{4}(9.5 \text{ mm})^2 = 70.88 \text{ mm}^2$$

Solving for maximum normal stress gives

$$\sigma_{max} = \frac{W_1 + W_2}{A} + \gamma(L_1 + L_2) = \frac{380 \text{ N} + 700 \text{ N}}{70.88 \text{ mm}^2} + 77.0\frac{\text{kN}}{\text{m}^3}(6 \text{ m} + 5 \text{ m}) = 16.08 \text{ MPa}$$

4. ***Finalize***: If the weight of the wire is ignored, the maximum normal stress is reduced to 15.24 MPa, which is a decrease of more than 5%. Although the stresses are low here, eliminating the self-weight of the wire from the stress calculation is not recommended.

1.5 Mechanical Properties of Materials

The design of machines and structures so that they function properly requires an understanding of the **mechanical behavior** of the materials being used. Ordinarily, the only way to determine how materials behave when they are subjected to loads is to perform experiments in the laboratory. The usual procedure is to place small specimens of the material in testing machines, apply the loads, and then measure the resulting deformations (such as changes in length and changes in diameter).

A typical **tensile-test machine** is shown in Fig. 1-28. The test specimen is installed between the two large grips of the testing machine and then loaded in tension. Measuring devices record the deformations, and the automatic control and data-processing systems tabulate and graph the results.

FIGURE 1-28

Tensile-test machine with automatic data-processing system (Courtesy of MTS Systems Corporation)

A more detailed view of a **tensile-test specimen** is shown in Fig. 1-29. The ends of the circular specimen are enlarged where they fit in the grips so that failure will not occur near the grips themselves. A failure at the ends would not produce the desired information about the material because the stress distribution near the grips is not uniform, as explained in Section 1.4. In a properly designed specimen, failure occurs in the prismatic portion of the specimen where the stress distribution is uniform and the bar is subjected only to pure tension. This situation is shown in Fig. 1-29, where the steel specimen has just fractured under load. The device at the right, which is attached by two arms to the specimen, is an **extensometer** that measures the elongation during loading.

In order that test results stay comparable, the dimensions of test specimens and the methods of applying loads must be standardized. One of the major standards organizations in the United States is the American Society for Testing and Materials (ASTM), which is a technical society that publishes specifications and standards for materials and testing. Other standardizing organizations include the American Standards Association (ASA), the National Institute of Standards and Technology (NIST), and Bureau International des Poids et Mesures (BIPM). Similar organizations exist in other countries.

The ASTM standard tension specimen has a diameter of 12.8 mm and a **gage length** of 50.8 mm between the gage marks, which are the points where the extensometer arms are attached to the specimen (see Fig. 1-29). As the

FIGURE 1-29

Typical tensile-test specimen with extensometer attached (Courtesy of MTS Systems Corporation)

specimen is pulled, the axial load is measured and recorded, either automatically or by reading from a dial. The elongation over the gage length is measured simultaneously, either by mechanical gages of the kind shown in Fig. 1-29 or by electrical-resistance strain gages.

In a **static test**, the load is applied slowly and the precise *rate* of loading is not of interest because it does not affect the behavior of the specimen. However, in a **dynamic test**, the load is applied rapidly and sometimes in a cyclical manner. Since the nature of a dynamic load affects the properties of the materials, the rate of loading must be measured.

Compression tests of metals are customarily made on small specimens in the shape of cubes or circular cylinders. For instance, cubes may be 50 mm on a side, and cylinders may have diameters of 25 mm and lengths from 25 to 300 mm. Both the load applied by the machine and the shortening of the specimen may be measured. The shortening should be measured over a gage length that is less than the total length of the specimen in order to eliminate end effects.

Concrete is tested in compression on important construction projects to ensure that the required strength has been obtained. One type of concrete test specimen is 152 mm in diameter, 305 mm in length, and 28 days old (the age of concrete is important because concrete gains strength as it cures). Similar but somewhat smaller specimens are used when performing compression tests of rock (see Fig. 1-30).

Stress-Strain Diagrams

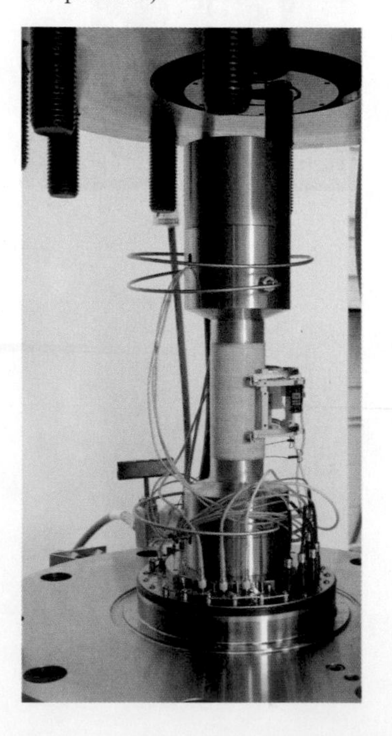

FIGURE 1-30

Rock sample being tested in compression to obtain compressive strength, elastic modulus and Poisson's ratio (Courtesy of MTS Systems Corporation)

Test results generally depend upon the dimensions of the specimen being tested. Since it is unlikely that you will design a structure having parts that are the same size as the test specimens, you need to express the test results in a form that can be applied to members of any size. A simple way to achieve this objective is to convert the test results to stresses and strains.

The axial stress σ in a test specimen is calculated by dividing the axial load P by the cross-sectional area A (Eq. 1-6). When the initial area of the specimen is used in the calculation, the stress is called the **nominal stress** (other names are *conventional stress* and *engineering stress*). A more exact value of the axial stress, called the **true stress**, can be calculated by using the actual area of the bar at the cross section where failure occurs. Since the actual area in a tension test is always less than the initial area (see Fig. 1-29), the true stress is larger than the nominal stress.

The average axial strain ε in the test specimen is found by dividing the measured elongation δ between the gage marks by the gage length L (see Fig. 1-29 and Eq. 1-7). If the initial gage length is used in the calculation (for instance, 50 mm), the **nominal strain** is obtained. Since the distance between the gage marks increases as the tensile load is applied, you can calculate the **true strain** (or *natural strain*) at any value of the load by using the actual distance between the gage marks. In tension, true strain is always smaller than nominal strain. However, for most engineering purposes, nominal stress and nominal strain are adequate.

After performing a tension or compression test and determining the stress and strain at various magnitudes of the load, you can plot a diagram of stress versus strain. Such a **stress-strain diagram** is a characteristic of the particular material being tested and conveys important information about the mechanical properties and type of behavior.[1]

[1]Stress-strain diagrams were originated by Jacob Bernoulli (1654–1705) and J. V. Poncelet (1788–1867) (Ref. 1-4).

Stresses and Strains for Structural Steel **Structural steel**, also known as *mild steel* or *low-carbon steel*, is one of the most widely used metals found in buildings, bridges, cranes, ships, towers, vehicles, and many other types of construction. A stress-strain diagram for a typical structural steel in tension is shown in Fig. 1-31. Strains are plotted on the horizontal axis and stresses on the vertical axis.

The diagram begins with a straight line from the origin O to point A, showing that the relationship between stress and strain in this initial region is not only *linear* but also *proportional*.[2] Beyond point A, the proportionality between stress and strain no longer exists; hence the stress at A is called the **proportional limit**. For low-carbon steels, this limit is in the range 210 to 350 MPa, but high-strength steels (with higher carbon content plus other alloys) can have proportional limits of more than 550 MPa. The slope of the straight line from O to A is called the **modulus of elasticity**. Because the slope has units of stress divided by strain, modulus of elasticity has the same units as stress. (See Section 1.6.)

With an increase in stress beyond the proportional limit, the strain begins to increase more rapidly for each increment in stress. Consequently, the stress-strain curve has a smaller and smaller slope, until, at point B, the curve becomes horizontal (see Fig. 1-31). Beginning at this point, considerable elongation of the test specimen occurs with no noticeable increase in the tensile force (from B to C). This phenomenon is known as **yielding** of the material, and point B is called the **yield point**. The corresponding stress is known as the **yield stress** of the steel.

In the region from B to C (see Fig. 1-31), the material becomes **perfectly plastic**, which means that it deforms without an increase in the applied load. The elongation of a mild-steel specimen in the perfectly plastic region is typically 10 to 15 times the elongation that occurs in the linear region (between the onset of loading and the proportional limit). The presence of very large

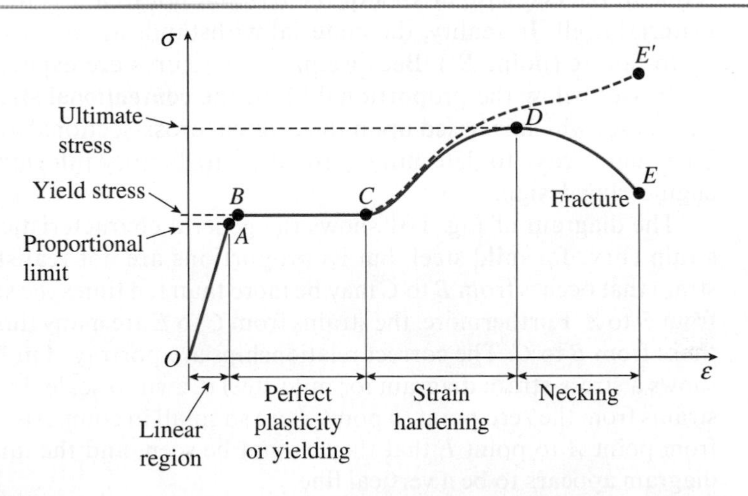

FIGURE 1-31
Stress-strain diagram for a typical structure steel in tension (not to scale)

[2]Two variables are said to be *proportional* if their ratio remains constant. Therefore, a proportional relationship may be represented by a straight line through the origin. Although a proportional relationship is linear, the converse is not necessarily true, because a relationship represented by a straight line that does *not* pass through the origin is linear but not proportional. "Directly proportional" is synonymous with "proportional" (Ref. 1-5).

FIGURE 1-32

Necking of a mild-steel bar in tension (© Barry Goodno)

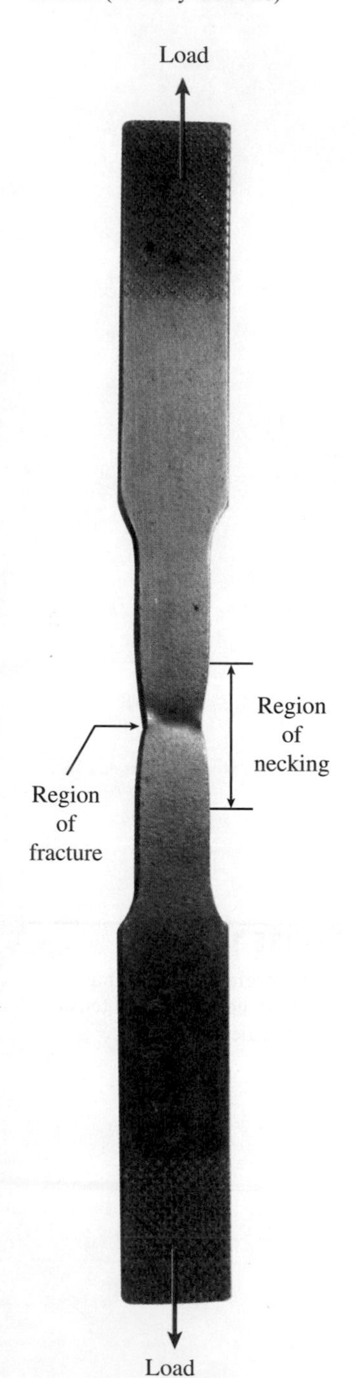

strains in the plastic region (and beyond) is the reason for not plotting this diagram to scale.

After undergoing the large strains that occur during yielding in the region BC, the steel begins to **strain harden**. During strain hardening, the material undergoes changes in its crystalline structure, resulting in increased resistance of the material to further deformation. Elongation of the test specimen in this region requires an increase in the tensile load, and therefore the stress-strain diagram has a positive slope from C to D. The load eventually reaches its maximum value, and the corresponding stress (at point D) is called the **ultimate stress**. Further stretching of the bar is actually accompanied by a reduction in the load, and fracture finally occurs at a point such as E in Fig. 1-31.

The yield stress and ultimate stress of a material are also called the **yield strength** and **ultimate strength**, respectively. **Strength** is a general term that refers to the capacity of a structure to resist loads. For instance, the yield strength of a beam is the magnitude of the load required to cause yielding in the beam, and the ultimate strength of a truss is the maximum load it can support, that is, the failure load. However, when conducting a tension test of a particular material, define load-carrying capacity by the stresses in the specimen rather than by the total loads acting on the specimen. As a result, the strength of a material is usually stated as a stress.

When a test specimen is stretched, **lateral contraction** occurs. The resulting decrease in cross-sectional area is too small to have a noticeable effect on the calculated values of the stresses up to about point C in Fig. 1-31, but beyond that point the reduction in area begins to alter the shape of the curve. In the vicinity of the ultimate stress, the reduction in area of the bar becomes clearly visible and a pronounced **necking** of the bar occurs (see Fig. 1-32).

If the actual cross-sectional area at the narrow part of the neck is used to calculate the stress, the **true stress-strain curve** (the dashed line CE' in Fig. 1-31) is obtained. The total load the bar can carry does indeed diminish after the ultimate stress is reached (as shown by curve DE), but this reduction is due to the decrease in area of the bar and not to a loss in strength of the material itself. In reality, the material withstands an increase in true stress up to failure (point E'). Because most structures are expected to function at stresses below the proportional limit, the **conventional stress-strain curve** $OABCDE$, which is based upon the original cross-sectional area of the specimen and is easy to determine, provides satisfactory information for use in engineering design.

The diagram of Fig. 1-31 shows the general characteristics of the stress-strain curve for mild steel, but its proportions are not realistic because the strain that occurs from B to C may be more than ten times the strain occurring from O to A. Furthermore, the strains from C to E are many times greater than those from B to C. The correct relationships are portrayed in Fig. 1-33, which shows a stress-strain diagram for mild steel drawn to scale. In this figure, the strains from the zero point to point A are so small in comparison to the strains from point A to point E that they cannot be seen, and the initial part of the diagram appears to be a vertical line.

Ductility The presence of a clearly defined yield point followed by large plastic strains is an important characteristic of structural steel that is sometimes utilized in practical design (see discussions of elastoplastic behavior in Sections 2.12

and 6.10). Metals such as structural steel that undergo large *permanent* strains before failure are classified as **ductile**. Ductility is the property that enables a bar of steel to be bent into a circular arc or drawn into a wire without breaking. A desirable feature of ductile materials is that visible distortions occur if the loads become too large, thus providing an opportunity to take remedial action before an actual fracture occurs. Also, materials exhibiting ductile behavior are capable of absorbing large amounts of strain energy prior to fracture.

Structural steel is an alloy of iron containing about 0.2% carbon, and therefore it is classified as a low-carbon steel. With increasing carbon content, steel becomes less ductile but stronger (higher yield stress and higher ultimate stress). The physical properties of steel are also affected by heat treatment, the presence of other metals, and manufacturing processes such as rolling. Other materials that behave in a ductile manner (under certain conditions) include aluminum, copper, magnesium, lead, molybdenum, nickel, brass, bronze, monel metal, nylon, and teflon.

FIGURE 1-33

Stress-strain diagram for a typical structural steel in tension (drawn to scale)

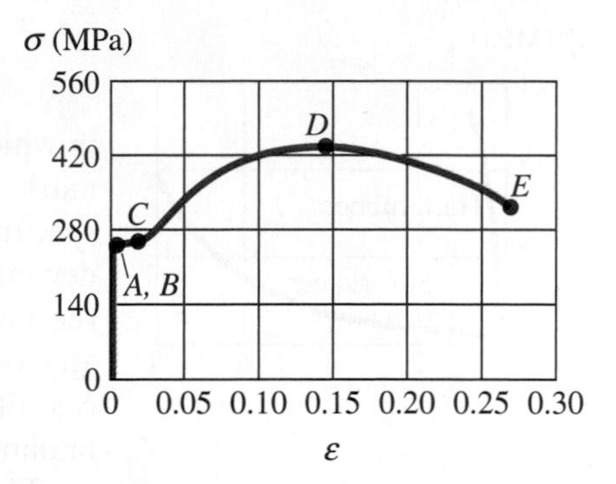

FIGURE 1-34

Typical stress-strain diagram for an aluminum alloy

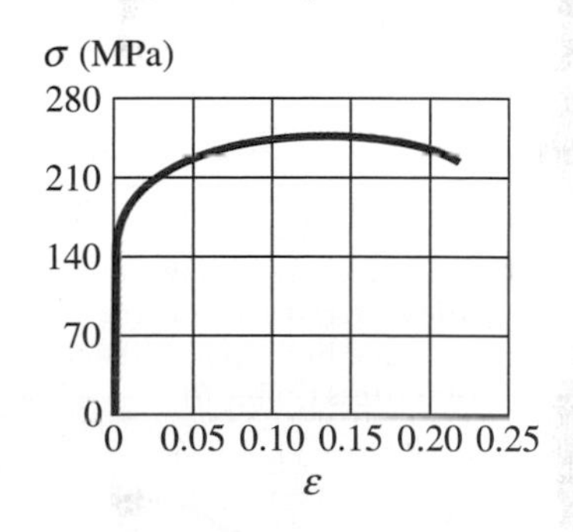

FIGURE 1-35

Arbitrary yields stress determined by the offset method

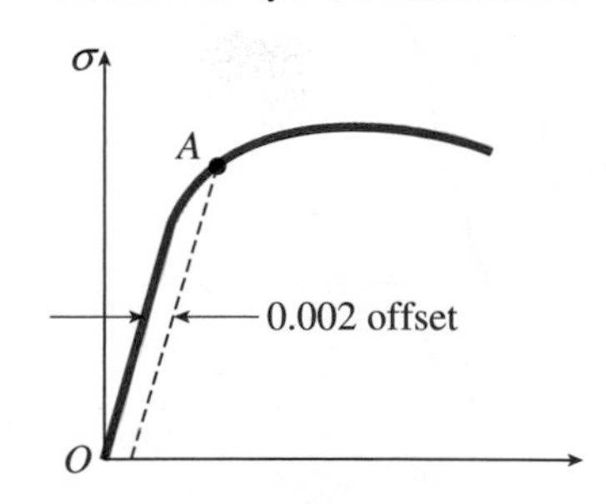

Aluminum Alloys Although they have considerable ductility, **aluminum alloys** typically do not have a clearly definable yield point, as shown by the stress-strain diagram of Fig. 1-34. However, they do have an initial linear region with a recognizable proportional limit. Alloys produced for structural purposes have proportional limits in the range 70 to 410 MPa and ultimate stresses in the range 140 to 550 MPa.

When a material such as aluminum does not have an obvious yield point yet undergoes large strains after the proportional limit is exceeded, an *arbitrary* yield stress may be determined by the **offset method**. A straight line is drawn on the stress-strain diagram parallel to the initial linear part of the curve (Fig. 1-35) but offset by some standard strain, such as 0.002 (or 0.2%). The intersection of the offset line and the stress-strain curve (point A in the figure) defines the yield stress. Because this stress is determined by an arbitrary rule and is not an inherent physical property of the material, it should be distinguished from a true yield stress by referring to it as the **offset yield stress**. For a material such as aluminum, the offset yield stress is slightly above the proportional limit. In the case of structural steel, with its abrupt transition from the linear region to the region of plastic stretching, the offset stress is essentially the same as both the yield stress and the proportional limit.

Rubber Rubber maintains a linear relationship between stress and strain up to relatively large strains (as compared to metals). The strain at the proportional limit may be as high as 0.1 or 0.2 (10 or 20%). Beyond the proportional limit, the behavior depends upon the type of rubber (Fig. 1-36). Some kinds of soft rubber stretch enormously without failure, reaching lengths several times their original lengths. The material eventually offers increasing resistance to the load, and the stress-strain curve turns markedly upward. You can easily sense this characteristic behavior by stretching a rubber band with your hands. (Note that although rubber exhibits very large strains, it is not a ductile material because the strains are not permanent. It is, of course, an elastic material; see Section 1.6.)

FIGURE 1-36

Stress-strain curves for two kinds of rubber in tension

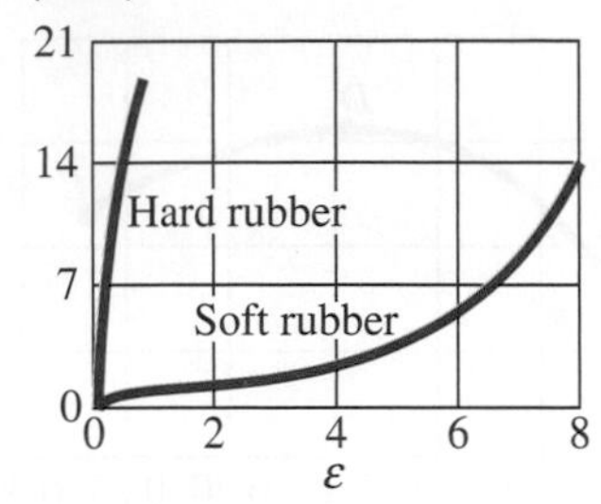

Ductility and Elongation The ductility of a material in tension can be characterized by its elongation and by the reduction in area at the cross section where fracture occurs. The **percent elongation** is defined as

$$\text{Percent elongation} = \frac{L_1 - L_0}{L_0}(100) \qquad \textbf{(1-10)}$$

in which L_0 is the original gage length and L_1 is the distance between the gage marks at fracture. Because the elongation is not uniform over the length of the specimen but is concentrated in the region of necking, the percent elongation depends upon the gage length. Therefore, when stating the percent elongation, the gage length should always be given. For a 50 mm gage length, steel may have an elongation in the range from 3 to 40%, depending upon composition; in the case of structural steel, values of 20 or 30% are common. The elongation of aluminum alloys varies from 1 to 45%, depending upon composition and treatment.

The **percent reduction in area** measures the amount of necking that occurs and is defined as

$$\text{Percent reduction} = \frac{A_0 - A_1}{A_0}(100) \qquad \textbf{(1-11)}$$

in which A_0 is the original cross-sectional area and A_1 is the final area at the fracture section. For ductile steels, the reduction is about 50%.

FIGURE 1-37

Typical stress-strain diagram for a brittle material showing the proportional limit (point A) and fracture stress (point B)

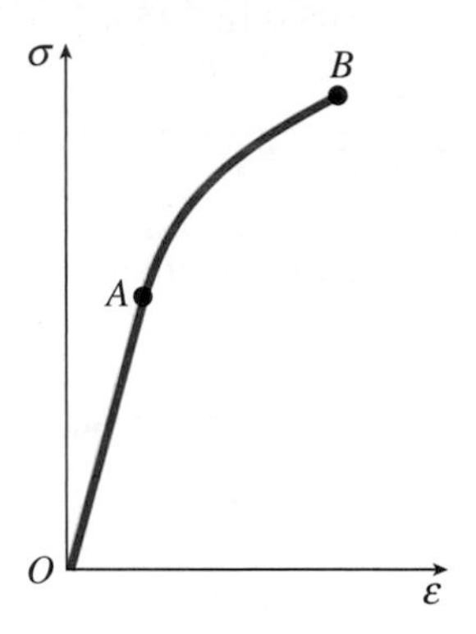

Brittle Materials Materials that fail in tension at relatively low values of strain are classified as **brittle**. Examples are concrete, stone, cast iron, glass, ceramics, and a variety of metallic alloys. Brittle materials fail with only little elongation after the proportional limit (the stress at point A in Fig. 1-37) is exceeded. Furthermore, the reduction in area is insignificant, and so the nominal fracture stress (point B) is the same as the true ultimate stress. High-carbon steels have very high yield stresses—over 700 MPa in some cases—but they behave in a brittle manner and fracture occurs at an elongation of only a few percent.

Ordinary glass is a nearly ideal brittle material because it exhibits almost no ductility. The stress-strain curve for glass in tension is essentially a straight line, with failure occurring before any yielding takes place. The ultimate stress is about 70 MPa for certain kinds of plate glass, but great variations exist, depending upon the type of glass, the size of the specimen, and the presence of microscopic defects. **Glass fibers** can develop enormous strengths, and ultimate stresses over 7 GPa have been attained.

Plastics Many types of **plastics** are used for structural purposes because of their light weight, resistance to corrosion, and good electrical insulation properties. Their mechanical properties vary tremendously, with some plastics being brittle and others ductile. When designing with plastics, it is important to realize that their properties are greatly affected by both temperature changes and the passage of time. For instance, the ultimate tensile stress of some plastics is cut in half merely by raising the temperature from 10°C to 50°C. Also, a loaded plastic may stretch gradually over time until it is no longer serviceable. For example, a bar of polyvinyl chloride subjected to a tensile load that initially produces a strain of 0.005 may have that strain doubled after one week, even though the load remains constant. (This phenomenon, known as *creep*, is discussed in the next section.)

Ultimate tensile stresses for plastics are generally in the range 14 to 350 MPa and weight densities vary from 8 to 14 kN/m^3. One type of nylon has an ultimate stress of 80 MPa and weighs only 11 kN/m^3, which is only 12% heavier than water. Because of its light weight, the strength-to-weight ratio for nylon is about the same as for structural steel (see Prob. 1.5-4).

Composites A **filament-reinforced material** consists of a base material (or *matrix*) in which high-strength filaments, fibers, or whiskers are embedded. The resulting composite material has much greater strength than the base material. As an example, the use of glass fibers can more than double the strength of a plastic matrix. Composites are widely used in aircraft, boats, rockets, and space vehicles where high strength and light weight are needed.

Compression

Stress-strain curves for materials in compression differ from those in tension. Ductile metals such as steel, aluminum, and copper have proportional limits in compression very close to those in tension, and the initial regions of their compressive and tensile stress-strain diagrams are about the same. However, after yielding begins, the behavior is quite different. In a tension test, the specimen is stretched, necking may occur, and fracture ultimately takes place. When the material is compressed, it bulges outward on the sides and becomes barrel shaped, because friction between the specimen and the end plates prevents lateral expansion. With increasing load, the specimen is flattened out and offers greatly increased resistance to further shortening (which means that the stress-strain curve becomes very steep). These characteristics are illustrated in Fig. 1-38, which shows a compressive stress-strain diagram for copper. Since the actual cross-sectional area of a specimen tested in compression is larger than the initial area, the true stress in a compression test is smaller than the nominal stress.

FIGURE 1-38

Stress-strain diagrams for copper in compression

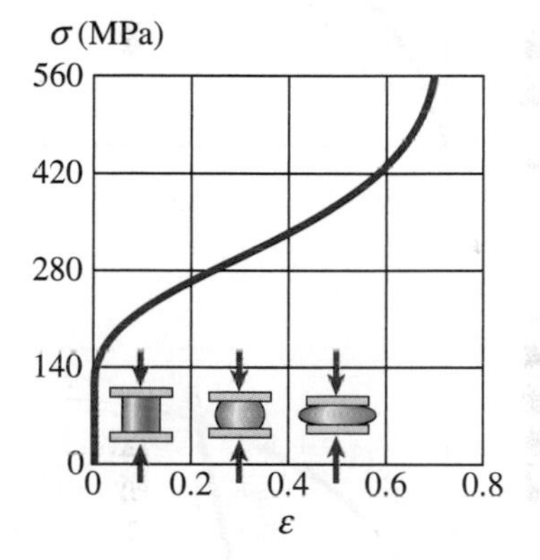

Brittle materials loaded in compression typically have an initial linear region followed by a region in which the shortening increases at a slightly higher rate than does the load. The stress-strain curves for compression and tension often have similar shapes, but the ultimate stresses in compression are much higher than those in tension. Also, unlike ductile materials, which flatten out when compressed, brittle materials actually break at the maximum load.

Tables of Mechanical Properties

Properties of materials are listed in the tables of Appendix I at the back of the book. The data in the tables are typical of the materials and are suitable for solving problems in this book. However, properties of materials and stress-strain curves vary greatly—even for the same material—because of different manufacturing processes, chemical composition, internal defects, temperature, and many other factors.

For these reasons, data obtained from Appendix I (or other tables of a similar nature) should not be used for specific engineering or design purposes. Instead, the manufacturers or materials suppliers should be consulted for information about a particular product.

1.6 Elasticity, Plasticity, and Creep

Stress-strain diagrams portray the behavior of engineering materials when the materials are loaded in tension or compression, as described in the preceding section. Now consider what happens when the load is removed and the material is *unloaded.*

Assume, for instance, that you apply a load to a tensile specimen so that the stress and strain go from the origin O to point A on the stress-strain curve of Fig. 1-39a. Suppose that when the load is removed, the material follows exactly the same curve back to the origin O. This property of a material, by which it returns to its original dimensions during unloading, is called **elasticity**, and the material itself is said to be *elastic. Note that the stress-strain curve from O to A need not be linear in order for the material to be elastic.*

FIGURE 1-39

Stress-strain diagrams illustrating (a) elastic behavior, and (b) partially elastic behavior

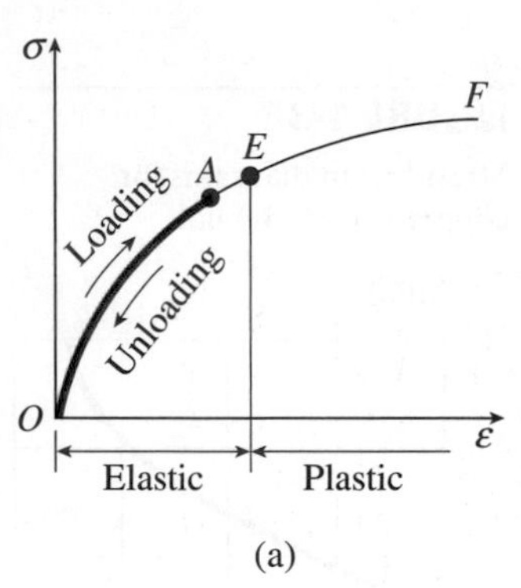

(a)

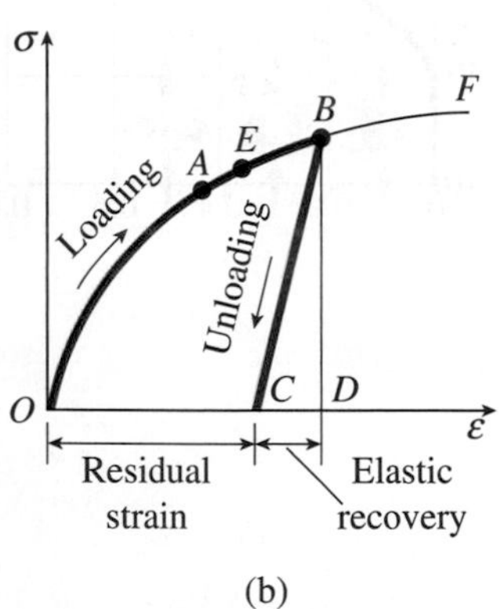

(b)

Now suppose that you load this same material to a higher level, so that point B is reached on the stress-strain curve (Fig. 1-39b). When unloading occurs from point B, the material follows line BC on the diagram. This unloading line is parallel to the initial portion of the loading curve; that is, line BC is parallel to a tangent to the stress-strain curve at the origin. When point C is reached, the load has been entirely removed, but a **residual strain**, or *permanent strain*, which is represented by line OC, remains in the material. As a consequence, the bar being tested is longer than it was before loading. This residual elongation of the bar is called the **permanent set**. Of the total strain OD developed during loading from O to B, the strain CD has been recovered elastically and the strain OC remains as a permanent strain. Thus, during unloading the bar returns partially to its original shape, and the material is said to be **partially elastic**.

Between points A and B on the stress-strain curve (Fig. 1-39b), there must be a point before which the material is elastic and beyond which the material is partially elastic. To find this point, load the material to some selected value of stress and then remove the load. If there is no permanent set (that is, if the elongation of the bar returns to zero), the material is fully elastic up to the selected value of the stress.

The process of loading and unloading can be repeated for successively higher values of stress. Eventually, a stress will be reached such that not all the strain is recovered during unloading. By this procedure, it is possible to determine the stress at the upper limit of the elastic region, for instance, the stress at point E in Figs. 1-39a and b. The stress at this point is known as the **elastic limit** of the material.

Many materials, including most metals, have linear regions at the beginning of their stress-strain curves (for example, see Figs. 1-31 and 1-34). The stress at the upper limit of this linear region is the proportional limit. The elastic limit is usually the same as, or slightly above, the proportional limit. Hence, for many materials the two limits are assigned the same numerical value. In the case of mild steel, the yield stress is also very close to the proportional limit, so for practical purposes, the yield stress, the elastic limit, and the proportional limit are assumed to be equal. Of course, this situation does not hold for all materials. Rubber is an outstanding example of a material that is elastic far beyond the proportional limit.

The characteristic of a material that undergoes inelastic strains beyond the strain at the elastic limit is known as **plasticity**. Thus, on the stress-strain curve of Fig. 1-39a, an elastic region is followed by a plastic region. When large deformations occur in a ductile material loaded into the plastic region, the material is said to undergo **plastic flow**.

Reloading of a Material

If the material remains within the elastic range, it can be loaded, unloaded, and loaded again without significantly changing its behavior. However, when loaded into the plastic range, the internal structure of the material is altered and its properties change. For instance, you have already observed that a permanent strain exists in the specimen after unloading from the plastic region (Fig. 1-39b). Now suppose that the material is reloaded after such an unloading (Fig. 1-40). The new loading begins at point C on the diagram and continues upward to point B, which is the point where unloading began during the first loading cycle. The material then follows the original stress-strain curve toward point F. Thus, for the second loading, imagine that there is a new stress-strain diagram with its origin at point C.

FIGURE 1-40

Reloading of a material and raising of the elastic and proportional limits

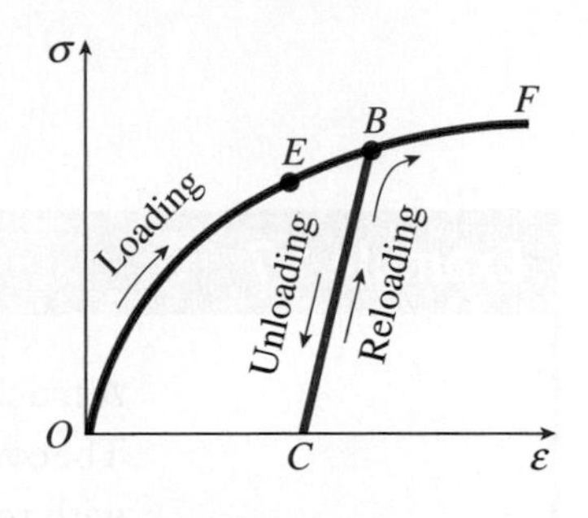

During the second loading, the material behaves in a linearly elastic manner from C to B, with the slope of line CB being the same as the slope of the tangent to the original loading curve at the origin O. The proportional limit is now at point B, which is at a higher stress than the original elastic limit (point E). Thus, by stretching a material such as steel or aluminum into the inelastic or plastic range, the *properties of the material are changed*—the linearly elastic region is increased, the proportional limit is raised, and the elastic limit is raised. However, the ductility is reduced because in the "new material" the amount of yielding beyond the elastic limit (from B to F) is less than in the original material (from E to F).[3]

FIGURE 1-41

Creep in a bar under constant load

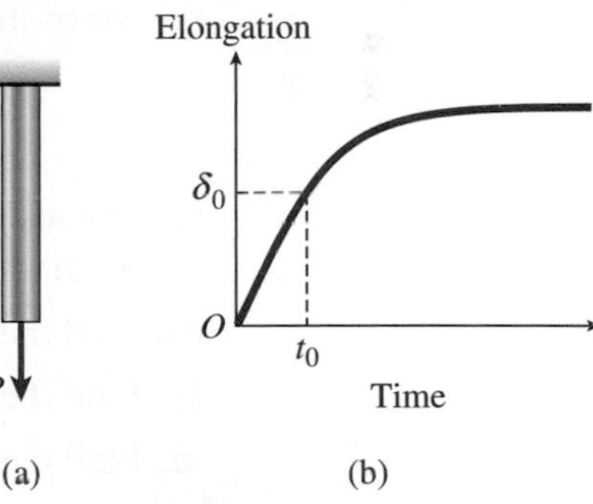

Creep

The stress-strain diagrams described previously were obtained from tension tests involving static loading and unloading of the specimens, and the passage of time did not enter the discussion. However, when loaded for long periods of time, some materials develop additional strains and are said to **creep**.

Creep can manifest itself in a variety of ways. For instance, suppose that a vertical bar (Fig. 1-41a) is loaded slowly by a force P, producing an elongation equal to δ_0. Assume that the loading and corresponding elongation take place during a time interval of duration t_0 (Fig. 1-41b). Subsequent to time t_0, the load remains constant. However, due to creep, the bar may gradually lengthen, as shown in Fig. 1-41b, even though the load does not change. This behavior occurs with many materials, although sometimes the change is too small to be of concern.

FIGURE 1-42

Relaxation of stress in a wire under constant strain

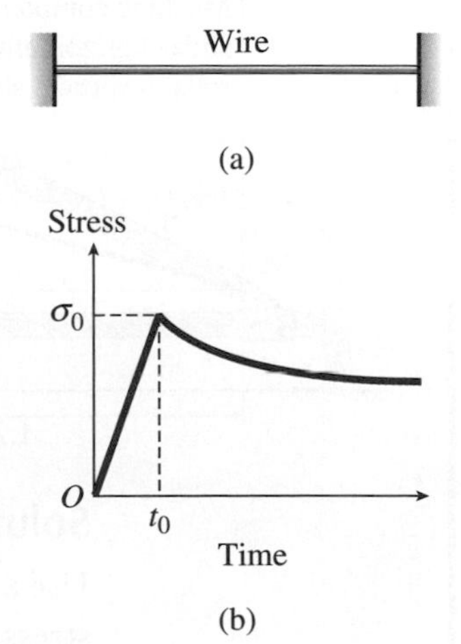

As another manifestation of creep, consider a wire that is stretched between two immovable supports so that it has an initial tensile stress σ_0 (Fig. 1-42). Again, denote the time during which the wire is initially stretched as t_0. With the elapse of time, the stress in the wire gradually diminishes, eventually reaching a constant value, even though the supports at the ends of the wire do not move. This process is called **relaxation** of the material.

Creep is usually more important at high temperatures than at ordinary temperatures, therefore it should always be considered in the design of engines, furnaces, and other structures that operate at elevated temperatures for long

[3]The study of material behavior under various environmental and loading conditions is an important branch of applied mechanics. For more detailed engineering information about materials, consult a textbook devoted solely to this subject.

periods of time. However, materials such as steel, concrete, and wood will creep slightly even at atmospheric temperatures. For example, creep of concrete over long periods of time can create undulations in bridge decks because of sagging between the supports. (One remedy is to construct the deck with an upward **camber**, which is an initial displacement above the horizontal, so that when creep occurs, the spans lower to the level position.)

Example 1-7

A machine component slides along a horizontal bar at A and moves in a vertical slot B. The component is represented as a rigid bar AB (length $L = 1.5$ m, weight $W = 4.5$ kN) with roller supports at A and B (neglect friction). When not in use, the machine component is supported by a single wire (diameter $d = 3.5$ mm) with one end attached at A and the other end supported at C (see Fig. 1-43). The wire is made of a copper alloy; the stress-strain relationship for the wire is

$$\sigma(\varepsilon) = \frac{124{,}000\varepsilon}{1 + 240\varepsilon} \qquad 0 \leq \varepsilon \leq 0.03 \quad (\sigma \text{ in MPa})$$

(a) Plot a stress-strain diagram for the material; What is the modulus of elasticity E (GPa)? What is the 0.2% offset yield stress (MPa)?

(b) Find the tensile force T (kN) in the wire.

(c) Find the normal axial strain ε and elongation δ (mm) of the wire.

(d) Find the permanent set of the wire if all forces are removed.

FIGURE 1-43

Example 1-7: Rigid bar supported by copper alloy wire

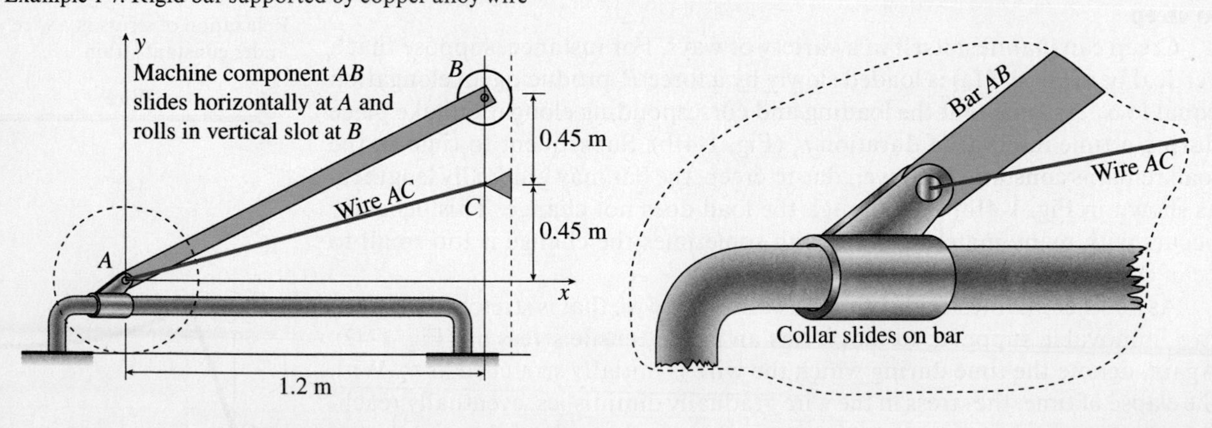

Solution:

Use a four-step problem-solving approach to find the modulus of elasticity, yield stress, tensile force, normal strain and elongation, and the permanent set of copper alloy wire AC.

1. ***Conceptualize***: The copper alloy has considerable ductility but will have a stress-strain curve without a well-defined yield point. Define the yield point

using an *offset method* as illustrated in Fig. 1-35. Find the residual strain and then the *permanent set* of the wire, as shown in Fig. 1-39.

2. *Categorize*: The given analytical expression for the stress-strain curve $\sigma(\varepsilon)$ is based on measured laboratory data for the copper alloy used to manufacture this wire. Hence, the analytical expression is an approximation of the actual behavior of this material and was formulated based on test data. Analytical representations of actual stress-strain curves are often used in computer programs to model and analyze structures of different materials under applied loads of various kinds.
3. *Analyze*:

Part (a): Plot a stress-strain diagram for the material; What is the modulus of elasticity *E* (GPa)? What is the 0.2% offset yield stress (MPa)?

Plot the function $\sigma(\varepsilon)$ for strain values between 0 and 0.03 (Fig. 1-44). The stress at strain $\varepsilon = 0.03$ is 454 MPa.

$$\sigma(\varepsilon) = \frac{124{,}000\varepsilon}{1 + 240\varepsilon} \qquad \varepsilon = 0, 0.001, \ldots, 0.03$$

$$\sigma(0) = 0 \qquad \sigma(0.03) = 454 \text{ MPa}$$

FIGURE 1-44

Stress-strain curve for copper alloy wire in Example 1-7

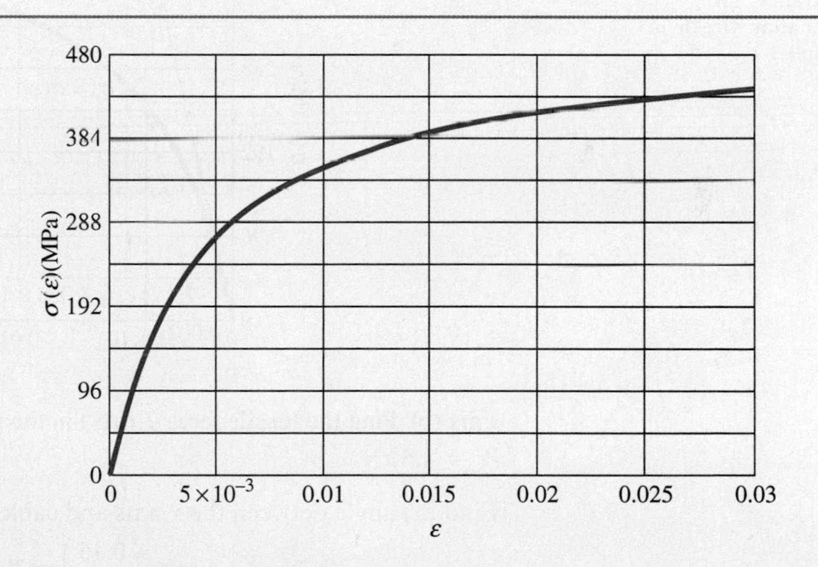

The slope of the tangent to the stress-strain curve at strain $\varepsilon = 0$ is the modulus of elasticity E (see Fig. 1-45). Take the derivative of $\sigma(\varepsilon)$ to get the slope of the tangent to the $\sigma(\varepsilon)$ curve, and evaluate the derivative at strain $\varepsilon = 0$ to find E:

$$E(\varepsilon) = \frac{d}{d\varepsilon}\sigma(\varepsilon) = \frac{124{,}000}{(240\varepsilon + 1)^2}$$

$$E = E(0) \qquad E = 124{,}000 \text{ MPa} = 124 \text{ GPa}$$

Next, find an expression for the yield strain ε_y, the point at which the 0.2% offset line crosses the stress-strain curve (see Fig. 1-45). Substitute the expression ε_y into the $\sigma(\varepsilon)$ expression and then solve for yield $\sigma(\varepsilon_y) = \sigma_y$:

$$\varepsilon_y = 0.002 + \frac{\sigma_y}{E} \text{ and } \sigma(\varepsilon_y) = \sigma_y \quad \text{or} \quad \sigma_y = \frac{124{,}000\varepsilon_y}{1 + 240\varepsilon_y}$$

Rearranging the equation in terms of σ_y gives

$$\sigma_y^2 + \left(\frac{E}{500}\right)\sigma_y - \frac{E^2}{120{,}000} = 0$$

Solving this quadratic equation for the 0.2% offset yield, stress σ_y gives $\sigma_y = 255$ MPa.

The yield strain is computed as

$$\varepsilon_y = 0.002 + \frac{\sigma_y}{E(\text{GPa})} = 4.056 \times 10^{-3}$$

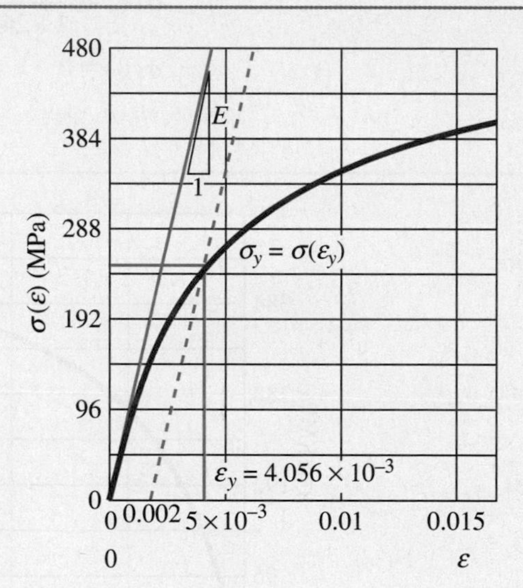

FIGURE 1-45
Modulus of elasticity E, 0.2% offset line, and yield stress σ_y and strain ε_y for copper alloy wire in Example 1-7

Part (b): Find the tensile force T (kN) in the wire. Recall that bar weight $W = 4.5$ kN.

Find the angle between the x-axis and cable attachment position at C:

$$\alpha_C = \arctan\left(\frac{0.45}{1.2}\right) = 20.556°$$

Sum the moments about A to obtain one equation and one unknown. The reaction B_x acts to the left:

$$B_x = \frac{-W(0.6\text{ m})}{0.9\text{ m}} = -3\text{ kN}$$

Next, sum the forces in the x direction to find the cable force T_C:

$$T_C = \frac{-B_x}{\cos(\alpha_C)} \qquad T_C = 3.2\text{ kN}$$

Part (c): Find the normal axial strain ε and elongation δ (mm) of the wire.

Compute the normal stress then find the associated strain from stress-strain plot (or from the $\sigma(\varepsilon)$ equation). The wire elongation is strain times wire length.

The wire diameter, cross-sectional area, and length are

$$d = 3.5 \text{ mm} \quad A = \frac{\pi}{4}d^2 = 9.6211 \text{ mm}^2$$

$$L_C = \sqrt{(1.2 \text{ m})^2 + (0.45 \text{ m})^2} = 1.282 \text{ m}$$

Now compute the stress and strain in the wire and the elongation of the wire as

$$\sigma_C = \frac{T_C}{A} = 333 \text{ MPa}$$

Note that the stress in the wire exceeds the 0.2% offset yield stress of 255 MPa. The corresponding normal strain is found from the $\sigma(\varepsilon)$ plot or by rearranging the $\sigma(\varepsilon)$ equation to give

$$\varepsilon(\sigma) = \frac{\sigma}{124{,}000 - 240\sigma}$$

Then,

$$\varepsilon(\sigma_C) = \varepsilon_C, \quad \text{or} \quad \varepsilon_C = \frac{\sigma_C}{124 \text{ GPa} - 240\sigma_C} = 7.556 \times 10^{-3}$$

Finally, the wire elongation is

$$\delta_C = \varepsilon_C L_C = 9.68 \text{ mm}$$

Part (d): Find the permanent set of the wire if all forces are removed.

If the load is removed from the wire, the stress in the wire will return to zero following the unloading line in Fig. 1-46 (see also Fig. 1-39b). The elastic recovery strain is

$$\varepsilon_{er} = \frac{\sigma_C}{E} = 3.895 \times 10^{-4}$$

The residual strain is the difference between the total strain (ε_C) and the elastic recovery strain (ε_{er})

$$\varepsilon_{res} = \varepsilon_C - \varepsilon_{er} = 7.166 \times 10^{-3}$$

Finally, the permanent set of the wire is the product of the residual strain and the length of the wire:

$$\varepsilon_{res}\, L_C = 9.184 \text{ mm}$$

4. ***Finalize***: This example presents an analytical model of the stress-strain relationship for a copper alloy. The computed values of modulus of elasticity E and yield stress σ_y are consistent with values listed in

FIGURE 1-46

Residual strain (ε_{res}) and elastic recovery strain (ε_{er}) for copper alloy wire in Example 1-7

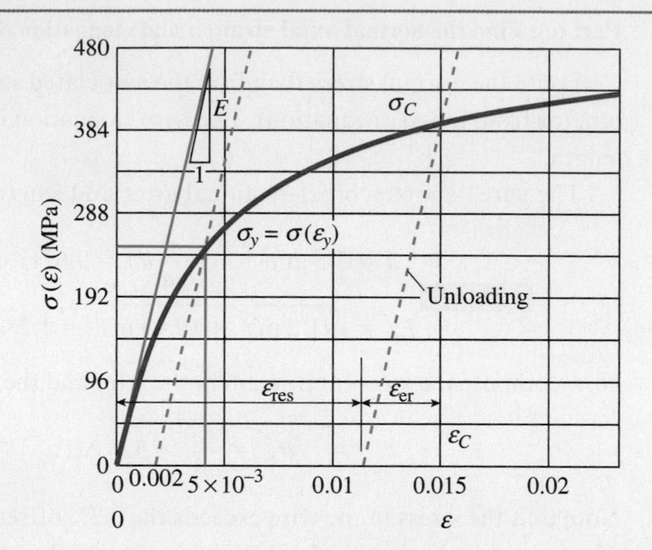

Appendix I. The tensile force, normal strain and elongation, and permanent set are computed for the wire when stressed beyond the apparent yield point of the material.

1.7 Linear Elasticity, Hooke's Law, and Poisson's Ratio

Many structural materials, including most metals, wood, plastics, and ceramics, behave both elastically and linearly when first loaded. Consequently, their stress-strain curves begin with a straight line passing through the origin. An example is the stress-strain curve for structural steel (Fig. 1-31), where the region from the origin O to the proportional limit (point A) is both linear and elastic. Other examples are the regions below *both* the proportional limits and the elastic limits on the diagrams for aluminum (Fig. 1-34), brittle materials (Fig. 1-37), and copper (Fig. 1-38).

When a material behaves elastically and also exhibits a linear relationship between stress and strain, it is said to be **linearly elastic**. By designing structures and machines to function in this region, engineers avoid permanent deformations due to yielding.

Hooke's Law

The linear relationship between stress and strain for a bar in simple tension or compression is expressed by the equation

$$\sigma = E\varepsilon \tag{1-12}$$

in which σ is the axial stress, ε is the axial strain, and E is a constant of proportionality known as the **modulus of elasticity** for the material. The modulus of elasticity is the slope of the stress-strain diagram in the linearly elastic region, as mentioned previously in Section 1.5. Since strain is dimensionless, the units of E are the same as the units of stress. Typical units of E are pascals (or multiples thereof) in SI units.

The equation $\sigma = E\varepsilon$ is commonly known as **Hooke's law**, named for the famous English scientist Robert Hooke (1635–1703). Hooke was the first person to investigate scientifically the elastic properties of materials, and he tested such diverse materials as metal, wood, stone, bone, and sinew. He measured the stretching of long wires supporting weights and observed that the elongations "always bear the same proportions one to the other that the weights do that made them." (Ref. 1-6). Thus, Hooke established the linear relationship between the applied loads and the resulting elongations.

Equation (1-12) is actually a very limited version of Hooke's law because it relates only to the longitudinal stresses and strains developed in simple tension or compression of a bar (*uniaxial stress*). To deal with more complicated states of stress, such as those found in most structures and machines, more extensive equations of Hooke's law are needed (see Sections 7.5 and 7.6).

The modulus of elasticity has relatively large values for materials that are very stiff, such as structural metals. Steel has a modulus of approximately 210 GPa; for aluminum, values around 73 GPa are typical. More flexible materials have a lower modulus—values for plastics range from 0.7 to 14 GPa. Some representative values of E are listed in Table I-2, Appendix I. For most materials, the value of E in compression is nearly the same as in tension.

Modulus of elasticity is often called **Young's modulus**, after another English scientist, Thomas Young (1773–1829). In connection with an investigation of tension and compression of prismatic bars, Young introduced the idea of a "modulus of the elasticity." However, his modulus was not the same as the one in use today because it involved properties of the bar as well as of the material (Ref. 1-7).

Poisson's Ratio

When a prismatic bar is loaded in tension, the axial elongation is accompanied by **lateral contraction** (that is, contraction normal to the direction of the applied load). This change in shape is pictured in Fig. 1-47, where part (a) shows the bar before loading and part (b) shows it after loading. In part (b), the dashed lines represent the shape of the bar prior to loading.

Lateral contraction is easily seen by stretching a rubber band, but in metals, the changes in lateral dimensions (in the linearly elastic region) are usually too small to be visible. However, they can be detected with sensitive measuring devices.

The **lateral strain** ε' at any point in a bar is proportional to the axial strain ε at that same point if the material is linearly elastic. The ratio of these strains is a property of the material known as **Poisson's ratio**. This dimensionless ratio, usually denoted by the Greek letter ν (nu), can be expressed by

$$\nu = -\frac{\text{lateral strain}}{\text{axial strain}} = -\frac{\varepsilon'}{\varepsilon} \quad \textbf{(1-13)}$$

The minus sign is inserted in the equation to compensate for the fact that the lateral and axial strains normally have opposite signs. For instance, the axial strain in a bar in tension is positive, and the lateral strain is negative (because the width of the bar decreases). The opposite is true for

FIGURE 1-47

Axial elongation and lateral contraction of a prismatic bar in tension: (a) bar before loading and (b) bar after loading (The deformations of the bar are highly exaggerated)

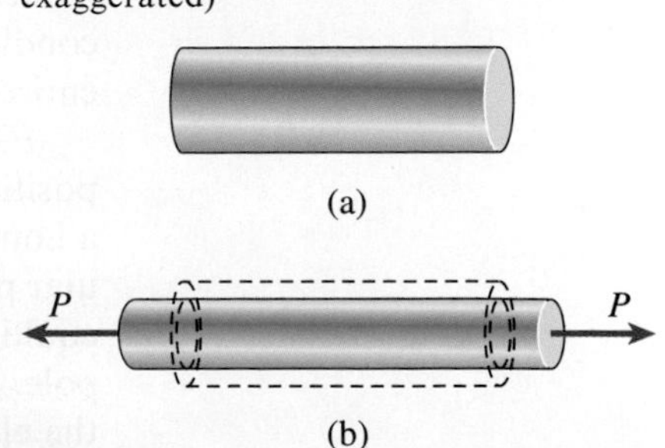

compression, with the bar becoming shorter (negative axial strain) and wider (positive lateral strain). Therefore, for ordinary materials, Poisson's ratio has a positive value. Some materials such as low-density open-cell polymer foams, however, can have a negative Poisson's ratio so, in the linear elastic range, Poisson's ratio lies between -1 and $+0.5$.

When Poisson's ratio for a material is known, you can obtain the lateral strain from the axial strain as

$$\varepsilon' = -\nu\varepsilon \tag{1-14}$$

When using Eqs. (1-13) and (1-14), always keep in mind that they apply only to a bar in uniaxial stress, that is, a bar for which the only stress is the normal stress σ in the axial direction.

Poisson's ratio is named for the famous French mathematician Siméon Denis Poisson (1781–1840), who attempted to calculate this ratio by a molecular theory of materials (Ref. 1-8). For isotropic materials, Poisson found $\nu = 1/4$. More recent calculations based upon better models of atomic structure give $\nu = 1/3$. Both of these values are close to actual measured values, which are in the range 0.25 to 0.35 for most metals and many other materials. Materials with an extremely low value of Poisson's ratio include cork, for which ν is practically zero, and concrete, for which ν is about 0.1 or 0.2. A theoretical upper limit for Poisson's ratio is 0.5, as explained later in Section 7.5. Rubber comes close to this limiting value. Note however, that Poisson's ratio may be as low as -1.0 for materials such as low-density open cell polymer foams. Hence, in the elastic range, Poisson's ratio varies between -1.0 and $+0.5$.

A table of Poisson's ratios for various materials in the linearly elastic range is given in Appendix I (see Table I-2). For most purposes, Poisson's ratio is assumed to be the same in both tension and compression.

When the strains in a material become large, Poisson's ratio changes. For instance, in the case of structural steel, the ratio becomes almost 0.5 when plastic yielding occurs. Thus, Poisson's ratio remains constant only in the linearly elastic range. When the material behavior is nonlinear, the ratio of lateral strain to axial strain is often called the *contraction ratio*. Of course, in the special case of linearly elastic behavior, the contraction ratio is the same as Poisson's ratio.

Limitations

For a particular material, Poisson's ratio remains constant throughout the linearly elastic range, as explained previously. Therefore, at any given point in the prismatic bar of Fig. 1-47, the lateral strain remains proportional to the axial strain as the load increases or decreases. However, for a given value of the load (which means that the axial strain is constant throughout the bar), additional conditions must be met if the lateral strains are to be the same throughout the entire bar.

First, the material must be **homogeneous**, that is, it must have the same composition (and hence the same elastic properties) at every point. However, having a homogeneous material does not mean that the elastic properties at a particular point are the same in all *directions*. For instance, the modulus of elasticity could be different in the axial and lateral directions, as in the case of a wood pole. Therefore, a second condition for uniformity in the lateral strains is that the elastic properties must be the same in all directions *perpendicular* to the longitudinal axis. When the preceding conditions are met, as is often the case with metals, the lateral strains in a prismatic bar subjected to uniform tension will be the same at every point in the bar and the same in all lateral directions.

Materials having the same properties in all directions (whether axial, lateral, or any other direction) are said to be **isotropic**. If the properties differ in various directions, the material is **anisotropic**.

In this book, all examples and problems are solved with the assumption that the material is linearly elastic, homogeneous, and isotropic—unless a specific statement is made to the contrary.

Example 1-8

A hollow plastic circular pipe (length L_p, inner and outer diameters d_1 and d_2, respectively; see Fig. 1-48) is inserted as a liner inside a cast iron pipe (length L_c, inner and outer diameters d_3 and d_4, respectively).

(a) Derive a formula for the required initial length L_p of the plastic pipe so that, when it is compressed by some force P, the final length of both pipes is the same and at the same time the final outer diameter of the plastic pipe is equal to the inner diameter of the cast iron pipe.

(b) Using the numerical data given, find the initial length L_p (m) and final thickness t_p (mm) for the plastic pipe.

(c) What is the required compressive force P (N)? What are the final normal stresses (MPa) in both pipes?

(d) Compare the initial and final volumes (mm^3) for the plastic pipe.

Numerical data and pipe cross-section properties are

$$L_c = 0.25 \text{ m} \quad E_c = 170 \text{ GPa} \quad E_p = 2.1 \text{ GPa} \quad \nu_c = 0.3 \quad \nu_p = 0.4$$

$$d_1 = 109.8 \text{ mm} \qquad d_2 = 110 \text{ mm} \qquad d_3 = 110.2 \text{ mm}$$

$$d_4 = 115 \text{ mm} \qquad t_p = \frac{d_2 - d_1}{2} = 0.1 \text{ mm}$$

FIGURE 1-48

Example 1-8: Plastic pipe compressed inside cast iron pipe

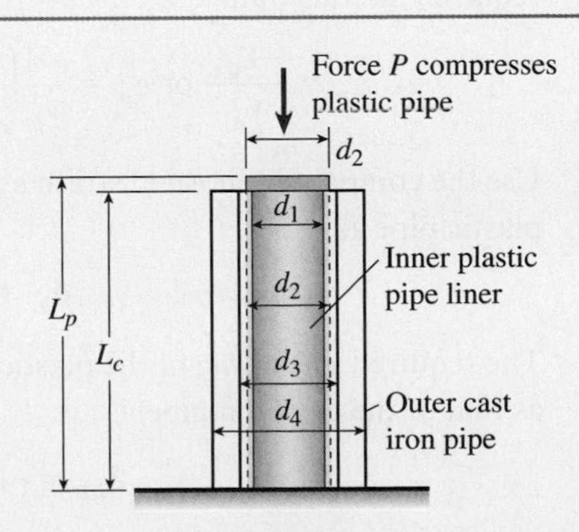

Solution:

Use the four-step problem-solving approach to find the dimensions and force required to fit a plastic liner into a cast iron pipe.

1. ***Conceptualize***: Application of a compressive force P results in compressive normal strains and extensional lateral strains in the plastic pipe, while the cast iron pipe is stress-free. The initial length of the plastic pipe (L_p) is greater than that of the cast iron pipe (L_c). With full application of force P, the lengths are made equal.

   The initial cross-sectional areas of the plastic and cast iron pipes are

$$A_p = \frac{\pi}{4}(d_2^2 - d_1^2) = 34.526 \text{ mm}^2 \quad A_c = \frac{\pi}{4}(d_4^2 - d_3^2) = 848.984 \text{ mm}^2$$

2. ***Categorize***: The two requirements are (a) compression of the plastic pipe must close the gap ($d_3 - d_2$) between the plastic pipe and the inner surface of the cast iron pipe and (b) the final lengths of the two pipes are the same. The first requirement depends on lateral strain and the second on normal strain. Each requirement leads to an expression for shortening of the plastic pipe. Equating the two expressions (i.e., enforcing *compatibility of displacements*) leads to a solution for the required length of the plastic pipe.

3. ***Analyze***:

   Part (a): Derive a formula for the required initial length L_p of the plastic pipe. The lateral strain resulting from compression of the plastic pipe must close the gap ($d_3 - d_2$) between the plastic pipe and the inner surface of the cast iron pipe. The required *extensional* lateral strain is positive (here, $\varepsilon_{\text{lat}} = \varepsilon'$):

$$\varepsilon_{\text{lat}} = \frac{d_3 - d_2}{d_2} = 1.818 \times 10^{-3}$$

   The accompanying *compressive* normal strain in the plastic pipe is obtained using Eq. 1-14, which requires Poisson's ratio for the plastic pipe and also the required lateral strain:

$$\varepsilon_p = \frac{-\varepsilon_{\text{lat}}}{\nu_p} \text{ or } \varepsilon_p = \frac{-1}{\nu_p}\left(\frac{d_3 - d_2}{d_2}\right) = -4.545 \times 10^{-3}$$

   Use the compressive normal strain ε_p to compute the *shortening* δ_{p1} of the plastic pipe as

$$\delta_{p1} = \varepsilon_p L_p$$

   The required *shortening* of the plastic pipe (so that it has the same final length as that of the cast iron pipe) is

$$\delta_{p2} = -(L_p - L_c)$$

Equating δ_{p1} and δ_{p2} leads to a formula for the required initial length L_p of the plastic pipe:

$$L_p = \frac{L_c}{1+\varepsilon_p} \quad \text{or} \quad L_p = \frac{L_c}{1 - \dfrac{d_3 - d_2}{\nu_p d_2}}$$

Part (b): Now substitute the numerical data to find the initial length L_p, change in thickness Δt_p, and final thickness t_{pf} for the plastic pipe.

As expected, L_p is greater than the length of the cast iron pipe, $L_c = 0.25$ m, and the thickness of the compressed plastic pipe increases by Δt_p:

$$L_p = \frac{L_c}{1 - \left(\dfrac{d_3 - d_2}{\nu_p d_2}\right)} = 0.25114 \text{ m}$$

$$\Delta t_p = \varepsilon_{\text{lat}} t_p = 1.818 \times 10^{-4} \text{ mm} \quad \text{so} \quad t_{pf} = t_p + \Delta t_p = 0.10018 \text{ mm}$$

Part (c): Next find the required compressive force P and the final normal stresses in both pipes.

A check on the normal compressive stress in the plastic pipe, computed using Hooke's law (Eq. 1-12), shows that it is well below the ultimate stress for selected plastics (see Table I-3, Appendix I); this is also the final normal stress in the plastic pipe:

$$\sigma_p = E_p \varepsilon_p = -9.55 \text{ MPa}$$

The required downward force to compress the plastic pipe is

$$P_{\text{reqd}} = \sigma_p A_p = -330 \text{ N}$$

Both the initial and final stresses in the cast iron pipe are zero because no force is applied to the cast iron pipe.

Part (d): Lastly, compare the initial and final volumes of the plastic pipe.

The initial cross-sectional area of the plastic pipe is

$$A_p = 34.526 \text{ mm}^2$$

The final cross-sectional area of the plastic pipe is

$$A_{pf} = \frac{\pi}{4}[d_3^2 - (d_3 - 2t_{pf})^2] = 34.652 \text{ mm}^2$$

The initial volume of the plastic pipe is

$$V_{\text{pinit}} = L_p A_p = 8671 \text{ mm}^3$$

and the final volume of the plastic pipe is

$$V_{pfinal} = L_c A_{pf} \text{ or } V_{pfinal} = 8663 \text{ mm}^3$$

4. ***Finalize***: The ratio of final to initial volume reveals little change in the volume of the plastic pipe:

$$\frac{V_{pfinal}}{V_{pinit}} = 0.99908$$

The numerical results obtained in this example illustrate that the dimensional changes in structural materials under normal loading conditions are extremely small. In spite of their smallness, changes in dimensions can be important in certain kinds of analysis (such as the analysis of statically indeterminate structures) and in the experimental determination of stresses and strains.

1.8 Shear Stress and Strain

Diagonal bracing for an elevated walkway showing a clevis and a pin in double shear

The preceding sections discussed the effects of normal stresses produced by axial loads acting on straight bars. These stresses are called "normal stresses" because they act in directions *perpendicular* to the surface of the material. Now consider another kind of stress, called a **shear stress**, that acts *tangential* to the surface of the material.

As an illustration of the action of shear stresses, consider the bolted connection shown in Fig. 1-49a. This connection consists of a flat bar A, a clevis C, and a bolt B that pass through holes in the bar and clevis. Under the action of the tensile loads P, the bar and clevis press against the bolt in **bearing**, and contact stresses, called **bearing stresses**, are developed. In addition, the bar and clevis tend to shear the bolt, that is, cut through it, and this tendency is resisted by shear stresses in the bolt. As an example, see the bracing for an elevated pedestrian walkway shown in the photograph.

To show more clearly the actions of the bearing and shear stresses, look at this type of connection in a schematic side view (Fig. 1-49b). With this view in mind, draw a free-body diagram of the bolt (Fig. 1-49c). The bearing stresses exerted by the clevis against the bolt appear on the left-hand side of the free-body diagram and are labeled 1 and 3. The stresses from the bar appear on the right-hand side and are labeled 2. The actual distribution of the bearing stresses is difficult to determine, so it is customary to assume that the stresses are uniformly distributed. Based upon the assumption of uniform distribution, calculate an **average bearing** stress σ_b by dividing the total bearing force F_b by the bearing area A_b:

$$\sigma_b = \frac{F_b}{A_b} \tag{1-15}$$

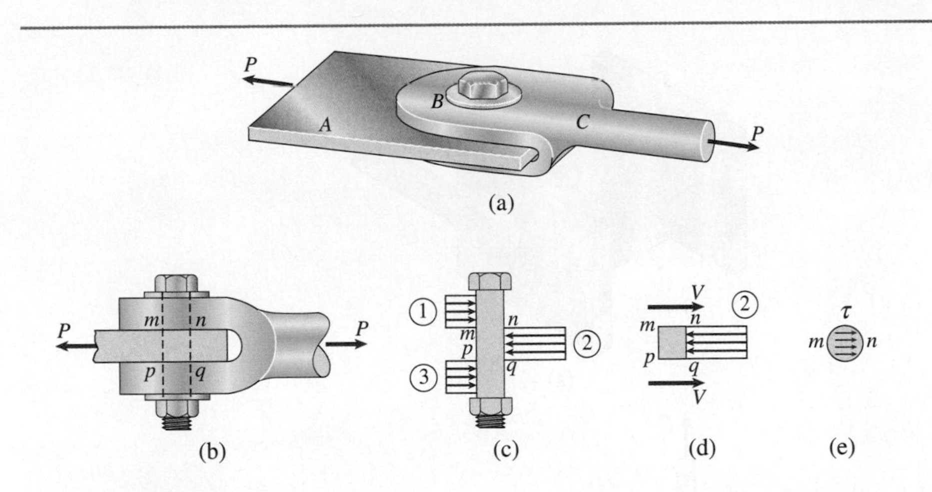

FIGURE 1-49

Bolted connection in which the bolt is loaded in double shear

The **bearing area** is defined as the projected area of the curved bearing surface. For instance, consider the bearing stresses labeled 1. The projected area A_b on which they act is a rectangle having a height equal to the thickness of the clevis and a width equal to the diameter of the bolt. Also, the bearing force F_b represented by the stresses labeled 1 is equal to $P/2$. The same area and the same force apply to the stresses labeled 3.

Now look at the bearing stresses between the flat bar and the bolt (the stresses labeled 2). For these stresses, the bearing area A_b is a rectangle with height equal to the thickness of the flat bar and width equal to the bolt diameter. The corresponding bearing force F_b is equal to the load P.

The free-body diagram of Fig. 1-49c shows that there is a tendency to shear the bolt along cross sections *mn* and *pq*. From a free-body diagram of the portion *mnpq* of the bolt (see Fig. 1-49d), note that shear forces V act over the cut surfaces of the bolt. There are two planes of shear (*mn* and *pq*), and the bolt is said to be in **double shear**. In double shear, each of the shear forces is equal to one-half of the total load transmitted by the bolt, that is, $V = P/2$.

The shear forces V are the resultants of the shear stresses distributed over the cross-sectional area of the bolt. For instance, the shear stresses acting on cross section *mn* are shown in Fig. 1-49e. These stresses act parallel to the cut surface. The exact distribution of the stresses is not known, but they are highest near the center and become zero at certain locations on the edges. As indicated in Fig. 1-49e, shear stresses are customarily denoted by the Greek letter τ (tau).

A bolted connection in single shear is shown in Fig. 1-50a, where the axial force P in the metal bar is transmitted to the flange of the steel column through a bolt. A cross-sectional view of the column (Fig. 1-50b) shows the connection in more detail. Also, a sketch of the bolt (Fig. 1-50c) shows the assumed distribution of the bearing stresses acting on the bolt. The actual distribution of these bearing stresses is much more complex than shown in the figure. Furthermore, bearing stresses are also developed against the inside surfaces of the bolt head and nut. Thus, Fig. 1-50c is *not* a free-body diagram—only the idealized bearing stresses acting on the shank of the bolt are shown in the figure.

Cutting through the bolt at section *mn* reveals the diagram shown in Fig. 1-50d. This diagram includes the shear force V (equal to the load P) acting

FIGURE 1-50

Bolted connection in which the bolt is loaded in single shear

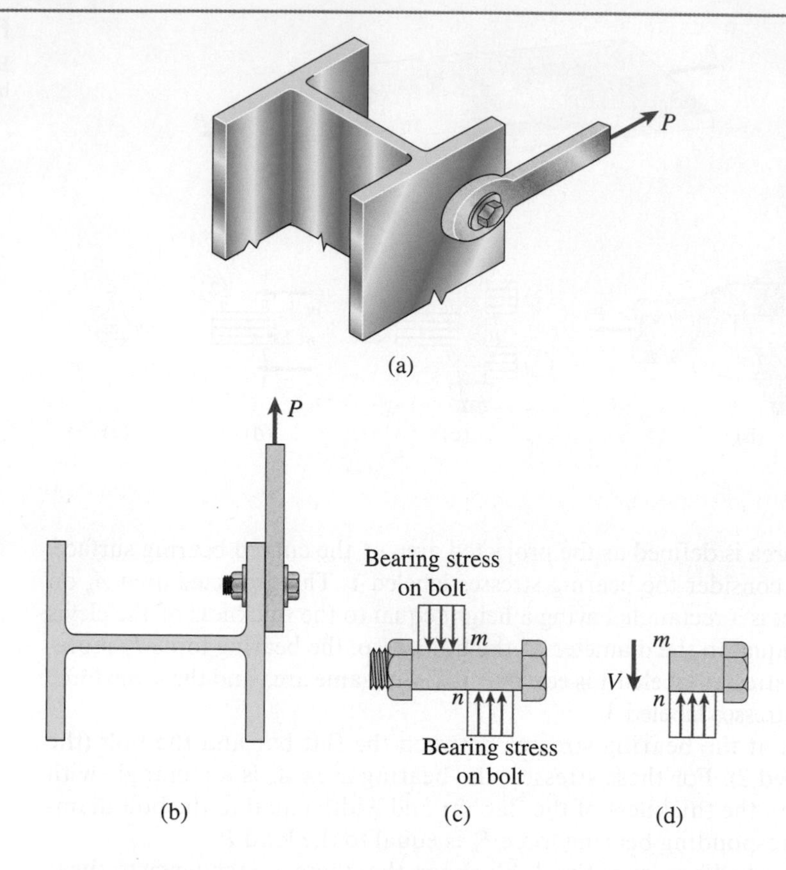

on the cross section of the bolt. This shear force is the resultant of the shear stresses that act over the cross-sectional area of the bolt.

The deformation of a bolt loaded almost to fracture in single shear is shown in Fig. 1-51 (compare with Fig. 1-50c).

The preceding discussion of bolted connections disregarded **friction** (produced by tightening of the bolts) between the connecting elements. The presence of friction means that part of the load is carried by friction forces, thereby reducing the loads on the bolts. Since friction forces are unreliable and difficult to estimate, it is common practice to err on the conservative side and omit them from the calculations.

FIGURE 1-51

Failure of a bolt in single shear (© Barry Goodno)

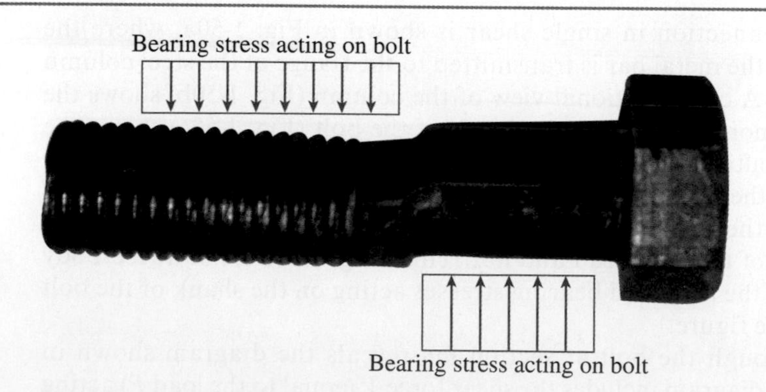

The **average shear stress** on the cross section of a bolt is obtained by dividing the total shear force V by the area A of the cross section on which it acts, as

$$\tau_{aver} = \frac{V}{A} \quad \textbf{(1-16)}$$

In the example of Fig. 1-50, which shows a bolt in *single shear*, the shear force V is equal to the load P, and the area A is the cross-sectional area of the bolt. However, in the example of Fig. 1-49, where the bolt is in *double shear*, the shear force V equals $P/2$.

Equation (1-16) shows that shear stresses, like normal stresses, represent intensity of force, or force per unit of area. Thus, the **units of** shear stress are the same as those for normal stress, namely, pascals or multiples thereof in SI units.

The loading arrangements shown in Figs. 1-49 and 1-50 are examples of **direct shear** (or *simple shear*) in which the shear stresses are created by the direct action of the forces trying to cut through the material. Direct shear arises in the design of bolts, pins, rivets, keys, welds, and glued joints.

Equality of Shear Stresses on Perpendicular Planes

To obtain a more complete picture of the action of shear stresses, consider a small element of material in the form of a rectangular parallelepiped having sides of lengths a, b, and c in the x, y, and z directions, respectively (Fig. 1-52).[4] The front and rear faces of this element are free of stress.

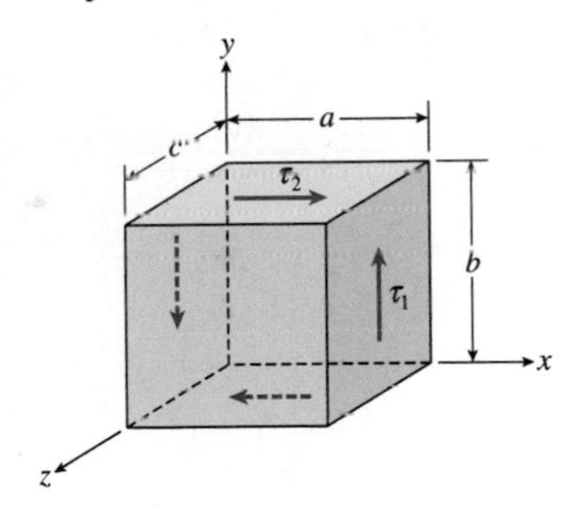

FIGURE 1-52
Small element of material subjected to shear stresses

Now assume that a shear stress τ_1 is distributed uniformly over the right-hand face, which has area bc. In order for the element to be in equilibrium in the y direction, the total shear force $\tau_1 bc$ acting on the right-hand face must be balanced by an equal but oppositely directed shear force on the left-hand face. Since the areas of these two faces are equal, it follows that the shear stresses on the two faces must be equal.

The forces $\tau_1 bc$ acting on the left- and right-hand side faces (Fig. 1-52) form a couple having a moment about the z axis of magnitude $\tau_1 abc$, acting counterclockwise in the figure.[5] Equilibrium of the element requires that this moment be balanced by an equal and opposite moment resulting from shear stresses acting on the top and bottom faces of the element. If the stresses on the top and bottom faces are labeled as τ_2, the corresponding horizontal shear forces equal $\tau_2 ac$. These forces form a clockwise couple of moment $\tau_2 abc$. From moment equilibrium of the element about the z axis, $\tau_1 abc$ equals $\tau_2 abc$, or

$$\tau_1 = \tau_2 \quad \textbf{(1-17)}$$

[4] A **parallelepiped** is a prism whose bases are parallelograms; thus, a parallelepiped has six faces, each of which is a parallelogram. Opposite faces are parallel and identical parallelograms. A **rectangular parallelepiped** has all faces in the form of rectangles.

[5] A **couple** consists of two parallel forces that are equal in magnitude and opposite in direction.

FIGURE 1-53

Element of material subjected to shear stresses and strains

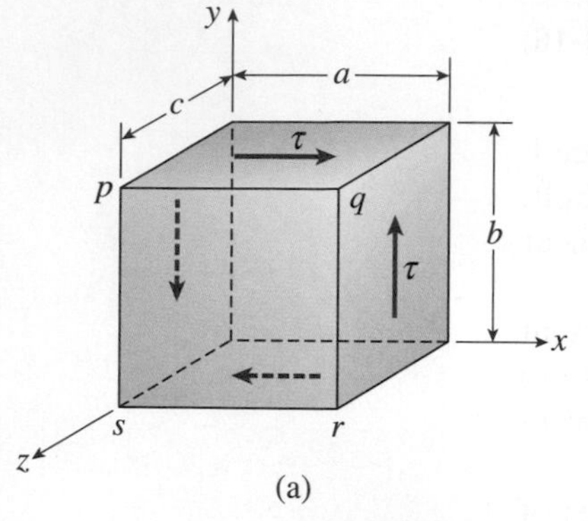

(a)

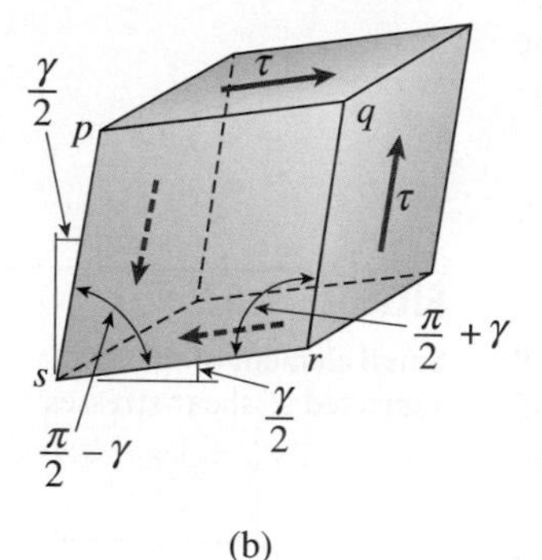

(b)

Therefore, the magnitudes of the four shear stresses acting on the element are equal, as shown in Fig. 1-53a.

The following are observations regarding shear stresses acting on a rectangular element:

1. Shear stresses on opposite (and parallel) faces of an element are equal in magnitude and opposite in direction.
2. Shear stresses on adjacent (and perpendicular) faces of an element are equal in magnitude and have directions such that both stresses point toward, or both point away from, the line of intersection of the faces.

These observations apply to an element subjected only to shear stresses (no normal stresses), as pictured in Figs. 1-52 and 1-53. This state of stress is called **pure shear**.

For most purposes, the preceding conclusions remain valid even when normal stresses act on the faces of the element. The reason is that the normal stresses on opposite faces of a small element usually are equal in magnitude and opposite in direction.

Shear Strain

Shear stresses acting on an element of material (Fig. 1-53a) are accompanied by *shear strains*. As an aid in visualizing these strains, note that the shear stresses have no tendency to elongate or shorten the element in the x, y, and z directions—in other words, the lengths of the sides of the element do not change. Instead, the shear stresses produce a change in the *shape* of the element (Fig. 1-53b). The original element, which is a rectangular parallelepiped, is deformed into an oblique parallelepiped, and the front and rear faces become rhomboids.[6]

Because of this deformation, the angles between the side faces change. For instance, the angles at points q and s, which were $\pi/2$ before deformation, are reduced by a small angle γ to $\pi/2 - \gamma$ (Fig. 1-53b). At the same time, the angles at points p and r are increased to $\pi/2 + \gamma$. The angle γ is a measure of the **distortion**, or change in shape, of the element and is called the shear strain. Because **shear strain** is an angle, it is usually measured in degrees or radians.

Sign Conventions for Shear Stresses and Strains

As an aid in establishing sign conventions for shear stresses and strains, you need a scheme for identifying the various faces of a stress element (Fig. 1-53a). Henceforth, the faces oriented toward the positive directions of the axes are referred to as the positive faces of the element. In other words, a positive face has its outward normal directed in the positive direction of a coordinate axis. The opposite faces are negative faces. Thus, in Fig. 1-53a, the right-hand, top, and front faces are the positive x, y, and z faces, respectively, and the opposite faces are the negative x, y, and z faces.

[6]An **oblique angle** can be either acute or obtuse, but it is *not* a right angle. A **rhomboid** is a parallelogram with oblique angles and adjacent sides *not* equal. (A *rhombus* is a parallelogram with oblique angles and all four sides equal, sometimes called a *diamond-shaped figure*.)

Using the terminology described in the preceding paragraph, the sign convention for shear stresses is as follows:

A shear stress acting on a positive face of an element is positive if it acts in the positive direction of one of the coordinate axes and negative if it acts in the negative direction of an axis. A shear stress acting on a negative face of an element is positive if it acts in the negative direction of an axis and negative if it acts in a positive direction.

Thus, all shear stresses shown in Fig. 1-53a are positive.

The sign convention for shear strains is as follows:

Shear strain in an element is positive when the angle between two positive faces (or two negative faces) is reduced. The strain is negative when the angle between two positive (or two negative) faces is increased.

Thus, the strains shown in Fig. 1-53b are positive, and the positive shear stresses are accompanied by positive shear strains.

Hooke's Law in Shear

The properties of a material in shear can be determined experimentally from direct-shear tests or from torsion tests. The latter tests are performed by twisting hollow, circular tubes, thereby producing a state of pure shear. The results of these tests are used to plot **shear stress-strain diagrams** (that is, diagrams of shear stress τ versus shear strain γ). These diagrams are similar in shape to tension-test diagrams (σ versus ε) for the same materials, although they differ in magnitudes.

From shear stress-strain diagrams, you can obtain material properties such as the proportional limit, modulus of elasticity, yield stress, and ultimate stress. Numerical values of these properties in shear are usually about half as large as those in tension. For instance, the yield stress for structural steel in shear is 0.5 to 0.6 times the yield stress in tension.

For many materials, the initial part of the shear stress-strain diagram is a straight line through the origin, just as it is in tension. For this linearly elastic region, the shear stress and shear strain are proportional, resulting in the following equation for **Hooke's law in shear**:

$$\tau = G\gamma \tag{1-18}$$

in which G is the **shear modulus of elasticity** (also called the *modulus of rigidity*).

The shear modulus G has the same units as the tension modulus E, namely, pascals (or multiples thereof) in SI units. For mild steel, typical values of G are 75 GPa; for aluminum alloys, typical values are 28 GPa. Additional values are listed in Table I-2, Appendix I.

The moduli of elasticity in tension and shear are related by

$$G = \frac{E}{2(1 + \nu)} \tag{1-19}$$

in which ν is Poisson's ratio. This relationship shows that E, G, and ν are not independent elastic properties of the material. Because the value of Poisson's ratio for ordinary materials is between zero and one-half, from Eq. (1-19) G must be from one-third to one-half of E.

The following examples illustrate some typical analyses involving the effects of shear. Example 1-9 is concerned with shear stresses in a plate, Example 1-10 involves finding shear stresses and shear strains in an elastomeric bearing pad subjected to a horizontal shear force, and Example 1-11 deals with normal and shear stresses in a bolted bracket.

Example 1-9

A punch for making holes in steel plates is shown in Fig. 1-54a. Assume that a punch having diameter $d = 20$ mm is used to punch a hole in an 8 mm plate, as shown in the cross-sectional view (Fig. 1-54b).

If a force $P = 110$ kN is required to create the hole, what is the average shear stress in the plate and the average compressive stress in the punch?

FIGURE 1-54

Example 1-9: Punching a hole in a steel plate

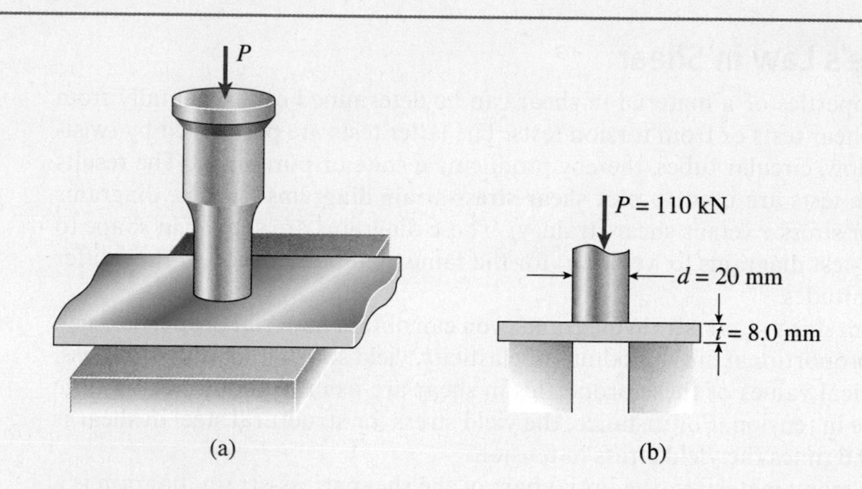

Solution:

Use the four-step problem-solving approach to find average shear stress in the plate and average compressive stress in the punch.

1. ***Conceptualize***: Assume that the shaft of the punch is in compression over its entire length due to applied load P. Force P acts downward on the plate and is applied as a uniformly distributed force along a circle of diameter d as the punch passes through the plate.
2. ***Categorize***: The average shear stress in the plate is obtained by dividing the force P by the shear area of the plate. The shear area is the cylindrical area of the plate that is exposed when the punch passes through the plate. The compressive stress of interest is the one acting on a circular cross section through the lower segment of the punch (Fig. 1-54).
3. ***Analyze***: The shear area A_s is equal to the circumference of the hole times the thickness of the plate, or

$$A_s = \pi dt = \pi(20 \text{ mm})(8.0 \text{ mm}) = 502.7 \text{ mm}^2$$

in which d is the diameter of the punch and t is the thickness of the plate. Therefore, the average shear stress in the plate is

$$\tau_{\text{aver}} = \frac{P}{A_s} = 110 \text{ kN}/502.7 \text{ mm}^2 = 219 \text{ MPa}$$

The average compressive stress in the punch is

$$\sigma_c = \frac{P}{A_{\text{punch}}} = \frac{P}{\pi d^2/4} = 110 \text{ kN}/\pi(20 \text{ mm})^2/4 = 350 \text{ MPa}$$

in which A_{punch} is the cross-sectional area of the lower segment of the punch.

4. ***Finalize***: The normal and shear stress distributions are not uniform due to stress concentration effects; hence, the calculations result in "average" stresses. In addition, this analysis is highly idealized because impact effects that occur when a punch is rammed through a plate are not part of this analysis.

Example 1-10

A bearing pad of the kind used to support machines and bridge girders consists (see photos) of a linearly elastic material (usually an elastomer, such as rubber) capped by a steel plate (Fig. 1-55a). Assume that the thickness of the elastomer is h, the dimensions of the plate are $a \times b$, and the pad is subjected to a horizontal shear force V.

Obtain formulas for the average shear stress τ_{aver} in the elastomer and the horizontal displacement d of the plate (Fig. 1-55b).

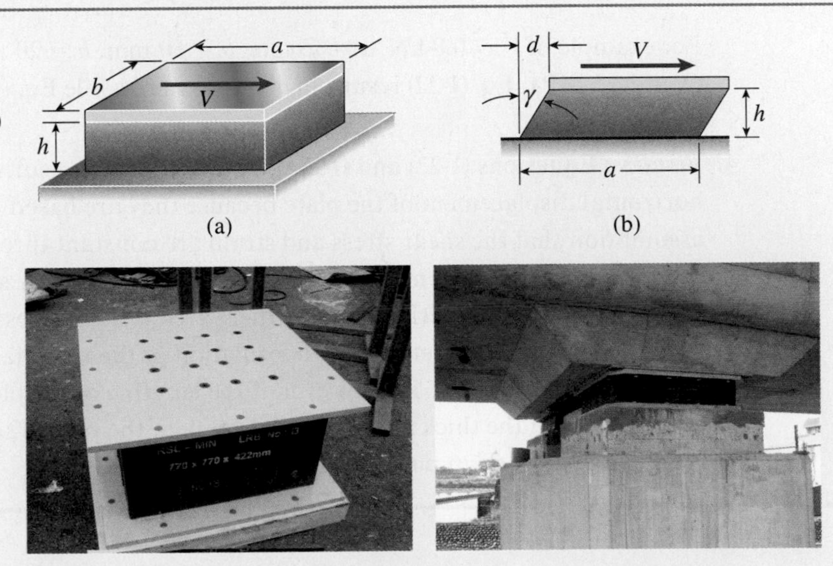

FIGURE 1-55
Example 1-10: Bearing pad in shear (Courtesy of Mageba)

Solution:

Use the following four-step problem-solving approach to find average shear stress in the elastomer and horizontal displacement d of the plate.

1. ***Conceptualize***: The distortion of the bearing pad under shear force V is assumed to be linear through the thickness h, as shown in Fig. 1-55b.
2. ***Categorize***: Assume that the shear stresses in the elastomer are uniformly distributed throughout its entire volume and that the shear strain γ is small.
3. ***Analyze***: The shear stress on any horizontal plane through the elastomer equals the shear force V divided by the area ab of the plane (Fig. 1-55a):

$$\tau_{\text{aver}} = \frac{V}{ab} \tag{1-20}$$

The corresponding shear strain [from Hooke's law in shear; Eq. (1-18)] is

$$\gamma = \frac{\tau_{\text{aver}}}{G_e} = \frac{V}{abG_e} \tag{1-21}$$

in which G_e is the shear modulus of the elastomeric material. Finally, the horizontal displacement d is equal to $h \tan \gamma$ (from Fig. 1-55b):

$$d = h\tan\gamma = h\tan\left(\frac{V}{abG_e}\right) \tag{1-22}$$

In most practical situations, the shear strain γ is a small angle, and in such cases, replace $\tan\gamma$ with γ and obtain

$$d = h\gamma = \frac{hV}{abG_e} \tag{1-23}$$

For example, if $V = 0.8$ kN, $a = 75$ mm, $b = 60$ mm, $h = 20$ mm, and $G_e = 1.25$ MPa, Eq. (1-22) results in $d = 2.86$ mm, while Eq. (1-23) gives $d = 2.84$ mm.

4. ***Finalize***: Equations (1-22) and (1-23) give approximate results for the horizontal displacement of the plate because they are based upon the assumption that the shear stress and strain are constant throughout the volume of the elastomeric material. In reality, the shear stress is zero at the edges of the material (because there are no shear stresses on the free vertical faces), and therefore, the deformation of the material is more complex than pictured in Fig. 1-55b. However, if the length a of the plate is large compared with the thickness h of the elastomer, the preceding results are satisfactory for design purposes.

Example 1-11

A basketball player hangs on the rim after dunking the ball. The player applies a downward force at point A with an estimated magnitude of $P = 1.75$ kN (Fig. 1-56a). Later, the player dunks again and hangs on the rim with two hands: one at A and one at B (see Fig. 1-56b, c). The rim and support bracket are bolted to the backboard using four bolts with washers (Fig. 1-56d). Find and compare the stresses in the bolted connection at bolt location 2 for the one-handed and two-handed load cases. Assume that the backboard is a fixed support and that bolt and washer diameters are $d_b = 10$ mm and $d_w = 16$ mm, respectively. The support bracket has thickness $t = 5$ mm.

FIGURE 1-56

(a) Load case 1—downward force on rim at A; (b) load case 2—forces applied at A and B; (c) top view of rim; (d) support bracket and bolt detail

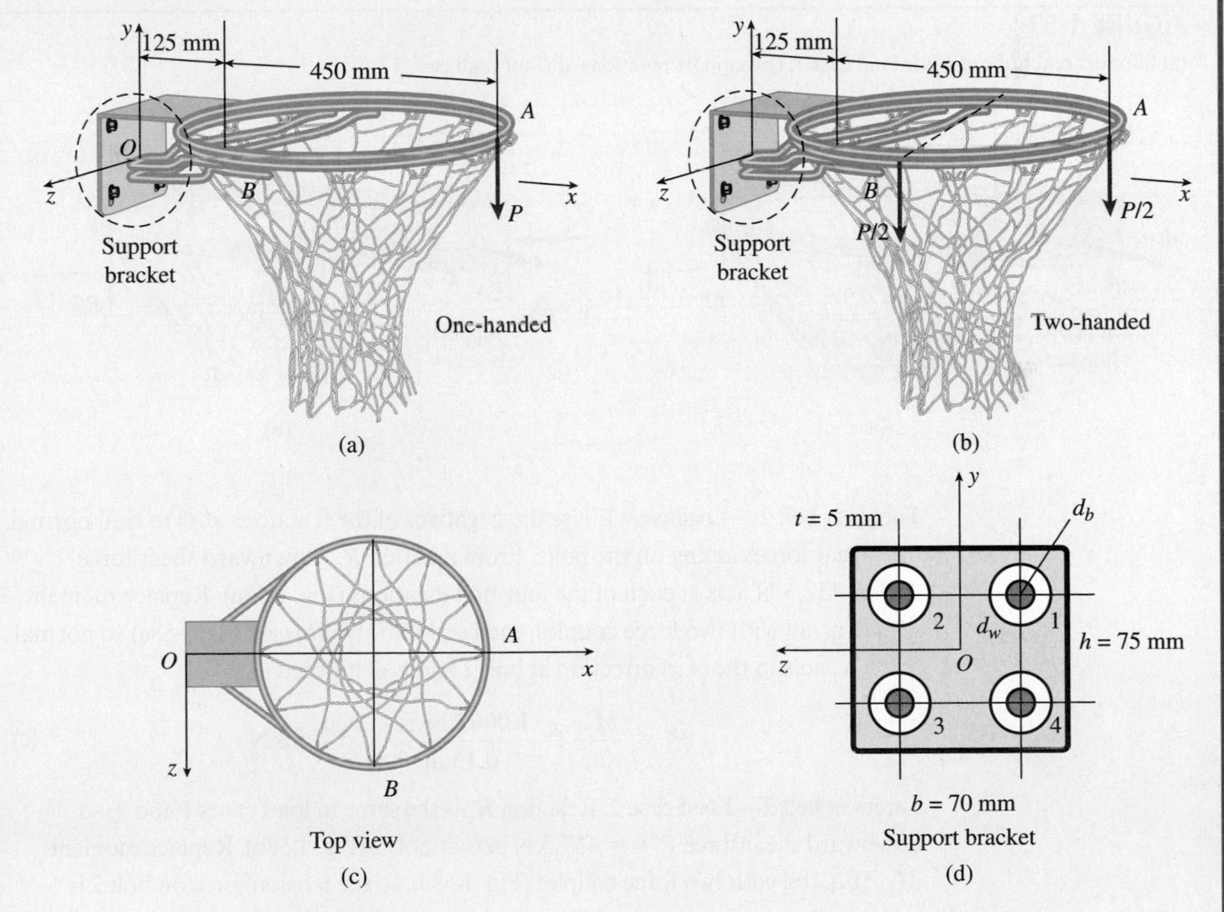

Solution:

Use the following four-step problem-solving approach.

1. ***Conceptualize***: Find reaction forces and moments at support point O and then distribute forces and moments to each bolt location. The rim and bracket act

as a cantilever beam. From the free-body diagrams in Fig. 1-57, the reactions for each load case are as follows.

Reactions—Load case 1: Load P is applied in the $(-y)$ direction at A. Sum forces and moments to find:

$$\Sigma F_y = 0 \quad R_y = P = 1.75\text{ kN}$$

$$\Sigma M_z = 0 \quad M_{z1} = P(125 + 450)\text{ mm} = (1.75)(0.575)\text{ kN}\cdot\text{m} = 1.006\text{ kN}\cdot\text{m} \tag{a}$$

Reactions—Load case 2: Loads $P/2$ are applied at both A and B. Reactions at O are

$$\Sigma F_y = 0 \quad R_y = P = 1.75\text{ kN}$$

$$\Sigma M_z = 0 \quad M_{z2} = \frac{P}{2}(125 + 225)\text{ mm} + \frac{P}{2}(575\text{ mm}) = 0.809\text{ kN}\cdot\text{m} \tag{b}$$

$$\Sigma M_x = 0 \quad T_x = -\frac{P}{2}(225\text{ mm}) = -0.1969\text{ kN}\cdot\text{m}$$

FIGURE 1-57

(a) Support reactions at O for load case 1; (b) support reactions at O for load case 2

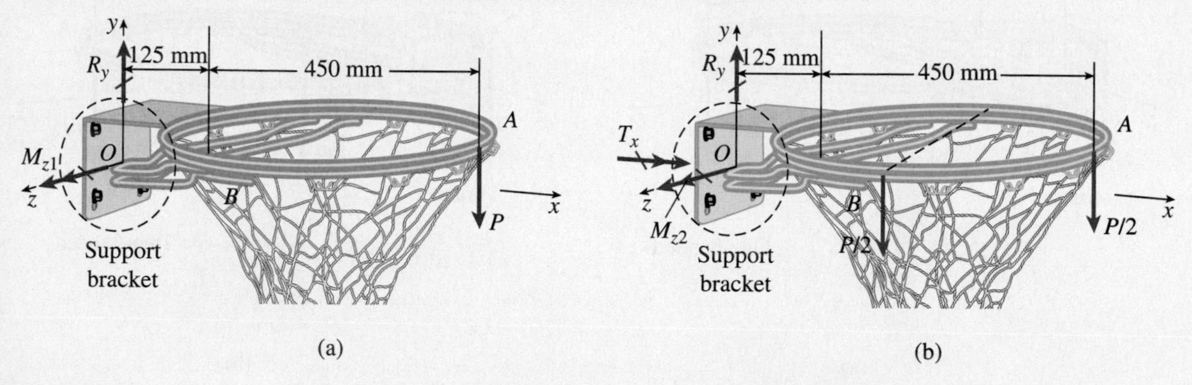

Forces at bolt 2—Load case 1: Use the negatives of the reactions at O to find normal and shear forces acting on the bolts. From reaction R_y, downward shear force $P/4 = 437.5$ N acts at each of the four bolt locations (Fig. 1-58a). Replace moment M_{z1} [Eq. (a)] with two force couples, each equal to $(N_1)(h)$ (see Fig. 1-58a) so normal force N_1 acts in the $(+x)$ direction at bolt 2 and is computed as

$$N_1 = \frac{M_{z1}}{2h} = \frac{1.006\text{ kN}\cdot\text{m}}{0.15\text{ m}} = 6.708\text{ kN} \tag{c}$$

Forces at bolt 2—Load case 2: Reaction R_y is the same in load cases 1 and 2, so downward shear force $P/4 = 437.5$ N acts at bolt 2 (Fig. 1-58b). Replace moment M_{z2} [Eq. (b)] with two force couples (Fig. 1-58b), so the tension force on bolt 2 is

$$N_2 = \frac{M_{z2}}{2h} = \frac{0.809\text{ kN}\cdot\text{m}}{0.15\text{ m}} = 5.396\text{ kN} \tag{d}$$

Load case 2 also creates a torsional reaction moment T_x [see Eq. (b)] that can be replaced by two counterclockwise force couples each equal to

FIGURE 1-58

(a) Bolt 2 forces for load case 1; (b) bolt 2 forces for load case 2

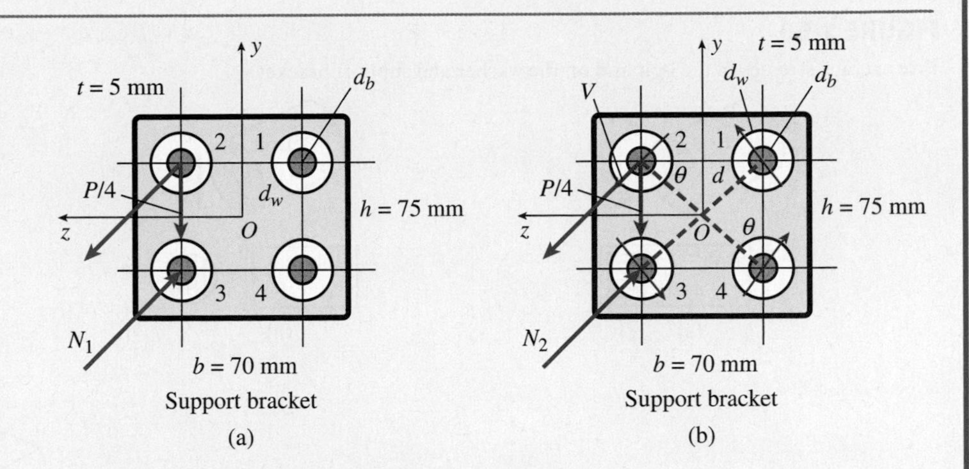

$(V)(d)$ (see Fig. 1-58b) where $d = \sqrt{b^2 + h^2} = 102.6$ mm. Compute the additional in-plane shear force on bolt 2 as

$$V = \frac{T_x}{2d} = \frac{196.9 \text{ N}\cdot\text{m}}{2(102.6 \text{ mm})} = 0.9595 \text{ kN} \quad \textbf{(e)}$$

The line of action of force V is shown in Fig. 1-58(b) at angle $\theta = \tan^{-1}\left(\frac{b}{h}\right) = 43.025°$. The total in-plane shear force on bolt 2 is the resultant R of forces V and $P/4$ computed as

$$R = \sqrt{(V\cos\theta)^2 + \left(\frac{P}{4} + V\sin\theta\right)^2} = \sqrt{(707.34)^2 + (437.5 + 648.39)^2}\text{ N} = 1.298 \text{ kN} \quad \textbf{(f)}$$

Resultant R also can be found using the parallelogram law, as shown in Fig. 1-59 with $\beta = 14.265°$ and $\gamma = 133.025°$.

FIGURE 1-59

Resultant R using parallelogram law

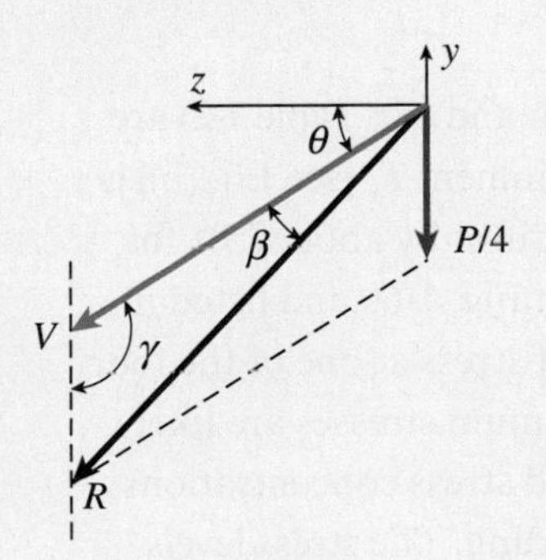

2. *Categorize*: Use the forces acting on bolt 2 in simple formulas to compute average stresses in the bolt and on the washer and support bracket at bolt location 2. The five connection stresses of interest are (a) normal stress in bolt (Fig. 1-60a); (b) shear stress in bolt (Fig. 60b); (c) bearing stress on shank of bolt (Fig. 1-60c); (d) bearing stress on washer (Fig. 1-60d); and (e) shear stress through bracket on periphery of washer (Fig. 1-60e).
3. *Analyze*: The five connection stresses at bolt location 2 are listed in Table 1-2 for load cases 1 and 2. Numerical dimensions for the bolt, washer, and bracket are $d_b = 10$ mm, $d_w = 16$ mm, and $t = 5$ mm. Areas needed in stress calculations are

Cross-sectional area of bolt:

$$A_b = \frac{\pi}{4} d_b^2 = 78.54 \text{ mm}^2$$

FIGURE 1-60

Five average stresses in the bolt and on the washer and support bracket

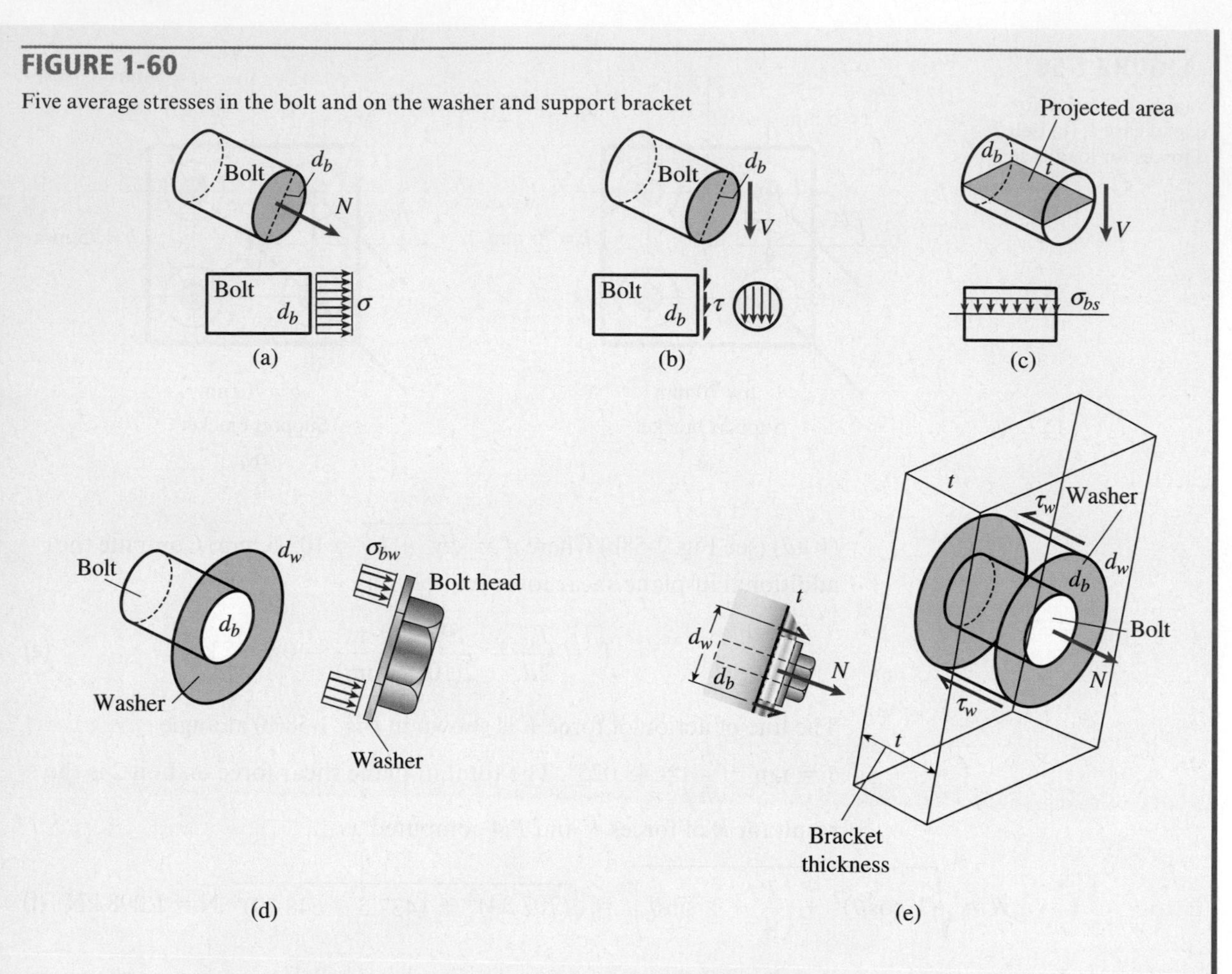

Surface area of washer:

$$A_w = \frac{\pi}{4}\left(d_w^2 - d_b^2\right) = 122.52\text{ mm}^2$$

Cylindrical area through bracket on periphery of washer:

$$A_p = \pi d_w t = 251.33\text{ mm}^2$$

4. ***Finalize***: Shear and bearing stresses on bolt 2 (items b and c in Table 1-2) are increased three-fold for load case 2 when torsional moment T_x [see Eq. (b)] is applied to the bracket. The other three bolt stresses differ by about 25% for the two load cases. The average stresses illustrated in Fig. 1-60 and listed in Table 1-2 are only approximations to the true state of stress at one of the four bolt locations on the support bracket. The true maximum stresses are likely to be higher for a variety of reasons, such as localized stress concentrations, pretensioning of bolts, and impact aspects of the loading. The stress levels computed here are low. If stress values are higher, a more detailed and sophisticated analysis using computer models employing the finite element method may be required.

Table 1-2 Five connection stresses at bolt location 2

Connection Stress	Load Case 1	Load Case 2
a. Normal stress in bolt	$\frac{N_1}{A_b} = 85.4$ MPa	$\frac{N_2}{A_b} = 68.7$ MPa
b. Shear stress in bolt	$\frac{P}{4A_b} = 5.57$ MPa	$\frac{R}{A_b} = 16.53$ MPa
c. Bearing stress on shank of bolt	$\frac{P}{4d_b t} = 8.75$ MPa	$\frac{R}{d_b t} = 26.0$ MPa
d. Bearing stress on washer	$\frac{N_1}{A_w} = 54.7$ MPa	$\frac{N_2}{A_w} = 44.0$ MPa
e. Shear stress through bracket on periphery of washer	$\frac{N_1}{A_p} = 26.7$ MPa	$\frac{N_2}{A_p} = 21.5$ MPa

1.9 Allowable Stresses and Allowable Loads

Engineers design a seemingly endless variety of objects to serve the basic needs of society. These needs include housing, agriculture, transportation, communication, and many other aspects of modern life. Factors to be considered in design include functionality, strength, appearance, economics, and environmental effects. However, when studying mechanics of materials, our principal design interest is **strength**, that is, *the capacity of the object to support or transmit loads*. Objects that must sustain loads include buildings, machines, containers, trucks, aircraft, ships, and the like. For simplicity, all such objects are referred to as **structures**; thus, a *structure is any object that must support or transmit loads*.

Factors of Safety

Strength is the ability of a structure to resist loads. *The actual strength of a structure must exceed the required strength.* The ratio of the actual strength to the required strength is called the **factor of safety** n:

$$\text{Factor of safety } n = \frac{\text{Actual strength}}{\text{Required strength}} \tag{1-24}$$

Of course, the factor of safety must be greater than 1.0 if failure is to be avoided. Depending upon the circumstances, factors of safety from slightly above 1.0 to as much as 10 are used.

The incorporation of factors of safety into design is not a simple matter, because both strength and failure have many different meanings. Strength may be measured by the load-carrying capacity of a structure, or it may be measured by the stress in the material. Failure may mean the fracture and complete collapse of a structure, or it may mean that the deformations have become so large that the structure can no longer perform its intended functions. The latter kind of failure may occur at loads much smaller than those that cause actual collapse.

The determination of a factor of safety must also take into account such matters as the following: probability of accidental overloading of the structure by loads that exceed the design loads; types of loads (static or dynamic); whether the loads are applied once or are repeated; how accurately the loads are known; possibilities for fatigue failure; inaccuracies in construction; variability in the quality of workmanship; variations in properties of materials; deterioration due to corrosion or other environmental effects; accuracy of the methods of analysis; whether failure is gradual (ample warning) or sudden (no warning); consequences of failure (minor damage or major catastrophe); and other such considerations. If the factor of safety is too low, the likelihood of failure will be high and the structure will be unacceptable; if the factor is too large, the structure will be wasteful of materials and perhaps unsuitable for its function (for instance, it may be too heavy).

Because of these complexities and uncertainties, factors of safety must be determined on a probabilistic basis. They usually are established by groups of experienced engineers who write the codes and specifications used by other designers, and sometimes they are even enacted into law. The provisions of codes and specifications are intended to provide reasonable levels of safety without unreasonable costs.

In aircraft design, it is customary to speak of the **margin of safety** rather than the factor of safety. The margin of safety is defined as the factor of safety minus one:

$$\text{Margin of safety} = n - 1 \tag{1-25}$$

Margin of safety is often expressed as a percent, in which case the value given above is multiplied by 100. Thus, a structure having an actual strength that is 1.75 times the required strength has a factor of safety of 1.75 and a margin of safety of 0.75 (or 75%). When the margin of safety is reduced to zero or less, the structure (presumably) will fail.

Allowable Stresses

Factors of safety are defined and implemented in various ways. For many structures, it is important that the material remain within the linearly elastic range in order to avoid permanent deformations when the loads are removed. Under these conditions, the factor of safety is established with respect to yielding of the structure. Yielding begins when the yield stress is reached at *any* point within the structure. Therefore, by applying a factor of safety with respect to the yield stress (or yield strength), you obtain an **allowable stress** (or *working stress*) that must not be exceeded anywhere in the structure. Thus,

$$\text{Allowable stress} = \frac{\text{Yield strength}}{\text{Factor of safety}} \tag{1-26}$$

or for tension and shear, respectively,

$$\sigma_{\text{allow}} = \frac{\sigma_Y}{n_1} \quad \text{and} \quad \tau_{\text{allow}} = \frac{\tau_Y}{n_2} \tag{1-27a,b}$$

in which σ_Y and τ_Y are the yield stresses and n_1 and n_2 are the corresponding factors of safety. In building design, a typical factor of safety with respect to yielding in tension is 1.67; thus, a mild steel having a yield stress of 250 MPa has an allowable stress of 150 MPa.

Sometimes the factor of safety is applied to the **ultimate stress** instead of the yield stress. This method is suitable for brittle materials, such as concrete and some plastics, and for materials without a clearly defined yield stress, such as wood and high-strength steels. In these cases, the allowable stresses in tension and shear are

$$\sigma_{\text{allow}} = \frac{\sigma_U}{n_3} \quad \text{and} \quad \tau_{\text{allow}} = \frac{\tau_U}{n_4} \tag{1-28}$$

in which σ_U and τ_U are the ultimate stresses (or ultimate strengths). Factors of safety with respect to the ultimate strength of a material are usually larger than those based upon yield strength. In the case of mild steel, a factor of safety of 1.67 with respect to yielding corresponds to a factor of approximately 2.8 with respect to the ultimate strength.

Allowable Loads

After the allowable stress has been established for a particular material and structure, the **allowable load** on that structure can be determined. The relationship between the allowable load and the allowable stress depends upon the type of structure. This chapter is concerned only with the most elementary kinds of structures, namely, bars in tension or compression and pins (or bolts) in direct shear and bearing.

In these kinds of structures, the stresses are uniformly distributed (or at least *assumed* to be uniformly distributed) over an area. For instance, in the case of a bar in tension, the stress is uniformly distributed over the cross-sectional area provided the resultant axial force acts through the centroid of the cross section. The same is true of a bar in compression provided the bar is not subject to buckling. In the case of a pin subjected to shear, consider only the average shear stress on the cross section, which is equivalent to assuming that the shear stress is uniformly distributed. Similarly, consider only an average value of the bearing stress acting on the projected area of the pin.

Therefore, in all four of the preceding cases, the **allowable load** (also called the *permissible load* or the *safe load*) is equal to the allowable stress times the area over which it acts:

$$\text{Allowable load} = (\text{Allowable stress})(\text{Area}) \tag{1-29}$$

For bars in direct *tension* and *compression* (no buckling), this equation becomes

$$P_{\text{allow}} = \sigma_{\text{allow}} A \tag{1-30}$$

in which σ_{allow} is the permissible normal stress and A is the cross-sectional area of the bar. If the bar has a hole through it, the *net area* is normally used when the bar is in tension. The **net area** is the gross cross-sectional area minus the area removed by the hole. For compression, the gross area may be used if the hole is filled by a bolt or pin that can transmit the compressive stresses.

For pins in *direct shear*, Eq. (1-29) becomes

$$P_{\text{allow}} = \tau_{\text{allow}} A \tag{1-31}$$

in which τ_{allow} is the permissible shear stress and A is the area over which the shear stresses act. If the pin is in single shear, the area is the cross-sectional area of the pin; in double shear, it is twice the cross-sectional area.

Finally, the permissible load based upon *bearing* is

$$P_{\text{allow}} = \sigma_b A_b \tag{1-32}$$

in which σ_b is the allowable bearing stress and A_b is the projected area of the pin or other surface over which the bearing stresses act.

The following example illustrates how allowable loads are determined when the allowable stresses for the material are known.

Example 1-12

A steel bar serving as a vertical hanger to support heavy machinery in a factory is attached to a support by the bolted connection shown in Fig. 1-61. Two clip angles (thickness $t_c = 9.5$ mm) are fastened to an upper support by bolts 1 and 2 each with a diameter of 12 mm; each bolt has a washer with a diameter of $d_w = 28$ mm. The main part of the hanger is attached to the clip angles by a single bolt (bolt 3 in Fig. 1-61a) with a diameter of $d = 25$ mm. The hanger has a rectangular cross section with a width of $b_1 = 38$ mm and thickness of $t = 13$ mm, but at the bolted connection, the hanger is enlarged to a width of $b_2 = 75$ mm. Determine the allowable value of the tensile load P in the hanger based upon the following considerations.

(a) The allowable tensile stress in the main part of the hanger is 110 MPa.

(b) The allowable tensile stress in the hanger at its cross section through the bolt 3 hole is 75 MPa. (The permissible stress at this section is lower because of the stress concentrations around the hole.)

(c) The allowable bearing stress between the hanger and the shank of bolt 3 is 180 MPa.

(d) The allowable shear stress in bolt 3 is 45 MPa.

(e) The allowable normal stress in bolts 1 and 2 is 160 MPa.

(f) The allowable bearing stress between the washer and the clip angle at either bolt 1 or 2 is 65 MPa.

(g) The allowable shear stress through the clip angle at bolts 1 and 2 is 35 MPa.

Solution:

Use a four-step problem-solving approach to find the allowable value of the tensile load P in the hanger based upon a variety of different allowable stresses in the different connection components.

1. ***Conceptualize***: Start by sketching a series of free-body diagrams to find the forces acting on each connection component. Express the force on each component in terms of an allowable stress times the associated area upon which it acts. This force is the allowable value of applied load P for that stress condition. Each of the seven stress states [(a)–(g) in the problem statement] and the associated applied load are illustrated in this example's figures; each is adjacent to the corresponding calculations in Step (3).

2. *Categorize*: Compute seven different values of the allowable load P, each based on an allowable stress and a corresponding area. The minimum value of load P will control.

Numerical data for the hanger connection design shown in Fig. 1-61 are as follows.

(a) Connection component dimensions:

$$t_c = 9.5 \text{ mm} \quad t = 13 \text{ mm} \quad b_1 = 38 \text{ mm} \quad b_2 = 75 \text{ mm}$$
$$d_1 = 12 \text{ mm} \quad d = 25 \text{ mm} \quad d_w = 28 \text{ mm}$$

(b) Allowable stresses:

$$\sigma_a = 110 \text{ MPa} \quad \sigma_{a3} = 75 \text{ MPa} \quad \sigma_{ba3} = 180 \text{ MPa} \quad \tau_{a3} = 45 \text{ MPa}$$
$$\tau_{a1} = 35 \text{ MPa} \quad \sigma_{a1} = 160 \text{ MPa} \quad \sigma_{ba1} = 65 \text{ MPa}$$

3. *Analyze*:

Part (a): Find the allowable load based upon the *stress in the main part of the hanger* (Fig. 1-61c). This is equal to the allowable stress in tension (110 MPa) times the cross-sectional area of the hanger (Eq. 1-30):

$$P_a = \sigma_a b_1 t = (110 \text{ MPa})(38 \text{ mm} \times 13 \text{ mm}) = 54.3 \text{ kN}$$

A load greater than this value will overstress the main part of the hanger (that is, the actual stress will exceed the allowable stress), thereby reducing the factor of safety.

FIGURE 1-61

Example 1-12: Vertical hanger subjected to a tensile load P: (a) front view of bolted connection and (b) side view of connection

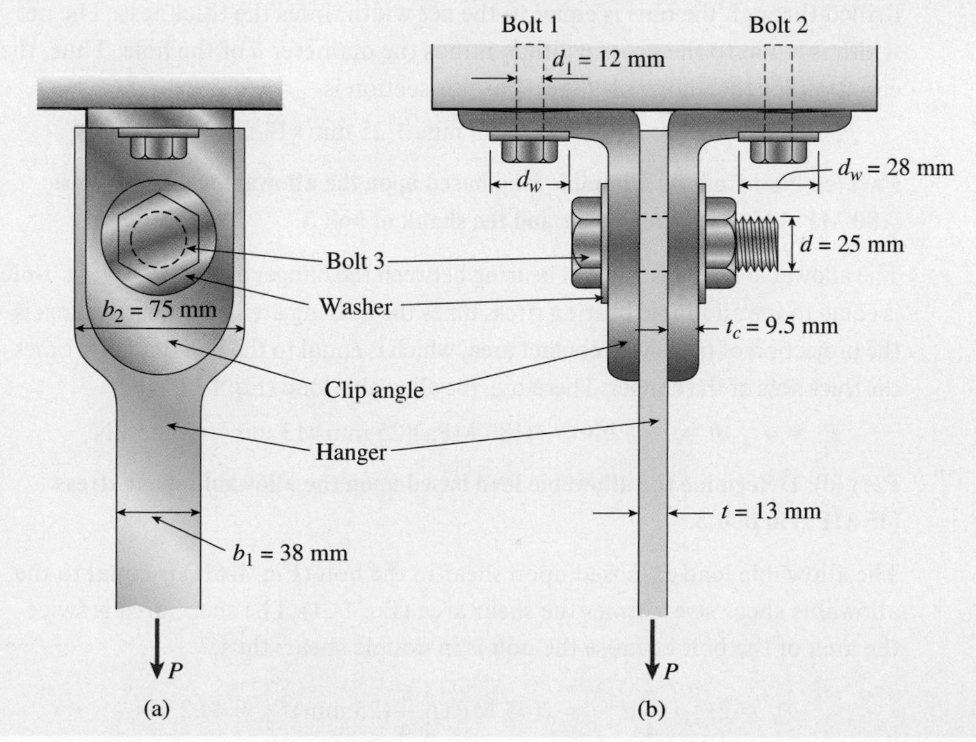

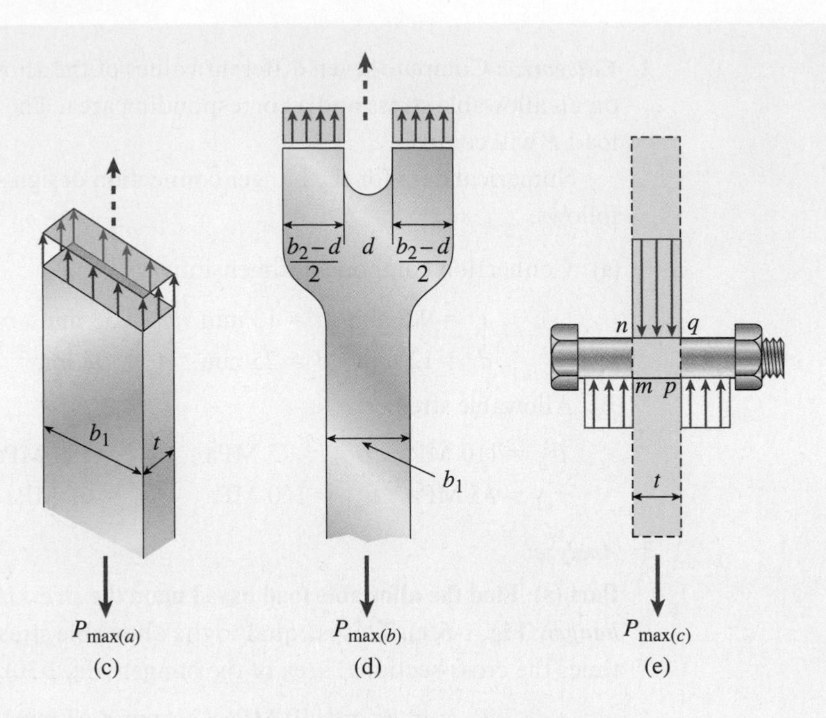

Part (b): Find the allowable load based upon the allowable tensile stress (75 MPa) in the hanger at its cross section through the bolt 3 hole.

At the cross section of the hanger through the bolt hole (Fig. 1-61d), make a similar calculation but with a different allowable stress and a different area. The net cross-sectional area (that is, the area that remains after the hole is drilled through the bar) is equal to the net width times the thickness. The net width is equal to the gross width b_2 minus the diameter d of the hole. Thus, the equation for the allowable load P_b at this section is

$$P_b = \sigma_{a3}(b_2 - d)t = (75 \text{ MPa})(75 \text{ mm} - 25 \text{ mm})(13 \text{ mm}) = 48.8 \text{ kN}$$

Part (c): Now find the allowable load based upon the allowable bearing stress (180 MPa) between the hanger and the shank of bolt 3.

The allowable load based upon bearing between the hanger and the bolt (Fig. 1-61e) is equal to the allowable bearing stress times the bearing area. The bearing area is the projection of the actual contact area, which is equal to the bolt diameter times the thickness of the hanger. Therefore, the allowable load (Eq. 1-32) is

$$P_c = \sigma_{ba3}dt = 58.5 \text{ kN} = (180 \text{ MPa})(25 \text{ mm})(13 \text{ mm}) = 58.5 \text{ kN}$$

Part (d): Determine the allowable load based upon the allowable shear stress (45 MPa) in bolt 3.

The allowable load P_d based upon shear in the bolt (Fig. 1-61f) is equal to the allowable shear stress times the shear area (Eq. 1-31). The shear area is twice the area of the bolt because the bolt is in double shear; thus,

$$P_d = 2\tau_{a3}\left(\frac{\pi}{4}d^2\right) = 2(45 \text{ MPa})\left[\frac{\pi}{4}(25 \text{ mm})^2\right] = 44.2 \text{ kN}$$

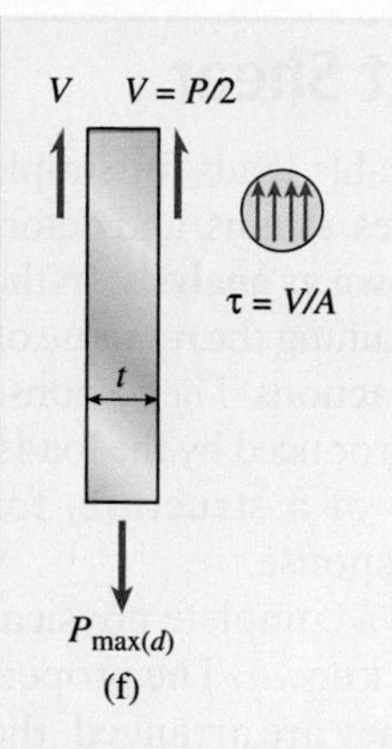

(f)

Part (e): Find the allowable load based upon the allowable normal stress (160 MPa) in bolts 1 and 2.

The allowable normal stress in bolts 1 and 2 is 160 MPa. Each bolt carries one half of the applied load P (see Fig. 1-61g). The allowable total load P_e is the product of the allowable normal stress in the bolt and the sum of the cross-sectional areas of bolts 1 and 2:

$$P_e = \sigma_{a1}(2)\left(\frac{\pi}{4}d_1^2\right) = (160 \text{ MPa})(2)\left[\frac{\pi}{4}(12 \text{ mm})^2\right] = 36.2 \text{ kN}$$

Part (f): Now find the allowable load based upon the allowable bearing stress (65 MPa) between the washer and the clip angle at either bolt 1 or 2.

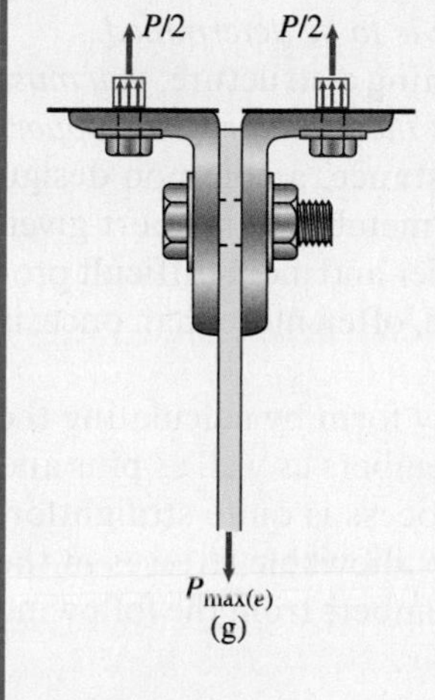

(g)

The allowable bearing stress between the washer and the clip angle at either bolt 1 or 2 is 65 MPa. Each bolt (1 or 2) carries one half of the applied load P (see Fig. 1-61h). The bearing area here is the ring-shaped circular area of the washer (the washer is assumed to fit snugly against the bolt). The allowable total load P_f is the allowable bearing stress on the washer times twice the area of the washer:

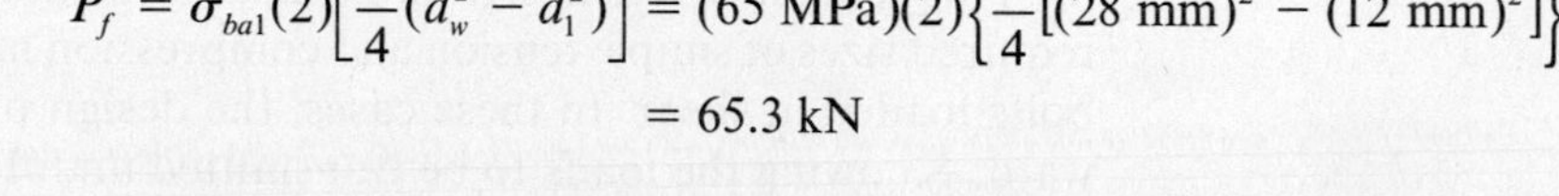

$$P_f = \sigma_{ba1}(2)\left[\frac{\pi}{4}(d_w^2 - d_1^2)\right] = (65 \text{ MPa})(2)\left\{\frac{\pi}{4}[(28 \text{ mm})^2 - (12 \text{ mm})^2]\right\}$$
$$= 65.3 \text{ kN}$$

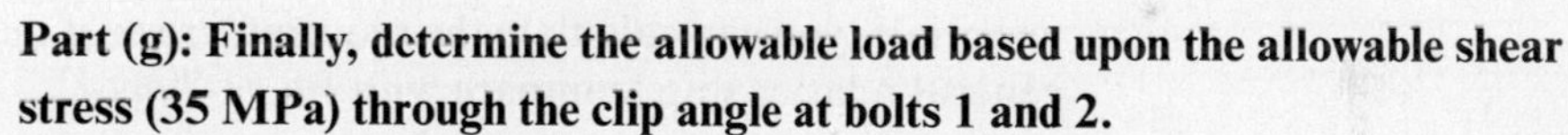

Part (g): Finally, determine the allowable load based upon the allowable shear stress (35 MPa) through the clip angle at bolts 1 and 2.

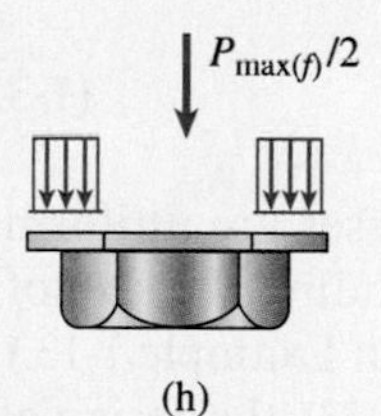

(h)

The allowable shear stress through the clip angle at bolts 1 and 2 is 35 MPa. Each bolt (1 or 2) carries one half of the applied load P (see Fig. 1-61i). The shear area at each bolt is equal to the circumference of the hole ($\pi \times d_w$) times the thickness of the clip angle (t_c).

The allowable total load P_g is the the allowable shear stress times twice the shear area:

$$P_g = \tau_{a1}(2)(\pi d_w t_c) = (35 \text{ MPa})(2)(\pi \times 28 \text{ mm} \times 9.5 \text{ mm}) = 58.5 \text{ kN}$$

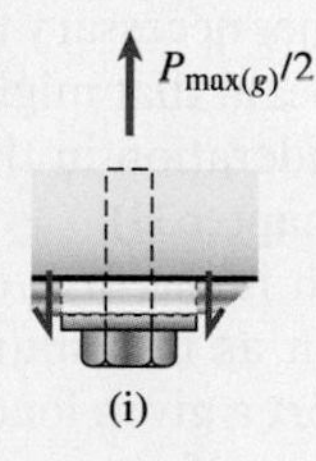

(i)

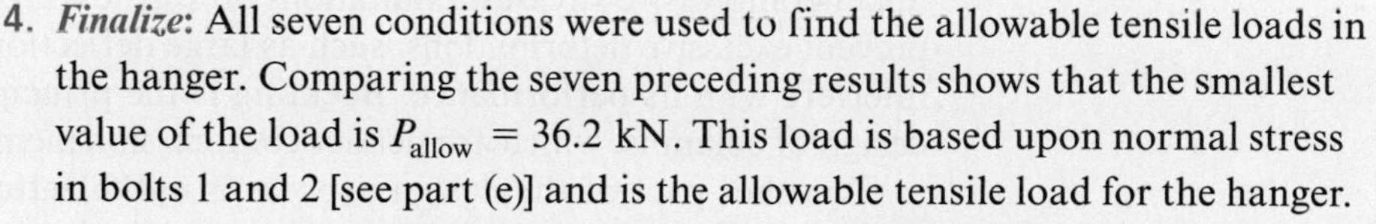

4. ***Finalize***: All seven conditions were used to find the allowable tensile loads in the hanger. Comparing the seven preceding results shows that the smallest value of the load is $P_{\text{allow}} = 36.2$ kN. This load is based upon normal stress in bolts 1 and 2 [see part (e)] and is the allowable tensile load for the hanger.

   A more refined analysis that includes the weight of the entire hanger assembly can be carried out as shown in Example 1-6. As in Example 1-11, note that these computed stresses are average values only and do not include localized effects such as stress concentrations around bolt holes.

1.10 Design For Axial Loads and Direct Shear

The preceding section discussed the determination of allowable loads for simple structures, and earlier sections showed how to find the stresses, strains, and deformations of bars. The determination of such quantities is known as **analysis**. In the context of mechanics of materials, analysis consists of determining the *response* of a structure to loads, temperature changes, and other physical actions. The response of a structure means the stresses, strains, and deformations produced by the loads.

Response also refers to the load-carrying capacity of a structure; for instance, the allowable load on a structure is a form of response.

A structure is said to be *known* (or *given*) when there is a complete physical description of the structure, that is, all of its *properties* are known. The properties of a structure include the types of members and how they are arranged, the dimensions of all members, the types of supports and where they are located, the materials used, and the properties of the materials. Thus, when analyzing a structure, *the properties are given, and the response is to be determined.*

The inverse process is called **design**. When designing a structure, *you must determine the properties of the structure in order that the structure will support the loads and perform its intended functions.* For instance, a common design problem in engineering is to determine the size of a member to support given loads. Designing a structure is usually a much lengthier and more difficult process than analyzing it—indeed, analyzing a structure, often more than once, is typically part of the design process.

This section covers design in its most elementary form by calculating the required sizes of simple tension and compression members as well as pins and bolts loaded in shear. In these cases, the design process is quite straightforward. Knowing the loads to be transmitted and the allowable stresses in the materials, you can calculate the required areas of members from the following general relationship [compare with Eq. (1-29)]:

$$\text{Required area} = \frac{\text{Load to be transmitted}}{\text{Allowable stress}} \tag{1-33}$$

Apply this equation to any structure in which the stresses are uniformly distributed over the area. (The use of this equation for finding the size of a cable in tension and the size of a pin in shear is illustrated in Example 1-13.)

In addition to **strength** considerations, as shown by Eq. (1-33), the design of a structure is likely to involve **stiffness** and **stability**. Stiffness refers to the ability of the structure to resist changes in shape (for instance, to resist stretching, bending, or twisting), and stability refers to the ability of the structure to resist buckling under compressive stresses. Limitations on stiffness are sometimes necessary to prevent excessive deformations, such as large deflections of a beam that might interfere with its performance. Buckling is the principal consideration in the design of columns, which are slender compression members (Chapter 11).

Another part of the design process is **optimization**, which is the task of designing the best structure to meet a particular goal, such as minimum weight. For instance, there are many structures that can support a given load, but in some circumstances the best structure is the lightest one. Of course, a goal such as minimum weight usually must be balanced against more general considerations, including the aesthetic, economic, environmental, political, and technical aspects of the particular design project.

When analyzing or designing a structure, the forces that act on it are **loads** and **reactions**. Loads are *active forces* that are applied to the structure by some external cause, such as gravity, water pressure, wind, and earthquake ground motion. Reactions are *passive forces* that are induced at the supports of the structure—their magnitudes and directions are determined by the nature of the structure itself. Thus, reactions must be calculated as part of the analysis, whereas loads are known in advance.

Example 1-13

Continuous cable ADB runs over a small frictionless pulley at D to support beam $OABC$, which is part of an entrance canopy for a building (see Fig. 1-62). Load $P = 4.5$ kN is applied at the end of the canopy at C. Assume that the canopy segment has weight $W = 7.5$ kN.

(a) Find cable force T and pin support reactions at O and D.

(b) Find the required cross-sectional area of cable ADB if the allowable normal stress is 125 MPa.

(c) Determine the required diameter of the pins at O, A, B, and D if the allowable stress in shear is 80 MPa.

(*Note*: The pins at O, A, B, and D are in double shear. Also, consider only load P and the weight W of the canopy; disregard the weight of cable ADB.)

FIGURE 1-62

(a) Inclined canopy at entrance to building; (b) two-dimensional model of one beam and supporting cable (© Archimage / Alamy Stock Photo)

(a)

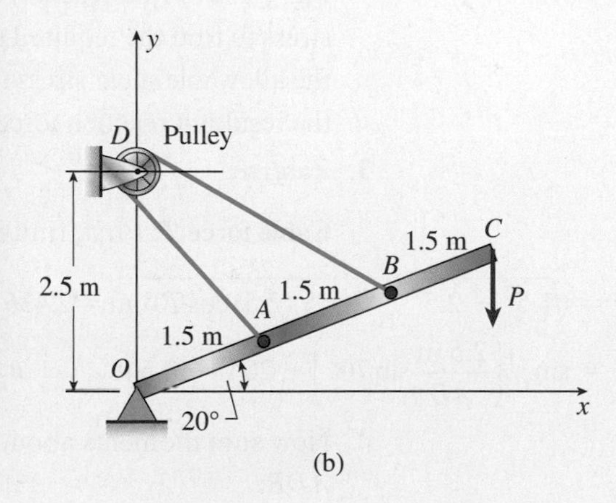

(b)

Solution:

Use the following four-step problem-solving approach.

1. ***Conceptualize***: Begin with a free-body diagram of beam $OABC$ (Fig. 1-63a). Also sketch free-body diagrams of the entire structure (Fig. 1-63b) and of joint D alone (Fig. 1-63c). Show cable force T and all applied and reaction force components.

FIGURE 1-63

Free-body diagrams of: (a) beam $OABC$; (b) beam and cable structure; (c) joint D

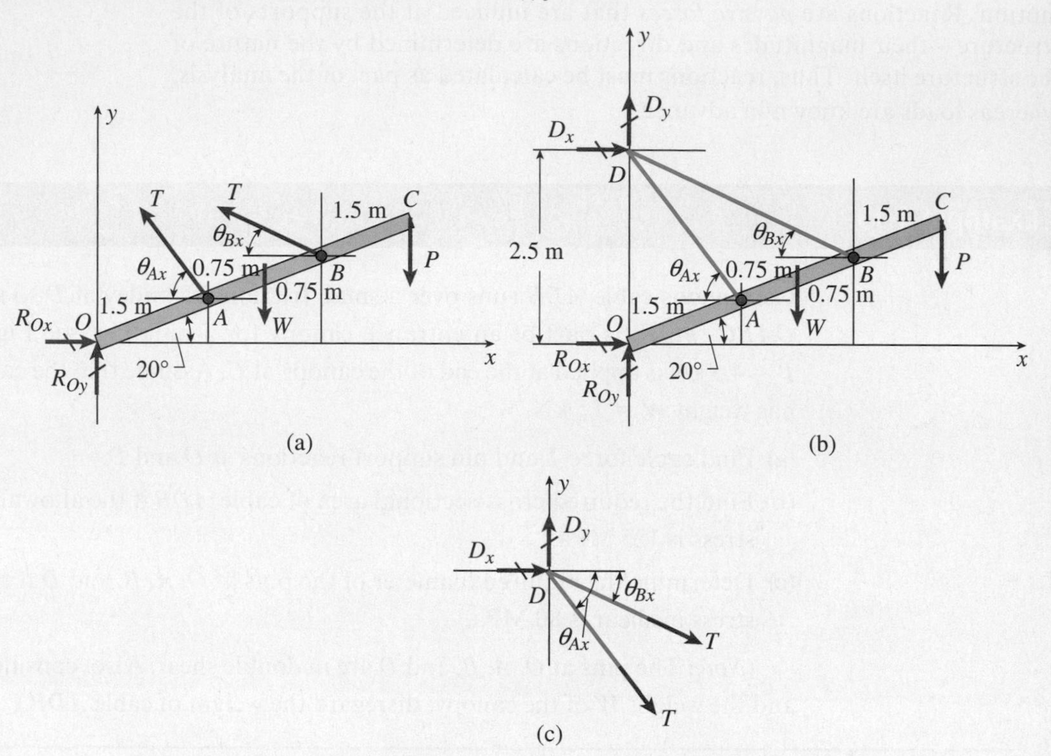

2. ***Categorize***: First, use the free-body diagram of beam $OABC$ (Fig. 1-63a) to find cable force T and reaction force components at O. Then use Fig. 1-63b or Fig. 1-63c to find reaction forces at D. Use cable force T and the allowable normal stress to find the required cross-sectional area of the cable. Also use force T and the allowable shear stress to find the required diameter of the pins at A and B. Use the resultant reaction forces at O and D to find pin diameters at these locations.

3. ***Analyze***:

Cable force T: First, find required distances and angles in Fig. 1-63a:

$$AD = \sqrt{1.5^2 + 2.5^2 - 2(1.5)(2.5)\cos 70^\circ}\ \text{m} = 2.436\ \text{m} \quad BD = \sqrt{3^2 + 2.5^2 - 2(3)(2.5)\cos 70^\circ}\ \text{m} = 3.181\ \text{m}$$

$$\theta_{Ax} = \sin^{-1}\left(\frac{2.5\ \text{m}}{AD}\sin 70^\circ\right) - 20^\circ = 54.648^\circ \qquad \theta_{Bx} = \sin^{-1}\left(\frac{2.5\ \text{m}}{BD}\sin 70^\circ\right) - 20^\circ = 27.603^\circ$$

Now sum moments about O in Fig. 1-63a to find tension T in continuous cable ADB:

$$T = \frac{W(2.25\ \text{m})(\cos 20^\circ) + P(4.5\ \text{m})(\cos 20^\circ)}{d_1 + d_2} = 9.527\ \text{kN} \qquad \text{(a)}$$

where

$$d_1 = \cos\theta_{Ax}(2.5\ \text{m} - AD\sin\theta_{Ax}) + \sin\theta_{Ax}(1.5\ \text{m}\ \cos 20^\circ) = 1.446\ \text{m}$$
$$d_2 = \cos\theta_{Bx}(2.5\ \text{m} - BD\sin\theta_{Bx}) + \sin\theta_{Bx}(3\ \text{m}\ \cos 20^\circ) = 2.215\ \text{m}$$

Reaction force at *O*: Sum forces in Fig. 1-63a to find reaction force components at *O*:

$$\Sigma F_x = 0 \quad R_{Ox} = T(\cos\theta_{Ax} + \cos\theta_{Bx}) = 13.95 \text{ kN}$$
$$\Sigma F_y = 0 \quad R_{Oy} = -T(\sin\theta_{Ax} + \sin\theta_{Bx}) + W + P = -0.184 \text{ kN}$$

The resultant reaction force at *O* is $R_{Ores} = \sqrt{R_{Ox}^2 + R_{Oy}^2} = 13.96 \text{ kN}$ **(b)**

Reaction force at *D*: Sum forces in Fig. 1-63c to find reaction force components at *D*:

$$\Sigma F_x = 0 \quad D_x = -T(\cos\theta_{Ax} + \cos\theta_{Bx}) = -13.95 \text{ kN}$$
$$\Sigma F_y = 0 \quad D_y = T(\sin\theta_{Ax} + \sin\theta_{Bx}) = 12.18 \text{ kN}$$

The resultant reaction force at *D* is

$$D_{res} = \sqrt{D_x^2 + D_y^2} = 18.53 \text{ kN} \qquad \textbf{(c)}$$

Cross-sectional area of cable *ADB*: Use the allowable normal stress of 125 MPa and cable force $T = 9.527$ kN [Eq. (a)] to find the required cross-sectional area of the cable:

$$A_{cable} = \frac{9.527 \text{ kN}}{125 \text{ MPa}} = 76.21 \text{ mm}^2$$

Required diameter of the pins at *O*, *A*, *B*, and *D*: All pins are in double shear. The allowable shear stress is $\tau_{allow} = 80$ MPa. Required diameters of each pin are computed as

$$\text{Pins } A, B: \quad A_{reqd} = \frac{T}{2\tau_{allow}} = \frac{9.527 \text{ kN}}{2(80 \text{ MPa})} = 59.54 \text{ mm}^2 \text{ so } d = \sqrt{\frac{4}{\pi}(59.54 \text{ mm}^2)} = 8.71 \text{ mm}$$

$$\text{Pin } O: \quad A_{reqd} = \frac{R_{Ores}}{2\tau_{allow}} = \frac{13.96 \text{ kN}}{2(80 \text{ MPa})} = 87.22 \text{ mm}^2 \text{ so } d = \sqrt{\frac{4}{\pi}(87.22 \text{ mm}^2)} = 10.54 \text{ mm}$$

$$\text{Pin } D: \quad A_{reqd} = \frac{D_{res}}{2\tau_{allow}} = \frac{18.53 \text{ kN}}{2(80 \text{ MPa})} = 115.8 \text{ mm}^2 \text{ so } d = \sqrt{\frac{4}{\pi}(115.8 \text{ mm}^2)} = 12.14 \text{ mm}$$

4. ***Finalize*:** In practice, other loads besides the weight of the canopy would have to be considered before making a final decision about the sizes of the cables and pins. Loads that could be important include wind loads, earthquake loads, and the weights of objects that might have to be supported temporarily by the structure. In addition, if cables *AD* and *BD* are *separate cables* (instead of one continuous cable *ADB*), the forces in the two cables are not equal in magnitude. The structure is now *statically indeterminate*, and the cable forces and the reactions at *O* and *D* cannot be determined using the equations of static equilibrium alone. Problems of this type are discussed in Chapter 2, Section 2.4 (see Example 2-7).

CHAPTER SUMMARY AND REVIEW

Chapter 1 covered mechanical properties of construction materials. Normal stresses and strains in bars loaded by centroidal axial loads were computed. Shear stresses and strains (as well as bearing stresses) in pin connections used to assemble simple structures such as trusses, were evaluated. Allowable levels of stress were calculated from appropriate factors of safety and used to set allowable loads that could be applied to the structure.

Some of the major concepts presented in this chapter are:

1. The principal objective of mechanics of materials is to determine the **stresses, strains, and displacements** in structures and their components due to the loads acting on them. These components include bars with axial loads, shafts in torsion, beams in bending, and columns in compression.

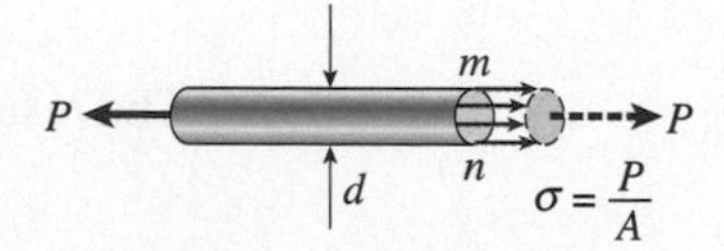

2. Prismatic bars subjected to tensile or compressive loads acting through the centroid of their cross section (to avoid bending) experience **normal stress** (σ) and **strain** (ε):

$$\sigma = \frac{P}{A} \qquad \varepsilon = \frac{\delta}{L}$$

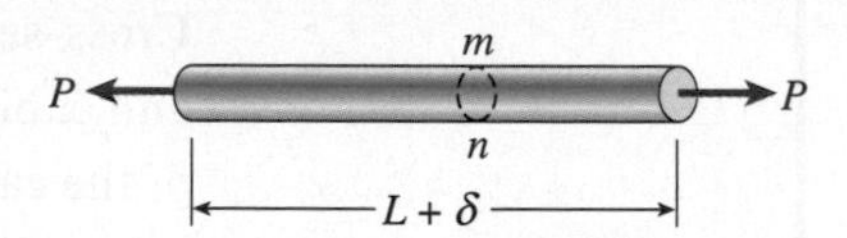

and either extension or contraction proportional to their lengths. These stresses and strains are **uniform** except near points of load application where high localized stresses, or **stress concentrations**, occur.

3. The **mechanical behavior** of various materials was displayed in a stress-strain diagram. **Ductile** materials such as mild steel have an initial linear relationship between normal stress and strain up to the **proportional limit** and are **linearly elastic** with stress and strain related by **Hooke's law**:

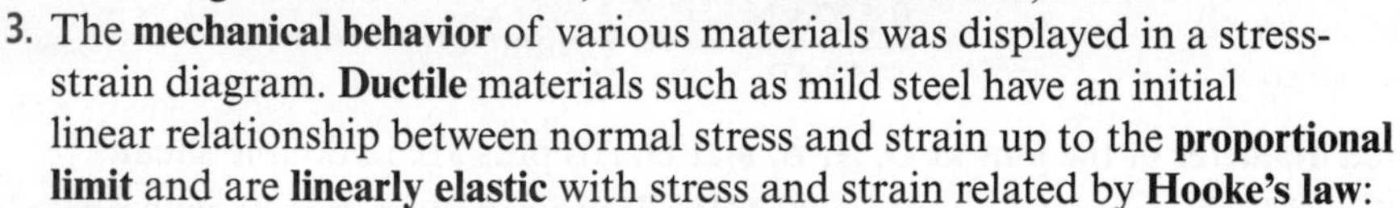

$$\sigma = E\varepsilon$$

They also have a well-defined yield point. Other ductile materials such as aluminum alloys typically do not have a clearly definable yield point, so an arbitrary yield stress is determined using the **offset method**.

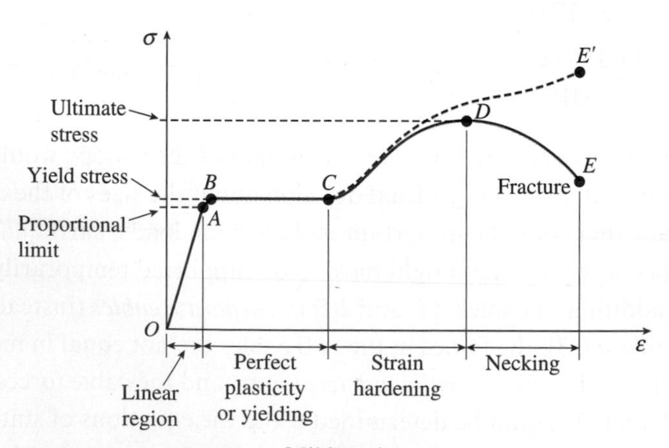

4. Materials that fail in tension at relatively low values of strain (such as concrete, stone, cast iron, glass ceramics, and a variety of metallic alloys) are classified as **brittle**. Brittle materials fail with only little elongation after the proportional limit.

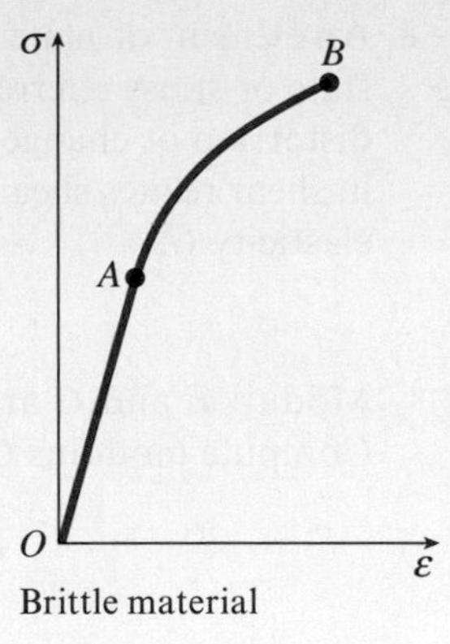

Brittle material

5. If a material remains within the elastic range, it can be loaded, unloaded, and loaded again without significantly changing its behavior. However, when loaded into the plastic range, the internal structure of the material is altered and its properties change. Loading and unloading behavior of materials depends on their **elasticity** and **plasticity** properties such as the **elastic limit** and the possibility of **permanent set** (residual strain). Sustained loading over time may lead to **creep** and **relaxation**.

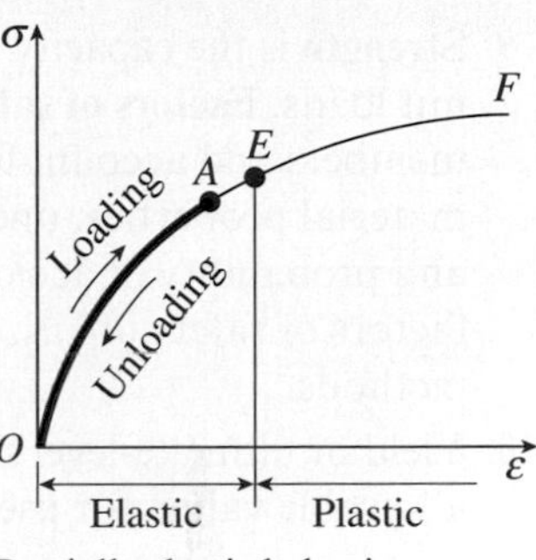

Partially elastic behavior

6. Axial elongation of bars loaded in tension is accompanied by lateral contraction; the ratio of lateral strain to normal strain is known as **Poisson's ratio** (ν):

$$\nu = -\frac{\text{lateral strain}}{\text{axial strain}} = -\frac{\varepsilon'}{\varepsilon}$$

Poisson's ratio remains constant throughout the linearly elastic range, provided the material is homogeneous and isotropic. Most of the examples and problems in the text are solved with the assumption that the material is linearly elastic, homogeneous, and **isotropic**.

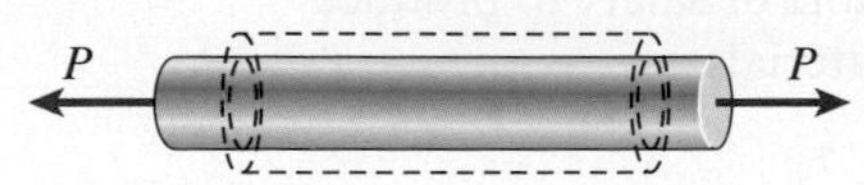

Lateral contraction

7. **Normal** stresses (σ) act perpendicular to the surface of the material, and **shear** stresses (τ) act tangential to the surface. In bolted connections between plates, the bolts are subjected to either single or double shear (τ_{aver}) where the average shear stress is

$$\tau_{aver} = \frac{V}{A}$$

Average **bearing** stresses (σ_b) act on the rectangular projected area (A_b) of the actual curved contact surface between a bolt and plate:

$$\sigma_b = \frac{F_b}{A_b}$$

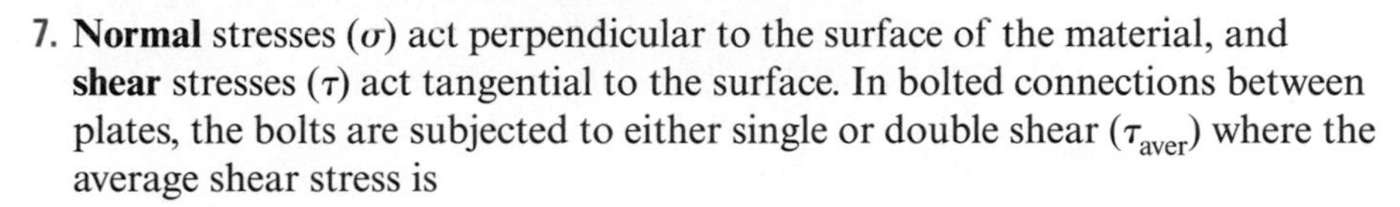

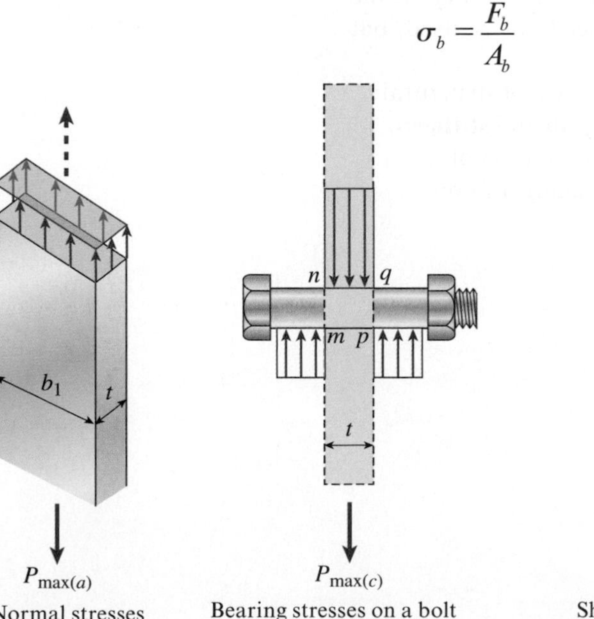

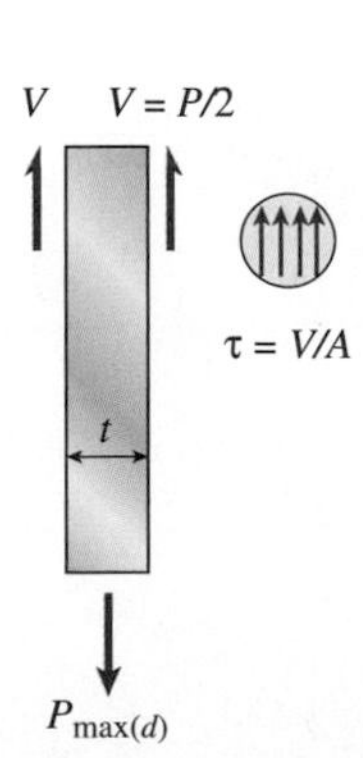

Normal stresses

Bearing stresses on a bolt passing through a bar

Shear stresses on a bolt passing through a bar

8. An element of material acted on by only shear stresses and strains is in a state of stress referred to as **pure shear**. Shear strain (γ) is a measure of the distortion or change in shape of the element in pure shear. Hooke's law in shear relates shear stress (τ) to shear strain by the shearing modulus of elasticity G:

$$\tau = G\gamma$$

Moduli E and G are not independent elastic properties of the material. Compute modulus G from E using Poisson's ratio:

$$G = \frac{E}{2(1+\nu)}$$

9. **Strength** is the capacity of a structure or component to support or transmit loads. **Factors of safety** relate actual to required strength of structural members and account for a variety of uncertainties, such as variations in material properties, uncertain magnitudes or distributions of loadings, and probability of accidental overload. Because of these uncertainties, factors of safety (n_1, n_2, n_3, and n_4) must be determined using probabilistic methods.
10. Yield or ultimate-level stresses are divided by factors of safety to produce allowable values for use in design. For **ductile** materials,

$$\sigma_{\text{allow}} = \frac{\sigma_Y}{n_1} \qquad \tau_{\text{allow}} = \frac{\tau_Y}{n_2}$$

while for **brittle** materials,

$$\sigma_{\text{allow}} = \frac{\sigma_U}{n_3} \qquad \tau_{\text{allow}} = \frac{\tau_U}{n_4}$$

A typical value of n_1 and n_2 is 1.67, while n_3 and n_4 might be 2.8.

For a pin-connected member in axial tension, the **allowable load** depends on the allowable stress times the appropriate area (such as the net cross-sectional area for bars acted on by centroidal tensile loads, cross-sectional area of pin for pins in shear, and projected area for bolts in bearing). If the bar is in compression, the net cross-sectional area need not be used, but buckling may be an important consideration.
11. **Design** is the iterative process by which the appropriate size of structural members is determined to meet a variety of both **strength** and **stiffness** requirements. Incorporation of factors of safety into design is not a simple matter because both strength and failure have many different meanings.

PROBLEMS Chapter 1

1.3 Statics Review

Introductory Problems

1.3-1 Find support reactions at A and B and then calculate the axial force N, shear force V, and bending moment M at mid-span of AB. Let $L = 4.25$ m, $q_0 = 175$ N/m, $P = 225$ N, and $M_0 = 410$ N·m.

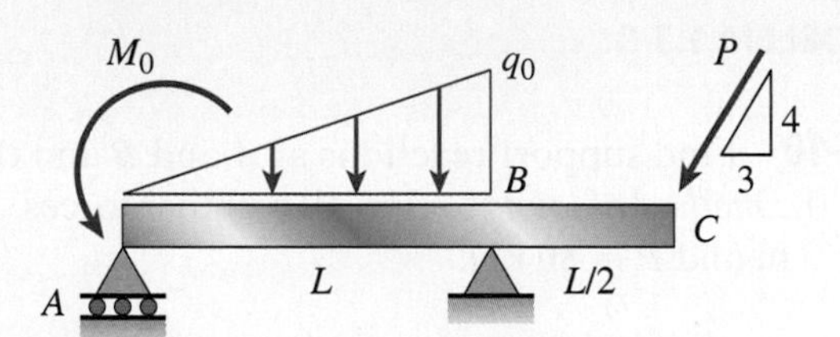

PROBLEM 1.3-1

1.3-2 Find support reactions at A and B and then calculate the axial force N, shear force V, and bending moment M at mid-span of AB. Let $L = 4$ m, $q_0 = 160$ N/m, $P = 200$ N, and $M_0 = 380$ N · m.

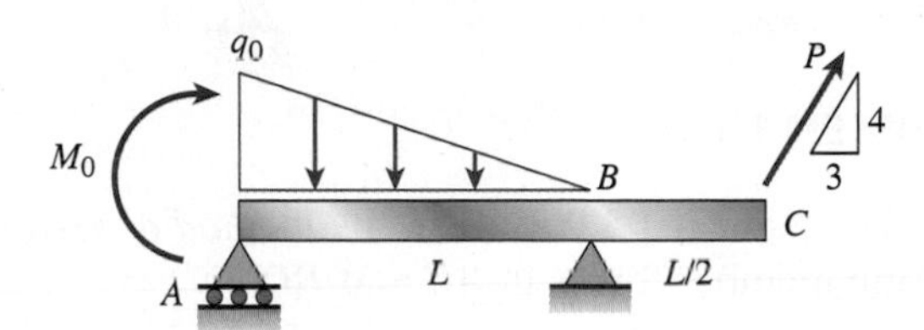

PROBLEM 1.3-2

1.3-3 Segments AB and BC of beam ABC are pin connected a small distance to the right of joint B (see figure). Axial loads act at A and at the mid-span of AB. A concentrated moment is applied at joint B.

(a) Find reactions at supports A, B, and C.

(b) Find internal stress resultants N, V, and M at 4.5 m.

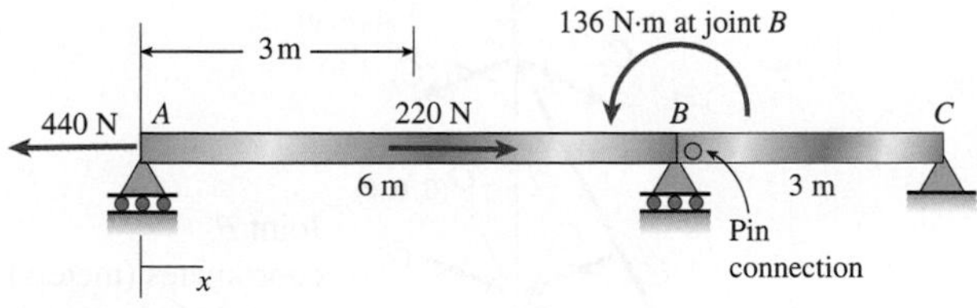

PROBLEM 1.3-3

1.3-4 Segments AB and BCD of beam $ABCD$ are pin connected at $x = 4$ m. The beam is supported by a sliding support at A and roller supports at C and D (see figure). A triangularly distributed load with peak intensity of 80 N/m acts on BC. A concentrated moment is applied at joint D.

(a) Find reactions at supports A, C, and D.

(b) Find internal stress resultants N, V, and M at $x = 5$ m.

(c) Repeat parts (a) and (b) for the case of the roller support at C replaced by a linear spring of stiffness $k_y = 200$ kN/m (see figure).

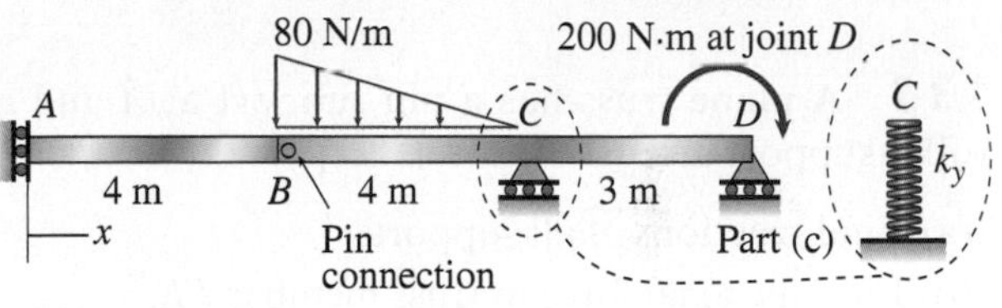

PROBLEM 1.3-4

1.3-5 Segments AB and BCD of beam $ABCD$ are pin connected at $x = 3$ m. The beam is supported by a pin support at A and roller supports at C and D; the roller at D is rotated by 30° from the x axis (see figure). A trapezoidal distributed load on BC varies in intensity from 74 N/m at B to 37 N/m at C. A concentrated moment is applied at joint A, and a 180 N inclined load is applied at the mid-span of CD.

(a) Find reactions at supports A, C, and D.

(b) Find the resultant force in the pin connection at B.

(c) Repeat parts (a) and (b) if a rotational spring ($k_r = 68$ N·m) is added at A *and* the roller at C is removed.

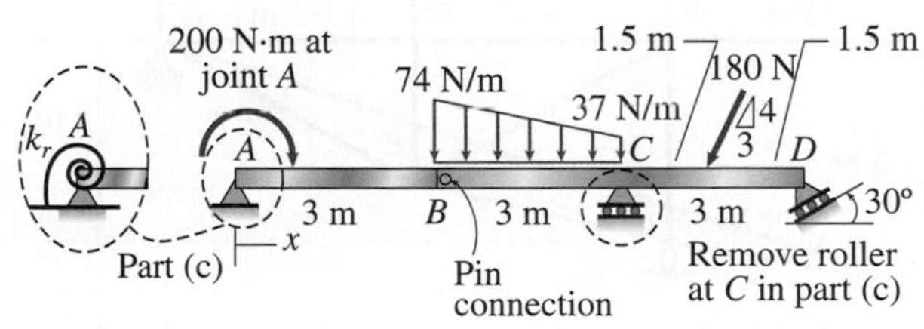

PROBLEM 1.3-5

1.3-6 Consider the plane truss with a pin support at joint 3 and a vertical roller support at joint 5 (see figure).

(a) Find reactions at support joints 3 and 5.

(b) Find axial forces in truss members 11 and 13.

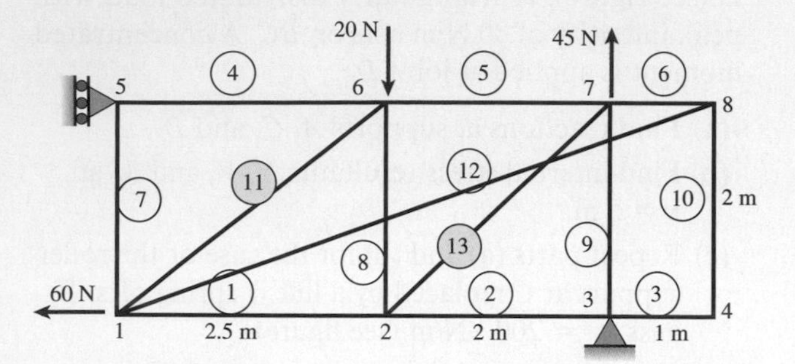

PROBLEM 1.3-6

1.3-7 A plane truss has a pin support at A and a roller support at E (see figure).

(a) Find reactions at all supports.

(b) Find the axial force in truss member FE.

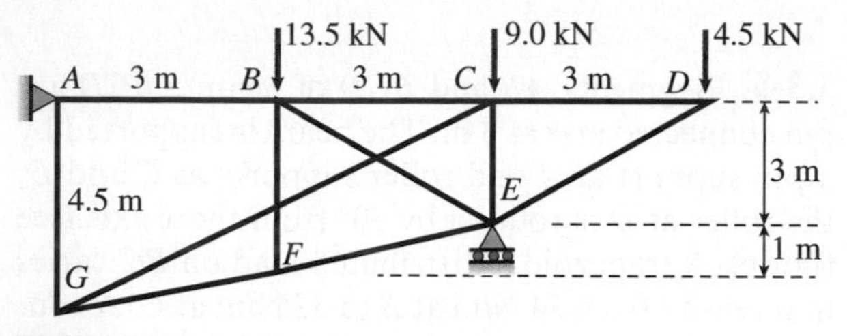

PROBLEM 1.3-7

1.3-8 A plane truss has a pin support at F and a roller support at D (see figure).

(a) Find reactions at both supports.

(b) Find the axial force in truss member FE.

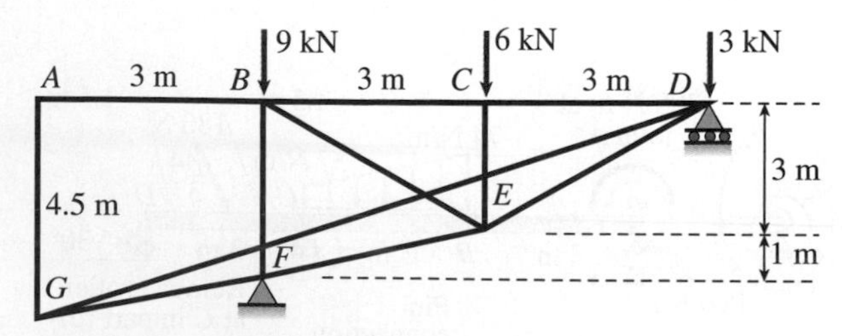

PROBLEM 1.3-8

1.3-9 Find support reactions at A and B and then use the *method of joints* to find all member forces. Let $c = 2.5$ m and $P = 90$ kN.

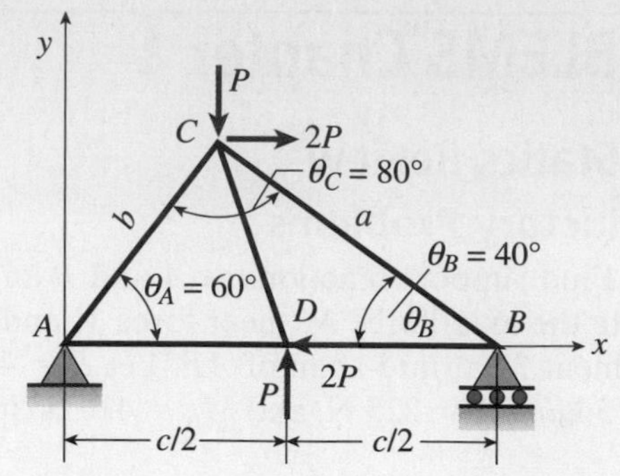

PROBLEM 1.3-9

1.3-10 Find support reactions at A and B and then use the *method of joints* to find all member forces. Let $b = 3$ m and $P = 80$ kN.

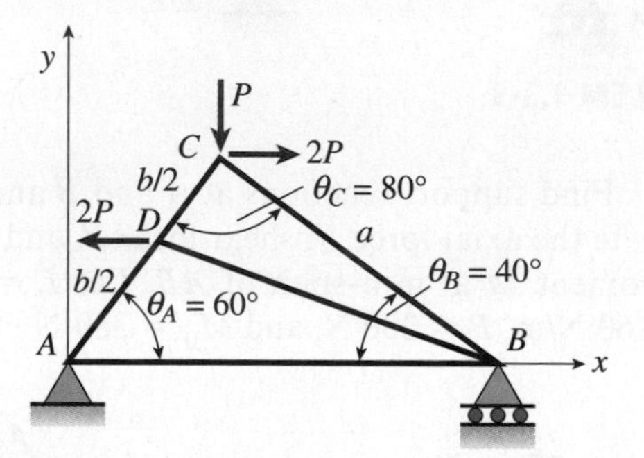

PROBLEM 1.3-10

1.3-11 Repeat 1.3-9 but use the *method of sections* to find member forces in AC and BD.

1.3-12 Repeat 1.3-10 but use the *method of sections* to find member forces in AB and DC.

1.3-13 A space truss has three-dimensional pin supports at joints O, B, and C. Load P is applied at joint A and acts toward point Q. Coordinates of all joints are given in meters (see figure).

(a) Find reaction force components B_x, B_z, and O_z.

(b) Find the axial force in truss member AC.

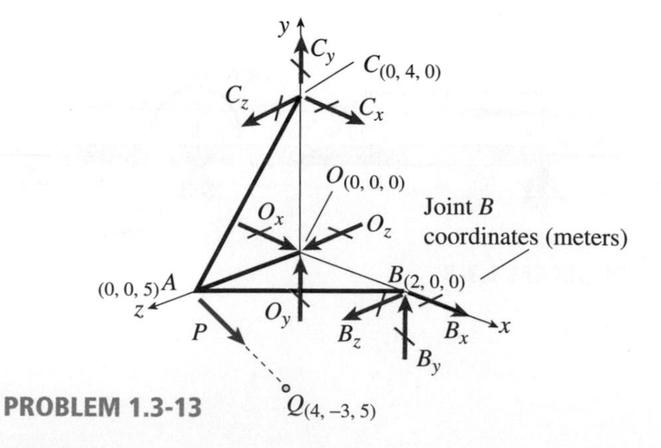

PROBLEM 1.3-13

1.3-14 A space truss is restrained at joints O, A, B, and C, as shown in the figure. Load P is applied at joint A and load $2P$ acts downward at joint C.

(a) Find reaction force components A_x, B_y, and B_z in terms of load variable P.

(b) Find the axial force in truss member AB in terms of load variable P.

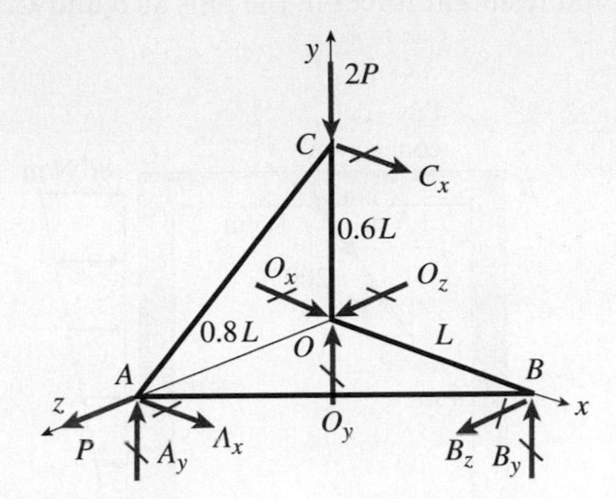

PROBLEM 1.3-14

1.3-15 A space truss is restrained at joints A, B, and C, as shown in the figure. Load $2P$ is applied at in the $-x$ direction at joint A, load $3P$ acts in the $+z$ direction at joint B, and load P is applied in the $+z$ direction at joint C. Coordinates of all joints are given in terms of dimension variable L (see figure).

(a) Find reaction force components A_y and A_z in terms of load variable P.

(b) Find the axial force in truss member AB in terms of load variable P.

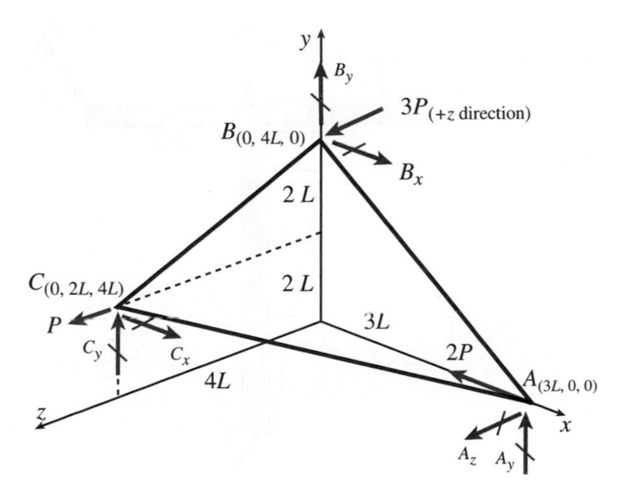

PROBLEM 1.3-15

1.3-16 A space truss is restrained at joints A, B, and C, as shown in the figure. Load P acts in the +z direction at joint B and in the −z directions at joint C. Coordinates of all joints are given in terms of dimension variable L (see figure). Let $P = 5$ kN and $L = 2$ m.

(a) Find the reaction force components A_z and B_x.

(b) Find the axial force in truss member AB.

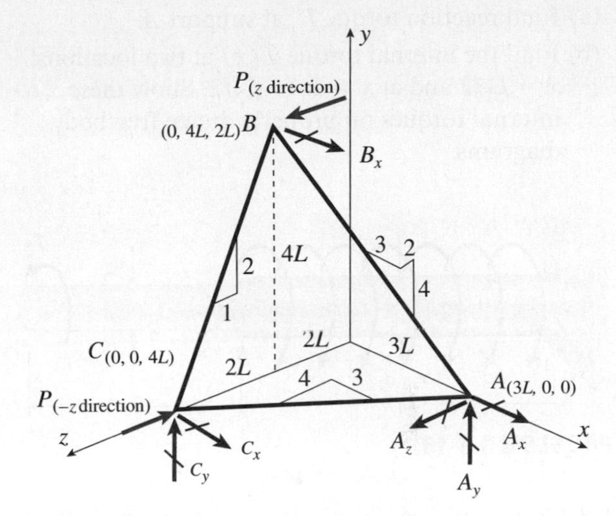

PROBLEM 1.3-16

1.3-17 A stepped shaft ABC consisting of two solid, circular segments is subjected to torques T_1 and T_2 acting in opposite directions, as shown in the figure. The larger segment of the shaft has a diameter of $d_1 = 58$ mm and a length $L_1 = 0.75$ m; the smaller segment has a diameter $d_2 = 44$ mm and a length $L_2 = 0.5$ m. The torques are $T_1 = 2400$ N·m and $T_2 = 1130$ N·m.

(a) Find reaction torque T_A at support A.

(b) Find the internal torque $T(x)$ at two locations: $x = L_1/2$ and $x = L_1 + L_2/2$. Show these internal torques on properly drawn free-body diagrams (FBDs).

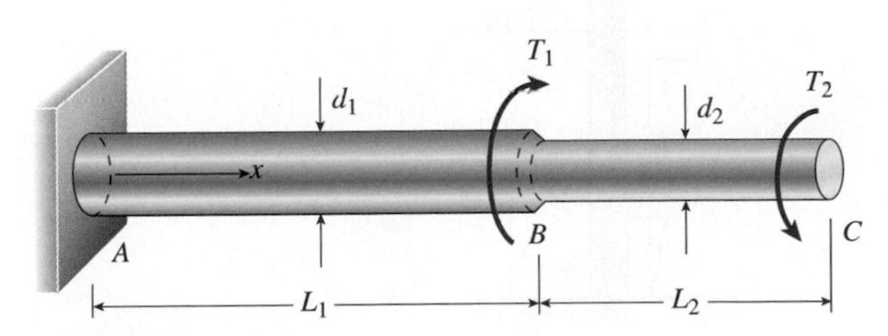

PROBLEM 1.3-17

1.3-18 A stepped shaft ABC consisting of two solid, circular segments is subjected to uniformly distributed torque t_1 acting over segment 1 and concentrated torque T_2 applied at C, as shown in the figure. Segment 1 of the shaft has a diameter of $d_1 = 57$ mm and length of $L_1 = 0.75$ m; segment 2 has a diameter $d_2 = 44$ mm and length $L_2 = 0.5$ m. Torque intensity $t_1 = 3100$ N · m/m and $T_2 = 1100$ N · m.

(a) Find reaction torque T_A at support A.

(b) Find the internal torque $T(x)$ at two locations: $x = L_1/2$ and at $x = L_1 + L_2/2$. Show these internal torques on properly drawn free-body diagrams.

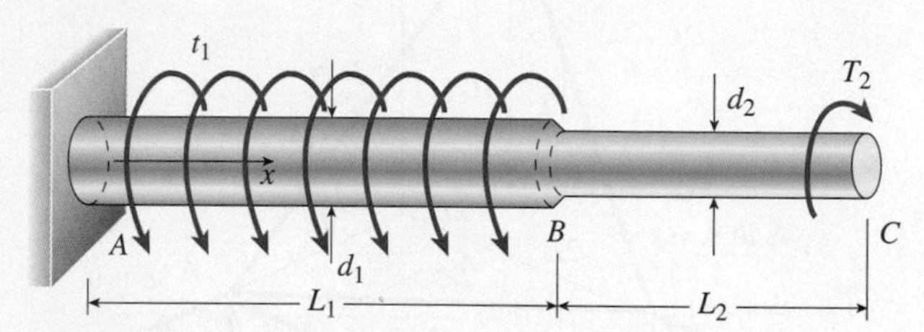

PROBLEM 1.3-18

1.3-19 A plane frame is restrained at joints A and C, as shown in the figure. Members AB and BC are pin connected at B. A triangularly distributed lateral load with a peak intensity of 1300 N/m acts on AB. A concentrated moment is applied at joint C.

(a) Find reactions at supports A and C.

(b) Find internal stress resultants N, V, and M at $x = 1.0$ m on column AB.

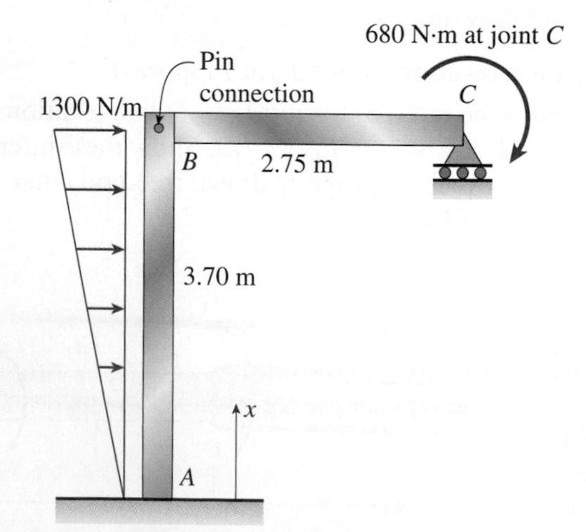

PROBLEM 1.3-19

1.3-20 A plane frame is restrained at joints A and D, as shown in the figure. Members AB and BCD are pin connected at B. A triangularly distributed lateral load with peak intensity of 80 N/m acts on CD. An inclined concentrated force of 200 N acts at the mid-span of BC.

(a) Find reactions at supports A and D.

(b) Find resultant forces in the pins at B and C.

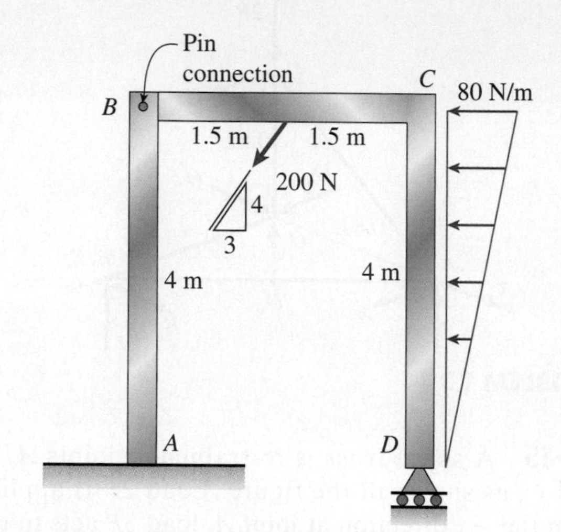

PROBLEM 1.3-20

1.3-21 Find support reactions at A and D and then calculate the axial force N, shear force V, and bending moment M at mid-span of AB. Let $L = 4.25$ m, $q_0 = 175$ N/m, $P = 225$ N, and $M_0 = 400$ N·m.

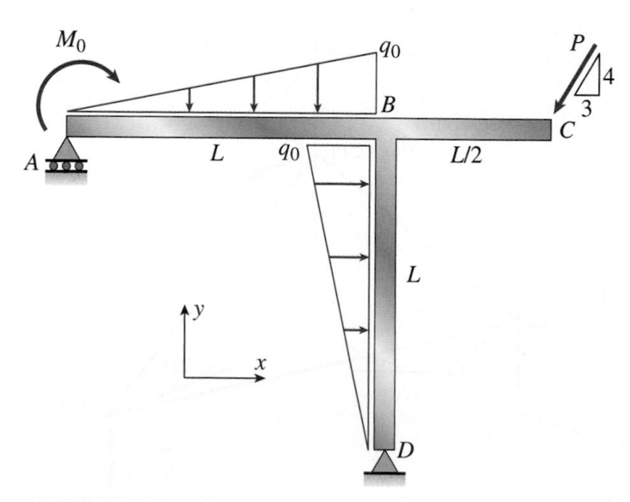

PROBLEM 1.3-21

1.3-22 Find support reactions at A and D and then calculate the axial force N, shear force V, and bending moment M at mid-span of column BD. Let $L = 4$ m, $q_0 = 160$ N/m, $P = 200$ N, and $M_0 = 380$ N · m.

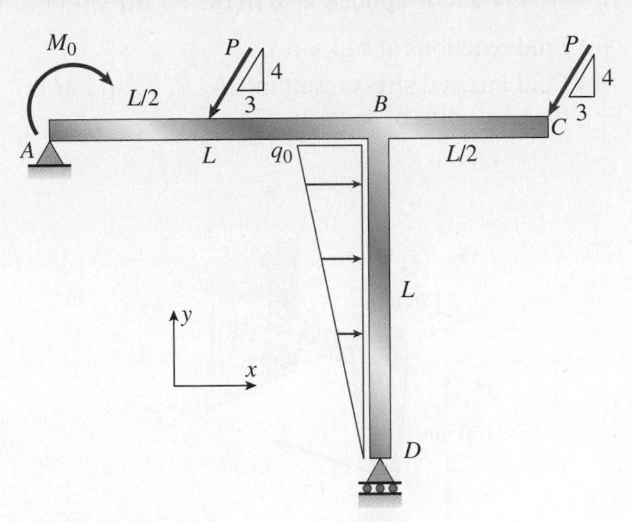

PROBLEM 1.3-22

1.3-23 A 1900 N trap door (AB) is supported by a strut (BC) which is pin connected to the door at B (see figure).

(a) Find reactions at supports A and C.
(b) Find internal stress resultants N, V, and M on the trap door at 0.5 m from A.

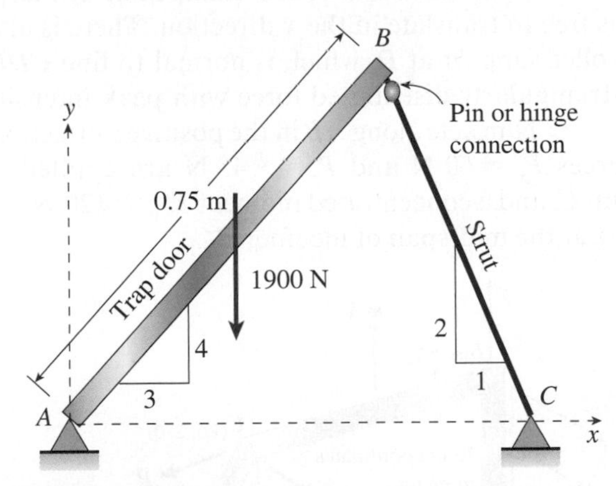

PROBLEM 1.3-23

1.3-24 A plane frame is constructed by using a pin connection between segments ABC and CDE. The frame has pin supports at A and E and joint loads at B and D (see figure).

(a) Find reactions at supports A and E.
(b) Find the resultant force in the pin at C.

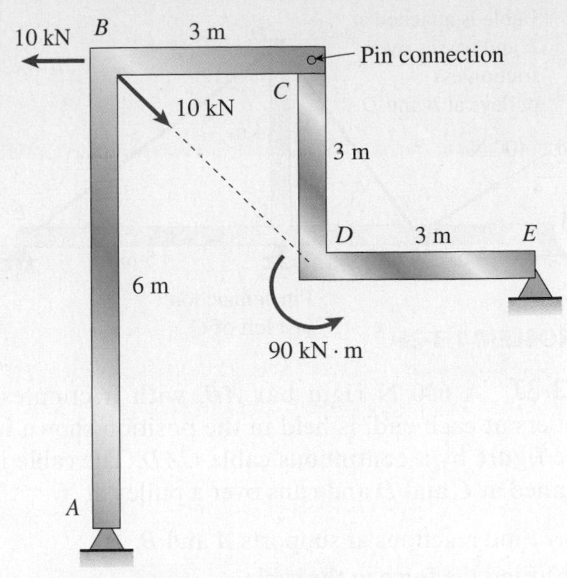

PROBLEM 1.3-24

1.3-25 A plane frame with pin supports at A and E has a cable attached at C, which runs over a frictionless pulley at F (see figure). The cable force is known to be 2.25 kN.

(a) Find reactions at supports A and E.
(b) Find internal stress resultants N, V, and M at point H.

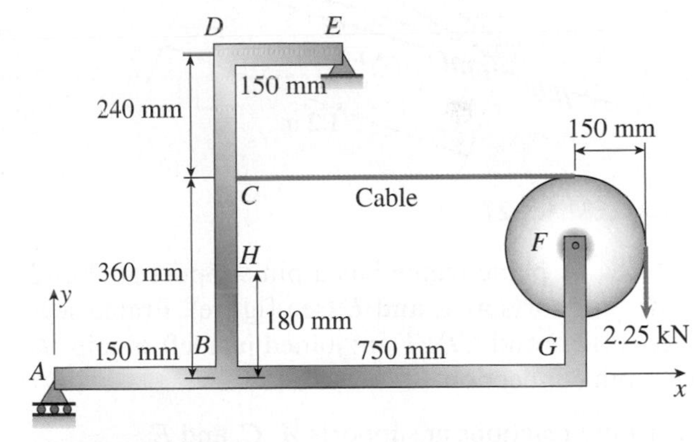

PROBLEM 1.3-25

1.3-26 A plane frame with a pin support at A and roller supports at C and E has a cable attached at E, which runs over frictionless pulleys at D and B (see figure). The cable force is known to be 400 N. There is a pin connection just to the left of joint C.

(a) Find reactions at supports A, C, and E.
(b) Find internal stress resultants N, V, and M just to the right of joint C.
(c) Find resultant force in the pin near C.

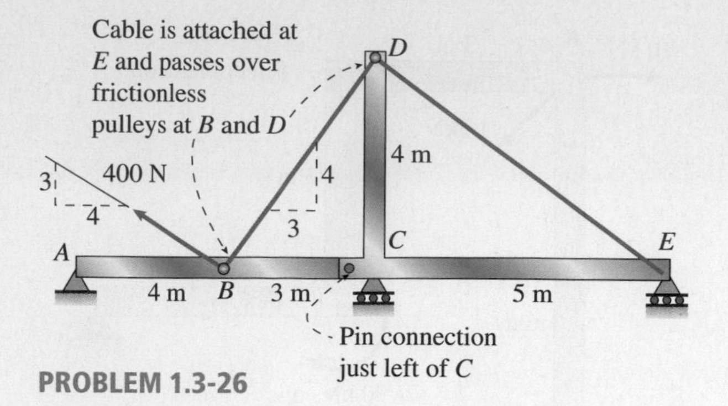

PROBLEM 1.3-26

1.3-27 A 650 N rigid bar AB, with frictionless rollers at each end, is held in the position shown in the figure by a continuous cable CAD. The cable is pinned at C and D and runs over a pulley at A.

(a) Find reactions at supports A and B.
(b) Find the force in the cable.

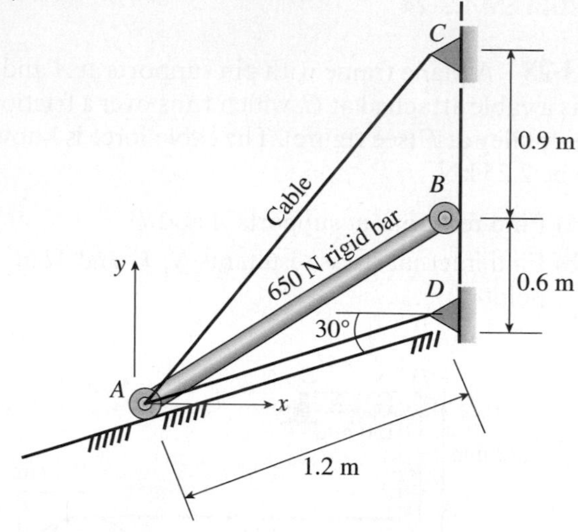

PROBLEM 1.3-27

1.3-28 A plane frame has a pin support at A and roller supports at C and E (see figure). Frame segments ABD and $CDEF$ are joined just left of joint D by a pin connection.

(a) Find reactions at supports A, C, and E.
(b) Find the resultant force in the pin just left of D.

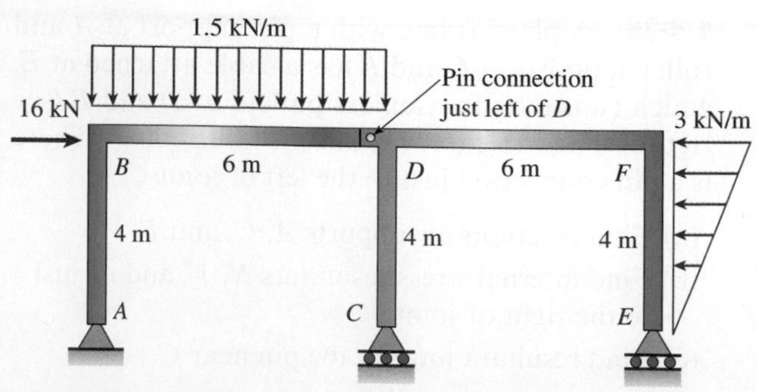

PROBLEM 1.3-28

1.3-29 A special vehicle brake is clamped at O when the brake force P_1 is applied (see figure). Force $P_1 = 220$ N and lies in a plane that is parallel to the x-z plane and is applied at C normal to line BC. Force $P_2 = 180$ N and is applied at B in the $-y$ direction.

(a) Find reactions at support O.
(b) Find internal stress resultants N, V, T, and M at the mid-point of segment OA.

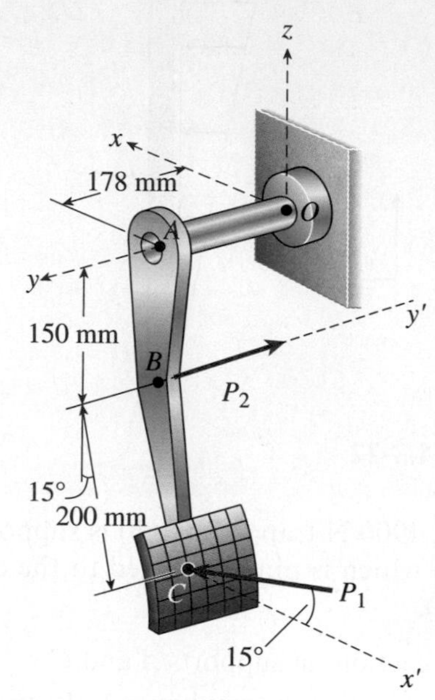

PROBLEM 1.3-29

1.3-30 Space frame $ABCD$ is clamped at A, except it is free to translate in the x direction. There is also a roller support at D, which is normal to line CDE. A triangularly distributed force with peak intensity $q_0 = 75$ N/m acts along AB in the positive z direction. Forces $P_x = 60$ N and $P_z = -45$ N are applied at joint C, and a concentrated moment $M_y = 120$ N · m acts at the mid-span of member BC.

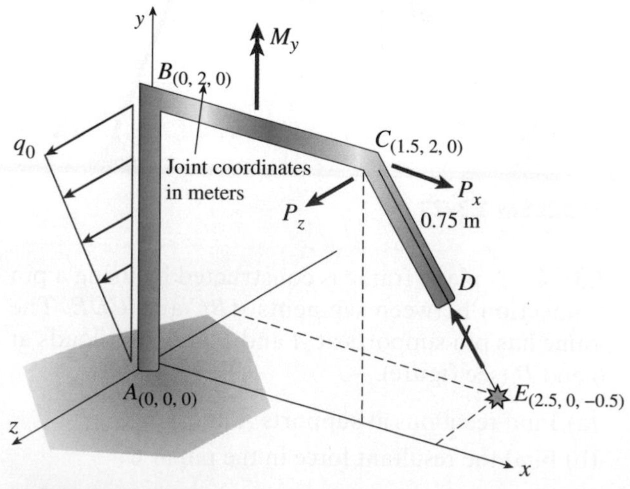

PROBLEM 1.3-30

(a) Find reactions at supports A and D.
(b) Find internal stress resultants N, V, T, and M at the mid-height of segment AB.

1.3-31 Space frame ABC is clamped at A, except it is free to rotate at A about the x and y axes. Cables DC and EC support the frame at C. Force $P_y = -220$ N is applied at the mid-span of AB, and a concentrated moment $M_x = -2.25$ N·m acts at joint B.

(a) Find reactions at support A.
(b) Find cable tension forces.

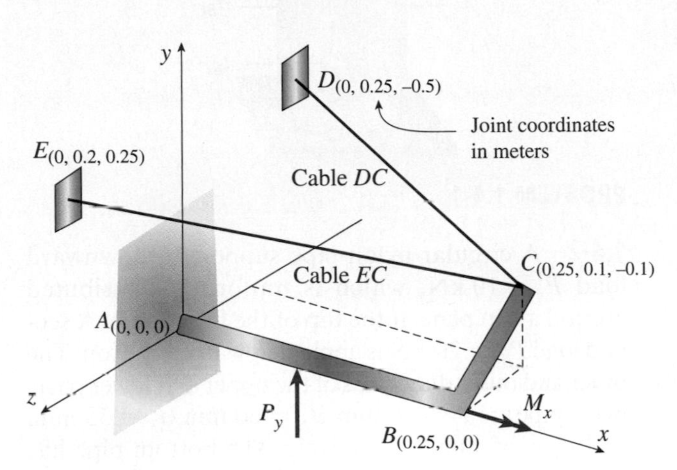

PROBLEM 1.3-31

1.3-32 A soccer goal is subjected to gravity loads (in the $-z$ direction, $w = 73$ N/m for DG, BG, and BC; $w = 29$ N/m for all other members; see figure) and a force $F = 200$ N applied eccentrically at the mid-height of member DG. Find reactions at supports C, D, and H.

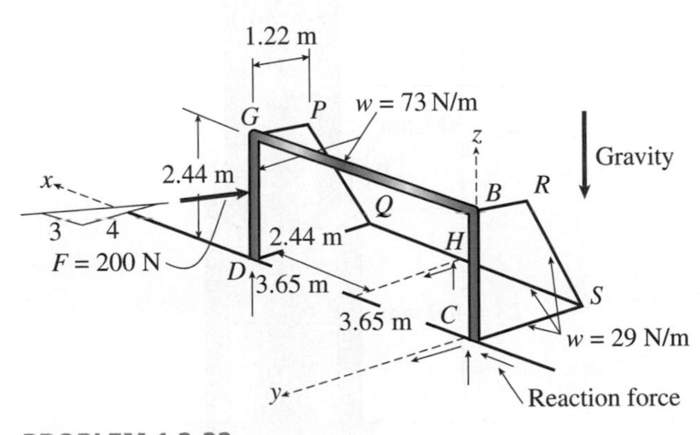

PROBLEM 1.3-32

1.3-33 An elliptical exerciser machine (see figure part a) is composed of front and back rails. A simplified plane-frame model of the back rail is shown in figure part b. Analyze the plane-frame model to find reaction forces at supports A, B, and C for the position and applied loads given in figure part b. Note that there are axial and moment releases at the base of member 2 so that member 2 can lengthen and shorten as the roller support at B moves along the 30° incline. (These releases indicate that the internal axial force N and moment M must be zero at this location.)

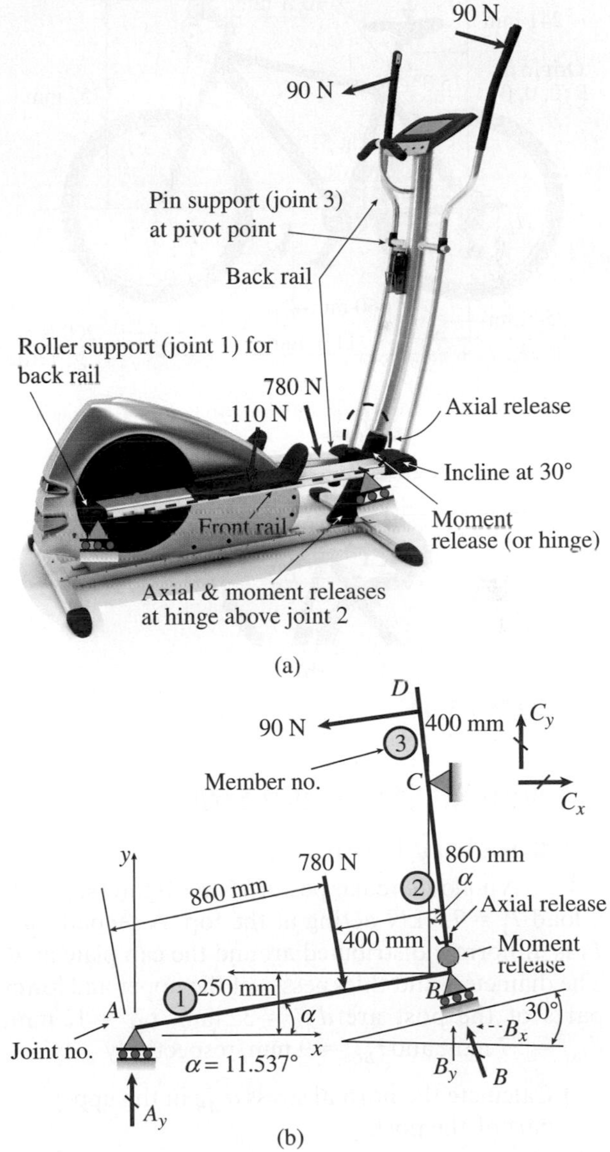

PROBLEM 1.3-33

1.3-34 A mountain bike is moving along a flat path at constant velocity. At some instant, the rider (weight = 670 N) applies pedal and hand forces, as shown in the figure part a.

(a) Find reaction forces at the front and rear hubs. (Assume that the bike is pin supported at the rear hub and roller supported at the front hub.)

(b) Find internal stress resultants N, V, and M in the inclined seat post (see figure part b).

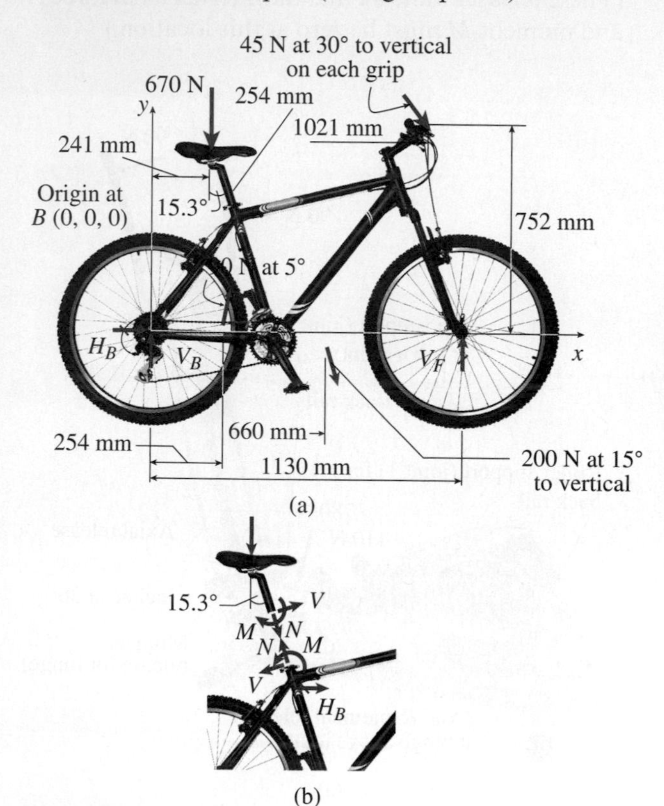

PROBLEM 1.3-34

1.4 Normal Stress and Strain

Introductory Problems

1.4-1 A hollow circular post ABC (see figure) supports a load $P_1 = 7.5$ kN acting at the top. A second load P_2 is uniformly distributed around the cap plate at B. The diameters and thicknesses of the upper and lower parts of the post are $d_{AB} = 32$ mm, $t_{AB} = 12$ mm, $d_{BC} = 57$ mm, and $t_{BC} = 9$ mm, respectively.

(a) Calculate the normal stress σ_{AB} in the upper part of the post.

(b) If you want the lower part of the post to have the same compressive stress as the upper part, what should be the required magnitude of load P_2?

(c) If P_1 remains at 7.5 kN and P_2 is set at 10 kN, what new thickness of BC will result in the same compressive stress in both parts?

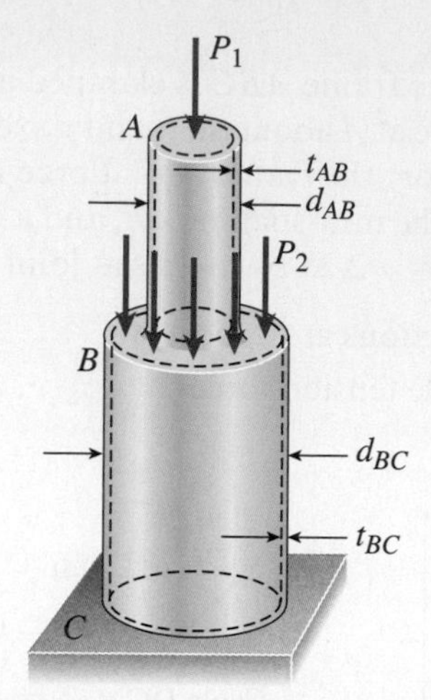

PROBLEM 1.4-1

1.4-2 A circular nylon pipe supports a downward load $P_A = 10$ kN, which is uniformly distributed around a cap plate at the top of the lower pipe. A second load $P_B = 20$ kN is applied upward at the top. The inner and outer diameters of the upper and lower parts of the pipe are $d_1 = 50$ mm, $d_2 = 60$ mm, $d_3 = 55$ mm, and $d_4 = 65$ mm, respectively. The bottom pipe has length 400 mm and the upper pipe has length 300 mm.

(a) Calculate the axial normal stress in each pipe segment.

(b) Calculate the strain in each pipe segment if the upward displacement at B on the upper pipe is 3.29 mm and the upward displacement at A on the bottom part is 1.25 mm.

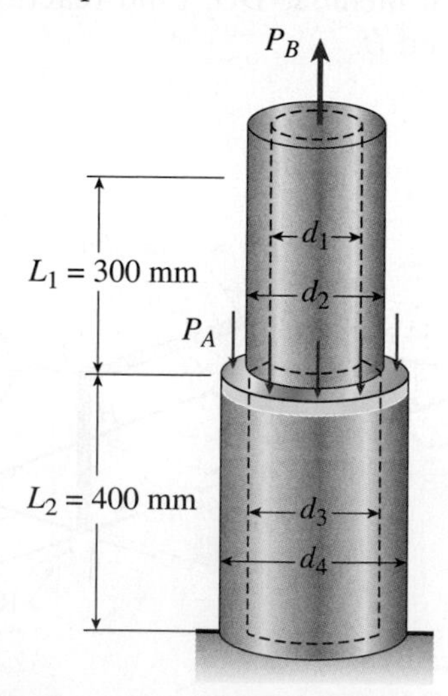

PROBLEM 1.4-2

1.4-3 A circular tube AB is fixed at one end and free at the other end. The tube is subjected to axial force at joint B. If the outer diameter of the tube is 75 mm and the thickness is 19 mm, calculate the maximum normal stress in the tube.

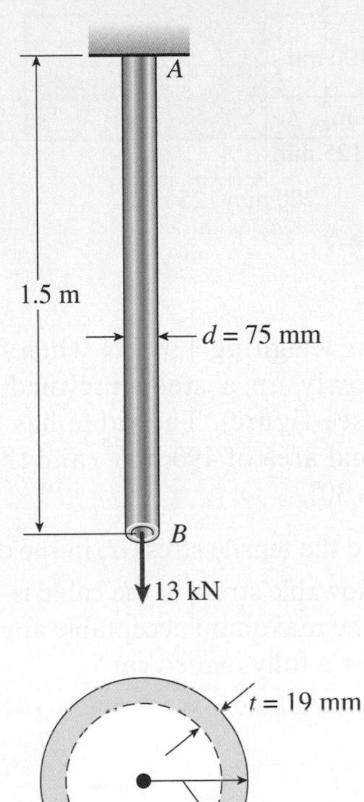

PROBLEM 1.4-3

1.4-4 A force P of 70 N is applied by a rider to the front hand brake of a bicycle (P is the resultant of an evenly distributed pressure). As the hand brake pivots at A, a tension T develops in the 460-mm long brake cable ($A_e = 1.075\ \text{mm}^2$), which elongates by $\delta = 0.214$ mm. Find the normal stress δ and strain ε in the brake cable.

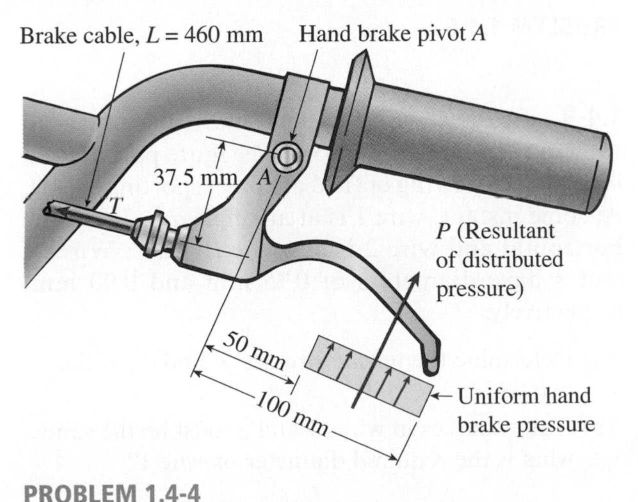

PROBLEM 1.4-4

Representative Problems

1.4-5 A bicycle rider wants to compare the effectiveness of cantilever hand brakes (see figure part a) versus V brakes (figure part b).

(a) Calculate the braking force R_B at the wheel rims for each of the bicycle brake systems shown. Assume that all forces act in the plane of the

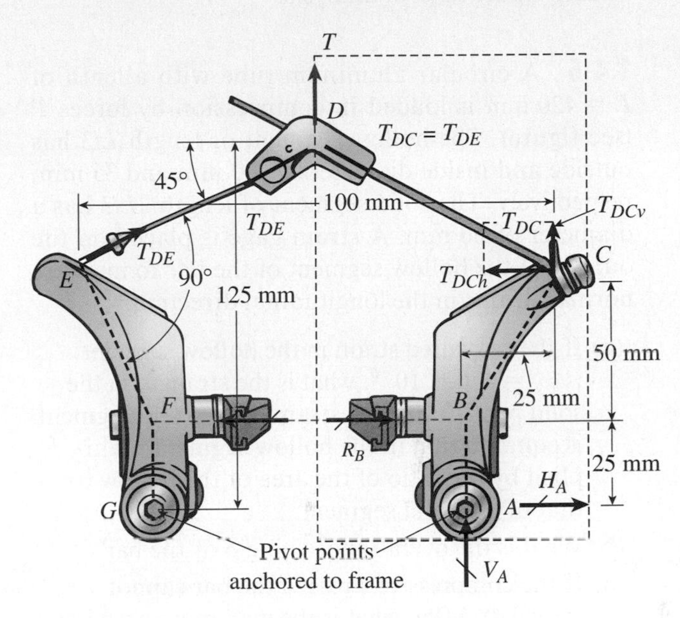

(a) Cantilever brakes

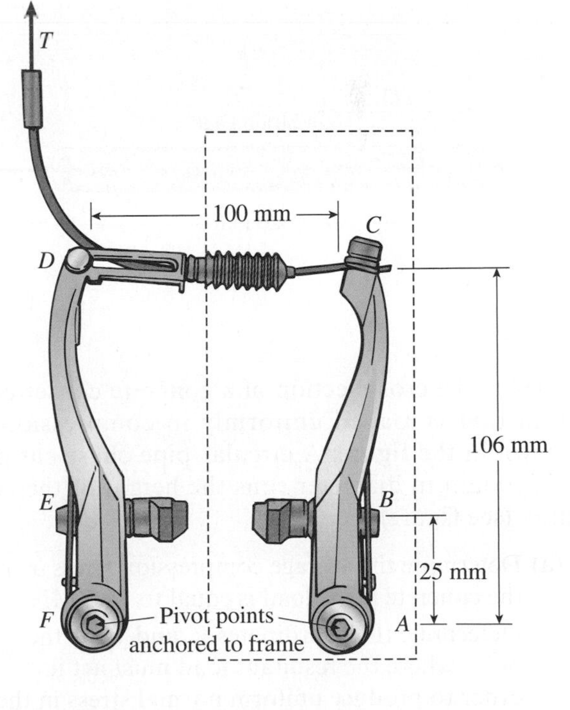

(b) V brakes

PROBLEM 1.4-5

figure and that cable tension $T = 200$ N. Also, what is the average compressive normal stress σ_c on the brake pad ($A = 4\ \text{cm}^2$)?

(b) For each braking system, what is the stress in the brake cable if the effective cross-sectional area is $1.077\ \text{mm}^2$?

Hint: Because of symmmetry, use only the right half of each figure in your analysis.

1.4-6 A circular aluminum tube with a lenth of $L = 420$ mm is loaded in compression by forces P (see figure). The hollow segment of length $L/3$ has outside and inside diameters of 60 mm and 35 mm, respectively. The solid segment of length $2L/3$ has a diameter of 60 mm. A strain gage is placed on the outside of the hollow segment of the bar to measure normal strains in the longitudinal direction.

(a) If the measured strain in the hollow segment is $\varepsilon_h = 470 \times 10^{-6}$, what is the strain ε_s in the solid part? *Hint:* The strain in the solid segment is equal to that in the hollow segment multiplied by the ratio of the area of the hollow to that of the solid segment.

(b) What is the overall shortening δ of the bar?

(c) If the compressive stress in the bar cannot exceed 48 MPa, what is the maximum permissible value of load P?

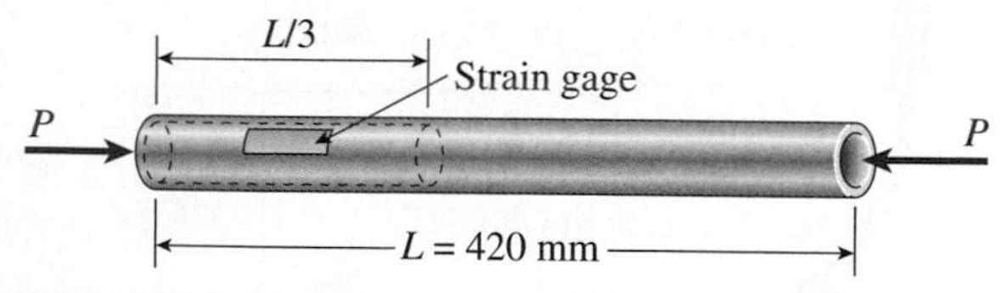

PROBLEM 1.4-6

1.4-7 The cross section of a concrete corner column that is loaded uniformly in compression is shown in the figure. A circular pipe chase cut-out of 250 mm in diameter runs the height of the column (see figure).

(a) Determine the average compression stress σ_c in the concrete if the load is equal to 14.5 MN.

(b) Determine the coordinates x_c and y_c of the point where the resultant load must act in order to produce uniform normal stress in the column.

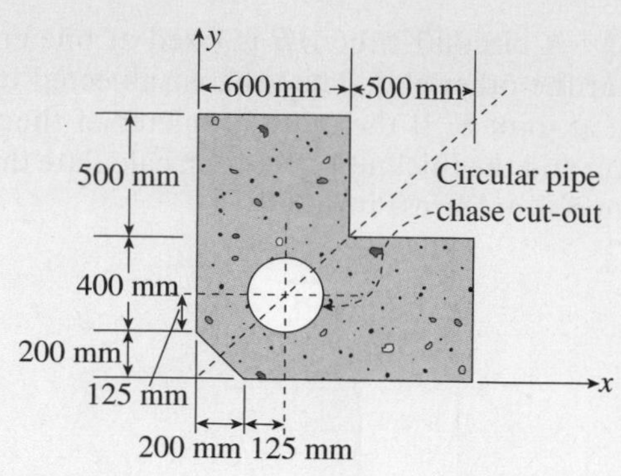

PROBLEM 1.4-7

1.4-8 A car weighing 130 kN when fully loaded is pulled slowly up a steep inclined track by a steel cable (see figure). The cable has an effective cross-sectional area of $490\ \text{mm}^2$, and the angle α of the incline is 30°.

(a) Calculate the tensile stress σ_t in the cable.

(b) If the allowable stress in the cable is 150 MPa, what is the maximum acceptable angle of the incline for a fully loaded car?

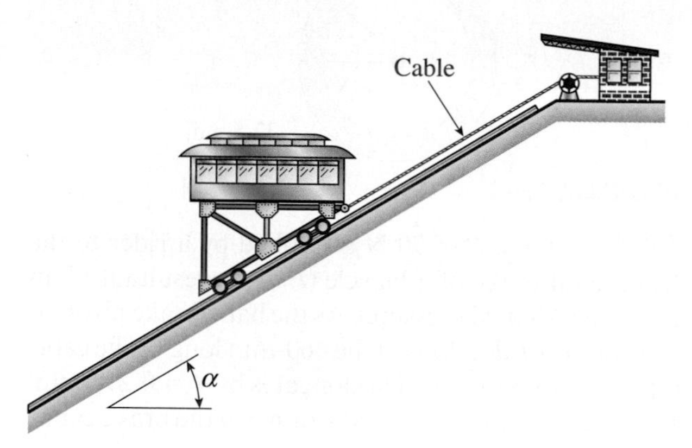

PROBLEM 1.4-8

1.4-9 Two steel wires support a moveable overhead camera weighing $W = 125$ N (see fgure part a) used for close-up viewing of field action at sporting events. At some instant, wire 1 is at an angle $\alpha = 22°$ to the horizontal and wire 2 is at angle $\beta = 40°$. Wires 1 and 2 have diameters of 0.75 mm and 0.90 mm, respectively.

(a) Determine the tensile stresses σ_1 and σ_2 in the two wires.

(b) If the stresses in wires 1 and 2 must be the same, what is the required diameter of wire 1?

(c) To stabilize the camera for windy outdoor conditions, a third wire is added (see figure part b). Assume the three wires meet at a common point coordinates (0, 0, 0) above the camera at the instant shown in figure part b. Wire 1 is attached to a support at coordinates (25 m, 16 m, 23 m). Wire 2 is supported at (−23 m, 18 m, 27 m). Wire 3 is supported at (−3 m, −28 m, 25 m). Assume that all three wires have a diameter of 0.8 mm. Find the tensile stresses in all three wires.

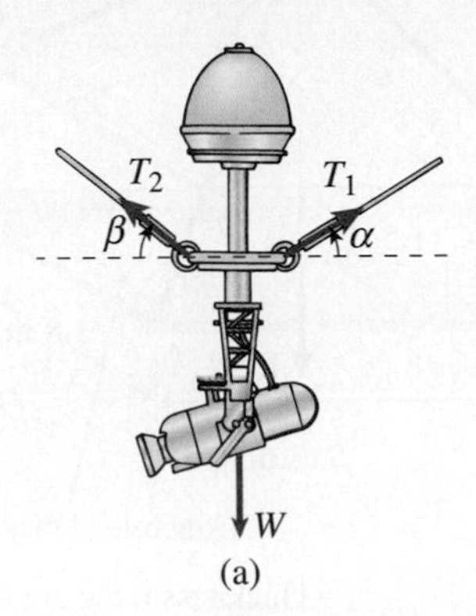

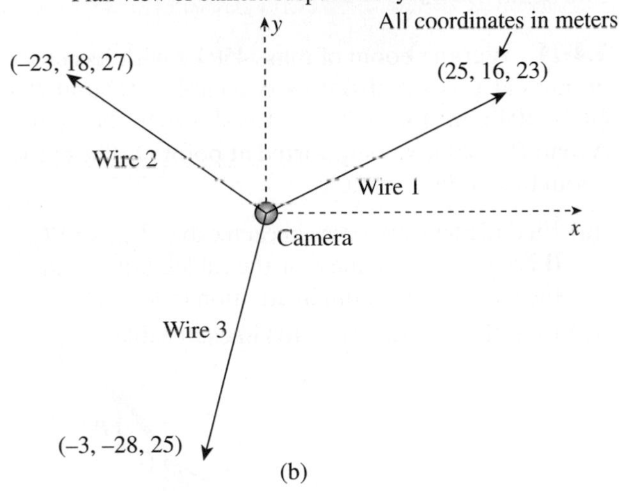

PROBLEM 1.4-9

1.4-10 A long retaining wall is braced by wood shores set at an angle of 30° and supported by concrete thrust blocks, as shown in the first part of the figure. The shores are evenly spaced at 3 m apart.

For analysis purposes, the wall and shores are idealized as shown in the second part of the figure. Note that the base of the wall and both ends of the shores are assumed to be pinned. The pressure of the soil against the wall is assumed to be triangularly distributed, and the resultant force acting on a 3-meter length of the walls is $F = 190$ kN.

If each shore has a 150 mm × 150 mm square cross section, what is the compressive stress σ_c in the shores?

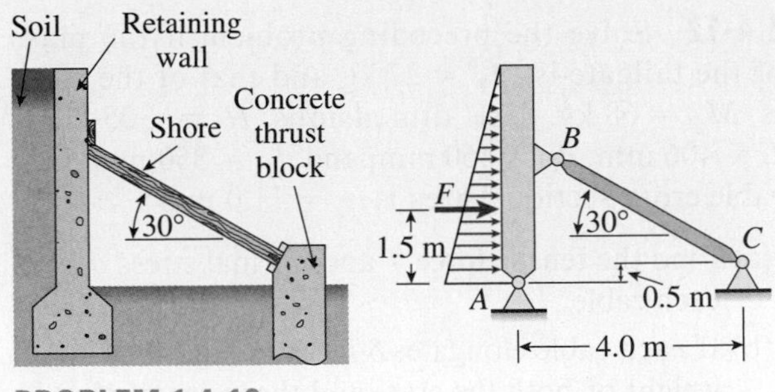

PROBLEM 1.4-10

1.4-11 A pickup truck tailgate supports a crate where $W_C = 900$ N, as shown in the figure. The tailgate weighs $W_T = 270$ N and is supported by two cables (only one is shown in the figure). Each cable has an effective cross-sectional area $A_e = 11\ \text{mm}^2$.

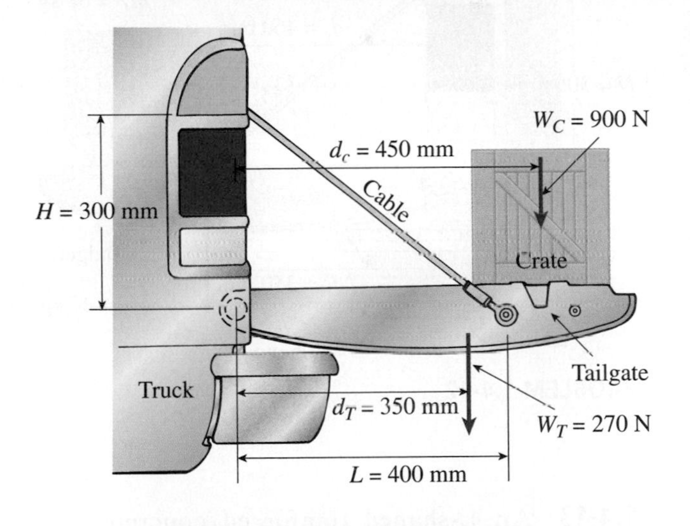

PROBLEM 1.4-11

(a) Find the tensile force T and normal stress σ in each cable.

(b) If each cable elongates $\delta = 0.42$ mm due to the weight of both the crate and the tailgate, what is the average strain in the cable?

1.4-12 Solve the preceding problem if the mass of the tailgate is $M_T = 27$ kg and that of the crate is $M_C = 68$ kg. Use dimensions $H = 305$ mm, $L = 406$ mm, $d_C = 460$ mm, and $d_T = 350$ mm. The cable cross-sectional area is $A_e = 11.0\ \text{mm}^2$.

(a) Find the tensile force T and normal stress σ in each cable.

(b) If each cable elongates $\delta = 0.25$ mm due to the weight of both the crate and the tailgate, what is the average strain in the cable?

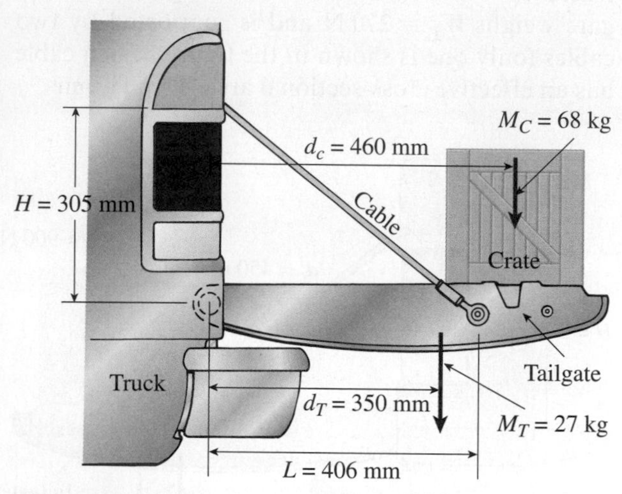

PROBLEM 1.4-12

1.4-13 An L-shaped reinforced concrete slab 3.6 m × 3.6 m, with a 1.8 m × 1.8 m cut-out and thickness $t = 230$ mm, is lifted by three cables attached at O, B, and D, as shown in the figure. The cables are are combined at point Q, which is 2.1 m above the top of the slab and directly above the center of mass at C. Each cable has an effective cross-sectional area of $A_e = 77\ \text{mm}^2$.

(a) Find the tensile force $T_i (i = 1, 2, 3)$ in each cable due to the weight W of the concrete slab (ignore weight of cables).

(b) Find the average stress σ_i in each cable. (See Table I-1 in Appendix I for the weight density of reinforced concrete.)

(c) Add cable AQ so that OQA is one continuous cable, with each segment having force T_1, which is connected to cables BQ and DQ at point Q. Repeat parts (a) and (b). *Hint*: There are now three forced equilibrium equations and one constraint equation, $T_1 = T_4$.

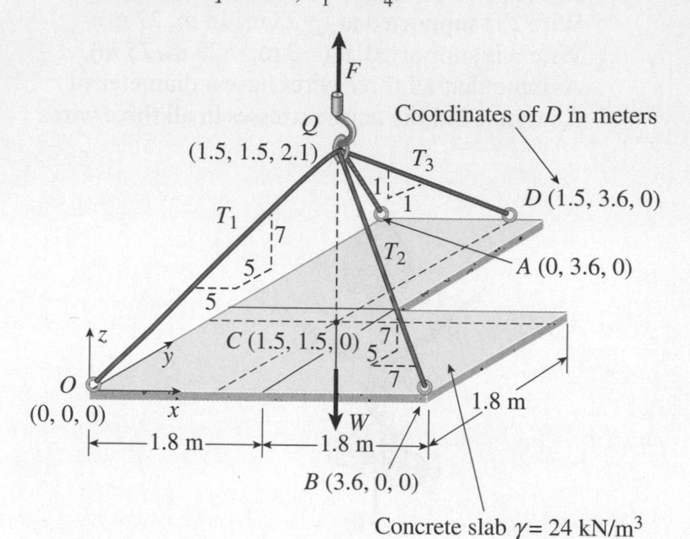

PROBLEM 1.4-13

1.4-14 A crane boom of mass 450 kg with its center of mass at C is stabilized by two cables AQ and BQ ($A_e = 304\ \text{mm}^2$ for each cable) as shown in the figure. A load $P = 20$ kN is supported at point D. The crane boom lies in the y-z plane.

(a) Find the tension forces in each cable: T_{AQ} and T_{BQ} (kN). Neglect the mass of the cables, but include the mass of the boom in addition to load P.

(b) Find the average stress (σ) in each cable.

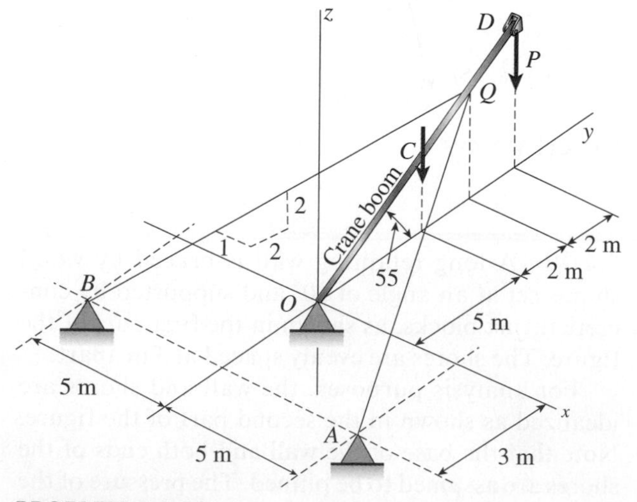

PROBLEM 1.4-14

1.4-15 Two gondolas on a ski lift are locked in the position shown in the figure while repairs are being made elsewhere. The distance between support towers is $L = 30.5$ m. The length of each cable segment under gondolas weighing $W_B = 2000$ N and $W_C = 2900$ N are $D_{AB} = 3.7$ m, $D_{BC} = 21.4$ m, and $D_{CD} = 6.1$ m. The cable sag at B is $\Delta_B = 1.3$ m and that at C is $\Delta_C = 2.3$ m. The effective cross-sectional area of the cables is $A_e = 77$ mm^2.

(a) Find the tension force in each segment; neglect the mass of the cable.

(b) Find the average stress (σ) in each cable segment.

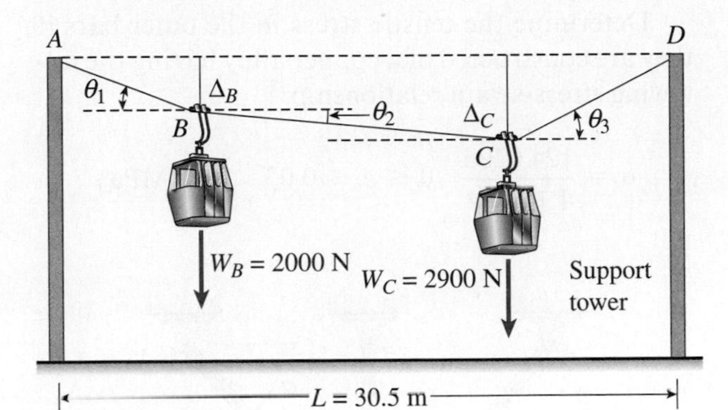

PROBLEM 1.4-15

1.4-16 A round bar ABC of length $2L$ (see figure) rotates about an axis through the midpoint C with constant angular speed ω (radians per second). The material of the bar has weight density γ.

(a) Derive a formula for the tensile stress σ_x in the bar as a function of the distance x from the midpoint C.

(b) What is the maximum tensile stress σ_{max}?

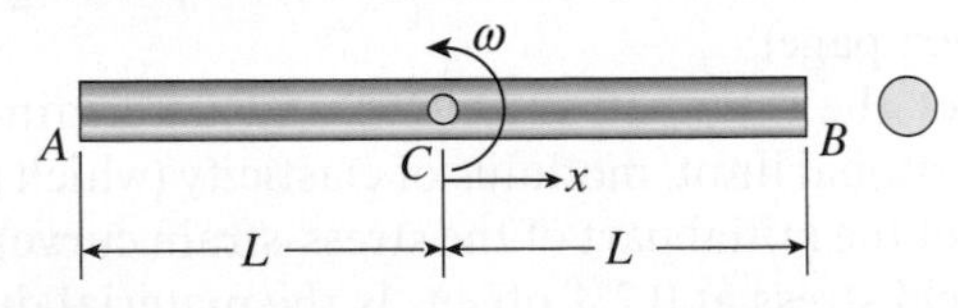

PROBLEM 1.4-16

1.4-17 Two separate cables AC and BC support a sign structure of weight $W = 7$ kN attached to a building. The sign is also supported by a pin support at O and a lateral restraint in the z-direction at D.

(a) Find the tension in each cable. Neglect the mass of the cables.

(b) Find the average stress in each cable if the area of each cable is $A_e = 300$ mm^2.

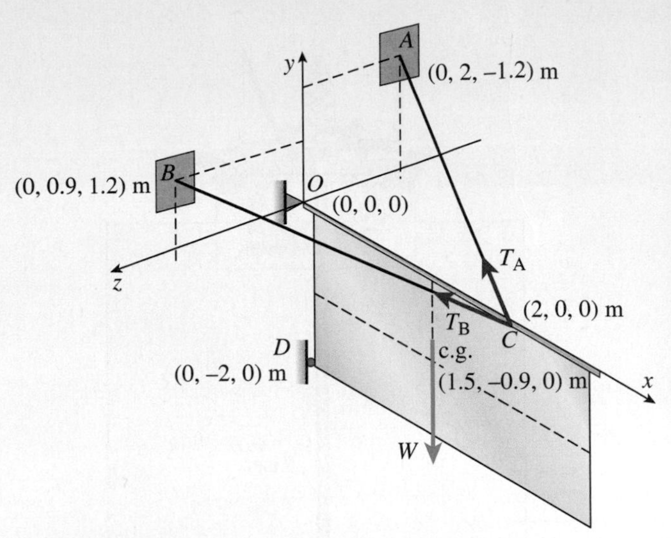

PROBLEM 1.4-17

1.5 Mechanical Properties of Materials

Introductory Problems

1.5-1 Imagine that a long steel wire hangs vertically from a high-altitude balloon.

(a) What is the greatest length (meters) it can have without yielding if the steel yields at 260 MPa?

(b) If the same wire hangs from a ship at sea, what is the greatest length? (Obtain the weight densities of steel and sea water from Table I-1, Appendix I.)

1.5-2 A steel riser pipe hangs from a drill rig located offshore in deep water (see figure).

(a) What is the greatest length (meters) it can have without breaking if the pipe is suspended in the air and the ultimate strength (or breaking strength) is 550 MPa?

(b) If the same riser pipe hangs from a drill rig at sea, what is the greatest length? (Obtain the weight densities of steel and sea water from Table I-1, Appendix I. Neglect the effect of buoyant foam casings on the pipe.)

1.5-3 Three different materials, designated A, B, and C, are tested in tension using test specimens having diameters of 12 mm and gage lengths of 50 mm (see figure). At failure, the distances between the gage marks are found to be 54.5 mm, 63.2 mm, and 69.4 mm, respectively. Also, at the failure cross sections, the diameters are found to be 11.46 mm, 9.48 mm, and 6.06 mm, respectively.

Determine the percent elongation and percent reduction in area of each specimen. Using your own judgment, classify each material as brittle or ductile.

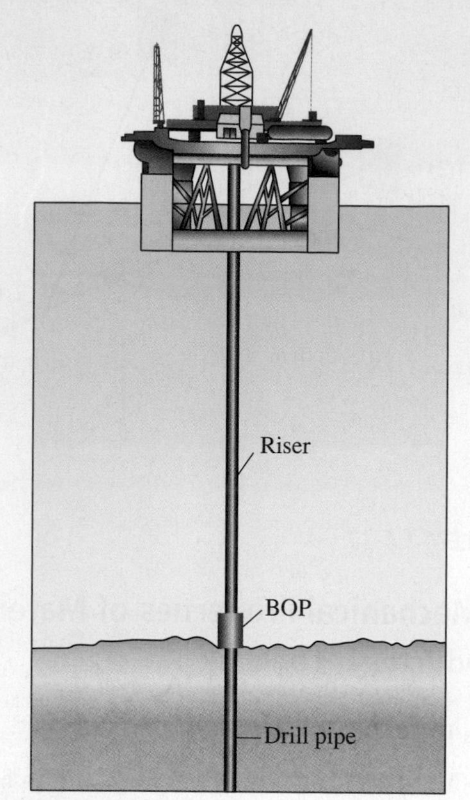

PROBLEM 1.5-2

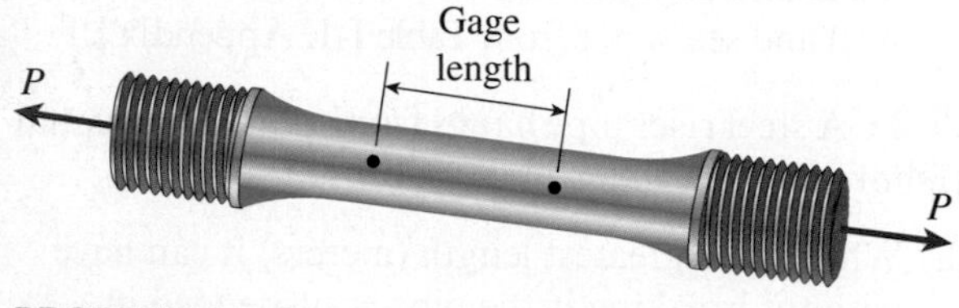

PROBLEM 1.5-3

Representative Problems

1.5-4 The *strength-to-weight ratio* of a structural material is defined as its load-carrying capacity divided by its weight. For materials in tension, use a characteristic tensile stress obtained from a stress-strain curve as a measure of strength. For instance, either the yield stress or the ultimate stress could be used, depending upon the particular application. Thus, the strength-to-weight ratio $R_{S/W}$ for a material in tension is defined as

$$R_{S/W} = \frac{\sigma}{\gamma}$$

in which σ is the characteristic stress and γ is the weight density. Note that the ratio has units of length.

Using the ultimate stress σ_U as the strength parameter, calculate the strength-to-weight ratio (in units of meters) for each of the following materials: aluminum alloy 6061-T6, Douglas fir (in bending), nylon, structural steel ASTM-A572, and a titanium alloy. Obtain the material properties from Tables I-1 and I-3 of Appendix I. When a range of values is given in a table, use the average value.

1.5-5 A symmetrical framework consisting of three pin-connected bars is loaded by a force P (see figure). The angle between the inclined bars and the horizontal is $\alpha = 52°$. The axial strain in the middle bar is measured as 0.027.

Determine the tensile stress in the outer bars if they are constructed of a copper alloy having the following stress-strain relationship:

$$\sigma = \frac{124{,}020\varepsilon}{1 + 300\varepsilon} \qquad 0 \le \varepsilon \le 0.03 \qquad (\sigma = \text{MPa})$$

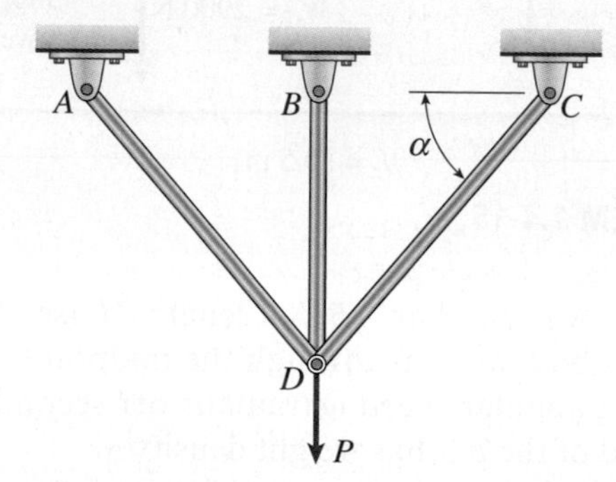

PROBLEM 1.5-5

1.5-6 A specimen of a methacrylate plastic is tested in tension at room temperature (see figure), producing the stress-strain data listed in the accompanying table (see next page).

Plot the stress-strain curve and determine the proportional limit, modulus of elasticity (which is the slope of the initial part of the stress-strain curve), and the yield stress at 0.2% offset. Is the material ductile or brittle?

PROBLEM 1.5-6

STRESS-STRAIN DATA FOR PROB. 1.5-6

Stress (MPa)	Strain
8.0	0.0032
17.5	0.0073
25.6	0.0111
31.1	0.0129
39.8	0.0163
44.0	0.0184
48.2	0.0209
53.9	0.0260
58.1	0.0331
60.2	0.0384
62.0	0.0429
62.1	Fracture

1.5-7 The data shown in the accompanying table are from a tensile test of high-strength steel. The test specimen has a diameter of 13 mm and a gage length of 50 mm (see figure for Prob. 1.5-3). At fracture, the elongation between the gage marks is 3.0 mm and the minimum diameter is 10.7 mm.

Plot the conventional stress-strain curve for the steel and determine the proportional limit, modulus of elasticity (the slope of the initial part of the stress-strain curve), yield stress at 0.1% offset, ultimate stress, percent elongation in 50 mm, and percent reduction in area.

TENSILE-TEST DATA FOR PROB. 1.5-7

Load (kN)	Elongation (mm)
5	0.005
10	0.015
30	0.048
50	0.084
60	0.099
64.5	0.109
67.0	0.119
68.0	0.137
69.0	0.160
70.0	0.229
72.0	0.259
76.0	0.330
84.0	0.584
92.0	0.853
100.0	1.288
112.0	2.814
113.0	Fracture

1.6 Elasticity, Plasticity, and Creep

Introductory Problems

1.6-1 A bar made of structural steel having the stress-strain diagram shown in the figure has a length of 1.5 m. The yield stress of the steel is 350 MPa, and the slope of the initial linear part of the stress-strain curve is 200 GPa.

(a) The bar is loaded axially until it elongates 5 mm and then the load is removed. How does the final length of the bar compare with its original length?

(b) If the bar has a circular cross section with a diameter $d = 40$ mm and is loaded by tensile forces $P = 350$ kN, what is the stress in the bar? What is the permanent set of the bar?

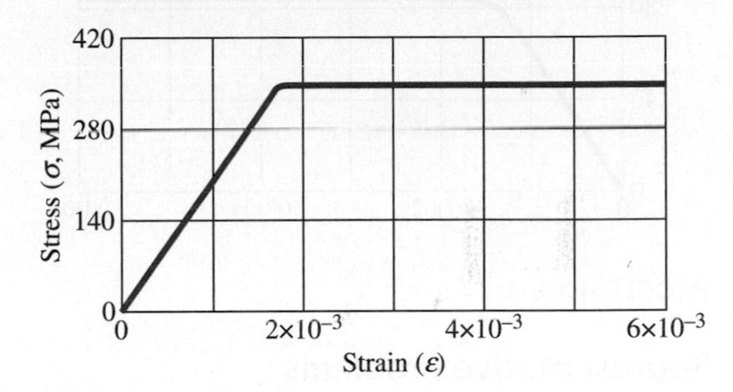

PROBLEM 1.6-1

1.6-2 A bar of length 2.0 m is made of a structural steel having the stress-strain diagram shown in the figure. The yield stress of the steel is 250 MPa, and the slope of the initial linear part of the stress-strain curve (modulus of elasticity) is 200 GPa. The bar is loaded axially until it elongates 6.5 mm, and then the load is removed.

How does the final length of the bar compare with its original length? *Hint:* Use the concepts illustrated in Fig. 1-39b.

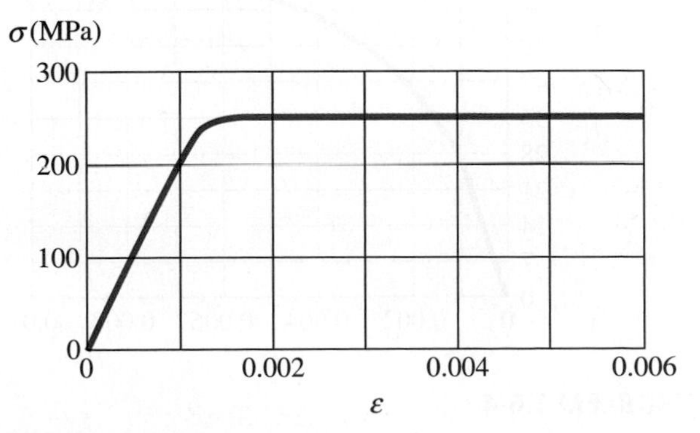

PROBLEM 1.6-2

1.6-3 A bar made of structural steel having the stress-strain diagram shown in the figure has a length of 1.5 m. The yield stress of the steel is 290 MPa, and the slope of the initial linear part of the stress-strain curve (modulus of elasticity) is 207 GPa. The bar is loaded axially until it elongates 7.6 mm, and then the load is removed.

How does the final length of the bar compare with its original length? *Hint:* Use the concepts illustrated in Fig. 1-39b.

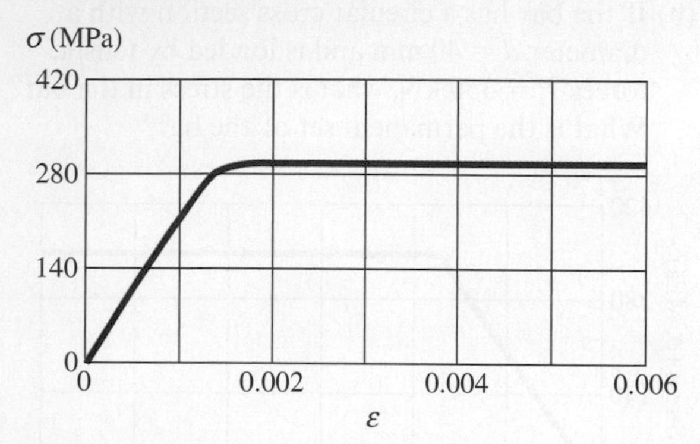

PROBLEM 1.6-3

Representative Problems

1.6-4 A circular bar of magnesium alloy is 750 mm long. The stress-strain diagram for the material is shown in the figure. The bar is loaded in tension to an elongation of 6.0 mm, and then the load is removed.

(a) What is the permanent set of the bar?

(b) If the bar is reloaded, what is the proportional limit?

Hint: Use the concepts illustrated in Figs. 1-39b and 1-40.

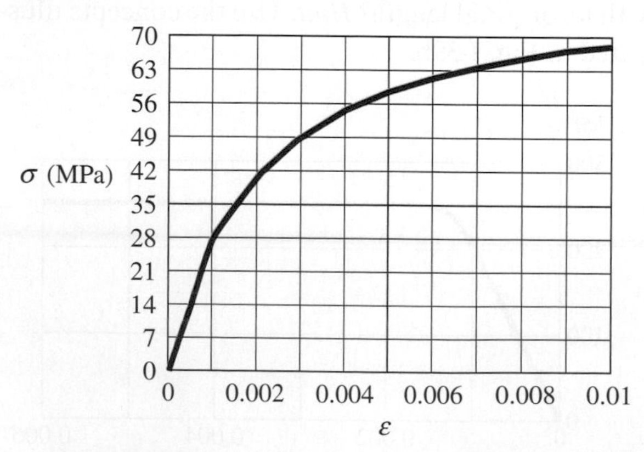

PROBLEM 1.6-4

1.6-5 An aluminum bar has length $L = 1.8$ m and diameter $d = 34$ mm. The stress-strain curve for the aluminum is shown in Fig. 1-34. The initial straight-line part of the curve has a slope (modulus of elasticity) of 73 GPa. The bar is loaded by tensile forces $P = 200$ kN and then unloaded.

(a) What is the permanent set of the bar?

(b) If the bar is reloaded, what is the proportional limit?

Hint: Use the concepts illustrated in Figs. 1-39b and 1-40.

1.6-6 A continuous cable (diameter 6 mm) with tension force T is attached to a horizontal frame member at B and C to support a sign structure. The cable passes over a small frictionless pulley at D. The wire is made of a copper alloy, and the stress-strain relationship for the wire is

$$\sigma(\varepsilon) = \frac{124{,}000\varepsilon}{1 + 300\varepsilon} \quad 0 \le \varepsilon \le 0.03 \quad (\sigma \text{ in MPa})$$

(a) Find the axial normal strain in the cable and its elongation due to the load $W = 6.8$ kN.

(b) If the forces are removed, what is the permanent set of the cable?

Hint: Start with constructing the stress-strain diagram and determine the modulus of elasticity, E, and the 0.2% offset yield stress.

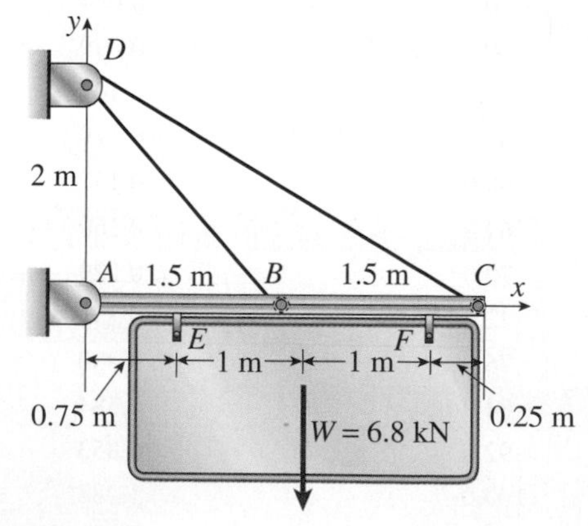

PROBLEM 1.6-6

1.6-7 A wire of length $L = 2.5$ m and diameter $d = 1.6$ mm is stretched by tensile forces $P = 600$ N. The wire is made of a copper alloy having a stress-strain relationship that may be described mathematically by

$$\sigma = \frac{124{,}020\varepsilon}{1 + 300\varepsilon} \quad 0 \le \varepsilon \le 0.03 \quad (\sigma = \text{MPa})$$

in which ε is nondimensional and σ has units of MPa.

(a) Construct a stress-strain diagram for the material.

(b) Determine the elongation of the wire due to the forces P.

(c) If the forces are removed, what is the permanent set of the bar?

(d) If the forces are applied again, what is the proportional limit?

1.7 Linear Elasticity, Hooke's Law, and Poisson's Ratio

When solving the problems for Section 1.7, assume that the material behaves linearly elastically.

Introductory Problems

1.7-1 A high-strength steel bar used in a large crane has a diameter $d = 50$ mm (see figure). The steel has a modulus of elasticity $E = 200$ GPa and Poisson's ratio is $\nu = 0.3$. Because of clearance requirements, the diameter of the bar is limited to 50.025 mm when it is compressed by axial forces.

What is the largest compressive load P_{max} that is permitted?

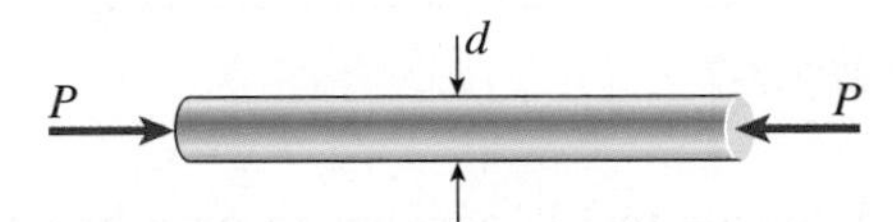

PROBLEM 1.7-1

1.7-2 A round bar of 10 mm diameter is made of aluminum alloy 7075-T6 (see figure). When the bar is stretched by axial forces P, its diameter decreases by 0.016 mm.

Find the magnitude of the load P. Obtain the material properties from Appendix I.

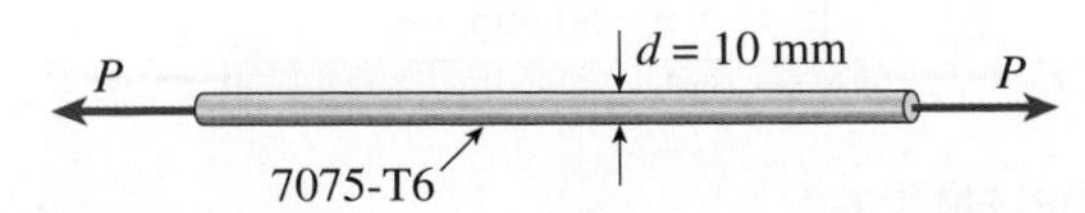

PROBLEM 1.7-2

1.7-3 A polyethylene bar with a diameter $d_1 = 70$ mm is placed inside a steel tube with an inner diameter $d_2 = 70.2$ mm (see figure). The polyethylene bar is then compressed by an axial force P.

At what value of the force P will the space between the polyethylene bar and the steel tube be closed? For polyethylene, assume $E = 1.4$ GPa and $\nu = 0.4$.

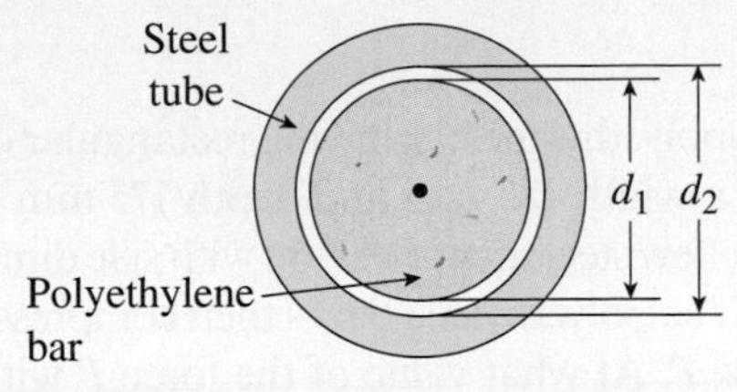

PROBLEM 1.7-3

1.7-4 A square plastic bar (length L_p, side dimension $s_p = 193$ mm) is inserted inside a hollow, square, cast iron tube (length $L_c = 400$ mm, side $s_c = 200$ mm, and thickness $t_c = 3$ mm).

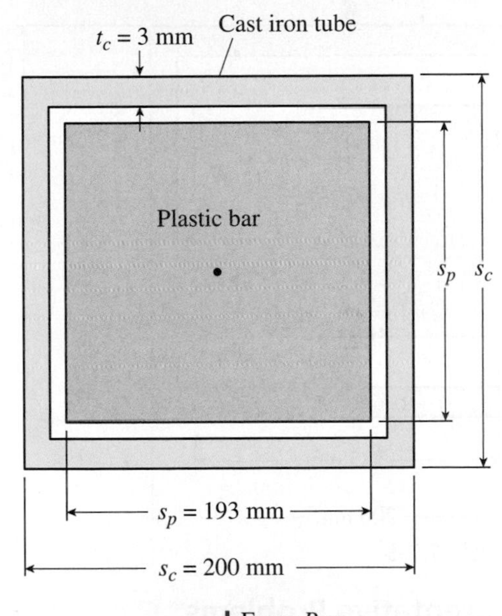

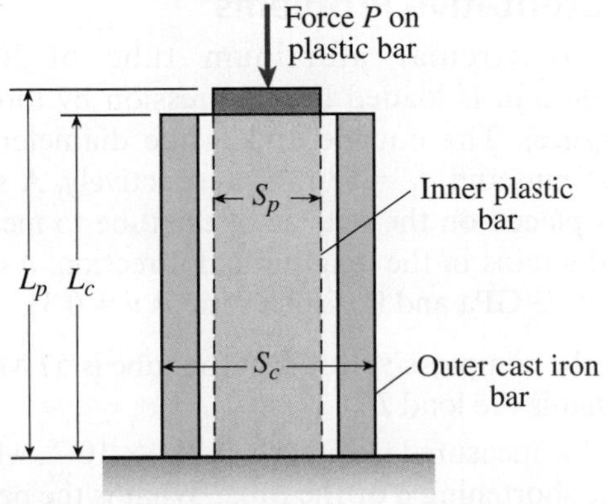

PROBLEM 1.7-4

(a) What is the required initial length L_p of the plastic bar so that, when it is compressed by some force P, the final length of bar and tube are equal to length L_c and, at the same time, the gap between plastic bar and cast iron tube is closed?

(b) Compare initial and final volumes for the plastic bar.

Assume that $E_c = 170$ GPa, $E_p = 2.1$ GPa, $\nu_c = 0.3$, and $\nu_p = 0.4$.

1.7-5 A polyethylene bar having rectangular cross section with a width 185 mm and depth 175 mm is placed inside a hollow steel square section with side dimension of 200 mm. The polyethylene bar is then compressed by an axial force P. At what value of the force P will the gap between the polyethylene bar and the steel tube be closed for the first time on one side? What is the remaining gap between the polyethylene bar and the steel tube on the other side? For polyethylene, assume $E = 1400$ MPa and $\nu = 0.4$.

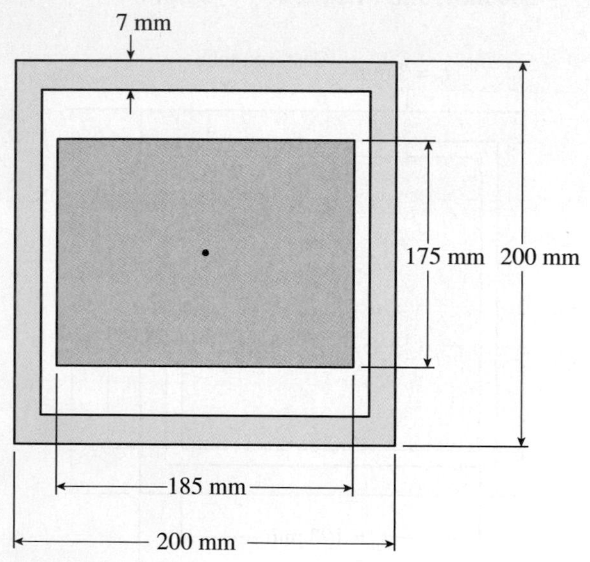

PROBLEM 1.7-5

Representative Problems

1.7-6 A circular aluminum tube of length $L = 600$ mm is loaded in compression by forces P (see figure). The outside and inside diameters are $d_2 = 75$ mm and $d_1 = 63$ mm, respectively. A strain gage is placed on the outside of the tube to measure normal strains in the longitudinal direction. Assume that $E = 73$ GPa and Poisson's ratio is $\nu = 0.33$.

(a) If the compressive stress in the tube is 57 MPa, what is the load P?

(b) If the measured strain is $\varepsilon = 781 \times 10^{-6}$, what is the shortening δ of the tube? What is the percent change in its cross-sectional area? What is the volume change of the tube?

(c) If the tube has a constant outer diameter of $d_2 = 75$ mm along its entire length L but now has increased *inner* diameter d_3 with a normal stress of 70 MPa over the middle third (see figure, part b) while the rest of the tube remains at normal stress of 57 MPa, what is the diameter d_3?

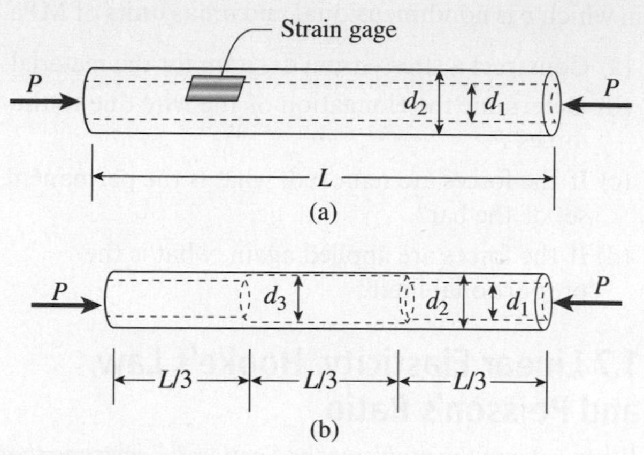

PROBLEM 1.7-6

1.7-7 A bar of monel metal with a length $L = 230$ mm and a diameter $d = 6$ mm is loaded axially by a tensile force P (see figure). If the bar elongates by 0.5 mm, what is the decrease in diameter d? What is the magnitude of the load P? Use the data in Table I-2, Appendix I.

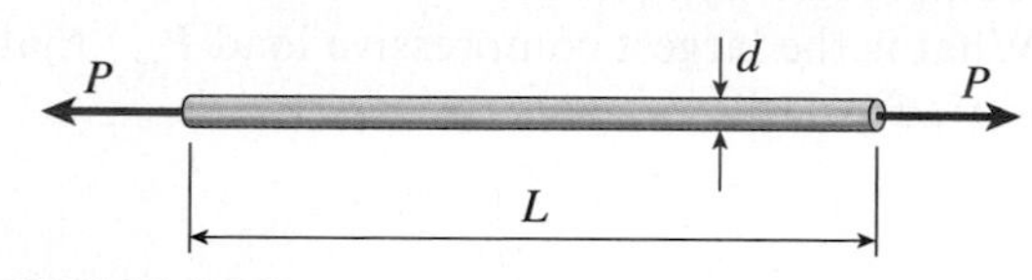

PROBLEM 1.7-7

1.7-8 A tensile test is performed on a brass specimen 10 mm in diameter using a gage length of 50 mm (see figure). When the tensile load P reaches a value of 20 kN, the distance between the gage marks has increased by 0.122 mm.

(a) What is the modulus of elasticity E of the brass?

(b) If the diameter decreases by 0.00830 mm, what is Poisson's ratio?

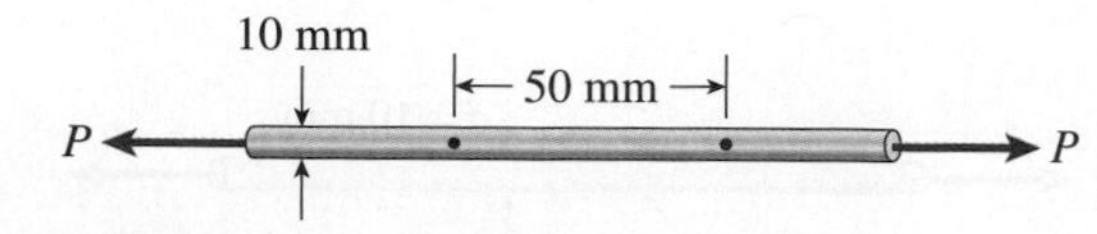

PROBLEM 1.7-8

1.7-9 A hollow, brass circular pipe ABC (see figure) supports a load $P_1 = 118$ kN acting at the top. A second load $P_2 = 98$ kN is uniformly distributed around the cap plate at B. The diameters and thicknesses of the upper and lower parts of the pipe are $d_{AB} = 31$ mm, $t_{AB} = 12$ mm, $d_{BC} = 57$ mm, and $t_{BC} = 9$ mm, respectively. The modulus of elasticity is 96 GPa. When both loads are fully applied, the wall thickness of pipe segment BC increases by 5×10^{-3} mm.

(a) Find the increase in the inner diameter of pipe segment BC.
(b) Find Poisson's ratio for the brass.
(c) Find the increase in the wall thickness of pipe segment AB and the increase in the inner diameter of segment AB.

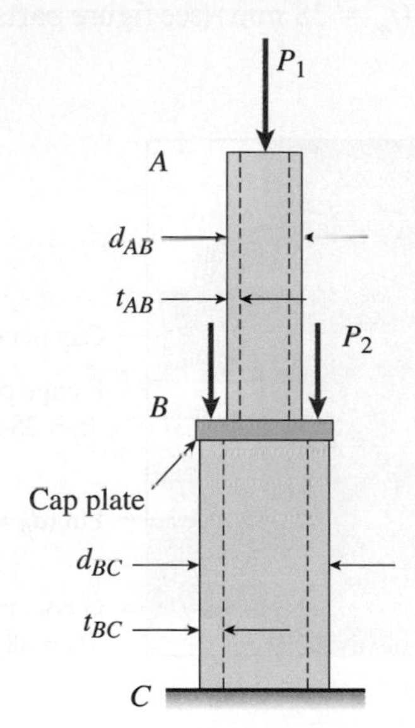

PROBLEM 1.7-9

1.7-10 Three round, copper alloy bars having the same length L but different shapes are shown in the figure. The first bar has a diameter d over its entire length, the second has a diameter d over one-fifth of its length, and the third has a diameter d over one-fifteenth of its length. Elsewhere, the second and third bars have a diameter $2d$. All three bars are subjected to the same axial load P.

Use the following numerical data: $P = 1400$ kN, $L = 5$ m, $d = 80$ mm, $E = 110$ GPa, and $\nu = 0.33$.

(a) Find the change in length of each bar.
(b) Find the change in volume of each bar.

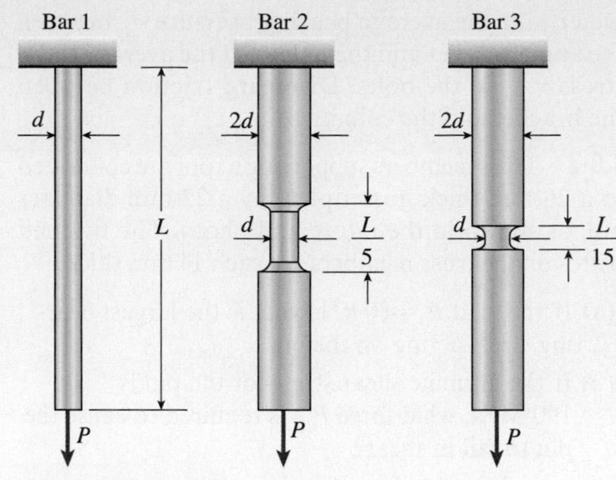

PROBLEM 1.7-10

1.8 Shear Stress and Strain

Introductory Problems

1.8-1 An angle bracket having a thickness $t = 19$ mm is attached to the flange of a column by two 16 mm diameter bolts (see figure). A uniformly distributed load from a floor joist acts on the top face of the bracket with a pressure $p = 1.9$ MPa. The top face of the bracket has a length $L = 200$ mm and width $b = 75$ mm.

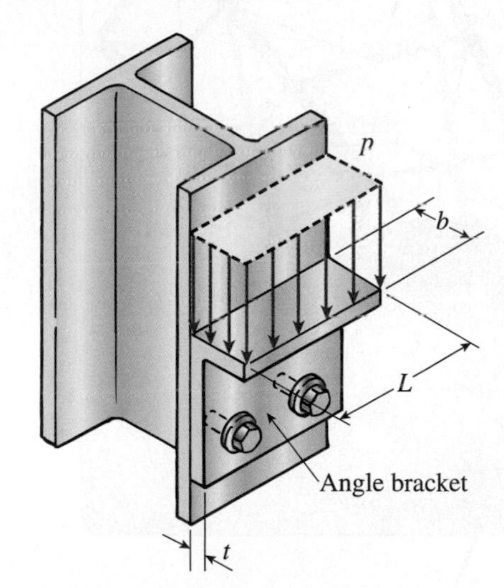

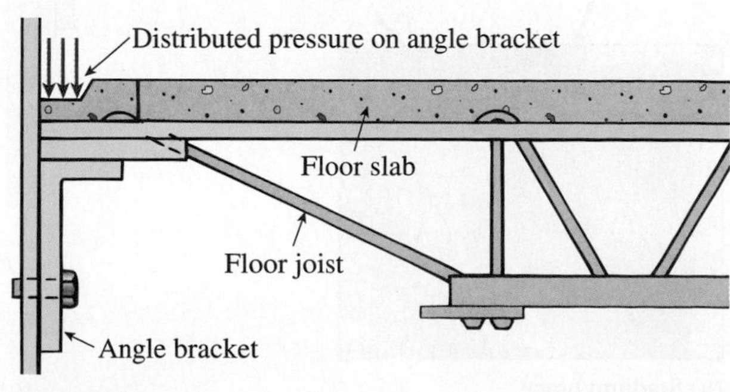

PROBLEM 1.8-1

Determine the average bearing pressure σ_b between the angle bracket and the bolts and the average shear stress τ_{aver} in the bolts. Disregard friction between the bracket and the column.

1.8-2 Truss members supporting a roof are connected to a 26-mm-thick gusset plate by a 22-mm diameter pin, as shown in the figure and photo. The two end plates on the truss members are each 14 mm thick.

(a) If the load $P = 80$ kN, what is the largest bearing stress acting on the pin?
(b) If the ultimate shear stress for the pin is 190 MPa, what force P_{ult} is required to cause the pin to fail in shear?

Disregard friction between the plates.

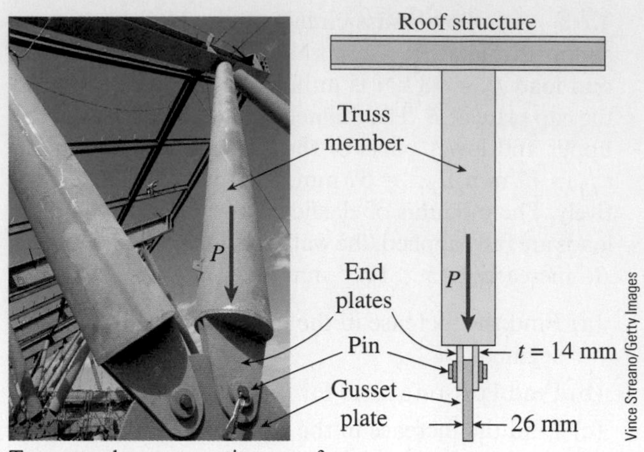

Truss members supporting a roof

PROBLEM 1.8-2

1.8-3 The upper deck of a football stadium is supported by braces, each of which transfers a load $P = 700$ kN to the base of a column (see figure part a). A cap plate at the bottom of the brace distributes the load P to four flange plates ($t_f = 25$ mm) through a pin ($d_p = 50$ mm) to two gusset plates ($t_g = 38$ mm) (see figure parts b and c).

(a) Stadium brace

(b) Detail at bottom of brace

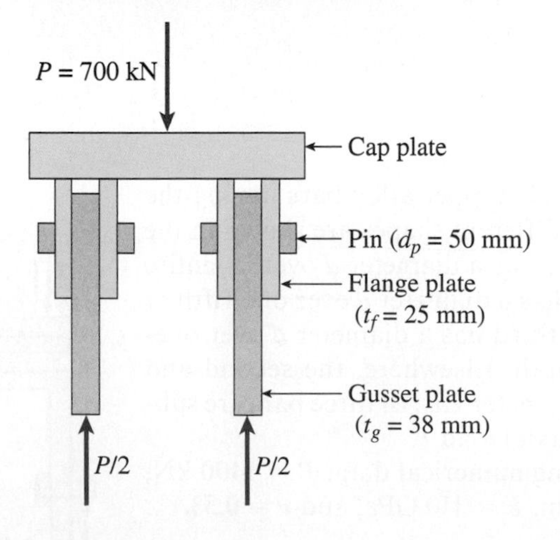

(c) Section through bottom of brace

PROBLEM 1.8-3

Determine the following quantities.

(a) The average shear stress τ_{aver} in the pin.

(b) The average bearing stress between the flange plates and the pin (σ_{bf}), and also between the gusset plates and the pin (σ_{bg}).

Disregard friction between the plates.

1.8-4 The inclined ladder AB supports a house painter (85 kg) at C and the weight ($q = 40$ N/m) of the ladder itself. Each ladder rail ($t_r = 4$ mm) is supported by a shoe ($t_s = 5$ mm) that is attached to the ladder rail by a bolt of diameter $d_p = 8$ mm.

(a) Find support reactions at A and B.

(b) Find the resultant force in the shoe bolt at A.

(c) Find maximum average shear τ and bearing (σ_b) stresses in the shoe bolt at A.

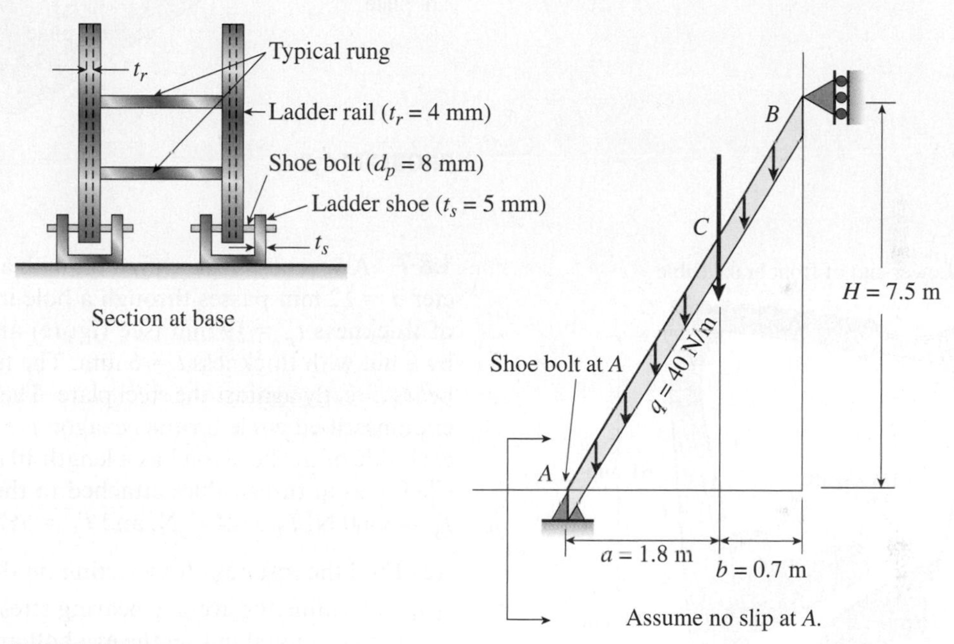

PROBLEM 1.8-4

1.8-5 The force in the brake cable of the V-brake system shown in the figure is $T = 200$ N. The pivot pin at A has a diameter $d_p = 6$ mm and length $L_p = 16$ mm.

Use the dimensions shown in the figure. Neglect the weight of the brake system.

(a) Find the average shear stress τ_{aver} in the pivot pin where it is anchored to the bicycle frame at B.

(b) Find the average bearing stress $\sigma_{b,\text{aver}}$ in the pivot pin over segment AB.

1.8-6 A steel plate of dimensions $2.5 \times 1.5 \times 0.08$ m and weighing 23.1 kN is hoisted by steel cables with lengths $L_1 = 3.2$ m and $L_2 = 3.9$ m that are each attached to the plate by a clevis and pin (see figure). The pins through the clevises are 18 mm in diameter and are located 2.0 m apart. The orientation angles are measured to be $\theta = 94.4°$ and $\alpha = 54.9°$.

For these conditions, first determine the cable forces T_1 and T_2, then find the average shear stress τ_{aver} in both pin 1 and pin 2, and then the average bearing stress σ_b between the steel plate and each pin. Ignore the weight of the cables.

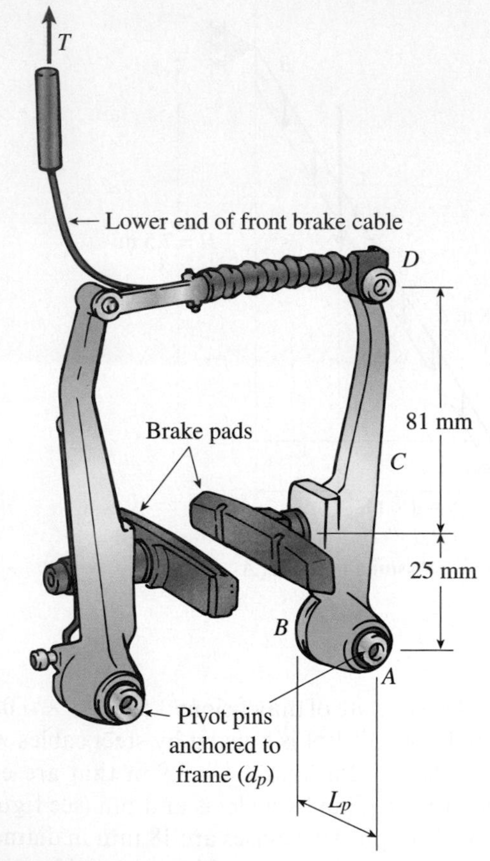

PROBLEM 1.8-5

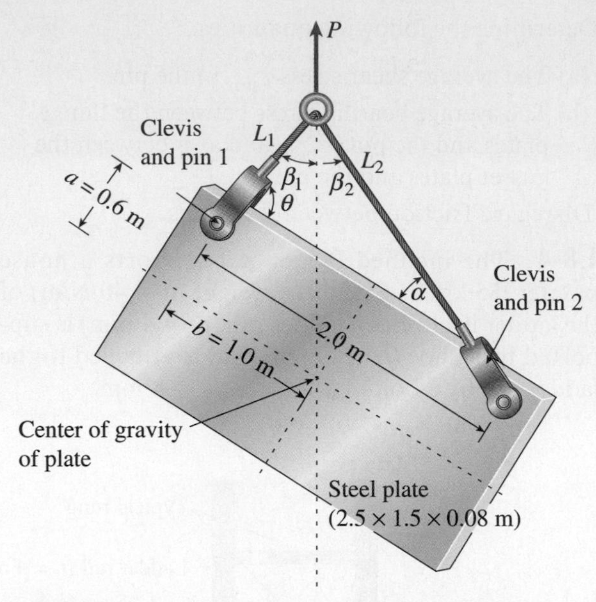

PROBLEM 1.8-6

1.8-7 A special-purpose eye bolt with a shank diameter $d = 12$ mm passes through a hole in a steel plate of thickness $t_p = 19$ mm (see figure) and is secured by a nut with thickness $t = 6$ mm. The hexagonal nut bears directly against the steel plate. The radius of the circumscribed circle for the hexagon is $r = 10$ mm, so each side of the hexagon has a length 10 mm. The tensile forces in three cables attached to the eye bolt are $T_1 = 3560$ N, $T_2 = 2448$ N, and $T_3 = 5524$ N.

(a) Find the resultant force acting on the eye bolt.

(b) Determine the average bearing stress σ_b between the hexagonal nut on the eye bolt and the plate.

(c) Determine the average shear stress τ_{aver} in the nut and also in the steel plate.

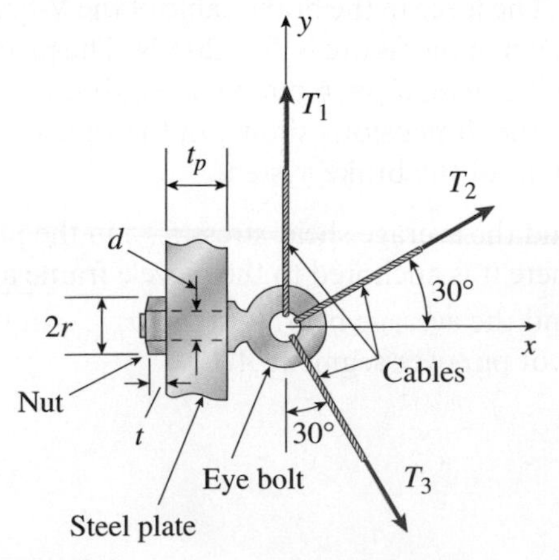

PROBLEM 1.8-7

Representative Problems

1.8-8 An elastomeric bearing pad consisting of two steel plates bonded to a chloroprene elastomer (an artificial rubber) is subjected to a shear force V during a static loading test (see figure). The pad has dimensions $a = 125$ mm and $b = 240$ mm, and the elastomer has a thickness $t = 50$ mm. When the force V equals 12 kN, the top plate is found to have displaced laterally 8.0 mm with respect to the bottom plate.

What is the shear modulus of elasticity G of the chloroprene?

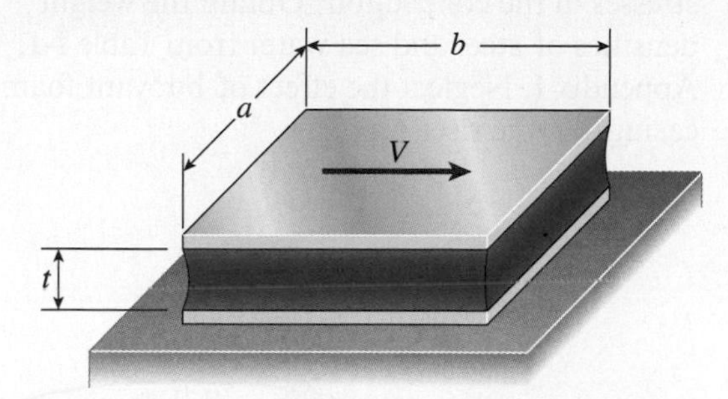

PROBLEM 1.8-8

1.8-9 A joint between two concrete slabs A and B is filled with a flexible epoxy that bonds securely to the concrete (see figure). The height of the joint is $h = 100$ mm, its length is $L = 1.0$ m, and its thickness is $t = 12$ mm. Under the action of shear forces V, the slabs displace vertically through the distance $d = 0.048$ mm relative to each other.

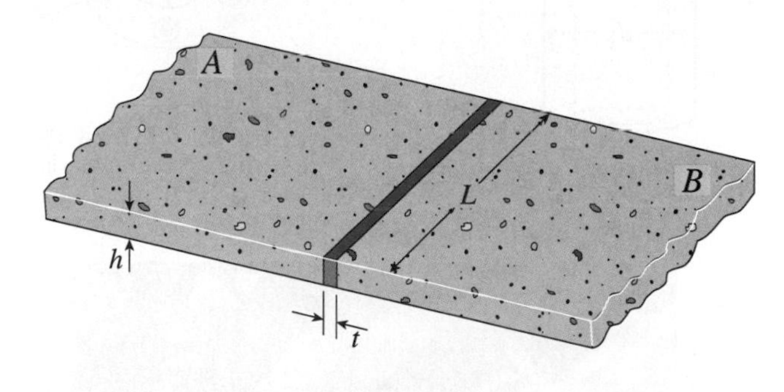

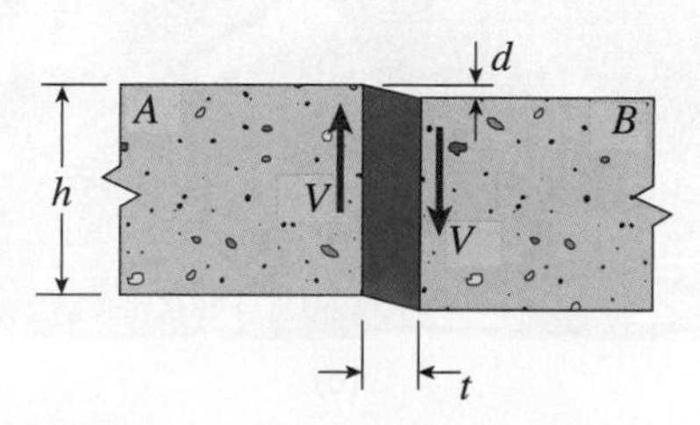

PROBLEM 1.8-9

(a) What is the average shear strain γ_{aver} in the epoxy?

(b) What is the magnitude of the forces V if the shear modulus of elasticity G for the epoxy is 960 MPa?

1.8-10 A steel punch consists of two shafts: upper shaft and lower shaft. Assume that the upper shaft has a diameter $d_1 = 24$ mm and the bottom shaft has a diameter $d_2 = 16$ mm. The punch is used to insert a hole in a 4 mm plate, as shown in the figure. If a force $P = 70$ kN is required to create the hole, what is the average shear stress in the plate and the average compressive stress in the upper and lower shaft of the punch?

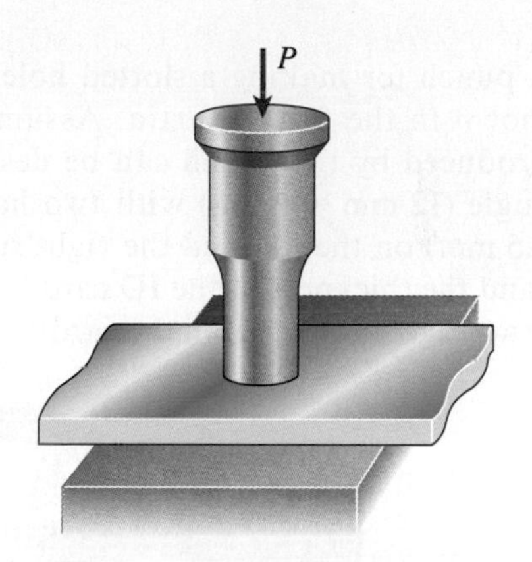

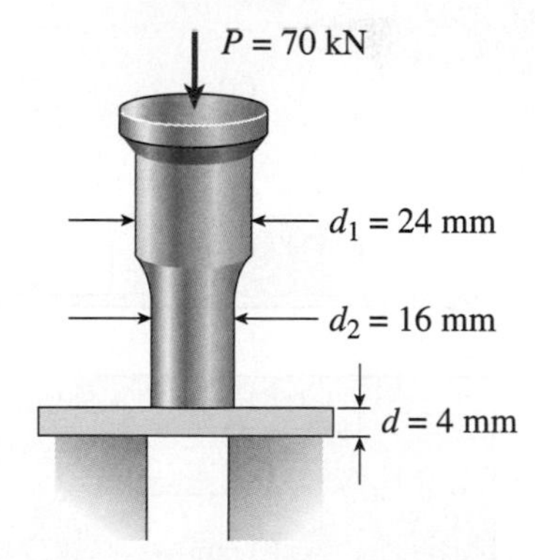

PROBLEM 1.8-10

1.8-11 A joint between two glass plates A and B is filled with a flexible epoxy that bonds securely to the glass. The height of the joint is $h = 13$ mm, its length is $L = 750$ mm, and its thickness is $t = 13$ mm. Shear force of $V = 110$ kN is applied to the joint. Calculate the displacement of the joint if the shear modulus of elasticity G of the epoxy is 690 MPa. Calculate the average shear strain in the epoxy.

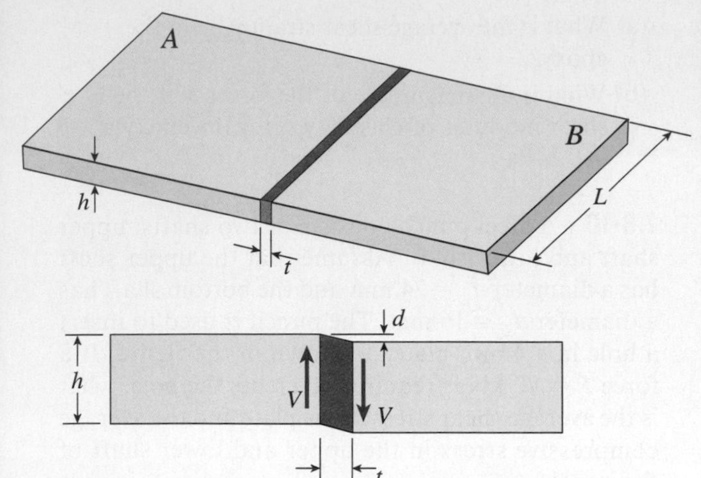

PROBLEM 1.8-11

1.8-12 A punch for making a slotted hole in ID cards is shown in the figure part a. Assume that the hole produced by the punch can be described as a rectangle (12 mm × 3 mm) with two half circles (r = 1.5 mm) on the left and the right sides. If P = 10 N and the thickness of the ID card is 1 mm, what is the average shear stress in the card?

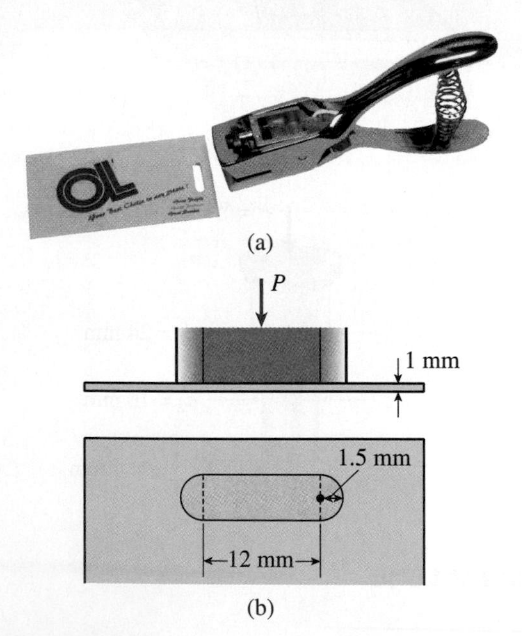

PROBLEM 1.8-12

1.8-13 A steel riser pipe hangs from a drill rig located offshore in deep water (see figure). Separate segments are joined using bolted flange plages (see figure part b and photo). Assume that there are six bolts at each pipe segment connection. Assume that the total length of the riser pipe is L = 1500 m; outer and inner diameters are d_2 = 405 mm and d_1 = 380 mm; flange plate thickness t_f = 44 mm; and bolt and washer diameters are d_b = 28 mm, and d_w = 47 mm, respectively.

(a) If the entire length of the riser pipe is suspended in air, find the average normal stress σ in each bolt, the average bearing stress σ_b beneath each washer, and the average shear stress τ through the flange plate at each bolt location for the top-most bolted connection.

(b) If the same riser pipe hangs from a drill rig at sea, what are the normal, bearing, and shear stresses in the connection? Obtain the weight densities of steel and sea water from Table I-1, Appendix I. Neglect the effect of buoyant foam casings on the riser pipe.

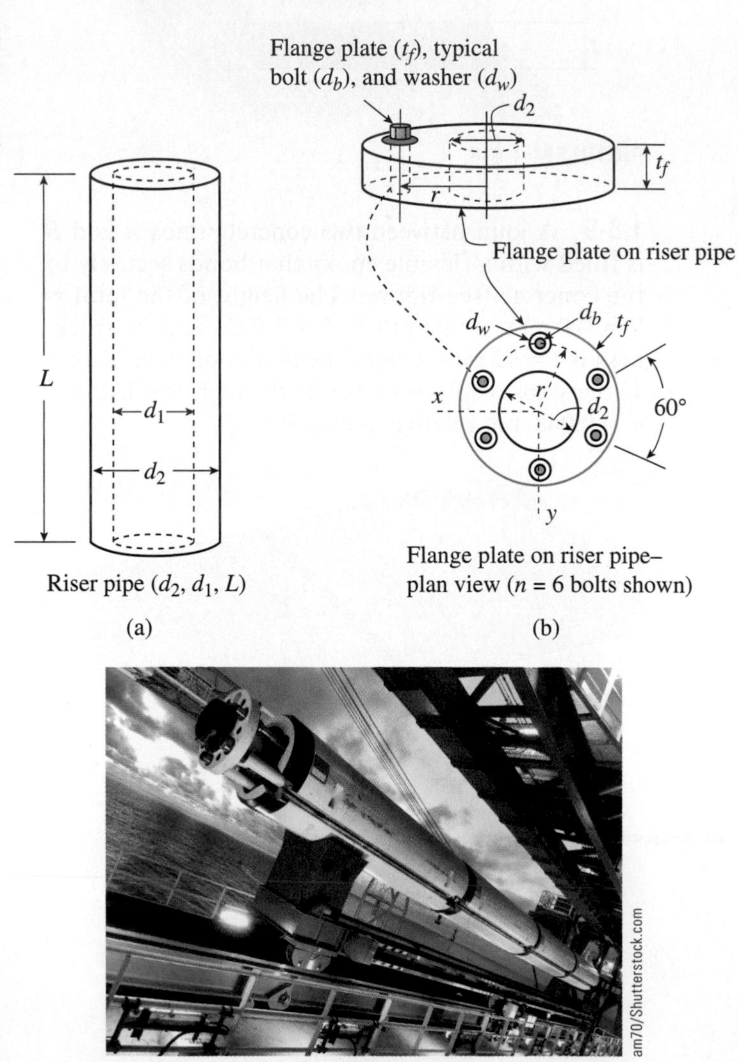

PROBLEM 1.8-13

1.8-14 A flexible connection consisting of rubber pads (thickness $t = 9$ mm) bonded to steel plates is shown in the figure. The pads are 160 mm long and 80 mm wide.

(a) Find the average shear strain γ_{aver} in the rubber if the force $P = 16$ kN and the shear modulus for the rubber is $G = 1250$ kPa.

(b) Find the relative horizontal displacement δ between the interior plate and the outer plates.

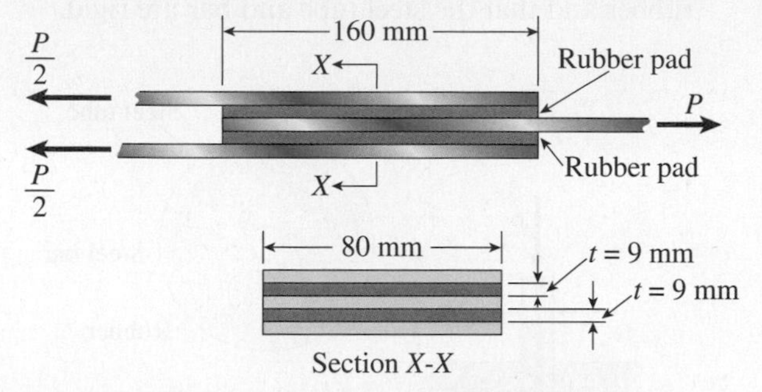

PROBLEM 1.8-14

1.8-15 A hitch-mounted bicycle rack is designed to carry up to four 135 N bikes mounted on and strapped to two arms GH (see bike loads in the figure part a). The rack is attached to the vehicle at A and is assumed to be like a cantilever beam $ABCDGH$ (figure part b). The weight of fixed segment AB is $W_1 = 45$ N, centered 225 mm from A (see figure part b) and the rest of the rack weighs $W_2 = 180$ N, centered 480 mm from A. Segment $ABCDG$ is a steel tube of 50 mm × 50 mm with a thickness $t = 3$ mm. Segment $BCDGH$ pivots about a bolt at B with a diameter $d_B = 6$ mm to allow access to the rear of the vehicle without removing the hitch rack. When in use, the rack is secured in an upright position by a pin at C (diameter of pin $d_p = 8$ mm) (see photo and figure part c). The overturning effect of the bikes on the rack is resisted by a force couple $F \cdot h$ at BC.

(a) Find the support reactions at A for the fully loaded rack.

(b) Find forces in the bolt at B and the pin at C.

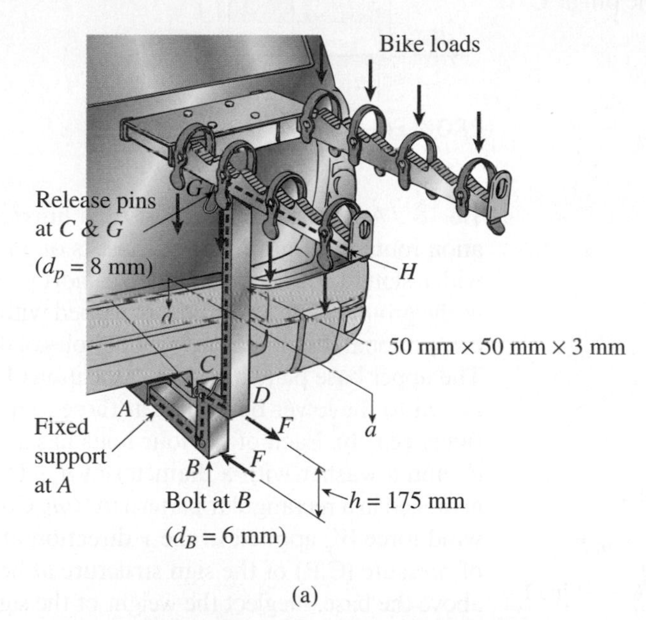

(a)

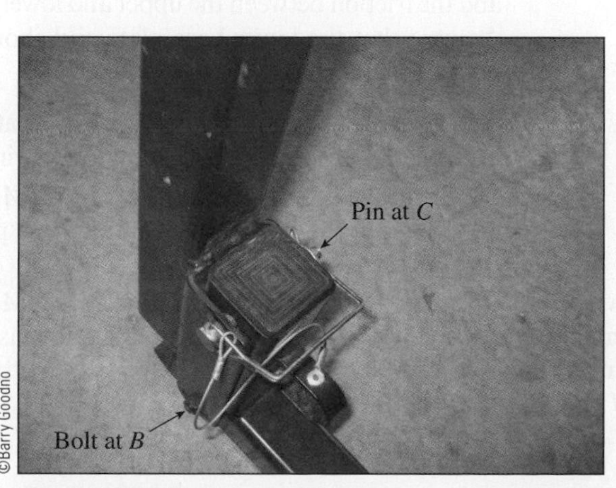

y
480 mm
4 bike loads
G
H
685 mm
3 @ 100 mm
W_2
150 mm
C
D
F
54 mm
h = 175 mm
W_1
A
B
F
x
230 mm 200 mm

(b)

Pin at C
C
54 mm
D
50 × 50 × 3 mm tube

(c) Section a–a

PROBLEM 1.8-15

(c) Find average shear stresses τ_{aver} in both the bolt at B and the pin at C.

(d) Find average bearing stresses σ_b in the bolt at B and the pin at C.

1.8-16 The clamp shown in the figure supports a load hanging from the lower flange of a steel beam. The clamp consists of two arms (A and B) joined by a pin at C. The pin has a diameter $d = 12$ mm. Because arm B straddles arm A, the pin is in double shear.

Line 1 in the figure defines the line of action of the resultant horizontal force H acting between the lower flange of the beam and arm B. The vertical distance from this line to the pin is $h = 250$ mm. Line 2 defines the line of action of the resultant vertical force V acting between the flange and arm B. The horizontal distance from this line to the centerline of the beam is $c = 100$ mm. The force conditions between arm A and the lower flange are symmetrical with those given for arm B.

Determine the average shear stress in the pin at C when the load $P = 18$ kN.

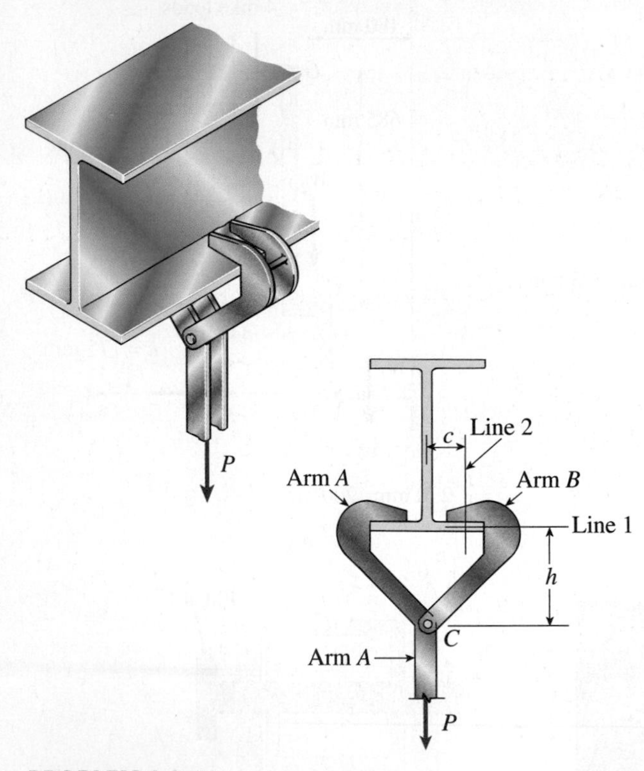

PROBLEM 1.8-16

1.8-17 A shock mount constructed as shown in the figure is used to support a delicate instrument. The mount consists of an outer steel tube with inside diameter b, a central steel bar of diameter d that supports the load P, and a hollow rubber cylinder (height h) bonded to the tube and bar.

(a) Obtain a formula for the shear stress τ in the rubber at a radial distance r from the center of the shock mount.

(b) Obtain a formula for the downward displacement δ of the central bar due to the load P, assuming that G is the shear modulus of elasticity of the rubber and that the steel tube and bar are rigid.

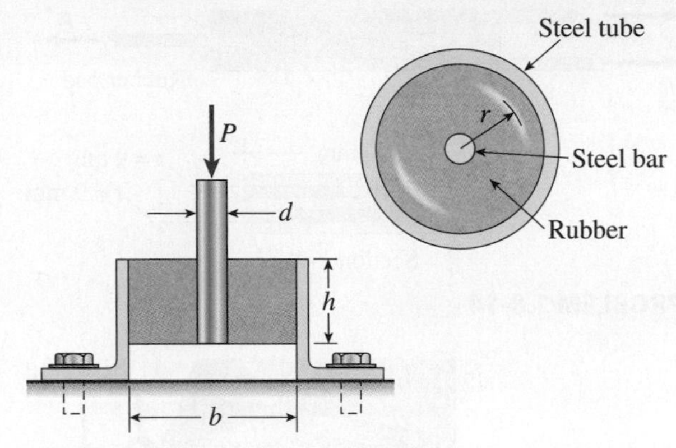

PROBLEM 1.8-17

1.8-18 A removable sign post on a hurricane evacuation route (see figure part a) consists of an upper pole with a slotted base plate bolted to a short post anchored in the ground. The lower post is capped with a separate conventional base plate having four holes of diameter d_b. The upper base plate has slots at locations 1 to 4 and is bolted to the lower base plate at these four points (see figure part b). Each of the four bolts has a diameter of d_b and a washer with a diameter of d_w. The bolts are arranged in a rectangular pattern ($b \times h$). Consider only wind force W_y applied in the y direction at the center of pressure (C.P.) of the sign structure at height $z = L$ above the base. Neglect the weight of the sign and post and the friction between the upper and lower base plates. Assume that the lower base plate and short anchored post are rigid.

(a) Find the average shear stress τ (MPa) at bolts 1 and 4 (see figure part c) due to the wind force W_y.

(b) Find the average bearing stress σ_b (MPa) between the bolt and the upper base plate (thickness t) at bolts 1 and 4.

(c) Find the average bearing stress σ_b (MPa) between the upper base plate and washer at bolt 4 due to the wind force W_y (assume the initial bolt pretension is zero).

(d) Find the average shear stress τ (MPa) through the upper base plate at bolt 4 due to the wind force W_y.

(e) Find an expression for the normal stress σ in bolt 3 due to the wind force W_y.

(See problem 1.9-17 for additional discussion of wind on a sign and the resulting forces acting on a conventional base plate.)

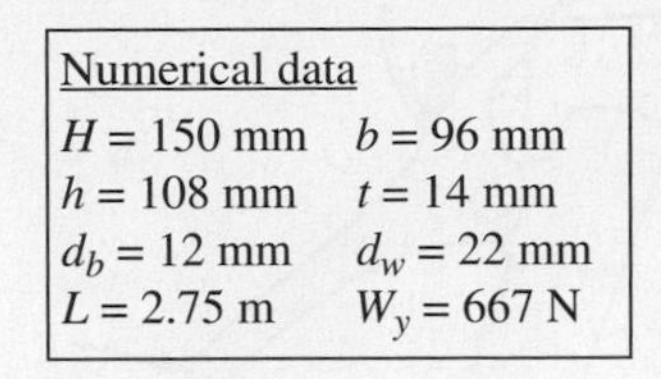

Numerical data	
$H = 150$ mm	$b = 96$ mm
$h = 108$ mm	$t = 14$ mm
$d_b = 12$ mm	$d_w = 22$ mm
$L = 2.75$ m	$W_y = 667$ N

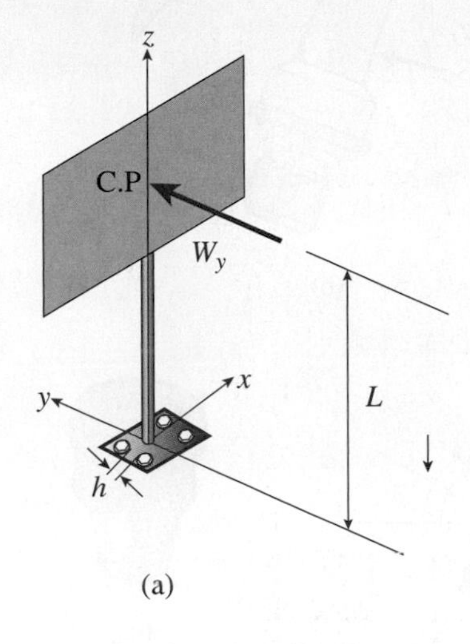

(a)

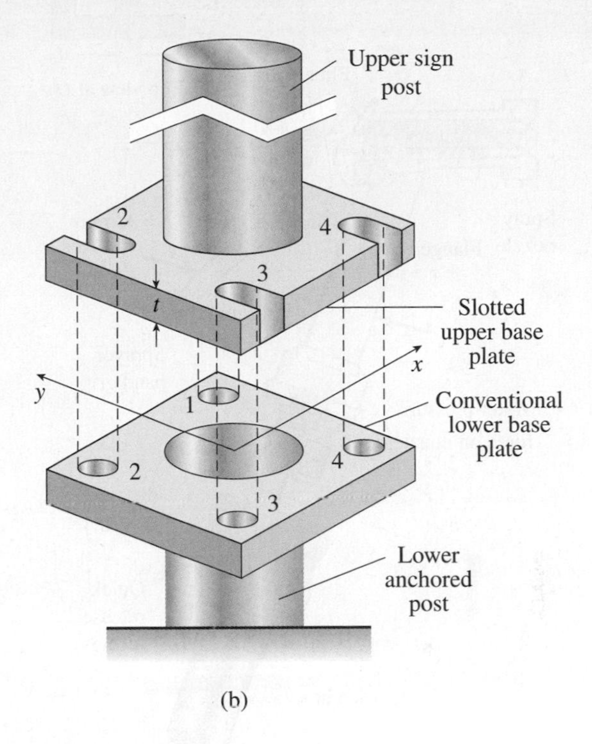

(b)

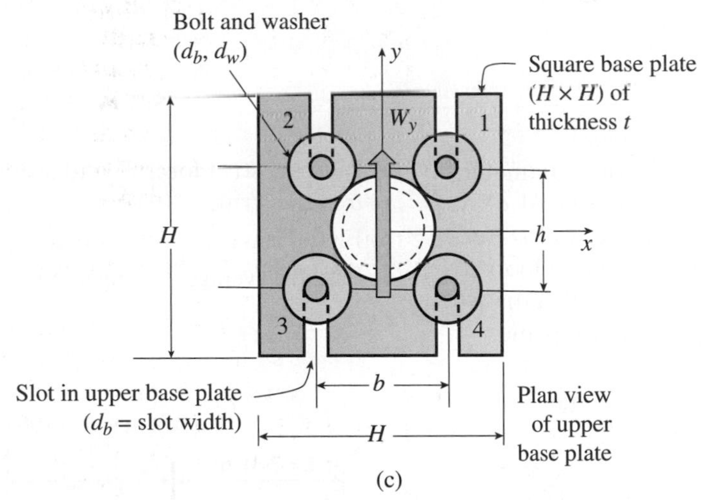

(c)

PROBLEM 1.8-18

1.8-19 A spray nozzle for a garden hose requires a force $F = 22$ N to open the spring-loaded spray chamber AB. The nozzle hand grip pivots about a pin through a flange at O. Each of the two flanges has a thickness $t = 1.5$ mm, and the pin has a diameter $d_p = 3$ mm (see figure part a). The spray nozzle is attached to the garden hose with a quick release fitting at B (see figure part b). Three brass balls (diameter $d_b = 4.5$ mm) hold the spray head in place under a water pressure force $f_p = 135$ N at C (see figure part c). Use dimensions given in figure part a.

(a) Find the force in the pin at O due to applied force F.

(b) Find average shear stress τ_{aver} and bearing stress σ_b in the pin at O.

(c) Find the average shear stress τ_{aver} in the brass retaining balls at C due to water pressure force f_p.

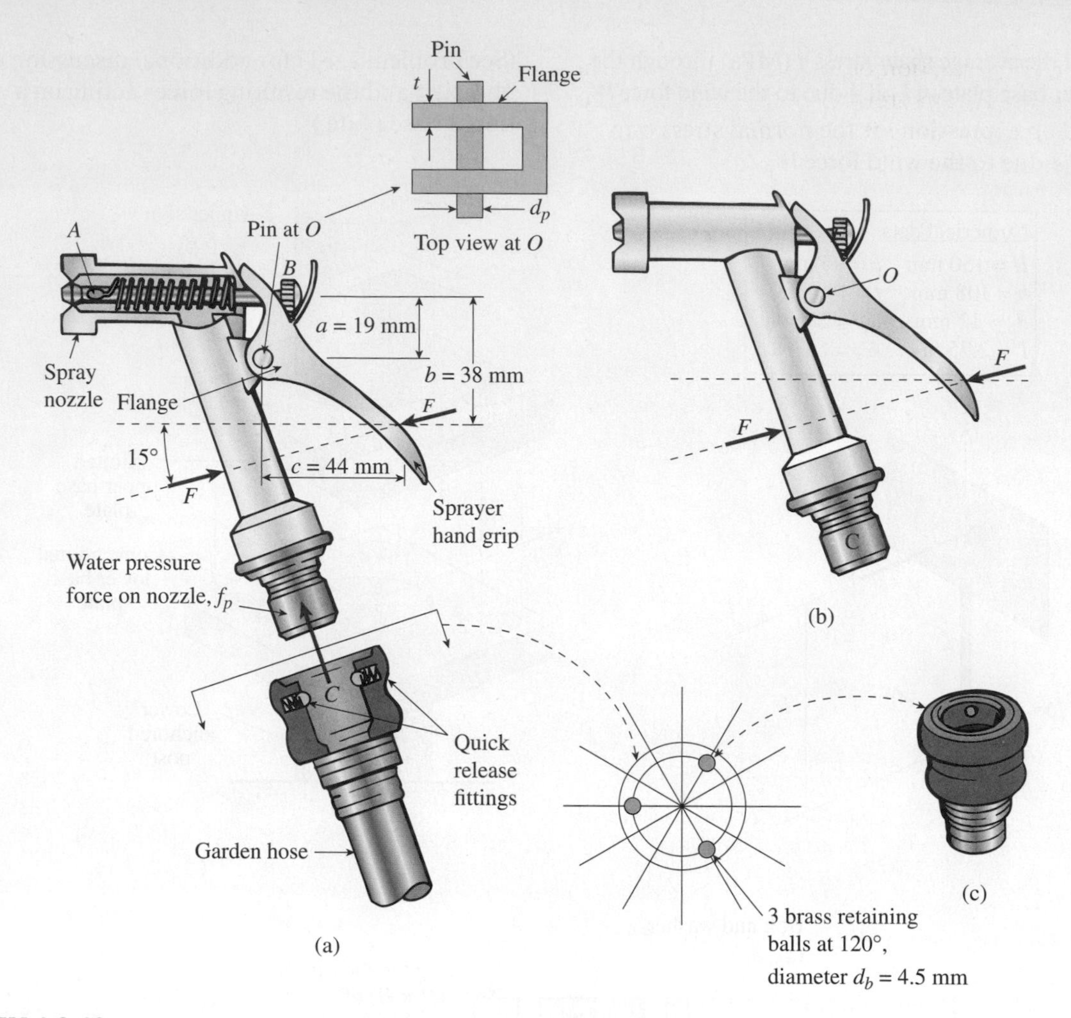

PROBLEM 1.8-19

1.8-20 A single steel strut AB with a diameter $d_s = 8$ mm supports the vehicle engine hood of a mass 20 kg that pivots about hinges at C and D (see figure parts a and b). The strut is bent into a loop at its end and then attached to a bolt at A with a diameter $d_b = 10$ mm. Strut AB lies in a vertical plane.

(a) Find the strut force F_s and average normal stress σ in the strut.

(b) Find the average shear stress τ_{aver} in the bolt at A.

(c) Find the average bearing stress σ_b on the bolt at A.

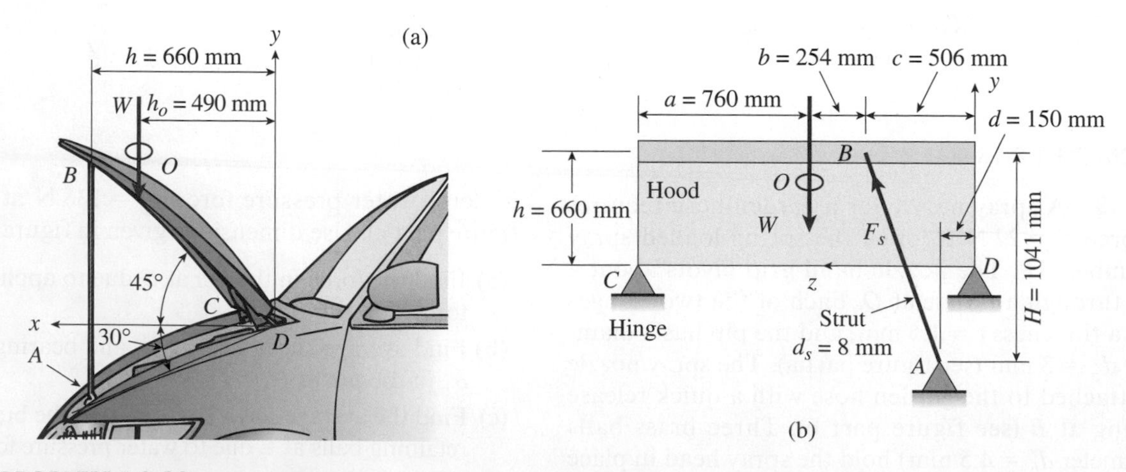

PROBLEM 1.8-20

1.8-21 The top portion of a pole saw used to trim small branches from trees is shown in the figure part a. The cutting blade BCD (see figure parts a and c) applies a force P at point D. Ignore the effect of the weak return spring attached to the cutting blade below B. Use properties and dimensions given in the figure.

(a) Find the force P on the cutting blade at D if the tension force in the rope is $T = 110$ N (see free-body diagram in figure part b).

(b) Find force in the pin at C.

(c) Find average shear stress τ_{aver} and bearing stress σ_b in the support pin at C (see section a–a through cutting blade in figure part c).

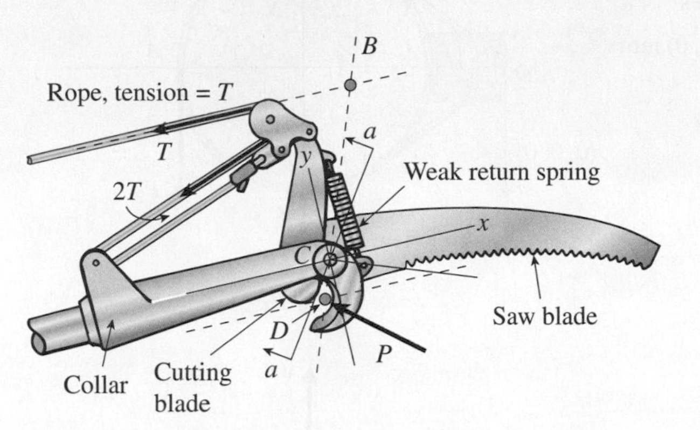

(a) Top part of pole saw

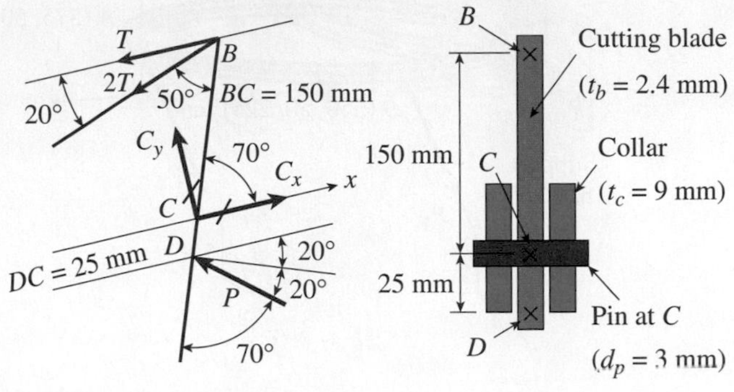

(b) Free-body diagram

(c) Section a–a

PROBLEM 1.8-21

1.8-22 A cargo ship is tied down to marine bollards at a number of points along its length while its cargo is unloaded by a container handling crane. Each bollard is fastened to the wharf using anchor bolts. Three cables having known tension force magnitudes $F_1 = 110$ kN, $F_2 = 85$ kN, and $F_3 = 90$ kN are secured to one bollard at a point A with coordinates (0, 0.45 m, 0) in the x-y-z coordinate system shown in the figure part b. Each cable force is directed at an attachment point on the ship. Force F_1 is directed from point A to a point on the ship having coordinates (3 m, 9 m, 0); force F_2 is directed at a point with coordinates (6.5 m, 8.5 m, 2 m); and force F_3 is directed at a point with coordinates (8 m, 9 m, 5 m). The diameter of each anchor bolts is $d_b = 24$ mm.

(a) Find the reaction forces and reaction moments at the base of the bollard.

(b) Calculate the average shear stress in the anchor bolts (in the x-z plane). Assume each bolt carries an equal share of the total force.

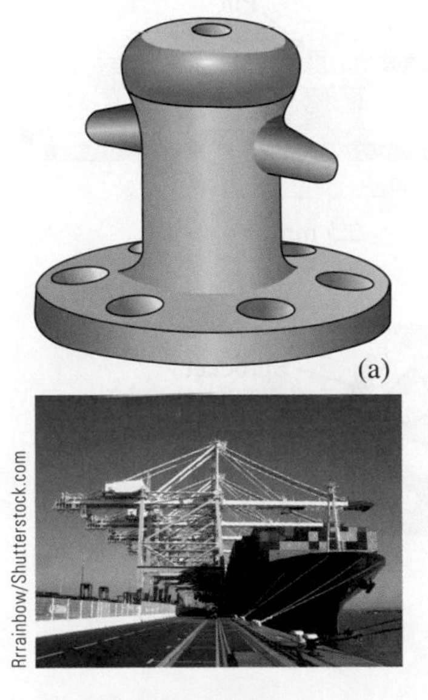

(a)

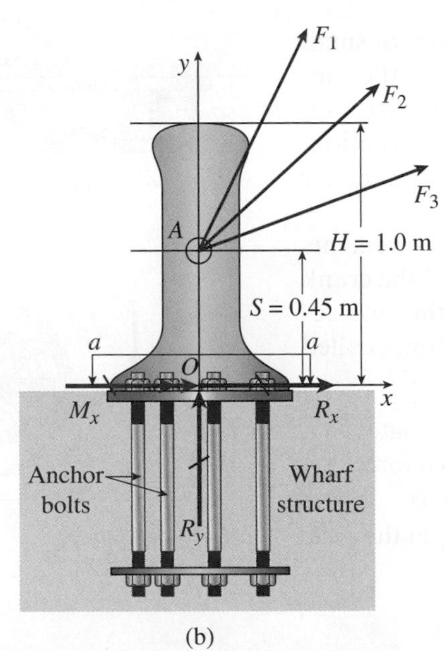

(b)

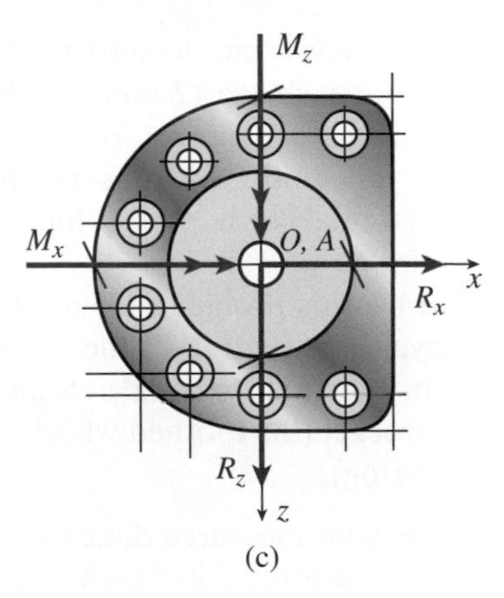

(c)

PROBLEM 1.8-22

1.8-23 A basketball player hangs on the rim after a dunk. He applies equal forces $P_1 = P_2 = 500$ N at both A and B (see joint coordinates in the figure). Forces P_1 *and* P_2 act parallel to the *y-z* plane.

(a) Find the *reactions* at the support bracket (assume that the bracket-rim assembly is a cantilever beam).

(b) Find connection shear stresses at bolt 2.

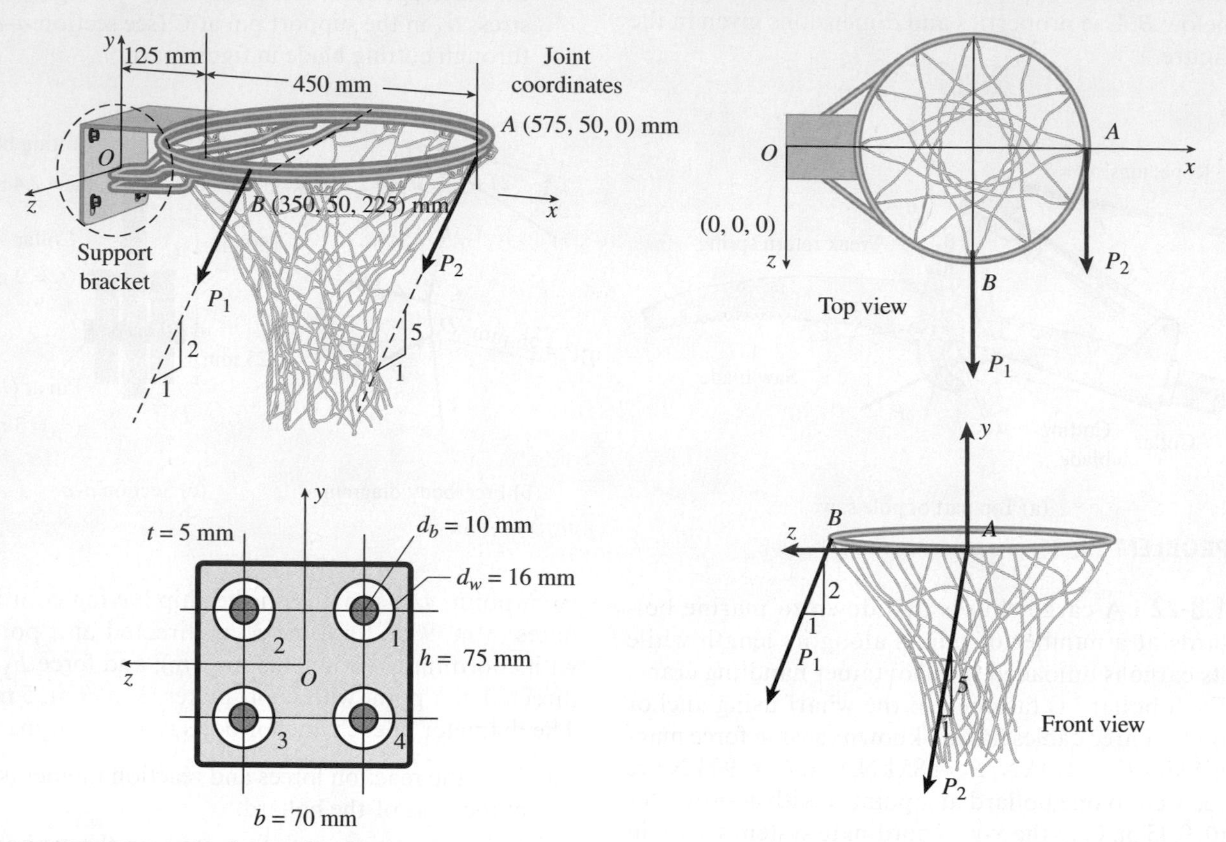

PROBLEM 1.8-23 Support bracket

1.8-24 A bicycle chain consists of a series of small links, where each are 12 mm long between the centers of the pins (see figure). You might wish to examine a bicycle chain and observe its construction. Note particularly the pins, which have a diameter of 2.5 mm.

To solve this problem, make two measurements on a bicycle (see figure): (1) the length L of the crank arm from main axle to pedal axle and (2) the radius R of the sprocket (the toothed wheel, sometimes called the chainring).

(a) Using your measured dimensions, calculate the tensile force T in the chain due to a force $F = 800$ N applied to one of the pedals.

(b) Calculate the average shear stress τ_{aver} in the pins.

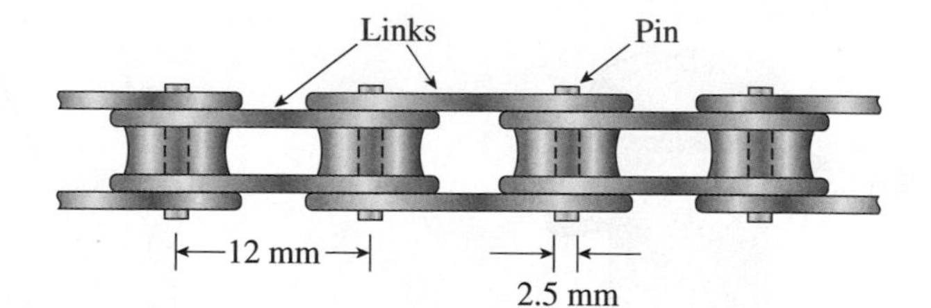

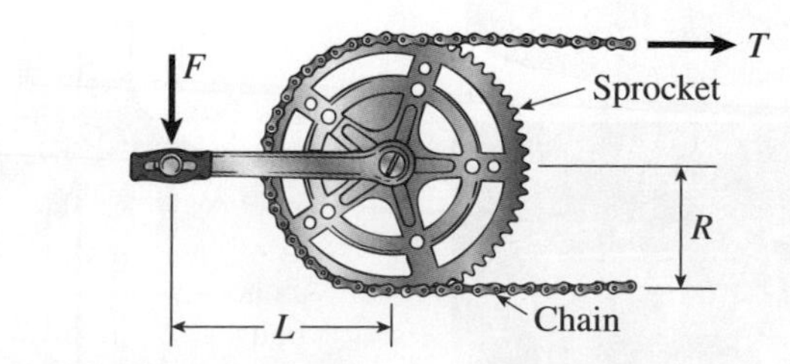

PROBLEM 1.8-24

1.9 Allowable Stresses and Allowable Loads

Introductory Problems

1.9-1 A bar of solid circular cross section is loaded in tension by forces P (see figure). The bar has a length $L = 380$ mm and diameter $d = 6$ mm. The material is a magnesium alloy having a modulus of elasticity $E = 42.7$ GPa. The allowable stress in tension is $\sigma_{allow} = 89.6$ MPa, and the elongation of the bar must not exceed 0.8 mm.

What is the allowable value of the forces P?

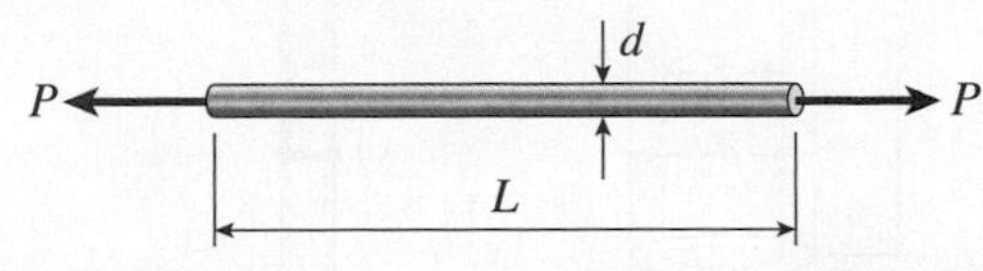

PROBLEM 1.9-1

1-9.2 A torque T_0 is transmitted between two flanged shafts by means of ten 20-mm bolts (see figure and photo). The diameter of the bolt circle is $d = 250$ mm.

If the allowable shear stress in the bolts is 85 MPa, what is the maximum permissible torque? (Disregard friction between the flanges.)

Drive shaft coupling on a ship propulsion motor

PROBLEM 1.9-2

1.9-3 A tie-down on the deck of a sailboat consists of a bent bar bolted at both ends, as shown in the figure. The diameter d_B of the bar is 6 mm, the diameter d_W of the washers is 22 mm, and the thickness t of the fiberglass deck is 10 mm.

If the allowable shear stress in the fiberglass is 2.1 MPa, and the allowable bearing pressure between the washer and the fiberglass is 3.8 MPa, what is the allowable load P_{allow} on the tie-down?

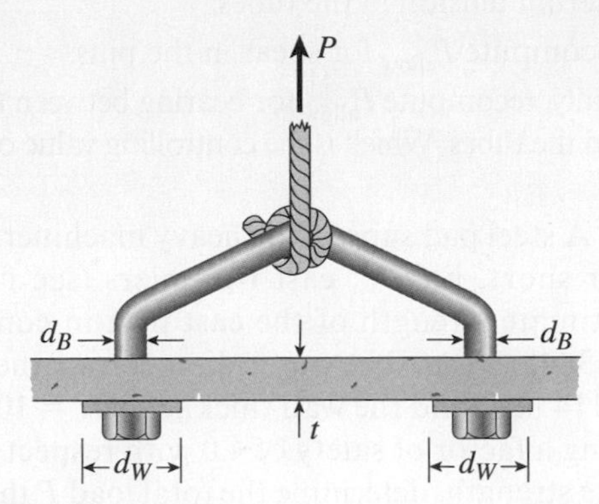

PROBLEM 1.9-3

1.9-4 Two steel tubes are joined at B by four pins ($d_p = 11$ mm), as shown in the cross section a–a in the figure. The outer diameters of the tubes are $d_{AB} = 41$ mm and $d_{BC} = 28$ mm. The wall thickness are $t_{AB} = 6.5$ mm and $t_{BC} = 7.5$ mm. The yield stress in tension for the steel is $\sigma_Y = 200$ MPa and the ultimate stress in *tension* is $\sigma_U = 340$ MPa. The corresponding yield and ultimate values in *shear* for the

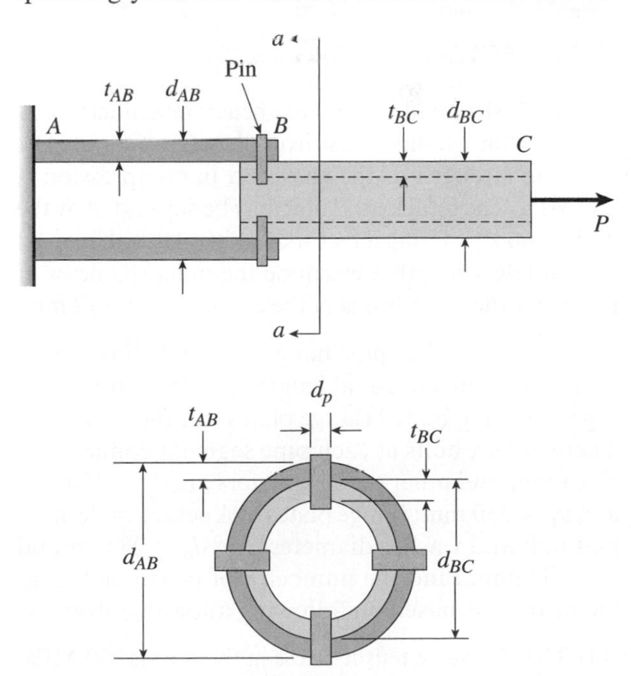

Section a–a

PROBLEM 1.9-4

pin are 80 MPa and 140 MPa, respectively. Finally, the yield and ultimate values in *bearing* between the pins and the tubes are 260 MPa and 450 MPa, respectively. Assume that the factors of safety with respect to yield stress and ultimate stress are 3.5 and 4.5, respectively.

(a) Calculate the allowable tensile force P_{allow} considering tension in the tubes.

(b) Recompute P_{allow} for shear in the pins.

(c) Finally, recompute P_{allow} for bearing between the pin and the tubes. Which is the controlling value of P?

1.9-5 A steel pad supporting heavy machinery rests on four short, hollow, cast iron piers (see figure). The ultimate strength of the cast iron in compression is 344.5 MPa. The outer diameter of the piers is $d = 114$ mm, and the wall thickness is $t = 10$ mm.

Using a factor of safety of 4.0 with respect to the ultimate strength, determine the total load P that can be supported by the pad.

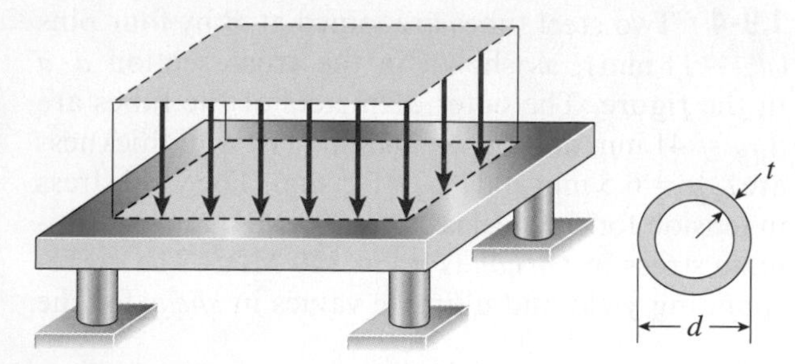

PROBLEMS 1.9-5 and 1.9-6

1.9-6 A steel pad supporting heavy machinery rests on four short, hollow, cast iron piers (see figure). The ultimate strength of the cast iron in compression is 400 MPa. The total load P that may be supported by the pad is 900 kN. Using a factor of safety 3.0 with respect to ultimate strength, determine the outer diameter of the pier if the thickness is of the cross section is 12 mm.

1.9-7 A steel riser pipe hangs from a drill rig. Individual segments of equal length $L = 15$ m are joined together using bolted flange plates (see figure part b). There are six bolts at each pipe segment connection. The outer and inner pipe diameters are $d_2 = 350$ mm and $d_1 = 330$ mm; flange plate thickness $t_f = 38$ mm; and bolt and washer diameters are $d_b = 28$ mm and $d_w = 48$ mm. Find the number n of permissible segments of pipe based on following allowable stresses.

(a) The allowable tensile stress in the pipe is 350 MPa.

(b) The allowable tensile stress in a bolt is 825 MPa.

Find number of segments n for two cases: pipe hanging in air and pipe hanging in seawater.

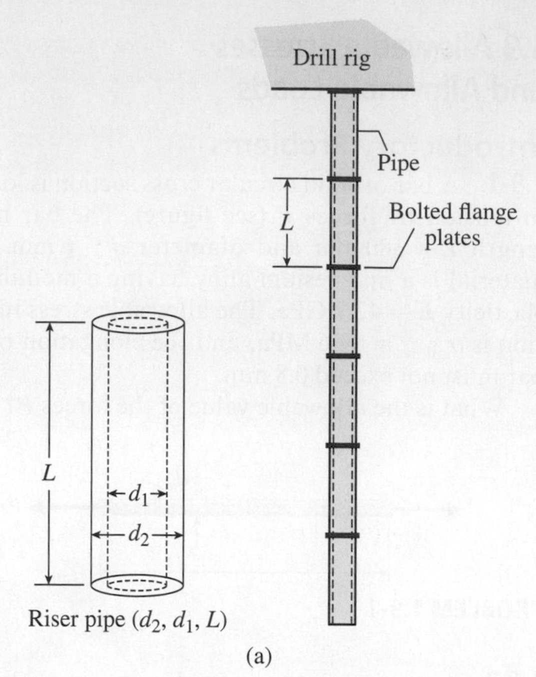

(a)

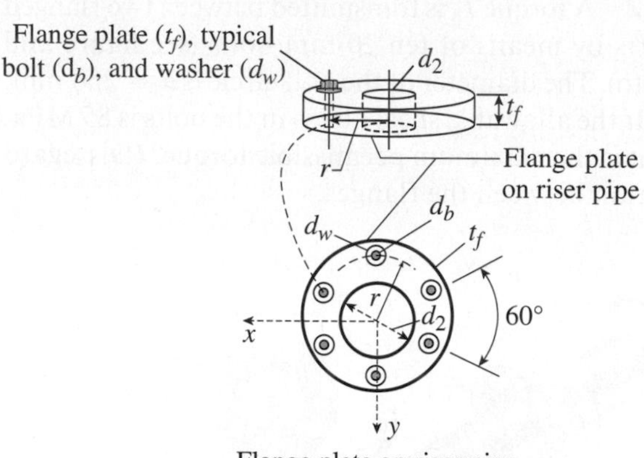

(b)

PROBLEM 1.9-7

Representative Problems

1.9-8 The rear hatch of a van ($BDCG$ in figure part a) is supported by two hinges at B_1 and B_2 and by two struts A_1B_1 and A_2B_2 (diameter $d_s = 10$ mm), as shown in figure part b. The struts are supported at A_1 and A_2 by pins, each with a diameter $d_p = 9$ mm and passing through an eyelet of thickness $t = 8$ mm at the end of the strut (figure part b). A closing force $P = 50$ N is applied at G, and the mass of the hatch $M_h = 43$ kg is concentrated at C.

(a) What is the force F in each strut? (Use the free-body diagram of one half of the hatch in the figure part c.)

(b) What is the maximum permissible force in the strut, F_{allow}, if the allowable stresses are compressive stress in the strut, 70 MPa; shear stress in the pin, 45 MPa; and bearing stress between the pin and the end of the strut, 110 MPa.

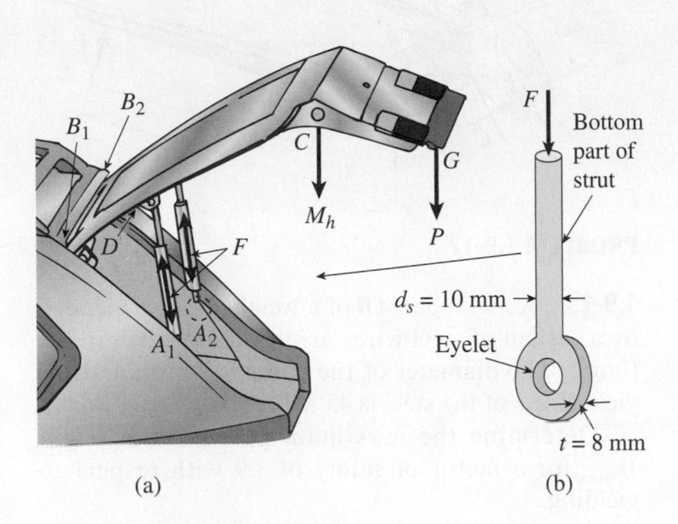

(a) (b)

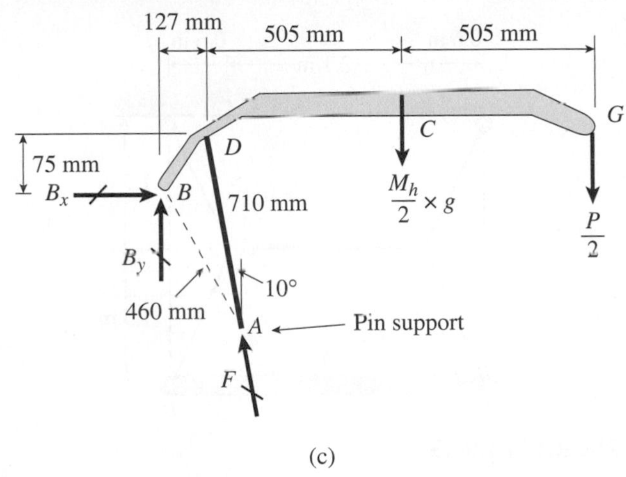

(c)

PROBLEM 1.9-8

1.9-9 A lifeboat hangs from two ship's davits, as shown in the figure. A pin of diameter $d = 20$ mm passes through each davit and supports two pulleys, one on each side of the davit.

Cables attached to the lifeboat pass over the pulleys and wind around winches that raise and lower the lifeboat. The lower parts of the cables are vertical and the upper parts make an angle $\alpha = 15°$ with the horizontal. The allowable tensile force in each cable is 8 kN, and the allowable shear stress in the pins is 27.5 MPa.

If the lifeboat weighs 6.7 kN, what is the maximum weight that can be carried in the lifeboat?

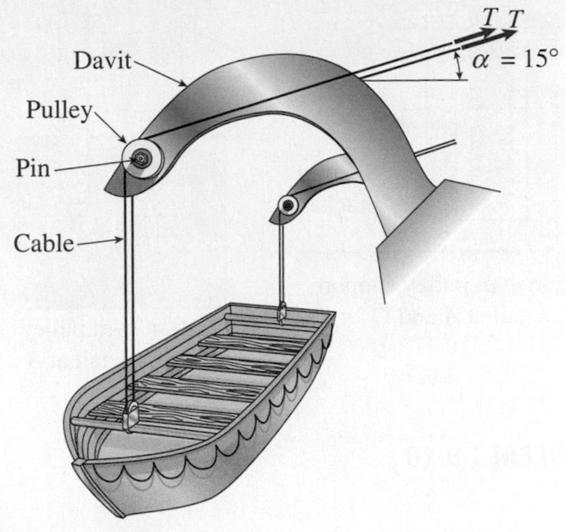

PROBLEM 1.9-9

1.9-10 A cable and pulley system in the figure part a supports a cage of a mass 300 kg at B. Assume that this includes the mass of the cables as well. The thickness of each of the three steel pulleys is $t = 40$ mm. The pin diameters are $d_{pA} = 25$ mm, $d_{pB} = 30$ mm, and $d_{pC} = 22$ mm (see figure part a and part b).

(a) Find expressions for the resultant forces acting on the pulleys at A, B, and C in terms of cable tension T.

(b) What is the maximum weight W that can be added to the cage at B based on the following allowable stresses? Shear stress in the pins is 50 MPa; bearing stress between the pin and the pulley is 110 MPa.

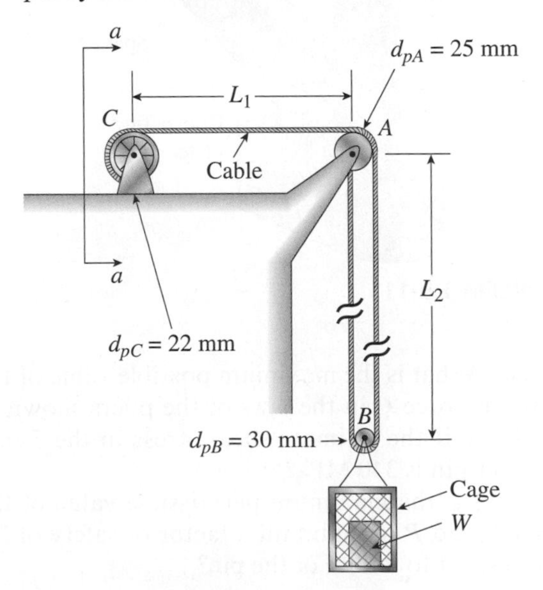

(a)

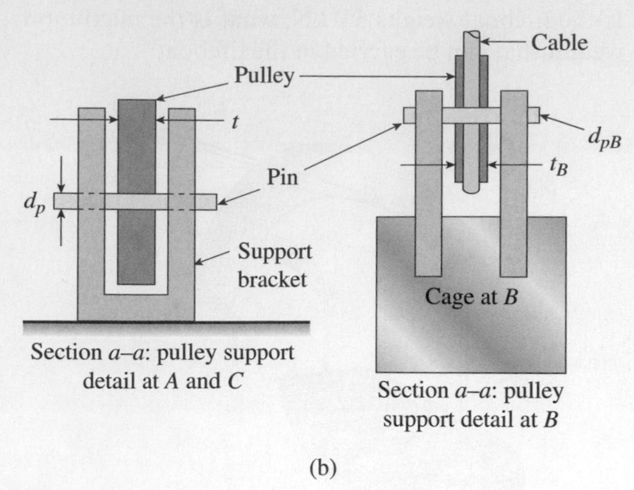

(b)

PROBLEM 1.9-10

1.9-11 A ship's spar is attached at the base of a mast by a pin connection (see figure). The spar is a steel tube of outer diameter $d_2 = 80$ mm and inner diameter $d_1 = 70$ mm. The steel pin has a diameter $d = 25$ mm, and the two plates connecting the spar to the pin have a thickness $t = 12$ mm. The allowable stresses are compressive stress in the spar, 75 MPa; shear stress in the pin, 50 MPa; and bearing stress between the pin and the connecting plates, 120 MPa.

Determine the allowable compressive force P_{allow} in the spar.

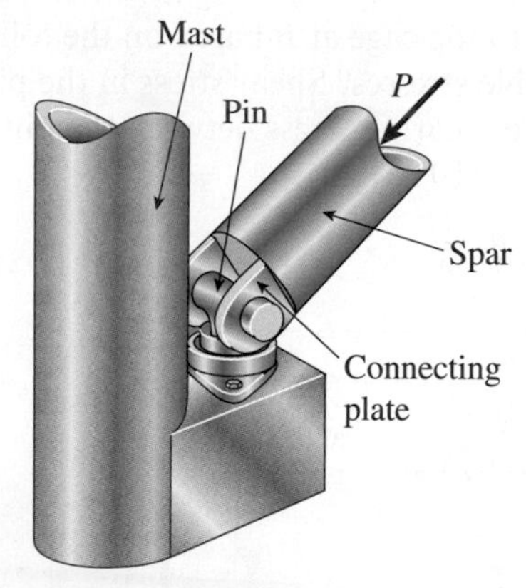

PROBLEM 1.9-11

1.9-12 What is the maximum possible value of the clamping force C in the jaws of the pliers shown in the figure if the ultimate shear stress in the 5-mm diameter pin is 340 MPa?

What is the maximum permissible value of the applied load P to maintain a factor of safety of 3.0 with respect to failure of the pin?

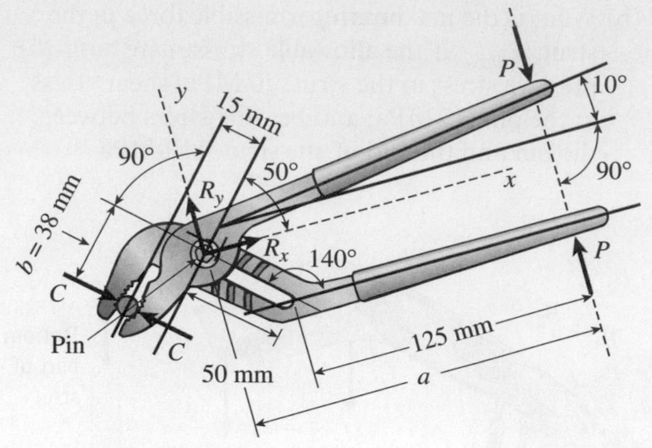

PROBLEM 1.9-12

1.9-13 A metal bar AB of a weight W is suspended by a system of steel wires arranged as shown in the figure. The diameter of the wires is 2 mm, and the yield stress of the steel is 45 MPa.

Determine the maximum permissible weight W_{max} for a factor of safety of 1.9 with respect to yielding.

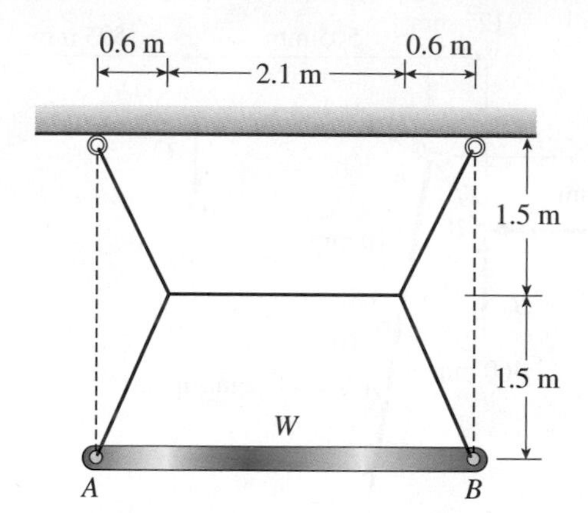

PROBLEM 1.9-13

1.9-14 A plane truss is subjected to loads $2P$ and P at joints B and C, respectively, as shown in the figure part a. The truss bars are made of two steel angles (cross-sectional area of the two angles, $A = 2180\ \text{mm}^2$, and thickness = 6.4 mm, see figure part b) having an ultimate stress in tension equal to 390 MPa. The angles are connected to a 12-mm-thick gusset plate at C (figure part c) with 16-mm diameter rivets; assume each rivet transfers an equal share of the member force to the gusset plate. The ultimate stresses in shear and bearing for the rivet steel are 190 MPa and 550 MPa, respectively.

Determine the allowable load P_{allow} if a safety factor of 2.5 is desired with respect to the ultimate load that can be carried. Consider tension in the bars, shear in the rivets, bearing between the rivets and the bars, and also bearing between the rivets and the gusset plate. Disregard friction between the plates and the weight of the truss itself.

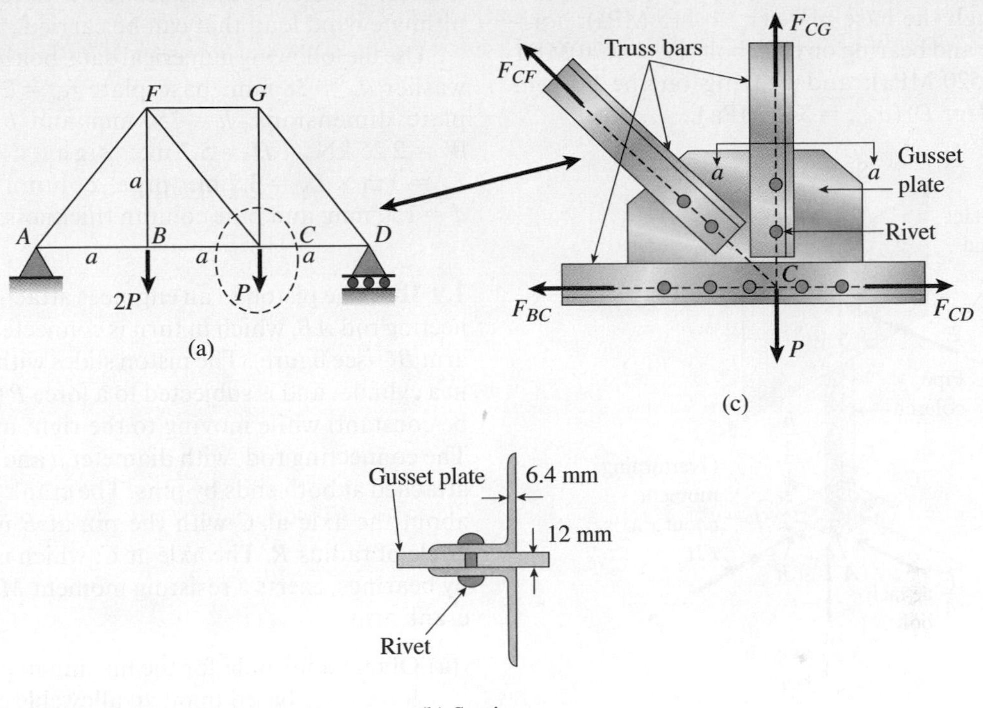

PROBLEM 1.9-14

1.9-15 A solid bar of circular cross section (diameter d) has a hole of diameter $d/5$ drilled laterally through the center of the bar (see figure). The allowable average tensile stress on the net cross section of the bar is σ_{allow}.

(a) Obtain a formula for the allowable load P_{allow} that the bar can carry in tension.

(b) Calculate the value of P_{allow} if the bar is made of brass with a diameter $d = 45$ mm and $\sigma_{allow} = 83$ MPa.

Hint: Use the formulas of Case 15, Appendix E.

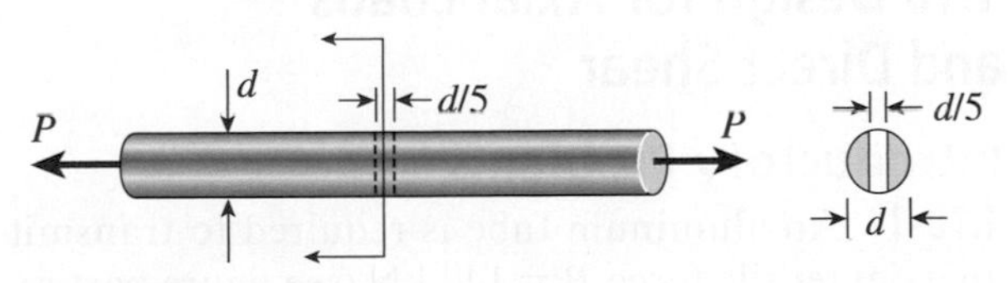

PROBLEM 1.9-15

1.9-16 A solid steel bar of a diameter $d_1 = 60$ mm has a hole of a diameter $d_2 = 32$ mm drilled through it (see figure). A steel pin of a diameter d_2 passes through the hole and is attached to supports.

Determine the maximum permissible tensile load P_{allow} in the bar if the yield stress for shear in the pin is $\tau_Y = 120$ MPa, the yield stress for tension in the bar is $\sigma_Y = 250$ MPa, and a factor of safety of 2.0 with respect to yielding is required. *Hint*: Use the formulas of Case 15, Appendix E.

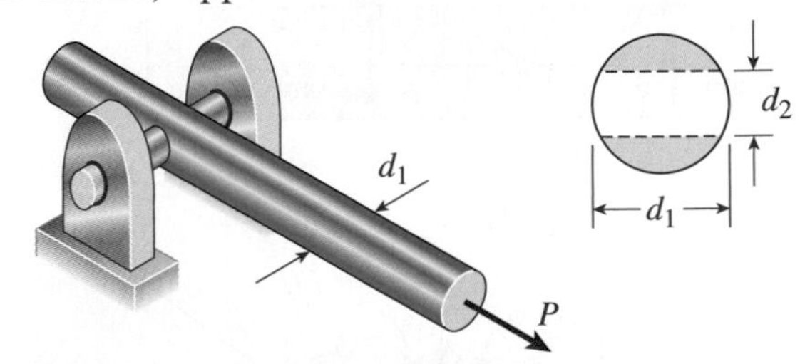

PROBLEM 1.9-16

1.9-17 A sign of weight W is supported at its base by four bolts anchored in a concrete footing. Wind pressure p acts normal to the surface of the sign; the resultant of the uniform wind pressure is force F at the center of pressure (C.P.). The wind force is assumed to create equal shear forces $F/4$ in the y direction at each bolt (see figure parts a and c). The overturning effect of the wind force also causes an uplift force R at bolts A and C and a downward force $(-R)$ at bolts B and D

(see figure part b). The resulting effects of the wind and the associated ultimate stresses for each stress condition are normal stress in each bolt ($\sigma_u = 410$ MPa); shear through the base plate ($\tau_u = 115$ MPa); horizontal shear and bearing on each bolt ($\tau_{hu} = 170$ MPa and $\sigma_{bu} = 520$ MPa); and bearing on the bottom washer at B (or D) ($\sigma_{bw} = 340$ MPa).

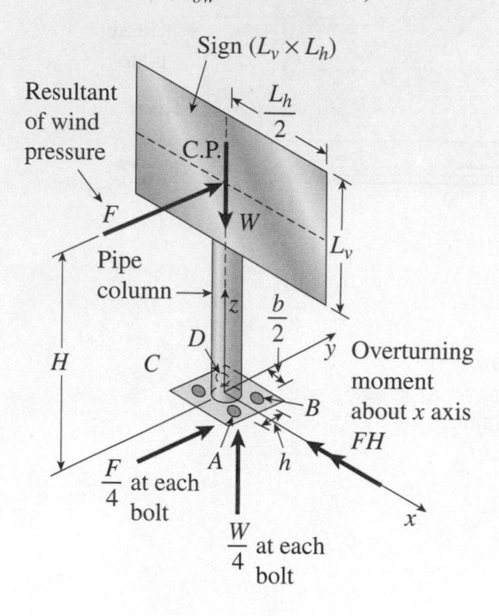

(a)

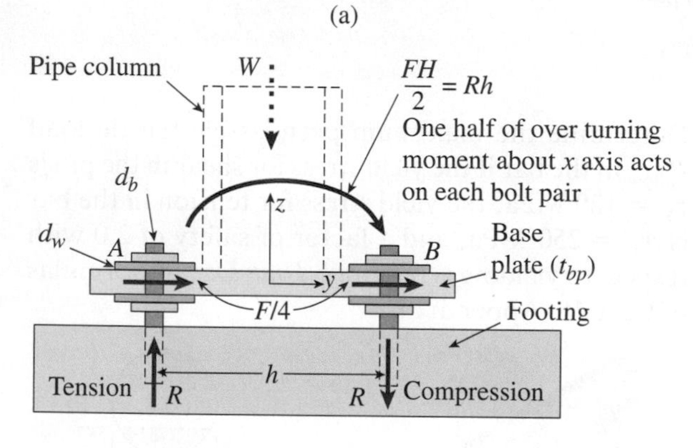

(b)

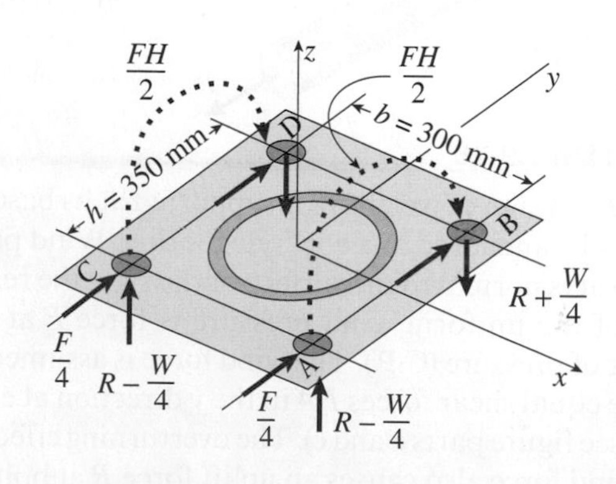

(c)

PROBLEM 1.9-17

Find the maximum wind pressure p_{max} (Pa) that can be carried by the bolted support system for the sign if a safety factor of 2.5 is desired with respect to the ultimate wind load that can be carried.

Use the following numerical data: bolt $d_b = 19$ mm; washer $d_w = 38$ mm; base plate $t_{bp} = 25$ mm; base plate dimensions: $h = 350$ mm and $b = 300$ mm; $W = 2.25$ kN; $H = 5.2$ m; sign dimensions: $L_v = 3$ m $\times$ $L_h = 3.7$ m; pipe column diameter: $d = 150$ mm; and pipe column thickness: $t = 10$ mm.

1.9-18 The piston in an engine is attached to a connecting rod AB, which in turn is connected to a crank arm BC (see figure). The piston slides without friction in a cylinder and is subjected to a force P (assumed to be constant) while moving to the right in the figure. The connecting rod, with diameter d and length L, is attached at both ends by pins. The crank arm rotates about the axle at C with the pin at B moving in a circle of radius R. The axle at C, which is supported by bearings, exerts a resisting moment M against the crank arm.

(a) Obtain a formula for the maximum permissible force P_{allow} based upon an allowable compressive stress σ_c in the connecting rod.
(b) Calculate the force P_{allow} for the following data: $\sigma_c = 160$ MPa, $d = 9.00$ mm, and $R = 0.28L$.

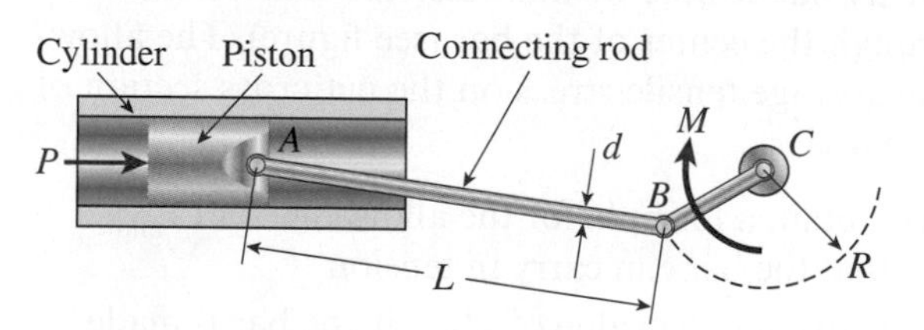

PROBLEM 1.9-18

1.10 Design for Axial Loads and Direct Shear

Introductory Problems

1.10-1 An aluminum tube is required to transmit an axial tensile force $P = 148$ kN (see figure part a). The thickness of the wall of the tube is 6 mm.

(a) What is the minimum required outer diameter d_{min} if the allowable tensile stress is 84 MPa?
(b) Repeat part (a) if the tube has a hole of a diameter $d/10$ at mid-length (see figure parts b and c).

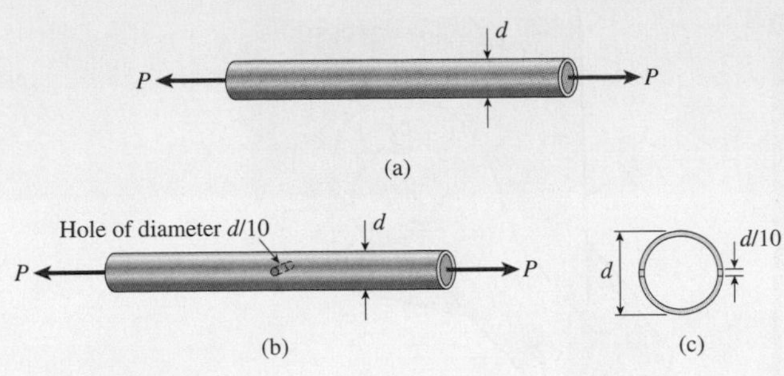

PROBLEM 1.10-1

1.10-2 A copper alloy pipe with a yield stress $\sigma_Y = 290$ MPa is to carry an axial tensile load $P = 1500$ kN (see figure part a). Use a factor of safety of 1.8 against yielding.

(a) If the thickness t of the pipe is one-eighth of its outer diameter, what is the minimum required outer diameter d_{min}?

(b) Repeat part (a) if the tube has a hole of diameter $d/10$ drilled through the entire tube, as shown in the figure part b.

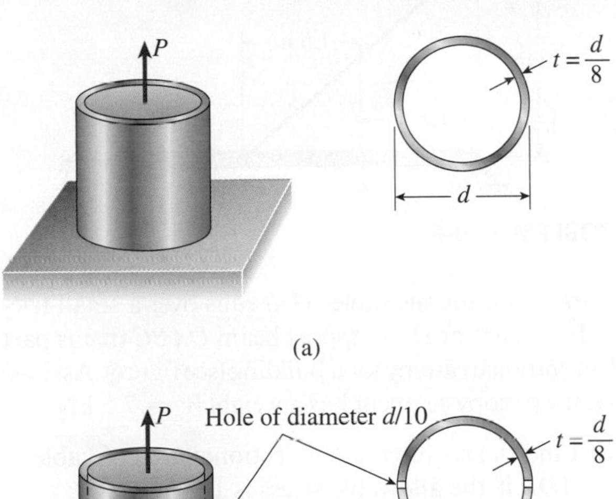

(a)

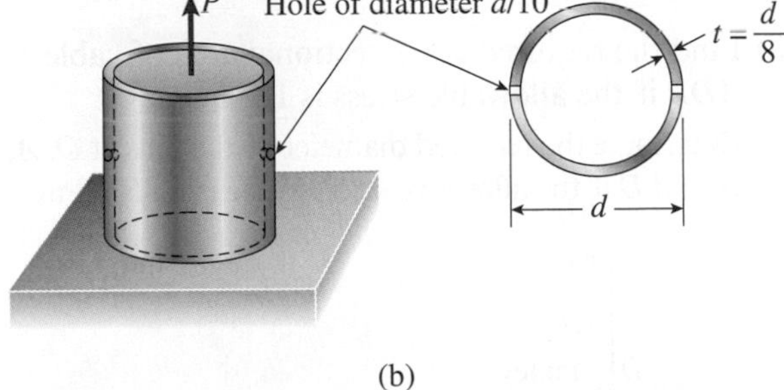

(b)

PROBLEM 1.10-2

1.10-3 A horizontal beam AB with cross-sectional dimensions $(b = 19\text{ mm}) \times (h = 200\text{ mm})$ is supported by an inclined strut CD and carries a load $P = 12$ kN at joint B (see figure part a). The strut, which consists of two bars each of thickness $5b/8$, is connected to the beam by a bolt passing through the three bars meeting at joint C (see figure part b).

(a) If the allowable shear stress in the bolt is 90 MPa, what is the minimum required diameter d_{min} of the bolt at C?

(b) If the allowable bearing stress in the bolt is 130 MPa, what is the minimum required diameter d_{min} of the bolt at C?

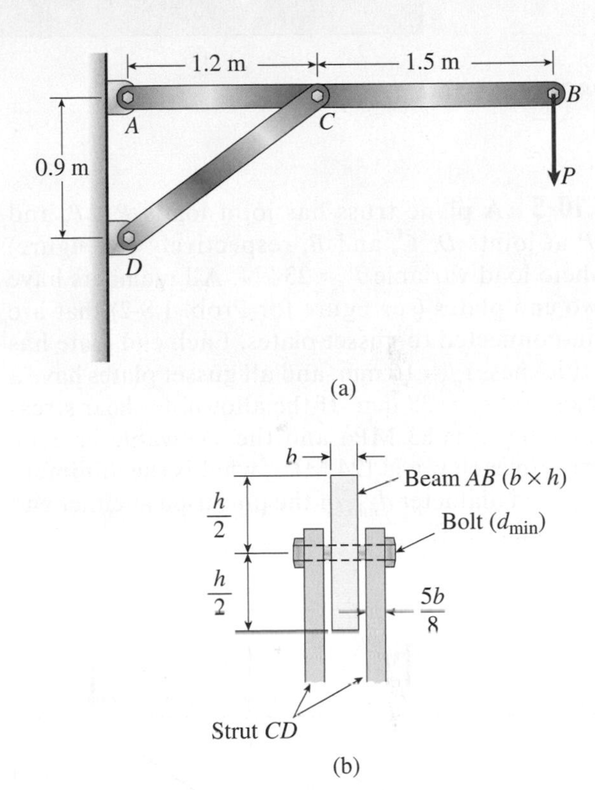

PROBLEM 1.10-3

Representative Problems

1.10-4 Lateral bracing for an elevated pedestrian walkway is shown in the figure part a. The thickness of the clevis plate $t_c = 16$ mm and the thickness of the gusset plate $t_g = 20$ mm (see figure part b). The maximum force in the diagonal bracing is expected to be $F = 190$ kN.

If the allowable shear stress in the pin is 90 MPa and the allowable bearing stress between the pin and both the clevis and gusset plates is 150 MPa, what is the minimum required diameter d_{min} of the pin?

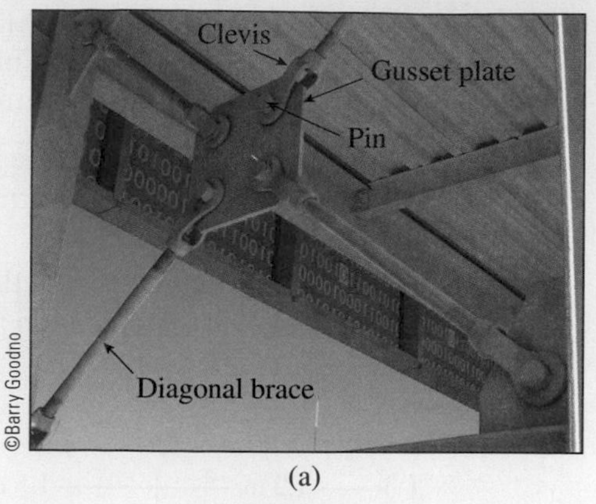

(a)

(b)

PROBLEM 1.10-4

1.10-5 A plane truss has joint loads P, $2P$, and $3P$ at joints D, C, and B, respectively (see figure) where load variable $P = 23$ kN. All members have two end plates (see figure for Prob. 1.8-2) that are pin-connected to gusset plates. Each end plate has a thickness $t_p = 16$ mm, and all gusset plates have a thickness $t_g = 28$ mm. If the allowable shear stress in each pin is 83 MPa and the allowable bearing stress in each pin is 124 MPa, what is the minimum required diameter d_{min} of the pins used at either end of member BE?

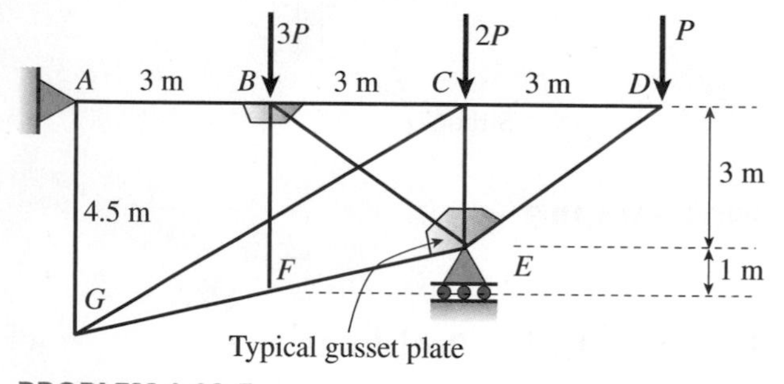

PROBLEM 1.10-5

1.10-6 Cable DB supports canopy beam $OABC$ as shown in the figure. Find the required cross-sectional area of cable BD if the allowable normal stress is 125 MPa. Determine the required diameter of the pins at O, B, and D if the allowable stress in shear is 80 MPa. Assume that canopy beam weight is $W = 8$ kN. *Note*: The pins at O, A, B, and D are in double shear. Consider only the weight of the canopy; disregard the weight of cable DB.

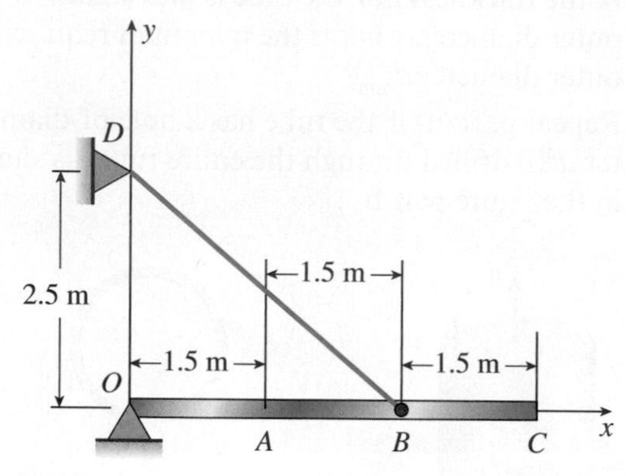

PROBLEM 1.10-6

1.10-7 Continuous cable ADB runs over a small frictionless pulley at D to support beam $OABC$ that is part of an entrance canopy for a building (see figure). Assume that the canopy segment has a weight $W = 7.5$ kN.

(a) Find the required cross-sectional area of cable ADB if the allowable stress is 125 MPa.

(b) Determine the required diameter of the pins at O, A, B, and D if the allowable stress in shear is 85 MPa.

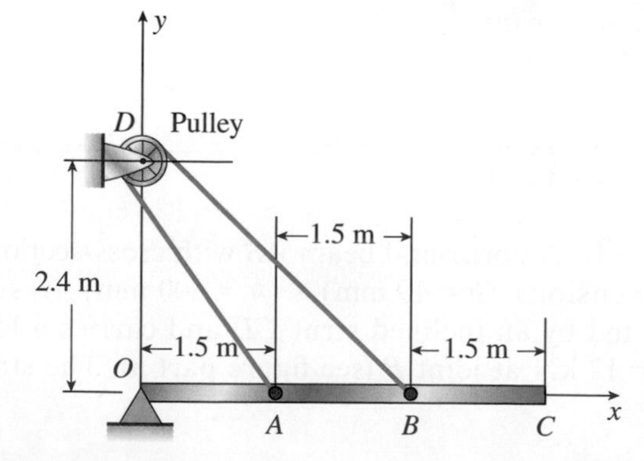

PROBLEM 1.10-7

1.10-8 A suspender on a suspension bridge consists of a cable that passes over the main cable (see figure) and supports the bridge deck, which is far below. The suspender is held in position by a metal tie that is prevented from sliding downward by clamps around the suspender cable.

Let P represent the load in each part of the suspender cable, and let θ represent the angle of the suspender cable just above the tie. Let σ_{allow} represent the allowable tensile stress in the metal tie.

(a) Obtain a formula for the minimum required cross-sectional area of the tie.
(b) Calculate the minimum area if $P = 130$ kN, $\theta = 75°$, and $\sigma_{allow} = 80$ MPa.

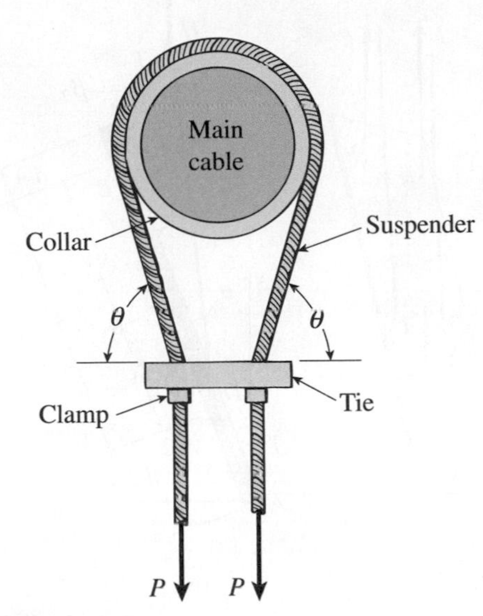

PROBLEM 1.10-8

1.10-9 A square steel tube of a length $L = 6$ m and width $b_2 = 250$ mm is hoisted by a crane (see figure). The tube hangs from a pin of diameter d that is held by the cables at points A and B. The cross section is a hollow square with an inner dimension $b_1 = 210$ mm and outer dimension $b_2 = 250$ mm. The allowable shear stress in the pin is 60 MPa, and the allowable bearing stress between the pin and the tube is 90 MPa.

Determine the minimum diameter of the pin in order to support the weight of the tube. *Note:* Disregard the rounded corners of the tube when calculating its weight.

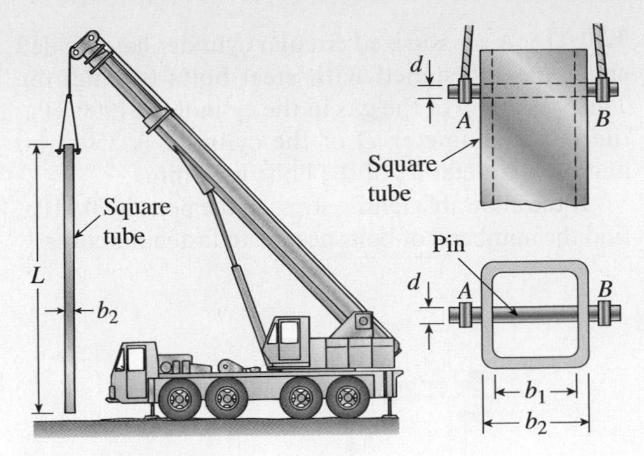

PROBLEM 1.10-9

1.10-10 A cable and pulley system at D is used to bring a 230-kg pole (ACB) to a vertical position, as shown in the figure part a. The cable has tensile force T and is attached at C. The length L of the pole is 6.0 m, the outer diameter is $d = 140$ mm, and the wall thickness is $t = 12$ mm. The pole pivots about a pin at A in figure part b. The allowable shear stress in the pin is 60 MPa and the allowable bearing stress is 90 MPa.

Find the minimum diameter of the pin at A in order to support the weight of the pole in the position shown in the figure part a.

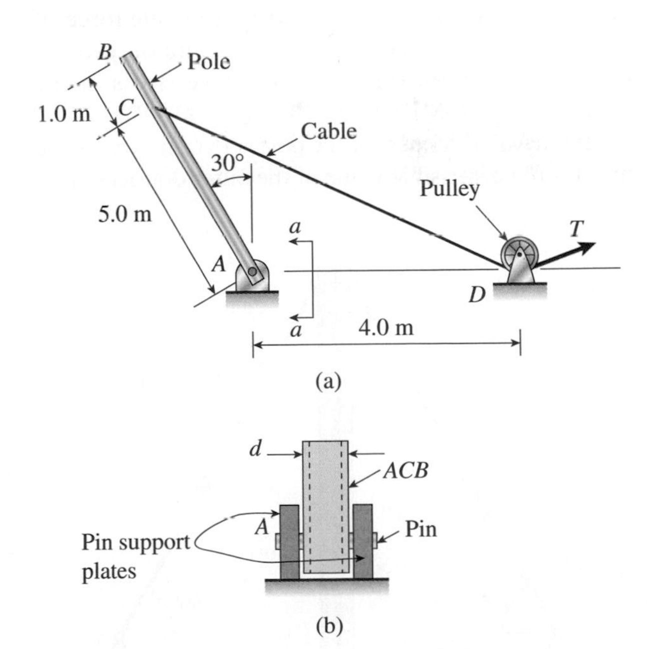

PROBLEM 1.10-10

1.10-11 A pressurized circular cylinder has a sealed cover plate fastened with steel bolts (see figure). The pressure p of the gas in the cylinder is 1900 kPa, the inside diameter D of the cylinder is 250 mm, and the diameter d_B of the bolts is 12 mm.

If the allowable tensile stress in the bolts is 70 MPa, find the number n of bolts needed to fasten the cover.

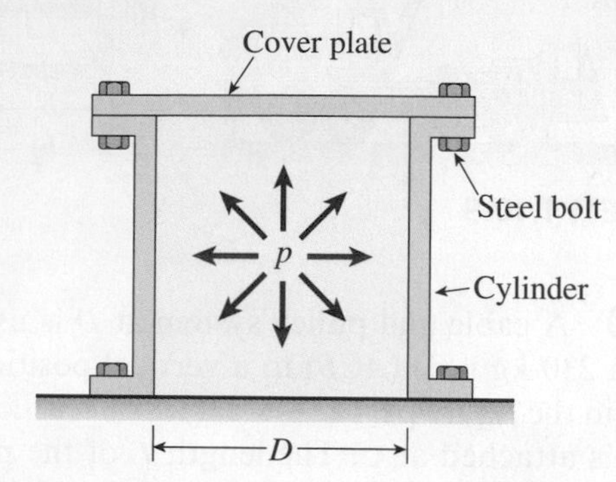

PROBLEM 1.10-11

1.10-12 A tubular post of outer diameter d_2 is guyed by two cables fitted with turnbuckles (see figure). The cables are tightened by rotating the turnbuckles, producing tension in the cables and compression in the post. Both cables are tightened to a tensile force of 110 kN. The angle between the cables and the ground is 60°, and the allowable compressive stress in the post is $\sigma_c = 35$ MPa.

If the wall thickness of the post is 15 mm, what is the minimum permissible value of the outer diameter d_2?

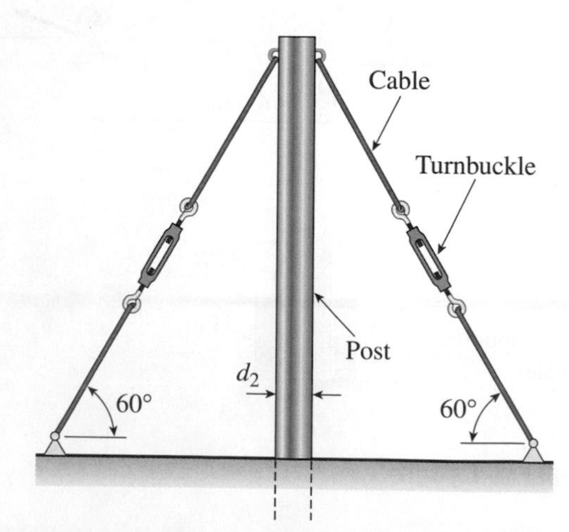

PROBLEM 1.10-12

1.10-13 A large precast concrete panel for a warehouse is raised using two sets of cables at two lift lines, as shown in the figure part a. Cable 1 has a length $L_1 = 6.7$ m, cable 2 has a length $L_2 = 2$ m, and the distance along the panel between lift points B and D is $d = 5.2$ m (see figure part b). The total weight of the panel is $W = 378$ kN. Assuming the cable lift forces F at each lift line are about equal, use the simplified model of one half of the panel in figure part b to perform your analysis for the lift position shown.

Find the required cross-sectional area AC of the cable if its breaking stress is 630 MPa and a factor of safety of 4 with respect to failure is desired.

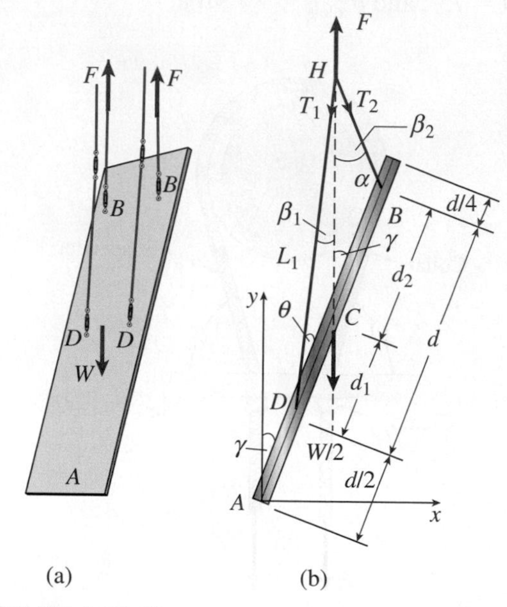

PROBLEM 1.10-13

1.10-14 A steel column of hollow circular cross section is supported on a circular, steel base plate and a concrete pedestal (see figure). The column has an outside diameter $d = 250$ mm and supports a load $P = 750$ kN.

(a) If the allowable stress in the column is 55 MPa, what is the minimum required thickness t? Based upon your result, select a thickness for the column. (Select a thickness that is an even integer, such as 10, 12, 14, . . ., in units of millimeters.)

(b) If the allowable bearing stress on the concrete pedestal is 11.5 MPa, what is the minimum required diameter D of the base plate if it is designed for the allowable load P_{allow} that the column with the selected thickness can support?

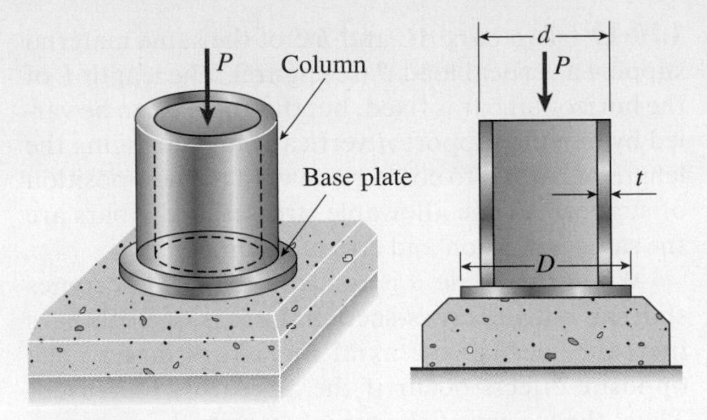

PROBLEM 1.10-14

1.10-15 An elevated jogging track is supported at intervals by a wood beam AB ($L = 2.3$ m) that is pinned at A and supported by steel rod BC and a steel washer at B. Both the rod ($d_{BC} = 5$ mm) and the washer ($d_B = 25$ mm) were designed using a rod tension force of $T_{BC} = 1890$ N. The rod was sized using a factor of safety of 3 against reaching the ultimate stress $\sigma_u = 410$ MPa. An allowable bearing stress $\sigma_{ba} = 3.9$ MPa was used to size the washer at B.

A small platform HF is suspended below a section of the elevated track to support some mechanical and electrical equipment. The equipment load is uniform load $q = 730$ N/m and concentrated load $W_E = 780$ N at mid-span of beam HF. The plan is to drill a hole through beam AB at D and install the same rod (d_{BC}) and washer (d_B) at both D and F to support beam HF.

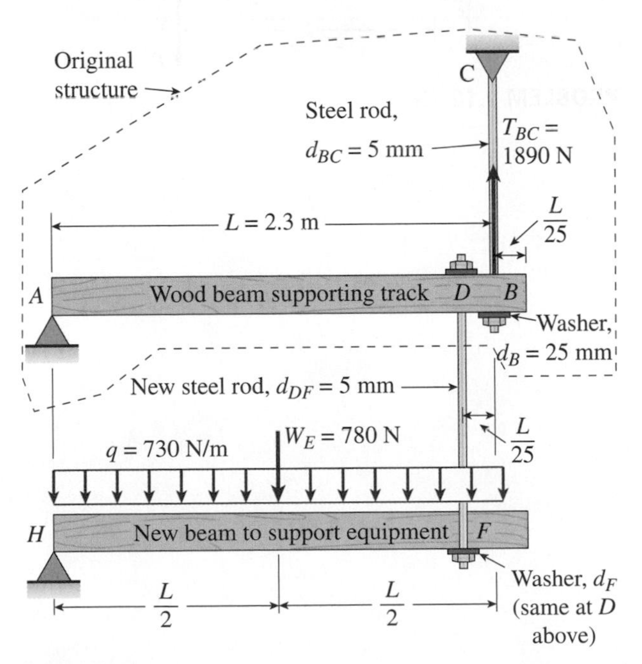

PROBLEM 1.10-15

(a) Use σ_u and σ_{ba} to check the proposed design for rod DF and washer d_F; are they acceptable?

(b) Re-check the normal tensile stress in rod BC and bearing stress at B; if either is inadequate under the additional load from platform HF, redesign them to meet the original design criteria.

1.10-16 A flat bar of a width $b = 60$ mm and thickness $t = 10$ mm is loaded in tension by a force P (see figure). The bar is attached to a support by a pin of a diameter d that passes through a hole of the same size in the bar. The allowable tensile stress on the net cross section of the bar is $\sigma_T = 140$ MPa, the allowable shear stress in the pin is $\tau_S = 80$ MPa, and the allowable bearing stress between the pin and the bar is $\sigma_B = 200$ MPa.

(a) Determine the pin diameter d_m for which the load P is a maximum.

(b) Determine the corresponding value P_{max} of the load.

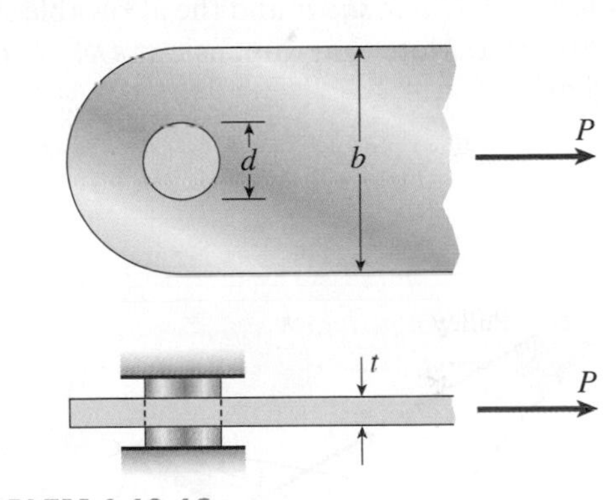

PROBLEM 1.10-16

1.10-17 Continuous cable ADB runs over a small frictionless pulley at D to support beam $OABC$, which is part of an entrance canopy for a building (see figure). The canopy segment has a weight $W = 7.5$ kN that acts as a concentrated load in the middle of segment AB.

(a) What is the maximum permissible value of load P at C if the allowable force in the cable is 18 kN?

(b) If $P = 10$ kN, what is the required diameter of pins A, B, and D? Assume that the pins are in *double shear* and the allowable shear stress in the pins is 70 MPa.

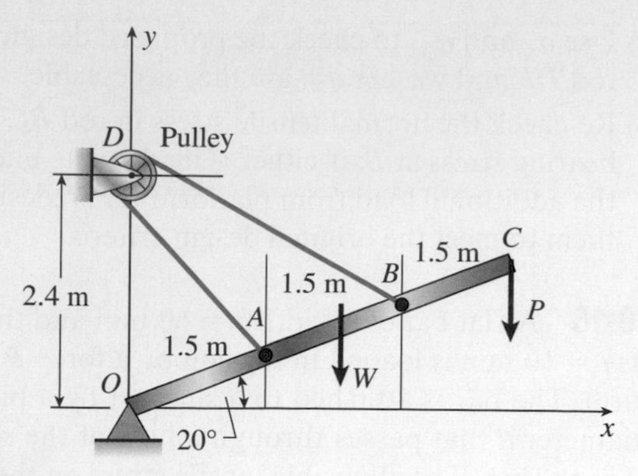

PROBLEM 1.10-17

1.10-18 Continuous cable ADB runs over a small frictionless pulley at D to support beam $OABC$, which is part of an entrance canopy for a building (see figure). A downward distributed load with peak intensity $q_0 = 5$ kN/m at O acts on the beam (see figure). Assume that canopy weight $W = 8$ kN and that the cable cross-sectional area is 100 mm^2.

What is the required diameter of pins A, B, and D if the pins are in *double shear* and the allowable shear stress is 80 MPa? Note that dimensions $OA = AB = BC = 1.5$ m.

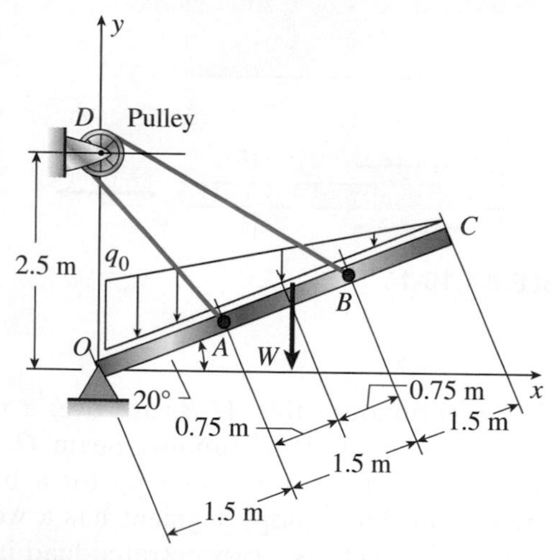

PROBLEM 1.10-18

1.10-19 Two bars AC and BC of the same material support a vertical load P (see figure). The length L of the horizontal bar is fixed, but the angle θ can be varied by moving support A vertically and changing the length of bar AC to correspond with the new position of support A. The allowable stresses in the bars are the same in tension and compression.

When the angle θ is reduced, bar AC becomes shorter, but the cross-sectional areas of both bars increase because the axial forces are larger. The opposite effects occur if the angle θ is increased. Thus, the weight of the structure (which is proportional to the volume) depends upon the angle θ.

Determine the angle θ so that the structure has minimum weight without exceeding the allowable stresses in the bars. *Note*: The weights of the bars are very small compared to the force P and may be disregarded.

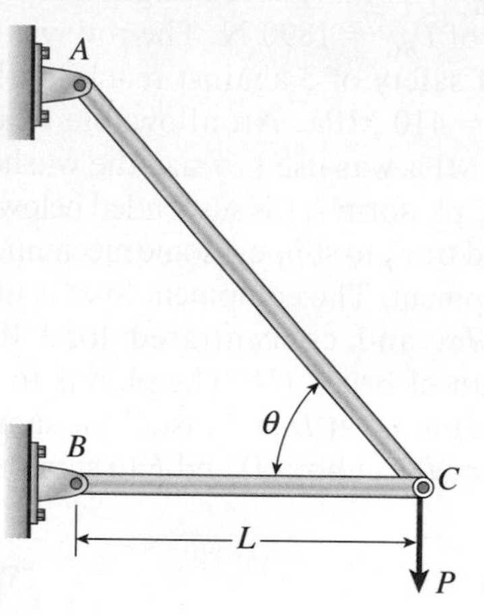

PROBLEM 1.10-19

SOME ADDITIONAL REVIEW PROBLEMS: Chapter 1

R-1.1 A plane truss has downward applied load P at joint 2 and another load P applied leftward at joint 5. The force in member 3–5 is:

(A) 0
(B) $-P/2$
(C) $-P$
(D) $+1.5\,P$

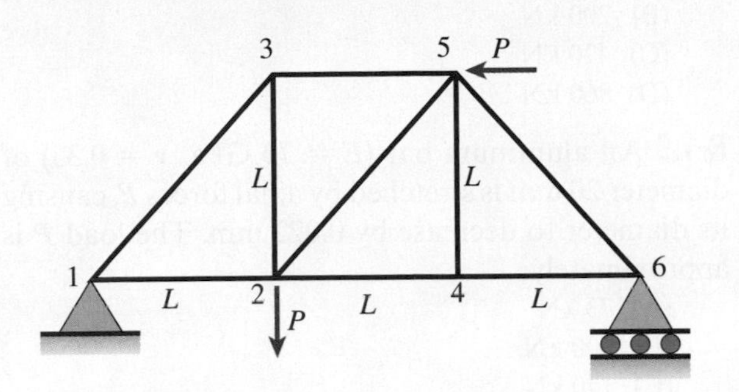

R-1.2 The force in member FE of the plane truss below is approximately:

(A) -1.5 kN
(B) -2.2 kN
(C) 3.9 kN
(D) 4.7 kN

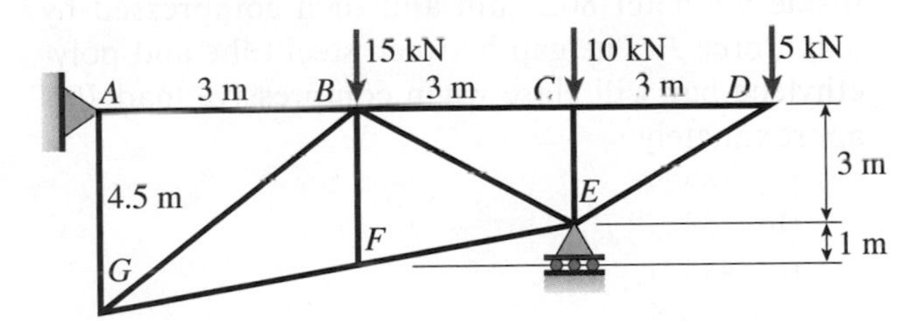

R-1.3 The moment reaction at A in the plane frame below is approximately:

(A) $+1400$ N·m
(B) -2280 N·m
(C) -3600 N·m
(D) $+6400$ N·m

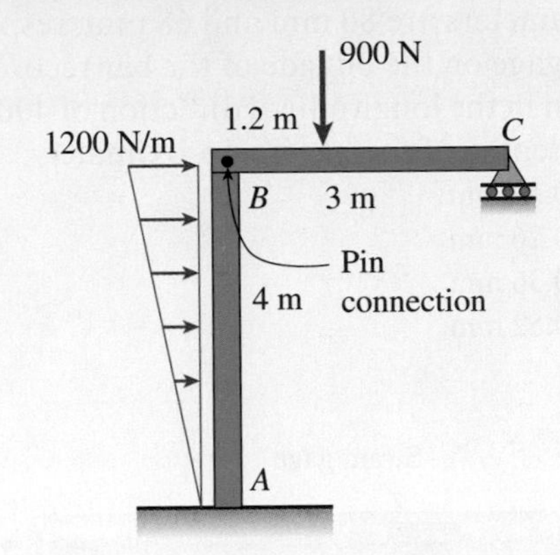

R-1.4 A hollow circular post ABC (see figure) supports a load $P_1 = 16$ kN acting at the top. A second load P_2 is uniformly distributed around the cap plate at B. The diameters and thicknesses of the upper and lower parts of the post are $d_{AB} = 30$ mm, $t_{AB} = 12$ mm, $d_{BC} = 60$ mm, and $t_{BC} = 9$ mm, respectively. The lower part of the post must have the same compressive stress as the upper part. The required magnitude of the load P_2 is approximately:

(A) 18 kN
(B) 22 kN
(C) 28 kN
(D) 46 kN

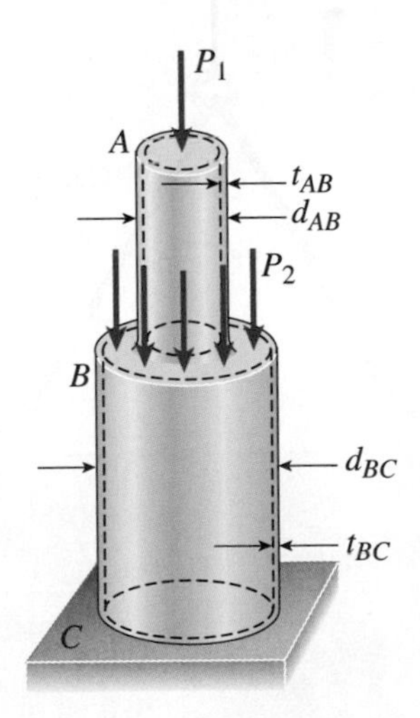

R-1.5 A circular aluminum tube of length $L = 650$ mm is loaded in compression by forces P. The outside and inside diameters are 80 mm and 68 mm, respectively. A strain gage on the outside of the bar records a normal strain in the longitudinal direction of 400×10^{-6}. The shortening of the bar is approximately:

(A) 0.12 mm
(B) 0.26 mm
(C) 0.36 mm
(D) 0.52 mm

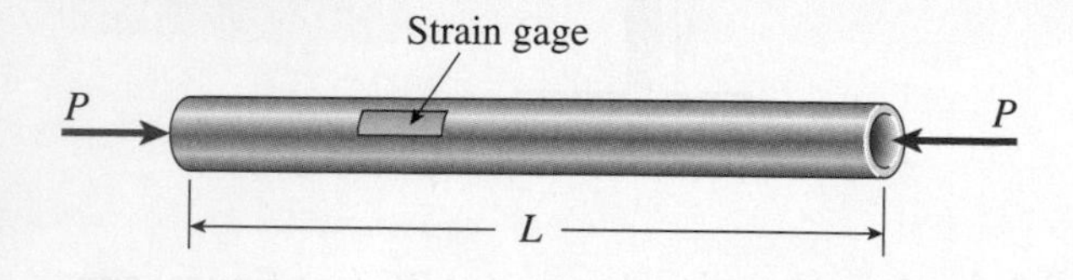

R-1.6 A steel plate weighing 27 kN is hoisted by a cable sling that has a clevis at each end. The pins through the clevises are 22 mm in diameter. Each half of the cable is at an angle of 35° to the vertical. The average shear stress in each pin is approximately:

(A) 22 MPa
(B) 28 MPa
(C) 40 MPa
(D) 48 MPa

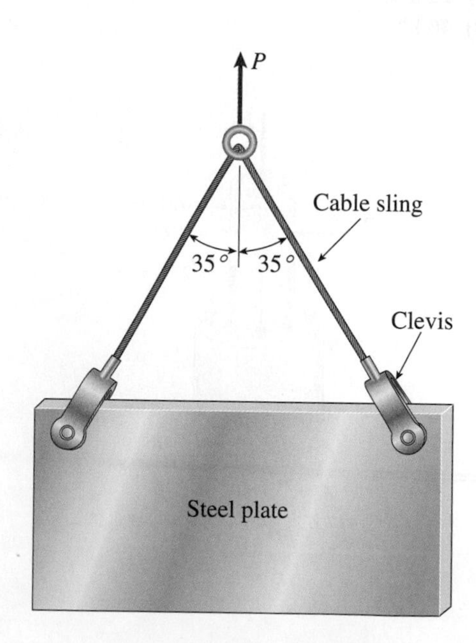

R-1.7 A steel wire hangs from a high-altitude balloon. The steel has unit weight 77 kN/m^3 and yield stress of 280 MPa. The required factor of safety against yield is 2.0. The maximum permissible length of the wire is approximately:

(A) 1800 m
(B) 2200 m
(C) 2600 m
(D) 3000 m

R-1.8 An aluminum bar (E = 72 GPa, v = 0.33) of diameter 50 mm cannot exceed a diameter of 50.1 mm when compressed by axial force P. The maximum acceptable compressive load P is approximately:

(A) 190 kN
(B) 200 kN
(C) 470 kN
(D) 860 kN

R-1.9 An aluminum bar (E = 70 GPa, v = 0.33) of diameter 20 mm is stretched by axial forces P, causing its diameter to decrease by 0.022 mm. The load P is approximately:

(A) 73 kN
(B) 100 kN
(C) 140 kN
(D) 339 kN

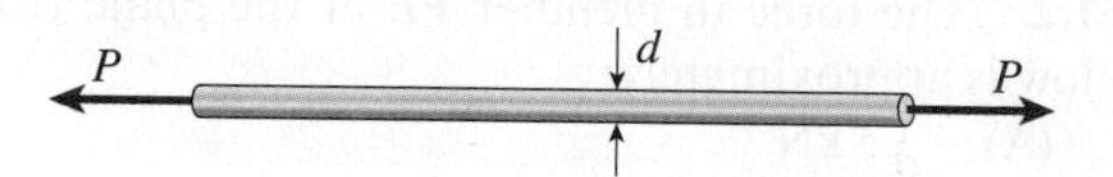

R-1.10 A polyethylene bar (E = 1.4 GPa, v = 0.4) of diameter 80 mm is inserted in a steel tube of inside diameter 80.2 mm and then compressed by axial force P. The gap between steel tube and polyethylene bar will close when compressive load P is approximately:

(A) 18 kN
(B) 25 kN
(C) 44 kN
(D) 60 kN

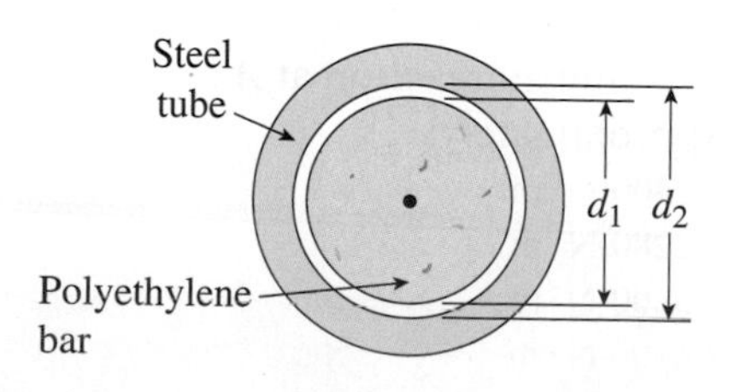

R-1.11 A pipe (E = 110 GPa) carries a load P_1 = 120 kN at A and a uniformly distributed load P_2 = 100 kN on the cap plate at B. Initial pipe diameters and thicknesses are d_{AB} = 38 mm, t_{AB} = 12 mm, d_{BC} = 70 mm, and t_{BC} = 10 mm. Under loads P_1 and P_2, wall thickness t_{BC} increases by 0.0036 mm. Poisson's ratio v for the pipe material is approximately:

(A) 0.27
(B) 0.30
(C) 0.31
(D) 0.34

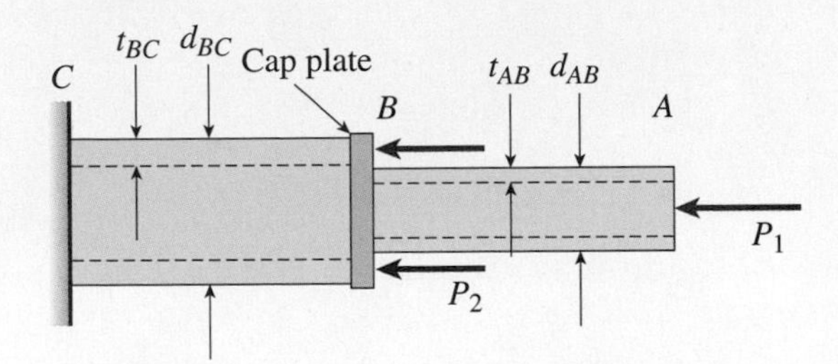

R-1.12 A titanium bar ($E = 100$ GPa, $v = 0.33$) with square cross section ($b = 75$ mm) and length $L = 3.0$ m is subjected to tensile load $P = 900$ kN. The increase in volume of the bar is approximately:

(A) 1400 mm^3
(B) 3500 mm^3
(C) 4800 mm^3
(D) 9200 mm^3

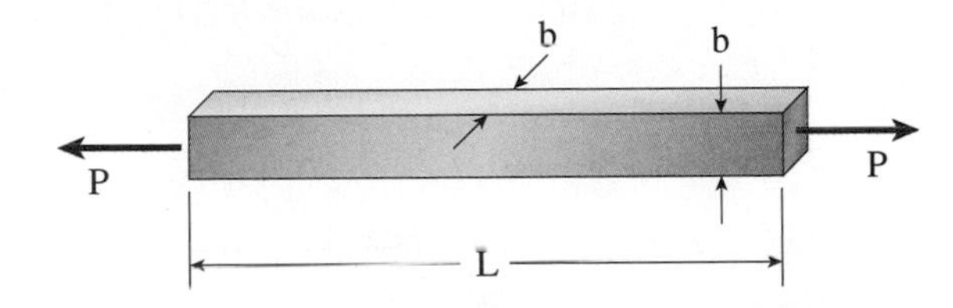

R-1.13 An elastomeric bearing pad is subjected to a shear force V during a static loading test. The pad has dimensions $a = 150$ mm and $b = 225$ mm, and thickness $t = 55$ mm. The lateral displacement of the top plate with respect to the bottom plate is 14 mm under a load $P = 16$ kN. The shear modulus of elasticity G of the elastomer is approximately:

(A) 1.0 MPa
(B) 1.5 MPa
(C) 1.7 MPa
(D) 1.9 MPa

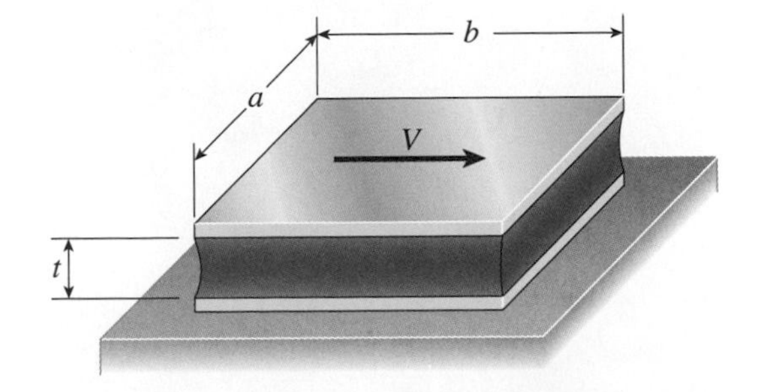

R-1.14 A bar of diameter $d = 18$ mm and length $L = 0.75$ m is loaded in tension by forces P. The bar has modulus $E = 45$ GPa and allowable normal stress of 180 MPa. The elongation of the bar must not exceed 2.7 mm. The allowable value of forces P is approximately:

(A) 41 kN
(B) 46 kN
(C) 56 kN
(D) 63 kN

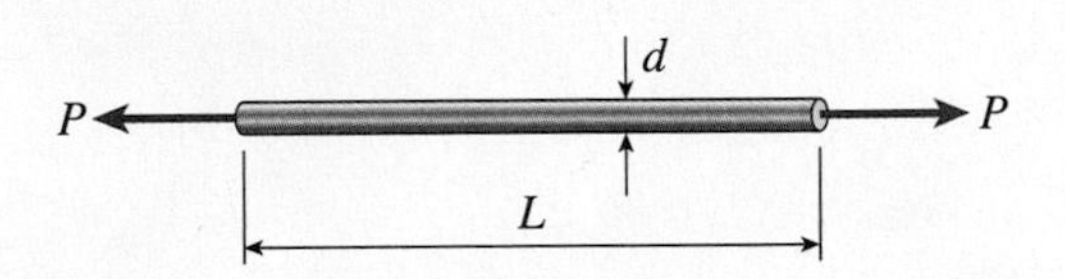

R-1.15 Two flanged shafts are connected by eight 18-mm bolts. The diameter of the bolt circle is 240 mm. The allowable shear stress in the bolts is 90 MPa. Ignore friction between the flange plates. The maximum value of torque T_0 is approximately:

(A) 19 kN·m
(B) 22 kN·m
(C) 29 kN·m
(D) 37 kN·m

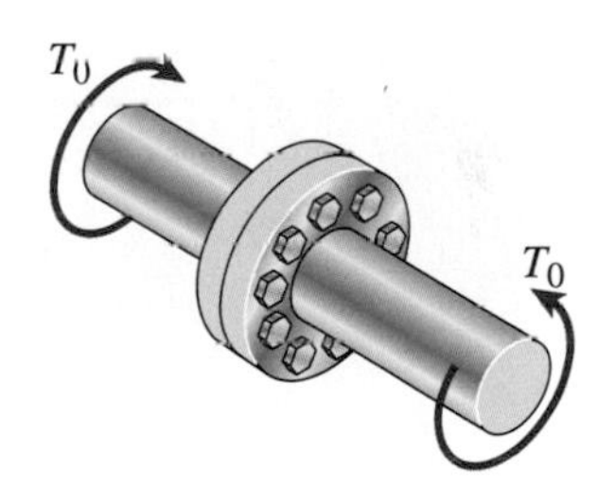

R-1.16 A copper tube with wall thickness of 8 mm must carry an axial tensile force of 175 kN. The allowable tensile stress is 90 MPa. The minimum required outer diameter is approximately:

(A) 60 mm
(B) 72 mm
(C) 85 mm
(D) 93 mm

that of a [illegible] 0.0005 and the [illegible] ratio [illegible] of the [illegible] is approximately:

(A) [illegible] kN
(B) [illegible] kN
(C) [illegible] kN
(D) [illegible] kN

(A) 1340 mm³
(B) 3660 mm³
(C) [illegible] mm³
(D) 9300 mm³

(A) [illegible]
(B) [illegible]
(C) [illegible]
(D) [illegible] MPa

(A) [illegible] mm
(B) [illegible] mm
(C) [illegible] mm
(D) [illegible] mm

CHAPTER 2

Axially Loaded Members

Roberto Giobbi/Shutterstock.com

An oil drilling rig is comprised of axially loaded members that must be designed for a variety of loading conditions, including self-weight, impact, and temperature effects.

Chapter Objectives

- Study changes in lengths of axially loaded members based on a force-displacement relation.
- Find support reactions in statically indeterminate bars acted on by concentrated and distributed axial forces.
- Find changes in lengths of bars due to temperature, misfit, and prestrain effects.
- Find both normal and shear stresses on inclined sections at points of interest on axially loaded bars.
- Study the effect of holes through axially loaded bars that cause localized stress concentrations.
- Study selected advanced topics such as strain energy, impact, fatigue, and nonlinear behavior.

Chapter Outline

*** Specialized and/or advanced topics**

2.1 Introduction

Structural components subjected only to tension or compression are known as **axially loaded members**. Solid bars with straight longitudinal axes are the most common type, although cables and coil springs also carry axial loads. Examples of axially loaded bars include truss members, connecting rods in engines, spokes in bicycle wheels, columns in buildings, and struts in aircraft engine mounts. Chapter 1 covered the stress-strain behavior of such members and also how to obtain equations for the stresses acting on cross sections ($\sigma = P/A$) and the strains in longitudinal directions ($\varepsilon = \delta/L$).

2.2 Changes in Lengths of Axially Loaded Members

When determining the changes in lengths of axially loaded members, it is convenient to begin with a **coil spring** (Fig. 2-1). Springs of this type are used in large numbers in many kinds of machines and devices—for instance, there are dozens of them in every automobile.

FIGURE 2-1
Spring subjected to an axial load P

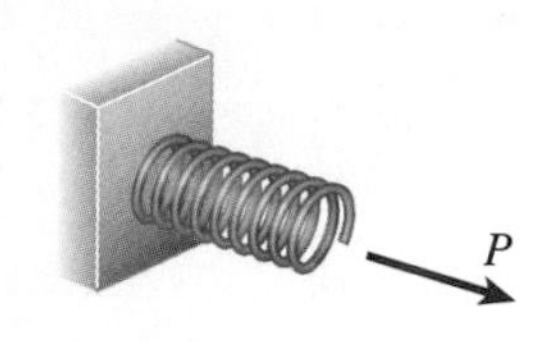

When a load is applied along the axis of a spring, as shown in Fig. 2-1, the spring gets longer or shorter depending upon the direction of the load. If the load acts away from the spring, the spring elongates, and the spring is loaded in *tension*. If the load acts toward the spring, the spring shortens, and it is in *compression*. However, it should not be inferred from this terminology that the individual coils of a spring are subjected to direct tensile or compressive stresses; rather, the coils act primarily in direct shear and torsion (or twisting). Nevertheless, the overall stretching or shortening of a spring is analogous to the behavior of a bar in tension or compression, and so the same terminology is used.

Springs

The elongation of a spring is pictured in Fig. 2-2, where the upper part of the figure shows a spring in its **natural length** L (also called its *unstressed length, relaxed length*, or *free length*), and the lower part of the figure shows the effects of applying a tensile load. Under the action of the force P, the spring lengthens by an amount δ and its final length becomes $L + \delta$. If the material of the spring is **linearly elastic**, the load and elongation will be proportional:

FIGURE 2-2
Elongation of an axially loaded spring

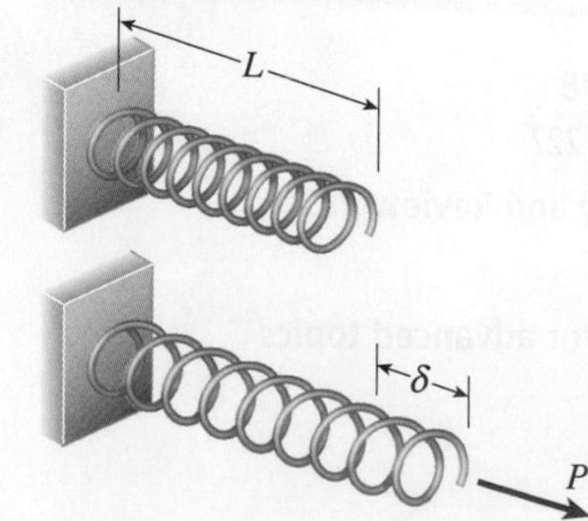

$$P = k\delta \qquad \delta = fP \tag{2-1a,b}$$

in which k and f are constants of proportionality.

The constant k is called the **stiffness** of the spring and is defined as the force required to produce a unit elongation, that is, $k = P/\delta$. Similarly, the constant f is known as the **flexibility** and is defined as the elongation produced by a load of unit value, that is, $f = \delta/P$. Although a spring in tension is used for this discussion, it should be obvious that Eqs. (2-1a and b) also apply to springs in compression.

From the preceding discussion, you can see that the stiffness and flexibility of a spring are the reciprocal of each other:

$$k = \frac{1}{f} \qquad f = \frac{1}{k} \tag{2-2a,b}$$

The flexibility of a spring easily can be determined by measuring the elongation produced by a known load, and then the stiffness can be calculated from Eq. (2-2a). Other terms for the stiffness and flexibility of a spring are the **spring constant** and **compliance**, respectively.

The spring properties given by Eqs. (2-1) and (2-2) can be used in the analysis and design of various mechanical devices involving springs, as illustrated later in Example 2-1.

Prismatic Bars

Axially loaded bars elongate under tensile loads and shorten under compressive loads, just as springs do. To analyze this behavior, let us consider the prismatic bar shown in Fig. 2-3. A **prismatic bar** is a structural member having a straight longitudinal axis and constant cross section throughout its length. Although circular bars are used in these illustrations, structural members may have a variety of cross-sectional shapes, such as those shown in Fig. 2-4.

FIGURE 2-3

Prismatic bar of circular cross section

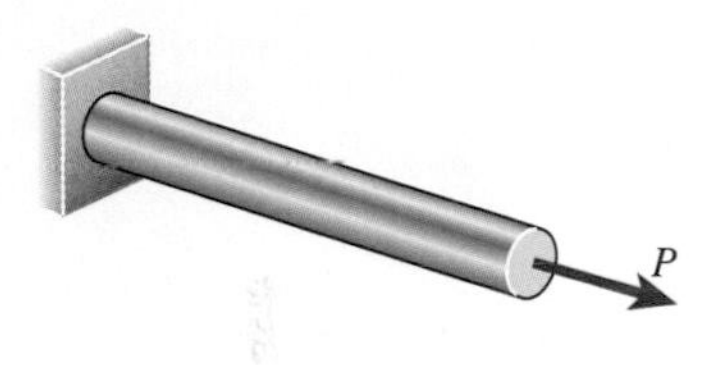

FIGURE 2-4

Typical cross sections of structural members

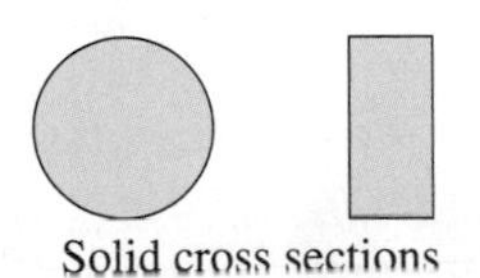

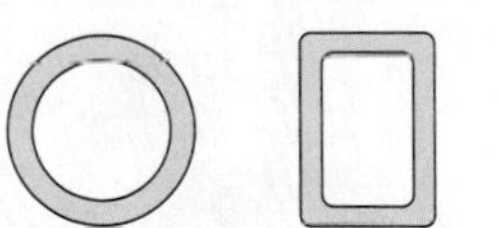

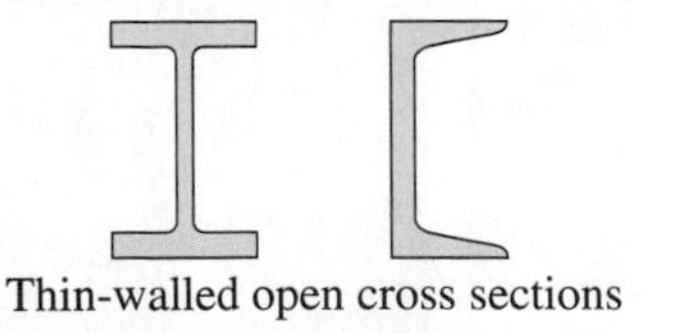

The **elongation** δ of a prismatic bar subjected to a tensile load P is shown in Fig. 2-5. If the load acts through the centroid of the end cross section, the uniform normal stress at cross sections away from the ends is given by the formula $\sigma = P/A$, where A is the cross-sectional area. Furthermore, if the bar is made of a homogeneous material, the axial strain is $\varepsilon = \delta/L$, where δ is the elongation and L is the length of the bar.

Assume that the material is **linearly elastic**, which means that it follows Hooke's law. Then the longitudinal stress and strain are related by the equation $\sigma = E\varepsilon$, where E is the modulus of elasticity. Combining these basic relationships results in the following equation for the elongation of the bar:

$$\delta = \frac{PL}{EA} \tag{2-3}$$

This equation shows that the elongation is directly proportional to the load P and the length L and inversely proportional to the modulus of elasticity E and the cross-sectional area A. The product EA is known as the **axial rigidity** of the bar.

Although Eq. (2-3) was derived for a member in tension, it applies equally well to a member in compression, in which case δ represents the shortening of the bar. Usually you know by inspection whether a member gets longer or shorter; however, there are occasions when a **sign convention** is needed (for

FIGURE 2-5

Elongation of a prismatic bar in tension

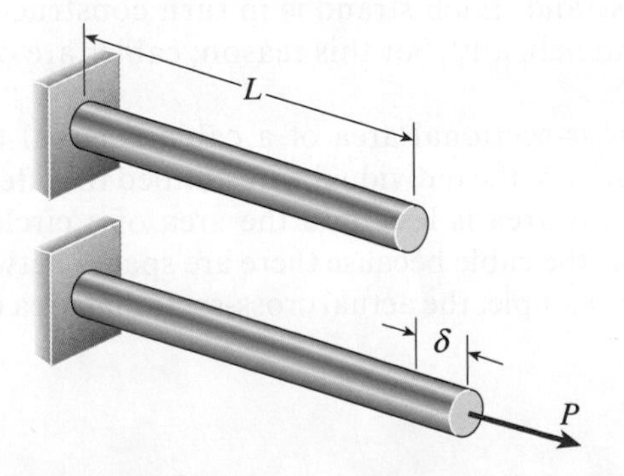

instance, when analyzing a statically indeterminate bar). When that happens, elongation is usually taken as positive and shortening as negative. This is known as a *deformation sign convention.*

The change in length of a bar is normally very small in comparison to its length, especially when the material is a structural metal, such as steel or aluminum. As an example, consider an aluminum strut that is 2 m long and subjected to a moderate compressive stress of 48 MPa. If the modulus of elasticity is 72 GPa, the shortening of the strut (from Eq. (2-3) with P/A replaced by σ) is $\delta = 0.0013$ m. Consequently, the ratio of the change in length to the original length is 0.0013/2, or 1/1500, and the final length is 0.999 times the original length. Under ordinary conditions similar to these, you can use the original length of a bar (instead of the final length) in calculations.

The stiffness and flexibility of a prismatic bar are defined in the same way as for a spring. The stiffness is the force required to produce a unit elongation, or P/δ, and the flexibility is the elongation due to a unit load, or δ/P. Thus, from Eq. (2-3) the **stiffness** and **flexibility** of a prismatic bar are, respectively,

$$k = \frac{EA}{L} \qquad f = \frac{L}{EA} \tag{2-4a,b}$$

Stiffnesses and flexibilities of structural members, including those given by Eqs. (2-4a and b), have a special role in the analysis of large structures by computer-oriented methods.

© Barsik/Dreamstime.com

Steel cables on a pulley

Cables

Cables are used to transmit large tensile forces, for example, when lifting and pulling heavy objects, raising elevators, guying towers, and supporting suspension bridges. Unlike springs and prismatic bars, cables cannot resist compression. Furthermore, they have little resistance to bending and therefore may be curved as well as straight. Nevertheless, a cable is considered to be an axially loaded member because it is subjected only to tensile forces. Because the tensile forces in a cable are directed along the axis, the forces may vary in both direction and magnitude, depending upon the configuration of the cable.

Cables are constructed from a large number of wires wound in some particular manner. While many arrangements are available depending upon how the cable will be used, a common type of cable, shown in Fig. 2-6, is formed by six *strands* wound helically around a central strand. Each strand is in turn constructed of many wires, also wound helically. For this reason, cables are often referred to as **wire rope**.

The cross-sectional area of a cable is equal to the total cross-sectional area of the individual wires, called the **effective area** or **metallic area**. This area is less than the area of a circle having the same diameter as the cable because there are spaces between the individual wires. For example, the actual cross-sectional area (effective area) of a particular 25-mm diameter cable is only 300 mm^2, whereas the area of a 25-mm diameter circle is 491 mm^2.

Under the same tensile load, the elongation of a cable is greater than the elongation of a solid bar of the same material and same metallic cross-sectional area, because the wires in a cable "tighten up" in the same manner as the fibers in a rope. Thus, the modulus of elasticity (called the **effective modulus**) of a cable is less than the modulus of the material of which it is made. The effective modulus of steel cables is about 140 GPa, whereas the steel itself has a modulus of about 210 GPa.

When determining the **elongation** of a cable from Eq. (2-3), the effective modulus should be used for E and the effective area should be used for A.

In practice, the cross-sectional dimensions and other properties of cables are obtained from the manufacturers. However, for use in solving problems in this book (and definitely *not* for use in engineering applications), Table 2-1 lists the properties of a particular type of cable. Note that the last column contains the *ultimate load*, which is the load that would cause the cable to break. The *allowable load* is obtained from the ultimate load by applying a safety factor that may range from 3 to 10, depending upon how the cable is to be used. The individual wires in a cable are usually made of high-strength steel, and the calculated tensile stress at the breaking load can be as high as 1400 MPa.

The following examples illustrate techniques for analyzing simple devices containing springs and bars. The solutions require the use of free-body diagrams, equations of equilibrium, and equations for changes in length. The problems at the end of the chapter provide many additional examples.

FIGURE 2-6

Typical arrangement of strands and wires in a steel cable

Tom Grundy/Shutterstock.com

Table 2-1

Properties of steel cables[1]

Nominal Diameter	Approximate Weight	Effective Area	Ultimate Load
mm	N/m	mm^2	kN
12	6.1	76.7	102
20	13.9	173	231
25	24.4	304	406
32	38.5	481	641
38	55.9	697	930
44	76.4	948	1260
50	99.8	1230	1650

[1]To be used solely for solving problems in this book.

Example 2-1

A machine component is modeled as two rigid bars connected to each other by a spring at *CD* and supported by linearly elastic springs at *B* and *E* (see Fig. 2-7). Before load *P* at *C* is applied, the bars are parallel and the springs are without stress. Ignore the weight of the bars.

(a) Derive formulas for downward displacement δ_C at *C* and rotation θ of bar *DEF* when load *P* is applied at *C*. (Assume that the bars rotate through very small angles under the action of the load *P*.)

(b) Find the forces in both translational springs and the moment in the rotational spring at *E*.

FIGURE 2-7

Two rigid bars supported by and connected by elastic springs

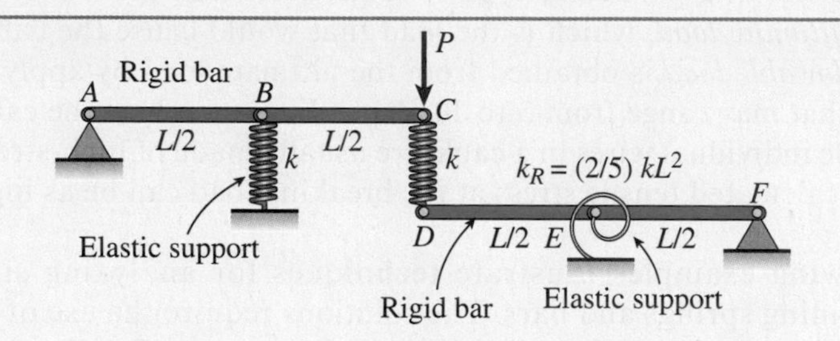

Solution:

Use a four-step problem-solving approach to find the rotations of both bars, the forces in translational springs, and the moment in the rotational spring.

1. ***Conceptualize*** [*hypothesize, sketch*]: Bar *ABC* is expected to displace downward under load *P*, and bar *DEF* is expected to rotate counterclockwise, as shown by the dashed lines in Fig. 2-8. As a result, spring forces F_B and F_{CD} and moment M_E in the rotational spring will act as shown in Fig. 2-8. Now sketch a free-body diagram of each of the bars by replacing the translational spring support at *B* with force $F_B = k\delta_B$ and the rotational spring support at *E* with moment $M_E = k_R\theta$. The spring connecting the two bars at *C* and *D* applies equal but opposite forces $F_{CD} = k(\delta_C - \delta_D)$ at points *C*

FIGURE 2-8

Free-body diagrams of upper and lower bars joined by a spring connecting points *C* and *D*

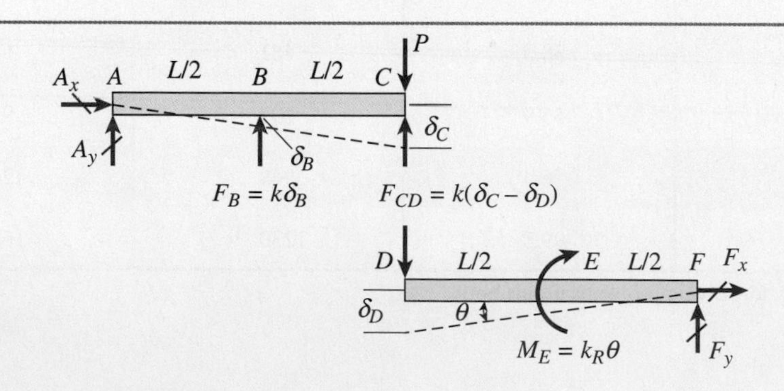

and D, which depend on the relative displacement $(\delta_C - \delta_D)$. Force P is applied downward at joint C so translation δ_C is larger than δ_D. A negative result means that the associated force or moment acts opposite to that shown in Fig. 2-8.

2. ***Categorize*** [*simplify, classify*]: The bars are rigid and the displacements and rotations are small, so it follows that $\delta_B = \delta_C/2$ and $\theta = \delta_D/L$.

3. ***Analyze*** [*evaluate; select relevant equations, carry out mathematical solution*]:

Equilibrium of bar *ABC*: Sum moments about point A in the upper free-body diagram in Fig. 2-8:

$$\Sigma M_A = 0 \quad F_B\frac{L}{2} + F_{CD}L - PL = 0 \quad \text{or} \quad \frac{k}{2}\delta_B + k(\delta_C - \delta_D) = P$$

Substitute $\delta_B = \delta_C/2$ and simplify to obtain

$$\frac{5k}{4}\delta_C - k\delta_D = P \tag{a}$$

Equilibrium of bar *DEF*: Next sum moments about point F in the lower free-body diagram in Fig. 2-8:

$$\Sigma M_F = 0 \quad F_{CD}L - M_E = 0 \quad \text{or} \quad k(\delta_C - \delta_D)L = k_R\theta$$

Substitute $\theta = \dfrac{\delta_D}{L}$ and $k_R = \dfrac{2}{5}kL^2$ and simplify to get

$$\delta_C = \frac{7}{5}\delta_D \tag{b}$$

Finally, substitute Eq. (b) into Eq. (a) and solve for displacements δ_C, δ_D, and rotation θ as

$$\delta_D = \frac{4}{3}\frac{P}{k}, \ \delta_C = \frac{28}{15}\frac{P}{k} \ \text{and} \ \theta = \frac{4}{3}\frac{P}{kL} \tag{c}$$

Forces in both translational springs and the moment in the rotational spring at *E*: Using the displacements in Eqs. (b) and (c), the forces in both translational springs and the moment in the rotational spring support at E are computed as

$$F_B = k\delta_B = \frac{14}{15}P, \ F_{CD} = k(\delta_C - \delta_D) = \frac{8}{15}P, \ M_E = k_R\theta = \frac{8}{15}PL$$

4. ***Finalize*** [*conclude; examine answer—Does it make sense? Are units correct? How does it compare to similar problem solutions?*]: The spring force and moment results are all positive, so the assumed directions of force and moment vectors in Fig. 2-8 are correct. Both translational springs are in compression, and the rotational spring at E acts to oppose counterclockwise rotation of bar DEF, as expected.

Example 2-2

Two frames are constructed using a rigid beam (ABC) and an elastic column (BD, see Fig. 2-9). Frame 1 has a sliding support at A and load P applied at C. Frame 2 has a pin support at A and moment $M = PL/3$ applied at C. For both frames, column BD has cross-sectional area $A = 8500 \text{ mm}^2$ and height $L = 725$ mm. Assume that the columns are made of steel with $E = 205$ GPa.

(a) For each frame, find the allowable value of load variable P if the displacement δ_C of point C is limited to 0.75 mm.

(b) If load variable $P = 1200$ kN, what is the required cross-sectional area of each column so that $\delta_C = 0.75$ mm?

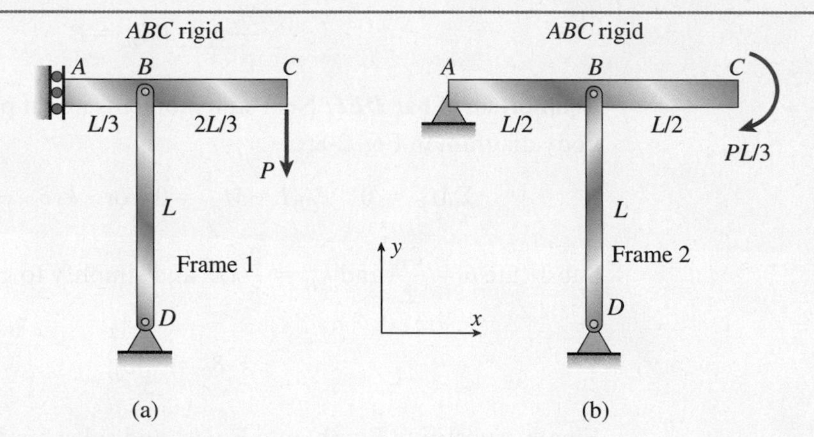

FIGURE 2-9
Two frames: rigid beam supported by an elastic column

Solution:

Use a four-step problem-solving approach to find the allowable load P and required cross-sectional areas.

1. ***Conceptualize***: The free-body diagrams of rigid beam ABC in Frames 1 and 2 are shown in Fig. 2-10. Use these diagrams to find the force F_{BD} in column BD for each frame.
2. ***Categorize***: Rigid beam ABC in Frame 1 cannot rotate because of the sliding support at A, so it will translate downward as shown in the displacement diagram in Fig. 2-11a. As a result, the downward displacement δ_B applied to

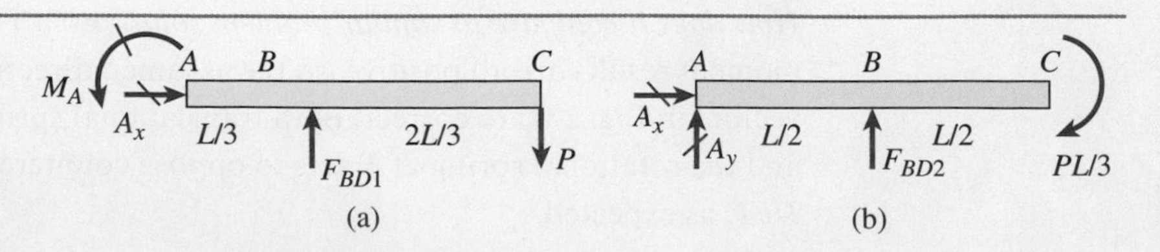

FIGURE 2-10
Free-body diagrams of rigid beam ABC in (a) Frame 1 and (b) Frame 2

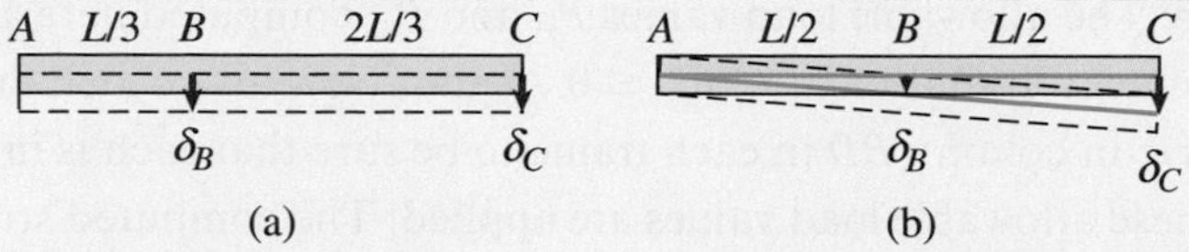

A L/3 B 2L/3 C A L/2 B L/2 C
δ_B δ_C δ_B δ_C
(a) (b)

FIGURE 2-11
Displacement diagrams for rigid beam *ABC* in (a) Frame 1 and (b) Frame 2

column *BD* is equal to δ_C. In Frame 2, rigid beam *ABC* must rotate about the pin support at *A* as shown in the displacement diagram in Fig. 2-11b. From similar triangles, the resulting downward displacement δ_B at the top of column *BD* is equal to $\delta_C/2$. All displacements in Fig. 2-11 are assumed to be small.

3. ***Analyze*:**

Allowable load variable *P* for Frame 1: Sum forces in the *y* direction in Fig. 2-10a to find that $F_{BD1} = P$. Use the *force-displacement relation* and $\delta_B = \delta_C$ to find an expression for allowable load P_{a1} on Frame 1 that satisfies the requirement that downward displacement at *C* cannot exceed δ_C. Substitute numerical values to obtain

$$P_{a1} = \frac{EA}{L}\delta_C = \frac{(205 \text{ GPa})(8500 \text{ mm}^2)}{725 \text{ mm}}(0.75 \text{ mm}) = 1803 \text{ kN}$$

Allowable load variable *P* for Frame 2: Sum moments about joint *A* in Fig. 2-10b to find that $F_{BD2} = 2P/3$. Use the *force-displacement relation* and $\delta_B = \delta_C/2$ to find an expression for allowable load P_{a2} on Frame 2 and then substitute numerical values:

$$\delta_B = \frac{F_{BD2}L}{EA} = \frac{(2P/3)L}{EA} = \frac{\delta_C}{2}$$

so

$$P_{a2} = \frac{3}{4}\frac{EA}{L}\delta_C = \frac{3}{4}\frac{(205 \text{ GPa})(8500 \text{ mm}^2)}{725 \text{ mm}}(0.75 \text{ mm}) = 1352 \text{ kN}$$

It follows that the allowable value of applied moment *M* at joint *C* is $P_{a2}L/3 = 327 \text{ kN}\cdot\text{m}$.

Required cross-sectional area of column *BD*: Now find the required cross-sectional area of column *BD* in each frame based on the limiting value of displacement δ_C but for a specified load variable $P = 1200$ kN. From the force-displacement relation,

$$A_{BD1} = \frac{F_{BD1}L}{E\delta_B} = \frac{(1200 \text{ kN})(725 \text{ mm})}{(205 \text{ GPa})(0.75 \text{ mm})} = 5659 \text{ mm}^2$$

$$A_{BD2} = \frac{F_{BD2}L}{E\delta_B} = \frac{\left(\frac{2}{3}1200 \text{ kN}\right)(725 \text{ mm})}{(205 \text{ GPa})\left(\frac{0.75 \text{ mm}}{2}\right)} = 7545 \text{ mm}^2$$

4. ***Finalize***: The allowable load values P_{a1} and P_{a2} computed here are based on a displacement limitation at C: $\delta_C \leq 0.75$ mm. Now also check the axial normal stresses σ in column BD in each frame to be sure that each is in the elastic range when these allowable load values are applied. The computed stresses in column BD in Frames 1 and 2 are

$$\sigma_{BD1} = \frac{F_{BD1}}{A} = \frac{1803 \text{ kN}}{8500 \text{ mm}^2} = 212 \text{ MPa and } \sigma_{BD2} = \frac{F_{BD2}}{A} = \frac{901 \text{ kN}}{8500 \text{ mm}^2} = 106 \text{ MPa}$$

Both stress values are acceptable, since the yield stress for steel is 250 MPa (see Table I-3 in Appendix I). Note also that both columns are in compression, so the possibility of *buckling instability* will have to be considered in future discussions.

2.3 Changes in Lengths under Nonuniform Conditions

When a prismatic bar of linearly elastic material is loaded only at the ends, you can obtain its change in length from the equation $\delta = PL/EA$, as described in the preceding section. In this section, you will see how this same equation can be used in more general situations.

Bars with Intermediate Axial Loads

Suppose, for instance, that a prismatic bar is loaded by one or more axial loads acting at intermediate points along the axis (Fig. 2-12a). You can determine the change in length of this bar algebraically by adding the elongations and shortenings of the individual segments. The procedure is as follows.

1. Identify the segments AB, BC, and CD of the bar as 1, 2, and 3, respectively.
2. Determine the internal axial forces N_1, N_2, and N_3 in segments 1, 2, and 3, respectively, from the free-body diagrams of Fig. 2-12b, c, and d. Note that the internal axial forces are denoted by the letter N to distinguish them from the external loads P. By summing forces in the vertical direction, you obtain the following expressions for the axial forces:

$$N_1 = -P_B + P_C + P_D \quad N_2 = P_C + P_D \quad N_3 = P_D$$

The deformation sign convention was used in writing these equations (internal axial forces are positive when in tension and negative when in compression). You can also sketch the *axial force diagram* (AFD) for this bar, which displays the variation of internal axial force $N(x)$ over the length of the bar (Fig. 2-12e).

3. Determine the changes in the lengths of the segments from Eq. (2-3):

$$\delta_1 = \frac{N_1 L_1}{EA} \quad \delta_2 = \frac{N_2 L_2}{EA} \quad \delta_3 = \frac{N_3 L_3}{EA}$$

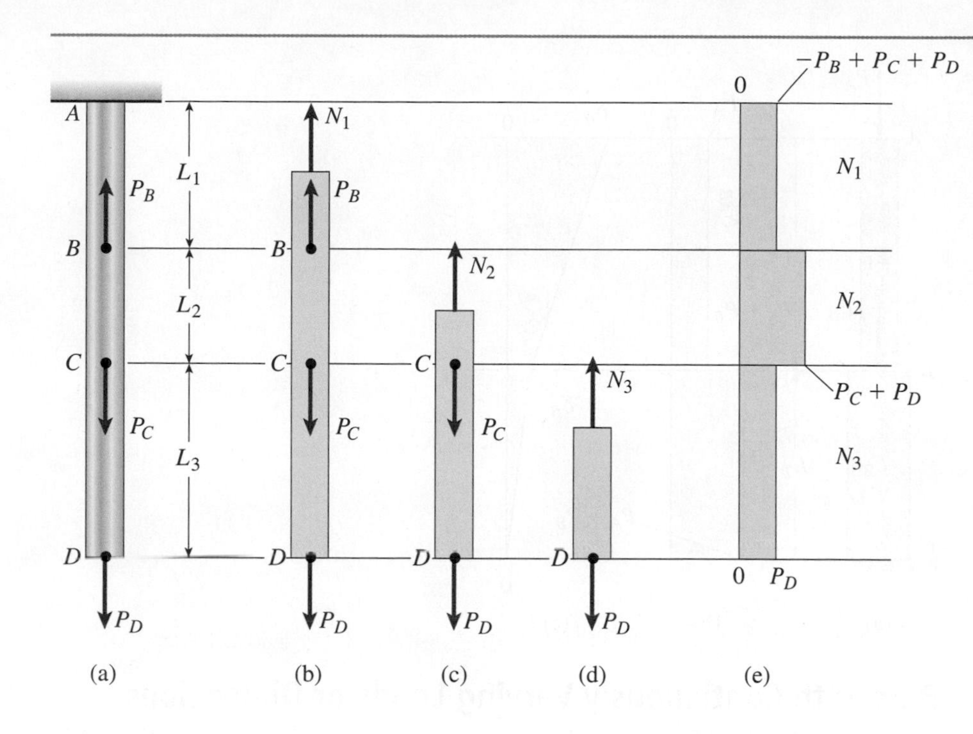

FIGURE 2-12

(a) Bar with external loads acting at intermediate points; (b), (c), and (d) free-body diagrams showing the internal axial forces N_1, N_2 and N_3; (e) Axial Force Diagram

in which L_1, L_2, and L_3 are the lengths of the segments and EA is the axial rigidity of the bar.

4. Add δ_1, δ_2, and δ_3 to obtain δ, which is the change in length of the entire bar:

$$\delta = \sum_{i=1}^{3} \delta_i = \delta_1 + \delta_2 + \delta_3$$

The changes in lengths must be added algebraically, with elongations being positive and shortenings negative. A plot of displacements at the ends of each bar segment is referred to as the *axial displacement diagram* (ADD) for the structure.

Bars Consisting of Prismatic Segments

This same general approach can be used when the bar consists of several prismatic segments, each having different axial forces, different dimensions, and different materials (Fig. 2-13a). The change in length is obtained from

$$\delta = \sum_{i=1}^{n} \frac{N_i L_i}{E_i A_i} \qquad \textbf{(2-5)}$$

in which the subscript i is a numbering index for the various segments of the bar and n is the total number of segments. Note especially that N_i is not an external load but is the internal axial force in segment i. A sketch of the AFD is given in Fig. 2-13b, and the ADD is shown in Fig. 2-13c. The displacement at A, δ_A, is the sum of segment shortenings $\delta_{AB} + \delta_{BC}$ and that at B, δ_B, is equal to the shortening of segment BC (δ_{BC}). Support C is restrained, so $\delta_C = 0$. The internal forces in each segment are constant within each segment, but internal deformations vary linearly within each segment (Fig. 2-13c).

FIGURE 2-13

(a) Bar consisting of prismatic segments having different axial forces, different dimensions, and different materials; (b) AFD; (c) ADD

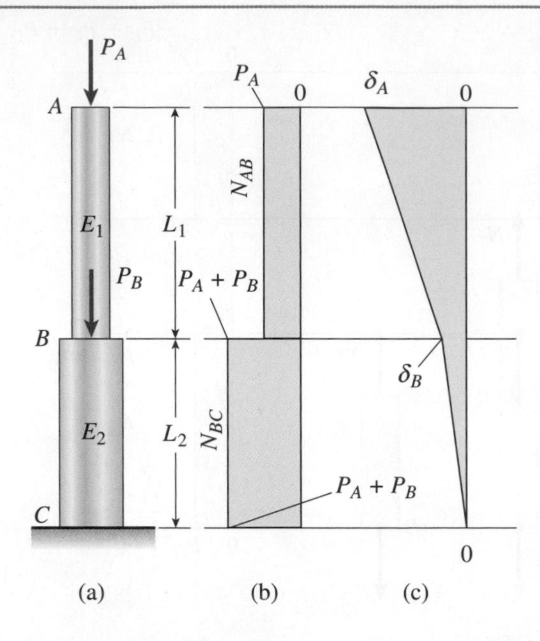

Bars with Continuously Varying Loads or Dimensions

Sometimes the axial force N and the cross-sectional area A vary continuously along the axis of a bar, as illustrated by the tapered bar of Fig. 2-14a. This bar not only has a continuously varying cross-sectional area but also a continuously varying axial force. In this illustration, the load consists of two parts: a single force P_B acting at end B of the bar and distributed forces $p(x)$ acting along the axis. (A distributed force has units of force per unit distance, such as newtons per meter.) A distributed axial load may be produced by such factors as centrifugal forces, friction forces, or the weight of a bar hanging in a vertical position.

Under these conditions, Eq. (2-5) is no longer useful to obtain the change in length. Instead, determine the change in length of a differential element of the bar and then integrate over the length of the bar.

Select a differential element at a distance x from the left-hand end of the bar (Fig. 2-14c). The internal axial force $N(x)$ acting at this cross section (Fig. 2-14b) may be determined from equilibrium using either segment AC or segment CB as a free body. In general, this force is a function of x. Also, knowing the dimensions of the bar, the cross-sectional area $A(x)$ is expressed as a function of x.

FIGURE 2-14

Bar with varying cross-sectional area and varying axial force

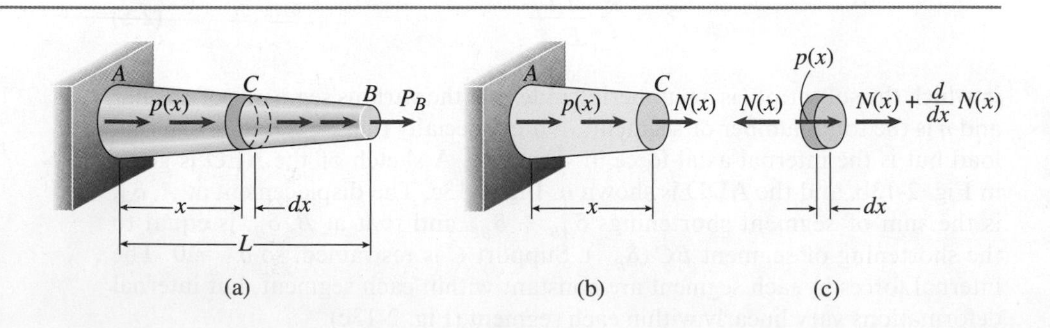

The elongation $d\delta$ of the differential element (Fig. 2-14c) may be obtained from $\delta = PL/EA$ by substituting $N(x)$ for P, dx for L, and $A(x)$ for A, as

$$d\delta = \frac{N(x)dx}{EA(x)} \tag{2-6}$$

The elongation of the entire bar is obtained by integrating over the length:

$$\delta = \int_0^L d\delta = \int_0^L \frac{N(x)dx}{EA(x)} \tag{2-7}$$

If the expressions for $N(x)$ and $A(x)$ are not too complicated, the integral can be evaluated analytically and a formula for δ or $\delta(x)$ can be obtained, as illustrated later in Example 2-4. However, if formal integration is either difficult or impossible, a numerical method for evaluating the integral should be used.

Limitations

Equations (2-5) through (2-7) apply only to bars made of linearly elastic materials, as shown by the presence of the modulus of elasticity E in the formulas. Also, the formula $\delta = PL/EA$ was derived using the assumption that the stress distribution is uniform over every cross section (because it is based on the formula $\sigma = P/A$). This assumption is valid for prismatic bars but not for tapered bars; therefore, Eq. (2-7) gives satisfactory results for a tapered bar only if angle between the sides of the bar is small.

As an illustration, if the angle between the sides of a bar is 20°, the stress calculated from the expression $\sigma = P/A$ (at an arbitrarily selected cross section) is 3% less than the exact stress for that same cross section (calculated by more advanced methods). For smaller angles, the error is even less. Consequently, Eq. (2-7) is satisfactory if the angle of taper is small. If the taper is large, more accurate methods of analysis are needed (Ref. 2-1).

Axial Displacement Diagram (ADD)

Equilibrium was used to develop a graphical display of the variation in internal axial force $N(x)$ over the length of a bar so critical regions of the bar, such as the location of maximum internal axial force N_{max} (which may also be the location of maximum axial normal stress), could be identified. This display is called the *axial force diagram* (AFD). Rules for construction of axial force diagrams are based on the differential and integral relationships between $q(x)$ and $N(x)$. Now use the *force-displacement relation* in Eq. (2-7) to create a display of the variation of axial displacement $\delta(x)$ over the length of the bar, referred to here as an *axial displacement diagram* (ADD).

To construct the ADD, first note in Eq. 2-7 that internal axial force $N(x)$ is part of the integrand in the expression for axial displacement $\delta(x)$. It follows that the ordinate $N(x)$ on the axial force diagram is *proportional to* the slope on the axial displacement diagram; the full expression for slope must include the term $EA(x)$ in the denominator in Eq. 2-7. Second, the integral expression indicates that the change in axial displacement between any two points A and B is proportional to the area under the AFD between those same two points:

$$\delta_B - \delta_A = \int_A^B \frac{N(x)}{EA(x)} dx.$$

Use these observations to develop the following rules or guidelines for constructing the ADD from the AFD:

- The slope at any point on the ADD is equal to the ordinate on the AFD at that same point divided by the axial rigidity of the bar EA at that same location.
- The change in axial displacement between any two points along a bar is equal to the area under the axial force diagram between those same two points divided by the axial rigidity of the bar EA over that same interval.

If the bar is piecewise prismatic (so that EA is constant in each bar segment) and the variation of $N(x)$ is constant or linear in x, the ADD can be constructed using simple geometry as illustrated below. Otherwise, analytical or numerical integration of Eq. (2-7) will be required to develop the ADD.

Axial force diagrams were plotted in Figs. 2-12e and 2-13b for several bars acted on by axial forces. Now consider the equilibrium of the bar in Figs. 2-15a and b but use the AFD in Fig. 2-15c to construct the ADD in Fig. 2-15d. The ADD in Fig. 2-15d is based on the force-displacement relation in Eq. 2-7 but take advantage of the graphical interpretations expressed in the two rules listed above. First, note that support displacement $\delta_A = 0$. Next, the ordinate on the AFD between A and B is constant ($+2P$) and the area under the AFD between A and B is $+PL/2EA$, so the ADD varies linearly with slope $+2P/EA$ and the change in ordinate over this segment is $\delta_B - \delta_A = \delta_B = +PL/2EA$. The change in ordinate from B to C is $-PL/4EA$, so $\delta_C = +PL/4EA$. Finally, the change in ordinate over segment CD is $+PL/2EA$, so $\delta_D = \delta_{max} = +3PL/4EA$. Use a statics sign convention in developing the ADD: Axial displacement $\delta(x)$ is considered positive (and plotted above the reference axis on the ADD) if the axial displacement vector points in the $+x$ direction. Examples 2-3 and 2-4 provide further application of these concepts.

FIGURE 2-15

(a) Prismatic bar subjected to intermediate axial centroidal loads; (b, c, d) Overall FBD, axial force diagram (AFD), and axial displacement diagram (ADD)

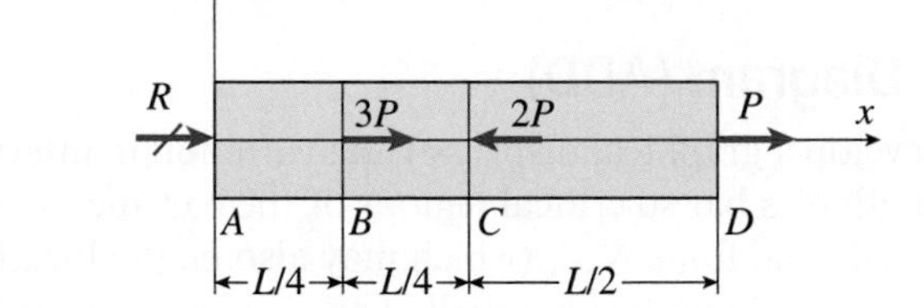

Example 2-3

A section of steel drill pipe hangs from an oil drilling rig just before the start of drilling operations (Fig. 2-16). The pipe has a length L, diameter d, and wall thickness t. Find expressions for the weight W of the pipe, support reaction R at the support point at the top of the pipe, and internal axial force $N(x)$ and displacement $\delta(x)$ at any point x along the pipe due to gravity. Plot both the axial force diagram (AFD) and the axial displacement diagram (ADD).

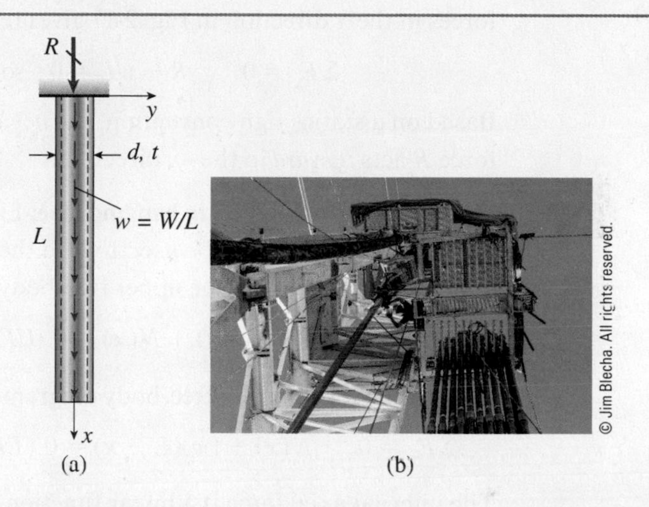

FIGURE 2-16

Prismatic pipe hanging from drill rig

Solution:

Use a four-step problem-solving approach.

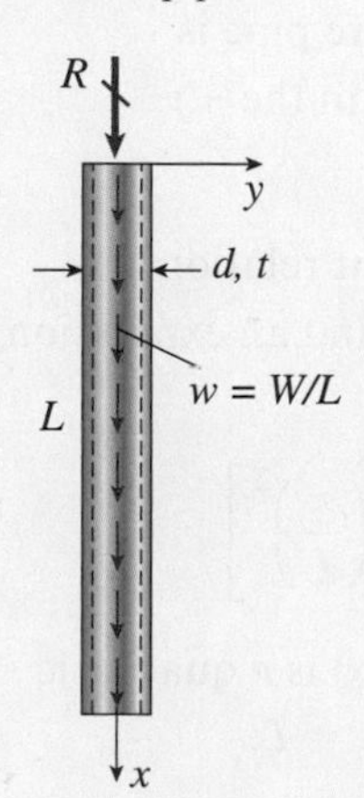

FIGURE 2-17

Free-body diagram of entire pipe

1. ***Conceptualize***: The drill pipe is prismatic and the total pipe weight is W, so the weight per unit length is a uniformly distributed axial load of constant intensity $w = W/L$. The x axis is the longitudinal centroidal axis of the pipe and runs from the support at the top to the free end of the pipe at the bottom. The top support is shown as a fixed support, but this represents axial restraint only. Start with a free-body diagram of the pipe as a whole (Fig. 2-17) to find the support reaction R at the top. A section cut through the pipe at distance x from the top and $L - x$ from the bottom breaks the pipe structure into upper and lower free-body diagrams (Fig. 2-18) and shows the internal axial force $N(x)$.
2. ***Categorize***: The drill pipe is subjected to axial forces only at the start of drilling operations. Only internal axial force $N(x)$ is present at any section cut along the pipe; no shear forces or moments are developed in the pipe at this stage. The pipe will increase in length due to uniformly distributed gravitational force $w(x)$.

FIGURE 2-18

Upper and lower free-body diagrams

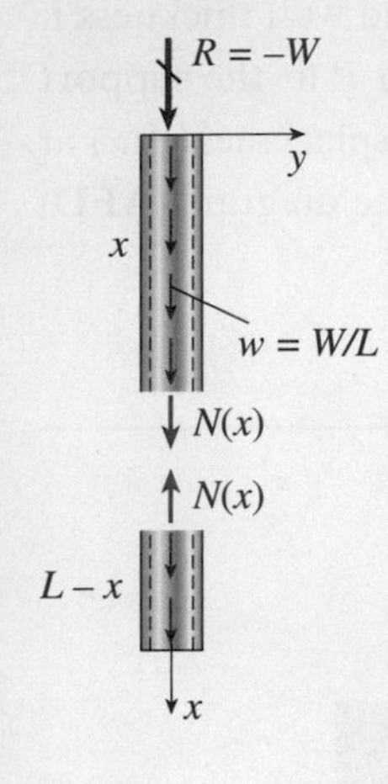

3. *Analyze:*

Pipe weight W and support reaction R: The cross-sectional area of the circular pipe is

$$A = \frac{\pi}{4}\left(d^2 - (d - 2t)^2\right) = \pi t(d - t)$$

The weight density of steel is γ (see Appendix Table I-1 for numerical values of weight density for various materials), so the pipe weight per unit length is $w = \gamma A$ and the total pipe weight is $W = wL$. The hanging pipe is statically determinate, and only axial centroidal forces are considered here. Summing forces in the x direction in Fig. 2-17 gives reaction R:

$$\Sigma F_x = 0 \qquad R + wL = 0 \quad \text{so} \quad R = -wL = -W$$

Based on a statics sign convention, the negative result indicates that reaction force R acts *upward* in the $-x$ direction.

Internal axial forces $N(x)$ in hanging pipe: Either the upper or lower free-body diagram in Fig. 2-18 can be used to find the internal axial force at any position x along the pipe. Using the upper free-body diagram gives

$$\Sigma F_x = 0 \quad R + (w)(x) + N(x) = 0 \ (UFBD) \quad \text{or} \quad N(x) = w(L - x)$$

Equilibrium of the lower free-body diagram in Fig. 2-18 yields the same result:

$$\Sigma F_x = 0 \quad -N(x) + (w)(L - x) = 0 \ (LFBD) \quad \text{or} \quad N(x) = w(L - x)$$

The internal axial force is a linear function of x and is tensile as assumed in Fig. 2-18 based on a deformation sign convention. Substituting values of x into the equation for $N(x)$ shows that maximum tensile force is $N(0) = N_{max} = W$ at the top support ($x = 0$), decreasing linearly to $N(L) = 0$ at the free end at the bottom of the pipe ($x = L$).

Axial force diagram (AFD): A plot of internal axial force $N(x)$ is shown in Fig. 2-19 and is referred to as an axial force diagram. The entire pipe is in tension, so $N(x)$ is plotted to the right of the reference axis (in the $+y$ direction). The plot has constant slope equal to $\frac{dN}{dx} = -w$.

Axial displacement diagram (ADD): Use the force-displacement relation [Eq. (2-7)] (with zeta as a dummy variable for integration) to find an expression for displacement $\delta(x)$ at any location x along the pipe:

$$\delta(x) = \int_0^x \frac{N(\zeta)d\zeta}{EA} = \int_0^x \frac{w(L-\zeta)d\zeta}{EA} = \frac{wL^2}{2EA}\left[\frac{2x}{L} - \left(\frac{x}{L}\right)^2\right]$$

Axial force $N(x)$ is a linear function of x, and displacement $\delta(x)$ is a quadratic function equal to zero at the top support and a maximum at $x = L$:

$$\delta(0) = 0, \quad \delta_{max} = \delta(L) = \frac{wL^2}{2EA} = \frac{WL}{2EA}$$

FIGURE 2-19

(a) Axial force diagram (AFD) and (b) axial displacement diagram (ADD) for entire pipe

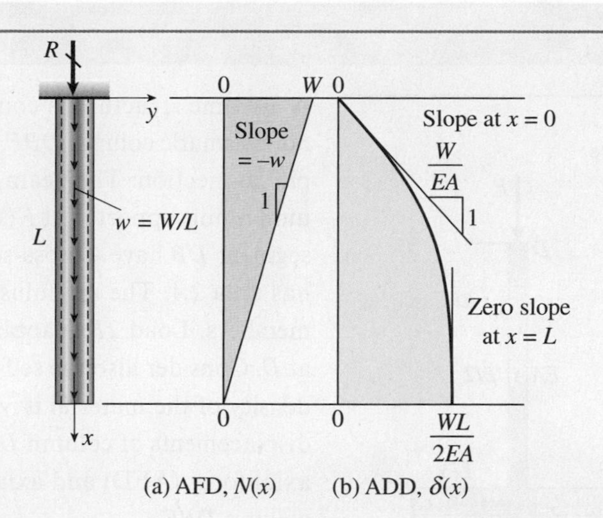

The maximum displacement at the lower end of the pipe is one-half of that which would occur if the entire pipe weight W had been applied as a concentrated force at $x = L$. A display of displacement $\delta(x)$ over the height of the pipe is referred to as an axial displacement diagram, as plotted in Fig. 2-19b.

4. ***Finalize***: From the axial force diagram (Fig. 2-19a), the maximum tensile internal force N_{max} is at the top support. As expected, N_{max} is equal to the pipe weight W. The pipe is prismatic, so maximum stress and strain also occur at the top support.

The distributed load intensity w is constant over the entire pipe. The differential relationship between load intensity w and internal axial force N is $\frac{dN}{dx} = -w$, as shown in Fig. 2-14(c) [after replacing load intensity $p(x)$ with $w(x)$ here]. Note that the constant slope of the AFD in the x-y coordinate system is $-w = -(W/L)$, which confirms that *the slope on the AFD is equal to the (−) ordinate on the distributed axial load diagram as w(x)*. In addition, *the area under the axial load diagram is −wL, which is equal to the change in internal force ordinate from start to end of the pipe on the AFD*. This is expressed as

$$N_{bottom} - N_{top} = -W = -\int_0^L w\,dx = -wL.$$

Similar geometric relationships between internal axial force $N(x)$ and displacement $\delta(x)$ are apparent if the AFD and ADD are compared in Fig. 2-19. Note in Eq. (2-7) that internal axial force $N(x)$ is part of the integrand in the expression for axial displacement $\delta(x)$, so it follows that the ordinate $N(x)$ on the axial force diagram is proportional to the slope on the axial displacement diagram [the full expression for slope must include the term $EA(x)$ in the denominator in Eq. (2-7)]. In addition, the integral expression means that the change in axial displacement between any two points is proportional to the *area* under the AFD between those same two points. If top ($x = 0$) and bottom ($x = L$) of pipe are selected as the two locations, the entire area under the AFD is $\frac{1}{2}\,WL$ and $\delta_{bottom} - \delta_{top} = \delta_{max} - 0 = \frac{WL}{2EA}$.

Example 2-4

FIGURE 2-20

T-frame structure

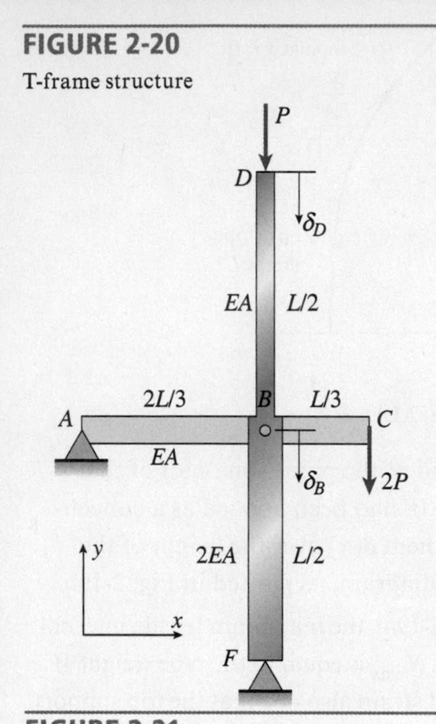

FIGURE 2-21

Free-body diagrams of (a) beam *ABC* and (b) column *DBF*

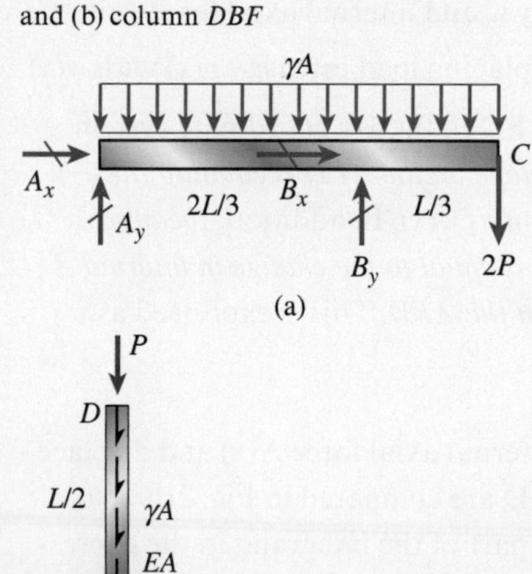

A T-frame structure is composed of prismatic beam ABC and nonprismatic column DBF; these are joined at B by a frictionless pin connection. The beam has a pin support at A, and the column is pin supported at F (see Fig. 2-20). Beam ABC and column segment DB have a cross-sectional area A; column segment BF has area $2A$. The modulus of elasticity E is the same for both members. Load $2P$ is applied downward at C, and load P acts at D. Consider also the self-weight of both members: The weight density of the material is γ. Find expressions for the downward displacements of column DBF at D (δ_D) and also at B (δ_B). Plot axial force (AFD) and axial displacement (ADD) diagrams for column DBF.

Solution:

Use a four-step problem-solving approach to find displacements at joints B and D for this T-frame structure.

1. ***Conceptualize***: Remove the pin at B and split the frame into two separate free-body diagrams: one for beam ABC and the other for column DBF (Fig. 2-21). Show reaction force components at A and F; equal but opposite pin force components at B; applied loads $2P$ and P at joints C and D, respectively; uniformly distributed self-weight transverse load γA on beam ABC; and uniformly distributed self-weight axial loads γA on DB and $2\gamma A$ on BF. Use forces in column segments DB and BF in the *force-displacement relations* to find the axial deformations of DB and BF and then joint displacements δ_B and δ_D.
2. ***Categorize***: Displacement δ_B at B results from the compression of column segment BF, while displacement δ_D at D is the sum of δ_B and the compression of column segment DB. The separate contributions to displacements from the loads shown in Fig. 2-21b are summarized in Table 2-2. Recall from Example 2-3 that the displacement due to a uniformly distributed axial load w is one-half of that due to a concentrated load or

$$\frac{wL^2}{2EA} = \frac{(\gamma A)L^2}{2EA}$$

Table 2-2 Contributions of loads in Fig. 2-21b to downward displacement at D

(1) Load Component in Figure 2-21b	(2) Column Segment DB	(3) Column Segment BF
Force P at D	$\frac{P(L/2)}{EA}$	$\frac{P(L/2)}{2EA}$
Uniform axial load γA on DB	$\frac{\gamma A(L/2)^2}{2EA}$	$\frac{\gamma A(L/2)(L/2)}{2EA}$
Force B_y at B	none	$\frac{B_y(L/2)}{2EA}$
Uniform axial load $2\gamma A$ on BF	none	$\frac{2\gamma A(L/2)^2}{2(2EA)}$

3. ***Analyze:***

Statics—pin force component B_y: Sum moments about joint A in Fig. 2-21a to find pin force B_y:

$$\Sigma M_A = 0 \quad B_y = \frac{1}{(2L/3)}\left(\gamma AL\left(\frac{L}{2}\right) + 2PL\right) = \frac{3}{4}\gamma AL + 3P \tag{a}$$

Summing forces in the y direction in Fig. 2-21b gives reaction force component F_y at the base of column DBF:

$$F_y = P + B_y + (\gamma A + 2\gamma A)\frac{L}{2} = 4P + \frac{9}{4}\gamma AL \tag{b}$$

Downward displacement at B: After substituting the expression for force B_y from Eq. (a), sum the elements in column 3 of Table 2-2 to find displacement δ_B:

$$\delta_B = \frac{7}{16}\frac{\gamma L^2}{E} + \frac{PL}{EA} \tag{c}$$

Downward displacement at D: To find displacement δ_D, sum the elements in column 2 of Table 2-2 and then add δ_B from Eq. (c):

$$\delta_D = \delta_B + \frac{\gamma L^2}{8E} + \frac{PL}{2EA} = \frac{9}{16}\frac{\gamma L^2}{E} + \frac{3}{2}\frac{PL}{EA} \tag{d}$$

Now, to simplify axial force and displacement plots, assume that γAL is equal to applied load P. Substitute this expression for γA into Eqs. (a) to (d) to obtain the simplified expressions for forces F_y and B_y and for displacements at B and D:

$$B_y = \frac{15}{4}P \text{ and } F_y = \frac{25}{4}P; \text{ also } \delta_B = \frac{23}{16}\frac{PL}{EA} \text{ and } \delta_D = \frac{33}{16}\frac{PL}{EA}$$

Axial force diagram (AFD): Start with the free body diagram (Fig. 2-22a) then plot the axial force diagram (Fig. 2-22b) which shows the variation of internal

FIGURE 2-22

(a) Free body diagram of DBF; (b) internal axial forces N; and (c) axial displacements δ

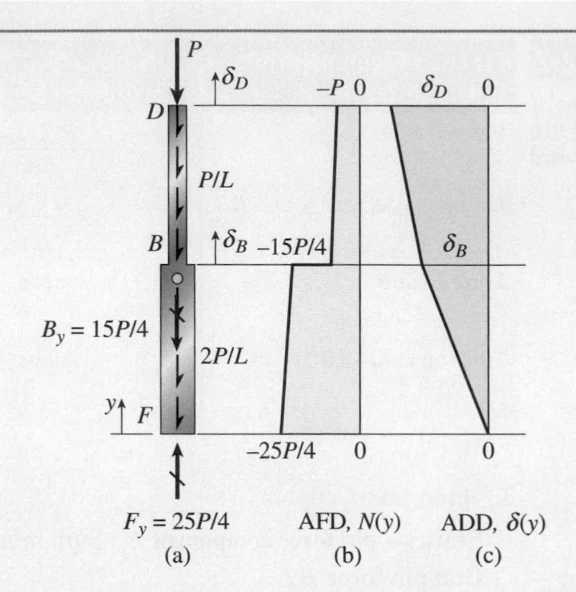

axial force $N(y)$ over the height of column DBF. Longitudinal centroidal axis y is measured upward from the base at joint F. Using a deformation sign convention, negative $N(y)$ indicates compressive internal force. At the base, internal force $N(0) = -F_y$. At the top of column DBF, $N(L) = -P$. Just below B, the compressive axial force increases by B_y. Between D and B and between D and F, $N(y)$ varies linearly, so the AFD consists of straight lines connecting known values of axial force N at $y = 0$, $L/2$, and L.

Axial displacement diagram (ADD): The axial displacement diagram (Fig. 2-22c) is a plot of axial displacement $\delta(y)$ over the height of column DBF. Note that $\delta(0) = 0$, $\delta(L/2) = \delta_B$, and $\delta(L) = \delta_D$. Internal force $N(y)$ varies linearly over the height of column DBF, so downward displacement $\delta(y)$ varies quadratically in segments DB and BF.

4. ***Finalize***: Plots of axial force and axial displacement identify locations of maximum force and displacement in the structure. For column DBF, the maximum compressive force is at the base and the maximum downward displacement is at the top as expected. Free-body diagrams like those in Fig. 2-21 are essential to the solution for internal forces $N(y)$. Internal forces are then used in the force-displacement relations to find the axial displacement distribution $\delta(y)$ over the height of the column.

The diagrams in Fig. 2-22 were drawn by inserting lines between points of known force (AFD) or displacement (ADD). A more rigorous approach is to cut column DBF into upper and lower free-body diagrams (see Fig. 2-23) and then write two functions for internal force $N(y)$ for segments BF and DB:

$$N(y) = -\frac{25}{4}P + \frac{2P}{L}y \text{ if } y \leq \frac{L}{2} \qquad N(y) = -P - \frac{P}{L}(L - y) \text{ if } \frac{L}{2} \leq y \leq L$$

FIGURE 2-23
Upper and lower free-body diagrams of column *DBF*

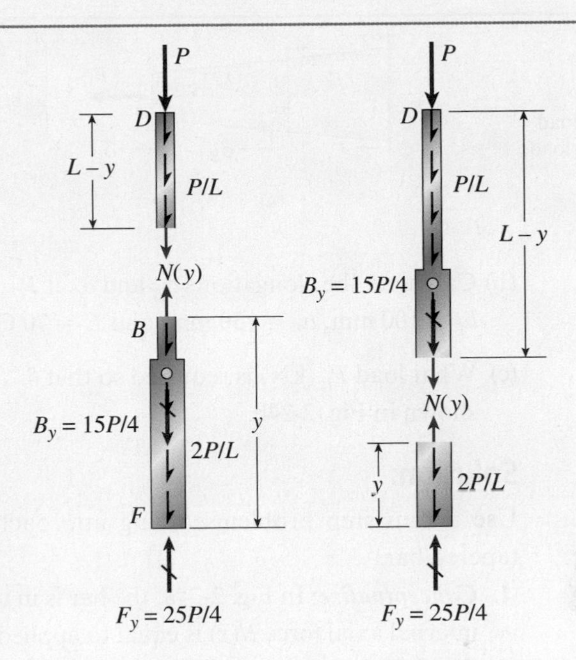

Internal forces $N(y)$ are negative because column *DBF* is in compression over its entire height. The axial displacement distribution $\delta(y)$ over the height of the column is obtained by inserting forces $N(y)$ into the force-displacement relations to obtain the following two functions:

$$\delta(y) = \int_0^y \frac{\left(-\frac{25}{4}P + \frac{2P}{L}\zeta\right)}{2EA}\,d\zeta = \frac{P}{EA}\left(\frac{y^2}{2L} - \frac{25}{8}y\right) \quad \text{if } y \le \frac{L}{2}$$

$$\delta(y) = -\frac{23}{16}\frac{PL}{EA} - \int_{L/2}^{y} \frac{P + \left(\frac{P}{L}\right)(L-\zeta)}{EA}\,d\zeta = \frac{P}{EA}\left(\frac{y^2}{2L} - 2y - \frac{9L}{16}\right) \quad \frac{L}{2} \le y \le L$$

Displacements $\delta(y)$ are downward and so are negative. Plots of functions $N(y)$ and $\delta(y)$ are shown in Figs. 2-22b and c, respectively.

Example 2-5

A flat bar has a length L, constant thickness t, and a rectangular cross section whose width varies linearly between b_2 at the fixed support to b_1 at the free end (see Fig. 2-24a). Assume that the taper of the bar is small. The bar has a modulus of elasticity E.

(a) Find an expression for the elongation δ_C of the bar due to axial centroidal load P_C at the free end. What is the elongation δ_B at B due to P_C alone?

FIGURE 2-24

Flat bar with rectangular cross section: (a) axial load P_C only; (b) axial loads P_B and P_C

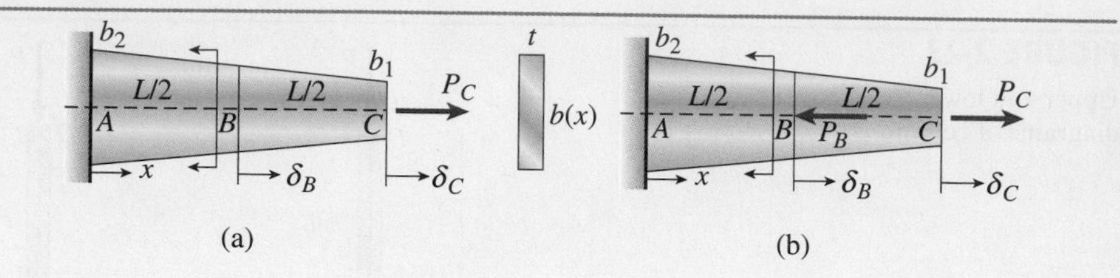

(b) Calculate the elongations δ_B and δ_C if $P_C = 275$ kN, $L = 1.5$ m, $t = 19$ mm, $b_1 = 100$ mm, $b_2 = 150$ mm, and $E = 70$ GPa.

(c) What load P_B (kN) is required so that $\delta_C = 0$ when both loads are applied as shown in Fig. 2-24b.

Solution:

Use a four-step problem-solving approach to find changes in length of this tapered bar.

1. ***Conceptualize***: In Fig. 2-24a, the bar is in tension over its entire length, and the internal axial force $N(x)$ is equal to applied force P_C (see free-body diagram in Fig. 2-25a). When load P_B is added in Fig. 2-24b, the internal tensile force $N(x)$ is

$$N(x) = P_C - P_B \quad \text{for} \quad 0 \le x \le L/2$$
$$N(x) = P_C \quad \text{for} \quad L/2 \le x \le L$$

Several free-body diagram sketches (Fig. 2-25b and 2-25c) are used as guides to obtain these equations. In all cases, internal force $N(x)$ is initially assumed to be tensile in accordance with a deformation sign convention, so arrow heads for $N(x)$ are directed away from each face of the cut section.

2. ***Categorize***: Bar ABC has constant axial force over each bar segment but continuously varying cross-sectional dimensions, so an integral form of the *force-displacement relation* [see Eq. (2-7)] must be used to find displacements

FIGURE 2-25

Internal axial tension force $N(x)$: (a) Free-body diagram for case of applied axial load P_C only; (b) two section cuts in AB and BC for case of both axial loads P_B and P_C

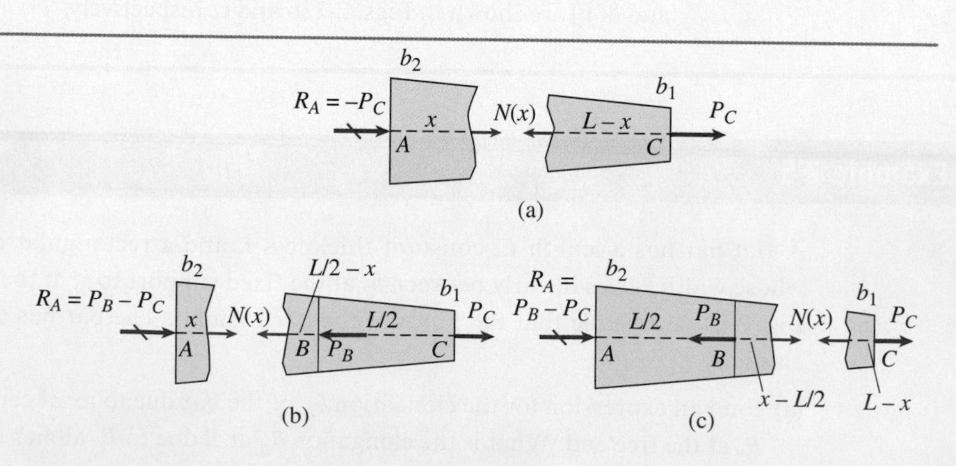

at B and C. The cross-sectional area varies linearly with x and depends on bar widths b_1 and b_2 and thickness t as $A(x) = t\left[b_2 - \frac{x}{L}(b_2 - b_1)\right]$.

3. ***Analyze:***

Part (a): Bar elongations at B and C due to load P_C only.

Apply load P_C at $x = L$ and use Eq. (2-7) to find the elongation of nonprismatic bar ABC:

$$\delta_C = \int_0^L \frac{P_C}{EA(x)}dx = \int_0^L \frac{P_C}{Et\left[b_2 - \frac{x}{L}(b_2 - b_1)\right]}dx = \frac{P_C L}{Et(b_2 - b_1)}\ln\left[\frac{b_2}{b_1}\right]$$

This same formula can also be used to find displacement δ_B at B due to P_C alone if L is replaced with $L/2$ and b_1 is replaced with the bar width at $x = L/2: b_B = (b_1 + b_2)/2$. The displacement at B is

$$\delta_B = \int_0^{L/2} \frac{P_C}{Et\left[b_2 - \frac{x}{L}(b_2 - b_B)\right]}dx = \frac{P_C L/2}{Et(b_2 - b_B)}\ln\left[\frac{b_2}{b_B}\right] = \frac{P_C L}{Et(b_2 - b_1)}\ln\left[\frac{2b_2}{b_1 + b_2}\right]$$

Part (b): Numerical calculations, bar elongations at B and C due to load P_C only. Bar elongations at B and C are obtained by substituting the numerical properties into the previous equations for δ_B and δ_C: $P_C = 275$ kN, $L = 1.5$ m, $t = 19$ mm, $b_1 = 100$ mm, $b_2 = 150$ mm, and $E = 70$ GPa:

$$\delta_B = \frac{275\text{ kN}(0.75\text{ m})}{70\text{ GPa}(19\text{ mm})(25\text{ mm})}\ln\left[\frac{6}{5}\right] = 1.131\text{ mm}$$

$$\delta_C = \frac{275\text{ kN}(1.5\text{ m})}{70\text{ GPa}(19\text{ mm})(50\text{ mm})}\ln\left[\frac{3}{2}\right] = 2.515\text{ mm}$$

Part (c): Required load P_B so net displacement at joint C is zero due to loads P_B and P_C. If load P_B *alone* is applied in the negative x direction, joints B and C will displace the same amount and are computed using the expression for δ_B given here except that P_C is replaced with $(-P_B)$.

Compatibility requires that displacements at C sum to zero when both P_B and P_C are applied, which leads to the following solution for the required load P_B:

$$P_B = \frac{Et(b_2 - b_B)}{\frac{L}{2}\ln\left[\frac{b_2}{b_B}\right]}\delta_C = \frac{70\text{ GPa}(19\text{ mm})(25\text{ mm})}{(0.75\text{ m})\ln\left[\frac{6}{5}\right]}(2.515\text{ mm}) = 612\text{ kN}$$

4. ***Finalize:***

Note 1: Computed average stresses at various locations x along the bar are $\sigma(x) = N(x)/A(x)$. For the combined loading in part (c), stresses are well below yield for this 7075-T6 aluminum alloy (see Table I-3: $\sigma_y = 480$ MPa):

$$\sigma_{\text{near}A} = \frac{(-612+275)\text{kN}}{(150\text{ mm})(19\text{ mm})} = -118.2\text{ MPa}, \sigma_{\text{near}B} = \frac{275\text{ kN}}{(125\text{ mm})(19\text{ mm})} = 115.8\text{ MPa},$$

$$\sigma_{\text{near}C} = \frac{275\text{ kN}}{(100\text{ mm})(19\text{ mm})} = 144.7\text{ MPa}$$

Note 2: In the limit, as width b_2 approaches width b_1 to produce a prismatic bar with constant cross-sectional area $A = (b_1)(t)$ (and constant thickness t), the force-displacement relation reduces to $\delta = PL/EA$ [see Eq. (2-3)]. To show this, express the above equation for δ_C in terms of $\xi = b_2/b_1$. Then, use L'Hospital's Rule with $f(\xi) = \ln(\xi)$ and $g(\xi) = b_1(\xi - 1)$ to find the displacement δ_C as $b_2 \to b_1$:

$$\lim_{x\to 1}\frac{f(\xi)}{g(\xi)} = \lim_{x\to 1}\frac{\ln(\xi)}{b_1(\xi-1)} = \lim_{x\to 1}\frac{f'(\xi)}{g'(\xi)} = \frac{1}{b_1} \text{ so } \delta_C = \frac{P_C L}{E(b_1 t)} = \frac{P_C L}{EA}$$

2.4 Statically Indeterminate Structures

The springs, bars, and cables covered in the preceding sections have one important feature in common—their reactions and internal forces can be determined solely from free-body diagrams and equations of equilibrium. Structures of this type are classified as **statically determinate**. Note especially that the forces in a statically determinate structure can be found without knowing the properties of the materials. Consider, for instance, the bar AB shown in Fig. 2-26. The calculations for the internal axial forces in both parts of the bar, as well as for the reaction R at the base, are independent of the material of which the bar is made.

FIGURE 2-26

Statically determinate bar

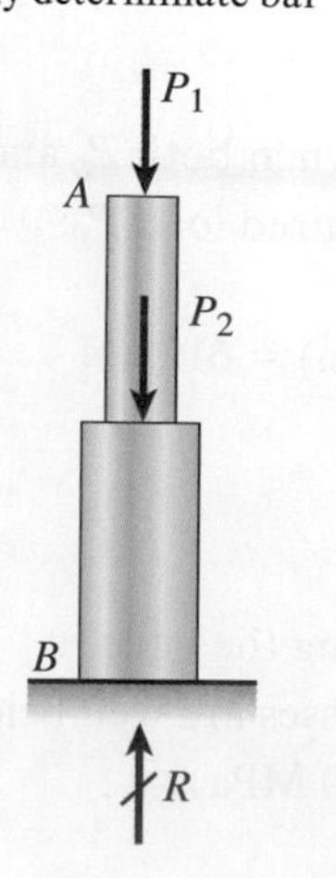

Most structures are more complex than the bar of Fig. 2-26, and their reactions and internal forces cannot be found by statics alone. This situation is illustrated in Fig. 2-27, which shows a bar AB fixed at *both* ends. There are now two vertical reactions (R_A and R_B) but only one useful equation of equilibrium—the equation for summing forces in the vertical direction. Since this equation contains two unknowns, it is not sufficient for finding the reactions. Structures of this kind are classified as **statically indeterminate**. To analyze such structures, you must supplement the equilibrium equations with additional equations pertaining to the displacements of the structure.

To see how a statically indeterminate structure is analyzed, consider the example of Fig. 2-28. The prismatic bar AB is attached to rigid supports at both ends and is axially loaded by a force P at an intermediate point C. As already discussed, the reactions R_A and R_B. cannot be found by statics alone, because only one **equation of equilibrium** is available:

$$\Sigma F_{\text{vert}} = 0 \quad R_A - P + R_B = 0 \tag{2-8}$$

An additional equation is needed in order to solve for the two unknown reactions.

The additional equation is based upon the observation that a bar with both ends fixed does not change in length. If you separate the bar from its supports

(Fig. 2-28), you obtain a bar that is free at both ends and loaded by the three forces, R_A, R_B, and P. These forces cause the bar to change in length by an amount δ_{AB}, which must be equal to zero:

$$\delta_{AB} = 0 \tag{2-9}$$

This equation, called an **equation of compatibility**, expresses the fact that the change in length of the bar must be compatible with the conditions at the supports.

In order to solve Eqs. (2-8) and (2-9), express the compatibility equation in terms of the unknown forces R_A and R_B. The relationships between the forces acting on a bar and its changes in length are known as **force-displacement relations**. These relations have various forms depending upon the properties of the material. If the material is linearly elastic, the equation $\delta = PL/EA$ can be used to obtain the force-displacement relations.

Assume that the bar of Fig. 2-28 has a cross-sectional area A and is made of a material with a modulus E. Then the changes in lengths of the upper and lower segments of the bar are, respectively,

$$\delta_{AC} = \frac{R_A a}{EA} \qquad \delta_{CB} = -\frac{R_B b}{EA} \tag{2-10a,b}$$

where the minus sign indicates a shortening of the bar. Equations (2-10a and b) are the force-displacement relations.

Now solve simultaneously the three sets of equations (the equation of *equilibrium*, the equation of *compatibility*, and the *force-displacement* relations). This illustration begins by combining the force-displacement relations with the equation of compatibility:

$$\delta_{AB} - \delta_{AC} + \delta_{CB} - \frac{R_A a}{EA} - \frac{R_B b}{EA} = 0 \tag{2-11}$$

Note that this equation contains the two reactions as unknowns.

FIGURE 2-27

Statically indeterminate bar

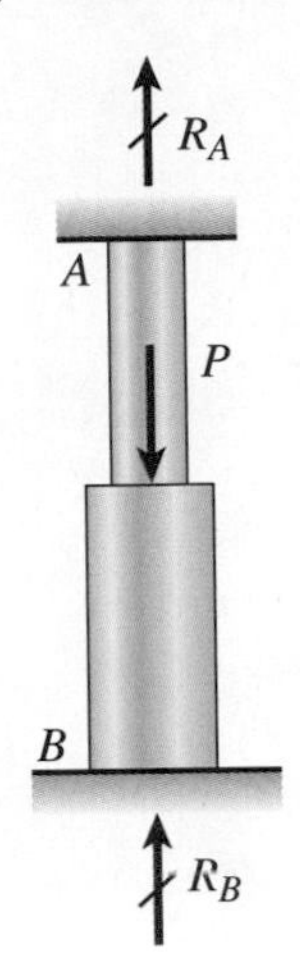

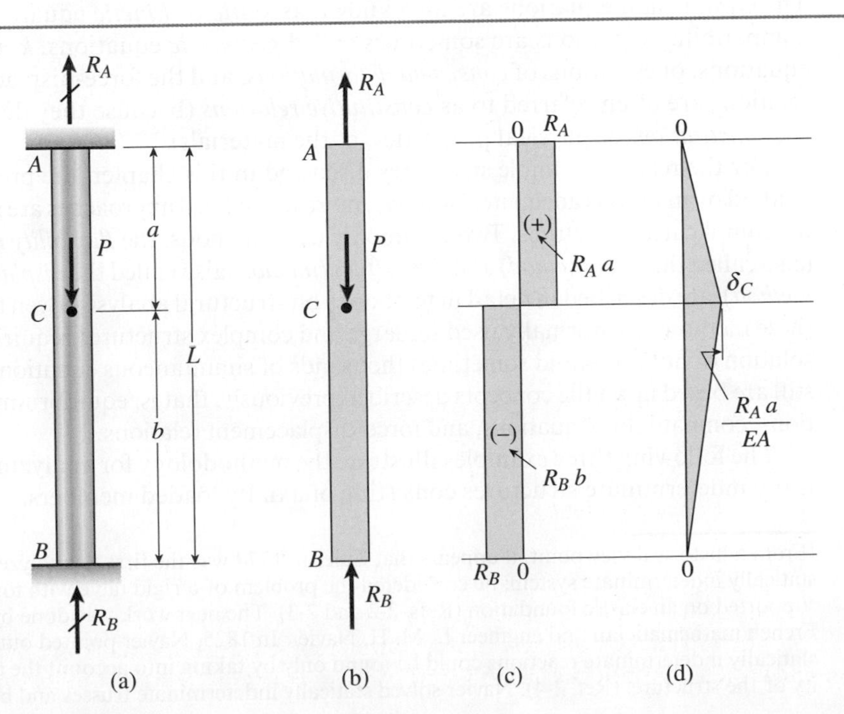

FIGURE 2-28

Analysis of a statically indeterminate bar: (a) statically indeterminate bar; (b) free-body diagram; (c) axial-force diagram; (d) axial-displacement diagram

The next step is to solve simultaneously the equation of equilibrium (Eq. 2-8) and the preceding equation [Eq. (2-11)]. The results are

$$R_A = \frac{Pb}{L} \qquad R_B = \frac{Pa}{L} \tag{2-12a,b}$$

With the reactions known, all other force and displacement quantities can be determined. Suppose, for instance, that you wish to find the downward displacement δ_C of point C. This displacement is equal to the elongation of segment AC:

$$\delta_C = \delta_{AC} = \frac{R_A a}{EA} = \frac{Pab}{LEA} \tag{2-13}$$

Also, you can find the stresses in the two segments of the bar directly from the internal axial forces (e.g., $\sigma_{AC} = R_A/A = Pb/AL$). A plot of the distribution of internal axial forces N (the AFD) is shown in Fig. 2-28c. The variation of axial displacements δ over the height of the bar (the ADD) is given in Fig. 2-28d. The bar is fully restrained at A and B, so displacements $\delta_A = \delta_B = 0$ as shown on the ADD.

General Comments

From the preceding discussion, the analysis of a statically indeterminate structure involves setting up and solving equations of *equilibrium*, equations of *compatibility*, and *force-displacement* relations. The equilibrium equations relate the loads acting on the structure to the unknown forces (which may be reactions or internal forces), and the compatibility equations express conditions on the displacements of the structure. The force-displacement relations are expressions that use the dimensions and properties of the structural members to relate the forces and displacements of those members. In the case of axially loaded bars that behave in a linearly elastic manner, the relations are based upon the equation $\delta = PL/EA$. Finally, all three sets of equations may be solved simultaneously for the unknown forces and displacements.

In the engineering literature, various terms are used for the conditions expressed by the equilibrium, compatibility, and force-displacement equations. The equilibrium equations are also known as *static* or *kinetic* equations; the compatibility equations are sometimes called *geometric* equations, *kinematic* equations, or equations of *consistent deformations*; and the force-displacement relations are often referred to as *constitutive relations* (because they deal with the *constitution*, or physical properties, of the materials).

For the relatively simple structures discussed in this chapter, the preceding method of analysis is adequate. However, more formalized approaches are needed for complicated structures. Two commonly used methods, the *flexibility method* (also called the *force method*) and the *stiffness method* (also called the *displacement method*), are described in detail in textbooks on structural analysis. Even though these methods are normally used for large and complex structures requiring the solution of hundreds and sometimes thousands of simultaneous equations, they still are based upon the concepts described previously, that is, equilibrium equations, compatibility equations, and force-displacement relations.[1]

The following three examples illustrate the methodology for analyzing statically indeterminate structures consisting of axially loaded members.

---

[1]From a historical viewpoint, it appears that Euler in 1774 was the first to analyze a statically indeterminate system; he considered the problem of a rigid table with four legs supported on an elastic foundation (Refs. 2-2 and 2-3). The next work was done by the French mathematician and engineer L. M. H. Navier. In 1825, Navier pointed out that statically indeterminate reactions could be found only by taking into account the elasticity of the structure. (Ref. 2-4). Navier solved statically indeterminate trusses and beams.

Example 2-6

Find expressions for all support reaction forces in the plane frame with load $2P$ applied at C shown in Fig. 2-29. Assume that member AC is a flat prismatic bar of length L, width b, and thickness t. Beam ABC is pinned to column DBF at mid-height (point B). Column DBF has constant thickness t and tapers linearly from width $b_1 = b$ at D to width $b_2 = 5b/4$ at F. Consider load $2P$ at C only; neglect the weights of all members. The modulus of elasticity E is the same for both members.

FIGURE 2-29
T-frame structure

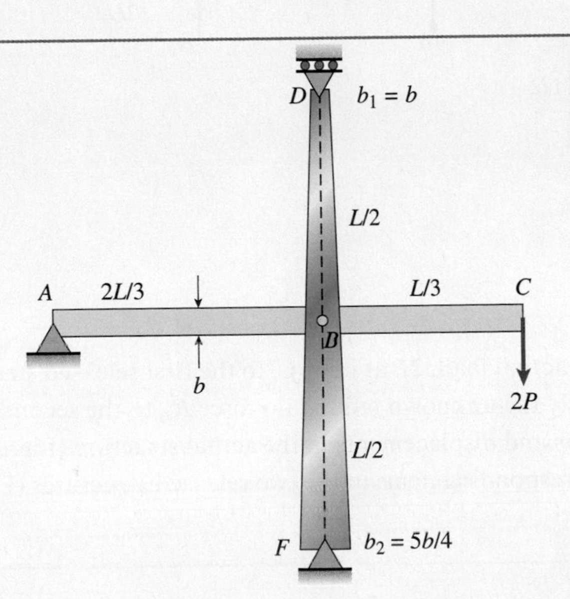

Solutions:

Use a four-step problem-solving approach to find support reactions for this T-frame structure.

1. ***Conceptualize***: The frame is one-degree statically indeterminate: There are five reaction force components (two each at joints A and F, one at D; see Fig. 2-30a) but only three overall equilibrium equations plus one additional equation due to the pin at B. Remove the pin at B and split the frame into two separate free-body diagrams: One of beam ABC and the other of column DBF (Fig. 2-30b). Show equal but opposite pin force components at B on each separate free-body diagram.
2. ***Categorize***: Select reaction R_D as the redundant force and use superposition to develop one additional equation based on the *compatibility* of displacements at joint D.

Superposition of released structures: Redundant force R_D is removed to create the released, or statically determinate, structure. Apply

FIGURE 2-30

(a) Overall free-body diagram; (b) free-body diagrams of beam ABC and column DBF

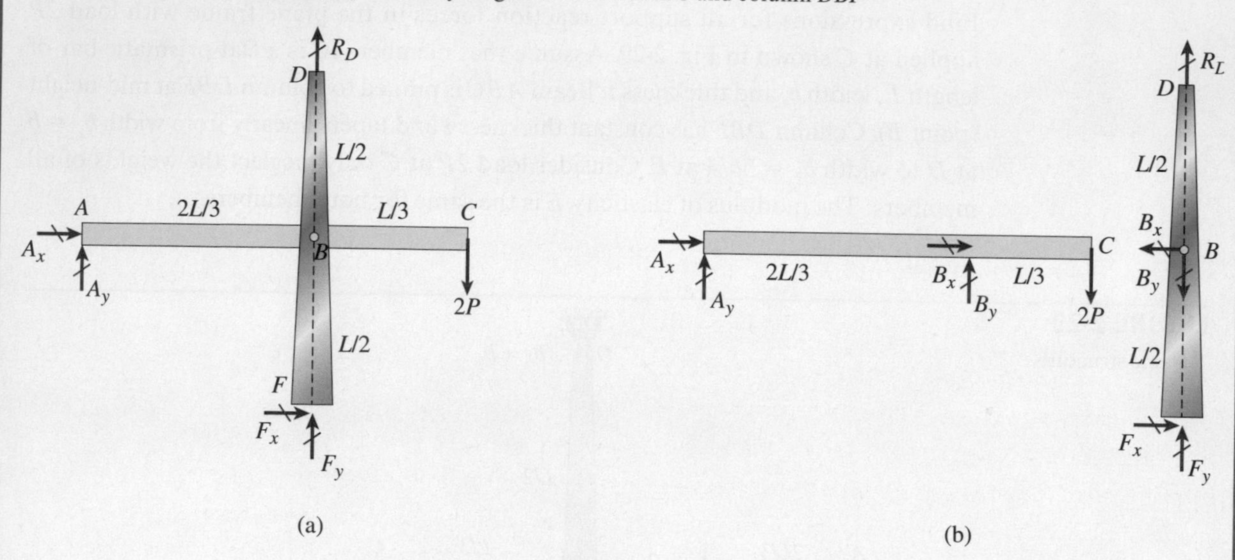

the actual load, $2P$ at joint C, to the first released structure (Fig. 2-31a) and apply the unknown redundant force R_D to the second (Fig. 2-31b). All reactions and displacements in the actual structure (Fig. 2-30a) must be the sum of corresponding items in the two released structures (Fig. 2-31a and b).

FIGURE 2-31

(a) Released structure 1 under actual load; (b) released structure 2 with redundant R_D applied

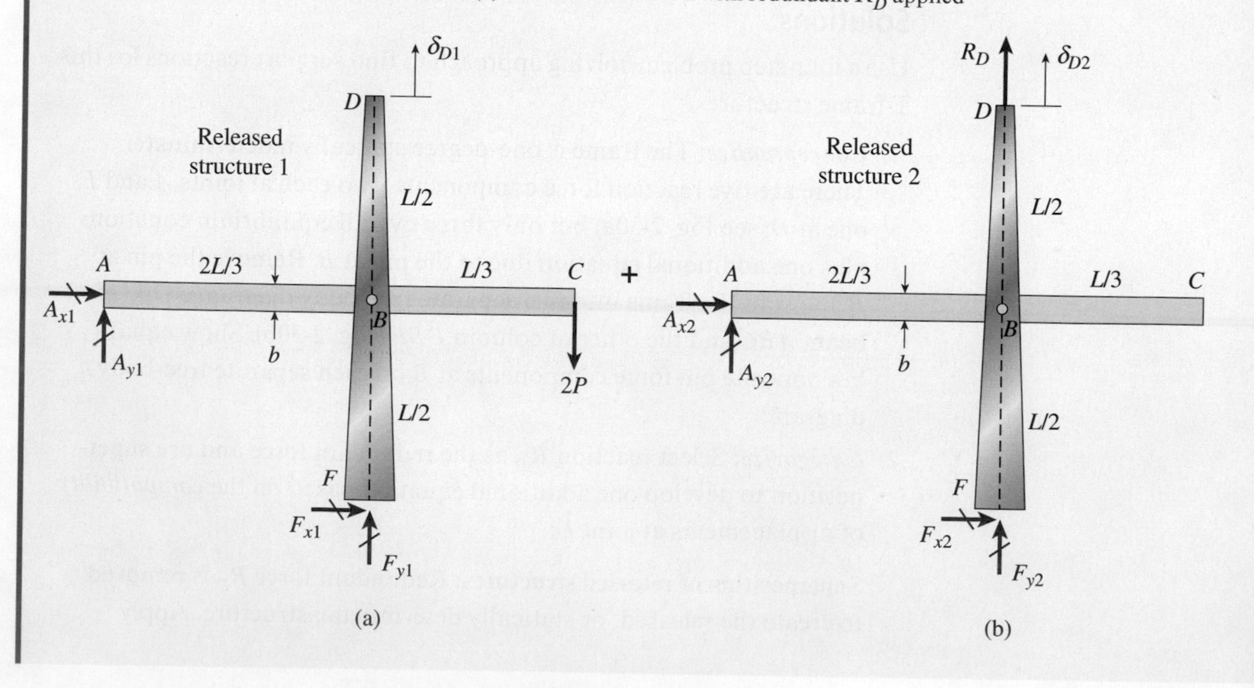

Compatibility equation: The vertical displacement at joint D in the actual structure (Fig. 2-30a) is zero. Superposition of displacements in Fig. 2-31a and b at joint D gives the compatibility equation:

$$\delta_D = 0 = \delta_{D1} + \delta_{D2} \tag{a}$$

Applying the force-displacement relation to the released structure in Fig. 2-31a will lead to an expression for δ_{D1} in terms of applied load $2P$ at joint C. Similarly, solution for displacement δ_{D2} in Fig. 2-31b will give an expression in terms of unknown reaction R_D. Substitution of both expressions into Eq. (a) will give reaction R_D. The final step is the application of static *equilibrium* to find all remaining reactions.

3. *Analyze*:

Find displacement δ_{D1} in released structure 1: Remove the pin at B and draw separate free-body diagrams for beam ABC and column DBF (Fig. 2-32a). Sum moments about joint A in beam ABC to find that $B_{y1} = 3P$. Apply B_{y1} to column DBF to see that the internal force in segment DB is zero and segment BF has a compressive internal force of $3P$. The width of the tapered bar is $b_B = 9b/8$ at B and $b_2 = 5b/4$ at F. The taper of bar DBF is small, so use the displacement expressions derived in Example 2-5 to find that the downward displacement at D in released structure 1 is

$$\delta_{D1} = \frac{(-3P)\left(\frac{L}{2}\right)}{Et(b_2 - b_B)}\ln\left(\frac{b_2}{b_B}\right) = -\frac{PL}{Ebt}(12)\ln\left(\frac{10}{9}\right) = -1.2643\frac{PL}{Ebt} \tag{b}$$

FIGURE 2-32

Beam and column free-body diagrams for (a) released structure 1 and (b) released structure 2

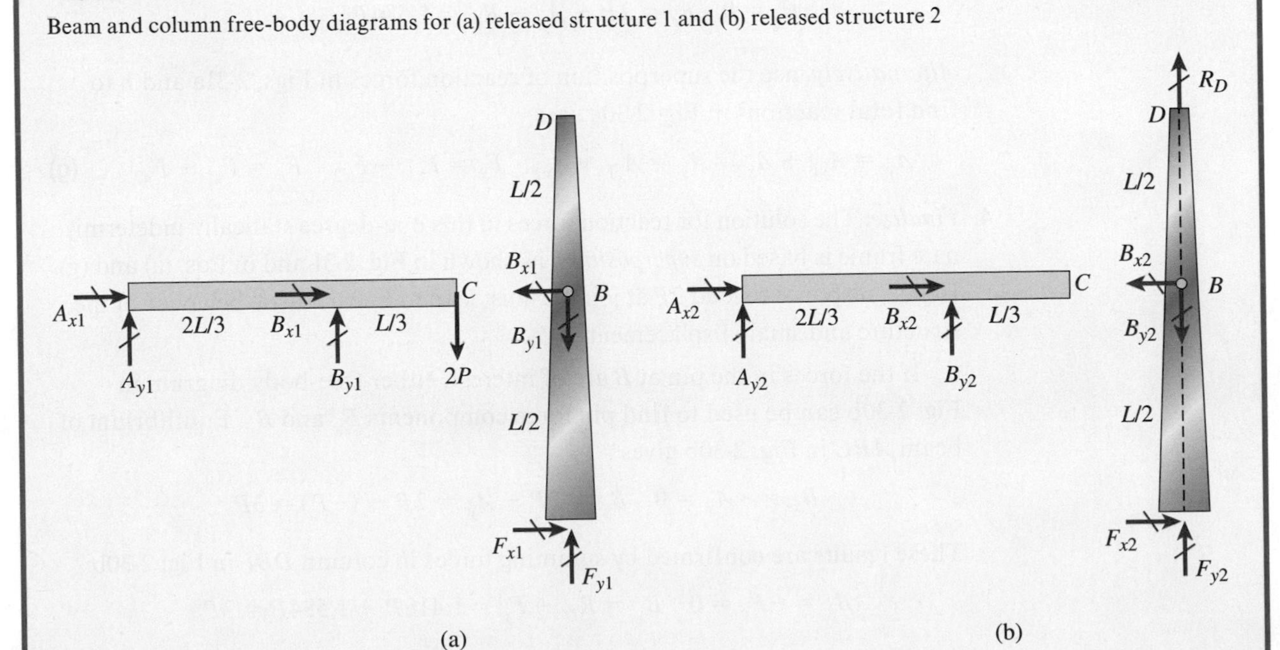

Find displacement δ_{D2} in released structure 2: Summing moments about joint A in Fig. 2-32b gives $B_{y2} = 0$. Next sum forces in the y direction in the free-body diagram of column DBF to find that $F_{y2} = -R_D$. Column DBF has a constant internal force equal to applied load R_D, and the width of bar DBF is $b_1 = b$ at D. Thus, the upward displacement at D in released structure 2 is

$$\delta_{D2} = \frac{(R_D L)}{Et(b_2 - b_1)} \ln\left(\frac{b_2}{b_1}\right) = \frac{R_D L}{Ebt}(4)\ln\left(\frac{5}{4}\right) = 0.8926\frac{R_D L}{Ebt} \quad \text{(c)}$$

Solve compatibility equation to find redundant R_D: Substitute expressions for δ_{D1} and δ_{D2} into Eq. (a) and solve for redundant reaction force R_D:

$$R_D = P\frac{1.2643}{0.8926} = 1.416P \quad \text{(d)}$$

Equilibrium to find remaining reaction forces: With redundant R_D now known, reaction force components at joints A and F are found using the free-body diagrams in Fig. 2-30. From beam ABC in Fig. 2-30b,

$$\Sigma M_B = 0 \quad A_y = \frac{-1}{\left(\frac{2L}{3}\right)}\left(2P\left(\frac{L}{3}\right)\right) = -P \quad \text{(e)}$$

Now use the overall free-body diagram in Fig. 2-30 to find all remaining reactions:

$$\Sigma M_F = 0 \quad A_x = \frac{-1}{\left(\frac{L}{2}\right)}\left(A_y\left(\frac{2L}{3}\right) + 2P\left(\frac{L}{3}\right)\right) = 0$$

$$\Sigma F_x = 0 \quad F_x = -A_x = 0 \quad \text{(f)}$$

$$\Sigma F_y = 0 \quad F_y = 2P - A_y - R_D = 1.584P$$

Alternatively, use the superposition of reaction forces in Figs. 2-31a and b to find total reactions in Fig. 2-30a as

$$A_x = A_{x1} + A_{x2} \quad A_y = A_{y1} + A_{y2} \quad F_x = F_{x1} + F_{x2} \quad F_y = F_{y1} + F_{y2} \quad \text{(g)}$$

4. ***Finalize***: The solution for reaction forces in this one-degree statically indeterminate frame is based on *superposition*, as shown in Fig. 2-31 and in Eqs. (a) and (g). Hence, response to load $2P$ at joint C must lead to linear elastic behavior of the structure and small displacements.

If the forces in the pin at B are of interest, either free-body diagram in Fig. 2-30b can be used to find pin force components B_x and B_y. Equilibrium of beam ABC in Fig. 2-30b gives

$$B_x = -A_x = 0 \quad B_y = 2P - A_y = 2P - (-P) = 3P$$

These results are confirmed by summing forces in column DBF in Fig. 2-30b:

$$B_x = -F_x = 0 \quad B_y = R_D + F_y = 1.416P + 1.584P = 3P$$

Example 2-7

Rigid beam $OABC$ is part of the support structure for a building entrance canopy (Fig. 2-33a). Beam $OABC$ is pinned at end O and supported by two cables (AD and BD) at points A and B (Fig. 2-33b). The combined canopy and cable weight of $W = 7.5$ kN is applied as a concentrated load between A and B and, vertical load P acts at joint C. The beam has a length of 4.5 m. Both cables have a nominal diameter of 12 mm, an effective cross-sectional area of 76.7 mm^2 (see Table 2-1), and an effective modulus of 135 GPa. Find the maximum load P that can act at C if the maximum permissible force in each cable is 20 kN.

FIGURE 2-33

(a) Inclined canopy at entrance to building; (b) two-dimensional model of one beam and two supporting cables

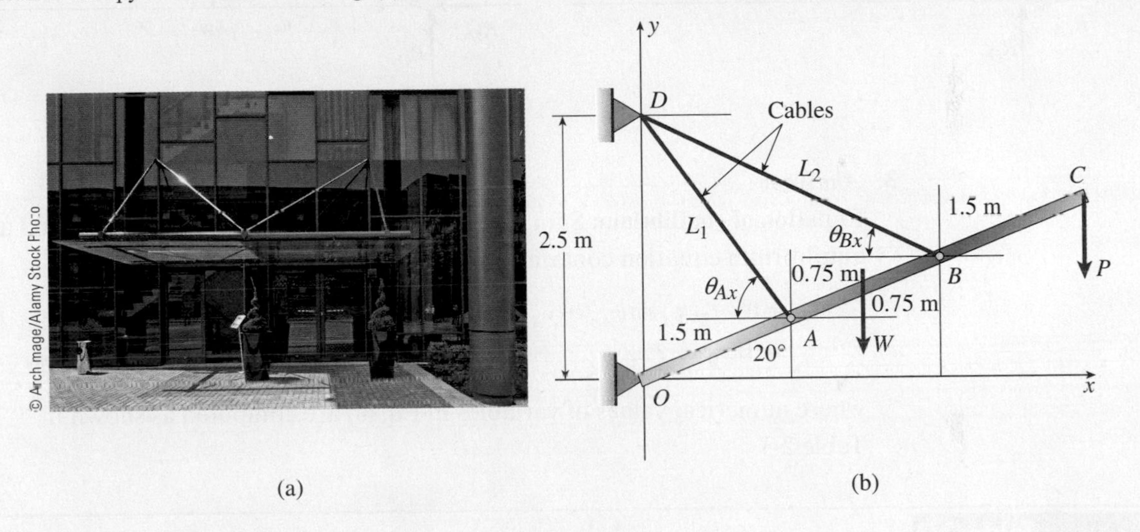

(a) (b)

Solution:

Use a four-step problem-solving approach.

1. ***Conceptualize***: Start with a free-body diagram of the entire structure (Fig. 2-34a) and then pass a horizontal plane through the cables to create upper and lower free-body diagrams. The lower free-body diagram in Fig. 2-34b will be of most use here because the forces T_1 and T_2 in cables AD and BD, respectively, are exposed.
2. ***Categorize***: This structure was studied in Example 1-13, but in that example, a continuous cable running over a pulley at D supported the canopy in place of the two separate cables AD and BD used here. Two unknown cable forces T_1 and T_2 replace the single unknown cable force T in Example 1-13, so beam OABC in Fig. 2-34b is *one degree statically indeterminate*: There are four unknowns (R_{Ox}, R_{Oy}, T_1, and T_2) but only three independent *equations of equilibrium*. A *compatibility equation* is needed to find another relationship between the two unknown cable forces.

FIGURE 2-34

Free-body diagrams of (a) beam and cable structure and (b) beam *OABC*

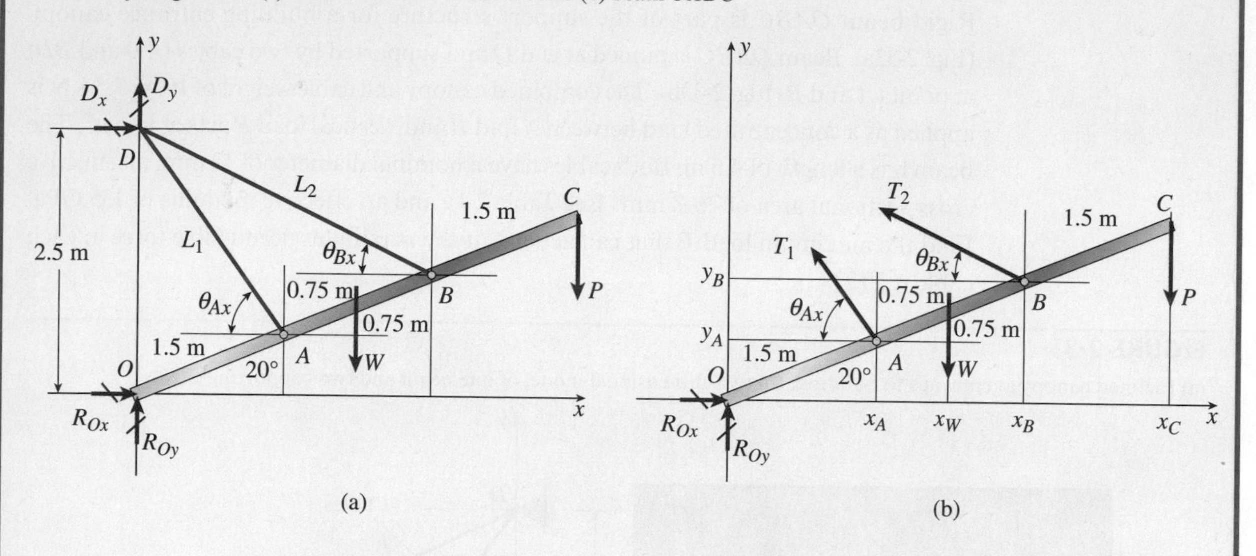

3. ***Analyze***:

Equation of equilibrium: Sum moments about point O in Fig. 2-34b to find an equilibrium equation containing cable forces T_1 and T_2:

$$\Sigma M_O = 0 \quad T_1(x_A \sin\theta_{Ax} + y_A \cos\theta_{Ax}) + T_2(x_B \sin\theta_{Bx} + y_B \cos\theta_{Bx}) = Wx_W + Px_C \qquad \text{(a)}$$

where numerical values of variables in Eq. (a) are computed as shown in Table 2-3.

Table 2-3

Numerical values of variables for use in Eq. (a)

Moment arm distances from point O	$x_A = (1.5\text{ m})\cos(20°) = 1.4095\text{ m}$ $y_A = (1.5\text{ m})\sin(20°) = 0.5130\text{ m}$ $x_B = 2.8191\text{ m},\ y_B = 1.0261\text{ m}$ $x_C = 4.2286\text{ m},\ x_W = 2.1143\text{ m}$
Initial cable lengths using law of cosines	$L_1 = \sqrt{1.5^2 + 2.5^2 - 2(1.5)(2.5)\cos 70°}\text{ m} = 2.436\text{ m}$ $L_2 = \sqrt{3^2 + 2.5^2 - 2(3)(2.5)\cos 70°}\text{ m} = 3.181\text{ m}$
Angles between cables and beam using law of sines	$\theta_A = \sin^{-1}\left(\dfrac{2.5\text{ m}}{L_1}\sin 70°\right) = 74.648°$ $\theta_B = \sin^{-1}\left(\dfrac{2.5\text{ m}}{L_2}\sin 70°\right) = 47.603°$ $\theta_{Ax} = \theta_A - 20° = 54.648°,\ \theta_{Bx} = \theta_B - 20° = 27.603°$

Inserting numerical values from Table 2-3 into moment equilibrium Eq. (a) using meter and kN units gives

$$1.4465T_1 + 2.2155T_2 = 15.8573 \text{ kN}\cdot\text{m} + 4.2286P \quad \textbf{(b)}$$

Equation of compatibility: Rigid beam *OABC* rotates clockwise about point *O* due to loads *W* and *P*. The displaced position of the beam is straight line *OA'B'C'* in Fig. 2-35 with the understanding that all displacements are very small. Displacements Δ_A and Δ_B of points *A* and *B*, respectively, are normal to line *OABC*. Use of similar triangles *OAA'* and *OBB'* gives the compatibility equation:

$$\Delta_B = 2\Delta_A \quad \textbf{(c)}$$

Force-displacement relations: Displacements in Fig. 2-35 are very small but are shown sufficiently large so that geometric relationships between displacements Δ_A and Δ_B of points *A* and *B* and cable elongations δ_1 for cable *AD* and δ_2 for cable *BD* can be written as

$$\Delta_A = \frac{\delta_1}{\sin\theta_A} \text{ and } \Delta_B = \frac{\delta_2}{\sin\theta_B}$$

Substitute these expressions for Δ_A and Δ_B into Eq. (c) to find:

$$\delta_2 = 2\left(\frac{\sin\theta_B}{\sin\theta_A}\right)\delta_1$$

The cables behave in a linearly elastic manner, so their elongations are written in terms of their flexibilities *f* in the *force-displacement relations*:

$$\delta_1 = \left(\frac{L_1}{EA}\right)T_1 = f_1T_1 \quad \text{and} \quad \delta_2 = \left(\frac{L_2}{EA}\right)T_2 = f_2T_2$$

It follows that

$$f_2T_2 = 2\left(\frac{\sin\theta_B}{\sin\theta_A}\right)f_1T_1 \quad \text{or} \quad T_2 = 2\left(\frac{\sin\theta_B}{\sin\theta_A}\right)\left(\frac{f_1}{f_2}\right)T_1$$

Inserting numerical values gives

$$T_2 = 2\left(\frac{\sin 47.603°}{\sin 74.648°}\right)\left(\frac{2.353(10^{-4})}{3.072(10^{-4})}\right)T_1 = 1.173\,T_1 \quad \textbf{(d)}$$

FIGURE 2-35

Displacement diagram

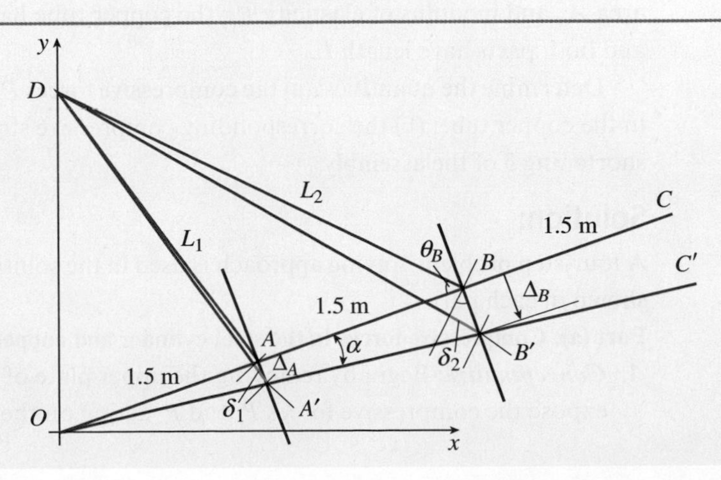

This relationship between cable forces T_1 and T_2 is a result of inserting the force-displacement relations into the compatibility equation [Eq. (c)].

Solution of equations: A constraint here is that neither cable force T_1 nor T_2 can exceed 20 kN. From Eq. (d), force T_2 is larger than T_1 and will reach the 20 kN limit first as force P at point C is increased in magnitude. Substitution of Eq. (d) into the equilibrium equation [Eq. (b)] and substitution of T_2 = 20 kN leads to a solution for P_{max}:

$$P_{max} = \frac{1}{4.2286}\left[1.4465\left(\frac{T_2}{1.173}\right) + 2.2155\,T_2 - 15.8573\right] \text{kN} = 12.56 \text{ kN} \quad \text{(e)}$$

4. ***Finalize***: When P_{max} = 12.56 kN at point C, cable BD reaches the limit force value of T_2 = 20 kN, and cable AD has tension force $T_1 = \dfrac{20 \text{ kN}}{1.173} = 17.05$ kN. Cable flexibilities f are known, so use the force-displacement relations to find cable elongations of $\delta_1 = f_1 T_1 = 4.012$ mm and $\delta_2 = f_2 T_2 = 6.144$ mm. Next, downward deflections at points A and B (normal to beam $OABC$ see Fig. 2-35) are $\Delta_A = \dfrac{\delta_1}{\sin\theta_A} = 4.16$ mm and $\Delta_B = \dfrac{\delta_2}{\sin\theta_B} = 8.32$ mm. So deflections in x and y directions at point B are for example, $\Delta_{Bx} = \Delta_B \sin(20°) = 2.85$ mm, $\Delta_{By} = -\Delta_B \cos(20°) = -7.82$ mm.

Using deflection Δ_B, the clockwise rotation of beam $OABC$ is computed as $\alpha = \Delta_B/3 \text{ m} = 2.77(10^{-3})$ radians = 0.159°, which confirms the assumption of small displacements. Finally, reactions at supports O and D (see Fig. 2-34a) can be computed using statics now that cable forces T_1 and T_2 are known.

Example 2-8

A solid circular steel cylinder S is encased in a hollow circular copper tube C (Fig. 2-36a and b). The cylinder and tube are compressed between the rigid plates of a testing machine by compressive forces P. The steel cylinder has cross-sectional area A_s and modulus of elasticity E_s, the copper tube has area A_c and modulus E_c, and both parts have length L.

Determine the quantities: (a) the compressive forces P_s in the steel cylinder and P_c in the copper tube; (b) the corresponding compressive stresses σ_s and σ_c; and (c) the shortening δ of the assembly.

Solution:

A four-step problem-solving approach is used in the solution; only essential steps are shown in each part.

Part (a): Compressive forces in the steel cylinder and copper tube.

1. ***Conceptualize***: Begin by removing the upper plate of the assembly in order to expose the compressive forces P_s and P_c acting on the steel cylinder and copper

FIGURE 2-36

Example 2-8: Analysis of a statically indeterminate structure

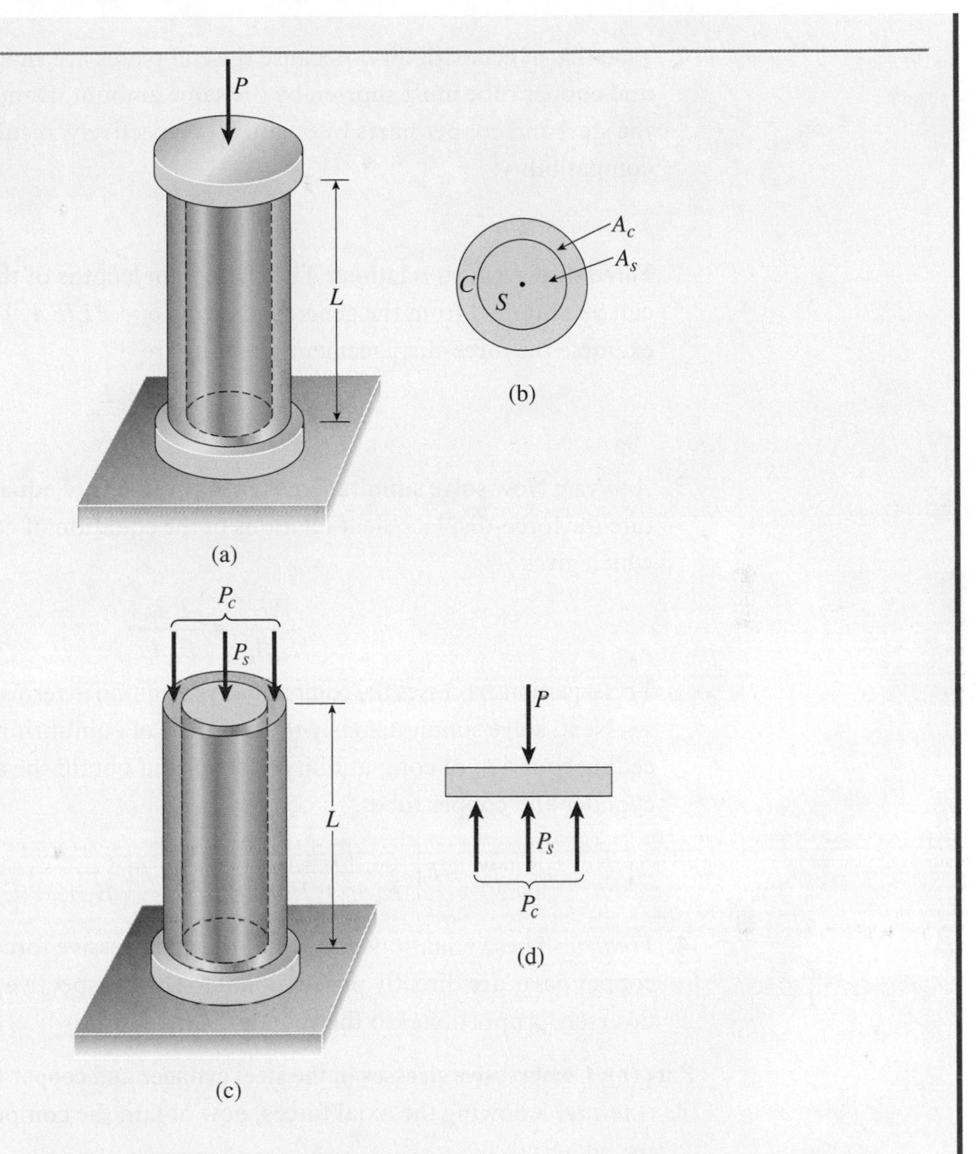

tube, respectively (Fig. 2-36c). The force P_s is the resultant of the uniformly distributed stresses acting over the cross section of the steel cylinder, and the force P_c is the resultant of the stresses acting over the cross section of the copper tube.

2. ***Categorize:***

Equation of equilibrium: A free body diagram of the upper plate is shown in Fig. 2-36d. This plate is subjected to the force P and to the unknown compressive forces P_s and P_c; thus, the equation of equilibrium is

$$\Sigma F_{\text{vert}} = 0 \qquad P_s + P_c - P = 0 \tag{a}$$

This equation, which is the only nontrivial equilibrium equation available, contains two unknowns. Therefore, the structure is statically indeterminate.

Equation of compatibility: Because the end plates are rigid, the steel cylinder and copper tube must shorten by the same amount. Denoting the shortening of the steel and copper parts by δ_s and δ_c, respectively, results in the equation of compatibility:

$$\delta_s = \delta_c \quad \text{(b)}$$

Force-displacement relations: The changes in lengths of the cylinder and tube can be obtained from the general equation $\delta = PL/EA$. Therefore, in this example the force-displacement relations are

$$\delta_s = \frac{P_s L}{E_s A_s} \qquad \delta_c = \frac{P_c L}{E_c A_c} \quad \text{(c,d)}$$

3. ***Analyze***: Now solve simultaneously the three sets of equations. First, substitute the force-displacement relations in the equation of compatibility [Eq.(b)], which gives

$$\frac{P_s L}{E_s A_s} = \frac{P_c L}{E_c A_c} \quad \text{(e)}$$

This equation expresses the compatibility condition in terms of the unknown forces.

Next, solve simultaneously the equation of equilibrium [Eq. (a)] and the preceding equation of compatibility [Eq. (e)] and obtain the axial forces in the steel cylinder and copper tube:

$$P_s = P\left(\frac{E_s A_s}{E_s A_s + E_c A_c}\right) \qquad P_c = P\left(\frac{E_c A_c}{E_s A_s + E_c A_c}\right) \quad \text{(2-14a,b)}$$

4. ***Finalize***: These equations show that the compressive forces in the steel and copper parts are directly proportional to their respective axial rigidities and inversely proportional to the sum of their rigidities.

Part (b): Compressive stresses in the steel cylinder and copper tube.

3. ***Analyze***: Knowing the axial forces, now obtain the compressive stresses in the two materials:

$$\sigma_s = \frac{P_s}{A_s} = \frac{PE_s}{E_s A_s + E_c A_c} \qquad \sigma_c = \frac{P_c}{A_c} = \frac{PE_c}{E_s A_s + E_c A_c} \quad \text{(2-15a,b)}$$

4. ***Finalize***: Note that the ratio σ_s/σ_c of the stresses is equal to the ratio E_s/E_c of the moduli of elasticity, showing that in general the "stiffer" material always has the larger stress.

Part (c): Shortening of the assembly.

2. ***Categorize***: The shortening δ of the entire assembly can be obtained from either Eq. (c) or (d). Substituting the forces [from Eqs. (2-14a and b)] gives

$$\delta = \frac{P_s L}{E_s A_s} = \frac{P_c L}{E_c A_c} = \frac{PL}{E_s A_s + E_c A_c} \quad \text{(2-16)}$$

4. *Finalize*: This result shows that the shortening of the assembly is equal to the total load divided by the sum of the stiffness of the two parts [recall from Eq. (2-4a) that the stiffness of an axially loaded bar is $k = EA/L$]

Alternative solution of the equations: Instead of substituting the force-displacement relations [Eqs. (c) and (d)] into the equation of compatibility, those relations can be written in the form

$$P_s = \frac{E_s A_s}{L}\delta_s \quad P_c = \frac{E_c A_c}{L}\delta_c \tag{f,g}$$

and substituted into the equation of equilibrium [Eq. (a)];

$$\frac{E_s A_s}{L}\delta_s + \frac{E_c A_c}{L}\delta_c = P \tag{h}$$

This equation expresses the equilibrium condition in terms of the unknown displacements. Then solve simultaneously the equation of compatibility [Eq. (b)] and the preceding equation, thus obtaining the displacements:

$$\delta_s = \delta_c = \frac{PL}{E_s A_s + E_c A_c} \tag{i}$$

which agrees with Eq. (2-16). Finally, substitute expression (i) into Eqs. (f) and (g) and obtain the compressive forces P_s and P_c [(see Eqs. (2-14a and b)].

Note: The alternative method of solving the equations is a simplified version of the **stiffness** (or **displacement**) **method of analysis**, and the first method of solving the equations is a simplified version of the **flexibility** (or **force**) **method**. The names of these two methods arise from the fact that Eq. (h) has displacements as unknowns and stiffnesses as coefficients [see Eq. (2-4a)], whereas Eq. (e) has forces as unknowns and flexibilities as coefficients [see Eq. (2-4b)]

2.5 Thermal Effects, Misfits, and Prestrains

External loads are not the only sources of stresses and strains in a structure. Other sources include *thermal effects* arising from temperature changes, *misfits* resulting from imperfections in construction, and *prestrains* that are produced by initial deformations. Still other causes are settlements (or movements) of supports, inertial loads resulting from accelerating motion, and natural phenomenon such as earthquakes.

Thermal effects, misfits, and prestrains are commonly found in both mechanical and structural systems and are described in this section. As a general rule, they are much more important in the design of statically indeterminate structures than in statically determinate ones.

Thermal Effects

FIGURE 2-37

Block of material subjected to an increase in temperature

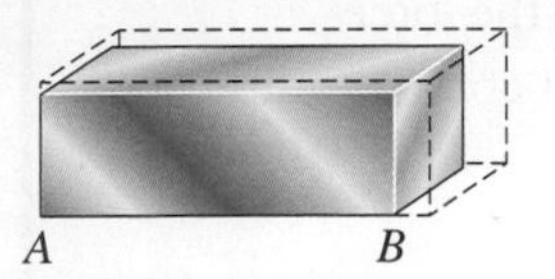

Changes in temperature produce expansion or contraction of structural materials, resulting in **thermal strains** and **thermal stresses**. A simple illustration of thermal expansion is shown in Fig. 2-37, where the block of material is unrestrained and therefore free to expand. When the block is heated, every element of the material undergoes thermal strains in all directions, and consequently the dimensions of the block increase. If corner A is taken as a fixed reference point and side AB maintains its original alignment, the block will have the shape shown by the dashed lines.

For most structural materials, thermal strain ε_T is proportional to the temperature change ΔT; that is,

$$\varepsilon_T = \alpha(\Delta T) \tag{2-17}$$

in which α is a property of the material called the **coefficient of thermal expansion**. Since strain is a dimensionless quantity, the coefficient of thermal expansion has units equal to the reciprocal of temperature change. In SI units, the dimensions of α can be expressed as either 1/K (the reciprocal of kelvins) or 1/°C (the reciprocal of degrees Celsius). The value of α is the same in both cases because a *change* in temperature is numerically the same in both kelvins and degrees Celsius.[2] Typical values of α are listed in Table I-4 of Appendix I.

When a **sign convention** is needed for thermal strains, assume that expansion is positive and contraction is negative.

To demonstrate the relative importance of thermal strains, compare thermal strains with load-induced strains in the following manner. Suppose you have an axially loaded bar with longitudinal strains given by the equation $\varepsilon = \sigma/E$, where σ is the stress and E is the modulus of elasticity. Then suppose you have an identical bar subjected to a temperature change ΔT, which means that the bar has thermal strains given by Eq. (2-17). Equating the two strains gives

$$\sigma = E\alpha(\Delta T)$$

Use this equation to calculate the axial stress σ that produces the same strain as does the temperature change ΔT. For instance, consider a stainless steel bar with $E = 210$ GPa and $\alpha = 17 \times 10^{-6}$ /°F. A quick calculation from the preceding equation for σ shows that a change in temperature of 60°C produces the same strain as a stress of 214 MPa. This stress is in the range of typical allowable stresses for stainless steel. Thus, a relatively modest change in temperature produces strains of the same magnitude as the strains caused by ordinary loads, which shows that temperature effects can be important in engineering design.

Ordinary structural materials expand when heated and contract when cooled, and therefore an increase in temperature produces a positive thermal strain. Thermal strains usually are reversible, in the sense that the member returns to its original shape when its temperature returns to the original value. However, a few special metallic alloys have recently been developed that do not behave in the customary manner. Instead, over certain temperature ranges, their dimensions decrease when heated and increase when cooled.

[2]For a discussion of temperature units and scales, see Section A.4 of Appendix A.

Water is also an unusual material from a thermal standpoint—it expands when heated at temperatures above 4°C and also expands when cooled below 4°C. Thus, water has its maximum density at 4°C.

Now return to the block of material shown in Fig. 2-37. Assume that the material is homogeneous and isotropic and that the temperature increase ΔT is uniform throughout the block. Calculate the increase in *any* dimension of the block by multiplying the original dimension by the thermal strain. For instance, if one of the dimensions is L, then that dimension will increase by the amount

$$\delta_T = \varepsilon_T L = \alpha(\Delta T)L \tag{2-18}$$

Equation (2-18) is a **temperature-displacement relation**, analogous to the force-displacement relations described in the preceding section. It can be used to calculate changes in lengths of structural members subjected to uniform temperature changes, such as the elongation δ_T of the prismatic bar shown in Fig. 2-38. (The transverse dimensions of the bar also change, but these changes are not shown in the figure since they usually have no effect on the axial forces being transmitted by the bar.)

FIGURE 2-38

Increase in length of a prismatic bar due to a uniform increase in temperature [Eq. (2-18)]

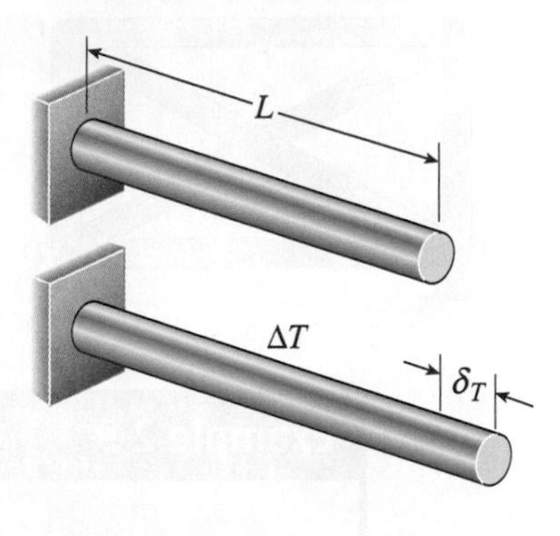

The preceding discussions of thermal strains assumed that the structure had no restraints and was able to expand or contract freely. These conditions exist when an object rests on a frictionless surface or hangs in open space. In such cases, no stresses are produced by a uniform temperature change throughout the object, although nonuniform temperature changes may produce internal stresses. However, many structures have supports that prevent free expansion and contraction, in which case **thermal stresses** will develop even when the temperature change is uniform throughout the structure.

FIGURE 2-39

Statically determinate truss with a uniform temperature change in each member

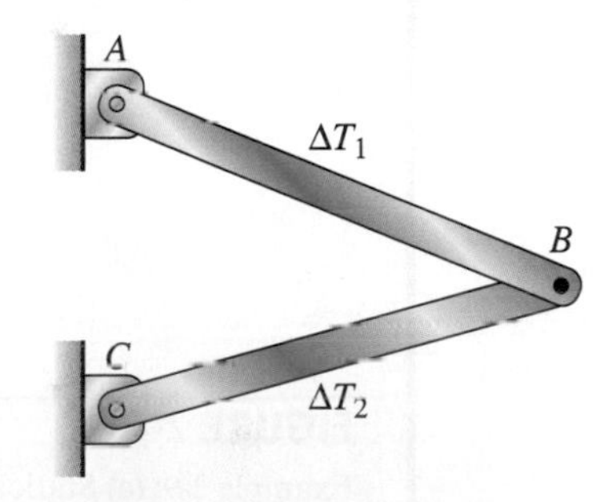

To illustrate some of these ideas about thermal effects, consider the two-bar truss ABC of Fig. 2-39 and assume that the temperature of bar AB is changed by ΔT_1 and the temperature of bar BC is changed by ΔT_2. Because the truss is statically determinate, both bars are free to lengthen or shorten, resulting in a displacement of joint B. However, there are no stresses in either bar and no reactions at the supports. This conclusion applies generally to **statically determinate structures**; that is, uniform temperature changes in the members produce thermal strains (and the corresponding changes in lengths) without producing any corresponding stresses.

A **statically indeterminate structure** may or may not develop temperature stresses, depending upon the character of the structure and the nature of the temperature changes. To illustrate some of the possibilities, consider the statically indeterminate truss shown in Fig. 2-40. Because the supports of this structure permit joint D to move horizontally, no stresses are developed when the *entire* truss is heated uniformly. All members increase in length in proportion to their original lengths, and the truss becomes slightly larger in size.

However, if some bars are heated and others are not, thermal stresses will develop because the statically indeterminate arrangement of the bars prevents free expansion. To visualize this condition, imagine that just one bar is heated. As this bar becomes longer, it meets resistance from the other bars, and therefore stresses develop in all members.

Forces can develop in statically indeterminate trusses due to temperature and prestrain

FIGURE 2-40

Statically indeterminate truss subjected to temperature changes

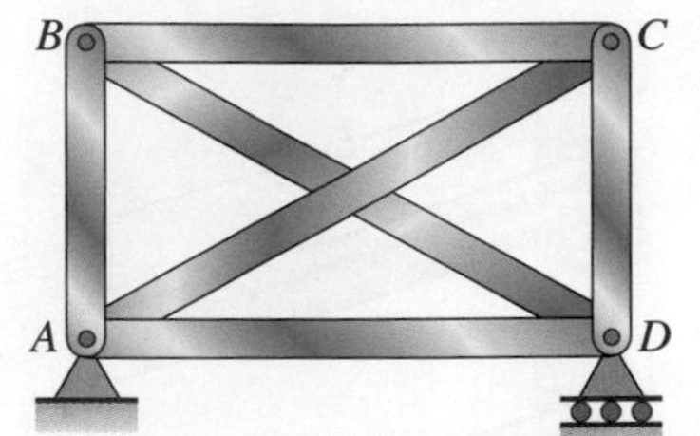

The analysis of a statically indeterminate structure with temperature changes is based upon the concepts discussed in the preceding section, namely equilibrium equations, compatibility equations, and force-displacement relations. The principal difference is that temperature-displacement relations [Eq. (2-18)] are used in addition to force-displacement relations (such as $\delta = PL/EA$) when performing the analysis. The following two examples illustrate the procedures in detail.

Example 2-9

A prismatic bar AB of length L is made of linearly elastic material and is held between immovable supports (Fig. 2-41a). The bar has a modulus of elasticity E and a coefficient of thermal expansion α.

(a) If the temperature of the bar is raised uniformly by an amount ΔT, derive a formula for the thermal stress σ_T developed in the bar.

(b) Modify the formula in part (a) if the rigid support at B is replaced by an elastic support having a spring constant k (Fig. 2-41b); assume that only bar AB is subject to the uniform temperature increase ΔT.

FIGURE 2-41

Example 2-9: (a) Statically indeterminate bar with uniform temperature increase ΔT, (b) statically indeterminate bar with elastic support and uniform temperature increase ΔT, and (c) statically indeterminate bar with elastic support and nonuniform temperature increase $\Delta T(x)$

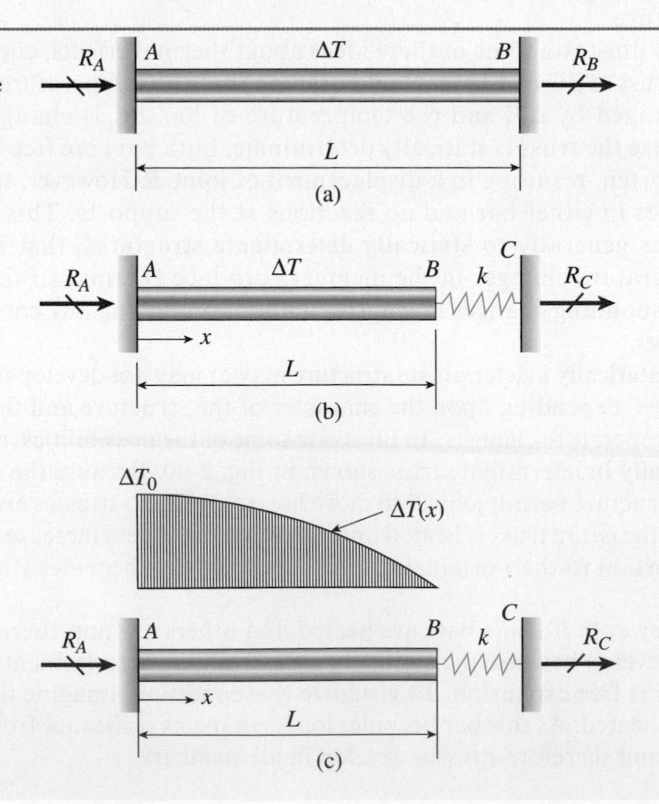

(c) Repeat part (b), but now assume that the bar is heated nonuniformly such that the temperature increase at distance x from A is given by $\Delta T(x) = \Delta T_0 \left(1 - x^2/L^2\right)$ (see Fig. 2-41c).

Solution:

A four-step problem-solving approach is used in the solution; only essential steps are shown in each part.

Part (a): Determine thermal stress in the bar fixed at A and B subjected to uniform temperature increase ΔT.

1. ***Conceptualize***: Because the temperature increases, the bar tends to elongate but is restrained by the rigid supports at A and B. Therefore, reactions R_A and R_B are developed at the supports, and the bar is subjected to uniform compressive stresses.
2. ***Categorize***: **Equation of Equilibrium:** The only nontrivial equation of static equilibrium is that reactions R_A and R_B must sum to zero. So there is one equation but two unknowns, which is a *one-degree statically indeterminate problem*:

$$\Sigma F_x = 0 \quad R_A + R_B = 0 \tag{a}$$

Select reaction R_B as the *redundant* and use the *superposition* of two statically determinate "released" structures (Fig. 2-41d) to develop an additional equation: an equation of *compatibility*. The first released structure is subjected to the temperature increase ΔT and hence elongates by amount δ_T. The second elongates δ_B under redundant R_B, which is applied as a load. Use a statics sign convention, so that forces and displacements in the x direction are assumed to be positive.

Equation of compatibility: The equation of compatibility expresses the fact that the net change in length of the bar is zero, because supports A and B are fully restrained:

$$\delta_T + \delta_B = 0 \tag{b}$$

Temperature-displacement and force-displacement relations: The increase in length of the bar due to temperature is [Eq. (2-18)]

$$\delta_T = \alpha(\Delta T)L \tag{c}$$

FIGURE 2-41 Continued

Example 2-9: (d) Statically determinate bars with support B removed (i.e., *released* structures)

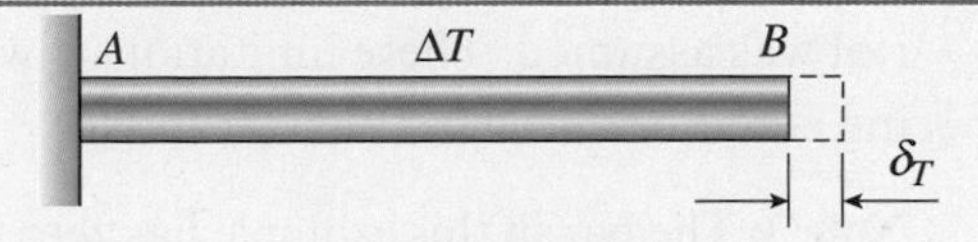

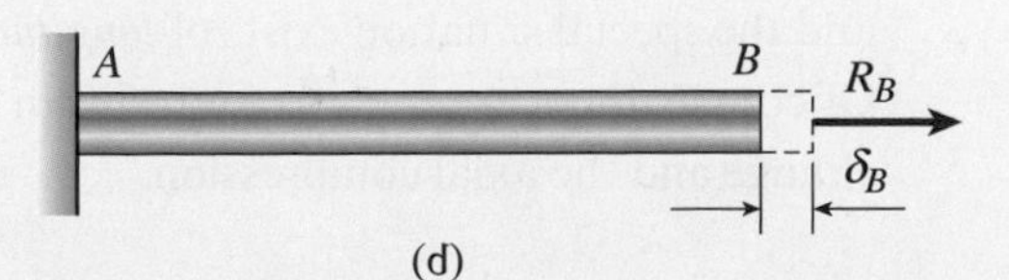

(d)

where α is the coefficient of thermal expansion of the material. The increase in bar length due to unknown applied force R_B is obtained from the force-displacement relation:

$$\delta_B = R_B\left(\frac{L}{EA}\right) = R_B f_{AB} \quad \text{(d)}$$

in which E is the modulus of elasticity, A is the bar cross-sectional area, and f_{AB} is the flexibility of the bar.

3. ***Analyze***: Substituting Eqs. (c) and (d) into the compatibility equation Eq. (b) and solving for redundant R_B gives

$$R_B = \frac{-\alpha(\Delta T)L}{f_{AB}} = -EA\alpha(\Delta T) \quad \text{(e)}$$

and the equilibrium equation Eq. (a) gives

$$R_A = -R_B = EA\alpha(\Delta T) \quad \text{(f)}$$

Based on a statics sign convention, R_B is in the negative x direction, while R_A is in the positive x direction. As a final step, compute the comprehensive stress in the bar (assuming that ΔT is positive and therefore an increase in temperature) to be

$$\sigma_T = \frac{R_A}{A} = E\alpha(\Delta T) \quad \text{(g)}$$

4. ***Finalize***:

Note 1: In this example, the reactions are independent of the length of the bar, and the stress is independent of both the length and the cross-sectional area [see Eqs. (f) and (g)]. Thus, again, the usefulness of a symbolic solution is demonstrated because these important features of the bar's behavior might not be noticed in a purely numerical solution.

Note 2: When determining the thermal elongation of the bar [Eq. (c)], the material was assumed to be homogeneous and the increase in temperature uniform throughout the volume of the bar. Also, when determining the increase in length due to the reactive force [Eq. (d)], linearly elastic behavior of the material was assumed. These limitations always should be kept in mind when writing equations, such as Eqs. (c) and (d).

Note 3: The bar in this example has zero longitudinal displacements, not only at the ends but also at every cross section. Thus, there are ***no axial strains in this bar***, and the special situation exists of *longitudinal stresses without longitudinal strains.* Of course, there are transverse strains in the bar from both the temperature change and the axial compression.

Part (b): Determine thermal stress in the bar fixed at A with elastic support at B and subjected to uniform temperature change ΔT.

2. *Categorize*: The structure in Fig. 2-41b is one-degree statically indeterminate, so select reaction R_C as the redundant and once again use the superposition of two released structures to solve the problem.
3. *Analyze*: First, *static equilibrium* of the original indeterminate structure requires that

$$R_A + R_C = 0 \tag{h}$$

while *compatibility of displacements* at joint C for the two released structures is expressed as

$$\delta_T + \delta_C = 0 \tag{i}$$

In the first released structure, apply uniform temperature change ΔT to bar AB only, so

$$\delta_T = \alpha(\Delta T)L \tag{c, repeated}$$

Note that the spring displaces in the positive x direction but is not deformed by the temperature change. Next, redundant R_C is applied to the end of the spring in the second released structure, resulting in displacement in the positive x direction. Both bar AB and the spring are subject to force R_C, so the total displacement at C is the sum of the elongations of bar and spring:

$$\delta_C = R_C\left(\frac{L}{EA}\right) + \frac{R_C}{k} = R_C(f_{AB} + f) \tag{j}$$

where $f = 1/k$ is the flexibility of the spring. Substituting the temperature-displacement equation [Eq. (c)] and force-displacement equation [Eq. (j)] into the compatibility equation [Eq. (i)], then solving for redundant R_C gives

$$R_C = \frac{-\alpha(\Delta T)L}{f_{AB} + f} = \frac{-\alpha(\Delta T)L}{\dfrac{L}{EA} + \dfrac{1}{k}} \quad \text{or} \quad R_C = -\left[\frac{EA\alpha(\Delta T)}{1 + \dfrac{EA}{kL}}\right] \tag{k}$$

Then equilibrium [Eq. (h)] leads to

$$R_A = -R_C = \frac{EA\alpha(\Delta T)}{1 + \dfrac{EA}{kL}} \tag{l}$$

Recall that, based on a statics sign convention, reaction force R_A is in the positive x direction, while reaction force R_C is in the negative x direction. Finally, the compressive stress in the bar is

$$\sigma_T = \frac{R_A}{A} = \frac{E\alpha(\Delta T)}{1 + \dfrac{EA}{kL}} \tag{m}$$

4. ***Finalize***: Note that *if the spring stiffness k goes to infinity*, Eq. (l) becomes Eq. (f) and Eq. (m) becomes Eq. (g). In effect, use of an infinitely stiff spring moves the rigid support from C back to B.

Part (c): Determine thermal stress in the bar fixed at A with elastic support at B and subjected to *nonuniform* temperature change.

2. ***Categorize***: The structure in Fig. 2-41c is one-degree statically indeterminate. So, once again, select reaction R_C as the redundant and, as in parts (a) and (b), use superposition of two released structures to solve the one-degree statically indeterminate problem (Fig. 2-41e, f).

FIGURE 2-41 Continued

Example 2.9: (e) Statically determinate bar with support C removed (i.e., *released* structure) under nonuniform temperature increase

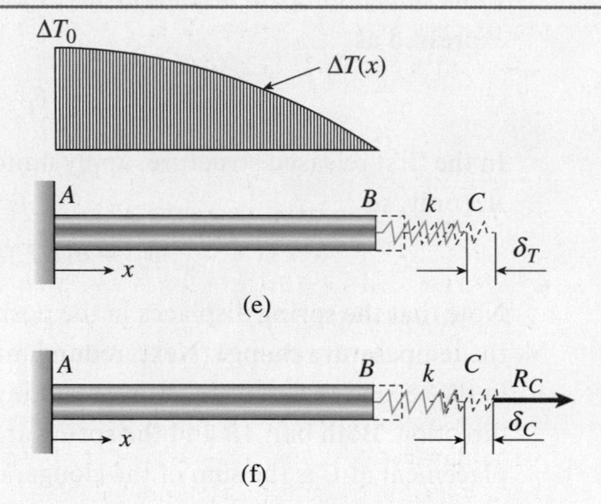

3. ***Analyze***: The equation of *static equilibrium* for the overall structure is Eq. (h), and the equation of *compatibility* is Eq. (i). First, solve for displacement δ_T in the released structure (Fig. 2-41e) as

$$\delta_T = \int_0^L \alpha[\Delta T(x)]dx = \int_0^L \alpha\left\{\Delta T_0\left[1 - \left(\frac{x}{L}\right)^2\right]\right\}dx = \frac{2}{3}\alpha(\Delta T_0)L \quad \textbf{(n)}$$

and δ_C for the second released structure (Fig. 2-41f) is the same as Eq. (j) giving

$$\delta_C = R_C(f_{AB} + f) \quad \textbf{(j, repeated)}$$

Substituting the temperature-displacement equation [Eq. (n)] and the force-displacement equation [Eq. (j)] into the compatibility equation [Eq. (i)] gives

$$R_C = \frac{\frac{-2}{3}\alpha(\Delta T_0)L}{f_{AB} + f} = \frac{-2\alpha(\Delta T_0)L}{3\left(\frac{L}{EA} + \frac{1}{k}\right)} \text{ or } R_C = -\left(\frac{2}{3}\right)\left[\frac{EA\alpha(\Delta T_0)}{1 + \frac{EA}{kL}}\right] \quad \textbf{(o)}$$

From the static equilibrium equation [Eq. (h)],

$$R_A = -R_C = \left(\frac{2}{3}\right)\left[\frac{EA\alpha(\Delta T_0)}{1 + \frac{EA}{kL}}\right] \quad \textbf{(p)}$$

Finally, the compressive stress in the bar under nonuniform temperature change $\Delta T(x) = \Delta T_0(1 - (x/L)^2)$ is

$$\sigma_T = \frac{R_A}{A} = \left(\frac{2}{3}\right)\left[\frac{E\alpha(\Delta T_0)}{1 + \dfrac{EA}{kL}}\right] \qquad \text{(q)}$$

4. ***Finalize***: Note once again that use of an infinitely stiff spring eliminates the EA/kL term from Eq. (q) and provides the solution for a prismatic bar fixed at A and B with nonuniform temperature change as $\Delta T(x) = \Delta T_0(1 - (x/L)^2)$.

Example 2-10

A sleeve in the form of a circular tube of length L is placed around a bolt and fitted between washers at each end (Fig. 2-42a). The nut is then turned until it is just snug. The sleeve and bolt are made of different materials and have different cross-sectional areas. (Assume that the coefficient of thermal expansion α_S of the sleeve is greater than the coefficient α_B of the bolt.)

(a) If the temperature of the entire assembly is raised by an amount ΔT, what stresses σ_S and σ_B are developed in the sleeve and bolt, respectively?

(b) What is the increase δ in the length L of the sleeve and bolt?

FIGURE 2-42

Example 2-10: Sleeve and bolt assembly with uniform temperature increase ΔT

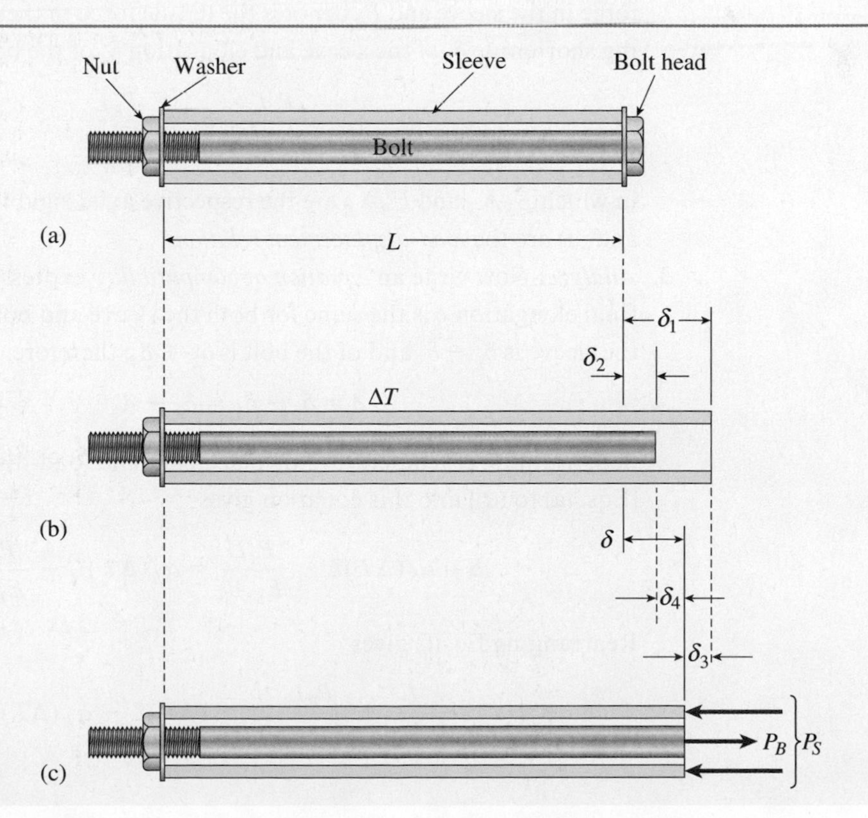

Solution:

A four-step problem-solving approach is used in the solution; only essential steps are shown in each part.

Part (a): Stresses in sleeve and bolt.

1. ***Conceptualize***: Because the sleeve and bolt are of different materials, they will elongate by different amounts when heated and allowed to expand freely. However, when they are held together by the assembly, free expansion cannot occur, and thermal stresses are developed in both materials. To find these stresses, use the same concepts as in any statically indeterminate analysis—equilibrium equations, compatibility equations, and displacement relations. However, you cannot formulate these equations until you disassemble the structure.
2. ***Categorize***: A simple way to cut the structure is to remove the head of the bolt, thereby allowing the sleeve and bolt to expand freely under the temperature change ΔT (Fig. 2-42b). The resulting elongations of the sleeve and bolt are denoted δ_1 and δ_2, respectively, and the corresponding *temperature-displacement relations* are

$$\delta_1 = \alpha_S(\Delta T)L \quad \delta_2 = \alpha_B(\Delta T)L \qquad \textbf{(a,b)}$$

Since α_S is greater than δ_B, the elongation δ_1 is greater than δ_2, as shown in Fig. 2-42b.

The axial forces in the sleeve and bolt must be such that they shorten the sleeve and stretch the bolt until the final lengths of the sleeve and bolt are the same. These forces are shown in Fig. 2-42c, where P_S denotes the compressive force in the sleeve and P_B denotes the tensile force in the bolt. The corresponding shortening δ_3 of the sleeve and elongation δ_4 of the bolt are

$$\delta_3 = \frac{P_S L}{E_S A_S} \quad \delta_4 = \frac{P_B L}{E_B A_B} \qquad \textbf{(c,d)}$$

in which $E_S A_S$ and $E_B A_B$ are the respective axial rigidities. Equations (c) and (d) are the *load-displacement relations*.

3. ***Analyze***: Now write an *equation of compatibility* expressing the fact that the final elongation δ is the same for both the sleeve and bolt. The elongation of the sleeve is $\delta_1 - \delta_3$ and of the bolt is $\delta_2 + \delta_4$; therefore,

$$\delta = \delta_1 - \delta_3 = \delta_2 + \delta_4 \qquad \textbf{(e)}$$

Substituting the temperature-displacement and load-displacement relations [Eqs. (a) to (d)] into this equation gives

$$\delta = \alpha_S(\Delta T)L - \frac{P_S L}{E_S A_S} = \alpha_B(\Delta T)L + \frac{P_B L}{E_B A_B} \qquad \textbf{(f)}$$

Rearranging Eq. (f) gives

$$\frac{P_S L}{E_S A_S} + \frac{P_B L}{E_B A_B} = \alpha_S(\Delta T)L - \alpha_B(\Delta T)L \qquad \textbf{(g)}$$

which is a modified form of the compatibility equation. Note that it contains the forces P_S and P_B as unknowns.

An *equation of equilibrium* is obtained from Fig. 2-42c, which is a free-body diagram of the part of the assembly remaining after the head of the bolt is removed. Summing forces in the horizontal direction gives

$$P_S = P_B \tag{h}$$

which expresses the obvious fact that the compressive force in the sleeve is equal to the tensile force in the bolt.

Now solve simultaneously Eqs. (g) and (h) and obtain the axial forces in the sleeve and bolt:

$$P_S = P_B = \frac{(\alpha_S - \alpha_B)(\Delta T)E_S A_S E_B A_B}{E_S A_S + E_B A_B} \tag{2-19}$$

4. ***Finalize***: This equation was derived assuming that the temperature increased and that the coefficient α_S was greater than the coefficient α_B. Under these conditions, P_S is the compressive force in the sleeve and P_B is the tensile force in the bolt.

The results will be quite different if the temperature increases, but the coefficient α_S is less than the coefficient α_B. Under these conditions, a gap will open between the bolt head and the sleeve, and there will be no stresses in either part of the assembly.

Stresses in the sleeve and bolt: Exprcssions for the stresses σ_S and σ_B in the sleeve and bolt, respectively, are obtained by dividing the corresponding forces by the appropriate areas:

$$\sigma_S = \frac{P_S}{A_S} = \frac{(\alpha_S - \alpha_B)(\Delta T)E_S E_B A_B}{E_S A_S + E_B A_B} \tag{2-20a}$$

$$\sigma_B = \frac{P_B}{A_B} = \frac{(\alpha_S - \alpha_B)(\Delta T)E_S A_S E_B}{E_S A_S + E_B A_B} \tag{2-20b}$$

Under the assumed conditions, the stress σ_S in the sleeve is compressive and the stress σ_B in the bolt is tensile. It is interesting to note that these stresses are independent of the length of the assembly and their magnitudes are inversely proportional to their respective areas (that is, $\sigma_S / \sigma_B = A_B / A_S$).

Part (b): Increase in length of the sleeve and bolt.

2, 3. ***Categorize, Analyze***: The elongation δ of the assembly can be found by substituting either P_S or P_B from Eq. (2-19) into Eq. (f), yielding

$$\delta = \frac{(\alpha_S E_S A_S + \alpha_B E_B A_B)(\Delta T)L}{E_S A_S + E_B A_B} \tag{2-21}$$

With the preceding formulas available, you can readily calculate the forces, stresses, and displacements of the assembly for any given set of numerical data.

4. ***Finalize***: As a partial check on the results, note that Eqs. (2-19), (2-20), and (2-21) reduce to known values in simplified cases. For instance, suppose that the bolt is rigid and therefore unaffected by temperature changes. You can represent this situation by setting $\alpha_B = 0$ and letting E_B become infinitely large, thereby creating an assembly in which the sleeve is held between rigid supports. Substituting these values into Eqs. (2-19), (2-20), and (2-21) gives

$$P_S = E_S A_S \alpha_S (\Delta T) \quad \sigma_S = E_S \alpha_S (\Delta T) \quad \delta = 0$$

These results agree with those of Example 2-9 for a bar held between rigid supports.

As a second special case, suppose that the sleeve and bolt are made of the same material. Then both parts will expand freely and will lengthen the same amount when the temperature changes. No forces or stresses will be developed. To see if the derived equations predict this behavior, substitute $\alpha_S = \alpha_B = \alpha$ into Eqs. (2-19), (2-20), and (2-21) and obtain

$$P_S = P_B = 0 \quad \sigma_S = \sigma_B = 0 \quad \delta = \alpha(\Delta T)L$$

which are the expected results.

Misfits and Prestrains

Suppose that a member of a structure is manufactured with its length slightly different from its prescribed length. Then the member will not fit into the structure in its intended manner, and the geometry of the structure will be different from what was planned. Situations of this kind are called **misfits**. Sometimes misfits are intentionally created in order to introduce strains into the structure at the time it is built. Because these strains exist before any loads are applied to the structure, they are called **prestrains**. Accompanying the prestrains are prestresses, and the structure is said to be **prestressed**. Common examples of prestressing are spokes in bicycle wheels (which would collapse if not prestressed), the pretensioned faces of tennis racquets, shrink-fitted machine parts, and prestressed concrete beams.

FIGURE 2-43

Statically determinate structure with a small misfit

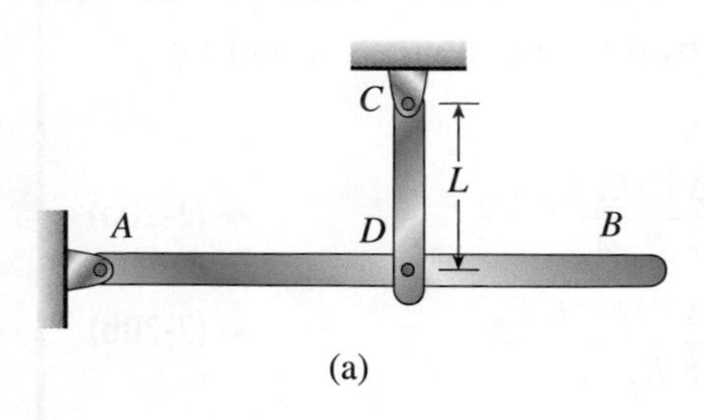

(a)

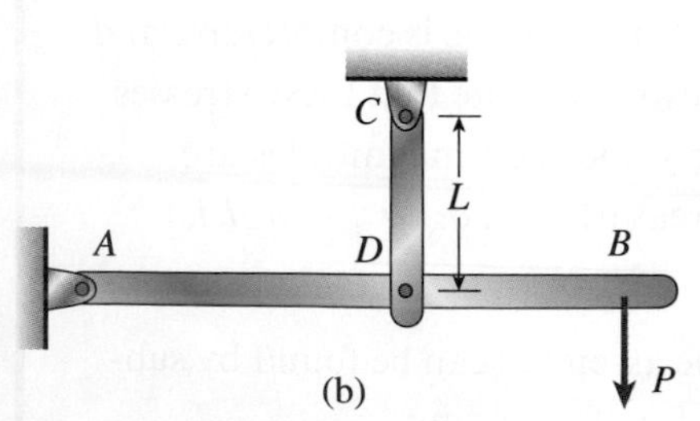

(b)

If a structure is **statically determinate**, small misfits in one or more members will not produce strains or stresses, although there will be departures from the theoretical configuration of the structure. To illustrate this statement, consider a simple structure consisting of a horizontal beam AB supported by a vertical bar CD (Fig. 2-43a). If bar CD has exactly the correct length L, the beam will be horizontal at the time the structure is built. However, if the bar is slightly longer than intended, the beam will make a small angle with the horizontal. Nevertheless, there will be no strains or stresses in either the bar or the beam attributable to the incorrect length of the bar. Furthermore, if a load P acts at the end of the beam (Fig. 2-43b), the stresses in the structure due to that load will be unaffected by the incorrect length of bar CD.

In general, if a structure is statically determinate, the presence of small misfits will produce small changes in geometry but no strains or stresses. Thus, the effects of a misfit are similar to those of a temperature change.

The situation is quite different if the structure is **statically indeterminate** because then the structure is not free to adjust to misfits (just as it is not free to adjust to certain kinds of temperature changes). To show this, consider a beam supported by two vertical bars (Fig. 2-44a). If both bars have exactly the correct length L, the structure can be assembled with no strains or stresses and the beam will be horizontal.

Suppose, however, that bar CD is slightly longer than the prescribed length. Then, in order to assemble the structure, bar CD must be compressed by external forces (or bar EF stretched by external forces), the bars must be fitted into place, and then the external forces must be released. As a result, the beam will deform and rotate, bar CD will be in compression, and bar EF will be in tension. In other words, prestrains will exist in all members and the structure will be prestressed, even though no external loads are acting. If a load P is now added (Fig. 2-44b), additional strains and stresses will be produced.

The analysis of a statically indeterminate structure with misfits and prestrains proceeds in the same general manner as described previously for loads and temperature changes. The basic ingredients of the analysis are equations of equilibrium, equations of compatibility, force-displacement relations, and (if appropriate) temperature-displacement relations. The methodology is illustrated in Example 2-12.

FIGURE 2-44

Statically indeterminate structure with a small misfit

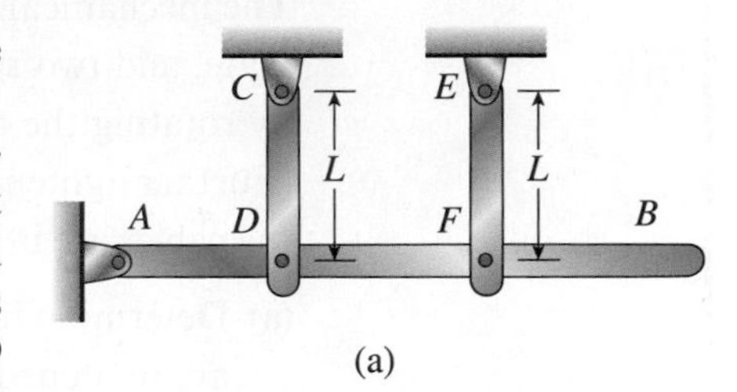

(a)

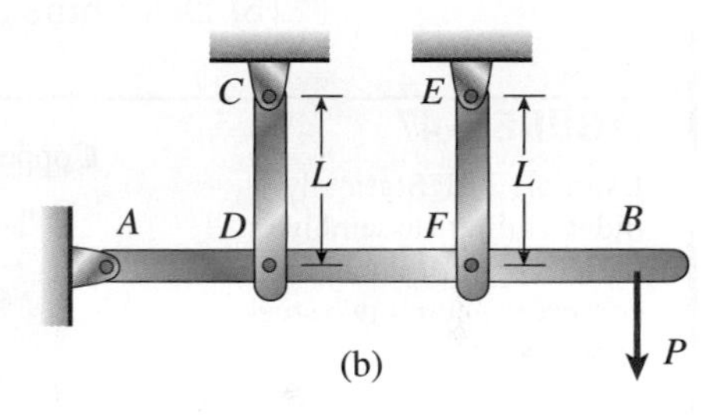

(b)

Bolts and Turnbuckles

Prestressing a structure requires that one or more parts of the structure be stretched or compressed from their theoretical lengths. A simple way to produce a change in length is to tighten a bolt or a turnbuckle. In the case of a **bolt** (Fig. 2-45) each turn of the nut will cause the nut to travel along the bolt a distance equal to the spacing p of the threads (called the *pitch* of the threads). Thus, the distance δ traveled by the nut is

$$\delta = np \qquad \textbf{(2-22)}$$

in which n is the number of revolutions of the nut (not necessarily an integer). Depending upon how the structure is arranged, turning the nut can stretch or compress a member.

FIGURE 2-45

The *pitch* of the threads is the distance from one thread to the next

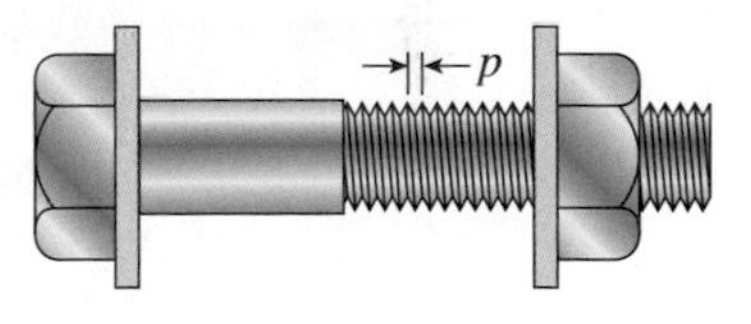

In the case of a **double-acting turnbuckle** (Fig. 2-46), there are two end screws. Because a right-hand thread is used at one end and a left-hand thread at the other, the device either lengthens or shortens when the buckle is rotated. Each full turn of the buckle causes it to travel a distance p along each screw, where again p is the pitch of the threads. Therefore, if the turnbuckle is tightened by one turn, the screws are drawn closer together by a distance $2p$, and the effect is to shorten the device by $2p$. For n turns, the distance is

$$\delta = 2np \qquad \textbf{(2-23)}$$

Turnbuckles are often inserted in cables and then tightened, thus creating initial tension in the cables, as illustrated in the following example.

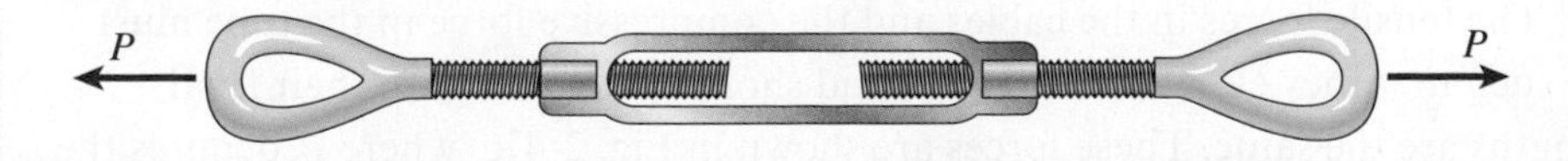

FIGURE 2-46

Double-acting turnbuckle. (Each full turn of the turnbuckle shortens or lengthens the cable by $2p$, where p is the pitch of the screw threads.)

Example 2-11

The mechanical assembly shown in Fig. 2-47a consists of a copper tube, a rigid end plate, and two steel cables with turnbuckles. The slack is removed from the cables by rotating the turnbuckles until the assembly is snug but with no initial stresses. (Further tightening of the turnbuckles will produce a prestressed condition in which the cables are in tension and the tube is in compression.)

(a) Determine the forces in the tube and cables (Fig. 2-47a) when the turnbuckles are tightened by n turns.

(b) Determine the shortening of the tube.

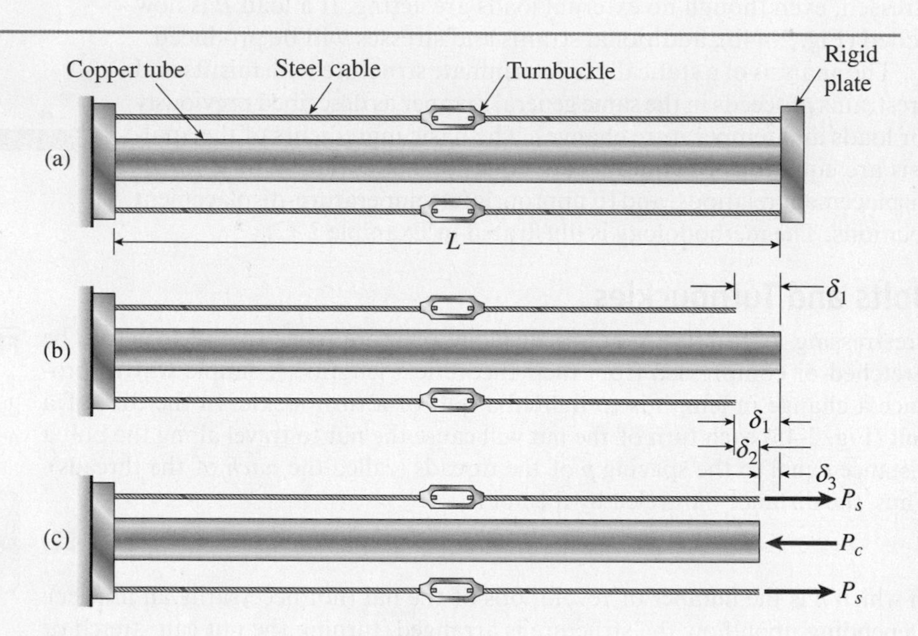

FIGURE 2-47

Example 2-11: Statically indeterminate assembly with a copper tube in compression and two steel cables in tension

Solution:

A four-step problem-solving approach is used in the solution; only essential steps are shown in each part.

Part (a): Forces in the cables and tube.

1. ***Conceptualize***: Begin the analysis by removing the plate at the right-hand end of the assembly so that the tube and cables are free to change in length (Fig. 2-47b). Rotating the turnbuckles through n turns will shorten the cables by a distance

$$\delta_1 = 2np \qquad \text{(a)}$$

as shown in Fig. 2-47b.

The tensile forces in the cables and the compressive force in the tube must be such that they elongate the cables and shorten the tube until their final lengths are the same. These forces are shown in Fig. 2-47c, where P_s denotes the tensile force in one of the steel cables and P_c denotes the compressive force in the copper tube. The elongation of a cable due to the force P_s is

$$\delta_2 = \frac{P_s L}{E_s A_s} \tag{b}$$

in which $E_s A_s$ is the axial rigidity and L is the length of a cable. Also, the compressive force P_c in the copper tube causes it to shorten by

$$\delta_3 = \frac{P_c L}{E_c A_c} \tag{c}$$

in which $E_c A_c$ is the axial rigidity of the tube. Equations (b) and (c) are the *load-displacement relations.*

2. ***Categorize*:** The final shortening of one of the cables is equal to the shortening δ_1 caused by rotating the turnbuckle minus the elongation δ_2 caused by the force P_s. This final shortening of the cable must equal the shortening δ_3 of the tube:

$$\delta_1 - \delta_2 = \delta_3 \tag{d}$$

which is the *equation of compatibility.*

Substituting the turnbuckle relation [Eq. (a)] and the load-displacement relations [Eqs. (b) and (c)] into the preceding equation yields

$$2np - \frac{P_s L}{E_s A_s} = \frac{P_c L}{E_c A_c} \tag{e}$$

or

$$\frac{P_s L}{E_s A_s} + \frac{P_c L}{E_c A_c} = 2np \tag{f}$$

which is a modified form of the compatibility equation. Note that it contains P_s and P_c as unknowns.

Using Fig. 2-47c, which is a free-body diagram of the assembly with the end plate removed, write the following equation of equilibrium:

$$2P_s = P_c \tag{g}$$

3. ***Analyze*: Forces in the cables and tube:** Now solve simultaneously Eqs. (f) and (g) and obtain the axial forces in the steel cables and copper tube, respectively:

$$P_s = \frac{2npE_c A_c E_s A_s}{L(E_c A_c + 2E_s A_s)} \qquad P_c = \frac{4npE_c A_c E_s A_s}{L(E_c A_c + 2E_s A_s)} \tag{2-24a,b}$$

Recall that the forces P_s are tensile forces and the force P_c is compressive. If desired, the stresses σ_s and σ_c in the steel and copper now can be obtained by dividing the forces P_s and P_c by the cross-sectional areas A_s and A_c, respectively.

Part (b): Shortening of the tube.

2, 3. ***Categorize, Analyze***: The decrease in length of the tube is the quantity δ_3 [see (Fig. 2-47) and Eq. (c)]:

$$\delta_3 = \frac{P_c L}{E_c A_c} = \frac{4npE_s A_s}{E_c A_c + 2E_s A_s} \tag{2-25}$$

4. ***Finalize***: With the preceding formulas available, you can readily calculate the forces, stresses, and displacements of the assembly for any given set of numerical data.

Example 2-12

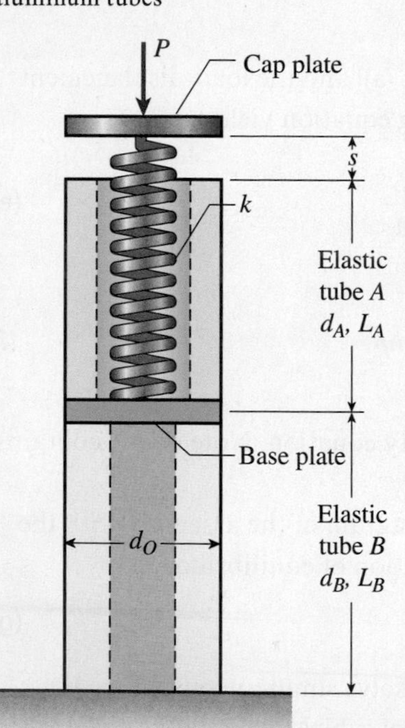

FIGURE 2-48
Elastic spring inside aluminum tubes

An elastic spring is enclosed by tube A but is longer than tube A by an amount s. The spring rests on a rigid base plate, and the base plate rests on top of tube B (see Fig. 2-48). Force P is applied to a rigid cap plate to compress the spring and the tubes. Tubes A and B are made of an aluminum alloy and have an outer diameter d_O and inner diameters d_A and d_B, respectively. If applied load $P = 140$ kN, find the axial normal compressive stresses in tubes A and B and plot the axial force and displacement diagrams for the tube structure. Assume that $E = 72$ GPa, spring stiffness $k = 200$ MN/m, $d_O = 76$ mm, $d_A = 62$ mm, $d_B = 58$ mm, and $s = 0.5$ mm. The lengths of tubes A and B are $L_A = 0.35$ m and $L_B = 0.4$ m.

Solution:

Use a four-step problem-solving approach.

1. ***Conceptualize***: The elastic spring does not fit in tube A, so force P acts to compress the spring to close the gap; the force in tube A is zero until the gap is closed. In Fig. 2-49a, an upper section cut through the spring and also through tube A creates an upper free-body diagram that shows the spring force $N_s = (k)(x) = P$ for downward cap displacement x. The internal force N_A in tube A is zero until the cap plate rests on tube A when the gap is closed. In Fig. 2-49b, force P now compresses both the spring and tube A when cap displacement x exceeds gap s. Figure 2-49 also gives the lower free-body diagram with a section cut through tube B and with the internal compressive force N_B in tube B equal to the applied load P. (Internal forces in tubes A and B are shown as two arrows, one at each tube

FIGURE 2-49

(a, b) Free-body diagrams ($x \le s$, $x > s$)

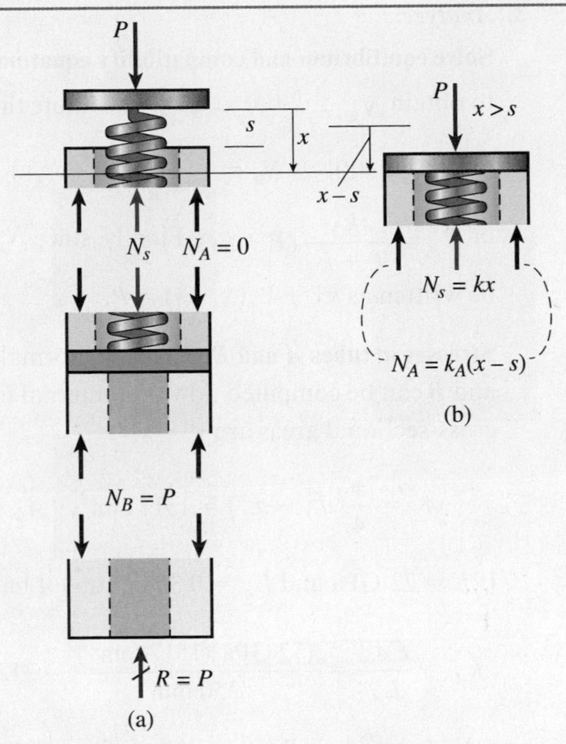

wall, indicating that N_A and N_B are actually uniformly distributed forces acting on the circular cross section of each tube.)

2. *Categorize*: If force P is large enough to close gap s, the structure becomes statically indeterminate; part of force P is carried by the spring and part by tube A, as shown in Fig. 2-49b. *Equilibrium* and *compatibility* equations are needed to determine the force in the spring N_s and internal force N_A in tube A. The force P required to close gap s is $(k)(s)$. After that, the spring and tube A deform together in parallel.

Equilibrium: Summing axial forces in the x direction in the lower free-body diagram in Fig. 2-49a gives the compressive force in tube B as $N_B = P$. Summing forces in the upper free-body diagram in Fig. 2-49b leads to

$$N_A + N_s = P \qquad \textbf{(a)}$$

This one equation has two unknowns, so the structure is *one degree statically indeterminate*.

Compatibility: Once the gap is closed by compression of the spring by force P, tube A and the spring begin to deform together. The compatibility equation is

$$\frac{N_A}{k_A} = \frac{N_s}{k} - s \qquad \textbf{(b)}$$

where the axial stiffness of tube A is $k_A = \dfrac{EA_A}{L_A}$ with A_A as the cross-sectional area of tube A.

3. ***Analyze***:

Solve equilibrium and compatibility equations: Solve for N_A in Eq. (b) to obtain $N_A = \frac{k_A}{k}N_s - k_A s$. Substitute this expression into Eq. (a) to find that the spring force is $N_s = \frac{k}{k + k_A}(P + k_A s)$. Then $N_A = P - N_s$ or $N_A = \frac{k_A}{k + k_A}(P - ks)$. Finally, since $N_s = kx$, equilibrium Eq. (a) also can be written as $kx + k_A(x - s) = P$.

Stresses in tubes *A* and *B*: The axial normal *compressive* stresses in tubes A and B can be computed now that internal forces N_A and N_B are known. Tube cross-sectional areas are

$$A_A = \frac{\pi}{4}\left(d_O^2 - d_A^2\right) = 1517\ \text{mm}^2, \quad A_B = \frac{\pi}{4}\left(d_O^2 - d_B^2\right) = 1894\ \text{mm}^2$$

If $E = 72$ GPa and $L_A = 0.35$ m, tube A has axial stiffness:

$$k_A = \frac{EA_A}{L_A} = \frac{(72\ \text{GPa})(1517\ \text{mm}^2)}{350\ \text{mm}} = 312\frac{\text{MN}}{\text{m}} \quad \text{and}$$

$$N_A = \frac{k_A}{k + k_A}(P - ks) = \frac{312}{200 + 312}\left(140\ \text{kN} - 200\frac{\text{MN}}{\text{m}}(0.5\ \text{mm})\right) = 24.4\ \text{kN}$$

The resulting stresses in tubes A and B are

$$\sigma_A = \frac{N_A}{A_A} = \frac{24.4\ \text{kN}}{1517\ \text{mm}^2} = 16.1\ \text{MPa} \text{ and } \sigma_B = \frac{N_B}{A_B} = \frac{140\ \text{kN}}{1894\ \text{mm}^2} = 73.9\ \text{MPa}$$

Axial force and displacement diagrams: From equilibrium, the force in the spring is $N_s = P - N_A = 140\ \text{kN} - 24.4\ \text{kN} = 115.6\ \text{kN}$, so the total compression of the spring is $x = \frac{N_s}{k} = \frac{115.6\ \text{kN}}{200\frac{\text{MN}}{\text{m}}} = 0.578$ mm. This confirms that force P is large enough to close gap s and compress tube A an amount equal to $x - s = 0.078$ mm.

The self-weight of the tubes is small, so it is not included in this analysis. Hence, forces N_A and N_B are taken as constant over, the height of each tube, as shown in the axial force diagram (AFD) in Fig. 2-50a. The AFD is composed of constant force segments, so the ADD (Fig. 2-50b) varies linearly over the height of the tube structure. The change in axial displacement between the base and the top of tube B is equal to the area under the AFD over distance L_B divided by EA_B: $\frac{(140\ \text{kN})L_B}{EA_B} = 0.411$ mm. Add to this the compression $(x - s)$ of tube A to find the total downward displacement 0.489 mm at the top.

FIGURE 2-50

(a, b) Axial force and displacement diagrams for tubes A and B

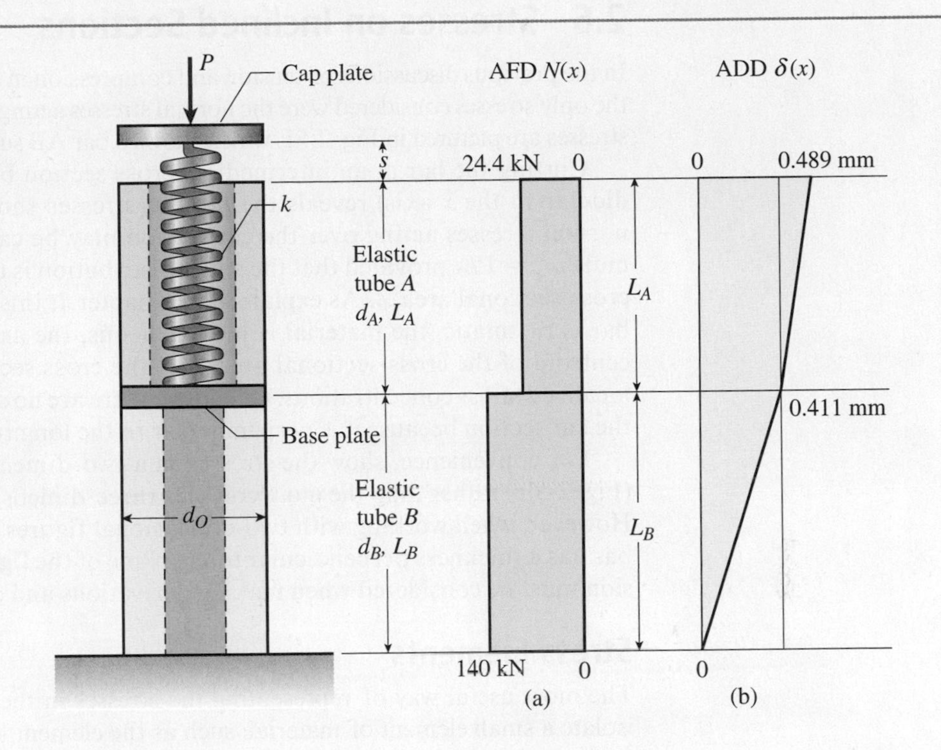

4. ***Finalize***: Once the gap s is closed, tube A and spring k become *springs in parallel*, and each carries part of force P in proportion to its axial stiffness (k or k_A) in relation to the total axial stiffness of the spring and tube A together ($k + k_A$). Here, as the gap closes, the spring alone carries $(k)(s) = 100$ kN; after that the remaining force $(P - ks)$ is distributed to tube A and the spring based on the ratios of axial stiffness: $\frac{k_A}{k + k_A}$ for tube A and $\frac{k}{k + k_A}$ for the elastic spring.

In the end, the spring carries $N_s = 100\text{ kN} + 15.6\text{ kN} = 115.6\text{ kN}$ (83% of applied load P) and tube A carries $N_A = 24.4$ kN (17% of applied load P).

Recall that Example 1-5 used a model of this structure in which tubes A and B are assumed to be *rigid*. The assumption of rigid tubes leads to a simpler analysis procedure and may be appropriate if spring stiffness k is much smaller that the axial stiffness of tube A; this is not true in this example, in which $k/k_A = 0.64$. In Example 1-5, the elastic spring is almost 50 times more flexible than that used here.

If tube A is assumed to be *rigid* in this example, the spring force is $N_s = (k)(s) = 100$ kN, which is 13.5% lower than computed previously. At the same time, the force in tube A is $N_A = P - N_s = (140 - 100)\text{ kN} = 40$ kN, which is approximately 64% larger than that obtained in the statically indeterminate analysis, leading to $N_A = 24.4$ kN. The assumption of rigid tubes and elastic spring in this example leads to erroneous results.

2.6 Stresses on Inclined Sections

In the previous discussions of tension and compression in axially loaded members, the only stresses considered were the normal stresses acting on cross sections. These stresses are pictured in Fig. 2-51, which shows a bar AB subjected to axial loads P.

Cutting the bar at an intermediate cross section by a plane mn (perpendicular to the x axis) reveals the normal stresses shown in Fig. 2-51b. The normal stresses acting over the cut section may be calculated from the formula $\sigma_x = P/A$ provided that the stress distribution is uniform over the entire cross-sectional area A. As explained in Chapter 1, this condition exists if the bar is prismatic, the material is homogeneous, the axial force P acts at the centroid of the cross-sectional area, and the cross section is away from any localized stress concentrations. Of course, there are no shear stresses acting on the cut section because it is perpendicular to the longitudinal axis of the bar.

For convenience, show the stresses in a two-dimensional view of the bar (Fig. 2-51c) rather than the more complex three-dimensional view (Fig. 2-51b). However, when working with two-dimensional figures, do not forget that the bar has a thickness perpendicular to the plane of the figure. This third dimension must be considered when making derivations and calculations.

Stress Elements

The most useful way of representing the stresses in the bar of Fig. 2-51 is to isolate a small element of material, such as the element labeled C in Fig. 2-51c, and then show the stresses acting on all faces of this element. An element of this

FIGURE 2-51

Prismatic bar in tension showing the stresses acting on cross section mn: (a) bar with axial forces P, (b) three-dimensional view of the cut bar showing the normal stresses, and (c) two-dimensional view

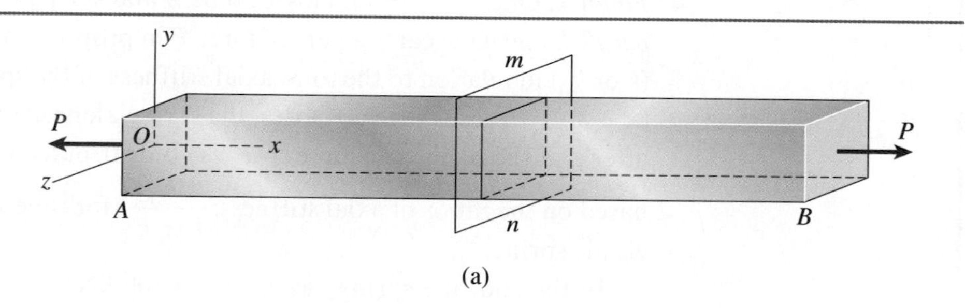

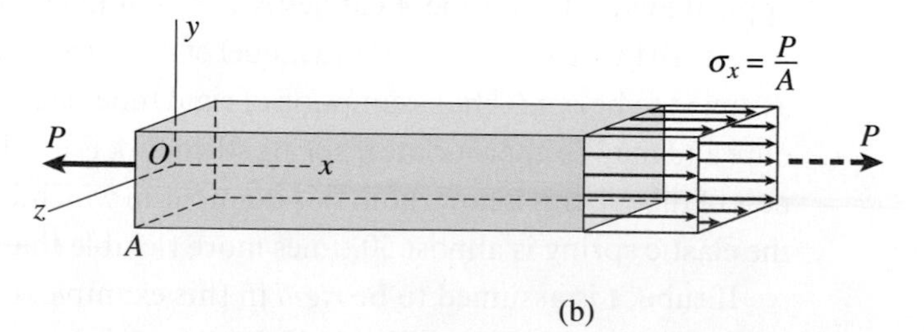

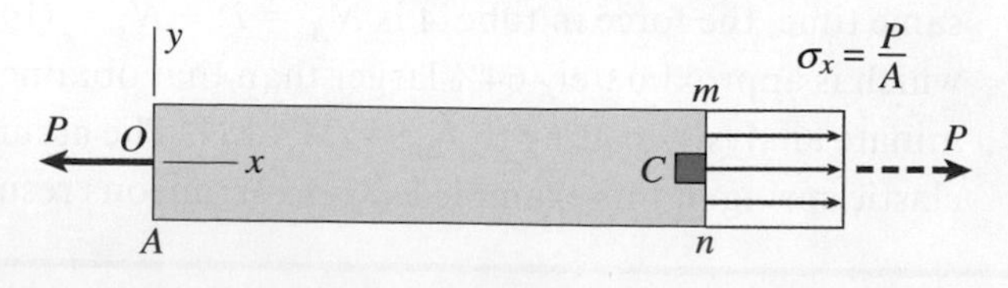

kind is called a **stress element**. The stress element at point C is a small rectangular block (it doesn't matter whether it is a cube or a rectangular parallelepiped) with its right-hand face lying in cross section mn.

The dimensions of a stress element are assumed to be infinitesimally small, but for clarity, draw the element to a large scale, as in Fig. 2-52a. In this case, the edges of the element are parallel to the x, y, and z axes, and the only stresses are the normal stresses σ_x acting on the x faces (recall that the x faces have their normals parallel to the x axis). Because it is more convenient, draw a two-dimensional view of the element (Fig. 2-52b) instead of a three-dimensional view.

Stresses on Inclined Sections

The stress element of Fig. 2-52 provides only a limited view of the stresses in an axially loaded bar. To obtain a more complete picture, you need to investigate the stresses acting on **inclined sections**, such as the section cut by the inclined plane pq in Fig. 2-53a. Because the stresses are the same throughout the entire bar, the stresses acting over the inclined section must be uniformly distributed, as pictured in the free-body diagrams of Fig. 2-53b (three-dimensional view) and Fig. 2-53c (two-dimensional view). From the equilibrium of the free body, the resultant of the stresses must be a horizontal force P. (The resultant is drawn with a dashed line in Figs. 2-53b and c.)

As a preliminary matter, the **orientation** of the inclined section pq must be specified. A standard method is to specify the angle θ between the x axis and the normal n to the section (see Fig. 2-54a). Thus, the angle θ for the inclined section shown in the figure is approximately 30°. By contrast, cross section mn (Fig. 2-51a) has an angle θ equal to zero (because the normal to the section is the x axis). For additional examples, consider the stress element of Fig. 2-52. The angle θ for the right-hand face is 0, for the top face is 90° (a longitudinal section of the bar), for the left-hand face is 180°, and for the bottom face is 270° (or −90°).

Now return to the task of finding the stresses acting on section pq (Fig. 2-54b). As already mentioned, the resultant of these stresses is a force P acting in the x direction. This resultant may be resolved into two components, a normal force N that is perpendicular to the inclined plane pq and a shear force V that is tangential to it. These force components are

$$N = P\cos\theta \quad V = P\sin\theta \qquad \text{(2-26a,b)}$$

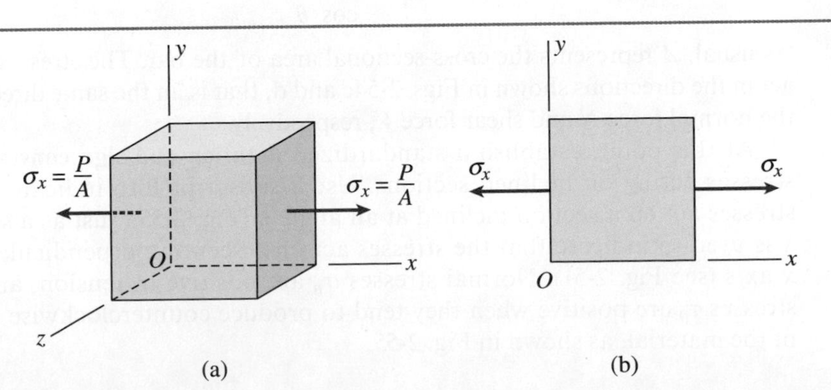

FIGURE 2-52

Stress element at point C of the axially loaded bar shown in Fig. 2-51c: (a) three-dimensional view of the element and (b) two-dimensional view of the element

FIGURE 2-53

Prismatic bar in tension showing the stresses acting on an inclined section pq: (a) bar with axial forces P, (b) three-dimensional view of the cut bar showing the stresses, and (c) two-dimensional view

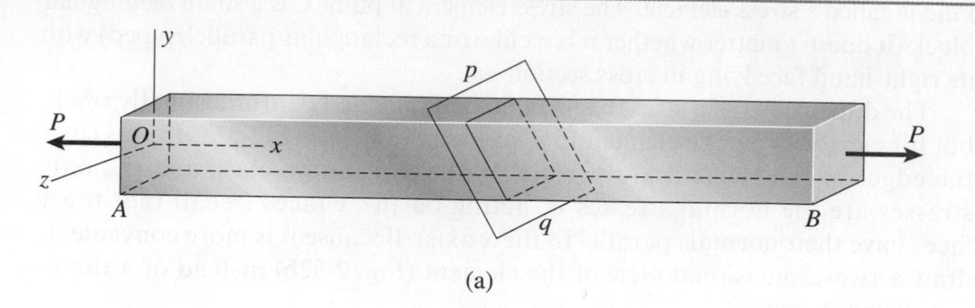

(a)

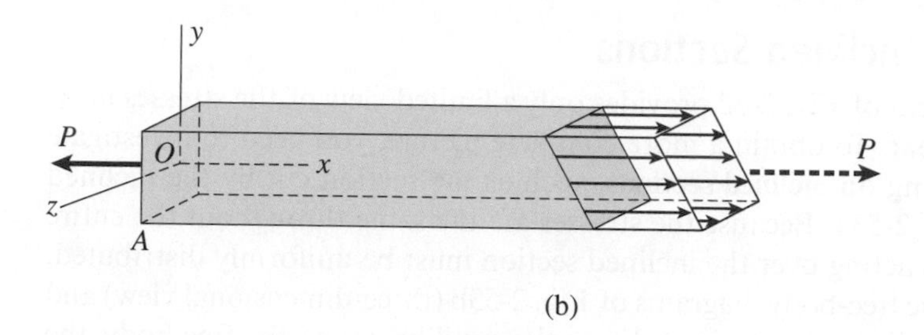

(b)

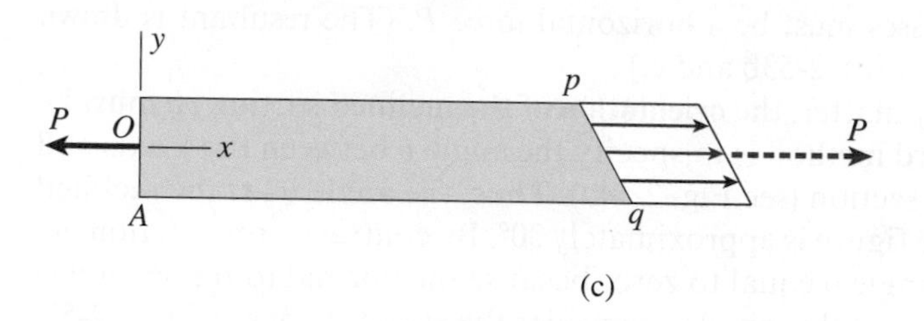

(c)

Associated with the forces N and V are normal and shear stresses that are uniformly distributed over the inclined section (Figs. 2-54c and d). The normal stress is equal to the normal force N divided by the area of the section, and the shear stress is equal to the shear force V divided by the area of the section. Thus, the stresses are

$$\sigma = \frac{N}{A_1} \qquad \tau = \frac{V}{A_1} \tag{2-27a,b}$$

in which A_1 is the area of the inclined section, as

$$A_1 = \frac{A}{\cos\theta} \tag{2-28}$$

As usual, A represents the cross-sectional area of the bar. The stresses σ and τ act in the directions shown in Figs. 2-54c and d, that is, in the same directions as the normal force N and shear force V, respectively.

At this point, establish a standardized **notation and sign convention** for stresses acting on inclined sections. Use a subscript θ to indicate that the stresses act on a section inclined at an angle θ (Fig. 2-55), just as a subscript x is used to indicate that the stresses act on a section perpendicular to the x axis (see Fig. 2-51). Normal stresses σ_θ are positive in tension, and shear stresses τ_θ are positive when they tend to produce counterclockwise rotation of the material, as shown in Fig. 2-55.

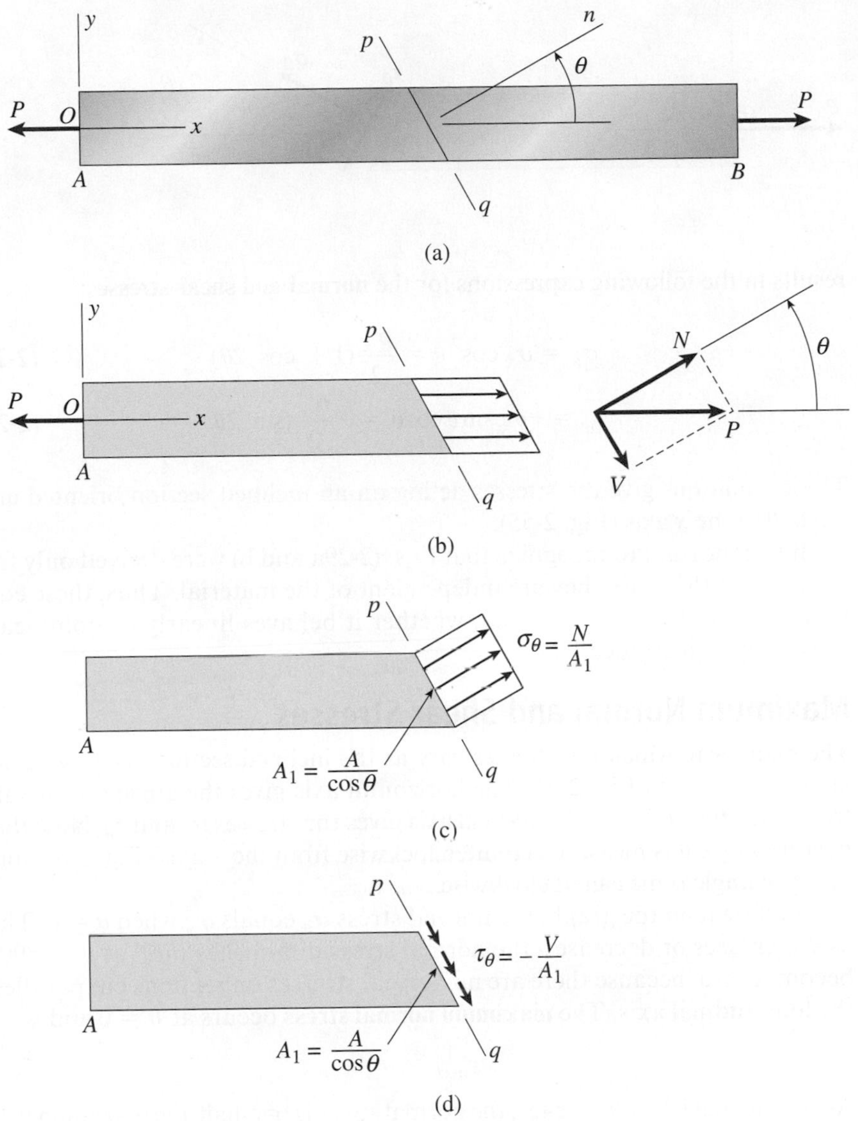

FIGURE 2-54

Prismatic bar in tension showing the stresses acting on an inclined section *pq*

For a bar in tension, the normal force N produces positive normal stresses σ_θ (see Figs. 2-54c), and the shear force V produces negative shear stresses τ_θ (see Fig. 2-54d). These stresses are given by [see Eqs. (2-26), (2-27), and (2-28)]

$$\sigma_\theta = \frac{N}{A_1} = \frac{P}{A}\cos^2\theta \quad \tau_\theta = -\frac{V}{A_1} = -\frac{P}{A}\sin\theta\cos\theta$$

Introducing the notation $\sigma_x = P/A$, in which σ_x is the normal stress on a cross section, and also using the trigonometric relations

$$\cos^2\theta = \frac{1}{2}(1 + \cos 2\theta) \quad \sin\theta\cos\theta = \frac{1}{2}(\sin 2\theta)$$

FIGURE 2-55

Sign convention for stresses acting on an inclined section (Normal stresses are positive when in tension, and shear stresses are positive when they tend to produce counterclockwise rotation)

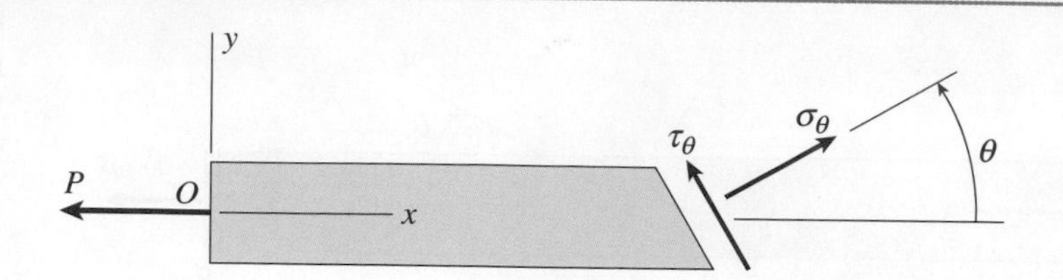

results in the following expressions for the **normal and shear stresses**:

$$\sigma_\theta = \sigma_x \cos^2\theta = \frac{\sigma_x}{2}(1 + \cos 2\theta) \tag{2-29a}$$

$$\tau_\theta = -\sigma_x \sin\theta\cos\theta = -\frac{\sigma_x}{2}(\sin 2\theta) \tag{2-29b}$$

These equations give the stresses acting on an inclined section oriented at an angle θ to the x axis (Fig. 2-55).

It is important to recognize that Eqs. (2-29a and b) were derived only from statics, and therefore they are independent of the material. Thus, these equations are valid for any material, whether it behaves linearly or nonlinearly, elastically or inelastically.

Maximum Normal and Shear Stresses

The manner in which the stresses vary as the inclined section is cut at various angles is shown in Fig. 2-56. The horizontal axis gives the angle θ as it varies from $-90°$ to $+90°$, and the vertical axis gives the stresses σ_θ and τ_θ. Note that a positive angle θ is measured counterclockwise from the x axis (Fig. 2-55) and a negative angle is measured clockwise.

As shown on the graph, the normal stress σ_θ equals σ_x when $\theta = 0$. Then, as θ increases or decreases, the normal stress diminishes until at $\theta = \pm 90°$ it becomes zero, because there are no normal stresses on sections cut parallel to the longitudinal axis. The **maximum normal stress** occurs at $\theta = 0$ and is

$$\sigma_{max} = \sigma_x \tag{2-30}$$

Also, note that when $\theta = \pm 45°$, the normal stress is one-half the maximum value.

The shear stress τ_θ is zero on cross sections of the bar ($\theta = 0$) as well as on longitudinal sections ($\theta = \pm 90°$). Between these extremes, the stress varies as shown on the graph, reaching the largest positive value when $\theta = -45°$ and the largest negative value when $\theta = +45°$. These **maximum shear stresses** have the same magnitude:

$$\tau_{max} = \frac{\sigma_x}{2} \tag{2-31}$$

but they tend to rotate the element in opposite directions.

The maximum stresses in a **bar in tension** are shown in Fig. 2-57. Two stress elements are selected—element A is oriented at $\theta = 0°$ and element B is oriented at $\theta = 45°$. Element A has the maximum normal stresses [Eq. (2-30)] and element B has the maximum shear stresses [Eq. (2-31)]. In the case of element A (Fig. 2-57b), the only stresses are the maximum normal stresses (no shear stresses exist on any of the faces).

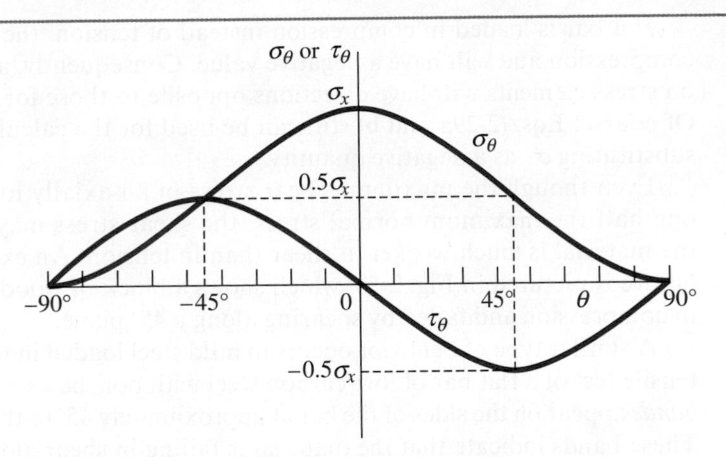

FIGURE 2-56

Graph of normal stress σ_θ and shear stress τ_θ versus angle θ of the inclined section [see Fig. 2-55 and Eqs. (2-29a and b)]

In the case of element B (Fig. 2-57c), both normal and shear stresses act on all faces (except, of course, the front and rear faces of the element). Consider, for instance, the face at 45° (the upper right-hand face). On this face, the normal and shear stresses [from Eqs. (2-29a and b)] are $\sigma_x/2$ and $-\sigma_x/2$, respectively. Hence, the normal stress is tension (positive) and the shear stress acts clockwise (negative) against the element. The stresses on the remaining faces are obtained in a similar manner by substituting $\theta = 135°, -45°$, and $-135°$ into Eqs. (2-29a and 29b).

Thus, in this special case of an element oriented at $\theta = 45°$, the normal stresses on all four faces are the same (equal to $\sigma_x/2$), and all four shear stresses have the maximum magnitude (equal to $\sigma_x/2$). Also, note that the shear stresses acting on perpendicular planes are equal in magnitude and have directions either toward, or away from, the line of intersection of the planes, as discussed in detail in Section 1.8.

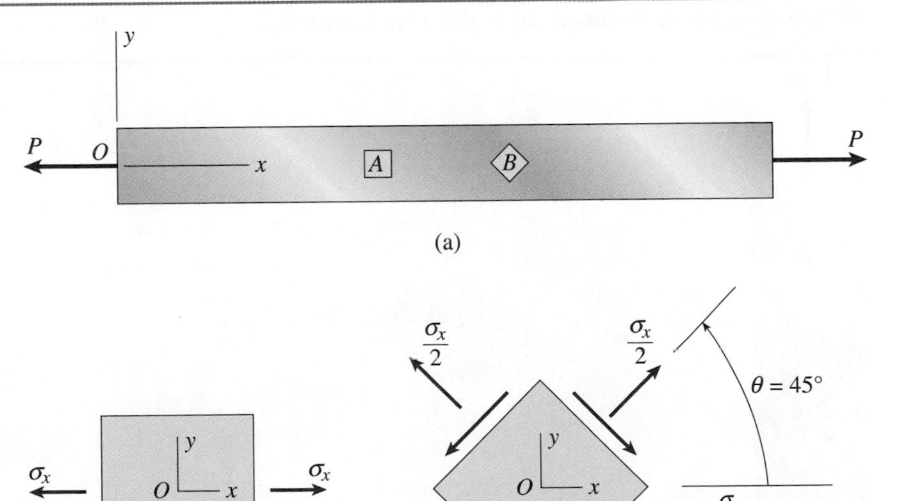

FIGURE 2-57

Normal and shear stresses acting on stress elements oriented at $\theta = 0°$ and $\theta = 45°$ for a bar in tension

FIGURE 2-58

Shear failure along a 45° plane of a wood block loaded in compression (©Jim Gere)

If a bar is loaded in compression instead of tension, the stress σ_x will be compression and will have a negative value. Consequently, all stresses acting on stress elements will have directions opposite to those for a bar in tension. Of course, Eqs. (2-29a and b) still can be used for the calculations simply by substituting σ_x as a negative quantity.

Even though the maximum shear stress in an axially loaded bar is only one-half the maximum normal stress, the shear stress may cause failure if the material is much weaker in shear than in tension. An example of a shear failure is pictured in Fig. 2-58, which shows a block of wood that was loaded in compression and failed by shearing along a 45° plane.

A similar type of behavior occurs in mild steel loaded in tension. During a tensile test of a flat bar of low-carbon steel with polished surfaces, visible *slip bands* appear on the sides of the bar at approximately 45° to the axis (Fig. 2-59). These bands indicate that the material is failing in shear along the planes on which the shear stress is maximum. Such bands were first observed by G. Piobert in 1842 and W. Lüders in 1860 (see Refs. 2-5 and 2-6), and today they are called either *Lüders' bands* or *Piobert's bands*. They begin to appear when the yield stress is reached in the bar (point B in Fig. 1-31 of Section 1.5).

Uniaxial Stress

The state of stress described throughout this section is called **uniaxial stress**, for the obvious reason that the bar is subjected to simple tension or compression in just one direction. The most important orientations of stress elements for uniaxial stress are $\theta = 0$ and $\theta = 45°$ (Figs. 2-57b and c); the former has the maximum normal stress and the latter has the maximum shear stress. If sections are cut through the bar at other angles, the stresses acting on the faces of the corresponding stress elements can be determined from Eqs. (2-29a and b), as illustrated in Examples 2-13 and 2-14.

Uniaxial stress is a special case of a more general stress state known as *plane stress*, which is described in detail in Chapter 7.

FIGURE 2-59

Slip bands (or Lüder's bands) in a polished steel specimen loaded in tension (©Jim Gere)

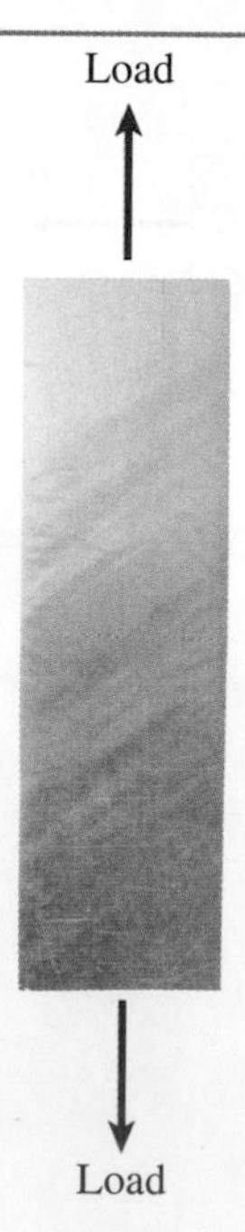

Example 2-13

A compression bar having a square cross section of width b must support a load $P = 35$ kN (Fig. 2-60a). The bar is constructed from two pieces of material that are connected by a glued joint (known as a *scarf joint*) along plane pq, which is at an angle $\alpha = 40°$ to the vertical. The material is a structural plastic for which the allowable stresses in compression and shear are 7.5 MPa and 4.0 MPa, respectively. Also, the allowable stresses in the glued joint are 5.2 MPa in compression and 3.4 MPa in shear.

Determine the minimum width b of the bar.

Solution:

Use a four-step problem-solving approach in the solution.

1. ***Conceptualize***: For convenience, rotate a segment of the bar to a horizontal position (Fig. 2-60b) that matches the figures used in deriving the equations for the stresses on an inclined section (see Figs. 2-54 and 2-55). With the bar in this position, observe that the normal n to the plane of the glued joint (plane pq) makes an angle $\beta = 90° - \alpha$, or 50°, with the axis of the bar. Since the angle θ is defined as positive when counterclockwise (Fig. 2-55), angle $\theta = -50°$ for the glued joint.

   The cross-sectional area of the bar is related to the load P and the stress σ_x acting on the cross sections by

$$A = \frac{P}{\sigma_x} \tag{a}$$

2. ***Categorize***: To find the required area, determine the value of σ_x corresponding to each of the four allowable stresses. Then the smallest value of σ_x will determine the required area. The values of σ_x are obtained by rearranging Eqs. (2-29a and b) as

$$\sigma_x = \frac{\sigma_\theta}{\cos^2\theta} \qquad \sigma_x = -\frac{\tau_\theta}{\sin\theta\cos\theta} \tag{2-32a,b}$$

   Now apply these equations to the glued joint and to the plastic.

3. ***Analyze***:

Part (a): Values of σ_x based upon the allowable stresses in the glued joint.

For compression in the glued joint, $\sigma_\theta = -5.2$ MPa and $\theta = -50°$. Substitute into Eq. (2-32a) to get

$$\sigma_x = \frac{-5.2\text{ MPa}}{(\cos(-50°))^2} = -12.6\text{ MPa} \tag{b}$$

For shear in the glued joint, the allowable stress is 3.4 MPa. However, it is not immediately evident whether τ_θ is +3.4 MPa or −3.4 MPa. One approach is to substitute both +3.4 MPa and −3.4 MPa into Eq. (2-32b) and then select the value of σ_x that is negative. The other value of σ_x will be positive (tension) and does not apply to this bar. Another approach is to inspect the bar itself (Fig. 2-60b) and observe from the directions of the loads that the shear stress will

FIGURE 2-60

Example 2-13: Stresses on an inclined section

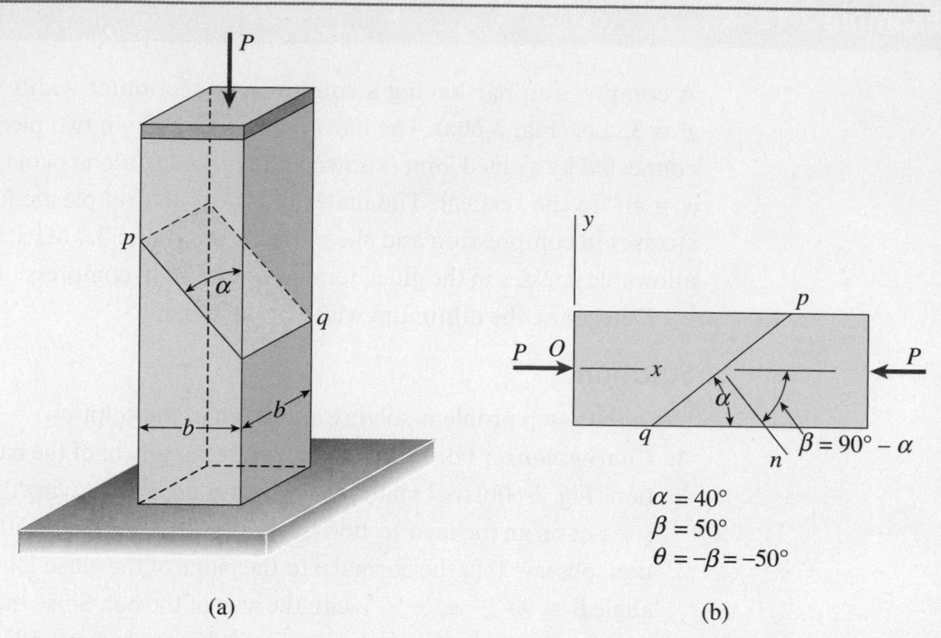

act clockwise against plane pq, which means that the shear stress is negative. Therefore, substitute $\tau_\theta = -3.4$ MPa and $\theta = -50°$ into Eq. (2-32b) and obtain

$$\sigma_x = -\frac{-3.4 \text{ MPa}}{(\sin(-50°))(\cos(-50°))} = -6.9 \text{ MPa} \quad \textbf{(c)}$$

Part (b): Values of σ_x based upon the allowable stresses in the plastic.

The maximum compressive stress in the plastic occurs on a cross section. Therefore, since the allowable stress in compression is 7.5 MPa, you know immediately that

$$\sigma_x = -7.5 \text{ MPa} \quad \textbf{(d)}$$

The maximum shear stress occurs on a plane at 45° and is numerically equal to $\sigma_x/2$ [see Eq. (2-31)]. Since the allowable stress in shear is 4 MPa, it follows that

$$\sigma_x = -8 \text{ MPa} \quad \textbf{(e)}$$

This same result can be obtained from Eq. (2-32b) by substituting $\tau_\theta = 4$ MPa and $\theta = 45°$.

Part (c): Minimum width of the bar.

Compare the four values of σ_x [Eqs. (b), (c), (d), and (e)] and note that the smallest is $\sigma_x = -6.9$ MPa. Therefore, this value governs the design. Substitute into Eq. (a) and use only numerical values to obtain the required area:

$$A = \frac{35 \text{ kN}}{6.9 \text{ MPa}} = 5072 \text{ mm}^2$$

4. *Finalize*: Since the bar has a square cross section ($A = b^2$), the minimum width is

$$b_{\min} = \sqrt{A} = \sqrt{5072 \text{ mm}^2} = 71.2 \text{ mm}$$

Any width larger than $b_{\min}$ will ensure that the allowable stresses are not exceeded.

Example 2-14

A prismatic brass bar with a length of $L = 0.5$ m and a cross-sectional area $A = 1200$ mm² is compressed by an axial load $P = 90$ kN (Fig. 2-61a).

(a) Determine the complete state of stress acting on an inclined section *pq* cut through the bar at an angle $\theta = 25°$, and show the stresses on a properly oriented stress element.

(b) If the bar is now fixed between supports *A* and *B* (Fig. 2-61b) and then subjected to a temperature increase of $\Delta T = 33°C$, the compressive stress on plane *rs* is known to be 65 MPa. Find the shear stress τ_θ on plane *rs*. What is angle θ? (Assume the modulus of elasticity is $E = 110$ GPa and coefficient of thermal expansion is $\alpha = 20 \times 10^{-6}/°C$.)

(c) If the allowable normal stress is ± 82 MPa and the allowable shear stress is ± 40 MPa, find the maximum permissible temperature *increase* (ΔT) in the bar if allowable stress values in the bar are not to be exceeded.

FIGURE 2-61

Example 2-14: (a) Stresses on inclined section *pq* through bar and (b) stresses on inclined section *rs* through bar

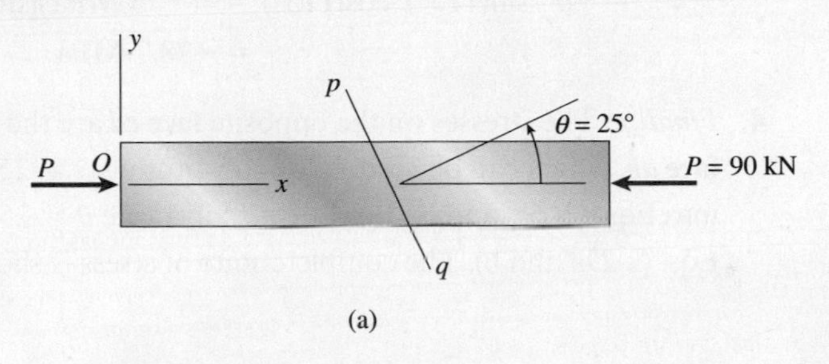

(a)

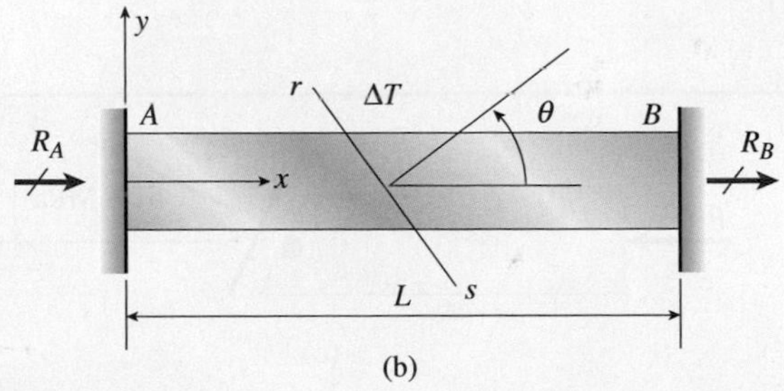

(b)

Solution:

A four-step problem-solving approach is used in the solution. Some steps are combined and only essential steps are shown in each part.

Part (a): Determine the complete state of stress on stress element aligned with inclined section *pq*.

1, 2. ***Conceptualize, Categorize*:** To find the state of stress on inclined section *pq*, start by finding compressive normal stress σ_x due to applied load *P*:

$$\sigma_x = \frac{-P}{A} = \frac{-90 \text{ kN}}{1200 \text{ mm}^2} = -75 \text{ MPa}$$

3. ***Analyze***: Next find normal and shear stresses from Eqs. (2-29a and b) with $\theta = 25°$ as

$$\sigma_\theta = \sigma_x \cos(\theta)^2 = (-75\text{ MPa})\cos(25°)^2 = -61.6\text{ MPa}$$
$$\tau_\theta = -\sigma_x \sin(\theta)\cos(\theta) = -(-75\text{ MPa})\sin(25°)\cos(25°)$$
$$= 28.7\text{ MPa}$$

These stresses are shown acting on the inclined section pq in Fig. 2-61c. Stress element **face *ab*** (Fig. 2-61d) is aligned with section pq. Note that the normal stress σ_θ is *negative* (compressive), and the shear stress τ_θ is *positive* (counterclockwise). Now use Eqs. (2-29a and b) to find normal and shear stresses on the remaining three faces of the stress element (see Fig. 2-61d).

The normal and shear stresses on **face *cb*** are computed using angle $\theta + 90° = 115°$ in Eqs. (2-29a and b):

$$\sigma_{cb} = \sigma_x \cos(115°)^2 = (-75\text{ MPa})\cos(115°)^2 = -13.4\text{ MPa}$$
$$\tau_{cb} = -\sigma_x \sin(115°)\cos(115°) = -(-75\text{ MPa})[\sin(115°)\cos(115°)]$$
$$= -28.7\text{ MPa}$$

4. ***Finalize***: The stresses on the opposite **face *cd*** are the same as those on **face *ab***, which can be verified by substituting $\theta = 25° + 180° = 205°$ into Eqs. (2-29a and b). For **face *ad*** substitute $\theta = 25° - 90° = -65°$ into Eqs. (2-29a and b). The complete state of stress is shown in Fig. (2-61d).

FIGURE 2-61 Continued

Example 2-14: (c) Stresses on element at inclined section pq through bar and (d) complete state of stress on element at inclined section pq through bar

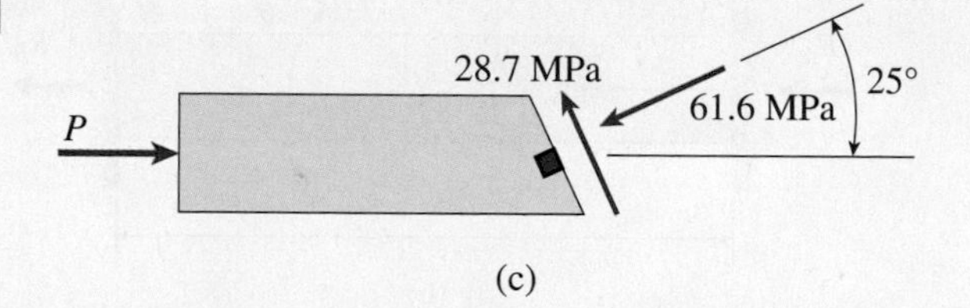

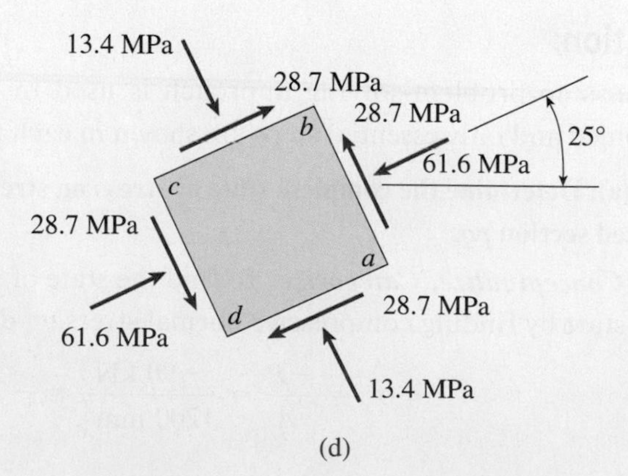

FIGURE 2-61 Continued

Example 2-14: (e) Normal and shear stresses on element at inclined section rs through bar

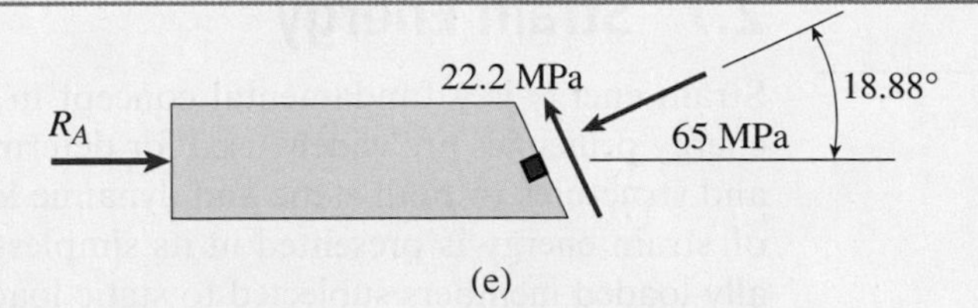

Part (b): Determine the normal and shear stresses due to temperature increase on the stress element aligned with inclined section rs.

2. ***Categorize***: From Example 2-9, reactions R_A and R_B (Fig. 2-61b) due to temperature increase $\Delta T = 33°$ are

$$R_A = -R_B = EA\alpha(\Delta T) \tag{a}$$

and the resulting axial compressive thermal stress is

$$\sigma_T = \frac{R_A}{A} = E\alpha(\Delta T) \tag{b}$$

So

$$\sigma_x = -(110 \text{ GPa})[20 \times 10^{-6}/°\text{C})](33°\text{C}) = -72.6 \text{ MPa}$$

3. ***Analyze***: Since the compressive stress on plane rs is known to be 65 MPa, angle θ for inclined plane rs from Eq. (2-29a) is

$$\theta_{rs} = \cos^{-1}\left(\sqrt{\frac{\sigma_\theta}{\sigma_x}}\right) = \cos^{-1}\left(\sqrt{\frac{-65 \text{ MPa}}{-72.6 \text{ MPa}}}\right) = 18.878°$$

and from Eq. (2-29b), shear stress τ_θ on inclined plane rs is

$$\tau_\theta = -\sigma_x(\sin(\theta_{rs})\cos(\theta_{rs})) = -(-72.6 \text{ MPa})\sin(18.878°)\cos(18.878°)$$
$$= 22.2 \text{ MPa}$$

Part (c): Determine the maximum permissible temperature *increase* (ΔT) in the bar based on allowable stress values.

2, 3. ***Categorize, Analyze***: The maximum normal stress σ_{max} occurs on a stress element inclined at $\theta = 0$ [Eq. (2-30)], so $\sigma_{\text{max}} = \sigma_x$. Equate *thermal stress* from Eq. (b) to allowable normal stress $\sigma_a = 82$ MPa to find the value of ΔT_{max} based on allowable normal stress:

$$\Delta T_{\text{max}1} = \frac{\sigma_a}{E\alpha} = \frac{82 \text{ MPa}}{(110 \text{ GPa})[20 \times 10^{-6}/°\text{C}]} = 37.3°\text{C} \tag{c}$$

From Eq. 2-31, maximum shear stress τ_{max} occurs at a section inclination of 45° for which $\tau_{\text{max}} = \sigma_x/2$. Using the given allowable shear stress value, $\tau_a = 40$ MPa, and the relationship between maximum normal and shear stresses in Eq. (2-31) compute a second value for ΔT_{max} as

$$\Delta T_{\text{max}2} = \frac{2\tau_a}{E\alpha} = \frac{2(40 \text{ MPa})}{(110 \text{ GPa})[20 \times 10^{-6}/°\text{C}]} = 36.4°\text{C}$$

4. ***Finalize***: The lower temperature increase value, based on not exceeding allowable shear stress τ_a, controls. You could have anticipated this because $\tau_{\text{allow}} < \sigma_{\text{allow}}/2$.

2.7 Strain Energy

FIGURE 2-62

Prismatic bar subjected to a statically applied load

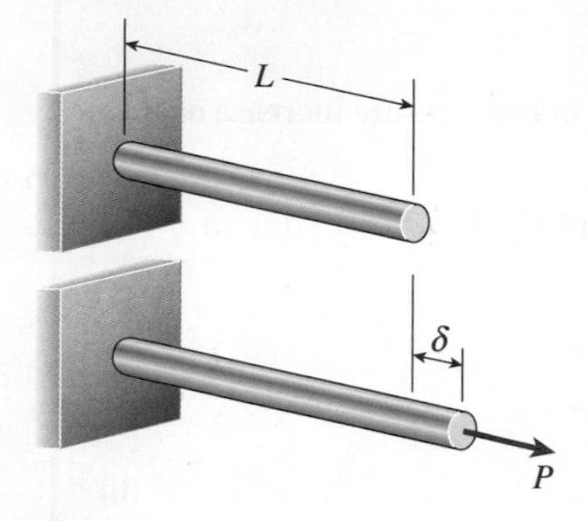

Strain energy is a fundamental concept in applied mechanics, and strain-energy principles are widely used for determining the response of machines and structures to both static and dynamic loads. In this section, the subject of strain energy is presented in its simplest form by considering only axially loaded members subjected to static loads. More complicated structural elements are discussed in later chapters—bars in torsion in Section 3.9 and beams in bending in Section 9.8. In addition, the use of strain energy in connection with dynamic loads is described in Sections 2.8 and 9.10.

To illustrate the basic ideas, consider a prismatic bar of length L subjected to a tensile force P (Fig. 2-62). Assume that the load is applied slowly, so that it gradually increases from zero to its maximum value P. Such a load is called a **static load** because there are no dynamic or inertial effects due to motion. The bar gradually elongates as the load is applied, eventually reaching its maximum elongation δ at the same time that the load reaches its full value P. Thereafter, the load and elongation remain unchanged.

During the loading process, the load P moves slowly through the distance δ and does a certain amount of **work**. To evaluate this work, recall from elementary mechanics that a constant force does work equal to the product of the force and the distance through which it moves. However, in this case, the force varies in magnitude from zero to its maximum value P. To find the work done by the load under these conditions, you need to know the manner in which the force varies. This information is supplied by a **load-displacement diagram**, such as the one plotted in Fig. 2-63. On this diagram, the vertical axis represents the axial load and the horizontal axis represents the corresponding elongation of the bar. The shape of the curve depends upon the properties of the material.

Denote by P_1 any value of the load between zero and the maximum value P, and denote the corresponding elongation of the bar by δ_1. Then an increment dP_1 in the load will produce an increment $d\delta_1$ in the elongation. The work done by the load during this incremental elongation is the product of the load and the distance through which it moves, that is, the work equals $P_1 d\delta_1$. This work is represented in the figure by the area of the shaded strip below the load-displacement curve. The total work done by the load as it increases from zero to the maximum value P is the summation of all such elemental strips:

$$W = \int_0^{\delta} P_1 d\delta_1 \tag{2-33}$$

In geometric terms, *the work done by the load is equal to the area below the load-displacement curve.*

FIGURE 2-63

Load-displacement diagram

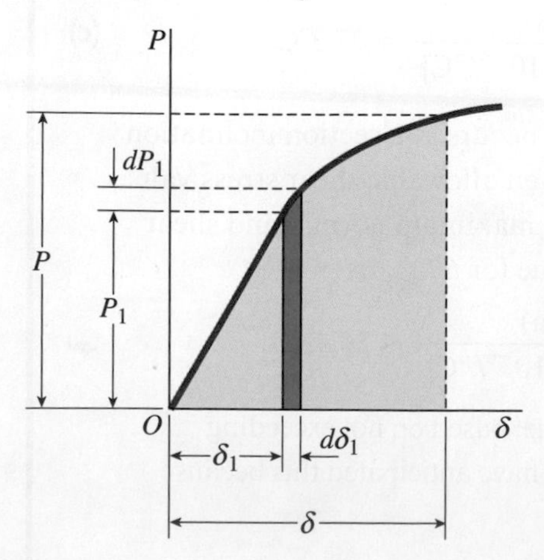

When the load stretches the bar, strains are produced. The presence of these strains increases the energy level of the bar itself. Therefore, a new quantity, called **strain energy**, is defined as the energy absorbed by the bar during the loading process. From the principle of conservation of energy, this strain energy is equal to the work done by the load provided no energy is added or subtracted in the form of heat. Therefore,

$$U = W = \int_0^{\delta} P_1 d\delta_1 \tag{2-34}$$

in which U is the symbol for strain energy. Sometimes strain energy is referred to as **internal work** to distinguish it from the external work done by the load.

Work and energy are expressed in the same **units**. In SI, the unit of work and energy is the joule (J), which is equal to one newton meter ($1 \text{ J} = 1 \text{ N} \cdot \text{m}$).

Elastic and Inelastic Strain Energy

If the force P (Fig. 2-62) is slowly removed from the bar, the bar will shorten. If the elastic limit of the material is not exceeded, the bar will return to its original length. If the limit is exceeded, *a permanent set* will remain (see Section 1.6). Thus, either all or part of the strain energy will be recovered in the form of work. This behavior is shown on the load-displacement diagram of Fig. 2-64. During loading, the work done by the load is equal to the area below the curve (area $OABCDO$). When the load is removed, the load-displacement diagram follows line BD if point B is beyond the elastic limit, and a permanent elongation OD remains. Thus, the strain energy recovered during unloading, called the **elastic strain energy**, is represented by the shaded triangle BCD. Area $OABDO$ represents energy that is lost in the process of permanently deforming the bar. This energy is known as the **inelastic strain energy**.

Most structures are designed with the expectation that the material will remain within the elastic range under ordinary conditions of service. Assume that the load at which the stress in the material reaches the elastic limit is represented by point A on the load-displacement curve (Fig. 2-64). As long as the load is below this value, all of the strain energy is recovered during unloading and no permanent elongation remains. Thus, the bar acts as an elastic spring, storing and releasing energy as the load is applied and removed.

FIGURE 2-64

Elastic and inelastic strain energy

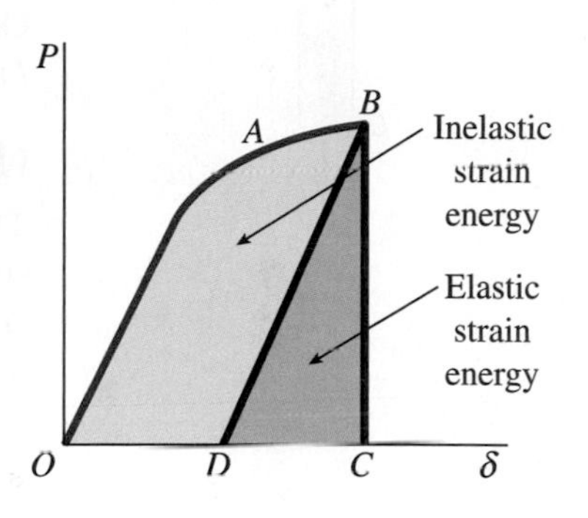

Linearly Elastic Behavior

Now assume that the material of the bar follows Hooke's law, so that the load-displacement curve is a straight line (Fig. 2-65). Then the strain energy U stored in the bar (equal to the work W done by the load) is

$$U = W = \frac{P\delta}{2} \tag{2-35}$$

which is the area of the shaded triangle OAB in the figure.[3]

The relationship between the load P and the elongation δ for a bar of linearly elastic material is given by the equation

$$\delta = \frac{PL}{EA} \tag{2-36}$$

Combine this equation with Eq. (2-35) to express the strain energy of a **linearly elastic bar** in either of the following forms:

$$U = \frac{P^2L}{2EA} \qquad U = \frac{EA\delta^2}{2L} \tag{2-37a,b}$$

The first equation expresses the strain energy as a function of the load and the second expresses it as a function of the elongation.

FIGURE 2-65

Load-displacement diagram for a bar of linearly elastic material

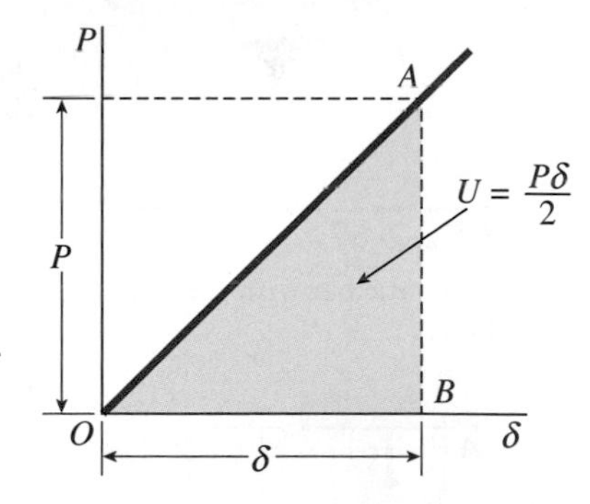

[3]The principle that the work of the external loads is equal to the strain energy (for the case of linearly elastic behavior) was first stated by the French engineer B. P. E Clapeyron (1799–1864) and is known as *Clapeyron's theorem* (Ref. 2-7).

From the first equation, note that increasing the length of a bar increases the amount of strain energy even though the load is unchanged (because more material is being strained by the load). On the other hand, increasing either the modulus of elasticity or the cross-sectional area decreases the strain energy because the strains in the bar are reduced. These ideas are illustrated in Examples 2-15 and 2-17.

Strain-energy equations analogous to Eqs. (2-37a and b) can be written for a **linearly elastic spring** by replacing the stiffness EA/L of the prismatic bar by the stiffness k of the spring. Thus,

$$U = \frac{P^2}{2k} \qquad U = \frac{k\delta^2}{2} \tag{2-38a,b}$$

Other forms of these equations can be obtained by replacing k by $1/f$, where f is the flexibility.

FIGURE 2-66

Bar consisting of prismatic segments having different cross-sectional areas and different axial forces

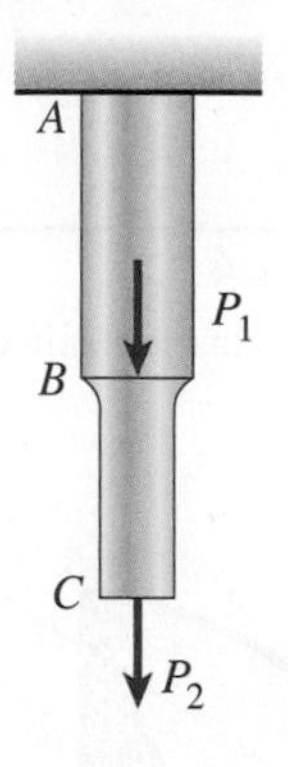

Nonuniform Bars

The total strain energy U of a bar consisting of several segments is equal to the sum of the strain energies of the individual segments. For instance, the strain energy of the bar pictured in Fig. 2-66 equals the strain energy of segment AB plus the strain energy of segment BC. This concept is expressed in general terms by

$$U = \sum_{i=1}^{n} U_i \tag{2-39}$$

in which U_i is the strain energy of segment i of the bar and n is the number of segments. (This relation holds whether the material behaves in a linear or nonlinear manner.)

Now assume that the material of the bar is linearly elastic and that the internal axial force is constant within each segment. Next use Eq. (2-37a) to obtain the strain energies of the segments, and Eq. (2-39) becomes

$$U = \sum_{i=1}^{n} \frac{N_i^2 L_i}{2E_i A_i} \tag{2-40}$$

in which N_i is the axial force acting in segment i and L_i, E_i, and A_i are properties of segment i. (The use of this equation is illustrated in Examples 2-15 and 2-17 at the end of the section.)

Obtain the strain energy of a nonprismatic bar with continuously varying axial force (Fig. 2-67) by applying Eq. (2-37a) to a differential element (shown shaded in the figure) and then integrate along the length of the bar:

$$U = \int_0^L \frac{[N(x)]^2\,dx}{2EA(x)} \tag{2-41}$$

FIGURE 2-67

Nonprismatic bar with varying axial force

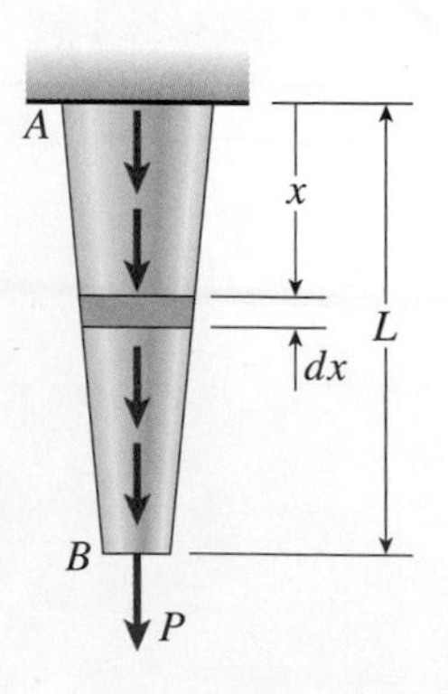

In this equation, $N(x)$ and $A(x)$ are the axial force and cross-sectional area at distance x from the end of the bar. (Example 2-16 illustrates the use of this equation.)

Comments

The preceding expressions for strain energy [Eqs. (2-37) through (2-41)] show that strain energy is *not* a linear function of the loads, not even when the material is linearly elastic. Thus, it is important to realize that *you cannot obtain the strain energy of a structure supporting more than one load by combining the strain energies obtained from the individual loads acting separately.*

In the case of the nonprismatic bar shown in Fig. 2-66, the total strain energy is *not* the sum of the strain energy due to load P_1 acting alone and the strain energy due to load P_2 acting alone. Instead, evaluate the strain energy with all of the loads acting simultaneously, as demonstrated later in Example 2-16.

Although only tension members are considered in the preceding discussions of strain energy, all of the concepts and equations apply equally well to members in **compression**. Since the work done by an axial load is positive regardless of whether the load causes tension or compression, it follows that strain energy is always a positive quantity. This fact is also evident in the expressions for strain energy of linearly elastic bars [such as Eqs. (2-37a and b)]. These expressions are always positive because the load and elongation terms are squared.

Strain energy is a form of **potential energy** (or "energy of position") because it depends upon the relative locations of the particles or elements that make up the member. When a bar or a spring is compressed, its particles are crowded more closely together; when it is stretched, the distances between particles increase. In both cases, the strain energy of the member increases as compared to its strain energy in the unloaded position.

Displacements Caused by a Single Load

The displacement of a linearly elastic structure supporting only one load can be determined from its strain energy. To illustrate the method, consider a two-bar truss (Fig. 2-68) loaded by a vertical force P. The objective is to determine the vertical displacement δ at joint B where the load is applied.

When applied slowly to the truss, the load P does work as it moves through the vertical displacement δ. However, it does no work as it moves laterally, that is, sideways. Therefore, since the load-displacement diagram is linear [see Fig. 2-65 and Eq. (2-35)], the strain energy U stored in the structure, equal to the work done by the load, is

$$U = W = \frac{P\delta}{2}$$

which leads to

$$\delta = \frac{2U}{P} \qquad \textbf{(2-42)}$$

This equation shows that under certain special conditions, as outlined in the following paragraph, the displacement of a structure can be determined directly from the strain energy.

In order to use Eq. (2-42): (1) the structure must behave in a linearly elastic manner, and (2) only one load may act on the structure. Furthermore, the only displacement that can be determined is the displacement corresponding to the load itself (that is, the displacement must be in the direction of the load and must be at the point where the load is applied). Therefore, this method for finding displacements is extremely limited in its application and is not a good indicator of the great importance of strain-energy principles in structural mechanics. However, the method does provide an introduction to the use of strain energy. (This method is illustrated later in Example 2-18.)

FIGURE 2-68

Structure supporting a single load P

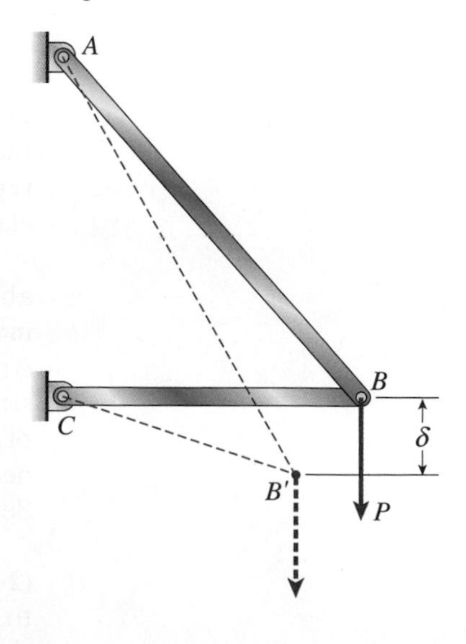

Strain-Energy Density

In many situations, it is convenient to use a quantity called **strain-energy density**, which is defined as the strain energy per unit volume of material. Expressions for strain-energy density in the case of linearly elastic materials can be obtained from the formulas for strain energy of a prismatic bar [Eqs. (2-37a and b)]. Since the strain energy of the bar is distributed uniformly throughout its volume, you can determine the strain-energy density by dividing the total strain energy U by the volume AL of the bar. Thus, the strain-energy density, denoted by the symbol u, can be expressed in either of these forms:

$$u = \frac{P^2}{2EA^2} \qquad u = \frac{E\delta^2}{2L^2} \tag{2-43a,b}$$

If P/A is replaced by the stress σ and δ/L by the strain ε,

$$u = \frac{\sigma^2}{2E} \qquad u = \frac{E\varepsilon^2}{2} \tag{2-44a,b}$$

These equations give the strain-energy density in a linearly elastic material in terms of either the normal stress σ or the normal strain ε.

The expressions in Eqs. (2-44a and b) have a simple geometric interpretation. They are equal to the area $\sigma\varepsilon/2$ of the triangle below the stress-strain diagram for a material that follows Hooke's law ($\sigma = E\varepsilon$). In more general situations where the material does not follow Hooke's law, the strain-energy density is still equal to the area below the stress-strain curve, but the area must be evaluated for each particular material.

Strain-energy density has **units** of energy divided by volume. The SI units are joules per cubic meter (J/m^3). Since all of these units reduce to units of stress (recall that 1 J = 1 N · m), you also can use units such as pascals (Pa) for strain-energy density.

The strain-energy density of the material when it is stressed to the proportional limit is called the **modulus of resilience** u_r. It is found by substituting the proportional limit σ_{pl} into Eq. (2-44a):

$$u_r = \frac{\sigma_{pl}^2}{2E} \tag{2-45}$$

For example, a mild steel having $\sigma_{pl} = 210$ MPa and $E = 210$ GPa has a modulus of resilience $u_r = 149$ kPa. Note that the modulus of resilience is equal to the area below the stress-strain curve up to the proportional limit. *Resilience* represents the ability of a material to absorb and release energy within the elastic range.

Another quantity, called *toughness*, refers to the ability of a material to absorb energy without fracturing. The corresponding modulus, called the **modulus of toughness** u_t, is the strain-energy density when the material is stressed to the point of failure. It is equal to the area below the entire stress-strain curve. The higher the modulus of toughness, the greater the ability of the material to absorb energy without failing. A high modulus of toughness is therefore important when the material is subject to impact loads (see Section 2.8).

The preceding expressions for strain-energy density [Eqs. (2-43) through (2-45)] were derived for *uniaxial stress*, that is, for materials subjected only to tension or compression. Formulas for strain-energy density in other stress states are presented in Chapters 3 and 7.

Example 2-15

Three round bars having the same length L but different shapes are shown in Fig. 2-69. The first bar has diameter d over its entire length, the second has diameter d over one-fifth of its length, and the third has diameter d over one-fifteenth of its length. Also, the second and third bars have diameter $2d$. All three bars are subjected to the same axial load P.

Compare the amounts of strain energy stored in the bars, assuming linearly elastic behavior. (Disregard the effects of stress concentrations and the weights of the bars.)

FIGURE 2-69

Example 2-15: Calculation of strain energy

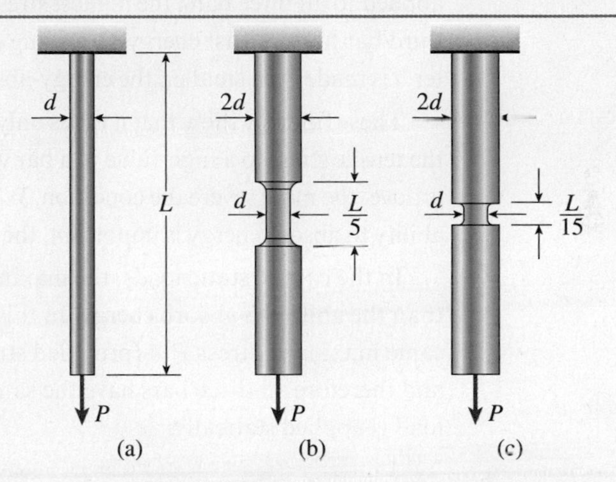

Solution:

A four-step problem-solving approach is used in the solution; only essential steps are shown in each part.

Part (a): Strain energy U_1 of the first bar.

1, 2, 3. ***Conceptualize, Categorize, Analyze***: The strain energy of the first bar is found directly from Eq. (2-37a):

$$U_1 = \frac{P^2 L}{2EA} \qquad \text{(a)}$$

in which $A = \pi d^2/4$.

Part (b): Strain energy U_2 of the second bar.

1, 2, 3. ***Conceptualize, Categorize, Analyze***: The strain energy is found by summing the strain energies in the three segments of the bar [see Eq. (2-40)]. Thus,

$$U_2 = \sum_{i=1}^{n} \frac{N_i^2 L_i}{2E_i A_i} = \frac{P^2(L/5)}{2EA} + \frac{P^2(4L/5)}{2E(4A)} = \frac{P^2 L}{5EA} = \frac{2U_1}{5} \qquad \text{(b)}$$

which is only 40% of the strain energy of the first bar. Thus, increasing the cross-sectional area over part of the length has greatly reduced the amount of strain energy that can be stored in the bar.

Part (c): Strain energy U_3 of the third bar.

1, 2, 3. ***Conceptualize, Categorize, Analyze***: Again using Eq. (2-40),

$$U_3 = \sum_{i=1}^{n} \frac{N_i^2 L_i}{2E_i A_i} = \frac{P^2(L/15)}{2EA} + \frac{P^2(14L/15)}{2E(4A)} = \frac{3P^2 L}{20EA} = \frac{3U_1}{10} \quad \text{(c)}$$

The strain energy has now decreased to 30% of the strain energy of the first bar.

4. ***Finalize***: Comparing the results shows that the strain energy decreases as the part of the bar with the larger area increases. If the same amount of work is applied to all three bars, the highest stress will be in the third bar, because the third bar has the least energy-absorbing capacity. If the region having diameter d is made even smaller, the energy-absorbing capacity will decrease further.

These findings show that it takes only a small amount of work to bring the tensile stress to a high value in a bar with a groove, and the narrower the groove, the more severe the condition. When the loads are dynamic and the ability to absorb energy is important, the presence of grooves is very damaging.

In the case of static loads, the maximum stresses are more important than the ability to absorb energy. In this example, all three bars have the same maximum stress P/A (provided stress concentrations are alleviated), and therefore all three bars have the same load-carrying capacity when the load is applied statically.

Example 2-16

Determine the strain energy of a prismatic bar suspended from its upper end (Fig. 2-70). Consider the following loads: (a) the weight of the bar itself and (b) the weight of the bar plus a load P at the lower end. (Assume linearly elastic behavior.)

FIGURE 2-70

Example 2-16: (a) Bar hanging under its own weight and (b) bar hanging under its own weight and also supporting a load P

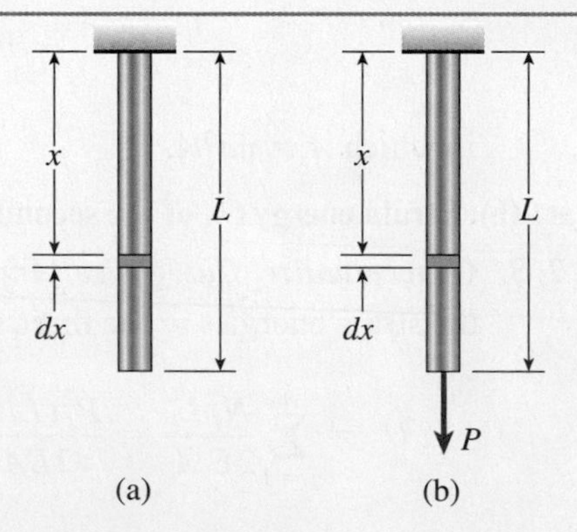

Solution:

A four-step problem-solving approach is used in the solution; only essential steps are shown in each part.

Part (a): Strain energy due to the weight of the bar itself (Fig. 2-70a).

1, 2, 3. ***Conceptualize, Categorize, Analyze*:** The bar is subjected to a varying axial force with the internal force being zero at the lower end and maximum at the upper end. To determine the axial force, consider an element of length dx (shown shaded in the figure) at distance x from the upper end. The internal axial force $N(x)$ acting on this element is equal to the weight of the bar below the element:

$$N(x) = \gamma A(L - x) \tag{a}$$

in which γ is the weight density of the material and A is the cross-sectional area of the bar. Substituting into Eq. (2-41) and integrating gives the total strain energy:

$$U = \int_0^L \frac{[N(x)]^2 dx}{2EA(x)} = \int_0^L \frac{[\gamma A(L - x)]^2 dx}{2EA} = \frac{\gamma^2 AL^3}{6E} \tag{2-46}$$

Part (b): Strain energy due to the weight of the bar plus the load *P* (Fig. 2-70b).

1, 2, 3. ***Conceptualize, Categorize, Analyze*:** In this case, the axial force $N(x)$ acting on the element is

$$N(x) = \gamma A(L - x) + P \tag{b}$$

[compare with Eq. (a)]. From Eq. (2-41), the total strain energy is

$$U = \int_0^L \frac{[\gamma A(L - x) + P]^2 dx}{2EA} = \frac{\gamma^2 AL^3}{6E} + \frac{\gamma PL^2}{2E} + \frac{P^2 L}{2EA} \tag{2-47}$$

4. ***Finalize*:** The first term in Eq. (2-47) is the same as the strain energy of a bar hanging under its own weight [Eq. (2-46)], and the last term is the same as the strain energy of a bar subjected only to an axial force P [Eq. (2-37a)]. However, the middle term contains both γ and P, showing that it depends upon both the weight of the bar and the magnitude of the applied load.

This example illustrates that the strain energy of a bar subjected to two loads is *not* equal to the sum of the strain energies produced by the individual loads acting separately.

Example 2-17

The cylinder for a compressed air machine is clamped by bolts that pass through the flanges of the cylinder (Fig. 2-71a). A detail of one of the bolts is shown in part (b) of the figure. The diameter d of the shank is 13 mm, and the root diameter d_r of the threaded portion is 10 mm. The grip g of the bolts is 40 mm, and the threads extend a distance $t = 6.5$ mm into the grip. Under the action of repeated cycles of high and low pressure in the chamber, the bolts may eventually break.

FIGURE 2-71

Example 2-17: (a) cylinder with piston and clamping bolts and (b) detail of one bolt

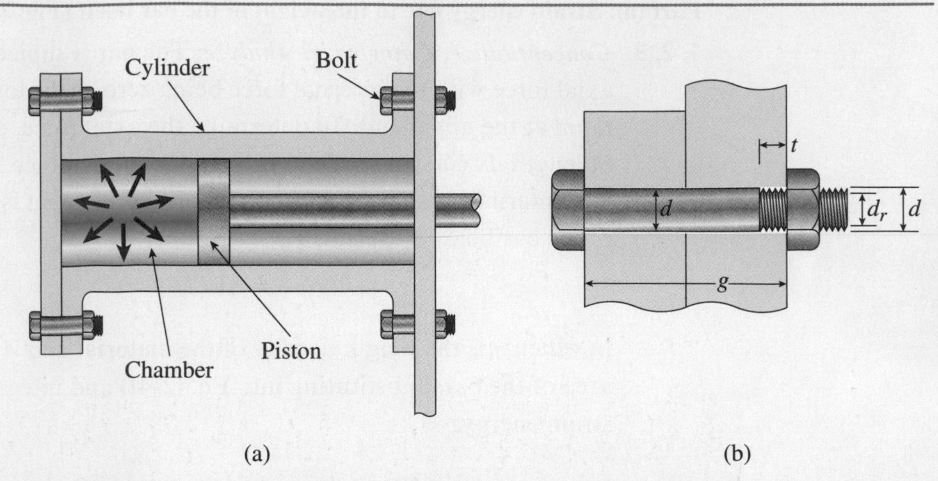

To reduce the likelihood of the bolts failing, the designers suggest two possible modifications: (1) Machine down the shanks of the bolts so that the shank diameter is the same as the thread diameter d_r, as shown in Fig. 2-72a. (2) Replace each pair of bolts by a single long bolt, as shown in Fig. 2-72b. The long bolts are similar to the original bolts (Fig. 2-71b) except that the grip is increased to the distance $L = 340$ mm.

Compare the energy-absorbing capacity of the three bolt configurations: (a) original bolts, (b) bolts with reduced shank diameter, and (c) long bolts. (Assume linearly elastic behavior and disregard the effects of stress concentrations.)

Solution:

A four-step problem-solving approach is used in the solution; only essential steps are shown in each part.

Part (a): Original bolts.

1, 2, 3. ***Conceptualize, Categorize, Analyze***: The original bolts are idealized as bars consisting of two segments (Fig. 2-71b). The left-hand segment has length $g - t$ and diameter d, and the right-hand segment has length t and diameter d_r. The strain energy of one bolt under a tensile load P is obtained by adding the strain energies of the two segments [Eq. (2-40)]:

$$U_1 = \sum_{i=1}^{n} \frac{N_i^2 L_i}{2E_i A_i} = \frac{P^2(g - t)}{2EA_s} + \frac{P^2 t}{2EA_r} \qquad \textbf{(a)}$$

in which A_s is the cross-sectional area of the shank and A_r is the cross-sectional area at the root of the threads; thus,

$$A_s = \frac{\pi d^2}{4} \qquad A_r = \frac{\pi d_r^2}{4} \qquad \textbf{(b)}$$

Substituting these expressions into Eq. (a) gives the formula for the strain energy of one of the original bolts:

$$U_1 = \frac{2P^2(g - t)}{\pi E d^2} + \frac{2P^2 t}{\pi E d_r^2} \qquad \textbf{(c)}$$

Part (b): Bolts with reduced shank diameter.

1, 2, 3. ***Conceptualize, Categorize, Analyze***: These bolts are idealized as prismatic bars having length g and diameter d_r (Fig. 2-72a). Therefore, the strain energy of one bolt [see Eq. (2-37a)] is

$$U_2 = \frac{P^2 g}{2EA_r} = \frac{2P^2 g}{\pi E d_r^2} \tag{d}$$

The ratio of the strain energies for cases (1) and (2) is

$$\frac{U_2}{U_1} = \frac{gd^2}{(g - t)d_r^2 + td^2} \tag{e}$$

or, upon substituting numerical values,

$$\frac{U_2}{U_1} = \frac{(40 \text{ mm})(13 \text{ mm})^2}{(40 \text{ mm} - 6.5 \text{ mm})(10 \text{ mm})^2 + (6.5 \text{ mm})(13 \text{ mm})^2} = 1.52$$

4. ***Finalize***: Using bolts with reduced shank diameters results in a 52% increase in the amount of strain energy that can be absorbed by the bolts. If implemented, this scheme should reduce the number of failures caused by the impact loads.

Part (c): Long bolts.

1, 2, 3. ***Conceptualize, Categorize, Analyze***: The calculations for the long bolts (Fig. 2-72b) are the same as for the original bolts except the grip g is changed to the grip L. Therefore, the strain energy of one long bolt [compare with Eq. (c)] is

$$U_3 = \frac{2P^2(L - t)}{\pi E d^2} + \frac{2P^2 t}{\pi E d_r^2} \tag{f}$$

FIGURE 2-72

Example 2-17: Proposed modifications to the bolts: (a) bolts with reduced shank diameter and (b) bolts with increased length

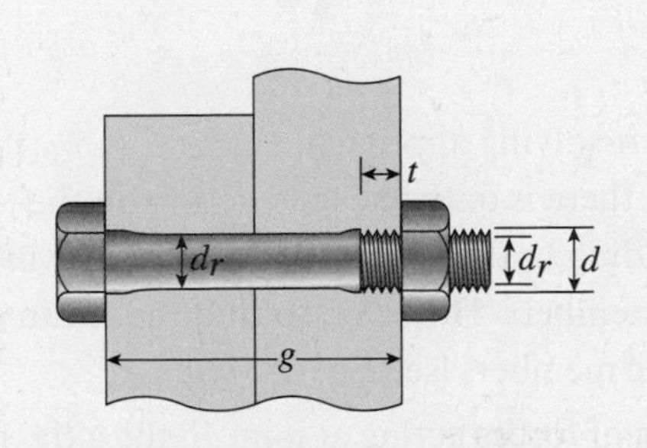

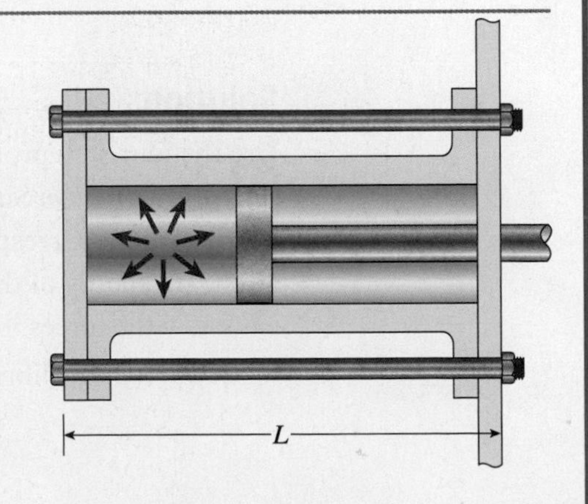

Since one long bolt replaces two of the original bolts, compare the strain energies by taking the ratio of U_3 to $2U_1$, as

$$\frac{U_3}{2U_1} = \frac{(L - t)d_r^2 + td^2}{2(g - t)d_r^2 + 2td^2} \tag{g}$$

Substituting numerical values gives

$$\frac{U_3}{2U_1} = \frac{(340 \text{ mm} - 6.5 \text{ mm})(10 \text{ mm})^2 + (6.5 \text{ mm})(13 \text{ mm})^2}{2(40 \text{ mm} - 6.5 \text{ mm})(10 \text{ mm})^2 + 2(6.5 \text{ mm})(13 \text{ mm})^2} = 3.87$$

4. ***Finalize:*** Using long bolts increases the energy-absorbing capacity by 318% and achieves the greatest safety from the standpoint of strain energy.

Note: When designing bolts, designers must also consider the maximum tensile stresses, maximum bearing stresses, stress concentrations, and many other matters.

Example 2-18

Determine the vertical displacement δ_B of joint B of the truss shown in Fig. 2-73. Note that the only load acting on the truss is a vertical load P at joint B. Assume that both members of the truss have the same axial rigidity EA.

FIGURE 2-73

Example 2-18: Displacement of a truss supporting a single load P

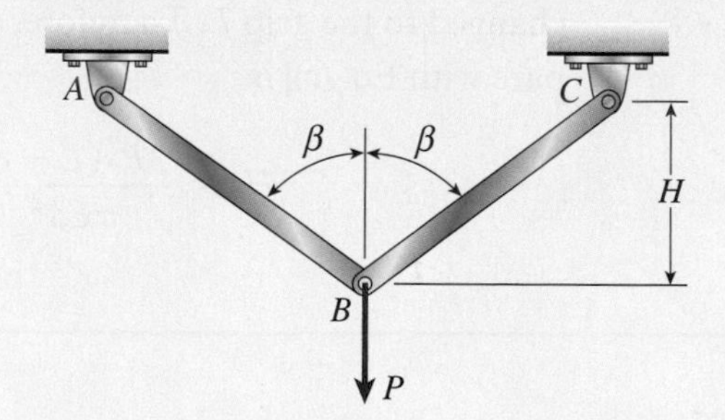

Solution:

Use the four-step problem-solving approach.

1. ***Conceptualize*:** Since there is only one load acting on the truss, find the displacement corresponding to that load by equating the work of the load to the strain energy of the members. However, to find the strain energy, you must know the forces in the members [see Eq. (2-37a)].

From the equilibrium of forces acting at joint B, the axial force F in either bar is

$$F = \frac{P}{2 \cos \beta} \tag{a}$$

in which β is the angle shown in the figure.

Also, from the geometry of the truss, the length of each bar is

$$L_1 = \frac{H}{\cos\beta} \tag{b}$$

in which H is the height of the truss.

2. ***Categorize***: Now obtain the strain energy of the two bars from Eq. (2-37a):

$$U = (2)\frac{F^2 L_1}{2EA} = \frac{P^2 H}{4EA\cos^3\beta} \tag{c}$$

Also, the work of the load P [from Eq. (2-35)] is

$$W = \frac{P\delta_B}{2} \tag{d}$$

where δ_B is the downward displacement of joint B.

3. ***Analyze***: Equating U and W and solving for δ_B,

$$\delta_B = \frac{PH}{2EA\cos^3\beta} \tag{2-48}$$

4. ***Finalize***: This displacement was found using only equilibrium and strain energy. There was no need to draw a displacement diagram at joint B.

*2.8 Impact Loading

Loads can be classified as static or dynamic depending upon whether they remain constant or vary with time. A **static load** is applied slowly, so that it causes no vibrational or dynamic effects in the structure. The load increases gradually from zero to its maximum value, and thereafter it remains constant.

A **dynamic load** may take many forms—some loads are applied and removed suddenly (*impact loads*), others persist for long periods of time and continuously vary in intensity (*fluctuating loads*). Impact loads are produced when two objects collide or when a falling object strikes a structure. Fluctuating loads are produced by rotating machinery, traffic, wind gusts, water waves, earthquakes, and manufacturing processes.

As an example of how structures respond to dynamic loads, consider now the impact of an object falling onto the lower end of a prismatic bar (Fig. 2-74). A collar of mass M, initially at rest, falls from a height h onto a flange at the end of bar AB. When the collar strikes the flange, the bar begins to elongate, creating axial stresses within the bar. In a very short interval of time, such as a few milliseconds, the flange will move downward and reach its position of maximum displacement. Thereafter, the bar shortens, then lengthens, then shortens again as the bar vibrates longitudinally and the end of the bar moves up and down. The vibrations are analogous to those that occur when a spring is stretched and then released, or when a person makes a bungee jump.

FIGURE 2-74

Impact load on a prismatic bar AB due to a falling object of mass M

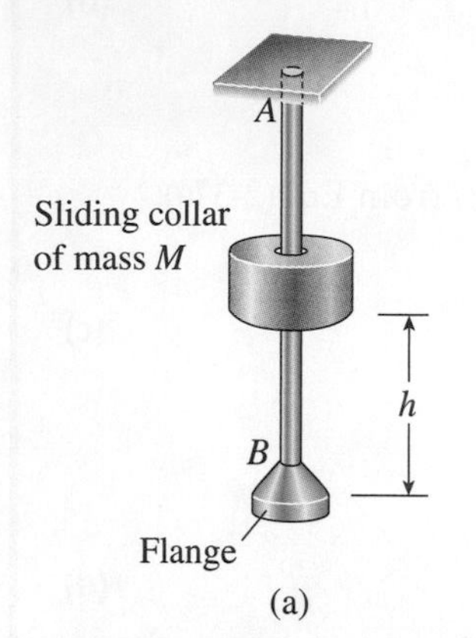

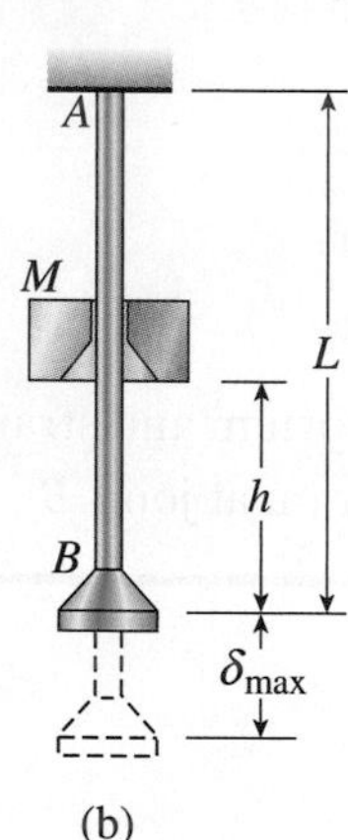

The vibrations of the bar soon cease because of various damping effects, and then the bar comes to rest with the mass M supported on the flange.

The response of the bar to the falling collar is obviously very complicated, and a complete and accurate analysis requires the use of advanced mathematical techniques. However, an approximate analysis can be made by using the concept of strain energy (Section 2.7) and making several simplifying assumptions.

Begin by considering the energy of the system just before the collar is released (Fig. 2-74a). The potential energy of the collar with respect to the elevation of the flange is Mgh, where g is the acceleration of gravity.[5] This potential energy is converted into kinetic energy as the collar falls. At the instant the collar strikes the flange, its potential energy with respect to the elevation of the flange is zero and its kinetic energy is $Mv^2/2$, where $v = \sqrt{2gh}$ is its velocity.[6]

During the ensuing impact, the kinetic energy of the collar is transformed into other forms of energy. Part of the kinetic energy is transformed into the strain energy of the stretched bar. Some of the energy is dissipated in the production of heat and in causing localized plastic deformations of the collar and flange. A small part remains as the kinetic energy of the collar, which either moves further downward (while in contact with the flange) or else bounces upward.

To make a simplified analysis of this very complex situation, the behavior is idealized by making the following assumptions. (1) Assume that the collar and flange are so constructed that the collar "sticks" to the flange and moves downward with it (that is, the collar does not rebound). This behavior is more likely to prevail when the mass of the collar is large compared to the mass of the bar. (2) Disregard all energy losses and assume that the kinetic energy of the falling mass is transformed entirely into strain energy of the bar. This assumption predicts larger stresses in the bar than would be predicted if we took energy losses into account. (3) Disregard any change in the potential energy of the bar itself (due to the vertical movement of elements of the bar), and ignore the existence of strain energy in the bar due to its own weight. Both of these effects are extremely small. (4) Assume that the stresses in the bar remain within the linearly elastic range. (5) Assume that the stress distribution throughout the bar is the same as when the bar is loaded statically by a force at the lower end, that is, assume the stresses are uniform throughout the volume of the bar. (In reality, longitudinal stress waves will travel through the bar, thereby causing variations in the stress distribution.)

On the basis of the preceding assumptions, you can calculate the maximum elongation and the maximum tensile stresses produced by the impact load. (Recall that the weight of the bar itself is ignored and the stresses are due solely to the falling collar.)

Maximum Elongation of the Bar

The maximum elongation δ_{max} (Fig. 2-74b) can be obtained from the principle of *conservation of energy* by equating the potential energy lost by the falling mass to the maximum strain energy acquired by the bar. The potential energy lost is $W(h + \delta_{max})$, where $W = Mg$ is the weight of the collar and $h + \delta_{max}$ is the distance through which it moves. The strain energy of the bar is $EA\delta_{max}^2/2L$,

[5]In SI units, the acceleration of gravity $g = 9.81 \text{ m/s}^2$. For more precise values of g, or for a discussion of mass and weight, see Appendix A.

[6]In engineering work, velocity is usually treated as a vector quantity. However, since kinetic energy is a scalar, here the word "velocity" refers to the *magnitude* of the velocity, or the *speed*.

where EA is the axial rigidity and L is the length of the bar [see Eq. (2-37b)]. The resulting equation is:

$$W(h + \delta_{max}) = \frac{EA\delta^2_{max}}{2L} \tag{2-49}$$

This equation is quadratic in δ_{max} and can be solved for the positive root; the result is

$$\delta_{max} = \frac{WL}{EA} + \left[\left(\frac{WL}{EA}\right)^2 + 2h\left(\frac{WL}{EA}\right)\right]^{1/2} \tag{2-50}$$

Note that the maximum elongation of the bar increases if either the weight of the collar or the height of fall is increased. The elongation diminishes if the stiffness EA/L is increased.

The preceding equation can be written in simpler form by introducing the notation

$$\delta_{st} = \frac{WL}{EA} = \frac{MgL}{EA} \tag{2-51}$$

in which δ_{st} is the elongation of the bar due to the weight of the collar under static loading conditions. Eq. (2-50) now becomes

$$\delta_{max} = \delta_{st} + (\delta_{st}^2 + 2h\delta_{st})^{1/2} \tag{2-52}$$

or

$$\delta_{max} = \delta_{st}\left[1 + \left(1 + \frac{2h}{\delta_{st}}\right)^{1/2}\right] \tag{2-53}$$

From this equation, note that the elongation of the bar under the impact load is much larger than it would be if the same load were applied statically. Suppose, for instance, that the height h is 40 times the static displacement δ_{st}; the maximum elongation would then be 10 times the static elongation.

When the height h is large compared to the static elongation, disregard the "ones" on the right-hand side of Eq. (2-53) and obtain

$$\delta_{max} = \sqrt{2h\delta_{st}} = \sqrt{\frac{Mv^2L}{EA}} \tag{2-54}$$

in which $M = W/g$ and $v = \sqrt{2gh}$ is the velocity of the falling mass when it strikes the flange. This equation also can be obtained directly from Eq. (2-49) by omitting δ_{max} on the left-hand side of the equation and then solving for δ_{max}. Because of the omitted terms, values of δ_{max} calculated from Eq. (2-54) are always less than those obtained from Eq. (2-53).

Maximum Stress in the Bar

The maximum stress can be calculated easily from the maximum elongation because the stress distribution is assumed to be uniform throughout the length of the bar. From the general equation $\delta = PL/EA = \sigma L/E$, so

$$\sigma_{max} = \frac{E\delta_{max}}{L} \tag{2-55}$$

Substitute from Eq. (2-50), to obtain the following equation for the maximum tensile stress:

$$\sigma_{max} = \frac{W}{A} + \left[\left(\frac{W}{A}\right)^2 + \frac{2WhE}{AL}\right]^{1/2} \quad \textbf{(2-56)}$$

Introducing the notation

$$\sigma_{st} = \frac{W}{A} = \frac{Mg}{A} = \frac{E\delta_{st}}{L} \quad \textbf{(2-57)}$$

in which σ_{st} is the stress when the load acts statically, you can write Eq. (2-56) in the form

$$\sigma_{max} = \sigma_{st} + \left(\sigma_{st}^2 + \frac{2hE}{L}\sigma_{st}\right)^{1/2} \quad \textbf{(2-58)}$$

or

$$\sigma_{max} = \sigma_{st}\left[1 + \left(1 + \frac{2hE}{L\sigma_{st}}\right)^{1/2}\right] \quad \textbf{(2-59)}$$

This equation is analogous to Eq. (2-53) and again shows that an impact load produces much larger effects than when the same load is applied statically.

Again consider the case where the height h is large compared to the elongation of the bar [compare with Eq. (2-54)] to obtain

$$\sigma_{max} = \sqrt{\frac{2hE\sigma_{st}}{L}} = \sqrt{\frac{Mv^2E}{AL}} \quad \textbf{(2-60)}$$

From this result, an increase in the kinetic energy $Mv^2/2$ of the falling mass will increase the stress, whereas an increase in the volume AL of the bar will reduce the stress. This situation is quite different from static tension of the bar, where the stress is independent of the length L and the modulus of elasticity E.

The preceding equations for the maximum elongation and maximum stress apply only at the instant when the flange of the bar is at its lowest position. After the maximum elongation is reached in the bar, the bar will vibrate axially until it comes to rest at the static elongation. From then on, the elongation and stress have the values given by Eqs. (2-51) and (2-57).

Although the preceding equations were derived for the case of a prismatic bar, they can be used for any linearly elastic structure subjected to a falling load, provided you know the appropriate stiffness of the structure. In particular, the equations can be used for a spring by substituting the stiffness k of the spring (see Section 2.2) for the stiffness EA/L of the prismatic bar.

Impact Factor

The ratio of the dynamic response of a structure to the static response (for the same load) is known as an **impact factor**. For instance, the impact factor for the elongation of the bar of Fig. 2-74 is the ratio of the maximum elongation to the static elongation:

$$\text{Impact factor} = \frac{\delta_{max}}{\delta_{st}} \quad \textbf{(2-61)}$$

This factor represents the amount by which the static elongation is amplified due to the dynamic effects of the impact.

Equations analogous to Eq. (2-61) can be written for other impact factors, such as the impact factor for the stress in the bar (the ratio of σ_{max} to σ_{st}). When the collar falls through a considerable height, the impact factor can be very large, such as 100 or more.

Suddenly Applied Load

A special case of impact occurs when a load is applied suddenly with no initial velocity. To explain this kind of loading, consider again the prismatic bar shown in Fig. 2-74 and assume that the sliding collar is lowered gently until it just touches the flange. Then the collar is suddenly released. Although in this instance no kinetic energy exists at the beginning of extension of the bar, the behavior is quite different from that of static loading of the bar. Under static loading conditions, the load is released gradually and equilibrium always exists between the applied load and the resisting force of the bar.

However, consider what happens when the collar is released suddenly from its point of contact with the flange. Initially, the elongation of the bar and the stress in the bar are zero, but then the collar moves downward under the action of its own weight. During this motion, the bar elongates and its resisting force gradually increases. The motion continues until at some instant the resisting force just equals W, which is the weight of the collar. At this particular instant, the elongation of the bar is δ_{st}. However, the collar now has a certain kinetic energy, which it acquired during the downward displacement δ_{st}. Therefore, the collar continues to move downward until its velocity is brought to zero by the resisting force in the bar. The maximum elongation for this condition is obtained from Eq. (2-53) by setting h equal to zero; thus,

$$\delta_{max} = 2\delta_{st} \tag{2-62}$$

From this equation, note that a suddenly applied load produces an elongation twice as large as the elongation caused by the same load applied statically. Thus, the impact factor is 2.

After the maximum elongation $2\delta_{st}$ has been reached, the end of the bar will move upward and begin a series of up and down vibrations, eventually coming to rest at the static elongation produced by the weight of the collar.[7]

Limitations

The preceding analyses were based upon the assumption that no energy losses occur during impact. In reality, energy losses always occur, with most of the lost energy being dissipated in the form of heat and localized deformation of the materials. Because of these losses, the kinetic energy of a system immediately after an impact is less than it was before the impact. Consequently, less energy is converted into strain energy of the bar than was previously assumed. As a result, the actual displacement of the end of the bar of Fig. 2-74 is less than that predicted by this simplified analysis.

The stresses in the bar were also assumed to remain within the proportional limit. If the maximum stress exceeds this limit, the analysis becomes more

[7]Equations (2-62) was first obtained by the French mathematician and scientist J. V. Poncelet (1788–1867) (see Ref. 2-8).

complicated because the elongation of the bar is no longer proportional to the axial force. Other factors to consider are the effects of stress waves, damping, and imperfections at the contact surfaces. Therefore, remember that all of the formulas in this section are based upon highly idealized conditions and give only a rough approximation of the true conditions (usually overestimating the elongation).

Materials that exhibit considerable ductility beyond the proportional limit generally offer much greater resistance to impact loads than do brittle materials. Also, bars with grooves, holes, and other forms of stress concentrations (see Sections 2.9 and 2.10) are very weak against impact—a slight shock may produce fracture, even when the material itself is ductile under static loading.

Example 2-19

A round, prismatic steel bar ($E = 210$ GPa) of length $L = 2$ m and diameter $d = 15$ mm hangs vertically from a support at its upper end (Fig. 2-75). A sliding collar of mass $M = 20$ kg drops from a height $h = 150$ mm onto the flange at the lower end of the bar without rebounding.

(a) Calculate the maximum elongation of the bar due to the impact and determine the corresponding impact factor.

(b) Calculate the maximum tensile stress in the bar and determine the corresponding impact factor.

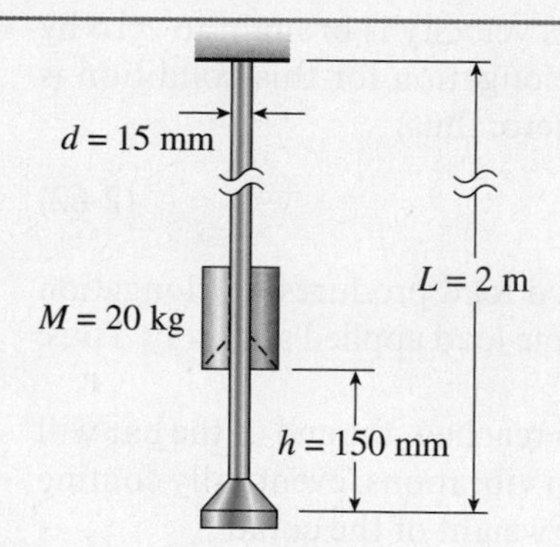

FIGURE 2-75
Example 2-19: Impact load on a vertical bar

Solution:

Use a four-step problem-solving approach.

1, 2. ***Conceptualize, Categorize***: Because the arrangement of the bar and collar in this example matches the arrangement shown in Fig. 2-74, you can use the equations derived previously [Eqs. (2-49) to (2-60)].

Part (a): Maximum elongation.

3. ***Analyze***: The elongation of the bar produced by the falling collar can be determined from Eq. (2-53). The first step is to determine the static elongation of the bar due to the weight of the collar. Since the weight of the collar is Mg

$$\delta_{st} = \frac{MgL}{EA} = \frac{(20 \text{ kg})\left(9.81 \text{ m/s}^2\right)(2 \text{ mm})}{210 \text{ GPa}\left(\dfrac{\pi}{4}\right)(15 \text{ mm})^2} = 0.0106 \text{ mm}$$

From this result,

$$\frac{h}{\delta_{st}} = \frac{150 \text{ mm}}{0.0106 \text{ mm}} = 14{,}150$$

The preceding numerical values may now be substituted into Eq. (2-53) to obtain the maximum elongation:

$$\delta_{max} = \delta_{st}\left[1 + 1\left(1 + \frac{2h}{\delta_{st}}\right)^{1/2}\right]$$

$$= (0.0106 \text{ mm})\left[1 + \sqrt{1 + 2(14{,}150)}\right]$$

$$= 1.79 \text{ mm}$$

Since the height of the fall is very large compared to the static elongation, nearly the same result is obtained by calculating the maximum elongation from Eq. (2-54):

$$\delta_{max} = \sqrt{2h\delta_{st}} = \sqrt{2(150 \text{ mm})[2(150 \text{ mm})(0.0106 \text{ mm})]^{1/2}} = 1.78 \text{ mm}$$

The impact factor is equal to the ratio of the maximum elongation to the static elongation:

$$\text{Impact factor} = \frac{\delta_{max}}{\delta_{st}} = \frac{1.79 \text{ mm}}{0.0106 \text{ mm}} = 169$$

4. ***Finalize***: This result shows that the effects of a dynamically applied load can be very large as compared to the effects of the same load acting statically.

Part (b): Maximum tensile stress.

3. ***Analyze***: The maximum stress produced by the falling collar is obtained from Eq. (2-55) as follows:

$$\sigma_{max} = \frac{E\delta_{max}}{L} = \frac{(210 \text{ GPa})(1.79 \text{ mm})}{2.0 \text{ m}} = 188 \text{ MPa}$$

This stress may be compared with the static stress [see Eq. (2-57)], which is

$$\sigma_{st} = \frac{W}{A} = \frac{Mg}{A} = \frac{(20 \text{ kg})(9.81 \text{ m/s}^2)}{(\pi/4)(15 \text{ mm})^2} = 1.11 \text{ MPa}$$

4. ***Finalize***: The ratio of σ_{max} to σ_{st} is 188/1.11 = 169, which is the same impact factor as for the elongations. This result is expected, because the stresses are directly proportional to the corresponding elongations [see Eqs. (2-55) and (2-57)].

Example 2-20

A horizontal bar AB of length L is struck at its free end by a heavy block of mass M moving horizontally with a velocity v (Fig. 2-76).

(a) Determine the maximum shortening δ_{max} of the bar due to the impact and determine the corresponding impact factor.

(b) Determine the maximum compressive stress σ_{max} and the corresponding impact factor. (Let EA represent the axial rigidity of the bar.)

FIGURE 2-76

Example 2-20: Impact load on a horizontal bar

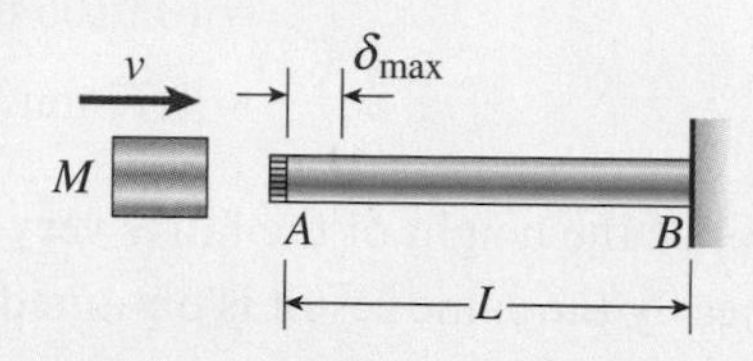

Solution:

Use a four-step problem-solving approach.

1, 2. ***Conceptualize, Categorize***: The loading on the bar in this example is quite different from the loads on the bars pictured in Figs. 2-74 and 2-75. Therefore, a new analysis is required based upon conservation of energy.

Part (a): Maximum shortening of the bar.

3. ***Analyze***: For this analysis, make the same assumptions as those described previously. Thus, disregard all energy losses and assume that the kinetic energy of the moving block is transformed entirely into strain energy of the bar.

The kinetic energy of the block at the instant of impact is $Mv^2/2$. The strain energy of the bar when the block comes to rest at the instant of maximum shortening is $EA\delta^2_{max}/2L$, as given by Eq. (2-37b). Therefore, the following equation of conservation of energy applies:

$$\frac{Mv^2}{2} = \frac{EA\delta^2{}_{max}}{2L} \tag{2-63}$$

Solving for δ_{max} gives

$$\delta_{max} = \sqrt{\frac{Mv^2L}{EA}} \tag{2-64}$$

As expected, this equation is the same as Eq. (2-54).

Now compute the static displacement of the end of the bar to find the impact factor. In this case, the static displacement is the shortening of the bar due to the weight of the block applied as a compressive load on the bar [see Eq. (2-51)]:

$$\delta_{st} = \frac{WL}{EA} = \frac{MgL}{EA}$$

Thus, the impact factor is

$$\text{Impact factor} = \frac{\delta_{max}}{\delta_{st}} = \sqrt{\frac{EAv^2}{Mg^2L}} \tag{2-65}$$

4. ***Finalize***: The value determined from this equation may be much larger than 1.

Part (b): Maximum compressive strength in the bar.

3. ***Analyze***: The maximum stress in the bar is found from the maximum shortening by means of Eq. (2-55):

$$\sigma_{max} = \frac{E\delta_{max}}{L} = \frac{E}{L}\sqrt{\frac{Mv^2L}{EA}} = \sqrt{\frac{Mv^2E}{AL}} \tag{2-66}$$

This equation is the same as Eq. (2-60).

4. ***Finalize***: The static stress σ_{st} in the bar is equal to W/A or Mg/A, which [in combination with Eq. (2-66)] leads to the same impact factor as before [Eq. (2-65)].

*2.9 Repeated Loading and Fatigue

The behavior of a structure depends not only upon the nature of the material but also upon the character of the loads. In some situations, the loads are static—they are applied gradually, act for long periods of time, and change slowly. Other loads are dynamic in character—examples are impact loads acting suddenly (Section 2.8) and repeated loads recurring for large numbers of cycles.

Some typical patterns for **repeated loads** are sketched in Fig. 2-77. The first graph (a) shows a load that is applied, removed, and applied again, while always acting in the same direction. The second graph (b) shows an alternating load that reverses direction during every cycle of loading, and the third graph (c) illustrates a fluctuating load that varies about an average value. Repeated loads are commonly associated with machinery, engines, turbines, generators, shafts, propellers, airplane parts, automobile parts, and the like. Some of these structures are subjected to millions (and even billions) of loading cycles during their useful life.

A structure subjected to dynamic loads is likely to fail at a lower stress than when the same loads are applied statically, especially when the loads are repeated for a large number of cycles. In such cases, failure is usually caused by **fatigue**, or **progressive fracture**. A familiar example of a fatigue failure is stressing a metal paper clip to the breaking point by repeatedly bending it back and forth. If the clip is bent only once, it does not break. But if the load is reversed by bending the clip in the opposite direction and if the entire loading cycle is repeated several times, the clip will finally break. *Fatigue* may be defined as the

FIGURE 2-77

Types of repeated loads: (a) load acting in one direction only, (b) alternating or reversed load, and (c) fluctuating load that varies about an average value

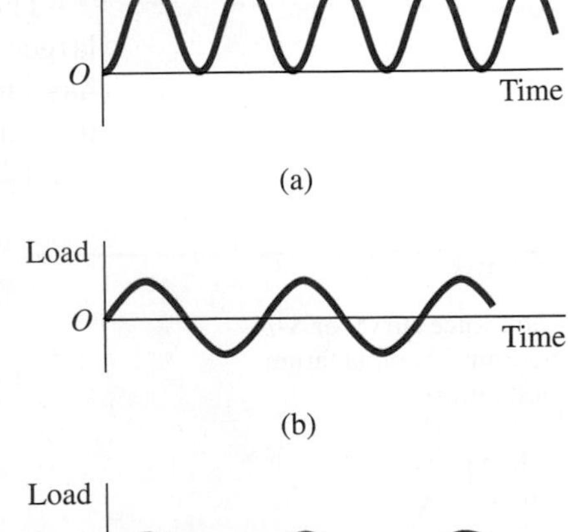

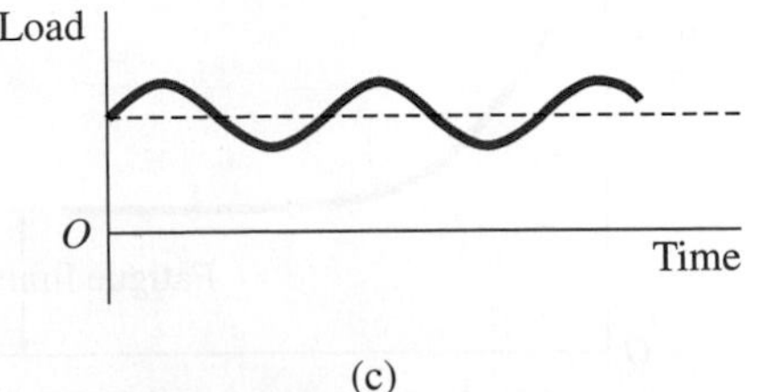

FIGURE 2-78

Fatigue failure of a bar loaded repeatedly in tension; the crack spread gradually over the cross section until fracture occurred suddenly (Courtesy of MTS System Corporation)

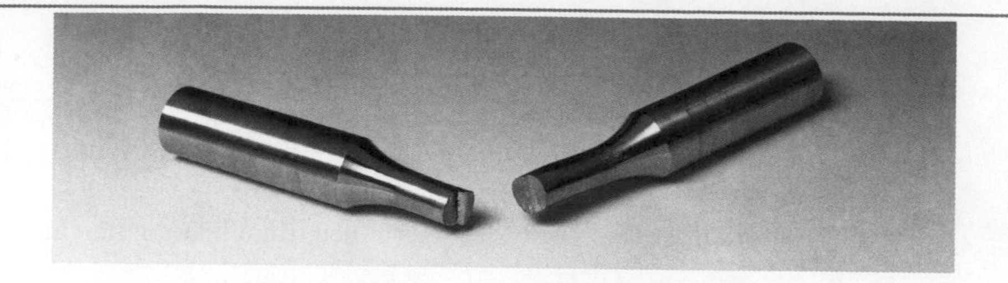

deterioration of a material under repeated cycles of stress and strain resulting in progressive cracking that eventually produces fracture.

In a typical fatigue failure, a microscopic crack forms at a point of high stress (usually at a *stress concentration*, as discussed in the next section) and gradually enlarges as the loads are applied repeatedly. When the crack becomes so large that the remaining material cannot resist the loads, a sudden fracture of the material occurs (Fig. 2-78). Depending upon the nature of the material, it may take anywhere from a few cycles of loading to hundreds of millions of cycles to produce a fatigue failure.

The magnitude of the load causing a fatigue failure is less than the load that can be sustained statically, as already pointed out. To determine the failure load, tests of the material must be performed. In the case of repeated loading, the material is tested at various stress levels and the number of cycles to failure is counted. For instance, a specimen of material is placed in a fatigue-testing machine and loaded repeatedly to a certain stress, say σ_1. The loading cycles are continued until failure occurs, and the number n of loading cycles to failure is noted. The test is then repeated for a different stress, say σ_2. If σ_2 is greater than σ_1, the number of cycles to failure will be smaller. If σ_2 is less than σ_1, the number will be larger. Eventually, enough data are accumulated to plot an **endurance curve**, or ***S-N* diagram**, in which failure stress (S) is plotted versus the number (N) of cycles to failure (Fig. 2-79). The vertical axis is usually a linear scale, and the horizontal axis is usually a logarithmic scale.

The endurance curve of Fig. 2-79 shows that the smaller the stress, the larger the number of cycles to produce failure. For some materials, the curve has a horizontal asymptote known as the **fatigue limit** or **endurance limit**. When it exists, this limit is the stress below which a fatigue failure will not occur regardless of how many times the load is repeated. The precise shape of an endurance curve depends upon many factors, including properties of the material, geometry of the test specimen, speed of testing, pattern of loading, and surface condition of the specimen. The results of numerous fatigue tests, made on a great variety of materials and structural components, have been reported in the engineering literature.

FIGURE 2-79

Endurance curve, or *S-N* diagram, showing fatigue limit failure

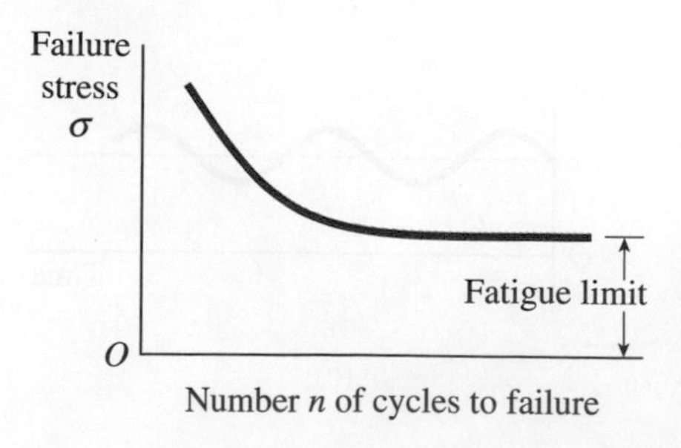

Typical *S-N* diagrams for steel and aluminum are shown in Fig. 2-80. The ordinate is the failure stress, expressed as a percentage of the ultimate stress for the material, and the abscissa is the number of cycles at which failure occurred. Note that the number of cycles is plotted on a logarithmic scale. The curve for steel becomes horizontal at about 10^7 cycles, and the fatigue limit is about 50% of the ultimate tensile stress for ordinary static loading. The fatigue limit for aluminum is not as clearly defined as that for steel, but a typical value of the fatigue limit is the stress at 5×10^8 cycles, or about 25% of the ultimate stress.

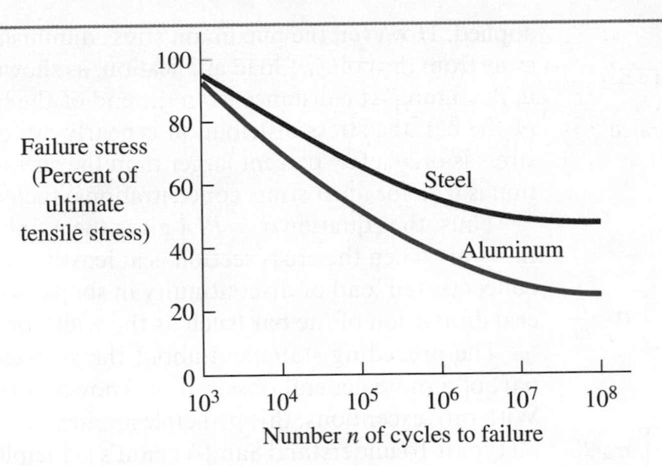

FIGURE 2-80

Typical endurance curves for steel and aluminum in alternating (reversed) loading

Since fatigue failures usually begin with a microscopic crack at a point of high localized stress (that is, at a stress concentration), the condition of the surface of the material is extremely important. Highly polished specimens have higher endurance limits. Rough surfaces, especially those at stress concentrations around holes or grooves, greatly lower the endurance limit. Corrosion, which creates tiny surface irregularities, has a similar effect. For steel, ordinary corrosion may reduce the fatigue limit by more than 50%.

*2.10 Stress Concentrations

When determining the stresses in axially loaded bars, it is customary to use the basic formula $\sigma = P/A$ in which P is the axial force in the bar and A is its cross-sectional area. This formula is based upon the assumption that the stress distribution is uniform throughout the cross section. In reality, bars often have holes, grooves, notches, keyways, shoulders, threads, or other abrupt changes in geometry that create a disruption in the otherwise uniform stress pattern. These discontinuities in geometry cause high stresses in very small regions of the bar, and these high stresses are known as **stress concentrations**. The discontinuities themselves are known as **stress raisers**.

Stress concentrations also appear at points of loading. For instance, a load may act over a very small area and produce high stresses in the region around its point of application. An example is a load applied through a pin connection, where the load is applied over the bearing area of the pin.

The stresses existing at stress concentrations can be determined either by experimental methods or by advanced methods of analysis, including the finite-element method. The results of such research for many cases of practical interest are readily available in the engineering literature (Ref. 2-9). Some typical stress-concentration data are given later in this section and also in Chapters 3 and 5.

Saint-Venant's Principle

To illustrate the nature of stress concentrations, consider the stresses in a bar of rectangular cross section (width b, thickness t) subjected to a concentrated load P at the end (Fig. 2-81). The peak stress directly under the load may be several times the average stress P/bt, depending upon the area over which the load is

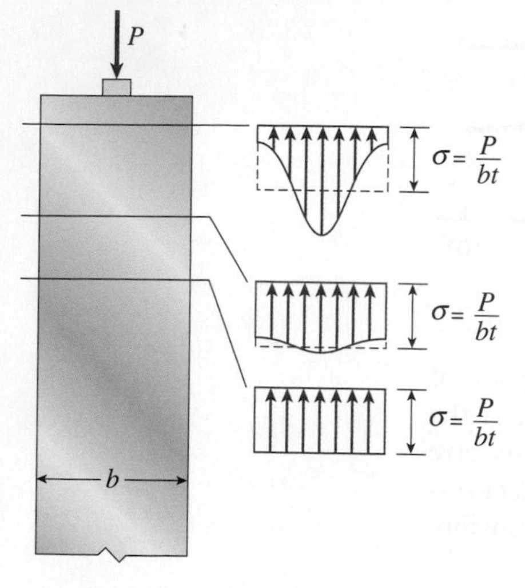

FIGURE 2-81

Stress distributions near the end of a bar of rectangular cross section (width b, thickness t) subjected to a concentrated load P acting over a small area

applied. However, the maximum stress diminishes rapidly as you move away from the point of load application, as shown by the stress diagrams in the figure. At a distance from the end of the bar equal to the width b of the bar, the stress distribution is nearly uniform, and the maximum stress is only a few percent larger than the average stress. This observation is true for most stress concentrations, such as holes and grooves.

Thus, the equation $\sigma = P/A$ gives the axial stresses on a cross section only when the cross section is at least a distance b away from any concentrated load or discontinuity in shape, where b is the largest lateral dimension of the bar (such as the width or diameter).

The preceding statement about the stresses in a prismatic bar is part of a more general observation known as **Saint-Venant's principle**. With rare exceptions, this principle applies to linearly elastic bodies of all types. To understand Saint-Venant's principle, consider a body with a system of loads acting over a small part of its surface such as a prismatic bar of width b subjected to a system of several concentrated loads acting at the end (Fig. 2-82a). For simplicity, assume that the loads are symmetrical and have only a vertical resultant.

Next, consider a different but statically equivalent load system acting over the same small region of the bar. ("Statically equivalent" means the two load systems have the same force resultant and same moment resultant.) For instance, the uniformly distributed load shown in Fig. 2-82b is statically equivalent to the system of concentrated loads shown in Fig. 2-82a. Saint-Venant's principle states that the stresses in the body caused by either of the two systems of loading are the same, provided you move away from the loaded region a distance at least equal to the largest dimension of the loaded region (distance b in this example). Thus, the stress distributions shown in Fig. 2-81 are an illustration of Saint-Venant's principle. Of course, this "principle" is not a rigorous law of mechanics but is a common-sense observation based upon theoretical and practical experience.

FIGURE 2-82

Illustration of Saint-Venant's principle: (a) system of concentrated loads acting over a small region of a bar and (b) statically equivalent system

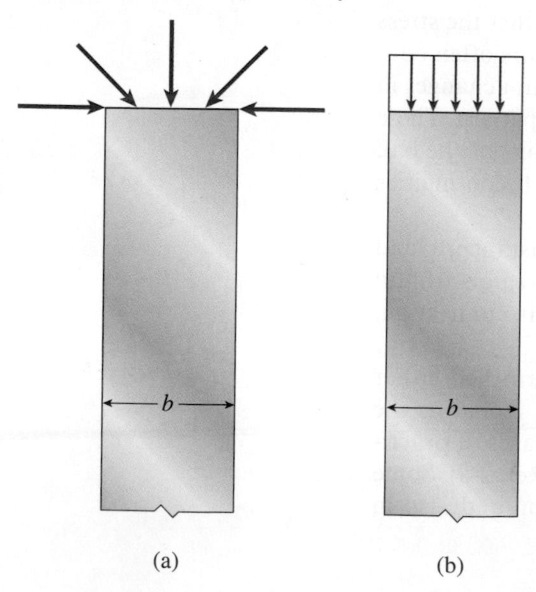

Saint-Venant's principle has great practical significance in the design and analysis of bars, beams, shafts, and other structures encountered in mechanics of materials. Because the effects of stress concentrations are localized, all of the standard stress formulas (such as $\sigma = P/A$) can be used at cross sections a sufficient distance away from the source of the concentration. Close to the source, the stresses depend upon the details of the loading and the shape of the member. Furthermore, formulas that are applicable to entire members, such as formulas for elongations, displacements, and strain energy, give satisfactory results even when stress concentrations are present. The explanation lies in the fact that stress concentrations are localized and have little effect on the overall behavior of a member.[8]

[8]Saint-Venant's principle is named for Barré de Saint-Venant (1797–1886), a famous French mathematician and elastician (Ref. 2-10). It appears that the principle applies generally to solid bars and beams but not to all thin-walled open sections. For a discussion of the limitations of Saint-Venant's principle, see Ref. 2-11.

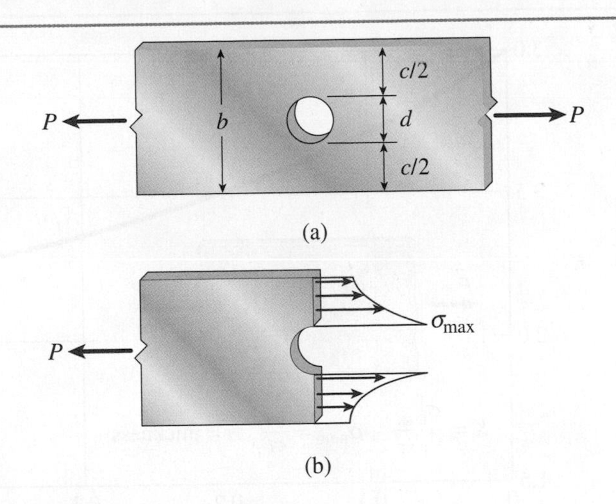

FIGURE 2-83

Stress distribution in a flat bar with a circular hole

Stress-Concentration Factors

Now consider some particular cases of stress concentrations caused by discontinuities in the shape of a bar. Begin with a bar of rectangular cross section having a circular hole and subjected to a tensile force P (Fig. 2-83). The bar is relatively thin, with its width b being much larger than its thickness t. Also, the hole has a diameter d.

The normal stress acting on the cross section through the center of the hole has the distribution shown in Fig. 2-83. The maximum stress σ_{max} occurs at the edges of the hole and may be significantly larger than the *nominal stress* $\sigma = P/ct$ at the same cross section. (Note that ct is the net area at the cross section through the hole.) The intensity of a stress concentration is usually expressed by the ratio of the maximum stress to the nominal stress, called the **stress-concentration factor** K:

$$K = \frac{\sigma_{max}}{\sigma_{nom}} \qquad \textbf{(2-67)}$$

For a bar in tension, the nominal stress is the average stress based upon the net cross-sectional area. In other cases, a variety of stresses may be used. Thus, whenever a stress concentration factor is used, it is important to note carefully how the nominal stress is defined.

A graph of the stress-concentration factor K for a bar with a hole is given in Fig. 2-84. If the hole is tiny, the factor K equals 3, which means that the maximum stress is three times the nominal stress. As the hole becomes larger in proportion to the width of the bar, K becomes smaller, and the effect of the concentration is not as severe.

From Saint-Venant's principle, observe that, at distances equal to the width b of the bar *away* from the hole in either axial direction, the stress distribution is practically uniform and equal to P divided by the gross cross-sectional area ($\sigma = P/bt$).

To reduce the stress-concentration effects (see Fig. 2-85), *fillets* are used to round off the re-entrant corners.[9] Stress-concentration factors for two other cases of practical interest are given in Figs. 2-86 and 2-87. These graphs

[9]A *fillet* is a curved concave surface formed where two other surfaces meet. Its purpose is to round off what would otherwise be a re-entrant corner.

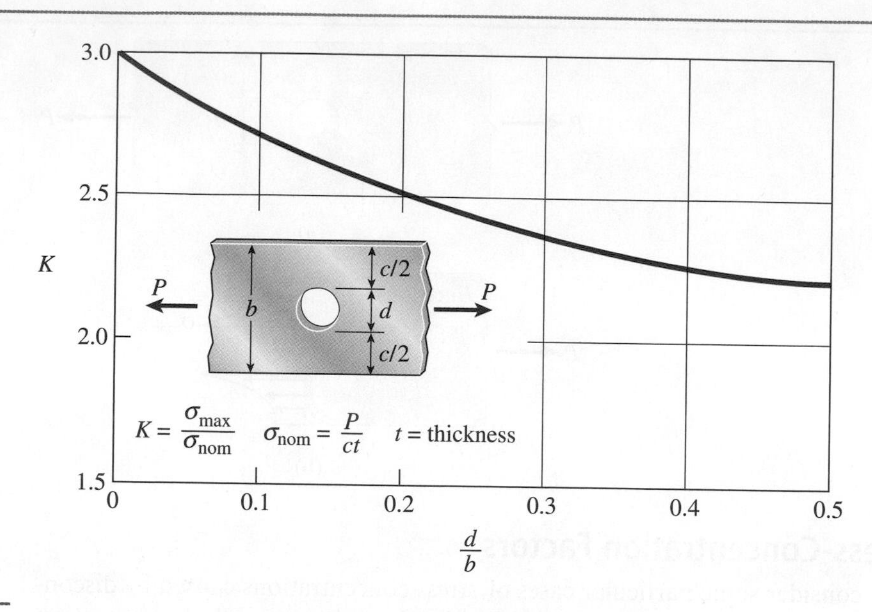

FIGURE 2-84

Stress-concentration factor K for flat bars with circular holes

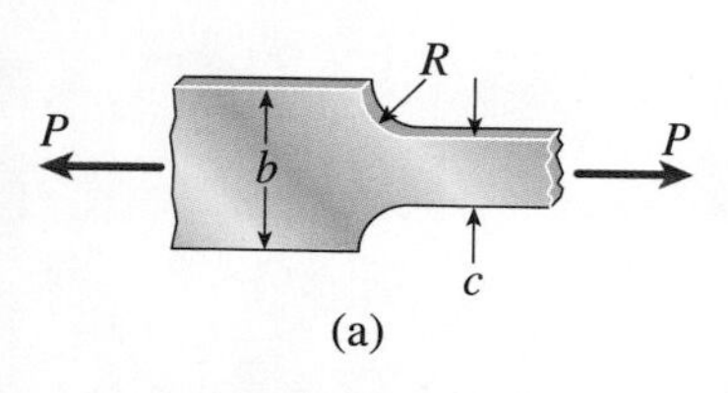

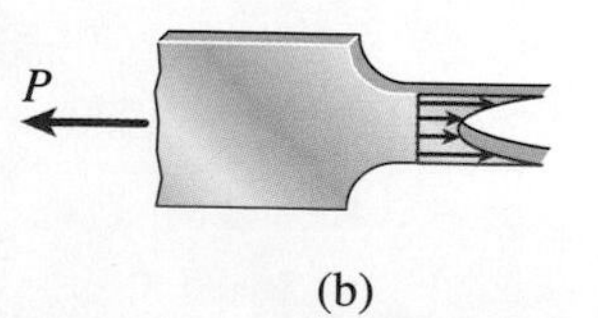

FIGURE 2-85

Stress distribution in a flat bar with shoulder fillets

are for flat bars and circular bars, respectively, that are stepped down in size, forming a *shoulder.* Without the fillets, the stress-concentration factors would be extremely large, as indicated at the left-hand side of each graph where K approaches infinity as the fillet radius R approaches zero. In both cases, the maximum stress occurs in the smaller part of the bar in the region of the fillet.[10]

Designing for Stress Concentrations

Because of the possibility of fatigue failures, stress concentrations are especially important when the member is subjected to repeated loading. As explained in the preceding section, cracks begin at the point of highest stress and then spread gradually through the material as the load is repeated. In practical design, the fatigue limit (Fig. 2-79) is considered to be the ultimate stress for the material

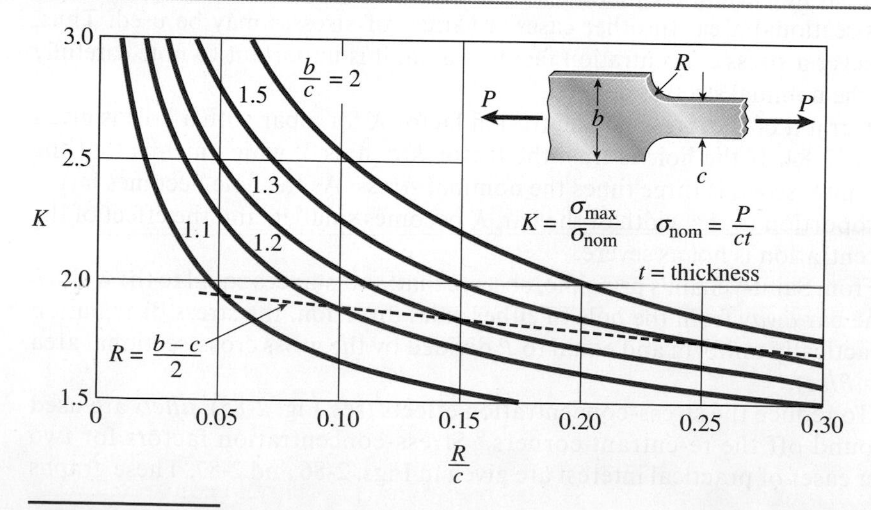

FIGURE 2-86

Stress-concentration factor K for flat bars with shoulder fillets; the dashed line is for a full quarter-circular fillet

[10]The stress-concentration factors given in the graphs are theoretical factors for bars of linearly elastic material. The graphs are plotted from the formulas given in Ref. 2-9.

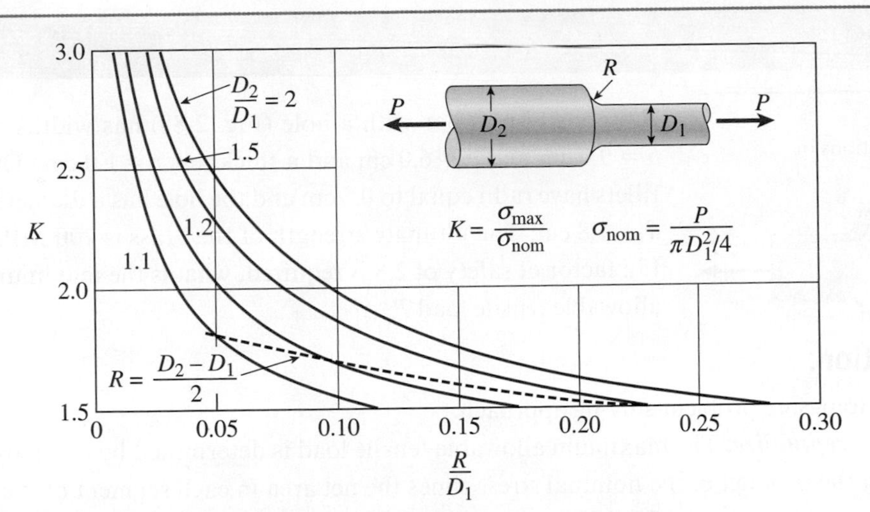

FIGURE 2-87
Stress-concentration factor K for round bars with shoulder fillets; the dashed line is for a full quarter-circular fillet

when the number of cycles is extremely large. The allowable stress is obtained by applying a factor of safety with respect to this ultimate stress. Then the peak stress at the stress concentration is compared with the allowable stress.

In many situations, the use of the full theoretical value of the stress-concentration factor is too severe. Fatigue tests usually produce failure at higher levels of the nominal stress than those obtained by dividing the fatigue limit by K. In other words, a structural member under repeated loading is not as sensitive to a stress concentration as the value of K indicates, and a reduced stress-concentration factor is often used.

Other kinds of dynamic loads, such as impact loads, also require that stress-concentration effects be taken into account. Unless better information is available, the full stress-concentration factor should be used. Members subjected to low temperatures also are highly susceptible to failures at stress concentrations, and therefore, special precautions should be taken in such cases.

The significance of stress concentrations when a member is subjected to static loading depends upon the kind of material. With ductile materials, such as structural steel, a stress concentration can often be ignored. The reason is that the material at the point of maximum stress (such as around a hole) will yield and plastic flow will occur, thus reducing the intensity of the stress concentration and making the stress distribution more nearly uniform. On the other hand, with brittle materials (such as glass), a stress concentration will remain up to the point of fracture. Therefore, the general observation is that, with static loads and a ductile material, the stress-concentration effect is not likely to be important, but with static loads and a brittle material, the full stress-concentration factor should be considered.

Stress concentrations can be reduced in intensity by properly proportioning the parts. Generous fillets reduce stress concentrations at re-entrant corners. Smooth surfaces at points of high stress, such as on the inside of a hole, inhibit the formation of cracks. Proper reinforcing around holes also can be beneficial. There are many other techniques for smoothing out the stress distribution in a structural member and thereby reducing the stress-concentration factor. These techniques, which are usually studied in engineering design courses, are extremely important in the design of aircraft, ships, and machines. Many unnecessary structural failures have occurred because designers failed to recognize the effects of stress concentrations and fatigue.

Example 2-21

FIGURE 2-88

Example 2-21: Stress concentrations in stepped bar with a hole

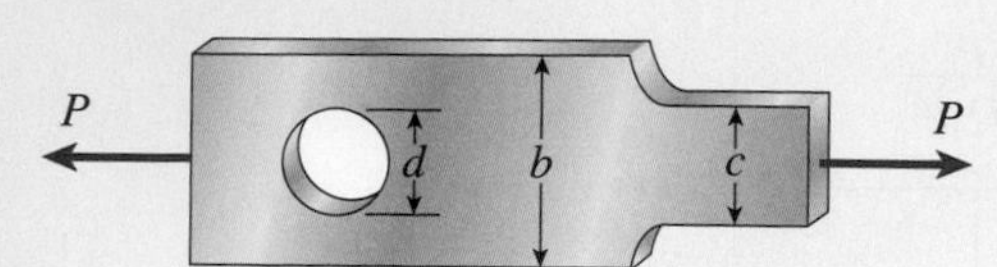

A stepped brass bar with a hole (Fig. 2-88) has widths of $b = 9.0$ cm and $c = 6.0$ cm and a thickness $t = 1.0$ cm. The fillets have radii equal to 0.5 cm and the hole has a diameter $d = 1.8$ cm. The ultimate strength of the brass is 200 MPa. If a factor of safety of 2.8 is required, what is the maximum allowable tensile load P_{max}?

Solution:

Use a four-step problem-solving approach.

1. ***Conceptualize***: The maximum allowable tensile load is determined by comparing the product of the nominal stress times the net area in each segment of the stepped bar (that is, the segment with a hole and the segment with fillets). Use the appropriate stress concentration factor K for each segment: for the segment with the hole, use Fig. 2-84, and for the segment with fillets, use Fig. 2-86.
2. ***Categorize***: For the segment of the bar of width b and thickness t and having a hole of diameter d, the net cross-sectional area is $(b - d)(t)$, and the nominal axial stress is computed as

$$\sigma_1 = \frac{P}{(b-d)t} \quad \text{and} \quad \sigma_1 = \frac{\sigma_{\text{allow}}}{K_{\text{hole}}} = \frac{\left(\dfrac{\sigma_U}{FS_U}\right)}{K_{\text{hole}}} \qquad \textbf{(a, b)}$$

where the maximum stress is set equal to the allowable stress and the stress concentration factor K_{hole} is obtained from Fig. 2-84. Now equate the nominal stress expressions in Eqs. (a) and (b) and solve for P_{max}:

$$\sigma_1 = \frac{P}{(b-d)t} = \frac{\left(\dfrac{\sigma_U}{FS_U}\right)}{K_{\text{hole}}} \quad \text{so} \quad P_{\text{maxh}} = \frac{\left(\dfrac{\sigma_U}{FS_U}\right)}{K_{\text{hole}}}(b-d)t \qquad \textbf{(c)}$$

Next investigate the tensile load-carrying capacity of the stepped bar in the segment having fillets of radius R. Following the procedure used to find P_{maxh} in Eqs. (a) through (c) leads to a second expression for P_{max}:

$$P_{\text{max f}} = \frac{\left(\dfrac{\sigma_U}{FS_U}\right)}{K_{\text{fillet}}}(ct) \qquad \textbf{(d)}$$

3. ***Analyze***: **Segment with hole:** Using the given numerical properties of the bar, ratio $d/b = 1.8/9.0 = 0.2$ so K_{hole} is approximately 2.51 from Fig. 2-84 (see Fig. 2-89). The allowable tensile load on the stepped bar *accounting for stress concentrations at the hole* using Eq. (c) is

$$P_{\text{maxh}} = \frac{\left(\dfrac{200\text{ MPa}}{2.8}\right)}{2.51} \cdot (9.0\text{ cm} - 1.8\text{ cm}) \cdot (1.0\text{ cm}) = 20.5\text{ kN} \qquad \textbf{(e)}$$

FIGURE 2-89
Selection of factor K_{hole} using Fig. 2-84

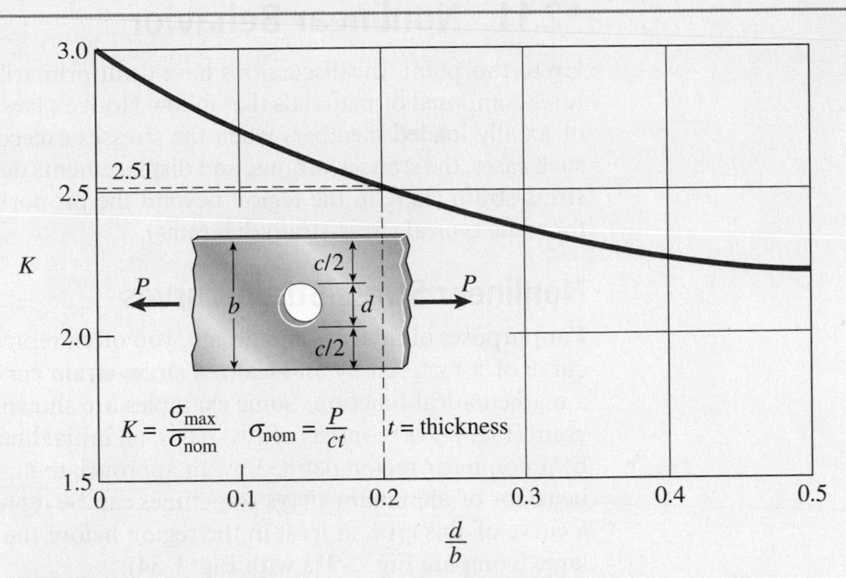

Segment with fillets: The stress concentration factor K_{fillet} is obtained from Fig. 2-86 using two parameters: the ratio of fillet radius to reduced width c ($R/c = 0.1$) and the ratio of bar full width to reduced width ($b/c = 1.5$). In the fillet segment, reduced width $c = b - 2R = 6.0$ cm. The stress concentration factor is approximately 2.35 (see Fig. 2-90), and so the maximum allowable tensile load based on stress concentrations in the fillet region of the bar is found using Eq. (d):

$$P_{maxf} = \frac{\left(\frac{200\text{ MPa}}{2.8}\right)}{2.35}(6.0\text{ cm})(1\text{ cm}) = 18.24\text{ kN} \tag{f}$$

FIGURE 2-90
Selection of factor K_{fillet} using Fig. 2-86

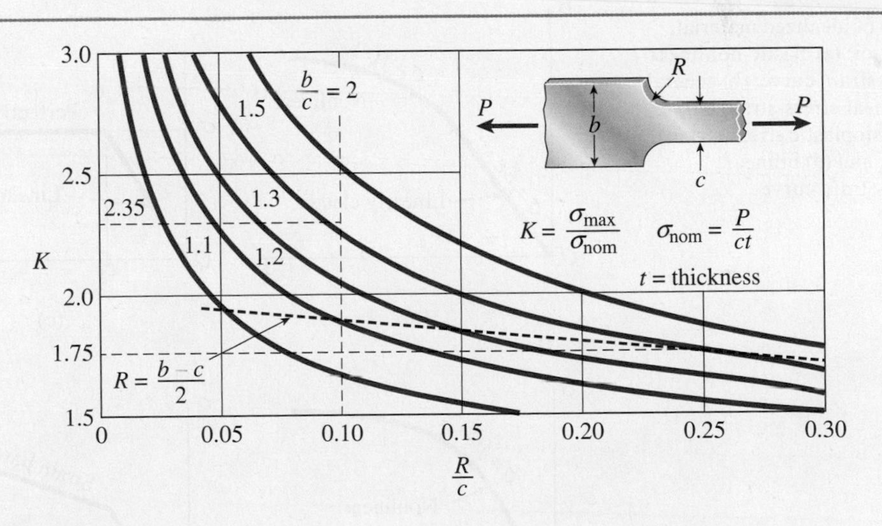

4. ***Finalize***: Comparing Eqs. (e) and (f), the lesser value of the maximum allowable tensile load, P_{maxf}, which is based on stress concentrations in the fillet region of the stepped bar, controls here.

*2.11 Nonlinear Behavior

Up to this point, the discussions have dealt primarily with members and structures composed of materials that follow Hooke's law. Now consider the behavior of axially loaded members when the stresses exceed the proportional limit. In such cases, the stresses, strains, and displacements depend upon the shape of the stress-strain curve in the region beyond the proportional limit (see Section 1.5 for some typical stress-strain diagrams).

Nonlinear Stress-Strain Curves

For purposes of analysis and design, you often represent the actual stress-strain curve of a material by an **idealized stress-strain curve** that can be expressed as a mathematical function. Some examples are shown in Fig. 2-91. The first diagram (Fig. 2-91a) consists of two parts: an initial linearly elastic region followed by a nonlinear region defined by an appropriate mathematical expression. The behavior of aluminum alloys sometimes can be represented quite accurately by a curve of this type, at least in the region before the strains become excessively large (compare Fig. 2-91a with Fig. 1-34).

In the second example (Fig. 2-91b), a single mathematical expression is used for the entire stress-strain curve. The best known expression of this kind is the Ramberg-Osgood stress-strain law, which is described later in more detail.

The stress-strain diagram frequently used for structural steel is shown in Fig. 2-91c. Because steel has a linearly elastic region followed by a region of considerable yielding (see the stress-strain diagrams of Figs. 1-31 and 1-33), its behavior can be represented by two straight lines. The material is assumed to follow Hooke's law up to the yield stress σ_y, after which

FIGURE 2-91

Types of idealized material behavior: (a) elastic-nonlinear stress-strain curve, (b) general nonlinear stress-strain curve, (c) elastoplastic stress-strain curve, and (d) bilinear stress-strain curve

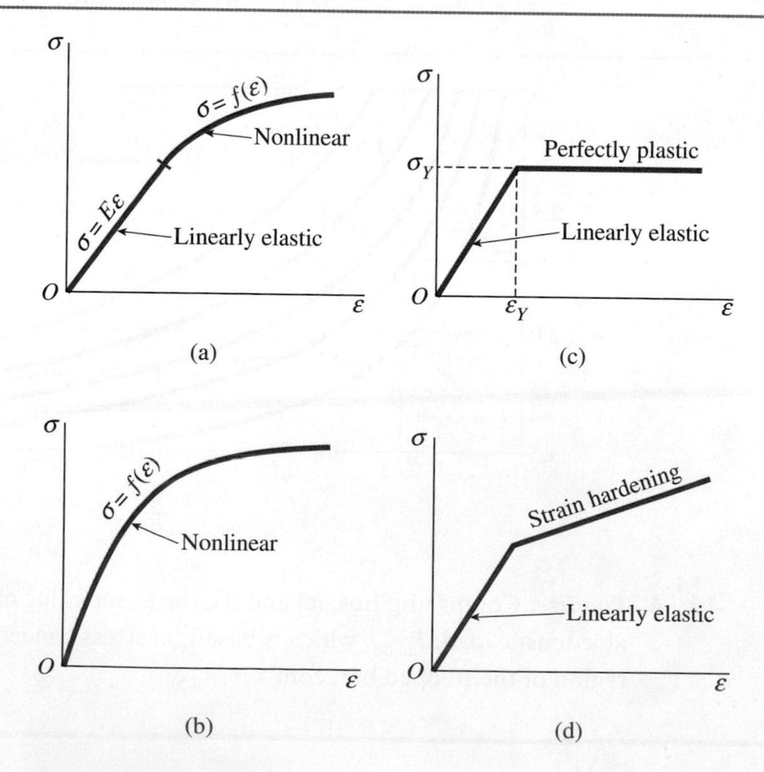

it yields under constant stress, the latter behavior being known as **perfect plasticity**. The perfectly plastic region continues until the strains are 10 or 20 times larger than the yield strain. A material having a stress-strain diagram of this kind is called an **elastoplastic material** (or *elastic-plastic material*).

Eventually, as the strain becomes extremely large, the stress-strain curve for steel rises above the yield stress due to strain hardening, as explained in Section 1.5. However, by the time strain hardening begins, the displacements are so large that the structure will have lost its usefulness. Consequently, it is common practice to analyze steel structures on the basis of the elastoplastic diagram shown in Fig. 2-91c, with the same diagram being used for both tension and compression. An analysis made with these assumptions is called an **elastoplastic analysis**, or simply **plastic analysis**, and is described in the next section.

Fig. 2-91d shows a stress-strain diagram consisting of two lines having different slopes, called a **bilinear stress-strain diagram**. Note that in both parts of the diagram the relationship between stress and strain is linear, but only in the first part is the stress proportional to the strain (Hooke's law). This idealized diagram may be used to represent materials with strain hardening, or it may be used as an approximation to diagrams of the general nonlinear shapes shown in Figs. 2-91a and b.

Changes in Lengths of Bars

The elongation or shortening of a bar can be determined if the stress-strain curve of the material is known. To illustrate the general procedure, consider the tapered bar AB shown in Fig. 2-92a. Both the cross-sectional area and the axial force vary along the length of the bar, and the material has a general nonlinear stress-strain curve (Fig. 2-92b). Because the bar is statically determinate, you can determine the internal axial forces at all cross sections from static equilibrium alone. Then find the stresses by dividing the forces by the cross-sectional areas, and find the strains from the stress-strain curve. Lastly, determine the change in length from the strains, as described in the following paragraph.

The change in length of an element dx of the bar (Fig. 2-92a) is εdx, where ε is the strain at distance x from the end. Integrate this expression from one end of the bar to the other to obtain the change in length of the entire bar:

$$\delta = \int_0^L \varepsilon dx \qquad \textbf{(2-68)}$$

where L is the length of the bar. If the strains are expressed analytically, that is, by algebraic formulas, it may be possible to integrate Eq. (2-68) by formal mathematical means and thus obtain an expression for the change in length. If the stresses and strains are expressed numerically, that is, by a series of numerical

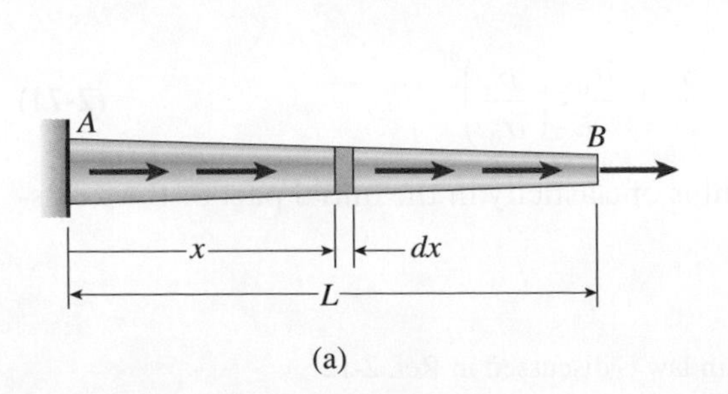

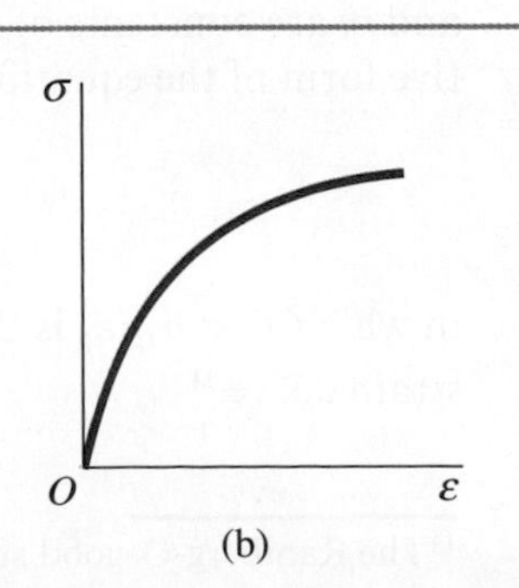

FIGURE 2-92

Change in length of a tapered bar consisting of a material having a nonlinear stress-strain curve

FIGURE 2-93

Stress-strain curve for an aluminum alloy using the Ramberg-Osgood equation [Eq. (2-72)]

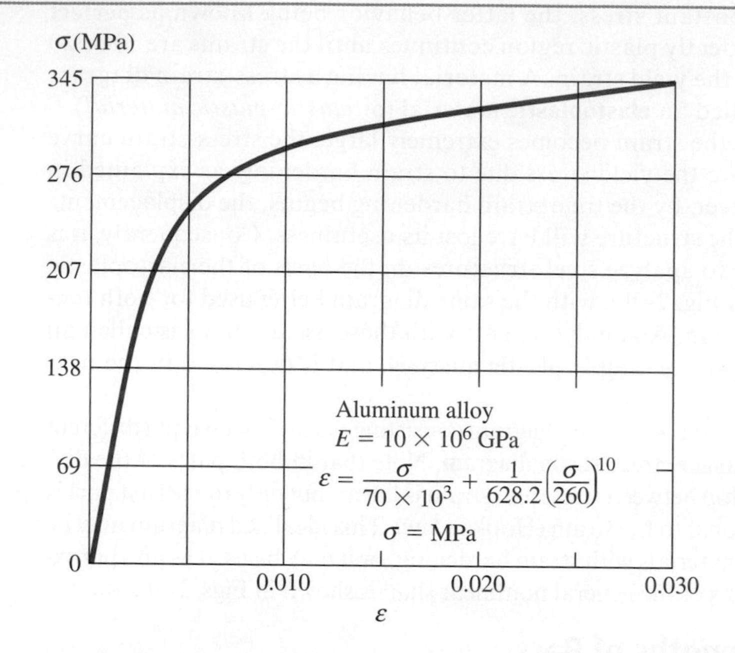

values, proceed as follows. Divide the bar into small segments of length Δx, determine the average stress and strain for each segment, and then calculate the elongation of the entire bar by summing the elongations for the individual segments. This process is equivalent to evaluating the integral in Eq. (2-68) by numerical methods instead of by formal integration.

If the strains are uniform throughout the length of the bar, as in the case of a prismatic bar with constant axial force, the integration of Eq. (2-68) is trivial and the change in length is

$$\delta = \varepsilon L \tag{2-69}$$

as expected [compare with Eq. (1-7) in Section 1.4].

Ramberg-Osgood Stress-Strain Law

Stress-strain curves for several metals, including aluminum and magnesium, can be accurately represented by the **Ramberg-Osgood equation**:

$$\frac{\varepsilon}{\varepsilon_0} = \frac{\sigma}{\sigma_0} + \alpha\left(\frac{\sigma}{\sigma_0}\right)^m \tag{2-70}$$

In this equation, σ and ε are the stress and strain, respectively, and ε_0, σ_0, α, and m are constants of the material (obtained from tension tests). An alternative form of the equation is

$$\varepsilon = \frac{\sigma}{E} + \frac{\sigma_0 \alpha}{E}\left(\frac{\sigma}{\sigma_0}\right)^m \tag{2-71}$$

in which $E = \sigma_0/\varepsilon_0$ is the modulus of elasticity in the initial part of the stress-strain curve.[11]

[11]The Ramberg-Osgood stress-strain law is discussed in Ref. 2-12.

A graph of Eq. (2-71) is given in Fig. 2-93 for an aluminum alloy for which the constants are as follows: $E = 70$ GPa, $\sigma_0 = 260$ MPa, $\alpha = 3/7$, and $m = 10$. The equation of this particular stress-strain curve is

$$\varepsilon = \frac{\sigma}{70{,}000} + \frac{1}{628.2}\left(\frac{\sigma}{260}\right)^{10} \tag{2-72}$$

where σ has units of megapascals (MPa).

The calculation of the change in length of a bar, using Eq. (2-71) for the stress-strain relationship, is illustrated in Example 2-22.

Statically Indeterminate Structures

If a structure is statically indeterminate and the material behaves nonlinearly, the stresses, strains, and displacements can be found by solving the same general equations as those described in Section 2.4 for linearly elastic structures, namely, equations of equilibrium, equations of compatibility, and force-displacement relations (or equivalent stress-strain relations). The principal difference is that the force-displacement relations are now nonlinear, which means that analytical solutions cannot be obtained except in very simple situations. Instead, the equations must be solved numerically, using a suitable computer program.

Example 2-22

A prismatic bar AB of length $L = 2.2$ m and cross-sectional area $A = 480$ mm^2 supports two concentrated loads $P_1 = 108$ kN and $P_2 = 27$ kN, as shown in Fig. 2-94. The material of the bar is an aluminum alloy having a nonlinear stress-strain curve described by the Ramberg-Osgood equation [Eq. (2-73)]:

$$\varepsilon = \frac{\sigma}{70{,}000} + \frac{1}{628.2}\left(\frac{\sigma}{260}\right)^{10}$$

in which σ has units of MPa. (The general shape of this stress-strain curve is shown in Fig. 2-93.)

Determine the displacement δ_B of the lower end of the bar under each of the following conditions: (a) the load P_1 acts alone, (b) the load P_2 acts alone, and (c) the loads P_1 and P_2 act simultaneously.

FIGURE 2-94

Example 2-22: Elongation of a bar of nonlinear material using the Ramberg-Osgood equation

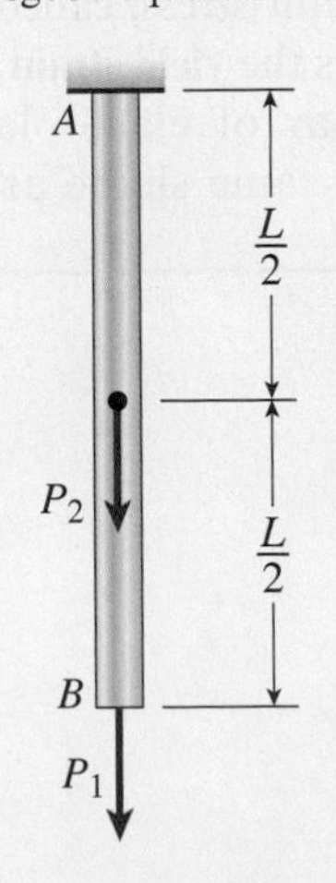

Solution:

Use a four-step problem-solving approach.

Part (a): Displacement due to the load P_1 acting alone.

1, 2, 3. ***Conceptualize, Categorize, Analyze***: The load P_1 produces a uniform tensile stress throughout the length of the bar equal to P_1/A, or 225 MPa. Substituting this value into the stress-strain relation gives $\varepsilon = 0.003589$. Therefore, the elongation of the bar equal to the displacement at point B is [see Eq. (2-69)]

$$\delta_B = \varepsilon L = (0.003589)(2.2 \text{ m}) = 7.90 \text{ mm}$$

Part (b): Displacement due to the load P_2 acting alone.

The stress in the upper half of the bar is P_2/A or 56.25 MPa, and there is no stress in the lower half. Proceeding as in part (a) leads to

$$\delta_B = \varepsilon L/2 = (0.0008036)(1.1 \text{ m}) = 0.884 \text{ mm}$$

Part (c): Displacement due to both loads acting simultaneously.

The stress in the lower half of the bar is P_1/A and in the upper half is $(P_1 + P_2)/A$. The corresponding stresses are 225 MPa and 281.25 MPa, and the corresponding strains are 0.003589 and 0.007510 (from the Ramberg-Osgood equation).

Therefore, the elongation of the bar is

$$\delta_B = (0.003589)(1.1 \text{ m}) + (0.007510)(1.1 \text{ m})$$
$$= 3.95 \text{ mm} + 8.26 \text{ mm} = 12.2 \text{ mm}$$

4. ***Finalize***: The three calculated values of δ_B illustrate an important principle pertaining to a structure made of a material that behaves nonlinearly:

In a nonlinear structure, the displacement produced by two (or more) loads acting simultaneously is not equal to the sum of the displacements produced by the loads acting separately.

*2.12 Elastoplastic Analysis

In the preceding section, the behavior of structures when the stresses in the material exceed the proportional limit was discussed. Now consider a material of considerable importance in engineering design—steel, the most widely used structural metal. Mild steel (or structural steel) can be modeled as an *elastoplastic* material with a stress-strain diagram as shown in Fig. 2-95. An elastoplastic material initially behaves in a linearly elastic manner with a modulus of elasticity E. After plastic yielding begins, the strains increase at a more-or-less constant stress, called the **yield stress** σ_Y. The strain at the onset of yielding is known as the **yield strain** ε_Y.

The load-displacement diagram for a prismatic bar of elastoplastic material subjected to a tensile load (Fig. 2-96) has the same shape as the

FIGURE 2-95

Idealized stress-strain diagram for an elastoplastic material, such as structural steel

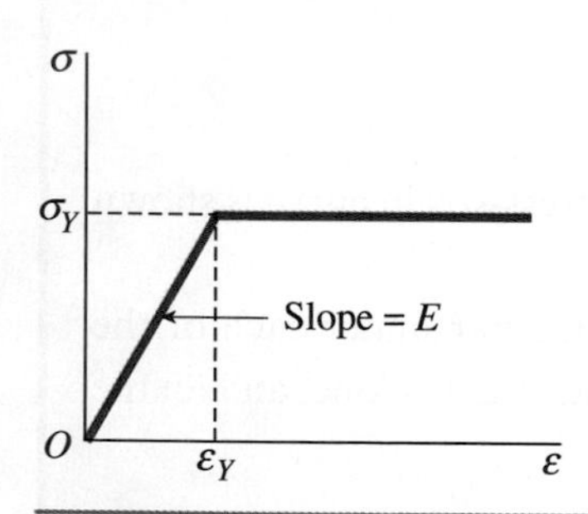

FIGURE 2-96

Load-displacement diagram for a prismatic bar of elastoplastic material

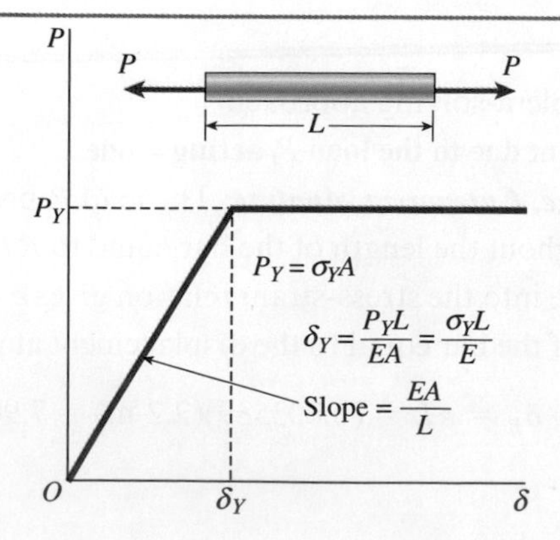

stress-strain diagram. Initially, the bar elongates in a linearly elastic manner and Hooke's law is valid. Therefore, in this region of loading, you can find the change in length from the familiar formula $\delta = PL/EA$. Once the yield stress is reached, the bar may elongate without an increase in load, and the elongation has no specific magnitude. The load at which yielding begins is called the **yield load** P_Y, and the corresponding elongation of the bar is called the **yield displacement** δ_Y. Note that for a single prismatic bar, the yield load P_Y equals $\sigma_Y A$ and the yield displacement δ_Y equals $P_Y L/EA$, or $\sigma_Y L/E$. (Similar comments apply to a bar in compression, provided buckling does not occur.)

FIGURE 2-97

Statically determinate structure consisting of axially loaded members

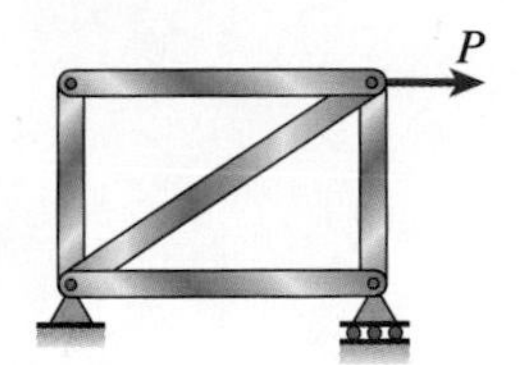

If a structure consisting only of axially loaded members is **statically determinate** (Fig. 2-97), its overall behavior follows the same pattern. The structure behaves in a linearly elastic manner until one of its members reaches the yield stress. Then that member will begin to elongate (or shorten) with no further change in the axial load in that member. Thus, the entire structure will yield, and its load-displacement diagram has the same shape as that for a single bar (Fig. 2-96).

Statically Indeterminate Structures

The situation is more complex if an elastoplastic structure is statically indeterminate. If one member yields, other members will continue to resist any increase in the load. However, eventually enough members will yield to cause the entire structure to yield.

To illustrate the behavior of a statically indeterminate structure, consider the simple arrangement shown in Fig. 2-98. This structure consists of three steel bars supporting a load P applied through a rigid plate. The two outer bars have length L_1, the inner bar has length L_2, and all three bars have the same cross-sectional area A. The stress-strain diagram for the steel is idealized as shown in Fig. 2-95, and the modulus of elasticity in the linearly elastic region is $E = \sigma_Y/\varepsilon_Y$.

As is normally the case with a statically indeterminate structure, the analysis begins with the equations of *equilibrium* and *compatibility*. From equilibrium of the rigid plate in the vertical direction,

$$2F_1 + F_2 = P \quad \textbf{(2-74)}$$

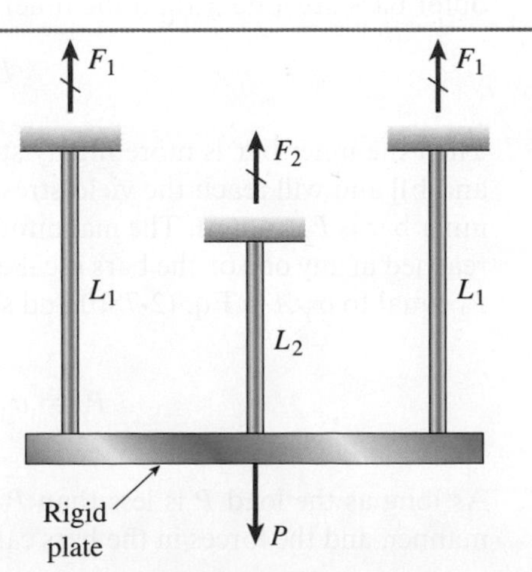

FIGURE 2-98

Elastoplastic analysis of a statically indeterminate structure where F_1 and F_2 are the axial forces in the outer and inner bars, respectively. Because the plate moves downward as a rigid body when the load is applied, the compatibility equation is

$$\delta_1 = \delta_2 \tag{2-75}$$

where δ_1 and δ_2 are the elongations of the outer and inner bars, respectively. Because they depend only upon equilibrium and geometry, the two preceding equations are valid at all levels of the load P; it does not matter whether the strains fall in the linearly elastic region or in the plastic region.

When the load P is small, the stresses in the bars are less than the yield stress σ_Y, and the material is stressed within the linearly elastic region. Therefore, the *force-displacement relations* between the bar forces and their elongations are

$$\delta_1 = \frac{F_1 L_1}{EA} \qquad \delta_2 = \frac{F_2 L_2}{EA} \tag{2-76a,b}$$

Substitute in the compatibility equation [Eq. (2-75)], to get

$$F_1 L_1 = F_2 L_2 \tag{2-77}$$

Solving simultaneously Eqs. (2-74) and (2-77),

$$F_1 = \frac{PL_2}{L_1 + 2L_2} \qquad F_2 = \frac{PL_1}{L_1 + 2L_2} \tag{2-78a,b}$$

Thus, knowing the forces in the bars in the linearly elastic region, the corresponding stresses are

$$\sigma_1 = \frac{F_1}{A} = \frac{PL_2}{A(L_1 + 2L_2)} \qquad \sigma_2 = \frac{F_2}{A} = \frac{PL_1}{A(L_1 + 2L_2)} \tag{2-79a,b}$$

These equations for the forces and stresses are valid provided the stresses in all three bars remain below the yield stress σ_Y.

As the load P gradually increases, the stresses in the bars increase until the yield stress is reached in either the inner bar or the outer bars. Assume that the outer bars are longer than the inner bar, as sketched in Fig. 2-98:

$$L_1 > L_2 \tag{2-80}$$

Then the inner bar is more highly stressed than the outer bars [see Eqs. (2-79a and b)] and will reach the yield stress first. When that happens, the force in the inner bar is $F_2 = \sigma_Y A$. The magnitude of the load P when the yield stress is first reached in any one of the bars is called the **yield load** P_Y. Determine P_Y by setting F_2 equal to $\sigma_Y A$ in Eq. (2-78b) and solving for the load:

$$P_Y = \sigma_Y A\left(1 + \frac{2L_2}{L_1}\right) \tag{2-81}$$

As long as the load P is less than P_Y, the structure behaves in a linearly elastic manner, and the forces in the bars can be determined from Eqs. (2-78a and b).

The downward displacement of the rigid bar at the yield load, called the **yield displacement** δ_Y, is equal to the elongation of the inner bar when its stress first reaches the yield stress σ_Y:

$$\delta_Y = \frac{F_2 L_2}{EA} = \frac{\sigma_2 L_2}{E} = \frac{\sigma_Y L_2}{E} \tag{2-82}$$

The relationship between the applied load P and the downward displacement δ of the rigid bar is portrayed in the load-displacement diagram of Fig. 2-99. The behavior of the structure up to the yield load P_Y is represented by line OA.

With a further increase in the load, the forces F_1 in the outer bars increase but the force F_2 in the inner bar remains constant at the value $\sigma_Y A$ because this bar is now perfectly plastic (see Fig. 2-96). When the forces F_1 reach the value $\sigma_Y A$, the outer bars also yield, and therefore the structure cannot support any additional load. Instead, all three bars will elongate plastically under this constant load, called the **plastic load** P_p. The plastic load is represented by point B on the load-displacement diagram (Fig. 2-99), and the horizontal line BC represents the region of continuous plastic deformation without any increase in the load.

The plastic load P_P can be calculated from static equilibrium [Eq. (2-74)] knowing that

$$F_1 = \sigma_Y A \quad F_2 = \sigma_Y A \tag{2-83a,b}$$

Thus, from equilibrium

$$P_P = 3\sigma_Y A \tag{2-84}$$

The **plastic displacement** δ_P at the instant the load just reaches the plastic load P_P is equal to the elongation of the outer bars at the instant they reach the yield stress. Therefore,

$$\delta_P = \frac{F_1 L_1}{EA} = \frac{\sigma_1 L_1}{E} = \frac{\sigma_Y L_1}{E} \tag{2-85}$$

Comparing δ_P with δ_Y, in this example the ratio of the plastic displacement to the yield displacement is

$$\frac{\delta_P}{\delta_Y} = \frac{L_1}{L_2} \tag{2-86}$$

Also, the ratio of the plastic load to the yield load is

$$\frac{P_P}{P_Y} = \frac{3L_1}{L_1 + 2L_2} \tag{2-87}$$

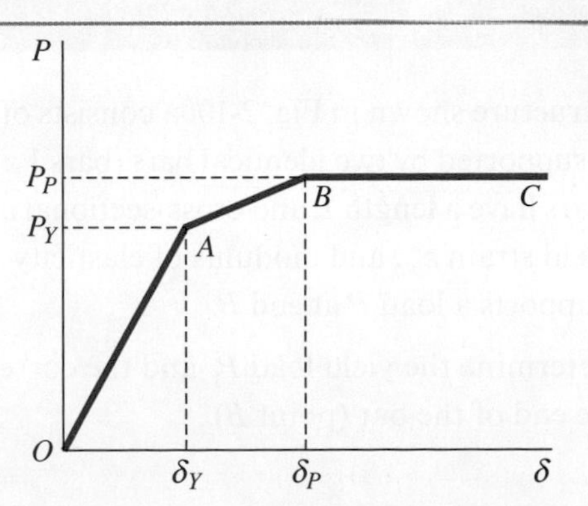

FIGURE 2-99
Load-displacement diagram for the statically indeterminate structure shown in Fig. 2-98

For example, if $L_1 = 1.5L_2$, the ratios are $\delta_P/\delta_Y = 1.5$ and $P_P/P_Y = 9/7 = 1.29$. In general, the ratio of the displacements is always larger than the ratio of the corresponding loads, and the partially plastic region AB on the load-displacement diagram (Fig. 2-99) always has a smaller slope than does the elastic region OA. Of course, the fully plastic region BC has the smallest slope (zero).

General Comments

To understand why the load-displacement graph is linear in the partially plastic region (line AB in Fig. 2-99) and has a slope that is less than in the linearly elastic region, consider the following. In the partially plastic region of the structure, the outer bars still behave in a linearly elastic manner. Therefore, their elongation is a linear function of the load. Since their elongation is the same as the downward displacement of the rigid plate, the displacement of the rigid plate must also be a linear function of the load. Consequently, there is a straight line between points A and B. However, the slope of the load-displacement diagram in this region is less than in the initial linear region because the inner bar yields plastically, and only the outer bars offer increasing resistance to the increasing load. In effect, the stiffness of the structure has diminished.

From the discussion associated with Eq. (2-84), the calculation of the plastic load P_P requires only the use of statics, because all members have yielded and their axial forces are known. In contrast, the calculation of the yield load P_Y requires a statically indeterminate analysis, which means that equilibrium, compatibility, and force-displacement equations must be solved.

After the plastic load P_P is reached, the structure continues to deform as shown by line BC on the load-displacement diagram (Fig. 2-99). Strain hardening occurs eventually, and then the structure is able to support additional loads. However, the presence of very large displacements usually means that the structure is no longer of use, and so the plastic load P_P is usually considered to be the failure load.

The preceding discussion has dealt with the behavior of a structure when the load is applied for the first time. If the load is removed before the yield load is reached, the structure will behave elastically and return to its original unstressed condition. However, if the yield load is exceeded, some members of the structure will retain a permanent set when the load is removed, thus creating a prestressed condition. Consequently, the structure will have *residual stresses* in it even though no external loads are acting. If the load is applied a second time, the structure will behave in a different manner.

Example 2-23

The structure shown in Fig. 2-100a consists of a horizontal beam AB (assumed to be rigid) supported by two identical bars (bars 1 and 2) made of an elastoplastic material. The bars have a length L and cross-sectional area A, and the material has yield stress σ_Y, yield strain ε_Y, and modulus of elasticity $E = \sigma_Y/\varepsilon_Y$. The beam has a length $3b$ and supports a load P at end B.

(a) Determine the yield load P_Y and the corresponding yield displacement δ_Y at the end of the bar (point B).

(b) Determine the plastic load P_P and the corresponding plastic displacement δ_P at point B.

(c) Construct a load-displacement diagram relating the load P to the displacement δ_B of point B.

FIGURE 2-100

Example 2-23: Elastoplastic analysis of a statically indeterminate structure

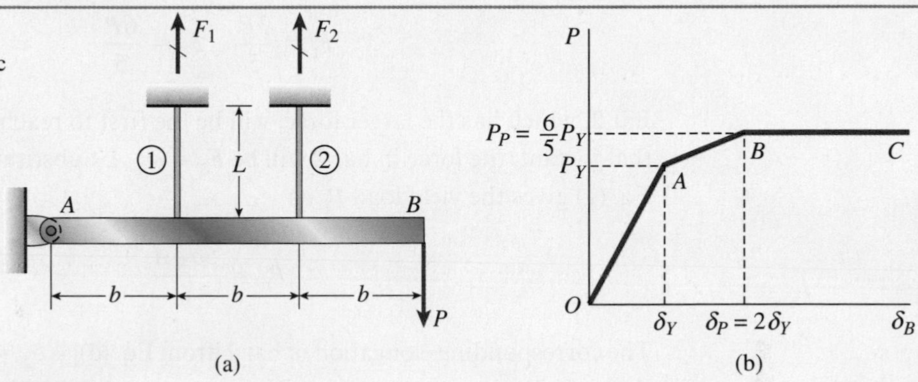

Solution:

Use a four-step problem-solving approach.

1, 2. ***Conceptualize, Categorize:***

Equation of equilibrium: Because the structure is statically indeterminate, begin with the equilibrium and compatibility equations. Considering the equilibrium of beam AB, take moments about point A and obtain

$$\Sigma M_A = 0 \quad F_1(b) + F_2(2b) - P(3b) = 0$$

in which F_1 and F_2 are the axial forces in bars 1 and 2, respectively. This equation simplifies to

$$F_1 + 2F_2 = 3P \tag{a}$$

Equation of compatibility: The compatibility equation is based upon the geometry of the structure. Under the action of load P, the rigid beam rotates about point A, and therefore the downward displacement at every point along the beam is proportional to its distance from point A. Thus, the compatibility equation is

$$\delta_2 = 2\delta_1 \tag{b}$$

where δ_2 is the elongation of bar 2 and δ_1 is the elongation of bar 1.

3. ***Analyze***: Consider yield load and displacement and then plastic load and displacement.

Part (a): Yield load and yield displacement.

When the load P is small and the stresses in the material are in the linearly elastic region, the force-displacement relations for the two bars are

$$\delta_1 = \frac{F_1 L}{EA} \quad \delta_2 = \frac{F_2 L}{EA} \tag{c, d}$$

Combining these equations with the compatibility condition [Eq. (b)] gives

$$\frac{F_2 L}{EA} = 2\frac{F_1 L}{EA} \quad \text{or} \quad F_2 = 2F_1 \qquad \textbf{(e)}$$

Substitute Eq. (e) into the equilibrium equation [Eq. (a)] to find

$$F_1 = \frac{3P}{5} \quad F_2 = \frac{6P}{5} \qquad \textbf{(f, g)}$$

Bar 2, which has the larger force, will be the first to reach the yield stress. At that instant, the force in bar 2 will be $F_2 = \sigma_Y A$. Substituting that value into Eq. (g) gives the yield load P_Y as

$$P_Y = \frac{5\sigma_Y A}{6} \qquad \textbf{(2-88)}$$

The corresponding elongation of bar 2 [from Eq. (d)] is $\delta_2 = \sigma_Y L/E$, and therefore the yield displacement at point B is

$$\delta_Y = \frac{3\delta_2}{2} = \frac{3\sigma_Y L}{2E} \qquad \textbf{(2-89)}$$

Both P_Y and δ_Y are indicated on the load-displacement diagram (Fig. 2-100b).

Part (b): Plastic load and plastic displacement.

When the plastic load P_P is reached, both bars will be stretched to the yield stress and both forces F_1 and F_2 will be equal to $\sigma_Y A$. It follows from equilibrium [Eq. (a)] that the plastic load is

$$P_P = \sigma_Y A \qquad \textbf{(2-90)}$$

At this load, the left-hand bar (bar 1) has just reached the yield stress; therefore, its elongation [from Eq. (c)] is $\delta_1 = \sigma_Y L/E$, and the plastic displacement of point B is

$$\delta_P = 3\delta_1 = \frac{3\sigma_Y L}{E} \qquad \textbf{(2-91)}$$

The ratio of the plastic load P_P to the yield load P_Y is 6/5, and the ratio of the plastic displacement δ_P to the yield displacement δ_Y is 2. These values are also shown on the load-displacement diagram.

4. ***Finalize:***

Part (c): Load-displacement diagram.

The complete load-displacement behavior of the structure is pictured in Fig. 2-100b. The behavior is linearly elastic in the region from O to A, partially plastic from A to B, and fully plastic from B to C.

CHAPTER SUMMARY AND REVIEW

Chapter 2 discussed the behavior of axially loaded bars acted on by distributed loads, such as self-weight, and also temperature changes and prestrains. Force-displacement relations were derived for use in computing changes in lengths of bars under both uniform and nonuniform conditions. Equilibrium and compatibility equations were developed for statically indeterminate structures. Equations were developed for normal and shear stresses on inclined sections. These are major concepts presented in this chapter.

1. The elongation or shortening (δ) of prismatic bars subjected to tensile or compressive centroidal loads is proportional to both the load (P) and the length (L) of the bar and inversely proportional to the axial rigidity (EA) of the bar; this relationship is called a **force-displacement relation.**

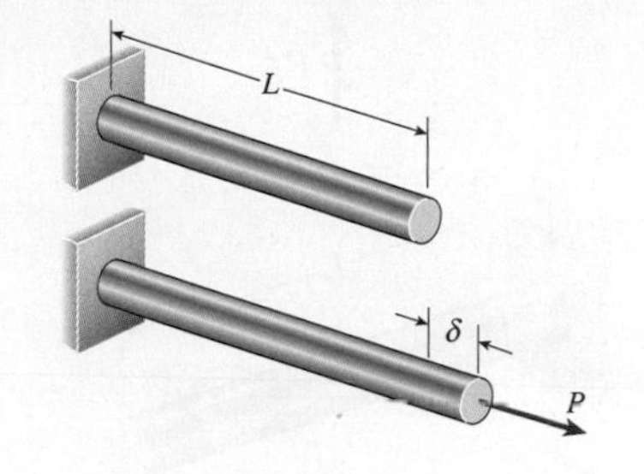

$$\delta = \frac{PL}{EA}$$

2. Cables are **tension-only elements**, and an effective modulus of elasticity (E_e) and effective cross-sectional area (A_e) should be used to account for the tightening effect that occurs when cables are placed under load.
3. The axial rigidity per unit length of a bar is referred to as its **stiffness** (k), and the inverse relationship is the **flexibility** (f) of the bar.

$$\delta = Pf = \frac{P}{k} \quad f = \frac{L}{EA} = \frac{1}{k}$$

4. The summation of the displacements of the individual segments of a nonprismatic bar equals the elongation or shortening of the entire bar (δ).

$$\delta = \sum_{i=1}^{n} \frac{N_i L_i}{E_i A_i}$$

A graphical display of the internal forces N_i is called an **axial force diagram** (AFD). The AFD is used along with the *force-displacement relation* to create the displaced shape of the bar referred to as the **axial displacement diagram** (ADD). These diagrams are useful visual aids that help to identify locations of maximum force and stress, displacement, and strain.

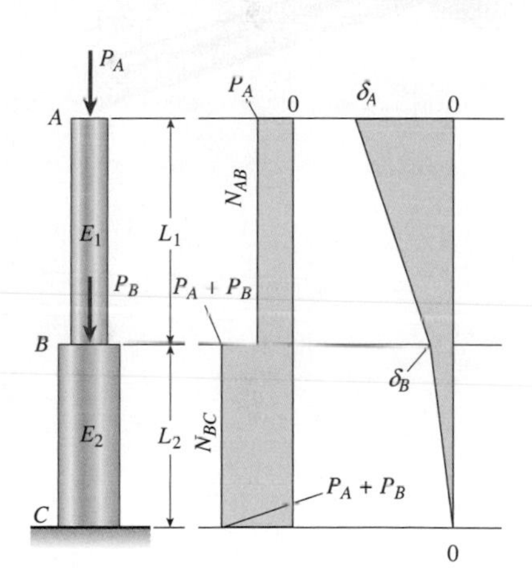

5. **Free-body diagrams** are used to find the axial force (N_i) in each segment i; if axial forces and/or cross-sectional areas vary continuously, an integral expression is required.

$$\delta = \int_0^L d\delta = \int_0^L \frac{N(x)dx}{EA(x)}$$

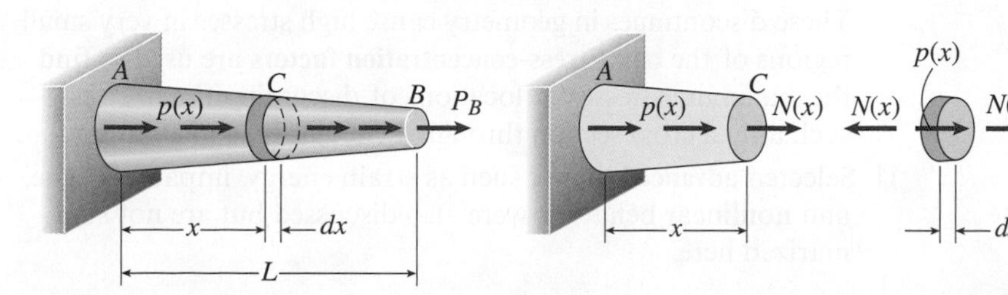

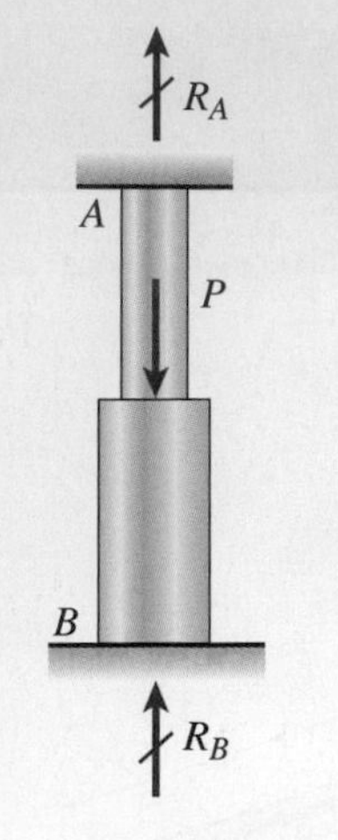

6. If the bar structure is **statically indeterminate**, additional equations beyond those available from statics are required to solve for unknown forces. **Compatibility equations** are used to relate bar displacements to support conditions and thereby generate additional relationships among the unknowns. It is convenient to use a **superposition** of "released" (or statically determinate) structures to represent the actual statically indeterminate bar structure.
7. **Thermal effects** result in displacements, proportional to the temperature change (ΔT) and the length (L) of the bar, but not stresses in statically determinate structures. The coefficient of thermal expansion (α) of the material also is required to compute axial strains (ε_T) and axial displacements (δ_T) due to thermal effects.

$$\varepsilon_T = \alpha(\Delta T) \quad \delta_T = \varepsilon_T L = \alpha(\Delta T)L$$

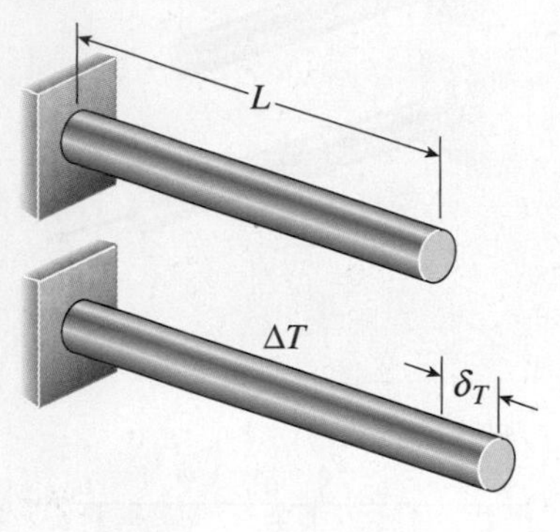

8. **Misfits** and **prestrains** induce axial forces only in statically indeterminate bars.
9. **Maximum normal** (σ_{max}) and **shear stresses** (τ_{max}) are obtained by considering an inclined stress element for a bar loaded by axial forces. The maximum normal stress occurs along the axis of the bar, but the maximum shear stress occurs at an inclination of 45° to the axis of the bar, and the maximum shear stress is one-half of the maximum normal stress.

$$\sigma_{max} = \sigma_x \quad \tau_{max} = \frac{\sigma_x}{2}$$

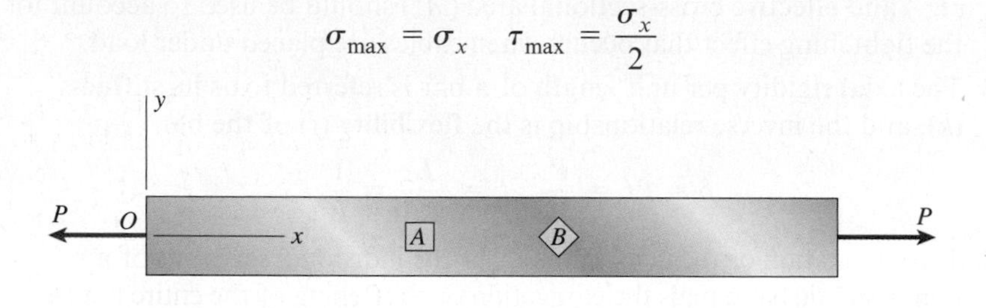

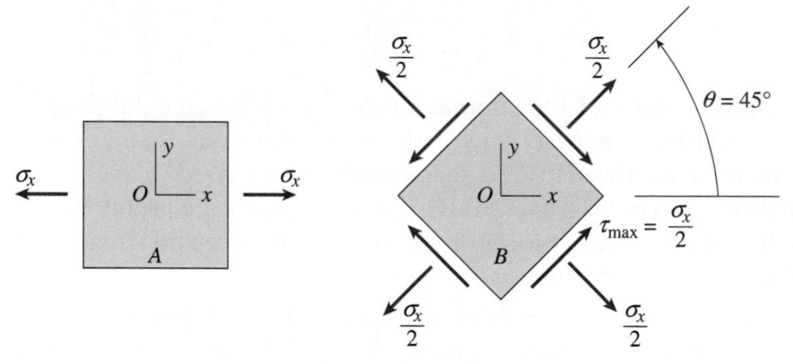

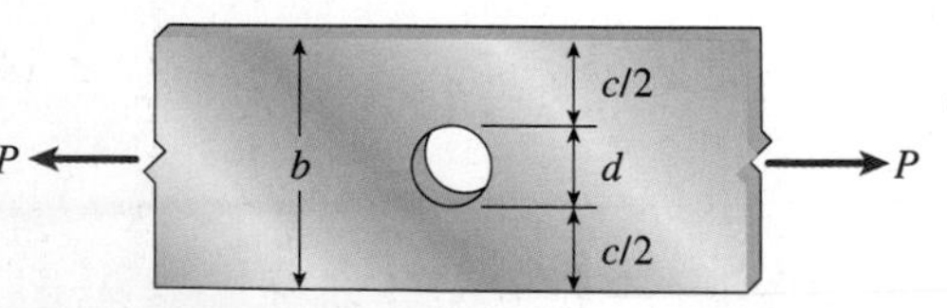

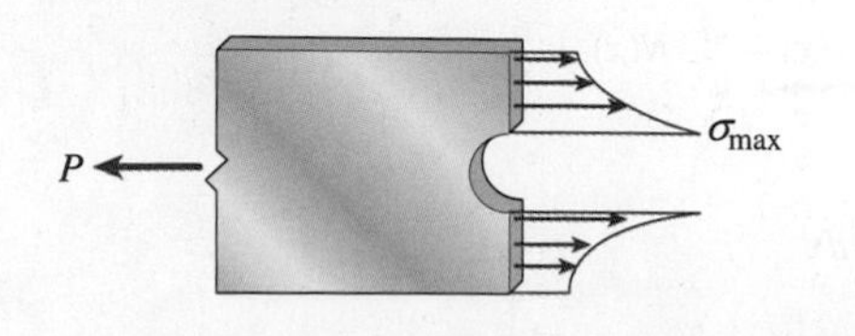

10. Axially loaded bars often have holes, grooves, notches, keyways, shoulders, threads, or other abrupt changes in geometry that create a disruption in the otherwise uniform stress pattern. These discontinues in geometry cause high stresses in very small regions of the bar. **Stress-concentration factors** are used to find the maximum stresses of locations of discontinuities in stress, such as at a cross section through a circular hole in a flat bar.
11. Selected **advanced topics**, such as strain energy, impact, fatigue, and nonlinear behavior, were also discussed but are not summarized here.

PROBLEMS Chapter 2

2.2 Changes in Lengths of Axially Loaded Members

Introductory Problems

2.2-1 A 3-m rigid bar AB is supported with a vertical translational spring at A and a pin at B. The bar is subjected to a linearly varying distributed load with maximum intensity q_0. Calculate the vertical deformation of the spring if the spring constant is 700 kN/m.

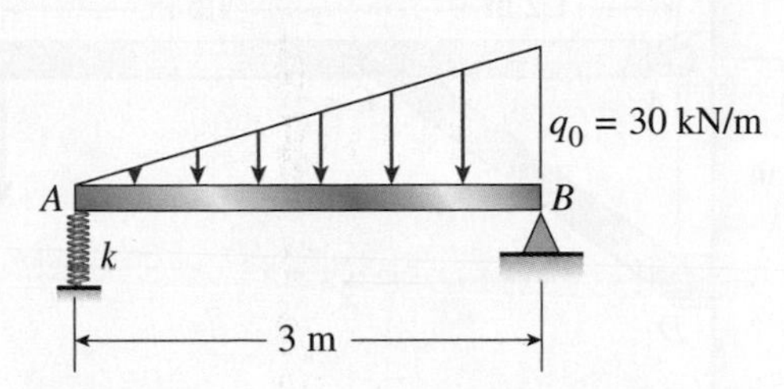

PROBLEM 2.2-1

2.2-2 Rigid bar ABC is supported with a pin at A and an elastic steel rod at C. The elastic rod has a diameter of 25 mm and modulus of elasticity $E = 200$ GPa. The bar is subjected to a uniform load q on span AC and a point load at B. Calculate the change in length of the elastic rod. What is the vertical displacement at point B?

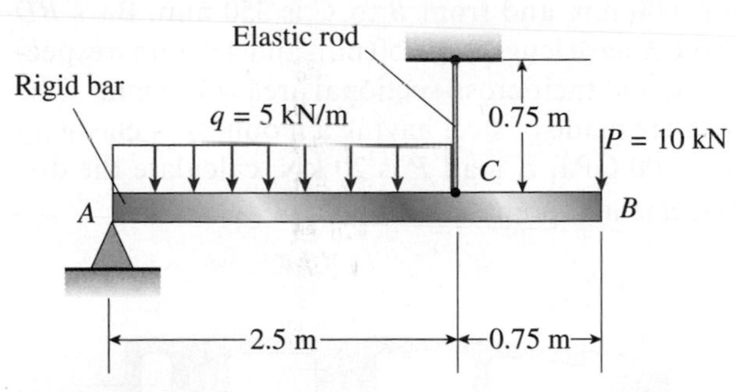

PROBLEM 2.2-2

2.2-3 The L-shaped arm $ABCD$ shown in the figure lies in a vertical plane and pivots about a horizontal pin at A. The arm has a constant cross-sectional area and total weight W. A vertical spring of stiffness k supports the arm at point B.

(a) Obtain a formula for the elongation of the spring due to the weight of the arm.

(b) Repeat part (a) if the pin support at A is moved to D.

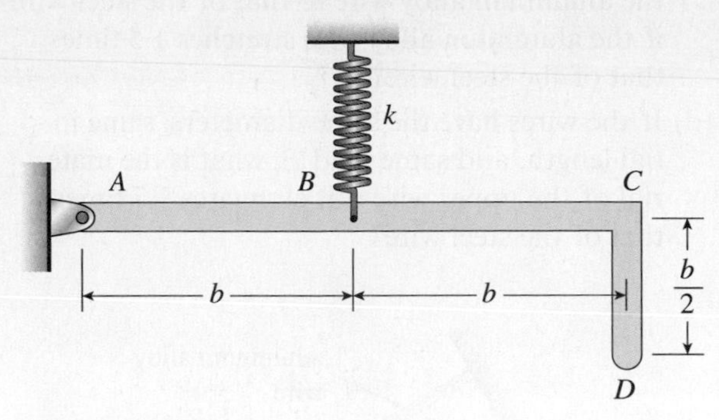

PROBLEM 2.2-3

2.2-4 A steel cable with a nominal diameter of 25 mm (see Table 2-1) is used in a construction yard to lift a bridge section weighing 38 kN, as shown in the figure. The cable has an effective modulus of elasticity $E = 140$ GPa.

(a) If the cable is 14 m long, how much will it stretch when the load is picked up?

(b) If the cable is rated for a maximum load of 70 kN, what is the factor of safety with respect to failure of the cable?

PROBLEM 2.2-4

2.2-5 A steel wire and an aluminum alloy wire have equal lengths and support equal loads P (see figure). The moduli of elasticity for the steel and aluminum alloy are $E_s = 206$ GPa and $E_a = 76$ GPa, respectively.

(a) If the wires have the same diameters, what is the ratio of the elongation of the aluminum alloy wire to the elongation of the steel wire?

(b) If the wires stretch the same amount, what is the ratio of the diameter of the aluminum alloy wire to the diameter of the steel wire?

(c) If the wires have the same diameters and same load P, what is the ratio of the initial length of the aluminum alloy wire to that of the steel wire if the aluminum alloy wire stretches 1.5 times that of the steel wire?

(d) If the wires have the same diameters, same initial length, and same load P, what is the material of the upper wire if it elongates 1.7 times that of the steel wire?

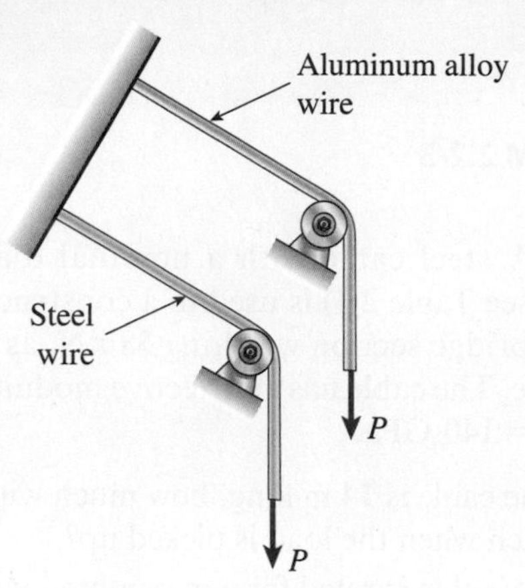

PROBLEM 2.2-5

2.2-6 By what distance h does the cage shown in the figure move downward when the weight W is placed inside it? (See the figure.)

Consider only the effects of the stretching of the cable, which has axial rigidity $EA = 10{,}700$ kN. The pulley at A has a diameter $d_A = 300$ mm and the pulley at B has a diameter $d_B = 150$ mm. Also, the distance $L_1 = 4.6$ m, the distance $L_2 = 10.5$ m, and the weight $W = 22$ kN. *Note:* When calculating the length of the cable, include the parts of the cable that go around the pulleys at A and B.

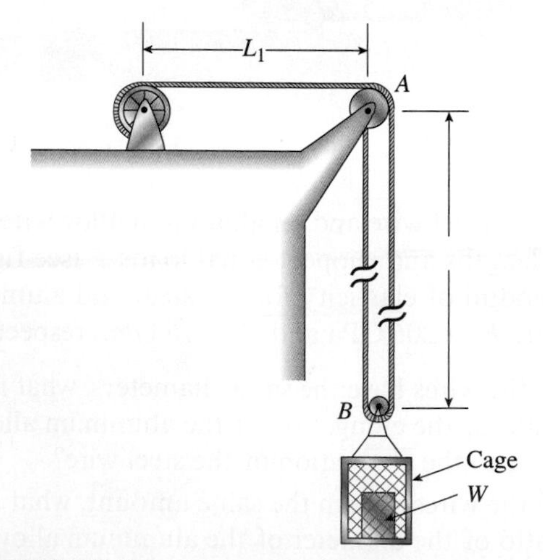

PROBLEM 2.2-6

2.2-7 Rigid bar ACB is supported by an elastic circular strut DC having an outer diameter of 380 mm and inner diameter of 365 mm. The strut is made of steel with a modulus elasticity of $E = 200$ GPa. Point load $P = 22$ kN is applied at B. Calculate the change in length of the circular strut DC. What is the vertical displacement of the rigid bar at point B?

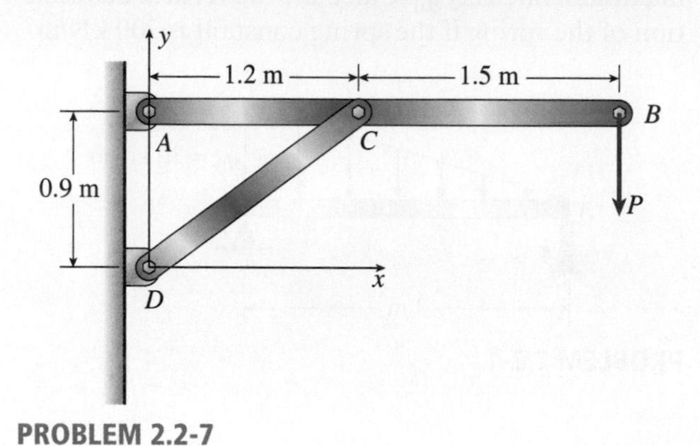

PROBLEM 2.2-7

2.2-8 A device consists of a horizontal beam ABC supported by two vertical bars BD and CE. Bar CE is pinned at both ends but bar BD is fixed to the foundation at its lower end. The distance from A to B is 600 mm and from B to C is 350 mm. Bars BD and CE have lengths of 350 mm and 450 mm, respectively, and their cross-sectional area is 720 mm^2. The bars are made of steel having a modulus of elasticity $E = 200$ GPa. If load P is 20 kN, calculate the displacement at point A.

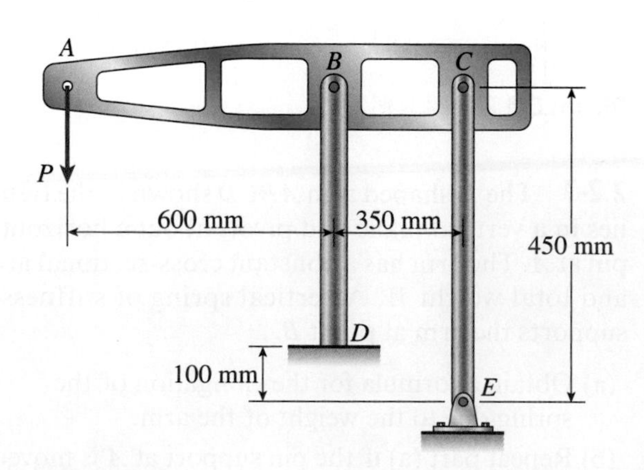

PROBLEM 2.2-8

Representative Problems

2.2-9 A safety valve on the top of a tank containing steam under pressure p has a discharge hole of diameter d (see figure). The valve is designed to release the steam when the pressure reaches the value p_{max}.

If the natural length of the spring is L and its stiffness is k, what should be the dimension h of the valve? (Express your result as a formula for h.)

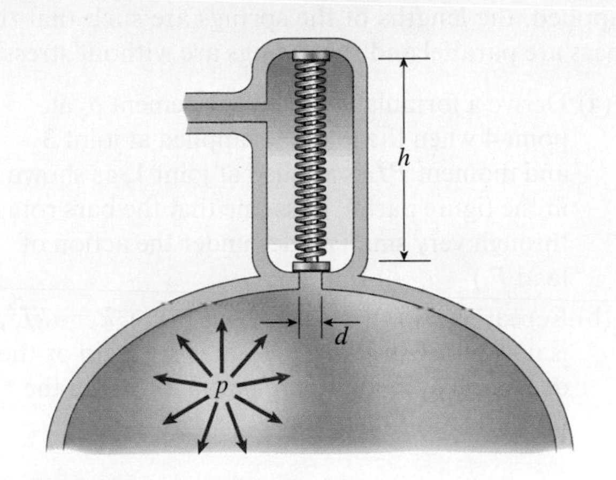

PROBLEM 2.2-9

2.2-10 The device shown in the figure consists of a prismatic rigid pointer ABC supported by a uniform translational spring of stiffness $k = 950$ N/m. The spring is positioned at distance $b = 165$ mm from the pinned end A of the pointer. The device is adjusted so that, when there is no load P, the pointer reads zero on the angular scale.

(a) If the load $P = 11$ N, at what distance x should the load be placed so that the pointer will read $\theta = 2.5°$ on the scale (see figure part a)?

(b) Repeat part (a) if a rotational spring $k_r = kb^2$ is added at A (see figure part b).

(c) Let $x = 7b/8$. What is P_{max} (N) if θ cannot exceed 2°? Include spring k_r in your analysis.

(d) Now, if the weight of the pointer ABC is known to be $W_p = 3$ N and the weight of the spring is $W_s = 2.75$ N, what initial angular position (i.e., θ in degrees) of the pointer will result in a zero reading on the angular scale once the pointer is released from rest? Assume $P = k_r = 0$.

(e) If the pointer is rotated to a vertical position (see figure part c), find the required load P applied at mid-height of the pointer that will result in a pointer reading of $\theta = 2.5°$ on the scale. Consider the weight of the pointer W_p in your analysis.

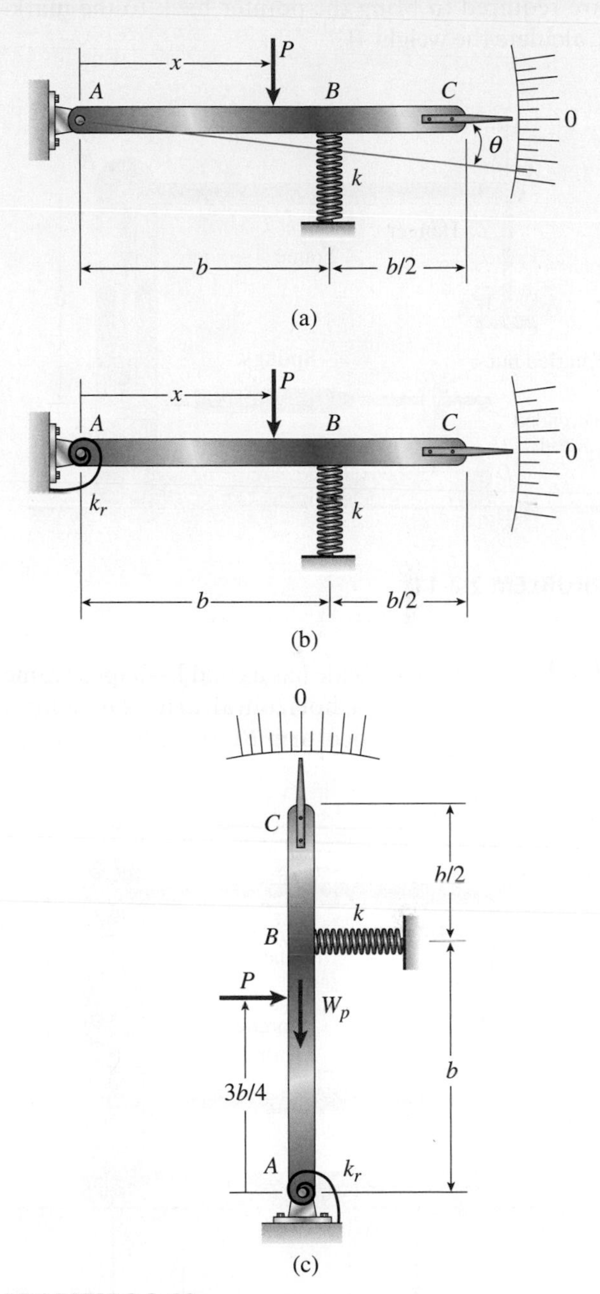

PROBLEM 2.2-10

2.2-11 A small lab scale has a rigid L-shaped frame ABC consisting of a horizontal arm AB (length $b = 250$ mm) and a vertical arm BC (length $c = 175$ mm) pivoted at point B. The pivot is attached to the outer frame BCD that stands on a laboratory bench. The position of the pointer at C is controlled by a spring, $k = 875$ N/m, that is attached to a threaded rod. The pitch of the threads is $p = 1.6$ mm. Under application of load W, 12 revolutions of the nut are required to bring the pointer back to the mark. Calculate the weight W.

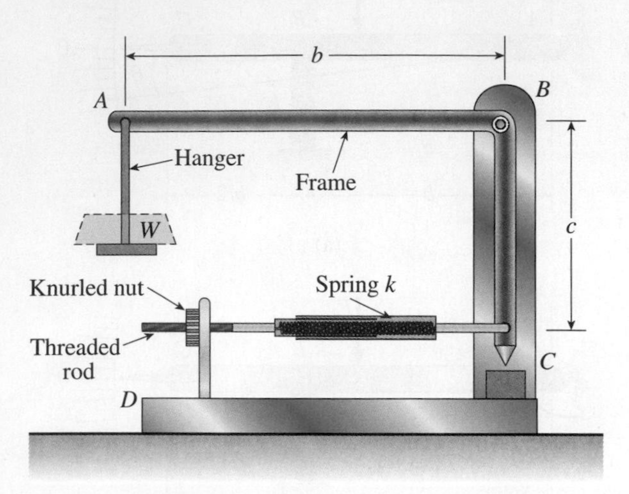

PROBLEM 2.2-11

2.2-12 A small lab scale has a rigid L-shaped frame ABC consisting of a horizontal arm AB (length $b = 30$ cm) and a vertical arm BC (length $c = 20$ cm) pivoted at point B. The pivot is attached to the outer frame BCD that stands on a laboratory bench. The position of the pointer at C is controlled by two parallel springs, each having a spring constant $k = 3650$ N/m, that are attached to a threaded rod. The pitch of the threads is $p = 1.5$ mm. If the weight is 65 N, how many revolutions of the nut are required to bring the pointer back to the mark?

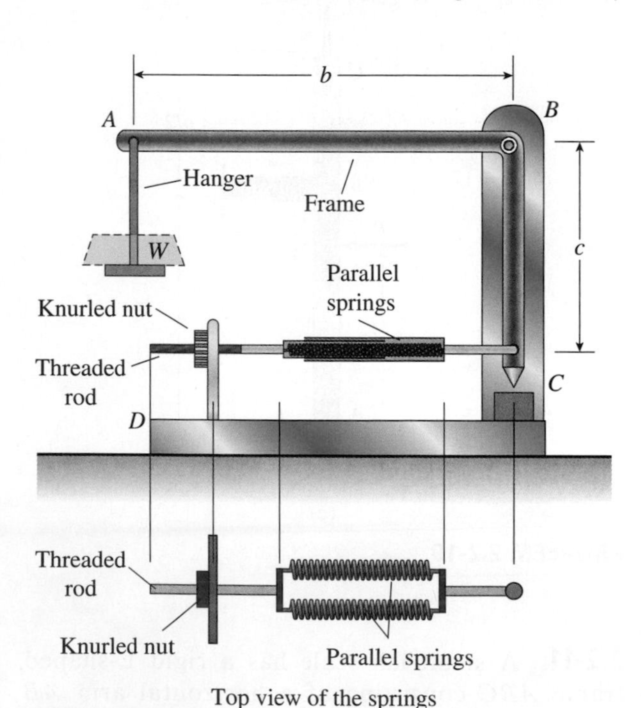

PROBLEM 2.2-12

2.2-13 Two rigid bars are connected to each other by two linearly elastic springs. Before loads are applied, the lengths of the springs are such that the bars are parallel and the springs are without stress.

(a) Derive a formula for the displacement δ_4 at point 4 when the load P is applied at joint 3 and moment PL is applied at joint 1, as shown in the figure part a. (Assume that the bars rotate through very small angles under the action of load P.)

(b) Repeat part (a) if a rotational spring, $k_r = kL^2$, is now added at joint 6. What is the ratio of the deflection δ_4 in the figure part a to that in the figure part b ?

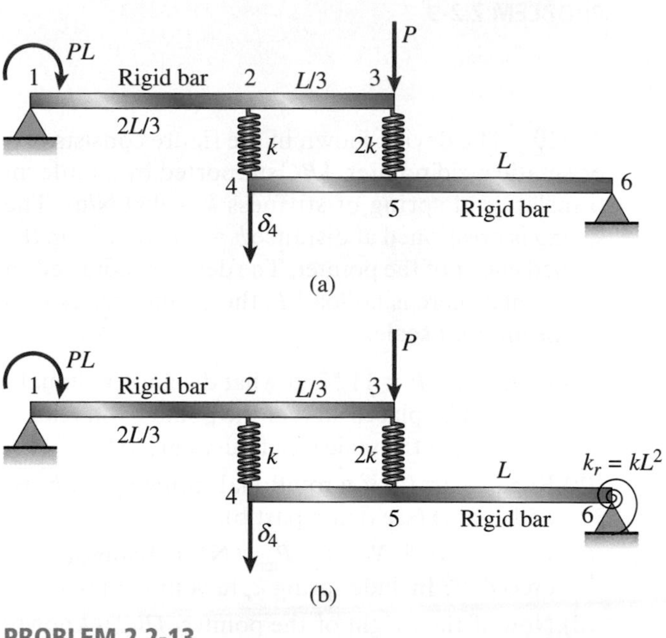

PROBLEM 2.2-13

2.2-14 The three-bar truss ABC shown in the figure part a has a span $L = 3$ m and is constructed of steel pipes having a cross-sectional area $A = 3900$ mm^2 and modulus of elasticity $E = 200$ GPa. Identical loads P act both vertically and horizontally at joint C, as shown.

(a) If $P = 475$ kN, what is the horizontal displacement of joint B?

(b) What is the maximum permissible load value P_{max} if the displacement of joint B is limited to 1.5 mm?

(c) Repeat parts (a) and (b) if the plane truss is replaced by a space truss (see figure part b).

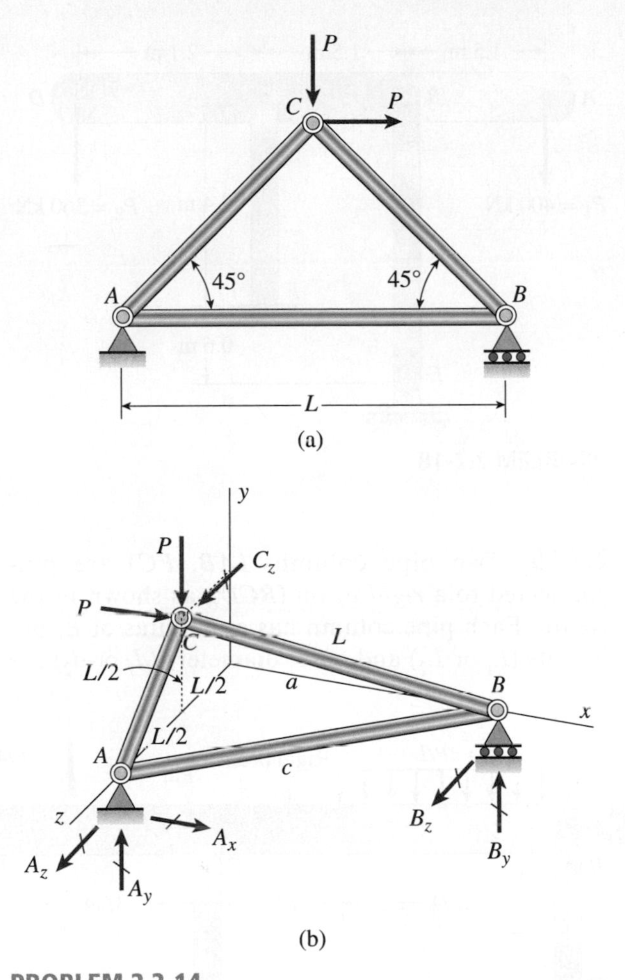

PROBLEM 2.2-14

2.2-15 An aluminum wire having a diameter $d = 2$ mm and length $L = 3.8$ m is subjected to a tensile load P (see figure). The aluminum has a modulus of elasticity $E = 75$ GPa.

If the maximum permissible elongation of the wire is 3 mm and the allowable stress in tension is 60 MPa, what is the allowable load P_{max}?

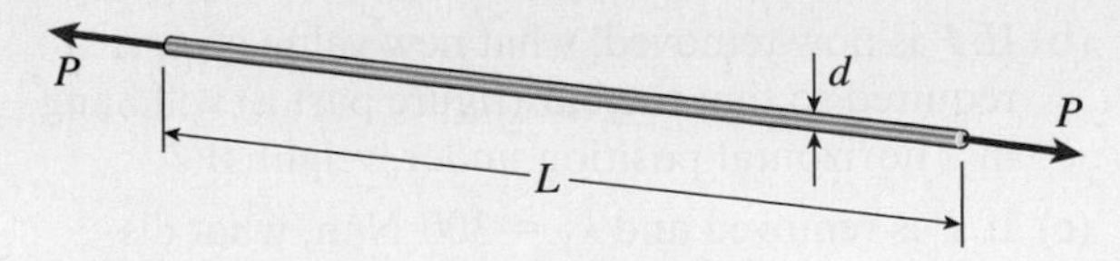

PROBLEM 2.2-15

2.2-16 A uniform bar AB of weight $W = 25$ N is supported by two springs, as shown in the figure. The spring on the left has a stiffness $k_1 = 300$ N/m and natural length $L_1 = 250$ mm. The corresponding quantities for the spring on the right are $k_2 = 400$ N/m and $L_2 = 200$ mm. The distance between the springs is $L = 350$ mm, and the spring on the right is suspended from a support that is a distance $h = 80$ mm below the point of support for the spring on the left. Neglect the weight of the springs.

(a) At what distance x from the left-hand spring (figure part a) should a load $P = 18$ N be placed in order to bring the bar to a horizontal position?

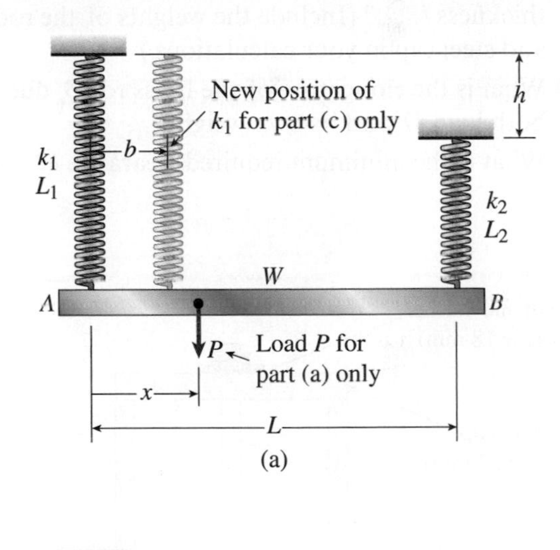

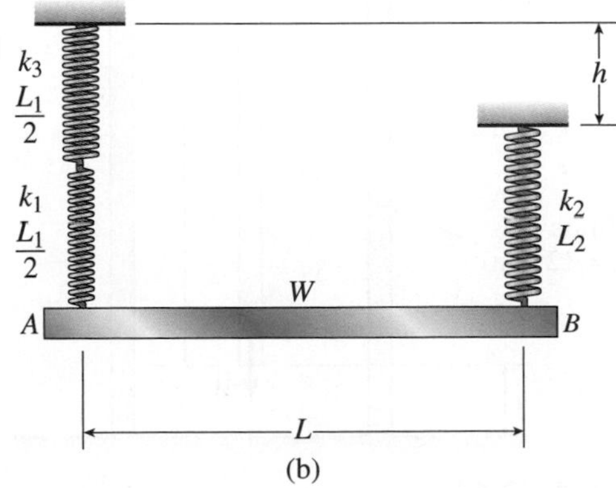

PROBLEM 2.2-16

(b) If P is now removed, what new value of k_1 is required so that the bar (figure part a) will hang in a horizontal position under weight W?

(c) If P is removed and $k_1 = 300$ N/m, what distance b should spring k_1 be moved to the right so that the bar (figure part a) will hang in a horizontal position under weight W?

(d) If the spring on the left is now replaced by two springs in series ($k_1 = 300$ N/m, k_3) with overall natural length $L_1 = 250$ mm (see figure part b), what value of k_3 is required so that the bar will hang in a horizontal position under weight W?

2.2-17 A hollow, circular, cast-iron pipe (E_c = 83 GPa) supports a brass rod (E_b = 96 GPa) and weight W = 9 kN, as shown. The outside diameter of the pipe is d_c = 150 mm.

(a) If the allowable compressive stress in the pipe is 35 MPa and the allowable shortening of the pipe is 0.5 mm, what is the minimum required wall thickness $t_{c,\min}$? (Include the weights of the rod and steel cap in your calculations.)

(b) What is the elongation of the brass rod δ_r due to both load W and its own weight?

(c) What is the minimum required clearance h?

Nut and washer (d_w = 18 mm)
Steel cap (t_s = 25 mm)
Cast iron pipe (d_c = 150 mm, t_c)
L_c = 1.25 m
Brass rod (d_r = 12 mm)
L_r = 1.1 m
h
W

PROBLEM 2.2-17

2.2-18 The horizontal rigid beam $ABCD$ is supported by vertical bars BE and CF and is loaded by vertical forces P_1 = 400 kN and P_2 = 360 kN acting at points A and D, respectively (see figure). Bars BE and CF are made of steel (E = 200 GPa) and have cross-sectional areas A_{BE} = 11,100 mm^2 and A_{CF} = 9280 mm^2. The distances between various points on the bars are shown in the figure.

Determine the vertical displacements δ_A and δ_D of points A and D, respectively.

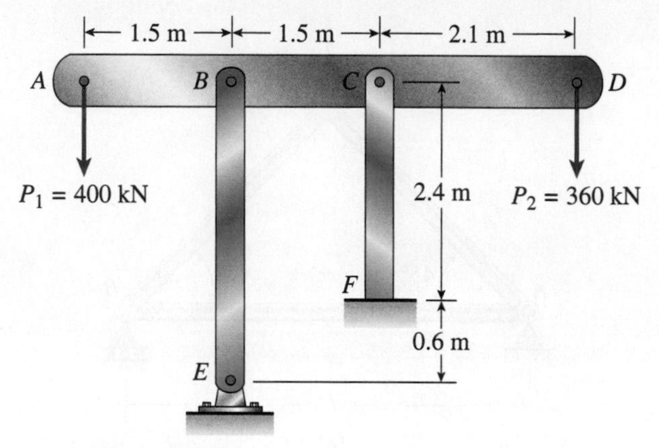

PROBLEM 2.2-18

2.2-19 Two pipe columns (AB, FC) are pin-connected to a *rigid* beam (BCD), as shown in the figure. Each pipe column has a modulus of E, but heights (L_1 or L_2) and outer diameters (d_1 or d_2) are

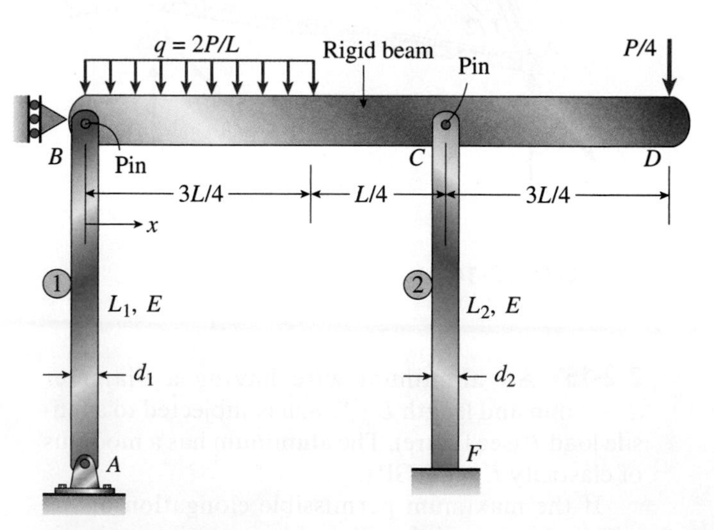

PROBLEM 2.2-19

different for each column. Assume the inner diameter of each column is 3/4 of outer diameter. Uniformly distributed downward load $q = 2P/L$ is applied over a distance of $3L/4$ along BC, and concentrated load $P/4$ is applied downward at D.

(a) Derive a formula for the displacement δ_D at point D in terms of P and column flexibilities f_1 and f_2.
(b) If $d_1 = (9/8)d_2$, find the L_1/L_2 ratio so that beam BCD displaces downward to a horizontal position under the load system in part (a).
(c) If $L_1 = 2L_2$, find the d_1/d_2 ratio so that beam BCD displaces downward to a horizontal position under the load system in part (a).
(d) If $d_1 = (9/8)d_2$ and $L_1/L_2 = 1.5$, at what horizontal distance x from B should load $P/4$ be placed so that beam BCD displaces downward to a horizontal position under the load system in part (a)?

2.2-20 A framework ABC consists of two rigid bars AB and BC, each having a length b (see the first part of the figure part a). The bars have pin connections at A, B, and C and are joined by a spring of stiffness k. The spring is attached at the midpoints of the bars. The framework has a pin support at A and a roller support at C, and the bars are at an angle α to the horizontal.

When a vertical load P is applied at joint B (see the second part of the figure part a) the roller support C moves to the right, the spring is stretched, and the angle of the bars decreases from α to the angle θ.

(a) Determine the angle θ and the increase δ in the distance between points A and C. Also find reactions at A and C. (Use the following data: $b = 200$ mm, $k = 3.2$ kN/m, $\alpha = 45°$, and $P = 50$ N.)
(b) Repeat part (a) if a translational spring $k_1 = k/2$ is added at C and a rotational spring $k_r = kb^2/2$ is added at A (see figure part b).

2.2-21 Solve the preceding problem for the following data: $b = 300$ mm, $k = 7.8$ kN/m, $\alpha = 55°$, and $P = 100$ N.

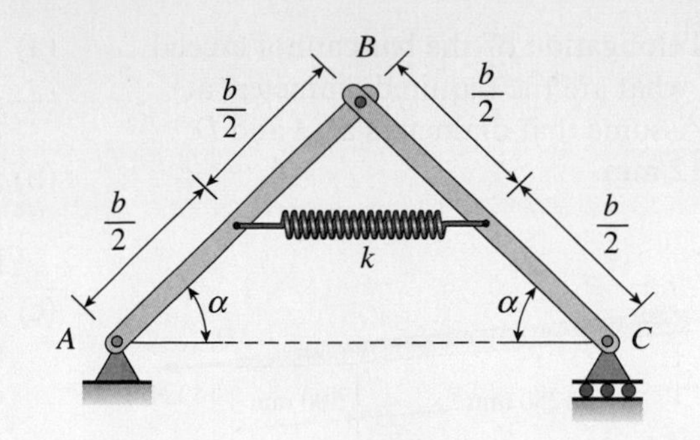

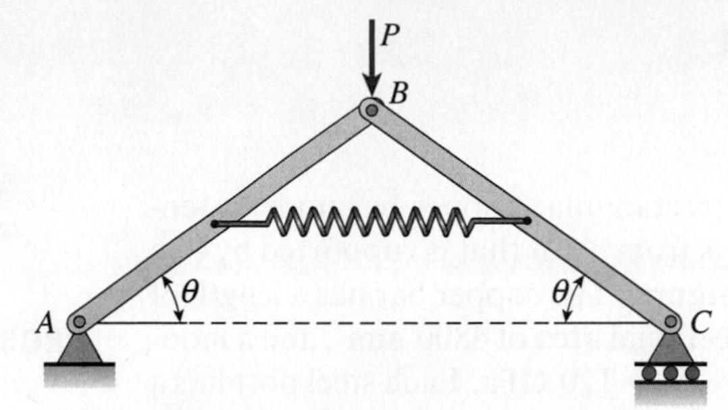

(a) Initial and displaced structures

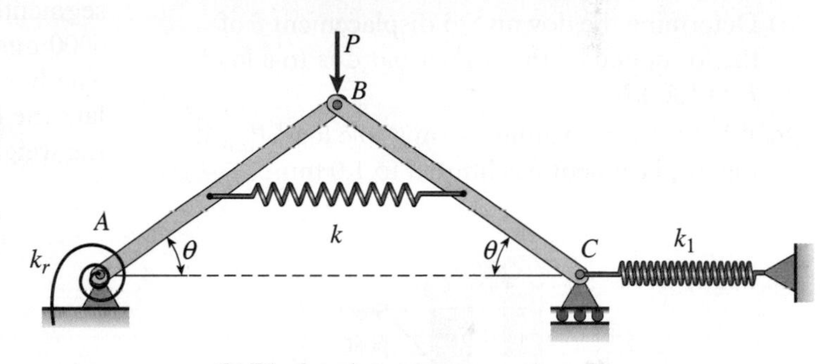

(b) Displaced structure

PROBLEMS 2.2-20 and 2.2-21

2.3 Changes in Lengths Under Nonuniform Conditions

Introductory Problems

2.3-1 The length of the end segments of the bar (see figure) is 500 mm and the length of the prismatic middle segment is 1250 mm. Also, the diameters at cross sections A, B, C, and D are 12, 24, 24, and 12 mm, respectively, and the modulus of elasticity is 120 GPa.

(a) Calculate the elongation of a copper bar of solid circular cross section with tapered ends when it is stretched by axial loads of magnitude 14 kN (see figure).

(b) If the total elongation of the bar cannot exceed 0.635 mm, what are the required diameters at B and C? Assume that diameters at A and D remain at 12 mm.

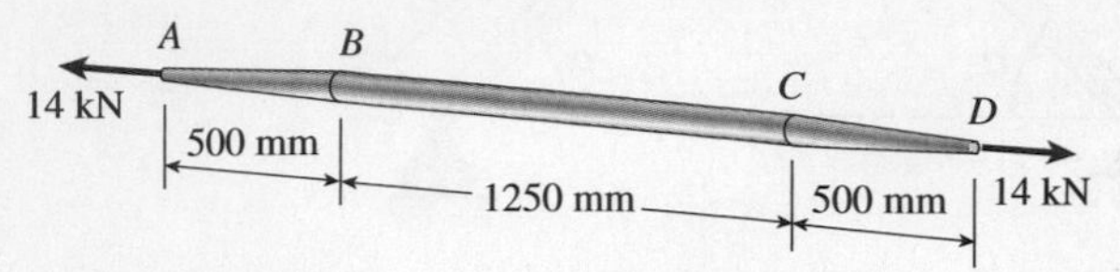

PROBLEM 2.3-1

2.3-2 A long, rectangular copper bar under a tensile load P hangs from a pin that is supported by two steel posts (see figure). The copper bar has a length of 2.0 m, a cross-sectional area of 4800 mm^2, and a modulus of elasticity $E_c = 120$ GPa. Each steel post has a height of 0.5 m, a cross-sectional area of 4500 mm^2, and a modulus of elasticity $E_s = 200$ GPa.

(a) Determine the downward displacement δ of the lower end of the copper bar due to a load $P = 180$ kN.

(b) What is the maximum permissible load P_{max} if the displacement δ is limited to 1.0 mm?

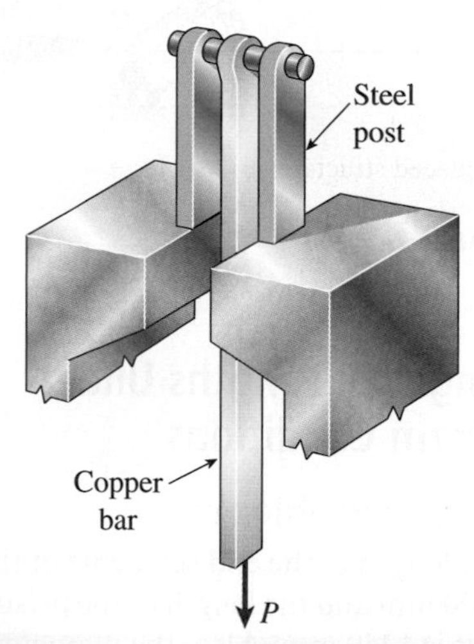

PROBLEM 2.3-2

2.3-3 An aluminum bar AD (see figure) has a cross-sectional area of 250 mm^2 and is loaded by forces $P_1 = 7560$ N, $P_2 = 5340$ N, and $P_3 = 5780$ N. The lengths of the segments of the bar are $a = 1525$ mm, $b = 610$ mm, and $c = 910$ mm.

(a) Assuming that the modulus of elasticity is $E = 72$ GPa, calculate the change in length of the bar. Does the bar elongate or shorten?

(b) By what amount P should the load P_3 be increased so that the bar does not change in length when the three loads are applied?

(c) If P_3 remains at 5780 N, what revised cross-sectional area for segment AB will result in no change of length when all three loads are applied?

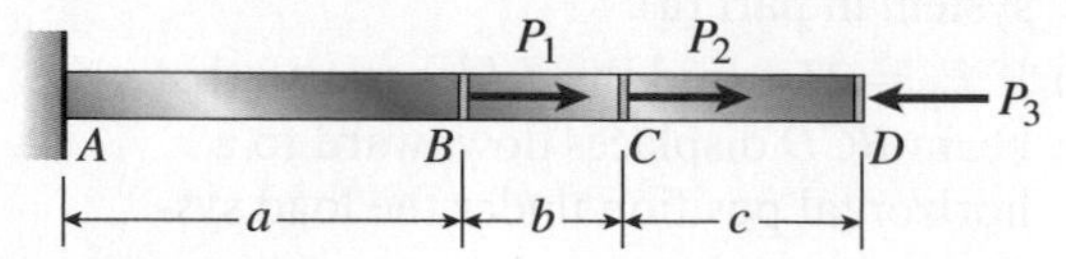

PROBLEM 2.3-3

2.3-4 A vertical bar consists of three prismatic segments A_1, A_2, and A_3 with cross-sectional areas of 6000 mm^2, 5000 mm^2, and 4000 mm^2, respectively. The bar is made of steel with $E = 200$ GPa. Calculate the displacements at points B, D, and E. Ignore the weight of the bar.

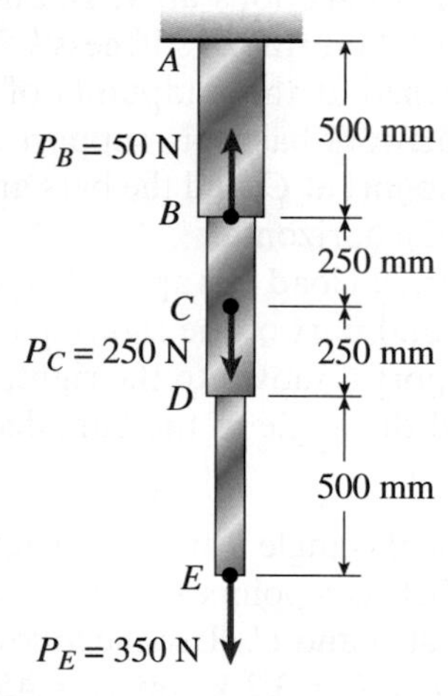

PROBLEM 2.3-4

2.3-5 A vertical bar is loaded with axial loads at points B, C, and D, as shown in the figure. The bar is made of steel with a modulus of elasticity $E = 200$ GPa. The bar has a cross-sectional area of 5300 mm^2. Calculate the displacements at points B, C, and D. Ignore the weight of the bar.

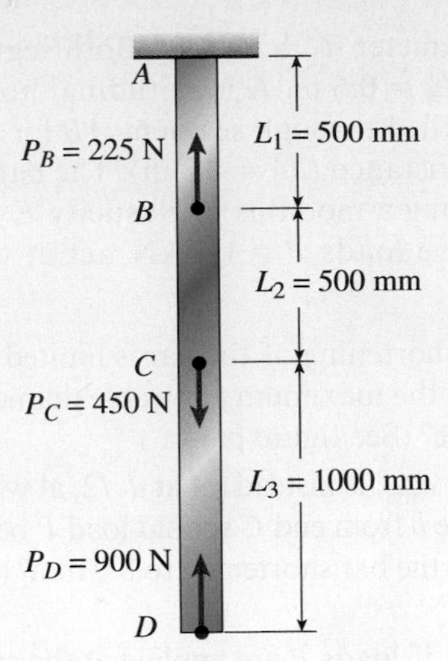

PROBLEM 2.3-5

2.3-6 Repeat Problem 2.3-4, but now include the weight of the bar. See Table I-1 in Appendix I for the weight density of steel.

2.3-7 Repeat Problem 2.3-5, but now include the weight of the bar. See Table I-1 in Appendix I for the weight density of steel.

Representative Problems

2.3-8 A rectangular bar of length L has a slot in the middle half of its length (see figure). The bar has width b, thickness t, and modulus of elasticity E. The slot has width $b/4$.

(a) Obtain a formula for the elongation δ of the bar due to the axial loads P.

(b) Calculate the elongation of the bar if the material is high-strength steel, the axial stress in the middle region is 160 MPa, the length is 750 mm, and the modulus of elasticity is 210 GPa.

(c) If the total elongation of the bar is limited to $\delta_{max} = 0.475$ mm, what is the maximum length of the slotted region? Assume that the axial stress in the middle region remains at 160 MPa.

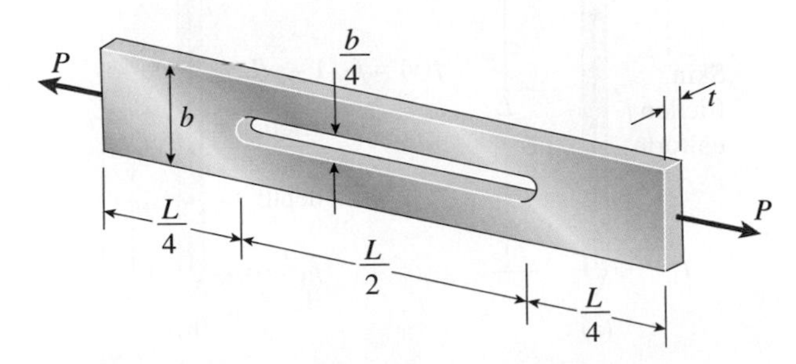

PROBLEMS 2.3-8 and 2.3-9

2.3-9 Solve the preceding problem if the axial stress in the middle region is 165 MPa, the length is 760 mm, and the modulus of elasticity is 207 GPa. In part (c), assume that $\delta_{max} = 0.5$ mm.

2.3-10 A two-story building has steel columns AB in the first floor and BC in the second floor, as shown in the figure. The roof load P_1 equals 400 kN, and the second-floor load P_2 equals 720 kN. Each column has a length $L = 3.75$ m. The cross-sectional areas of the first- and second-floor columns are 11,000 mm^2 and 3900 mm^2, respectively.

(a) Assuming that $E = 206$ GPa, determine the total shortening δ_{AC} of the two columns due to the combined action of the loads P_1 and P_2.

(b) How much additional load P_0 can be placed at the top of the column (point C) if the total shortening δ_{AC} is not to exceed 4.0 mm?

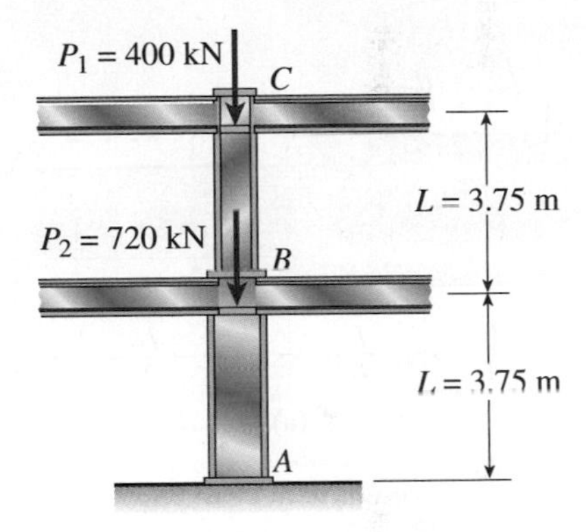

PROBLEM 2.3-10

2.3-11 A steel bar is 2.4 m long and has a circular cross section of diameter $d_1 = 20$ mm over one-half of its length and diameter $d_2 = 12$ mm over the other half (see figure on following page part a). The modulus of elasticity is $E = 205$ GPa.

(a) How much will the bar elongate under a tensile load $P = 22$ kN?

(b) If the same volume of material is made into a bar of constant diameter d and length 2.4 m, what will be the elongation under the same load P?

(c) If the uniform axial centroidal load $q = 18.33$ kN/m is applied to the left over segment 1 (see figure part b), find the ratio of the total elongation of the bar to that in parts (a) and (b).

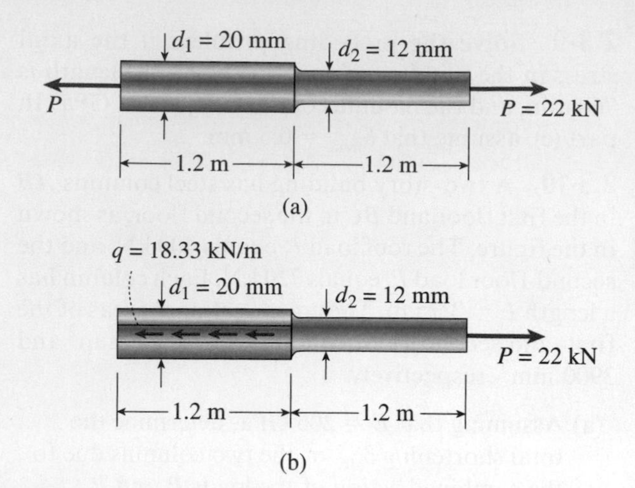

PROBLEM 2.3-11

2.3-12 A bar ABC of length L consists of two parts of equal lengths but different diameters. Segment AB has diameter $d_1 = 100$ mm, and segment BC has diameter $d_2 = 60$ mm. Both segments have a length $L/2 = 0.6$ m. A longitudinal hole of diameter d is drilled through segment AB for one-half of its length (distance $L/4 = 0.3$ m). The bar is made of plastic having a modulus of elasticity $E = 4.0$ GPa. Compressive loads $P = 110$ kN act at the ends of the bar.

(a) If the shortening of the bar is limited to 8.0 mm, what is the maximum allowable diameter d_{max} of the hole? (See figure part a.)

(b) Now, if d_{max} is instead set at $d_2/2$, at what distance b from end C should load P be applied to limit the bar shortening to 8.0 mm? (See figure part b.)

(c) Finally, if loads P are applied at the ends and $d_{max} = d_2/2$, what is the permissible length x of the hole if shortening is to be limited to 8.0 mm? (See figure part c.)

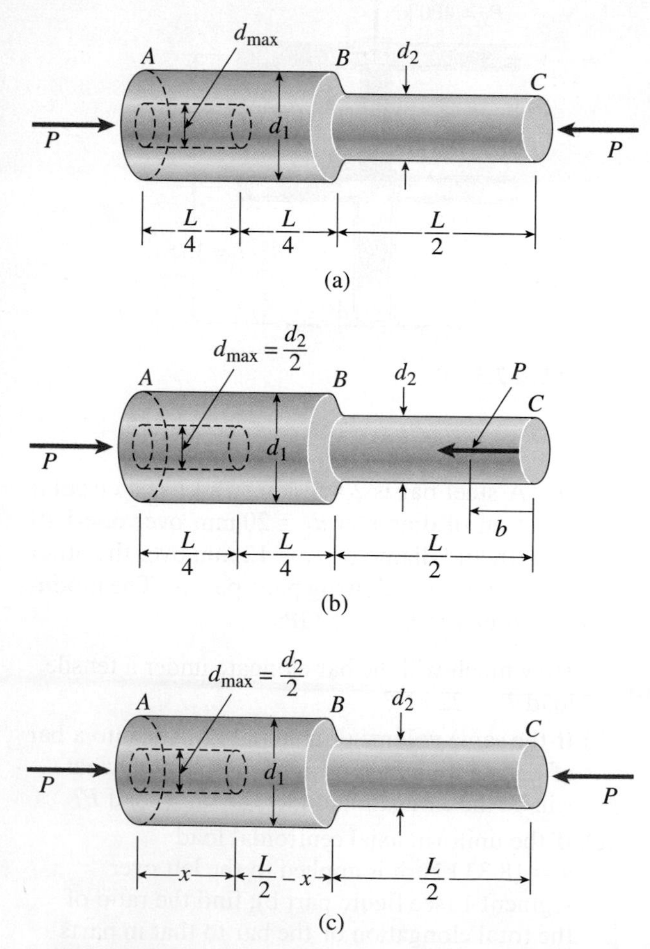

PROBLEM 2.3-12

2.3-13 A wood pile, driven into the earth, supports a load P entirely by friction along its sides (see figure part a). The friction force f per unit length of the pile is assumed to be uniformly distributed over the surface of the pile. The pile has a length L, cross-sectional area A, and modulus of elasticity E.

(a) Derive a formula for the shortening δ of the pile in terms of P, L, E, and A.

(b) Draw a diagram showing how the compressive stress σ_c varies throughout the length of the pile.

(c) Repeat parts (a) and (b) if skin friction f varies linearly with depth (see figure part b).

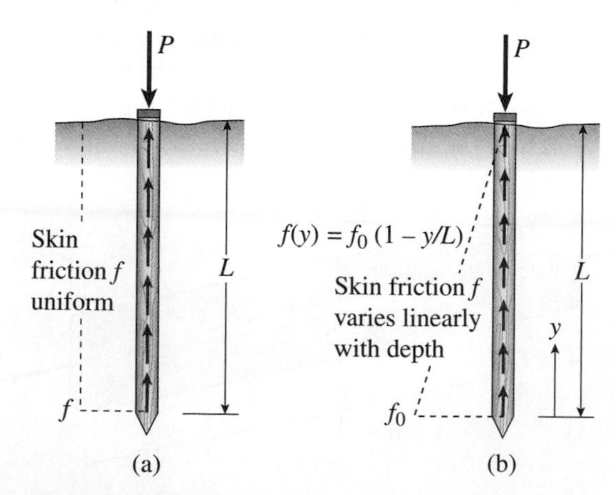

PROBLEM 2.3-13

2.3-14 Consider the copper tubes joined in the figure using a "sweated" joint. Use the properties and dimensions given.

(a) Find the total elongation of segment 2-3-4 (δ_{2-4}) for an applied tensile force of $P = 5$ kN. Use $E_c = 120$ GPa.

(b) If the yield strength in shear of the tin-lead solder is $\tau_y = 30$ MPa and the tensile yield strength of the copper is $\sigma_y = 200$ MPa, what is the maximum load P_{max} that can be applied to the joint if the desired factor of safety in shear is $FS_\tau = 2$ and in tension is $FS_\sigma = 1.7$?

(c) Find the value of L_2 at which tube and solder capacities are equal.

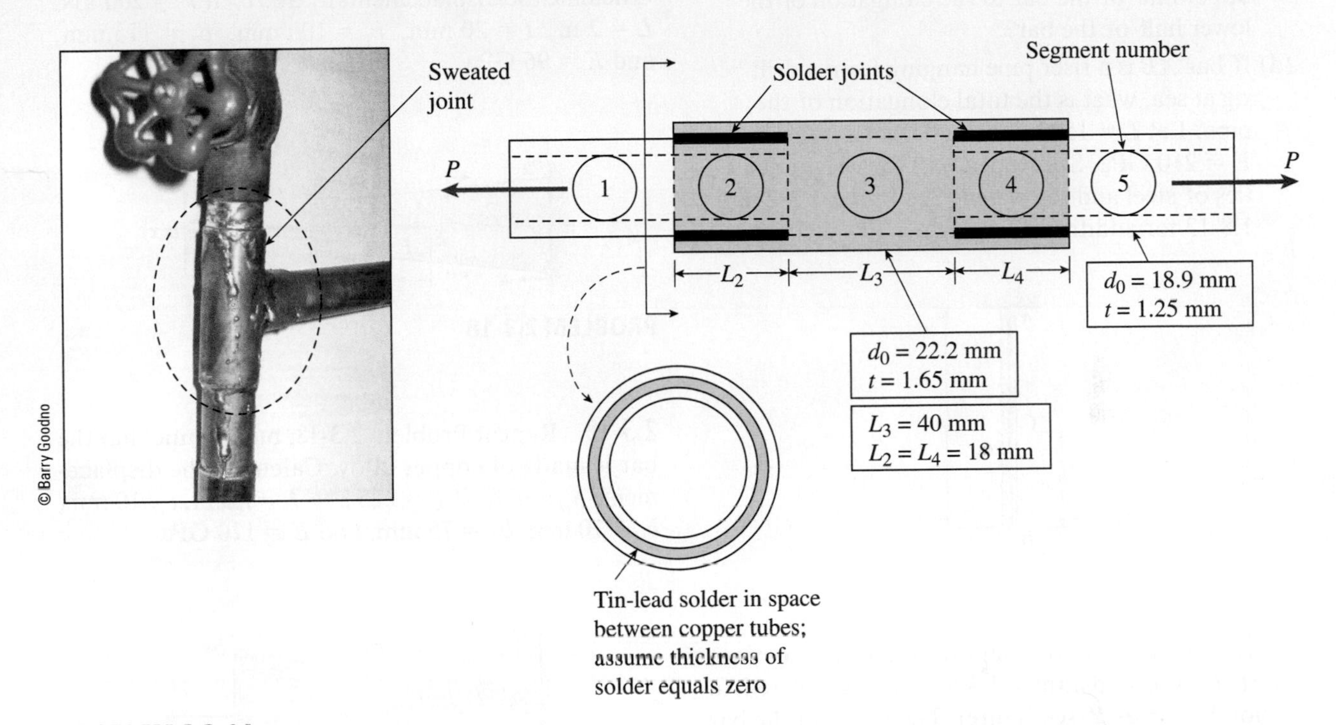

PROBLEM 2.3-14

2.3-15 The nonprismatic cantilever circular bar shown has an internal cylindrical hole of diameter $d/2$ from 0 to x, so the net area of the cross section for segment 1 is $(3/4)A$. Load P is applied at x, and load $P/2$ is applied at $x = L$. Assume that E is constant.

(a) Find reaction force R_1.

(b) Find internal axial forces N_i in segments 1 and 2.

(c) Find x required to obtain axial displacement at joint 3 of $\delta_3 = PL/EA$.

(d) In part (c), what is the displacement at joint 2, δ_2?

(e) If P acts at $x = 2L/3$ and $P/2$ at joint 3 is replaced by βP, find β so that $\delta_3 = PL/EA$.

(f) Draw the *axial force* (AFD: $N(x)$, $0 \leq x \leq L$) and *axial displacement diagrams* (ADD: $\delta(x)$, $0 \leq x \leq L$) using results from parts (b) through (d).

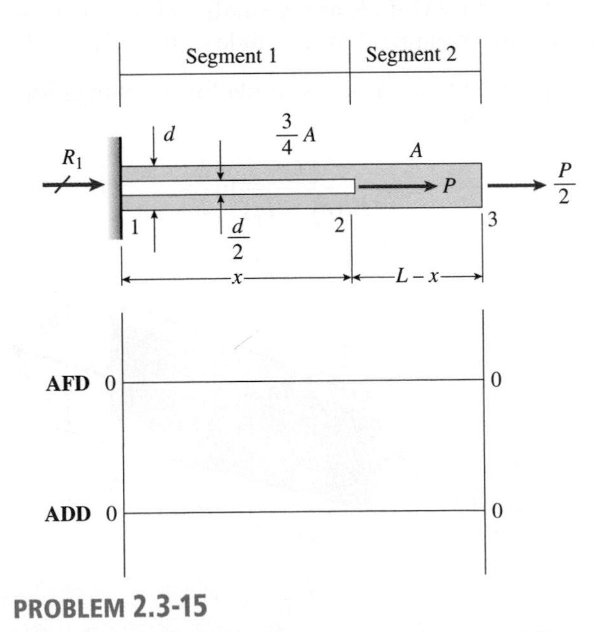

PROBLEM 2.3-15

2.3-16 A prismatic bar AB of length L, cross-sectional area A, modulus of elasticity E, and weight W hangs vertically under its own weight (see figure).

(a) Derive a formula for the downward displacement δ_C of point C, located at distance h from the lower end of the bar.

(b) What is the elongation δ_B of the entire bar?

(c) What is the ratio β of the elongation of the upper half of the bar to the elongation of the lower half of the bar?

(d) If bar AB is a riser pipe hanging from a drill rig at sea, what is the total elongation of the pipe? Let $L = 1500$ m, $A = 0.0157$ m^2, and $E = 210$ GPa. See Appendix I for weight densities of steel and sea water. (See Probs. 1.5-2 and 1.8-13 for additional figures.)

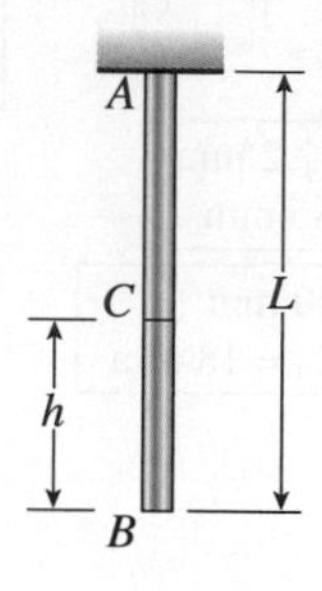

PROBLEM 2.3-16

2.3-17 A flat bar of rectangular cross section, length L, and constant thickness t is subjected to tension by forces P (see figure). The width of the bar varies linearly from b_1 at the smaller end to b_2 at the larger end. Assume that the angle of taper is small.

(a) Derive the following formula for the elongation of the bar:

$$\delta = \frac{PL}{Et(b_2 - b_1)} \ln \frac{b_2}{b_1}$$

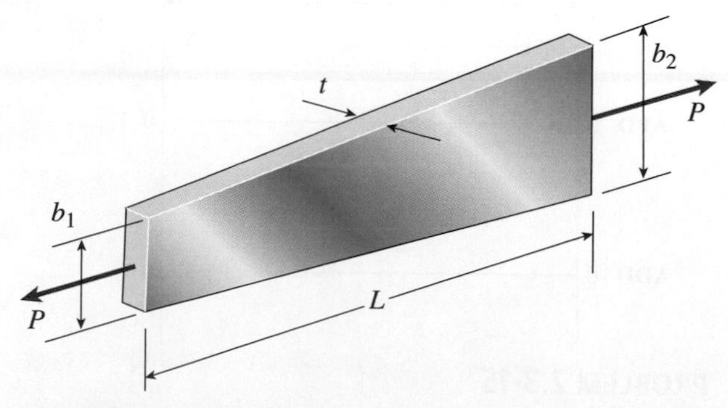

PROBLEM 2.3-17

(b) Calculate the elongation, assuming $L = 1.5$ m, $t = 25$ mm, $P = 125$ kN, $b_1 = 100$ mm, $b_2 = 150$ mm, and $E = 200$ GPa.

2.3-18 A flat brass bar has length L, constant thickness t, and a rectangular cross section whose width varies linearly between b_2 at the fixed support to b_1 at the free end (see figure). Assume that the taper of the bar is small. The bar has modulus of elasticity E. Calculate the displacements δ_B and δ_C if $P = 200$ kN, $L = 2$ m, $t = 20$ mm, $b_1 = 100$ mm, $b_2 = 115$ mm, and $E = 96$ GPa.

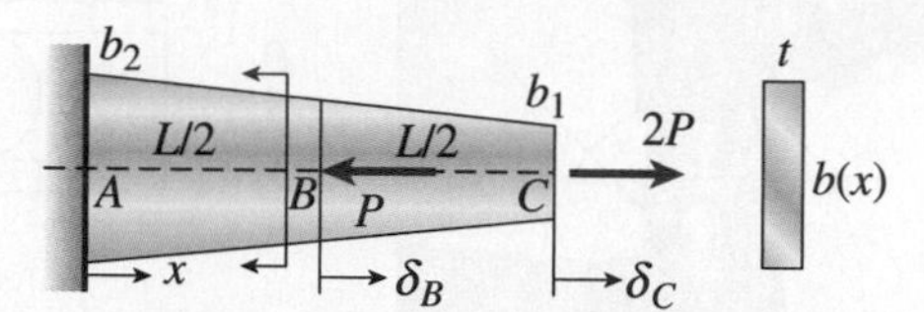

PROBLEM 2.3-18

2.3-19 Repeat Problem 2.3-18, but assume that the bar is made of copper alloy. Calculate the displacements δ_B and δ_C if $P = 225$ kN, $L = 1.5$ m, $t = 10$ mm, $b_1 = 70$ mm, $b_2 = 75$ mm, and $E = 110$ GPa.

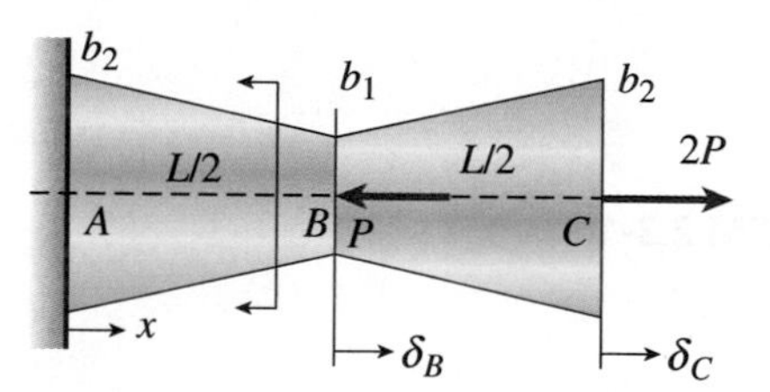

PROBLEM 2.3-19

2.3-20 Repeat Problem 2.3-18, but assume that the bar is made of aluminum alloy. If $P_2 = 200$ kN, what is P_1 so that displacement $\delta_C = 0$? What is displacement δ_B? Assume that $L = 2$ m, $t = 20$ mm, $b_1 = 100$ mm, $b_2 = 115$ mm, and $E = 72$ GPa.

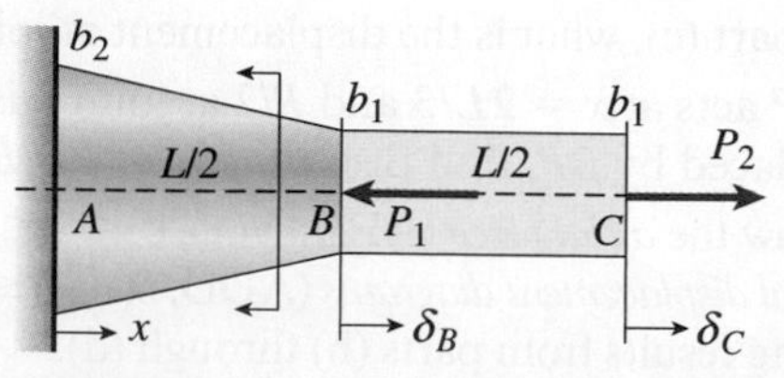

PROBLEM 2.3-20

2.3-21 A slightly tapered bar AB of solid circular cross section and length L is supported at end B and subjected to a tensile load P at the free end A. The diameters of the bar at ends A and B are d_A and d_B, respectively. Determine the length of the bar if the elongation of the bar due to the load $P = 200$ kN is 0.5 mm. Assume that $E = 72$ GPa.

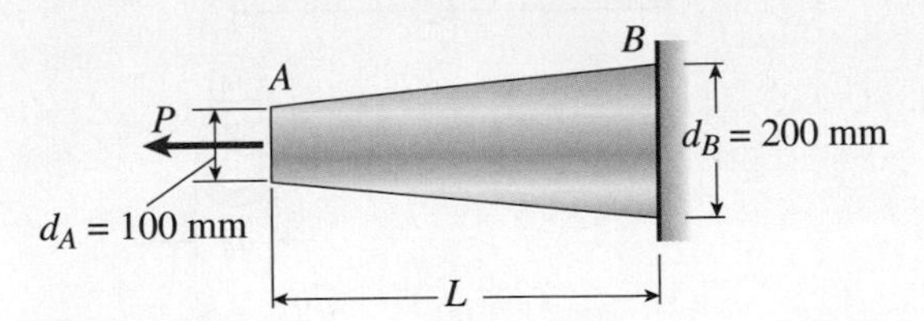

PROBLEM 2.3-21

2.3-22 A circular aluminum alloy bar of length $L = 1.8$ m has a slot in the middle half of its length (see figure). The bar has a radius $r = 36$ mm and modulus of elasticity $E = 72$ GPa. The slot has a height $2a = r/4$. Calculate the elongation of the bar due to forces P applied at the ends if the axial stress in the middle region is known to be 180 MPa.

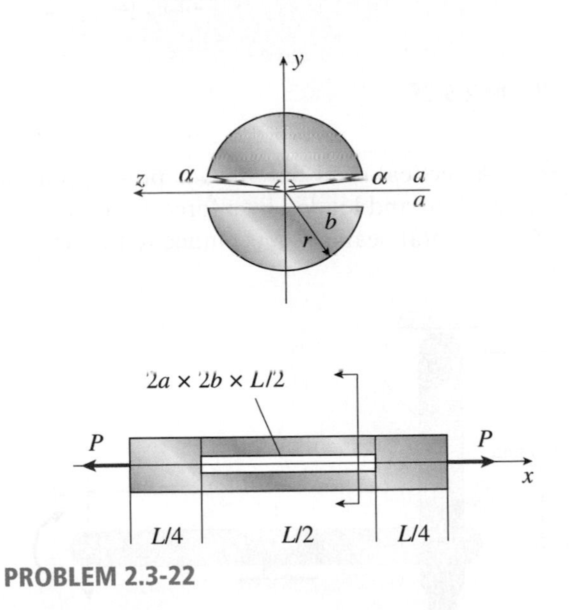

PROBLEM 2.3-22

2.3-23 A long, slender bar in the shape of a right circular cone with length L and base diameter d hangs vertically under the action of its own weight (see figure). The weight of the cone is W and the modulus of elasticity of the material is E.

Derive a formula for the increase δ in the length of the bar due to its own weight. (Assume that the angle of taper of the cone is small.)

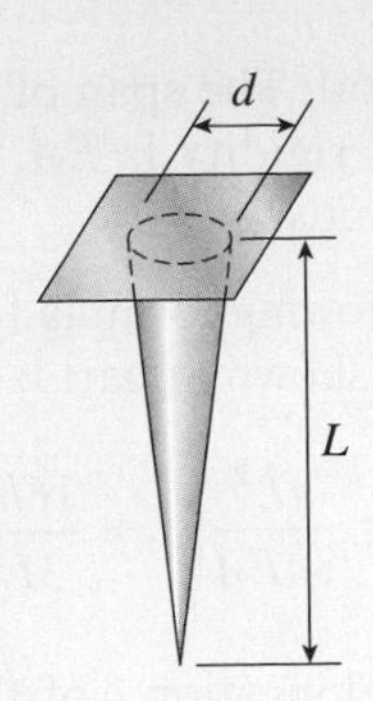

PROBLEM 2.3-23

2.3-24 A post AB supporting equipment in a laboratory is tapered uniformly throughout its height H (see figure). The cross sections of the post are square, with dimensions $b \times b$ at the top and $1.5b \times 1.5b$ at the base.

Derive a formula for the shortening δ of the post due to the compressive load P acting at the top. (Assume that the angle of taper is small and disregard the weight of the post itself.)

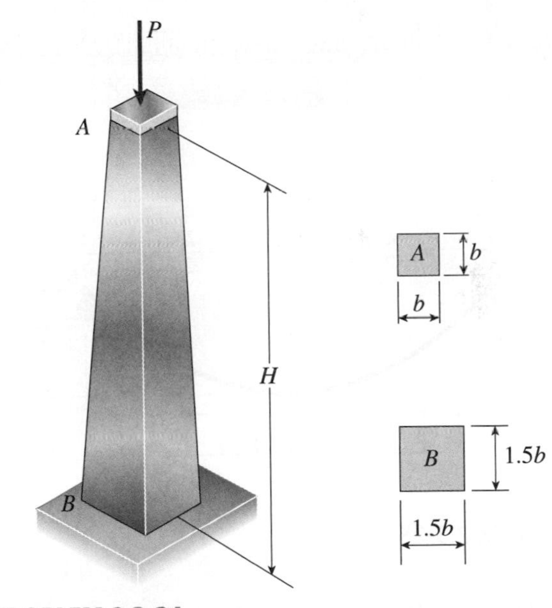

PROBLEM 2.3-24

2.3-25 The main cables of a suspension bridge (see figure part a) follow a curve that is nearly parabolic because the primary load on the cables is the weight of the bridge deck, which is uniform in intensity along the horizontal. Therefore, represent the central region AOB of one of the main cables (see part b of the figure) as a parabolic cable supported at points A and B and carrying a uniform load of intensity q

along the horizontal. The span of the cable is L, the sag is h, the axial rigidity is EA, and the origin of coordinates is at midspan.

(a) Derive the following formula for the elongation of cable AOB shown in part b of the figure:

$$\delta = \frac{qL^3}{8hEA}\left(1 + \frac{16h^2}{3L^2}\right)$$

(b) Calculate the elongation δ of the central span of one of the main cables of the Golden Gate Bridge for which the dimensions and properties are $L = 1300$ m, $h = 140$ m, $q = 185$ kN/m, and $E = 200$ GPa. The cable consists of 27,572 parallel wires of diameter 5 mm.

Hint: Determine the tensile force T at any point in the cable from a free-body diagram of part of the cable; then determine the elongation of an element of the cable of length ds; finally, integrate along the curve of the cable to obtain an equation for the elongation δ.

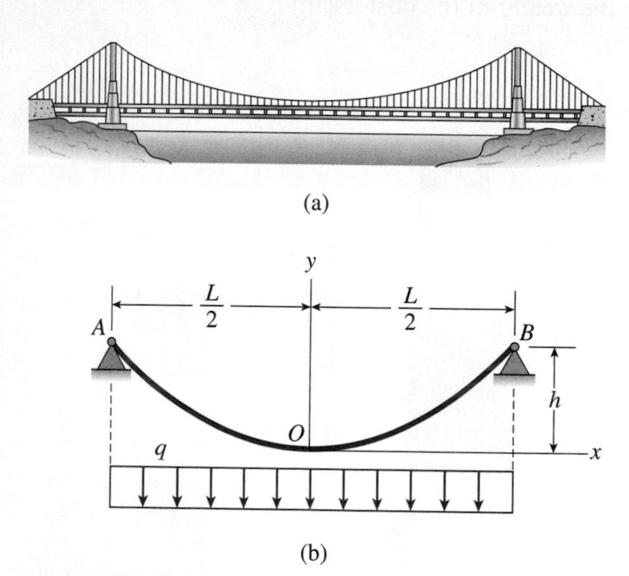

PROBLEM 2.3-25

2.3-26 A uniformly tapered tube AB of circular cross section and length L is shown in the figure. The average diameters at the ends are d_A and $d_B = 2d_A$. Assume E is constant. Find the elongation δ of the tube when it is subjected to loads P acting at the ends. Use the following numerical data: $d_A = 35$ mm, $L = 300$ mm, $E = 2.1$ GPa, and $P = 25$ kN. Consider the following cases.

(a) A hole of *constant* diameter d_A is drilled from B toward A to form a hollow section of length $x = L/2$.

(b) A hole of *variable* diameter $d(x)$ is drilled from B toward A to form a hollow section of length $x = L/2$ and constant thickness $t = d_A/20$.

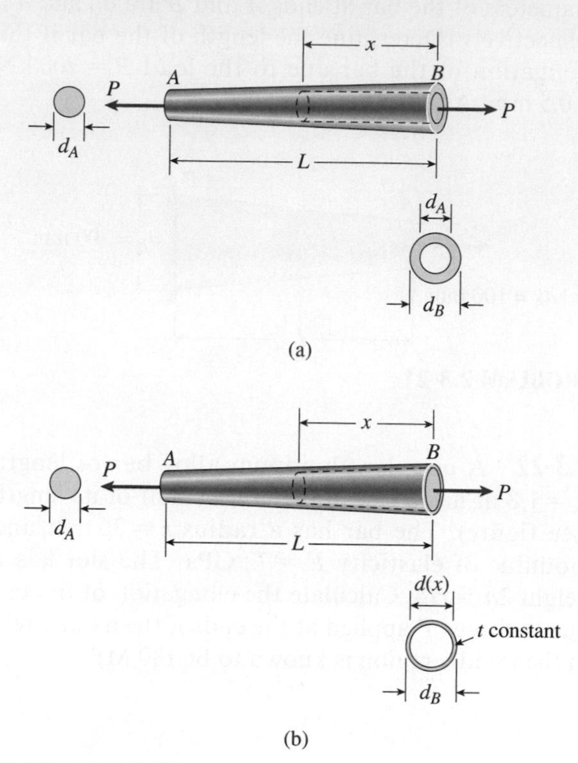

PROBLEM 2.3-26

2.3-27 A vertical steel bar ABC is pin-supported at its upper end and loaded by a force P_1 at its lower end. A horizontal beam BDE is pinned to the vertical

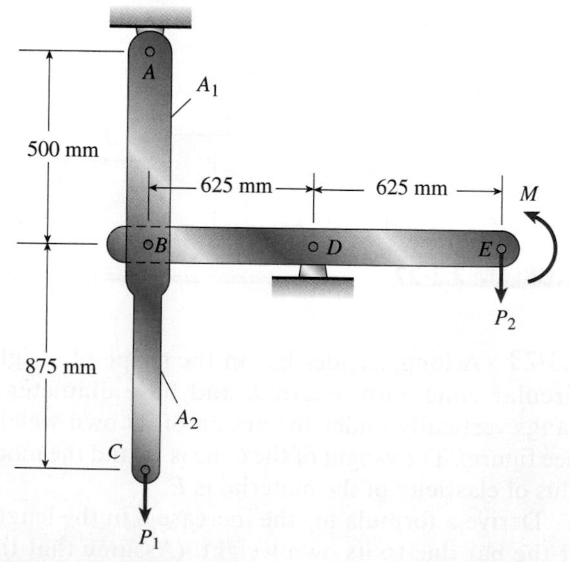

PROBLEM 2.3-27

bar at joint B and supported at point D. Load P_2 and moment M are applied at end E. Calculate the vertical displacement δ_C at point C if the loads are $P_1 = 11$ kN, $P_2 = 4.5$ kN, and $M = 2.8$ kN·m. The modulus of elasticity is $E = 200$ GPa and cross-sectional areas are $A_1 = 160$ mm^2 and $A_2 = 100$ mm^2. Ignore the weight of the bar.

2.3-28 A T-frame structure is composed of a prismatic beam ABC and a nonprismatic column DBF. The beam and the column have a pin support at A and D, respectively. Both members are connected with a pin at B. The lengths and properties of the members are shown in the figure. Find the vertical displacement of the column at points F and B. Plot axial force (AFD) and axial displacement (ADD) diagrams for column DBF.

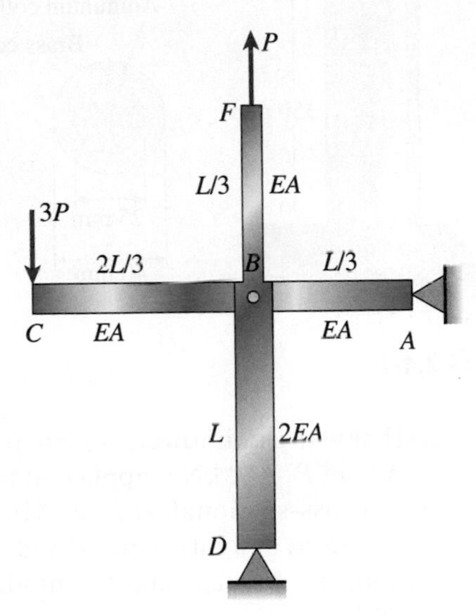

PROBLEM 2.3-28

2.3-29 A T-frame structure is composed of prismatic beam ABC and nonprismatic column DBF that are joined at B by a frictionless pin connection. The beam has a sliding support at A and the column is pin supported at F (see figure). Beam ABC and column segment DB have cross-sectional area A; column segment BF has area $2A$. The modulus of elasticity E is the same for both members. Load $2P$ is applied downward at C, and load P acts at D. Find expressions for the downward displacements of column DBF at D (δ_D) and also at B (δ_B). Plot axial force (AFD) and axial displacement (ADD) diagrams for column DBF.

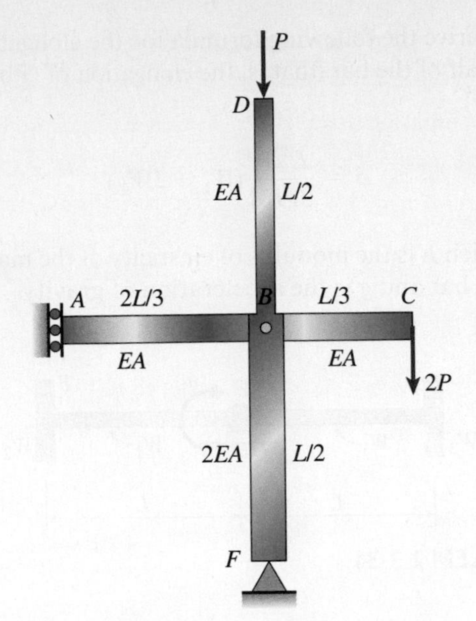

PROBLEM 2.3-29

2.3-30 Repeat Problem 2.3-29 if vertical load P at D is replaced by a horizontal load P at D (see figure).

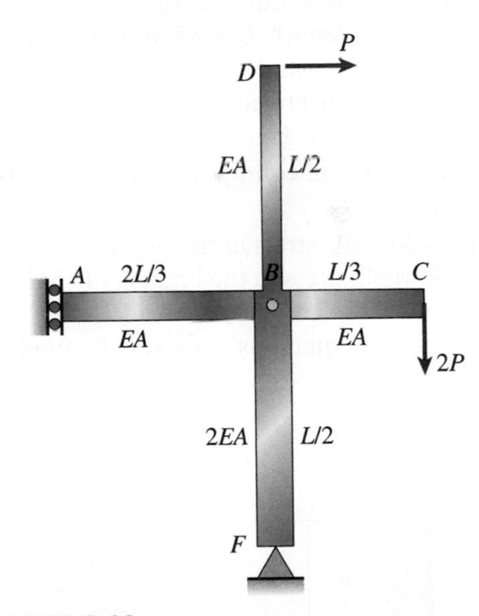

PROBLEM 2.3-30

2.3-31 A bar ABC revolves in a horizontal plane about a vertical axis at the midpoint C (see figure). The bar, which has a length $2L$ and cross-sectional area A, revolves at constant angular speed ω. Each half of the bar (AC and BC) has a weight W_1 and supports a weight W_2 at its end.

Derive the following formula for the elongation of one-half of the bar (that is, the elongation of either AC or BC):

$$\delta = \frac{L^2\omega^2}{3gEA}(W_1 + 3W_2)$$

in which E is the modulus of elasticity of the material of the bar and g is the acceleration of gravity.

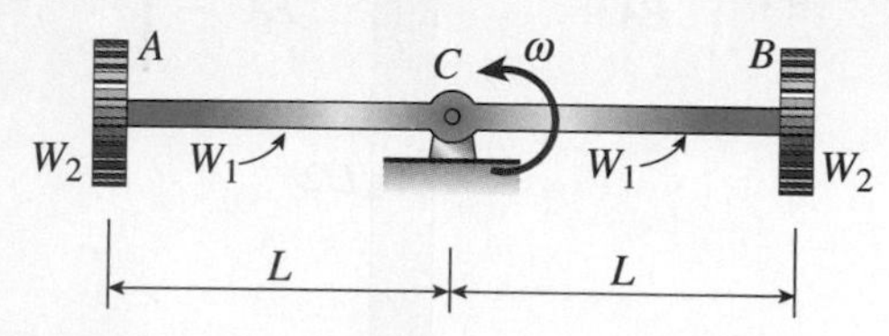

PROBLEM 2.3-31

2.4 Statically Indeterminate Structures

Introductory Problems

2.4-1 The assembly shown in the figure consists of a brass core (diameter $d_1 = 6$ mm) surrounded by a steel shell (inner diameter $d_2 = 7$ mm, outer diameter $d_3 = 9$ mm). A load P compresses the core and shell that both have a length $L = 85$ mm. The moduli of elasticity of the brass and steel are $E_b = 100$ GPa and $E_s = 200$ GPa, respectively.

(a) What load P will compress the assembly by 0.1 mm?

(b) If the allowable stress in the steel is 180 MPa and the allowable stress in the brass is 140 MPa, what is the allowable compressive load P_{allow}? (*Suggestion:* Use the equations derived in Example 2-8.)

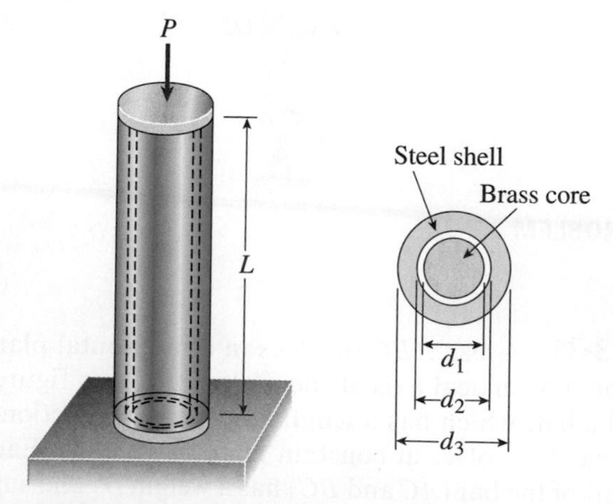

PROBLEM 2.4-1

2.4-2 A cylindrical assembly consisting of a brass core and an aluminum collar is compressed by a load P (see figure). The length of the aluminum collar and brass core is 350 mm, the diameter of the core is 25 mm, and the outside diameter of the collar is 40 mm. Also, the moduli of elasticity of the aluminum and brass are 72 GPa and 100 GPa, respectively.

(a) If the length of the assembly decreases by 0.1% when the load P is applied, what is the magnitude of the load?

(b) What is the maximum permissible load P_{max} if the allowable stresses in the aluminum and brass are 80 MPa and 120 MPa, respectively?

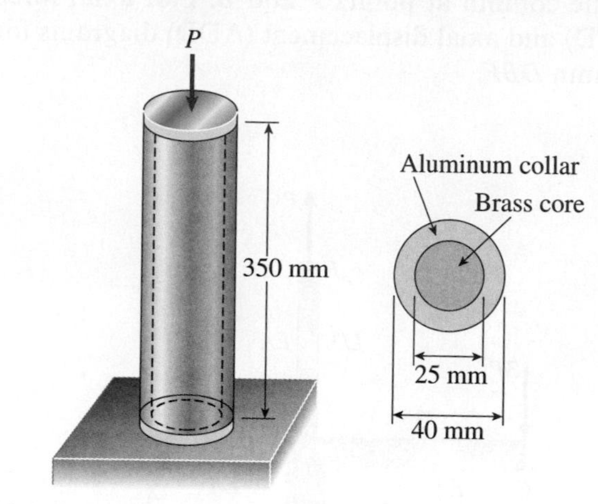

PROBLEM 2.4-2

2.4-3 A steel bar with a uniform cross section is fixed at both ends. A load $P = 10$ kN is applied at point C. The bar has a cross-sectional area of 5100 mm^2. Calculate the reactions at joints A and B and the displacement at joint C. Assume that the modulus of elasticity $E = 200$ GPa.

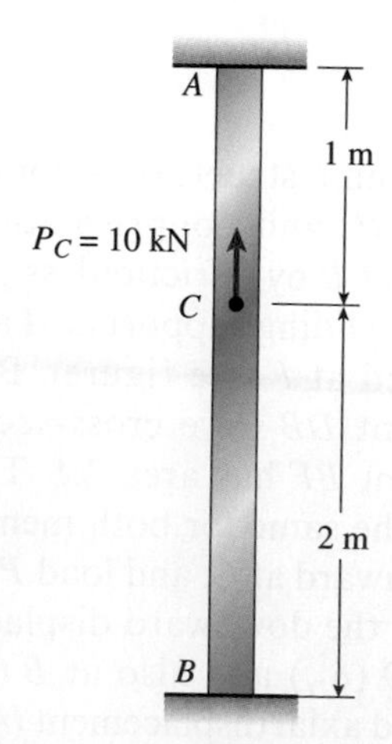

PROBLEM 2.4-3

2.4-4 A horizontal rigid bar *ABC* is pinned at end *A* and supported by two cables at points *B* and *C*. A vertical load $P = 10$ kN acts at end *C* of the bar. The two cables are made of steel with a modulus elasticity $E = 200$ GPa and have the same cross-sectional area. Calculate the minimum cross-sectional area of each cable if the yield stress of the cable is 400 MPa and the factor of safety is 2.0. Consider load *P* only; ignore the weight of bar *ABC* and the cables.

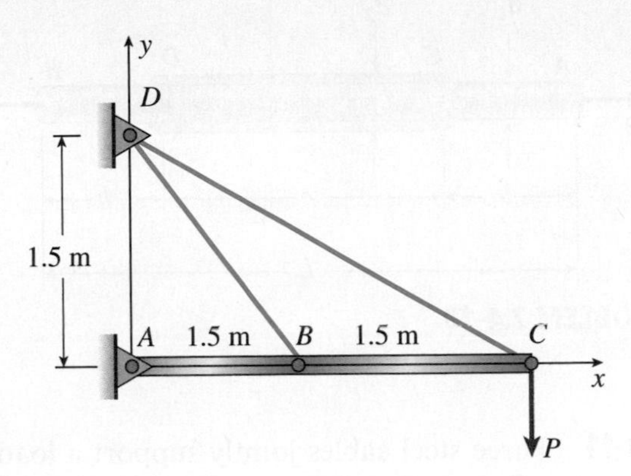

PROBLEM 2.4-4

2.4-5 A solid circular steel cylinder *S* is encased in a hollow circular aluminum tube *A*. The cylinder and tube are compressed between the rigid plates of a testing machine which applies forces *P*. Calculate the allowable value of the compressive force if the yield stresses of steel and aluminum are $\sigma_{yS} = 340$ MPa and $\sigma_{yA} = 410$ MPa, respectively. Assume that $A_S = 7700\ \text{mm}^2$, $A_A = 3800\ \text{mm}^2$, $L = 500\ \text{mm}$, $E_S = 200$ GPa, and $E_A = 73$ GPa.

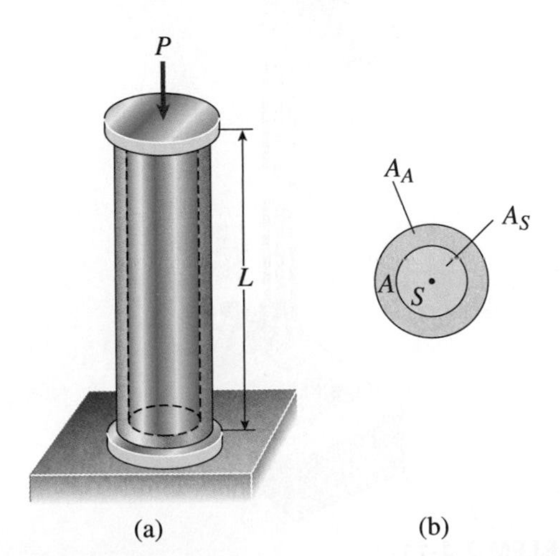

PROBLEM 2.4-5

2.4-6 Three prismatic bars, two of material *A* and one of material *B*, transmit a tensile load *P* (see figure). The two outer bars (material *A*) are identical. The cross-sectional area of the middle bar (material *B*) is 50% larger than the cross-sectional area of one of the outer bars. Also, the modulus of elasticity of material *A* is twice that of material *B*.

(a) What fraction of the load *P* is transmitted by the middle bar?

(b) What is the ratio of the stress in the middle bar to the stress in the outer bars?

(c) What is the ratio of the strain in the middle bar to the strain in the outer bars?

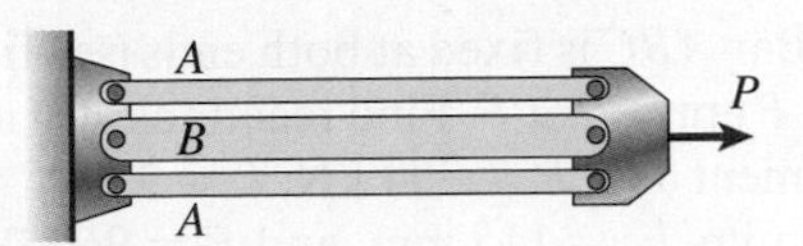

PROBLEM 2.4-6

Representative Problems

2.4-7 A circular bar *ACB* of a diameter *d* having a cylindrical hole of length *x* and diameter *d*/2 from *A* to *C* is held between rigid supports at *A* and *B*.

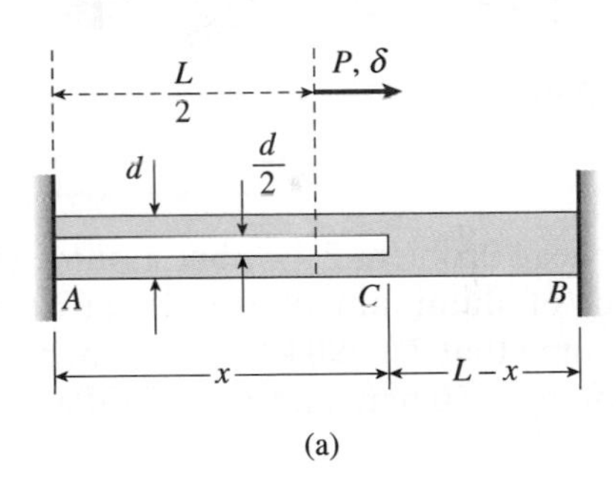

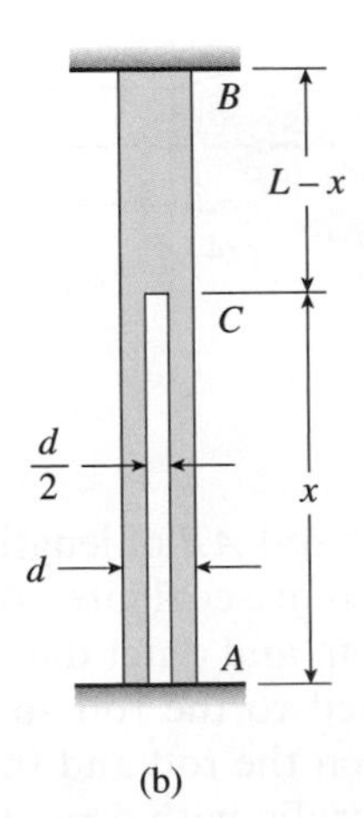

PROBLEM 2.4-7

A load P acts at $L/2$ from ends A and B. Assume E is constant.

(a) Obtain formulas for the reactions R_A and R_B at supports A and B, respectively, due to the load P (see figure part a).

(b) Obtain a formula for the displacement δ at the point of load application (see figure part a).

(c) For what value of x is $R_B = (6/5)R_A$? (See figure part a.)

(d) Repeat part (a) if the bar is now rotated to a vertical position, load P is removed, and the bar is hanging under its own weight (assume mass density $= \rho$). (See figure part b.) Assume that $x = L/2$.

2.4-8 Bar *ABC* is fixed at both ends (see figure) and has load P applied at B. Find reactions at A and C and displacement δ_B if $P = 200$ kN, $L = 2$ m, $t = 20$ mm, $b_1 = 100$ mm, $b_2 = 115$ mm, and $E = 96$ GPa.

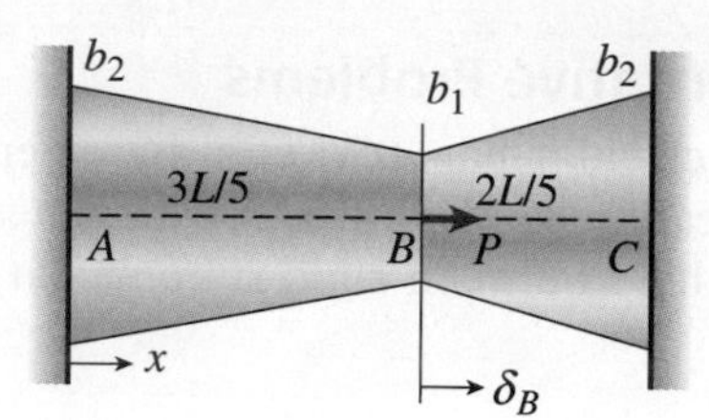

PROBLEM 2.4-8

2.4-9 Repeat Problem 2.4-8, but assume that the bar is made of aluminum alloy and that *BC* is prismatic. Assume that $P = 90$ kN, $L = 1$ m, $t = 6$ mm, $b_1 = 50$ mm, $b_2 = 60$ mm, and $E = 72$ GPa.

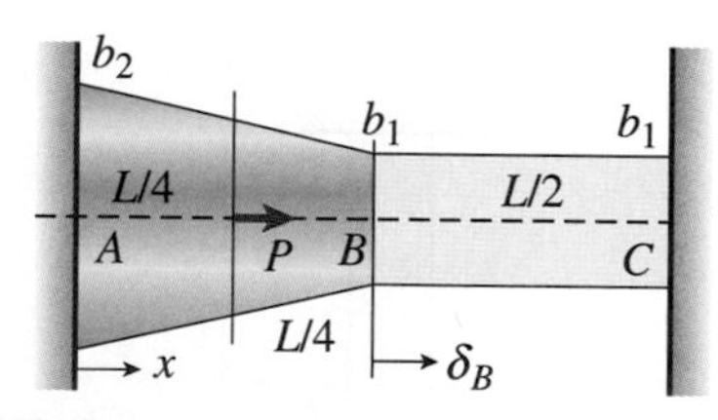

PROBLEM 2.4-9

2.4-10 A plastic rod *AB* of length $L = 0.5$ m has a diameter $d_1 = 30$ mm (see figure). A plastic sleeve *CD* of length $c = 0.3$ m and outer diameter $d_2 = 45$ mm is securely bonded to the rod so that no slippage can occur between the rod and the sleeve. The rod is made of an acrylic with a modulus of elasticity $E_1 = 3.1$ GPa, and the sleeve is made of a polyamide with $E_2 = 2.5$ GPa.

(a) Calculate the elongation δ of the rod when it is pulled by axial forces $P = 12$ kN.

(b) If the sleeve is extended for the full length of the rod, what is the elongation?

(c) If the sleeve is removed, what is the elongation?

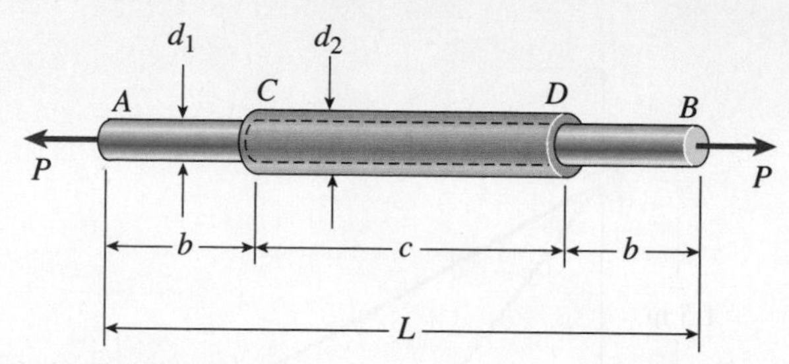

PROBLEM 2.4-10

2.4-11 Three steel cables jointly support a load of 60 kN (see figure). The diameter of the middle cable is 20 mm and the diameter of each outer cable is 12 mm. The tensions in the cables are adjusted so that each cable carries one-third of the load (i.e., 20 kN). Later, the load is increased by 40 kN to a total load of 100 kN.

(a) What percent of the total load is now carried by the middle cable?

(b) What are the stresses σ_M and σ_O in the middle and outer cables, respectively? *Note:* See Table 2-1 in Section 2.2 for properties of cables.

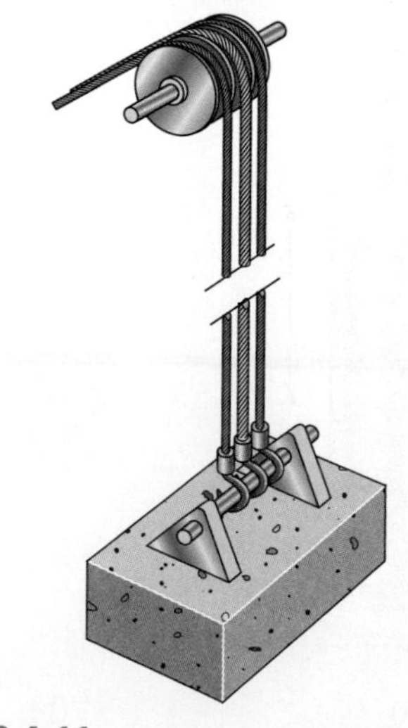

PROBLEM 2.4-11

2.4-12 The fixed-end bar *ABCD* consists of three prismatic segments, as shown in the figure. The end segments have a cross-sectional area $A_1 = 840\ \text{mm}^2$ and length $L_1 = 200$ mm. The middle segment has a cross-sectional area $A_2 = 1260\ \text{mm}^2$ and length $L_2 = 250$ mm. Loads P_B and P_C are equal to 25.5 kN and 17.0 kN, respectively.

(a) Determine the reactions R_A and R_D at the fixed supports.

(b) Determine the compressive axial force F_{BC} in the middle segment of the bar.

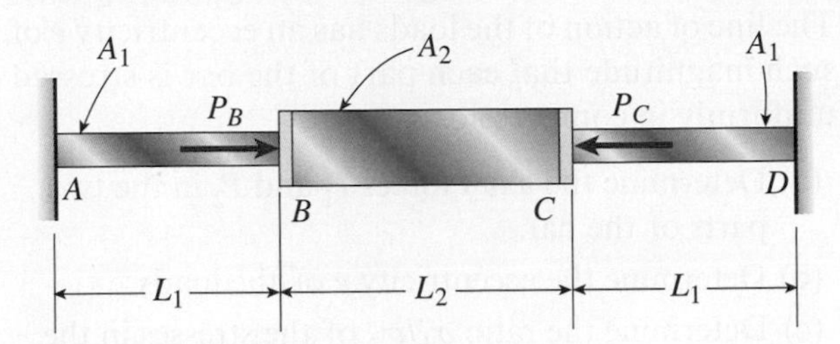

PROBLEM 2.4-12

2.4-13 A tube structure is acted on by loads at *B* and *D*, as shown in the figure. The tubes are joined using two flange plates at *C* that are bolted together using six 12.5 mm diameter bolts.

(a) Derive formulas for the reactions R_A and R_E at the ends of the bar.

(b) Determine the axial displacements δ_B, δ_C, and δ_D at points *B*, *C*, and *D*, respectively.

(c) Draw an axial-displacement diagram (ADD) in which the abscissa is the distance *x* from support *A* to any point on the bar and the ordinate is the horizontal displacement δ at that point.

(d) Find the maximum value of the load variable *P* if allowable normal stress in the bolts is 96 MPa.

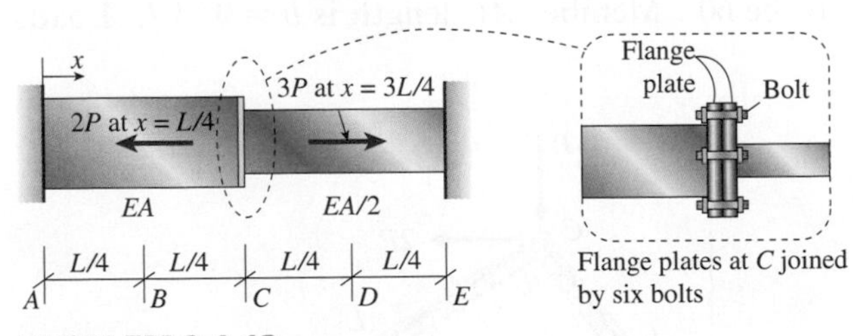

PROBLEM 2.4-13

2.4-14 A hollow circular pipe (see figure) supports a load *P* that is uniformly distributed around a cap plate at the top of the lower pipe. The inner and outer diameters of the upper and lower parts of the pipe are $d_1 = 50$ mm, $d_2 = 60$ mm, $d_3 = 57$ mm, and $d_4 = 64$ mm, respectively. Pipe lengths are $L_1 = 2$ m and $L_2 = 3$ m. Neglect the self-weight of the pipes. Assume that cap plate thickness is small compared to L_1 and L_2. Let $E = 110$ MPa.

(a) If the tensile stress in the upper part is $\sigma_1 = 10.5$ MPa, what is load *P*? Also, what are reactions R_1 at the upper support and R_2 at the lower support? What is the stress σ_2 (MPa) in the lower part?

(b) Find displacement δ (mm) at the cap plate. Plot the axial force diagram (AFD) [$N(x)$] and axial displacement diagram (ADD) [$\delta(x)$].

(c) Add the uniformly distributed load *q* along the centroidal axis of pipe segment 2. Find *q* (kN/m) so that $R_2 = 0$. Assume that load *P* from part (a) is also applied.

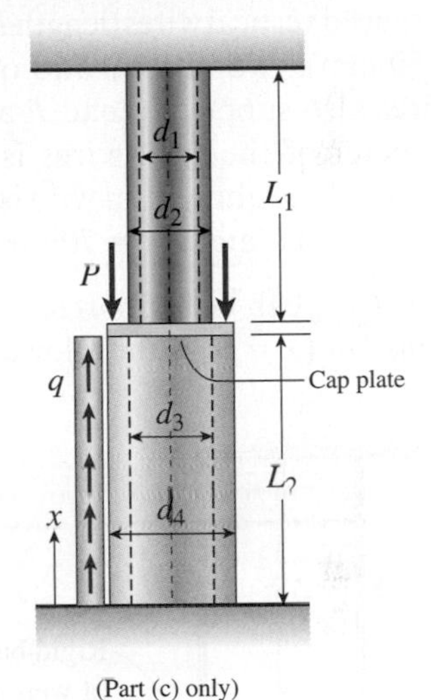

PROBLEM 2.4-14

2.4-15 The aluminum and steel pipes shown in the figure are fastened to rigid supports at ends *A* and *B* and to a rigid plate *C* at their junction. The aluminum pipe is twice as long as the steel pipe. Two equal and symmetrically placed loads *P* act on the plate at *C*.

(a) Obtain formulas for the axial stresses σ_a and σ_s in the aluminum and steel pipes, respectively.

(b) Calculate the stresses for the following data:
$P = 50$ kN,
$A_a = 6000\ \text{mm}^2$, cross-sectional area of aluminum pipe,
$A_s = 600\ \text{mm}^2$, cross-sectional area of steel pipe,
$E_a = 70$ GPa, modulus of elasticity of aluminum, and
$E_s = 200$ GPa, modulus of elasticity of steel.

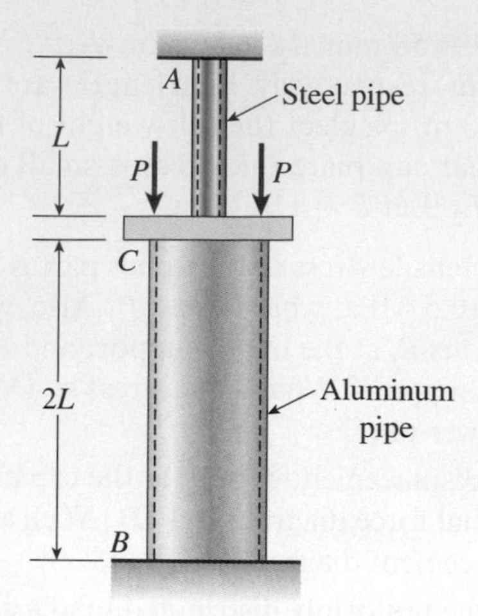

PROBLEM 2.4-15

2.4-16 A rigid bar of weight $W = 800$ N hangs from three equally spaced vertical wires (length $L = 150$ mm, spacing $a = 50$ mm): two of steel and one of aluminum. The wires also support a load P acting on the bar. The diameter of the steel wires is $d_s = 2$ mm, and the diameter of the aluminum wire is $d_a = 4$ mm. Assume $E_s = 210$ GPa and $E_a = 70$ GPa.

(a) What load P_{allow} can be supported *at the midpoint of the bar* ($x = a$) if the allowable stress in the steel wires is 220 MPa and in the aluminum wire is 80 MPa? (See figure part a.)

(b) What is P_{allow} if the load is positioned at $x = a/2$? (See figure part a.)

(c) Repeat part (b) if the second and third wires are *switched* as shown in the figure part b.

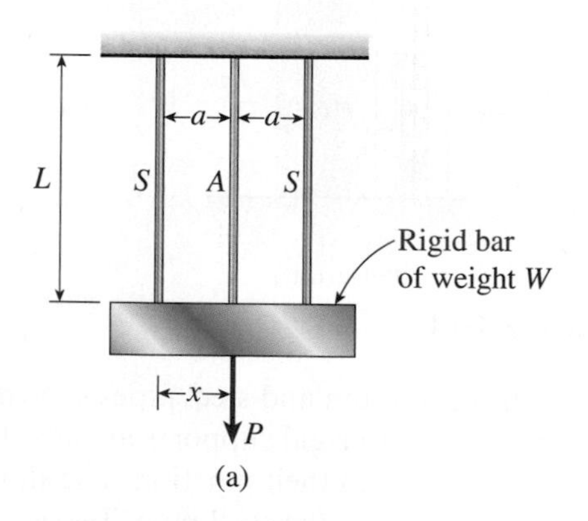

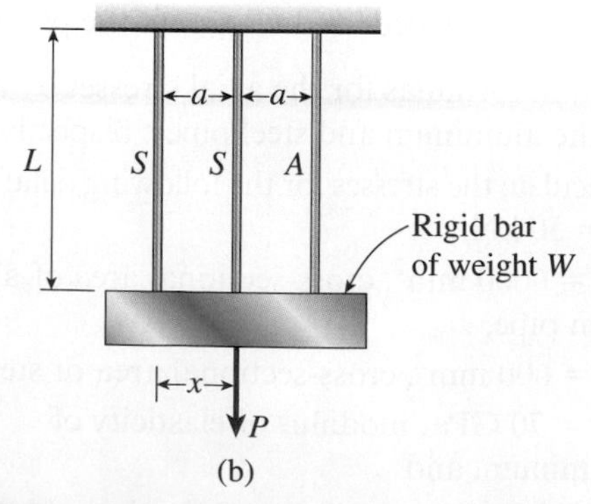

PROBLEM 2.4-16

2.4-17 A *bimetallic* bar (or composite bar) of square cross section with dimensions $2b \times 2b$ is constructed of two different metals having moduli of elasticity E_1 and E_2 (see figure). The two parts of the bar have the same cross-sectional dimensions. The bar is compressed by forces P acting through rigid end plates. The line of action of the loads has an eccentricity e of such magnitude that each part of the bar is stressed uniformly in compression.

(a) Determine the axial forces P_1 and P_2 in the two parts of the bar.

(b) Determine the eccentricity e of the loads.

(c) Determine the ratio σ_1/σ_2 of the stresses in the two parts of the bar.

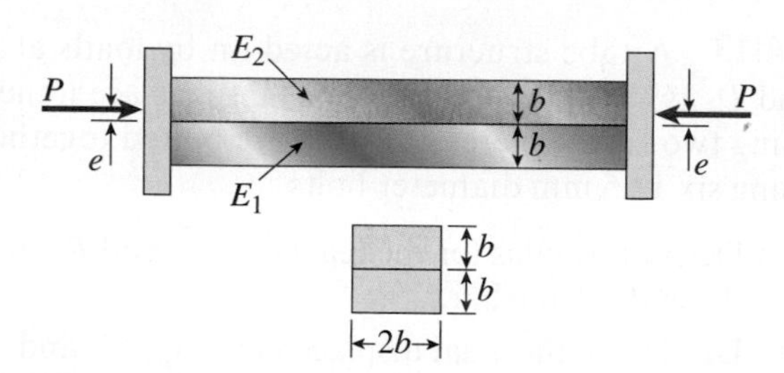

PROBLEM 2.4-17

2.4-18 Three-bar truss ABC (see figure) is constructed of steel pipes having a cross-sectional area $A = 3500$ mm^2 and a modulus of elasticity $E = 210$ GPa. Member BC is of length $L = 2.5$ m, and the angle between members AC and AB is known to be 60°. Member AC length is $b = 0.71\,L$. Loads

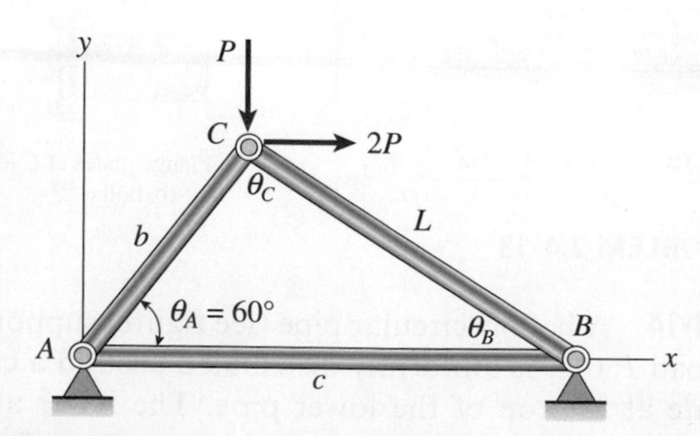

PROBLEM 2.4-18

$P = 185$ kN and $2P = 370$ kN act vertically and horizontally at joint C, as shown. Joints A and B are pinned supports. (Use the law of sines and law of cosines to find missing dimensions and angles in the figure.)

(a) Find the support reactions at joints A and B. Use horizontal reaction B_x as the redundant.

(b) What is the maximum permissible value of load variable P if the allowable normal stress in each truss member is 150 MPa?

2.4-19 A horizontal rigid bar of weight $W = 32$ kN is supported by three slender circular rods that are equally spaced (see figure). The two outer rods are made of aluminum ($E_1 = 70$ GPa) with diameter $d_1 = 10$ mm and length $L_1 = 1$ m. The inner rod is magnesium ($E_2 = 42$ GPa) with diameter d_2 and length L_2. The allowable stresses in the aluminum and magnesium are 165 MPa and 90 MPa, respectively.

If it is desired to have all three rods loaded to their maximum allowable values, what should be the diameter d_2 and length L_2 of the middle rod?

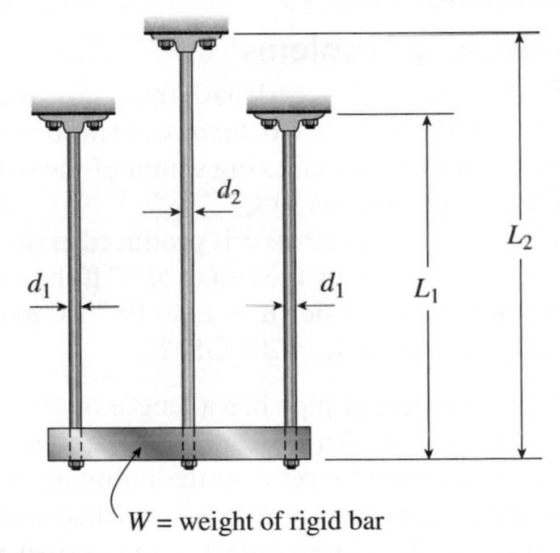

PROBLEM 2.4-19

2.4-20 A rigid bar $ABCD$ is pinned at point B and supported by springs at A and D (see figure). The springs at A and D have stiffnesses $k_1 = 10$ kN/m and $k_2 = 25$ kN/m, respectively, and the dimensions a, b, and c are 250 mm, 500 mm, and 200 mm, respectively. A load P acts at point C.

If the angle of rotation of the bar due to the action of the load P is limited to 3°, what is the maximum permissible load P_{max}?

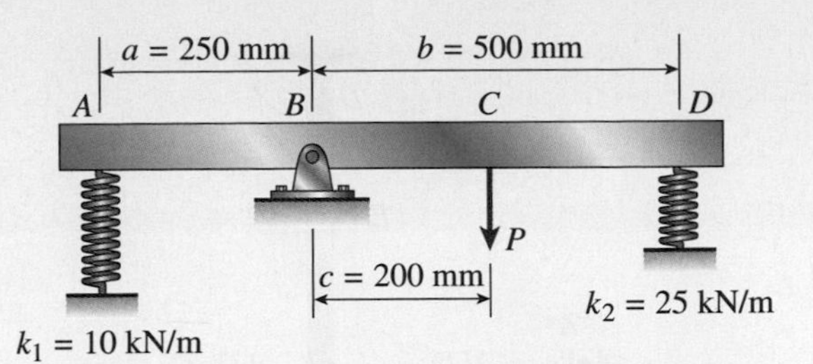

PROBLEM 2.4-20

2.4-21 A rigid bar AB of a length $L = 1600$ mm is hinged to a support at A and supported by two vertical wires attached at points C and D (see figure). Both wires have the same cross-sectional area ($A = 16$ mm^2) and are made of the same material (modulus $E = 200$ GPa). The wire at C has a length $h = 0.4$ m and the wire at D has a length twice that amount. The horizontal distances are $c = 0.5$ m and $d = 1.2$ m.

(a) Determine the tensile stresses σ_C and σ_D in the wires due to the load $P = 970$ N acting at end B of the bar.

(b) Find the downward displacement δ_B at end B of the bar.

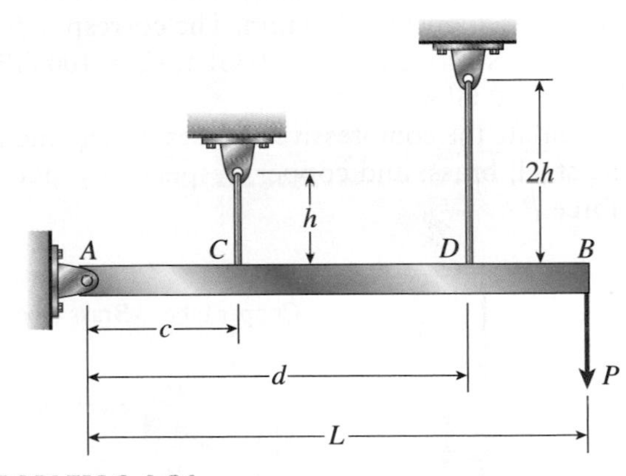

PROBLEM 2.4-21

2.4-22 Find expressions for all support reaction forces in the plane frame with load $2P$ applied at C, as shown in the figure. Joint A is a sliding support, joint D is pinned, and joint F is a roller support. Assume that member AC is a flat prismatic bar of length L, width b, and thickness t. Beam ABC is pinned to column DBF at mid-height (point B). Column DBF has constant thickness t and width b for DB but width $2b$ for BF. Consider load $2P$ at C only; neglect the weights of all members. The modulus of elasticity E is the same for both members. Select reaction R_F as the *redundant*.

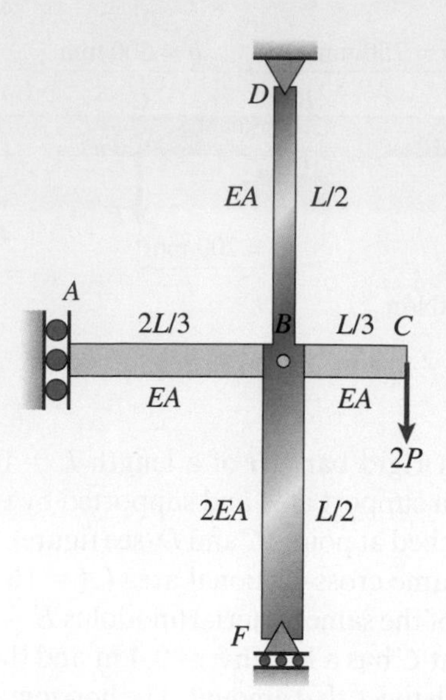

PROBLEM 2.4-22

2.4-23 A trimetallic bar is uniformly compressed by an axial force $P = 12$ kN applied through a rigid end plate (see figure). The bar consists of a circular steel core surrounded by brass and copper tubes. The steel core has a diameter of 10 mm, the brass tube has an outer diameter of 15 mm, and the copper tube has an outer diameter of 20 mm. The corresponding moduli of elasticity are $E_s = 210$ GPa, $E_b = 100$ GPa, and $E_c = 120$ GPa.

Calculate the compressive stresses σ_s, σ_b, and σ_c in the steel, brass, and copper, respectively, due to the force P.

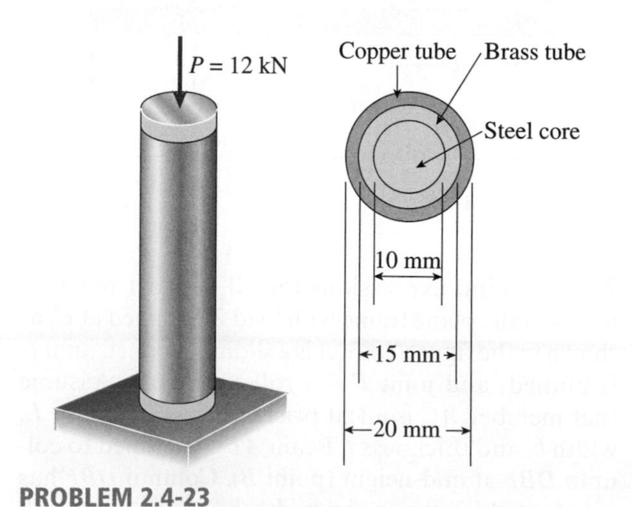

PROBLEM 2.4-23

2.4-24 Find expressions for all support reaction forces in the plane frame with load $3P$ applied at C as shown in the figure. Joints A and D are pin supported, and there is a roller support at joint F. The lengths and the properties of the members are shown in the figure. Neglect the weights of all members. Select R_F as the redundant.

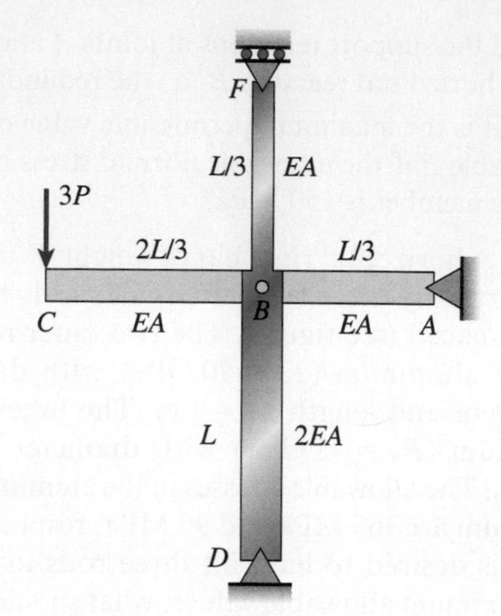

PROBLEM 2.4-24

2.5 Thermal Effects

Introductory Problems

2.5-1 The rails of a railroad track are welded together at their ends (to form continuous rails and thus eliminate the clacking sound of the wheels) when the temperature is 10°C.

What compressive stress σ is produced in the rails when they are heated by the sun to 52°C if the coefficient of thermal expansion $\alpha = 12 \times 10^{-6}$/°C and the modulus of elasticity $E = 200$ GPa?

2.5-2 An aluminum pipe has a length of 60 m at a temperature of 10°C. An adjacent steel pipe at the same temperature is 5 mm longer than the aluminum pipe.

At what temperature (degrees Celsius) will the aluminum pipe be 15 mm longer than the steel pipe? (Assume that the coefficients of thermal expansion of aluminum and steel are $\alpha_a = 23 \times 10^{-6}$/°C and as $\alpha_S = 12 \times 10^{-6}$/°C, respectively.)

2.5-3 A rigid bar of weight $W = 3560$ N hangs from three equally spaced wires: two of steel and one of aluminum (see figure). The diameter of the wires is 3.2 mm. Before they were loaded, all three wires had the same length.

What temperature increase ΔT in all three wires will result in the entire load being carried by the steel wires? (Assume $E_S = 205$ GPa, $\alpha_S = 12 \times 10^{-6}$/°C, and $\alpha_a = 24 \times 10^{-6}$/°C.)

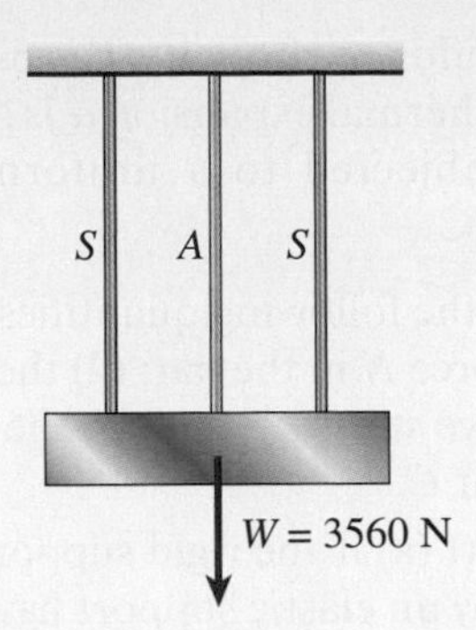

PROBLEM 2.5-3

2.5-4 A steel rod of 15-mm diameter is held snugly (but without any initial stresses) between rigid walls by the arrangement shown in the figure part a. (For the steel rod, use $\alpha = 12 \times 10^{-6}$ /°C and $E = 200$ GPa.)

(a) Calculate the temperature drop ΔT (degrees Celsius) at which the average shear stress in the 12-mm diameter bolt becomes 45 MPa. Also, what is the normal stress in the rod?

(b) What are the average bearing stresses in the bolt and clevis at A and between the washer ($d_w = 20$ mm) and wall ($t = 18$ mm) at B?

(c) If the connection to the wall at B is changed to an end plate with two bolts (see figure part b), what is the required diameter d_b of each bolt if the temperature drop is $\Delta T = 38$°C and the allowable bolt stress is 90 MPa?

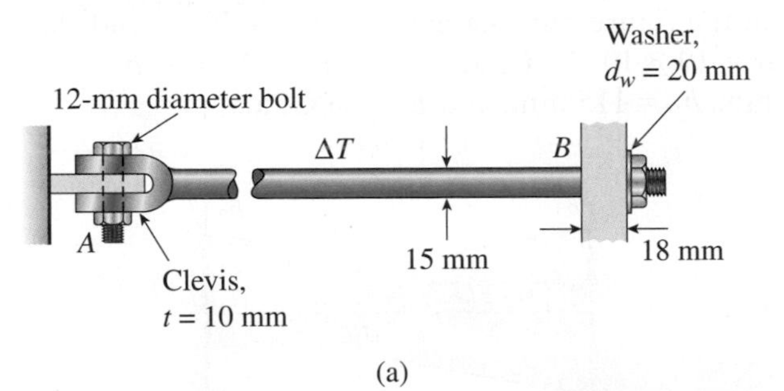

(a)

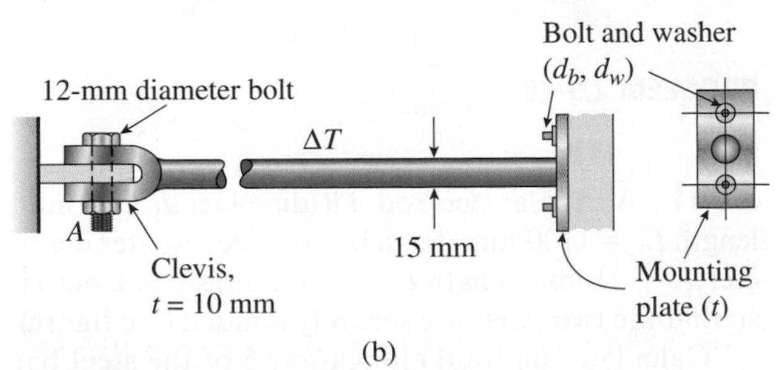

(b)

PROBLEM 2.5-4

Representative Problems

2.5-5 A bar AB of length L is held between rigid supports and heated nonuniformly in such a manner that the temperature increase ΔT at a distance x from end A is given by the expression $\Delta T = \Delta T_B x^3/L^3$, where ΔT_B is the increase in temperature at end B of the bar (see figure part a).

(a) Derive a formula for the compressive stress σ_c in the bar. (Assume that the material has modulus of elasticity E and coefficient of thermal expansion α).

(b) Now modify the formula in part (a) if the rigid support at A is replaced by an elastic support at A having a spring constant k (see figure part b). Assume that only bar AB is subject to the temperature increase.

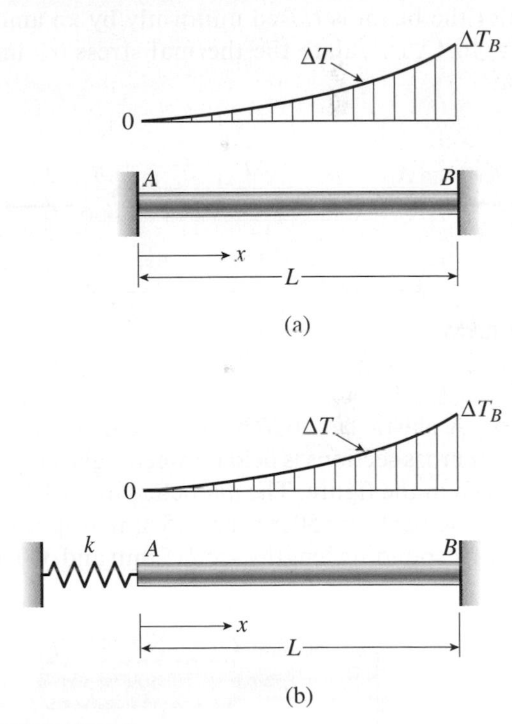

PROBLEM 2.5-5

2.5-6 A beam is constructed using two angle sections (L120 × 80 × 10) arranged back to back, as shown in the figure. The beam is fixed at joint A and attached to an elastic support having a spring constant $k = 1750$ kN/m at joint B. Assume only the beam is subjected to temperature increase $\Delta T = 45$°C. Calculate the thermal stress developed in the beam and the displacement at point B. Assume that $\alpha = 12 \times 10^{-6}$/°C. Let $E = 205$ GPa.

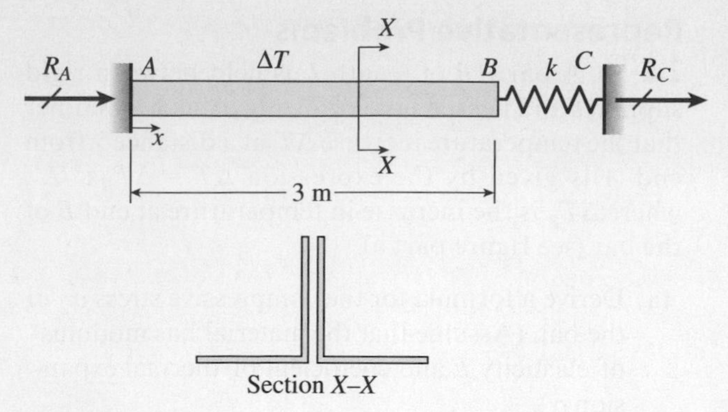

PROBLEM 2.5-6

2.5-7 An IPN 200 beam of a length 3 m is held between immoveable supports. The beam has a modulus of elasticity $E = 200$ GPa and coefficient of thermal expansion $\alpha = 12 \times 10^{-6}$/°C. If the temperature of the beam is raised uniformly by an amount $\Delta T = 10$°C, calculate the thermal stress σ_T in the beam.

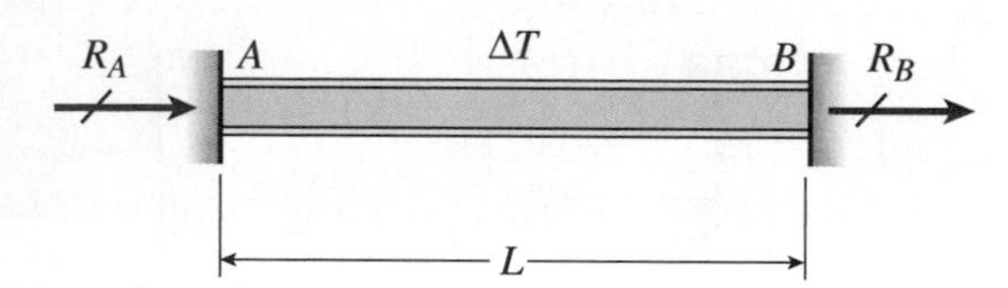

PROBLEM 2.5-7

2.5-8 A plastic bar ACB having two different solid circular cross sections is held between rigid supports, as shown in the figure. The diameters in the left- and right-hand parts are 50 mm and 75 mm, respectively. The corresponding lengths are 225 mm and 300 mm. Also, the modulus of elasticity E is 6.0 GPa, and the coefficient of thermal expansion α is 100×10^{-6} /°C. The bar is subjected to a uniform temperature increase of 30°C.

(a) Calculate the following quantities: (1) the compressive force N in the bar; (2) the maximum compressive stress σ_c; and (3) the displacement δ_C of point C.

(b) Repeat part (a) if the rigid support at A is replaced by an elastic support having spring constant $k = 50$ MN/m (see figure part b; assume that only the bar ACB is subject to the temperature increase).

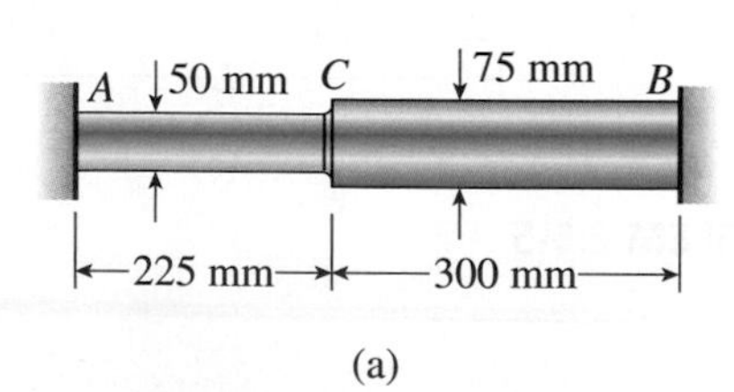

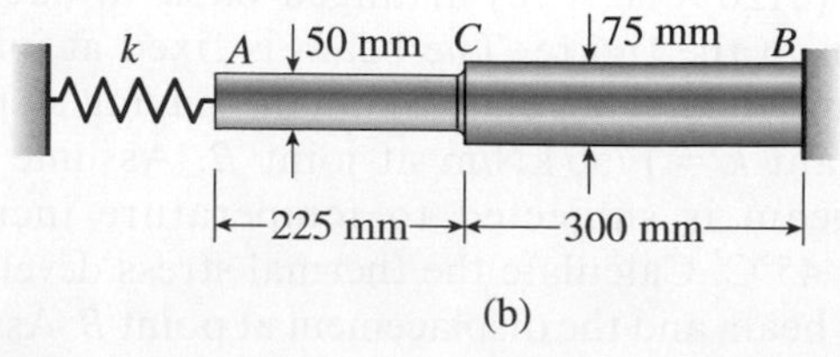

PROBLEM 2.5-8

2.5-9 A flat aluminum alloy bar is fixed at both ends. Segment AB has a slight taper. If the temperature of the bar is raised uniformly by an amount $\Delta T = 10$°C, find reactions at A and C. What is the displacement at B? Assume that $L = 1$ m, $t = 6$ mm, $b_1 = 50$ mm, $b_2 = 60$ mm, $E = 72$ GPa, and the coefficient of thermal expansion $\alpha = 23 \times 10^{-6}$ /°C.

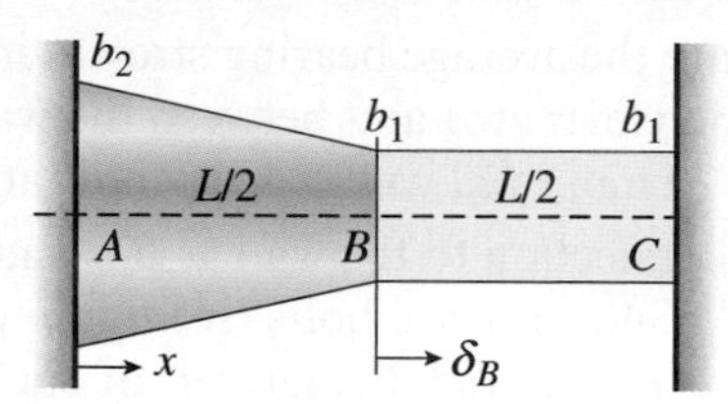

PROBLEM 2.5-9

2.5-10 Repeat Problem 2.5-9 for the flat bar shown in the figure but assume that $\Delta T = 30$°C, and that $\alpha = 19 \times 10^{-6}$ /°C, $L = 2$ m, $t = 20$ mm, $b_1 = 100$ mm, $b_2 = 115$ mm, and $E = 96$ GPa.

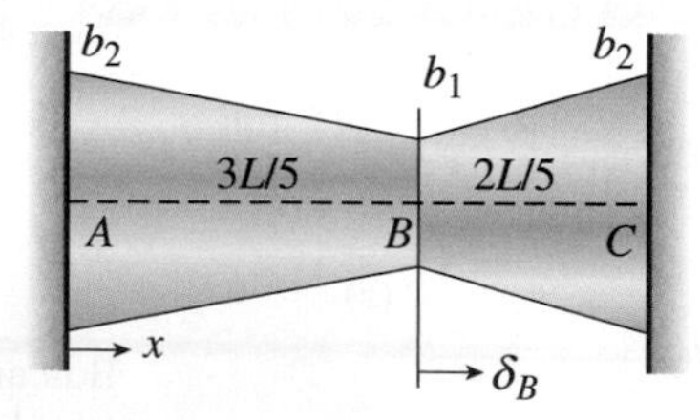

PROBLEM 2.5-10

2.5-11 A circular steel rod AB (diameter $d_1 = 15$ mm, length $L_1 = 1100$ mm) has a bronze sleeve (outer diameter $d_2 = 21$ mm, length $L_2 = 400$ mm) shrunk onto it so that the two parts are securely bonded (see figure).

Calculate the total elongation δ of the steel bar due to a temperature rise $\Delta T = 350$°C. (Material properties are as follows: for steel, $E_S = 210$ GPa and $\alpha_S = 12 \times 10^{-6}$ /°C; for bronze, $E_b = 110$ GPa and $\alpha_b = 20 \times 10^{-6}$/°C.)

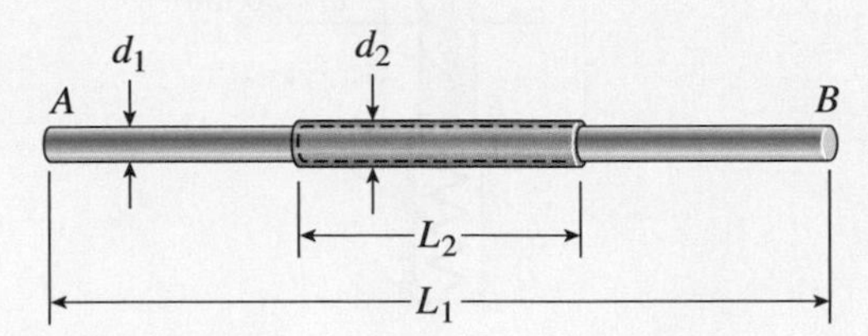

PROBLEM 2.5-11

2.5-12 A circular, aluminum alloy bar of a length $L = 1.8$ m has a slot in the middle half of its length (see figure). The bar has a radius $r = 36$ mm and modulus of elasticity $E = 72$ GPa. The slot has a height $2a = r/4$. If the temperature of the beam is raised uniformly by an amount $\Delta T = 15$°C, calculate the thermal stress σ_T developed in the bar. Assume that $\alpha = 23 \times 10^{-6}$/°C.

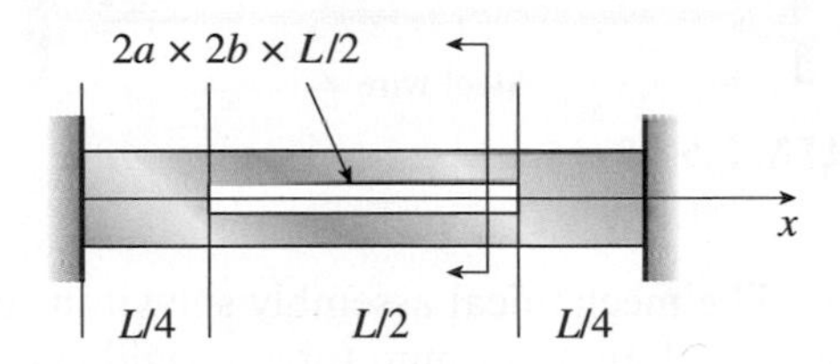

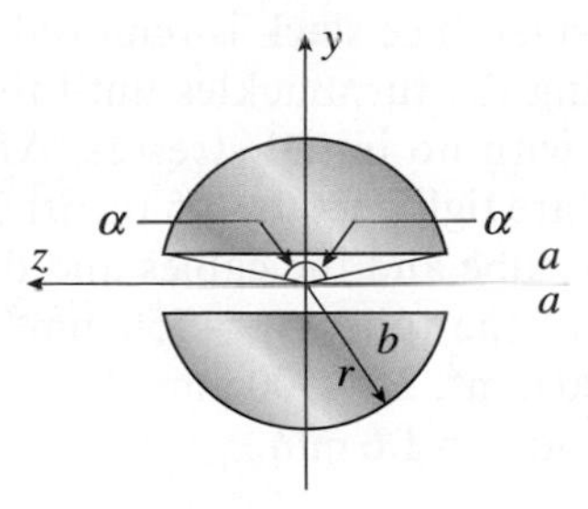

PROBLEM 2.5-12

2.5-13 Rectangular bars of copper and aluminum are held by pins at their ends, as shown in the figure. Thin spacers provide a separation between the bars. The copper bars have cross-sectional dimensions 12 mm × 50 mm, and the aluminum bar has dimensions 25 mm × 50 mm.

Determine the shear stress in the 11 mm diameter pins if the temperature is raised by 40°C. (For copper, $E_c = 124$ GPa and $\alpha_c = 20 \times 10^{-6}$/°C; for aluminum, $E_a = 69$ GPa and $\alpha_a = 26 \times 10^{-6}$/°C.) *Suggestion:* Use the results of Example 2-10.

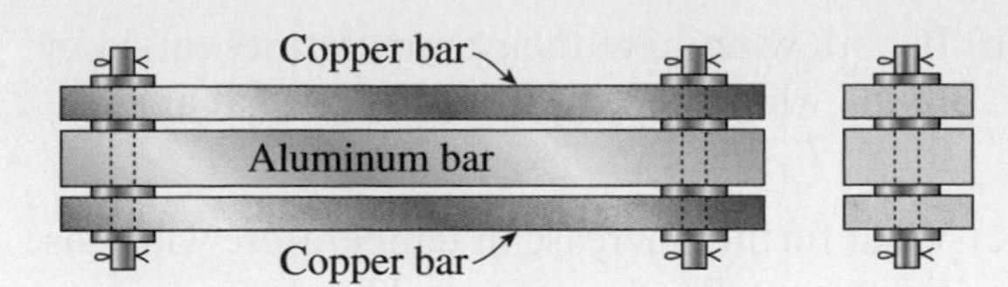

PROBLEM 2.5-13

2.5-14 A brass sleeve S is fitted over a steel bolt B (see figure), and the nut is tightened until it is just snug. The bolt has a diameter $d_B = 25$ mm, and the sleeve has inside and outside diameters $d_1 = 26$ mm and $d_2 = 36$ mm, respectively.

Calculate the temperature rise ΔT that is required to produce a compressive stress of 25 MPa in the sleeve. (Use material properties as follows: for the sleeve, $\alpha_S = 21 \times 10^{-6}$ /°C and $E_S = 100$ GPa; for the bolt, $\alpha_B = 10 \times 10^{-6}$ /°C and $E_B = 200$ GPa.)

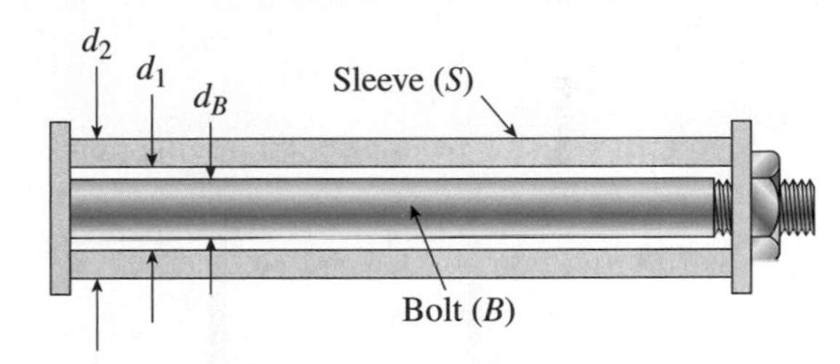

PROBLEM 2.5-14

2.5-15 A rigid triangular frame is pivoted at C and held by two identical horizontal wires at points A and B (see figure). Each wire has an axial rigidity $EA = 540$ kN and coefficient of thermal expansion $\alpha = 23 \times 10^{-6}$/°C.

(a) If a vertical load $P = 2.2$ kN acts at point D, what are the tensile forces T_A and T_B in the wires at A and B, respectively?

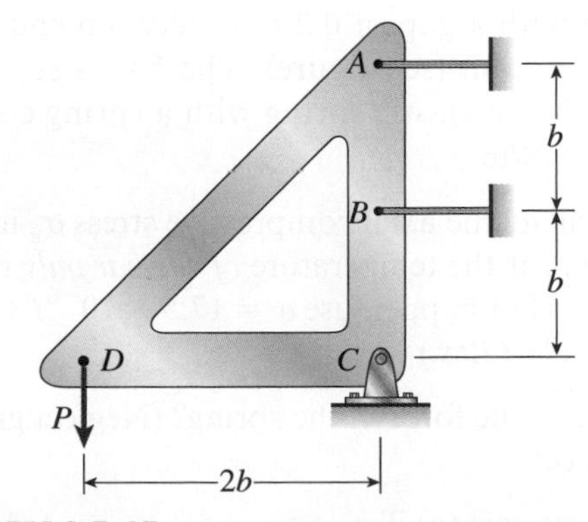

PROBLEM 2.5-15

(b) If both wires have their temperatures raised by 100°C while the load P is acting, what are the forces T_A and T_B?

(c) What further increase in temperature will cause the wire at B to become slack?

2.5-16 A rigid bar $ABCD$ is pinned at end A and supported by two cables at points B and C (see figure). The cable at B has a nominal diameter $d_B = 12$ mm and the cable at C has a nominal diameter $d_C = 20$ mm. A load P acts at end D of the bar.

What is the allowable load P if the temperature rises by 60°C and each cable is required to have a factor of safety of at least 5 against its ultimate load?

Note: The cables have an effective modulus of elasticity $E = 140$ GPa and coefficient of thermal expansion $\alpha = 12 \times 10^{-6}$/°C. Other properties of the cables can be found in Table 2-1, Section 2.2.

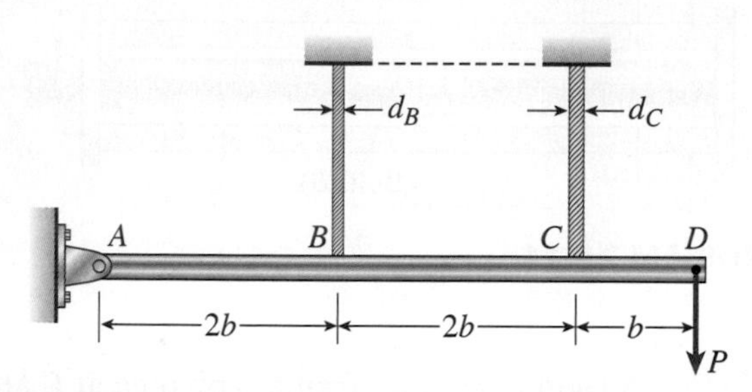

PROBLEM 2.5-16

2.5 Misfits and Prestrains

Introductory Problems

2.5-17 A copper bar AB with a length 0.635 m and diameter 50 mm is placed in position at room temperature with a gap of 0.2 mm between end A and a rigid restraint (see figure). The bar is supported at end B by an elastic spring with a spring constant $k = 210$ MN/m.

(a) Calculate the axial compressive stress σ_c in the bar if the temperature *of the bar only* rises 27°C. (For copper, use $\alpha = 17.5 \times 10^{-6}$/°C and $E = 110$ GPa.)

(b) What is the force in the spring? (Neglect gravity effects.)

(c) Repeat part (a) if $k \to \infty$.

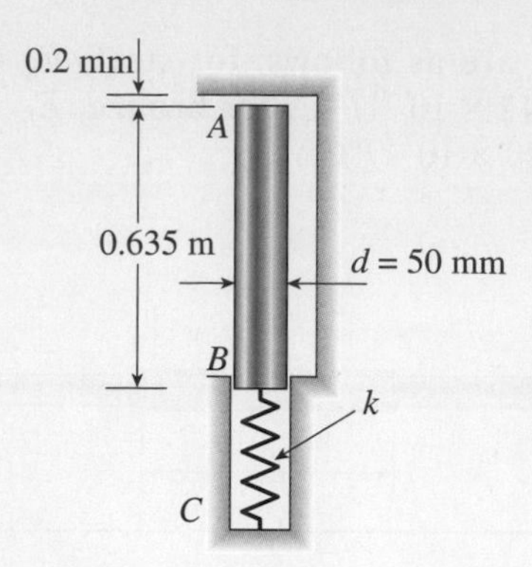

PROBLEM 2.5-17

2.5-18 A steel wire AB is stretched between rigid supports (see figure). The initial prestress in the wire is 42 MPa when the temperature is 20°C.

(a) What is the stress σ in the wire when the temperature drops to 0°C?

(b) At what temperature T will the stress in the wire become zero? (Assume $\alpha = 14 \times 10^{-6}$/°C and $E = 200$ GPa.)

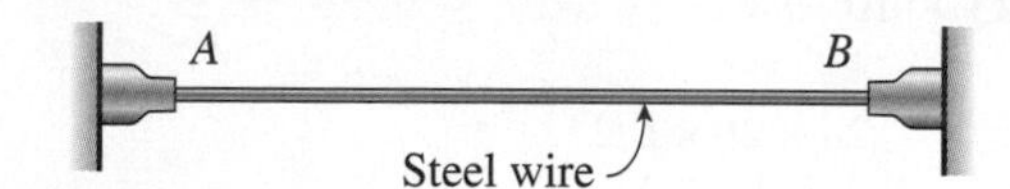

PROBLEM 2.5-18

2.5-19 The mechanical assembly shown in the figure consists of an aluminum tube, a rigid end plate, and two steel cables. The slack is removed from the cables by rotating the turnbuckles until the assembly is snug but with no initial stresses. Afterward, the turnbuckles are tightened by 1.5 turns. Calculate the forces in the tube and the cables and determine the shortening of the tube. $A_s = 550$ mm^2 for each cable, $A_A = 2900$ mm^2, $L = 500$ mm, $E_s = 200$ GPa, $E_A = 73$ GPa, and $p = 1.6$ mm.

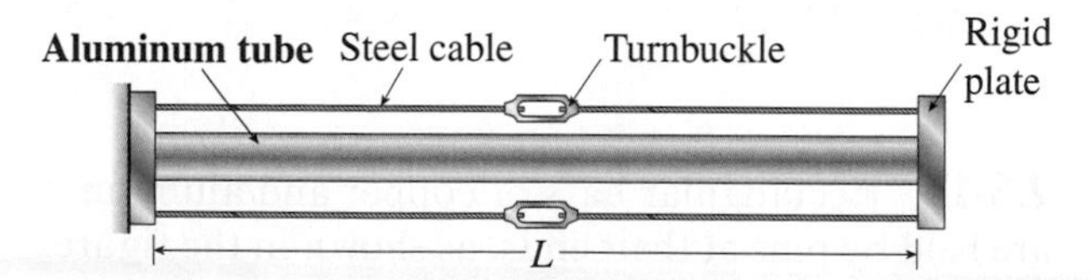

PROBLEM 2.5-19

2.5-20 A bar AB having a length L and axial rigidity EA is fixed at end A (see figure). At the other end, a small gap of dimension s exists between the end of the bar and a rigid surface. A load P acts on the bar at point C, which is two-thirds of the length from the fixed end.

If the support reactions produced by load P are to be equal in magnitude, what should be the size s of the gap?

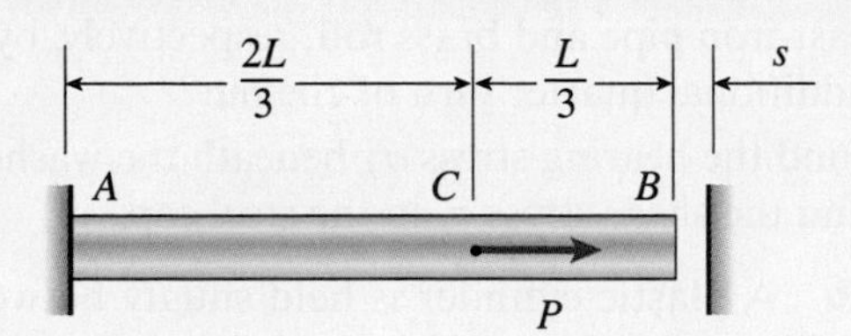

PROBLEM 2.5-20

2.5-21 Pipe 2 has been inserted snugly into Pipe 1, but the holes for a connecting pin do not line up; there is a gap s. The user decides to apply *either* force P_1 to Pipe 1 *or* force P_2 to Pipe 2, whichever is smaller. Determine the following using the numerical properties in the box.

(a) If only P_1 is applied, find P_1 (kN) required to close gap s; if a pin is then inserted and P_1 removed, what are reaction forces R_A and R_B for this load case?

(b) If only P_2 is applied, find P_2 (kN) required to close gap s; if a pin is inserted and P_2 removed, what are reaction forces R_A and R_B for this load case?

(c) What is the maximum *shear* stress in the pipes, for the loads in parts (a) and (b)?

(d) If a temperature increase ΔT is to be applied to the entire structure to close gap s (*instead of applying forces P_1 and P_2*), find the ΔT required to close the gap. If a pin is inserted after the gap has closed, what are reaction forces R_A and R_B for this case?

(e) Finally, if the structure (with pin inserted) then cools to the *original* ambient temperature, what are reaction forces R_A and R_B?

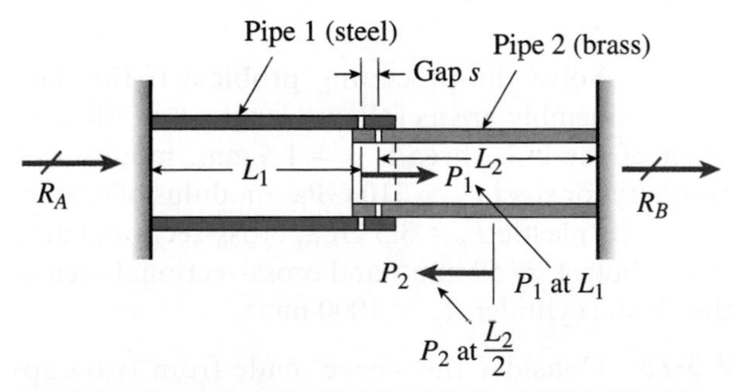

Numerical properties:
$E_1 = 210$ GPa, $E_2 = 96$ GPa
$\alpha_1 = 12 \times 10^{-6}$/°C, $\alpha_2 = 21 \times 10^{-6}$/°C
Gap $s = 1.25$ mm
$L_1 = 1.4$ m, $d_1 = 152$ mm, $t_1 = 12.5$ mm, $A_1 = 5478$ mm^2
$L_2 = 0.9$ m, $d_2 = 127$ mm, $t_2 = 6.5$ mm, $A_2 = 2461$ mm^2

PROBLEM 2.5-21

2.5-22 A nonprismatic bar ABC made up of segments AB (length L_1, cross-sectional area A_1) and BC (length L_2, cross-sectional area A_2) is fixed at end A and free at end C (see figure). The modulus of elasticity of the bar is E. A small gap of dimension s exists between the end of the bar and an elastic spring of length L_3 and spring constant k_3. If bar ABC only (*not the spring*) is subjected to temperature increase ΔT, determine the following.

(a) Write an expression for reaction forces R_A and R_D if the elongation of ABC exceeds gap length s.

(b) Find expressions for the displacements of points B and C if the elongation of ABC exceeds gap length s.

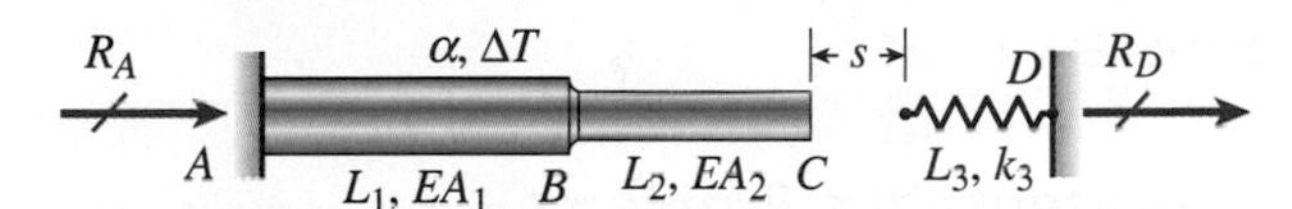

PROBLEM 2.5-22

2.5-23 Wires B and C are attached to a support at the left-hand end and to a pin-supported rigid bar at the right-hand end (see figure). Each wire has cross-sectional area $A = 19.3$ mm^2 and modulus of elasticity $E = 210$ GPa. When the bar is in a vertical position, the length of each wire is $L = 2.032$ m. However, before being attached to the bar, the length of wire B was 2.031 m and wire C was 2.030 m.

Find the tensile forces T_B and T_C in the wires under the action of a force $P = 3.115$ kN acting at the upper end of the bar.

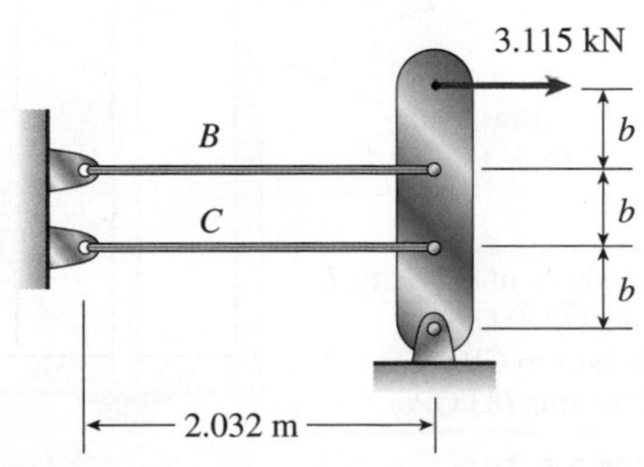

PROBLEM 2.5-23

2.5-24 A rigid steel plate is supported by three posts of high-strength concrete each having an effective cross-sectional area $A = 40{,}000\ \text{mm}^2$ and length $L = 2\ \text{m}$ (see figure). Before the load P is applied, the middle post is shorter than the others by an amount $s = 1.0\ \text{mm}$.

Determine the maximum allowable load P_{allow} if the allowable compressive stress in the concrete is $\sigma_{\text{allow}} = 20\ \text{MPa}$. (Use $E = 30\ \text{GPa}$ for concrete.)

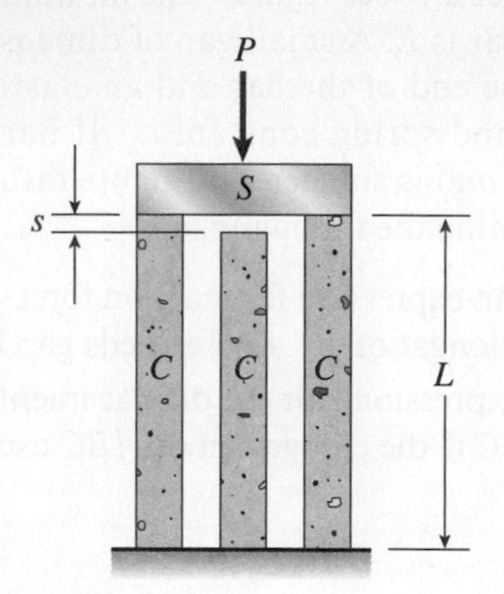

PROBLEM 2.5-24

Representative Problems

2.5-25 A capped cast-iron pipe is compressed by a brass rod, as shown. The nut is turned until it is just snug, then add an additional quarter turn to pre-compress the cast-iron pipe. The pitch of the threads of the bolt is $p = 1.3\ \text{mm}$. Use the numerical properties provided.

(a) What stresses σ_p and σ_r will be produced in the cast-iron pipe and brass rod, respectively, by the additional quarter turn of the nut?

(b) Find the bearing stress σ_b beneath the washer and the shear stress τ_c in the steel cap.

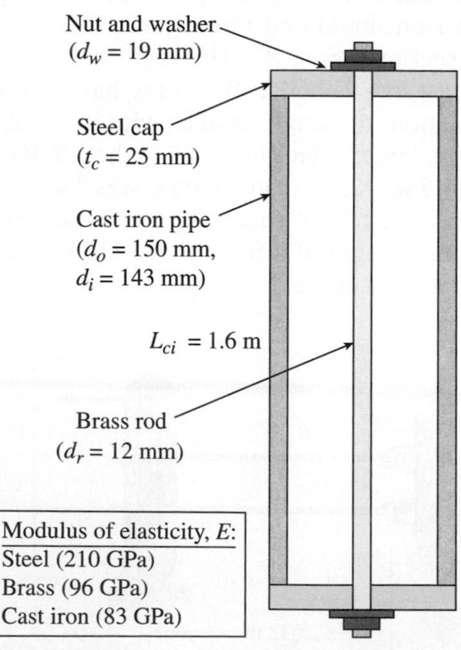

PROBLEM 2.5-25

2.5-26 A plastic cylinder is held snugly between a rigid plate and a foundation by two steel bolts (see figure).

Determine the compressive stress σ_P in the plastic when the nuts on the steel bolts are tightened by one complete turn.

Data for the assembly are as follows: length $L = 200\ \text{mm}$, pitch of the bolt threads $p = 1.0\ \text{mm}$, modulus of elasticity for steel $E_s = 200\ \text{GPa}$, modulus of elasticity for the plastic $E_P = 7.5\ \text{GPa}$, cross-sectional area of one bolt $A_s = 36.0\ \text{mm}^2$, and cross-sectional area of the plastic cylinder $A_P = 960\ \text{mm}^2$.

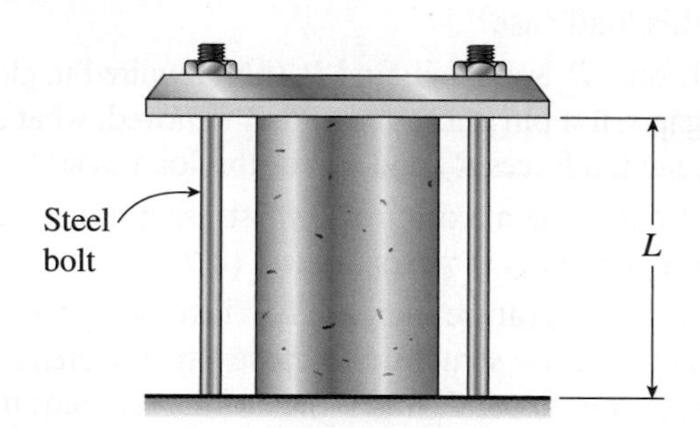

PROBLEMS 2.5-26 and 2.5-27

2.5-27 Solve the preceding problem if the data for the assembly are as follows: length $L = 300\ \text{mm}$, pitch of the bolt threads $p = 1.5\ \text{mm}$, modulus of elasticity for steel $E_s = 210\ \text{GPa}$, modulus of elasticity for the plastic $E_P = 3.5\ \text{GPa}$, cross-sectional area of one bolt $A_s = 50\ \text{mm}^2$, and cross-sectional area of the plastic cylinder $A_P = 1000\ \text{mm}^2$.

2.5-28 Consider the sleeve made from two copper tubes joined by tin-lead solder over distance s. The sleeve has brass caps at both ends that are held in place by a steel bolt and washer with the nut turned just snug at the outset. Then, two "loadings" are applied: $n = 1/2$ turn applied to the nut; at the same time, the internal temperature is raised by $\Delta T = 30°\text{C}$.

(a) Find the forces in the sleeve and bolt, P_S and P_B, due to both the prestress in the bolt and the temperature increase. For copper, use $E_c = 120$ GPa and $\alpha_c = 17 \times 10^{-6}/°C$; for steel, use $E_s = 200$ GPa and $\alpha_s = 12 \times 10^{-6}/°C$. The pitch of the bolt threads is $p = 1.0$ mm. Assume $s = 26$ mm and bolt diameter $d_B = 5$ mm.

(b) Find the required length of the solder joint, s, if shear stress in the sweated joint cannot exceed the allowable shear stress $\tau_{aj} = 18.5$ MPa.

(c) What is the final elongation of the entire assemblage due to both temperature change ΔT and the initial prestress in the bolt?

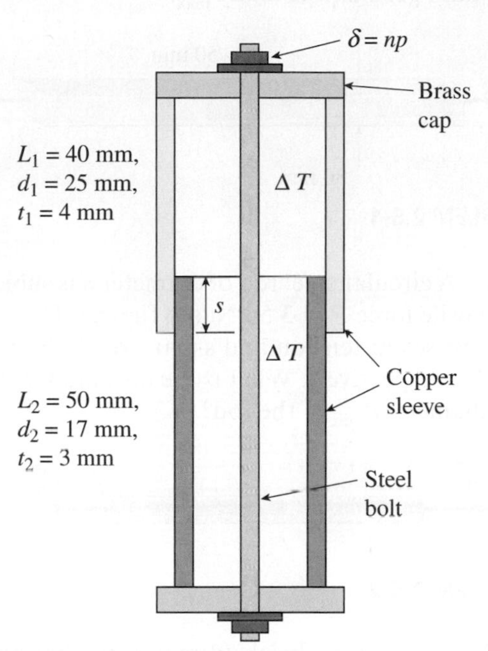

PROBLEM 2.5-28

2.5-29 A polyethylene tube (length L) has a cap that when installed compresses a spring (with undeformed length $L_1 > L$) by an amount $\delta = (L_1 - L)$. Ignore deformations of the cap and base. Use the force at the base of the spring as the redundant. Use numerical properties given in the boxes.

(a) What is the resulting force in the spring, F_k?

(b) What is the resulting force in the tube, F_t?

(c) What is the final length of the tube, L_f?

(d) What temperature change ΔT inside the tube will result in zero force in the spring?

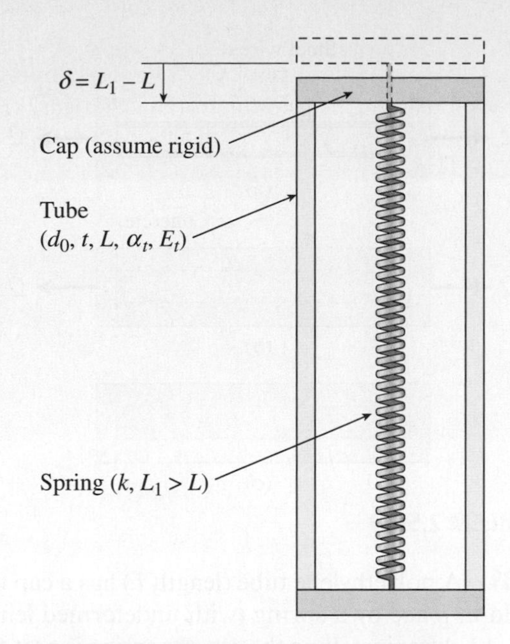

Modulus of elasticity:
Polyethylene tube ($E_t = 0.7$ GPa)

Coefficients of thermal expansion:
$\alpha_t = 140 \times 10^{-6}/°C$, $\alpha_k = 12 \times 10^{-6}/°C$

Properties and dimensions:
$d_0 = 150$ mm $t = 3$ mm
$L_1 = 308$ mm $> L = 305$ mm $k = 262.5$ kN/m

PROBLEM 2.5-29

2.5-30 Prestressed concrete beams are sometimes manufactured in the following manner. High-strength steel wires are stretched by a jacking mechanism that applies a force Q, as represented schematically in part a of the figure. Concrete is then poured around the wires to form a beam, as shown in the figure part b.

After the concrete sets properly, the jacks are released, and the force Q is removed (see part c of the figure). Thus, the beam is left in a prestressed condition with the wires in tension and the concrete in compression.

Assume that the prestressing force Q produces in the steel wires an initial stress $\sigma_0 = 620$ MPa. If the moduli of elasticity of the steel and concrete are in the ratio 12:1 and the cross-sectional areas are in the ratio 1:50, what are the final stresses σ_s and σ_c in the two materials?

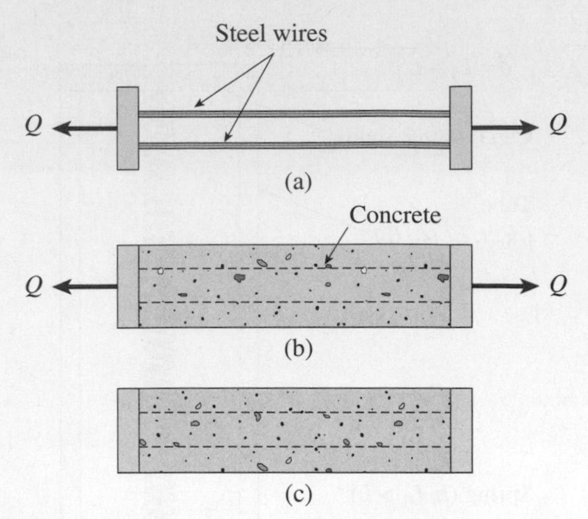

PROBLEM 2.5-30

2.5-31 A polyethylene tube (length L) has a cap that is held in place by a spring (with undeformed length $L_1 < L$). After installing the cap, the spring is post-tensioned by turning an adjustment screw by an amount δ. Ignore deformations of the cap and base. Use the force at the base of the spring as the redundant. Use numerical properties in the boxes below the figure.

(a) What is the resulting force in the spring, F_k?
(b) What is the resulting force in the tube, F_t?
(c) What is the final length of the tube, L_f?
(d) What temperature change ΔT inside the tube will result in zero force in the spring?

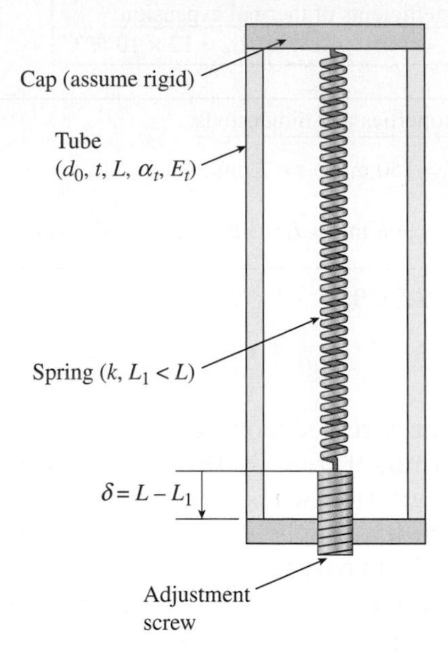

Modulus of elasticity:
Polyethylene tube (E_t = 0.7 GPa)
Coefficients of thermal expansion:
$\alpha_t = 140 \times 10^{-6}$/°C, $\alpha_k = 12 \times 10^{-6}$/°C

Properties and dimensions:
d_0 = 150 mm t = 3 mm
L = 305 mm L_1 = 302 mm k = 262.5 kN/m

PROBLEM 2.5-31

2.6 Stresses on Inclined Sections

Introductory Problems

2.6-1 A steel bar of rectangular cross section (50 mm × 50 mm) carries a tensile load P (see figure). The allowable stresses in tension and shear are 125 MPa and 76 MPa, respectively. Determine the maximum permissible load P_{max}.

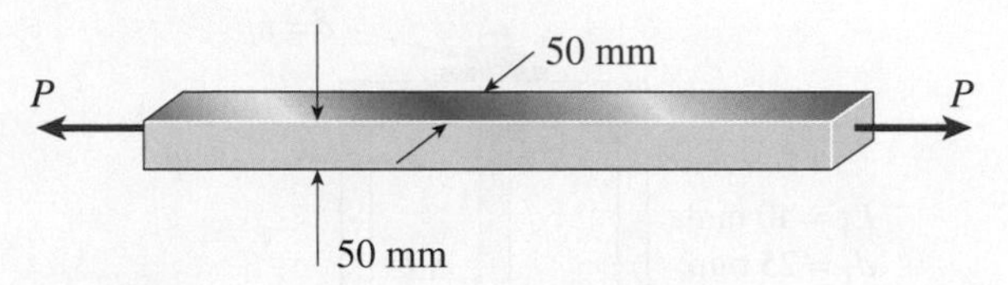

PROBLEM 2.6-1

2.6-2 A circular steel rod of diameter d is subjected to a tensile force P = 3.5 kN (see figure). The allowable stresses in tension and shear are 118 MPa and 48 MPa, respectively. What is the minimum permissible diameter d_{min} of the rod?

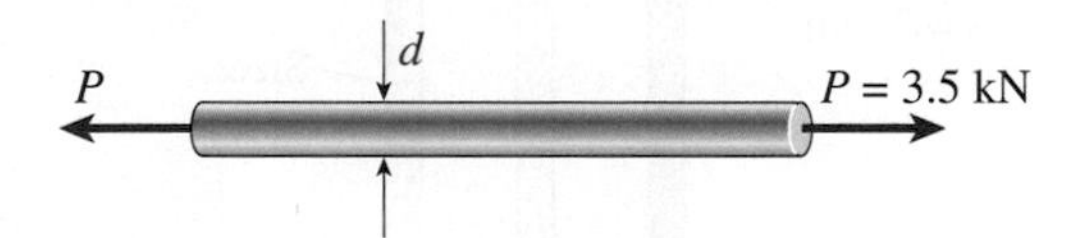

PROBLEM 2.6-2

2.6-3 A standard brick (dimensions 200 mm × 100 mm × 65 mm) is compressed lengthwise by a force P, as shown in the figure. If the ultimate shear stress for brick is 8 MPa and the ultimate compressive stress is 26 MPa, what force P_{max} is required to break the brick?

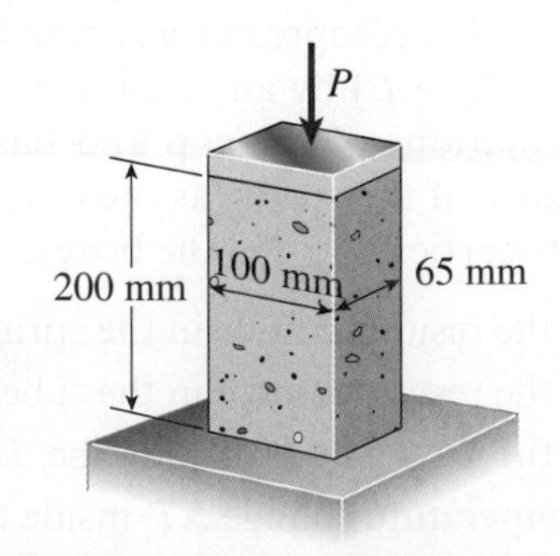

PROBLEM 2.6-3

2.6-4 A brass wire of diameter $d = 2.42$ mm is stretched tightly between rigid supports so that the tensile force is $T = 98$ N (see figure). The coefficient of thermal expansion for the wire is 19.5×10^{-6}/°C, and the modulus of elasticity is $E = 110$ GPa.

(a) What is the maximum permissible temperature drop ΔT if the allowable shear stress in the wire is 60 MPa?

(b) At what temperature change does the wire go slack?

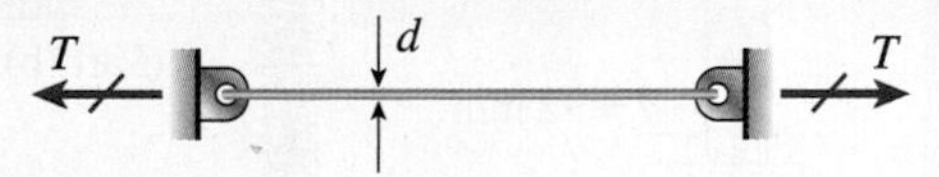

PROBLEMS 2.6-4 and 2.6-5

2.6-5 A brass wire of diameter $d = 1.6$ mm is stretched between rigid supports with an initial tension T of 200 N (see figure). Assume that the coefficient of thermal expansion is 21.2×10^{-6}/°C and the modulus of elasticity is 110 GPa.

(a) If the temperature is lowered by 30°C, what is the maximum shear stress τ_{max} in the wire?

(b) If the allowable shear stress is 70 MPa, what is the maximum permissible temperature drop?

(c) At what temperature change ΔT does the wire go slack?

2.6-6 A steel bar with a diameter $d = 12$ mm is subjected to a tensile load $P = 9.5$ kN (see figure).

(a) What is the maximum normal stress σ_{max} in the bar?

(b) What is the maximum shear stress τ_{max}?

(c) Draw a stress element oriented at 45° to the axis of the bar and show all stresses acting on the faces of this element.

(d) Repeat part (c) for a stress element oriented at 22.5° to the axis of the bar.

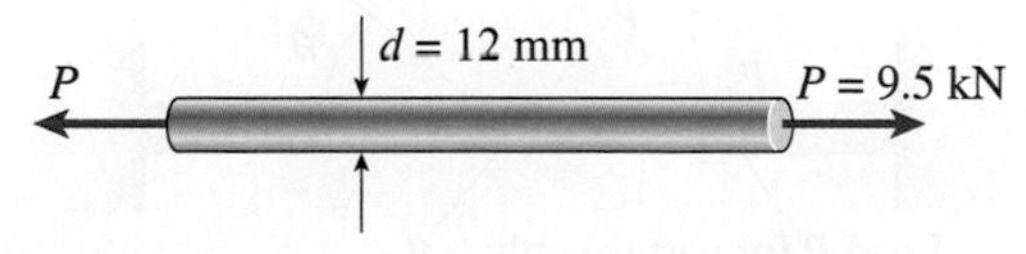

PROBLEM 2.6-6

2.6-7 During a tension test of a mild-steel specimen (see figure), the extensometer shows an elongation of 0.004 mm with a gage length of 50 mm. Assume that the steel is stressed below the proportional limit and that the modulus of elasticity $E = 210$ GPa.

(a) What is the maximum normal stress σ_{max} in the specimen?

(b) What is the maximum shear stress τ_{max}?

(c) Draw a stress element oriented at an angle of 45° to the axis of the bar, and show all stresses acting on the faces of this element.

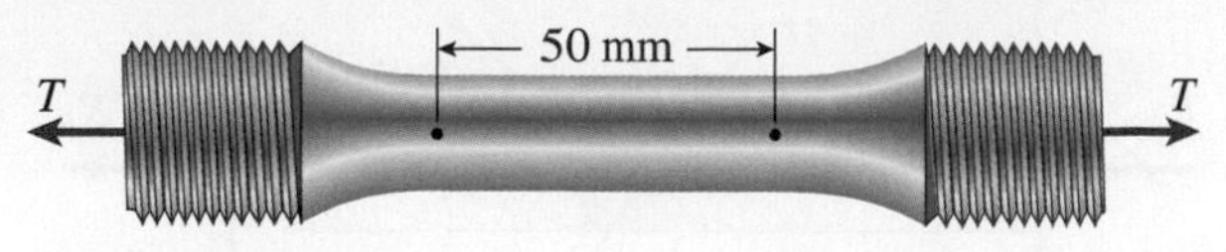

PROBLEM 2.6-7

Representative Problems

2.6-8 A copper bar with a rectangular cross section is held without stress between rigid supports (see figure). Subsequently, the temperature of the bar is raised 50°C.

(a) Determine the stresses on all faces of the elements A and B, and show these stresses on sketches of the elements. (Assume $\alpha = 17.5 \times 10^{-6}$/°C and $E = 120$ GPa.)

(b) If the shear stress at B is known to be 48 MPa at some inclination θ, find angle θ and show the stresses on a sketch of a properly oriented element.

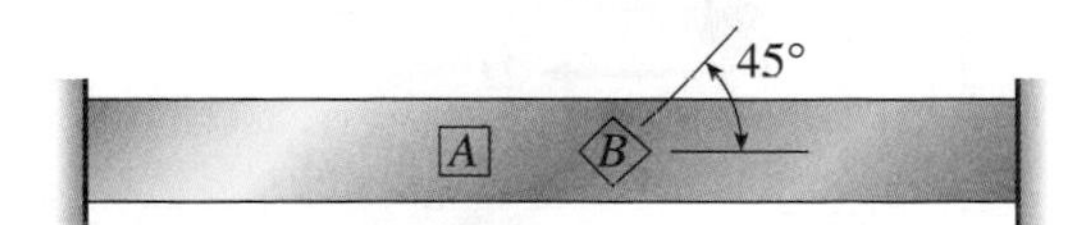

PROBLEM 2.6-8

2.6-9 A prismatic bar with a length $L = 1$ m and cross-sectional area $A = 5200$ mm^2 is compressed by an axial centroidal load $P = 45$ kN. Determine the complete state of stress acting on an inclined section pq that is cut through the bar at an angle $\theta = 35°$, and show the stresses on a properly oriented stress element.

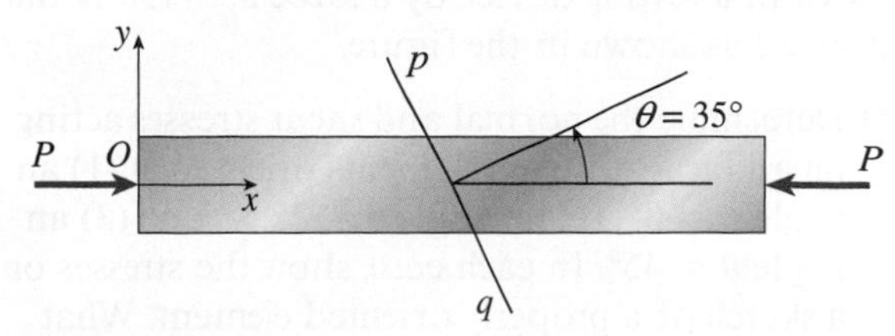

PROBLEM 2.6-9

2.6-10 A prismatic bar with a length $L = 1$ m and cross-sectional area $A = 1200\ \text{mm}^2$ is supported at the ends. The bar is then subjected to a temperature increase of $\Delta T = 25°\text{C}$. Calculate the complete state of stress acting on an inclined section rs that is cut through the bar at an angle $\theta = 45°$. Use $E = 200$ GPa and the coefficient of thermal expansion $\alpha = 12 \times 10^{-6}$ /°C.

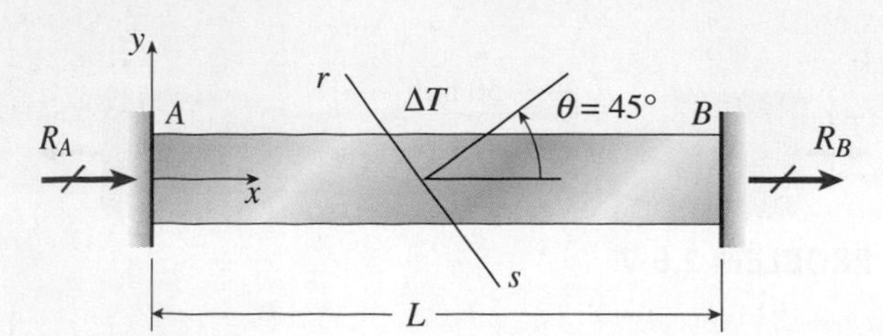

PROBLEM 2.6-10

2.6-11 The plane truss in the figure is assembled from steel UPN 220 shapes (see Table 3 in Appendix F). Assume that $L = 3$ m and $b = 0.71L$.

(a) If load variable $P = 220$ kN, what is the maximum shear stress τ_{max} in each truss member?

(b) What is the maximum permissible value of load variable P if the allowable normal stress is 96 MPa and the allowable shear stress is 52 MPa?

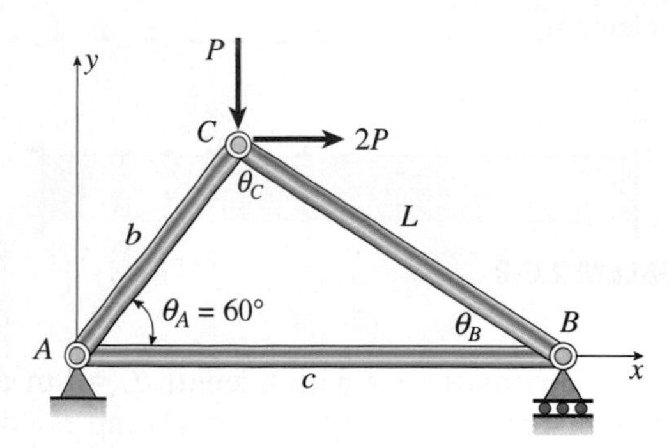

PROBLEM 2.6-11

2.6-12 A plastic bar of diameter $d = 32$ mm is compressed in a testing device by a force $P = 190$ N that is applied as shown in the figure.

(a) Determine the normal and shear stresses acting on all faces of stress elements oriented at (1) an angle $\theta = 0°$, (2) an angle $\theta = 22.5°$, and (3) an angle $\theta = 45°$. In each case, show the stresses on a sketch of a properly oriented element. What are σ_{max} and τ_{max}?

(b) Find σ_{max} and τ_{max} in the plastic bar if a re-centering spring of stiffness k is inserted into the testing device, as shown in the figure. The spring stiffness is 1/6 of the axial stiffness of the plastic bar.

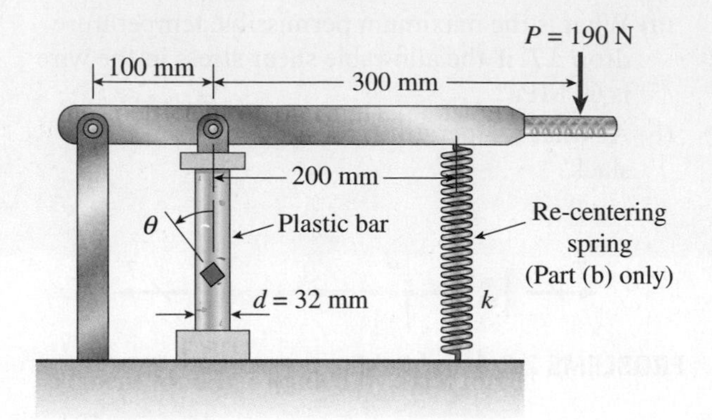

PROBLEM 2.6-12

2.6-13 A plastic bar of rectangular cross section ($b = 38$ mm and $h = 75$ mm) fits snugly between rigid supports at room temperature (20°C) but with no initial stress (see figure). When the temperature of the bar is raised to 70°C, the compressive stress on an inclined plane pq at mid-span becomes 8.7 MPa.

(a) What is the shear stress on plane pq? (Assume $\alpha = 95 \times 10^{-6}$/°C and $E = 2.4$ GPa.)

(b) Draw a stress element oriented to plane pq and show the stresses acting on all faces of this element.

(c) If the allowable normal stress is 23 MPa and the allowable shear stress is 11.3 MPa, what is the maximum load P (*in the positive x direction*), which can be added at the quarter point (in addition to thermal effects given) without exceeding allowable stress values in the bar?

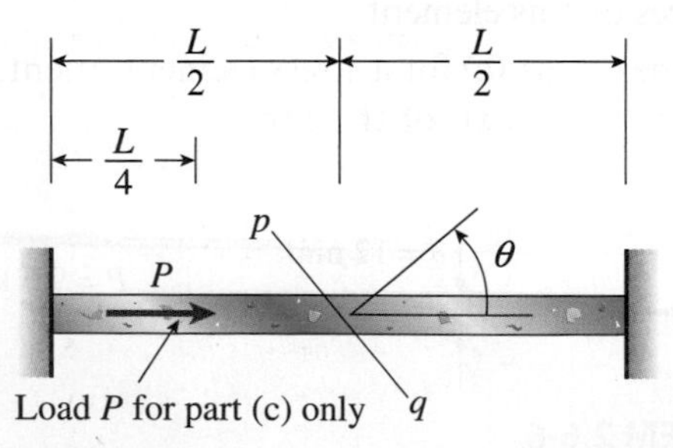

PROBLEM 2.6-13

2.6-14 A copper bar of rectangular cross section ($b = 18$ mm and $h = 40$ mm) is held snugly (but without any initial stress) between rigid supports (see figure). The allowable stresses on the inclined plane *pq* at midspan, for which $\theta = 55°$, are specified as 60 MPa in compression and 30 MPa in shear.

(a) What is the maximum permissible temperature rise ΔT if the allowable stresses on plane *pq* are not to be exceeded? (Assume $\alpha = 17 \times 10^{-6}$/°C and $E = 120$ GPa.)

(b) If the temperature increases by the maximum permissible amount, what are the stresses on plane *pq*?

(c) If the temperature rises $\Delta T = 28$°C, how far to the right of end *A* (distance βL, which is expressed as a fraction of length *L*) can load $P = 15$ kN be applied without exceeding allowable stress values in the bar? Assume that $\sigma_a = 75$ MPa and $\tau_a = 35$ MPa.

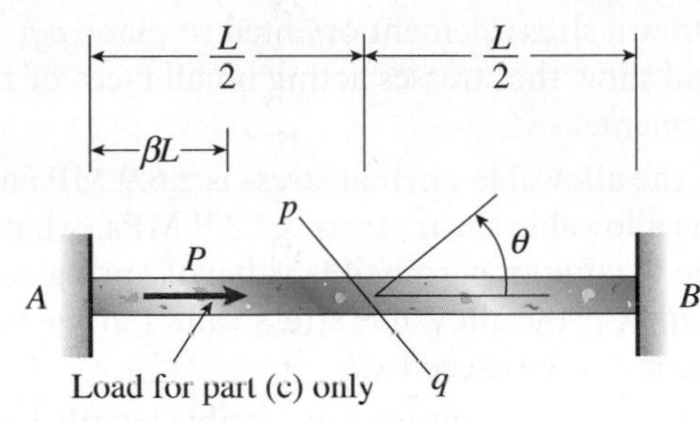

PROBLEM 2.6-14

2.6-15 A circular brass bar with a diameter *d* is member *AC* in truss *ABC* that has load $P = 30$ kN applied at joint *C*. Bar *AC* is composed of two segments brazed together on a plane *pq*, making an angle $\alpha = 36°$ with the axis of the bar (see figure). The allowable stresses in the brass are 90 MPa in tension and 48 MPa in shear. On the brazed joint, the allowable stresses are 40 MPa in tension and 20 MPa in shear. What is the tensile force N_{AC} in bar *AC*? What is the minimum required diameter d_{min} of bar *AC*?

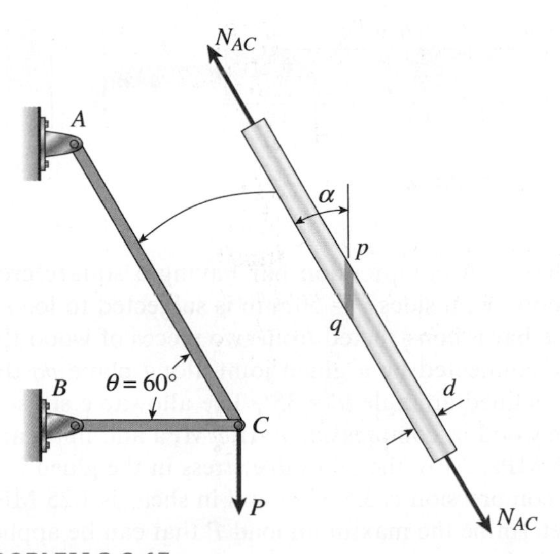

PROBLEM 2.6-15

2.6-16 Two boards are joined by gluing along a scarf joint, as shown in the figure. For purposes of cutting and gluing, the angle α between the plane of the joint and the faces of the boards must be between 10° and 40°. Under a tensile load *P*, the normal stress in the boards is 4.9 MPa.

(a) What are the normal and shear stresses acting on the glued joint if $\alpha = 20°$?

(b) If the allowable shear stress on the joint is 2.25 MPa, what is the largest permissible value of the angle α?

(c) For what angle α will the shear stress on the glued joint be numerically equal to twice the normal stress on the joint?

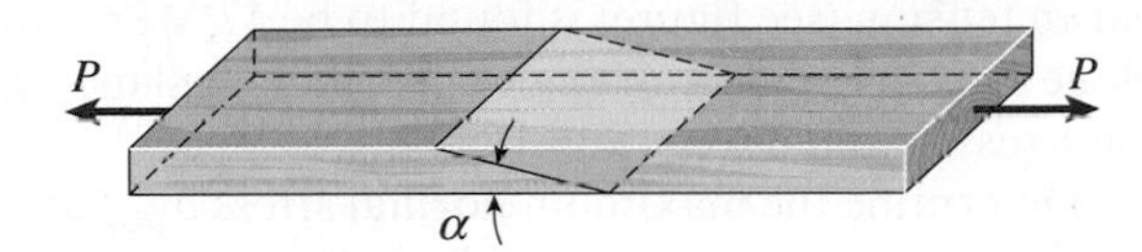

PROBLEM 2.6-16

2.6-17 Acting on the sides of a stress element cut from a bar in uniaxial stress are tensile stresses of 60 MPa and 20 MPa, as shown in the figure.

(a) Determine the angle θ and the shear stress τ_θ, and show all stresses on a sketch of the element.

(b) Determine the maximum normal stress σ_{max} and the maximum shear stress τ_{max} in the material.

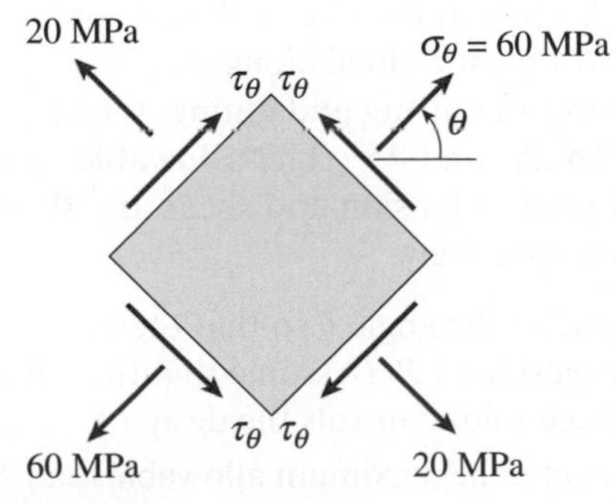

PROBLEM 2.6-17

2.6-18 A prismatic bar is subjected to an axial force that produces a tensile stress $\sigma_\theta = 65$ MPa and a shear stress $\tau_\theta = 23$ MPa on a certain inclined plane (see figure). Determine the stresses acting on all faces of a stress element oriented at $\theta = 30°$, and show the stresses on a sketch of the element.

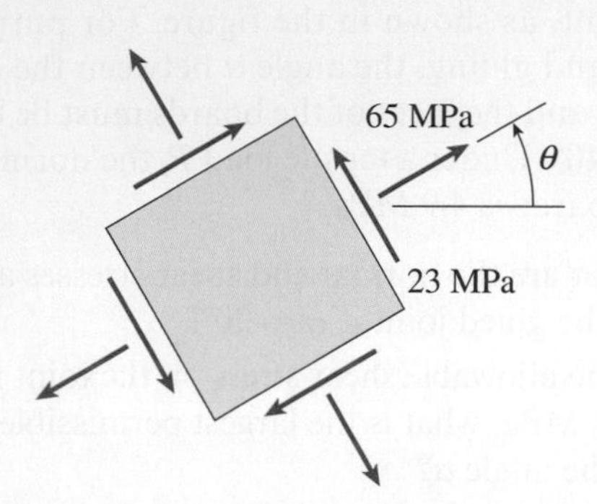

PROBLEM 2.6-18

2.6-19 The normal stress on plane *pq* of a prismatic bar in tension (see figure) is found to be 57 MPa. On plane *rs*, which makes an angle $\beta = 30°$ with plane *pq*, the stress is found to be 23 MPa.

Determine the maximum normal stress σ_{max} and maximum shear stress τ_{max} in the bar.

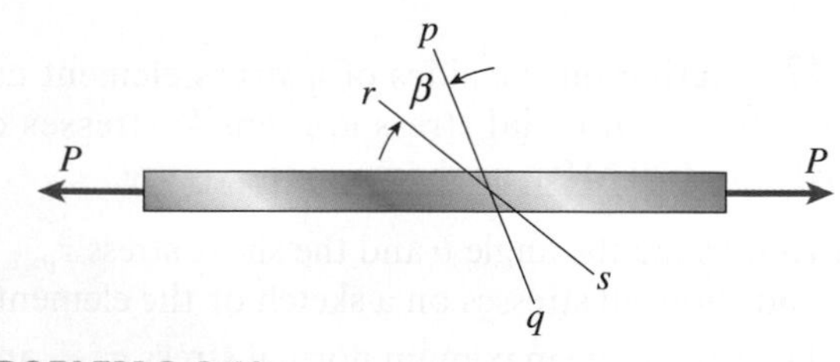

PROBLEM 2.6-19

2.6-20 A tension member is to be constructed of two pieces of plastic glued along plane *pq* (see figure). For purposes of cutting and gluing, the angle θ must be between 25° and 45°. The allowable stresses on the glued joint in tension and shear are 5.0 MPa and 3.0 MPa, respectively

(a) Determine the angle θ so that the bar will carry the largest load *P*. (Assume that the strength of the glued joint controls the design.)

(b) Determine the maximum allowable load P_{max} if the cross-sectional area of the bar is 225 mm^2.

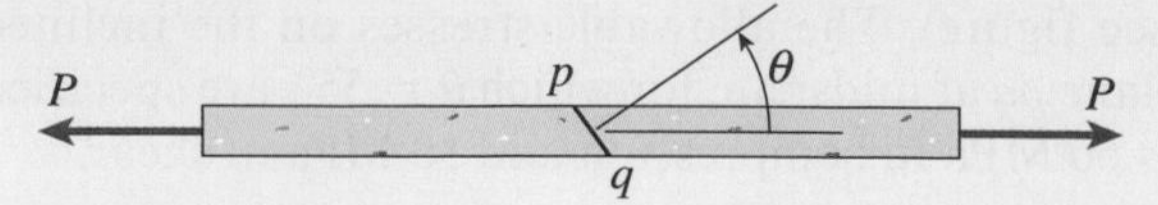

PROBLEM 2.6-20

2.6-21 Plastic bar *AB* of rectangular cross section ($b = 19$ mm and $h = 38$ mm) and length $L = 0.6$ m is fixed at *A* and has a spring support ($k = 3150$ kN/m) at *C* (see figure). Initially, the bar and spring have no stress. When the temperature of the bar is *raised* by 48°C, the *compressive* stress on an inclined plane *pq* at $L_\theta = 0.46$ m becomes 5.3 MPa. Assume the spring is massless and is unaffected by the temperature change. Let $\alpha = 95 \times 10^{-6}$/°C and $E = 2.8$ GPa.

(a) What is the shear stress τ_θ on plane *pq*? What is angle θ?

(b) Draw a stress element oriented to plane *pq*, and show the stresses acting on all faces of this element.

(c) If the allowable normal stress is ±6.9 MPa and the allowable shear stress is ±3.9 MPa, what is the maximum permissible value of spring constant *k* if the allowable stress values in the bar are not to be exceeded?

(d) What is the maximum permissible length *L* of the bar if the allowable stress values in the bar are not be exceeded? (Assume $k = 3150$ kN/m.)

(e) What is the maximum permissible temperature increase (ΔT) in the bar if the allowable stress values in the bar are not to be exceeded? (Assume $L = 0.6$ m and $k = 3150$ kN/m.)

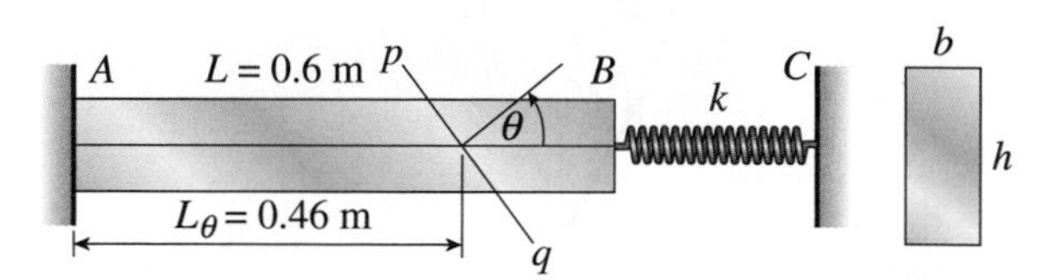

PROBLEM 2.6-21

2.6-22 A compression bar having a square cross section with sides $b = 50$ mm is subjected to load *P*. The bar is constructed from two pieces of wood that are connected by a glued joint along plane *pq* that is inclined at angle $\alpha = 35°$. The allowable stress in the wood in compression is 11.5 MPa and in shear is 4.5 MPa. Also, the allowable stress in the glued joint in compression is 3.5 MPa and in shear is 1.25 MPa. Determine the maximum load *P* that can be applied to the bar.

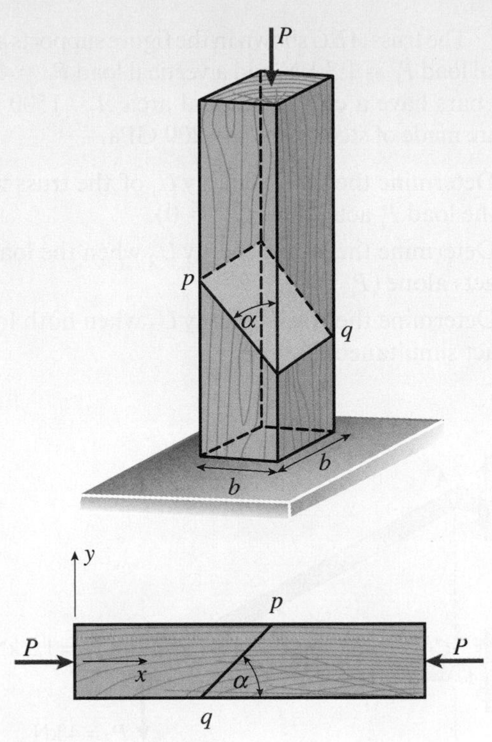

PROBLEM 2.6-22

2.7 Strain Energy

When solving the problems for Section 2.7, assume that the material behaves linearly elastically.

2.7-1 A prismatic bar *AD* of length *L*, cross-sectional area *A*, and modulus of elasticity *E* is subjected to loads 5*P*, 3*P*, and *P* acting at points *B*, *C*, and *D*, respectively (see figure). Segments *AB*, *BC*, and *CD* have lengths *L*/6, *L*/2, and *L*/3, respectively.

(a) Obtain a formula for the strain energy *U* of the bar.
(b) Calculate the strain energy if $P = 27$ kN, $L = 130$ cm, $A = 18\ \text{cm}^2$, and the material is aluminum with 72 GPa.

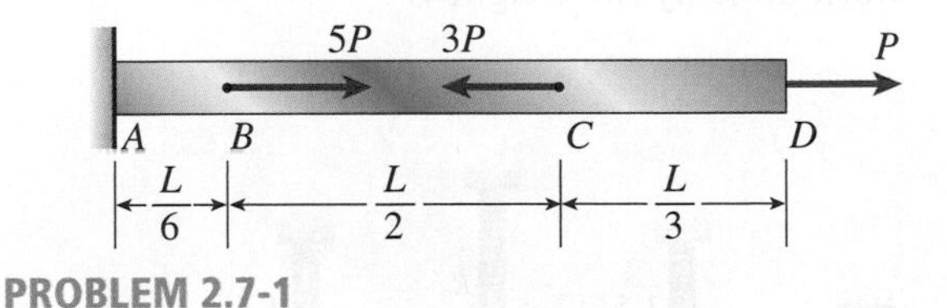

PROBLEM 2.7-1

2.7-2 A bar with a circular cross section having two different diameters *d* and 2*d* is shown in the figure. The length of each segment of the bar is *L*/2, and the modulus of elasticity of the material is *E*.

(a) Obtain a formula for the strain energy *U* of the bar due to the load *P*.
(b) Calculate the strain energy if the load $P = 27$ kN, the length $L = 600$ mm, the diameter $d = 40$ mm, and the material is brass with $E = 105$ GPa.

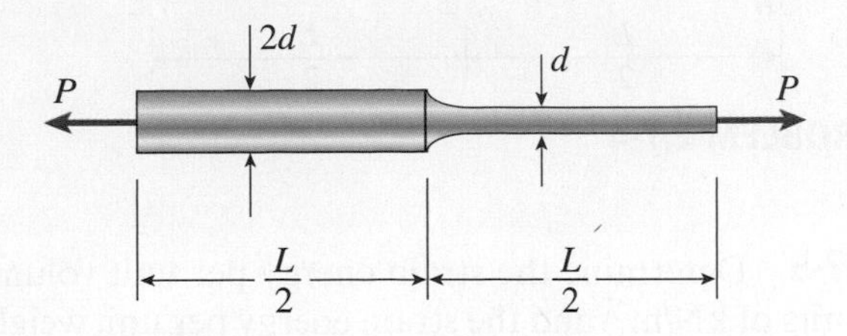

PROBLEM 2.7-2

2.7-3 A three-story steel column in a building supports roof and floor loads as shown in the figure. The story height *H* is 3 m, the cross-sectional area *A* of the column is 7500 mm², and the modulus of elasticity *E* of the steel is 200 GPa.

Calculate the strain energy *U* of the column assuming $P_1 = 150$ kN and $P_2 = P_3 = 300$ kN.

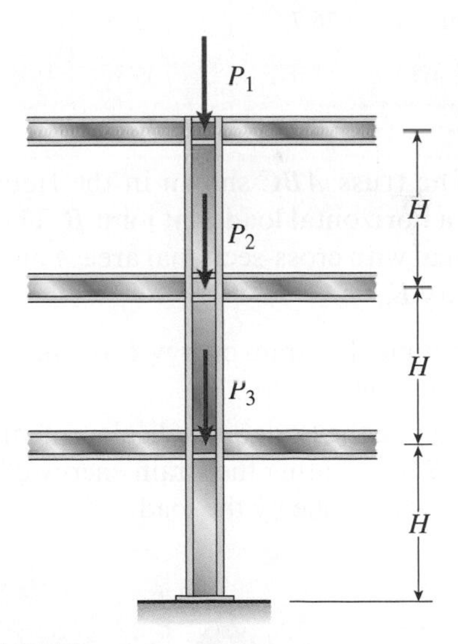

PROBLEM 2.7-3

2.7-4 The bar *ABC* shown in the figure is loaded by a force *P* acting at end *C* and by a force *Q* acting at the midpoint *B*. The bar has a constant axial rigidity *EA*.

(a) Determine the strain energy U_1 of the bar when the force *P* acts alone ($Q = 0$).

(b) Determine the strain energy U_2 when the force Q acts alone ($P = 0$).

(c) Determine the strain energy U_3 when the forces P and Q act simultaneously upon the bar.

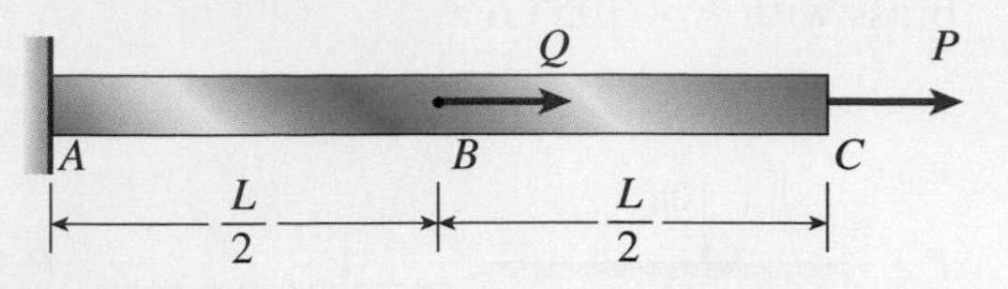

PROBLEM 2.7-4

2.7-5 Determine the strain energy per unit volume (units of kN/m²) and the strain energy per unit weight (units of m) that can be stored in each of the materials listed in the accompanying table, assuming that the material is stressed to the proportional limit.

DATA FOR PROBLEM 2.7-5

Material	Weight Density (kN/m³)	Modulus of Elasticity (GPa)	Proportional Limit (MPa)
Mild steel	77.1	207	248
Tool steel	77.1	207	827
Aluminum	26.7	72	345
Rubber (soft)	11.0	2	1.38

2.7-6 The truss ABC shown in the figure is subjected to a horizontal load P at joint B. The two bars are identical with cross-sectional area A and modulus of elasticity E.

(a) Determine the strain energy U of the truss if the angle $\beta = 60°$.

(b) Determine the horizontal displacement δ_B of joint B by equating the strain energy of the truss to the work done by the load.

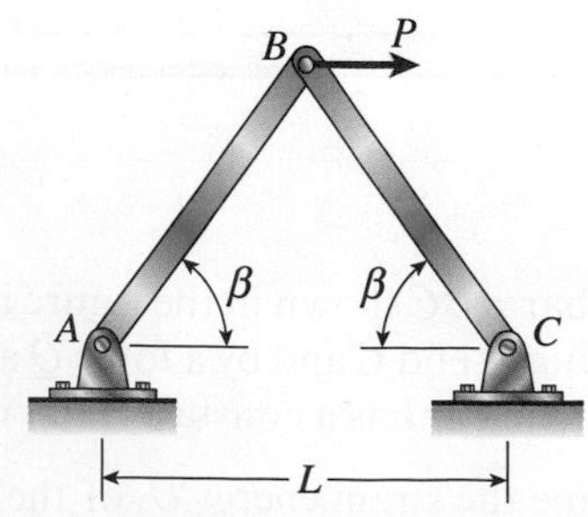

PROBLEM 2.7-6

2.7-7 The truss ABC shown in the figure supports a horizontal load $P_1 = 1.3$ kN and a vertical load $P_2 = 4$ kN. Both bars have a cross-sectional area $A = 1500$ mm² and are made of steel with $E = 200$ GPa.

(a) Determine the strain energy U_1 of the truss when the load P_1 acts alone ($P_2 = 0$).

(b) Determine the strain energy U_2 when the load P_2 acts alone ($P_1 = 0$).

(c) Determine the strain energy U_3 when both loads act simultaneously.

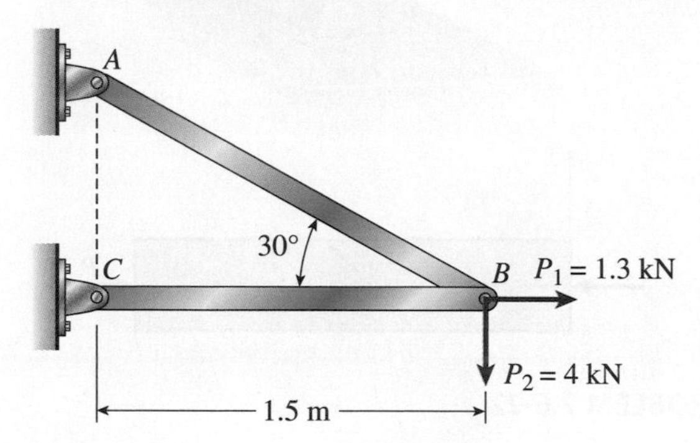

PROBLEM 2.7-7

2.7-8 The statically indeterminate structure shown in the figure consists of a horizontal rigid bar AB supported by five equally spaced springs. Springs 1, 2, and 3 have stiffnesses $3k$, $1.5k$, and k, respectively. When unstressed, the lower ends of all five springs lie along a horizontal line. Bar AB, which has weight W, causes the springs to elongate by an amount δ.

(a) Obtain a formula for the total strain energy U of the springs in terms of the downward displacement δ of the bar.

(b) Obtain a formula for the displacement δ by equating the strain energy of the springs to the work done by the weight W.

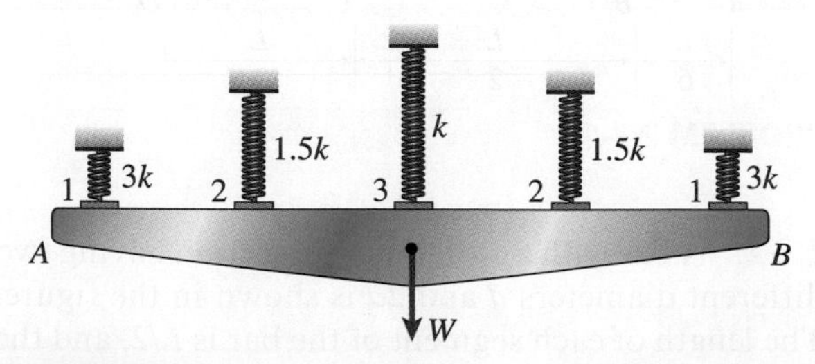

PROBLEM 2.7-8

(c) Determine the forces F_1, F_2, and F_3 in the springs.

(d) Evaluate the strain energy U, the displacement δ, and the forces in the springs if $W = 600$ N and $k = 7.5$ N/mm.

2.7-9 A slightly tapered bar AB of rectangular cross section and length L is acted upon by a force P (see figure). The width of the bar varies uniformly from b_2 at end A to b_1 at end B. The thickness t is constant.

(a) Determine the strain energy U of the bar.

(b) Determine the elongation δ of the bar by equating the strain energy to the work done by the force P.

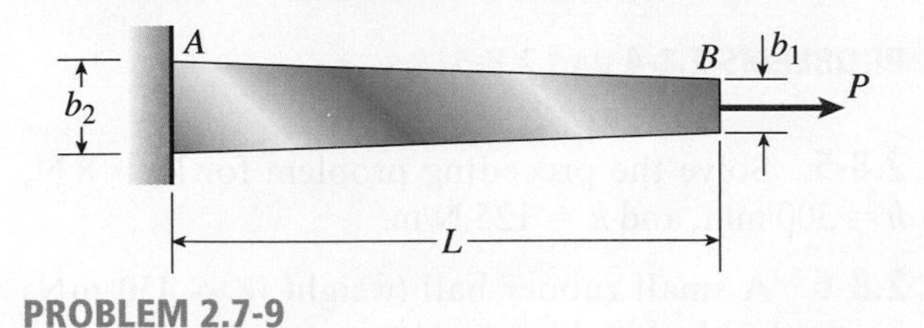

PROBLEM 2.7-9

2.7-10 A compressive load P is transmitted through a rigid plate to three magnesium-alloy bars that are identical except that initially the middle bar is slightly shorter than the other bars (see figure). The dimensions and properties of the assembly are as follows: length $L = 1.0$ m, cross-sectional area of each bar $A = 3000\ \text{mm}^2$, modulus of elasticity $E = 45$ GPa, and the gap $s = 1.0$ mm.

(a) Calculate the load P_1 required to close the gap.

(b) Calculate the downward displacement δ of the rigid plate when $P = 400$ kN.

(c) Calculate the total strain energy U of the three bars when $P = 400$ kN.

(d) Explain why the strain energy U is *not* equal to $P\delta/2$. *Hint:* Draw a load-displacement diagram.

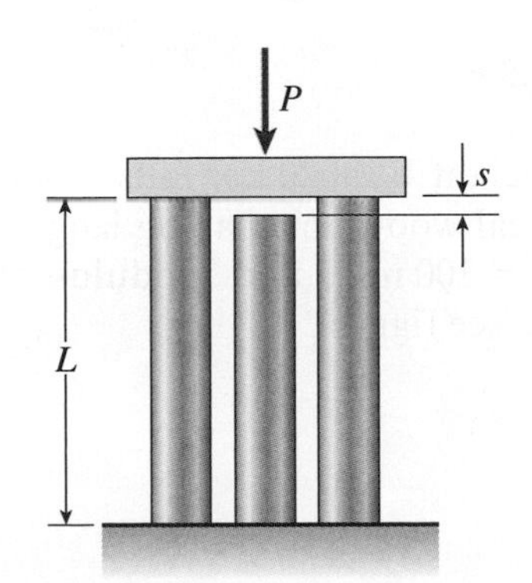

PROBLEM 2.7-10

2.7-11 A block B is pushed against three springs by a force P (see figure). The middle spring has a stiffness k_1, and the outer springs each have stiffness k_2. Initially, the springs are unstressed, and the middle spring is longer than the outer springs (the difference in length is denoted s).

(a) Draw a force-displacement diagram with the force P as ordinate and the displacement x of the block as abscissa.

(b) From the diagram, determine the strain energy U_1 of the springs when $x = 2s$.

(c) Explain why the strain energy U_1 is not equal to $P\delta/2$, where $\delta = 2s$.

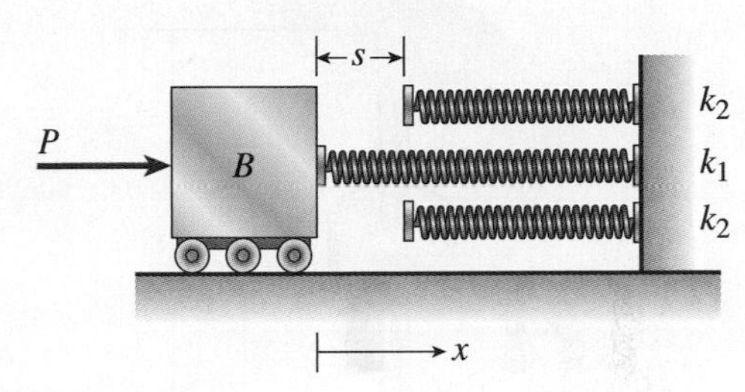

PROBLEM 2.7-11

2.7-12 A bungee cord that behaves linearly elastically has an unstressed length $L_0 = 760$ mm and a stiffness $k = 140$ N/m. The cord is attached to two pegs, distance $b = 380$ mm apart, and is pulled at its midpoint by a force $P = 80$ N (see figure).

(a) How much strain energy U is stored in the cord?

(b) What is the displacement δ_C of the point where the load is applied?

(c) Compare the strain energy U with the quantity $P\delta_C/2$.

Note: The elongation of the cord is *not* small compared to its original length.

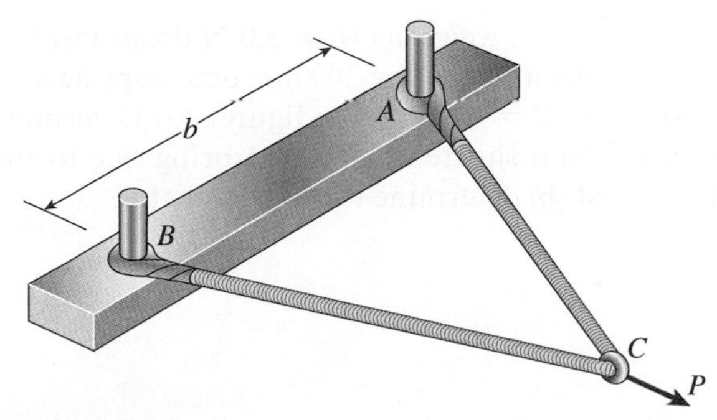

PROBLEM 2.7-12

2.8 Impact Loading

Solve the problems for Section 2.8 by assuming that the material behaves linearly elastically and no energy is lost during the impact.

2.8-1 A sliding collar of weight $W = 650$ N falls from a height $h = 50$ mm onto a flange at the bottom of a slender vertical rod (see figure). The rod has a length $L = 1.2$ m, cross-sectional area $A = 5\ \text{cm}^2$, and modulus of elasticity $E = 210$ GPa.

Calculate the following quantities: (a) the maximum downward displacement of the flange, (b) the maximum tensile stress in the rod, and (c) the impact factor.

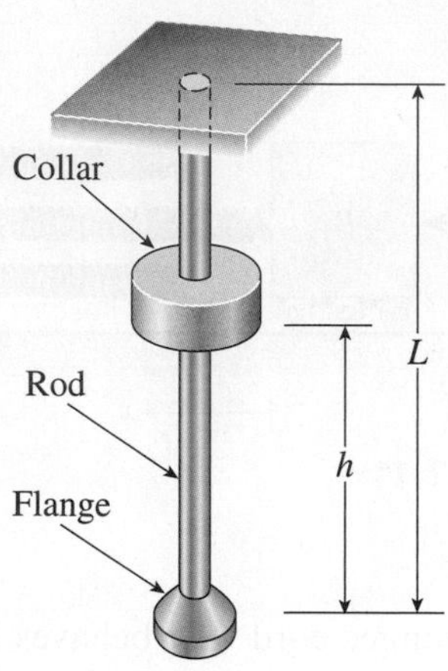

PROBLEMS 2.8-1, 2.8-2, and 2.8-3

2.8-2 Solve the preceding problem if the collar has mass $M = 80$ kg, the height $h = 0.5$ m, the length $L = 3.0$ m, the cross-sectional area $A = 350\ \text{mm}^2$, and the modulus of elasticity $E = 170$ GPa.

2.8-3 Solve Prob. 2.8-1 if the collar has weight $W = 200$ N, the height $h = 50$ mm, the length $L = 0.9$ m, the cross-sectional area $A = 1.5\ \text{cm}^2$, and the modulus of elasticity $E = 210$ GPa.

2.8-4 A block weighing $W = 5.0$ N drops inside a cylinder from a height $h = 200$ mm onto a spring having stiffness $k = 90$ N/m (see figure). (a) Determine the maximum shortening of the spring due to the impact and (b) determine the impact factor.

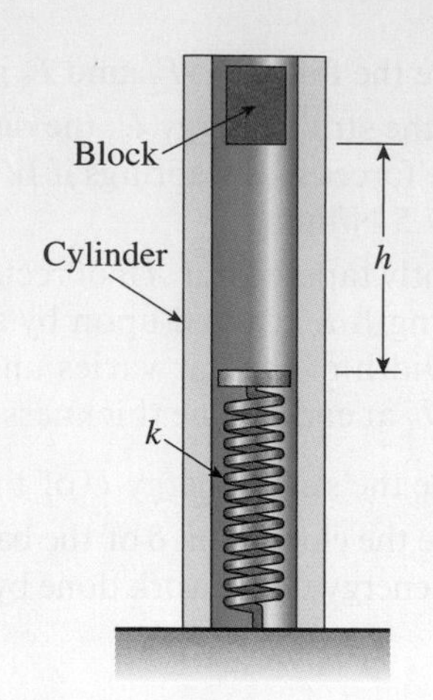

PROBLEMS 2.8-4 and 2.8-5

2.8-5 Solve the preceding problem for $W = 8$ N, $h = 300$ mm, and $k = 125$ N/m.

2.8-6 A small rubber ball (weight $W = 450$ mN) is attached by a rubber cord to a wood paddle (see figure). The natural length of the cord is $L_0 = 200$ mm, its cross-sectional area is $A = 1.6\ \text{mm}^2$, and its modulus of elasticity is $E = 2.0$ MPa. After being struck by the paddle, the ball stretches the cord to a total length $L_1 = 900$ mm.

What was the velocity v of the ball when it left the paddle? (Assume linearly elastic behavior of the rubber cord, and disregard the potential energy due to any change in elevation of the ball.)

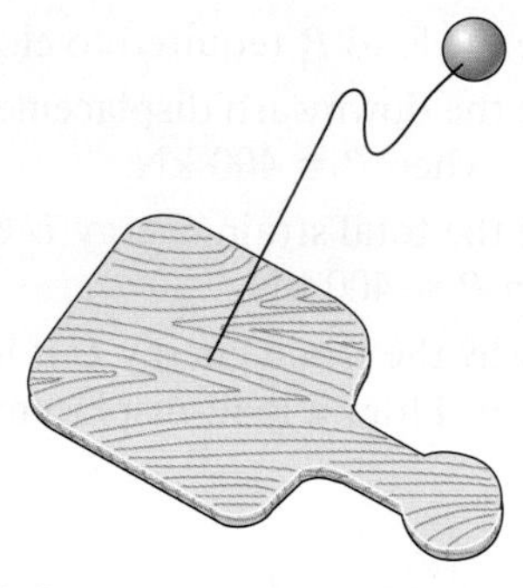

PROBLEM 2.8-6

2.8-7 A weight $W = 20$ kN falls from a height h onto a vertical wood pole having length $L = 5.5$ m, diameter $d = 300$ mm, and modulus of elasticity $E = 10$ GPa (see figure).

If the allowable stress in the wood under an impact load is 17 MPa, what is the maximum permissible height h?

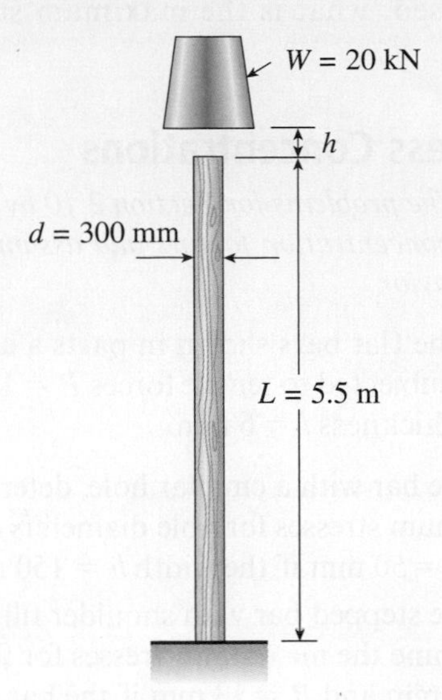

PROBLEM 2.8-7

2.8-8 A cable with a restrainer at the bottom hangs vertically from its upper end (see figure). The cable has an effective cross-sectional area $A = 40\ \text{mm}^2$ and an effective modulus of elasticity $E = 130$ GPa. A slider of mass $M = 35$ kg drops from a height $h = 1.0$ m onto the restrainer.

If the allowable stress in the cable under an impact load is 500 MPa, what is the minimum permissible length L of the cable?

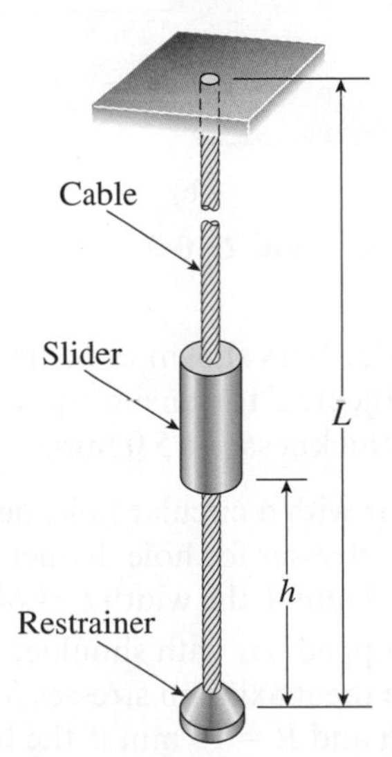

PROBLEMS 2.8-8 and 2.8-9

2.8-9 Solve the preceding problem if the slider has $W = 145$ N, $h = 120$ cm, $A = 0.5\ \text{cm}^2$, $E = 150$ GPa, and the allowable stress is 480 MPa.

2.8-10 A bumping post at the end of a track in a railway yard has a spring constant $k = 8.0$ MN/m (see figure). The maximum possible displacement d of the end of the striking plate is 450 mm.

What is the maximum velocity v_{max} that a railway car of weight $W = 545$ kN can have without damaging the bumping post when it strikes it?

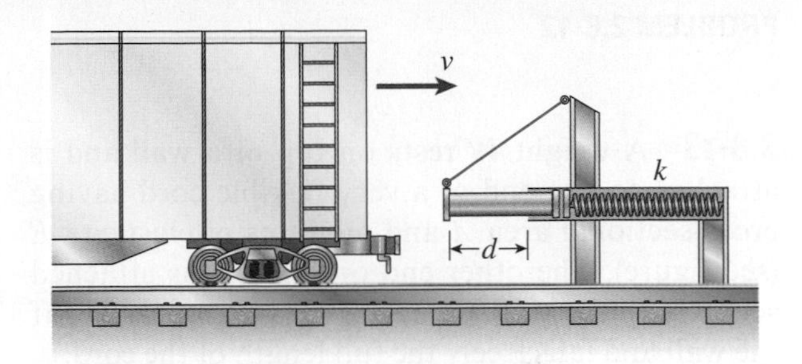

PROBLEM 2.8-10

2.8-11 A bumper for a mine car is constructed with a spring of stiffness $k = 176$ kN/m (see figure). If a car weighing 14 kN is traveling at velocity $v = 8$ km/h when it strikes the spring, what is the maximum shortening of the spring?

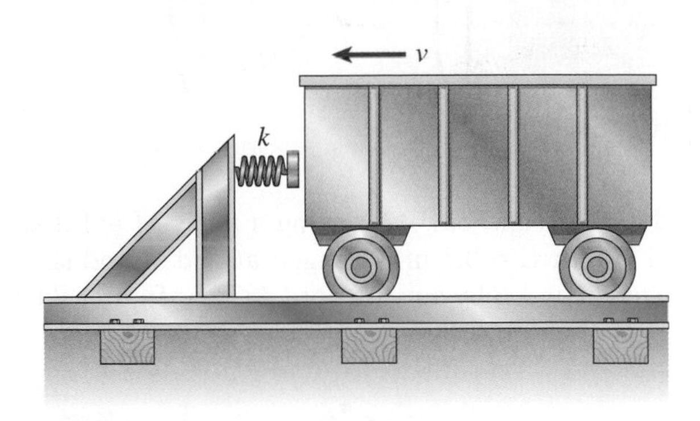

PROBLEM 2.8-11

2.8-12 A bungee jumper having a mass of 55 kg leaps from a bridge, braking her fall with a long elastic shock cord having axial rigidity $EA = 2.3$ kN (see figure).

If the jumpoff point is 60 m above the water, and if it is desired to maintain a clearance of 10 m between the jumper and the water, what length L of cord should be used?

PROBLEM 2.8-12

2.8-13 A weight W rests on top of a wall and is attached to one end of a very flexible cord having cross-sectional area A and modulus of elasticity E (see figure). The other end of the cord is attached securely to the wall. The weight is then pushed off the wall and falls freely the full length of the cord.

(a) Derive a formula for the impact factor.
(b) Evaluate the impact factor if the weight, when hanging statically, elongates the band by 2.5% of its original length.

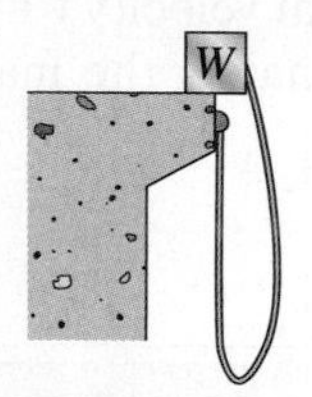

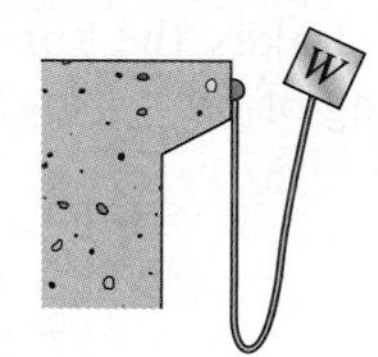

PROBLEM 2.8-13

2.8-14 A rigid bar AB having a mass $M = 1.0$ kg and length $L = 0.5$ m is hinged at end A and supported at end B by a nylon cord BC (see figure). The cord has cross-sectional area $A = 30\ \text{mm}^2$, length $b = 0.25$ m, and modulus of elasticity $E = 2.1$ GPa.

If the bar is raised to its maximum height and then released, what is the maximum stress in the cord?

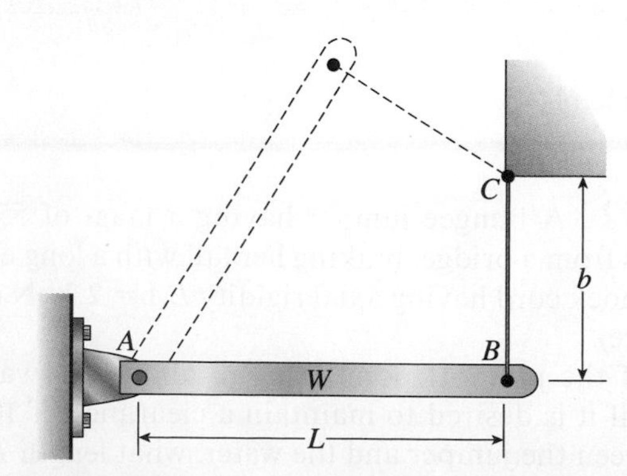

PROBLEM 2.8-14

2.10 Stress Concentrations

Solve the The problems for Section 2.10 by considering the stress-concentration factors and assuming linearly elastic behavior.

2.10-1 The flat bars shown in parts a and b of the figure are subjected to tensile forces $P = 13$ kN. Each bar a has thickness $t = 6$ mm.

(a) For the bar with a circular hole, determine the maximum stresses for hole diameters $d = 25$ mm and $d = 50$ mm if the width $b = 150$ mm.
(b) For the stepped bar with shoulder fillets, determine the maximum stresses for fillet radii $R = 6$ mm and $R = 13$ mm if the bar widths are $b = 100$ mm and $c = 65$ mm.

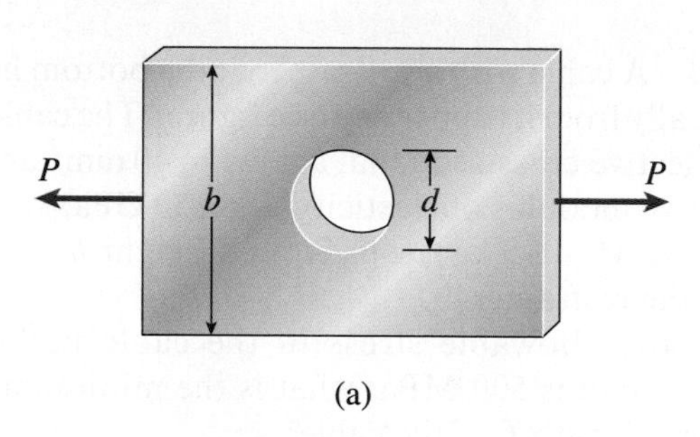

(a)

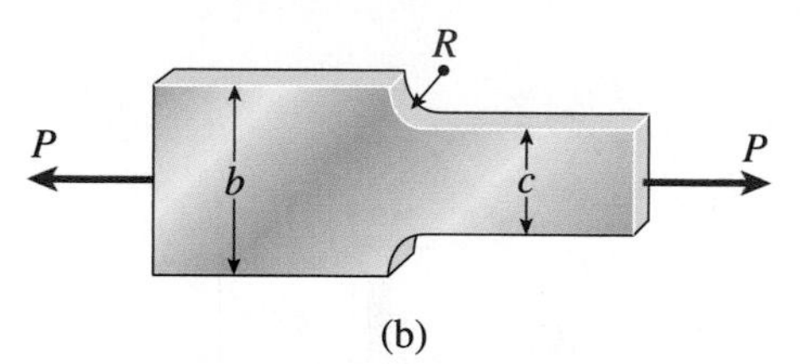

(b)

PROBLEMS 2.10-1 and 2.10-2

2.10-2 The flat bars shown in parts a and b of the figure are subjected to tensile forces $P = 2.5$ kN. Each bar has thickness $t = 5.0$ mm.

(a) For the bar with a circular hole, determine the maximum stresses for hole diameters $d = 12$ mm and $d = 20$ mm if the width $b = 60$ mm.
(b) For the stepped bar with shoulder fillets, determine the maximum stresses for fillet radii $R = 6$ mm and $R = 10$ mm if the bar widths are $b = 60$ mm and $c = 40$ mm.

2.10-3 A flat bar of width b and thickness t has a hole of diameter d drilled through it (see figure). The hole may have any diameter that will fit within the bar.

What is the maximum permissible tensile load P_{max} if the allowable tensile stress in the material is σ_t?

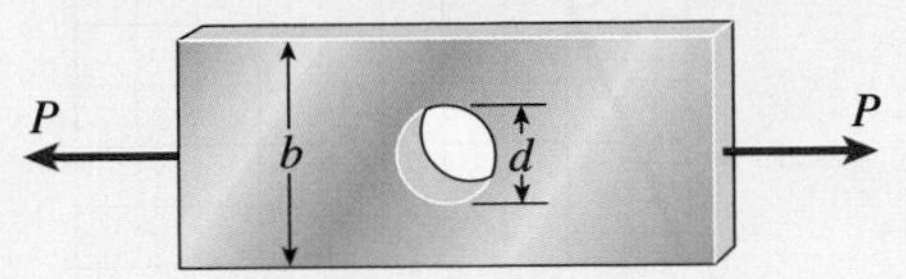

PROBLEM 2.10-3

2.10-4 A round brass bar of a diameter $d_1 = 20$ mm has upset ends each with a diameter $d_2 = 26$ mm (see figure). The lengths of the segments of the bar are $L_1 = 0.3$ m and $L_2 = 0.1$ m. Quarter-circular fillets are used at the shoulders of the bar, and the modulus of elasticity of the brass is $E = 100$ GPa.

If the bar lengthens by 0.12 mm under a tensile load P, what is the maximum stress σ_{max} in the bar?

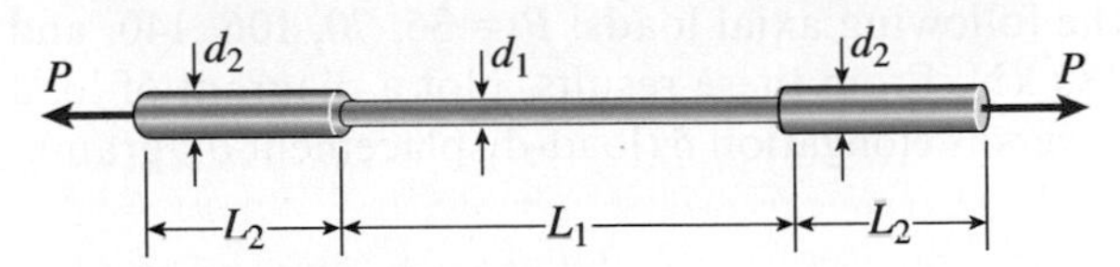

PROBLEMS 2.10-4 and 2.10-5

2.10-5 Solve the preceding problem for a bar of monel metal having the following properties: $d_1 = 25$ mm, $d_2 = 36$ mm, $L_1 = 500$ mm, $L_2 = 125$ mm, and $E = 170$ GPa. Also, the bar lengthens by 0.1 mm when the tensile load is applied.

2.10-6 A prismatic bar with a diameter $d_0 = 20$ mm is being compared with a stepped bar of the same diameter ($d_1 = 20$ mm) that is enlarged in the middle region to a diameter $d_2 = 25$ mm (see figure). The radius of the fillets in the stepped bar is 2.0 mm.

(a) Does enlarging the bar in the middle region make it stronger than the prismatic bar? Demonstrate your answer by determining the maximum permissible load P_1 for the prismatic bar and the maximum permissible load P_2 for the enlarged bar, assuming that the allowable stress for the material is 80 MPa.

(b) What should be the diameter d_0 of the prismatic bar if it is to have the same maximum permissible load as does the stepped bar?

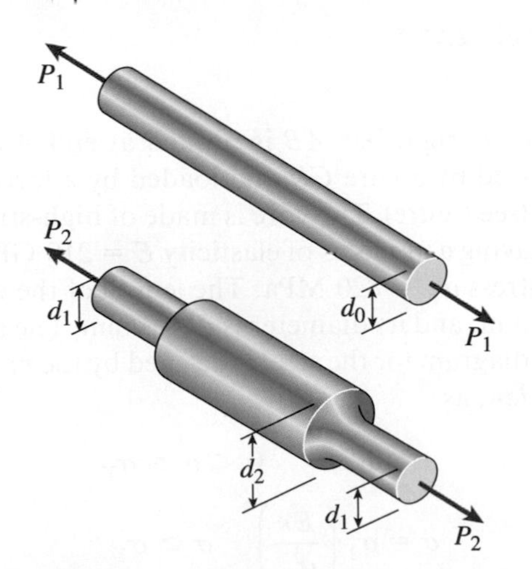

PROBLEM 2.10-6

2.10-7 A stepped bar with a hole (see figure) has widths $b = 60$ mm and $c = 40$ mm. The fillets have radii equal to 5 mm.

What is the diameter d_{max} of the largest hole that can be drilled through the bar without reducing the load-carrying capacity?

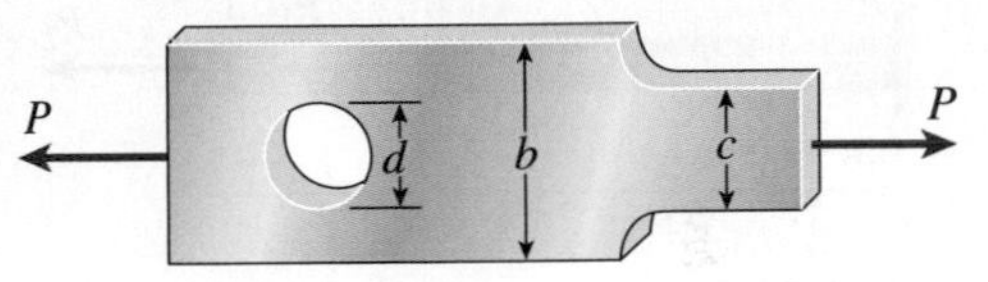

PROBLEM 2.10-7

2.11 Nonlinear Behavior

2.11-1 A bar AB of length L and weight density γ hangs vertically under its own weight (see figure). The stress-strain relation for the material is given by the Ramberg-Osgood equation [Eq. (2-71)]:

$$\varepsilon = \frac{\sigma}{E} + \frac{\sigma_0 \alpha}{E}\left(\frac{\sigma}{\sigma_0}\right)^m$$

Derive the formula

$$\delta = \frac{\gamma L^2}{2E} + \frac{\sigma_0 \alpha L}{(m+1)E}\left(\frac{\gamma L}{\sigma_0}\right)^m$$

for the elongation of the bar.

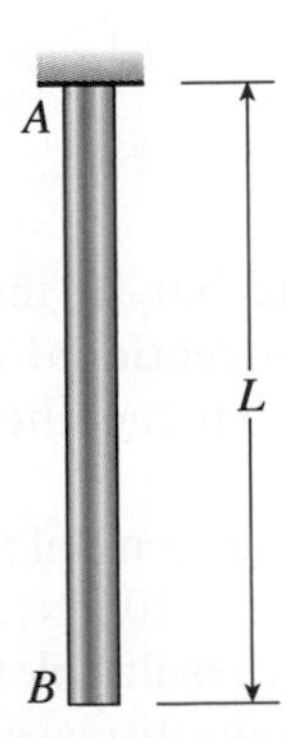

PROBLEM 2.11-1

2.11-2 A prismatic bar of length $L = 1.8$ m and cross-sectional area $A = 480\text{ mm}^2$ is loaded by forces $P_1 = 30$ kN and $P_2 = 60$ kN (see figure). The bar is constructed of magnesium alloy having a stress-strain curve described by the Ramberg-Osgood equation:

$$\varepsilon = \frac{\sigma}{45{,}000} + \frac{1}{618}\left(\frac{\sigma}{170}\right)^{10}$$

in which σ has units of megapascals (MPa).

(a) Calculate the displacement δ_C of the end of the bar when the load P_1 acts alone.

(b) Calculate the displacement when the load P_2 acts alone.

(c) Calculate the displacement when both loads act simultaneously.

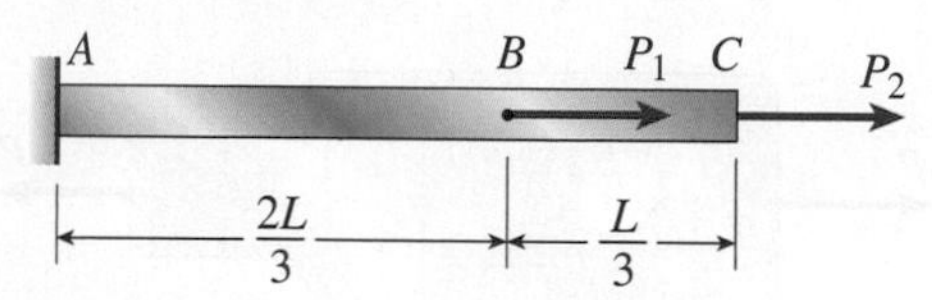

PROBLEM 2.11-2

2.11-3 A circular bar of length $L = 810$ mm and diameter $d = 19$ mm is subjected to tension by forces P (see figure). The wire is made of a copper alloy having the *hyperbolic stress-strain relationship*:

$$\sigma = \frac{18{,}000\varepsilon}{1 + 300\varepsilon} \times 6.809 \quad 0 \le \varepsilon \le 0.03 \quad (\sigma = \text{MPa})$$

(a) Draw a stress-strain diagram for the material.

(b) If the elongation of the wire is limited to 6 mm and the maximum stress is limited to 275 MPa, what is the allowable load P?

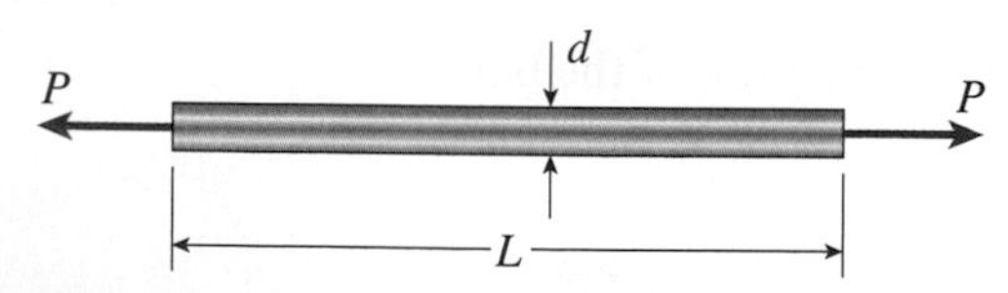

PROBLEM 2.11-3

2.11-4 A prismatic bar in tension has a length $L = 2.0$ m and cross-sectional area $A = 249\text{ mm}^2$. The material of the bar has the stress-strain curve shown in the figure.

Determine the elongation δ of the bar for each of the following axial loads: $P = 10$ kN, 20 kN, 30 kN, 40 kN, and 45 kN. From these results, plot a diagram of load P versus elongation δ (load-displacement diagram).

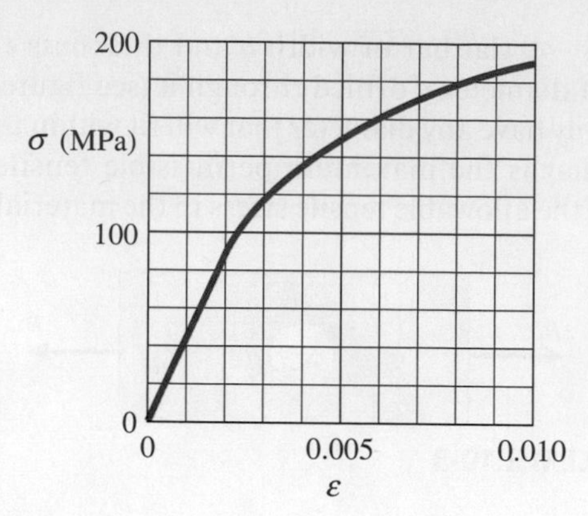

PROBLEM 2.11-4

2.11-5 An aluminum bar subjected to tensile forces P has a length $L = 3.8$ m and cross-sectional area $A = 1290\text{ mm}^2$. The stress-strain behavior of the aluminum may be represented approximately by the bilinear stress-strain diagram shown in the figure.

Calculate the elongation δ of the bar for each of the following axial loads: $P = 35$, 70, 106, 140, and 180 kN. From these results, plot a diagram of load P versus elongation δ (load-displacement diagram).

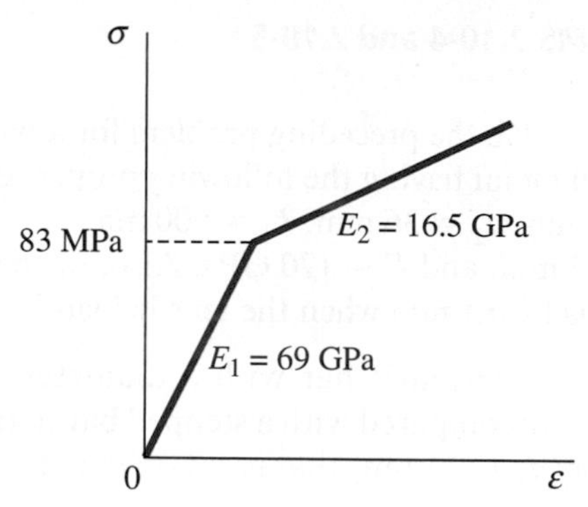

PROBLEM 2.11-5

2.11-6 A rigid bar AB is pinned at end A and is supported by a wire CD and loaded by a force P at end B (see figure). The wire is made of high-strength steel having a modulus of elasticity $E = 210$ GPa and yield stress $\sigma_Y = 820$ MPa. The length of the wire is $L = 1.0$ m, and its diameter is $d = 3$ mm. The stress-strain diagram for the steel is defined by the *modified power law*, as

$$\sigma = E\varepsilon \qquad 0 < \sigma \le \sigma_Y$$

$$\sigma = \sigma_Y\left(\frac{E\varepsilon}{\sigma_Y}\right)^n \qquad \sigma \ge \sigma_Y$$

(a) Assuming $n = 0.2$, calculate the displacement δ_B at the end of the bar due to the load P. Take values of P from 2.4 kN to 5.6 kN in increments of 0.8 kN.

(b) Plot a load-displacement diagram showing P versus δ_B.

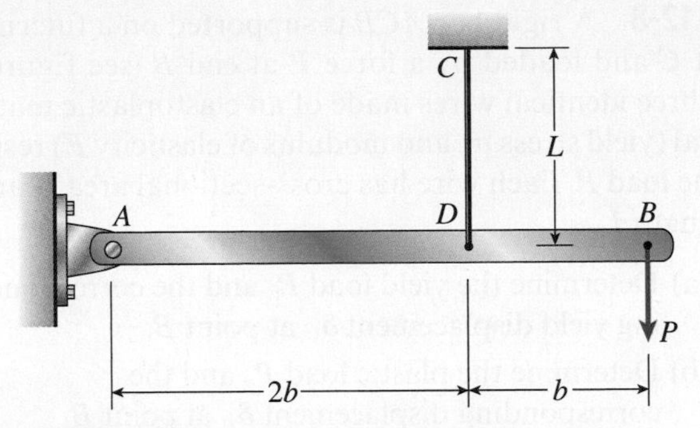

PROBLEM 2.11-6

2.12 Elastoplastic Analysis

Solve the problems for Section 2.12 assuming that the material is elastoplastic with yield stress σ_Y, yield strain ε_Y, and modulus of elasticity E in the linearly elastic region (see Fig. 2-95).

2.12-1 Two identical bars AB and BC support a vertical load P (see figure). The bars are made of steel having a stress-strain curve that may be idealized as elastoplastic with yield stress σ_Y. Each bar has cross-sectional area A.

Determine the yield load P_Y and the plastic load P_P.

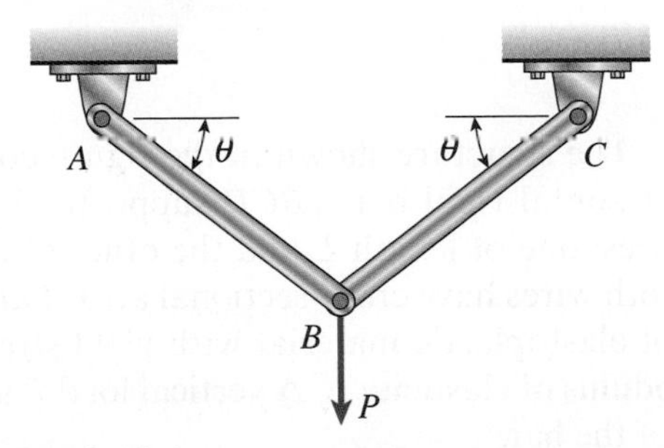

PROBLEM 2.12-1

2.12-2 A stepped bar ACB with circular cross sections is held between rigid supports and loaded by an axial force P at midlength (see figure). The diameters for the two parts of the bar are $d_1 = 20$ mm and $d_2 = 25$ mm, and the material is elastoplastic with yield stress $\sigma_Y = 250$ MPa.

Determine the plastic load P_P.

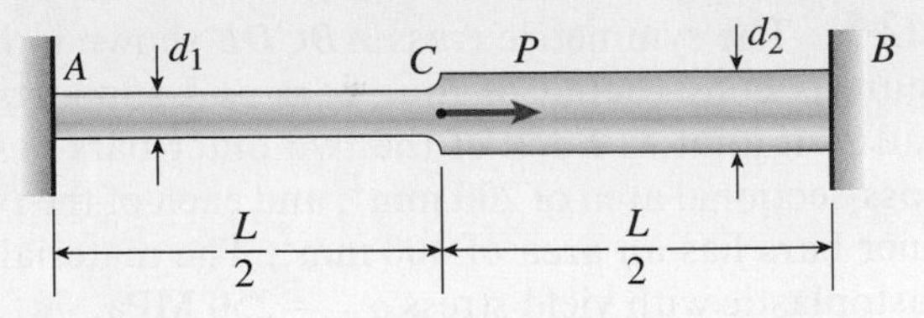

PROBLEM 2.12-2

2.12-3 A horizontal rigid bar AB supporting a load P is hung from five symmetrically placed wires, each of cross-sectional area A (see figure). The wires are fastened to a curved surface of radius R.

(a) Determine the plastic load P_P if the material of the wires is elastoplastic with yield stress σ_Y.

(b) How is P_P changed if bar AB is flexible instead of rigid?

(c) How is P_P changed if the radius R is increased?

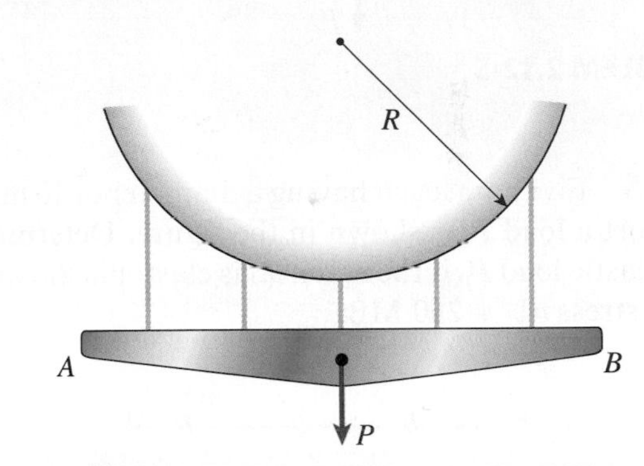

PROBLEM 2.12-3

2.12-4 A load P acts on a horizontal beam that is supported by four rods arranged in the symmetrical pattern shown in the figure. Each rod has a cross-sectional area A, and the material is elastoplastic with a yield stress σ_Y. Determine the plastic load P_P.

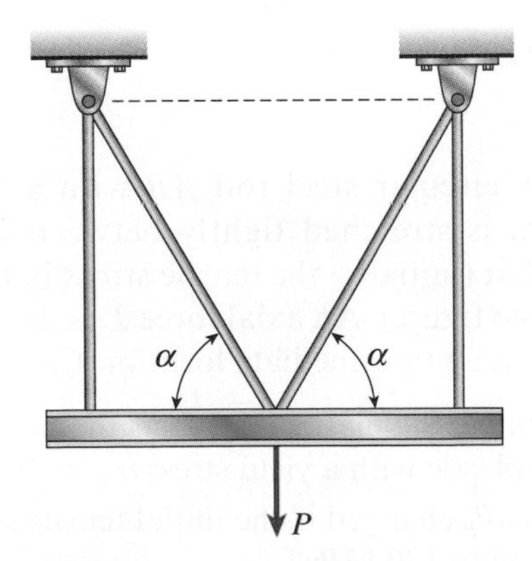

PROBLEM 2.12-4

2.12-5 The symmetric truss $ABCDE$ shown in the figure is constructed of four bars and supports a load P at joint E. Each of the two outer bars has a cross-sectional area of 200 mm^2, and each of the two inner bars has an area of 400 mm^2. The material is elastoplastic with yield stress $\sigma_Y = 250$ MPa.

Determine the plastic load P_P.

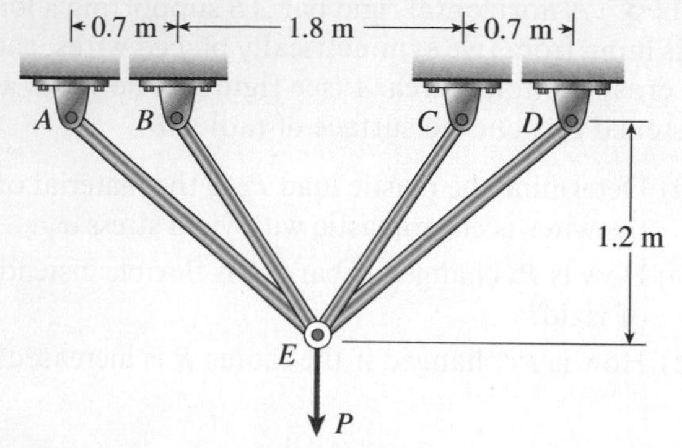

PROBLEM 2.12-5

2.12-6 Five bars, each having a diameter of 10 mm, support a load P as shown in the figure. Determine the plastic load P_P if the material is elastoplastic with yield stress $\sigma_Y = 250$ MPa.

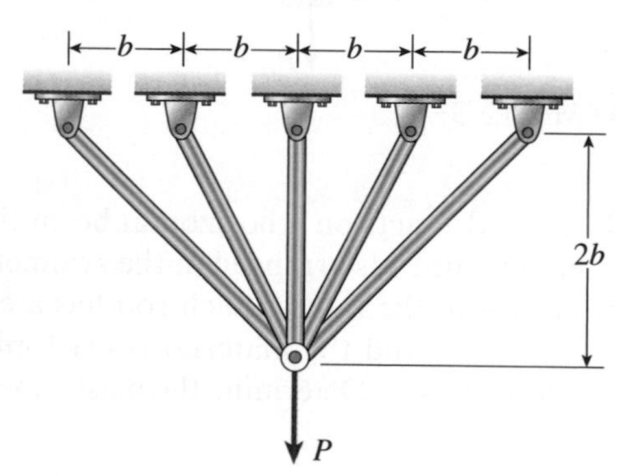

PROBLEM 2.12-6

2.12-7 A circular steel rod AB with a diameter $d = 15$ mm is stretched tightly between two supports so that (initially) the tensile stress in the rod is 60 MPa (see figure). An axial force P is then applied to the rod at an intermediate location C.

(a) Determine the plastic load P_P if the material is elastoplastic with a yield stress $\sigma_Y = 290$ MPa.

(b) How is P_P changed if the initial tensile stress is doubled to 120 MPa?

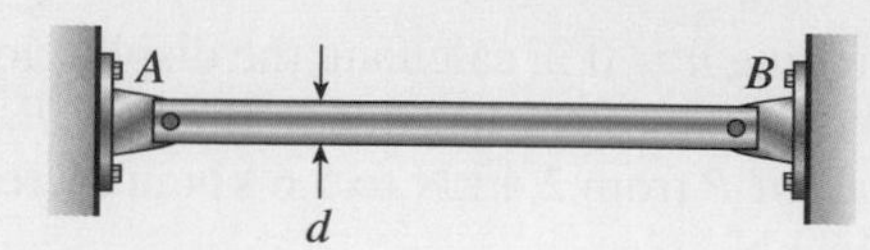

PROBLEM 2.12-7

2.12-8 A rigid bar ACB is supported on a fulcrum at C and loaded by a force P at end B (see figure). Three identical wires made of an elastoplastic material (yield stress σ_Y and modulus of elasticity E) resist the load P. Each wire has cross-sectional area A and length L.

(a) Determine the yield load P_Y and the corresponding yield displacement δ_Y at point B.

(b) Determine the plastic load P_P and the corresponding displacement δ_P at point B when the load just reaches the value P_P.

(c) Draw a load-displacement diagram with the load P as ordinate and the displacement δ_B of point B as abscissa.

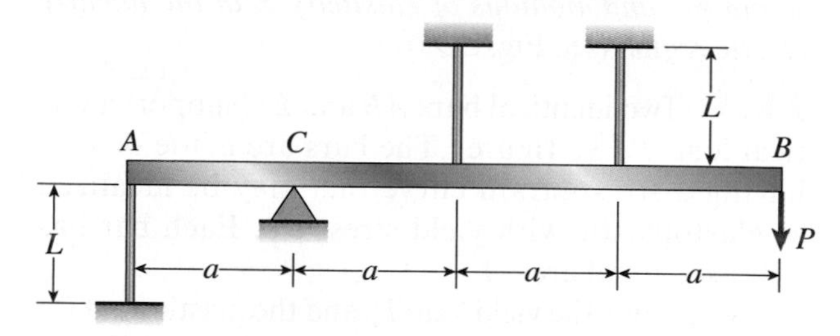

PROBLEM 2.12-8

2.12-9 The structure shown in the figure consists of a horizontal rigid bar $ABCD$ supported by two steel wires: one of length L and the other of length $3L/4$. Both wires have cross-sectional area A and are made of elastoplastic material with yield stress σ_Y and modulus of elasticity E. A vertical load P acts at end D of the bar.

(a) Determine the yield load P_Y and the corresponding yield displacement δ_Y at point D.

(b) Determine the plastic load P_P and the corresponding displacement δ_Y at point D when the load just reaches the value P_P.

(c) Draw a load-displacement diagram with the load P as ordinate and the displacement δ_D of point D as abscissa.

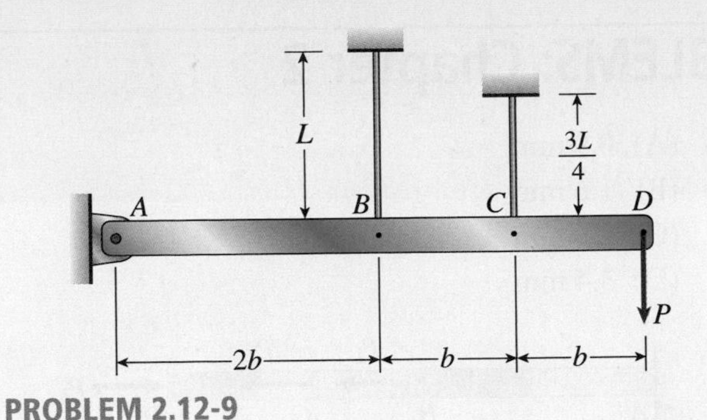

PROBLEM 2.12-9

2.12-10 Two cables, each having a length L of approximately 40 m, support a loaded container of weight W (see figure). The cables, which have an effective cross-sectional area $A = 48.0 \text{ mm}^2$ and effective modulus of elasticity $E = 160$ GPa, are identical except that one cable is longer than the other when they are hanging separately and unloaded. The difference in lengths is $d = 100$ mm. The cables are made of steel having an elastoplastic stress-strain diagram with $\sigma_Y = 500$ MPa. Assume that the weight W is initially zero and is slowly increased by the addition of material to the container.

(a) Determine the weight W_Y that first produces yielding of the shorter cable. Also, determine the corresponding elongation δ_Y of the shorter cable.

(b) Determine the weight W_P that produces yielding of both cables. Also, determine the elongation δ_P of the shorter cable when the weight W just reaches the value W_P.

(c) Construct a load-displacement diagram showing the weight W as ordinate and the elongation δ of the shorter cable as abscissa. *Hint*: The load displacement diagram is not a single straight line in the region $0 \leq W \leq W_{Y'}$

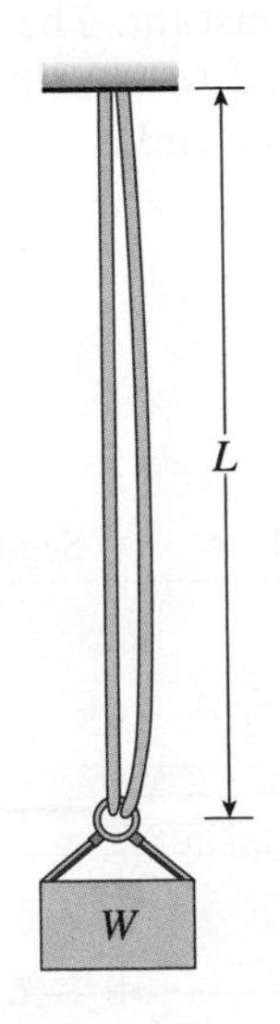

PROBLEM 2.12-10

2.12-11 A hollow circular tube T of a length $L = 380$ mm is uniformly compressed by a force P acting through a rigid plate (see figure). The outside and inside diameters of the tube are 76 and 70 mm, respectively. A concentric solid circular bar B of 38 mm diameter is mounted inside the tube. When no load is present, there is a clearance $c = 0.26$ mm between the bar B and the rigid plate. Both bar and tube are made of steel having an elastoplastic stress-strain diagram with $E = 200$ GPa and $\sigma_Y = 250$ MPa.

(a) Determine the yield load P_Y and the corresponding shortening δ_Y of the tube.

(b) Determine the plastic load P_P and the corresponding shortening δ_P of the tube.

(c) Construct a load-displacement diagram showing the load P as ordinate and the shortening δ of the tube as abscissa. *Hint:* The load-displacement diagram is not a single straight line in the region $0 \leq P \leq P_{Y'}$.

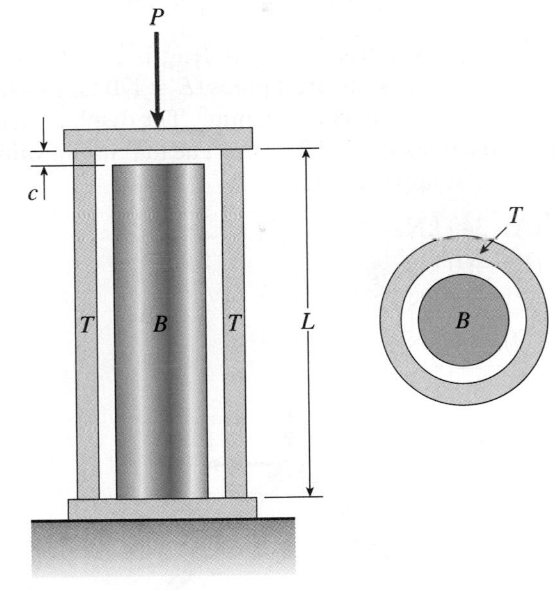

PROBLEM 2.12-11

SOME ADDITIONAL REVIEW PROBLEMS: Chapter 2

R-2.1 Two wires, one copper and the other steel, of equal length stretch the same amount under an applied load P. The moduli of elasticity for each is $E_s = 210$ GPa and $E_c = 120$ GPa. The ratio of the diameter of the copper wire to that of the steel wire is approximately:

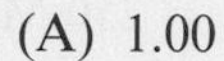

(A) 1.00
(B) 1.08
(C) 1.19
(D) 1.32

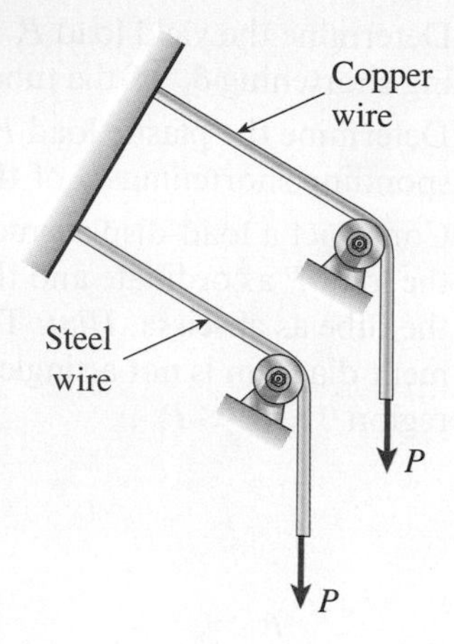

R-2.2 A plane truss with span length $L = 4.5$ m is constructed using cast iron pipes ($E = 170$ GPa) with a cross-sectional area of 4500 mm². The displacement of joint B cannot exceed 2.7 mm. The maximum value of loads P is approximately:

(A) 340 kN
(B) 460 kN
(C) 510 kN
(D) 600 kN

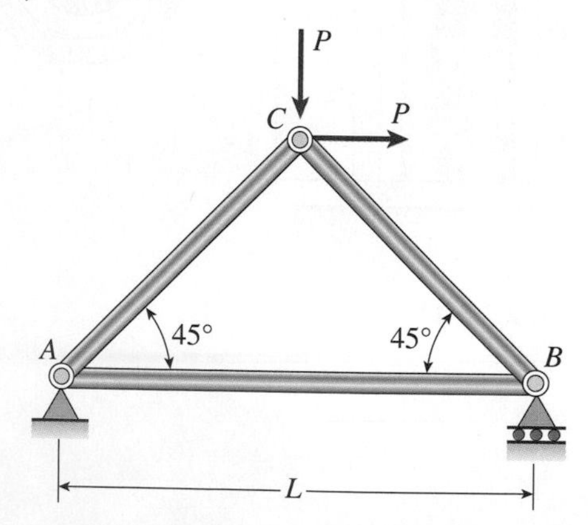

R-2.3 A brass rod ($E = 110$ GPa) with a cross-sectional area of 250 mm² is loaded by forces $P_1 = 15$ kN, $P_2 = 10$ kN, and $P_3 = 8$ kN. Segment lengths of the bar are $a = 2.0$ m, $b = 0.75$ m, and $c = 1.2$ m. The change in length of the bar is approximately:

(A) 0.9 mm
(B) 1.6 mm
(C) 2.1 mm
(D) 3.4 mm

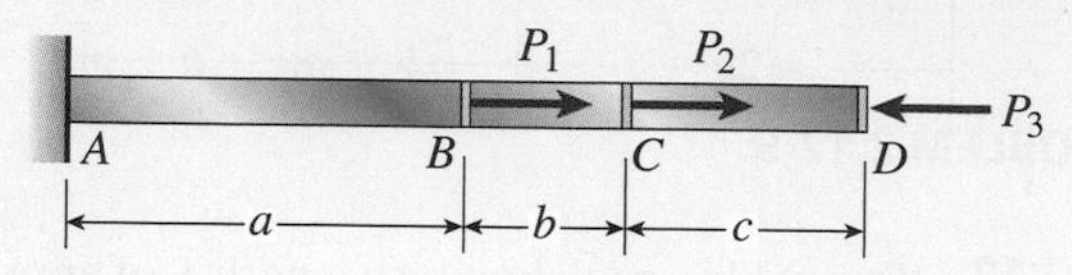

R-2.4 A brass bar ($E = 110$ MPa) of length $L = 2.5$ m has diameter $d_1 = 18$ mm over one-half of its length and diameter $d_2 = 12$ mm over the other half. Compare this nonprismatic bar to a prismatic bar of the same volume of material with constant diameter d and length L. The elongation of the prismatic bar under the same load $P = 25$ kN is approximately:

(A) 3 mm
(B) 4 mm
(C) 5 mm
(D) 6 mm

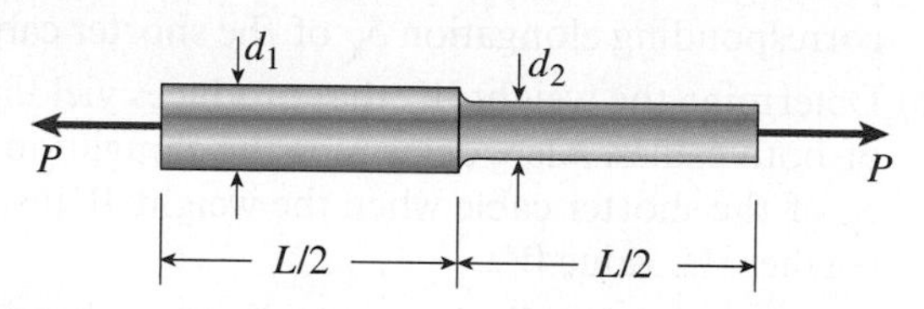

R-2.5 A nonprismatic cantilever bar has an internal cylindrical hole of diameter $d/2$ from 0 to x, so the net area of the cross section for Segment 1 is $(3/4)A$. Load $2P$ is applied at x, and load $-P/2$ is applied at $x = L$. Assume that E is constant. The length of the hollow segment, x, required to obtain axial displacement $\delta = PL/EA$ at the free end is:

(A) $x = L/5$
(B) $x = L/4$
(C) $x = L/3$
(D) $x = 3L/5$

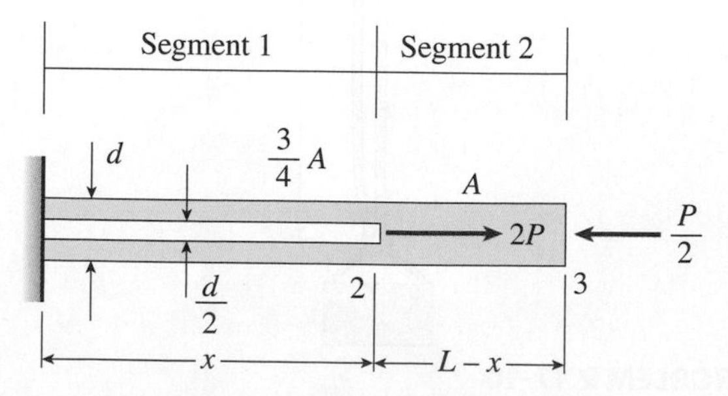

R-2.6 A nylon bar (E = 2.1 GPa) with diameter 12 mm, length 4.5 m, and weight 5.6 N hangs vertically under its own weight. The elongation of the bar at its free end is approximately:

(A) 0.05 mm
(B) 0.07 mm
(C) 0.11 mm
(D) 0.17 mm

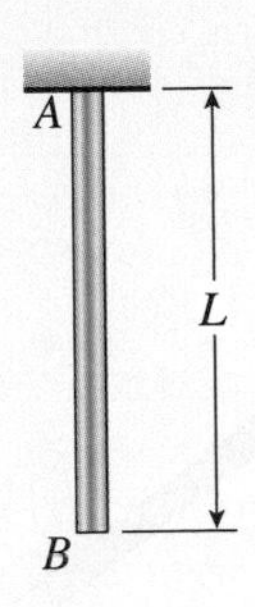

R-2.7 A monel shell (E_m = 170 GPa, d_3 = 12 mm, d_2 = 8 mm) encloses a brass core (E_b = 96 GPa, d_1 = 6 mm). Initially, both shell and core are a length of 100 mm. A load P is applied to both shell and core through a cap plate. The load P required to compress both shell and core by 0.10 mm is approximately:

(A) 10.2 kN
(B) 13.4 kN
(C) 18.5 kN
(D) 21.0 kN

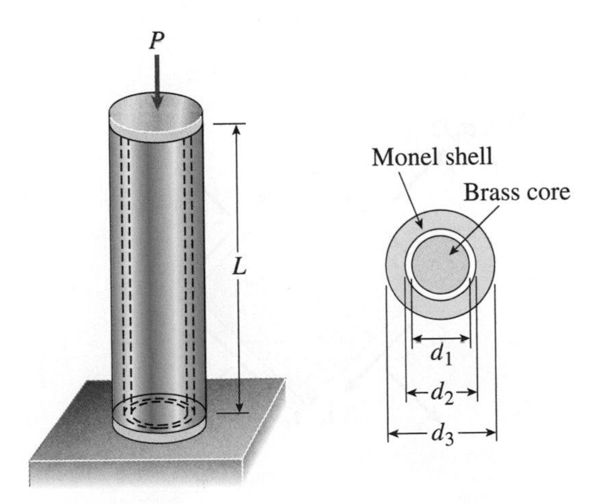

R-2.8 A steel rod (E_s = 210 GPa, d_r = 12 mm, α_s = 12 × 10^{-6}/°C) is held stress-free between rigid walls by a clevis and pin (d_p = 15 mm) assembly at each end. If the allowable shear stress in the pin is 45 MPa and the allowable normal stress in the rod is 70 MPa, the maximum permissible temperature drop ΔT is approximately:

(A) 14°C
(B) 20°C
(C) 28°C
(D) 40°C

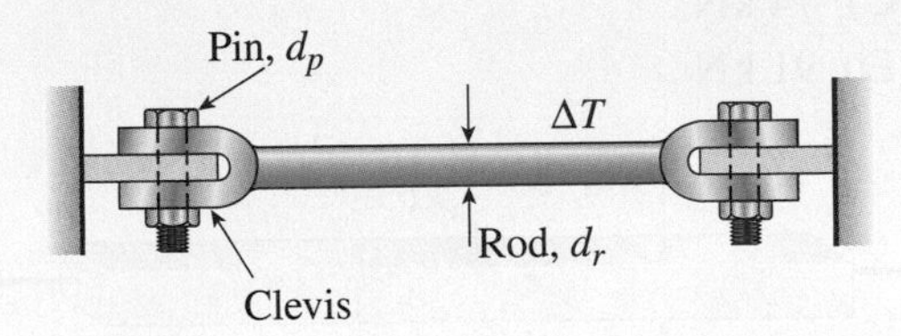

R-2.9 A threaded steel rod (E_s = 210 GPa, d_r = 15 mm, α_s = 12 × 10^{-6}/°C) is held stress-free between rigid walls by a nut and washer (d_w = 22 mm) assembly at each end. If the allowable bearing stress between the washer and wall is 55 MPa and the allowable normal stress in the rod is 90 MPa, the maximum permissible temperature drop ΔT is approximately:

(A) 25°C
(B) 30°C
(C) 38°C
(D) 46°C

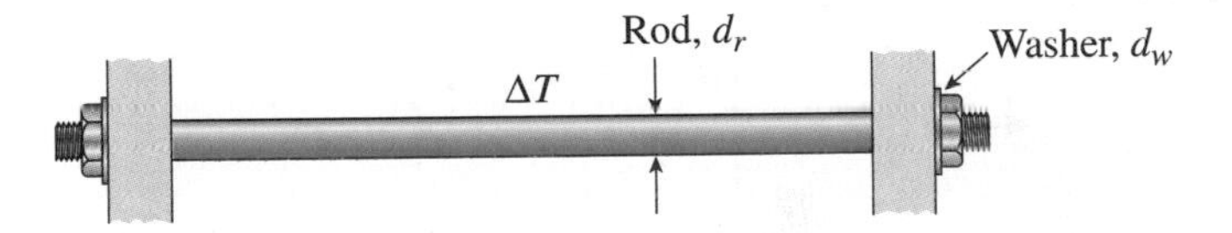

R-2.10 A steel bolt (area = 130 mm^2, E_s = 210 GPa) is enclosed by a copper tube (length = 0.5 m, area = 400 mm^2, E_c = 110 GPa) and the end nut is turned until it is just snug. The pitch of the bolt threads is 1.25 mm. The bolt is now tightened by a quarter turn of the nut. The resulting stress in the bolt is approximately:

(A) 56 MPa
(B) 62 MPa
(C) 74 MPa
(D) 81 MPa

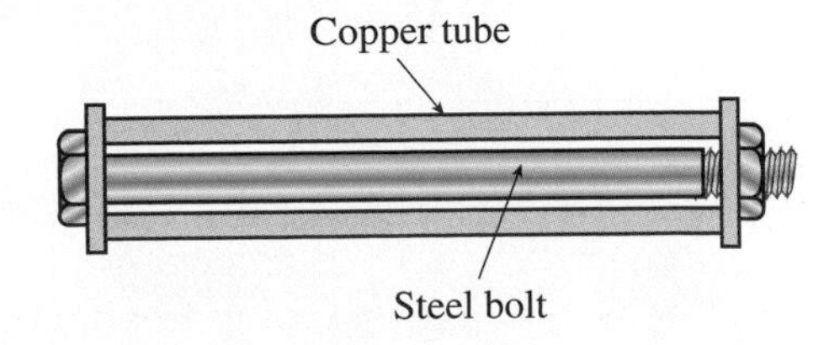

R-2.11 A steel bar of rectangular cross section (a = 38 mm, b = 50 mm) carries a tensile load P. The allowable stresses in tension and shear are 50 MPa and 24 MPa, respectively. The maximum permissible load P_{max} is approximately:

(A) 56 kN
(B) 62 kN
(C) 74 kN
(D) 91 kN

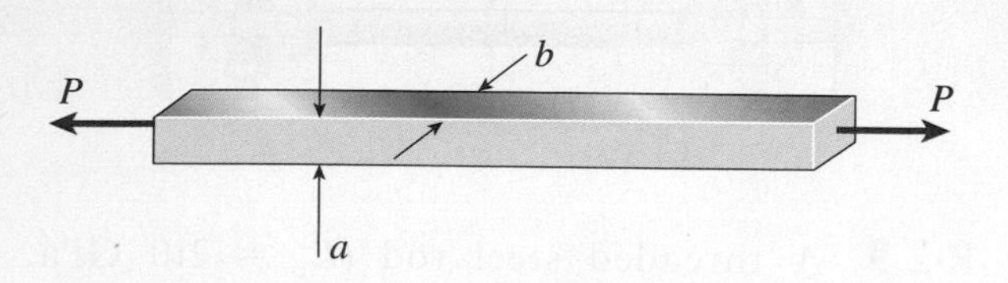

R-2.12 A brass wire (d = 2.0 mm, E = 110 GPa) is pretensioned to T = 85 N. The coefficient of thermal expansion for the wire is 19.5 = 10^{-6}/°C. The temperature change at which the wire goes slack is approximately:

(A) +5.7°C
(B) −12.6°C
(A) +12.6°C
(B) −18.2°C

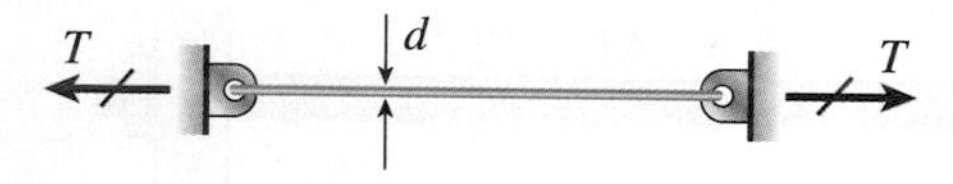

R-2.13 A copper bar (d = 10 mm, E = 110 GPa) is loaded by tensile load P = 11.5 kN. The maximum shear stress in the bar is approximately:

(A) 73 MPa
(B) 87 MPa
(C) 145 MPa
(D) 150 MPa

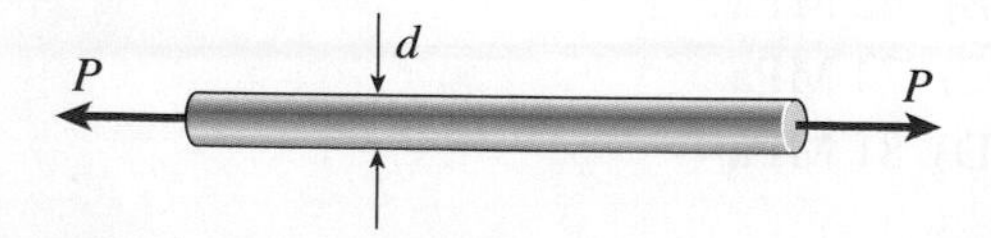

R-2.14 A steel plane truss is loaded at B and C by forces P = 200 kN. The cross-sectional area of each member is A = 3970 mm^2. Truss dimensions are H = 3 m and L = 4 m. The maximum shear stress in bar AB is approximately:

(A) 27 MPa
(B) 33 MPa
(C) 50 MPa
(D) 69 MPa

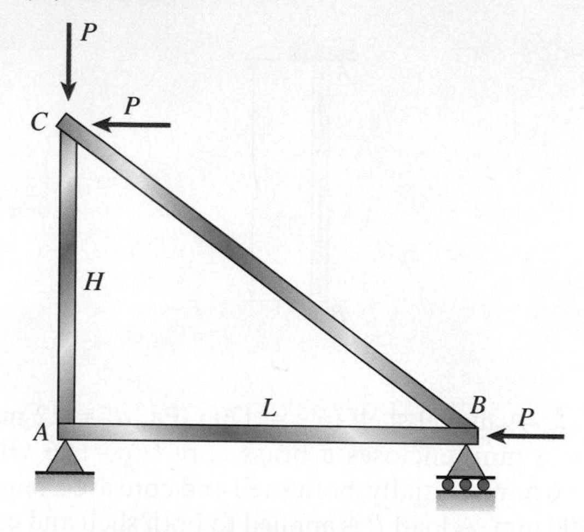

R-2.15 A plane stress element on a bar in uniaxial stress has a tensile stress of σ_θ = 78 MPa (see fig.). The maximum shear stress in the bar is approximately:

(A) 29 MPa
(B) 37 MPa
(C) 50 MPa
(D) 59 MPa

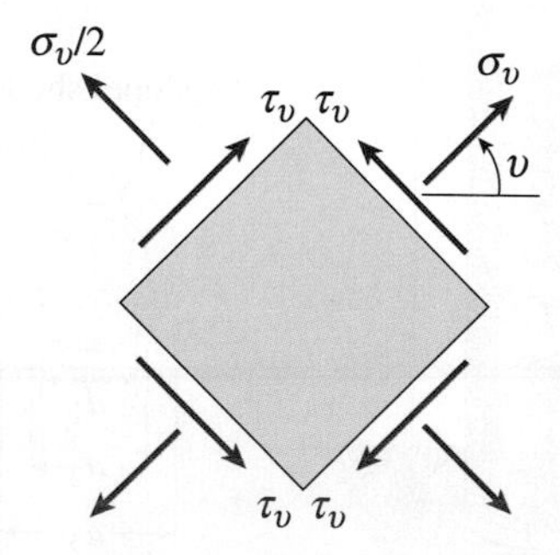

R-2.16 A prismatic bar (diameter $d_0 = 18$ mm) is loaded by force P_1. A stepped bar (diameters $d_1 = 20$ mm and $d_2 = 25$ mm with radius of fillets $R = 2$ mm) is loaded by force P_2. The allowable axial stress in the material is 75 MPa. The ratio P_1/P_2 of the maximum permissible loads that can be applied to the bars, considering stress concentration effects in the stepped bar, is:

(A) 0.9
(B) 1.2
(C) 1.4
(D) 2.1

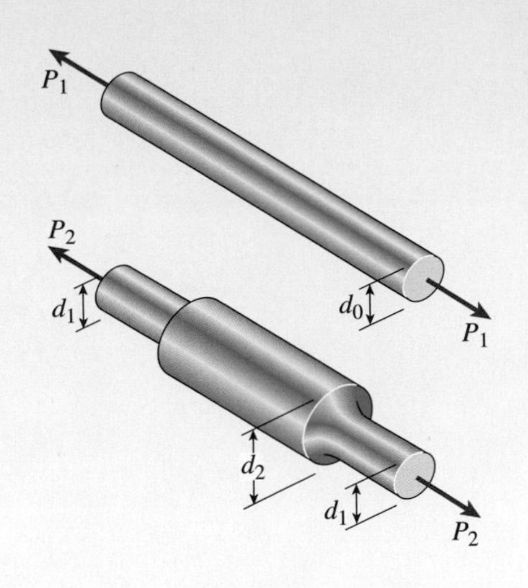

FIGURE 2-87

Stress-concentration factor K for round bars with shoulder fillets. The dashed line is for a full quarter-circular fillet.

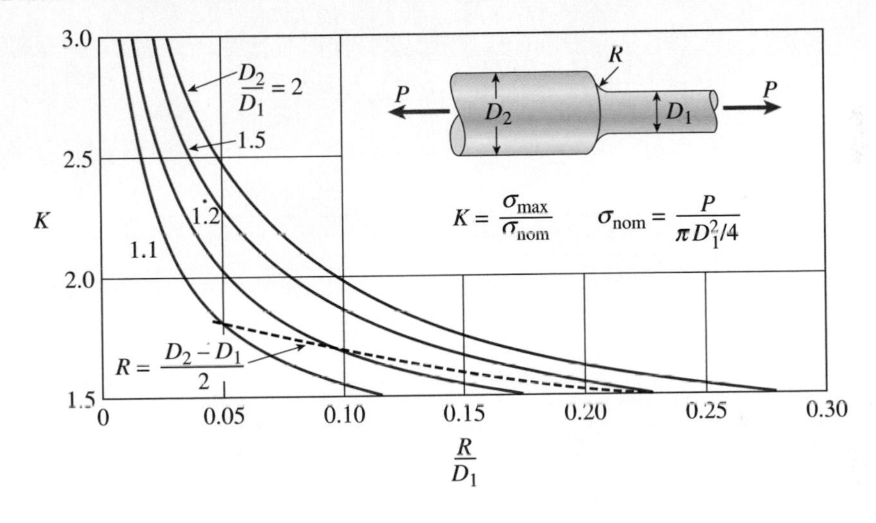

CHAPTER 3

Torsion

RGB Ventures/SuperStock/Alamy Stock Photo

A large crankshaft is needed to turn the propeller shaft to power large ships. The propeller shaft develops torsional stresses and twisting deformations during rotation.

Chapter Objectives

- Study twisting of circular bars and hollow shafts acted on by torsional moments.
- Develop the torsion formula, which relates the variation of shear stress with increasing radial distance in the cross section.
- Define the torque-displacement relation, which shows that the angle of twist is proportional to the product of the internal torsional moment and the torsional flexibility of a circular bar.
- Find both normal and shear stresses at points of interest on inclined sections of circular shafts.
- Define the relationship between modulus of elasticity E and shearing modulus G.
- Study power transmission by circular shafts to find a relation among power P, torque T, and rotational speed n.
- Find support reaction moments in statically indeterminate shafts acted on by torsional moments.
- Study maximum shears stresses in noncircular shafts and stress concentrations in shafts at locations of abrupt changes in geometry.

Chapter Objectives

***Specialized and/or advanced topics**

3.1 Introduction

Chapters 1 and 2 discussed the behavior of the simplest type of structural member—namely, a straight bar subjected to axial loads. Now a slightly more complex type of behavior known as **torsion** is considered. Torsion refers to the twisting of a straight bar when it is loaded by moments (or torques) that tend to produce rotation about the longitudinal axis of the bar. For instance, when you turn a screwdriver (Fig. 3-1a), your hand applies a torque T to the handle (Fig. 3-1b) and twists the shank of the screwdriver. Other examples of bars in torsion are drive shafts in automobiles, axles, propeller shafts, steering rods, and drill bits.

FIGURE 3-1

Torsion of a screwdriver due to a torque T applied to the handle

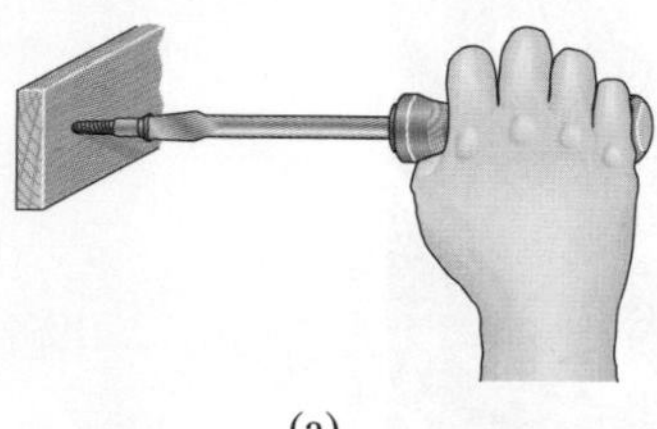

(a)

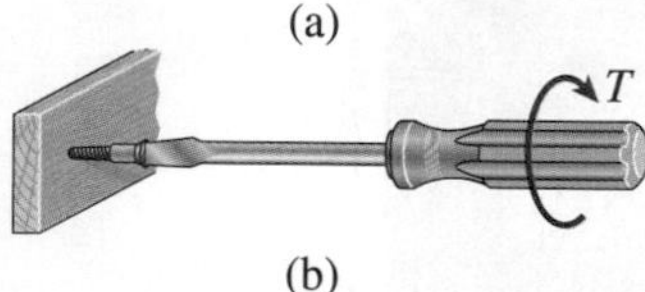

(b)

An idealized case of torsional loading is pictured in Fig. 3-2a, which shows a straight bar supported at one end and loaded by two pairs of equal and opposite forces. The first pair consists of the forces P_1 acting near the midpoint of the bar and the second pair consists of the forces P_2 acting at the end. Each pair of forces forms a **couple** that tends to twist the bar about its longitudinal axis. As you know from statics, the **moment of a couple** is equal to the product of one of the forces and the perpendicular distance between the lines of action of the forces; thus, the first couple has a moment $T_1 = P_1d_1$ and the second has a moment $T_2 = P_2d_2$.

The SI unit for moment is the newton meter (N·m).

The moment of a couple may be represented by a **vector** in the form of a double-headed arrow (Fig. 3-2b). The arrow is perpendicular to the plane containing the couple, and in this case, both arrows are parallel to the axis of the bar. The direction (or *sense*) of the moment is indicated by the *right-hand rule* for moment vectors—namely, using your right hand, let your fingers curl in the direction of the moment, and then your thumb will point in the direction of the vector.

An alternative representation of a moment is a curved arrow acting in the direction of rotation (Fig. 3-2c). Both the curved arrow and vector representations are in common use and are used in this book. The choice depends upon convenience and personal preference.

FIGURE 3-2

Circular bar subjected to torsion by torques T_1 and T_2

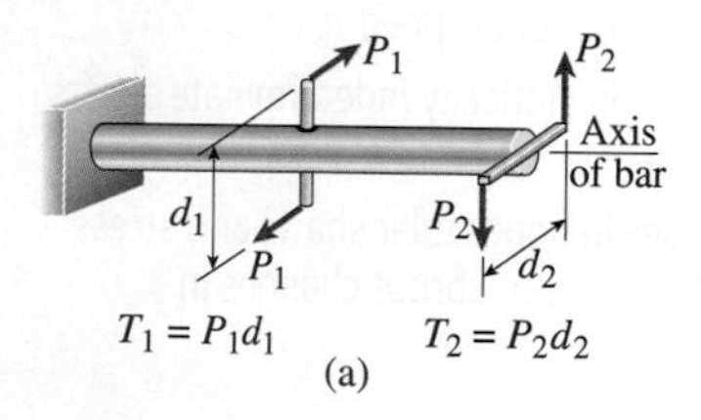

(a)

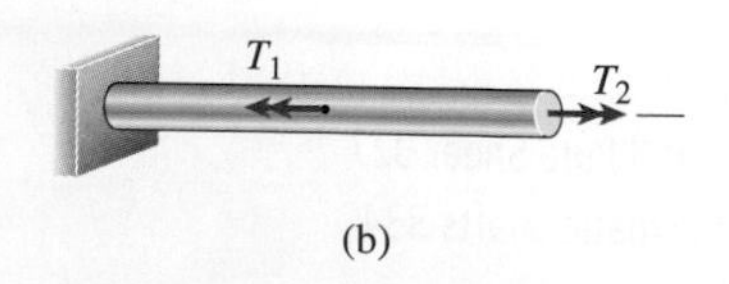

(b)

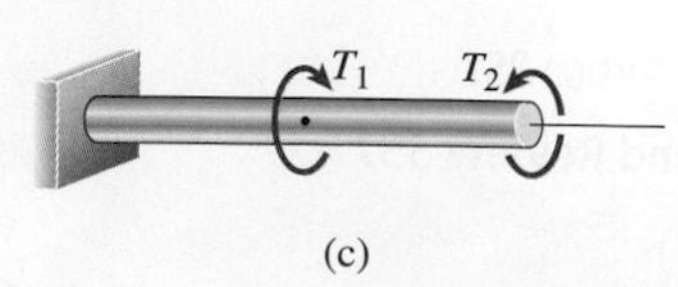

(c)

Moments that produce the twisting of a bar, such as the moments T_1 and T_2 in Fig. 3-2, are called **torques** or **twisting moments**. Cylindrical members that are subjected to torques and transmit power through rotation are called **shafts**; for instance, the drive shaft of an automobile or the propeller shaft of a ship. Most shafts have circular cross sections, either solid or tubular.

This chapter begins by developing formulas for the deformations and stresses in circular bars subjected to torsion. Then analysis of the state of stress known as *pure shear* and the relationship between the moduli of elasticity E and G in tension and shear, respectively, are presented. Next, rotating shafts are analyzed to determine the power they transmit. Finally, several additional topics related to torsion are covered, namely, statically indeterminate members, noncircular prismatic shafts, and stress concentrations.

3.2 Torsional Deformations of a Circular Bar

First consider a prismatic bar of circular cross section twisted by torques T acting at the ends (Fig. 3-3a). Since every cross section of the bar is identical, and since every cross section is subjected to the same internal torque T, the bar is in **pure torsion**. From considerations of symmetry, it can be proved that cross sections of the bar do not change in shape as they rotate about the longitudinal axis. In other words, all cross sections remain plane and circular, and all radii remain straight. Furthermore, if the angle of rotation between one end of the bar and the other is small, neither the length of the bar nor its radius will change.

To aid in visualizing the deformation of the bar, imagine that the left-hand end of the bar (Fig. 3-3a) is fixed in position. Then, under the action of the torque T, the right-hand end will rotate (with respect to the left-hand end) through a small angle ϕ, known as the **angle of twist** (or *angle of rotation*). Because of this rotation, a straight longitudinal line pq on the surface of the bar will become a helical curve pq', where q' is the position of point q after the end cross section has rotated through the angle ϕ (Fig. 3-3b).

The angle of twist changes along the axis of the bar and, at intermediate cross sections, will have a value $\phi(x)$ that is between zero at the left-hand end and ϕ at the right-hand end. If every cross section of the bar has the same radius and is subjected to the same torque (pure torsion), the angle $\phi(x)$ will vary linearly between the ends.

Shear Strains at the Outer Surface

Now consider an element of the bar between two cross sections distance dx apart (see Fig. 3-4a). This element is shown enlarged in Fig. 3-4b. On its outer surface, a small element $abcd$ is identified with sides ab and cd that initially are parallel to the longitudinal axis. During twisting of the bar, the right-hand cross section rotates with respect to the left-hand cross section through a small angle of twist $d\phi$, so that points b and c move to b' and c', respectively. The lengths of the sides of the element, which is now element $ab'c'd$, do not change during this small rotation.

However, the angles at the corners of the element (Fig. 3-4b) are no longer equal to 90°. The element is therefore in a state of **pure shear**, which means that the element is subjected to shear strains but no normal strains (see Fig. 1-53 of Section 1.8). The magnitude of the shear strain at the outer surface of the bar, denoted γ_{max}, is equal to the decrease in the angle at point a, that is, the decrease in angle bad. From Fig. 3-4b, the decrease in this angle is

$$\gamma_{max} = \frac{bb'}{ab} \tag{3-1}$$

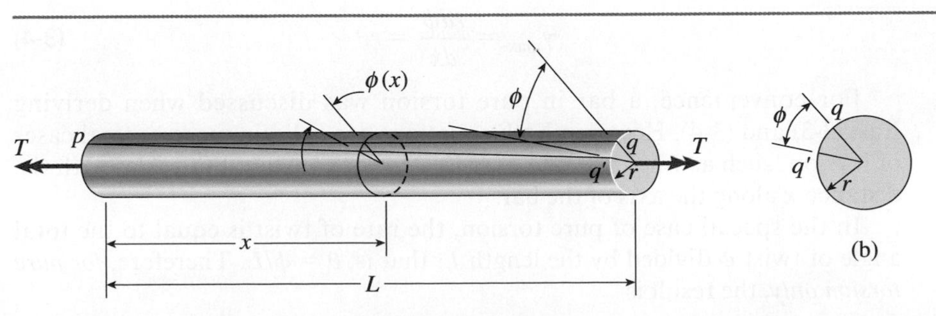

FIGURE 3-3
Deformations of a circular bar in pure torsion

FIGURE 3-4

Deformation of an element of length dx cut from a bar in torsion

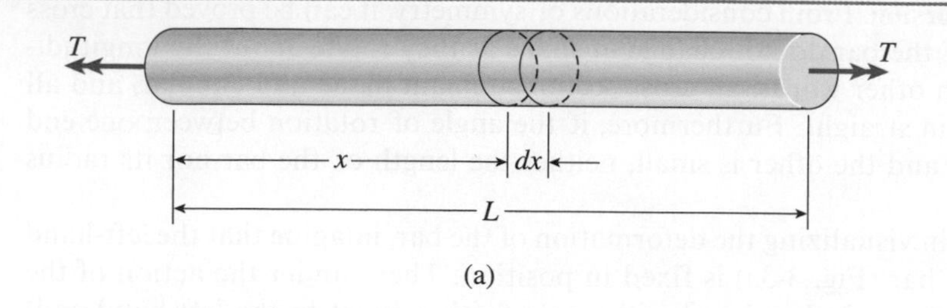

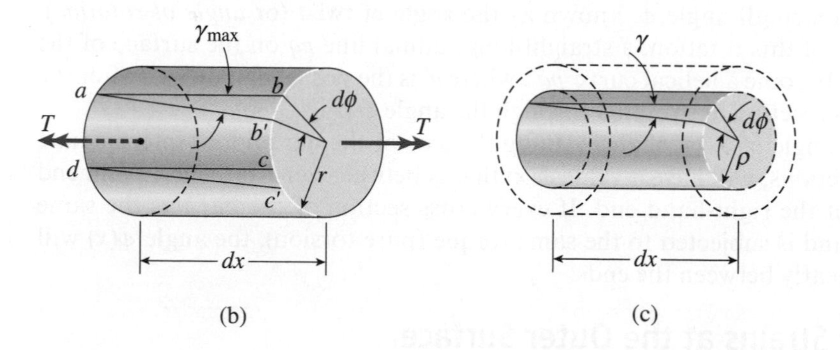

where γ_{max} is measured in radians, bb' is the distance through which point b moves, and ab is the length of the element (equal to dx). With r denoting the radius of the bar, express the distance bb' as $rd\phi$, where $d\phi$ also is measured in radians. Thus, the preceding equation becomes

$$\gamma_{max} = \frac{rd\phi}{dx} \tag{3-2}$$

This equation relates the shear strain at the outer surface of the bar to the angle of twist.

The quantity $d\phi/dx$ is the rate of change of the angle of twist ϕ with respect to the distance x measured along the axis of the bar. Denote $d\phi/dx$ by the symbol θ and refer to it as the **rate of twist**, or the **angle of twist per unit length**:

$$\theta = \frac{d\phi}{dx} \tag{3-3}$$

With this notation, now write the equation for the shear strain at the outer surface [Eq. (3-2)] as

$$\gamma_{max} = \frac{rd\phi}{dx} = r\theta \tag{3-4}$$

For convenience, a bar in pure torsion was discussed when deriving Eqs. (3-3) and (3-4). However, both equations are valid in more general cases of torsion, such as when the rate of twist θ is not constant but varies with the distance x along the axis of the bar.

In the special case of pure torsion, the rate of twist is equal to the total angle of twist ϕ divided by the length L, that is, $\theta = \phi/L$. Therefore, *for pure torsion only*, the result is

$$\gamma_{max} = r\theta = \frac{r\phi}{L} \tag{3-5}$$

This equation can be obtained directly from the geometry of Fig. 3-3a by noting that γ_{max} is the angle between lines pq and pq', that is, γ_{max} is the angle qpq'. Therefore, $\gamma_{max}L$ is equal to the distance qq' at the end of the bar. But since the distance qq' also equals $r\phi$ (Fig. 3-3b), the relation $r\phi = \gamma_{max}L$ is produced, which agrees with Eq. (3-5).

Shear Strains within the Bar

The shear strains within the interior of the bar can be found by the same method used to find the shear strain γ_{max} at the surface. Because radii in the cross sections of a bar remain straight and undistorted during twisting, the preceding discussion for an element *abcd* at the outer surface (Fig. 3-4b) will also hold for a similar element situated on the surface of an interior cylinder of radius ρ (Fig. 3-4c). Thus, interior elements are also in pure shear with the corresponding shear strains given by the equation [compare with Eq. (3-4)]:

$$\gamma = \rho\theta = \frac{\rho}{r}\gamma_{max} \qquad \textbf{(3-6)}$$

This equation shows that the shear strains in a circular bar vary linearly with the radial distance ρ from the center with the strain being zero at the center and reaching a maximum value γ_{max} at the outer surface.

Circular Tubes

A review of the preceding discussions will show that the equations for the shear strains [Eqs. (3-2) to (3-4)] apply to **circular tubes** (Fig. 3-5) as well as to solid circular bars. Figure 3-5 shows the linear variation in shear strain between the maximum strain at the outer surface and the minimum strain at the interior surface. The equations for these strains are

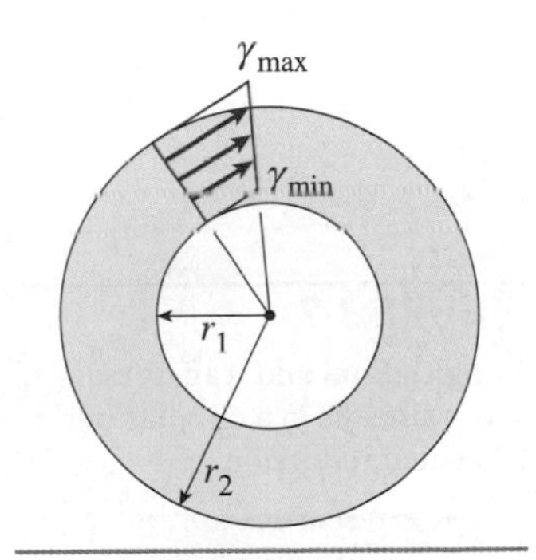

FIGURE 3-5

Shear strains in a circular tube

$$\gamma_{max} = \frac{r_2\phi}{L} \qquad \gamma_{min} = \frac{r_1}{r_2}\gamma_{max} = \frac{r_1\phi}{L} \qquad \textbf{(3-7a,b)}$$

in which r_1 and r_2 are the inner and outer radii, respectively, of the tube.

All of the preceding equations for the strains in a circular bar are based upon geometric concepts and do not involve the material properties. Therefore, the equations are valid for any material, whether it behaves elastically or inelastically, linearly or nonlinearly. However, the equations are limited to bars having small angles of twist and small strains.

3.3 Circular Bars of Linearly Elastic Materials

Now that the shear strains in a circular bar in torsion have been investigated (see Figs. 3-3 to 3-5), it is time to determine the directions and magnitudes of the corresponding shear stresses. The directions of the stresses can be determined by inspection, as illustrated in Fig. 3-6a. Observe that the torque T tends to rotate the right-hand end of the bar counterclockwise when viewed from the right. Therefore, the shear stresses τ acting on a stress element located on the surface of the bar will have the directions shown in the figure.

For clarity, the stress element shown in Fig. 3-6a is enlarged in Fig. 3-6b, where both the shear strain and the shear stresses are shown. As explained previously in Section 2.6, stress elements are customarily drawn in two

FIGURE 3-6

Shear stresses in a circular bar in torsion

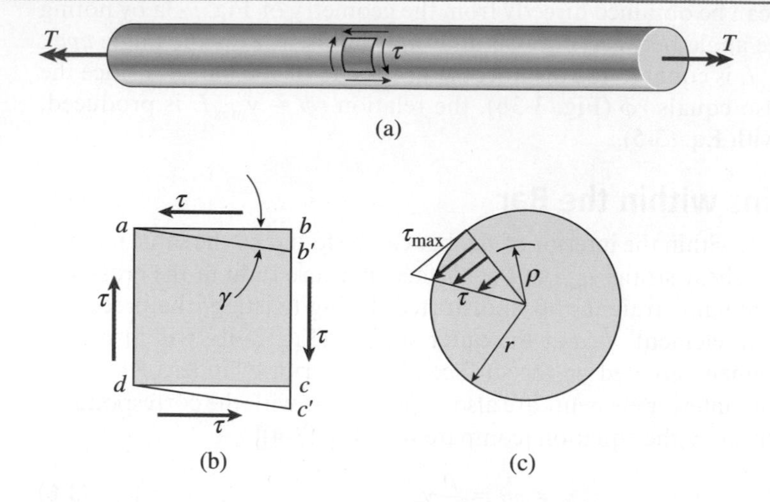

dimensions, as in Fig. 3-6b. But always remember that stress elements are actually three-dimensional objects with a thickness perpendicular to the plane of the figure.

The magnitudes of the shear stresses can be determined from the strains by using the stress-strain relation for the material of the bar. If the material is linearly elastic, use **Hooke's law in shear** [Eq. (1-18)]:

$$\tau = G\gamma \tag{3-8}$$

in which G is the shear modulus of elasticity and γ is the shear strain in radians. Combine this equation with the equations for the shear strains [Eqs. (3-2) and (3-4)] to get

$$\tau_{max} = Gr\theta \qquad \tau = G\rho\theta = \frac{\rho}{r}\tau_{max} \qquad \textbf{(3-9a,b)}$$

in which τ_{max} is the shear stress at the outer surface of the bar (radius r), τ is the shear stress at an interior point (radius ρ), and θ is the rate of twist. (In these equations, θ has units of radians per unit of length.)

Equations (3-9a and b) show that the shear stresses vary linearly with the distance from the center of the bar, as illustrated by the triangular stress diagram in Fig. 3-6c. This linear variation of stress is a consequence of Hooke's law. If the stress-strain relation is nonlinear, the stresses will vary nonlinearly, and other methods of analysis will be needed.

FIGURE 3-7

Longitudinal and transverse shear stresses in a circular bar subjected to torsion

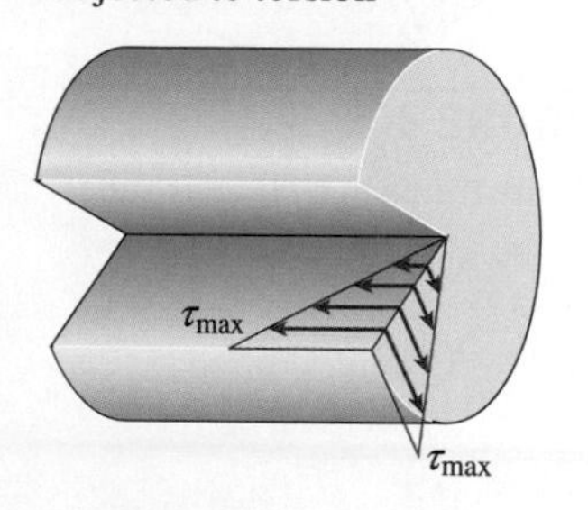

The shear stresses acting on a cross-sectional plane are accompanied by shear stresses of the same magnitude acting on longitudinal planes (Fig. 3-7). This is because equal shear stresses always exist on mutually perpendicular planes, as explained in Section 1.8. If the material of the bar is weaker in shear on longitudinal planes than on cross-sectional planes, as is typical of wood when the grain runs parallel to the axis of the bar, the first cracks due to torsion will appear on the surface in the longitudinal direction.

FIGURE 3-8

Tensile and compressive stresses acting on a stress element oriented at 45° to the longitudinal axis

The state of pure shear at the surface of a bar (Fig. 3-6b) is equivalent to equal tensile and compressive stresses acting on an element oriented at an angle of 45°, as explained later in Section 3.5. Therefore, a rectangular element with sides at 45° to the axis of the shaft will be subjected to tensile and compressive stresses, as shown in Fig. 3-8. If a torsion bar is made of a material that is weaker in tension than in shear, failure will occur in tension along a helix inclined at 45° to the axis, as you can demonstrate by twisting a piece of classroom chalk.

The Torsion Formula

The next step in this analysis is to determine the relationship between the shear stresses and the torque T. Once this is accomplished, you will be able to calculate the stresses and strains in a bar due to any set of applied torques.

The distribution of the shear stresses acting on a cross section is pictured in Figs. 3-6c and 3-7. Because these stresses act continuously around the cross section, they have a resultant in the form of a moment—a moment equal to the torque T acting on the bar. To determine this resultant, consider an element of area dA located at radial distance ρ from the axis of the bar (Fig. 3-9). The shear force acting on this element is equal to $\tau\, dA$, where τ is the shear stress at radius ρ. The moment of this force about the axis of the bar is equal to the force times its distance from the center, or $\tau\rho dA$. Substitute for the shear stress τ from Eq. (3-9b) to express this elemental moment as

FIGURE 3-9

Determination of the resultant of the shear stresses acting on a cross section

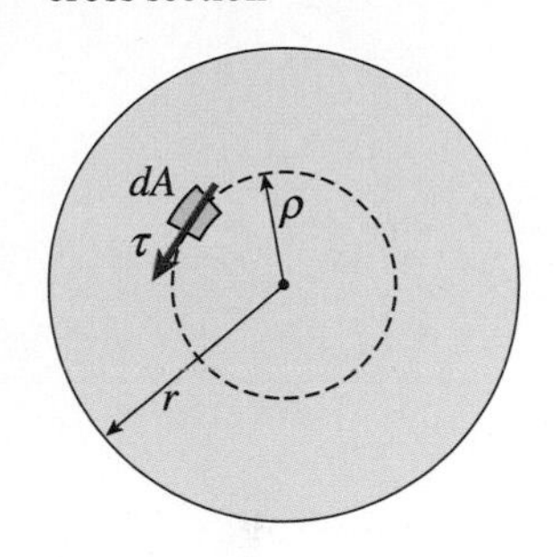

$$dM = \tau\rho dA = \frac{\tau_{\max}}{r}\rho^2 dA$$

The resultant moment (equal to the torque T) is the summation over the entire cross-sectional area of all such elemental moments:

$$T = \int_A dM = \frac{\tau_{\max}}{r}\int_A \rho^2 dA = \frac{\tau_{\max}}{r} I_p \quad \textbf{(3-10)}$$

in which

$$I_p = \int_A \rho^2 dA \quad \textbf{(3-11)}$$

is the **polar moment of inertia** of the circular cross section.

For a **circle** of radius r and diameter d, the polar moment of inertia is

$$I_p = \frac{\pi r^4}{2} = \frac{\pi d^4}{32} \quad \textbf{(3-12)}$$

as given in Appendix E, Case 9. Note that moments of inertia have units of length to the fourth power.[1]

An expression for the maximum shear stress is obtained by rearranging Eq. (3-10) as

$$\tau_{\max} = \frac{Tr}{I_p} \quad \textbf{(3-13)}$$

This equation, known as the **torsion formula**, shows that the maximum shear stress is proportional to the applied torque T and inversely proportional to the polar moment of inertia I_p.

Typical **units** used with the torsion formula are as follows. In SI, the torque T is usually expressed in newton meters (N·m), the radius r in meters (m), the

[1]Polar moments of inertia are discussed in Section D.5 of Appendix D.

polar moment of inertia I_p in meters to the fourth power (m^4), and the shear stress τ in pascals (Pa).

Substitute $r = d/2$ and $I_p = \pi d^4/32$ into the torsion formula to get the following equation for the maximum stress:

$$\tau_{max} = \frac{16T}{\pi d^3} \tag{3-14}$$

This equation applies only to bars of *solid circular cross section*, whereas the torsion formula itself [Eq. (3-13)] applies to both solid bars and circular tubes, as explained later. Equation (3-14) shows that the shear stress is inversely proportional to the cube of the diameter. Thus, if the diameter is doubled, the stress is reduced by a factor of eight.

The shear stress at distance ρ from the center of the bar is

$$\tau = \frac{\rho}{r}\tau_{max} = \frac{T\rho}{I_p} \tag{3-15}$$

which is obtained by combining Eq. (3-9b) with the torsion formula [Eq. (3-13)]. Equation (3-15) is a *generalized torsion formula*, and once again, the shear stresses vary linearly with the radial distance from the center of the bar.

Angle of Twist

The angle of twist of a bar of linearly elastic material now can be related to the applied torque T. Combine Eq. (3-9a) with the torsion formula to get

$$\theta = \frac{T}{GI_p} \tag{3-16}$$

in which θ has units of radians per unit of length. This equation shows that the rate of twist θ is directly proportional to the torque T and inversely proportional to the product GI_p, which is known as the **torsional rigidity** of the bar.

For a bar in **pure torsion**, the total angle of twist ϕ, which is equal to the rate of twist times the length of the bar (that is, $\phi = \theta L$), is

$$\phi = \frac{TL}{GI_p} \tag{3-17}$$

where ϕ is measured in radians. The use of the preceding equations in both analysis and design is illustrated later in Examples 3-1 and 3-2.

The quantity GI_p/L, called the **torsional stiffness** of the bar, is the torque required to produce a unit angle of rotation. The **torsional flexibility** is the reciprocal of the stiffness, or L/GI_p, and is defined as the angle of rotation produced by a unit torque. Thus, the expressions for torsional stiffness and flexibility are

$$k_T = \frac{GI_p}{L} \qquad f_T = \frac{L}{GI_p} \tag{3-18a,b}$$

These quantities are analogous to the axial stiffness $k = EA/L$ and axial flexibility $f = L/EA$ of a bar in tension or compression [compare with

Eqs. (2-4a and b)]. Stiffnesses and flexibilities have important roles in structural analysis.

The equation for the angle of twist [Eq. (3-17)] provides a convenient way to determine the shear modulus of elasticity G for a material. Conduct a torsion test on a circular bar and measure the angle of twist ϕ produced by a known torque T. Then the value of G can be calculated from Eq. (3-17).

Circular Tubes

Circular tubes are more efficient than solid bars in resisting torsional loads. The shear stresses in a solid circular bar are maximum at the outer boundary of the cross section and zero at the center. Therefore, most of the material in a solid shaft is stressed significantly below the maximum shear stress. Furthermore, the stresses near the center of the cross section have a smaller moment arm ρ for use in determining the torque [see Fig. 3-9 and Eq. (3-10)].

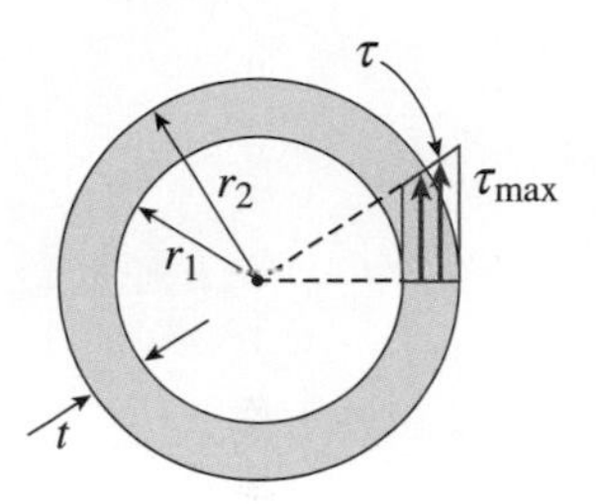

FIGURE 3-10

Circular tube in torsion

By contrast, in a typical hollow tube, most of the material is near the outer boundary of the cross section where both the shear stresses and the moment arms are highest (Fig. 3-10). Thus, if weight reduction and savings of material are important, it is advisable to use a circular tube. For instance, large drive shafts, propeller shafts, and generator shafts usually have hollow circular cross sections.

The analysis of the torsion of a circular tube is almost identical to that for a solid bar. The same basic expressions for the shear stresses may be used [for instance, Eqs. (3-9a and b)]. Of course, the radial distance ρ is limited to the range r_1 to r_2, where r_1 is the inner radius and r_2 is the outer radius of the bar (Fig. 3-10).

The relationship between the torque T and the maximum stress is given by Eq. (3-10), but the limits on the integral for the polar moment of inertia [Eq. (3-11)] are $\rho = r_1$ and $\rho = r_2$. Therefore, the polar moment of inertia of the cross-sectional area of a tube is

$$I_p = \frac{\pi}{2}(r_2^4 - r_1^4) = \frac{\pi}{32}(d_2^4 - d_1^4) \tag{3-19}$$

The preceding expressions also can be written in the forms:

$$I_p = \frac{\pi r t}{2}(4r^2 + t^2) = \frac{\pi d t}{4}(d^2 + t^2) \tag{3-20}$$

where r is the *average radius* of the tube equal to $(r_1 + r_2)/2$; d is the *average diameter* equal to $(d_1 + d_2)/2$; and t is the *wall thickness* (Fig. 3-10) equal to $r_2 - r_1$. Of course, Eqs. (3-19) and (3-20) give the same results, but sometimes the latter is more convenient.

If the tube is relatively thin so that the wall thickness t is small compared to the average radius r, disregard the terms t^2 in Eq. (3-20). This simplification leads to the following *approximate formulas* for the polar moment of inertia:

$$I_p \approx 2\pi r^3 t = \frac{\pi d^3 t}{4} \tag{3-21}$$

These expressions are given in Case 22 of Appendix E.

Reminders: In Eqs. 3-20 and 3-21, the quantities r and d are the average radius and diameter, not the maximums. Also, Eqs. 3-19 and 3-20 are exact; Eq. 3-21 is approximate.

The torsion formula [Eq. (3-13)] may be used for a circular tube of linearly elastic material provided I_p is evaluated according to Eq. (3-19), Eq. (3-20), or, if appropriate, Eq. (3-21). The same comment applies to the general equation for shear stress [Eq. (3-15)], the equations for rate of twist and angle of twist [Eqs. (3-16) and Eq. (3-17)], and the equations for stiffness and flexibility [Eqs. (3-18a and b)].

The shear stress distribution in a tube is pictured in Fig. 3-10. From the figure, note that the average stress in a thin tube is nearly as great as the maximum stress. This means that a hollow bar is more efficient in the use of material than is a solid bar, as explained previously and as demonstrated later in Examples 3-2 and 3-3.

When designing a circular tube to transmit a torque, be sure that the thickness t is large enough to prevent wrinkling or buckling of the wall of the tube. For instance, a maximum value of the radius to thickness ratio, such as $(r_2/t)_{max} = 12$, may be specified. Other design considerations include environmental and durability factors, which also may impose requirements for minimum wall thickness. These topics are discussed in courses and textbooks on mechanical design.

Limitations

The equations derived in this section are limited to bars of circular cross section (either solid or hollow) that behave in a linearly elastic manner. In other words, the loads must be such that the stresses do not exceed the proportional limit of the material. Furthermore, the equations for stresses are valid only in parts of the bars away from stress concentrations (such as holes and other abrupt changes in shape) and away from cross sections where loads are applied. (Stress concentrations in torsion are discussed later in Section 3.12.)

Finally, it is important to emphasize that the equations for the torsion of circular bars and tubes cannot be used for bars of other shapes. Noncircular bars, such as rectangular bars and bars having I-shaped cross sections, behave quite differently than do circular bars. For instance, their cross sections do *not* remain plane, and their maximum stresses are *not* located at the farthest distances from the midpoints of the cross sections. Thus, these bars require more advanced methods of analysis, such as those presented in books on theory of elasticity and advanced mechanics of materials.[2] (A brief overview of torsion of noncircular prismatic shafts is presented in Section 3.10.)

[2]The torsion theory for circular bars originated with the work of the famous French scientist C. A. de Coulomb (1736–1806); further developments were due to Thomas Young and A. Duleau (Ref. 3-1). The general theory of torsion (for bars of any shape) is due to the most famous elastician of all time, Barré de Saint-Venant (1797–1886); see (Ref. 2-10).

Example 3-1

A solid steel bar of circular cross section (Fig. 3-11) has a diameter $d = 40$ mm, length $L = 1.3$ m, and shear modulus of elasticity $G = 80$ GPa. The bar is subjected to torques T acting at the ends.

(a) If the torques have a magnitude $T = 340$ N·m, what is the maximum shear stress in the bar? What is the angle of twist between the ends?

(b) If the allowable shear stress is 42 MPa and the allowable angle of twist is 2.5°, what is the maximum permissible torque?

FIGURE 3-11

Example 3-1: Bar in pure torsion

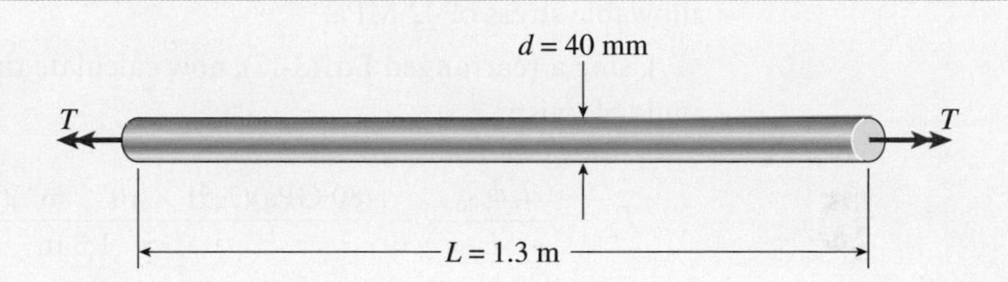

Solution:

Use a four-step problem-solving approach. Combine steps as needed for an efficient solution.

Part (a): Maximum shear stress and angle of twist.

1, 2. ***Conceptualize*** *[hypothesize, sketch]*, ***Categorize*** *[simplify, classify]*: Because the bar has a solid circular cross section, compute the maximum shear stress from Eq. (3-14) as

$$\tau_{max} = \frac{16T}{\pi d^3} = \frac{16(340\ \text{N}\cdot\text{m})}{\pi(0.04\ \text{m})^3} = 27.1\ \text{MPa}$$

In a similar manner, the angle of twist is obtained from Eq. (3-17) with the polar moment of inertia given by Eq. (3-12).

3. ***Analyze*** *[evaluate; select relevant equations, carry out mathematical solution]*:

$$I_p = \frac{\pi d^4}{32} = \frac{\pi(0.04\ \text{m})^4}{32} = 2.51 \times 10^{-7}\ \text{m}^4$$

$$\phi = \frac{TL}{GI_p} = \frac{(340\ \text{N}\cdot\text{m})(1.3\ \text{m})}{(80\ \text{GPa})(2.51 \times 10^{-7}\ \text{m}^4)} = 0.02198\ \text{rad} = 1.26°$$

4. ***Finalize*** *[conclude; examine answer—Does it make sense? Are units correct? How does it compare to similar problem solutions?]*: The formulas used here apply to bars of circular cross section only. The twist distortion of the bar is very small as expected.

Part (b): Maximum permissible torque.

1, 2. ***Conceptualize, Categorize*:** The maximum permissible torque is determined either by the allowable shear stress or by the allowable angle of twist.

3. ***Analyze*:** Beginning with the shear stress, rearrange Eq. (3-14) and calculate as

$$T_1 = \frac{\pi d^3 \tau_{\text{allow}}}{16} = \frac{\pi}{16}(0.04 \text{ m})^3(42 \text{ MPa}) = 528 \text{ N}\cdot\text{m}$$

Any torque larger than this value will result in a shear stress that exceeds the allowable stress of 42 MPa.

Using a rearranged Eq. (3-17), now calculate the torque based upon the angle of twist:

$$T_2 = \frac{GI_p\phi_{\text{allow}}}{L} = \frac{(80 \text{ GPa})(2.51 \times 10^{-7} \text{ m}^4)(2.5°)(\pi \text{rad}/180°)}{1.3 \text{ m}}$$
$$= 674 \text{ N}\cdot\text{m}$$

4. ***Finalize*:** Any torque larger than T_2 will exceed the allowable angle of twist. The maximum permissible torque is the smaller of T_1 and T_2:

$$T_{\text{max}} = 528 \text{ N}\cdot\text{m}$$

In this example, the allowable shear stress provides the limiting condition.

Example 3-2

A steel shaft is to be manufactured either as a solid circular bar or as a circular tube (Fig. 3-12). The shaft is required to transmit a torque of 1200 N·m without exceeding an allowable shear stress of 40 MPa nor an allowable rate of twist of 0.75°/m. (The shear modulus of elasticity of the steel is 78 GPa.)

(a) Determine the required diameter d_0 of the solid shaft.

(b) Determine the required outer diameter d_2 of the hollow shaft if the thickness t of the shaft is specified as one-tenth of the outer diameter.

FIGURE 3-12

Example 3-2: Torsion of a steel shaft

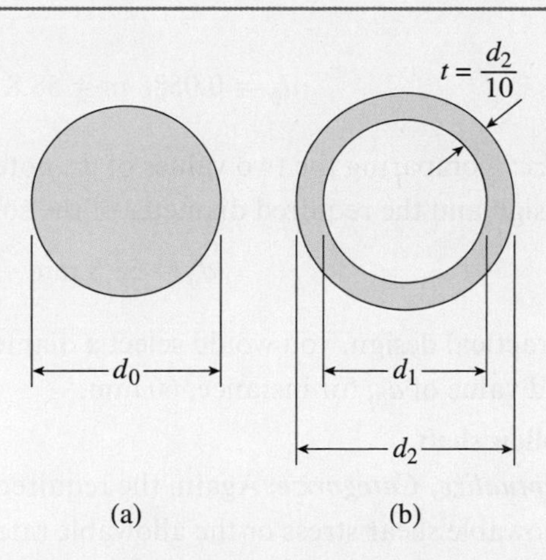

(c) Determine the ratio of diameters (that is, the ratio d_2/d_0) and the ratio of weights of the hollow and solid shafts.

Solution:

A four-step problem-solving approach is used in the solution; only essential steps are shown in each part.

Part (a): Solid shaft.

1, 2. ***Conceptualize, Categorize:*** The required diameter d_0 is determined either from the allowable shear stress or from the allowable rate of twist.

3. ***Analyze:*** In the case of the allowable shear stress, rearrange Eq. (3-14) and obtain

$$d_0^3 = \frac{16T}{\pi\tau_{\text{allow}}} = \frac{16(1200 \text{ N}\cdot\text{m})}{\pi(40 \text{ MPa})} = 152.8 \times 10^{-6} \text{ m}^3$$

so the required diameter is

$$d_0 = 0.0535 \text{ m} = 53.5 \text{ mm}$$

In the case of the allowable rate of twist, start by finding the required polar moment of inertia [see Eq. (3-16)]:

$$I_p = \frac{T}{G\theta_{\text{allow}}} = \frac{1200 \text{ N}\cdot\text{m}}{(78 \text{ GPa})(0.75°/\text{m})(\pi \text{rad}/180°)} = 1175 \times 10^{-9} \text{ m}^4$$

Since the polar moment of inertia is equal to $\pi d^4/32$, the required diameter is

$$d_0^4 = \frac{32I_p}{\pi} = \frac{32(1175 \times 10^{-9} \text{ m}^4)}{\pi} = 11.97 \times 10^{-6} \text{ m}^4$$

or

$$d_0 = 0.0588 \text{ m} = 58.8 \text{ mm}$$

4. ***Finalize***: Comparing the two values of d_0, note that the rate of twist governs the design and the required diameter of the solid shaft is

$$d_0 = 58.8 \text{ mm}$$

In a practical design, you would select a diameter slightly larger than the calculated value of d_0; for instance, 60 mm.

Part (b): Hollow shaft.

1, 2. ***Conceptualize, Categorize***: Again, the required diameter is based upon either the allowable shear stress or the allowable rate of twist.

3. ***Analyze***: Begin by noting that the outer diameter of the bar is d_2 and the inner diameter is

$$d_1 = d_2 - 2t = d_2 - 2(0.1d_2) = 0.8d_2$$

Thus, the polar moment of inertia [Eq. (3-19)] is

$$I_p = \frac{\pi}{32}(d_2^4 - d_1^4) = \frac{\pi}{32}\left[d_2^4 - (0.8d_2)^4\right] = \frac{\pi}{32}(0.5904d_2^4) = 0.05796d_2^4$$

In the case of the allowable shear stress, use the torsion formula [Eq. (3-13)] as

$$\tau_{\text{allow}} = \frac{Tr}{I_p} = \frac{T(d_2/2)}{0.05796d_2^4} = \frac{T}{0.1159d_2^3}$$

Rearrange to get

$$d_2^3 = \frac{T}{0.1159\tau_{\text{allow}}} = \frac{1200 \text{ N} \cdot \text{m}}{0.1159(40 \text{ MPa})} = 258.8 \times 10^{-6} \text{ m}^3$$

Solving for d_2 gives

$$d_2 = 0.0637 \text{ m} = 63.7 \text{ mm}$$

which is the required outer diameter based upon the shear stress.

In the case of the allowable rate of twist, use Eq. (3-16) with θ replaced by θ_{allow} and I_p replaced by the previously obtained expression; thus,

$$\theta_{\text{allow}} = \frac{T}{G(0.05796d_2^4)}$$

from which

$$d_2^4 = \frac{T}{0.05796G\theta_{allow}}$$

$$= \frac{1200 \text{ N} \cdot \text{m}}{0.05796(78 \text{ GPa})(0.75°/\text{m})(\pi \text{rad}/180°)} = 20.28 \times 10^{-6} \text{ m}^4$$

Solving for d_2 gives

$$d_2 = 0.0671 \text{ m} = 67.1 \text{ mm}$$

which is the required diameter based upon the rate of twist.

4. ***Finalize***: Compare the two values of d_2 and note that the rate of twist governs the design. The required outer diameter of the hollow shaft is

$$d_2 = 67.1 \text{ mm}$$

The inner diameter d_1 is equal to $0.8d_2$, or 53.7 mm. (As practical values, you might select $d_2 = 70$ mm and $d_1 = 0.8d_2 = 56$ mm.)

Part (c): Ratios of diameters and weights.

1, 2. ***Conceptualize, Categorize***: The ratio of the outer diameter of the hollow shaft to the diameter of the solid shaft (using the calculated values) is

$$\frac{d_2}{d_0} = \frac{67.1 \text{ mm}}{58.8 \text{ mm}} = 1.14$$

3. ***Analyze***: Since the weights of the shafts are proportional to their cross-sectional areas, express the ratio of the weight of the hollow shaft to the weight of the solid shaft as

$$\frac{W_H}{W_S} = \frac{A_H}{A_S} = \frac{\pi(d_2^2 - d_1^2)/4}{\pi d_0^2/4} = \frac{d_2^2 - d_1^2}{d_0^2}$$

$$= \frac{(67.1 \text{ mm})^2 - (53.7 \text{ mm})^2}{(58.8 \text{ mm})^2} = 0.47$$

4. ***Finalize***: These results show that the hollow shaft uses only 47% as much material as does the solid shaft, while its outer diameter is only 14% larger.

Note: This example illustrates how to determine the required sizes of both solid bars and circular tubes when allowable stresses and allowable rates of twist are known. It also illustrates the fact that circular tubes are more efficient in the use of materials than are solid circular bars.

Example 3-3

A hollow shaft and a solid shaft constructed of the same material have the same length and the same outer radius R (Fig. 3-13). The inner radius of the hollow shaft is $0.6R$.

(a) Assuming that both shafts are subjected to the same torque, compare their shear stresses, angles of twist, and weights.

(b) Determine the strength-to-weight ratios for both shafts.

FIGURE 3-13

Example 3-3: Comparison of hollow and solid shafts

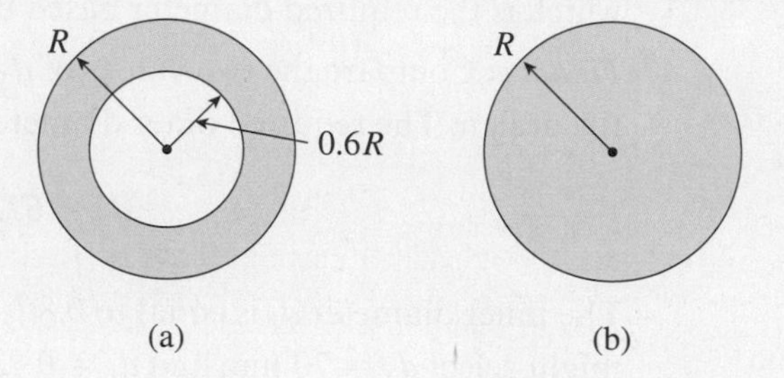

Solution:

Use a four-step problem-solving approach. Combine steps as needed for an efficient solution.

Part (a): Comparison of shear stresses.

2, 3. ***Categorize, Analyze***: The maximum shear stresses given by the torsion formula [Eq. (3-13)] are proportional to $1/I_p$ inasmuch as the torques and radii are the same. For the hollow shaft,

$$I_p = \frac{\pi R^4}{2} - \frac{\pi(0.6R)^4}{2} = 0.4352\pi R^4$$

and for the solid shaft,

$$I_p = \frac{\pi R^4}{2} = 0.5\pi R^4$$

Therefore, the ratio β_1 of the maximum shear stress in the hollow shaft to that in the solid shaft is

$$\beta_1 = \frac{\tau_H}{\tau_S} = \frac{0.5\pi R^4}{0.4352\pi R^4} = 1.15$$

where the subcripts H and S refer to the hollow shaft and the solid shaft, respectively.

Comparison of angles of twist.

2, 3. ***Categorize, Analyze***: The angles of twist [Eq. (3-17)] are also proportional to $1/I_p$ because the torque T, length L, and shearing modulus G are the same for both shafts. Therefore, the ratio of angles is the same as for the shear stresses:

$$\beta_2 = \frac{\phi_H}{\phi_S} = \frac{0.5\pi R^4}{0.4352\pi R^4} = 1.15$$

Comparison of weights.

2, 3. ***Categorize, Analyze*:** The weights of the shafts are proportional to their cross-sectional areas; consequently, the weight of the solid shaft is proportional to πR^2 and the weight of the hollow shaft is proportional to

$$\pi R^2 - \pi(0.6R)^2 = 0.64\pi R^2$$

Therefore, the ratio of the weight of the hollow shaft to the weight of the solid shaft is

$$\beta_3 = \frac{W_H}{W_S} = \frac{0.64\pi R^2}{\pi R^2} = 0.64$$

4. ***Finalize*:** From the preceding ratios, note again the inherent advantage of hollow shafts. In this example, the hollow shaft has 15% greater stress and 15% greater angle of rotation than the solid shaft but 36% less weight.

Part (b): Strength-to-weight ratios.

2. ***Categorize*:** The relative efficiency of a structure is sometimes measured by its *strength-to-weight ratio*, which is defined for a bar in torsion as the allowable torque divided by the weight.

3. ***Analyze*:** The allowable torque for the hollow shaft of Fig. 3-13a (from the torsion formula) is

$$T_H = \frac{\tau_{max} I_p}{R} = \frac{\tau_{max}(0.4352\pi R^4)}{R} = 0.4352\pi R^3 \tau_{max}$$

and for the solid shaft is

$$T_S = \frac{\tau_{max} I_p}{R} = \frac{\tau_{max}(0.5\pi R^4)}{R} = 0.5\pi R^3 \tau_{max}$$

The weights of the shafts are equal to the cross-sectional areas times the length L times the weight density γ of the material:

$$W_H = 0.64\pi R^2 L\gamma \qquad W_S = \pi R^2 L\gamma$$

Thus, the strength-to-weight ratios S_H and S_S for the hollow and solid bars, respectively, are

$$S_H = \frac{T_H}{W_H} = 0.68\frac{\tau_{max} R}{\gamma L} \qquad S_S = \frac{T_S}{W_S} = 0.5\frac{\tau_{max} R}{\gamma L}$$

4. ***Finalize*:** In this example, the strength-to-weight ratio of the hollow shaft is 36% greater than the strength-to-weight ratio for the solid shaft, demonstrating once again the relative efficiency of hollow shafts. For a thinner shaft, the percentage will increase; for a thicker shaft, it will decrease.

3.4 Nonuniform Torsion

As explained in Section 3.2, *pure torsion* refers to torsion of a prismatic bar subjected to torques acting only at the ends. **Nonuniform torsion** differs from pure torsion in that the bar need not be prismatic, and the applied torques may act anywhere along the axis of the bar. Bars in nonuniform torsion can be analyzed by applying the formulas of pure torsion to finite segments of the bar and then adding the results or by applying the formulas to differential elements of the bar and then integrating.

Consider three cases of nonuniform torsion. Other cases can be handled by techniques similar to those described here.

Case 1. *Bar consisting of prismatic segments with constant torque throughout each segment* (Fig. 3-14). The bar shown in part (a) of the figure has two different diameters and is loaded by torques acting at points A, B, C, and D. Consequently, divide the bar into segments in such a way that each segment is prismatic and subjected to a constant torque. In this example, there are three such segments, AB, BC, and CD. Each segment is in pure torsion; therefore, all of the formulas derived in the preceding section may be applied to each part separately.

FIGURE 3-14

Bar in nonuniform torsion (Case 1)

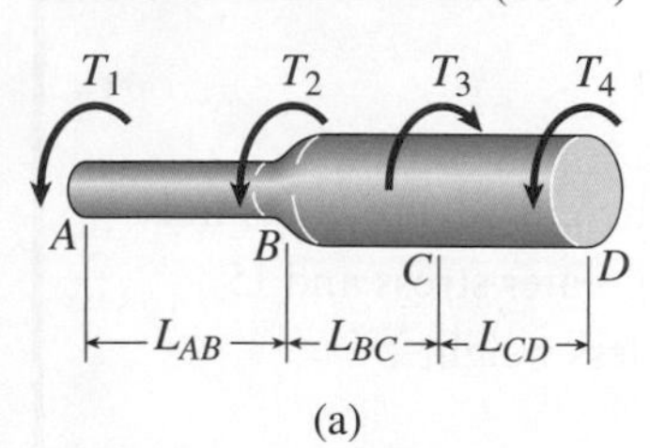

(a)

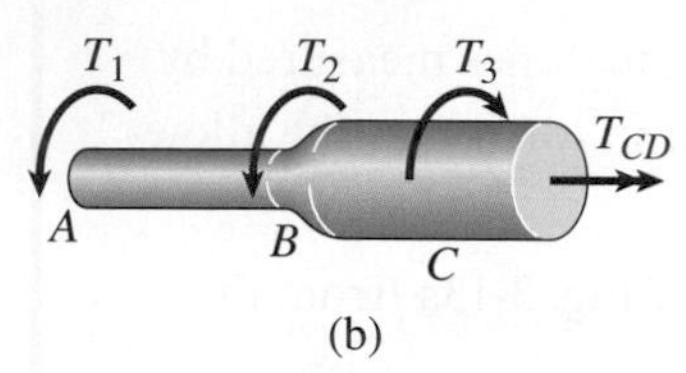

(b)

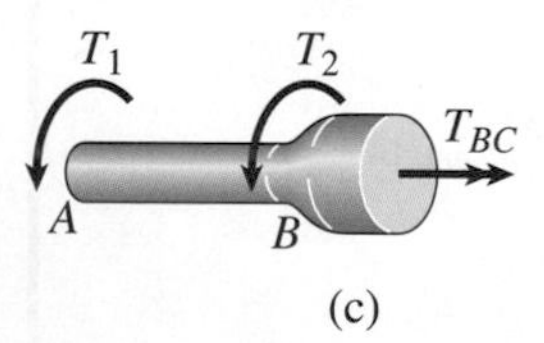

(c)

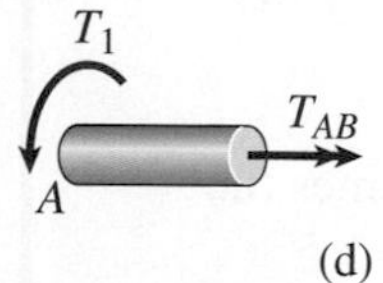

(d)

The first step in the analysis is to determine the magnitude and direction of the internal torque in each segment. Usually, the torques can be determined by inspection, but if necessary, they can be found by cutting sections through the bar, drawing free-body diagrams, and solving equations of equilibrium. This process is illustrated in Figs. 3-14b, c, and d. The first cut is made anywhere in segment CD, thereby exposing the internal torque T_{CD}. From the free-body diagram (Fig. 3-14b), T_{CD} is equal to $-T_1 - T_2 + T_3$. From the next diagram, T_{BC} equals to $-T_1 - T_2$, and from the last, T_{AB} equals $-T_1$. Thus,

$$T_{CD} = -T_1 - T_2 + T_3 \qquad T_{BC} = -T_1 - T_2 \qquad T_{AB} = -T_1 \qquad \textbf{(3-22a,b,c)}$$

Each of these torques is constant throughout the length of its segment.

To find the shear stresses in each segment, only the magnitudes of these internal torques are needed, since the directions of the stresses are not of interest. However, when finding the angle of twist for the entire bar, the direction of twist in each segment must be known in order to combine the angles of twist correctly. Therefore, establish a *sign convention* for the internal torques. A convenient rule in many cases is

> *An internal torque is positive when its vector points away from the cut section and negative when its vector points toward the section.*

Thus, all of the internal torques shown in Figs. 3-14b, c, and d are pictured in their positive directions. If the calculated torque [from (Eq. 3-22a, b, or c)] turns out to have a positive sign, it means that the torque acts in the assumed direction; if the torque has a negative sign, it acts in the opposite direction.

The maximum shear stress in each segment of the bar is readily obtained from the torsion formula [Eq. (3-13)] using the appropriate cross-sectional dimensions and internal torque. For instance, the maximum stress in segment BC (Fig. 3-14) is found using the diameter of that segment and the torque T_{BC} calculated from Eq. (3-22b). The maximum stress in the entire bar is the largest stress from among the stresses calculated for each of the three segments.

The angle of twist for each segment is found from Eq. (3-17), again using the appropriate dimensions and torque. The total angle of twist of one end of the bar with respect to the other is then obtained by algebraic summation, as

$$\phi = \phi_1 + \phi_2 + \ldots + \phi_n \tag{3-23}$$

where ϕ_1 is the angle of twist for segment 1, ϕ_2 is the angle for segment 2, and so on, while n is the total number of segments. Since each angle of twist is found from Eq. (3-17), the general formula used to find the total angle of twist is

$$\phi = \sum_{i=1}^{n} \phi_i = \sum_{i=1}^{n} \frac{T_i L_i}{G_i (I_p)_i} \tag{3-24}$$

in which the subscript i is a numbering index for the various segments. For segment i of the bar, T_i is the internal torque (found from equilibrium, as illustrated in Fig. 3-14), L_i is the length, G_i is the shear modulus, and $(I_p)_i$ is the polar moment of inertia. Some of the torques (and the corresponding angles of twist) may be positive, and some may be negative. By summing *algebraically* the angles of twist for all segments, you obtain the total angle of twist ϕ between the ends of the bar. This process is illustrated later in Example 3-4.

Case 2. *Bar with continuously varying cross sections and constant torque* (Fig. 3-15). When the torque is constant, the maximum shear stress in a solid bar always occurs at the cross section having the smallest diameter, as shown by Eq. (3-14). Furthermore, this observation usually holds for tubular bars. If this is the case, you only need to investigate the smallest cross section in order to calculate the maximum shear stress. Otherwise, it may be necessary to evaluate the stresses at more than one location in order to obtain the maximum.

FIGURE 3-15

Bar in nonuniform torsion (Case 2)

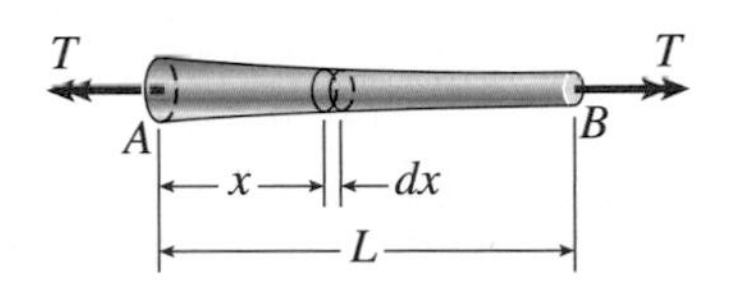

To find the angle of twist, consider an element of length dx at distance x from one end of the bar (Fig. 3-15). The differential angle of rotation $d\phi$ for this element is

$$d\phi = \frac{T dx}{GI_p(x)} \tag{3-25}$$

in which $I_p(x)$ is the polar moment of inertia of the cross section at distance x from the end. The angle of twist for the entire bar is the summation of the differential angles of rotation:

$$\phi = \int_0^L d\phi = \int_0^L \frac{T dx}{GI_p(x)} \tag{3-26}$$

If the expression for the polar moment of inertia $I_p(x)$ is not too complex, this integral can be evaluated analytically. In other cases, it must be evaluated numerically.

Case 3. *Bar with continuously varying cross sections and continuously varying torque* (Fig. 3-16). The bar shown in part (a) of the figure is subjected to a *distributed torque* of intensity t per unit distance along the axis of the bar. As a result, the internal torque $T(x)$ varies continuously along the axis (Fig. 3-16b). The internal torque can be evaluated with the aid of a free-body diagram and an equation of equilibrium. As in Case 2, the polar moment of inertia $I_p(x)$ can be evaluated from the cross-sectional dimensions of the bar.

FIGURE 3-16

Bar in nonuniform torsion (Case 3)

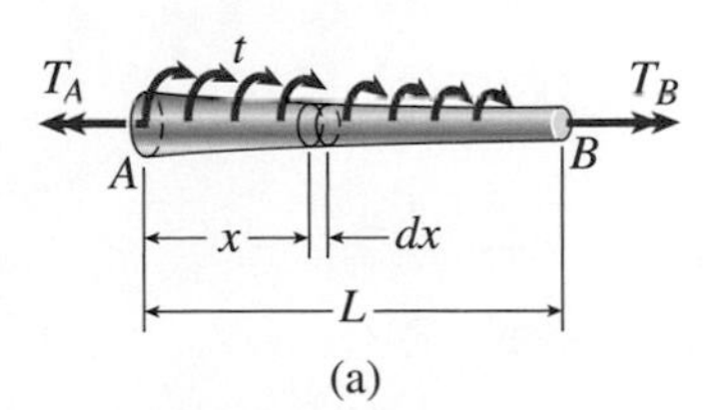

(a)

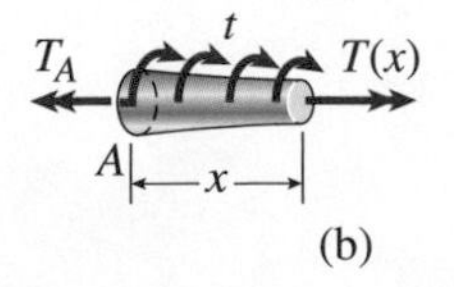

(b)

Knowing both the torque and polar moment of inertia as functions of x, use the torsion formula to determine how the shear stress varies along the axis of the bar. The cross section of maximum shear stress can then be identified, and the maximum shear stress can be determined.

The angle of twist for the bar of Fig. 3-16a can be found in the same manner as described for Case 2. The only difference is that the torque, like the polar moment of inertia, also varies along the axis. Consequently, the equation for the angle of twist becomes

$$\phi = \int_0^L d\phi = \int_0^L \frac{T(x)dx}{GI_p(x)} \tag{3-27}$$

This integral can be evaluated analytically in some cases, but usually it must be evaluated numerically.

Limitations

The analyses described in this section are valid for bars made of linearly elastic materials with circular cross sections (either solid or hollow). Also, the stresses determined from the torsion formula are valid in regions of the bar *away* from stress concentrations, which are highly localized stresses that occur wherever the diameter changes abruptly and wherever concentrated torques are applied (see Section 3.12). However, stress concentrations have relatively little effect on the angle of twist; therefore, the equations for ϕ are generally valid.

Finally, keep in mind that the torsion formula and the formulas for angles of twist were derived for prismatic bars with circular cross sections (see Section 3.10 for a brief discussion of noncircular bars in torsion). You can safely apply them to bars with varying cross sections only when the changes in diameter are small and gradual. As a rule of thumb, the formulas given here are satisfactory as long as the angle of taper (the angle between the sides of the bar) is less than 10°.

Torsional Displacement Diagram (TDD)

A graphical display of the variation in internal torsional moment $T(x)$ over the length of a shaft is presented here, so critical regions of the shaft (such as the location of maximum internal torsional moment T_{max} that also may be the location of maximum shear stress) can be identified. This display is referred to as the *torsional moment diagram* (TMD). Rules for construction of torsional moment diagrams are based on the differential and integral relationships between the applied torque of intensity $t(x)$ and the internal torsional moment $T(x)$. Now the *torque-displacement relation* in Eq. (3-27) is used to create a display of the variation of torsional displacement (or twist) $\phi(x)$ over the length of the shaft, referred to here as a *torsional displacement diagram* (TDD).

To construct the TDD, first note in Eq. (3-27) that the internal torsional moment $T(x)$ is part of the integrand in the expression for torsional displacement $\phi(x)$. It follows that the slope on the torsional displacement diagram is *proportional to* the ordinate on the torsional moment diagram; the full expression for slope must include the torsional rigidity term $GI_p(x)$ in the denominator in Eq. (3-27). Second, the integral expression indicates that the change in torsional displacement between any two points is proportional to the area under the TMD between those same two points:

$$\phi_B - \phi_A = \int_A^B \frac{T(x)}{GI_p(x)} dx$$

These above observations lead to the following rules or guidelines to be used in developing the TDD from the TMD:

- The slope at any point on the TDD is equal to the ordinate on the TMD at the same point divided by the torsional rigidity of the shaft (GI_p) at the same location.
- The change in torsional displacement between any two points along a shaft is equal to the area under the torsional moment diagram between those same two points divided by the torsional rigidity of the shaft (GI_p) over that same interval.

If the shaft is piecewise prismatic (so that GI_p is constant in each shaft segment) and the internal torsional moment $T(x)$ is constant, the TDD can be constructed using simple geometry. For instance, if the internal torsional moment is constant within each segment, the twist will vary linearly within that segment. Otherwise, analytical or numerical integration of Eq. (3-27) will be required to develop the TDD.

Construction of the TMD is illustrated using a cantilever shaft subjected to triangularly distributed torque $t(x)$ (see Figs. 3-17a and b). The torsional loading has linear variation, so the internal torque $T(x)$ has a quadratic variation as shown on the TMD in Fig. 3-17c. It follows that the twist displacement $\phi(x)$ has a cubic variation on the TDD (Fig. 3-17d). Comparison of the TMD and TDD plots in Figs. 3-17c and d leads to the following observations: (a) the slope of the tangent to the TDD curve at $x = 0$ is equal to the TMD ordinate $t_0L/2$ divided by the torsional rigidity GI_p; (b) the TMD ordinate is zero at $x - L$, so the TDD has zero slope at that location, and (c) the area A_m under

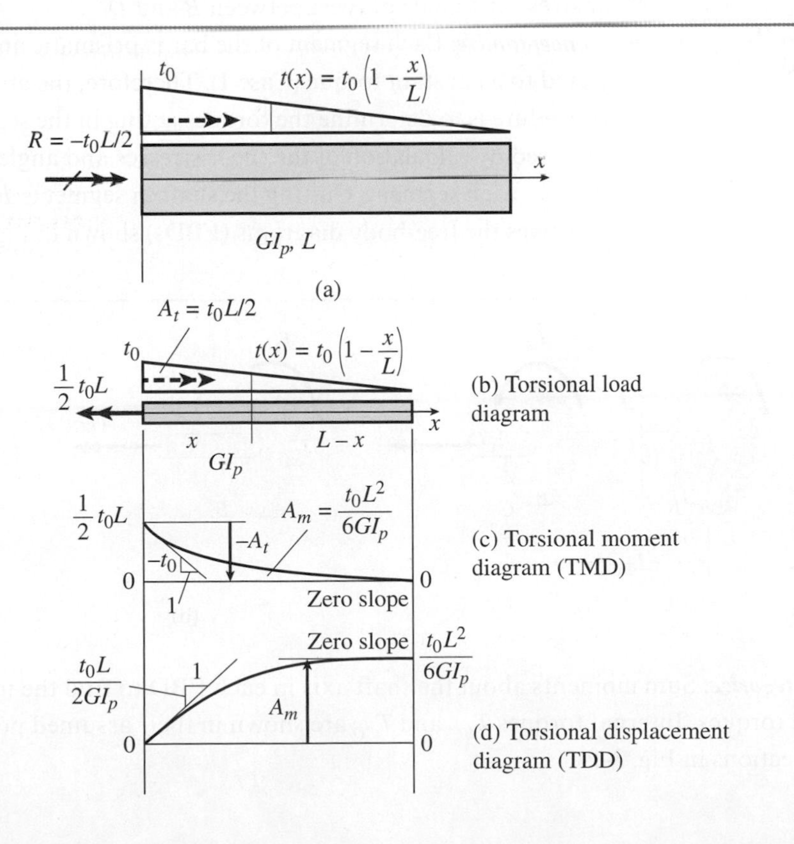

FIGURE 3-17

(a) Prismatic shaft subject to triangularly distributed torque (b, c, d) Overall FBD, torsional moment diagram (TMD), and torsional displacement diagram (TDD)

the entire TMD is proportional to the change in twist from $x = 0$ to $x = L$, so A_m/GI_p is equal to the maximum twist at $x = L$. Use a *statics sign convention* in developing the TDD: twist displacement $\phi(x)$ is considered positive (and plotted above the reference axis on the TDD) if the twist displacement vector points in the $+x$ direction. Examples 3-4 and 3-5 provide further application of these concepts.

Example 3-4

FIGURE 3-18

Example 3-4: Steel shaft in torsion

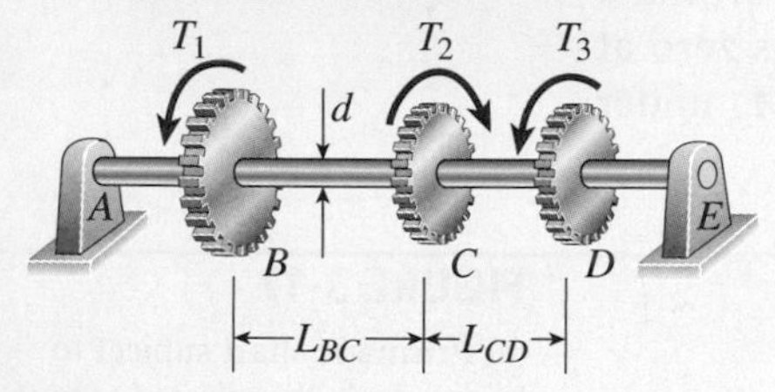

A solid steel shaft $ABCDE$ (Fig. 3-18) having a diameter $d = 30$ mm turns freely in bearings at points A and E. The shaft is driven by a gear at C, which applies a torque $T_2 = 450$ N·m in the direction shown in the figure. Gears at B and D are driven by the shaft and have resisting torques $T_1 = 275$ N·m and $T_3 = 175$ N·m, respectively, acting in the opposite direction to the torque T_2. Segments BC and CD have lengths $L_{BC} = 500$ mm and $L_{CD} = 400$ mm, respectively, and the shear modulus is $G = 80$ GPa.

Determine the maximum shear stress in each part of the shaft and the angle of twist between gears B and D.

Solution:

Use a four-step problem-solving approach to find the maximum shear stress and angle of twist between B and D.

1. ***Conceptualize***: Each segment of the bar is prismatic and subjected to a constant torque (Case 1). Therefore, the analysis procedure is to determine the torques acting in the segments, followed by calculation of the shear stresses and angles of twist in each segment. Cutting the shaft in segments BC and CD gives the free-body diagrams (FBDs) shown in Fig. 3-19.

FIGURE 3-19

Free-body diagrams for Example 3-4

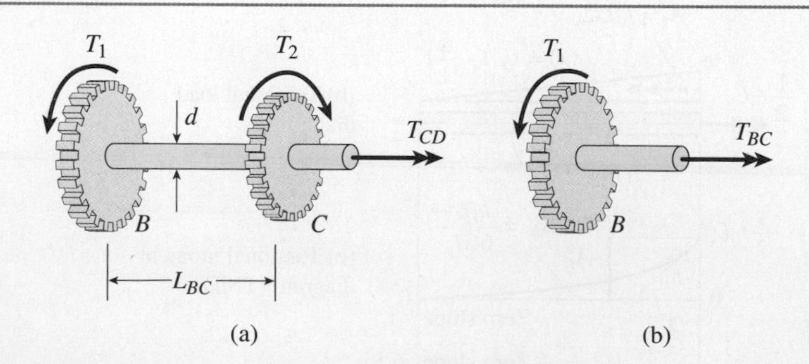

2. ***Categorize***: Sum moments about the shaft axis in each FBD to find the internal torques. Internal torques T_{BC} and T_{CD} are shown in their assumed positive directions in Fig. 3-19.

Torques acting in the segments: The torques in the end segments (AB and DE) are zero, since any friction in the bearings at the supports is assumed to be too small to consider. Therefore, the end segments have no stresses and no angles of twist.

The torque T_{CD} in segment CD is found by cutting a section through the segment and constructing a free-body diagram, as in Fig. 3-19a. The torque is assumed to be positive; therefore, its vector points away from the cut section. From equilibrium of the free body, torque T_{CD} is

$$T_{CD} = T_2 - T_1 = 450 \text{ N} \cdot \text{m} - 275 \text{ N} \cdot \text{m} = 175 \text{ N} \cdot \text{m}$$

The positive sign in the result means that T_{CD} acts in the assumed positive direction.

The torque in segment BC is found in a similar manner, using the free-body diagram of Fig. 3-19b:

$$T_{BC} = -T_1 = -275 \text{ N} \cdot \text{m}$$

Note that this torque has an negative sign, which means that its direction is opposite to the direction shown in the figure.

With all internal torques known, plot the TMD as shown in Fig. 3-20a.

FIGURE 3-20

TMD and TDD for steel shaft in torsion

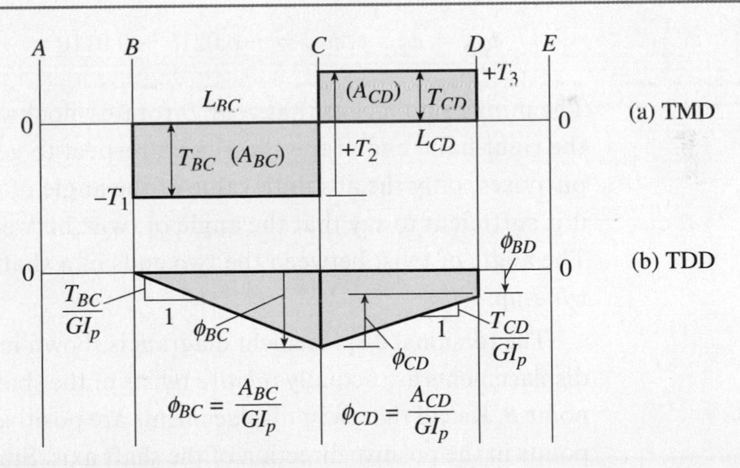

3. ***Analyze:***

Shear stresses: The maximum shear stresses in segments BC and CD are found from the modified form of the torsion formula [Eq. (3-14)]; thus,

$$\tau_{BC} = \frac{16T_{BC}}{\pi d^3} = \frac{16(275 \text{ N} \cdot \text{m})}{\pi(30 \text{ mm})^3} = 51.9 \text{ MPa}$$

$$\tau_{CD} = \frac{16T_{CD}}{\pi d^3} = \frac{16(175 \text{ N} \cdot \text{m})}{\pi(30 \text{ mm})^3} = 33.0 \text{ MPa}$$

Since the direction of shear stresses are not of the interest of this example, only absolute values of the torques are used in the preceding calculations.

Angles of twist: The angle of twist ϕ_{BD} between gears B and D is the algebraic sum of the angles of twist for the intervening segments of the bar, as given by Eq. (3-23); thus,

$$\phi_{BD} = \phi_{BC} + \phi_{CD}$$

When calculating the individual angles of twist, use the polar moment of inertia of the cross section, which is computed as

$$I_p = \frac{\pi d^4}{32} = \frac{\pi(30 \text{ mm})^4}{32} = 79{,}520 \text{ mm}^4$$

Now find the angles of twist, as

$$\phi_{BC} = \frac{T_{BC}L_{BC}}{GI_p} = \frac{(-275 \text{ N}\cdot\text{m})(500 \text{ mm})}{(80 \text{ GPa})(79{,}520 \text{ mm}^4)} = -0.0216 \text{ rad}$$

and

$$\phi_{CD} = \frac{T_{CD}L_{CD}}{GI_p} = \frac{(175 \text{ N}\cdot\text{m})(400 \text{ mm})}{(80 \text{ GPa})(79{,}520 \text{ mm}^4)} = 0.0110 \text{ rad}$$

Note that in this example the angles of twist have opposite directions. Add algebraically to obtain the total angle of twist:

$$\phi_{BD} = \phi_{BC} + \phi_{CD} = -0.0216 + 0.0110 = -0.0106 \text{ rad} = -0.61°$$

The minus sign means that gear D rotates clockwise (when viewed from the right-hand end of the shaft) with respect to gear B. However, for most purposes, only the absolute value of the angle of twist is needed; therefore, it is sufficient to say that the angle of twist between gears B and D is 0.61°. The angle of twist between the two ends of a shaft is sometimes called the *wind-up*.

The torsional displacement diagram is shown in Fig. 3-20b. Here the twist displacements are actually *relative* twists of the shaft segments with respect to point B. Recall that twist displacements are positive if the displacement vector points in the positive direction of the shaft axis. Study Figs. 3-20a and b to confirm the geometric relationships labeled on the figures.

4. ***Finalize:*** The procedures illustrated in this example can be used for shafts having segments of different diameters or of different materials, as long as the dimensions and properties remain constant within each segment.

Only the effects of torsion are considered in this example and in the problems at the end of the chapter. Bending effects are considered later, beginning with Chapter 4.

Example 3-5

Two sections (AB, BC) of steel drill pipe, joined by bolted flange plates at B, are being tested to assess the adequacy of both the pipe and the bolted connection (see Fig. 3-21). In the test, the pipe structure is fixed at A, a concentrated torque $2T_0$ is applied at $x = 2L/5$, and a uniformly distributed torque with intensity $t_0 = 3T_0/L$ is applied on pipe BC.

(a) Find the maximum shear stress τ_{max} in the pipes and its location. Assume that load variable $T_0 = 226$ kN·m. Let $G = 81$ GPa, and assume that both pipes have the same inner diameter $d = 250$ mm. Pipe AB has a thickness of $t_{AB} = 19$ mm, while pipe BC has a thickness of $t_{BC} = 16$ mm.

(b) Find the expressions for twist rotation $\phi(x)$ over the length of the pipe structure. If the maximum allowable twist of the pipe structure is $\phi_{allow} = 0.5°$, find the maximum permissible value of *load variable* T_0 (kN·m). Let $L = 3$ m. Also, plot the torsional moment diagram (TMD) and the torsional displacement diagram (TDD).

(c) Use maximum T_0 from part (b) to find the number of $d_b = 22$ mm diameter bolts at radius $r = 380$ mm required in the flange plate connection at B. Assume that the allowable shear stress for the bolts is $\tau_a = 190$ MPa.

Courtesy of Subsea Riser Products

©Can Stock Photo Inc./Nostalgie

Solution:

Use a four-step problem-solving approach to find maximum shear stress, maximum twist, and the number of connection bolts at B.

1. ***Conceptualize***: Make section cuts along the shaft to find internal torques $T(x)$ and then plot the torsional moment diagram (TMD) as shown in Fig. 3-21c. From the TMD, the maximum internal torque is $6T_0/5$ in the central region of the structure. It appears that the maximum shear stress may be in the thinner pipe segment BC just right of the connection at B.

2. ***Categorize***: Use the torsion formula [Eq. (3-13)] to find shear stresses on the surface of each segment. Use the torque-displacement relation [Eq. (3-17)] to find the twist rotations over the length of the pipe that are displayed as the TDD. Finally, examine shear stresses in the bolted connection to find the required number of bolts.

FIGURE 3-21

Example 3-5: (a, b) Two pipes in nonuniform torsion; (c) Torsional moment diagram (TMD) for entire shaft

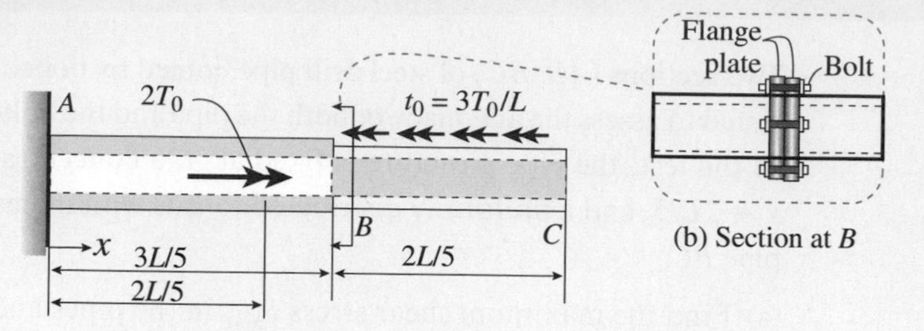

(a) Nonprismatic pipe

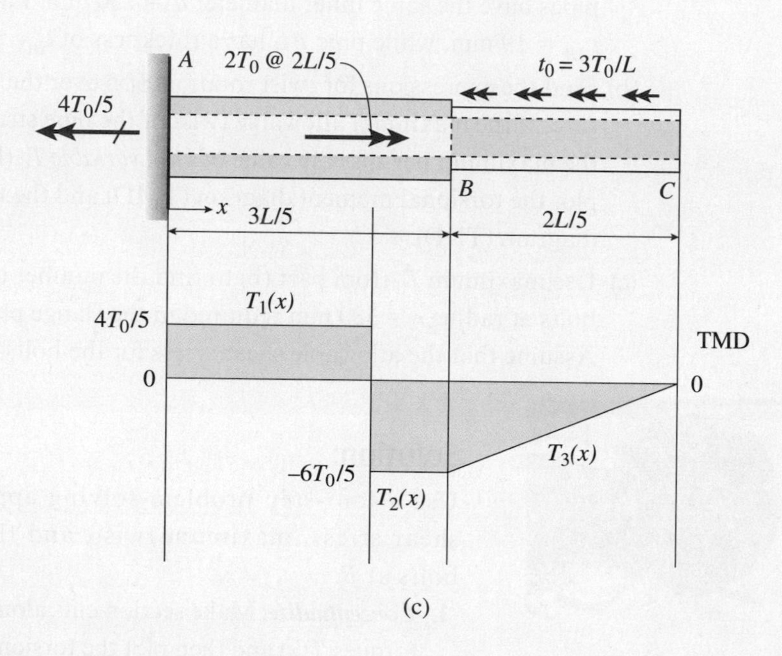

(c)

3. *Analyze*:

Part (a): Maximum shear stress in pipe τ_{max}.

Use the torsion formula [Eq. (3-13)] to compute the shear stress in the pipe. The maximum shear stress is on the surface of the pipe. The polar moment of inertia of each pipe is computed as

$$I_{pAB} = \frac{\pi}{32}\left[(d + 2t_{AB})^4 - (d)^4\right]$$

$$= \frac{\pi}{32}\left[[250 \text{ mm} + 2(19 \text{ mm})]^4 - (250 \text{ mm})^4\right] = 2.919 \times 10^{-4} \text{ m}^4$$

and

$$I_{pBC} = \frac{\pi}{32}\left[(d + 2t_{BC})^4 - (d)^4\right]$$

$$= \frac{\pi}{32}\left[[250 \text{ mm} + 2(16 \text{ mm})]^4 - (250 \text{ mm})^4\right] = 2.374 \times 10^{-4} \text{ m}^4$$

The shear modulus G is constant, so the torsional rigidity of AB is 1.23 times that of BC. From the TMD (Fig. 3-21c), the maximum torsional moments in both AB and BC (each equal to $6T_0/5$) are near joint B. Applying the torsion formula to pipes AB and BC near B gives

$$\tau_{\max AB} = \frac{\left(\frac{6}{5}T_0\right)\left(\frac{d + 2t_{AB}}{2}\right)}{I_{pAB}}$$

$$= \frac{\left(\frac{6}{5}226\text{ kN}\cdot\text{m}\right)\left[\frac{250\text{ mm} + 2(19\text{ mm})}{2}\right]}{2.919 \times 10^{-4}\text{ m}^4} = 133.8\text{ MPa}$$

$$\tau_{\max BC} = \frac{\left(\frac{6}{5}T_0\right)\left(\frac{d + 2t_{BC}}{2}\right)}{I_{pBC}}$$

$$= \frac{\left(\frac{6}{5}226\text{ kN}\cdot\text{m}\right)\left[\frac{250\text{ mm} + 2(16\text{ mm})}{2}\right]}{2.374 \times 10^{-4}\text{ m}^4} = 161.1\text{ MPa}$$

So the maximum shear stress in the pipe is *just to the right of* the flange plate connection at joint B. "Just to the right of" means an appropriate distance away from the connection to avoid any stress-concentration effects at the point of attachment of the two pipes in accordance with St. Venant's principle (see Section 3.12).

Part (b): Twist rotations $\phi(x)$.

Next, use the *torque-displacement relation* from Eqs. (3-24) through (3-27) to find the variation of twist rotation ϕ over the length of the pipe structure. Support A is fixed, so $\phi_A = \phi(0) = 0$. The internal torque from $x = 0$ to $x = 2L/5$ (segment 1) is constant, so use Eq. (3-24) to find twist rotation $\phi_1(x)$, which varies linearly from $x = 0$ to $x = 2L/5$:

$$\phi_1(x) = \frac{T_1(x)(x)}{GI_{pAB}} = \frac{\left(\frac{4T_0}{5}\right)(x)}{GI_{pAB}} = \frac{4T_0 x}{5GI_{pAB}} \qquad 0 \le x \le \frac{2L}{5} \tag{a}$$

Evaluate Eq. (a) at $x = 2L/5$ to find the twist rotation at the point of application of torque $2T_0$ as

$$\phi_1\left(\frac{2L}{5}\right) = \frac{T_1\left(\frac{2L}{5}\right)\left(\frac{2L}{5}\right)}{GI_{pAB}} = \frac{\left(\frac{4T_0}{5}\right)\left(\frac{2L}{5}\right)}{GI_{pAB}} = \frac{8T_0 L}{25GI_{pAB}} = \frac{0.32T_0 L}{GI_{pAB}} \tag{b}$$

Next, find an expression for the variation of twist angle $\phi_2(x)$ from $x = 2L/5$ to $x = 3L/5$ (point B). As with $\phi_1(x)$, twist $\phi_2(x)$ varies linearly over segment 2, because torque $T_2(x)$ is constant (Fig. 3-21c). Use Eq. (3-24) to get

$$\phi_2(x) = \phi_1\left(\frac{2L}{5}\right) + \frac{T_2(x)\left(x - \frac{2L}{5}\right)}{GI_{pAB}} = \frac{8T_0L}{25GI_{pAB}} + \frac{\left(\frac{-6}{5}T_0\right)\left(x - \frac{2L}{5}\right)}{GI_{pAB}}$$

$$= \frac{2T_0(2L - 3x)}{5GI_{pAB}} \quad \frac{2L}{5} \le x \le \frac{3L}{5} \tag{c}$$

Finally, develop an expression for twist over segment 3 (or pipe BC). The internal torsional moment now has a linear variation (Fig. 3-21c), so an integral form of the torque-displacement relation [Eq. (3-27)] is required. Insert an expression for $T_3(x)$ and add the torsional displacement at B to get a formula for the variation of twist in BC (note that variable ζ is a dummy variable of integration):

$$\phi_3(x) = \phi_2\left(\frac{3L}{5}\right) + \int_{\frac{3L}{5}}^{x} \frac{\left[3T_0\left(\frac{\zeta}{L} - 1\right)\right]}{GI_{pBC}}\,d\zeta$$

$$= \frac{2T_0\left[2L - 3\left(\frac{3L}{5}\right)\right]}{5GI_{pAB}} + \int_{\frac{3L}{5}}^{x} \frac{\left[3T_0\left(\frac{\zeta}{L} - 1\right)\right]}{GI_{pBC}}\,d\zeta$$

Torque $T_3(x)$ has a linear variation, so evaluating the integral yields a quadratic expression for variation of twist in BC:

$$\phi_3(x) = \frac{2T_0L}{25GI_{pAB}} + \frac{3T_0(21L^2 - 50Lx + 25x^2)}{50GI_{pBC}L} \quad \frac{3L}{5} \le x \le L \tag{d}$$

Substitute $x = 3L/5$ to find the twist at B:

$$\phi_3\left(\frac{3L}{5}\right) = \frac{2T_0L}{25GI_{pAB}}$$

At $x = L$, the twist at C is

$$\phi_3(L) = \frac{2T_0L}{25GI_{pAB}} - \frac{6T_0L}{25GI_{pBC}} = -0.219\frac{T_0L}{GI_{pAB}}$$

Now assume that $I_{pAB} = 1.23I_{pBC}$ (based on the numerical properties here), and plot the variation of twist over the length of the pipe structure (Fig. 3-22), noting that ϕ_{max} occurs at $x = 2L/5$ [see Eq. (b)].

Finally, restrict ϕ_{max} to the allowable value of 0.5° and solve for the maximum permissible value of *load variable* T_0 (kN·m) using the numerical properties given previously:

$$T_{0\,max} = \frac{GI_{pAB}}{0.32L}(\phi_{allow}) = \frac{(81\text{ GPa})(2.919 \times 10^{-4}\text{ m}^4)}{0.32(3\text{ m})}(0.5°)$$

$$= 215\text{ kN·m} \tag{e}$$

FIGURE 3-22

Example 3-5: Torsional displacement diagram (TDD)

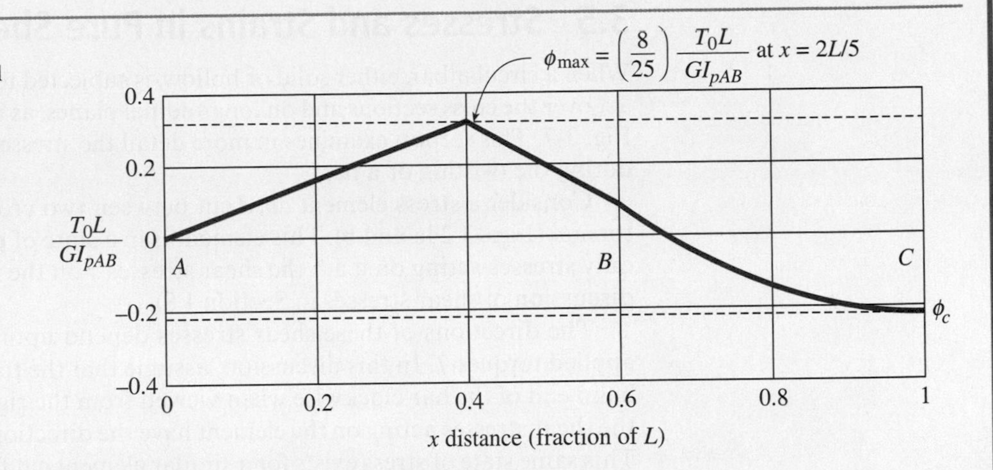

Part (c): Number of bolts required in flange plate.

Now use $T_{0,max}$ from Eq. (e) to find the required number of $d_b = 22$ mm diameter bolts at radius $r = 380$ mm in the flange plate connection at B. The allowable shear stress in the bolts is $\tau_a = 190$ MPa. Assume that each bolt carries an equal share of the torque at B, so each of n bolts carries shear force F_b at a distance r from the centroid of the cross section (Fig. 3-23).

The maximum shear force F_b per bolt is τ_a times the bolt cross-sectional area A_b, and the total torque at B is $6T_{0,max}/5$ (see TMD in Fig. 3-21c), so

$$nF_br = \frac{6}{5}T_{0max} \text{ or } n = \frac{\frac{6}{5}T_{0max}}{\tau_aA_br} = \frac{\frac{6}{5}(215\text{ kN}\cdot\text{m})}{(190\text{ MPa})\left[\frac{\pi}{4}(22\text{ mm})^2\right](380\text{ mm})} = 9.4$$

Use ten 22 mm diameter bolts at a radius of 380 mm in the flange plate connection at B.

4. ***Finalize:*** Confirm the TDD in Fig. 3-22 using the TMD in Fig. 3-21c along with the rules presented in Section 3.4. Values and locations of maximum torsional moment and maximum twist are readily visible on these diagrams and are essential information for use in an engineering evaluation of this pipe structure.

FIGURE 3-23

Example 3-5: Flange plate bolts at B

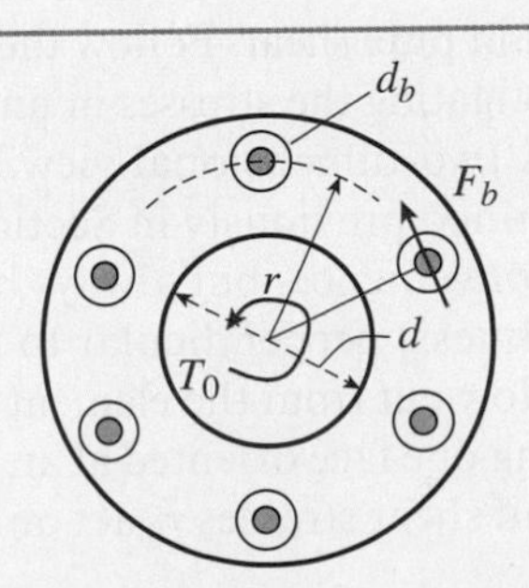

3.5 Stresses and Strains in Pure Shear

When a circular bar, either solid or hollow, is subjected to torsion, shear stresses act over the cross sections and on longitudinal planes, as illustrated previously in Fig. 3-7. This section examines in more detail the stresses and strains produced during the twisting of a bar.

Consider a stress element *abcd* cut between two cross sections of a bar in torsion (Figs. 3-24a and b). This element is in a state of **pure shear**, because the only stresses acting on it are the shear stresses τ on the four side faces (see the discussion of shear stresses in Section 1.8).

The directions of these shear stresses depend upon the directions of the applied torques *T*. In this discussion, assume that the torques rotate the right-hand end of the bar clockwise when viewed from the right (Fig. 3-24a); hence, the shear stresses acting on the element have the directions shown in the figure. This same state of stress exists for a similar element cut from the interior of the bar, except that the magnitudes of the shear stresses are smaller because the radial distance to the element is smaller.

The directions of the torques shown in Fig. 3-24a are intentionally chosen so that the resulting shear stresses (Fig. 3-24b) are positive according to the sign convention for shear stresses described previously in Section 1.8. This **sign convention** is repeated here:

A shear stress acting on a positive face of an element is positive if it acts in the positive direction of one of the coordinate axes and negative if it acts in the negative direction of an axis. Conversely, a shear stress acting on a negative face of an element is positive if it acts in the negative direction of one of the coordinate axes and negative if it acts in the positive direction of an axis.

Apply this sign convention to the shear stresses acting on the stress element of Fig. 3-24b, and see that all four shear stresses are positive. For instance, the stress on the right-hand face (which is a positive face because the x axis is directed to the right) acts in the positive direction of the y axis; therefore, it is a positive shear stress. Also, the stress on the left-hand face (which is a negative face) acts in the negative direction of the y axis; therefore, it is a positive shear stress. Analogous comments apply to the remaining stresses.

FIGURE 3-24

Stresses acting on a stress element cut from a bar in torsion (pure shear)

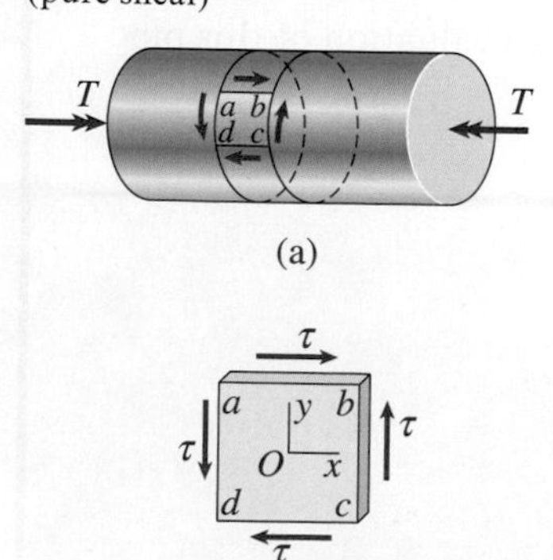

Stresses on Inclined Planes

Now determine the stresses acting on *inclined planes* cut through the stress element in pure shear. Follow the same approach as the one used in Section 2.6 for investigating the stresses in uniaxial stress.

A two-dimensional view of the stress element is shown in Fig. 3-25a. As explained previously in Section 2.6, a two-dimensional view is usually drawn for convenience, but always be aware that the element has a third dimension (thickness) perpendicular to the plane of the figure.

Now cut from the element a wedge-shaped (or "triangular") stress element having one face oriented at an angle θ to the x axis (Fig. 3-25b). Normal stresses σ_θ and shear stresses τ_θ act on this inclined face and are shown in their positive directions in the figure. The **sign convention** for stresses σ_θ and τ_θ was described previously in Section 2.6 and is repeated here:

Normal stresses σ_θ are positive in tension, and shear stresses τ_θ are positive when they tend to produce counterclockwise rotation of the material. (Note that this sign convention for the shear stress τ_θ acting on an inclined plane is different from the sign convention for ordinary shear stresses τ that act on the sides of rectangular elements oriented to a set of *x-y* axes.)

The horizontal and vertical faces of the triangular element (Fig. 3-25b) have positive shear stresses τ acting on them, and the front and rear faces of the element are free of stress. Therefore, all stresses acting on the element are visible in this figure.

The stresses σ_θ and τ_θ now may be determined from the equilibrium of the triangular element. The *forces* acting on its three side faces can be obtained by multiplying the stresses by the areas over which they act. For instance, the force on the left-hand face is equal to τA_0, where A_0 is the area of the vertical face. This force acts in the negative *y* direction and is shown in the *free-body diagram* of Fig. 3-25c. Because the thickness of the element in the *z* direction is constant, the area of the bottom face is $A_0 \tan\theta$ and the area of the inclined face is $A_0 \sec\theta$. Multiplying the stresses acting on these faces by the corresponding areas enables us to obtain the remaining forces and thereby complete the free-body diagram (Fig. 3-25c).

Now write two equations of equilibrium for the triangular element: one in the direction of σ_θ and the other in the direction of τ_θ. When writing these equations, the forces acting on the left-hand and bottom faces must be resolved into components in the directions of σ_θ and τ_θ. Thus, the first equation, obtained by summing forces in the direction of σ_θ, is

$$\sigma_\theta A_0 \sec\theta = \tau A_0 \sin\theta + \tau A_0 \tan\theta \cos\theta$$

or

$$\sigma_\theta = 2\tau \sin\theta \cos\theta \tag{3-28a}$$

The second equation is obtained by summing forces in the direction of τ_θ:

$$\tau_\theta A_0 \sec\theta = \tau A_0 \cos\theta - \tau A_0 \tan\theta \sin\theta$$

or

$$\tau_\theta = \tau(\cos^2\theta - \sin^2\theta) \tag{3-28b}$$

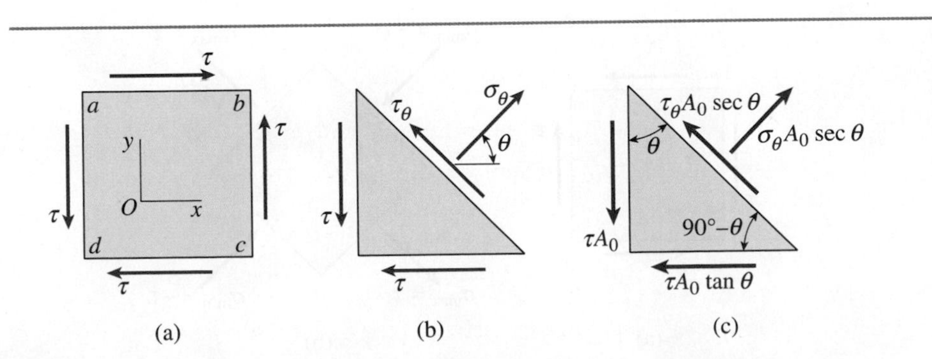

FIGURE 3-25
Analysis of stresses on inclined planes: (a) element in pure shear, (b) stresses acting on a triangular stress element, and (c) forces acting on the triangular stress element (free-body diagram)

These equations can be expressed in simpler forms by introducing the following trigonometric identities (see Appendix C):

$$\sin 2\theta = 2 \sin\theta \cos\theta \qquad \cos 2\theta = \cos^2\theta - \sin^2\theta$$

Then the equations for σ_θ and τ_θ become

$$\sigma_\theta = \tau \sin 2\theta \qquad \tau_\theta = \tau \cos 2\theta \qquad \textbf{(3-29a,b)}$$

Equations (3-29a and b) give the normal and shear stresses acting on any inclined plane in terms of the shear stresses τ acting on the x and y planes (Fig. 3-25a) and the angle θ defining the orientation of the inclined plane (Fig. 3-25b).

The manner in which the stresses σ_θ and τ_θ vary with the orientation of the inclined plane is shown by the graph in Fig. 3-26, which is a plot of Eqs. (3-29a and b). For $\theta = 0$, which is the right-hand face of the stress element in Fig. 3-25a, the graph gives $\sigma_\theta = 0$ and $\tau_\theta = \tau$. This latter result is expected, because the shear stress τ acts counterclockwise against the element and therefore produces a positive shear stress τ_θ.

FIGURE 3-26

Graph of normal stresses σ_θ and shear stresses τ_θ versus angle θ of the inclined plane

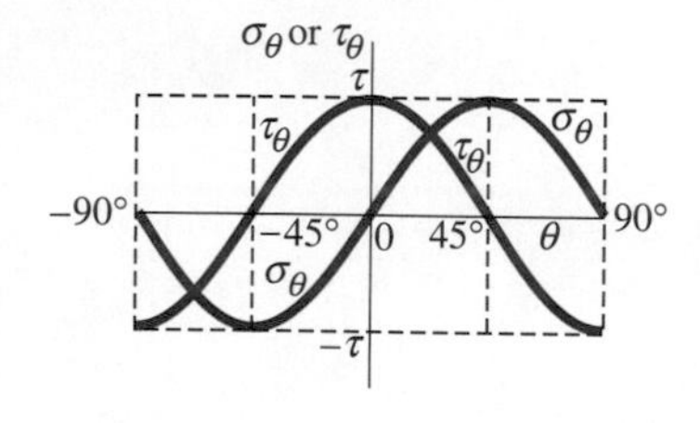

For the top face of the element ($\theta = 90°$), $\sigma_\theta = 0$ and $\tau_\theta = -\tau$. The minus sign for τ_θ means that it acts clockwise against the element, that is, to the right on face ab (Fig. 3-25a), which is consistent with the direction of the shear stress τ. Note that the numerically largest shear stresses occur on the planes for which $\theta = 0$ and $90°$, as well as on the opposite faces ($\theta = 180°$ and $270°$).

In the graph, the normal stress σ_θ reaches a maximum value at $\theta = 45°$. At that angle, the stress is positive (tension) and equal numerically to the shear stress τ. Similarly, σ_θ has its minimum value (which is compressive) at $\theta = -45°$. At both of these 45° angles, the shear stress τ_θ is equal to zero. These conditions are pictured in Fig. 3-27, which shows stress elements oriented at $\theta = 0$ and $\theta = 45°$. The element at 45° is acted upon by equal tensile and compressive stresses in perpendicular directions with no shear stresses.

Note that the normal stresses acting on the 45° element (Fig. 3-27b) correspond to an element subjected to shear stresses τ acting in the directions shown in Fig. 3-27a. If the shear stresses acting on the element of Fig. 3-27a are reversed in direction, the normal stresses acting on the 45° planes also will change directions.

FIGURE 3-27

Stress elements oriented at $\theta = 0$ and $\theta = 45°$ for pure shear

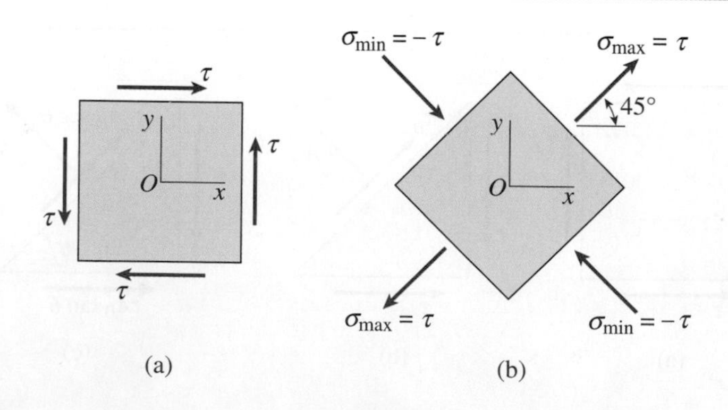

If a stress element is oriented at an angle other than 45°, both normal and shear stresses will act on the inclined faces [see Eqs. (3-29a and b) and Fig. 3-26]. Stress elements subjected to these more general conditions are discussed in detail in Chapter 7.

The equations derived in this section are valid for a stress element in pure shear regardless of whether the element is cut from a bar in torsion or from some other structural element. Also, since Eqs. (3-29) were derived from equilibrium only, they are valid for any material, whether or not it behaves in a linearly elastic manner.

The existence of maximum tensile stresses on planes at 45° to the x axis (Fig. 3-27b) explains why bars in torsion that are made of materials that are brittle and weak in tension fail by cracking along a 45° helical surface (Fig. 3-28). As mentioned in Section 3.3, this type of failure is readily demonstrated by twisting a piece of classroom chalk.

Strains in Pure Shear

Now consider the strains that exist in an element in pure shear. For instance, consider the element in pure shear shown in Fig. 3-27a. The corresponding shear strains are shown in Fig. 3-29a, where the deformations are highly exaggerated. The shear strain γ is the change in angle between two lines that were originally perpendicular to each other, as discussed previously in Section 1.8. Thus, the decrease in the angle at the lower left-hand corner of the element is the shear strain γ (measured in radians). This same change in angle occurs at the upper right-hand corner (where the angle decreases) and at the other two corners (where the angles increase). However, the lengths of the sides of the element, including the thickness perpendicular to the plane of the paper, do not change when these shear deformations occur. Therefore, the element changes its shape from a rectangular parallelepiped (Fig. 3-27a) to an oblique parallelepiped (Fig. 3-29a). This change in shape is called a **shear distortion**.

If the material is linearly elastic, the shear strain for the element oriented at $\theta = 0$ (Fig. 3-29a) is related to the shear stress by Hooke's law in shear:

$$\gamma = \frac{\tau}{G} \tag{3-30}$$

where, as usual, the symbol G represents the shear modulus of elasticity.

Next, consider the strains that occur in an element oriented at $\theta = 45°$ (Fig. 3-29b). The tensile stresses acting at 45° tend to elongate the element in that direction. Because of the Poisson effect, they also tend to shorten it in the perpendicular direction (the direction where $\theta = 135°$ or $-45°$). Similarly, the

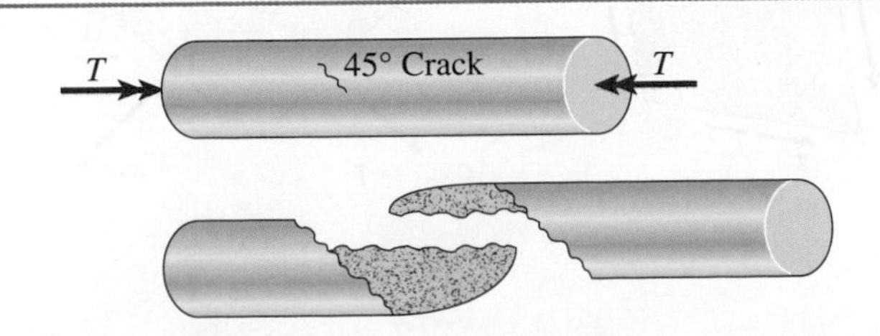

FIGURE 3-28

Torsion failure of a brittle material by tension cracking along a 45° helical surface

compressive stresses acting at 135° tend to shorten the element in that direction and elongate it in the 45° direction. These dimensional changes are shown in Fig. 3-29b, where the dashed lines show the deformed element. Since there are no shear distortions, the element remains a rectangular parallelepiped even though its dimensions have changed.

If the material is linearly elastic and follows Hooke's law, you can obtain an equation relating strain to stress for the element at $\theta = 45°$ (Fig. 3-29b). The tensile stress σ_{max} acting at $\theta = 45°$ produces a positive normal strain in that direction equal to σ_{max}/E. Since $\sigma_{max} = \tau$, this strain also can be expressed as τ/E. The stress σ_{max} also produces a negative strain in the perpendicular direction equal to $-\nu\tau/E$, where ν is Poisson's ratio. Similarly, the stress $\sigma_{min} = -\tau$ (at $\theta = 135°$) produces a negative strain equal to $-\tau/E$ in that direction and a positive strain in the perpendicular direction (the 45° direction) equal to $\nu\tau/E$. Therefore, the normal strain in the 45° direction is

$$\varepsilon_{max} = \frac{\tau}{E} + \frac{\nu\tau}{E} = \frac{\tau}{E}(1 + \nu) \qquad \textbf{(3-31)}$$

which is positive, representing elongation. The strain in the perpendicular direction is a negative strain of the same amount. In other words, pure shear produces elongation in the 45° direction and shortening in the 135° direction. These strains are consistent with the shape of the deformed element of Fig. 3-29a, because the 45° diagonal has lengthened and the 135° diagonal has shortened.

The next section uses the geometry of the deformed element to relate the shear strain γ (Fig. 3-29a) to the normal strain ε_{max} in the 45° direction (Fig. 3-29b). In so doing, the following relationship is derived:

$$\varepsilon_{max} = \frac{\gamma}{2} \qquad \textbf{(3-32)}$$

This equation, in conjunction with Eq. (3-30), can be used to calculate the maximum shear strains and maximum normal strains in pure torsion when the shear stress τ is known.

FIGURE 3-29

Strains in pure shear: (a) shear distortion of an element oriented at $\theta = 0$ and (b) distortion of an element oriented at $\theta = 45°$

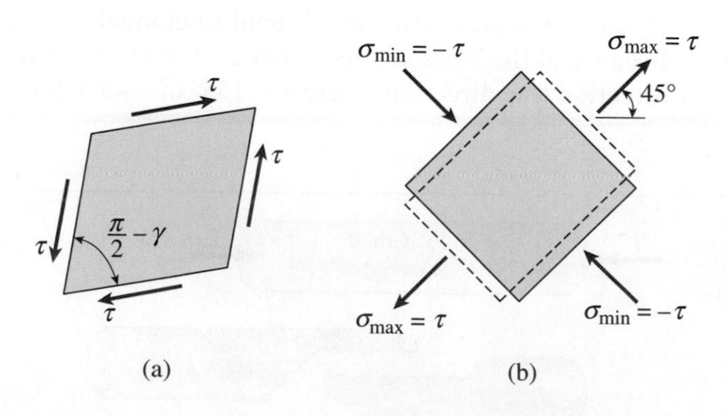

Example 3-6

A circular tube with an outside diameter of 80 mm and an inside diameter of 60 mm is subjected to a torque $T = 4.0$ kN·m (Fig. 3-30). The tube is made of aluminum alloy 7075-T6.

(a) Determine the maximum shear, tensile, and compressive stresses in the tube and show these stresses on sketches of properly oriented stress elements.

(b) Determine the corresponding maximum strains in the tube and show these strains on sketches of the deformed elements.

(c) What is the maximum permissible torque T_{max} if the allowable normal strain is $\varepsilon_a = 0.9 \times 10^{-3}$?

(d) If $T = 4.0$ kN·m and $\varepsilon_a = 0.9 \times 10^{-3}$, what new outer diameter is required so that the tube can carry the required torque T (assume that the inner diameter of the tube remains at 60 mm)?

FIGURE 3-30

Example 3-6: Circular tube in torsion

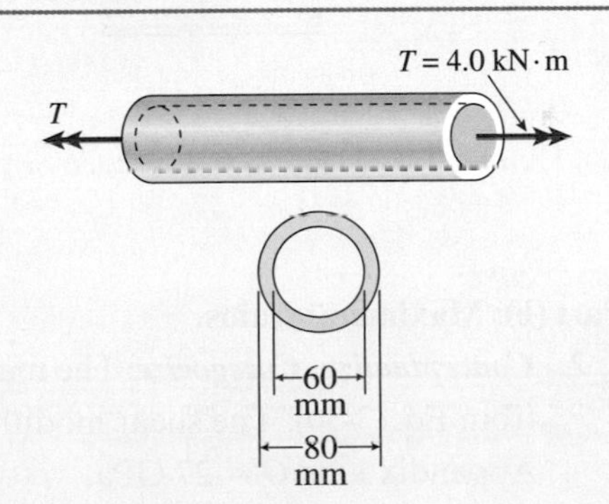

Solution:

A four-step problem-solving approach is used in the solution; only essential steps are shown in each part.

Part (a): Maximum stresses.

1, 2. *Conceptualize, Categorize*: The maximum values of all three stresses (shear, tensile, and compressive) are equal numerically, although they act on different planes. Their magnitudes are found from the torsion formula:

$$\tau_{max} = \frac{Tr}{I_p} = \frac{(4000 \text{ N} \cdot \text{m})(0.040 \text{ m})}{\frac{\pi}{32}\left[(0.080 \text{ m})^4 - (0.060 \text{ m})^4\right]} = 58.2 \text{ MPa}$$

3. *Analyze*: The maximum shear stresses act on cross-sectional and longitudinal planes, as shown by the stress element in Fig. 3-31a, where the x axis is parallel to the longitudinal axis of the tube.

The maximum tensile and compressive stresses are

$$\sigma_t = 58.2 \text{ MPa} \quad \sigma_c = -58.2 \text{ MPa}$$

4. *Finalize*: These stresses act on planes at 45° to the axis (Fig. 3-31b).

FIGURE 3-31

Stress and strain elements for the tube of Example 3-6: (a) maximum shear stresses, (b) maximum tensile and compressive stresses, (c) maximum shear strains, and (d) maximum tensile and compressive strains

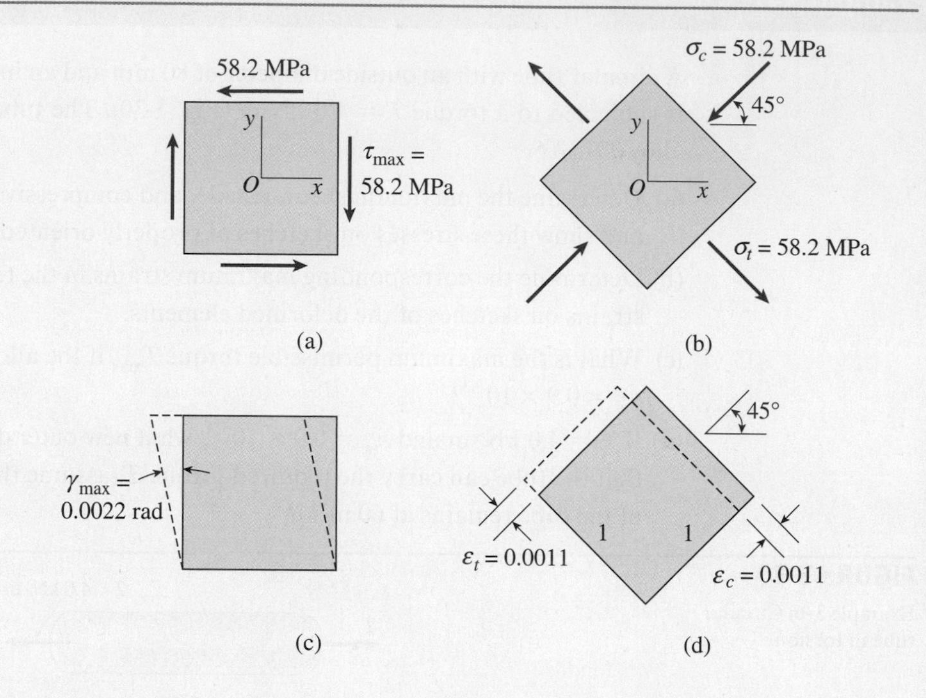

Part (b): Maximum strains.

1, 2. ***Conceptualize, Categorize***: The maximum shear strain in the tube is obtained from Eq. (3-30). The shear modulus of elasticity is obtained from Table I-2, Appendix I, as $G = 27$ GPa.

3. ***Analyze***: Therefore, the maximum shear strain is

$$\gamma_{max} = \frac{\tau_{max}}{G} = \frac{58.2 \text{ MPa}}{27 \text{ GPa}} = 0.0022 \text{ rad}$$

The deformed element is shown by the dashed lines in Fig. 3-31c.

The magnitude of the maximum normal strains [from Eq. (3-32)] is

$$\varepsilon_{max} = \frac{\gamma_{max}}{2} = 0.0011$$

Thus, the maximum tensile and compressive strains are

$$\varepsilon_t = 0.0011 \qquad \varepsilon_c = -0.0011$$

4. ***Finalize***: The deformed element is shown by the dashed lines in Fig. 3-31d for an element with sides of unit length.

Part (c): Maximum permissible torque.

1, 2. ***Conceptualize, Categorize***: The tube is in *pure shear*, so the allowable shear strain is twice the allowable normal strain [see Eq. (3-32)]:

$$\gamma_a = 2\varepsilon_a = 2(0.9 \times 10^{-3}) = 1.8 \times 10^{-3}$$

3. ***Analyze***: From the shear formula [Eq. (3-13)],

$$\tau_{max} = \frac{T\left(\frac{d_2}{2}\right)}{I_p} \quad \text{so} \quad T_{max} = \frac{\tau_a I_p}{\left(\frac{d_2}{2}\right)} = \frac{2(G\gamma_a)I_p}{d_2}$$

where d_2 is the outer diameter. Substituting numerical values gives

$$T_{max} = \frac{2(27\text{ GPa})(1.8 \times 10^{-3})\left[\frac{\pi}{32}[(0.08\text{ m})^4 - (0.06\text{ m})^4]\right]}{0.08\text{ m}}$$

$$= 3.34\text{ kN} \cdot \text{m}$$

4. ***Finalize***: The allowable shear strain is less than γ_{max} in part (b), so T_{max} is less than 4.0 kN·m, as expected.

Part (d): New outer diameter of tube.

1, 2. ***Conceptualize, Categorize***: Use the previous equation but with $T = 4.0\text{ kN} \cdot \text{m}$ to find the required outer diameter d_2:

$$\frac{I_p}{d_2} = \frac{T}{2G\gamma_a} \quad \text{or} \quad \frac{d_2^4 - (0.06\text{ m})^4}{d_2} = \frac{\left(\frac{32}{\pi}\right)4\text{ kN} \cdot \text{m}}{2(27\text{ GPa})(1.8 \times 10^{-3})} = 4.192 \times 10^{-4}\text{ m}^3$$

3. ***Analyze***: Solving for the required outer diameter d_2 numerically gives

$$d_2 = 83.2\text{ mm}$$

4. ***Finalize***: Outer diameter d_2 is larger than that in part (c), as expected, since the applied torque here is larger than T_{max} in part (c).

3.6 Relationship Between Moduli of Elasticity *E* and *G*

An important relationship between the moduli of elasticity E and G can be obtained from the equations derived in the preceding section. For this purpose, consider the stress element *abcd* shown in Fig. 3-32a. The front face of the element is assumed to be square with the length of each side denoted as h. When this element is subjected to pure shear by stresses τ, the front face distorts into a rhombus (Fig. 3-32b) with sides of length h and with shear strain $\gamma = \tau/G$. Because of the distortion, diagonal *bd* is lengthened, and diagonal *ac* is shortened. The length of diagonal *bd* is equal to its initial length $\sqrt{2}h$ times the factor $1 + \varepsilon_{max}$, where ε_{max} is the normal strain in the 45° direction; thus,

$$L_{bd} = \sqrt{2}h(1 + \varepsilon_{max}) \qquad \text{(3-33)}$$

This length can be related to the shear strain γ by considering the geometry of the deformed element.

FIGURE 3-32

Geometry of deformed element in pure shear

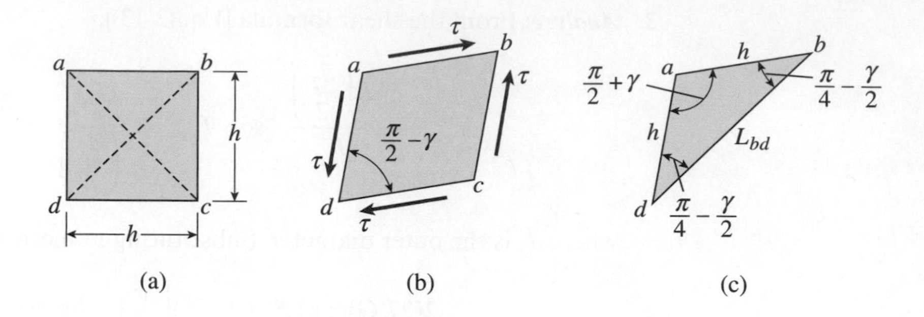

To obtain the required geometric relationships, consider triangle *abd* (Fig. 3-32c), which represents one-half of the rhombus pictured in Fig. 3-32b. Side *bd* of this triangle has a length L_{bd} [Eq. (3-33)], and the other sides each have a length *h*. Angle *adb* of the triangle is equal to one-half of angle *adc* of the rhombus, or $\pi/4 - \gamma/2$. The angle *abd* in the triangle is the same. Therefore, angle *dab* of the triangle equals $\pi/2 + \gamma$. Now use the law of cosines (see Appendix C) for triangle *abd* to get

$$L_{bd}^2 = h^2 + h^2 - 2h^2 \cos\left(\frac{\pi}{2} + \gamma\right)$$

Substitute for L_{bd} from Eq. (3-33) and simplify to get

$$(1 + \varepsilon_{max})^2 = 1 - \cos\left(\frac{\pi}{2} + \gamma\right)$$

Expand the term on the left-hand side, and also observe that $\cos(\pi/2 + \gamma) = -\sin\gamma$ to obtain

$$1 + 2\varepsilon_{max} + \varepsilon_{max}^2 = 1 + \sin\gamma$$

Because ε_{max} and γ are very small strains, disregard ε_{max}^2 in comparison with $2\varepsilon_{max}$ and replace $\sin\gamma$ by γ. The resulting expression is

$$\varepsilon_{max} = \frac{\gamma}{2} \qquad \textbf{(3-34)}$$

which establishes the relationship already presented in Section 3.5 as Eq. (3-32).

The shear strain γ appearing in Eq. (3-34) is equal to τ/G by Hooke's law [Eq. (3-30)], and the normal strain ε_{max} is equal to $\tau(1 + \nu)/E$ by Eq. (3-31). Making both of these substitutions in Eq. (3-34) yields

$$G = \frac{E}{2(1 + \nu)} \qquad \textbf{(3-35)}$$

So *E*, *G*, and ν are not independent properties of a linearly elastic material. Instead, if any two of them are known, the third can be calculated from Eq. (3-35).

Typical values of *E*, *G*, and ν are listed in Table I-2, Appendix I.

3.7 Transmission of Power by Circular Shafts

The most important use of circular shafts is to transmit mechanical power from one device or machine to another, as in the drive shaft of an automobile, the propeller shaft of a ship, or the axle of a bicycle. The power is transmitted through the rotary motion of the shaft, and the amount of power transmitted depends upon the magnitude of the torque and the speed of rotation. A common design problem is to determine the required size of a shaft so that it will transmit a specified amount of power at a specified rotational speed without exceeding the allowable stresses for the material.

FIGURE 3-33

Shaft transmitting a constant torque T at an angular speed ω

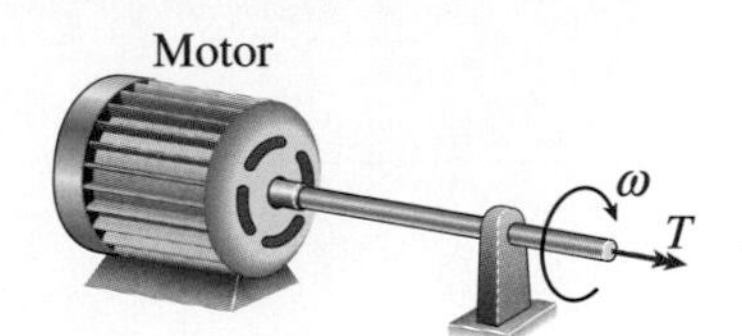

Suppose that a motor-driven shaft (Fig. 3-33) is rotating at an angular speed ω, measured in radians per second (rad/s). The shaft transmits a torque T to a device (not shown in the figure) that is performing useful work. The torque applied by the shaft to the external device has the same sense as the angular speed ω, that is, its vector points to the left. However, the torque shown in the figure is the torque exerted *on the shaft* by the device, and so its vector points in the opposite direction.

In general, the work W done by a torque of constant magnitude is equal to the product of the torque and the angle through which it rotates; that is,

$$W = T\psi \tag{3-36}$$

where ψ is the angle of rotation in radians.

Power is the *rate* at which work is done, or

$$P = \frac{dW}{dt} = T\frac{d\psi}{dt} \tag{3-37}$$

in which P is the symbol for power and t represents time. The rate of change $d\psi/dt$ of the angular displacement ψ is the angular speed ω, and therefore, the preceding equation becomes

$$P = T\omega \quad (\omega = \text{rad/s}) \tag{3-38}$$

This formula, which is familiar from elementary physics, gives the power transmitted by a rotating shaft transmitting a constant torque T.

The **units** to be used in Eq. (3-38) are as follows. If the torque T is expressed in newton meters, then the power is expressed in watts (W). One watt is equal to one newton meter per second (or one joule per second).

Angular speed is often expressed as the frequency f of rotation, which is the number of revolutions per unit of time. The unit of frequency is the hertz (Hz), which is equal to one revolution per second (s^{-1}). One revolution equals 2π radians, so

$$\omega = 2\pi f \quad (\omega = \text{rad/s},\ f = \text{Hz} = \text{s}^{-1}) \tag{3-39}$$

The expression for power [Eq. (3-3)] then becomes

$$P = 2\pi f T \quad (f = \text{Hz} = \text{s}^{-1}) \tag{3-40}$$

Another commonly used unit is the number of revolutions per minute (rpm), denoted by the letter n. It follows that

$$n = 60f \tag{3-41}$$

and

$$P = \frac{2\pi nT}{60} \quad (n = \text{rpm}) \tag{3-42}$$

In Eqs. (3-40) and (3-42), the quantities P and T have the same units as in Eq. (3-38); that is, P has units of watts if T has units of newton meters.

In U.S. engineering practice, power is sometimes expressed in horsepower (hp), which is a unit equal to 550 ft-lb/s. Therefore, the horsepower H being transmitted by a rotating shaft is

$$H = \frac{2\pi nT}{60(550)} = \frac{2\pi nT}{33{,}000} \quad (n = \text{rpm},\ T = \text{lb-ft},\ H = \text{hp}) \tag{3-43}$$

One horsepower is approximately 746 watts.

The preceding equations relate the torque acting in a shaft to the power transmitted by the shaft. Once the torque is known, you can determine the shear stresses, shear strains, angles of twist, and other desired quantities by the methods described in Sections 3.2 through 3.5.

The following examples illustrate some of the procedures for analyzing rotating shafts.

Example 3-7

A motor driving a solid circular steel shaft transmits 30 kW to a gear at B (Fig. 3-34). The allowable shear stress in the steel is 42 MPa.

(a) What is the required diameter d of the shaft if it is operated at 500 rpm?
(b) What is the required diameter d if it is operated at 3000 rpm?

FIGURE 3-34

Example 3-7: Steel shaft in torsion

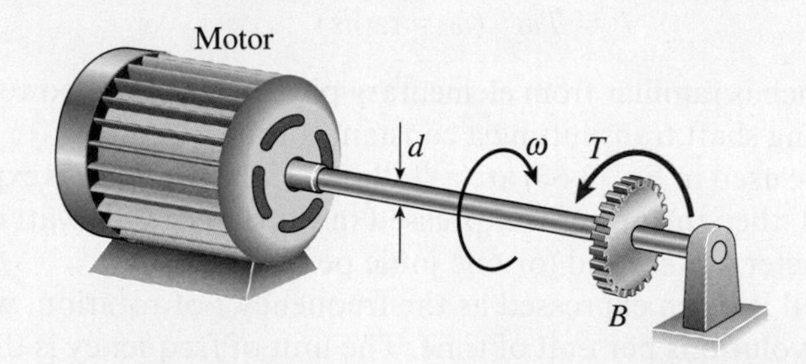

Solution:

Use a four-step problem-solving approach. Combine steps as needed for an efficient solution.

Part (a): Motor operating at 500 rpm.

1, 2. ***Conceptualize, Categorize***: Knowing the power and the speed of rotation, the torque T acting on the shaft is found from Eq. (3-43). Solve that equation for T to get

$$T = \frac{60\ P}{2\pi n} = \frac{60(30\ \text{kW})}{2\pi(500\ \text{rpm})} = 573\ \text{N}\cdot\text{m}$$

This torque is transmitted by the shaft from the motor to the gear.

The maximum shear stress in the shaft can be obtained from the modified torsion formula [Eq. (3-14)]:

$$\tau_{max} = \frac{16T}{\pi d^3}$$

3. ***Analyze***: Solve that equation for the diameter d, and also substitute τ_{allow} for τ_{max} to get

$$d^3 = \frac{16T}{\pi \tau_{allow}} = \frac{16(573\ \text{N}\cdot\text{m})}{\pi(42\ \text{MPa})} = 6.948 \times 10^{-5}\ \text{m}^3$$

from which

$$d = 41.1\ \text{mm}$$

4. ***Finalize***: The diameter of the shaft must be at least this large if the allowable shear stress is not to be exceeded.

Part (b): Motor operating at 3000 rpm.

1, 2. ***Conceptualize, Categorize***: Follow the same procedure as in part (a).

3. ***Analyze***: Torque T and diameter d are now

$$T = \frac{60\ P}{2\pi n} = \frac{60(30\ \text{kW})}{2\pi(3000\ \text{rpm})} = 95.49\ \text{N}\cdot\text{m}$$

$$d^3 = \frac{16T}{\pi \tau_{allow}} = \frac{16(95.49\ \text{N}\cdot\text{m})}{\pi(42\ \text{MPa})} = 1.158 \times 10^{-5}\ \text{m}^3$$

$$d = 22.6\ \text{mm}$$

which is less than the diameter found in part (a).

4. ***Finalize***: This example illustrates that the higher the speed of rotation, the smaller the required size of the shaft (for the same power and the same allowable stress).

Example 3-8

A solid steel shaft ABC with a 50 mm diameter (Fig. 3-35a) is driven at A by a motor that transmits 50 kW to the shaft at 10 Hz. The gears at B and C drive machinery requiring power equal to 35 kW and 15 kW, respectively.

Compute the maximum shear stress τ_{max} in the shaft and the angle of twist ϕ_{AC} between the motor at A and the gear at C. (Use $G = 80$ GPa.)

Solution:

Use a four-step problem-solving approach.

1. ***Conceptualize***:

Torques acting on the shaft: Begin the analysis by determining the torques applied to the shaft by the motor and the two gears. Since the motor supplies

FIGURE 3-35

Example 3-8: Steel shaft in torsion

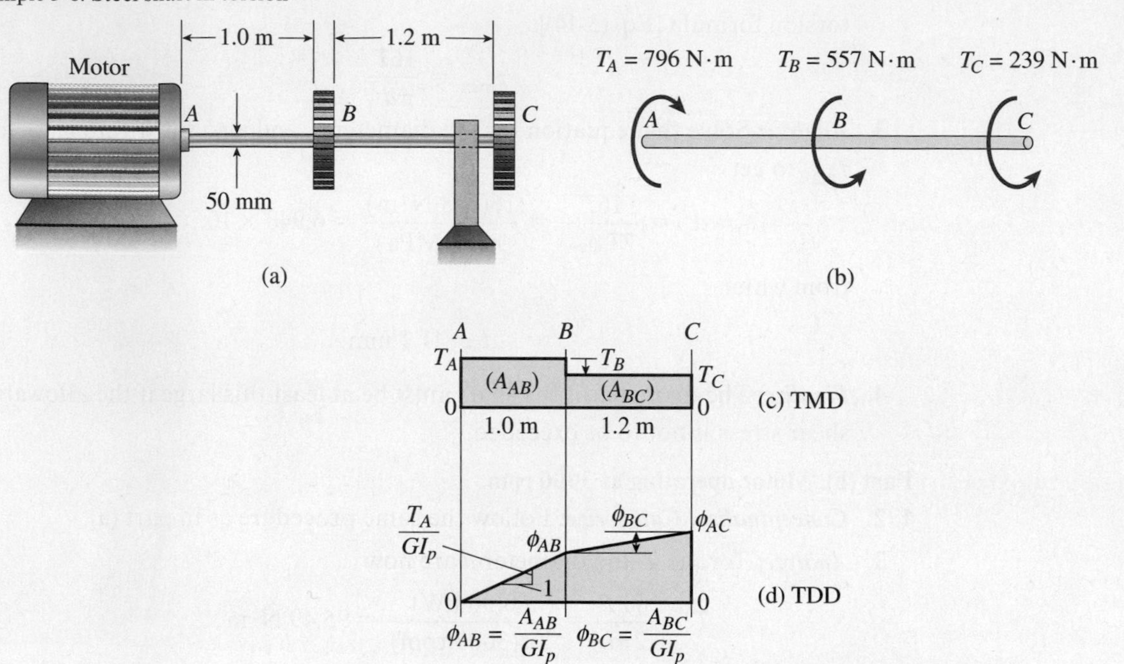

50 kW at 10 Hz, it creates a torque T_A at end A of the shaft (Fig. 3-35b) calculated from Eq. (3-40):

$$T_A = \frac{P}{2\pi f} = \frac{50 \text{ kW}}{2\pi(10 \text{ Hz})} = 796 \text{ N} \cdot \text{m}$$

In a similar manner, calculate the torques T_B and T_C applied by the gears to the shaft:

$$T_B = \frac{P}{2\pi f} = \frac{35 \text{ kW}}{2\pi(10 \text{ Hz})} = 557 \text{ N} \cdot \text{m}$$

$$T_C = \frac{P}{2\pi f} = \frac{15 \text{ kW}}{2\pi(10 \text{ Hz})} = 239 \text{ N} \cdot \text{m}$$

These torques are shown in the free-body diagram of the shaft (Fig. 3-35b). Note that the torques applied by the gears are opposite in direction to the torque applied by the motor. (Think of T_A as the "load" applied to the shaft by the motor, then the torques T_B and T_C are the "reactions" of the gears.)

The internal torques in the two segments of the shaft are now found (by inspection) from the free-body diagram of Fig. 3-35b:

$$T_{AB} = 796 \text{ N} \cdot \text{m} \quad T_{BC} = 239 \text{ N} \cdot \text{m}$$

The TMD is shown in Fig. 3-35c.

2. ***Categorize:*** Both internal torques act in the same direction; therefore, the angles of twist in segments AB and BC are additive when finding the total angle of

twist. (To be specific, both torques are positive according to the sign convention adopted in Section 3.4.)

3. ***Analyze:***

Shear stresses and angles of twist: The shear stress and angle of twist in segment AB of the shaft are found in the usual manner from Eqs. (3-14) and (3-17):

$$\tau_{AB} = \frac{16T_{AB}}{\pi d^3} = \frac{16(796 \text{ N} \cdot \text{m})}{\pi(50 \text{ mm})^3} = 32.4 \text{ MPa}$$

$$\phi_{AB} = \frac{T_{AB}L_{AB}}{GI_p} = \frac{(796 \text{ N} \cdot \text{m})(1.0 \text{ m})}{(80 \text{ GPa})\left(\frac{\pi}{32}\right)(50 \text{ mm})^4} = 0.0162 \text{ rad}$$

The corresponding quantities for segment BC are

$$\tau_{BC} = \frac{16T_{BC}}{\pi d^3} = \frac{16(239 \text{ N} \cdot \text{m})}{\pi(50 \text{ mm})^3} = 9.7 \text{ MPa}$$

$$\phi_{BC} = \frac{T_{BC}L_{BC}}{GI_p} = \frac{(239 \text{ N} \cdot \text{m})(1.2 \text{ m})}{(80 \text{ GPa})\left(\frac{\pi}{32}\right)(50 \text{ mm})^4} = 0.0058 \text{ rad}$$

4. ***Finalize:*** Thus, the maximum shear stress in the shaft occurs in segment AB and is

$$\tau_{max} = 32.4 \text{ MPa}$$

Also, the total angle of twist between the motor at A and the gear at C is

$$\phi_{AC} = \phi_{AB} + \phi_{BC} = 0.0162 \text{ rad} + 0.0058 \text{ rad} = 0.0220 \text{ rad} = 1.26°$$

The TDD is shown in Fig. 3-35d. As explained previously, both parts of the shaft twist in the same direction; therefore, the angles of twist are added.

3.8 Statically Indeterminate Torsional Members

The bars and shafts described in the preceding sections of this chapter are *statically determinate* because all internal torques and all reactions can be obtained from free-body diagrams and equations of equilibrium. However, if additional restraints, such as fixed supports, are added to the bars, the equations of equilibrium will no longer be adequate for determining the torques. The bars are then classified as **statically indeterminate**. Torsional members of this kind can be analyzed by supplementing the equilibrium equations with compatibility equations pertaining to the rotational displacements. Thus, the general method for analyzing statically indeterminate torsional members is the same as described in Section 2.4 for statically indeterminate bars with axial loads.

The first step in the analysis is to write **equations of equilibrium**, obtained from free-body diagrams of the given physical situation. The unknown quantities in the equilibrium equations are torques, either internal torques or reaction torques.

FIGURE 3-36
Statically indeterminate bar in torsion

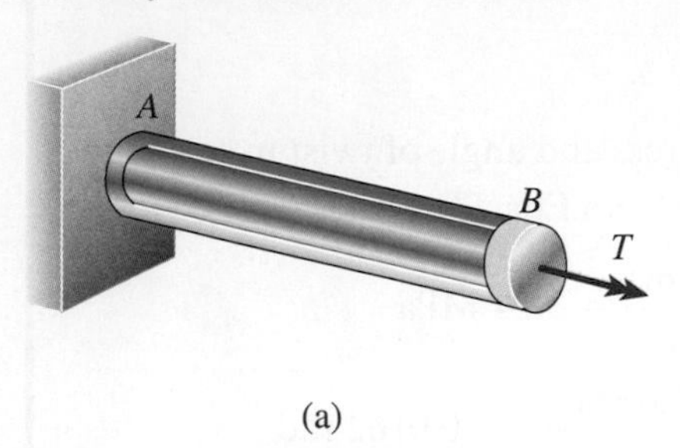

(a)

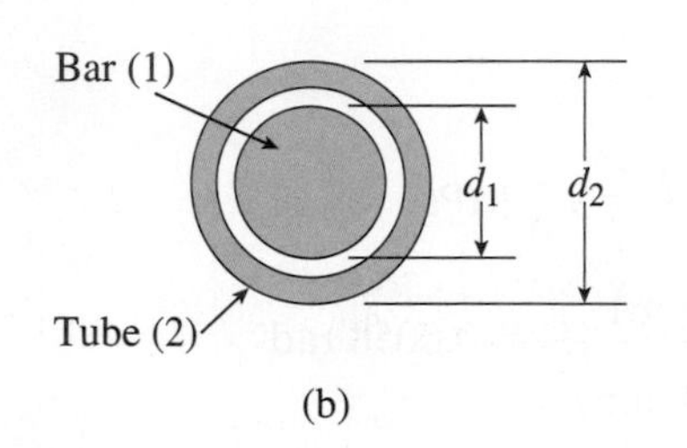

(b)

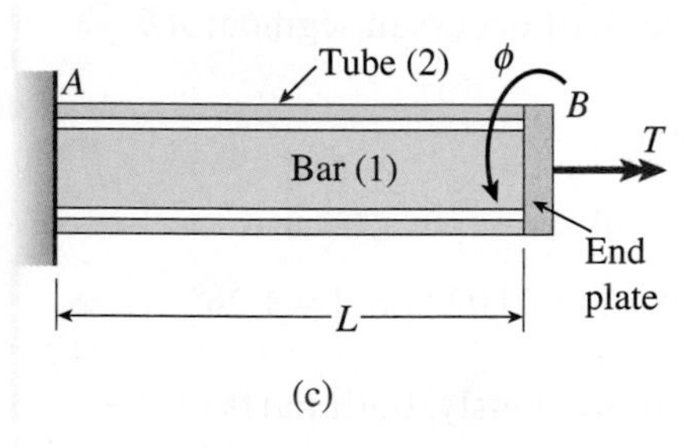

(c)

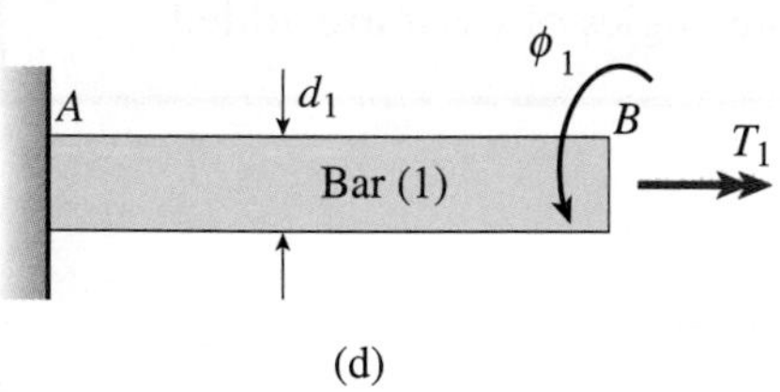

(d)

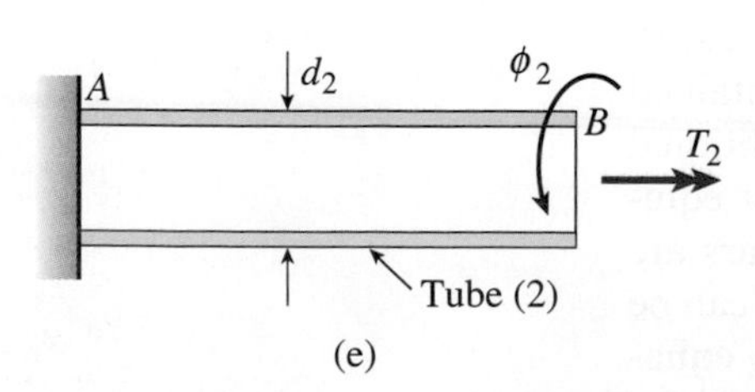

(e)

The second step in the analysis is to formulate **equations of compatibility** based upon physical conditions pertaining to the angles of twist. As a consequence, the compatibility equations contain angles of twist as unknowns.

The third step is to relate the angles of twist to the torques by **torque-displacement relations**, such as $\phi = TL/GI_p$. After introducing these relations into the compatibility equations, they too become equations containing torques as unknowns. Therefore, the last step is to obtain the unknown torques by solving simultaneously the equations of equilibrium and compatibility.

To illustrate the method of solution, analyze the composite bar AB shown in Fig. 3-36a. The bar is attached to a fixed support at end A and loaded by a torque T at end B. Furthermore, the bar consists of two parts: a solid bar and a tube (Figs. 3-36b and c) with both the solid bar and the tube joined to a rigid end plate at B.

For convenience, identify the solid bar and tube (and their properties) by the numerals 1 and 2, respectively. For instance, the diameter of the solid bar is denoted d_1 and the outer diameter of the tube is denoted d_2. A small gap exists between the bar and the tube; therefore, the inner diameter of the tube is slightly larger than the diameter d_1 of the bar.

When the torque T is applied to the composite bar, the end plate rotates through a small angle ϕ (Fig. 3-36c), and torques T_1 and T_2 are developed in the solid bar and the tube, respectively (Figs. 3-36d and e). From equilibrium, the sum of these torques equals the applied load, so the *equation of equilibrium* is

$$T_1 + T_2 = T \quad \textbf{(3-44)}$$

Because this equation contains two unknowns (T_1 and T_2), the composite bar is statically indeterminate.

To obtain a second equation, consider the rotational displacements of both the solid bar and the tube. Note that the angle of twist of the solid bar (Fig. 3-36d) is ϕ_1, and the angle of twist of the tube is ϕ_2 (Fig. 3-36e). These angles of twist must be equal because the bar and tube are securely joined to the end plate and rotate with it; consequently, the *equation of compatibility* is

$$\phi_1 = \phi_2 \quad \textbf{(3-45)}$$

The angles ϕ_1 and ϕ_2 are related to the torques T_1 and T_2 by the *torque-displacement relations*, which in the case of linearly elastic materials are obtained from the equation $\phi = TL/GI_p$. Thus,

$$\phi_1 = \frac{T_1L}{G_1I_{p1}} \qquad \phi_2 = \frac{T_2L}{G_2I_{p2}} \quad \textbf{(3-46a,b)}$$

in which G_1 and G_2 are the shear moduli of elasticity of the materials and I_{p1} and I_{p2} are the polar moments of inertia of the cross sections.

When the preceding expressions for ϕ_1 and ϕ_2 are substituted into Eq. (3-45), the equation of compatibility becomes

$$\frac{T_1L}{G_1I_{p1}} = \frac{T_2L}{G_2I_{p2}} \quad \textbf{(3-47)}$$

The two equations [Eqs. (3-44) and (3-47)] have two unknowns, so solving for the torques T_1 and T_2 gives

$$T_1 = T\left(\frac{G_1 I_{p1}}{G_1 I_{p1} + G_2 I_{p2}}\right) \quad T_2 = T\left(\frac{G_2 I_{p2}}{G_1 I_{p1} + G_2 I_{p2}}\right) \qquad \textbf{(3-48a,b)}$$

With these torques known, the essential part of the statically indeterminate analysis is completed. All other quantities, such as stresses and angles of twist, now can be found from the torques.

The preceding discussion illustrates the general methodology for analyzing a statically indeterminate system in torsion. In the following example, this same approach is used to analyze a bar that is fixed against rotation at both ends. In the example and in the problems, assume that the bars are made of linearly elastic materials. However, the general methodology is also applicable to bars of nonlinear materials—the only change is in the torque-displacement relations.

Example 3-9

The bar ACB shown in Figs. 3-37a and b is fixed at both ends and loaded by a torque T_0 at point C. Segments AC and CB of the bar have diameters d_A and d_B, lengths L_A and L_B, and polar moments of inertia I_{PA} and I_{PB}, respectively. The material of the bar is the same throughout both segments.

Obtain formulas for (a) the reactive torques T_A and T_B at the ends, (b) the maximum shear stresses τ_{AC} and τ_{CB} in each segment of the bar, and (c) the angle of rotation ϕ_C at the cross section where the load T_0 is applied.

Solution:

Use a four-step problem-solving approach. Combine steps as needed for an efficient solution.

1, 2. ***Conceptualize, Categorize:***

Equation of equilibrium: The load T_0 produces reactions T_A and T_B at the fixed ends of the bar, as shown in Figs. 3-37a and b. Equilibrium of the bar gives

$$T_A + T_B = T_0 \qquad \textbf{(a)}$$

Because there are two unknowns in this equation (and no other useful equations of equilibrium), the bar is statically indeterminate.

Equation of compatibility: Now separate the bar from its support at end B and obtain a bar that is fixed at end A and free at end B (Figs. 3-37c and d). When the load T_0 acts alone (Fig. 3-37c), it produces an angle of twist at end B denoted as ϕ_1. Similarly, when the reactive torque T_B acts alone, it produces an angle ϕ_2 (Fig. 3-37d). The angle of twist at end B in the original bar, which is equal to the sum of ϕ_1 and ϕ_2, is zero. Therefore, the equation of compatibility is

$$\phi_1 + \phi_2 = 0 \qquad \textbf{(b)}$$

Note that ϕ_1 and ϕ_2 are assumed to be positive in the direction shown in the figure.

FIGURE 3-37

Example 3-9: Statically indeterminate bar in torsion

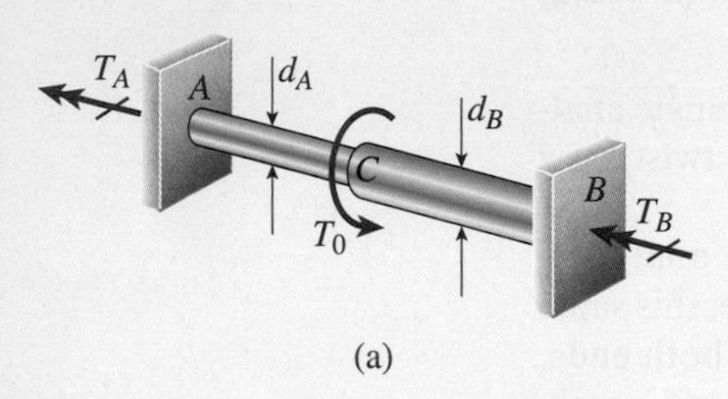

(a)

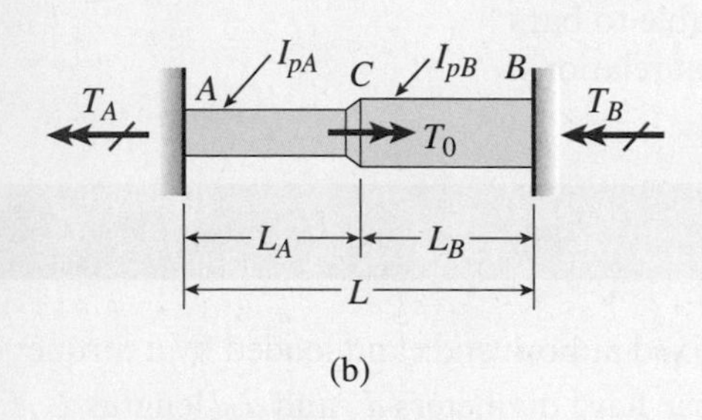

(b)

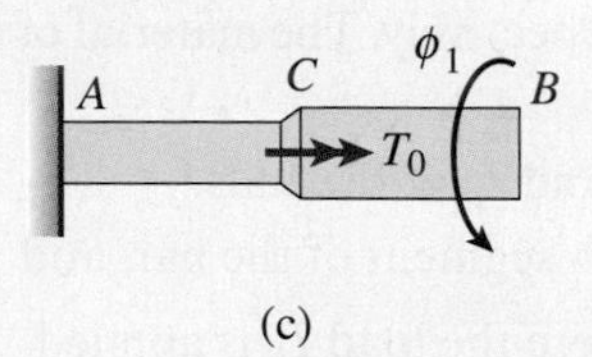

(c)

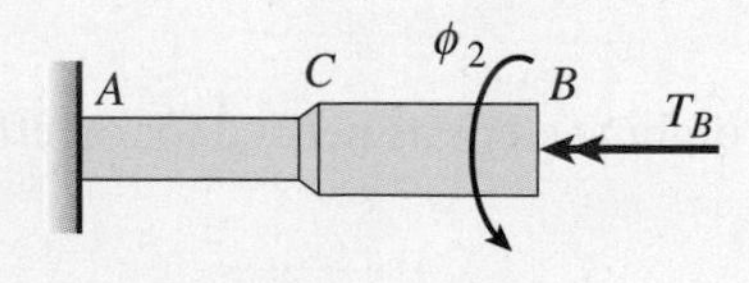

(d)

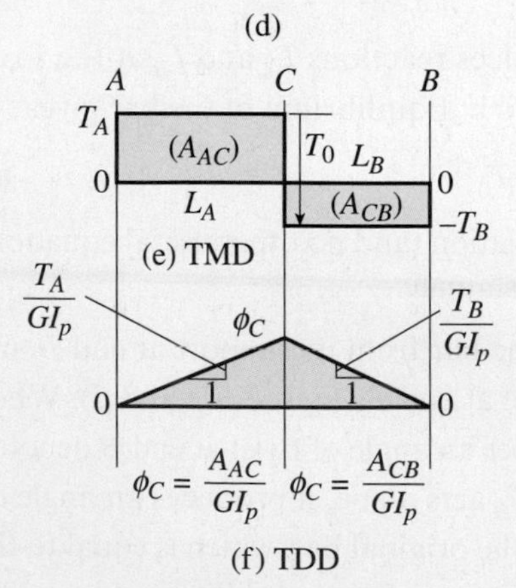

Torque-displacement equations: The angles of twist ϕ_1 and ϕ_2 can be expressed in terms of the torques T_0 and T_B by referring to Figs. 3-37c and d and using the equation $\phi = TL/GI_p$. The equations are

$$\phi_1 = \frac{T_0 L_A}{GI_{pA}} \quad \phi_2 = -\frac{T_B L_A}{GI_{pA}} - \frac{T_B L_B}{GI_{pB}} \tag{c,d}$$

The minus signs appear in Eq. (d) because T_B produces a rotation that is opposite in direction to the positive direction of ϕ_2 (Fig. 3-37d).

Now substitute the angles of twist [Eqs. (c) and (d)] into the compatibility equation [Eq. (b)] and obtain

$$\frac{T_0 L_A}{GI_{pA}} - \frac{T_B L_A}{GI_{pA}} - \frac{T_B L_B}{GI_{pB}} = 0$$

or

$$\frac{T_B L_A}{I_{pA}} + \frac{T_B L_B}{I_{pB}} = \frac{T_0 L_A}{I_{pA}} \tag{e}$$

3. *Analyze:*

Solution of equations: The preceding equation can be solved for the torque T_B, which then can be substituted into the equation of equilibrium [Eq. (a)] to obtain the torque T_A. The results are

$$T_A = T_0\left(\frac{L_B I_{pA}}{L_B I_{pA} + L_A I_{pB}}\right) \quad T_B = T_0\left(\frac{L_A I_{pB}}{L_B I_{pA} + L_A I_{pB}}\right) \tag{3-49a,b}$$

Thus, the reactive torques at the ends of the bar have been found, and the statically indeterminate part of the analysis is completed.

As a special case, note that if the bar is prismatic ($I_{pA} = I_{pB} = I_p$), the preceding results simplify to

$$T_A = \frac{T_0 L_B}{L} \quad T_B = \frac{T_0 L_A}{L} \tag{3-50a,b}$$

where L is the total length of the bar. These equations are analogous to those for the reactions of an axially loaded bar with fixed ends [see Eqs. (2-12a and b)]. The TMD for this special case is shown in Fig. 3-37e.

Maximum shear stresses: The maximum shear stresses in each part of the bar are obtained directly from the torsion formula:

$$\tau_{AC} = \frac{T_A d_A}{2I_{pA}} \quad \tau_{CB} = \frac{T_B d_B}{2I_{pB}}$$

Substituting from Eqs. (3-49a and b) gives

$$\tau_{AC} = \frac{T_0 L_B d_A}{2(L_B I_{pA} + L_A I_{pB})} \qquad \tau_{CB} = \frac{T_0 L_A d_B}{2(L_B I_{pA} + L_A I_{pB})} \qquad \text{(3-51a,b)}$$

By comparing the product $L_B d_A$ with the product $L_A d_B$, you can immediately determine which segment of the bar has the larger stress.

Angle of rotation: The angle of rotation ϕ_C at section C is equal to the angle of twist of either segment of the bar, since both segments rotate through the same angle at section C. Therefore, the angle of rotation is

$$\phi_C = \frac{T_A L_A}{G I_{pA}} = \frac{T_B L_B}{G I_{pB}} = \frac{T_0 L_A L_B}{G(L_B I_{pA} + L_A I_{pB})} \qquad \text{(3-52)}$$

In the special case of a prismatic bar ($I_{pA} = I_{pB} = I_p$), the angle of rotation at the section where the load is applied is

$$\phi_C = \frac{T_0 L_A L_B}{G L I_p} \qquad \text{(3-53)}$$

The TDD for the case of a prismatic bar is shown in Fig. 3-37f.

4. ***Finalize***: This example illustrates not only the analysis of a statically indeterminate bar but also the techniques for finding stresses and angles of rotation. In addition, note that the results obtained in this example are valid for a bar consisting of either solid or tubular segments.

3.9 Strain Energy in Torsion and Pure Shear

When a load is applied to a structure, work is performed by the load and strain energy is developed in the structure, as described in detail in Section 2.7 for a bar subjected to axial loads. This section uses the same basic concepts to determine the strain energy of a bar in torsion.

Consider a prismatic bar AB in **pure torsion** under the action of a torque T (Fig. 3-38). When the load is applied statically, the bar twists and the free end rotates through an angle ϕ. Assuming that the material of the bar is linearly elastic and follows Hooke's law, then the relationship between the applied torque and the angle of twist will also be linear, as shown by the torque-rotation diagram of Fig. 3-39 and as given by the equation $\phi = TL/GI_p$.

FIGURE 3-38

Prismatic bar in pure torsion

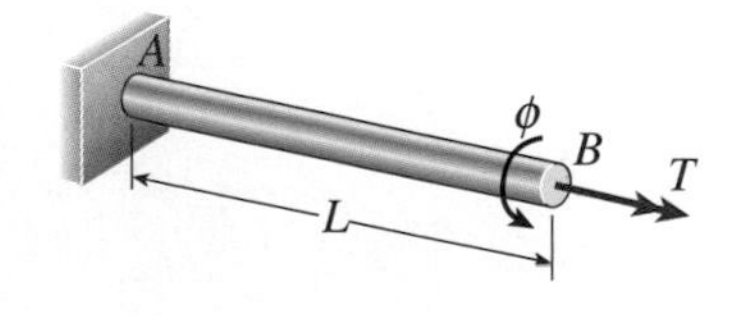

The work W done by the torque as it rotates through the angle ϕ is equal to the area below the torque-rotation line OA, that is, it is equal to the area of the shaded triangle in Fig. 3-39. Furthermore, from the principle of conservation of energy, the strain energy of the bar is equal to the work done by the load, provided no energy is gained or lost in the form of heat. Therefore, the equation for the strain energy U of the bar is

$$U = W = \frac{T\phi}{2} \qquad \text{(3-54)}$$

FIGURE 3-39

Torque-rotation diagram for a bar in pure torsion (linearly elastic material)

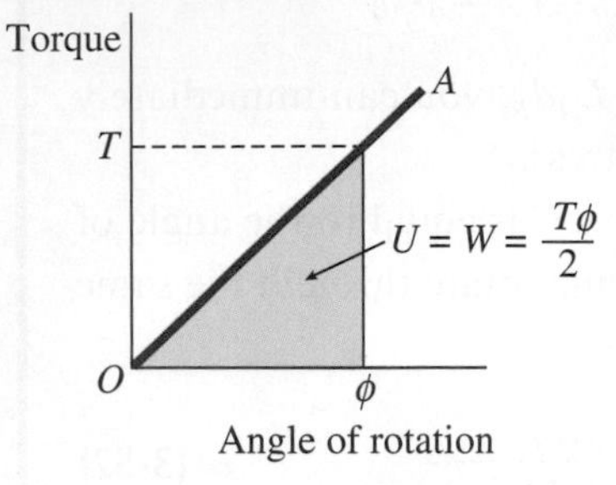

This equation is analogous to the equation $U = W = P\delta/2$ for a bar subjected to an axial load [see Eq. (2-35)].

Using the equation $\phi = TL/GI_p$, the strain energy is expressed in the following forms

$$U = \frac{T^2L}{2GI_p} \qquad U = \frac{GI_p\phi^2}{2L} \tag{3-55a,b}$$

The first expression is in terms of the load, and the second is in terms of the angle of twist. Again, note the analogy with the corresponding equations for a bar with an axial load [see Eqs. (2-37a and b)].

The SI unit for both work and energy is the joule (J), which is equal to one newton meter (1 J = 1 N·m).

Nonuniform Torsion

If a bar is subjected to nonuniform torsion (described in Section 3.4), additional formulas are needed for the strain energy. In those cases where the bar consists of prismatic segments with constant torque in each segment (see Fig. 3-14a of Section 3.4), find the strain energy of each segment and then add to obtain the total energy of the bar:

$$U = \sum_{i=1}^{n} U_i \tag{3-56}$$

in which U_i is the strain energy of segment i and n is the number of segments. For instance, if Eq. (3-55a) is used to obtain the individual strain energies, the preceding equation becomes

$$U = \sum_{i=1}^{n} \frac{T_i^2 L_i}{2G_i(I_p)_i} \tag{3-57}$$

in which T_i is the internal torque in segment i and L_i, G_i, and $(I_p)_i$ are the torsional properties of the segment.

If either the cross section of the bar or the internal torque varies along the axis, as illustrated in Figs. 3-15 and 3-16 of Section 3.4, obtain the total strain energy by first determining the strain energy of an element and then integrating along the axis. For an element of a length dx, the strain energy is [see (Eq. 3-55a)]

$$dU = \frac{[T(x)]^2\,dx}{2GI_p(x)}$$

in which T(x) is the internal torque acting on the element and $I_p(x)$ is the polar moment of inertia of the cross section at the element. Therefore, the total strain energy of the bar is

$$U = \int_0^L \frac{[T(x)]^2\,dx}{2GI_p(x)} \tag{3-58}$$

Once again, the similarities of the expressions for strain energy in torsion and axial load should be noted [compare Eqs. (3-57) and (3-58) with (2-40) and (2-41) of Section 2.7].

The use of the preceding equations for nonuniform torsion is illustrated in the examples that follow. In Example 3-10, the strain energy is found for a bar in pure torsion with prismatic segments, and in Examples 3-11 and 3-12, the strain energy is found for bars with varying torques and varying cross-sectional dimensions.

In addition, Example 3-12 shows how, under very limited conditions, the angle of twist of a bar can be determined from its strain energy. (For a more detailed discussion of this method, including its limitations, see the subsection "Displacements Caused by a Single Load" in Section 2.7.)

Limitations

When evaluating strain energy, keep in mind that the equations derived in this section apply only to bars of linearly elastic materials with small angles of twist. Also, remember the important observation stated previously in Section 2.7, namely, *the strain energy of a structure supporting more than one load cannot be obtained by adding the strain energies obtained for the individual loads acting separately*. This observation is demonstrated in Example 3-10.

Strain-Energy Density in Pure Shear

Because the individual elements of a bar in torsion are stressed in pure shear, it is useful to obtain expressions for the strain energy associated with the shear stresses. Begin the analysis by considering a small element of material subjected to shear stresses τ on its side faces (Fig. 3-40a). For convenience, assume that the front face of the element is square, with each side having a length h. Although the figure shows only a two-dimensional view of the element, recognize that the element is actually three dimensional with the thickness t perpendicular to the plane of the figure.

Under the action of the shear stresses, the element is distorted, so the front face becomes a rhombus, as shown in Fig. 3-40b. The change in angle at each corner of the element is the shear strain γ.

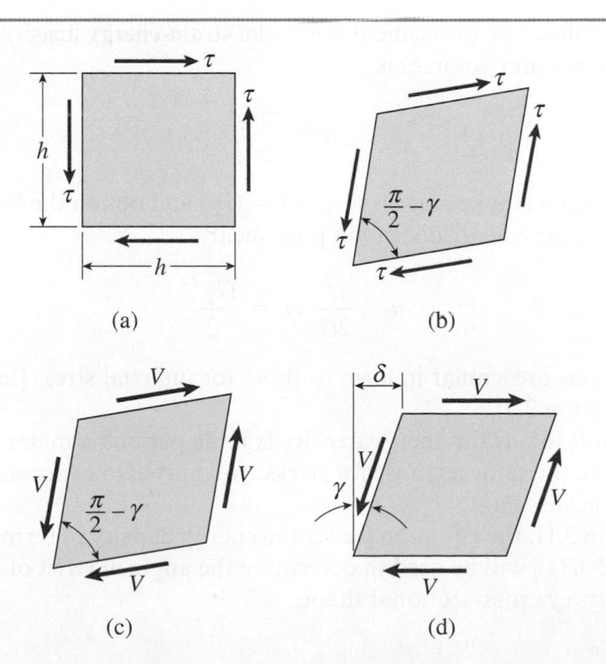

FIGURE 3-40
Element in pure shear

The shear forces V acting on the side faces of the element (Fig. 3-40c) are found by multiplying the stresses by the areas ht over which they act:

$$V = \tau ht \tag{3-59}$$

These forces produce work as the element deforms from its initial shape (Fig. 3-40a) to its distorted shape (Fig. 3-40b). To calculate this work, determine the relative distances through which the shear forces move. This task is made easier if the element in Fig. 3-40c is rotated as a rigid body until two of its faces are horizontal, as in Fig. 3-40d. During the rigid-body rotation, the net work done by the forces V is zero because the forces occur in pairs that form two equal and opposite couples.

As can be seen in Fig. 3-40d, the top face of the element is displaced horizontally through a distance δ (relative to the bottom face) as the shear force is gradually increased from zero to its final value V. The displacement δ is equal to the product of the shear strain γ (which is a small angle) and the vertical dimension of the element:

$$\delta = \gamma h \tag{3-60}$$

Now assume that the material is linearly elastic and follows Hooke's law, so that work done by the forces V is equal to $V\delta / 2$, which is also the strain energy stored in the element:

$$U = W = \frac{V\delta}{2} \tag{3-61}$$

Note that the forces acting on the side faces of the element (Fig. 3-40d) do not move along their lines of action—hence they do no work.

Substitute from Eqs. (3-59) and (3-60) into Eq. (3-61) to get the total strain energy of the element:

$$U = \frac{\tau \gamma h^2 t}{2}$$

Because the volume of the element is $h^2 t$, the **strain-energy density** u (that is, the strain energy per unit volume) is

$$u = \frac{\tau \gamma}{2} \tag{3-62}$$

Finally, substitute Hooke's law in shear ($\tau = G\gamma$) and obtain the following equations for the strain-energy density in pure shear:

$$u = \frac{\tau^2}{2G} \qquad u = \frac{G\gamma^2}{2} \tag{3-63a,b}$$

These equations are similar in form to those for uniaxial stress [see Eqs. (2-44a and b) of Section 2.7].

The SI unit for strain-energy density is joule per cubic meter (J/m^3). Since these units are the same as those for stress, you may also express strain-energy density in pascals (Pa).

In Section 3.11, the equation for strain-energy density in terms of the shear stress [Eq. (3-63a)] will be used to determine the angle of twist of a thin-walled tube of arbitrary cross-sectional shape.

Example 3-10

FIGURE 3-41

Example 3-10: Strain energy produced by two loads

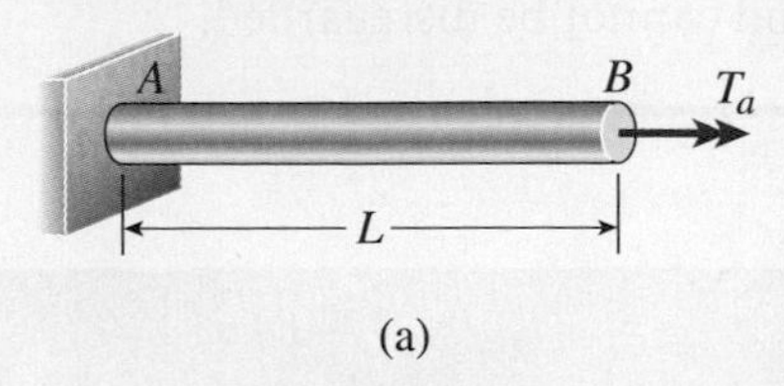

(a)

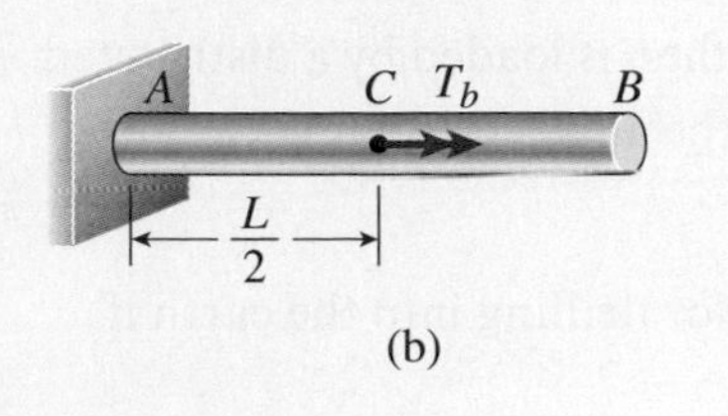

(b)

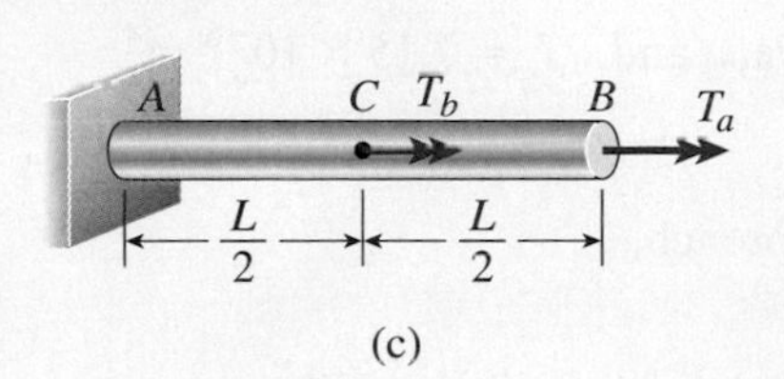

(c)

A solid circular bar AB of length L is fixed at one end and free at the other (Fig. 3-41). Three different loading conditions are to be considered: (a) torque T_a acting at the free end; (b) torque T_b acting at the midpoint of the bar; and (c) torques T_a and T_b acting simultaneously.

For each case of loading, obtain a formula for the strain energy stored in the bar. Then evaluate the strain energy for the following data: $T_a = 100\text{ N}\cdot\text{m}$, $T_b = 150\text{ N}\cdot\text{m}$, $L = 1.6\text{ m}$, $G = 80\text{ GPa}$, and $I_p = 79.52 \times 10^3\text{ mm}^4$.

Solution:

Use a four-step problem-solving approach. Combine steps as needed for an efficient solution.

1. ***Conceptualize, Categorize:***

Part (a): Torque T_a acting at the free end (Fig. 3-41a).

In this case, the strain energy is obtained directly from Eq. (3-55a):

$$U_a = \frac{T_a^2 L}{2GI_p} \quad \text{(a)}$$

Part (b): Torque T_b acting at the midpoint (Fig. 3-41b).

When the torque acts at the midpoint, apply Eq. (3-55a) to segment AC of the bar:

$$U_b = \frac{T_b^2(L/2)}{2GI_p} = \frac{T_b^2 L}{4GI_p} \quad \text{(b)}$$

Part (c): Torques T_a and T_b acting simultaneously (Fig. 3-41c).

When both loads act on the bar, the torque in segment CB is T_a and the torque in segment AC is $T_a + T_b$. Thus, the strain energy [from Eq. (3-57)] is

$$U_c = \sum_{i=1}^{n} \frac{T_i^2 L_i}{2G(I_p)_i} = \frac{T_a^2(L/2)}{2GI_p} + \frac{(T_a + T_b)^2(L/2)}{2GI_p}$$

$$= \frac{T_a^2 L}{2GI_p} + \frac{T_a T_b L}{2GI_p} + \frac{T_b^2 L}{4GI_p} \quad \text{(c)}$$

A comparison of Eqs. (a), (b), and (c) shows that the strain energy produced by the two loads acting simultaneously is *not* equal to the sum of the strain energies produced by the loads acting separately. As pointed out in Section 2.7, the reason is that strain energy is a quadratic function of the loads, not a linear function.

3. ***Analyze:***

Part (d): Numerical results.

Substitute the given data into Eq. (a) to obtain

$$U_a = \frac{T_a^2 L}{2GI_p} = \frac{(100\text{ N}\cdot\text{m})^2(1.6\text{ m})}{2(80\text{ GPa})(79.52 \times 10^3\text{ mm}^4)} = 1.26\text{ J}$$

Recall that one joule is equal to one newton-meter ($1\text{ J} = 1\text{ N}\cdot\text{m}$).

Proceed in the same manner for Eqs. (b) and (c) to find

$$U_b = 1.41\text{ J}$$
$$U_c = 1.26\text{ J} + 1.89\text{ J} + 1.41\text{ J} = 4.56\text{ J}$$

4. ***Finalize***: Note that the middle term, involving the product of the two loads, contributes significantly to the strain energy and cannot be disregarded.

Example 3-11

A prismatic bar AB, fixed at one end and free at the other, is loaded by a distributed torque of constant intensity t per unit distance along the axis of the bar (Fig. 3-42).

(a) Derive a formula for the strain energy of the bar.

(b) Evaluate the strain energy of a hollow shaft used for drilling into the earth if the data are

$$t = 2100\text{ N}\cdot\text{m/m},\quad L = 3.7\text{ m},\quad G = 80\text{ GPa},\quad \text{and}\quad I_p = 7.15 \times 10^{-6}\text{ m}^4$$

FIGURE 3-42

Example 3-11: Strain energy produced by a distributed torque

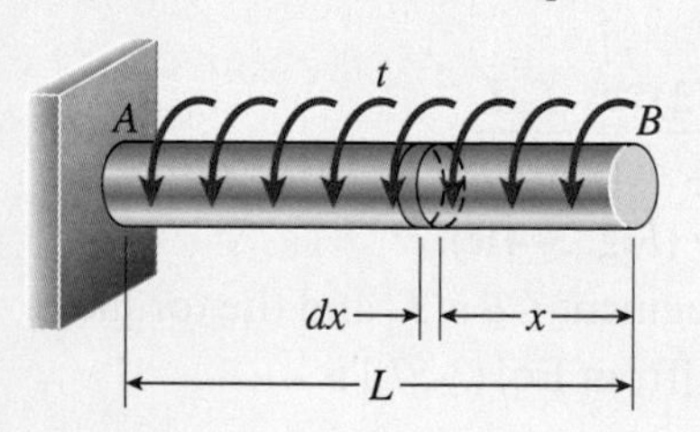

Solution:

Use a four-step problem-solving approach.

1,2. ***Conceptualize, Categorize***:

Part (a): Strain energy of the bar.

The first step in the solution is to determine the internal torque $T(x)$ acting at distance x from the free end of the bar (Fig. 3-42). This internal torque is equal to the total torque acting on the part of the bar between $x = 0$ and $x = x$. This latter torque is equal to the intensity t of torque times the distance x over which it acts:

$$T(x) = tx \tag{a}$$

Substitute into Eq. (3-58) to obtain

$$U = \int_0^L \frac{[T(x)]^2\,dx}{2GI_p} = \frac{1}{2GI_p}\int_0^L (tx)^2\,dx = \frac{t^2L^3}{6GI_p} \tag{3-64}$$

This expression gives the total strain energy stored in the bar.

3. ***Analyze***:

Part (b): Numerical results.

To evaluate the strain energy of the hollow shaft, substitute the given data into Eq. (3-64):

$$U = \frac{t^2L^3}{6GI_p} = \frac{(2100\text{ N}\cdot\text{m/m})^2(3.7\text{ m})^3}{6(80\text{ GPa})(7.15 \times 10^{-6}\text{ m}^4)} = 65.1\text{ N}\cdot\text{m}$$

4. ***Finalize***: This example illustrates the use of integration to evaluate the strain energy of a bar subjected to a distributed torque.

Example 3-12

FIGURE 3-43

Example 3-12: Tapered bar in torsion

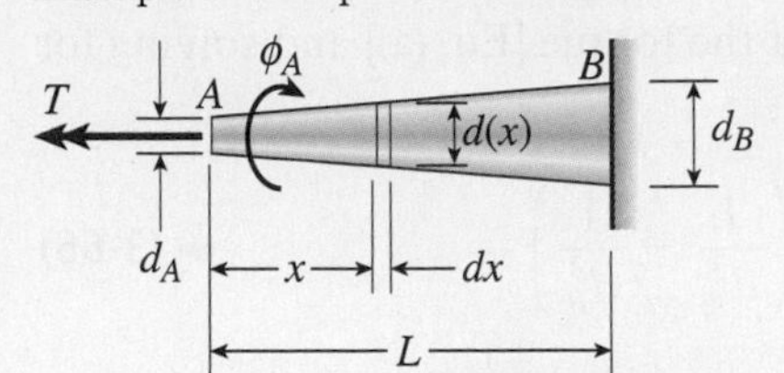

A tapered bar AB of solid circular cross section is supported at the right-hand end and loaded by a torque T at the other end (Fig. 3-43). The diameter of the bar varies linearly from d_A at the left-hand end to d_B at the right-hand end.

Determine the angle of rotation ϕ_A at end A of the bar by equating the strain energy to the work done by the load.

Solution:

Use a four-step problem-solving approach. Combine steps as needed for an efficient solution.

1, 2. ***Conceptualize, Categorize***: From the principle of conservation of energy, the work done by the applied torque equals the strain energy of the bar; thus, $W = U$. The work is given by

$$W = \frac{T\phi_A}{2} \tag{a}$$

and the strain energy U can be found from Eq. (3-58).

3. ***Analyze***: To use Eq. (3-58), first find expressions for the torque $T(x)$ and the polar moment of inertia $I_p(x)$. The torque is constant along the axis of the bar and equal to the load T, and the polar moment of inertia is

$$I_p(x) = \frac{\pi}{32}[d(x)]^4$$

in which $d(x)$ is the diameter of the bar at distance x from end A. From the geometry of the figure, diameter $d(x)$ is

$$d(x) = d_A + \frac{d_B - d_A}{L}x \tag{b}$$

and therefore

$$I_p(x) = \frac{\pi}{32}\left(d_A + \frac{d_B - d_A}{L}x\right)^4 \tag{c}$$

Now substitute into Eq. (3-58)

$$U = \int_0^L \frac{[T(x)]^2\,dx}{2GI_p(x)} = \frac{16T^2}{\pi G}\int_0^L \frac{dx}{\left(d_A + \dfrac{d_B - d_A}{L}x\right)^4}$$

The integral in this expression can be integrated with the aid of a table of integrals (see Appendix C) with the result:

$$\int_0^L \frac{dx}{\left(d_A + \dfrac{d_B - d_A}{L}x\right)^4} = \frac{L}{3(d_B - d_A)}\left(\frac{1}{d_A^3} - \frac{1}{d_B^3}\right)$$

Therefore, the strain energy of the tapered bar is

$$U = \frac{16T^2L}{3\pi G(d_B - d_A)}\left(\frac{1}{d_A^3} - \frac{1}{d_B^3}\right) \tag{3-65}$$

Equating the strain energy to the work of the torque [Eq. (a)] and solving for ϕ_A gives

$$\phi_A = \frac{32TL}{3\pi G(d_B - d_A)}\left(\frac{1}{d_A^3} - \frac{1}{d_B^3}\right) \tag{3-66}$$

4. **Finalize:** This equation gives the angle of rotation at end A of the tapered bar. [*Note:* This is the same angle of twist expression obtained in the solution of Prob. 3.4-10(a).]

Note especially that the method used in this example for finding the angle of rotation is suitable only when the bar is subjected to a single load, and then only when the desired angle corresponds to that load. Otherwise, angular displacements must be obtained by the usual methods described in Sections 3.3, 3.4, and 3.8.

3.10 Torsion of Noncircular Prismatic Shafts

Sections 3.1 through 3.9 of this chapter covered the twisting of circular shafts. Shafts with circular cross sections (either solid or hollow) do not *warp* when torsional moments are applied. Plane cross sections remain plane (as shown in Fig. 3-4), and shearing stresses and strains vary linearly with a distance ρ from the longitudinal axis of the shaft to the outer surface ($\rho = r$). Now consider prismatic shafts of length L acted upon by torsional moments T at either end but having *noncircular* cross sections. These cross sections could be solid (such as the elliptical, triangular, and rectangular shapes shown in Fig. 3-44), or they could

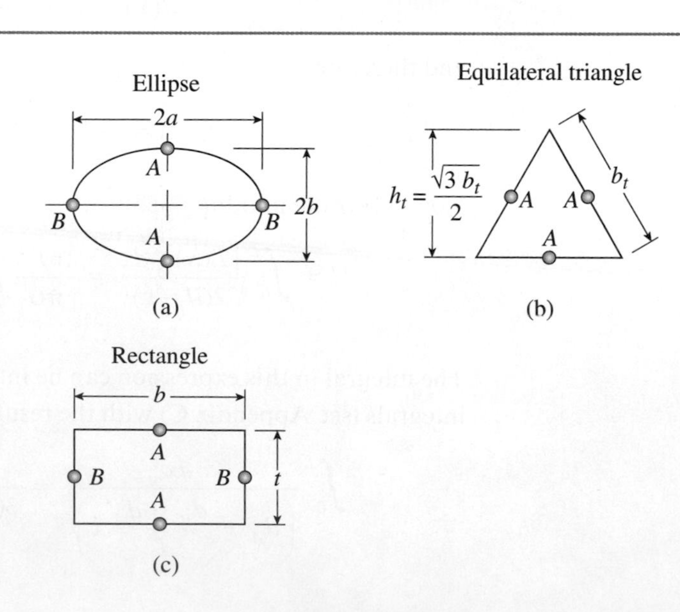

FIGURE 3-44
Solid elliptical, triangular, and rectangular cross-sectional shapes

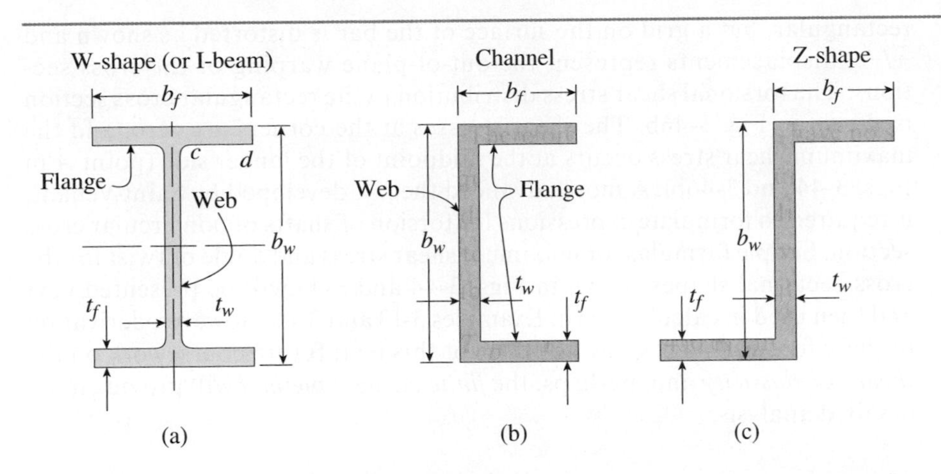

FIGURE 3-45

Thin-walled *open* cross sections: I-beam, channel, and Z-shape

be thin-walled *open* cross sections such as the I-beam, channel, and Z-shaped cross sections depicted in Fig. 3-45.

These noncircular cross sections *warp* under the action of torsional moments, and this warping alters the shear stress and strain distributions in the cross section. The simple *torsion formula* of Eq. 3-13 can no longer be used to compute shear stresses, and the *torque-displacement relation* of Eq. 3-17 cannot be used to find the angle of twist of the shaft. For example, warping distortions of a rectangular bar of length L acted upon by torques T at either end are shown in Fig. 3-46a; the cross sections remain

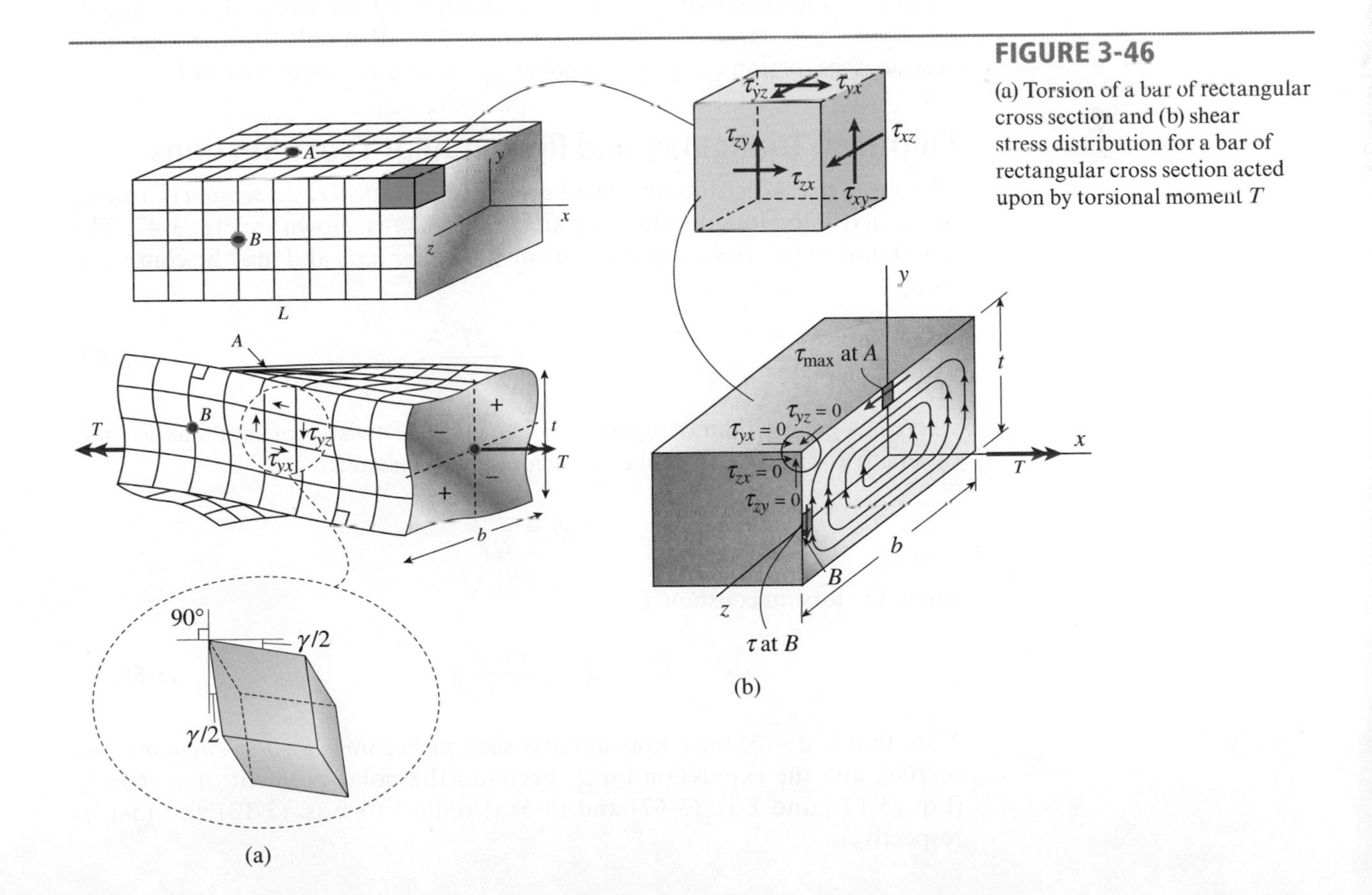

FIGURE 3-46

(a) Torsion of a bar of rectangular cross section and (b) shear stress distribution for a bar of rectangular cross section acted upon by torsional moment T

rectangular, but a grid on the surface of the bar is distorted as shown and +/−x displacements represent the out-of-plane warping of the cross sections. The torsional shear stress distribution in the rectangular cross section is shown in Fig. 3-46b. The shear stresses at the corners are zero, and the maximum shear stress occurs at the midpoint of the longer side (point A in Figs. 3-44 and 3-46b). A more advanced theory, developed by Saint-Venant, is required to formulate expressions for torsion of shafts of noncircular cross section. Simple formulas for maximum shear stress and angle of twist for the cross-sectional shapes shown in Figs. 3-44 and 3-45 will be presented next and then used in calculations in Examples 3-13 and 3-14. However, derivation of these formulas is beyond the scope of this text; future coursework on the *theory of elasticity* and, perhaps, the *finite element method* will provide more detailed analysis.

Shear Stress Distribution and Angle of Twist

The following discussion presents only the basic relations between applied torsional moment T and three key items of interest for a variety of noncircular cross sections.

1. The location and value of the maximum shear stress τ_{max} in the cross section
2. The torsional rigidity GJ
3. The angle of twist ϕ of a prismatic bar of length L

Constant G is the shearing modulus of elasticity of the material, and variable J is the *torsion constant* for the cross section. Note that only for a circular cross section does torsion constant J become the polar moment of inertia I_p.

Elliptical, Triangular, and Rectangular Cross Sections

The shear stress distribution for a bar with an *elliptical cross section* ($2a$ along major axis, $2b$ along minor axis, area $A = \pi ab$) is shown in Fig. 3-47. The maximum shear stress is at the ends of the *minor axis* and may be computed using

$$\tau_{max} = \frac{2T}{\pi ab^2} \quad \textbf{(3-67)}$$

where a is greater than or equal to b. The angle of twist ϕ of a prismatic shaft of length L with an elliptical cross section is expressed as

$$\phi = \frac{TL}{GJ_e}$$

where the torsion constant J_e is

$$J_e = \frac{\pi a^3 b^3}{a^2 + b^2} \quad \textbf{(3-68a,b)}$$

Note that if $a = b$, the elliptical cross section becomes a *solid circular cross section*, and the expression for J_e becomes the polar moment of inertia I_p [Eq. (3-12)] and Eqs. (3-67) and (3-68a) reduce to Eqs. (3-13) and (3-17), respectively.

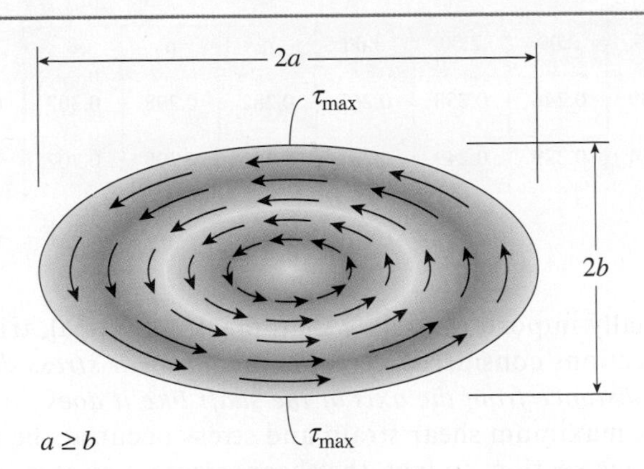

FIGURE 3-47
Shear stress distribution in an elliptical cross section

Next consider an *equilateral triangular cross section* (Fig. 3-44b) for the shaft of length L acted upon by torques T at each end. Each side has dimension b_t, and the triangle height is h_t. The torsion constant J_t is

$$J_t = \frac{h_t^4}{15\sqrt{3}} \tag{3-69}$$

The maximum shear stress occurs on the surface at the *midpoint of each side* (points A in Fig. 3-44b). The maximum shear stress and the angle of twist ϕ of a prismatic shaft of length L with an *equilateral triangular cross section* are expressed as

$$\tau_{max} = \frac{T\left(\frac{h_t}{2}\right)}{J_t} = \frac{15\sqrt{3}T}{2h_t^3} \tag{3-70}$$

where

$$J_t = \frac{h_t^4}{15\sqrt{3}}$$

and

$$\phi = \frac{TL}{GJ_t} = \frac{15\sqrt{3}TL}{Gh_t^4} \tag{3-71}$$

Finally, consider a *rectangular cross section* ($b \times t$, $b/t \geq 1$) (see Figs. 3-44c and 3-46). Theory of elasticity solutions provide expressions for *maximum shear stress at point A* in the cross section and the angle of twist for a variety of aspect ratios b/t as

$$\tau_{max} = \frac{T}{k_1 bt^2} \tag{3-72}$$

$$\phi = \frac{TL}{(k_2 bt^3)G} = \frac{TL}{GJ_r} \tag{3-73}$$

where

$$J_r = k_2 bt^3$$

and dimensionless coefficients k_1 and k_2 are listed in Table 3-1.

Table 3-1 Dimensionless coefficients for rectangular bars

b/t	1.00	1.50	1.75	2.00	2.50	3.00	4	6	8	10	∞
k_1	0.208	0.231	0.239	0.246	0.258	0.267	0.282	0.298	0.307	0.312	0.333
k_2	0.141	0.196	0.214	0.229	0.249	0.263	0.281	0.298	0.307	0.312	0.333

It is especially important to note that, for the elliptical, triangular, and rectangular sections considered here, *maximum shear stress does not occur at the largest distance from the axis of the shaft like it does for circular sections.* Instead, maximum shear strain and stress occur at the midpoints of the sides for each section. In fact, the shear stresses are zero in the corners of the triangular and rectangular sections (as indicated by the appearance of zero shear strain at the corners of the recangular section in Fig. 3-46a, for example).

Thin-Walled Open Cross Sections: I-beam, Angle, Channel, and Z-shape

Metal structural shapes of open cross section (see Fig. 3-45) can be represented as assemblages of rectangles for purposes of computing their torsional properties and response to applied torsional moments. Torsion constants for typical structural steel shapes are tabulated in the American Institute of Steel Construction (AISC) manual and may be up to 10% higher than properties based on use of rectangles to represent flanges and web. Hence, maximum shear stress values and twist angles computed using the formulas presented here may be somewhat conservative.

The total torque is assumed to be equal to the sum of the torques carried by the flanges and web. First compute the flange b_f/t_f ratio (see Fig. 3-45 for cross-sectional dimensions). Then find constant k_2 from Table 3-1 (interpolation between values may be necessary). For the web, use the ratio $(b_w - 2t_f)/t_w$ in Table 3-1 to find a new constant k_2 for the web. The separate torsion constants for both flanges and the web are expressed as

$$J_f = k_2 b_f t_f^3 \quad J_w = k_2(b_w - 2t_f)\left(t_w^3\right) \tag{3-74a,b}$$

The total torsion constant for the thin, open cross section is obtained (assuming two flanges) as

$$J = J_w + 2J_f \tag{3-75}$$

The maximum shear stress and angle of twist then can be computed as

$$\tau_{\max} = \frac{2T\left(\dfrac{t}{2}\right)}{J} \quad \text{and} \quad \phi = \frac{TL}{GJ} \tag{3-76a,b}$$

where the *larger* of t_f and t_w is used in the formula for $\tau_{\max}$.

Examples 3-13 and 3-14 illustrate the application of these formulas to obtain maximum shear stress and angle of twist values for prismatic bars with noncircular cross sections, such as those presented in Figs. 3-44 and 3-45.

Example 3-13

A shaft with a length $L = 1.8$ m is subjected to torques $T = 5$ kN·m at either end (Fig. 3-48). Segment AB($L_1 = 900$ mm) is made of brass ($G_b = 41$ GPa) and has a square cross section ($a = 75$ mm). Segment BC($L_2 = 900$ mm) is made of steel ($G_s = 74$ GPa) and has a circular cross section ($d = a = 75$ mm). Ignore stress concentrations near B.

(a) Find the maximum shear stress and angle of twist for each segment of the shaft.

(b) Find a new value for the dimension a of bar AB if the maximum shear stress in AB and BC are to be equal.

(c) Repeat part (b) if the angles of twist of segments AB and BC are to be equal.

(d) If dimension a is reset to $a = 75$ mm and bar BC is now a hollow pipe with an outer diameter $d_2 = a$, find the inner diameter d_1 so that the angles of twist of segments AB and BC are equal.

FIGURE 3-48

Example 3-13: Torsion of shaft with noncircular cross section

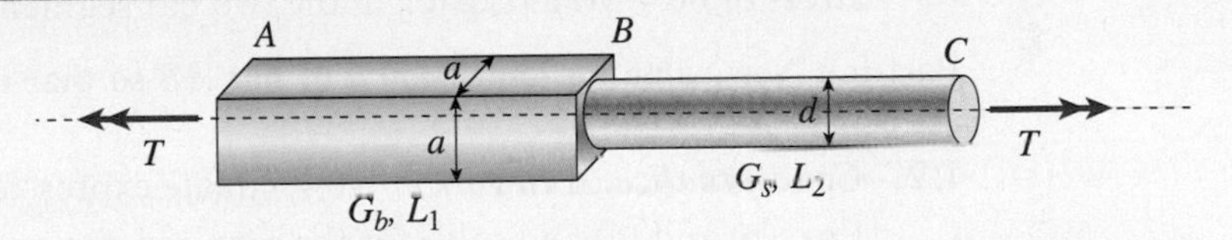

Solution:

Use a four-step problem-solving approach. Combine steps as needed for an efficient solution.

Part (a): Maximum shear stress and angles of twist for each segment.

1, 2. ***Conceptualize, Categorize:*** Both segments of the shaft have internal torque equal to the applied torque T. For square segment AB, obtain torsion coefficients k_1 and k_2 from Table 3-1.

3. ***Analyze:*** Use Eqs. (3-72) and (3-73) to compute the maximum shear stress and angle of twist as

$$\tau_{\max 1} = \frac{T}{k_1 b t^2} = \frac{T}{k_1 a^3} = \frac{(5\ \text{kN}\cdot\text{m})}{0.208(75\ \text{mm})^3} = 57\ \text{MPa} \quad \text{(a)}$$

$$\phi_1 = \frac{T L_1}{(k_2 b t^3) G_b} = \frac{T L_1}{k_2 a^4 G_b} = \frac{(5\ \text{kN}\cdot\text{m})(900\ \text{mm})}{0.141(75\ \text{mm})^4 (41\ \text{GPa})}$$
$$= 2.46 \times 10^{-2}\ \text{radians} \quad \text{(b)}$$

The maximum shear stress in AB occurs at the midpoint of each side of the square cross section.

Segment BC is a solid, circular cross section, so use Eqs. (3-14) and (3-17) to compute the maximum shear stress and angle of twist for segment BC:

$$\tau_{\max 2} = \frac{16T}{\pi d^3} = \frac{16(5\ \text{kN}\cdot\text{m})}{\pi(75\ \text{mm})^3} = 60.4\ \text{MPa} \quad \text{(c)}$$

$$\phi_2 = \frac{T L_2}{G_s I_p} = \frac{(5\ \text{kN}\cdot\text{m})(900\ \text{mm})}{74\ \text{GPa}\left[\frac{\pi}{32}(75\ \text{mm})^4\right]} = 1.958 \times 10^{-2} \quad \text{(d)}$$

4. ***Finalize***: Compare the shear stress and angle of twist values for square segment *AB* and circular segment *BC*. Steel pipe *BC* has 6% greater maximum shear stress but 20% less twist rotation than the brass bar *AB*.

Part (b): New value for dimension *a* of bar *AB* so that maximum shear stresses in *AB* and *BC* are equal.

1, 2. ***Conceptualize, Categorize***: Equate expressions for τ_{max1} and τ_{max2} in Eqs. (a) and (c) and solve for the required new value of dimension *a* of bar *AB*.

3. ***Analyze***:

$$\tau_{max1} = \tau_{max2} \quad \text{so} \quad \frac{16}{\pi d^3} = \frac{1}{k_1 a_{new}^3} \quad \text{or} \quad a_{new} = \left(\frac{\pi d^3}{16\, k_1}\right)^{\frac{1}{3}} = 73.6 \text{ mm} \qquad \text{(e)}$$

4. ***Finalize***: The diameter of bar *BC* remains unchanged at $d = 75$ mm, so a slight reduction in dimension *a* for bar *AB* leads to the same maximum shear stress of 60.4 MPa [Eq. (c)] in the two bar segments.

Part (c): New value for dimension *a* of bar *AB* so that twist rotations in *AB* and *BC* are equal.

1, 2. ***Conceptualize, Categorize***: Now, equate expressions for ϕ_1 and ϕ_2 in Eqs. (b) and (d) and solve for the required new value of dimension *a* of bar *AB*.

3. ***Analyze***:

$$\phi_1 = \phi_2$$

so

$$\frac{L_1}{k_2 a_{new}^4 G_b} = \frac{L_2}{G_s I_p}$$

or

$$a_{new} = \left[\frac{L_1}{L_2}\left(\frac{G_s I_p}{k_2 G_b}\right)\right]^{\frac{1}{4}} = 79.4 \text{ mm} \qquad \text{(f)}$$

4. ***Finalize***: The diameter of bar *BC* remains unchanged at $d = 75$ mm, so a slight increase in dimension *a* for brass bar *AB* leads to the same twist rotation of 0.01958 radians, as in Eq. (d) in each of the two bar segments.

FIGURE 3-49

Hollow pipe cross section for segment *BC*

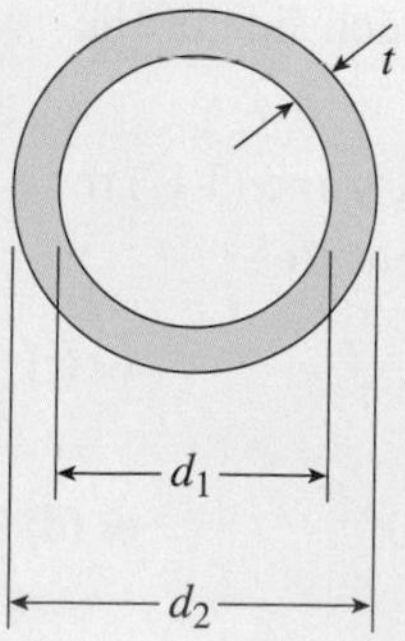

Part (d): Change segment *BC* to hollow pipe; find inner diameter d_1 so that twist rotations in *AB* and *BC* are equal.

1, 2. ***Conceptualize, Categorize***: Side dimension *a* of square segment *AB* is equal to 75 mm, and outer diameter $d_2 = 75$ mm (Fig. 3-49). Using Eq. (3-19) for the polar moment of inertia of segment *BC*, twist angle ϕ_2 is

$$\phi_2 = \frac{TL_2}{G_s\left[\frac{\pi}{32}\left(d_2^4 - d_1^4\right)\right]} \qquad \text{(g)}$$

3. ***Analyze***: Once again, equate expressions for ϕ_1 and ϕ_2 but now use Eqs. (b) and (g). Solve for d_1 to get

$$d_1 = \left[d_2^4 - 32\left(\frac{L_2}{L_1}\right)\left(\frac{G_b}{G_s}\right)\left(\frac{a^4 k_2}{\pi}\right)\right]^{\frac{1}{4}}$$

$$= \left[(75\text{ mm})^4 - 32\left(\frac{900\text{ mm}}{900\text{ mm}}\right)\left(\frac{41\text{ GPa}}{74\text{ GPa}}\right)\left[\frac{(75\text{ mm})^4(0.141)}{\pi}\right]\right]^{\frac{1}{4}} \quad \textbf{(h)}$$

$$= 50.4\text{ mm}$$

4. ***Finalize***: So the square, solid brass pipe AB ($a \times a$, $a = 75$ mm) and hollow steel pipe BC ($d_2 = 75$ mm, $d_1 = 50.4$ mm) are each 900 mm in length and have the same twist rotation (0.0246 radians) due to applied torque T. However, additional calculations will show that the maximum shear stress in segment BC is now increased from 60.4 MPa [Eq. (c)] to 75.9 MPa by using a hollow rather than solid bar for BC.

Note that by deriving the formula for inner diameter d_1 in Eq. (h) (rather than finding a numerical solution alone), you can also investigate other solutions of possible interest using different values of the key variables. For example, if bar AB is increased in length to $L_1 = 1100$ mm, inner diameter d_1 for BC can be increased to 57.6 mm, and the angles of twist for AB and BC will be the same.

Example 3-14

FIGURE 3-50

Example 3-14: Wide-flange shape and angle steel sections in torsion

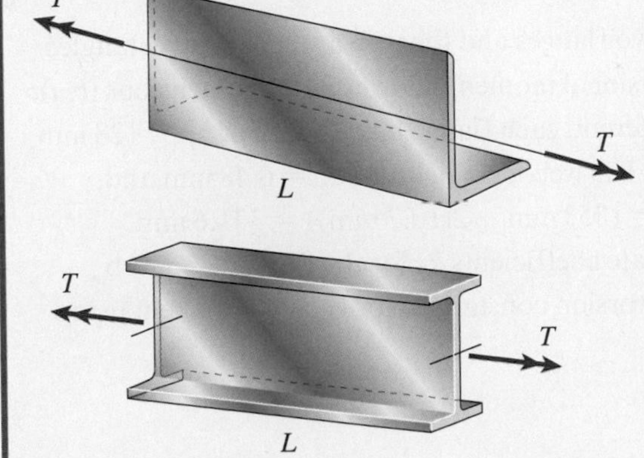

A steel angle and a steel wide-flange beam, each of length $L = 3.5$ m, are subjected to torque T (see Fig. 3-50). The allowable shear stress is 45 MPa, and the maximum permissible twist rotation is 5°. Find the value of the maximum torque T than can be applied to each section. Assume that $G = 80$ GPa and ignore stress concentration effects. Use the cross-sectional properties and dimensions given below. [See Table F-1 for wide flange beams and Table F-5 for angles with similar cross-sectional properties and dimensions.]

Angle and beam cross sectional properties and dimensions:

Angle: A = 49.6 cm^2 , total leg length = b_L = 280 mm, leg thickness = t_L = 19 mm

Beam: A = 49.6 cm^2, depth of web = d_w = 353 mm, web thickness = t_w = 6.48 mm, flange width = b_f = 128 mm, flange thickness = t_f = 10.7 mm

Solution:

Use a four-step problem-solving approach. Combine steps as needed for an efficient solution.

The angle and wide-flange steel shapes have the same cross-sectional area but the thicknesses of flange and web components of each section are quite different. First, consider the angle section.

Part (a): Steel angle section.

1, 2. ***Conceptualize, Categorize***: Approximate the unequal leg angle as one long rectangle with length $b_L = 280$ mm and constant thickness $t_L = 19$ mm, so $b_L / t_L = 14.7$. From Table 3-1, estimate coefficients $k_1 = k_2$ to be approximately 0.319.

3. ***Analyze***: The maximum allowable torques can be obtained from Eqs. (3-72) and (3-73) based on the given allowable shear stress and allowable twist rotation, respectively, as

$$T_{\max 1} = \tau_a k_1 b_L t_L^2 = 45\text{ MPa}(0.319)(280\text{ mm})[(19\text{ mm})^2] = 1451\text{ N}\cdot\text{m} \quad \textbf{(a)}$$

$$T_{\max 2} = \phi_a (k_2 b_L t_L^3)\frac{G}{L} = \left(\frac{5\pi}{180}\text{ rad}\right)(0.319)(280\text{ mm})[(19\text{ mm})^3]\frac{80\text{ GPa}}{3500\text{ mm}}$$
$$= 1222\text{ N}\cdot\text{m} \quad \textbf{(b)}$$

Alternatively, compute the torsion constant for the angle J_L as

$$J_L = k_2 b_L t_L^3 = 6.128 \times 10^5\text{ mm}^4 \quad \textbf{(c)}$$

then use Eqs. (3-74) and (3-76) to find the maximum allowable torque values. From Eq. (3-76a), find $T_{\max 1}$, and from Eq. (3-76b), obtain $T_{\max 2}$:

$$T_{\max 1} = \frac{\tau_a J_L}{t_L} = 1451\text{ N}\cdot\text{m} \quad \text{and} \quad T_{\max 2} = \frac{GJ_L}{L}\phi_a = 1222\text{ N}\cdot\text{m}$$

4. ***Finalize***: For the angle, the lesser value controls, so $T_{\max} = 1222\text{ N}\cdot\text{m}$.

Part (b): Steel Wide-flange shape.

1, 2. ***Conceptualize, Categorize***: The two flanges and the web are separate rectangles that together resist the applied torsional moment. However, the dimensions (b, t) of each of these rectangles are different: each flange has a width of $b_f = 128$ mm and a thickness of $t_f = 10.7$ mm. The web has thickness $t_w = 6.48$ mm and, conservatively, $b_w = (d_w - 2t_f) = (353\text{ mm} - 2(10.7\text{ mm})) = 331.6$ mm. Based on the b/t ratios, find separate coefficients k_2 for the flanges and web from Table 3-1, then compute the torsion constants J for each component using Eqs. (3-74) as

For the flanges:

$$\frac{b_f}{t_f} = 11.963$$

so an estimated value for $k_{2f} = 0.316$. Thus,

$$J_f = k_{2f} b_f t_f^3 = 0.316(128 \text{ mm})[(10.7 \text{ mm})^3] = 4.955 \times 10^4 \text{ mm}^4 \quad \text{(d)}$$

For the web:

$$\frac{d_w - 2t_f}{t_w} = 51.173$$

and k_{2w} is estimated as $k_{2w} = 0.329$, so

$$J_w = k_{2w}(d_w - 2t_f)(t_w^3) = 0.329[353 \text{ mm} - 2(10.7 \text{ mm})][(6.48 \text{ mm})^3]$$

$$= 2.968 \times 10^4 \text{ mm}^4 \quad \text{(e)}$$

The torsion constant for the entire wide flange section is obtained by adding web and flange contributions [Eqs. (d) and (e)]:

$$J_W = 2J_f + J_w = [2(4.955) + 2.968](10^4) \text{ mm}^4 = 1.288 \times 10^5 \text{ mm}^4 \quad \text{(f)}$$

3. ***Analyze***: Now, use Eq. (3-76a) and the allowable shear stress τ_a to compute the maximum allowable torque based on both flange and web maximum shear stresses:

$$T_{\max f} = \tau_a \frac{J_W}{t_f} = 45 \text{ MPa}\left(\frac{1.288 \times 10^5 \text{ mm}^4}{10.7 \text{ mm}}\right) = 542 \text{ N} \cdot \text{m} \quad \text{(g)}$$

$$T_{\max w} = \tau_a \frac{J_W}{t_w} = 45 \text{ MPa}\left(\frac{1.288 \times 10^5 \text{ mm}^4}{6.48 \text{ mm}}\right) = 894 \text{ N} \cdot \text{m} \quad \text{(h)}$$

Note that since the flanges have greater thickness than the web, the maximum shear stress will be in the flanges. So a calculation of $T_{\max}$ based on the maximum web shear stress using Eq. (h) is not necessary.

Finally, use Eq. (3-76b) to compute $T_{\max}$ based on the allowable angle of twist:

$$T_{\max \phi} = \frac{GJ_W}{L}\phi_a = \frac{80 \text{ GPa}(1.288 \times 10^5 \text{ mm}^4)}{3500 \text{ mm}}\left(\frac{5\pi}{180}\text{rad}\right)$$

$$= 257 \text{ N} \cdot \text{m} \quad \text{(i)}$$

4. ***Finalize***: For the wide-flange shape, the most restrictive requirement is the allowable twist rotation, so $T_{\max} = 257 \text{ N} \cdot \text{m}$ governs [Eq. (i)].

It is interesting to note that, even though both angle and wide-flange shapes have the same cross-sectional area, the wide flange shape is considerably weaker in torsion, because its component rectangles are much thinner ($t_w = 6.48$ mm, $t_f = 10.7$ mm) than the angle section ($t_L = 19$ mm). However, Chapter 5 shows that, although weak in torsion, the wide flange shape has a considerable advantage in resisting bending and *transverse* shear stresses.

3.11 Thin-Walled Tubes

With the exception of Section 3.10, which covered torsion of bars of noncircular cross section, the torsion theory described in the preceding sections is applicable only to solid or hollow bars of circular cross section. Circular shapes are the most efficient shapes for resisting torsion and consequently are the most commonly used. However, in lightweight structures, such as aircraft and spacecraft, thin-walled tubular members with noncircular closed cross sections are often required to resist torsion. In this section, structural members of this kind are analyzed.

To obtain formulas that are applicable to a variety of shapes, consider a thin-walled tube of arbitrary cross section (Fig. 3-51a). The tube is cylindrical in shape—that is, all cross sections are identical and the longitudinal axis is a straight line. The thickness t of the wall is not necessarily constant but may vary around the cross section. However, the thickness must be small in comparison with the total width of the tube. The tube is subjected to pure torsion by torques T acting at the ends.

Shear Stresses and Shear Flow

The shear stresses τ acting on a cross section of the tube are pictured in Fig. 3-51b, which shows an element of the tube cut out between two cross sections that are distance dx apart. The stresses act parallel to the boundaries of the cross section and "flow" around the cross section. Also, the intensity of

FIGURE 3-51

Thin-walled tube of arbitrary cross-sectional shape

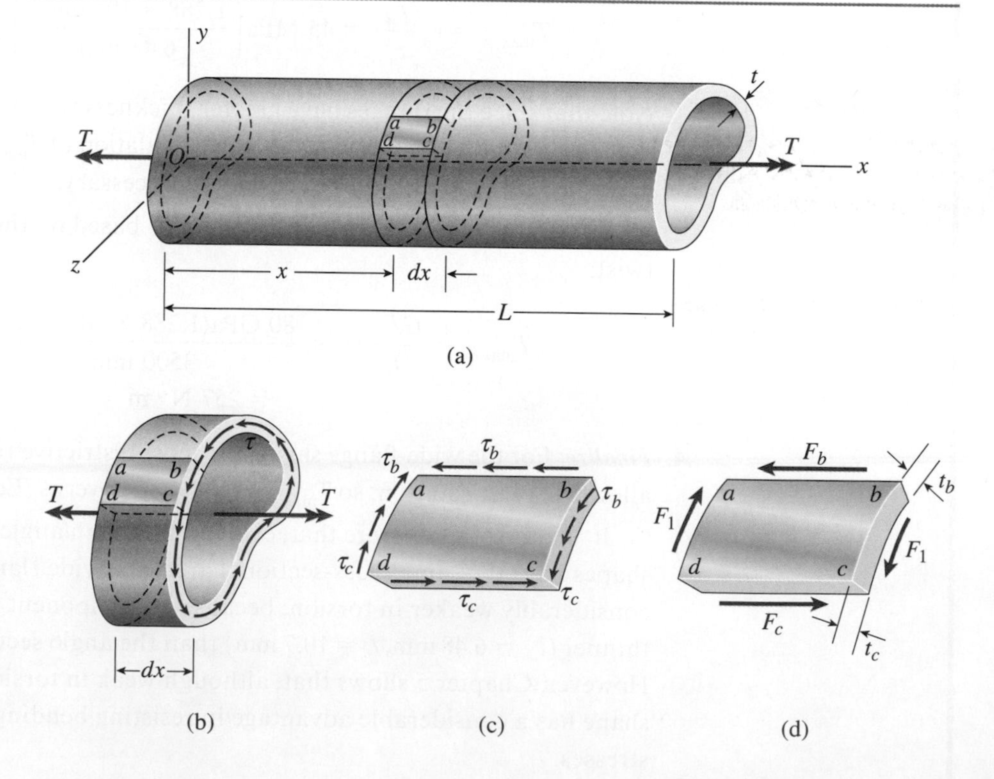

the stresses varies so slightly *across* the thickness of the tube (because the tube is assumed to be thin) that τ may be assumed to be constant in that direction. However, if the thickness t is not constant, the stresses will vary in intensity as we go *around* the cross section, and the manner in which they vary must be determined from equilibrium.

To determine the magnitude of the shear stresses, consider a rectangular element *abcd* obtained by making two longitudinal cuts *ab* and *cd* (Figs. 3-51a and b). This element is isolated as a free body in Fig. 3-51c. Acting on the cross-sectional face *bc* are the shear stresses τ shown in Fig. 3-51b. Assume that these stresses vary in intensity along the cross section from b to c; therefore, the shear stress at b is denoted τ_b and the stress at c is denoted τ_c (see Fig. 3-51c).

From equilibrium, identical shear stresses act in the opposite direction on the opposite cross-sectional face *ad*, and shear stresses of the same magnitude also act on the longitudinal faces *ab* and *cd*. Thus, the constant shear stresses acting on faces *ab* and *cd* are equal to τ_b and τ_c, respectively.

The stresses acting on the longitudinal faces *ab* and *cd* produce forces F_b and F_c (Fig. 3-51d). These forces are obtained by multiplying the stresses by the areas on which they act:

$$F_b = \tau_b t_b dx \quad F_c = \tau_c t_c dx$$

in which t_b and t_c represent the thicknesses of the tube at points b and c, respectively (Fig. 3-51d).

In addition, forces F_1 are produced by the stresses acting on faces *bc* and *ad*. From the equilibrium of the element in the longitudinal direction (the x direction), it follows that $F_b = F_c$, or

$$\tau_b t_b = \tau_c t_c$$

Because the locations of the longitudinal cuts *ab* and *cd* were selected arbitrarily, it follows from the preceding equation that the product of the shear stress τ and the thickness t of the tube is the same at every point in the cross section. This product is known as the **shear flow** and is denoted by the letter f:

$$f = \tau t = \text{constant} \qquad \textbf{(3-77)}$$

This relationship shows that the largest shear stress occurs where the thickness of the tube is smallest, and vice versa. In regions where the thickness is constant, the shear stress is constant. Note that shear flow is the shear force per unit distance along the cross section.

Torsion Formula for Thin-Walled Tubes

The next step in the analysis is to relate the shear flow f (and hence the shear stress τ) to the torque T acting on the tube. For that purpose, examine the cross section of the tube, as pictured in Fig. 3-52. The **median line** (also called the *centerline* or the *midline*) of the wall of the tube is shown as a dashed line in the figure. Consider an element of area of length ds (measured along the median line) and thickness t. The distance s defining the location of the element is measured along the median line from some arbitrarily chosen reference point.

FIGURE 3-52

Cross section of thin-walled tube

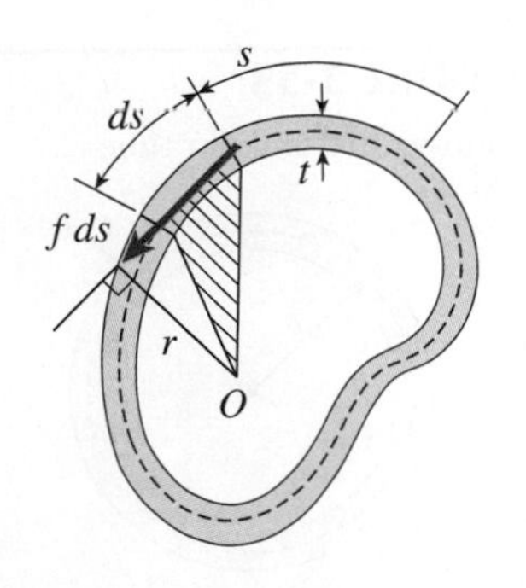

The total shear force acting on the element of area is fds, and the moment of this force about any point O within the tube is

$$dT = rfds$$

in which r is the perpendicular distance from point O to the line of action of the force fds. (Note that the line of action of the force fds is tangent to the median line of the cross section at the element ds.) The total torque T produced by the shear stresses is obtained by integrating along the median line of the cross section:

$$T = f\int_0^{L_m} rds \tag{3-78}$$

in which L_m denotes the length of the median line.

The integral in Eq. (3-78) can be difficult to integrate by formal mathematical means, but fortunately it can be evaluated easily by giving it a simple geometric interpretation. The quantity rds represents twice the area of the shaded triangle shown in Fig. 3-52. (Note that the triangle has base length ds and height equal to r.) Therefore, the integral represents twice the area A_m enclosed by the median line of the cross section:

$$\int_0^{L_m} rds = 2A_m \tag{3-79}$$

It follows from Eq. (3-78) that $T = 2fA_m$, and therefore the **shear flow** is

$$f = \frac{T}{2A_m} \tag{3-80}$$

Now eliminate the shear flow f between Eqs. (3-77) and (3-80) and obtain a **torsion formula for thin-walled tubes:**

$$\tau = \frac{T}{2tA_m} \tag{3-81}$$

Since t and A_m are properties of the cross section, the shear stresses τ can be calculated from Eq. (3-81) for any thin-walled tube subjected to a known torque T. (*Reminder:* The area A_m is the area *enclosed* by the median line—it is *not* the cross-sectional area of the tube.)

To illustrate the use of the torsion formula, consider a thin-walled **circular tube** (Fig. 3-53) of thickness t and radius r to the median line. The area enclosed by the median line is

$$A_m = \pi r^2 \tag{3-82}$$

and therefore the shear stress (constant around the cross section) is

$$\tau = \frac{T}{2\pi r^2 t} \tag{3-83}$$

This formula agrees with the stress obtained from the standard torsion formula [Eq. (3-13)] when the standard formula is applied to a circular tube with thin walls using the approximate expression $I_p \approx 2\pi r^3 t$ for the polar moment of inertia [Eq. (3-21)].

FIGURE 3-53

Thin-walled circular tube

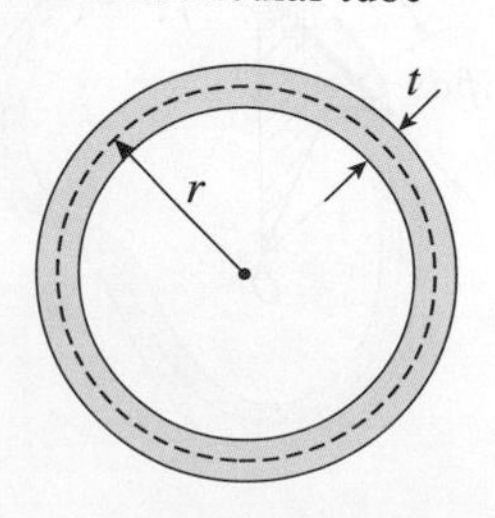

As a second illustration, consider a thin-walled **rectangular tube** (Fig. 3-54) having thickness t_1 on the sides and thickness t_2 on the top and bottom. Also, the height and width (measured to the median line of the cross section) are h and b, respectively. The area within the median line is

$$A_m = bh \tag{3-84}$$

and thus, the shear stresses in the vertical and horizontal sides, respectively, are

$$\tau_{\text{vert}} = \frac{T}{2t_1bh} \qquad \tau_{\text{horiz}} = \frac{T}{2t_2bh} \tag{3-85a,b}$$

If t_2 is larger than t_1, the maximum shear stress will occur in the vertical sides of the cross section.

FIGURE 3-54

Thin-walled rectangular tube

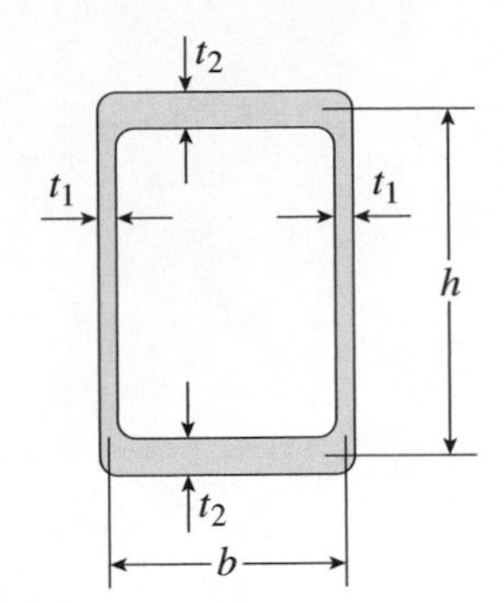

Strain Energy and Torsion Constant

The strain energy of a thin-walled tube can be determined by first finding the strain energy of an element and then integrating throughout the volume of the bar. Consider an element of the tube having area *tds* in the cross section (see the element in Fig. 3-52) and length *dx* (see the element in Fig. 3-51). The volume of such an element, which is similar in shape to the element *abcd* shown in Fig. 3-51a, is *tdsdx*. Because elements of the tube are in pure shear, the strain-energy density of the element is $\tau^2/2G$, as given by Eq. (3-63a). The total strain energy of the element is equal to the strain-energy density times the volume:

$$dU = \frac{\tau^2}{2G}tdsdx = \frac{\tau^2t^2}{2G}\frac{ds}{t}dx = \frac{f^2}{2G}\frac{ds}{t}dx \tag{3-86}$$

in which is replaced τt by the shear flow f (a constant).

Obtain the total strain energy of the tube by integrating dU throughout the volume of the tube, that is, ds is integrated from 0 to L_m around the median line and dx is integrated along the axis of the tube from 0 to L, where L is the length. Thus,

$$U = \int dU = \frac{f^2}{2G}\int_0^{L_m}\frac{ds}{t}\int_0^L dx \tag{3-87}$$

Note that the thickness t may vary around the median line and must remain with ds under the integral sign. Since the last integral is equal to the length L of the tube, the equation for the strain energy becomes

$$U = \frac{f^2L}{2G}\int_0^{L_m}\frac{ds}{t} \tag{3-88}$$

Substitute for the shear flow from Eq. (3-80) to obtain

$$U = \frac{T^2L}{8GA_m^2}\int_0^{L_m}\frac{ds}{t} \tag{3-89}$$

as the equation for the strain energy of the tube in terms of the torque T.

The preceding expression for strain energy can be written in simpler form by introducing a new property of the cross section, called the **torsion constant**. For a thin-walled tube, the torsion constant (denoted by the letter J) is defined as

$$J = \frac{4A_m^2}{\displaystyle\int_0^{L_m} \frac{ds}{t}} \tag{3-90}$$

With this notation, the equation for **strain energy** [Eq. (3-89)] becomes

$$U = \frac{T^2 L}{2GJ} \tag{3-91}$$

which has the same form as the equation for strain energy in a circular bar [see Eq. (3-55a)]. The only difference is that the torsion constant J has replaced the polar moment of inertia I_p. Note that the torsion constant has units of length to the fourth power.

In the special case of a cross section having constant thickness t, the expression for J [Eq. (3-90)] simplifies to

$$J = \frac{4tA_m^2}{L_m} \tag{3-92}$$

For each shape of cross section, evaluate J from either Eq. (3-90) or Eq. (3-92).

FIGURE 3-53 (Repeated)

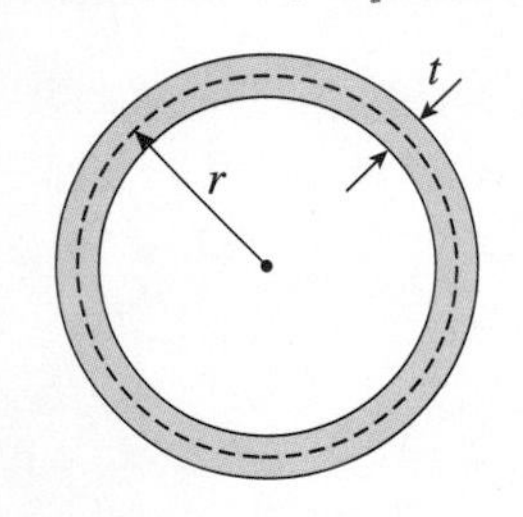

As an illustration, consider again the thin-walled **circular tube** of Fig. 3-53. Since the thickness is constant, use Eq. (3-92) and substitute $L_m = 2\pi r$ and $A_m = \pi r^2$; the result is

$$J = 2\pi r^3 t \tag{3-93}$$

which is the approximate expression for the polar moment of inertia [Eq. (3-21)]. Thus, in the case of a thin-walled circular tube, the polar moment of inertia is the same as the torsion constant.

FIGURE 3-54 (Repeated)

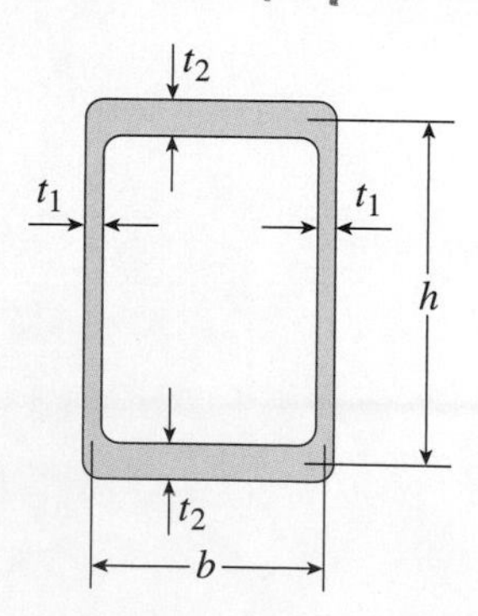

As a second illustration, use the **rectangular tube** of Fig. 3-54. For this cross section, $A_m = bh$. Also, the integral in Eq. (3-90) is

$$\int_0^{L_m} \frac{ds}{t} = 2\int_0^{h} \frac{ds}{t_1} + 2\int_0^{b} \frac{ds}{t_2} = 2\left(\frac{h}{t_1} + \frac{b}{t_2}\right)$$

Thus, the torsion constant [Eq. (3-90)] is

$$J = \frac{2b^2h^2t_1t_2}{bt_1 + ht_2} \tag{3-94}$$

Torsion constants for other thin-walled cross sections can be found in a similar manner.

Angle of Twist

The angle of twist ϕ for a thin-walled tube of arbitrary cross-sectional shape (Fig. 3-55) may be determined by equating the work W done by the applied torque T to the strain energy U of the tube. Thus,

$$W = U \quad \text{or} \quad \frac{T\phi}{2} = \frac{T^2 L}{2GJ}$$

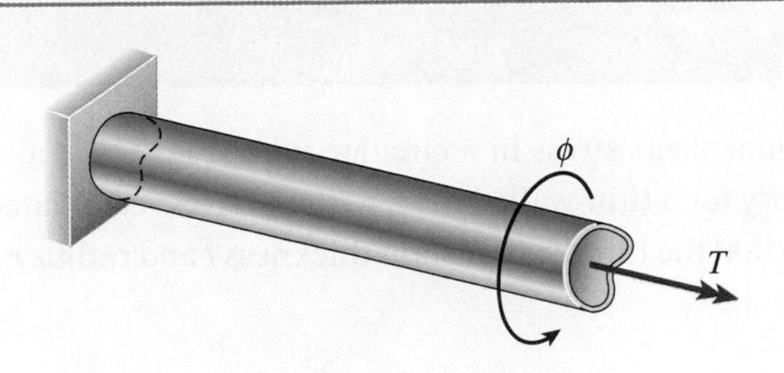

FIGURE 3-55

Angle of twist ϕ for a thin-walled tube

from which follows the equation for the angle of twist:

$$\phi = \frac{TL}{GJ}$$

Again, observe that the equation has the same form as the corresponding equation for a circular bar [Eq. (3-17)] but with the polar moment of inertia replaced by the torsion constant. The quantity GJ is called the **torsional rigidity** of the tube.

Limitations

The formulas developed in this section apply to prismatic members having *closed* tubular shapes with thin walls. If the cross section is thin walled but *open*, as in the case of I-beams and channel sections, the theory given here does not apply. To emphasize this point, take a thin-walled tube and slit it lengthwise—then the cross section becomes an open section, the shear stresses and angles of twist increase, the torsional resistance decreases, and the formulas given in this section cannot be used. Recall that the torsion of prismatic bars with noncircular cross sections was reviewed briefly in Section 3.10. This included solid rectangular, triangular, and elliptical cross sections, as well as thin-walled open sections (such as I-beams and channels). An advanced theory is required to derive formulas for shear stress and the angle of twist of such bars, so only key formulas and their application were presented.

Some of the formulas given in this section on thin-walled tubes are restricted to linearly elastic materials—for instance, any equation containing the shear modulus of elasticity G is in this category. However, the equations for shear flow and shear stress [Eqs. (3-80) and (3-81)] are based only upon equilibrium and are valid regardless of the material properties. The entire theory is approximate because it is based upon centerline dimensions, and the results become less accurate as the wall thickness t increases.[3]

An important consideration in the design of any thin-walled member is the possibility that the walls will buckle. The thinner the walls and the longer the tube, the more likely it is that buckling will occur. In the case of noncircular tubes, stiffeners and diaphragms are often used to maintain the shape of the tube and prevent localized buckling. In all discussions and problems, assume that buckling is prevented.

[3]The torsion theory for thin-walled tubes described in this section was developed by R. Bredt, a German engineer who presented it in 1896 (Ref. 3-2). It is often called *Bredt's theory of torsion.*

Example 3-15

FIGURE 3-56

Example 3-15: Comparison of approximate and exact theories of torsion

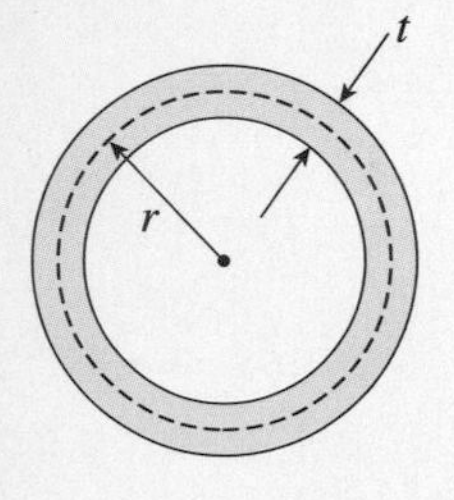

Compare the maximum shear stress in a circular tube (Fig. 3-56), as calculated by the approximate theory for a thin-walled tube, with the stress calculated by the exact torsion theory. (Note that the tube has constant thickness t and radius r to the median line of the cross section.)

Solution:

Use a four-step problem-solving approach.

1, 2. *Conceptualize, Categorize:*

Approximate theory: The shear stress obtained from the approximate theory for a thin-walled tube [Eq. (3-83)] is

$$\tau_1 = \frac{T}{2\pi r^2 t} = \frac{T}{2\pi t^3 \beta^2} \tag{3-95}$$

in which the relation

$$\beta = \frac{r}{t} \tag{3-96}$$

is introduced.

Torsion formula: The maximum stress obtained from the more accurate torsion formula [Eq. (3-13)] is

$$\tau_2 = \frac{T(r + t/2)}{I_p} \tag{a}$$

where

$$I_p = \frac{\pi}{2}\left[\left(r + \frac{t}{2}\right)^4 - \left(r - \frac{t}{2}\right)^4\right] \tag{b}$$

After expansion, this expression simplifies to

$$I_p = \frac{\pi r t}{2}(4r^2 + t^2) \tag{3-97}$$

and the expression for the shear stress [Eq. (a)] becomes

$$\tau_2 = \frac{T(2r + t)}{\pi r t(4r^2 + t^2)} = \frac{T(2\beta + 1)}{\pi t^3 \beta(4\beta^2 + 1)} \tag{3-98}$$

3. *Analyze:*

Ratio: The ratio τ_1/τ_2 of the shear stresses is

$$\frac{\tau_1}{\tau_2} = \frac{4\beta^2 + 1}{2\beta(2\beta + 1)} \tag{3-99}$$

which depends only on the ratio β.

4. ***Finalize:*** Using values of β equal to 5, 10, and 20 in Eq. (3-99) results in values $\tau_1/\tau_2 = 0.92$, 0.95, and 0.98, respectively. Thus, the approximate formula for

the shear stresses gives results that are slightly less than those obtained from the exact formula. The accuracy of the approximate formula increases as the wall of the tube becomes thinner. In the limit, as the thickness approaches zero and β approaches infinity, the ratio τ_1/τ_2 becomes 1.

Example 3-16

A circular tube and a square tube (Fig. 3-57) are constructed of the same material and subjected to the same torque. Both tubes have the same length, same wall thickness, and same cross-sectional area.

What are the ratios of their shear stresses and angles of twist? (Disregard the effects of stress concentrations at the corners of the square tube.)

FIGURE 3-57

Example 3-16: Comparison of circular and square tubes

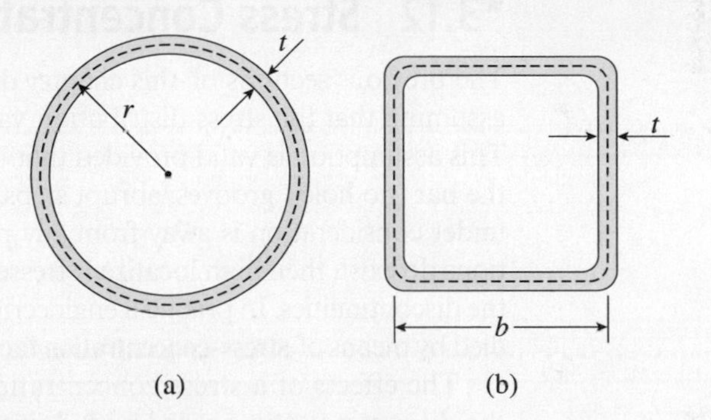

Solution:

Use a four-step problem-solving approach.

1, 2. ***Conceptualize, Categorize:***

Circular tube: For the circular tube, the area A_{m1} enclosed by the median line of the cross section is

$$A_{m1} = \pi r^2 \tag{a}$$

where r is the radius to the median line. Also, the torsion constant [Eq. (3-93)] and cross-sectional area are

$$J_1 = 2\pi r^3 t \qquad A_1 = 2\pi rt \tag{b,c}$$

Square tube: For the square tube, the cross-sectional area is

$$A_2 = 4bt \tag{d}$$

where b is the length of one side measured along the median line. Inasmuch as the areas of the tubes are the same, length b is $b = \pi r/2$. Also, the torsion constant [Eq. (3-94)] and area enclosed by the median line of the cross section are

$$J_2 = b^3 t = \frac{\pi^3 r^3 t}{8} \qquad A_{m2} = b^2 = \frac{\pi^2 r^2}{4} \tag{e,f}$$

3. ***Analyze***:

Ratios: The ratio τ_1/τ_2 of the shear stress in the circular tube to the shear stress in the square tube [from Eq. (3-81)] is

$$\frac{\tau_1}{\tau_2} = \frac{A_{m2}}{A_{m1}} = \frac{\pi^2 r^2/4}{\pi r^2} = \frac{\pi}{4} = 0.79 \qquad \text{(g)}$$

From the torque-displacement relation $\phi = \frac{TL}{GJ}$, the ratio of the angles of twist is

$$\frac{\phi_1}{\phi_2} = \frac{J_2}{J_1} = \frac{\pi^3 r^3 t/8}{2\pi r^3 t} = \frac{\pi^2}{16} = 0.62 \qquad \text{(j)}$$

4. ***Finalize***: These results show that the circular tube not only has a 21% lower shear stress than does the square tube but also a greater stiffness against rotation.

*3.12 Stress Concentrations in Torsion

The previous sections of this chapter discussed the stresses in torsional members, assuming that the stress distribution varied in a smooth and continuous manner. This assumption is valid provided that there are no abrupt changes in the shape of the bar (no holes, grooves, abrupt steps, and the like) and provided that the region under consideration is away from any points of loading. If such disruptive conditions do exist, then high localized stresses will be created in the regions surrounding the discontinuities. In practical engineering work, these **stress concentrations** are handled by means of **stress-concentration factors**, as explained previously in Section 2.10.

The effects of a stress concentration are confined to a small region around the discontinuity in accord with Saint-Venant's principle (see Section 2.10). For instance, consider a stepped shaft consisting of two segments having different diameters (Fig. 3-58). The larger segment has a diameter D_2, and the smaller segment has a diameter D_1. The junction between the two segments forms a "step" or "shoulder" that is machined with a fillet of radius R. Without the fillet, the theoretical stress-concentration factor would be infinitely large because of the abrupt 90° reentrant corner. Of course, infinite stresses cannot occur. Instead, the material at the reentrant corner would deform and partially relieve the high stress concentration. However, such a situation is very dangerous under dynamic loads, and in good design a fillet is always used. The larger the radius of the fillet, the lower the stresses.

At a distance from the shoulder approximately equal to the diameter D_2 (for instance, at cross section A–A in Fig. 3-58a), the torsional shear stresses are practically unaffected by the discontinuity. Therefore, the maximum stress τ_2 at a sufficient distance to the left of the shoulder can be found from the torsion formula using D_2 as the diameter (Fig. 3-58b). The same general comments apply at section C–C, which is a distance D_1 (or greater) from the toe of the fillet. Because the diameter D_1 is less than the diameter D_2, the maximum stress τ_1 at section C–C (Fig. 3-58d) is larger than the stress τ_2.

The stress-concentration effect is greatest at section B–B, which cuts through the toe of the fillet. At this section, the maximum stress is

$$\tau_{max} = K\tau_{nom} = K\frac{Tr}{I_p} = K\left(\frac{16T}{\pi D_1^3}\right) \qquad \text{(3-100)}$$

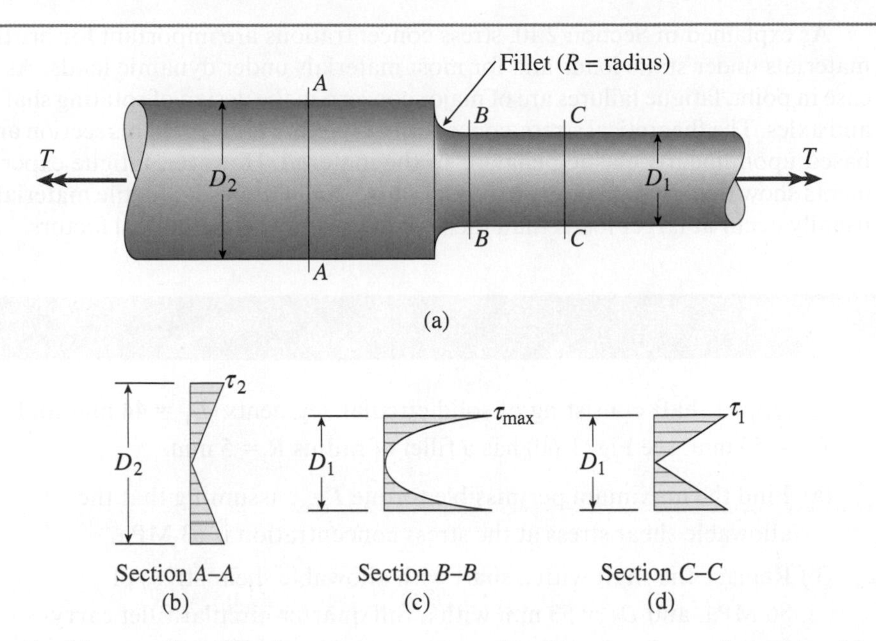

FIGURE 3-58

Stepped shaft in torsion

In this equation, K is the stress-concentration factor and τ_{nom} (equal to τ_1) is the nominal shear stress, that is, the shear stress in the smaller part of the shaft.

Values of the factor K are plotted in Fig. 3-59 as a function of the ratio R/D_1. Curves are plotted for various values of the ratio D_2/D_1. Note that when the fillet radius R becomes very small and the transition from one diameter to the other is abrupt, the value of K becomes quite large. Conversely, when R is large, the value of K approaches 1.0 and the effect of the stress concentration disappears. The dashed curve in Fig. 3-59 is for the special case of a full quarter-circular fillet, which means that $D_2 = D_1 + 2R$. (*Note:* Probs. 3.12-1 through 3.12-5 provide practice in obtaining values of K from Fig. 3-59.)

Many other cases of stress concentrations for circular shafts, such as a shaft with a keyway and a shaft with a hole, are available in the engineering literature. (see, for example, Ref. 2-9).

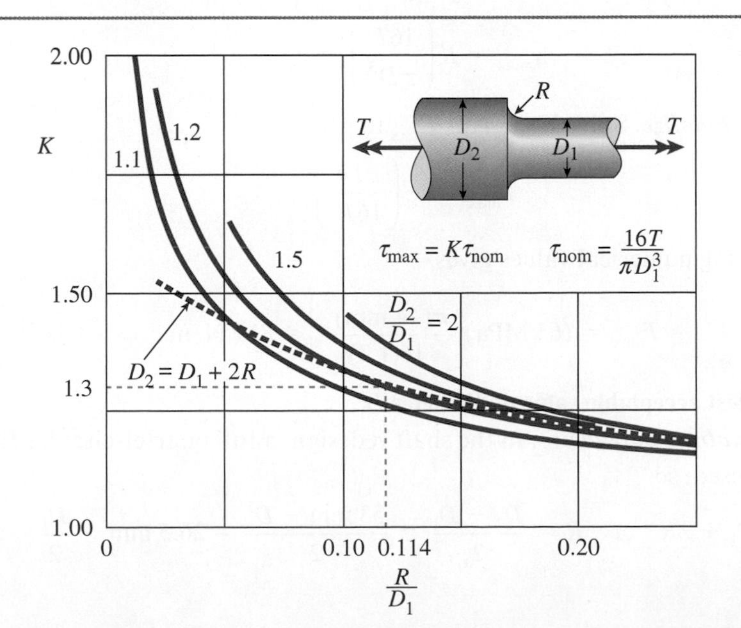

FIGURE 3-59

Stress-concentration factor K for a stepped shaft in torsion (the dashed line is for a full quarter-circular fillet)

As explained in Section 2.10, stress concentrations are important for brittle materials under static loads and for most materials under dynamic loads. As a case in point, fatigue failures are of major concern in the design of rotating shafts and axles. The theoretical stress-concentration factors K given in this section are based upon linearly elastic behavior of the material. However, fatigue experiments show that these factors are conservative, and failures in ductile materials usually occur at larger loads than those predicted by the theoretical factors.

Example 3-17

FIGURE 3-60

Example 3-17: Stepped circular shaft in torsion

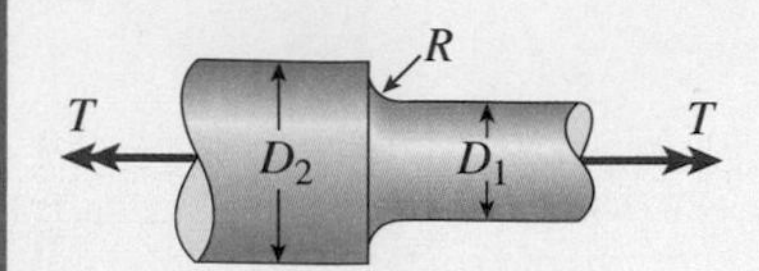

A stepped shaft consisting of solid circular segments ($D_1 = 44$ mm and $D_2 = 53$ mm, see Fig. 3-60) has a fillet of radius $R = 5$ mm.

(a) Find the maximum permissible torque T_{max}, assuming that the allowable shear stress at the stress concentration is 63 MPa.

(b) Replace the shaft with a shaft with allowable shear stress of 86 MPa, and $D_2 = 53$ mm with a full quarter-circular fillet carrying a torque of $T = 960$ N·m. Find the smallest acceptable value of diameter D_1.

Solution:

Use a four-step problem-solving approach. Combine steps as needed for an efficient solution.

Part (a): Maximum permissible torque.

1, 2. ***Conceptualize, Categorize***: Compute the ratio of the shaft diameters ($D_2/D_1 = 1.2$) and the ratio of the fillet radius R to diameter D_1 ($R/D_1 = 0.114$) to find the stress concentration factor K to be approximately 1.3 from Fig. 3-59 (repeated on next page). Then, equate the maximum shear stress in the smaller shaft to the allowable shear stress τ_a to get

$$\tau_{max} = K\left(\frac{16T}{\pi D_1^3}\right) = \tau_a \tag{a}$$

3, 4. ***Analyze, Finalize***: Solve Eq. (a) for T_{max} to get

$$T_{max} = \tau_a\left(\frac{\pi D_1^3}{16K}\right) \tag{b}$$

Substituting numerical values gives

$$T_{max} = (63 \text{ MPa})\left[\frac{\pi(44 \text{ mm})^3}{16(1.3)}\right] = 811 \text{ N·m}$$

Part (b): Smallest acceptable value of diameter D_1.

1, 2. ***Conceptualize, Categorize***: In the shaft redesign, a full quarter-circular fillet is being used, so

$$D_2 = D_1 + 2R \quad \text{or} \quad R = \frac{D_2 - D_1}{2} = \frac{53 \text{ mm} - D_1}{2} = 26.5 \text{ mm} - \frac{D_1}{2} \tag{c}$$

FIGURE 3-59 (Repeated)

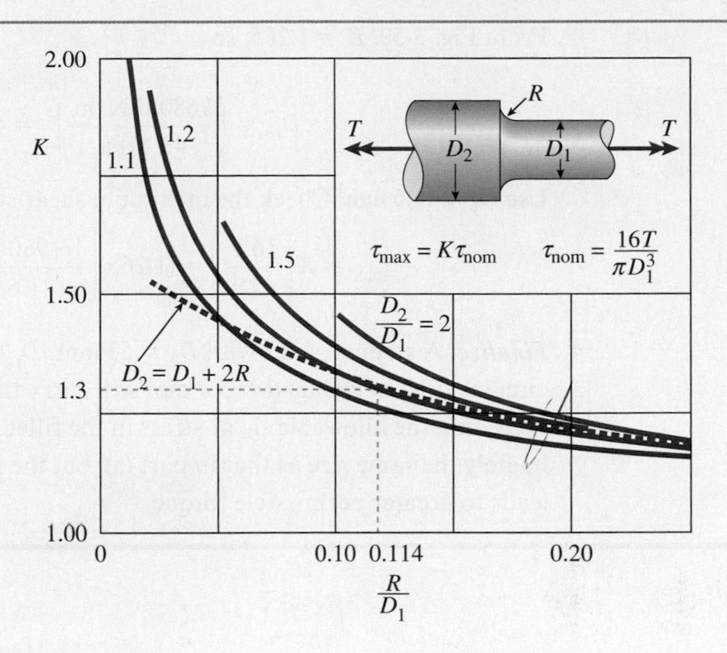

Next, solve Eq. (a) for diameter D_1 in terms of the unknown stress concentration factor K:

$$D_1 = \left[K\left(\frac{16T}{\pi\tau_a}\right)\right]^{\frac{1}{3}} = \left[K\left[\frac{16(960\ \text{N}\cdot\text{m})}{\pi(86\ \text{MPa})}\right]\right]^{\frac{1}{3}} = \left(\frac{7680K\ \text{N}\cdot\text{m}}{43\ \text{MPa}}\right)^{\frac{1}{3}} \quad \textbf{(d)}$$

3. ***Analyze***: Solving Eqs. (c) and (d) *using trial and error* and using Fig. 3-59 to obtain K produces the following results:

Trial 1:

$$D_{1a} = 38\ \text{mm} \quad R = 26.5\ \text{mm} - \frac{D_{1a}}{2} = 7.5\ \text{mm} \quad \frac{R}{D_{1a}} = 0.197$$

From Fig. 3-59, $K = 1.24$, so

$$D_{1b} = \left(\frac{7680K\ \text{N}\cdot\text{m}}{43\ \text{MPa}}\right)^{\frac{1}{3}} = 41.31\ \text{mm}$$

Trial 2:

$$D_{1a} = 41.3\ \text{mm} \quad R = 26.5\ \text{mm} - \frac{D_{1a}}{2} = 5.85\ \text{mm} \quad \frac{R}{D_{1a}} = 0.142$$

From Fig. 3-59, $K = 1.26$, so

$$D_{1b} = \left(\frac{7680K\ \text{N}\cdot\text{m}}{43\ \text{MPa}}\right)^{\frac{1}{3}} = 41.53\ \text{mm}$$

Trial 3:

$$D_{1a} = 41.6\ \text{mm} \quad R = 26.5\ \text{mm} - \frac{D_{1a}}{2} = 5.7\ \text{mm} \quad \frac{R}{D_{1a}} = 0.137$$

From Fig. 3-59, $K = 1.265$, so

$$D_{1b} = \left(\frac{7680K\ \mathrm{N \cdot m}}{43\ \mathrm{MPa}}\right)^{\frac{1}{3}} = 41.59\ \mathrm{mm}$$

Use $D_1 = 41.6$ mm. Check the maximum shear stress:

$$\tau_{max} = K\left(\frac{16T}{\pi D_1^3}\right) = (1.265)\left[\frac{16(960\ \mathrm{N \cdot m})}{\pi(41.6\ \mathrm{mm})^3}\right] = 86\ \mathrm{MPa}$$

4. ***Finalize***: A stepped shaft with $D_2 = 53$ mm, $D_1 = 41.6$ mm, and a full quarter-circular fillet of radius $R = 5.7$ mm will carry the required torque T without exceeding the allowable shear stress in the fillet region. The shaft is approximately the same size as that in part (a), but the higher allowable shear stress leads to greater permissible torque.

CHAPTER SUMMARY AND REVIEW

Chapter 3 discussed the behavior of bars and hollow tubes acted on by concentrated torques or distributed torsional moments as well as prestrain effects. Torque-displacement relations were derived for use in computing angles of twist of bars under both uniform and nonuniform conditions. Then, equilibrium and compatibility equations were developed for statically indeterminate structures in a superposition procedure leading to solution for all unknown torques, rotational displacements and, stresses. Equations were developed for normal and shear stresses on inclined sections. Here are the major concepts presented in this chapter.

1. For circular bars and tubes, the **shearing stress** (τ) and **strain** (γ) vary linearly with radial distance from the center of the cross-section.

$$\tau = (\rho/r)\tau_{max} \quad \gamma = (\rho/r)\gamma_{max}$$

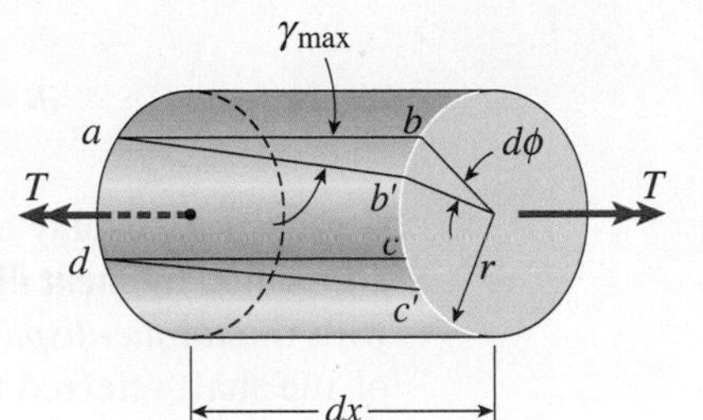

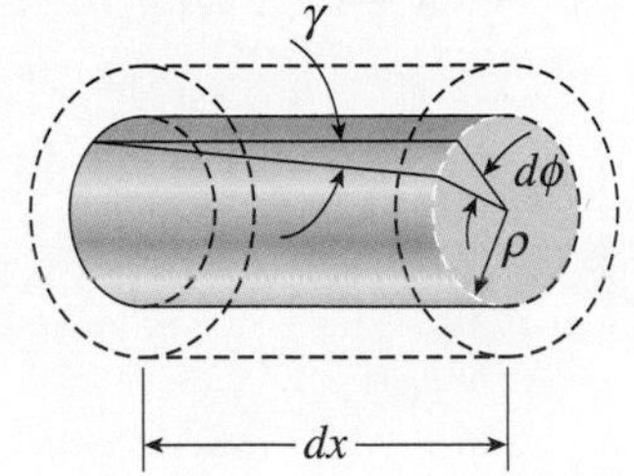

2. The **torsion formula** defines the relation between shear stress and torsional moment. Maximum shear stress τ_{max} occurs on the outer surface of the bar or tube and depends on torsional moment T, radial distance r, and polar moment of inertia I_p for circular cross sections. Thin-walled tubes are more efficient in torsion because the available material is more uniformly stressed than solid circular bars.

$$\tau_{max} = \frac{Tr}{I_p}$$

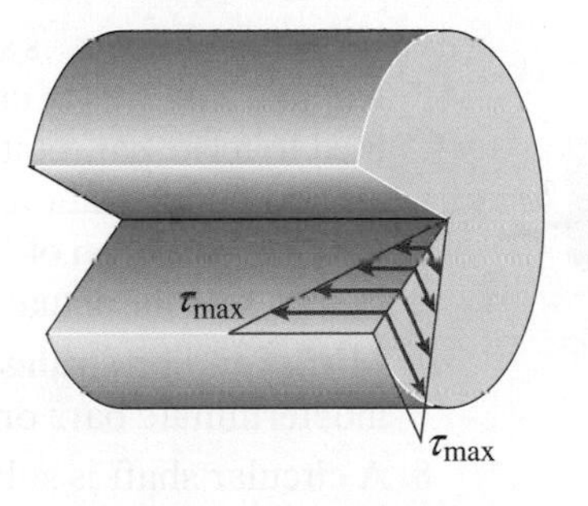

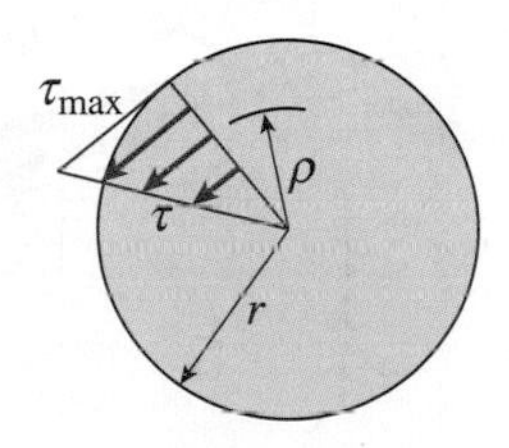

3. The angle of twist ϕ of prismatic circular bars subjected to torsional moment(s) is proportional to both the torque T and the length of the bar L, and it is inversely proportional to the torsional rigidity (GI_p) of the bar; this relationship is called the **torque-displacement relation**.

$$\phi = \frac{TL}{GI_p}$$

4. The angle of twist per unit length of a bar is referred to as its **torsional flexibility** (f_T), and the inverse relationship is the **torsional stiffness** ($k_T = 1/f_T$) of the bar or shaft.

$$k_T = \frac{GI_p}{L} \qquad f_T = \frac{L}{GI_p}$$

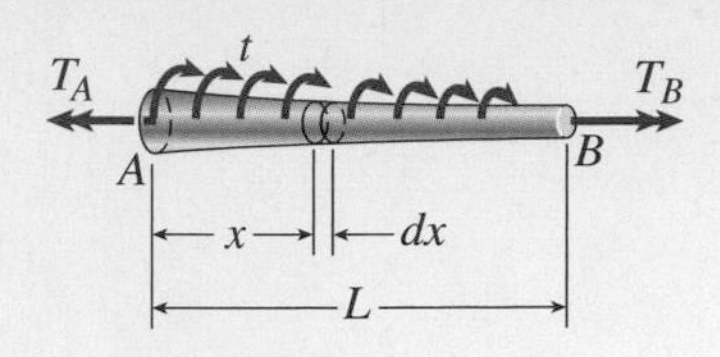

5. The summation of the twisting deformations of the individual segments of a nonprismatic shaft equals the twist of the entire bar (ϕ). Free-body diagrams are used to find the torsional moments (T_i) in each segment i.

$$\phi = \sum_{i=1}^{n} \phi_i = \sum_{i=l}^{n} \frac{T_i L_i}{G_i (I_p)_i}$$

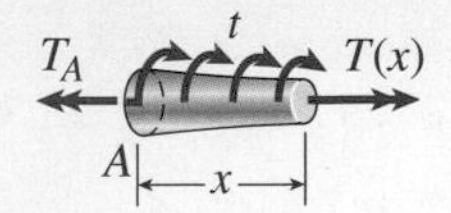

If torsional moments and/or cross sectional properties (I_p) vary continuously, an integral expression is required.

$$\phi = \int_0^L d\phi = \int_0^L \frac{T(x)dx}{GI_p(x)}$$

A graphical display of the internal torsional moments T_i is called a **torsional moment diagram** (TMD). The TMD can be used along with the *torque-displacement relation* to create the displaced shape of the shaft referred to as the **torsional displacement diagram** (TDD). These diagrams are useful visual aids that help to identify locations of maximum moment and shear stress, twist displacement, and shear strain.

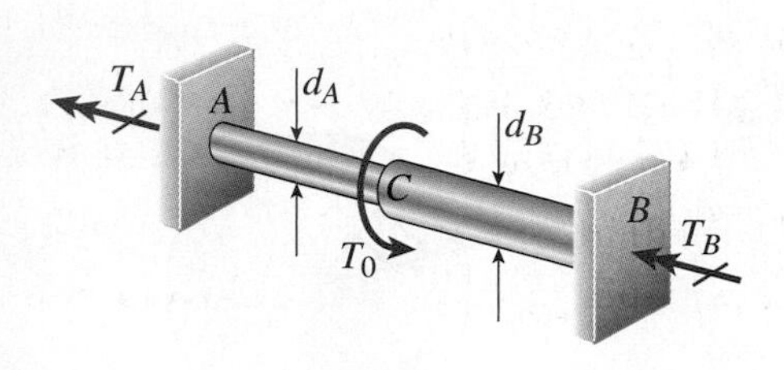

6. If the bar structure is **statically indeterminate**, additional equations are required to solve for unknown moments. **Compatibility equations** are used to relate bar rotations to support conditions and thereby generate additional relationships among the unknowns. It is convenient to use a **superposition** of "released" (or statically determinate) structures to represent the actual statically indeterminate bar structure.
7. **Misfits** and **prestrains** induce torsional moments only in statically indeterminate bars or shafts.

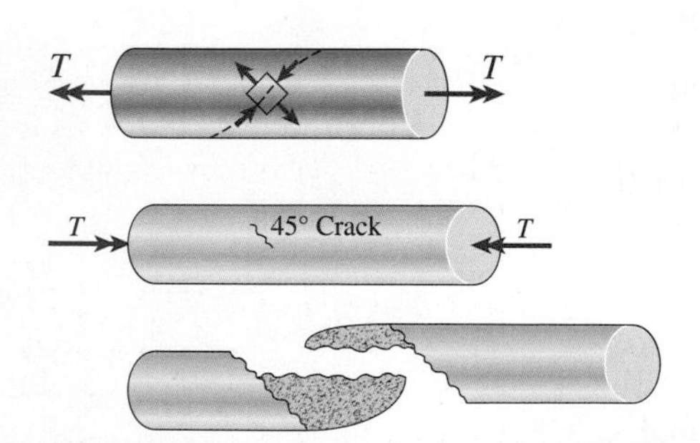

8. A circular shaft is subjected to **pure shear** due to torsional moments. **Maximum normal** and **shear stresses** are obtained by considering an inclined stress element. The maximum shear stress occurs on an element aligned with the axis of the bar, but the maximum normal stress occurs at an inclination of 45° to the axis of the bar, and the maximum normal stress is equal to the maximum shear stress

$$\sigma_{\text{max}} = \tau$$

The relationship between the maximum shear and normal strains for the case of pure shear is:

$$\varepsilon_{\text{max}} = \gamma_{\text{max}}/2$$

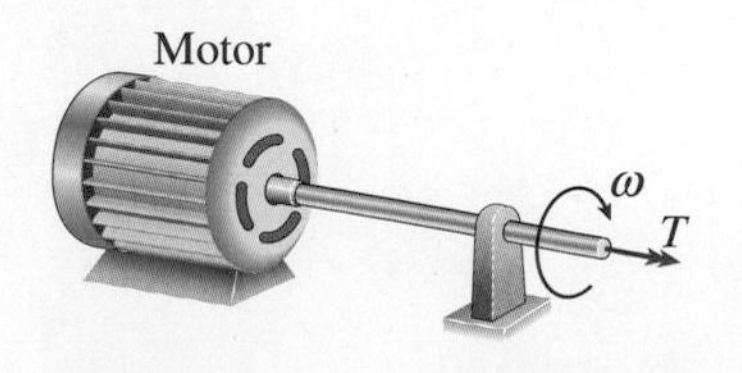

9. Circular shafts are commonly used to transmit mechanical power from one device or machine to another. If the torque T is expressed in newton meters and n is the shaft rpm, the power P is expressed in watts as

$$P = \frac{2\pi n T}{60}$$

In U.S. customary units, torque T is given in ft-lb and power is given in horsepower (hp), H, as

$$H = \frac{2\pi nT}{33{,}000}$$

10. The torsion formula and torque-displacement relation do not apply to prismatic **noncircular shafts.** Shafts with elliptical, triangular, and rectangular cross sections *warp* under the action of torsional moments, and the *maximum shear stress does not occur at the largest distance from the axis of the shaft like it does for circular sections.*

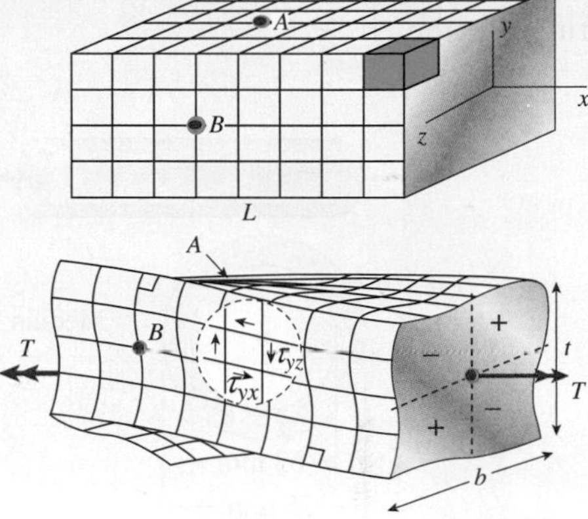

11. If the shaft has grooves, notches, keyways, shoulders, threads, or other abrupt changes in geometry, **stress concentrations** occur to disrupt the otherwise uniform stress pattern. These discontinuities in geometry cause high stresses in very small regions of the bar. Stress concentration factors are used to find the maximum stresses at these locations.

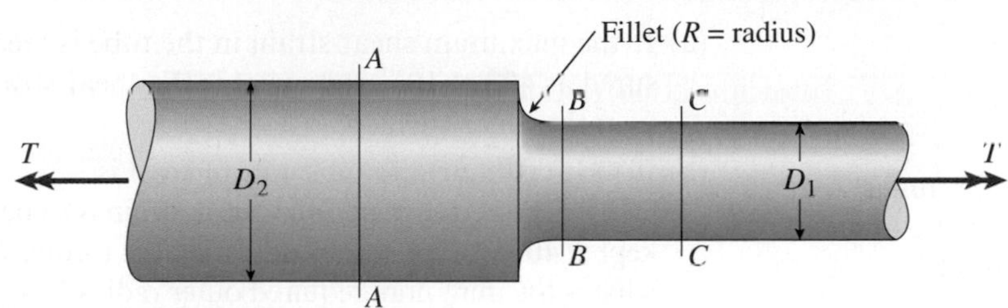

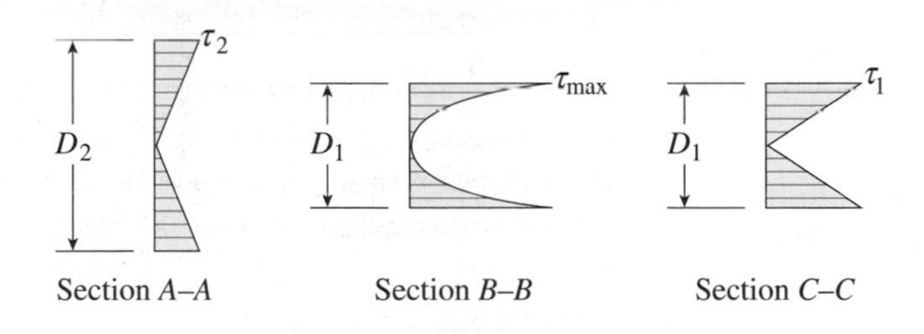

12. **Additional topics** such as strain energy and behavior of thin-walled tubes were also discussed but are not summarized here.

PROBLEMS Chapter 3

3.2 Torsional Deformations of a Circular Bar

Introductory Problems

3.2-1 A circular tube is subjected to torque T at its ends. The resulting maximum shear strain in the tube is 0.005. Calculate the minimum shear strain in the tube and the shear strain at the median line of the tube section.

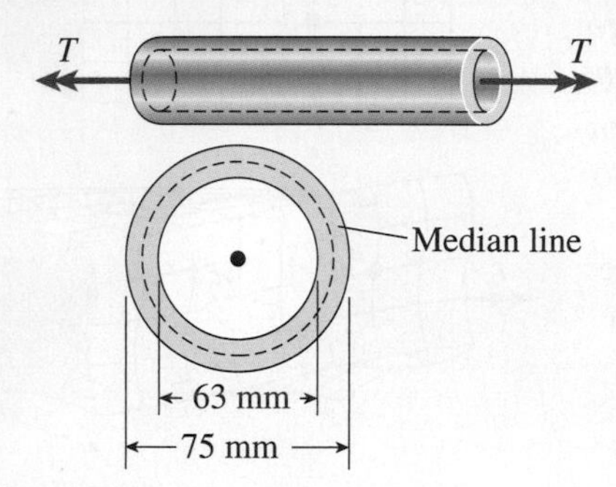

PROBLEM 3.2-1

3.2-2 A plastic bar of diameter $d = 56$ mm is to be twisted by torques T (see figure) until the angle of rotation between the ends of the bar is 4.0°.

(a) If the allowable shear strain in the plastic is 0.012 rad, what is the minimum permissible length of the bar?

(b) If the length of the bar is 200 mm, what is the maximum permissible diameter of the bar?

3.2-3 A copper rod of length $L = 460$ mm is to be twisted by torques T (see figure) until the angle of rotation between the ends of the rod is 3.0°.

(a) If the allowable shear strain in the copper is 0.0006 rad, what is the maximum permissible diameter of the rod?

(b) If the rod diameter is 12.5 mm, what is the minimum permissible length of the rod?

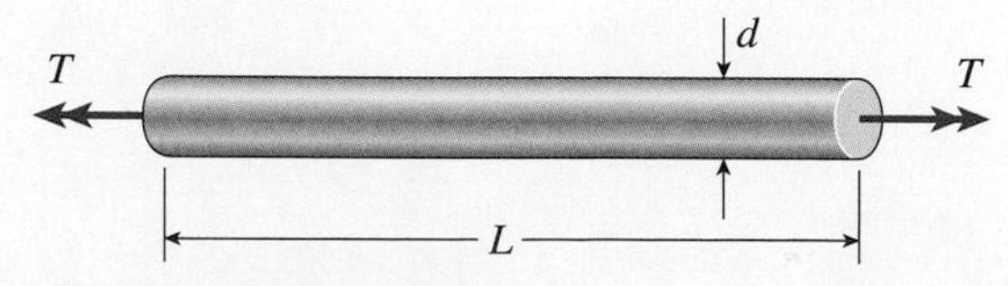

PROBLEMS 3.2-2 and 3.2-3

Representative Problems

3.2-4 A circular steel tube of length $L = 1.0$ m is loaded in torsion by torques T (see figure).

(a) If the inner radius of the tube is $r_1 = 45$ mm and the measured angle of twist between the ends is 0.5°, what is the shear strain γ_1 (in radians) at the inner surface?

(b) If the maximum allowable shear strain is 0.0004 rad and the angle of twist is to be kept at 0.45° by adjusting the torque T, what is the maximum permissible outer radius $(r_2)_{max}$?

3.2-5 Solve the preceding problem if the length $L = 1420$ mm, the inner radius $r_1 = 32$ mm, the angle of twist is 0.5°, and the allowable shear strain is 0.0004 rad.

3.2-6 A circular aluminum tube subjected to pure torsion by torques T (see figure) has an outer radius r_2 equal to 1.5 times the inner radius r_1.

(a) If the maximum shear strain in the tube is measured as 400×10^{-6} rad, what is the shear strain γ_1 at the inner surface?

(b) If the maximum allowable rate of twist is 0.125 °/m and the maximum shear strain is to be kept at 400×10^{-6} rad by adjusting the torque T, what is the minimum required outer radius $(r_2)_{min}$?

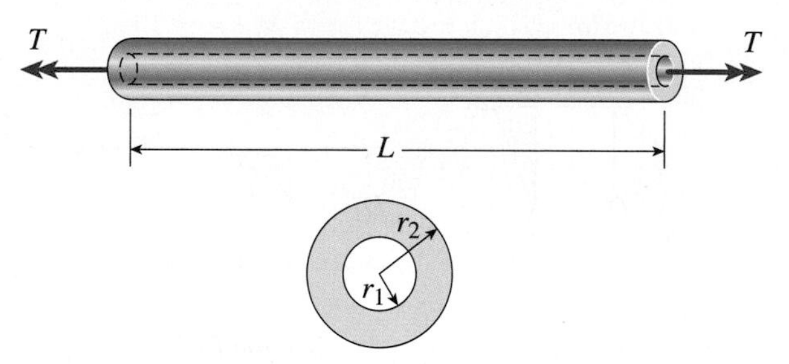

PROBLEMS 3.2-4, 3.2-5, and 3.2-6

3.3 Circular Bars of Linearly Elastic Materials

Introductory Problems

3.3-1 A solid steel bar of circular cross section has diameter $d = 63$ mm, $L = 1.5$ m, and shear modulus of elasticity $G = 80$ GPa. The bar is subjected to torques $T = 400$ N·m at the ends. Calculate the angle of twist between the ends. What is the maximum shear stress and the shear stress at a distance $r_A = 25$ mm measured from the center of the bar?

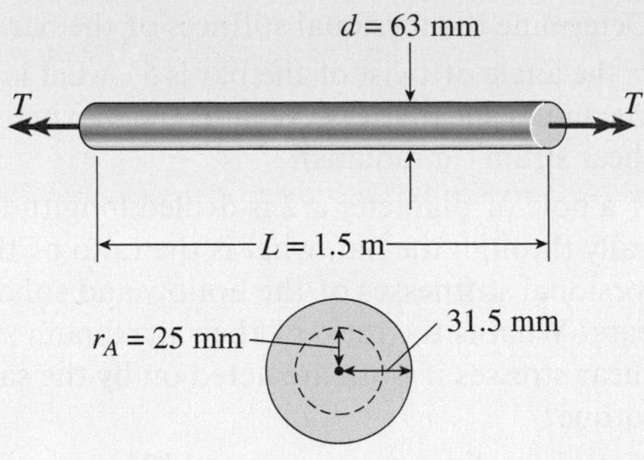

PROBLEM 3.3-1

3.3-2 A solid copper bar of circular cross section has length $L = 1.25$ m and shear modulus of elasticity $G = 45$ GPa. The bar is designed to carry a 250 N · m torque acting at the ends. If the allowable shear stress is 30 MPa and the allowable angle of twist between the ends is 2.5°, what is the minimum required diameter?

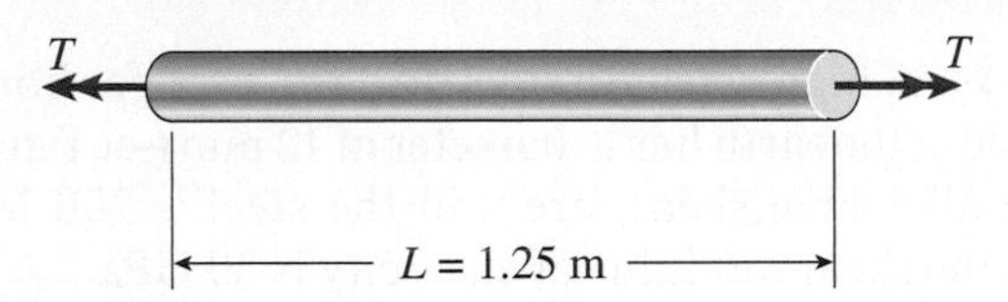

PROBLEM 3.3-2

3.3-3 Repeat Problem 3.3-1, but now use a circular tube with outer diameter $d_o = 63$ mm and inner diameter $d_i = 38$ mm.

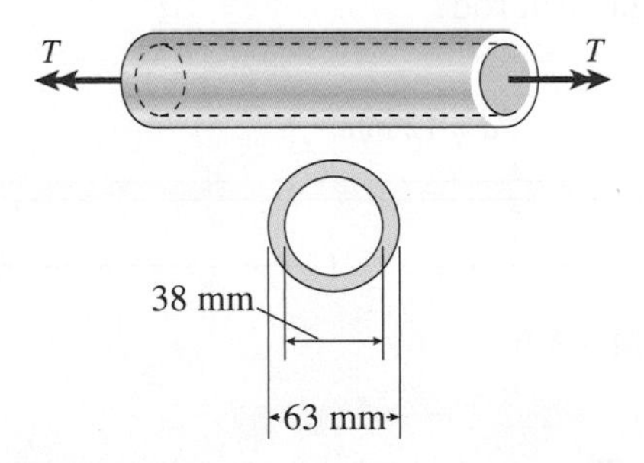

PROBLEM 3.3-3

3.3-4 A copper tube with circular cross section has length $L = 1.25$ m, thickness $t = 2$ mm, and shear modulus of elasticity $G = 45$ GPa. The bar is designed to carry a 300 N · m torque acting at the ends. If the allowable shear stress is 25 MPa and the allowable angle of twist between the ends is 2.5°, what is the minimum required outer diameter d?

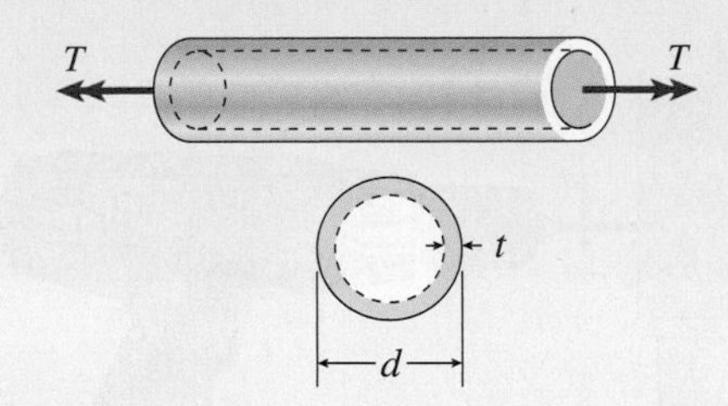

PROBLEM 3.3-4

Representative Problems

3.3-5 A prospector uses a hand-powered winch (see figure) to raise a bucket of ore in his mine shaft. The axle of the winch is a steel rod of diameter $d = 15$ mm. Also, the distance from the center of the axle to the center of the lifting rope is $b = 100$ mm.

(a) If the weight of the loaded bucket is $W = 400$ N, what is the maximum shear stress in the axle due to torsion?

(b) If the maximum bucket load is 510 N and the allowable shear stress in the axle is 65 MPa, what is the minimum permissible axle diameter?

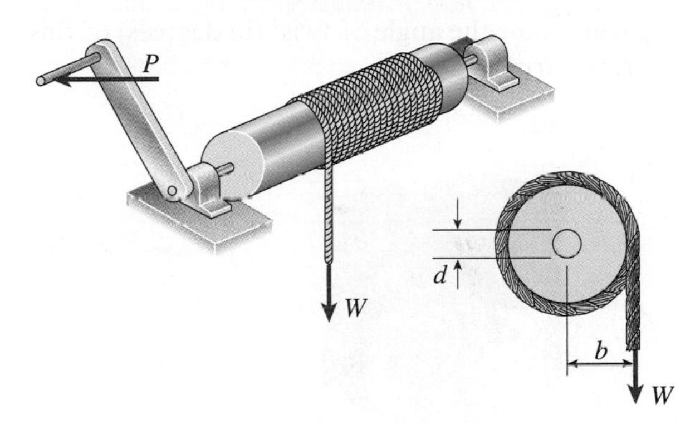

PROBLEM 3.3-5

3.3-6 When drilling a hole in a table leg, a furniture maker uses a hand-operated drill (see figure) with a bit of diameter $d = 4.0$ mm.

(a) If the resisting torque supplied by the table leg is equal to 0.3 N · m, what is the maximum shear stress in the drill bit?

(b) If the allowable shear stress in the drill bit is 32 MPa, what is the maximum resisting torque before the drill binds up?

(c) If the shear modulus of elasticity of the steel is $G = 75$ GPa, what is the rate of twist of the drill bit (degrees per meter)?

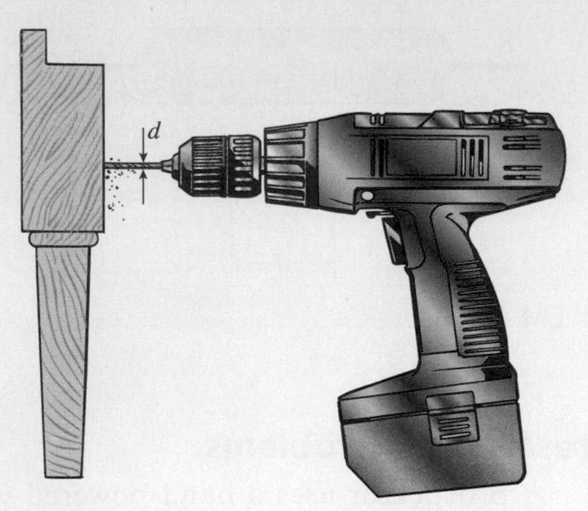

PROBLEM 3.3-6

3.3-7 While removing a wheel to change a tire, a driver applies forces $P = 100$ N at the ends of two of the arms of a lug wrench (see figure). The wrench is made of steel with shear modulus of elasticity $G = 78$ GPa. Each arm of the wrench is 225 mm long and has a solid circular cross section of diameter $d = 12$ mm.

(a) Determine the maximum shear stress in the arm that is turning the lug nut (arm A).

(b) Determine the angle of twist (in degrees) of this same arm.

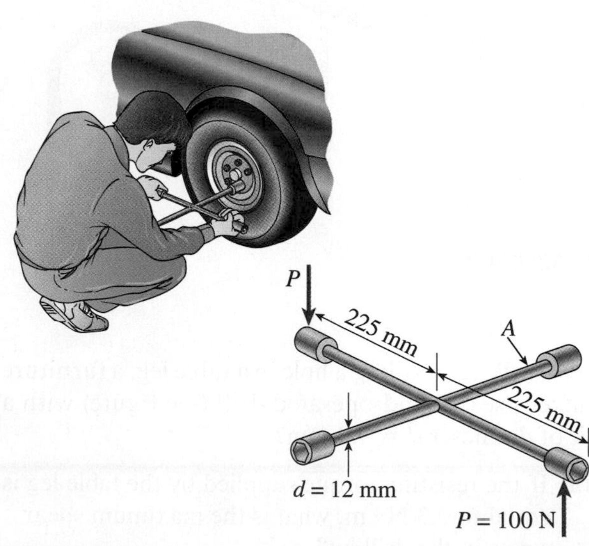

PROBLEM 3.3-7

3.3-8 An aluminum bar of solid circular cross section is twisted by torques T acting at the ends (see figure). The dimensions and shear modulus of elasticity are $L = 1.4$ m, $d = 32$ mm, and $G = 28$ GPa.

(a) Determine the torsional stiffness of the bar.

(b) If the angle of twist of the bar is 5°, what is the maximum shear stress? What is the maximum shear strain (in radians)?

(c) If a hole of diameter $d/2$ is drilled longitudinally through the bar, what is the ratio of the torsional stiffnesses of the hollow and solid bars? What is the ratio of their maximum shear stresses if both are acted on by the same torque?

(d) If the hole diameter remains at $d/2$, what new outside diameter d_2 will result in equal stiffnesses of the hollow and solid bars?

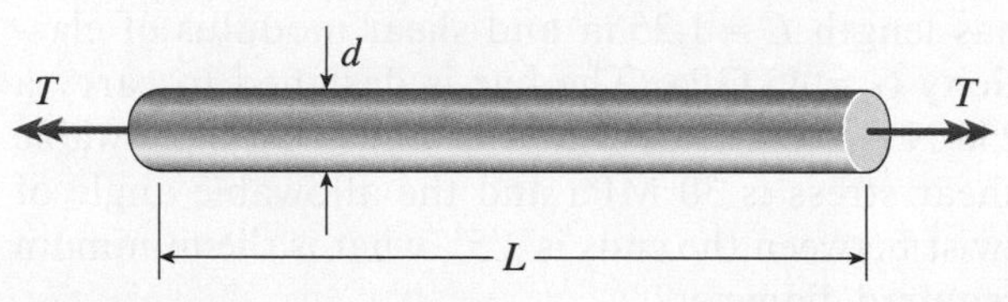

PROBLEM 3.3-8

3.3-9 A high-strength steel drill rod used for boring a hole in the earth has a diameter of 12 mm (see figure). The allowable shear stress in the steel is 300 MPa and the shear modulus of elasticity is 80 GPa.

(a) What is the minimum required length of the rod so that one end of the rod can be twisted 30° with respect to the other end without exceeding the allowable stress?

(b) If the shear strain in part (a) is limited to 3.2×10^{-3}, what is the minimum required length of the drill rod?

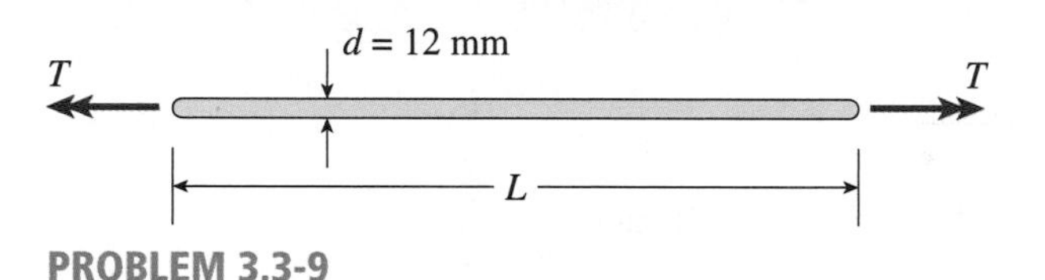

PROBLEM 3.3-9

3.3-10 The steel shaft of a socket wrench has a diameter of 8.0 mm and a length of 200 mm (see figure).

If the allowable stress in shear is 60 MPa, what is the maximum permissible torque T_{max} that may be exerted with the wrench?

Through what angle ϕ (in degrees) will the shaft twist under the action of the maximum torque? (Assume $G = 78$ GPa and disregard any bending of the shaft.)

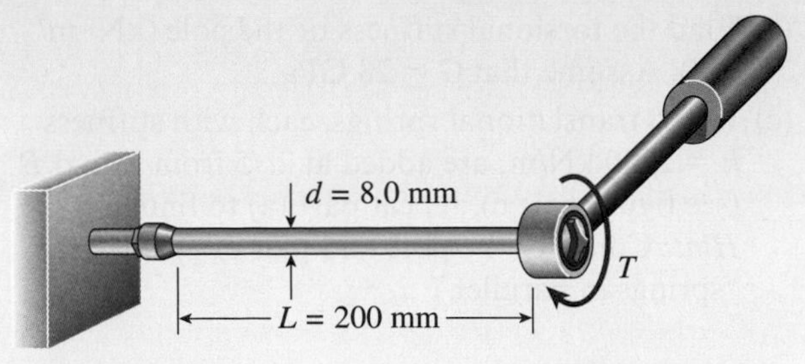

PROBLEM 3.3-10

3.3-11 A circular tube of aluminum is subjected to torsion by torques T applied at the ends (see figure). The bar is 0.75 m long, and the inside and outside diameters are 28 mm and 45 mm, respectively. It is determined by measurement that the angle of twist is 4° when the torque is 700 N·m.

(a) Calculate the maximum shear stress τ_{max} in the tube, the shear modulus of elasticity G, and the maximum shear strain γ_{max} (in radians).

(b) If the maximum shear strain in the tube is limited to 2.2×10^{-3} and the inside diameter is increased to 35 mm, what is the maximum permissible torque?

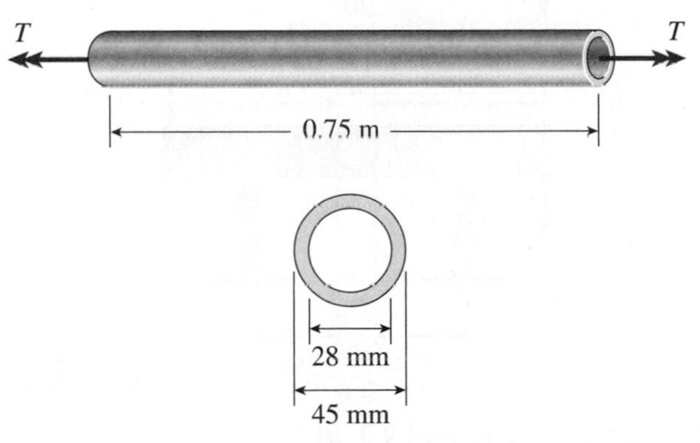

PROBLEM 3.3-11

3.3-12 A propeller shaft for a small yacht is made of a solid steel bar 104 mm in diameter. The allowable stress in shear is 48 MPa, and the allowable rate of twist is 2.0° in 3.5 meters.

(a) Assuming that the shear modulus of elasticity is $G = 80$ GPa, determine the maximum torque T_{max} that can be applied to the shaft.

(b) Repeat part (a) if the shaft is now hollow with an inner diameter of $5d/8$. Compare T_{max} values to corresponding values from part (a).

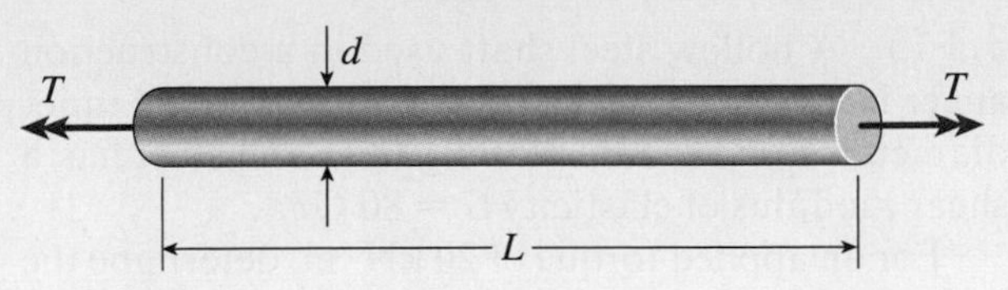

PROBLEM 3.3-12

3.3-13 Three identical circular disks A, B, and C are welded to the ends of three identical solid circular bars (see figure). The bars lie in a common plane and the disks lie in planes perpendicular to the axes of the bars. The bars are welded at their intersection D to form a rigid connection. Each bar has diameter $d_1 = 10$ mm and each disk has diameter $d_2 = 75$ mm.

Forces P_1, P_2, and P_3 act on disks A, B, and C, respectively, thus subjecting the bars to torsion. If $P_1 = 100$ N, what is the maximum shear stress τ_{max} in any of the three bars?

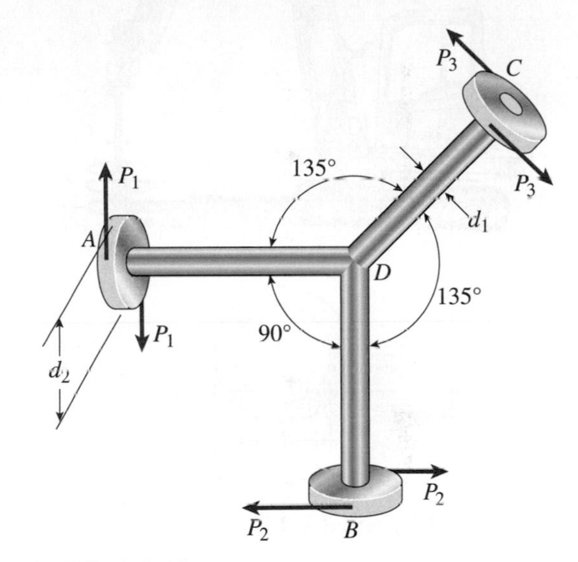

PROBLEM 3.3-13

3.3-14 The steel axle of a large winch on an ocean liner is subjected to a torque of 1.65 kN · m (see figure).

(a) What is the minimum required diameter d_{min} if the allowable shear stress is 48 MPa and the allowable rate of twist is 0.75°/m? (Assume that the shear modulus of elasticity is 80 GPa.)

(b) Repeat part (a) if the shaft is now hollow with an inner diameter of $5d/8$. Compare d_{min} values to corresponding values from part (a).

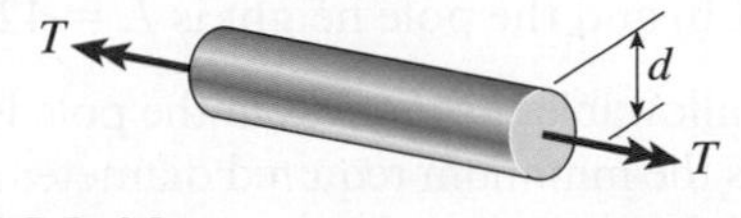

PROBLEM 3.3-14

3.3-15 A hollow steel shaft used in a construction auger has an outer diameter $d_2 = 175$ mm and inner diameter $d_1 = 125$ mm (see figure). The steel has a shear modulus of elasticity $G = 80$ GPa.

For an applied torque of 20 kN · m, determine the following quantities:

(a) shear stress τ_2 at the outer surface of the shaft,
(b) shear stress τ_1 at the inner surface, and
(c) rate of twist θ (degrees per unit of length).

Also, draw a diagram showing how the shear stresses vary in magnitude along a radial line in the cross section.

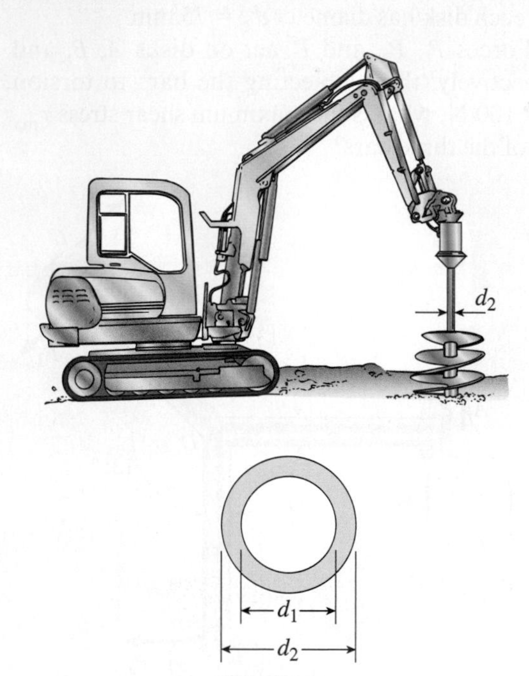

PROBLEMS 3.3-15 and 3.3-16

3.3-16 Solve the preceding problem if the shaft has an outer diameter $d_2 = 150$ mm and inner diameter $d_1 = 100$ mm. Also, the steel has a shear modulus of elasticity $G = 75$ GPa, and the applied torque is 16 kN · m.

3.3-17 A vertical pole of solid, circular cross section is twisted by horizontal forces $P = 12$ kN acting at the ends of a rigid horizontal arm AB (see figure part a). The distance from the outside of the pole to the line of action of each force is $c = 212$ mm (see figure part b) and the pole height is $L = 425$ mm.

(a) If the allowable shear stress in the pole is 32 MPa, what is the minimum required diameter $d_{min.}$ of the pole?
(b) Find the torsional stiffness of the pole (kN · m/rad). Assume that $G = 28$ GPa.
(c) If two translational springs, each with stiffness $k = 2700$ kN/m, are added at $2c/5$ from A and B (see figure part c), repeat part (a) to find d_{min}. *Hint:* Consider the pole and pair of springs as "springs in parallel."

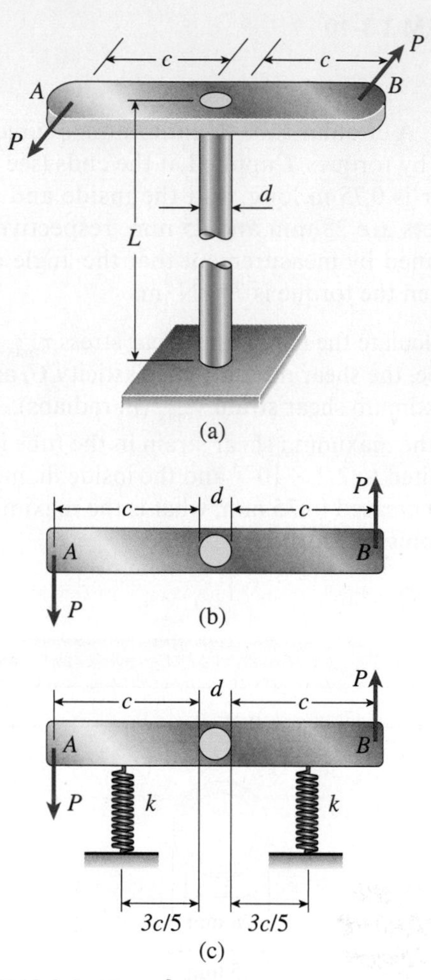

PROBLEMS 3.3-17 and 3.3-18

3.3-18 A vertical pole of solid, circular cross section is twisted by horizontal forces $P = 5$ kN acting at the ends of a rigid horizontal arm AB (see figure part a). The distance from the outside of the pole to the line of action of each force is $c = 125$ mm (see figure part b) and the pole height $L = 350$ mm.

(a) If the allowable shear stress in the pole is 30 MPa, what is the minimum required diameter d_{min} of the pole?
(b) What is the torsional stiffness of the pole (kN · m/rad)? Assume that $G = 28$ GPa.

(c) If two translational springs, each with stiffness $k = 2550$ kN/m, are added at $2c/5$ from A and B (see figure part c), repeat part (a) to find d_{min}. *Hint*: Consider the pole and pair of springs as "springs in parallel."

3.3-19 A solid brass bar of diameter $d = 30$ mm is subjected to torques T_1, as shown in part a of the figure. The allowable shear stress in the brass is 80 MPa.

(a) What is the maximum permissible value of the torques T_1?
(b) If a hole of diameter 15 mm is drilled longitudinally through the bar, as shown in part b of the figure, what is the maximum permissible value of the torques T_2?
(c) What is the percent decrease in torque and the percent decrease in weight due to the hole?

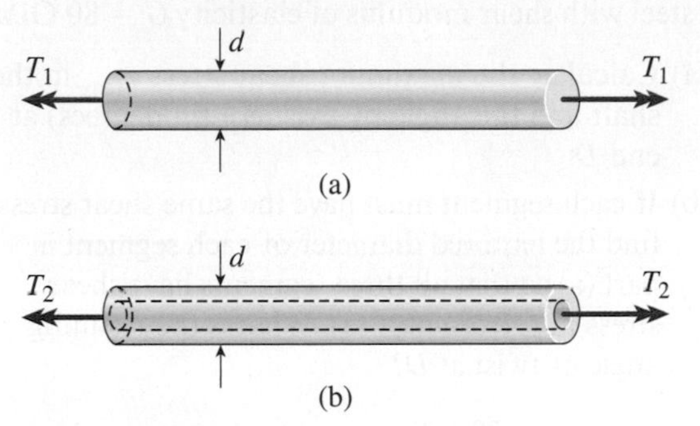

PROBLEM 3.3-19

3.3-20 A hollow aluminum tube used in a roof structure has an outside diameter $d_2 = 104$ mm and an inside $d_1 = 82$ mm (see figure). The tube is 2.75 m long, and the aluminum has shear modulus $G = 28$ GPa.

(a) If the tube is twisted in pure torsion by torques acting at the ends, what is the angle of twist (in degrees) when the maximum shear stress is 48 MPa?
(b) What diameter d is required for a solid shaft (see figure) to resist the same torque with the same maximum stress?
(c) What is the ratio of the weight of the hollow tube to the weight of the solid shaft?

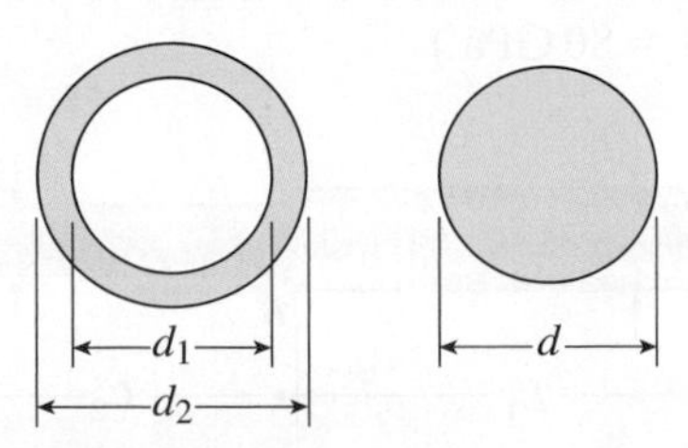

PROBLEM 3.3-20

3.3-21 A circular tube of inner radius r_1 and outer radius r_2 is subjected to a torque produced by forces $P = 4000$ N (see figure part a). The forces have their lines of action at a distance $b = 140$ mm from the outside of the tube.

(a) If the allowable shear stress in the tube is 43 MPa and the inner radius $r_1 = 30$ mm, what is the minimum permissible outer radius r_2?
(b) If a torsional spring of stiffness $k_R = 50$ kN·m/rad is added at the end of the tube (see figure part b), what is the maximum value of forces P if the allowable shear stress is not to be exceeded? Assume that the tube has a length of $L = 450$ mm, outer radius of $r_2 = 37$ mm, and shear modulus $G = 74$ GPa. *Hint*: Consider the tube and torsional spring as "springs in parallel."

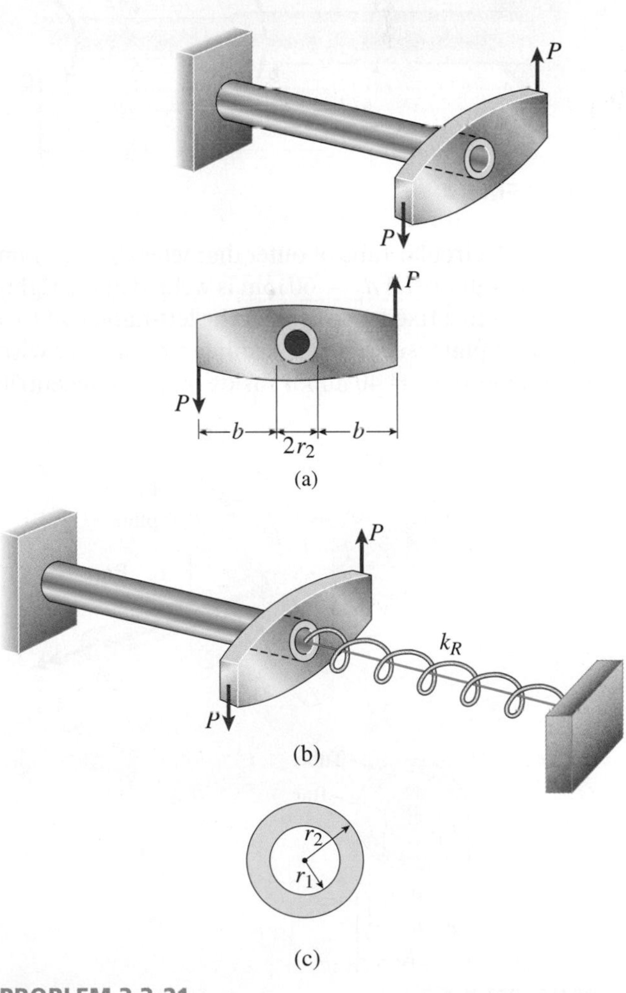

PROBLEM 3.3-21

3.4 Nonuniform Torsion

Introductory Problems

3.4-1 A stepped shaft ABC consisting of two solid circular segments is subjected to torques T_1 and T_2 acting in opposite directions, as shown in the figure. The larger segment of the shaft has a diameter of $d_1 = 58$ mm and length $L_1 = 760$ mm; the smaller segment has a diameter of $d_2 = 45$ mm and length of $L_2 = 510$ mm. The material is steel with shear modulus $G = 76$ GPa, and the torques are $T_1 = 2300$ N·m and $T_2 = 900$ N·m.

(a) Calculate the maximum shear stress τ_{max} in the shaft and the angle of twist ϕ_c (in degrees) at end C.

(b) If the maximum shear stress in BC must be the same as that in AB, what is the required diameter of segment BC? What is the resulting twist at end C?

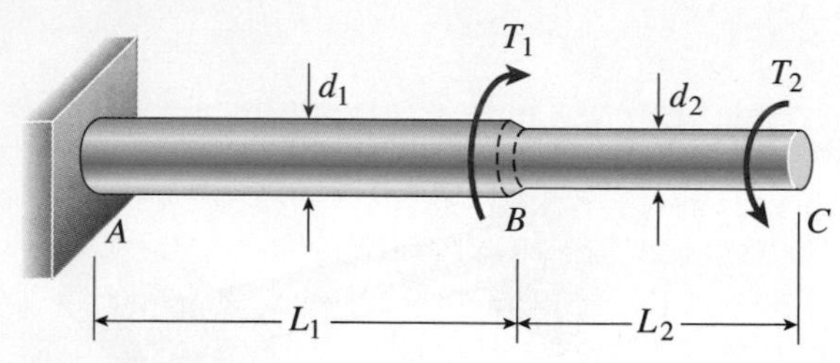

PROBLEM 3.4-1

3.4-2 A circular tube of outer diameter $d_3 = 70$ mm and inner diameter $d_2 = 60$ mm is welded at the right-hand end to a fixed plate and at the left-hand end to a rigid end plate (see figure). A solid, circular bar with a diameter of $d_1 = 40$ mm is inside of, and concentric with, the tube. The bar passes through a hole in the fixed plate and is welded to the rigid end plate.

The bar is 1.0 m long and the tube is half as long as the bar. A torque $T = 1000$ N·m acts at end A of the bar. Also, both the bar and tube are made of an aluminum alloy with a shear modulus of elasticity $G = 27$ GPa.

(a) Determine the maximum shear stresses in both the bar and tube.

(b) Determine the angle of twist (in degrees) at end A of the bar.

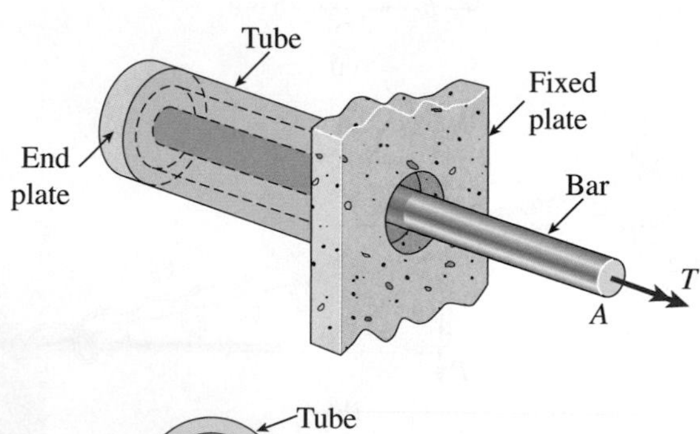

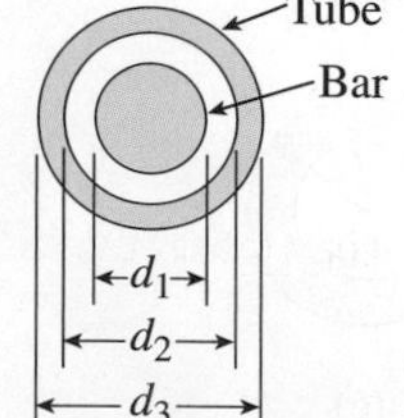

PROBLEM 3.4-2

3.4-3 A stepped shaft $ABCD$ consisting of solid circular segments is subjected to three torques, as shown in the figure. The torques have magnitudes of 3000 N·m, 2000 N·m, and 800 N·m. The length of each segment is 0.5 m and the diameters of the segments are 80 mm, 60 mm, and 40 mm. The material is steel with shear modulus of elasticity $G = 80$ GPa.

(a) Calculate the maximum shear stress τ_{max} in the shaft and the angle of twist ϕ_D (in degrees) at end D.

(b) If each segment must have the same shear stress, find the required diameter of each segment in part (a) so that all three segments have shear stress τ_{max} from part (a). What is the resulting angle of twist at D?

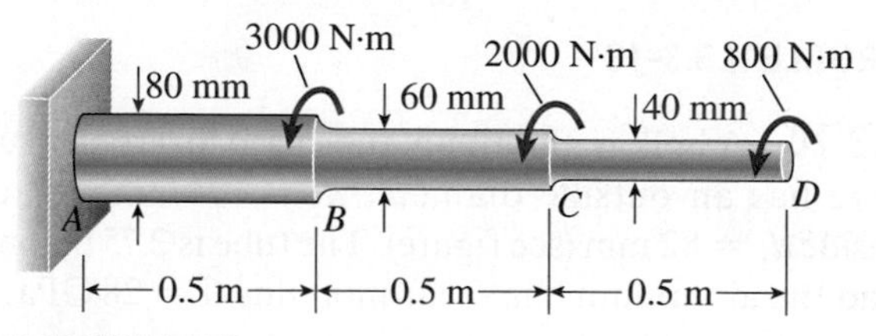

PROBLEM 3.4-3

3.4-4 A solid, circular bar ABC consists of two segments, as shown in the figure. One segment has a diameter of $d_1 = 56$ mm and length of $L_1 = 1.45$ m; the other segment has a diameter of $d_2 = 48$ mm and length of $L_2 = 1.2$ m.

What is the allowable torque T_{allow} if the shear stress is not to exceed 30 MPa and the angle of twist between the ends of the bar is not to exceed 1.25°? (Assume $G = 80$ GPa.)

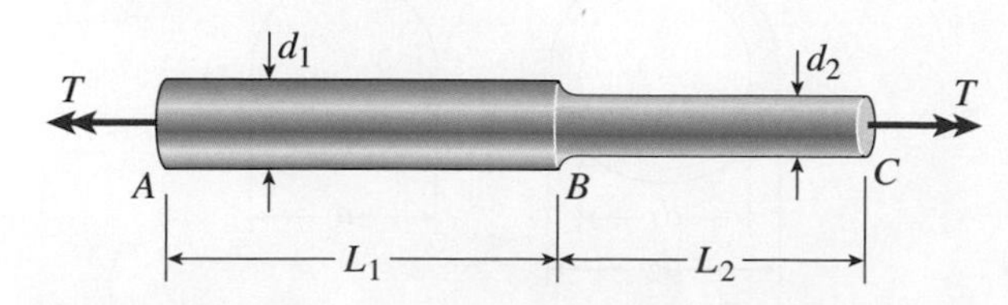

PROBLEM 3.4-4

3.4-5 A hollow tube *ABCDE* constructed of monel metal is subjected to five torques acting in the directions shown in the figure. The magnitudes of the torques are $T_1 = 100$ N·m, $T_2 = T_4 = 50$ N·m, and $T_3 = T_5 = 80$ N·m. The tube has an outside diameter of $d_2 = 25$ mm. The allowable shear stress is 80 MPa and the allowable rate of twist is 6°/m.

Determine the maximum permissible inside diameter d_1 of the tube.

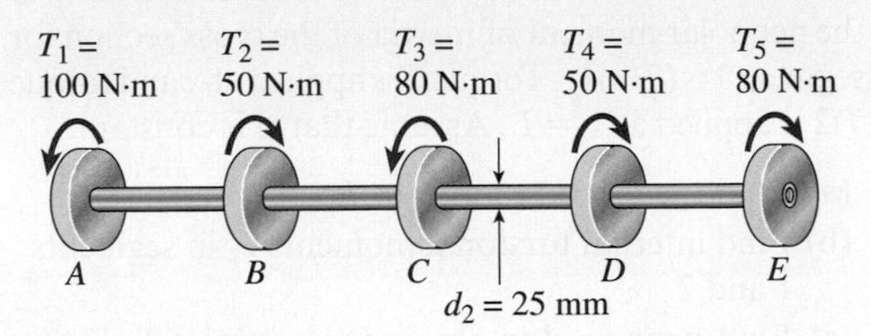

PROBLEM 3.4-5

3.4-6 A shaft with a solid, circular cross section consisting of two segments is shown in part a of the figure. The left-hand segment has a diameter of 80 mm and length of 1.2 m; the right-hand segment has a diameter of 60 mm and length of 0.9 m.

Shown in part b of the figure is a hollow shaft made of the same material and having the same length. The thickness *t* of the hollow shaft is *d*/10, where *d* is the outer diameter. Both shafts are subjected to the same torque.

(a) If the hollow shaft is to have the same torsional stiffness as the solid shaft, what should be its outer diameter *d*?

(b) If torque *T* is applied at either end of both shafts and the hollow shaft is to have the same maximum shear stress as the solid shaft, what should be its outer diamteter *d*?

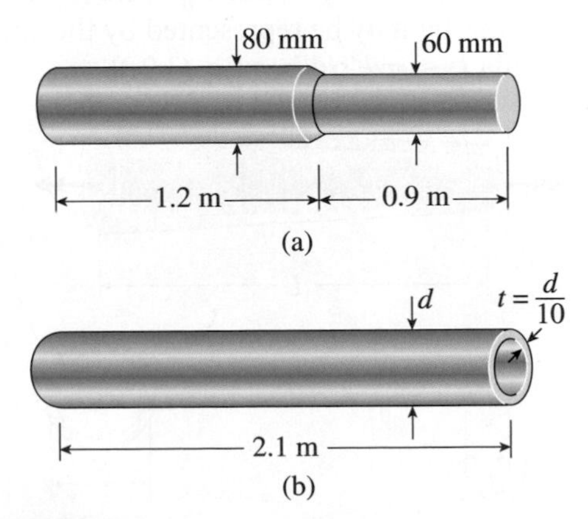

PROBLEM 3.4-6

Representative Problems

3.4-7 A solid steel shaft *ABCDE* turns freely in bearings at points *A* and *E*. The shaft is driven by the gear at *C*, which applies a torque $T_2 = 440$ N·m. Gears at *B* and *D* are driven by the shaft and have resisting torques $T_1 = 270$ N·m and $T_3 = 170$ N·m, respectively. Segments *BC* and *CD* have lengths $L_{BC} = 500$ mm and $L_{CD} = 380$ mm and the shear modulus $G = 80$ GPa. Determine the minimum required diameter (*d*) of the shaft if the allowable shear stress $\tau_a = 40$ MPa. Also calculate the angle of twist between gears *B* and *D*.

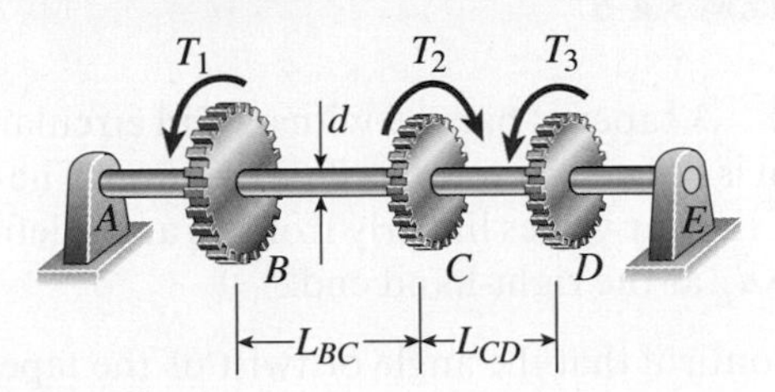

PROBLEM 3.4-7

3.4-8 Two sections of steel drill pipe, joined by bolted flange plates at *B*, are being tested to assess the adequacy of both the pipes. In the test, the pipe structure is fixed at *A*, a concentrated torque of 500 kN · m is applied at $x = 0.5$ m, and uniformly distributed torque intensity $t_0 = 250$ kN · m/m is applied on pipe *BC*. Both pipes have the same inner diameter $d = 200$ mm. Pipe *AB* has thickness $t_{AB} = 15$ mm, while pipe *BC* has thickness $t_{BC} = 12$ mm. Find the maximum shear stress and maximum twist of the pipe and their locations along the pipe. Assume $G = 75$ GPa.

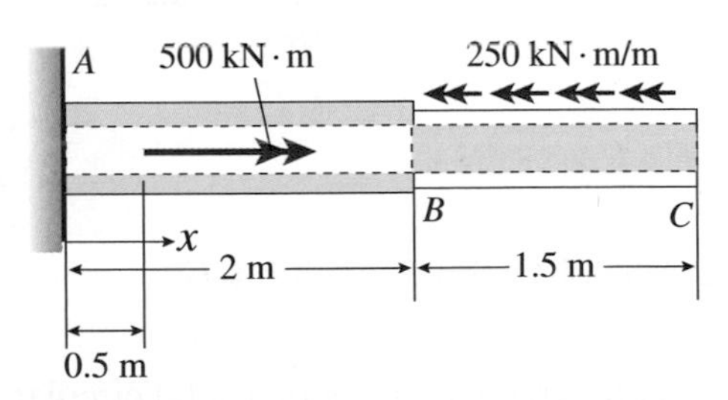

PROBLEM 3.4-8

3.4-9 Four gears are attached to a circular shaft and transmit the torques shown in the figure. The allowable shear stress in the shaft is 70 MPa.

(a) What is the required diameter *d* of the shaft if it has a solid cross section?

(b) What is the required outside diameter *d* if the shaft is hollow with an inside diameter of 40 mm?

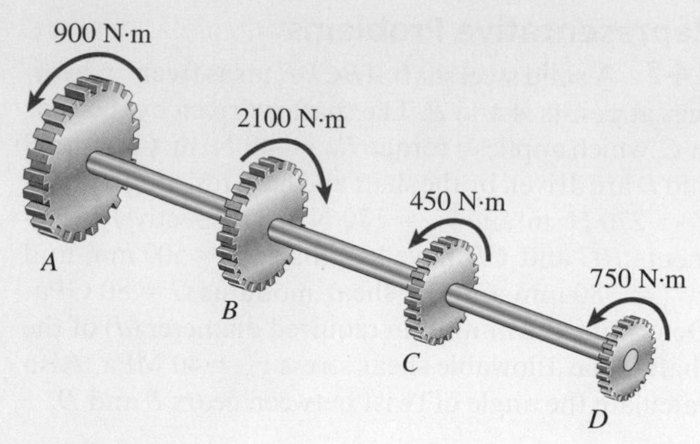

PROBLEM 3.4-9

3.4-10 A tapered bar AB with a solid circular cross section is twisted by torques T (see figure). The diameter of the bar varies linearly from d_A at the left-hand end to d_B at the right-hand end.

(a) Confirm that the angle of twist of the tapered bar is

$$\phi = \frac{32TL}{3\pi G(d_B - d_A)}\left(\frac{1}{d_A^3} - \frac{1}{d_B^3}\right)$$

(b) For what ratio d_B/d_A will the angle of twist of the tapered bar be one-half the angle of twist of a prismatic bar of diameter d_A? (The prismatic bar is made of the same material, has the same length, and is subjected to the same torque as the tapered bar.)

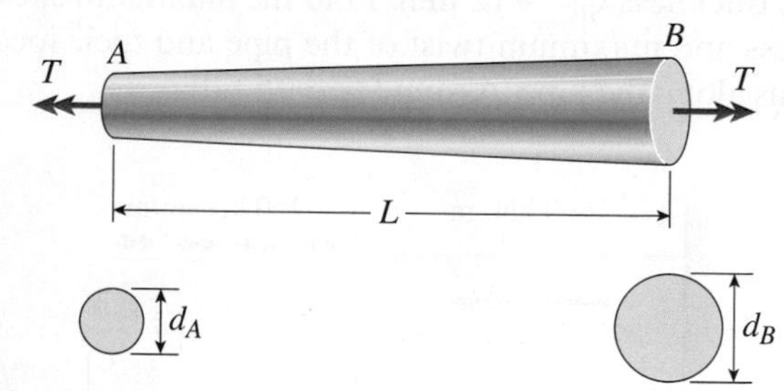

PROBLEMS 3.4-10, 3.4-11, and 3.4-12

3.4-11 A tapered bar AB with a solid circular cross section is twisted by torques $T = 2035$ N·m (see figure). The diameter of the bar varies linearly from d_A at the left-hand end to d_B at the right-hand end. The bar has length $L = 2.4$ m and is made of an aluminum alloy having shear modulus of elasticity $G = 27$ GPa. The allowable shear stress in the bar is 52 MPa and the allowable angle of twist is 3.0°.

If the diameter at end B is 1.5 times the diameter at end A, what is the minimum required diameter d_A at end A?

3.4-12 The bar shown in the figure is tapered linearly from end A to end B and has a solid circular cross section. The diameter at the smaller end of the bar is $d_A = 25$ mm and the length is $L = 300$ mm. The bar is made of steel with shear modulus of elasticity $G = 82$ GPa.

If the torque $T = 180$ N · m and the allowable angle of twist is 0.3°, what is the minimum allowable diameter d_B at the larger end of the bar?

3.4-13 The nonprismatic, cantilever circular bar shown has an internal cylindrical hole from 0 to x, so the net polar moment of inertia of the cross section for segment 1 is $(7/8)I_p$. Torque T is applied at x and torque $T/2$ is applied at $x = L$. Assume that G is constant.

(a) Find the reaction moment R_1.
(b) Find internal torsional moments T_i in segments 1 and 2.
(c) Find x required to obtain twist at joint 3 of $\phi_3 = TL/GI_p$.
(d) What is the rotation at joint 2, ϕ_2?
(e) Draw the torsional moment ($\text{TMD}: T(x), 0 \le x \le L$) and displacement ($\text{TDD}: \phi(x), 0 \le x \le L$) diagrams.

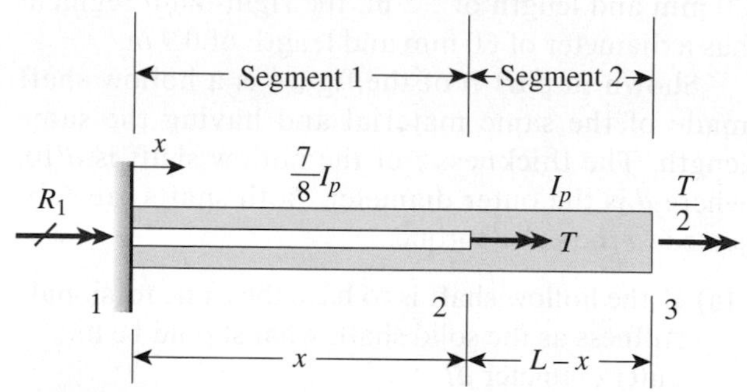

PROBLEM 3.4-13

3.4-14 A uniformly tapered tube AB with a hollow circular cross section is shown in the figure. The tube has constant wall thickness t and length L. The average diameters at the ends are d_A and $d_B = 2d_A$. The polar moment of inertia may be represented by the approximate formula $I_p \approx \pi d^3 t/4$ [see Eq. (3-21)].

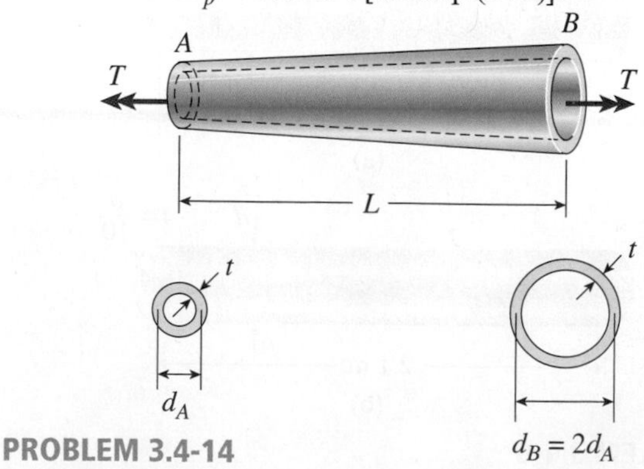

PROBLEM 3.4-14

Derive a formula for the angle of twist ϕ of the tube when it is subjected to torques T acting at the ends.

3.4-15 A uniformly tapered aluminum-alloy tube AB with a circular cross section and length L is shown in the figure. The outside diameters at the ends are d_A and $d_B = 2d_A$. A hollow section of length $L/2$ and constant thickness $t = d_A/10$ is cast into the tube and extends from B halfway toward A.

(a) Find the angle of twist ϕ of the tube when it is subjected to torques T acting at the ends. Use numerical values: $d_A = 65$ mm, $L = 1.2$ m, $G = 27$ GPa, and $T = 4.5$ kN·m.

(b) Repeat part (a) if the hollow section has constant diameter d_A (see figure part b).

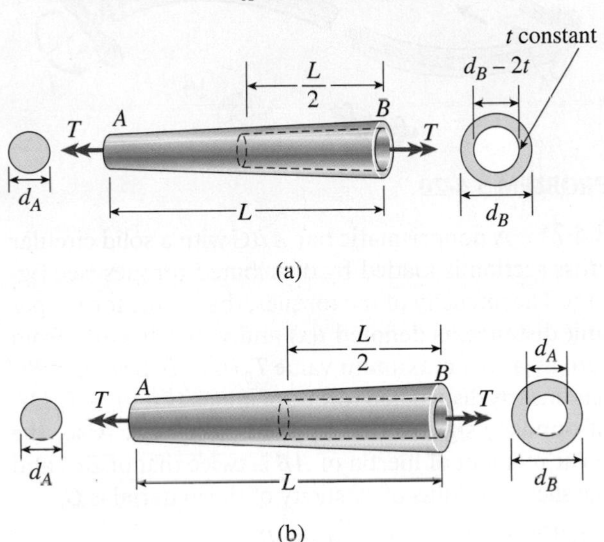

PROBLEM 3.4-15

3.4-16 For the *thin* nonprismatic steel pipe of constant thickness t and variable diameter d shown with applied torques at joints 2 and 3, determine the following.

(a) Find the reaction moment R_1.

(b) Find an expression for twist rotation ϕ_3 at joint 3. Assume that G is constant.

(c) Draw the torsional moment diagram (TMD: $T(x), 0 \le x \le L$).

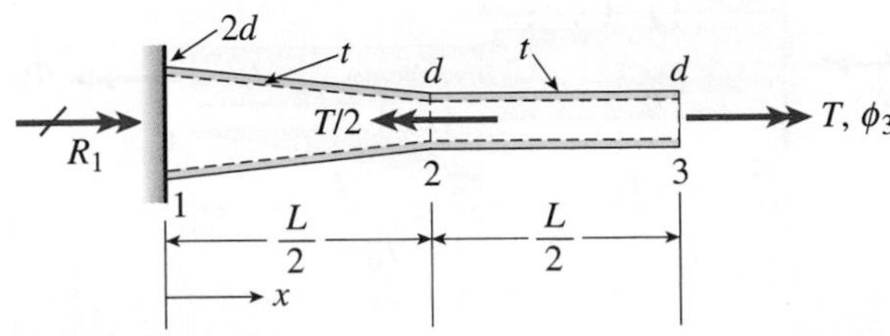

PROBLEM 3.4-16

3.4-17 A mountain-bike rider going uphill applies torque $T = Fd$ ($F = 65$ N, $d = 100$ mm) to the end of the handlebars $ABCD$ by pulling on the handlebar extenders DE. Consider the right half of the handlebar assembly only (assume the bars are fixed at the fork at A). Segments AB and CD are prismatic with lengths $L_1 = 50$ mm and $L_3 = 210$ mm and with outer diameters and thicknesses $d_{01} = 40$ mm, $t_{01} = 3$ mm and $d_{03} = 22$ mm, $t_{03} = 2.8$ mm respectively as shown. Segment BC of length $L_2 = 38$ mm however, is tapered, and outer diameter and thickness vary linearly between dimensions at B and C.

Consider torsion effects only. Assume $G = 28$ GPa is constant.

Derive an integral expression for the angle of twist ϕ_D of half of the handlebar tube when it is subjected to torque $T = Fd$ acting at the end. Evaluate ϕ_D for the given numerical values.

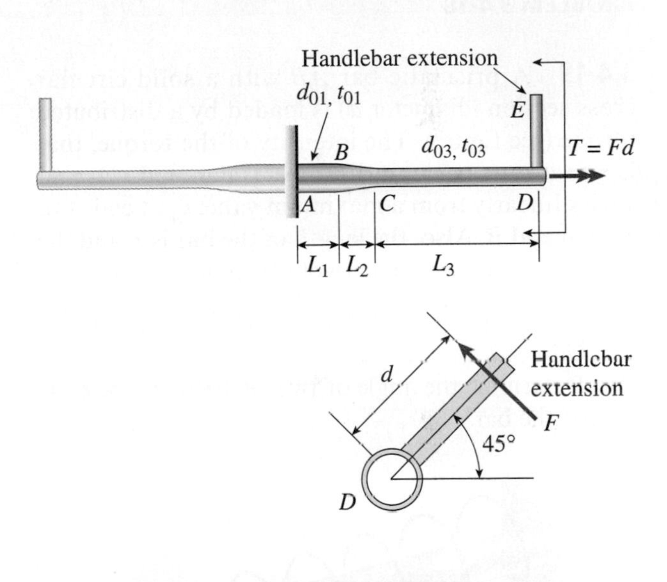

Bontrager Race XXX Lite Flat Handlebar, used Courtesy of Bontrager

PROBLEM 3.4-17

3.4-18 A prismatic bar AB of length L and solid circular cross section (diameter d) is loaded by a distributed torque of constant intensity t per unit distance (see figure).

(a) Determine the maximum shear stress τ_{max} in the bar.

(b) Determine the angle of twist ϕ between the ends of the bar.

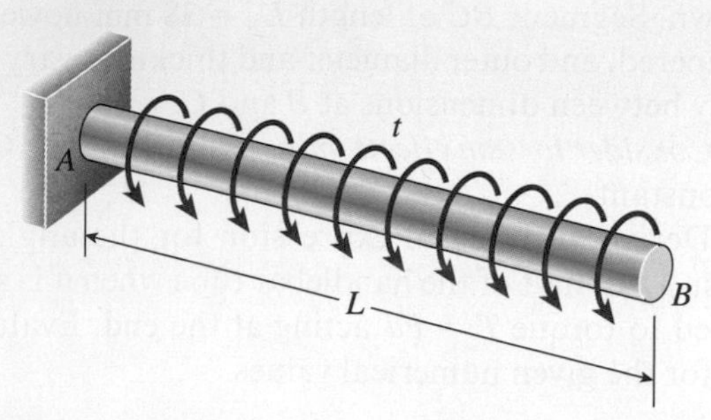

PROBLEM 3.4-18

3.4-19 A prismatic bar AB with a solid circular cross section (diameter d) is loaded by a distributed torque (see figure). The intensity of the torque, that is, the torque per unit distance, is denoted $t(x)$ and varies linearly from a maximum value t_A at end A to zero at end B. Also, the length of the bar is L and the shear modulus of elasticity of the material is G.

(a) Determine the maximum shear stress τ_{max} in the bar.

(b) Determine the angle of twist ϕ between the ends of the bar.

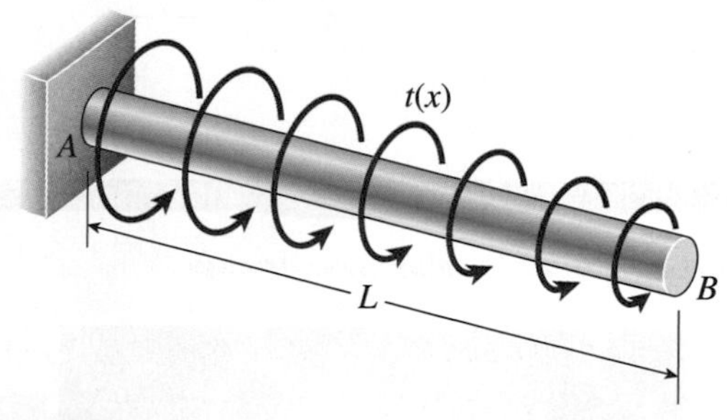

PROBLEM 3.4-19

3.4-20 A magnesium-alloy wire of diameter $d = 4$ mm and length L rotates inside a flexible tube in order to open or close a switch from a remote location (see figure). A torque T is applied manually (either clockwise or counterclockwise) at end B, thus twisting the wire inside the tube. At the other end A, the rotation of the wire operates a handle that opens or closes the switch.

A torque $T_0 = 0.2\ \text{N} \cdot \text{m}$ is required to operate the switch. The torsional stiffness of the tube, combined with friction between the tube and the wire, induces a distributed torque of constant intensity $t = 0.04\ \text{N} \cdot \text{m/m}$ (torque per unit distance) acting along the entire length of the wire.

(a) If the allowable shear stress in the wire is $\tau_{allow} = 30$ MPa, what is the longest permissible length L_{max} of the wire?

(b) If the wire has length $L = 4.0$ m and the shear modulus of elasticity for the wire is $G = 15$ GPa, what is the angle of twist ϕ (in degrees) between the ends of the wire?

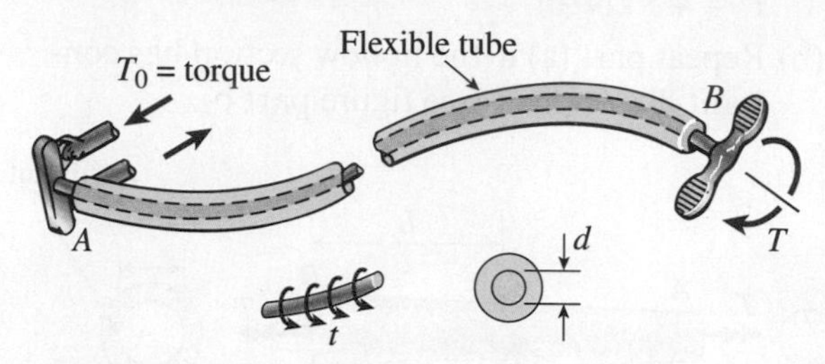

PROBLEM 3.4-20

3.4-21 A nonprismatic bar ABC with a solid circular cross section is loaded by distributed torques (see figure). The intensity of the torques, that is, the torque per unit distance, is denoted $t(x)$ and varies linearly from zero at A to a maximum value T_0/L at B. Segment BC has linearly distributed torque of intensity $t(x) = T_0/3L$ of opposite sign to that applied along AB. Also, the polar moment of inertia of AB is twice that of BC, and the shear modulus of elasticity of the material is G.

(a) Find the reaction torque R_A.

(b) Find internal torsional moments $T(x)$ in segments AB and BC.

(c) Find the rotation ϕ_C.

(d) Find the maximum shear stress τ_{max} and its location along the bar.

(e) Draw the torsional moment diagram (TMD: $T(x), 0 \leq x \leq L$).

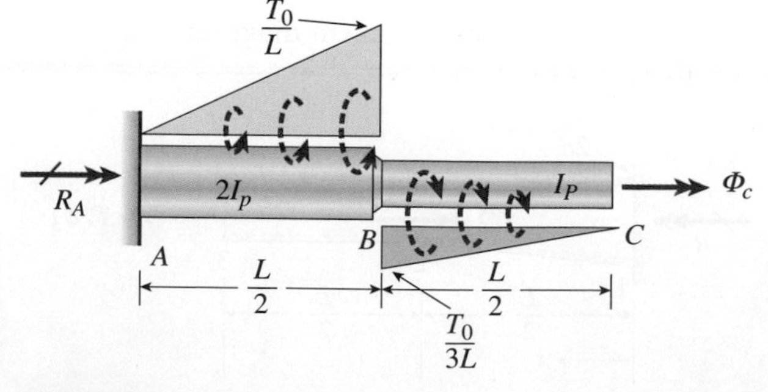

PROBLEM 3.4-21

3.4-22 Two tubes (AB, BC) of the same material are connected by three pins (pin diameter $= d_p$) just left of B as shown in the figure. Properties and dimensions for each tube are given in the figure. Torque $2T$ is applied at $x = 2L/5$ and uniformly distributed torque intensity $t_0 = 3T/L$ is applied on tube BC.

(a) Find the maximum value of load variable T(N·m) based on allowable shear (τ_a) and bearing (σ_{ba}) stresses in the three pins which connect the two tubes at B. Use the following numerical properties: $L = 1.5$ m, $E = 74$ GPa, $v = 0.33$, $d_p = 18$ mm, $\tau_a = 45$ MPa, $\sigma_{ba} = 90$ MPa, $d_1 = 85$ mm, $d_2 = 73$ mm, and $d_3 = 60$ mm.

(b) What is the maximum shear stress in the tubes for the applied torque in part (a)?

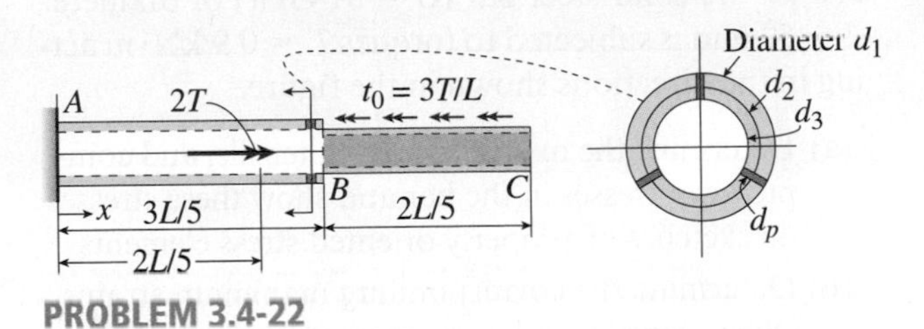

PROBLEM 3.4-22

3.5 Stresses and Strains in Pure Shear

Introductory Problems

3.5-1 A circular copper bar with diameter $d = 75$ mm is subjected to torques $T = 3.4$ kN·m at its ends. Find the maximum shear, tensile, and compressive stresses in the tube and their corresponding strains. Assume that $G = 41$ GPa.

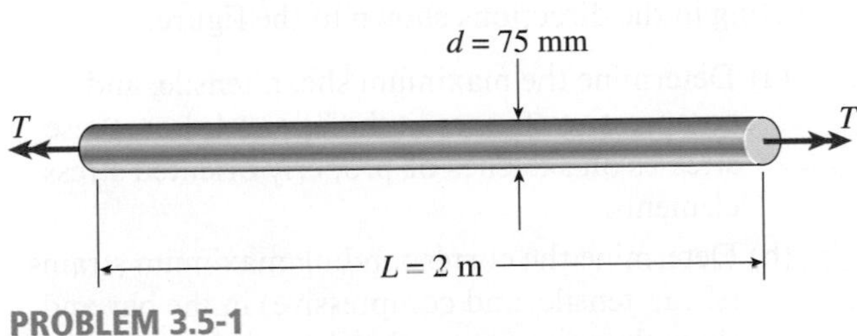

PROBLEM 3.5-1

3.5-2 A circular steel tube with an outer diameter of 75 mm and inner diameter of 65 mm is subjected to torques T at its ends. Calculate the maximum permissible torque T_{max} if the allowable normal strain is $\varepsilon_a = 5 \times 10^{-4}$. Assume that $G = 75$ GPa.

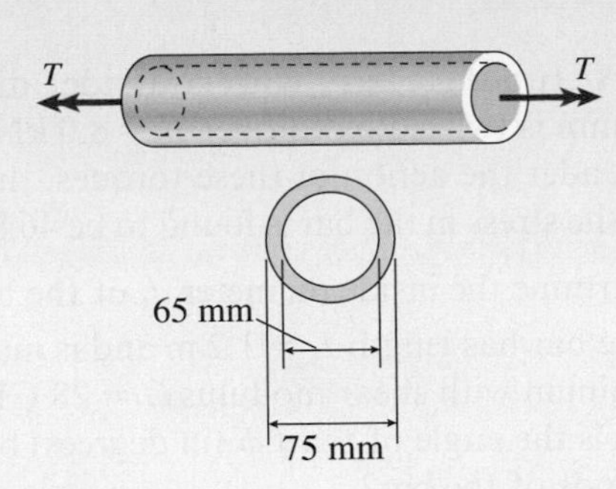

PROBLEM 3.5-2

3.5-3 A hollow aluminum shaft (see figure) has an outside diameter $d_2 = 100$ mm and inside diameter $d_1 = 50$ mm. When twisted by torques T, the shaft has an angle of twist per unit distance equal to 2°/m. The shear modulus of elasticity of the aluminum is $G = 27.5$ GPa.

(a) Determine the maximum tensile stress σ_{max} in the shaft.

(b) Determine the magnitude of the applied torques T.

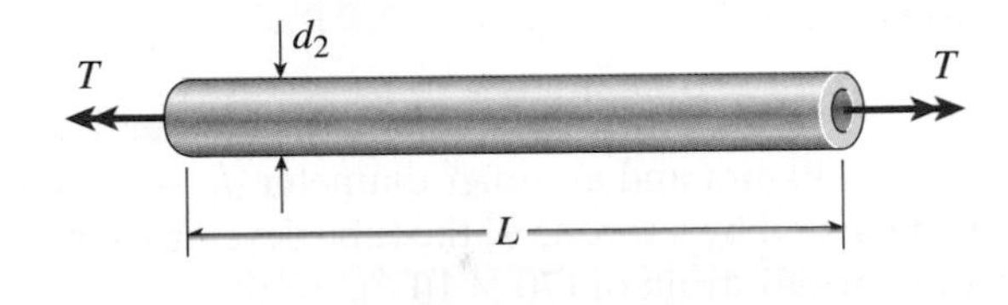

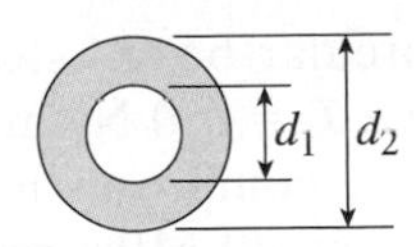

PROBLEMS 3.5-3, 3.5-4, and 3.5-5

3.5-4 A hollow steel bar ($G = 80$ GPa) is twisted by torques T (see figure). The twisting of the bar produces a maximum shear strain $\gamma_{max} = 640 \times 10^{-6}$ rad. The bar has outside and inside diameters of 150 mm and 120 mm, respectively.

(a) Determine the maximum tensile strain in the bar.

(b) Determine the maximum tensile stress in the bar.

(c) What is the magnitude of the applied torques T?

3.5-5 A tubular bar with outside diameter $d_2 = 100$ mm is twisted by torques $T = 8.0$ kN·m (see figure). Under the action of these torques, the maximum tensile stress in the bar is found to be 46.8 MPa.

(a) Determine the inside diameter d_1 of the bar.

(b) If the bar has length $L = 1.2$ m and is made of aluminum with shear modulus $G = 28$ GPa, what is the angle of twist ϕ (in degrees) between the ends of the bar?

(c) Determine the maximum shear strain γ_{max} (in radians)?

Representative Problems

3.5-6 A solid circular bar of diameter $d = 50$ mm (see figure) is twisted in a testing machine until the applied torque reaches the value $T = 500$ N · m. At this value of torque, a strain gage oriented at 45° to the axis of the bar gives a reading $\varepsilon = 339 \times 10^{-6}$.

What is the shear modulus G of the material?

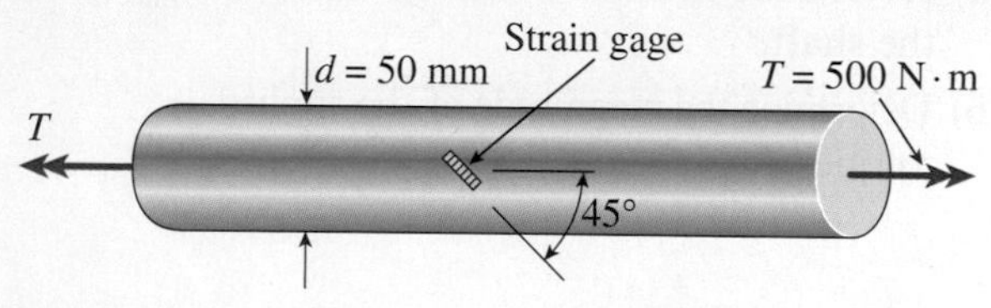

PROBLEM 3.5-6

3.5-7 A steel tube ($G = 80$ GPa) has an outer diameter $d_2 = 40$ mm and an inner diameter $d_1 = 30$ mm. When twisted by a torque T, the tube develops a maximum normal strain of 170×10^{-6}.

What is the magnitude of the applied torque T?

3.5-8 A solid circular bar of steel ($G = 78$ GPa) transmits a torque $T = 360$ N · m. The allowable stresses in tension, compression, and shear are 90 MPa, 70 MPa, and 40 MPa, respectively. Also, the allowable tensile strain is 220×10^{-6}.

(a) Determine the minimum required diameter d of the bar.

(b) If the bar diameter $d = 40$ mm, what is T_{max}?

3.5-9 The normal strain in the 45° direction on the surface of a circular tube (see figure) is 880×10^{-6} when the torque $T = 85$ N·m. The tube is made of copper alloy with $G = 42$ GPa and $\nu = 0.35$.

(a) If the outside diameter d_2 of the tube is 20 mm, what is the inside diameter d_1?

(b) If the allowable normal stress in the tube is 96 MPa, what is the maximum permissible inside diameter d_1?

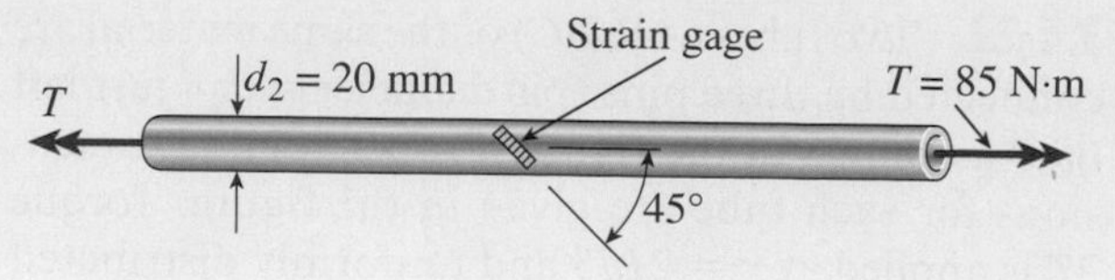

PROBLEM 3.5-9

3.5-10 An aluminum tube has inside diameter $d_1 = 50$ mm, shear modulus of elasticity $G = 27$ GPa, $\nu = 0.33$, and torque $T = 4.0$ kN · m. The allowable shear stress in the aluminum is 50 MPa, and the allowable normal strain is 900×10^{-6}.

(a) Determine the required outside diameter d_2.

(b) Re-compute the required outside diameter d_2 if allowable normal stress is 62 MPa and allowable shear strain is 1.7×10^{-3}.

3.5-11 A solid steel bar ($G = 81$ GPa) of diameter $d = 50$ mm is subjected to torques $T = 0.9$ kN·m acting in the directions shown in the figure.

(a) Determine the maximum shear, tensile, and compressive stresses in the bar and show these stresses on sketches of properly oriented stress elements.

(b) Determine the corresponding maximum strains (shear, tensile, and compressive) in the bar and show these strains on sketches of the deformed elements.

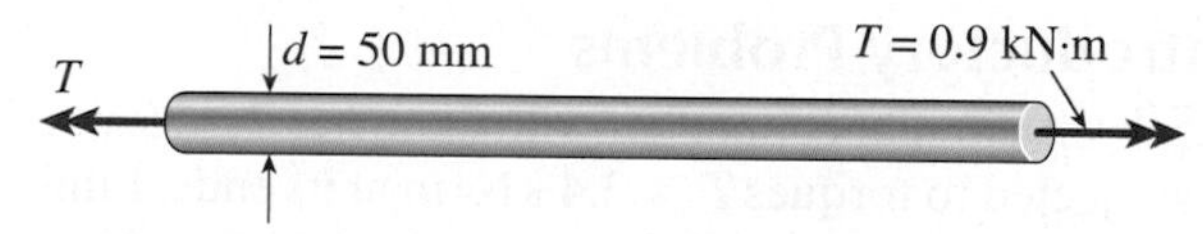

PROBLEM 3.5-11

3.5-12 A solid aluminum bar ($G = 27$ GPa) of diameter $d = 40$ mm is subjected to torques $T = 300$ N · m acting in the directions shown in the figure.

(a) Determine the maximum shear, tensile, and compressive stresses in the bar and show these stresses on sketches of properly oriented stress elements.

(b) Determine the corresponding maximum strains (shear, tensile, and compressive) in the bar and show these strains on sketches of the deformed elements.

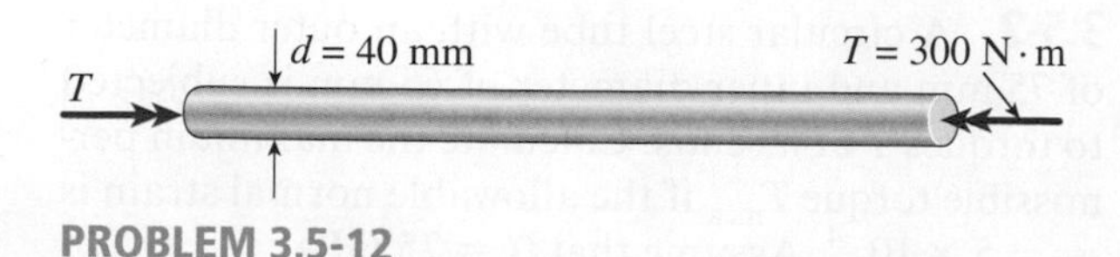

PROBLEM 3.5-12

3.5-13 Two circular aluminum pipes of equal length $L = 610$ mm are loaded by torsional moments T (see figure). Pipe 1 has outside and inside diameters $d_2 = 76$ mm and $d_1 = 64$ mm, respectively. Pipe 2 has a constant outer diameter of d_2 along its entire length L and an inner diameter of d_1 but has an increased inner diameter of $d_3 = 67$ mm over the middle third.

Assume that $E = 72$ GPa, $v = 0.33$, and allowable shear stress $\tau_a = 45$ MPa.

(a) Find the maximum acceptable torques that can be applied to Pipe 1; repeat for Pipe 2.

(b) If the maximum twist ϕ of Pipe 2 cannot exceed 5/4 of that of Pipe 1, what is the maximum acceptable length of the middle segment? Assume both pipes have total length L and the same applied torque T.

(c) Find the new value of inner diameter d_3 of Pipe 2 if the maximum torque carried by Pipe 2 is to be 7/8 of that for Pipe 1.

(d) If the maximum normal strain in each pipe is known to be $\varepsilon_{max} = 811 \times 10^{-6}$, what is the applied torque on each pipe? Also, what is the maximum twist of each pipe? Use the original properties and dimensions.

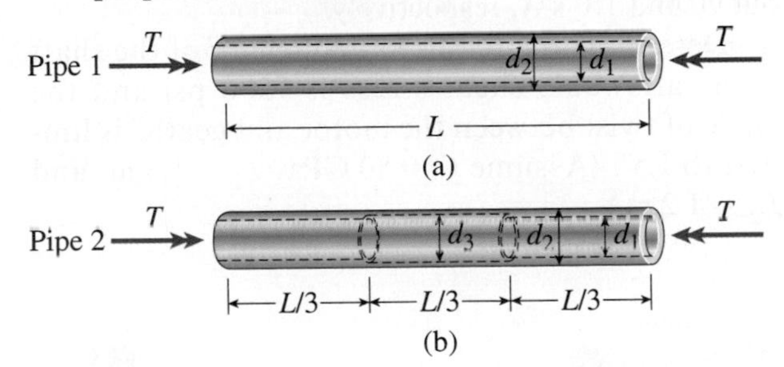

PROBLEM 3.5-13

3.7 Transmission of Power by Circular Shafts

Introductory Problems

3.7-1 A generator shaft in a small hydroelectric plant turns at 120 rpm and delivers 38 kW (see figure).

(a) If the diameter of the shaft is $d = 75$ mm, what is the maximum shear τ_{max} in the shaft?

(b) If the shear stress is limited to 28 MPa, what is the minimum permissible diameter d_{min} of the shaft?

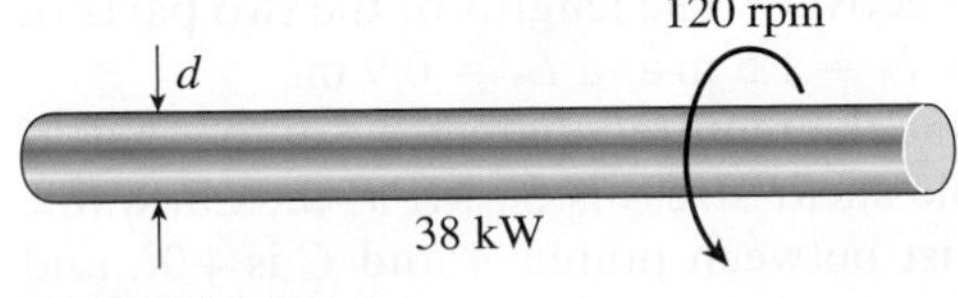

PROBLEM 3.7-1

3.7-2 A motor drives a shaft at 12 Hz and delivers 20 kW of power (see figure).

(a) If the shaft has a diameter of 30 mm, what is the maximum shear stress τ_{max} in the shaft?

(b) If the maximum allowable shear stress is 40 MPa, what is the maximum permissible diameter d_{min} of the shaft?

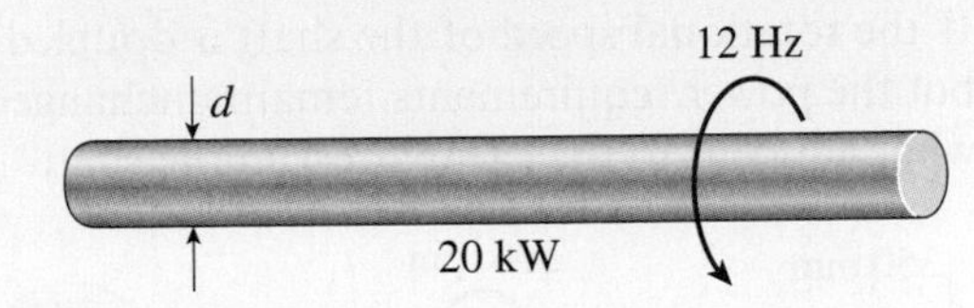

PROBLEM 3.7-2

3.7-3 A motor driving a solid circular steel shaft with diameter $d = 38$ mm transmits 37 kW to a gear at B. The allowable shear stress in the steel is 40 MPa. Calculate the required speed of rotation (number of revolutions per minute) so that the shear stress in the shaft does not exceed the allowable limit.

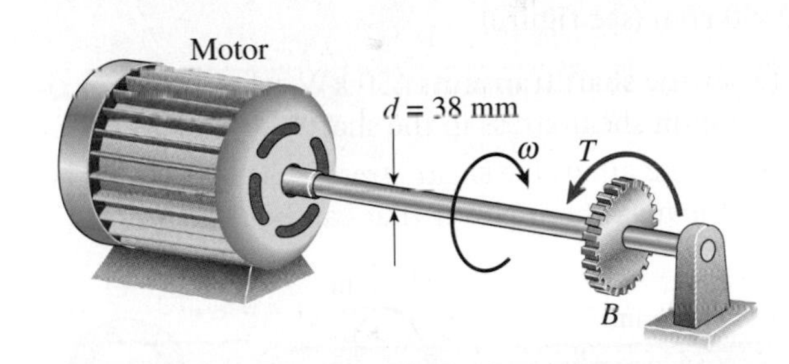

PROBLEM 3.7-3

3.7-4 A solid steel shaft ABC with diameter $d = 40$ mm is driven at A by a motor that transmits 75 kW to the shaft at 15 Hz. The gears at B and C drive machinery requiring power equal to 50 kW and 25 kW, respectively. Compute the maximum shear stress and angle of twist in the shaft between the motor at A and the gear at C. Assume that $G = 75$ GPa.

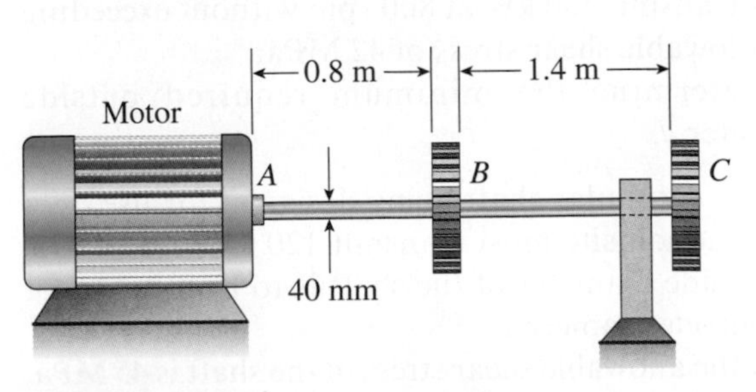

PROBLEM 3.7-4

Representative Problems

3.7-5 The propeller shaft of a large ship has an outside diameter 350 mm and inside diameter 250 mm, as shown in the figure. The shaft is rated for a maximum shear stress of 62 MPa.

(a) If the shaft is turning at 500 rpm, what is the maximum horsepower that can be transmitted without exceeding the allowable stress?

(b) If the rotational speed of the shaft is doubled but the power requirements remain unchanged, what happens to the shear stress in the shaft?

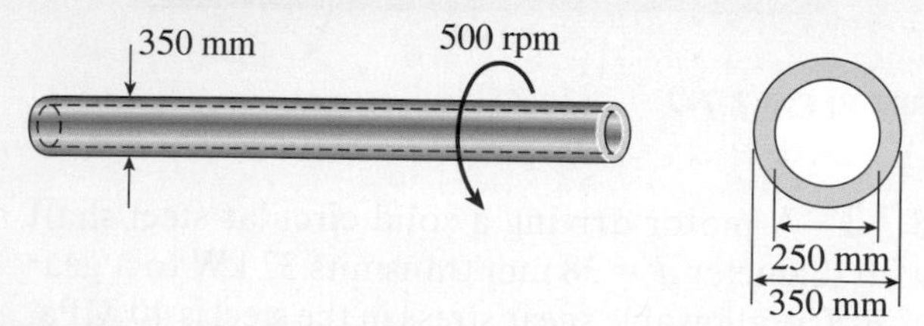

PROBLEM 3.7-5

3.7-6 The drive shaft for a truck (outer diameter 60 mm and inner diameter 40 mm) is running at 2500 rpm (see figure).

(a) If the shaft transmits 150 kW, what is the maximum shear stress in the shaft?

(b) If the allowable shear stress is 30 MPa, what is the maximum power that can be transmitted?

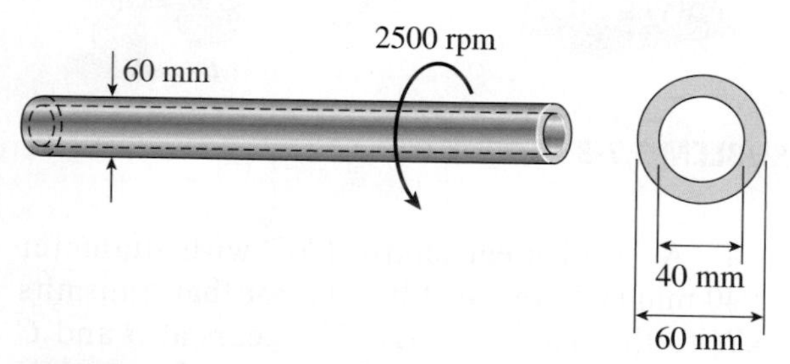

PROBLEM 3.7-6

3.7-7 A hollow circular shaft for use in a pumping station is being designed with an inside diameter equal to 0.8 times the outside diameter. The shaft must transmit 300 kW at 800 rpm without exceeding the allowable shear stress of 42 MPa.

Determine the minimum required outside diameter d.

3.7-8 A tubular shaft being designed for use on a construction site must transmit 120 kW at 1.75 Hz. The inside diameter of the shaft is to be one-half of the outside diameter.

If the allowable shear stress in the shaft is 45 MPa, what is the minimum required outside diameter d?

3.7-9 A propeller shaft of solid circular cross section and diameter d is spliced by a collar of the same material (see figure). The collar is securely bonded to both parts of the shaft.

What should be the minimum outer diameter d_1 of the collar in order that the splice can transmit the same power as the solid shaft?

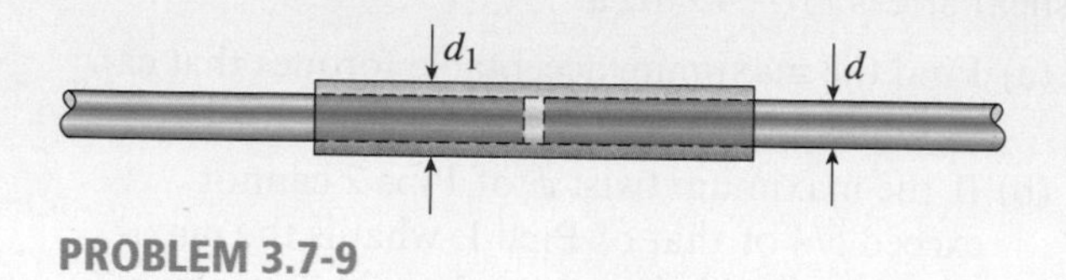

PROBLEM 3.7-9

3.7-10 What is the maximum power that can be delivered by a hollow propeller shaft (outside diameter 50 mm, inside diameter 40 mm, and shear modulus of elasticity 80 GPa) turning at 600 rpm if the allowable shear stress is 100 MPa and the allowable rate of twist is 3.0°/m?

3.7-11 A motor delivers 200 kW at 1000 rpm to the end of a shaft (see figure). The gears at B and C take out 90 and 110 kW, respectively.

Determine the required diameter d of the shaft if the allowable shear stress is 7500 psi and the angle of twist between the motor and gear C is limited to 1.5°. (Assume $G = 80$ GPa, $L_1 = 1.8$ m, and $L_2 = 1.2$ m.)

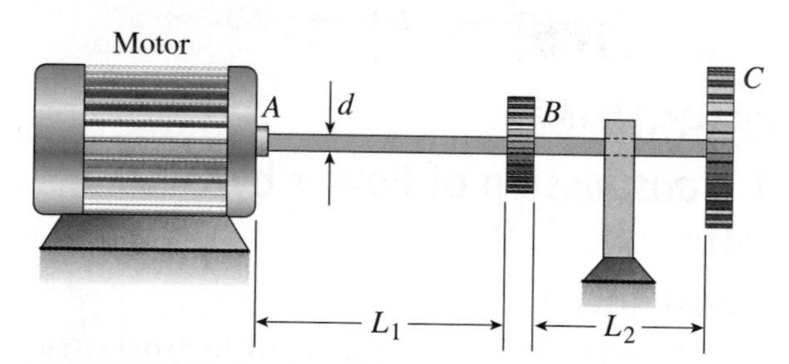

PROBLEMS 3.7-11 and 3.7-12

3.7-12 The shaft ABC shown in the figure is driven by a motor that delivers 300 kW at a rotational speed of 32 Hz. The gears at B and C take out 120 kW and 180 kW, respectively. The lengths of the two parts of the shaft are $L_1 = 1.5$ m and $L_2 = 0.9$ m.

Determine the required diameter d of the shaft if the allowable shear stress is 50 MPa, the allowable angle of twist between points A and C is 4.0°, and $G = 75$ GPa.

3.8 Statically Indeterminate Torsional Members

Introductory Problems

3.8-1 A solid circular bar *ABCD* with fixed supports is acted upon by torques T_0 and $2T_0$ at the locations shown in the figure.

(a) Obtain a formula for the maximum angle of twist ϕ_{max} of the bar.

(b) What is ϕ_{max} if the applied torque T_0 at *B* is reversed in direction?

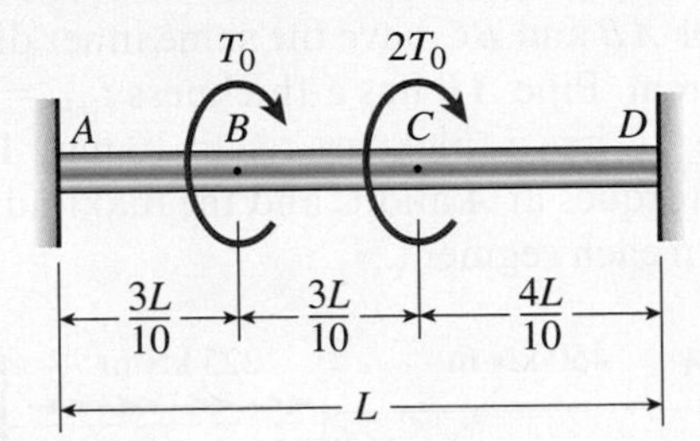

PROBLEM 3.8-1

3.8-2 A solid circular bar *ABCD* with fixed supports at ends *A* and *D* is acted upon by two equal and oppositely directed torques T_0, as shown in the figure. The torques are applied at points *B* and *C*, each of which is located at distance *x* from one end of the bar. (The distance *x* may vary from zero to *L*/2.)

(a) For what distance *x* will the angle of twist at points *B* and *C* be a maximum?

(b) What is the corresponding angle of twist ϕ_{max}?

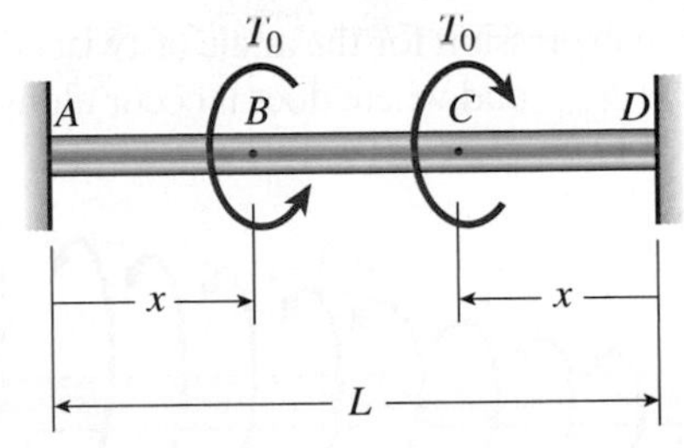

PROBLEM 3.8-2

3.8-3 A solid circular shaft *AB* of diameter *d* is fixed against rotation at both ends (see figure). A circular disk is attached to the shaft at the location shown.

What is the largest permissible angle of rotation ϕ_{max} of the disk if the allowable shear stress in the shaft is τ_{allow}? [Assume that $a > b$. Also, use Eqs. (3-50a and b) of Example 3-9 to obtain the reactive torques.]

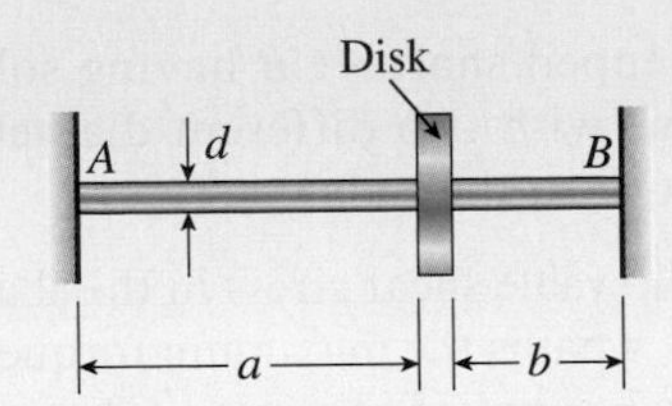

PROBLEM 3.8-3

3.8-4 A hollow steel shaft *ACB* of outside diameter 50 mm and inside diameter 40 mm is held against rotation at ends *A* and *B* (see figure). Horizontal forces *P* are applied at the ends of a vertical arm that is welded to the shaft at point *C*.

Determine the allowable value of the forces *P* if the maximum permissible shear stress in the shaft is 45 MPa.

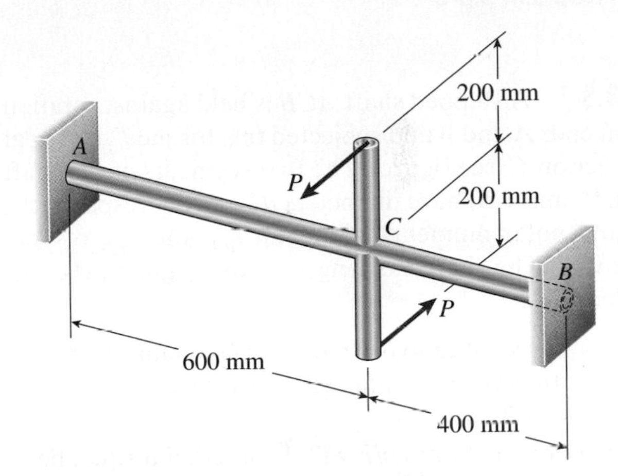

PROBLEM 3.8-4

3.8-5 A stepped shaft *ACB* having solid circular cross sections with two different diameters is held against rotation at the ends (see figure).

(a) If the allowable shear stress in the shaft is 42 MPa, what is the maximum torque $(T_0)_{max}$ that may be applied at section C?

(b) Find $(T_0)_{max}$ if the maximum angle of twist is limited to 0.55°. Let $G = 73$ GPa.

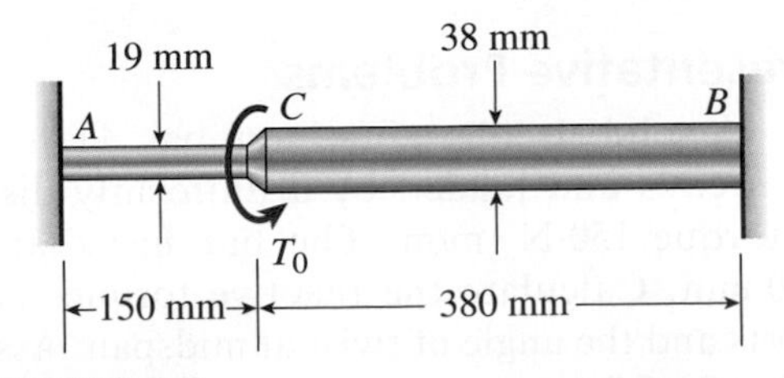

PROBLEM 3.8-5

3.8-6 A stepped shaft ACB having solid circular cross sections with two different diameters is held against rotation at the ends (see figure).

(a) If the allowable shear stress in the shaft is 43 MPa, what is the maximum torque $(T_0)_{max}$ that may be applied at section C?

(b) Find $(T_0)_{max}$ if the maximum angle of twist is limited to 1.85°. Let $G = 28$ GPa.

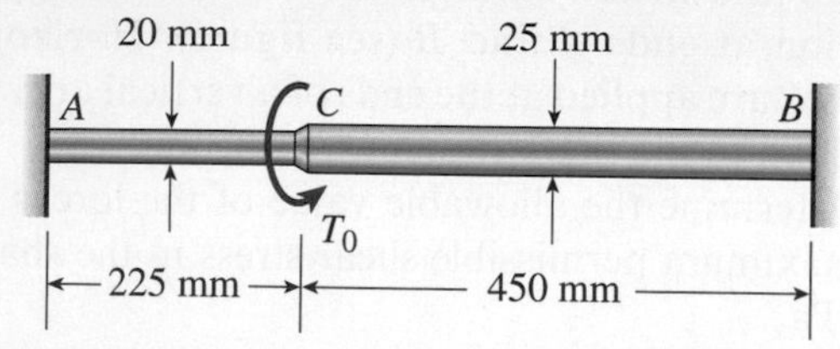

PROBLEM 3.8-6

3.8-7 A stepped shaft ACB is held against rotation at ends A and B and subjected to a torque T_0 acting at section C (see figure). The two segments of the shaft (AC and CB) have diameters d_A and d_B, respectively, and polar moments of inertia I_{pA} and I_{pB}, respectively. The shaft has length L and segment AC has length a.

(a) For what ratio a/L will the maximum shear stresses be the same in both segments of the shaft?

(b) For what ratio a/L will the internal torques be the same in both segments of the shaft?

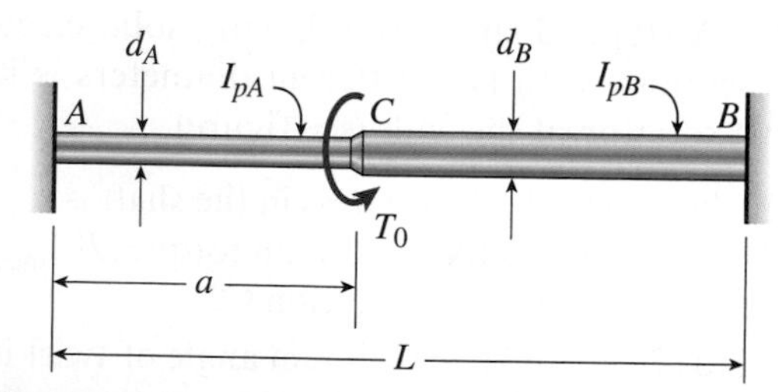

PROBLEM 3.8-7

Representative Problems

3.8-8 A solid circular aluminum bar AB is fixed at both ends and loaded by a uniformly distributed torque 150 N · m/m. The bar has diameter $d = 30$ mm. Calculate the reactive torques at the supports and the angle of twist at midspan. Assume that $G = 28$ GPa.

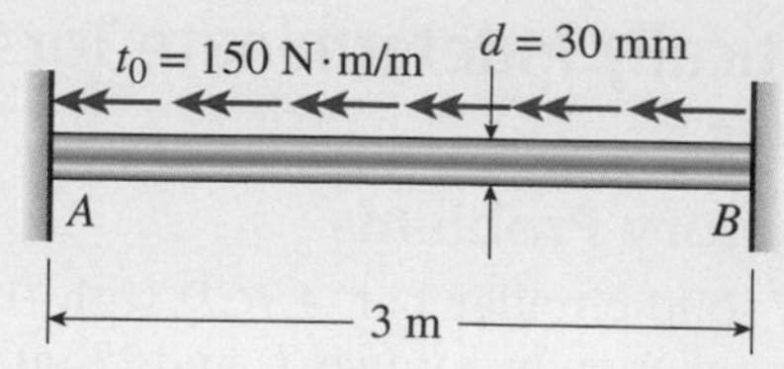

PROBLEM 3.8-8

3.8-9 Two sections of steel drill pipe, joined by bolted flange plates at B, are subjected to a concentrated torque 450 kN · m at $x = 0.9$ m, and a uniformly distributed torque $t_0 = 225$ kN·m/m is applied on pipe BC. Let $G = 81$ GPa and assume that pipes AB and BC have the same inner diameter, $d = 300$ mm. Pipe AB has a thickness $t_{AB} = 19$ mm, and pipe BC has a thickness $t_{BC} = 16$ mm. Find the reactive torques at A and C and the maximum shear stresses in each segment.

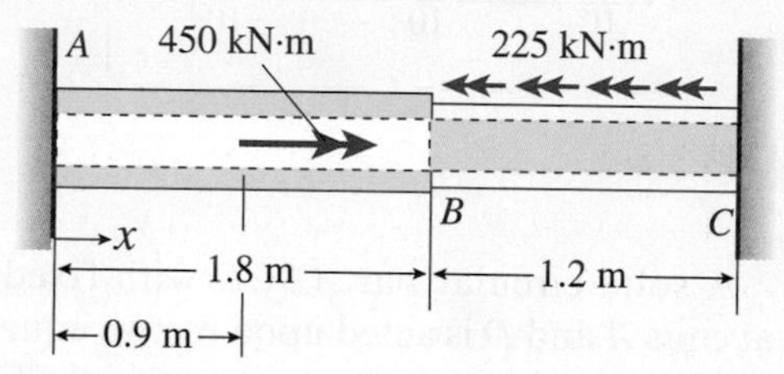

PROBLEM 3.8-9

3.8-10 A circular bar AB of length L is fixed against rotation at the ends and loaded by a distributed torque $t(x)$ that varies linearly in intensity from zero at end A to t_0 at end B (see figure).

(a) Obtain formulas for the fixed-end torques T_A and T_B.

(b) Find an expression for the angle of twist $\phi(x)$. What is ϕ_{max}, and where does it occur along the bar?

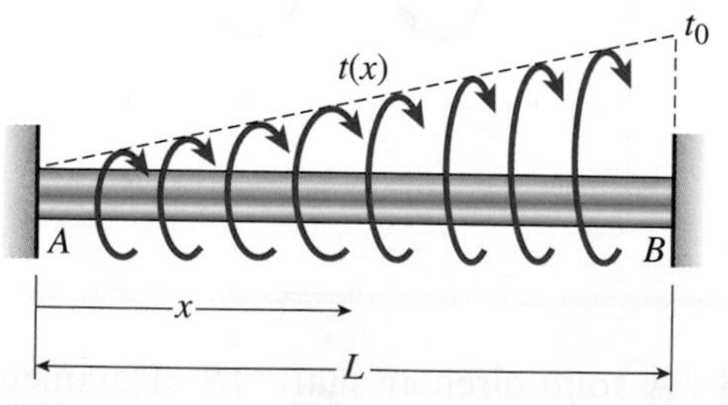

PROBLEM 3.8-10

3.8-11 A circular bar AB with ends fixed against rotation has a hole extending for half of its length (see figure). The outer diameter of the bar is $d_2 = 76$ mm, and the diameter of the hole is $d_1 = 61$ mm. The total length of the bar is $L = 1270$ mm.

(a) At what distance x from the left-hand end of the bar should a torque T_0 be applied so that the reactive torques at the supports will be equal?

(b) Based on the solution for x in part (a), what is ϕ_{max}, and where does it occur? Assume that $T_0 = 10$ kN·m and $G = 73$ GPa.

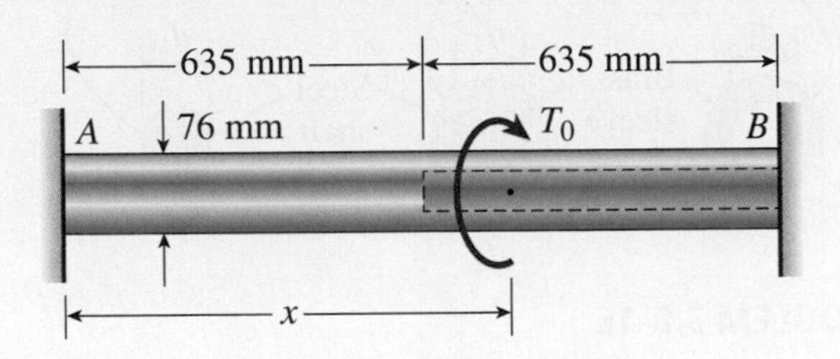

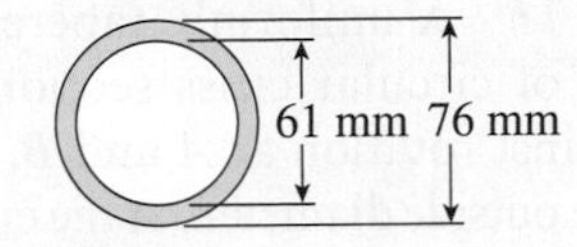

PROBLEM 3.8-11

3.8-12 A solid steel bar of diameter $d_1 = 25.0$ mm is enclosed by a steel tube of outer diameter $d_3 = 37.5$ mm and inner diameter $d_2 = 30.0$ mm (see figure). Both bar and tube are held rigidly by a support at end A and joined securely to a rigid plate at end B. The composite bar, which has a length $L = 550$ mm, is twisted by a torque $T = 400$ N · m acting on the end plate.

(a) Determine the maximum shear stresses τ_1 and τ_2 in the bar and tube, respectively.

(b) Determine the angle of rotation ϕ (in degrees) of the end plate, assuming that the shear modulus of the steel is $G = 80$ GPa.

(c) Determine the torsional stiffness k_T of the composite bar.

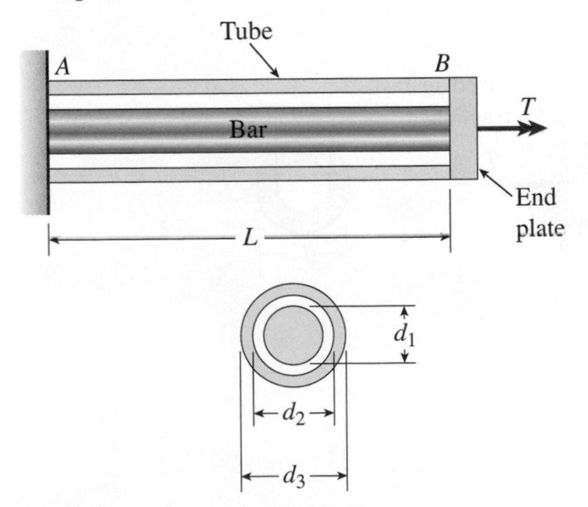

PROBLEMS 3.8-12 and 3.8-13

3.8-13 A solid steel bar of diameter $d_1 = 50$ mm is enclosed by a steel tube of outer diameter $d_3 = 75$ mm and inner diameter $d_2 = 65$ mm (see figure). Both bar and tube are held rigidly by a support at end A and joined securely to a rigid plate at end B. The composite bar, which has length $L = 660$ mm, is twisted by a torque $T = 2$ kN·m acting on the end plate.

(a) Determine the maximum shear stresses τ_1 and τ_2 in the bar and tube, respectively.

(b) Determine the angle of rotation ϕ (in degrees) of the end plate, assuming that the shear modulus of the steel is $G = 80$ GPa.

(c) Determine the torsional stiffness k_T of the composite bar.

3.8-14 The composite shaft shown in the figure is manufactured by shrink-fitting a steel sleeve over a brass core so that the two parts act as a single solid bar in torsion. The outer diameters of the two parts are $d_1 = 40$ mm for the brass core and $d_2 = 50$ mm for the steel sleeve. The shear moduli of elasticity are $G_b = 36$ GPa for the brass and $G_s = 80$ GPa for the steel.

(a) Assuming that the allowable shear stresses in the brass and steel are $\tau_b = 48$ MPa and $\tau_s = 80$ MPa, respectively, determine the maximum permissible torque T_{max} that may be applied to the shaft.

(b) If the applied torque $T = 2500$ kN · m, find the required diameter d_2 so that allowable shear stress τ_s is reached in the steel.

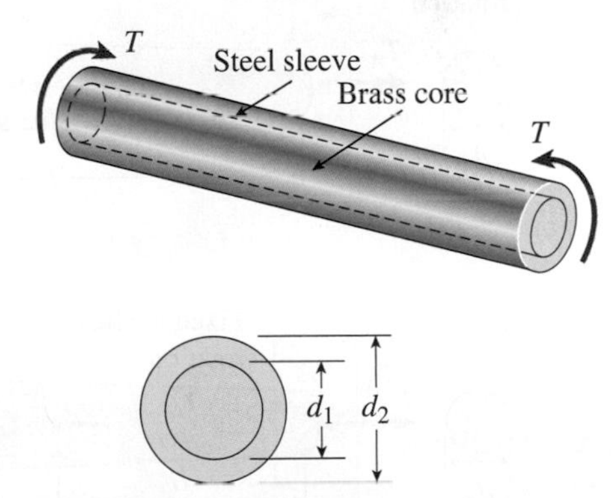

PROBLEMS 3.8-14 and 3.8-15

3.8-15 The composite shaft shown in the figure is manufactured by shrink-fitting a steel sleeve over a brass core so that the two parts act as a single solid bar in torsion. The outer diameters of the two parts are $d_1 = 41$ mm for the brass core and $d_2 = 51$ mm for the steel sleeve. The shear moduli of elasticity are $G_b = 37$ GPa for the brass and $G_s = 83$ GPa for the steel.

(a) Assuming that the allowable shear stresses in the brass and steel are $\tau_b = 31$ MPa and $\tau_s = 52$ MPa, respectively, determine the maximum permissible torque T_{max} that may be applied to the shaft.

(b) If the applied torque $T = 1250$ N·m, find the required diameter d_2 so that allowable shear stress τ_s is reached in the steel.

3.8-16 A steel shaft ($G_s = 80$ GPa) of total length $L = 3.0$ m is encased for one-third of its length by a brass sleeve ($G_b = 40$ GPa) that is securely bonded to the steel (see figure). The outer diameters of the shaft and sleeve are $d_1 = 70$ mm and $d_2 = 90$ mm, respectively.

(a) Determine the allowable torque T_1 that may be applied to the ends of the shaft if the angle of twist between the ends is limited to 8.0°.

(b) Determine the allowable torque T_2 if the shear stress in the brass is limited to $\tau_b = 70$ MPa.

(c) Determine the allowable torque T_3 if the shear stress in the steel is limited to $\tau_s = 110$ MPa.

(d) What is the maximum allowable torque T_{max} if all three of the preceding conditions must be satisfied?

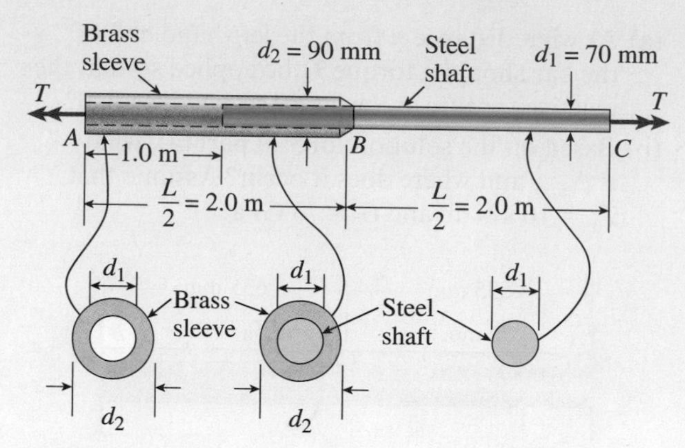

PROBLEM 3.8-16

3.8-17 A uniformly tapered aluminum-alloy tube AB of circular cross section and length L is fixed against rotation at A and B, as shown in the figure. The outside diameters at the ends are d_A and $d_B = 2d_A$. A hollow section of length $L/2$ and constant thickness $t = d_A/10$ is cast into the tube and extends from B halfway toward A. Torque T_0 is applied at $L/2$.

(a) Find the reactive torques at the supports, T_A and T_B. Use numerical values as follows: $d_A = 64$ mm, $L = 1.2$ m, $G = 27$ GPa, and $T_0 = 4.5$ kN·m.

(b) Repeat part (a) if the hollow section has constant diameter d_A.

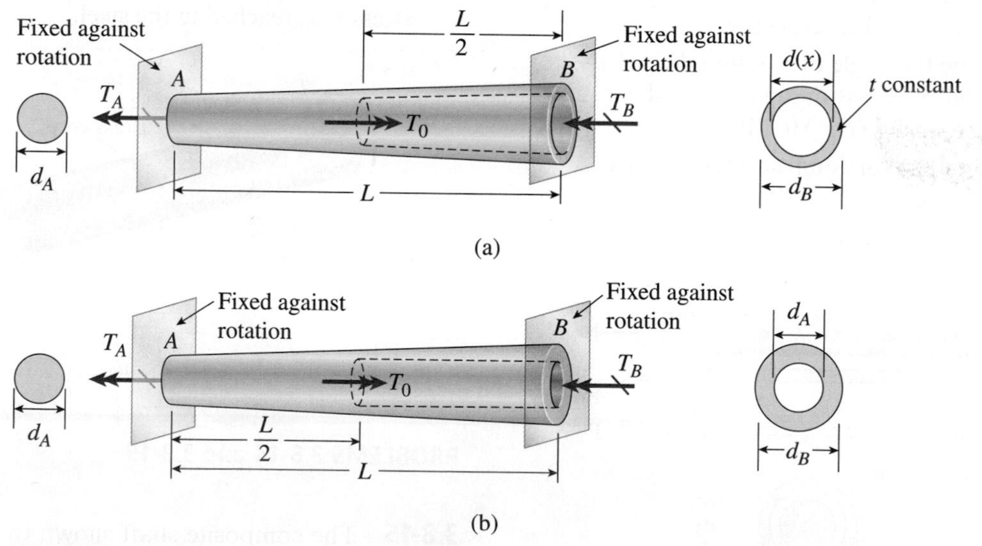

PROBLEM 3.8-17

3.8-18 Two pipes ($L_1 = 2.5$ m and $L_2 = 1.5$ m) are joined at B by flange plates (thickness $t_f = 14$ mm) with five bolts ($d_{bf} = 13$ mm) arranged in a circular pattern (see figure). Also, each pipe segment is attached to a wall (at A and C, see figure) using a base plate ($t_b = 15$ mm) and four bolts ($d_{bb} = 16$ mm). All bolts are tightened until just snug. Assume $E_1 = 110$ GPa, $E_2 = 73$ GPa, $\nu_1 = 0.33$, and $\nu_2 = 0.25$. Neglect the self-weight of the pipes, and assume the pipes are in a stress-free state initially. The cross-sectional areas of the pipes are $A_1 = 1500$ mm^2 and $A_2 = (3/5)A_1$. The outer diameter of Pipe 1 is 60 mm. The outer diameter of Pipe 2 is equal to the inner diameter of Pipe 1. The bolt radius $r = 64$ mm for both base and flange plates.

(a) If torque T is applied at $x = L_1$, find an expression for reactive torques R_1 and R_2 in terms of T.

(b) Find the maximum load variable T (i.e., T_{max}) if allowable torsional stress in the two pipes is $\tau_{allow} = 65$ MPa.

(c) Draw torsional moment (TMD) and torsional displacement (TDD) diagrams. Label all key ordinates. What is ϕ_{max}?

(d) Find T_{max} if allowable shear and bearing stresses in the base plate and flange bolts cannot be exceeded. Assume allowable stresses in shear and bearing for all bolts are $\tau_{allow} = 45$ MPa and $\sigma_{allow} = 90$ MPa.

(e) Remove torque T at $x = L_1$. Now assume the flange-plate bolt holes are misaligned by some angle β (see figure). Find the expressions for reactive torques R_1 and R_2 if the pipes are twisted to align the flange-plate bolt holes, bolts are then inserted, and the pipes released.

(f) What is the maximum permissible misalignment angle β_{max} if allowable stresses in shear and bearing for all bolts [from part (d)] are not to be exceeded?

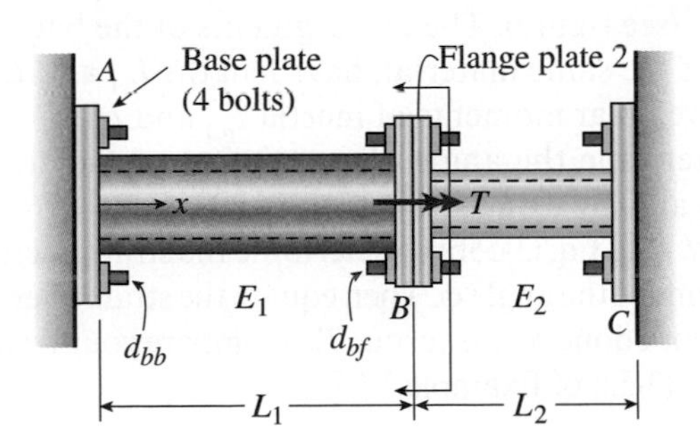

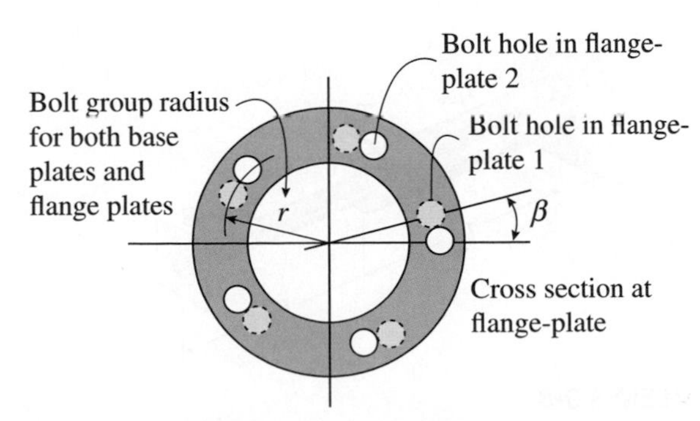

PROBLEM 3.8-18

3.9 Strain Energy in Torsion and Pure Shear

Introductory Problems

3.9-1 A solid circular bar of steel ($G = 80$ GPa) with length $L = 1.5$ m and diameter $d = 75$ mm is subjected to pure torsion by torques T acting at the ends (see figure).

(a) Calculate the amount of strain energy U stored in the bar when the maximum shear stress is 45 MPa.

(b) From the strain energy, calculate the angle of twist ϕ (in degrees).

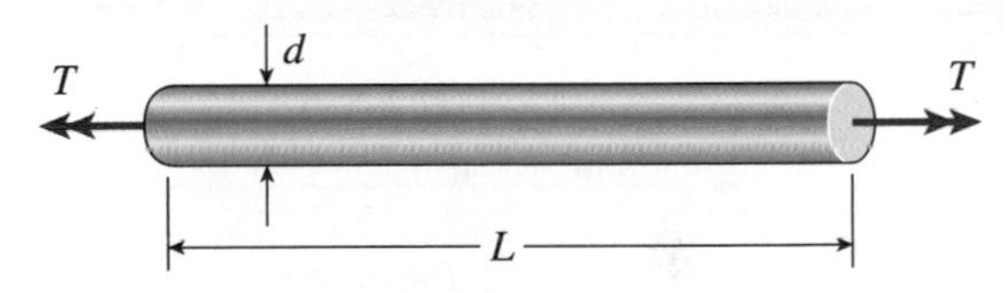

PROBLEMS 3.9-1 and 3.9-2

3.9-2 A solid circular bar of copper ($G = 45$ GPa) with length $L = 0.75$ m and diameter $d = 40$ mm is subjected to pure torsion by torques T acting at the ends (see figure).

(a) Calculate the amount of strain energy U stored in the bar when the maximum shear stress is 32 MPa.

(b) From the strain energy, calculate the angle of twist ϕ (in degrees).

3.9-3 A stepped shaft of solid circular cross sections (see figure) has length $L = 2.6$ m, diameter $d_2 = 50$ mm, and diameter $d_1 = 40$ mm. The material is brass with $G = 81$ GPa.

Determine the strain energy U of the shaft if the angle of twist is 3.0°.

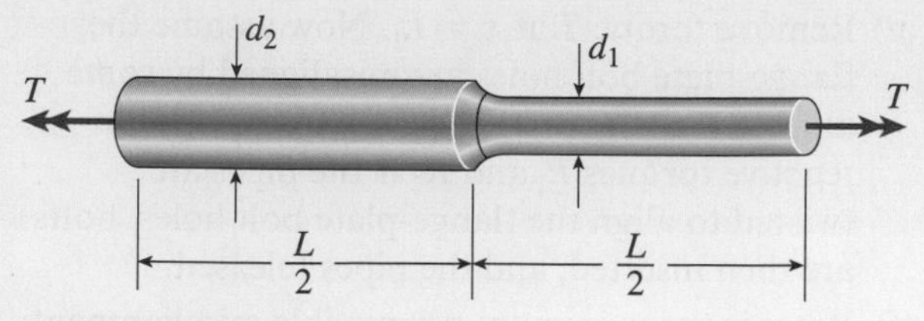

PROBLEMS 3.9-3 and 3.9-4

3.9-4 A stepped shaft of solid circular cross sections (see figure) has length $L = 0.80$ m, diameter $d_2 = 40$ mm, and diameter $d_1 = 30$ mm. The material is steel with $G = 80$ GPa.

Determine the strain energy U of the shaft if the angle of twist is 1.0°.

Representative Problems

3.9-5 A circular tube AB is fixed at one end and free at the other. The tube is subjected to concentrated torques as shown in the figure. If the outer radius of the tube is 37.5 mm and the thickness is 19 mm, calculate the strain energy stored in the tube. Let $G = 81$ GPa.

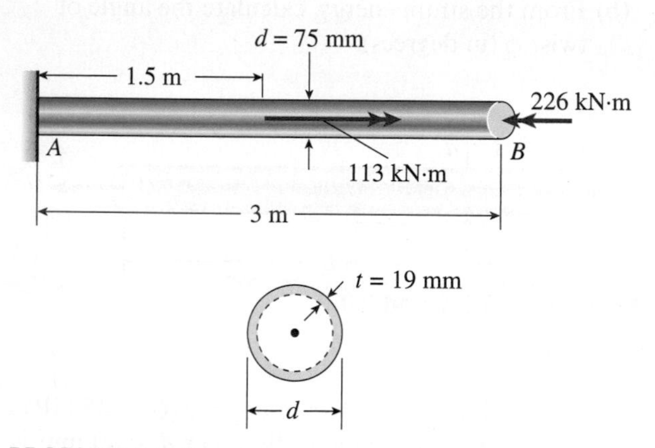

PROBLEM 3.9-5

3.9-6 A cantilever bar of circular cross section and length L is fixed at one end and free at the other (see figure). The bar is loaded by a torque T at the free end and by a distributed torque of constant intensity t per unit distance along the length of the bar.

(a) What is the strain energy U_1 of the bar when the load T acts alone?

(b) What is the strain energy U_2 when the load t acts alone?

(c) What is the strain energy U_3 when both loads act simultaneously?

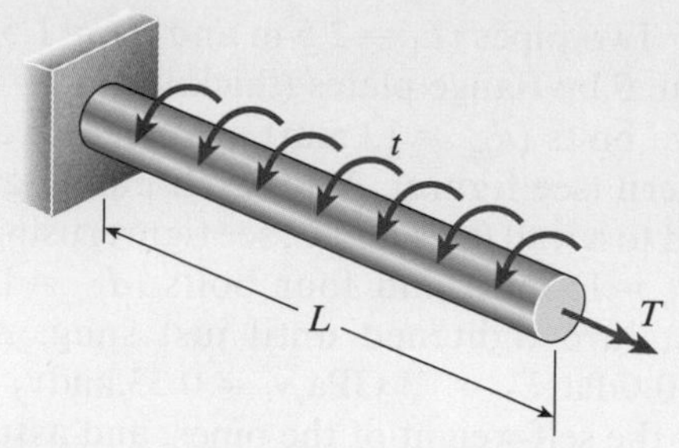

PROBLEM 3.9-6

3.9-7 Obtain a formula for the strain energy U of the statically indeterminate circular bar shown in the figure. The bar has fixed supports at ends A and B and is loaded by torques $2T_0$ and T_0 at points C and D, respectively.

Hint: Use Eqs. (3-50a and b) of Example 3-9 to obtain the reactive torques.

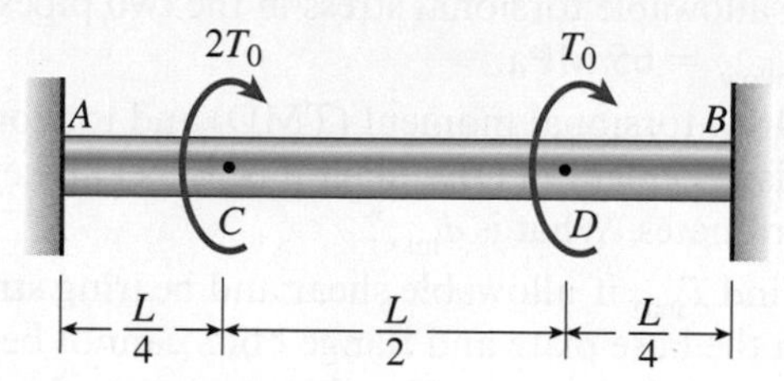

PROBLEM 3.9-7

3.9-8 A statically indeterminate stepped shaft ACB is fixed at ends A and B and loaded by a torque T_0 at point C (see figure). The two segments of the bar are made of the same material, have lengths L_A and L_B, and have polar moments of inertia I_{pA} and I_{pB}.

Determine the angle of rotation ϕ of the cross section at C by using strain energy.

Hint: Use Eq. (3-55b) to determine the strain energy U in terms of the angle ϕ. Then equate the strain energy to the work done by the torque T_0. Compare your result with Eq. (3-52) of Example 3-9.

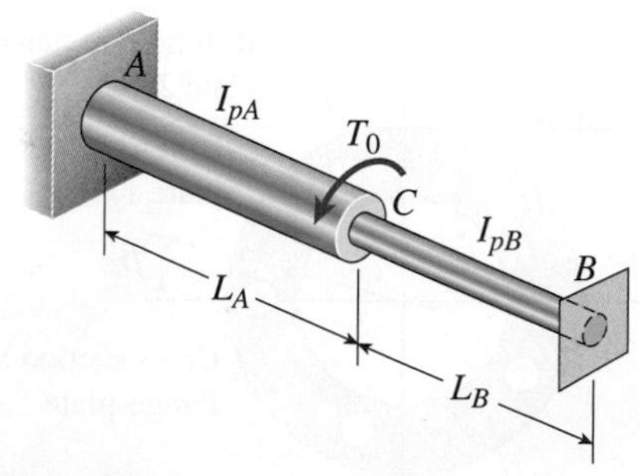

PROBLEM 3.9-8

3.9-9 Derive a formula for the strain energy U of the cantilever bar shown in the figure.

The bar has circular cross sections and length L. It is subjected to a distributed torque of intensity t per unit distance. The intensity varies linearly from $t = 0$ at the free end to a maximum value $t = t_0$ at the support.

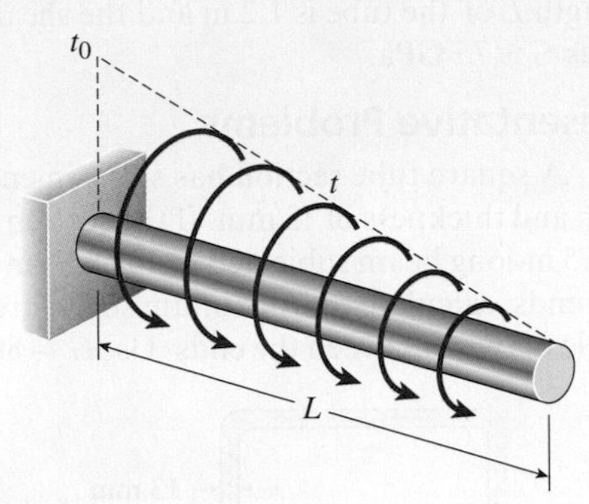

PROBLEM 3.9-9

3.9-10 A thin-walled hollow tube AB of conical shape has constant thickness t and average diameters d_A and d_B at the ends (see figure).

(a) Determine the strain energy U of the tube when it is subjected to pure torsion by torques T.

(b) Determine the angle of twist ϕ of the tube.

Note: Use the approximate formula $I_p \approx \pi d^3 t/4$ for a thin circular ring; see Case 22 of Appendix E.

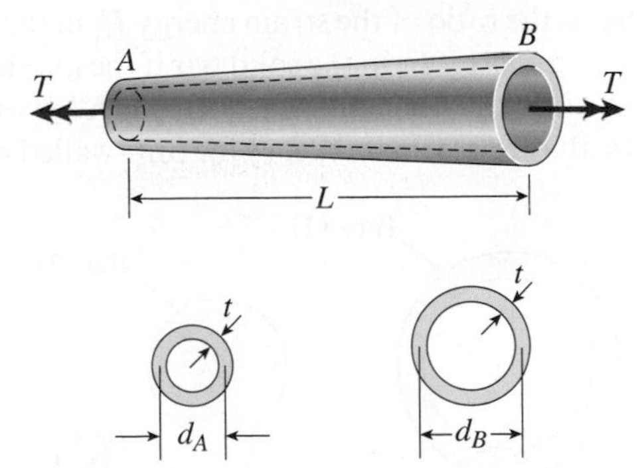

PROBLEM 3.9-10

3.9-11 A hollow circular tube A fits over the end of a solid circular bar B, as shown in the figure. The far ends of both bars are fixed. Initially, a hole through bar B makes an angle β with a line through two holes in tube A. Then bar B is twisted until the holes are aligned, and a pin is placed through the holes.

When bar B is released and the system returns to equilibrium, what is the total strain energy U of the two bars? (Let I_{pA} and I_{pB} represent the polar moments of inertia of bars A and B, respectively. The length L and shear modulus of elasticity G are the same for both bars.)

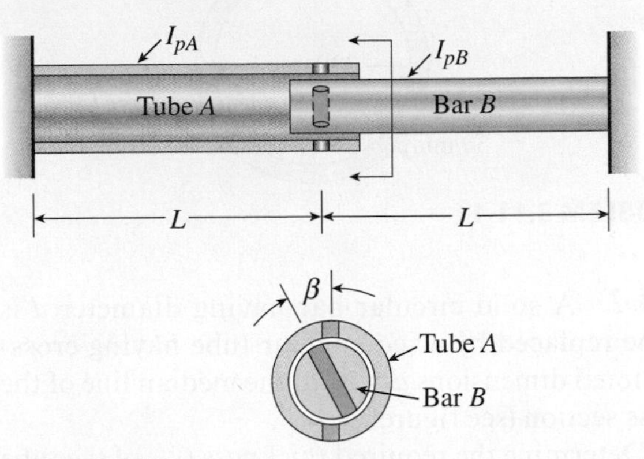

PROBLEM 3.9-11

3.9-12 A heavy flywheel rotating at n revolutions per minute is rigidly attached to the end of a shaft of diameter d (see figure). If the bearing at A suddenly freezes, what will be the maximum angle of twist ϕ of the shaft? What is the corresponding maximum shear stress in the shaft?

(Let L = length of the shaft, G = shear modulus of elasticity, and I_m = mass moment of inertia of the flywheel about the axis of the shaft. Also, disregard friction in the bearings at B and C and disregard the mass of the shaft.)

Hint: Equate the kinetic energy of the rotating flywheel to the strain energy of the shaft.

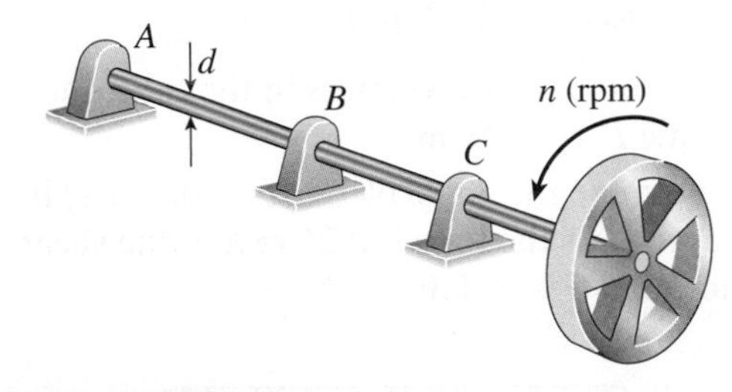

PROBLEM 3.9-12

3.11 Thin-Walled Tubes

Introductory Problems

3.11-1 A hollow circular tube having an inside diameter of 230 mm and a wall thickness of 30 mm (see figure) is subjected to a torque $T = 136$ kN·m.

Determine the maximum shear stress in the tube using (a) the approximate theory of thin-walled tubes, and (b) the exact torsion theory. Does the approximate theory give conservative or nonconservative results?

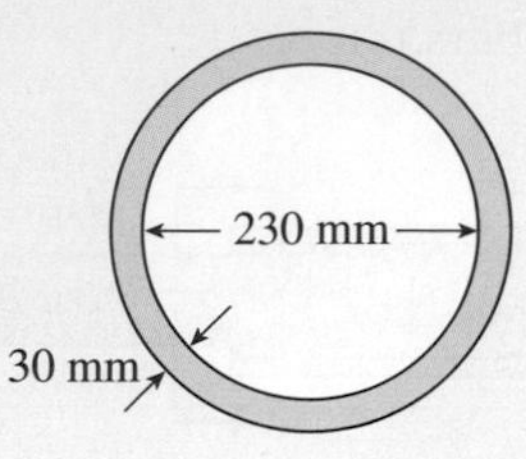

PROBLEM 3.11-1

3.11-2 A solid circular bar having diameter d is to be replaced by a rectangular tube having cross-sectional dimensions $d \times 2d$ to the median line of the cross section (see figure).

Determine the required thickness t_{min} of the tube so that the maximum shear stress in the tube will not exceed the maximum shear stress in the solid bar.

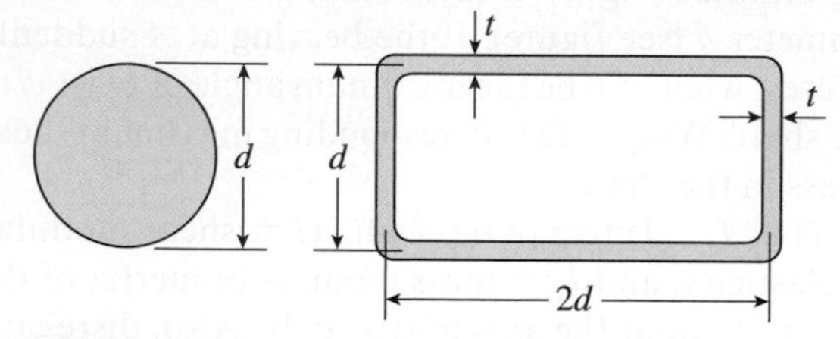

PROBLEM 3.11-2

3.11-3 A thin-walled aluminum tube of rectangular cross section (see figure) has a centerline dimensions $b = 50$ mm and $h = 20$ mm. The wall thickness t is constant and equal to 3 mm.

(a) Determine the shear stress in the tube due to a torque $T = 90$ N·m.

(b) Determine the angle of twist (in degrees) if the length L of the tube is 0.25 m and the shear modulus G is 26 GPa.

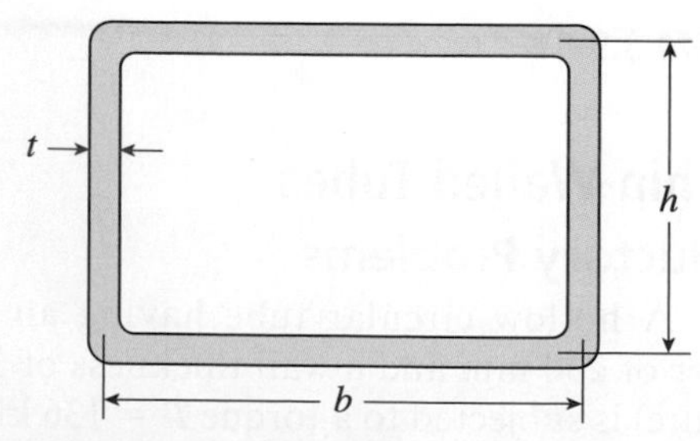

PROBLEMS 3.11-3 and 3.11-4

3.11-4 A thin-walled steel tube of rectangular cross section (see figure) has centerline dimensions $b = 150$ mm and $h = 100$ mm. The wall thickness t is constant and equal to 6.0 mm.

(a) Determine the shear stress in the tube due to a torque $T = 1650$ N · m.

(b) Determine the angle of twist (in degrees) if the length L of the tube is 1.2 m and the shear modulus G is 75 GPa.

Representative Problems

3.11-5 A square tube section has side dimension of 500 mm and thickness of 13 mm. If the section is used for a 2.75 m-long beam subjected to 140 kN·m torque at both ends, calculate the maximum shear stress and the angle of twist between the ends. Use $G = 80$ GPa.

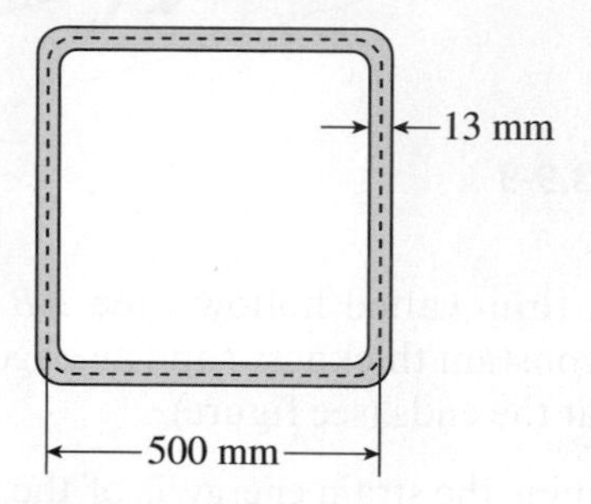

PROBLEM 3.11-5

3.11-6 A thin-walled circular tube and a solid circular bar of the same material (see figure) are subjected to torsion. The tube and bar have the same cross-sectional area and the same length.

What is the ratio of the strain energy U_1 in the tube to the strain energy U_2 in the solid bar if the maximum shear stresses are the same in both cases? (For the tube, use the approximate theory for thin-walled bars.)

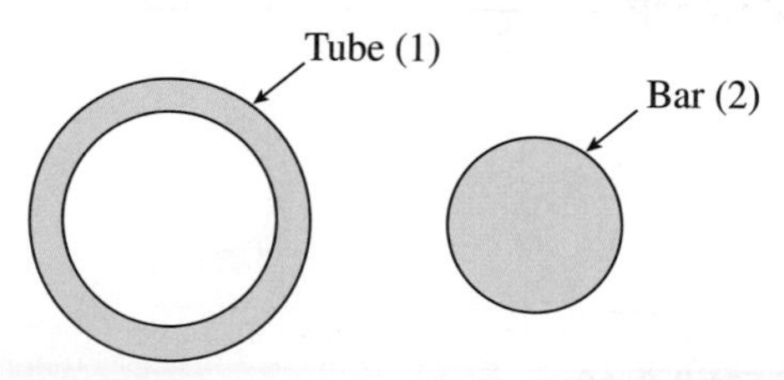

PROBLEM 3.11-6

3.11-7 A thin-walled steel tube having an elliptical cross section with constant thickness t (see figure) is subjected to a torque $T = 5.5$ kN·m.

Determine the shear stress τ and the rate of twist θ (in degrees per meter) if $G = 83$ GPa, $t = 5$ mm, $a = 75$ mm, and $b = 50$ mm. *Note:* See Appendix E, Case 16, for the properties of an ellipse.

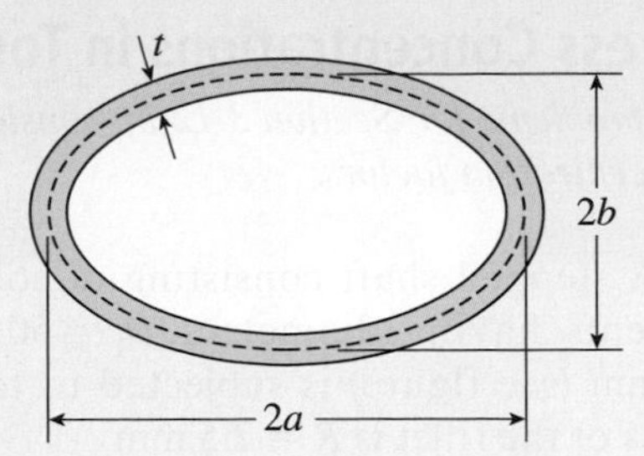

PROBLEM 3.11-7

3.11-8 Calculate the shear stress τ and the angle of twist ϕ (in degrees) for a steel tube ($G = 76$ GPa) having the cross section shown in the figure. The tube has length $L = 1.5$ m and is subjected to a torque $T = 10$ kN · m.

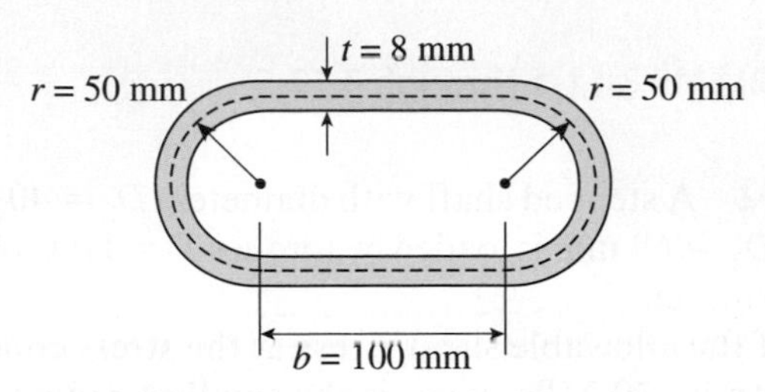

PROBLEM 3.11-8

3.11-9 A torque T is applied to a thin-walled tube having a cross section in the shape of a regular hexagon with constant wall thickness t and side length b (see figure). Obtain formulas for the shear stress τ and the rate of twist θ.

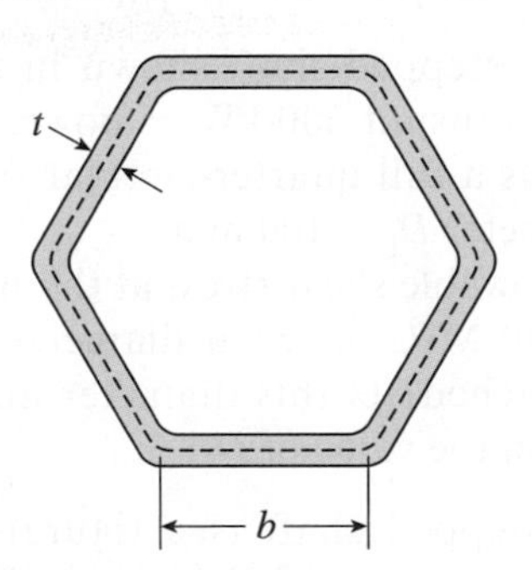

PROBLEM 3.11-9

3.11-10 Compare the angle of twist ϕ_1 for a thin-walled circular tube (see figure) calculated from the approximate theory for thin-walled bars with the angle of twist ϕ_2 calculated from the exact theory of torsion for circular bars.

(a) Express the ratio ϕ_1/ϕ_2 in terms of the nondimensional ratio $\beta = r/t$.

(b) Calculate the ratio of angles of twist for $\beta = 5$, 10, and 20. What conclusion about the accuracy of the approximate theory do you draw from these results?

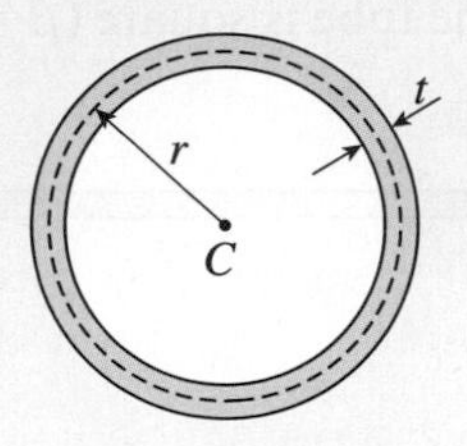

PROBLEM 3.11-10

3.11-11 A tubular aluminum bar ($G = 28$ GPa) of square cross section (see figure) with outer dimensions 50 mm × 50 mm must resist a torque $T = 300$ N·m.

Calculate the minimum required wall thickness t_{min} if the allowable shear stress is 20 MPa and the allowable rate of twist is 0.025 rad/m.

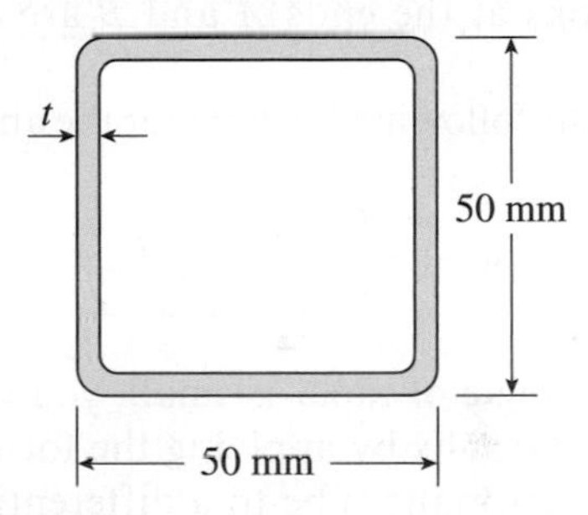

PROBLEM 3.11-11

3.11-12 A thin tubular shaft with a circular cross section (see figure) and with inside diameter 100 mm is subjected to a torque of 5000 N · m.

If the allowable shear stress is 42 MPa, determine the required wall thickness t by using (a) the approximate theory for a thin-walled tube and (b) the exact torsion theory for a circular bar.

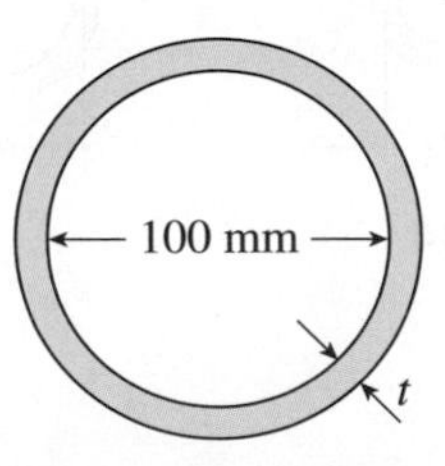

PROBLEM 3.11-12

3.11-13 A thin-walled rectangular tube has uniform thickness t and dimensions $a \times b$ to the median line of the cross section (see figure).

How does the shear stress in the tube vary with the ratio $\beta = a/b$ if the total length L_m of the median line of the cross section and the torque T remain constant?

From your results, show that the shear stress is smallest when the tube is square ($\beta = 1$).

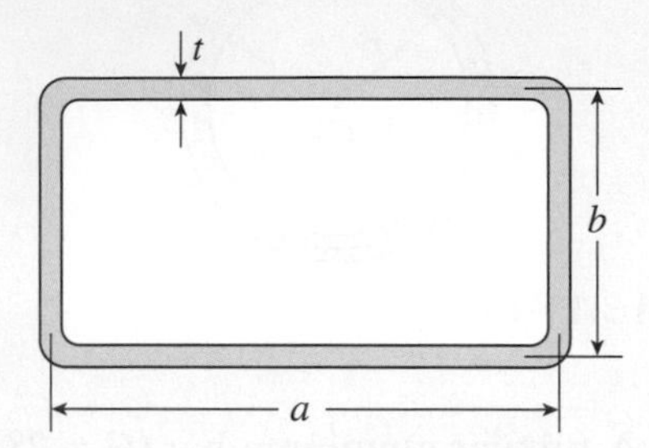

PROBLEM 3.11-13

3.11-14 A long, thin-walled tapered tube AB with a circular cross section (see figure) is subjected to a torque T. The tube has length L and constant wall thickness t. The diameter to the median lines of the cross sections at the ends A and B are d_A and d_B, respectively.

Derive the following formula for the angle of twist of the tube:

$$\phi = \frac{2TL}{\pi Gt}\left(\frac{d_A + d_B}{d_A^2 d_B^2}\right)$$

Hint: If the angle of taper is small, you may obtain approximate results by applying the formulas for a thin-walled prismatic tube to a differential element of the tapered tube and then integrating along the axis of the tube.

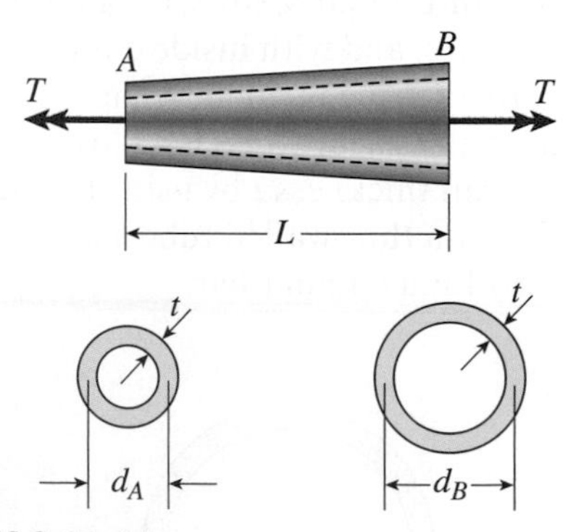

PROBLEM 3.11-14

3.12 Stress Concentrations in Torsion

Solve the problems for Section 3.12 by considering the stress-concentration factors.

3.12-1 A stepped shaft consisting of solid circular segments having diameters $D_1 = 50$ mm and $D_2 = 60$ mm (see figure) is subjected to torques T. The radius of the fillet is $R = 2.5$ mm.

If the allowable shear stress at the stress concentration is 110 MPa, what is the maximum permissible torque T_{max}?

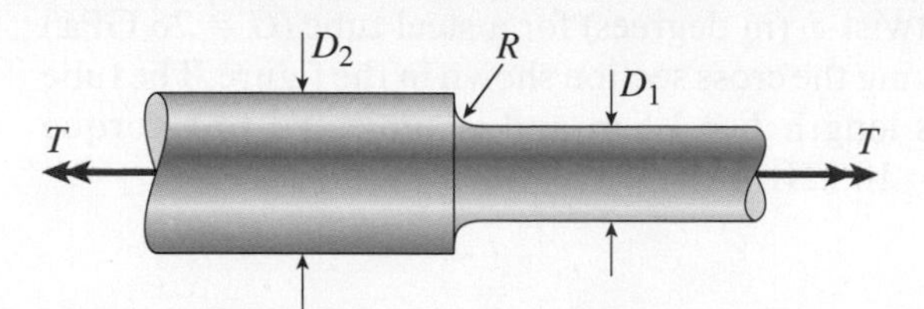

PROBLEMS 3.12-1 through 3.12-5

3.12-2 A stepped shaft with diameters $D_1 = 40$ mm and $D_2 = 60$ mm is loaded by torques $T = 1100$ N · m (see figure).

If the allowable shear stress at the stress concentration is 120 MPa, what is the smallest radius R_{min} that may be used for the fillet?

3.12-3 A full quarter-circular fillet is used at the shoulder of a stepped shaft having diameter $D_2 = 25$ mm (see figure). A torque $T = 115$ N·m acts on the shaft.

Determine the shear stress τ_{max} at the stress concentration for values as follows: $D_1 = 18$, 20 and 22 mm. Plot a graph showing τ_{max} versus D_1.

3.12-4 The stepped shaft shown in the figure is required to transmit 600 kW of power at 400 rpm. The shaft has a full quarter-circular fillet, and the smaller diameter $D_1 = 100$ mm.

If the allowable shear stress at the stress concentration is 100 MPa, at what diameter D_2 will this stress be reached? Is this diameter an upper or a lower limit on the value of D_2?

3.12-5 A stepped shaft (see figure) has diameter $D_2 = 40$ mm and a full quarter-circular fillet. The allowable shear stress is 100 MPa and the load $T = 540$ N·m.

What is the smallest permissible diameter D_1?

SOME ADDITIONAL REVIEW PROBLEMS: Chapter 3

R-3.1 A brass rod of length $L = 0.75$ m is twisted by torques T until the angle of rotation between the ends of the rod is 3.5°. The allowable shear strain in the copper is 0.0005 rad. The maximum permissible diameter of the rod is approximately:

(A) 6.5 mm
(B) 8.6 mm
(C) 9.7 mm
(D) 12.3 mm

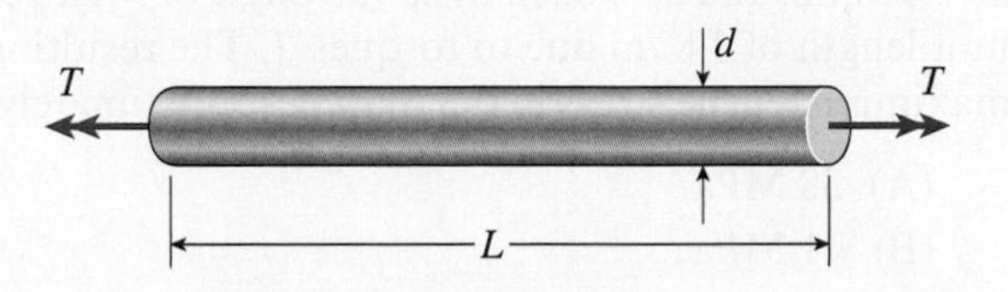

R-3.2 The angle of rotation between the ends of a nylon bar is 3.5°. The bar diameter is 70 mm and the allowable shear strain is 0.014 rad. The minimum permissible length of the bar is approximately:

(A) 0.15 m
(B) 0.27 m
(C) 0.40 m
(D) 0.55 m

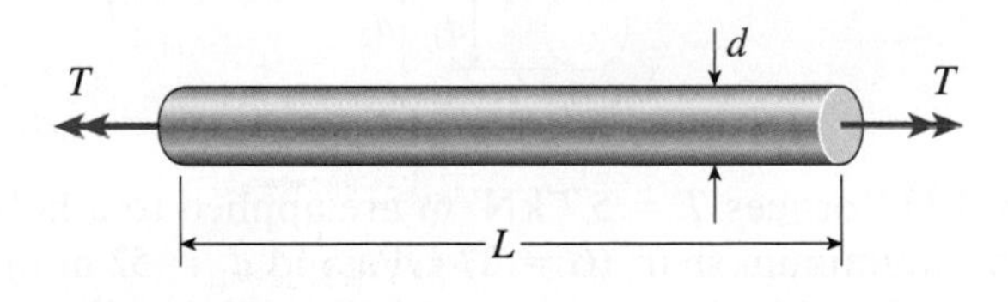

R-3.3 A brass bar twisted by torques T acting at the ends has the following properties: $L = 2.1$ m, $d = 38$ mm, and $G = 41$ GPa. The torsional stiffness of the bar is approximately:

(A) 1200 N·m
(B) 2600 N·m
(C) 4000 N·m
(D) 4800 N·m

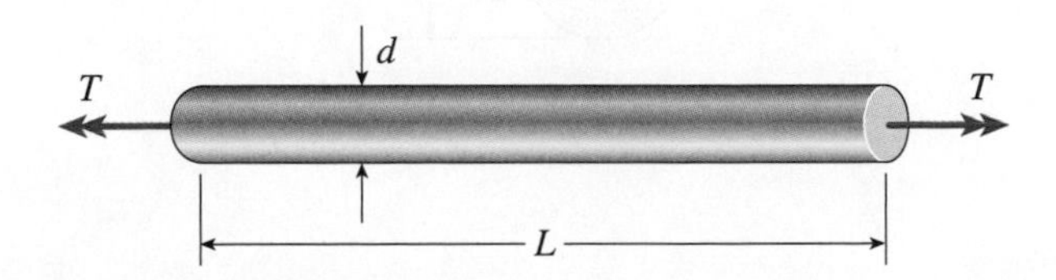

R-3.4 A brass pipe is twisted by torques $T = 800$ N·m acting at the ends causing an angle of twist of 3.5°. The pipe has the following properties: $L = 2.1$ m, $d_1 = 38$ mm, and $d_2 = 56$ mm. The shear modulus of elasticity G of the pipe is approximately:

(A) 36.1 GPa
(B) 37.3 GPa
(C) 38.7 GPa
(D) 40.6 GPa

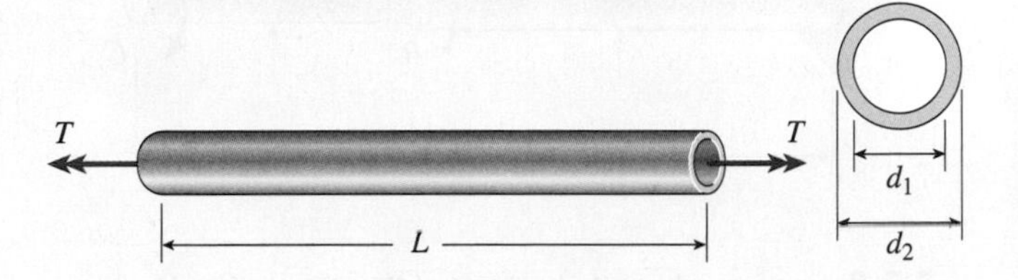

R-3.5 An aluminum bar of diameter $d = 52$ mm is twisted by torques T_1 at the ends. The allowable shear stress is 65 MPa. The maximum permissible torque T_1 is approximately:

(A) 1450 N·m
(B) 1675 N·m
(C) 1710 N·m
(D) 1800 N·m

R-3.6 A steel tube with diameters $d_2 = 86$ mm and $d_1 = 52$ mm is twisted by torques at the ends. The diameter of a solid steel shaft that resists the same torque at the same maximum shear stress is approximately:

(A) 56 mm
(B) 62 mm
(C) 75 mm
(D) 82 mm

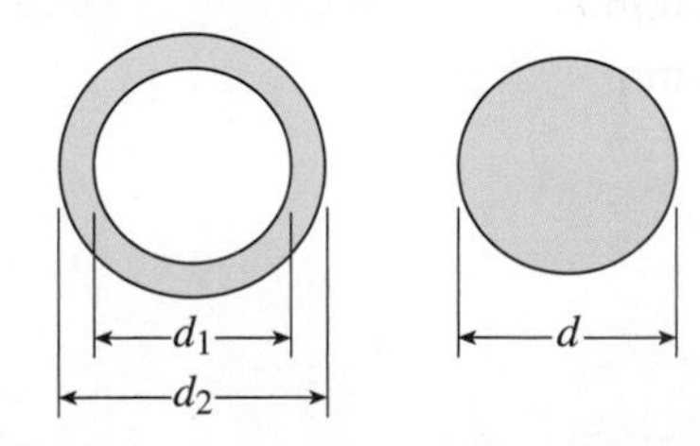

R-3.7 A stepped steel shaft with diameters $d_1 = 56$ mm and $d_2 = 52$ mm is twisted by torques $T_1 = 3.5$ kN·m and $T_2 = 1.5$ kN·m acting in opposite directions. The maximum shear stress is approximately:

(A) 54 MPa
(B) 58 MPa
(C) 62 MPa
(D) 79 MPa

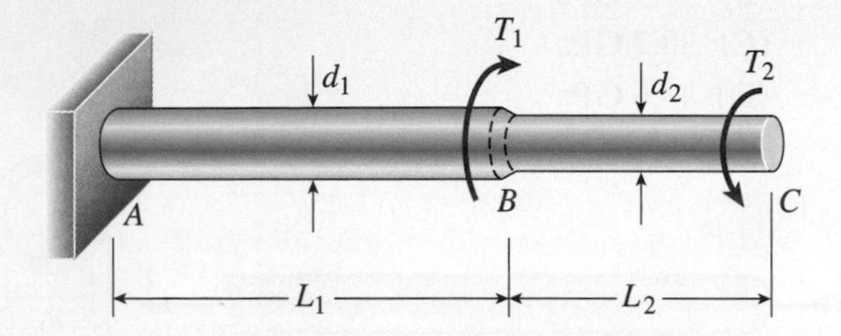

R-3.8 A stepped steel shaft ($G = 75$ GPa) with diameters $d_1 = 36$ mm and $d_2 = 32$ mm is twisted by torques T at each end. Segment lengths are $L_1 = 0.9$ m and $L_2 = 0.75$ m. If the allowable shear stress is 28 MPa and maximum allowable twist is 1.8°, the maximum permissible torque is approximately:

(A) 142 N·m
(B) 180 N·m
(C) 185 N·m
(D) 257 N·m

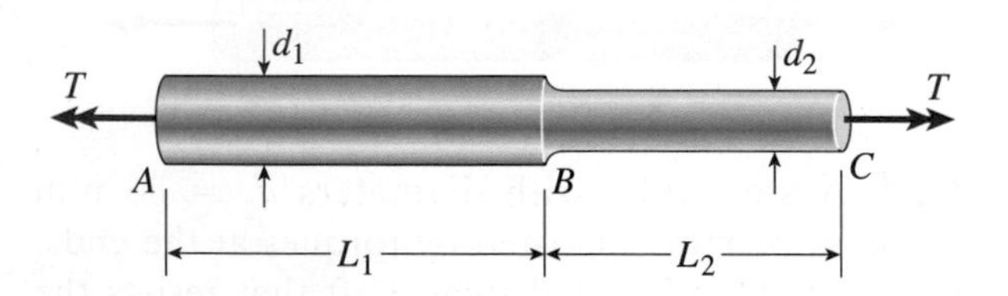

R-3.9 A gear shaft transmits torques $T_A = 975$ N·m, $T_B = 1750$ N·m, $T_C = 650$ N·m, and $T_D = 125$ N·m. If the allowable shear stress is 50 MPa, the required shaft diameter is approximately:

(A) 38 mm
(B) 44 mm
(C) 46 mm
(D) 48 mm

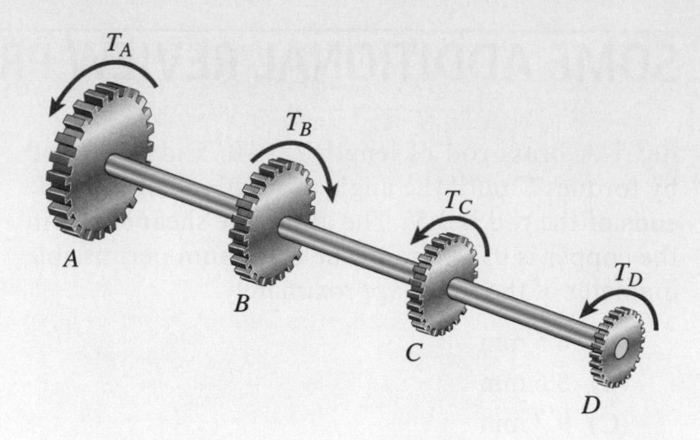

R-3.10 A hollow aluminum shaft $G = 27$ GPa, $d_2 = 96$ mm, and $d_1 = 52$ mm) has an angle of twist per unit length of 1.8°/m due to torques T. The resulting maximum tensile stress in the shaft is approximately:

(A) 38 MPa
(B) 41 MPa
(C) 49 MPa
(D) 58 MPa

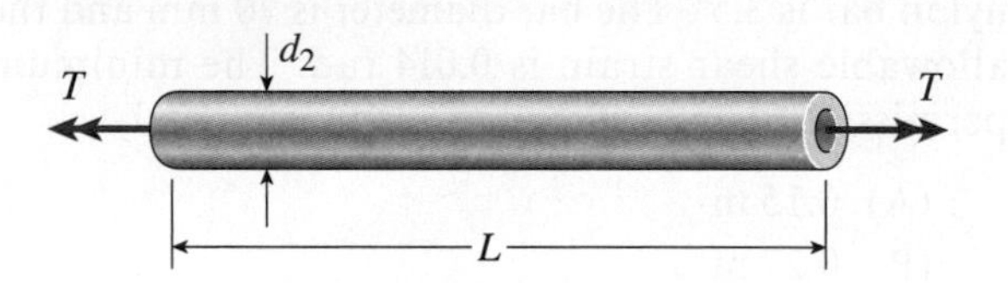

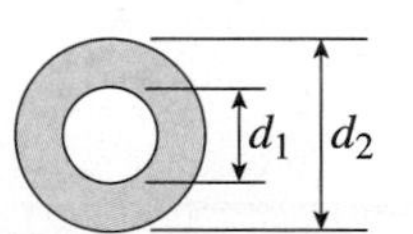

R-3.11 Torques $T = 5.7$ kN·m are applied to a hollow aluminum shaft ($G = 27$ GPa and $d_1 = 52$ mm). The allowable shear stress is 45 MPa and the allowable normal strain is 8.0×10^{-4}. The required outside diameter d_2 of the shaft is approximately:

(A) 38 mm
(B) 56 mm
(C) 87 mm
(D) 91 mm

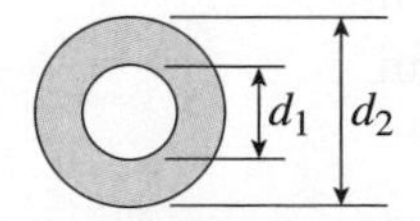

R-3.12 A motor drives a shaft with diameter d = 46 mm at f = 5.25 Hz and delivers P = 25 kW of power. The maximum shear stress in the shaft is approximately:

(A) 32 MPa
(B) 40 MPa
(C) 83 MPa
(D) 91 MPa

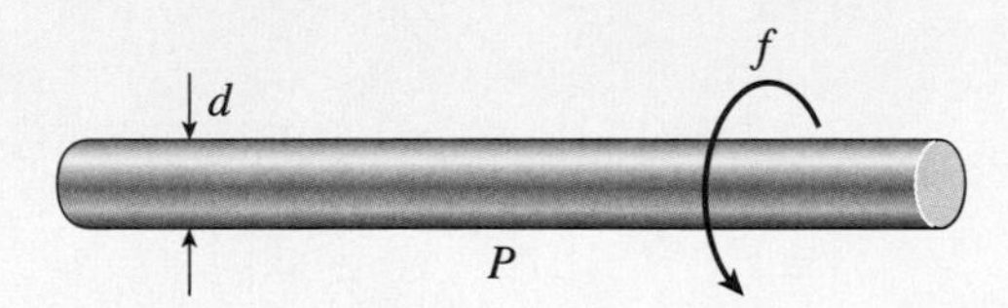

R-3.13 A motor drives a shaft at f = 10 Hz and delivers P = 35 kW of power. The allowable shear stress in the shaft is 45 MPa. The minimum diameter of the shaft is approximately:

(A) 35 mm
(B) 40 mm
(C) 47 mm
(D) 61 mm

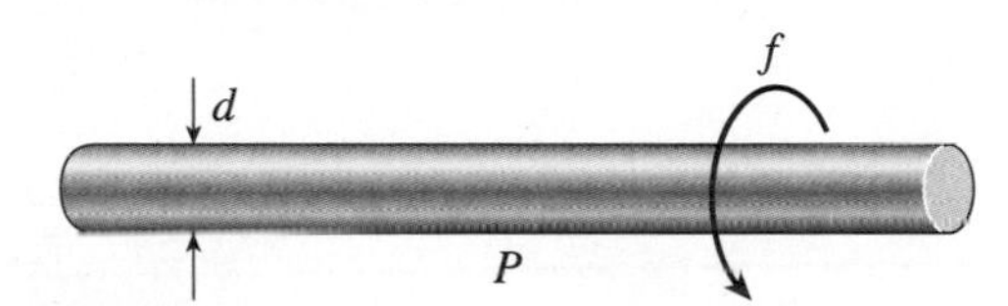

R-3.14 A drive shaft running at 2500 rpm has outer diameter 60 mm and inner diameter 40 mm. The allowable shear stress in the shaft is 35 MPa. The maximum power that can be transmitted is approximately:

(A) 220 kW
(B) 240 kW
(C) 288 kW
(D) 312 kW

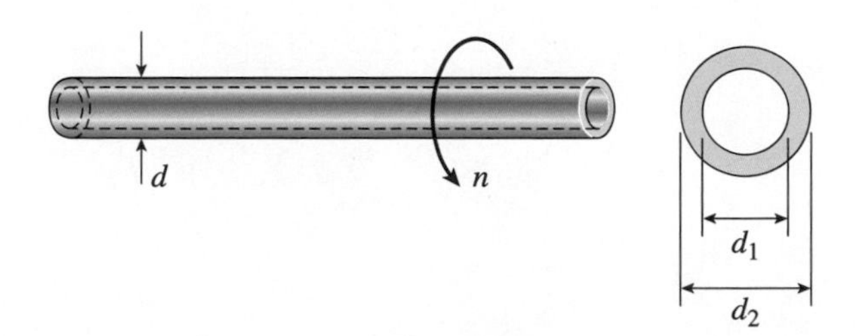

R-3.15 A prismatic shaft (diameter d_0 = 19 mm) is loaded by torque T_1. A stepped shaft (diameters d_1 = 20 mm and d_2 = 25 mm with radius of fillets R = 2 mm) is loaded by torque T_2. The allowable shear stress in the material is 42 MPa. The ratio T_1/T_2 of the maximum permissible torques that can be applied to the shafts, considering stress concentration effects in the stepped shaft is:

(A) 0.9
(B) 1.2
(C) 1.4
(D) 2.1

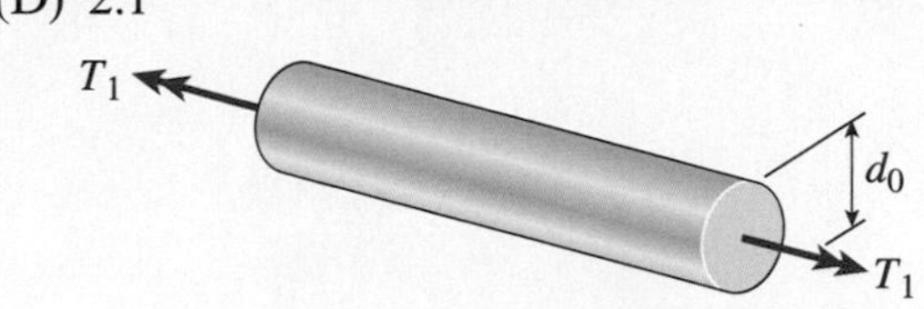

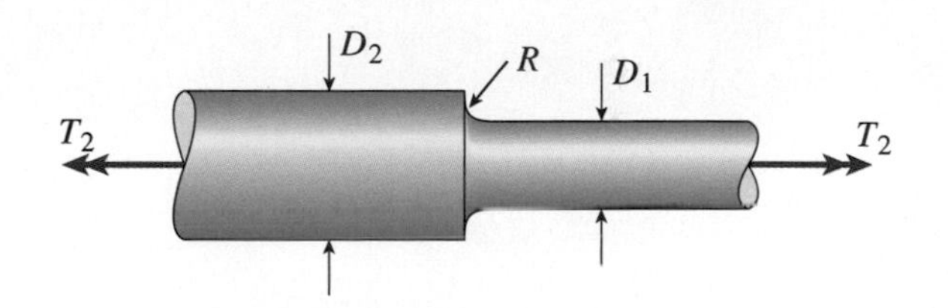

FIGURE 3-59

Stress-concentration factor K for a stepped shaft in torsion. (The dashed line is for a full quarter-circular fillet.)

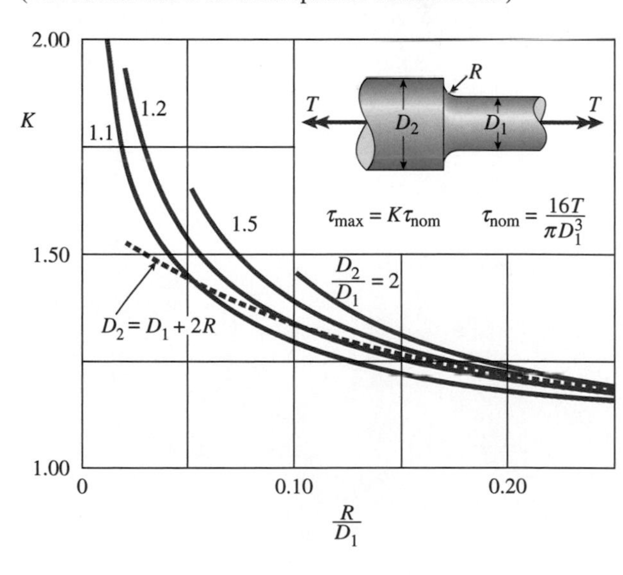

(A) [illegible] mm

(B) [illegible] mm

(C) [illegible] mm

(D) [illegible] mm

(A) [illegible] kW

(B) [illegible] kW

(C) [illegible] kW

(D) [illegible] kW

CHAPTER 4

Shear Forces and Bending Moments

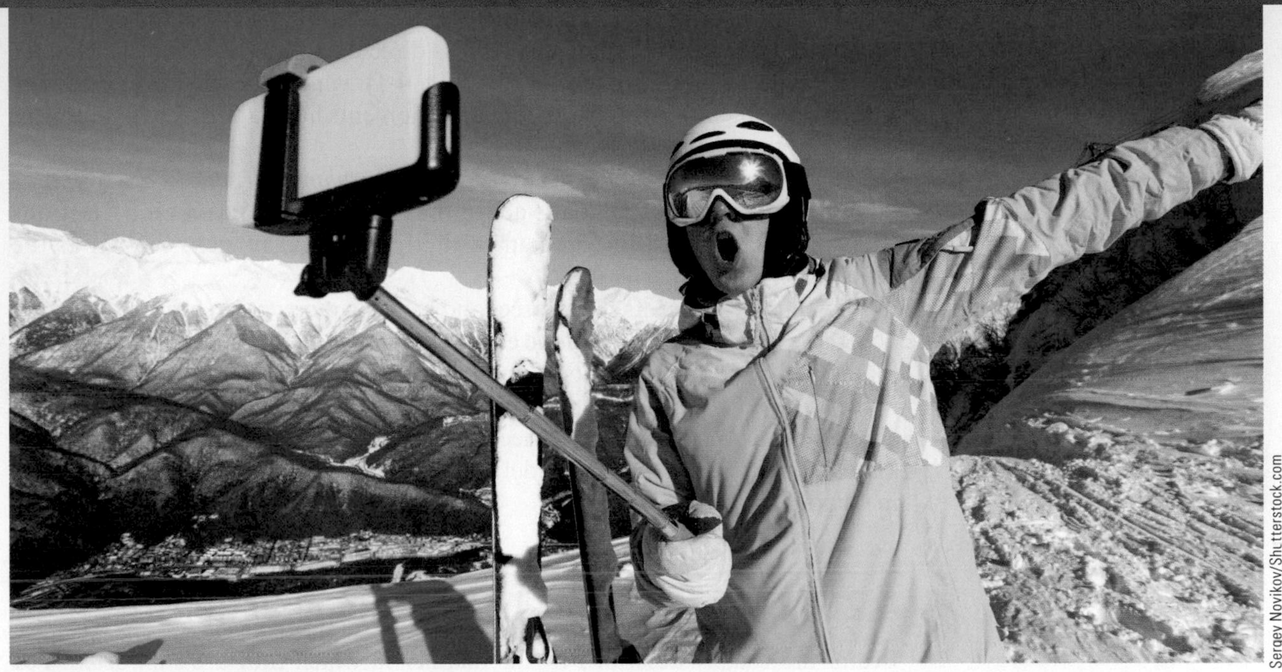
Sergey Novikov/Shutterstock.com

The selfie stick acts as a cantilever beam and is subjected to shear force and bending moment along its length.

Chapter Objectives

- Study internal forces and moments in beams and frames.
- Define different types of beams, various types of loads that act on beams and frames, and different support conditions for beams and frames.
- Describe possible internal releases that can exist in beams and frames and must be included in an accurate model of the beam member.
- Derive differential and integral relationships among distributed transverse load $q(x)$, internal shear force $V(x)$, and bending moment $M(x)$ for beams and frames.
- Draw axial force diagrams (N) and shear (V) and bending moment (M) diagrams for beams and frames.
- Show how axial force and shear and bending moment diagrams help to identify values and locations of maximum axial force and shear and moment in beams and frames which are important in their design.

Chapter Outline

4.1 Introduction

Structural members are usually classified according to the types of loads that they support. For instance, an *axially loaded bar* supports forces having their vectors directed along the axis of the bar, and a *bar in torsion* supports torques (or couples) having their moment vectors directed along the axis. This chapter is concerned with the study of **beams** (Fig. 4-1), which are structural members subjected to lateral loads, that is, forces or moments having their vectors perpendicular to the axis of the bar.

FIGURE 4-1

Examples of beams subjected to lateral loads

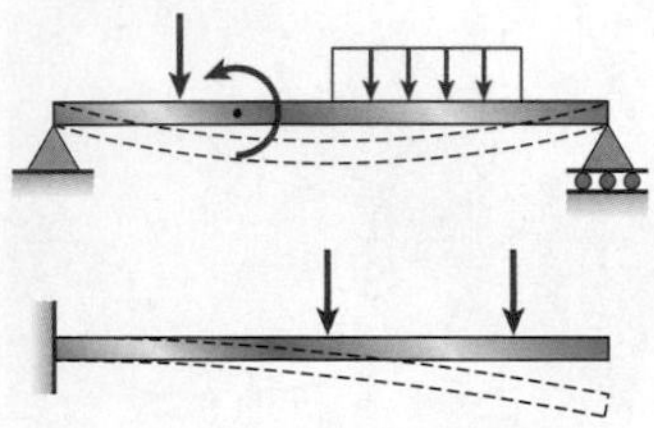

The beams shown in Fig. 4-1 are classified as *planar structures* because they lie in a single plane. If all loads act in that same plane, and if all deflections (shown by the dashed lines) occur in that plane, that plane is known as the **plane of bending**.

This chapter presents shear forces and bending moments in beams and shows how these quantities are related to each other and to the loads. Finding the shear forces and bending moments is an essential step in the design of any beam. Maximum values of these quantities, and also the manner in which they vary along the axis, are of interest. Once the shear forces and bending moments are known, the stresses, strains, and deflections can be computed, as discussed later in Chapters 5, 6, and 9.

4.2 Types of Beams, Loads, and Reactions

Beams are usually described by the manner in which they are supported. For instance, a beam with a pin support at one end and a roller support at the other (Fig. 4-2a) is called a **simply supported beam** or a **simple beam**. The essential feature of a **pin support** is that it prevents translation at the end of a beam but does not prevent rotation. Thus, end A of the beam of Fig. 4-2a cannot move horizontally or vertically but the axis of the beam can rotate in the plane of the figure. Consequently, a pin support is capable of developing a force reaction with both horizontal and vertical components (H_A and R_A), but it cannot develop a moment reaction.

At end B of the beam (Fig. 4-2a) the **roller support** prevents translation in the vertical direction but not in the horizontal direction; hence this support can resist a vertical force (R_B) but not a horizontal force. Of course, the axis of the beam is free to rotate at B just as it is at A. The vertical reactions at roller supports and pin supports may act *either* upward or downward, and the horizontal reaction at a pin support may act either to the left or to the right. In the figures, reactions are indicated by slashes across the arrows in order to distinguish them from loads, as explained previously in Section 1.3.

The beam shown in Fig. 4-2b, which is fixed at one end and free at the other, is called a **cantilever beam**. At the **fixed support** (or *clamped support*) the beam can neither translate nor rotate, whereas at the free end it may do both. Consequently, both force and moment reactions may exist at the fixed support. (The selfie stick in the chapter opener photo is a cantilever beam and the skier's hand provides the fixed support).

The third example is a **beam with an overhang** (Fig. 4-2c). This beam is simply supported at points A and B (that is, it has a pin support at A and a roller support at B), but it also projects beyond the support at B. The overhanging segment BC is similar to a cantilever beam except that the beam axis may rotate at point B.

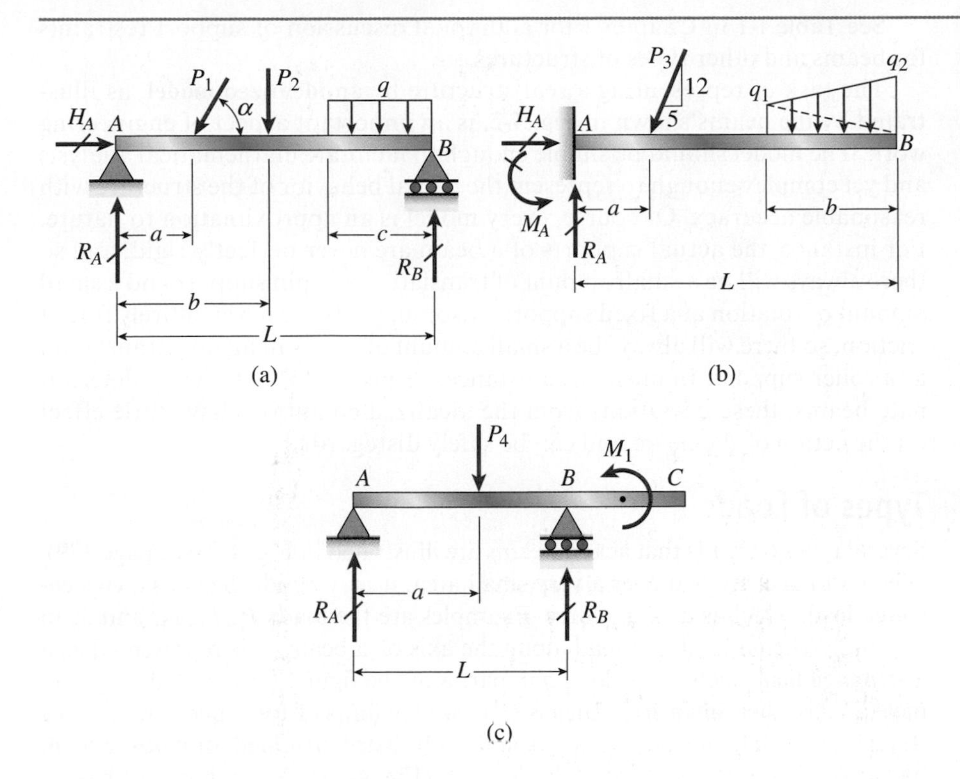

FIGURE 4-2
Types of beams: (a) simple beam, (b) cantilever beam, and (c) beam with an overhang

When drawing sketches of beams, the supports are identified by **conventional symbols**, such as those shown in Fig. 4-2. These symbols indicate the manner in which the beam is restrained, and therefore they also indicate the nature of the reactive forces and moments. However, *the symbols do not represent the actual physical construction*. For instance, consider the examples shown in Fig. 4-3. Part a of the figure shows a wide-flange beam supported on a concrete wall and held down by anchor bolts that pass through slotted holes in the lower flange of the beam. This connection restrains the beam against vertical movement (either upward or downward) but does not prevent horizontal movement. Also, any restraint against rotation of the longitudinal axis of the beam is small and ordinarily may be disregarded. Consequently, this type of support is usually represented by a roller, as shown in part b of the figure.

The second example (Fig. 4-3c) is a beam-to-column connection in which the beam is attached to the column flange by bolted angles. (See the photo on the next page.) This type of support is usually assumed to restrain the beam against horizontal and vertical movement but not against rotation (restraint against rotation is slight because both the angles and the column can bend). Thus, this connection is usually represented as a pin support for the beam (Fig. 4-3d).

The last example (Fig. 4-3e and f on the next page) is a metal pole welded to a base plate that is anchored to a concrete pier embedded deep in the ground. Since the base of the pole is fully restrained against both translation and rotation, it is represented as a fixed support (Fig. 4-3f).

FIGURE 4-3
Beam supported on a wall: (a) actual construction and (b) representation as a roller support; Beam-to-column connection: (c) actual construction and (d) representation as a pin support

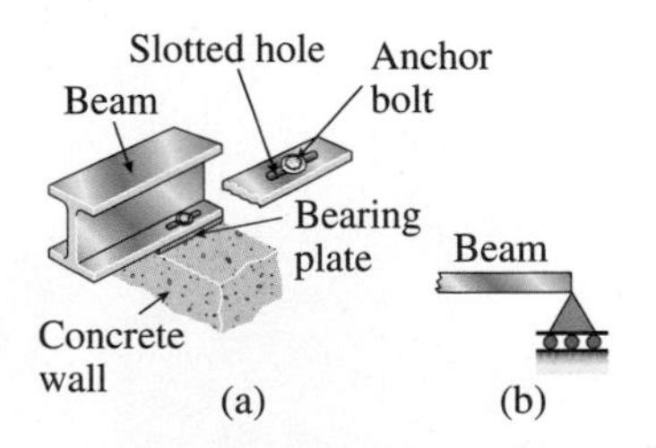

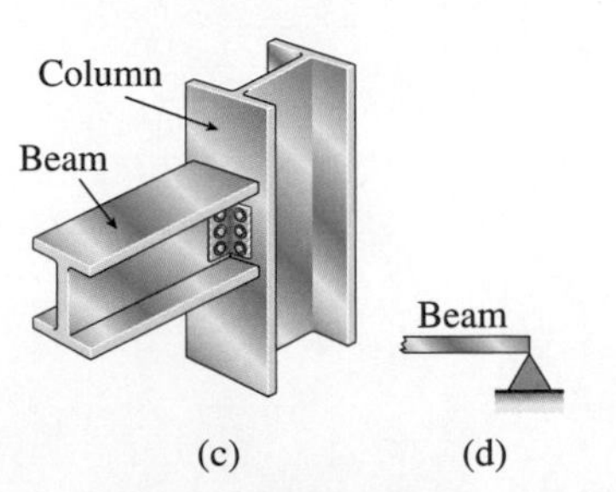

Beam-to-column connection with one beam attached to a column flange and other attached to a column web

FIGURE 4-3 (*Continued*)
Pole anchored to a concrete pier: (e) actual construction and (f) representation as a fixed support

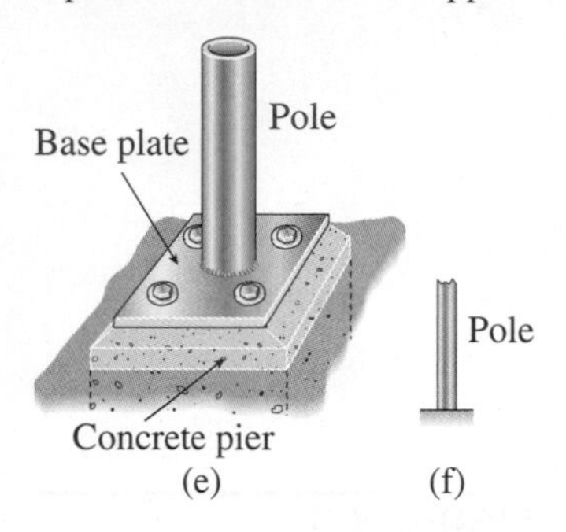

See Table 1-1 in Chapter 1 for additional discussion of support restraints for beams and other types of structures.

The task of representing a real structure by an **idealized model**, as illustrated by the beams shown in Fig. 4-2, is an important aspect of engineering work. The model should be simple enough to facilitate mathematical analysis and yet complex enough to represent the actual behavior of the structure with reasonable accuracy. Of course, every model is an approximation to nature. For instance, the actual supports of a beam are never perfectly rigid, and so there always will be a small amount of translation at a pin support and a small amount of rotation at a fixed support. Also, supports are never entirely free of friction, so there will always be a small amount of restraint against translation at a roller support. In most circumstances, especially for statically determinate beams, these deviations from the idealized conditions have little effect on the action of the beam and can be safely disregarded.

Types of Loads

Several types of loads that act on beams are illustrated in Fig. 4-2 (see page 379). When a load is applied over a very small area, it may be idealized as a **concentrated load**, which is a single force. Examples are the loads P_1, P_2, P_3, and P_4 in Fig. 4-2. When a load is spread along the axis of a beam, it is represented as a **distributed load**, such as the load q in part a of the figure. Distributed loads are measured by their **intensity**, which is expressed in units of force per unit distance (for example, newtons per meter). A **uniformly distributed load**, or **uniform load**, has a constant intensity q per unit distance (Fig. 4-2a). A varying load has an intensity that changes with distance along the axis; for instance, the **linearly varying load** of Fig. 4-2b has an intensity that varies linearly from q_1 to q_2. Another kind of load is a **couple**, as illustrated by the couple of moment M_1 acting on the overhanging beam (Fig. 4-2c).

As mentioned in Section 4.1, assume in this discussion that the loads act in the plane of the figure, which means that all forces have their vectors in the plane of the figure and all couples have their moment vectors perpendicular to the plane of the figure. Furthermore, assume that the beam itself is symmetric about that plane, which means that every cross section of the beam has a vertical axis of symmetry. Under these conditions, the beam will deflect only in the *plane of bending* (the plane of the figure).

Reactions

Finding the reactions is usually the first step in the analysis of a beam. Once the reactions are known, the shear forces and bending moments can be found, as described later in this chapter. If a beam is supported in a statically determinate manner, all reactions can be found from free-body diagrams and equations of equilibrium.

Internal Releases

In some instances, it may be necessary to add internal releases into the beam or frame model to better represent actual conditions of construction that may have an important effect on overall structure behavior. Consider the four types of releases shown in Fig. 4-4. The first is an **axial release** (Fig. 4-4a) illustrated by the insertion of pipe 2 into pipe 1. Force P applied to pipe 1 cannot be

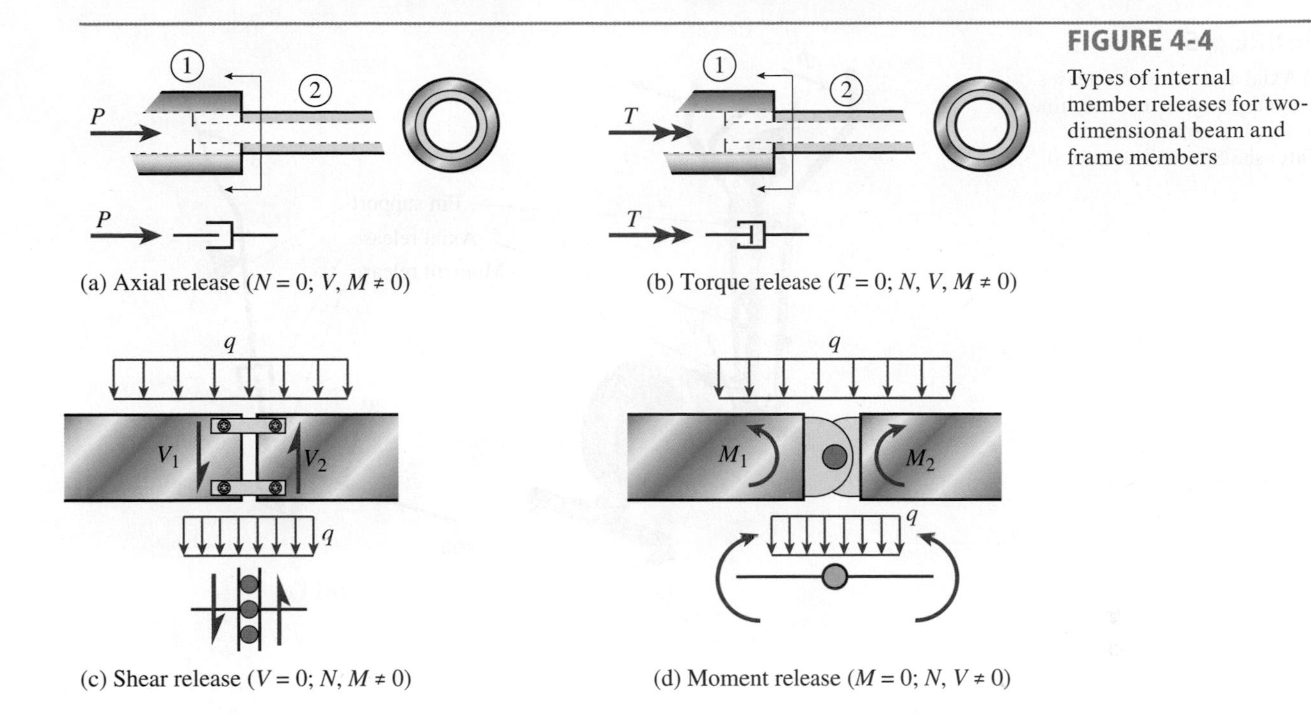

(a) Axial release ($N = 0$; $V, M \neq 0$)

(b) Torque release ($T = 0$; $N, V, M \neq 0$)

(c) Shear release ($V = 0$; $N, M \neq 0$)

(d) Moment release ($M = 0$; $N, V \neq 0$)

FIGURE 4-4

Types of internal member releases for two-dimensional beam and frame members

transmitted to pipe 2 because the internal axial force N (in the absence of friction) must be zero at this connection. Shear V and bending moment M can be carried by this connection, however. The symbol for an axial release is shown at the bottom of Fig. 4-4a. The same pipe-in-pipe connection also functions as a **torque release** (see Fig. 4-4b) because applied torsional moment T cannot be transferred from pipe 1 to pipe 2. Of more interest to beams and frames are the shear and moment releases shown in the Figs. 4-4c and d, respectively. A **shear release** is represented by a pair of links at the top and bottom of the beam, which can transfer axial tension or compression forces (and hence, moment), but transverse shear forces must be zero in this connection. The **moment release** in Fig. 4-4d is usually referred to as a **pin** connection; this connection carries axial and shear forces, but the moment must be zero at that location on a beam or frame.

An illustration of axial and moment releases (in addition to pin and roller supports) is presented in Fig. 4-5a, which shows one type of elliptical trainer. The trainer machine has vertical arms that are pin connected to a vertical center post, and these arms vary in length with the front and back motion applied during exercise. The connection between the vertical arm and the weight-bearing step is a moment release, so these two components can rotate independently. A simple frame model that can be used for analysis of the machine in a variety of different positions is shown in Fig. 4-5b. (See Problem 1.3-33 in Chapter 1 for a sample analysis.)

As Examples 4-1, 4-2 and 4-3 show, if axial, shear, or moment **releases** are present in the structure model, the structure must be broken into separate free-body diagrams by cutting through the release; an additional equation of equilibrium is then available for use in solving for the unknown support reactions included in that free-body diagram.

FIGURE 4-5

(a) Axial and moment releases in an elliptical trainer machine; (b) simple frame model (Tatuasha/Shutterstock.com)

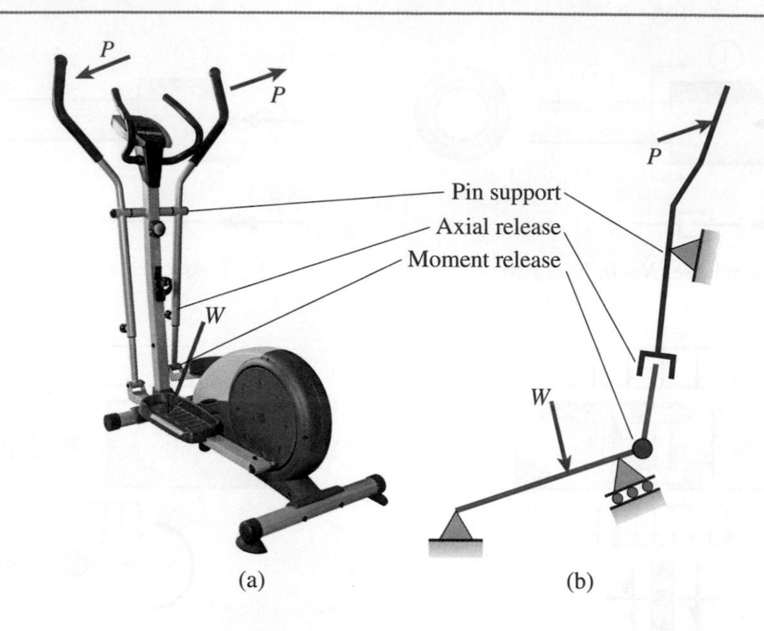

Example 4-1

Find the support reactions for the two simply supported beams in Fig. 4-6. Beam (a) has a pin support at A and a roller support at B. Beam (b) has pin supports at both A and B and an internal axial release just to the left of the point of application of inclined load P_1.

FIGURE 4-6

Simply supported beams with uniform and concentrated loads

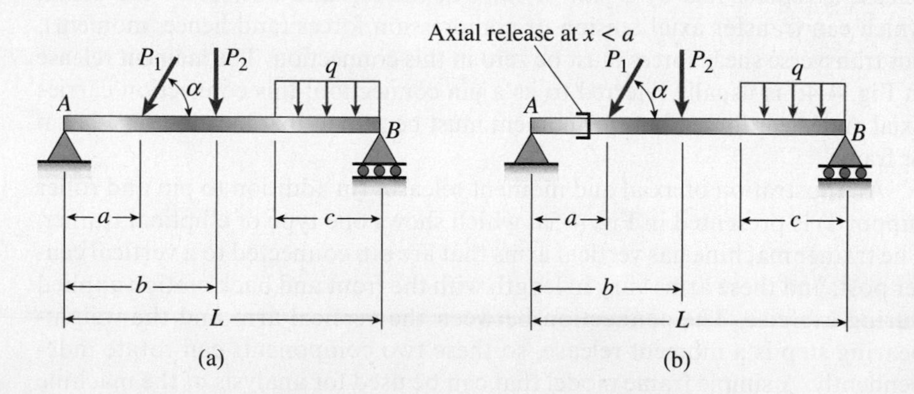

Solution:

Use a four-step problem-solving approach for these beams.

1. ***Conceptualize*** [*hypothesize, sketch*]: Start by drawing overall free-body diagrams of each beam (Figs. 4-7 and 4-8). Beam (a) has two reaction force components at A and one at B, while beam (b) has two reaction components at both A and B.

FIGURE 4-7

Free-body diagram of beam (a)

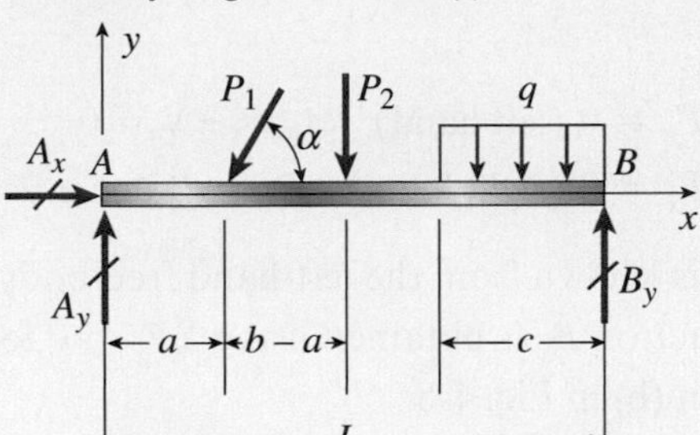

FIGURE 4-8

Overall free-body diagram of beam (b)

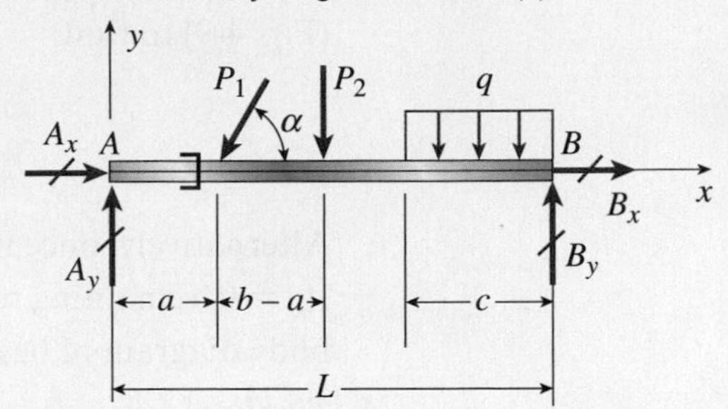

2. ***Categorize*** [*simplify, classify*]: Beam (a) is statically determinate; there are three reaction force unknowns and three static equilibrium equations. Beam (b) has four reaction force unknowns, so it is necessary to cut the beam into separate left- and right-hand free-body diagrams (Fig. 4-9) to find one additional equation of equilibrium. Because of the axial release, internal axial force N is known to be zero at the section cut. This is the additional equilibrium equation needed to solve for all four reaction components.

FIGURE 4-9

Left-hand and right-hand free-body diagrams of beam (b)

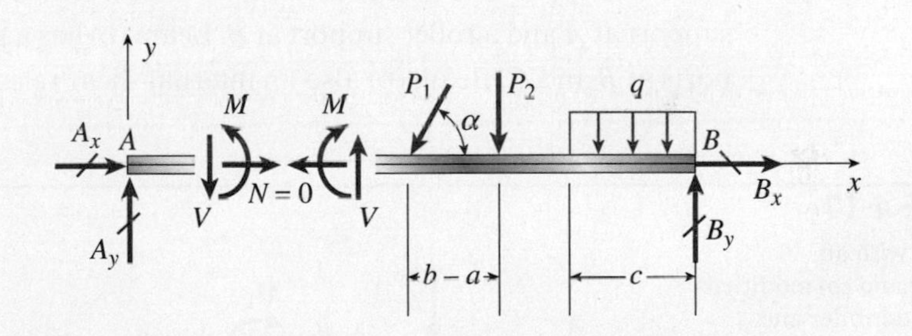

3. ***Analyze*** [*evaluate; select relevant equations, carry out mathematical solution*]:

External reactions—beam (a): One possible solution approach to find reactions in beam (a) proceeds as follows. A *statics sign convention* is used in the solution.

$$\Sigma F_x = 0 \quad A_x - P_1 \cos\alpha = 0 \quad \text{so} \quad A_x = P_1 \cos\alpha$$

$$\Sigma M_B = 0 \quad -A_y L + (P_1 \sin\alpha)(L - a) + P_2(L - b) + \frac{qc^2}{2} = 0$$

$$\text{so} \quad A_y = P_1 \sin\alpha\left(1 - \frac{a}{L}\right) + P_2\left(1 - \frac{b}{L}\right) + \frac{qc^2}{2L}$$

$$\Sigma M_A = 0 \quad B_y L - (P_1 \sin\alpha)(a) - P_2 b - qc\left(L - \frac{c}{2}\right) = 0$$

$$\text{so} \quad B_y = P_1 \sin\alpha\left(\frac{a}{L}\right) + P_2\left(\frac{b}{L}\right) + qc\left(1 - \frac{c}{2L}\right)$$

External reactions—beam (b): The previous solutions for A_y and B_y are still valid for beam (b) but now use left- and right-hand free-body diagrams (Fig. 4-9) to find

$$\Sigma F_x = 0 \text{ (left hand)} \quad A_x = -N = 0$$
$$\Sigma F_x = 0 \text{ (right hand)} \quad B_x = P_1 \cos\alpha$$

Alternatively, once it is known from the left-hand free-body diagram that $A_x = 0$, remaining reaction B_x is obtained using $\Sigma F_x = 0$ for the overall free-body diagram of beam (b) in Fig. 4-8.

4. ***Finalize*** *[conclude; examine answer—Does it make sense? Are units correct? How does it compare to similar problem solutions?]*: If beam (a) is modified to replace the roller support at B with a pin support, it is now one degree statically indeterminate. However, if an axial force release is inserted into the model to create beam (b), the beam can be analyzed using the laws of statics alone because the release provides one additional equilibrium equation.

Example 4-2

Find the support reactions for the two beams shown in Fig. 4-10. Beam (a) has a pin support at A and a roller support at B. Beam (b) has a pin support at A and roller supports at B and C. Beam (b) also an internal shear release just to the left of support B.

FIGURE 4-10

(a) Beam with an overhang and (b) modified beam—add roller and shear release

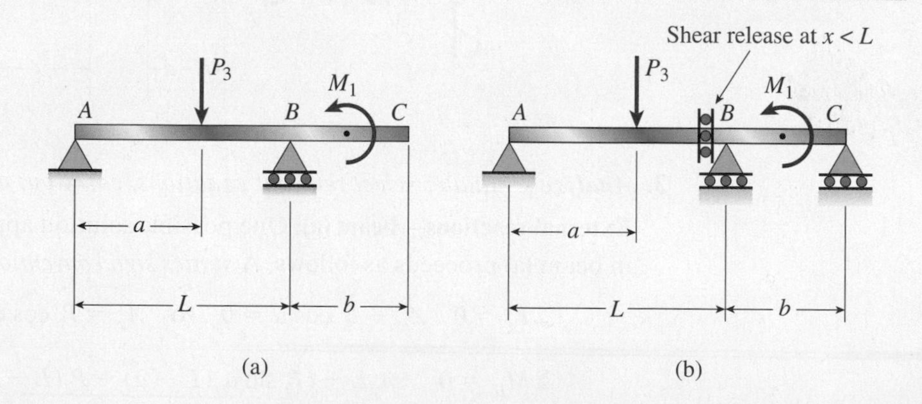

Solution:

Use a four-step problem-solving approach for these beams.

1. ***Conceptualize***: The overall free-body diagrams for each beam are shown in Figs. 4-11 and 4-12. Beam (a) has three reaction force components: two at A and one at B. Beam (b), however, has two reaction components at A and one each at B and C, giving a total of four reactions.

FIGURE 4-11
Free-body diagram of beam (a)

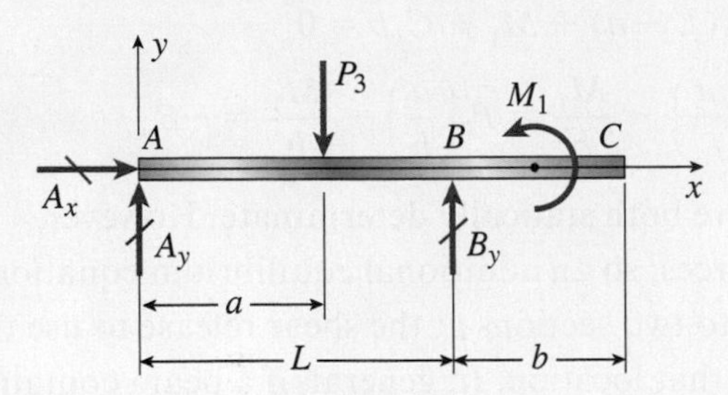

FIGURE 4-12
Overall free-body diagram of beam (b)

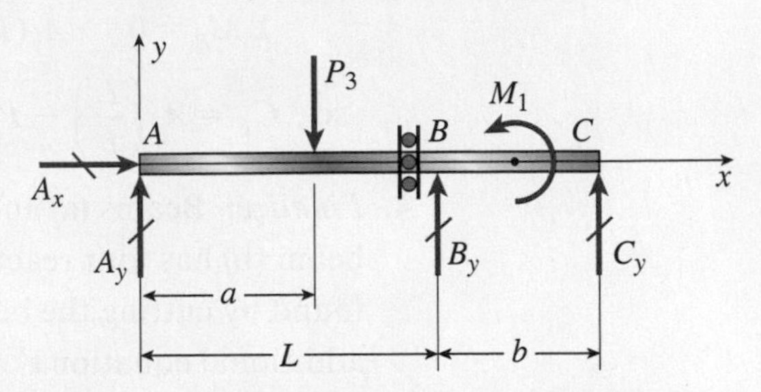

2. *Categorize*: Beam (a) is statically determinate; there are three reaction force unknowns and three static equilibrium equations. Beam (b) has four reaction force unknowns, so it appears to be statically indeterminate. It will be necessary to cut the beam into separate left- and right-hand free-body diagrams (Fig. 4-13) to make use of the fact that the shear force V is zero at the shear release. This additional equilibrium equation is needed to solve for all four reaction components in beam (b).

FIGURE 4-13
Left-hand and right-hand free-body diagrams of beam (b)

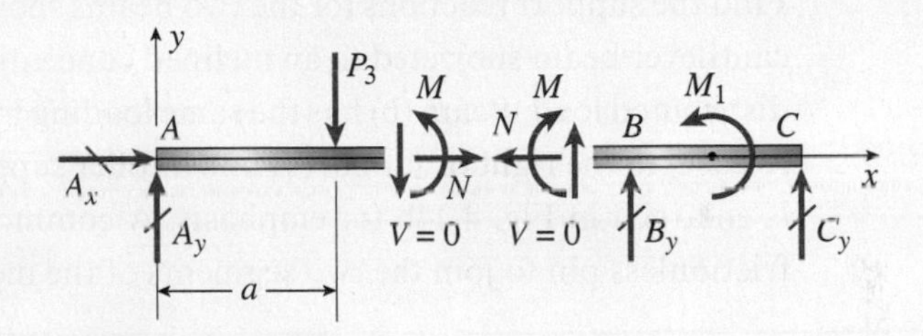

3. *Analyze*:

External reactions—beam (a): Sum forces in the x direction in Fig. 4-11 to find that $A_x = 0$. Summing moments about A to find B_y and summing moments about B to find A_y offers a direct path to a solution for each vertical reaction force:

$$\Sigma F_x = 0 \quad A_x = 0$$

$$\Sigma M_A = 0 \quad -P_3 a + B_y L + M_1 = 0 \quad \text{so} \quad B_y = P_3\left(\frac{a}{L}\right) - \frac{M_1}{L} \quad \text{(a)}$$

$$\Sigma M_B = 0 \quad -A_y L + P_3(L - a) + M_1 = 0 \quad \text{so} \quad A_y = P_3\left(1 - \frac{a}{L}\right) + \frac{M_1}{L}$$

External reactions—beam (b): Now use left- and right-hand free-body diagrams (Fig. 4-13) to find

$$\Sigma F_y = 0 \text{ (left hand)} \quad A_y = P_3$$

$$\Sigma F_y = 0 \text{ (right hand)} \quad B_y + C_y = 0 \quad \text{so} \quad B_y = -C_y \quad \text{(b)}$$

Use the overall free-body diagram of beam (b) in Fig. 4-12; either sum moments about joint C to find B_y or sum moments about joint B to find C_y:

$$\Sigma M_B = 0 \quad -A_y(L) + P_3(L - a) + M_1 + C_y b = 0$$

$$\text{so} \quad C_y = A_y\left(\frac{L}{b}\right) - P_3\left(\frac{L-a}{b}\right) - \frac{M_1}{b} = P_3\left(\frac{a}{b}\right) - \frac{M_1}{b} = -B_y \qquad \text{(c)}$$

4. ***Finalize***: Beams (a) and (b) are both statically determinate. However, beam (b) has four reaction forces, so an additional equilibrium equation was found by cutting the beam into two sections at the shear release to use the additional equation $V = 0$ at that location. In general, if a beam contains an internal release, it is necessary to cut the beam at the location of the release to create separate free-body diagrams in order to find all of the support reactions.

Example 4-3

Find the support reactions for the two beams shown in Fig. 4-14a and b. Beam (a) is a cantilever beam subjected to an inclined concentrated load P_4 and a linearly varying distributed load. Beam (b) has the same loading but has a pin connection (or moment release) to the right of support A and a roller support at joint B. The moment release is enlarged in Fig. 4-14b for emphasis. A common alternate representation using a frictionless pin to join the two segments of the member is shown in Fig. 4-14c.

FIGURE 4-14

(a) Cantilever beam (a); (b) propped cantilever beam (b); and (c) alternate representation of propped cantilever beam (b)

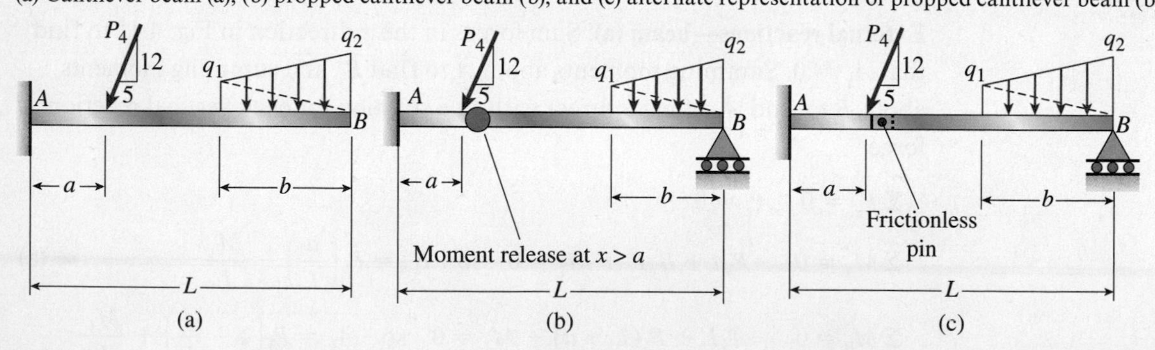

Solution:

Use a four-step problem-solving approach for these beams.

1. ***Conceptualize***: Overall free-body diagrams of each beam are shown in Figs. 4-15 and 4-16. Both beam (a) and beam (b) have three reactions at A (two force components and a moment) and, beam (b) has an additional force reaction at roller support B.

FIGURE 4-15

Free-body diagram of beam (a)

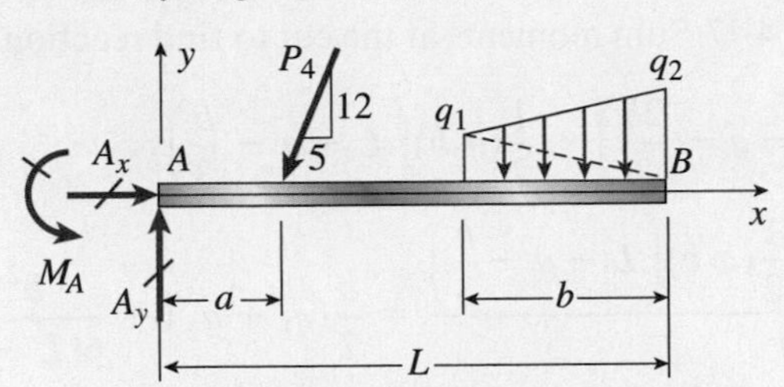

FIGURE 4-16

Overall free-body diagram of beam (b)

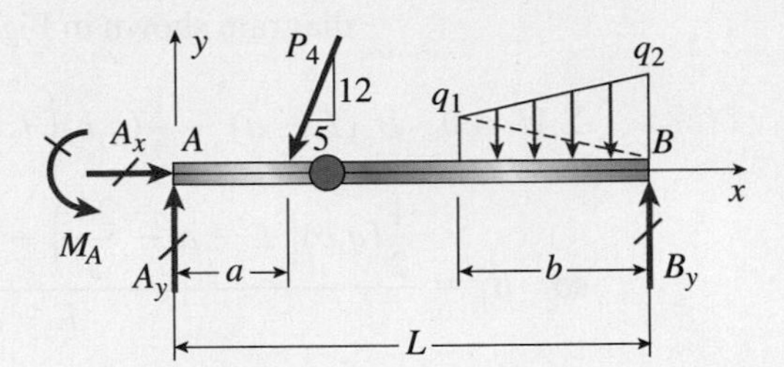

2. *Categorize*: Beam (a) is readily identified as statically determinate because there are three reaction force unknowns and three static equilibrium equations. Beam (b) has four reaction force unknowns, so as in Examples 4-1 and 4-2, it is necessary to cut the beam into separate left- and right-hand free-body diagrams (Fig. 4-17) to find one additional equation of equilibrium: At the moment release the moment is assumed to be equal to zero. This additional equilibrium equation is needed to solve for all four reaction components.

FIGURE 4-17

Left-hand and right-hand free-body diagrams of beam (b)

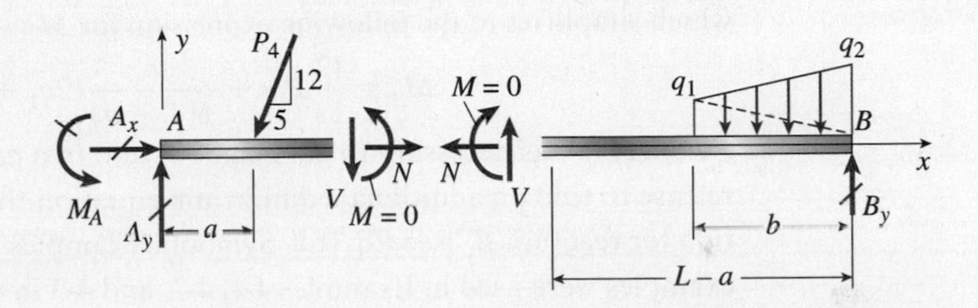

3. *Analyze*:

External reactions—beam (a): Summing forces in x and y directions and moments about A in Fig. 4-15 lead to a solution for reactions at A. Divide the distributed load trapezoid into two triangles, as shown by the dashed line in Fig. 4-16 [or use a uniform load of intensity q_1 plus a triangular load of peak intensity $(q_2 - q_1)$]. Each load triangle can be replaced by its resultant, which is a force with magnitude equal to the area of the triangle and line of action through the centroid of the triangle. Use a statics sign convention to find

$$\Sigma F_x = 0 \quad A_x = \frac{5}{13}P_4$$

$$\Sigma F_y = 0 \quad A_y = \frac{12}{13}P_4 + \frac{1}{2}(q_1 + q_2)b$$

$$\Sigma M_A = 0 \quad M_A = \frac{12}{13}P_4(a) + \frac{1}{2}q_1 b\left(L - \frac{2}{3}b\right) + \frac{1}{2}q_2 b \qquad \text{(a)}$$

or

$$M_A = \frac{12}{13}P_4(a) + \frac{bL}{2}(q_1 + q_2) - \frac{b^2}{6}(2q_1 + q_2)$$

External reactions—beam (b): First, note that reaction A_x is the same for beams (a) and (b). Now cut through the moment release and use the right-hand free-body diagram shown in Fig. 4-17. Sum moments at the cut to find reaction B_y as

$$\Sigma M = 0 \quad B_y(L-a) - \frac{1}{2}(q_1 b)\left(L - a - \frac{2b}{3}\right) - \frac{1}{2}(q_2 b)\left(L - a - \frac{b}{3}\right)$$

$$\text{so} \quad B_y = \frac{\frac{1}{2}(q_1 b)\left(L - a - \frac{2b}{3}\right) + \frac{1}{2}(q_2 b)\left(L - a - \frac{b}{3}\right)}{L-a} = \frac{b}{2}(q_1 + q_2) - \frac{b^2}{6(L-a)}(2q_1 + q_2) \qquad \text{(b)}$$

Next, sum forces in the y direction for the entire free-body diagram (Fig. 4-16) to find reaction force A_y:

$$\Sigma F_y = 0 \quad A_y = \frac{12}{13}P_4 + \frac{1}{2}(q_1 + q_2)(b) - B_y = \frac{12}{13}P_4 + \frac{b^2}{6(L-a)}(2q_1 + q_2) \qquad \text{(c)}$$

Finally, sum moments about A for the entire free-body diagram (Fig. 4-16) to get reaction moment M_A:

$$\Sigma M_A = 0 \quad M_A = \frac{12}{13}P_4 a + \frac{1}{2}q_1 b\left(L - \frac{2b}{3}\right) + \frac{1}{2}q_2 b\left(L - \frac{b}{3}\right) - B_y L$$

which simplifies to the following expression for M_A:

$$M_A = \frac{12}{13}P_4 a + \frac{ab^2}{6(L-a)}(2q_1 + q_2) \qquad \text{(d)}$$

4. ***Finalize:*** It was necessary to cut beam (b) into two parts through the moment release to find an additional equilibrium equation that resulted in a solution for reaction B_y [see Eq. (b)]. Symbolic examples rather than numerical examples were used in Examples 4-1, 4-2, and 4-3 in order to show how the individual steps are carried out. In each case, free-body diagrams like those in Figs. 4-15, 4-16, and 4-17 were essential to an efficient solution for all reactions.

4.3 Shear Forces and Bending Moments

When a beam is loaded by forces or couples, stresses and strains are created throughout the interior of the beam. To determine these stresses and strains, start by finding the internal forces and internal couples that act on cross sections of the beam.

As an illustration of how these internal quantities are found, consider a cantilever beam AB loaded by a force P at its free end (Fig. 4-18a). Cut through the beam at a cross section mn located at distance x from the free end and isolate the left-hand part of the beam as a free body (Fig. 4-18b). The free body is held in equilibrium by the force P and by the stresses that act over the cut cross section. These stresses represent the action of the right-hand part of the beam on the left-hand part. At this stage, the distribution of the stresses acting over the cross section is unknown, but the resultant of these stresses must be such as to maintain equilibrium of the free body.

From statics, the resultant of the stresses acting on the cross section can be reduced to a **shear force** V and a **bending moment** M (Fig. 4-18b). Because the load P is transverse to the axis of the beam, no axial force exists at the cross section. Both the shear force and the bending moment act in the plane of the beam, that is, the vector for the shear force lies in the plane of the figure and the vector for the moment is perpendicular to the plane of the figure.

Shear forces and bending moments, like axial forces in bars and internal torques in shafts, are the resultants of stresses distributed over the cross section. Therefore, these quantities are known collectively as **stress resultants.**

The stress resultants in statically determinate beams can be calculated from equations of equilibrium. In the case of the cantilever beam of Fig. 4-18a, use the free-body diagram of Fig. 4-18b. Summing forces in the vertical direction and also taking moments about the cut section gives

$$\Sigma F_{\text{vert}} = 0 \qquad P - V = 0 \quad \text{or} \quad V = P$$
$$\Sigma M = 0 \qquad M - Px = 0 \quad \text{or} \quad M = Px$$

where x is the distance from the free end of the beam to the cross section where V and M are being determined. Thus, through the use of a free-body diagram and equations of equilibrium, the shear force and bending moment values and their variation with distance x can be found without difficulty.

FIGURE 4-18

Shear force V and bending moment M in a beam

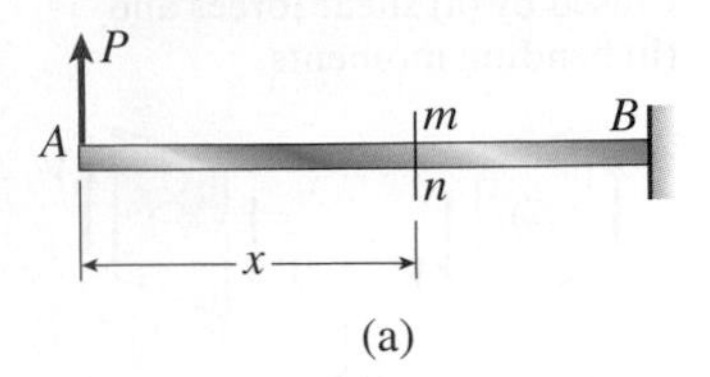

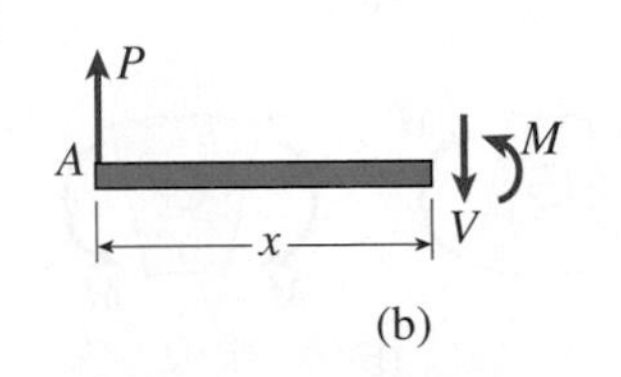

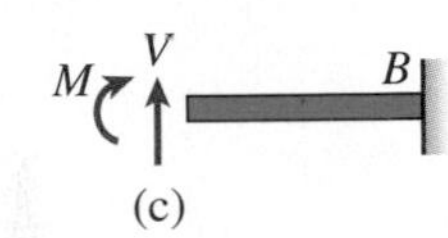

Sign Conventions

Now consider the sign conventions for shear forces and bending moments. It is customary to assume that shear forces and bending moments are positive when they act in the directions shown in Fig. 4-18b. Note that the shear force tends to rotate the material clockwise, and the bending moment tends to compress the upper part of the beam and elongate the lower part. Also, in this instance, the shear force acts downward and the bending moment acts counterclockwise.

The action of these *same* stress resultants against the right-hand part of the beam is shown in Fig. 4-18c. The directions of both quantities are now reversed—the shear force acts upward and the bending moment acts clockwise. However, the shear force still tends to rotate the material clockwise, and the bending moment still tends to compress the upper part of the beam and elongate the lower part.

Therefore, recognize that the algebraic sign of a stress resultant is determined by how it deforms the material on which it acts, rather than by its direction in space. In the case of a beam, *a positive shear force acts clockwise against the material* (Figs. 4-18b and c) *and a negative shear force acts counterclockwise against the material. Also, a positive bending moment compresses the upper part of the beam* (Figs. 4-18b and c) *and a negative bending moment compresses the lower part.*

To make these conventions clear, both positive and negative shear forces and bending moments are shown in Fig. 4-19. The forces and moments are shown acting on an element of a beam cut out between two cross sections that are a small distance apart.

FIGURE 4-19

Sign convention for shear force V and bending moment M

The *deformations* of an element caused by both positive and negative shear forces and bending moments are sketched in Fig. 4-20. Observe that a positive shear force tends to deform the element by causing the right-hand face to move downward with respect to the left-hand face, and, as already mentioned, a positive bending moment compresses the upper part of a beam and elongates the lower part.

FIGURE 4-20

Deformations (highly exaggerated) of a beam element caused by (a) shear forces and (b) bending moments

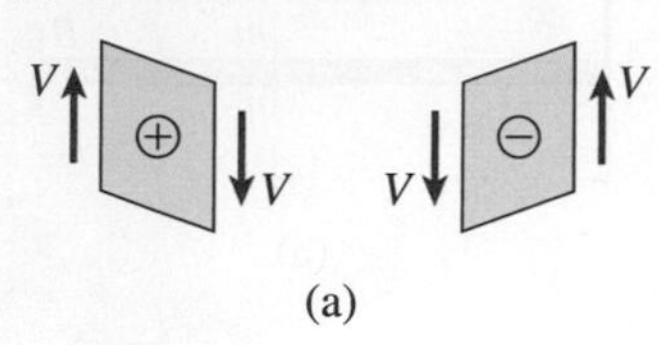

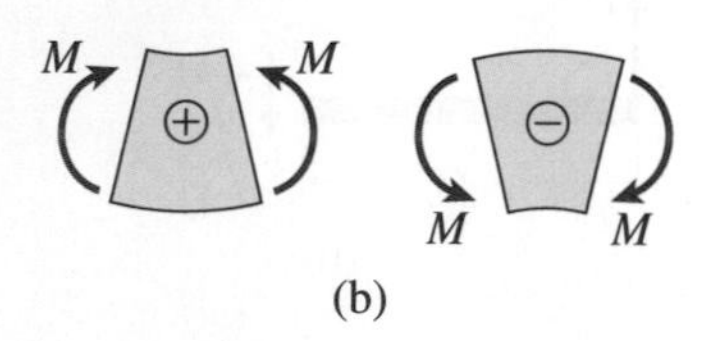

Sign conventions for stress resultants are called **deformation sign conventions** because they are based upon how the material is deformed. For instance, a deformation sign convention was used in dealing with axial forces in a bar. An axial force producing elongation (or tension) in a bar is positive, and an axial force producing shortening (or compression) is negative. Thus, the sign of an axial force depends upon how it deforms the material, not upon its direction in space.

By contrast, when writing equations of equilibrium, use **static sign conventions** in which forces are positive or negative according to their directions along the coordinate axes. For instance, if summing forces in the y direction, forces acting in the positive direction of the y axis are taken as positive and forces acting in the negative direction are taken as negative.

As an example, consider Fig. 4-18b, which is a left hand free-body diagram of part of the cantilever beam. Suppose that forces are summed in the vertical direction and that the y axis is positive upward. Then the load P is given a positive sign in the equation of equilibrium because it acts upward. However, the shear force V (which is a *positive* shear force) is given a negative sign because it acts downward (that is, in the negative direction of the y axis). This example shows the distinction between the deformation sign convention used for the shear force and the static sign convention used in the equation of equilibrium.

The following examples illustrate the techniques for handling sign conventions and determining shear forces and bending moments in beams. The general procedure consists of constructing free-body diagrams and solving equations of equilibrium.

Example 4-4

A simple beam AB supports two loads, a force P and a couple M_0, acting as shown in Fig. 4-21a. Find the shear force V and bending moment M in the beam at cross sections located: (a) a small distance to the left of the midpoint of the beam and (b) a small distance to the right of the midpoint of the beam.

FIGURE 4-21

(a) Simple beam with loads P and M_0 and (b) free-body diagram of simple beam

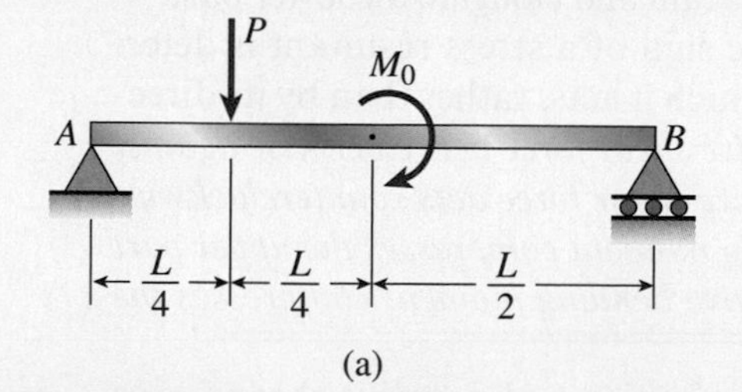

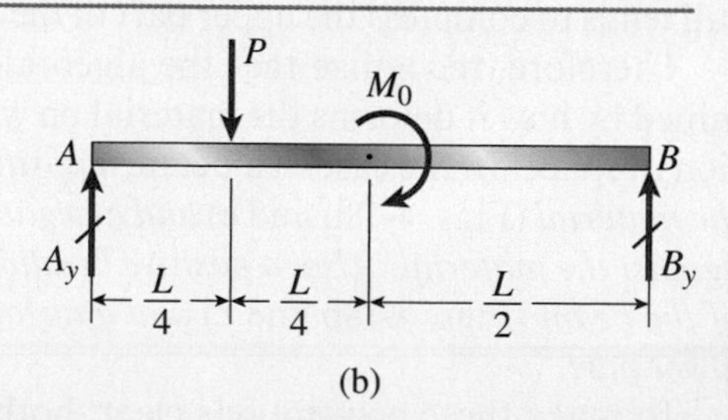

Solution:

Use a four-step problem-solving approach.

1. ***Conceptualize***: Draw the free-body diagram of the entire beam (Fig. 4-21b). Horizontal reaction component A_x is zero because only vertical loading is applied to the beam. Make a section cut through the beam just left of and also just right of mid-span where moment M_0 is applied. The resulting two free-body diagrams are shown in Figs. 4-22a and b.

FIGURE 4-22

(a) Left-hand free-body diagram—Case (a); (b) left-hand free-body diagram—Case (b)

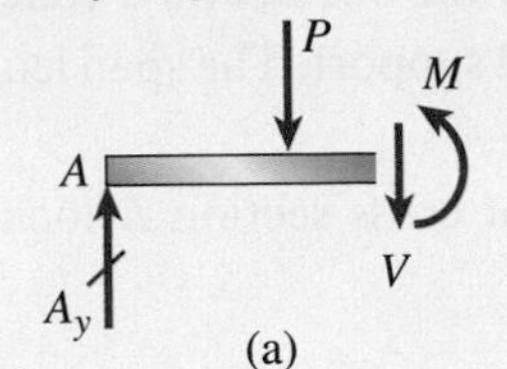

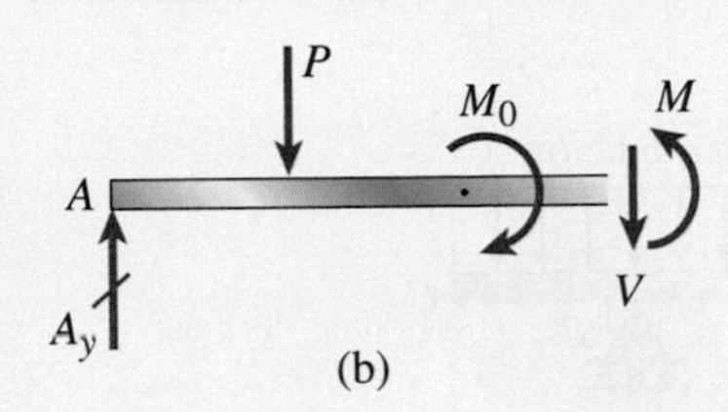

2. ***Categorize***: The beam in Fig. 4-21 is statically determinate. Sum moments about B to find expressions for reaction force A_y in terms of known loads P and M_0. Use the free-body diagrams in Fig. 4-22 to find shear force V and moment M at the two locations of interest.

3. ***Analyze***:

Reaction force A_y: Sum moments about B and use a statics sign convention to find

$$\Sigma M_B = 0 \quad A_y = \frac{1}{L}\left(P\left(\frac{3L}{4}\right) - M_0\right) = \frac{3}{4}P - \frac{M_0}{L} \tag{a}$$

Only reaction force A_y is needed here for the left-hand free-body diagrams in Fig. 4-22. If a right-hand free-body diagram is selected instead, reaction force B_y is needed and is computed as

$$\Sigma F_y = 0 \qquad A_y + B_y = P \quad \text{so} \quad B_y = P - A_y = \frac{1}{4}P + \frac{M_0}{L} \tag{b}$$

Shear force and bending moment just left of mid-span: Use the free-body diagram in Fig. 4-22a. Shear force V and moment M are shown in their assumed positive directions. Sum forces and moments at the cut section to find V and M as

$$\Sigma F_y = 0 \quad V = A_y - P = -\frac{P}{4} - \frac{M_0}{L}$$

$$\Sigma M = 0 \quad M = A_y\left(\frac{L}{2}\right) - P\left(\frac{L}{4}\right) = \frac{PL}{8} - \frac{M_0}{2} \tag{c}$$

Shear force V is negative, so it acts upward—opposite to that assumed in Fig. 4-22a. Bending moment M may be either positive or negative, depending on the relative magnitudes of loads P and M_0.

Shear force and bending moment just right of mid-span: Repeat the analysis using the free-body diagram in Fig. 4-22b, which now includes applied moment M_0:

$$\Sigma F_y = 0 \quad V = A_y - P = -\frac{P}{4} - \frac{M_0}{L}$$

$$\Sigma M = 0 \quad M = A_y\left(\frac{L}{2}\right) - P\left(\frac{L}{4}\right) + M_0 = \frac{PL}{8} + \frac{M_0}{2} \tag{d}$$

4. ***Finalize***: These results show that, when the cut section is shifted from the left to the right of the couple M_0, the shear force does not change (because the vertical forces acting on the free body do not change) but the bending moment increases algebraically by an amount equal to M_0 [compare the moment expressions in Eqs. (c) and (d)].

If the right-hand free-body diagrams are used (instead of the left-hand free-body diagrams in Fig. 4-22) to find shear force V and moment M near the mid-span, the same expressions as those in Eqs. (c) and (d) will be obtained.

Example 4-5

A simple beam with an overhang is supported at points A and B (Fig. 4-23). A uniform load of intensity $q = 6$ kN/m acts throughout the length of the beam, and a concentrated load $P = 28$ kN acts at a point 3 m from the left-hand support. The span length is 8 m and the length of the overhang is 2 m.

Calculate the shear force V and bending moment M at cross section D located 5 m from the left-hand support

FIGURE 4-23

Beam with an overhang and uniform and concentrated loads

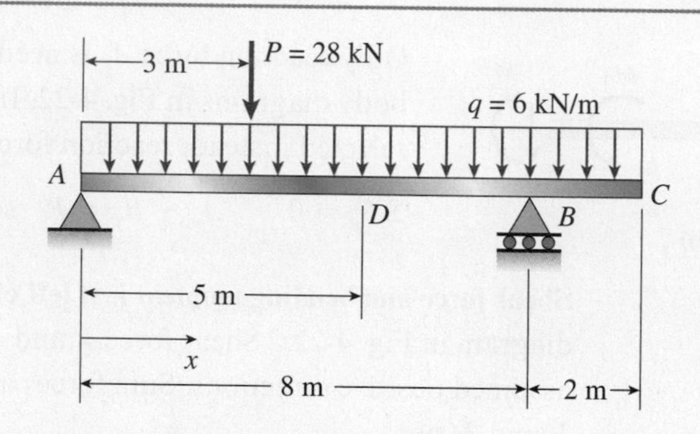

Solution:

Find external reactions at supports; then find the internal shear force and bending moment at point D. Use a four-step problem-solving approach.

1. ***Conceptualize***: Find the reaction forces R_A and R_B using the free-body diagram of the overall structure shown in Fig. 4-24. The horizontal reaction component at the pin at A is zero because only vertical loading is applied to the beam. Obtain internal shear force V_D and bending moment M_D at point D (Fig. 4-25) by cutting the beam at D. Either the left-hand or right-hand free-body diagram in Fig. 4-25 may be used to find V_D and M_D.

FIGURE 4-24

Free-body diagram of beam

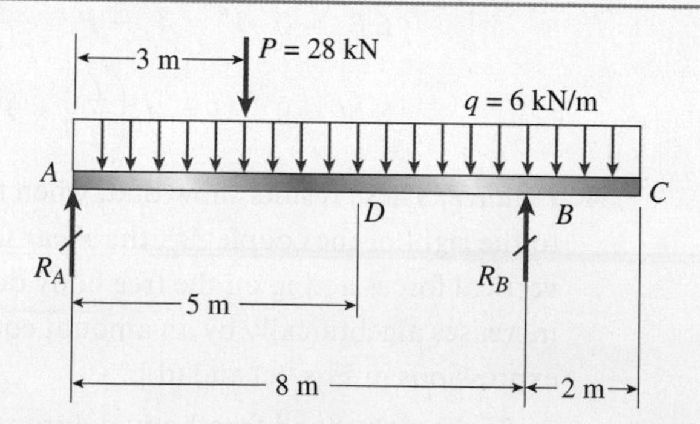

2. ***Categorize***: The free body diagrams in Fig. 4-25 show internal shear force V_D and bending moment M_D in their assumed positive directions based on a deformation sign convention. Start by finding reaction forces R_A and R_B for this statically determinate beam.

FIGURE 4-25

Left- and right-hand free-body diagrams from section cut at D

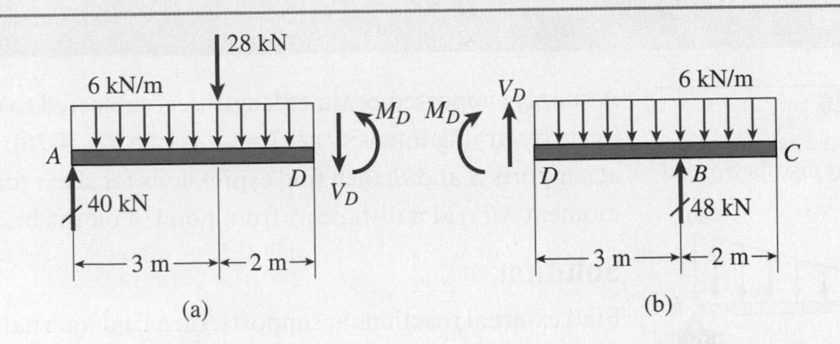

3. ***Analyze***:

Solution for external reactions: Sum forces in the y direction and moments about point A and use a *statics sign convention* to find

$$\Sigma F_y = 0 \quad R_A + R_B = P + qL = 28\text{ kN} + (6\text{ kN/m})(10\text{ m}) = 88\text{ kN}$$
$$\Sigma M_A = 0 \quad R_B(8\text{ m}) = (28\text{ kN})(3\text{ m}) + (6\text{ kN/m})(10\text{ m})(5\text{ m}) = 384\text{ kN}\cdot\text{m} \tag{a}$$

In Eq. (a), the distributed load $q(x)$ has been replaced by its resultant acting at $x = (8 + 2)/2\text{ m} = 5\text{ m}$. Solve Eq. (a) to obtain $R_B = 48\text{ kN}$; then solve for $R_A = 40\text{ kN}$. These reaction forces are shown in Fig. 4-25.

Solution for internal force V and moment M at point D: Use the left-hand free-body diagram in Fig. 4-25a to find

$$\Sigma F_y = 0 \quad V_D = 40\text{ kN} - 28\text{ kN} - (6\text{ kN/m})(3 + 2)\text{ m} = -18\text{ kN }(\uparrow)$$
$$\Sigma M_D = 0 \tag{b}$$
$$M_D = 40\text{ kN}(5\text{ m}) - 28\text{ kN}(2\text{ m}) - (6\text{ kN/m})(5\text{ m})(5\text{ m}/2)$$
$$= 69\text{ kN}\cdot\text{m}$$

4. ***Finalize***: An alternative solution approach is to use the right-hand free-body diagram (Fig. 4-25b) to find shear V_D and moment M_D. First sum vertical forces:

$$V_D = -48\text{ kN} + (6\text{ kN/m})(5\text{ m}) = -18\text{ kN}$$

Next, sum moments about the cut at D:

$$M_D = (48\text{ kN})(3\text{ m}) - (6\text{ kN/m})(5\text{ m})(5\text{ m}/2) = 69\text{ kN}\cdot\text{m}$$

These results agree with those in Eq. (b), so either the left-hand or right-hand free-body diagram can be used in the solution.

Example 4-6

FIGURE 4-26

Simply supported beam with a linearly varying distributed load

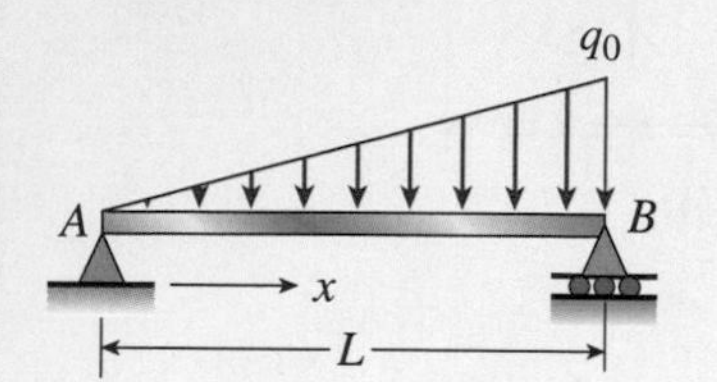

A simply supported beam of length L is subjected to a distributed load of linearly varying intensity $q(x) = \frac{x}{L}q_0$ (see Fig. 4-26). Find reaction forces at supports A and B; then find expressions for shear force $V(x)$ and bending moment $M(x)$ at a distance x from point A on the beam.

Solution:

Find external reactions at supports; then find internal shear force $V(x)$ and bending moment $M(x)$. Use a four-step problem-solving approach.

1. ***Conceptualize***: Find the reaction forces R_A and R_B using the free-body diagram of the overall structure shown in Fig. 4-27. Internal shear force V and bending moment M can be displayed by cutting the beam at some distance x from support A (Fig. 4-28). This cut produces a left-hand and a right-hand free-body diagram, but only the left-hand free-body diagram is shown.

FIGURE 4-27

Free-body diagram of beam

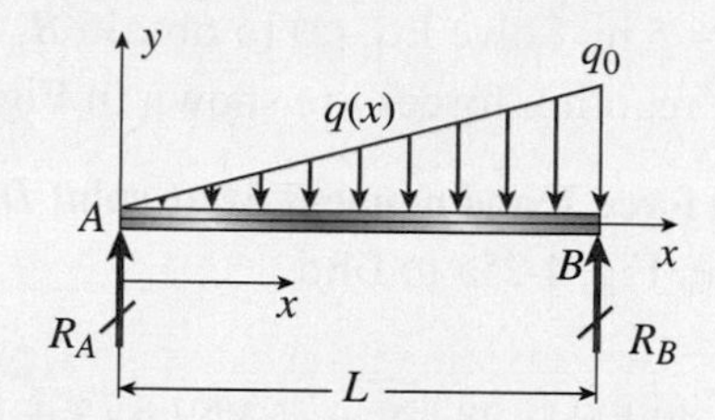

FIGURE 4-28

Left-hand free-body diagram

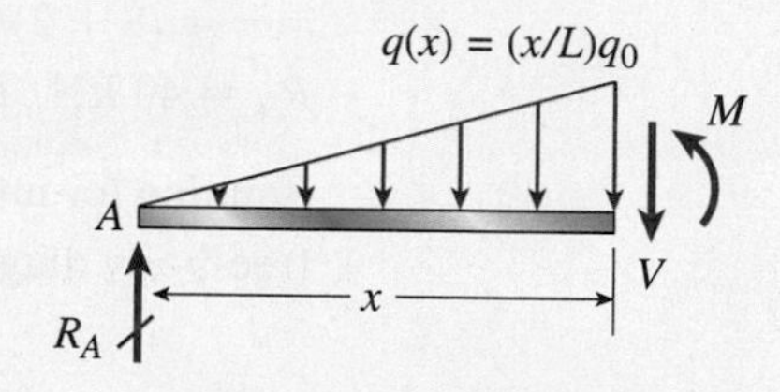

2. ***Categorize***: The left-hand free-body diagram (Fig. 4-28) shows internal shear force V and bending moment M in their assumed positive directions based on a deformation sign convention. Summing forces and then moments at the cut section provides functions $V(x)$ and $M(x)$. First find reaction force R_A for this statically determinate beam.
3. ***Analyze***:

Solution for external reactions: Sum forces in the y direction and moments about point A and use a *statics sign convention* to find

$$\Sigma F_y = 0 \quad R_A + R_B - \frac{1}{2}q_0 L = 0$$

$$\Sigma M_A = 0 \quad R_B L - \frac{1}{2}(q_0 L)\left(\frac{2}{3}L\right) = 0 \tag{a}$$

$$R_B = \frac{1}{3}q_0 L$$

Substituting R_B in the first equation gives

$$R_A = -R_B + \frac{1}{2}q_0 L = \frac{1}{6}q_0 L \tag{b}$$

The reaction at A carries one-third of the applied load and at B carries two-thirds of the load. Note that the x-direction reaction at pin A is zero by inspection because no horizontal load or load component is applied.

Solution for internal force $V(x)$ and moment $M(x)$: Sum forces and moments for the free-body diagram in Fig. 4-28 to get the following expressions, which show the variation in V and M over the length of the beam:

$$\Sigma F_y = 0 \quad V(x) = R_A - \frac{1}{2}\left(\frac{x}{L}q_0\right)(x) = \frac{q_0 L}{6} - \frac{1}{2}\left(\frac{x}{L}q_0\right)(x)$$

$$V(x) = \frac{q_0}{6L}(L^2 - 3x^2) \tag{c}$$

$$\Sigma M = 0 \quad M(x) = R_A(x) - \frac{1}{2}\left(\frac{x}{L}q_0\right)(x)\left(\frac{x}{3}\right) = \frac{q_0 L}{6}(x) - \frac{1}{6}\left(\frac{x^3}{L}q_0\right)$$

$$M(x) = \frac{q_0 x}{6L}(L^2 - x^2) \tag{d}$$

The numerically largest shear is at B, where R_B is twice the value of R_A:

$$V_{\max} = \frac{-q_0 L}{3} \tag{e}$$

It is not readily apparent where along the beam the maximum moment occurs. However, differentiate the expression for $M(x)$, equate it to zero, solve for x, and find the point (x_m) at which a local maxima or minima occurs in the function $M(x)$. Solving for x_m and then substituting x_m into the moment expression gives

$$\frac{d}{dx}(M(x)) = \frac{d}{dx}\left[\frac{q_0 x(L^2 - x^2)}{6L}\right] = \frac{q_0(L^2 - 3x^2)}{6L} = 0$$

This results in

$$x_m = \frac{L}{\sqrt{3}}$$

and so

$$M_{\max} = M(x_m) = \frac{\sqrt{3}}{27}q_0 L^2 \tag{f}$$

4. ***Finalize***: When $x = 0$, $V(0) = R_A$ and $M(0) = 0$ at joint A. When $x = L$, $V(L) = -R_B$ and once again $M(L) = 0$ because no moment is applied at the roller support at joint B.

Note that the expression that results from $\frac{d}{dx}(M(x))$ is the same as that for $V(x)$ in Eq. (c). The functions for internal shear V and moment M will be plotted in the next section to create shear and moment diagrams. The diagrams will show the variation of shear and moment over the length of the beam. Locations and values of maximum shear and moment are readily identified on these diagrams for use in the design of the beam.

4.4 Relationships Among Loads, Shear Forces, and Bending Moments

Now consider some important relationships among loads, shear forces, and bending moments in beams. These relationships are quite useful when investigating the shear forces and bending moments throughout the entire length of a beam, and they are especially helpful when constructing shear-force and bending-moment diagrams (Section 4.5).

As a means of obtaining the relationships, an element of a beam is cut out between two cross sections that are a distance dx apart (Fig. 4-29). The load acting on the top surface of the element may be a distributed load, a concentrated load, or a couple, as shown in Figs. 4-29a, b, and c, respectively. The **sign conventions** for these loads are as follows.

Distributed loads and concentrated loads are positive when they act downward on the beam and negative when they act upward. A couple acting as a load on a beam is positive when it is counterclockwise and negative when it is clockwise.

If other sign conventions are used, changes may occur in the signs of the terms appearing in the equations derived in this section.

The shear forces and bending moments acting on the sides of the element are shown in their positive directions in Fig. 4-20. In general, the shear forces and bending moments vary along the axis of the beam. Therefore, their values on the right-hand face of the element may be different from their values on the left-hand face.

In the case of a distributed load (Fig. 4-29a), the increments in V and M are infinitesimal, so denote them by dV and dM, respectively. The corresponding stress resultants on the right-hand face are $V + dV$ and $M + dM$.

In the case of a concentrated load (Fig. 4-29b) or a couple (Fig. 4-29c), the increments may be finite, so they are denoted V_1 and M_1. The corresponding stress resultants on the right-hand face are $V + V_1$ and $M + M_1$.

For each type of loading, two equations of equilibrium can be written for the element—one equation for the equilibrium of forces in the vertical direction and one for the equilibrium of moments. The first of these equations gives the relationship between the load and the shear force, and the second gives the relationship between the shear force and the bending moment.

FIGURE 4-29
Element of a beam used in deriving the relationships among loads, shear forces, and bending moments (All loads and stress resultants are shown in their positive directions)

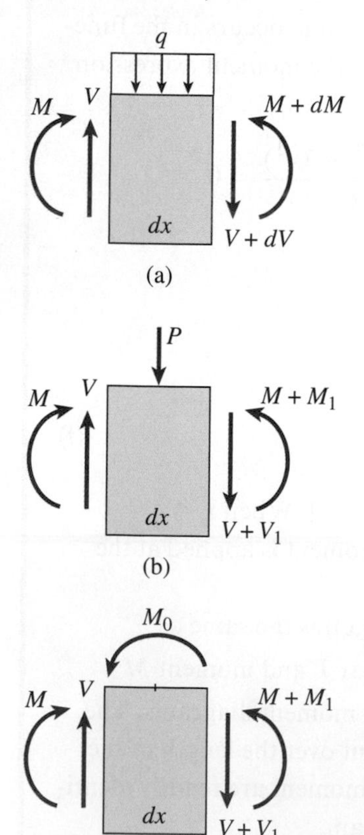

Distributed Loads (Fig. 4-29a)

The first type of loading is a distributed load of intensity q, as shown in Fig. 4-29a. Consider first its relationship to the shear force and second its relationship to the bending moment.

Shear Force Equilibrium of forces in the vertical direction (upward forces are positive) gives

$$\Sigma F_{\text{vert}} = 0 \qquad V - q\,dx - (V + dV) = 0$$

or

$$\frac{dV}{dx} = -q \tag{4-1}$$

From this equation, observe that the rate of change of the shear force at any point on the axis of the beam is equal to the *negative* of the intensity of the distributed load at that same point. (*Note:* If the sign convention for the distributed load is reversed so that q is positive upward instead of downward, the minus sign is omitted in the preceding equation.)

Some useful relations are immediately obvious from Eq. (4-1). For instance, if there is no distributed load on a segment of the beam (that is, if $q = 0$), then $dV/dx = 0$ and the shear force is constant in that part of the beam. Also, if the distributed load is uniform along part of the beam (q = constant), then dV/dx is also constant and the shear force varies linearly in that part of the beam.

As a demonstration of Eq. (4-1), consider the simply supported beam with a linearly varying load discussed in Example 4-6 of the preceding section (see Fig. 4-27). The load on the beam is

$$q = \frac{q_0 x}{L}$$

which is positive because it acts downward. Also, the shear force is (see Eq. c in Example 4-6)

$$V(x) = \frac{q_0}{6L}(L^2 - 3x^2)$$

Taking the derivative dV/dx gives

$$\frac{dV}{dx} = \frac{d}{dx}\left[\frac{q_0}{6L}(L^2 - 3x^2)\right] = -\frac{q_0 x}{L} = -q$$

which agrees with Eq. (4-1).

A useful relationship pertaining to the shear forces at two different cross sections of a beam can be obtained by integrating Eq. (4-1) along the axis of the beam. To obtain this relationship, multiply both sides of Eq. (4-1) by dx and then integrate between any two points A and B on the axis of the beam; thus,

$$\int_A^B dV = -\int_A^B q dx \tag{4-2}$$

where x increases as you move from point A to point B. The left-hand side of this equation equals the difference $(V_B - V_A)$ of the shear forces at B and A. The integral on the right-hand side represents the area of the loading diagram between A and B, which in turn is equal to the magnitude of the resultant of the distributed load acting between points A and B. Thus, Eq. (4-2), gives

$$V_B - V_A = -\int_A^B q dx \tag{4-3}$$
$$= -(\text{area of the loading diagram between } A \text{ and } B)$$

In other words, the change in shear force between two points along the axis of the beam is equal to the negative of the total downward load between those points. The area of the loading diagram may be positive (if q acts downward) or negative (if q acts upward).

Because Eq. (4-1) was derived for an element of the beam subjected *only* to a distributed load (or to no load), Eq. (4-1) cannot be used at a point where a concentrated load is applied (because the *intensity* of load is not defined for a concentrated load). For the same reason, do not use Eq. (4-3) if a concentrated load P acts on the beam between points A and B.

FIGURE 4-29a (Repeated)

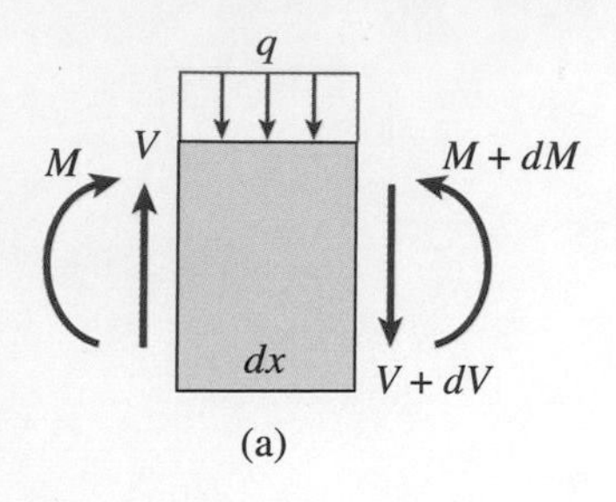

(a)

Bending Moment Now consider the moment equilibrium of the beam element shown in Fig. 4-29a. Summing moments about an axis at the left-hand side of the element (the axis is perpendicular to the plane of the figure) and taking counterclockwise moments as positive gives

$$\Sigma M = 0 \quad -M - qdx\left(\frac{dx}{2}\right) - (V + dV)dx + M + dM = 0$$

Discarding products of differentials (because they are negligible compared to the other terms) results in the relationship:

$$\frac{dM}{dx} = V \tag{4-4}$$

This equation shows that the rate of change of the bending moment at any point on the axis of a beam is equal to the shear force at that same point. For instance, if the shear force is zero in a region of the beam, then the bending moment is constant in that same region.

Equation (4-4) applies only in regions where distributed loads (or no loads) act on the beam. At a point where a concentrated load acts, a sudden change (or discontinuity) in the shear force occurs, and the derivative dM/dx is undefined at that point.

Again using the simply supported beam of Example 4-6, recall that the bending moment is (see Eq. d in Example 4-6)

$$M(x) = \frac{q_0 x}{6L}(L^2 - x^2)$$

Therefore, the derivative dM/dx is

$$\frac{dM}{dx} = \frac{d}{dx}\left[\frac{q_0 x}{6L}\left(L^2 - x^2\right)\right] = \frac{q_0}{6L}(L^2 - 3x^2)$$

which is equal to the shear force in the beam.

Integrating Eq. (4-4) between two points A and B on the beam axis gives

$$\int_A^B dM = \int_A^B V dx \tag{4-5}$$

The integral on the left-hand side of this equation is equal to the difference $(M_B - M_A)$ of the bending moments at points B and A. To interpret the integral on the right-hand side, visualize a shear-force diagram showing the variation of V with x. Then the integral on the right-hand side represents the area below the shear-force diagram between A and B. Therefore, Eq. (4-5) is expressed as

$$\begin{aligned} M_B - M_A &= \int_A^B V dx \\ &= (\text{area of the shear-force diagram between } A \text{ and } B) \end{aligned} \tag{4-6}$$

This equation is valid even when concentrated loads act on the beam between points A and B. However, it is not valid if a couple acts between A and B. A couple produces a sudden change in the bending moment, and the left-hand side of Eq. (4-5) cannot be integrated across such a discontinuity.

Concentrated Loads (Fig. 4-29b)

Now consider a concentrated load P acting on the beam element (Fig. 4-29b). Equilibrium of forces in the vertical direction gives

$$V - P - (V + V_1) = 0 \quad \text{or} \quad V_1 = -P \qquad \textbf{(4-7)}$$

This result means that an abrupt change in the shear force occurs at any point where a concentrated load acts. Passing from left to right through the point of the load application, the shear force decreases by an amount equal to the magnitude of the downward load P.

Equilibrium of moments about the left-hand face of the element (Fig. 4-29b) gives

$$-M - P\left(\frac{dx}{2}\right) - (V + V_1)dx + M + M_1 = 0$$

or

$$M_1 = P\left(\frac{dx}{2}\right) + Vdx + V_1dx \qquad \textbf{(4-8)}$$

FIGURE 4-29b (Repeated)

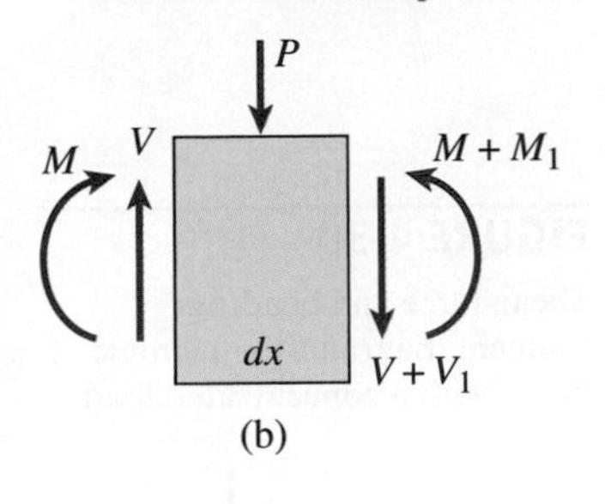

(b)

Since the length dx of the element is infinitesimally small, the increment M_1 in the bending moment is also infinitesimally small. *Thus, the bending moment does not change when passing through the point of application of a concentrated load.*

Even though the bending moment M does not change at a concentrated load, its rate of change dM/dx undergoes an abrupt change. At the left-hand side of the element (Fig. 4-29b), the rate of change of the bending moment [see Eq. (4-4)] is $dM/dx = V$. At the right-hand side, the rate of change is $dM/dx = V + V_1 = V - P$. *Therefore, at the point of application of a concentrated load P, the rate of change dM/dx of the bending moment decreases abruptly by an amount equal to P.*

Loads in the Form of Couples (Fig. 4-29c)

The last case to be considered is a load in the form of a couple M_0 (Fig. 4-29c). Equilibrium of the element in the vertical direction leads to $V_1 = 0$, which shows that *the shear force does not change at the point of application of a couple.*

Equilibrium of moments about the left-hand side of the element gives

$$-M + M_0 - (V + V_1)dx + M + M_1 = 0$$

Disregarding terms that contain differentials (because they are negligible compared to the finite terms) results in

$$M_1 = -M_0 \qquad \textbf{(4-9)}$$

FIGURE 4-29c (Repeated)

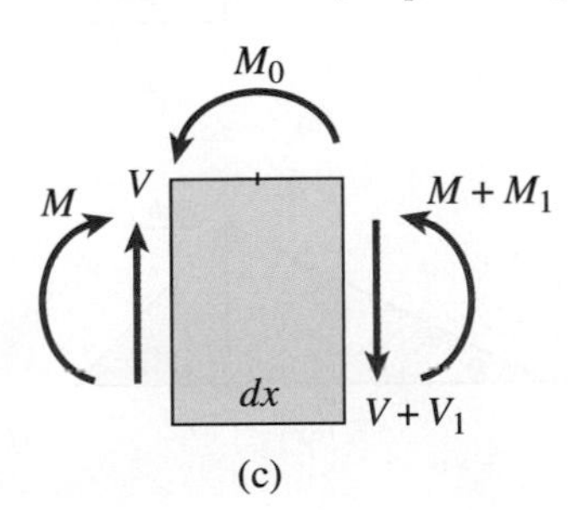

(c)

This equation shows that the bending moment decreases by M_0 when moving from left to right through the point of load application. *Thus, the bending moment changes abruptly at the point of application of a couple.*

Equations (4-1) through (4-9) are useful when making a complete investigation of the shear forces and bending moments in a beam, as discussed in the next section.

4.5 Shear-Force and Bending-Moment Diagrams

When designing a beam, you usually need to know how the shear forces and bending moments vary throughout the length of the beam. Of special importance are the maximum and minimum values of these quantities. Information of this kind is usually provided by graphs in which the shear force and bending moment are plotted as ordinates and the distance x along the axis of the beam is plotted as the abscissa. Such graphs are called **shear-force and bending-moment diagrams.**

To provide a clear understanding of these diagrams, consider how they are constructed and interpreted for three basic loading conditions—a single concentrated load, a uniform load, and several concentrated loads. In addition, Examples 4-7 to 4-13 at the end of the section provide detailed illustration of the techniques for handling various kinds of loads, including the case of a couple acting as a load on a beam.

FIGURE 4-30

Shear-force and bending-moment diagrams for a simple beam with a concentrated load

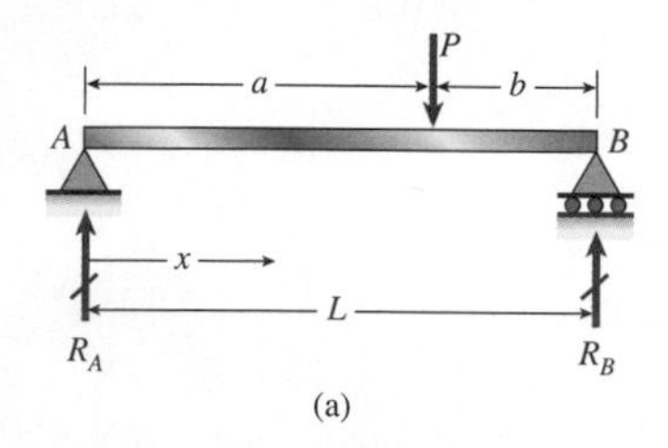

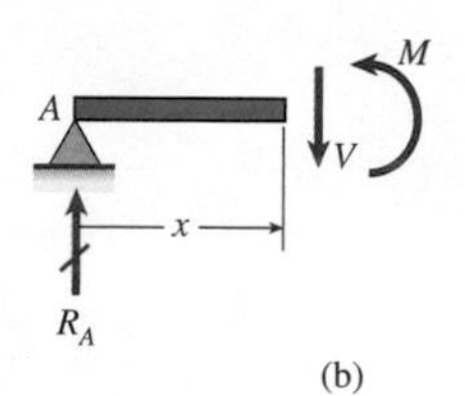

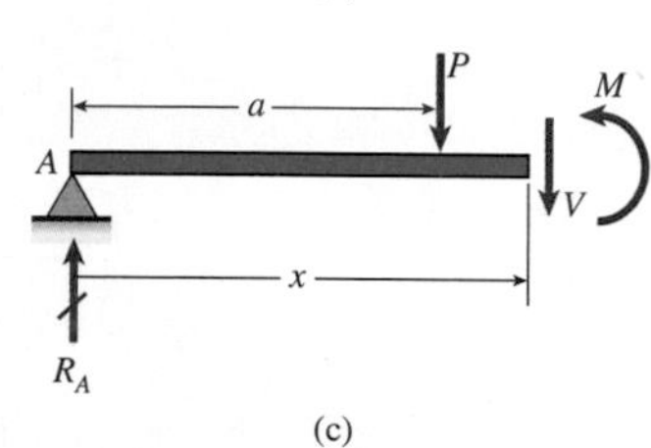

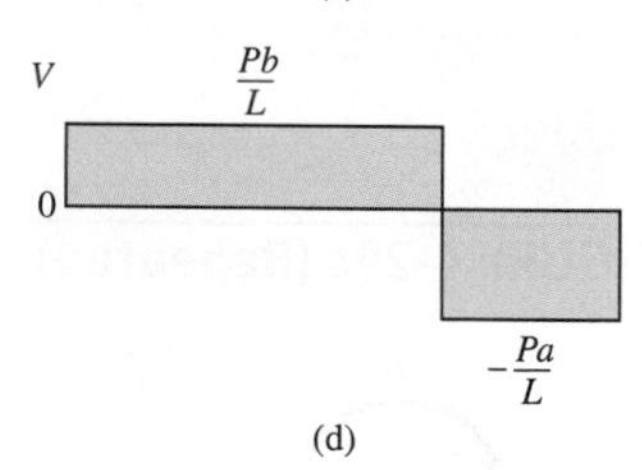

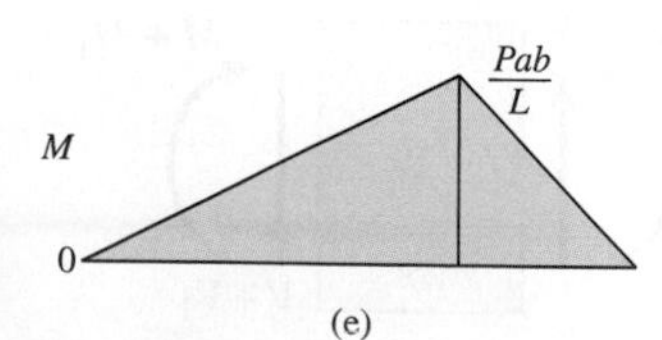

Concentrated Load

Begin with a simple beam AB supporting a concentrated load P (Fig. 4-30a). The load P acts at distance a from the left-hand support and distance b from the right-hand support. Considering the entire beam as a free body, the reactions of the beam from equilibrium are

$$R_A = \frac{Pb}{L} \quad R_B = \frac{Pa}{L} \tag{4-10a, b}$$

Now cut through the beam at a cross section to the left of the load P and at a distance x from the support at A; then draw a free-body diagram of the left-hand part of the beam (Fig. 4-30b). From the equations of equilibrium for this free body, the shear force V and bending moment M at distance x from the support are

$$V = R_A = \frac{Pb}{L} \quad M = R_A x = \frac{Pbx}{L} \quad (0 < x < a) \tag{4-11a, b}$$

These expressions are valid only for the part of the beam to the left of the load P.

Next, cut through the beam to the right of load P (that is, in the region $a < x < L$) and again draw a free-body diagram of the left-hand part of the beam (Fig. 4-30c). From the equations of equilibrium for this free-body, the expressions for the shear force and bending moment are

$$V = R_A - P = \frac{Pb}{L} - P = -\frac{Pa}{L} \quad (a < x < L) \tag{4-12a}$$

and

$$\begin{aligned} M &= R_A x - P(x - a) = \frac{Pbx}{L} - P(x - a) \\ &= \frac{Pa}{L}(L - x) \quad (a < x < L) \end{aligned} \tag{4-12b}$$

Note that these equations are valid only for the right-hand part of the beam.

The equations for the shear forces and bending moments [Eqs. (4-11) and (4-12)] are plotted in Fig. 4-30. Figure 4-30d is the *shear-force diagram*, and Fig. 4-30e is the *bending-moment diagram*.

From the shear-force diagram, observe that the shear force at end A of the beam ($x = 0$) is equal to the reaction R_A. Then it remains constant to the point of application of the load P. At that point, the shear force decreases abruptly by an amount equal to the load P. In the right-hand part of the beam, the shear force is again constant but equal numerically to the reaction at B.

As shown in Fig. 4-30e, the bending moment in the left-hand part of the beam increases linearly from zero at the support to Pab/L at the concentrated load ($x = a$). In the right-hand part, the bending moment is again a linear function of x, varying from Pab/L at $x = a$ to zero at the support ($x = L$). Thus, the maximum bending moment is

$$M_{max} = \frac{Pab}{L} \tag{4-13}$$

and occurs under the concentrated load.

The expressions for the shear force and bending moment to the right of the load P [Eqs. (4-12a and b)] result from equilibrium of the left-hand part of the beam (Fig. 4-30c). This free-body is acted upon by the forces R_A and P in addition to V and M. It is slightly simpler in this particular example to consider the right-hand portion of the beam as a free-body, because then only one force (R_B) appears in the equilibrium equations (in addition to V and M). Of course, the final results are unchanged.

Certain characteristics of the shear-force and bending moment diagrams (Figs. 4-30d and e) may now be seen. Note first that the slope dV/dx of the shear-force diagram is zero in the regions $0 < x < a$ and $a < x < L$, which is in accord with the equation $dV/dx = -q$ [Eq. (4-1)]. Also, in these same regions, the slope dM/dx of the bending moment diagram is equal to V [Eq. (4-4)]. To the left of load P, the slope of the moment diagram is positive and equal to Pb/L; to the right, it is negative and equal to $-Pa/L$. Thus, at the point of application of the load P, there is an abrupt change in the shear-force diagram (equal to the magnitude of load P) and a corresponding change in the slope of the bending-moment diagram.

Now consider the *area* of the shear-force diagram. Moving from $x = 0$ to $x = a$, the area of the shear-force diagram is $(Pb/L)a$, or Pab/L. This quantity represents the increase in bending moment between these same two points [see Eq. (4-6)]. From $x = a$ to $x = L$, the area of the shear-force diagram is $-Pab/L$, which means that in this region the bending moment decreases by that amount. Consequently, the bending moment is zero at end B of the beam, as expected.

If the bending moments at both ends of a beam are zero, as is usually the case with a simple beam, then the area of the shear-force diagram between the ends of the beam must be zero—provided no couples act on the beam [see the discussion in Section 4.4 following Eq. (4-6)].

Recall that the maximum and minimum values of the shear forces and bending moments are needed when designing beams. For a simple beam with a single concentrated load, the maximum shear force occurs at the end of the beam nearest to the concentrated load and the maximum bending moment occurs under the load itself.

FIGURE 4-31

Shear-force and bending-moment diagrams for a simple beam with a uniform load

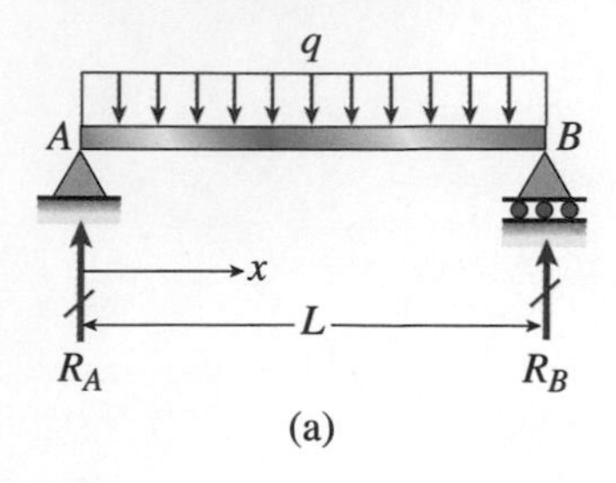

(a)

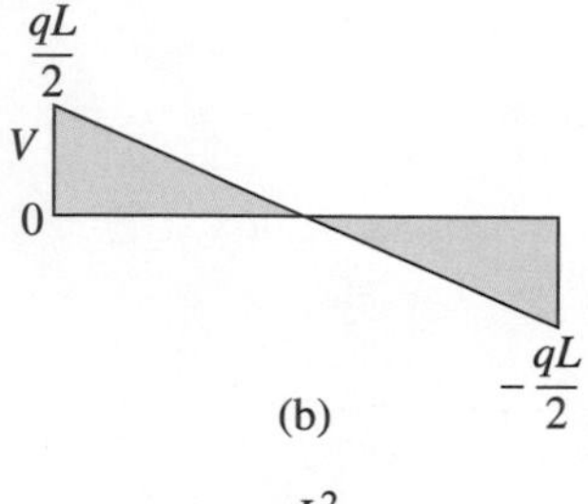

(b)

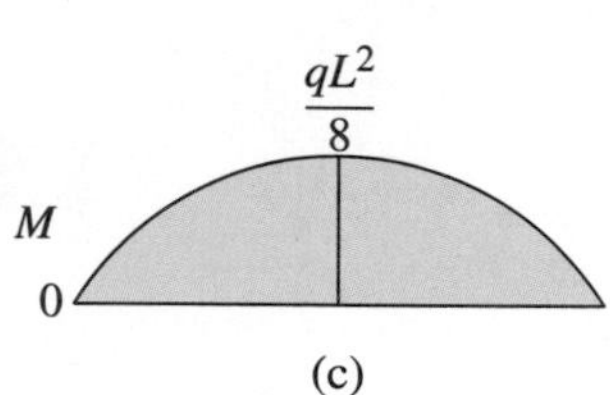

(c)

Uniform Load

A simple beam with a uniformly distributed load of constant intensity q is shown in Fig. 4-31a. Because the beam and its loading are symmetric, each of the reactions (R_A and R_B) is equal to $qL/2$. Therefore, the shear force and bending moment at distance x from the left-hand end are

$$V = R_A - qx = \frac{qL}{2} - qx \tag{4-14a}$$

and

$$M = R_A x - qx\left(\frac{x}{2}\right) = \frac{qLx}{2} - \frac{qx^2}{2} \tag{4-14b}$$

These equations, which are valid throughout the length of the beam, are plotted as shear-force and bending-moment diagrams in Figs. 4-31b and c, respectively.

The shear-force diagram consists of an inclined straight line having ordinates at $x = 0$ and $x = L$, which are equal numerically to the reactions. The slope of the line is $-q$, as expected from Eq. (4-1). The bending-moment diagram is a parabolic curve that is symmetric about the midpoint of the beam. At each cross section, the slope of the bending-moment diagram is equal to the shear force [see Eq. (4-4)]:

$$\frac{dM}{dx} = \frac{d}{dx}\left(\frac{qLx}{2} - \frac{qx^2}{2}\right) = \frac{qL}{2} - qx = V$$

The maximum value of the bending moment occurs at the midpoint of the beam where both dM/dx and the shear force V are equal to zero. Therefore, substitute $x = L/2$ into the expression for M and obtain

$$M_{\text{max}} = \frac{qL^2}{8} \tag{4-15}$$

as shown on the bending-moment diagram.

The diagram of load intensity (Fig. 4-31a) has area qL, and according to Eq. (4-3), the shear force V must decrease by this amount moving along the beam from A to B. This is indeed the case, because the shear force decreases from $qL/2$ to $-qL/2$.

The area of the shear-force diagram between $x = 0$ and $x = L/2$ is $qL^2/8$, and this area represents the increase in the bending moment between those same two points [Eq. (4-6)]. In a similar manner, the bending moment decreases by $qL^2/8$ in the region from $x = L/2$ to $x = L$.

Several Concentrated Loads

If several concentrated loads act on a simple beam (Fig. 4-32a), expressions for the shear forces and bending moments may be determined for each segment of the beam between the points of load application. Again, using free-body diagrams of the left-hand part of the beam and measuring the distance x from end A results in the following equations for the first segment of the beam:

$$V = R_A \qquad M = R_A x \qquad (0 < x < a_1) \tag{4-16a,b}$$

For the second segment, the shear and moment are

$$V = R_A - P_1 \quad M = R_A x - P_1(x - a_1) \quad (a_1 < x < a_2) \qquad \textbf{(4-17a,b)}$$

For the third segment of the beam, it is advantageous to consider the right-hand part of the beam rather than the left, because fewer loads act on the corresponding free body. Here shear and moment are

$$V = -R_B + P_3 \qquad \textbf{(4-18a)}$$

$$M = R_B(L - x) - P_3(L - b_3 - x) \quad (a_2 < x < a_3) \qquad \textbf{(4-18b)}$$

Finally, for the fourth segment of the beam

$$V = -R_B \quad M = R_B(L - x) \quad (a_3 < x < L) \qquad \textbf{(4-19a,b)}$$

Equations (4-16) through (4-19) can be used to construct the shear-force and bending-moment diagrams (Figs. 4-32b and c).

From the shear-force diagram, note that the shear force is constant in each segment of the beam and changes abruptly at every load point with the amount of each change being equal to the load. Also, the bending moment in each segment is a linear function of x; therefore, the corresponding part of the bending-moment diagram is an inclined straight line. To assist in drawing these lines, obtain the bending moments under the concentrated loads by substituting $x = a_1$, $x = a_2$, and $x = a_3$ into Eqs. (4-16b), (4-17b), and (4-18b), respectively. In this manner, the bending moments are

$$M_1 = R_A a_1 \quad M_2 = R_A a_2 - P_1(a_2 - a_1) \quad M_3 = R_B b_3 \qquad \textbf{(4-20a,b,c)}$$

Knowing these values, the bending-moment diagram is constructed by connecting the points with straight lines.

At each discontinuity in the shear force, there is a corresponding change in the slope dM/dx of the bending-moment diagram. Also, the change in bending moment between two load points equals the area of the shear-force diagram between those same two points [see Eq. (4-6)]. For example, the change in bending moment between loads P_1 and P_2 is $M_2 - M_1$. Substituting from Eqs. (4-20a and b) gives

$$M_2 - M_1 = (R_A - P_1)(a_2 - a_1)$$

which is the area of the rectangular shear-force diagram between $x = a_1$ and $x = a_2$.

The maximum bending moment in a beam having only concentrated loads *must* occur under one of the loads or at a reaction. To show this, recall that the slope of the bending-moment diagram is equal to the shear force. Therefore, whenever the bending moment has a maximum or minimum value, the derivative dM/dx (and hence the shear force) must change sign. However, in a beam with only concentrated loads, the shear force can change sign only under a load.

If, in moving along the x axis, the shear force changes from positive to negative (as in Fig. 4-32b), then the slope in the bending moment diagram also changes from positive to negative. Therefore, there must be a maximum bending moment at this cross section. Conversely, a change in shear force from a negative to a positive value indicates a minimum bending moment. Theoretically, the shear-force diagram can intersect the horizontal axis at several

FIGURE 4-32

Shear-force and bending-moment diagrams for a simple beam with several concentrated loads

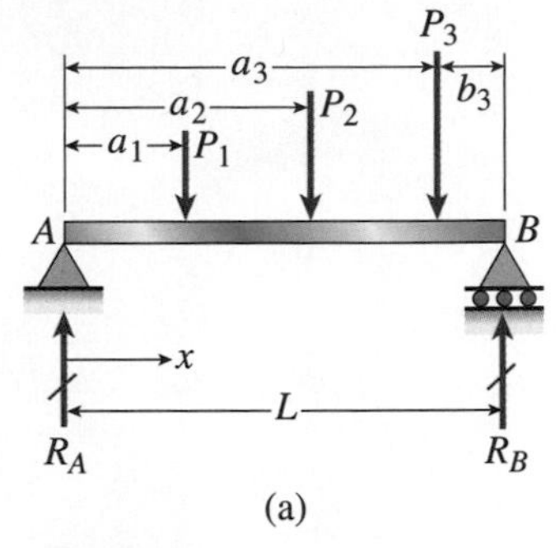

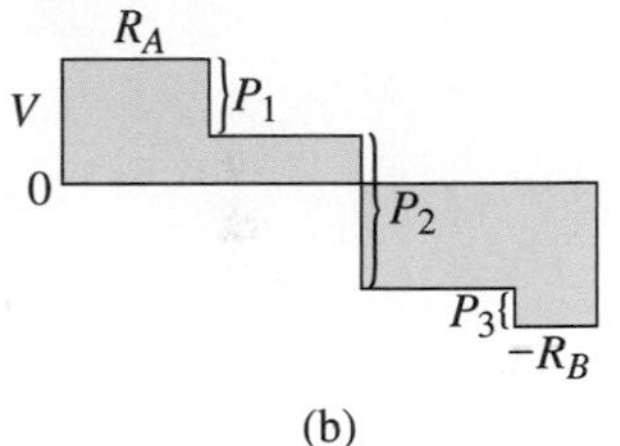

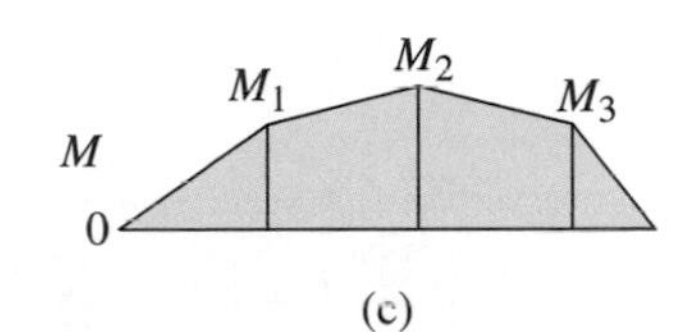

points, although this is quite unlikely. Corresponding to each such intersection point, there is a local maximum or minimum in the bending-moment diagram. The values of all local maximums and minimums must be determined in order to find the maximum positive and negative bending moments in a beam.

General Comments

The terms "maximum" and "minimum" are frequently used with their common meanings of "largest" and "smallest." Consequently, refer to "the maximum bending moment in a beam" regardless of whether the bending-moment diagram is described by a smooth, continuous function (as in Fig. 4-31c) or by a series of lines (as in Fig. 4-32c).

Furthermore, there is often need to distinguish between positive and negative quantities. Therefore, use expressions such as "maximum positive moment" and "maximum negative moment." In both of these cases, the expression refers to the numerically largest quantity, that is, the term "maximum negative moment" really means "numerically largest negative moment." Analogous comments apply to other beam quantities, such as shear forces and deflections.

The maximum positive and negative bending moments in a beam may occur at the following places: (1) a cross section where a concentrated load is applied and the shear force changes sign (see Figs. 4-30 and 4-32), (2) a cross section where the shear force equals zero (see Fig. 4-31), (3) a point of support where a vertical reaction is present, and (4) a cross section where a couple is applied. The preceding discussions and the following examples illustrate all of these possibilities.

When several loads act on a beam, the shear-force and bending-moment diagrams can be obtained by superposition (or summation) of the diagrams obtained for each of the loads acting separately. For instance, the shear-force diagram of Fig. 4-32b is actually the sum of three separate diagrams, each of the type shown in Fig. 4-30d for a single concentrated load. The same is true for the bending-moment diagram of Fig. 4-32c. Superposition of shear-force and bending-moment diagrams is permissible because shear forces and bending moments in statically determinate beams are linear functions of the applied loads.

Computer programs are readily available for drawing shear-force and bending-moment diagrams. After you have developed an understanding of the nature of the diagrams by constructing them manually, you should feel secure in using computer programs to plot the diagrams and obtain numerical results. For convenient reference, the differential relationships used in drawing shear-force and bending-moment diagrams are summarized in the Chapter 4 Summary & Review following Example 4-13.

Example 4-7

Draw the shear-force and bending-moment diagrams for a simple beam with a uniform load of intensity q acting over part of the span (Fig. 4-33a).

Solution:

Use a four-step problem-solving approach. Combine steps as needed for an efficient solution.

FIGURE 4-33

Example 4-7: Simple beam with a uniform load over part of the span

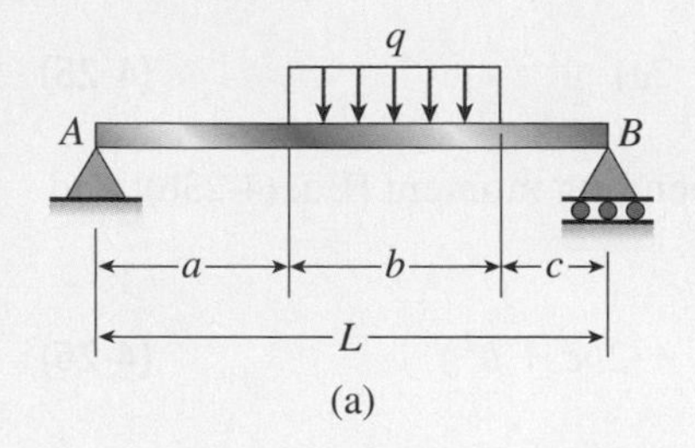

(a)

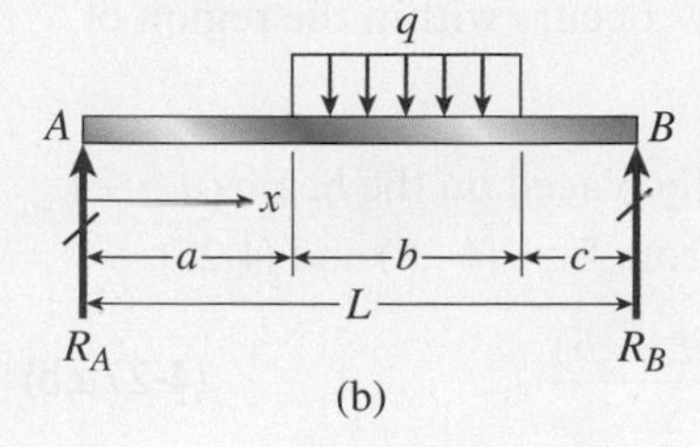

(b)

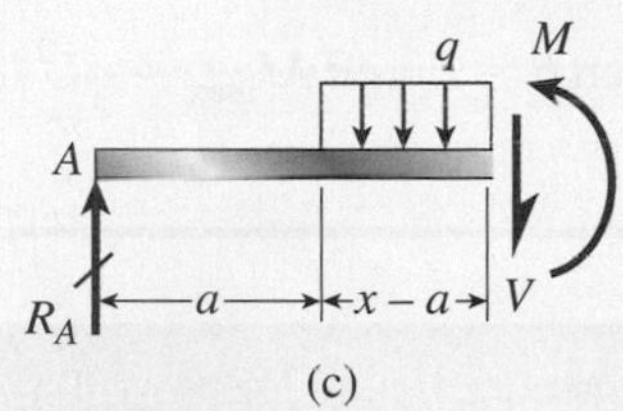

(c)

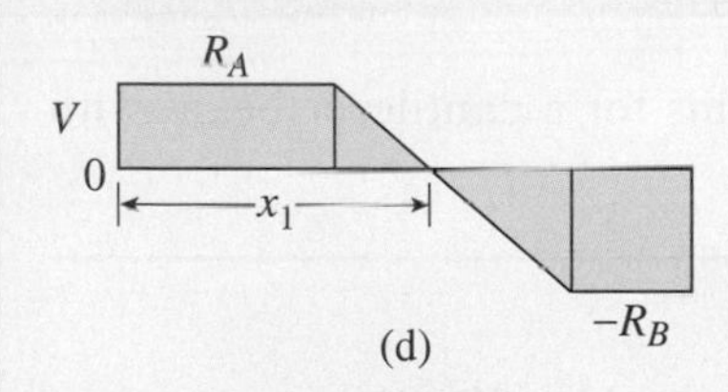

(d)

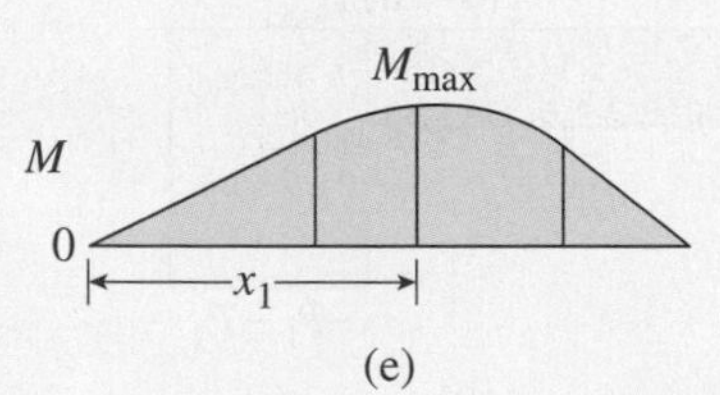

(e)

1,2. ***Conceptualize, Categorize*****:** Begin the analysis by determining the reactions of the beam from a free-body diagram of the entire beam (Fig. 4-33b). The results are

$$R_A = \frac{qb(b + 2c)}{2L} \qquad R_B = \frac{qb(b + 2a)}{2L} \tag{4-21a,b}$$

3. ***Analyze*****:**

Shear forces and bending moments: To obtain the shear forces and bending moments for the entire beam, consider the three segments of the beam individually. For each segment, cut through the beam to expose the shear force V and bending moment M. Then draw a free-body diagram containing V and M as unknown quantities. For example, the left-hand free-body diagram for the middle segment of the beam is shown in Fig. 4-33c. Lastly, sum forces in the vertical direction to obtain the shear force and take moments about the cut section to obtain the bending moment. The results for all three segments are

$$V = R_A \quad M = R_A x \quad (0 < x < a) \tag{4-22a,b}$$

$$V = R_A - q(x - a) \quad M = R_A x - \frac{q(x - a)^2}{2} \quad (a < x < a + b) \tag{4-23a,b}$$

$$V = -R_B \quad M = R_B(L - x) \quad (a + b < x < L) \tag{4-24a,b}$$

These equations give the shear force and bending moment at every cross section of the beam and are expressed in terms of a deformation sign convention. As a partial check on these results, apply Eq. (4-1) to the shear forces and Eq. (4-4) to the bending moments and verify that the equations are satisfied.

Now construct the shear-force and bending-moment diagrams (Figs. 4-33d and e) from Eqs. (4-22) through (4-24). The shear-force diagram consists of horizontal straight lines in the unloaded regions of the beam and an inclined straight line with a negative slope in the loaded region, as expected from the equation $dV/dx = -q$ (Eq. (4-1)).

The bending-moment diagram consists of two inclined straight lines in the unloaded portions of the beam and a parabolic curve in the loaded portion. The inclined lines have slopes equal to R_A and $-R_B$, respectively, as expected from the equation $dM/dx = V$ (Eq. (4-4)). Also, each of these inclined lines is tangent to the parabolic curve at the point where it meets the curve. Note that, with a deformation sign convention, the bending-moment diagram is plotted on the compression side of the beam. So, the entire top surface of beam AB is in compression as expected.

Maximum bending moment: The maximum moment occurs where the shear force equals zero. This point can be found by setting the shear force V [from Eq. (4-23a)] equal to zero and solving for the value of x, denoted here by x_1:

$$x_1 = a + \frac{b}{2L}(b + 2c) \tag{4-25}$$

Now substitute x_1 into the expression for the bending moment [Eq. (4-23b)] and solve for the maximum moment:

$$M_{max} = \frac{qb}{8L^2}(b + 2c)(4aL + 2bc + b^2) \tag{4-26}$$

4. ***Finalize***: The maximum bending moment always occurs within the region of the uniform load, as shown by Eq. (4-25).

Special cases: If the uniform load is symmetrically placed on the beam ($a = c$), then the following simplified expressions come from Eqs. (4-25) and (4-26):

$$x_1 = \frac{L}{2} \qquad M_{max} = \frac{qb(2L - b)}{8} \tag{4-27a,b}$$

If the uniform load extends over the entire span, then $b = L$ and $M_{max} = qL^2/8$, which agrees with Fig. 4-31 and Eq. (4-15).

Example 4-8

Draw the shear-force and bending-moment diagrams for a cantilever beam with two concentrated loads (Fig. 4-34a).

FIGURE 4-34

Example 4-8:
(a) Cantilever beam with two concentrated loads
(b) shear diagram
(c) moment diagram

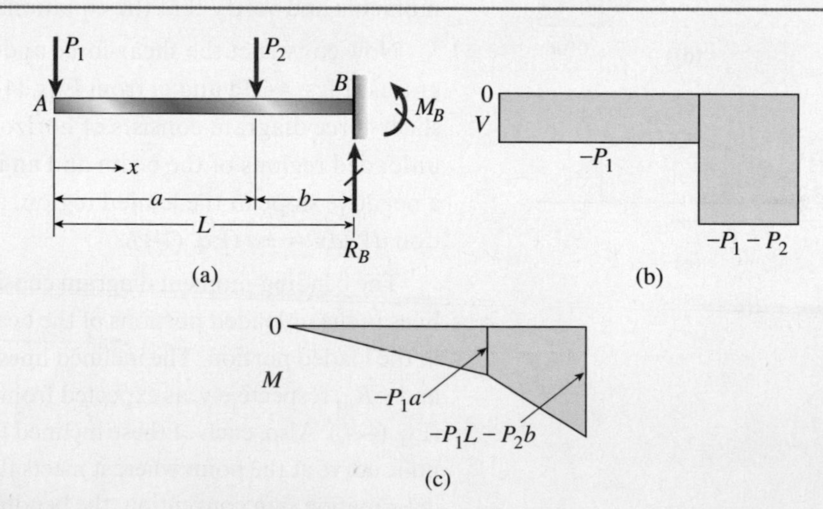

Solution:

Use a four-step problem-solving approach. Combine steps as needed for an efficient solution.

1,2. ***Conceptualize, Categorize:***

Reactions: From a free-body diagram of the entire beam, find the vertical reaction R_B (positive when upward) and the moment reaction M_B (positive when counterclockwise in a statics sign convention):

$$R_B = P_1 + P_2 \quad M_B = -(P_1 L + P_2 b) \tag{4-28a,b}$$

Shear forces and bending moments: Find the shear forces and bending moments by cutting through the beam in each of the two segments, drawing the corresponding free-body diagrams, and solving the equations of equilibrium to obtain

$$V = -P_1 \quad M = -P_1 x \quad (0 < x < a) \tag{4-29a,b}$$

$$V = -P_1 - P_2 \quad M = -P_1 x - P_2(x - a) \quad (a < x < L) \tag{4-30a,b}$$

3. ***Analyze:*** The corresponding shear-force and bending-moment diagrams are shown in Figs. 4-34b and c. The shear force is constant between the loads and reaches its maximum numerical value at the support, where it is equal numerically to the vertical reaction R_B [Eq. (4-28a)].

The bending-moment diagram consists of two inclined straight lines, each having a slope equal to the shear force in the corresponding segment of the beam. The maximum bending moment occurs at the support and is equal numerically to the moment reaction M_B [Eq. (4-28b)]. It is also equal to the area of the entire shear-force diagram, as expected from Eq. (4-6).

4. ***Finalize:*** The shear diagram is constant between applied loads, so the moment diagram must be linear between these same load points. The moment diagram is plotted on the compression side of the beam, so the entire top of this cantilever is in tension as expected.

Example 4-9

The simple beam with an overhang from Example 4-5 is shown in Fig. 4-35. A uniform load of intensity $q = 6$ kN/m acts throughout the length of the beam, and a concentrated load $P = 28$ kN acts at a point 3 m from the left-hand support.

Find the shear force V and bending moment M as functions of x from the left-hand support, then plot the shear and moment diagrams.

Solution:

Use a four-step problem-solving approach. Find internal shear force and bending moment as functions of x. Plot the shear force and bending moment diagrams.

FIGURE 4-35

Beam with an overhang and uniform and concentrated loads

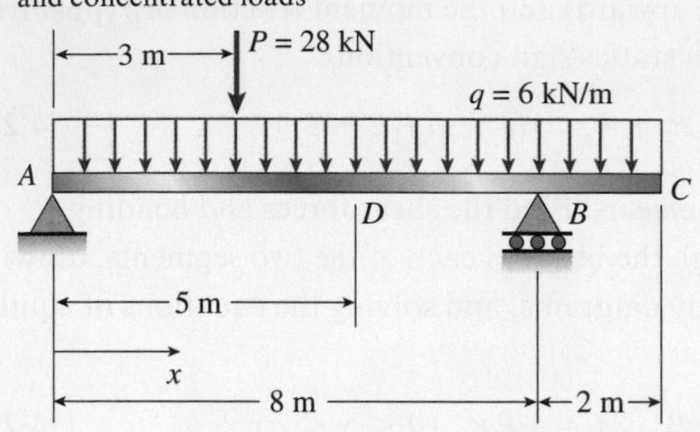

FIGURE 4-36

Free-body diagram of beam

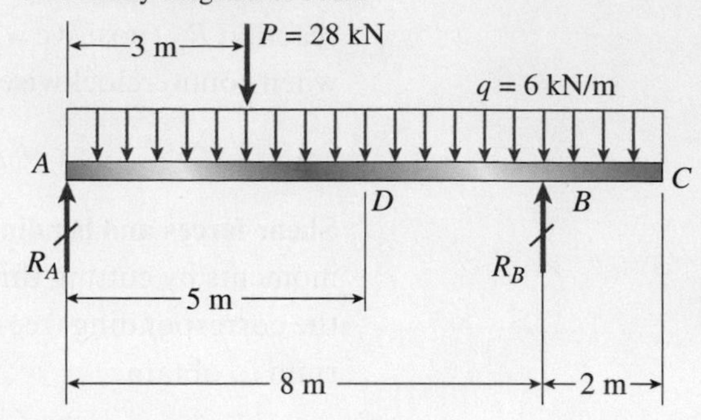

1. *Conceptualize*: Reaction forces R_A and R_B were computed in Example 4-5 using the free-body diagram of the overall structure shown in Fig. 4-36. Then internal shear force V_D and bending moment M_D at point D were obtained by cutting the beam at D. Now the variation in shear V and moment M over the entire length of the beam will be computed and plotted. Shear and moment functions $V(x)$ and $M(x)$ are discontinuous due to concentrated load P and reaction force R_B, so three separate cuts must be made to find $V(x)$ and $M(x)$ over the entire length of the beam (Fig. 4-37).

FIGURE 4-37

Three section cuts required to find functions $V(x)$ and $M(x)$

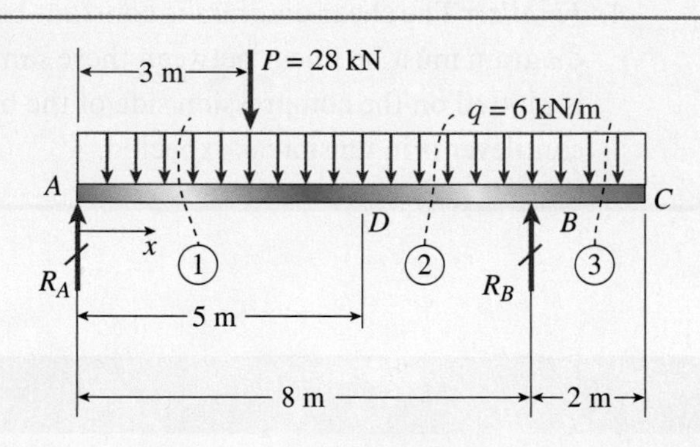

2. *Categorize*: From Example 4-5, sum forces in the y direction and moments about point A to find reaction forces R_A and R_B for this statically determinate beam:

$$\Sigma F_y = 0 \quad R_A + R_B = P + qL = 28\text{ kN} + (6\text{ kN/m})(10\text{ m}) = 88\text{ kN}$$

$$\Sigma M_A = 0 \quad R_B(8\text{ m}) = (28\text{ kN})(3\text{ m}) + (6\text{ kN/m})(10\text{ m})(5\text{ m}) = 384\text{ kN}\cdot\text{m} \tag{a}$$

Solve the second equation to find $R_B = 48$ kN. Then solve for $R_A = 40$ kN from the first equation. Use these reaction forces when summing forces and moments at section cuts in Fig. 4-37 to find functions $V(x)$ and $M(x)$.

3. ***Analyze:***

Solution for internal force $V(x)$ and moment $M(x)$: Cut the beam into sections at locations 1 through 3 (Fig. 4-38) and then use either the left-hand (LHFB) or right-hand (RHFB) free-body diagrams to find $V(x)$ and $M(x)$:

Shear $V(x)$:

Sec. 1 (LHFB) $V(x) = R_A - qx = 40 - 6x$ (kN) $\quad 0 \le x \le 3$ m

Sec. 2 (LHFB) $V(x) = R_A - qx - P = 12 - 6x$ (kN) $\quad 3\text{ m} < x \le 8$ m **(b)**

Sec. 3 (RHFB) $V(x) = q(10\text{ m} - x) = (6\text{ kN/m})(10\text{ m} - x)$ $\quad x \ge 8$ m

FIGURE 4-38

(a) Section 1 left-hand free-body diagram; (b) Section 2 left-hand free-body diagram; and (c) Section 3 right-hand free-body diagram

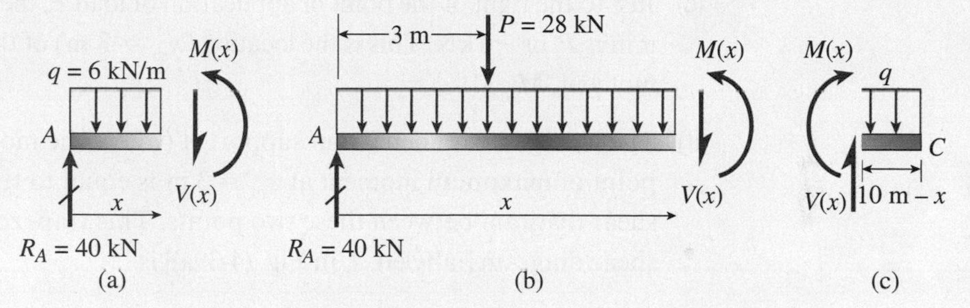

For Section 3, it is easier to find $V(x)$ if the RHFB is used rather than the LHFB. If the LHFB had been used, the expression for $V(x)$ would include R_A and P but would simplify to that shown in Eq. (b) after numerical values were substituted. Next find the bending moment functions.

Moment $M(x)$:

Sec. 1 (LHFB) $M(x) = R_A x - q\dfrac{x^2}{2} = 40x - 3x^2$ (kN · m) $\quad 0 \le x \le 3$ m

Sec. 2 (LHFB) $M(x) = R_A x - q\dfrac{x^2}{2} - P(x - 3\text{ m}) = 12x - 3x^2 + 84$ (kN · m) $\quad 3\text{ m} \le x \le 8$ m **(c)**

Sec. 3 (RHFB) $M(x) = \dfrac{-q}{2}(10\text{ m} - x)^2 = -3(10 - x)^2$ (kN · m) $\quad x \ge 8$ m

The maximum value of the shear force occurs to the right of support A and to the left of support B, and the maximum value of the moment occurs at the point at which $V = 0$. The point of maximum moment (at $x = x_m$) is located by setting the expression for $V(x)$ [Eq. (b)] equal to zero, then solving for $x_m = 3$ m. Solving for $M(x_m)$ [Eq. (c)] gives $M_{max} = 93$ kN · m.

Shear and moment diagrams: One option to plot shear and moment diagrams is to find the functions $V(x)$ and $M(x)$ over the entire length of the beam [Eqs. (b) and (c)] and then use computer software to create the V and M diagram plots. The preferred alternative approach for constructing the shear and moment diagrams relies on graphical interpretation of the differential and integral relationships among load, shear, and moment. The resulting diagrams are presented in Fig. 4-39. Based on this graphical approach, observe the following from these diagrams.

(a) The shear force V a small distance to the right of A is $+R_A$. Moving further to the right of A, the shear force decreases by load intensity q per meter. At distance $x = 3$ m, shear V has dropped from 40 kN to [40 kN − 3 m (6 kN/m)] = 22 kN.

(b) The slope of the tangent to the shear curve is equal to the negative ordinate on the load curve, since $\frac{dV}{dx} = -q(x)$. Load $q(x)$ is constant, so the shear decreases linearly.

(c) Just to the right of the point of application of load P, the shear is 22 kN minus P, or −6 kN. This is the location ($x_m = 3$ m) of the maximum positive moment, M_{max}.

(d) The change in moment from support A (where the moment is zero) to the point of maximum moment at $x_m = 3$ m is equal to the area under the shear diagram between these two points. This trapezoidal area under the shear diagram [labeled A_1 in Fig. (4-39a)] is

$$M_{max} = \frac{1}{2}[40 \text{ kN} + 22 \text{ kN}](x_m) = 93 \text{ kN} \cdot \text{m}$$

(e) The integral relationships between load and shear and between shear and moment mean that, since the load curve $q(x)$ has constant magnitude, the shear diagram $V(x)$ is a linear plot and moment $M(x)$ diagram is a quadratic plot. [This is shown in Eqs. (b) and (c).]

(f) The shear at D is $(-2 \text{ m})(q) = -12$ kN less than that at location $x = x_m$. The moment at D is M_{max} minus the trapezoidal area under the shear curve from $x = x_m$ to $x = 5$ m, or

$$M_D = M_{max} - \frac{1}{2}(6+18)\text{kN (2 m)} = [93 - \frac{1}{2}(24 \text{ kN})(2 \text{ m})] \text{ kN} \cdot \text{m} = 69 \text{ kN} \cdot \text{m}$$

(g) By subtracting the entire trapezoidal area under the shear curve from $x = x_m$ to $x = 8$ m, the moment at B is

$$M_B = M_{max} - \frac{1}{2}(42 \text{ kN})(5 \text{ m}) = (93 - 105) \text{ kN} \cdot \text{m} = -12 \text{ kN} \cdot \text{m}$$

(h) Moving past B on the shear diagram, the shear increases from −36 kN to +12 kN, which is an increase of +48 kN due to reaction force R_B.

(i) The moment is zero just to the left of B at the *inflection point.* Equate the two areas labeled A_1 on the shear diagram [the trapezoidal area A_1 is computed in observation (d) and can be equated to the trapezoidal area labeled $-A_1$] to find that the inflection point is 0.343 m to the left of B. The area under the shear curve from B to C, labeled A_2, is equal to 12 kN · m; adding this to moment M_B takes the moment diagram back to zero at C.

FIGURE 4-39

(a, b) Shear and moment diagrams

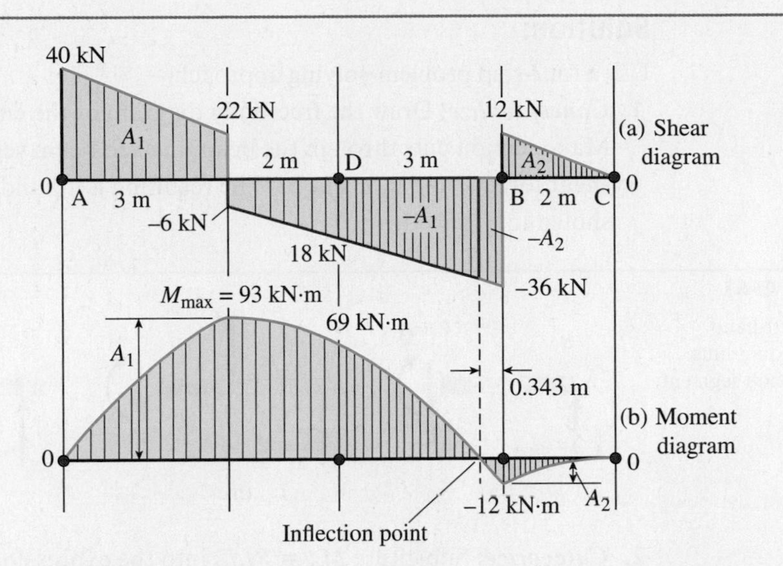

4. ***Finalize***: Graphical interpretation of the differential and integral relationships among load, shear, and moment was used to construct the diagrams shown in Fig. 4-39. The functions $V(x)$ and $M(x)$ [Eqs. (b) and (c)] can be plotted using widely available computer software (such as Maple, Mathcad, Mathematica, MATLAB, etc.) to confirm these plots.

   Note that the moment diagram is plotted on the compression side of the beam. Hence, most of the top of the beam from A to B (i.e., from point A to the inflection point) is in compression, while the top of the beam from the inflection point to C is in tension. This is important information needed in the design of beams made of certain materials.

Example 4-10

Draw shear and moment diagrams for the simple beam AB from Example 4-4. The beam supports two loads: a force P and a couple M_0, which act as shown in Fig. 4-40a. Assume that applied moment $M_0 = PL/2$ in this example.

FIGURE 4-40

(a) Simple beam with loads P and M_0 and (b) free-body diagram of simple beam

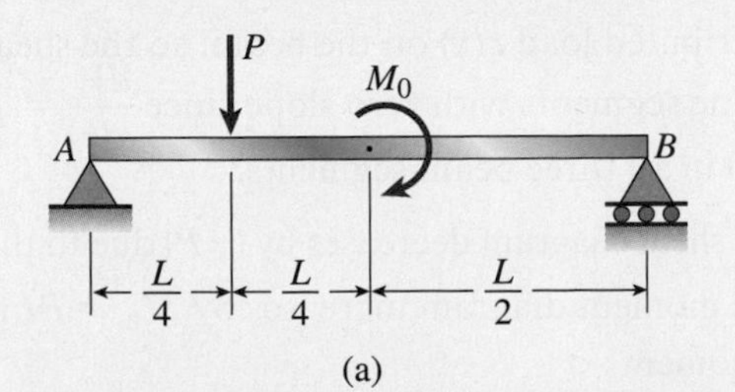

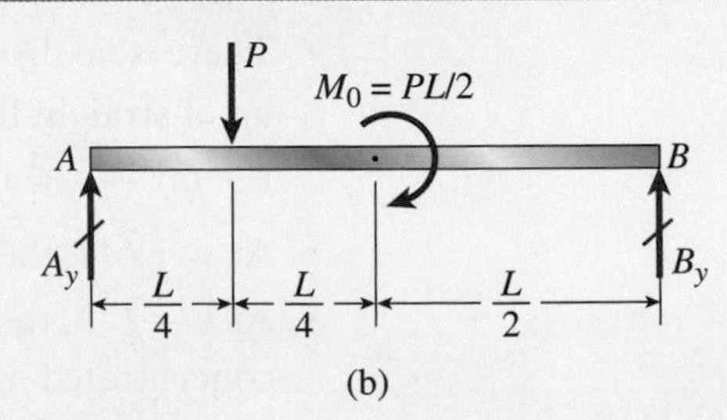

Solution:

Use a four-step problem-solving approach.

1. ***Conceptualize***: Draw the free-body diagram of the entire beam (Fig. 4-40b). Make section cuts through the beam in each beam segment to show internal shear force V and moment M. The resulting left-hand free-body diagrams are shown in Fig. 4-41.

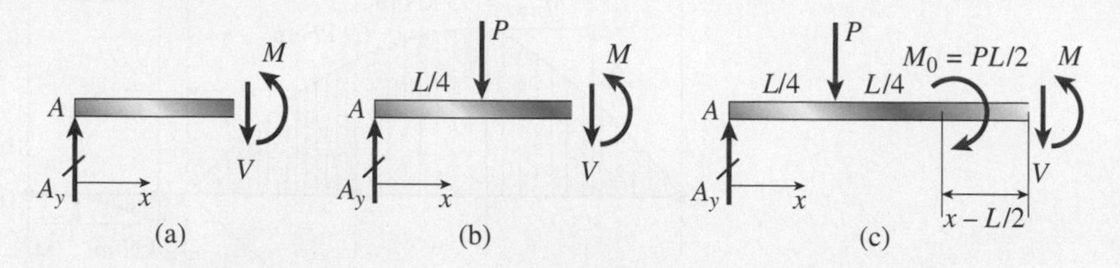

FIGURE 4-41

(a, b, c) Left-hand free-body diagrams for each beam segment

2. ***Categorize***: Substitute $M_0 = PL/2$ into the expressions for reactions A_y and B_y in Example 4-4 to find $A_y = P/4$ and $B_y = 3P/4$. Use the free-body diagrams in Fig. 4-41 to find shear force V and moment M in each of the three segments.

3. ***Analyze***:

Shear and moment functions: Use the left-hand free-body diagrams in Fig. 4-41 to find functions $V(x)$ and $M(x)$:

$$0 \le x \le \frac{L}{4} \quad V(x) = A_y = \frac{P}{4} \qquad M(x) = A_y x = \frac{Px}{4}$$

$$\frac{L}{4} \le x \le \frac{L}{2} \quad V(x) = A_y - P = \frac{-3P}{4} \qquad M(x) = A_y x - P\left(x - \frac{L}{4}\right) = \frac{-3Px}{4} + \frac{PL}{4} \qquad \text{(a)}$$

$$\frac{L}{2} \le x \le L \quad V(x) = A_y - P = \frac{-3P}{4} \qquad M(x) = A_y x - P\left(x - \frac{L}{4}\right) + M_0 = \frac{3P}{4}(L - x)$$

Note that the V and M expressions for segment 3 are more easily obtained from the right-hand free body than from the left-hand free body shown in Fig. 4-41c.

Shear-force and bending-moment diagrams: Use a graphical interpretation of the differential and integral relationships among load, shear, and moment to draw shear and moment diagrams; the resulting diagrams are presented in Fig. 4-42. The diagrams are in agreement with the expressions in Eq. (a). Review these diagrams for the following observations.

- There is no distributed load $q(x)$ on the beam, so the shear diagram is made up of straight line segments with zero slope since $\frac{dV}{dx} = q$. Note that in Eq. (a), $\frac{dV}{dx} = 0$ in all three beam segments.
- At $x = L/4$, the shear diagram decreases by $(-P)$ due to the concentrated load.
- At $x = L/2$, the moment diagram increases by $M_0 = PL/2$ due to the concentrated moment.

- In segment 1, the slope of the moment diagram is $+P/4$, which is the ordinate on the shear diagram, and in segments 2 and 3, the moment diagram has a slope of $(-3P/4)$; this follows from the differential relationship $\frac{dM}{dx} = V$. Each slope can be obtained by differentiating $M(x)$ in Eq. (a).
- Areas A_V on the shear diagram are the corresponding changes in ordinate for each beam segment on the moment diagram. For example, the moment changes from zero at joint A to $A_{V1} = PL/16$ at $x = L/4$.

FIGURE 4-42

(a, b) Shear and moment diagrams

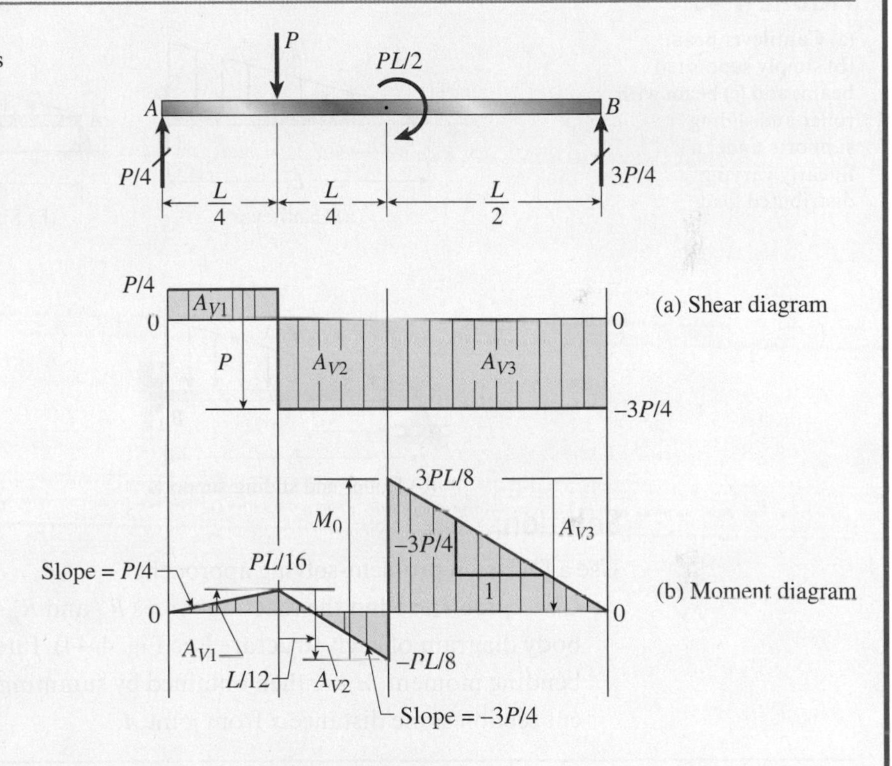

4. ***Finalize***: Confirm that if the *right-hand free-body diagrams* are used to find shear force V and moment M for each segment of the beam (instead of the left-hand free-body diagrams in Fig. 4-41), the same expressions as those in Eq. (a) are obtained.

The shear and moment expressions just left and right of mid-span, as computed from Eq. (a) or the diagrams in Fig. 4-42, are in agreement with the expressions obtained in Example 4-4. Most of the top surface of the beam is in compression because the moment diagram is plotted on the compression side of the beam. There is an *inflection point* a distance $L/12$ to the right of the point of application of the concentrated load P (at $x = L/4 + L/12 = L/3$). Observe from the second moment equation in Eq. (a) that $M(L/3) = 0$, which identifies an inflection point.

Example 4-11

Plot shear and bending moment diagrams for three beams acted on by a distributed load of linearly varying intensity $q(x) = \frac{x}{L} q_0$ (see Fig. 4-43). The three different support cases are (a) cantilever beam, (b) simply supported beam, and (c) beam with a roller support at A and a sliding support at B. Use a graphical interpretation of the differential and integral relationships among load, shear, and moment to construct these diagrams.

FIGURE 4-43

(a) Cantilever beam; (b) simply supported beam; and (c) beam with roller and sliding supports under a linearly varying distributed load

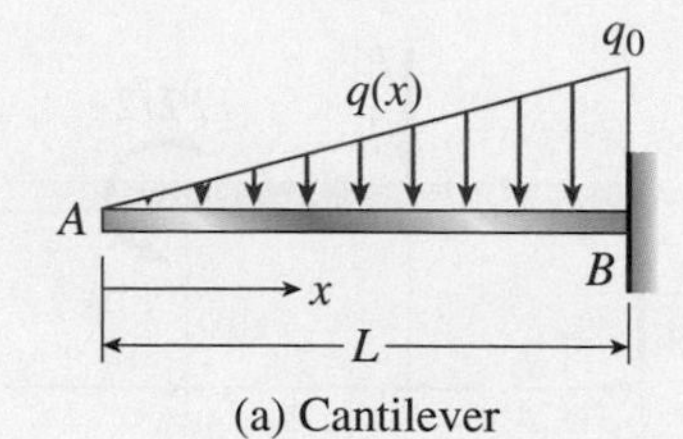

(a) Cantilever

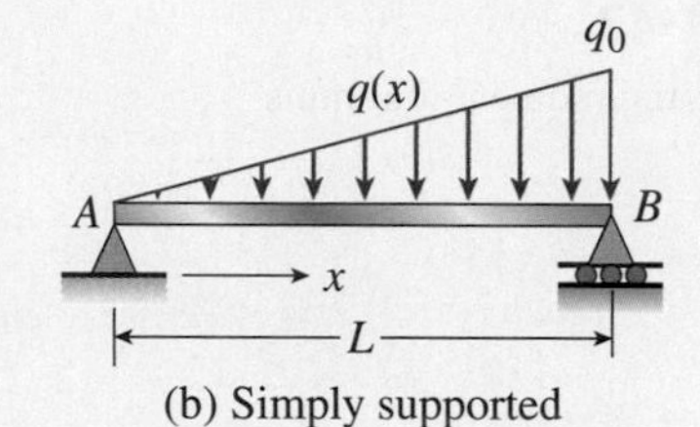

(b) Simply supported

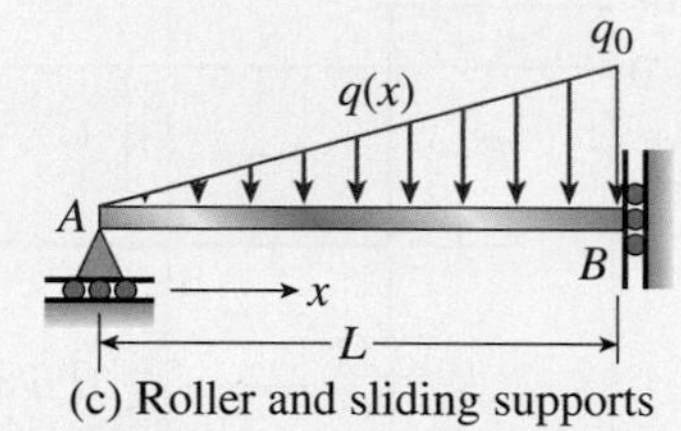

(c) Roller and sliding supports

Solution:

Use a four-step problem-solving approach.

1. ***Conceptualize***: Find the reaction forces R_A and R_B for each beam using a free-body diagram of each structure (see Fig. 4-44). Internal shear force V and bending moment M are then obtained by summing forces and moments at a cut section some distance x from joint A.

FIGURE 4-44

Free-body diagrams of the three beams: (a) cantilever beam; (b) simply supported beam; and (c) beam with roller and sliding supports

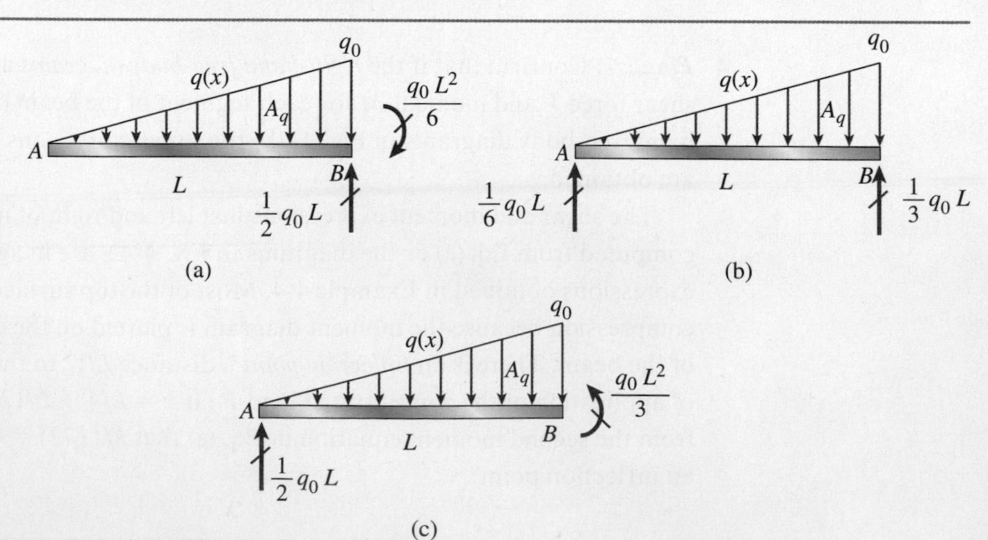

2. *Categorize*: In Example 4-6, shear and moment functions $V(x)$ and $M(x)$ were developed for the simply supported beam in Fig. 4-43b. A left-hand free-body diagram was created by cutting the beam at some distance x from joint A and then summing forces and moments at the cut section to obtain functions $V(x)$ and $M(x)$. The same approach can be used to find functions $V(x)$ and $M(x)$ for the beams in Figs. 4-43a and c. However, a more direct approach is employed here: Use the differential and integral relationships among load, shear, and moment to construct V and M diagrams in a more efficient manner.
3. *Analyze*:

Cantilever beam: Find the shear forces and bending moments by integrating the differential relationships among load, shear force, and bending moment. The shear force V at distance x from the free end at A is obtained from the load by integrating Eq. (4-3) as

$$V - V_A = V - 0 = V = -\int_0^x q(x)dx \tag{a}$$

Integrate over the entire length of the beam to find that the change in shear from A to B is equal to the negative value of the area under the distributed load diagram $(-A_q)$, as shown in Fig. 4-45a. In addition, the slope of a tangent to the shear diagram at any point x is equal to the negative of the corresponding ordinate on the distributed load curve at that same point. Since the load curve is *linear*, the shear diagram is *quadratic*. The bending moment M at distance x from point A is obtained from the shear force by integrating Eq. (4-6):

$$M - M_A = M - 0 = M = \int_0^x V(x)dx \tag{b}$$

Integrate over the entire length of the beam to find that the change in moment from A to B is equal to the value of the area under the shear diagram (A_V), as shown in Fig. 4-45a. Also, the slope of a tangent to the moment diagram at any point x is equal to the value of the corresponding ordinate on the shear-force diagram at that same point. Since the shear-force diagram is *quadratic*, the bending-moment diagram is *cubic*. The maximum values of the shear force and bending moment occur at the fixed support where $x = L$.

Simply supported beam: As with the cantilever beam, the slope of the shear-force diagram at any point x along the beam is equal to negative $q(x)$, and the slope of the bending-moment diagram at any point x is equal to $V(x)$. The maximum value of the shear force occurs at support B where $x = L$, and the maximum value of the moment occurs at the point at which $V = 0$ $(x_m = L/\sqrt{3})$ [see Eq. (c) in Example 4-6]. Now, the maximum moment is obtained from the moment function as $M(x_m)$ [see Eq. (f) in Example 4-6]. In general, for the simply supported beam, the change in shear from A to B is equal to the negative value of the area under the distributed-load diagram $(-A_q)$, as shown in Figure 4-45b, and the change in moment from A to B is equal to the value of the area under the shear-force diagram (A_V).

FIGURE 4-45

Shear and moment diagrams for the three beams: (a) cantilever beam; (b) simply supported beam; and (c) beam with roller and sliding supports

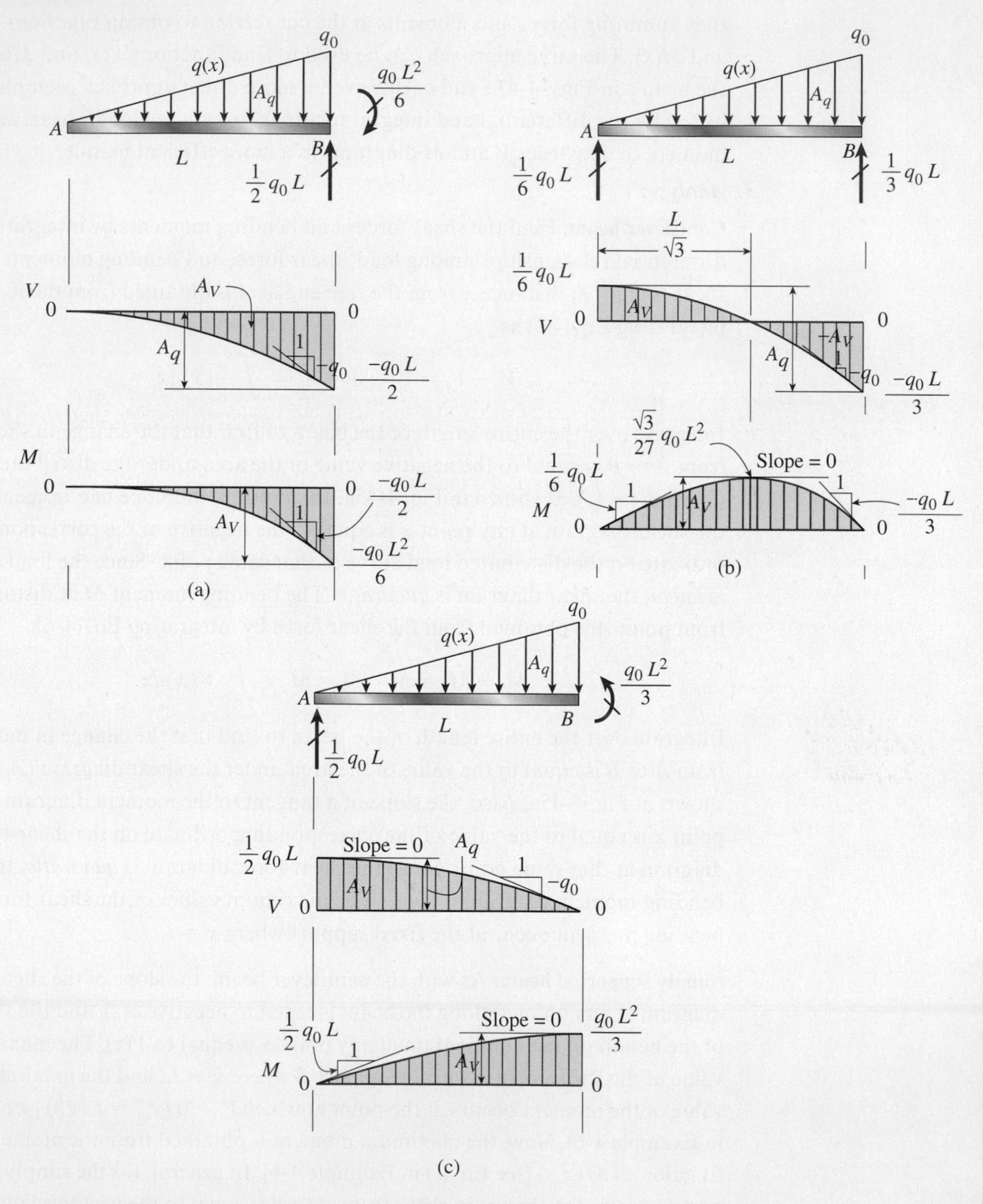

Beam with roller and sliding supports: The entire distributed load $\left(A_q = \frac{1}{2}q_0 L\right)$ is carried by the support at joint A, so the maximum shear is at A (Fig. 4-45c). The shear drop follows a second-degree curve from A_q at joint A to zero at the sliding support at joint B, (a sliding support cannot resist any vertical force). The maximum moment is at B and is equal to $A_V = \frac{1}{3}q_0 L^2$.

4. ***Finalize***: Integrating the differential relationships is straightforward in this example because the loading pattern is linear and continuous and there are no concentrated loads or couples in the regions of integration. If concentrated loads or couples are present, discontinuities in the V and M diagrams will exist, and the differential relationships must be applied separately over each region between concentrated loads.

Example 4-12

The plane frame shown in Fig. 4-46 is loaded by both concentrated and distributed loads. Moment M_A is applied at pin support joint A, and load F_B is applied at joint B. A uniform load with intensity q_1 acts on member BC, and a linearly distributed load with peak intensity q_0 is applied downward on member ED. Find the support reactions at joints A and D; then solve for internal forces at the top of member BC. Plot the axial force, shear, and moment diagrams for all members. Use the numerical properties given below.

FIGURE 4-46

Plane frame with both concentrated and distributed loads

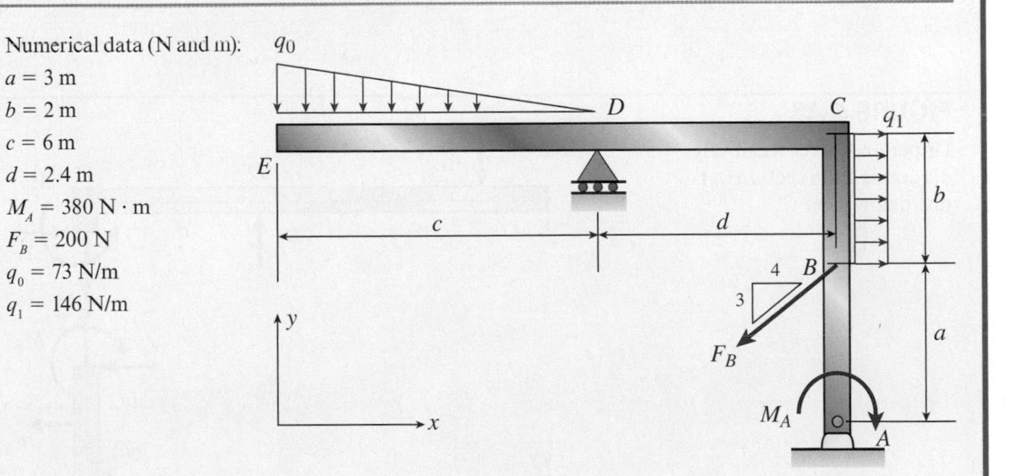

Solution:

Use a four-step problem-solving approach. Solve for external reactions at supports and internal forces and moments in all members. Plot the axial-force, shear-force, and bending-moment diagrams.

1. ***Conceptualize***: Find the reaction forces at supports A and D using the free-body diagram of the overall structure shown in Fig. 4-47. Distributed loads on member ED and BC can be replaced by their static equivalents (F_{q0} and F_{q1}) with each acting at the centroid of its distributed load. The statically equivalent forces are equal in magnitude to the area under the distributed load curves. Internal axial force N_C, shear force V_C, and bending moment M_C just below joint C can be obtained by cutting the frame as shown in Fig. 4-48. Either the upper or lower free-body diagram in Fig. 4-48 may be used to find the internal forces and moment at that location.

FIGURE 4-47

Free-body diagram of frame

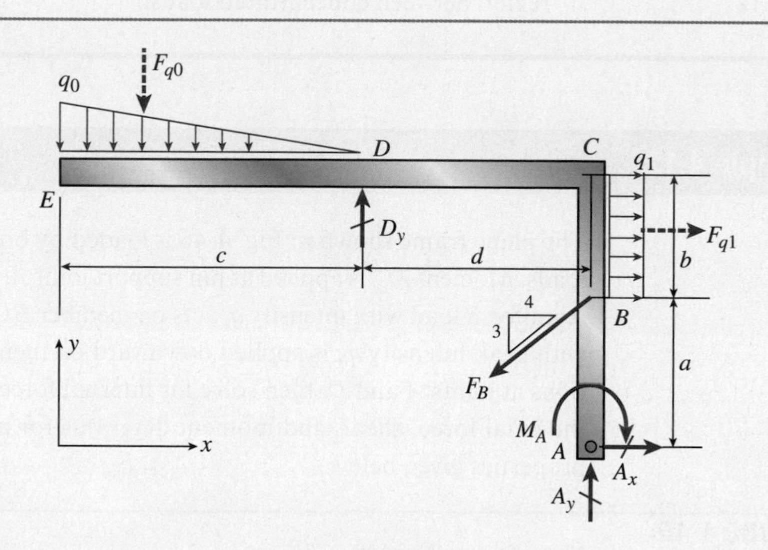

FIGURE 4-48

Upper and lower free-body diagrams from section cut just below joint C

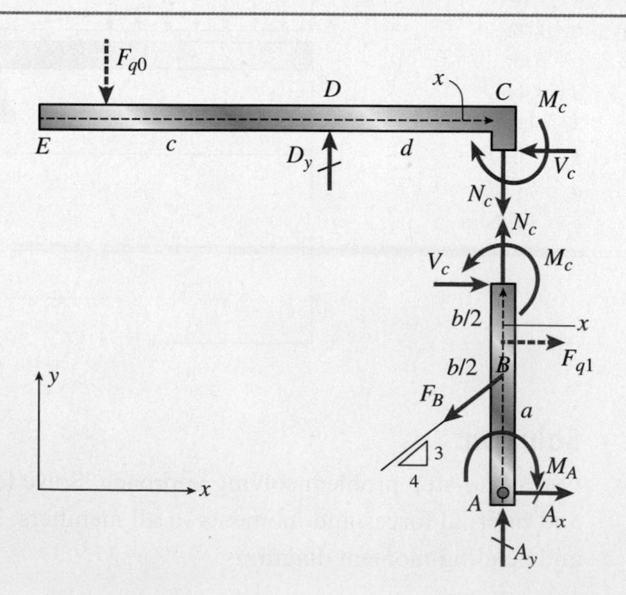

2. ***Categorize***: The overall frame is statically determinate. First solve for the reaction forces at A and D using the overall free-body diagram in Fig. 4-47; then use either free-body diagram in Fig. 4-48 to find the internal forces and moment below joint C. Treat members ABC and EDC as separate beams with each having its own longitudinal centroidal axis x (see Fig. 4-48); then find axial-force, shear-force, and moment functions [$N(x)$, $V(x)$, and $M(x)$] and the associated diagrams for each member.

3. ***Analyze***:

Statically equivalent forces and force components: Distributed forces are replaced by their static equivalents (F_{q0} and F_{q1}). Load F_{q0} acts downward at $c/3 = 2$ m to the right of E, and load F_{q1} acts at the center of BC. The static equivalents and the components of the inclined concentrated force at B are computed as

$$F_{q0} = \frac{1}{2}q_0 c = \frac{1}{2}(73 \text{ N/m})(6 \text{ m}) = 219 \text{ N} \quad F_{q1} = q_1 b = (146 \text{ N/m})(2 \text{ m}) = 292 \text{ N}$$

$$F_{Bx} = \frac{4}{5}F_B = \frac{4}{5}(200 \text{ N}) = 160 \text{ N} \qquad F_{By} = \frac{3}{5}F_B = \frac{3}{5}(200 \text{ N}) = 120 \text{ N} \qquad \text{(a)}$$

External reactions: Use a *statics sign convention* when solving for reaction forces:

$$\Sigma F_x = 0 \quad A_x = -F_{q1} + F_{Bx} = -132 \text{ N}$$

$$\Sigma M_A = 0 \quad D_y = \frac{1}{d}[F_{q0}(d + \frac{2c}{3}) - F_{q1}(a + \frac{b}{2}) + F_{Bx}(a) - M_A] = 139 \text{ N} \qquad \text{(b)}$$

$$\Sigma F_y = 0 \quad A_y = F_{q0} - D_y + F_{By} = 200 \text{ N}$$

Internal forces N and V and moment M just below point C: Use a *deformation sign convention* to find internal quantities N_C, V_C, and M_C, which are shown in their assumed positive directions in Fig. 4-48. Equilibrium of the upper free-body diagram in Fig. 4-48 gives

$$N_C = D_y - F_{q0} = -80 \text{ N} (\uparrow)$$

$$V_C = 0 \qquad \text{(c)}$$

$$M_C = F_{q0}(d + \frac{2c}{3}) - D_y(d) = 1068 \text{ N}\cdot\text{m(CW)}$$

Moment M_C is clockwise (CW) as assumed in Fig. 4-48, so the result is positive.

Axial force, shear force, and moment functions [$N(x)$, $V(x)$, and $M(x)$]: Treat each member as a beam and use the centroidal axis x for each member to write axial force, shear, and moment functions over the entire length of the member.

Beam EDC (Fig. 4-48, Upper free-body diagram): Split the triangular load on ED into triangular load segments (see dashed line in Fig. 4-49).

FIGURE 4-49
Triangular load on ED

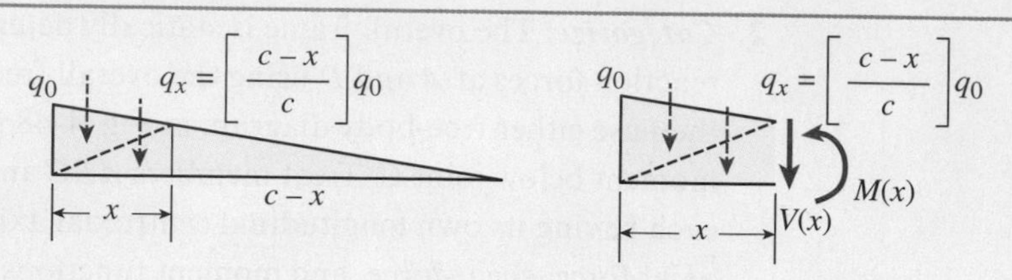

to make it easier to write expressions for $V(x)$ and $M(x)$ for the two parts of EDC as

$$N(x) = 0 \qquad 0 \le x \le c + d$$

$$V(x) = -\frac{1}{2}(q_0 + q_x)(x) \text{ with } q_x = \left[\frac{c-x}{c}\right]q_0 \qquad 0 \le x \le c$$

$$= -\frac{1}{2}(q_0)(c) + D_y \qquad c \le x \le c + d \tag{d}$$

$$M(x) = -\frac{1}{2}(q_0 x)\left(\frac{2x}{3}\right) - \frac{1}{2}(q_x)\left(\frac{x^2}{3}\right) \qquad 0 \le x \le c$$

$$= -\frac{1}{2}(q_0 c)\left(x - \frac{c}{3}\right) + D_y(x - c) \qquad c \le x \le c + d$$

Beam ABC (Fig. 4-48, lower free-body diagram): Treat column ABC as a beam for the purpose of writing the axial-force, shear-force, and moment expressions for the two segments, AB and BC. The origin of member axis x is at A (Fig. 4-48); member ABC may be more conveniently viewed if rotated clockwise to a horizontal position. Functions $N(x)$, $V(x)$, and $M(x)$ for the two parts of ABC are:

$$N(x) = -A_y \quad 0 \le x \le a, \quad N(x) = -A_y + F_{By} \quad a \le x \le a + b$$

$$V(x) = -A_x \quad 0 \le x \le a, \quad V(x) = -A_x + F_{Bx} - q_1(x - a) \quad a \le x \le a + b$$

$$M(x) = M_A - A_x x \qquad 0 \le x \le a \tag{e}$$

$$= M_A - A_x x + F_{Bx}(x - a) - \frac{q_1}{2}(x - a)^2 \quad a \le x \le a + b$$

As a check on the functions in Eq. (e), note that $N(a+b) = N_C$, $V(a+b) = V_C$, and $M(a+b) = M_C$ where internal forces and moment at C are given in Eq. (c).

Axial force, shear-force, and bending-moment diagrams: An alternative approach for constructing the axial-force, shear-force, and moment diagrams is based on the differential and integral relationships among load, shear, and moment. The resulting diagrams are presented in Fig. 4-50. Based on this graphical approach, observe the following from these diagrams.

(a) The axial force is constant and compressive in the two column segments AB and BC [see axial-force diagram, Fig. (4-50a)]. In AB, $N(x) = -A_y$; in BC, $N(x) = N_C$ [from Eq. (c)].

(b) The slope of the tangent to the shear curve at E (Fig. 4-50b) is equal to $-q_0$, since q_0 is downward and $\frac{dV}{dx} = -q(x)$; at D, the load ordinate $q(c) = 0$, so the slope is zero on shear curve. The area under the load curve from E to D is F_{q0}, so the change in shear from E to D is $-F_{q0}$. The change in shear at D is $+D_y = 139$ N, and the shear remains constant from D to C. The shear just left of joint C is equal to N_C (Fig. 4-48).

(c) The area under the shear curve from E to D, A_{VED}, is equal to the change in moment from E to D (Figs. 4-50b and c). The change in moment from D to C equals the negative rectangular area under the shear curve from D to C [i.e., $(-80\ \text{N})(d) = -192\ \text{N}\cdot\text{m}$)].

(d) The shear from A to B is constant, since the slope is equal to the distributed load magnitude, but $q = 0$ in this segment. At B, the shear increases by the magnitude of the load component $F_{Bx} = 160$ N [Eq. (a)] and then varies linearly with slope of $(-q_1)$ to a value of $V_C = 0$ below joint C. The decrease in shear from B to C equals $(-F_{q1})$, which is the area under the load curve from B to C.

(e) The change in moment from support A (where moment M_A is applied) to point B is equal to the positive area under the shear curve from A to B, so $A_{VAB} = (132\ \text{N})(3\ \text{m}) = 396\ \text{N}\cdot\text{m}$. The moment from B to C increases further by the triangular area under the shear curve, which is equal to $\frac{1}{2}(292\ \text{N})(b) = 292\ \text{N}\cdot\text{m}$.

(f) From the diagrams in Fig. 4-50: (1) maximum axial compressive force occurs on segment AB; (2) maximum shear force is just above joint B on member AB; and (3) maximum moment can be found on either side of joint C.

FIGURE 4-50

(a) Axial-force, (b) shear-force, and (c) bending-moment diagrams

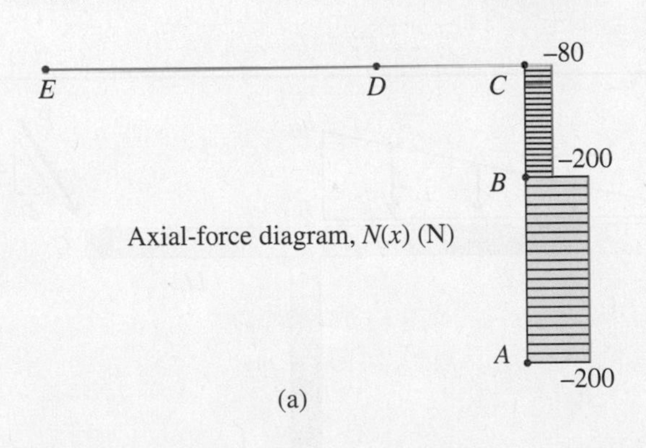

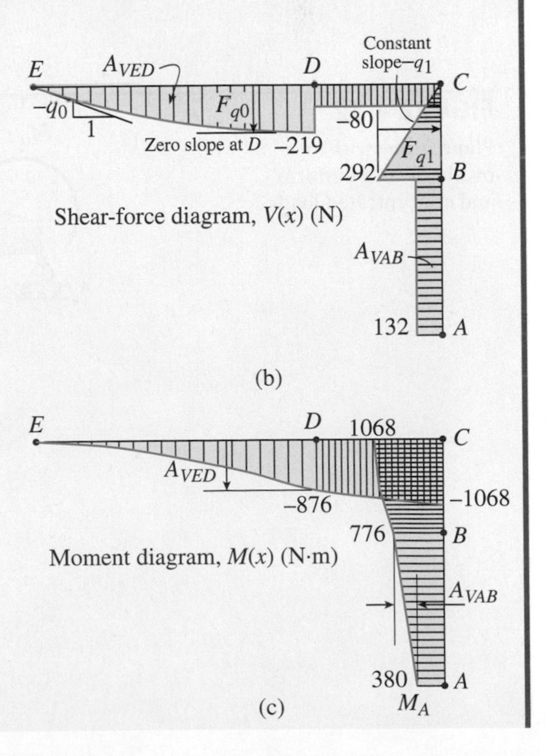

FIGURE 4-51

Internal forces and moments on either side of joint C

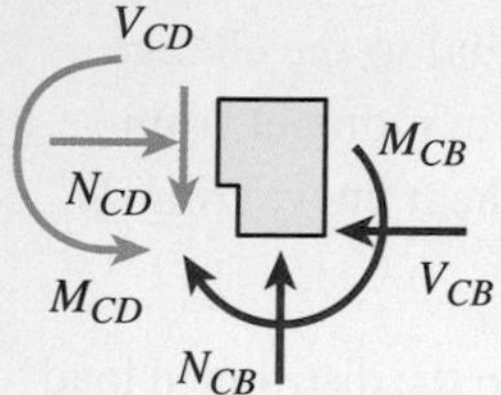

4. ***Finalize***: Axial-force, shear-force, and moment diagrams for frames are frequently plotted on top of the frame, as shown in Fig. 4-50. There is no axial force in beam EDC, but column ABC is in compression over its entire height. The moment diagram shows that the left side of column AB and the entire underside of beam EDC are in compression.

The connection between beam EDC and column ABC is an important design feature of this frame. Use the N, V, and M diagrams and a free-body diagram of joint C to find forces and moments on this connection, which are needed to find stresses in the connection components. From the free-body diagram of the connection at C (Fig. 4-51), observe that force and moment equilibrium are satisfied for this joint. Shear to the left of C (V_{CD}) equals the axial force in BC (N_{CB}), while the shear below joint C (V_{CB}) equals the axial force (N_{CD}) at the end of member DC.

Example 4-13

A plane frame with an overhang is supported at points A and D (Fig. 4-52). (This is the frame from Example 1-4.) A linearly varying distributed load of peak intensity $q_0 = 160$ N/m acts on span AB. Concentrated moment $M_0 = 380$ N · m is applied at A, and an inclined concentrated load $P = 200$ N acts at C. Force P also acts at mid-height of column BD. The lengths of segments AB and BD are $L = 4$ m, and the length of the overhang BC is 2 m. Draw the axial-force, shear-force, and bending-moment diagrams for this frame.

FIGURE 4-52

Plane frame with an overhang and uniform and concentrated loads

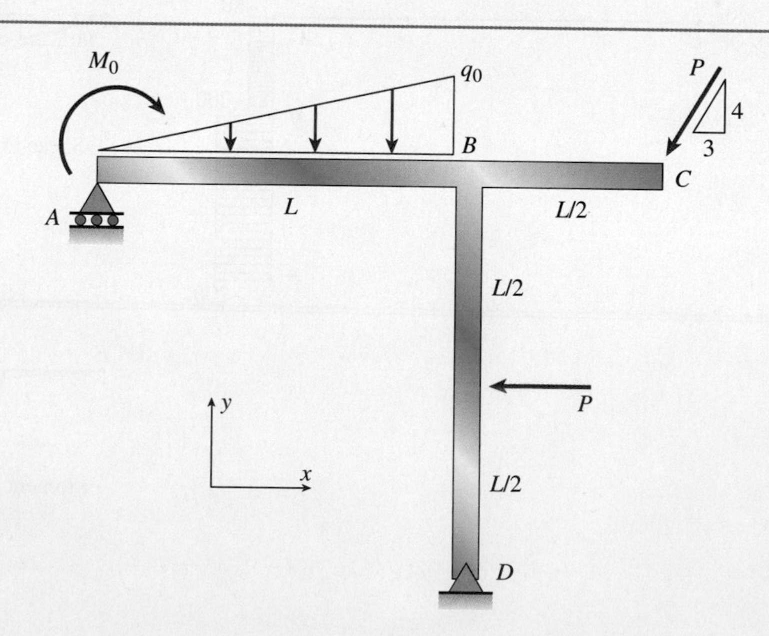

Solution:

Use a four-step problem-solving approach for this plane frame.

1. ***Conceptualize***: Start by drawing the free-body diagram of the entire frame (Fig. 4-53); then solve for reaction forces A_y, D_x, and D_y (as in Example 1-4). Make section cuts in each member to expose internal axial force N, shear force V, and bending moment M in each member (Fig. 4-54). (The section cut is not shown for BC but is like that for AB.) Rotate the frame 90 degrees clockwise to view column BD with joint D on the left and B on the right to establish the assumed positive directions of N, V, and M on either side of the cut section, as shown in Fig. 4-54.

FIGURE 4-53

Overall free-body diagram of plane frame

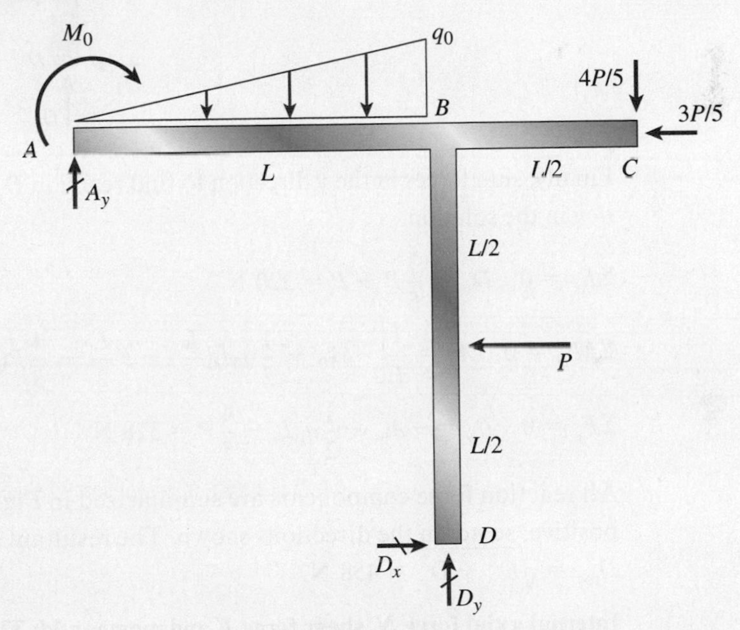

2. ***Categorize***: Use either functions $N(x)$, $V(x)$, and $M(x)$, as in Example 4-6, or the differential relationships between V and M, as in Example 4-10, to plot the axial-force diagram, shear-force diagram, and moment diagram. The free-body diagrams in Fig. 4-54 show internal axial force N, shear force V, and bending moment M in their assumed positive directions based on a deformation sign convention. Each section cut divides the structure into two parts. Use either part to plot N, V, and M values at that location.
3. ***Analyze***:

   Solution for external reactions: Sum forces in the x direction to find reaction force component D_x. Next sum moments about D to find reaction component A_y.

FIGURE 4-54

Section cuts in each member to show internal forces N and V and moment M

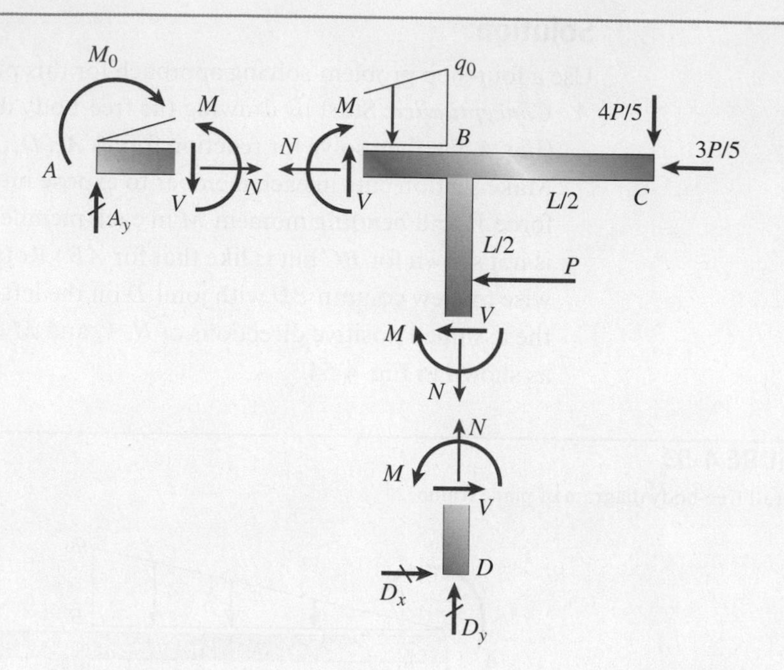

Finally, sum forces in the y direction to find reaction D_y. Use a *statics sign convention* in the solution

$$\Sigma F_x = 0 \quad D_x = \frac{3}{5}P + P = 320 \text{ N}$$

$$\Sigma M_D = 0 \quad A_y = \frac{1}{L}\left[-M_0 + \frac{1}{2}q_0 L\left(\frac{L}{3}\right) + P\frac{L}{2} - \frac{4}{5}P\left(\frac{L}{2}\right) + \frac{3}{5}PL\right] = 152 \text{ N} \quad \textbf{(a)}$$

$$\Sigma F_y = 0 \quad D_y = -A_y + \frac{1}{2}q_0 L + \frac{4}{5}P = 328 \text{ N}$$

All reaction force components are summarized in Fig. 4-55. Reactions are positive, so act in the directions shown. The resultant reaction force at D is $D_{\text{res}} = \sqrt{D_x^2 + D_y^2} = 458$ N.

Internal axial force N, shear force V and moment M: The first option is to use Fig. 4-55 and write functions for $N(x)$, $V(x)$, and $M(x)$ at some distance x from the start of each member. Alternatively, differential and geometric relationships between load and shear and between shear and moment may be used.

Member AB: Use the left-hand free-body in Figs. 4-54 and Fig. 4-55 and measure x from point A to find

$$\Sigma F_x = 0 \quad N(x) = 0$$

$$\Sigma F_y = 0 \quad V(x) = A_y - \frac{1}{2}\left(\frac{x}{L}q_0\right)x = \left(152 - 20x^2\right) \text{ N} \qquad \textbf{(b)}$$

$$\Sigma M = 0 \quad M(x) = M_0 + A_y x - \frac{1}{2}\left(\frac{x}{L}q_0\right)(x)\left(\frac{x}{3}\right) = \left(380 + 152x - \frac{20}{3}x^3\right) \text{N}\cdot\text{m}$$

FIGURE 4-55

Free-body diagram of plane frame segments

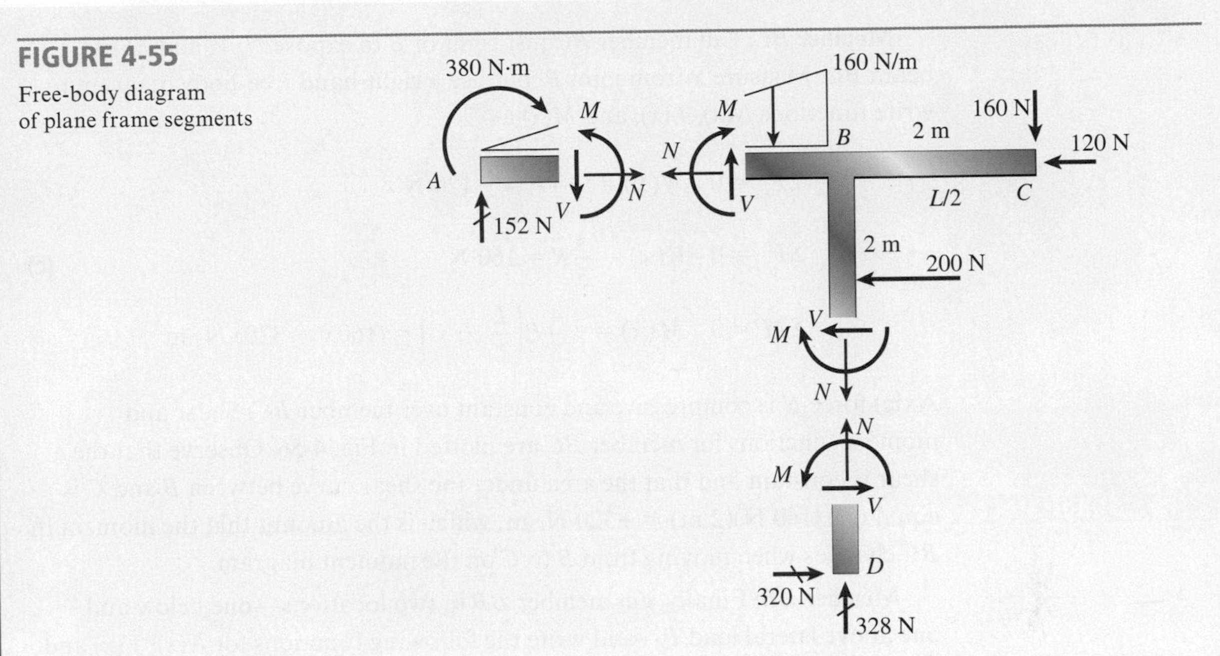

The shear diagram $V(x)$ is plotted in Fig. 4-56a and the moment diagram $M(x)$ in Fig. 4-56b. The shear just right of A is equal to $A_y = 152$ N and decreases by 320 N moving from A to B, which is an amount equal to the area under the triangular loading on AB. The moment at A is equal to $M_0 = 380$ N · m, and the moment increases by 279 N · m (the area under the shear curve between these two points) to a maximum value of 659 N · m at $x = 2.76$ m to the right of point A where the shear is equal to zero. The moment decreases by 98 N · m between the maximum moment location and joint B.

FIGURE 4-56

(a, b) Shear and moment diagrams for plane frame

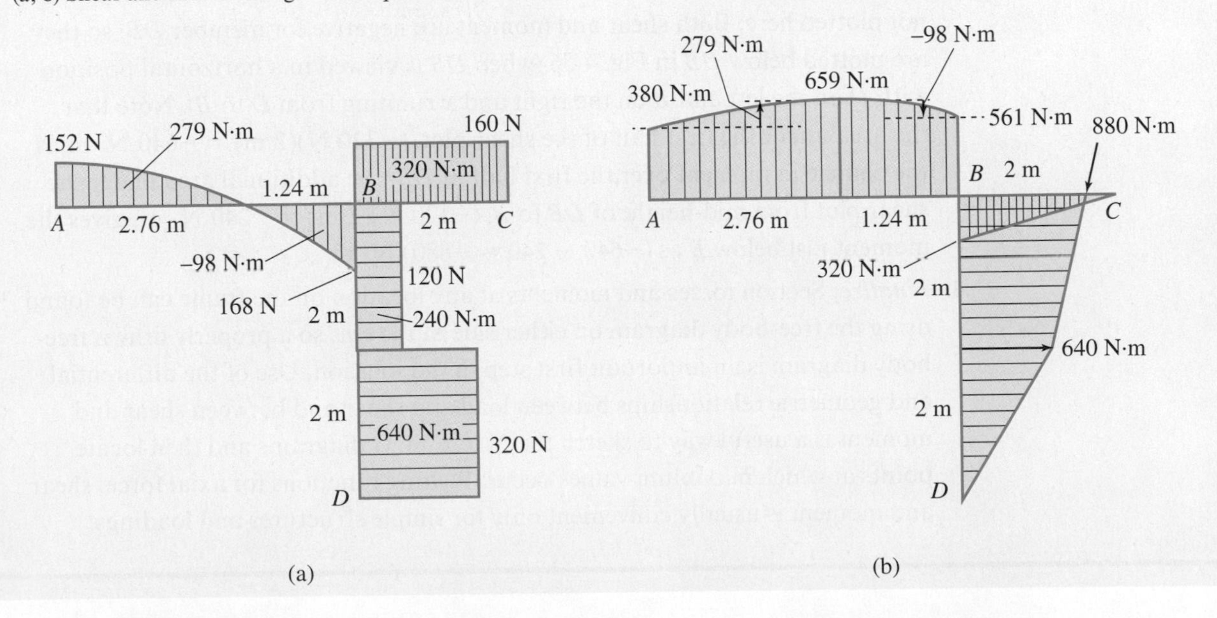

Member *BC*: Cut member *BC* just right of *B* to expose *N, V,* and *M* in beam *BC*. Measure *x* from joint *B*, but use a right-hand free-body diagram to write functions $N(x)$, $V(x)$, and $M(x)$ as

$$\Sigma F_x = 0 \quad N(x) = -\frac{3}{5}P = -120 \text{ N}$$

$$\Sigma F_y = 0 \quad V(x) = \frac{4}{5}P = 160 \text{ N} \qquad \textbf{(c)}$$

$$\Sigma M = 0 \quad M(x) = -\frac{4}{5}P\left(\frac{L}{2} - x\right) = (160x - 320) \text{ N} \cdot \text{m}$$

Axial force *N* is compressive and constant over member *BC*. Shear and moment functions for member *BC* are plotted in Fig. 4-56. Observe that the shear is constant and that the area under the shear curve between *B* and *C* is equal to +(160 N)(2 m) = +320 N · m, which is the amount that the moment in *BC* changes when moving from *B* to *C* on the moment diagram.

Member *DB*: Finally, cut member *DB* in two locations—one below and one above lateral load *P*—and write the following functions for $N(x)$, $V(x)$ and $M(x)$ for column *DB* with *x* measured upward from joint *D*:

$$N(x) = -D_y = -328 \text{ N} \qquad 0 \le x \le 4 \text{ m}$$

$$V(x) = -D_x = -320 \text{ N} \qquad 0 \le x \le 2 \text{ m}$$

$$= P - D_x = -120 \text{ N} \quad 2 \text{ m} \le x \le 4 \text{ m} \qquad \textbf{(d)}$$

$$M(x) = -D_x(x) = -320x \text{ N} \cdot \text{m} \quad 0 \le x \le 2 \text{ m}$$

$$= -D_x(x) + P\left(x - \frac{L}{2}\right) = (-400 - 120x) \text{ N} \cdot \text{m} \quad 2 \text{ m} \le x \le 4 \text{ m}$$

As for *BC*, axial force *N* is compressive and constant over member *DB*, so it is not plotted here. Both shear and moment are negative for member *DB*, so they are plotted below *DB* in Fig. 4-56 (when *DB* is viewed in a horizontal position with *D* on the left and *B* on the right and *x* running from *D* to *B*). Note that the area under the first half of the shear plot, (−320 N)(2 m) = −640 N · m, is the change in moment over the first half of *DB*; the additional area under the shear plot from mid-height of *DB* to *B*, (−120 N)(2 m) = −240 N · m, gives the moment just below *B* as (−640 − 240 = −880) N · m.

4. ***Finalize*:** Section forces and moments at any location on the frame can be found using the free-body diagram on either side of the cut, so a properly drawn free-body diagram is an important first step in the solution. Use of the differential and geometric relationships between load and shear and between shear and moment is a useful way to sketch the *N, V,* and *M* diagrams and then locate points at which maximum values occur. Writing functions for axial force, shear, and moment is usually convenient only for simple structures and loadings.

CHAPTER SUMMARY AND REVIEW

Chapter 4 reviewed the analysis of statically determinate beams and simple frames to find support reactions, internal forces, and moments (N, V, and M). Then **axial-force, shear-force**, and **bending-moment diagrams** were drawn to show the variation of these quantities throughout the structure. A variety of different support conditions, and both concentrated and distributed loadings, were considered in assembling models of structures for analysis. A **statics sign convention** was used when solving for external forces and moments, and a **deformation sign convention** was used when finding internal quantities at selected points along a member.

Here are some of the major concepts presented in Chapter 4.

1. If the structure is **statically determinate**, the laws of statics alone are sufficient to solve for all values of support reaction forces and moments, as well as the magnitude of the internal axial force (N), shear force (V), and bending moment (M) at any location in a beam or frame member.

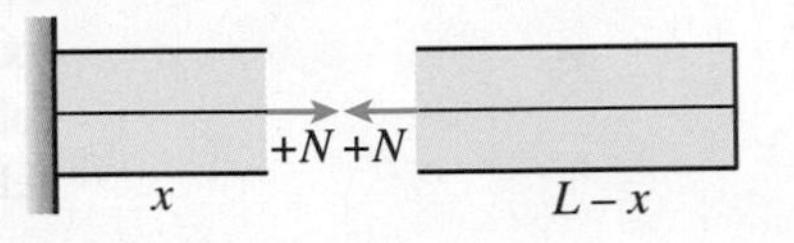

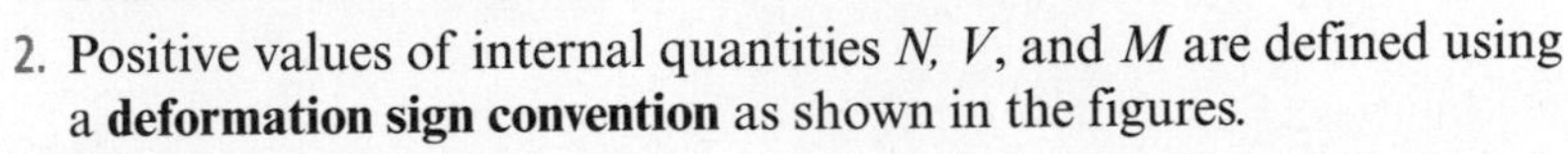

2. Positive values of internal quantities N, V, and M are defined using a **deformation sign convention** as shown in the figures.

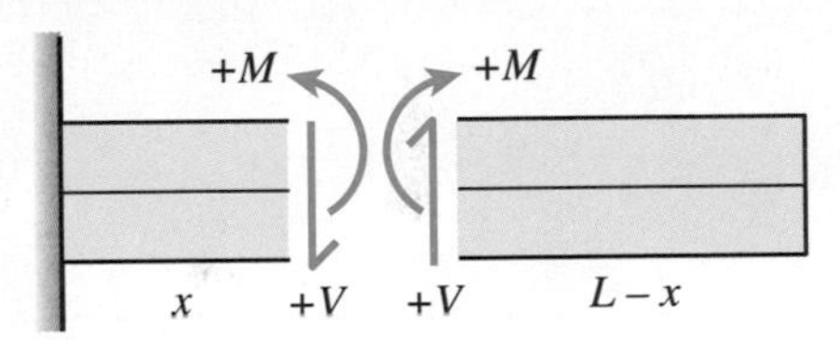

3. Graphical displays or **diagrams** showing the variation of N, V, and M over a structure are useful in design because they readily show the location of maximum values of N, V, and M needed in **design**.

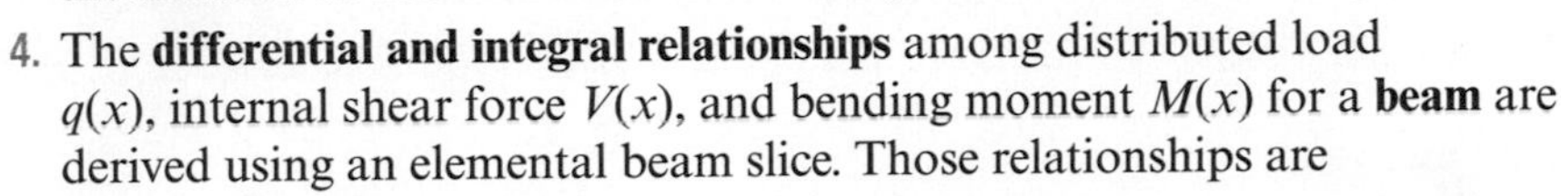

4. The **differential and integral relationships** among distributed load $q(x)$, internal shear force $V(x)$, and bending moment $M(x)$ for a **beam** are derived using an elemental beam slice. Those relationships are

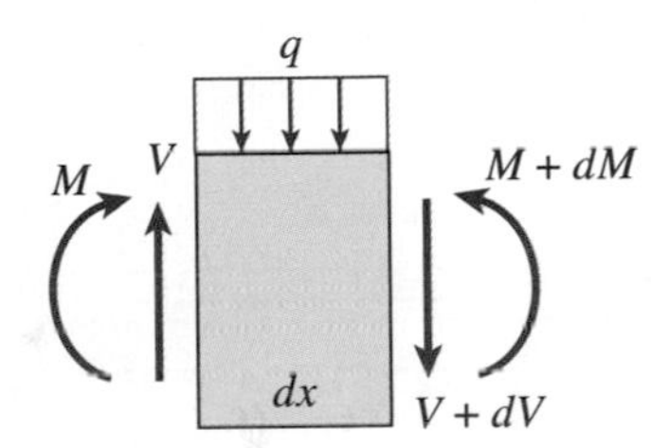

$$\frac{dV}{dx} = -q \text{ and } V_B - V_A = -\int_A^B q(x)dx, \text{ also } \frac{dM}{dx} = V \text{ and } M_B - M_A = \int_A^B V(x)\, dx$$

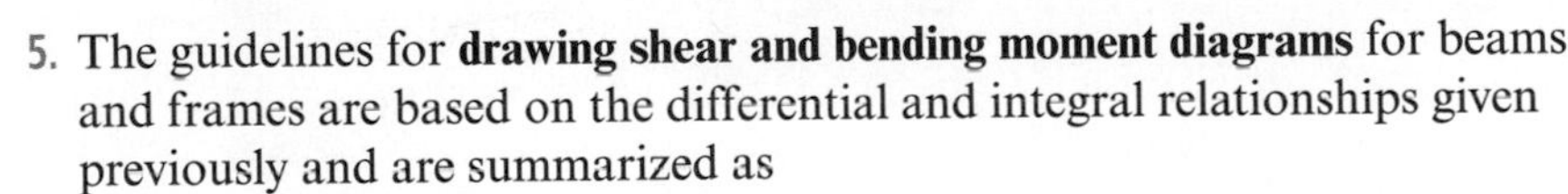

5. The guidelines for **drawing shear and bending moment diagrams** for beams and frames are based on the differential and integral relationships given previously and are summarized as

   - The ordinate on the distributed load curve (q) is equal to the negative of the slope $\frac{dV}{dx}$ on the shear diagram.
   - The difference in shear values between any two points A and B on the shear diagram is equal to the (−) area under the distributed load curve between those same two points.
   - The ordinate on the shear diagram (V) is equal to the slope $\frac{dM}{dx}$ on the bending-moment diagram.
   - The difference in values between any two points on the moment diagram is equal to the area under the shear diagram between those same two points.
   - At those points where the shear curve crosses the reference axis ($V = 0$), the value of the moment on the bending-moment diagram is a local maximum or minimum.
   - At points of application of concentrated loads or bending moments, there are changes in ordinates (or discontinuities) on the V and M diagrams.

6. With a deformation sign convention, a positive moment causes compression at the top and tension at the bottom of the beam. As a result, the bending-moment diagram is plotted on the **compression side** of a structural member or portion of a member.
7. The point at which the bending-moment diagram crosses the reference axis ($M = 0$) is called an **inflection point**.
8. **Axial-force, shear-force**, and **bending-moment diagrams** help to identify values and locations of maximum axial force, shear, and moment so that **maximum stresses and strains** can be found in members.
9. The ordinate on the axial-force diagram (N) is equal to zero at an **axial-force release**; the ordinate on the shear diagram (V) is zero at a **shear release**; and the ordinate on the moment diagram (M) is zero at a **moment release**.

PROBLEMS Chapter 4

4.3 Shear Forces and Bending Moments

Introductory Problems

4.3-1 Calculate the shear force V and bending moment M at a cross section just to the right of the 3.5 kN load acting on the simple beam AB shown in the figure.

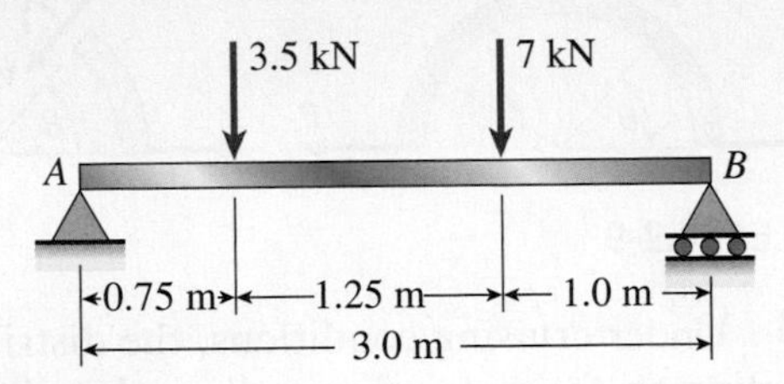

PROBLEM 4.3-1

4.3-2 Determine the shear force V and bending moment M just right of the 6 kN load on the simple beam AB shown in the figure.

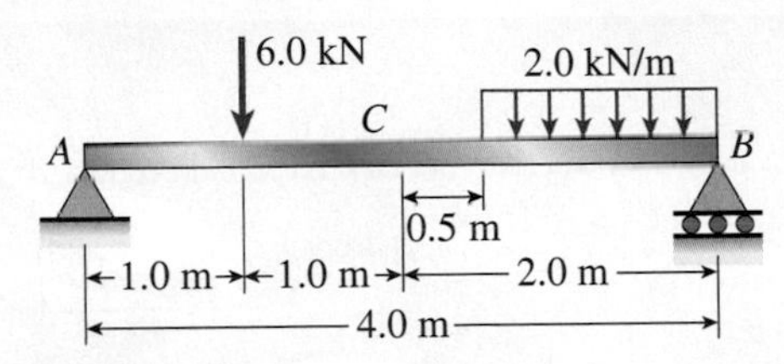

PROBLEM 4.3-2

4.3-3 Determine the shear force V and bending moment M at the midpoint of the beam with overhangs (see figure). Note that one load acts downward and the other upward, and clockwise moments Pb are applied at each support. Repeat if moments Pb are moved to the ends of the beam (Fig. b).

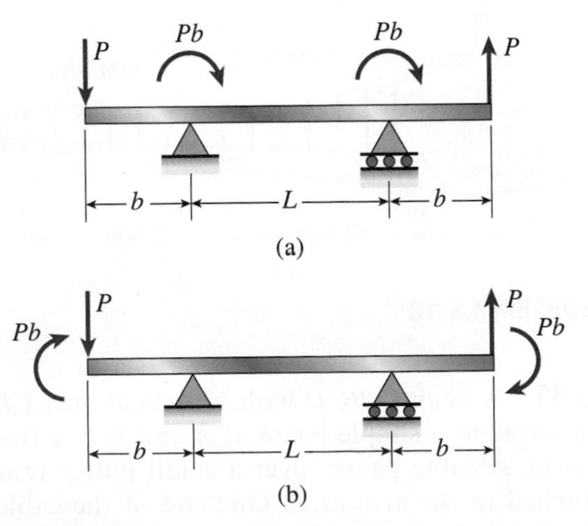

PROBLEM 4.3-3

4.3-4 Calculate the shear force V and bending moment M at a cross section located just right of the 4 kN load on the cantilever beam AB shown in the figure.

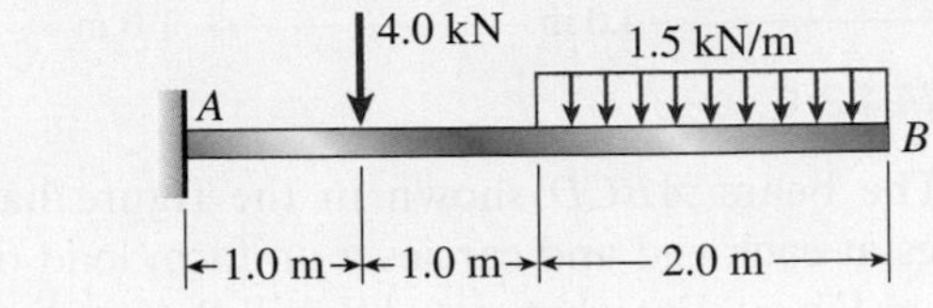

PROBLEM 4.3-4

Representative Problems

4.3-5 Consider the beam with an overhang shown in the figure.

(a) Determine the shear force V and bending moment M at a cross section located 5.5 m from the left-hand end A.

(b) Find the required magnitude of load intensity q acting on the right half of member BC that will result in a zero shear force on the cross section 5.5 m from A.

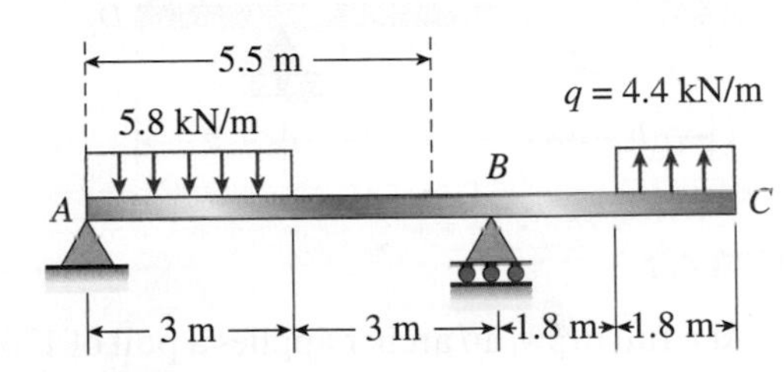

PROBLEM 4.3-5

4.3-6 The beam ABC shown in the figure is simply supported at A and B and has an overhang from B to C. The loads consist of a horizontal force $P_1 = 4.0$ kN acting at the end of a vertical arm and a vertical force $P_2 = 8.0$ kN acting at the end of the overhang.

(a) Determine the shear force V and bending moment M at a cross section located 3.0 m from the left-hand support. *Note:* Disregard the widths of the beam and vertical arm and use centerline dimensions when making calculations.

(b) Find the value of load P_2 that results in $V = 0$ at a cross section located 2.0 m from the left-hand support.

(c) If $P_2 = 8.0$ kN, find the value of load P_1 that results in $M = 0$ at a cross section located 2.0 m from the left-hand support.

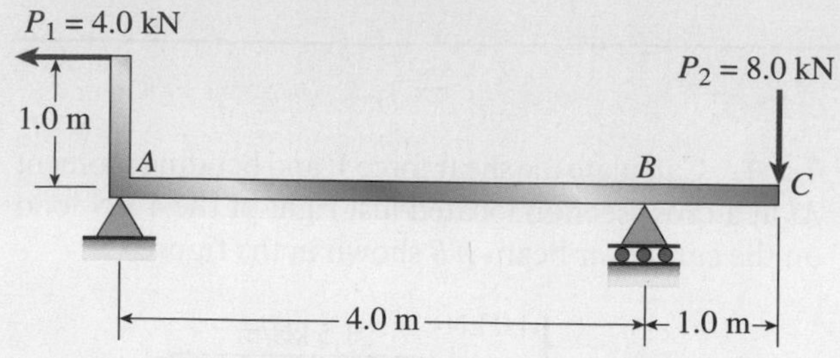

PROBLEM 4.3-6

4.3-7 The beam *ABCD* shown in the figure has overhangs at each end and carries a uniform load of intensity q (Fig. a). For what ratio b/L will the bending moment at the midpoint of the beam be zero?

Repeat for a triangular load with peak intensity q_0 at $L/2$ (Fig. b).

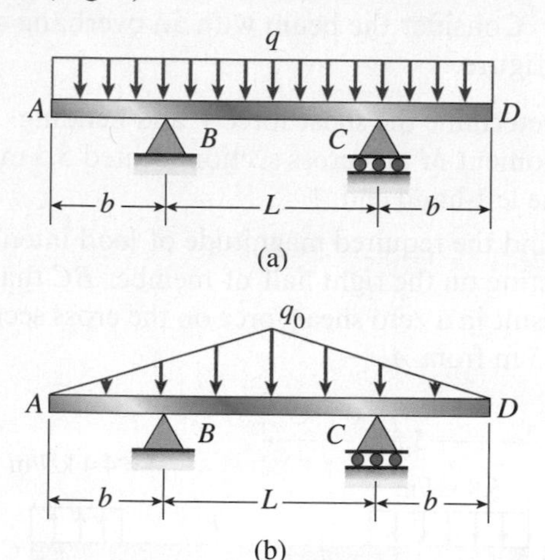

PROBLEM 4.3-7

4.3-8 At a full draw, an archer applies a pull of 130 N to the bowstring of the bow shown in the figure. Determine the bending moment at the midpoint of the bow.

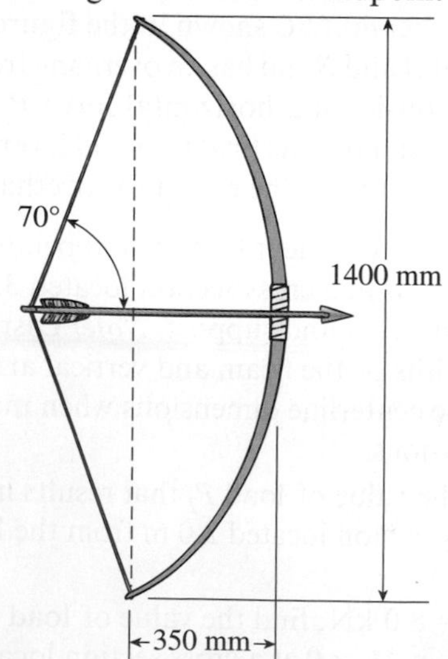

PROBLEM 4.3-8

4.3-9 A curved bar *ABC* is subjected to loads in the form of two equal and opposite forces *P*, as shown in the figure. The axis of the bar forms a semicircle of radius *r*.

Determine the axial force *N*, shear force *V*, and bending moment *M* acting at a cross section defined by the angle θ.

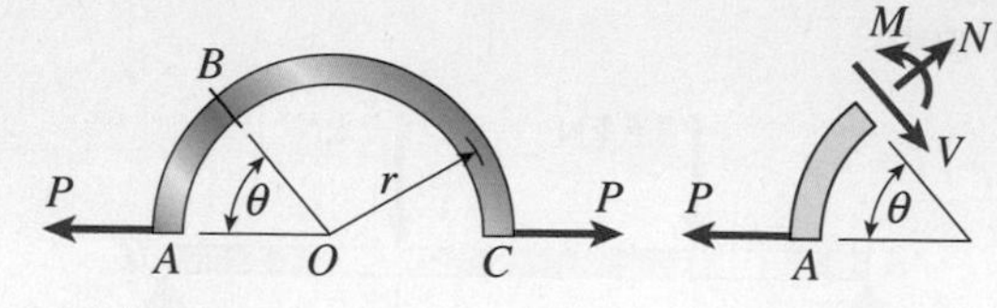

PROBLEM 4.3-9

4.3-10 Under cruising conditions, the distributed load acting on the wing of a small airplane has the idealized variation shown in the figure.

Calculate the shear force *V* and bending moment *M* at 4 m from the tip of the wing.

(a)

Wings of a small airplane have distributed uplift loads.

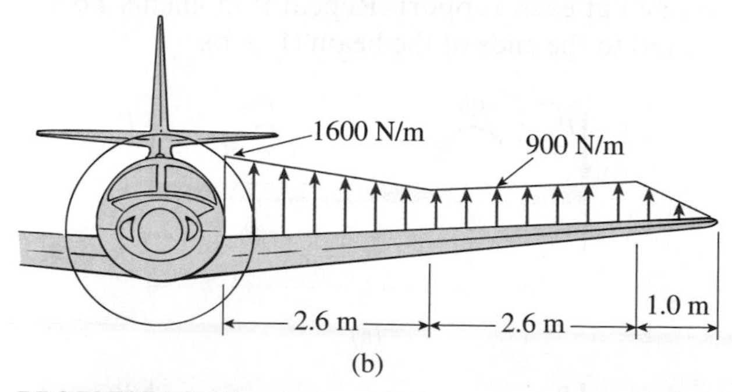

(b)

PROBLEM 4.3-10

4.3-11 A beam *ABCD* with a vertical arm *CE* is supported as a simple beam at *A* and *D* (see figure part a). A cable passes over a small pulley that is attached to the arm at *E*. One end of the cable is attached to the beam at point *B*.

(a) What is the force P in the cable if the bending moment in the beam just to the left of point C is equal numerically to 7.5 kN·m? *Note:* Disregard the widths of the beam and vertical arm and use centerline dimensions when making calculations.

(b) Repeat part (a) if a roller support is added at C and a shear release is inserted just left of C (see figure part b).

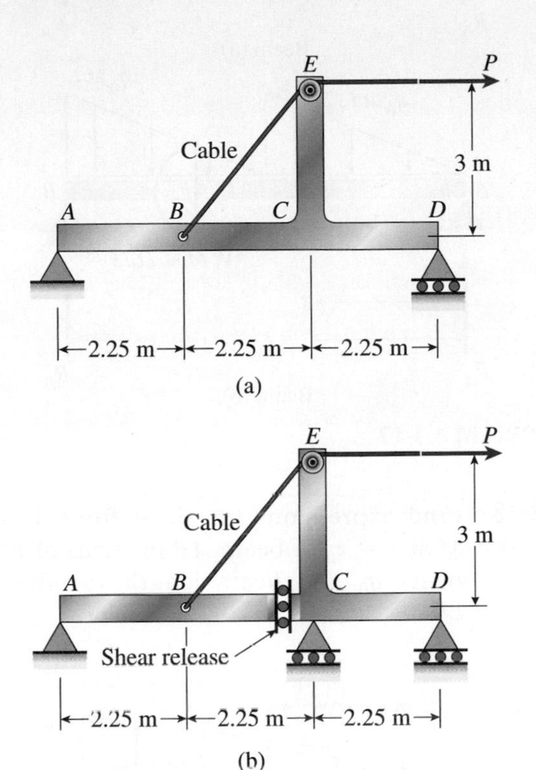

PROBLEM 4.3-11

4.3-12 A simply supported beam AB supports a trapezoidally distributed load (see figure). The intensity of the load varies linearly from 50 kN/m at support A to 25 kN/m at support B.

Calculate the shear force V and bending moment M at the midpoint of the beam.

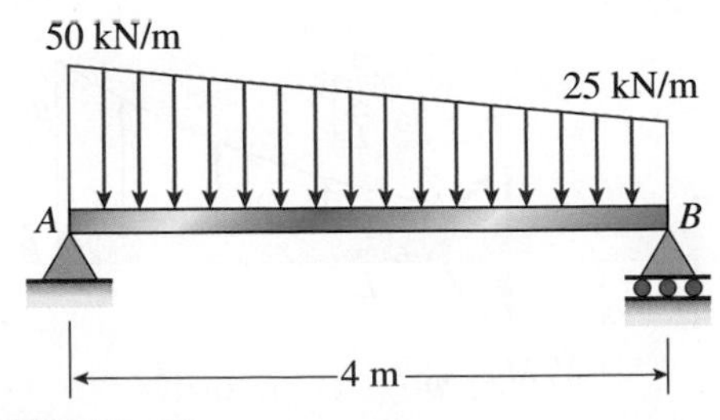

PROBLEM 4.3-12

4.3-13 Beam $ABCD$ represents a reinforced-concrete foundation beam that supports a uniform load of intensity $q_1 = 40$ kN/m (see figure). Assume that the soil pressure on the underside of the beam is uniformly distributed with intensity q_2.

(a) Find the shear force V_B and bending moment M_B at point B.

(b) Find the shear force V_m and bending moment M_m at the midpoint of the beam.

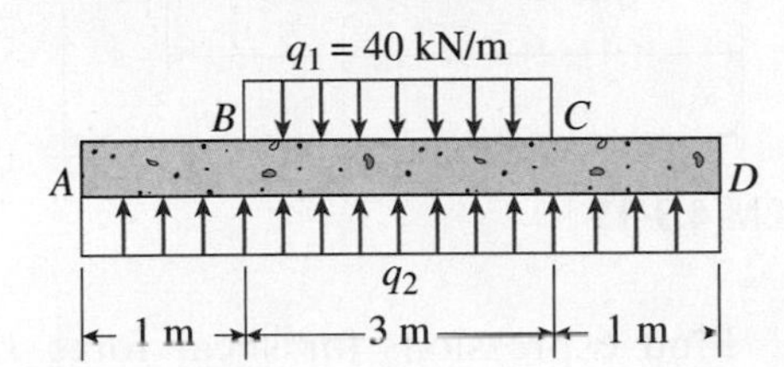

PROBLEM 4.3-13

4.3-14 Find shear (V) and moment (M) at $x = 3L/4$ for the beam shown in Fig. a. Let $M_A = 24$ kN·m, $P = 48$ kN, $L = 6$ m, and $q_0 = 8$ kN/m. Repeat for the beam in Fig. b (first solve for the reaction moment at fixed support A).

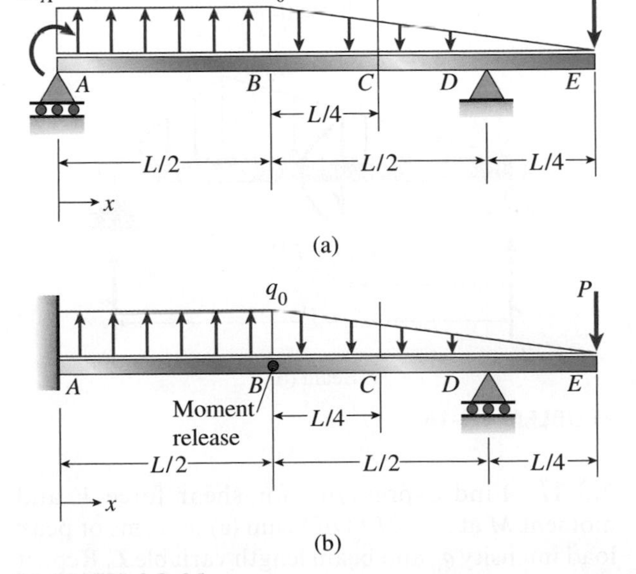

PROBLEM 4.3-14

4.3-15 Find expressions for shear force V and moment M at mid-span of beam AB in terms of peak load intensity q_0 and beam length variables a and L Let $a = 5L/6$.

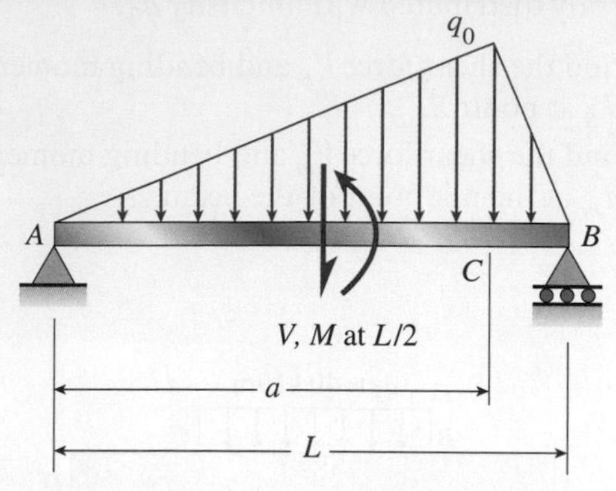

PROBLEM 4.3-15

4.3-16 Find expressions for shear force V and moment M at $x = 2L/3$ of beam (a) in terms of peak load intensity q_0 and beam length variable L. Repeat for beam (b) but at $x = L/2$.

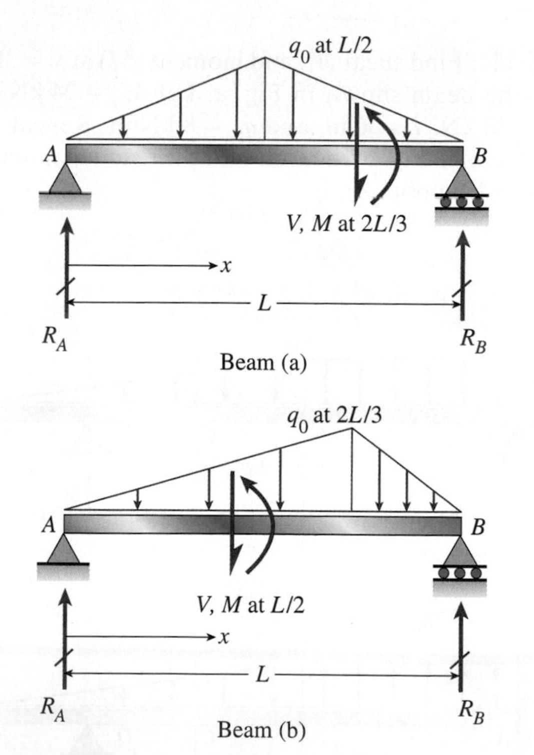

PROBLEM 4.3-16

4.3-17 Find expressions for shear force V and moment M at $x = 2L/3$ of beam (a) in terms of peak load intensity q_0 and beam length variable L. Repeat for beam (b).

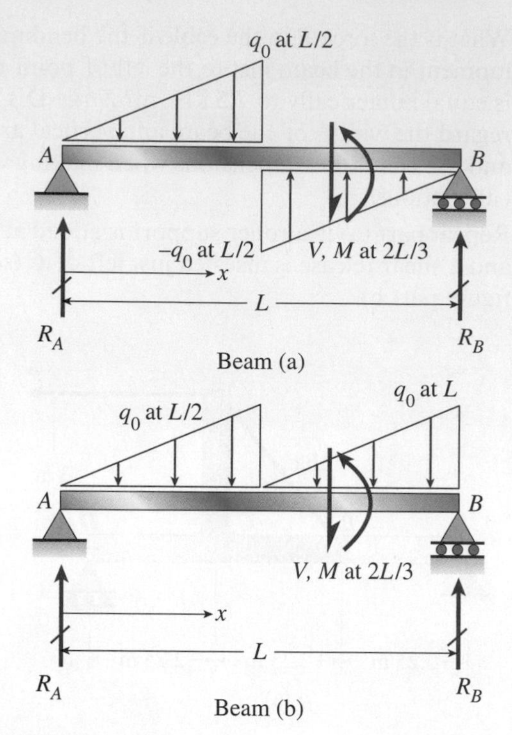

PROBLEM 4.3-17

4.3-18 Find expressions for shear force V and moment M at $x = x_0$ of beam AB in terms of peak load intensity q_0 and beam length variable L. Let $x_0 = L/2$.

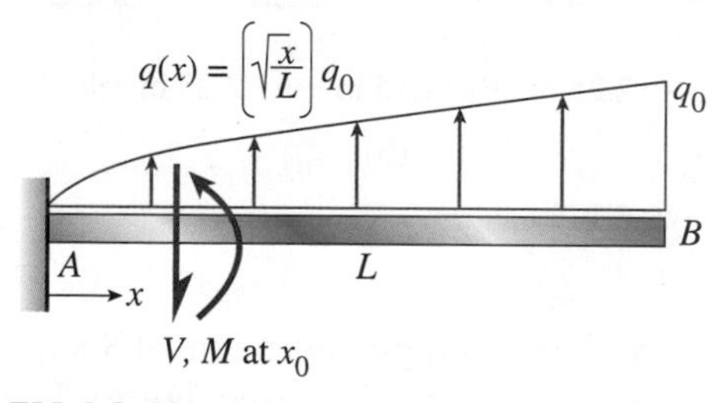

PROBLEM 4.3-18

4.3-19 Find expressions for shear force V and moment M at $x = x_0$ of beam AB in terms of peak load intensity q_0 and beam length variable L. Let $x_0 = 2L/3$.

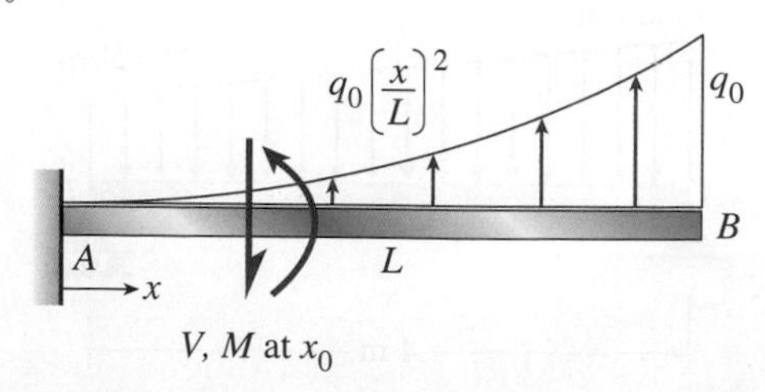

PROBLEM 4.3-19

4.3-20 Find expressions for shear force V and moment M at $x = L/2$ of beam BC. Express V and M in terms of peak load intensity q_0 and beam length variable L.

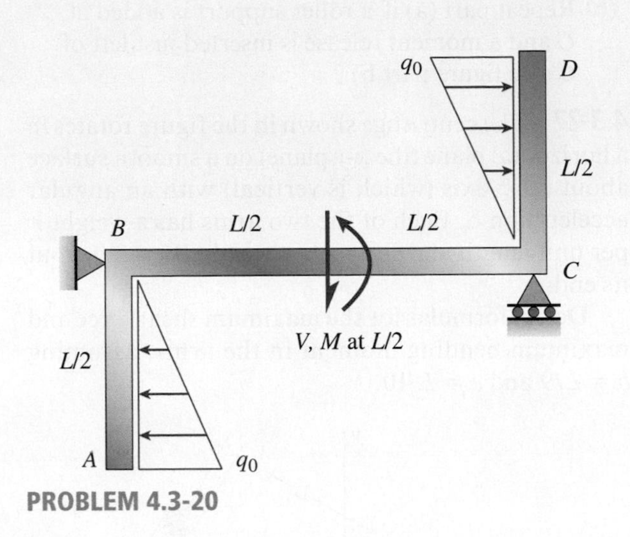

PROBLEM 4.3-20

4.3-21 A cable with force P is attached to a frame at A and runs over a frictionless pulley at D. Find expressions for shear force V and moment M at $x = L/2$ of beam BC.

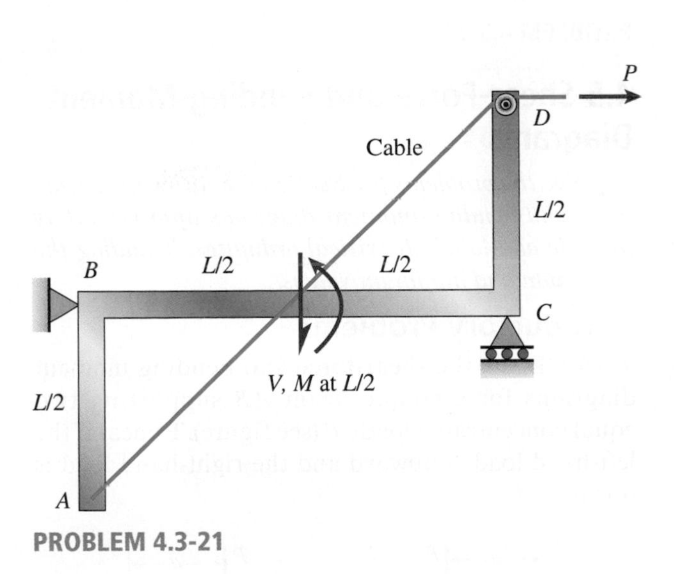

PROBLEM 4.3-21

4.3-22 Find expressions for shear force V and moment M at $x = L/2$ of beam AB in structure (a). Express V and M in terms of peak load intensity q_0 and beam length variable L. Repeat for structure (b) but find V and M at mid-span of member BC.

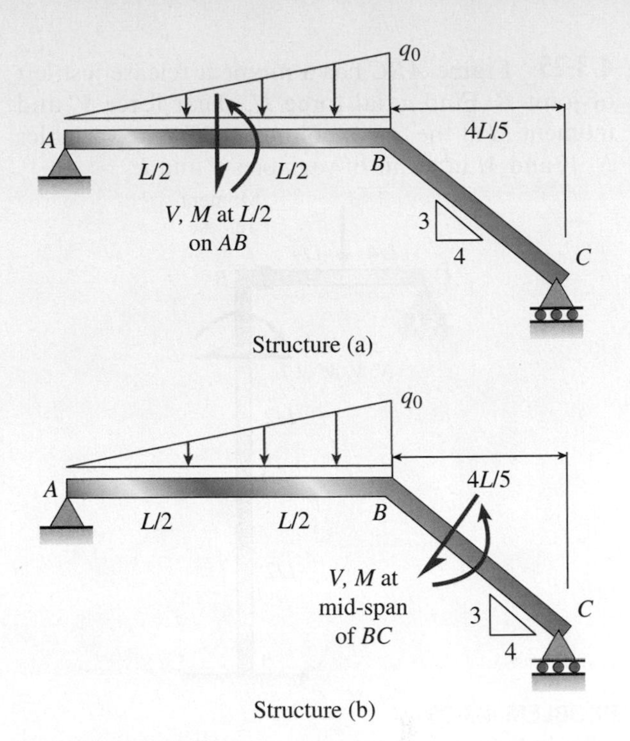

PROBLEM 4.3-22

4.3-23 A cable with force P is attached to a frame at D and runs over a frictionless pulley at B. Find expressions for shear force V and moment M at $x = L/3$ of beam AB.

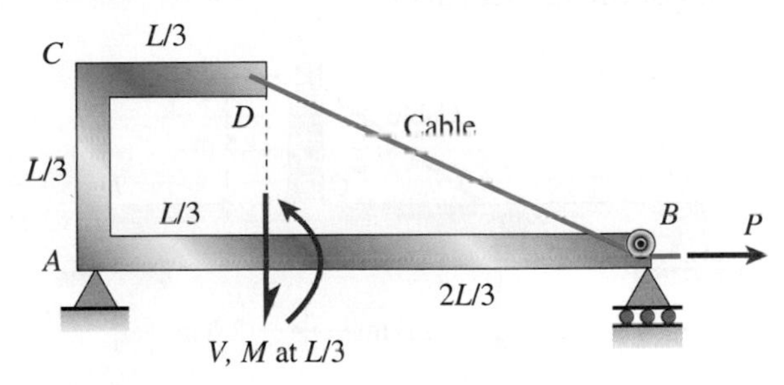

PROBLEM 4.3-23

4.3-24 Frame $ABCD$ carries two concentrated loads ($2P$ at C and P at D, see figure) and also a linearly varying distributed load on AB. Find expressions for shear force V and moment M at $x = L/3$ of beam AB in terms of peak load intensity q_0, force P, and beam length variable L. Let $q_0 = P/L$.

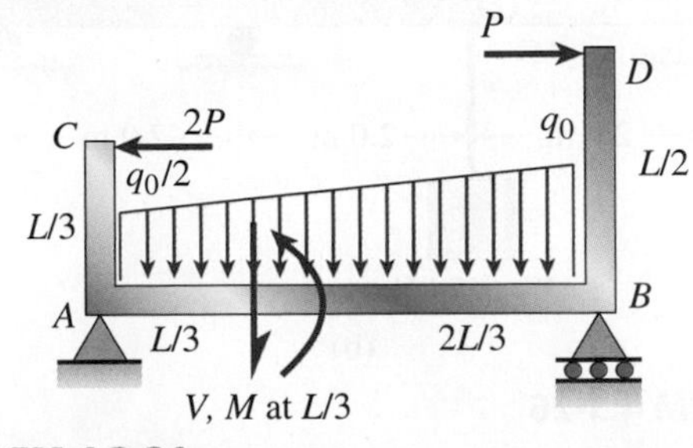

PROBLEM 4.3-24

4.3-25 Frame ABC has a moment release just left of joint B. Find axial force N, shear force V, and moment M at the top of column AB. Write variables N, V, and M in terms of variables P and L.

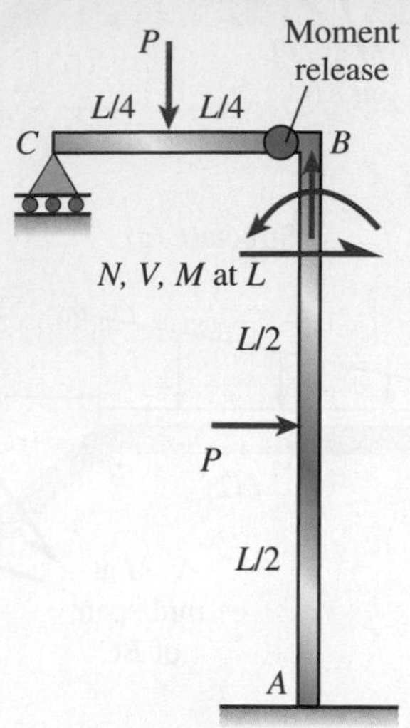

PROBLEM 4.3-25

4.3-26 The simply supported beam $ABCD$ is loaded by a weight $W = 27$ kN through the arrangement shown in the figure part a. The cable passes over a small frictionless pulley at B and is attached at E to the end of the vertical arm.

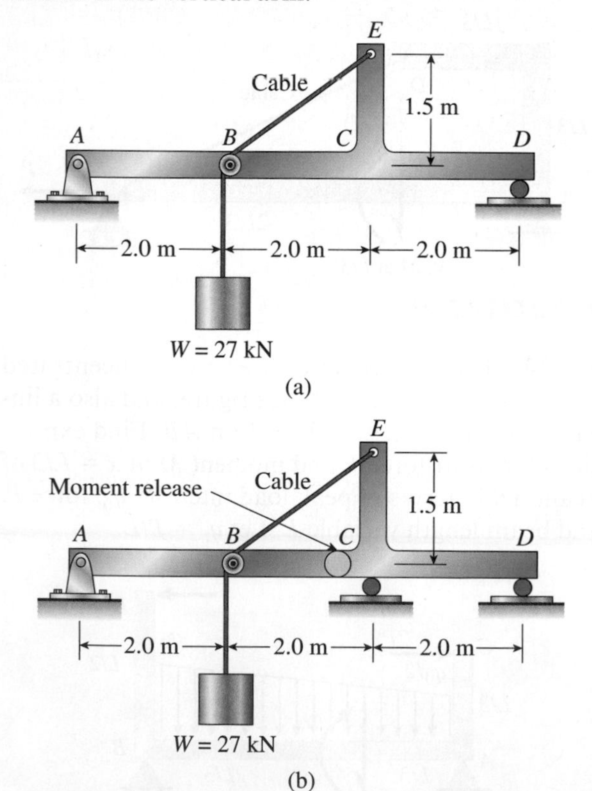

PROBLEM 4.3-26

(a) Calculate the axial force N, shear force V, and bending moment M at section C, which is just to the left of the vertical arm. *Note:* Disregard the widths of the beam and vertical arm and use centerline dimensions when making calculations.

(b) Repeat part (a) if a roller support is added at C and a moment release is inserted just left of C (see figure part b).

4.3-27 The centrifuge shown in the figure rotates in a horizontal plane (the x-y plane) on a smooth surface about the z axis (which is vertical) with an angular acceleration α. Each of the two arms has a weight w per unit length and supports a weight $W = 2.0wL$ at its end.

Derive formulas for the maximum shear force and maximum bending moment in the arms, assuming $b = L/9$ and $c = L/10$.

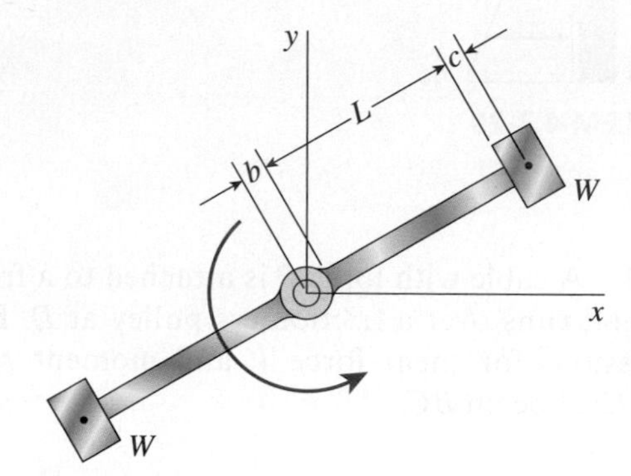

PROBLEM 4.3-27

4.5 Shear-Force and Bending-Moment Diagrams

To solve the problems for Section 4.5, draw the shear-force and bending-moment diagrams approximately to scale and label all critical ordinates, including the maximum and minimum values.

Introductory Problems

4.5-1 Draw the shear-force and bending-moment diagrams for a simple beam AB supporting two equal concentrated loads P (see figure). Repeat if the left-hand load is upward and the right-hand load is downward.

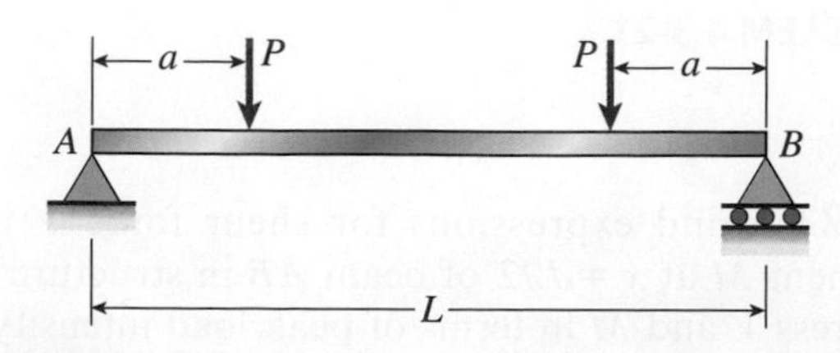

PROBLEM 4.5-1

4.5-2 A simple beam AB is subjected to a counterclockwise couple of moment M_0 acting at distance a from the left-hand support (see figure). Draw the shear-force and bending-moment diagrams for this beam.

Also draw the shear-force and bending-moment diagrams if a second moment M_0 is added at distance a from support B.

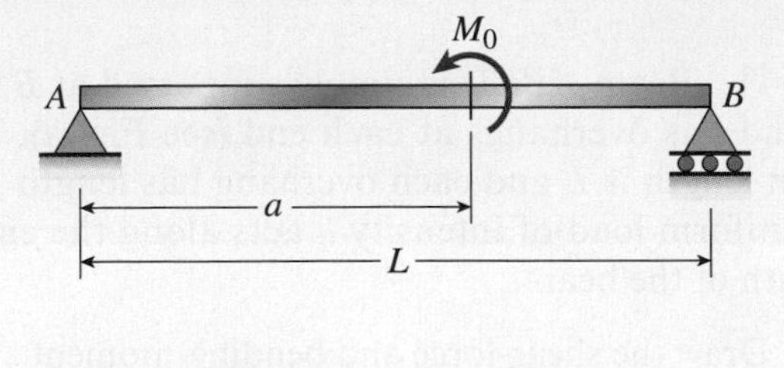

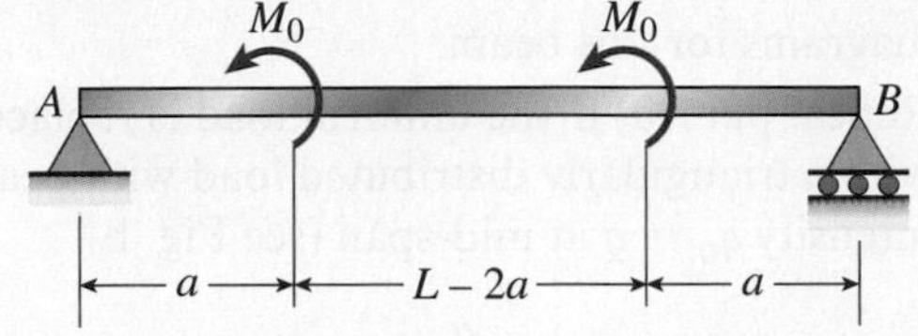

PROBLEM 4.5-2

4.5-3 Draw the shear-force and bending-moment diagrams for a cantilever beam AB carrying a uniform load of intensity q over one-half of its length (see figure).

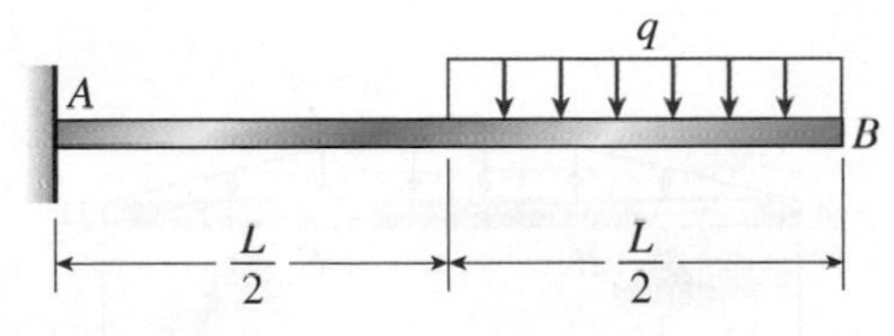

PROBLEM 4.5-3

4.5-4 The cantilever beam AB shown in the figure is subjected to a concentrated load P at the midpoint and a counterclockwise couple of moment $M_1 = PL/4$ at the free end.

Draw the shear-force and bending-moment diagrams for this beam.

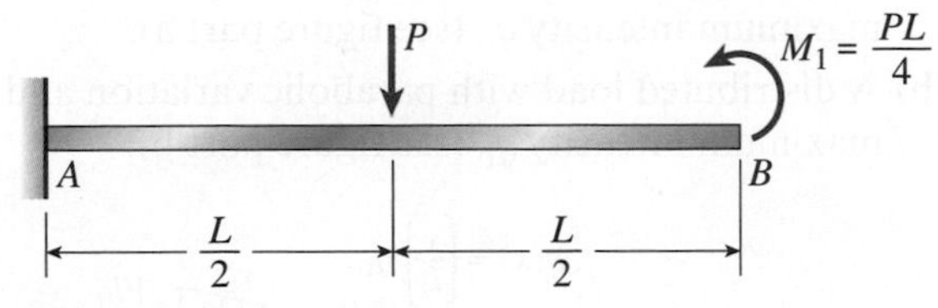

PROBLEM 4.5-4

4.5-5 Cantilever beam AB carries an upward uniform load of intensity q_1 from $x = 0$ to $L/2$ (see Fig. a) and a downward uniform load of intensity q from $x = L/2$ to L.

(a) Find q_1 in terms of q if the resulting moment at A is zero. Draw V and M diagrams for the case of both q and q_1 as applied loadings.

(b) Repeat part (a) for the case of an upward triangularly distributed load with peak intensity q_0 (see Fig. b). For part (b), find q_0 instead of q_1.

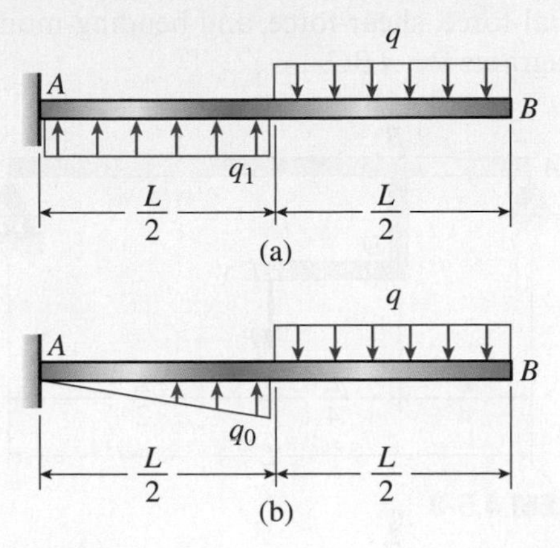

PROBLEM 4.5-5

4.5-6 The simple beam AB shown in the figure is subjected to a concentrated load P and a clockwise couple $M_1 = PL/3$ acting at the third points.

Draw the shear-force and bending-moment diagrams for this beam.

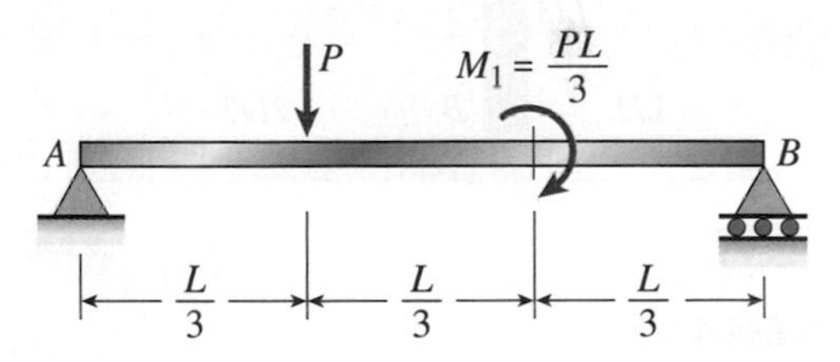

PROBLEM 4.5-6

4.5-7 A simple beam AB subjected to couples M_1 and $3M_1$ acting at the third points is shown in the figure.

Draw the shear-force and bending-moment diagrams for this beam.

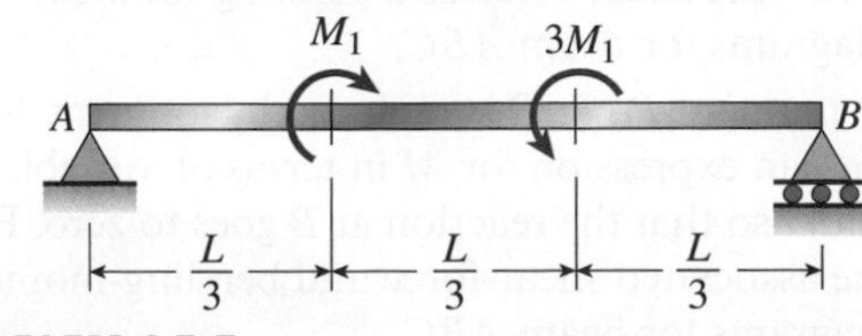

PROBLEM 4.5-7

4.5-8 A simply supported beam ABC is loaded by a vertical load P acting at the end of a bracket BDE (see figure).

(a) Draw the shear-force and bending-moment diagrams for beam ABC.

(b) Now assume that load P at E is directed to the right. The vertical dimension BD is $L/5$. Draw axial-force, shear-force, and bending-moment diagrams for ABC.

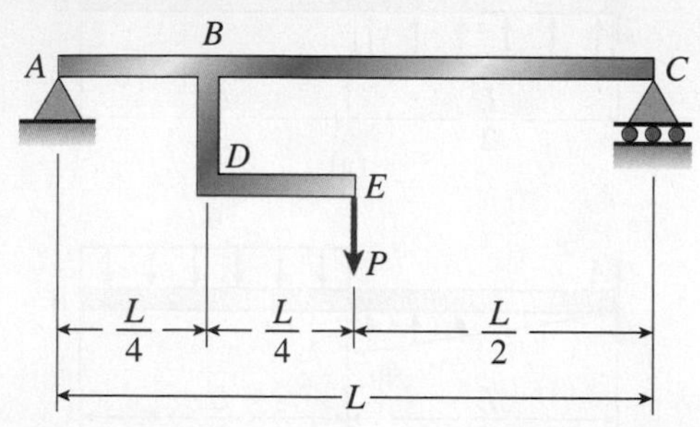

PROBLEM 4.5-8

4.5-9 A simply supported beam ABC is loaded at the end of a bracket BDE (see figure). Draw axial-force, shear-force, and bending-moment diagrams for ABC.

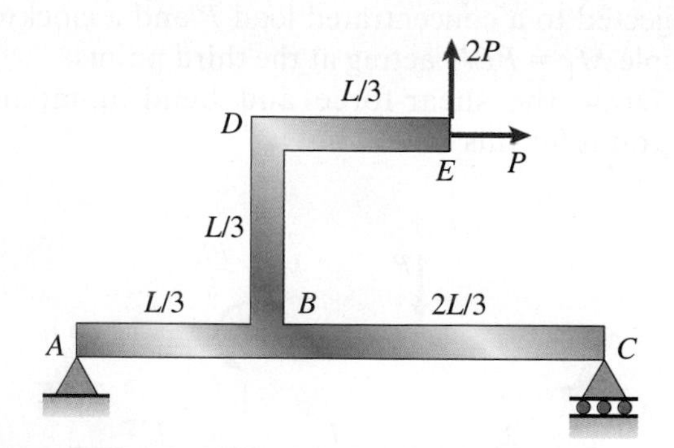

PROBLEM 4.5-9

Representative Problems

4.5-10 A beam ABC is simply supported at A and B and has an overhang BC (see figure). The beam is loaded by two forces P and a clockwise couple of moment Pa at D that act through the arrangement shown.

(a) Draw the shear-force and bending-moment diagrams for beam ABC.

(b) If moment Pa at D is replaced by moment M, find an expression for M in terms of variables P and a so that the reaction at B goes to zero. Plot the associated shear-force and bending-moment diagrams for beam ABC.

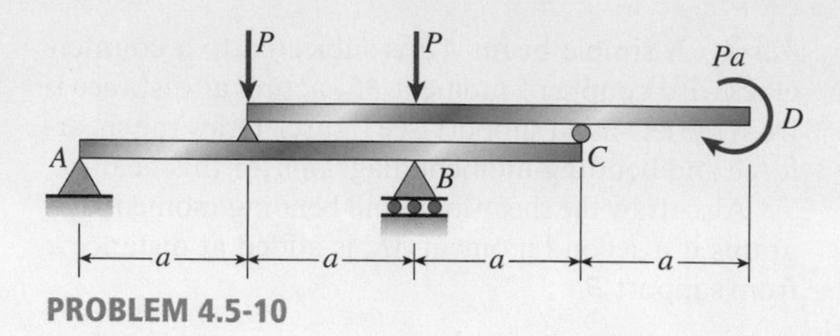

PROBLEM 4.5-10

4.5-11 Beam $ABCD$ is simply supported at B and C and has overhangs at each end (see Fig. a). The span length is L and each overhang has length $L/3$. A uniform load of intensity q acts along the entire length of the beam.

(a) Draw the shear-force and bending-moment diagrams for this beam.

(b) Repeat part (a) if the uniform load is replaced with a triangularly distributed load with peak intensity $q_0 = q$ at mid-span (see Fig. b).

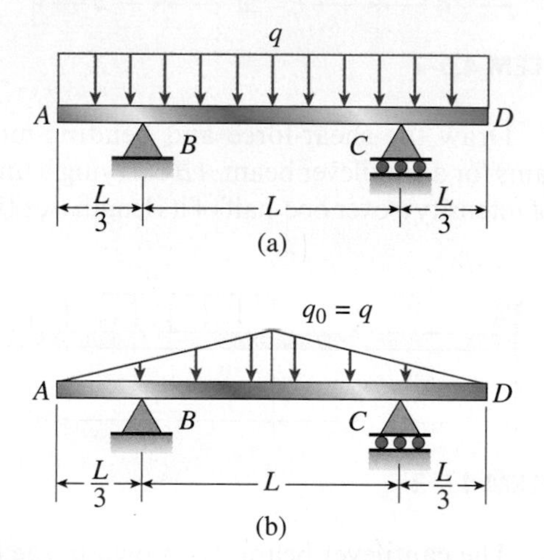

PROBLEM 4.5-11

4.5-12 Draw the shear-force and bending-moment diagrams for a cantilever beam AB acted upon by two different load cases.

(a) A distributed load with linear variation and maximum intensity q_0 (see figure part a).

(b) A distributed load with parabolic variation and maximum intensity q_0 (see figure part b).

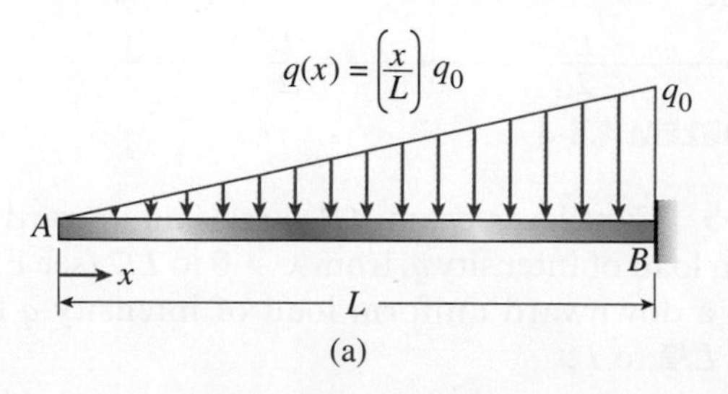

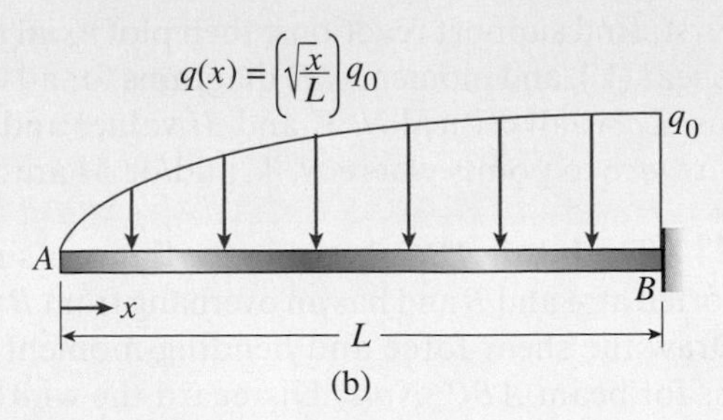

(b)

PROBLEM 4.5-12

4.5-13 The simple beam AB supports a triangular load of maximum intensity $q_0 = 1.75$ kN/m acting over one-half of the span and a concentrated load $P = 350$ N acting at midspan (see figure). Draw the shear-force and bending-moment diagrams for this beam.

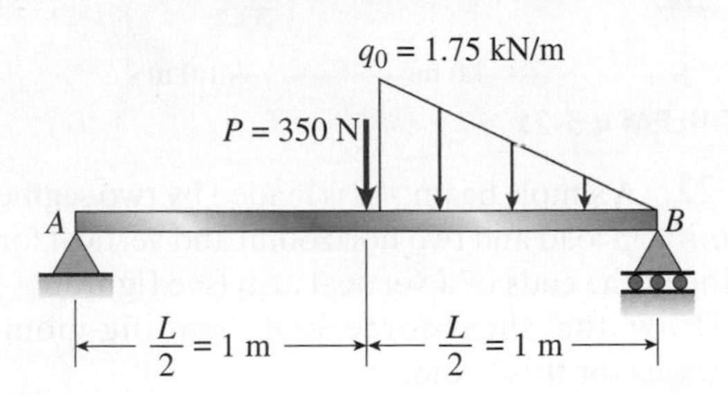

PROBLEM 4.5-13

4.5-14 The beam AB shown in the figure supports a uniform load of intensity 3000 N/m acting over half the length of the beam. The beam rests on a foundation that produces a uniformly distributed load over the entire length.

(a) Draw the shear-force and bending-moment diagrams for this beam.

(b) Repeat part (a) for the distributed load variation shown in Fig. b.

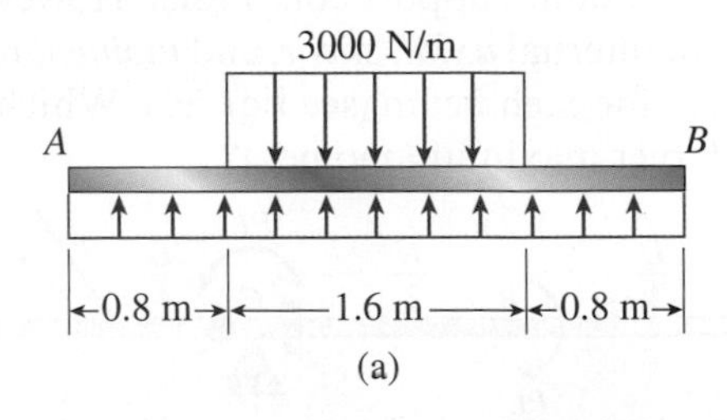

(a)

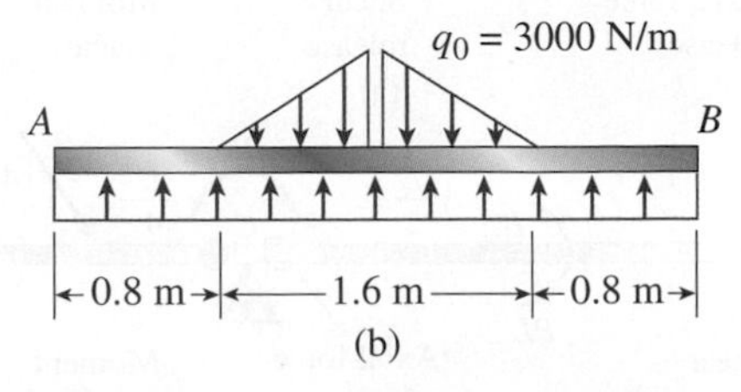

(b)

PROBLEM 4.5-14

4.5-15 A cantilever beam AB supports a couple and a concentrated load, as shown in the figure. Draw the shear-force and bending-moment diagrams for this beam.

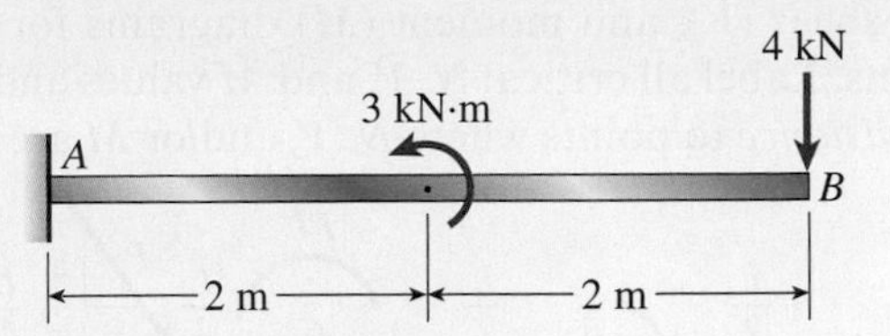

PROBLEM 4.5-15

4.5-16 The cantilever beam AB shown in the figure is subjected to a triangular load acting over one-half of its length and a concentrated load acting at the free end.

Draw the shear-force and bending-moment diagrams for this beam.

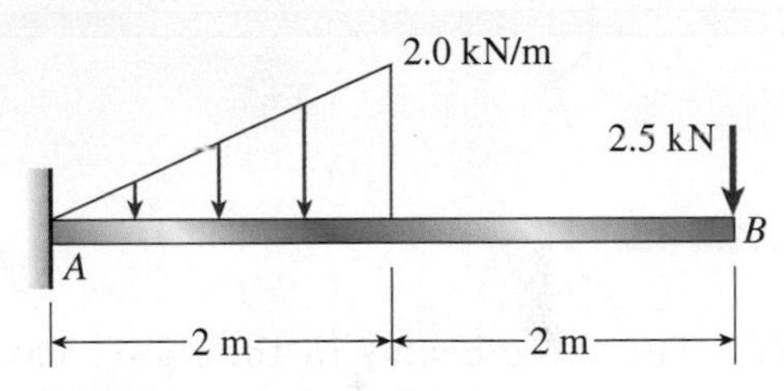

PROBLEM 4.5-16

4.5-17 Beam ABC has simple supports at A and B, an overhang BC and the distributed loading shown in the figure.

Draw the shear-force and bending-moment diagrams for this beam.

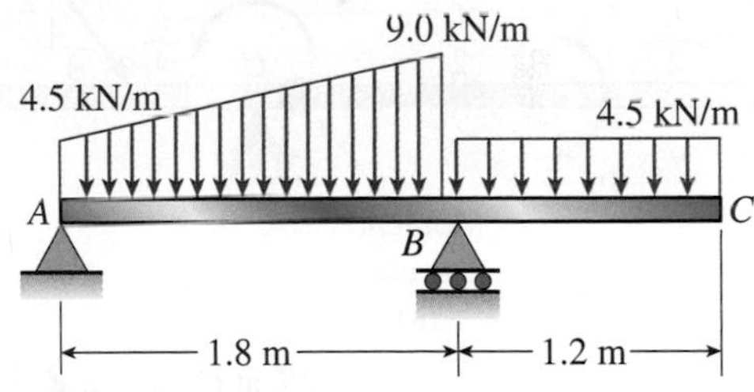

PROBLEM 4.5-17

4.5-18 Beam ABC with an overhang at one end supports a partial uniform load of intensity 12 kN/m and a concentrated moment of magnitude 4 kN·m at C (see figure).

Draw the shear-force and bending-moment diagrams for this beam.

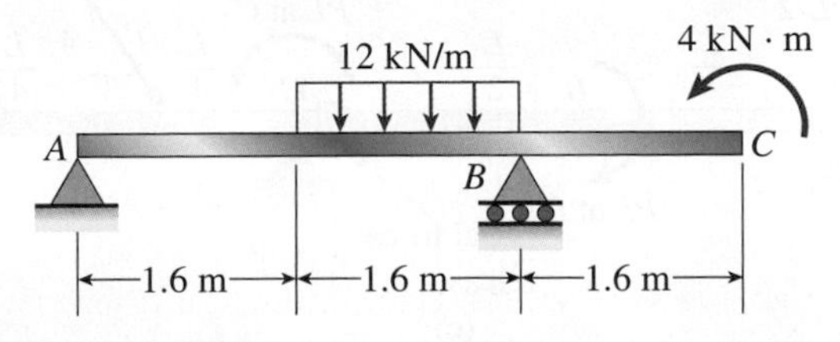

PROBLEM 4.5-18

4.5-19 Consider the two beams shown in the figures. Which beam has the larger maximum moment?

First, find support reactions; then plot axial force (N), shear (V), and moment (M) diagrams for both beams. Label all critical N, V, and M values and also the *distance* to points where N, V, and/or M are zero.

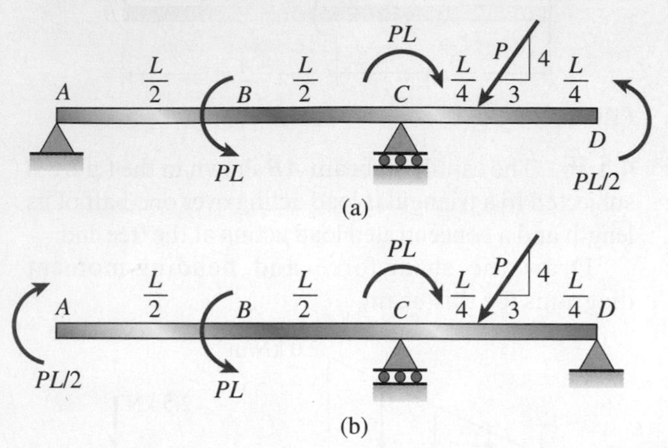

PROBLEM 4.5-19

4.5-20 The three beams in the figure have the same loading. However, one has a *moment release* just to the left of C, the second has a *shear release* just to the right of C, and the third has an *axial release* just to the left of C. Which beam has the largest maximum moment?

First, find support reactions; then plot axial force (N), shear (V), and moment (M) diagrams for all three beams. *Label* all critical N, V, and M values and also the *distance* to points where N, V, and/or M are zero.

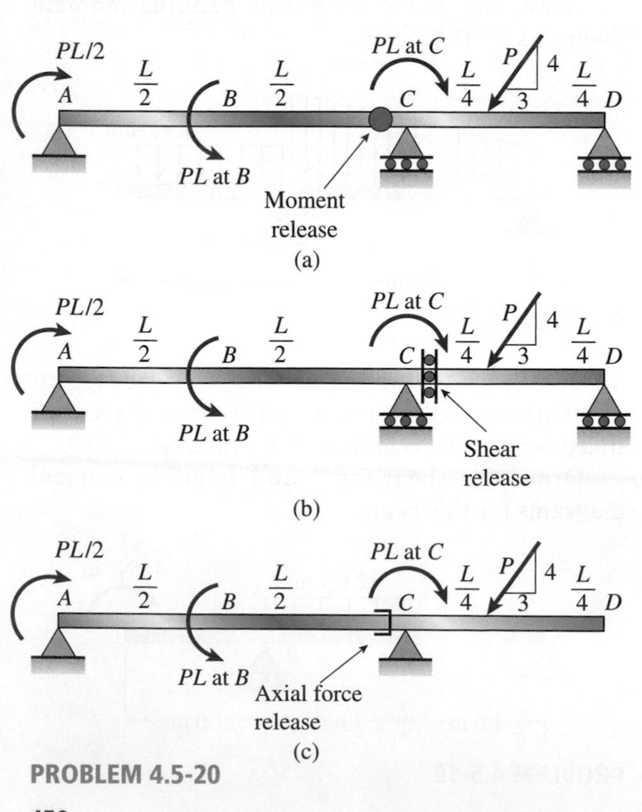

PROBLEM 4.5-20

4.5-21 The beam ABC shown in the figure is simply supported at A and B and has an overhang from B to C.

Draw the shear-force and bending-moment diagrams for beam ABC. *Note:* Disregard the widths of the beam and vertical arm and use centerline dimensions when making calculations.

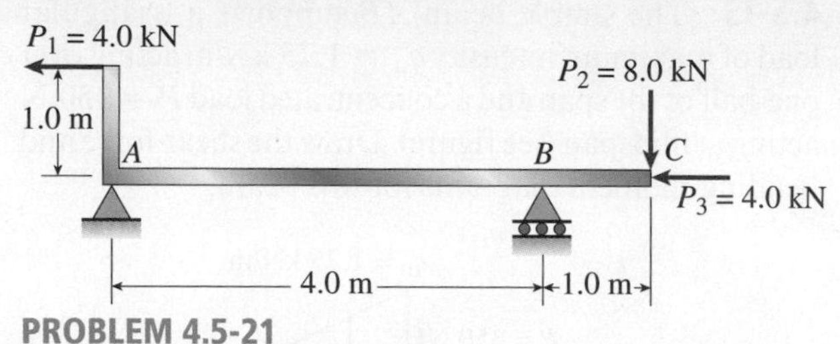

PROBLEM 4.5-21

4.5-22 A simple beam AB is loaded by two segments of uniform load and two horizontal and vertical forces acting at the ends of a vertical arm (see figure).

Draw the shear-force and bending-moment diagrams for this beam.

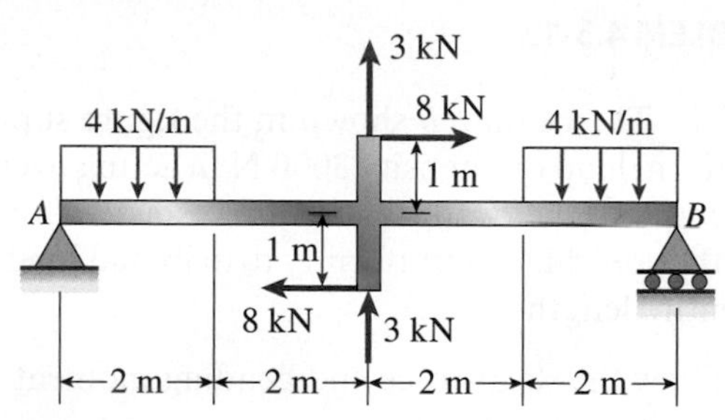

PROBLEM 4.5-22

4.5-23 Two beams (see figure) are loaded the same and have the same support conditions. However, the location of internal *axial, shear*, and *moment releases* is different for each beam (see figures). Which beam has the larger maximum moment?

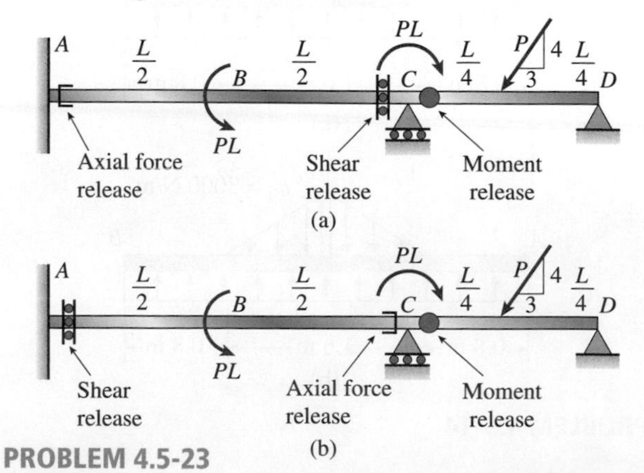

PROBLEM 4.5-23

First, find support reactions; then plot axial force (N), shear (V), and moment (M) diagrams for both beams. *Label* all critical N, V, and M values and also the *distance* to points where N, V, and/or M are zero.

4.5-24 The beam $ABCD$ shown in the figure has overhangs that extend in both directions for a distance of 4.2 m from the supports at B and C, which are 1.2 m apart.

Draw the shear-force and bending-moment diagrams for this overhanging beam.

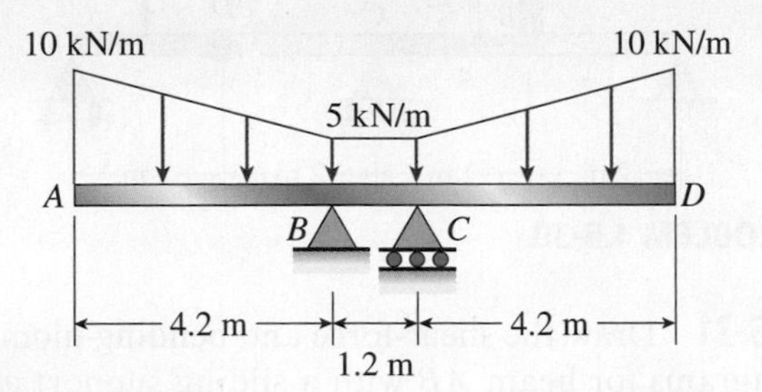

PROBLEM 4.5-24

4.5-25 A beam $ABCD$ with a vertical arm CE is supported as a simple beam at A and D (see figure). A cable passes over a small pulley that is attached to the arm at E. One end of the cable is attached to the beam at point B. The tensile force in the cable is 8 kN.

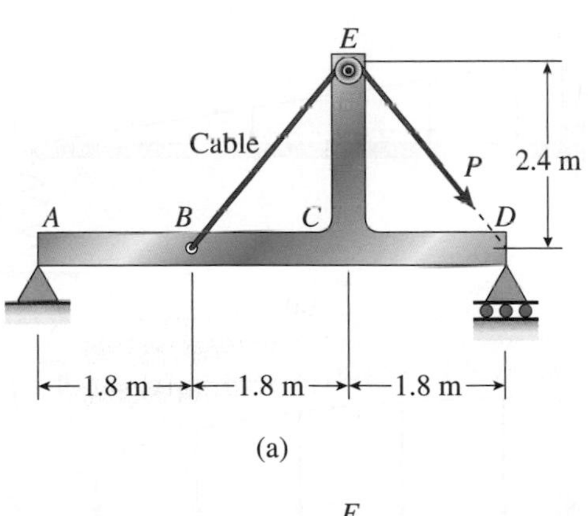

(a)

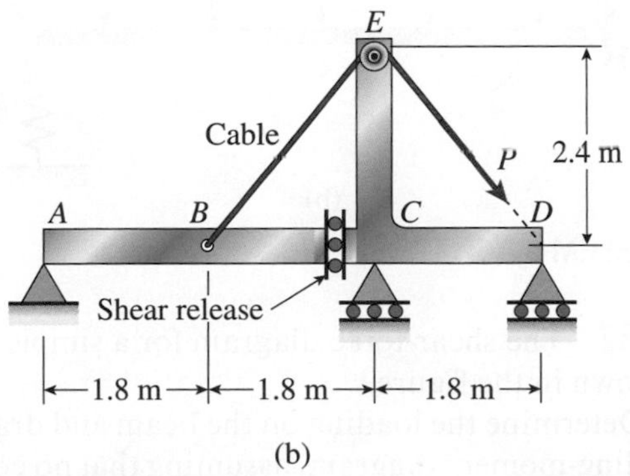

(b)

PROBLEM 4.5-25

(a) Draw the shear-force and bending-moment diagrams for beam $ABCD$. *Note:* Disregard the widths of the beam and vertical arm and use centerline dimensions when making calculations.

(b) Repeat part (a) if a roller support is added at C and a shear release is inserted just left of C (see figure part b).

4.5-26 Beams ABC and CD are supported at A, C, and D and are joined by a hinge (or *moment release*) just to the left of C. The support at A is a sliding support (hence reaction $A_y = 0$ for the loading shown below). Find all support reactions; then plot shear (V) and moment (M) diagrams. *Label* all critical V and M values and also the *distance* to points where either V and/or M are zero.

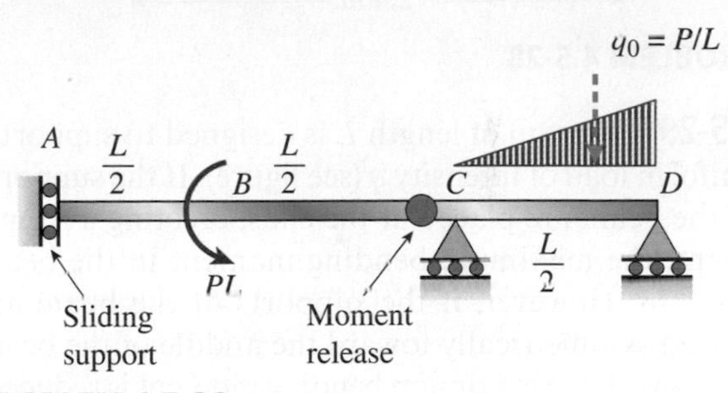

PROBLEM 4.5-26

4.5-27 The simple beam ACB shown in the figure is subjected to a triangular load of maximum intensity $q_0 = 2.6$ kN/m at $a = 1.8$ m and a concentrated moment $M = 400$ N·m at A.

(a) Draw the shear-force and bending-moment diagrams for this beam.

(b) Find the value of distance a that results in the maximum moment occurring at $L/2$. Draw the shear-force and bending-moment diagrams for this case.

(c) Find the value of distance a for which M_{max} is the largest possible value.

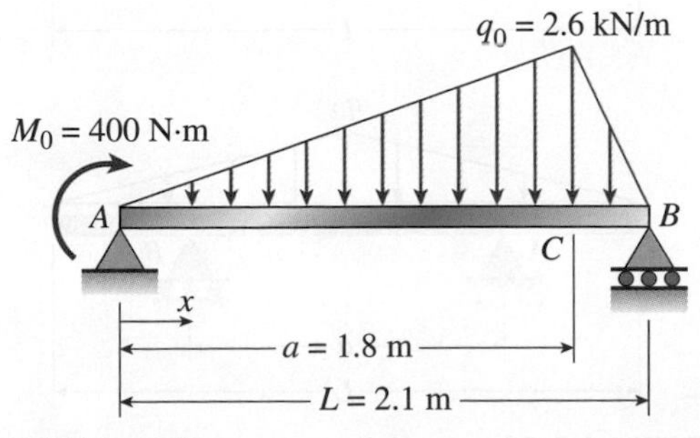

PROBLEM 4.5-27

4.5-28 A beam with simple supports is subjected to a trapezoidally distributed load (see figure). The intensity of the load varies from 1.0 kN/m at support A to 2.5 kN/m at support B.

(a) Draw the shear-force and bending-moment diagrams for this beam. Assume that M_0 at B is zero.

(b) Find the required moment M_0 at B so that the maximum moment in the beam does not exceed 1.0 kN · m.

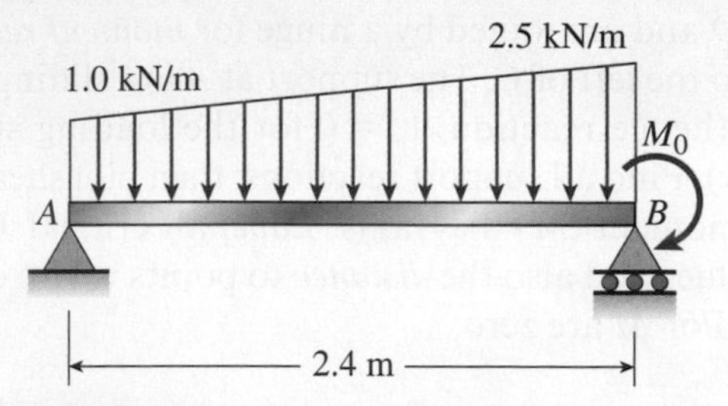

PROBLEM 4.5-28

4.5-29 A beam of length L is designed to support a uniform load of intensity q (see figure). If the supports of the beam are placed at the ends, creating a simple beam, the maximum bending moment in the beam is $qL^2/8$. However, if the supports of the beam are moved symmetrically toward the middle of the beam (as shown), the maximum bending moment is reduced.

(a) Determine the distance a between the supports so that the maximum bending moment in the beam has the smallest possible numerical value. Draw the shear-force and bending-moment diagrams for this condition.

(b) Repeat part (a) if the uniform load is replaced with a triangularly distributed load with peak intensity $q_0 = q$ at mid-span (see Fig. b).

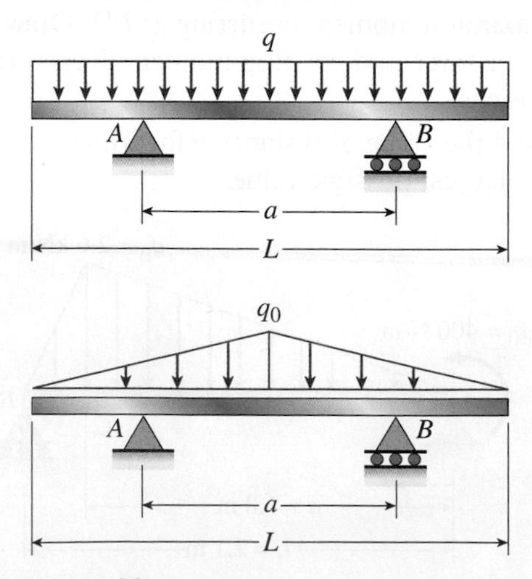

PROBLEM 4.5-29

4.5-30 The compound beam $ABCDE$ shown in the figure consists of two beams (AD and DE) joined by a hinged connection (or moment release) at D. The moment release can transmit a shear force but not a bending moment. Draw the shear-force and bending-moment diagrams for this compound beam.

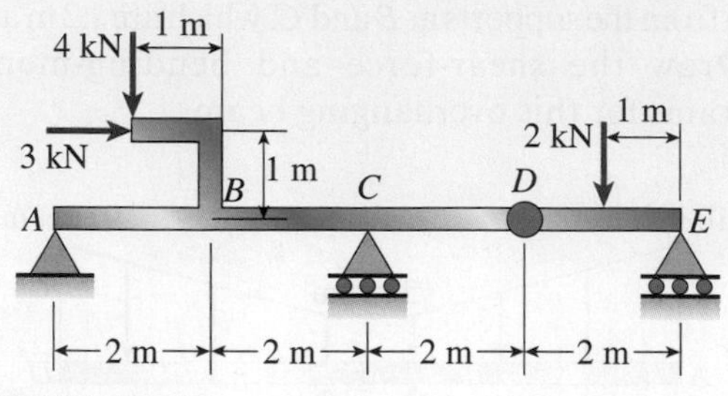

PROBLEM 4.5-30

4.5-31 Draw the shear-force and bending-moment diagrams for beam AB with a sliding support at A and an elastic support with spring constant k at B acted upon by two different load cases:

(a) A distributed load with linear variation and maximum intensity q_0 (see figure part a).

(b) A distributed load with parabolic variation with maximum intensity q_0 (see figure part b).

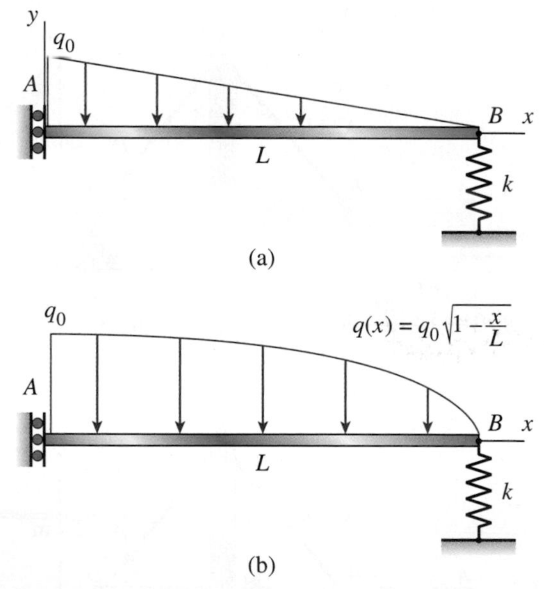

PROBLEM 4.5-31

4.5-32 The shear-force diagram for a simple beam is shown in the figure.

Determine the loading on the beam and draw the bending-moment diagram, assuming that no couples act as loads on the beam.

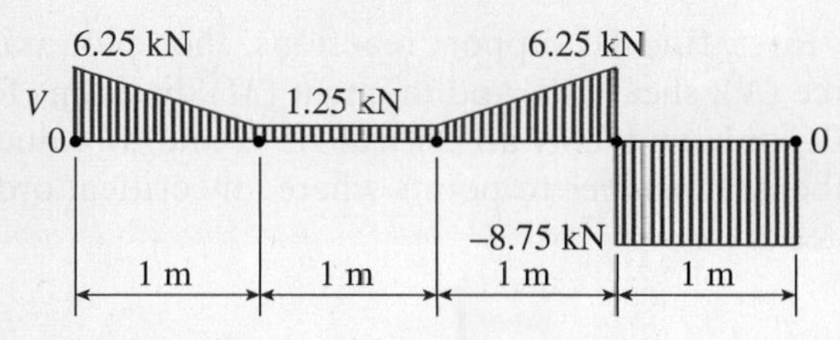

PROBLEM 4.5-32

4.5-33 The shear-force diagram for a beam is shown in the figure. Assuming that no couples act as loads on the beam, determine the forces acting on the beam and draw the bending-moment diagram.

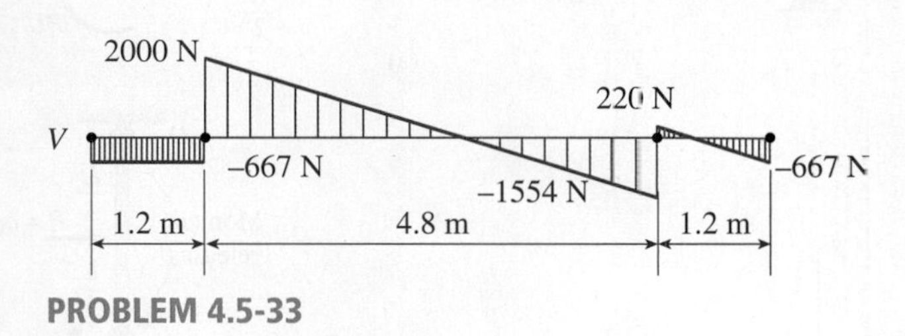

PROBLEM 4.5-33

4.5-34 A compound beam (see figure) has an internal *moment release* just to the left of *B* and a *shear release* just to the right of *C*. Reactions have been computed at *A*, *C*, and *D* and are shown in the figure.

First, confirm the reaction expressions using statics; then plot shear (*V*) and moment (*M*) diagrams. *Label* all critical *V* and *M* values and also the *distance* to points where either *V* and/or *M* are zero.

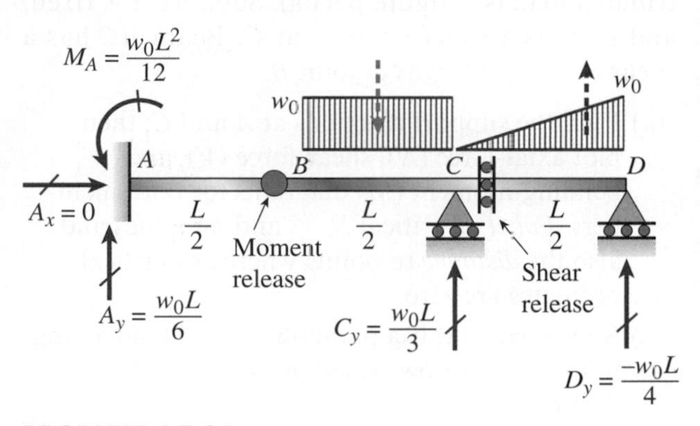

PROBLEM 4.5-34

4.5-35 A compound beam (see figure) has an *shear release* just to the left of *C* and a *moment release* just to the right of *C*. A plot of the moment diagram is provided below the beam for applied load *P* at *B* and triangular distributed loads *w*(*x*) on segments *BC* and *CD*.

First, solve for reactions using statics; then plot axial force (*N*) and shear force (*V*) diagrams. Confirm that the moment diagram is that shown below. *Label* all critical *N*, *V*, and *M* values and also the *distance* to points where *N*, *V*, and/or *M* are zero.

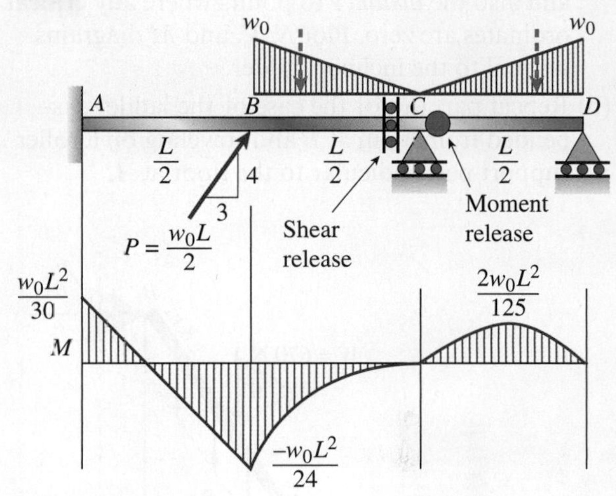

PROBLEM 4.5-35

4.5-36 A simple beam *AB* supports two connected wheel loads 3*P* and 2*P* that are a distance *d* apart (see figure). The wheels may be placed at any distance *x* from the left-hand support of the beam. (Assume $P = 12$ kN, $d = 2$ m, and $L = 15$ m.)

(a) Determine the distance *x* that will produce the maximum shear force in the beam, and also determine the maximum shear force V_{max}.

(b) Determine the distance *x* that will produce the maximum bending moment in the beam, and also draw the corresponding bending-moment diagram.

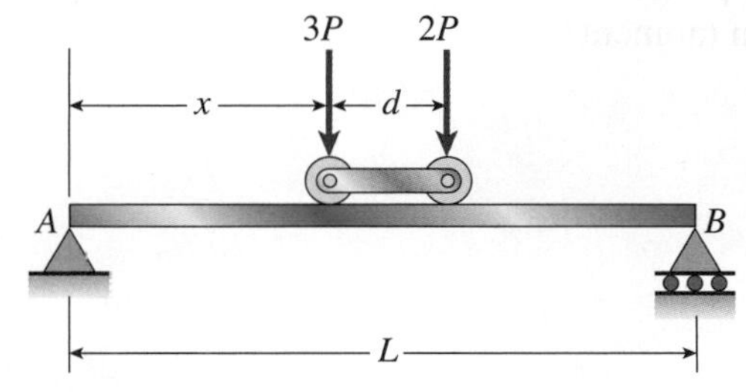

PROBLEM 4.5-36

4.5-37 The inclined beam represents a ladder with the following applied loads: the weight (W) of the house painter and the distributed weight (w) of the ladder itself.

(a) Find support reactions at A and B; then plot axial force (N), shear (V), and moment (M) diagrams. *Label* all critical N, V, and M values and also the *distance* to points where any critical ordinates are zero. Plot N, V, and M diagrams normal to the inclined ladder.

(b) Repeat part (a) for the case of the ladder suspended from a pin at B and traveling on a roller support perpendicular to the floor at A.

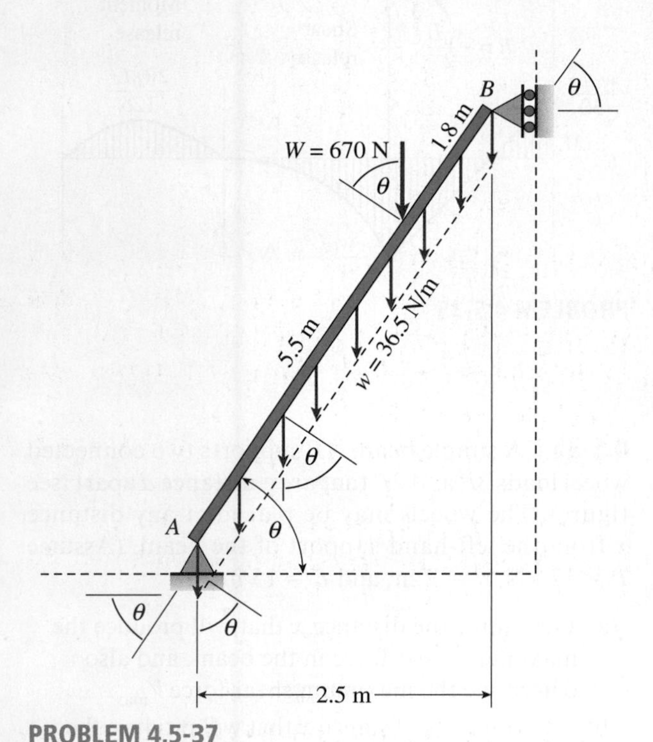

PROBLEM 4.5-37

4.5-38 Beam ABC is supported by a tie rod CD as shown. Two configurations are possible: pin support at A and downward triangular load on AB or pin at B and upward load on AB. Which has the larger maximum moment?

First, find all support reactions; then plot axial force (N), shear (V), and moment (M) diagrams for ABC only and *label* all critical N, V, and M values. Label the *distance* to points where any critical ordinates are zero.

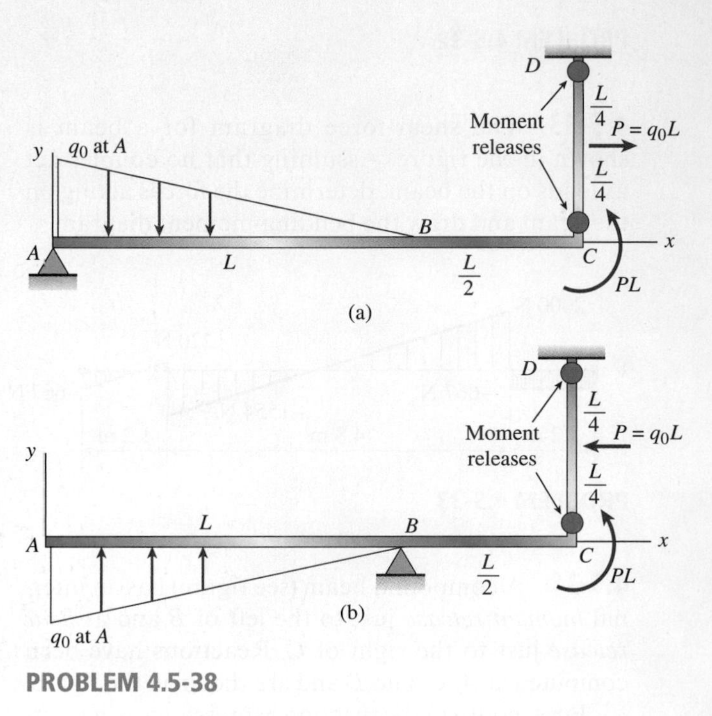

PROBLEM 4.5-38

4.5-39 A plane frame (see figure) consists of column AB and beam BC that carries a triangular distributed load (see figure part a). Support A is fixed, and there is a roller support at C. Beam BC has a shear release just right of joint B.

(a) Find the support reactions at A and C; then plot axial-force (N), shear-force (V), and bending-moment (M) diagrams for both members. *Label* all critical N, V, and M values and also the *distance* to points where any critical ordinates are zero.

(b) Repeat part (a) if a parabolic lateral load acting to the right is now added on column AB (figure part b).

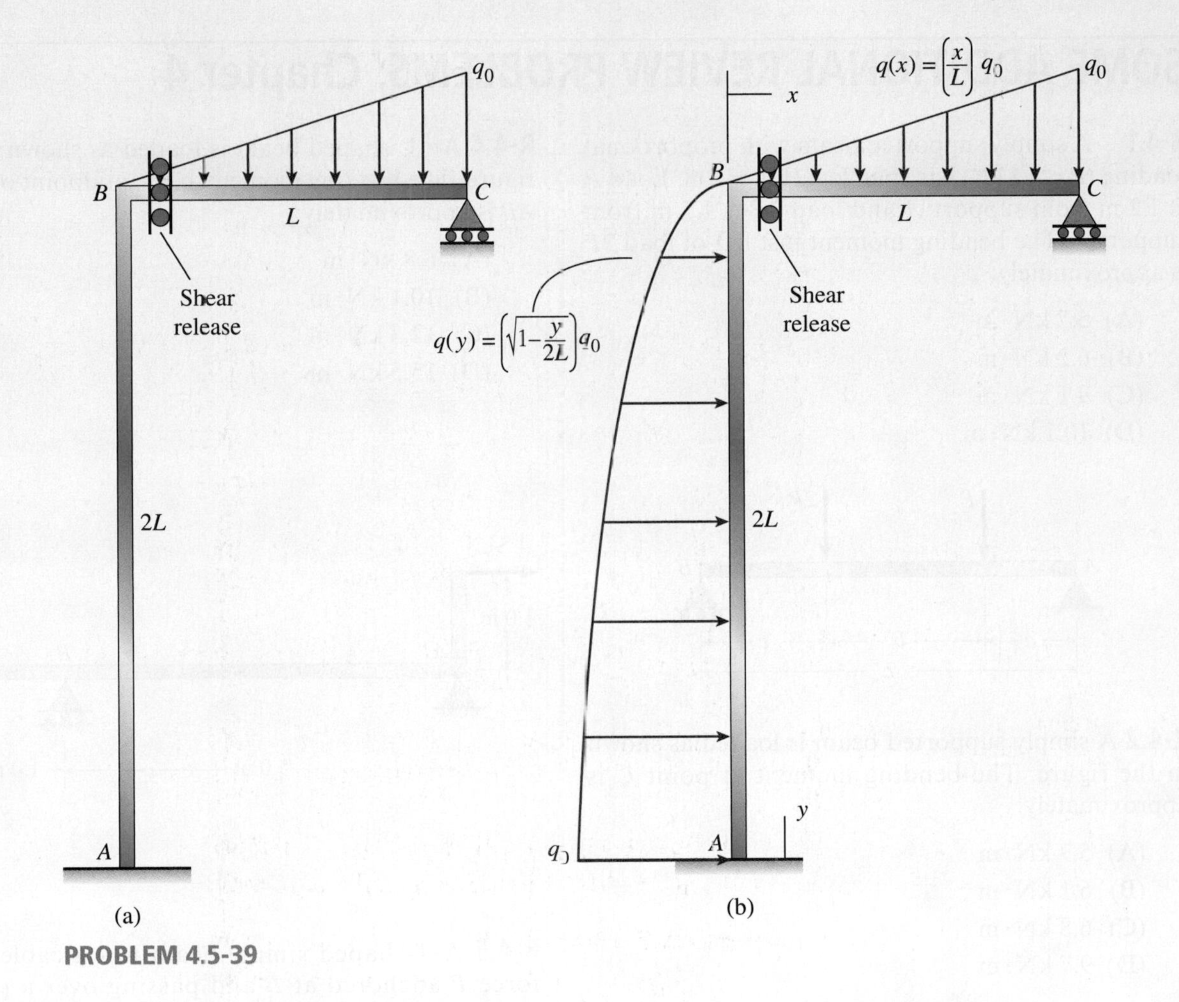

PROBLEM 4.5-39

4.5-40 The plane frame shown in the figure is part of an elevated freeway system. Supports at A and D are fixed, but there are *moment releases* at the base of both columns (AB and DE) as well as in column BC and at the end of beam BE.

Find all support reactions; then plot axial-force (N), shear (V), and moment (M) diagrams for all beam and column members. *Label* all critical N, V, and M values and also the *distance* to points where any critical ordinates are zero.

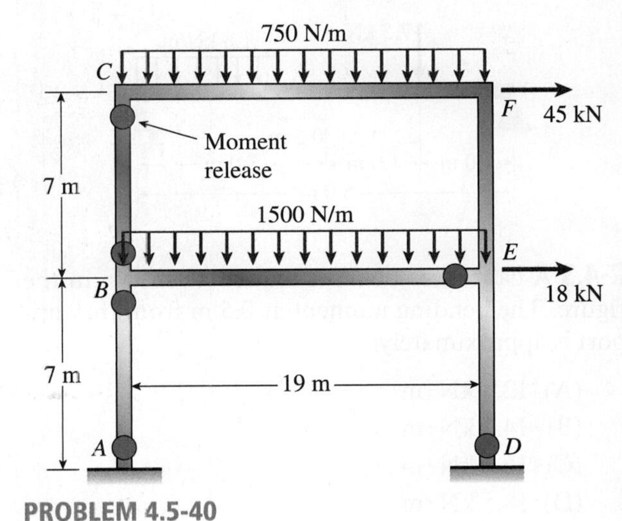

PROBLEM 4.5-40

SOME ADDITIONAL REVIEW PROBLEMS: Chapter 4

R-4.1 A simply supported beam with proportional loading ($P = 4.1$ kN) has span length $L = 5$ m. Load P is 1.2 m from support A and load $2P$ is 1.5 m from support B. The bending moment just left of load $2P$ is approximately:

(A) 5.7 kN·m
(B) 6.2 kN·m
(C) 9.1 kN·m
(D) 10.1 kN·m

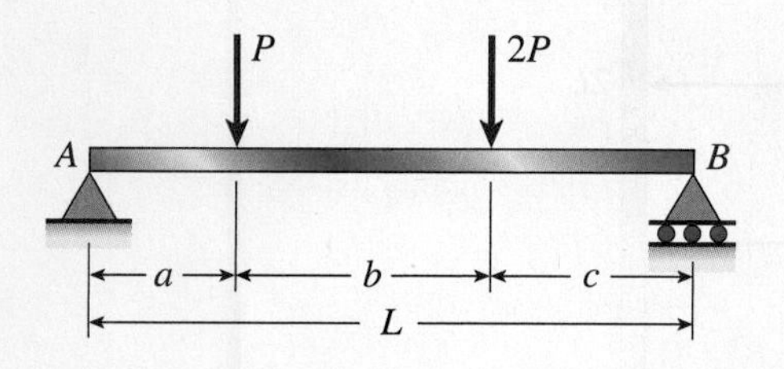

R-4.2 A simply supported beam is loaded as shown in the figure. The bending moment at point C is approximately:

(A) 5.7 kN·m
(B) 6.1 kN·m
(C) 6.8 kN·m
(D) 9.7 kN·m

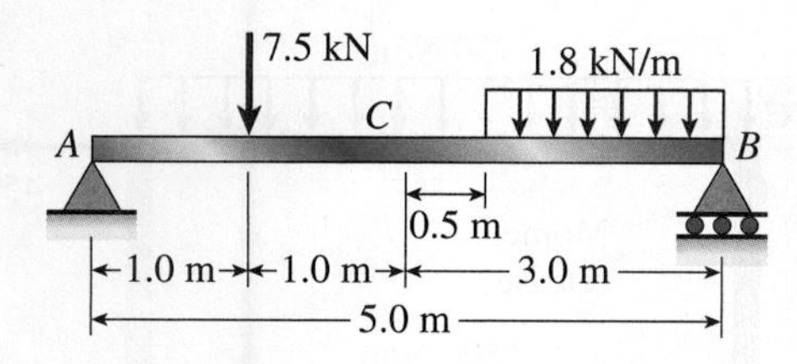

R-4.3 A cantilever beam is loaded as shown in the figure. The bending moment at 0.5 m from the support is approximately:

(A) 12.7 kN·m
(B) 14.2 kN·m
(C) 16.1 kN·m
(D) 18.5 kN·m

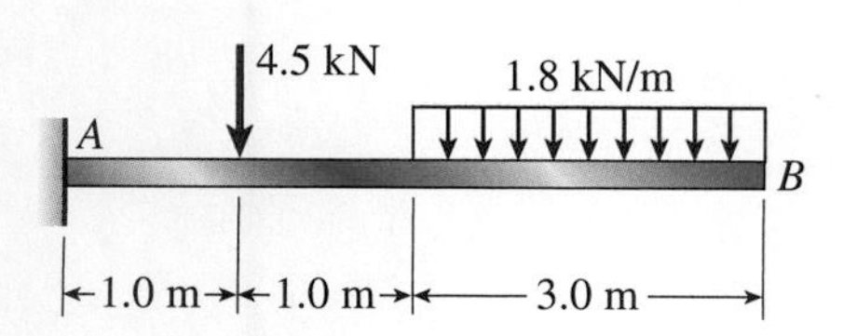

R-4.4 An L-shaped beam is loaded as shown in the figure. The bending moment at the midpoint of span AB is approximately:

(A) 6.8 kN·m
(B) 10.1 kN·m
(C) 12.3 kN·m
(D) 15.5 kN·m

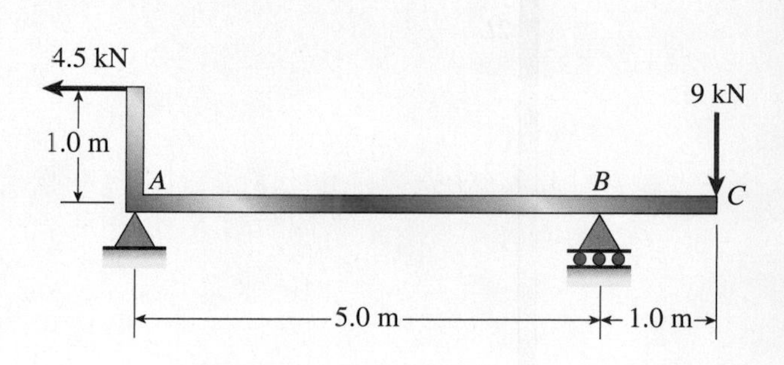

R-4.5 A T-shaped simple beam has a cable with force P anchored at B and passing over a pulley at E, as shown in the figure. The bending moment just left of C is 1.25 kN·m. The cable force P is approximately:

(A) 2.7 kN
(B) 3.9 kN
(C) 4.5 kN
(D) 6.2 kN

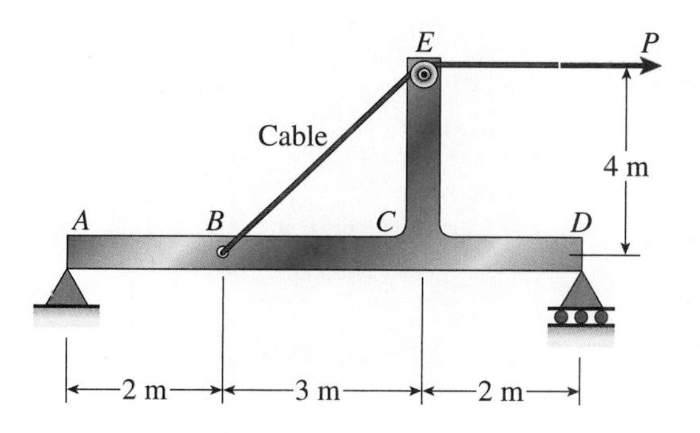

R-4.6 A simple beam ($L = 9$ m) with attached bracket BDE has force $P = 5$ kN applied downward at E. The bending moment just right of B is approximately:

(A) 6 kN·m
(B) 10 kN·m
(C) 19 kN·m
(D) 22 kN·m

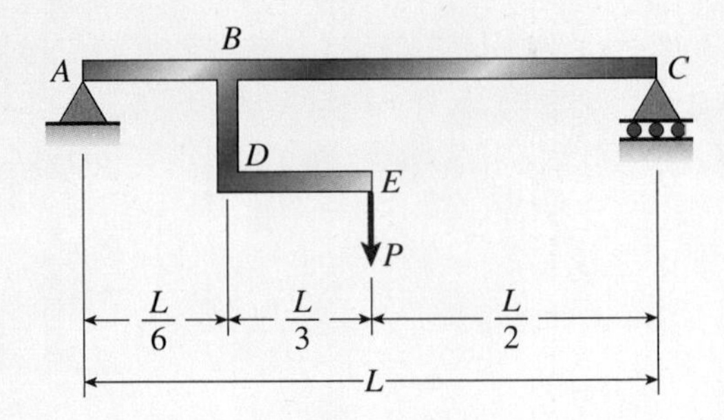

R-4.7 A simple beam AB with an overhang BC is loaded as shown in the figure. The bending moment at the midspan of AB is approximately:

(A) 8 kN·m
(B) 12 kN·m
(C) 17 kN·m
(D) 21 kN·m

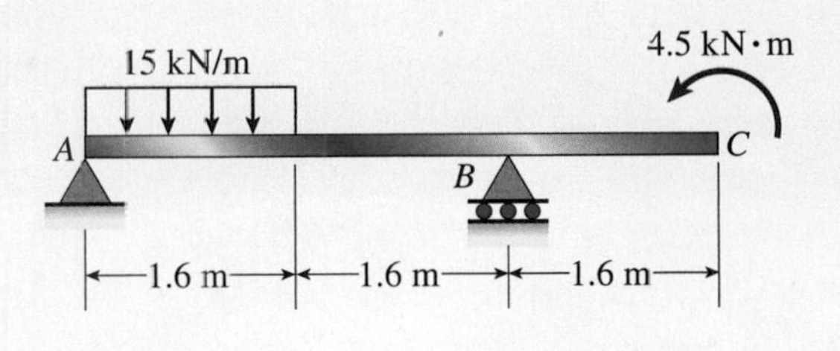

CHAPTER 5

Stresses in Beams (Basic Topics)

Jeff_Hu/iStock/Getty Images Plus/Getty Images

Beams are essential load carrying components in a wide variety of modern structures.

Chapter Objectives

- Develop a relationship between moment and curvature for a beam loaded by transverse applied loads and bending moments.
- Define the flexure formula, which shows that normal stresses vary linearly over the depth of a beam and are proportional to the bending moment and inversely proportional to the moment of inertia of the cross section.
- Define the section modulus of a beam and use it to design beams made of steel, wood, or other materials based upon an allowable stress for the material.
- Investigate shear stresses in beams of different shapes and study the variation of shear stress over the depth of a beam using a shear formula.
- Design the glued or nailed connections between the parts of built-up beams to ensure that the connections are strong enough to transmit the horizontal shear forces acting between the parts of the beam.
- Superpose bending and axial stresses for structural members subjected to simultaneous action of transverse and axial loads.
- Evaluate normal stresses in beams at locations of holes or other abrupt changes in cross section where stress concentrations occur.

Chapter Outline

***Specialized and/or advanced topics**

5.1 Introduction

In the preceding chapter, you saw how the loads acting on a beam create internal actions (or *stress resultants*) in the form of shear forces and bending moments. This chapter goes one step further and investigates the *stresses* and *strains* associated with those shear forces and bending moments. Knowing the stresses and strains, you will analyze and design beams subjected to a variety of loading conditions.

The loads acting on a beam cause the beam to bend (or *flex*), thereby deforming its axis into a curve. As an example, consider a cantilever beam AB subjected to a load P at the free end (Fig. 5-1a). The initially straight axis is bent into a curve (Fig. 5-1b), called the **deflection curve** of the beam.

FIGURE 5-1

Bending of a cantilever beam: (a) beam with load and (b) deflection curve

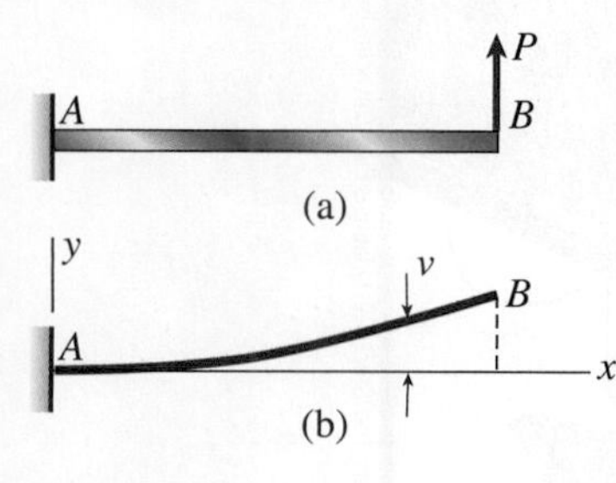

For reference purposes, a system of **coordinate axes** (Fig. 5-1b) is constructed with the origin located at a suitable point on the longitudinal axis of the beam. In this illustration, the origin is placed at the fixed support. The positive x axis is directed to the right, and the positive y axis is directed upward. The z axis, not shown in the figure, is directed outward (that is, toward the viewer), so that the three axes form a right-handed coordinate system.

The beams considered in this chapter are assumed to be symmetric about the x-y plane, which means that the y axis is an axis of symmetry of the cross section. In addition, all loads must act in the x-y plane. As a consequence, the bending deflections occur in this same plane, known as the **plane of bending**. Thus, the deflection curve shown in Fig. 5-1b is a plane curve lying in the plane of bending.

The **deflection** of the beam at any point along its axis is the *displacement* of that point from its original position measured in the y direction. Denote the deflection by the letter v to distinguish it from the coordinate y itself (see Fig. 5-1b).[1]

5.2 Pure Bending and Nonuniform Bending

When analyzing beams, it is often necessary to distinguish between pure bending and nonuniform bending. **Pure bending** refers to the flexure of a beam under a constant bending moment. Therefore, pure bending occurs only in regions of a beam where the shear force is zero (because $V = dM/dx$). In contrast, **nonuniform bending** refers to flexure in the presence of shear forces, which means that the bending moment changes as you move along the axis of the beam.

FIGURE 5-2

Simple beam in pure bending ($M = M_1$)

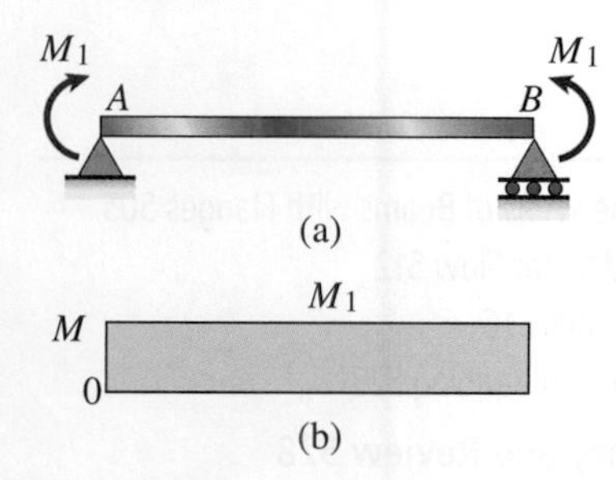

As an example of pure bending, consider a simple beam AB loaded by two couples M_1 having the same magnitude but acting in opposite directions (Fig. 5-2a). These loads produce a constant bending moment $M = M_1$ throughout the length of the beam, as shown by the bending moment diagram in (Fig. 5-2b). Note that the shear force V is zero at all cross sections of the beam.

Figure 5-3a shows pure bending, where the cantilever beam AB is subjected to a clockwise couple M_2 at the free end. There are no shear forces in this beam, and the bending moment M is constant throughout its length. The bending moment is negative ($M = -M_2$), as shown by the bending moment diagram in Fig. 5-3b.

The symmetrically loaded simple beam of Fig. 5-4a is an example of a beam that is partly in pure bending and partly in nonuniform bending, as seen from the shear-force and bending-moment diagrams (Figs. 5-4b and c).

[1]In applied mechanics, the traditional symbols for displacements in the x, y, and z directions are u, v, and w, respectively.

The central region of the beam is in pure bending because the shear force is zero and the bending moment is constant. The parts of the beam near the ends are in nonuniform bending because shear forces are present and the bending moments vary.

In the following two sections, the strains and stresses in beams subjected only to pure bending are investigated. Fortunately, the results obtained for pure bending can be used even when shear forces are present, as explained in Section 5.8.

FIGURE 5-3

Cantilever beam in pure bending ($M = -M_2$)

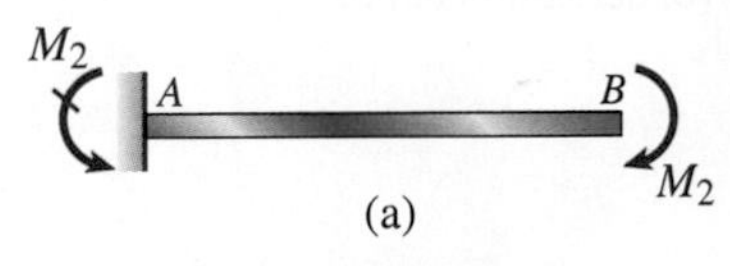

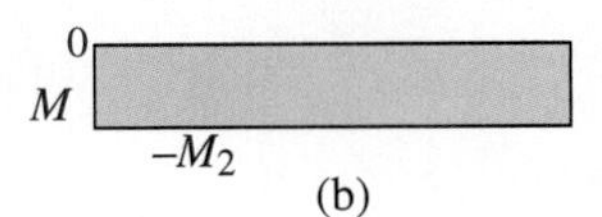

5.3 Curvature of a Beam

When loads are applied to a beam, its longitudinal axis is deformed into a curve, as illustrated previously in Fig. 5-1. The resulting strains and stresses in the beam are directly related to the **curvature** of the deflection curve.

To illustrate the concept of curvature, consider again a cantilever beam subjected to a load P acting at the free end (see Fig. 5-5a on the next page). The deflection curve of this beam is shown in Fig. 5-5b. For purposes of analysis, identify two points m_1 and m_2 on the deflection curve. Point m_1 is selected at an arbitrary distance x from the y axis, and point m_2 is located a small distance ds further along the curve. At each of these points, draw a line normal to the *tangent* to the deflection curve, that is, normal to the curve itself. These normals intersect at point O', which is the **center of curvature** of the deflection curve. Because most beams have very small deflections and nearly flat deflection curves, point O' is usually located much farther from the beam than is indicated in the figure.

The distance m_1O' from the curve to the center of curvature is called the **radius of curvature** ρ (rho), and the **curvature** κ (kappa) is defined as the reciprocal of the radius of curvature. Thus,

$$\kappa = \frac{1}{\rho} \tag{5-1}$$

Curvature is a measure of how sharply a beam is bent. If the load on a beam is small, the beam will be nearly straight, the radius of curvature will be very large, and the curvature will be very small. If the load is increased, the amount of bending will increase—the radius of curvature will become smaller, and the curvature will become larger.

The geometry of triangle $O'm_1m_2$ (Fig. 5-5b) leads to

$$\rho d\theta = ds \tag{5-2}$$

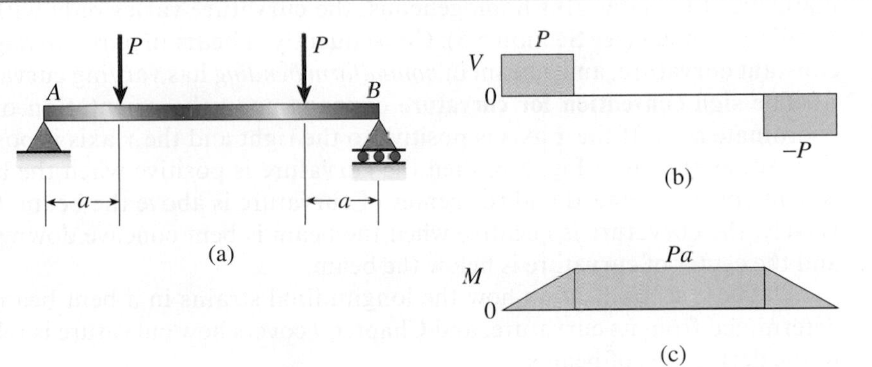

FIGURE 5-4

Simple beam with central region in pure bending and end regions in nonuniform bending

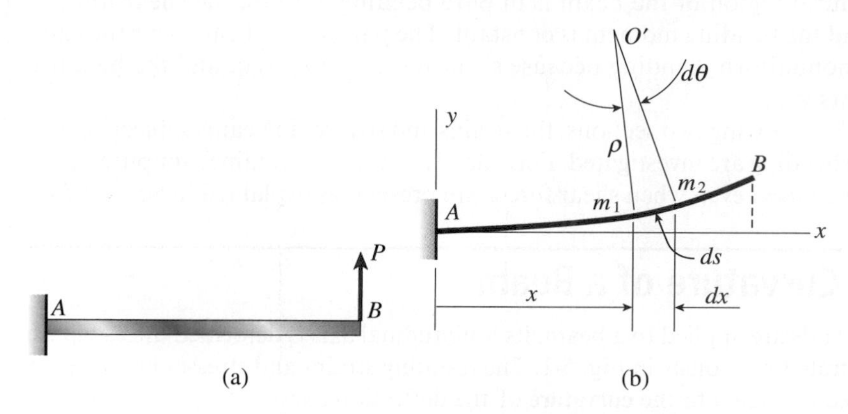

FIGURE 5-5

Curvature of a bent beam: (a) beam with load and (b) deflection curve

in which $d\theta$ (measured in radians) is the infinitesimal angle between the normals and ds is the infinitesimal distance along the curve between points m_1 and m_2. Combine Eq. (5-2) with Eq. (5-1) to get

$$\kappa = \frac{1}{\rho} = \frac{d\theta}{ds} \tag{5-3}$$

This equation for **curvature** is derived in textbooks on calculus and holds for any curve, regardless of the amount of curvature. If the curvature is *constant* throughout the length of a curve, the radius of curvature also will be constant, and the curve will be an arc of a circle.

The deflections of a beam are usually very small compared to its length (consider, for instance, the deflections of the structural frame of an automobile or a beam in a building). Small deflections mean that the deflection curve is nearly flat. Consequently, the distance ds along the curve may be set equal to its horizontal projection dx (see Fig. 5-5b). Under these special conditions of **small deflections**, the equation for the curvature becomes

$$\kappa = \frac{1}{\rho} = \frac{d\theta}{dx} \tag{5-4}$$

Both the curvature and the radius of curvature are functions of the distance x measured along the x axis. It follows that the position O' of the center of curvature also depends upon the distance x.

The curvature at a particular point on the axis of a beam depends upon the bending moment at that point and upon the properties of the beam itself (shape of cross section and type of material). Therefore, if the beam is prismatic and the material is homogeneous, the curvature varies only with the bending moment (see Section 5.5). Consequently, a beam in *pure bending* has constant curvature, and a beam in *nonuniform bending* has varying curvature.

The **sign convention for curvature** depends upon the orientation of the coordinate axes. If the x axis is positive to the right and the y axis is positive upward, as shown in Fig. 5-6, then the curvature is positive when the beam is bent concave upward and the center of curvature is above the beam. Conversely, the curvature is negative when the beam is bent concave downward, and the center of curvature is below the beam.

FIGURE 5-6

Sign convention for curvature

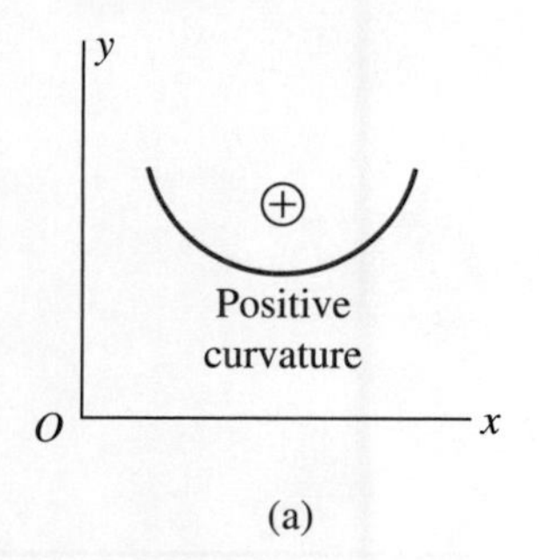

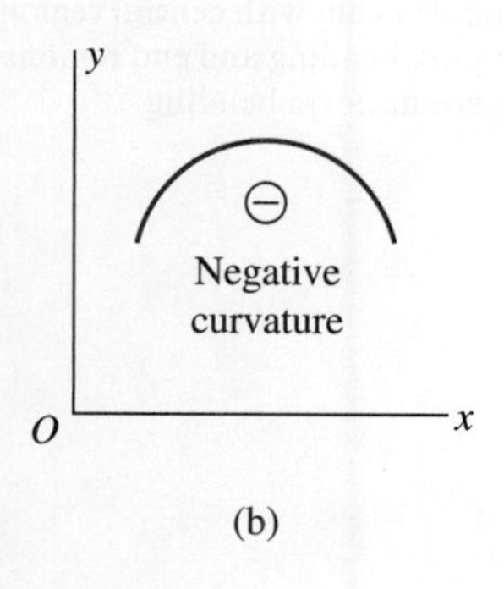

The next section shows how the longitudinal strains in a bent beam are determined from its curvature, and Chapter 9 covers how curvature is related to the deflections of beams.

5.4 Longitudinal Strains in Beams

The longitudinal strains in a beam can be found by analyzing the curvature of the beam and the associated deformations. For this purpose, consider a portion AB of a beam in pure bending subjected to positive bending moments M (Fig. 5-7a). Assume that the beam initially has a straight longitudinal axis (the x axis in the figure) and that its cross section is symmetric about the y axis, as shown in Fig. 5-7b.

Under the action of the bending moments, the beam deflects in the x-y plane (the plane of bending) and its longitudinal axis is bent into a circular curve (curve s–s in Fig. 5-7c). The beam is bent concave upward, which is positive curvature (Fig. 5-6a).

Cross sections of the beam, such as sections mn and pq in Fig. 5-7a, remain plane and normal to the longitudinal axis (Fig. 5-7c). The fact that cross sections of a beam in pure bending remain plane is so fundamental to beam theory that it is often called an assumption. However, it also could be called a theorem because it can be proved rigorously using only rational arguments based upon symmetry (Ref. 5-1). The basic point is that the symmetry of the beam and its loading (Figs. 5-7a and b) means that all elements of the beam (such as element $mpqn$) must deform in an identical manner, which is possible only if cross sections remain plane during bending (Fig. 5-7c). This conclusion is valid for beams of any material, whether the material is elastic or inelastic, linear or nonlinear. Of course, the material properties, like the dimensions, must be symmetric about the plane of bending. *Note:* Even though a plane cross section in pure bending remains plane, there still may be deformations in the plane

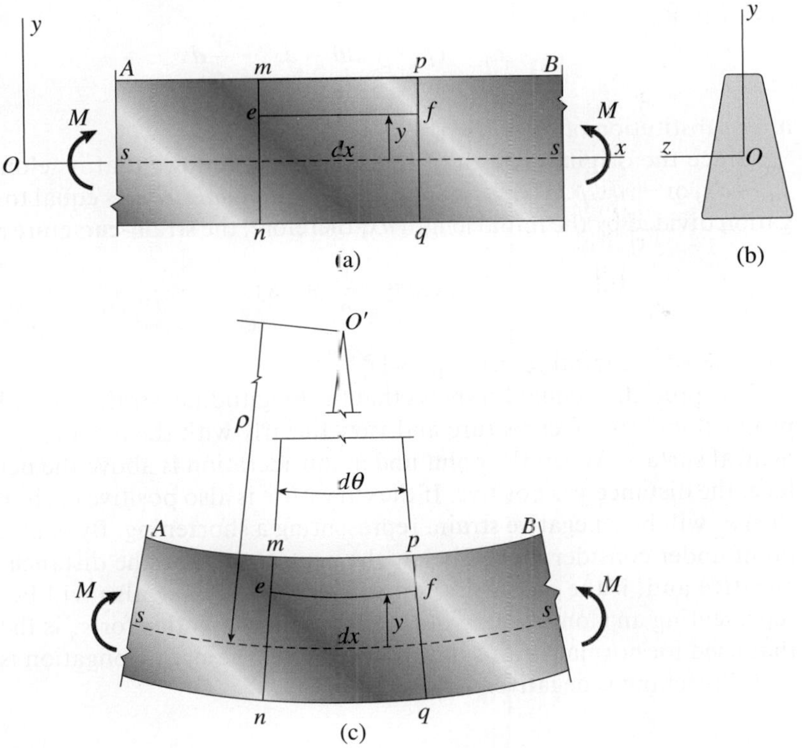

FIGURE 5-7

Deformations of a beam in pure bending: (a) side view of beam, (b) cross section of beam, and (c) deformed beam

itself. Such deformations are due to the effects of Poisson's ratio, as explained at the end of this discussion.

Because of the bending deformations shown in Fig. 5-7c, cross sections *mn* and *pq* rotate with respect to each other about axes perpendicular to the *x-y* plane. Longitudinal lines on the lower part of the beam are elongated, whereas those on the upper part are shortened. Thus, the lower part of the beam is in tension and the upper part is in compression. Somewhere between the top and bottom of the beam is a surface in which longitudinal lines do not change in length. This surface, indicated by the dashed line *s–s* in Figs. 5-7a and c, is called the **neutral surface** of the beam. Its intersection with any cross-sectional plane is called the **neutral axis** of the cross section; for instance, the z axis is the neutral axis for the cross section of Fig. 5-7b.

The planes containing cross sections *mn* and *pq* in the deformed beam (Fig. 5-7c) intersect in a line through the center of curvature O'. The angle between these planes is denoted $d\theta$, and the distance from O' to the neutral surface *s–s* is the radius of curvature ρ. The initial distance dx between the two planes (Fig. 5-7a) is unchanged at the neutral surface (Fig. 5-7c), hence $\rho d\theta = dx$. However, all other longitudinal lines between the two planes either lengthen or shorten, thereby creating **normal strains** ε_x.

To evaluate these normal strains, consider a typical longitudinal line *ef* located within the beam between planes *mn* and *pq* (Fig. 5-7a). Identify line *ef* by its distance y from the neutral surface in the initially straight beam. Now assume that the x axis lies along the neutral surface of the *undeformed* beam. Of course, when the beam deflects, the neutral surface moves with the beam, but the x axis remains fixed in position. Nevertheless, the longitudinal line *ef* in the deflected beam (Fig. 5-7c) is still located at the same distance y from the neutral surface. Thus, the length L_1 of line *ef* after bending takes place is

$$L_1 = (\rho - y)d\theta = dx - \frac{y}{\rho}dx$$

after substitution of $d\theta = dx/\rho$.

Since the original length of line *ef* is dx, it follows that its elongation is $L_1 - dx$, or $-ydx/\rho$. The corresponding *longitudinal strain* is equal to the elongation divided by the initial length dx; therefore, the **strain-curvature relation** is

$$\varepsilon_x = -\frac{y}{\rho} = -\kappa y \qquad \textbf{(5-5)}$$

where κ is the curvature [see Eq. (5-1)].

The preceding equation shows that the longitudinal strains in the beam are proportional to the curvature and vary linearly with the distance y from the neutral surface. When the point under consideration is above the neutral surface, the distance y is positive. If the curvature is also positive (as in Fig. 5-7c), then ε_x will be a negative strain, representing a shortening. By contrast, if the point under consideration is below the neutral surface, the distance y will be negative and, if the curvature is positive, the strain ε_x also will be positive, representing an elongation. Note that the **sign convention** for ε_x is the same as that used for normal strains in earlier chapters, namely, elongation is positive and shortening is negative.

Equation (5-5) for the normal strains in a beam was derived solely from the geometry of the deformed beam—the properties of the material did not enter into the discussion. Therefore, *the strains in a beam in pure bending vary linearly with distance from the neutral surface regardless of the shape of the stress-strain curve of the material.*

The next step in the analysis, namely, finding the stresses from the strains, requires the use of the *stress-strain curve*. This step is described in the next section for linearly elastic materials and in Section 6.10 for elastoplastic materials.

The longitudinal strains in a beam are accompanied by *transverse strains* (that is, normal strains in the y and z directions) because of the effects of Poisson's ratio. However, there are no accompanying transverse stresses because beams are free to deform laterally. This stress condition is analogous to that of a prismatic bar in tension or compression, and therefore, *longitudinal elements in a beam in pure bending are in a state of uniaxial stress.*

Example 5-1

FIGURE 5-8

Example 5-1: Beam in pure bending: (a) beam with loads and (b) deflection curve

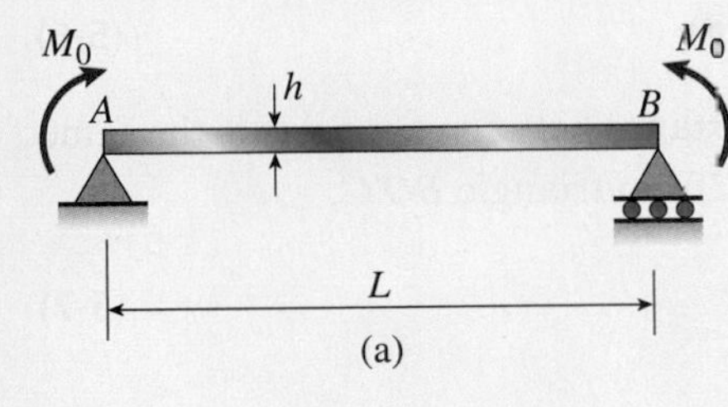

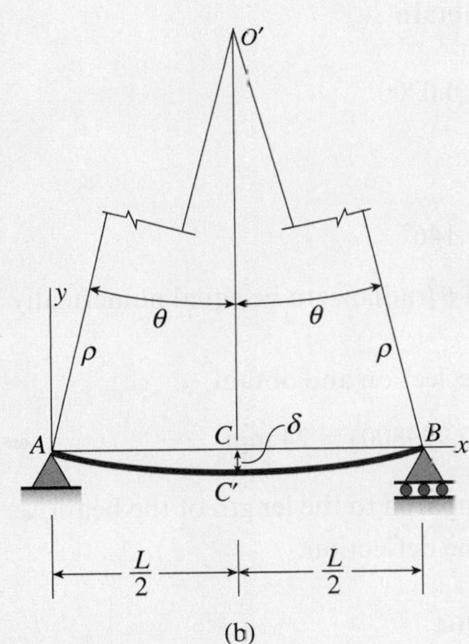

A simply supported steel beam AB (Fig. 5-8a) of a length $L = 4.9$ m and height $h = 300$ mm is bent by couples M_0 into a circular arc with a downward deflection δ at the midpoint (Fig. 5-8b). The longitudinal normal strain (elongation) on the bottom surface of the beam is 0.00125, and the distance from the neutral surface to the bottom surface of the beam is 150 mm.

Determine the radius of curvature ρ, the curvature κ, and the deflection δ of the beam.

Note: This beam has a relatively large deflection because its length is large compared to its height ($L/h = 16.33$), and the strain of 0.00125 is also large. (This is about the same as the yield strain for ordinary structural steel.)

Solution:

Use a four-step problem-solving approach. Combine steps as needed for an efficient solution.

Part (a): Curvature.

1, 2. ***Conceptualize*** *[hypothesize, sketch]*, ***Categorize*** *[simplify, classify]*: Since the longitudinal strain at the bottom surface of the beam ($\varepsilon_x = 0.00125$) and the distance from the neutral surface to the bottom surface ($y = -150$ mm) are known, use Eq. (5-5) to calculate both the radius of curvature and the curvature.

3. ***Analyze*** [*evaluate; select relevant equations, carry out mathematical solution*]: Rearrange Eq. (5-5) and substitute numerical values to get

$$\rho = -\frac{y}{\varepsilon_x} = -\frac{-150 \text{ mm}}{0.00125} = 120 \text{ m} \quad \kappa = \frac{1}{\rho} = 8.33 \times 10^{-3} \text{m}^{-1}$$

4. ***Finalize*** [*conclude; examine answer—Does it make sense? Are units correct? How does it compare to similar problem solutions?*]: These results show that the radius of curvature is extremely large compared to the length of the beam even when the strain in the material is large. If, as usual, the strain is less, the radius of curvature is even larger.

Part (b): Deflection.

1, 2. ***Conceptualize, Categorize***: As pointed out in Section 5.3, a constant bending moment (pure bending) produces constant curvature throughout the length of a beam. Therefore, the deflection curve is a circular arc. From Fig. 5-8b, the distance from the center of curvature O' to the midpoint C' of the deflected beam is the radius of curvature ρ, and the distance from O' to point C on the x axis is $\rho \cos \theta$, where θ is angle $BO'C$. This leads to the expression for the deflection at the midpoint of the beam:

$$\delta = \rho(1 - \cos\ \theta) \tag{5-6}$$

For a nearly flat curve, assume that the distance between supports is the same as the length of the beam itself. Therefore, from triangle $BO'C$,

$$\sin\ \theta = \frac{L/2}{\rho} \tag{5-7}$$

3. ***Analyze***: Substitute numerical values to obtain

$$\sin\ \theta = \frac{4.9 \text{ m}}{2(120 \text{ m})} = 0.0200$$

and

$$\theta = 0.0200 \text{ rad} = 1.146°$$

For practical purposes, consider $\sin \theta$ and θ (radians) to be equal numerically because θ is a very small angle.

Now substitute into Eq. (5-6) for the deflection and obtain

$$\delta = \rho(1 - \cos\ \theta) = (120 \text{ m})(1 - 0.999800) = 24 \text{ mm}$$

4. ***Finalize***: This deflection is very small compared to the length of the beam, as shown by the ratio of the span length to the deflection:

$$\frac{L}{\delta} = \frac{4.9 \text{ m}}{24 \text{ mm}} = 204$$

This confirms that the deflection curve is nearly flat in spite of the large strains. Of course, in Fig. 5-8b, the deflection of the beam is highly exaggerated for clarity.

Note: The purpose of this example is to show the relative magnitudes of the radius of curvature, length of the beam, and deflection of the beam. However, the method used for finding the deflection has little practical value because it is limited to pure bending, which produces a circular deflected shape. More useful methods for finding beam deflections are presented in Chapter 9.

5.5 Normal Stresses in Beams (Linearly Elastic Materials)

Longitudinal strains ε_x in a beam in pure bending were investigated in the preceding section [see Eq. (5-5) and Fig. 5-7]. Since longitudinal elements of a beam are subjected only to tension or compression, now use the **stress-strain curve** for the material to determine the stresses from the strains. The stresses act over the entire cross section of the beam and vary in intensity, depending upon the shape of the stress-strain diagram and the dimensions of the cross section. Since the x direction is longitudinal (Fig. 5-7a), use the symbol σ_x to denote these stresses.

The most common stress-strain relationship encountered in engineering is the equation for a **linearly elastic material**. For such materials, substitute Hooke's law for uniaxial stress ($\sigma = E\varepsilon$) into Eq. (5-5) and obtain

$$\sigma_x = E\varepsilon_x = -\frac{Ey}{\rho} = -E\kappa y \quad \textbf{(5-8)}$$

This equation shows that the normal stresses acting on the cross section vary linearly with the distance y from the neutral surface. This stress distribution is pictured in Fig. 5-9a for the case in which the bending moment M is positive and the beam bends with positive curvature.

When the curvature is positive, the stresses σ_x are negative (compression) above the neutral surface and positive (tension) below it. In the figure, compressive stresses are indicated by arrows pointing *toward* the cross section and tensile stresses are indicated by arrows pointing *away* from the cross section.

In order for Eq. (5-8) to be of practical value, locate the origin of the coordinates so that you can determine the distance y. In other words, locate the neutral axis of the cross section. You also need to obtain a relationship between the curvature and the bending moment—so that you can substitute into Eq. (5-8) and obtain an equation relating the stresses to the bending moment. These two objectives can be accomplished by determining the resultant of the stresses σ_x acting on the cross section.

In general, the **resultant of the normal stresses** consists of two stress resultants: (1) a force acting in the x direction and (2) a bending couple acting about the z axis. However, the axial force is zero when a beam is in pure bending. Therefore, write the following equations of statics: (1) The resultant force in the x direction is equal to zero, and (2) the resultant moment is equal to the bending moment M. The first equation gives the location of the neutral axis, and the second gives the moment-curvature relationship.

FIGURE 5-9

Normal stresses in a beam of linearly elastic material: (a) side view of beam showing distribution of normal stresses and (b) cross section of beam showing the z axis as the neutral axis of the cross section

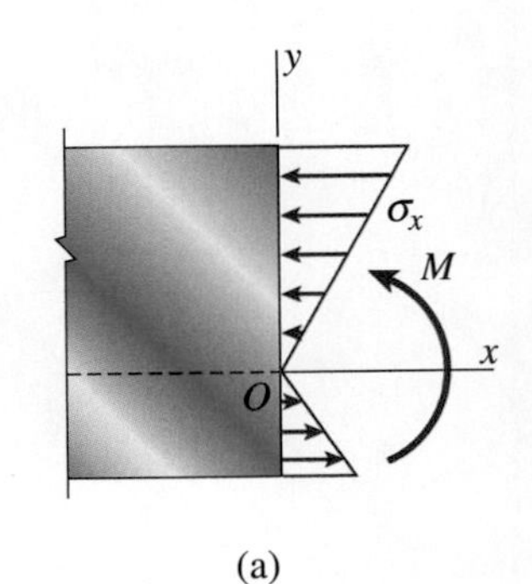

(a)

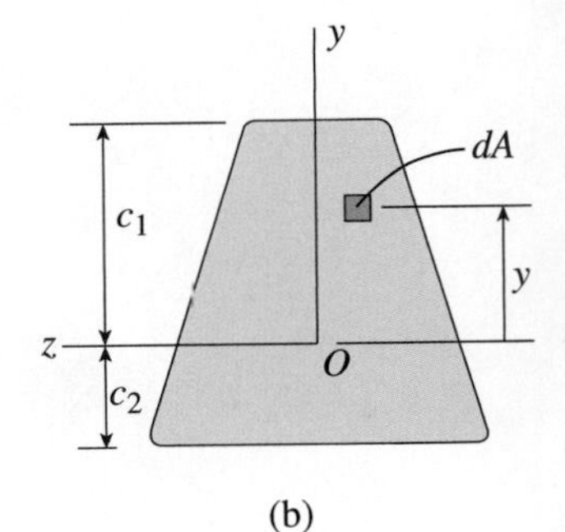

(b)

Location of Neutral Axis

To obtain the first equation of statics, consider an element of area dA in the cross section (Fig. 5-9b). The element is located at a distance y from the neutral axis; therefore, the stress σ_x acting on the element is given by Eq. (5-8). The *force* acting on the element is equal to $\sigma_x dA$ and is compressive when y is positive. Because there is no resultant force acting on the cross section, the integral of $\sigma_x dA$ over the area A of the entire cross section must vanish; thus, the *first equation of statics* is

$$\int_A \sigma_x dA = -\int_A E\kappa y\, dA = 0 \tag{5-9a}$$

Because the curvature κ and modulus of elasticity E are nonzero constants at any given cross section of a bent beam, they are not involved in the integration over the cross-sectional area. Therefore, drop them from the equation and obtain

$$\int_A y\, dA = 0 \tag{5-9b}$$

This equation states that the first moment of the area of the cross section, when evaluated with respect to the z axis, is zero. In other words, the z axis must pass through the centroid of the cross section.[2]

The z axis is also the neutral axis, so

The neutral axis passes through the centroid of the cross-sectional area when the material follows Hooke's law and there is no axial force acting on the cross section.

This observation makes it relatively simple to determine the position of the neutral axis.

As explained in Section 5.1, this discussion is limited to beams for which the y axis is an axis of symmetry. Consequently, the y axis also passes through the centroid. Therefore,

The origin O of coordinates (Fig. 5-9b) is located at the centroid of the cross-sectional area.

Because the y axis is an axis of symmetry of the cross section, the y axis is a *principal axis* (see Appendix D, Section D.8, for a discussion of principal axes). Since the z axis is perpendicular to the y axis, it too is a principal axis. Thus, when a beam of linearly elastic material is subjected to pure bending, *the y and z axes are principal centroidal axes.*

Moment-Curvature Relationship

The *second equation of statics* expresses the fact that the moment resultant of the normal stresses σ_x acting over the cross section is equal to the bending moment M (Fig. 5-9a). The element of force $\sigma_x dA$ acting on the element of area dA (Fig. 5-9b) is in the positive direction of the x axis when σ_x is positive and in the negative direction when σ_x is negative. Since the element dA is located above the neutral axis, a positive stress σ_x acting on that element produces an element of moment equal to $\sigma_x y dA$. This element of moment acts opposite in direction to the positive bending moment M shown in Fig. 5-9a. Therefore, the elemental moment is

$$dM = -\sigma_x y\, dA$$

[2]Centroids and first moments of areas are discussed in Appendix D, Sections D.1 and D.2.

The integral of all such elemental moments over the entire cross-sectional area A must equal the bending moment:

$$M = -\int_A \sigma_x y \, dA \tag{5-10a}$$

or, upon substituting for σ_x from Eq. (5-9),

$$M = \int_A \kappa E y^2 dA = \kappa E \int_A y^2 dA \tag{5-10b}$$

This equation relates the curvature of the beam to the bending moment M.

Since the integral in the preceding equation is a property of the cross-sectional area, it is convenient to rewrite the equation as

$$M = \kappa EI \tag{5-11}$$

in which

$$I = \int_A y^2 dA \tag{5-12}$$

This integral is the **moment of inertia** of the cross-sectional area with respect to the z axis (that is, with respect to the neutral axis). Moments of inertia are always positive and have dimensions of length to the fourth power; for instance, typical SI units are mm^4 when performing beam calculations.[3]

Equation (5-11) now can be rearranged to express the *curvature* in terms of the bending moment in the beam:

$$\kappa = \frac{1}{\rho} = \frac{M}{EI} \tag{5-13}$$

Known as the **moment-curvature equation**, Eq. (5-13) shows that the curvature is directly proportional to the bending moment M and inversely proportional to the quantity EI, which is called the **flexural rigidity** of the beam. Flexural rigidity is a measure of the resistance of a beam to bending, that is, the larger the flexural rigidity, the smaller the curvature for a given bending moment.

Comparing the **sign convention** for bending moments (Fig. 4-19) with that for curvature (Fig. 5-6), note that *a positive bending moment produces positive curvature and a negative bending moment produces negative curvature* (see Fig. 5-10).

FIGURE 5-10

Relationships between signs of bending moments and signs of curvatures

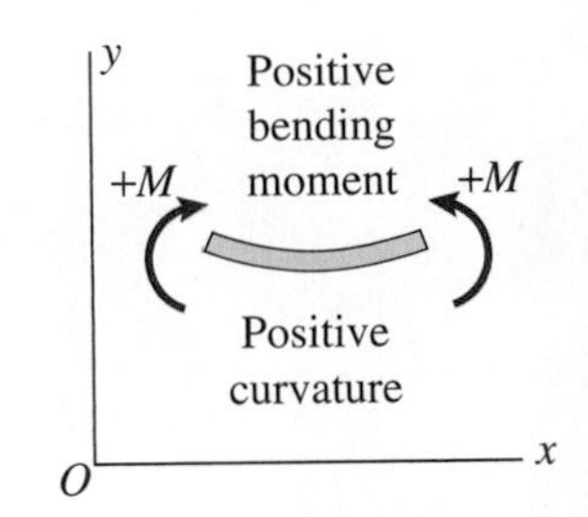

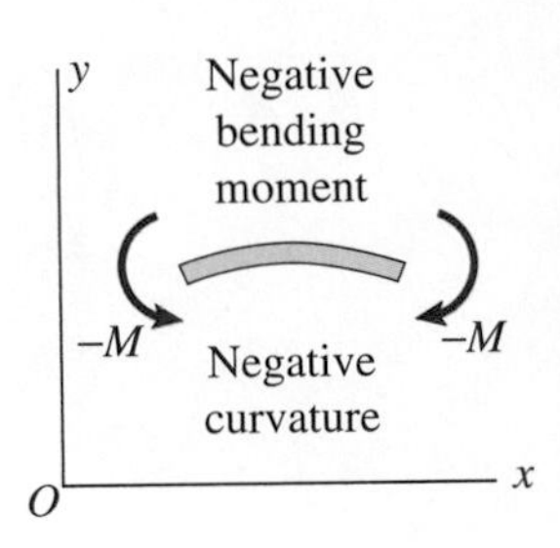

Flexure Formula

Now that the neutral axis has been located and the moment-curvature relationship has been derived, determine the stresses in terms of the bending moment. Substitute the expression for curvature [Eq. (5-13)] into the expression for the stress σ_x [Eq. (5-8)] to get

$$\sigma_x = -\frac{My}{I} \tag{5-14}$$

This equation, called the **flexure formula**, shows that the stresses are directly proportional to the bending moment M and inversely proportional to the moment of inertia I of the cross section. Also, the stresses vary linearly with the distance y from the neutral axis, as previously observed. Stresses calculated from the flexure formula are called **bending stresses** or **flexural stresses**.

[3]Moments of inertia of areas are discussed in Appendix D, Section D.3.

FIGURE 5-11

Relationships between signs of bending moments and directions of normal stresses: (a) positive bending moment and (b) negative bending moment

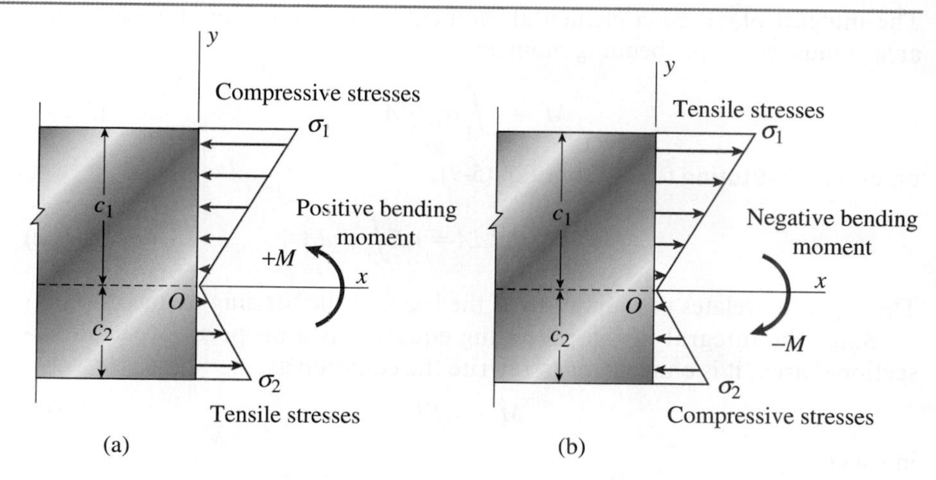

If the bending moment in the beam is positive, the bending stresses will be positive (tension) over the part of the cross section where y is negative, that is, over the lower part of the beam. The stresses in the upper part of the beam will be negative (compression). If the bending moment is negative, the stresses will be reversed. These relationships are shown in Fig. 5-11.

Maximum Stresses at a Cross Section

The maximum tensile and compressive bending stresses acting at any given cross section occur at points located farthest from the neutral axis. Denote by c_1 and c_2 the distances from the neutral axis to the extreme elements in the positive and negative y directions, respectively (see Fig. 5-9b and Fig. 5-11). Then the corresponding **maximum normal stresses** σ_1 and σ_2 (from the flexure formula) are

$$\sigma_1 = -\frac{Mc_1}{I} = -\frac{M}{S_1} \qquad \sigma_2 = \frac{Mc_2}{I} = \frac{M}{S_2} \tag{5-15a,b}$$

in which

$$S_1 = \frac{I}{c_1} \qquad S_2 = \frac{I}{c_2} \tag{5-16a,b}$$

The quantities S_1 and S_2 are known as the **section moduli** of the cross-sectional area. From [Eqs. (5-16a and b)], note that each section modulus has dimensions of a length to the third power (for example, mm^3). Also note that the distances c_1 and c_2 to the top and bottom of the beam are always taken as positive quantities.

The advantage of expressing the maximum stresses in terms of section moduli arises from the fact that each section modulus combines the beam's relevant cross-sectional properties into a single quantity. Then this quantity can be listed in tables and handbooks as a property of the beam, which is a convenience to designers. (Design of beams using section moduli is explained in the next section.)

Doubly Symmetric Shapes

If the cross section of a beam is symmetric with respect to the z axis as well as the y axis (*doubly symmetric cross section*), then $c_1 = c_2 = c$, and the maximum tensile and compressive stresses are equal numerically:

$$\sigma_1 = -\sigma_2 = -\frac{Mc}{I} = -\frac{M}{S} \quad \text{or} \quad \sigma_{max} = \frac{M}{S} \tag{5-17a,b}$$

in which

$$S = \frac{I}{c} \tag{5-18}$$

is the only section modulus for the cross section.

For a beam of **rectangular cross section** with width b and height h (Fig. 5-12a), the moment of inertia and section modulus are

$$I = \frac{bh^3}{12} \qquad S = \frac{bh^2}{6} \tag{5-19a,b}$$

For a **circular cross section** of diameter d (Fig. 5-12b), these properties are

$$I = \frac{\pi d^4}{64} \qquad S = \frac{\pi d^3}{32} \tag{5-20a,b}$$

Properties of other doubly symmetric shapes, such as hollow tubes (either rectangular or circular) and wide-flange shapes, can be readily obtained from the preceding formulas.

FIGURE 5-12

Doubly symmetric cross-sectional shapes

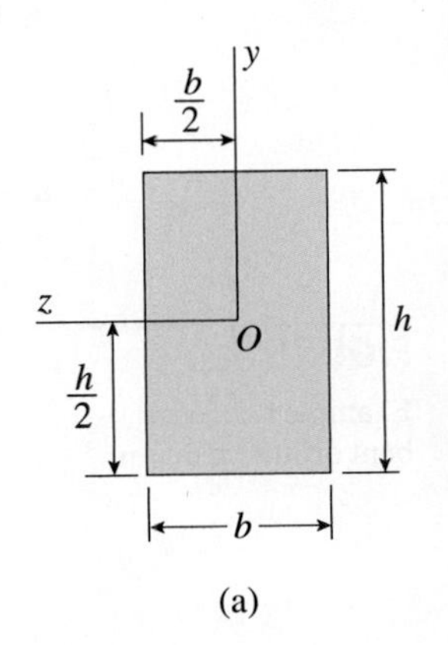

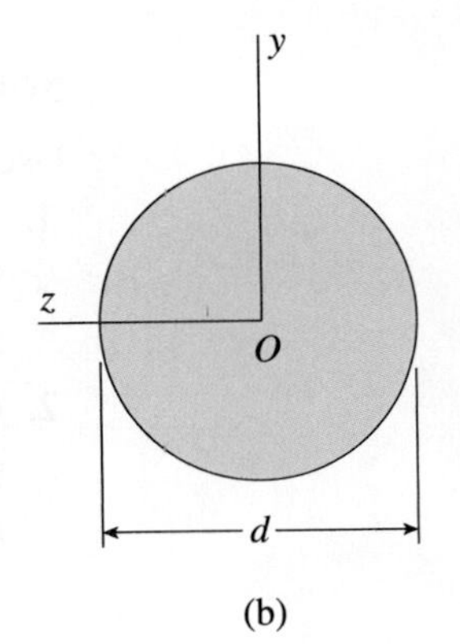

Properties of Beam Cross Sections

Moments of inertia of many plane figures are listed in Appendix E for convenient reference. Also, the dimensions and properties of standard sizes of steel and wood beams are listed in Appendixes F and G and in many engineering handbooks, as explained in more detail in the next section.

For other cross-sectional shapes, determine the location of the neutral axis, the moment of inertia, and the section moduli by direct calculation, using the techniques described in Appendix D. This procedure is illustrated later in Example 5-4.

Limitations

The analysis presented in this section is for the pure bending of prismatic beams composed of homogeneous, linearly elastic materials. If a beam is subjected to nonuniform bending, the shear forces will produce *warping* (or out-of-plane distortion) of the cross sections. Thus, a cross section that was plane before bending is no longer plane after bending. Warping due to shear deformations greatly complicates the behavior of the beam. However, detailed investigations show that the normal stresses calculated from the flexure formula are not significantly altered by the presence of shear stresses and the associated warping (Ref. 2-1, pp. 42 and 48). Thus, you may justifiably use the theory of pure bending for calculating normal stresses in beams subjected to nonuniform bending.[4]

The flexure formula gives results that are accurate only in regions of the beam where the stress distribution is not disrupted by changes in the shape of the beam or by discontinuities in loading. For instance, the flexure formula is not applicable near the supports of a beam or close to a concentrated load. Such irregularities produce localized stresses, or *stress concentrations*, that are much greater than the stresses obtained from the flexure formula (see Section 5.13).

[4]Beam theory began with Galileo Galilei (1564–1642), who investigated the behavior of various types of beams. His work in mechanics of materials is described in his famous book *Two New Sciences*, first published in 1638 (Ref. 5-2). Although Galileo made many important discoveries regarding beams, he did not obtain the stress distribution used today. Further progress in beam theory was made by Mariotte, Jacob Bernoulli, Euler, Parent, Saint-Venant, and others (Ref. 5-3).

Example 5-2

A high-strength steel wire with a diameter d is bent around a cylindrical drum of radius R_0 (Fig. 5-13).

Determine the bending moment M and maximum bending stress σ_{max} in the wire, assuming $d = 4$ mm and $R_0 = 0.5$ m. (The steel wire has a modulus of elasticity $E = 200$ GPa and a proportional limit $\sigma_{pl} = 1200$ MPa.)

FIGURE 5-13

Example 5-2: Wire bent around a drum

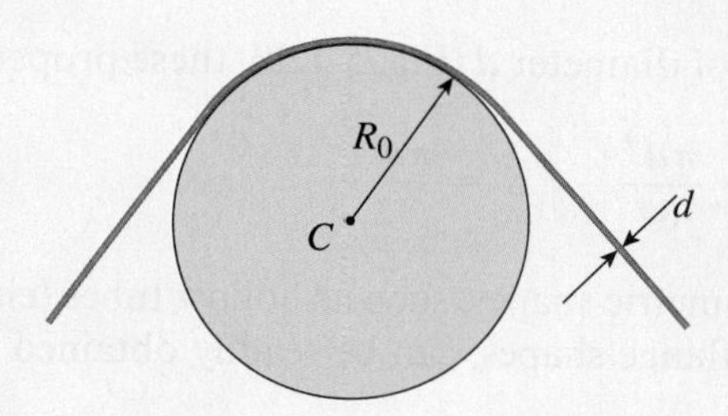

Solution:

Use a four-step problem-solving approach.

1. ***Conceptualize:*** The first step in this example is to determine the radius of curvature ρ of the bent wire. Knowing ρ, then find the bending moment and maximum stresses.
2. ***Categorize:***

Radius of curvature: The radius of curvature of the bent wire is the distance from the center of the drum to the neutral axis of the cross section of the wire:

$$\rho = R_0 + \frac{d}{2} \tag{5-21}$$

Bending moment: The bending moment in the wire may be found from the moment-curvature relationship (Eq. 5-13):

$$M = \frac{EI}{\rho} = \frac{2EI}{2R_0 + d} \tag{5-22}$$

in which I is the moment of inertia of the cross-sectional area of the wire. Substitute for I in terms of the diameter d of the wire [Eq. (5-20a)] to get

$$M = \frac{\pi E d^4}{32(2R_0 + d)} \tag{5-23}$$

This result was obtained without regard to the *sign* of the bending moment, since the direction of bending is obvious from the figure.

Maximum bending stresses: The maximum tensile and compressive stresses, which are equal numerically, are obtained from the flexure formula as given by Eq. (5-17b):

$$\sigma_{max} = \frac{M}{S}$$

in which S is the section modulus for a circular cross section. Substitute for M from Eq. (5-23) and for S from Eq. (5-20b) to get

$$\sigma_{max} = \frac{Ed}{2R_0 + d} \tag{5-24}$$

This same result can be obtained directly from Eq. (5-8) by replacing y with $d/2$ and substituting for ρ from Eq. (5-21).

Inspection of Fig. 5-13 reveals that the stress is compressive on the lower (or inner) part of the wire and tensile on the upper (or outer) part.

3. ***Analyze***:

Numerical results: Now substitute the given numerical data into Eqs. (5-23) and (5-24) and obtain

$$M = \frac{\pi E d^4}{32(2R_0 + d)} = \frac{\pi(200 \text{ GPa})(4 \text{ mm})^4}{32[2(0.5 \text{ m}) + 4 \text{ mm}]} = 5.01 \text{ N} \cdot \text{m}$$

$$\sigma_{max} = \frac{Ed}{2R_0 + d} = \frac{(200 \text{ GPa})(4 \text{ mm})}{2(0.5 \text{ m}) + 4 \text{ mm}} = 797 \text{ MPa}$$

4. ***Finalize***: Maximum stress σ_{max} is less than the proportional limit of the steel wire; therefore, the calculations are valid.

Note: Because the radius of the drum is large compared to the diameter of the wire, d in comparison with $2R_0$ in the denominators of the expressions for M and σ_{max} can be safely disregarded. Then Eqs. (5-23) and (5-24) give

$$M = 5.03 \text{ N} \cdot \text{m} \qquad \sigma_{max} = 800 \text{ MPa}$$

These results are on the conservative side and differ by less than 1% from the more precise values.

Example 5-3

A simple beam with an overhang (from Examples 4-5 and 4-9) is shown in Fig. 5-14. A uniform load with an intensity $q = 6$ kN/m acts throughout the length of the beam, and a concentrated load $P = 28$ kN acts at a point 3 m from the left-hand support. Uniform load q includes the weight of the beam. The beam is constructed of structural glued and laminated timber, has a cross section width of $b = 140$ mm, and has a height of $h = 580$ mm (Fig. 5-15).

(a) Determine the maximum tensile and compressive stresses in the beam due to bending.

(b) If load q is unchanged, find the maximum permissible value of load P if the allowable normal stress in tension and compression is $\sigma_a = 13$ MPa.

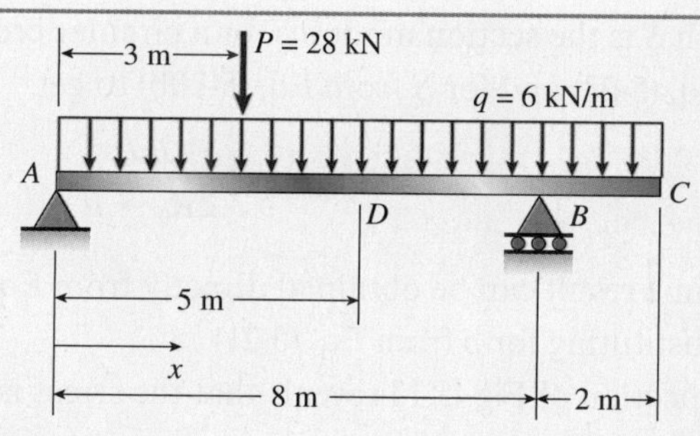

FIGURE 5-14
Beam with an overhang and uniform and concentrated loads

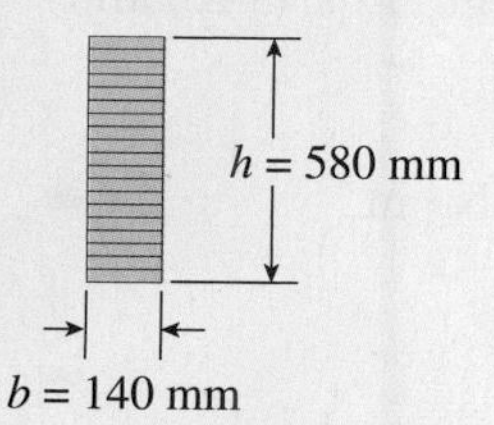

FIGURE 5-15
Beam cross section

Solution:

Use a four-step problem-solving approach. Combine steps as needed for an efficient solution.

Part (a): Maximum normal stresses.

1, 2. ***Conceptualize, Categorize:*** Begin the analysis by drawing the shear-force and bending-moment diagrams (Fig. 5-16); then determine the maximum bending moment, which occurs under the concentrated load. This is detailed in Example 4-9, and the resulting diagrams are shown in Fig. 5-16. The moment diagram shows that $M_{max} = 93 \text{ kN}\cdot\text{m}$ at 3 m to the right of support A. The maximum bending stresses in the beam occur at the cross section of the maximum moment.

Section modulus: The section modulus for the rectangular cross-sectional area in Fig. 5-15 is from Eq. (5-19b):

$$S = \frac{bh^2}{6} = \frac{1}{6}(140 \text{ mm})(580 \text{ mm})^2 = 7849.3 \text{ cm}^3 \quad \text{(a)}$$

3. ***Analyze:***

Maximum stresses: The maximum tensile and compressive stresses are obtained from Eq. (5-17):

$$\sigma_t = \frac{M_{max}}{S} = \frac{(93 \text{ kN}\cdot\text{m})}{7849.3 \text{ cm}^3} = 11.85 \text{ MPa}$$

$$\sigma_c = -\frac{M_{max}}{S} = -11.85 \text{ MPa} \quad \text{(b)}$$

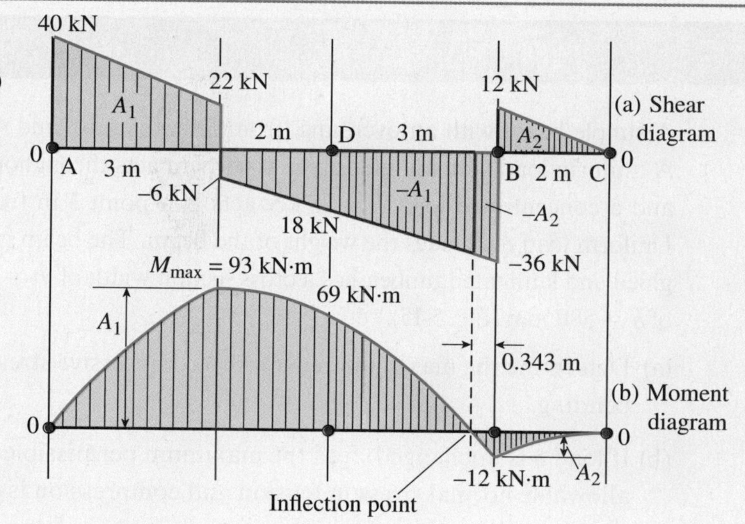

FIGURE 5-16
(a, b) Shear and moment diagrams (from Example 4-9)

4. ***Finalize***: The moment diagram is plotted on the compression side of the beam, so most of span AB has compressive stress on the top and tension stress on the bottom of the beam. The reverse is true for the portion of the beam to the right of the inflection point, which includes overhang segment BC.

Part (b): Maximum permissible load *P*.

1, 2. ***Conceptualize, Categorize***: The normal stresses in Eq. (b) at the location of the maximum moment are well below the allowable value of 13 MPa, so the beam can carry a much larger value of load P than that applied in part (a). Let the distance from support A to load P be $a = 3$ m, span AB length $L = 8$ m, overhang BC length = L/4, and the uniform load be unchanged at $q = 6$ kN/m.

3. ***Analyze***: Apply concentrated load P and uniform load q and solve for the reaction at A:

$$R_A = P\left(\frac{L-a}{L}\right) + \frac{15}{32}qL \qquad \text{(c)}$$

The maximum moment is at distance a from support A and is written as

$$M_{max} = R_A a - \frac{qa^2}{2} \qquad \text{(d)}$$

Equate M_{max} to $(\sigma_a)(S) = 102.04$ kN·m, insert numerical values in Eqs. (c) and (d), and solve for $P_{max} = 32.8$ kN.

Alternate solution: Apply additional load ΔP to increase the maximum moment from 93 kN·m to 102.04 kN·m, that is, $\Delta M = 9.04$ kN·m. The required additional load ΔP is computed using Eq. (4-13), which gives the moment at the location of a concentrated load:

$$\Delta P = \frac{L}{a(L-a)}\Delta M = \frac{8\text{ m}}{3\text{ m}(8\text{ m} - 3\text{ m})}(9.04\text{ kN}\cdot\text{m}) = 4.82\text{ kN} \qquad \text{(e)}$$

Add ΔP to the load $P = 28$ kN from part (a) to get

$$P_{max} = P + \Delta P = 28\text{ kN} + 4.8\text{ kN} = 32.8\text{ kN} \qquad \text{(f)}$$

4. ***Finalize***: Check that the maximum permissible value of P produces normal stresses at the allowable level at the point of maximum moment. Substitute P_{max} into Eqs. (c) and (d) to find that $R_A = 43.01$ kN and $M_{max} = 102.04$ kN·m. Using these values, the stresses at the point of load P_{max} application are

$$\sigma_t = -\sigma_c = \frac{M_{max}}{S} = \frac{(102.04\text{ kN}\cdot\text{m})}{7849.3\text{ cm}^3} = 13\text{ MPa} \qquad \text{(g)}$$

Example 5-4

The beam ABC shown in Fig. 5-17a has simple supports at A and B and an overhang from B to C. The length of the span is $L = 3.0$ m, and the length of the overhang is $L/2 = 1.5$ m. A uniform load of intensity $q = 3.2$ kN/m acts throughout the entire length of the beam (4.5 m).

The beam has a cross section of channel shape with a width of $b = 300$ mm and height of $h = 80$ mm (Fig. 5-18). The web thickness is $t = 12$ mm, and the average thickness of the sloping flanges is the same. For the purpose of calculating the properties of the cross section, assume that the cross section consists of three rectangles, as shown in Fig. 5-18b.

(a) Determine the maximum tensile and compressive stresses in the beam due to the uniform load.

(b) Find the maximum permissible value of uniform load q (in kN/m) if allowable stresses in tension and compression are $\sigma_{aT} = 110$ MPa and $\sigma_{aC} = 92$ MPa, respectively.

FIGURE 5-17

Example 5-4: Stresses in a beam with an overhang

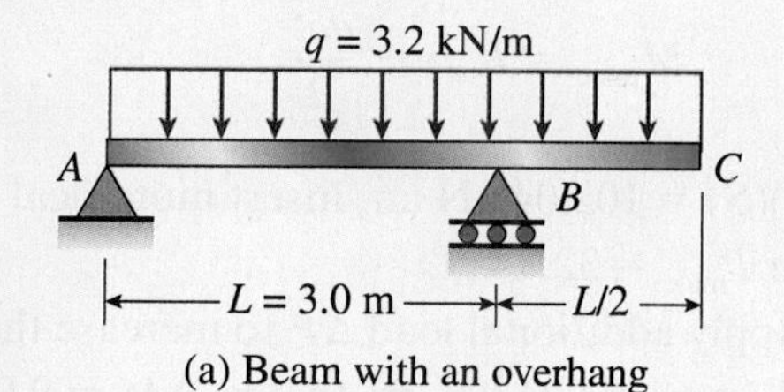

(a) Beam with an overhang

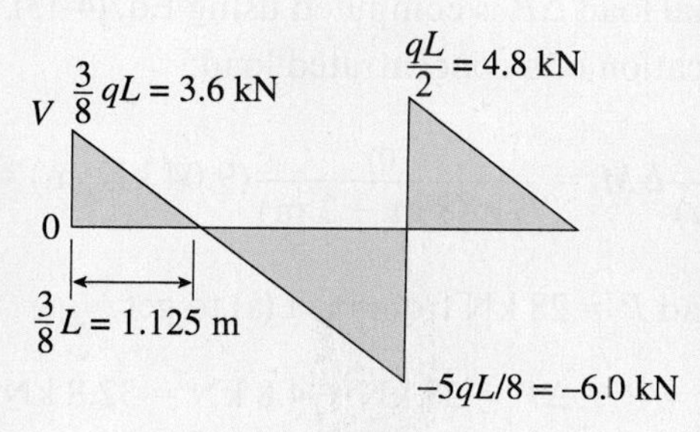

(b) Shear diagram

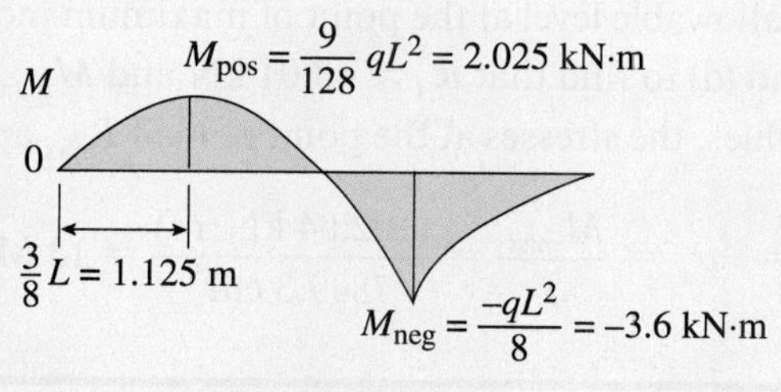

(c) Moment diagram

Solution:

Use a four-step problem-solving approach. Combine steps as needed for an efficient solution.

Part (a): Maximum tensile and compressive stresses.

1, 2. ***Conceptualize, Categorize:*** Reactions, shear forces, and bending moments are computed in the analysis of this beam. First, find the reactions at supports A and B using statics, as described in Chapter 4. The results are

$$R_A = \frac{3}{8}qL = 3.6 \text{ kN} \qquad R_B = \frac{9}{8}qL = 10.8 \text{ kN}$$

From these values, construct the shear-force diagram (Fig. 5-17b). Note that the shear force changes sign and is equal to zero at two locations: (1) at a distance of 1.125 m from the left-hand support and (2) at the right-hand reaction.

Next, draw the bending-moment diagram shown in Fig. 5-17c. Both the maximum positive and maximum negative bending moments occur at the cross sections where the shear force changes sign. These maximum moments are

$$M_{\text{pos}} = \frac{9}{128}qL^2 = 2.025 \text{ kN} \cdot \text{m} \qquad M_{\text{neg}} = \frac{-qL^2}{8} = -3.6 \text{ kN} \cdot \text{m}$$

respectively.

Neutral axis of the cross section (Fig. 5-18b): The origin O of the y-z coordinates is placed at the centroid of the cross-sectional area; therefore, the z axis becomes the neutral axis of the cross section. The centroid is located by using the techniques described in Appendix D, Section D.2, as follows.

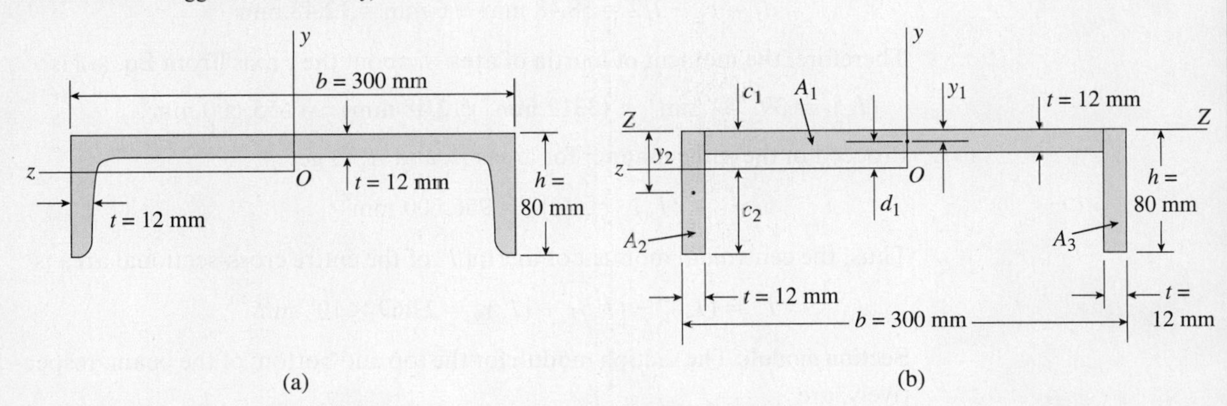

FIGURE 5-18

Cross section of beam discussed in Example 5-4: (a) actual shape and (b) idealized shape for use in analysis (the thickness of the beam is exaggerated for clarity)

First, divide the area into three rectangles (A_1, A_2, and A_3). Second, establish a reference axis Z–Z across the upper edge of the cross section, and let y_1 and y_2 be the distances from the Z–Z axis to the centroids of areas A_1 and A_2, respectively. Then the calculations for locating the centroid of the entire channel section (distances c_1 and c_2) are

Area 1: $y_1 = t/2 = 6 \text{ mm}$

$$A_1 = (b - 2t)(t) = (276 \text{ mm})(12 \text{ mm}) = 3312 \text{ mm}^2$$

Area 2: $y_2 = h/2 = 40 \text{ mm}$

$$A_2 = ht = (80 \text{ mm})(12 \text{ mm}) = 960 \text{ mm}^2$$

Area 3: $y_3 = y_2 \quad A_3 = A_2$

$$c_1 = \frac{\Sigma y_i A_i}{\Sigma A_i} = \frac{y_1 A_1 + 2 y_2 A_2}{A_1 + 2A_2}$$

$$= \frac{(6 \text{ mm})(3312 \text{ mm}^2) + 2(40 \text{ mm})(960 \text{ mm}^2)}{3312 \text{ mm}^2 + 2(960 \text{ mm}^2)} = 18.48 \text{ mm}$$

$$c_2 = h - c_1 = 80 \text{ mm} - 18.48 \text{ mm} = 61.52 \text{ mm}$$

Thus, the position of the neutral axis (the z axis) is determined.

Moment of inertia: In order to calculate the stresses from the flexure formula, determine the moment of inertia of the cross-sectional area with respect to the neutral axis. These calculations require the use of the parallel axis theorem (see Appendix D, Section D.4).

Beginning with area A_1, obtain its moment of inertia $(I_z)_1$ about the z axis from the equation

$$(I_z)_1 = (I_c)_1 + A_1 d_1^2 \qquad \textbf{(a)}$$

In this equation, $(I_c)_1$ is the moment of inertia of area A_1 about its own centroidal axis:

$$(I_c)_1 = \frac{1}{12}(b - 2t)(t)^3 = \frac{1}{12}(276 \text{ mm})(12 \text{ mm})^3 = 39{,}744 \text{ mm}^4$$

and d_1 is the distance from the centroidal axis of area A_1 to the z axis:

$$d_1 = c_1 - t/2 = 18.48 \text{ mm} - 6 \text{ mm} = 12.48 \text{ mm}$$

Therefore, the moment of inertia of area A_1 about the z axis [from Eq. (a)] is

$$(I_z)_1 = 39{,}744 \text{ mm}^4 + (3312 \text{ mm}^2)(12.48 \text{ mm})^2 = 555{,}600 \text{ mm}^4$$

Proceed in the same manner for areas A_2 and A_3 to get

$$(I_z)_2 = (I_z)_3 = 956{,}600 \text{ mm}^4$$

Thus, the centroidal moment of inertia I_z of the entire cross-sectional area is

$$I_z = (I_z)_1 + (I_z)_2 + (I_z)_3 = 2.469 \times 10^6 \text{ mm}^4$$

Section moduli: The section moduli for the top and bottom of the beam, respectively, are

$$S_1 = \frac{I_z}{c_1} = 133{,}600 \text{ mm}^3 \qquad S_2 = \frac{I_z}{c_2} = 40{,}100 \text{ mm}^3$$

[see Eqs. (5-16a and b)]. With the cross-sectional properties determined, now calculate the maximum stresses from Eqs. (5-15a and b).

3. ***Analyze:***

Maximum stresses: At the cross section of maximum positive bending moment, the largest tensile stress occurs at the bottom of the beam (σ_2) and the largest compressive stress occurs at the top (σ_1). Thus, from Eqs. (5-15b) and (5-15a), respectively, you get

$$\sigma_t = \sigma_2 = \frac{M_{pos}}{S_2} = \frac{2.025 \text{ kN} \cdot \text{m}}{40{,}100 \text{ mm}^3} = 50.5 \text{ MPa}$$

$$\sigma_c = \sigma_1 = -\frac{M_{pos}}{S_1} = -\frac{2.025 \text{ kN} \cdot \text{m}}{133{,}600 \text{ mm}^3} = -15.2 \text{ MPa}$$

Similarly, the largest stresses at the section of maximum negative moment are

$$\sigma_t = \sigma_1 = -\frac{M_{neg}}{S_1} = -\frac{-3.6 \text{ kN} \cdot \text{m}}{133{,}600 \text{ mm}^3} = 26.9 \text{ MPa}$$

$$\sigma_c = \sigma_2 = \frac{M_{neg}}{S_2} = \frac{-3.6 \text{ kN} \cdot \text{m}}{40{,}100 \text{ mm}^3} = -89.8 \text{ MPa}$$

4. ***Finalize:*** A comparison of these four stresses shows that the largest tensile stress in the beam is 50.5 MPa and occurs at the bottom of the beam at the cross section of maximum positive bending moment; thus,

$$(\sigma_t)_{max} = 50.5 \text{ MPa}$$

The largest compressive stress is −89.8 MPa and occurs at the bottom of the beam at the section of maximum negative moment:

$$(\sigma_c)_{max} = -89.8 \text{ MPa}$$

Recall that these are the maximum bending stresses due to the uniform load acting on the beam.

Part (b): Maximum permissible value of uniform load q.

1, 2. ***Conceptualize, Categorize:*** Next, find q_{max} based on the given allowable normal stresses, which are different for tension and compression. The allowable compression stress is σ_{aC} lower than that for tension, σ_{aT}, to account for the possibility of local buckling of the flanges of the C shape (if they are in compression).

Use the flexure formula to compute potential values of q_{max} at four locations: at the top and bottom of the beam at the location of the maximum positive moment (M_{pos}) and at the top and bottom of the beam at the location of the maximum negative moment (M_{neg}). In each case, be sure to use the proper value of allowable stress. Assume that the C shape is used in the orientation shown in Fig. 5-18 (flanges downward), so at the location of M_{pos}, the top of the beam is in compression and the bottom is in tension, while the opposite is true at point B. Using the expressions for M_{pos} and M_{neg} and equating each to the appropriate product of allowable stress and section modulus, solve for possible values of q_{max} as given here.

3. ***Analyze:*** In beam segment AB at the *top* of beam,

$$M_{\text{pos}} = \frac{9}{128}q_1L^2 = \sigma_{aC}S_1 \qquad \text{so} \qquad q_1 = \frac{128}{9L^2}(\sigma_{aC}S_1) = 19.42 \text{ kN/m}$$

In beam segment AB at the *bottom* of beam,

$$M_{\text{pos}} = \frac{9}{128}q_2L^2 = \sigma_{aT}S_2 \qquad \text{so} \qquad q_2 = \frac{128}{9L^2}(\sigma_{aT}S_2) = 6.97 \text{ kN/m}$$

At joint B at the top of beam,

$$M_{\text{pos}} = \frac{1}{8}q_3L^2 = \sigma_{aT}S_1 \qquad \text{so} \qquad q_3 = \frac{8}{L^2}(\sigma_{aT}S_1) = 13.06 \text{ kN/m}$$

At joint B at *bottom* of the beam,

$$M_{\text{pos}} = \frac{1}{8}q_4L^2 = \sigma_{aC}S_2 \qquad \text{so} \qquad q_4 = \frac{8}{L^2}(\sigma_{aC}S_2) = 3.28 \text{ kN/m}$$

4. ***Finalize:*** From these calculations, the bottom of the beam near joint B (where the flange tips are in compression) does indeed control the maximum permissible value of uniform load q. Hence,

$$q_{\max} = 3.28 \text{ kN/m}$$

5.6 Design of Beams for Bending Stresses

The process of designing a beam requires that many factors be considered, including the type of structure (airplane, automobile, bridge, building, or whatever), the materials to be used, the loads to be supported, the environmental conditions to be encountered, and the costs to be paid. However, from the standpoint of strength, the task eventually reduces to selecting a shape and size of beam such that the actual stresses in the beam do not exceed the allowable stresses for the material. This section considers only the bending stresses [that is, the stresses obtained from the flexure formula, Eq. (5-14)].

When designing a beam to resist bending stresses, begin by calculating the **required section modulus**. For instance, if the beam has a doubly symmetric cross section and the allowable stresses are the same for both tension and compression, calculate the required modulus by dividing the maximum bending moment by the allowable bending stress for the material [see Eq. (5-17)]:

$$S = \frac{M_{\max}}{\sigma_{\text{allow}}} \tag{5-25}$$

The allowable stress is based upon the properties of the material and the desired factor of safety. To ensure that this stress is not exceeded, choose a beam that provides a section modulus at least as large as that obtained from Eq. (5-25).

If the cross section is not doubly symmetric, or if the allowable stresses are different for tension and compression, it may be necessary to determine two required section moduli—one based upon tension and the other based upon compression. Then provide a beam that satisfies both criteria.

To minimize weight and save material, select a beam that has the least cross-sectional area while still providing the required section moduli (and also meeting any other design requirements that may be imposed).

Beams are constructed in a great variety of shapes and sizes to suit a myriad of purposes. For instance, very large steel beams are fabricated by welding (Fig. 5-19), aluminum beams are extruded as round or rectangular tubes, wood beams are cut and glued to fit special requirements, and reinforced concrete beams are cast in any desired shape by proper construction of the forms.

FIGURE 5-19

Welder fabricating a large wide flange steel beam

(Courtesy of AISC)

In addition, beams of steel, aluminum, plastic, and wood can be ordered in **standard shapes and sizes** from catalogs supplied by dealers and manufacturers. Readily available shapes include wide-flange beams, I-beams, angles, channels, rectangular beams, and tubes.

Beams of Standardized Shapes and Sizes

The dimensions and properties of many kinds of beams are listed in engineering handbooks and manuals. For instance, in the UK, the British Constructional Steelwork Association publishes the *National Structural Steelwork Specification*, and in the United States the shapes and sizes of structural-steel beams are standardized by the American Institute of Steel Construction (AISC). AISC publishes the *Steel Construction Manual* which lists their properties in both USCS and SI units. The tables in these manuals give cross-sectional dimensions and properties such as mass, cross-sectional area, moment of inertia, and section modulus. Properties of structural steel shapes used in many parts of the world are readily available online as well. In Europe, design of steel structures is governed by *Eurocode 3* (Ref. 5-4).

Properties of aluminum beams are tabulated in a similar manner and are available in publications of the Aluminum Association (Ref. 5-5) (see the *Aluminum Design Manual, Part 6*, for dimensions and section properties). In Europe, design of aluminum structures is governed by *Eurocode 9* and properties of available shapes may be found online at manufacturers' web sites (Ref. 5-5). Finally, design of timber beams in Europe is covered in *Eurocode 5*; in the United States, the *National Design Specification for Wood Construction (ASD/LRFD)* is used (Ref. 5-6). Abridged tables of steel beams and wood beams are given later in this book for use in solving problems (see Appendices F and G).

Structural-steel sections are given a designation such as HE 600A, which means that the section is a wide-flange shape with a nominal depth of 600 mm; as Table F-1 (Appendix F) shows, its width is 300 mm, its cross-sectional area is 226.5 cm^2, and its mass is 178 kilograms per meter of length. Table F-2 lists similar properties for European Standard Beams (IPN shapes); Table F-3 provides properties of European Standard Channels (UPN shapes); and Tables F-4 and F-5 give properties of European equal and unequal angles, respectively. All of the standardized steel sections described above are manufactured by *rolling*, a process in which a billet of hot steel is passed back and forth between rolls until it is formed into the desired shape.

Aluminum structural sections are usually made by the process of *extrusion*, in which a hot billet is pushed, or extruded, through a shaped die. Since dies are relatively easy to make and the material is workable, aluminum beams can be extruded in almost any desired shape. Standard shapes of wide-flange beams, I-beams, channels, angles, tubes, and other sections are listed in *Part 6* of the *Aluminum Design Manual* and structural aluminum shapes available in Europe may be found online (Ref. 5-5). In addition, custom-made shapes can be ordered.

Most **wood beams** have rectangular cross sections and are designated by nominal dimensions, such as 50 mm × 75 mm. These dimensions represent the rough-cut size of the lumber. The net dimensions (or actual dimensions) of a wood beam are smaller than the nominal dimensions if the sides of the rough lumber have been planed, or *surfaced*, to make them smooth. Thus, a 50 mm × 75 mm wood beam has actual dimensions 47 × 72 mm after it has been surfaced. Of course, the net dimensions of surfaced lumber should be used in all engineering computations. Therefore, net dimensions and the corresponding properties are given in Appendix G.

Relative Efficiency of Various Beam Shapes

One of the objectives in designing a beam is to use the material as efficiently as possible within the constraints imposed by function, appearance, manufacturing costs, and the like. From the standpoint of strength alone, efficiency in bending depends primarily upon the shape of the cross section. In particular, the most efficient beam is one in which the material is located as far as practical from the neutral axis. The farther a given amount of material is from the neutral axis, the larger the section modulus becomes—and the larger the section modulus, the larger the bending moment that can be resisted (for a given allowable stress).

As an illustration, consider a cross section in the form of a **rectangle** of width b and height h (Fig. 5-20a). The section modulus [from Eq. (5-19b)] is

$$S = \frac{bh^2}{6} = \frac{Ah}{6} = 0.167Ah \qquad \textbf{(5-26)}$$

where A denotes the cross-sectional area. This equation shows that a rectangular cross section of given area becomes more efficient as the height h is increased (and the width b is decreased to keep the area constant). Of course, there is a practical limit to the increase in height, because the beam becomes laterally unstable when the ratio of height to width becomes too large. Thus, a beam of very narrow rectangular section will fail due to lateral (sideways) buckling rather than to insufficient strength of the material.

FIGURE 5-20

Cross-sectional shapes of beams

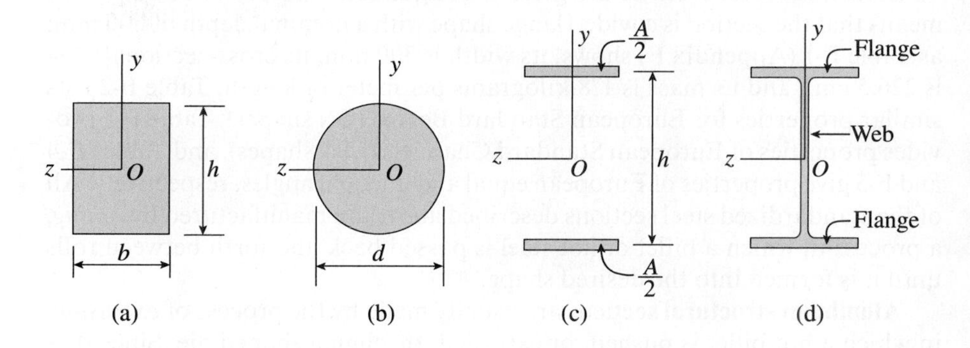

Next, compare a **solid circular cross section** of diameter d (Fig. 5-20b) with a square cross section of the same area. The side h of a square having the same area as the circle is $h = (d/2)\sqrt{\pi}$. The corresponding section moduli [from Eqs. (5-19b) and (5-20b)] are

$$S_{\text{square}} = \frac{h^3}{6} = \frac{\pi\sqrt{\pi}d^3}{48} = 0.1160d^3 \quad \textbf{(5-27a)}$$

$$S_{\text{circle}} = \frac{\pi d^3}{32} = 0.0982d^3 \quad \textbf{(5-27b)}$$

which gives

$$\frac{S_{\text{square}}}{S_{\text{circle}}} = 1.18 \quad \textbf{(5-28)}$$

This result shows that a beam of square cross section is more efficient in resisting bending than is a circular beam of the same area. The reason, of course, is that a circle has a relatively larger amount of material located near the neutral axis. This material is less highly stressed; therefore, it does not contribute as much to the strength of the beam.

The **ideal cross-sectional shape** for a beam of given cross-sectional area A and height h would be obtained by placing one-half of the area at a distance $h/2$ above the neutral axis and the other half at distance $h/2$ below the neutral axis, as shown in Fig. 5-20c. For this ideal shape, obtain

$$I = 2\left(\frac{A}{2}\right)\left(\frac{h}{2}\right)^2 = \frac{Ah^2}{4} \qquad S = \frac{I}{h/2} = 0.5Ah \quad \textbf{(5-29a,b)}$$

These theoretical limits are approached in practice by wide-flange sections and I-sections, which have most of their material in the flanges (Fig. 5-20d). For standard wide-flange beams, the section modulus is approximately

$$S \approx 0.35Ah \quad \textbf{(5-30)}$$

which is less than the ideal but much larger than the section modulus for a rectangular cross section of the same area and height [see Eq. (5-26)].

Another desirable feature of a wide-flange beam is its greater width; hence, its greater stability with respect to sideways buckling when compared to a rectangular beam of the same height and section modulus. On the other hand, there are practical limits to how thin the web can be for a wide-flange beam. A web that is too thin is susceptible to localized buckling or it may be overstressed in shear (see Section 5.10).

The following four examples illustrate the process of selecting a beam on the basis of the allowable stresses. Only the effects of bending stresses (obtained from the flexure formula) are considered.

Note: When solving examples and problems that require the selection of a steel or wood beam from the tables in the appendixes, use the following rule: *If several choices are available in a table, select the lightest beam that provides the required section modulus.*

Example 5-5

A simply supported wood beam with a span length $L = 3$ m carries a uniform load $q = 4$ kN/m (Fig. 5-21). The allowable bending stress is 12 MPa, the wood weighs 5.4 kN/m^3, and the beam is supported laterally against sideways buckling and tipping.

Select a suitable size for the beam from the table in Appendix G.

FIGURE 5-21

Example 5-5: Design of a simply supported wood beam

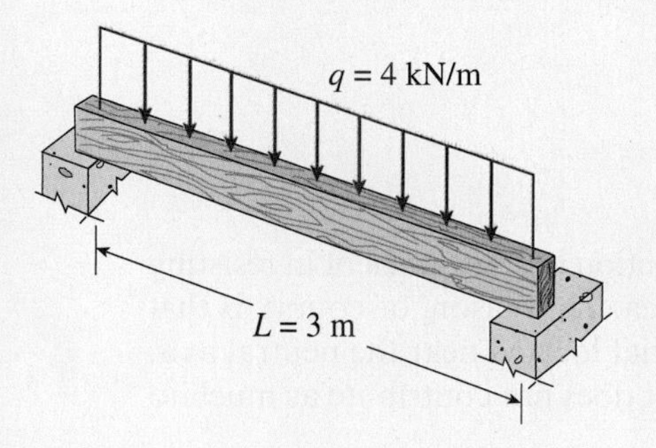

Solution:

Use a four-step problem-solving approach. Combine steps as needed for an efficient solution.

1, 2. ***Conceptualize, Categorize***: Since the beam weight is not known in advance, proceed by trial-and-error:

i. Calculate the required section modulus based upon the given uniform load.

ii. Select a trial size for the beam.

iii. Add the weight of the beam to the uniform load and calculate a new required section modulus.

Check to see that the selected beam is still satisfactory. If it is not, select a larger beam and repeat the process.

3. ***Analyze***:

i. The maximum bending moment in the beam occurs at the midpoint:

$$M_{\max} = \frac{qL^2}{8} = \frac{(4 \text{ kN/m})(3 \text{ m})^2}{8} = 4.5 \text{ kN} \cdot \text{m}$$

The required section modulus [Eq. (5-25)] is

$$S = \frac{M_{\max}}{\sigma_{\text{allow}}} = \frac{4.5 \text{ kN} \cdot \text{m}}{12 \text{ MPa}} = 0.375 \times 10^6 \text{ mm}^3$$

ii. From the table in Appendix G, the lightest beam that supplies a section modulus of at least 0.375×10^6 mm^3 about axis 1–1 is a 75 mm × 200 mm beam (nominal dimensions). This beam has a section modulus equal to

0.456×10^6 mm^3 and weighs 77.11 N/m. (Note that Appendix G gives weights of beams based upon a density of 5.4 kN/m^3.)

iii. The uniform load on the beam now becomes 4.077 kN/m, and the corresponding required section modulus is

$$S = (0.375 \times 10^6 \text{ mm}^3)\left(\frac{4.077}{4.0}\right) = 0.382 \times 10^6 \text{ mm}^3$$

4. ***Finalize***: The previously selected beam has a section modulus of 0.456×10^6 mm^3, which is larger than the required modulus of 0.382×10^6 mm^3.

   Therefore, a 75 mm × 100 mm. beam is satisfactory.

   Note: If the weight density of the wood is other than 5.4 kN/m^3, compute the weight of the beam per linear meter by multiplying the value in the last column in Appendix G by the ratio of the actual weight density to 5.4 kN/m^3.

Example 5-6

A vertical post 2.5-meters high must support a lateral load $P = 12$ kN at its upper end (Fig. 5-22). Two plans are proposed—a solid wood post and a hollow aluminum tube.

(a) What is the minimum required diameter d_1 of the wood post if the allowable bending stress in the wood is 15 MPa?

(b) What is the minimum required outer diameter d_2 of the aluminum tube if its wall thickness is to be one-eighth of the outer diameter and the allowable bending stress in the aluminum is 50 MPa?

FIGURE 5-22

Example 5-6: (a) Solid wood post and (b) aluminum tube

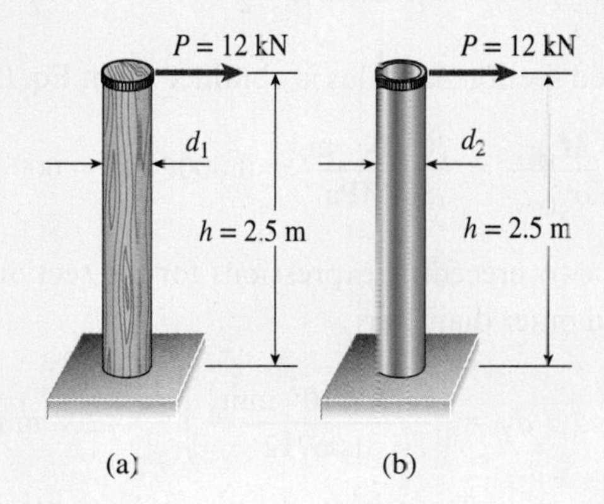

Solution:

Use a four-step problem-solving approach. Combine steps as needed for an efficient solution.

1. ***Conceptualize***:

Maximum bending moment: The maximum moment occurs at the base of the post and is equal to the load P times the height h; thus,

$$M_{max} = Ph = (12 \text{ kN})(2.5 \text{ m}) = 30 \text{ kN} \cdot \text{m}$$

Part (a): Wood post.

2, 3. ***Categorize, Analyze***: The required section modulus S_1 for the wood post [see Eqs. (5-20b and 5-25)] is

$$S_1 = \frac{\pi d_1^3}{32} = \frac{M_{max}}{\sigma_{allow}} = \frac{30 \text{ kN} \cdot \text{m}}{15 \text{ MPa}} = 0.0020 \text{ m}^3 = 2 \times 10^6 \text{ mm}^3$$

Solving for the diameter gives

$$d_1 = 273 \text{ mm}$$

4. ***Finalize***: The diameter selected for the wood post must be equal to or larger than 273 mm if the allowable stress is not to be exceeded.

Part (a): Aluminum tube.

2, 3. ***Categorize, Analyze***: To determine the section modulus S_2 for the tube, first find the moment of inertia I_2 of the cross section. The wall thickness of the tube is $d_2/8$; therefore, the inner diameter is $d_2 - d_2/4$, or $0.75d_2$. Thus, the moment of inertia [see Eq. (5-20a)] is

$$I_2 = \frac{\pi}{64}[d_2^4 - (0.75d_2)^4] = 0.03356d_2^4$$

The section modulus of the tube is now obtained from Eq. (5-18) as

$$S_2 = \frac{I_2}{c} = \frac{0.03356d_2^4}{d_2/2} = 0.06712d_2^3$$

The required section modulus is obtained from Eq. (5-25):

$$S_2 = \frac{M_{max}}{\sigma_{allow}} = \frac{30 \text{ kN} \cdot \text{m}}{50 \text{ MPa}} = 0.0006 \text{ m}^3 = 600 \times 10^3 \text{ mm}^3$$

Equate the two preceding expressions for the section modulus, then solve for the required outer diameter:

$$d_2 = \left(\frac{600 \times 10^3 \text{ mm}^3}{0.06712}\right)^{1/3} = 208 \text{ mm}$$

4. ***Finalize***: The corresponding inner diameter is 0.75(208 mm), or 156 mm.

Example 5-7

A simple beam AB of span length 7 m must support a uniform load $q = 60$ kN/m distributed along the beam in the manner shown in Fig. 5-23a.

Considering both the uniform load and the weight of the beam, and also using an allowable bending stress of 110 MPa, select a structural steel beam of wide-flange shape to support the loads.

FIGURE 5-23

Example 5-7: Design of a simple beam with partial uniform loads

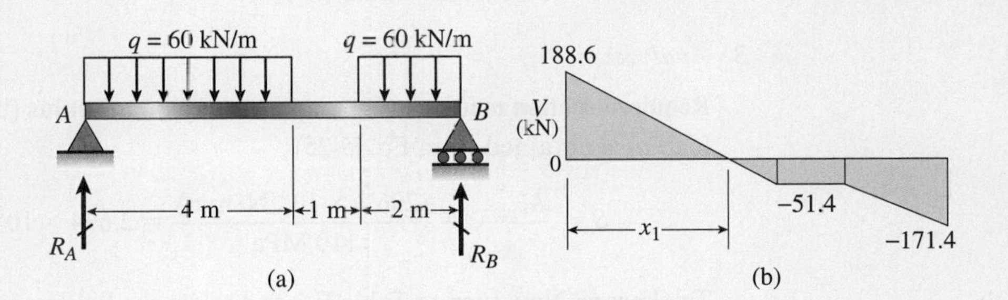

Solution:

Use a four-step problem-solving approach. Combine steps as needed for an efficient solution.

1, 2. ***Conceptualize, Categorize***: In this example, proceed as follows:

i. Find the maximum bending moment in the beam due to the uniform load.

ii. Knowing the maximum moment, find the required section modulus.

iii. Select a trial wide-flange beam from Table F-1 in Appendix F and obtain the weight of the beam.

iv. With the weight known, calculate a new value of the bending moment and a new value of the section modulus.

Determine whether the selected beam is still satisfactory. If it is not, select a new beam size and repeat the process until a satisfactory size of beam has been found.

Maximum bending moment: To assist in locating the cross section of maximum bending moment, construct the shear-force diagram (Fig. 5-23b) using the methods described in Chapter 4. As part of that process, determine the reactions at the supports:

$$R_A = 188.6 \text{ kN} \qquad R_B = 171.4 \text{ kN}$$

The distance x_1 from the left-hand support to the cross section of zero shear force is obtained from

$$V = R_A - qx_1 = 0$$

which is valid in the range $0 \le x \le 4$ m. Solve for x_1 to get

$$x_1 = \frac{R_A}{q} = \frac{188.6 \text{ kN}}{60 \text{ kN/m}} = 3.14 \text{ m}$$

which is less than 4 m; therefore, the calculation is valid.

The maximum bending moment occurs at the cross section where the shear force is zero; therefore,

$$M_{max} = R_A x_1 - \frac{q x_1^2}{2} = 296.3 \text{ kN} \cdot \text{m}$$

3. *Analyze*:

Required section modulus: The required section modulus (based only upon the load q) is obtained from Eq. (5-25):

$$S = \frac{M_{max}}{\sigma_{allow}} = \frac{(296.3 \times 10^6 \text{ N} \cdot \text{mm})}{110 \text{ MPa}} = 2.694 \times 10^6 \text{ mm}^3$$

Trial beam: Now turn to Table F-1 and select the lightest wide-flange beam having a section modulus greater than 2694 cm^3. The lightest beam that provides this section modulus is HE 450A with $S = 2896$ cm^3. This beam weighs 140 kg/m, or 1.373 kN/m. (Recall that the tables in Appendix F are abridged, so a lighter beam may actually be available.)

Now recalculate the reactions, maximum bending moment, and required section modulus with the beam loaded by both the uniform load q and its own weight. Under these combined loads the reactions are

$$R_A = 193.4 \text{ kN} \qquad R_B = 176.2 \text{ kN}$$

and the distance to the cross section of zero shear becomes

$$x_1 = \frac{193.4 \text{ kN}}{(60 \text{ kN/m} + 1.373 \text{ kN/m})} = 3.151 \text{ m}$$

The maximum bending moment increases to 304.7 kN · m, and the new required section modulus is

$$S = \frac{M_{max}}{\sigma_{allow}} = \frac{(304.7 \times 10^6 \text{N} \cdot \text{mm})}{110 \text{ MPa}} = 2770 \text{ cm}^3$$

4. *Finalize*: Thus, the HE 450A beam with section modulus $S = 2896$ cm^3 is still satisfactory.

Note: If the new required section modulus exceeded that of the HE 450A beam, a new beam with a larger section modulus would be selected and the process repeated.

Example 5-8

FIGURE 5-24

Example 5-8: Wood dam with horizontal planks A supported by vertical posts B

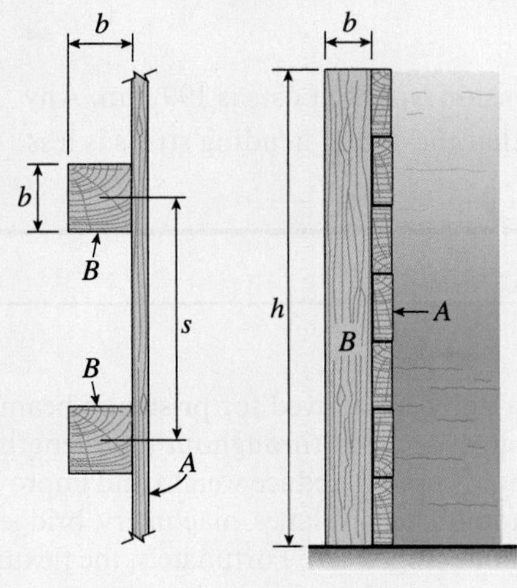

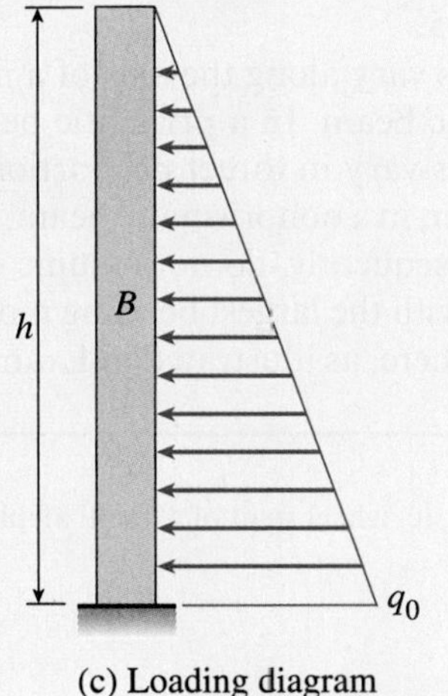

A temporary wood dam is constructed of horizontal planks A supported by vertical wood posts B that are sunk into the ground so that they act as cantilever beams (Fig. 5-24). The posts are of square cross section (dimensions $b \times b$) and spaced at distance $s = 0.8$ m, center to center. Assume that the water level behind the dam is at its full height $h = 2.0$ m.

Determine the minimum required dimension b of the posts if the allowable bending stress in the wood is $\sigma_{\text{allow}} = 8.0$ MPa.

Solution:

Use a four-step problem-solving approach.

1. *Conceptualize:*

Loading diagram: Each post is subjected to a triangularly distributed load produced by the water pressure acting against the planks. Consequently, the loading diagram for each post is triangular (Fig. 5-24c). The maximum intensity q_0 of the load on the posts is equal to the water pressure at depth h times the spacing s of the posts:

$$q_0 = \gamma h s \tag{a}$$

in which γ is the specific weight of water. Note that q_0 has units of force per unit distance, γ has units of force per unit volume, and both h and s have units of length.

2. *Categorize:*

Section modulus: Since each post is a cantilever beam, the maximum bending moment occurs at the base and is given by

$$M_{\text{max}} = \frac{q_0 h}{2}\left(\frac{h}{3}\right) = \frac{\gamma h^3 s}{6} \tag{b}$$

Therefore, the required section modulus [Eq. (5-25)] is

$$S = \frac{M_{\text{max}}}{\sigma_{\text{allow}}} = \frac{\gamma h^3 s}{6\sigma_{\text{allow}}} \tag{c}$$

3. *Analyze:* For a beam of square cross section, the section modulus is $S = b^3/6$ [see Eq. (5-19b)]. Substitute this expression for S into Eq. (c) to get a formula for the cube of the minimum dimension b of the posts:

$$b^3 = \frac{\gamma h^3 s}{\sigma_{\text{allow}}} \tag{d}$$

Numerical values: Now substitute numerical values into Eq. (d) and obtain

$$b^3 = \frac{(9.81\ \text{kN/m}^3)(2.0\ \text{m})^3(0.8\ \text{m})}{8.0\ \text{MPa}} = 0.007848\ \text{m}^3 = 7.848 \times 10^6\ \text{mm}^3$$

from which

$$b = 199\ \text{mm}$$

4. ***Finalize***: Thus, the minimum required dimension b of the posts is 199 mm. Any larger dimension, such as 200 mm, ensures that the actual bending stress is less than the allowable stress.

5.7 Nonprismatic Beams

The beam theories described in this chapter were derived for prismatic beams, that is, straight beams having the same cross sections throughout their lengths. However, nonprismatic beams are commonly used to reduce weight and improve appearance. Such beams are found in automobiles, airplanes, machinery, bridges, buildings, tools, and many other applications (Fig. 5-25). Fortunately, the flexure formula [Eq. (5-13)] gives reasonably accurate values for the bending stresses in nonprismatic beams whenever the changes in cross-sectional dimensions are gradual, as in the examples shown in Fig. 5-25.

The manner in which the bending stresses vary along the axis of a nonprismatic beam is not the same as for a prismatic beam. In a prismatic beam, the section modulus S is constant, so the stresses vary in direct proportion to the bending moment (because $\sigma = M/S$). However, in a nonprismatic beam, the section modulus also varies along the axis. Consequently, do not assume that the maximum stresses occur at the cross section with the largest bending moment—sometimes the maximum stresses occur elsewhere, as illustrated in Example 5-9.

FIGURE 5-25

Examples of nonprismatic beams: (a) street lamp, (b) bridge with tapered girders and piers, (c) wheel strut of a small airplane, and (d) wrench handle

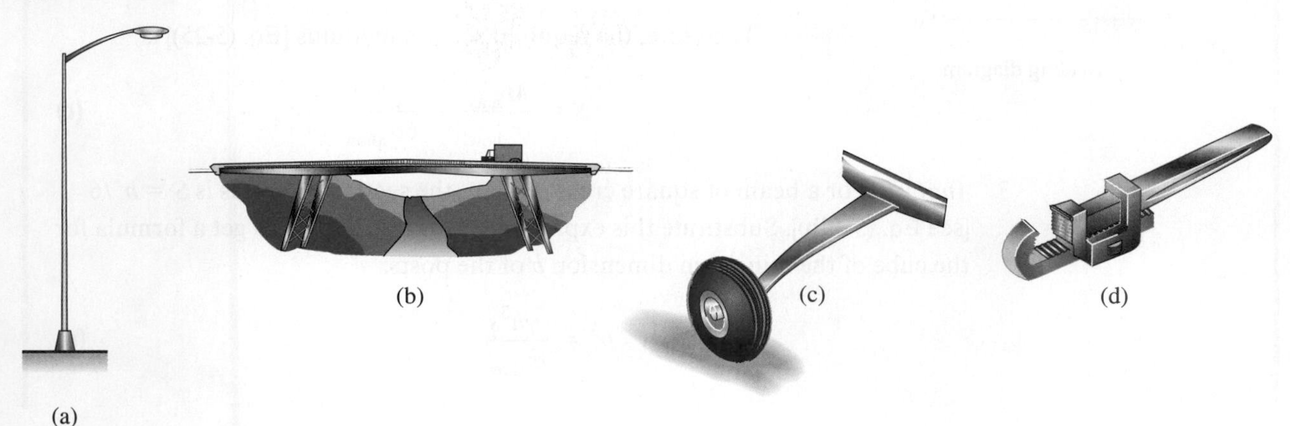

Fully Stressed Beams

To minimize the amount of material and thereby have the lightest possible beam, vary the dimensions of the cross sections to have the maximum allowable bending stress at every section. A beam in this condition is called a **fully stressed beam**, or a *beam of constant strength*.

Of course, these ideal conditions are seldom attained because of practical problems in constructing the beam and the possibility of the loads being different from those assumed in design. Nevertheless, knowing the properties of a fully stressed beam can be an important aid when designing structures for minimum weight. Familiar examples of structures designed to maintain nearly constant maximum stress are leaf springs in automobiles, bridge girders that are tapered, and some of the structures shown in Fig. 5-25.

The determination of the shape of a fully stressed beam is illustrated in Example 5-10.

Example 5-9

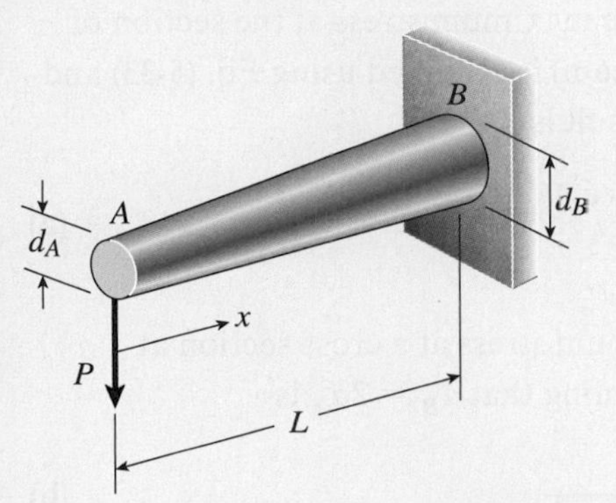

FIGURE 5-26
Example 5-9: Tapered cantilever beam of circular cross section

A tapered cantilever beam AB with a solid circular cross section supports a load P at the free end (Fig. 5-26). The diameter d_B at the large end is twice the diameter d_A at the small end:

$$\frac{d_B}{d_A} = 2$$

Determine the bending stress σ_B at the fixed support and the maximum bending stress σ_{max}.

Solution:

Use a four-step problem-solving approach. Combine steps as needed for an efficient solution.

1, 2. ***Conceptualize, Categorize***: If the angle of taper of the beam is small, the bending stresses obtained from the flexure formula differ only slightly from the exact values. As a guideline concerning accuracy, note that if the angle between line AB (Fig. 5-26) and the longitudinal axis of the beam is about 20°, the error in calculating the normal stresses from the flexure formula is about 10%. Of course, as the angle of taper decreases, the error becomes smaller.

3. ***Analyze***:

Section modulus: The section modulus at any cross section of the beam can be expressed as a function of the distance x measured along the axis of the beam. Since the section modulus depends upon the diameter, first express the diameter in terms of x, as

$$d_x = d_A + (d_B - d_A)\frac{x}{L} \tag{5-31}$$

in which d_x is the diameter at distance x from the free end. Therefore, the section modulus at distance x from the end [Eq. (5-20b)] is

$$S_x = \frac{\pi d_x^3}{32} = \frac{\pi}{32}\left[d_A + (d_B - d_A)\frac{x}{L}\right]^3 \tag{5-32}$$

Bending stresses: Since the bending moment equals Px, the maximum normal stress at any cross section is given by

$$\sigma_1 = \frac{M_x}{S_x} = \frac{32Px}{\pi[d_A + (d_B - d_A)(x/L)]^3} \tag{5-33}$$

The stress σ_1 is tensile at the top of the beam and compressive at the bottom.

Note that Eqs. (5-31), (5-32), and (5-33) are valid for any values of d_A and d_B, provided the angle of taper is small. In the following, consider only the case where $d_B = 2d_A$.

Maximum stress at the fixed support: The maximum stress at the section of largest bending moment (end B of the beam) is obtained using Eq. (5-33) and substituting $x = L$ and $d_B = 2d_A$; the result is

$$\sigma_B = \frac{4PL}{\pi d_A^3} \tag{a}$$

Maximum stress in the beam: The maximum stress at a cross section at distance x from the end [Eq. (5-33)] assuming that $d_B = 2d_A$ is

$$\sigma_1 = \frac{32Px}{\pi d_A^3(1 + x/L)^3} \tag{b}$$

To determine the location of the cross section having the largest bending stress in the beam, find the value of x that makes σ_1 a maximum. Take the derivative $d\sigma_1/dx$ and equate it to zero, then solve for the value of x that makes σ_1 a maximum; the result is

$$x = \frac{L}{2} \tag{c}$$

The corresponding maximum stress, obtained by substituting $x = L/2$ into Eq. (b), is

$$\sigma_{max} = \frac{128PL}{27\pi d_A^3} = \frac{4.741PL}{\pi d_A^3} \tag{d}$$

4. *Finalize*: In this particular example, the maximum stress occurs at the midpoint of the beam and is 19% greater than the stress σ_B at the built-in end.

Note: If the taper of the beam is reduced, the cross section of maximum normal stress moves from the midpoint toward the fixed support. For small angles of taper, the maximum stress occurs at end *B*.

Example 5-10

FIGURE 5-27

Example 5-10: Fully stressed beam having constant maximum normal stress (theoretical shape with shear stresses disregarded)

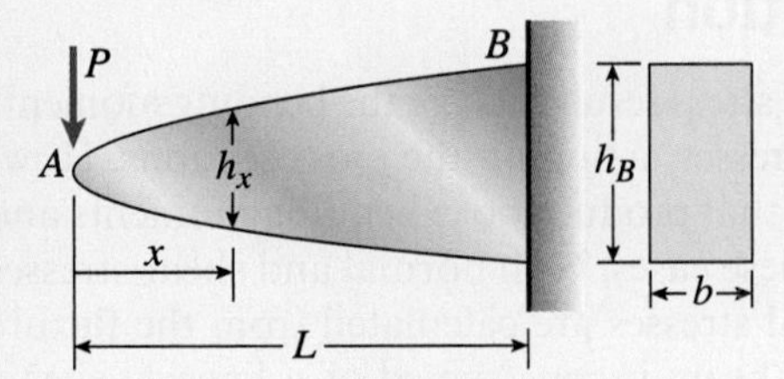

A cantilever beam *AB* of length *L* is being designed to support a concentrated load *P* at the free end (Fig. 5-27). The cross sections of the beam are rectangular with a constant width *b* and varying height *h*. To assist in designing this beam, the designers want to know how the height of an idealized beam should vary in order that the maximum normal stress at every cross section will be equal to the allowable stress σ_{allow}.

Considering only the bending stresses obtained from the flexure formula, determine the height of the fully stressed beam.

Solution:

Use a four-step problem-solving approach. Combine steps as needed for an efficient solution.

1, 2. *Conceptualize, Categorize*: The bending moment and section modulus at distance *x* from the free end of the beam are

$$M = Px \qquad S = \frac{bh_x^2}{6}$$

where h_x is the height of the beam at distance *x*. Substitute in the flexure formula to obtain

$$\sigma_{\text{allow}} = \frac{M}{S} = \frac{Px}{bh_x^2/6} = \frac{6Px}{bh_x^2} \tag{a}$$

3. *Analyze*: Solve for the height of the beam to find

$$h_x = \sqrt{\frac{6Px}{b\sigma_{\text{allow}}}} \tag{b}$$

At the fixed end of the beam ($x = L$), the height h_B is

$$h_B = \sqrt{\frac{6PL}{b\sigma_{\text{allow}}}} \tag{c}$$

therefore, the height h_x is expressed as

$$h_x = h_B \sqrt{\frac{x}{L}} \quad \text{(d)}$$

4. ***Finalize***: This last equation shows that the height of the fully stressed beam varies with the square root of x. Consequently, the idealized beam has the parabolic shape shown in Fig. 5-27.

Note: At the loaded end of the beam ($x = 0$), the theoretical height is zero because there is no bending moment at that point. A beam of this shape is not practical because it is incapable of supporting the shear forces near the end of the beam. Nevertheless, the idealized shape can provide a useful starting point for a realistic design in which shear stresses and other effects are considered.

5.8 Shear Stresses in Beams of Rectangular Cross Section

When a beam is in *pure bending*, the only stress resultants are the bending moments and the only stresses are the normal stresses acting on the cross sections. However, most beams are subjected to loads that produce both bending moments and shear forces (*nonuniform bending*). In these cases, both normal and shear stresses are developed in the beam. The normal stresses are calculated from the flexure formula (see Section 5.5), provided the beam is constructed of a linearly elastic material. The shear stresses are discussed in this and the following two sections.

Vertical and Horizontal Shear Stresses

Consider a beam of rectangular cross section (width b and height h) subjected to a positive shear force V (Fig. 5-28a). It is reasonable to assume that the shear stresses τ acting on the cross section are parallel to the shear force, that is, parallel to the vertical sides of the cross section. It is also reasonable to assume that the shear stresses are uniformly distributed across the width of the beam, although they may vary over the height. Using these two assumptions, you can determine the intensity of the shear stress at any point on the cross section.

For purposes of analysis, isolate a small element mn of the beam (Fig. 5-28a) by cutting between two adjacent cross sections and between two horizontal planes. Assume the shear stresses τ acting on the front face of this element are vertical and uniformly distributed from one side of the beam to the other. Also, from the discussion of shear stresses in Section 1.8, shear stresses acting on one side of an element are accompanied by shear stresses of equal magnitude acting on perpendicular faces of the element (Figs. 5-28b and c). Thus, there are horizontal shear stresses acting between horizontal layers of the beam as well as vertical shear stresses acting on the cross sections. At any point in the beam, these complementary shear stresses are equal in magnitude.

The equality of the horizontal and vertical shear stresses acting on an element leads to an important conclusion regarding the shear stresses at the top and bottom of the beam. If you imagine that the element mn (Fig. 5-28a) is located at either the top or the bottom, it follows that the horizontal shear stresses must vanish, because there are no stresses on the outer surfaces of the beam. It follows that the vertical shear stresses must also vanish at those locations; in other words, $\tau = 0$ where $y = \pm h/2$.

The existence of horizontal shear stresses in a beam can be demonstrated by a simple experiment. Place two identical rectangular beams on simple supports and load them by a force P, as shown in Fig. 5-29a. If friction between the beams is small, the beams bend independently (Fig. 5-29b). Each beam is in compression above its own neutral axis and in tension below its neutral axis; therefore, the bottom surface of the upper beam slides with respect to the top surface of the lower beam.

Now suppose that the two beams are glued along the contact surface, so they become a single solid beam. When this beam is loaded, horizontal shear stresses must develop along the glued surface in order to prevent the sliding shown in Fig. 5-29b. Because of the presence of these shear stresses, the single solid beam is much stiffer and stronger than the two separate beams.

FIGURE 5-28

Shear stresses in a beam of rectangular cross section

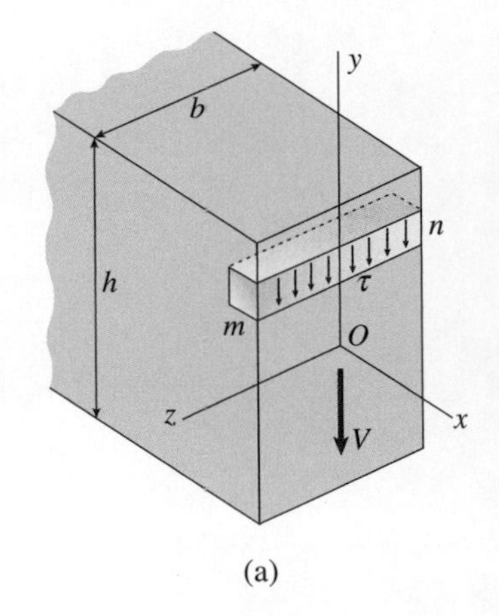

(a)

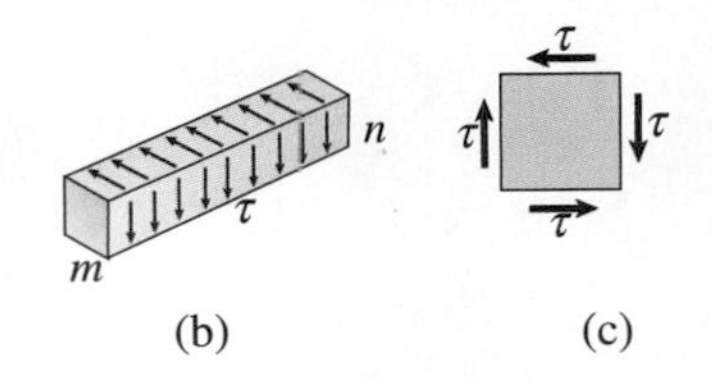

(b) (c)

Derivation of Shear Formula

Now derive a formula for the shear stresses τ in a rectangular beam. However, instead of evaluating the vertical shear stresses acting on a cross section, it is easier to evaluate the horizontal shear stresses acting between layers of the beam. Of course, the vertical shear stresses have the same magnitudes as the horizontal shear stresses.

Now consider a beam in nonuniform bending (Fig. 5-30a). Take two adjacent cross sections mn and m_1n_1 at a distance dx apart, and consider the **element** mm_1n_1n. The bending moment and shear force acting on the left-hand face of this element are denoted M and V, respectively. Since both the bending moment and shear force may change when moving along the axis of the beam, the corresponding quantities on the right-hand face (Fig. 5-30a) are denoted $M + dM$ and $V + dV$.

Because of the presence of the bending moments and shear forces, the element shown in Fig. 5-30a is subjected to normal and shear stresses on both cross-sectional faces. However, only the normal stresses are needed in the following derivation, so only the normal stresses are shown in Fig. 5-30b. On cross sections mn and m_1n_1, the normal stresses are, respectively,

$$\sigma_1 = -\frac{My}{I} \quad \text{and} \quad \sigma_2 = -\frac{(M + dM)y}{I} \qquad \textbf{(5-34a,b)}$$

as given by the flexure formula [Eq. (5-14)]. In these expressions, y is the distance from the neutral axis and I is the moment of inertia of the cross-sectional area about the neutral axis.

Next, isolate a **subelement** mm_1p_1p by passing a horizontal plane pp_1 through element mm_1n_1n (Fig. 5-30b). The plane pp_1 is at distance y_1 from the neutral surface of the beam. The subelement is shown separately in Fig. 5-30c. Note that its top face is part of the upper surface of the beam and thus is free from stress. Its bottom face (which is parallel to the neutral surface and distance y_1 from it) is acted upon by the horizontal shear stresses τ existing at this level in the beam. Its cross-sectional faces mp and m_1p_1 are acted upon by the bending stresses σ_1 and σ_2, respectively, which are produced by the bending moments. Vertical shear stresses also act on the cross-sectional faces; however, these stresses do not affect the equilibrium of the subelement in the horizontal direction (the x direction), so they are not shown in Fig. 5-30c.

If the bending moments at cross sections mn and m_1n_1 (Fig. 5-30b) are equal (that is, if the beam is in pure bending), the normal stresses σ_1 and σ_2 acting

FIGURE 5-29

Bending of two separate beams

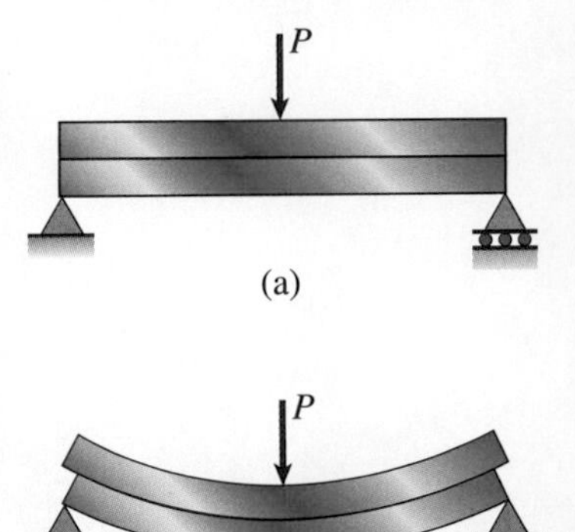

(b)

over the sides mp and m_1p_1 of the subelement (Fig. 5-30c) also are equal. Under these conditions, the subelement is in equilibrium under the action of the normal stresses alone; therefore, the shear stresses τ acting on the bottom face pp_1 vanish. This conclusion is obvious inasmuch as a beam in pure bending has no shear force and hence no shear stresses.

If the bending moments vary along the x axis (nonuniform bending), the shear stress τ acting on the bottom face of the subelement (Fig. 5-30c) can be determined by considering the equilibrium of the subelement in the x direction.

Begin by identifying an element of area dA in the *cross section* at a distance y from the neutral axis (Fig. 5-30d). The force acting on this element is σdA, in which σ is the normal stress obtained from the flexure formula. If the element of area is located on the left-hand face mp of the subelement (where the bending moment is M), the normal stress is given by Eq. (5-34a); therefore, the element of force is

$$\sigma_1 dA = \frac{My}{I} dA$$

Note that only absolute values are used in this equation because the directions of the stresses are obvious from Fig. 5-30. Summing these elements of force over the area of face mp of the subelement (Fig. 5-30c) gives the total horizontal force F_1 acting on that face:

$$F_1 = \int \sigma_1 dA = \int \frac{My}{I} dA \qquad \textbf{(5-35a)}$$

Note that this integration is performed over the area of the shaded part of the cross section shown in Fig. 5-30d, that is, over the area of the cross section from $y = y_1$ to $y = h/2$.

The force F_1 is shown in Fig. 5-31 on a partial free-body diagram of the subelement. (Vertical forces have been omitted.)

FIGURE 5-30

Shear stresses in a beam of rectangular cross section

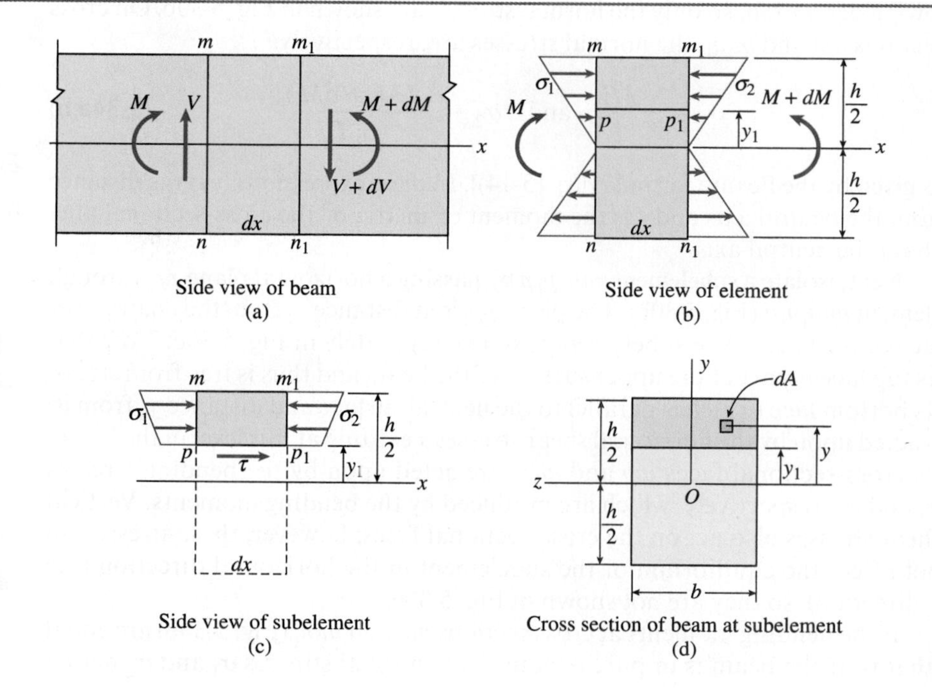

In a similar manner, the total force F_2 acting on the right-hand face m_1p_1 of the subelement (Fig. 5-31 and Fig. 5-30c) is

$$F_2 = \int \sigma_2 dA = \int \frac{(M + dM)y}{I} dA \quad \text{(5-35b)}$$

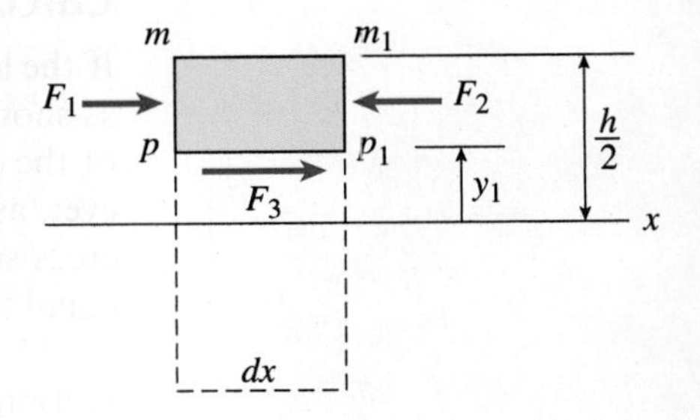

FIGURE 5-31

Partial free-body diagram of subelement showing all horizontal forces (compare with Fig. 5-30c)

Knowing the forces F_1 and F_2, now determine the horizontal force F_3 acting on the bottom face of the subelement.

Since the subelement is in equilibrium, sum forces in the x direction and obtain

$$F_3 = F_2 - F_1 \quad \text{(5-35c)}$$

or

$$F_3 = \int \frac{(M + dM)y}{I} dA - \int \frac{My}{I} dA = \int \frac{(dM)y}{I} dA$$

The quantities dM and I in the last term can be moved outside the integral sign because they are constants at any given cross section and are not involved in the integration. Thus, the expression for the force F_3 becomes

$$F_3 = \frac{dM}{I} \int y dA \quad \text{(5-36)}$$

If the shear stresses τ are uniformly distributed across the width b of the beam, the force F_3 is also equal to

$$F_3 = \tau b dx \quad \text{(5-37)}$$

in which bdx is the area of the bottom face of the subelement.

Combine Eqs. (5-36) and (5-37) and solve for the shear stress τ to get

$$\tau = \frac{dM}{dx}\left(\frac{1}{Ib}\right) \int y dA \quad \text{(5-38)}$$

The quantity dM/dx is equal to the shear force V (see Eq. 4-4), so the preceding expression becomes

$$\tau = \frac{V}{Ib} \int y dA \quad \text{(5-39)}$$

The integral in this equation is evaluated over the shaded part of the cross section (Fig. 5-30d), as already explained. Thus, the integral is the first moment of the shaded area with respect to the neutral axis (the z axis). In other words, *the integral is the first moment of the cross-sectional area above the level at which the shear stress τ is being evaluated.* This first moment is usually denoted by the symbol Q:

$$Q = \int y dA \quad \text{(5-40)}$$

With this notation, the equation for the shear stress becomes

$$\tau = \frac{VQ}{Ib} \quad \text{(5-41)}$$

This equation, known as the **shear formula**, can be used to determine the shear stress τ at any point in the cross section of a rectangular beam. Note that for a specific cross section, the shear force V, moment of inertia I, and width b are constants. However, the first moment Q (and hence the shear stress τ) varies with the distance y_1 from the neutral axis.

Calculation of the First Moment Q

If the level at which the shear stress is to be determined is above the neutral axis, as shown in Fig. 5-30d, it is natural to obtain Q by calculating the first moment of the cross-sectional area *above* that level (the shaded area in the figure). However, as an alternative, you could calculate the first moment of the remaining cross-sectional area, that is, the area *below* the shaded area. Its first moment is equal to the negative of Q.

The explanation lies in the fact that the first moment of the entire cross-sectional area with respect to the neutral axis is equal to zero (because the neutral axis passes through the centroid). Therefore, the value of Q for the area below the level y_1 is the negative of Q for the area above that level. Use the area above the level y_1 when the point where the shear stress is computed is in the upper part of the beam, and use the area below the level y_1 when the point is in the lower part of the beam.

Furthermore, don't bother with sign conventions for V and Q. Instead, treat all terms in the shear formula as positive quantities and determine the direction of the shear stresses by inspection, since the stresses act in the same direction as the shear force V itself. This procedure for determining shear stresses is illustrated in Example 5-11.

FIGURE 5-32

Distribution of shear stresses in a beam of rectangular cross section: (a) cross section of beam and (b) diagram showing the parabolic distribution of shear stresses over the height of the beam

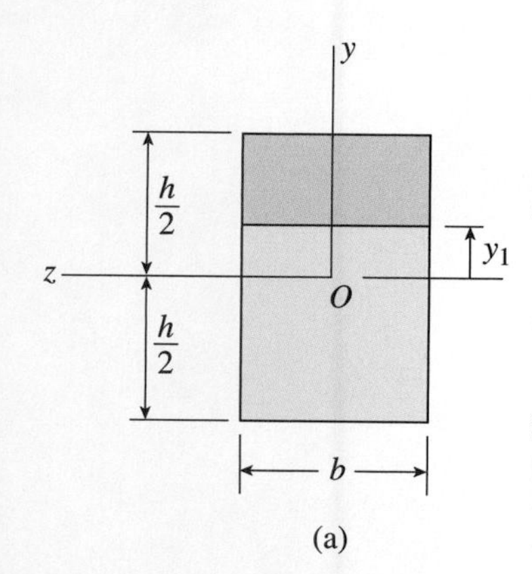

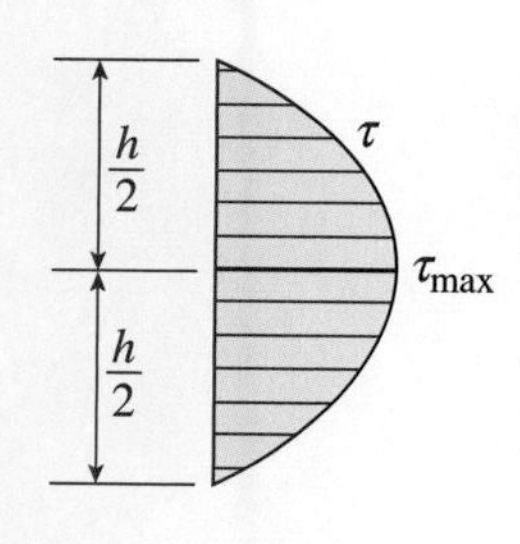

Distribution of Shear Stresses in a Rectangular Beam

Now find the distribution of the shear stresses in a beam of rectangular cross section (Fig. 5-32). Obtain the first moment Q of the shaded part of the cross-sectional area by multiplying the area by the distance from its own centroid to the neutral axis:

$$Q = b\left(\frac{h}{2} - y_1\right)\left(y_1 + \frac{h/2 - y_1}{2}\right) = \frac{b}{2}\left(\frac{h^2}{4} - y_1^2\right) \quad \textbf{(5-42a)}$$

This same result can be obtained by integration using Eq. (5-40):

$$Q = \int y\,dA = \int_0^{h/2} ybdy = \frac{b}{2}\left(\frac{h^2}{4} - y_1^2\right) \quad \textbf{(5-42b)}$$

Substitute the expression for Q into the shear formula [Eq. (5-41)] to get

$$\tau = \frac{V}{2I}\left(\frac{h^2}{4} - y_1^2\right) \quad \textbf{(5-43)}$$

This equation shows that the shear stresses in a rectangular beam vary quadratically with the distance y_1 from the neutral axis. Thus, when plotted along the height of the beam, τ varies as shown in Fig. 5-32b. Note that the shear stress is zero when $y_1 = \pm h/2$.

The maximum value of the shear stress occurs at the neutral axis ($y_1 = 0$) where the first moment Q has its maximum value. Substitute $y_1 = 0$ into Eq. (5-43) to get

$$\tau_{max} = \frac{Vh^2}{8I} = \frac{3V}{2A} \quad \textbf{(5-44)}$$

in which $A = bh$ is the cross-sectional area. Thus, the maximum shear stress in a beam of rectangular cross section is 50% larger than the average shear stress V/A.

Note again that the preceding equations for the shear stresses can be used to calculate either the vertical shear stresses acting on the cross sections or the horizontal shear stresses acting between horizontal layers of the beam.[5]

Limitations

The formulas for shear stresses in this section are subject to the same restrictions as the flexure formula from which they are derived. Thus, they are valid only for beams of linearly elastic materials with small deflections.

In the case of rectangular beams, the accuracy of the shear formula depends upon the height-to-width ratio of the cross section. The formula may be considered as exact for very narrow beams (height h much larger than the width b). However, it becomes less accurate as b increases relative to h. For instance, when the beam is square ($b = h$), the true maximum shear stress is about 13% larger than the value given by Eq. (5-44). (For a more complete discussion of the limitations of the shear formula, see Ref. 5-9.)

A common error is to apply the shear formula [(Eq. (5-41)] to cross-sectional shapes for which it is not applicable. For instance, it is not applicable to sections of triangular or semicircular shapes. To avoid misusing the formula, keep in mind the following assumptions that underlie the derivation: (1) The edges of the cross section must be parallel to the y axis (so that the shear stresses act parallel to the y axis), and (2) the shear stresses must be uniform across the width of the cross section. These assumptions are fulfilled only in certain cases, such as those discussed in this and the next two sections.

Finally, the shear formula applies only to prismatic beams. If a beam is nonprismatic (for instance, if the beam is tapered), the shear stresses are quite different from those predicted by the formulas given here (see Refs. 5-9 and 5-10).

Effects of Shear Strains

Because the shear stress τ varies parabolically over the height of a rectangular beam, it follows that the shear strain $\gamma = \tau/G$ also varies parabolically. As a result of these shear strains, cross sections of the beam that were originally plane surfaces become warped. This warping is shown in Fig. 5-33, where cross sections mn and pq, originally plane, have become curved surfaces m_1n_1 and p_1q_1, with the maximum shear strain occurring at the neutral surface. At points m_1, p_1, n_1, and q_1, the shear strain is zero, and therefore the curves m_1n_1 and p_1q_1 are perpendicular to the upper and lower surfaces of the beam.

If the shear force V is constant along the axis of the beam, warping is the same at every cross section. Therefore, stretching and shortening of longitudinal elements due to the bending moments is unaffected by the shear strains, and the distribution of the normal stresses is the same as in pure bending. Moreover, detailed investigations using advanced methods of analysis show that the warping of cross sections due to shear strains does not substantially affect the longitudinal strains even when the shear force varies continuously along the length. Thus, under most conditions, it is justifiable to use the flexure formula [Eq. (5-14)] for nonuniform bending, even though the formula was derived for pure bending.

FIGURE 5-33

Warping of the cross sections of a beam due to shear strains

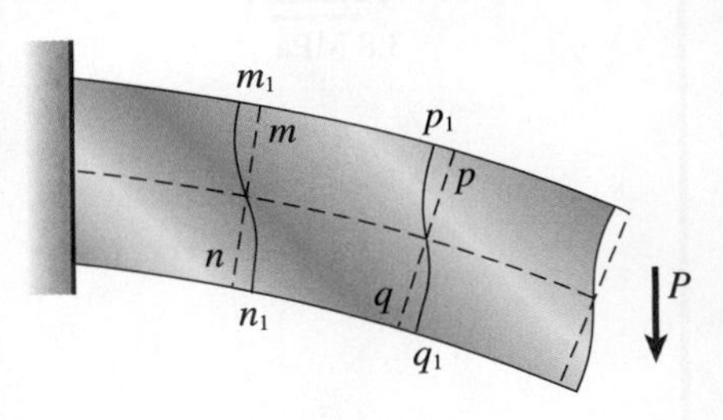

[5]The shear-stress analysis presented in this section was developed by the Russian engineer D. J. Jourawski; see Refs. 5-7 and 5-8.

Example 5-11

FIGURE 5-34

Example 5-11: (a) Simple beam with uniform load, (b) cross section of beam, and (c) stress element showing the normal and shear stresses at point C

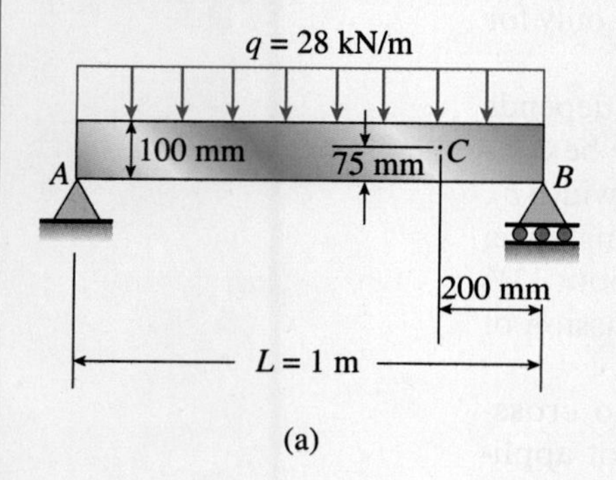

(a)

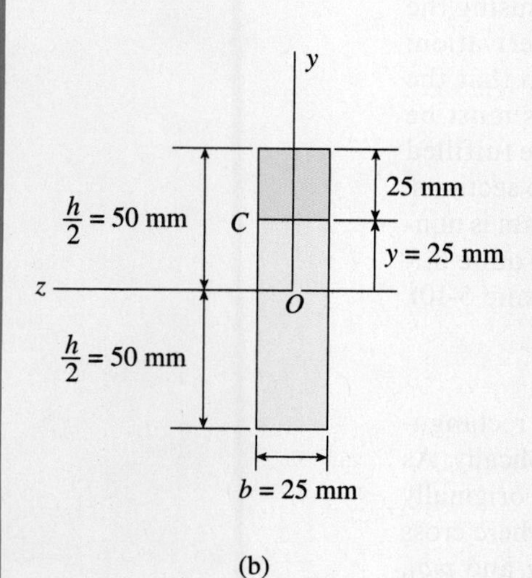

(b)

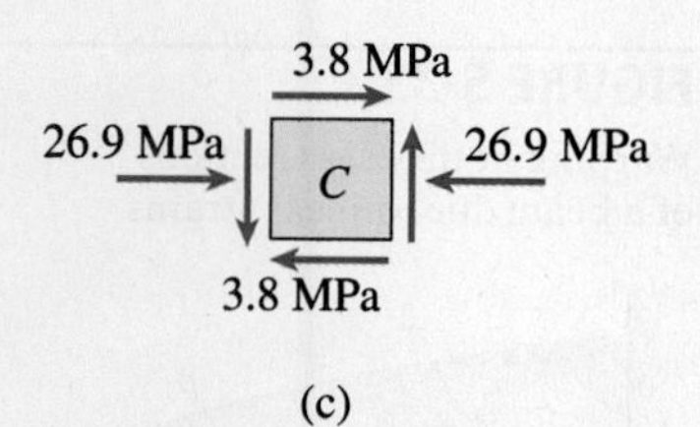

(c)

A metal beam with a span $L = 1$ m is simply supported at points A and B (Fig. 5-34a). The uniform load on the beam (including its own weight) is $q = 28$ kN/m. The cross section of the beam is rectangular (Fig. 5-34b) with width $b = 25$ mm and height $h = 100$ mm. The beam is adequately supported against sideways buckling.

Determine the normal stress σ_C and shear stress τ_C at point C, which is located 25 mm below the top of the beam and 200 mm from the right-hand support. Show these stresses on a sketch of a stress element at point C.

Solution:

Use a four-step problem-solving approach. Combine steps as needed for an efficient solution.

1, 2. ***Conceptualize, Categorize:***

Shear force and bending moment: The shear force V_C and bending moment M_C at the cross section through point C are found as described in Chapter 4. The results are

$$M_C = 2.22 \text{ kN}\cdot\text{m} \qquad V_C = -8.4 \text{ kN}$$

The signs of these quantities are based upon the standard sign conventions for bending moments and shear forces (see Fig. 4-19).

Moment of inertia: The moment of inertia of the cross-sectional area about the neutral axis (the z axis in Fig. 5-34b) is

$$I = \frac{bh^3}{12} = \frac{1}{12}(25 \text{ mm})(100 \text{ mm})^3 = 2083 \times 10^3 \text{ mm}^4$$

3. ***Analyze:***

Normal stress at point C: The normal stress at point C is found from the flexure formula [Eq. (5-14)] with the distance y from the neutral axis equal to 25 mm; thus,

$$\sigma_C = -\frac{My}{I} = -\frac{(2.24 \times 10^6 \text{ N}\cdot\text{mm})(25 \text{ mm})}{2083 \times 10^3 \text{ mm}^4} = -26.9 \text{ MPa}$$

The minus sign indicates that the stress is compressive, as expected.

Shear stress at point C: To obtain the shear stress at point C, evaluate the first moment Q_C of the cross-sectional area above point C (Fig. 5-34b). This first moment is equal to the product of the area and its centroidal distance (denoted y_C) from the z axis; thus,

$$A_C = (25 \text{ mm})(25 \text{ mm}) = 625 \text{ mm}^2 \quad y_C = 37.5 \text{ mm} \quad Q_C = A_C y_C = 23{,}440 \text{ mm}^3$$

Now substitute numerical values into the shear formula [Eq. (5-41)] and obtain the magnitude of the shear stress:

$$\tau_C = \frac{V_C Q_C}{Ib} = \frac{(8400\ \text{N})(23{,}440\ \text{mm}^3)}{(2083 \times 10^3\ \text{mm}^4)(25\ \text{mm})} = 3.8\ \text{MPa}$$

4. *Finalize*: The direction of this stress can be established by inspection because it acts in the same direction as the shear force. In this example, the shear force acts upward on the part of the beam to the left of point C and downward on the part of the beam to the right of point C. The best way to show the directions of both the normal and shear stresses is to draw a stress element.

   Stress element at point C: The stress element, shown in Fig. 5-34c, is cut from the side of the beam at point C (Fig. 5-34a). Compressive stresses $\sigma_C = 26.9$ MPa act on the cross-sectional faces of the element and shear stresses $\tau_C = 3.8$ MPa act on the top and bottom faces as well as the cross-sectional faces.

Example 5-12

FIGURE 5-35

Example 5-12: Wood beam with concentrated loads

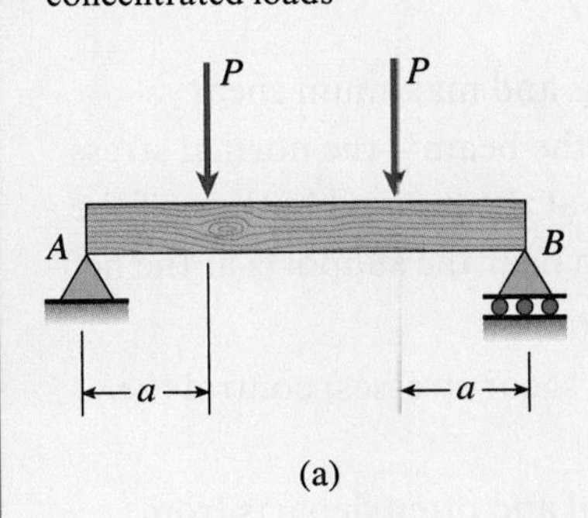

(a)

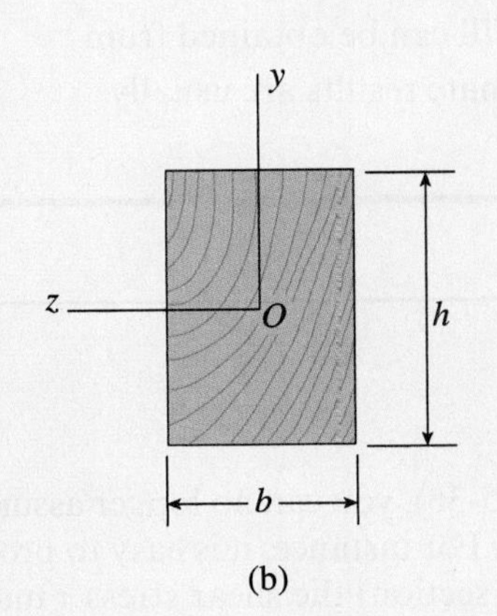

(b)

A wood beam AB supporting two concentrated loads P (Fig. 5-35a) has a rectangular cross section of width $b = 100$ mm and height $h = 150$ mm (Fig. 5-35b). The distance from each end of the beam to the nearest load is $a = 0.5$ m.

Determine the maximum permissible value P_{max} of the loads if the allowable stress in bending is $\sigma_{allow} = 11$ MPa (for both tension and compression) and the allowable stress in horizontal shear is $\tau_{allow} = 1.2$ MPa. (Disregard the weight of the beam itself.)

Note: Wood beams are much weaker in *horizontal shear* (shear parallel to the longitudinal fibers in the wood) than in *cross-grain shear* (shear on the cross sections). Consequently, the allowable stress in horizontal shear is usually considered in design.

Solution:

Use a four-step problem-solving approach.

1. *Conceptualize*: The maximum shear force occurs at the supports, and the maximum bending moment occurs throughout the region between the loads. Their values are

$$V_{max} = P \quad M_{max} = Pa$$

Also, the section modulus S and cross-sectional area A are

$$S = \frac{bh^2}{6} \quad A = bh$$

2. ***Categorize***: The maximum normal and shear stresses in the beam are obtained from the flexure and shear formulas [Eqs. (5-17) and (5-44)]:

$$\sigma_{max} = \frac{M_{max}}{S} = \frac{6Pa}{bh^2} \qquad \tau_{max} = \frac{3V_{max}}{2A} = \frac{3P}{2bh}$$

Therefore, the maximum permissible values of the load P in bending and shear, respectively, are

$$P_{bending} = \frac{\sigma_{allow}bh^2}{6a} \qquad P_{shear} = \frac{2\tau_{allow}bh}{3}$$

3. ***Analyze***: Substitute numerical values into these formulas to get

$$P_{bending} = \frac{(11\text{ MPa})(100\text{ mm})(150\text{ mm})^2}{6(0.5\text{ m})} = 8.25\text{ kN}$$

$$P_{shear} = \frac{2(1.2\text{ MPa})(100\text{ mm})(150\text{ mm})}{3} = 12.0\text{ kN}$$

Thus, the bending stress governs the design, and the maximum permissible load is

$$P_{max} = 8.25\text{ kN}$$

4. ***Finalize***: A more complete analysis of this beam would require that the weight of the beam be taken into account, thus reducing the permissible load.

Notes:

i. In this example, the maximum normal stresses and maximum shear stresses do not occur at the same locations in the beam—the normal stress is maximum in the middle region of the beam at the top and bottom of the cross section, and the shear stress is maximum near the supports at the neutral axis of the cross section.

ii. For most beams, the bending stresses (not the shear stresses) control the allowable load, as in this example.

iii. Although wood is not a homogeneous material and often departs from linearly elastic behavior, approximate results still can be obtained from the flexure and shear formulas. These approximate results are usually adequate for designing wood beams.

5.9 Shear Stresses in Beams of Circular Cross Section

When a beam has a **circular cross section** (Fig. 5-36), you can no longer assume that the shear stresses act parallel to the y axis. For instance, it is easy to prove that at point m (on the boundary of the cross section) the shear stress τ must act *tangent* to the boundary. This observation follows from the fact that the outer surface of the beam is free of stress, and the shear stress acting on the cross section can have no component in the radial direction.

Although there is no simple way to find the shear stresses acting throughout the entire cross section, the shear stresses at the neutral axis (where the stresses are the largest) are found by making some reasonable assumptions about the stress distribution. Assume that the stresses act parallel to the y axis and have a constant intensity across the width of the beam (from point p to point q in Fig. 5-36). Since these assumptions are the same as those used in deriving the shear formula $\tau = VQ/Ib$ [Eq. (5-41)], use the shear formula to calculate the stresses at the neutral axis.

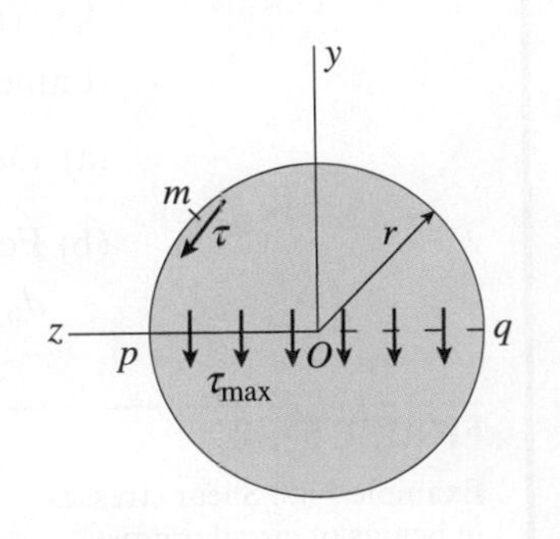

FIGURE 5-36
Shear stresses acting on the cross section of a circular beam

For use in the shear formula, the following properties pertaining to a circular cross section having radius r are needed:

$$I = \frac{\pi r^4}{4} \quad Q = A\bar{y} = \left(\frac{\pi r^2}{2}\right)\left(\frac{4r}{3\pi}\right) = \frac{2r^3}{3} \quad b = 2r \qquad \textbf{(5-45a,b)}$$

The expression for the moment of inertia I is taken from Case 9 of Appendix E, and the expression for the first moment Q is based upon the formulas for a semicircle (Case 10, Appendix E). Substitute these expressions into the shear formula to obtain

$$\tau_{\max} = \frac{VQ}{Ib} = \frac{V(2r^3/3)}{(\pi r^4/4)(2r)} = \frac{4V}{3\pi r^2} = \frac{4V}{3A} \qquad \textbf{(5-46)}$$

in which $A = \pi r^2$ is the area of the cross section. This equation shows that the maximum shear stress in a circular beam is equal to 4/3 times the average vertical shear stress V/A.

For a beam with a **hollow circular cross section** (Fig. 5-37), again assume with reasonable accuracy that the shear stresses at the neutral axis are parallel to the y axis and uniformly distributed across the section. Consequently, the shear formula is used to find the maximum stresses. The required properties for a hollow circular section are

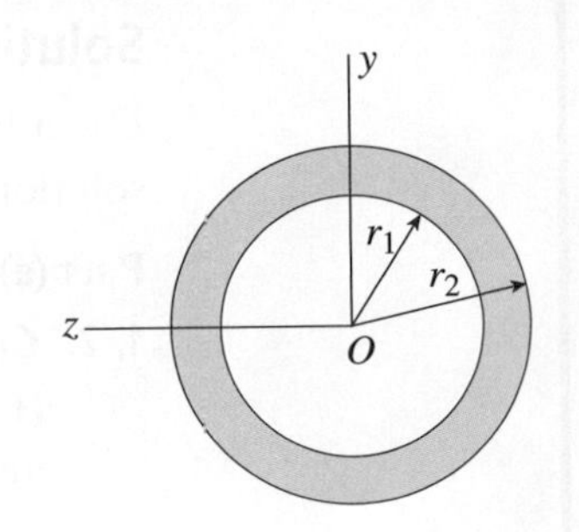

FIGURE 5-37
Hollow circular cross section

$$I = \frac{\pi}{4}(r_2^4 - r_1^4) \quad Q = \frac{2}{3}(r_2^3 - r_1^3) \quad b = 2(r_2 - r_1) \qquad \textbf{(5-47a,b,c)}$$

in which r_1 and r_2 are the inner and outer radii of the cross section. Therefore, the maximum stress is

$$\tau_{\max} = \frac{VQ}{Ib} = \frac{4V}{3A}\left(\frac{r_2^2 + r_2 r_1 + r_1^2}{r_2^2 + r_1^2}\right) \qquad \textbf{(5-48)}$$

in which

$$A = \pi(r_2^2 - r_1^2)$$

is the area of the cross section. Note that if $r_1 = 0$, Eq. (5-48) reduces to Eq. (5-46) for a solid circular beam.

Although the preceding theory for shear stresses in beams of circular cross section is approximate, it gives results differing by only a few percent from those obtained using the exact theory of elasticity (Ref. 5-9). Consequently, Eqs. (5-46) and (5-48) can be used to determine the maximum shear stresses in circular beams under ordinary circumstances.

Example 5-13

A vertical pole consisting of a circular tube of outer diameter $d_2 = 100$ mm and inner diameter $d_1 = 80$ mm is loaded by a horizontal force $P = 6675$ N (Fig. 5-38a).

(a) Determine the maximum shear stress in the pole.

(b) For the same load P and the same maximum shear stress, what is the diameter d_0 of a solid circular pole (Fig. 5-38b)?

FIGURE 5-38

Example 5-13: Shear stresses in beams of circular cross section

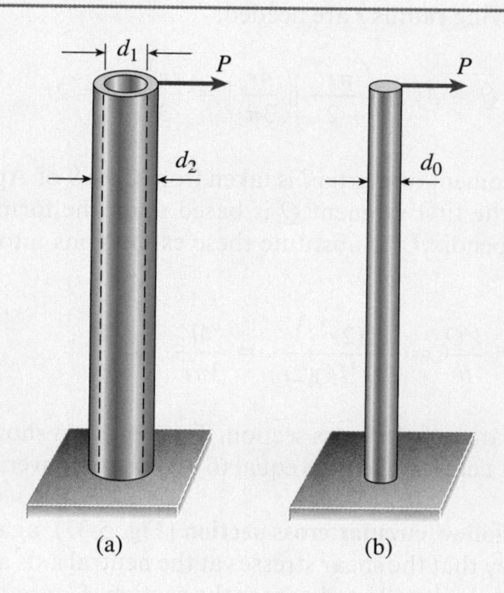

Solution:

Use a four-step problem-solving approach. Combine steps as needed for an efficient solution.

Part (a): Maximun shear stress.

1, 2. ***Conceptualize, Categorize***: For the pole having a hollow circular cross section (Fig. 5-38a), use Eq. (5-48) with the shear force V replaced by the load P and the cross-sectional area A replaced by the expression $\pi(r_2^2 - r_1^2)$; thus,

$$\tau_{max} = \frac{4P}{3\pi}\left(\frac{r_2^2 + r_2 r_1 + r_1^2}{r_2^4 - r_1^4}\right) \qquad \textbf{(a)}$$

3, 4. ***Analyze, Finalize***: Next, substitute numerical values, namely,

$$P = 6675 \text{ N} \qquad r_2 = d_2/2 = 50 \text{ mm} \qquad r_1 = d_1/2 = 40 \text{ mm}$$

to obtain

$$\tau_{max} = 4.68 \text{ MPa}$$

which is the maximum shear stress in the pole.

Part (b): Diameter of solid circular pole.

1, 2. ***Conceptualize, Categorize***: For the pole having a solid circular cross section (Fig. 5-36b), use Eq. (5-46) with V replaced by P and r replaced by $d_0/2$:

$$\tau_{max} = \frac{4P}{3\pi(d_0/2)^2} \qquad \text{(b)}$$

3. ***Analyze***: Solve for d_0 to obtain

$$d_0^2 = \frac{16P}{3\pi\tau_{max}} = \frac{16(6675\ \text{N})}{3\pi(4.68\ \text{MPa})} = 2.42 \times 10^{-3}\ \text{m}^2$$

that produces

$$d_0 = 49.21\ \text{mm}$$

4. ***Finalize***: In this particular example, the solid circular pole has a diameter approximately one-half that of the tubular pole.

Note: Shear stresses rarely govern the design of either circular or rectangular beams made of metals such as steel and aluminum. In these kinds of materials, the allowable shear stress is usually in the range 25 to 50% of the allowable tensile stress. In the case of the tubular pole in this example, the maximum shear stress is only 4.68 MPa. In contrast, the maximum bending stress obtained from the flexure formula is 69 MPa for a relatively short pole of length 600 mm. Thus, as the load increases, the allowable tensile stress will be reached long before the allowable shear stress is reached.

The situation is quite different for materials that are weak in shear, such as wood. For a typical wood beam, the allowable stress in horizontal shear is in the range of 4 to 10% of the allowable bending stress. Consequently, even though the maximum shear stress is relatively low in value, it sometimes governs the design.

5.10 Shear Stresses in the Webs of Beams with Flanges

When a beam of wide-flange shape (Fig. 5-39a) is subjected to shear forces as well as bending moments (nonuniform bending), both normal and shear stresses are developed on the cross sections. The distribution of the shear stresses in a wide-flange beam is more complicated than in a rectangular beam. For instance, the shear stresses in the flanges of the beam act in both vertical and horizontal directions (the y and z directions), as shown by the small arrows in Fig. 5-39b. The horizontal shear stresses are much larger than the vertical shear stresses in the flanges and are discussed later in Section 6.8.

FIGURE 5-39

(a) Beam of wide-flange shape and (b) directions of the shear stresses acting on a cross section

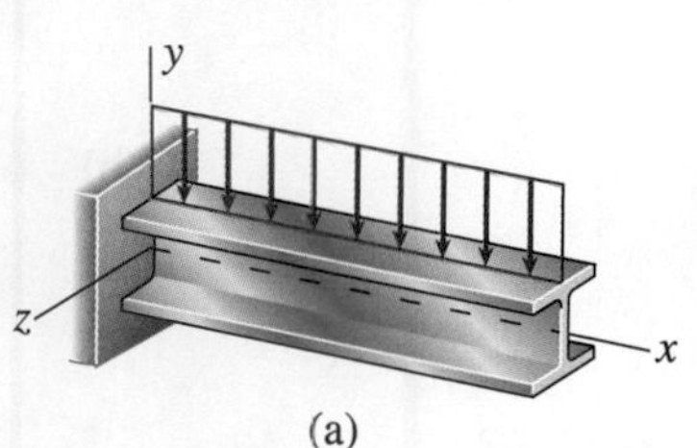

(a)

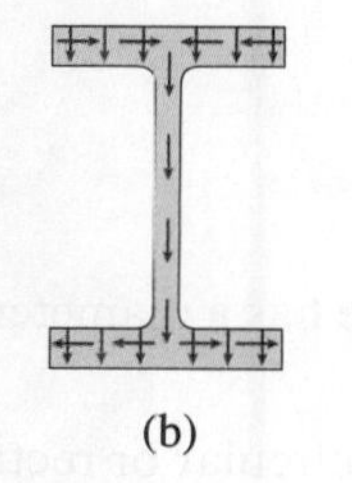

(b)

The shear stresses in the web of a wide-flange beam act only in the vertical direction and are larger than the stresses in the flanges. These stresses can be found by the same techniques used for finding shear stresses in rectangular beams.

Shear Stresses in the Web

Begin the analysis by determining the shear stresses at line *ef* in the web of a wide-flange beam (Fig. 5-40a). Make the same assumptions as those made for a rectangular beam; that is, assume that the shear stresses act parallel to the y axis and are uniformly distributed across the thickness of the web. Then the shear formula $\tau = VQ/Ib$ will still apply. However, the width b is now the thickness t of the web, and the area used in calculating the first moment Q is the area between line *ef* and the top edge of the cross section (indicated by the shaded area of Fig. 5-40a).

When finding the first moment Q of the shaded area, disregard the effects of the small fillets at the juncture of the web and flange (points b and c in Fig. 5-40a). The error in ignoring the areas of these fillets is very small. Then divide the shaded area into two rectangles. The first rectangle is the upper flange itself, which has the area

$$A_1 = b\left(\frac{h}{2} - \frac{h_1}{2}\right) \qquad \textbf{(5-49a)}$$

in which b is the width of the flange, h is the overall height of the beam, and h_1 is the distance between the insides of the flanges. The second rectangle is the part of the web between *ef* and the flange, that is, rectangle *efcb*, which has the area

$$A_2 = t\left(\frac{h_1}{2} - y_1\right) \qquad \textbf{(5-49b)}$$

in which t is the thickness of the web and y_1 is the distance from the neutral axis to line *ef*.

The first moments of areas A_1 and A_2, evaluated about the neutral axis, are obtained by multiplying these areas by the distances from their respective

FIGURE 5-40

Shear stresses in the web of a wide-flange beam: (a) cross section of beam and (b) distribution of vertical shear stresses in the web

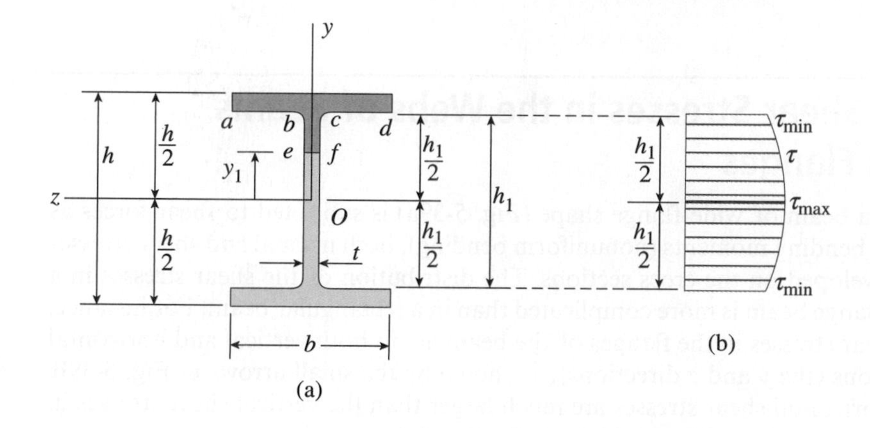

(a)

(b)

centroids to the z axis. Adding these first moments gives the first moment Q of the combined area:

$$Q = A_1\left(\frac{h_1}{2} + \frac{h/2 - h_1/2}{2}\right) + A_2\left(y_1 + \frac{h_1/2 - y_1}{2}\right)$$

Substituting for A_1 and A_2 from Eqs. (5-49a and b) and then simplifying gives

$$Q = \frac{b}{8}(h^2 - h_1^2) + \frac{t}{8}(h_1^2 - 4y_1^2) \quad \textbf{(5-50)}$$

Therefore, the shear stress τ in the web of the beam at distance y_1 from the neutral axis is

$$\tau = \frac{VQ}{It} = \frac{V}{8It}\left[b(h^2 - h_1^2) + t(h_1^2 - 4y_1^2)\right] \quad \textbf{(5-51)}$$

in which the moment of inertia of the cross section is

$$I = \frac{bh^3}{12} - \frac{(b-t)h_1^3}{12} = \frac{1}{12}(bh^3 - bh_1^3 + th_1^3) \quad \textbf{(5-52)}$$

Since all quantities in Eq. (5-51) are constants except y_1, note that τ varies quadratically throughout the height of the web, as shown by the graph in Fig. 5-40b. The graph is drawn only for the web and does not include the flanges. The reason is simple enough—Eq. (5-51) cannot be used to determine the vertical shear stresses in the flanges of the beam (see the discussion titled "Limitations" later in this section).

Maximum and Minimum Shear Stresses

The maximum shear stress in the web of a wide-flange beam occurs at the neutral axis where $y_1 = 0$. The minimum shear stress occurs where the web meets the flanges ($y_1 = \pm h_1/2$). These stresses, found from Eq. (5-51), are

$$\tau_{max} = \frac{V}{8It}(bh^2 - bh_1^2 + th_1^2) \qquad \tau_{min} = \frac{Vb}{8It}(h^2 - h_1^2) \quad \textbf{(5-53a,b)}$$

Both τ_{max} and τ_{min} are labeled on the graph of Fig. 5-40b. For typical wide-flange beams, the maximum stress in the web is from 10 to 60% greater than the minimum stress.

Although it may not be apparent from the preceding discussion, the stress τ_{max} given by Eq. (5-53a) not only is the largest shear stress in the web but also is the largest shear stress anywhere in the cross section.

Shear Force in the Web

The vertical shear force carried by the web alone may be determined by multiplying the area of the shear-stress diagram (Fig. 5-40b) by the thickness t of the web. The shear-stress diagram consists of two parts: a rectangle of area $h_1\tau_{min}$ and a parabolic segment of area

$$\frac{2}{3}(h_1)(\tau_{max} - \tau_{min})$$

FIGURE 5-40 (Repeated)

Shear stresses in the web of a wide-flange beam: (a) cross section of beam and (b) distribution of vertical shear stresses in the web

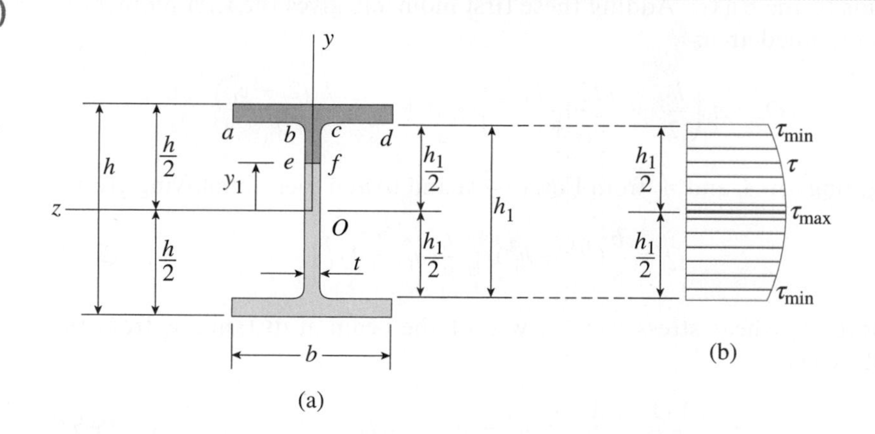

Adding these two areas, multiplying by the thickness t of the web, and then combining terms gives the total shear force in the web:

$$V_{web} = \frac{th_1}{3}(2\tau_{max} + \tau_{min}) \qquad \textbf{(5-54)}$$

For beams of typical proportions, the shear force in the web is 90 to 98% of the total shear force V acting on the cross section; the remainder is carried by shear in the flanges.

Since the web resists most of the shear force, designers often calculate an approximate value of the maximum shear stress by dividing the total shear force by the area of the web. The result is the average shear stress in the web, assuming that the web carries *all* of the shear force:

$$\tau_{aver} = \frac{V}{th_1} \qquad \textbf{(5-55)}$$

For typical wide-flange beams, the average stress calculated in this manner is within 10% (plus or minus) of the maximum shear stress calculated from Eq. (5-53a). Thus, Eq. (5-55) provides a simple way to estimate the maximum shear stress.

Limitations

The elementary shear theory presented in this section is suitable for determining the vertical shear stresses in the web of a wide-flange beam. However, when investigating vertical shear stresses in the flanges, you can no longer assume that the shear stresses are constant across the width of the section, that is, across the width b of the flanges (Fig. 5-40a). Hence, you cannot use the shear formula to determine these stresses.

To emphasize this point, consider the junction of the web and upper flange ($y_1 = h_1/2$), where the width of the section changes abruptly from t to b. The shear stresses on the free surfaces *ab* and *cd* (Fig. 5-40a) must be zero, whereas the shear stress across the web at line *bc* is τ_{min}. These observations indicate that the distribution of shear stresses at the junction of the web and the flange is quite complex and cannot be investigated by elementary methods. The stress analysis is further complicated by the use of fillets at the re-entrant corners (corners b and c). The fillets are necessary to prevent the stresses from becoming dangerously large, but they also alter the stress distribution across the web.

Thus, the shear formula cannot be used to determine the vertical shear stresses in the flanges. However, the shear formula does give good results for the shear stresses acting *horizontally* in the flanges (Fig. 5-39b), as discussed later in Section 6.8.

This method for determining shear stresses in the webs of wide-flange beams also can be used for other sections having thin webs. For instance, Example 5-15 illustrates the procedure for a T-beam.

Example 5-14

A beam of wide-flange shape (Fig. 5-41a) is subjected to a vertical shear force $V = 45$ kN. The cross-sectional dimensions of the beam are $b = 165$ mm, $t = 7.5$ mm, $h = 320$ mm, and $h_1 = 290$ mm.

Determine the maximum shear stress, minimum shear stress, and total shear force in the web. (Disregard the areas of the fillets when making calculations.)

FIGURE 5-41

Example 5-14: Shear stresses in the web of a wide-flange beam

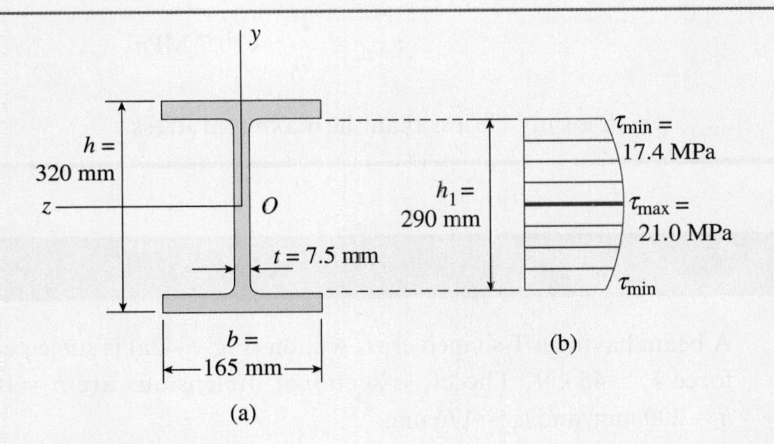

Solution:

Use a four-step problem-solving approach. Combine steps as needed for an efficient solution.

1, 2. *Conceptualize, Categorize*:

Maximum and minimum shear stresses: The maximum and minimum shear stresses in the web of the beam are given by Eqs. (5-53a and b). Before substituting into those equations, calculate the moment of inertia of the cross-sectional area from Eq. (5-52):

$$I = \frac{1}{12}(bh^3 - bh_1^3 + th_1^3) = 130.45 \times 10^6 \text{ mm}^4$$

3. ***Analyze***: Now substitute this value for I, as well as the numerical values for the shear force V and the cross-sectional dimensions, into Eqs. (5-53a and b):

$$\tau_{max} = \frac{V}{8It}(bh^2 - bh_1^2 + th_1^2) = 21.0 \text{ MPa}$$

$$\tau_{min} = \frac{Vb}{8It}(h^2 - h_1^2) = 17.4 \text{ MPa}$$

In this case, the ratio of τ_{max} to τ_{min} is 1.21, that is, the maximum stress in the web is 21% larger than the minimum stress. The variation of the shear stresses over the height h_1 of the web is shown in Fig. 5-41b.

Total shear force: The shear force in the web is calculated from Eq. (5-54) as

$$V_{web} = \frac{th_1}{3}(2\tau_{max} + \tau_{min}) = 43.0 \text{ kN}$$

4. ***Finalize***: From this result, note that the web of this particular beam resists 96% of the total shear force.

Note: The average shear stress in the web of the beam [from Eq. (5-55)] is

$$\tau_{aver} = \frac{V}{th_1} = 20.7 \text{ MPa}$$

which is only 1% less than the maximum stress.

Example 5-15

A beam having a T-shaped cross section (Fig. 5-42a) is subjected to a vertical shear force $V = 45$ kN. The cross-sectional dimensions are $b = 100$ mm, $t = 24$ mm, $h = 200$ mm, and $h_1 = 176$ mm.

Determine the shear stress τ_1 at the top of the web (level *nn*) and the maximum shear stress τ_{max}. (Disregard the areas of the fillets.)

Solution:

Use a four-step problem-solving approach. Combine steps as needed for an efficient solution.

1, 2. ***Conceptualize, Categorize***:

Location of neutral axis: The neutral axis of the T-beam is located by calculating the distances c_1 and c_2 from the top and bottom of the beam to the centroid of the cross section (Fig. 5-42a). First, divide the cross section into two rectangles: the flange and the web (see the dashed line in Fig. 5-42a). Then calculate the first moment Q_{aa} of these two rectangles with respect to line *aa* at the bottom of the beam. The distance c_2 is equal to Q_{aa} divided by the area

FIGURE 5-42

Example 5-15: Shear stresses in web of T-shaped beam

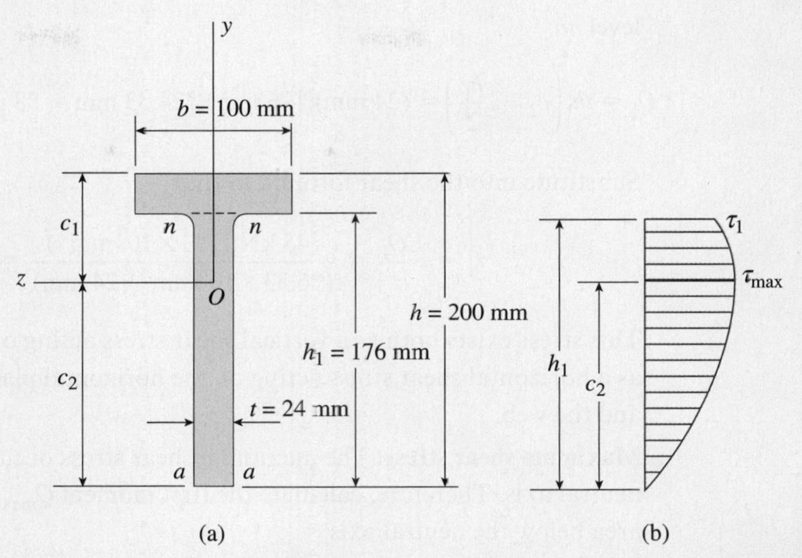

A of the entire cross section (see Appendix D, Section D.2, for methods for locating centroids of composite areas). The calculations are

$$A = \Sigma A_i = b(h - h_1) + th_1 = 6624 \text{ mm}^2$$

$$Q_{aa} = \Sigma y_i A_i = \left(\frac{h + h_1}{2}\right)(b)(h - h_1) + \frac{h_1}{2}(th_1) = 822{,}912 \text{ mm}^3$$

$$c_2 = \frac{Q_{aa}}{A} = \frac{822{,}912 \text{ mm}^3}{6624 \text{ mm}^2} = 124.23 \text{ mm} \quad c_1 = h - c_2 = 75.77 \text{ mm}$$

Moment of inertia: Find the moment of inertia I of the entire cross-sectional area (with respect to the neutral axis) by determining the moment of inertia I_{aa} about line aa at the bottom of the beam and then use the parallel-axis theorem (see Section D.4, Appendix D):

$$I = I_{aa} - Ac_2^2$$

The calculations are

$$I_{aa} = \frac{bh^3}{3} - \frac{(b - t)h_1^3}{3} = 128.56 \times 10^6 \text{ mm}^4 \quad Ac_2^2 = 102.23 \times 10^6 \text{ mm}^4 \quad I = 26.33 \times 10^6 \text{ mm}^4$$

3. ***Analyze***:

Shear stress at top of web: To find the shear stress τ_1 at the top of the web (along line nn) calculate the first moment Q_1 of the area above level nn. This first moment is equal to the area of the flange times the distance from the neutral axis to the centroid of the flange:

$$Q_1 = b(h - h_1)\left(c_1 - \frac{h - h_1}{2}\right)$$

$$= (100 \text{ mm})(24 \text{ mm})(75.77 \text{ mm} - 12 \text{ mm}) = 153.0 \times 10^3 \text{ mm}^3$$

You get the same result if you calculate the first moment of the area below level *nn*:

$$Q_1 = th_1\left(c_2 - \frac{h_1}{2}\right) = (24\text{ mm})(176\text{ mm})(124.33\text{ mm} - 88\text{ mm}) = 153 \times 10^3\text{ mm}^3$$

Substitute into the shear formula to find

$$\tau_1 = \frac{VQ_1}{It} = \frac{(45\text{ kN})(153 \times 10^3\text{ mm}^3)}{(26.33 \times 10^6\text{ mm}^4)(24\text{ mm})} = 10.9\text{ MPa}$$

This stress exists both as a vertical shear stress acting on the cross section and as a horizontal shear stress acting on the horizontal plane between the flange and the web.

Maximum shear stress: The maximum shear stress occurs in the web at the neutral axis. Therefore, calculate the first moment Q_{max} of the cross-sectional area below the neutral axis:

$$Q_{max} = tc_2\left(\frac{c_2}{2}\right) = (24\text{ mm})(124.23\text{ mm})\left(\frac{124.23\text{ mm}}{2}\right) = 185 \times 10^3\text{ mm}^3$$

The same result is obtained if the first moment of the area above the neutral axis is computed, but those calculations would be slighter longer.

Substitute into the shear formula to obtain

$$\tau_{max} = \frac{VQ_{max}}{It} = \frac{(45\text{ kN})(185 \times 10^3\text{ mm}^3)}{(26.33 \times 10^6\text{ mm}^4)(24\text{ mm})} = 13.2\text{ MPa}$$

which is the maximum shear stress in the beam.

4. ***Finalize***: The parabolic distribution of shear stresses in the web is shown in Fig. 5-42b.

*5.11 Built-Up Beams and Shear Flow

Built-up beams are fabricated from two or more pieces of material joined together to form a single beam. Such beams can be constructed in a great variety of shapes to meet special architectural or structural needs and to provide larger cross sections than are ordinarily available.

Figure 5-43 shows some typical cross sections of built-up beams. A wood **box beam** (Fig. 5-43a) is constructed of two planks that serve as flanges and two plywood webs. The pieces are joined together with nails, screws, or glue in such a manner that the entire beam acts as a single unit. Box beams are also constructed of other materials, including steel, plastics, and composites.

The second example (Fig. 5-43b) is a glued laminated beam (called a **glulam beam**) made of boards glued together to form a much larger beam than could be cut from a tree as a single member. Glulam beams are widely used in the construction of small buildings.

The third example (Fig. 5-43c) is a steel **plate girder** of the type commonly used in bridges and large buildings. These girders, consisting of three steel plates joined by welding, can be fabricated in much larger sizes than are available with ordinary wide-flange or I-beams.

Built-up beams must be designed so that the beam behaves as a single member. Consequently, the design calculations involve two phases. In the first phase, the beam is designed as though it were made of one piece, taking into account both bending and shear stresses. In the second phase, the *connections* between the parts (such as nails, bolts, welds, and glue) are designed to ensure that the beam does indeed behave as a single entity. In particular, the connections must be strong enough to transmit the horizontal shear forces acting between the parts of the beam. To obtain these forces, make use of the concept of *shear flow*.

FIGURE 5-43

Cross sections of typical built-up beams: (a) wood box beam, (b) glulam beam, and (c) plate girder

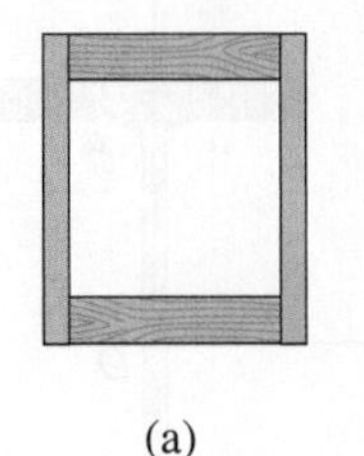

(a)

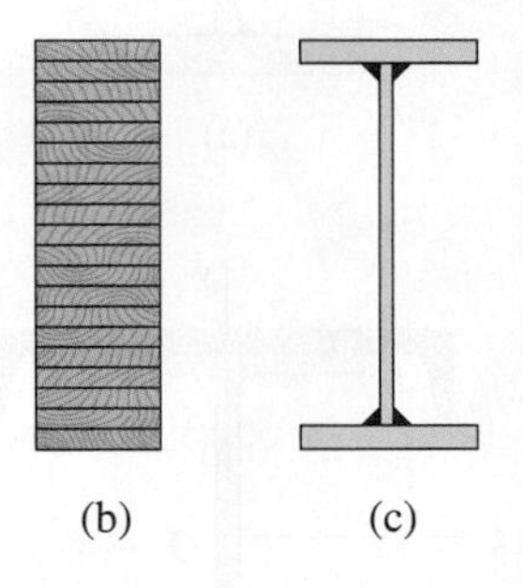

(b) (c)

Shear Flow

To obtain a formula for the horizontal shear forces acting between parts of a beam, return to the derivation of the shear formula (see Figs. 5-30 and 5-31 of Section 5.8). In that derivation, element mm_1n_1n was cut from a beam (Fig. 5-44a) and horizontal equilibrium of a subelement mm_1p_1p was investigated (Fig. 5-44b). From the horizontal equilibrium of the subelement, the force F_3 (Fig. 5-44c) acting on its lower surface was found to be

$$F_3 = \frac{dM}{I}\int y\,dA \tag{5-56}$$

This equation is repeated from Eq. (5-36) of Section 5.8.

FIGURE 5-44

Horizontal shear stresses and shear forces in a beam (*Note*: These figures are repeated from Figs. 5-30 and 5-31)

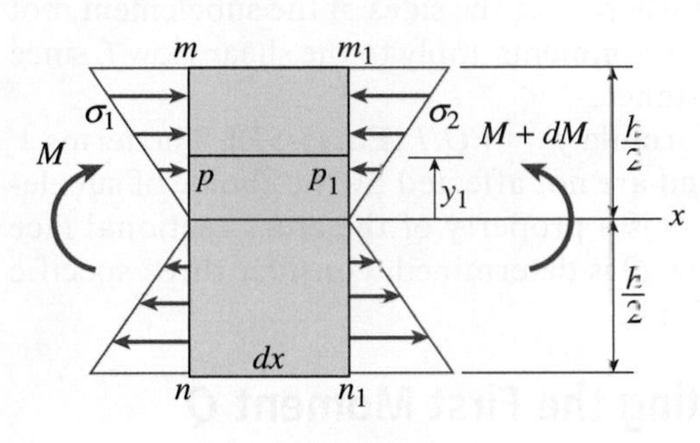

Side view of element

(a)

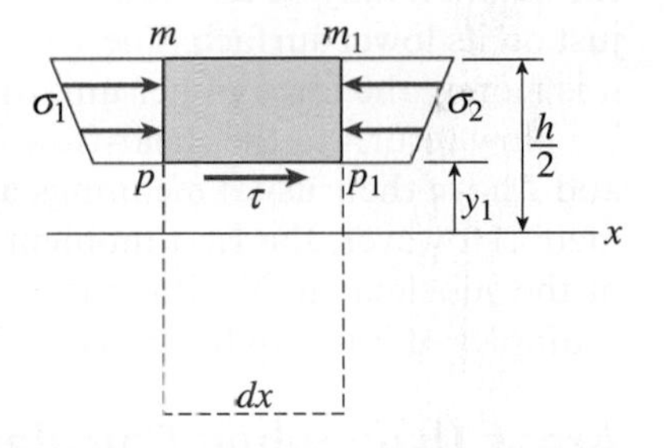

Side view of subelement

(b)

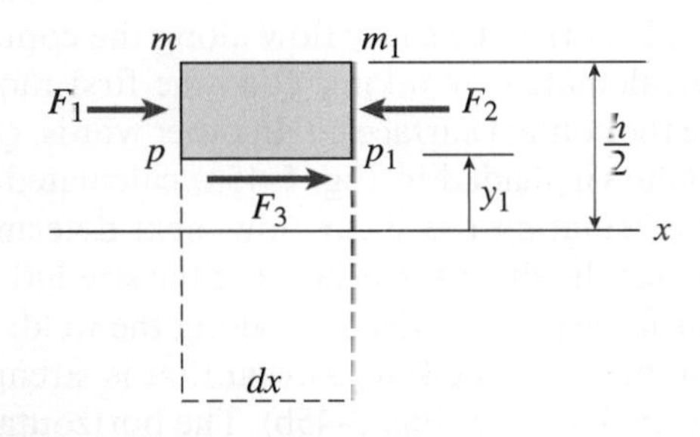

Forces on of subelement

(c)

FIGURE 5-45

Areas used when calculating the first moment Q

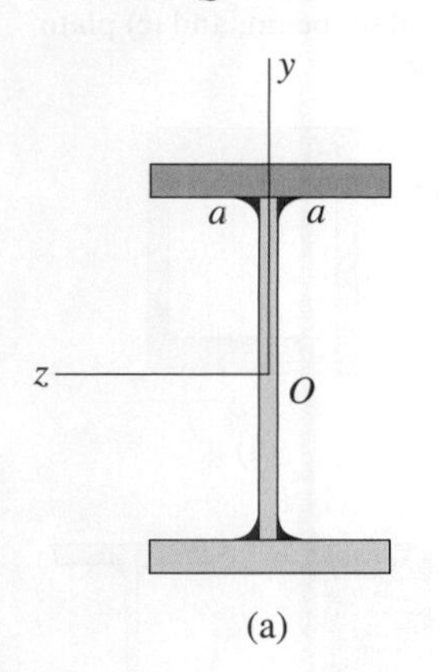

(a)

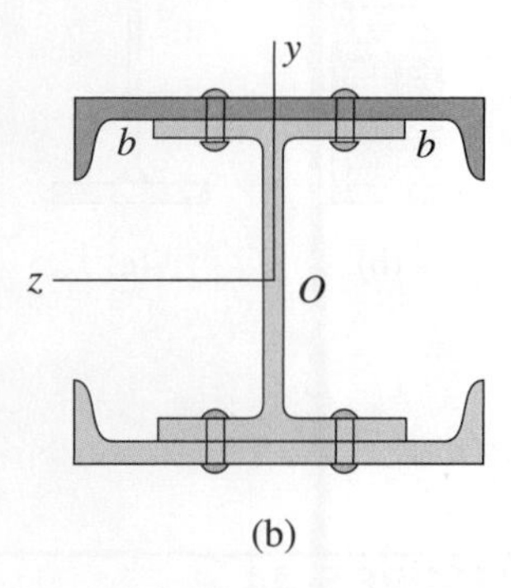

(b)

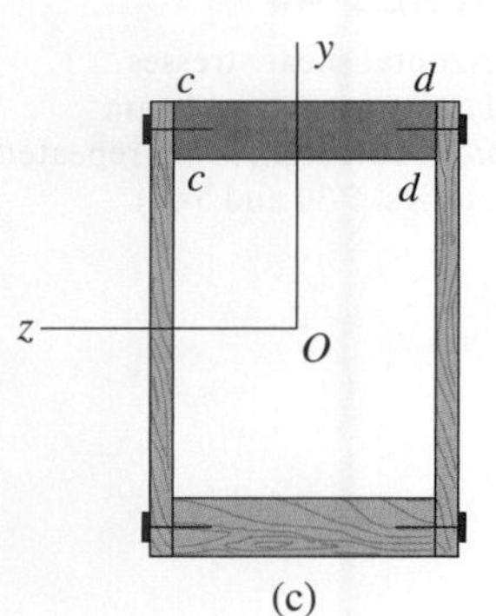

(c)

Now define a new quantity called the **shear flow** f. Shear flow is the *horizontal shear force per unit distance along the longitudinal axis of the beam*. Since the force F_3 acts along the distance dx, the shear force per unit distance is equal to F_3 divided by dx; thus,

$$f = \frac{F_3}{dx} = \frac{dM}{dx}\left(\frac{1}{I}\right)\int y\,dA$$

Replacing dM/dx by the shear force V and denoting the integral by Q leads to the **shear-flow formula**:

$$f = \frac{VQ}{I} \tag{5-57}$$

This equation gives the shear flow acting on the horizontal plane pp_1 shown in Fig. 5-44a. The terms V, Q, and I have the same meanings as in the shear formula [Eq. (5-41)].

If the shear stresses on plane pp_1 are uniformly distributed, as assumed for rectangular beams and wide-flange beams, the shear flow f equals τb. In that case, the shear-flow formula reduces to the shear formula. However, the derivation of Eq. (5-56) for the force F_3 does not involve any assumption about the distribution of shear stresses in the beam. Instead, the force F_3 is found solely from the horizontal equilibrium of the subelement (Fig. 5-44c). Therefore, the subelement and the force F_3 can be interpreted in more general terms than before.

The subelement may be *any* prismatic block of material between cross sections mn and m_1n_1 (Fig. 5-44a). It does not have to be obtained by making a single horizontal cut (such as pp_1) through the beam. Also, since the force F_3 is the total horizontal shear force acting between the subelement and the rest of the beam, it may be distributed anywhere over the sides of the subelement, not just on its lower surface. These same comments apply to the shear flow f, since it is merely the force F_3 per unit distance.

Now return to the shear-flow formula $f = VQ/I$ [Eq. (5-57)]. The terms V and I have their usual meanings and are not affected by the choice of subelement. However, the first moment Q is a property of the cross-sectional face of the subelement. To illustrate how Q is determined, consider three specific examples of built-up beams (Fig. 5-45).

Areas Used when Calculating the First Moment Q

The first example of a built-up beam is a welded steel **plate girder** (Fig. 5-45a). The welds must transmit the horizontal shear forces that act between the flanges and the web. At the upper flange, the horizontal shear force (per unit distance along the axis of the beam) is the shear flow along the contact surface *aa*. This shear flow may be calculated by taking Q as the first moment of the cross-sectional area above the contact surface *aa*. In other words, Q is the first moment of the flange area (shown shaded in Fig. 5-45a) calculated with respect to the neutral axis. After calculating the shear flow, next determine the amount of welding needed to resist the shear force, because the strength of a weld is usually specified in terms of force per unit distance along the weld.

The second example is a **wide-flange beam** that is strengthened by riveting a channel section to each flange (Fig. 5-45b). The horizontal shear force acting between each channel and the main beam must be transmitted by the rivets. This force is calculated from the shear-flow formula using Q as the first moment of the area of the entire channel (shown shaded in the figure). The resulting shear flow is the longitudinal force per unit distance acting along the contact surface *bb*, and the rivets must be of adequate size and longitudinal spacing to resist this force.

The last example is a **wood box beam** with two flanges and two webs that are connected by nails or screws (Fig. 5-45c). The total horizontal shear force between the upper flange and the webs is the shear flow acting along *both* contact surfaces *cc* and *dd*, and therefore the first moment Q is calculated for the upper flange (the shaded area). In other words, the shear flow calculated from the formula $f = VQ/I$ is the total shear flow along all contact surfaces that surround the area for which Q is computed. In this case, the shear flow f is resisted by the combined action of the nails on *both* sides of the beam, that is, at both *cc* and *dd*, as illustrated in the following example.

Example 5-16

A wood box beam (Fig. 5-46) is constructed of two boards, each 40 × 180 mm in cross section, that serve as flanges and two plywood webs, each 15 mm thick. The total height of the beam is 280 mm. The plywood is fastened to the flanges by wood screws having an allowable load in shear of $F = 800$ N each.

If the shear force V acting on the cross section is 10.5 kN, determine the maximum permissible longitudinal spacing s of the screws (Fig. 5-46b).

FIGURE 5-46

Example 5-16: Wood box beam

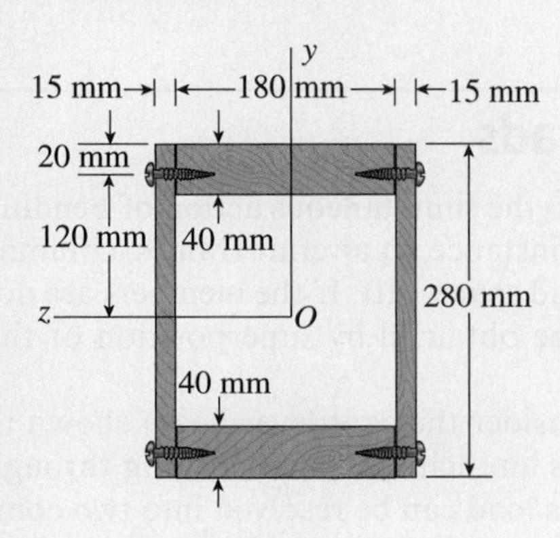

(a) Cross section

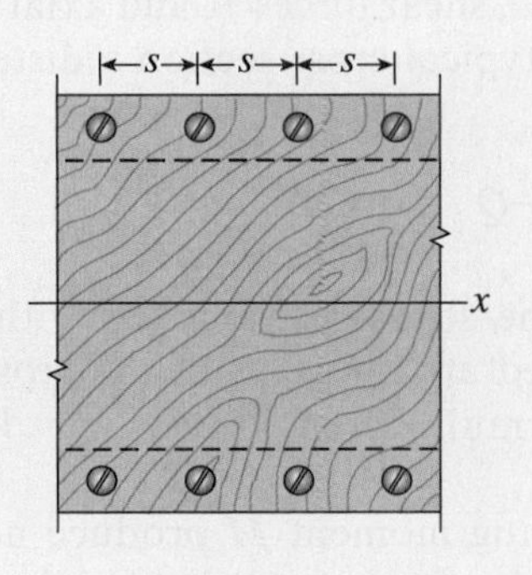

(b) Side view

Solution:

Use a four-step problem-solving approach. Combine steps as needed for an efficient solution.

1, 2. ***Conceptualize, Categorize:***

Shear flow: The horizontal shear force transmitted between the upper flange and the two webs can be found from the shear-flow formula $f = VQ/I$, in which Q is the first moment of the cross-sectional area of the flange. To find this first moment, multiply the area A_f of the flange by the distance d_f from its centroid to the neutral axis:

$$A_f = 40 \text{ mm} \times 180 \text{ mm} = 7200 \text{ mm}^2 \quad d_f = 120 \text{ mm}$$

$$Q = A_f d_f = (7200 \text{ mm}^2)(120 \text{ mm}) = 864 \times 10^3 \text{ mm}$$

The moment of inertia of the entire cross-sectional area about the neutral axis is equal to the moment of inertia of the outer rectangle minus the moment of inertia of the "hole" (the inner rectangle):

$$I = \frac{1}{12}(210 \text{ mm})(280 \text{ mm})^3 - \frac{1}{12}(180 \text{ mm})(200 \text{ mm})^3$$

$$= 264.2 \times 10^6 \text{ mm}^4$$

Substituting V, Q, and I into the shear-flow formula [Eq. (5-57)] gives

$$f = \frac{VQ}{I} = \frac{(10{,}500 \text{ N})(864 \times 10^3 \text{ mm}^3)}{264.2 \times 10^6 \text{ mm}^4} = 34.3 \text{ N/mm}$$

which is the horizontal shear force per millimeter of length that must be transmitted between the flange and the two webs.

3. *Analyze*:

Spacing of screws: Since the longitudinal spacing of the screws is s, and since there are two lines of screws (one on each side of the flange), the load capacity of the screws is $2F$ per distance s along the beam. Therefore, the capacity of the screws per unit distance along the beam is $2F/s$. Equating $2F/s$ to the shear flow f and solving for the spacing s gives

$$s = \frac{2F}{f} = \frac{2(800 \text{ N})}{34.3 \text{ N/mm}} = 46.6 \text{ mm}$$

4. *Finalize*: This value of s is the maximum permissible spacing of the screws based upon the allowable load per screw. Any spacing greater than 46.6 mm would overload the screws. For convenience in fabrication (and to be on the safe side), a spacing such as $s = 45$ mm should be selected.

*5.12 Beams with Axial Loads

Structural members are often subjected to the simultaneous action of bending loads and axial loads. This happens, for instance, in aircraft frames, columns in buildings, machinery, parts of ships, and spacecraft. If the members are not too slender, the combined stresses can be obtained by superposition of the bending stresses and the axial stresses.

To see how this is accomplished, consider the cantilever beam shown in Fig. 5-47a. The only load on the beam is an inclined force P acting through the centroid of the end cross section. This load can be resolved into two components, a lateral load Q and an axial load S. These loads produce **stress resultants** in the form of bending moments M, shear forces V, and axial forces N throughout the beam (Fig. 5-47b). On a typical cross section a distance x from the support, these stress resultants are

$$M = Q(L - x) \quad V = -Q \quad N = S$$

in which L is the length of the beam. The stresses associated with each of these stress resultants can be determined at any point in the cross section by means of the appropriate formula ($\sigma = -My/I$, $\tau = VQ/Ib$, and $\sigma = N/A$).

Since both the axial force N and bending moment M produce normal stresses, combine those stresses to obtain the final stress distribution. The **axial force** (when acting alone) produces a uniform stress distribution $\sigma = N/A$

over the entire cross section, as shown by the stress diagram in Fig. 5-47c. In this particular example, the stress σ is tensile, as indicated by the plus signs attached to the diagram.

The **bending moment** produces a linearly varying stress $\sigma = -My/I$ (Fig. 5-47d) with compression on the upper part of the beam and tension on the lower part. The distance y is measured from the z axis, which passes through the centroid of the cross section.

The final distribution of normal stresses is obtained by superposing the stresses produced by the axial force and the bending moment. Thus, the equation for the **combined stresses** is

$$\sigma = \frac{N}{A} - \frac{My}{I} \quad \textbf{(5-58)}$$

Note that N is positive when it produces tension and M is positive, according to the bending-moment sign convention (positive bending moment produces compression in the upper part of the beam and tension in the lower part). Also, the y axis is positive upward. As long as these sign conventions are used in Eq. (5-58), the normal stress σ is positive for tension and negative for compression.

The final stress distribution depends upon the relative algebraic values of the terms in Eq. (5-58). For this example, the three possibilities are shown in Figs. 5-47e, f, and g. If the bending stress at the top of the beam (Fig. 5-47d) is numerically less than the axial stress (Fig. 5-47c), the entire cross section is in tension, as shown in Fig. 5-47e. If the bending stress at the top equals the axial stress, the distribution is triangular (Fig. 5-47f), and if the bending stress is numerically larger than the axial stress, the cross section is partially in compression and partially in tension (Fig. 5-47g). Of course, if the axial force is a compressive force, or if the bending moment is reversed in direction, the stress distributions change accordingly.

Whenever bending and axial loads act simultaneously, the neutral axis (that is, the line in the cross section where the normal stress is zero) no longer passes through the centroid of the cross section. As shown in Figs. 5-47e, f, and g, respectively, the neutral axis may be outside the cross section, at the edge of the section, or within the section.

The use of Eq. (5-58) to determine the stresses in a beam with axial loads is illustrated later in Example 5-17.

FIGURE 5-47

Normal stresses in a cantilever beam subjected to both bending and axial loads: (a) beam with load P acting at the free end, (b) stress resultants N, V, and M acting on a cross section at distance x from the support, (c) tensile stresses due to the axial force N acting alone, (d) tensile and compressive stresses due to the bending moment M acting alone, and (e), (f), (g) are possible final stress distributions due to the combined effects of N and M

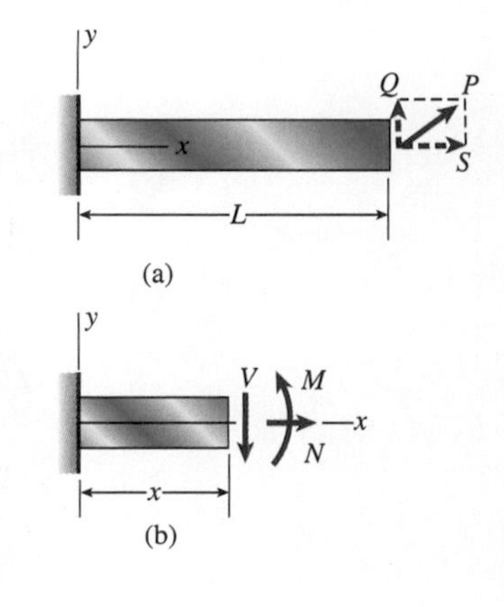

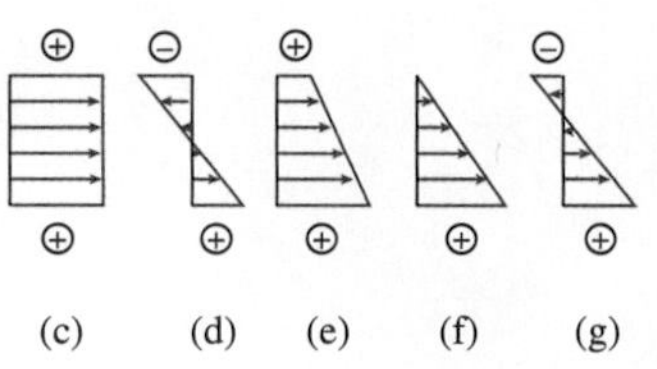

Eccentric Axial Loads

An **eccentric axial load** is an axial force that does *not* act through the centroid of the cross section. An example is shown in Fig. 5-48a, where the cantilever beam AB is subjected to a tensile load P acting at distance e from the x axis (the x axis passes through the centroids of the cross sections). The distance e, called the *eccentricity* of the load, is positive in the positive direction of the y axis.

The eccentric load P is statically equivalent to an axial force P acting along the x axis and a bending moment Pe acting about the z axis (Fig. 5-48b). Note that the moment Pe is a negative bending moment.

A cross-sectional view of the beam (Fig. 5-48c) shows the y and z axes passing through the centroid C of the cross section. The eccentric load P intersects the y axis, which is an axis of symmetry.

(Lester Lefkowitz/Getty Images)

Bending due to self-weight of beam and axial compression due to horizontal component of cable lifting force

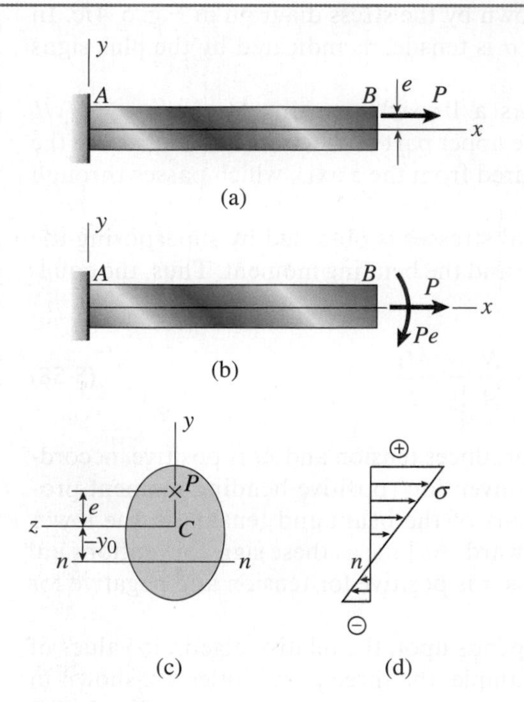

FIGURE 5-48

(a) Cantilever beam with an eccentric axial load P, (b) equivalent loads P and Pe, (c) cross section of beam, and (d) distribution of normal stresses over the cross section

Since the axial force N at any cross section is equal to P, and since the bending moment M is equal to $-Pe$, the **normal stress** at any point in the cross section [from Eq. (5-58)] is

$$\sigma = \frac{P}{A} + \frac{Pey}{I} \tag{5-59}$$

in which A is the area of the cross section and I is the moment of inertia about the z axis. The stress distribution obtained from Eq. (5-59), for the case where both P and e are positive, is shown in Fig. 5-48d.

The position of the **neutral axis** nn (Fig. 5-48c) can be obtained from Eq. (5-59) by setting the stress σ equal to zero and solving for the coordinate y, denoted as y_0. The result is

$$y_0 = -\frac{I}{Ae} \tag{5-60}$$

The coordinate y_0 is measured from the z axis (which is the neutral axis under pure bending) to the line nn of zero stress (the neutral axis under combined bending and axial load). Because y_0 is positive in the direction of the y axis (upward in Fig. 5-48c), it is labeled $-y_0$ when it is shown downward in the figure.

From Eq. (5-60), note that the neutral axis lies below the z axis when e is positive and above the z axis when e is negative. If the eccentricity is reduced, the distance y_0 increases and the neutral axis moves away from the centroid. In the limit, as e approaches zero, the load acts at the centroid, the neutral axis is at an infinite distance, and the stress distribution is uniform. If the eccentricity is increased, the distance y_0 decreases and the neutral axis moves toward the centroid. In the limit, as e becomes extremely large, the load acts at an infinite distance, the neutral axis passes through the centroid, and the stress distribution is the same as in pure bending.

Eccentric axial loads are analyzed in some of the problems at the end of this chapter.

Limitations

The preceding analysis of beams with axial loads is based upon the assumption that the bending moments can be calculated without considering the deflections of the beams. In other words, when determining the bending moment M for use in Eq. (5-58), you must be able to use the original dimensions of the beam—that is, the dimensions *before* any deformations or deflections occur. The use of the original dimensions is valid provided the beams are relatively stiff in bending, so that the deflections are very small.

Thus, when analyzing a beam with axial loads, it is important to distinguish between a **stocky beam**, which is relatively short and therefore highly resistant to bending, and a **slender beam**, which is relatively long and therefore very flexible. In the case of a stocky beam, the lateral deflections are so small as to have no significant effect on the line of action of the axial forces. As a consequence, the bending moments will not depend upon the deflections, and the stresses can be found from Eq. (5-58).

In the case of a slender beam, the lateral deflections (even though small in magnitude) are large enough to alter significantly the line of action of the axial forces. When that happens, an additional bending moment equal to the product of the axial force and the lateral deflection is created at every cross section. In other words, there is an interaction, or coupling, between the axial effects and the bending effects. This type of behavior is discussed in Chapter 11 on **columns**.

The distinction between a stocky beam and a slender beam is obviously not a precise one. In general, the only way to know whether interaction effects are important is to analyze the beam with and without the interaction and notice whether the results differ significantly. However, this procedure may require considerable calculating effort. Therefore, as a guideline for practical use, consider a beam with a length-to-height ratio of 10 or less to be a stocky beam. Only stocky beams are considered in the problems pertaining to this section.

Example 5-17

A tubular beam ACB with a length of $L = 1.5$ m is pin-supported at its ends, A and B. A powered winch at E lifts load W below C using a cable which passes over a frictionless pulley at midlength (point D, Fig. 5-49a). The distance from the center of the pulley to the longitudinal axis of the tube is $d = 140$ mm. The cross section of the tube is square (Fig. 5-49b) with an outer dimension of $b = 150$ mm, area of $A = 125\ \text{cm}^2$, and moment of inertia of $I = 3385\ \text{cm}^4$.

(a) Determine the maximum tensile and compressive stresses in the beam due to a load $W = 13.5$ kN.

(b) If the allowable normal stress in the tube is 24 MPa, find the maximum permissible load W. Assume that the cable, pulley, and bracket CD are adequate to carry load W_{max}.

FIGURE 5-49

Example 5-17: Tubular beam subjected to combined bending and axial load

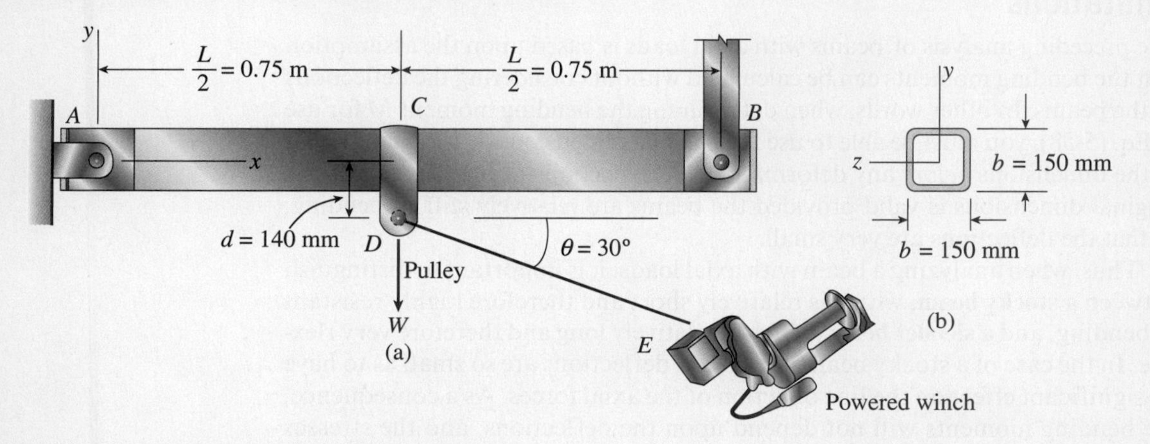

Solution:

Use a four-step problem-solving approach. Combine steps as needed for an efficient solution.

Part (a): Maximum tensile and compressive stresses in the beam.

1, 2. ***Conceptualize, Categorize:***

Beam and loading: Begin by representing the beam and its load in idealized form for the purposes of analysis (Fig. 5-50a). Since the support at end A resists both horizontal and vertical displacement, it is represented as a pin support. The support at B prevents vertical displacement but offers no resistance to horizontal displacement, so it is shown as a roller support.

Replace the cable forces at D with statically equivalent forces F_H and F_V and moment M_O, all of which are applied on the axis of the beam at C (see Fig. 5-50a):

$$F_H = W\cos(\theta) = 11.691 \text{ kN} \qquad F_V = W[1 + \sin(\theta)] = 20.25 \text{ kN}$$

$$M_0 W\cos(\theta)d = 1636.8 \text{ N}\cdot\text{m}$$

FIGURE 5-50

Solution of Example 5-17: (a) Idealized beam and loading, (b) axial-force diagram, (c) shear-force diagram, and (d) bending-moment diagram

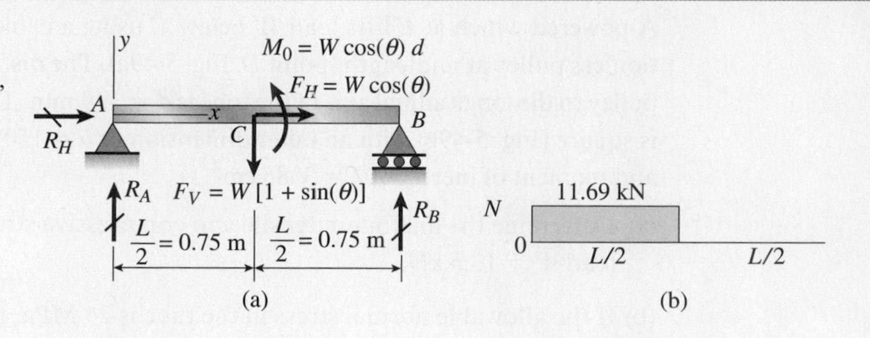

FIGURE 5-50 (Continued)

Solution of Example 5-17:
(a) Idealized beam and loading,
(b) axial-force diagram,
(c) shear-force diagram,
and (d) bending-moment diagram

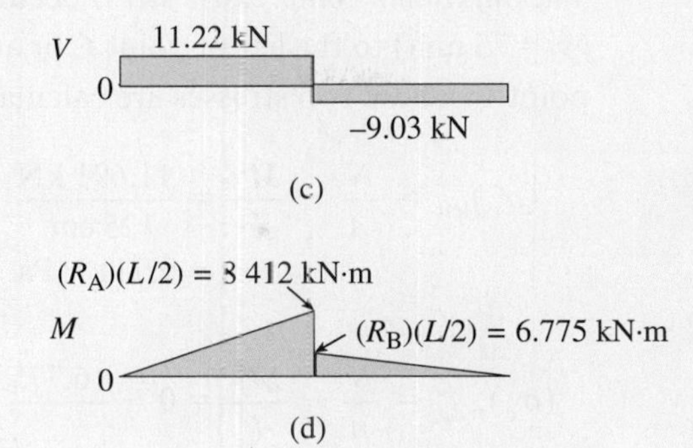

Reactions and stress resultants: The reactions of the beam (R_H, R_A, and R_B) are labeled in Fig. 5-50a. Also, the diagrams of axial force N, shear force V, and bending moment M are shown in Figs. 5-50b, c, and d, respectively. All of these quantities are found from free-body diagrams and equations of equilibrium using the techniques described in Chapter 4. For example, use equations of statics to find that

$$\Sigma F_H = 0: \quad R_H = -F_H = -W\cos(\theta) = -(13.5\text{ kN})\cos(30°) = -11.691\text{ kN} \qquad \textbf{(a)}$$

$$\Sigma M_A = 0: \quad R_B = \frac{1}{L}\left(F_V\frac{L}{2} - M_C\right) = \frac{W}{2}[1+\sin(\theta)] - W\frac{d}{L}[\cos(\theta)]$$

$$R_B = (13.5\text{ kN})\left[\frac{1+\sin(30°)}{2} - \left(\frac{140\text{ mm}}{1.5\text{ m}}\right)\cos(30°)\right] = 9.034\text{ kN} \qquad \textbf{(b)}$$

$$\Sigma F_V = 0: \quad R_A = F_V - R_B = (13.5\text{ kN})(1+\sin(30°)) - 9.034\text{ kN} = 11.216\text{ kN} \qquad \textbf{(c)}$$

Next, use the axial-force (N), shear-force (V), and bending-moment (M) diagrams (Figs. 5-50b, c, and d, respectively) to find the combined stresses in beam ACB using Eq. (5-58).

3. ***Analyze:***

Stresses in the beam: The *maximum tensile stress* in the beam occurs at the bottom of the beam ($y = -75$ mm) just to the left of the midpoint C. Note that at this point in the beam the tensile stress due to the axial force adds to the tensile stress produced by the largest bending moment. Thus, from Eq. (5-58),

$$(\sigma_t)_{\max} = \frac{N}{A} - \frac{My}{I} = \frac{11.691\text{ kN}}{125\text{ cm}^2} - \frac{(8.412\text{ kN}\cdot\text{m})(-75\text{ mm})}{3385\text{ cm}^4}$$
$$= 0.935\text{ MPa} + 18.638\text{ MPa} = 19.57\text{ MPa}$$

The *maximum compressive stress* occurs either at the top of the beam ($y = 75$ mm) to the left of point C or at the top of the beam to the right of point C. These two stresses are calculated as

$$(\sigma_c)_{\text{left}} = \frac{N}{A} - \frac{My}{I} = \frac{11.691\text{ kN}}{125\text{ cm}^2} - \frac{(8.412\text{ kN}\cdot\text{m})(75\text{ mm})}{3385\text{ cm}^4}$$
$$= 0.934\text{ MPa} - 18.638\text{ MPa} = -17.7\text{ MPa}$$

$$(\sigma_c)_{\text{right}} = \frac{N}{A} - \frac{My}{I} = 0 - \frac{(6.775\text{ kN}\cdot\text{m})(75\text{ mm})}{3385\text{ cm}^4} = -15.01\text{ MPa}$$

Thus, the maximum compressive stress is

$$(\sigma_c)_{\text{max}} = -17.7\text{ MPa}$$

and occurs at the top of the beam to the left of point C.

Part (b): Maximum permissible load W.

1, 2. ***Conceptualize, Categorize***: From Eq. (a), the *tensile stress at the bottom of the beam* just left of C (equal to 19.57 MPa for a load $W = 13.5$ kN) will reach allowable normal stress $\sigma_a = 24$ MPa first and thus will be the determining factor in finding W_{max}. Using expressions for reactions [Eqs. (a), (b), and (c)], the axial tension force in beam segment AC and the positive moment just left of C are

$$N = W\cos(\theta) \qquad M = R_A\frac{L}{2} = W\left(\frac{1+\sin(\theta)}{2} + \frac{d}{L}\cos(\theta)\right)\left(\frac{L}{2}\right)$$

From Eq. (5-58), the combined normal stress is

$$\sigma_a = \frac{W\cos(\theta)}{A} - \frac{W\left(\frac{1+\sin(\theta)}{2} + \frac{d}{L}\cos(\theta)\right)\left(\frac{L}{2}\right)\left(\frac{-b}{2}\right)}{I}$$

3. ***Analyze***: Solving for $W = W_{\text{max}}$ gives

$$W_{\text{max}} = \frac{\sigma_a}{\dfrac{\cos(\theta)}{A} + \dfrac{bL[1+\sin(\theta)]}{8I} + \dfrac{bd\cos(\theta)}{4I}} = 16.56\text{ kN}$$

4. ***Finalize***: This example shows how the normal stresses in a beam due to combined bending and axial load can be determined. The shear stresses acting on cross sections of the beam (due to the shear forces V) can be determined independently of the normal stresses, as described earlier in this chapter. Later, in Chapter 7, stresses on inclined planes are computed when both the normal and shear stresses acting on cross-sectional planes are known.

*5.13 Stress Concentrations in Bending

The flexure and shear formulas discussed in earlier sections of this chapter are valid for beams without holes, notches, or other abrupt changes in dimensions. Whenever such discontinuities exist, high localized stresses are produced. These **stress concentrations** can be extremely important when a member is made of brittle material or is subjected to dynamic loads. (See Chapter 2, Section 2.10, for a discussion of the conditions under which stress concentrations are important.)

For illustrative purposes, two cases of stress concentrations in beams are described in this section. The first case is a beam of rectangular cross section with a **hole at the neutral axis** (Fig. 5-51). The beam has a height h and thickness b (perpendicular to the plane of the figure) and is in pure bending under the action of bending moments M.

FIGURE 5-51

Stress distributions in a beam in pure bending with a circular hole at the neutral axis (The beam has a rectangular cross section with height h and thickness b)

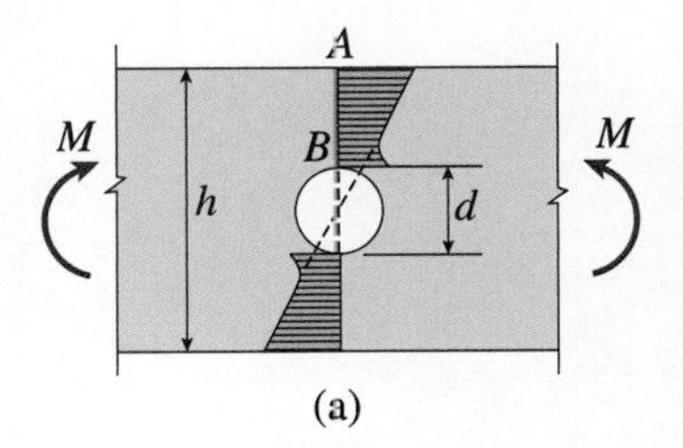

(a)

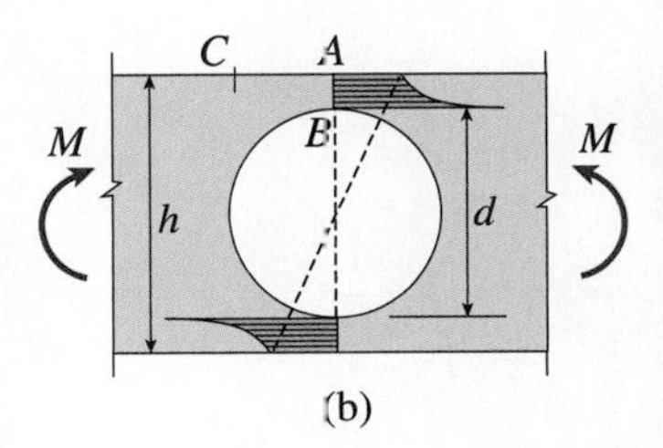

(b)

When the diameter d of the hole is small compared to the height h, the stress distribution on the cross section through the hole is approximately as shown by the diagram in Fig. 5-51a. At point B on the edge of the hole, the stress is much larger than the stress that would exist at that point if the hole were not present. (The dashed line in the figure shows the stress distribution with no hole.) However, moving toward the outer edges of the beam (toward point A), the stress distribution varies linearly with distance from the neutral axis and is only slightly affected by the presence of the hole.

When the hole is relatively large, the stress pattern is approximately as shown in Fig. 5-51b. There is a large increase in stress at point B and only a small change in stress at point A, as compared to the stress distribution in the beam without a hole (again shown by the dashed line). The stress at point C is larger than the stress at A but smaller than the stress at B.

Extensive investigations have shown that the stress at the edge of the hole (point B) is approximately twice the *nominal stress* at that point. The nominal stress is calculated from the flexure formula in the standard way, that is, $\sigma = My/I$, in which y is the distance $d/2$ from the neutral axis to point B and I is the moment of inertia of the net cross section at the hole. Thus, the following approximate formula can be used to find the stress at point B:

$$\sigma_B \approx 2\frac{My}{I} = \frac{12Md}{b(h^3 - d^3)} \tag{5-61}$$

At the outer edge of the beam (at point C), the stress is approximately equal to the *nominal stress* (not the actual stress) at point A (where $y = h/2$):

$$\sigma_C \approx \frac{My}{I} = \frac{6Mh}{b(h^3 - d^3)} \tag{5-62}$$

From the last two equations, the ratio σ_B/σ_C is approximately $2d/h$. Hence, when the ratio d/h of hole diameter to height of beam exceeds 1/2, the largest stress occurs at point B. When d/h is less than 1/2, the largest stress is at point C.

The second case is a **rectangular beam with notches** (Fig. 5-52). The beam shown in the figure is subjected to pure bending and has a height h and thickness b (perpendicular to the plane of the figure). Also, the net height of the beam (that is, the distance between the bases of the notches) is h_1, and the radius at the base of each notch is R. The maximum stress in this beam occurs at the base

FIGURE 5-52

Stress-concentration factor K for a notched beam of rectangular cross section in pure bending (h = height of beam; b = thickness of beam, perpendicular to the plane of the figure), where the dashed line is for semicircular notches ($h = h_1 + 2R$)

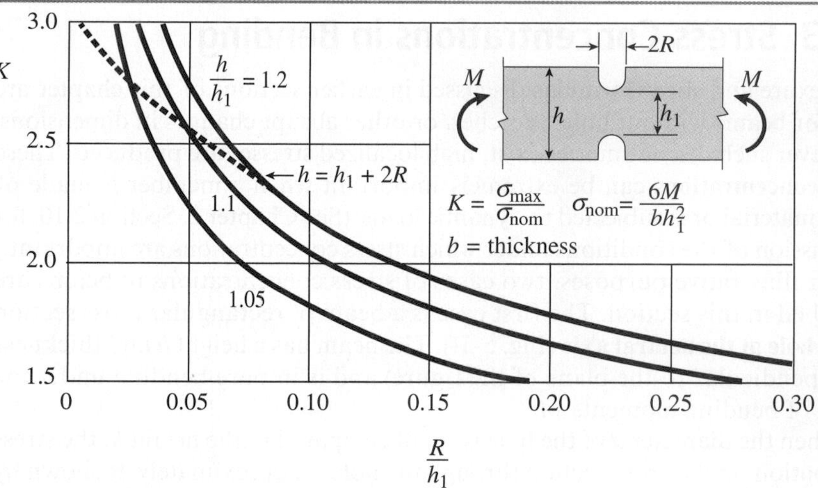

of the notches and may be much larger than the nominal stress at that same point. The nominal stress is calculated from the flexure formula with $y = h_1/2$ and $I = bh_1^3/12$; thus,

$$\sigma_{nom} = \frac{My}{I} = \frac{6M}{bh_1^2} \tag{5-63}$$

The maximum stress is equal to the stress-concentration factor K times the nominal stress:

$$\sigma_{max} = K\sigma_{nom} \tag{5-64}$$

The stress-concentration factor K is plotted in Fig. 5-52 for a few values of the ratio h/h_1. Note that when the notch becomes "sharper," that is, the ratio R/h_1 becomes smaller, the stress-concentration factor increases. (Fig. 5-52 is plotted from the formulas given in Ref. 2-9.)

The effects of the stress concentrations are confined to small regions around the holes and notches, as explained in the discussion of Saint-Venant's principle in Section 2.10. At a distance equal to h or greater from the hole or notch, the stress-concentration effect is negligible and the ordinary formulas for stresses may be used.

Example 5-18

A simple beam AB with rectangular cross section ($b \times h$) has a hole with a diameter of d at its centerline and two notches on either side and equidistant from the beam centerline. Beam AB is simply supported, and loads P are applied at $L/5$ from each end of the beam. Assume that dimensions given in Fig. 5-53 are $L = 4.5$ m, $b = 50$ mm, $h = 144$ mm, $h_1 = 120$ mm, $d = 85$ mm, and $R = 10$ mm. Assume that the allowable bending stress is $\sigma_a = 150$ MPa.

(a) Find the maximum permissible value of applied load P.

(b) If $P = 11$ kN, find the smallest acceptable radius of the notches, R_{min}.

(c) If $P = 11$ kN, find the maximum acceptable diameter of the hole at mid-height of beam.

FIGURE 5-53

Example 5-18: Rectangular steel beam with notches and a hole

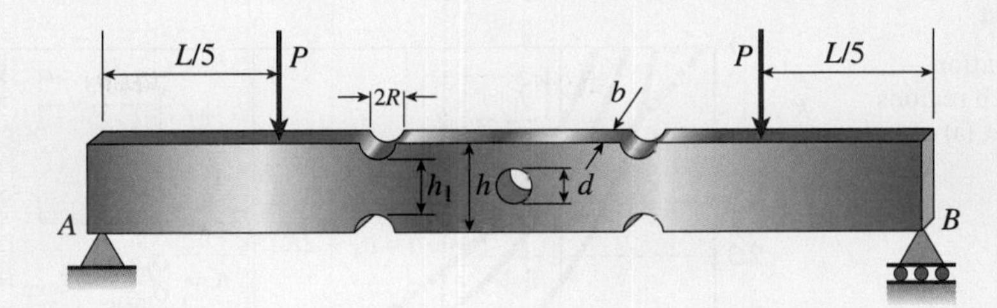

Solution:

Use a four-step problem-solving approach. Combine steps as needed for an efficient solution.

Part (a): Maximum permissible load *P*.

1, 2. ***Conceptualize, Categorize*:** The central part of the beam between the loads P ($x = L/5$ to $x = 4L/5$) is in pure bending, and the maximum moment in this region is $M = PL/5$. To find P_{max}, compare the maximum bending stress (at midspan around the hole and in the notch regions) to the allowable stress value of $\sigma_a = 150$ MPa.

First, check the *maximum stresses around the hole*. The hole diameter-to-beam depth ratio $d/h = 85 \text{ mm}/144 \text{ mm} = 0.59$ exceeds 1/2, so the stress at B rather than at C (Fig. 5-51) will govern. Setting σ_B equal to σ_a and substituting $PL/5$ for M in Eq. (5-61) gives the expression for P_{max}:

$$M_{max} = \sigma_a\left[\frac{b(h^3 - d^3)}{12d}\right] \quad \text{and} \quad P_{max1} = \frac{5}{L}\left\{\sigma_a\left[\frac{b(h^3 - d^3)}{12d}\right]\right\}$$

3. ***Analyze*:** Use this expression to compute

$$P_{max1} = \frac{5}{4.5 \text{ m}}\left\{150 \text{ MPa}\left[\frac{50 \text{ mm}[(144 \text{ mm}^3) - (85 \text{ mm})^3]}{12(85 \text{ mm})}\right]\right\} = 19.38 \text{ kN}$$

Next, check the peak stresses at the base of the two notches to get a second value of P_{max}. The ratio of notch radius R to height h_1 is equal to 0.083, and the ratio $h/h_1 = 1.2$. So from Fig. 5-52, the stress concentration factor K is approximately equal to 2.3 (see Fig. 5-54).

Use Eqs. (5-63) and (5-64) to get the expressions:

$$\sigma_{max} = K\sigma_{nom} = K\left(\frac{6M}{bh_1^2}\right) = K\left[\frac{6}{bh_1^2}\left(\frac{PL}{5}\right)\right]$$

so

$$P_{max2} = \sigma_a\left(\frac{5bh_1^2}{6KL}\right) = 150 \text{ MPa}\left[\frac{5(50 \text{ mm})(120 \text{ mm})^2}{6(2.3)(4.5 \text{ m})}\right] = 8.7 \text{ kN}$$

FIGURE 5-54

Stress concentration factor K in notch regions of beam for part (a) of Example 5-18

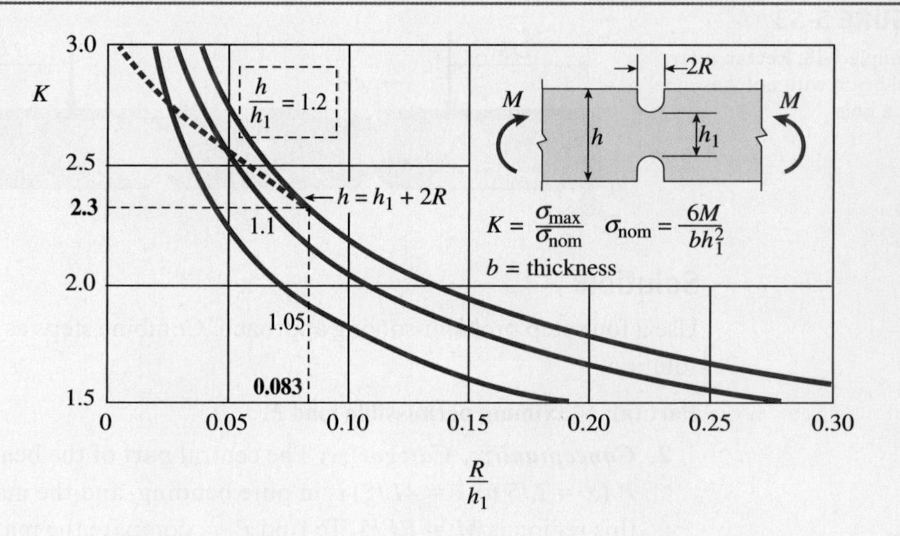

4. ***Finalize***: Compare P_{max1} and P_{max2}, to see that the peak stress at the base of the notches controls, so

$$P_{max} = 8.7 \text{ kN}$$

Part (b): Smallest acceptable radius R of the notches.

1, 2. ***Conceptualize, Categorize***: The stress concentration factor K in Fig. 5-52 increases as the ratio of the notch radius R to dimension h_1 decreases.

3, 4. ***Analyze, Finalize***: Compute the nominal stress using Eq. (5-63) as

$$\sigma_{nom} = \frac{6\left(\frac{PL}{5}\right)}{bh_1^2} = \frac{6(11 \text{ kN})(4.5 \text{ m})}{5(50 \text{ mm})(120 \text{ mm})^2} = 82.5 \text{ MPa}$$

Then set the maximum bending stress σ_{max} equal to the allowable stress $\sigma_a = 150$ MPa to find the stress concentration factor K:

$$K = \frac{\sigma_a}{\sigma_{nom}} = \frac{150 \text{ MPa}}{82.5 \text{ MPa}} = 1.82$$

From Fig. 5-55, with $h/h_1 = 1.2$ and $K = 1.82$, obtain

$$\frac{R}{h_1} = 0.16 \quad \text{so} \quad R_{min} = 0.16(120 \text{ mm}) = 19.2 \text{ mm}$$

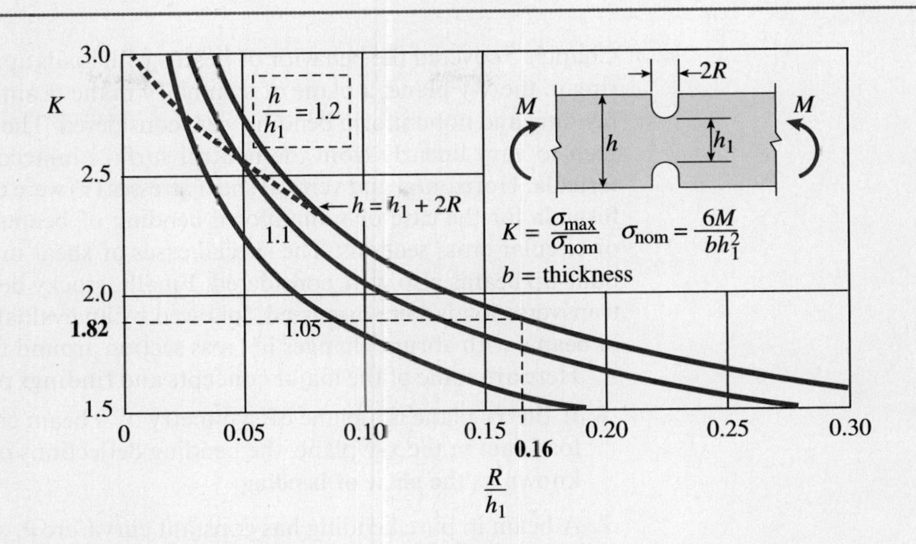

FIGURE 5-55 Stress concentration factor K in notch regions of beam for part (b) of Example 5-18

Part (c): Maximum acceptable diameter of the hole.

1, 2. *Conceptualize, Categorize*: Begin by assuming that ratio $d/h > 1/2$, and start with Eq. (5-61) (which assumes that maximum bending stress is at B, as in Fig. 5-51) to find d_{max}. If d/h turns out to be less than 1/2, use Eq. (5-62), which means that maximum bending stress is in fact at point C. If the peak stress is at B, write Eq. (5-61) as

$$\frac{12\left(\frac{PL}{5}\right)d}{b(h^3 - d^3)} = \sigma_a$$

3. *Analyze*: Solve the previous equation numerically to find that $d_{max} = 108.3$ mm.

4. *Finalize*: The original assumption about the d/h ratio is confirmed, since $d_{max}/h = 0.752$ exceeds 1/2, so the peak stress is indeed at B and not at C.

CHAPTER SUMMARY AND REVIEW

Chapter 5 covered the behavior of beams with loads applied and bending occurring in the x-y plane: a plane of symmetry in the beam cross section. Both pure bending and nonuniform bending were considered. The normal stresses (σ) were seen to vary linearly from the neutral surface in accordance with the **flexure formula**. Horizontal and vertical shear stresses (τ) were computed using the **shear formula** for the case of nonuniform bending of beams with either rectangular or circular cross sections. The special cases of shear in beams with flanges and built-up beams also were considered. Finally, stocky beams with both axial and transverse loads were discussed, followed by an evaluation of localized stresses in beams with abrupt changes in cross section around notches or holes.

Here are some of the major concepts and findings presented in this chapter.

1. If the xy plane is a plane of symmetry of a beam cross section and applied loads act in the x-y plane, the bending deflections occur in this same plane, known as the **plane of bending**.

2. A beam in pure bending has constant curvature κ, and a beam in nonuniform bending has varying curvature. Longitudinal strains (ε_x) in a bent beam are proportional to its curvature, and the strains in a beam in pure bending vary linearly with distance from the neutral surface, regardless of the shape of the stress-strain curve of the material, as

$$\varepsilon_x = -\kappa y$$

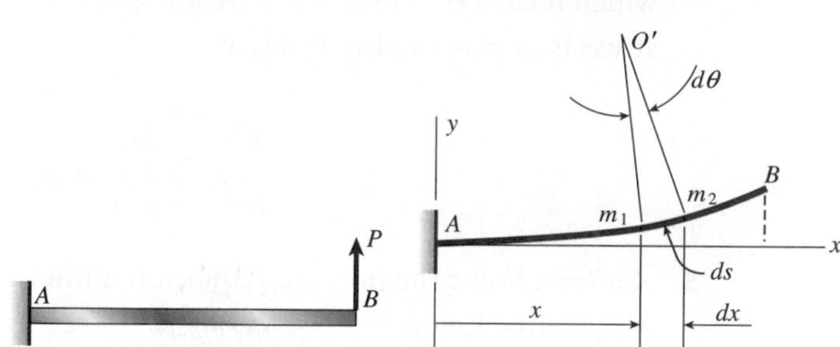

3. The neutral axis passes through the centroid of the cross-sectional area when the material follows Hooke's law and there is no axial force acting on the cross section. When a beam of linearly elastic material is subjected to pure bending, the y and z axes are **principal centroidal** axes.

4. If the material of a beam is linearly elastic and follows Hooke's law, the **moment-curvature equation** shows that the curvature is directly proportional to the bending moment M and inversely proportional to the quantity EI, the **flexural rigidity** of the beam. The moment curvature relation is

$$\kappa = \frac{M}{EI}$$

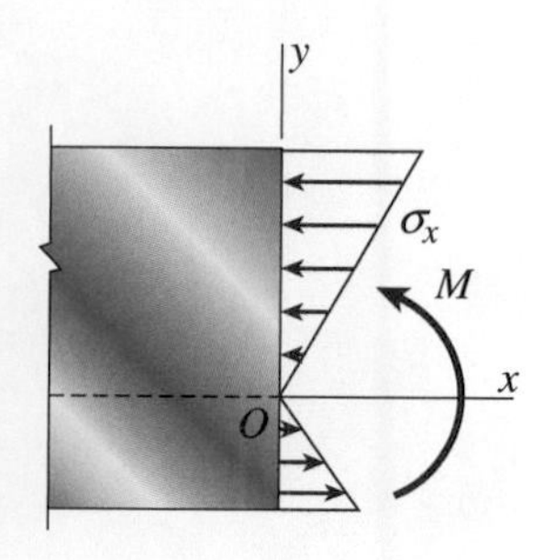

5. The **flexure formula** shows that the normal stresses σ_x are directly proportional to the bending moment M and inversely proportional to the moment of inertia I of the cross section:

$$\sigma_x = -\frac{My}{I}$$

The maximum tensile and compressive bending stresses acting at any given cross section occur at points located farthest from the neutral axis. Thus,

$$(y = c_1,\ y = -c_2)$$

6. The normal stresses calculated from the flexure formula are not significantly altered by the presence of shear stresses and the associated warping of the cross section for the case of nonuniform bending. However, the flexure formula is not applicable near the supports of a beam or close to a concentrated load, because such irregularities produce **stress concentrations** that are much greater than the stresses obtained from the flexure formula.
7. To **design** a beam to resist bending stresses, calculate the required **section modulus** S from the maximum moment and allowable normal stress as

$$S = \frac{M_{\text{max}}}{\sigma_{\text{allow}}}$$

   To minimize weight and save material, select a beam from a material design manual (see sample tables in Appendixes F and G for steel and wood) that has the least cross-sectional area while still providing the required section modulus; wide-flange sections and I-sections have most of their material in the flanges, and the width of their flanges helps to reduce the likelihood of sideways buckling.
8. **Nonprismatic beams** (found in automobiles, airplanes, machinery, bridges, buildings, tools, and many other applications) commonly are used to reduce weight and improve appearance. The flexure formula gives reasonably accurate values for the bending stresses in nonprismatic beams, provided that the changes in cross-sectional dimensions are gradual. However, in a nonprismatic beam, the section modulus also varies along the axis, so do not assume that the maximum stresses occur at the cross section with the largest bending moment.
9. Beams subjected to loads that produce both bending moments (M) and shear forces (V) (**nonuniform bending**) develop both normal and shear stresses in the beam. Normal stresses are calculated from the **flexure formula** (provided the beam is constructed of a linearly elastic material), and shear stresses are computed using the **shear formula**

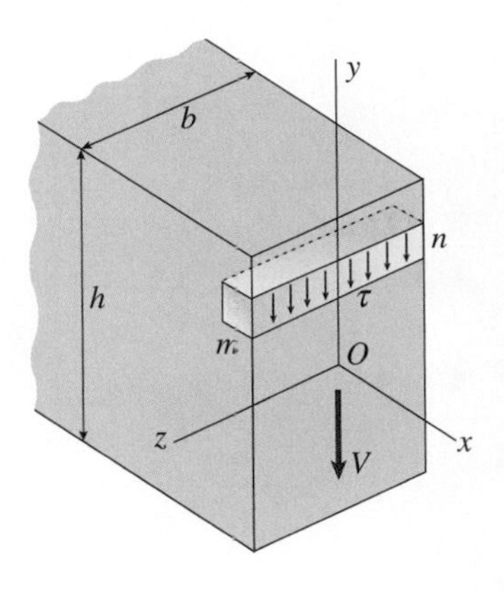

$$\tau = \frac{VQ}{Ib}$$

   Shear stress varies parabolically over the height of a rectangular beam, and shear strain also varies parabolically; these shear strains cause cross sections of the beam that were originally plane surfaces to become warped. The maximum values of the shear stress and strain (τ_{max}, γ_{max}) occur at the neutral axis, and the shear stress and strain are zero on the top and bottom surfaces of the beam.
10. The shear formula applies only to prismatic beams and is valid only for beams of linearly elastic materials with small deflections; also, the edges of the cross section must be **parallel** to the y axis. For **rectangular** beams, the accuracy of the shear formula depends upon the height-to-width ratio of the cross section: The formula may be considered exact for very narrow beams but becomes less accurate as width b increases relative to height h.

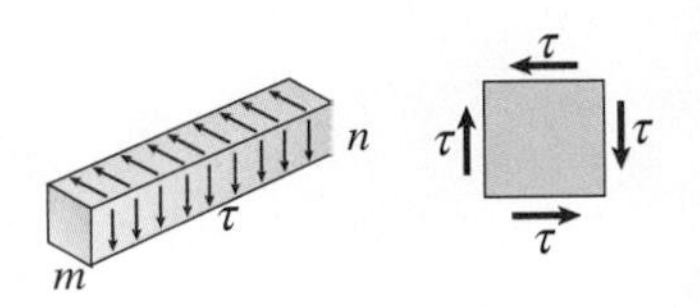

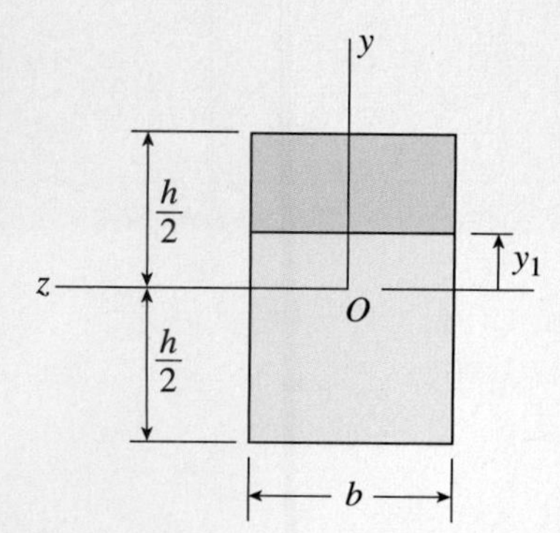

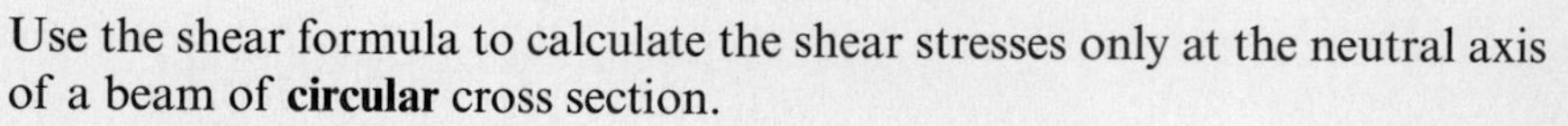

Use the shear formula to calculate the shear stresses only at the neutral axis of a beam of **circular** cross section.

For rectangular cross sections,

$$\tau_{max} = \frac{3}{2}\frac{V}{A}$$

and for solid circular cross sections

$$\tau_{max} = \frac{4}{3}\frac{V}{A}$$

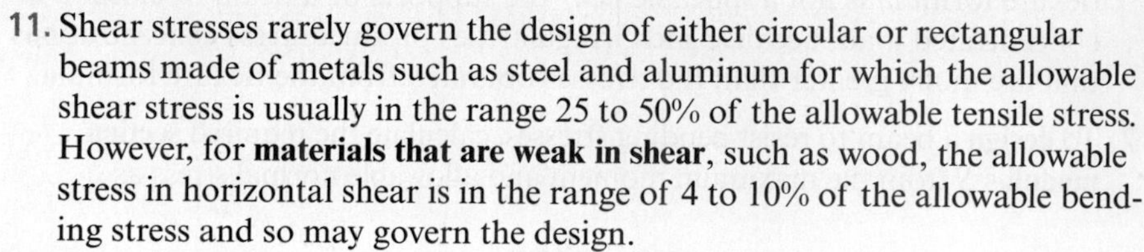

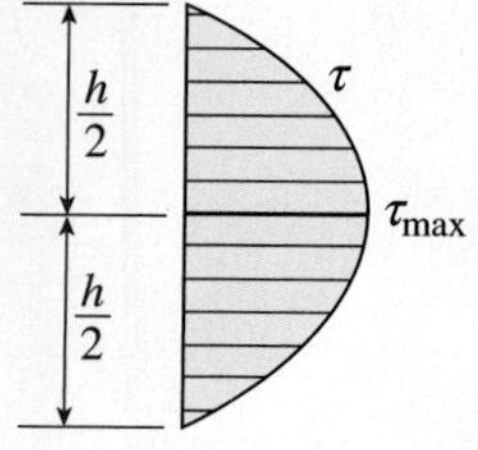

11. Shear stresses rarely govern the design of either circular or rectangular beams made of metals such as steel and aluminum for which the allowable shear stress is usually in the range 25 to 50% of the allowable tensile stress. However, for **materials that are weak in shear**, such as wood, the allowable stress in horizontal shear is in the range of 4 to 10% of the allowable bending stress and so may govern the design.

12. Shear stresses in the flanges of **wide-flange beams** act in both vertical and horizontal directions. The horizontal shear stresses are much larger than the vertical shear stresses in the flanges. The shear stresses in the **web of a wide-flange beam** act only in the vertical direction, are larger than the stresses in the flanges, and may be computed using the shear formula. The maximum shear stress in the web of a wide-flange beam occurs at the neutral axis, and the minimum shear stress occurs where the web meets the flanges. For beams of typical proportions, the shear force in the web is 90 to 98% of the total shear force V acting on the cross section; the remainder is carried by shear in the flanges.

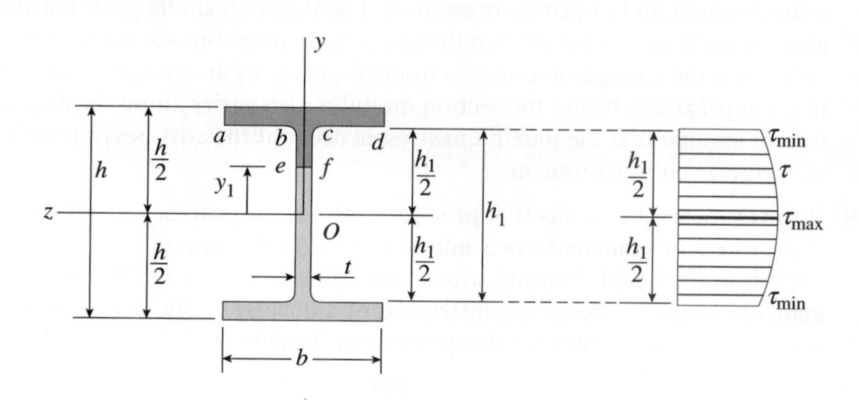

13. Connections between the parts in **built-up beams** (such as nails, bolts, welds, and glue) must be strong enough to transmit the horizontal shear forces acting between the parts of the beam. The connections are designed using the **shear flow formula**

$$f = \frac{VQ}{I}$$

to ensure that the beam behaves as a single entity. **Shear flow** f is defined as horizontal shear force per unit distance along the longitudinal axis of the beam.

14. Normal stresses in **beams with axial loads** are obtained by superposing the stresses produced by the axial force N and the bending moment M as

$$\sigma = \frac{N}{A} - \frac{My}{I}$$

Whenever bending and axial loads act simultaneously, the neutral axis no longer passes through the centroid of the cross section and may be outside the cross section, at the edge of the section, or within the section. This discussion applies only to **stocky beams** for which the lateral deflections are so small as to have no significant effect on the line of action of the axial forces.

15. Stress distributions in beams are altered by holes, notches, or other abrupt changes in dimensions leading to high localized stresses or **stress concentrations**. These are especially important to consider when the beam material is brittle or the member is subjected to dynamic loads. The maximum stress values may be several times larger than the nominal stress.

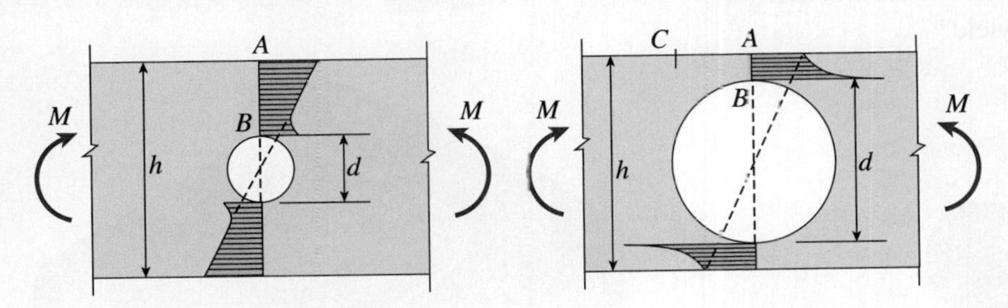

PROBLEMS Chapter 5

5.4 Longitudinal Strains in Beams

Introductory Problems

5.4-1 A steel wire with a diameter of $d = 1.6$ mm is bent around a cylindrical drum with a radius of $R = 0.9$ m (see figure).

(a) Determine the maximum normal strain ε_{max}.
(b) What is the minimum acceptable radius of the drum if the maximum normal strain must remain below yield? Assume $E = 210$ GPa and $\sigma_Y = 690$ MPa.
(c) If $R = 0.9$ m, what is the maximum acceptable diameter of the wire if the maximum normal strain must remain below yield?

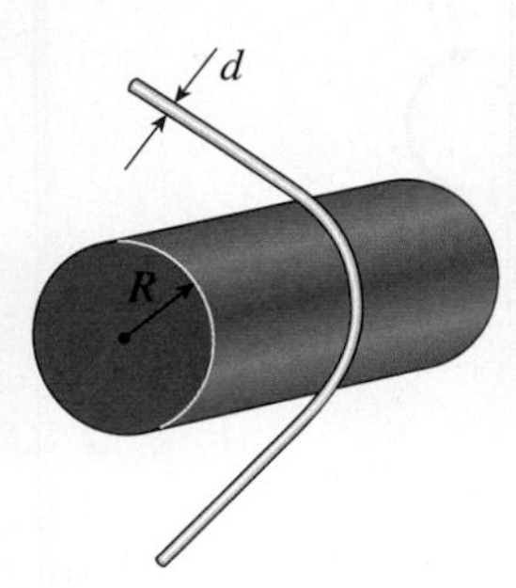

PROBLEM 5.4-1

5.4-2 A copper wire having a diameter of $d = 4$ mm is bent into a circle and held with the ends just touching (see figure).

(a) If the maximum permissible strain in the copper is $\varepsilon_{max} = 0.0024$, what is the shortest length L of wire that can be used?
(b) If $L = 5.5$ m, what is the maximum acceptable diameter of the wire if the maximum normal strain must remain below yield? Assume $E = 120$ GPa and $\sigma_Y = 300$ MPa.

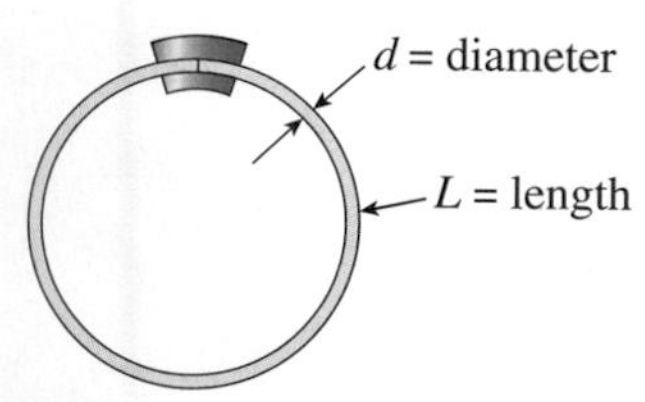

PROBLEM 5.4-2

5.4-3 A 120 mm outside diameter polyethylene pipe designed to carry chemical waste is placed in a trench and bent around a quarter-circular 90° bend (see figure). The bent section of the pipe is 16 m long.

(a) Determine the maximum compressive strain ε_{max} in the pipe.
(b) If the normal strain cannot exceed 6.1×10^{-3}, what is the maximum diameter of the pipe?
(c) If $d = 120$ mm, what is the minimum acceptable length of the bent section of the pipe?

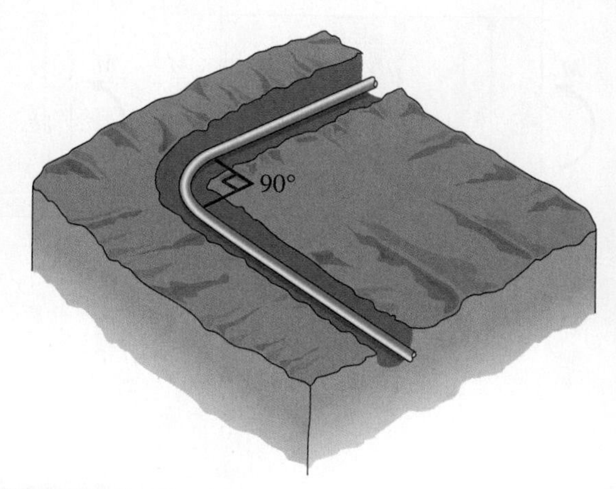

PROBLEM 5.4-3

Representative Problems

5.4-4 A cantilever beam AB is loaded by a couple M_0 at its free end (see figure). The length of the beam is $L = 2.0$ m, and the longitudinal normal strain at the top surface is $\varepsilon = 0.0010$. The distance from the top surface of the beam to the neutral surface is $c = 85$ mm.

(a) Calculate the radius of curvature ρ, the curvature κ, and the vertical deflection δ at the end of the beam.
(b) If allowable strain $\varepsilon_a = 0.0008$, what is the maximum acceptable depth of the beam? [Assume that the curvature is unchanged from part(a)].
(c) If allowable strain $\varepsilon_a = 0.0008$, $c = 85$ mm, and $L = 4$ m, what is deflection δ?

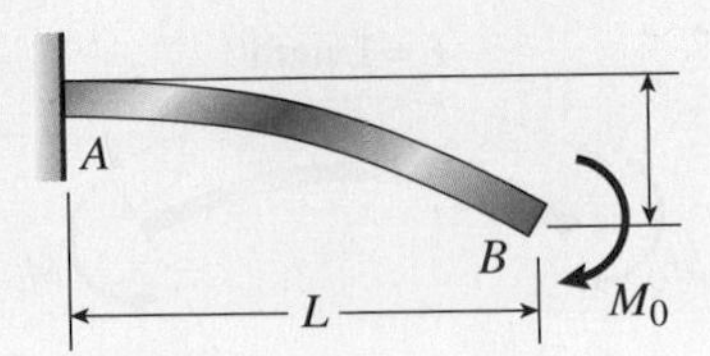

PROBLEM 5.4-4

5.4-5 A thin strip of steel with a length of $L = 0.5$ m and thickness of $t = 7$ mm is bent by couples M_0 (see figure). The deflection at the midpoint of the strip (measured from a line joining its end points) is found to be 7.5 mm.

(a) Determine the longitudinal normal strain ε at the top surface of the strip.
(b) If allowable strain $\varepsilon_a = 0.0008$, what is the maximum acceptable thickness of the strip?
(c) If allowable strain $\varepsilon_a = 0.0008$, $t = 7$ mm, and $L = 0.8$ m, what is deflection δ?
(d) If allowable strain $\varepsilon_a = 0.0008$, $t = 7$ mm, and the deflection cannot exceed 25 mm, what is the maximum permissible length of the strip?

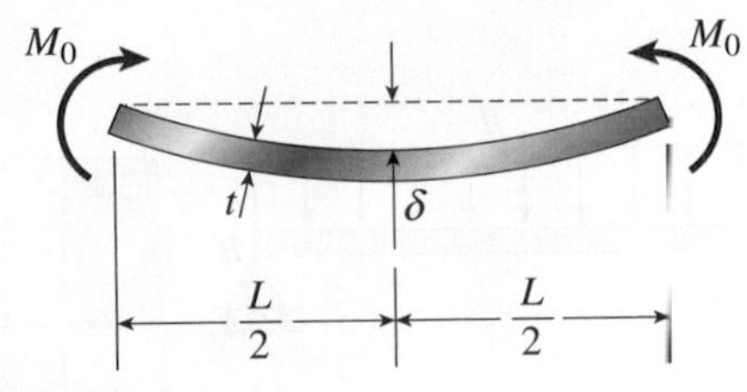

PROBLEM 5.4-5

5.4-6 A bar of rectangular cross section is loaded and supported as shown in the figure. The distance between supports is $L = 1.75$ m, and the height of the bar is $h = 140$ mm. The deflection at the midpoint is measured as 2.5 mm.

(a) What is the maximum normal strain ε at the top and bottom of the bar?

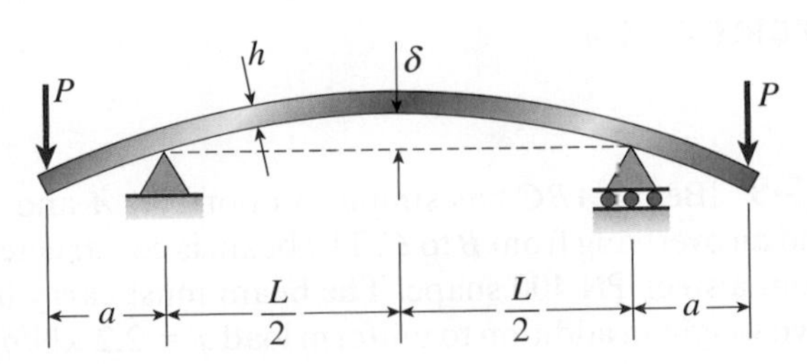

PROBLEM 5.4-6

(b) If allowable strain $\varepsilon_a = 0.0006$ and the deflection cannot exceed 4.3 mm, what is the maximum permissible length of the bar?

5.4-7 A simply supported beam with a length $L = 3$ m and height 175 mm is bent by couples M_0 into a circular arc with downward deflection δ at the midpoint. If the curvature of the beam is 9.8×10^{-3} m^{-1}, calculate the deflection, δ, at the mid-span of the beam and the longitudinal strain at the bottom fiber given that the distance between the neutral surface and the bottom surface is 87.5 mm.

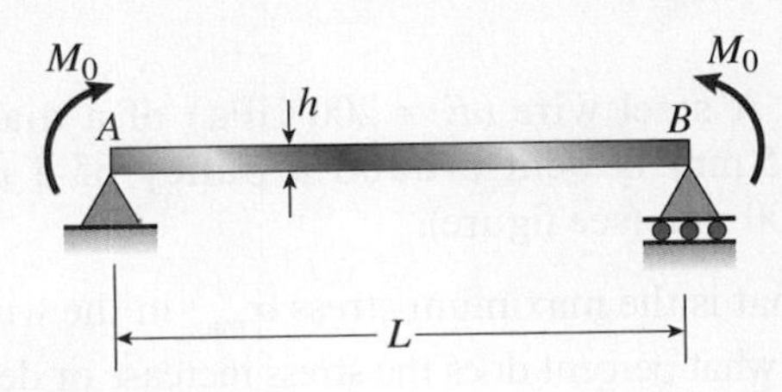

PROBLEM 5.4-7

5.4-8 A cantilever beam is subjected to a concentrated moment at B. The length of the beam $L = 3$ m and the height $h = 600$ mm. The longitudinal strain at the top of the beam is 0.0005 and the distance from the neutral surface to the bottom surface of the beam is 300 mm. Find the radius of curvature, the curvature, and the deflection of the beam at B.

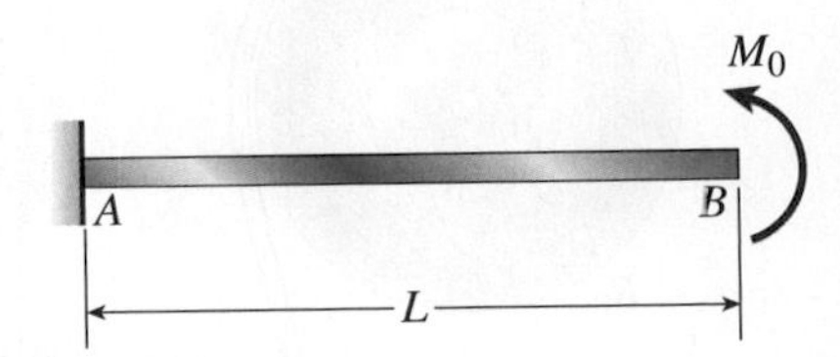

PROBLEM 5.4-8

5.5 Normal Stresses in Beams (Linearly Elastic Materials)

Introductory Problems

5.5-1 A thin strip of hard copper ($E = 110$ GPa) having length $L = 2.3$ m and thickness $t = 2.4$ mm is bent into a circle and held with the ends just touching (see figure).

(a) Calculate the maximum bending stress σ_{max} in the strip.
(b) By what percent does the stress increase or decrease if the thickness of the strip is increased by 0.8 mm?

(c) Find the new length of the strip so that the stress in part (b) ($t = 3.2$ m and $L = 2.3$ m) is equal to that in part (a) ($t = 2.4$ mm and $L = 2.3$ m).

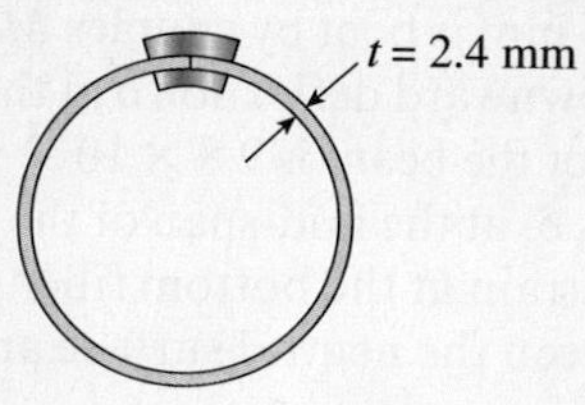

PROBLEM 5.5-1

5.5-2 A steel wire ($E = 200$ GPa) of a diameter $d = 1.25$ mm is bent around a pulley of a radius $R_0 = 500$ mm (see figure).

(a) What is the maximum stress σ_{max} in the wire?
(b) By what percent does the stress increase or decrease if the radius of the pulley is increased by 25%?
(c) By what percent does the stress increase or decrease if the diameter of the wire is increased by 25% while the pulley radius remains at $R_0 = 500$ mm?

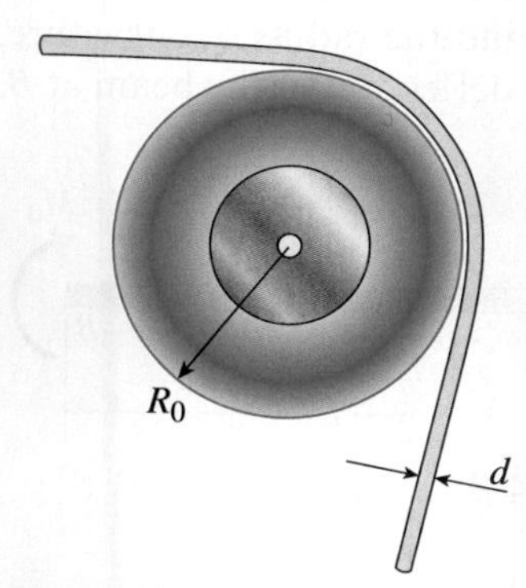

PROBLEM 5.5-2

5.5-3 A thin, high-strength steel rule ($E = 200$ GPa) having a thickness $t = 4$ mm and length $L = 1.5$ m is bent by couples M_0 into a circular arc subtending a central angle $\alpha = 40°$ (see figure).

(a) What is the maximum bending stress σ_{max} in the rule?
(b) By what percent does the stress increase or decrease if the central angle is increased by 10%?
(c) What percent increase or decrease in rule thickness will result in the maximum stress reaching the allowable value of 290 MPa?

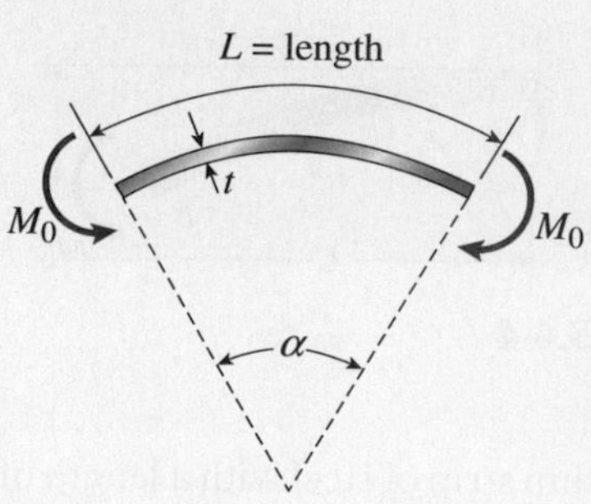

PROBLEM 5.5-3

Representative Problems

5.5-4 A simply supported wood beam AB with a span length $L = 4$ m carries a uniform load of intensity $q = 5.8$ kN/m (see figure).

(a) Calculate the maximum bending stress σ_{max} due to the load q if the beam has a rectangular cross section with width $b = 140$ mm and height $h = 240$ mm.
(b) Repeat part (a) but use the trapezoidal distributed load shown in the figure part b.

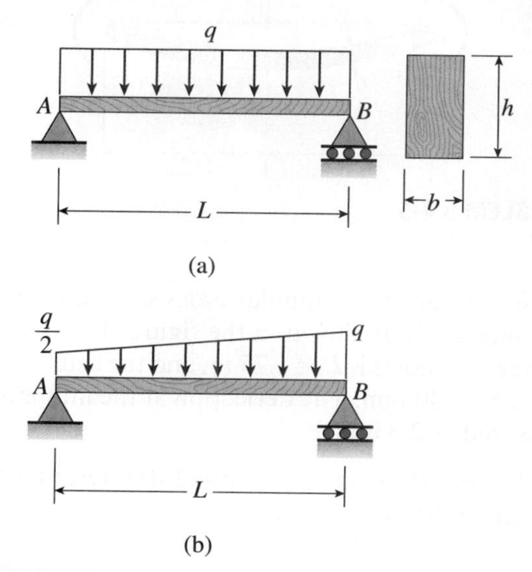

PROBLEM 5.5-4

5.5-5 Beam ABC has simple supports at A and B and an overhang from B to C. The beam is constructed from a steel IPN 400 shape. The beam must carry its own weight in addition to uniform load $q = 2.2$ kN/m. Determine the maximum tensile and compressive stresses in the beam.

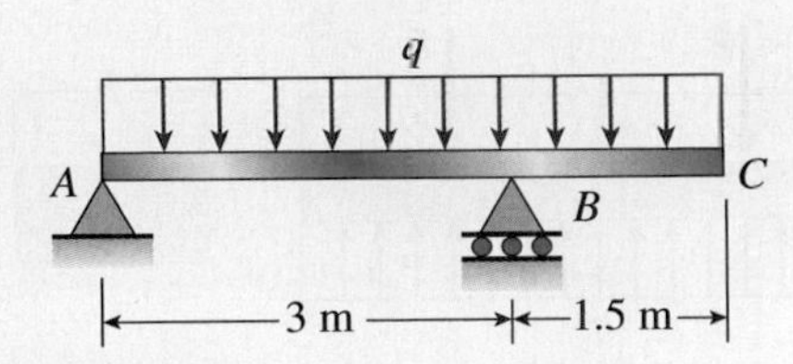

PROBLEM 5.5-5

5.5-6 A simply supported beam is subjected to a linearly varying distributed load $q(x) = \frac{x}{L} q_0$ with maximum intensity q_0 at B. The beam has a length $L = 4$ m and rectangular cross section with a width of 200 mm and height of 300 mm. Determine the maximum permissible value for the maximum intensity, q_0, if the allowable normal stresses in tension and compression are 120 MPa.

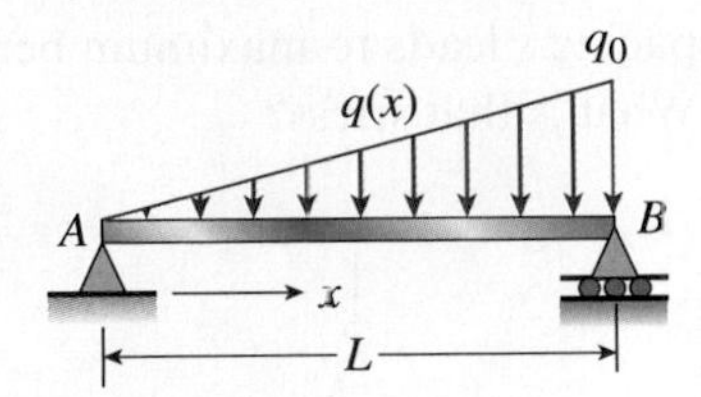

PROBLEM 5.5-6

5.5-7 Each girder of the lift bridge (see figure) is 50 m long and simply supported at the ends. The design load for each girder is a uniform load of intensity 18 kN/m. The girders are fabricated by welding three steel plates to form an I-shaped cross section (see figure) having section modulus $S = 46{,}000$ cm^3.

What is the maximum bending stress σ_{max} in a girder due to the uniform load?

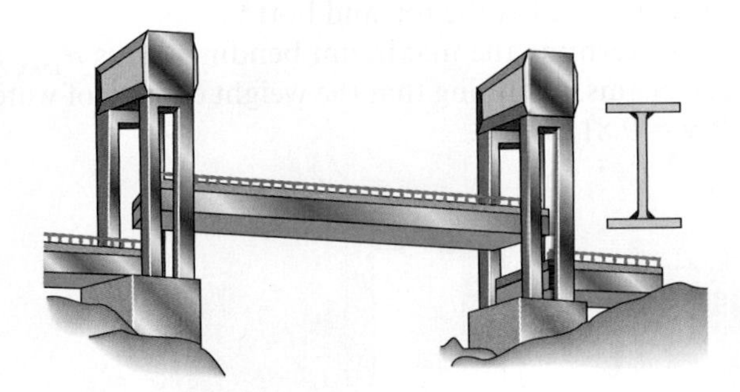

PROBLEM 5.5-7

5.5-8 A freight-car axle AB is loaded approximately as shown in the figure, with the forces P representing the car loads (transmitted to the axle through the axle boxes) and the forces R representing the rail loads (transmitted to the axle through the wheels). The diameter of the axle is $d = 82$ mm, the distance between centers of the rails is L, and the distance between the forces P and R is $b = 220$ mm.

Calculate the maximum bending stress σ_{max} in the axle if $P = 50$ kN.

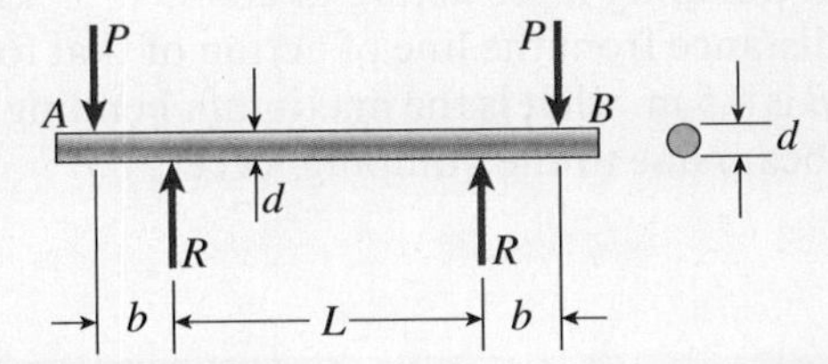

PROBLEM 5.5-8

5.5-9 A seesaw weighing 45 N/m of length is occupied by two children, each weighing 400 N (see figure). The center of gravity of each child is 2.5 m from the fulcrum. The board is 6 m long, 200 mm wide, and 40 mm thick.

What is the maximum bending stress in the board?

PROBLEM 5.5-9

5.5-10 During construction of a highway bridge, the main girders are cantilevered outward from one pier toward the next (see figure). Each girder has a cantilever length of 48 m and an I-shaped cross

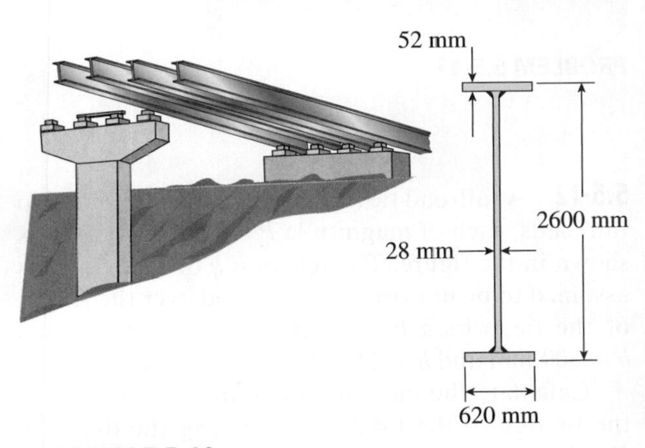

PROBLEM 5.5-10

section with dimensions shown in the figure. The load on each girder (during construction) is assumed to be 9.5 kN/m, which includes the weight of the girder.

Determine the maximum bending stress in a girder due to this load.

5.5-11 The horizontal beam ABC of an oil-well pump has the cross section shown in the figure. If the vertical pumping force acting at end C is 39 kN and if the distance from the line of action of that force to point B is 4.5 m, what is the maximum bending stress in the beam due to the pumping force?

Gabriel M. Covian/Getty Images

Horizontal beam transfers loads as part of oil well pump

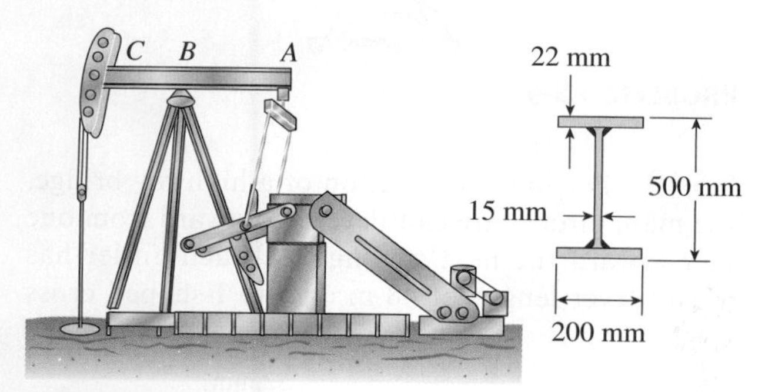

PROBLEM 5.5-11

5.5-12 A railroad tie (or *sleeper*) is subjected to two rail loads, each of magnitude $P = 175$ kN, acting as shown in the figure. The reaction q of the ballast is assumed to be uniformly distributed over the length of the tie, which has cross-sectional dimensions $b = 300$ mm and $h = 250$ mm.

Calculate the maximum bending stress σ_{max} in the tie due to the loads P, assuming the distance $L = 1500$ mm and the overhang length $a = 500$ mm.

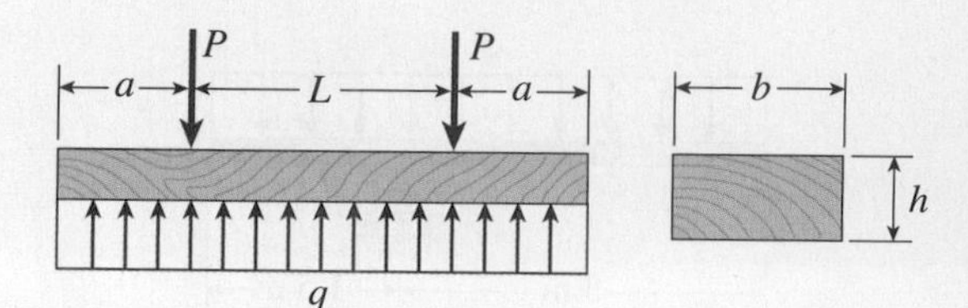

PROBLEM 5.5-12

5.5-13 A fiberglass pipe is lifted by a sling, as shown in the figure. The outer diameter of the pipe is 150 mm, its thickness is 6 mm, and its weight density is 18 kN/m^3. The length of the pipe is $L = 13$ m and the distance between lifting points is $s = 4$ m.

(a) Determine the maximum bending stress in the pipe due to its own weight.

(b) Find the spacing s between lift points which minimizes the bending stress. What is the minimum bending stress?

(c) What spacing s leads to maximum bending stress? What is that stress?

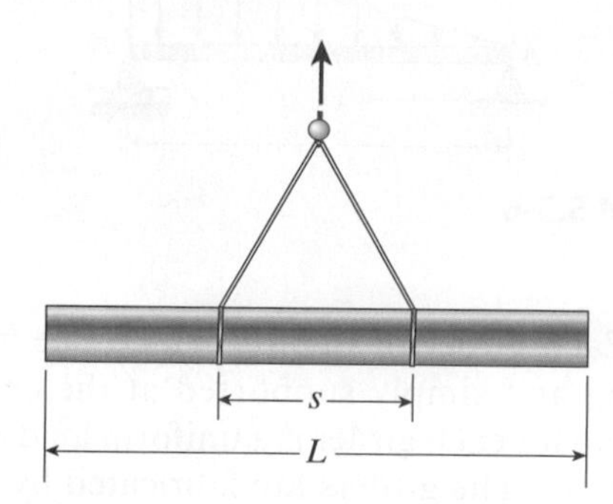

PROBLEM 5.5-13

5.5-14 A small dam of height $h = 2.0$ m is constructed of vertical wood beams AB of thickness $t = 120$ mm, as shown in the figure. Consider the beams to be simply supported at the top and bottom.

Determine the maximum bending stress σ_{max} in the beams, assuming that the weight density of water is $\gamma = 9.81$ kN/m^3.

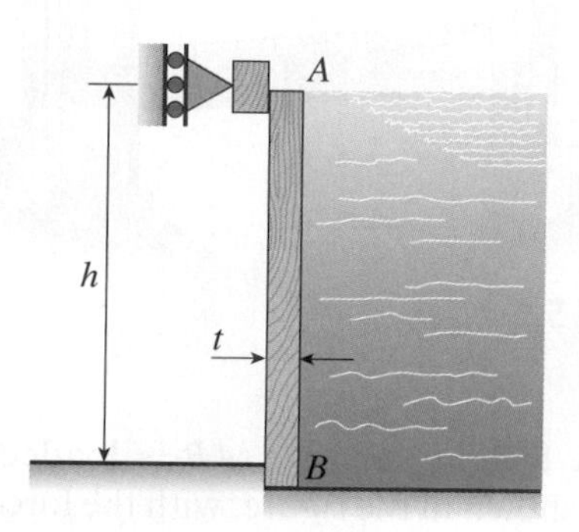

PROBLEM 5.5-14

5.5-15 Determine the maximum tensile stress σ_t (due to pure bending about a horizontal axis through C by positive bending moments M) for beams having cross sections as follows (see figure).

(a) A semicircle of diameter d.
(b) An isosceles trapezoid with bases $b_1 = b$ and $b_2 = 4b/3$ and altitude h.
(c) A circular sector with $\alpha = \pi/3$ and $r = d/2$

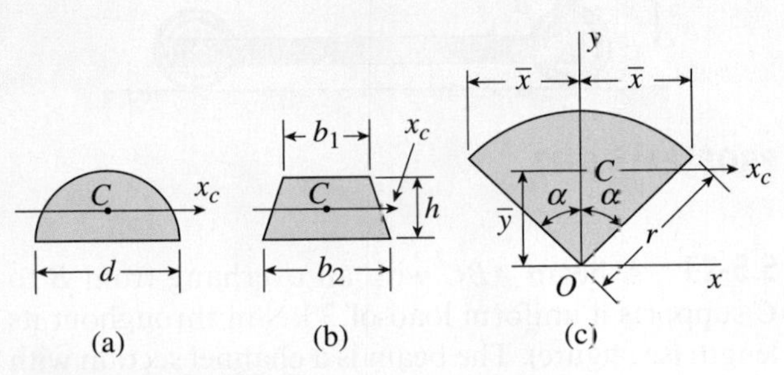

PROBLEM 5.5-15

5.5-16 Determine the maximum bending stress σ_{max} (due to pure bending by a moment M) for a beam having a cross section in the form of a circular core (see figure). The circle has diameter d and the angle $\beta = 60°$. *Hint:* Use the formulas given in Appendix E, Cases 9 and 15.

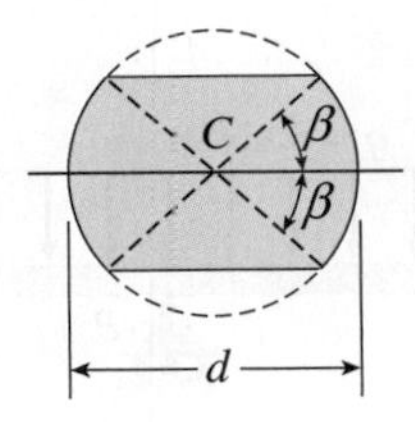

PROBLEM 5.5-16

5.5-17 A simple beam AB of a span length $L = 7$ m is subjected to two wheel loads acting at a distance $d = 1.5$ m apart (see figure). Each wheel transmits a load $P = 14$ kN, and the carriage may occupy any position on the beam.

(a) Determine the maximum bending stress σ_{max} due to the wheel loads if the beam is an I-beam having section modulus $S = 265\ \text{cm}^3$.
(b) If $d = 1.5$ m, find the required span length L to reduce the maximum stress in part (a) to 124 MPa.
(c)If $L = 7$ m, find the required wheel spacing s to reduce the maximum stress in part (a) to 124 MPa.

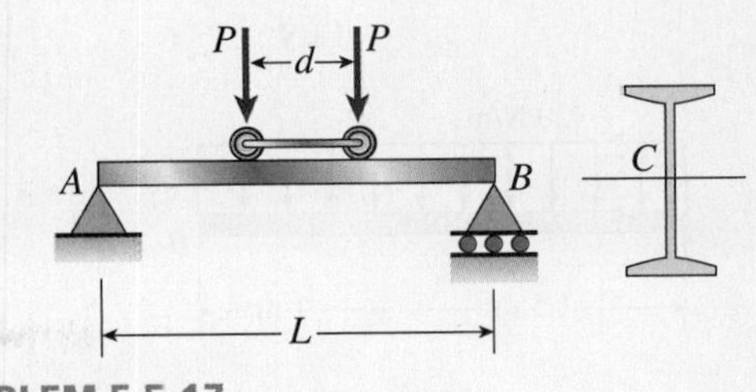

PROBLEM 5.5-17

5.5-18 Determine the maximum tensile stress σ_t and maximum compressive stress σ_c due to the load P acting on the simple beam AB (see figure).

(a) Data are $P = 6.2$ kN, $L = 3.2$ m, $d = 1.25$ m, $b = 80$ mm, $t = 25$ mm, $h = 120$ mm, and $h_1 = 90$ mm.
(b) Find the value of d for which tensile and compressive stresses are the largest. What are these stresses?

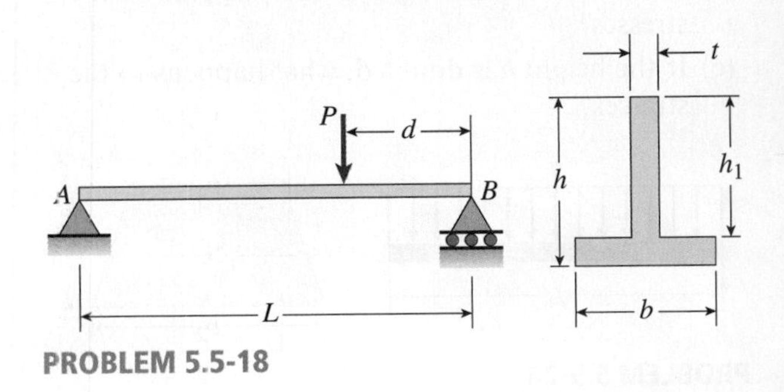

PROBLEM 5.5-18

5.5-19 A cantilever beam AB, loaded by a uniform load and a concentrated load (see figure), is constructed of a channel section.

(a) Find the maximum tensile stress σ_t and maximum compressive stress σ_c if the cross section has the dimensions indicated and the moment of inertia about the z axis (the neutral axis) is $I = 130\ \text{cm}^4$. *Note:* The uniform load represents the weight of the beam.
(b) Find the maximum value of the concentrated load if the maximum tensile stress cannot exceed 27 MPa and the maximum compressive stress is limited to 100 MPa.
(c) How far from A can load $P = 1$ kN be positioned if the maximum tensile stress cannot exceed 27 MPa and the maximum compressive stress is limited to 100 MPa?

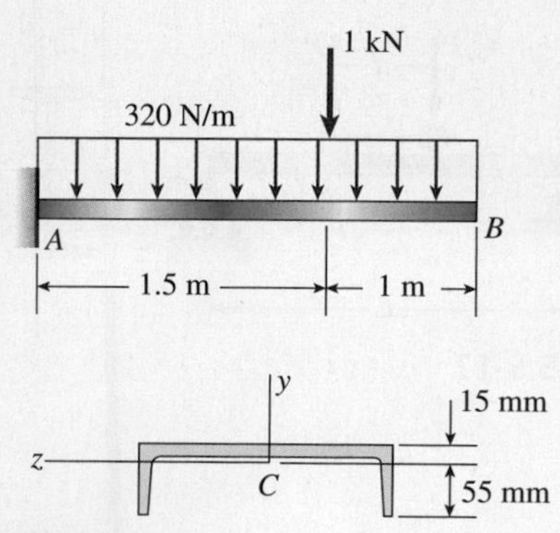

PROBLEM 5.5-19

5.5-20 A cantilever beam AB of an isosceles trapezoidal cross section has a length $L = 0.8$ m, dimensions $b_1 = 80$ mm and $b_2 = 90$ mm, and height $h = 110$ mm (see figure). The beam is made of brass weighing 85 kN/m^3.

(a) Determine the maximum tensile stress σ_t and maximum compressive stress σ_c due to the beam's own weight.

(b) If the width b_1 is doubled, what happens to the stresses?

(c) If the height h is doubled, what happens to the stresses?

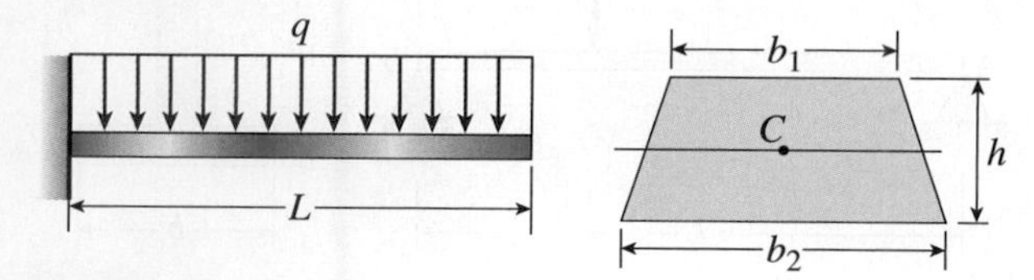

PROBLEM 5.5-20

5.5-21 A cantilever beam, a UPN 300 section, is subjected to its own weight and a point load at B. Find the maximum permissible value of load P at B (kN) if the allowable stress in tension and compression is $\sigma_a = 125$ MPa.

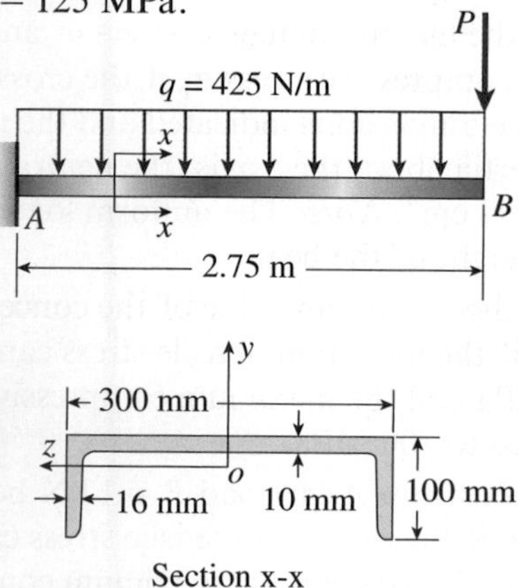

PROBLEM 5.5-21

5.5-22 A frame ABC travels horizontally with an acceleration a_0 (see figure). Obtain a formula for the maximum stress σ_{max} in the vertical arm AB, which has length L, thickness t, and mass density ρ.

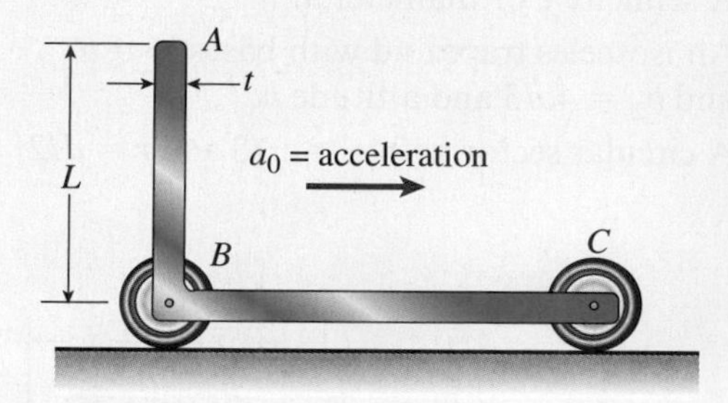

PROBLEM 5.5-22

5.5-23 A beam ABC with an overhang from B to C supports a uniform load of 3 kN/m throughout its length (see figure). The beam is a channel section with dimensions as shown in the figure. The moment of inertia about the z axis (the neutral axis) equals 210 cm^4.

(a) Calculate the maximum tensile stress σ_t and maximum compressive stress σ_c due to the uniform load.

(b) Find the required span length a that results in the ratio of larger to smaller compressive stress being equal to the ratio of larger to smaller tensile stress for the beam. Assume that the total length $L = a + b = 6$ m remains unchanged.

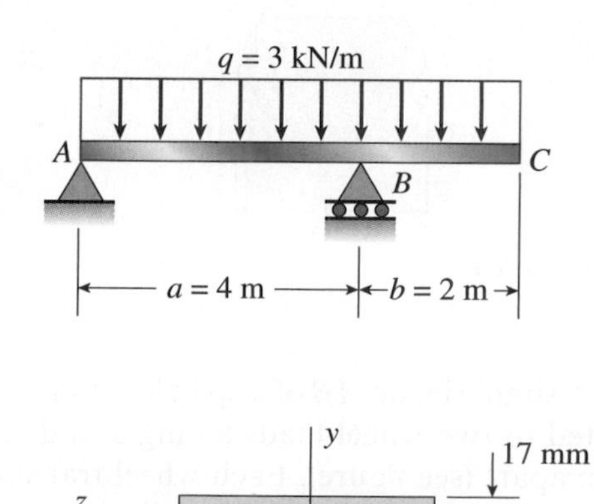

PROBLEM 5.5-23

5.5-24 A cantilever beam AB with a rectangular cross section has a longitudinal hole drilled throughout its length (see figure). The beam supports a load $P = 600$ N. The cross section is 25 mm wide and 50 mm high, and the hole has a diameter of 10 mm.

Find the bending stresses at the top of the beam, at the top of the hole, and at the bottom of the beam.

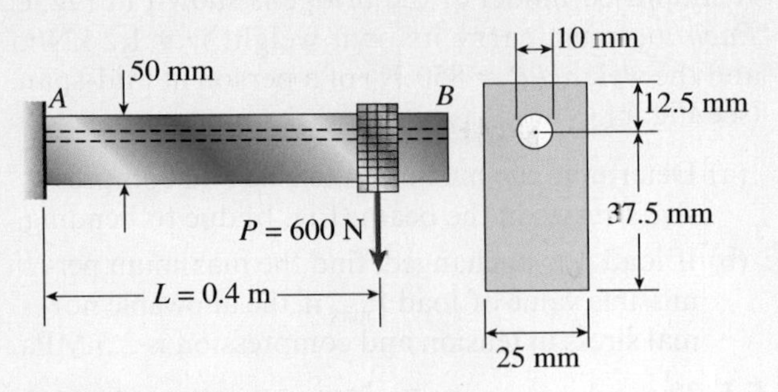

PROBLEM 5.5-24

5.5-25 A beam with a T-section is supported and loaded as shown in the figure. The cross section has width $b = 65$ mm, height $h = 75$ mm, and thickness $t = 13$ mm.

(a) Determine the maximum tensile and compressive stresses in the beam.

(b) If the allowable stresses in tension and compression are 124 MPa and 82 MPa, respectively, what is the required depth h of the beam? Assume that thickness t remains at 13 mm and that flange width $b = 65$ mm.

(c) Find the new values of loads P and q so that the allowable tension (124 MPa) and compression (82 MPa) stresses are reached simultaneously for the beam. Use the beam cross section in part (a) (see figure) and assume that L_1, L_2, and L_3 are unchanged.

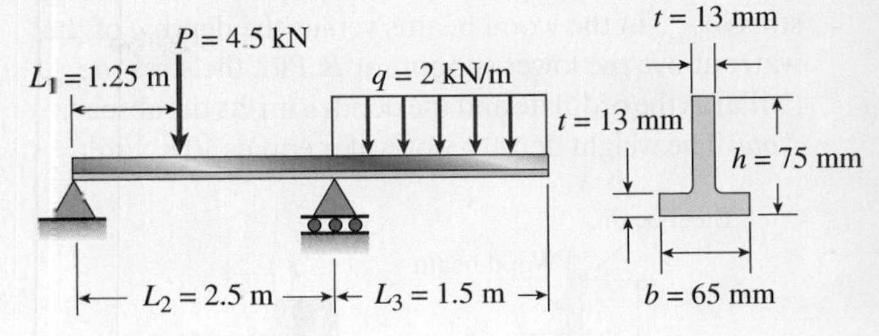

PROBLEM 5.5-25

5.5-26 Consider the compound beam with segments AB and BCD joined by a pin connection (moment release) just right of B (see figure part a). The beam cross section is a double-T made up from three 50 mm × 150 mm wood members (actual dimensions, see figure part b).

(a) Find the centroid C of the double-T cross section (c_1, c_2); then compute the moment of inertia, $[I_z\,(\text{mm}^4)]$.

(b) Find the maximum *tensile* normal stress σ_t and maximum *compressive* normal stress σ_c (kPa) for the loading shown. (Ignore the weight of the beam.)

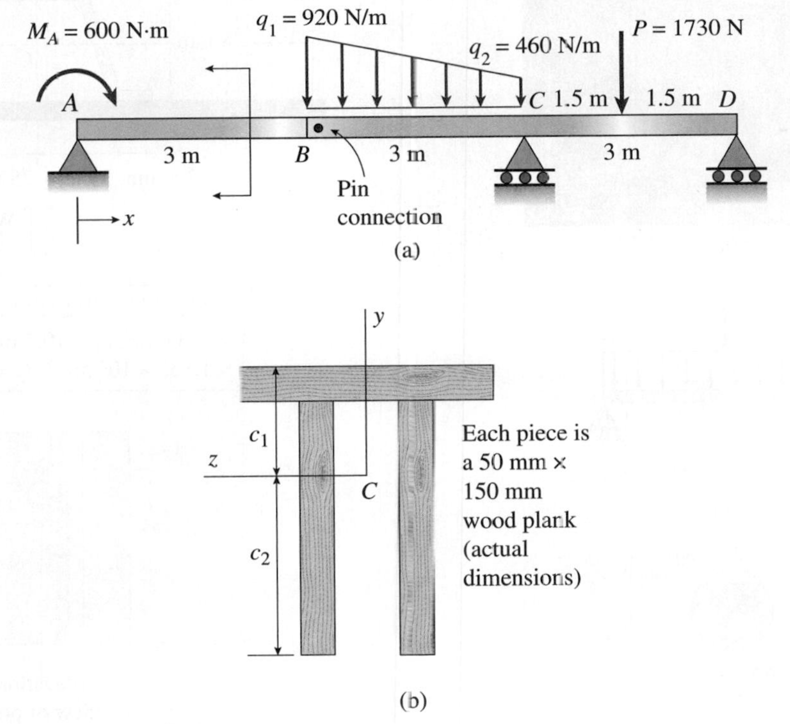

PROBLEM 5.5-26

5.5-27 A small dam of a height $h = 2$ m is constructed of vertical wood beams AB, as shown in the figure. The wood beams, which have a thickness $t = 64$ mm, are simply supported by horizontal steel beams at A and B.

Construct a graph showing the maximum bending stress σ_{max} in the wood beams versus the depth d of the water above the lower support at B. Plot the stress σ_{max} (MPa) as the ordinate and the depth d (m) as the abscissa. *Note:* The weight density γ of water equals 10 kN/m^3.

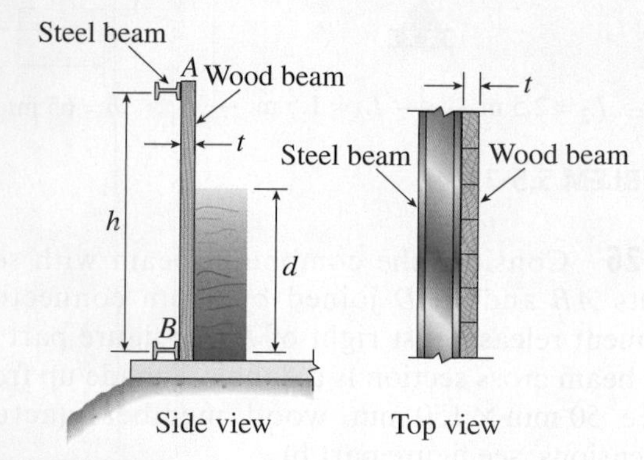

PROBLEM 5.5-27

5.5-28 A foot bridge on a hiking trail is constructed using two timber logs each having a diameter $d = 0.5$ m (see figure a). The bridge is simply supported and has a length $L = 4$ m. The top of each log is trimmed to form the walking surface (see Fig. b). A simplified model of the bridge is shown in Fig. c. *Each log* must carry its own weight $w = 1.2$ kN/m and the weight ($P = 850$ N) of a person at mid-span. (see Fig. b).

(a) Determine the maximum tensile and compressive stresses in the beam (Fig. b) due to bending.

(b) If load w is unchanged, find the maximum permissible value of load P_{max} if the allowable normal stress in tension and compression is 2.5 MPa.

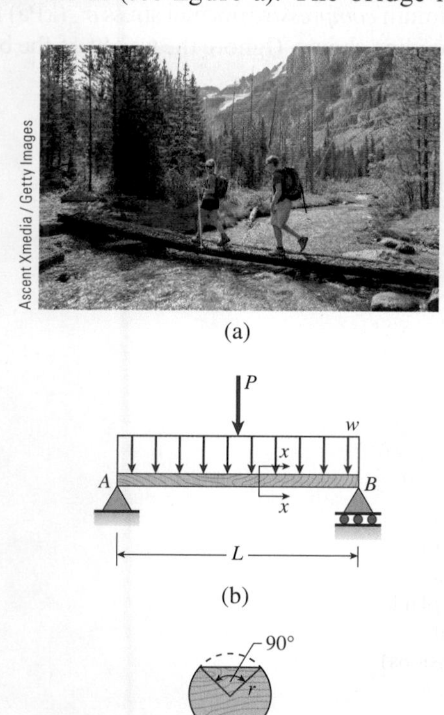

(a)

(b)

(c)

PROBLEM 5.5-28

5.5-29 A steel post ($E = 200$ GPa) having thickness $t = 3$ mm and height $L = 2$ m supports a stop

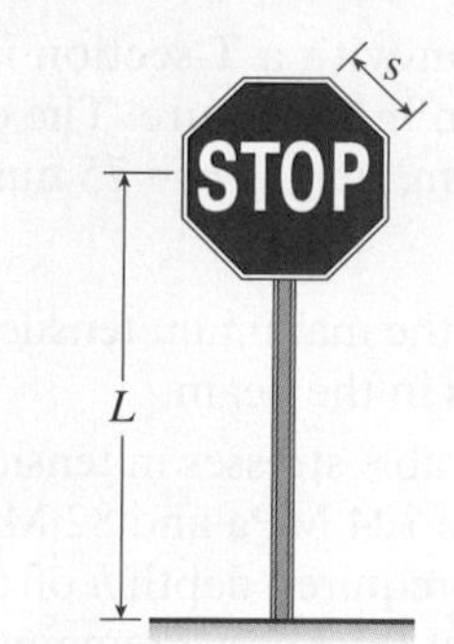

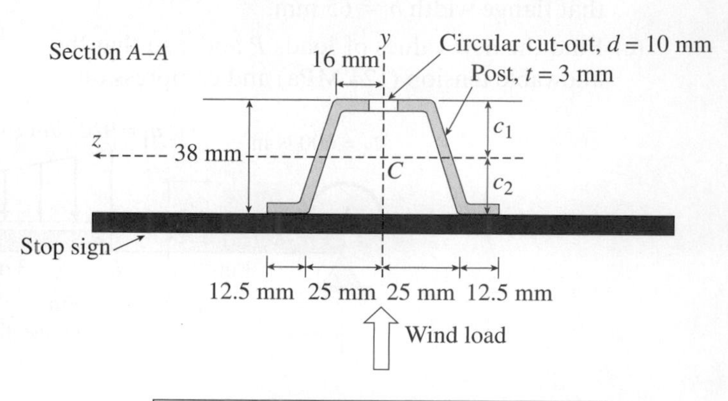

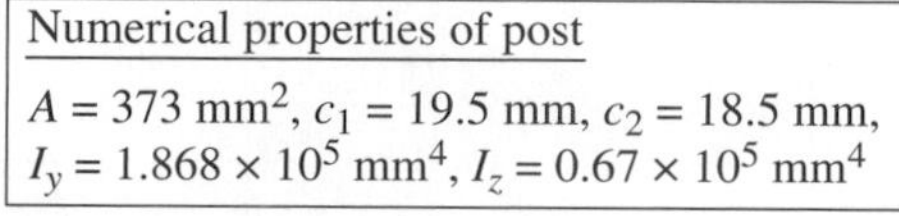

Numerical properties of post
A = 373 mm², c1 = 19.5 mm, c2 = 18.5 mm,
Iy = 1.868 × 10⁵ mm⁴, Iz = 0.67 × 10⁵ mm⁴

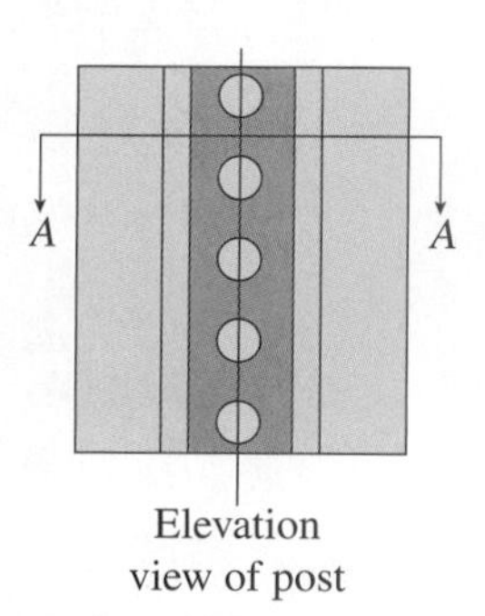

PROBLEM 5.5-29

sign (see figure), where $s = 310$ mm. The height of the post L is measured from the base to the centroid of the sign. The stop sign is subjected to wind pressure $p = 0.95$ kPa normal to its surface. Assume that the post is fixed at its base.

(a) What is the resultant load on the sign? (See Appendix E, Case 25, for properties of an octagon, $n = 8$.)
(b) What is the maximum bending stress σ_{max} in the post?
(c) Repeat part (b) if the circular cut-outs are eliminated over the height of the post.

5.5-30 Beam $ABCDE$ has a moment release just right of joint B and has concentrated moment loads at D and E. In addition, a cable with tension P is attached at F and runs over a pulley at C (Fig. a). The beam is constructed using two steel plates, which are welded to form a T cross section (see Fig. b). *Consider flexural stresses only.* Find the maximum permissible value of load *variable* P if the allowable bending stress is 130 MPa. Ignore the self-weight of the frame members and let length *variable* $L = 0.75$ m.

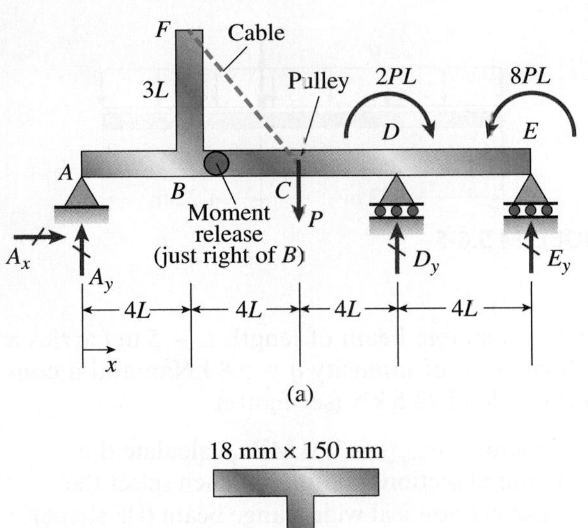

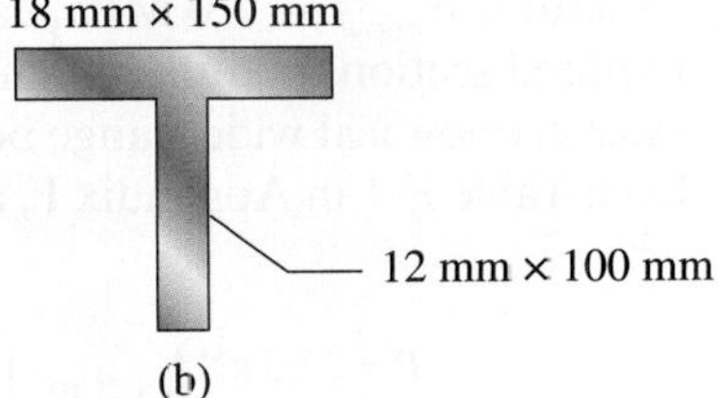

PROBLEM 5.5-30

5.6 Design of Beams for Bending Stresses

Introductory Problems

5.6-1 A simply supported wood beam having a span length $L = 3.6$ m is subjected to unsymmetrical point loads, as shown in the figure. Select a suitable size for the beam from the table in Appendix G. The allowable bending stress is 12 MPa and the wood weighs 5.5 kN/m^3.

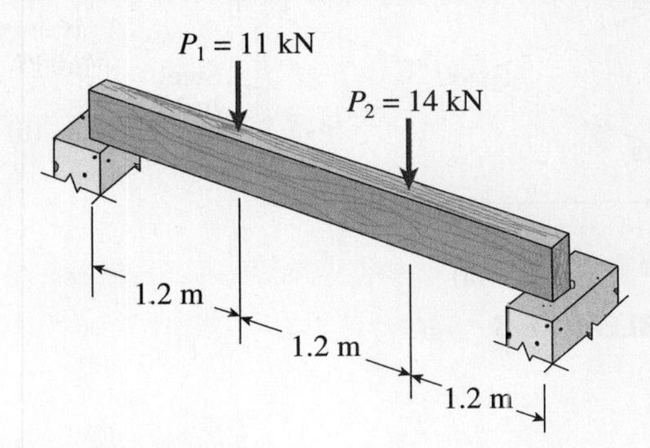

PROBLEM 5.6-1

5.6-2 A simply supported beam ($L = 4.5$ m) must support mechanical equipment represented as a distributed load with intensity $q = 30$ kN/m acting over the middle segment of the beam (see figure). Select the most economical IPN shape steel beam from Table F-2 to support the loads. Consider both the distributed force q *and the weight of the beam.* Use an allowable bending stress of 140 MPa.

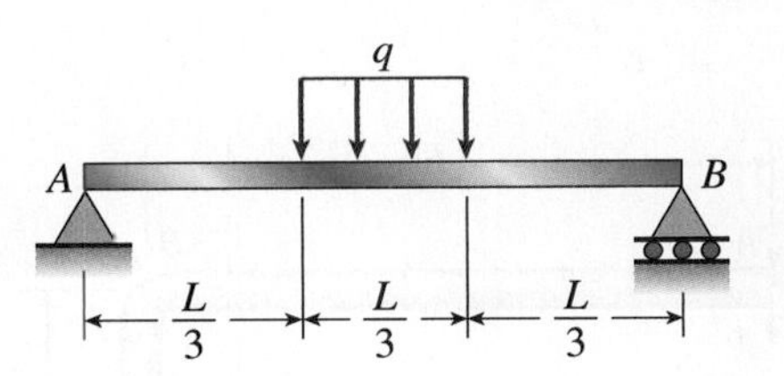

PROBLEM 5.6-2

5.6-3 The cross section of a narrow-gage railway bridge is shown in part a of the figure. The bridge is constructed with longitudinal steel girders that support the wood cross ties. The girders are restrained against lateral buckling by diagonal bracing, as indicated by the dashed lines.

The spacing of the girders is $s_1 = 0.8$ m and the spacing of the rails is $s_2 = 0.6$ m. The load transmitted by each rail to a single tie is $P = 16$ kN. The cross section of a tie, shown in part b of the figure, has a width $b = 120$ mm and depth d.

Determine the minimum value of d based upon an allowable bending stress of 8 MPa in the wood tie. (Disregard the weight of the tie itself.)

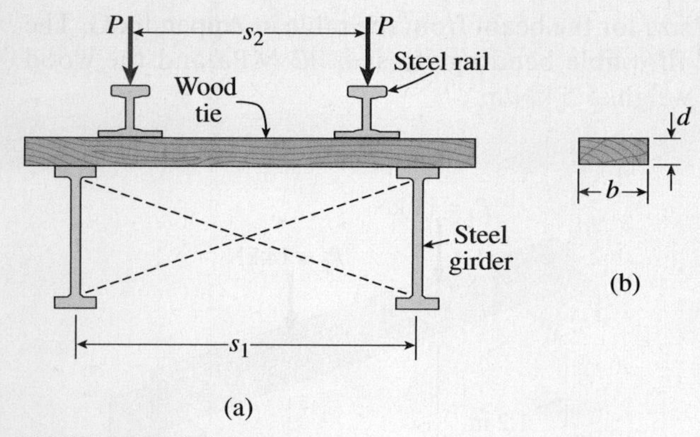

PROBLEM 5.6-3

5.6-4 A fiberglass bracket $ABCD$ with a solid circular cross section has the shape and dimensions shown in the figure. A vertical load $P = 40$ N acts at the free end D.

(a) Determine the minimum permissible diameter d_{min} of the bracket if the allowable bending stress in the material is 30 MPa and $b = 37$ mm. *Note:* Disregard the weight of the bracket itself.

(b) If $d = 10$ mm, $b = 37$ mm, and $\sigma_{allow} = 30$ MPa, what is the maximum value of load P if vertical load P at D is replaced with horizontal loads P at B and D (see figure part b)?

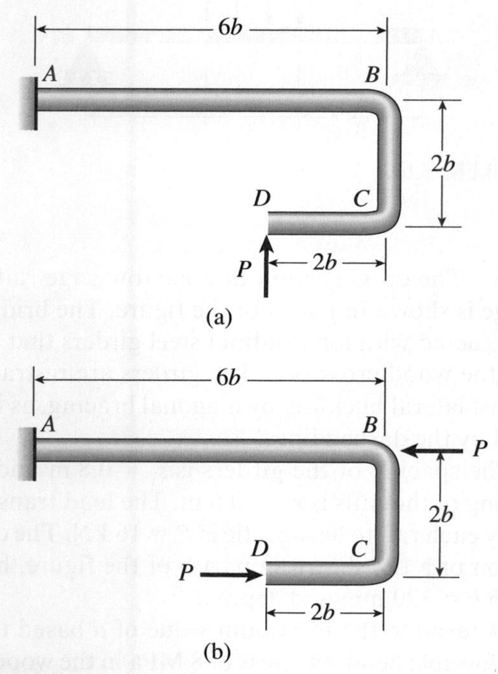

PROBLEM 5.6-4

Representative Problems

5.6-5 A cantilever beam AB is loaded by a uniform load q and a concentrated load P, as shown in the figure.

(a) Select the most economical steel UPN shape from Table F-3 in Appendix F; use $q = 292$ N/m and $P = 1.33$ kN (assume allowable normal stress is $\sigma_a = 124$ MPa).

(b) Select the most economical steel IPN shape from Table F-2 in Appendix F; use $q = 657$ N/m and $P = 9$ kN (assume allowable normal stress is $\sigma_a = 138$ MPa).

(c) Select the most economical steel HE shape from Table F-1 in Appendix F; use $q = 657$ N/m and $P = 9$ kN (assume allowable normal stress is $\sigma_a = 138$ MPa). *However*, assume that the design requires that the HE shape must be used in weak axis bending, i.e., it must bend about the 2–2 (or y) axis of the cross section.

Note: For parts (a), (b), and (c), revise your initial beam selection as needed to include the distributed weight of the beam in addition to uniform load q.

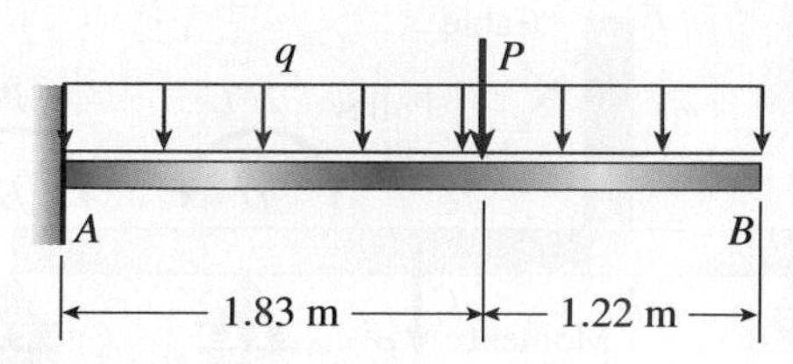

PROBLEM 5.6-5

5.6-6 A simple beam of length $L = 5$ m carries a uniform load of intensity $q = 5.8$ kN/m and a concentrated load 22.5 kN (see figure).

(a) Assuming $\sigma_{allow} = 110$ MPa, calculate the required section modulus S. Then select the most economical wide-flange beam (HE shape) from Table F-1 in Appendix F, and recalculate

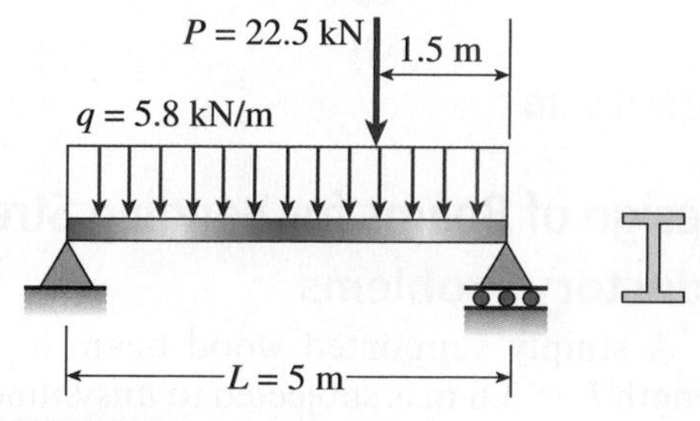

PROBLEM 5.6-6

S, taking into account the weight of the beam. Select a new beam if necessary.

(b) Repeat part (a), but now assume that the design requires that the HE shape must be used in weak axis bending (i.e., it must bend about the 2–2 (or y) axis of the cross section).

5.6-7 A simple beam AB is loaded as shown in the figure.

(a) Calculate the required section modulus S if $\sigma_{allow} = 124$ MPa, $L = 9.75$ m, $P = 13$ kN, and $q = 6.6$ kN/m. Then select a suitable I-beam (IPN shape) from Table F-2, Appendix F, and recalculate S taking into account the weight of the beam. Select a new beam size if necessary.

(b) What is the maximum load P that can be applied to your final beam selection in part (a)?

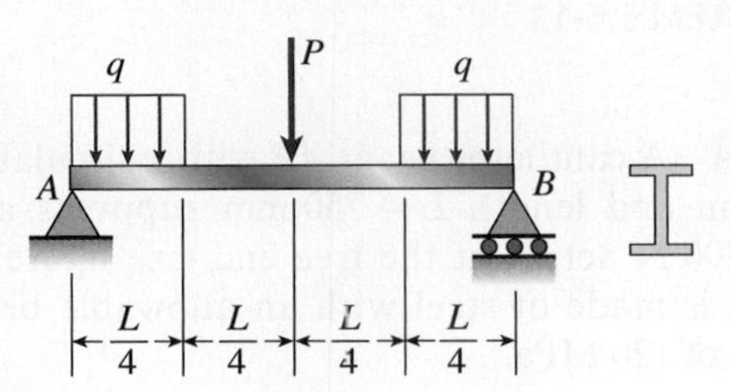

PROBLEM 5.6-7

5.6-8 A pontoon bridge (see figure) is constructed of two longitudinal wood beams, known as *balks*, that span between adjacent pontoons and support the transverse floor beams, which are called *chesses*. For purposes of design, assume that a uniform floor load of 7.5 kPa acts over the chesses. (This load includes an allowance for the weights of the chesses and balks.) Also, assume that the chesses are 2.5 m long and that the balks are simply supported with a span of 3.0 m. The allowable bending stress in the wood is 15 MPa.

(a) If the balks have a square cross section, what is their minimum required width b_{min}?

(b) Repeat part (a) if the balk width is 1.5 b and the balk depth is b; compare the cross-sectional areas of the two designs.

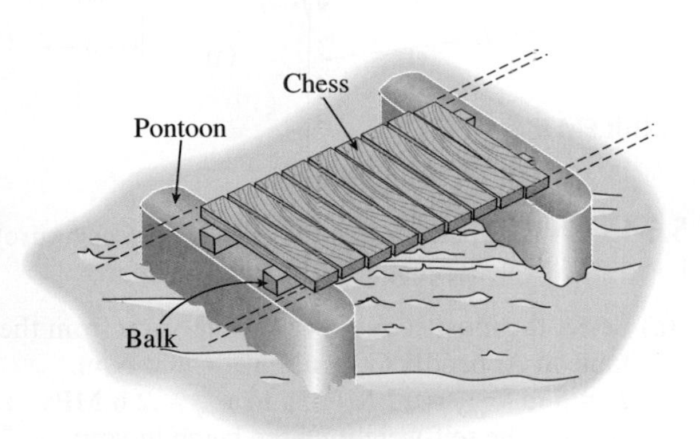

PROBLEM 5.6-8

5.6-9 A floor system in a small building consists of wood planks supported by 50 mm (nominal width) joists spaced at distance s and measured from center to center (see figure). The span length L of each joist is 3 m, the spacing s of the joists is 400 mm, and the allowable bending stress in the wood is 8 MPa. The uniform floor load is 6 kN/m^2, which includes an allowance for the weight of the floor system itself.

(a) Calculate the required section modulus S for the joists, and then select a suitable joist size (surfaced lumber) from Appendix G, assuming that each joist may be represented as a simple beam carrying a uniform load.

(b) What is the maximum floor load that can be applied to your final beam selection in part (a)?

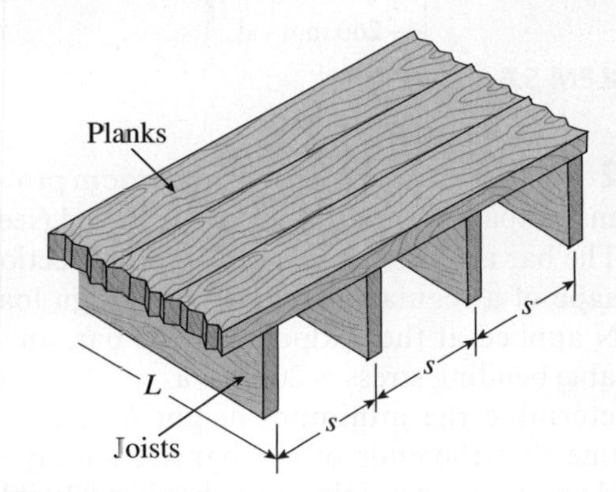

PROBLEMS 5.6-9 and 5.6-10

5.6-10 The wood joists supporting a plank floor (see figure) are 38 mm × 220 mm in cross section (actual dimensions) and have a span length of $L = 4.0$ m. The floor load is 5.0 kPa, which includes the weight of the joists and the floor.

(a) Calculate the maximum permissible spacing s of the joists if the allowable bending stress is 14 MPa. (Assume that each joist may be represented as a simple beam carrying a uniform load.)

(b) If spacing $s = 406$ mm, what is the required depth h of the joist? Assume all other variables remain unchanged.

5.6-11 A beam ABC with an overhang from B to C is constructed of a UPN 260 channel section with flanges facing upward (see figure). The beam supports its own weight (372 N/m) plus a *triangular* load of maximum intensity q_0 acting on the overhang. The allowable stresses in tension and compression are 138 MPa and 75 MPa, respectively.

(a) Determine the allowable *triangular* load intensity $q_{0,allow}$ if the distance L equals 1.2 m.

(b) What is the allowable triangular load intensity $q_{0,allow}$ if the beam is rotated 180° about its longitudinal centroidal axis so that the flanges are downward?

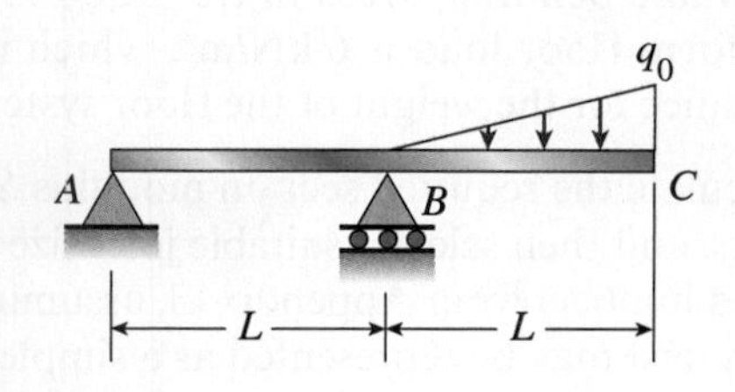

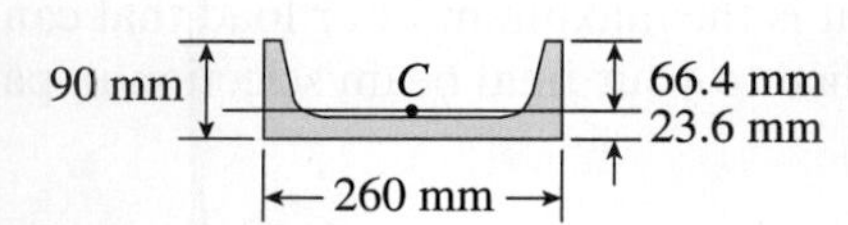

PROBLEM 5.6-11

5.6-12 A "trapeze bar" in a hospital room provides a means for patients to exercise while in bed (see figure). The bar is 2.1 m long and has a cross section in the shape of a regular octagon. The design load is 1.2 kN applied at the midpoint of the bar, and the allowable bending stress is 200 MPa.

Determine the minimum height h of the bar. (Assume that the ends of the bar are simply supported and that the weight of the bar is negligible.)

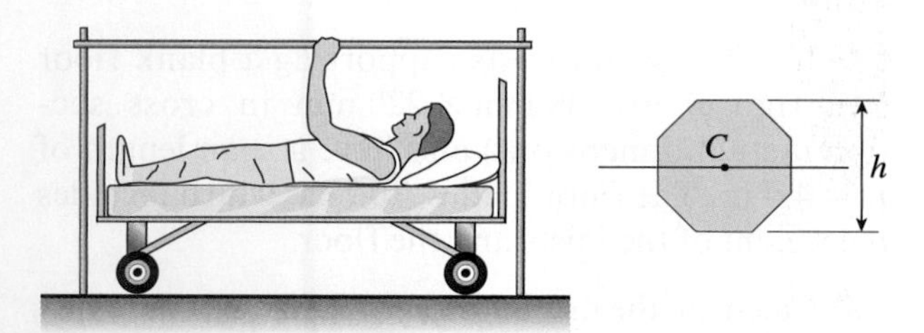

PROBLEM 5.6-12

5.6-13 A two-axle carriage that is part of an overhead traveling crane in a testing laboratory moves slowly across a simple beam AB (see figure). The load transmitted to the beam from the front axle is 9 kN and from the rear axle is 18 kN. The weight of the beam itself may be disregarded.

(a) Determine the minimum required section modulus S for the beam if the allowable bending stress is 110 MPa, the length of the beam is 5 m, and the wheelbase of the carriage is 1.5 m.

(b) Select the most economical beam (IPN shape) from Table F-2, Appendix F.

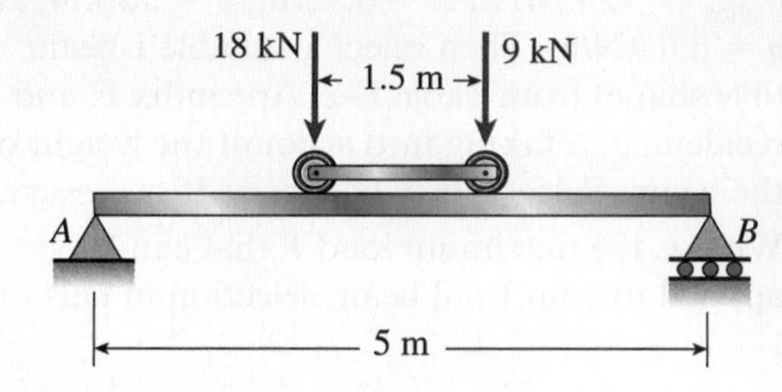

PROBLEM 5.6-13

5.6-14 A cantilever beam AB with a circular cross section and length $L = 750$ mm supports a load $P = 800$ N acting at the free end (see figure). The beam is made of steel with an allowable bending stress of 120 MPa.

(a) Determine the required diameter d_{min} (figure part a) of the beam, considering the effect of the beam's own weight.

(b) Repeat part (a) if the beam is hollow with wall thickness $t = d/8$ (figure part b); compare the cross-sectional areas of the two designs.

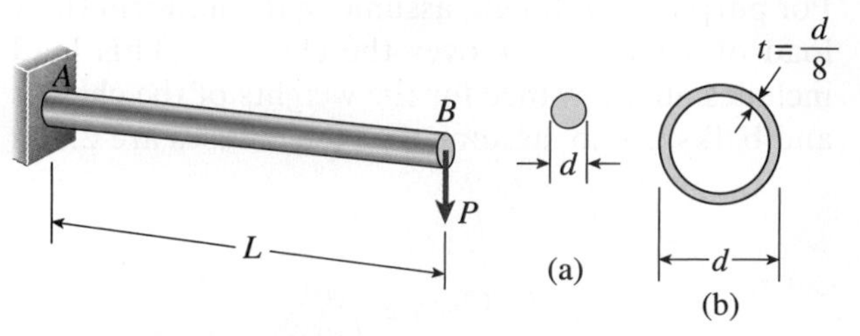

PROBLEM 5.6-14

5.6-15 A propped cantilever beam ABC (see figure) has a shear release just right of the mid-span.

(a) Select the most economical wood beam from the table in Appendix G; assume $q = 800$ N/m, $L = 5$ m, $\sigma_{aw} = 12$ MPa, and $\tau_{aw} = 2.6$ MPa. Include the self-weight of the beam in your design.

(b) If a UPN 180 steel beam is now used for beam ABC, what is the maximum permissible value of load variable q? Assume $\sigma_{as} = 110$ MPa and $L = 3$ m. Include the self-weight of the beam in your analysis.

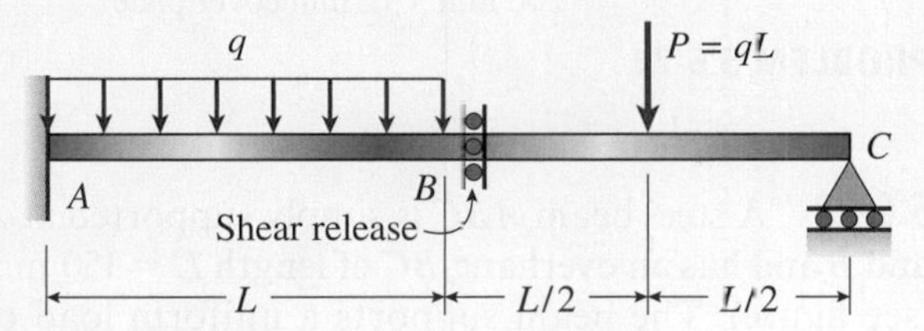

PROBLEM 5.6-15

5.6-16 A small balcony constructed of wood is supported by three identical cantilever beams (see figure). Each beam has length $L_1 = 2.1$ m, width b, and height $h = 4b/3$. The dimensions of the balcony floor are $L_1 \times L_2$, where $L_2 = 2.5$ m. The design load is 5.5 kPa acting over the entire floor area. (This load accounts for all loads except the weights of the cantilever beams, which have a weight density $\gamma = 5.5$ kN/m^3.) The allowable bending stress in the cantilevers is 15 MPa.

Assuming that the middle cantilever supports 50% of the load and each outer cantilever supports 25% of the load, determine the required dimensions b and h.

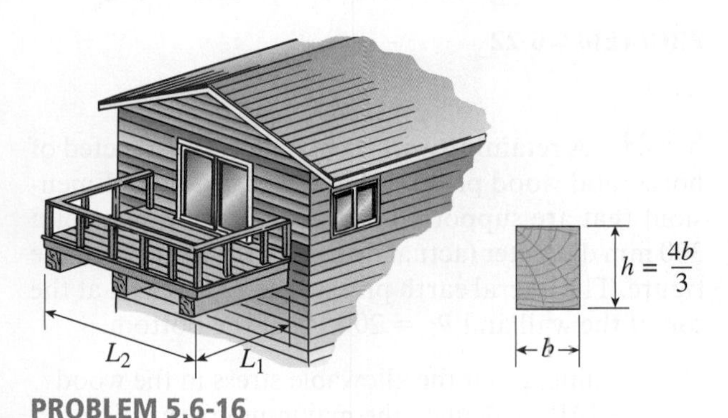

PROBLEM 5.6-16

5.6-17 A beam having a cross section in the form of an unsymmetric wide-flange shape (see figure) is subjected to a negative bending moment acting about the z axis.

Determine the width b of the top flange in order that the stresses at the top and bottom of the beam will be in the ratio 4:3, respectively.

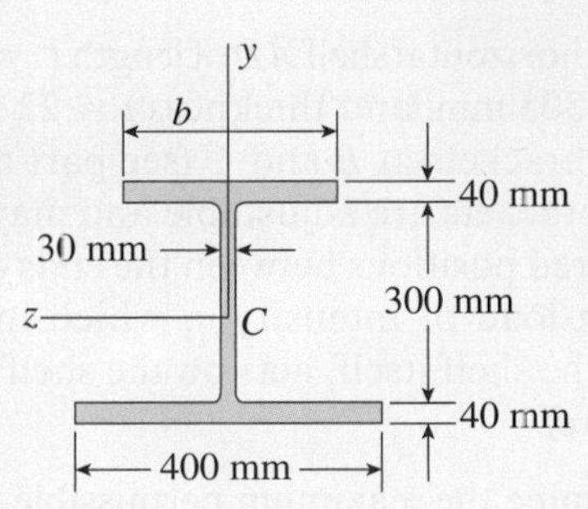

PROBLEM 5.6-17

5.6-18 A beam having a cross section in the form of a channel (see figure) is subjected to a bending moment acting about the z axis.

Calculate the thickness t of the channel in order that the bending stresses at the top and bottom of the beam will be in the ratio 7:3, respectively.

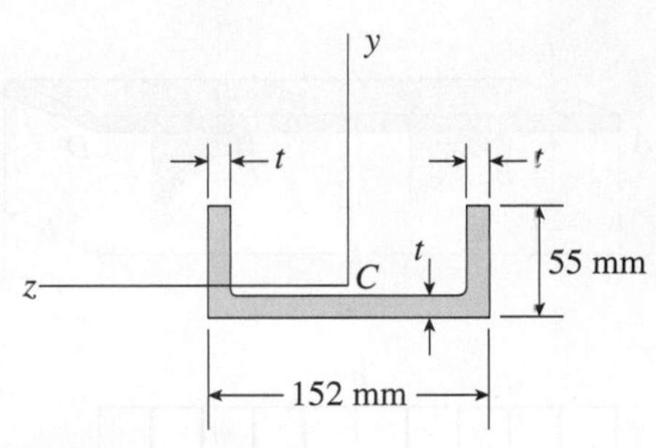

PROBLEM 5.6-18

5.6-19 Determine the ratios of the weights of four beams that have the same length, are made of the same material, are subjected to the same maximum bending moment, and have the same maximum bending stress if their cross sections are (1) a rectangle with height equal to twice the width, (2) a square, (3) a circle, and (4) a pipe with outer diameter d and wall thickness $t = d/8$ (see figures).

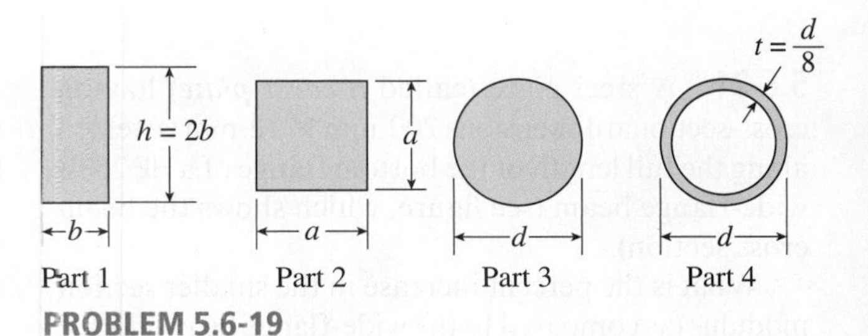

PROBLEM 5.6-19

5.6-20 A horizontal shelf AD of length $L = 1215$ mm, width $b = 305$ mm, and thickness $t = 22$ mm is supported by brackets at B and C (see part a of the figure). The brackets are adjustable and may be placed in any desired positions between the ends of the shelf. A uniform load of intensity q, which includes the weight of the shelf itself, acts on the shelf (see part b of the figure).

(a) Determine the maximum permissible value of the load q if the allowable bending stress in the shelf is $\sigma_{allow} = 8.5$ MPa and the position of the supports is adjusted for maximum load carrying capacity.

(b) The bookshelf owner decides to reinforce the shelf with a bottom wood plate $b/2 \times t/2$ along its entire length (see figure part c). Find the new maximum permissible value of the load q if the allowable bending stress in the shelf remains at $\sigma_{allow} = 8.5$ MPa.

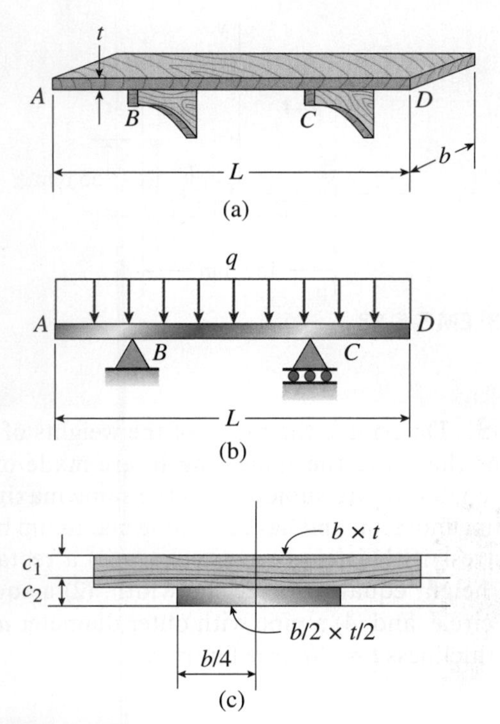

PROBLEM 5.6-20

5.6-21 A steel plate (called a *cover plate*) having cross-sectional dimensions 200 mm × 12 mm is welded along the full length of the bottom flange of a HE 260B wide-flange beam (see figure, which shows the beam cross section).

What is the percent increase in the smaller section modulus (as compared to the wide-flange beam alone)?

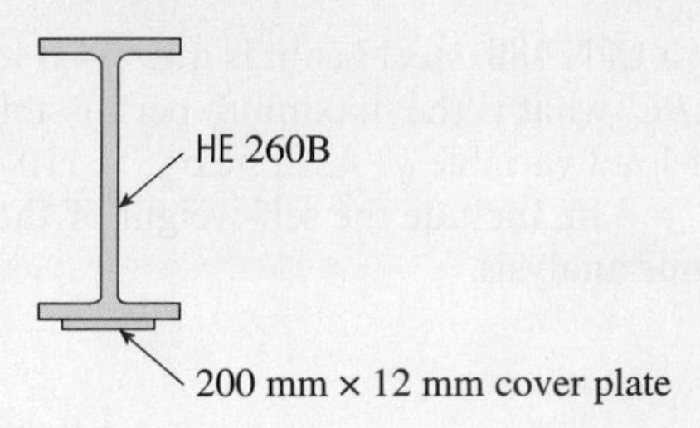

PROBLEM 5.6-21

5.6-22 A steel beam ABC is simply supported at A and B and has an overhang BC of length $L = 150$ mm (see figure). The beam supports a uniform load of intensity $q = 4.0$ kN/m over its entire span AB and $1.5q$ over BC. The cross section of the beam is rectangular with width b and height $2b$. The allowable bending stress in the steel is $\sigma_{allow} = 60$ MPa, and its weight density is $\gamma = 77.0$ kN/m^3.

(a) Disregarding the weight of the beam, calculate the required width b of the rectangular cross section.

(b) Taking into account the weight of the beam, calculate the required width b.

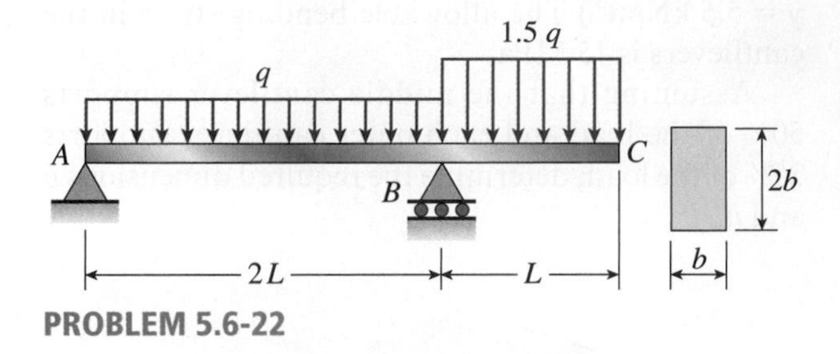

PROBLEM 5.6-22

5.6-23 A retaining wall 1.5 m high is constructed of horizontal wood planks 75 mm thick (actual dimension) that are supported by vertical wood piles of a 300 mm diameter (actual dimension), as shown in the figure. The lateral earth pressure is $p_1 = 5$ kPa at the top of the wall and $p_2 = 20$ kPa at the bottom.

(a) Assuming that the allowable stress in the wood is 8 MPa, calculate the maximum permissible spacing s of the piles.

(b) Find the required diameter of the wood piles so that piles and planks ($t = 75$ mm) reach the allowable stress at the same time.

Hint: Observe that the spacing of the piles may be governed by the load-carrying capacity of either the planks or the piles. Consider the piles to act as cantilever beams subjected to a trapezoidal distribution of load, and consider the planks to act as simple beams between the piles. To be on the safe side, assume that the pressure on the bottom plank is uniform and equal to the maximum pressure.

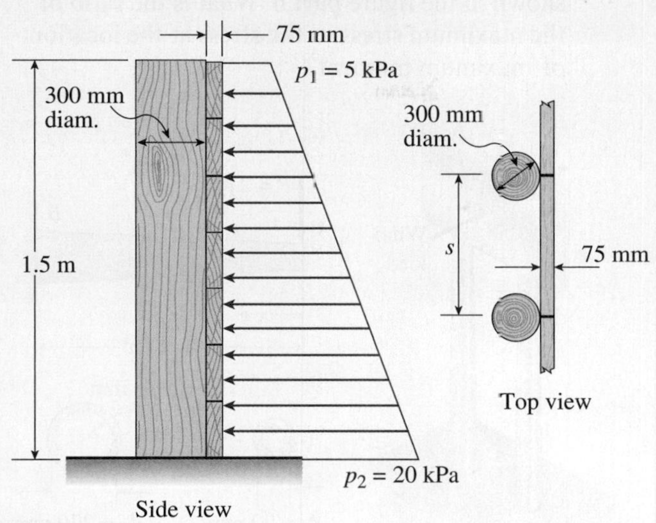

PROBLEM 5.6-23

5.6-24 A retaining wall (Fig. a) is constructed using steel IPN shape columns and concrete panel infill (Fig. b). Each column is subjected to lateral soil pressure with peak intensity q_0 (Figs. b and c). The tensile and compressive strength of the beam is 600 MPa. Select the most economical IPN section from Table F-2 based on safety factor of 3.0.

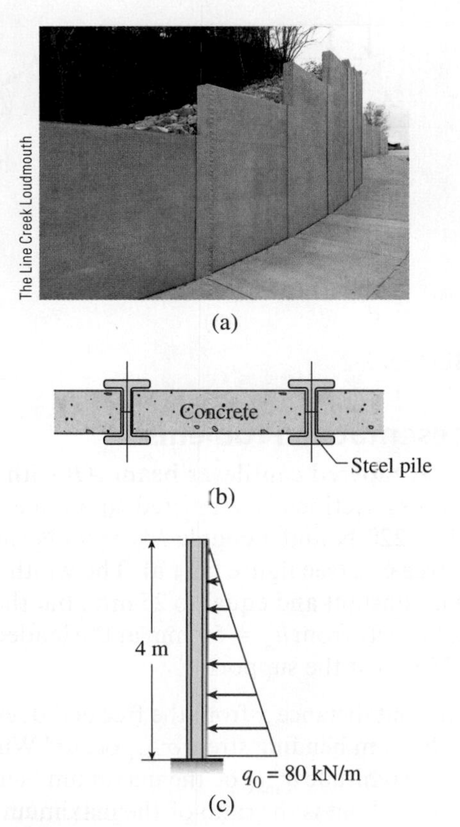

PROBLEM 5.6-24

5.6-25 A beam of square cross section (a = length of each side) is bent in the plane of a diagonal (see figure). By removing a small amount of material at the top and bottom corners, as shown by the shaded triangles in the figure, you can increase the section modulus and obtain a stronger beam, even though the area of the cross section is reduced.

(a) Determine the ratio β defining the areas that should be removed in order to obtain the strongest cross section in bending.
(b) By what percent is the section modulus increased when the areas are removed?

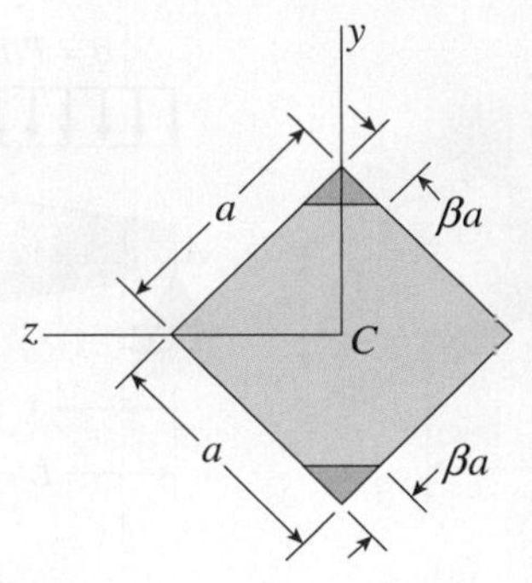

PROBLEM 5.6-25

5.6-26 The cross section of a rectangular beam having a width b and height h is shown in part a of the figure. For reasons unknown to the beam designer, it is planned to add structural projections of width $b/9$ and height d to the top and bottom of the beam (see part b of the figure).

For what values of d is the bending-moment capacity of the beam increased? For what values is it decreased?

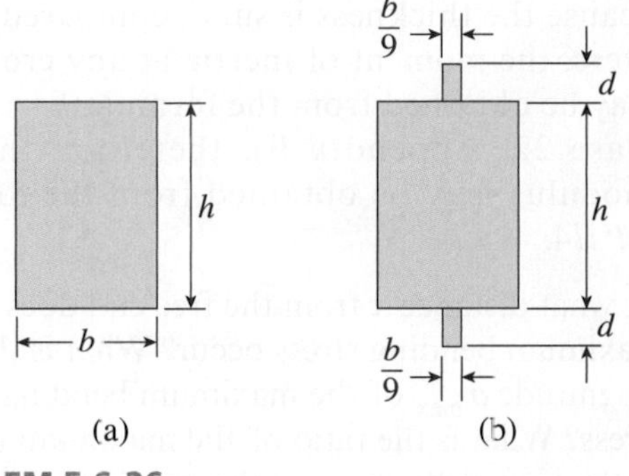

PROBLEM 5.6-26

5.7 Nonprismatic Beams

Introductory Problems

5.7-1 A tapered cantilever beam AB of length L has square cross sections and supports a concentrated load P at the free end (see figure part a). The width and height of the beam vary linearly from h_A at the free end to h_B at the fixed end.

Determine the distance x from the free end A to the cross section of maximum bending stress if $h_B = 3h_A$.

(a) What is the magnitude σ_{max} of the maximum bending stress? What is the ratio of the maximum stress to the largest stress B at the support?

(b) Repeat part (a) if load P is now applied as a uniform load of intensity $q = P/L$ over the entire beam, A is restrained by a roller support, and B is a sliding support (see figure part b).

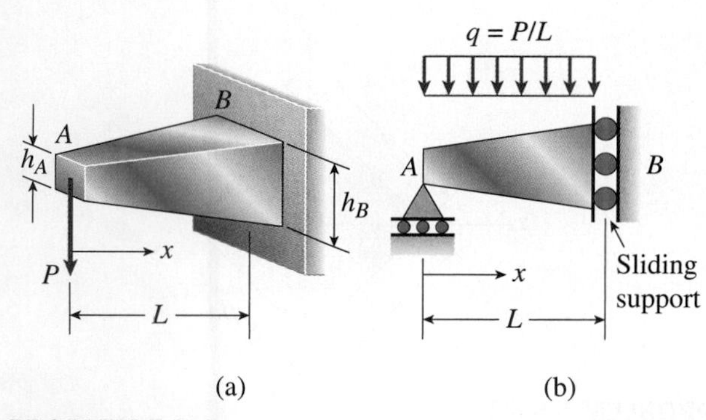

PROBLEM 5.7-1

5.7-2 A tall signboard is supported by two vertical beams consisting of thin-walled, tapered circular tubes (see figure part a). For purposes of this analysis, each beam may be represented as a cantilever AB of length $L = 8.0$ m subjected to a lateral load $P = 2.4$ kN at the free end. The tubes have a constant thickness $t = 10.0$ mm and average diameters $d_A = 90$ mm and $d_B = 270$ mm at ends A and B, respectively.

Because the thickness is small compared to the diameters, the moment of inertia at any cross section may be obtained from the formula $I = \pi d^3 t/8$ (see Case 22, Appendix E); therefore, the section modulus may be obtained from the formula $S = \pi d^2 t/4$.

(a) At what distance x from the free end does the maximum bending stress occur? What is the magnitude σ_{max} of the maximum bending stress? What is the ratio of the maximum stress to the largest stress σ_B at the support?

(b) Repeat part (a) if concentrated load P is applied upward at A and downward uniform load $q(x) = 2P/L$ is applied over the entire beam as shown in the figure part b. What is the ratio of the maximum stress to the stress at the location of maximum moment?

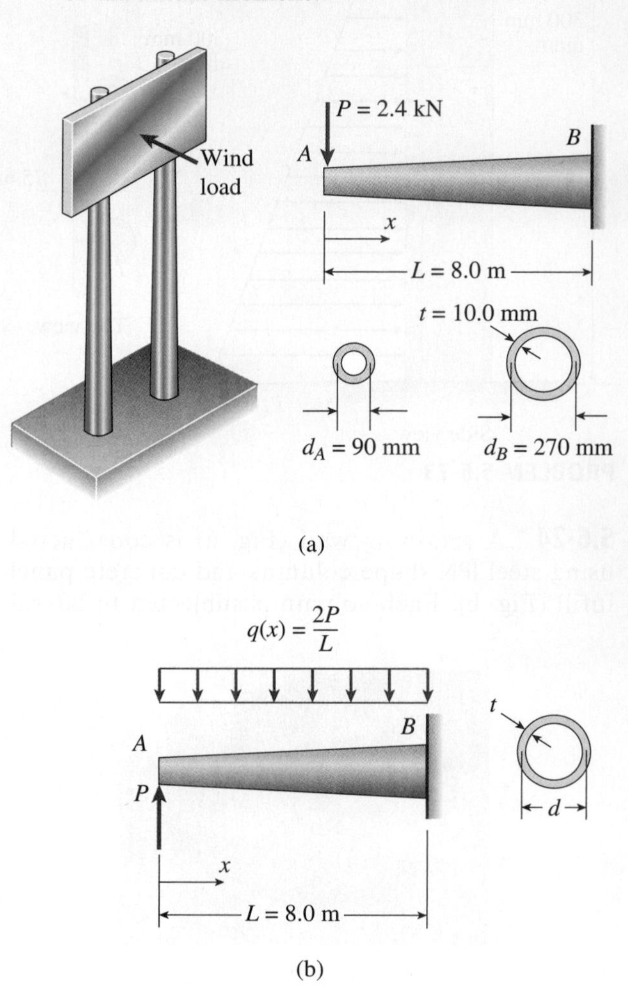

PROBLEM 5.7-2

Representative Problems

5.7-3 A tapered cantilever beam AB with rectangular cross sections is subjected to a concentrated load $P = 220$ N and a couple $M_0 = 90$ N·m acting at the free end (see figure part a). The width b of the beam is constant and equal to 25 mm, but the height varies linearly from $h_A = 50$ mm at the loaded end to $h_B = 75$ mm at the support.

(a) At what distance x from the free end does the maximum bending stress σ_{max} occur? What is the magnitude σ_{max} of the maximum bending stress? What is the ratio of the maximum stress to the largest stress σ_B at the support?

(b) Repeat part a if, in addition to P and M_0, a triangular distributed load with peak intensity $q_0 = 3P/L$ acts upward over the entire beam as shown in the figure part b. What is the ratio of the maximum stress to the stress at the location of maximum moment?

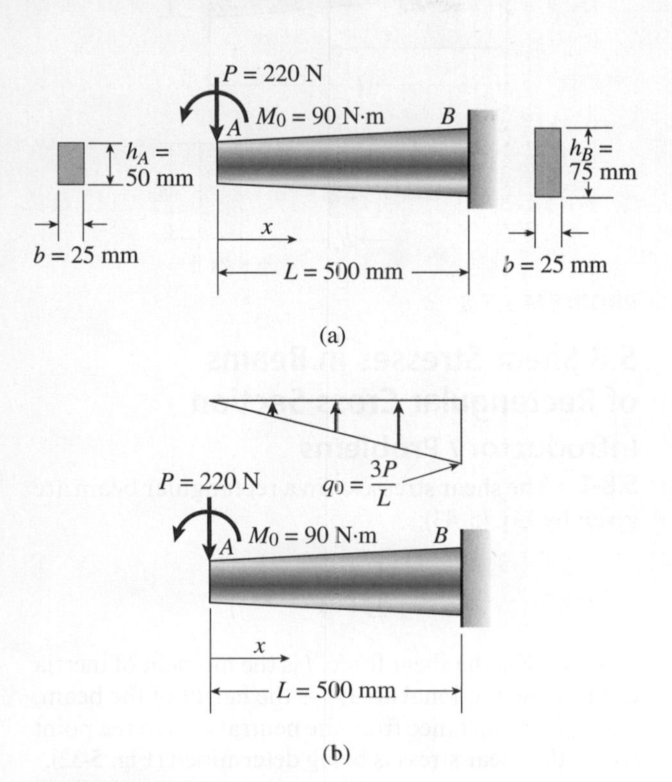

PROBLEM 5.7-3

5.7-4 The spokes in a large flywheel are modeled as beams fixed at one end and loaded by a force P and a couple M_0 at the other (see figure). The cross sections of the spokes are elliptical with major and minor axes (height and width, respectively) having the lengths shown in the figure part a. The cross-sectional dimensions vary linearly from end A to end B.

Considering only the effects of bending due to the loads P and M_0, determine the following quantities.

(a) The largest bending stress σ_A at end A.

(b) The largest bending stress σ_B at end B.

(c) The distance x to the cross section of maximum bending stress.

(d) The magnitude σ_{max} of the maximum bending stress.

(e) Repeat part d if a uniform load $q(x) = 10P/3L$ is added to loadings P and M_0, as shown in the figure part b.

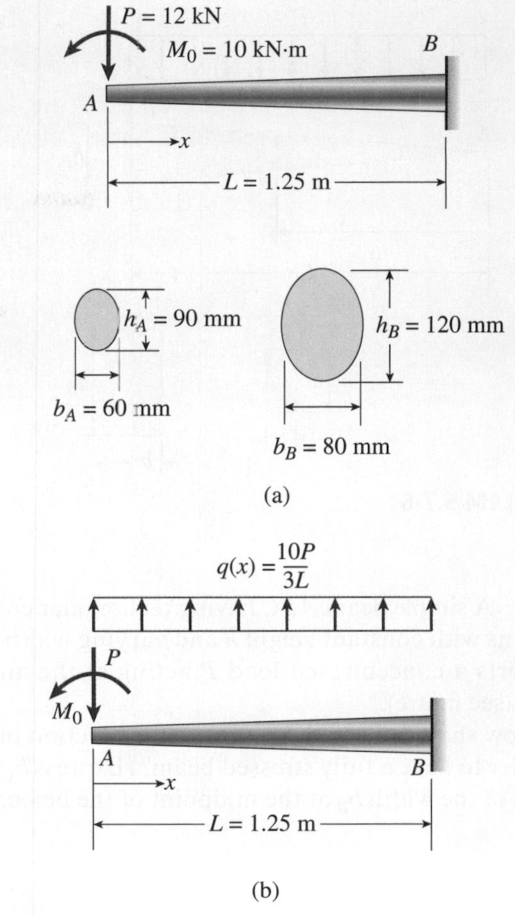

PROBLEM 5.7-4

5.7-5 Refer to the tapered cantilever beam of solid circular cross section shown in Fig. 5-26 of Example 5-9.

(a) Considering only the bending stresses due to the load P, determine the range of values of the ratio d_B/d_A for which the maximum normal stress occurs at the support.

(b) What is the maximum stress for this range of values?

Fully Stressed Beams

Problems 5.7-6 to 5.7-8 pertain to fully stressed beams of rectangular cross section. Consider only the bending stresses obtained from the flexure formula and disregard the weights of the beams.

5.7-6 A cantilever beam AB with rectangular cross sections of a constant width b and varying height h_x is subjected to a uniform load of intensity q (see figure).

How should the height h_x vary as a function of x (measured from the free end of the beam) in order to have a fully stressed beam? (Express h_x in terms of the height h_B at the fixed end of the beam.)

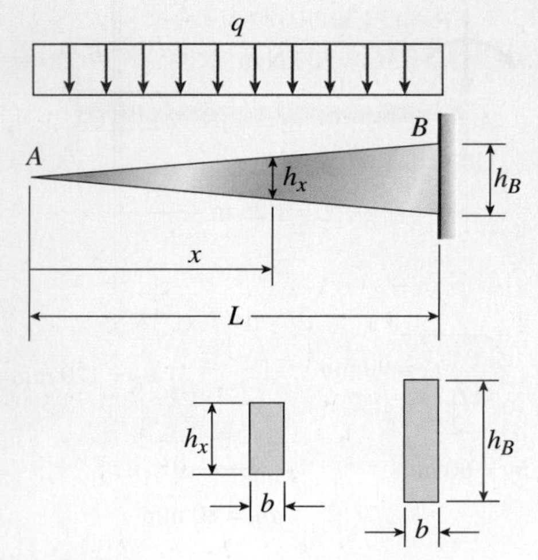

PROBLEM 5.7-6

5.7-7 A simple beam *ABC* having rectangular cross sections with constant height *h* and varying width b_x supports a concentrated load *P* acting at the midpoint (see figure).

How should the width b_x vary as a function of *x* in order to have a fully stressed beam? (Express b_x in terms of the width b_B at the midpoint of the beam.)

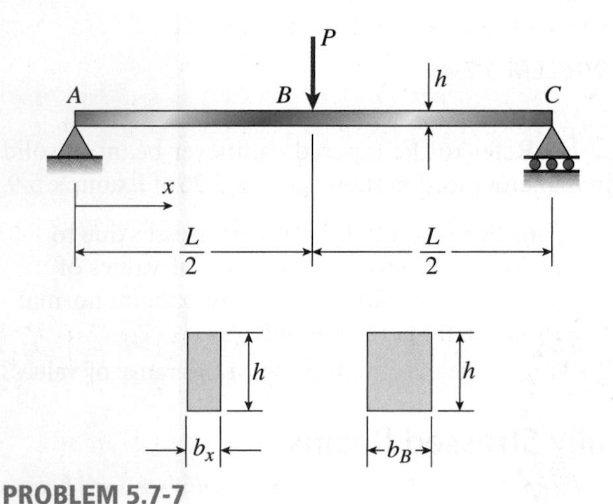

PROBLEM 5.7-7

5.7-8 A cantilever beam *AB* having rectangular cross sections with varying width b_x and varying height h_x is subjected to a uniform load of intensity *q* (see figure). If the width varies linearly with *x* according to the equation $b_x = b_B x/L$, how should the height h_x vary as a function of *x* in order to have a fully stressed beam? (Express h_x in terms of the height h_B at the fixed end of the beam.)

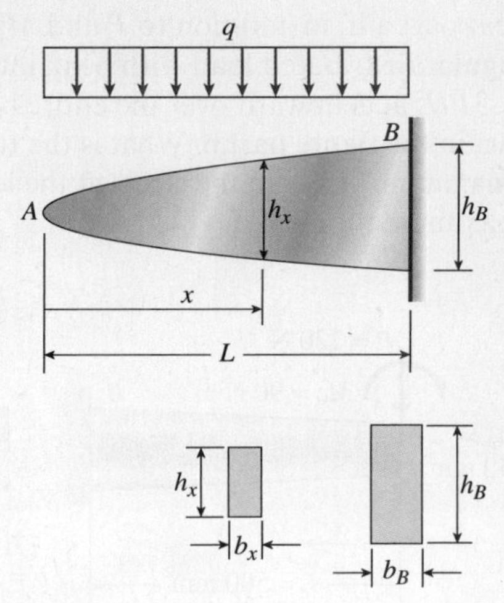

PROBLEM 5.7-8

5.8 Shear Stresses in Beams of Rectangular Cross Section

Introductory Problems

5.8-1 The shear stresses τ in a rectangular beam are given by Eq. (5-43):

$$\tau = \frac{V}{2I}\left(\frac{h^2}{4} - y_1^2\right)$$

in which *V* is the shear force, *I* is the moment of inertia of the cross-sectional area, *h* is the height of the beam, and y_1 is the distance from the neutral axis to the point where the shear stress is being determined (Fig. 5-32).

By integrating over the cross-sectional area, show that the resultant of the shear stresses is equal to the shear force *V*.

5.8-2 Calculate the maximum shear stress τ_{max} and the maximum bending stress σ_{max} in a wood beam

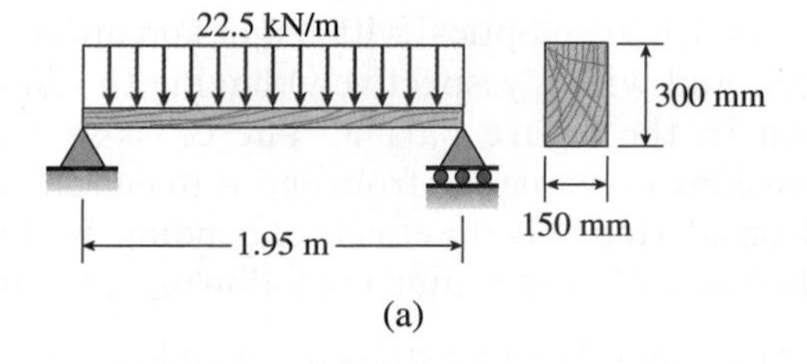

(a)

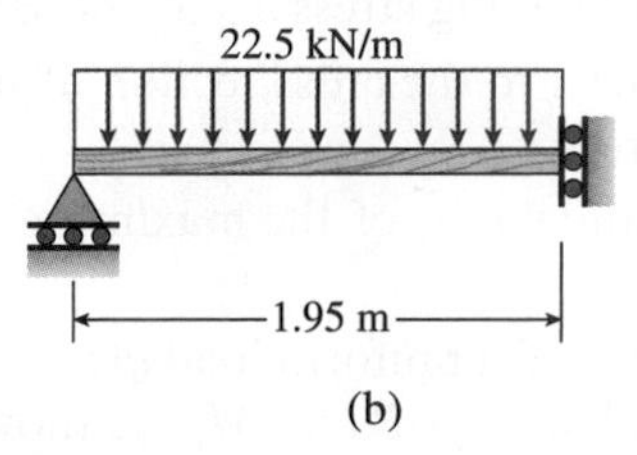

(b)

PROBLEM 5.8-2

(see figure) carrying a uniform load of 22.5 kN/m (which includes the weight of the beam) if the length is 1.95 m and the cross section is rectangular with width 150 mm and height 300 mm, and the beam is either (a) simply supported as in the figure part a, or b has a sliding support at right as in the figure part b.

5.8-3 A simply supported wood beam is subjected to uniformly distributed load q. The width of the beam is 150 mm and the height is 200 mm. Determine the normal stress and the shear stress at point C. Show these stresses on a sketch of a stress element at point C.

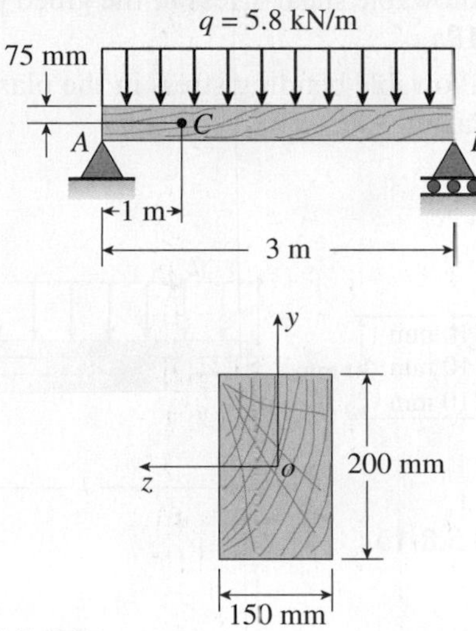

PROBLEM 5.8-3

5.8-4 A simply supported wood beam with overhang is subjected to uniformly distributed load q. The beam has a rectangular cross section with width $b = 200$ mm and height $h = 250$ mm. Determine the maximum permissible value q if the allowable bending stress is $\sigma_{all} = 11$ MPa, and the allowable shear stress is $\tau_{all} = 1.2$ MPa.

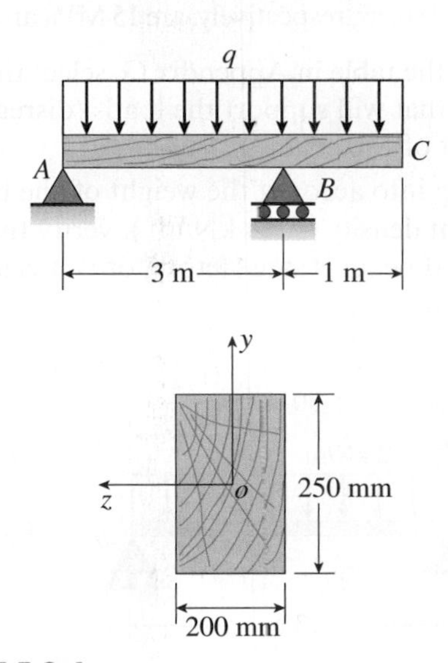

PROBLEM 5.8-4

5.8-5 Two wood beams, each of rectangular cross section (100 mm × 90 mm, actual dimensions), are glued together to form a solid beam with dimensions 200 mm × 90 mm (see figure). The beam is simply supported with a span of 2.5 m.

(a) What is the maximum moment M_{max} that may be applied at the left support if the allowable shear stress in the glued joint is 1.4 MPa? (Include the effects of the beam's own weight, assuming that the wood weighs 5.4 kN/m^3.)

(b) Repeat part (a) if M_{max} is based on allowable bending stress of 17.25 MPa.

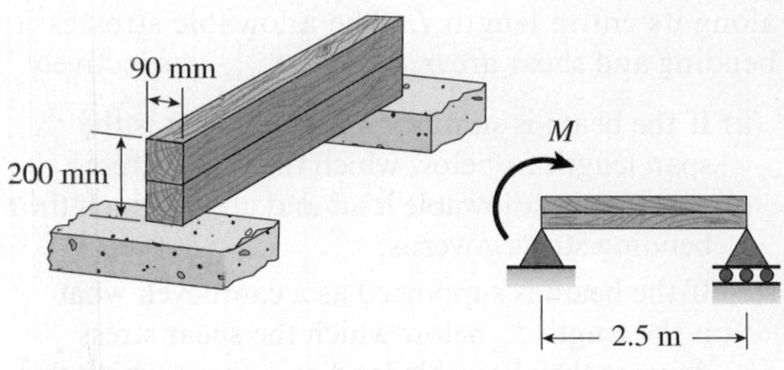

PROBLEM 5.8-5

5.8-6 A cantilever beam of length $L = 2$ m supports a load $P = 8.0$ kN (see figure). The beam is made of wood with cross-sectional dimensions 120 mm × 200 mm.

Calculate the shear stresses due to the load P at points located 25 mm, 50 mm, 75 mm, and 100 mm from the top surface of the beam. From these results, plot a graph showing the distribution of shear stresses from top to bottom of the beam.

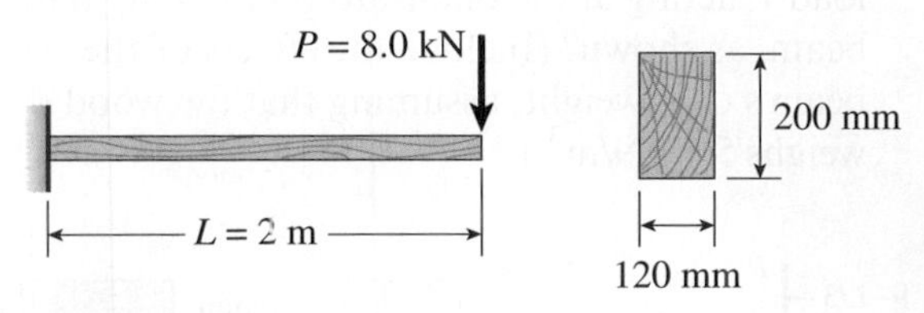

PROBLEM 5.8-6

5.8-7 A steel beam of length $L = 400$ mm and cross-sectional dimensions $b = 12$ mm and $h = 50$ mm (see figure) supports a uniform load of intensity $q = 45$ kN/m, which includes the weight of the beam.

Calculate the shear stresses in the beam (at the cross section of maximum shear force) at points located 6.25 mm, 12.5 mm, 18.75 mm, and 25 mm from the top surface of the beam. From these calculations, plot a graph showing the distribution of shear stresses from top to bottom of the beam.

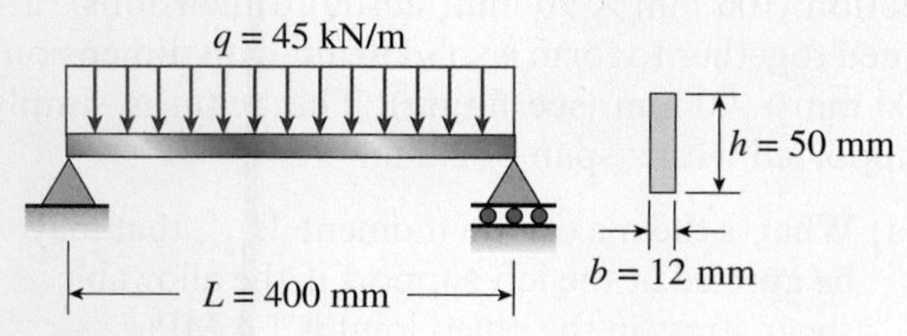

PROBLEM 5.8-7

Representative Problems

5.8-8 A beam of rectangular cross section (width b and height h) supports a uniformly distributed load along its entire length L. The allowable stresses in bending and shear are σ_{allow} and τ_{allow}, respectively.

(a) If the beam is simply supported, what is the span length L_0 below which the shear stress governs the allowable load and above which the bending stress governs?

(b) If the beam is supported as a cantilever, what is the length L_0 below which the shear stress governs the allowable load and above which the bending stress governs?

5.8-9 A laminated wood beam on simple supports (figure part a) is built up by gluing together four 50 mm × 100 mm boards (actual dimensions) to form a solid beam 100 mm × 200 mm in cross section, as shown in the figure part b. The allowable shear stress in the glued joints is 425 kPa, the allowable shear stress in the wood is 1.2 MPa, and the allowable bending stress in the wood is 11.4 MPa.

(a) If the beam is 3.6 m long, what is the allowable load P acting at the one-third point along the beam, as shown? (Include the effects of the beam's own weight, assuming that the wood weighs 5.5 kN/m^3.)

(b) Repeat part (a) if the beam is assembled by gluing together two 75 mm × 100 mm boards and a 50 mm × 100 mm board (see figure part c).

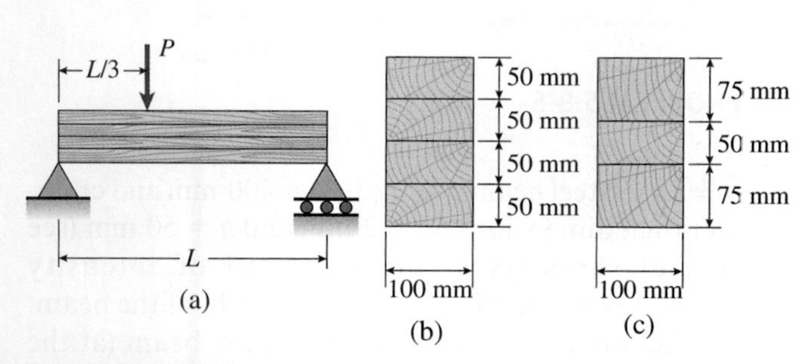

PROBLEM 5.8-9

5.8-10 A laminated plastic beam of square cross section is built up by gluing together three strips, each 10 mm × 30 mm in cross section (see figure). The beam has a total weight of 3.6 N and is simply supported with span length L = 360 mm.

Considering the weight of the beam (q), calculate the maximum permissible CCW moment M that may be placed at the right support.

(a) The allowable shear stress in the glued joints is 0.3 MPa.

(b) The allowable bending stress in the plastic is 8 MPa.

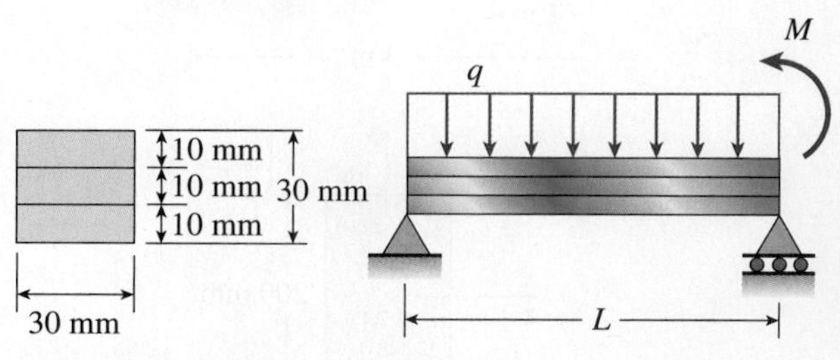

PROBLEM 5.8-10

5.8-11 A wood beam AB on simple supports with span length equal to 3 m is subjected to a uniform load of intensity 2 kN/m acting along the entire length of the beam, a concentrated load of magnitude 30 kN acting at a point 1 m from the right-hand support, and a moment at A of 26 kN · m (see figure). The allowable stresses in bending and shear, respectively, are 15 MPa and 1.1 MPa.

(a) From the table in Appendix G, select the lightest beam that will support the loads (disregard the weight of the beam).

(b) Taking into account the weight of the beam (weight density = 5.4 kN/m^3), verify that the selected beam is satisfactory, or if it is not, select a new beam.

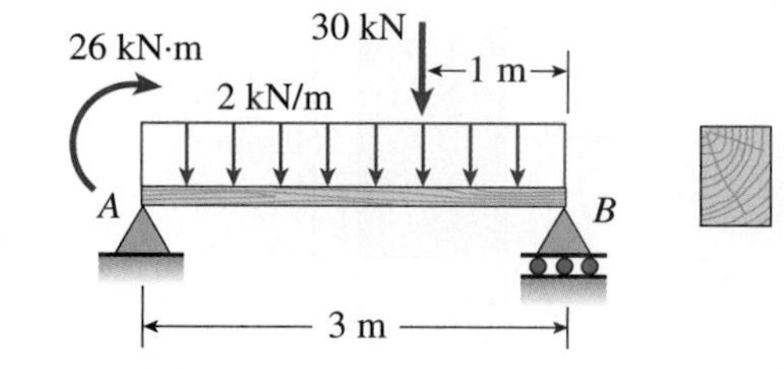

PROBLEM 5.8-11

5.8-12 A simply supported wood beam of rectangular cross section and span length 1.2 m carries a concentrated load P at midspan in addition to its own weight (see figure). The cross section has width 140 mm and height 240 mm. The weight density of the wood is 5.4 kN/m^3.

Calculate the maximum permissible value of the load P if (a) the allowable bending stress is 8.5 MPa and (b) the allowable shear stress is 0.8 MPa.

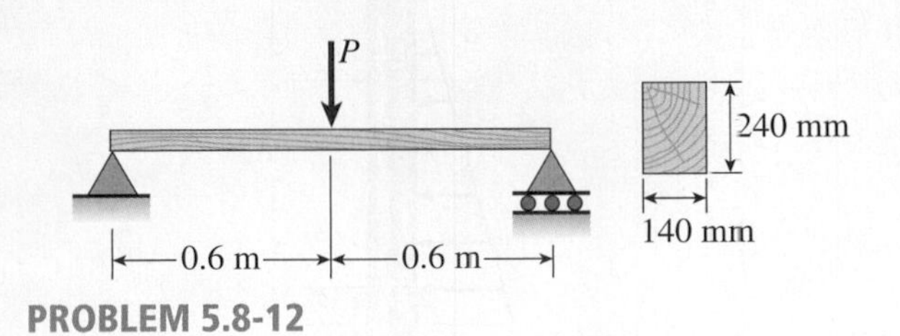

PROBLEM 5.8-12

5.8-13 A square wood platform is 2.4 m × 2.4 m in area and rests on masonry walls (see figure). The deck of the platform is constructed of 50 mm nominal thickness tongue-and-groove planks (actual thickness 47 mm; see Appendix G) supported on two 2.4 m long beams. The beams have 100 mm × 150 mm nominal dimensions (actual dimensions 97 mm × 147 mm).

The planks are designed to support a uniformly distributed load w (kN/m^2) acting over the entire top surface of the platform. The allowable bending stress for the planks is 17 MPa and the allowable shear stress is 0.7 MPa. When analyzing the planks, disregard their weights and assume that their reactions are uniformly distributed over the top surfaces of the supporting beams.

(a) Determine the allowable platform load w_1 (kN/m^2) based upon the bending stress in the planks.

(b) Determine the allowable platform load w_2 (kN/m^2) based upon the shear stress in the planks.

(c) Which of the preceding values becomes the allowable load w_{allow} on the platform?

Hints: Use care in constructing the loading diagram for the planks, noting especially that the reactions are distributed loads instead of concentrated loads. Also, note that the maximum shear forces occur at the inside faces of the supporting beams.

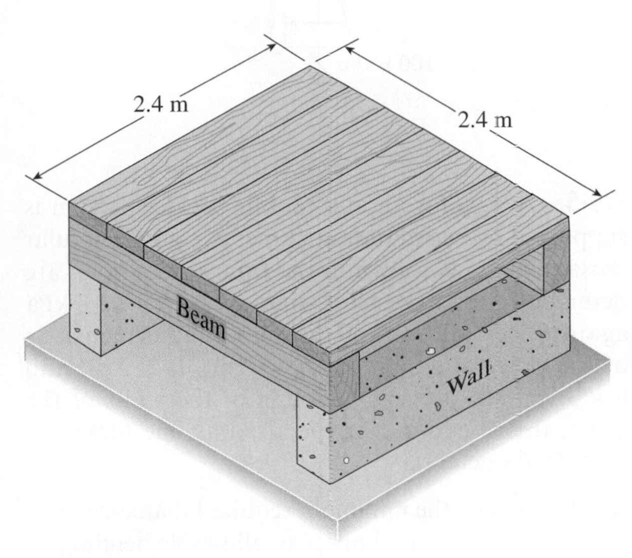

PROBLEM 5.8-13

5.8-14 A wood beam ABC with simple supports at A and B and an overhang BC has height h = 300 mm (see figure). The length of the main span of the beam is L = 3.6 m and the length of the overhang is $L/3$ = 1.2 m. The beam supports a concentrated load $3P$ = 18 kN at the midpoint of the main span and a moment $PL/2$ = 10.8 kN · m at the free end of the overhang. The wood has a weight density γ = 5.5 kN/m^3.

(a) Determine the required width b of the beam based upon an allowable bending stress of 8.2 MPa.

(b) Determine the required width based upon an allowable shear stress of 0.7 MPa.

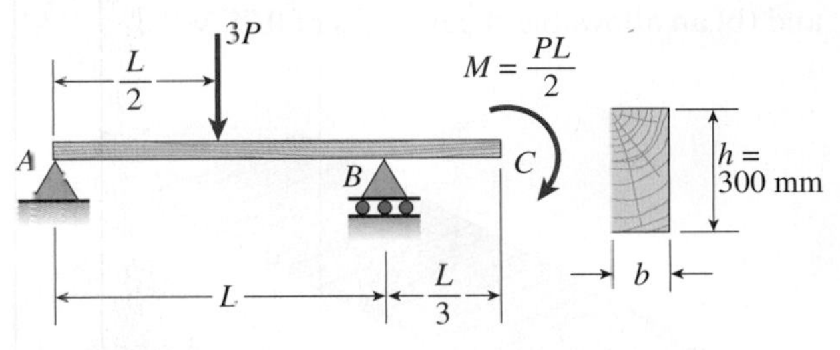

PROBLEM 5.8-14

5.9 Shear Stresses in Beams of Circular Cross Section

Introductory Problems

5.9-1 A wood pole with a solid circular cross section (d = diameter) is subjected to a triangular distributed horizontal force of peak intensity q_0 = 3.75 kN/m (see figure). The length of the pole is L = 2 m, and the allowable stresses in the wood are 13 MPa in bending and 820 kPa in shear.

Determine the minimum required diameter of the pole based upon (a) the allowable bending stress, and (b) the allowable shear stress.

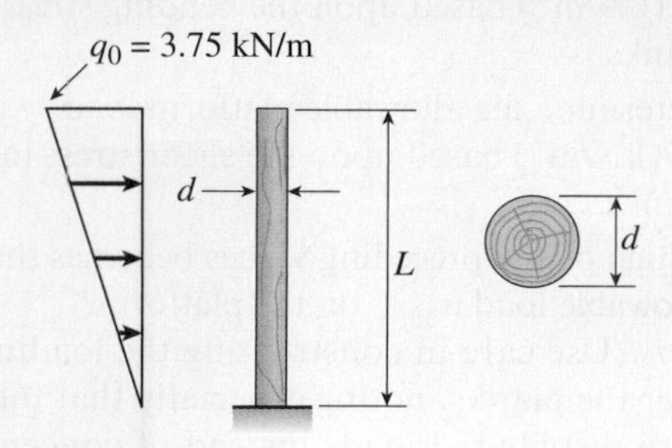

PROBLEM 5.9-1

5.9-2 A simple log bridge in a remote area consists of two parallel logs with planks across them (see figure). The logs are Douglas fir with an average diameter 300 mm. A truck moves slowly across the bridge, which spans 2.5 m. Assume that the weight of the truck is equally distributed between the two logs.

Because the wheelbase of the truck is greater than 2.5 m, only one set of wheels is on the bridge at a time. Thus, the wheel load on one log is equivalent to a concentrated load W acting at any position along the span. In addition, the weight of one log and the planks it supports is equivalent to a uniform load of 850 N/m acting on the log.

Determine the maximum permissible wheel load W based upon (a) an allowable bending stress of 7.0 MPa and (b) an allowable shear stress of 0.75 MPa.

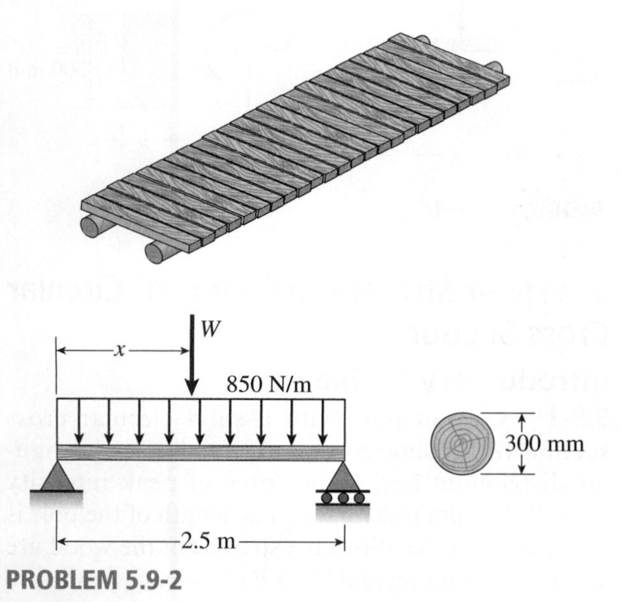

PROBLEM 5.9-2

Representative Problems

5.9-3 A vertical pole consisting of a circular tube of outer diameter 127 mm and inner diameter 115 mm is loaded by a linearly varying distributed force with maximum intensity of q_0. Find the maximum shear stress in the pole.

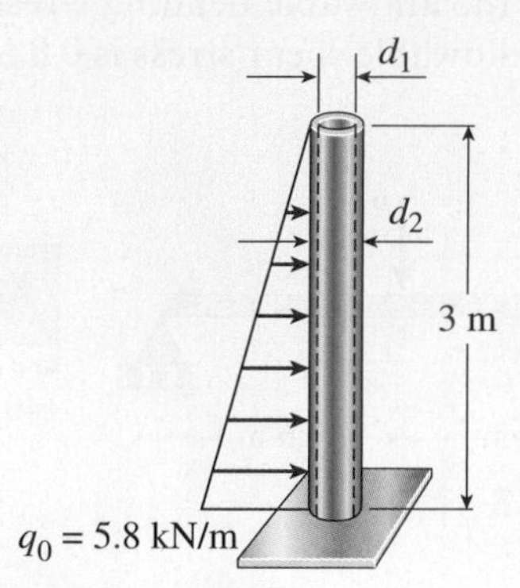

PROBLEM 5.9-3

5.9-4 A circular pole is subjected to linearly varying distributed force with maximum intensity q_0. Calculate the diameter d_0 of the pole if the maximum allowable shear stress for the pole is 75 MPa.

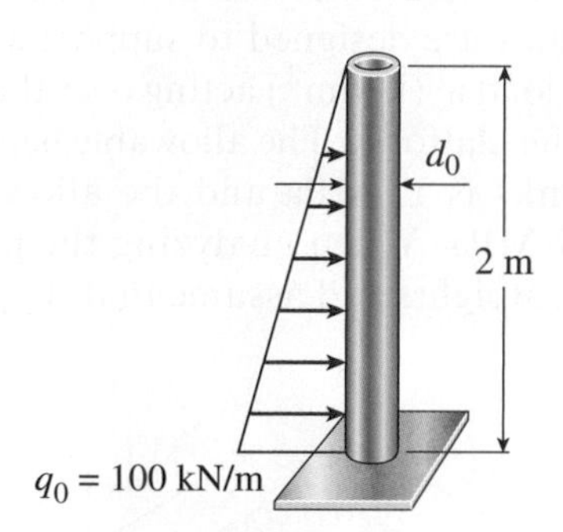

PROBLEM 5.9-4

5.9-5 A sign for an automobile service station is supported by two aluminum poles of hollow circular cross section, as shown in the figure. The poles are being designed to resist a wind pressure of 3.8 kPa against the full area of the sign. The dimensions of the poles and sign are $h_1 = 7$ m, $h_2 = 2$ m, and $b = 3.5$ m. To prevent buckling of the walls of the poles, the thickness t is specified as one-tenth the outside diameter d.

(a) Determine the minimum required diameter of the poles based upon an allowable bending stress of 52 MPa in the aluminum.

(b) Determine the minimum required diameter based upon an allowable shear stress of 14 MPa.

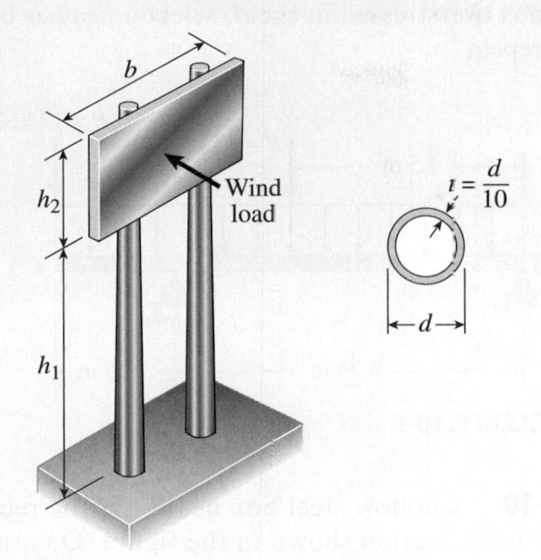

PROBLEM 5.9-5

5.9-6 A steel pipe is subjected to a quadratic distributed load over its height with the peak intensity q_0 at the base (see figure). Assume the following pipe properties and dimensions: height L, outside diameter $d = 200$ mm, and wall thickness $t = 10$ mm. Allowable stresses for flexure and shear are $\sigma_a = 125$ MPa and $\tau_a = 30$ MPa.

(a) If $L = 2.6$ m, find $q_{0,max}$ (kN/m), assuming that allowable flexure and shear stresses in the pipe are not to be exceeded.

(b) If $q_0 = 60$ kN/m, find the maximum height L_{max} (m) of the pipe if the allowable flexure and shear stresses in the pipe are not to be exceeded.

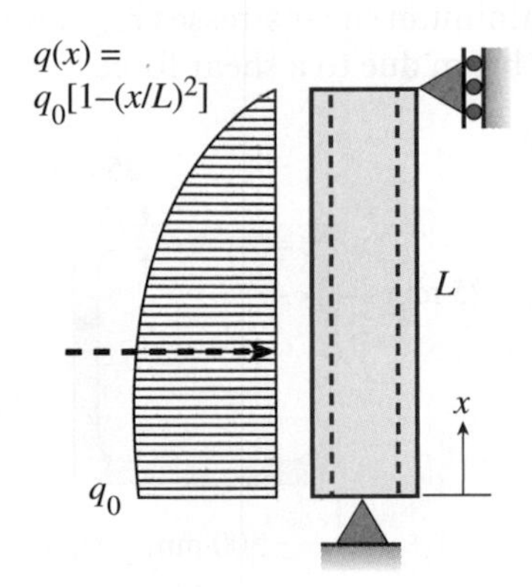

PROBLEM 5.9-6

5.10 Shear Stresses in the Webs of Beams with Flanges

Introductory Problems

5.10-1 through 5.10-6 A wide-flange beam (see figure) is subjected to a shear force V. Using the dimensions of the cross section, calculate the moment of inertia and then determine the following quantities:

(a) The maximum shear stress τ_{max} in the web.

(b) The minimum shear stress τ_{min} in the web.

(c) The average shear stress τ_{aver} (obtained by dividing the shear force by the area of the web) and the ratio τ_{max}/τ_{aver}.

(d) The shear force V_{web} carried in the web and the ratio V_{web}/V.

Note: Disregard the fillets at the junctions of the web and flanges and determine all quantities, including the moment of inertia, by considering the cross section to consist of three rectangles.

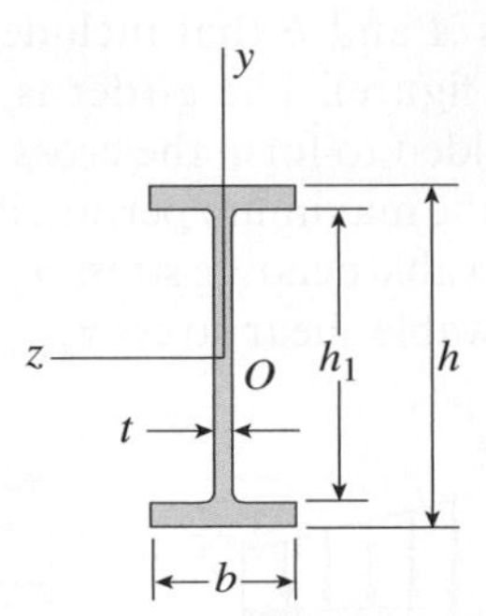

PROBLEMS 5.10-1 through 5.10-6

5.10-1 Dimensions of cross section: $b = 150$ mm, $t = 12$ mm, $h = 300$ mm, $h_1 = 270$ mm, and $V = 130$ kN.

5.10-2 Dimensions of cross section: $b = 180$ mm, $t = 12$ mm, $h = 420$ mm, $h_1 = 380$ mm, and $V = 125$ kN.

5.10-3 Wide-flange shape, HE 160B (see Table F-1, Appendix F); $V = 45$ kN.

5.10-4 Dimensions of cross section: $b = 220$ mm, $t = 12$ mm, $h = 600$ mm, $h_1 = 570$ mm, and $V = 200$ kN.

5.10-5 Wide-flange shape, HE 450A (see Table F-1, Appendix F); $V = 90$ kN.

5.10-6 Dimensions of cross section: $b = 120$ mm, $t = 7$ mm, $h = 350$ mm, $h_1 = 330$ mm, and $V = 60$ kN.

Representative Problems

5.10-7 A cantilever beam AB of length $L = 2$ m supports a trapezoidal distributed load of peak intensity q, and minimum intensity $q/2$, that includes the weight of the beam (see figure). The beam is a steel HE 340B wide-flange shape (see Table F-1, Appendix F).

Calculate the maximum permissible load q based upon (a) an allowable bending stress $\sigma_{allow} = 124$ MPa and (b) an allowable shear stress $\tau_{allow} = 52$ MPa. *Note:* Obtain the moment of inertia and section modulus of the beam from Table F-1.

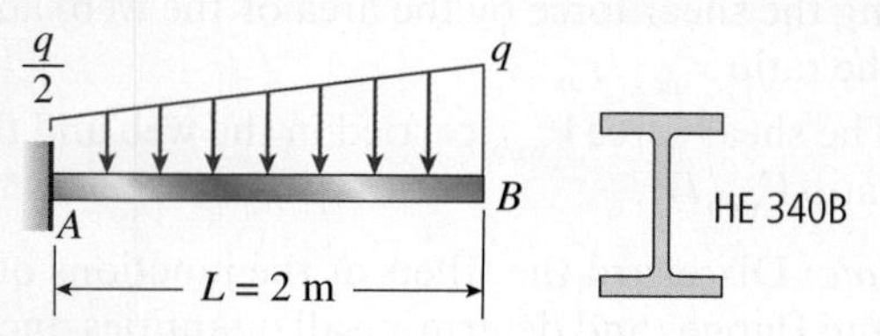

PROBLEM 5.10-7

5.10-8 A bridge girder AB on a simple span of length $L = 14$ m supports a distributed load of maximum intensity q at mid-span and minimum intensity $q/2$ at supports A and B that includes the weight of the girder (see figure). The girder is constructed of three plates welded to form the cross section shown.

Determine the maximum permissible load q based upon (a) an allowable bending stress $\sigma_{allow} = 110$ MPa and (b) an allowable shear stress $\tau_{allow} = 50$ MPa.

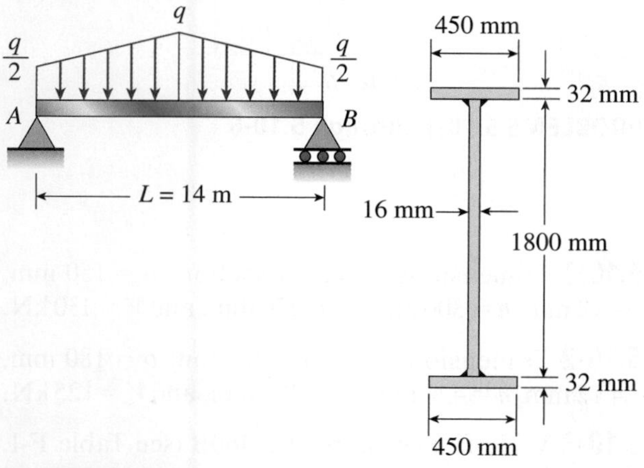

PROBLEM 5.10-8

5.10-9 A simple beam with an overhang supports a uniform load of intensity $q = 17.5$ kN/m and a concentrated $P = 13$ kN load at 2.5 m to the right of A and also at C (see figure). The uniform load includes an allowance for the weight of the beam. The allowable stresses in bending and shear are 124 MPa and 76 MPa, respectively.

Select from Table F-2, Appendix F, the lightest I-beam (IPN shape) that will support the given loads.

Hint: Select a beam based upon the bending stress and then calculate the maximum shear stress. If the beam is overstressed in shear, select a heavier beam and repeat.

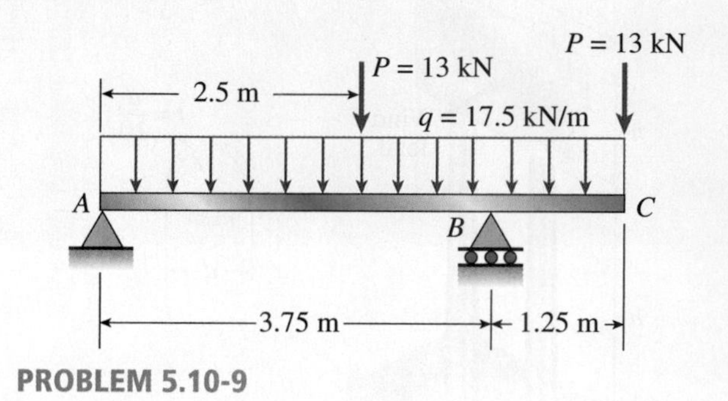

PROBLEM 5.10-9

5.10-10 A hollow steel box beam has the rectangular cross section shown in the figure. Determine the maximum allowable shear force V that may act on the beam if the allowable shear stress is 36 MPa.

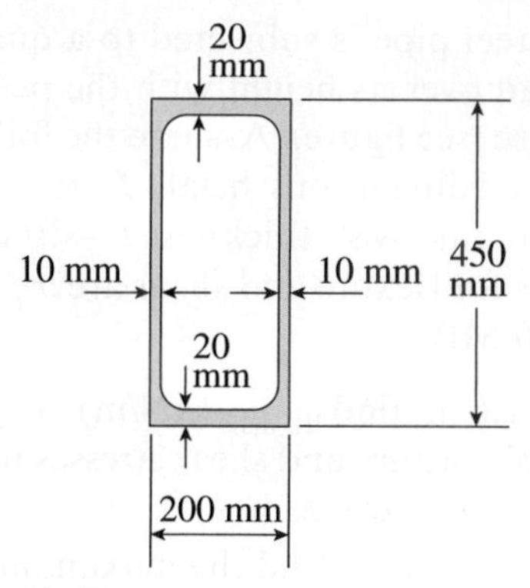

PROBLEM 5.10-10

5.10-11 A hollow aluminum box beam has the square cross section shown in the figure. Calculate the maximum and minimum shear stresses τ_{max} and τ_{min} in the webs of the beam due to a shear force $V = 125$ kN.

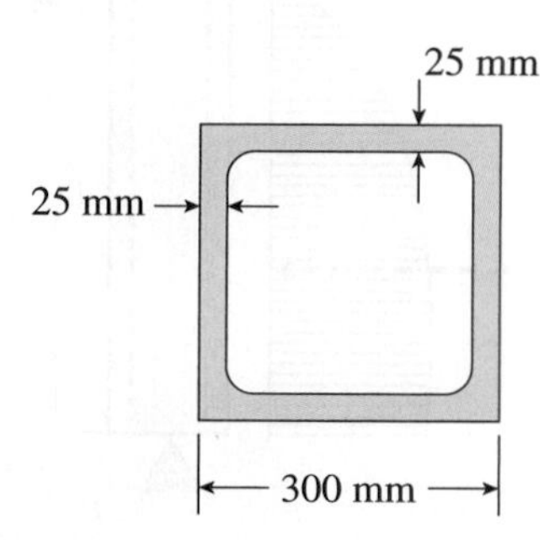

PROBLEM 5.10-11

5.10-12 The T-beam shown in the figure has cross-sectional dimensions: $b = 210$ mm, $t = 16$ mm, $h = 300$ mm, and $h_1 = 280$ mm. The beam is subjected to a shear force $V = 68$ kN.

Determine the maximum shear stress τ_{max} in the web of the beam.

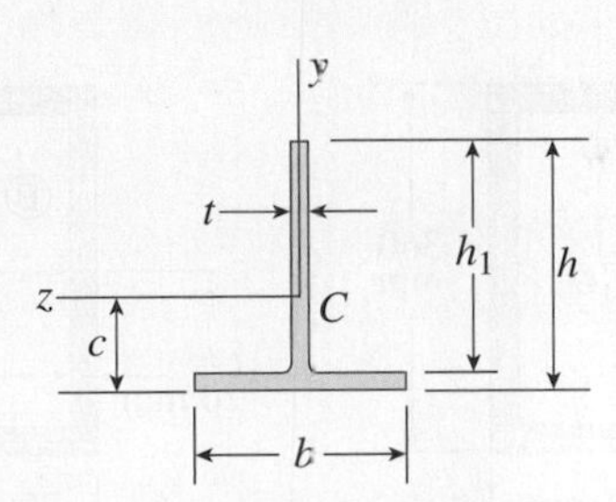

PROBLEMS 5.10-12 and 5.10-13

5.10-13 Calculate the maximum shear stress τ_{max} in the web of the T-beam shown in the figure which is one half of an HE 450A (see Appendix F-1). Assume the shear force $V = 24$ kN.

5.11 Built-Up Beams and Shear Flow

Introductory Problems

5.11-1 A prefabricated wood I-beam serving as a floor joist has the cross section shown in the figure. The allowable load in shear for the glued joints between the web and the flanges is 12 kN/m in the longitudinal direction.

Determine the maximum allowable shear force V_{max} for the beam.

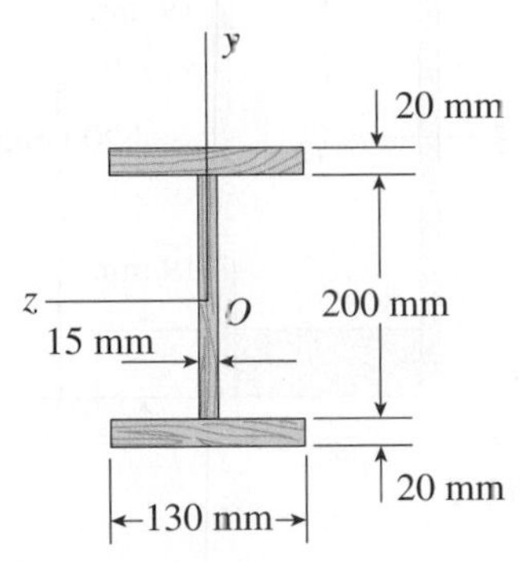

PROBLEM 5.11-1

5.11-2 A welded steel girder having the cross section shown in the figure is fabricated of two 300 mm × 25 mm flange plates and a 800 mm × 16 mm web plate. The plates are joined by four fillet welds that run continuously for the length of the girder. Each weld has an allowable load in shear of 920 kN/m.

Calculate the maximum allowable shear force V_{max} for the girder.

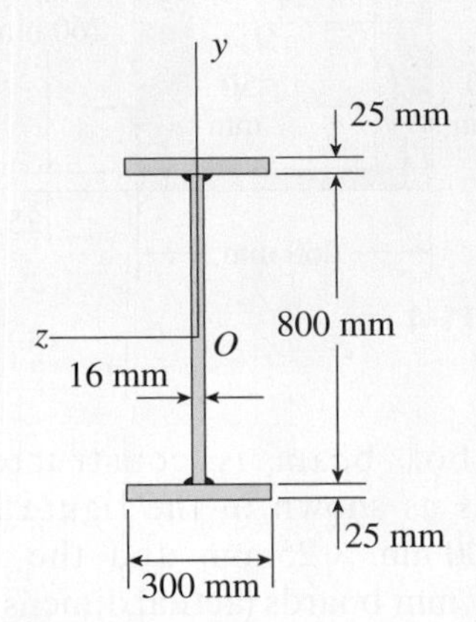

PROBLEM 5.11-2

5.11-3 A welded steel girder having the cross section shown in the figure is fabricated of two 450 mm × 24 mm flange plates and a 1.6 mm × 10 mm web plate. The plates are joined by four longitudinal fillet welds that run continuously throughout the length of the girder.

If the girder is subjected to a shear force of 1300 kN, what force F (per meter of length of weld) must be resisted by each weld?

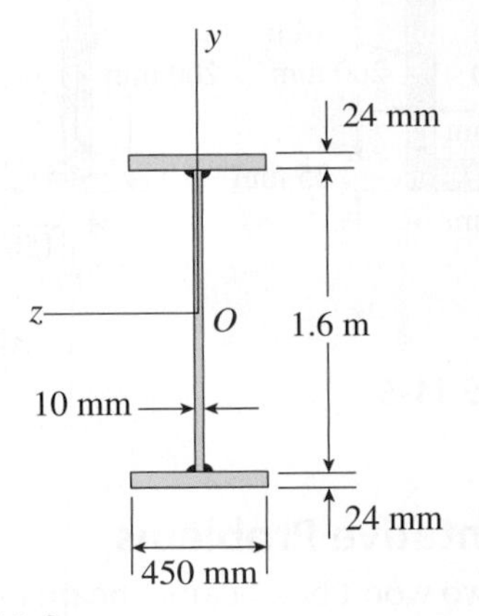

PROBLEM 5.11-3

5.11-4 A wood box beam is constructed of two 260 mm × 50 mm boards and two 260 mm × 25 mm boards (see figure). The boards are nailed at a longitudinal spacing $s = 100$ mm.

If each nail has an allowable shear force $F = 1200$ N, what is the maximum allowable shear force V_{max}?

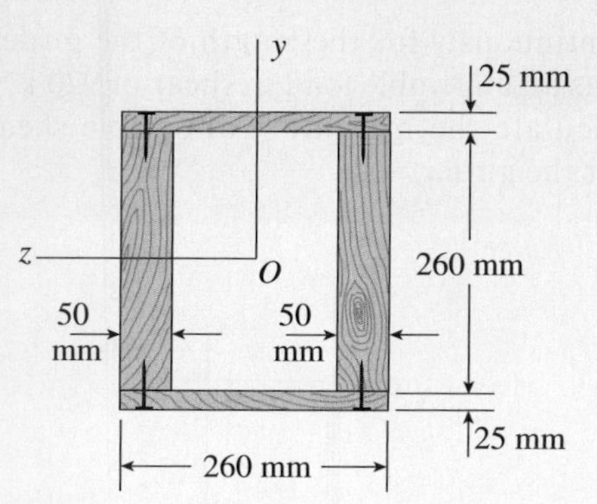

PROBLEM 5.11-4

5.11-5 A box beam is constructed of four wood boards as shown in the figure part a. The webs are 200 mm × 25 mm and the flanges are 150 mm × 25 mm boards (actual dimensions), joined by screws for which the allowable load in shear is $F = 1.1$ kN per screw.

(a) Calculate the maximum permissible longitudinal spacing s_{max} of the screws if the shear force V is 5.3 kN.

(b) Repeat part (a) if the flanges are attached to the webs using a *horizontal arrangement* of screws as shown in the figure part b.

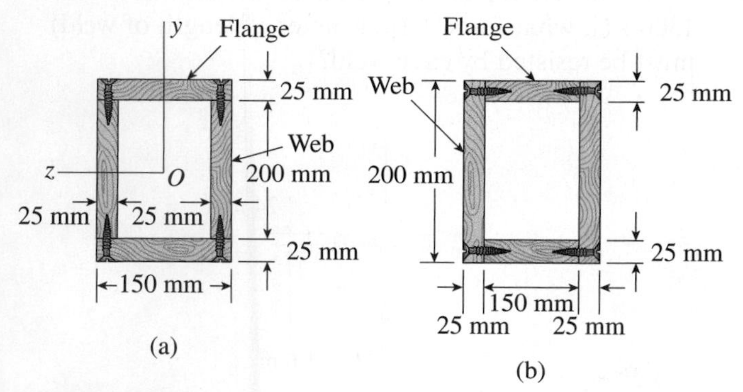

PROBLEM 5.11-5

Representative Problems

5.11-6 Two wood box beams (beams A and B) have the same outside dimensions (200 mm × 360 mm) and the same thickness ($t = 20$ mm) throughout, as shown in the figure. Both beams are formed by nailing, with each nail having an allowable shear load of 250 N. The beams are designed for a shear force $V = 3.2$ kN.

(a) What is the maximum longitudinal spacing s_A for the nails in beam A?

(b) What is the maximum longitudinal spacing s_B for the nails in beam B?

(c) Which beam is more efficient in resisting the shear force?

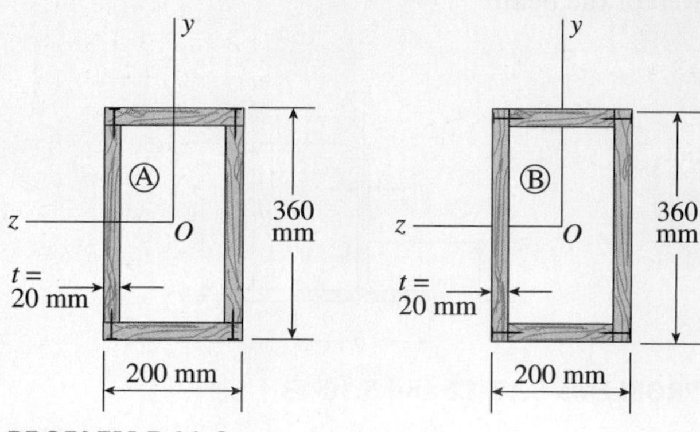

PROBLEM 5.11-6

5.11-7 A hollow wood beam with plywood webs has the cross-sectional dimensions shown in the figure. The plywood is attached to the flanges by means of small nails. Each nail has an allowable load in shear of 130 N.

Find the maximum allowable spacing s of the nails at cross sections where the shear force V is equal to (a) 900 N and (b) 1350 N.

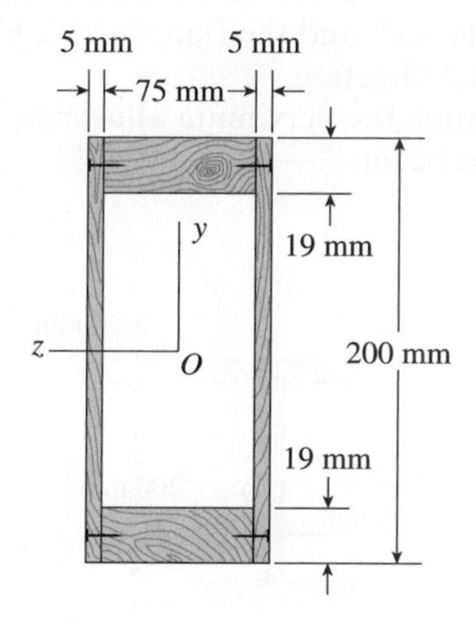

PROBLEM 5.11-7

5.11-8 A beam of a T cross section is formed by nailing together two boards having the dimensions shown in the figure.

If the total shear force V acting on the cross section is 1500 N and each nail may carry 760 N in shear, what is the maximum allowable nail spacing s?

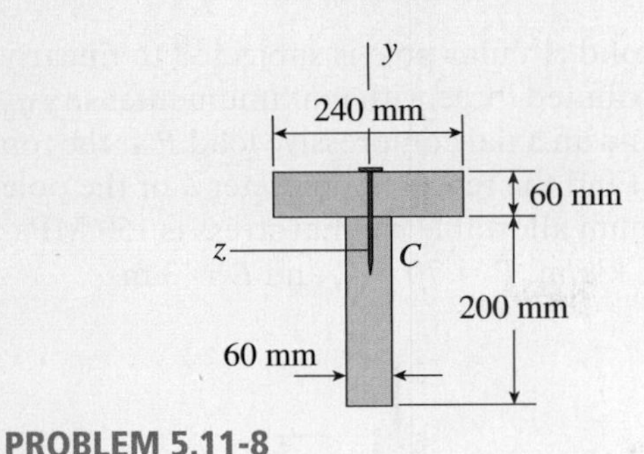

PROBLEM 5.11-8

5.11-9 The T-beam shown in the figure is fabricated by welding together two steel plates. If the allowable load for each weld is 400 kN/m in the longitudinal direction, what is the maximum allowable shear force V?

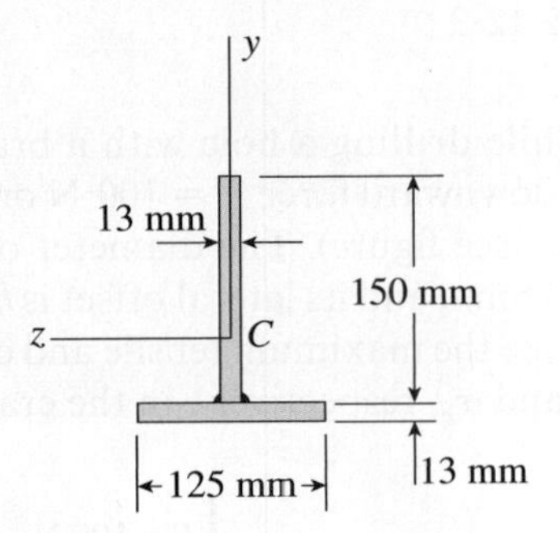

PROBLEM 5.11-9

5.11-10 A steel beam is built up from a HE 180B wide flange beam and two 180 mm × 9 mm cover plates (see figure). The allowable load in shear on each bolt is 9.8 kN. What is the required bolt spacing s in the longitudinal direction if the shear force $V = 110$ kN *Note:* Obtain the dimensions and moment of inertia of the HE shape from Table F-1.

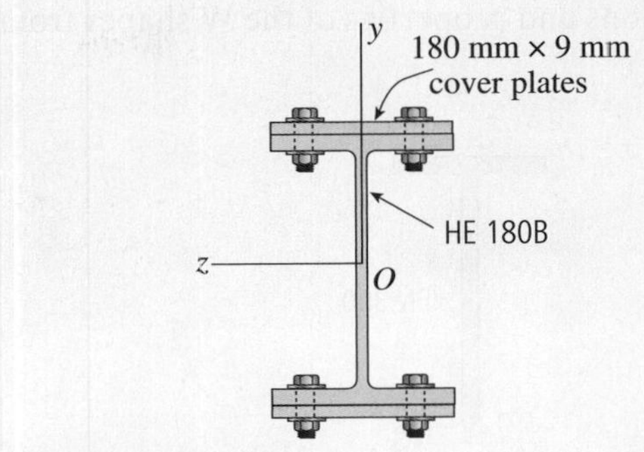

PROBLEM 5.11-10

5.11-11 The three beams shown have approximately the same cross-sectional area. Beam 1 is an IPN 400 with flange plates; beam 2 consists of a web plate with four angles; and beam 3 is constructed of 2 UPN shapes with flange plates.

(a) Which design has the largest moment capacity?
(b) Which has the largest shear capacity?
(c) Which is the most economical in bending?
(d) Which is the most economical in shear?

Assume allowable stress values are: $\sigma_a = 120$ MPa and $\tau_a = 75$ MPa. The most economical beam is that having the largest capacity-to-weight ratio. Neglect fabrication costs in answering parts (c) and (d) above. *Note:* Obtain the dimensions and properties of all rolled shapes from tables in Appendix F.

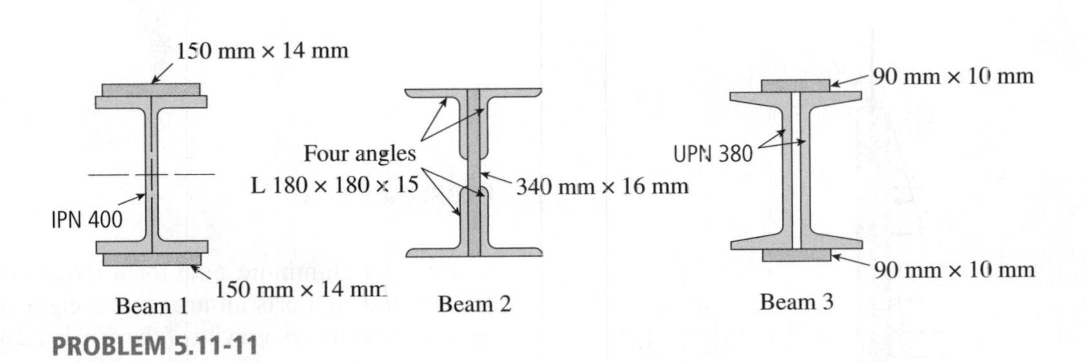

PROBLEM 5.11-11

5.11-12 Two IPN 300 steel beams are bolted together to form a built-up beam as shown in the figure. What is the maximum permissible bolt spacing s if the shear force $V = 80$ kN and the allowable load in shear on each bolt is $F = 13.5$ kN. *Note:* Obtain the dimensions and properties of the W shapes from Table F-2.

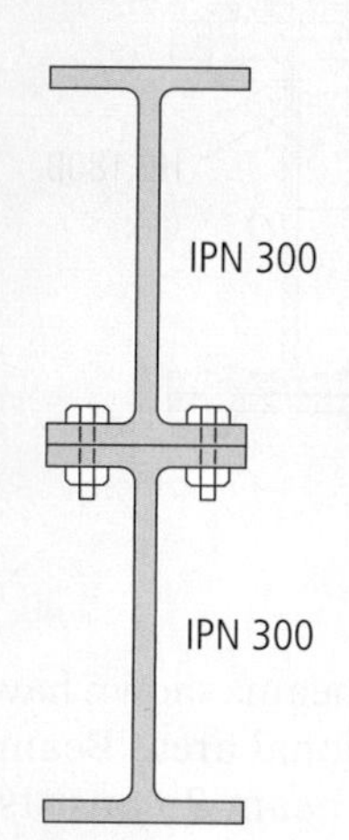

PROBLEM 5.11-12

5.12 Beams with Axial Loads

When solving the problems for Section 5.12, assume that the bending moments are not affected by the presence of lateral deflections.

Introductory Problems

5.12-1 A pole is fixed at the base and is subjected to a linearly varying distributed force with maximum intensity of q_0 *and* an axial compressive load $P = 90$ kN at the top (see figure). The pole has a circular cross section with an outer diameter of 127 mm and an inner diameter of 115 mm. Find the *normal* stresses on the surface of the pole at the base at locations A and B.

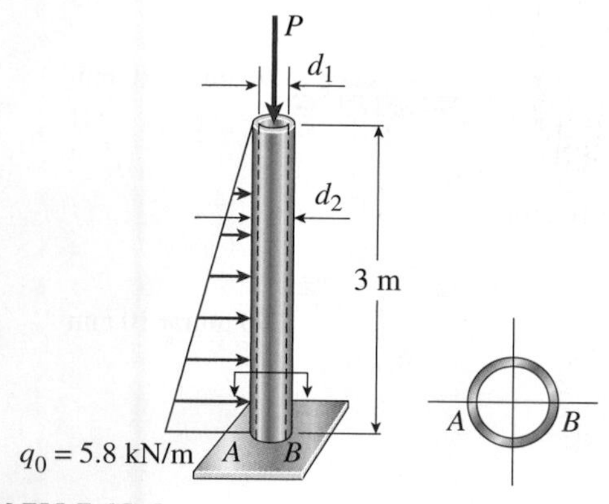

PROBLEM 5.12-1

5.12-2 A solid circular pole is subjected to linearly varying distributed force with maximum intensity q_0 at the base and an axial compressive load P at the top (see figure). Find the required diameter d of the pole if the maximum allowable normal stress is 150 MPa. Let $q_0 = 6.5$ kN/m, $P = 70$ kN, and $L = 3$ m.

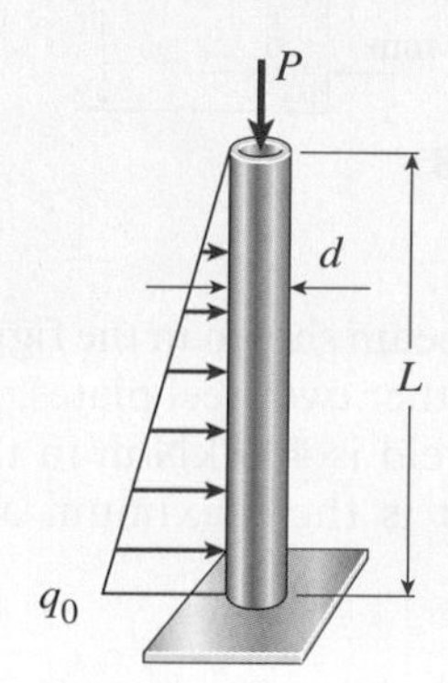

PROBLEM 5.12-2

5.12-3 While drilling a hole with a brace and bit, you exert a downward force $P = 100$ N on the handle of the brace (see figure). The diameter of the crank arm is $d = 11$ mm and its lateral offset is $b = 125$ mm

Determine the maximum tensile and compressive stresses σ_t and σ_c, respectively, in the crank.

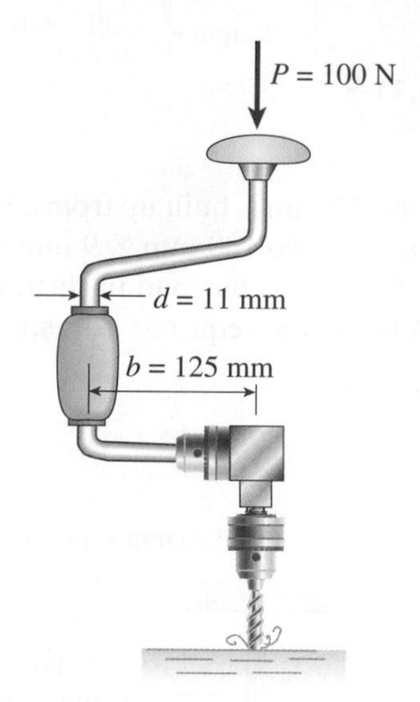

PROBLEM 5.12-3

5.12-4 An aluminum pole for a street light weighs 4600 N and supports an arm that weighs 660 N (see figure). The center of gravity of the arm is 1.2 m from the axis of the pole. A wind force of 300 N also acts in the $(-y)$ direction at 9 m above the base. The outside diameter of the pole (at its base) is 225 mm, and its thickness is 18 mm.

Determine the maximum tensile and compressive stresses σ_t and σ_c, respectively, in the pole (at its base) due to the weights and the wind force.

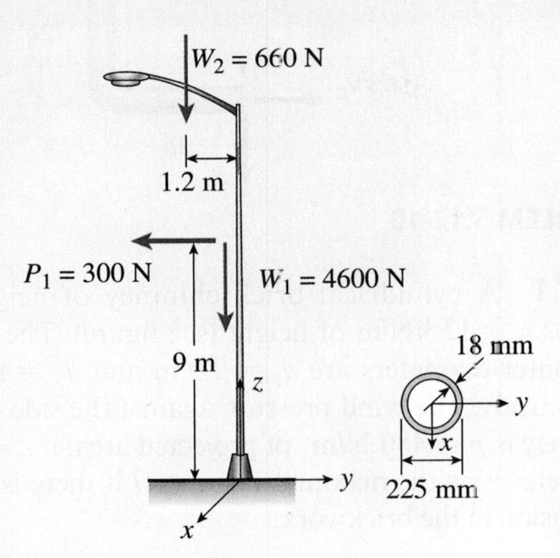

PROBLEM 5.12-4

Representative Problems

5.12-5 A curved bar ABC having a circular axis (radius $r = 300$ mm) is loaded by forces $P = 1.6$ kN (see figure). The cross section of the bar is rectangular with height h and thickness t.

If the allowable tensile stress in the bar is 80 MPa and the height $h = 30$ mm, what is the minimum required thickness t_{min}?

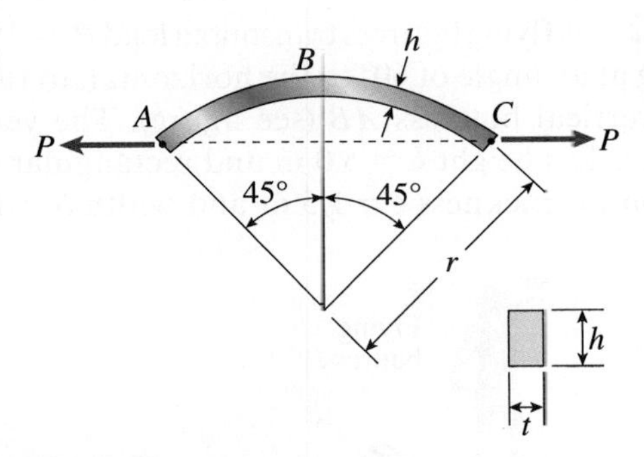

PROBLEM 5.12-5

5.12-6 A rigid frame ABC is formed by welding two steel pipes at B (see figure). Each pipe has cross-sectional area $A = 11.31 \times 10^3$ mm^2, moment of inertia $I = 46.37 \times 10^6$ mm^4, and outside diameter $d = 200$ mm.

Find the maximum tensile and compressive stresses σ_t and σ_c, respectively, in the frame due to the load $P = 8.0$ kN if $L = H = 1.4$ m.

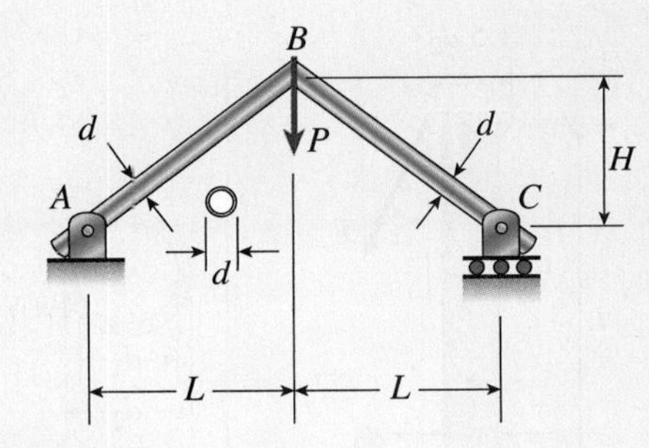

PROBLEM 5.12-6

5.12-7 A palm tree weighing 5 kN is inclined at an angle of 60° (see figure). The weight of the tree may be resolved into two resultant forces: a force $P_1 = 4.5$ kN acting at a point 3.6 m from the base and a force $P_2 = 500$ N acting at the top of the tree, which is 9 m long. The diameter at the base of the tree is 350 mm.

Calculate the maximum tensile and compressive stresses σ_t and σ_c, respectively, at the base of the tree due to its weight.

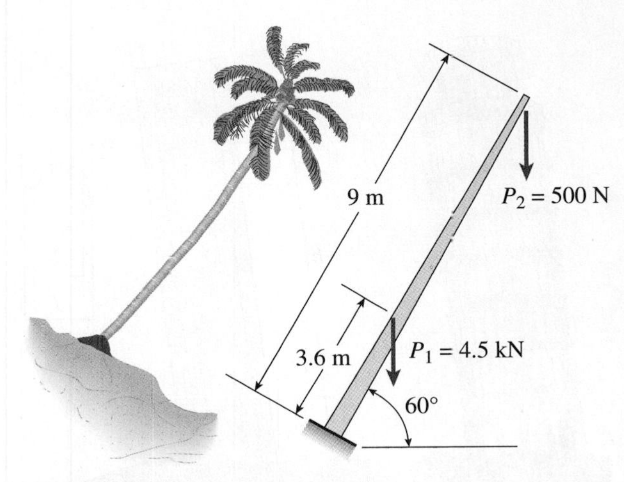

PROBLEM 5.12-7

5.12-8 A vertical pole of aluminum is fixed at the base and pulled at the top by a cable having a tensile force T (see figure). The cable is attached at the outer edge of a stiffened cover plate on top of the pole and makes an angle $\alpha = 20°$ at the point of attachment. The pole has length $L = 2.5$ m and a hollow circular cross section with an outer diameter $d_2 = 280$ mm and inner diameter $d_1 = 220$ mm. The circular cover plate has diameter $1.5d_2$.

Determine the allowable tensile force T_{allow} in the cable if the allowable compressive stress in the aluminum pole is 90 MPa.

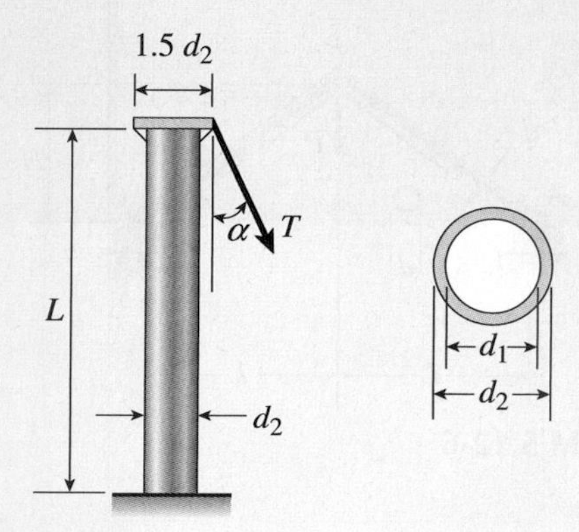

PROBLEM 5.12-8

5.12-9 Because of foundation settlement, a circular tower is leaning at an angle α to the vertical (see figure). The structural core of the tower is a circular cylinder of height h, outer diameter d_2, and inner diameter d_1. For simplicity in the analysis, assume that the weight of the tower is uniformly distributed along the height.

Obtain a formula for the maximum permissible angle α if there is to be no tensile stress in the tower.

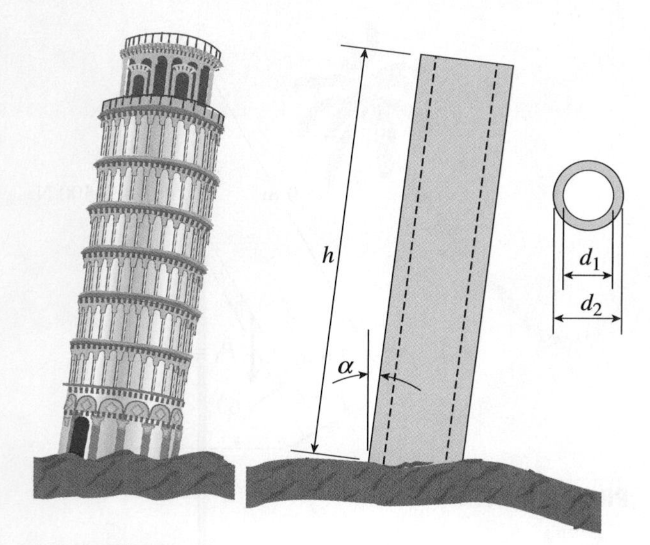

PROBLEM 5.12-9

5.12-10 A steel bracket of solid circular cross section is subjected to two loads, each of which is $P = 4.5$ kN at D (see figure). Let the dimension variable be $b = 240$ mm.

(a) Find the minimum permissible diameter d_{min} of the bracket if the allowable normal stress is 110 MPa.

(b) Repeat part (a), including the weight of the bracket. The weight density of steel is 77.0 kN/m^3.

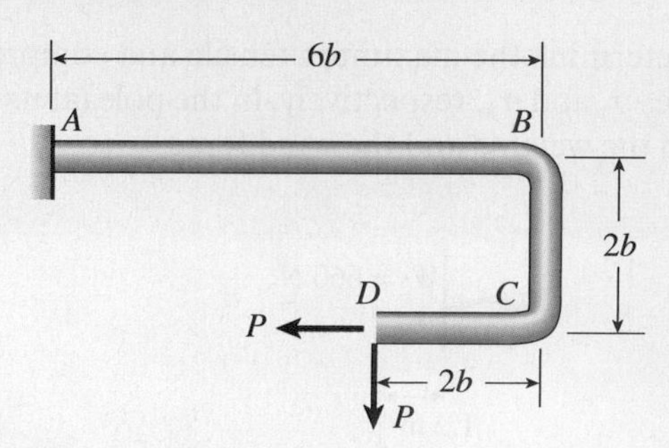

PROBLEM 5.12-10

5.12-11 A cylindrical brick chimney of height H weighs $w = 12$ kN/m of height (see figure). The inner and outer diameters are $d_1 = 0.9$ m and $d_2 = 1.2$ m, respectively. The wind pressure against the side of the chimney is $p = 480$ N/m^2 of projected area.

Determine the maximum height H if there is to be no tension in the brickwork.

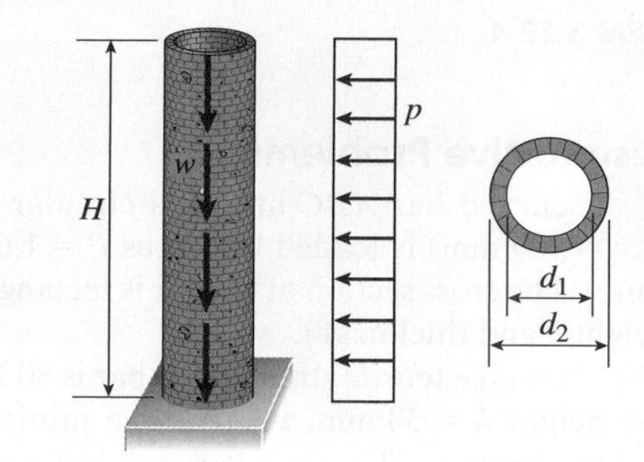

PROBLEM 5.12-11

5.12-12 A flying buttress transmits a load $P = 25$ kN, acting at an angle of 60° to the horizontal, to the top of a vertical buttress AB (see figure). The vertical buttress has height $h = 5.0$ m and rectangular cross section of thickness $t = 1.5$ m and width $b = 1.0$ m

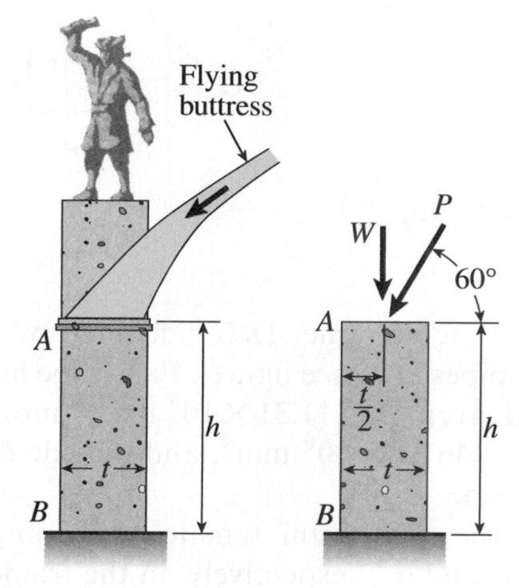

PROBLEM 5.12-12

(perpendicular to the plane of the figure). The stone used in the construction weighs $\gamma = 26\ \text{kN/m}^3$.

What is the required weight W of the pedestal and statue above the vertical buttress (that is, above section A) to avoid any tensile stresses in the vertical buttress?

5.12-13 A plain concrete wall (i.e., a wall with no steel reinforcement) rests on a secure foundation and serves as a small dam on a creek (see figure). The height of the wall is $h = 2$ m and the thickness of the wall is $t = 0.3$ m.

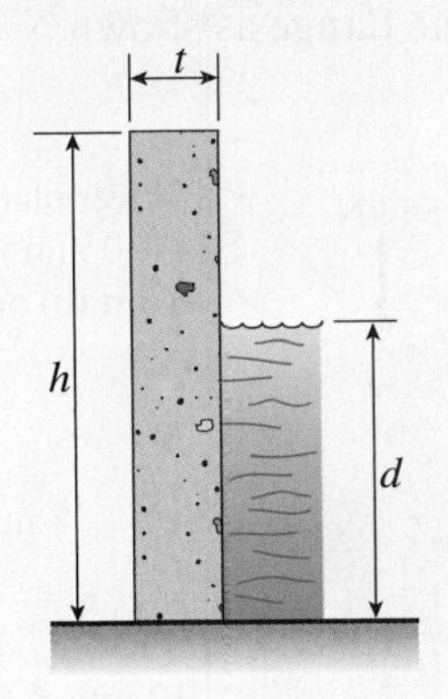

PROBLEM 5.12-13

(a) Determine the maximum tensile and compressive stresses σ_t and σ_c, respectively, at the base of the wall when the water level reaches the top ($d = h$). Assume plain concrete has weight density $\gamma_c = 23\ \text{kN/m}^3$.

(b) Determine the maximum permissible depth d_{max} of the water if there is to be no tension in the concrete.

Eccentric Axial Loads

5.12-14 A circular post, a rectangular post, and a post of cruciform cross section are each compressed by loads that produce a resultant force P acting at the edge of the cross section (see figure). The diameter of the circular post and the depths of the rectangular and cruciform posts are the same.

(a) For what width b of the rectangular post will the maximum tensile stresses be the same in the circular and rectangular posts?

(b) Repeat part (a) for the post with cruciform cross section.

(c) Under the conditions described in parts (a) and (b), which post has the largest compressive stress?

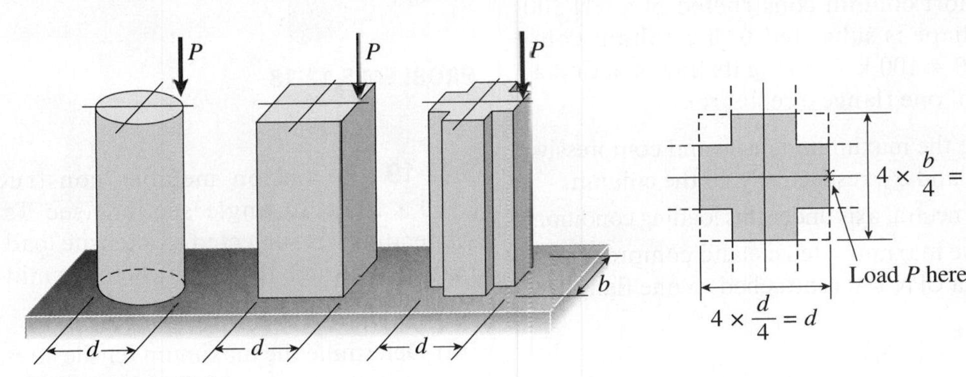

PROBLEM 5.12-14

5.12-15 Two cables, each carrying a tensile force $P = 5.5$ kN, are bolted to a block of steel (see figure). The block has thickness $t = 25$ mm and width $b = 75$ mm.

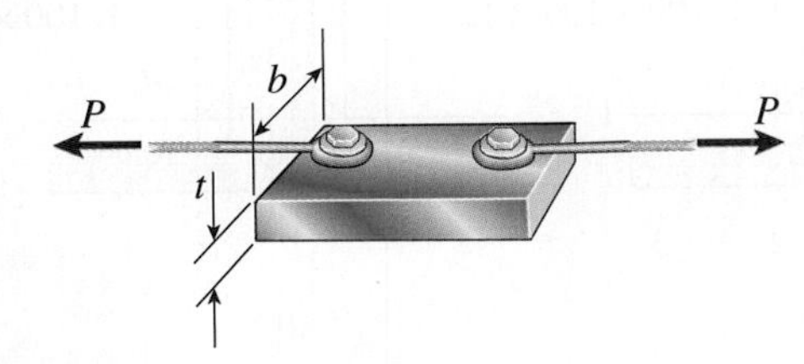

PROBLEM 5.12-15

(a) If the diameter d of the cable is 7 mm, what are the maximum tensile and compressive stresses σ_t and σ_c, respectively, in the block?

(b) If the diameter of the cable is increased (without changing the force P), what happens to the maximum tensile and compressive stresses?

5.12-16 A bar AB supports a load P acting at the centroid of the end cross section (see figure). In the middle region of the bar the cross-sectional area is reduced by removing one-half of the bar.

(a) If the end cross sections of the bar are square with sides of length b, what are the maximum tensile and compressive stresses σ_t and σ_c, respectively, at cross section *mn* within the reduced region?

(b) If the end cross sections are circular with diameter *b*, what are the maximum stresses σ_t and σ_c?

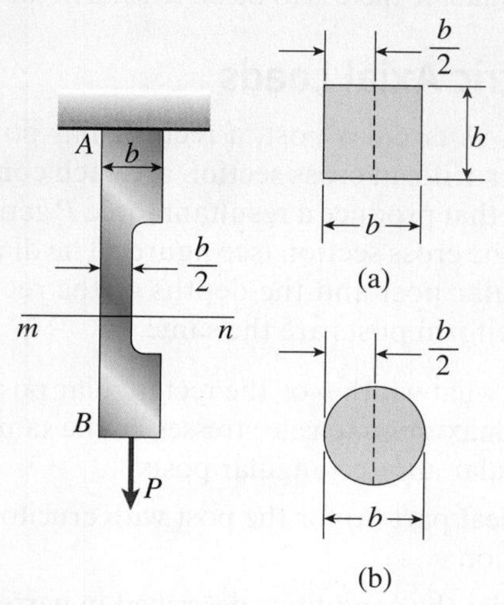

PROBLEM 5.12-16

5.12-17 A short column constructed of a IPN 300 wide-flange shape is subjected to a resultant compressive load $P = 100$ kN having its line of action at the midpoint of one flange (see figure).

(a) Determine the maximum tensile and compressive stresses σ_t and σ_c, respectively, in the column.

(b) Locate the neutral axis under this loading condition.

(c) Recompute maximum tensile and compressive stresses if a UPN 180 is attached to one flange, as shown.

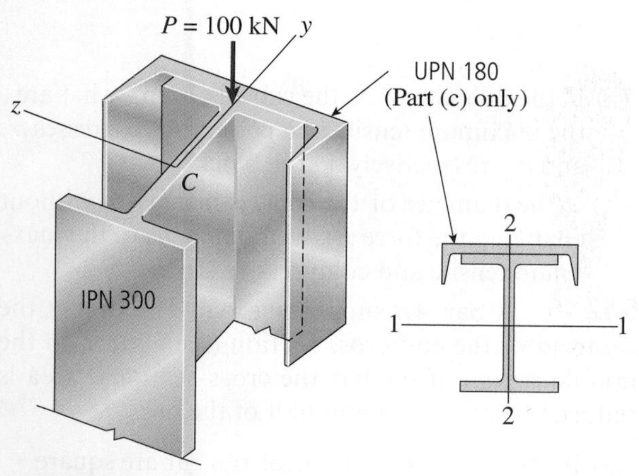

PROBLEM 5.12-17

5.12-18 A short column with a wide-flange shape is subjected to a compressive load that produces a resultant force $P = 55$ kN acting at the midpoint of one flange (see figure).

(a) Determine the maximum tensile and compressive stresses σ_t and σ_c, respectively, in the column.

(b) Locate the neutral axis under this loading condition.

(c) Recompute maximum tensile and compressive stresses if a 120 mm × 10 mm cover plate is added to one flange as shown.

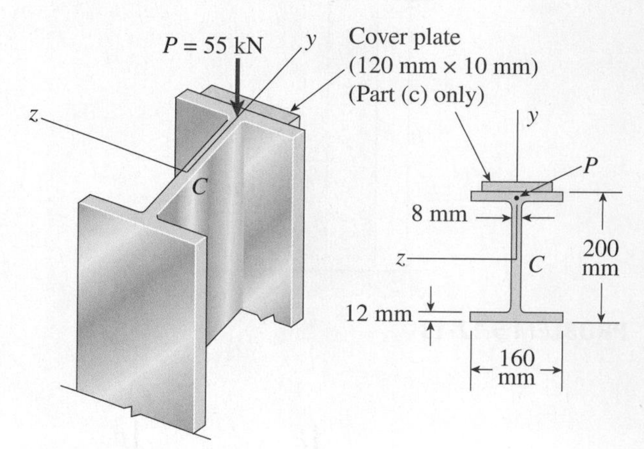

PROBLEM 5.12-18

5.12-19 A tension member constructed of an L 100 × 100 × 12 angle section (see Table F-4 in Appendix F) is subjected to a tensile load $P = 56$ kN that acts through the point where the mid-lines of the legs intersect (see figure part a).

(a) Determine the maximum tensile stress σ_t in the angle section.

(b) Recompute the maximum tensile stress if two angles are used and *P* is applied as shown in the figure part b.

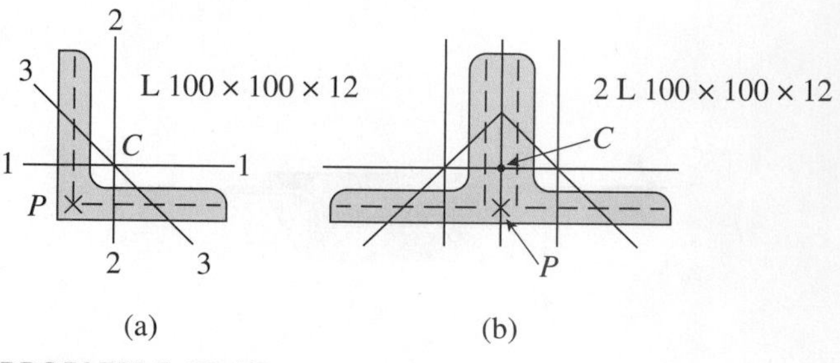

PROBLEM 5.12-19

5.12-20 A short length of a UPN 200 channel is subjected to an axial compressive force P that has its line of action through the midpoint of the web of the channel (see figure part a).

(a) Determine the equation of the neutral axis under this loading condition.
(b) If the allowable stresses in tension and compression are 76 MPa and 52 MPa respectively, find the maximum permissible load P_{max}.
(c) Repeat parts (a) and (b) if two L 90 × 90 × 7 angles are added to the channel as shown in the figure part b.

See Table F-3 in Appendix F for channel properties and Table F-4 for angle properties.

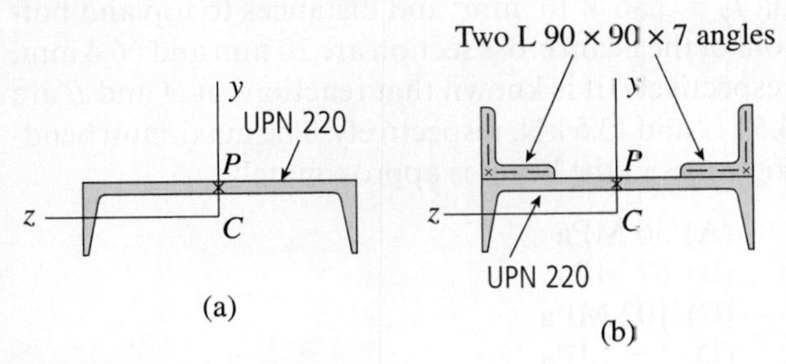

PROBLEM 5.12-20

5.13 Stress Concentrations in Bending

The problems for Section 5.13 are to be solved considering the stress-concentration factors.

5.13-1 The beams shown in the figure are subjected to bending moments $M = 250$ N · m. Each beam has a rectangular cross section with height $h = 40$ mm and width $b = 10$ mm (perpendicular to the plane of the figure).

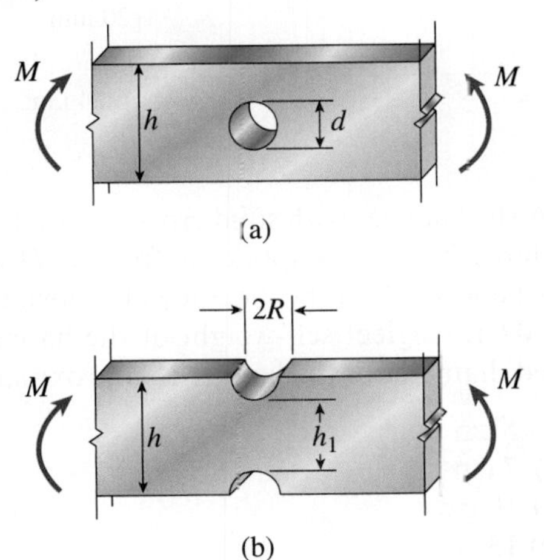

PROBLEMS 5.13-1 through 5.13-4

(a) For the beam with a hole at midheight, determine the maximum stresses for hole diameters $d = 6$, 12, 18, 24 mm.
(b) For the beam with two identical notches (inside height $h_1 = 30$ mm), determine the maximum stresses for notch radii $R = 1.25$, 2.5, 3.75, and 5.0 mm.

5.13-2 The beams shown in the figure are subjected to bending moments $M = 250$ N · m. Each beam has a rectangular cross section with height $h = 44$ mm and width $b = 10$ mm (perpendicular to the plane of the figure).

(a) For the beam with a hole at midheight, determine the maximum stresses for hole diameters $d = 10$, 16, 22, and 28 mm.
(b) For the beam with two identical notches (inside height $h_1 = 40$ mm), determine the maximum stresses for notch radii $R = 2$, 4, 6, and 8 mm.

5.13-3 A rectangular beam with semicircular notches, as shown in part b of the figure, has dimensions $h = 22$ mm and $h_1 = 20$ mm. The maximum allowable bending stress in the metal beam is $\sigma_{max} = 410$ MPa, and the bending moment is $M = 68$ N · m.

Determine the minimum permissible width b_{min} of the beam.

5.13-4 A rectangular beam with semicircular notches, as shown in part b of the figure, has dimensions $h = 120$ mm and $h_1 = 100$ mm. The maximum allowable bending stress in the plastic beam is $\sigma_{max} = 6$ MPa, and the bending moment is $M = 150$ N · m.

Determine the minimum permissible width b_{min} of the beam.

5.13-5 A rectangular beam with notches and a hole (see figure) has dimensions $h = 140$ mm, $h_1 = 128$ mm, and width $b = 40$ mm. The beam is subjected to a bending moment $M = 15$ kN · m, and the maximum allowable bending stress in the material (steel) is $\sigma_{max} = 290$ MPa.

(a) What is the smallest radius R_{min} that should be used in the notches?
(b) What is the diameter d_{max} of the largest hole that should be drilled at the midheight of the beam?

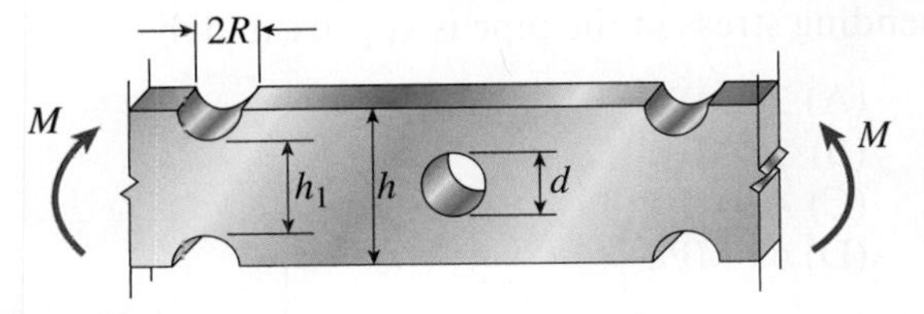

PROBLEM 5.13-5

SOME ADDITIONAL REVIEW PROBLEMS: Chapter 5

R-5.1 A copper wire ($d = 1.5$ mm) is bent around a tube of radius R = 0.6 m. The maximum normal strain in the wire is approximately:

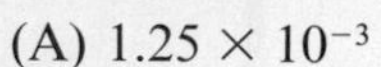

(A) 1.25×10^{-3}
(B) 1.55×10^{-3}
(C) 1.76×10^{-3}
(D) 1.92×10^{-3}

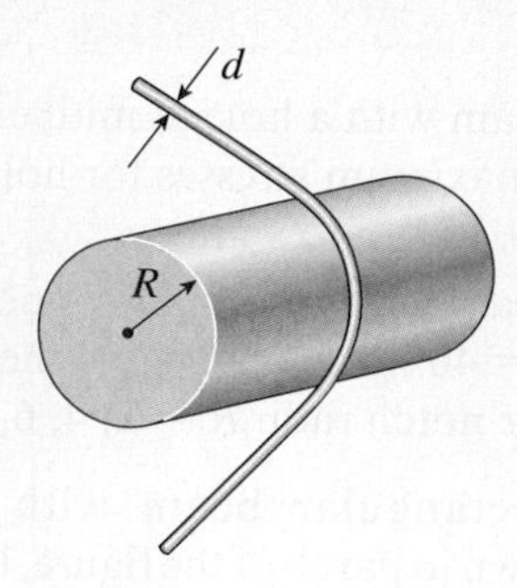

R-5.2 A simply supported wood beam ($L = 5$ m) with rectangular cross section ($b = 200$ mm, $h = 280$ mm) carries uniform load $q = 9.5$ kN/m includes the weight of the beam. The maximum flexural stress is approximately:

(A) 8.7 MPa
(B) 10.1 MPa
(C) 11.4 MPa
(D) 14.3 MPa

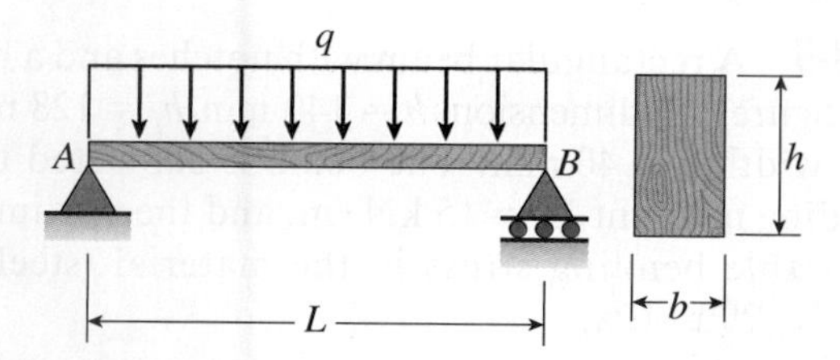

R-5.3 A cast iron pipe ($L = 12$ m, weight density = 72 kN/m³, $d_2 = 100$ mm, and $d_1 = 75$ mm) is lifted by a hoist. The lift points are 4 m apart. The maximum bending stress in the pipe is approximately:

(A) 28 MPa
(B) 33 MPa
(C) 47 MPa
(D) 59 MPa

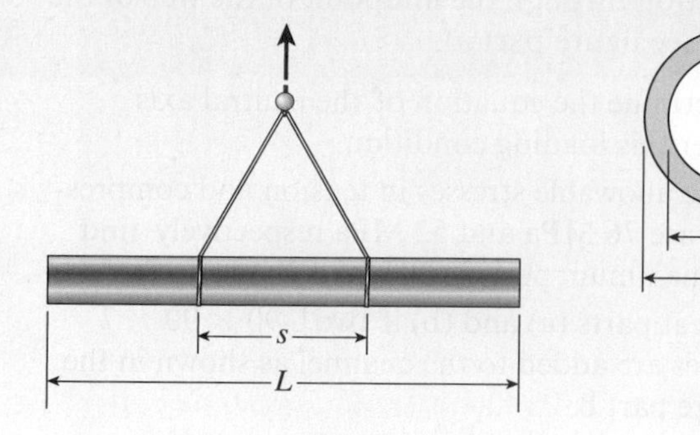

R-5.4 A beam with an overhang is loaded by a uniform load of 3 kN/m over its entire length. Moment of inertia $I_z = 3.36 \times 10^6$ mm⁴ and distances to top and bottom of the beam cross section are 20 mm and 66.4 mm, respectively. It is known that reactions at A and B are 4.5 kN and 13.5 kN, respectively. The maximum bending stress in the beam is approximately:

(A) 36 MPa
(B) 67 MPa
(C) 102 MPa
(D) 119 MPa

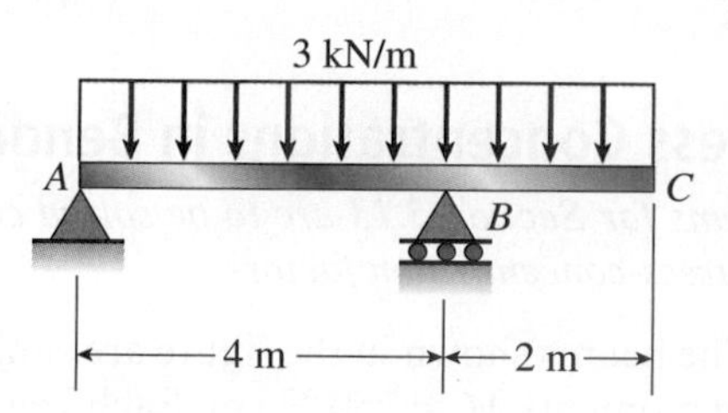

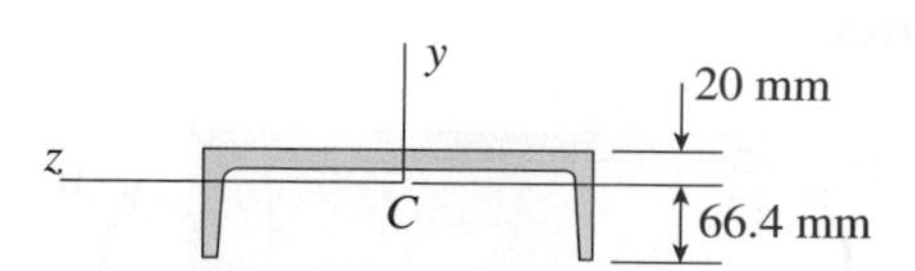

R-5.5 A steel hanger with solid cross section has horizontal force $P = 5.5$ kN applied at free end D. Dimension variable $b = 175$ mm and allowable normal stress is 150 MPa. Neglect self-weight of the hanger. The required diameter of the hanger is approximately:

(A) 5 cm
(B) 7 cm
(C) 10 cm
(D) 13 cm

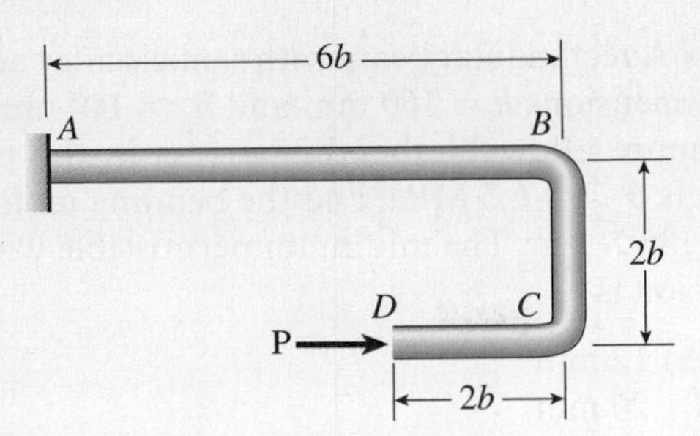

R-5.6 A cantilever wood pole carries force $P = 300$ N applied at its free end, as well as its own weight (weight density $= 6$ kN/m^3). The length of the pole is $L = 0.75$ m and the allowable bending stress is 14 MPa. The required diameter of the pole is approximately:

(A) 4.2 cm
(B) 5.5 cm
(C) 6.1 cm
(D) 8.5 cm

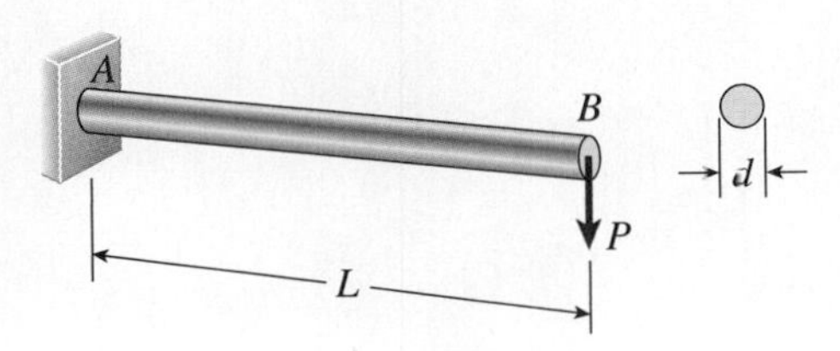

R-5.7 A simply supported steel beam of length $L =$ 1.5 m and rectangular cross section ($h = 75$ mm, $b =$ 20 mm) carries a uniform load of $q = 48$ kN/m that includes its own weight. The maximum transverse shear stress on the cross section at 0.25 m from the left support is approximately:

(A) 20 MPa
(B) 24 MPa
(C) 30 MPa
(D) 36 MPa

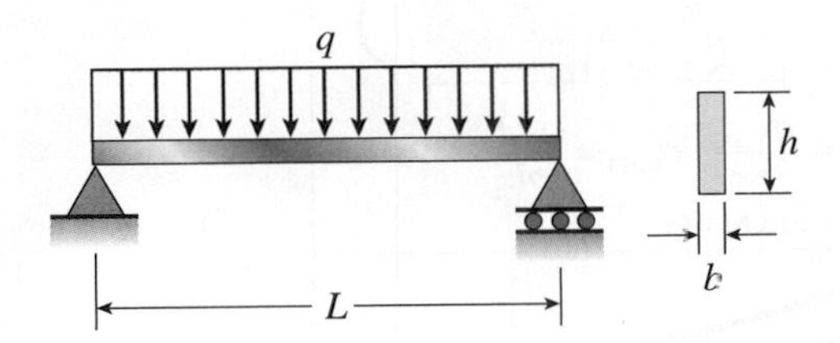

R-5.8 A simply supported laminated beam of length $L = 0.5$ m and square cross section weighs 4.8 N. Three strips are glued together to form the beam, with the allowable shear stress in the glued joint equal to 0.3 MPa. Considering also the weight of the beam, the maximum load P that can be applied at $L/3$ from the left support is approximately:

(A) 240 N
(B) 360 N
(C) 434 N
(D) 510 N

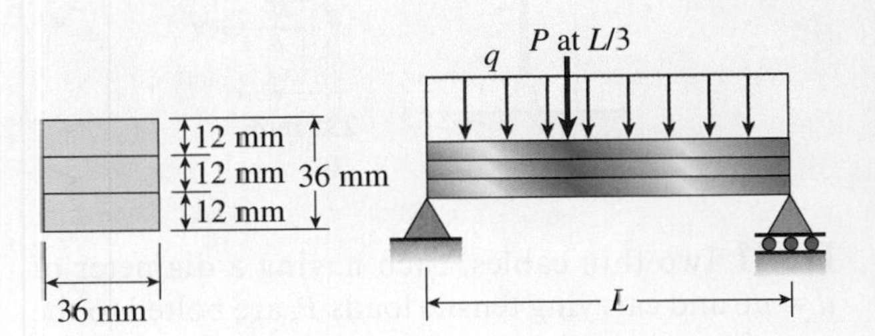

R-5.9 An aluminum cantilever beam of length $L =$ 0.65 m carries a distributed load, which includes its own weight, of intensity $q/2$ at A and q at B. The beam cross section has a width of 50 mm and height of 170 mm. Allowable bending stress is 95 MPa and allowable shear stress is 12 MPa. The permissible value of load intensity q is approximately:

(A) 110 kN/m
(B) 122 kN/m
(C) 130 kN/m
(D) 139 kN/m

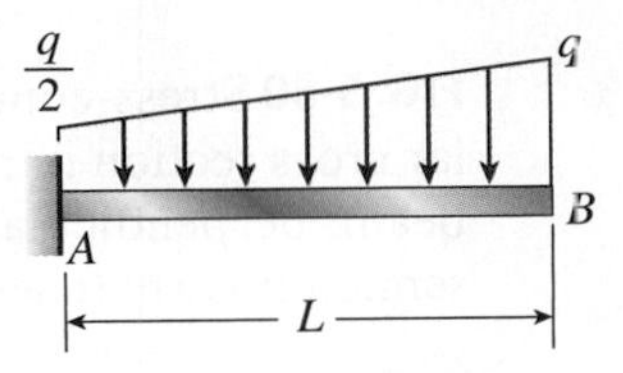

R-5.10 An aluminum light pole weights 4300 N and supports an arm of weight 700 N, with the arm center of gravity at 1.2 m left of the centroidal axis of the pole. A wind force of 1500 N acts to the right at 7.5 m above the base. The pole cross section at the base has an outside diameter of 235 mm and thickness of 20 mm. The maximum compressive stress at the base is approximately:

(A) 16 MPa
(B) 18 MPa
(C) 21 MPa
(D) 24 MPa

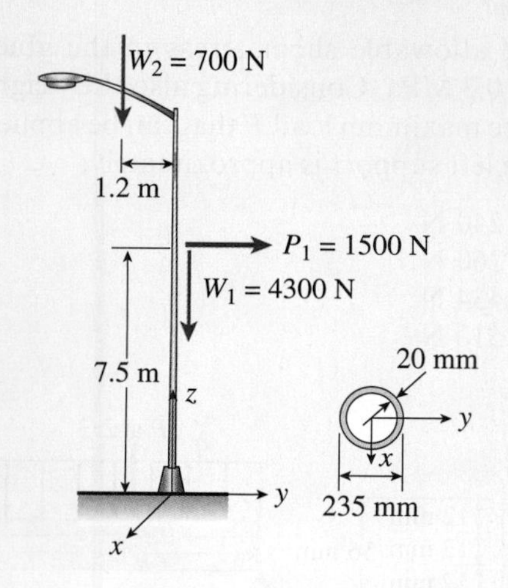

R-5.11 Two thin cables, each having a diameter of $d = t/6$ and carrying tensile loads P, are bolted to the top of a rectangular steel block with cross-sectional dimensions $b \times t$. The ratio of the maximum tensile to compressive stress in the block due to loads P is:

(A) 1.5
(B) 1.8
(C) 2.0
(D) 2.5

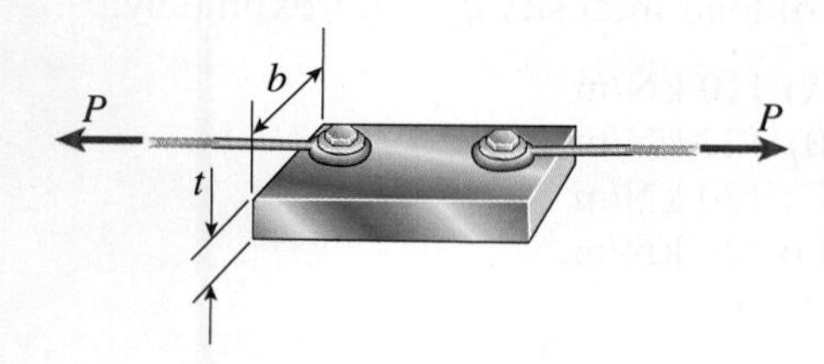

R-5.12 A rectangular beam with semicircular notches has dimensions $h = 160$ mm and $h_1 = 140$ mm. The maximum allowable bending stress in the plastic beam is $\sigma_{max} = 6.5$ MPa, and the bending moment is $M = 185$ N · m. The minimum permissible width of the beam is:

(A) 12 mm
(B) 20 mm
(C) 28 mm
(D) 32 mm

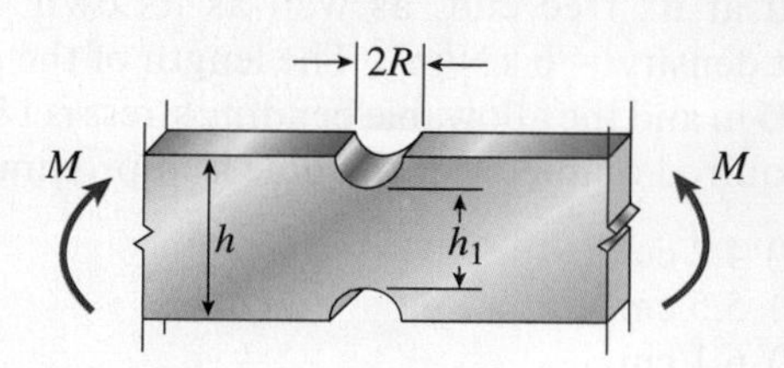

FIG. 5-50 Stress-concentration factor K for a notched beam of rectangular cross section in pure bending (h = height of beam; b = thickness of beam, perpendicular to the plane of the figure). The dashed line is for semicircular notches ($h = h_1 + 2R$).

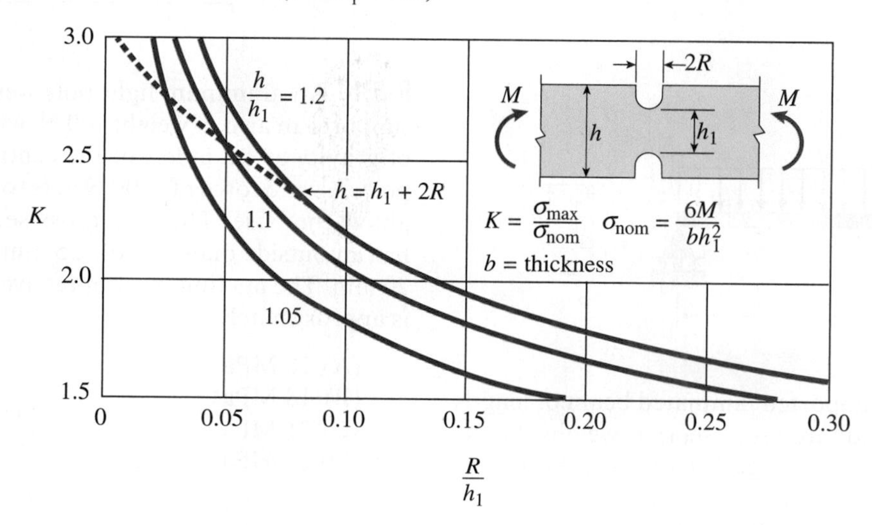

CHAPTER 6

Stresses in Beams (Advanced Topics)

A more advanced theory is required for analysis and design of composite beams, thin-wall beams, and beams with unsymmetric cross sections.

Chapter Objectives

- Develop a modified flexure formula for computation of normal stresses in beams made up of two different materials, referred to as composite beams.
- Study bending of doubly symmetric beams acted on by inclined loads having a line of action through the centroid of the cross section.
- Develop a general procedure for analyzing an unsymmetric beam subjected to any bending moment M resolved into components along the principal centroidal axes of the cross section or a generalized flexure formula for use with the original axes of the cross section.
- Locate the shear center of the cross section: that point through which loads must act so that twisting of the beam about a longitudinal axis can be avoided.
- Find distributions of shear stresses in the elements of the cross sections of a number of beams of thin-walled open section, such as channels, angles, and Z shapes.
- Study the bending of elastoplastic beams in which the normal stresses go beyond the linear elastic range of behavior and define *yield* and *plastic* moments for such beams.

Chapter Outline

***Specialized or Advanced Topics**

6.1 Introduction

Chapter 6 continues the study of the bending of beams by examining several specialized topics. These subjects are based upon the fundamental topics discussed previously in Chapter 5—topics such as curvature, normal stresses in beams (including the flexure formula), and shear stresses in beams. However, beams need not be composed of one material only; and the beams are not required to have a plane of symmetry in which transverse loads must be applied. Finally, the performance will be extended into the inelastic range of behavior for beams made of elastoplastic materials.

Later, two additional subjects of fundamental importance in beam design—deflections of beams and statically indeterminate beams—are discussed in Chapters 9 and 10.

FIGURE 6-1

Examples of composite beams: (a) bimetallic beam, (b) plastic-coated steel pipe, and (c) wood beam reinforced with a steel plate

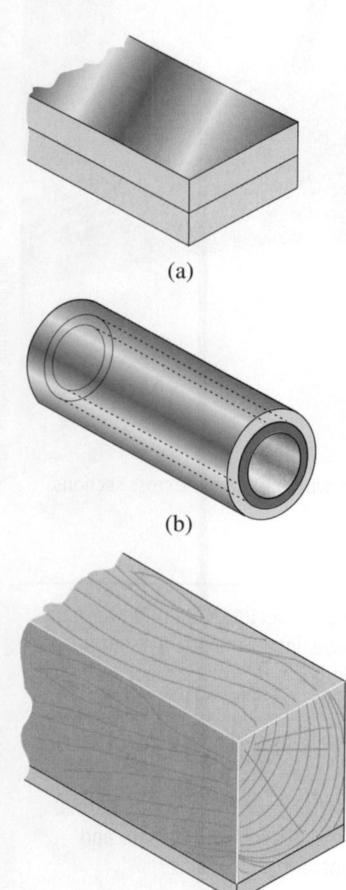

6.2 Composite Beams

Beams that are fabricated of more than one material are called **composite beams**. Examples are bimetallic beams (such as those used in thermostats), plastic-coated pipes, and wood beams with steel reinforcing plates (see Fig. 6-1).

Many other types of composite beams have been developed in recent years, primarily to save material and reduce weight. For instance, **sandwich beams** are widely used in the aviation and aerospace industries, where light weight plus high strength and rigidity are required. Such familiar objects as skis, doors, wall panels, book shelves, and cardboard boxes are also manufactured in sandwich style.

A typical sandwich beam (Fig. 6-2) consists of two thin *faces* of relatively high-strength material (such as aluminum) separated by a thick *core* of lightweight, low-strength material. Since the faces are at the greatest distance from the neutral axis (where the bending stresses are highest), they function somewhat like the flanges of an I-beam. The core serves as a filler and provides support for the faces, stabilizing them against wrinkling or buckling. Lightweight plastics and foams, as well as honeycombs and corrugations, are often used for cores.

Strains and Stresses

The strains in composite beams are determined from the same basic axiom used for finding the strains in beams of one material, namely, cross sections remain plane during bending. This axiom is valid for pure bending regardless of the nature of the material (see Section 5.4). Therefore, the longitudinal strains ε_x in a composite beam vary linearly from top to bottom of the beam, as expressed by Eq. (5-5), which is repeated here:

$$\varepsilon_x = -\frac{y}{\rho} = -\kappa y \tag{6-1}$$

In this equation, y is the distance from the neutral axis, ρ is the radius of curvature, and κ is the curvature.

Begin with the linear strain distribution represented by Eq. (6-1) to determine the strains and stresses in any composite beam. To show how this is accomplished, consider the composite beam shown in Fig. 6-3. This beam

Roof structure: composite timber-steel I-beam and rafters

consists of two materials, labeled 1 and 2 in the figure, which are securely bonded so that they act as a single solid beam.

As in previous discussions of beams (Chapter 5), assume that the x-y plane is a plane of symmetry and that the x-z plane is the neutral plane of the beam. However, the neutral axis (the z axis in Fig. 6-3b) does *not* pass through the centroid of the cross-sectional area when the beam is made of two different materials.

If the beam is bent with positive curvature, the strains ϵ_x will vary as shown in Fig. 6-3c, where ε_A is the compressive strain at the top of the beam, ε_B is the tensile strain at the bottom, and ε_C is the strain at the contact surface of the two materials. Of course, the strain is zero at the neutral axis (the z axis).

The normal stresses acting on the cross section can be obtained from the strains by using the stress-strain relationships for the two materials. Assume that both materials behave in a linearly elastic manner so that Hooke's law for uniaxial stress is valid. Then the stresses in the materials are obtained by multiplying the strains by the appropriate modulus of elasticity.

Denoting the moduli of elasticity for materials 1 and 2 as E_1 and E_2, respectively, and also assuming that $E_2 > E_1$, obtain the stress diagram shown in Fig. 6-3d. The compressive stress at the top of the beam is $\sigma_A = E_1\varepsilon_A$ and the tensile stress at the bottom is $\sigma_B = E_2\varepsilon_B$.

At the contact surface (C) the stresses in the two materials are different because their moduli are different. In material 1, the stress is $\sigma_{1C} = E_1\varepsilon_C$, and in material 2, it is $\sigma_{2C} = E_2\varepsilon_C$.

Using Hooke's law and Eq. (6-1), the normal stresses at distance y from the neutral axis in terms of the curvature are

$$\sigma_{x1} = -E_1\kappa y \qquad \sigma_{x2} = -E_2\kappa y \tag{6-2a,b}$$

in which σ_{x1} is the stress in material 1 and σ_{x2} is the stress in material 2. With the aid of these equations, the next step is to locate the neutral axis and obtain the moment-curvature relationship.

FIGURE 6-2

Sandwich beams: (a) plastic core, (b) honeycomb core, and (c) corrugated core (© Barry Goodno)

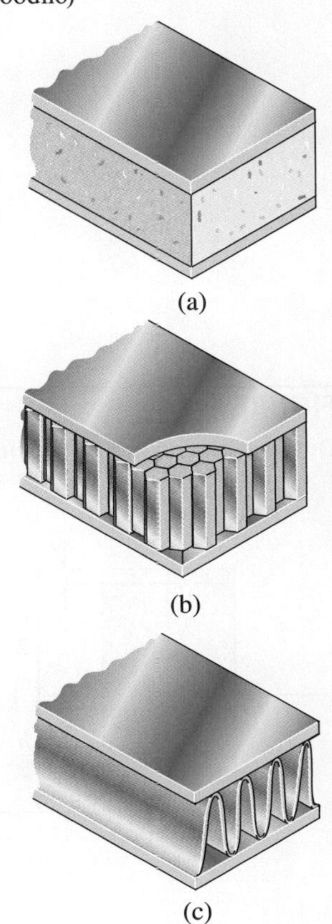

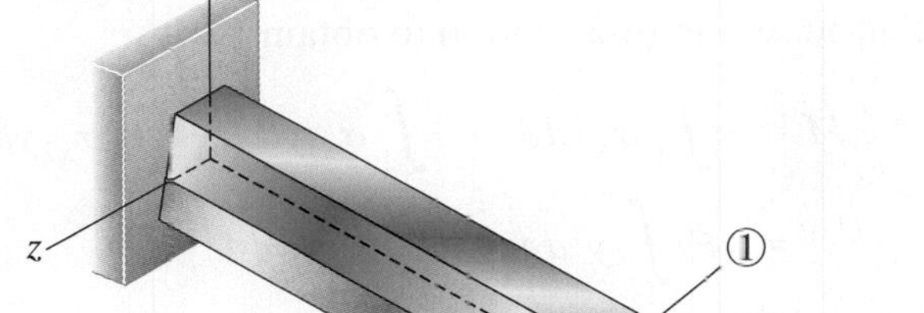

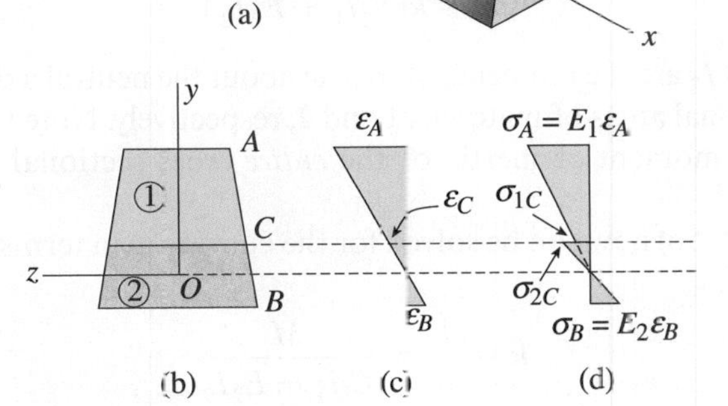

FIGURE 6-3

(a) Composite beam of two materials, (b) cross section of beam, (c) distribution of strains ε_x throughout the height of the beam, and (d) distribution of stresses σ_x in the beam for the case where $E_2 > E_1$

Neutral Axis

The position of the neutral axis (the z axis) is found from the condition that the resultant axial force acting on the cross section is zero (see Section 5.5); therefore,

$$\int_1 \sigma_{x1} dA + \int_2 \sigma_{x2} dA = 0 \tag{6-3}$$

where it is understood that the first integral is evaluated over the cross-sectional area of material 1 and the second integral is evaluated over the cross-sectional area of material 2. Replace σ_{x1} and σ_{x2} in the preceding equation by their expressions from Eqs. (6-2a and b) to get

$$-\int_1 E_1 \kappa y dA - \int_2 E_2 \kappa y dA = 0$$

Since the curvature is a constant at any given cross section, it is not involved in the integrations and can be cancelled from the equation; thus, the equation for locating the **neutral axis** becomes

$$E_1 \int_1 y dA + E_2 \int_2 y dA = 0 \tag{6-4}$$

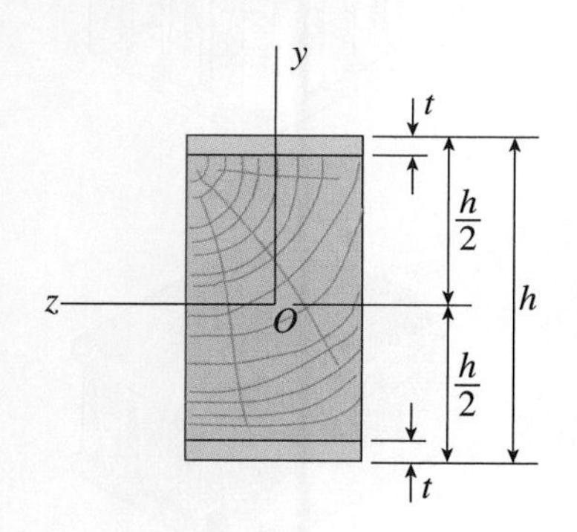

FIGURE 6-4
Doubly symmetric cross section

The integrals in this equation represent the first moments of the two parts of the cross-sectional area with respect to the neutral axis. (If there are more than two materials—a rare condition—additional terms are required in the equation.)

Equation (6-4) is a generalized form of the analogous equation for a beam of one material (Eq. 5-9). The details of the procedure for locating the neutral axis with the aid of Eq. (6-4) are illustrated later in Example 6-1.

If the cross section of a beam is **doubly symmetric**, as in the case of a wood beam with steel cover plates on the top and bottom (Fig. 6-4), the neutral axis is located at the mid-height of the cross section and Eq. (6-4) is not needed.

Moment-Curvature Relationship

The moment-curvature relationship for a composite beam of two materials (Fig. 6-3) may be determined from the condition that the moment resultant of the bending stresses is equal to the bending moment M acting at the cross section. Follow the same steps as for a beam of one material [see Eqs. (5-10) through (5-13)], and also use Eqs. (6-2a and b) to obtain

$$M = -\int_A \sigma_x y dA = -\int_1 \sigma_{x1} y dA - \int_2 \sigma_{x2} y dA$$

$$= \kappa E_1 \int_1 y^2 dA + \kappa E_2 \int_2 y^2 dA \tag{6-5a}$$

This equation can be written in the simpler form

$$M = \kappa(E_1 I_1 + E_2 I_2) \tag{6-5b}$$

in which I_1 and I_2 are the moments of inertia about the neutral axis (the z axis) of the cross-sectional areas of materials 1 and 2, respectively. Note that $I = I_1 + I_2$, where I is the moment of inertia of the *entire* cross-sectional area about the neutral axis.

Equation (6-5b) now can be solved for the curvature in terms of the bending moment:

$$\kappa = \frac{1}{\rho} = \frac{M}{E_1 I_1 + E_2 I_2} \tag{6-6}$$

This equation is the **moment-curvature relationship** for a beam of two materials [compare with Eq. (5-13) for a beam of one material]. The denominator on the right-hand side is the **flexural rigidity** of the composite beam.

Normal Stresses (Flexure Formulas)

The normal stresses (or bending stresses) in the beam are obtained by substituting the expression for curvature [Eq. (6-6)] into the expressions for σ_{x1} and σ_{x2} [Eqs. (6-2a and b)]; thus,

$$\sigma_{x1} = -\frac{MyE_1}{E_1I_1 + E_2I_2} \qquad \sigma_{x2} = -\frac{MyE_2}{E_1I_1 + E_2I_2} \tag{6-7a,b}$$

These expressions, known as the **flexure formulas for a composite beam**, give the normal stresses in materials 1 and 2, respectively. If the two materials have the same modulus of elasticity ($E_1 = E_2 = E$), both equations reduce to the flexure formula for a beam of one material [Eq. (5-14)].

The analysis of composite beams, using Eqs. (6-4) through (6-7), is illustrated in Examples 6-1 and 6-2 at the end of this section.

Approximate Theory for Bending of Sandwich Beams

Sandwich beams having doubly symmetric cross sections and composed of two linearly elastic materials (Fig. 6-5) can be analyzed for bending using Eqs. (6-6) and (6-7), as described previously. However, an approximate theory for bending of sandwich beams can be developed by introducing some simplifying assumptions.

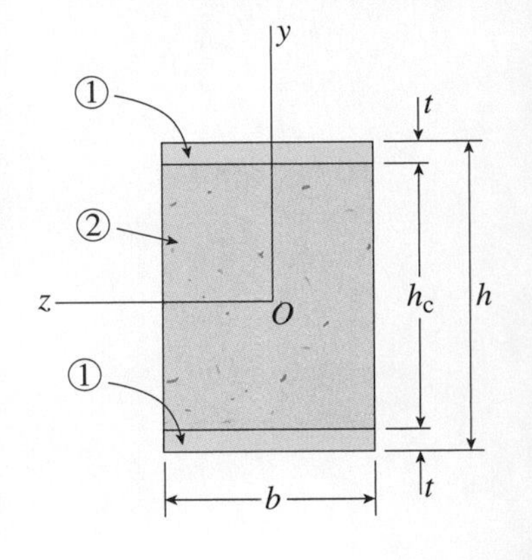

FIGURE 6-5
Cross section of a sandwich beam having two axes of symmetry (doubly symmetric cross section)

If the material of the faces (material 1) has a much larger modulus of elasticity than does the material of the core (material 2), it is reasonable to disregard the normal stresses in the core and assume that the faces resist all of the longitudinal bending stresses. This assumption is equivalent to saying that the modulus of elasticity E_2 of the core is zero. Under these conditions, the flexure formula for material 2 [Eq. (6-7b)] gives $\sigma_{x2} = 0$ (as expected), and the flexure formula for material 1 [Eq. (6-7a)] gives

$$\sigma_{x1} = -\frac{My}{I_1} \tag{6-8}$$

which is similar to the ordinary flexure formula [Eq. (5-14)]. The quantity I_1 is the moment of inertia of the two faces evaluated with respect to the neutral axis; thus,

$$I_1 = \frac{b}{12}\left(h^3 - h_c^3\right) \tag{6-9}$$

in which b is the width of the beam, h is the overall height of the beam, and h_c is the height of the core. Note that $h_c = h - 2t$ where t is the thickness of the faces.

The maximum normal stresses in the sandwich beam occur at the top and bottom of the cross section where $y = h/2$ and $-h/2$, respectively. Thus, from Eq. (6-8),

$$\sigma_{top} = -\frac{Mh}{2I_1} \qquad \sigma_{bottom} = \frac{Mh}{2I_1} \tag{6-10a,b}$$

If the bending moment M is positive, the upper face is in compression and the lower face is in tension. [These equations are conservative because they give stresses in the faces that are higher than those obtained from Eqs. (6-7a and b).]

If the faces are thin compared to the thickness of the core (that is, if t is small compared to h_c), disregard the shear stresses in the faces and assume that the core carries all of the shear stresses. Under these conditions the average shear stress and average shear strain in the core are, respectively,

$$\tau_{aver} = \frac{V}{bh_c} \qquad \gamma_{aver} = \frac{V}{bh_c G_c} \tag{6-11a,b}$$

in which V is the shear force acting on the cross section and G_c is the shear modulus of elasticity for the core material. (Although the maximum shear stress and maximum shear strain are larger than the average values, the average values are often used for design purposes.)

Limitations

The preceding discussion of composite beams assumed that both materials followed Hooke's law and that the two parts of the beam were adequately bonded so that they acted as a single unit. Thus, the analysis is highly idealized and represents only a first step in understanding the behavior of composite beams and composite materials. Methods for dealing with nonhomogeneous and nonlinear materials, bond stresses between the parts, shear stresses on the cross sections, buckling of the faces, and other such matters are treated in reference books dealing specifically with composite construction.

Reinforced concrete beams are one of the most complex types of composite construction (Fig. 6-6), and their behavior differs significantly from that of the composite beams discussed in this section. Concrete is strong in compression but extremely weak in tension. Consequently, its tensile strength is usually disregarded entirely. Under those conditions, *the formulas given in this section do not apply*. Working stress design in which the portion of the beam in tension is removed from the composite beam cross section is used in evaluating deflections of reinforced concrete beams, so an allowable stress approach is presented in Example 6-4 to show the general analysis procedure. Example 6-4 uses a "cracked section analysis" to demonstrate this reinforced-concrete analysis procedure.

Note that most reinforced concrete beams are not designed on the basis of linearly elastic behavior—instead, more realistic design methods based upon load-carrying capacity instead of allowable stresses are used. The design of reinforced concrete members is highly specialized and is presented in courses and textbooks devoted solely to that subject.

FIGURE 6-6

Reinforced concrete beam with longitudinal reinforcing bars and vertical stirrups

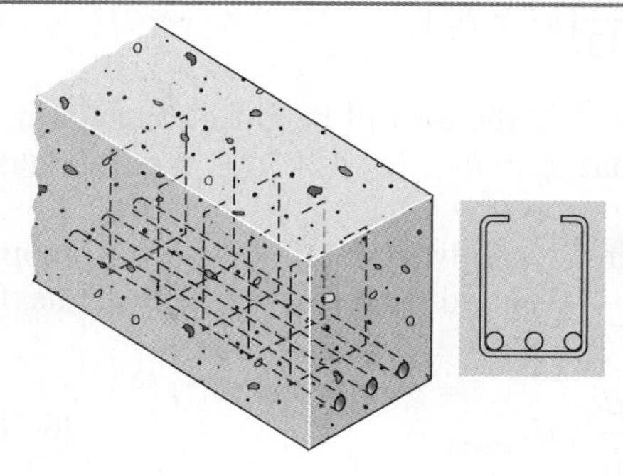

Example 6-1

FIGURE 6-7

Example 6-1: Cross section of a composite beam of wood and steel

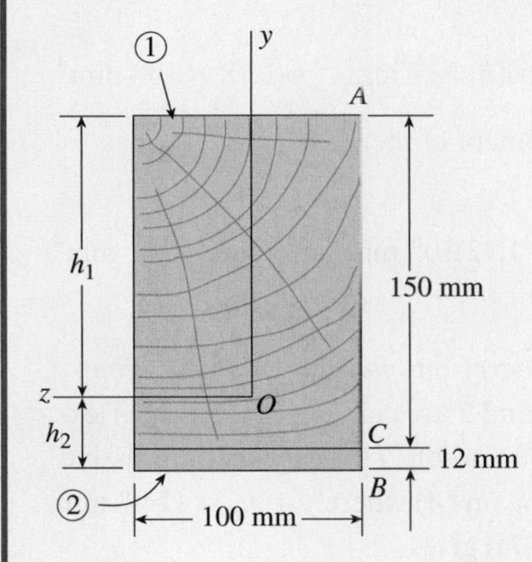

A composite beam (Fig. 6-7) is constructed from a wood beam (100 mm × 150 mm actual dimensions) and a steel reinforcing plate (100 mm wide and 12 mm thick). The wood and steel are securely fastened to act as a single beam. The beam is subjected to a positive bending moment $M = 6$ kN·m.

Calculate the largest tensile and compressive stresses in the wood (material 1) and the maximum and minimum tensile stresses in the steel (material 2) if $E_1 = 10.5$ GPa and $E_2 = 210$ GPa.

Solution:

Use a four-step problem-solving approach.

1. ***Conceptualize*** [*hypothesize, sketch*]: Use the general theory of flexure for composite beams.
2. ***Categorize*** [*simplify, classify*]:

Neutral axis: The first step in the analysis is to locate the neutral axis of the cross section. For that purpose, denote the distances from the neutral axis to the top and bottom of the beam as h_1 and h_2, respectively. To obtain these distances, use Eq. (6-4). Evaluate the integrals in that equation by taking the first moments of areas 1 and 2 about the z axis, as

$$\int_1 ydA = \bar{y}_1 A_1 = (h_1 - 75 \text{ mm})(100 \text{ mm} \times 150 \text{ mm}) = (h_1 - 75 \text{ mm})(15000 \text{ mm}^2)$$

$$\int_2 ydA = \bar{y}_2 A_2 = -(156 \text{ mm} - h_1)(100 \text{ mm} \times 12 \text{ mm}) = (h_1 - 75 \text{ mm})(1200 \text{ mm}^2)$$

in which A_1 and A_2 are the areas of parts 1 and 2 of the cross section, $\bar{y}_1$ and $\bar{y}_2$ are the y coordinates of the centroids of the respective areas, and h_1 has units of millimeters.

Substitute the preceding expressions into Eq. (6-4) to get the equation for locating the neutral axis as

$$E_1 \int_1 ydA + E_2 \int_2 ydA = 0$$

or

$$(10.5 \text{ GPa})(h_1 - 75 \text{ mm})(15000 \text{ mm}^2) + (210 \text{ GPa})(h_1 - 75 \text{ mm})(1200 \text{ mm}^2) = 0$$

Solve this equation to obtain the distance h_1 from the neutral axis to the top of the beam:

$$h_1 = 124.8 \text{ mm}$$

Also, the distance h_2 from the neutral axis to the bottom of the beam is

$$h_2 = 162 \text{ mm} - h_1 = 37.2 \text{ mm}$$

Thus, the position of the neutral axis is established.

Moments of inertia: The moments of inertia I_1 and I_2 of areas A_1 and A_2 with respect to the neutral axis are found by using the parallel-axis theorem (see Section D.4 of Appendix D). Beginning with area 1 (Fig. 6-7),

$$I_1 = \frac{1}{12}(100 \text{ mm})(150 \text{ mm})^3 + (100 \text{ mm})(150 \text{ mm})(h_1 - 75 \text{ mm})^2 = 65.33 \times 10^6 \text{ mm}^4$$

Similarly, for area 2,

$$I_2 = \frac{1}{12}(100 \text{ mm})(12 \text{ mm})^3 + (100 \text{ mm})(12 \text{ mm})(h_2 - 6 \text{ mm})^2 = 1.18 \times 10^6 \text{ mm}^4$$

To check these calculations, compute the moment of inertia I of the entire cross-sectional area about the z axis as

$$I = \frac{1}{3}(100 \text{ mm})h_1^3 + \frac{1}{3}(100 \text{ mm})h_2^3 = (64.79 + 1.72)10^6 \text{ mm}^4 = 66.51 \times 10^6 \text{ mm}^4$$

which agrees with the sum of I_1 and I_2.

3. ***Analyze*** [*evaluate; select relevant equations, carry out mathematical solution*]:
Normal stresses: The stresses in materials 1 and 2 are calculated from the flexure formulas for composite beams [Eqs. (6-7a and b)]. The largest compressive stress in material 1 occurs at the top of the beam (A) where $y = h_1 = 124.8$ mm Denoting this stress by σ_{1A} and using Eq. (6-7a) gives

$$\sigma_{1A} = -\frac{Mh_1E_1}{E_1I_1 + E_2I_2}$$

$$= -\frac{(6 \text{ kN}\cdot\text{m})(124.8 \text{ mm})(10.5 \text{ GPa})}{(10.5 \text{ GPa})(65.33 \times 10^6 \text{ mm}^4) + (210 \text{ GPa})(1.18 \times 10^6 \text{ mm}^4)} = -8.42 \text{ MPa}$$

The largest tensile stress in material 1 occurs at the contact plane between the two materials (C) where $y = -(h_2 - 12 \text{ mm}) = -25.2$ mm. Proceed as in the previous calculation to obtain

$$\sigma_{1C} = -\frac{(6 \text{ kN}\cdot\text{m})(-25.2 \text{ mm})(10.5 \text{ GPa})}{(10.5 \text{ GPa})(65.33 \times 10^6 \text{ mm}^4) + (210 \text{ GPa})(1.18 \times 10^6 \text{ mm}^4)} = 1.7 \text{ MPa}$$

These are the largest compressive and tensile stresses in the wood.

The steel plate (material 2) is located below the neutral axis; therefore, it is entirely in tension. The maximum tensile stress occurs at the bottom of the beam (B) where $y = -h_2 = -37.2$ mm. Hence, from Eq. (6-7b),

$$\sigma_{2B} = -\frac{M(-h_2)E_2}{E_1I_1 + E_2I_2}$$

$$= -\frac{(6 \text{ kN}\cdot\text{m})(-37.2 \text{ mm})(210 \text{ GPa})}{(10.5 \text{ GPa})(65.33 \times 10^6 \text{ mm}^4) + (210 \text{ GPa})(1.18 \times 10^6 \text{ mm}^4)} = 50.2 \text{ MPa}$$

The minimum tensile stress in material 2 occurs at the contact plane (C) where $y = -25.2$ mm. Thus,

$$\sigma_{2C} = -\frac{(6 \text{ kN}\cdot\text{m})(-25.2 \text{ mm})(210 \text{ GPa})}{(10.5 \text{ GPa})(65.33 \times 10^6 \text{ mm}^4) + (210 \text{ GPa})(1.18 \times 10^6 \text{ mm}^4)} = 34 \text{ MPa}$$

These stresses are the maximum and minimum tensile stresses in the steel.

FIGURE 6-8

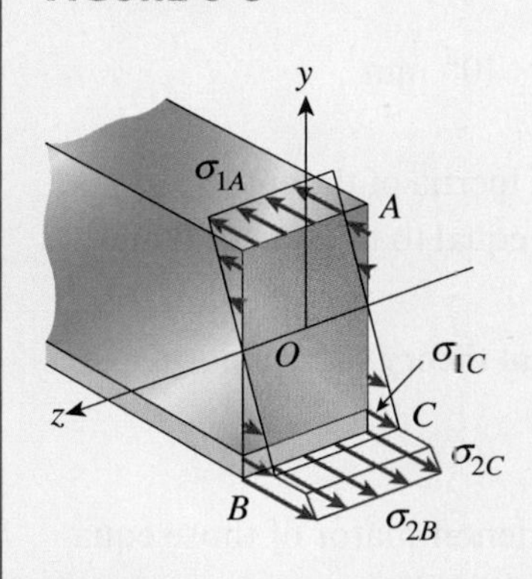

4. ***Finalize*** [*conclude; examine answer—Does it make sense? Are units correct? How does it compare to similar problem solutions?*]: The stress distribution over the cross section of the composite wood-steel beam is shown in Fig. 6-8.

   Note: At the contact plane, the ratio of the stress in the steel to the stress in the wood is

$$\sigma_{2C}/\sigma_{1C} = 34 \text{ MPa}/1.7 \text{ MPa} = 20$$

   which is equal to the ratio E_2/E_1 of the moduli of elasticity (as expected). Although the strains in the steel and wood are equal at the contact plane, the stresses are different because of the different moduli.

Example 6-2

A sandwich beam having aluminum-alloy faces enclosing a plastic core (Fig. 6-9) is subjected to a bending moment $M = 3.0 \text{ kN} \cdot \text{m}$. The thickness of the faces is $t = 5$ mm, and their modulus of elasticity is $E_1 = 72$ GPa. The height of the plastic core is $h_c = 150$ mm, and its modulus of elasticity is $E_2 = 800$ MPa. The overall dimensions of the beam are $h = 160$ mm and $b = 200$ mm.

Determine the maximum tensile and compressive stresses in the faces and the core using (a) the general theory for composite beams and (b) the approximate theory for sandwich beams.

FIGURE 6-9

Example 6-2: Cross section of sandwich beam having aluminum-alloy faces and a plastic core

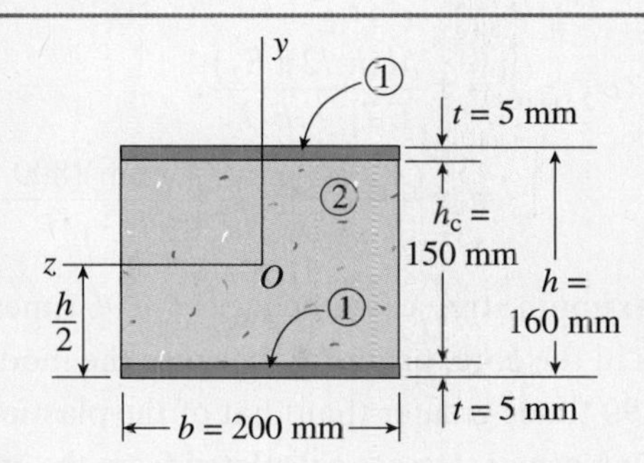

Solution:

Use a four-step problem-solving approach. Combine steps as needed for an efficient solution.

1. ***Conceptualize***: Use the general theory of flexure for composite beams.
2. ***Categorize***:

   Neutral axis: Because the cross section is doubly symmetric, the neutral axis (the z axis in Fig. 6-9) is located at mid-height.

   Moments of inertia: The moment of inertia I_1 of the cross-sectional areas of the faces (about the z axis) is

$$I_1 = \frac{b}{12}(h^3 - h_c^3) = \frac{200 \text{ mm}}{12}\left[(160 \text{ mm})^3 - (150 \text{ mm})^3\right]$$

$$= 12.017 \times 10^6 \text{ mm}^4$$

and the moment of inertia I_2 of the plastic core is

$$I_2 = \frac{b}{12}(h_c^3) = \frac{200 \text{ mm}}{12}(150 \text{ mm})^3 = 56.250 \times 10^6 \text{ mm}^4$$

As a check on these results, note that the moment of inertia of the entire cross-sectional area about the z axis ($I = bh^3/12$) is equal to the sum of I_1 and I_2.

3, 4. ***Analyze, Finalize:***

Part (a): Normal stresses calculated from the general theory for composite beams.

To calculate these stresses, use Eqs. (6-7a and b).

As a preliminary matter, evaluate the term in the denominator of those equations (that is, the flexural rigidity of the composite beam):

$$E_1 I_1 + E_2 I_2 = (72 \text{ GPa})(12.017 \times 10^6 \text{ mm}^4) + (800 \text{ MPa})(56.250 \times 10^6 \text{ mm}^4)$$
$$= 910{,}200 \text{ N} \cdot \text{m}^2$$

The maximum tensile and compressive stresses in the aluminum faces are found from Eq. (6-7a):

$$(\sigma_1)_{\max} = \pm \frac{M(h/2)(E_1)}{E_1 I_1 + E_2 I_2}$$
$$= \pm \frac{(3.0 \text{ kN} \cdot \text{m})(80 \text{ mm})(72 \text{ GPa})}{910{,}200 \text{ N} \cdot \text{m}^2} = \pm 19.0 \text{ MPa}$$

The corresponding quantities for the plastic core (from Eq. 6-7b) are

$$(\sigma_2)_{\max} = \pm \frac{M(h_c/2)(E_2)}{E_1 I_1 + E_2 I_2}$$
$$= \pm \frac{(3.0 \text{ kN} \cdot \text{m})(75 \text{ mm})(800 \text{ MPa})}{910{,}200 \text{ N} \cdot \text{m}^2} = \pm 0.198 \text{ MPa}$$

The maximum stresses in the faces are 96 times greater than the maximum stresses in the core, primarily because the modulus of elasticity of the aluminum is 90 times greater than that of the plastic.

Part (b): Normal stresses calculated from the approximate theory for sandwich beams:

The approximate theory disregards the normal stresses in the core and assumes that the faces transmit the entire bending moment. Then the maximum tensile and compressive stresses in the faces is found from Eqs. (6-10a and b), as

$$(\sigma_1)_{\max} = \pm \frac{Mh}{2I_1} = \pm \frac{(3.0 \text{ kN} \cdot \text{m})(80 \text{ mm})}{12.017 \times 10^6 \text{ mm}^4} = \pm 20.0 \text{ MPa}$$

As expected, the approximate theory gives slightly higher stresses in the faces than does the general theory for composite beams.

6.3 Transformed-Section Method

The transformed-section method is an alternative procedure for analyzing the bending stresses in a composite beam. The method is based upon the theories and equations developed in the preceding section; therefore, it is subject to the same limitations (for instance, it is valid only for linearly elastic materials) and gives the same results. Although the transformed-section method does not reduce the calculating effort, many designers find that it provides a convenient way to visualize and organize the calculations.

The method consists of transforming the cross section of a composite beam into an equivalent cross section of an imaginary beam that is composed of only one material. This new cross section is called the **transformed section**. Then the imaginary beam with the transformed section is analyzed in the customary manner for a beam of one material. As a final step, the stresses in the transformed beam are converted to those in the original beam.

Neutral Axis and Transformed Section

If the transformed beam is to be equivalent to the original beam, *its neutral axis must be located in the same place and its moment-resisting capacity must be the same*. To show how these two requirements are met, consider again a composite beam of two materials (Fig. 6-10a). The **neutral axis** of the cross section is obtained from Eq. (6-4), which is repeated here:

$$E_1 \int_1 y dA + E_2 \int_2 y dA = 0 \tag{6-12}$$

In this equation, the integrals represent the first moments of the two parts of the cross section with respect to the neutral axis.

Now introduce the notation

$$n = \frac{E_2}{E_1} \tag{6-13}$$

where n is the **modular ratio**. With this notation, rewrite Eq. (6-12) in the form

$$\int_1 y dA + \int_2 y n dA = 0 \tag{6-14}$$

Since Eqs. (6-12) and (6-14) are equivalent, the preceding equation shows that the neutral axis is unchanged if each element of area dA in material 2 is multiplied by the factor n, provided that the y coordinate for each such element of area is not changed.

Therefore, create a new cross section consisting of two parts: (1) area 1 with its dimensions unchanged and (2) area 2 with its *width* (that is, its dimension parallel to the neutral axis) multiplied by n. This new cross section (the transformed section) is shown in Fig. 6-10b for the case where $E_2 > E_1$ (and therefore $n > 1$). Its neutral axis is in the same position as the neutral axis of the original beam. (Note that all dimensions perpendicular to the neutral axis remain the same.)

Since the stress in the material (for a given strain) is proportional to the modulus of elasticity ($\sigma = E\varepsilon$), multiplying the width of material 2 by $n = E_2/E_1$ is equivalent to transforming it to material 1. For instance, suppose that $n = 10$. Then the area of part 2 of the cross section is now 10 times wider than before. If you imagine that this part of the beam is now material 1, it will

FIGURE 6-10

Composite beam of two materials: (a) actual cross section and (b) transformed section consisting only of material 1

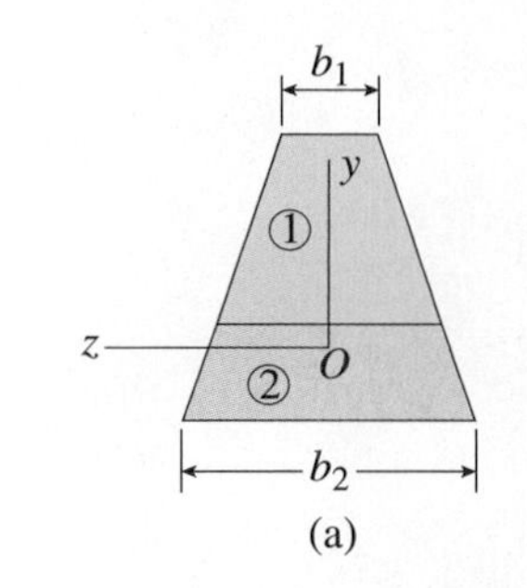

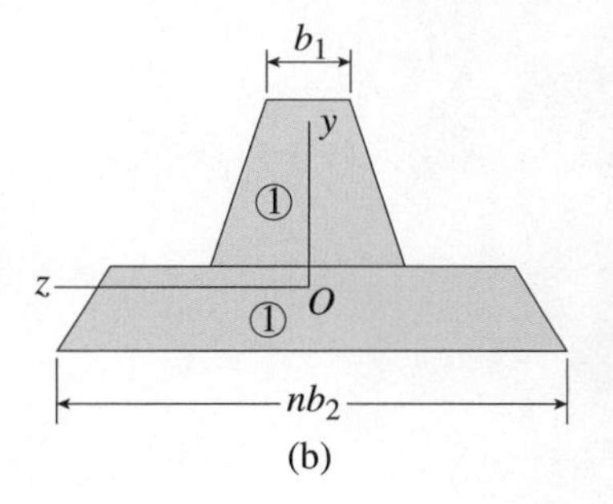

carry the same force as before because its modulus is *reduced* by a factor of 10 (from E_2 to E_1) at the same time that its area is *increased* by a factor of 10. Thus, the new section (the transformed section) consists only of material 1.

Moment-Curvature Relationship

The *moment-curvature relationship* for the transformed beam must be the same as for the original beam. To show that this is indeed the case, note that the stresses in the transformed beam (since it consists only of material 1) are given by Eq. (5-8) of Section 5.5:

$$\sigma_x = -E_1 \kappa y$$

Using this equation, and also following the same procedure as for a beam of one material (see Section 5.5), the moment-curvature relation for the transformed beam is

$$M = -\int_A \sigma_x y dA = -\int_1 \sigma_x y dA - \int_2 \sigma_x y dA$$

$$= E_1 \kappa \int_1 y^2 dA + E_1 \kappa \int_2 y^2 dA = \kappa(E_1 I_1 + E_1 n I_2)$$

or

$$M = \kappa(E_1 I_1 + E_2 I_2) \quad \textbf{(6-15)}$$

This equation is the same as Eq. (6-5), thereby demonstrating that the moment-curvature relationship for the transformed beam is the same as for the original beam.

Normal Stresses

Since the transformed beam consists of only one material, the *normal stresses* (or *bending stresses*) can be found from the standard flexure formula [Eq. (5-14)]. Thus, the normal stresses in the beam transformed to material 1 (Fig. 6-10b) are

$$\sigma_{x1} = -\frac{My}{I_T} \quad \textbf{(6-16)}$$

where I_T is the moment of inertia of the transformed section with respect to the neutral axis. By substituting into this equation, you can calculate the stresses at any point in the *transformed* beam. (As explained later, the stresses in the transformed beam match those in the part of the original beam consisting of material 1; however, in the part of the original beam consisting of material 2, the stresses are different from those in the transformed beam.)

Equation (6-16) can be verified by noting that the moment of inertia of the transformed section (Fig. 6-10b) is related to the moment of inertia of the original section (Fig. 6-10a) by the relation:

$$I_T = I_1 + nI_2 = I_1 + \frac{E_2}{E_1} I_2 \quad \textbf{(6-17)}$$

Substituting this expression for I_T into Eq. (6-16) gives

$$\sigma_{x1} = -\frac{MyE_1}{E_1 I_1 + E_2 I_2} \quad \textbf{(6-18a)}$$

which is the same as Eq. (6-7a), demonstrating that the stresses in material 1 in the original beam are the same as the stresses in the corresponding part of the transformed beam.

As mentioned previously, the stresses in material 2 in the original beam are *not* the same as the stresses in the corresponding part of the transformed beam. Instead, the stresses in the transformed beam [Eq. (6-16)] must be multiplied by the modular ratio n to obtain the stresses in material 2 of the original beam:

$$\sigma_{x2} = -\frac{My}{I_T}n \tag{6-18b}$$

Verify this formula by substituting Eq. (6-17) for I_T into Eq. (6-18b) to get

$$\sigma_{x2} = -\frac{MynE_1}{E_1I_1 + E_2I_2} = -\frac{MyE_2}{E_1I_1 + E_2I_2} \tag{6-18c}$$

which is the same as Eq. (6-7b).

General Comments

In this discussion of the transformed-section method, the original beam was transformed to a beam consisting entirely of material 1. It is also possible to transform the beam to material 2. In that case, the stresses in the original beam in material 2 will be the same as the stresses in the corresponding part of the transformed beam. However, the stresses in material 1 in the original beam must be obtained by multiplying the stresses in the corresponding part of the transformed beam by the modular ratio n, which in this case is defined as $n = E_1/E_2$.

It is also possible to transform the original beam into a material having any arbitrary modulus of elasticity E, in which case all parts of the beam must be transformed to the fictitious material. Of course, the calculations are simpler if you transform to one of the original materials. Finally, with a little ingenuity, it is possible to extend the transformed-section method to composite beams of more than two materials.

Example 6-3

The composite beam shown in Fig. 6-11a is formed of a wood beam (100 mm × 150 mm actual dimensions) and a steel reinforcing plate (100 mm wide and 12 mm thick). The beam is subjected to a positive bending moment $M = 6\ \text{kN} \cdot \text{m}$.

Using the transformed-section method, calculate the largest tensile and compressive stresses in the wood (material 1) and the maximum and minimum tensile stresses in the steel (material 2) if $E_1 = 10.5$ GPa and $E_2 = 210$ GPa.

Note: This same beam was analyzed previously in Example 6-1 of Section 6.2.

FIGURE 6-11

Example 6-3: Composite beam of Example 6-1 analyzed by the transformed-section method: (a) cross section of original beam and (b) transformed section (material 1)

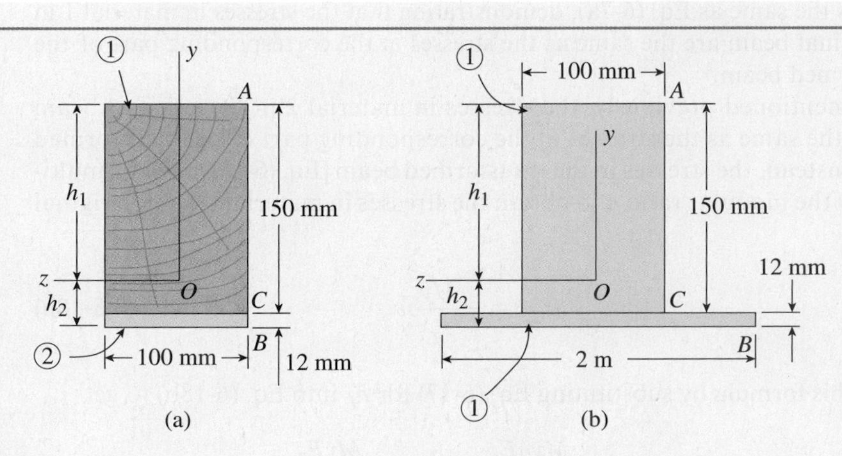

Solution:

Use a four-step problem-solving approach. Combine steps as needed for an efficient solution.

1, 2. ***Conceptualize, Categorize*****:** Use the transformed-section approach and compare results to those found in Example 6-1.

3. ***Analyze*****:**

Transformed section: Transform the original beam into a beam of material 1, which means that the modular ratio is defined as

$$n = \frac{E_2}{E_1} = \frac{210 \text{ GPa}}{10.5 \text{ GPa}} = 20$$

The part of the beam made of wood (material 1) is not altered, but the part made of steel (material 2) has its width multiplied by the modular ratio. Thus, the width of this part of the beam becomes

$$n(100 \text{ mm}) = 20(100 \text{ mm}) = 2 \text{ m}$$

in the transformed section (Fig. 6-11b).

Neutral axis: Because the transformed beam consists of only one material, the neutral axis passes through the centroid of the cross-sectional area. Therefore, with the top edge of the cross section serving as a reference line and the distance y_i measured positive downward, calculate the distance h_1 to the centroid as

$$h_1 = \frac{\Sigma y_i A_i}{\Sigma A_i} = \frac{(75 \text{ mm})(100 \text{ mm})(150 \text{ mm}) + (156 \text{ mm})(2000 \text{ mm})(12 \text{ mm})}{(100 \text{ mm})(150 \text{ mm}) + (2000 \text{ mm})(12 \text{ mm})}$$

$$= \frac{4869 \times 10^3 \text{ mm}^3}{39 \times 10^3 \text{ mm}^2} = 124.8 \text{ mm}$$

Also, the distance h_2 from the lower edge of the section to the centroid is

$$h_2 = 162 \text{ mm} - h_1 = 37.2 \text{ mm}$$

Thus, the location of the neutral axis is determined.

Moment of inertia of the transformed section: Using the parallel-axis theorem (see Section D.4 of Appendix D), calculate the moment of inertia I_T of the entire cross-sectional area with respect to the neutral axis as

$$I_T = \frac{1}{12}(100 \text{ mm})(150 \text{ mm})^3 + (100 \text{ mm})(150 \text{ mm})(h_1 - 75 \text{ mm})^2$$
$$+\frac{1}{12}(2000 \text{ mm})(12 \text{ mm})^3 + (2000 \text{ mm})(12 \text{ mm})(h_2 - 6 \text{ mm})^2$$
$$= 65.3 \times 10^6 \text{ mm}^4 + 23.7 \times 10^6 \text{ mm}^4 = 89.0 \times 10^6 \text{ mm}^4$$

Normal stresses in the wood (material 1): The stresses in the transformed beam (Fig. 6-11b) at the top of the cross section (A) and at the contact plane between the two parts (C) are the same as in the original beam (Fig. 6-11a). These stresses can be found from the flexure formula in Eq. (6-16) as

$$\sigma_{1A} = -\frac{My}{I_T} = -\frac{(6 \times 10^6 \text{ N}\cdot\text{mm})(124.8 \text{ mm})}{89 \times 10^6 \text{ mm}^4} = -8.42 \text{ MPa}$$

$$\sigma_{1C} = -\frac{My}{I_T} = -\frac{(6 \times 10^6 \text{ N}\cdot\text{mm})(-25.2 \text{ mm})}{89 \times 10^6 \text{ mm}^4} = 1.7 \text{ MPa}$$

These are the largest tensile and compressive stresses in the wood (material 1) in the original beam. The stress σ_{1A} is compressive and the stress σ_{1C} is tensile.

Normal stresses in the steel (material 2): The maximum and minimum stresses in the steel plate are found by multiplying the corresponding stresses in the transformed beam by the modular ratio n in Eq. (6-18b). The maximum stress occurs at the lower edge of the cross section (B) and the minimum stress occurs at the contact plane (C):

$$\sigma_{2B} = -\frac{My}{I_T}n = -\frac{(6 \times 10^6 \text{ N}\cdot\text{mm})(-37.2 \text{ mm})}{89 \times 10^6 \text{ mm}^4}(20) = 50.2 \text{ MPa}$$

$$\sigma_{2C} = -\frac{My}{I_T}n = -\frac{(6 \times 10^6 \text{ N}\cdot\text{mm})(-25.2 \text{ mm})}{89 \times 10^6 \text{ mm}^4}(20) = 34 \text{ MPa}$$

Both of these stresses are tensile.

4. ***Finalize***: Note that the stresses calculated by the transformed-section method agree with those found in Example 6-1 by direct application of the formulas for a composite beam.

Balanced design: As a final evaluation of the wood-steel composite beam considered here and in Example 6-1, note that neither wood nor steel has reached typical allowable stress levels. Perhaps some redesigning of this beam would be of interest; consider only the steel plate here (you could also re-size the wood beam).

A *balanced design* is one in which wood and steel reach their allowable stress values at the same time under the design moment; this could be regarded as a more efficient design of this beam. *First*, holding the steel plate thickness at $t_s = 12$ mm, find the required width b_s of the steel plate, so the wood and steel reach allowable stress values simultaneously under design moment M_D. Then

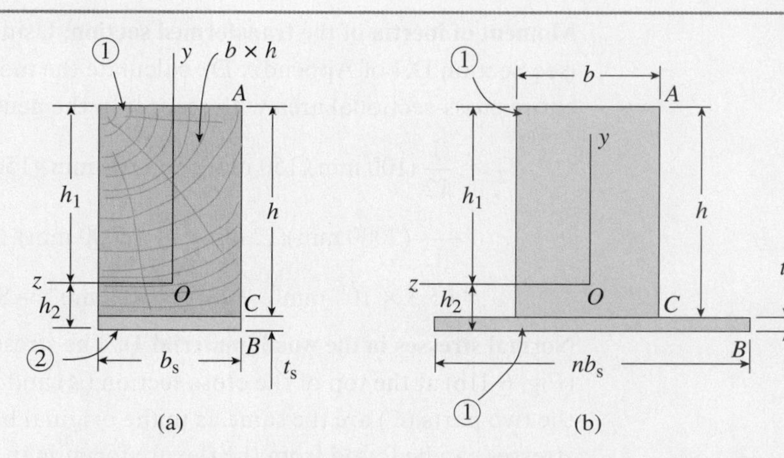

FIGURE 6-12
Example 6-3: Balanced design of composite beam: (a) original beam and (b) transformed beam

setting $b_s = 100$ mm, repeat the previous process but also find required plate thickness t_s to achieve the same objective. Assume that the allowable stress values for wood and steel are $\sigma_{aw} = 12.7$ MPa and $\sigma_{as} = 96$ MPa, respectively. Also assume that the wood beam dimensions are unchanged.

Using the transformed-section approach, write the expressions for the stresses at the top of the wood and bottom of the steel. Equate each to its allowable value as

$$\sigma_{aw} = \frac{-M_D h_1}{I_T} \quad \text{and} \quad \sigma_{as} = \frac{-M_D h_2 n}{I_T} \qquad \textbf{(a,b)}$$

Next, solve each of Eqs. (a) and (b) for ratio M_D/I_T; then equate the two expressions to find the h_1/h_2 ratio for which allowable stress levels are reached in both materials:

$$\frac{h_1}{h_2} = n\frac{\sigma_{aw}}{\sigma_{as}} \qquad \textbf{(c)}$$

Expressions for h_1 and h_2 can be obtained in terms of the transformed-section dimensions b, h, b_s, and t_s (Figs. 6-12a and b) by taking first moments about the z axis to get

$$h_1 = \frac{h}{2} + \frac{(b_s n t_s^2) + (b_s h n t_s)}{(2bh) + (2b_s n t_s)} \quad \text{and} \quad h_2 = \frac{n b_s t_s\left(\frac{t_s}{2}\right) + bh\left(t_s + \frac{h}{2}\right)}{(n b_s t_s) + (bh)} \qquad \textbf{(d)}$$

With some effort (and perhaps with computer assistance), rewrite Eq. (c) as

$$\frac{bh^2 + (2b_s n h t_s) + (b_s n t_s^2)}{bh^2 + (2bht_s) + (b_s n t_s^2)} = n\frac{\sigma_{aw}}{\sigma_{as}} \qquad \textbf{(e)}$$

Collecting terms and solving for the required width of b_s for the steel plate (with thickness t_s unchanged) then substituting numerical values, width b_s (instead of the original width of 100 mm) is

$b = 100$ mm $\quad h = 150$ mm $\quad t_s = 12$ mm $\quad n = 20 \quad \sigma_{aw} = 12.7$ MPa $\quad \sigma_{as} = 96$ MPa

$$b_s = \frac{\left(n\frac{\sigma_{aw}}{\sigma_{as}}\right)(bh^2 + 2bht_s) - bh^2}{(2nht_s) + nt_s^2\left(1 - n\frac{\sigma_{aw}}{\sigma_{as}}\right)} = 69.2 \text{ mm} \qquad \textbf{(f)}$$

So for a 100 mm × 150 mm wood beam reinforced by a 69.2 mm × 12 mm steel plate (Fig. 6-12a) under any applied moment M that is less than or equal to M_D, the stress ratio σ_{1A}/σ_{2B} will be equal to σ_{aw}/σ_{as}. If $M = M_D$, then $\sigma_{1A} = \sigma_{aw}$ and $\sigma_{2B} = \sigma_{as}$.

Alternatively, reformulate Eq. (e) to get a quadratic equation for the steel plate thickness t_s (with the original width $b_s = 100$ mm) to obtain

$$t_s^2\left[nb_s\left(1 - n\frac{\sigma_{aw}}{\sigma_{as}}\right)\right] + t_s\left\{2h\left[nb_s - b\left(n\frac{\sigma_{aw}}{\sigma_{as}}\right)\right]\right\} + bh^2\left(1 - n\frac{\sigma_{aw}}{\sigma_{as}}\right) = 0 \quad \textbf{(g)}$$

The solution of Eq. (g) results in a reduced steel plate thickness leading to a *balanced design* of the wood-steel composite beam as $t_s = 7.46$ mm. Once again, for $b_s = 100$ mm and $t_s = 7.46$ mm, the stress ratio $\sigma_{1A}/\sigma_{2B} = \sigma_{aw}/\sigma_{as}$ for applied moments M which are less than or equal to M_D.

Example 6-4

FIGURE 6-13

Example 6-4: Cross section of a singly reinforced concrete inverted T-beam

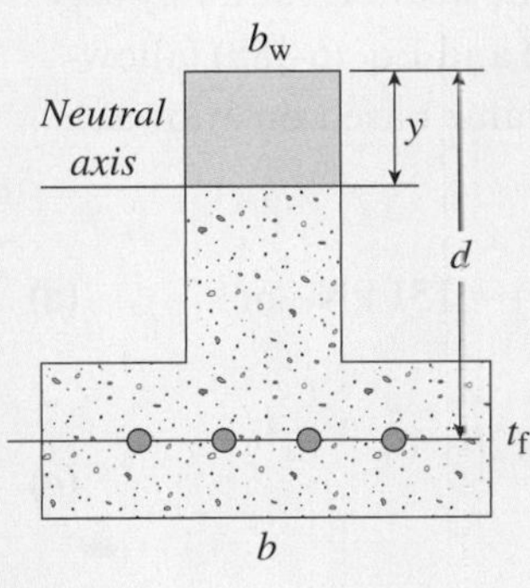

An inverted precast concrete T-beam is used to support precast double-tee floor beams in a parking deck (see Fig. 6-13 and photo). Beam dimensions are $b = 500$ mm, $b_w = 300$ mm, $d = 600$ mm, and $t_f = 100$ mm. Steel reinforcement consists of four bars each with a 25-mm diameter. The modulus of elasticity for the concrete is $E_c = 25$ GPa, while that of the steel is $E_s = 200$ GPa. Allowable stresses for concrete and steel are $\sigma_{ac} = 9.3$ MPa and $\sigma_{as} = 137$ MPa, respectively.

(a) Use the transformed section in Fig. 6-14 (in which the concrete in tension is neglected and the steel reinforcing bars are converted to the equivalent concrete) to find the maximum permissible moment that can be applied to this beam.

(b) Repeat part (a) if the beam is rotated 180°, as shown in Fig. 6-15, and if the steel reinforcement remains in the bottom tension zone.

FIGURE 6-14

Transformed section for singly reinforced concrete inverted T-beam

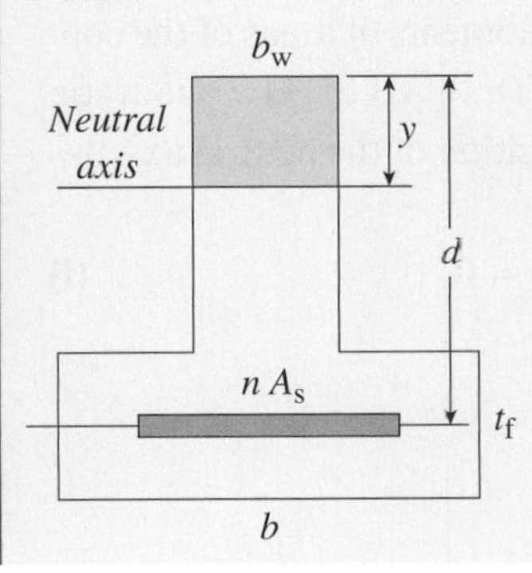

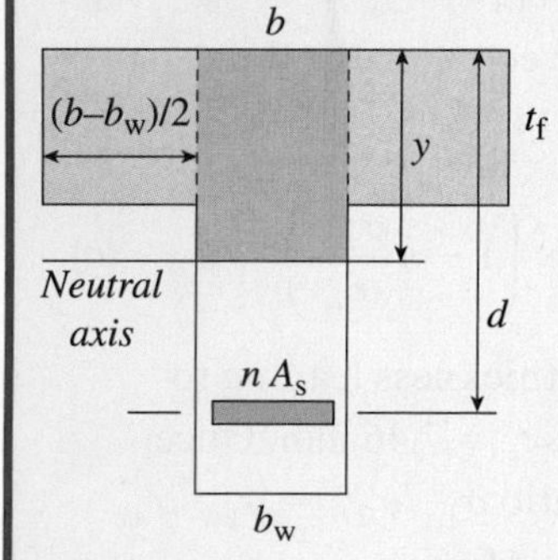

FIGURE 6-15
Transformed section for singly reinforced concrete T-beam

Solution:

Use a four-step problem-solving approach. Combine steps as needed for an efficient solution.

Part (a): Inverted T-beam.

1, 2. *Conceptualize, Categorize*: Start by finding the *neutral axis* (at some distance y down from the top of the beam) for the transformed section shown in Fig. 6-14. Equating the first moments of areas of concrete in compression ($b_w \times y$) and the transformed area of steel in tension ($n \times A_s$) leads to a quadratic equation. The solution for y gives the position of the neutral axis as

$$b_w y \frac{y}{2} - nA_s(d - y) = 0 \quad \text{where} \quad n = \frac{E_s}{E_c} = 8 \tag{a}$$

$$y = \sqrt{\left(\frac{nA_s}{b_w}\right)^2 + 2d\left(\frac{nA_s}{b_w}\right)} - \left(\frac{nA_s}{b_w}\right) = 0.204 \text{ m} \tag{b}$$

Now use Eq. (6-17) to compute the moment of inertia of the transformed section:

$$I_T = \frac{b_w y^3}{3} + nA_s[(d - y)^2] = 3.312 \times 10^{-3} \text{ m}^4 \tag{c}$$

3, 4. *Analyze, Finalize*: Finally, the moment capacity of the beam is found by solving Eqs. (6-16) (allowable stress in concrete controls) and Eq. (6-18b) (allowable stress in steel controls) for M, where the lower value based on allowable stress in the steel governs:

$$M_c = \frac{\sigma_{ac}}{y} I_T = \left(\frac{9.3 \text{ MPa}}{0.204 \text{ m}}\right)(3.312 \times 10^{-3} \text{ m}^4) = 151 \text{ kN} \cdot \text{m} \tag{d}$$

$$M_s = \frac{\sigma_{as}}{n(d - y)} I_T = \frac{137 \text{ MPa}}{8(0.6 \text{ m} - 0.204 \text{ m})}(3.312 \times 10^{-3} \text{ m}^4) = 143.2 \text{ kN} \cdot \text{m} \tag{e}$$

Part (b): T-beam.

1, 2. *Conceptualize, Categorize*: Now the flange of the T-beam with a thickness of t_f is on top, so start by assuming that neutral axis location distance y is greater than t_f. Divide the compression area of the concrete for the transformed section (Fig. 6-15) into three rectangles, then equate the first moments of areas of the concrete in compression and transformed area of the steel ($n \times A_s$) to get a quadratic equation for distance y. The solution for y gives the position of the neutral axis as

$$(b - b_w)t_f\left(y - \frac{t_f}{2}\right) + b_w y \frac{y}{2} - nA_s(d - y) = 0 \tag{f}$$

Solve Eq. (f) for y to get

$$y = 0.1702 \text{ m}$$

which exceeds t_f as assumed. The moment of inertia of the transformed section is now

$$I_T = \frac{b_w y^3}{3} + \frac{(b - b_w)t_f^3}{12} + (b - b_w)t_f\left(y - \frac{t_f}{2}\right)^2 + nA_s(d - y)^2$$
$$= 3.7 \times 10^{-3}\ \text{m}^4$$

3. ***Analyze***: Finally, repeat the solutions for maximum permissible moment M in Eqs. (d) and (e) as

$$M_c = \frac{\sigma_{ac}}{y} I_T = \frac{(9.3\ \text{MPa})}{0.1702\ \text{m}}(3.7 \times 10^{-3}\ \text{m}^4) = 202\ \text{kN} \cdot \text{m} \quad \textbf{(g)}$$

$$M_s = \frac{\sigma_{as}}{n(d - y)} I_T = \frac{137\ \text{MPa}}{8(0.6\ \text{m} - 0.1702\ \text{m})}(3.7 \times 10^{-3}\ \text{m}^4) \quad \textbf{(h)}$$
$$= 147.4\ \text{kN} \cdot \text{m}$$

4. ***Finalize***: Once again, the lower value of moment M based on allowable stress in the steel governs. Since the allowable stress in the reinforcing steel bars controls both beams, their moment capacities [Eqs. (e) and (h)] are approximately the same.

6.4 Doubly Symmetric Beams with Inclined Loads

Previous discussions of bending dealt with beams possessing a longitudinal plane of symmetry (the x-y plane in Fig. 6-16) and supporting lateral loads acting in that same plane. Under these conditions, the bending stresses are obtained from the flexure formula [Eq. (5-14)], provided that the material is homogeneous and linearly elastic.

FIGURE 6-16

Beam with a lateral load acting in a plane of symmetry

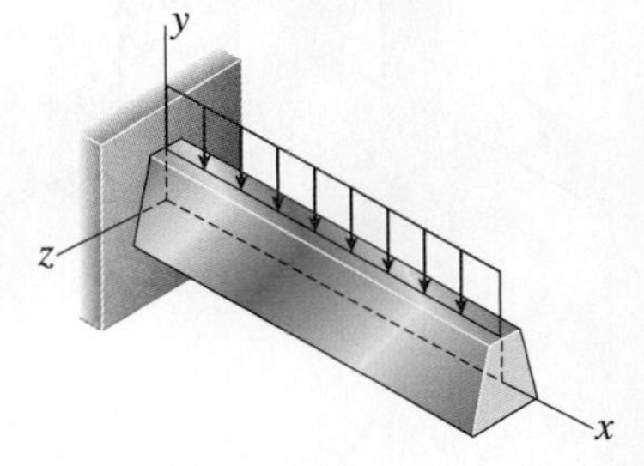

In this section, those ideas are extended to consider what happens when the beam is subjected to loads that do not act in the plane of symmetry, that is, **inclined loads** (Fig. 6-17). The discussion is limited to beams that have a doubly symmetric cross section, that is, both the x-y and x-z planes are planes of symmetry. Also, the inclined loads must act through the centroid of the cross section to avoid twisting the beam about the longitudinal axis.

The bending stresses in the beam shown in Fig. 6-17 are found by resolving the inclined load into two components, one acting in each plane of symmetry. Then the bending stresses are obtained from the flexure formula for each load component acting separately, and the final stresses are obtained by superposing the separate stresses.

Sign Conventions for Bending Moments

As a preliminary matter, first establish sign conventions for the bending moments acting on cross sections of a beam.[1] For this purpose, cut through the beam and consider a typical cross section (Fig. 6-18). The bending moments M_y and M_z acting about the y and z axes, respectively, are represented as vectors using

[1] The directions of the normal and shear stresses in a beam are usually apparent from an inspection of the beam and its loading; therefore, stresses are often calculated by ignoring sign conventions and using only absolute values. However, when deriving general formulas, maintain rigorous sign conventions to avoid ambiguity in the equations.

FIGURE 6-17

Doubly symmetric beam with an inclined load

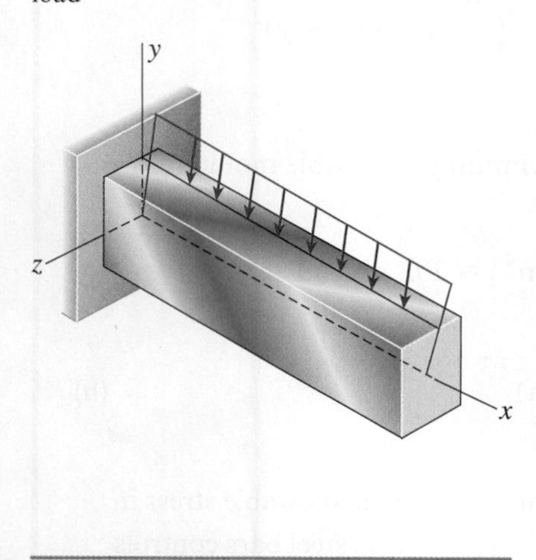

double-headed arrows. The moments are positive when their vectors point in the positive directions of the corresponding axes, and the right-hand rule for vectors gives the direction of rotation (indicated by the curved arrows in the figure).

From Fig. 6-18, note that a positive bending moment M_y produces compression on the right-hand side of the beam (the negative z side) and tension on the left-hand side (the positive z side). Similarly, a positive moment M_z produces compression on the upper part of the beam (where y is positive) and tension on the lower part (where y is negative). Also, note that the bending moments shown in Fig. 6-18 act on the positive x face of a segment of the beam, that is, on a face having its outward normal in the positive direction of the x axis.

Normal Stresses (Bending Stresses)

The normal stresses associated with the individual bending moments M_y and M_z are obtained from the flexure formula [Eq. (5-14)]. These stresses are then superposed to give the stresses produced by both moments acting simultaneously. For instance, consider the stresses at a point in the cross section having positive coordinates y and z (point A in Fig. 6-19). A positive moment M_y produces tension at this point, and a positive moment M_z produces compression; thus, the **normal stress** at point A is

$$\sigma_x = \frac{M_y z}{I_y} - \frac{M_z y}{I_z} \qquad \textbf{(6-19)}$$

FIGURE 6-18

Sign conventions for bending moments M_y and M_z

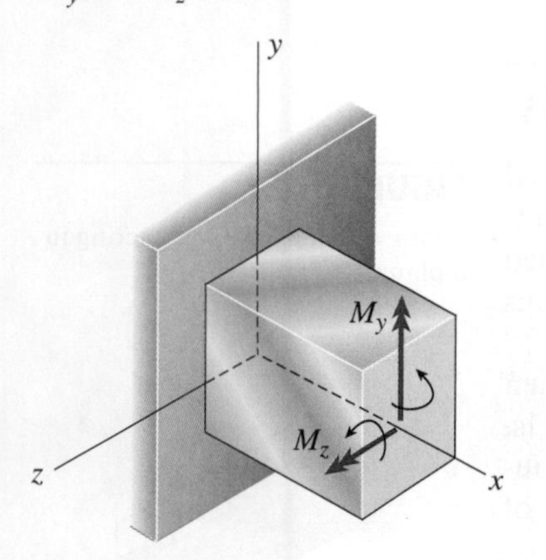

in which I_y and I_z are the moments of inertia of the cross-sectional area with respect to the y and z axes, respectively. Use this equation to find the normal stress at any point in the cross section by substituting the appropriate algebraic values of the moments and the coordinates.

Neutral Axis

The equation of the neutral axis is determined by equating the normal stress σ_x [Eq. (6-19)] to zero:

$$\frac{M_y}{I_y} z - \frac{M_z}{I_z} y = 0 \qquad \textbf{(6-20)}$$

This equation shows that the neutral axis nn is a straight line passing through the centroid C (Fig. 6-19). The angle β between the neutral axis and the z axis is determined as

$$\tan \beta = \frac{y}{z} = \frac{M_y I_z}{M_z I_y} \qquad \textbf{(6-21)}$$

FIGURE 6-19

Cross section of beam subjected to bending moments M_y and M_z

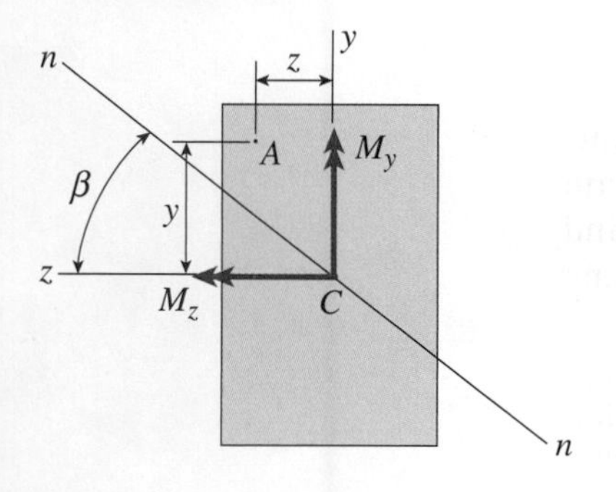

Depending upon the magnitudes and directions of the bending moments, the angle β may vary from $-90°$ to $+90°$. Knowing the orientation of the neutral axis is useful when determining the points in the cross section where the normal stresses are the largest. (Since the stresses vary linearly with distance from the neutral axis, the maximum stresses occur at points located farthest from the neutral axis.)

Relationship Between the Neutral Axis and the Inclination of the Loads

As noted previously, the orientation of the neutral axis with respect to the z axis is determined by the bending moments and the moments of inertia [Eq. (6-21)]. The next step is to determine the orientation of the neutral axis relative to the angle of inclination of the loads acting on the beam. For this purpose, use the cantilever beam shown in Fig. 6-20a as an example. The beam is loaded by a force P acting in the plane of the end cross section and inclined at an angle θ to the positive y axis. This particular orientation of the load is selected because it means that both bending moments (M_y and M_z) are positive when θ is between 0 and 90°.

The load P can be resolved into components $P \cos \theta$ in the positive y direction and $P \sin \theta$ in the negative z direction. Therefore, the bending moments M_y and M_z (Fig. 6-20b) acting on a cross section located at distance x from the fixed support are

$$M_y = (P \sin \theta)(L - x) \quad M_z = (P \cos \theta)(L - x) \qquad \textbf{(6-22a,b)}$$

in which L is the length of the beam. The ratio of these moments is

$$\frac{M_y}{M_z} = \tan \theta \qquad \textbf{(6-23)}$$

which shows that the resultant moment vector M is at the angle θ from the z axis (Fig. 6-20b). Consequently, the resultant moment vector is perpendicular to the longitudinal plane containing the force P.

The angle β between the neutral axis nn and the z axis (Fig. 6-20b) is obtained from Eq. (6-21):

$$\tan \beta = \frac{M_y I_z}{M_z I_y} = \frac{I_z}{I_y} \tan \theta \qquad \textbf{(6-24)}$$

which shows that the angle β is generally *not* equal to the angle θ. Thus, *except in special cases, the neutral axis is not perpendicular to the longitudinal plane containing the load.*

Exceptions to this general rule occur in three **special cases**:

1. When the load lies in the x-y plane ($\theta = 0$ or 180°), which means that the z axis is the neutral axis.
2. When the load lies in the x-z plane ($\theta = \pm 90°$), which means that the y axis is the neutral axis.
3. When the principal moments of inertia are equal, that is, when $I_y = I_z$.

In the third case, all axes through the centroid are principal axes and all have the same moment of inertia. The plane of loading, no matter what its direction, is always a principal plane, and the neutral axis is always perpendicular to it. (This situation occurs with square, circular, and certain other cross sections, as described in Section D.8 of Appendix D.)

The fact that the neutral axis is not necessarily perpendicular to the plane of the load can greatly affect the stresses in a beam, especially if the ratio of the principal moments of inertia is very large. Under these conditions, the stresses in the beam are very sensitive to slight changes in the direction of the load and to irregularities in the alignment of the beam itself. This characteristic of certain beams is illustrated next in Example 6-5.

FIGURE 6-20

Doubly symmetric beam with an inclined load P acting at an angle θ to the positive y axis

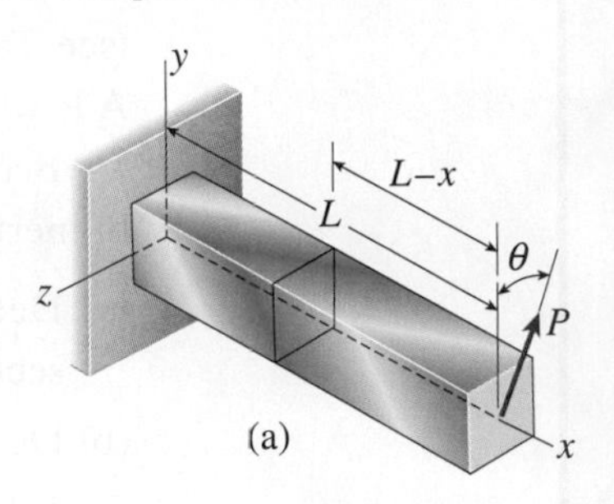

(a)

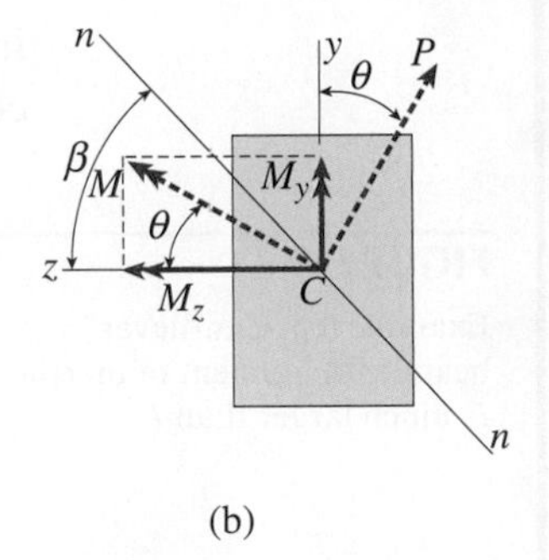

(b)

Example 6-5

A 4-m long cantilever beam (Fig. 6-21a) is constructed from an IPN 500 section (see Table F-2 of Appendix F for the dimensions and properties of this beam). A load $P = 45$ kN acts in the vertical direction at the end of the beam.

Because the beam is very narrow compared to its height (Fig. 6-21b), its moment of inertia about the z axis is much larger than its moment of inertia about the y axis.

(a) Determine the maximum bending stresses in the beam if the y axis of the cross section is vertical and therefore aligned with the load P (Fig. 6-21a).

(b) Determine the maximum bending stresses if the beam is inclined at a small angle $\alpha = 1°$ to the load P (Fig. 6-21b). (A small inclination can be caused by imperfections in the fabrication of the beam, misalignment of the beam during construction, or movement of the supporting structure.)

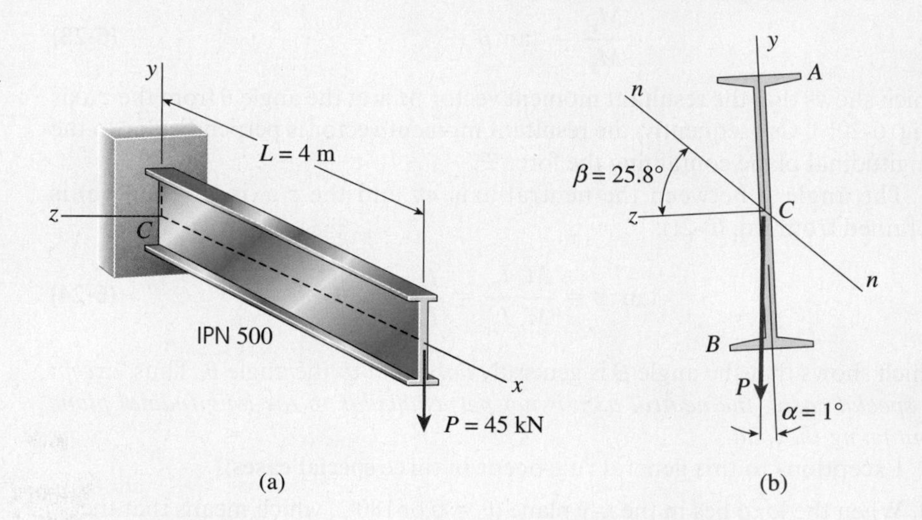

FIGURE 6-21

Example 6-5: Cantilever beam with moment of inertia I_z much larger than I_y

Solution:

Use a four-step problem-solving approach. Combine steps as needed for an efficient solution.

Part (a): Maximum bending stresses when the load is aligned with the y axis.

1, 2. ***Conceptualize, Categorize***: If the beam and load are in perfect alignment, the z axis is the neutral axis, and the maximum stresses in the beam (at the support) are obtained from the flexure formula:

$$\sigma_{max} = \frac{My}{I_z} = \frac{PL(h/2)}{I_z}$$

in which $M_z = -M = -PL$ and $M_y = 0$ so $M = PL$ is the bending moment at the support, h is the height of the beam, and I_z is the moment of inertia about the z axis.

3. ***Analyze***: Substituting numerical values gives

$$\sigma_{max} = \frac{(45 \text{ kN})(4000 \text{ mm})(250 \text{ mm})}{68740 \text{ cm}^4} = 65.5 \text{ MPa}$$

4. ***Finalize***: This stress is tensile at the top of the beam and compressive at the bottom of the beam.

Part (b): Maximum bending stresses when the load is inclined to the *y* axis.

1, 2. ***Conceptualize, Categorize***: Now assume that the beam has a small inclination (Fig. 6-21b), so that the angle between the y axis and the load is $\alpha = 1°$.

The components of the load P are $P \cos \alpha$ in the negative y direction and $P \sin \alpha$ in the positive z direction.

3. ***Analyze***: The bending moments at the support are

$$M_y = -(P \sin \alpha)L = -(45 \text{ kN})(\sin 1°)(4000 \text{ mm}) = -3.14 \text{ kN}\cdot\text{m}$$

$$M_z = -(P \cos \alpha)L = -(45 \text{ kN})(\cos 1°)(4000 \text{ mm}) = -180 \text{ kN}\cdot\text{m}$$

The angle β giving the orientation of the neutral axis nn (Fig. 6-21b) is obtained from Eq. (6-21):

$$\tan \beta = \frac{y}{z} = \frac{M_y I_z}{M_z I_y} = \frac{(-3.14 \text{ kN}\cdot\text{m})(68740 \text{ cm}^4)}{(-180 \text{ kN}\cdot\text{m})(2480 \text{ cm}^4)} = 0.878 \quad \beta = 25.8°$$

This calculation shows that the neutral axis is inclined at an angle of 25.8° from the z axis even though the plane of the load is inclined only 1° from the y axis. The sensitivity of the position of the neutral axis to the angle of the load is a consequence of the large I_z/I_y ratio.

From the position of the neutral axis (Fig. 6-21b), note that the maximum stresses in the beam occur at points A and B, which are located at the farthest distances from the neutral axis. The coordinates of point A are

$$z_A = -92.5 \text{ mm} \quad y_A = 250 \text{ mm}$$

Therefore, the tensile stress at point A [see Eq. (6-19)] is

$$\sigma_A = \frac{M_y z_A}{I_y} - \frac{M_z y_A}{I_z}$$

$$= \frac{(-3.14 \text{ kN}\cdot\text{m})(-92.5 \text{ mm})}{2480 \text{ cm}^4} - \frac{(-180 \text{ kN}\cdot\text{m})(250 \text{ mm})}{68740 \text{ cm}^4}$$

$$= 11.7 \text{ MPa} + 65.5 \text{ MPa} = 77.2 \text{ MPa}$$

The stress at B has the same magnitude but is a compressive stress:

$$\sigma_B = -77.2 \text{ MPa}$$

4. ***Finalize***: These stresses are 18% larger than the stress $\sigma_{max} = 65.5$ MPa for the same beam with a perfectly aligned load. Furthermore, the inclined load produces a lateral deflection in the z direction, whereas the perfectly aligned load does not.

This example shows that beams with I_z much larger than I_y may develop large stresses if the beam or its loads deviate even a small amount from their planned alignment. Therefore, such beams should be used with caution, because they are highly susceptible to overstress and to lateral (that is, sideways) bending and buckling. The remedy is to provide adequate lateral support for the beam, thereby preventing sideways bending. For instance, wood floor joists in buildings are supported laterally by installing bridging or blocking between the joists.

Example 6-6

A wood beam AB of rectangular cross section serving as a roof purlin (Figs. 6-22a and b) is simply supported by the top chords of two adjacent roof trusses. The beam supports the weight of the roof sheathing and the roofing material, plus its own weight and any additional loads that affect the roof (such as wind, snow, and earthquake loads).

FIGURE 6-22

Example 6-6: Wood beam of rectangular cross section serving as a roof purlin

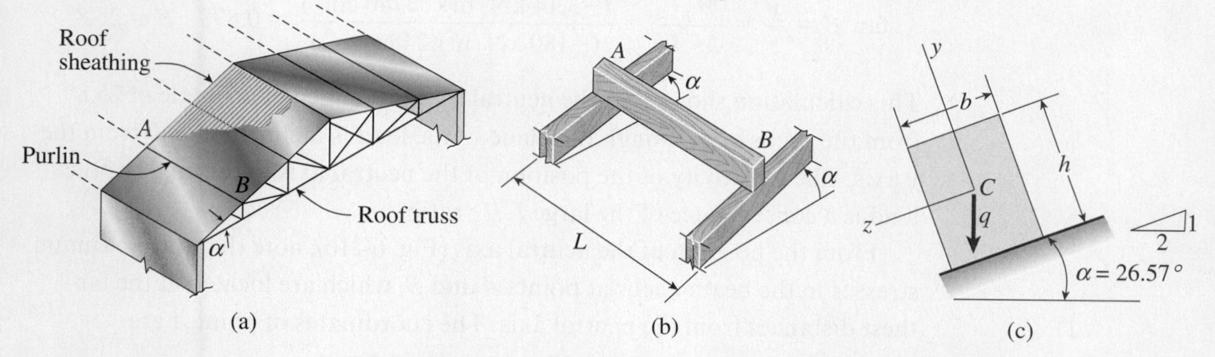

In this example, consider only the effects of a uniformly distributed load of intensity $q = 3.0$ kN/m acting in the vertical direction through the centroids of the cross sections (Fig. 6-22c). The load acts along the entire length of the beam and includes the weight of the beam. The top chords of the trusses have a slope of 1 on 2 ($\alpha = 26.57°$), and the beam has width $b = 100$ mm, height $h = 150$ mm, and span $L = 1.6$ m.

Determine the maximum tensile and compressive stresses in the beam and locate the neutral axis.

Solution:

Use a four-step problem-solving approach. Combine steps as needed for an efficient solution.

1, 2. *Conceptualize, Categorize*:

Loads and bending moments: The uniform load q acting in the vertical direction can be resolved into components in the y and z directions (Fig. 6-23a):

$$q_y = q \cos \alpha \quad q_z = q \sin \alpha \tag{6-25a,b}$$

FIGURE 6-23

Solution to Example 6-6: (a) Components of the uniform load, (b) bending moments acting on a cross section, and (c) normal stress distribution

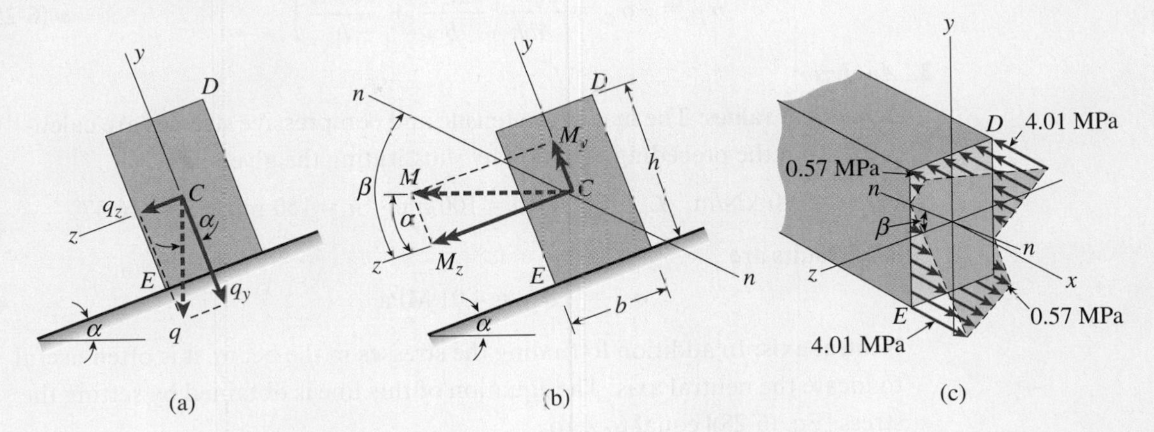

The maximum bending moments occur at the midpoint of the beam and are found from the general formula $M = qL^2/8$; hence,

$$M_y = \frac{q_z L^2}{8} = \frac{qL^2 \sin\alpha}{8} \qquad M_z = \frac{q_y L^2}{8} = \frac{qL^2 \cos\alpha}{8} \tag{6-26a,b}$$

Both of these moments are positive because their vectors are in the positive directions of the y and z axes (Fig. 6-23b).

Moments of inertia: The moments of inertia of the cross-sectional area with respect to the y and z axes are

$$I_y = \frac{hb^3}{12} \qquad I_z = \frac{bh^3}{12} \tag{6-27a,b}$$

Bending stresses: The stresses at the mid-section of the beam are obtained from Eq. (6-19) with the bending moments given by Eqs. (6-26) and the moments of inertia given by Eqs. (6-27):

$$\sigma_x = \frac{M_y z}{I_y} - \frac{M_z y}{I_z} = \frac{qL^2 \sin\alpha}{8hb^3/12} z - \frac{qL^2 \cos\alpha}{8bh^3/12} y$$
$$= \frac{3qL^2}{2bh}\left(\frac{\sin\alpha}{b^2} z - \frac{\cos\alpha}{h^2} y\right) \tag{6-28}$$

The stress at any point in the cross section can be obtained from this equation by substituting the coordinates y and z of the point.

From the orientation of the cross section and the directions of the loads and bending moments (Fig. 6-23), it is apparent that the maximum compressive stress occurs at point D (where $y = h/2$ and $z = -b/2$) and the maximum tensile stress occurs at point E (where $y = -h/2$ and $z = b/2$). Substitute these coordinates into Eq. (6-28) and then simplify to obtain expressions for the maximum and minimum stresses in the beam:

$$\sigma_E = -\sigma_D = \frac{3qL^2}{4bh}\left(\frac{\sin\,\alpha}{b} + \frac{\cos\,\alpha}{h}\right) \tag{6-29}$$

3. *Analyze*:

Numerical values: The maximum tensile and compressive stresses are calculated from the preceding equation by substituting the given data:

$$q = 3.0 \text{ kN/m} \quad L = 1.6 \text{ m} \quad b = 100 \text{ mm} \quad h = 150 \text{ mm} \quad \alpha = 26.57°$$

The results are

$$\sigma_E = -\sigma_D = 4.01 \text{ MPa}$$

Neutral axis: In addition to finding the stresses in the beam, it is often useful to locate the neutral axis. The equation of this line is obtained by setting the stress [Eq. (6-28)] equal to zero:

$$\frac{\sin\,\alpha}{b^2}z - \frac{\cos\,\alpha}{h^2}y = 0 \tag{6-30}$$

The neutral axis is shown in Fig. 6-23b as line *nn*. The angle β from the z axis to the neutral axis is obtained from Eq. (6-30) as

$$\tan\,\beta = \frac{y}{z} = \frac{h^2}{b^2}\tan\,\alpha \tag{6-31}$$

Substituting numerical values gives

$$\tan\,\beta = \frac{h^2}{b^2}\tan\,\alpha = \frac{(150 \text{ mm})^2}{(100 \text{ mm})^2}(\tan 26.57°) = 1.125 \quad \beta = 48.4°$$

Since the angle β is not equal to the angle α, the neutral axis is inclined to the plane of loading (which is vertical).

4. *Finalize*: From the orientation of the neutral axis (Fig. 6-23b), note that points D and E are the farthest from the neutral axis, thus confirming the assumption that the maximum stresses occur at those points. The part of the beam above and to the right of the neutral axis is in compression, and the part to the left and below the neutral axis is in tension.

6.5 Bending of Unsymmetric Beams

In previous discussions of bending, the beams were assumed to have cross sections with at least one axis of symmetry. Now that restriction is removed, and beams may have unsymmetric cross sections. First beams in pure bending are considered; then in later sections (Sections 6.6 through 6.9) the effects of lateral loads are investigated. As in earlier discussions, it is assumed that the beams are made of linearly elastic materials.

Suppose that a beam having an unsymmetric cross section is subjected to a bending moment M acting at the end cross section (Fig. 6-24a). Of interest here are the stresses in the beam and the position of the neutral axis. Unfortunately, at this stage of the analysis, there is no direct way of determining these quantities. Therefore, an indirect approach must be used—instead of starting with a bending moment and trying to find the neutral axis, start with an assumed neutral axis and find the associated bending moment.

FIGURE 6-24

Unsymmetric beam subjected to a bending moment M

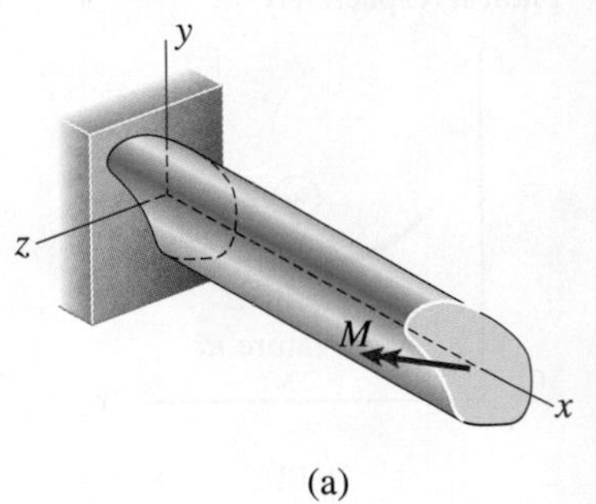

(a)

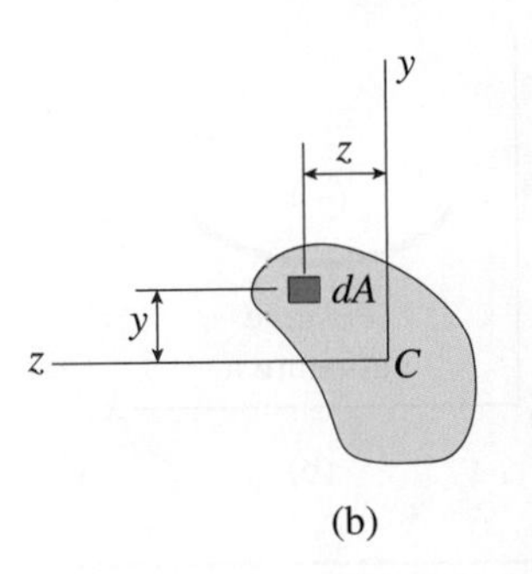

(b)

Neutral Axis

Begin by constructing two perpendicular axes (the y and z axes) at an arbitrarily selected point in the plane of the cross section (Fig. 6-24b). The axes may have any orientation, but for convenience, orient them horizontally and vertically. Next, *assume* that the beam is bent in such a manner that the z axis is the neutral axis of the cross section. Consequently, the beam deflects in the x-y plane, which becomes the plane of bending. Under these conditions, the normal stress acting on an element of area dA located at distance y from the neutral axis [see Fig. 6-24b and Eq. (5-8) of Chapter 5] is

$$\sigma_x = -E\kappa_y y \tag{6-32}$$

The minus sign is needed because the part of the beam above the z axis (the neutral axis) is in compression when the curvature is positive. (The sign convention for curvature when the beam is bent in the x-y plane is shown in Fig. 6-25a.)

The force acting on the element of area dA is $\sigma_x dA$, and the resultant force acting on the entire cross section is the integral of this elemental force over the cross-sectional area A. Since the beam is in pure bending, the resultant force must be zero; hence,

$$\int_A \sigma_x dA = -\int_A E\kappa_y y dA = 0$$

The modulus of elasticity and the curvature are constants at any given cross section, and therefore

$$\int_A y dA = 0 \tag{6-33}$$

This equation shows that the z axis (the neutral axis) passes through the centroid C of the cross section.

Now assume that the beam is bent in such a manner that the y axis is the neutral axis and the x-z plane is the plane of bending. Then the normal stress acting on the element of area dA (Fig. 6-24b) is

$$\sigma_x = -E\kappa_z z \tag{6-34}$$

The sign convention for the curvature κ_z in the x-z plane is shown in Fig. 6-25b. The minus sign is needed in Eq. (6-34) because positive curvature in the x-z plane produces compression on the element dA. The resultant force for this case is

$$\int_A \sigma_x dA = -\int_A E\kappa_z z dA = 0$$

which gives

$$\int_A z dA = 0 \tag{6-35}$$

and again note that the neutral axis must pass through the centroid. Thus, it follows that *the origin of the y and z axes for an unsymmetric beam must be placed at the centroid C.*

Unsymmetric composite beam made up from channel section and old wood beam

FIGURE 6-25

Sign conventions for curvatures κ_y and κ_z in the *x-y* and *x-z* planes, respectively

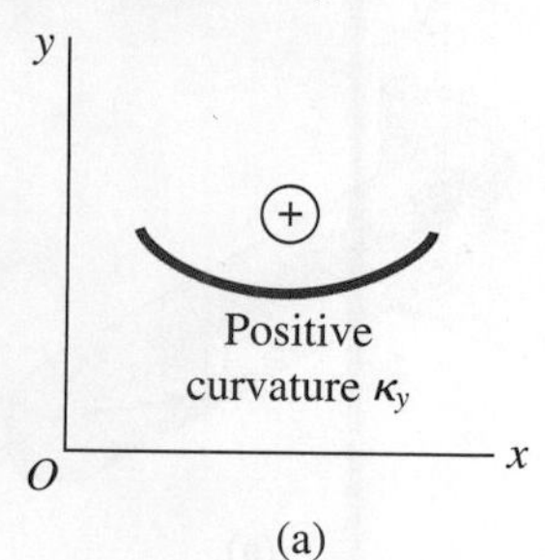

(a)

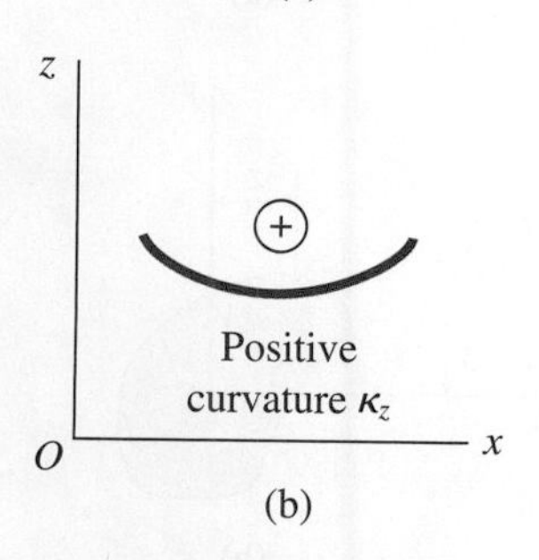

(b)

FIGURE 6-26

Bending moments M_y and M_z acting about the *y* and *z* axes, respectively

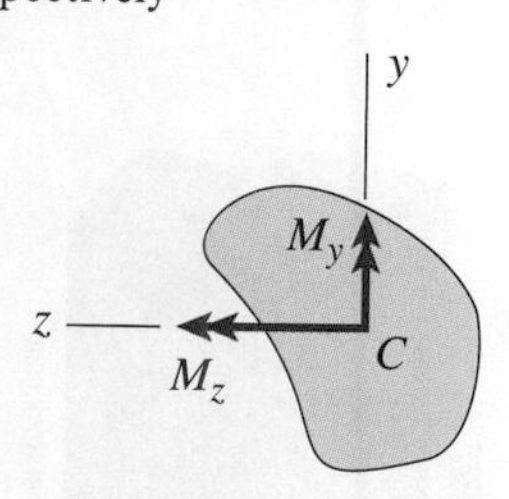

Now consider the moment resultant of the stresses σ_x. Once again assume that bending takes place with the *z* axis as the neutral axis, in which case the stresses σ_x are given by Eq. (6-32). The corresponding bending moments M_z and M_y about the *z* and *y* axes, respectively (Fig. 6-26), are

$$M_z = -\int_A \sigma_x y dA = \kappa_y E \int_A y^2 dA = \kappa_y E I_z \qquad \textbf{(6-36a)}$$

$$M_y = \int_A \sigma_x z dA = -\kappa_y E \int_A yz dA = -\kappa_y E I_{yz} \qquad \textbf{(6-36b)}$$

In these equations, I_z is the moment of inertia of the cross-sectional area with respect to the *z* axis and I_{yz} is the *product of inertia* with respect to the *y* and *z* axes.[2]

Equations (6-36a and b) lead to the following conclusions: (1) If the *z* axis is selected in an arbitrary direction through the centroid, it will be the neutral axis *only* if moments M_y and M_z act about the *y* and *z* axes and *only* if these moments are in the ratio established by Eqs. (6-36a and b). (2) If the *z* axis is selected as a *principal axis*, then the product of inertia I_{yz} equals zero and the only bending moment is M_z. In that case, the *z* axis is the neutral axis, bending takes place in the *x-y* plane, and the moment M_z acts in that same plane. Thus, bending occurs in a manner analogous to that of a symmetric beam.

In summary, an unsymmetric beam bends in the same general manner as a symmetric beam provided the *z* axis is a *principal centroidal axis* and the only bending moment is the moment M_z acting about that same axis.

Now assume that the *y* axis is the neutral axis to arrive at similar conclusions. The stresses σ_x are given by Eq. (6-34) and the bending moments are

$$M_y = \int_A \sigma_x z dA = -\kappa_z E \int_A z^2 dA = -\kappa_z E I_y \qquad \textbf{(6-37a)}$$

$$M_z = -\int_A \sigma_x y dA = \kappa_z E \int_A yz dA = \kappa_z E I_{yz} \qquad \textbf{(6-37b)}$$

in which I_y is the moment of inertia with respect to the *y* axis. Again observe that if the neutral axis (the *y* axis in this case) is oriented arbitrarily, moments M_y and M_z must exist. However, if the *y* axis is a principal axis, the only moment is M_y and there is ordinary bending in the *x-z* plane. Therefore, an unsymmetric beam bends in the same general manner as a symmetric beam when the *y* axis is a *principal centroidal axis* and the only bending moment is the moment M_y acting about that same axis.

One further observation—since the *y* and *z* axes are orthogonal, if *either* axis is a principal axis, then the other axis is automatically a principal axis.

This discussion leads to the following important conclusion: *When an unsymmetric beam is in pure bending, the plane in which the bending moment acts is perpendicular to the neutral surface only if the y and z axes are principal centroidal axes of the cross section and the bending moment acts in one of the two principal planes (the x-y plane or the x-z plane).* In such a case, the principal plane in which the bending moment acts becomes the plane of bending, and the usual bending theory (including the flexure formula) is valid.

This conclusion leads to a direct method for finding the stresses in an unsymmetric beam subjected to a bending moment acting in an arbitrary direction.

[2]Products of inertia are discussed in Section D.6 of Appendix D.

Procedure for Analyzing an Unsymmetric Beam

A general procedure is now presented for analyzing an unsymmetric beam subjected to any bending moment M (Fig. 6-27). Begin by locating the centroid C of the cross section and constructing a set of principal axes at that point (the y and z axes in the figure).[3] Next, the bending moment M is resolved into components M_y and M_z, which are positive in the directions shown in the figure. These components are

$$M_y = M \sin \theta \qquad M_z = M \cos \theta \tag{6-38a,b}$$

in which θ is the angle between the moment vector M and the z axis (Fig. 6-27). Since each component acts in a principal plane, it produces pure bending in that same plane. Thus, the usual formulas for pure bending apply, and the stresses are computed using the moments M_y and M_z acting separately. The bending stresses obtained from the moments acting separately are then superposed to obtain the stresses produced by the original bending moment M. (Note that this general procedure is similar to that described in the preceding section for analyzing doubly symmetric beams with inclined loads.)

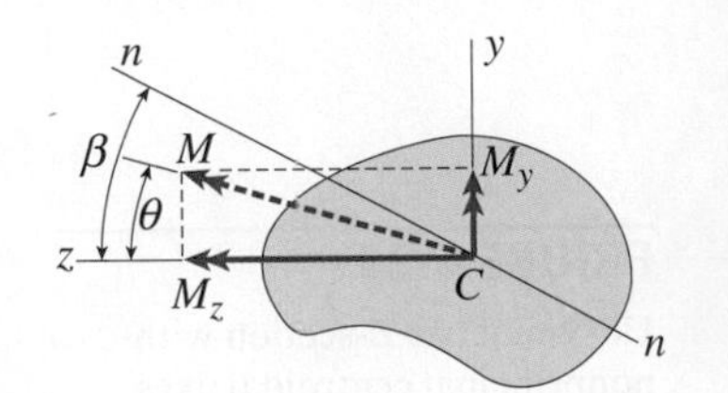

FIGURE 6-27

Unsymmetric cross section with the bending moment M resolved into components M_y and M_z acting about the principal centroidal axes

The superposition of the bending stresses in order to obtain the resultant stress at any point in the cross section is given by Eq. (6-19):

$$\sigma_x = \frac{M_y z}{I_y} - \frac{M_z y}{I_z} = \frac{(M \sin \theta)z}{I_y} - \frac{(M \cos \theta)y}{I_z} \tag{6-39}$$

in which y and z are the coordinates of the point under consideration.

Also, the equation of the neutral axis nn (Fig. 6-27) is obtained by setting σ_x equal to zero and simplifying:

$$\frac{\sin \theta}{I_y} z - \frac{\cos \theta}{I_z} y = 0 \tag{6-40}$$

The angle β between the neutral axis and the z axis can be obtained from the preceding equation, as

$$\tan \beta = \frac{y}{z} = \frac{I_z}{I_y} \tan \theta \tag{6-41}$$

This equation shows that in general the angles β and θ are not equal; hence the neutral axis is generally not perpendicular to the plane in which the applied couple M acts. The only exceptions are the three special cases described in the preceding section in the paragraph following Eq. (6-24).

This section has focused attention on unsymmetric beams. Of course, symmetric beams are special cases of unsymmetric beams; therefore, the discussions of this section also apply to symmetric beams. If a beam is singly symmetric, the axis of symmetry is one of the centroidal principal axes of the cross section; the other principal axis is perpendicular to the axis of symmetry at the centroid. If a beam is doubly symmetric, the two axes of symmetry are centroidal principal axes.

In a strict sense, the discussions of this section apply only to pure bending, which means that no shear forces act on the cross sections. *When shear forces do exist, the possibility arises that the beam will twist about the longitudinal axis.*

[3]Principal axes are discussed in Sections D.7 and D.8 of Appendix D.

However, twisting is avoided when the shear forces act through the *shear center*, which is described in the next section.

Example 6-7 illustrates the analysis of a beam having one axis of symmetry. The calculations for an unsymmetric beam having no axes of symmetry (see Example 6-8) proceed in the same general manner, except that the determination of the various cross-sectional properties is much more complex.

Alternate Procedure for Analyzing an Unsymmetric Beam

FIGURE 6-28

Unsymmetric Z-section with nonprincipal centroidal axes

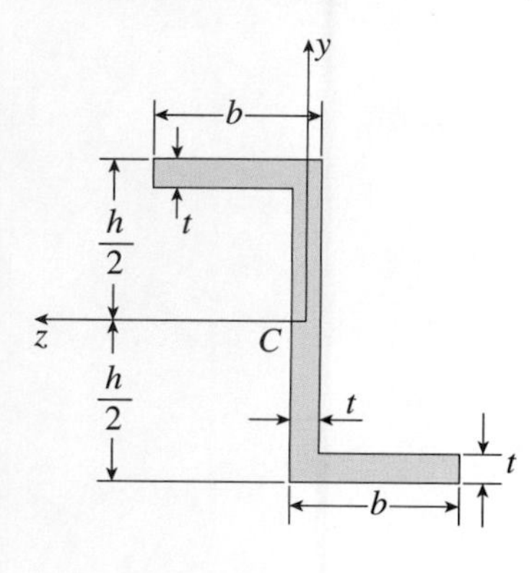

In the preceding discussion, an unsymmetric beam was analyzed by locating the principal centroidal axes of the cross section and then resolving the bending moment into components in those directions. The advantage of this method is that all standard formulas for stresses and deflections apply because bending takes place in a principal plane. However, the method is inconvenient when the orientation of the principal axes cannot be obtained by inspection (as when an axis of symmetry exists) or from tables (as in the case of standard angle sections). If the orientation of the principal axes and the magnitudes of the principal moments of inertia must be calculated, such as for the Z-section in Fig. 6-28, it may be easier to work with the **nonprincipal centroidal axes** that are aligned with the sides of the cross section.

To derive the equations of a generalized bending theory for nonprincipal axes, consider the unsymmetric cross section of Fig. 6-29. The *y* and *z* axes have their origin at the centroid, but they are not principal axes. Bending moments M_y and M_z act on the cross section, and bending of the beam occurs in both the *x-y* and *x-z* planes, neither of which is a principal plane. If the curvatures in the *x-y* and *x-z* planes are κ_y and κ_z, respectively (see Fig. 6-25), the normal stress at point *A* is

FIGURE 6-29

Unsymmetric cross section with the bending moment M resolved into components M_y and M_z that act about the nonpricipal centroidal axes

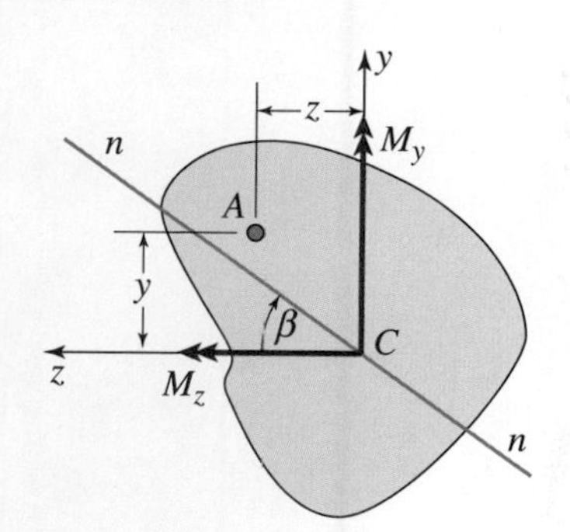

$$\sigma_x = -\kappa_y E y - \kappa_z E z \tag{6-42}$$

The resultant (axial) force acting on the cross section in the *x* direction is equated to zero to obtain

$$\int \sigma_x dA = 0 \quad \text{or} \quad \kappa_y E \int y dA + \kappa_z E \int z dA = 0 \tag{6-43}$$

This equation is satisfied automatically because the origin of the axes is at the centroid of the cross section.

The moment M_y is the moment stress resultant about the *y* axis:

$$M_y = \int \sigma_x z dA = -\kappa_y E \int yz dA - \kappa_z E \int z^2 dA \tag{6-44a}$$

or

$$M_y = -\kappa_y E I_{yz} - \kappa_z E I_y \tag{6-44b}$$

in which I_{yz} is the product of inertia of the cross-sectional area with respect to the *y* and *z* axes. In a similar way, the expression for moment about the *z* axis is

$$M_z = -\int \sigma_x y dA = \kappa_y E \int y^2 dA + \kappa_z E \int yz dA \tag{6-45a}$$

or

$$M_z = \kappa_y E I_z + \kappa_z E I_{yz} \tag{6-45b}$$

Solve Eqs. (6-44) and (6-45) simultaneously to obtain the expressions for curvatures in terms of the bending moments:

$$\kappa_y = \frac{M_z I_y + M_y I_{yz}}{E(I_y I_z - I_{yz}^2)} \quad \text{and} \quad \kappa_z = -\frac{M_y I_z + M_z I_{yz}}{E(I_y I_z - I_{yz}^2)} \tag{6-46}$$

Now substitute these expressions for curvatures into Eq. (6-42) to obtain the *normal stress* σ_x at any point in an unsymmetric beam when moments M_y and M_z are known:

$$\sigma_x = \frac{(M_y I_z + M_z I_{yz})z - (M_z I_y + M_y I_{yz})y}{I_y I_z - I_{yz}^2} \tag{6-47}$$

Equation (6-47) is the *generalized flexure formula* for an unsymmetric beam acted on by moments about perpendicular centroidal axes that are not necessarily principal axes. Note that if the y and z axes are principal centroidal axes, Eq. (6-47) reduces to Eq. (6-39), since $I_{yz} = 0$ for principal axes.

The orientation of the *neutral axis nn* (Fig. 6-29) is obtained by equating σ_x to zero to find an expression for angle β between the z axis and the neutral axis:

$$\tan \beta = \frac{y}{z} = \frac{M_y I_z + M_z I_{yz}}{M_z I_y + M_y I_{yz}} \tag{6-48}$$

The plane of bending is perpendicular to the neutral axis.

See Example 6-8 for calculation of bending stresses in a Z-section using both the flexure formula for principal centroidal axes and the generalized flexure formula.

Example 6-7

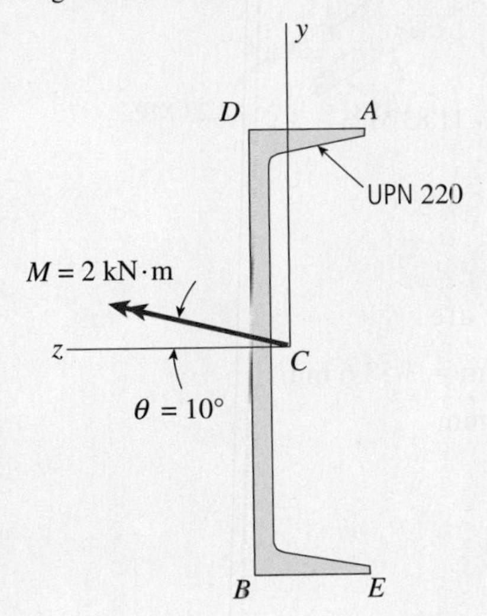

FIGURE 6-30

Example 6-7: Channel section subjected to a bending moment M acting at an angle θ to the z axis

A channel section UPN 220 is subjected to a bending moment $M = 2\ \text{kN}\cdot\text{m}$ oriented at an angle $\theta = 10°$ to the z axis (Fig. 6-30).

Calculate the bending stresses σ_A and σ_B at points A and B, respectively, and determine the position of the neutral axis.

Solution:

Use a four-step problem-solving approach. Combine steps as needed for an efficient solution.

1, 2. *Conceptualize, Categorize*:

Properties of the cross section: The centroid C is located on the axis of symmetry (the z axis) at a distance

$$c = 2.14\ \text{cm}$$

from the back of the channel (Fig. 6-31).[4] The y and z axes are principal centroidal axes with moments of inertia:

$$I_y = 197\ \text{cm}^4 \quad I_z = 2690\ \text{cm}^4$$

[4]See Table F-3, Appendix F, for dimensions and properties of channel sections.

FIGURE 6-31

Solution to Example 6-7: (a) Cross section properties and (b) normal stress distribution in channel section

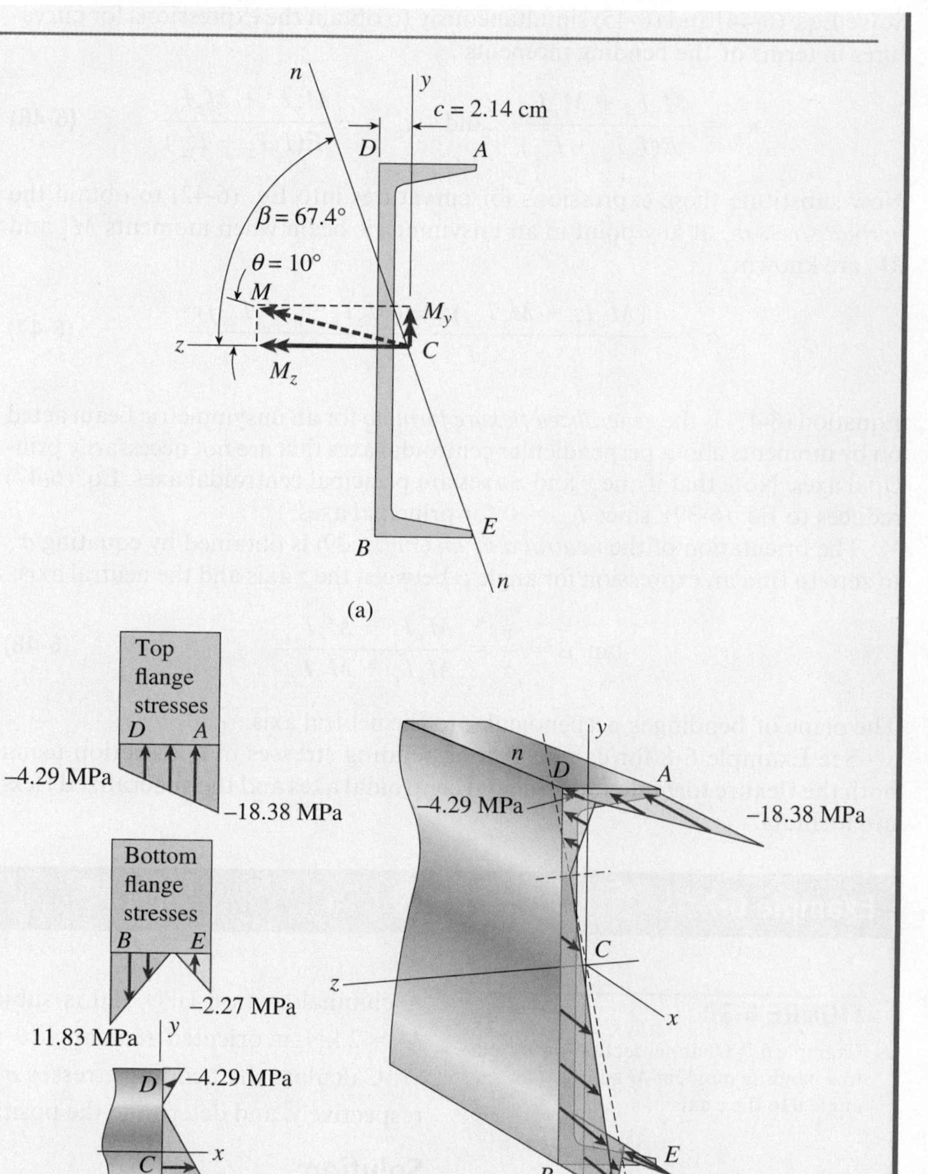

Also, the coordinates of points A, B, D, and E are

$$y_A = 110 \text{ mm} \quad z_A = -80 \text{ mm} + 21.4 \text{ mm} = -58.6 \text{ mm}$$

$$y_B = -110 \text{ mm} \quad z_B = 21.4 \text{ mm}$$

$$y_D = y_A,\ z_D = z_B$$

$$y_E = y_B,\ z_E = z_A$$

Bending moments: The bending moments about the y and z axes (Fig. 6-31a) are

$$M_y = M \sin\theta = (2\ \text{kN}\cdot\text{m})(\sin 10°) = 0.347\ \text{kN}\cdot\text{m}$$
$$M_z = M \cos\theta = (2\ \text{kN}\cdot\text{m})(\cos 10°) = 1.97\ \text{kN}\cdot\text{m}$$

3. ***Analyze*:**

Bending stresses: Now calculate the stress at point A from Eq. (6-39):

$$\sigma_A = \frac{M_y z_A}{I_y} - \frac{M_z y_A}{I_z}$$
$$= \frac{(0.347\ \text{kN}\cdot\text{m})(-0.0586\ \text{m})}{1.97 \times 10^{-6}\ \text{m}^4} - \frac{(1.97\ \text{kN}\cdot\text{m})(0.110\ \text{m})}{2.69 \times 10^{-5}\ \text{m}^4}$$
$$= -10.32\ \text{MPa} - 8.06\ \text{MPa} = -18.38\ \text{MPa}$$

By a similar calculation, the stress at point B is

$$\sigma_B = \frac{M_y z_B}{I_y} - \frac{M_z y_B}{I_z}$$
$$= \frac{(0.347\ \text{kN}\cdot\text{m})(0.0214\ \text{m})}{1.97 \times 10^{-6}\ \text{m}^4} - \frac{(1.97\ \text{kN}\cdot\text{m})(-0.110\ \text{m})}{2.69 \times 10^{-5}\ \text{m}^4}$$
$$= 3.77\ \text{MPa} + 8.06\ \text{MPa} = 11.83\ \text{MPa}$$

These stresses are the maximum compressive and tensile stresses in the beam.

The normal stresses at points D and E also can be computed using the procedure shown. Thus,

$$\sigma_D = -4.29\ \text{MPa}, \quad \sigma_E = -2.27\ \text{MPa}$$

The normal stresses acting on the cross section are shown in Fig. 6-31b.

Neutral axis: The angle β that locates the neutral axis [Eq. (6-41)] is found as

$$\tan\beta = \frac{I_z}{I_y}\tan\theta = \frac{2690\ \text{cm}^4}{197\ \text{cm}^4}\tan 10° = 2.408 \quad \beta = 67.4°$$

The neutral axis nn is shown in Fig. 6-31, and note that points A and B are located at the farthest distances from the neutral axis, thus confirming that σ_A and σ_B are the largest stresses in the beam.

4. ***Finalize*:** In this example, the angle β between the z axis and the neutral axis is much larger than the angle θ (Fig. 6-31) because the ratio I_z/I_y is large. The angle β varies from 0 to 67.4° as the angle θ varies from 0 to 10°. As discussed previously in Example 6-5 of Section 6.4, beams with large I_z/I_y ratios are very sensitive to the direction of loading. Thus, beams of this kind should be provided with lateral support to prevent excessive lateral deflections.

Example 6-8

A Z-section is subjected to bending moment $M = 3$ kN · m at an angle $\theta = -20°$ to the z axis, as shown in Fig. 6-32. Find the normal stresses at A, B, D, and E (σ_A, σ_B, σ_D and σ_E, respectively) and also find the position of the neutral axis. Compare the results obtained from use of the flexure formula for principal centroidal axes to those obtained from the generalized flexure formula. Use the following numerical data: $h = 200$ mm, $b = 90$ mm, and thickness $t = 15$ mm.

Solution:

Use a four-step problem-solving approach. Combine steps as needed for an efficient solution.

1, 2. ***Conceptualize, Categorize:***

Properties of the cross section: Use the results of Example D-7 in Section D.8 of Appendix D. Moments of inertia and orientation of principal centroidal axes are

$$I_z = 32.6 \times 10^6 \text{ mm}^4 \quad I_y = 2.4 \times 10^6 \text{ mm}^4$$

$$\theta_{p1} = 19.2° \qquad \theta_{p1} = (19.2)\frac{\pi}{180} \text{ radians}$$

Coordinates (y, z) of points A, B, D, D', E, and E' for the rotated principal coordinate axes (labeled y and z in Fig. 6-32) are

$$\theta = -20\left(\frac{\pi}{180}\right) \text{ radians}$$

$$y_A = \frac{h}{2}\cos(\theta_{p1}) + \left(b - \frac{t}{2}\right)\sin(\theta_{p1}) \qquad y_A = 121.569 \text{ mm}$$

$$y_B = -y_A \qquad y_B = -121.569 \text{ mm}$$

$$y_D = \frac{h}{2}\cos(\theta_{p1}) - \frac{t}{2}\sin(\theta_{p1}) \qquad y_D = 91.971 \text{ mm}$$

$$y_D{}' = \frac{h}{2}\cos(\theta_{p1}) \qquad y_D{}' = 94.438 \text{ mm}$$

$$y_E{}' = -y_D{}' \qquad y_E{}' = -94.438 \text{ mm}$$

$$y_E = -y_D \qquad y_E = -91.971 \text{ mm}$$

$$z_A = \left(b - \frac{t}{2}\right)\cos(\theta_{p1}) - \frac{h}{2}\sin(\theta_{p1}) \qquad z_A = 45.024 \text{ mm}$$

$$z_B = -z_A \qquad z_B = -45.024 \text{ mm}$$

$$z_D = \frac{-h}{2}\sin(\theta_{p1}) - \frac{t}{2}\cos(\theta_{p1}) \qquad z_D = -39.969 \text{ mm}$$

$$z_D{}' = \frac{-h}{2}\sin(\theta_{p1}) \qquad z_D{}' = -32.887 \text{ mm}$$

$$z_E{}' = -z_D{}' \qquad z_E{}' = 32.887 \text{ mm}$$

$$z_E = -z_D \qquad z_E = 39.969 \text{ mm}$$

FIGURE 6-32

(a) Z-section subjected to bending moment M at angle θ to principal z axis and (b) normal stress distribution in Z-section

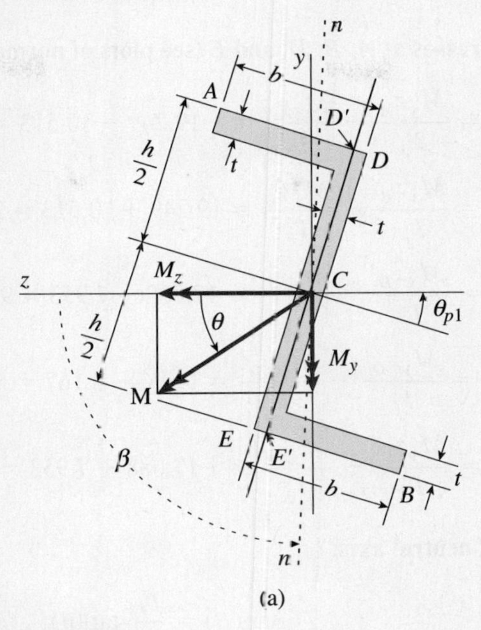

(a)

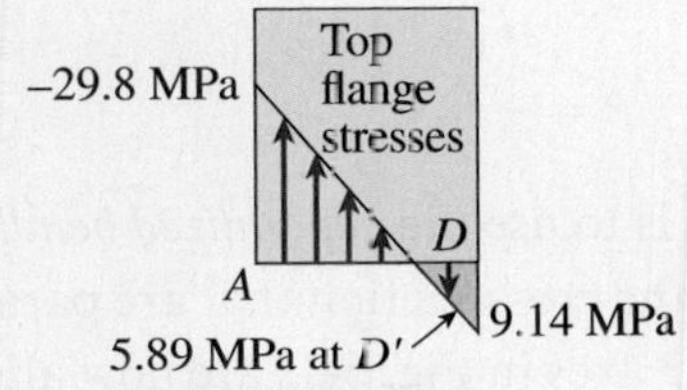

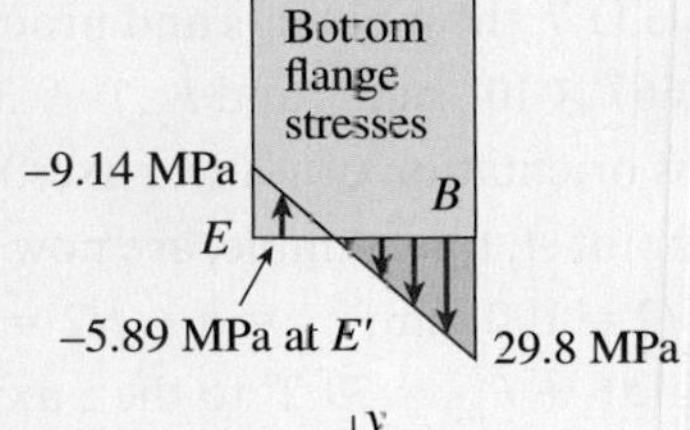

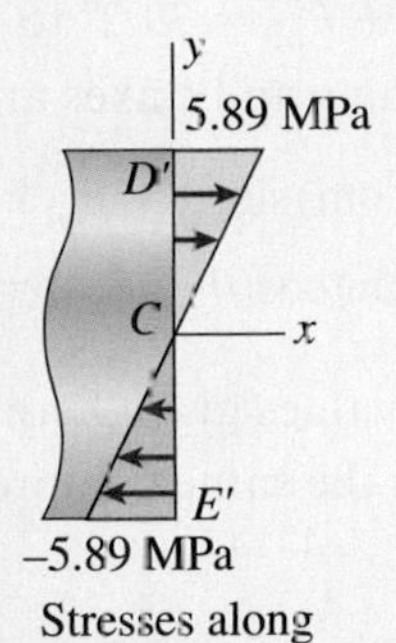

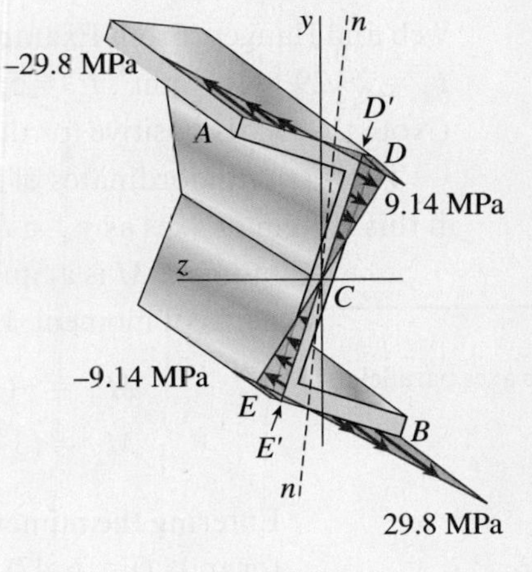

(b)

Bending moments (kN · m) $M = 3$ kN · m:

$$M_y = M\ \sin(\theta) \qquad M_y = -1.026\ \text{kN} \cdot \text{m}$$

$$M_z = M\ \cos(\theta) \qquad M_z = 2.819\ \text{kN} \cdot \text{m}$$

3. ***Analyze***:

Bending stresses at A, B, D, and E (see plots of normal stresses in Fig. 6-32b):

$$\sigma_A = \frac{M_y z_A}{I_y} - \frac{M_z y_A}{I_z} = -19.249 - 10.513 = -29.8 \text{ MPa}$$

$$\sigma_B = \frac{M_y z_B}{I_y} - \frac{M_z y_B}{I_z} = 19.249 + 10.513 = 29.8 \text{ MPa}$$

$$\sigma_D = \frac{M_y z_D}{I_y} - \frac{M_z y_D}{I_z} = 17.088 - 7.953 = 9.14 \text{ MPa}$$

$$\sigma_D' = \frac{M_y z_D'}{I_y} - \frac{M_z y_D'}{I_z} = 14.06 - 8.167 = 5.89 \text{ MPa} = -\sigma_E'$$

$$\sigma_E = \frac{M_y z_E}{I_y} - \frac{M_z y_E}{I_z} = -17.088 + 7.953 = -9.14 \text{ MPa}$$

Location of neutral axis:

$$\tan(\beta) = \frac{I_z}{I_y} \tan(\theta)$$

$$\beta = -78.6°$$

4. ***Finalize***: An alternate approach is to use the *generalized bending theory* for *nonprincipal centroidal axes* in the cross section that are parallel to the sides of the cross section. The y and z axes in Fig. 6-33 are now aligned with the web and flanges. From Example D-7, the moments and product of inertia are $I_z = 29.29 \times 10^6 \text{ mm}^4$, $I_y = 5.667 \times 10^6 \text{ mm}^4$, and $I_{yz} = 9.366 \times 10^6 \text{ mm}^4$. (Note that I_{yz} is positive for this orientation of y and z axes.)

The y and z coordinates of point A, for example, are now easily computed in this system of axes as $y_A = h/2 = 100$ mm, $z_A = b - t/2 = 82.5$ mm. Applied moment M is at angle $\theta + \theta_{p1} = 39.2°$ to the z axis, so the components of moment M along y and z axes are

$$M_y = -(3 \text{ kN} \cdot \text{m})\sin(\theta + \theta_{p1}) = -1.896 \text{ kN} \cdot \text{m}$$

$$M_z = (3 \text{ kN} \cdot \text{m})\cos(\theta + \theta_{p1}) = 2.325 \text{ kN} \cdot \text{m}$$

Entering the numerical values given here in the general flexure formula (Eq. 6-47) gives the same compressive stress value as that reported previously:

$$\sigma_A = \frac{(M_y I_z + M_z I_{yz}) z_A - (M_z I_y + M_y I_{yz}) y_A}{I_y I_z - I_{yz}^2} = -29.8 \text{ MPa}$$

Use of y and z coordinates of other points of interest on the cross section will give the normal stresses at B, D, and E listed in step 3.

FIGURE 6-33

Z-section with y and z axes parallel to sides of cross section

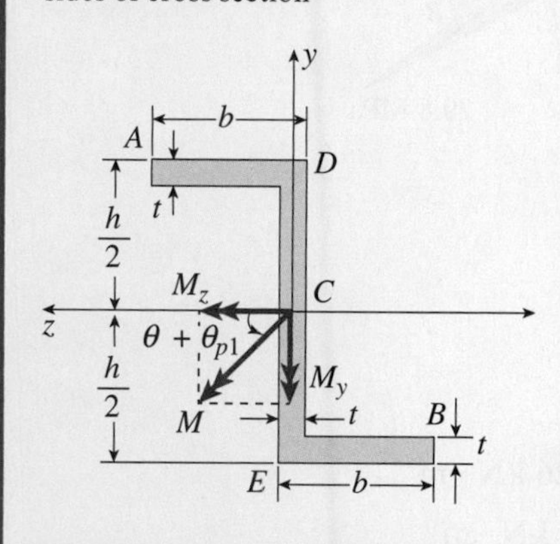

6.6 The Shear-Center Concept

Preceding sections of this chapter were concerned with determining the bending stresses in beams under a variety of special conditions. For instance, Section 6.4 considered symmetric beams with inclined loads, and Section 6.5 considered unsymmetric beams. However, lateral loads acting on a beam produce shear forces as well as bending moments, so in this and the next three sections the effects of shear are examined.

Chapter 5 discussed how to determine the shear stresses in beams when the loads act in a plane of symmetry and presented the derivation of the shear formula for calculating these stresses for certain shapes of beams. Now consider the shear stresses in beams when the lateral loads act in a plane that is *not* a plane of symmetry. You will see that the loads must be applied at a particular point in the cross section, called the *shear center*, if the beam is to bend without twisting.

Consider a cantilever beam of singly symmetric cross section supporting a load P at the free end (see Fig. 6-34a). A beam having the cross section shown in Fig. 6-34b is called an *unbalanced I-beam*. Beams of I-shape, whether balanced or unbalanced, are usually loaded in the plane of symmetry (the x-z plane), but in this case, the line of action of the force P is perpendicular to that plane. Since the origin of coordinates is taken at the centroid C of the cross section, and since the z axis is an axis of symmetry of the cross section, both the y and z axes are principal centroidal axes.

FIGURE 6-34

Cantilever beam with singly symmetric cross section: (a) beam with load and (b) intermediate cross section of beam showing stress resultants P and M_0, centroid C, and shear center S

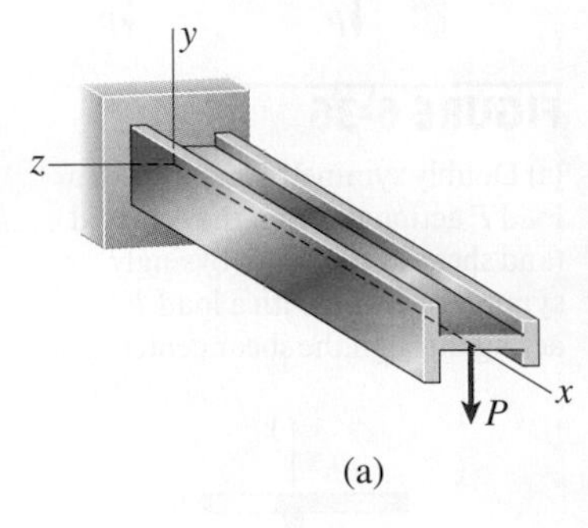

(a)

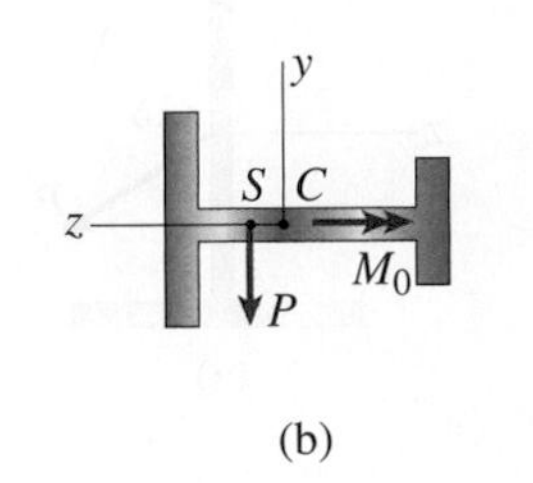

(b)

Assume that under the action of the load P the beam bends with the x-z plane as the neutral plane, which means that the x-y plane is the plane of bending. Under these conditions, two stress resultants exist at intermediate cross sections of the beam (Fig. 6-34b): a bending moment M_0 acting about the z axis and having its moment vector in the negative direction of the z axis and a shear force of magnitude P acting in the negative y direction. For a given beam and loading, both M_0 and P are known quantities.

The normal stresses acting on the cross section have a resultant that is the bending moment M_0, and the shear stresses have a resultant that is the shear force (equal to P). If the material follows Hooke's law, the normal stresses vary linearly with the distance from the neutral axis (the z axis) and can be calculated from the flexure formula. Since the shear stresses acting on a cross section are determined from the normal stresses solely from equilibrium considerations (see the derivation of the shear formula in Section 5.8), it follows that the distribution of shear stresses over the cross section is also determined. The resultant of these shear stresses is a vertical force equal in magnitude to the force P and having its line of action through some point S lying on the z axis (Fig. 6-34b). This point is known as the **shear center** (also called the *center of flexure*) of the cross section.

In summary, by assuming that the z axis is the neutral axis, not only the distribution of the normal stresses but also the distribution of the shear stresses and the position of the resultant shear force can be determined. Therefore, now recognize that a load P applied at the end of the beam (Fig. 6-34a) must act through a particular point (the shear center) if bending is to occur with the z axis as the neutral axis.

If the load is applied at some other point on the z axis (say, at point A in Fig. 6-35), it can be replaced by a statically equivalent system consisting of a force P acting at the shear center and a torque T. The force acting at the shear center produces bending about the z axis, and the torque produces torsion. It

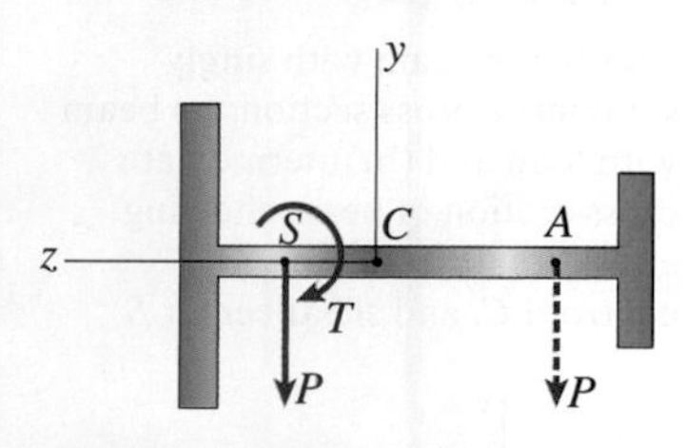

FIGURE 6-35

Singly symmetric beam with load P applied at point A

follows that *a lateral load acting on a beam will produce bending without twisting only if it acts through the shear center.*

The shear center (like the centroid) lies on any axis of symmetry; therefore, the shear center S and the centroid C coincide for a **doubly symmetric cross section** (Fig. 6-36a). A load P acting through the centroid produces bending about the y and z axes without torsion, and the corresponding bending stresses can be found by the method described in Section 6.4 for doubly symmetric beams.

If a beam has a **singly symmetric cross section** (Fig. 6-36b), both the centroid and the shear center lie on the axis of symmetry. A load P acting through the shear center can be resolved into components in the y and z directions. The component in the y direction produces bending in the x-y plane with the z axis as the neutral axis, and the component in the z direction produces bending (without torsion) in the x-z plane with the y axis as the neutral axis. The bending stresses produced by these components can be superposed to obtain the stresses caused by the original load.

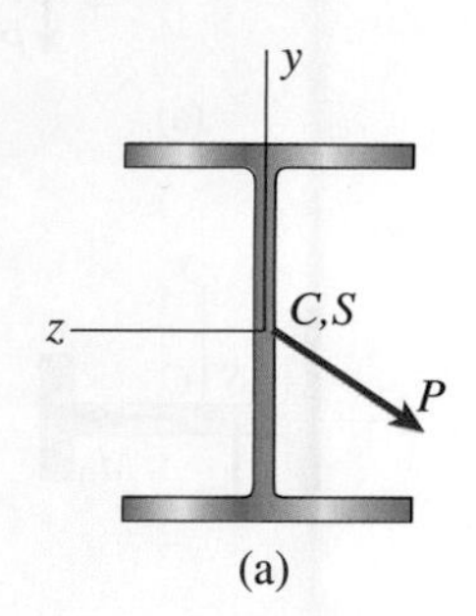

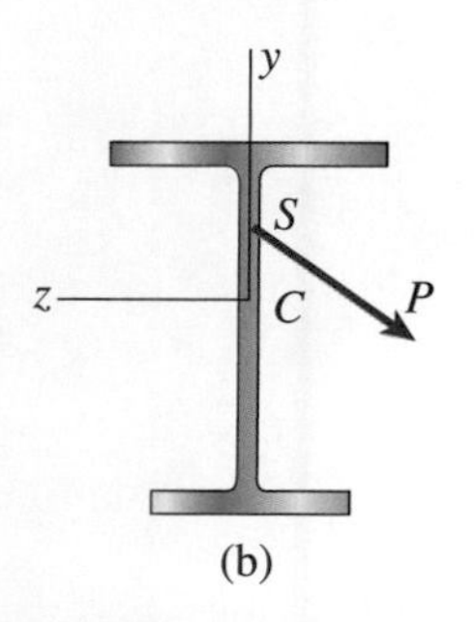

FIGURE 6-36

(a) Doubly symmetric beam with a load P acting through the centroid (and shear center) and (b) singly symmetric beam with a load P acting through the shear center

Finally, if a beam has an **unsymmetric cross section** (Fig. 6-37), the bending analysis proceeds as follows (provided the load acts through the shear center). First, locate the centroid C of the cross section and determine the orientation of the principal centroidal axes y and z. Then resolve the load into components (acting at the shear center) in the y and z directions and determine the bending moments M_y and M_z about the principal axes. Lastly, calculate the bending stresses using the method described in Section 6.5 for unsymmetric beams.

Now seeing the *significance* of the shear center and its use in beam analysis, it is natural to ask, "How do you *locate* the shear center?" For doubly symmetric shapes, the answer, of course, is simple—it is at the centroid. For singly symmetric shapes, the shear center lies on the axis of symmetry, but the precise location on that axis may not be easy to determine. Locating the shear center is even more difficult if the cross section is unsymmetric (Fig. 6-37). In such cases, the task requires more advanced methods than are appropriate for this book. (A few engineering handbooks give formulas for locating shear centers; e.g., see Ref. 2-9.)

Beams of **thin-walled open cross sections**, such as wide-flange beams, channels, angles, T-beams, and Z-sections, are a special case. Not only are they in common use for structural purposes, they also are very weak in torsion. Consequently, it is especially important to locate their shear centers. Cross sections of this type are considered in the following three sections—Sections 6.7 and 6.8 discuss how to find the shear stresses in such beams, and Section 6.9 shows how to locate their shear centers.

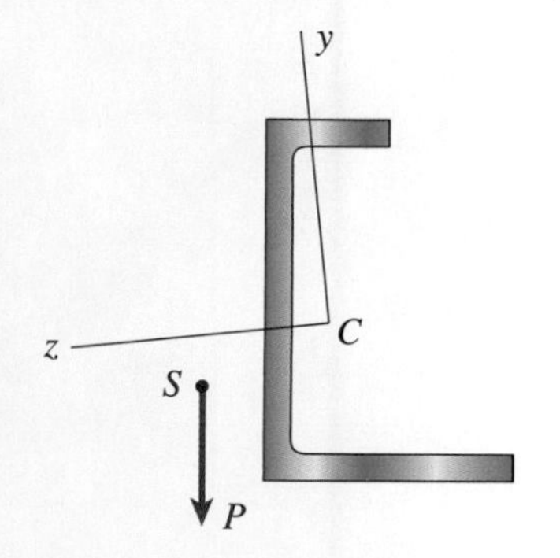

FIGURE 6-37

Unsymmetric beam with a load P acting through the shear center S

6.7 Shear Stresses in Beams of Thin-Walled Open Cross Sections

The distribution of shear stresses in rectangular beams, circular beams, and in the webs of beams with flanges was described previously in Sections 5.8, 5.9, and 5.10, where the shear formula (Eq. 5-41) for calculating the stresses was derived:

$$\tau = \frac{VQ}{Ib} \tag{6-47}$$

In this formula, V represents the shear force acting on the cross section, I is the moment of inertia of the cross-sectional area (with respect to the neutral

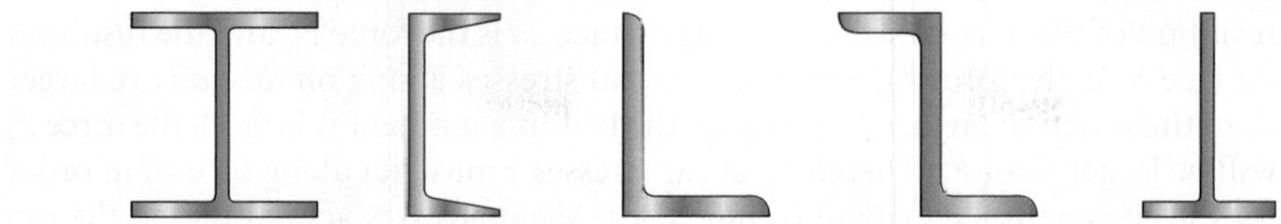

FIGURE 6-38

Typical beams of thin-walled open cross section (wide-flange beam or I-beam, channel beam, angle section, Z-section, and T-beam)

axis), b is the width of the beam at the location where the shear stress is to be determined, and Q is the first moment of the cross-sectional area outside of the location where the stress is being found.

Now consider the shear stresses in a special class of beams known as beams of **thin-walled open cross section.** Beams of this type are distinguished by two features: (1) The wall thickness is small compared to the height and width of the cross section, and (2) the cross section is open, as in the case of an I-beam or channel beam, rather than closed, as in the case of a hollow box beam. Examples are shown in Fig. 6-38. Beams of this type are also called **structural sections** or **profile sections**.

To determine the shear stresses in thin-walled beams of open cross section, use the same techniques as those used when deriving the shear formula [Eq. (6-47)]. To keep the derivation as general as possible, consider a beam having its cross-sectional centerline *mm* of arbitrary shape (Fig. 6-39a). The y and z axes are principal centroidal axes of the cross section, and the load P acts parallel to the y axis through the shear center S (Fig. 6-39b). Therefore, bending will occur in the x-y plane with the z axis as the neutral axis.

Under these conditions, the normal stress at any point in the beam from the flexure formula is

$$\sigma_x = -\frac{M_z y}{I_z} \tag{6-48}$$

where M_z is the bending moment about the z axis (positive as defined in Fig. 6-18) and y is a coordinate of the point under consideration.

Now consider a volume element *abcd* cut out between two cross sections at a distance dx apart (Fig. 6-39a). Note that the element begins at the edge of the cross section and has a length s measured along the centerline *mm* (Fig. 6-39b).

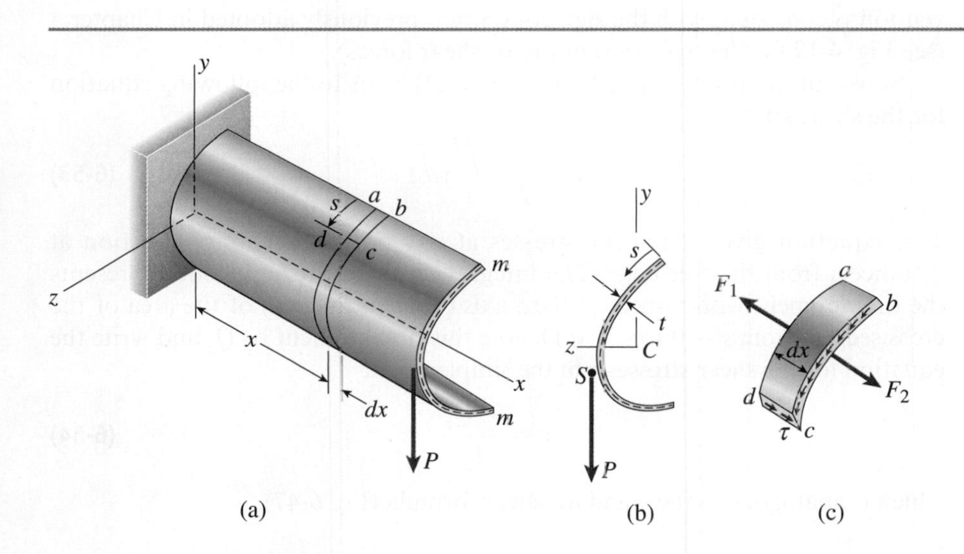

FIGURE 6-39

Shear stresses in a beam of thin-walled open cross section, where the y and z axes are principal centroidal axes

To determine the shear stresses, isolate the element as shown in Fig. 6-39c. The resultant of the normal stresses acting on face *ad* is the force F_1, and the resultant on face *bc* is the force F_2. Since the normal stresses acting on face *ad* are larger than those acting on face *bc* (because the bending moment is larger), the force F_1 will be larger than F_2. Therefore, shear stresses τ must act along face *cd* in order for the element to be in equilibrium. These shear stresses act parallel to the top and bottom surfaces of the element and must be accompanied by complementary shear stresses acting on the cross-sectional faces *ad* and *bc*, as shown in the figure.

To evaluate these shear stresses, sum forces in the x direction for element *abcd* (Fig. 6-39c); thus,

$$\tau t dx + F_2 - F_1 = 0 \quad \text{or} \quad \tau t dx = F_1 - F_2 \tag{6-49}$$

where t is the thickness of the cross section at face *cd* of the element. In other words, t is the thickness of the cross section at distance s from the free edge (Fig. 6-39b). Next, obtain an expression for the force F_1 by using Eq. (6-48):

$$F_1 = \int_0^s \sigma_x dA = -\frac{M_{z1}}{I_z}\int_0^s y dA \tag{6-50a}$$

where dA is an element of area on side *ad* of the volume element *abcd*, y is a coordinate to the element dA, and M_{z1} is the bending moment at the cross section. An analogous expression is obtained for the force F_2:

$$F_2 = \int_0^s \sigma_x dA = -\frac{M_{z2}}{I_z}\int_0^s y dA \tag{6-50b}$$

Substituting these expressions for F_1 and F_2 into Eq. (6-49) gives

$$\tau = \left(\frac{M_{z2} - M_{z1}}{dx}\right)\frac{1}{I_z t}\int_0^s y dA \tag{6-51}$$

The quantity $(M_{z2} - M_{z1})/dx$ is the rate of change dM/dx of the bending moment and is equal to the shear force acting on the cross section [see (Eq. 4-4)]:

$$\frac{dM}{dx} = \frac{M_{z2} - M_{z1}}{dx} = V_y \tag{6-52}$$

The shear force V_y is parallel to the y axis and positive in the negative direction of the y axis, that is, positive in the direction of the force P (Fig. 6-39). This convention is consistent with the sign convention previously adopted in Chapter 4 (see Fig. 4-19 for the sign convention for shear forces).

Substituting from Eq. (6-52) into Eq. (6-51) leads to the following equation for the shear stress τ:

$$\tau = \frac{V_y}{I_z t}\int_0^s y dA \tag{6-53}$$

This equation gives the shear stresses at any point in the cross section at distance s from the free edge. The integral on the right-hand side represents the first moment with respect to the z axis (the neutral axis) of the area of the cross section from $s = 0$ to $s = s$. Denote this first moment by Q_z and write the equation for the **shear stresses** τ in the simpler form

$$\tau = \frac{V_y Q_z}{I_z t} \tag{6-54}$$

which is analogous to the standard shear formula (Eq. 6-47).

The shear stresses are directed along the centerline of the cross section and act parallel to the edges of the section. It is tacitly assumed that these stresses have constant intensity across the thickness t of the wall, which is a valid assumption when the thickness is small. (Note that the wall thickness need not be constant but may vary as a function of the distance s.)

The **shear flow** at any point in the cross section, equal to the product of the shear stress and the thickness at that point, is

$$f = \tau t = \frac{V_y Q_z}{I_z} \tag{6-55}$$

Because V_y and I_z are constants, the shear flow is directly proportional to Q_z. At the top and bottom edges of the cross section, Q_z is zero; hence, the shear flow is also zero. The shear flow varies continuously between these end points and reaches its maximum value where Q_z is maximum, which is at the neutral axis.

Now suppose that the beam shown in Fig. 6-39 is bent by loads that act parallel to the z axis and through the shear center. Then the beam bends in the x-z plane and the y axis becomes the neutral axis. In this case, repeat the same type of analysis and arrive at the following equations for the shear stresses and shear flow [compare with Eqs. (6-54) and (6-55)]:

$$\tau = \frac{V_z Q_y}{I_y t} \qquad f = \tau t = \frac{V_z Q_y}{I_y} \tag{6-56a,b}$$

In these equations, V_z is the shear force parallel to the z axis and Q_y is the first moment with respect to the y axis.

In summary, expressions were derived for the shear stresses in beams of thin-walled open cross sections with the stipulations that the shear force must act through the shear center and must be parallel to one of the principal centroidal axes. If the shear force is inclined to the y and z axes (but still acts through the shear center), it can be resolved into components parallel to the principal axes. Then two separate analyses can be made, and the results can be superimposed.

To illustrate the use of the shear-stress equations, the shear stresses in a wide-flange beam are considered in the next section. Later, in Section 6.9, the shear-stress equations are used to locate the shear centers of several thin-walled beams with open cross sections.

6.8 Shear Stresses in Wide-Flange Beams

Now the concepts and equations discussed in the preceding section are used to investigate the shear stresses in wide-flange beams. For discussion purposes, consider the wide-flange beam of Fig. 6-40a. This beam is loaded by a force P acting in the plane of the web, that is, through the shear center, which coincides with the centroid of the cross section. The cross-sectional dimensions are shown in Fig. 6-40b; note that b is the flange width, h is the height between *centerlines* of the flanges, t_f is the flange thickness, and t_w is the web thickness.

Shear Stresses in the Upper Flange

First consider the shear stresses at section bb in the right-hand part of the upper flange (Fig. 6-40b). Since the distance s has its origin at the edge of the section (point a), the cross-sectional area between point a and section bb is st_f. Also, the distance from

FIGURE 6-40

Shear stresses in a wide-flange beam

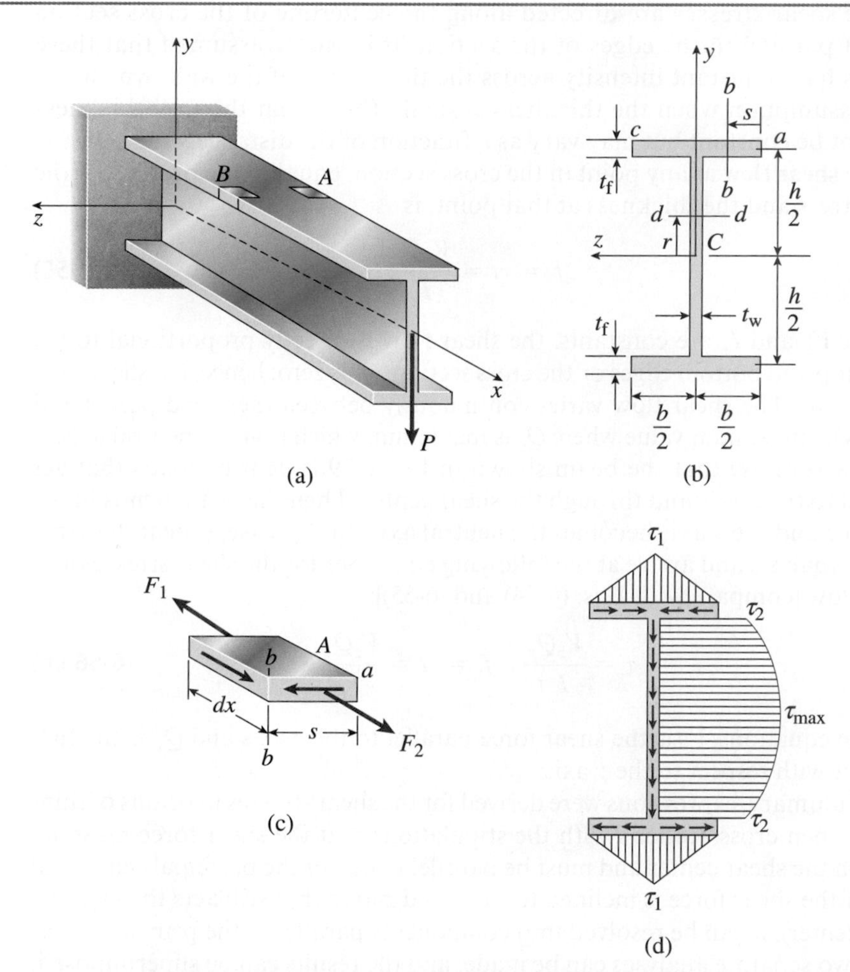

the centroid of this area to the neutral axis is $h/2$. Therefore, its first moment Q_z is equal to $st_f h/2$. Thus, the shear stress τ_f in the flange at section bb [from Eq. (6-54)] is

$$\tau_f = \frac{V_y Q_z}{I_z t} = \frac{P(st_f h/2)}{I_z t_f} = \frac{shP}{2I_z} \qquad \textbf{(6-57)}$$

The direction of this stress can be determined by examining the forces acting on element A, which is cut out of the flange between point a and section bb (see Figs. 6-40a and b).

The element is drawn to a larger scale in Fig. 6-40c in order to show clearly the forces and stresses acting on it. Note that the tensile force F_1 is larger than the force F_2, because the bending moment is larger on the rear face of the element than it is on the front face. It follows that the shear stress on the left-hand face of element A must act toward you if the element is to be in equilibrium. From this observation, it follows that the shear stresses on the front face of element A must act toward the left.

Returning now to Fig. 6-40b, note that the magnitude and direction of the shear stress at section bb, which may be located anywhere between point a and the junction of the top flange and the web, are now known. Thus, the shear stresses throughout the right-hand part of the flange are horizontal, act to the left, and have a magnitude given by Eq. (6-57). As seen from that equation, the shear stresses increase linearly with the distance s.

The variation of the stresses in the upper flange is shown graphically in Fig. 6-40d; observe that the stresses vary from zero at point a (where $s = 0$) to a maximum value τ_1 at $s = b/2$:

$$\tau_1 = \frac{bhP}{4I_z} \quad \text{(6-58)}$$

The corresponding shear flow is

$$f_1 = \tau_1 t_f = \frac{bht_f P}{4I_z} \quad \text{(6-59)}$$

Note that the shear stress and shear flow have been calculated at the junction of the *centerlines* of the flange and web, using only centerline dimensions of the cross section in the calculations. This approximate procedure simplifies the calculations and is satisfactory for thin-walled cross sections.

By beginning at point c on the left-hand part of the top flange (Fig. 6-40b) and measuring s toward the right, repeat the same type of analysis. It follows that the magnitude of the shear stresses is again given by Eqs. (6-57) and (6-58). However, by cutting out an element B (Fig. 6-40a) and considering its equilibrium, note that the shear stresses on the cross section now act toward the right, as shown in Fig. 6-40d.

Shear Stresses in the Web

The next step is to determine the shear stresses acting in the web. Considering a horizontal cut at the top of the web (at the junction of the flange and web) results in the first moment about the neutral axis to be $Q_z = bt_f h/2$, so that the corresponding shear stress is

$$\tau_2 = \frac{bht_f P}{2I_z t_w} \quad \text{(6-60)}$$

The associated shear flow is

$$f_2 = \tau_2 t_w = \frac{bht_f P}{2I_z} \quad \text{(6-61)}$$

Now the shear flow f_2 is equal to twice the shear flow f_1, which is expected, since the shear flows in the two halves of the upper flange combine to produce the shear flow at the top of the web.

The shear stresses in the web act downward and increase in magnitude until the neutral axis is reached. At section dd, located at distance r from the neutral axis (Fig. 6-40b), the shear stress τ_w in the web is calculated as

$$Q_z = \frac{bt_f h}{2} + \left(\frac{h}{2} - r\right)(t_w)\left(\frac{h/2 + r}{2}\right) = \frac{bt_f h}{2} + \frac{t_w}{2}\left(\frac{h^2}{4} - r^2\right)$$

$$\tau_w = \left(\frac{bt_f h}{t_w} + \frac{h^2}{4} - r^2\right)\frac{P}{2I_z} \quad \text{(6-62)}$$

When $r = h/2$, this equation reduces to Eq. (6-60), and when $r = 0$, it gives the maximum shear stress:

$$\tau_{max} = \left(\frac{bt_f}{t_w} + \frac{h}{4}\right)\frac{Ph}{2I_z} \quad \text{(6-63)}$$

Again note that all calculations were made on the basis of the centerline dimensions of the cross section. For this reason, the shear stresses in the web of a wide-flange beam calculated from Eq. (6-62) may be slightly different from those obtained by the more exact analysis made in Chapter 5 [see Eq. (5-51) of Section (5.10)].

The shear stresses in the web vary parabolically, as shown in Fig. 6-40d, although the variation is not large. The ratio of τ_{max} to τ_2 is

$$\frac{\tau_{max}}{\tau_2} = 1 + \frac{ht_w}{4bt_f} \tag{6-64}$$

For instance, if $h = 2b$ and $t_f = 2t_w$, the ratio is $\tau_{max}/\tau_2 = 1.25$.

Shear Stresses in the Lower Flange

As the final step in the analysis, consider the shear stresses in the lower flange using the same methods as those used for the top flange. The result is that the magnitudes of the stresses are the same as in the top flange, but the directions are as shown in Fig. 6-40d.

General Comments

From Fig. 6-40d, observe that the shear stresses on the cross section "flow" inward from the outermost edges of the top flange, then down through the web, and finally outward to the edges of the bottom flange. Because this flow is always continuous in any structural section, it serves as a convenient method for determining the directions of the stresses. For instance, if the shear force acts downward on the beam of Fig. 6-40a, it follows that the shear flow in the web must also be downward. Knowing the direction of the shear flow in the web gives the directions of the shear flows in the flanges because of the required continuity in the flow. Using this simple technique to get the directions of the shear stresses is easier than visualizing the directions of the forces acting on elements such as A (Fig. 6-40c) cut out from the beam.

The resultant of all the shear stresses acting on the cross section is clearly a vertical force, because the horizontal stresses in the flanges produce no resultant. The shear stresses in the web have a resultant R, which can be found by integrating the shear stresses over the height of the web, as

$$R = \int \tau dA = 2\int_0^{h/2} \tau t_w dr$$

Substituting from Eq. (6-62) gives

$$R = 2t_w \int_0^{h/2} \left(\frac{bt_f h}{t_w} + \frac{h^2}{4} - r^2\right)\left(\frac{P}{2I_z}\right) dr = \left(\frac{bt_f}{t_w} + \frac{h}{6}\right)\frac{h^2 t_w P}{2I_z} \tag{6-65}$$

The moment of inertia I_z can be calculated as (using centerline dimensions):

$$I_z = \frac{t_w h^3}{12} + \frac{bt_f h^2}{2} \tag{6-66}$$

in which the first term is the moment of inertia of the web and the second term is the moment of inertia of the flanges. Substitute this expression for I_z in Eq. (6-65) to get $R = P$, which demonstrates that the resultant of the shear stresses acting on the cross section is equal to the load. Furthermore, the line of action of the resultant is in the plane of the web, so the resultant passes through the shear center.

The preceding analysis provides a more complete picture of the shear stresses in a wide-flange or I-beam because it includes the flanges (recall that Chapter 5 considered only the shear stresses in the web). Furthermore, this analysis illustrates the general techniques for finding shear stresses in beams of thin-walled open cross section. Other illustrations can be found in the next section, where the shear stresses in a channel section and an angle section are determined as part of the process of locating their shear centers.

6.9 Shear Centers of Thin-Walled Open Sections

Sections 6.7 and 6.8 presented methods for finding the shear stresses in beams of thin-walled open cross section. Now those methods are used to locate the shear centers of several shapes of beams. Only beams with singly symmetric or unsymmetric cross sections are considered, because the shear center of a doubly symmetric cross section is known to be located at the centroid.

The procedure for locating the shear center consists of two principal steps: first, evaluate the shear stresses acting on the cross section when bending occurs about one of the principal axes, and second, determine the resultant of those stresses. The shear center is located on the line of action of the resultant. Consider bending about *both* principal axes to determine the position of the shear center.

As in Sections 6.7 and 6.8, use only centerline dimensions when deriving formulas and making calculations. This procedure is satisfactory if the beam is thin-walled, that is, if the thickness of the beam is small compared to the other dimensions of the cross section.

Channel Section

The first beam to be analyzed is a singly symmetric channel section (Fig. 6-41a). Based on general discussion in Section 6.6, the shear center is located on the axis of symmetry (the z axis). To find the position of the shear center on the z axis, assume that the beam is bent about the z axis as the neutral axis, and then determine the line of action of the resultant shear force V_y acting parallel to the y axis. The shear center is located where the line of action of V_y intersects the z axis. (Note that the origin of axes is at the centroid C, so that both the y and z axes are principal centroidal axes.)

Based upon the discussions in Section 6.8, the shear stresses in a channel vary linearly in the flanges and parabolically in the web (Fig. 6-41b). The resultant of those stresses can be found if the maximum stress τ_1 in the flange, the stress τ_2 at the top of the web, and the maximum stress τ_{max} in the web are known.

To find the stress τ_1 in the flange, use Eq. (6-54) with Q_z equal to the first moment of the flange area about the z axis:

$$Q_z = \frac{bt_f h}{2} \tag{6-67}$$

in which b is the flange width, t_f is the flange thickness, and h is the height of the beam. (Note again that the dimensions b and h are measured along the centerline of the section.) Thus, the stress τ_1 in the flange is

$$\tau_1 = \frac{V_y Q_z}{I_z t_f} = \frac{bhV_y}{2I_z} \tag{6-68}$$

where I_z is the moment of inertia about the z axis.

FIGURE 6-41

Shear center S of a channel section

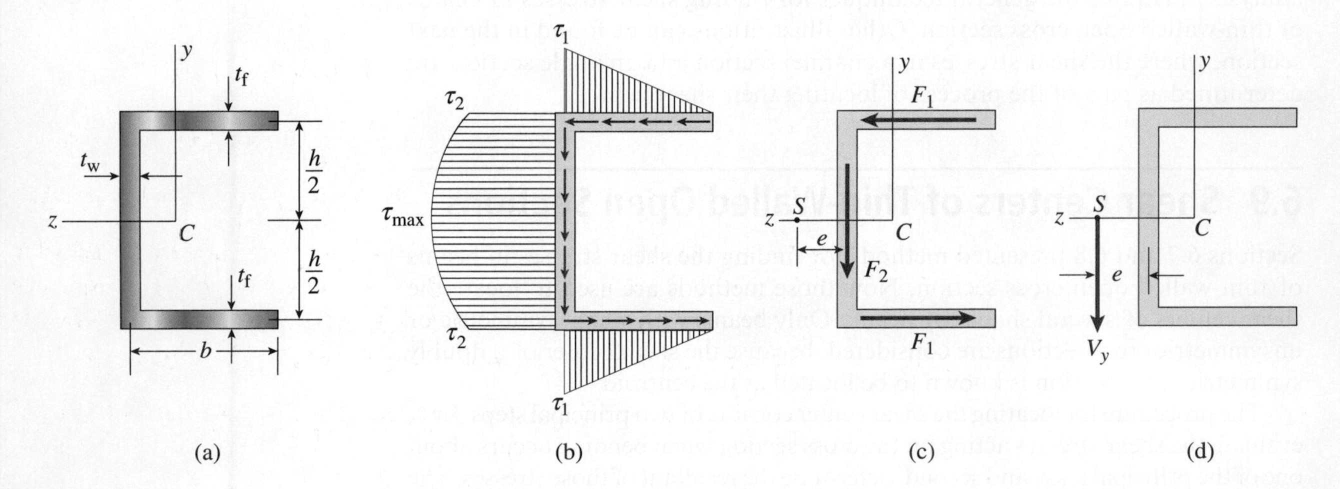

The stress τ_2 at the top of the web is obtained in a similar manner but with the thickness equal to the web thickness instead of the flange thickness:

$$\tau_2 = \frac{V_y Q_z}{I_z t_w} = \frac{b t_f h V_y}{2 t_w I_z} \tag{6-69}$$

Also, at the neutral axis the first moment of area is

$$Q_z = \frac{b t_f h}{2} + \frac{h t_w}{2}\left(\frac{h}{4}\right) = \left(b t_f + \frac{h t_w}{4}\right)\frac{h}{2} \tag{6-70}$$

Therefore, the maximum stress is

$$\tau_{max} = \frac{V_y Q_z}{I_z t_w} = \left(\frac{b t_f}{t_w} + \frac{h}{4}\right)\frac{h V_y}{2 I_z} \tag{6-71}$$

The stresses τ_1 and τ_2 in the lower half of the beam are equal to the corresponding stresses in the upper half (Fig. 6-41b).

The horizontal shear force F_1 in either flange (Fig. 6-41c) can be found from the triangular stress diagrams. Each force is equal to the area of the stress triangle multiplied by the thickness of the flange:

$$F_1 = \left(\frac{\tau_1 b}{2}\right)(t_f) = \frac{h b^2 t_f V_y}{4 I_z} \tag{6-72}$$

The vertical force F_2 in the web must be equal to the shear force V_y, since the forces in the flanges have no vertical components. As a check, verify that $F_2 = V_y$ by considering the parabolic stress diagram of Fig. 6-41b. The diagram is made up of two parts—a rectangle of area $\tau_2 h$ and a parabolic segment of area

$$\frac{2}{3}(\tau_{max} - \tau_2)h$$

Thus, the shear force F_2, which is equal to the area of the stress diagram times the web thickness t_w, is

$$F_2 = \tau_2 h t_w + \frac{2}{3}(\tau_{max} - \tau_2) h t_w$$

Substitute the expressions for τ_2 and τ_{max} [Eqs. (6-69) and (6-71)] into the preceding equation to get

$$F_2 = \left(\frac{t_w h^3}{12} + \frac{bh^2 t_f}{2}\right)\frac{V_y}{I_z} \qquad \text{(6-73)}$$

Finally, note that the expression for the moment of inertia is

$$I_z = \frac{t_w h^3}{12} + \frac{bh^2 t_f}{2} \qquad \text{(6-74)}$$

in which the calculations are based upon centerline dimensions. Substitute this expression for I_z into Eq. (6-73) for F_2 to obtain

$$F_2 = V_y \qquad \text{(6-75)}$$

as expected.

The three forces acting on the cross section (Fig. 6-41c) have a resultant V_y that intersects the z axis at the shear center S (Fig. 6-41d). Hence, the moment of the three forces about any point in the cross section must be equal to the moment of the force V_y about that same point. This moment relationship provides an equation from which the position of the shear center may be found.

As an illustration, select the shear center itself as the center of moments. In that case, the moment of the three forces (Fig. 6-41c) is $F_1 h - F_2 e$, where e is the distance from the centerline of the web to the shear center, and the moment of the resultant force V_y is zero (Fig. 6-41d). Equating these moments gives

$$F_1 h - F_2 e = 0 \qquad \text{(6-76)}$$

Substitute for F_1 from Eq. (6-72) and for F_2 from Eq. (6-75) and then solve for e to get

$$e = \frac{b^2 h^2 t_f}{4 I_z} \qquad \text{(6-77)}$$

When the expression for I_z (Eq. 6-74) is substituted, Eq. (6-77) becomes

$$e = \frac{3b^2 t_f}{h t_w + 6 b t_f} \qquad \text{(6-78)}$$

Thus, the position of the shear center of a channel section has been determined.

As explained in Section 6.6, a channel beam undergoes bending without twisting whenever it is loaded by forces acting through the shear center. If the loads act parallel to the y axis but through some point other than the shear center (for example, if the loads act in the plane of the web), they can be replaced by a statically equivalent force system consisting of loads through the shear center and twisting couples. This leads to a combination of bending and torsion of the beam. If the loads act along the z axis, the result is simple bending about the y axis. If the loads act in skew directions through the shear center, they can be replaced by statically equivalent loads acting parallel to the y and z axes.

Angle Section

The next shape to be considered is an equal-leg angle section (Fig. 6-42a), in which each leg of the angle has length b and thickness t. The z axis is an axis of symmetry and the origin of coordinates is at the centroid C; therefore, both the y and z axes are principal centroidal axes.

To locate the shear center, follow the same general procedure as that described for a channel section to determine the distribution of the shear stresses as part of the analysis. However, as shown later, the shear center of an angle section can be determined by inspection.

Begin by assuming that the section is subjected to a shear force V_y acting parallel to the y axis. Then use Eq. (6-54) to find the corresponding shear stresses in the legs of the angle. For this purpose, find the first moment of the cross-sectional area between point a at the outer edge of the beam (Fig. 6-42b) and section bb located at distance s from point a. The area is equal to st and its centroidal distance from the neutral axis is

$$\frac{b - s/2}{\sqrt{2}}$$

Thus, the first moment of the area is

$$Q_z = st\left(\frac{b - s/2}{\sqrt{2}}\right) \tag{6-79}$$

FIGURE 6-42
Shear center of an equal-leg angle section

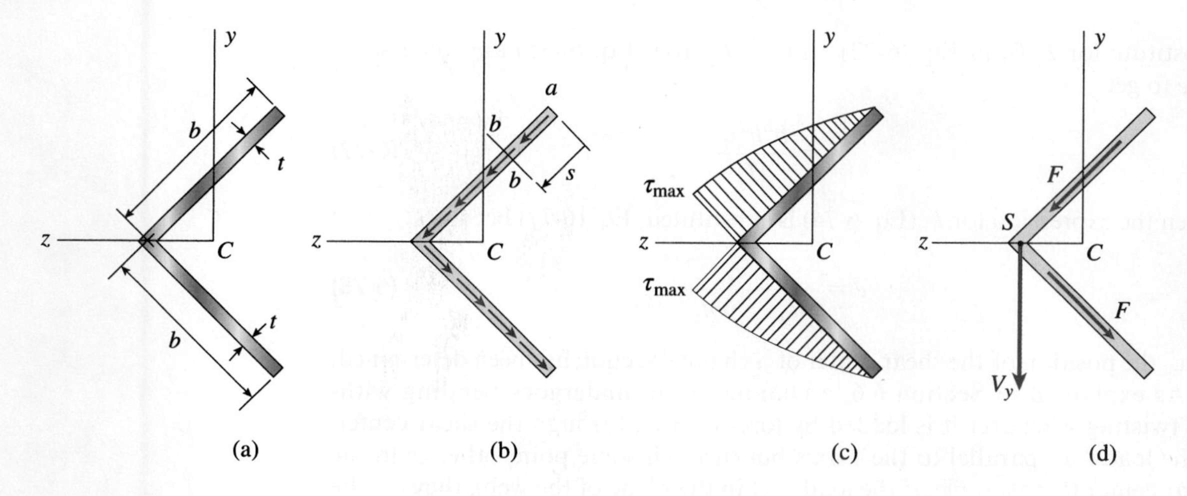

Substitute into Eq. (6-54) to get the following expression for the shear stress at distance s from the edge of the cross section:

$$\tau = \frac{V_y Q_z}{I_z t} = \frac{V_y s}{I_z \sqrt{2}}\left(b - \frac{s}{2}\right) \tag{6-80}$$

The moment of inertia I_z can be obtained from Case 24 of Appendix E with $\beta = 45°$:

$$I_z = 2I_{BB} = 2\left(\frac{tb^3}{6}\right) = \frac{tb^3}{3} \tag{6-81}$$

Substitute this expression for I_z into Eq. (6-80) to get

$$\tau = \frac{3V_y s}{b^3 t\sqrt{2}}\left(b - \frac{s}{2}\right) \tag{6-82}$$

This equation gives the shear stress at any point along the leg of the angle. The stress varies quadratically with s, as shown in Fig. 6-42c.

The maximum value of the shear stress occurs at the intersection of the legs of the angle and is obtained from Eq. (6-82) by substituting $s = b$:

$$\tau_{max} = \frac{3V_y}{2bt\sqrt{2}} \tag{6-83}$$

The shear force F in each leg (Fig. 6-42d) is equal to the area of the parabolic stress diagram (Fig. 6-42c) times the thickness t of the legs:

$$F = \frac{2}{3}(\tau_{max} b)(t) = \frac{V_y}{\sqrt{2}} \tag{6-84}$$

Since the horizontal components of the forces F cancel each other, only the vertical components remain. Each vertical component is equal to $F/\sqrt{2}$, or $V_y/2$, so the resultant vertical force is equal to the shear force V_y, as expected.

Since the resultant force passes through the intersection point of the lines of action of the two forces F (Fig. 6-42d), note that the shear center S is located at the junction of the two legs of the angle.

Sections Consisting of Two Intersecting Narrow Rectangles

In the preceding discussion of an angle section, the shear stresses and the forces in the legs were evaluated to illustrate the general methodology for analyzing thin-walled open sections. However, if the sole objective had been to locate the shear center, it would not have been necessary to evaluate the stresses and forces.

Since the shear stresses are parallel to the centerlines of the legs (Fig. 6-42b), their resultants are two forces F (Fig. 6-42d). The resultant of those two forces is a single force that passes through their point of intersection. Consequently, this point is the shear center. Thus, the shear center of an equal-leg angle section can be found by a simple line of reasoning (without making *any* calculations).

The same line of reasoning is valid for all cross sections consisting of two thin, intersecting rectangles (Fig. 6-43). In each case, the resultants of the shear stresses are forces that intersect at the junction of the rectangles. Therefore, the shear center S is located at that point.

FIGURE 6-43

Shear centers of sections consisting of two intersecting narrow rectangles

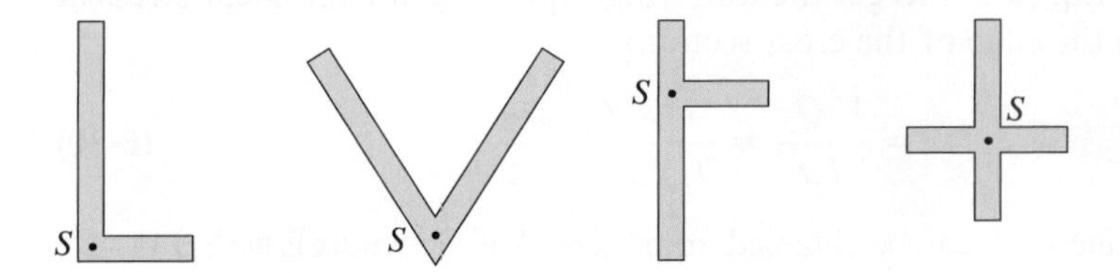

Z-Section

Now determine the location of the shear center of a Z-section having thin walls (Fig. 6-44a). The section has no axes of symmetry but is symmetric about the centroid C (see Section D.1 of Appendix D for a discussion of *symmetry about a point*). The y and z axes are principal axes through the centroid.

Begin by assuming that a shear force V_y acts parallel to the y axis and causes bending about the z axis as the neutral axis. Then the shear stresses in the flanges and web will be directed as shown in Fig. 6-44a. From symmetry note that the forces F_1 in the two flanges must be equal to each other (Fig. 6-44b). The resultant of the three forces acting on the cross section (F_1 in the flanges and F_2 in the web) must be equal to the shear force V_y. The forces F_1 have a resultant $2F_1$ acting through the centroid and parallel to the flanges. This force intersects the force F_2 at the centroid C; therefore, it follows that the line of action of the shear force V_y must be through the centroid.

If the beam is subjected to a shear force V_z parallel to the z axis, the shear force acts through the centroid. Since the shear center is located at the intersection of the lines of action of the two shear forces, the shear center of the Z-section coincides with the centroid.

This conclusion applies to any Z-section that is symmetric about the centroid, that is, any Z-section having identical flanges (same width and same thickness). Note, however, that the thickness of the web does not have to be the same as the thickness of the flanges.

The locations of the shear centers of many other structural shapes are given in the problems at the end of this chapter.[5]

FIGURE 6-44

Shear center of a thin-walled Z-section

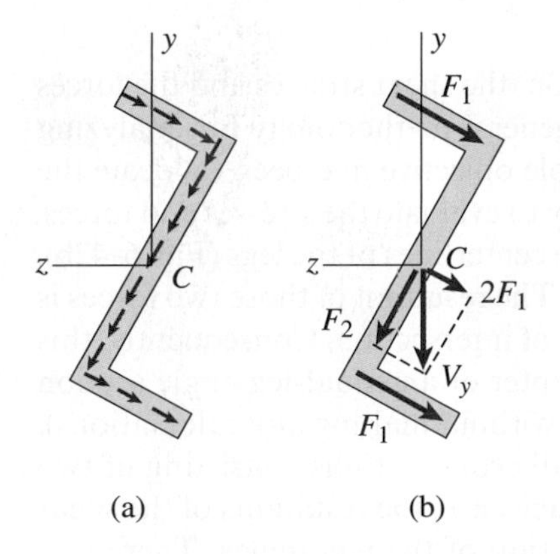

[5]The first determination of a shear center was made by S. P. Timoshenko in 1913 (Ref. 6-1).

Example 6-9

A thin-walled semicircular cross section of radius r and thickness t is shown in Fig. 6-45a. Determine the distance e from the center O of the semicircle to the shear center S.

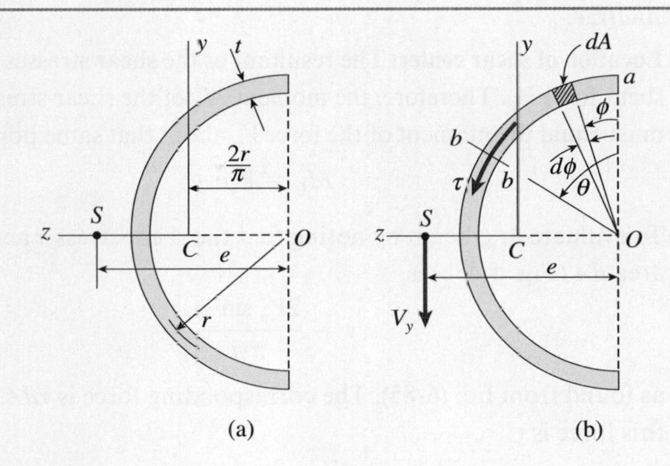

FIGURE 6-45

Example 6-9: Shear center of a thin-walled semicircular section

Solution:

Use a four-step problem-solving approach. Combine steps as needed for an efficient solution.

1. *Conceptualize*: The shear center is located somewhere on the axis of symmetry (the z axis). To determine the exact position, assume that the beam is bent by a shear force V_y acting parallel to the y axis and producing bending about the z axis as the neutral axis (Fig. 6-45b).

2. *Categorize*:

Shear stresses: The first step is to determine the shear stresses τ acting on the cross section (Fig. 6-45b). Consider a section bb defined by the distance s measured along the centerline of the cross section from point a. The central angle subtended between point a and section bb is denoted θ. Therefore, the distance s equals $r\theta$, where r is the radius of the centerline and θ is measured in radians.

To evaluate the first moment of the cross-sectional area between point a and section bb, identify an element of area dA (shown shaded in the figure) and integrate as

$$Q_z = \int y dA = \int_0^\theta (r\cos\phi)(tr\, d\phi) = r^2 t \sin\theta \tag{a}$$

in which ϕ is the angle to the element of area and t is the thickness of the section. Thus, the shear stress τ at section bb is

$$\tau = \frac{V_y Q_z}{I_z t} = \frac{V_y r^2 \sin\theta}{I_z} \tag{b}$$

Substituting $I_z = \pi r^3 t/2$ (see Case 22 or Case 23 of Appendix E) gives

$$\tau = \frac{2V_y \sin\theta}{\pi r t} \tag{6-85}$$

When $\theta = 0$ or $\theta = \pi$, this expression gives $\tau = 0$, as expected. When $\theta = \pi/2$, it gives the maximum shear stress.

3. *Analyze:*

Location of shear center: The resultant of the shear stresses must be the vertical shear force V_y. Therefore, the moment M_0 of the shear stresses about the center O must equal the moment of the force V_y about that same point:

$$M_0 = V_y e \tag{c}$$

To evaluate M_0, begin by noting that the shear stress τ acting on the element of area dA (Fig. 6-45b) is

$$\tau = \frac{2V_y \sin\phi}{\pi r t}$$

as found from Eq. (6-85). The corresponding force is τdA, and the moment of this force is

$$dM_0 = r(\tau dA) = \frac{2V_y \sin\phi \, dA}{\pi t}$$

Since $dA = trd\phi$, this expression becomes

$$dM_0 = \frac{2rV_y \sin\phi \, d\phi}{\pi}$$

Therefore, the moment produced by the shear stresses is

$$M_0 = \int dM_0 = \int_0^{\pi} \frac{2rV_y \sin\phi \, d\phi}{\pi} = \frac{4rV_y}{\pi} \tag{d}$$

It follows from Eq. (c) that the distance e to the shear center is

$$e = \frac{M_0}{V_y} = \frac{4r}{\pi} \approx 1.27r \tag{6-86}$$

4. ***Finalize***: This result shows that the shear center S is located outside of the semicircular section.

Note: The distance from the center O of the semicircle to the centroid C of the cross section (Fig. 6-45a) is $2r/\pi$ (from Case 23 of Appendix E), which is one-half of the distance e. Thus, the centroid is located midway between the shear center and the center of the semicircle.

The location of the shear center in a more general case of a thin-walled circular section is determined in Prob. 6.9-13.

*6.10 Elastoplastic Bending

Previous discussions of bending assumed that the beams were made of materials that followed Hooke's law (linearly elastic materials). Now consider the bending of elastoplastic beams when the material is strained beyond the linear region. When that happens, the distribution of the stresses is no longer linear but varies according to the shape of the stress-strain curve.

Elastoplastic materials were considered earlier in the discussion of axially loaded bars in Section 2.12. As explained in that section, elastoplastic materials follow Hooke's law up to the yield stress σ_Y and then yield plastically under constant stress (see the stress-strain diagram of Fig. 6-46). From the figure, note that an elastoplastic material has a region of linear elasticity between regions of perfect plasticity. Throughout this section, the material is assumed to have the same yield stress σ_Y and same yield strain ε_Y in both tension and compression.

FIGURE 6-46

Idealized stress-strain diagram for an elastoplastic material

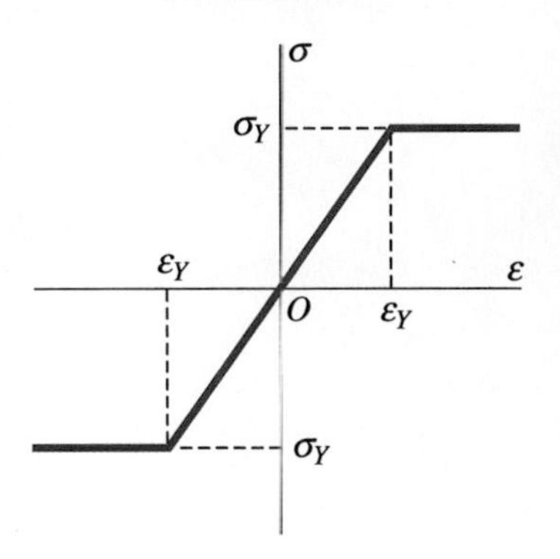

Structural steels are excellent examples of elastoplastic materials because they have sharply defined yield points and undergo large strains during yielding. Eventually, the steels begin to strain harden, and then the assumption of perfect plasticity is no longer valid. However, strain hardening provides an increase in strength; therefore, the assumption of perfect plasticity is on the side of safety.

Yield Moment

Consider a beam of elastoplastic material subjected to a bending moment M that causes bending in the x-y plane (Fig. 6-47). When the bending moment is small, the maximum stress in the beam is less than the yield stress σ_Y, so the beam is in the same condition as a beam in ordinary elastic bending with a linear stress distribution, as shown in Fig. 6-48b. Under these conditions, the neutral axis passes through the centroid of the cross section, and the normal stresses are obtained from the flexure formula ($\sigma = -My/I$). Since the bending moment is positive, the stresses are compressive above the z axis and tensile below it.

FIGURE 6-47

Beam of elastoplastic material subjected to a positive bending moment M

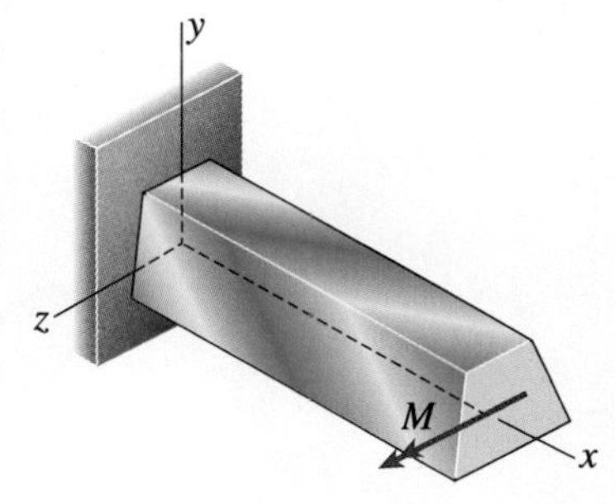

The preceding conditions exist until the stress in the beam at the point farthest from the neutral axis reaches the yield stress σ_Y either in tension or in compression (Fig. 6-48c). The bending moment in the beam when the maximum stress just reaches the yield stress, called the **yield moment** M_Y, can be obtained from the flexure formula:

$$M_Y = \frac{\sigma_Y I}{c} = \sigma_Y S \tag{6-87}$$

in which c is the distance to the point farthest from the neutral axis and S is the corresponding section modulus.

Plastic Moment and Neutral Axis

If the bending moment is now increased above the yield moment M_Y, the strains in the beam continue to increase and the maximum strain will exceed the yield strain ε_Y. However, because of perfectly plastic yielding, the maximum stress remains constant and equal to σ_Y, as pictured in Fig. 6-48d. Note that the outer regions of the beam have become fully plastic while a central core (called the **elastic core**) remains linearly elastic.

FIGURE 6-48

Stress distributions in a beam of elastoplastic material

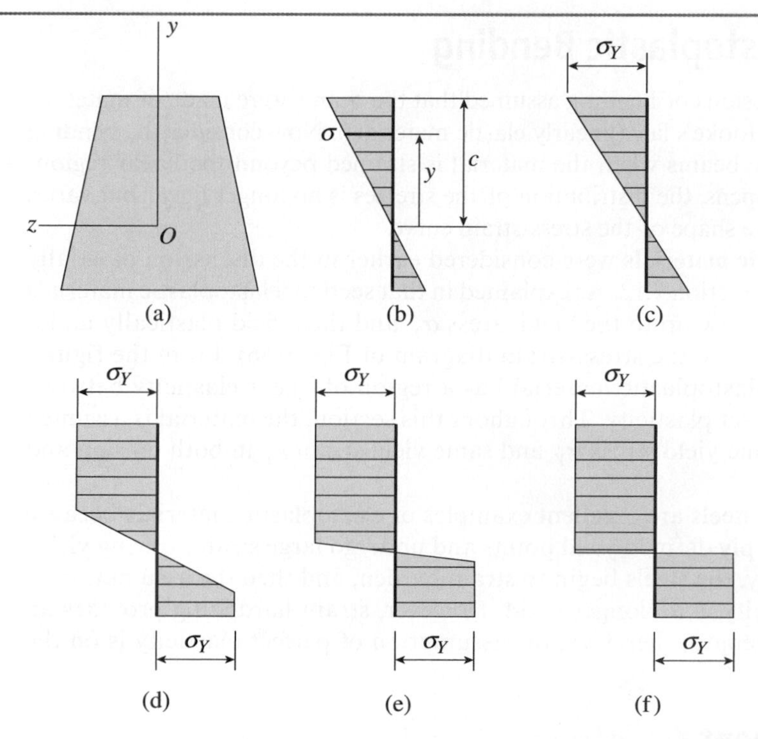

If the z axis is not an axis of symmetry (singly symmetric cross section), the neutral axis moves away from the centroid when the yield moment is exceeded. This shift in the location of the neutral axis is not large, and in the case of the trapezoidal cross section of Fig. 6-48, it is too small to be seen in the figure. If the cross section is doubly symmetric, the neutral axis passes through the centroid even when the yield moment is exceeded.

As the bending moment increases still further, the plastic region enlarges and moves inward toward the neutral axis until the condition shown in Fig. 6-48e is reached. At this stage, the maximum strain in the beam (at the farthest distance from the neutral axis) is perhaps 10 or 15 times the yield strain ε_Y and the elastic core has almost disappeared. Thus, for practical purposes, the beam has reached its ultimate moment-resisting capacity, and the ultimate stress distribution is idealized as consisting of two rectangular parts (Fig. 6-48f). The bending moment corresponding to this idealized stress distribution, called the **plastic moment** M_P, represents the maximum moment that can be sustained by a beam of elastoplastic material.

To find the plastic moment M_P, begin by locating the **neutral axis** of the cross section under fully plastic conditions. For this purpose, consider the cross section shown in Fig. 6-49a and let the z axis be the neutral axis. Every point in the cross section above the neutral axis is subjected to a compressive stress σ_Y (Fig. 6-49b), and every point below the neutral axis is subjected to a tensile stress σ_Y. The resultant compressive force C is equal to σ_Y times the cross-sectional area A_1 above the neutral axis (Fig. 6-49a), and the resultant tensile force T equals σ_Y times the area A_2 below the neutral axis. Since the resultant force acting on the cross section is zero, it follows that

$$T = C \quad \text{or} \quad A_1 = A_2 \qquad \textbf{(6-88a,b)}$$

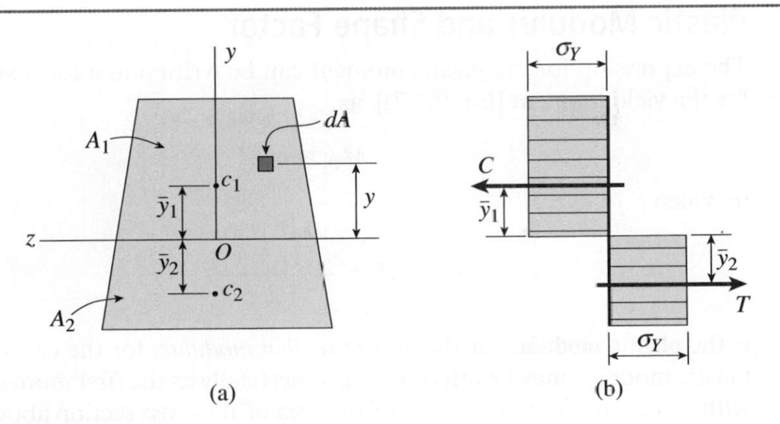

FIGURE 6-49

Location of the neutral axis and determination of the plastic moment M_P under fully plastic conditions

Because the total area A of the cross section is equal to $A_1 + A_2$, note that

$$A_1 = A_2 = \frac{A}{2} \tag{6-89}$$

Therefore, under fully plastic conditions, *the neutral axis divides the cross section into two equal areas.*

As a result, the location of the neutral axis for the plastic moment M_P may be different from its location for linearly elastic bending. For instance, in the case of a trapezoidal cross section that is narrower at the top than at the bottom (Fig. 6-49a), the neutral axis for fully plastic bending is slightly below the neutral axis for linearly elastic bending.

Since the plastic moment M_P is the moment resultant of the stresses acting on the cross section, it can be found by integrating over the cross-sectional area A (Fig. 6-49a):

$$\begin{aligned} M_P &= -\int_A \sigma y dA = -\int_{A_1} (-\sigma_Y) y dA - \int_{A_2} \sigma_Y y dA \\ &= \sigma_Y(\bar{y}_1 A_1) - \sigma_Y(-\bar{y}_2 A_2) = \frac{\sigma_Y A(\bar{y}_1 + \bar{y}_2)}{2} \end{aligned} \tag{6-90}$$

in which y is the coordinate (positive upward) of the element of area dA and $\bar{y}_1$ and $\bar{y}_2$ are the distances from the neutral axis to the centroids c_1 and c_2 of areas A_1 and A_2, respectively.

An easier way to obtain the plastic moment is to evaluate the moments about the neutral axis of the forces C and T (Fig. 6-49b):

$$M_P = C\bar{y}_1 + T\bar{y}_2 \tag{6-91}$$

Replace T and C by $\sigma_Y A/2$ to get

$$M_P = \frac{\sigma_Y A(\bar{y}_1 + \bar{y}_2)}{2} \tag{6-92}$$

which is the same as Eq. (6-90).

The procedure for obtaining the **plastic moment** is to divide the cross section of the beam into two equal areas, locate the centroid of each half, and then use Eq. (6-92) to calculate M_P.

Plastic Modulus and Shape Factor

The expression for the plastic moment can be written in a form similar to that for the yield moment [Eq. (6-87)], as

$$M_P = \sigma_Y Z \tag{6-93}$$

in which

$$Z = \frac{A(\bar{y}_1 + \bar{y}_2)}{2} \tag{6-94}$$

is the **plastic modulus** (or the *plastic section modulus*) for the cross section. The plastic modulus may be interpreted geometrically as the first moment (evaluated with respect to the neutral axis) of the area of the cross section above the neutral axis plus the first moment of the area below the neutral axis.

The ratio of the plastic moment to the yield moment is solely a function of the shape of the cross section and is called the **shape factor** f:

$$f = \frac{M_P}{M_Y} = \frac{Z}{S} \tag{6-95}$$

This factor is a measure of the reserve strength of the beam after yielding first begins. It is highest when most of the material is located near the neutral axis (for instance, a beam having a solid circular section) and lowest when most of the material is away from the neutral axis (for instance, a beam having a wide-flange section). Values of f for cross sections of rectangular, wide-flange, and circular shapes are given in the remainder of this section. Other shapes are considered in the problems at the end of the chapter.

Beams of Rectangular Cross Section

FIGURE 6-50

Rectangular cross section

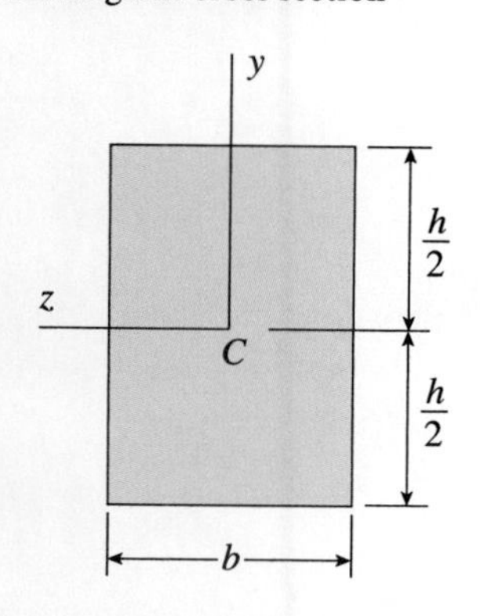

Now determine the properties of a beam of rectangular cross section (Fig. 6-50) when the material is elastoplastic. The section modulus is $S = bh^2/6$; therefore, the **yield moment** [Eq. (6-87)] is

$$M_Y = \frac{\sigma_Y bh^2}{6} \tag{6-96}$$

in which b is the width and h is the height of the cross section.

Because the cross section is doubly symmetric, the neutral axis passes through the centroid even when the beam is loaded into the plastic range. Consequently, the distances to the centroids of the areas above and below the neutral axis are

$$\bar{y}_1 = \bar{y}_2 = \frac{h}{4} \tag{6-97}$$

Therefore, the **plastic modulus** [Eq. (6-94)] is

$$Z = \frac{A(\bar{y}_1 + \bar{y}_2)}{2} = \frac{bh}{2}\left(\frac{h}{4} + \frac{h}{4}\right) = \frac{bh^2}{4} \tag{6-98}$$

and the **plastic moment** (Eq. 6-93) is

$$M_P = \frac{\sigma_Y bh^2}{4} \tag{6-99}$$

Finally, the **shape factor** for a rectangular cross section is

$$f = \frac{M_P}{M_Y} = \frac{Z}{S} = \frac{3}{2} \tag{6-100}$$

which means that the plastic moment for a rectangular beam is 50% greater than the yield moment.

Next, consider the stresses in a rectangular beam when the bending moment M is greater than the yield moment but has not yet reached the plastic moment. The outer parts of the beam are at the yield stress σ_Y, and the inner part (the **elastic core**) has a linearly varying stress distribution (Figs. 6-51a and b). The fully plastic zones are shaded in Fig. 6-51a, and the distances from the neutral axis to the inner edges of the plastic zones (or the outer edges of the elastic core) are denoted by e.

The stresses acting on the cross section have the force resultants C_1, C_2, T_1, and T_2, as shown in Fig. 6-51c. The forces C_1 and T_1 in the plastic zones are each equal to the yield stress times the cross-sectional area of the zone:

$$C_1 = T_1 = \sigma_Y b\left(\frac{h}{2} - e\right) \tag{6-101}$$

The forces C_2 and T_2 in the elastic core are each equal to the area of the stress diagram times the width b of the beam:

$$C_2 = T_2 = \frac{\sigma_Y e}{2} b \tag{6-102}$$

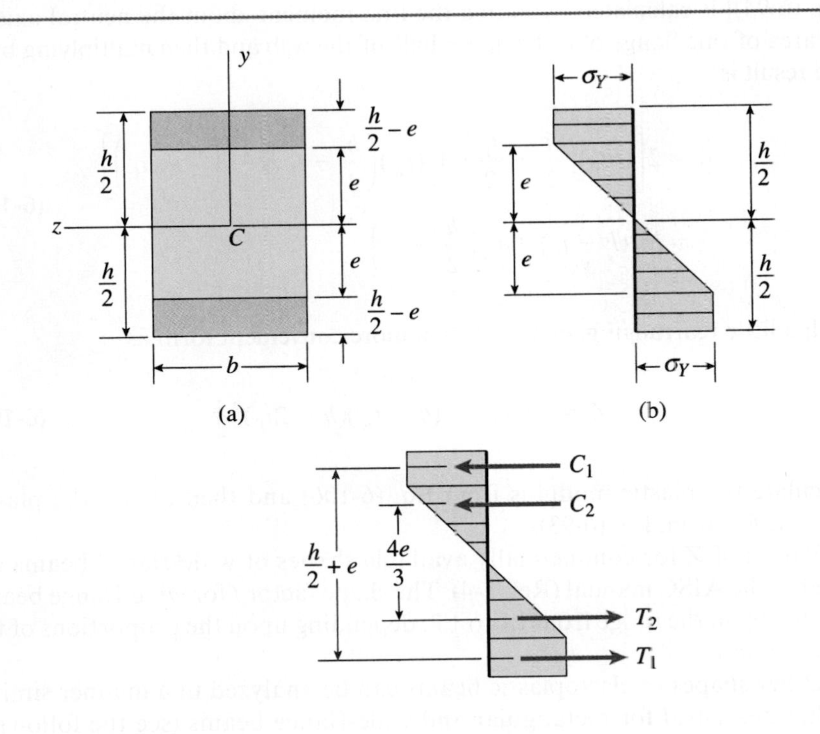

FIGURE 6-51

Stress distribution in a beam of rectangular cross section with an elastic core ($M_Y \le M \le M_P$)

Thus, the **bending moment** (see Fig. 6-51c) is

$$M = C_1\left(\frac{h}{2} + e\right) + C_2\left(\frac{4e}{3}\right) = \sigma_Y b\left(\frac{h}{2} - e\right)\left(\frac{h}{2} + e\right) + \frac{\sigma_Y be}{2}\left(\frac{4e}{3}\right)$$

$$= \frac{\sigma_Y bh^2}{6}\left(\frac{3}{2} - \frac{2e^2}{h^2}\right) = M_Y\left(\frac{3}{2} - \frac{2e^2}{h^2}\right) \quad M_Y \le M \le M_P \tag{6-103}$$

Note that when $e = h/2$, the equation gives $M = M_Y$, and when $e = 0$, it gives $M = 3M_Y/2$, which is the plastic moment M_P.

Equation (6-103) can be used to determine the bending moment when the dimensions of the elastic core are known. However, a more common requirement is to determine the **size of the elastic core** when the bending moment is known. Therefore, solve Eq. (6-103) for e in terms of the bending moment:

$$e = h\sqrt{\frac{1}{2}\left(\frac{3}{2} - \frac{M}{M_Y}\right)} \quad M_Y \le M \le M_P \tag{6-104}$$

Again note the limiting conditions: When $M = M_Y$, the equation gives $e = h/2$, and when $M = M_P = 3M_Y/2$, it gives $e = 0$, which is the fully plastic condition.

Beams of Wide-Flange Shape

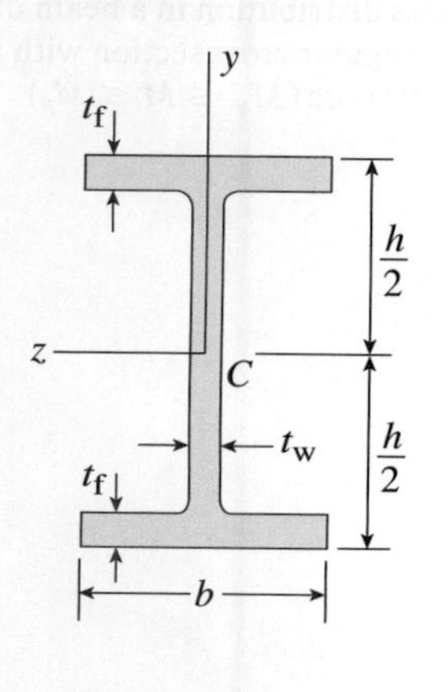

FIGURE 6-52
Cross section of a wide-flange beam

For a doubly symmetric wide-flange beam (Fig. 6-52), the plastic modulus Z [Eq. (6-94)] is calculated by taking the first moment about the neutral axis of the area of one flange plus the upper half of the web and then multiplying by 2. The result is

$$Z = 2\left[(bt_f)\left(\frac{h}{2} - \frac{t_f}{2}\right) + (t_w)\left(\frac{h}{2} - t_f\right)\left(\frac{1}{2}\right)\left(\frac{h}{2} - t_f\right)\right]$$
$$= bt_f(h - t_f) + t_w\left(\frac{h}{2} - t_f\right)^2 \tag{6-105}$$

With a little rearranging, express Z in a more convenient form as

$$Z = \frac{1}{4}\left[bh^2 - (b - t_w)(h - 2t_f)^2\right] \tag{6-106}$$

Calculate the plastic modulus from Eq. (6-106) and then obtain the plastic moment M_P from Eq. (6-93).

Values of Z for commercially available shapes of wide-flange beams are listed in the AISC manual (Ref. 5-4). The shape factor f for wide-flange beams is typically in the range from 1.1 to 1.2, depending upon the proportions of the cross section.

Other shapes of elastoplastic beams can be analyzed in a manner similar to that described for rectangular and wide-flange beams (see the following examples and the problems at the end of the chapter).

Example 6-10

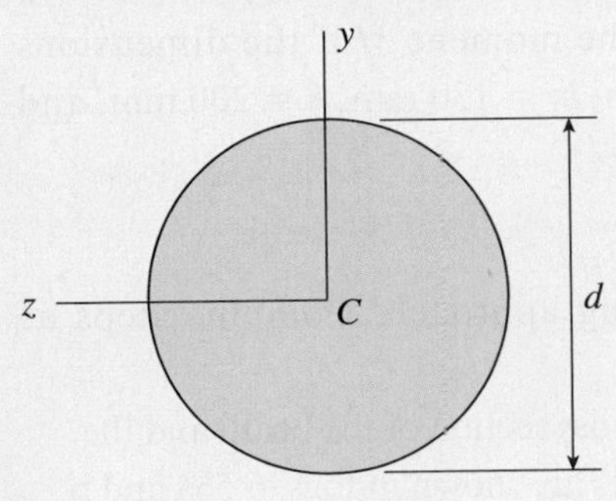

FIGURE 6-53

Example 6-10: Cross section of a circular beam (elastoplastic material)

Determine the yield moment, plastic modulus, plastic moment, and shape factor for a beam of circular cross section with diameter d (Fig. 6-53).

Solution:

Use a four-step problem-solving approach. Combine steps as needed for an efficient solution.

1, 2. ***Conceptualize, Categorize***: As a preliminary matter, note that since the cross section is doubly symmetric, the neutral axis passes through the center of the circle for both linearly elastic and elastoplastic behavior.

3. ***Analyze***: The yield moment M_Y is found from the flexure formula [Eq. (6-87)] as

$$M_Y = \frac{\sigma_Y I}{c} = \frac{\sigma_Y(\pi d^4/64)}{d/2} = \sigma_Y\left(\frac{\pi d^3}{32}\right) \tag{6-107}$$

The plastic modulus Z is found from Eq. (6-94), in which A is the area of the circle and $\overline{y}_1$ and $\overline{y}_2$ are the distances to the centroids c_1 and c_2 of the two halves of the circle (Fig. 6-54). Use Cases 9 and 10 of Appendix E to get

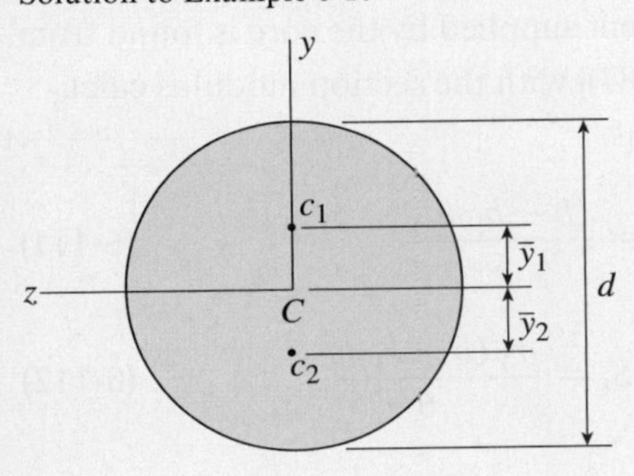

FIGURE 6-54

Solution to Example 6-10

$$A = \frac{\pi d^2}{4} \qquad \overline{y}_1 = \overline{y}_2 = \frac{2d}{3\pi}$$

Now substitute into Eq. (6-94) for the plastic modulus to find

$$Z = \frac{A(\overline{y}_1 + \overline{y}_2)}{2} = \frac{d^3}{6} \tag{6-108}$$

Therefore, the plastic moment M_P [Eq. (6-93)] is

$$M_P = \sigma_Y Z = \frac{\sigma_Y d^3}{6} \tag{6-109}$$

and the shape factor f [Eq. (6-95)] is

$$f = \frac{M_P}{M_Y} = \frac{16}{3\pi} \approx 1.70 \tag{6-110}$$

4. ***Finalize***: This result shows that the maximum bending moment for a circular beam of elastoplastic material is about 70% larger than the bending moment when the beam first begins to yield.

Example 6-11

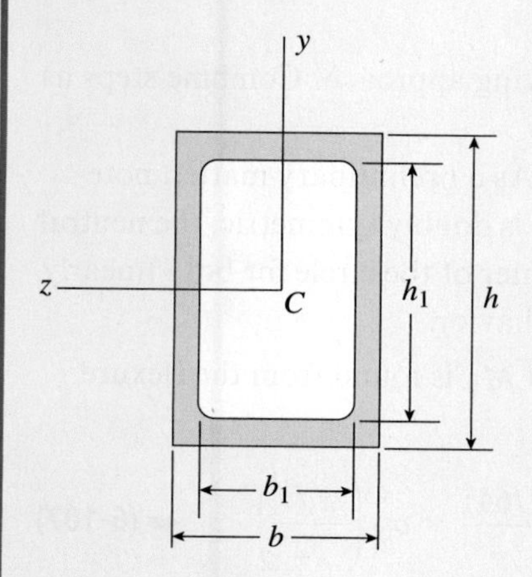

FIGURE 6-55
Example 6-11: Cross section of a hollow box beam (elastoplastic material)

A doubly symmetric hollow box beam (Fig. 6-55) of elastoplastic material ($\sigma_Y = 220$ MPa) is subjected to a bending moment M of such magnitude that the flanges yield but the webs remain linearly elastic.

Determine the magnitude of the moment M if the dimensions of the cross section are $b = 150$ mm, $b_1 = 130$ mm, $h = 200$ mm, and $h_1 = 160$ mm.

Solution:

Use a four-step problem-solving approach. Combine steps as needed for an efficient solution.

1, 2. ***Conceptualize, Categorize:*** The cross section of the beam and the distribution of the normal stresses are shown in Figs. 6-56a and b, respectively. From the figure, note that the stresses in the webs increase linearly with distance from the neutral axis and the stresses in the flanges equal the yield stress σ_Y. Therefore, the bending moment M acting on the cross section consists of two parts:

(i) A moment M_1 corresponding to the elastic core.

(ii) A moment M_2 produced by the yield stresses σ_Y in the flanges.

3. ***Analyze:*** The bending moment supplied by the core is found from the flexure formula [Eq. (6-87)] with the section modulus calculated for the webs alone; thus,

$$S_1 = \frac{(b - b_1)h_1^2}{6} \quad \textbf{(6-111)}$$

and

$$M_1 = \sigma_Y S_1 = \frac{\sigma_Y (b - b_1)h_1^2}{6} \quad \textbf{(6-112)}$$

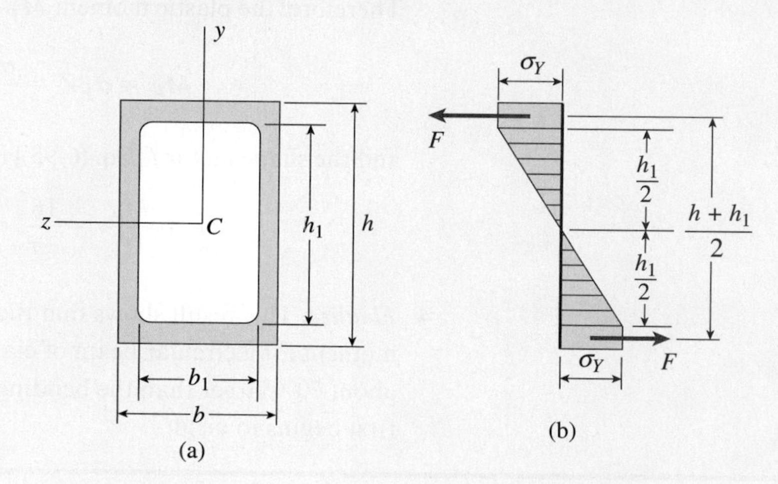

FIGURE 6-56
Solution to Example 6-11

To find the moment supplied by the flanges, note that the resultant force F in each flange (Fig. 6-56b) is equal to the yield stress multiplied by the area of the flange:

$$F = \sigma_Y b\left(\frac{h - h_1}{2}\right) \tag{a}$$

The force in the top flange is compressive, and the force in the bottom flange is tensile if the bending moment M is positive. Together, the two forces create the bending moment M_2:

$$M_2 = F\left(\frac{h + h_1}{2}\right) = \frac{\sigma_Y b(h^2 - h_1^2)}{4} \tag{6-113}$$

Therefore, the total moment acting on the cross section, after some rearranging, is

$$M = M_1 + M_2 = \frac{\sigma_Y}{12}\left[3bh^2 - (b + 2b_1)h_1^2\right] \tag{6-114}$$

Substitute the given numerical values to obtain

$$M = 138 \text{ kN}\cdot\text{m}$$

4. *Finalize*: *Note:* The yield moment M_Y and the plastic moment M_P for the beam in this example have the following values (determined in Prob. 6.10-13):

$$M_Y = 122 \text{ kN}\cdot\text{m} \quad M_P = 147 \text{ kN}\cdot\text{m}$$

The bending moment M is between these values, as expected.

CHAPTER SUMMARY AND REVIEW

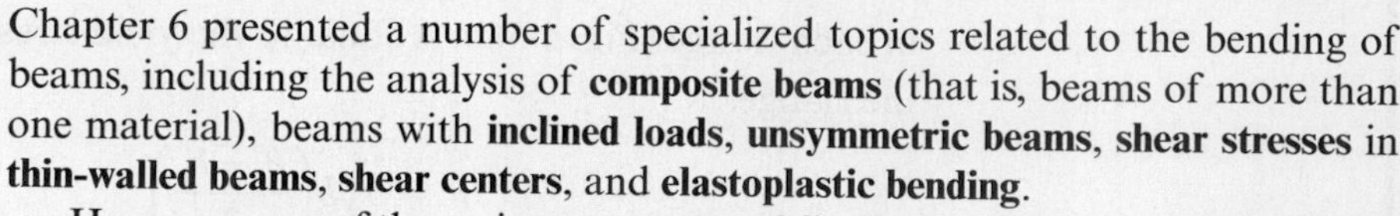

Chapter 6 presented a number of specialized topics related to the bending of beams, including the analysis of **composite beams** (that is, beams of more than one material), beams with **inclined loads**, **unsymmetric beams**, **shear stresses** in **thin-walled beams**, **shear centers**, and **elastoplastic bending**.

Here are some of the major concepts and findings presented in this chapter.

1. In the introduction to **composite beams**, specialized moment-curvature relationship and flexure formulas for composite beams of two materials were developed:

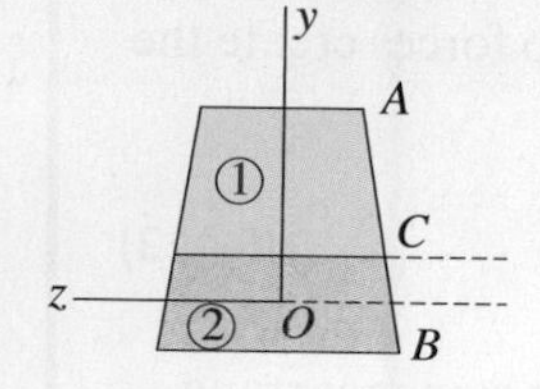

$$\kappa = \frac{1}{\rho} = \frac{M}{E_1 I_1 + E_2 I_2}$$

$$\sigma_{x1} = -\frac{M_Y E_1}{E_1 I_1 + E_2 I_2} \qquad \sigma_{x2} = -\frac{M_Y E_2}{E_1 I_1 + E_2 I_2}$$

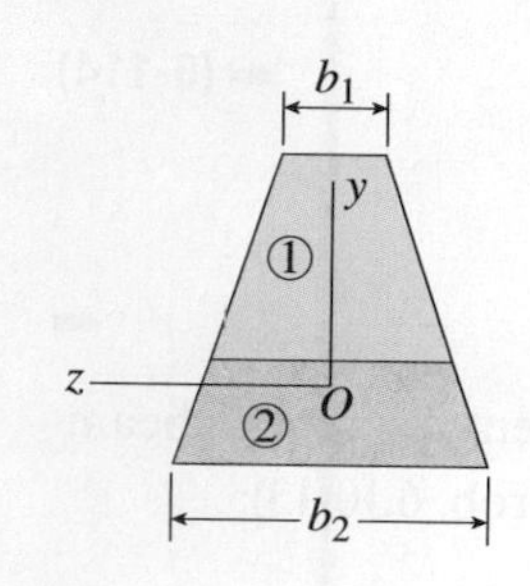

Both materials must follow Hooke's law, and the two parts of the beam must be adequately bonded so that they act as a single unit. The formulas here do not apply to reinforced concrete beams, which are not designed on the basis of linearly elastic behavior. However, a transformed-section approach (see Example 6-4) can be used as part of a cracked section analysis of reinforced concrete beams.

2. The **transformed-section method** offers a convenient way of transforming the cross section of a composite beam into an equivalent cross section of an imaginary beam that is composed of only one material. The ratio of the modulus of elasticity of material 2 to that of material 1 is known as the **modular ratio**, $n = E_2/E_1$. The moment of inertia of the transformed section is defined as

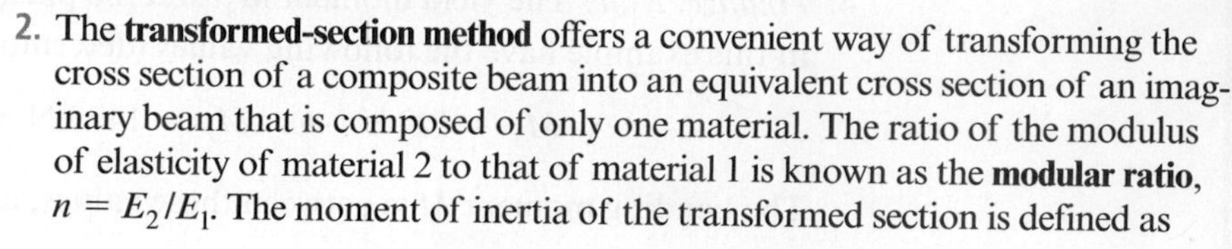

$$I_T = I_1 + nI_2 = I_1 + \frac{E_2}{E_1} I_2$$

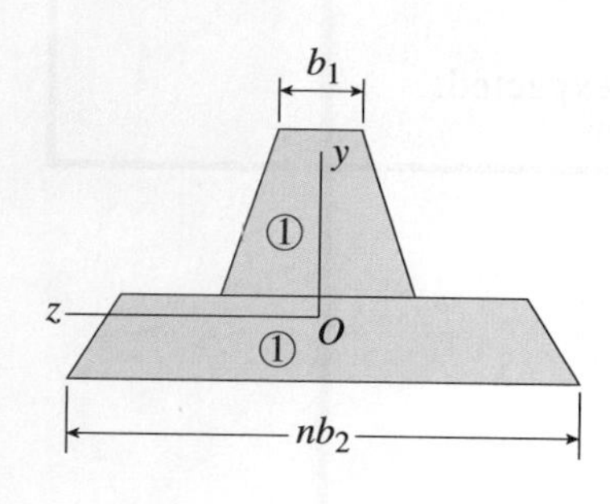

Normal stresses in the beam transformed to material 1 are computed using the simplified flexure formula:

$$\sigma_{x1} = -\frac{My}{I_T}$$

while those in material 2 are computed as

$$\sigma_{x2} = -\frac{My}{I_T} n$$

3. If **inclined loads** act through the centroid of the cross section of beams with two axes of symmetry in the cross section, there will be no twisting of the beam about the longitudinal axis. For these beams, compute the bending stresses by resolving the inclined load into two components—one acting in each plane of symmetry. The bending stresses were obtained from the flexure formula for each load component acting separately, and the final stresses obtained by superposing the separate stresses. The stresses in the beam are very sensitive to slight changes in the direction of the load and to irregularities in the alignment of the beam itself.

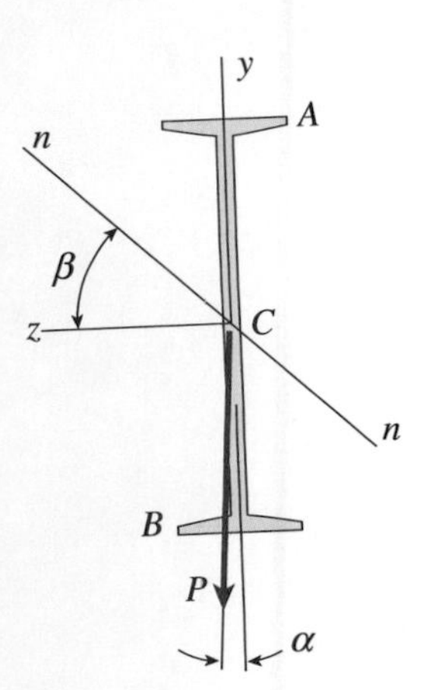

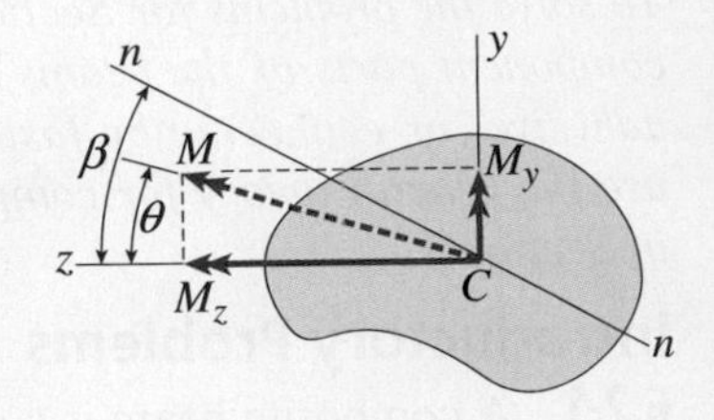

4. When the restriction of symmetry about at least one axis of the cross section was removed for pure bending, the plane in which the bending moment acts is perpendicular to the neutral surface only if the y and z axes are principal centroidal axes of the cross section, and the bending moment acts in one of the two principal planes (the x-y plane or the x-z plane). A **general procedure for computing normal stresses in *unsymmetric beams*** acted on by any moment M is (a) find the centroid and (b) superpose results of the flexure formula about the two principal centroidal axes to find normal stresses. An alternate approach is to use the **generalized flexure formula** with the original centroidal axes aligned with the beam cross section.

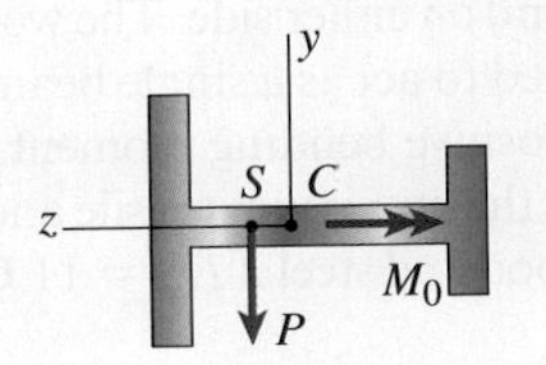

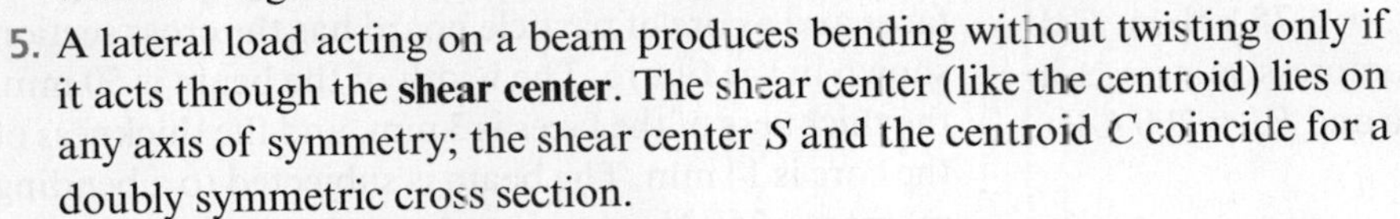

5. A lateral load acting on a beam produces bending without twisting only if it acts through the **shear center**. The shear center (like the centroid) lies on any axis of symmetry; the shear center S and the centroid C coincide for a doubly symmetric cross section.
6. Beams of **thin-walled open cross sections** (such as wide-flange beams, channels, angles, T-beams, and Z-sections) are in common use for structural purposes, but are **very weak in torsion**.

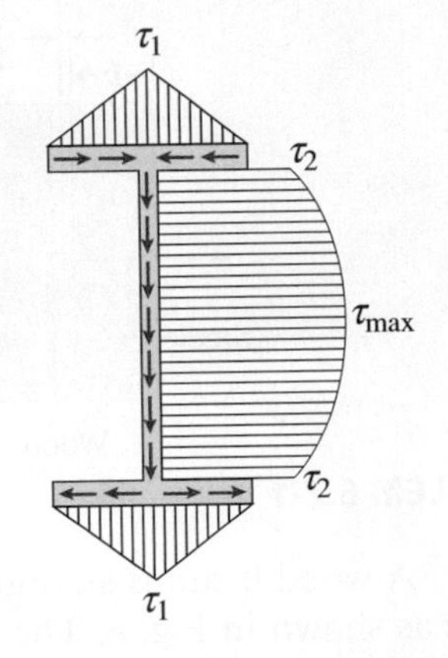

7. The **shear stresses** in beams of thin-walled open cross sections *flow* inward from the outermost edges, then down through the web, and finally outward to the edges of the bottom flange.
8. Any Z-section that is symmetric about the centroid (i.e., any Z-section having identical flanges—same width and same thickness) has its shear center at the centroid of the cross section. The locations of the shear centers of many other structural shapes are given in both the examples and the problems at the end of the chapter.

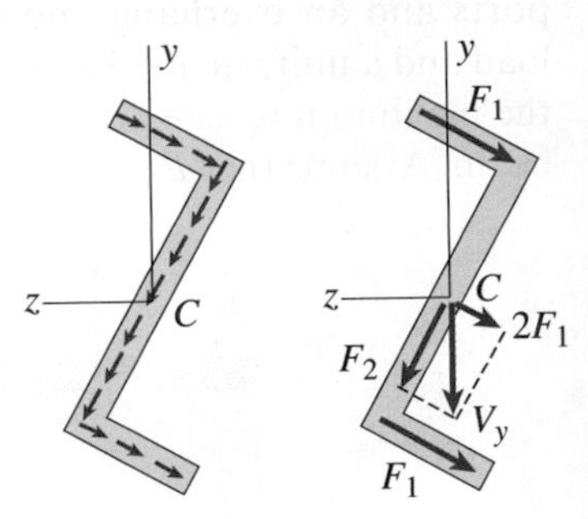

9. **Elastoplastic materials** follow Hooke's law up to the yield stress σ_Y and then yield plastically under constant stress. Structural steels are excellent examples of elastoplastic materials, because they have sharply defined yield points and undergo large strains during yielding. First the *yield moment* is computed as

$$M_Y = \sigma_Y \times S$$

using the flexure formula. Then, the *plastic moment* is

$$M_P = \sigma_Y \times Z$$

where S and Z are the section modulus and plastic section modulus of the cross section, respectively. M_Y is the bending moment in the beam when the maximum stress just reaches the yield stress, and M_P is the maximum moment that can be sustained by a beam of elastoplastic material. The shape factor f is defined as

$$f = M_P / M_Y = Z/S$$

and is a measure of the reserve strength of the beam after yielding first begins.

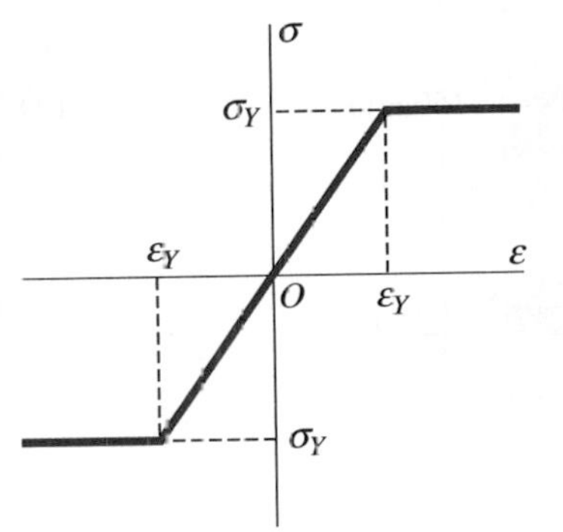

PROBLEMS Chapter 6

6.2 Composite Beams

To solve the problems for Section 6.2, assume that the component parts of the beams are securely bonded by adhesives or connected by fasteners. Also, be sure to use the general theory for composite beams described in Section 6.2.

Introductory Problems

6.2-1 A composite beam is constructed using a steel plate (13 mm × 150 mm) with two wood beams (75 mm × 150 mm) on either side. The wood and steel are securely fastened to act as a single beam. The beam is subjected to a positive bending moment $M_z = 6.75$ kN·m. Calculate the maximum tensile and compressive stresses in the wood and steel if $E_w = 11$ GPa and $E_s = 200$ GPa.

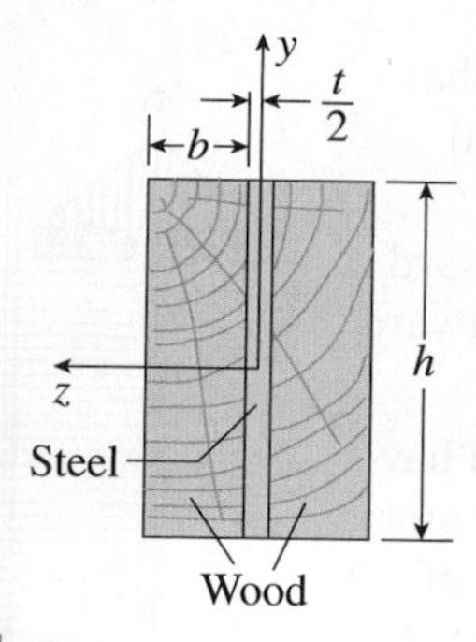

PROBLEM 6.2-1

6.2-2 A wood beam is strengthened using two steel plates as shown in Fig. a. The beam has simple supports and an overhang and is subjected to a point load and a uniform load as shown in Fig. b. Calculate the maximum tensile and compressive stresses of the beam. Assume that $E_w = 11$ GPa and $E_s = 200$ GPa.

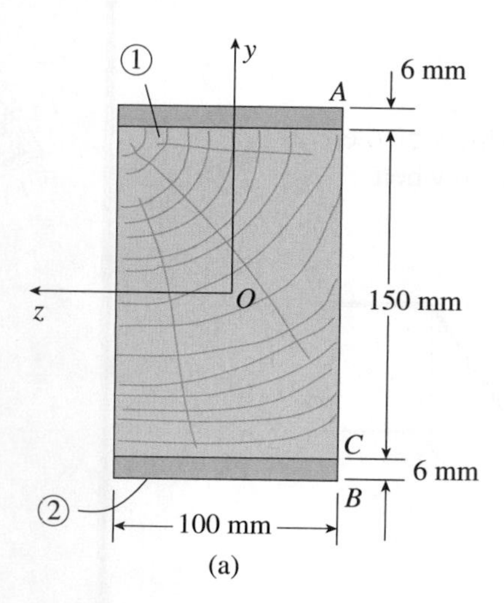

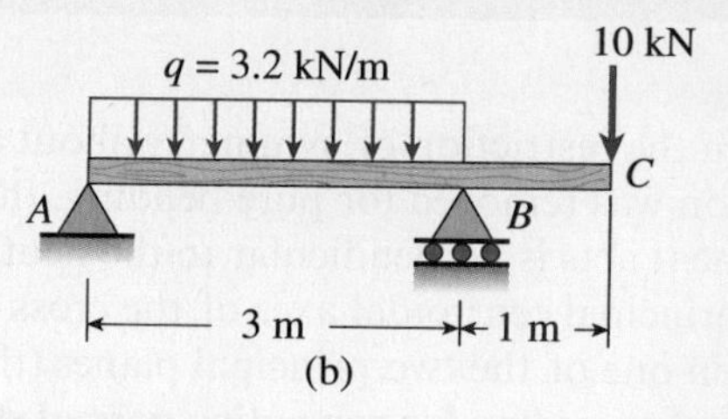

PROBLEM 6.2-2

6.2-3 A composite beam consisting of fiberglass faces and a core of particle board has the cross section shown in the figure. The width of the beam is 50 mm, the thickness of the faces is 3 mm, and the thickness of the core is 14 mm. The beam is subjected to a bending moment of 55 N·m acting about the z axis.

Find the maximum bending stresses σ_f and σ_c in the faces and the core, respectively, if their respective moduli of elasticity are 28 GPa and 10 GPa.

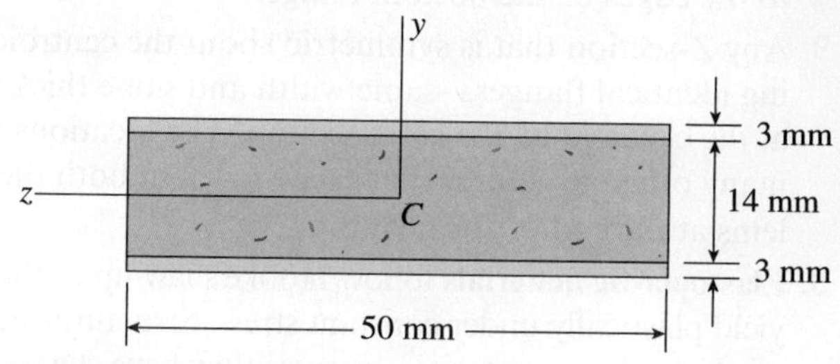

PROBLEM 6.2-3

6.2-4 A wood beam with cross-sectional dimensions 200 mm × 300 mm is reinforced on its sides by steel plates 12 mm thick (see figure). The moduli of elasticity for the steel and wood are $E_s = 190$ GPa and $E_w = 11$ GPa, respectively. Also, the corresponding allowable stresses are $\sigma_s = 110$ MPa and $\sigma_w = 7.5$ MPa.

(a) Calculate the maximum permissible bending moment M_{max} when the beam is bent about the z axis.

(b) Repeat part (a) if the beam is now bent about its y axis.

(c) Find the required thickness of the steel plates on the beam bent about the y axis so that M_{max} is the same for both beam orientations.

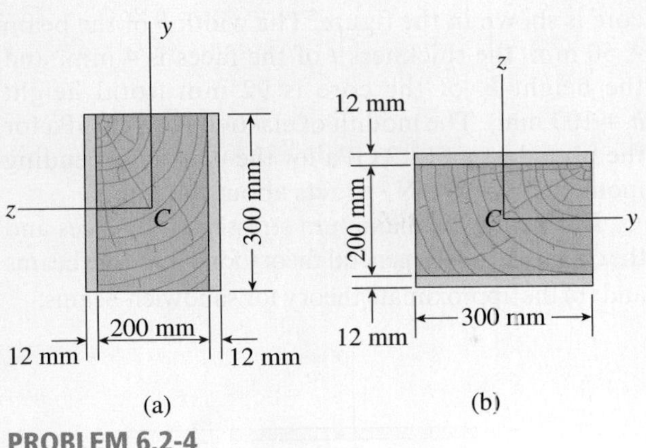

PROBLEM 6.2-4

6.2-5 A hollow box beam is constructed with webs of Douglas-fir plywood and flanges of pine, as shown in the figure in a cross-sectional view. The plywood is 24 mm thick and 300 mm wide; the flanges are 50 mm × 100 mm (actual size). The modulus of elasticity for the plywood is 11 GPa and for the pine is 8 GPa.

(a) If the allowable stresses are 14 MPa for the plywood and 12 MPa for the pine, find the allowable bending moment M_{max} when the beam is bent about the z axis.

(b) Repeat part (a) if the beam is now bent about its y axis.

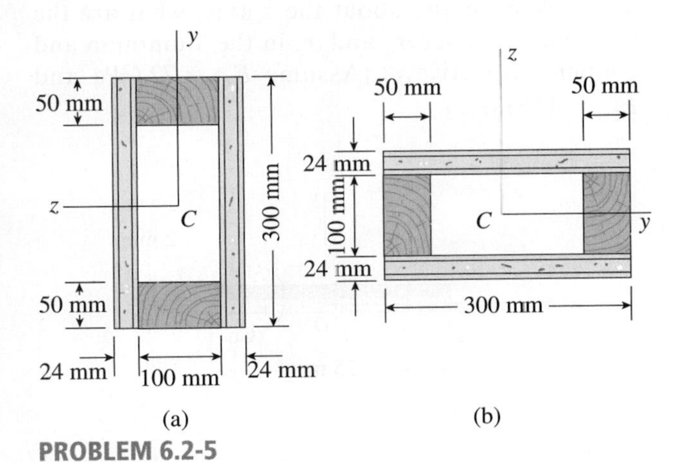

PROBLEM 6.2-5

Representative Problems

6.2-6 A round *titanium* tube of outside diameter d_2 and a *copper* core of diameter d_1 are bonded to form a composite beam, as shown in the figure.

(a) Derive formulas for the allowable bending moment M that can be carried by the beam based upon an allowable stress σ_{Ti} in the titanium and an allowable stress σ_{Cu} in the copper (Assume that the moduli of elasticity for the titanium and copper are E_{Ti} and E_{Cu}, respectively.)

(b) If $d_2 = 40$ mm, $d_1 = 36$ mm, $E_{Ti} = 120$ GPa, $E_{Cu} = 110$ GPa, $\sigma_{Ti} = 840$ MPa, and $\sigma_{Cu} = 700$ MPa, what is the maximum bending moment M?

(c) What new value of copper diameter d_1 will result in a balanced design? (i.e., a balanced design is that in which titanium and copper reach allowable stress values at the same time).

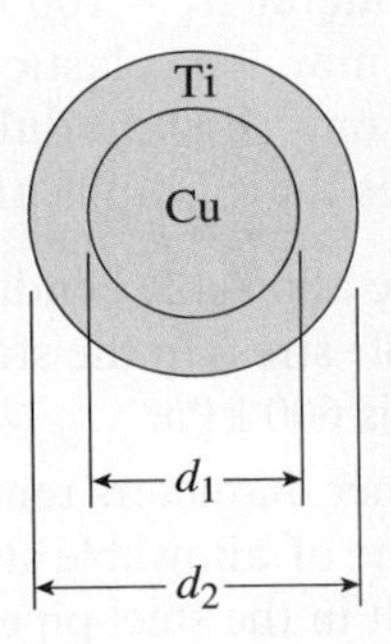

PROBLEM 6.2-6

6.2-7 A beam with a guided support and 4 m span supports a distributed load of intensity $q = 4$ kN/m over its first half (see figure part a) and a moment $M_0 = 5$ kN·m at joint B. The beam consists of a wood member (actual dimensions 97 mm × 295 mm in cross section, as shown in the figure part b) that is reinforced by 7-mm thick steel plates on top and bottom. The moduli of elasticity for the steel and wood are $E_s = 210$ GPa and $E_w = 10$ GPa, respectively.

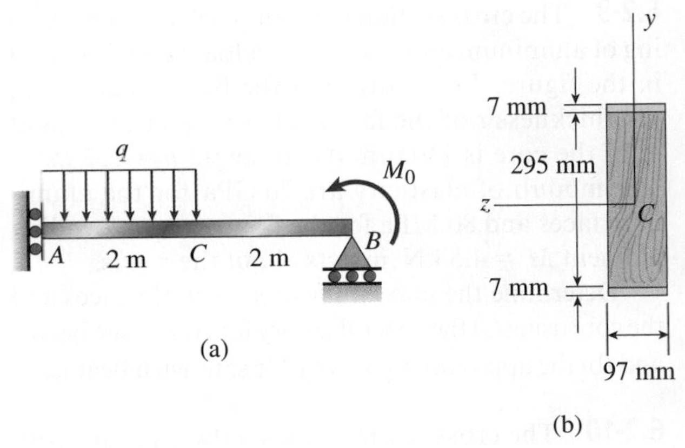

PROBLEM 6.2-7

(a) Calculate the maximum bending stresses σ_s in the steel plates and σ_w in the wood member due to the applied loads.

(b) If the allowable bending stress in the steel plates is $\sigma_{as} = 100$ MPa and that in the wood is $\sigma_{aw} = 6.5$ MPa, find q_{max}. (Assume that the moment at B, M_0, remains at 5 kN·m.)

(c) If $q = 4$ kN/m and allowable stress values in part (b) apply, what is $M_{0,max}$ at B?

6.2-8 A plastic-lined steel pipe has the cross-sectional shape shown in the figure. The steel pipe has an outer diameter $d_3 = 100$ mm and an inner diameter $d_2 = 94$ mm. The plastic liner has an inner diameter $d_1 = 82$ mm. The modulus of elasticity of the steel is 75 times the modulus of the plastic.

(a) Determine the allowable bending moment M_{allow} if the allowable stress in the steel is 35 MPa and in the plastic is 600 kPa.

(b) If pipe and liner diameters remain unchanged, what new value of allowable stress for the steel pipe will result in the steel pipe and plastic liner reaching their allowable stress values under the same maximum moment (i.e., a balanced design)? What is the new maximum moment?

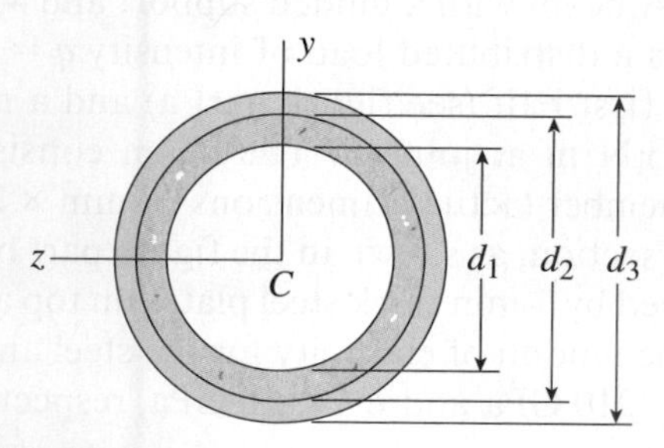

PROBLEM 6.2-8

6.2-9 The cross section of a sandwich beam consisting of aluminum alloy faces and a foam core is shown in the figure. The width b of the beam is 200 mm, the thickness t of the faces is 6 mm, and the height h_c of the core is 140 mm (total height $h = 152$ mm). The moduli of elasticity are 70 GPa for the aluminum faces and 80 MPa for the foam core. A bending moment $M = 4.5$ kN·m acts about the z axis.

Determine the maximum stresses in the faces and the core using (a) the general theory for composite beams and (b) the approximate theory for sandwich beams.

6.2-10 The cross section of a sandwich beam consisting of fiberglass faces and a lightweight plastic core is shown in the figure. The width b of the beam is 50 mm, the thickness t of the faces is 4 mm, and the height h_c of the core is 92 mm (total height $h = 100$ mm). The moduli of elasticity are 75 GPa for the fiberglass and 1.2 GPa for the plastic. A bending moment $M = 275$ N·m acts about the z axis.

Determine the maximum stresses in the faces and the core using (a) the general theory for composite beams and (b) the approximate theory for sandwich beams.

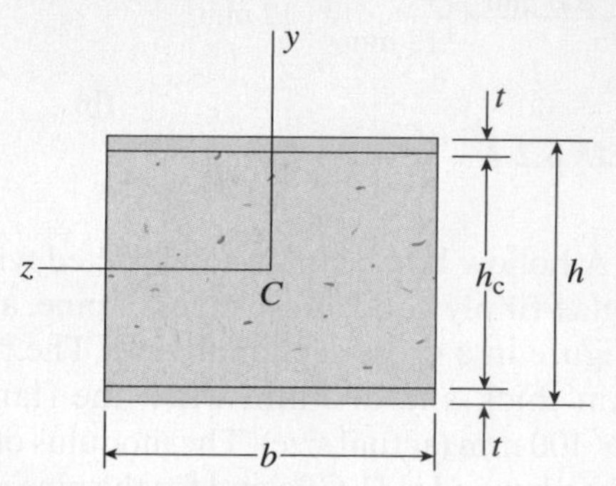

PROBLEMS 6.2-9 and 6.2-10

6.2-11 A bimetallic beam used in a temperature-control switch consists of strips of aluminum and copper bonded together as shown in the figure, which is a cross-sectional view. The width of the beam is 25 mm, and each strip has a thickness of 2 mm.

Under the action of a bending moment $M = 2$ N·m acting about the z axis, what are the maximum stresses σ_a and σ_c in the aluminum and copper, respectively? (Assume $E_{Al} = 72$ GPa and $E_{Cu} = 115$ GPa.)

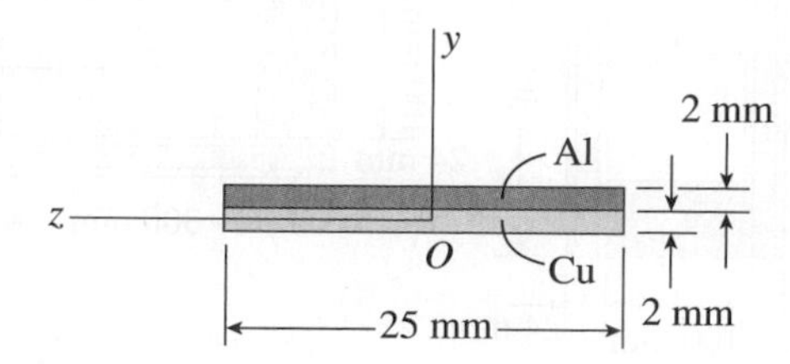

PROBLEM 6.2-11

6.2-12 A simply supported composite beam 3 m long carries a uniformly distributed load of intensity $q = 30$ kN/m (see figure). The beam is constructed of a wood member, 100 mm wide by 150 mm deep, and is reinforced on its lower side by a steel plate 8 mm thick and 100 mm wide.

(a) Find the maximum bending stresses σ_w and σ_s in the wood and steel, respectively, due to the uniform load if the moduli of elasticity are $E_w = 10$ GPa for the wood and $E_s = 210$ GPa for the steel.

(b) Find the required thickness of the steel plate so that the steel plate and wood reach their allowable stress values, $\sigma_{as} = 100$ MPa and $\sigma_{aw} = 8.5$ MPa, simultaneously under the maximum moment.

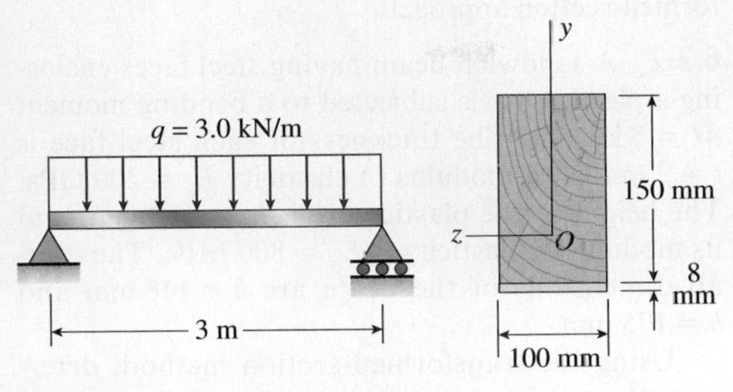

PROBLEM 6.2-12

6.2-13 A simply supported wooden I-beam with a 3.6 m span supports a distributed load of intensity $q = 1.3$ kN/m over its length (see figure part a). The beam is constructed with a web of Douglas-fir plywood and flanges of pine glued to the web, as shown in the figure part b. The plywood is 10 mm thick; the flanges are 50 mm × 50 mm (actual size). The modulus of elasticity for the plywood is 11 GPa and for the pine is 8.3 GPa.

(a) Calculate the maximum bending stresses in the pine flanges and in the plywood web.

(b) What is q_{max} if allowable stresses are 11 MPa in the flanges and 8 MPa in the web?

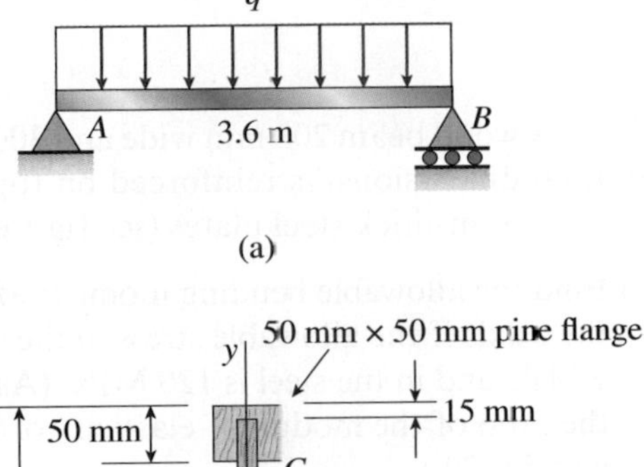

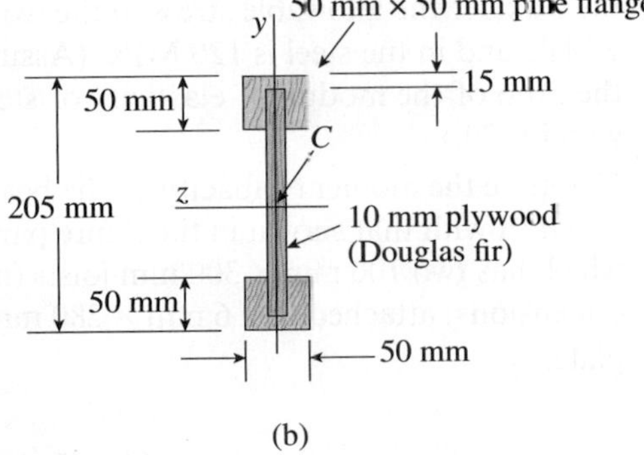

PROBLEM 6.2-13

6.2-14 A simply supported composite beam with a 3.6 m span supports a triangularly distributed load of peak intensity q_0 at mid-span (see figure part a). The beam is constructed of two wood joists, each 50 mm × 280 mm, fastened to two steel plates, one of dimensions 6 mm × 80 mm and the lower plate of dimensions 6 mm × 120 mm (see figure part b). The modulus of elasticity for the wood is 11 GPa and for the steel is 210 GPa.

If the allowable stresses are 7 MPa for the wood and 120 MPa for the steel, find the allowable peak load intensity $q_{0,max}$ when the beam is bent about the z axis. Neglect the weight of the beam.

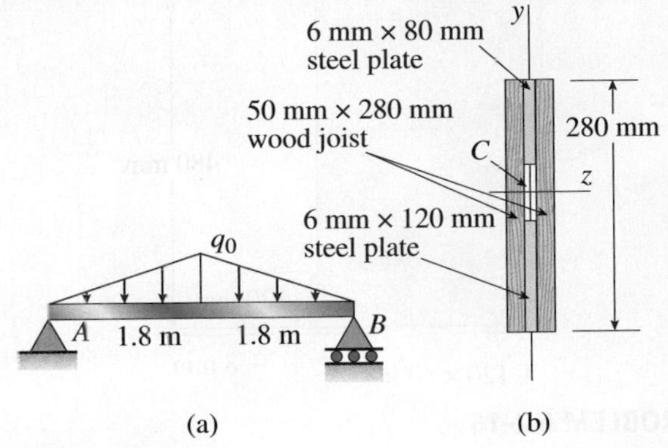

PROBLEM 6.2-14

6.2-15 A composite beam is constructed from a wood beam (75 mm × 150 mm) and a steel plate (75 mm). The wood and the steel are securely fastened to act as a single beam. The beam is subjected to a positive bending moment $M_z = 8.5$ kN·m. Calculate the required thickness of the steel plate based on the following limit states:

(a) Allowable compressive stress in the wood = 14 MPa

(b) Allowable tensile stress in the wood = 14 MPa

(c) Allowable tensile stress in the steel plate = 110 MPa Assume that $E_w = 11$ GPa and $E_s = 200$ GPa.

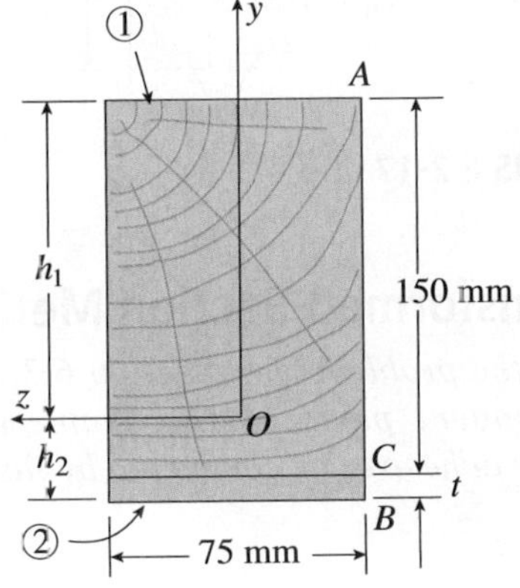

PROBLEM 6.2-15

6.2-16 A wood beam in a historic theater is reinforced with two angle sections at the outside lower corners (see figure). If the allowable stress in the wood is 12 MPa and that in the steel is 140 MPa, what is the ratio of the maximum permissible moments for the beam before and after reinforcement with the angle sections? See Appendix F Table F-5 for angle section properties. Assume that $E_w = 12$ GPa and $E_s = 210$ GPa.

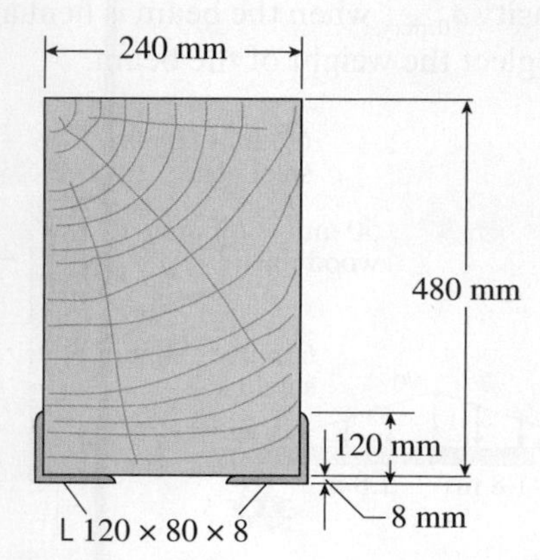

PROBLEM 6.2-16

6.2-17 Repeat Problem 6.2-1 but now assume that the steel plate is smaller (13 mm × 130 mm) and is aligned with the top of the beam as shown in the figure.

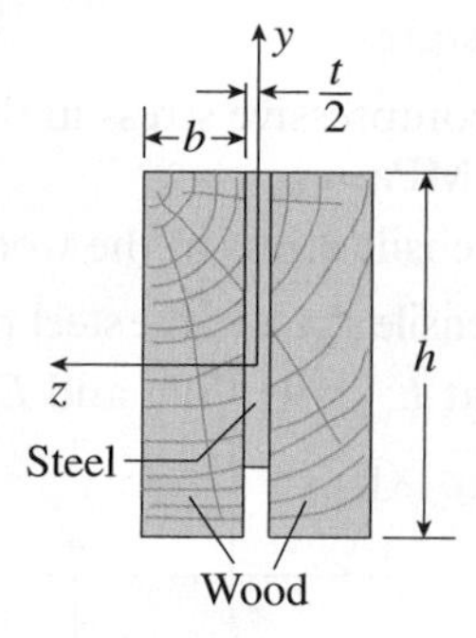

PROBLEMS 6.2-17 and 6.3-1

6.3 Transformed-Section Method

To solve the problems for Section 6.3, assume that the component parts of the beams are securely bonded by adhesives or connected by fasteners. Also, be sure to use the transformed-section method in the solutions.

Introductory Problems

6.3-1 Repeat Problem 6.2-17 but now use a transformed-section approach.

6.3-2 A sandwich beam having steel faces enclosing a plastic core is subjected to a bending moment $M = 5$ kN · m. The thickness of each steel face is $t = 3$ mm with modulus of elasticity $E_s = 200$ GPa. The height of the plastic core is $h_p = 140$ mm, and its modulus of elasticity is $E_p = 800$ MPa. The overall dimensions of the beam are $h = 146$ mm and $b = 175$ mm.

Using the transformed-section method, determine the maximum tensile and compressive stresses in the faces and the core.

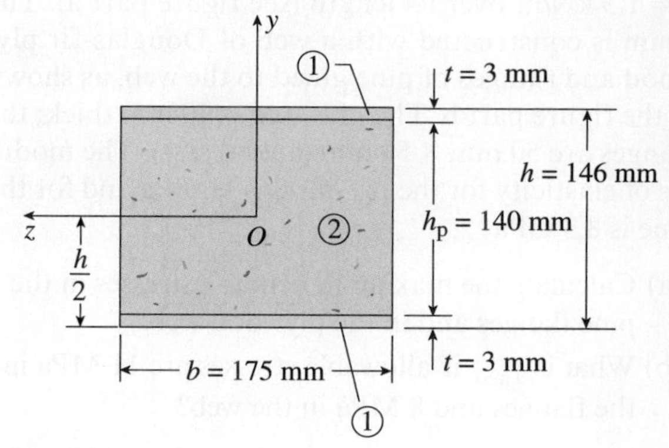

PROBLEM 6.3-2

6.3-3 A wood beam 200 mm wide and 300 mm deep (nominal dimensions) is reinforced on top and bottom by 12 mm-thick steel plates (see figure part a).

(a) Find the allowable bending moment M_{max} about the z axis if the allowable stress in the wood is 7 MPa and in the steel is 120 MPa. (Assume that the ratio of the moduli of elasticity of steel and wood is 20.)

(b) Compare the moment capacity of the beam in part a with that shown in the figure part b which has two 100 mm × 300 mm joists (nominal dimensions) attached to a 6 mm × 280 mm steel plate.

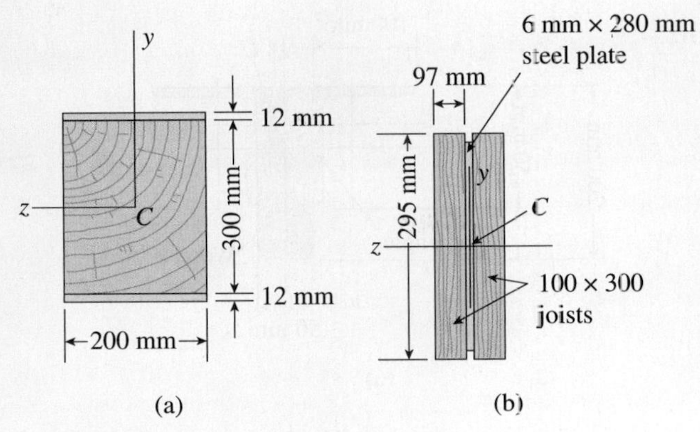

PROBLEM 6.3-3

6.3-4 A simple beam of span length 3.2 m carries a uniform load of intensity 48 kN/m. The cross section of the beam is a hollow box with wood flanges and steel side plates, as shown in the figure. The wood flanges are 75 mm × 100 mm in cross section, and the steel plates are 300 mm deep.

What is the required thickness t of the steel plates if the allowable stresses are 120 MPa for the steel and 6.5 MPa for the wood? (Assume that the moduli of elasticity for the steel and wood are 210 GPa and 10 GPa, respectively, and disregard the weight of the beam.)

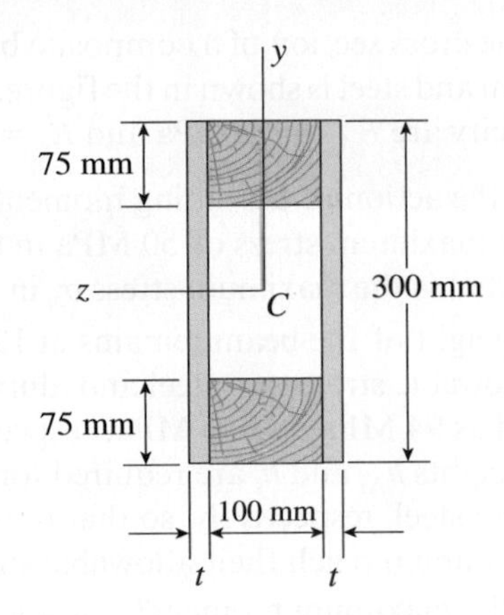

PROBLEM 6.3-4

6.3-5 A simple beam that is 5.5 m long supports a uniform load of intensity q. The beam is constructed of two UPN 200 sections on either side of a 97 mm × 195 mm (actual dimensions) wood beam (see the cross section shown in the figure part a). The modulus of elasticity of the steel (E_s = 210 GPa) is 20 times that of the wood (E_w).

(a) If the allowable stresses in the steel and wood are 110 MPa and 8.2 MPa, respectively, what is the allowable load q_{allow}? *Note:* Disregard the weight of the beam, and see Table F-3 of Appendix F for the dimensions and properties of the UPN-shape beam.

(b) If the beam is rotated 90° to bend about its y axis (see figure part b) and uniform load q = 3.6 kN/m is applied, find the maximum stresses σ_s and σ_w in the steel and wood, respectively. Include the weight of the beam. (Assume weight densities of 5.5 kN/m^3 and 77 kN/m^3 for the wood and steel, respectively.)

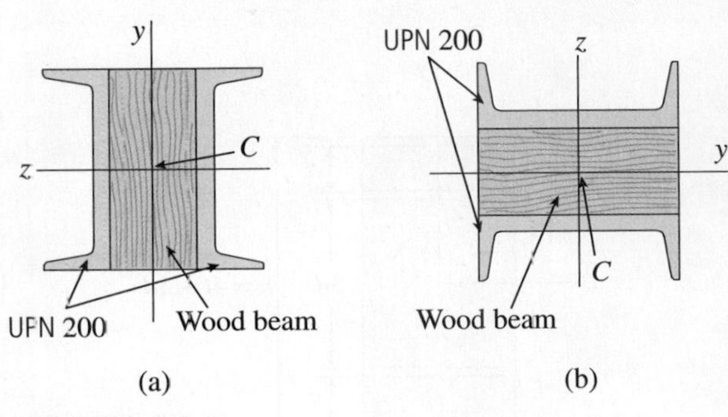

PROBLEM 6.3-5

6.3-6 The composite beam shown in the figure is simply supported and carries a total uniform load of 40 kN/m on a span length of 4.0 m. The beam is built of a southern pine wood member having cross-sectional dimensions of 150 mm × 250 mm and two brass plates of cross-sectional dimensions 30 mm × 150 mm.

(a) Determine the maximum stresses σ_B and σ_w in the brass and wood, respectively, if the moduli of elasticity are E_B = 96 GPa and E_w = 14 GPa. (Disregard the weight of the beam.)

(b) Find the required thickness of the brass plates so that the plate and wood reach their allowable stress values of σ_{aB} = 70 MPa and σ_{aw} = 8.5 MPa simultaneously under the maximum moment. What is the maximum moment?

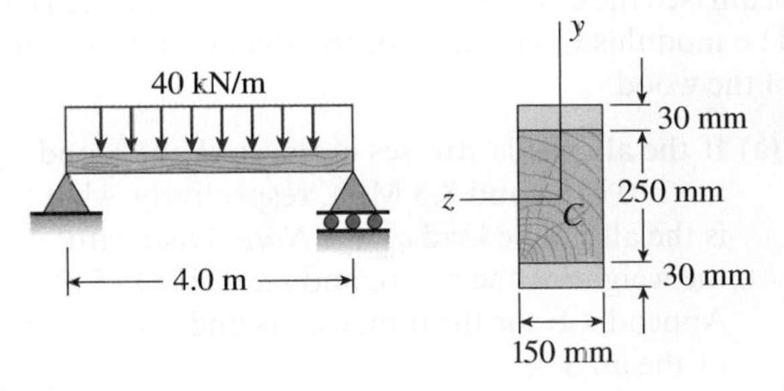

PROBLEM 6.3-6

Representative Problems

6.3-7 The cross section of a beam made of thin strips of aluminum separated by a lightweight plastic is shown in the figure. The beam has width $b = 75$ mm, the aluminum strips have thickness $t = 2.5$ mm, and the plastic segments have heights $d = 30$ mm and $3d = 90$ mm. The total height of the beam is $h = 160$ mm.

The moduli of elasticity for the aluminum and plastic are $E_{Al} = 75$ GPa and $E_p = 3$ GPa, respectively.

Determine the maximum stresses σ_{Al} and σ_p in the aluminum and plastic, respectively, due to a bending moment of 1.2 kN · m.

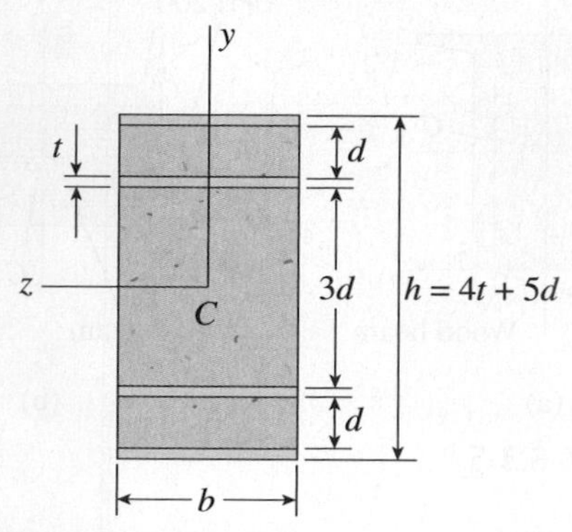

PROBLEMS 6.3-7 and 6.3-8

6.3-8 Consider the preceding problem if the beam has width $b = 75$ mm, the aluminum strips have thickness $t = 3$ mm, the plastic segments have heights $d = 40$ mm and $3d = 120$ mm, and the total height of the beam is $h = 212$ mm. Also, the moduli of elasticity are $E_{Al} = 75$ GPa and $E_p = 3$ GPa, respectively.

Determine the maximum stresses σ_{Al} and σ_p in the aluminum and plastic, respectively, due to a bending moment of 1.0 kN · m.

6.3-9 A simple beam that is 5.5 m long supports a uniform load of intensity q. The beam is constructed of two angle sections, each L 150 × 100 × 10, on either side of a 50 mm × 200 mm (actual dimensions) wood beam (see the cross section shown in the figure part a). The modulus of elasticity of the steel is 20 times that of the wood.

(a) If the allowable stresses in the steel and wood are 110 MPa and 8.3 MPa, respectively, what is the allowable load q_{allow}? *Note:* Disregard the weight of the beam, and see Table F-5 of Appendix F for the dimensions and properties of the angles.

(b) Repeat part (a) if a 25 mm × 250 mm wood flange (actual dimensions) is added (see figure part b).

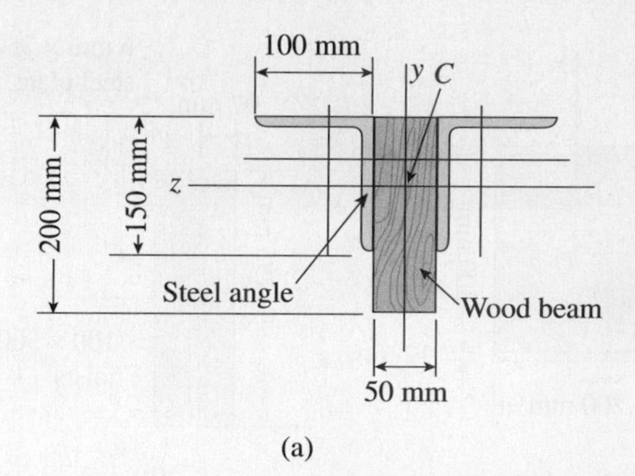

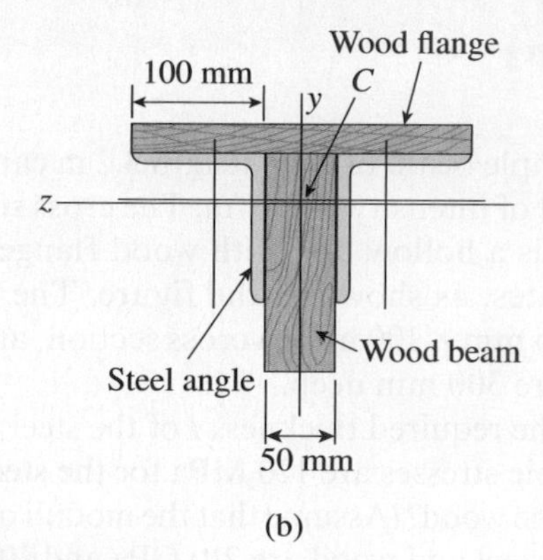

PROBLEM 6.3-9

6.3-10 The cross section of a composite beam made of aluminum and steel is shown in the figure. The moduli of elasticity are $E_{Al} = 75$ GPa and $E_s = 200$ GPa.

(a) Under the action of a bending moment that produces a maximum stress of 50 MPa in the aluminum, what is the maximum stress σ_s in the steel?

(b) If the height of the beam remains at 120 mm and allowable stresses in steel and aluminum are defined as 94 MPa and 40 MPa, respectively, what heights h_{Al} and h_s are required for aluminum and steel, respectively, so that both steel and aluminum reach their allowable stress values under the maximum moment?

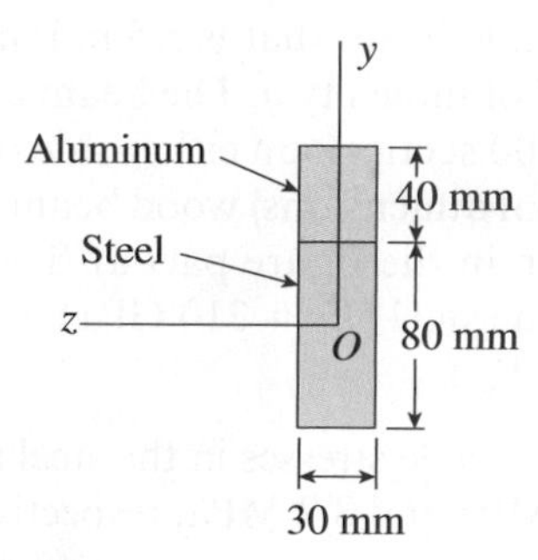

PROBLEM 6.3-10

6.3-11 A beam is constructed of two angle sections, each L 120 × 80 × 12, that reinforce a 50 mm × 200 mm (actual dimensions) wood plank (see the cross section shown in the figure). The modulus of elasticity for the wood is $E_w = 8$ GPa and for the steel is $E_s = 200$ GPa.

Find the allowable bending moment M_{allow} for the beam if the allowable stress in the wood is $\sigma_w = 10$ MPa and in the steel is $\sigma_s = 110$ MPa. *Note:* Disregard the weight of the beam, and see Table F-5 of Appendix F for the dimensions and properties of the angles.

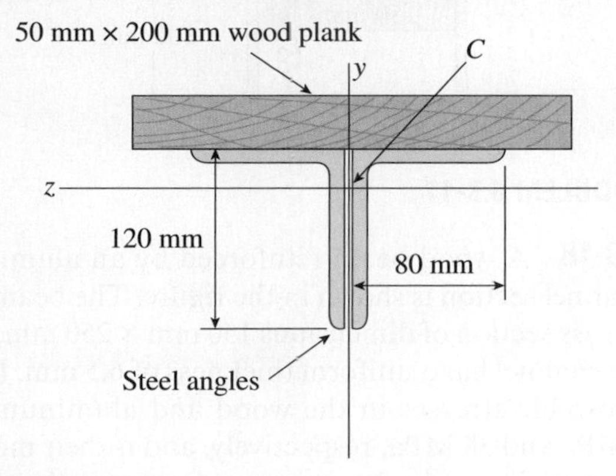

PROBLEM 6.3-11

6.3-12 The cross section of a bimetallic strip is shown in the figure. Assuming that the moduli of elasticity for metals *A* and *B* are $E_A = 168$ GPa and $E_B = 90$ GPa, respectively, determine the smaller of the two section moduli for the beam. (Recall that section modulus is equal to bending moment divided by maximum bending stress.) In which material does the maximum stress occur?

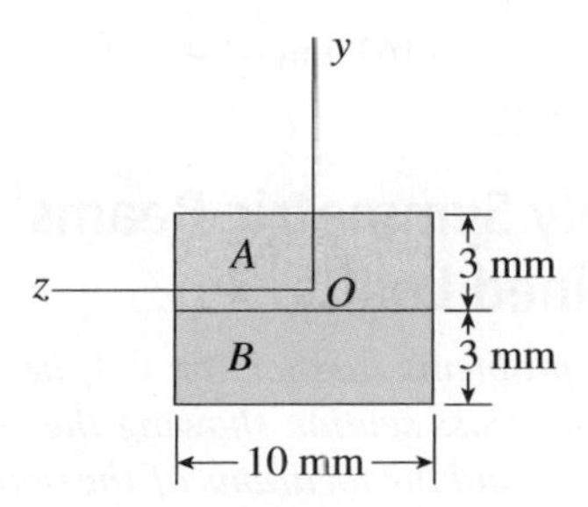

PROBLEM 6.3-12

6.3-13 An HE 260B steel wide-flange beam and a segment of a 100-mm thick concrete slab (see figure) jointly resist a positive bending moment of 130 kN · m. The beam and slab are joined by shear connectors that are welded to the steel beam. (These connectors resist the horizontal shear at the contact surface.) The moduli of elasticity of the steel and the concrete are in the ratio 12 to 1.

Determine the maximum stresses σ_s and σ_c in the steel and concrete, respectively. *Note:* See Table F-1 of Appendix F for the dimensions and properties of the steel beam.

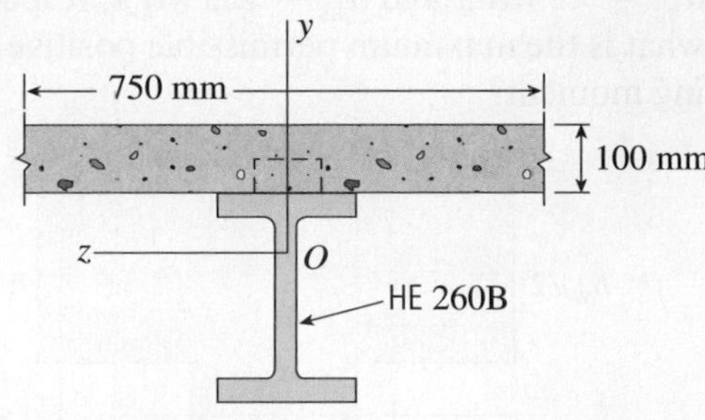

PROBLEM 6.3-13

6.3-14 A reinforced concrete beam (see figure) is acted on by a positive bending moment of $M = 160$ kN · m. Steel reinforcement consists of 4 bars of 28 mm diameter. The modulus of elasticity for the concrete is $E_c = 25$ GPa while that of the steel is $E_s = 200$ GPa.

(a) Find the maximum stresses in steel and concrete.

(b) If *allowable* stresses for concrete and steel are $\sigma_{ac} = 9.2$ MPa and $\sigma_{as} = 135$ MPa, respectively, what is the maximum permissible positive bending moment?

(c) What is the required area of steel reinforcement, A_s, if a balanced condition must be achieved? What is the allowable positive bending moment? (Recall that in a balanced design, both steel and concrete reach allowable stress values simultaneously under the design moment.)

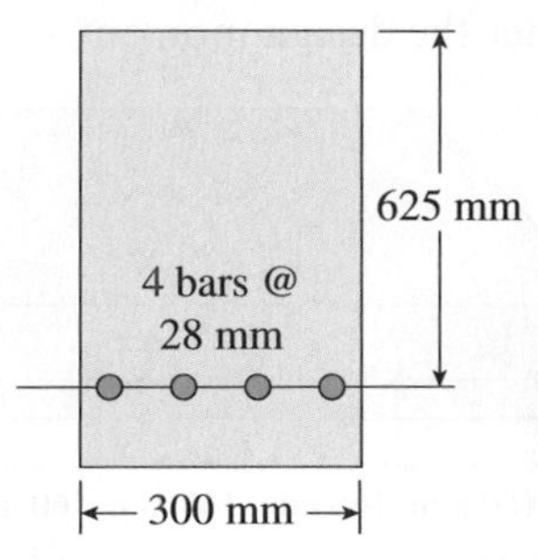

PROBLEM 6.3-14

6.3-15 A reinforced concrete T-beam (see figure) is acted on by a positive bending moment of $M = 240$ kN·m. Steel reinforcement consists of four bars of 40 mm diameter. The modulus of elasticity for the concrete is $E_c = 20$ GPa while that of the steel is $E_s = 210$ GPa. Let $b = 1200$ mm, $t_f = 100$ mm, $b_w = 380$ mm, and $d = 610$ mm.

(a) Find the maximum stresses in steel and concrete.

(b) If *allowable* stresses for concrete and steel are $\sigma_{ac} = 9.5$ MPa and $\sigma_{as} = 125$ MPa, respectively, what is the maximum permissible positive bending moment?

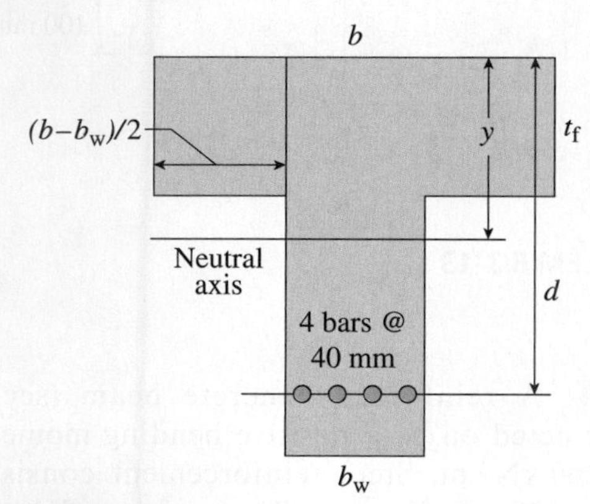

PROBLEM 6.3-15

6.3-16 A reinforced concrete slab (see figure) is reinforced with 13-mm bars spaced 160 mm apart at $d = 105$ mm from the top of the slab. The modulus of elasticity for the concrete is $E_c = 25$ GPa, while that of the steel is $E_s = 200$ GPa. Assume that *allowable* stresses for concrete and steel are $\sigma_{ac} = 9.2$ MPa and $\sigma_{as} = 135$ MPa.

(a) Find the maximum permissible positive bending moment for a 1-m wide strip of the slab.

(b) What is the required area of steel reinforcement, A_s, if a *balanced condition* must be achieved? What is the allowable positive bending moment? (Recall that in a balanced design, both steel and concrete reach allowable stress values simultaneously under the design moment.)

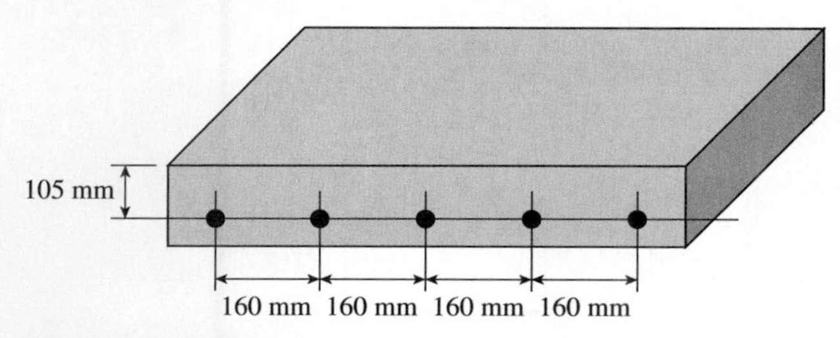

PROBLEM 6.3-16

6.3-17 A wood beam reinforced using two channels is subjected to a positive bending moment $M_z = 34$ kN·m. Calculate the largest tensile and compressive stresses in the wood and steel if $E_w = 11$ GPa and $E_s = 200$ GPa.

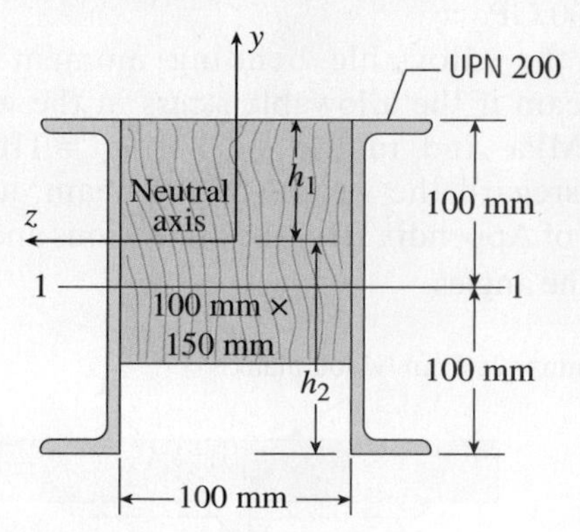

PROBLEM 6.3-17

6.3-18 A wood beam reinforced by an aluminum channel section is shown in the figure. The beam has a cross section of dimensions 150 mm × 250 mm, and the channel has a uniform thickness of 6.5 mm. If the allowable stresses in the wood and aluminum are 8 MPa and 38 MPa, respectively, and if their moduli of elasticity are in the ratio 1 to 6, what is the maximum allowable bending moment for the beam?

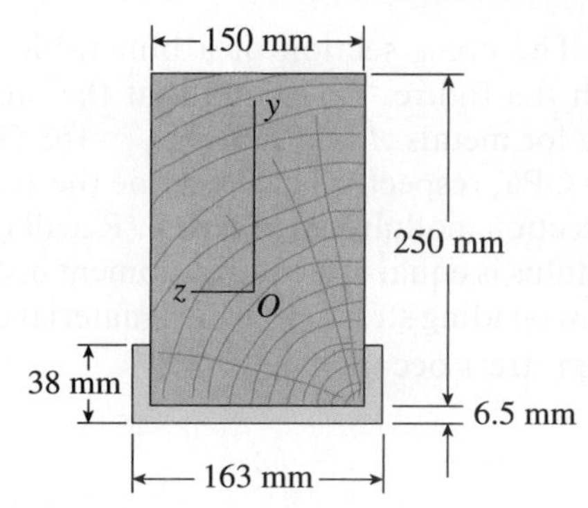

PROBLEM 6.3-18

6.4 Doubly Symmetric Beams with Inclined Loads

To solve the problems for Section 6.4, be sure to draw a sketch of the cross section showing the orientation of the neutral axis and the locations of the points where the stresses are being found.

Introductory Problems

6.4-1 A beam with a rectangular cross section supports an inclined load P having its line of action along a diagonal of the cross section (see figure). Show that the neutral axis lies along the other diagonal.

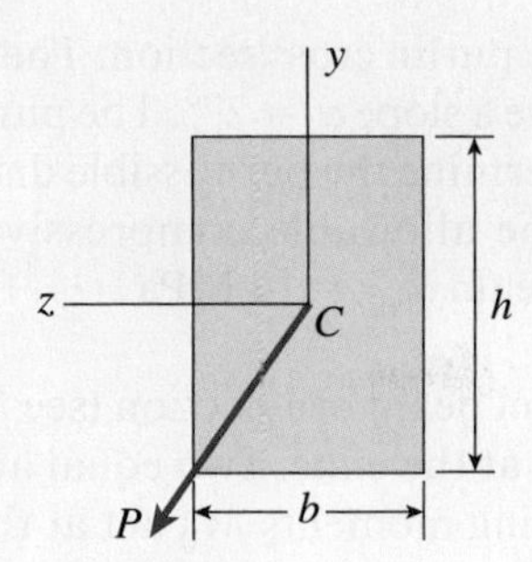

PROBLEM 6.4-1

6.4-2 A wood beam with a rectangular cross section (see figure) is simply supported on a span of length L. The longitudinal axis of the beam is horizontal, and the cross section is tilted at an angle α. The load on the beam is a vertical uniform load of intensity q acting through the centroid C.

Determine the orientation of the neutral axis and calculate the maximum tensile stress σ_{max} if $b = 80$ mm, $h = 140$ mm, $L = 1.75$ m, $\alpha = 22.5°$, and $q = 7.5$ kN/m.

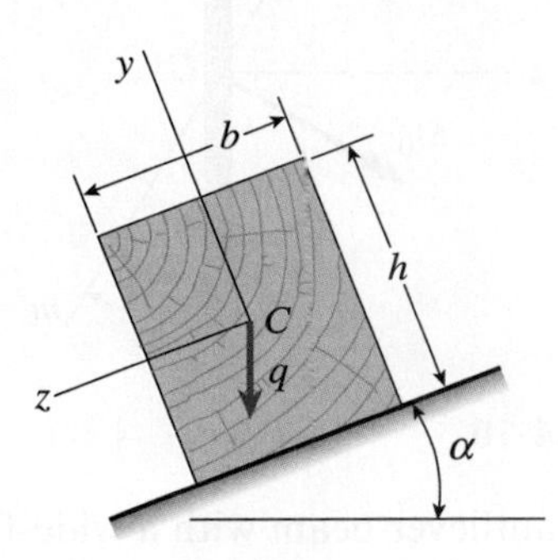

PROBLEMS 6.4-2 and 6.4-3

6.4-3 Solve the preceding problem for the following data: $b = 100$ mm, $h = 200$ mm, $L = 3$ m, $\tan\,\alpha = 1/3$, and $q = 3$ kN/m.

6.4-4 A simply supported beam of span length L carries a vertical concentrated load P acting through the centroid C at the midpoint of the span (see figure). The beam is attached to supports inclined at an angle α to the horizontal.

Determine the orientation of the neutral axis and calculate the maximum stresses at the outside corners of the cross section (points A, B, D, and E) due to the load P. Data for the beam are IPN 280 section, $L = 3.5$ m, $P = 18$ kN, and $\alpha = 26.57°$. *Note:* See Table F-2 of Appendix F for the dimensions and properties of the beam.

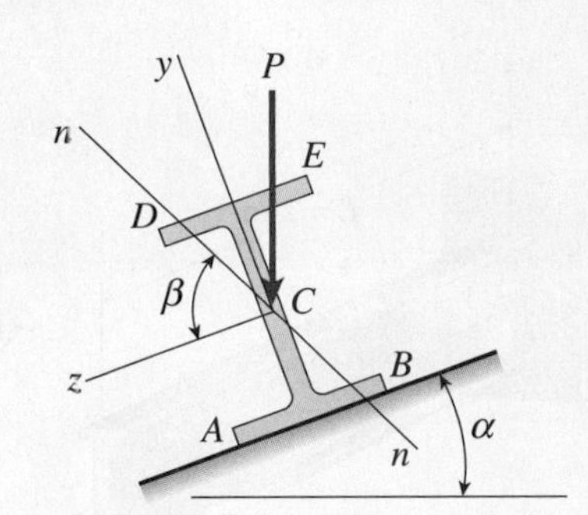

PROBLEMS 6.4-4 and 6.4-5

6.4-5 Solve the preceding problem using the following data: HE 140A section, $L = 2.5$ m, $P = 20$ kN, and $\alpha = 22.5°$.

Representative Problems

6.4-6 A wood cantilever beam with a rectangular cross section and length L supports an inclined load P at its free end (see figure).

Determine the orientation of the neutral axis and calculate the maximum tensile stress σ_{max} due to the load P. Data for the beam are $b = 80$ mm, $h = 140$ mm, $L = 2.0$ m, $P = 575$ N, and $\alpha = 30°$.

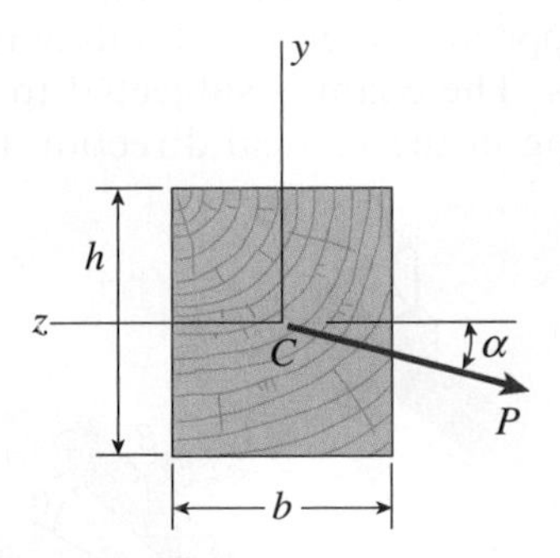

PROBLEMS 6.4-6 and 6.4-7

6.4-7 Solve the preceding problem for a cantilever beam with data as $b = 100$ mm, $h = 200$ mm, $L = 2$ m, $P = 2$ kN, and $\alpha = 45°$.

6.4-8 A 2-m-long cantilever beam is constructed using an IPN 320 section. Load P acts in an inclined direction at the free end (see figure). Determine the allowable load P that can be carried by the beam if the maximum permissible tensile and compressive stresses are 150 MPa. Include the weight of the beam in the calculations.

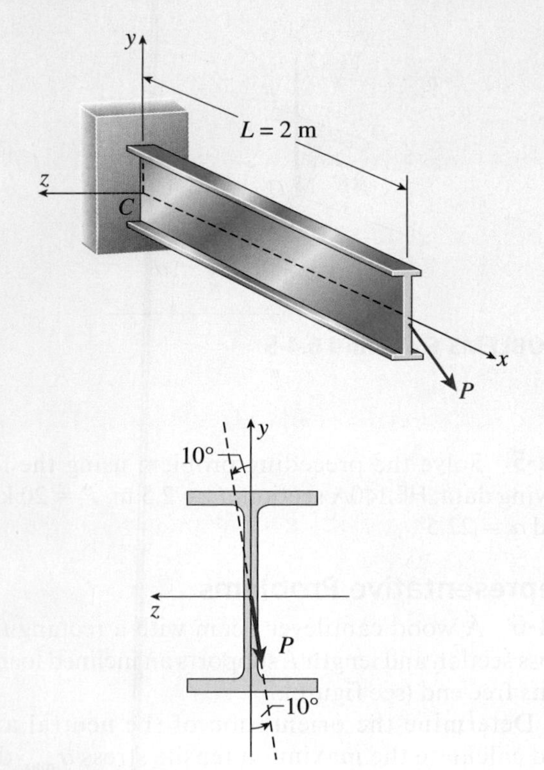

PROBLEM 6.4-8

6.4-9 A wood beam AB with a rectangular cross section (100 mm × 150 mm) serving as a roof purlin is simply supported by the top chords of two adjacent roof trusses. The beam is subjected to distributed load q acting in the vertical direction through the centroid of the purlin cross section. The top chords of the trusses have a slope $\alpha = 27°$. The purlin has length $L = 2$ m. Determine the permissible distributed load q based on the allowable compressive and tensile stress in the beam $\sigma_{all} = 14$ MPa.

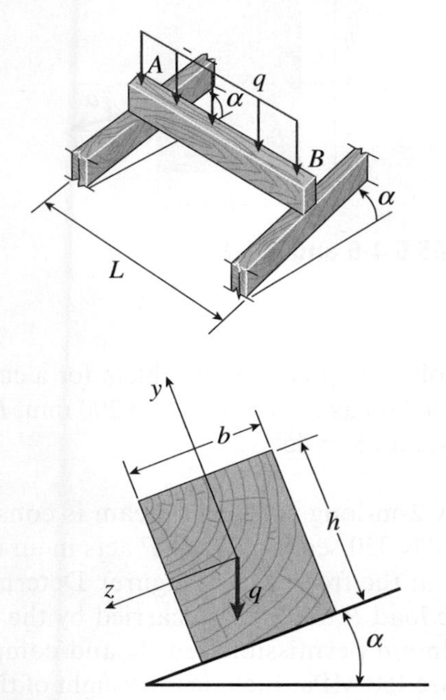

PROBLEM 6.4-9

6.4-10 A steel beam of I-section (see figure) is simply supported at the ends. Two equal and oppositely directed bending moments M_0 act at the ends of the beam, so the beam is in pure bending. The moments act in plane mm, which is oriented at an angle α to the x-y plane.

Determine the orientation of the neutral axis and calculate the maximum tensile stress σ_{max} due to the moments M_0. Data for the beam are IPN 220 section, $M_0 = 4$ kN · m, and $\alpha = 24°$. *Note:* See Table F-2 of Appendix F for the dimensions and properties of the beam.

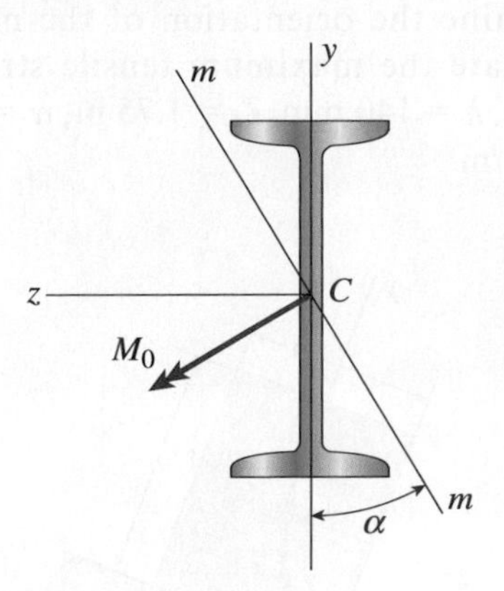

PROBLEM 6.4-10

6.4-11 A cantilever beam with a wide-flange cross section and length L supports an inclined load P at its free end (see figure).

Determine the orientation of the neutral axis and calculate the maximum tensile stress σ_{max} due to the load P.

Data for the beam are HE 650B section, $L = 2.5$ m, $P = 16.7$ kN, and $\alpha = 55°$. *Note:* See Table F-1 of Appendix F for the dimensions and properties of the beam.

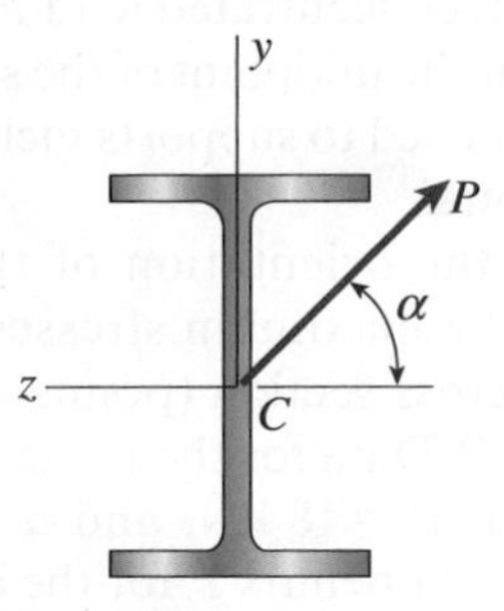

PROBLEMS 6.4-11 and 6.4-12

6.4-12 Solve the preceding problem using a HE 320B section, $L = 1.8$ m, $P = 9.5$ kN, and $\alpha = 60°$. See Table F-1 of Appendix F for the dimensions and properties of the beam.

6.4-13 A cantilever beam of IPN 300 section and length $L = 3$ m supports a slightly inclined load $P = 2.5$ kN at the free end (see figure).

(a) Plot a graph of the stress σ_A at point A as a function of the angle of inclination α.

(b) Plot a graph of the angle β, which locates the neutral axis nn, as a function of the angle α. (When plotting the graphs, let α vary from 0 to 10°.) See Table F-2 of Appendix F for the dimensions and properties of the beam.

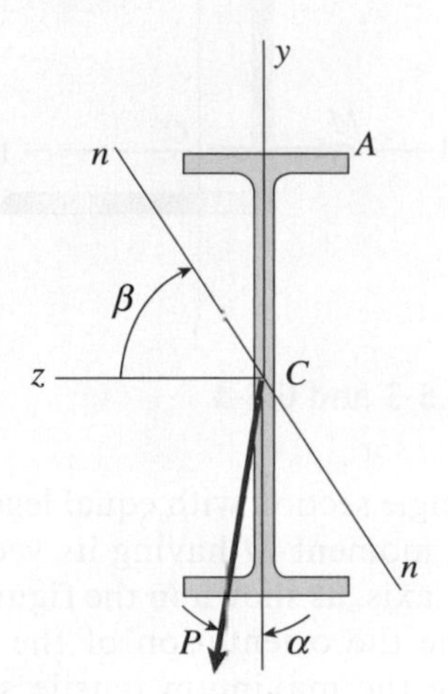

PROBLEM 6.4-13

6.4-14 A cantilever beam built up from two channel shapes, each UPN 180 and of length L, supports an inclined load P at its free end (see figure).

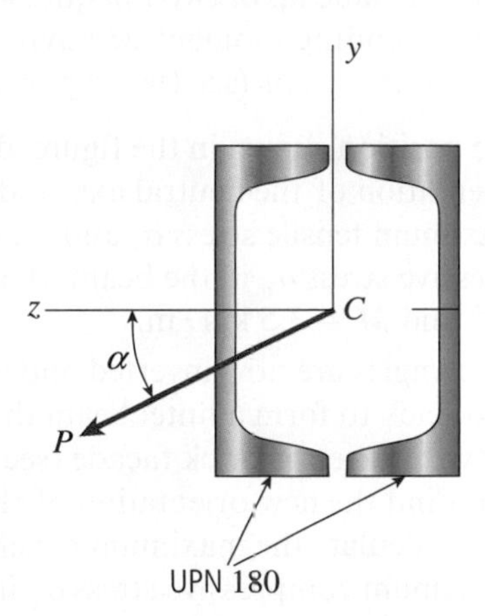

PROBLEM 6.4-14

Determine the orientation of the neutral axis and calculate the maximum tensile stress σ_{max} due to the load P. Data for the beam are $L = 4.5$ m, $P = 500$ N, and $\alpha = 30°$.

6.4-15 A built-up I-section steel beam with channels attached to the flanges (see figure part a) is simply supported at the ends. Two equal and oppositely directed bending moments M_0 act at the ends of the beam, so the beam is in pure bending. The moments act in plane mm, which is oriented at an angle α to the x-y plane.

(a) Determine the orientation of the neutral axis and calculate the maximum tensile stress σ_{max} due to the moments M_0.

(b) Repeat part (a) if the channels now have their flanges pointing away from the beam flange, as shown in the figure part b. Data for the beam are IPN 160 section with UPN 100 sections attached to the flanges, $M_0 = 5$ kN·m, and $\alpha = 40°$. See Tables F-2 and F-3 of Appendix F for the dimensions and properties of the IPN and UPN shapes.

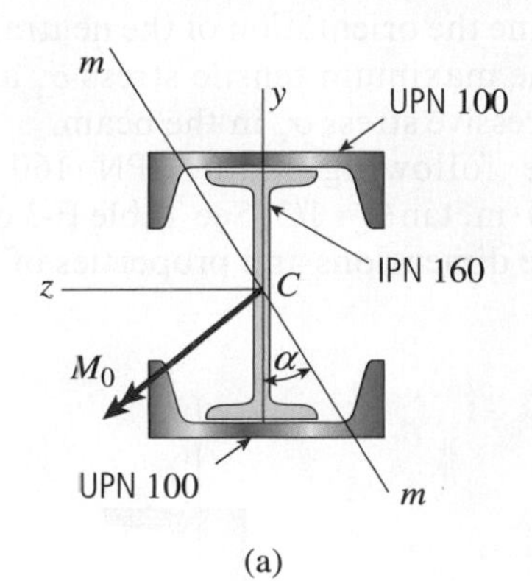

(a)

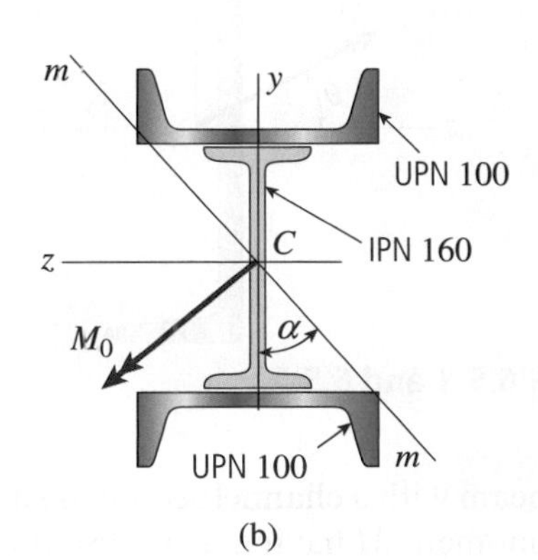

(b)

PROBLEM 6.4-15

6.4-16 Repeat Problem 6.4-14 but use the configuration of UPN 200 channel shapes and loading shown in the figure. Use $P = 250$ N.

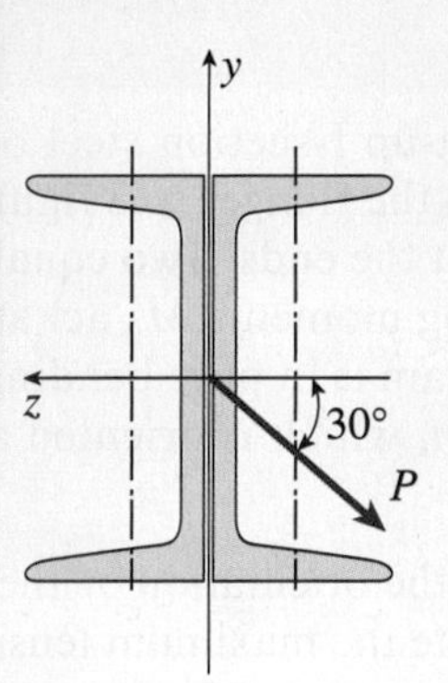

PROBLEM 6.4-16

6.5 Bending of Unsymmetric Beams

To solve the problems for Section 6.5, be sure to draw a sketch of the cross section showing the orientation of the neutral axis and the locations of the points where the stresses are being found.

6.5-1 A beam with a channel section is subjected to a bending moment M having its vector at an angle θ to the z axis (see figure).

Determine the orientation of the neutral axis and calculate the maximum tensile stress σ_t and maximum compressive stress σ_c in the beam.

Use the following data: UPN 160 section, $M = 2.5$ kN·m, $\tan\theta = 1/3$. See Table F-3 of Appendix F for the dimensions and properties of the channel section.

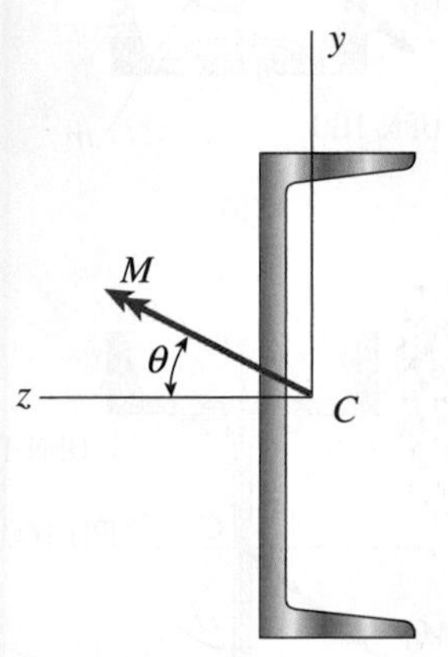

PROBLEMS 6.5-1 and 6.5-2

6.5-2 A beam with a channel section is subjected to a bending moment M having its vector at an angle θ to the z axis (see figure).

Determine the orientation of the neutral axis and calculate the maximum tensile stress σ_t and maximum compressive stress σ_c in the beam. Use a UPN 200 channel section with $M = 0.75$ kN · m and $\theta = 20°$.

6.5-3 An angle section with equal legs is subjected to a bending moment M having its vector directed along the 1–1 axis, as shown in the figure.

Determine the orientation of the neutral axis and calculate the maximum tensile stress σ_t and maximum compressive stress σ_c if the angle is an L 150 × 150 × 14 section and $M = 2.5$ kN·m. See Table F-4 of Appendix F for the dimensions and properties of the angle section.

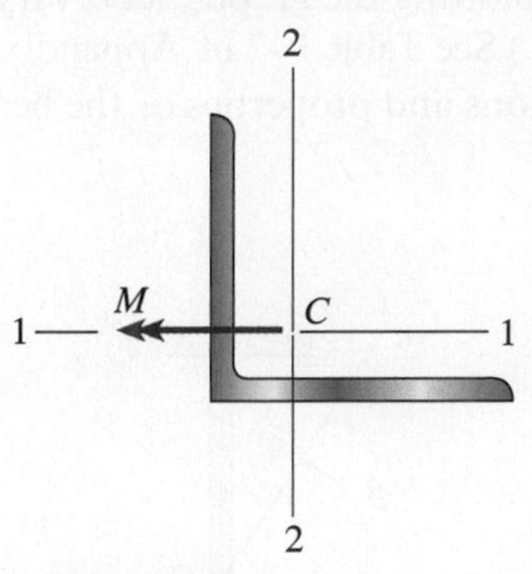

PROBLEMS 6.5-3 and 6.5-4

6.5-4 An angle section with equal legs is subjected to a bending moment M having its vector directed along the 1–1 axis, as shown in the figure.

Determine the orientation of the neutral axis and calculate the maximum tensile stress σ_t and maximum compressive stress σ_c if the section is an L 200 × 200 × 19 section and $M = 4.5$ kN · m. See Table F-4 of Appendix F for the dimensions and properties of the angle section.

6.5-5 A beam made up of two unequal leg angles is subjected to a bending moment M having its vector at an angle θ to the z axis (see figure part a).

(a) For the position shown in the figure, determine the orientation of the neutral axis and calculate the maximum tensile stress σ_t and maximum compressive stress σ_c in the beam. Assume that $\theta = 30°$ and $M = 3.5$ kN · m.

(b) The two angles are now inverted and attached back-to-back to form a lintel beam that supports two courses of brick façade (see figure part b). Find the new orientation of the neutral axis and calculate the maximum tensile stress σ_t and maximum compressive stress σ_c in the beam using $\theta = 30°$ and $M = 3.5$ kN · m.

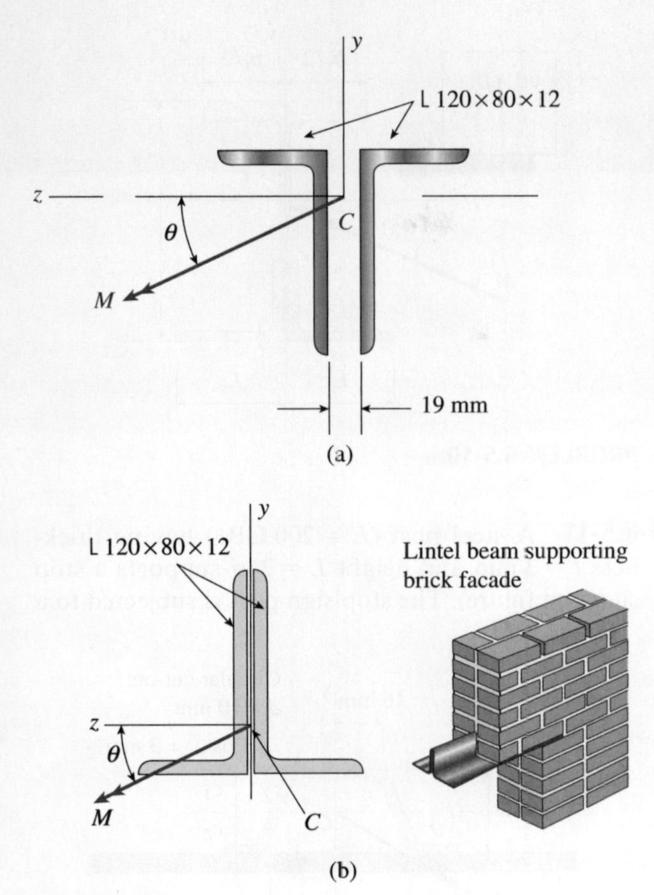

PROBLEM 6.5-5

6.5-6 The Z-section of Example D-7 is subjected to $M = 5 \text{ kN} \cdot \text{m}$, as shown.

Determine the orientation of the neutral axis and calculate the maximum tensile stress σ_t and maximum compressive stress σ_c in the beam. Use the following numerical data: height $h = 200$ mm, width $b = 90$ mm, constant thickness $t = 15$ mm, and $\theta_p = 19.2°$. Use $I_1 = 32.6 \times 10^6 \text{ mm}^4$ and $I_2 = 2.4 \times 10^6 \text{ mm}^4$ from Example D-7.

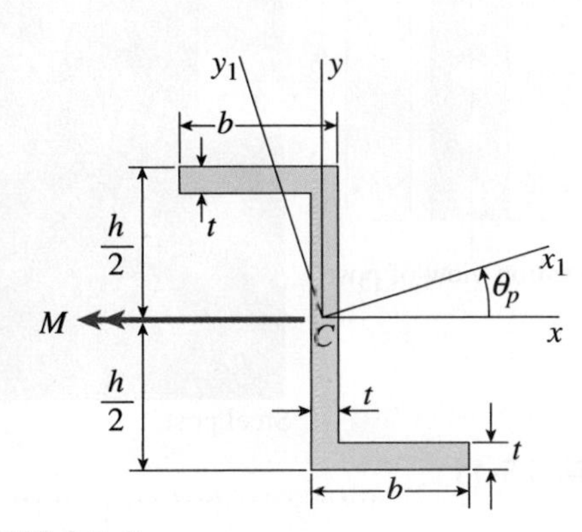

PROBLEM 6.5-6

6.5-7 The cross section of a steel beam is constructed of a HE 450A wide-flange section with a 25 cm × 1.5 cm cover plate welded to the top flange and a UPN 320 channel section welded to the bottom flange. This beam is subjected to a bending moment M having its vector at an angle θ to the z axis (see figure).

Determine the orientation of the neutral axis and calculate the maximum tensile stress σ_t and maximum compressive stress σ_c in the beam. Assume that $\theta = 30°$ and $M = 18.5 \text{ kN} \cdot \text{m}$. *Note:* The cross-sectional properties of this beam were computed in Examples D-2 and D-5.

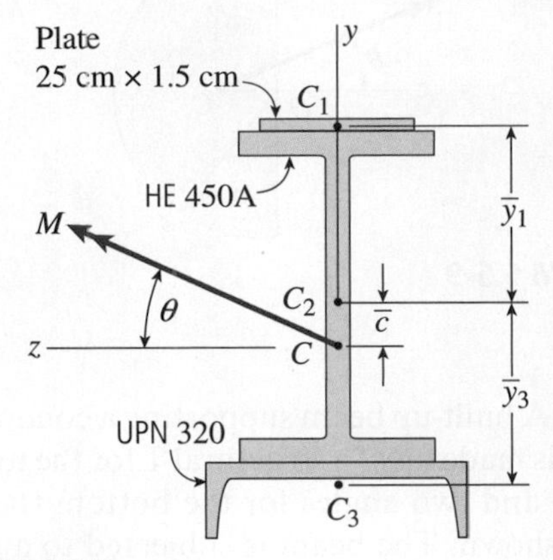

PROBLEM 6.5-7

6.5-8 The cross section of a steel beam is shown in the figure. This beam is subjected to a bending moment M having its vector at an angle θ to the z axis.

Determine the orientation of the neutral axis and calculate the maximum tensile stress σ_t and maximum compressive stress σ_c in the beam. Assume that $\theta = 22.5°$ and $M = 4.5 \text{ kN} \cdot \text{m}$. Use cross-sectional properties $I_{x_1} = 93.14 \times 10^6 \text{ mm}^4$, $I_{y_1} = 152.7 \times 10^6 \text{ mm}^4$, and $\theta_p = 27.3°$.

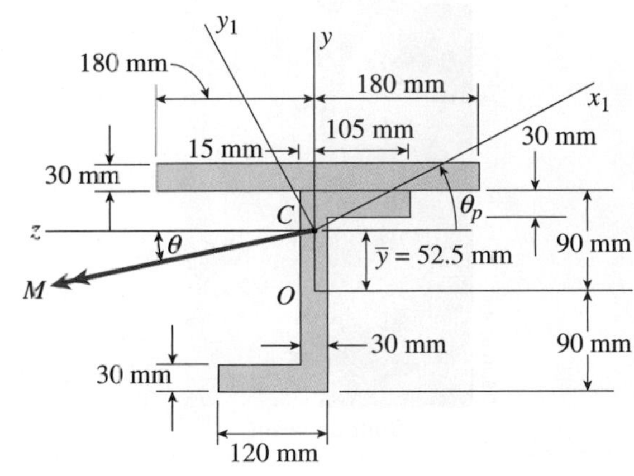

PROBLEM 6.5-8

6.5-9 A beam with a semicircular cross section of radius r is subjected to a bending moment M having its vector at an angle θ to the z axis (see figure).

Derive formulas for the maximum tensile stress σ_t and the maximum compressive stress σ_c in the beam for $\theta = 0, 45°$, and $90°$. Express the results in the form $\alpha\ M/r^3$, where α is a numerical value.

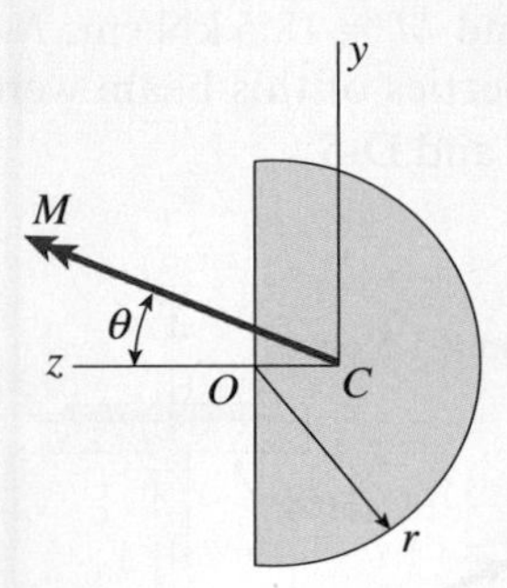

PROBLEM 6.5-9

6.5-10 A built-up beam supporting a condominium balcony is made up of a structural T for the top flange and web and two angles for the bottom flange and web, as shown. The beam is subjected to a bending moment M having its vector at an angle θ to the z axis (see figure).

Determine the orientation of the neutral axis and calculate the maximum tensile stress σ_t and maximum compressive stress σ_c in the beam. Assume that $\theta = 30°$ and $M = 10\ \text{kN} \cdot \text{m}$.

Use the numerical properties: $c_1 = 4.111$ mm, $c_2 = 4.169$ mm, $b_f = 134$ mm, $L_s = 76$ mm, $A = 4144$ mm^2, $I_y = 3.88 \times 10^6$ mm^4, and $I_z = 34.18 \times 10^6$ mm^4.

Built-up beam

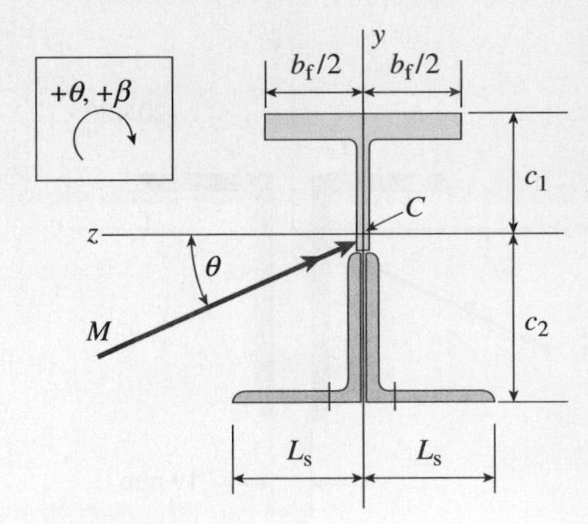

PROBLEM 6.5-10

6.5-11 A steel post ($E = 200$ GPa) having thickness $t = 3$ mm and height $L = 2$ m supports a stop sign (see figure). The stop sign post is subjected to a

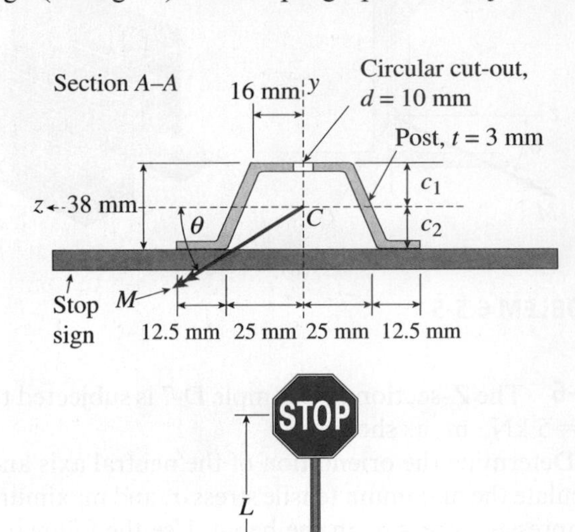

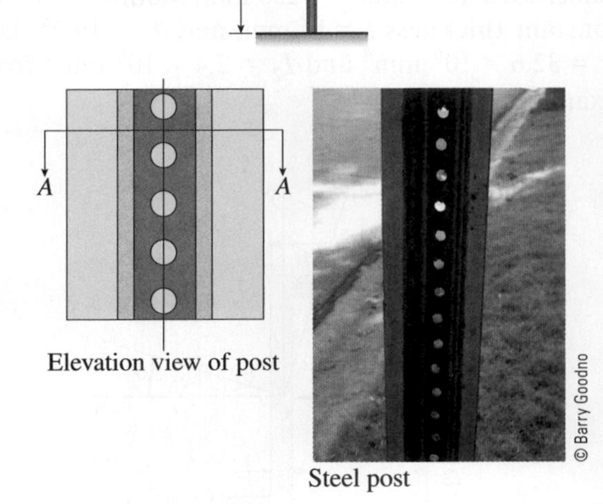

Elevation view of post

Steel post

PROBLEM 6.5-11

bending moment M having its vector at an angle θ to the z axis.

Determine the orientation of the neutral axis and calculate the maximum tensile stress σ_t and maximum compressive stress σ_c in the beam. Assume that $\theta = 30°$ and $M = 350\ \text{N} \cdot \text{m}$.

Use the following numerical properties for the post: $A = 373\ \text{mm}^2$, $c_1 = 19.5\ \text{mm}$, $c_2 = 18.5\ \text{mm}$, $I_y = 1.868 \times 10^5\ \text{mm}^4$, and $I_z = 0.67 \times 10^5\ \text{mm}^4$.

6.5-12 A UPN 220 channel section has an angle with equal legs attached as shown; the angle serves as a lintel beam. The combined steel section is subjected to a bending moment M having its vector directed along the z axis, as shown in the figure. The centroid C of the combined section is located at distances x_c and y_c from the centroid (C_1) of the channel alone. Principal axes x_1 and y_1 are also shown in the figure and properties I_{x1}, I_{y1}, and θ_p are given.

Find the orientation of the neutral axis and calculate the maximum tensile stress σ_t and maximum compressive stress σ_c if the angle is an L 90 × 90 × 7 section and $M = 3.5\ \text{kN} \cdot \text{m}$. Use the following properties for principal axes for the combined section: $I_{x1} = 35.14 \times 10^6\ \text{mm}^4$, $I_{y1} = 4.265 \times 10^6\ \text{mm}^4$, $\theta_p = 7.826°$ (CW), $x_c = 11.32$ mm, and $y_c = 21.08$ mm.

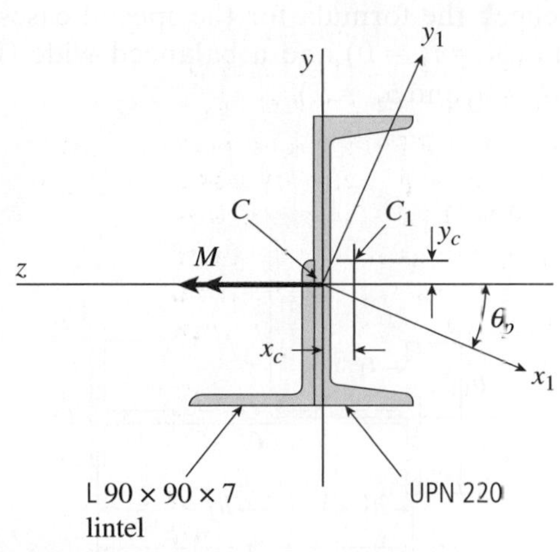

PROBLEM 6.5-12

6.5-13 A cold-formed steel section is made by folding a steel plate to form a structural section such as that shown in the figure. This beam is subjected to bending moment $M = 225\ \text{N}\cdot\text{m}$ at angle $\theta = 10°$ to the z axis. Find the centroid and the orientation of the neutral axis. Find flexural normal stresses at points A and B (see figure). Use the flexure formula based on principal axes and then confirm your solution using the general flexure formula.

Hint: the plate is thin so use centerline dimensions.

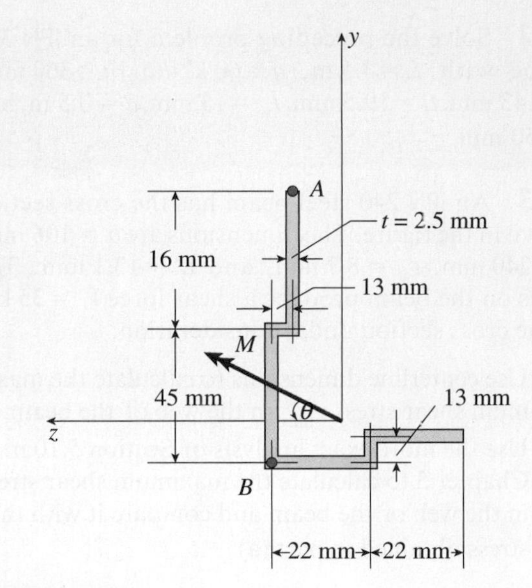

PROBLEM 6.5-13

6.8 Shear Stresses in Wide-Flange Beams

To solve the problems for Section 6.8, assume that the cross sections are thin-walled. Use centerline dimensions for all calculations and derivations, unless otherwise specified.

6.8-1 A simple beam with a HE 220B wide-flange cross section supports a uniform load of intensity $q = 45$ kN/m on a span of length $L = 3.8$ m (see figure). The dimensions of the cross section are $h = 220$ mm, $b = 220$ mm, $t_f = 16$ mm, and $t_w = 9.5$ mm.

(a) Calculate the maximum shear stress τ_{max} on cross section A–A located at distance $d = 0.75$ m from the end of the beam.

(b) Calculate the shear stress τ at point B on the cross section. Point B is located at a distance $a = 38$ mm from the edge of the lower flange.

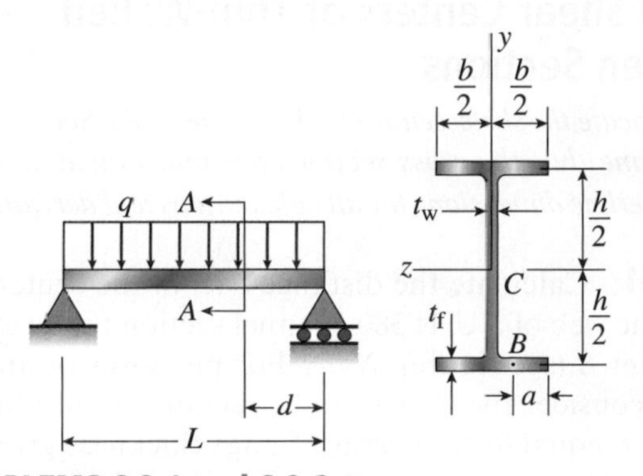

PROBLEMS 6.8-1 and 6.8-2

6.8-2 Solve the preceding problem for an IPN 360 shape with $L = 3.5$ m, $q = 60$ kN/m, $h = 360$ mm, $b = 143$ mm, $t_f = 19.5$ mm, $t_w = 13$ mm, $d = 0.5$ m, and $a = 50$ mm.

6.8-3 An IPN 240 steel beam has the cross section shown in the figure. The dimensions are $b = 106$ mm, $h = 240$ mm, $t_w = 8.7$ mm, and $t_f = 13.1$ mm. The loads on the beam produce a shear force $V = 35$ kN at the cross section under consideration.

(a) Use centerline dimensions to calculate the maximum shear stress τ_{max} in the web of the beam.
(b) Use the more exact analysis of Section 5.10 in Chapter 5 to calculate the maximum shear stress in the web of the beam and compare it with the stress obtained in part (a).

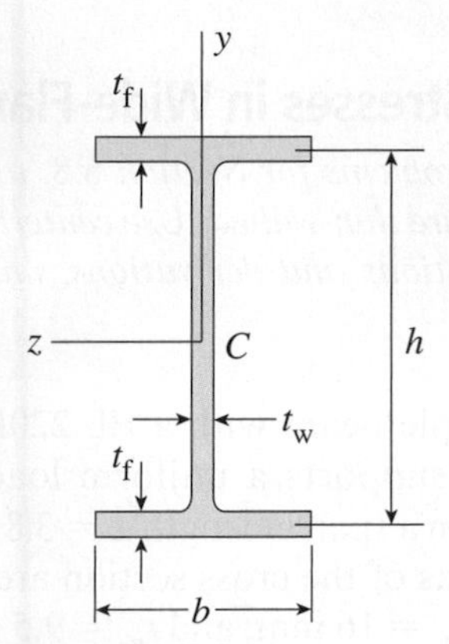

PROBLEMS 6.8-3 and 6.8-4

6.8-4 Solve the preceding problem for a HE 550B shape with $b = 300$ mm, $h = 550$ mm, $t_w = 15$ mm, $t_f = 29$ mm, and $V = 115$ kN.

6.9 Shear Centers of Thin-Walled Open Sections

To locate the shear centers in the problems for Section 6.9, assume that the cross sections are thin-walled and use centerline dimensions for all calculations and derivations.

6.9-1 Calculate the distance e from the centerline of the web of a UPN 380 channel section to the shear center S (see figure). *Note:* For purposes of analysis, consider the flanges to be rectangles with thickness t_f equal to the average flange thickness given in Table F-3 in Appendix F.

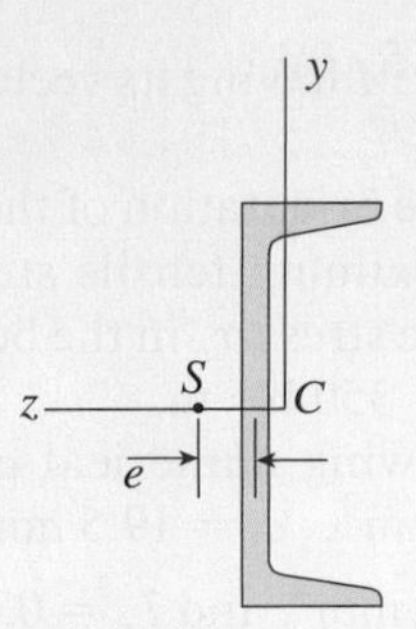

PROBLEMS 6.9-1 and 6.9-2

6.9-2 Calculate the distance e from the centerline of the web of a UPN 100 channel section to the shear center S (see figure). *Note:* For purposes of analysis, consider the flanges to be rectangles with thickness t_f equal to the average flange thickness given in Table F-3 in Appendix F.

6.9-3 The cross section of an unbalanced wide-flange beam is shown in the figure. Derive the following formula for the distance h_1 from the centerline of one flange to the shear center S:

$$h_1 = \frac{t_2 b_2^3 h}{t_1 b_1^3 + t_2 b_2^3}$$

Also, check the formula for the special cases of a T-beam ($b_2 = t_2 = 0$) and a balanced wide-flange beam ($t_2 = t_1$ and $b_2 = b_1$).

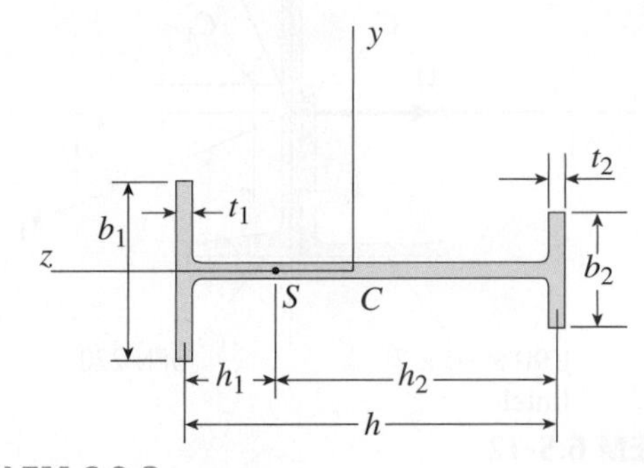

PROBLEM 6.9-3

6.9-4 The cross section of an unbalanced wide-flange beam is shown in the figure. Derive the following formula for the distance e from the centerline of the web to the shear center S:

$$e = \frac{3t_f(b_2^2 - b_1^2)}{ht_w + 6t_f(b_1 + b_2)}$$

Also, check the formula for the special cases of a channel section ($b_1 = 0$ and $b_2 = b$) and a doubly symmetric beam ($b_1 = b_2 = b/2$).

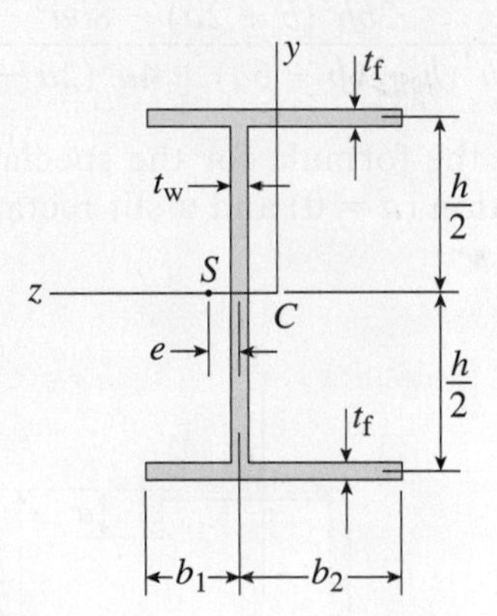

PROBLEM 6.9-4

6.9-5 The cross section of a channel beam with double flanges and constant thickness throughout the section is shown in the figure.

Derive the following formula for the distance e from the centerline of the web to the shear center S:

$$e = \frac{3b^2(h_1^2 + h_2^2)}{h_2^3 + 6b(h_1^2 + h_2^2)}$$

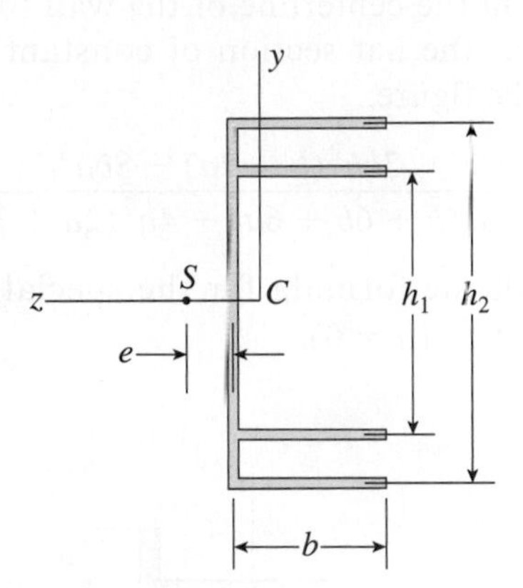

PROBLEM 6.9-5

6.9-6 The cross section of a slit circular tube of constant thickness is shown in the figure.

(a) Show that the distance e from the center of the circle to the shear center S is equal to $2r$ in the figure part a.

(b) Find an expression for e if flanges with the same thickness as that of the tube are added, as shown in the figure part b.

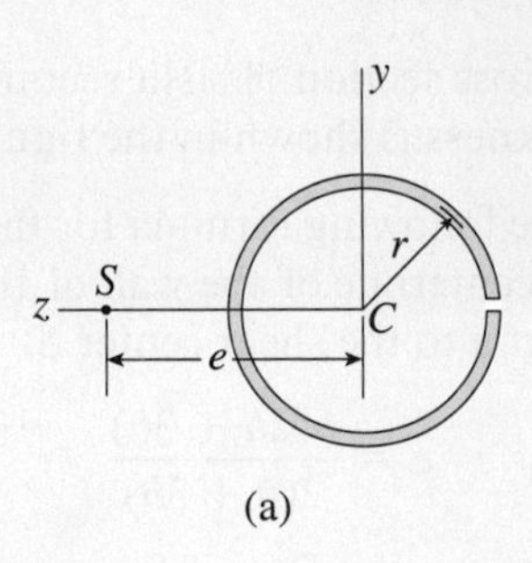

(a)

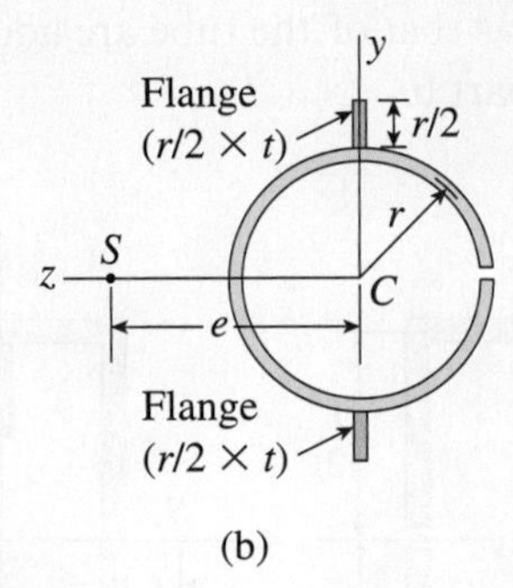

(b)

PROBLEM 6.9-6

6.9-7 The cross section of a slit square tube of constant thickness is shown in the figure. Derive the following formula for the distance e from the corner of the cross section to the shear center S:

$$e = \frac{b}{2\sqrt{2}}$$

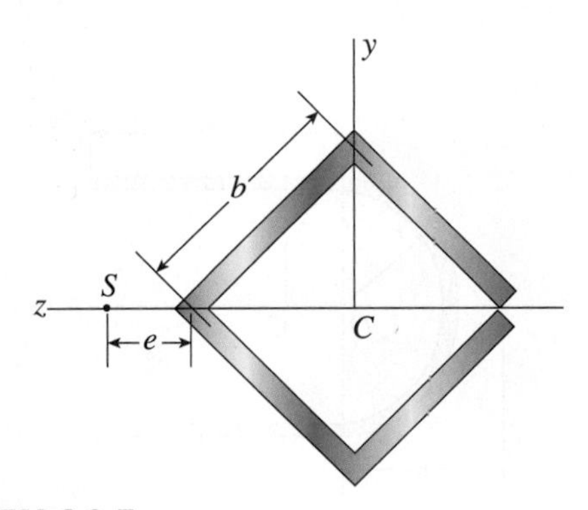

PROBLEM 6.9-7

6.9-8 The cross section of a slit rectangular tube of constant thickness is shown in the figures.

(a) Derive the following formula for the distance e from the centerline of the wall of the tube in the figure part a to the shear center S:

$$e = \frac{b(2h + 3b)}{2(h + 3b)}$$

(b) Find an expression for e if flanges with the same thickness as that of the tube are added as shown in figure part b.

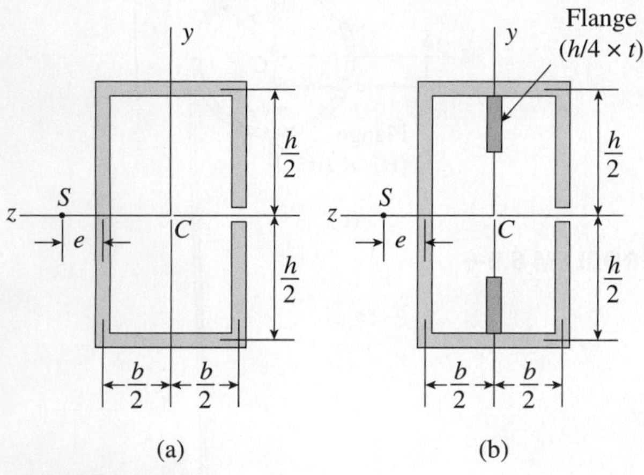

PROBLEM 6.9-8

6.9-9 A U-shaped cross section of constant thickness is shown in the figure. Derive the following formula for the distance e from the center of the semicircle to the shear center S:

$$e = \frac{2(2r^2 + b^2 + \pi br)}{4b + \pi r}$$

Also, plot a graph showing how the distance e (expressed as the nondimensional ratio e/r) varies as a function of the ratio b/r. (Let b/r range from 0 to 2.)

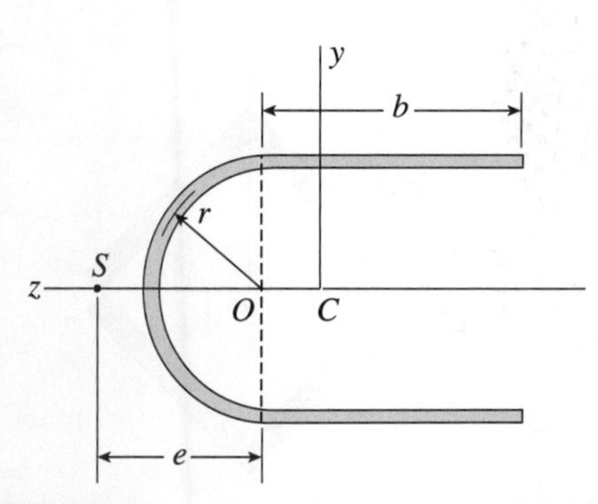

PROBLEM 6.9-9

6.9-10 Derive the following formula for the distance e from the centerline of the wall to the shear center S for the C-section of constant thickness shown in the figure:

$$e = \frac{3bh^2(b + 2a) - 8ba^3}{h^2(h + 6b + 6a) + 4a^2(2a - 3h)}$$

Also, check the formula for the special cases of a channel section ($a = 0$) and a slit rectangular tube ($a = h/2$).

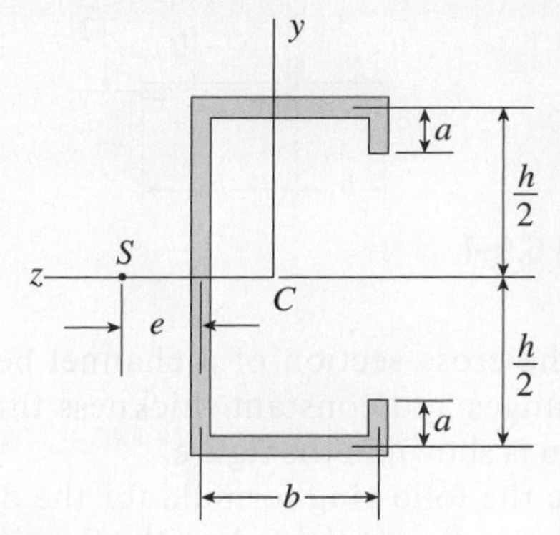

PROBLEM 6.9-10

6.9-11 Derive the following formula for the distance e from the centerline of the wall to the shear center S for the hat section of constant thickness shown in the figure:

$$e = \frac{3bh^2(b + 2a) - 8ba^3}{h^2(h + 6b + 6a) + 4a^2(2a + 3h)}$$

Also, check the formula for the special case of a channel section ($a = 0$).

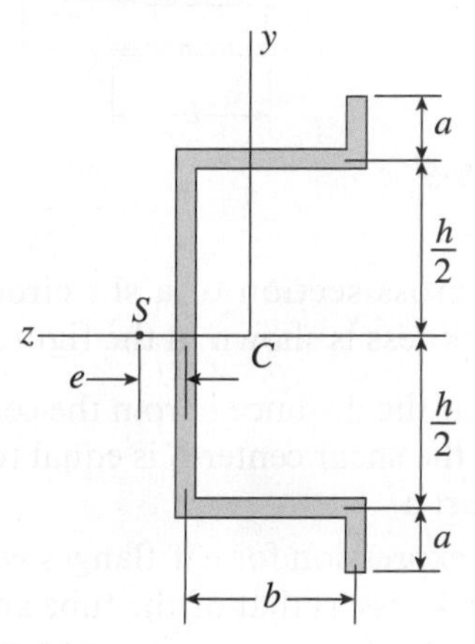

PROBLEM 6.9-11

6.9-12 The cross section of a sign post of constant thickness is shown in the figure.

Derive the formula for the distance e from the centerline of the wall of the post to the shear center S:

$$e = \frac{1}{3} t\, ba[4a^2 + 3ab \sin(\beta) + 3ab$$
$$+ 2 \sin(\beta) b^2] \frac{\cos(\beta)}{I_z}$$

where I_z = moment of inertia about the z axis.

Also, compare this formula with that given in Problem 6.9-11 for the special case of $\beta = 0$ here and $a = h/2$ in both formulas.

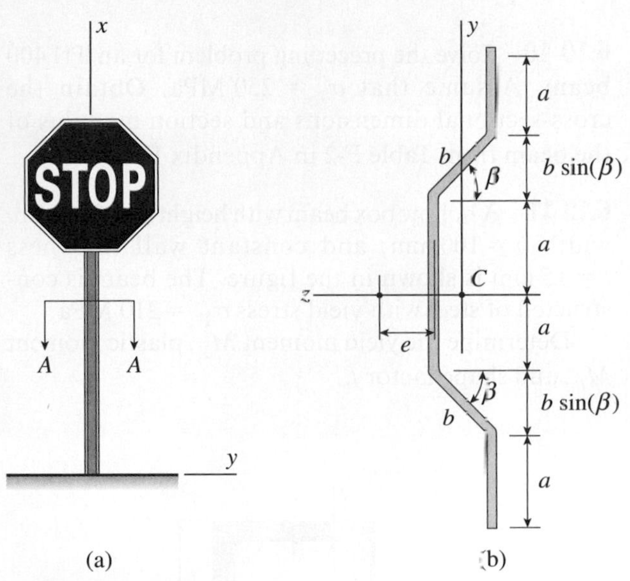

PROBLEM 6.9-12

6.9-13 A cross section in the shape of a circular arc of constant thickness is shown in the figure. Derive the following formula for the distance e from the center of the arc to the shear center S:

$$e = \frac{2r(\sin(\beta) - \beta \cos(\beta))}{\beta - \sin(\beta)\cos(\beta)}$$

in which β is in radians. Also, plot a graph showing how the distance e varies as β varies from 0 to π.

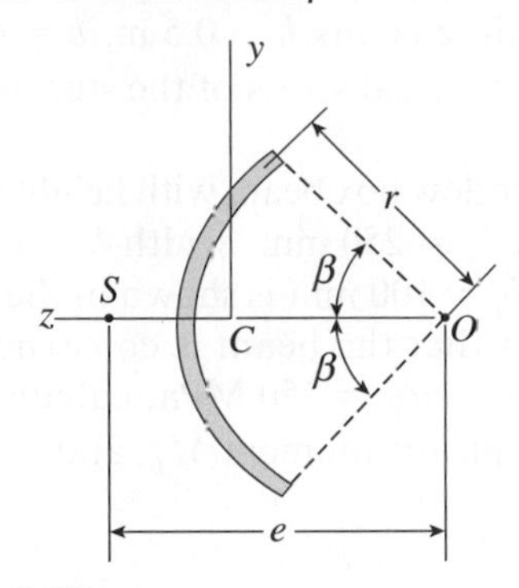

PROBLEM 6.9-13

6.10 Elastoplastic Bending

Solve the problems for Section 6.10 using the assumption that the material is elastoplastic with yield stress σ_Y.

6.10-1 Determine the shape factor f for a cross section in the shape of a double trapezoid having the dimensions shown in the figure.

Also, check your result for the special cases of a rhombus ($b_1 = 0$) and a rectangle ($b_1 = b_2$).

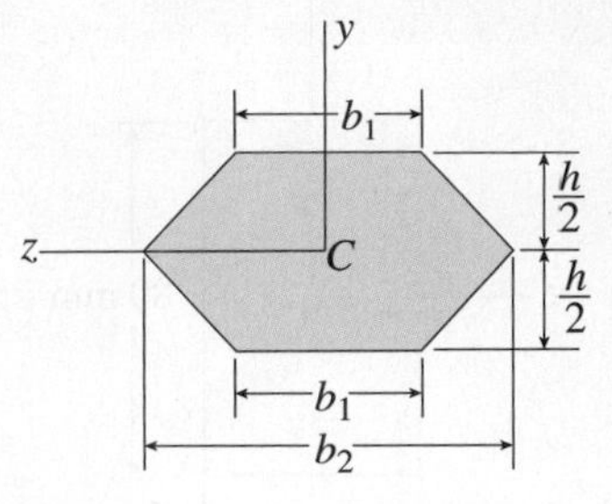

PROBLEM 6.10-1

6.10-2 (a) Determine the shape factor f for a hollow circular cross section having inner radius r_1 and outer radius r_2 (see figure). (b) If the section is very thin, what is the shape factor?

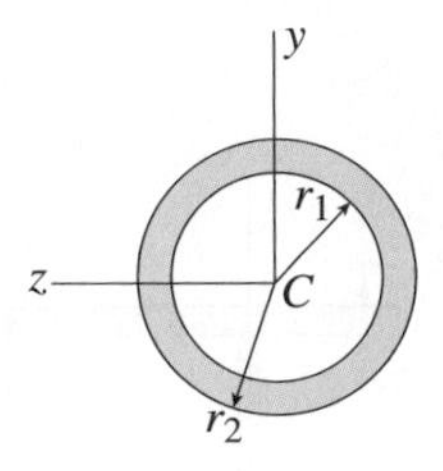

PROBLEM 6.10-2

6.10-3 A propped cantilever beam of length $L = 1.5$ m with a sliding support supports a uniform load of intensity q (see figure). The beam is made of steel ($\sigma_Y = 250$ MPa) and has a rectangular cross section of width $b = 100$ mm and height $h = 150$ mm.

What load intensity q will produce a fully plastic condition in the beam?

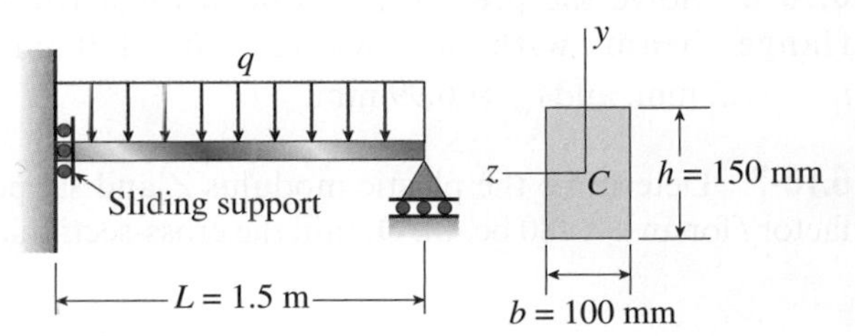

PROBLEM 6.10-3

6.10-4 A steel beam of rectangular cross section is 40 mm wide and 80 mm high (see figure). The yield stress of the steel is 210 MPa.

(a) What percent of the cross-sectional area is occupied by the elastic core if the beam is subjected to a bending moment of 12.0 kN · m acting about the z axis?

(b) What is the magnitude of the bending moment that will cause 50% of the cross section to yield?

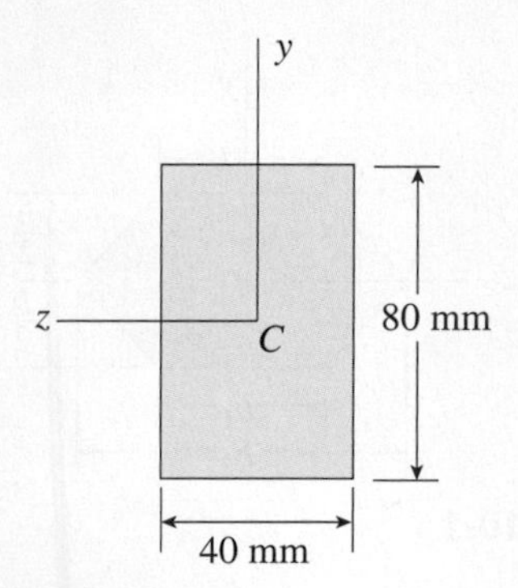

PROBLEM 6.10-4

6.10-5 Calculate the shape factor f for the wide-flange beam shown in the figure if $h = 310$ mm, $b = 300$ mm, $t_f = 15.5$ mm, and $t_w = 9$ mm.

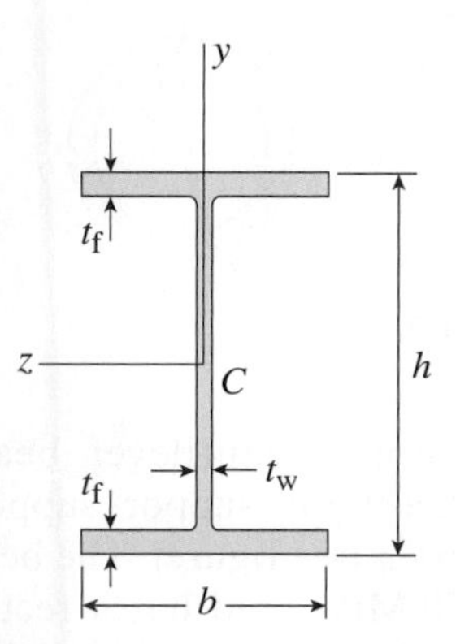

PROBLEMS 6.10-5 and 6.10-6

6.10-6 Solve the preceding problem for a wide-flange beam with $h = 404$ mm, $b = 140$ mm, $t_f = 11.2$ mm, and $t_w = 6.99$ mm.

6.10-7 Determine the plastic modulus Z and shape factor f for an IPN 180 beam. Obtain the cross-sectional dimensions and section modulus of the beam from Table F-2 in Appendix F.

6.10-8 Solve the preceding problem for a HE 260B wide-flange beam. Obtain the cross-sectional dimensions and section modulus of the beam from Table F-1 in Appendix F.

6.10-9 Determine the yield moment M_Y, plastic moment M_P, and shape factor f for an IPN 500 beam if $\sigma_Y = 250$ MPa. Obtain the cross-sectional dimensions and section modulus of the beam from Table F-2 in Appendix F.

6.10-10 Solve the preceding problem for an IPN 400 beam. Assume that $\sigma_Y = 250$ MPa. Obtain the cross-sectional dimensions and section modulus of the beam from Table F-2 in Appendix F.

6.10-11 A hollow box beam with height $h = 300$ mm, width $b = 100$ mm, and constant wall thickness $t = 15$ mm is shown in the figure. The beam is constructed of steel with yield stress $\sigma_Y = 210$ MPa.

Determine the yield moment M_Y, plastic moment M_P, and shape factor f.

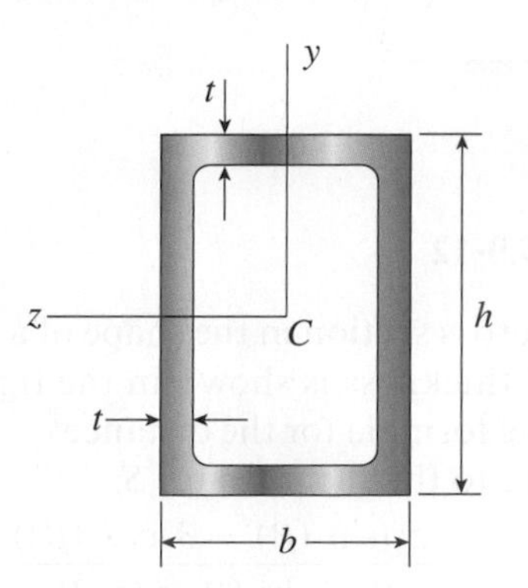

PROBLEMS 6.10-11 and 6.10-12

6.10-12 Solve the preceding problem for a box beam with dimensions $h = 0.5$ m, $b = 0.18$ m, and $t = 22$ mm. The yield stress of the steel is 210 MPa.

6.10-13 A hollow box beam with height $h = 300$ mm, inside height $h_1 = 250$ mm, width $b = 150$ mm, and inside width $b_1 = 100$ mm is shown in the figure.

Assuming that the beam is constructed of steel with yield stress $\sigma_Y = 250$ MPa, calculate the yield moment M_Y, plastic moment M_P, and shape factor f.

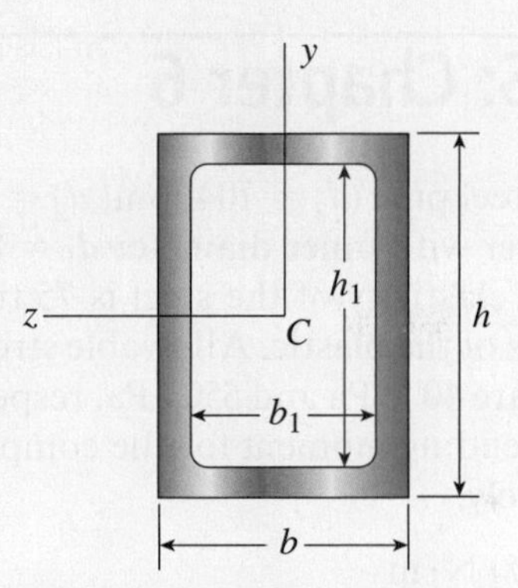

PROBLEMS 6.10-13 through 6.10-16

6.10-14 Solve the preceding problem for a box beam with dimensions $h = 200$ mm, $h_1 = 160$ mm, $b = 150$ mm, and $b_1 = 130$ mm. Assume that the beam is constructed of steel with yield stress $\sigma_Y = 220$ MPa.

6.10-15 The hollow box beam shown in the figure is subjected to a bending moment M of such magnitude that the flanges yield but the webs remain linearly elastic.

(a) Calculate the magnitude of the moment M if the dimensions of the cross section are $h = 350$ mm, $h_1 = 310$ mm, $b = 200$ mm, and $b_1 = 175$ mm. Also, the yield stress is $\sigma_Y = 220$ MPa.

(b) What percent of the moment M is produced by the elastic core?

6.10-16 Solve the preceding problem for a box beam with dimensions $h = 400$ mm, $h_1 = 360$ mm, $b = 200$ mm, and $b_1 = 160$ mm, and with yield stress $\sigma_Y = 220$ MPa.

6.10-17 A HE 260A wide-flange beam is subjected to a bending moment M of such magnitude that the flanges yield but the web remains linearly elastic.

(a) Calculate the magnitude of the moment M if the yield stress is $\sigma_Y = 250$ MPa.

(b) What percent of the moment M is produced by the elastic core?

6.10-18 A singly symmetric beam with a T-section (see figure) has cross-sectional dimensions $b = 140$ mm, $a = 190.8$ mm, $t_w = 6.99$ mm, and $t_f = 11.2$ mm.

Calculate the plastic modulus Z and the shape factor f.

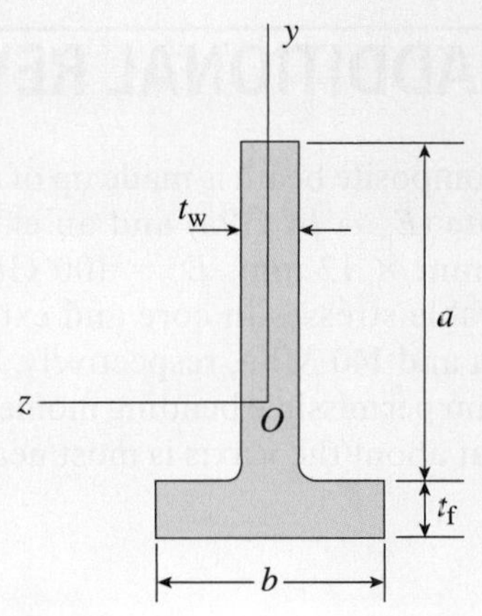

PROBLEM 6.10-18

6.10-19 A wide-flange beam with an unbalanced cross section has the dimensions shown in the figure.

Determine the plastic moment M_P if $\sigma_Y = 250$ MPa.

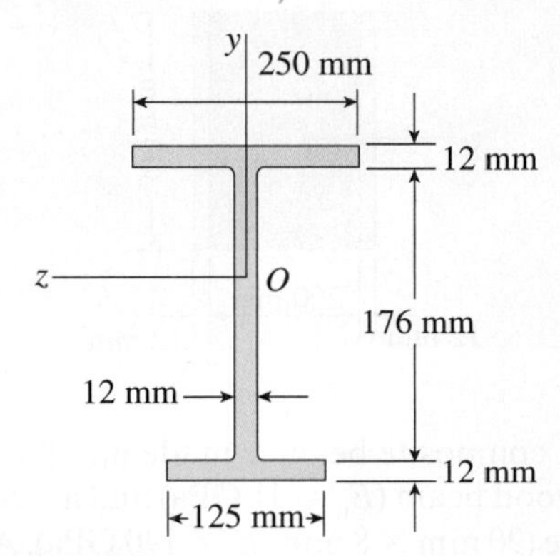

PROBLEM 6.10-19

6.10-20 Determine the plastic moment M_P for a beam having the cross section shown in the figure if $\sigma_Y = 210$ MPa.

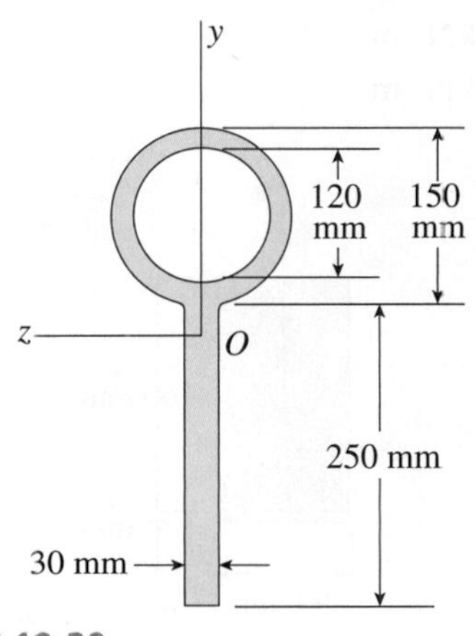

PROBLEM 6.10-20

SOME ADDITIONAL REVIEW PROBLEMS: Chapter 6

R-6.1 A composite beam is made up of a 200 mm × 300 mm core (E_c = 14 GPa) and an exterior cover sheet (300 mm × 12 mm, E_e = 100 GPa) on each side. Allowable stresses in core and exterior sheets are 9.5 MPa and 140 MPa, respectively. The ratio of the maximum permissible bending moment about the z axis to that about the y axis is most nearly:

(A) 0.5
(B) 0.7
(C) 1.2
(D) 1.5

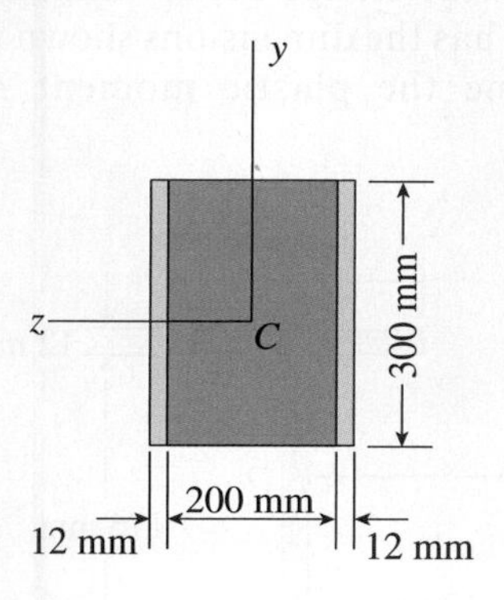

R-6.2 A composite beam is made up of a 90 mm × 160 mm wood beam (E_w = 11 GPa) and a steel bottom cover plate (90 mm × 8 mm, E_s = 190 GPa). Allowable stresses in wood and steel are 6.5 MPa and 110 MPa, respectively. The allowable bending moment about the z axis of the composite beam is most nearly:

(A) 2.9 kN·m
(B) 3.5 kN·m
(C) 4.3 kN·m
(D) 9.9 kN·m

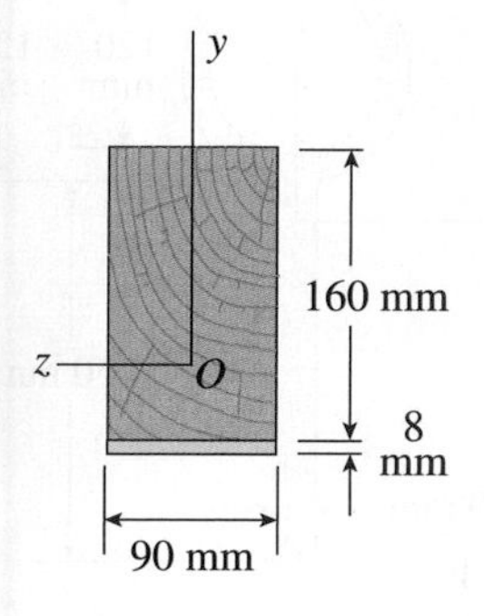

R-6.3 A steel pipe (d_3 = 104 mm, d_2 = 96 mm) has a plastic liner with inner diameter d_1 = 82 mm. The modulus of elasticity of the steel is 75 times that of the modulus of the plastic. Allowable stresses in steel and plastic are 40 MPa and 550 kPa, respectively. The allowable bending moment for the composite pipe is approximately:

(A) 1100 N·m
(B) 1230 N·m
(C) 1370 N·m
(D) 1460 N·m

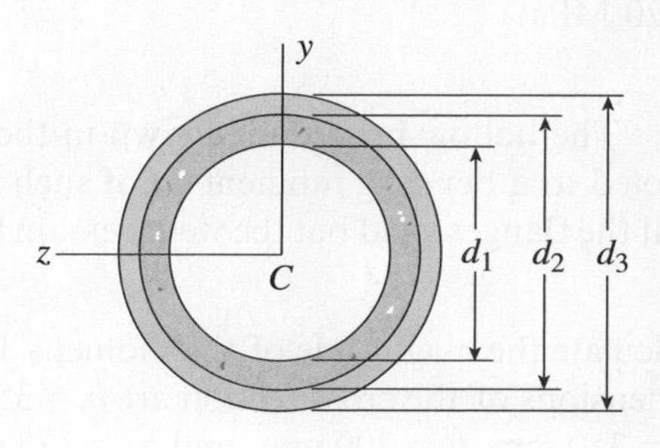

R-6.4 A bimetallic beam of aluminum (E_a = 70 GPa) and copper (E_c = 110 GPa) strips has a width of b = 25 mm; each strip has a thickness t = 1.5 mm. A bending moment of 1.75 N·m is applied about the z axis. The ratio of the maximum stress in the aluminum to that in the copper is approximately:

(A) 0.6
(B) 0.8
(C) 1.0
(D) 1.5

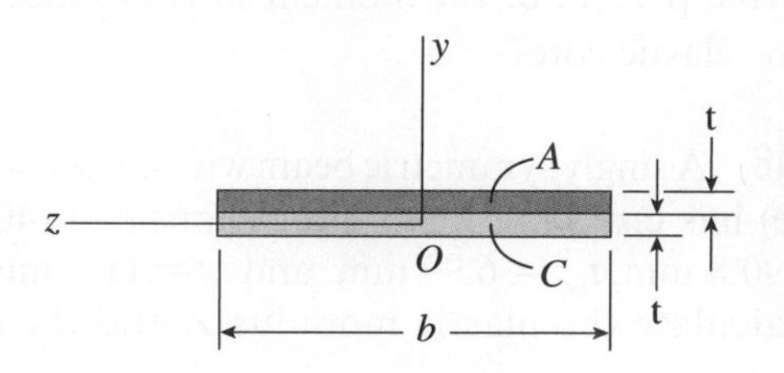

R-6.5 A composite beam of aluminum ($E_a = 72$ GPa) and steel ($E_s = 190$ GPa) has a width $b = 25$ mm and heights $h_a = 42$ mm and $h_s = 68$ mm, respectively. A bending moment is applied about the z axis resulting in a maximum stress in the aluminum of 55 MPa. The maximum stress in the steel is approximately:

(A) 86 MPa
(B) 90 MPa
(C) 94 MPa
(D) 98 MPa

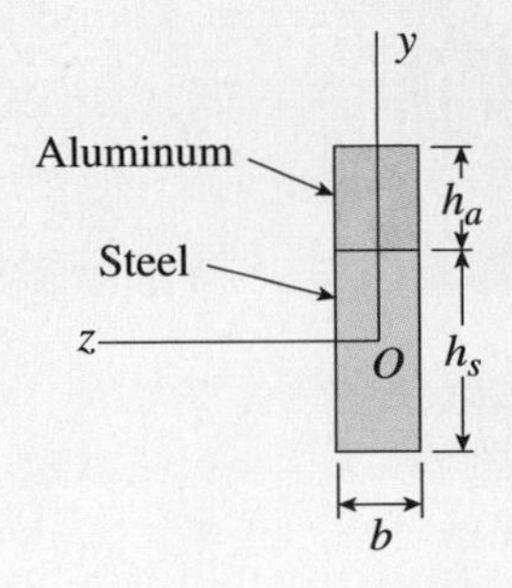

CHAPTER 7

Analysis of Stress and Strain

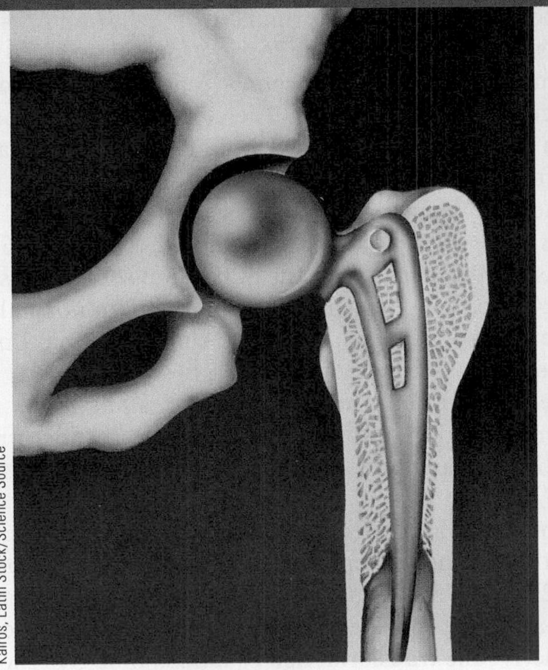

Kairos, Latin Stock/Science Source

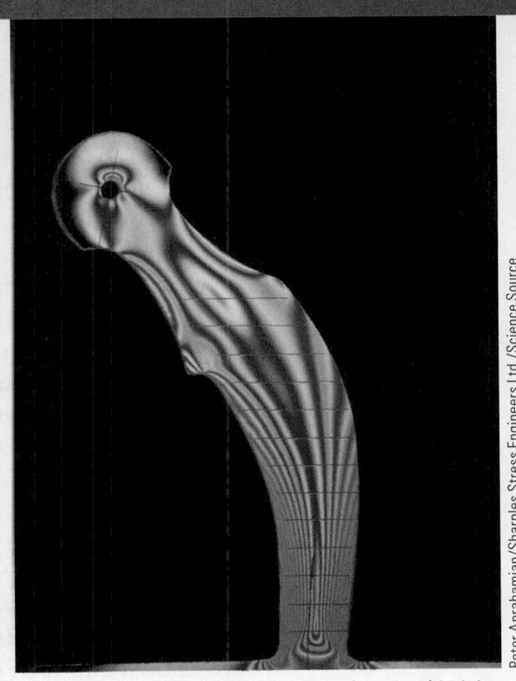

Peter Aprahamian/Sharples Stress Engineers Ltd./Science Source

Photoelasticity is an experimental method that can be used to find the complex state of stress in this prosthetic implant in a hip joint.

Chapter Objectives

- Define the state of stress at a point on a body using a plane stress element.
- Develop plane stress transformation equations that give equivalent descriptions of the state of stress at a point for various rotated positions of the stress element.
- Use the transformation equations to find the maximum and minimum normal stresses, known as principal normal stresses, and also the maximum shear stress at any point of interest.
- Display the transformation equations in graphical form on a plot known as Mohr's circle, which shows the relationship between normal and shear stresses acting on inclined planes at a point in a stressed body.
- Define Hooke's law for plane stress, which provides the relationship between normal strains and stresses for homogeneous and isotropic materials.
- Study states of stress in the three dimensions and define triaxial states of stress known as spherical stress and hydrostatic stress.
- Develop transformation equations for plane strain for use in evaluation of experimental measurements made with strain gages; use Mohr's circle for plane strain transformations.

Chapter Outline

7.1 Introduction

Normal and shear stresses in beams, shafts, and bars can be calculated from the basic formulas discussed in the preceding chapters. For instance, the stresses in a beam are given by the flexure and shear formulas ($\sigma = My/I$ and $\tau = VQ/Ib$), and the stresses in a shaft are given by the torsion formula $\tau = T\rho / I_p$. The stresses calculated from these formulas act on cross sections of the members, but larger stresses may occur on **inclined sections**. Therefore, the analysis of stresses and strains begins with a discussion of methods for finding the normal and shear stresses acting on inclined sections cut through a member.

Expressions for the normal and shear stresses acting on inclined sections in both *uniaxial stress* and *pure shear* were presented in Sections 2.6 and 3.5, respectively. In the case of uniaxial stress, the maximum shear stresses occur on planes inclined at 45° to the axis, whereas the maximum normal stresses occur on the cross sections. In the case of pure shear, the maximum tensile and compressive stresses occur on 45° planes. In an analogous manner, the stresses on inclined sections cut through a beam may be larger than the stresses acting on a cross section. To calculate such stresses, determine the stresses acting on inclined planes under a more general stress state known as **plane stress** (Section 7.2).

In this discussion of plane stress, **stress elements** represent the state of stress at a point in a body. Stress elements were discussed previously in a specialized context (see Sections 2.6 and 3.5), but now they are used in a more formalized manner. Begin the analysis by considering an element on which the stresses are known and then derive the **transformation equations** that give the stresses acting on the sides of an element oriented in a different direction.

When working with stress elements, keep in mind that only one intrinsic **state of stress** exists at a point in a stressed body, regardless of the orientation of the element being used to portray that state of stress. When there are two elements with different orientations at the same point in a body, the stresses acting on the faces of the two elements are different, but they still represent the same state of stress, namely, the stress at the point under consideration. This situation is analogous to the representation of a force vector by its components—although the components are different when the coordinate axes are rotated to a new position, the force itself is the same.

Furthermore, keep in mind that stresses are *not* vectors. This fact can sometimes be confusing, because engineers customarily represent stresses by arrows just as they represent force vectors by arrows. *Although the arrows used to represent stresses have magnitude and direction, they are not vectors because they do not combine according to the parallelogram law of addition.* Instead, stresses are much more complex quantities than are vectors, and in mathematics, they are called **tensors**. Other tensor quantities in mechanics are strains and moments of inertia.

7.2 Plane Stress

The stress conditions encountered in earlier chapters when analyzing bars in tension and compression, shafts in torsion, and beams in bending are examples of a state of stress called **plane stress**. To explain plane stress, consider the stress element shown in Fig. 7-1a. This element is infinitesimal in size and can be sketched either as a cube or as a rectangular parallelepiped. The xyz axes are parallel to the edges of the element, and the faces of the element are designated by the directions of their outward normals, as explained previously in Section 1.8. For instance, the right-hand face of the element is referred to as the positive x face, and the left-hand face (hidden from the viewer) is referred to as the negative x face. Similarly, the top face is the positive y face, and the front face is the positive z face.

When the material is in plane stress in the x-y plane, only the x and y faces of the element are subjected to stresses, and all stresses act parallel to the x and y axes, as shown in Fig. 7-1a. This stress condition is very common because it exists at the surface of any stressed body, except at points where external loads act on the surface. When the element shown in Fig. 7-1a is located at the free surface of a body, the z axis is normal to the surface and the z face is in the plane of the surface.

The symbols for the stresses shown in Fig. 7-1a have the following meanings. A **normal stress** σ has a subscript that identifies the face on which the stress acts; for instance, the stress σ_x acts on the x face of the element and the stress σ_y acts on the y face of the element. Since the element is infinitesimal in size, equal normal stresses act on the opposite faces. The **sign convention for normal stresses** is the familiar one, namely, tension is positive and compression is negative.

A **shear stress** τ has two subscripts—the first subscript denotes the face on which the stress acts, and the second gives the direction on that face. Thus, the stress τ_{xy} acts on the x face in the direction of the y axis (Fig. 7-1a), and the stress τ_{yx} acts on the y face in the direction of the x axis.

The **sign convention for shear stresses** is as follows. A shear stress is positive when it acts on a positive face of an element in the positive direction of an axis, and it is negative when it acts on a positive face of an element in the negative direction of an axis. Therefore, the stresses τ_{xy} and τ_{yx} shown on the positive x and y faces in Fig. 7-1a are positive shear stresses. Similarly, on a negative face of the element, a shear stress is positive when it acts in the negative direction of an axis. Hence, the stresses τ_{xy} and τ_{yx} shown on the negative x and y faces of the element are also positive.

This sign convention for shear stresses is easy to remember if stated as follows:

A shear stress is positive when the directions associated with its subscripts are plus-plus or minus-minus; the stress is negative when the directions are plus-minus or minus-plus.

The preceding sign convention for shear stresses is consistent with the equilibrium of the element, because shear stresses on opposite faces of an infinitesimal element must be equal in magnitude and opposite in direction. Hence, according to this sign convention, a positive stress τ_{xy} acts upward on the positive face (Fig. 7-1a) and downward on the negative face. In a similar manner, the stresses τ_{yx} acting on the top and bottom faces of the element are positive—although they have opposite directions.

Shear stresses on perpendicular planes are equal in magnitude and have directions such that both stresses point toward, or both point away from, the line of intersection of the faces. Inasmuch as τ_{xy} and τ_{yx} are positive in the directions shown in the figure, they are consistent with this observation. Therefore, note that

$$\tau_{xy} = \tau_{yx} \tag{7-1}$$

FIGURE 7-1

Elements in plane stress: (a) three-dimensional view of an element oriented to the xyz axes, (b) two-dimensional view of the same element, and (c) two-dimensional view of an element oriented to the $x_1y_1z_1$ axes

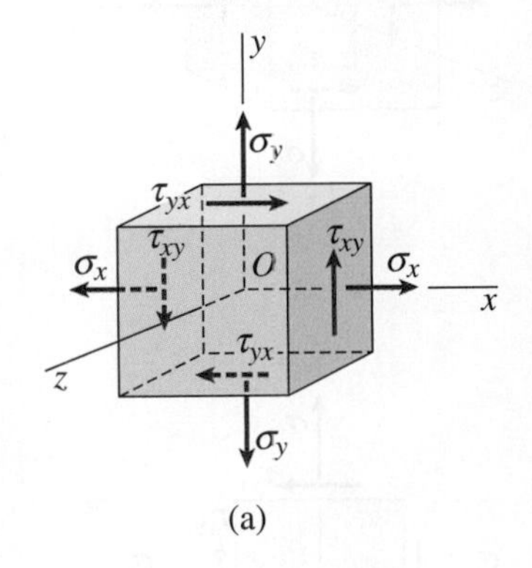

(a)

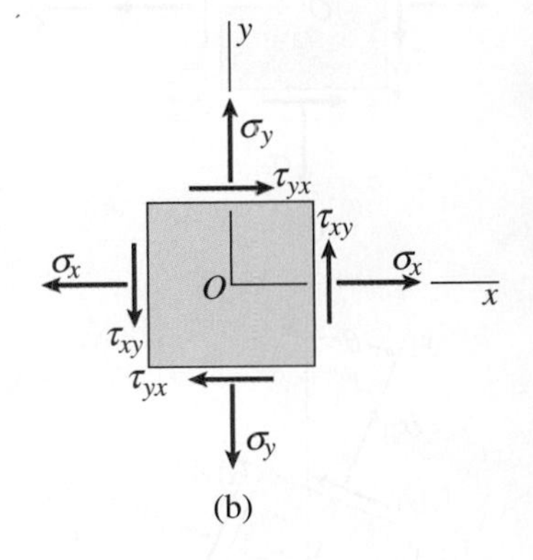

(b)

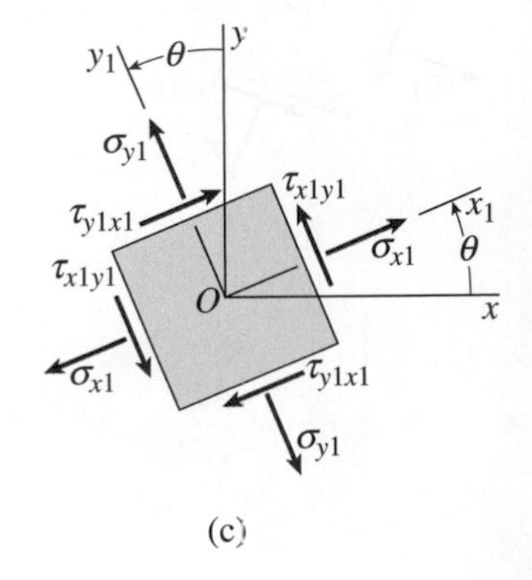

(c)

FIGURE 7-1 (Repeated)

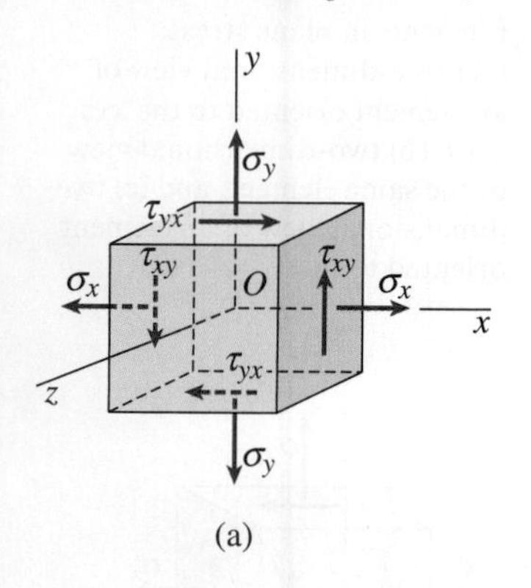

(a)

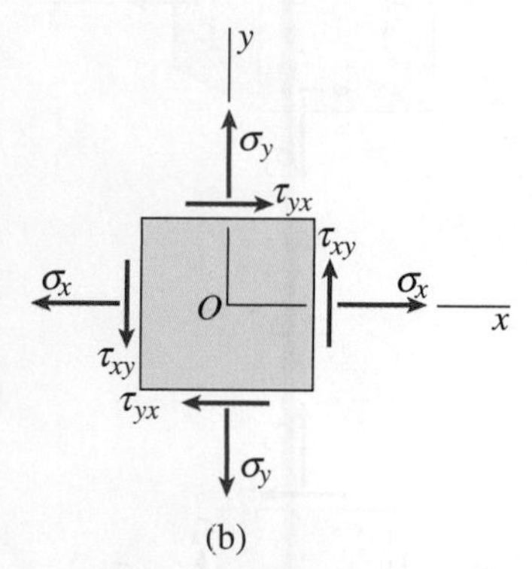

(b)

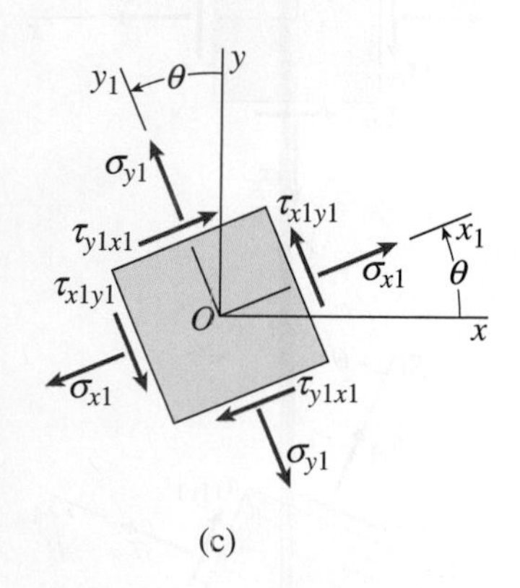

(c)

This relationship was derived previously from equilibrium of the element (see Section 1.8).

For convenience in sketching plane-stress elements, draw only a two-dimensional view of the element, as shown in Fig. 7-1b. Although a figure of this kind is adequate for showing all stresses acting on the element, keep in mind that the element is a solid body with a thickness perpendicular to the plane of the figure.

Stresses on Inclined Sections

Now consider the stresses acting on inclined sections, assuming that the stresses σ_x, σ_y, and τ_{xy} (Figs. 7-1a and b) are known. To portray the stresses acting on an inclined section, consider a new stress element (Fig. 7-1c) that is located at the same point in the material as the original element (Fig. 7-1b). However, the new element has faces that are parallel and perpendicular to the inclined direction. Associated with this new element are axes x_1, y_1, and z_1, such that the z_1 axis coincides with the z axis and the x_1y_1 axes are rotated counterclockwise through an angle θ with respect to the x-y axes.

The normal and shear stresses acting on this new element are denoted σ_{x1}, σ_{y1}, τ_{x1y1}, and τ_{y1x1}, using the same subscript designations and sign conventions described previously for the stresses acting on the x-y element. The previous conclusions regarding the shear stresses still apply, so that

$$\tau_{x1y1} = \tau_{y1x1} \tag{7-2}$$

From this equation and the equilibrium of the element, observe that *the shear stresses acting on all four side faces of an element in plane stress are known if you determine the shear stress acting on any one of those faces.*

The stresses acting on the inclined x_1y_1 element (Fig. 7-1c) can be expressed in terms of the stresses on the x-y element (Fig. 7-1b) by using equations of equilibrium. For this purpose, choose a **wedge-shaped stress element** (Fig. 7-2a) having an inclined face that is the same as the x_1 face of the inclined element shown in Fig. 7-1c. The other two side faces of the wedge are parallel to the x and y axes.

In order to write equations of equilibrium for the wedge, construct a free-body diagram showing the forces acting on the faces. Denote the area of the left-hand side face (that is, the negative x face) as A_0. Then the normal and shear forces acting on that face are $\sigma_x A_0$ and $\tau_{xy} A_0$, as shown in the free-body diagram of Fig. 7-2b. The area of the bottom face (or negative y face) is $A_0 \tan\theta$, and the area of the inclined face (or positive x_1 face) is $A_0 \sec\theta$. Thus, the normal and shear forces acting on these faces have the magnitudes and directions shown in Fig. 7-2b.

The forces acting on the left-hand and bottom faces can be resolved into orthogonal components acting in the x_1 and y_1 directions. Then obtain two equations of equilibrium by summing forces in those directions. The first equation, obtained by summing forces in the x_1 direction, is

$$\sigma_{x1} A_0 \sec\theta - \sigma_x A_0 \cos\theta - \tau_{xy} A_0 \sin\theta$$
$$- \sigma_y A_0 \tan\theta \sin\theta - \tau_{yx} A_0 \tan\theta \cos\theta = 0$$

In the same manner, summation of forces in the y_1 direction gives

$$\tau_{x1y1} A_0 \sec\theta + \sigma_x A_0 \sin\theta - \tau_{xy} A_0 \cos\theta$$
$$- \sigma_y A_0 \tan\theta \cos\theta + \tau_{yx} A_0 \tan\theta \sin\theta = 0$$

Use the relationship $\tau_{xy} = \tau_{yx}$ and also simplify and rearrange to obtain the following two equations:

$$\sigma_{x1} = \sigma_x \cos^2\theta + \sigma_y \sin^2\theta + 2\tau_{xy} \sin\theta\cos\theta \quad \textbf{(7-3a)}$$

$$\tau_{x1y1} = -(\sigma_x - \sigma_y)\sin\theta\cos\theta + \tau_{xy}(\cos^2\theta - \sin^2\theta) \quad \textbf{(7-3b)}$$

Equations (7-3a) and (7-3b) give the normal and shear stresses acting on the x_1 plane in terms of the angle θ and the stresses σ_x, σ_y, and τ_{xy} acting on the x and y planes.

For the special case when $\theta = 0$, note that Eqs. (7-3a and b) give $\sigma_{x1} = \sigma_x$ and $\tau_{x1y1} = \tau_{xy}$, as expected. Also, when $\theta = 90°$, the equations give $\sigma_{x1} = \sigma_y$ and $\tau_{x1y1} = -\tau_{xy} = -\tau_{yx}$. In the latter case, since the x_1 axis is vertical when $\theta = 90°$, the stress τ_{x1y1} will be positive when it acts to the left. However, the stress τ_{yx} acts to the right, so $\tau_{x1y1} = -\tau_{yx}$.

FIGURE 7-2

Wedge-shaped stress element in plane stress: (a) stresses acting on the element and (b) forces acting on the element (free-body diagram)

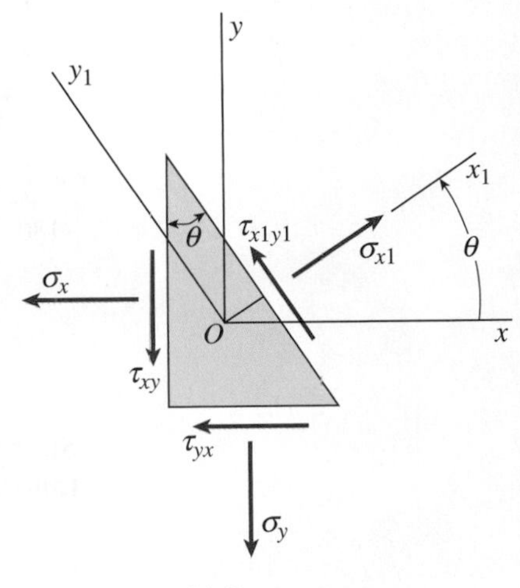

(a) Stresses

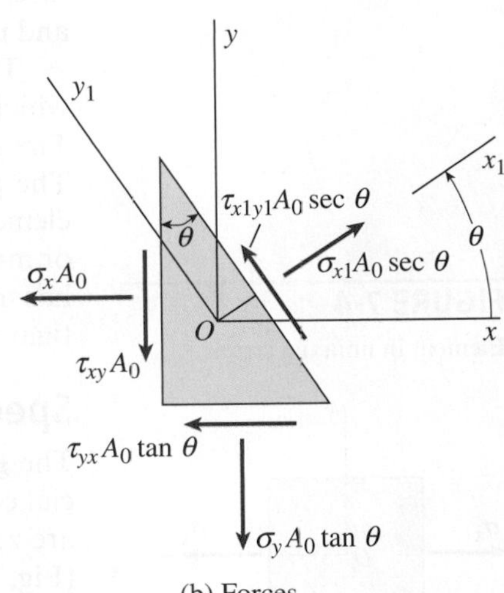

(b) Forces

Transformation Equations for Plane Stress

Equations (7-3a and b) for the stresses on an inclined section can be expressed in a more convenient form by introducing the following trigonometric identities (see Appendix C):

$$\cos^2\theta = \frac{1}{2}(1 + \cos 2\theta) \quad \sin^2\theta = \frac{1}{2}(1 - \cos 2\theta)$$

$$\sin\theta\cos\theta = \frac{1}{2}\sin 2\theta$$

When these substitutions are made, the equations become

$$\sigma_{x1} = \frac{\sigma_x + \sigma_y}{2} + \frac{\sigma_x - \sigma_y}{2}\cos 2\theta + \tau_{xy}\sin 2\theta \quad \textbf{(7-4a)}$$

$$\tau_{x1y1} = -\frac{\sigma_x - \sigma_y}{2}\sin 2\theta + \tau_{xy}\cos 2\theta \quad \textbf{(7-4b)}$$

These equations are usually called the **transformation equations for plane stress** because they transform the stress components from one set of axes to another. However, the intrinsic state of stress at the point under consideration is the same whether represented by stresses acting on the xy element (Fig. 7-1b) or by stresses acting on the inclined x_1y_1 element (Fig. 7-1c).

Since the transformation equations were derived solely from equilibrium of an element, they are applicable to stresses in any kind of material, whether linear or nonlinear, elastic or inelastic.

An important observation concerning the normal stresses can be obtained from the transformation equations. As a preliminary matter, note that the normal stress σ_{y1} acting on the y_1 face of the inclined element (Fig. 7-1c) can be obtained from Eq. (7-4a) by substituting $\theta + 90°$ for θ. The result is the following equation for σ_{y1}:

$$\sigma_{y1} = \frac{\sigma_x + \sigma_y}{2} - \frac{\sigma_x - \sigma_y}{2}\cos 2\theta - \tau_{xy}\sin 2\theta \quad \textbf{(7-5)}$$

FIGURE 7-3

Graph of normal stress σ_{x1} and shear stress τ_{x1y1} versus the angle θ (for $\sigma_y = 0.2\sigma_x$ and $\tau_{xy} = 0.8\sigma_x$)

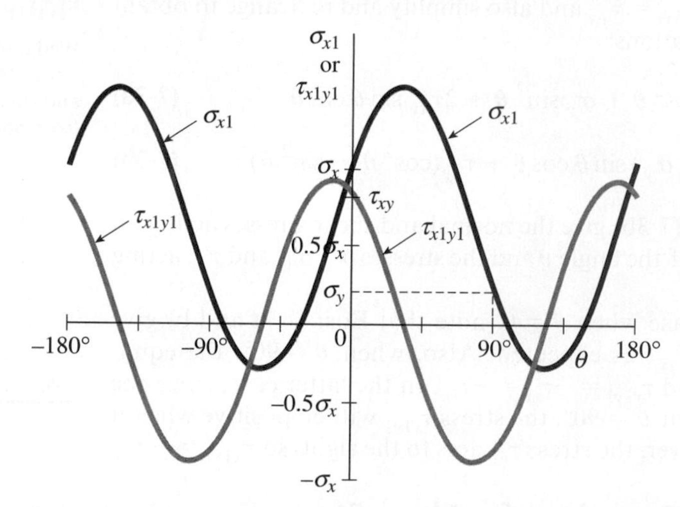

Summing the expressions for σ_{x1} and σ_{y1} [Eqs. (7-4a) and (7-5)] produces the following equation for plane stress:

$$\sigma_{x1} + \sigma_{y1} = \sigma_x + \sigma_y \tag{7-6}$$

This equation shows that the sum of the normal stresses acting on perpendicular faces of plane-stress elements (at a given point in a stressed body) is constant and independent of the angle θ.

The manner in which the normal and shear stresses vary is shown in Fig. 7-3, which is a graph of σ_{x1} and τ_{x1y1} versus the angle θ [from Eqs. (7-4a and b)]. The graph is plotted for the particular case of $\sigma_y = 0.2\sigma_x$ and $\tau_{xy} = 0.8\sigma_x$. The plot shows that the stresses vary continuously as the orientation of the element is changed. At certain angles, the normal stress reaches a maximum or minimum value; at other angles, it becomes zero. Similarly, the shear stress has maximum, minimum, and zero values at certain angles. A detailed investigation of these maximum and minimum values is made in Section 7.3.

FIGURE 7-4

Element in uniaxial stress

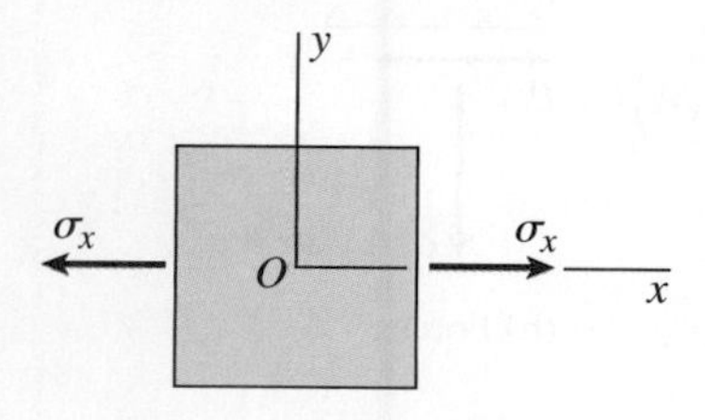

Special Cases of Plane Stress

The general case of plane stress reduces to simpler states of stress under special conditions. For instance, if all stresses acting on the xy element (Fig. 7-1b) are zero except for the normal stress σ_x, then the element is in **uniaxial stress** (Fig. 7-4). The corresponding transformation equations, obtained by setting σ_y and τ_{xy} equal to zero in Eqs. (7-4a and b), are

$$\sigma_{x1} = \frac{\sigma_x}{2}(1 + \cos 2\theta) \quad \tau_{x1y1} = -\frac{\sigma_x}{2}(\sin 2\theta) \tag{7-7a,b}$$

FIGURE 7-5

Element in pure shear

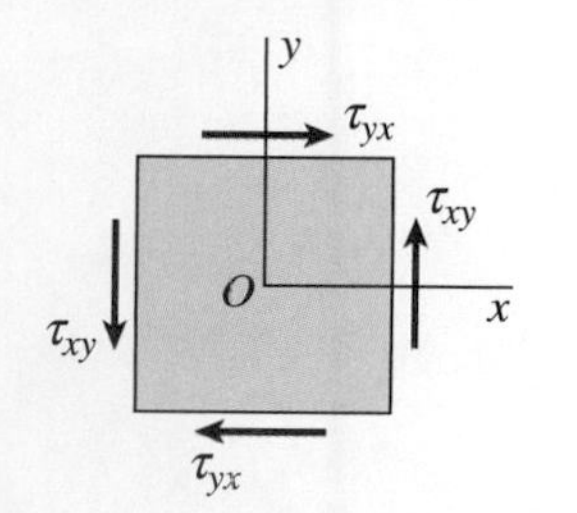

These equations agree with the equations derived previously in Section 2.6 [see Eqs. (2-29a and b)], except that now a more generalized notation for the stresses acting on an inclined plane is used.

Another special case is **pure shear** (Fig. 7-5), for which the transformation equations are obtained by substituting $\sigma_x = 0$ and $\sigma_y = 0$ into Eqs. (7-4a and b):

$$\sigma_{x1} = \tau_{xy} \sin 2\theta \quad \tau_{x1y1} = \tau_{xy} \cos 2\theta \tag{7-8a,b}$$

Again, these equations correspond to those derived earlier [see Eqs. (3-29a and 3-29b) in Section 3.5].

Finally, note the special case of **biaxial stress**, in which the xy element is subjected to normal stresses in both the x and y directions but without any shear stresses (Fig. 7-6). The equations for biaxial stress are obtained from Eqs. (7-4a and b) simply by dropping the terms containing τ_{xy}, as

$$\sigma_{x1} = \frac{\sigma_x + \sigma_y}{2} + \frac{\sigma_x - \sigma_y}{2} \cos 2\theta \tag{7-9a}$$

$$\tau_{x1y1} = -\frac{\sigma_x - \sigma_y}{2} \sin 2\theta \tag{7-9b}$$

Biaxial stress occurs in many kinds of structures, including thin-walled pressure vessels (see Sections 8.2 and 8.3).

FIGURE 7-6

Element in biaxial stress

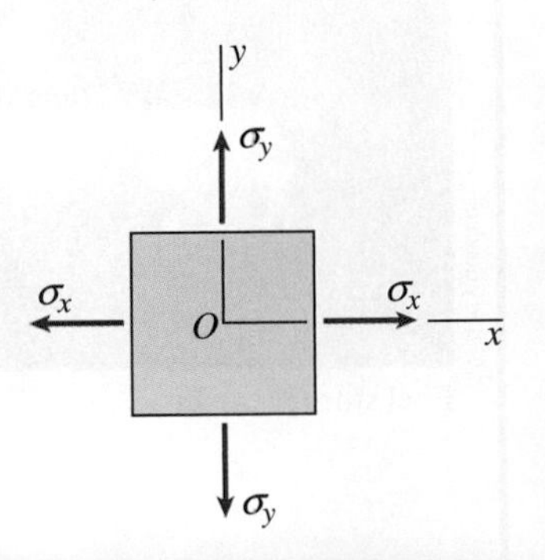

Example 7-1

FIGURE 7-7

Example 7-1: (a) Cylindrical pressure vessel with stress element at C, (b) element C in plane stress, and (c) element C inclined at an angle $\theta = 45°$

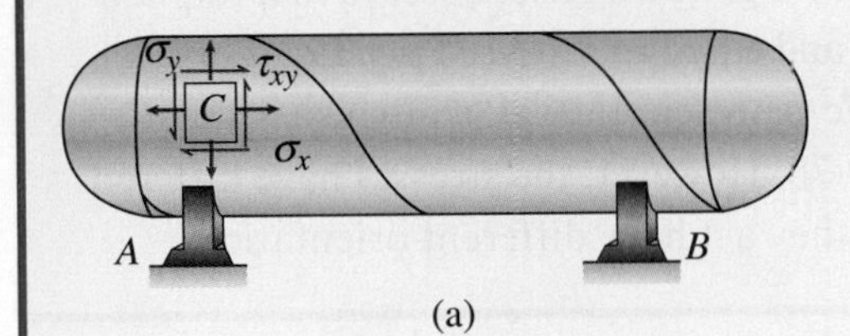

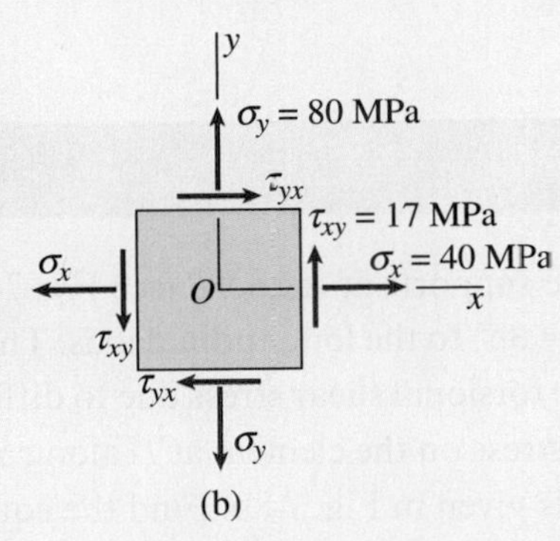

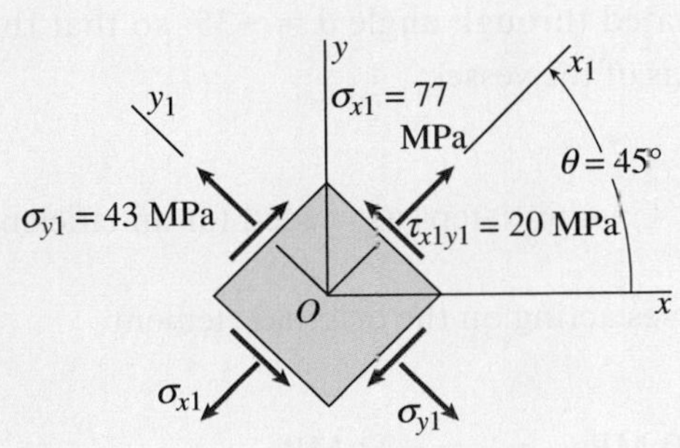

A cylindrical pressure vessel rests on simple supports at A and B (see Fig. 7-7). The vessel is under internal pressure resulting in longitudinal stress $\sigma_x = 40$ MPa and circumferential stress $\sigma_y = 80$ MPa on a stress element at point C on the wall of the vessel. In addition, differential settlement after an earthquake has caused the support at B to rotate, which applies a torsional moment to the vessel leading to shear stress $\tau_{xy} = 17$ MPa. Find the stresses acting on the element at C when rotated through angle $\theta = 45°$.

Solution:

Use a four-step problem-solving approach. Combine steps as needed for an efficient solution.

1, 2. *Conceptualize* [*hypothesize, sketch*], ***Categorize,*** [*simplify, classify*]:

Transformation equations: To determine the stresses acting on an inclined element, use the transformation equations given in Eqs. (7-4a and b). From the given numerical data, obtain the following values for substitution into those equations:

$$\frac{\sigma_x + \sigma_y}{2} = 60 \text{ MPa} \qquad \frac{\sigma_x - \sigma_y}{2} = -20 \text{ MPa} \qquad \tau_{xy} = 17 \text{ MPa}$$

$$\sin 2\theta = \sin 90° = 1 \qquad \cos 2\theta = \cos 90° = 0$$

3. *Analyze* [*evaluate; select relevant equations, carry out mathematical solution*]: Substitute these values into Eqs. (7-4a and b) to get

© Barry Goodno
Fuel storage tanks

$$\sigma_{x1} = \frac{\sigma_x + \sigma_y}{2} + \frac{\sigma_x - \sigma_y}{2}\cos 2\theta + \tau_{xy}\sin 2\theta$$
$$= 60 \text{ MPa} + (-20 \text{ MPa})(0) + (17 \text{ MPa})(1) = 77 \text{ MPa}$$

$$\tau_{x1y1} = -\frac{\sigma_x - \sigma_y}{2}\sin 2\theta + \tau_{xy}\cos 2\theta$$
$$= -(-20 \text{ MPa})(1) + (17 \text{ MPa})(0) = 20 \text{ MPa}$$

In addition, the stress σ_{y1} may be obtained from Eq. (7-5):

$$\sigma_{y1} = \frac{\sigma_x + \sigma_y}{2} - \frac{\sigma_x - \sigma_y}{2}\cos 2\theta - \tau_{xy}\sin 2\theta$$
$$= 60 \text{ MPa} - (-20 \text{ MPa})(0) - (17 \text{ MPa})(1) = 43 \text{ MPa}$$

4. ***Finalize*** [*conclude; examine answer—Does it make sense? Are units correct? How does it compare to similar problem solutions?*]:

Stress elements: From these results, you can readily obtain the stresses acting on all sides of an element oriented at $\theta = 45°$, as shown in Fig. 7-7c. The arrows show the true directions in which the stresses act. Note especially the directions of the shear stresses, all of which have the same magnitude. Also, observe that the sum of the normal stresses remains constant and equal to 120 MPa [see Eq. (7-6)].

Note: The stresses shown in Fig. 7-7c represent the same intrinsic state of stress as do the stresses shown in Fig. 7-7b. However, the stresses have different values, because the elements on which they act have different orientations.

Example 7-2

A cylindrical pressure vessel rests on simple supports at A and B (see Fig. 7-8). The vessel has a helical weld joint oriented at $\theta = 35°$ to the longitudinal axis. The vessel is under internal pressure and also has some torsional shear stress due to differential settlement of the support at B. The state of stress on the element at D along and perpendicular to the weld seam is known and is given in Fig. 7-8b. Find the equivalent stress state for the element at D when rotated through angle $\theta = -35°$ so that the element is aligned with the longitudinal axis of the vessel.

Solution:

Use a four-step problem-solving approach. Combine steps as needed for an efficient solution.

1, 2. ***Conceptualize, Categorize*:** The stresses acting on the original element (Fig. 7-8b) have the values:

$$\sigma_x = 40 \text{ MPa} \quad \sigma_y = 80 \text{ MPa} \quad \tau_{xy} = -22 \text{ MPa}$$

FIGURE 7-8

Example 7-2: (a) Cylindrical pressure vessel with stress element at D, (b) element D in plane stress, and (c) element D inclined at an angle $\theta = -35°$

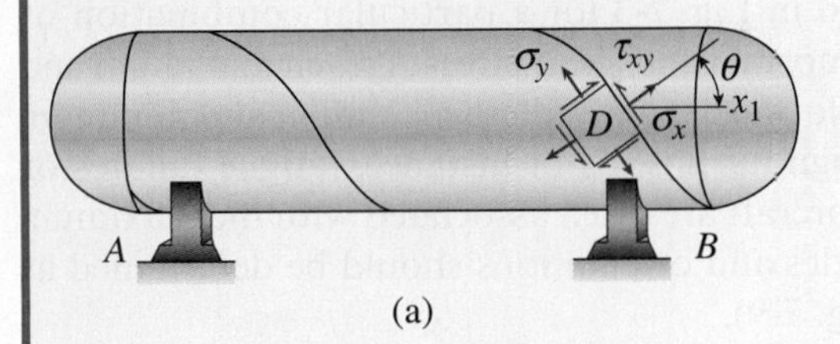

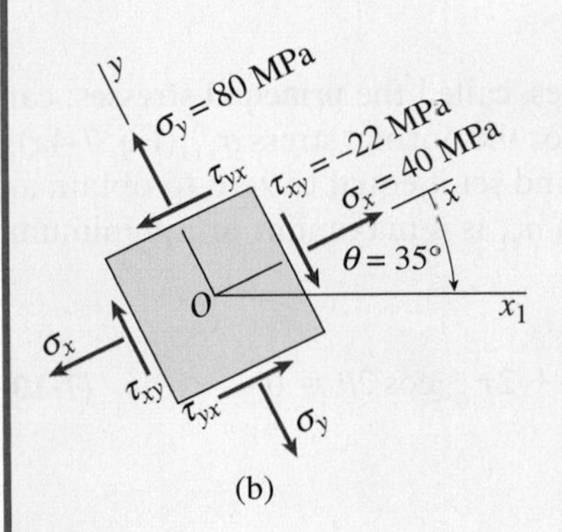

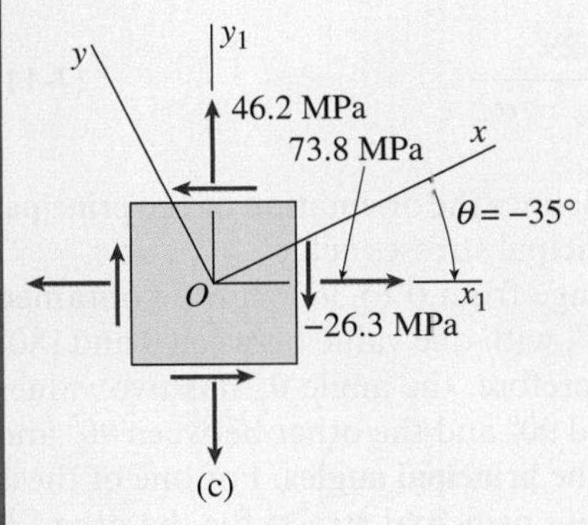

Fuel storage tank supported on pedestals

An element oriented at a clockwise angle of $-35°$ is shown in Fig. 7-8c, where the x_1 axis is at an angle $\theta = -35°$ with respect to the x axis.

Stress transformation equations: You can readily calculate the stresses on the x_1 face of the element oriented at $\theta = -35°$ by using the transformation equations given in Eqs. (7-4a and b). The calculation results are

$$\frac{\sigma_x + \sigma_y}{2} = 60 \text{ MPa} \qquad \frac{\sigma_x - \sigma_y}{2} = -20 \text{ MPa}$$

$$\sin 2\theta = \sin(-70°) = -0.94 \qquad \cos 2\theta = \cos(-70°) = 0.342$$

3. ***Analyze***: Substitute into the transformation equations to get

$$\begin{aligned}\sigma_{x1} &= \frac{\sigma_x + \sigma_y}{2} + \frac{\sigma_x - \sigma_y}{2}\cos(2\theta) + \tau_{xy}\sin(2\theta)\\ &= 60 \text{ MPa} + (-20 \text{ MPa})(0.342) + (-22 \text{ MPa})(-0.94)\\ &= 73.8 \text{ MPa}\end{aligned}$$

$$\begin{aligned}\tau_{x1y1} &= -\left(\frac{\sigma_x - \sigma_y}{2}\right)\sin(2\theta) + \tau_{xy}\cos(2\theta)\\ &= -(-20 \text{ MPa})(-0.94) + (-22 \text{ MPa})(0.342) = -26.3 \text{ MPa}\end{aligned}$$

The normal stress acting on the y_1 face [see Eq. (7-5)] is

$$\begin{aligned}\sigma_{y1} &= \frac{\sigma_x + \sigma_y}{2} - \frac{\sigma_x - \sigma_y}{2}\cos(2\theta) - \tau_{xy}\sin(2\theta)\\ &= 60 \text{ MPa} - (-20 \text{ MPa})(0.342) - (-22 \text{ MPa})(-0.94)\\ &= 46.2 \text{ MPa}\end{aligned}$$

4. ***Finalize***: As a check on the results, note that

$$\sigma_{x1} + \sigma_{y1} = \sigma_x + \sigma_y.$$

The stresses acting on the rotated element are shown in Fig. 7-8c, where the arrows indicate the true directions of the stresses. Note that both stress elements shown in Fig. 7-8 represent the same state of stress.

FIGURE 7-9

Photoelastic fringe pattern displays principal stresses in a model of a crane-hook: (a) Photo of a crane-hook, (b) Photoelastic fringe pattern

Frans Lemmens/Getty Images

(a)

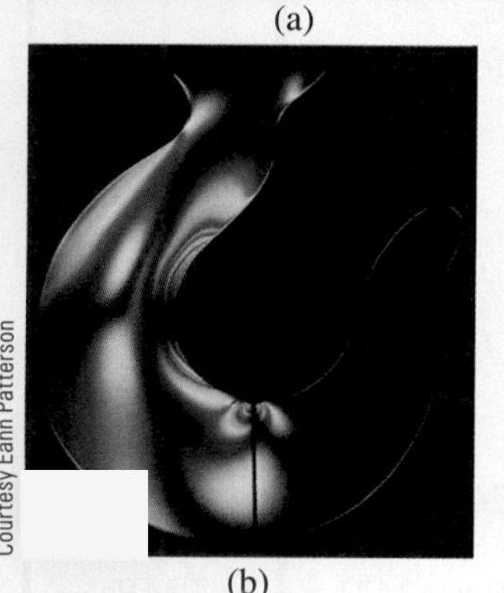

Courtesy Eann Patterson

(b)

7.3 Principal Stresses and Maximum Shear Stresses

The transformation equations for plane stress show that the normal stresses σ_{x1} and the shear stresses τ_{x1y1} vary continuously as the axes are rotated through the angle θ. This variation is pictured in Fig. 7-3 for a particular combination of stresses. From the figure, both the normal and shear stresses reach maximum and minimum values at 90° intervals. Not surprisingly, these maximum and minimum values are usually needed for design purposes. For instance, fatigue failures of structures such as machines and aircraft are often associated with the maximum stresses, and hence their magnitudes and orientations should be determined as part of the design process (see Fig. 7-9).

Principal Stresses

The maximum and minimum normal stresses, called the **principal stresses**, can be found from the transformation equation for the normal stress σ_{x1} [(Eq. 7-4a)]. Take the derivative of σ_{x1} with respect to θ and set it equal to zero to obtain an equation that gives the values of θ at which σ_{x1} is a maximum or a minimum. The equation for the derivative is

$$\frac{d\sigma_{x1}}{d\theta} = -(\sigma_x - \sigma_y)\sin 2\theta + 2\tau_{xy}\cos 2\theta = 0 \quad \textbf{(7-10)}$$

which now gives

$$\tan 2\theta_p = \frac{2\tau_{xy}}{\sigma_x - \sigma_y} \quad \textbf{(7-11)}$$

The subscript p indicates that the angle θ_p defines the orientation of the **principal planes**, that is, the planes on which the principal stresses act.

Two values of the angle $2\theta_p$ in the range from 0 to 360° can be obtained from Eq. (7-11). These values differ by 180°, with one value between 0 and 180° and the other between 180° and 360°. Therefore, the angle θ_p has two values that differ by 90°, one value between 0 and 90° and the other between 90° and 180°. The two values of θ_p are known as the **principal angles**. For one of these angles, the normal stress σ_{x1} is a *maximum* principal stress; for the other, it is a *minimum* principal stress. Because the principal angles differ by 90°, *the principal stresses occur on mutually perpendicular planes.*

The principal stresses can be calculated by substituting each of the two values of θ_p into the first stress-transformation equation [Eq. (7-4a)] and solving for σ_{x1}. By determining the principal stresses in this manner, you not only obtain the values of the principal stresses but you also learn which principal stress is associated with which principal angle.

FIGURE 7-10

Geometric representation of Eq. (7-11)

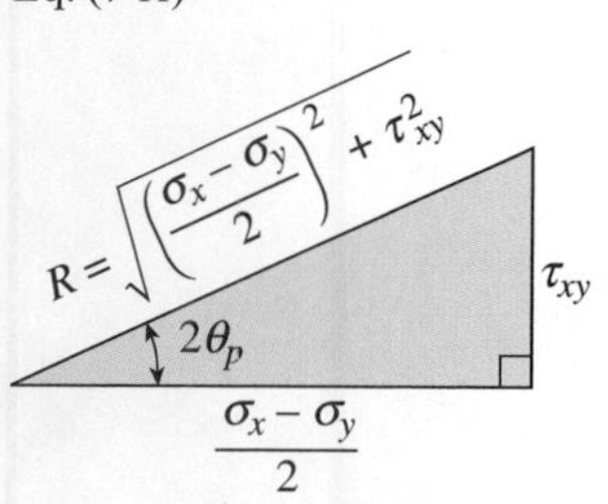

You can also obtain general formulas for the principal stresses. To do so, refer to the right triangle in Fig. 7-10, which is constructed from Eq. (7-11). Note that the hypotenuse of the triangle, obtained from the Pythagorean theorem, is

$$R = \sqrt{\left(\frac{\sigma_x - \sigma_y}{2}\right)^2 + \tau_{xy}^2} \quad \textbf{(7-12)}$$

The quantity R is always a positive number and, like the other two sides of the triangle, has units of stress. Two additional relations from the triangle are

$$\cos 2\theta_p = \frac{\sigma_x - \sigma_y}{2R} \qquad \sin 2\theta_p = \frac{\tau_{xy}}{R} \tag{7-13a,b}$$

Now substitute these expressions for $\cos 2\theta_p$ and $\sin 2\theta_p$ into Eq. (7-4a) and obtain the algebraically larger of the two principal stresses, denoted by σ_1:

$$\sigma_1 = \sigma_{x1} = \frac{\sigma_x + \sigma_y}{2} + \frac{\sigma_x - \sigma_y}{2}\cos 2\theta_p + \tau_{xy}\sin 2\theta_p$$

$$= \frac{\sigma_x + \sigma_y}{2} + \frac{\sigma_x - \sigma_y}{2}\left(\frac{\sigma_x - \sigma_y}{2R}\right) + \tau_{xy}\left(\frac{\tau_{xy}}{R}\right)$$

Substitute for R from Eq. (7-12) and perform some algebraic manipulations to obtain

$$\sigma_1 = \frac{\sigma_x + \sigma_y}{2} + \sqrt{\left(\frac{\sigma_x - \sigma_y}{2}\right)^2 + \tau_{xy}^2} \tag{7-14}$$

The smaller of the principal stresses, denoted by σ_2, may be found from the condition that the sum of the normal stresses on perpendicular planes is constant [see Eq. (7-6)]:

$$\sigma_1 + \sigma_2 = \sigma_x + \sigma_y \tag{7-15}$$

Substitute the expression for σ_1 into Eq. (7-15) and solve for σ_2 to get

$$\sigma_2 = \sigma_x + \sigma_y - \sigma_1$$

$$= \frac{\sigma_x + \sigma_y}{2} - \sqrt{\left(\frac{\sigma_x - \sigma_y}{2}\right)^2 + \tau_{xy}^2} \tag{7-16}$$

This equation has the same form as the equation for σ_1 but differs by the presence of the minus sign before the square root.

The preceding formulas for σ_1 and σ_2 can be combined into a single formula for the **principal stresses**:

$$\sigma_{1,2} = \frac{\sigma_x + \sigma_y}{2} \pm \sqrt{\left(\frac{\sigma_x - \sigma_y}{2}\right)^2 + \tau_{xy}^2} \tag{7-17}$$

The plus sign gives the algebraically larger principal stress, and the minus sign gives the algebraically smaller principal stress.

Principal Angles

Now denote the two angles defining the principal planes as θ_{p1} and θ_{p2}, corresponding to the principal stresses σ_1 and σ_2, respectively. Both angles can be determined from the equation for $\tan 2\theta_p$ [Eq. (7-11)]. However, you cannot tell from that equation which angle is θ_{p1} and which is θ_{p2}. A simple procedure for making this determination is to take one of the values and substitute it into the equation for σ_{x1} [Eq. (7-4a)]. The resulting value of σ_{x1} will be

FIGURE 7-11

Elements in uniaxial and biaxial stress

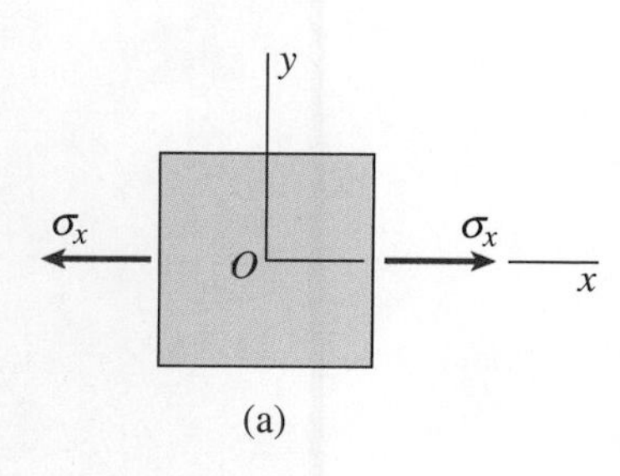

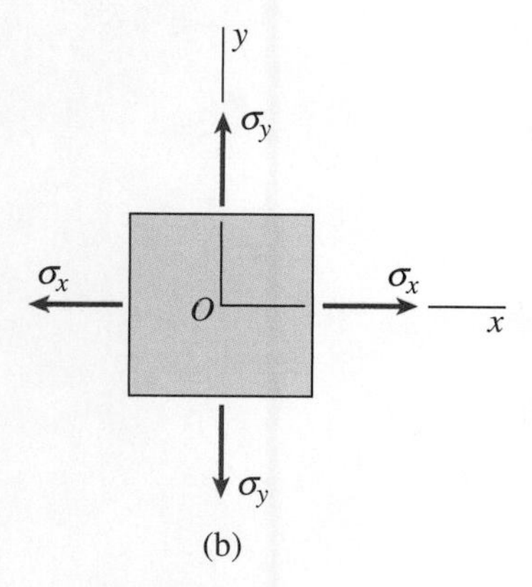

recognized as either σ_1 or σ_2 [assuming you have already found σ_1 and σ_2 from Eq. (7-17)], thus correlating the two principal angles with the two principal stresses.

Another method for correlating the principal angles and principal stresses is to use Eqs. (7-13a and b) to find θ_p, since the only angle that satisfies *both* of those equations is θ_{p1}. Thus, rewrite those equations as

$$\cos 2\theta_{p1} = \frac{\sigma_x - \sigma_y}{2R} \qquad \sin 2\theta_{p1} = \frac{\tau_{xy}}{R} \qquad \textbf{(7-18a,b)}$$

Only one angle exists between 0 and 360° that satisfies both of these equations. Thus, the value of θ_{p1} can be determined uniquely from Eqs. (7-18a and b). The angle θ_{p2}, corresponding to σ_2, defines a plane that is perpendicular to the plane defined by θ_{p1}. Therefore, θ_{p2} can be taken as 90° larger or 90° smaller than θ_{p1}.

Shear Stresses on the Principal Planes

An important characteristic of the principal planes can be obtained from the transformation equation for the shear stresses [Eq. (7-4b)]. If the shear stress τ_{x1y1} is set equal to zero, the resulting equation is the same as Eq. (7-10). Therefore, if you solve that equation for the angle 2θ, you get the same expression for $\tan 2\theta$ as before [Eq. (7-11)]. In other words, the angles to the planes of zero shear stress are the same as the angles to the principal planes.

This leads to the following important observation: *The shear stresses are zero on the principal planes.*

Special Cases

The principal planes for elements in **uniaxial stress** and **biaxial stress** are the x and y planes themselves (Fig. 7-11) because $\tan 2\theta_p = 0$ [see Eq. (7-11)] and the two values of θ_p are 0 and 90°. The x and y planes are the principal planes based on the fact that the shear stresses are zero on those planes.

For an element in **pure shear** (Fig. 7-12a), the principal planes are oriented at 45° to the x axis (Fig. 7-12b), because $\tan 2\theta_p$ is infinite, and the two values of θ_p are 45° and 135°. If τ_{xy} is positive, the principal stresses are $\sigma_1 = \tau_{xy}$ and $\sigma_2 = -\tau_{xy}$ (see Section 3.5 for a discussion of pure shear).

FIGURE 7-12

(a) Element in pure shear and (b) principal stresses

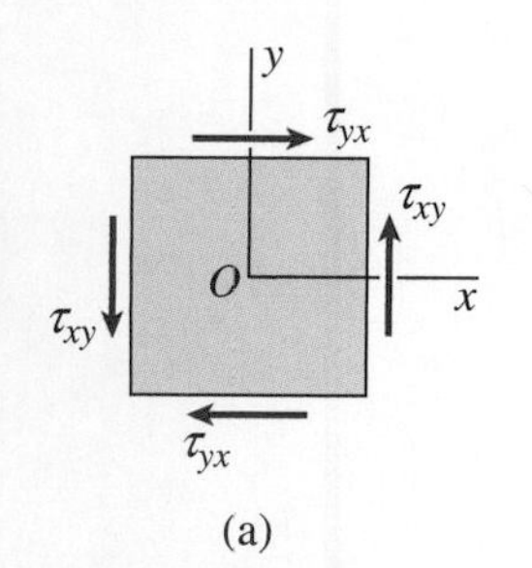

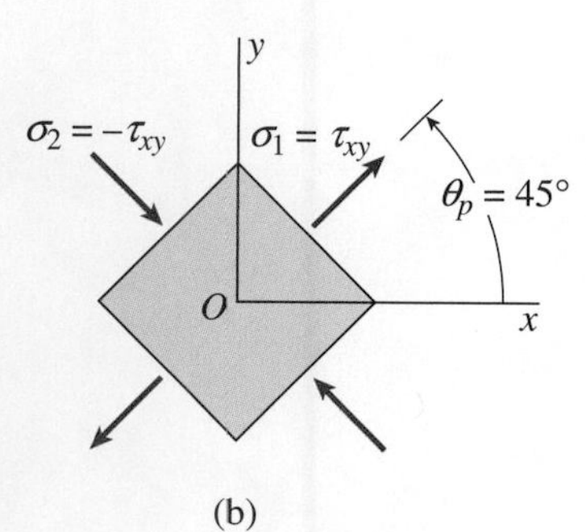

The Third Principal Stress

The preceding discussion of principal stresses refers only to rotation of axes in the x-y plane, that is, rotation about the z axis (Fig. 7-13a). Therefore, the two principal stresses determined from Eq. (7-17) are called the **in-plane principal stresses**. However, the stress element is actually three-dimensional and has three (not two) principal stresses acting on three mutually perpendicular planes.

A more complete three-dimensional analysis shows that the three principal planes for a plane-stress element are the two principal planes already described plus the z face of the element. These principal planes are shown in Fig. 7-13b, where a stress element has been oriented at the principal angle θ_{p1}, which corresponds to the principal stress σ_1. The principal stresses σ_1 and σ_2 are given by Eq. (7-17), and the third principal stress (σ_3) equals zero.

By definition, σ_1 is algebraically larger than σ_2, but σ_3 may be algebraically larger than, between, or smaller than σ_1 and σ_2. Of course, it is also possible

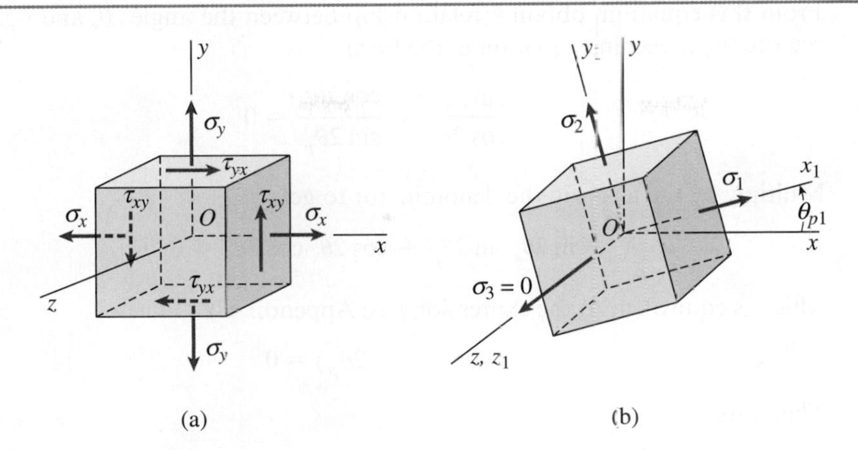

FIGURE 7-13

Elements in plane stress: (a) original element and (b) element oriented to the three principal planes and three principal stresses

for some or all of the principal stresses to be equal. Note again that there are no shear stresses on any of the principal planes.[1]

Maximum Shear Stresses

Having found the principal stresses and their directions for an element in plane stress, now consider the determination of the maximum shear stresses and the planes on which they act. The shear stresses τ_{x1y1} acting on inclined planes are given by the second transformation equation [Eq. (7-4b)]. Take the derivative of τ_{x1y1} with respect to θ and set it equal to zero to obtain

$$\frac{d\tau_{x1y1}}{d\theta} = -(\sigma_x - \sigma_y)\cos 2\theta - 2\tau_{xy}\sin 2\theta = 0 \tag{7-19}$$

from which

$$\tan 2\theta_s = -\frac{\sigma_x - \sigma_y}{2\tau_{xy}} \tag{7-20}$$

The subscript s indicates that the angle θ_s defines the orientation of the planes of maximum positive and negative shear stresses.

Equation (7-20) yields one value of θ_s between 0 and 90° and another between 90° and 180°. Furthermore, these two values differ by 90°; therefore, the maximum shear stresses occur on perpendicular planes. Because shear stresses on perpendicular planes are equal in absolute value, the maximum positive and negative shear stresses differ only in sign.

Comparing Eq. (7-20) for θ_s with Eq. (7-11) for θ_p shows that

$$\tan 2\theta_s = -\frac{1}{\tan 2\theta_p} = -\cot 2\theta_p \tag{7-21}$$

[1]The determination of principal stresses is an example of a type of mathematical analysis known as *eigenvalue analysis*, which is described in books on matrix algebra. The stress-transformation equations and the concept of principal stresses are due to the French mathematicians A. L. Cauchy (1789–1857) and Barré de Saint-Venant (1797–1886) and to the Scottish scientist and engineer W. J. M Rankine (1820–1872); see Refs. 7-1, 7-2, and 7-3, respectively.

From this equation, obtain a relationship between the angles θ_s and θ_p. First, rewrite the preceding equation in the form

$$\frac{\sin 2\theta_s}{\cos 2\theta_s} + \frac{\cos 2\theta_p}{\sin 2\theta_p} = 0$$

Multiply by the terms in the denominator to get

$$\sin 2\theta_s \sin 2\theta_p + \cos 2\theta_s \cos 2\theta_p = 0$$

which is equivalent to the expression (see Appendix C):

$$\cos(2\theta_s - 2\theta_p) = 0$$

Therefore,

$$2\theta_s - 2\theta_p = \pm 90°$$

and

$$\theta_s = \theta_p \pm 45° \tag{7-22}$$

This equation shows that *the planes of maximum shear stress occur at 45° to the principal planes.*

The plane of the maximum positive shear stress τ_{max} is defined by the angle θ_{s1}, for which the following equations apply:

$$\cos 2\theta_{s1} = \frac{\tau_{xy}}{R} \quad \sin 2\theta_{s1} = -\frac{\sigma_x - \sigma_y}{2R} \tag{7-23a,b}$$

in which R is given by Eq. (7-12). Also, the angle θ_{s1} is related to the angle θ_{p1} [see Eqs. (7-18a and b)] as

$$\theta_{s1} = \theta_{p1} - 45° \tag{7-24}$$

The corresponding maximum shear stress is obtained by substituting the expressions for $\cos 2\theta_{s1}$ and $\sin 2\theta_{s1}$ into the second transformation equation [Eq. (7-4b)], yielding

$$\tau_{max} = \sqrt{\left(\frac{\sigma_x - \sigma_y}{2}\right)^2 + \tau_{xy}^2} \tag{7-25}$$

The maximum negative shear stress τ_{min} has the same magnitude but opposite sign.

Another expression for the maximum shear stress can be obtained from the principal stresses σ_1 and σ_2, both of which are given by Eq. (7-17). Subtract the expression for σ_2 from that for σ_1, and then compare with Eq. (7-25) to get

$$\tau_{max} = \frac{\sigma_1 - \sigma_2}{2} \tag{7-26}$$

Thus, *the maximum shear stress is equal to one-half the difference of the principal stresses.*

The planes of maximum shear stress also contain normal stresses. The **normal stress** acting on the planes of maximum positive shear stress can be determined by substituting the expressions for the angle θ_{s1} [Eqs. (7-23a and b)] into the equation for σ_{x1} [Eq. (7-4a)]. The resulting stress is equal to the average of the normal stresses on the x and y planes:

$$\sigma_{\text{aver}} = \frac{\sigma_x + \sigma_y}{2} \tag{7-27}$$

This same normal stress acts on the planes of maximum negative shear stress.

In the particular cases of **uniaxial stress** and **biaxial stress** (Fig. 7-11), the planes of maximum shear stress occur at 45° to the x and y axes. In the case of **pure shear** (Fig. 7-12), the maximum shear stresses occur on the x and y planes.

In-Plane and Out-of-Plane Shear Stresses

The preceding analysis of shear stresses dealt only with **in-plane shear stresses**, that is, stresses acting in the x-y plane. To obtain the maximum in-plane shear stresses [Eqs. (7-25) and (7-26)], elements were considered that were obtained by rotating the xyz axes about the z axis, which is a principal axis (Fig. 7-13a). The result was that maximum shear stresses occur on planes at 45° to the principal planes. The principal planes for the element of Fig. 7-13a are shown in Fig. 7-13b, where σ_1 and σ_2 are the principal stresses. Therefore, the maximum in-plane shear stresses are found on an element obtained by rotating the $x_1y_1z_1$ axes (Fig. 7-13b) about the z_1 axis through an angle of 45°. These stresses are given by Eq. (7-25) or Eq. (7-26).

Maximum shear stresses also can be obtained by 45° rotations about the other two principal axes (the x_1 and y_1 axes in Fig. 7-13b). As a result, three sets of **maximum positive** and **maximum negative shear stresses** are found [compare with Eq. (7-26)]:

$$(\tau_{\max})_{x1} = \pm\frac{\sigma_2}{2} \qquad (\tau_{\max})_{y1} = \pm\frac{\sigma_1}{2} \qquad (\tau_{\max})_{z1} = \pm\frac{\sigma_1 - \sigma_2}{2} \tag{7-28a,b,c}$$

in which the subscripts indicate the principal axes about which the 45° rotations take place. The stresses obtained by rotations about the x_1 and y_1 axes are called **out-of-plane shear stresses**.

The algebraic values of σ_1 and σ_2 determine which of the preceding expressions gives the numerically largest shear stress. If σ_1 and σ_2 have the same sign, then one of the first two expressions is numerically largest; if they have opposite signs, the last expression is largest.

FIGURE 7-13 (Repeated)

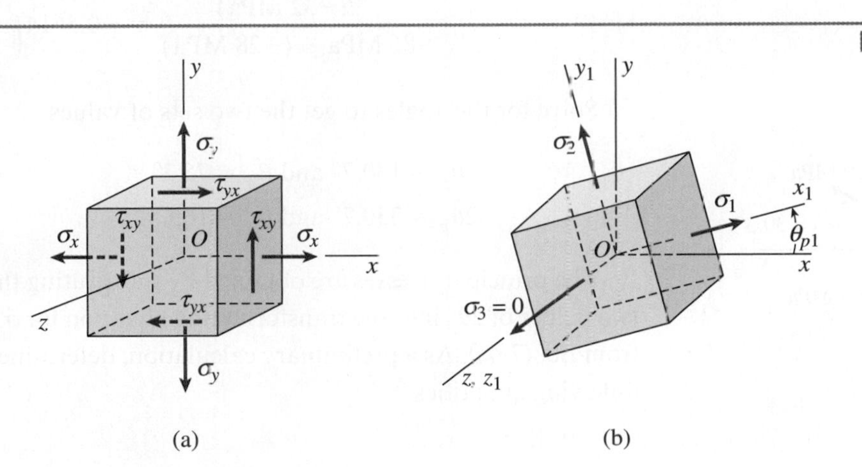

Example 7-3

FIGURE 7-14

Example 7-3: (a) Beam structure, (b) element at C in plane stress, (c) principal stresses, and (d) maximum shear stresses

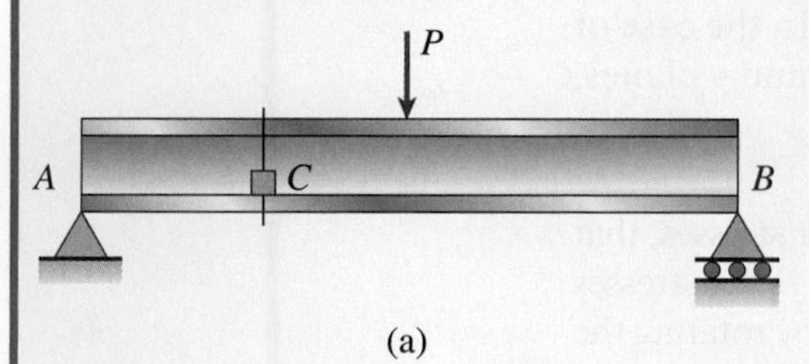

(a)

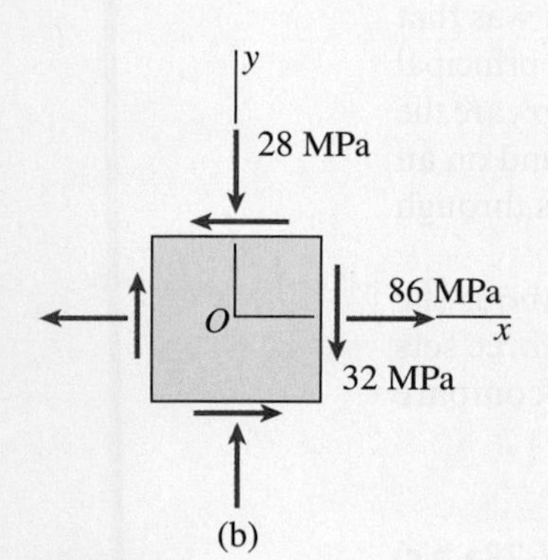

(b)

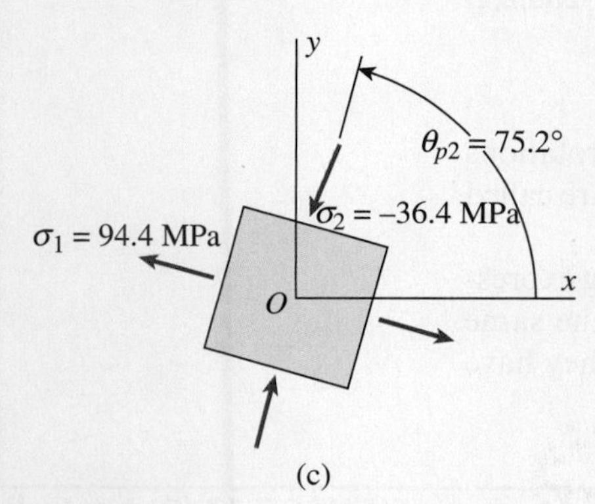

(c)

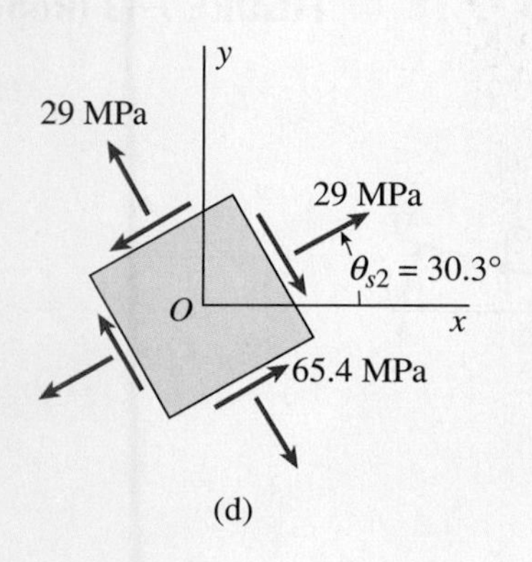

(d)

A simply supported, wide-flange beam has a concentrated load P applied at mid-span (Fig. 7-14a). The state of stress in the beam web at element C is known (Fig. 7-14b) to be $\sigma_x = 86$ MPa, $\sigma_y = -28$ MPa, and $\tau_{xy} = -32$ MPa.

(a) Determine the principal stresses and show them on a sketch of a properly oriented element.

(b) Determine the maximum shear stresses and show them on a sketch of a properly oriented element. (Consider only the in-plane stresses.)

Solution:

Use a four-step problem-solving approach. Combine steps as needed for an efficient solution.

Part (a): Principal stresses.

1, 2. *Conceptualize, Categorize*: The normal and shear stresses for the stress element at C (Fig. 7-14b) are aligned with the axis of the beam. These stresses were computed using the flexure and shear formulas presented in Sections 5.5 and 5.10, respectively. First, find the orientation of principal axes; then use stress-transformation formulas to find the principal normal stresses. Next, find the orientation angle for the plane on which maximum shear stress acts and the value of that shear stress using Eq. (7-25).

3. *Analyze*: The principal angles θ_p that locate the principal planes can be obtained from Eq. (7-11):

$$\tan 2\theta_p = \frac{2\tau_{xy}}{\sigma_x - \sigma_y}$$

$$= \frac{2(-32 \text{ MPa})}{86 \text{ MPa} - (-28 \text{ MPa})} = -0.5614$$

Solve for the angles to get the two sets of values

$$2\theta_p = 150.7° \text{ and } \theta_p = 75.3°$$
$$2\theta_p = 330.7° \text{ and } \theta_p = 165.3°$$

The principal stresses are obtained by substituting the two values of $2\theta_p$ into the transformation equation for σ_{x1} from Eq. (7-4a). As a preliminary calculation, determine the following quantities:

$$\frac{\sigma_x + \sigma_y}{2} = \frac{86 \text{ MPa} - 28 \text{ MPa}}{2} = 29 \text{ MPa}$$

$$\frac{\sigma_x - \sigma_y}{2} = \frac{86 \text{ MPa} + 28 \text{ MPa}}{2} = 57 \text{ MPa}$$

Now substitute the first value of $2\theta_p$ into Eq. (7-4a) and obtain

$$\sigma_{x1} = \frac{\sigma_x + \sigma_y}{2} + \frac{\sigma_x - \sigma_y}{2} \cos 2\theta + \tau_{xy} \sin 2\theta$$

$$= 29 \text{ MPa} + (57 \text{ MPa})(\cos 150.7°) - (32 \text{ MPa})(\sin 150.7°)$$

$$= -36.4 \text{ MPa}$$

In a similar manner, substitute the second value of $2\theta_p$ and obtain $\sigma_{x1} = 94.4$ MPa. Thus, the principal stresses and their corresponding principal angles are

$$\sigma_1 = 94.4 \text{ MPa} \quad \text{and} \quad \theta_{p1} = 165.3°$$

$$\sigma_2 = -36.4 \text{ MPa} \quad \text{and} \quad \theta_{p2} = 75.3°$$

4. ***Finalize***: Note that θ_{p1} and θ_{p2} differ by 90° and that $\sigma_1 + \sigma_2 = \sigma_x + \sigma_y$.

The principal stresses are shown on a properly oriented element in Fig. 7-14c. Of course, no shear stresses act on the principal planes.

Alternative solution for the principal stresses: The principal stresses also may be calculated directly from Eq. (7-17):

$$\sigma_{1,2} = \frac{\sigma_x + \sigma_y}{2} \pm \sqrt{\left(\frac{\sigma_x - \sigma_y}{2}\right)^2 + \tau_{xy}^2}$$

$$= 29 \text{ MPa} \pm \sqrt{(57 \text{ MPa})^2 + (-32 \text{ MPa})^2}$$

$$\sigma_{1,2} = 29 \text{ MPa} \pm 65.4 \text{ MPa}$$

Therefore,

$$\sigma_1 = 94.4 \text{ MPa} \quad \sigma_2 = -36.4 \text{ MPa}$$

The angle θ_{p1} to the plane on which σ_1 acts is obtained from Eqs. (7-18a and b):

$$\cos 2\theta_{p1} = \frac{\sigma_x - \sigma_y}{2R} = \frac{57 \text{ MPa}}{65.4 \text{ MPa}} = 0.872$$

$$\sin 2\theta_{p1} = \frac{\tau_{xy}}{R} = \frac{-32 \text{ MPa}}{65.4 \text{ MPa}} = -0.489$$

in which R is given by Eq. (7-12) and is equal to the square-root term in the preceding calculation for the principal stresses σ_1 and σ_2.

The only angle between 0 and 360° having the specified sine and cosine is $2\theta_{p1} = 330.7°$, hence, $\theta_{p1} = 165.3°$. This angle is associated with the algebraically larger principal stress $\sigma_1 = 94.4$ MPa. The other angle is 90° larger or smaller than θ_{p1}; hence, $\theta_{p2} = 75.3°$. This angle corresponds to the smaller principal stress $\sigma_2 = -36.4$ MPa. Note that these results for the principal stresses and principal angles agree with those found previously.

FIGURE 7-14 (Repeated)

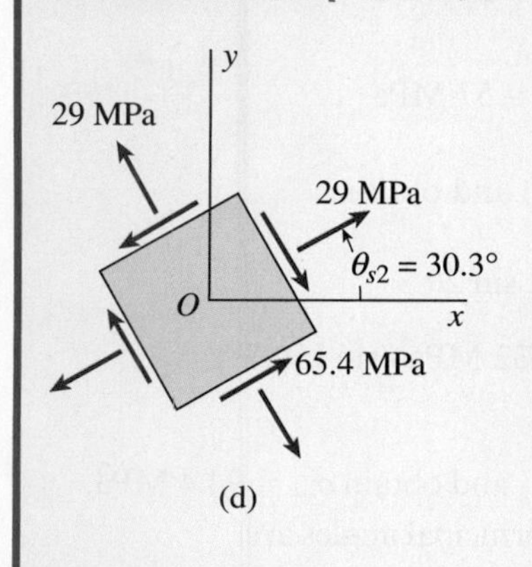

Part (b): Maximum shear stresses.

2, 3. ***Categorize, Analyze***: The maximum in-plane shear stresses are given by Eq. (7-25):

$$\tau_{max} = \sqrt{\left(\frac{\sigma_x - \sigma_y}{2}\right)^2 + \tau_{xy}^2}$$

$$= \sqrt{(57\text{ MPa})^2 + (-32\text{ MPa})^2} = 65.4\text{ MPa}$$

The angle θ_{s1} to the plane having the maximum positive shear stress is calculated from Eq. (7-24):

$$\theta_{s1} = \theta_{p1} - 45° = 165.3° - 45° = 120.3°$$

It follows that the maximum negative shear stress acts on the plane for which $\theta_{s2} = 120.3° - 90° = 30.3°$.

The normal stresses acting on the planes of maximum shear stresses are calculated from Eq. (7-27):

$$\sigma_{aver} = \frac{\sigma_x + \sigma_y}{2} = 29\text{ MPa}$$

4. ***Finalize***: Finally, the maximum shear stresses and associated normal stresses are shown on the stress element of Fig. 7-14d.

As an alternative approach to finding the maximum shear stresses, use Eq. (7-20) to determine the two values of the angles θ_s, and then use the second transformation equation (Eq. 7-4b) to obtain the corresponding shear stresses.

7.4 Mohr's Circle for Plane Stress

The transformation equations for plane stress can be represented in graphical form by a plot known as **Mohr's circle**. This graphical representation is extremely useful because it displays the relationships between the normal and shear stresses acting on various inclined planes at a point in a stressed body. It also provides a means for calculating principal stresses, maximum shear stresses, and stresses on inclined planes. Furthermore, Mohr's circle is valid not only for stresses but also for other quantities of a similar mathematical nature, including strains and moments of inertia.[2]

Equations of Mohr's Circle

The equations of Mohr's circle can be derived from the transformation equations for plane stress in Eqs. (7-4a and b). The two equations are repeated here, but with a slight rearrangement of the first equation:

$$\sigma_{x1} - \frac{\sigma_x + \sigma_y}{2} = \frac{\sigma_x - \sigma_y}{2}\cos 2\theta + \tau_{xy}\sin 2\theta \qquad \textbf{(7-29a)}$$

[2]Mohr's circle is named after the famous German civil engineer Otto Christian Mohr (1835–1918), who developed the circle in 1882 (Ref. 7-4).

$$\tau_{x1y1} = -\frac{\sigma_x - \sigma_y}{2} \sin 2\theta + \tau_{xy} \cos 2\theta \tag{7-29b}$$

From analytic geometry, you might recognize that these two equations are the equations of a circle in parametric form. The angle 2θ is the parameter and the stresses σ_{x1} and τ_{x1y1} are the coordinates. However, it is not necessary to recognize the nature of the equations at this stage—if you eliminate the parameter, the significance of the equations will become apparent.

To eliminate the parameter 2θ, square both sides of each equation and then add the two equations. The equation that results is

$$\left(\sigma_{x1} - \frac{\sigma_x + \sigma_y}{2}\right)^2 + \tau_{x1y1}^2 = \left(\frac{\sigma_x - \sigma_y}{2}\right)^2 + \tau_{xy}^2 \tag{7-30}$$

This equation can be written in simpler form by using the following notation from Section 7.3 [see Eqs. (7-27) and (7-12), respectively]:

$$\sigma_{\text{aver}} = \frac{\sigma_x + \sigma_y}{2} \qquad R = \sqrt{\left(\frac{\sigma_x - \sigma_y}{2}\right)^2 + \tau_{xy}^2} \tag{7-31a,b}$$

Equation (7-30) now becomes

$$(\sigma_{x1} - \sigma_{\text{aver}})^2 + \tau_{x1y1}^2 = R^2 \tag{7-32}$$

which is the equation of a circle in standard algebraic form. The coordinates are σ_{x1} and τ_{x1y1}, the radius is R, and the center of the circle has coordinates $\sigma_{x1} = \sigma_{\text{aver}}$ and $\tau_{x1y1} = 0$.

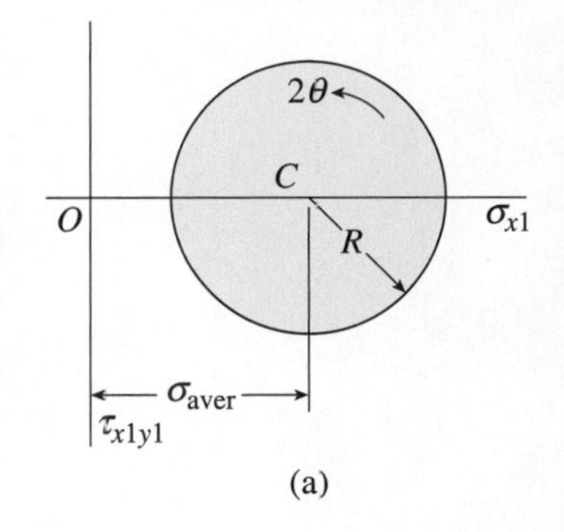

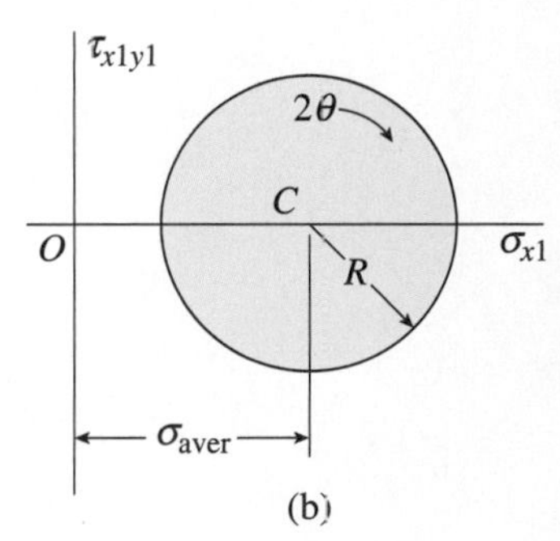

FIGURE 7-15

Two forms of Mohr's circle: (a) τ_{x1y1} is positive downward and the angle 2θ is positive counterclockwise and (b) τ_{x1y1} positive upward and the angle 2θ is positive clockwise (*Note:* The first form is used in this book)

Two Forms of Mohr's Circle

Mohr's circle can be plotted from Eqs. (7-29) and (7-32) in either of two forms. In the first form of Mohr's circle, plot the normal stress σ_{x1} positive to the right and the shear stress τ_{x1y1} positive downward, as shown in Fig. 7-15a. The advantage of plotting shear stresses positive downward is that the angle 2θ on Mohr's circle will be positive when counterclockwise, which agrees with the positive direction of 2θ in the derivation of the transformation equations (see Figs. 7-1 and 7-2).

In the second form of Mohr's circle, τ_{x1y1} is plotted positive upward but the angle 2θ is now positive clockwise (Fig. 7-15b), which is opposite to its usual positive direction.

Both forms of Mohr's circle are mathematically correct, and either one can be used. However, it is easier to visualize the orientation of the stress element if the positive direction of the angle 2θ is the same in Mohr's circle as it is for the element itself. Furthermore, a counterclockwise rotation agrees with the customary right-hand rule for rotation.

Therefore, the first form of Mohr's circle (Fig. 7-15a) is used here, where *a positive shear stress is plotted downward and a positive angle 2θ is plotted counterclockwise.*

Construction of Mohr's Circle

Mohr's circle can be constructed in a variety of ways, depending upon which stresses are known and which are to be found. To show the basic properties of the circle, assume that you know the stresses σ_x, σ_y, and τ_{xy} acting on the x and y planes of an element in plane stress (Fig. 7-16a). This information is sufficient

FIGURE 7-16

Construction of Mohr's circle for plane stress

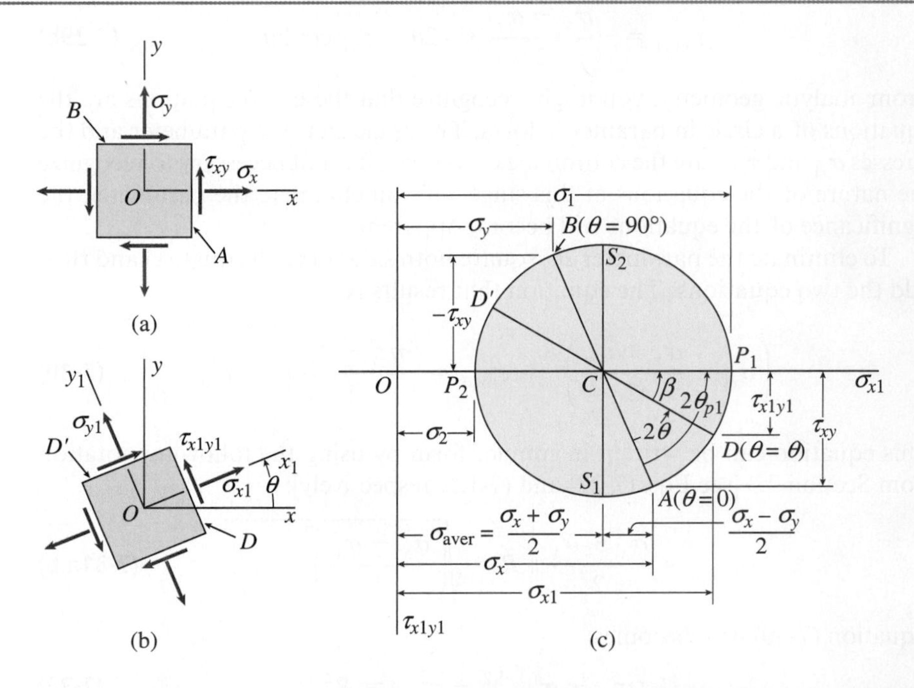

to construct the circle. Then, with the circle drawn, determine the stresses σ_{x1}, σ_{y1}, and τ_{x1y1} acting on an inclined element (Fig. 7-16b). You can also obtain the principal stresses and maximum shear stresses from the circle.

With σ_x, σ_y, and τ_{xy} known, the **procedure for constructing Mohr's circle** is as follows (see Fig. 7-16c).

1. Draw a set of coordinate axes with σ_{x1} as abscissa (positive to the right) and τ_{x1y1} as ordinate (positive downward).
2. Locate the center C of the circle at the point having coordinates and $\sigma_{x1} = \sigma_{\text{aver}}$ and $\tau_{x1y1} = 0$ [see Eqs. (7-31a and b)].
3. Locate point A, representing the stress conditions on the x face of the element shown in Fig. 7-16a, by plotting its coordinates $\sigma_{x1} = \sigma_x$ and $\tau_{x1y1} = \tau_{xy}$. Note that point A on the circle corresponds to $\theta = 0$. Also, note that the x face of the element (Fig. 7-16a) is labeled "A" to show its correspondence with point A on the circle.
4. Locate point B, representing the stress conditions on the y face of the element shown in Fig. 7-16a, by plotting its coordinates $\sigma_{x1} = \sigma_y$ and $\tau_{x1y1} = -\tau_{xy}$. Note that point B on the circle corresponds to $\theta = 90°$. In addition, the y face of the element (Fig. 7-16a) is labeled "B" to show its correspondence with point B on the circle.
5. Draw a line from point A to point B. This line is a diameter of the circle and passes through the center C. Points A and B, representing the stresses on planes at 90° to each other (Fig. 7-16a), are at opposite ends of the diameter (and therefore are 180° apart on the circle).
6. Using point C as the center, draw Mohr's circle through points A and B. The circle drawn in this manner has radius R (Eq. 7-31b), as shown in the next paragraph.

With the circle drawn, verify by geometry that lines CA and CB are radii and have lengths equal to R. Note that the abscissas of points C and

A are $(\sigma_x + \sigma_y)/2$ and σ_x, respectively. The difference in these abscissas is $(\sigma_x - \sigma_y)/2$, as dimensioned in the figure. Also, the ordinate to point A is τ_{xy}. Therefore, line CA is the hypotenuse of a right triangle having one side of length $(\sigma_x - \sigma_y)/2$ and the other side of length τ_{xy}. Taking the square root of the sum of the squares of these two sides gives the radius R:

$$R = \sqrt{\left(\frac{\sigma_x - \sigma_y}{2}\right)^2 + \tau_{xy}^2}$$

which is the same as Eq. (7-31b). By a similar procedure, you can show that the length of line CB is also equal to the radius R of the circle.

Stresses on an Inclined Element

Now consider the stresses σ_{x1}, σ_{y1}, and τ_{x1y1} acting on the faces of a plane-stress element oriented at an angle θ from the x axis (Fig. 7-16b). If the angle θ is known, these stresses can be determined from Mohr's circle. The procedure is as follows.

On the circle (Fig. 7-16c), measure an angle 2θ counterclockwise from radius CA, because point A corresponds to $\theta = 0$ and is the reference point from which angles are measured. The angle 2θ locates point D on the circle, which (as shown in the next paragraph) has coordinates σ_{x1} and τ_{x1y1}. Therefore, point D represents the stresses on the x_1 face of the element of Fig. 7-16b. Consequently, this face of the element is labeled "D" in Fig. 7-16b.

Note that an angle 2θ on Mohr's circle corresponds to an angle θ on a stress element. For instance, point D on the circle is at an angle 2θ from point A, but the x_1 face of the element shown in Fig. 7-16b (the face labeled "D") is at an angle θ from the x face of the element shown in Fig. 7-16a (the face labeled "A"). Similarly, points A and B are 180° apart on the circle, but the corresponding faces of the element (Fig. 7-16a) are 90° apart.

To show that the coordinates σ_{x1} and τ_{x1y1} of point D on the circle are indeed given by the stress-transformation equations given in Eqs. (7-4a and b), again use the geometry of the circle. Let β be the angle between the radial line CD and the σ_{x1} axis. Then, from the geometry of the figure, the following expressions are obtained for the coordinates of point D:

$$\sigma_{x1} = \frac{\sigma_x + \sigma_y}{2} + R\cos\beta \qquad \tau_{x1y1} = R\sin\beta \tag{7-33a,b}$$

Note that the angle between the radius CA and the horizontal axis is $2\theta + \beta$, so

$$\cos(2\theta + \beta) = \frac{\sigma_x - \sigma_y}{2R} \qquad \sin(2\theta + \beta) = \frac{\tau_{xy}}{R}$$

Expanding the cosine and sine expressions (see Appendix C) gives

$$\cos 2\theta \cos\beta - \sin 2\theta \sin\beta = \frac{\sigma_x - \sigma_y}{2R} \tag{7-34a}$$

$$\sin 2\theta \cos\beta + \cos 2\theta \sin\beta = \frac{\tau_{xy}}{R} \tag{7-34b}$$

Multiply the first of these equations by $\cos 2\theta$ and the second by $\sin 2\theta$ and then add to obtain

$$\cos\beta = \frac{1}{R}\left(\frac{\sigma_x - \sigma_y}{2}\cos 2\theta + \tau_{xy}\sin 2\theta\right) \tag{7-34c}$$

FIGURE 7-16 (Repeated)

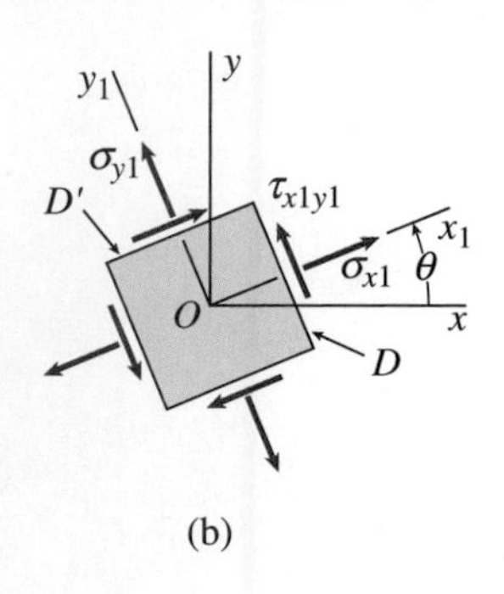

(b)

Also, multiply Eq. (7-34a) by $\sin 2\theta$ and Eq. (7-34b) by $\cos 2\theta$ and then subtract to get

$$\sin\beta = \frac{1}{R}\left(\frac{\sigma_x - \sigma_y}{2}\sin 2\theta + \tau_{xy}\cos 2\theta\right) \qquad \textbf{(7-34d)}$$

Substitute these expressions for $\cos\beta$ and $\sin\beta$ into Eqs. (7-33a and b) to obtain the stress-transformation equations for σ_{x1} and τ_{x1y1} [Eqs. (7-4a and b)]. Thus, point D on Mohr's circle, defined by the angle 2θ, represents the stress conditions on the x_1 face of the stress element defined by the angle θ (Fig. 7-16b).

Point D', which is diametrically opposite point D on the circle, is located by an angle 2θ (measured from line CA) that is 180° greater than the angle 2θ to point D. Therefore, point D' on the circle represents the stresses on a face of the stress element (Fig. 7-16b) at 90° from the face represented by point D. Thus, point D' on the circle gives the stresses σ_{y1} and $-\tau_{x1y1}$ on the y_1 face of the stress element (the face labeled "D'" in Fig. 7-16b).

The stresses represented by points on Mohr's circle are related to the stresses acting on an element. The stresses on an inclined plane defined by the angle θ (Fig. 7-16b) are found on the circle at the point where the angle from the reference point (point A) is 2θ. Thus, as the x_1y_1 axes rotate counterclockwise through an angle θ (Fig. 7-16b), the point on Mohr's circle corresponding to the x_1 face moves counterclockwise through an angle 2θ. Similarly, if the axes rotate clockwise through an angle, the point on the circle moves clockwise through an angle twice as large.

Principal Stresses

The determination of principal stresses is probably the most important application of Mohr's circle. Note that as you move around Mohr's circle (Fig. 7-16c), you encounter point P_1 where the normal stress reaches its algebraically largest value and the shear stress is zero. Hence, point P_1 represents a **principal stress** and a **principal plane**. The abscissa σ_1 of point P_1 gives the algebraically larger principal stress, and its angle $2\theta_{p1}$ from the reference point A (where $\theta = 0$) gives the orientation of the principal plane. The other principal plane, associated with the algebraically smallest normal stress, is represented by point P_2, which is diametrically opposite point P_1.

From the geometry of the circle, the algebraically larger principal stress is

$$\sigma_1 = \overline{OC} + \overline{CP_1} = \frac{\sigma_x + \sigma_y}{2} + R$$

which, upon substitution of the expression for R [Eq. (7-31b)], agrees with the earlier equation for this stress [Eq. (7-14)]. In a similar manner, you can verify the expression for the algebraically smaller principal stress σ_2.

The principal angle θ_{p1} between the x axis (Fig. 7-16a) and the plane of the algebraically larger principal stress is one-half the angle $2\theta_{p1}$, which is the angle on Mohr's circle between radii CA and CP_1. The cosine and sine of the angle $2\theta_{p1}$ can be obtained by inspection from the circle:

$$\cos 2\theta_{p1} = \frac{\sigma_x - \sigma_y}{2R} \qquad \sin 2\theta_{p1} = \frac{\tau_{xy}}{R}$$

These equations agree with Eqs. (7-18a and b), and once again the geometry of the circle matches the equations derived earlier. On the circle, the angle $2\theta_{p2}$ to the other principal point (point P_2) is 180° larger than $2\theta_{p1}$; hence, $\theta_{p2} = \theta_{p1} + 90°$, as expected.

Maximum Shear Stresses

Points S_1 and S_2, representing the planes of maximum positive and maximum negative shear stresses, respectively, are located at the bottom and top of Mohr's circle (Fig. 7-16c). These points are at angles $2\theta = 90°$ from points P_1 and P_2, which agrees with the fact that the planes of maximum shear stress are oriented at 45° to the principal planes.

The maximum shear stresses are numerically equal to the radius R of the circle [compare Eq. (7-31b) for R with Eq. (7-25) for τ_{max}]. Also, the normal stresses on the planes of maximum shear stress are equal to the abscissa of point C, which is the average normal stress σ_{aver} [see Eq. (7-31a)].

Alternative Sign Convention for Shear Stresses

An alternative sign convention for shear stresses is sometimes used when constructing Mohr's circle. In this convention, the direction of a shear stress acting on an element of the material is indicated by the sense of the rotation that it tends to produce (Figs. 7-17a and b). If the shear stress τ tends to rotate the stress element clockwise, it is called a *clockwise shear stress*, and if it tends to rotate it counterclockwise, it is called a *counterclockwise stress*. Then, when constructing Mohr's circle, clockwise shear stresses are plotted upward and counterclockwise shear stresses are plotted downward (Fig. 7-17c).

It is important to realize that *the alternative sign convention produces a circle that is identical to the circle already described* (Fig. 7-16c). The reason is that a positive shear stress τ_{x1y1} is also a counterclockwise shear stress, and both are plotted downward. Also, a negative shear stress τ_{x1y1} is a clockwise shear stress, and both are plotted upward.

Thus, the alternative sign convention merely provides a different point of view. Instead of thinking of the vertical axis as having negative shear stresses plotted upward and positive shear stresses plotted downward (which is a bit awkward), you can think of the vertical axis as having clockwise shear stresses plotted upward and counterclockwise shear stresses plotted downward (Fig. 7-17c).

FIGURE 7-17

Alternative sign convention for shear stresses: (a) clockwise shear stress, (b) counterclockwise shear stress, and (c) axes for Mohr's circle (Note that clockwise shear stresses are plotted upward and counterclockwise shear stresses are plotted downward)

(a)

(b)

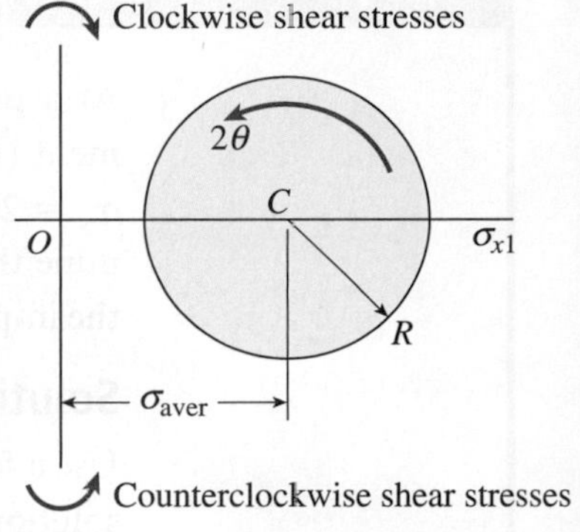

(c)

General Comments about the Circle

From the preceding discussions in this section, it is apparent that the stresses acting on any inclined plane, as well as the principal stresses and maximum shear stresses, can be found from Mohr's circle. However, only rotations of axes in the x-y plane (that is, rotations about the z axis) are considered; therefore, *all stresses on Mohr's circle are in-plane stresses.*

For convenience, the circle of Fig. 7-16 was drawn with σ_x, σ_y, and τ_{xy} as positive stresses, but the same procedures may be followed if one or more of the stresses is negative. If one of the normal stresses is negative, part or all of the circle will be located to the left of the origin, as illustrated in Example 7-6 that follows.

Point A in Fig. 7-16c, representing the stresses on the plane $\theta = 0$, may be situated anywhere around the circle. However, the angle 2θ is always measured counterclockwise from the radius CA, regardless of where point A is located.

In the special cases of *uniaxial stress*, *biaxial stress*, and *pure shear*, the construction of Mohr's circle is simpler than in the general case of plane stress. These special cases are illustrated in Example 7-4 and in Problems 7.4-1 through 7.4-9.

Besides using Mohr's circle to obtain the stresses on inclined planes when the stresses on the x and y planes are known, you can also use the circle in the opposite manner. If you know the stresses σ_{x1}, σ_{y1}, and τ_{x1y1} acting on an inclined element oriented at a known angle θ, you can easily construct the circle and determine the stresses σ_x, σ_y, and τ_{xy} for the angle $\theta = 0$. The procedure is to locate points D and D' from the known stresses and then draw the circle using line DD' as a diameter. Measure the angle 2θ in a negative sense from radius CD to locate point A, corresponding to the x face of the element. Then locate point B by constructing a diameter from A. Finally, determine the coordinates of points A and B, and thereby obtain the stresses acting on the element for which $\theta = 0$.

If desired, construct Mohr's circle to scale and measure values of stress from the drawing. However, it is usually preferable to obtain the stresses by numerical calculations—either directly from the various equations or by using trigonometry and the geometry of the circle.

Mohr's circle makes it possible to visualize the relationships between stresses acting on planes at various angles, and it also serves as a simple memory device for calculating stresses. Although many graphical techniques are no longer used in engineering work, Mohr's circle remains valuable because it provides a simple and clear picture of an otherwise complicated analysis.

Mohr's circle is also applicable to the transformations for plane strain and moments of inertia of plane areas because these quantities follow the same transformation laws as do stresses (see Section 7.7, and Sections D.7, and D.8 in Appendix D).

Example 7-4

At a point on the surface of a hydraulic ram on a piece of construction equipment (Fig. 7-18a), the material is subjected to biaxial stresses $\sigma_x = 90$ MPa and $\sigma_y = 20$ MPa, as shown on the stress element of Fig. 7-18b. Using Mohr's circle, determine the stresses acting on an element inclined at an angle $\theta = 30°$. (Consider only the in-plane stresses, and show the results on a sketch of a properly oriented element.)

Solution:

Use a four-step problem-solving approach. Combine steps as needed for an efficient solution.

1, 2. ***Conceptualize, Categorize:***

Construction of Mohr's circle: Begin by setting up the axes for the normal and shear stresses, with σ_{x1} positive to the right and τ_{x1y1} positive downward, as shown in Fig. 7-18c. Then place the center C of the circle on the σ_{x1} axis at the point where the stress equals the average normal stress [Eq. (7-31a)]:

$$\sigma_{\text{aver}} = \frac{\sigma_x + \sigma_y}{2} = \frac{90 \text{ MPa} + 20 \text{ MPa}}{2} = 55 \text{ MPa}$$

Point A, representing the stresses on the x face of the element ($\theta = 0$), has coordinates

$$\sigma_{x1} = 90 \text{ MPa} \qquad \tau_{x1y1} = 0$$

Similarly, the coordinates of point B, representing the stresses on the y face ($\theta = 90°$), are

$$\sigma_{x1} = 20 \text{ MPa} \qquad \tau_{x1y1} = 0$$

FIGURE 7-18

Example 7-4: (a) Hydraulic cylinder on construction equipment, (b) element on hydraulic ram in plane stress, and (c) the corresponding Mohr's circle (*Note:* All stresses on the circle have units of MPa)

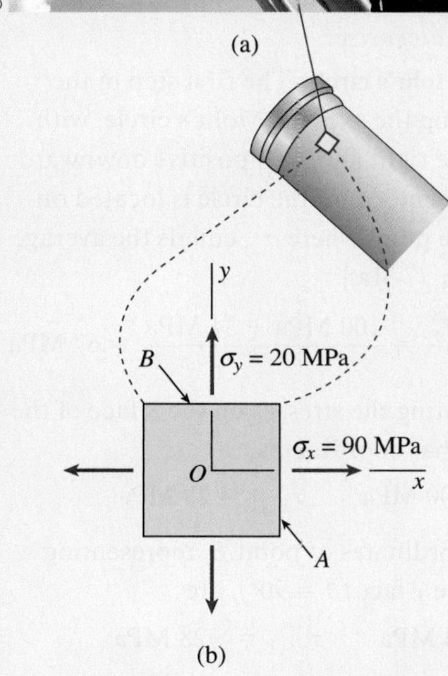

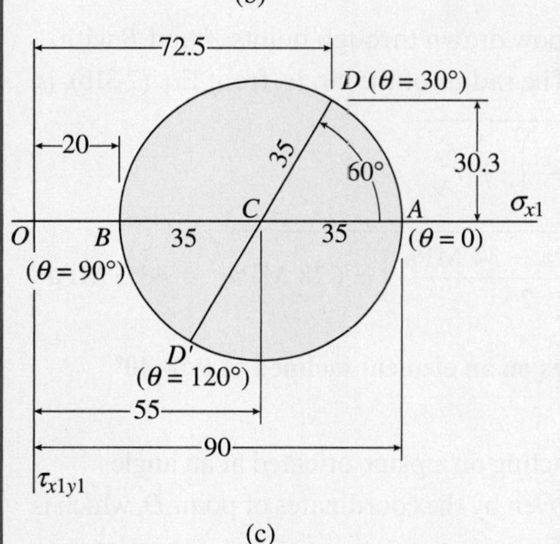

Now draw the circle through points A and B with the center at C and radius R [see Eq. (7-31b)] equal to

$$R = \sqrt{\left(\frac{\sigma_x - \sigma_y}{2}\right)^2 + \tau_{xy}^2} = \sqrt{\left(\frac{90 \text{ MPa} - 20 \text{ MPa}}{2}\right)^2 + 0}$$

$$= 35 \text{ MPa}$$

3. *Analyze:*

Stresses on an element inclined at $\theta = 30°$: The stresses acting on a plane oriented at an angle $\theta = 30°$ are given by the coordinates of point D, which is at an angle $2\theta = 60°$ from point A (Fig. 7-18c). By inspection of the circle, note that the coordinates of point D are

$$\text{(Point } D) \quad \sigma_{x1} = \sigma_{\text{aver}} + R\cos 60°$$
$$= 55 \text{ MPa} + (35 \text{ MPa})(\cos 60°) = 72.5 \text{ MPa}$$
$$\tau_{x1y1} = -R\sin 60° = -(35 \text{ MPa})(\sin 60°)$$
$$= -30.3 \text{ MPa}$$

In a similar manner, find the stresses represented by point D', which corresponds to an angle $\theta = 120°$ (or $2\theta = 240°$):

$$\text{(Point } D') \quad \sigma_{x1} = \sigma_{\text{aver}} - R\cos 60°$$
$$= 55 \text{ MPa} - (35 \text{ MPa})(\cos 60°) = 37.5 \text{ MPa}$$
$$\tau_{x1y1} = R\sin 60° = (35 \text{ MPa})(\sin 60°) = 30.3 \text{ MPa}$$

4. **Finalize:** These results are shown in Fig. 7-19 on a sketch of an element oriented at an angle $\theta = 30°$ with all stresses shown in their true directions. Note that the sum of the normal stresses on the inclined element is equal to $\sigma_x + \sigma_y$ or 110 MPa.

FIGURE 7-19

Example 7-4: Stresses acting on an element oriented at an angle $\theta = 30°$

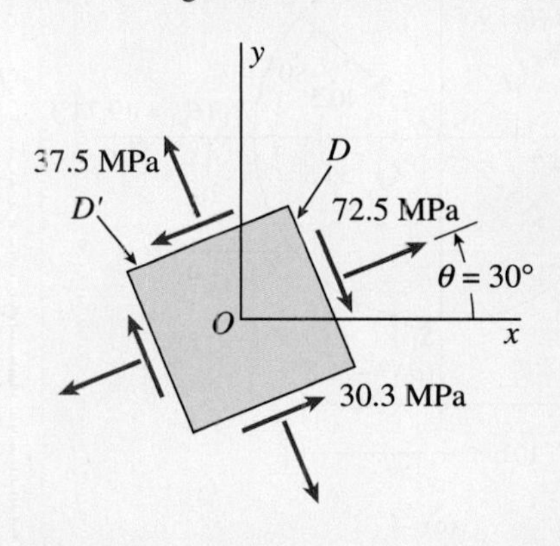

Example 7-5

FIGURE 7-20

Example 7-5: (a) Oil drilling pumps, (b) element in plane stress, and (c) the corresponding Mohr's circle (*Note:* All stresses on the circle have units of MPa)

(a)

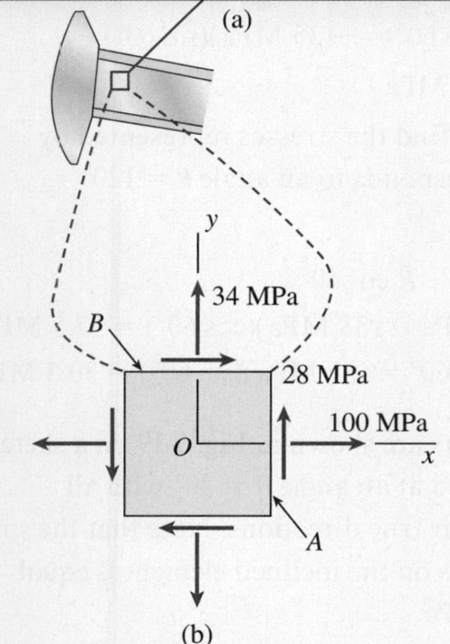

(b)

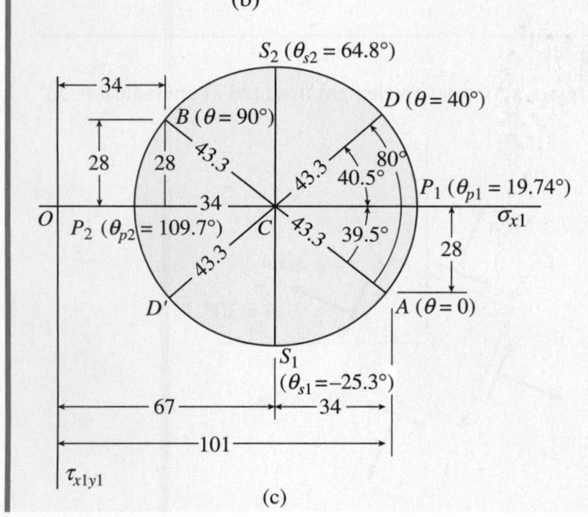

(c)

An element in plane stress on the surface of an oil-drilling pump arm (Fig. 7-20a) is subjected to stresses $\sigma_x = 100$ MPa, $\sigma_y = 34$ MPa, and $\tau_{xy} = 28$ MPa, as shown in Fig. 7-20b.

Using Mohr's circle, determine (a) the stresses acting on an element inclined at an angle $\theta = 40°$, (b) the principal stresses, and (c) the maximum shear stresses. (Consider only the in-plane stresses, and show all results on sketches of properly oriented elements.)

Solution:

Use a four-step problem-solving approach. Combine steps as needed for an efficient solution.

1, 2. ***Conceptualize, Categorize:***

Construction of Mohr's circle: The first step in the solution is to set up the axes for Mohr's circle, with σ_{x1} positive to the right and τ_{x1y1} positive downward (Fig. 7-20c). The center C of the circle is located on the σ_{x1} axis at the point where σ_{x1} equals the average normal stress [Eq. (7-31a)]:

$$\sigma_{\text{aver}} = \frac{\sigma_x + \sigma_y}{2} = \frac{100\text{ MPa} + 34\text{ MPa}}{2} = 67\text{ MPa}$$

Point A, representing the stresses on the x face of the element ($\theta = 0$), has coordinates

$$\sigma_{x1} = 100\text{ MPa} \qquad \tau_{x1y1} = 28\text{ MPa}$$

Similarly, the coordinates of point B, representing the stresses on the y face ($\theta = 90°$), are

$$\sigma_{x1} = 34\text{ MPa} \qquad \tau_{x1y1} = -28\text{ MPa}$$

The circle is now drawn through points A and B with center at C. The radius of the circle, from Eq. (7-31b), is

$$R = \sqrt{\left(\frac{\sigma_x - \sigma_y}{2}\right)^2 + \tau_{xy}^2}$$

$$= \sqrt{\left(\frac{101\text{ MPa} - 34\text{ MPa}}{2}\right)^2 + (28\text{ MPa})^2} = 43.3\text{ MPa}$$

Part (a): Stresses on an element inclined at $\boldsymbol{\theta = 40°}$.

3. ***Analyze:***

The stresses acting on a plane oriented at an angle $\theta = 40°$ are given by the coordinates of point D, which is

FIGURE 7-21

Example 7-5: (a) Stresses acting on an element oriented at $\theta = 40°$, (b) principal stresses, and (c) maximum shear stresses

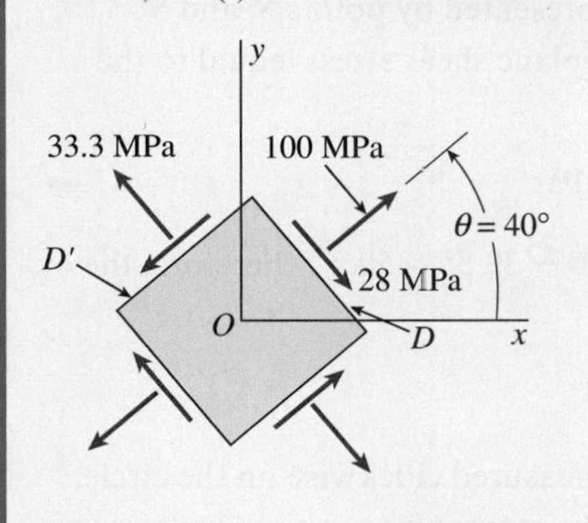

(a)

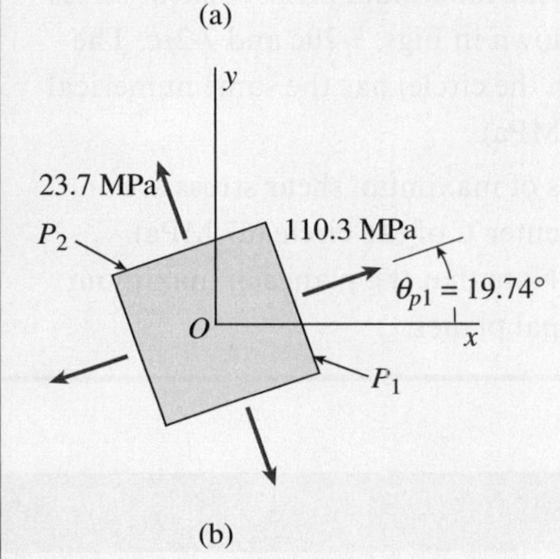

(b)

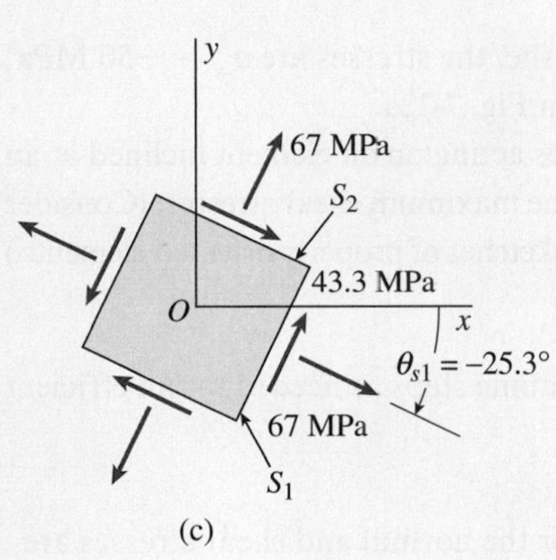

(c)

at an angle $2\theta = 80°$ from point A (Fig. 7-20c). To evaluate these coordinates, you need to know the angle between line CD and the σ_{x1} axis (that is, angle DCP_1), which in turn requires that you know the angle between line CA and the σ_{x1} axis (angle ACP_1). These angles are found from the geometry of the circle, as

$$\tan \overline{ACP_1} = \frac{28 \text{ MPa}}{34 \text{ MPa}} = 0.824 \qquad \overline{ACP_1} = 39.5°$$

$$\overline{DCP_1} = 80° - \overline{ACP_1} = 80° - 39.5° = 40.5°$$

Knowing these angles, determine the coordinates of point D directly from the Fig. 7-20c:

$$(\text{Point } D) \qquad \sigma_{x1} = 67 \text{ MPa} + (43.3 \text{ MPa})(\cos 40.5°)$$
$$= 100 \text{ MPa}$$
$$\tau_{x1y1} = -(43.3 \text{ MPa})(\sin 40.5°) = -28.1 \text{ MPa}$$

In an analogous manner, find the stresses represented by point D', which corresponds to a plane inclined at an angle $\theta = 130°$ (or $2\theta = 260°$):

$$(\text{Point } D') \quad \sigma_{x1} = 67 \text{ MPa} - (43.3 \text{ MPa})(\cos 40.5°)$$
$$= 33.3 \text{ MPa}$$
$$\tau_{x1y1} = (43.3 \text{ MPa})(\sin 40.5°) = 28 \text{ MPa}$$

4. ***Finalize***: These stresses are shown in Fig. 7-21a on a sketch of an element oriented at an angle $\theta = 40°$ (all stresses are shown in their true directions). Also, note that the sum of the normal stresses is equal to $\sigma_x + \sigma_y$ or 134 MPa.

Part (b): Principal stresses.

3. ***Analyze***: The principal stresses are represented by points P_1 and P_2 on Mohr's circle (Fig. 7-20c). The algebraically larger principal stress (point P_1) is

$$\sigma_1 = 67 \text{ MPa} + 43.3 \text{ MPa} = 110.3 \text{ MPa}$$

as seen by inspection of the circle. The angle $2\theta_{p1}$ to point P_1 from point A is the angle ACP_1 on the circle, that is,

$$\overline{ACP_1} = 2\theta_{p1} = 39.5° \qquad \theta_{p1} = 19.74°$$

Thus, the plane of the algebraically larger principal stress is oriented at an angle $\theta_{p1} = 19.74°$, as shown in Fig. 7-21b.

The algebraically smaller principal stress (represented by point P_2) is obtained from the circle in a similar manner:

$$\sigma_2 = 67 \text{ MPa} - 43.3 \text{ MPa} = 23.7 \text{ MPa}$$

4. ***Finalize***: The angle $2\theta_{p2}$ to point P_2 on the circle is $39.5° + 180° = 219.5°$; thus, the second principal plane is defined by the angle $\theta_{p2} = 109.7°$.

The principal stresses and principal planes are shown in Fig. 7-21b. Note that the sum of the normal stresses is equal to 134 MPa.

Part (c): Maximum shear stresses.

3. ***Analyze***: The maximum shear stresses are represented by points S_1 and S_2 on Mohr's circle; therefore, the maximum in-plane shear stress (equal to the radius of the circle) is

$$\tau_{max} = 43.3 \text{ MPa}$$

The angle ACS_1 from point A to point S_1 is $90° - 39.5° = 50.5°$; therefore, the angle $2\theta_{s1}$ for point S_1 is

$$2\theta_{s1} = -50.5°$$

4. ***Finalize***: This angle is negative because it is measured clockwise on the circle. The corresponding angle θ_{s1} to the plane of the maximum positive shear stress is one-half that value, or $\theta_{s1} = -25.3°$, as shown in Figs. 7-20c and 7-21c. The maximum negative shear stress (point S_2 on the circle) has the same numerical value as the maximum positive stress (43.3 MPa).

The normal stresses acting on the planes of maximum shear stress are equal to σ_{aver}, which is the abscissa of the center C of the circle (67 MPa). These stresses are also shown in Fig. 7-21c. Note that the planes of maximum shear stress are oriented at 45° to the principal planes.

Example 7-6

At a point on the surface of a metal-working lathe, the stresses are $\sigma_x = -50$ MPa, $\sigma_y = 10$ MPa, and $\tau_{xy} = -40$ MPa, as shown in Fig. 7-22a.

Using Mohr's circle, determine (a) the stresses acting on an element inclined at an angle $\theta = 45°$, (b) the principal stresses, and (c) the maximum shear stresses. (Consider only the in-plane stresses, and show all results on sketches of properly oriented elements.)

Solution:

Use a four-step problem-solving approach. Combine steps as needed for an efficient solution.

1, 2. ***Conceptualize, Categorize***:

Construction of Mohr's circle: The axes for the normal and shear stresses are shown in Fig. 7-22b, with σ_{x1} positive to the right and τ_{x1y1} positive downward. The center C of the circle is located on the σ_{x1} axis at the point where the stress equals the average normal stress [Eq. (7-31a)]:

$$\sigma_{aver} = \frac{\sigma_x + \sigma_y}{2} = \frac{-50 \text{ MPa} + 10 \text{ MPa}}{2} = -20 \text{ MPa}$$

FIGURE 7-22

Example 7-6: (a) Element in plane stress and (b) the corresponding Mohr's circle (*Note*: All stresses on the circle have units of MPa)

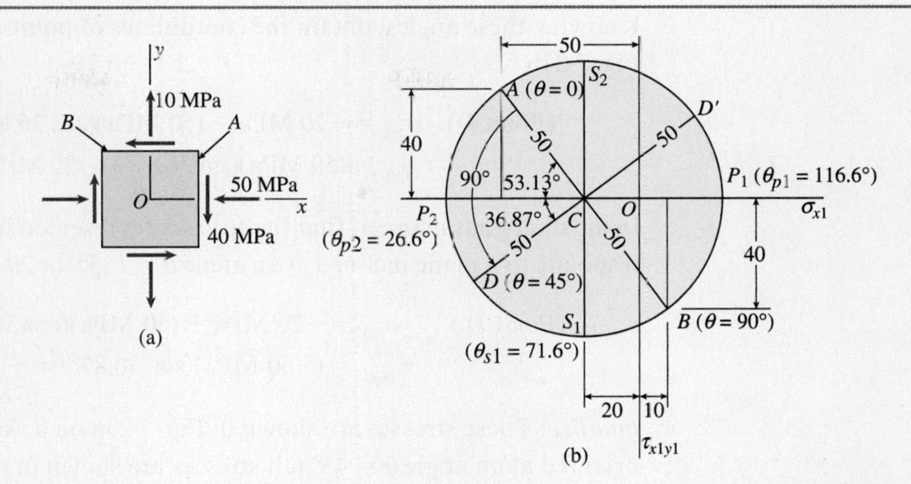

Point A, representing the stresses on the x face of the element ($\theta = 0$), has coordinates

$$\sigma_{x1} = -50 \text{ MPa} \qquad \tau_{x1y1} = -40 \text{ MPa}$$

Similarly, the coordinates of point B, representing the stresses on the y face ($\theta = 90°$), are

$$\sigma_{x1} = 10 \text{ MPa} \qquad \tau_{x1y1} = 40 \text{ MPa}$$

The circle is now drawn through points A and B with center at C and radius R equal to [from Eq. (7-31b)]

$$R = \sqrt{\left(\frac{\sigma_x - \sigma_y}{2}\right)^2 + \tau_{xy}^2}$$

$$= \sqrt{\left(\frac{-50 \text{ MPa} - 10 \text{ MPa}}{2}\right)^2 + (-40 \text{ MPa})^2} = 50 \text{ MPa}$$

Part (a): Stresses on an element inclined at $\theta = 45°$.

3. *Analyze*: The stresses acting on a plane oriented at an angle $\theta = 45°$ are given by the coordinates of point D, which is at an angle $2\theta = 90°$ from point A (Fig. 7-22b). To evaluate these coordinates, you need to know the angle between line CD and the negative σ_{x1} axis (that is, angle DCP_2), which in turn requires that you know the angle between line CA and the negative σ_{x1} axis (angle ACP_2). These angles are found from the geometry of the circle

$$\tan \overline{ACP_2} = \frac{40 \text{ MPa}}{30 \text{ MPa}} = \frac{4}{3} \qquad \overline{ACP_2} = 53.13°$$

$$\overline{DCP_2} = 90° - \overline{ACP_2} = 90° - 53.13° = 36.87°$$

Knowing these angles, obtain the coordinates of point D directly from Fig. 7-22b:

$$\text{(Point } D) \quad \sigma_{x1} = -20 \text{ MPa} - (50 \text{ MPa})(\cos 36.87°) = -60 \text{ MPa}$$
$$\tau_{x1y1} = (50 \text{ MPa})(\sin 36.87°) = 30 \text{ MPa}$$

In an analogous manner, find the stresses represented by point D', which corresponds to a plane inclined at an angle $\theta = 135°$ (or $2\theta = 270°$):

$$\text{(Point } D') \quad \sigma_{x1} = -20 \text{ MPa} + (50 \text{ MPa})(\cos 36.87°) = 20 \text{ MPa}$$
$$\tau_{x1y1} = (-50 \text{ MPa})(\sin 36.87°) = -30 \text{ MPa}$$

4. ***Finalize***: These stresses are shown in Fig. 7-23a on a sketch of an element oriented at an angle $\theta = 45°$ (all stresses are shown in their true directions). Also, note that the sum of the normal stresses is equal to $\sigma_x + \sigma_y$ or −40 MPa.

FIGURE 7-23

Example 7-6: (a) Stresses acting on an element oriented at $\theta = 45°$, (b) principal stresses, and (c) maximum shear stresses

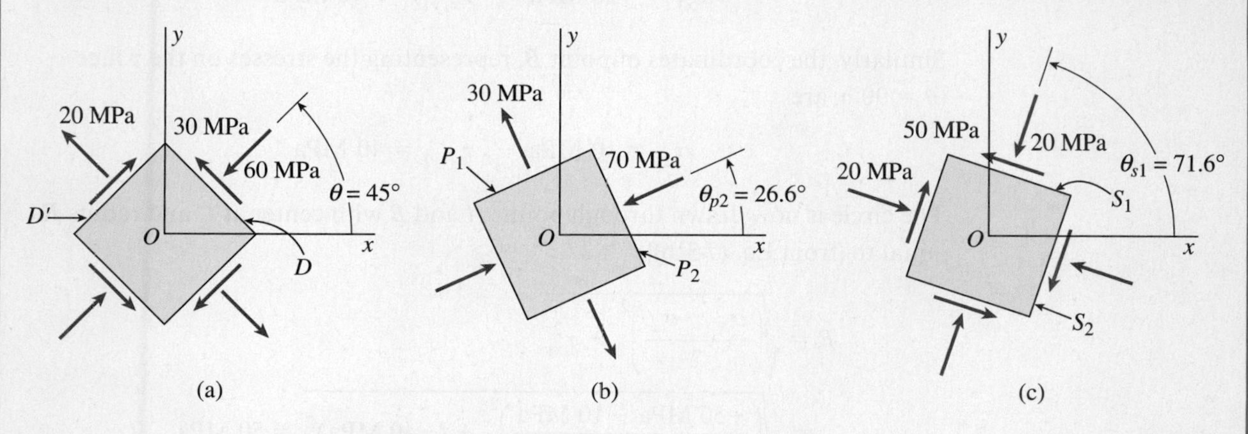

Part (b): Principal stresses.

3. ***Analyze***: The principal stresses are represented by points P_1 and P_2 on Mohr's circle (Fig. 7-22b). The algebraically larger principal stress (represented by point P_1) is

$$\sigma_1 = -20 \text{ MPa} + 50 \text{ MPa} = 30 \text{ MPa}$$

as seen by inspection of the circle. The angle $2\theta_{p1}$ to point P_1 from point A is the angle ACP_1 measured counterclockwise on the circle, that is,

$$\overline{ACP_1} = 2\theta_{p1} = 53.13° + 180° = 233.13° \quad \theta_{p1} = 116.6°$$

Thus, the plane of the algebraically larger principal stress is oriented at an angle $\theta_{p1} = 116.16°$.

The algebraically smaller principal stress (point P_2) is obtained from the circle in a similar manner:

$$\sigma_2 = -20 \text{ MPa} - 50 \text{ MPa} = -70 \text{ MPa}$$

The angle $2\theta_{p2}$ to point P_2 on the circle is 53.13°; thus, the second principal plane is defined by the angle $\theta_{p2} = 26.6°$.

4. ***Finalize***: The principal stresses and principal planes are shown in Fig. 7-23b. Note that the sum of the normal stresses is equal to $\sigma_x + \sigma_y$ or −40 MPa.

Part (c): Maximum shear stresses.

3. ***Analyze***: The maximum positive and negative shear stresses are represented by points S_1 and S_2 on Mohr's circle (Fig. 7-22b). Their magnitudes, equal to the radius of the circle, are

$$\tau_{\max} = 50 \text{ MPa}$$

The angle ACS_1 from point A to point S_1 is 90° + 53.13° = 143.13°; therefore, the angle $2\theta_{s1}$ for point S_1 is

$$2\theta_{s1} = 143.13°$$

4. ***Finalize***: The corresponding angle θ_{s1} to the plane of the maximum positive shear stress is one-half that value, or $\theta_{s1} = 71.6°$, as shown in Fig. 7-23c. The maximum negative shear stress (point S_2 on the circle) has the same numerical value as the positive stress (50 MPa).

The normal stresses acting on the planes of maximum shear stress are equal to σ_{aver}, which is the coordinate of the center C of the circle (−20 MPa). These stresses are also shown in Fig. 7-23c. Note that the planes of maximum shear stress are oriented at 45° to the principal planes.

7.5 Hooke's Law for Plane Stress

The stresses acting on inclined planes when the material is subjected to plane stress (Fig. 7-24) were discussed in Sections 7.2, 7.3, and 7.4. The stress-transformation equations derived in those discussions were obtained solely from equilibrium, so the properties of the materials were not needed. Now, in this section, the *strains* in the material are investigated, which means that the material properties must be considered. However, the discussion is limited to materials that meet two important conditions: First, *the material is uniform throughout the body and has the same properties in all directions* (homogeneous and isotropic material), and second, *the material follows Hooke's law* (linearly elastic material). Under these conditions, the relationships between the stresses and strains in the body are readily obtained.

FIGURE 7-24

Element of material in plane stress ($\sigma_z = 0$)

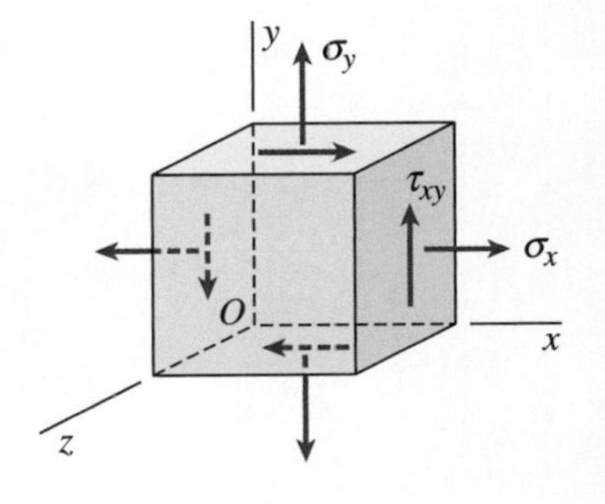

FIGURE 7-25

Element of material subjected to normal strains ε_x, ε_y, and ε_z

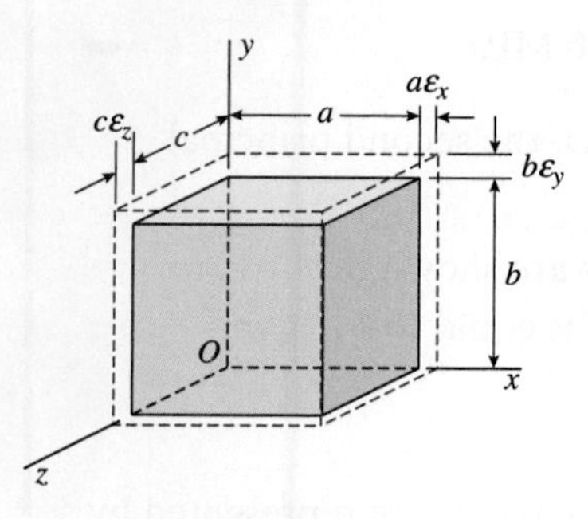

Begin by considering the **normal strains** ε_x, ε_y, and ε_z in plane stress. The effects of these strains are pictured in Fig. 7-25, which shows the changes in dimensions of a small element having edges of lengths a, b, and c. All three strains are shown positive (elongation) in the figure. The strains can be expressed in terms of the stresses (Fig. 7-24) by superimposing the effects of the individual stresses.

For instance, the strain ε_x in the x direction due to the stress σ_x is equal to σ_x/E, where E is the modulus of elasticity. Also, the strain ε_x due to the stress σ_y is equal to $-\nu\sigma_y/E$, where ν is Poisson's ratio (see Section 1.7). Of course, the shear stress τ_{xy} produces no normal strains in the x, y, or z directions. Thus, the resultant strain in the x direction is

$$\varepsilon_x = \frac{1}{E}(\sigma_x - \nu\sigma_y) \qquad \textbf{(7-35a)}$$

In a similar manner, the strains in the y and z directions are obtained:

$$\varepsilon_y = \frac{1}{E}(\sigma_y - \nu\sigma_x) \qquad \varepsilon_z = -\frac{\nu}{E}(\sigma_x + \sigma_y) \qquad \textbf{(7-35b,c)}$$

These equations may be used to find the normal strains (in plane stress) when the stresses are known.

FIGURE 7-26

Shear strain γ_{xy}

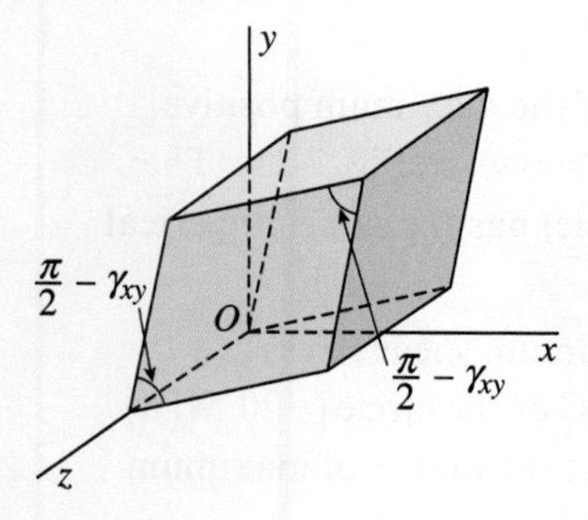

The shear stress τ_{xy} (Fig. 7-24) causes a distortion of the element such that each z face becomes a rhombus (Fig. 7-26). The **shear strain** γ_{xy} is the *decrease* in angle between the x and y faces of the element and is related to the shear stress by Hooke's law in shear, as

$$\gamma_{xy} = \frac{\tau_{xy}}{G} \qquad \textbf{(7-36)}$$

where G is the shear modulus of elasticity. Note that the normal stresses σ_x and σ_y have no effect on the shear strain γ_{xy}. Consequently, Eqs. (7-35) and (7-36) give the strains (in plane stress) when all stresses (σ_x, σ_y, and τ_{xy}) act simultaneously.

The first two equations (Eqs. 7-35a and b) give the strains ε_x and ε_y in terms of the stresses. These equations can be solved simultaneously for the stresses in terms of the strains:

$$\sigma_x = \frac{E}{1-\nu^2}(\varepsilon_x + \nu\varepsilon_y) \qquad \sigma_y = \frac{E}{1-\nu^2}(\varepsilon_y + \nu\varepsilon_x) \qquad \textbf{(7-37a,b)}$$

In addition, the following equation expresses the shear stress in terms of the shear strain:

$$\tau_{xy} = G\gamma_{xy} \qquad \textbf{(7-38)}$$

Equations (7-37) and (7-38) may be used to find the stresses (in plane stress) when the strains are known. Of course, the normal stress σ_z in the z direction is equal to zero.

Equations (7-35) through (7-38) are known collectively as **Hooke's law for plane stress**. They contain three material constants (E, G, and ν), but only two are independent because of the relationship

$$G = \frac{E}{2(1+\nu)} \qquad \textbf{(7-39)}$$

that was derived previously in Section 3.6.

FIGURE 7-11 (Repeated)

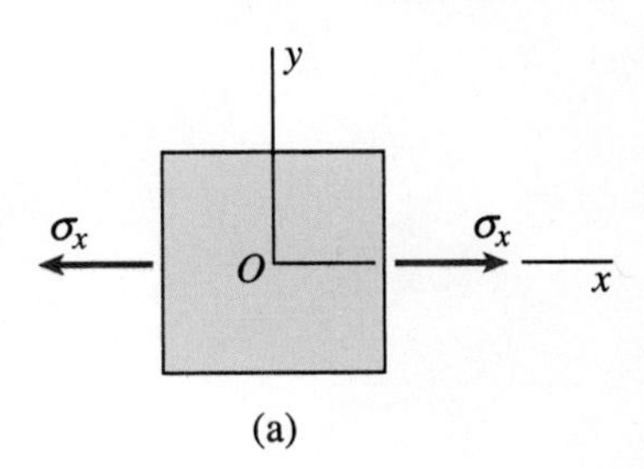

(a)

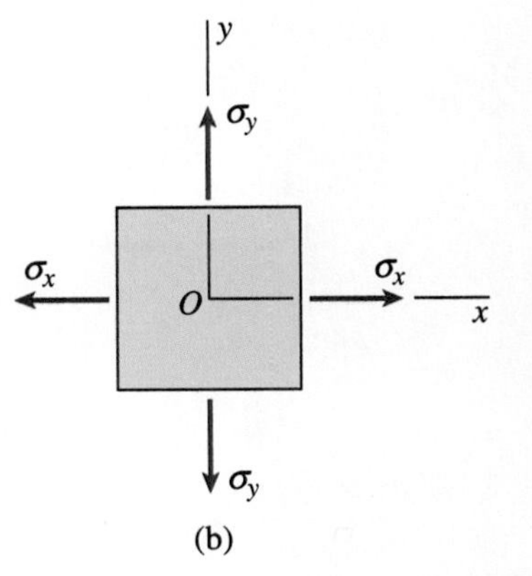

(b)

Special Cases of Hooke's Law

In the special case of **biaxial stress** (Fig. 7-11b), the shear stress $\tau_{xy} = 0$; therefore, Hooke's law for plane stress simplifies to

$$\varepsilon_x = \frac{1}{E}(\sigma_x - \nu\sigma_y) \quad \varepsilon_y = \frac{1}{E}(\sigma_y - \nu\sigma_x)$$

$$\varepsilon_z = -\frac{\nu}{E}(\sigma_x + \sigma_y) \quad \textbf{(7-40a,b,c)}$$

$$\sigma_x = \frac{E}{1-\nu^2}(\varepsilon_x + \nu\varepsilon_y) \quad \sigma_y = \frac{E}{1-\nu^2}(\varepsilon_y + \nu\varepsilon_x) \quad \textbf{(7-41a,b)}$$

These equations are the same as Eqs. (7-35) and (7-37) because the effects of normal and shear stresses are independent of each other.

For **uniaxial stress**, with $\sigma_y = 0$ (Fig. 7-11a), the equations of Hooke's law simplify even further:

$$\varepsilon_x = \frac{\sigma_x}{E} \quad \varepsilon_y = \varepsilon_z = -\frac{\nu\sigma_x}{E} \quad \sigma_x = E\varepsilon_x \quad \textbf{(7-42a,b,c)}$$

Finally, consider **pure shear** (Fig. 7-12a), which means that $\sigma_x = \sigma_y = 0$, so

$$\varepsilon_x = \varepsilon_y = \varepsilon_z = 0 \quad \gamma_{xy} = \frac{\tau_{xy}}{G} \quad \textbf{(7-43a,b)}$$

In all three of these special cases, the normal stress σ_z is equal to zero.

Volume Change

When a solid object undergoes strains, both its dimensions and its volume will change. The change in volume can be determined if the normal strains in three perpendicular directions are known. To show how this is accomplished, consider the small element of material shown in Fig. 7-25. The original element is a rectangular parallelepiped having sides of lengths a, b, and c in the x, y, and z directions, respectively. The strains ε_x, ε_y, and ε_z produce the changes in dimensions shown by the dashed lines. Thus, the increases in the lengths of the sides are $a\varepsilon_x$, $b\varepsilon_y$, and $c\varepsilon_z$.

The original volume of the element is

$$V_0 = abc \quad \textbf{(7-44a)}$$

and its final volume is

$$\begin{aligned} V_1 &= (a + a\varepsilon_x)(b + b\varepsilon_y)(c + c\varepsilon_z) \\ &= abc(1 + \varepsilon_x)(1 + \varepsilon_y)(1 + \varepsilon_z) \end{aligned} \quad \textbf{(7-44b)}$$

By referring to Eq. (7-44a), express the final volume of the element [Eq. (7-44b)] in the form

$$V_1 = V_0(1 + \varepsilon_x)(1 + \varepsilon_y)(1 + \varepsilon_z) \quad \textbf{(7-45a)}$$

Expand the terms on the right-hand side to obtain the following equivalent expression:

$$V_1 = V_0(1 + \varepsilon_x + \varepsilon_y + \varepsilon_z + \varepsilon_x\varepsilon_y + \varepsilon_x\varepsilon_z + \varepsilon_y\varepsilon_z + \varepsilon_x\varepsilon_y\varepsilon_z) \quad \textbf{(7-45b)}$$

The preceding equations for V_1 are valid for both large and small strains.

Limiting the discussion to structures having only very *small strains* (as is usually the case), you can disregard the terms in Eq. (7-45b) that consist of products of small strains. Such products are themselves small in comparison to the individual strains ε_x, ε_y, and ε_z. Then the expression for the final volume simplifies to

$$V_1 = V_0(1 + \varepsilon_x + \varepsilon_y + \varepsilon_z) \tag{7-46}$$

and the **volume change** is

$$\Delta V = V_1 - V_0 = V_0(\varepsilon_x + \varepsilon_y + \varepsilon_z) \tag{7-47}$$

This expression can be used for any volume of material *provided the strains are small and remain constant throughout the volume*. Note also that the material does not have to follow Hooke's law. Furthermore, the expression is not limited to plane stress, but is valid for any stress conditions. (As a final note, shear strains produce no change in volume.)

The **unit volume change** e, also known as the **dilatation**, is defined as the change in volume divided by the original volume; thus,

$$e = \frac{\Delta V}{V_0} = \varepsilon_x + \varepsilon_y + \varepsilon_z \tag{7-48}$$

By applying this equation to a differential element of volume and then integrating, you can obtain the change in volume of a body even when the normal strains vary throughout the body.

The preceding equations for volume changes apply to both tensile and compressive strains, inasmuch as the strains ε_x, ε_y, and ε_z are algebraic quantities (positive for elongation and negative for shortening). With this sign convention, positive values for ΔV and e represent increases in volume, and negative values represent decreases.

FIGURE 7-24 (Repeated)

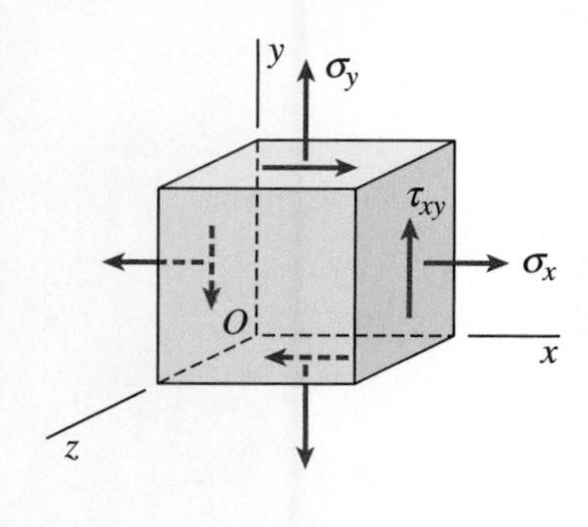

Now return to materials that follow **Hooke's law** and are subjected only to **plane stress** (Fig. 7-24). In this case, the strains ε_x, ε_y, and ε_z are given by Eqs. (7-35a, b, and c). Substitute those relationships into Eq. (7-48) to obtain the unit volume change in terms of stresses:

$$e = \frac{\Delta V}{V_0} = \frac{1 - 2\nu}{E}(\sigma_x + \sigma_y) \tag{7-49}$$

Note that this equation also applies to **biaxial stress**.

In the case of a prismatic bar in tension, that is, **uniaxial stress**, Eq. (7-49) simplifies to

$$e = \frac{\Delta V}{V_0} = \frac{\sigma_x}{E}(1 - 2\nu) \tag{7-50}$$

From this equation, note that the maximum possible value of Poisson's ratio for common materials is 0.5, because a larger value means that the volume decreases when the material is in tension, which is contrary to ordinary physical behavior.[3]

Strain-Energy Density in Plane Stress

The strain-energy density u is the strain energy stored in a unit volume of the material (see the discussions in Sections 2.7 and 3.9). For an element in plane stress, the strain-energy density is obtained by referring to the elements pictured

[3]See discussion in Section 1.7 about materials that have a negative Poisson's ratio.

in Figs. 7-25 and 7-26. Because the normal and shear strains occur independently, add the strain energies from these two elements to obtain the total energy.

Begin by finding the strain energy associated with the normal strains (Fig. 7-25). Since the stress acting on the x face of the element is σ_x (see Fig. 7-24), the force acting on the x face of the element (Fig. 7-25) is equal to $\sigma_x bc$. Of course, as the loads are applied to the structure, this force increases gradually from zero to its maximum value. At the same time, the x face of the element moves through the distance $a\varepsilon_x$. Therefore, the work done by this force is

FIGURE 7-25 (Repeated)

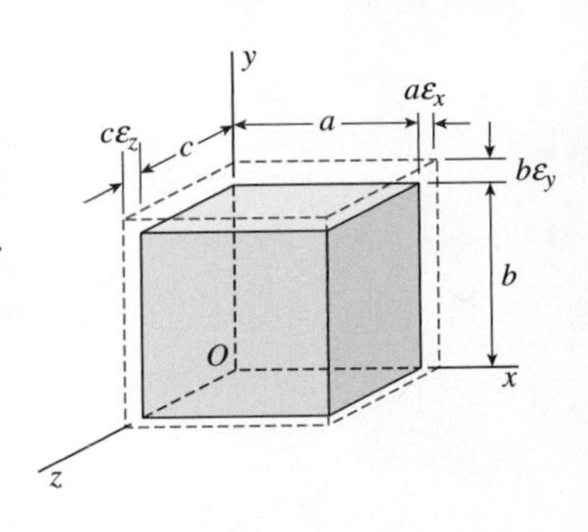

$$\frac{1}{2}(\sigma_x bc)(a\varepsilon_x)$$

provided Hooke's law holds for the material. Similarly, the force $\sigma_y ac$ acting on the y face does work equal to

$$\frac{1}{2}(\sigma_y ac)(b\varepsilon_y)$$

The sum of these two terms gives the strain energy stored in the element:

$$\frac{abc}{2}(\sigma_x\varepsilon_x + \sigma_y\varepsilon_y)$$

Thus, the strain-energy density (strain energy per unit volume) due to the normal stresses and strains is

$$u_1 = \frac{1}{2}(\sigma_x\varepsilon_x + \sigma_y\varepsilon_y) \quad \textbf{(7-51a)}$$

The strain-energy density associated with the shear strains (Fig. 7-26) was evaluated previously in Section 3.9 [see Eq. (3-62) of that section]:

$$u_2 = \frac{\tau_{xy}\gamma_{xy}}{2} \quad \textbf{(7-51b)}$$

FIGURE 7-26 (Repeated)

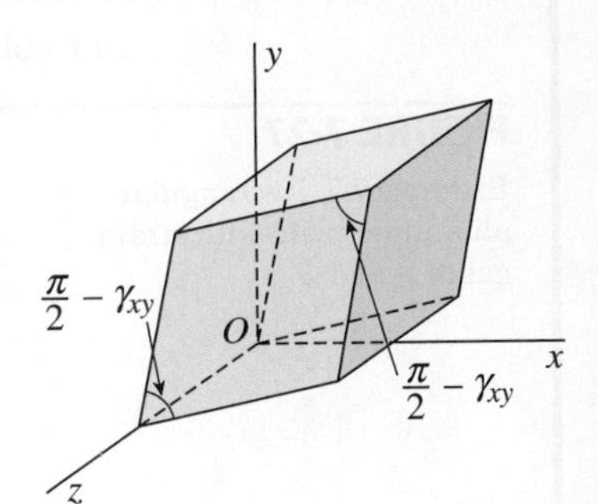

Combining the strain-energy densities for the normal and shear strains results in the following formula for the **strain-energy density in plane stress**:

$$u = \frac{1}{2}(\sigma_x\varepsilon_x + \sigma_y\varepsilon_y + \tau_{xy}\gamma_{xy}) \quad \textbf{(7-52)}$$

Substituting for the strains from Eqs. (7-35) and (7-36) gives the strain-energy density in terms of stresses alone:

$$u = \frac{1}{2E}(\sigma_x^2 + \sigma_y^2 - 2\nu\sigma_x\sigma_y) + \frac{\tau_{xy}^2}{2G} \quad \textbf{(7-53)}$$

In a similar manner, substitute for the stresses from Eqs. (7-37) and (7-38) and obtain the strain-energy density in terms of strains alone:

$$u = \frac{E}{2(1-\nu^2)}(\varepsilon_x^2 + \varepsilon_y^2 + 2\nu\varepsilon_x\varepsilon_y) + \frac{G\gamma_{xy}^2}{2} \quad \textbf{(7-54)}$$

To obtain the strain-energy density in the special case of **biaxial stress**, simply drop the shear terms in Eqs. (7-52), (7-53), and (7-54).

For the special case of **uniaxial stress**, substitute the following values

$$\sigma_y = 0 \quad \tau_{xy} = 0 \quad \varepsilon_y = -\nu\varepsilon_x \quad \gamma_{xy} = 0$$

into Eqs. (7-53) and (7-54) and obtain, respectively,

$$u = \frac{\sigma_x^2}{2E} \qquad u = \frac{E\varepsilon_x^2}{2} \qquad \textbf{(7-55a,b)}$$

These equations agree with Eqs. (2-44a) and (2-44b) of Section 2.7.

Also, for **pure shear** substitute

$$\sigma_x = \sigma_y = 0 \qquad \varepsilon_x = \varepsilon_y = 0$$

into Eqs. (7-53) and (7-54) and obtain

$$u = \frac{\tau_{xy}^2}{2G} \qquad u = \frac{G\gamma_{xy}^2}{2} \qquad \textbf{(7-56a,b)}$$

These equations agree with Eqs. (3-63a and b) of Section 3.9.

Example 7-7

Strain gages A and B (oriented in the x and y directions, respectively) are attached to a rectangular aluminum plate with a thickness of $t = 7$ mm. The plate is subjected to uniform normal stresses σ_x and σ_y, as shown in Fig. 7-27, and the gage readings for normal strains are $\varepsilon_x = -0.00075$ (shortening, gage A) and $\varepsilon_y = 0.00125$ (elongation, gage B). The modulus of elasticity is $E = 73$ GPa, and Poisson's ratio is $\nu = 0.33$. Find the stresses σ_x and σ_y and the change Δt in the thickness of the plate. Also, find the unit volume change (or dilatation) e and the strain energy density u for the plate.

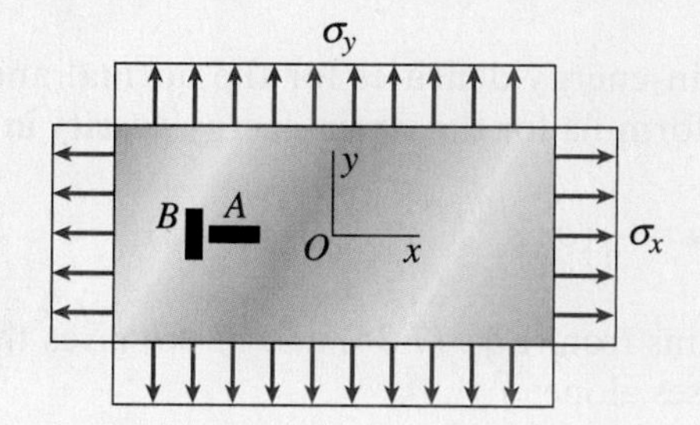

FIGURE 7-27
Example 7-7: Rectangular aluminum plate with strain gages A and B

Solution:

Use a four-step problem-solving approach. Combine steps as needed for an efficient solution.

1, 2, 3. ***Conceptualize, Categorize, Analyze***: For a plate in *biaxial stress*, use Eqs. (7-41a and b) to find the normal stresses σ_x and σ_y in the x and y directions, respectively, based upon the measured normal strains ε_x and ε_y:

$$\sigma_x = \frac{E}{1-\nu^2}(\varepsilon_x + \nu\varepsilon_y) = \frac{73\text{ GPa}}{1-0.33^2}[-0.00075 + (0.33)(0.00125)]$$
$$= -27.6\text{ MPa}$$

$$\sigma_y = \frac{E}{1-\nu^2}(\varepsilon_y + \nu\varepsilon_x) = \frac{73 \text{ GPa}}{1-0.33^2}[0.00125 + (0.33)(-0.00075)]$$
$$= 82.1 \text{ MPa}$$

The normal strain in the z direction is then computed from Eq. (7-40c) as

$$\varepsilon_z = \frac{-\nu}{E}(\sigma_x + \sigma_y) = \frac{-(0.33)}{73 \text{ GPa}}(-27.6 \text{ MPa} + 82.1 \text{ MPa})$$
$$= -2.464 \times 10^{-4}$$

The change (here a decrease) in the thickness of the plate is then

$$\Delta t = \varepsilon_z t = [-2.464(10^{-4})](7 \text{ mm}) = -1.725 \times 10^{-3} \text{ mm}$$

Use Eq. (7-49) to find the dilatation or unit volume change e of the plate as

$$e = \frac{1-2\nu}{E}(\sigma_x + \sigma_y) = 2.538 \times 10^{-4}$$

4. ***Finalize***: The positive sign for e means that the plate under biaxial stress increases in volume (although the increase is very small) while decreasing in thickness. Finally, compute the strain-energy density of the plate (substitute the normal stresses in Eq. (7-53), deleting the shear term):

$$u = \frac{1}{2E}(\sigma_x^2 + \sigma_y^2 - 2\nu\sigma_x\sigma_y)$$
$$= \frac{1}{2(73 \text{ GPa})}[(-27.6 \text{ MPa})^2 - (82.1 \text{ MPa})^2 - 2(0.33)(-27.6 \text{ MPa})(82.1 \text{ MPa})]$$
$$= 61.6 \text{ kPa}$$

7.6 Triaxial Stress

An element of material subjected to normal stresses σ_x, σ_y, and σ_z acting in three mutually perpendicular directions is said to be in a state of **triaxial stress** (Fig. 7-28a). Since there are no shear stresses on the x, y, and z faces, the stresses σ_x, σ_y, and σ_z are the *principal stresses* in the material.

If an inclined plane parallel to the z axis is cut through the element (Fig. 7-28b), the only stresses on the inclined face are the normal stress σ and shear stress τ, both of which act parallel to the x-y plane. These stresses are analogous to the stresses σ_{x1} and τ_{x1y1} encountered in earlier discussions of plane stress (see, for instance, Fig. 7-2a). Because the stresses σ and τ (Fig. 7-28b) are found from equations of force equilibrium in the x-y plane, they are independent of the normal stress σ_z. Therefore, use the transformation equations of plane stress, as well as Mohr's circle for plane stress, when determining the stresses σ and τ in triaxial stress. The same general conclusion holds for the normal and shear stresses acting on inclined planes cut through the element parallel to the x and y axes.

FIGURE 7-28
Element in triaxial stress

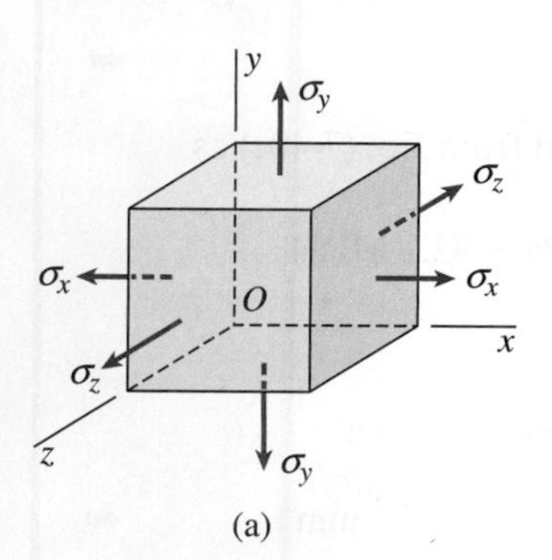

(a)

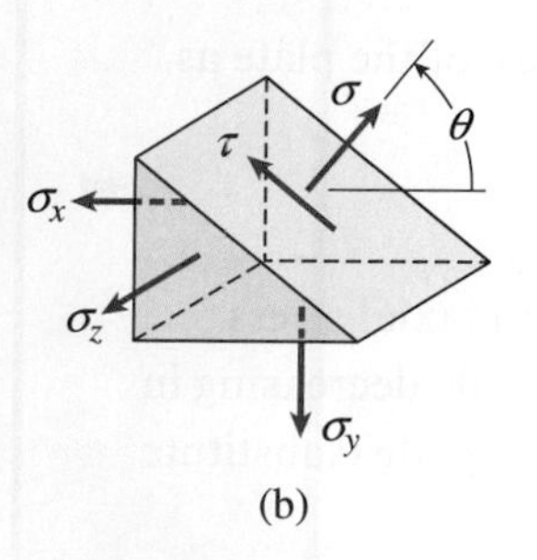

(b)

Maximum Shear Stresses

From the previous discussions of plane stress, the maximum shear stresses occur on planes oriented at 45° to the principal planes. Therefore, for a material in triaxial stress (Fig. 7-28a), the maximum shear stresses occur on elements oriented at angles of 45° to the *x*, *y*, and *z* axes. For example, consider an element obtained by a 45° rotation about the *z* axis. The maximum positive and negative shear stresses acting on this element are

$$(\tau_{\max})_z = \pm\frac{\sigma_x - \sigma_y}{2} \qquad \textbf{(7-57a)}$$

Similarly, by rotating about the *x* and *y* axes through angles of 45°, the following maximum shear stresses are obtained:

$$(\tau_{\max})_x = \pm\frac{\sigma_y - \sigma_z}{2} \qquad (\tau_{\max})_y = \pm\frac{\sigma_x - \sigma_z}{2} \qquad \textbf{(7-57b,c)}$$

The absolute maximum shear stress is the numerically largest of the stresses determined from Eqs. (7-57a, b, and c). It is equal to one-half the difference between the algebraically largest and algebraically smallest of the three principal stresses.

The stresses acting on elements oriented at various angles to the *x*, *y*, and *z* axes can be visualized with the aid of **Mohr's circles**. For elements oriented by rotations about the *z* axis, the corresponding circle is labeled *A* in Fig. 7-29. Note that this circle is drawn for the case in which $\sigma_x > \sigma_y$ and both σ_x and σ_y are tensile stresses.

FIGURE 7-29
Mohr's circles for an element in triaxial stress

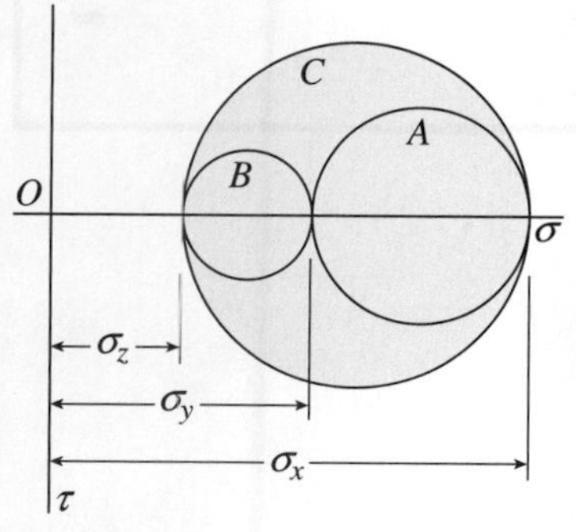

In a similar manner, construct circles *B* and *C* for elements oriented by rotations about the *x* and *y* axes, respectively. The radii of the circles represent the maximum shear stresses given by Eqs. (7-57a, b, and c), and the absolute maximum shear stress is equal to the radius of the largest circle. The normal stresses acting on the planes of maximum shear stresses have magnitudes given by the abscissas of the centers of the respective circles.

The preceding discussion of triaxial stress only considered stresses acting on planes obtained by rotating about the *x*, *y*, and *z* axes. Thus, every plane considered is parallel to one of the axes. For instance, the inclined plane of Fig. 7-28b is parallel to the *z* axis, and its normal is parallel to the *x*-*y* plane. Of course, you can also cut through the element in **skew directions**, so that the resulting inclined planes are skew to all three coordinate axes. The normal and shear stresses acting on such planes can be obtained by a more complicated three-dimensional analysis. However, the normal stresses acting on skew planes are intermediate in value between the algebraically maximum and minimum principal stresses, and the shear stresses on those planes are smaller (in absolute value) than the absolute maximum shear stress obtained from Eqs. (7-57a, b, and c).

Hooke's Law for Triaxial Stress

If the material follows Hooke's law, you can obtain the relationships between the normal stresses and normal strains by using the same procedure as for plane stress (see Section 7.5). The strains produced by the stresses σ_x, σ_y, and σ_z

acting independently are superimposed to obtain the resultant strains. Thus, the following equations pertain to the **strains in triaxial stress**:

$$\varepsilon_x = \frac{\sigma_x}{E} - \frac{\nu}{E}(\sigma_y + \sigma_z) \quad \text{(7-58a)}$$

$$\varepsilon_y = \frac{\sigma_y}{E} - \frac{\nu}{E}(\sigma_z + \sigma_x) \quad \text{(7-58b)}$$

$$\varepsilon_z = \frac{\sigma_z}{E} - \frac{\nu}{E}(\sigma_x + \sigma_y) \quad \text{(7-58c)}$$

In these equations, the standard sign conventions are used; that is, tensile stress σ and extensional strain ε are positive.

The preceding equations can be solved simultaneously for the **stresses in terms of the strains**:

$$\sigma_x = \frac{E}{(1+\nu)(1-2\nu)}[(1-\nu)\varepsilon_x + \nu(\varepsilon_y + \varepsilon_z)] \quad \text{(7-59a)}$$

$$\sigma_y = \frac{E}{(1+\nu)(1-2\nu)}[(1-\nu)\varepsilon_y + \nu(\varepsilon_z + \varepsilon_x)] \quad \text{(7-59b)}$$

$$\sigma_z = \frac{E}{(1+\nu)(1-2\nu)}[(1-\nu)\varepsilon_z + \nu(\varepsilon_x + \varepsilon_y)] \quad \text{(7-59c)}$$

Equations (7-58) and (7-59) represent **Hooke's law for triaxial stress**.

In the special case of **biaxial stress** (Fig. 7-11b), the equations of Hooke's law are obtained by substituting $\sigma_z = 0$ into the preceding equations. The resulting equations reduce to Eqs. (7-40) and (7-41) of Section 7.5.

Unit Volume Change

The unit volume change (or *dilatation*) for an element in triaxial stress is obtained in the same manner as for plane stress (see Section 7.5). If the element is subjected to strains ε_x, ε_y, and ε_z, use Eq. (7-48) for the unit volume change:

$$e = \varepsilon_x + \varepsilon_y + \varepsilon_z \quad \text{(7-60)}$$

This equation is valid for any material provided the strains are small.

If Hooke's law holds for the material, substitute for the strains ε_x, ε_y, and ε_z from Eqs. (7-58a, b, and c) and obtain

$$e = \frac{1-2\nu}{E}(\sigma_x + \sigma_y + \sigma_z) \quad \text{(7-61)}$$

Equations (7-60) and (7-61) give the unit volume change in triaxial stress in terms of the strains and stresses, respectively.

Strain-Energy Density

The strain-energy density for an element in triaxial stress is obtained by the same method used for plane stress. When stresses σ_x and σ_y act alone (biaxial stress), the strain-energy density [from Eq. (7-52) with the shear term discarded] is

$$u = \frac{1}{2}(\sigma_x\varepsilon_x + \sigma_y\varepsilon_y)$$

When the element is in triaxial stress and subjected to stresses σ_x, σ_y, and σ_z, the expression for strain-energy density becomes

$$u = \frac{1}{2}(\sigma_x \varepsilon_x + \sigma_y \varepsilon_y + \sigma_z \varepsilon_z) \tag{7-62a}$$

Substituting for the strains from Eqs. (7-58a, b, and c) gives the strain-energy density in terms of the stresses:

$$u = \frac{1}{2E}(\sigma_x^2 + \sigma_y^2 + \sigma_z^2) - \frac{\nu}{E}(\sigma_x \sigma_y + \sigma_x \sigma_z + \sigma_y \sigma_z) \tag{7-62b}$$

In a similar manner, using Eqs. (7-59a, b, and c) leads to the strain-energy density in terms of the strains:

$$u = \frac{E}{2(1+\nu)(1-2\nu)}[(1-\nu)(\varepsilon_x^2 + \varepsilon_y^2 + \varepsilon_z^2) + 2\nu(\varepsilon_x \varepsilon_y + \varepsilon_x \varepsilon_z + \varepsilon_y \varepsilon_z)] \tag{7-62c}$$

When calculating from these expressions, be sure to substitute the stresses and strains with their proper algebraic signs.

Spherical Stress

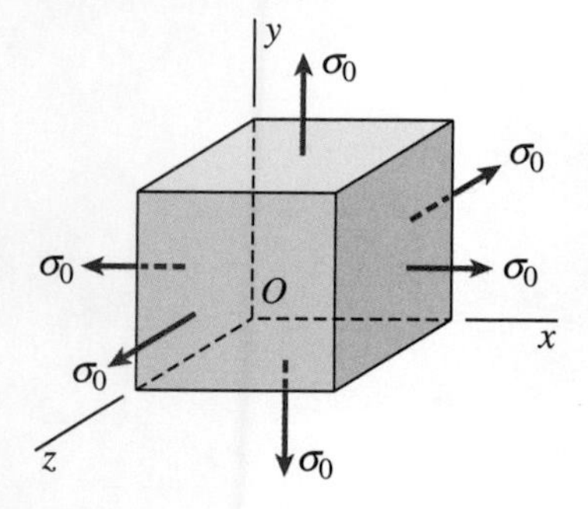

FIGURE 7-30
Element in spherical stress

A special type of triaxial stress, called **spherical stress**, occurs whenever all three normal stresses are equal (Fig. 7-30):

$$\sigma_x = \sigma_y = \sigma_z = \sigma_0 \tag{7-63}$$

Under these stress conditions, *any* plane cut through the element will be subjected to the same normal stress σ_0 and will be free of shear stress. Thus, there are equal normal stresses in every direction and no shear stresses anywhere in the material. Every plane is a principal plane, and the three Mohr's circles shown in Fig. 7-29 reduce to a single point.

The normal strains in spherical stress are also the same in all directions, provided the material is homogeneous and isotropic. If Hooke's law applies, the normal strains are

$$\varepsilon_0 = \frac{\sigma_0}{E}(1 - 2\nu) \tag{7-64}$$

as obtained from Eqs. (7-58a, b, and c).

Since there are no shear strains, an element in the shape of a cube changes in size but remains a cube. In general, any body subjected to spherical stress maintains its relative proportions but expands or contracts in volume depending upon whether σ_0 is tensile or compressive.

The expression for the unit volume change can be obtained from Eq. (7-60) by substituting for the strains from Eq. (7-64). The result is

$$e = 3\varepsilon_0 = \frac{3\sigma_0(1-2\nu)}{E} \tag{7-65}$$

Equation (7-65) is usually expressed in more compact form by introducing a new quantity K called the **volume modulus of elasticity**, or **bulk modulus of elasticity**, which is defined as

$$K = \frac{E}{3(1-2\nu)} \tag{7-66}$$

With this notation, the expression for the unit volume change becomes

$$e = \frac{\sigma_0}{K} \tag{7-67}$$

and the volume modulus is

$$K = \frac{\sigma_0}{e} \tag{7-68}$$

Thus, the volume modulus can be defined as the ratio of the spherical stress to the volumetric strain, which is analogous to the definition of the modulus E in uniaxial stress. Note that the preceding formulas for e and K are based upon the assumptions that *the strains are small and Hooke's law holds for the material.*

Equation (7-68) for K shows that if Poisson's ratio ν equals 1/3, the moduli K and E are numerically equal. If $\nu = 0$, then K has the value $E/3$, and if $\nu = 0.5$, K becomes infinite, which corresponds to a rigid material having no change in volume (that is, the material is incompressible).

The preceding formulas for spherical stress were derived for an element subjected to uniform tension in all directions, but of course, the formulas also apply to an element in uniform compression. In the case of uniform compression, the stresses and strains have negative signs. Uniform compression occurs when the material is subjected to uniform pressure in all directions; for example, an object submerged in water or rock deep within the earth. This state of stress is often called **hydrostatic stress.**

Although uniform compression is relatively common, a state of uniform tension is difficult to achieve. It can be realized by suddenly and uniformly heating the outer surface of a solid metal sphere, so that the outer layers are at a higher temperature than the interior. The tendency of the outer layers to expand produces uniform tension in all directions at the center of the sphere.

7.7 Plane Strain

The strains at a point in a loaded structure vary according to the orientation of the axes in a manner similar to that for stresses. This section presents the derivation of the transformation equations that relate the strains in inclined directions to the strains in the reference directions. These transformation equations are widely used in laboratory and field investigations involving measurements of strains.

Strains are customarily measured by *strain gages*; for example, gages are placed in aircraft to measure structural behavior during flight, and gages are placed in buildings to measure the effects of earthquakes. Since each gage measures the strain in one particular direction, it is usually necessary to calculate the strains in other directions by means of the transformation equations.

Plane Strain versus Plane Stress

Consider what is meant by plane strain and how it relates to plane stress. A small element of material has sides of lengths a, b, and c in the x, y, and z directions, respectively (Fig. 7-31a). If the only deformations are those in the x-y plane, then three strain components may exist—the normal strain ε_x in the x direction (Fig. 7-31b), the normal strain ε_y in the y direction (Fig. 7-31c), and the shear strain γ_{xy} (Fig. 7-31d). An element of material subjected to these strains (and *only* these strains) is said to be in a state of **plane strain**.

It follows that an element in plane strain has no normal strain ε_z in the z direction and no shear strains γ_{xz} and γ_{yz} in the x-z and y-z planes, respectively. Thus, plane strain is defined by the following conditions:

$$\varepsilon_z = 0 \quad \gamma_{xz} = 0 \quad \gamma_{yz} = 0 \qquad \textbf{(7-69a,b,c)}$$

The remaining strains (ε_x, ε_y, and γ_{xy}) may have nonzero values.

The preceding definition shows that plane strain occurs when the front and rear faces of an element of material (Fig. 7-31a) are fully restrained against displacement in the z direction—an idealized condition that is seldom reached in actual structures. However, this does not mean that the transformation equations of plane strain are not useful. It turns out that they are extremely useful because they also apply to the strains in plane stress, as explained in the following paragraphs.

The definition of plane strain [Eqs. (7-69a, b, and c)] is analogous to that for plane stress. In plane stress, the following stresses must be zero:

$$\sigma_z = 0 \quad \tau_{xz} = 0 \quad \tau_{yz} = 0 \qquad \textbf{(7-70a,b,c)}$$

whereas the remaining stresses (σ_x, σ_y, and τ_{xy}) may have nonzero values. A comparison of the stresses and strains in plane stress and plane strain is given in Fig. 7-32.

It should not be inferred from the similarities in the definitions of plane stress and plane strain that both occur simultaneously. In general, an element

FIGURE 7-31

Strain components ε_x, ε_y, and γ_{xy} in the x-y plane (plane strain)

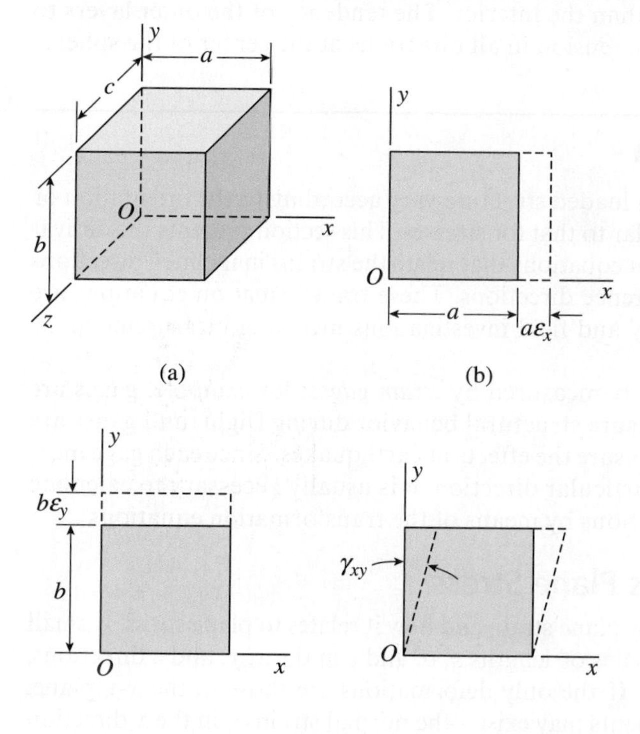

FIGURE 7-32

Comparison of plane stress and plane strain

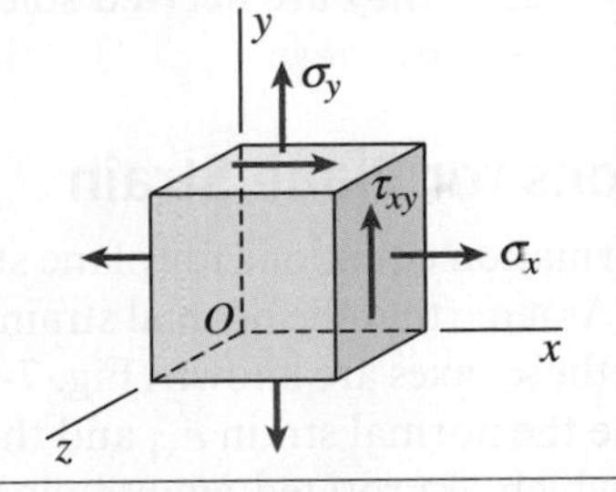

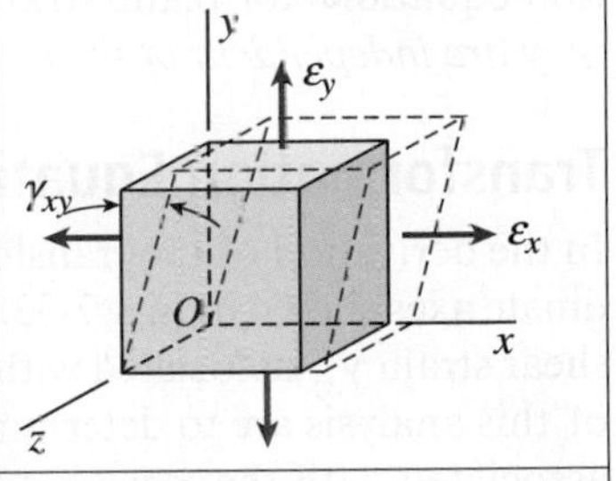

	Plane stress	Plane strain
Stresses	$\sigma_z = 0$ $\tau_{xz} = 0$ $\tau_{yz} = 0$ σ_x, σ_y, and τ_{xy} may have nonzero values	$\tau_{xz} = 0$ $\tau_{yz} = 0$ σ_x, σ_y, σ_z, and τ_{xy} may have nonzero values
Strains	$\gamma_{xz} = 0$ $\gamma_{yz} = 0$ ε_x, ε_y, ε_z, and γ_{xy} may have nonzero values	$\varepsilon_z = 0$ $\gamma_{xz} = 0$ $\gamma_{yz} = 0$ ε_x, ε_y, and γ_{xy} may have nonzero values

in plane stress undergoes a strain in the z direction (Fig. 7-32); hence, it is *not* in plane strain. Also, an element in plane strain usually has stresses σ_z acting on it because of the requirement that $\varepsilon_z = 0$; therefore, it is *not* in plane stress. Thus, under ordinary conditions plane stress and plane strain do not occur simultaneously.

An exception occurs when an element in plane stress is subjected to equal and opposite normal stresses (that is, when $\sigma_x = -\sigma_y$) and Hooke's law holds for the material. In this special case, there is no normal strain in the z direction, as shown by Eq. (7-35c); therefore, the element is in a state of plane strain as well as plane stress. Another special case, albeit a hypothetical one, is when a material has Poisson's ratio equal to zero ($\nu = 0$); then every plane stress element is also in plane strain because $\varepsilon_z = 0$ [Eq. (7-35c)].[4]

Application of the Transformation Equations

The stress-transformation equations derived for plane stress in the x-y plane [Eqs. (7-4a and b)] are valid even when a normal stress σ_z is present. The explanation lies in the fact that the stress σ_z does not enter the equations of equilibrium used in deriving Eqs. (7-4a and b). Therefore, *the transformation equations for plane stress also can be used for the stresses in plane strain.*

An analogous situation exists for plane strain. Although the strain-transformation equations are derived for the case of plane strain in the x-y plane, the equations are valid even when a strain ε_z exists. The reason is simple enough—the strain ε_z does not affect the geometric relationships used in the derivations. Therefore, *the transformation equations for plane strain also can be used for the strains in plane stress.*

[4]The discussions of this chapter omit the effects of temperature changes and prestrains, both of which produce additional deformations that may alter some of these conclusions.

Finally, recall that the transformation equations for plane stress were derived solely from equilibrium and therefore are valid for any material, whether linearly elastic or not. The same conclusion applies to the transformation equations for plane strain—since they are derived solely from geometry, *they are independent of the material properties.*

FIGURE 7-33

Axes x_1 and y_1 rotated through an angle θ from the x-y axes

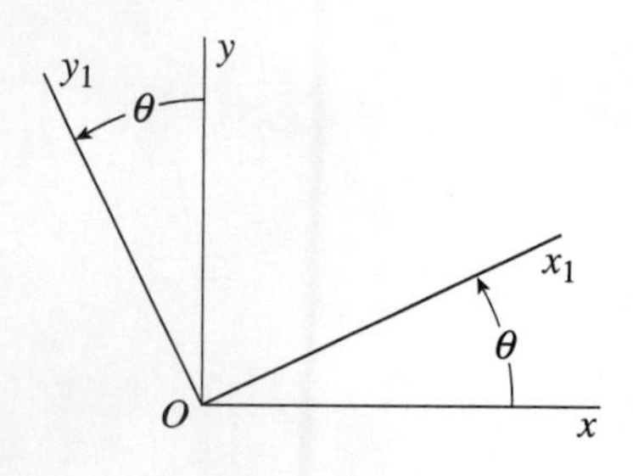

Transformation Equations for Plane Strain

In the derivation of the transformation equations for plane strain, use the coordinate axes shown in Fig. 7-33. Assume that the normal strains ε_x and ε_y and the shear strain γ_{xy} associated with the xy axes are known (Fig. 7-31). The objectives of this analysis are to determine the normal strain ε_{x1} and the shear strain γ_{x1y1} associated with the x_1y_1 axes, which are rotated counterclockwise through an angle θ from the xy axes. (It is not necessary to derive a separate equation for the normal strain ε_{y1} because it can be obtained from the equation for ε_{x1} by substituting $\theta + 90°$ for θ.)

Normal strain ε_{x1}. To determine the normal strain ε_{x1} in the x_1 direction, consider a small element of material selected so that the x_1 axis is along a diagonal of the z face of the element and the x and y axes are along the sides of the element (Fig. 7-34a). The figure shows a two-dimensional view of the element with the z axis toward the viewer. Of course, the element is actually three-dimensional, as in Fig. 7-31a, with a dimension in the z direction.

Consider first the strain ε_x in the x direction (Fig. 7-34a). This strain produces an elongation in the x direction equal to $\varepsilon_x dx$, where dx is the length of the corresponding side of the element. As a result of this elongation, the diagonal of the element increases in length by an amount

$$\varepsilon_x dx \cos\theta \tag{7-71a}$$

as shown in Fig. 7-34a.

Next, consider the strain ε_y in the y direction (Fig. 7-34b). This strain produces an elongation in the y direction equal to $\varepsilon_y dy$, where dy is the length of the side of the element parallel to the y axis. As a result of this elongation, the diagonal of the element increases in length by an amount

$$\varepsilon_y dy \sin\theta \tag{7-71b}$$

which is shown in Fig. 7-34b.

Finally, consider the shear strain γ_{xy} in the x-y plane (Fig. 7-34c). This strain produces a distortion of the element such that the angle at the lower left corner of the element decreases by an amount equal to the shear strain. Consequently, the upper face of the element moves to the right (with respect to the lower face) by an amount $\gamma_{xy}dy$. This deformation results in an increase in the length of the diagonal equal to

$$\gamma_{xy}dy \cos\theta \tag{7-71c}$$

as shown in Fig. 7-34c.

The total increase Δd in the length of the diagonal is the sum of the preceding three expressions; thus,

$$\Delta d = \varepsilon_x dx \cos\theta + \varepsilon_y dy \sin\theta + \gamma_{xy}dy \cos\theta \tag{7-72}$$

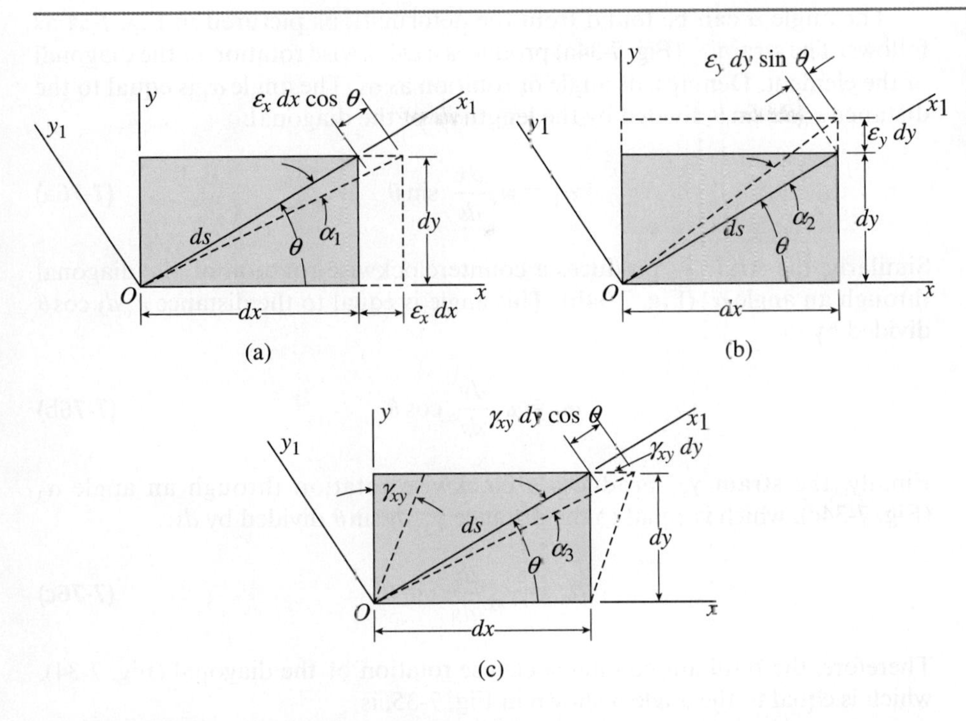

FIGURE 7-34

Deformations of an element in plane strain due to (a) normal strain ε_x, (b) normal strain ε_y, and (c) shear strain γ_{xy}

The normal strain ε_{x1} in the x_1 direction is equal to this increase in length divided by the initial length *ds* of the diagonal:

$$\varepsilon_{x1} = \frac{\Delta d}{ds} = \varepsilon_x \frac{dx}{ds} \cos\theta + \varepsilon_y \frac{dy}{ds} \sin\theta + \gamma_{xy} \frac{dy}{ds} \cos\theta \qquad \textbf{(7-73)}$$

The two relations $dx/ds = \cos\theta$ and $dy/ds = \sin\theta$ lead to the following equation for the **normal strain**:

$$\varepsilon_{x1} = \varepsilon_x \cos^2\theta + \varepsilon_y \sin^2\theta + \gamma_{xy} \sin\theta\cos\theta \qquad \textbf{(7-74)}$$

Thus, Eq. (7-74) represents an expression for the normal strain in the x_1 direction in terms of the strains ε_x, ε_y, and γ_{xy} associated with the *x-y* axes.

As mentioned previously, the normal strain ε_{y1} in the y_1 direction is obtained from the preceding equation by substituting $\theta + 90°$ for θ.

Shear strain γ_{x1y1}. Now turn to the shear strain γ_{x1y1} associated with the x_1y_1 axes. This strain is equal to the decrease in angle between lines in the material that were initially along the x_1 and x_1 axes. To clarify this idea, consider Fig. 7-35, which shows both the *xy* and x_1y_1 axes, with the angle θ between them. Let line *Oa* represent a line in the material that *initially* was along the x_1 axis (that is, along the diagonal of the element in Fig. 7-34). The deformations caused by the strains ε_x, ε_y, and γ_{xy} (Fig. 7-34) cause line *Oa* to rotate through a counterclockwise angle α from the x_1 axis to the position shown in Fig. 7-35. Similarly, line *Ob* was originally along the y_1 axis, but because of the deformations, it rotates through a clockwise angle β. The shear strain γ_{x1y1} is the decrease in angle between the two lines that originally were at right angles; therefore,

$$\gamma_{x1y1} = \alpha + \beta \qquad \textbf{(7-75)}$$

Thus, to find the shear strain γ_{x1y1}, find the angles α and β.

FIGURE 7-35

Shear strain γ_{x1y1} associated with the x_1y_1 axes

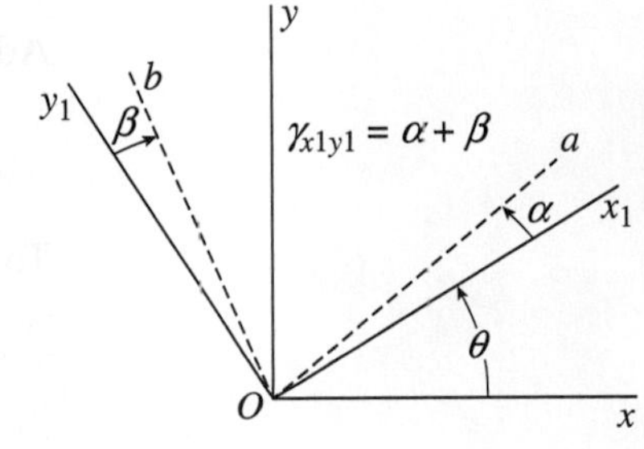

The angle α can be found from the deformations pictured in Fig. 7-34 as follows. The strain ε_x (Fig. 7-34a) produces a clockwise rotation of the diagonal of the element. Denote this angle of rotation as α_1. The angle α_1 is equal to the distance $\varepsilon_x dx \sin\theta$ divided by the length ds of the diagonal:

$$\alpha_1 = \varepsilon_x \frac{dx}{ds} \sin\theta \tag{7-76a}$$

Similarly, the strain ε_y produces a counterclockwise rotation of the diagonal through an angle α_2 (Fig. 7-34b). This angle is equal to the distance $\varepsilon_y dy \cos\theta$ divided by ds:

$$\alpha_2 = \varepsilon_y \frac{dy}{ds} \cos\theta \tag{7-76b}$$

Finally, the strain γ_{xy} produces a clockwise rotation through an angle α_3 (Fig. 7-34c), which is equal to the distance $\gamma_{xy} dy \sin\theta$ divided by ds:

$$\alpha_3 = \gamma_{xy} \frac{dy}{ds} \sin\theta \tag{7-76c}$$

Therefore, the resultant counterclockwise rotation of the diagonal (Fig. 7-34), which is equal to the angle α shown in Fig. 7-35, is

$$\begin{aligned} \alpha &= -\alpha_1 + \alpha_2 - \alpha_3 \\ &= -\varepsilon_x \frac{dx}{ds} \sin\theta + \varepsilon_y \frac{dy}{ds} \cos\theta - \gamma_{xy} \frac{dy}{ds} \sin\theta \end{aligned} \tag{7-77}$$

Again observe that $dx/ds = \cos\theta$ and $dy/ds = \sin\theta$ so

$$\alpha = -(\varepsilon_x - \varepsilon_y) \sin\theta \cos\theta - \gamma_{xy} \sin^2\theta \tag{7-78}$$

The rotation of line Ob (Fig. 7-35), which initially was at 90° to line Oa, can be found by substituting $\theta + 90°$ for θ in the expression for α. The resulting expression is counterclockwise when positive (because α is counterclockwise when positive); hence, it is equal to the negative of the angle β (because β is positive when clockwise). Thus,

$$\begin{aligned} \beta &= (\varepsilon_x - \varepsilon_y) \sin(\theta + 90°) \cos(\theta + 90°) + \gamma_{xy} \sin^2(\theta + 90°) \\ &= -(\varepsilon_x - \varepsilon_y) \sin\theta \cos\theta + \gamma_{xy} \cos^2\theta \end{aligned} \tag{7-79}$$

Adding α and β gives the shear strain γ_{x1y1} [see Eq. (7-75)]:

$$\gamma_{x1y1} = -2(\varepsilon_x - \varepsilon_y) \sin\theta \cos\theta + \gamma_{xy}(\cos^2\theta - \sin^2\theta) \tag{7-80}$$

To put the equation in a more useful form, divide each term by 2:

$$\frac{\gamma_{x1y1}}{2} = -(\varepsilon_x - \varepsilon_y) \sin\theta \cos\theta + \frac{\gamma_{xy}}{2}(\cos^2\theta - \sin^2\theta) \tag{7-81}$$

The result is an expression for the **shear strain** γ_{x1y1} associated with the x_1-y_1 axes in terms of the strains ε_x, ε_y, and γ_{xy} associated with the x-y axes.

Transformation equations for plane strain. The equations for plane strain [Eqs. (7-74) and (7-81)] can be expressed in terms of the angle 2θ by using the following trigonometric identities:

$$\cos^2\theta = \frac{1}{2}(1 + \cos 2\theta) \quad \sin^2\theta = \frac{1}{2}(1 - \cos 2\theta)$$

$$\sin\theta\cos\theta = \frac{1}{2}\sin 2\theta$$

Thus, the transformation equations for plane strain become

$$\varepsilon_{x1} = \frac{\varepsilon_x + \varepsilon_y}{2} + \frac{\varepsilon_x - \varepsilon_y}{2}\cos 2\theta + \frac{\gamma_{xy}}{2}\sin 2\theta \tag{7-82a}$$

and

$$\frac{\gamma_{x1y1}}{2} = -\frac{\varepsilon_x - \varepsilon_y}{2}\sin 2\theta + \frac{\gamma_{xy}}{2}\cos 2\theta \tag{7-82b}$$

These equations are the counterparts of Eqs. (7-4a and b) for plane stress.

When comparing the two sets of equations, note that ε_{x1} corresponds to σ_{x1}, $\gamma_{x1y1}/2$ corresponds to τ_{x1y1}, ε_x corresponds to σ_x, ε_y corresponds to σ_y, and $\gamma_{xy}/2$ corresponds to τ_{xy}. The corresponding variables in the two sets of transformation equations are listed in Table 7-1.

Table 7-1

Corresponding variables in the transformation equations for plane stress [Eqs. (7-4a and b)]and plane strain [Eqs. (7-82a and b)]

Stresses	Strains
σ_x	ε_x
σ_y	ε_y
τ_{xy}	$\gamma_{xy}/2$
σ_{x1}	ε_{x1}
τ_{x1y1}	$\gamma_{x1y1}/2$

The analogy between the transformation equations for plane stress and those for plane strain shows that all of the observations made in Sections 7.2, 7.3, and 7.4 concerning plane stress, principal stresses, maximum shear stresses, and Mohr's circle have their counterparts in plane strain. For instance, the sum of the normal strains in perpendicular directions is a constant [compare with Eq. (7-6)]:

$$\varepsilon_{x1} + \varepsilon_{y1} = \varepsilon_x + \varepsilon_y \tag{7-83}$$

This equality can be verified easily by substituting the expressions for ε_{x1} [from Eq. (7-82a)] and ε_{y1} [from Eq. (7-82a) with θ replaced by $\theta + 90°$].

Principal Strains

Principal strains exist on perpendicular planes with the principal angles θ_p calculated from the following equation [compare with Eq. (7-11)]:

$$\tan 2\theta_p = \frac{\gamma_{xy}}{\varepsilon_x - \varepsilon_y} \tag{7-84}$$

The principal strains can be calculated from the equation

$$\varepsilon_{1,2} = \frac{\varepsilon_x + \varepsilon_y}{2} \pm \sqrt{\left(\frac{\varepsilon_x - \varepsilon_y}{2}\right)^2 + \left(\frac{\gamma_{xy}}{2}\right)^2} \tag{7-85}$$

which corresponds to Eq. (7-17) for the principal stresses. The two principal strains (in the x-y plane) can be correlated with the two principal directions using the technique described in Section 7.3 for the principal stresses. (This technique is illustrated later in Example 7-8.) Finally, note that in plane strain the third principal strain is $\varepsilon_z = 0$. Also, the shear strains are zero on the principal planes.

Maximum Shear Strains

The maximum shear strains in the x-y plane are associated with axes at 45° to the directions of the principal strains. The algebraically maximum shear strain (in the x-y plane) is given by the following equation [compare with Eq. (7-25)]:

$$\frac{\gamma_{max}}{2} = \sqrt{\left(\frac{\varepsilon_x - \varepsilon_y}{2}\right)^2 + \left(\frac{\gamma_{xy}}{2}\right)^2} \quad \textbf{(7-86)}$$

The minimum shear strain has the same magnitude but is negative. In the directions of maximum shear strain, the normal strains are

$$\varepsilon_{aver} = \frac{\varepsilon_x + \varepsilon_y}{2} \quad \textbf{(7-87)}$$

which is analogous to Eq. (7-27) for stresses. The maximum out-of-plane shear strains, that is, the shear strains in the x-z and y-z planes, can be obtained from equations analogous to Eq. (7-86).

An element in plane stress that is oriented to the principal directions of stress (see Fig. 7-13b) has no shear stresses acting on its faces. Therefore, the shear strain γ_{x1y1} for this element is zero. It follows that the normal strains in this element are the principal strains. Thus, at a given point in a stressed body, *the principal strains and principal stresses occur in the same directions.*

Mohr's Circle for Plane Strain

Mohr's circle for plane strain is constructed in the same manner as the circle for plane stress, as illustrated in Fig. 7-36. Normal strain ε_{x1} is plotted as the abscissa (positive to the right), and one-half the shear strain ($\gamma_{x1y1}/2$) is plotted

FIGURE 7-36

Mohr's circle for plane strain

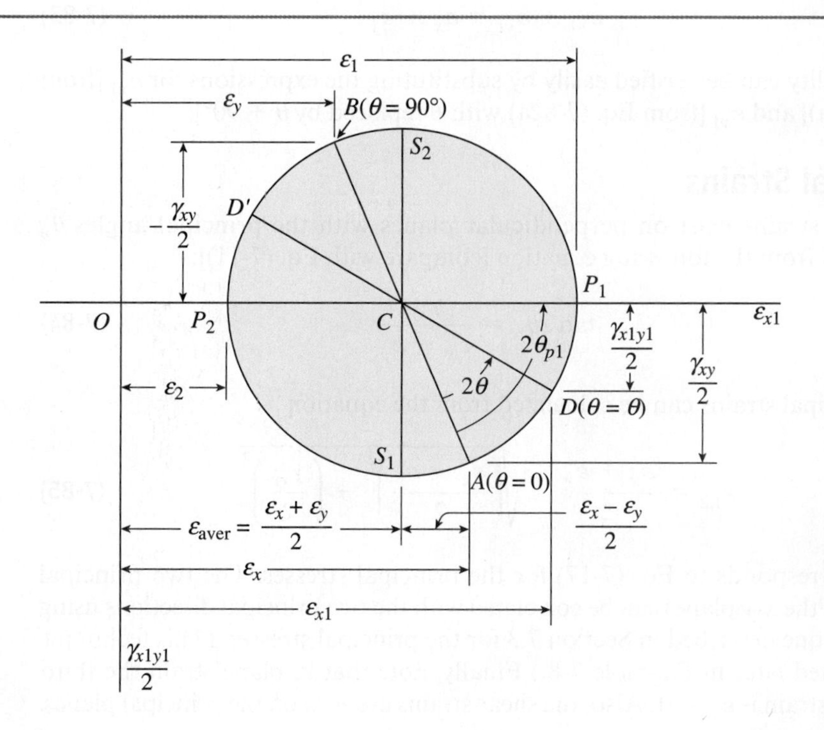

as the ordinate (positive downward). The center C of the circle has an abscissa equal to ε_{aver} [Eq. (7-87)].

Point A, representing the strains associated with the x direction ($\theta = 0$), has coordinates ε_x and $\gamma_{xy}/2$. Point B, at the opposite end of a diameter from A, has coordinates ε_y and $-\gamma_{xy}/2$, representing the strains associated with a pair of axes rotated through an angle $\theta = 90°$.

The strains associated with axes rotated through an angle θ are given by point D, which is located on the circle by measuring an angle 2θ counterclockwise from radius CA. The principal strains are represented by points P_1 and P_2, and the maximum shear strains by points S_1 and S_2. All of these strains can be determined from the geometry of the circle or from the transformation equations.

Strain Measurements

An electrical-resistance **strain gage** is a device for measuring normal strains on the surface of a stressed object. These gages are quite small, with lengths typically in the range from 3 mm to 12 mm. The gages are bonded securely to the surface of the object so that they change in length in proportion to the strains in the object itself.

Each gage consists of a fine metal grid that is stretched or shortened when the object is strained at the point where the gage is attached. The grid is equivalent to a continuous wire that goes back and forth from one end of the grid to the other, thereby effectively increasing its length (Fig. 7-37). The electrical resistance of the wire is altered when it stretches or shortens—then this change in resistance is converted into a measurement of strain. The gages are extremely sensitive and can measure strains as small as 1×10^{-6}.

Since each gage measures the normal strain in only one direction, and since the directions of the principal stresses are usually unknown, it is necessary to use three gages in combination, with each gage measuring the strain in a different direction. From three such measurements, it is possible to calculate the strains in any direction, as illustrated in Example 7-9.

A group of three gages arranged in a particular pattern is called a **strain rosette**. Because the rosette is mounted on the surface of the body, where the material is in plane stress, use the transformation equations for plane strain to

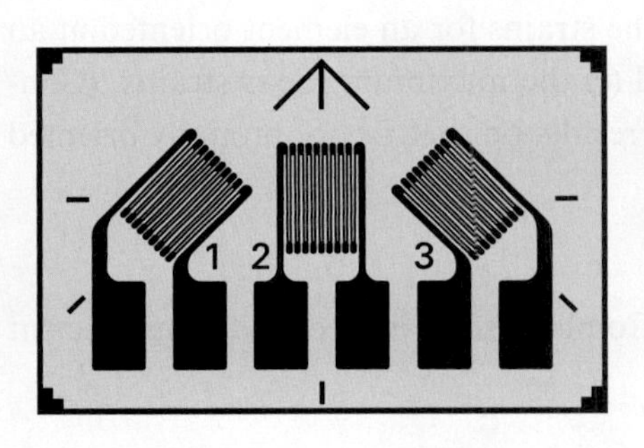

(a) 45° strain-gage three-element rosette

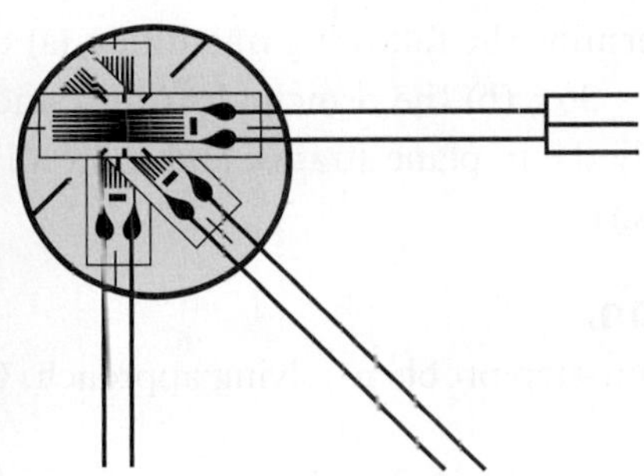

(b) Three-element strain-gage rosettes prewired

FIGURE 7-37

Three electrical-resistance strain gages arranged as a 45° strain rosette (magnified view) (Courtesy of Micro-Measurements Division of Vishay Precision Group, Raleigh, NC, USA)

calculate the strains in various directions. (As explained earlier in this section, the transformation equations for plane strain can also be used for the strains in plane stress.)

Calculation of Stresses from the Strains

The strain equations presented in this section are derived solely from geometry, as already pointed out. Therefore, the equations apply to any material, whether linear or nonlinear, elastic or inelastic. However, if it is desired to determine the stresses from the strains, the material properties must be taken into account.

If the material follows Hooke's law, find the stresses using the appropriate stress-strain equations from either Section 7.5 (for plane stress) or Section 7.6 (for triaxial stress).

As a first example, suppose that the material is in plane stress and that are known the strains ε_x, ε_y, and γ_{xy}, perhaps from strain-gage measurements. Then use the stress-strain equations for plane stress [Eqs. (7-37) and (7-38)] to obtain the stresses in the material.

Now consider a second example. Suppose you have determined the three principal strains ε_1, ε_2, and ε_3 for an element of material (if the element is in plane strain, then $\varepsilon_3 = 0$). Knowing these strains, you can find the principal stresses using Hooke's law for triaxial stress [see Eqs. (7-59a, b, and c)]. Once the principal stresses are known, the stresses on inclined planes are computed using the transformation equations for plane stress (see the discussion at the beginning of Section 7.6).

Example 7-8

An element of material in plane strain undergoes the following strains:

$$\varepsilon_x = 340 \times 10^{-6} \quad \varepsilon_y = 110 \times 10^{-6} \quad \gamma_{xy} = 180 \times 10^{-6}$$

These strains are shown highly exaggerated in Fig. 7-38a, which shows the deformations of an element of unit dimensions. Since the edges of the element have unit lengths, the changes in linear dimensions have the same magnitudes as the normal strains ε_x and ε_y. The shear strain γ_{xy} is the decrease in angle at the lower-left corner of the element.

Determine the following quantities: (a) the strains for an element oriented at an angle $\theta = 30°$, (b) the principal strains, and (c) the maximum shear strains. (Consider only the in-plane strains, and show all results on sketches of properly oriented elements.)

Solution:

Use a four-step problem-solving approach. Combine steps as needed for an efficient solution.

FIGURE 7-38

Example 7-8: Element of material in plane strain: (a) element oriented to the x and y axes, (b) element oriented at an angle $\theta = 30°$, (c) principal strains, and (d) maximum shear strains (*Note:* The edges of the elements have unit lengths)

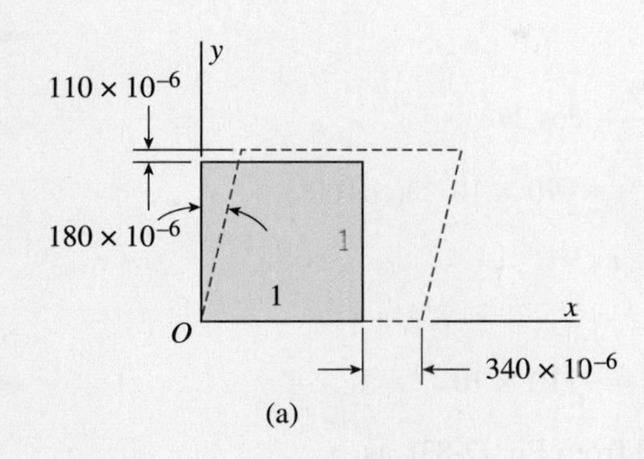

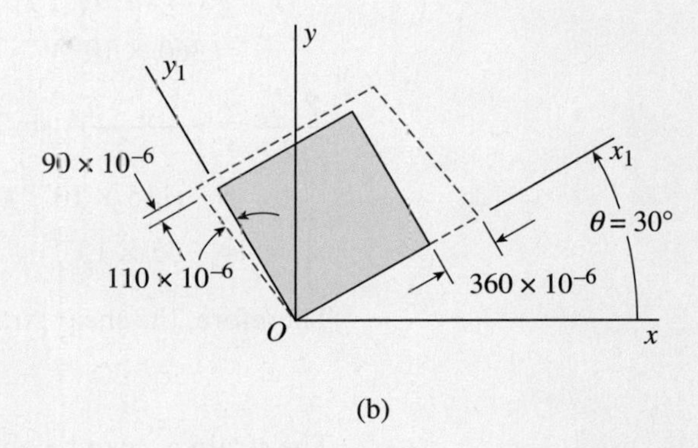

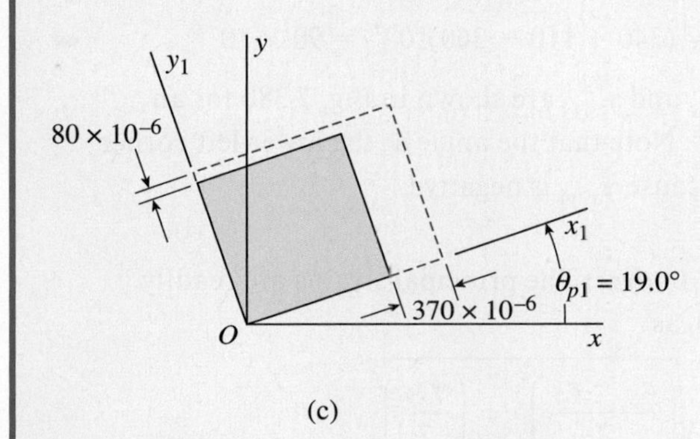

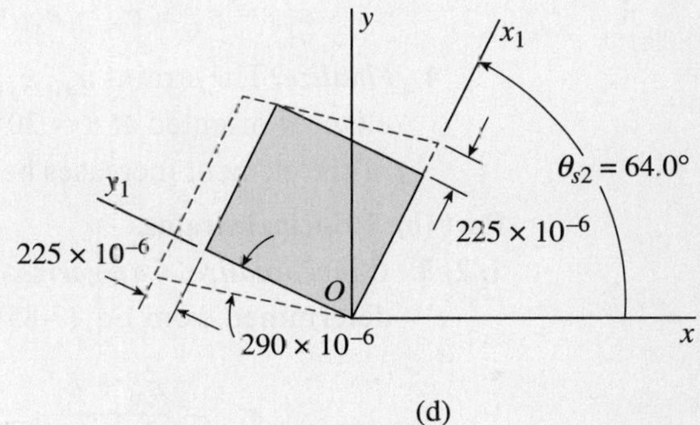

Part (a): Element oriented at an angle $\theta = 30°$.

1, 2, 3. ***Conceptualize, Categorize, Analyze***: The strains for an element oriented at an angle θ to the x axis can be found from the transformation equations of Eqs. (7-82a and b). As a preliminary matter, make the following calculations:

$$\frac{\varepsilon_x + \varepsilon_y}{2} = \frac{(340 + 110)10^{-6}}{2} = 225 \times 10^{-6}$$

$$\frac{\varepsilon_x - \varepsilon_y}{2} = \frac{(340 - 110)10^{-6}}{2} = 115 \times 10^{-6}$$

$$\frac{\gamma_{xy}}{2} = 90 \times 10^{-6}$$

Now substitute into Eqs. (7-82a and b) to get

$$\varepsilon_{x1} = \frac{\varepsilon_x + \varepsilon_y}{2} + \frac{\varepsilon_x - \varepsilon_y}{2} \cos 2\theta + \frac{\gamma_{xy}}{2} \sin 2\theta$$
$$= (225 \times 10^{-6}) + (115 \times 10^{-6})(\cos 60°) + (90 \times 10^{-6})(\sin 60°)$$
$$= 360 \times 10^{-6}$$

$$\frac{\gamma_{x1y1}}{2} = -\frac{\varepsilon_x - \varepsilon_y}{2} \sin 2\theta + \frac{\gamma_{xy}}{2} \cos 2\theta$$
$$= -(115 \times 10^{-6})(\sin 60°) + (90 \times 10^{-6})(\cos 60°)$$
$$= -55 \times 10^{-6}$$

Therefore, the shear strain is

$$\gamma_{x1y1} = -110 \times 10^{-6}$$

The strain ε_{y1} can be obtained from Eq. (7-83), as

$$\varepsilon_{y1} = \varepsilon_x + \varepsilon_y - \varepsilon_{x1} = (340 + 110 - 360)10^{-6} = 90 \times 10^{-6}$$

4. ***Finalize***: The strains ε_{x1}, ε_{y1}, and γ_{x1y1} are shown in Fig. 7-38b for an element oriented at $\theta = 30°$. Note that the angle at the lower-left corner of the element increases because γ_{x1y1} is negative.

Part (b): Principal strains.

1, 2, 3. ***Conceptualize, Categorize, Analyze***: The principal strains are readily determined from Eq. (7-85), as

$$\varepsilon_{1,2} = \frac{\varepsilon_x + \varepsilon_y}{2} \pm \sqrt{\left(\frac{\varepsilon_x - \varepsilon_y}{2}\right)^2 + \left(\frac{\gamma_{xy}}{2}\right)^2}$$
$$= 225 \times 10^{-6} \pm \sqrt{(115 \times 10^{-6})^2 + (90 \times 10^{-6})^2}$$
$$= 225 \times 10^{-6} \pm 146 \times 10^{-6}$$

Thus, the principal strains are

$$\varepsilon_1 = 370 \times 10^{-6} \qquad \varepsilon_2 = 80 \times 10^{-6}$$

FIGURE 7-38c,d (Repeated)

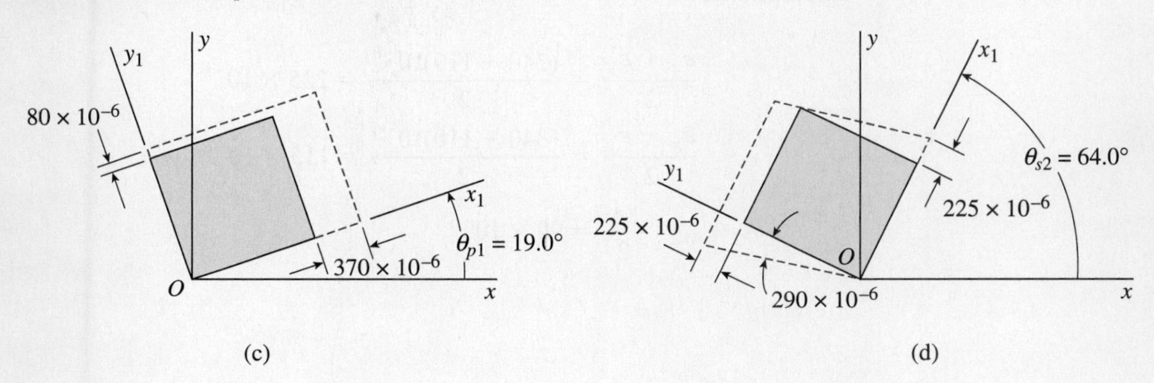

in which ε_1 denotes the algebraically larger principal strain and ε_2 denotes the algebraically smaller principal strain. (Recall that only in-plane strains are considered in this example.)

The angles to the principal directions can be obtained from Eq. (7-84):

$$\tan 2\theta_p = \frac{\gamma_{xy}}{\varepsilon_x - \varepsilon_y} = \frac{180}{340 - 110} = 0.7826$$

The values of $2\theta_p$ between 0 and 360° are 38.0° and 218.0°; therefore, the angles to the principal directions are

$$\theta_p = 19.0° \text{ and } 109.0°$$

To determine the value of θ_p associated with each principal strain, substitute $\theta_p = 19.0°$ into the first transformation equation [Eq. (7-82a)] and solve for the strain:

$$\varepsilon_{x1} = \frac{\varepsilon_x + \varepsilon_y}{2} + \frac{\varepsilon_x + \varepsilon_y}{2}\cos 2\theta + \frac{\gamma_{xy}}{2}\sin 2\theta$$
$$= (225 \times 10^{-6}) + (115 \times 10^{-6})(\cos 38.0°) + (90 \times 10^{-6})(\sin 38.0°)$$
$$= 370 \times 10^{-6}$$

4. ***Finalize***: This result shows that the larger principal strain ε_1 is at the angle $\theta_{p1} = 19.0°$. The smaller strain ε_2 acts at 90° from that direction ($\theta_{p2} = 109.0°$). Thus,

$$\varepsilon_1 = 370 \times 10^{-6} \quad \text{and} \quad \theta_{p1} = 19.0°$$
$$\varepsilon_2 = 80 \times 10^{-6} \quad \text{and} \quad \theta_{p2} = 109.0°$$

Note that $\varepsilon_1 + \varepsilon_2 = \varepsilon_x + \varepsilon_y$.

The principal strains are portrayed in Fig. 7-38c. There are, of course, no shear strains on the principal planes.

Part (c): Maximum shear strain.

1, 2, 3. ***Conceptualize, Categorize, Analyze***: The maximum shear strain is calculated from Eq. (7-86):

$$\frac{\gamma_{max}}{2} = \sqrt{\left(\frac{\varepsilon_x - \varepsilon_y}{2}\right)^2 + \left(\frac{\gamma_{xy}}{2}\right)^2} = 146 \times 10^{-6} \quad \gamma_{max} = 290 \times 10^{-6}$$

The element having the maximum shear strains is oriented at 45° to the principal directions; therefore, $\theta_s = 19.0° + 45° = 64.0°$ and $2\theta_s = 128.0°$. Substitute this value of $2\theta_s$ into the second transformation equation

[Eq. (7-82b)] to determine the sign of the shear strain associated with this direction. The calculations are

$$\frac{\gamma_{x1y1}}{2} = -\frac{\varepsilon_x - \varepsilon_y}{2} \sin 2\theta + \frac{\gamma_{xy}}{2} \cos 2\theta$$

$$= -(115 \times 10^{-6})(\sin 128.0°) + (90 \times 10^{-6})(\cos 128.0°)$$

$$= -146 \times 10^{-6}$$

4. ***Finalize***: This result shows that an element oriented at an angle $\theta_{s2} = 64.0°$ has the maximum negative shear strain.

You can arrive at the same result by observing that the angle θ_{s1} to the direction of maximum positive shear strain is always 45° less than θ_{p1}. Hence,

$$\theta_{s1} = \theta_{p1} - 45° = 19.0° - 45° = -26.0°$$

$$\theta_{s2} = \theta_{s1} + 90° = 64.0°$$

The shear strains corresponding to θ_{s1} and θ_{s2} are $\gamma_{max} = 290 \times 10^{-6}$, and $\gamma_{min} = -290 \times 10^{-6}$, respectively.

The normal strains on the element having the maximum and minimum shear strains are

$$\varepsilon_{aver} = \frac{\varepsilon_x + \varepsilon_y}{2} = 225 \times 10^{-6}$$

A sketch of the element having the maximum in-plane shear strains is shown in Fig. 7-38d.

In this example, the strains were computed by using the transformation equations. However, all of the results can be obtained just as easily from Mohr's circle, as illustrated in Fig. 7-39.

FIGURE 7-39

Example 7-8: Mohr's circle for element of material in plane strain

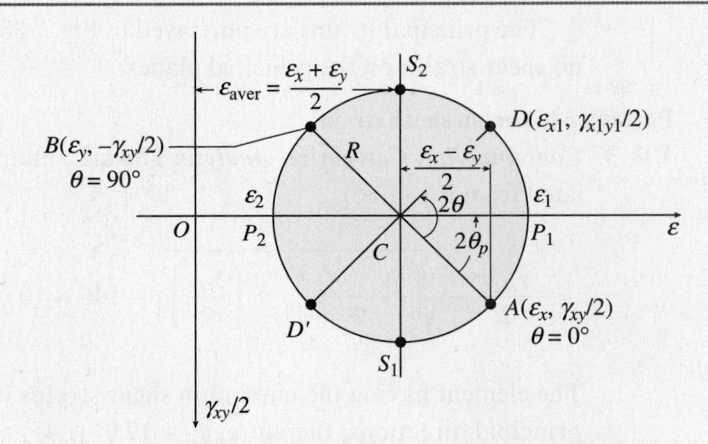

Example 7-9

FIGURE 7-40
Example 7-9: (a) 45° strain rosette and (b) element oriented at an angle θ to the x-y axes

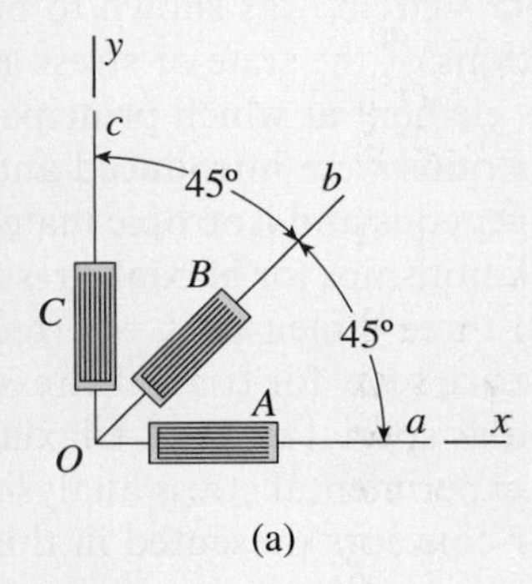

(a)

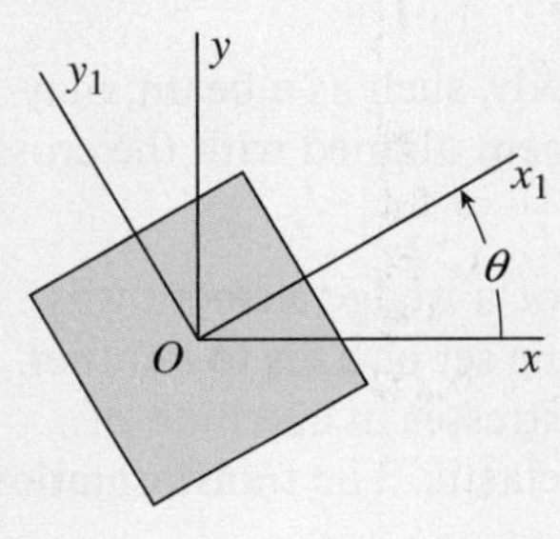

(b)

A 45° strain rosette (also called a *rectangular rosette*) consists of three electrical-resistance strain gages arranged to measure strains in two perpendicular directions and also at a 45° angle between them, as shown in Fig. 7-40a. The rosette is bonded to the surface of the structure before it is loaded. Gages A, B, and C measure the normal strains ε_a, ε_b, and ε_c in the directions of lines Oa, Ob, and Oc, respectively.

Explain how to obtain the strains ε_{x1}, ε_{y1}, and γ_{x1y1} associated with an element oriented at an angle θ to the xy axes (Fig. 7-40b).

Solution:

Use a four-step problem-solving approach. Combine steps as needed for an efficient solution.

1, 2. ***Conceptualize, Categorize:*** At the surface of the stressed object, the material is in plane stress. Since the strain-transformation equations [Eqs. (7-82a and b)] apply to plane stress as well as to plane strain, use those equations to determine the strains in any desired direction.

3. ***Analyze:***

Strains associated with the x-y axes: Begin by determining the strains associated with the x-y axes. Because gages A and C are aligned with the x and y axes, respectively, they give the strains ε_x and ε_y directly:

$$\varepsilon_x = \varepsilon_a \qquad \varepsilon_y = \varepsilon_c \tag{7-88a,b}$$

To obtain the shear strain γ_{xy}, use the transformation equation for normal strains [Eq. (7-82a)]:

$$\varepsilon_{x1} = \frac{\varepsilon_x + \varepsilon_y}{2} + \frac{\varepsilon_x - \varepsilon_y}{2}\cos 2\theta + \frac{\gamma_{xy}}{2}\sin 2\theta$$

For an angle $\theta = 45°$, $\varepsilon_{x1} = \varepsilon_b$ (Fig. 7-40a); therefore, the preceding equation gives

$$\varepsilon_b = \frac{\varepsilon_a + \varepsilon_c}{2} + \frac{\varepsilon_a - \varepsilon_c}{2}(\cos 90°) + \frac{\gamma_{xy}}{2}(\sin 90°)$$

Solve for γ_{xy} to get

$$\gamma_{xy} = 2\varepsilon_b - \varepsilon_a - \varepsilon_c \tag{7-89}$$

Thus, the strains ε_x, ε_y, and γ_{xy} are easily determined from the given strain-gage readings.

4. ***Finalize:***

Strains associated with the x_1y_1 axes: Using the strains ε_x, ε_y, and γ_{xy}, calculate the strains for an element oriented at any angle θ (Fig. 7-40b) from the strain-transformation equations [Eqs. (7-82a and b)] or from Mohr's circle. The principal strains and the maximum shear strains are computed using from Eqs. (7-85) and (7-86), respectively.

CHAPTER SUMMARY AND REVIEW

Chapter 7 discussed the **state of stress** at a point on a stressed body and then displayed it on a stress element. In two dimensions, **plane stress** was defined and expressed in terms of transformation equations that gave different, but equivalent, expressions of the state of normal and shear stresses at that point. **Principal normal stresses** and **maximum shear stress**, and their orientations, were seen to be the most important information for design. A graphical representation of the transformation equations, **Mohr's circle**, was shown to be a convenient way of exploring various representations of the state of stress at a point, including those orientations of the stress element at which principal stresses and maximum shear stress occur. Later, strains were introduced and **Hooke's law for plane stress** was derived for homogeneous and isotropic materials and then specialized to obtain stress-strain relationships for **biaxial stress**, uniaxial stress, and pure shear. The stress state in three dimensions, referred to as **triaxial stress**, was introduced along with Hooke's law for triaxial stress. **Spherical stress** and **hydrostatic stress** were defined as special cases of triaxial stress. Finally, **plane strain** was defined for use in experimental stress analysis and compared to **plane stress**. Here are the major concepts presented in this chapter.

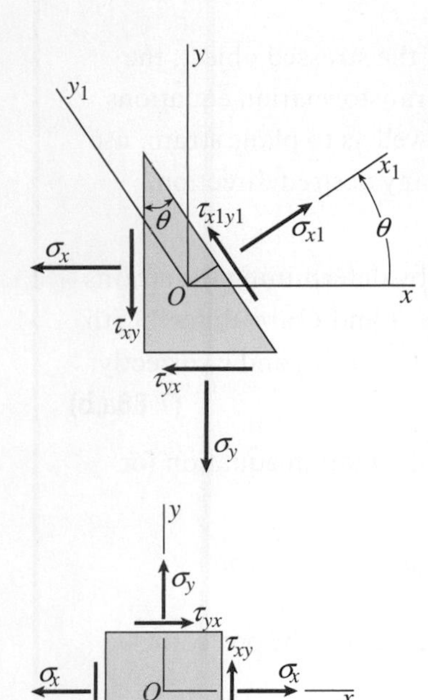

1. The **stresses on inclined sections** cut through a body, such as a beam, may be larger than the stresses acting on a stress element aligned with the cross section.
2. Stresses are tensors, not vectors, so equilibrium of a wedge element was used to transform the stress components from one set of axes to another. The transformation equations are applicable to stresses in any kind of material, whether linear, nonlinear, elastic, or inelastic. The **transformation equations for plane stress** are

$$\sigma_{x1} = \frac{\sigma_x + \sigma_y}{2} + \frac{\sigma_x - \sigma_y}{2}\cos 2\theta + \tau_{xy}\sin 2\theta$$

$$\tau_{x1y1} = -\frac{\sigma_x - \sigma_y}{2}\sin 2\theta + \tau_{xy}\cos 2\theta$$

$$\sigma_{y1} = \frac{\sigma_x + \sigma_y}{2} - \frac{\sigma_x - \sigma_y}{2}\cos 2\theta - \tau_{xy}\sin 2\theta$$

3. If two elements with different orientations are used to display the **state of plane stress** at the same point in a body, the stresses acting on the faces of the two elements are different, but they still represent the same intrinsic state of stress at that point.
4. The shear stresses acting on all four side faces of a stress element in plane stress are known if the shear stress acting on any one of those faces is known.

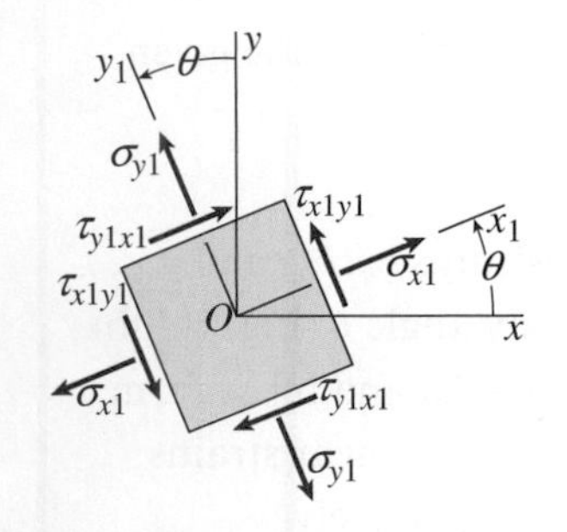

5. The sum of the normal stresses acting on perpendicular faces of plane-stress elements at a given point in a stressed body is constant and independent of the angle θ:

$$\sigma_{x1} + \sigma_{y1} = \sigma_x + \sigma_y$$

6. The maximum and minimum normal stresses, called the **principal stresses** σ_1 and σ_2, can be found from the transformation equation for normal stress as

$$\sigma_{1,2} = \frac{\sigma_x + \sigma_y}{2} \pm \sqrt{\left(\frac{\sigma_x - \sigma_y}{2}\right)^2 + \tau_{xy}^2}$$

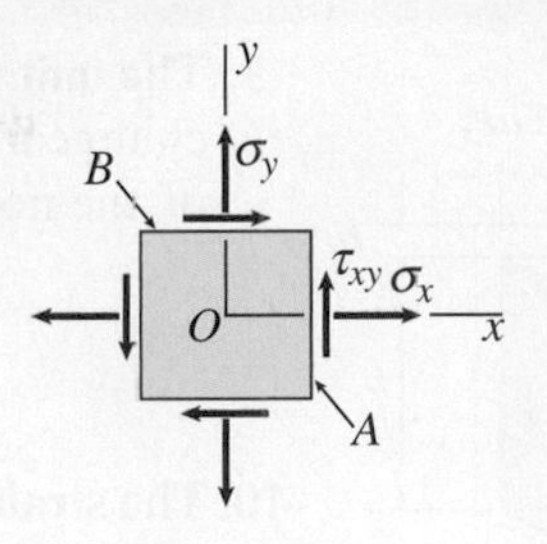

The principal planes, on which they act are at orientation angle θ_p. The shear stresses are zero on the principal planes, the planes of maximum shear stress occur at 45° to the principal planes, and the maximum shear stress is equal to one-half the difference of the principal stresses. Maximum shear stress can be computed from the normal and shear stresses on the original element, or from the principal stresses as

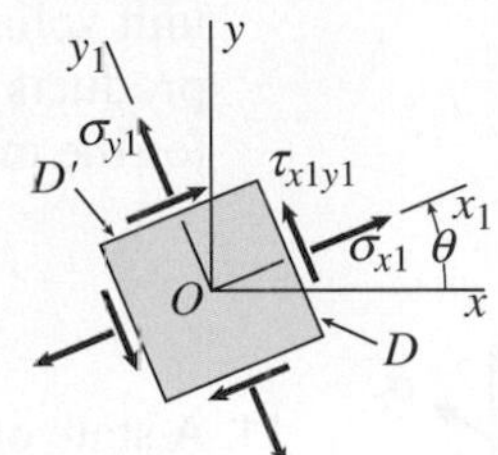

$$\tau_{max} = \sqrt{\left(\frac{\sigma_x - \sigma_y}{2}\right)^2 + \tau_{xy}^2}$$

$$\tau_{max} = \frac{\sigma_1 - \sigma_2}{2}$$

7. The transformation equations for plane stress are represented in graphical form as **Mohr's circle**, which displays the relationship between normal and shear stresses acting on various inclined planes at a point in a stressed body. It also is used for calculating principal stresses, maximum shear stresses, and the orientations of the elements on which they act.

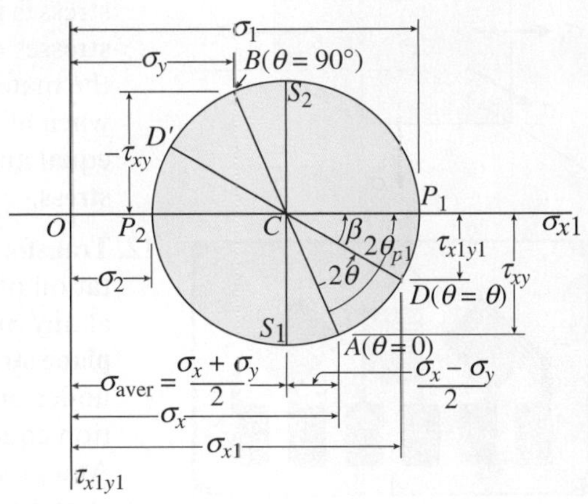

8. **Hooke's law for plane stress** provides the relationships between normal strains and stresses for homogeneous and isotropic materials which follow Hooke's law. These relationships contain three material constants (E, G, and ν). When the normal stresses in plane stress are known, the normal strains in the x, y and z directions are

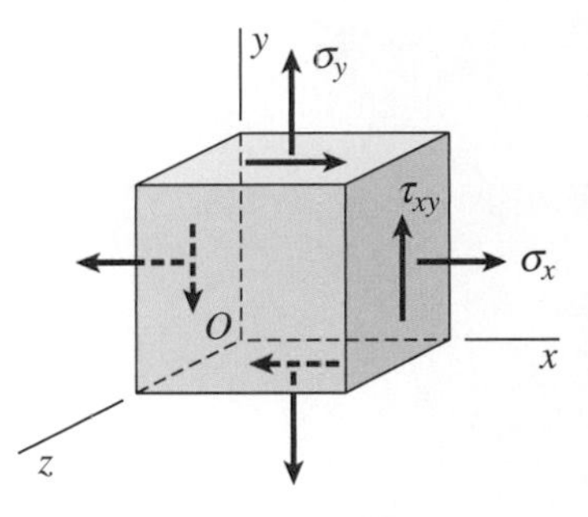

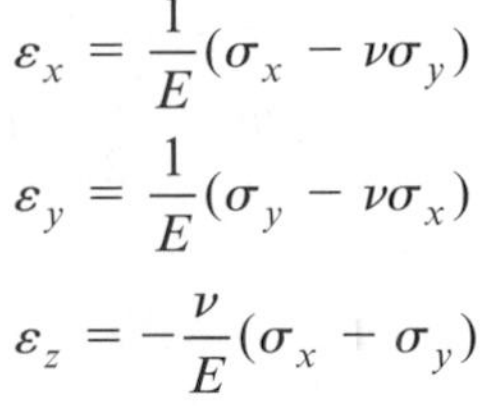

$$\varepsilon_x = \frac{1}{E}(\sigma_x - \nu\sigma_y)$$

$$\varepsilon_y = \frac{1}{E}(\sigma_y - \nu\sigma_x)$$

$$\varepsilon_z = -\frac{\nu}{E}(\sigma_x + \sigma_y)$$

Solution of these equations gives the x and y normal stresses in terms of the strains:

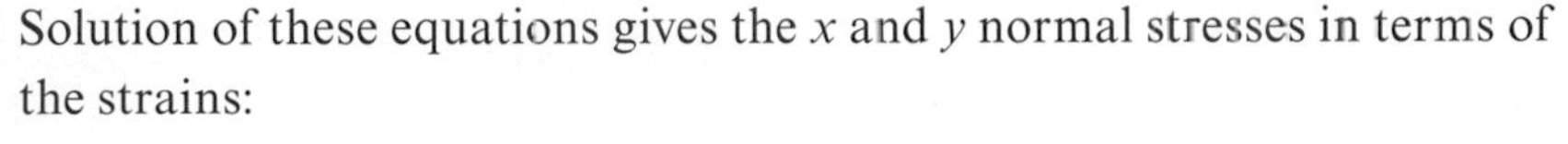

$$\sigma_x = \frac{E}{1 - \nu^2}(\varepsilon_x + \nu\varepsilon_y)$$

$$\sigma_y = \frac{E}{1 - \nu^2}(\varepsilon_y + \nu\varepsilon_x)$$

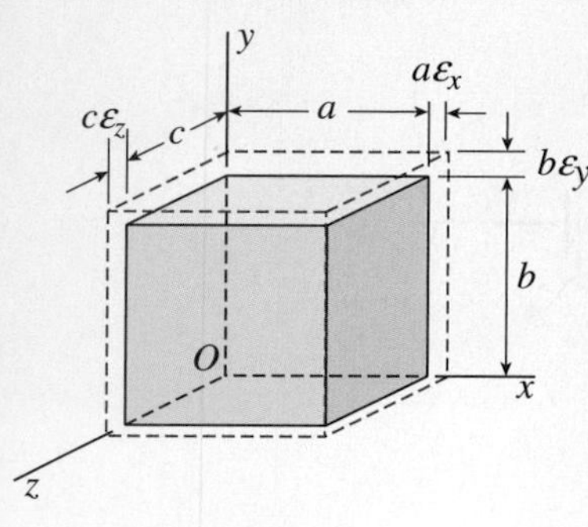

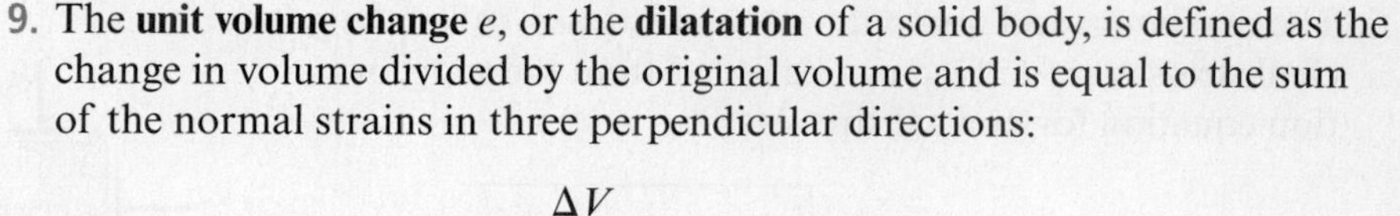

9. The **unit volume change** e, or the **dilatation** of a solid body, is defined as the change in volume divided by the original volume and is equal to the sum of the normal strains in three perpendicular directions:

$$e = \frac{\Delta V}{V_0} = \varepsilon_x + \varepsilon_y + \varepsilon_z$$

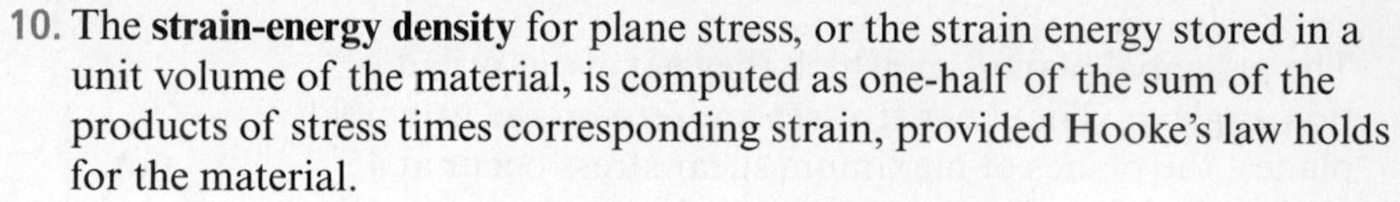

10. The **strain-energy density** for plane stress, or the strain energy stored in a unit volume of the material, is computed as one-half of the sum of the products of stress times corresponding strain, provided Hooke's law holds for the material.

$$u = \frac{1}{2}(\sigma_x \varepsilon_x + \sigma_y \varepsilon_y + \sigma_z \varepsilon_z)$$

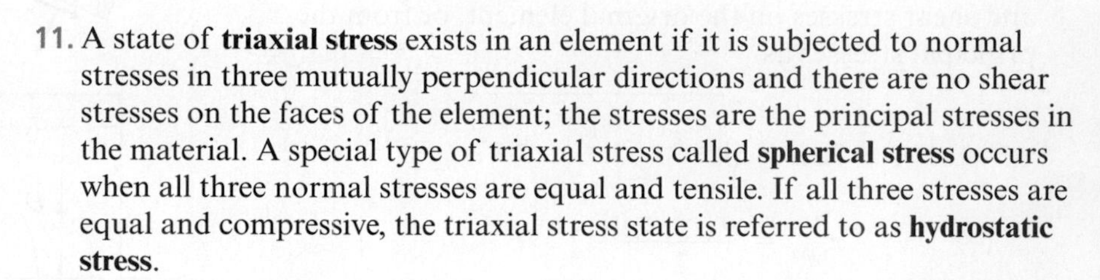

11. A state of **triaxial stress** exists in an element if it is subjected to normal stresses in three mutually perpendicular directions and there are no shear stresses on the faces of the element; the stresses are the principal stresses in the material. A special type of triaxial stress called **spherical stress** occurs when all three normal stresses are equal and tensile. If all three stresses are equal and compressive, the triaxial stress state is referred to as **hydrostatic stress**.

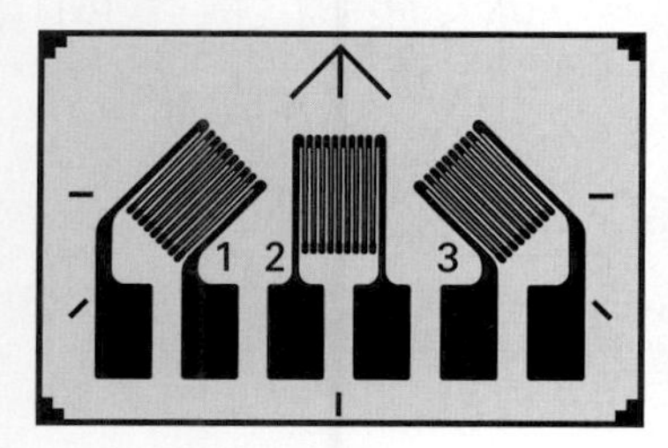

12. **Transformation equations for plane strain** are derived for use in the interpretation of experimental measurements made with strain gages. Plane strains at any orientation are represented in graphical form using **Mohr's circle for plane strain**. Plane stress and plane strain are compared in Fig. 7-32, and under ordinary conditions do not occur simultaneously. The transformation equations for plane strain are independent of the material properties. At a given point in a stressed body, the **principal strains** and **principal stresses** occur in the same directions. Last, the transformation equations for plane stress also can be used for the stresses in plane strain, and the transformation equations for plane strain can be used for the strains in plane stress. The transformation equations for plane strain are

$$\varepsilon_{x1} = \frac{\varepsilon_x + \varepsilon_y}{2} + \frac{\varepsilon_x - \varepsilon_y}{2}\cos 2\theta + \frac{\gamma_{xy}}{2}\sin 2\theta$$

$$\frac{\gamma_{x1y1}}{2} = -\frac{\varepsilon_x - \varepsilon_y}{2}\sin 2\theta + \frac{\gamma_{xy}}{2}\cos 2\theta$$

PROBLEMS Chapter 7

7.2 Plane Stress

Introductory Problems

7.2-1 The stresses on the bottom surface of a fuel tanker (figure part a) are known to be $\sigma_x = 50$ MPa, $\sigma_y = 8$ MPa, and $\tau_{xy} = 6.5$ MPa (figure part b).

Determine the stresses acting on an element oriented at an angle $\theta = 52°$ from the x axis, where the angle θ is positive when counterclockwise. Show these stresses on a sketch of an element oriented at the angle θ.

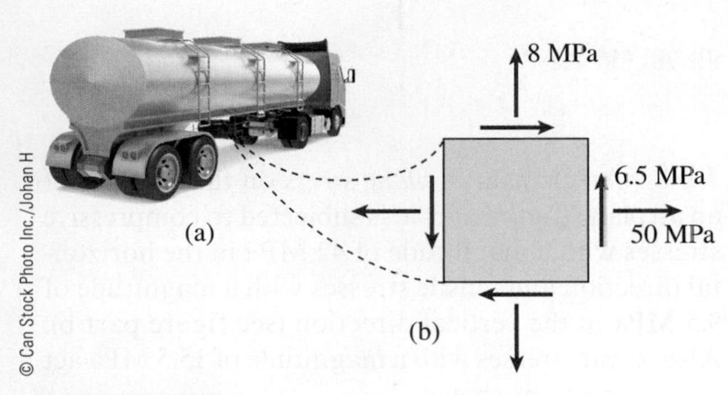

PROBLEM 7.2-1

7.2-2 Solve the preceding problem for an element in *plane stress* on the bottom surface of a fuel tanker (figure part a); stresses are $\sigma_x = 105$ MPa, $\sigma_y = 75$ MPa, and $\tau_{xy} = 25$ MPa.

Determine the stresses acting on an element oriented at an angle $\theta = 40°$ from the x axis, where the angle θ is positive when counterclockwise. Show these stresses on a sketch of an element oriented at the angle θ.

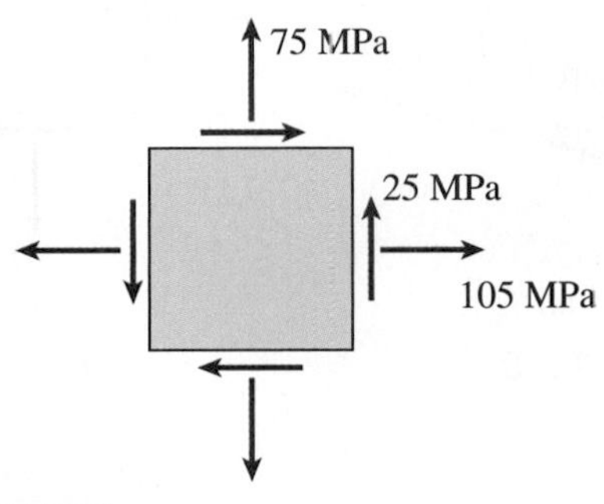

PROBLEM 7.2-2

7.2-3 The stresses on an element are $\sigma_x = 7$ MPa, $\sigma_y = 3.5$ MPa, and $\tau_{xy} = 2.4$ MPa. Find the stresses acting on an element oriented at an angle $\theta = 25°$. Show these stresses on the rotated element.

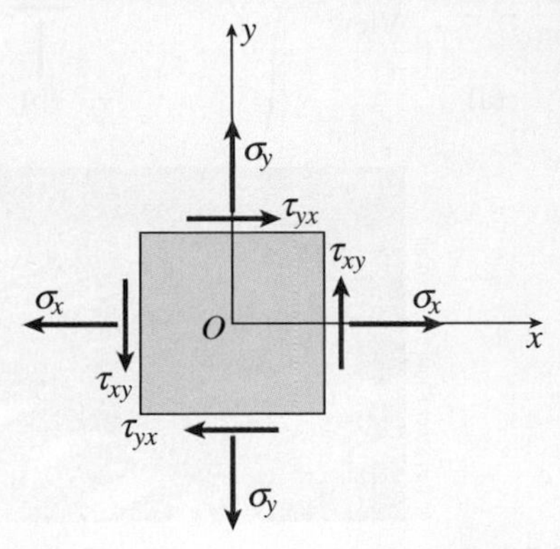

PROBLEM 7.2-3

7.2-4 The stresses on an element are known to be $\sigma_x = 120$ MPa, $\sigma_y = 100$ MPa, and $\tau_{xy} = 75$ MPa. Find the stresses on an inclined section through the element at an angle $\theta = 45°$.

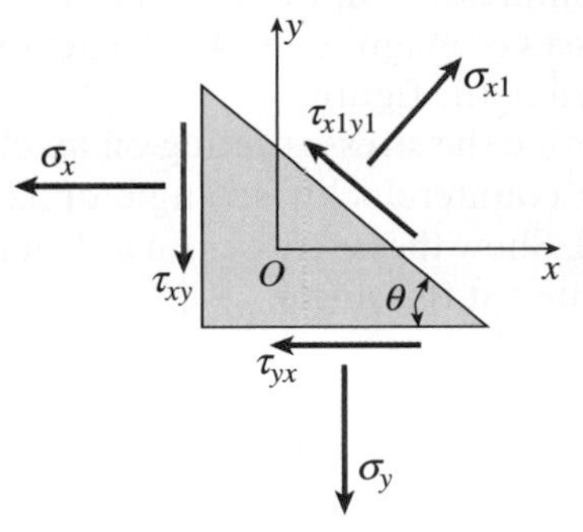

PROBLEM 7.2-4

7.2-5 The stresses acting on element A on the web of a train rail (see figure part a) are found to be 45 MPa tension in the horizontal direction and 120 MPa compression in the vertical direction (see figure part b). Also, shear stresses with a magnitude of 25 MPa act in the directions shown.

Determine the stresses acting on an element oriented at a counterclockwise angle of 32° from the horizontal. Show these stresses on a sketch of an element oriented at this angle.

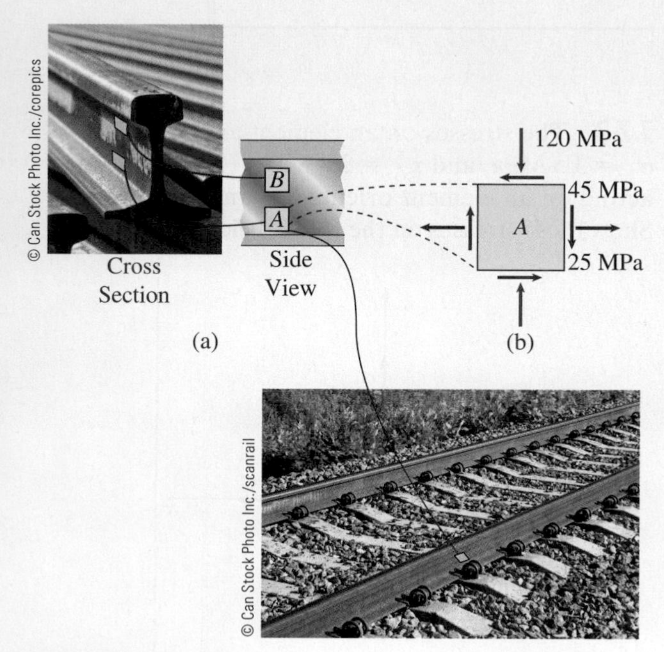

PROBLEM 7.2-5

7.2-6 Solve the preceding problem if the stresses acting on element *A* on the web of a train rail (see figure part a of Problem 7.2-5) are found to be 40 MPa in tension in the horizontal direction and 160 MPa in compression in the vertical direction. Also, shear stresses of magnitude 54 MPa act in the directions shown in the figure.

Determine the stresses acting on an element oriented at a counterclockwise angle of 52° from the horizontal. Show these stresses on a sketch of an element oriented at this angle.

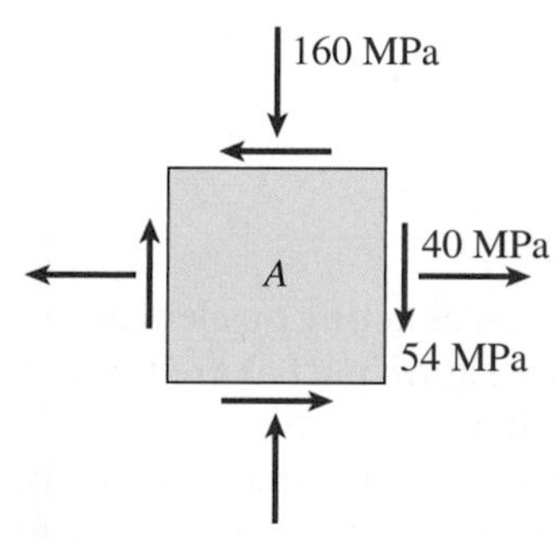

PROBLEM 7.2-6

7.2-7 The stresses acting on element *B* on the web of a train rail (see figure part a of Problem 7.2-5) are found to be 40 MPa in compression in the horizontal direction and 16 MPa in compression in the vertical direction (see figure). Also, shear stresses of magnitude 17 MPa act in the directions shown.

Determine the stresses acting on an element oriented at a counterclockwise angle of 48° from the horizontal. Show these stresses on a sketch of an element oriented at this angle.

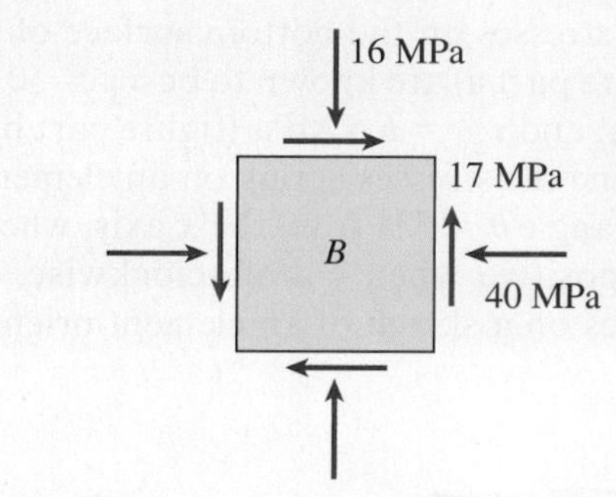

PROBLEM 7.2-7

7.2-8 An element in *plane stress* on the fuselage of an airplane (figure part a) is subjected to compressive stresses with a magnitude of 42 MPa in the horizontal direction and tensile stresses with a magnitude of 9.5 MPa in the vertical direction (see figure part b). Also, shear stresses with a magnitude of 15.5 MPa act in the directions shown.

Determine the stresses acting on an element oriented at a clockwise angle of 40° from the horizontal. Show these stresses on a sketch of an element oriented at this angle.

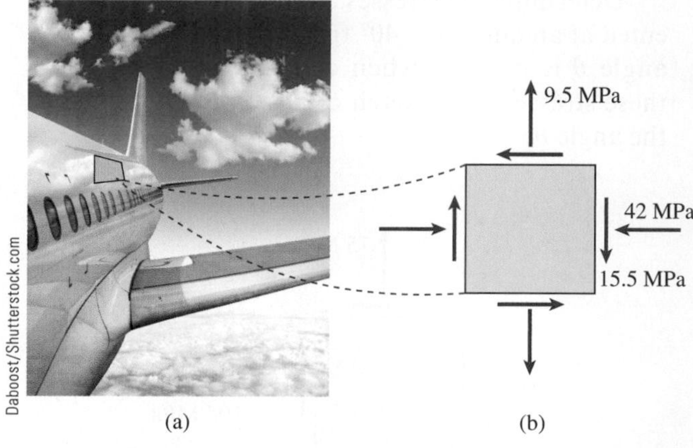

PROBLEM 7.2-8

7.2-9 The stresses acting on element *B* (see figure part a) on the web of a wide-flange beam are found to be 100 MPa in compression in the horizontal direction and 17 MPa in compression in the vertical direction (see figure part b). Also, shear stresses with a magnitude of 24 MPa act in the directions shown.

Determine the stresses acting on an element oriented at a counterclockwise angle of 36° from the horizontal. Show these stresses on a sketch of an element oriented at this angle.

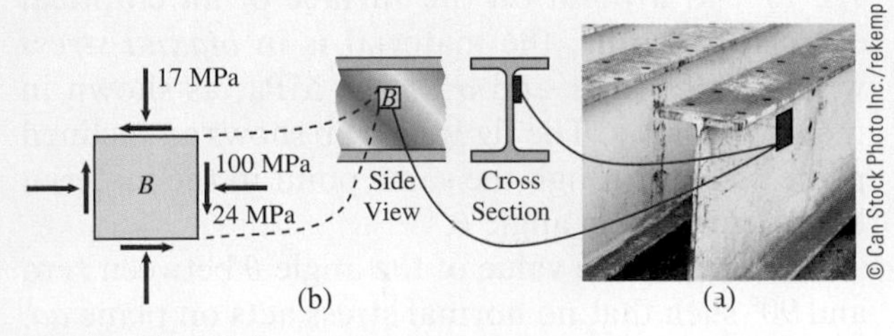

PROBLEM 7.2-9

7.2-10 Solve the preceding problem if the normal and shear stresses acting on element B are 56 MPa, 17 MPa, and 27 MPa (in the directions shown in the figure) and the angle is 40° (clockwise).

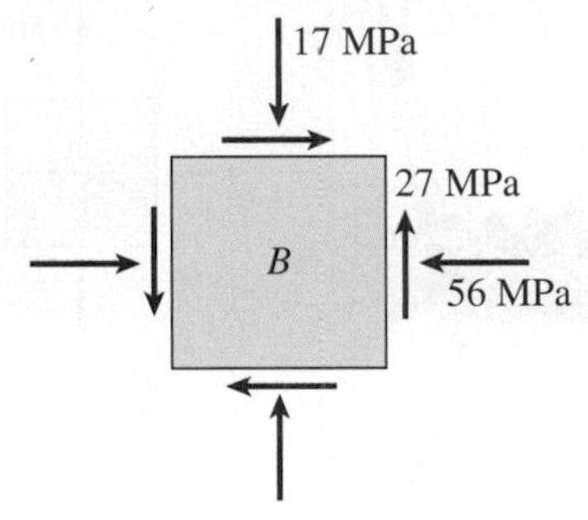

PROBLEM 7.2-10

Representative Problems

7.2-11 The polyethylene liner of a settling pond is subjected to stresses $\sigma_x = 2.5$ MPa, $\sigma_y = 0.75$ MPa, and $\tau_{xy} = -0.8$ MPa, as shown by the plane-stress element in the figure part a.

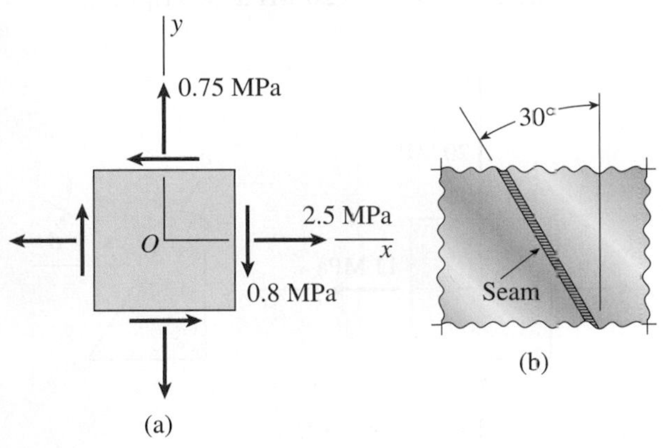

PROBLEM 7.2-11

Determine the normal and shear stresses acting on a seam oriented at an angle of 30° to the element, as shown in the figure part b. Show these stresses on a sketch of an element having its sides parallel and perpendicular to the seam.

7.2-12 Solve the preceding problem if the normal and shear stresses acting on the element are $\sigma_x = 2100$ kPa, $\sigma_y = 300$ kPa, and $\tau_{xy} = -560$ kPa, and the seam is oriented at an angle of 22.5° to the element.

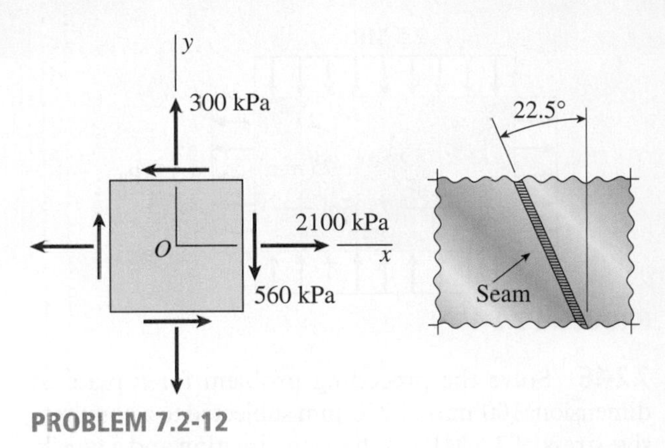

PROBLEM 7.2-12

7.2-13 Two steel rods are welded together (see figure); the seam is oriented at angle $\theta = 50°$. The stresses on the rotated element are $\sigma_{x1} = 70$ MPa, $\sigma_{y1} = -83$ MPa, and $\tau_{x1y1} = -35$ MPa. Find the state of plane stress on the element if it is rotated clockwise to align the x_1 axis with the longitudinal axis of the rods.

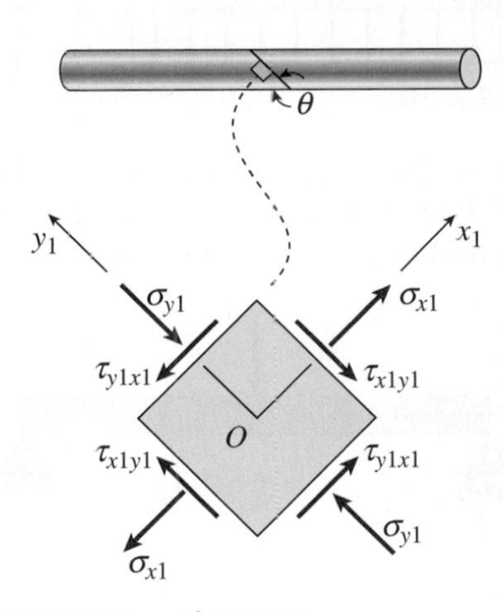

PROBLEMS 7.2-13 and 7.2-14

7.2-14 Repeat the previous problem using $\theta = 30°$ and stresses on the rotated element: $\sigma_{x1} = 70$ MPa, $\sigma_{y1} = -82$ MPa, and $\tau_{x1y1} = -35$ MPa.

7.2-15 A rectangular plate of dimensions 75 mm × 125 mm is formed by welding two triangular plates (see figure). The plate is subjected to a tensile stress of 3.5 MPa in the long direction and a compressive stress of 2.5 MPa in the short direction.

Determine the normal stress σ_w acting perpendicular to the line of the weld and the shear stress τ_w acting parallel to the weld. (Assume that the normal stress σ_w is positive when it acts in tension against the weld and the shear stress τ_w is positive when it acts counterclockwise against the weld.)

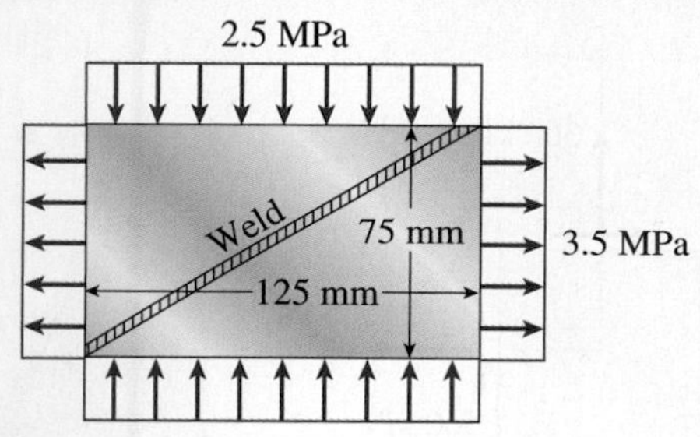

PROBLEM 7.2-15

7.2-16 Solve the preceding problem for a plate of dimensions 100 mm × 250 mm subjected to a compressive stress of 2.5 MPa in the long direction and a tensile stress of 12.0 MPa in the short direction (see figure).

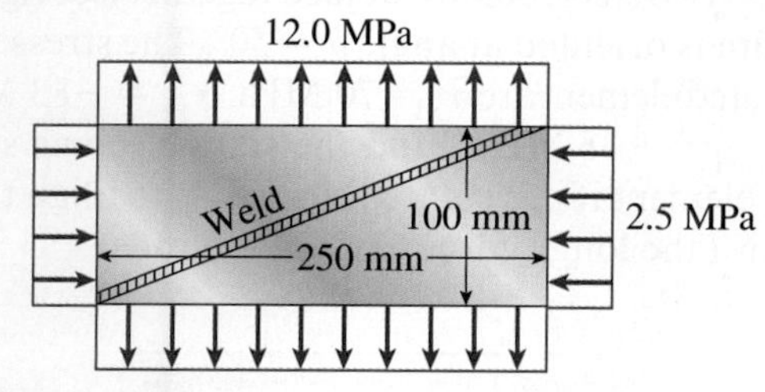

PROBLEM 7.2-16

7.2-17 A simply supported beam is subjected to point load P at mid-span. The normal stress on an element at mid-span is known to be $\sigma_x = 10$ MPa.

Determine the element stresses if it is rotated through angle $\theta = -45°$. Show these stresses on a sketch of an element oriented at that angle.

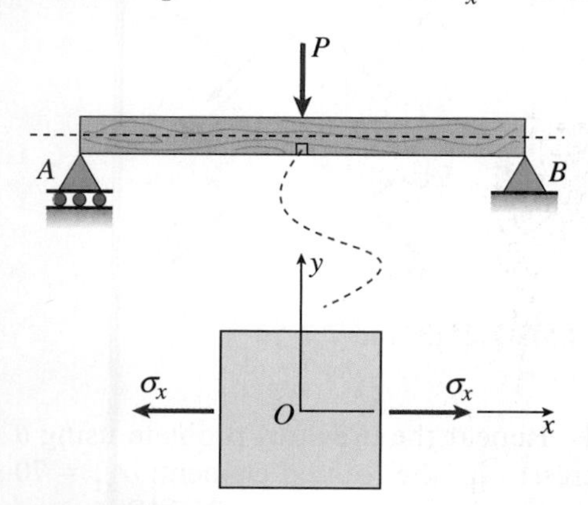

PROBLEMS 7.2-17 and 7.2-18

7.2-18 Repeat the previous problem using $\sigma_x = 12$ MPa and $\theta = -30°$.

7.2-19 At a point on the surface of an elliptical exercise machine, the material is in *biaxial stress* with $\sigma_x = 9.7$ MPa and $\sigma_y = -6$ MPa, as shown in the figure part a. The figure part b shows an inclined plane *aa* cut through the same point in the material but oriented at an angle θ.

Determine the value of the angle θ between zero and 90° such that no normal stress acts on plane *aa*. Sketch a stress element having plane *aa* as one of its sides and show all stresses acting on the element.

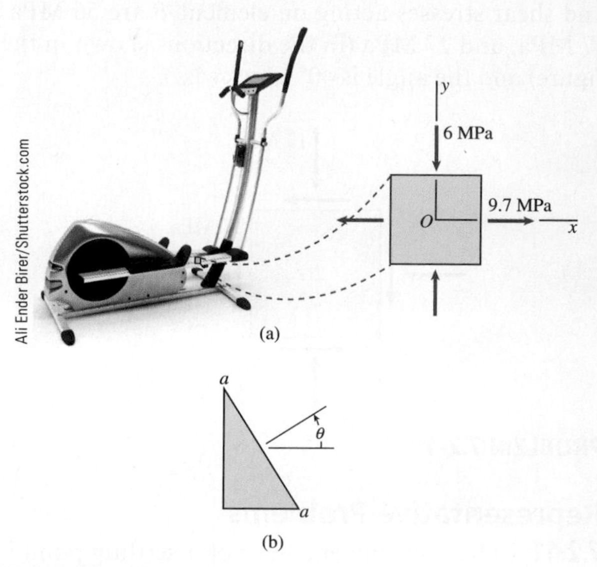

PROBLEM 7.2-19

7.2-20 Solve the preceding problem for $\sigma_x = 11$ MPa and $\sigma_y = -20$ MPa (see figure).

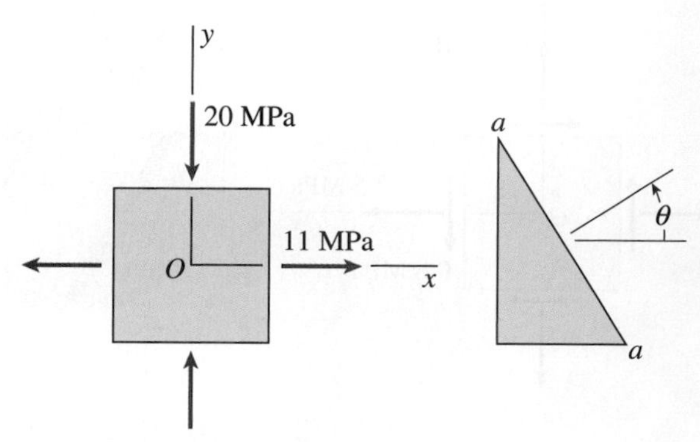

PROBLEM 7.2-20

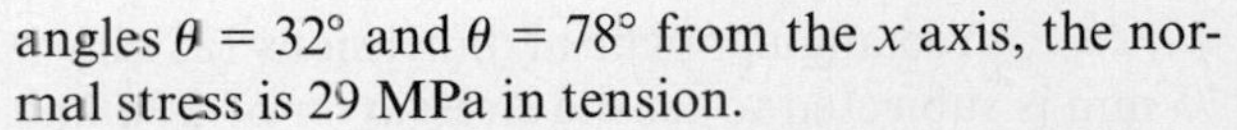

7.2-21 An element in *plane stress* from the frame of a racing car is oriented at a known angle θ (see figure). On this inclined element, the normal and shear stresses have the magnitudes and directions shown in the figure.

Determine the normal and shear stresses acting on an element whose sides are parallel to the xy axes, that is, determine σ_x, σ_y, and τ_{xy}. Show the results on a sketch of an element oriented at $\theta = 0°$.

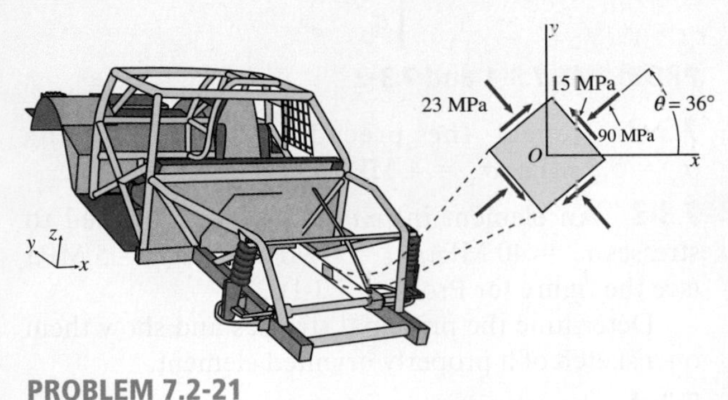

PROBLEM 7.2-21

7.2-22 Solve the preceding problem for the element shown in the figure.

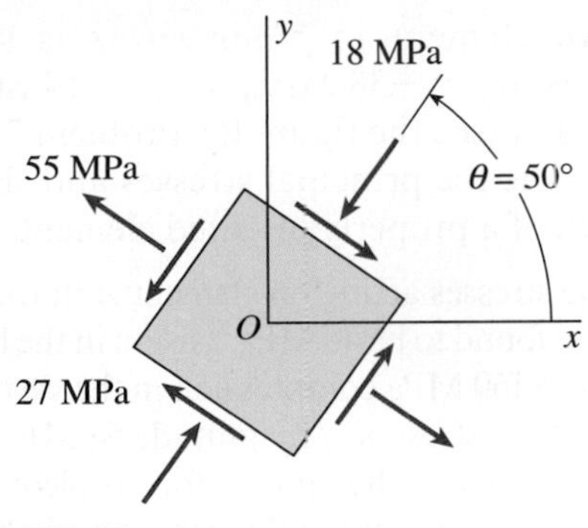

PROBLEM 7.2-22

7.2-23 A gusset plate on a truss bridge is in *plane stress* with normal stresses σ_x and σ_y and *shear stress* τ_{xy}, as shown in the figure. At counterclockwise angles $\theta = 32°$ and $\theta = 78°$ from the x axis, the normal stress is 29 MPa in tension.

If the stress σ_x equals 18 MPa in tension, what are the stresses σ_y and τ_{xy}?

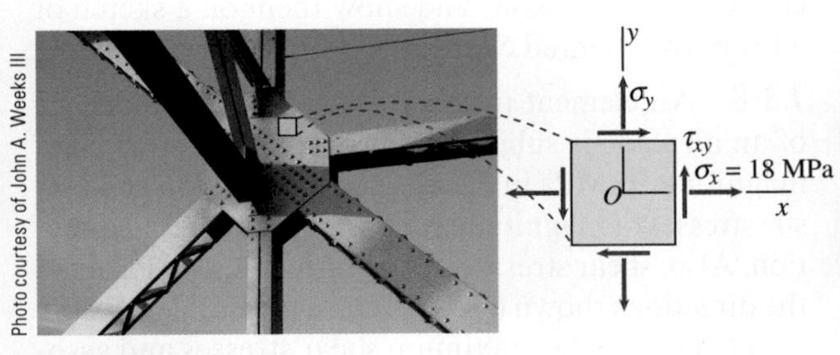

PROBLEM 7.2-23

7.2-24 The surface of an airplane wing is subjected to plane stress with normal stresses σ_x and σ_y and shear stress τ_{xy}, as shown in the figure. At a counterclockwise angle $\theta = 32°$ from the x axis, the normal stress is 29 MPa in tension, and at an angle $\theta = 46°$, it is 17 MPa in compression.

If the stress σ_x equals 105 MPa in tension, what are the stresses σ_y and τ_{xy}?

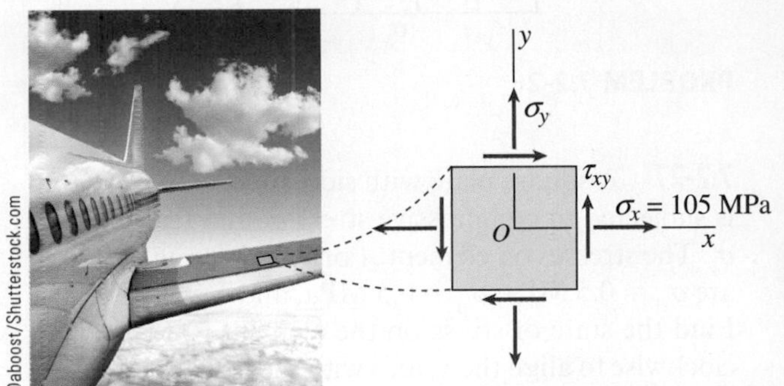

PROBLEM 7.2-24

7.2-25 At a point on the web of a girder on an overhead bridge crane in a manufacturing facility, the stresses are known to be $\sigma_x = -30$ MPa, $\sigma_y = 12$ MPa, and $\tau_{xy} = 21$ MPa (the sign convention for these stresses is shown in Fig. 7-1). A stress element located at the same point in the structure (but oriented at a counterclockwise angle θ_1 with respect to the x axis) is subjected to the stresses shown in the figure (σ_b, τ_b, and 14 MPa).

Assuming that the angle θ_1 is between zero and 90°, calculate the normal stress σ_b, the shear stress τ_b, and the angle θ_1.

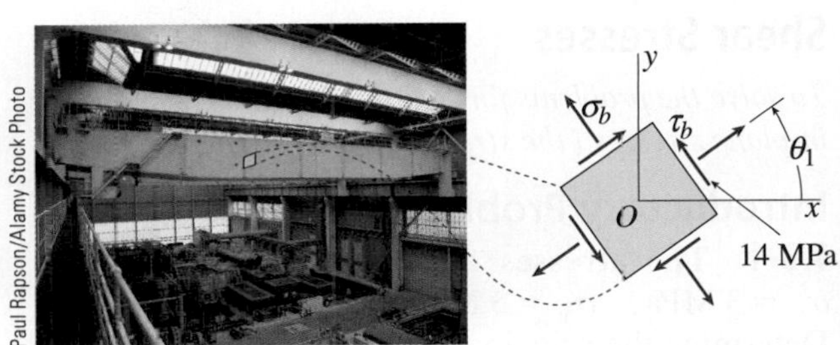

PROBLEM 7.2-25

7.2-26 A rectangular plate of dimensions 125 mm × 75 mm is subjected to tensile stress σ_x = 67 kPa and compressive stress σ_y. If it is known that the normal stress along the diagonal t–t is σ_t = −6.57 kPa, find stress σ_y on element A.

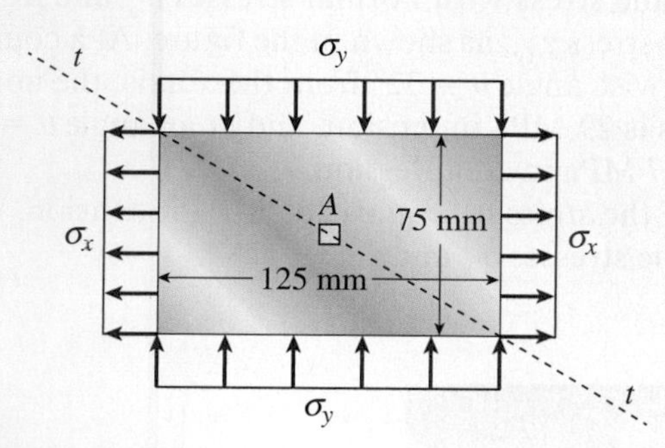

PROBLEM 7.2-26

7.2-27 A square plate with side dimension of 50 mm is subjected to compressive stress σ_x and tensile stress σ_y. The stresses on element A oriented at angle θ = 45° are σ_{x1} = 0.5 MPa, σ_{y1} = 0.5 MPa, and τ_{x1y1} = 1.9 MPa. Find the state of stress on the element if it is rotated clockwise to align the x_1 axis with the horizontal x axis.

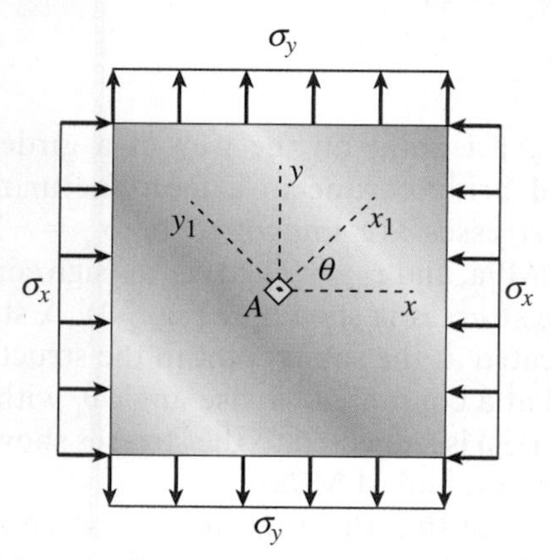

PROBLEM 7.2-27

7.3 Principal Stresses and Maximum Shear Stresses

To solve the problems for Section 7.3, consider only the in-plane stresses (the stresses in the x-y plane).

Introductory Problems

7.3-1 The stresses acting on an element are σ_x = 3 MPa, σ_y = 5 MPa, and τ_{xy} = 2.8 MPa. Determine the principal stresses and show them on a sketch of a properly oriented element.

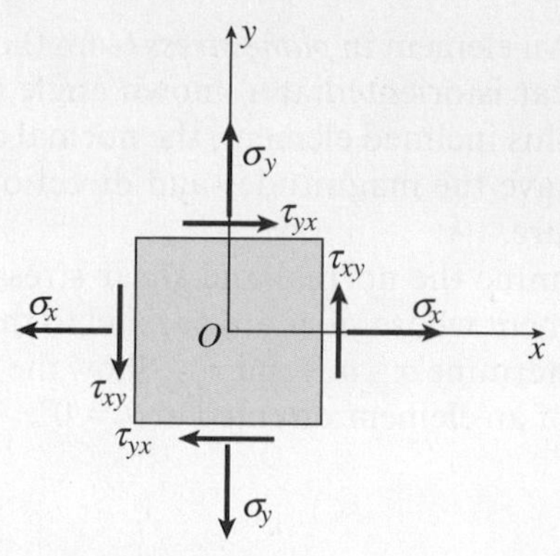

PROBLEMS 7.3-1 and 7.3-2

7.3-2 Repeat the preceding problem using σ_x = 5.5 MPa, σ_y = 4 MPa, and τ_{xy} = 3.2 MPa.

7.3-3 An element in *plane stress* is subjected to stresses σ_x = 40 MPa, σ_y = 8 MPa, and τ_{xy} = 5 MPa (see the figure for Problem 7.3-1).

Determine the principal stresses and show them on a sketch of a properly oriented element.

7.3-4 An element in *plane stress* is subjected to stresses σ_x = 105 MPa, σ_y = 75 MPa, and τ_{xy} = 25 MPa (see the figure for Problem 7.3-1).

Determine the principal stresses and show them on a sketch of a properly oriented element.

7.3-5 An element in *plane stress* is subjected to stresses σ_x = −38 MPa, σ_y = −14 MPa, and τ_{xy} = 13 MPa (see the figure for Problem 7.3-1).

Determine the principal stresses and show them on a sketch of a properly oriented element.

7.3-6 The stresses acting on element A in the web of a train rail are found to be 40 MPa tension in the horizontal direction and 160 MPa compression in the vertical direction. Also, shear stresses of magnitude 54 MPa act in the directions shown (see the figure b for Problem 7.2-5).

Determine the principal stresses and show them on a sketch of a properly oriented element.

7.3-7 The normal and shear stresses acting on element A are 45 MPa, 119 MPa, and 20 MPa (see the figure b for Problem 7.2-5).

Determine the maximum shear stresses and associated normal stresses and show them on a sketch of a properly oriented element.

7.3-8 An element in *plane stress* from the fuselage of an airplane is subjected to compressive stresses of magnitude 35 MPa in the horizontal direction and tensile stresses of magnitude 6.5 MPa in the vertical direction. Also, shear stresses of magnitude 12.5 MPa act in the directions shown (see the figure for Problem 7.2-8).

Determine the maximum shear stresses and associated normal stresses and show them on a sketch of a properly oriented element.

7.3-9 The stresses acting on element B in the web of a wide-flange beam are found to be 97 MPa compression in the horizontal direction and 18 MPa compression in the vertical direction. Also, shear stresses of magnitude 26 MPa act in the directions shown (see the figure for Problem 7.2-9).

Determine the maximum shear stresses and associated normal stresses and show them on a sketch of a properly oriented element.

7.3-10 The normal and shear stresses acting on element B are $\sigma_x = -46$ MPa, $\sigma_y = -13$ MPa, and $\tau_{xy} = 21$ MPa (see figure for Problem 7.2-10).

Determine the maximum shear stresses and associated normal stresses and show them on a sketch of a properly oriented element.

Representative Problems

7.3-11 The stresses on an element are $\sigma_x = -2$ MPa and $\sigma_y = -4$ MPa. Find the maximum shear stresses on the element and show them on a sketch of a properly oriented element.

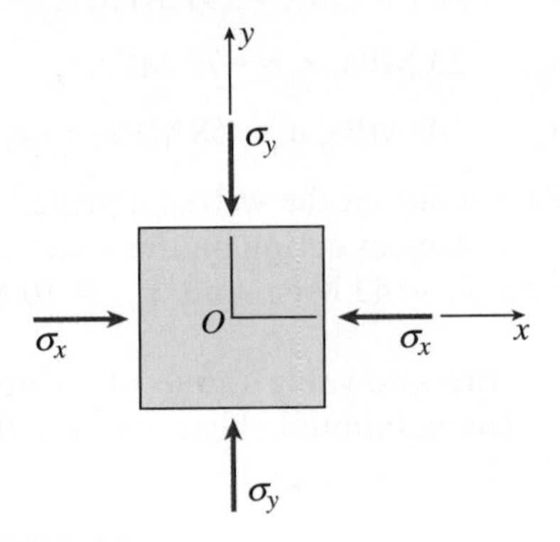

PROBLEM 7.3-11

7.3-12 A simply supported beam is subjected to two point loads as shown in the figure. The stresses on element A are $\tau_{xy} = -20$ kPa. Find the principal stresses on element A and show them on a sketch of a properly oriented element.

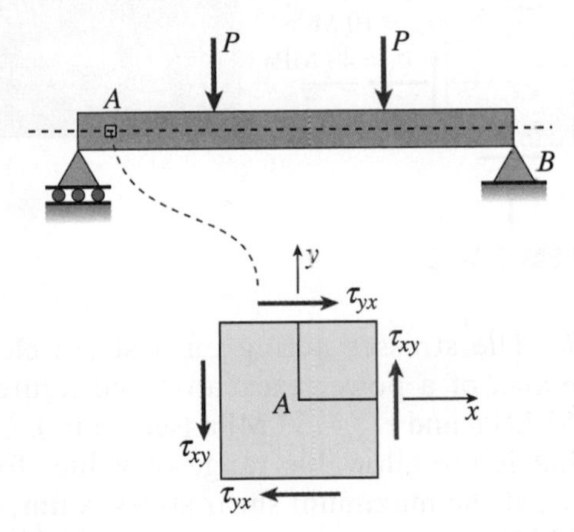

PROBLEM 7.3-12

7.3-13 A shear wall in a reinforced concrete building is subjected to a vertical uniform load of intensity q and a horizontal force H, as shown in the first part of the figure. (The force H represents the effects of wind and earthquake loads.) As a consequence of these loads, the stresses at point A on the surface of the wall have the values shown in the second part of the figure (compressive stress equal to 8 MPa and shear stress equal to 3 MPa).

(a) Determine the principal stresses and show them on a sketch of a properly oriented element.

(b) Determine the maximum shear stresses and associated normal stresses and show them on a sketch of a properly oriented element.

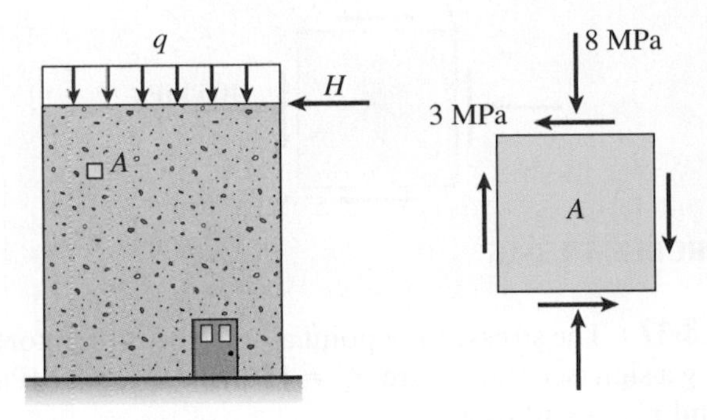

PROBLEM 7.3-13

7.3-14 The state of stress on an element along the hydraulic lift cylinder on a truck is $\sigma_y = -5$ MPa. Find the maximum shear stress on the element and show the state of stress on a sketch of a properly oriented element.

PROBLEMS 7.3-14 and 7.3-15

7.3-15 Repeat the preceding problem using $\sigma_y = -7.5$ MPa.

7.3-16 A propeller shaft subjected to combined torsion and axial thrust is designed to resist a shear stress of 57 MPa and a compressive stress of 105 MPa (see figure).

(a) Determine the principal stresses and show them on a sketch of a properly oriented element.
(b) Determine the maximum shear stresses and associated normal stresses and show them on a sketch of a properly oriented element.

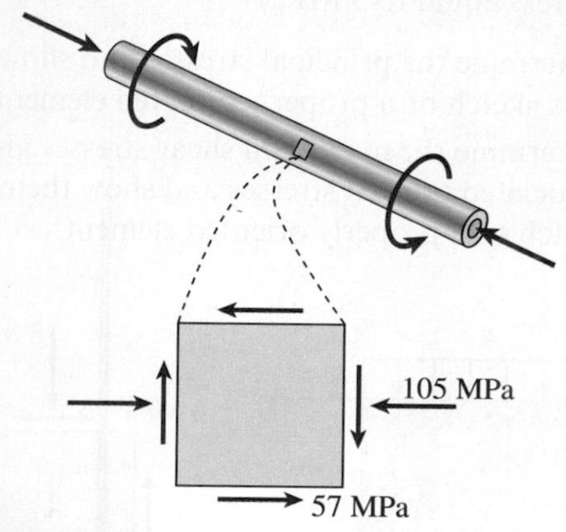

PROBLEM 7.3-16

7.3-17 The stresses at a point along a beam supporting a sign (see figure) are $\sigma_x = 15$ MPa, $\sigma_y = 8$ MPa, and $\tau_{xy} = -6$ MPa.

(a) Find the principal stresses. Show them on a sketch of a properly oriented element.
(b) Find the maximum shear stresses and associated normal stresses. Show them on a sketch of a properly oriented element.

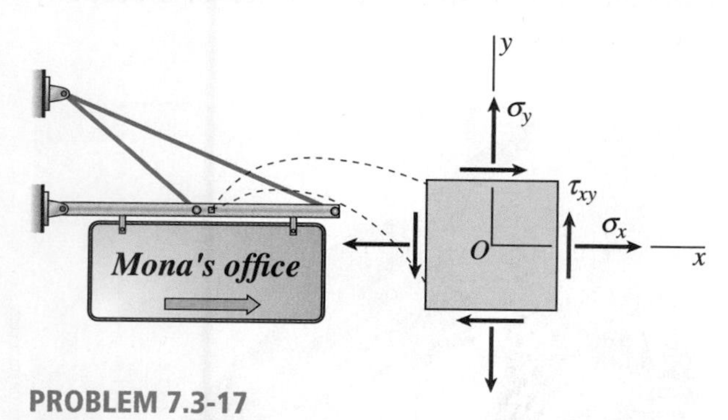

PROBLEM 7.3-17

7.3-18 through 7.3-22 An element in *plane stress* (see figure) is subjected to stresses σ_x, σ_y, and τ_{xy}.

(a) Determine the principal stresses and show them on a sketch of a properly oriented element.
(b) Determine the maximum shear stresses and associated normal stresses and show them on a sketch of a properly oriented element.

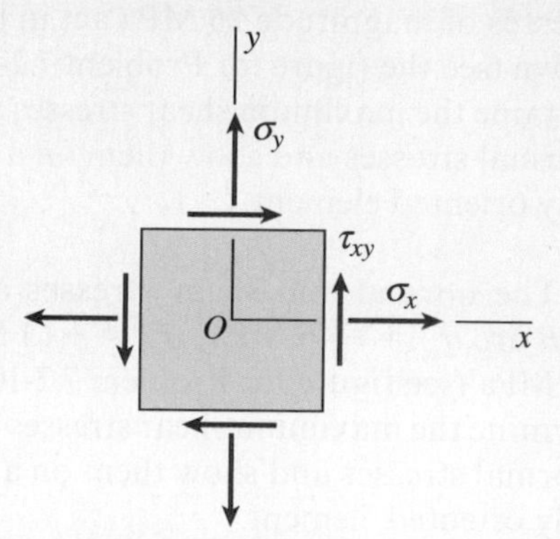

PROBLEMS 7.3-18 through 7.3-22

7.3-18 $\sigma_x = 2150$ kPa, $\sigma_y = 375$ kPa, $\tau_{xy} = -460$ kPa

7.3-19 $\sigma_x = 100$ MPa, $\sigma_y = 7.5$ MPa, $\tau_{xy} = 13$ MPa

7.3-20 $\sigma_x = 16.5$ MPa, $\sigma_y = -91$ MPa, $\tau_{xy} = -39$ MPa

7.3-21 $\sigma_x = -23$ MPa, $\sigma_y = -76$ MPa, $\tau_{xy} = 31$ MPa

7.3-22 $\sigma_x = -108$ MPa, $\sigma_y = 58$ MPa, $\tau_{xy} = -58$ MPa

7.3-23 At a point on the web of a girder on a gantry crane, the stresses acting on the x face of a stress element are $\sigma_x = 43$ MPa and $\tau_{xy} = 10$ MPa (see figure).

What is the allowable range of values for the stress σ_y if the maximum shear stress is limited to $\tau_0 = 15$ MPa?

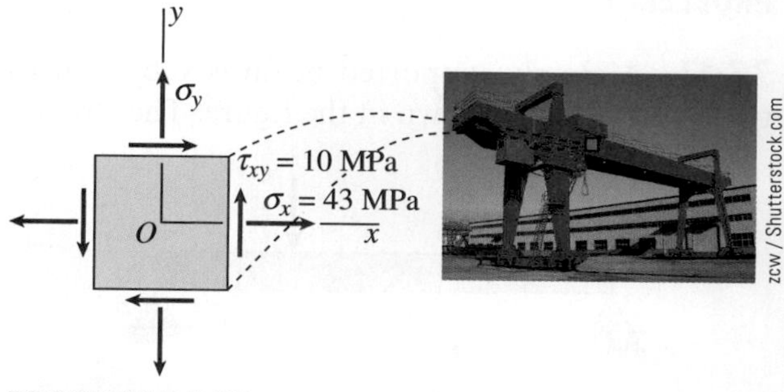

PROBLEM 7.3-23

7.3-24 The stresses acting on a stress element on the arm of a power excavator (see figure) are $\sigma_x = 52$ MPa and $\tau_{xy} = 33$ MPa (see figure).

What is the allowable range of values for the stress σ_y if the maximum shear stress is limited to $\tau_0 = 37$ MPa?

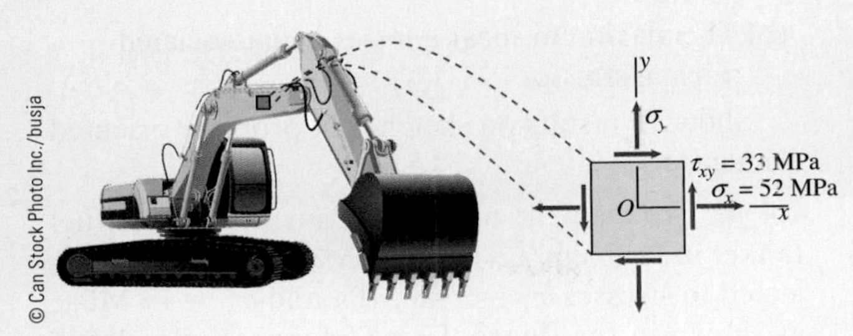

PROBLEM 7.3-24

7.3-25 The stresses at a point on the down tube of a bicycle frame are $\sigma_x = 33$ MPa and $\tau_{xy} = -13$ MPa (see figure). It is known that one of the principal stresses equals 44 MPa in tension.

(a) Determine the stress σ_y.

(b) Determine the other principal stress and the orientation of the principal planes, then show the principal stresses on a sketch of a properly oriented element.

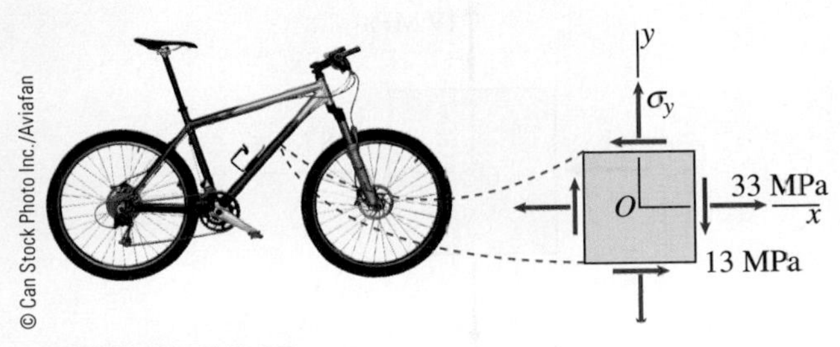

PROBLEM 7.3-25

7.3-26 An element in *plane stress* on the surface of an automobile drive shaft (see figure) is subjected to stresses of $\sigma_x = -45$ MPa and $\tau_{xy} = 39$ MPa (see figure). It is known that one of the principal stresses equals 41 MPa in tension.

(a) Determine the stress σ_y.

(b) Determine the other principal stress and the orientation of the principal planes, then show the principal stresses on a sketch of a properly oriented element.

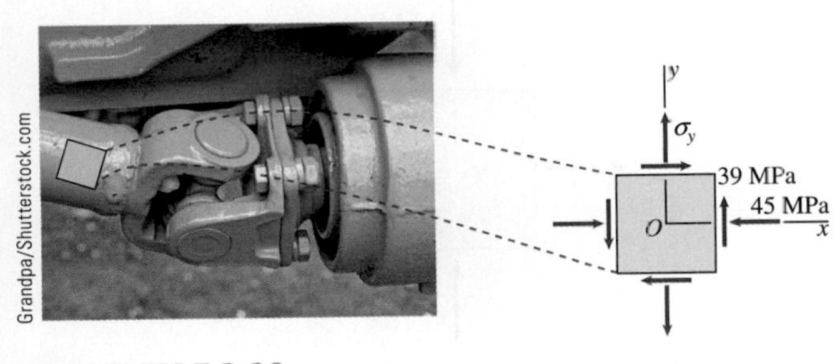

PROBLEM 7.3-26

7.3-27 A simply supported wood beam is subjected to point load P at mid-span. The stresses on element C are known to be $\sigma_x = -0.634$ MPa and $\tau_{xy} = -0.048$ MPa. Find the principal stresses on the element and show them on a sketch of a properly oriented element.

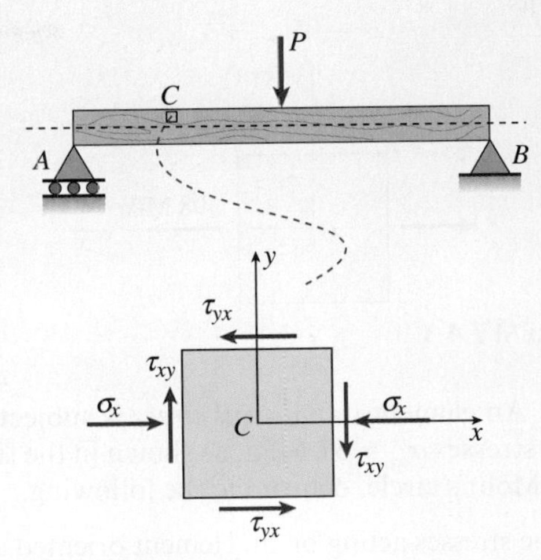

PROBLEM 7.3-27

7.3-28 A simply supported wood beam is subjected to point load P at mid-span. The normal stress on element C is known to be $\sigma_x = 12$ MPa. Find the maximum shear stress on the element and show the state of stress on a sketch of a properly oriented element.

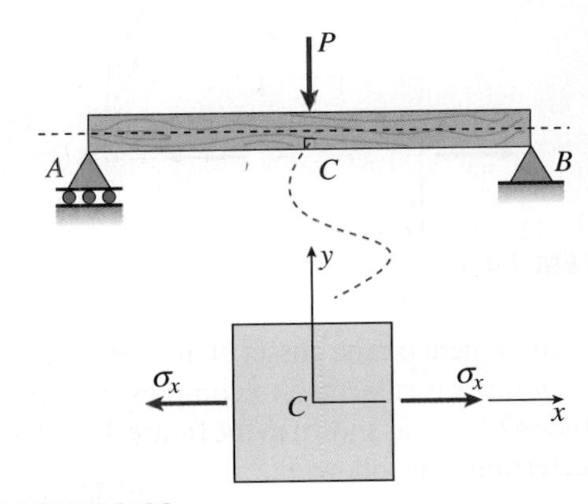

PROBLEM 7.3-28

7.4 Mohr's Circle

Solve the problems for Section 7.4 using Mohr's circle. Consider only the in-plane stresses (the stresses in the x-y plane).

Introductory Problems

7.4-1 An element in *uniaxial stress* is subjected to tensile stresses $\sigma_x = 98$ MPa, as shown in the figure. Using Mohr's circle, determine the following.

(a) The stresses acting on an element oriented at a counterclockwise angle $\theta = 29°$ from the x axis.

(b) The maximum shear stresses and associated normal stresses.

Show all results on sketches of properly oriented elements.

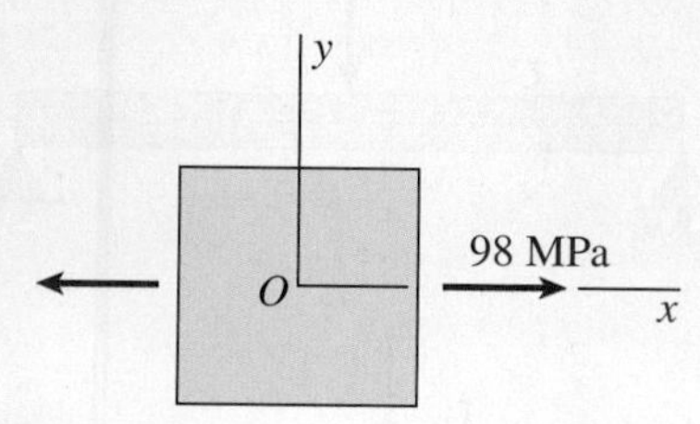

PROBLEM 7.4-1

7.4-2 An element in *uniaxial stress* is subjected to tensile stresses $\sigma_x = 57$ MPa, as shown in the figure. Using Mohr's circle, determine the following.

(a) The stresses acting on an element oriented at an angle $\theta = -33°$ from the x axis (minus means clockwise).

(b) The maximum shear stresses and associated normal stresses.

Show all results on sketches of properly oriented elements.

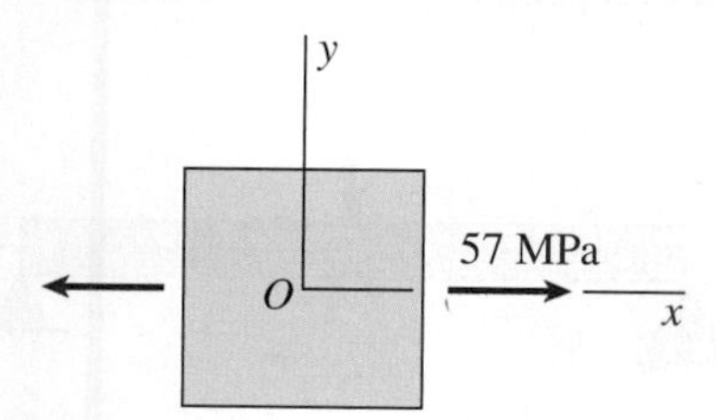

PROBLEM 7.4-2

7.4-3 An element on the gusset plate in Problem 7.2-23 in *uniaxial stress* is subjected to compressive stresses of magnitude 47 MPa, as shown in the figure. Using Mohr's circle, determine the following.

(a) The stresses acting on an element oriented at a slope of 1 on 2 (see figure).

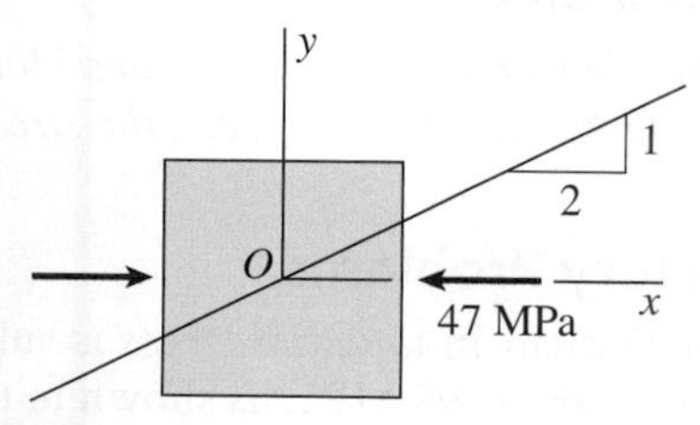

PROBLEM 7.4-3

(b) The maximum shear stresses and associated normal stresses.

Show all results on sketches of properly oriented elements.

7.4-4 An element on the top surface of the fuel tanker in Problem 7.2-1 is in *biaxial stress* and is subjected to stresses $\sigma_x = -48$ MPa and $\sigma_y = 19$ MPa, as shown in the figure. Using Mohr's circle, determine the following.

(a) The stresses acting on an element oriented at a counterclockwise angle $\theta = 25°$ from the x axis.

(b) The maximum shear stresses and associated normal stresses.

Show all results on sketches of properly oriented elements.

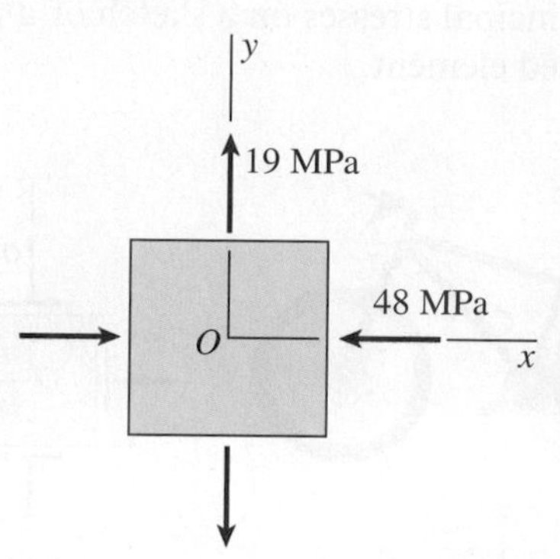

PROBLEM 7.4-4

7.4-5 An element on the top surface of the fuel tanker in Problem 7.2-1 is in *biaxial stress* and is subjected to stresses $\sigma_x = 43$ MPa and $\sigma_y = -12$ MPa, as shown in the figure. Using Mohr's circle, determine the following:

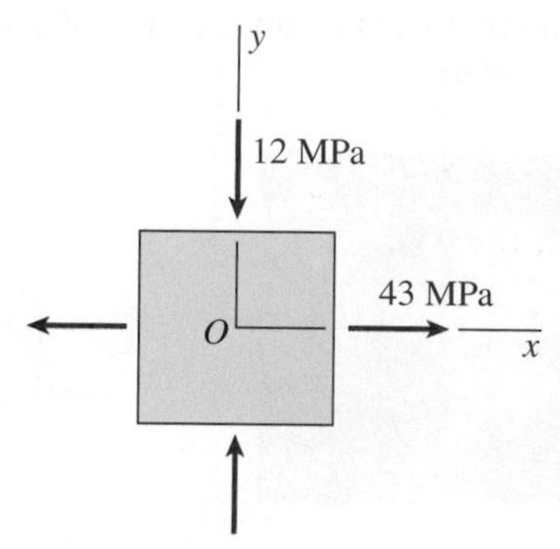

PROBLEM 7.4-5

(a) The stresses acting on an element oriented at a counterclockwise angle $\theta = 55°$ from the x axis.

(b) The maximum shear stresses and associated normal stresses.

Show all results on sketches of properly oriented elements.

7.4-6 An element in *biaxial stress* is subjected to stresses $\sigma_x = -29$ MPa and $\sigma_y = 57$ MPa, as shown in the figure. Using Mohr's circle, determine the following.

(a) The stresses acting on an element oriented at a slope of 1 on 2.5 (see figure).

(b) The maximum shear stresses and associated normal stresses.

Show all results on sketches of properly oriented elements.

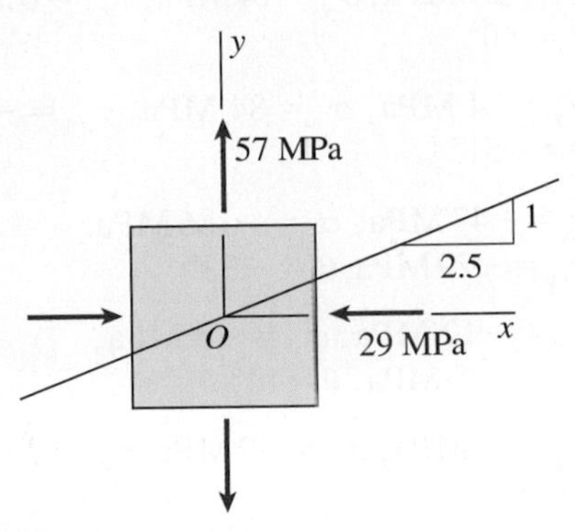

PROBLEM 7.4-6

7.4-7 An element on the surface of a drive shaft is in *pure shear* and is subjected to stresses $\tau_{xy} = 19$ MPa, as shown in the figure. Using Mohr's circle, determine the following.

(a) The stresses acting on an element oriented at a counterclockwise angle $\theta = 52°$ from the x axis.

(b) The principal stresses.

Show all results on sketches of properly oriented elements.

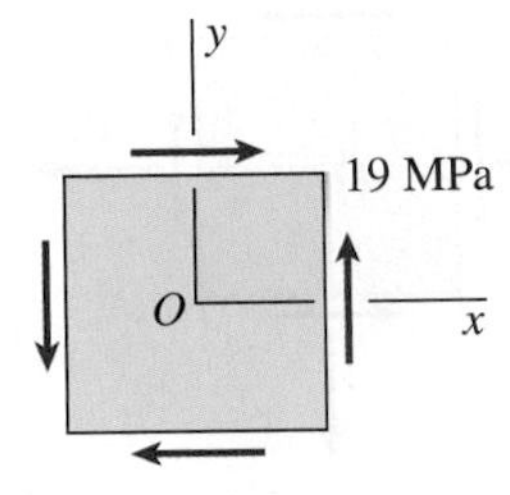

PROBLEM 7.4-7

Representative Problems

7.4-8 A specimen used in a coupon test has normal stress $\sigma_y = 15$ MPa (see figure). Using Mohr's circle, find the state of stress on the element oriented at angle $\theta = 20°$ and show the full stress state on a sketch of a properly oriented element.

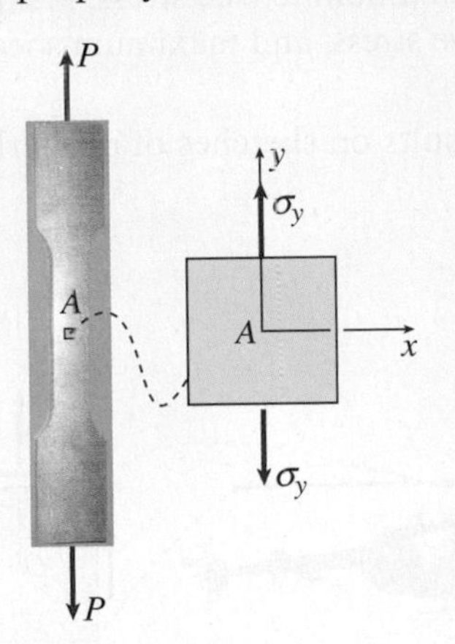

PROBLEM 7.4-8

7.4-9 A specimen used in a coupon test is shown in the figure. The stresses on element A are known to be $\sigma_y = -10$ MPa. Use Mohr's circle to:

(a) Find the stresses acting on the element oriented at an angle $\theta = -35°$.

(b) Find maximum normal and shear stresses and show them on sketches of properly oriented elements.

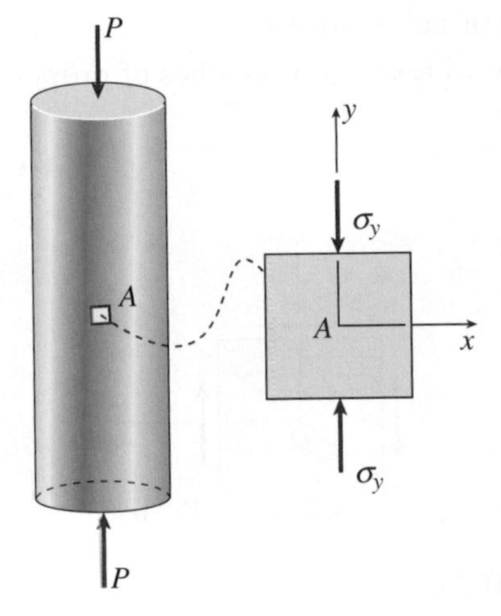

PROBLEM 7.4-9

7.4-10 The rotor shaft of a helicopter (see figure part a) drives the rotor blades that provide the lifting force and is subjected to a combination of torsion and axial loading (see figure part b).

It is known that normal stress $\sigma_y = 68$ MPa and shear stress $\tau_{xy} = -100$ MPa. Using Mohr's circle, determine the following:

(a) The stresses acting on an element oriented at a counterclockwise angle $\theta = 22.5°$ from the x axis.

(b) Find the maximum tensile stress, maximum compressive stress, and maximum shear stress in the shaft.

Show all results on sketches of properly oriented elements.

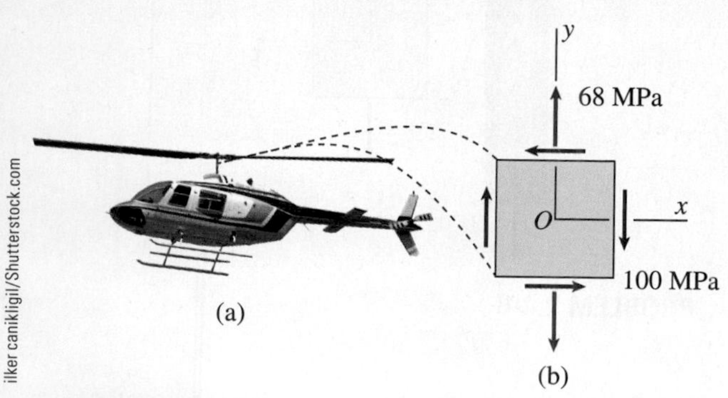

PROBLEM 7.4-10

7.4-11 An element in *pure shear* is subjected to stresses $\tau_{xy} = 26$ MPa, as shown in the figure. Using Mohr's circle, determine the following:

(a) The stresses acting on an element oriented at a slope of 3 on 4 (see figure).

(b) The principal stresses.

Show all results on sketches of properly oriented elements.

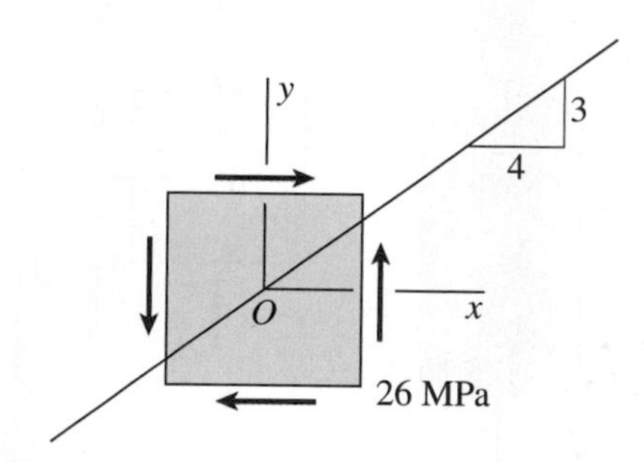

PROBLEM 7.4-11

7.4-12 through 7.4-17 An element in *plane stress* is subjected to stresses σ_x, σ_y, and τ_{xy} (see figure).

Using Mohr's circle, determine the stresses acting on an element oriented at an angle θ from the x axis. Show these stresses on a sketch of an element oriented at the angle θ. *Note:* The angle θ is positive when counterclockwise and negative when clockwise.

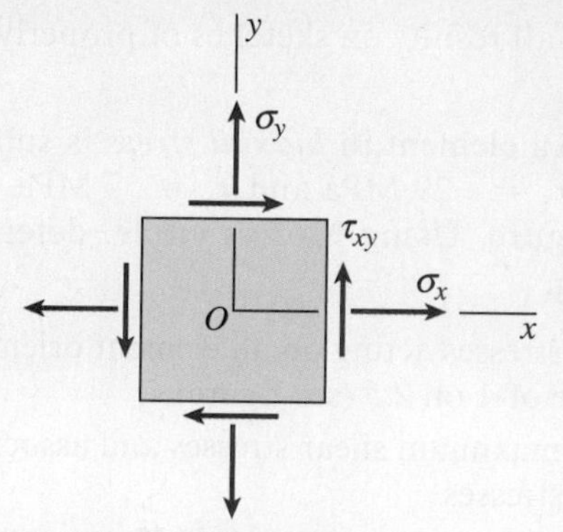

PROBLEMS 7.4-12 through 7.4-17

7.4-12 $\sigma_x = 27$ MPa, $\sigma_y = 14$ MPa, $\tau_{xy} = 6$ MPa, $\theta = 40°$

7.4-13 $\sigma_x = 24$ MPa, $\sigma_y = 84$ MPa, $\tau_{xy} = -23$ MPa, $\theta = -51°$

7.4-14 $\sigma_x = -47$ MPa, $\sigma_y = -186$ MPa, $\tau_{xy} = -29$ MPa, $\theta = -33°$

7.4-15 $\sigma_x = -12$ MPa, $\sigma_y = -5$ MPa, $\tau_{xy} = 2.5$ MPa, $\theta = 14°$

7.4-16 $\sigma_x = 33$ MPa, $\sigma_y = -9$ MPa, $\tau_{xy} = 29$ MPa, $\theta = 35°$

7.4-17 $\sigma_x = -39$ MPa, $\sigma_y = 7$ MPa, $\tau_{xy} = -15$ MPa, $\theta = 65°$

7.4-18 through 7.4-25 An element in *plane stress* is subjected to stresses σ_x, σ_y, and τ_{xy} (see figure).

Using Mohr's circle, determine (a) the principal stresses and (b) the maximum shear stresses and associated normal stresses. Show all results on sketches of properly oriented elements.

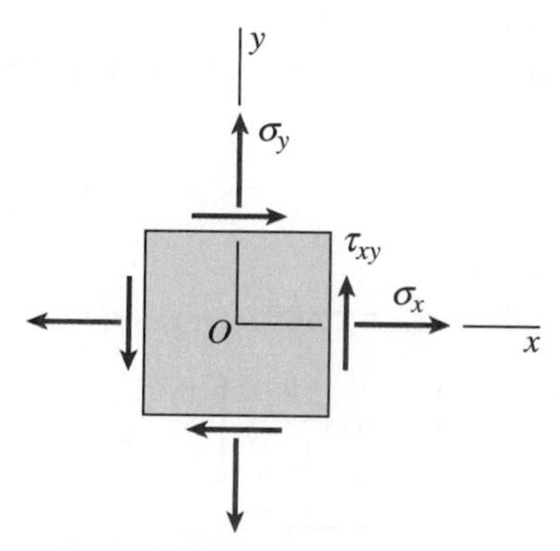

PROBLEMS 7.4-18 through 7.4-25

7.4-18 $\sigma_x = 2900\,\text{kPa}, \sigma_y = 9100\,\text{kPa}, \tau_{xy} = -3750\,\text{kPa}$

7.4-19 $\sigma_x = 5.5\ \text{MPa}, \sigma_y = -15\ \text{MPa}, \tau_{xy} = 20\ \text{MPa}$

7.4-20 $\sigma_x = -3.3\,\text{MPa}, \sigma_y = 8.9\,\text{MPa}, \tau_{xy} = -14.1\,\text{MPa}$

7.4-21 $\sigma_x = -80\ \text{MPa}, \sigma_y = -125\ \text{MPa},$
$\tau_{xy} = -50\ \text{MPa}$

7.4-22 $\sigma_x = -29.5\,\text{MPa}, \sigma_y = 29.5\,\text{MPa}, \tau_{xy} = 27\,\text{MPa}$

7.4-23 $\sigma_x = 14\ \text{MPa}, \sigma_y = 42\ \text{MPa}, \tau_{xy} = 19\ \text{MPa}$

7.4-24 $\sigma_x = 0\,\text{MPa}, \sigma_y = -23.4\,\text{MPa}, \tau_{xy} = -9.6\,\text{MPa}$

7.4-25 $\sigma_x = 50\ \text{MPa}, \sigma_y = 0, \tau_{xy} = 9\ \text{MPa}$

7.5 Hooke's Law for Plane Stress

To solve the problems for Section 7.5, assume that the material is linearly elastic with modulus of elasticity E and Poisson's ratio ν.

Introductory Problems

7.5-1 A rectangular steel plate with thickness $t = 16$ mm is subjected to uniform normal stresses σ_x and σ_y, as shown in the figure. Strain gages A and B, oriented in the x and y directions, respectively, are attached to the plate. The gage readings give normal strains $\varepsilon_x = 0.00065$ (elongation) and $\varepsilon_y = 0.00040$ (elongation).

Knowing that $E = 207$ GPa and $\nu = 0.3$, determine the stresses σ_x and σ_y and the change Δt in the thickness of the plate.

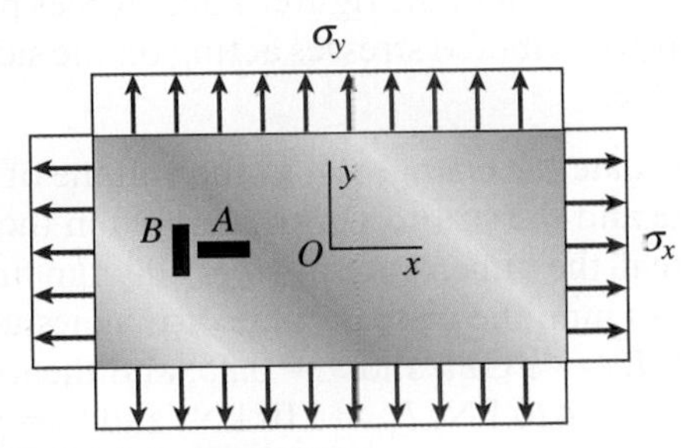

PROBLEM 7.5-1 and 7.5-2

7.5-2 Solve the preceding problem if the thickness of the steel plate is $t = 12$ mm, the gage readings are $\varepsilon_x = 530 \times 10^{-6}$ (elongation) and $\varepsilon_y = -210 \times 10^{-6}$ (shortening), the modulus is $E = 200$ GPa, and Poisson's ratio is $\nu = 0.30$.

7.5-3 The state of stress on an element of material is shown in the figure. Calculate the unit volume change of the element if the stresses σ_x and σ_y are –138 MPa and 70 MPa, respectively. Assume $E = 73$ GPa and $\nu = 0.33$.

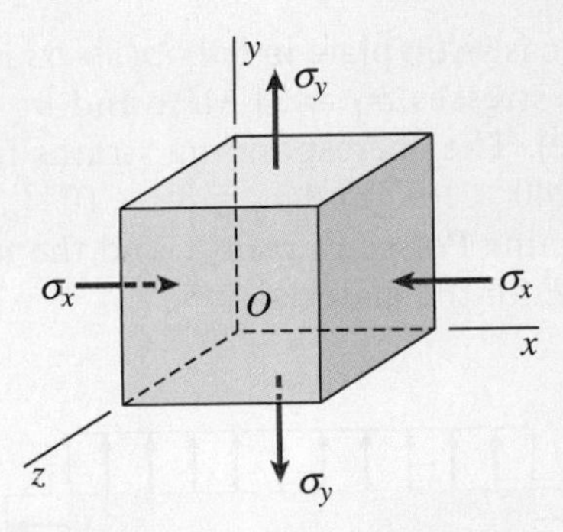

PROBLEM 7.5-3

7.5-4 An element of a material is subjected to plane stresses as shown in the figure. The stresses σ_x, σ_y, and τ_{xy} are 10 MPa, –15 MPa, and 5 MPa, respectively. Assume $E = 200$ GPa and $\nu = 0.3$.

(a) Calculate the normal strain in the x, y, and z directions and the shear strain.
(b) Calculate the strain-energy density of the element.

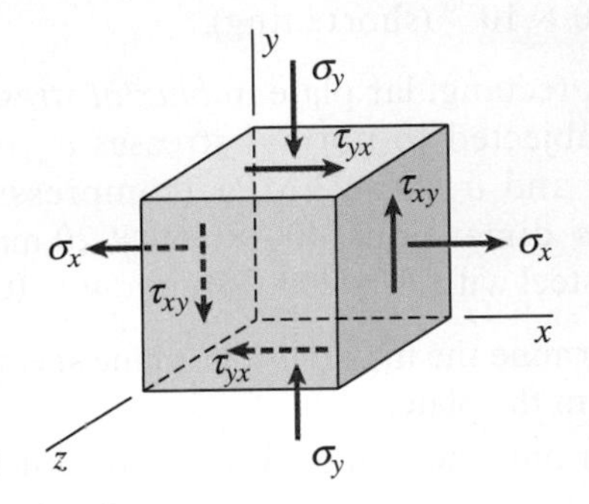

PROBLEM 7.5-4

7.5-5 Assume that the normal strains ε_x and ε_y for an element in plane stress (see figure) are measured with strain gages.

(a) Obtain a formula for the normal strain ε_z in the z direction in terms of ε_x, ε_y, and Poisson's ratio ν.
(b) Obtain a formula for the dilatation e in terms of ε_x, ε_y, and Poisson's ratio ν.

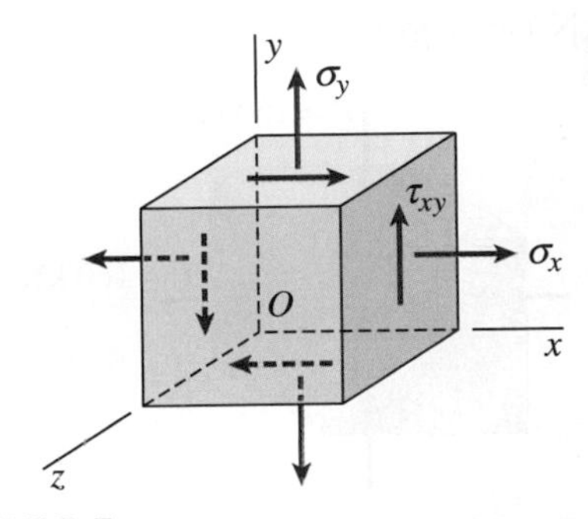

PROBLEM 7.5-5

7.5-6 A cast-iron plate in *biaxial stress* is subjected to tensile stresses $\sigma_x = 31$ MPa and $\sigma_y = 17$ MPa (see figure). The corresponding strains in the plate are $\varepsilon_x = 240 \times 10^{-6}$ and $\varepsilon_y = 85 \times 10^{-6}$.

Determine Poisson's ratio ν and the modulus of elasticity E for the material.

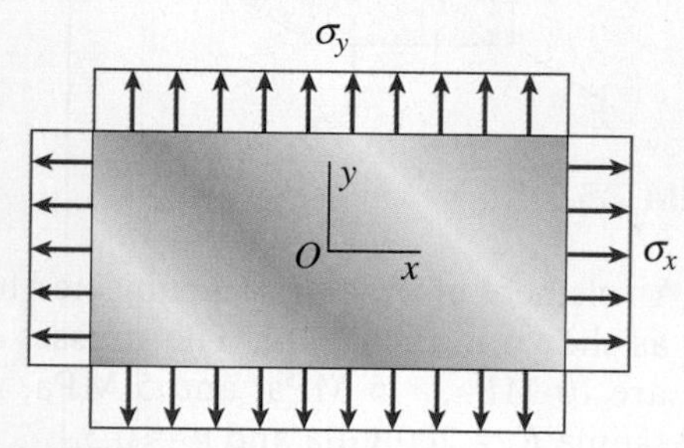

PROBLEMS 7.5-6 through 7.5-9

7.5-7 Solve the preceding problem for a steel plate with $\sigma_x = 80$ MPa (tension), $\sigma_y = -39$ MPa (compression), $\varepsilon_x = 450 \times 10^{-6}$ (elongation), and $\varepsilon_y = -310 \times 10^{-6}$ (shortening).

7.5-8 A rectangular plate in *biaxial stress* (see figure) is subjected to normal stresses $\sigma_x = 67$ MPa (tension) and $\sigma_x = -23$ MPa (compression). The plate has dimensions 400 × 550 × 20 mm and is made of steel with $E = 200$ GPa and $\nu = 0.30$.

(a) Determine the maximum in-plane shear strain γ_{max} in the plate.
(b) Determine the change Δt in the thickness of the plate.
(c) Determine the change ΔV in the volume of the plate.

7.5-9 Solve the preceding problem for an aluminum plate with $\sigma_x = 83$ MPa (tension), $\sigma_y = -21$ MPa (compression), dimensions 500 mm × 750 mm × 12.5 mm, $E = 72$ GPa, and $\nu = 0.33$.

7.5-10 A brass cube of 48 mm on each edge is compressed in two perpendicular directions by forces $P = 160$ kN (see figure).

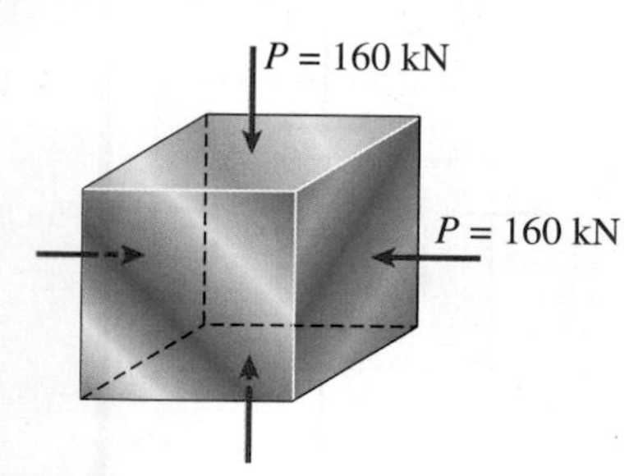

PROBLEM 7.5-10

(a) Calculate the change ΔV in the volume of the cube and the strain energy U stored in the cube, assuming $E = 100$ GPa and $\nu = 0.34$.
(b) Repeat part (a) if the cube is made of an aluminum alloy with $E = 73$ GPa and $\nu = 0.33$.

Representative Problems

7.5-11 A 100 mm cube of concrete ($E = 31$ GPa, $\nu = 0.2$) is compressed in *biaxial stress* by means of a framework that is loaded as shown in the figure.

Assuming that each load F equals 110 kN, determine the change ΔV in the volume of the cube and the strain energy U stored in the cube.

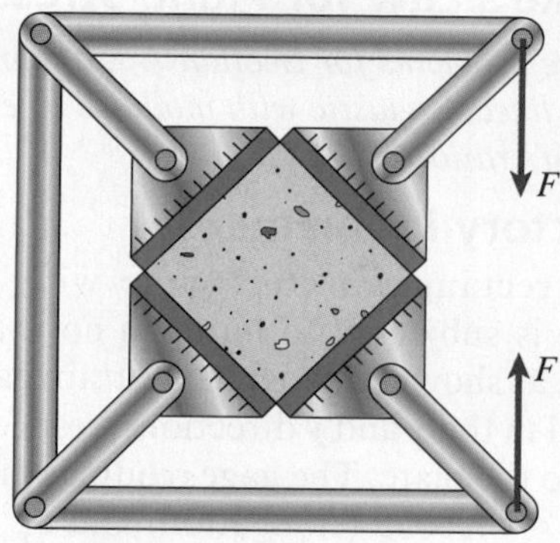

PROBLEM 7.5-11

7.5-12 A square plate of a width b and thickness t is loaded by normal forces P_x and P_y and by shear forces V, as shown in the figure. These forces produce uniformly distributed stresses acting on the side faces of the plate.

(a) Calculate the change ΔV in the volume of the plate and the strain energy U stored in the plate if the dimensions are $b = 600$ mm and $t = 40$ mm; the plate is made of magnesium with $E = 41$ GPa and $\nu = 0.35$; and the forces are $P_x = 420$ kN, $P_y = 210$ kN, and $V = 96$ kN.

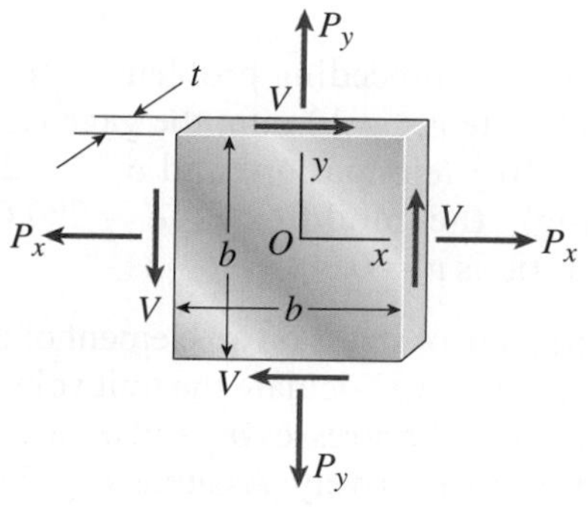

PROBLEMS 7.5-12 and 7.5-13

(b) Find the maximum permissible thickness of the plate when the strain energy U must be at least 62 J. [Assume that all other numerical values in part (a) are unchanged.]

(c) Find the minimum width b of the square plate of thickness $t = 40$ mm when the change in volume of the plate cannot exceed 0.018% of the original volume.

7.5-13 Solve the preceding problem for an aluminum plate with $b = 250$ mm, $t = 19$ mm, $E = 73$ GPa, $\nu = 0.33$, $P_x = 425$ kN, $P_y = 110$ kN, and $V = 80$ kN.

For part (b) of Problem 7.5-12, assume that the required strain energy stored is 72 J. In part (c), the change in volume cannot exceed 0.05%.

7.5-14 A circle of a diameter $d = 200$ mm is etched on a brass plate (see figure). The plate has dimensions of $400 \times 400 \times 20$ mm. Forces are applied to the plate, producing uniformly distributed normal stresses $\sigma_x = 59$ MPa and $\sigma_y = -17$ MPa. Calculate the following quantities: (a) the change in length Δac of diameter ac; (b) the change in length Δbd of diameter bd; (c) the change Δt in the thickness of the plate; (d) the change ΔV in the volume of the plate; (e) the strain energy U stored in the plate; (f) the maximum permissible thickness of the plate when strain energy U must be at least 78.4 J; and (g) the maximum permissible value of normal stress σ_x when the change in volume of the plate cannot exceed 0.015% of the original volume. (Assume $E = 100$ GPa and $\nu = 0.34$.)

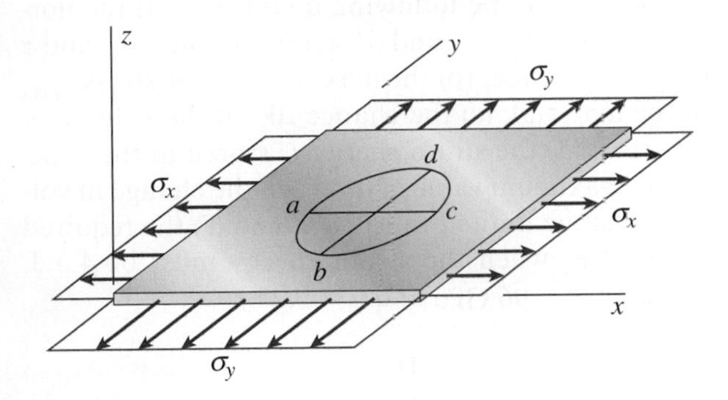

PROBLEM 7.5-14

7.5-15 The normal stress on an elastomeric rubber pad in a test machine is $\sigma_y = -0.7$ MPa (see figure). Assume $E = 2.2$ MPa and shear modulus $G = 0.7$ MPa.

(a) Calculate the strains in the pad in the x, y, and z directions.

(b) Calculate the unit volume change of the rubber.

(Dr. Mark E. Barkey, Department of Aerospace Engineering and Mechanics, University of Alabama)

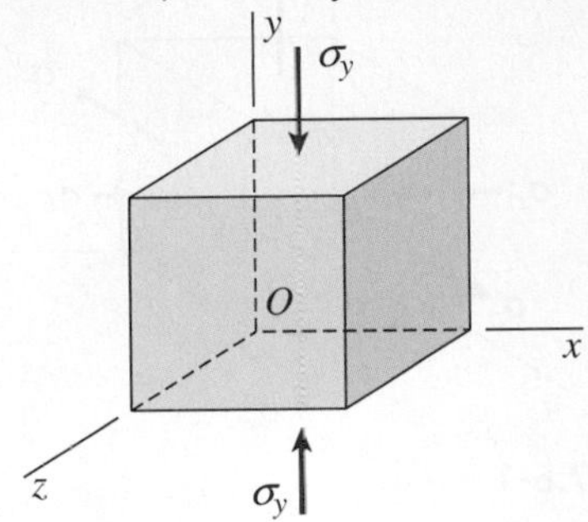

PROBLEM 7.5-15

7.5-16 A rubber sheet in biaxial stress is subjected to tensile stresses $\sigma_x = 270$ Pa and $\sigma_y = 144$ Pa. The corresponding strains in the sheet are $\varepsilon_x = 0.0002$ and $\varepsilon_y = 0.000015$. Determine Poisson's ratio and the modulus elasticity of the material.

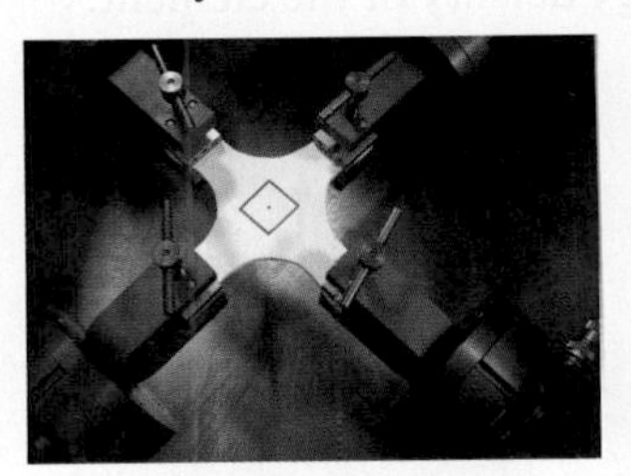

(Courtesy of TestResources, http://www.testresources.net)

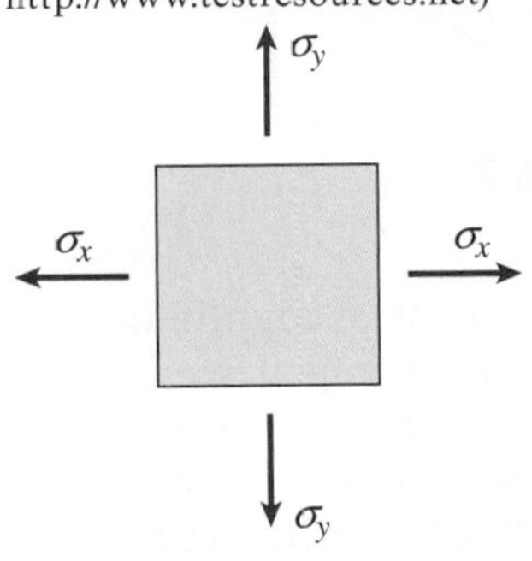

PROBLEM 7.5-16

7.6 Triaxial Stress

To solve the problems for Section 7.6, assume that the material is linearly elastic with modulus of elasticity E and Poisson's ratio ν.

Introductory Problems

7.6-1 An element of aluminum is subjected to triaxial stresses. Calculate the strains in the element in x, y, and z directions if the stresses σ_x, σ_y, and σ_z are –14 MPa, 24 MPa, and 18 MPa, respectively. Assume $E = 70$ GPa and $\nu = 0.33$. Also, find the unit volume change of the element.

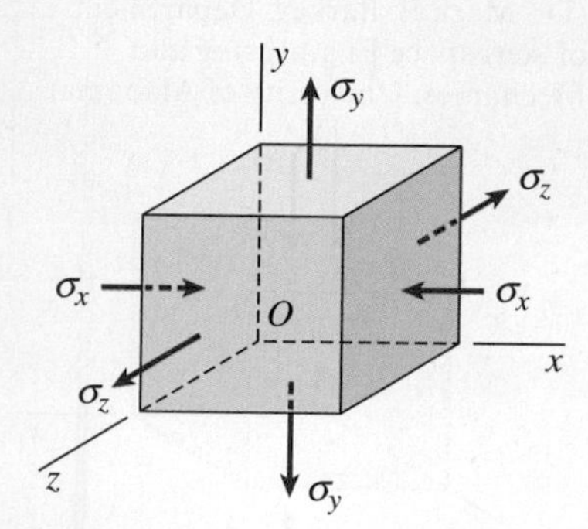

PROBLEM 7.6-1

7.6-2 An element of aluminum is subjected to triaxial stresses. Calculate the strains in the element in x, y, and z directions if the stresses σ_x, σ_y, and σ_z are –20 MPa, 28 MPa, and –18 MPa, respectively. Assume $E = 70$ GPa and $\nu = 0.33$. Also, find the strain energy density of the element.

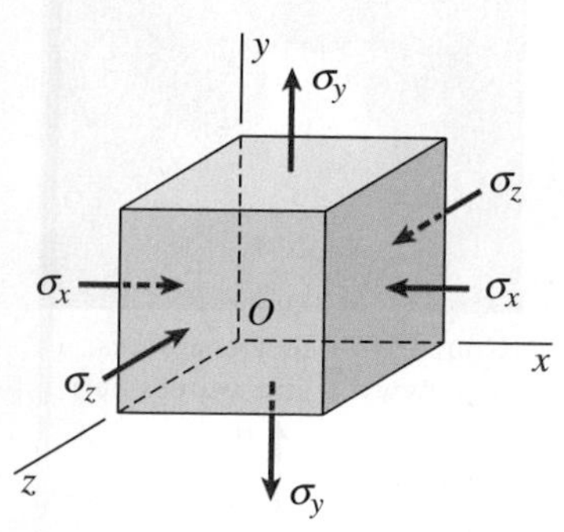

PROBLEM 7.6-2

7.6-3 An element of aluminum in the form of a rectangular parallelepiped (see figure) of dimensions $a = 140$ mm, $b = 115$ mm, and $c = 90$ mm is subjected to *triaxial stresses* $\sigma_x = 86$ MPa, $\sigma_y = -34$ MPa, and $\sigma_z = -10$ MPa acting on the x, y, and z faces, respectively.

Determine the following quantities: (a) the maximum shear stress τ_{max} in the material; (b) the changes Δa, Δb, and Δc in the dimensions of the element; (c) the change ΔV in the volume; (d) the strain energy U stored in the element; (e) the maximum value of σ_x when the change in volume must be limited to 0.021%; and (f) the required value of σ_x when the strain energy must be 102 J. (Assume $E = 72$ GPa and $\nu = 0.33$.)

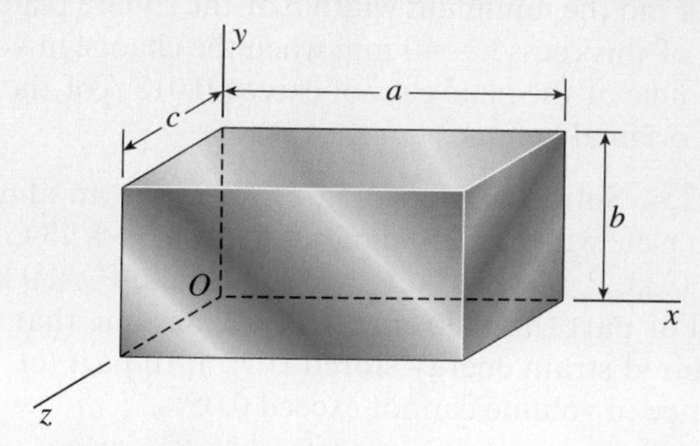

PROBLEMS 7.6-3 and 7.6-4

7.6-4 Solve the preceding problem if the element is steel ($E = 200$ GPa, $\nu = 0.30$) with dimensions $a = 300$ mm, $b = 150$ mm, and $c = 150$ mm and with the stresses $\sigma_x = -62$ MPa, $\sigma_y = -45$ MPa, and $\sigma_z = -45$ MPa.

For part (e) of Problem 7.6-3, find the maximum value of σ_x if the change in volume must be limited to −0.028%. For part (f), find the required value of σ_x if the strain energy must be 60 J.

7.6-5 A cube of cast iron with sides of length $a = 100$ mm (see figure) is tested in a laboratory under *triaxial stress*. Gages mounted on the testing machine show that the compressive strains in the material are $\varepsilon_x = -225 \times 10^{-6}$ and $\varepsilon_y = \varepsilon_z = -37.5 \times 10^{-6}$.

Determine the following quantities: (a) the normal stresses σ_x, σ_y, and σ_z acting on the x, y, and z faces of the cube; (b) the maximum shear stress τ_{max} in the material; (c) the change ΔV in the volume of the cube; (d) the strain energy U stored in the cube; (e) the maximum value of σ_x when the change in volume must be limited to 0.028%; and (f) the required value of ε_x when the strain energy must be 4.3 J. (Assume $E = 96$ GPa and $\nu = 0.25$.)

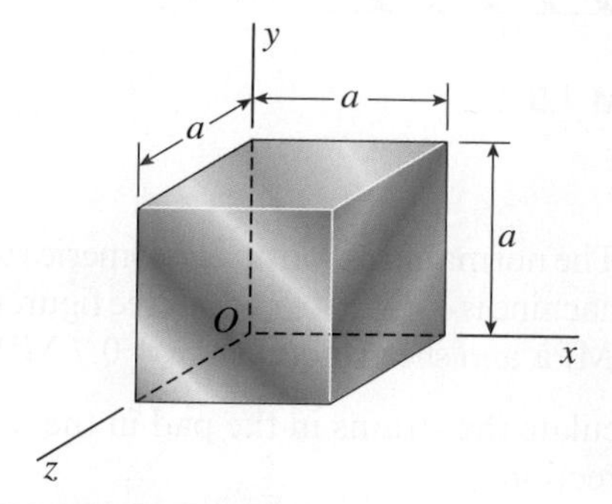

PROBLEMS 7.6-5 and 7.6-6

7.6-6 Solve the preceding problem if the cube is granite ($E = 80$ GPa, $\nu = 0.25$) with dimensions $a = 89$ mm and compressive strains $\varepsilon_x = 690 \times 10^{-6}$ and $\varepsilon_y = \varepsilon_z = 255 \times 10^{-6}$. For part (e) of Problem 7.6-5, find the maximum value of σ_x when the change in volume must be limited to 0.11%. For part (f), find the required value of ε_x when the strain energy must be 33 J.

Representative Problems

7.6-7 An element of aluminum is subjected to *triaxial* stress (see figure).

(a) Find the bulk modulus K for the aluminum if the following stress and strain data are known: normal stresses are $\sigma_x = 36$ MPa (tension), $\sigma_y = -33$ MPa (compression); and $\sigma_z = -21$ MPa (compression) and normal strains in the x and y directions are $\varepsilon_x = 713.8 \times 10^{-6}$ (elongation) and $\varepsilon_y = -502.3 \times 10^{-6}$ (shortening).

(b) If the element is replaced by one of magnesium, find the modulus of elasticity E and Poisson's ratio ν if the following data is given: bulk modulus $K = 47$ GPa; normal stresses are $\sigma_x = 31$ MPa (tension), $\sigma_y = -12$ MPa (compression), and $\sigma_z = -7.5$ MPa (compression); and normal strain in the x direction is $\varepsilon_x = 900 \times 10^{-6}$ (elongation).

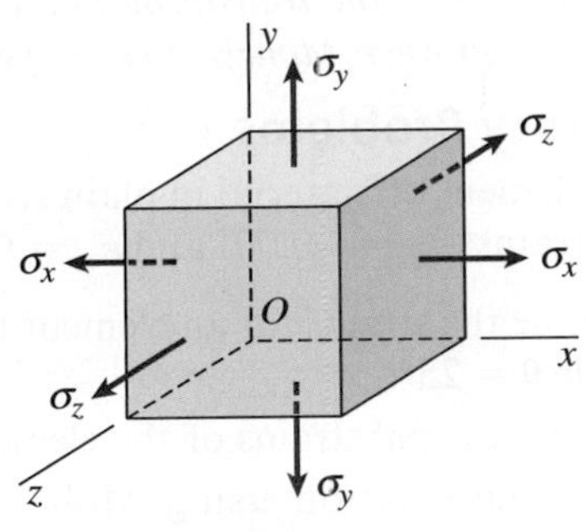

PROBLEMS 7.6-7 and 7.6-8

7.6-8 Solve the preceding problem if the material is nylon.

(a) Find the bulk modulus K for the nylon if the following stress and strain data is known: normal stresses are $\sigma_x = -3.9$ MPa, $\sigma_y = -3.2$ MPa, and $\sigma_z = -1.8$ MPa; and normal strains in the x and y directions are $\varepsilon_x = -640 \times 10^{-6}$ (shortening) and $\varepsilon_y = -310 \times 10^{-6}$ (shortening).

(b) If the element is replaced by one of polyethylene, find the modulus of elasticity E and Poisson's ratio ν if the following data is given: bulk modulus $K = 2162$ MPa; normal stresses are $\sigma_x = -3.6$ MPa (compression), $\sigma_y = -2.1$ MPa (compression), and $\sigma_z = -2.1$ MPa (compression); and normal strain in the x direction is $\varepsilon_x = -1480 \times 10^{-6}$ (shortening).

7.6-9 A rubber cylinder R of length L and cross-sectional area A is compressed inside a steel cylinder S by a force F that applies a uniformly distributed pressure to the rubber (see figure).

(a) Derive a formula for the lateral pressure p between the rubber and the steel. (Disregard friction between the rubber and the steel, and assume that the steel cylinder is rigid when compared to the rubber.)

(b) Derive a formula for the shortening δ of the rubber cylinder.

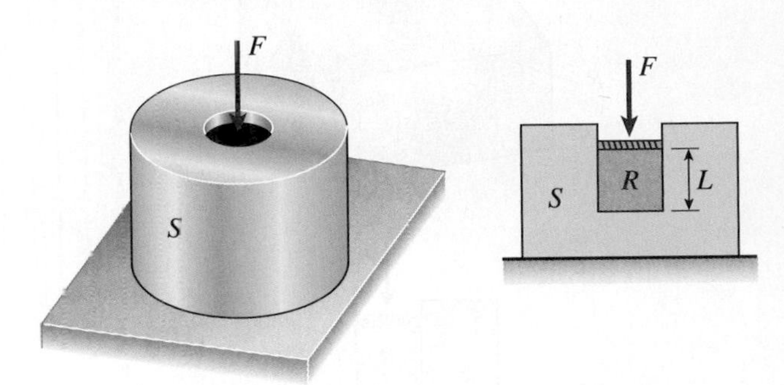

PROBLEM 7.6-9

7.6-10 A block R of rubber is confined between plane parallel walls of a steel block S (see figure). A uniformly distributed pressure p_0 is applied to the top of the rubber block by a force F.

(a) Derive a formula for the lateral pressure p between the rubber and the steel. (Disregard friction between the rubber and the steel, and

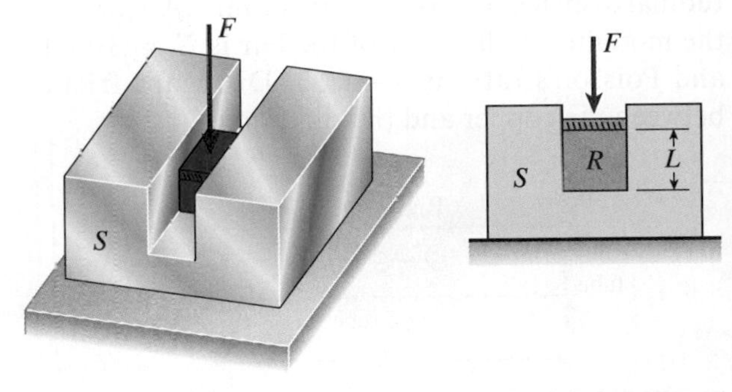

PROBLEM 7.6-10

assume that the steel block is rigid when compared to the rubber.)

(b) Derive a formula for the dilatation e of the rubber.

(c) Derive a formula for the strain-energy density u of the rubber.

7.6-11 A rubber cube R of a side $L = 75$ mm and cross-sectional area $A = 5800\ \text{mm}^2$ is compressed inside a steel cube S by a force $F = 22$ N that applies uniformly distributed pressure to the rubber. Assume $E = 2$ MPa and $\nu = 0.45$.

(a) Calculate the lateral pressure between the rubber and steel (disregard friction between the rubber and the steel, and assume that the steel block is rigid when compared to the rubber).

(b) Calculate the change in volume of the rubber.

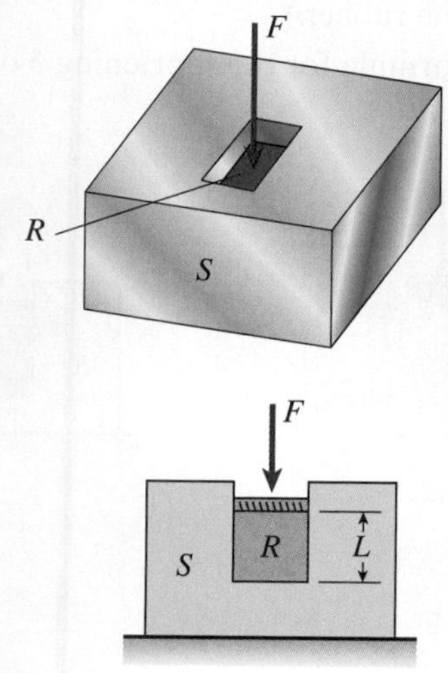

PROBLEM 7.6-11

7.6-12 A copper bar with a square cross section is inserted into a square rigid tube as shown in the figure. The length of the copper bar is 1.2 m and the area of the cross section is 300 mm². The bar is subjected to a force P that applies a uniformly distributed pressure to the copper. Calculate the force P if the longitudinal displacement of the bar is 2 mm. Assume that the modulus of elasticity of the bar is $E = 110$ GPa and Poisson's ratio is $\nu = 0.33$. Disregard friction between the copper and the rigid tube.

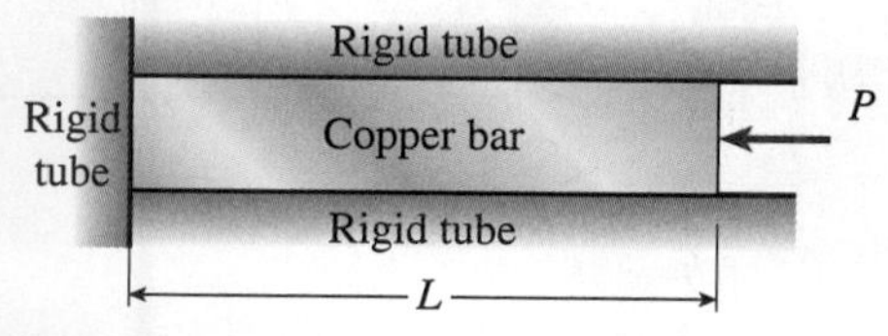

PROBLEM 7.6-12

7.6-13 A solid spherical ball of magnesium alloy ($E = 45$ GPa, $\nu = 0.35$) is lowered into the ocean to a depth of 2400 m. The diameter of the ball is 225 mm.

(a) Determine the decrease Δd in diameter, the decrease ΔV in volume, and the strain energy U of the ball.

(b) At what depth will the volume change be equal to 0.0324% of the original volume?

7.6-14 A solid steel sphere ($E = 210$ GPa, $\nu = 0.3$) is subjected to hydrostatic pressure p such that its volume is reduced by 0.4%.

(a) Calculate the pressure p.

(b) Calculate the volume modulus of elasticity K for the steel.

(c) Calculate the strain energy U stored in the sphere if its diameter is $d = 150$ mm.

7.6-15 A solid bronze sphere (volume modulus of elasticity $K = 100$ GPa) is suddenly heated around its outer surface. The tendency of the heated part of the sphere to expand produces uniform tension in all directions at the center of the sphere.

If the stress at the center is 83 MPa, what is the strain? Also, calculate the unit volume change e and the strain-energy density u at the center.

7.7 Plane Strain

To solve the problems for Section 7.7, consider only the in-plane strains (the strains in the x-y plane) unless stated otherwise. Use the transformation equations of plane strain except when Mohr's circle is specified.

Introductory Problems

7.7-1 An element of material in plain strain has the following strains: $\varepsilon_x = -0.001$ and $\varepsilon_y = 0.0015$.

(a) Determine the strains for an element oriented at an angle $\theta = 25°$.

(b) Find the principal strains of the element.

Confirm the solution using Mohr's circle for plane strain.

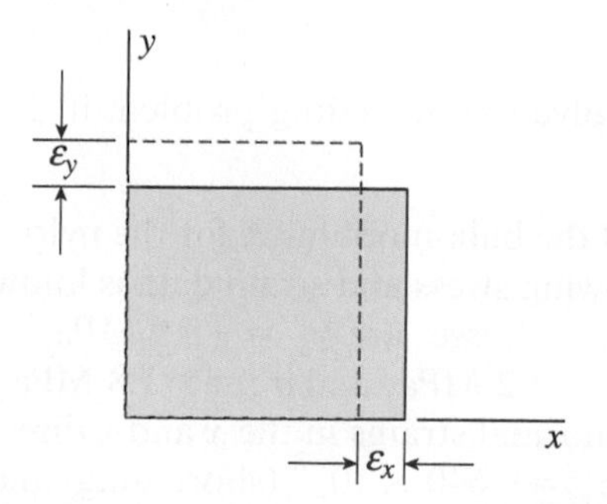

PROBLEM 7.7-1

7.7-2 An element of material in plain strain has the following strains: $\varepsilon_x = 0.002$ and $\varepsilon_y = 0.0015$.

(a) Determine the principal strains of the element.

(b) Determine the maximum shear strain of the element.

Confirm the solution using Mohr's circle for plane strain.

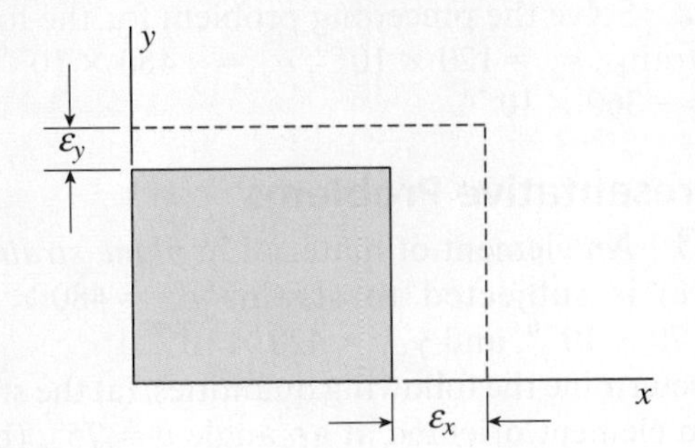

PROBLEM 7.7-2

7.7-3 An element of material in plain strain is subjected to shear strain $\gamma_{xy} = 0.0003$.

(a) Determine the strains for an element oriented at an angle $\theta = 30°$.

(b) Determine the principal strains of the element.

Confirm the solution using Mohr's circle for plane strain.

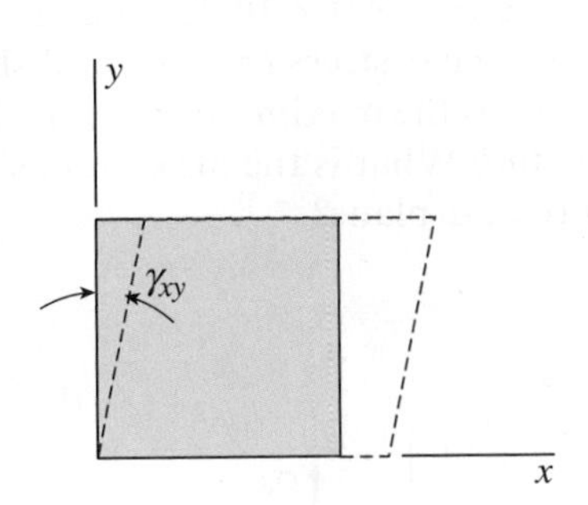

PROBLEM 7.7-3

7.7-4 An element of material in plain strain is subjected to strains $\varepsilon_x = 0.0015$, $\varepsilon_y = -0.0002$, and $\gamma_{xy} = 0.0003$.

(a) Determine the strains for an element oriented at an angle $\theta = 20°$.

(b) Determine the principal strains of the element.

Confirm the solution using Mohr's circle for plane strain.

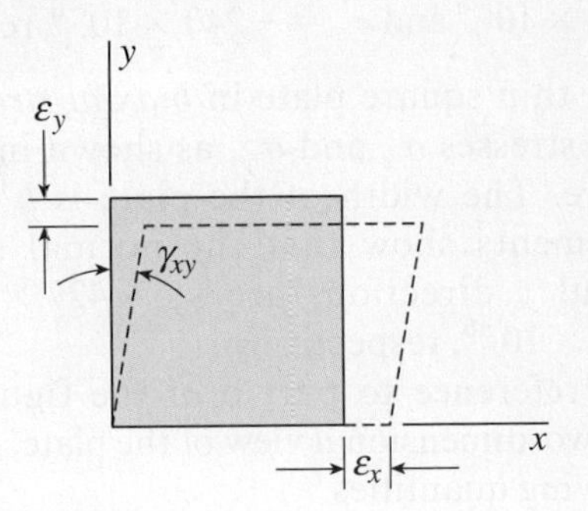

PROBLEM 7.7-4

7.7-5 A thin rectangular plate in *biaxial stress* is subjected to stresses σ_x and σ_y, as shown in part a of the figure. The width and height of the plate are $b = 190$ mm and $h = 63$ mm, respectively. Measurements show that the normal strains in the x and y directions are $\varepsilon_x = 285 \times 10^{-6}$ and $\varepsilon_y = -190 \times 10^{-6}$, respectively.

With reference to part b of the figure, which shows a two-dimensional view of the plate, determine the following quantities.

(a) The increase Δd in the length of diagonal Od.

(b) The change $\Delta\phi$ in the angle ϕ between diagonal Od and the x axis.

(c) The change $\Delta\psi$ in the angle ψ between diagonal Od and the y axis.

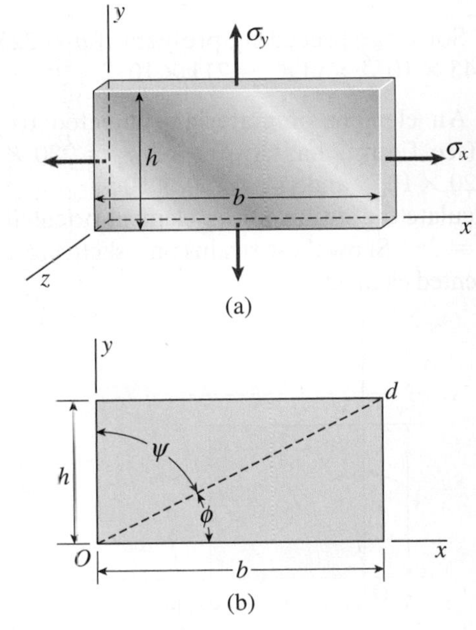

PROBLEMS 7.7-5 and 7.7-6

7.7-6 Solve the preceding problem if $b = 180$ mm and $h = 70$ mm, respectively. Measurements show that the normal strains in the x and y directions are $\varepsilon_x = 390 \times 10^{-6}$ and $\varepsilon_y = -240 \times 10^{-6}$, respectively.

7.7-7 A thin square plate in *biaxial stress* is subjected to stresses σ_x and σ_y, as shown in part a of the figure. The width of the plate is $b = 300$ mm Measurements show that the normal strains in the x and y directions are $\varepsilon_x = 427 \times 10^{-6}$ and $\varepsilon_y = 113 \times 10^{-6}$, respectively.

With reference to part b of the figure, which shows a two-dimensional view of the plate, determine the following quantities.

(a) The increase Δd in the length of diagonal Od.

(b) The change $\Delta\phi$ in the angle ϕ between diagonal Od and the x axis.

(c) The shear strain γ associated with diagonals Od and cf (that is, find the decrease in angle ced).

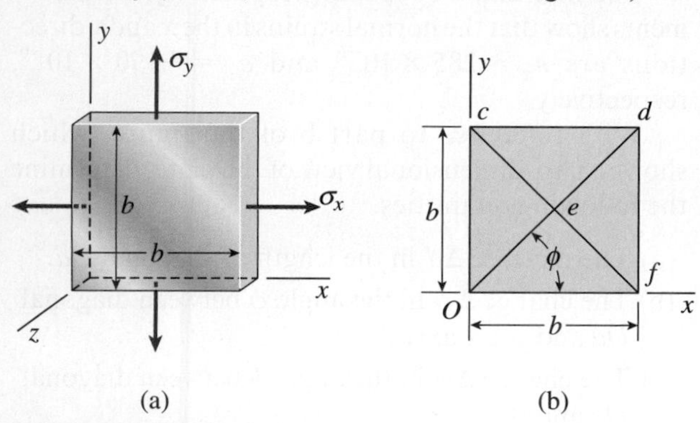

PROBLEMS 7.7-7 and 7.7-8

7.7-8 Solve the preceding problem if $b = 225$ mm, $\varepsilon_x = 845 \times 10^{-6}$, and $\varepsilon_y = 211 \times 10^{-6}$.

7.7-9 An element of material subjected to *plane strain* (see figure) has strains of $\varepsilon_x = 280 \times 10^{-6}$, $\varepsilon_y = 420 \times 10^{-6}$, and $\gamma_{xy} = 150 \times 10^{-6}$.

Calculate the strains for an element oriented at an angle $\theta = 35°$. Show these strains on a sketch of a properly oriented element.

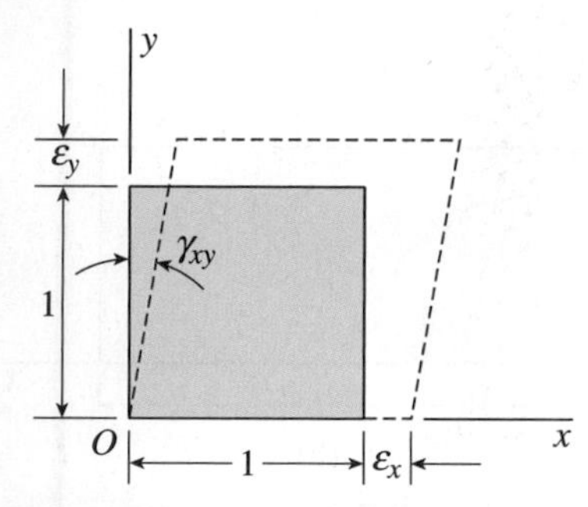

PROBLEMS 7.7-9 through 7.7-14

7.7-10 Solve the preceding problem for the following data: $\varepsilon_x = 190 \times 10^{-6}$, $\varepsilon_y = -230 \times 10^{-6}$, $\gamma_{xy} = 160 \times 10^{-6}$, and $\theta = 40°$.

7.7-11 The strains for an element of material in *plane strain* (see figure) are as follows: $\varepsilon_x = 480 \times 10^{-6}$, $\varepsilon_y = 140 \times 10^{-6}$, and $\gamma_{xy} = -350 \times 10^{-6}$.

Determine the principal strains and maximum shear strains, and show these strains on sketches of properly oriented elements.

7.7-12 Solve the preceding problem for the following strains: $\varepsilon_x = 120 \times 10^{-6}$, $\varepsilon_y = -450 \times 10^{-6}$, and $\gamma_{xy} = -360 \times 10^{-6}$.

Representative Problems

7.7-13 An element of material in *plane strain* (see figure) is subjected to strains $\varepsilon_x = 480 \times 10^{-6}$, $\varepsilon_y = 70 \times 10^{-6}$, and $\gamma_{xy} = 420 \times 10^{-6}$.

Determine the following quantities: (a) the strains for an element oriented at an angle $\theta = 75°$, (b) the principal strains, and (c) the maximum shear strains. Show the results on sketches of properly oriented elements.

7.7-14 Solve the preceding problem for the following data: $\varepsilon_x = -1120 \times 10^{-6}$, $\varepsilon_y = -430 \times 10^{-6}$, $\gamma_{xy} = 780 \times 10^{-6}$, and $\theta = 45°$.

7.7-15 A brass plate with a modulus of elasticity $E = 110$ GPa and Poisson's ratio $\nu = 0.34$ is loaded in *biaxial stress* by normal stresses σ_x and σ_y (see figure). A strain gage is bonded to the plate at an angle $\phi = 35°$.

If the stress σ_x is 74 MPa and the strain measured by the gage is $\varepsilon = 390 \times 10^{-6}$, what is the maximum in-plane shear stress $(\tau_{max})_{xy}$ and shear strain $(\gamma_{max})_{xy}$? What is the maximum shear strain $(\gamma_{max})_{xz}$ in the x-z plane? What is the maximum shear strain $(\gamma_{max})_{yz}$ in the y-z plane?

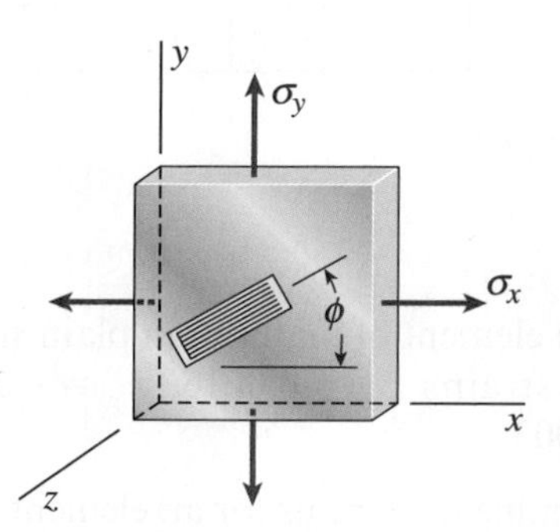

PROBLEMS 7.7-15 and 7.7-16

7.7-16 Solve the preceding problem if the plate is made of aluminum with $E = 72$ GPa and Poisson's ratio $\nu = 0.33$. The plate is loaded in *biaxial stress* with normal stress $\sigma_x = 79$ MPa, angle $\phi = 18°$, and the strain measured by the gage is $\varepsilon = 925 \times 10^{-6}$.

7.7-17 An element in *plane stress* is subjected to stresses $\sigma_x = -58$ MPa, $\sigma_y = 7.5$ MPa, and $\tau_{xy} = -12$ MPa (see figure). The material is aluminum with modulus of elasticity $E = 69$ GPa and Poisson's ratio $\nu = 0.33$.

Determine the following quantities: (a) the strains for an element oriented at an angle $\theta = 30°$, (b) the principal strains, and (c) the maximum shear strains. Show the results on sketches of properly oriented elements.

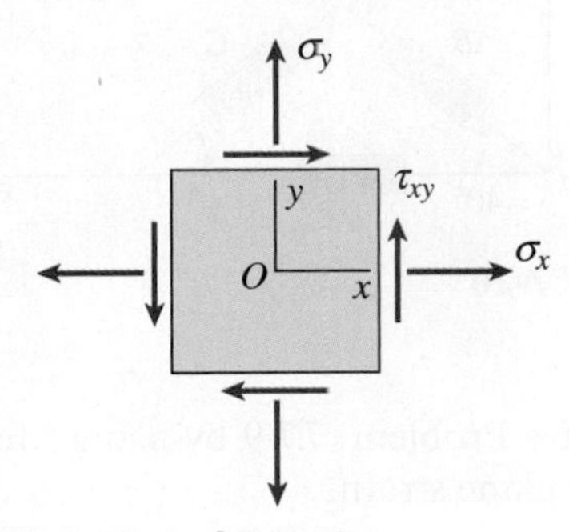

PROBLEMS 7.7-17 and 7.7-18

7.7-18 Solve the preceding problem for the following data: $\sigma_x = -150$ MPa, $\sigma_y = -210$ MPa, $\tau_{xy} = -16$ MPa, and $\theta = 50°$. The material is brass with $E = 100$ GPa and $\nu = 0.34$.

7.7-19 During a test of an airplane wing, the strain gage readings from a 45° rosette (see figure) are as follows: gage A, 520×10^{-6}; gage B, 360×10^{-6}; and gage C, -80×10^{-6}.

Determine the principal strains and maximum shear strains, and show them on sketches of properly oriented elements.

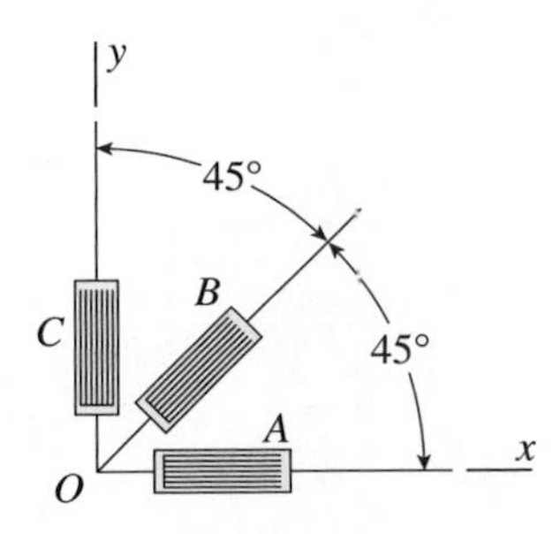

PROBLEMS 7.7-19 and 7.7-20

7.7-20 A 45° strain rosette (see figure) mounted on the surface of an automobile frame gives the following readings: gage A, 310×10^{-6}; gage B, 180×10^{-6}; and gage C, -160×10^{-6}.

Determine the principal strains and maximum shear strains, and show them on sketches of properly oriented elements.

7.7-21 A solid circular bar with a diameter of $d = 32$ mm is subjected to an axial force P and a torque T (see figure). Strain gages A and B mounted on the surface of the bar give readings $\varepsilon_A = 140 \times 10^{-6}$ and $\varepsilon_B = -60 \times 10^{-6}$. The bar is made of steel having $E = 210$ GPa and $\nu = 0.29$.

(a) Determine the axial force P and the torque T.

(b) Determine the maximum shear strain γ_{max} and the maximum shear stress τ_{max} in the bar.

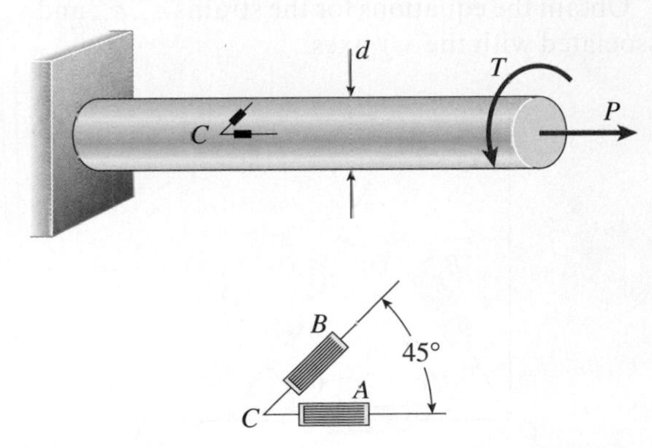

PROBLEM 7.7-21

7.7-22 A cantilever beam with a rectangular cross section (width $b = 20$ mm, height $h = 175$ mm) is loaded by a force P that acts at the mid-height of the beam and is inclined at an angle α to the vertical (see figure). Two strain gages are placed at point C, which also is at the mid-height of the beam. Gage A

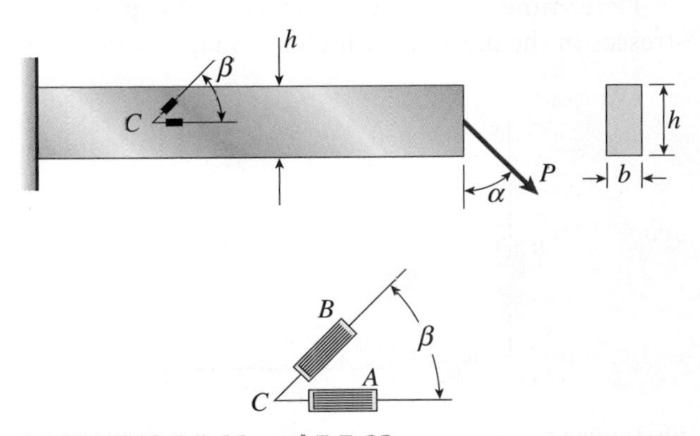

PROBLEMS 7.7-22 and 7.7-23

measures the strain in the horizontal direction, and gage B measures the strain at an angle $\beta = 60°$ to the horizontal. The measured strains are $\varepsilon_A = 145 \times 10^{-6}$ and $\varepsilon_B = -165 \times 10^{-6}$.

Determine the force P and the angle α, assuming the material is steel with $E = 200$ GPa and $\nu = 1/3$.

7.7-23 Solve the preceding problem if the cross-sectional dimensions are $b = 38$ mm and $h = 125$ mm, the gage angle is $\beta = 75°$, the measured strains are $\varepsilon_A = 209 \times 10^{-6}$ and $\varepsilon_B = -110 \times 10^{-6}$, and the material is a magnesium alloy with modulus $E = 43$ GPa and Poisson's ratio $\nu = 0.35$.

7.7-24 A 60° strain rosette, or *delta rosette*, consists of three electrical-resistance strain gages arranged as shown in the figure. Gage A measures the normal strain ε_a in the direction of the x axis. Gages B and C measure the strains ε_b and ε_c in the inclined directions shown.

Obtain the equations for the strains ε_x, ε_y, and γ_{xy} associated with the x-y axes.

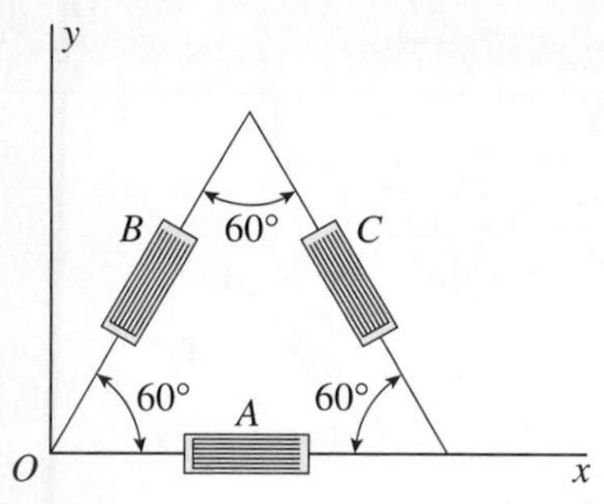

PROBLEM 7.7-24

7.7-25 On the surface of a structural component in a space vehicle, the strains are monitored by means of three strain gages arranged as shown in the figure. During a certain maneuver, the following strains were recorded: $\varepsilon_a = 1100 \times 10^{-6}$, $\varepsilon_b = 200 \times 10^{-6}$, and $\varepsilon_c = 200 \times 10^{-6}$.

Determine the principal strains and principal stresses in the material, which is a magnesium alloy for which $E = 41$ GPa and $\nu = 0.35$. Show the principal strains and principal stresses on sketches of properly oriented elements.

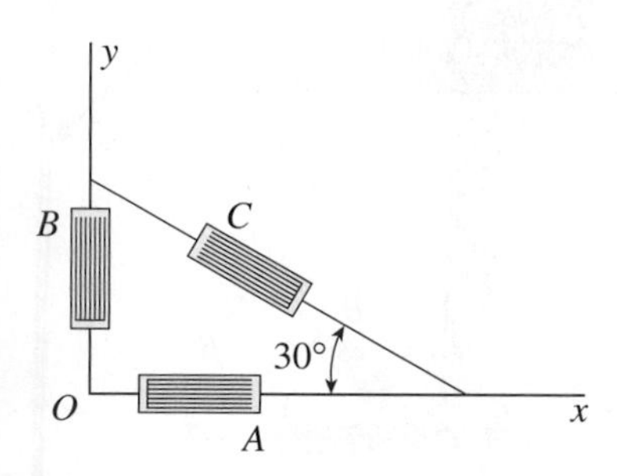

PROBLEM 7.7-25

7.7-26 The strains on the surface of an experimental device made of pure aluminum ($E = 70$ GPa, $\nu = 0.33$) and tested in a space shuttle were measured by means of strain gages. The gages were oriented as shown in the figure, and the measured strains were $\varepsilon_a = 1100 \times 10^{-6}$, $\varepsilon_b = 1496 \times 10^{-6}$, and $\varepsilon_c = -39.44 \times 10^{-6}$.

What is the stress σ_x in the x direction?

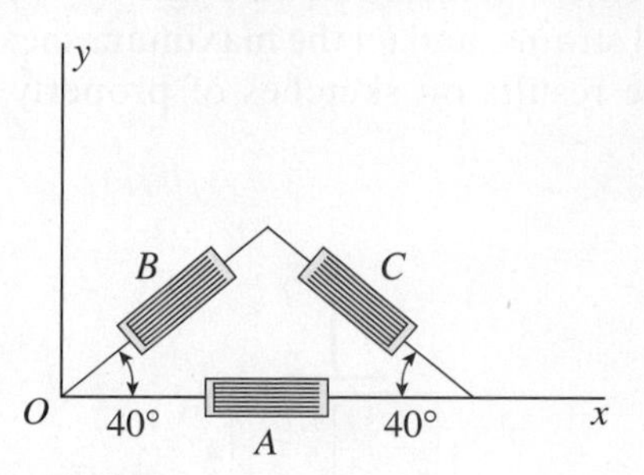

PROBLEM 7.7-26

7.7-27 Solve Problem 7.7-9 by using Mohr's circle for plane strain.

7.7-28 Solve Problem 7.7-10 by using Mohr's circle for plane strain.

7.7-29 Solve Problem 7.7-11 by using Mohr's circle for plane strain.

7.7-30 Solve Problem 7.7-12 by using Mohr's circle for plane strain.

7.7-31 Solve Problem 7.7-13 by using Mohr's circle for plane strain.

7.7-32 Solve Problem 7.7-14 by using Mohr's circle for plane strain.

SOME ADDITIONAL REVIEW PROBLEMS: Chapter 7

R-7.1 A rectangular plate ($a = 120$ mm, $b = 160$ mm) is subjected to compressive stress $\sigma_x = -4.5$ MPa and tensile stress $\sigma_y = 15$ MPa. The ratio of the normal stress acting perpendicular to the weld to the shear stress acting along the weld is approximately:

(A) 0.27
(B) 0.54
(C) 0.85
(D) 1.22

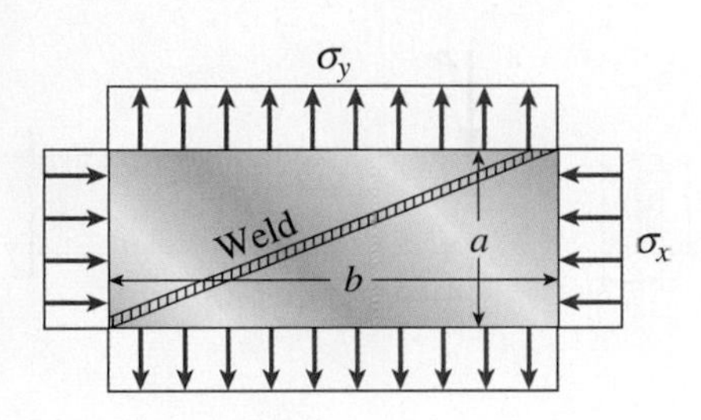

R-7.2 A rectangular plate in plane stress is subjected to normal stresses σ_x and σ_y and shear stress τ_{xy}. Stress σ_x is known to be 15 MPa, but σ_y and τ_{xy} are unknown. However, the normal stress is known to be 33 MPa at counterclockwise angles of 35° and 75° from the x axis. Based on this, the normal stress σ_y on the element in the figure is approximately:

(A) 14 MPa
(B) 21 MPa
(C) 26 MPa
(D) 43 MPa

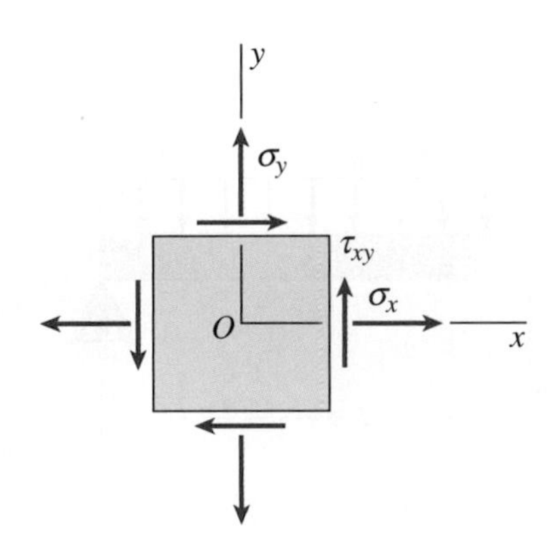

R-7.3 A rectangular plate in plane stress is subjected to normal stresses $\sigma_x = 35$ MPa, $\sigma_y = 26$ MPa, and shear stress $\tau_{xy} = 14$ MPa. The ratio of the magnitudes of the principal stresses (σ_1/σ_2) is approximately:

(A) 0.8
(B) 1.5
(C) 2.1
(D) 2.9

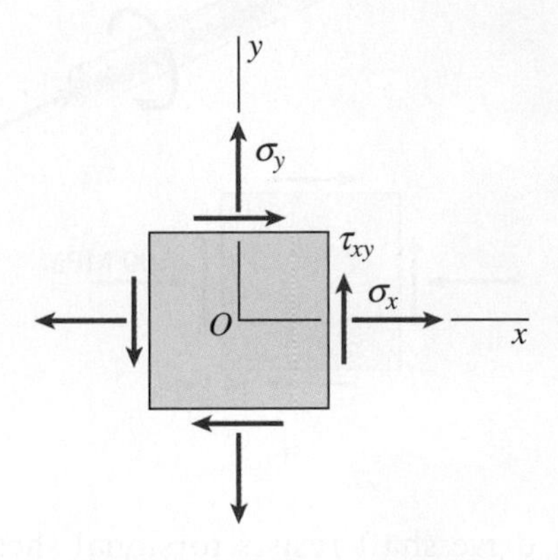

R-7.4 A drive shaft resists torsional shear stress of 45 MPa and axial compressive stress of 100 MPa. The ratio of the magnitudes of the principal stresses (σ_1/σ_2) is approximately:

(A) 0.15
(B) 0.55
(C) 1.2
(D) 1.9

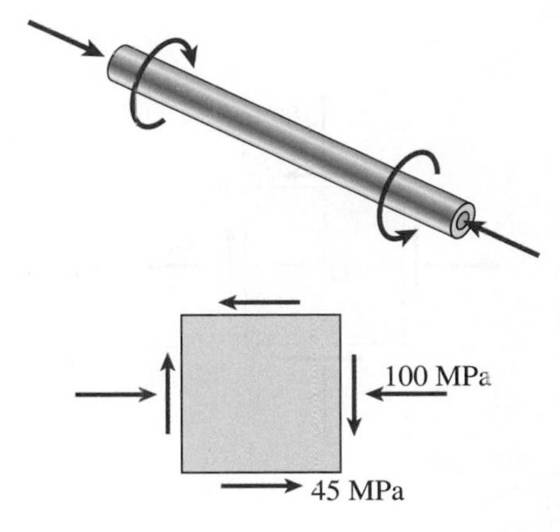

R-7.5 A drive shaft resists torsional shear stress of 45 MPa and axial compressive stress of 100 MPa. The maximum shear stress is approximately:

(A) 42 MPa
(B) 67 MPa
(C) 71 MPa
(D) 93 MPa

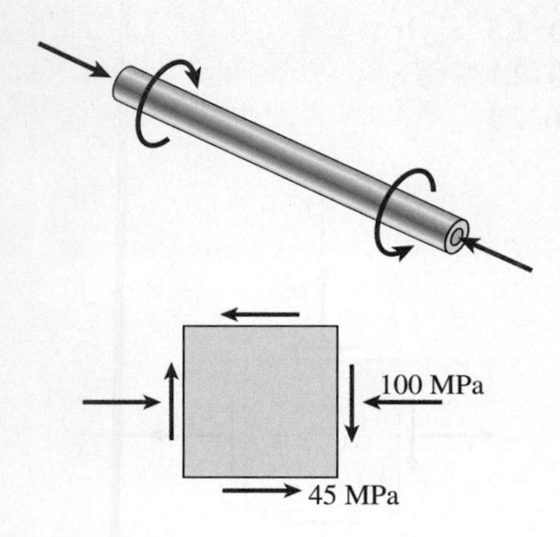

R-7.6 A drive shaft resists torsional shear stress of $\tau_{xy} = 40$ MPa and axial compressive stress $\sigma_x = -70$ MPa. One principal normal stress is known to be 38 MPa (tensile). The stress σ_y is approximately:

(A) 23 MPa
(B) 35 MPa
(C) 62 MPa
(D) 75 MPa

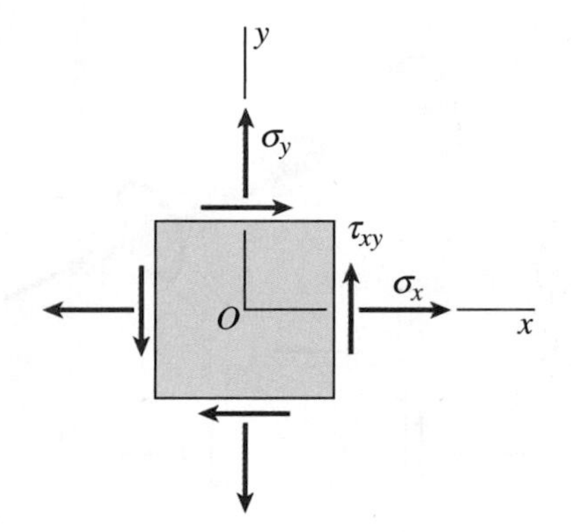

R-7.7 A cantilever beam with rectangular cross section ($b = 95$ mm, $h = 300$ mm) supports load $P = 160$ kN at its free end. The ratio of the magnitudes of the principal stresses (σ_1/σ_2) at point A (at distance $c = 0.8$ m from the free end and distance $d = 200$ mm up from the bottom) is approximately:

(A) 5
(B) 12
(C) 18
(D) 25

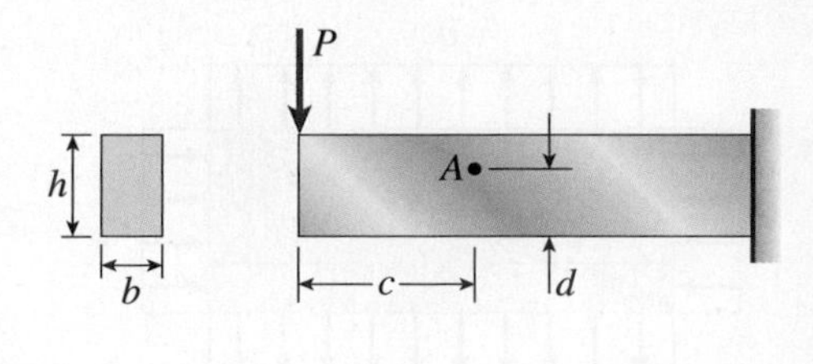

R-7.8 A simply supported beam ($L = 4.5$ m) with rectangular cross section ($b = 95$ mm, $h = 280$ mm) supports uniform load $q = 25$ kN/m. The ratio of the magnitudes of the principal stresses (σ_1/σ_2) at a point $a = 1.0$ m from the left support and distance $d = 100$ mm up from the bottom of the beam is approximately:

(A) 9
(B) 17
(C) 31
(D) 41

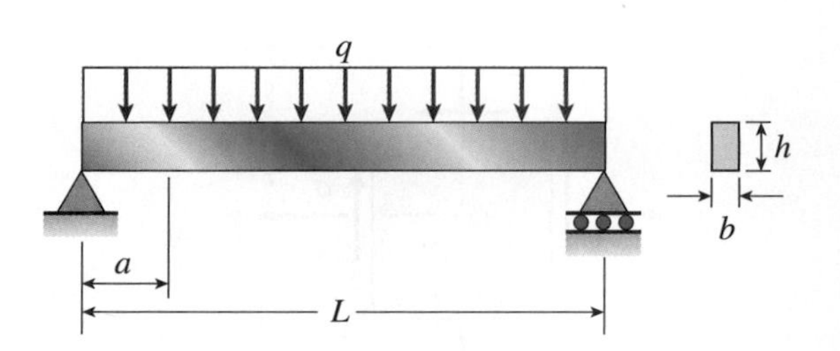

CHAPTER 8

Applications of Plane Stress
(Pressure Vessels, Beams, and Combined Loadings)

andrea crisante/Shutterstock.com

Airships rely on internal pressure to maintain their shape using a gas lighter than air for buoyant lift.

Chapter Objectives

- Study the condition of stress known as plane stress and its application in such structures as spherical and cylindrical pressure vessels.
- Investigate the distribution of principal stresses and maximum shear stresses in beams and plot either stress trajectories or stress contours to display the variation of these stresses over the length of the beam.
- Evaluate combined loadings such as axial, torsional, shear, bending, and possibly internal pressure at key locations on structures to define the state of plane stress at different points of interest.
- Present a detailed approach for analysis of critical points in a structure acted on by combined loadings.

Chapter Outline

8.1 Introduction

This chapter presents the analysis of some practical examples of structures and components in states of plane stress or strain, building upon the concepts presented in Chapter 7. First, stresses and strains in the walls of thin pressure vessels are examined. Then structures acted upon by combined loadings are evaluated to find the maximum normal and shear stresses that govern their design.

Wayne Eastep/Getty Images

Thin-walled spherical pressure vessel used for storage of propane in this oil refinery

8.2 Spherical Pressure Vessels

Pressure vessels are closed structures containing liquids or gases under pressure. Familiar examples include tanks, pipes, and pressurized cabins in aircraft and space vehicles. When pressure vessels have walls that are thin in comparison to their overall dimensions, they are included within a more general category known as **shell structures**. Examples of shell structures are roof domes, airplane wings, and submarine hulls.

This section considers thin-walled pressure vessels of spherical shape, like the compressed-air tank shown in Fig. 8-1. The term **thin-walled** is not precise, but as a general rule, pressure vessels are considered to be thin-walled when the ratio of radius r to wall thickness t (Fig. 8-2) is greater than 10. When this condition is met, the stresses in the walls can be determined with reasonable accuracy using statics alone.

FIGURE 8-1

Spherical pressure vessel

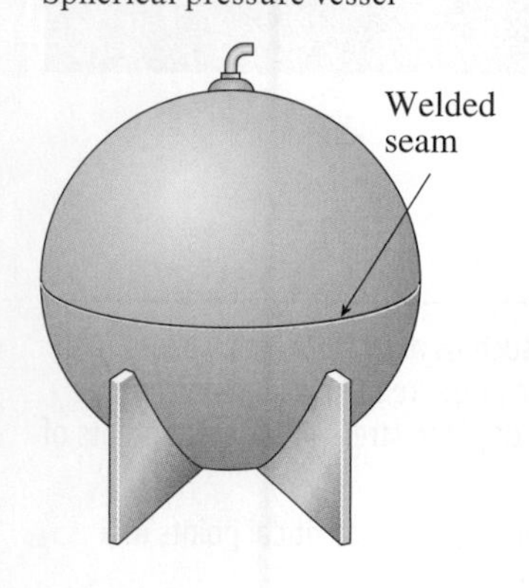

Assume in the following discussions that the internal pressure p (Fig. 8-2) exceeds the pressure acting on the outside of the shell. Otherwise, the vessel may collapse inward due to buckling.

A sphere is the theoretically ideal shape for a vessel that resists internal pressure. You only need to contemplate the familiar soap bubble to recognize that a sphere is the "natural" shape for this purpose. To determine the stresses in a spherical vessel, cut through the sphere on a vertical diametral plane (Fig. 8-3a) and isolate half of the shell *and its fluid contents* as a single free body (Fig. 8-3b). Acting on this free body are the tensile stresses σ in the wall of the vessel and the fluid pressure p. This pressure acts horizontally against the plane circular area of fluid remaining inside the hemisphere. Since the pressure is uniform, the resultant pressure force P (Fig. 8-3b) is

$$P = p(\pi r^2) \tag{8-1}$$

where r is the inner radius of the sphere.

FIGURE 8-2

Cross section of spherical pressure vessel showing inner radius r, wall thickness t, and internal pressure p

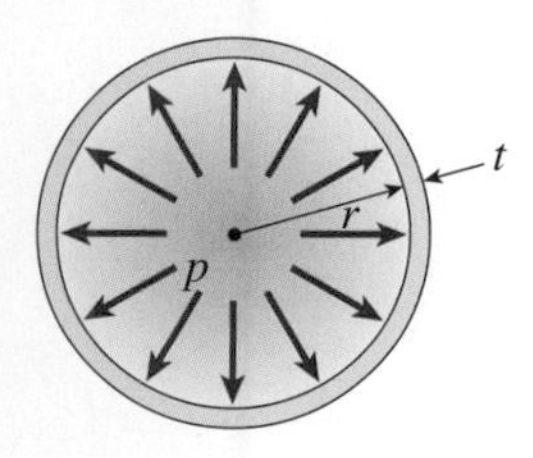

Note that the pressure p is not the absolute pressure inside the vessel but is the net internal pressure, or the **gage pressure**. Gage pressure is the internal pressure *above* the pressure acting on the outside of the vessel. If the internal and external pressures are the same, no stresses are developed in the wall of the vessel—only the excess of internal pressure over external pressure has any effect on these stresses.

Because of the symmetry of the vessel and its loading (Fig. 8-3b), the tensile stress σ is uniform around the circumference. Furthermore, since the wall is thin, you can assume with good accuracy that the stress is uniformly distributed across the thickness t. The accuracy of this approximation increases as the shell becomes thinner and decreases as it becomes thicker.

The resultant of the tensile stresses σ in the wall is a horizontal force equal to the stress σ times the area over which it acts, or

$$\sigma(2\pi r_m t)$$

where t is the thickness of the wall and r_m is its mean radius:

$$r_m = r + \frac{t}{2} \tag{8-2}$$

Thus, equilibrium of forces in the horizontal direction (Fig. 8-3b) gives

$$\Sigma F_{\text{horiz}} = 0: \quad \sigma(2\pi r_m t) - p(\pi r^2) = 0 \tag{8-3}$$

from which the *tensile stresses* in the wall of the vessel are derived:

$$\sigma = \frac{pr^2}{2r_m t} \tag{8-4}$$

Since this analysis is valid only for thin shells, disregard the small difference between the two radii appearing in Eq. (8-4) and replace r by r_m or replace r_m by r. While either choice is satisfactory for this approximate analysis, the stresses are closer to the theoretically exact stresses if the inner radius r is used instead of the mean radius r_m. Therefore, the following formula is used for calculating the **tensile stresses in the wall of a spherical shell**:

$$\sigma = \frac{pr}{2t} \tag{8-5}$$

As is evident from the symmetry of a spherical shell, you obtain the same equation for the tensile stresses when you cut a plane through the center of the sphere in any direction whatsoever. Thus, *the wall of a pressurized spherical vessel is subjected to uniform tensile stresses σ in all directions*. This stress condition is represented in Fig. 8-3c by the small stress element with stresses σ acting in mutually perpendicular directions.

Stresses that act tangentially to the curved surface of a shell, such as the stresses σ shown in Fig. 8-3c, are known as **membrane stresses**. The name arises from the fact that these are the only stresses that exist in true membranes, such as soap films.

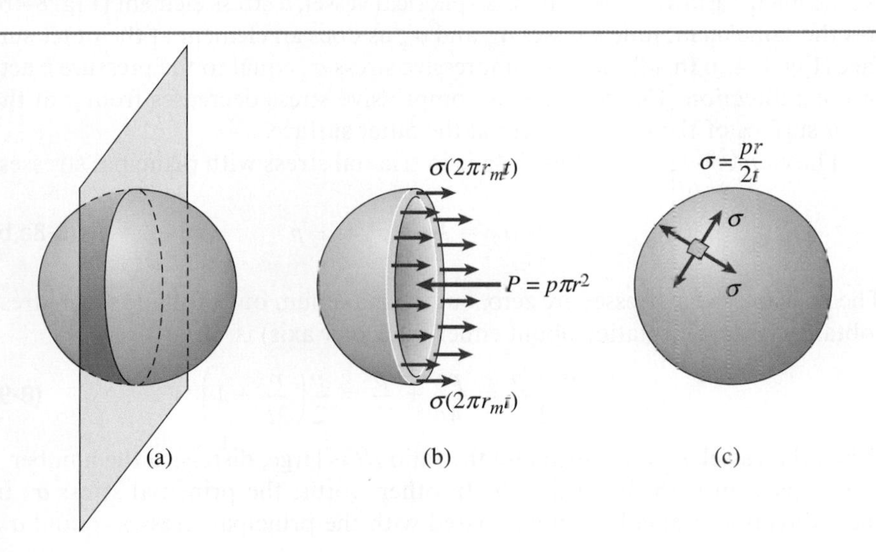

FIGURE 8-3

Tensile stresses σ in the wall of a spherical pressure vessel

FIGURE 8-4

Stresses in a spherical pressure vessel at (a) the outer surface and (b) the inner surface

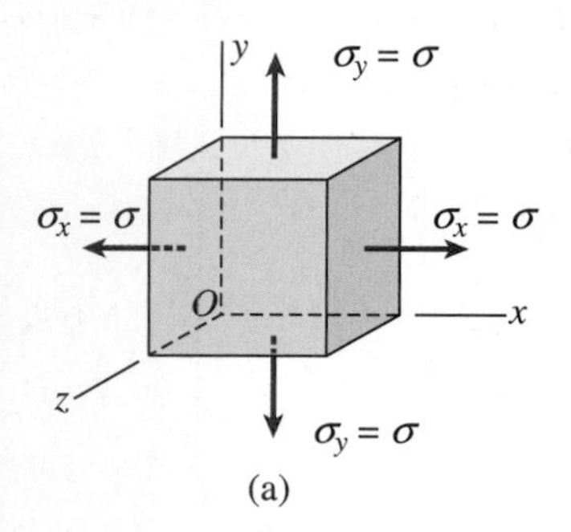

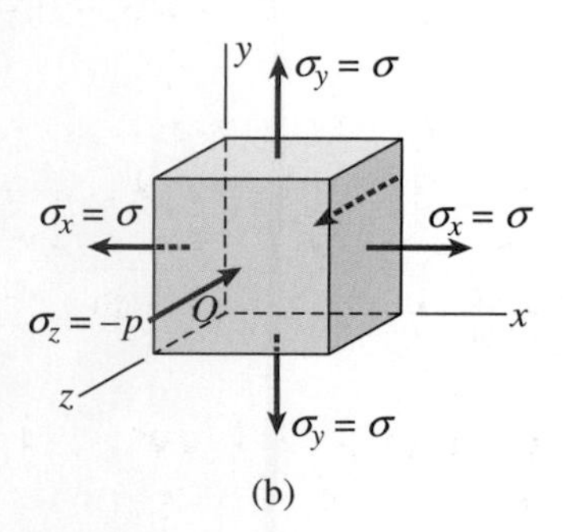

Stresses at the Outer Surface

The outer surface of a spherical pressure vessel is usually free of any loads. Therefore, the element shown in Fig. 8-3c is in *biaxial stress*. To aid in analyzing the stresses acting on this element, it appears again in Fig. 8-4a, where a set of coordinate axes is oriented parallel to the sides of the element. The x and y axes are tangential to the surface of the sphere, and the z axis is perpendicular to the surface. Thus, the normal stresses σ_x and σ_y are the same as the membrane stresses σ, and the normal stress σ_z is zero. No shear stresses act on the sides of this element.

Analyzing the element of Fig. 8-4a by using the transformation equations for plane stress [see Fig. 7-1 and Eqs. (7-4a and b) of Section 7.2] gives

$$\sigma_{x1} = \sigma \quad \text{and} \quad \tau_{x1y1} = 0$$

as expected. In other words, when you consider elements obtained by rotating the axes about the z axis, the normal stresses remain constant and there are no shear stresses. *Every plane is a principal plane and every direction is a principal direction.* Thus, the **principal stresses** for the element are

$$\sigma_1 = \sigma_2 = \frac{pr}{2t} \qquad \sigma_3 = 0 \tag{8-6a,b}$$

The stresses σ_1 and σ_2 lie in the x-y plane, and the stress σ_3 acts in the z direction.

To obtain the **maximum shear stresses**, consider out-of-plane rotations, that is, rotations about the x and y axes (because all in-plane shear stresses are zero). Elements oriented by making 45° rotations about the x and y axes have maximum shear stresses equal to $\sigma/2$ and normal stresses equal to $\sigma/2$. Therefore,

$$\tau_{max} = \frac{\sigma}{2} = \frac{pr}{4t} \tag{8-7}$$

These stresses are the largest shear stresses in the element.

Stresses at the Inner Surface

At the inner surface of the wall of a spherical vessel, a stress element (Fig. 8-4b) has the same membrane stresses σ_x and σ_y as does an element at the outer surface (Fig. 8-4a). In addition, a compressive stress σ_z equal to the pressure p acts in the z direction (Fig. 8-4b). This compressive stress decreases from p at the inner surface of the sphere to zero at the outer surface.

The element shown in Fig. 8-4b is in triaxial stress with principal stresses

$$\sigma_1 = \sigma_2 = \frac{pr}{2t} \qquad \sigma_3 = -p \tag{8-8a,b}$$

The in-plane shear stresses are zero, but the maximum out-of-plane shear stress (obtained by a 45° rotation about either the x or y axis) is

$$\tau_{max} = \frac{\sigma + p}{2} = \frac{pr}{4t} + \frac{p}{2} = \frac{p}{2}\left(\frac{r}{2t} + 1\right) \tag{8-9}$$

When the vessel is thin-walled and the ratio r/t is large, disregard the number 1 in comparison with the term $r/2t$. In other words, the principal stress σ_3 in the z direction is small when compared with the principal stresses σ_1 and σ_2.

Consequently, consider the stress state at the inner surface to be the same as at the outer surface (biaxial stress). This approximation is consistent with the approximate nature of thin-shell theory, so use Eqs. (8-5), (8-6a,b), and (8-7) to obtain the stresses in the wall of a spherical pressure vessel.

General Comments

Pressure vessels usually have openings in their walls (to serve as inlets and outlets for the fluid contents) as well as fittings and supports that exert forces on the shell (Fig. 8-1). These features result in nonuniformities in the stress distribution, or *stress concentrations*, that cannot be analyzed by the elementary formulas given here. Instead, more advanced methods of analysis are needed. Other factors that affect the design of pressure vessels include corrosion, accidental impacts, and temperature changes.

Some of the limitations of thin-shell theory as applied to pressure vessels are listed here.

1. The wall thickness must be small in comparison to the other dimensions (the ratio r/t should be 10 or more).
2. The internal pressure must exceed the external pressure (to avoid inward buckling).
3. The analysis presented in this section is based only on the effects of internal pressure (the effects of external loads, reactions, the weight of the contents, and the weight of the structure are not considered).
4. The formulas derived in this section are valid throughout the wall of the vessel *except* near points of stress concentrations.

The following example illustrates how the principal stresses and maximum shear stresses are used in the analysis of a spherical shell.

Example 8-1

A compressed-air tank having an inner diameter of 5.5 m and a wall thickness of 45 mm is formed by welding two steel hemispheres (Fig. 8-5).

(a) If the allowable tensile stress in the steel is 93 MPa, what is the maximum permissible air pressure p_a in the tank?

(b) If the allowable shear stress in the steel is 42 MPa, what is the maximum permissible pressure p_b?

(c) If the normal strain at the outer surface of the tank is not to exceed 0.0003, what is the maximum permissible pressure p_c? (Assume that Hooke's law is valid and that the modulus of elasticity for the steel is 210 GPa and Poisson's ratio is 0.28.)

(d) Tests on the welded seam show that failure occurs when the tensile load on the welds exceeds 7.5 MN/m of weld. If the required factor of safety against failure of the weld is 2.5, what is the maximum permissible pressure p_d?

(e) Considering the four preceding factors, what is the allowable pressure p_{allow} in the tank?

FIGURE 8-5

Example 8-1: Spherical pressure vessel. (Attachments and supports are shown in photo.)

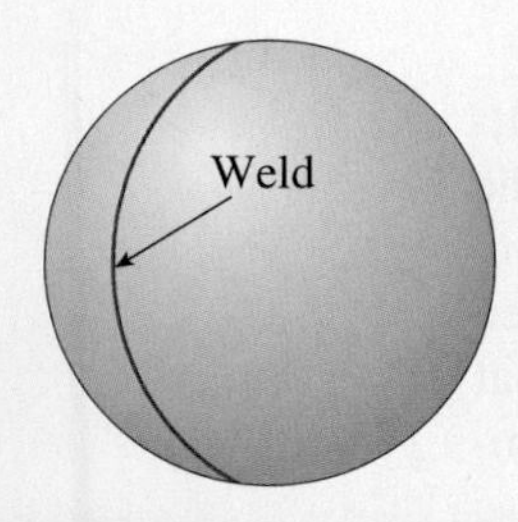

Spherical tanks at oil refinery

Solution:

Use a four-step problem-solving approach. Combine steps as needed for an efficient solution.

1. ***Conceptualize*** [*hypothesize, sketch*]: The compressed-air tank is a thin-walled pressure vessel. Evaluate membrane stresses at the outer surface of the tank. Tensile stresses in the wall are evaluated using Eq. (8-5). Computed stresses will be closer to theoretically exact values if the inner radius of the shell rather than the mean radius is used.

2,3. ***Categorize*** [*simplify, classify*], ***Analyze*** [*evaluate; select relevant equations, carry out mathematical solution*]:

Part (a): Allowable pressure based upon the tensile stress in the steel.

The maximum tensile stress in the wall of the tank is given by the formula $\sigma = pr/2t$ [see Eq. (8-5)]. Solve this equation for the pressure in terms of the allowable stress to get

$$p_a = \frac{2t\sigma_{\text{allow}}}{r} = \frac{2(45 \text{ mm})(93 \text{ MPa})}{2.75 \text{ m}} = 3.04 \text{ MPa}$$

Thus, the maximum allowable pressure based upon tension in the wall of the tank is $p_a = 3.04$ MPa. (Note that in a calculation of this kind, you round downward, not upward.)

Part (b): Allowable pressure based upon the shear stress in the steel.

The maximum shear stress in the wall of the tank is given by Eq. (8-7), from which the pressure is obtained:

$$p_b = \frac{4t\tau_{\text{allow}}}{r} = \frac{4(45 \text{ mm})(42 \text{ MPa})}{2.75 \text{ m}} = 2.75 \text{ MPa}$$

Therefore, the allowable pressure based upon shear is $p_b = 2.75$ MPa.

Part (c): Allowable pressure based upon the normal strain in the steel.

The normal strain is obtained from Hooke's law for biaxial stress [Eq. (7-40a)]:

$$\varepsilon_x = \frac{1}{E}(\sigma_x - \nu\sigma_y) \tag{a}$$

Substitute $\sigma_x = \sigma_y = \sigma = pr/2t$ (see Fig. 8-4a) to obtain

$$\varepsilon_x = \frac{\sigma}{E}(1 - \nu) = \frac{pr}{2tE}(1 - \nu) \tag{8-10}$$

This equation can be solved for the pressure p_c:

$$p_c = \frac{2tE\varepsilon_{\text{allow}}}{r(1 - \nu)} = \frac{2(45 \text{ mm})(210 \text{ GPa})(0.0003)}{2.75 \text{ m}(1 - 0.28)} = 2.86 \text{ MPa}$$

Thus, the allowable pressure based upon the normal strain in the wall is $p_c = 2.86$ MPa.

Part (d): Allowable pressure based upon the tension in the welded seam.
The allowable tensile load on the welded seam is equal to the failure load divided by the factor of safety:

$$T_{\text{allow}} = \frac{T_{\text{failure}}}{n} = \frac{7.5 \text{ MN/m}}{2.5} = 3 \text{ MN/m}$$

The corresponding allowable tensile stress is equal to the allowable load on a one-meter length of weld divided by the cross-sectional area of a one-meter length of weld:

$$\sigma_{\text{allow}} = \frac{T_{\text{allow}}(1.0 \text{ m})}{(1.0 \text{ m})(t)} = \frac{3 \text{ MN/m}(1.0 \text{ m})}{(1.0 \text{ m})(45 \text{ mm})} = 66.67 \text{ MPa}$$

Finally, solve for the internal pressure by using Eq. (8-5):

$$p_d = \frac{2t\sigma_{\text{allow}}}{r} = \frac{2(45 \text{ mm})(66.67 \text{ MPa})}{2.75 \text{ m}} = 2.18 \text{ MPa}$$

This result gives the allowable pressure based upon tension in the welded seam.

Part (e): Allowable pressure.
Comparing the preceding results for p_a, p_b, p_c, and p_d, note that tension in the welded seam governs and the allowable pressure in the tank is

$$p_{\text{allow}} = 2.18 \text{ MPa}$$

4. ***Finalize*** [*conclude; examine answer—Does it make sense? Are units correct? How does it compare to similar problem solutions?*]: This example illustrates how various stresses and strains enter into the design of a spherical pressure vessel.

Note: When the internal pressure is at its maximum allowable value (2.18 MPa), the tensile stresses in the shell are

$$\sigma = \frac{pr}{2t} = \frac{2.18 \text{ MPa}(2.75 \text{ m})}{2(45 \text{ mm})} = 66.6 \text{ MPa}$$

Thus, at the inner surface of the shell (Fig. 8-4b), the ratio of the principal stress in the z direction (2.18 MPa) to the in-plane principal stresses (66.6 MPa) is only 0.033. Therefore, the earlier assumption that the principal stress σ_3 in the z direction can be disregarded and the entire shell to be in biaxial stress is justified.

FIGURE 8-6

Cylindrical pressure vessels with circular cross sections

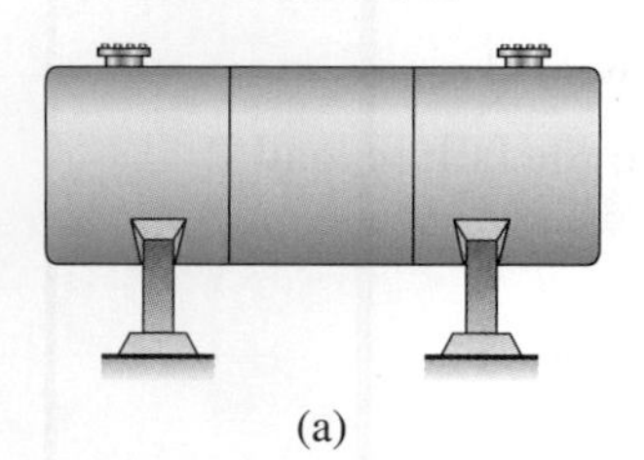

(a)

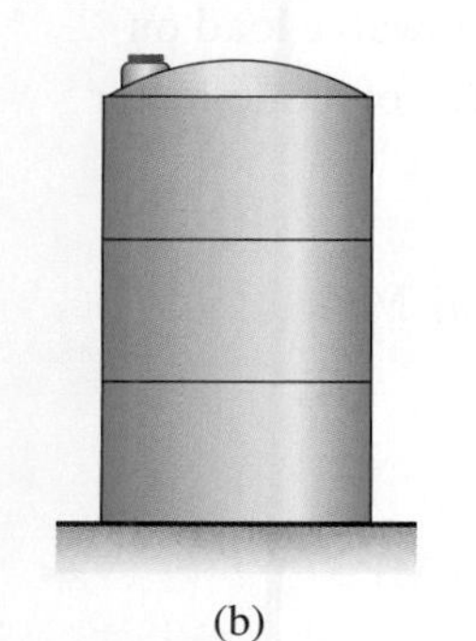

(b)

Cylindrical storage tanks in a petrochemical plant

8.3 Cylindrical Pressure Vessels

Cylindrical pressure vessels with a circular cross section (Fig. 8-6) are found in industrial settings (compressed air tanks and rocket motors), in homes (fire extinguishers and spray cans), and in the countryside (propane tanks and grain silos). Pressurized pipes, such as water-supply pipes and penstocks, are also classified as cylindrical pressure vessels.

The analysis of cylindrical vessels begins by determining the normal stresses in a *thin-walled circular tank AB* subjected to internal pressure (Fig. 8-7a). A *stress element* with its faces parallel and perpendicular to the axis of the tank is shown on the wall of the tank. The normal stresses σ_1 and σ_2 acting on the side faces of this element are the membrane stresses in the wall. No shear stresses act on these faces because of the symmetry of the vessel and its loading. Therefore, the stresses σ_1 and σ_2 are principal stresses.

Because of their directions, the stress σ_1 is called the **circumferential stress** or the **hoop stress**, and the stress σ_2 is called the **longitudinal stress** or the **axial stress**. Each of these stresses can be calculated from equilibrium by using appropriate free-body diagrams.

Circumferential Stress

To determine the circumferential stress σ_1, make two cuts (*mn* and *pq*) perpendicular to the longitudinal axis and a distance *b* apart (Fig. 8-7a). Then make a third cut in a vertical plane through the longitudinal axis of the tank, resulting in the free body shown in Fig. 8-7b. This free body consists not only of the half-circular piece of the tank but also of the fluid contained within the cuts. Acting on the longitudinal cut (plane *mpqn*) are the circumferential stresses σ_1 and the internal pressure *p*.

Stresses and pressures also act on the left-hand and right-hand faces of the free body. However, these stresses and pressures are not shown in the figure because they do not enter the equation of equilibrium used here. As in the analysis of a spherical vessel, disregard the weight of the tank and its contents.

FIGURE 8-7

Stresses in a circular cylindrical pressure vessel

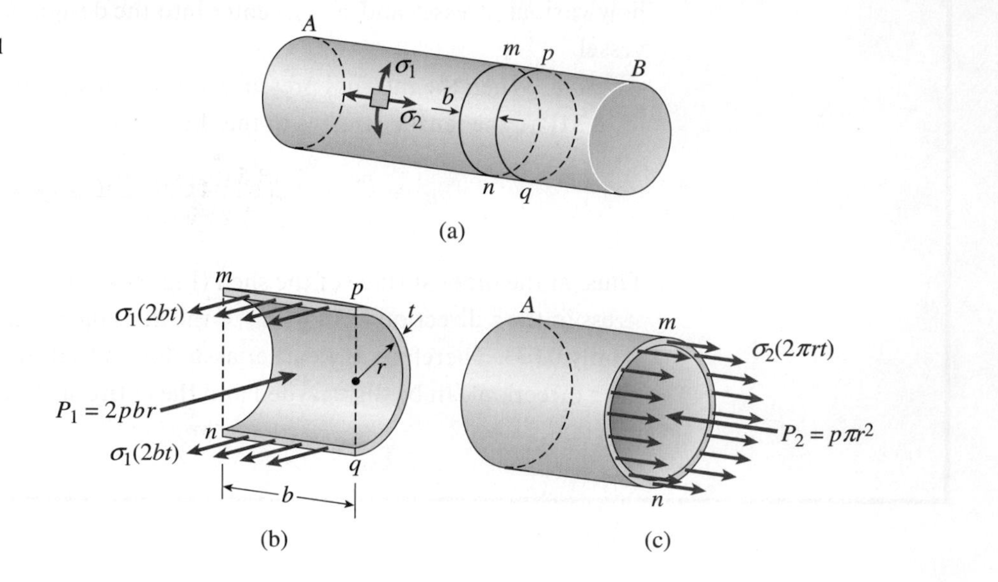

The circumferential stresses σ_1 acting in the wall of the vessel have a resultant equal to $\sigma_1(2bt)$, where t is the thickness of the wall. Also, the resultant force P_1 of the internal pressure is equal to $2pbr$, where r is the inner radius of the cylinder. The resulting equation of equilibrium is

$$\sigma_1(2bt) - 2pbr = 0$$

This equation leads to the formula for the *circumferential stress in a pressurized cylinder*:

$$\sigma_1 = \frac{pr}{t} \quad \textbf{(8-11)}$$

This stress is uniformly distributed over the thickness of the wall, provided the thickness is small compared to the radius.

Longitudinal Stress

The longitudinal stress σ_2 is obtained from the equilibrium of a free body of the part of the vessel to the left of cross section *mn* (Fig. 8-7c). Again, the free body includes not only part of the tank but also its contents. The stresses σ_2 act longitudinally and have a resultant force equal to $\sigma_2(2\pi rt)$. Note that the inner radius of the shell is used in place of the mean radius, as explained in Section 8.2.

The resultant force P_2 of the internal pressure is a force equal to $p\pi r^2$. Thus, the equation of equilibrium for the free body is

$$\sigma_2(2\pi rt) - p\pi r^2 = 0$$

Solve this equation for σ_2 to obtain the formula for the *longitudinal stress* in a cylindrical pressure vessel:

$$\sigma_2 = \frac{pr}{2t} \quad \textbf{(8-12)}$$

This stress is equal to the membrane stress in a spherical vessel [Eq. (8-5)].

Compare Eqs. (8-11) and (8-12) to see that the circumferential stress in a cylindrical vessel is equal to twice the longitudinal stress:

$$\sigma_1 = 2\sigma_2 \quad \textbf{(8-13)}$$

From this result, note that a longitudinal welded seam in a pressurized tank must be twice as strong as a circumferential seam.

Stresses at the Outer Surface

The principal stresses σ_1 and σ_2 at the outer surface of a cylindrical vessel are shown on the stress element of Fig. 8-8a. Since the third principal stress (acting in the z direction) is zero, the element is in *biaxial stress*.

The maximum *in-plane shear stresses* occur on planes that are rotated 45° about the z axis; these stresses are

$$(\tau_{max})_z = \frac{\sigma_1 - \sigma_2}{2} = \frac{\sigma_1}{4} = \frac{pr}{4t} \quad \textbf{(8-14)}$$

FIGURE 8-8

Stresses in a circular cylindrical pressure vessel at (a) the outer surface and (b) the inner surface

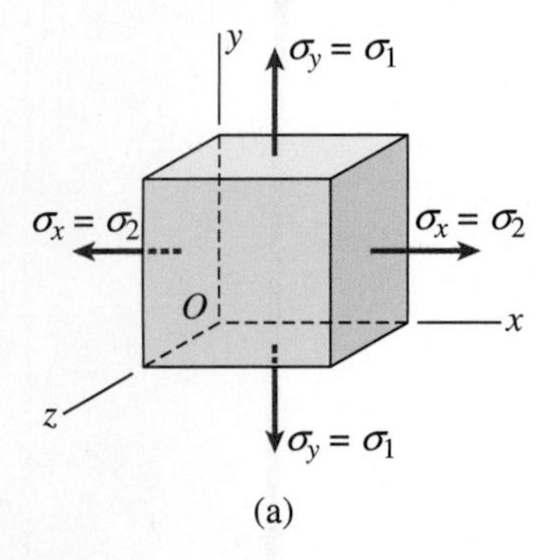

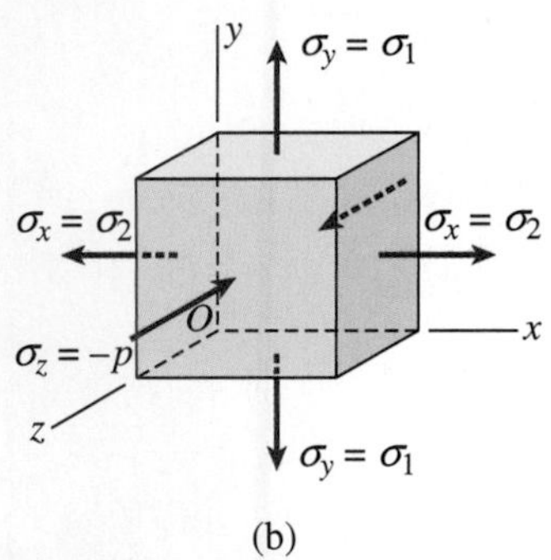

The maximum *out-of-plane shear stresses* are obtained by 45° rotations about the x and y axes, respectively; thus,

$$(\tau_{\max})_x = \frac{\sigma_1}{2} = \frac{pr}{2t} \qquad (\tau_{\max})_y = \frac{\sigma_2}{2} = \frac{pr}{4t} \tag{8-15a,b}$$

The preceding results show that the *absolute maximum shear stress* is

$$\tau_{\max} = \frac{\sigma_1}{2} = \frac{pr}{2t} \tag{8-16}$$

This stress occurs on a plane that has been rotated 45° about the x axis.

Stresses at the Inner Surface

The stress conditions at the inner surface of the wall of the vessel are shown in Fig. 8-8b. The principal stresses are

$$\sigma_1 = \frac{pr}{t} \quad \sigma_2 = \frac{pr}{2t} \quad \sigma_3 = -p \tag{8-17a,b,c}$$

The three maximum shear stresses, obtained by 45° rotations about the x, y, and z axes, are

$$(\tau_{\max})_x = \frac{\sigma_1 - \sigma_3}{2} = \frac{pr}{2t} + \frac{p}{2} \tag{8-18a}$$

$$(\tau_{\max})_y = \frac{\sigma_2 - \sigma_3}{2} = \frac{pr}{4t} + \frac{p}{2} \tag{8-18b}$$

$$(\tau_{\max})_z = \frac{\sigma_1 - \sigma_2}{2} = \frac{pr}{4t} \tag{8-18c}$$

The first of these three stresses is the largest. However, as explained in the discussion of shear stresses in a spherical shell, you may disregard the additional term $p/2$ in Eqs. (8-18a and b) when the shell is thin-walled. Equations (8-18a, b, and c) then become the same as Eqs. (8-15) and (8-14), respectively.

Therefore, in all examples and problems pertaining to cylindrical pressure vessels, *disregard the presence of the compressive stress in the z direction*. (This compressive stress varies from p at the inner surface to zero at the outer surface.) With this approximation, the stresses at the inner surface become the same as the stresses at the outer surface (biaxial stress). As explained in the discussion of spherical pressure vessels, this procedure is satisfactory when you consider the numerous other approximations in this theory.

General Comments

The preceding formulas for stresses in a circular cylinder are valid in parts of the cylinder away from any discontinuities that cause stress concentrations, as discussed previously for spherical shells. An obvious discontinuity exists at the ends of the cylinder where the heads are attached, because the geometry of the structure changes abruptly. Other stress concentrations occur at openings, at points of support, and wherever objects or fittings are attached to the cylinder. The stresses at such points cannot be determined solely from equilibrium equations; instead, more advanced methods of analysis (such as shell theory and finite-element analysis) must be used.

Some of the limitations of the elementary theory for thin-walled shells are listed in Section 8.2.

Example 8-2

FIGURE 8-9

Example 8-2: Cylindrical pressure vessel with a helical weld

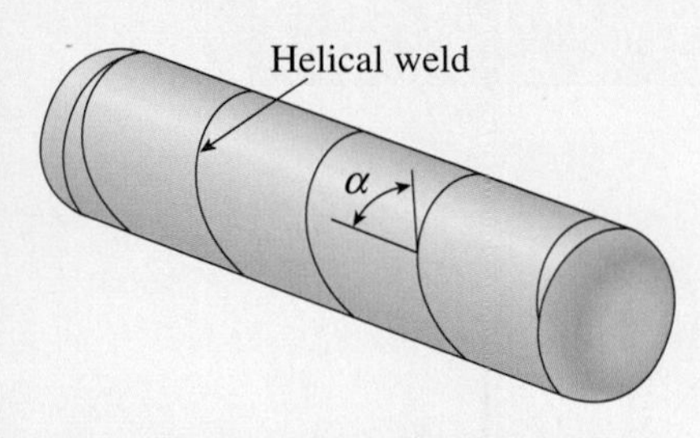

Cylindrical pressure vessel on simple supports

A cylindrical pressure vessel is constructed from a long, narrow steel plate by wrapping the plate around a mandrel and then welding along the edges of the plate to make a helical joint (Fig. 8-9). The helical weld makes an angle $\alpha = 55°$ with the longitudinal axis. The vessel has inner radius $r = 1.8$ m and wall thickness $t = 20$ mm. The material is steel with modulus $E = 200$ GPa and Poisson's ratio $\nu = 0.30$. The internal pressure p is 800 kPa.

Calculate the following quantities for the cylindrical part of the vessel: (a) the circumferential and longitudinal stresses σ_1 and σ_2, respectively; (b) the maximum in-plane and out-of-plane shear stresses; (c) the circumferential and longitudinal strains ε_1 and ε_2, respectively; and (d) the normal stress σ_w and shear stress τ_w acting perpendicular and parallel, respectively, to the welded seam.

Solution:

Use a four-step problem-solving approach. Combine steps as needed for an efficient solution.

1. ***Conceptualize***: The circular cylindrical tank is a thin-walled pressure vessel. Evaluate membrane stresses at the outer surface of the tank. Evaluate circumferential stresses in the wall using Eq. (8-11) and longitudinal stresses using Eq. (8-12). Computed stresses will be closer to theoretically exact values if you use the inner radius of the shell rather than the mean radius.

2, 3. ***Categorize, Analyze***:

Part (a): Circumferential and longitudinal stresses.

The circumferential and longitudinal stresses σ_1 and σ_2, respectively, are pictured in Fig. 8-10a, where they are shown acting on a stress element at point A on the wall of the vessel. Calculate the magnitudes of the stresses from Eqs. (8-11) and (8-12):

$$\sigma_1 = \frac{pr}{t} = \frac{(800 \text{ kPa})(1.8 \text{ m})}{20 \text{ mm}} = 72 \text{ MPa} \quad \sigma_2 = \frac{pr}{2t} = \frac{\sigma_1}{2} = 36 \text{ MPa}$$

The stress element at point A is shown again in Fig. 8-10b, where the x axis is in the longitudinal direction of the cylinder and the y axis is in the circumferential direction. Since there is no stress in the z direction ($\sigma_3 = 0$), the element is in biaxial stress.

Note that the ratio of the internal pressure (800 kPa) to the smaller in-plane principal stress (36 MPa) is 0.022. Therefore, the assumption that any stresses in the z direction can be disregarded and all elements in the cylindrical shell, even those at the inner surface, are in biaxial stress is justified.

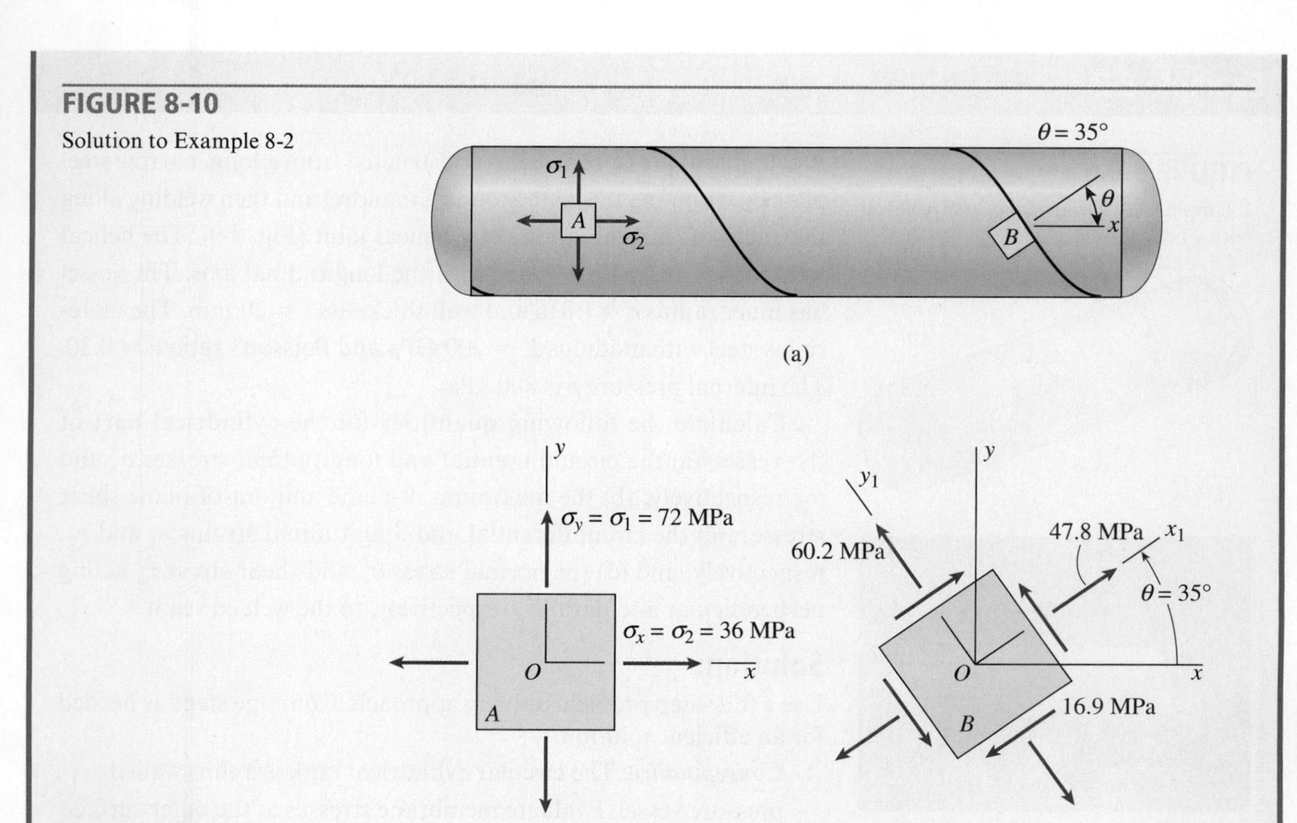

FIGURE 8-10
Solution to Example 8-2

Part (b): Maximum shear stresses.
The largest in-plane shear stress is obtained from Eq. (8-14):

$$(\tau_{\max})_z = \frac{\sigma_1 - \sigma_2}{2} = \frac{\sigma_1}{4} = \frac{pr}{4t} = 18 \text{ MPa}$$

Because the normal stress in the z direction is disregarded the largest out-of-plane shear stress is obtained from Eq. (8-15a):

$$\tau_{\max} = \frac{\sigma_1}{2} = \frac{pr}{2t} = 36 \text{ MPa}$$

This last stress is the absolute maximum shear stress in the wall of the vessel.

Part (c): Circumferential and longitudinal strains.
Since the largest stresses are well below the yield stress of steel (see Table I-3, Appendix I), assume that Hooke's law applies to the wall of the vessel. Then obtain the strains in the x and y directions (Fig. 8-10b) from Eqs. (7-40a and b) for biaxial stress:

$$\varepsilon_x = \frac{1}{E}(\sigma_x - \nu\sigma_y) \qquad \varepsilon_y = \frac{1}{E}(\sigma_y - \nu\sigma_x) \qquad \textbf{(a,b)}$$

Note that the strain ε_x is the same as the principal strain ε_2 in the longitudinal direction and that the strain ε_y is the same as the principal strain ε_1 in the circumferential direction. Also, the stress σ_x is the same as the stress σ_2, and the stress σ_y is the same as the stress σ_1. Therefore, the preceding two equations can be written in the forms:

$$\varepsilon_2 = \frac{\sigma_2}{E}(1 - 2\nu) = \frac{pr}{2tE}(1 - 2\nu) \tag{8-19a}$$

$$\varepsilon_1 = \frac{\sigma_1}{2E}(2 - \nu) = \frac{pr}{2tE}(2 - \nu) \tag{8-19b}$$

Substituting numerical values:

$$\varepsilon_2 = \frac{\sigma_2}{E}(1 - 2\nu) = \frac{(36 \text{ MPa})[1 - 2(0.30)]}{200 \text{ GPa}} = 72 \times 10^{-6}$$

$$\varepsilon_1 = \frac{\sigma_1}{2E}(2 - \nu) = \frac{(72 \text{ MPa})(2 - 0.30)}{2(200 \text{ GPa})} = 306 \times 10^{-6}$$

These are the longitudinal and circumferential strains in the cylinder.

Part (d): Normal and shear stresses acting on the welded seam.

The stress element at point B in the wall of the cylinder (Fig. 8-10a) is oriented so that its sides are parallel and perpendicular to the weld. The angle θ for the element is

$$\theta = 90° - \alpha = 35°$$

as shown in Fig. 8-10c. Either the stress-transformation equations or Mohr's circle may be used to obtain the normal and shear stresses acting on the side faces of this element.

Stress-transformation equations: The normal stress σ_{x1} and the shear stress τ_{x1y1} acting on the x_1 face of the element (Fig. 8-10c) are obtained from Eqs. (7-4a and b), which are repeated here:

$$\sigma_{x1} = \frac{\sigma_x + \sigma_y}{2} - \frac{\sigma_x - \sigma_y}{2} \cos 2\theta + \tau_{xy} \sin 2\theta \tag{8-20a}$$

$$\tau_{x1y1} = -\frac{\sigma_x - \sigma_y}{2} \sin 2\theta + \tau_{xy} \cos 2\theta \tag{8-20b}$$

Substitute $\sigma_x = \sigma_2 = pr/2t$, $\sigma_y = \sigma_1 = pr/t$, and $\tau_{xy} = 0$, to obtain

$$\sigma_{x1} = \frac{pr}{4t}(3 - \cos 2\theta) \quad \tau_{x1y1} = \frac{pr}{4t} \sin 2\theta \tag{8-21a,b}$$

These equations give the normal and shear stresses acting on an inclined plane oriented at an angle θ with the longitudinal axis of the cylinder.

Substitute $pr/4t = 18$ MPa and $\theta = 35°$ into Eqs. (8-21a and b) to obtain

$$\sigma_{x1} = 47.8 \text{ MPa} \quad \tau_{x1y1} = 16.9 \text{ MPa}$$

These stresses are shown on the stress element of Fig. 8-10c.

To complete the stress element, calculate the normal stress σ_{y1} acting on the y_1 face of the element from the sum of the normal stresses on perpendicular faces [Eq. (7-6)]:

$$\sigma_1 + \sigma_2 = \sigma_{x1} + \sigma_{y1} \tag{8-22}$$

FIGURE 8-11

Mohr's circle for the biaxial stress element of Figure 8-10b (*Note:* All stresses on the circle have units of MPa)

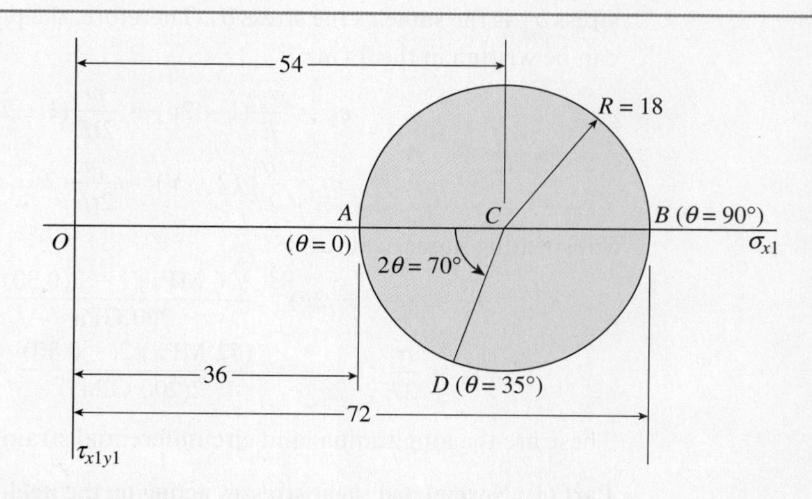

Substitute numerical values to get

$$\sigma_{y1} = \sigma_1 + \sigma_2 - \sigma_{x1} = 72 \text{ MPa} + 36 \text{ MPa} - 47.8 \text{ MPa} = 60.2 \text{ MPa}$$

as shown in Fig. 8-10c.

From the figure, the normal and shear stresses acting perpendicular and parallel, respectively, to the welded seam are

$$\sigma_w = 47.8 \text{ MPa} \quad \tau_w = 16.9 \text{ MPa}$$

3. ***Finalize***:

Mohr's circle: The Mohr's circle construction for the biaxial stress element of Fig. 8-10b is shown in Fig. 8-11. Point *A* represents the stress $\sigma_2 = 36$ MPa on the *x* face ($\theta = 0$) of the element, and point *B* represents the stress $\sigma_1 = 72$ MPa on the *y* face ($\theta = 90°$). The center *C* of the circle is at a stress of 54 MPa, and the radius of the circle is

$$R = \frac{72 \text{ MPa} - 36 \text{ MPa}}{2} = 18 \text{ MPa}$$

A counterclockwise angle $2\theta = 70°$ (measured on the circle from point *A*) locates point *D*, which corresponds to the stresses on the x_1 face ($\theta = 35°$) of the element. The coordinates of point *D* (from the geometry of the circle) are

$$\sigma_{x1} = 54 \text{ MPa} - R\cos 70° = 54 \text{ MPa} - (18 \text{ MPa})(\cos 70°) = 47.8 \text{ MPa}$$

$$\tau_{x1y1} = R\sin 70° = (18 \text{ MPa})(\sin 70°) = 16.9 \text{ MPa}$$

These results are the same as those found earlier from the stress-transformation equations.

Note: When seen in a side view, a **helix** follows the shape of a sine curve (Fig. 8-12). The pitch of the helix is

$$p = \pi d \tan\theta \quad \textbf{(8-23)}$$

FIGURE 8-12
Side view of a helix

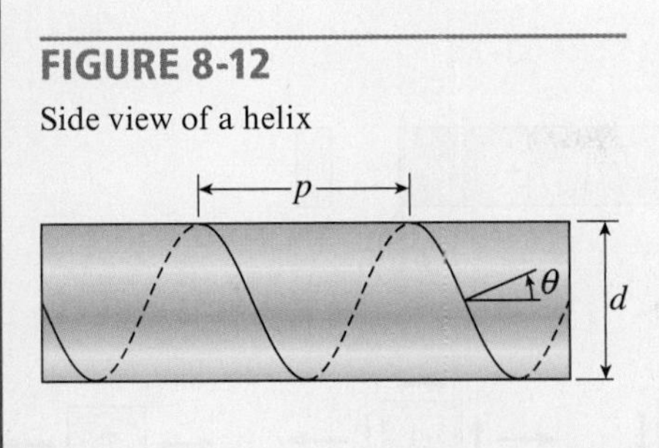

where d is the diameter of the circular cylinder and θ is the angle between a normal to the helix and a longitudinal line. The width of the flat plate that wraps into the cylindrical shape is

$$w = \pi d \sin \theta \quad \textbf{(8-24)}$$

Thus, if the diameter of the cylinder and the angle θ are given, both the pitch and the plate width are established. For practical reasons, the angle θ is usually in the range from 20° to 35°.

8.4 Maximum Stresses in Beams

The stress analysis of a beam usually begins by finding the normal and shear stresses acting on cross sections. For instance, when Hooke's law holds, the normal and shear stresses from the **flexure and shear formulas** [Eqs. (5-14) and (5-41), respectively, of Chapter 5] are

$$\sigma = -\frac{My}{I} \qquad \tau = \frac{VQ}{Ib} \quad \textbf{(8-25a,b)}$$

In the flexure formula, σ is the normal stress acting on the cross section, M is the bending moment, y is the distance from the neutral axis, and I is the moment of inertia of the cross-sectional area with respect to the neutral axis. (The sign conventions for M and y in the flexure formula are shown in Figs. 5-9 and 5-10 of Chapter 5.)

In the case of the shear formula, τ is the shear stress at any point in the cross section, V is the shear force, Q is the first moment of the cross-sectional area outside of the point in the cross section where the stress is being found, and b is the width of the cross section. (The shear formula is usually written without regard to signs because the directions of the shear stresses are apparent from the directions of the loads.)

The normal stresses obtained from the flexure formula have their maximum values at the farthest distances from the neutral axis, whereas the shear stresses obtained from the shear formula usually have their highest values at the neutral axis. The normal stresses are calculated at the cross section of the maximum bending moment, and the shear stresses are calculated at the cross section of maximum shear force. In most circumstances, these are the only stresses that are needed for design purposes.

However, to obtain a more complete picture of the stresses in a beam, it is desirable to find the principal stresses and maximum shear stresses at various points in the beam. First, consider the stresses in a rectangular beam.

Beams of Rectangular Cross Section

To better understand how the stresses in a beam vary, consider the simple beam of rectangular cross section shown in Fig. 8-13a. For the purposes of this discussion, choose a cross section to the left of the load and then select five points (A, B, C, D, and E) on the side of the beam. Points A and E are at the top and bottom of the beam, respectively, point C is at the mid-height of the beam, and points B and D are in between.

FIGURE 8-13

Stresses in a beam of rectangular cross section: (a) simple beam with points A, B, C, D, and E on the side of the beam; (b) normal and shear stresses acting on stress elements at points A, B, C, D, and E; (c) principal stresses; and (d) maximum shear stresses

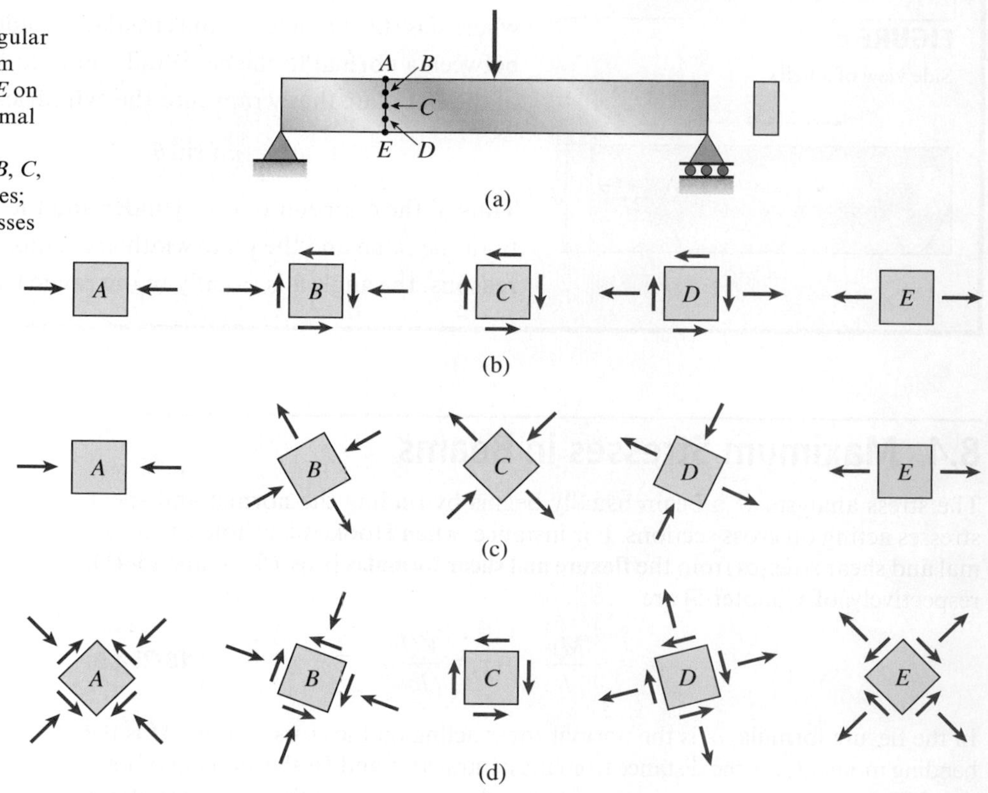

If Hooke's law applies, the normal and shear stresses at each of these five points can be readily calculated from the flexure and shear formulas. Since these stresses act on the cross section, picture them on stress elements having vertical and horizontal faces, as shown in Fig. 8-13b. Note that all elements are in plane stress, because there are no stresses acting perpendicular to the plane of the figure.

At point A, the normal stress is compressive, and there are no shear stresses. Similarly, at point E, the normal stress is tensile, and again there are no shear stresses. Thus, the elements at these locations are in uniaxial stress. At the neutral axis (point C), the element is in pure shear. At the other two locations (points B and D), both normal and shear stresses act on the stress elements.

To find the principal stresses and maximum shear stresses at each point, use either the transformation equations of plane stress or Mohr's circle. The directions of the principal stresses are shown in Fig. 8-13c, and the directions of the maximum shear stresses are shown in Fig. 8-13d. (Note that only the in-plane stresses are considered here.)

Now examine the **principal stresses** in more detail. From the sketches in Fig. 8-13c, observe how the principal stresses change when moving from top to bottom of the beam. Begin with the compressive principal stress. At point A, the compressive stress acts in the horizontal direction, and the other principal stress is zero. Moving toward the neutral axis, the compressive principal stress becomes inclined, and at the neutral axis (point C), it acts at 45° to the horizontal. At point D, the compressive principal stress is further inclined from the horizontal, and at the bottom of the beam, its direction becomes vertical (except that its magnitude is now zero).

Thus, the direction and magnitude of the compressive principal stress vary continuously from top to bottom of the beam. If the chosen cross-section is located in a region of large bending moment, the largest compressive principal stress occurs at the top of the beam (point A), and the smallest compressive principal stress (zero) occurs at the bottom of the beam (point E). If the cross section is located in a region of small bending moment and large shear force, then the largest compressive principal stress is at the neutral axis.

Analogous comments apply to the tensile principal stress, which also varies in both magnitude and direction from point A to point E. At point A, the tensile stress is zero, and at point E, it has its maximum value. (Graphs showing how the principal stresses vary in magnitude for a particular beam and particular cross section are given later in Example 8-3.)

The **maximum shear stresses** (Fig. 8-13d) at the top and bottom of the beam occur on 45° planes (because the elements are in uniaxial stress). At the neutral axis, the maximum shear stresses occur on horizontal and vertical planes (because the element is in pure shear). At all points, the maximum shear stresses occur on planes oriented at 45° to the principal planes. In regions of high bending moment, the largest shear stresses occur at the top and bottom of the beam; in regions of low bending moment and high shear force, the largest shear stresses occur at the neutral axis.

Investigation of the stresses at many cross sections of the beam shows how the principal stresses vary throughout the beam. This variation is displayed using two systems of orthogonal curves called **stress trajectories**, which give the directions of the principal stresses. Examples of stress trajectories for rectangular beams are shown in Fig. 8-14. Part a of the figure shows a cantilever beam with a load acting at the free end, and part b shows a simple beam with a uniform load. Solid lines are used for tensile principal stresses and dashed lines for compressive principal stresses. The curves for tensile and compressive principal stresses always intersect at right angles, and every trajectory crosses the longitudinal axis at 45°. At the top and bottom surfaces of the beam, where the shear stress is zero, the trajectories are either horizontal or vertical.[1]

Another type of curve that may be plotted from the principal stresses is a **stress contour**, which is a curve connecting points of equal principal stress. Stress contours for a cantilever beam of rectangular cross section are shown in Fig. 8-15 (for tensile principal stresses only). The contour of largest stress is at the upper left part of the figure. Moving downward in the figure, the tensile stresses represented by the contours become smaller and smaller. The contour line of zero tensile stress is at the lower edge of the beam. Thus, the largest tensile stress occurs at the support, where the bending moment has its largest value.

Note that stress trajectories (Fig. 8-14) give the directions of the principal stresses but give no information about the magnitudes of the stresses. In general, the magnitudes of the principal stresses vary as you move along a

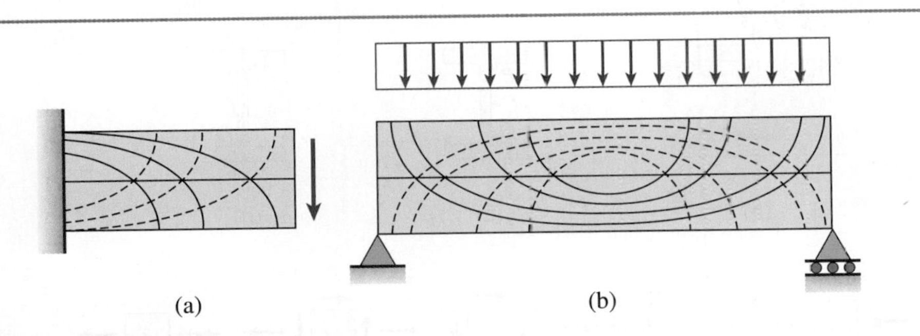

FIGURE 8-14

Principal-stress trajectories for beams of rectangular cross section; (a) cantilever beam and (b) simple beam (Solid lines represent tensile principal stresses and dashed lines represent compressive principal stresses)

[1]Stress trajectories were originated by the German engineer Karl Culmann (1821–1881); see Ref. 8-1.

FIGURE 8-15

Stress contours for a cantilever beam (tensile principal stresses only)

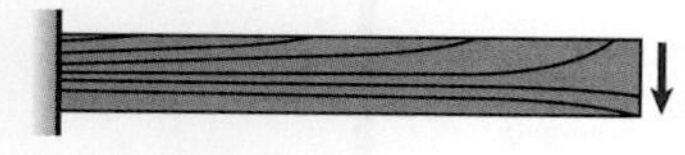

trajectory. In contrast, the magnitudes of the principal stresses are constant as you move along a stress contour (Fig. 8-15), but the contours give no information about the directions of the stresses. In particular, the principal stresses are neither parallel nor perpendicular to a stress contour.

The stress trajectories and contours of Figs. 8-14 and 8-15 were plotted from the flexure and shear formulas [Eqs. (8-25a and b)]. Stress concentrations near the supports and near the concentrated loads, as well as the direct compressive stresses caused by the uniform load bearing on the top of the beam (Fig. 8-14b), were disregarded in plotting these figures.

Wide-Flange Beams

FIGURE 8-13 (Repeated)

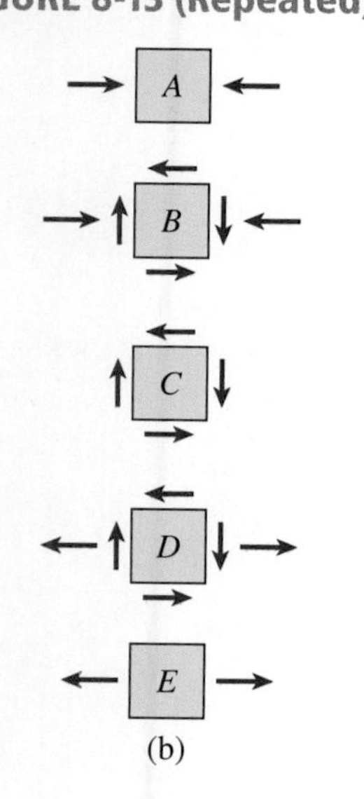

(b)

Beams having other cross-sectional shapes, such as wide-flange beams, can be analyzed for the principal stresses in a manner similar to that described previously for rectangular beams. For instance, consider the simply supported wide-flange beam shown in Fig. 8-16a. Proceeding as for a rectangular beam, identify points *A*, *B*, *C*, *D*, and *E* from top to bottom of the beam (Fig. 8-16b). Points *B* and *D* are in the web where it meets the flange, and point *C* is at the neutral axis. You can think of these points as being located either on the side of the beam (Figs. 8-16b and c) or inside the beam along a vertical axis of symmetry (Fig. 8-16d). The stresses determined from the flexure and shear formulas are the same at both sets of points.

Stress elements at points *A*, *B*, *C*, *D*, and *E* (as seen in a side view of the beam) are shown in parts (e) through (i) of Fig. 8-16. These elements have the same general appearance as those for a rectangular beam (Fig. 8-13b).

The largest **principal stresses** usually occur at the top and bottom of the beam (points *A* and *E*) where the stresses obtained from the flexure formula have their largest values. However, depending upon the relative magnitudes of the bending moment and shear force, the largest stresses sometimes occur in the web where it meets the flange (points *B* and *D*). The explanation lies in the fact that the normal stresses at points *B* and *D* are only slightly smaller than those at points *A* and *E*, whereas the shear stresses (which are zero at points *A* and *E*)

FIGURE 8-16

Stresses in a wide-flange beam

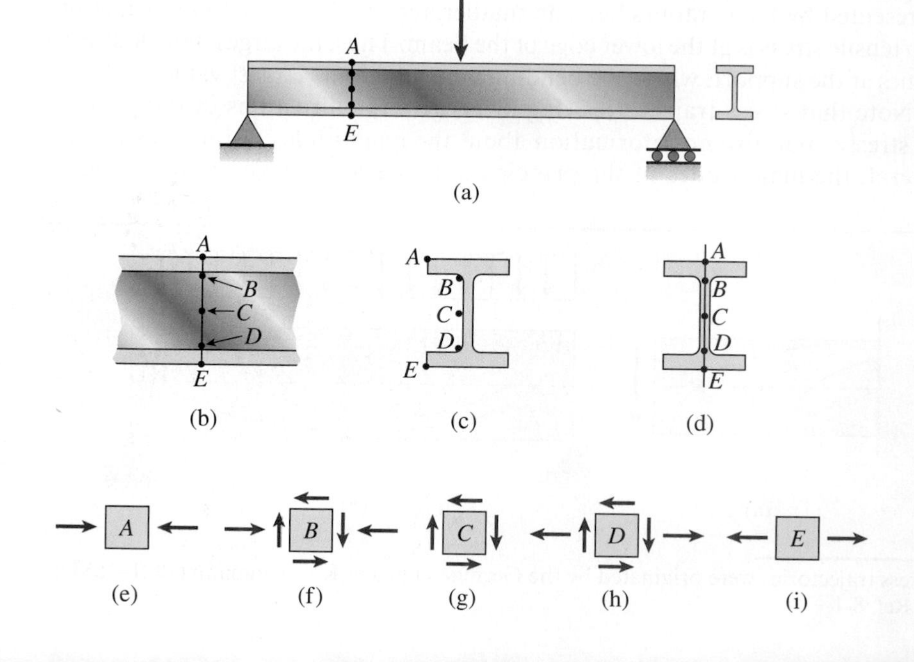

may be significant at points B and D because of the thin web. (*Note*: Fig. 5-40b in Chapter 5 shows how the shear stresses vary in the web of a wide-flange beam.)

The **maximum shear stresses** acting on a cross section of a wide-flange beam always occur at the neutral axis, as shown by the shear formula of Eq. (8-25b). However, the maximum shear stresses acting on inclined planes usually occur either at the top and bottom of the beam (points A and E) or in the web where it meets the flange (points B and D) because of the presence of normal stresses.

When analyzing a wide-flange beam for the maximum stresses, remember that high stresses may exist near supports, points of loading, fillets, and holes. Such *stress concentrations* are confined to the region very close to the discontinuity and cannot be calculated by elementary beam formulas.

The following example illustrates the procedure for determining the principal stresses and maximum shear stresses at a selected cross section in a rectangular beam. The procedures for a wide-flange beam are similar.

Example 8-3

A simple beam AB with a span length $L = 1.8$ m supports a concentrated load $P = 48$ kN acting at distance $c = 0.6$ m from the right-hand support (Fig. 8-17). The beam is made of steel and has a rectangular cross section of width $b = 50$ mm and height $h = 150$ mm.

Investigate the principal stresses and maximum shear stresses at cross section mn, located at distance $x = 230$ mm from end A of the beam. (Consider only the in-plane stresses.)

FIGURE 8-17

Example 8-3: Beam of rectangular cross section

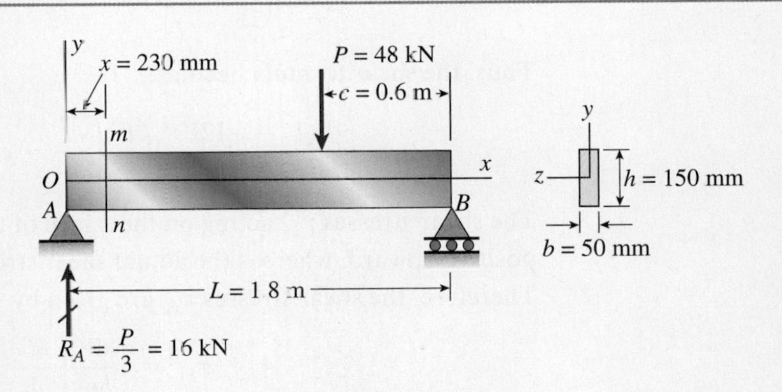

Solution:

Use a four-step problem-solving approach. Combine steps as needed for an efficient solution.

1. ***Conceptualize***: Begin by using the flexure and shear formulas to calculate the stresses acting on cross section mn. Once those stresses are known, the principal stresses and maximum shear stresses can be found from the equations of plane stress. Finally, plot graphs of these stresses to show how they vary over the height of the beam.

As a preliminary matter, note that the reaction of the beam at support A is $R_A = P/3 = 16$ kN; therefore, the bending moment and shear force at section mn are

$$M = R_A x = (16 \text{ kN})(230 \text{ mm}) = 3680 \text{ kN} \cdot \text{m} \quad V = R_A = 16 \text{ kN}$$

2. *Categorize:*

Normal stresses on cross section *mn*: Find these stresses from the flexure formula [Eq. (8-25a)] as

$$\sigma_x = -\frac{My}{I} = -\frac{12My}{bh^3} = -\frac{12(3680 \text{ kN}\cdot\text{m})y}{(50 \text{ mm})(150 \text{ mm})^3} = -271.7 \times 10^3 y \quad \textbf{(a)}$$

in which y has units of millimeters (mm) and σ_x has units of Newtons per square meter (Pa). The stresses calculated from Eq. (a) are positive when in tension and negative when in compression. For instance, note that a positive value of y (upper half of the beam) gives a negative stress, as expected.

FIGURE 8-18

Example 8-3: Plane-stress element at cross section *mn* of the beam of Figure 8-17

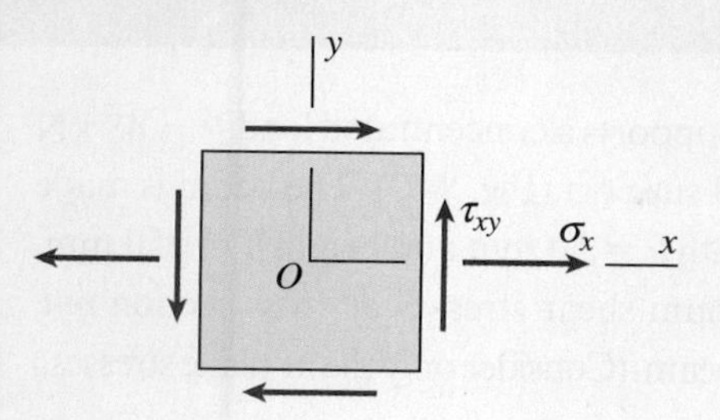

A stress element cut from the side of the beam at cross section *mn* (Fig. 8-17) is shown in Fig. 8-18. For reference purposes, a set of xy axes is associated with the element. The normal stress σ_x and the shear stress τ_{xy} are shown acting on the element in their positive directions. (Note that in this example there is no normal stress σ_y acting on the element.)

Shear stresses on cross section *mn*: The shear stresses are given by the shear formula [Eq. (8-25b)] in which the first moment Q for a rectangular cross section is

$$Q = b\left(\frac{h}{2} - y\right)\left(y + \frac{h/2 - y}{2}\right) = \frac{b}{2}\left(\frac{h^2}{4} - y^2\right) \quad \textbf{(8-26)}$$

Thus, the shear formula becomes

$$\tau = \frac{VQ}{Ib} = \frac{12V}{(bh^3)(b)}\left(\frac{b}{2}\right)\left(\frac{h^2}{4} - y^2\right) = \frac{6V}{bh^3}\left(\frac{h^2}{4} - y^2\right) \quad \textbf{(8-27)}$$

The shear stresses τ_{xy} acting on the x face of the stress element (Fig. 8-18) are positive upward, whereas the actual shear stresses τ [Eq. (8-27)] act downward. Therefore, the shear stresses τ_{xy} are given by

$$\tau_{xy} = -\frac{6V}{bh^3}\left(\frac{h^2}{4} - y^2\right) \quad \textbf{(8-28)}$$

Substitute numerical values into this equation to get

$$\tau_{xy} = -\frac{6(16 \text{ kN})}{(50 \text{ mm})(150 \text{ mm})^3}\left(\frac{(150 \text{ mm})^2}{4} - y^2\right) = -569(5625 - y^2) \quad \textbf{(b)}$$

in which y has units of millimeters (mm) and τ_{xy} has units of Newtons per square meter (Pa).

3. *Analyze:*

Calculation of stresses: In order to calculate the stresses at cross section *mn*, divide the height of the beam into six equal intervals and label the corresponding points from A to G, as shown in the side view of the beam (Fig. 8-19a). The y coordinates of these points are listed in column 2 of Table 8-1, and the corresponding stresses σ_x and τ_{xy} [calculated from Eqs. (a) and (b), respectively] are listed in columns 3 and 4. These stresses are plotted in Figs. 8-19b and c. The normal stresses vary linearly from a compressive stress of −19.63 MPa at the top of the beam (point A) to a tensile stress of 19.63 MPa at the bottom of the beam (point G). The shear stresses have a parabolic distribution with the maximum stress at the neutral axis (point D).

Principal stresses and maximum shear stresses: The principal stresses at each of the seven points A through G may be determined from Eq. (7-17):

$$\sigma_{1,2} = \frac{\sigma_x + \sigma_y}{2} \pm \sqrt{\left(\frac{\sigma_x - \sigma_y}{2}\right)^2 + \tau_{xy}^2} \tag{8-29}$$

Since there is no normal stress in the y direction (Fig. 8-18), this equation simplifies to

$$\sigma_{1,2} = \frac{\sigma_x}{2} \pm \sqrt{\left(\frac{\sigma_x}{2}\right)^2 + \tau_{xy}^2} \tag{8-30}$$

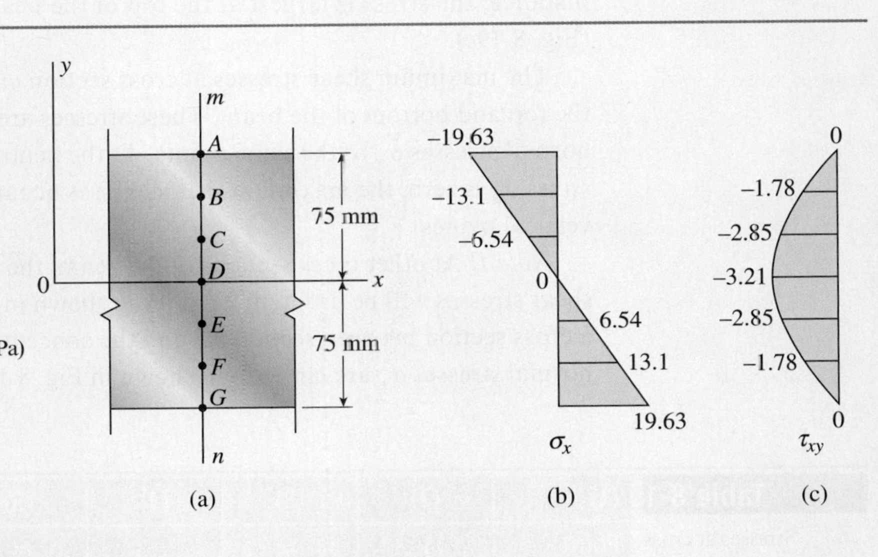

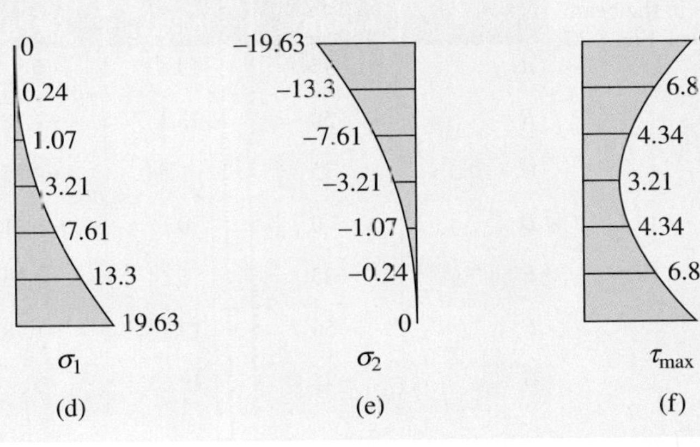

FIGURE 8-19

Example 8-3: Stresses in the beam of Figure 8-17 (a) points A, B, C, D, E, F, and G at cross section mn; (b) normal stresses σ_x acting on cross section mn; (c) shear stresses τ_{xy} acting on cross section mn; (d) principal tensil stresses σ_1; (e) principal compressive stresses σ_2; and (f) maximum shear stresses τ_{max} (*Note:* All stresses have units of MPa)

Also, the maximum shear stresses [from Eq. (7-25)] are

$$\tau_{max} = \sqrt{\left(\frac{\sigma_x - \sigma_y}{2}\right)^2 + \tau_{xy}^2} \quad \textbf{(8-31)}$$

which simplifies to

$$\tau_{max} = \sqrt{\left(\frac{\sigma_x}{2}\right)^2 + \tau_{xy}^2} \quad \textbf{(8-32)}$$

Thus, substitute the values of σ_x and τ_{xy} (from Table 8-1) into Eqs. (8-30) and (8-32), to calculate the principal stresses σ_1 and σ_2 and the maximum shear stress τ_{max}. These quantities are listed in the last three columns of Table 8-1 and are plotted in Figs. 8-19d, e, and f.

4. ***Finalize***: The tensile principal stresses σ_1 increase from zero at the top of the beam to a maximum of 19.63 MPa at the bottom (Fig. 8-19d). The directions of the stresses also change, varying from vertical at the top to horizontal at the bottom. At mid-height, the stress σ_1 acts on a 45° plane. Similar comments apply to the compressive principal stress σ_2, except in reverse. For instance, the stress is largest at the top of the beam and zero at the bottom (Fig. 8-19e).

The maximum shear stresses at cross section *mn* occur on 45° planes at the top and bottom of the beam. These stresses are equal to one-half of the normal stresses σ_x at the same points. At the neutral axis, where the normal stress σ_x is zero, the maximum shear stresses occur on the horizontal and vertical planes.

Note 1: At other cross sections of the beam, the maximum normal and shear stresses will be different from those shown in Fig. 8-19. For instance, at a cross section between section *mn* and the concentrated load (Fig. 8-17), the normal stresses σ_x are larger than shown in Fig. 8-19b because the bending

Table 8-1 Stresses at cross section *mn* in the beam of Fig. 8-17

(1)	(2)	(3)	(4)	(5)	(6)	(7)
Point	y (mm)	σ_x (MPa)	τ_{xy} (MPa)	σ_1 (MPa)	σ_2 (MPa)	τ_{max} (MPa)
A	75	−19.63	0	0	−19.63	9.82
B	50	−13.1	−1.78	0.24	−13.3	6.80
C	25	−6.54	−2.85	1.07	−7.61	4.34
D	0	0	−3.21	3.21	−3.21	3.21
E	−25	6.54	−2.85	7.61	−1.07	4.34
F	−50	13.1	−1.78	13.3	−0.24	6.80
G	−75	19.63	0	19.63	0	9.82

moment is larger. However, the shear stresses τ_{xy} are the same as those shown in Fig. 8-19c because the shear force doesn't change in that region of the beam. Consequently, the principal stresses σ_1 and σ_2 and maximum shear stresses τ_{max} will vary in the same general manner as shown in Figs. 8-19d, e, and f but with different numerical values.

The largest tensile stress anywhere in the beam is the normal stress at the bottom of the beam at the cross section of maximum bending moment. This stress is

$$(\sigma_{tens})_{max} = 102.4 \text{ MPa}$$

The largest compressive stress has the same numerical value and occurs at the top of the beam at the same cross section.

The largest shear stress τ_{xy} acting on a cross section of the beam occurs to the right of the load P (Fig. 8-17) because the shear force is larger in that region of the beam ($V = R_B = 32$ kN). Therefore, the largest value of τ_{xy} that occurs at the neutral axis is

$$(\tau_{xy})_{max} = 6.4 \text{ MPa}$$

The largest shear stress anywhere in the beam occurs on 45° planes at either the top or bottom of the beam at the cross section of maximum bending moment:

$$\tau_{max} = \frac{102.4 \text{ MPa}}{2} = 51.2 \text{ MPa}$$

Note 2: In the practical design of ordinary beams, the principal stresses and maximum shear stresses are rarely calculated. Instead, the tensile and compressive stresses to be used in design are calculated from the flexure formula at the cross section of maximum bending moment, and the shear stress to be used in design is calculated from the shear formula at the cross section of maximum shear force.

8.5 Combined Loadings

Previous chapters presented the analysis of structural members subjected to a single type of loading. For instance, axially loaded bars were analyzed in Chapters 1 and 2, shafts in torsion in Chapter 3, and beams in bending in Chapters 4, 5, and 6. Analysis of pressure vessels was presented earlier in this chapter. For each type of loading, methods were developed for finding stresses, strains, and deformations.

However, in many structures, the members are required to resist more than one kind of loading. For example, a beam may be subjected to the simultaneous action of bending moments and axial forces (Fig. 8-20a), a pressure vessel may be supported so that it also functions as a beam (Fig. 8-20b), or a shaft in

FIGURE 8-20

Example of structures subjected to combined loading: (a) wide-flange beam supported by a cable (combined bending and axial load), (b) cylindrical pressure vessel supported as a beam, and (c) shaft in combined torsion and bending

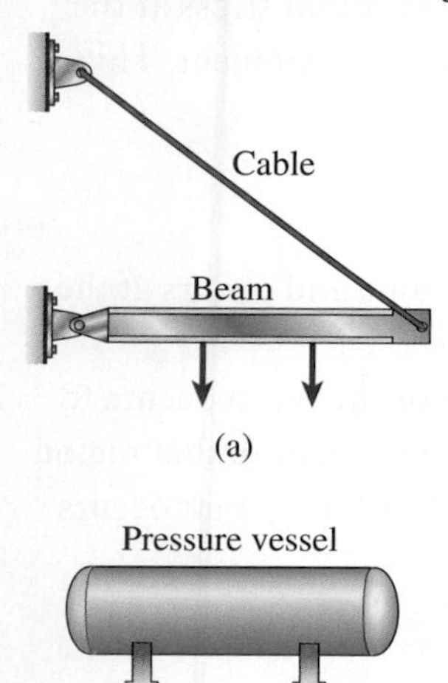

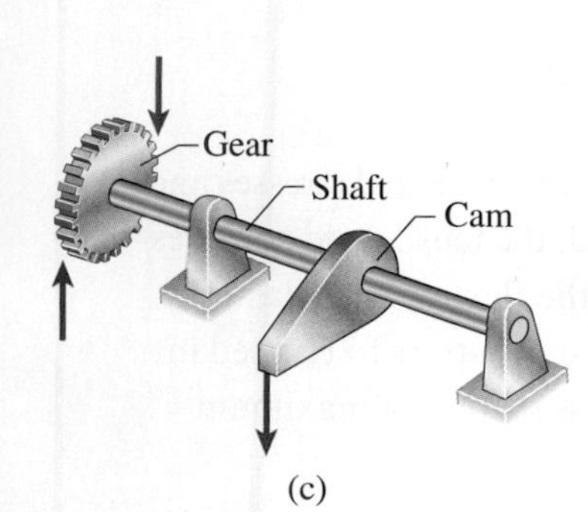

torsion may carry a bending load (Fig. 8-20c). Known as **combined loadings**, situations similar to those shown in Fig. 8-20 occur in a great variety of machines, buildings, vehicles, tools, equipment, and many other kinds of structures.

A structural member subjected to combined loadings often can be analyzed by superimposing the stresses and strains caused by each load acting separately. However, superposition of both stresses and strains is permissible only under certain conditions, as explained in earlier chapters. One requirement is that the stresses and strains must be linear functions of the applied loads, which in turn requires that the material follow Hooke's law and the displacements remain small.

A second requirement is that there must be no interaction between the various loads, that is, the stresses and strains due to one load must not be affected by the presence of the other loads. Most ordinary structures satisfy these two conditions; therefore, the use of superposition is very common in engineering work.

Method of Analysis

While there are many ways to analyze a structure subjected to more than one type of load, the procedure usually includes the following steps:

1. Select a point in the structure where the stresses and strains are to be determined. The point is usually selected at a cross section where the stresses are large, such as at a cross section where the bending moment has its maximum value.
2. For each load on the structure, determine the stress resultants at the cross section containing the selected point. The possible stress resultants are an axial force, a twisting moment, a bending moment, and a shear force.
3. Calculate the normal and shear stresses at the selected point due to each of the stress resultants. If the structure is a pressure vessel, determine the stresses due to the internal pressure. The stresses are found from the stress formulas derived previously; for instance, $\sigma = P/A$, $\tau = T\rho/I_p$, $\sigma = My/I$, $\tau = VQ/Ib$, and $\sigma = pr/t$.
4. Combine the individual stresses to obtain the resultant stresses at the selected point. In other words, obtain the stresses σ_x, σ_y, and τ_{xy} acting on a stress element at the point. (Note that this chapter deals only with elements in plane stress.)
5. Determine the principal stresses and maximum shear stresses at the selected point using either the stress-transformation equations or Mohr's circle. If required, determine the stresses acting on other inclined planes.
6. Determine the strains at the point with the aid of Hooke's law for plane stress.
7. Select additional points and repeat the process. Continue until enough stress and strain information is available to satisfy the purposes of the analysis.

Illustration of the Method

To illustrate the procedure for analyzing a member subjected to combined loadings, consider the stresses in the cantilever bar of circular cross section shown in Fig. 8-21a. This bar is subjected to two types of load—a torque T and a vertical load P, both acting at the free end of the bar.

Begin by arbitrarily selecting two points A and B for investigation (Fig. 8-21a). Point A is located at the top of the bar and point B is located on the side. Both points are located at the same cross section.

The stress resultants acting at the cross section (Fig. 8-21b) are a twisting moment equal to the torque T, a bending moment M equal to the load P times the distance b from the free end of the bar to the cross section, and a shear force V equal to the load P.

The stresses acting at points A and B are shown in Fig. 8-21c. The twisting moment T produces torsional shear stresses

$$\tau_1 = \frac{Tr}{I_p} = \frac{2T}{\pi r^3} \tag{8-33}$$

in which r is the radius of the bar and $I_p = \pi r^4/2$ is the polar moment of inertia of the cross-sectional area. The stress τ_1 acts horizontally to the left at point A and vertically downward at point B, as shown in the figure.

The bending moment M produces a tensile stress at point A:

$$\sigma_A = \frac{Mr}{I} = \frac{4M}{\pi r^3} \tag{8-34}$$

in which $I = \pi r^4/4$ is the moment of inertia about the neutral axis. However, the bending moment produces no stress at point B, because B is located on the neutral axis.

The shear force V produces no shear stress at the top of the bar (point A), but at point B, the shear stress is [see Eq. (5-46) in Chapter 5]:

$$\tau_2 = \frac{4V}{3A} = \frac{4V}{3\pi r^2} \tag{8-35}$$

in which $A = \pi r^2$ is the cross-sectional area.

The stresses σ_A and τ_1 acting at point A (Fig. 8-21c) are shown acting on a stress element in Fig. 8-22a. This element is cut from the top of the bar at point A. A two-dimensional view of the element, obtained by looking vertically downward on the element, is shown in Fig. 8-22b. For the purpose of determining the principal stresses and maximum shear stresses, construct x and y axes through the element. The x axis is parallel to the longitudinal axis of the circular bar (Fig. 8-21a) and the y axis is horizontal. Note that the element is in plane stress with $\sigma_x = \sigma_A$, $\sigma_y = 0$, and $\tau_{xy} = -\tau_1$.

A stress element at point B (also in plane stress) is shown in Fig. 8-23a. The only stresses acting on this element are the shear stresses, equal to $\tau_1 + \tau_2$ (see Fig. 8-21c). A two-dimensional view of the stress element is shown in Fig. 8-23b, with the x axis parallel to the longitudinal axis of the bar and the y axis in the vertical direction. The stresses acting on the element are $\sigma_x = \sigma_y = 0$ and $\tau_{xy} = -(\tau_1 + \tau_2)$.

Now that the stresses acting at points A and B have been determined and shown on the corresponding stress elements, you can use the transformation equations of plane stress (Sections 7.2 and 7.3) or Mohr's circle (Section 7.4) to determine principal stresses, maximum shear stresses, and stresses acting in inclined directions. You can also use Hooke's law (Section 7.5) to determine the strains at points A and B.

The procedure described previously for analyzing the stresses at points A and B (Fig. 8-21a) can be used at other points in the bar. Of particular interest are the points where the stresses calculated from the flexure and shear formulas have maximum or minimum values, called **critical points**. For instance, the normal stresses due to bending are largest at the cross section of maximum bending moment, which is at the support. Therefore, points C and D at the top and bottom of the beam at the fixed end (Fig. 8-21a) are critical points where the stresses

FIGURE 8-21

Cantilever bar subjected to combined torsion and bending: (a) loads acting on the bar, (b) stress resultants at a cross section, and (c) stresses at point A and B

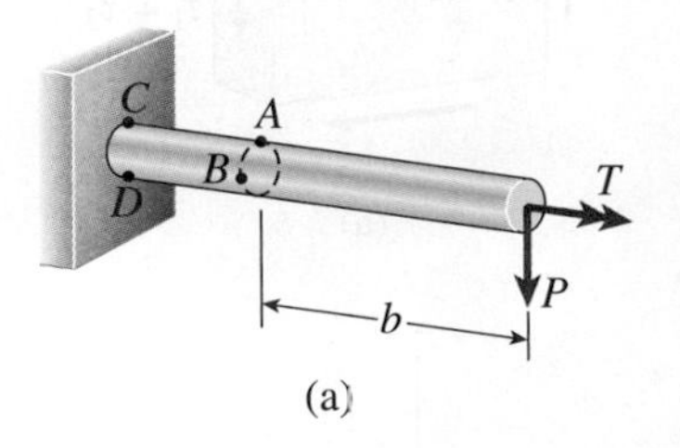

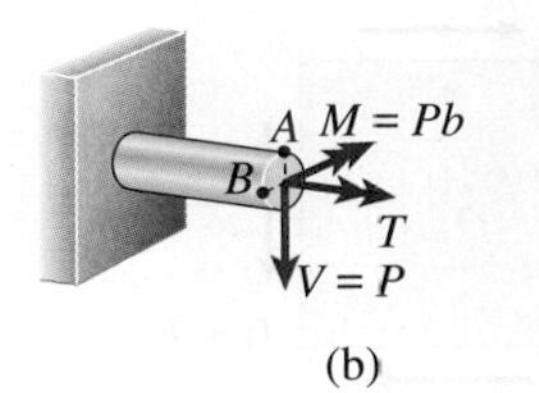

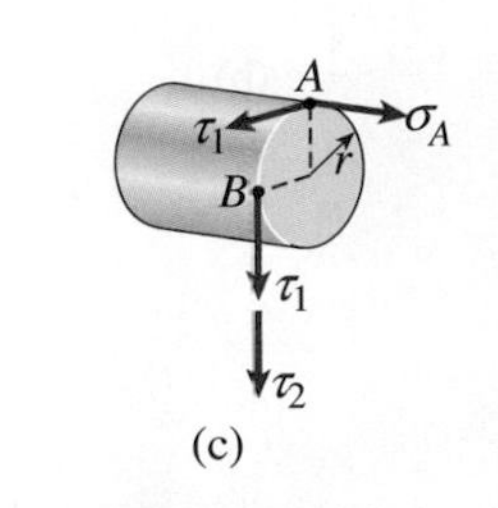

FIGURE 8-22

Stress element at point A

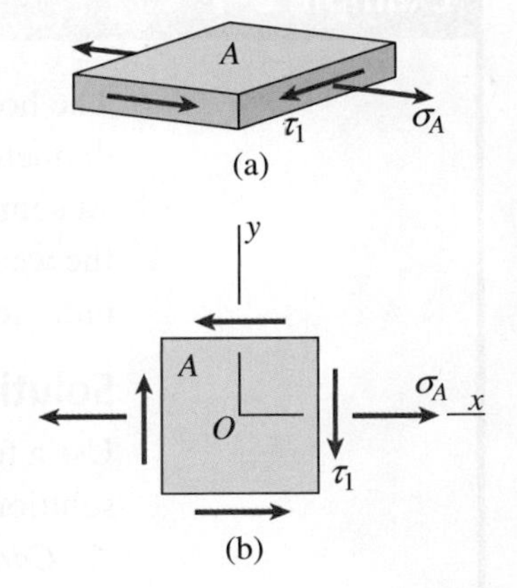

FIGURE 8-23

Stress element at point B

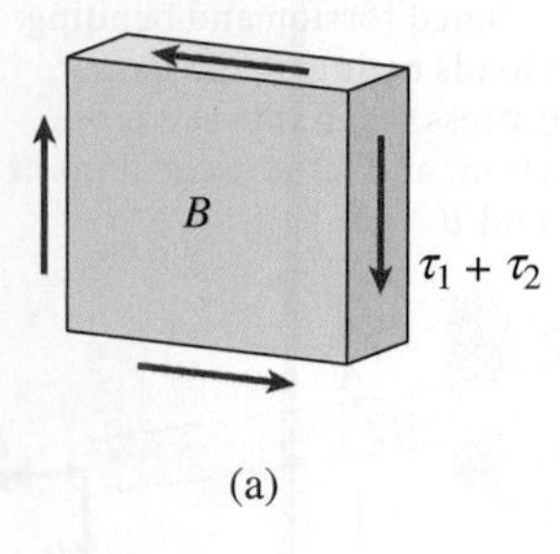

(a)

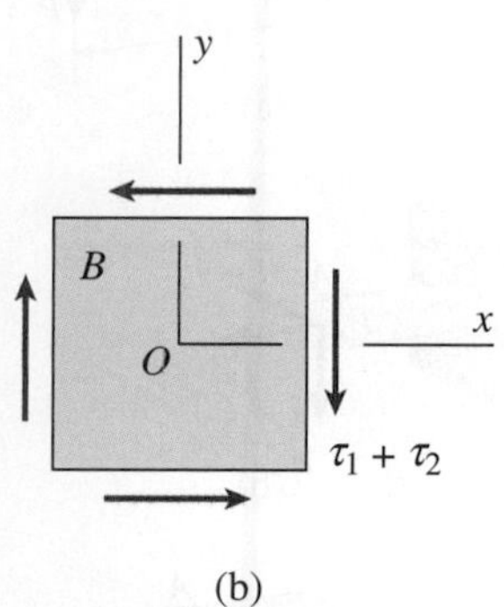

(b)

should be calculated. Another critical point is point B itself because the shear stresses are a maximum at this point. (Note that, in this example, the shear stresses do not change if point B is moved along the bar in the longitudinal direction.)

As a final step, the principal stresses and maximum shear stresses at the critical points can be compared with one another in order to determine the absolute maximum normal and shear stresses in the bar.

This example illustrates the general procedure for determining the stresses produced by combined loadings. Note that no new theories are involved—only applications of previously derived formulas and concepts. Since the variety of practical situations seems to be endless, it is not practical to derive general formulas for calculating the maximum stresses. Instead, treat each structure as a special case.

Selection of Critical Points

If the objective of the analysis is to determine the largest stresses *anywhere* in the structure, then the critical points should be selected at cross sections where the stress resultants have their largest values. Furthermore, within those cross sections, the points should be selected where either the normal stresses or the shear stresses have their largest values. By using good judgment in the selection of the points, you often can be reasonably certain of obtaining the absolute maximum stresses in the structure.

However, it is sometimes difficult to recognize in advance where the maximum stresses in the member are to be found. Then it may be necessary to investigate the stresses at a large number of points, perhaps even using trial-and-error in the selection of points. Other strategies may also prove fruitful—such as deriving equations specific to the problem at hand or making simplifying assumptions to facilitate an otherwise difficult analysis.

The following examples illustrate the methods used to calculate stresses in structures subjected to combined loadings.

Example 8-4

The hollow pipe casing for a production oil well (see Fig. 8-24) is 200 mm in outer diameter and 18 mm in thickness. The internal pressure due to gas and oil is 15 MPa. At some point above the blowout preventer, the compressive force in the pipe (due to the weight of the pipe) is 175 kN, and the torque is 14 kN · m. Determine the maximum tensile, compressive, and shear stresses in the pipe casing.

Solution:

Use a four-step problem-solving approach. Combine steps as needed for an efficient solution.

1. ***Conceptualize***: The stresses in the well casing are produced by the combined action of the axial force P, the torque T, and *internal pressure p* (Fig. 8-24b). Therefore, the stresses at any point on the surface of the shaft at some depth consist of circumferential stress σ_x, longitudinal stress σ_y, and shear stresses τ_{xy}, as shown on the stress element on the surface of the casing in Fig. 8-24b. Note that the y axis is parallel to the longitudinal axis of the casing.

FIGURE 8-24

Example 8-4: Production oil well casing (combined torsion and axial force and internal pressure) (Courtesy of EMNRD)

Production oil well

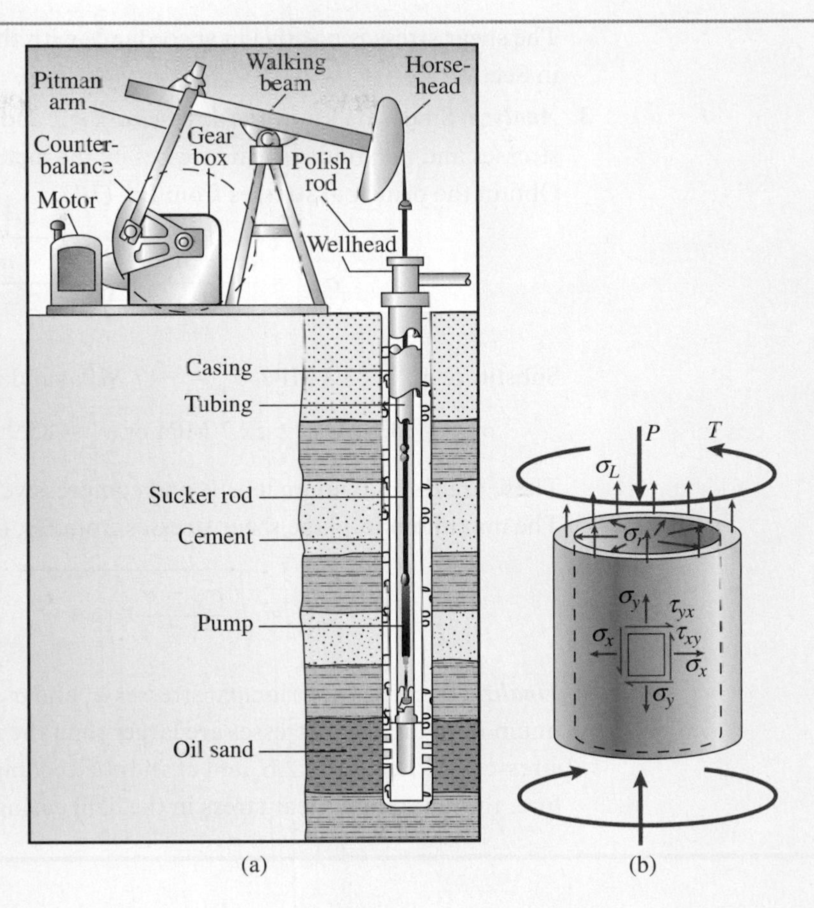

2. ***Categorize***: The circumferential stress σ_x is due to the internal pressure of oil and gas and is computed using Eq. (8-11) as

$$\sigma_x = \frac{pr}{t} = \frac{[15 \text{ MPa} \times (100 \text{ mm})]}{18 \text{ mm}} = 83.3 \text{ MPa}$$

The longitudinal stress σ_y is caused by the axial compressive force P (due to self-weight) and is divided by casing cross-sectional area A. The longitudinal tensile stress σ_L is due to internal pressure [see Eq. (8-12) to find σ_L, which is nonzero when the well is capped and not operational]. You can assume that oil and gas are flowing, so σ_L is zero and σ_y is computed as

$$\sigma_y = \frac{-P}{A} = \frac{-(175 \text{ kN})}{\pi\left[r^2 - (r-t)^2\right]} = -17 \text{ MPa}$$

Find the shear stress τ_{xy} from the torsion formula [see Eq. (3-13) of Section 3.3]:

$$\tau_{xy} = \frac{Tr}{I_p} = \frac{(14 \text{ kN} \cdot \text{m}) \times (100 \text{ mm})}{8.606(10^{-5}) \text{ m}^4} = 16.3 \text{ MPa}$$

The shear stress is positive in accordance with the sign convention established in Section 1.8.

3. ***Analyze***: Since you know the stresses σ_x, σ_y, and τ_{xy}, compute the principal stresses and maximum shear stresses by the methods described in Section 7.3. Obtain the principal stresses from Eq. (7-17):

$$\sigma_{1,2} = \frac{\sigma_x + \sigma_y}{2} \pm \sqrt{\left(\frac{\sigma_x - \sigma_y}{2}\right)^2 + \tau_{xy}^2}$$

Substitute $\sigma_x = 83.3$ MPa, $\sigma_y = -17$ MPa, and $\tau_{xy} = 16.3$ MPa to get

$$\sigma_{1,2} = 33.2 \text{ MPa} \pm 52.7 \text{ MPa or } \sigma_1 = 85.9 \text{ MPa} \quad \sigma_2 = -19.5 \text{ MPa}$$

These are the maximum tensile and compressive stresses in the drill casing. The maximum in-plane shear stresses from Eq. (7-25) are

$$\tau_{max} = \sqrt{\left(\frac{\sigma_x - \sigma_y}{2}\right)^2 + \tau_{xy}^2} = 52.7 \text{ MPa}$$

4. ***Finalize***: Because the principal stresses σ_1 and σ_2 have opposite signs, the maximum in-plane shear stresses are larger than the maximum out-of-plane shear stresses [see Eqs. (7-28a, b, and c) and the accompanying discussion]. Therefore, the maximum shear stress in the drill casing is 52.7 MPa.

Example 8-5

The cylindrical pressure vessel from Example 8-2 (see photo) is now placed on simple supports and is acted on by a uniformly distributed load $q = 150$ kN/m, which includes the weight of the tank and its contents. The 6-m-long tank has an inner radius of $r = 1.2$ m and a wall thickness of $t = 19$ mm. The material is steel with a modulus of $E = 200$ GPa and the internal pressure $p = 720$ kPa.

Example 8-2 investigated the longitudinal and circumferential stresses and strains, as well as the maximum in-plane and out-of-plane shear stresses. Now, the effect of distributed load q is considered to *find states of stress at element locations A and B* (see Fig. 8-25) due to the combined effects of internal pressure and transverse shear and bending moment (shear-force and bending-moment diagrams are given in Figs. 8-25c and d). (a) Element A is on the outer surface of the vessel, just to the right of the left-hand support. (b) Element B is located on the bottom surface of the tank at the mid-span.

Solution:

Use a four-step problem-solving approach. Combine steps as needed for an efficient solution.

FIGURE 8-25

Example 8-5: Cylindrical pressure vessel subjected to combined internal pressure p and transverse load q

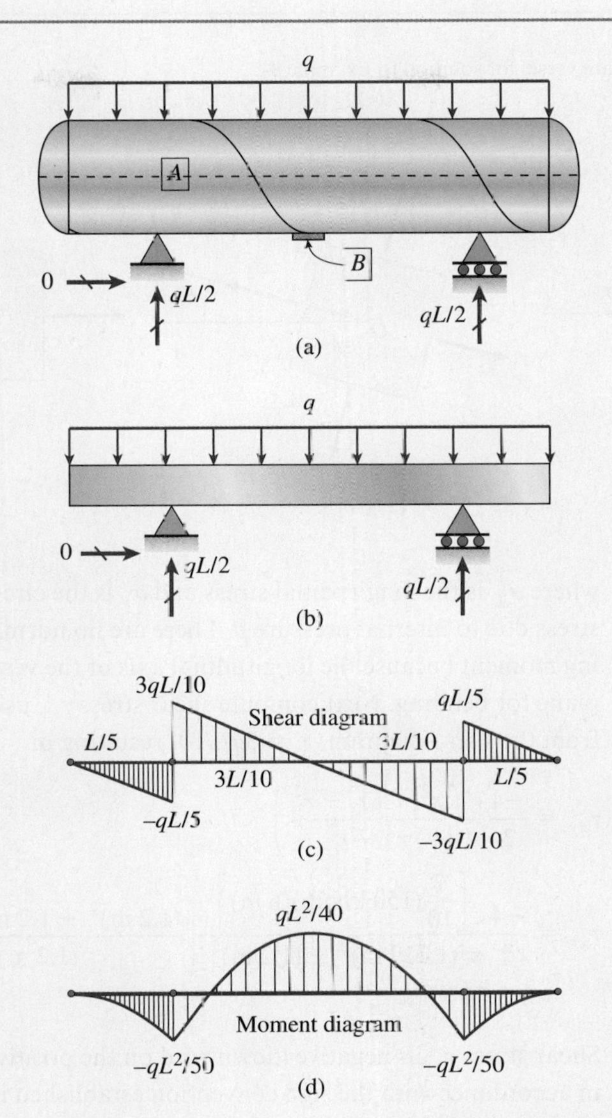

1. ***Conceptualize***: The stresses in the wall of the pressure vessel are caused by the *combined action* of internal pressure and transverse shear and bending.

Part (a): State of stress at point *A*.

2, 3. ***Categorize, Analyze***: Isolate a stress element similar to that shown in Fig. 8-26a. The x axis is parallel to the longitudinal axis of the pressure vessel, and the y axis is circumferential. There are shear stresses acting on element A due to load q (assume that element A is a sufficient distance from the support so that any stress concentration effects are negligible). The stresses are computed as

$$\sigma_x = \sigma_L = \frac{pr}{2t} = \frac{720 \text{ kPa} \times (1.2 \text{ m})}{2(19 \text{ mm})} = 22.7 \text{ MPa}$$

$$\sigma_y = \sigma_r = \frac{pr}{t} = \frac{720 \text{ kPa} \times (1.2 \text{ m})}{19 \text{ mm}} = 45.5 \text{ MPa}$$

FIGURE 8-26

Stresses in a cylindrical pressure vessel for solution to Example 8-5

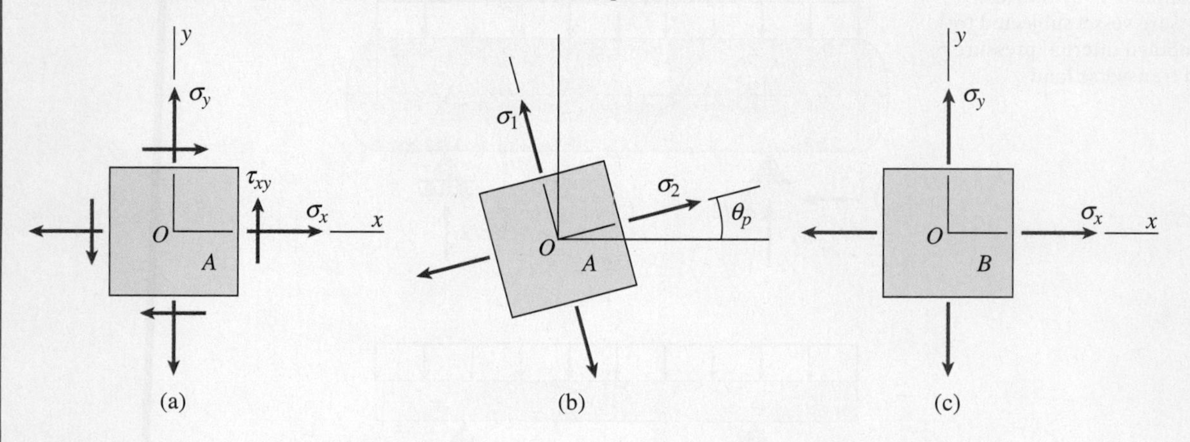

where σ_L is the longitudinal stress and σ_r is the circumferential (or radial) stress due to internal pressure p. There are no normal stresses due to bending moment because the longitudinal axis of the vessel lies in the neutral plane for bending. Next compute shear stress τ_{xy} using Eq. (5-48), where, from the shear diagram, $V = 3qL/10$, resulting in

$$\tau_{xy} = \frac{-4}{3}\frac{V}{A}\left(\frac{r_1^2 + r_1 r_2 + r_2^2}{r_1^2 + r_2^2}\right)$$

$$\tau_{xy} = \frac{-4}{3}\frac{\left[\frac{3}{10}(150\text{ kN/m})(6\text{ m})\right]}{\pi[(1.219\text{ m})^2 - (1.2\text{ m})^2]}\left[\frac{(1.2\text{ m})^2 + 1.2\text{ m}(1.219\text{ m}) + (1.219\text{ m})^2}{(1.2\text{ m})^2 + (1.219\text{ m})^2}\right]$$
$$= -3.74\text{ MPa}$$

Shear stress τ_{xy} is negative (downward on the positive x face of the element) in accordance with the sign convention established in Section 1.8.

Principal stresses and maximum shear stresses at point A: The principal stresses are obtained from Eq. (7-17), which is repeated here:

$$\sigma_{1,2} = \frac{\sigma_x + \sigma_y}{2} \pm \sqrt{\left(\frac{\sigma_x - \sigma_y}{2}\right)^2 + \tau_{xy}^2}$$

so

$$\sigma_1 = 34.1\text{ MPa} + 11.96\text{ MPa} = 46.1\text{ MPa}$$

$$\sigma_2 = 34.1\text{ MPa} - 11.96\text{ MPa} = 22.1\text{ MPa}$$

4. ***Finalize*:** The principal stresses at point A are shown on an element rotated through $\theta_p = 9.11°$ in Fig. 8-26b.

The *maximum in-plane shear stress* is computed using Eq. (7-28c):

$$\tau_{max} = \frac{\sigma_1 - \sigma_2}{2} = 12 \text{ MPa}$$

but the *maximum out-of-plane shear stress* controls using Eqs. (7-28b):

$$\tau_{max} = \frac{\sigma_1}{2} = 23.1 \text{ MPa}$$

Because the principal stresses have the same signs, one of the out-of-plane shear stresses is the largest shear stress [see the discussion following Eqs. (7-28a, b, and c)].

Part (b): State of stress at point *B*.

2, 3. ***Categorize, Analyze***: The stress element is located on the bottom surface of the vessel and (as you look up at it from the bottom of the tank) is oriented as shown in Fig. 8-26c. The *x* axis is parallel to the longitudinal axis of the pressure vessel, and the *y* axis is circumferential. There are no shear stresses acting on element *B* due to load *q*, because element *B* is on the bottom free surface, but normal tensile stress is maximum due to bending. The stresses are computed as

$$\sigma_x = \sigma_L + \frac{Mr}{I_z}$$

where I_z for the vessel is

$$I_z = \frac{\pi}{4}\left[(r+t)^4 - r^4\right] = 0.10562 \text{ m}^4$$

so

$$\sigma_x = \frac{pr}{2t} + \frac{\left(\frac{qL^2}{40}\right)(r+t)}{I_z}$$

$$= \frac{720 \text{ kPa}(1.2 \text{ m})}{2(19 \text{ mm})} + \frac{\left[150 \text{ kN/m}\frac{(6 \text{ m})^2}{40}\right](1.219 \text{ m})}{0.10562 \text{ m}^4}$$

$$\sigma_x = 22.74 \text{ MPa} + 1.558 \text{ MPa} = 24.3 \text{ MPa}$$

$$\sigma_y = \sigma_r = \frac{pr}{t} = \frac{720 \text{ kPa}(1.2 \text{ m})}{19 \text{ mm}} = 45.5 \text{ MPa}$$

4. ***Finalize***: Because there are no shear stresses acting at *B*, normal stresses σ_x and σ_y are the principal normal stresses, (i.e., $\sigma_x = \sigma_2$ and $\sigma_y = \sigma_1$). The maximum in-plane and out-of-plane shear stresses can be found from Eqs. (7-28a, b, and c).

The *maximum in-plane shear stress* is computed using Eq. (7-28c) as

$$\tau_{max} = \frac{\sigma_1 - \sigma_2}{2} = 10.6 \text{ MPa}$$

but the *maximum out-of-plane shear stress* controls using Eqs. (7-28b) for

$$\tau_{max} = \frac{\sigma_1}{2} = 23 \text{ MPa}$$

Example 8-6

FIGURE 8-27

Example 8-6: Wind pressure against a sign (combined bending, torsion, and shear of the pole)

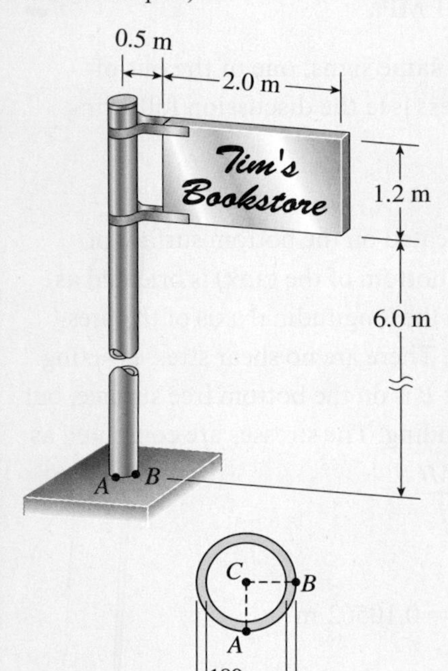

A sign with dimensions of 2.0 m × 1.2 m is supported by a hollow circular pole having an outer diameter of 220 mm and an inner diameter of 180 mm (Fig. 8-27). The sign is offset 0.5 m from the centerline of the pole, and its lower edge is 6.0 m above the ground.

(a) Determine the principal stresses and maximum shear stresses at points A and B at the base of the pole due to a wind pressure of 2.0 kPa against the sign.

(b) Compare the circular pole stresses at the base and twist at the top to stresses and twist of a square tube with the same height, same wall thickness, and same cross-sectional area.

Solution:

Use a four-step problem-solving approach. Combine steps as needed for an efficient solution.

Part (a): Circular pole stress resultants.

1. ***Conceptualize:*** The wind pressure against the sign produces a resultant force W that acts at the midpoint of the sign (Fig. 8-28a) and is equal to the pressure p times the area A over which it acts:

$$W = pA = (2.0\ \text{kPa})(2.0\ \text{m} \times 1.2\ \text{m}) = 4.8\ \text{kN}$$

The line of action of this force is at height $h = 6.6$ m above the ground and at distance $b = 1.5$ m from the centerline of the pole.

The wind force acting on the sign is statically equivalent to a lateral force W and a torque T acting on the pole (Fig. 8-28b). The torque is equal to the force W times the distance b:

$$T = Wb = (4.8\ \text{kN})(1.5\ \text{m}) = 7.2\ \text{kN·m}$$

The stress resultants at the base of the pole (Fig. 8-28c) consist of a bending moment M, a torque T, and a shear force V. Their magnitudes are

$$M = Wh = (4.8\ \text{kN})(6.6\ \text{m}) = 31.68\ \text{kN·m}$$

$$T = 7.2\ \text{kN·m} \quad V = W = 4.8\ \text{kN}$$

Examination of these stress resultants shows that maximum bending stresses occur at point A and maximum shear stresses at point B. Therefore, A and B are critical points where the stresses should be determined. (Another critical point is diametrically opposite point A, as explained in the note at the end of this example.)

FIGURE 8-28

Solution to Example 8-6

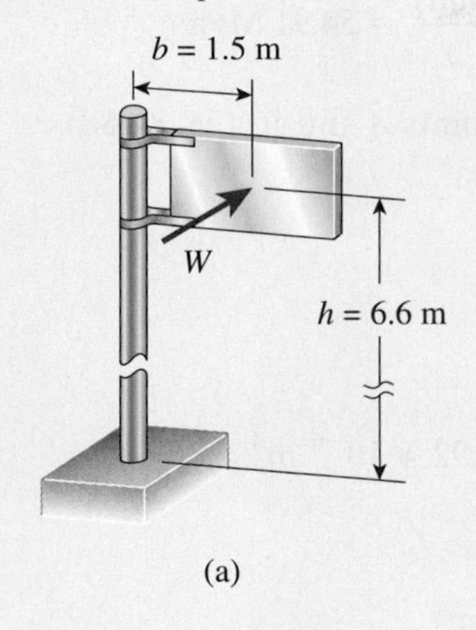

(a)

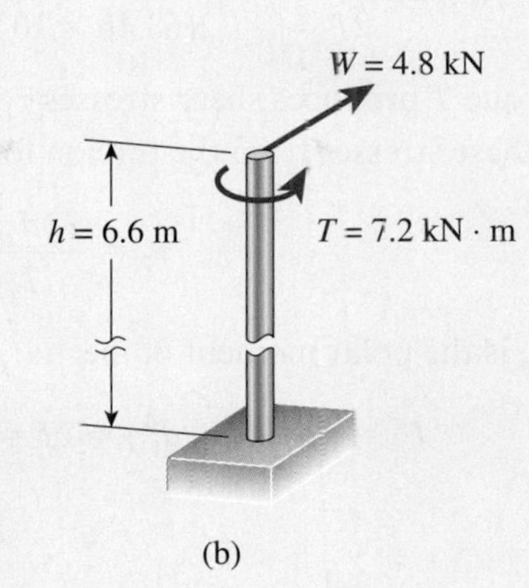

(b)

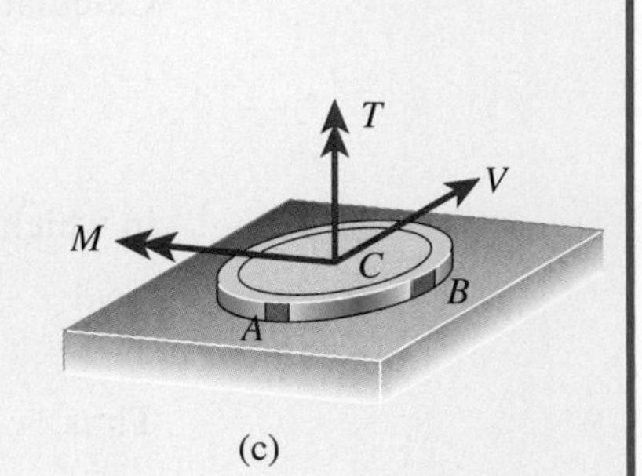

(c)

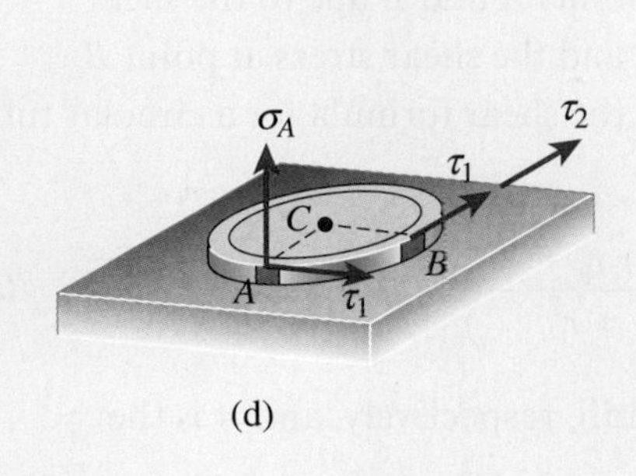

(d)

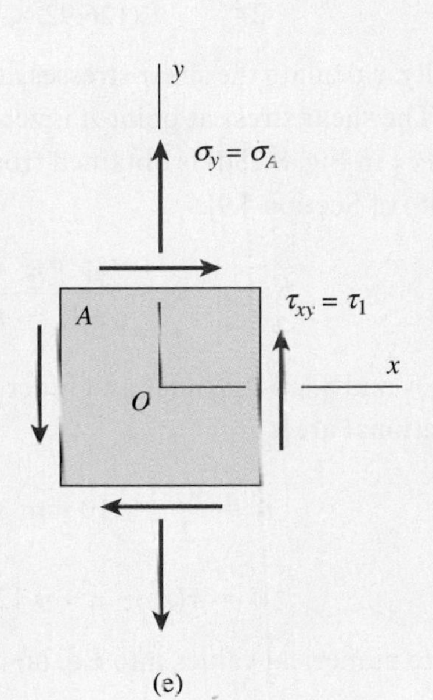

(e)

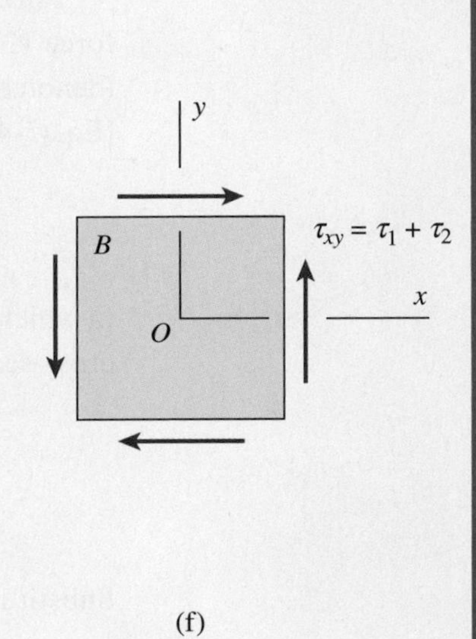

(f)

2, 3. ***Categorize, Analyze*:**

Stresses at points *A* and *B*: The bending moment M produces a tensile stress σ_A at point A (Fig. 8-28d) but no stress at point B (which is located on the neutral axis). The stress σ_A is obtained from the flexure formula:

$$\sigma_A = \frac{M(d_2/2)}{I}$$

in which d_2 is the outer diameter (220 mm) and I is the moment of inertia of the cross section. The moment of inertia is

$$I = \frac{\pi}{64}(d_2^4 - d_1^4) = \frac{\pi}{64}\left[(220\text{ mm})^4 - (180\text{ mm})^4\right] = 63.46 \times 10^{-6}\text{ m}^4$$

in which d_1 is the inner diameter. Therefore, the stress σ_A is

$$\sigma_A = \frac{Md_2}{2I} = \frac{(31.68 \text{ kN}\cdot\text{m})(220 \text{ mm})}{2(63.46 \times 10^{-6} \text{ m}^4)} = 54.91 \text{ MPa}$$

The torque T produces shear stresses τ_1 at points A and B (Fig. 8-28d). Calculate these stresses from the torsion formula:

$$\tau_1 = \frac{T(d_2/2)}{I_p}$$

in which I_p is the polar moment of inertia:

$$I_p = \frac{\pi}{32}(d_2^4 - d_1^4) = 2I = 126.92 \times 10^{-6} \text{ m}^4$$

Thus,

$$\tau_1 = \frac{Td_2}{2I_p} = \frac{(7.2 \text{ kN}\cdot\text{m})(220 \text{ mm})}{2(126.92 \times 10^{-6} \text{ m}^4)} = 6.24 \text{ MPa}$$

Finally, calculate the shear stresses at points A and B due to the shear force V. The shear stress at point A is zero, and the shear stress at point B (denoted τ_2 in Fig. 8-28d) is obtained from the shear formula for a circular tube [Eq. (5-48) of Section 5.9]:

$$\tau_2 = \frac{4V}{3A}\left(\frac{r_2^2 + r_2 r_1 + r_1^2}{r_2^2 + r_1^2}\right) \quad \text{(a)}$$

in which r_2 and r_1 are the outer and inner radii, respectively, and A is the cross-sectional area:

$$r_2 = \frac{d_2}{2} = 110 \text{ mm} \quad r_1 = \frac{d_1}{2} = 90 \text{ mm}$$

$$A = \pi(r_2^2 - r_1^2) = 12{,}570 \text{ mm}^2$$

Substitute numerical values into Eq. (a) to obtain

$$\tau_2 = 0.76 \text{ MPa}$$

The stresses acting on the cross section at points A and B now have been calculated.

Stress elements: The next step is to show these stresses on stress elements (Figs. 8-28e and f). For both elements, the y axis is parallel to the longitudinal axis of the pole, and the x axis is horizontal. At point A, the stresses acting on the element are

$$\sigma_x = 0 \quad \sigma_y = \sigma_A = 54.91 \text{ MPa} \quad \tau_{xy} = \tau_1 = 6.24 \text{ MPa}$$

At point B, the stresses are

$$\sigma_x = \sigma_y = 0 \quad \tau_{xy} = \tau_1 + \tau_2 = 6.24 \text{ MPa} + 0.76 \text{ MPa} = 7.00 \text{ MPa}$$

Since there are no normal stresses acting on the element, point B is in pure shear.

Now that all stresses acting on the stress elements (Figs. 8-28e and f) are known, use the equations given in Section 7.3 to determine the principal stresses and maximum shear stresses.

Principal stresses and maximum shear stresses at point A: The principal stresses are obtained from Eq. (7-17), which is repeated here:

$$\sigma_{1,2} = \frac{\sigma_x + \sigma_y}{2} \pm \sqrt{\left(\frac{\sigma_x - \sigma_y}{2}\right)^2 + \tau_{xy}^2} \quad \text{(b)}$$

Substitute $\sigma_x = 0, \sigma_y = 54.91$ MPa, and $\tau_{xy} = 6.24$ MPa to get

$$\sigma_{1,2} = 27.5 \text{ MPa} \pm 28.2 \text{ MPa}$$

or

$$\sigma_1 = 55.7 \text{ MPa} \quad \sigma_2 = -0.7 \text{ MPa}$$

The maximum in-plane shear stresses may be obtained from Eq. (7-25):

$$\tau_{max} = \sqrt{\left(\frac{\sigma_x - \sigma_y}{2}\right)^2 + \tau_{xy}^2} \quad \text{(c)}$$

This term was evaluated previously, so it follows that

$$\tau_{max} = 28.2 \text{ MPa}$$

Because the principal stresses σ_1 and σ_2 have opposite signs, the maximum in-plane shear stresses are larger than the maximum out-of-plane shear stresses [see Eqs. (7-28a, b, and c) and the accompanying discussion]. Therefore, the maximum shear stress at point A is 28.2 MPa.

Principal stresses and maximum shear stresses at point B: The stresses at this point are $\sigma_x = 0, \sigma_y = 0$, and $\tau_{xy} = 7.0$ MPa. Since the element is in pure shear, the principal stresses are

$$\sigma_1 = 7.0 \text{ MPa} \quad \sigma_2 = -7.0 \text{ MPa}$$

and the maximum in-plane shear stress is

$$\tau_{max} = 7.0 \text{ MPa}$$

The maximum out-of-plane shear stresses are half this value.

4. ***Finalize***: If the largest stresses anywhere in the pole are needed, then also determine the stresses at the critical point diametrically opposite point A, because at that point the compressive stress due to bending has its largest value. The principal stresses at that point are

$$\sigma_1 = 0.7 \text{ MPa} \quad \sigma_2 = -55.7 \text{ MPa}$$

FIGURE 8-29

Square tube for Example 8-6

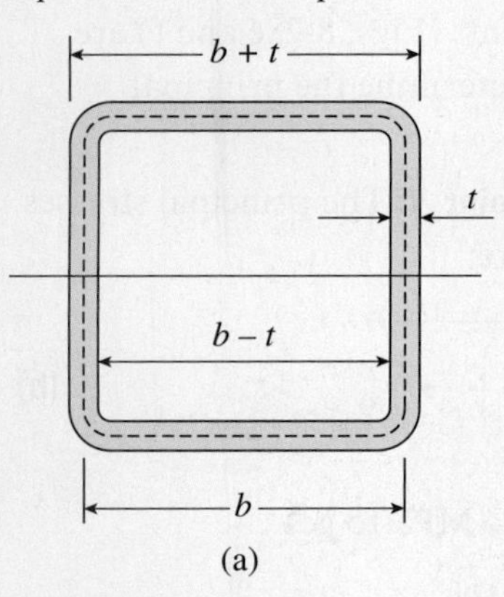

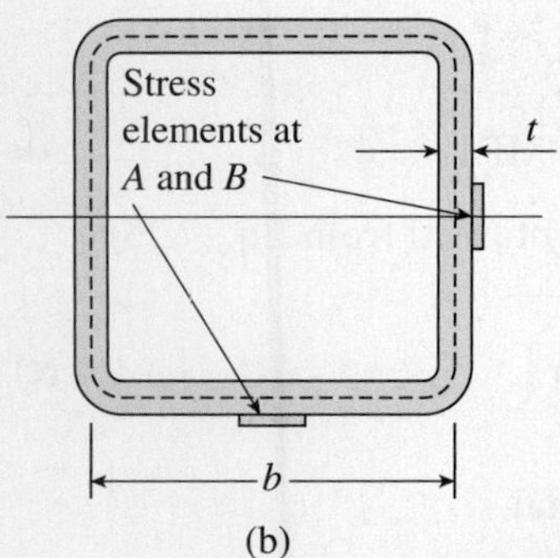

and the maximum shear stress is 28.2 MPa. Therefore, the largest tensile stress in the pole is 55.7 MPa, the largest compressive stress is −55.7 MPa, and the largest shear stress is 28.2 MPa. (Keep in mind that only the effects of the wind pressure are considered in this analysis. Other loads, such as the weight of the structure, also produce stresses at the base of the pole.)

Part (b): Square tube stress resultants.

1. ***Conceptualize***: The square tube has the same height ($h = 6.6$ m to the center of pressure on the sign), same wall thickness ($t = 20$ mm), and same cross-sectional area ($A = 12{,}570\ \text{mm}^2$) as that of the circular pole. Now compute tube dimension b (along the median line of the tube, see Fig. 8-29a) as

$$(b + t)^2 - (b - t)^2 = 12{,}570\ \text{mm}^2 \quad \text{so} \quad b = 157.125\ \text{mm}$$

The torsion constant J of the tube [see Eq. (3-94)] and the area A_m enclosed by the median line of the tube are

$$J = b^3 t = 7.758 \times 10^{-5}\ \text{m}^4 \qquad A_m = b^2 = 2.469 \times 10^4\ \text{mm}^2$$

(Assume that the formulas for thin-walled tubes in Section 3.11 apply here and disregard the effects of stress concentrations at the corners of the tube.)

For normal and transverse shear stress calculations, find the moment of inertia I_{tube} with respect to the neutral axis of the cross section for use in the flexure formula [from Eq. (5-14)] and the first moment Q_{tube} of the area with respect to the neutral axis for use in the shear formula [from Eq. (5-41)]. These properties are computed as

$$I_{\text{tube}} = \frac{1}{12}\left[(b + t)^4 - (b - t)^4\right] = 5.256 \times 10^{-5}\ \text{m}^4$$

$$Q_{\text{tube}} = (b + t)\left(\frac{b + t}{2}\right)\left(\frac{b + t}{4}\right) - (b - t)\left(\frac{b - t}{2}\right)\left(\frac{b - t}{4}\right)$$

$$= 3.723 \times 10^{-4}\ \text{m}^3$$

2, 3. ***Categorize, Analyze***:

Stresses at *A* and *B* on tube: Compute the normal tensile stress at A (see Fig. 8-29b) using the flexure formula with $M = 31.68\ \text{kN} \cdot \text{m}$, to get

$$\sigma_A = \frac{M(b + t)}{2I_{\text{tube}}} = 53.38\ \text{MPa}$$

The normal stress at B is zero, because it is located on the neutral axis. The transverse shear stress at A is zero, and the shear stress at B is obtained from the shear formula as

$$\tau_2 = \frac{VQ_{\text{tube}}}{I_{\text{tube}}(2t)} = 0.85\ \text{MPa}$$

The torque $T = 7.2\text{ kN}\cdot\text{m}$ produces shear stress at both A and B. Using Eq. (3-81), shear stress τ_1 is

$$\tau_1 = \frac{T}{2tA_m} = 7.29\text{ MPa}$$

The resulting stress states at locations A and B on the square tube in Fig. 8-29b are the same as that shown in Figs. 8-28e and f, where

$$\sigma_x = 0 \quad \sigma_y = \sigma_A = 53.38\text{ MPa}$$
$$\tau_{xy} = \tau_1 + \tau_2 = 8.14\text{ MPa}$$

Principal stresses and maximum in-plane shear stress at point A: Repeat the calculations for the square tube using Eq. (b) to get

$$\sigma_1 = \frac{\sigma_x + \sigma_y}{2} + \sqrt{\left(\frac{\sigma_x - \sigma_y}{2}\right)^2 + \tau_{xy}^2} = 54.6\text{ MPa}$$

$$\sigma_2 = \frac{\sigma_x + \sigma_y}{2} - \sqrt{\left(\frac{\sigma_x - \sigma_y}{2}\right)^2 + \tau_{xy}^2} = -1.2\text{ MPa}$$

$$\tau_{max} = \sqrt{\left(\frac{\sigma_x - \sigma_y}{2}\right)^2 + \tau_{xy}^2} = 27.9\text{ MPa}$$

Principal stresses and maximum in-plane shear stress at point B: The stress element at point B is in pure shear, so the principal stresses and maximum in-plane shear stress are

$$\sigma_1 = \tau_{xy} = 8.1\text{ MPa}$$
$$\sigma_2 = -\tau_{xy} = -8.1\text{ MPa}$$
$$\tau_{max} = \tau_{xy} = 8.1\text{ MPa}$$

4. ***Finalize***: These stresses at A and B on the square tube are comparable to those for the circular pole. As a final comparison, look at the twist displacement on each pole at the level of the sign center of pressure ($h = 6.6$ m). The twist rotation for the circular pole is computed using Eq. (3-17) (assuming that $G = 80$ GPa for steel):

$$\phi_c = \frac{Th}{GI_p} = 4.68 \times 10^{-3}\text{ radians}$$

and that for the square tube is computed using Eq. (3-73):

$$\phi_t = \frac{Th}{GJ} = 7.656 \times 10^{-3}\text{ radians}$$

The twist rotation for the circular pole is 39% lower than that of the square tube. (See Example 3-16 for more discussion of square and circular tube stresses and twist rotations.) Both circular and square tube sign posts also displace in the direction of the wind force, but calculation of bending displacements must be delayed until beam deflections are discussed in Chapter 9.

Example 8-7

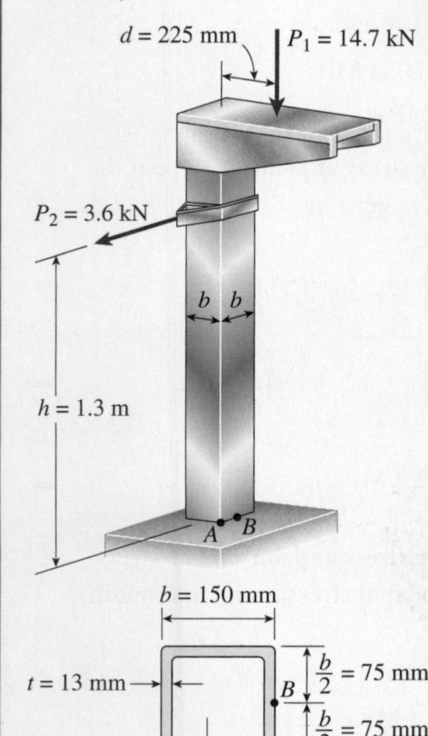

FIGURE 8-30

Example 8-7: Loads on a post (combined axial load, bending, and shear)

A tubular post of square cross section supports a horizontal platform (Fig. 8-30). The tube has an outer dimension $b = 150$ mm and wall thickness $t = 13$ mm. The platform has dimensions 175 mm $\times$ 600 mm and supports a uniformly distributed load of 140 kPa acting over its upper surface. The resultant of this distributed load is a vertical force P_1:

$$P_1 = (140 \text{ kPa})(175 \text{ mm} \times 600 \text{ mm}) = 14.7 \text{ kN}$$

This force acts at the midpoint of the platform, which is at distance $d = 225$ mm from the longitudinal axis of the post. A second load $P_2 = 3.6$ kN acts horizontally on the post at a height $h = 1.3$ m above the base.

Determine the principal stresses and maximum shear stresses at points A and B at the base of the post due to the loads P_1 and P_2.

Solution:

Use a four-step problem-solving approach. Combine steps as needed for an efficient solution.

1. *Conceptualize*:

Stress resultants: The force P_1 acting on the platform (Fig. 8-30) is statically equivalent to a force P_1 and a moment $M_1 = P_1 d$ acting at the centroid of the cross section of the post (Fig. 8-31a). The load P_2 is also shown in this figure.

The stress resultants at the base of the post due to the loads P_1 and P_2 and the moment M_1 are shown in Fig. 8-31b. These stress resultants are

i. An axial compressive force $P_1 = 14.7$ kN.

ii. A bending moment M_1 produced by the force P_1:

$$M_1 = P_1 d = (14.7 \text{ kN})(225 \text{ mm}) = 3307.5 \text{ N}\cdot\text{m}$$

iii. A shear force $P_2 = 3.6$ kN.

iv. A bending moment M_2 produced by the force P_2:

$$M_2 = P_2 h = (3.6 \text{ kN})(1.3 \text{ m}) = 4.68 \text{ kN}\cdot\text{m}$$

Examination of these stress resultants (Fig. 8-31b) shows that both M_1 and M_2 produce maximum compressive stresses at point A and the shear force produces maximum shear stresses at point B. Therefore, A and B are critical points where the stresses should be determined. (Another critical point is diagonally opposite point A, as explained in the note at the end of this example.)

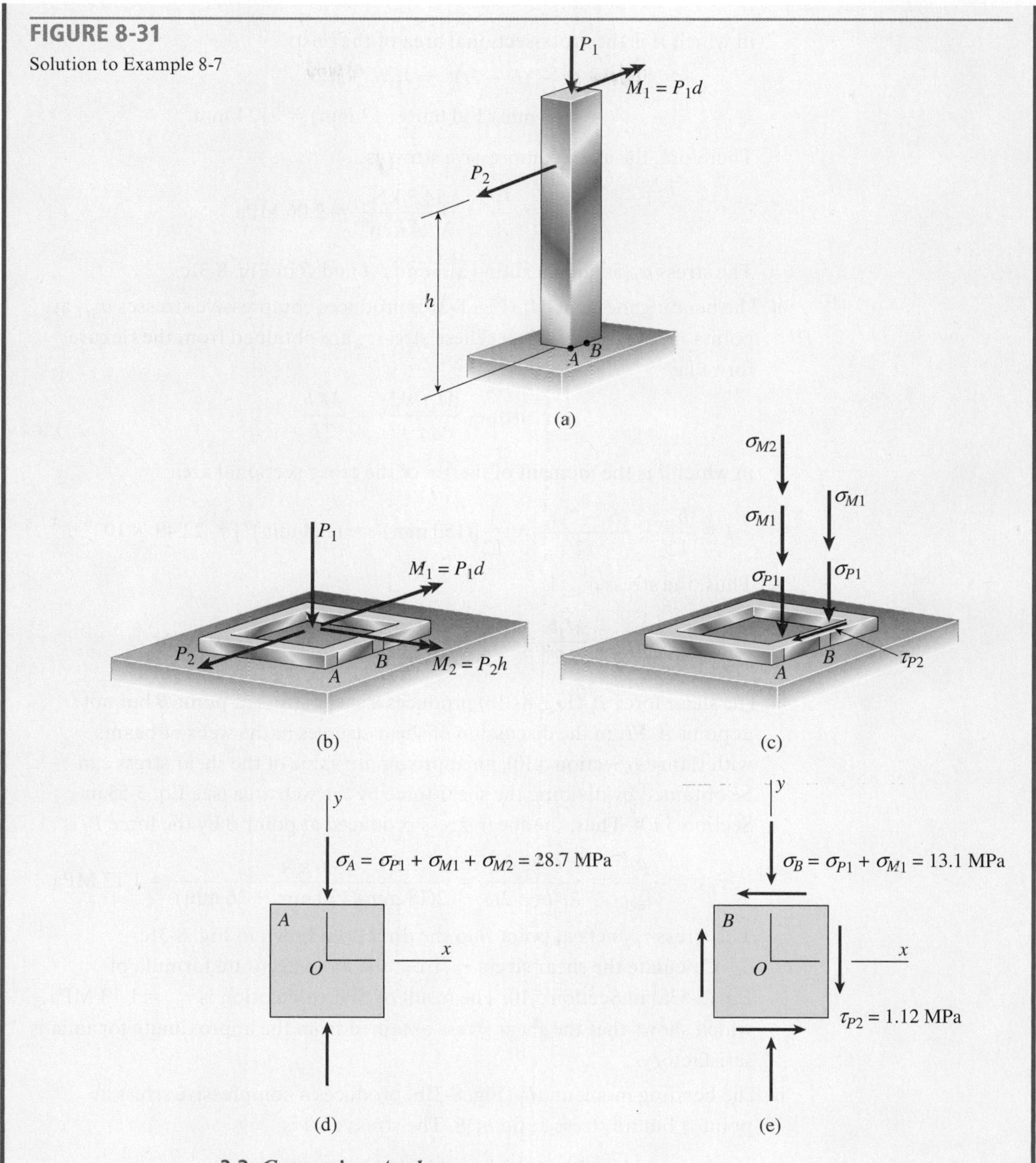

FIGURE 8-31
Solution to Example 8-7

2,3. *Categorize, Analyze:*

Stresses at points *A* and *B*:

i. The axial force P_1 (Fig. 8-31b) produces uniform compressive stresses throughout the post. These stresses are

$$\sigma_{P1} = \frac{P_1}{A}$$

in which A is the cross-sectional area of the post:

$$A = b^2 - (b - 2t)^2 = 4t(b - t)$$
$$= 4(13 \text{ mm})(150 \text{ mm} - 13 \text{ mm}) = 7124 \text{ mm}^2$$

Therefore, the axial compressive stress is

$$\sigma_{P1} = \frac{P_1}{A} = \frac{14.7 \text{ kN}}{7124 \text{ mm}^2} = 2.06 \text{ MPa}$$

The stress σ_{P1} is shown acting at points A and B in Fig. 8-31c.

ii. The bending moment M_1 (Fig. 8-31b) produces compressive stresses σ_{M1} at points A and B (Fig. 8-31c). These stresses are obtained from the flexure formula:

$$\sigma_{M1} = \frac{M_1(b/2)}{I} = \frac{M_1 b}{2I}$$

in which I is the moment of inertia of the cross-sectional area:

$$I = \frac{b^4}{12} - \frac{(b - 2t)^4}{12} = \frac{1}{12}[(150 \text{ mm})^4 - (124 \text{ mm})^4] = 22.49 \times 10^{-6} \text{ m}^4$$

Thus, the stress σ_{M1} is

$$\sigma_{M1} = \frac{M_1 b}{2I} = \frac{(3307.5 \text{ N}\cdot\text{m})(150 \text{ mm})}{2(22.49 \times 10^{-6} \text{ m}^4)} = 11.03 \text{ MPa}$$

iii. The shear force P_2 (Fig. 8-31b) produces a shear stress at point B but not at point A. From the discussion of shear stresses in the webs of beams with flanges (Section 5.10), an approximate value of the shear stress can be obtained by dividing the shear force by the web area (see Eq. 5-55 in Section 5.10). Thus, the shear stress produced at point B by the force P_2 is

$$\tau_{P2} = \frac{P_2}{A_{\text{web}}} = \frac{P_2}{2t(b - 2t)} = \frac{3.6 \text{ kN}}{2(13 \text{ mm})(150 \text{ mm} - 26 \text{ mm})} = 1.12 \text{ MPa}$$

The stress τ_{P2} acts at point B in the direction shown in Fig. 8-31c.

Calculate the shear stress τ_{P2} from the more accurate formula of Eq. (5-53a) in Section 5.10. The result of that calculation is $\tau_{P2} = 1.13$ MPa, which shows that the shear stress obtained from the approximate formula is satisfactory.

iv. The bending moment M_2 (Fig. 8-31b) produces a compressive stress at point A but no stress at point B. The stress at A is

$$\sigma_{M2} = \frac{M_2(b/2)}{I} = \frac{M_2 b}{2I} = \frac{(4.68 \text{ kN}\cdot\text{m})(150 \text{ mm})}{2(22.49 \times 10^{-6} \text{ m}^4)} = 15.61 \text{ MPa}$$

This stress is also shown in Fig. 8-31c.

Stress elements: The next step is to show the stresses acting on stress elements at points A and B (Figs. 8-31d and e). Each element is oriented so that

FIGURE 8-31 (Repeated)

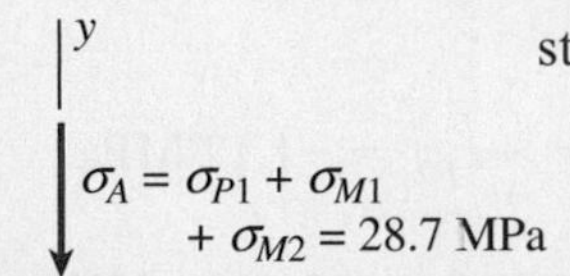

the y axis is vertical (that is, parallel to the longitudinal axis of the post) and the x axis is horizontal. At point A, the only stress is a compressive stress σ_A in the y direction (Fig. 8-31d):

$$\sigma_A = \sigma_{P1} + \sigma_{M1} + \sigma_{M2}$$
$$= 2.06 \text{ MPa} + 11.03 \text{ MPa} + 15.61 \text{ MPa} = 28.7 \text{ MPa (compression)}$$

Thus, this element is in uniaxial stress.

At point B, the compressive stress in the y direction (Fig. 8-31e) is

$$\sigma_B = \sigma_{P1} + \sigma_{M1} = 2.06 \text{ MPa} + 11.03 \text{ MPa} = 13.1 \text{ MPa (compression)}$$

and the shear stress is

$$\tau_{P2} = 1.12 \text{ MPa}$$

The shear stress acts leftward on the top face of the element and downward on the x face of the element.

Principal stresses and maximum shear stresses at point A: Using the standard notation for an element in plane stress (Fig. 8-32), write the stresses for element A (Fig. 8-31d) as

$$\sigma_x = 0 \quad \sigma_y = -\sigma_A = -28.7 \text{ MPa} \quad \tau_{xy} = 0$$

Since the element is in uniaxial stress, the principal stresses are

$$\sigma_1 = 0 \quad \sigma_2 = -28.7 \text{ MPa}$$

and the maximum in-plane shear stress [Eq. (7-26)] is

$$\tau_{max} = \frac{\sigma_1 - \sigma_2}{2} = \frac{28.7 \text{ MPa}}{2} = 14.4 \text{ MPa}$$

The maximum out-of-plane shear stress [Eq. (7-28a)] has the same magnitude.

FIGURE 8-32

Notation for an element in plane stress

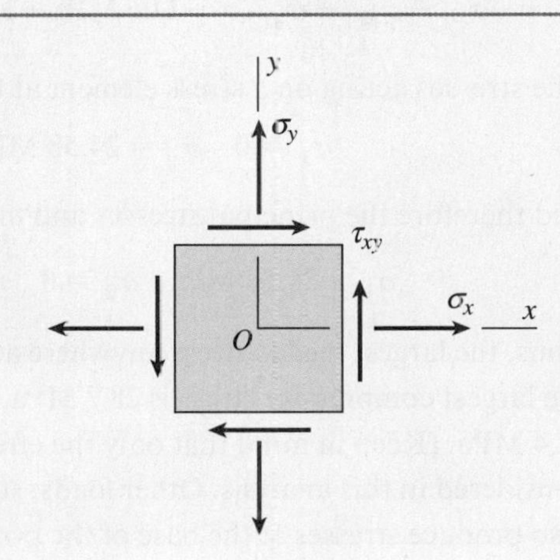

Principal stresses and maximum shear stresses at point *B*: Again using the standard notation for plane stress (Fig. 8-32), the stresses at point *B* (Fig. 8-31e) are

$$\sigma_x = 0 \quad \sigma_y = -\sigma_B = -13.1 \text{ MPa} \quad \tau_{xy} = -\tau_{P2} = -1.12 \text{ MPa}$$

To obtain the principal stresses, use Eq. (7-17), which is repeated here:

$$\sigma_{1,2} = \frac{\sigma_x + \sigma_y}{2} \pm \sqrt{\left(\frac{\sigma_x - \sigma_y}{2}\right)^2 + \tau_{xy}^2} \qquad \textbf{(a)}$$

Substitute for σ_x, σ_y, and τ_{xy} to get

$$\sigma_{1,2} = -6.55 \text{ MPa} \pm 6.65 \text{ MPa}$$

or

$$\sigma_1 = 0.1 \text{ MPa} \quad \sigma_2 = -13.2 \text{ MPa}$$

The maximum in-plane shear stresses may be obtained from Eq. (7-25):

$$\tau_{max} = \sqrt{\left(\frac{\sigma_x - \sigma_y}{2}\right)^2 + \tau_{xy}^2} \qquad \textbf{(b)}$$

This term was evaluated previously, so

$$\tau_{max} = 6.65 \text{ MPa}$$

Because the principal stresses σ_1 and σ_2 have opposite signs, the maximum in-plane shear stresses are larger than the maximum out-of-plane shear stresses [see Eqs. (7-28a, b, and c) and the accompanying discussion]. Therefore, the maximum shear stress at point *B* is 6.65 MPa.

4. ***Finalize***: If the largest stresses anywhere at the base of the post are needed, then you must also determine the stresses at the critical point diagonally opposite point *A* (Fig. 8-31c) because at that point each bending moment produces the maximum tensile stress. Thus, the tensile stress acting at that point is

$$\sigma_y = -\sigma_{P1} + \sigma_{M1} + \sigma_{M2} = -2.06 \text{ MPa} + 11.03 \text{ MPa} + 15.61 \text{ MPa} = 24.58 \text{ MPa}$$

The stresses acting on a stress element at that point (see Fig. 8-32) are

$$\sigma_x = 0 \quad \sigma_y = 24.58 \text{ MPa} \quad \tau_{xy} = 0$$

and therefore the principal stresses and maximum shear stress are

$$\sigma_1 = 24.58 \text{ MPa} \quad \sigma_2 = 0 \quad \tau_{max} = 12.3 \text{ MPa}$$

Thus, the largest tensile stress anywhere at the base of the post is 24.58 MPa, the largest compressive stress is 28.7 MPa, and the largest shear stress is 14.4 MPa. (Keep in mind that only the effects of the loads P_1 and P_2 are considered in this analysis. Other loads, such as the weight of the structure, also produce stresses at the base of the post.)

Example 8-8

A jib crane at a marina carries a load at the end of the jib beam with components 12 kN in the x direction, 22 kN in the $-y$ direction, and 18 kN in the z direction (Fig. 8-33a). The crane is fixed at the base at O. Find the state of plane stress on the surface of the vertical post at points such as A and B just above the base (see Fig. 8-33b). The post has an outer radius $r_2 = 165$ mm and wall thickness $t = 20$ mm. Neglect the weight of the crane.

Solution:

Use a four-step problem-solving approach. Combine steps as needed for an efficient solution.

1. ***Conceptualize***: Use statics to find reaction forces and moments at support point O or to find the load resultants just above the base of the post (Fig. 8-33b) (the resultants are the negatives of the reactions). Use the load resultants to find normal and shear stresses on each plane stress element. View each element by walking around the base of the post; in each case, the local x axis for the element (at A, B, or elsewhere) is to your right, and the y axis is upward (parallel to the y axis for the entire crane structure).
2. ***Categorize***: Start with axial force resultant F_y at the base (Fig. 8-33b) and find the axial compressive normal stress σ_y acting on each stress element. Next, use force resultants F_x and F_z in the *shear formula* for a hollow circular cross section [Eq. (5-48)] to find the transverse shear stress τ_V on the $+y$ face of each stress element. Use torsional moment M_y in the *torsion formula* [Eq. (3-13)] to compute the shear stress τ_T on the $+y$ face due to torsion. Finally, use moments M_x and M_z in the *flexure formula* [Eq. (5-14)] to find additional

FIGURE 8-33

(a) Load components at end of beam on jib crane; (b) load resultants and plane stress elements A and B on surface of post at base

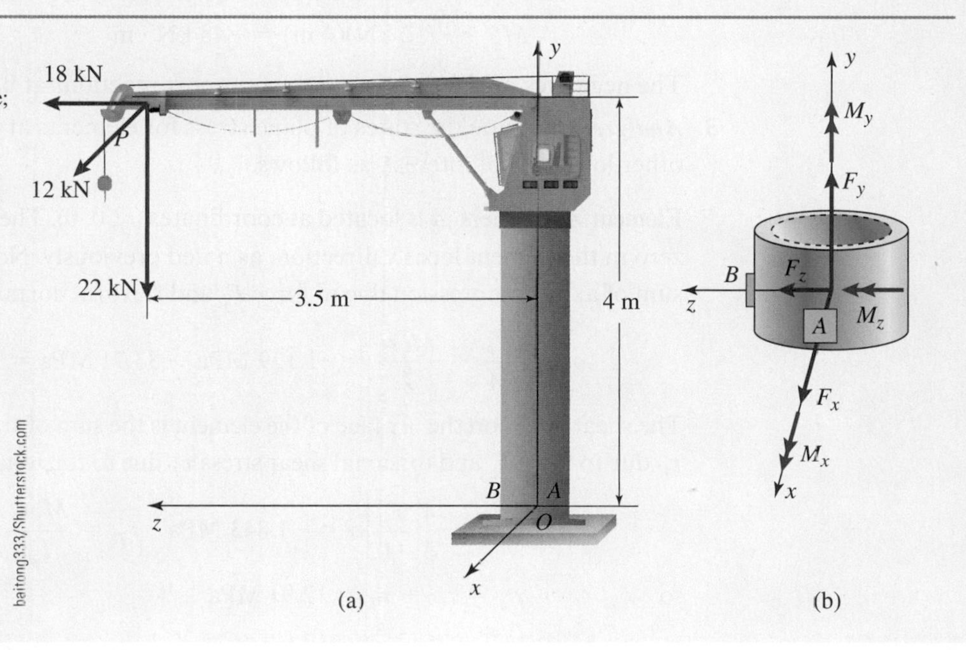

normal stresses σ_y on the $+y$ face of each element. Note that elements not aligned with the x and z axes are subjected to *biaxial bending* [see (Eq. 6-19)] and that every axis through the centroid of the cross section of the post is a *principal axis*. Also observe that the normal stress σ_x in the x direction on each element is zero. Assume that all stress elements are a sufficient distance from support O so that stress-concentration effects are negligible.

Cross-sectional properties of post: Compute the cross-sectional area A, polar moment of inertia I_p, and moment of inertia for flexure I for use in stress calculations. Use the inner and outer radii and thickness of the post, $r_2 = 165$ mm, $t = 20$ mm, and $r_1 = r_2 - t = 145$ mm, as

$$A = \pi(r_2^2 - r_1^2) = 1.948 \times 10^4 \text{ mm}^2$$
$$I_p = \frac{\pi}{2}(r_2^4 - r_1^4) = 4.699 \times 10^8 \text{ mm}^4 \quad \text{and} \quad I = I_p/2 = 2.350 \times 10^8 \text{ mm}^4 \qquad \text{(a)}$$

In addition, the following term is needed in the calculation of maximum shear stress at the neutral axis for a hollow circular cross section [see Eq. (5-48)]:

$$\alpha = \frac{r_2^2 + r_2 r_1 + r_1^2}{r_2^2 + r_1^2} = 1.496 \qquad \text{(b)}$$

Stress resultants (forces, moments) at base of post: Summing forces in x, y, and z directions and also moments about the x, y, and z axes leads to the values for the force and moment resultants shown in Fig. 8-33b:

$$F_x = 12 \text{ kN} \quad F_y = -22 \text{ kN} \quad F_z = 18 \text{ kN}$$
$$M_x = (22 \text{ kN})(3.5 \text{ m}) + (18 \text{ kN})(4 \text{ m}) = 149 \text{ kN} \cdot \text{m}$$
$$M_y = (12 \text{ kN})(3.5 \text{ m}) = 42 \text{ kN} \cdot \text{m} \qquad \text{(c)}$$
$$M_z = -(12 \text{ kN})(4 \text{ m}) = -48 \text{ kN} \cdot \text{m}$$

The negatives of these values are the support reactions at base point O.

3. ***Analyze***: Compute the states of plane stress for elements at points A and B, and other locations of interest, as follows.

Element A: Element A is located at coordinates $(r_2, 0, 0)$. The normal stress σ_x is zero in the element local x direction, as noted previously. Normal stress σ_y is the sum of axial compression due to force F_y and flexural normal stress due to M_z:

$$\sigma_y = \frac{F_y}{A} + \frac{M_z r_2}{I} = -1.129 \text{ MPa} - 33.71 \text{ MPa} = -34.8 \text{ MPa}$$

The shear stress on the $+y$ face of the element is the sum of transverse shear stress τ_V due to force F_z and torsional shear stress τ_T due to torsional moment M_y:

$$\tau_V = -\frac{4}{3}\left[\frac{F_z}{A}\right]\alpha = -1.843 \text{ MPa} \quad \tau_T = \frac{M_y r_2}{I_p} = 14.75 \text{ MPa}$$

so
$$\tau_A = \tau_V + \tau_T = 12.91 \text{ MPa}$$

FIGURE 8-34

(a) Plane stress state at element A; (b) principal normal stresses; (c) maximum shear stress; and (d) Mohr's circle for element A

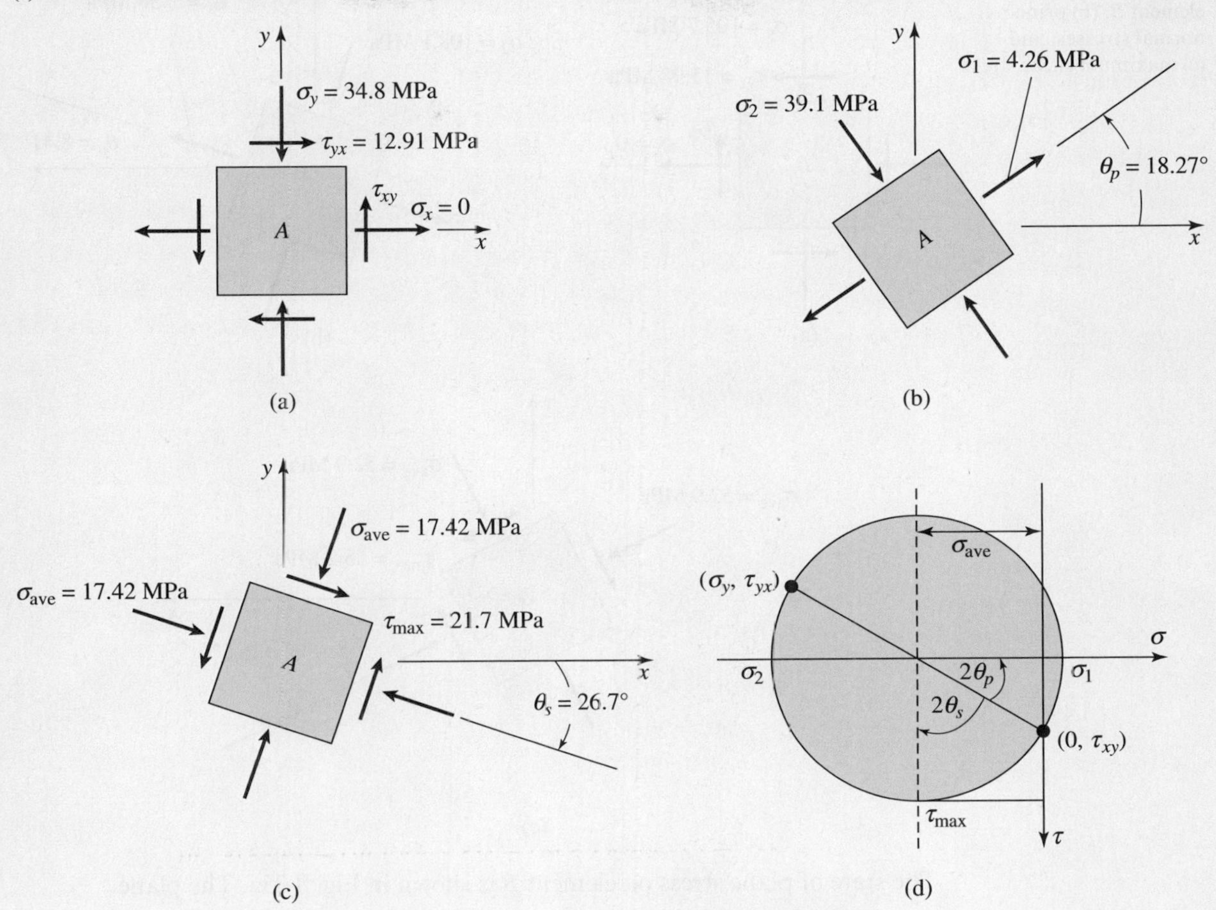

The state of plane stress at element A is shown in Fig. 8-34a. Use the plane stress-transformation equations [Eqs. (7-11), (7-17), and (7-25)] or Mohr's circle (see Fig. 8-34d) to find the principal normal and maximum shear stress states shown in Figs. 8-34b and c.

Element B: Element B is located at coordinates $(0, 0, r_2)$. Normal stress σ_y is the sum of axial compression due to force F_y and flexural normal stress due to M_x:

$$\sigma_y = \frac{F_y}{A} - \frac{M_x r_2}{I} = -1.129 \text{ MPa} - 104.6 \text{ MPa} = -105.7 \text{ MPa}$$

Similar to element A, the shear stress on the $+y$ face of the element is the sum of transverse shear stress τ_V due to force F_x and torsional shear stress τ_T due to torsional moment M_y:

$$\tau_B = \tau_V + \tau_T = \frac{4}{3}\left[\frac{F_x}{A}\right]\alpha + \frac{M_y r_2}{I_p} = 1.229 \text{ MPa} + 14.75 \text{ MPa} = 15.98 \text{ MPa}$$

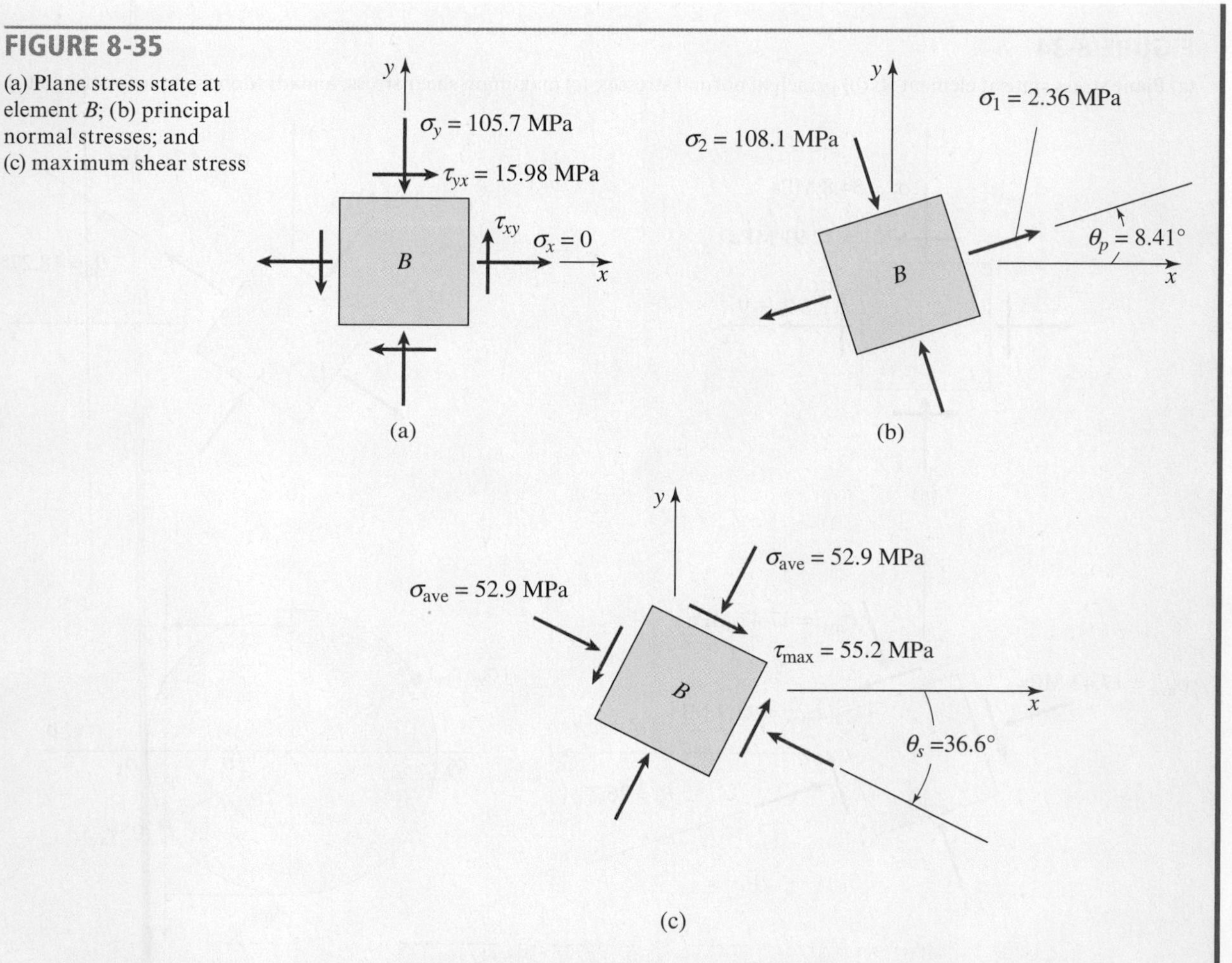

FIGURE 8-35

(a) Plane stress state at element B; (b) principal normal stresses; and (c) maximum shear stress

The state of plane stress on element B is shown in Fig. 8-35a. The plane stress-transformation equations (or Mohr's circle) provide the principal normal and maximum shear stress states and rotated element orientations shown in Figs. 8-35b and c, respectively.

Element C: Since $I_x = I_z$, the neutral axis nn for flexure only is defined by the angle between moment components M_x and M_z, that is, $\beta = \tan^{-1}\left[\dfrac{M_z}{M_x}\right] = 17.86°$, as shown in Fig. 8-36. (Recall that the neutral axis no longer passes through the centroid of the cross section when both axial and flexural effects are considered, as described in Section 5.12.) As a result, the maximum compressive normal stress is at point C and is the sum of uniform axial compressive stress due to force F_y and flexural normal stress due to the resultant of moment components M_x and M_z. Hence $\sigma_x = 0$, and the maximum normal stress σ_y is

$$\sigma_y = \frac{F_y}{A} - \frac{\left(\sqrt{M_x^2 + M_z^2}\right)r_2}{I} = \frac{-22\text{ kN}}{A} - \frac{(156.5\text{ kN}\cdot\text{m})r_2}{I}$$
$$= -1.129\text{ MPa} - 109.9\text{ MPa} = -111.1\text{ MPa}$$

FIGURE 8-36

(a) Resultant moment for flexural stress at C; (b) force components for transverse shear stress at element C

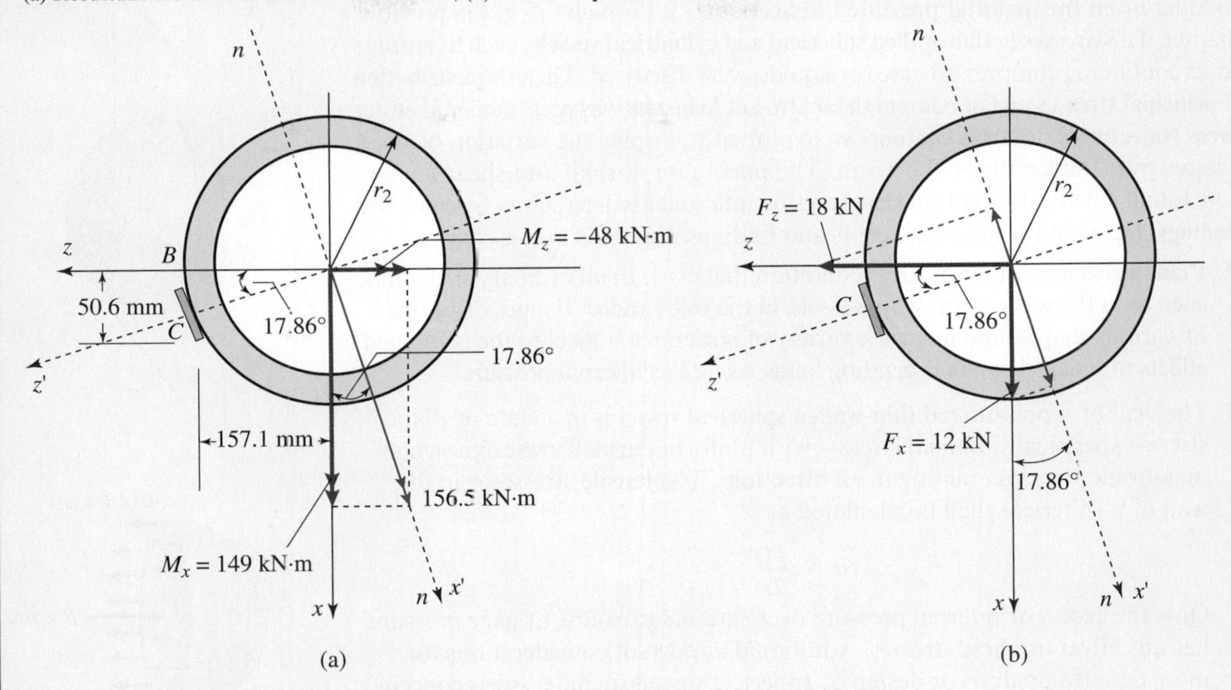

Alternatively, the flexural portion of the normal stress at point C can be computed using biaxial bending [Eq. (6-19)] as

$$(-M_x)\frac{r_2\cos(17.86°)}{I} + (M_z)\frac{r_2\sin(17.86°)}{I} = -99.6\text{ MPa} - 10.3\text{ MPa}$$
$$= -109.9\text{ MPa}$$

Additional analysis at point C leads to shear stress τ_V due to transverse shear force. Use components of forces F_x and F_z (Fig. 8-36c) to find transverse shear stress τ_V and then add uniform torsional shear stress as

$$\tau_C = \tau_V + \tau_T = \frac{4}{3}\left[\frac{F_x\cos(17.86°) - F_z\sin(17.86°)}{A}\right]\alpha + \frac{M_y r_2}{I_p}$$
$$= 0.60\text{ MPa} + 14.75\text{ MPa} = 15.35\text{ MPa}$$

Finally, use plane stress transformations to find principal normal stresses and the maximum shear stress at point C for points A and B. The principal normal stresses and maximum shear stress at point C are $\sigma_1 = 2.08$ MPa, $\sigma_2 = -113.2$ MPa, and $\tau_{max} = 57.6$ MPa.

4. ***Finalize***: As expected, the normal stress σ_y at point C is the largest axial compressive stress at the base of the post for this loading. If the weight of the crane itself is considered in this analysis, load resultants F_y and M_x will increase, resulting in larger normal compressive stress values at the base.

CHAPTER SUMMARY AND REVIEW

Chapter 8 presented some practical examples of structures in states of plane stress, building upon the material presented in Sections 7.2 through 7.5 in the previous chapter. The **stresses in thin-walled spherical and cylindrical vessels**, such as storage tanks containing compressed gases or liquids, were discussed. Then the **distribution of principal stresses and maximum shear stresses in beams** was presented and either **stress trajectories** or **stress contours** were plotted to display the variation of these stresses over the length of the beam. The maximum normal and shear stresses were found at various points in structures or components acted upon by **combined loadings**. Here are the major concepts and findings presented in this chapter.

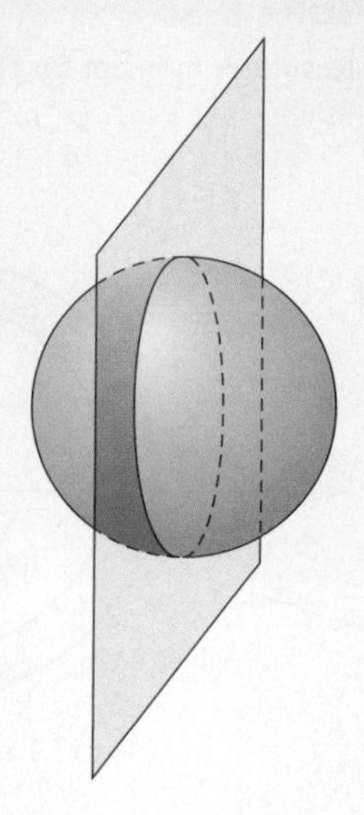

1. **Plane stress** is a common stress condition that exists in all ordinary structures, such as in the walls of pressure vessels, in the webs and/or flanges of beams of various shapes, and in a wide variety of structures subject to the combined effects of axial, shear, and bending loads, as well as internal pressure.

2. The wall of a pressurized **thin-walled spherical vessel** is in a state of plane stress—specifically, biaxial stress—with uniform tensile stresses known as membrane stresses σ acting in all directions. The tensile stresses σ in the wall of a spherical shell is calculated as

$$\sigma = \frac{pr}{2t}$$

Only the excess of internal pressure over external pressure, or gage pressure, has any effect on these stresses. Additional important considerations for more detailed analysis or design of spherical vessels include: stress concentrations around openings, effects of external loads and self weight, including contents, and influence of corrosion, impacts, and temperature changes.

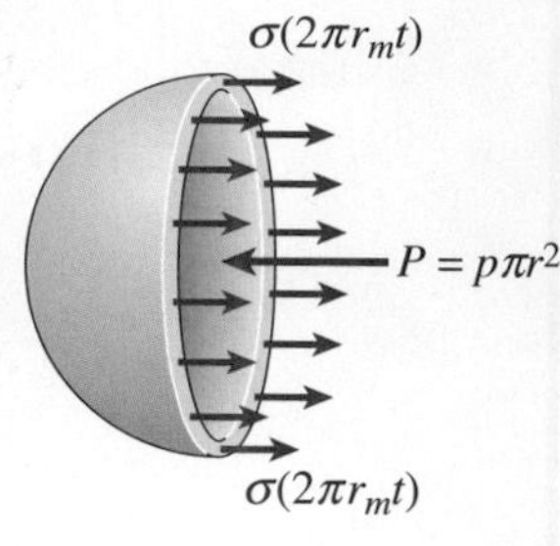

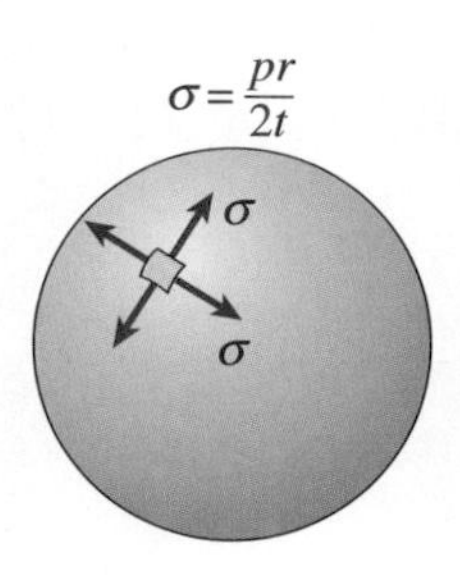

3. The walls of **thin-walled cylindrical pressure vessels** with circular cross sections are also in a state of biaxial stress. The circumferential stress σ_1 is referred to as the hoop stress, and the stress parallel to the axis of the tank is called the longitudinal stress or the axial stress σ_2. The circumferential stress is equal to twice the longitudinal stress. Both are principal stresses. The formulas for σ_1 and σ_2 are

$$\sigma_1 = \frac{pr}{t} \qquad \sigma_2 = \frac{pr}{2t}$$

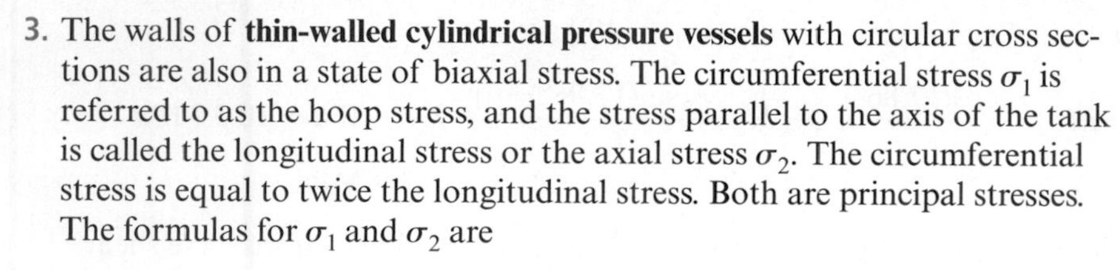

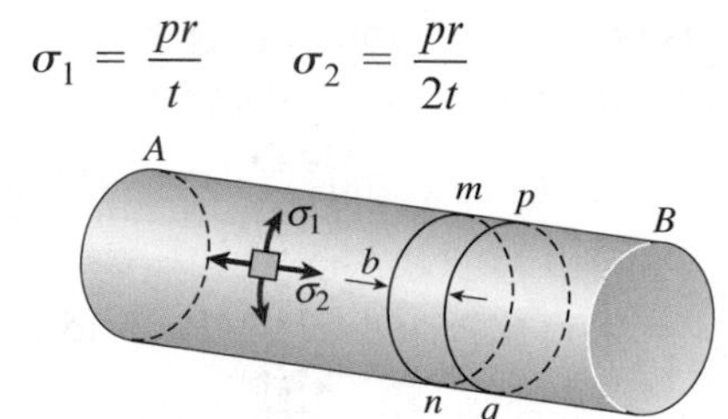

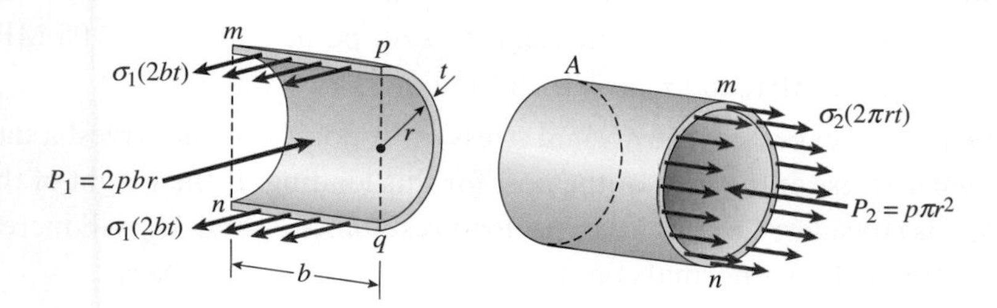

These formulas were derived using elementary theory for thin-walled shells and are only valid in parts of the cylinder away from any discontinuities that cause stress concentrations.

4. If Hooke's law applies, the flexure and shear formulas (Chapter 5) are used to find normal and shear stresses at points of interest along a beam. Investigation of the stresses at many cross sections of the beam for a given loading gives the variation in principal stresses throughout the beam and provides two systems of orthogonal curves (called **stress trajectories**) that give the directions of the principal stresses. Curves connecting points of equal principal stress, are known as **stress contours**.

5. **Stress trajectories** give the directions of the principal stresses but give no information about the magnitudes of the stresses. In contrast, the magnitudes of the principal stresses are constant along a **stress contour**, but the contours give no information about the directions of the stresses.

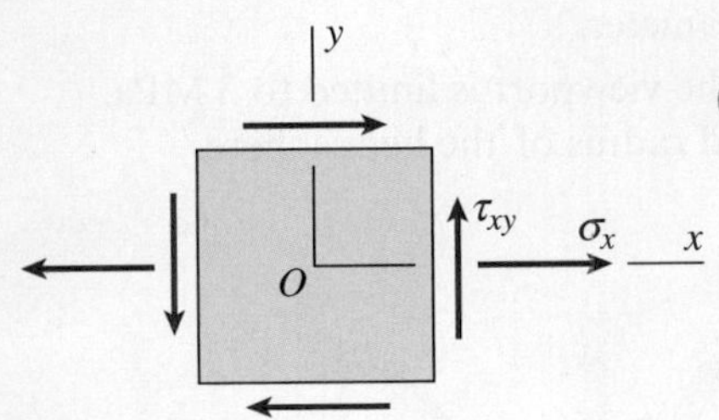

6. In the practical design of ordinary beams, the principal stresses and maximum shear stresses are rarely calculated. Instead, the tensile and compressive stresses used in design are calculated from the **flexure formula** at the cross section of the maximum bending moment

$$\sigma = -\frac{My}{I}$$

and the shear stress used in design is calculated from the **shear formula** at the cross section of the maximum shear force

$$\tau = \frac{VQ}{Ib}$$

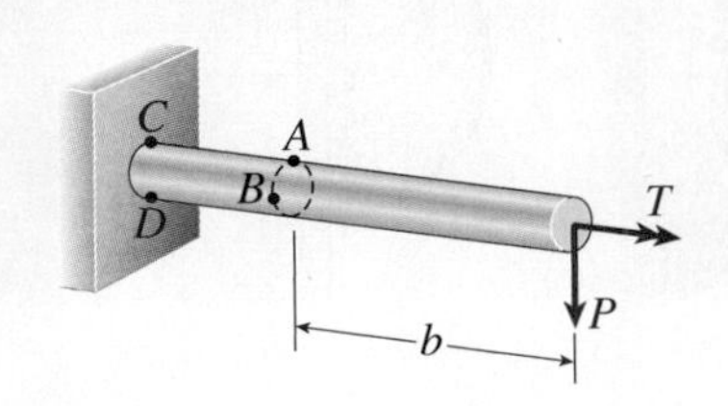

7. A structural member subjected to **combined loadings** often is analyzed by superimposing the stresses and strains caused by each load acting separately. However, the stresses and strains must be linear functions of the applied loads, which in turn require that the material follow Hooke's law and the displacements remain small. There must be no interaction between the various loads, that is, the stresses and strains due to one load must not be affected by the presence of the other loads.

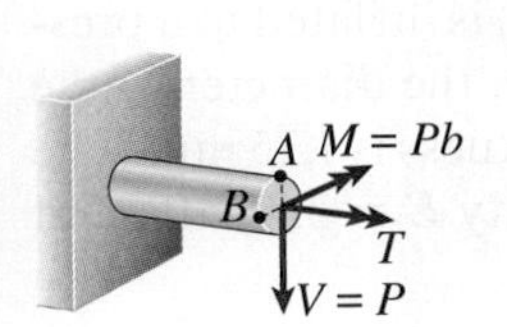

8. A **detailed approach for analysis** of critical points in a structure or component subjected to more than one type of load was presented in Section 8.5.

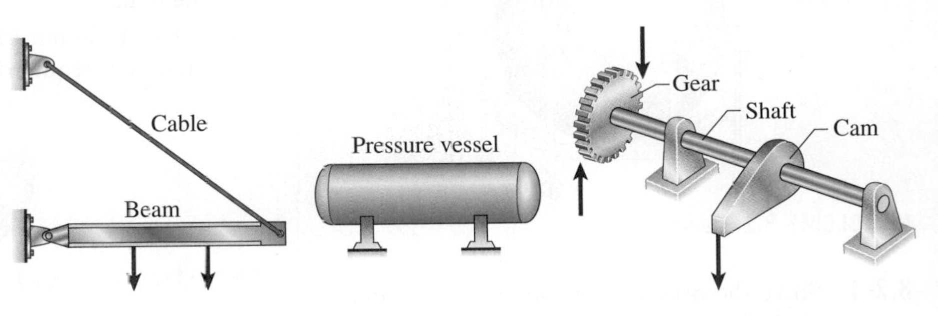

PROBLEMS Chapter 8

8.2 Spherical Pressure Vessels

To solve the problems for Section 8.2, assume that the given radius or diameter is an inside dimension and that all internal pressures are gage pressures.

Introductory Problems

8.2-1 A spherical balloon is filled with a gas. The outer diameter of the balloon is 500 mm and the thickness is 0.3 mm. Calculate the maximum permissible pressure in the balloon if the allowable tensile stress and the allowable shear stress in the balloon are 7 MPa and 2 MPa, respectively.

8.2-2 A spherical balloon with an outer diameter of 500 mm and thickness 0.3 mm is filled with a gas. Calculate maximum permissible pressure in the balloon if the allowable normal strain at the outer surface of the balloon is 0.1. Assume $E = 4$ MPa and $\nu = 0.45$.

8.2-3 A large spherical tank (see figure) contains gas at a pressure of 3.5 MPa. The tank is 20 m in diameter and is constructed of high-strength steel having a yield stress in tension of 550 MPa.

(a) Determine the required thickness of the wall of the tank if a factor of safety of 3.5 with respect to yielding is required.

(b) If the tank wall thickness is 100 mm, what is the maximum permissible internal pressure?

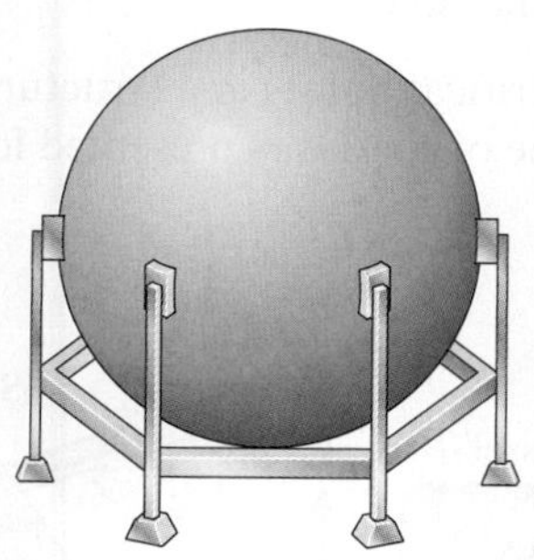

PROBLEMS 8.2-3 and 8.2-4

8.2-4 Solve the preceding problem if the internal pressure is 3.85 MPa, the diameter is 20 m, the yield stress is 590 MPa, and the factor of safety is 3.0.

(a) Determine the required thickness to the nearest millimeter.

(b) If the tank wall thickness is 85 mm, what is the maximum permissible internal pressure?

Representative Problems

8.2-5 A hemispherical window (or *viewport*) in a decompression chamber (see figure) is subjected to an internal air pressure of 575 kPa. The window is attached to the wall of the chamber by 14 bolts.

(a) Find the tensile force F in each bolt and the tensile stress σ in the viewport if the radius of the hemisphere is 350 mm, and its thickness is 32 mm.

(b) If the yield stress for each of the 14 bolts is 345 MPa, and the factor of safety is 3.0, find the required bolt diameter.

(c) If the stress in the viewport is limited to 3 MPa, find the required radius of the hemisphere.

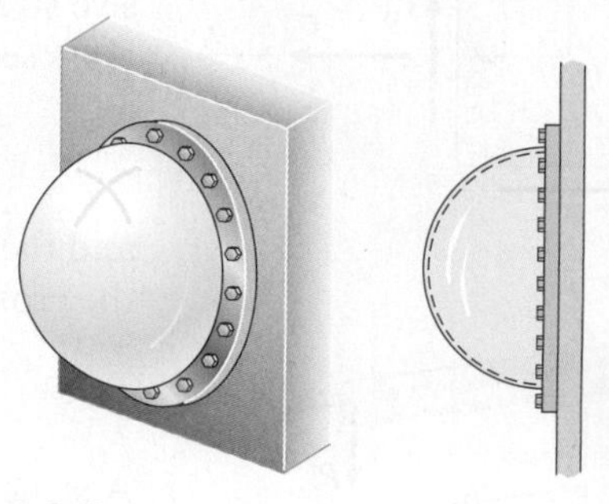

PROBLEM 8.2-5

8.2-6 A rubber ball (see figure) is inflated to a pressure of 65 kPa. At that pressure, the diameter of the ball is 240 mm and the wall thickness is 1.25 mm. The rubber has a modulus of elasticity $E = 3.7$ MPa and Poisson's ratio $\nu = 0.48$.

(a) Determine the maximum stress and strain in the ball.

(b) If the strain must be limited to 0.425, find the minimum required wall thickness of the ball.

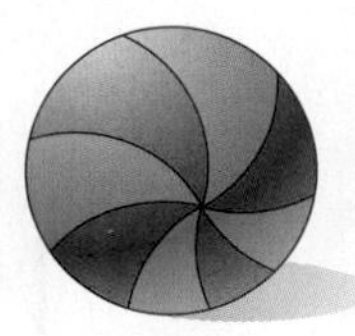

PROBLEMS 8.2-6 and 8.2-7

8.2-7 (a) Solve part (a) of the preceding problem if the pressure is 100 kPa, the diameter is 250 mm, the wall thickness is 1.5 mm, the modulus of elasticity is 3.5 MPa, and Poisson's ratio is 0.45.

(b) If the strain must be limited to 0.85, find the maximum acceptable inflation pressure.

8.2-8 A spherical steel pressure vessel (diameter 500 mm, thickness 10 mm) is coated with brittle lacquer that cracks when the strain reaches 150×10^{-6} (see figure).

(a) What internal pressure p will cause the lacquer to develop cracks? (Assume $E = 205$ GPa and $\nu = 0.30$.)

(b) If the strain is measured at 125×10^{-6}, what is the internal pressure at that point?

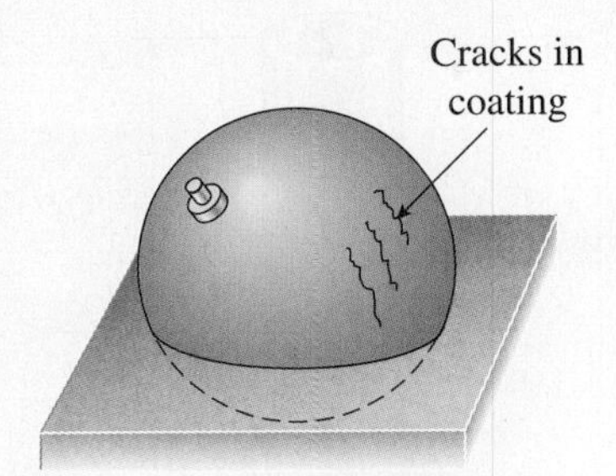

PROBLEM 8.2-8

8.2-9 A spherical tank of diameter 1.2 m and wall thickness 50 mm contains compressed air at a pressure of 17 MPa. The tank is constructed of two hemispheres joined by a welded seam (see figure).

(a) What is the tensile load f (N per mm of length of weld) carried by the weld?

(b) What is the maximum shear stress τ_{max} in the wall of the tank?

(c) What is the maximum normal strain ε in the wall? (For steel, assume $E = 210$ GPa and $\nu = 0.29$.)

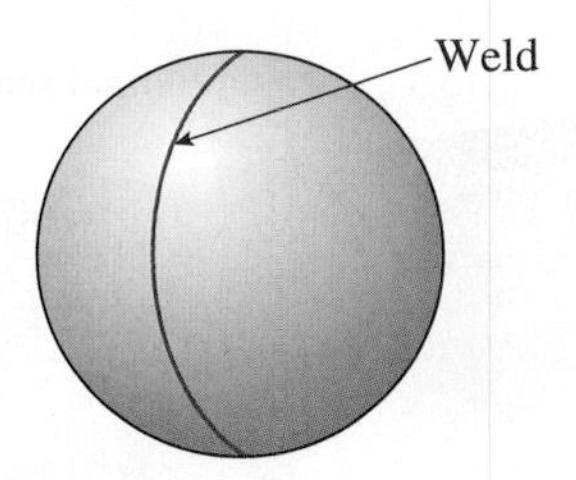

PROBLEMS 8.2-9 and 8.2-10

8.2-10 Solve the preceding problem for the following data: diameter 1.0 m, thickness 48 mm, pressure 22 MPa, modulus 210 GPa, and Poisson's ratio 0.29.

8.2-11 A spherical stainless-steel tank having a diameter of 500 mm is used to store propane gas at a pressure of 30 MPa. The properties of the steel are as follows: yield stress in tension, 950 MPa; yield stress in shear, 450 MPa; modulus of elasticity, 210 GPa; and Poisson's ratio, 0.28. The desired factor of safety with respect to yielding is 2.8. Also, the normal strain must not exceed 1250×10^{-6}.

(a) Determine the minimum permissible thickness t_{min} of the tank.

(b) If the tank thickness is 7 mm and normal strain is measured at 1000×10^{-6}, what is the internal pressure in the tank at that point?

8.2-12 Solve the preceding problem if the diameter is 480 mm, the pressure is 20 MPa, the yield stress in tension is 975 MPa, the yield stress in shear is 460 MPa, the factor of safety is 2.75, the modulus of elasticity is 210 GPa, Poisson's ratio is 0.28, and the normal strain must not exceed 1190×10^{-6}. For part (b), assume that the tank thickness is 8 mm and the measured normal strain is 990×10^{-6}.

8.2-13 A hollow, pressurized sphere having a radius $r = 150$ mm and wall thickness $t = 13$ mm is lowered into a lake (see figure). The compressed air in the tank is at a pressure of 140 kPa (gage pressure when the tank is out of the water).

At what depth D_0 will the wall of the tank be subjected to a compressive stress of 700 kPa?

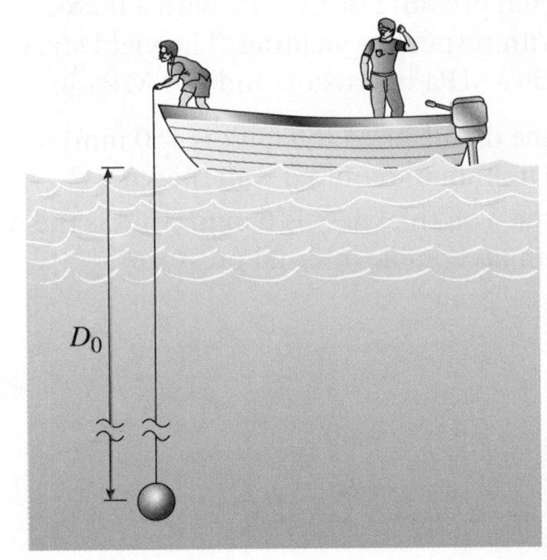

PROBLEM 8.2-13

8.3 Cylindrical Pressure Vessels

To solve the problems for Section 8.3, assume that the given radius or diameter is an inside dimension and that all internal pressures are gage pressures.

Introductory Problems

8.3-1 A fire extinguisher tank is designed for an internal pressure of 5.5 MPa. The tank has an outer diameter of 115 mm and thickness of 2 mm. Calculate the longitudinal stress, the circumferential stress, and the maximum shear stresses (out-of-plane and in-plane) at the outer surface of the tank.

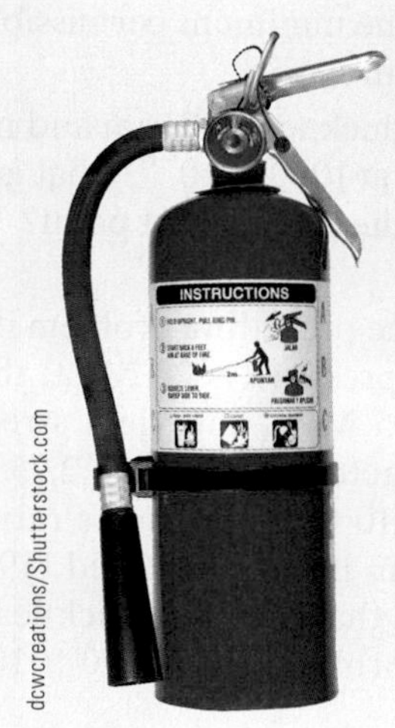

PROBLEMS 8.3-1 and 8.3-2

8.3-2 Repeat Problem 8.3-1 for a fire extinguisher tank with an internal pressure of 1.8 MPa, diameter of 130 mm, and thickness 1.5 mm.

8.3-3 A scuba tank (see figure) is being designed for an internal pressure of 12 MPa with a factor of safety of 2.0 with respect to yielding. The yield stress of the steel is 300 MPa in tension and 140 MPa in shear.

(a) If the diameter of the tank is 150 mm, what is the minimum required wall thickness?

(b) If the wall thickness is 6 mm, what is the maximum acceptable internal pressure?

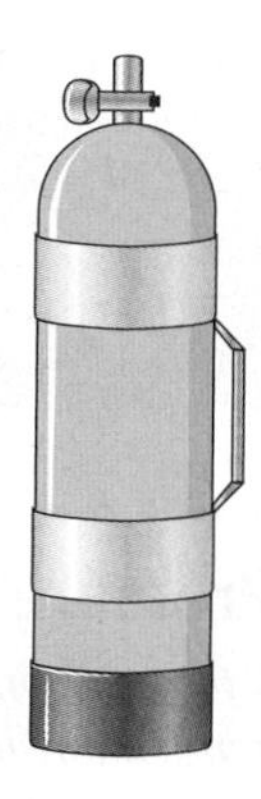

PROBLEM 8.3-3

8.3-4 A tall standpipe with an open top (see figure) has diameter $d = 2.2$ m and wall thickness $t = 20$ mm.

(a) What height h of water will produce a circumferential stress of 12 MPa in the wall of the standpipe?

(b) What is the axial stress in the wall of the tank due to the water pressure?

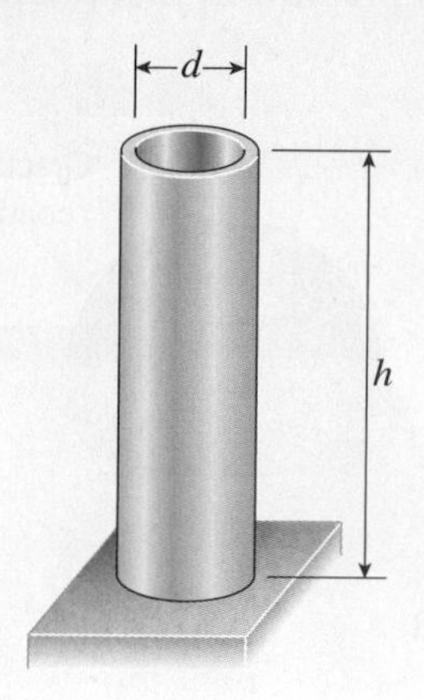

PROBLEM 8.3-4

Representative Problems

8.3-5 An inflatable structure used by a traveling circus has the shape of a half-circular cylinder with closed ends (see figure). The fabric and plastic structure is inflated by a small blower and has a radius of 12 m when fully inflated. A longitudinal seam runs the entire length of the "ridge" of the structure.

If the longitudinal seam along the ridge tears open when it is subjected to a tensile load of 100 N/mm of seam, what is the factor of safety n against tearing when the internal pressure is 3.5 kPa and the structure is fully inflated?

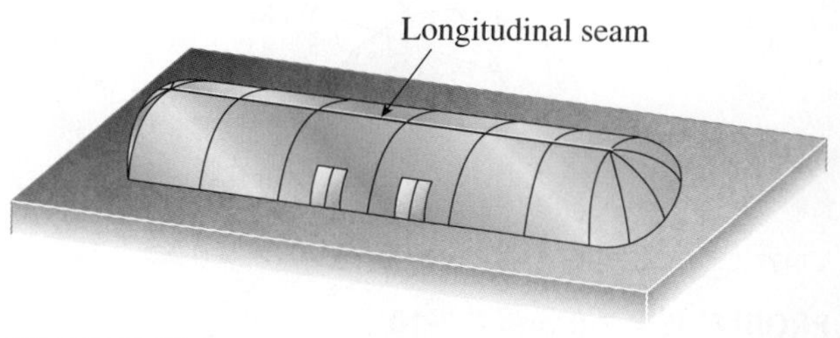

PROBLEM 8.3-5

8.3-6 A thin-walled cylindrical pressure vessel of a radius r is subjected simultaneously to internal gas pressure p and a compressive force F acting at the ends (see figure).

(a) What should be the magnitude of the force F in order to produce pure shear in the wall of the cylinder?

(b) If force $F = 190$ kN, internal pressure $p = 12$ MPa, inner diameter = 200 mm, and allowable normal and shear stresses are 110 MPa and 60 MPa, respectively, what is the required thickness of the vessel?

PROBLEM 8.3-6

8.3-7 A strain gage is installed in the longitudinal direction on the surface of an aluminum beverage can (see figure). The radius-to-thickness ratio of the can is 200. When the lid of the can is popped open, the strain changes by $\varepsilon_0 = 170 \times 10^{-6}$.

(a) What was the internal pressure p in the can? (Assume $E = 70$ GPa and $\nu = 0.33$.)

(b) What is the change in strain in the radial direction when the lid is opened?

PROBLEM 8.3-7

8.3-8 A circular cylindrical steel tank (see figure) contains a volatile fuel under pressure. A strain gage at point A records the longitudinal strain in the tank and transmits this information to a control room. The ultimate shear stress in the wall of the tank is 98 MPa, and a factor of safety of 2.8 is required.

(a) At what value of the strain should the operators take action to reduce the pressure in the tank? (Data for the steel are modulus of elasticity $E = 210$ GPa and Poisson's ratio $\nu = 0.30$.)

(b) What is the associated strain in the radial direction?

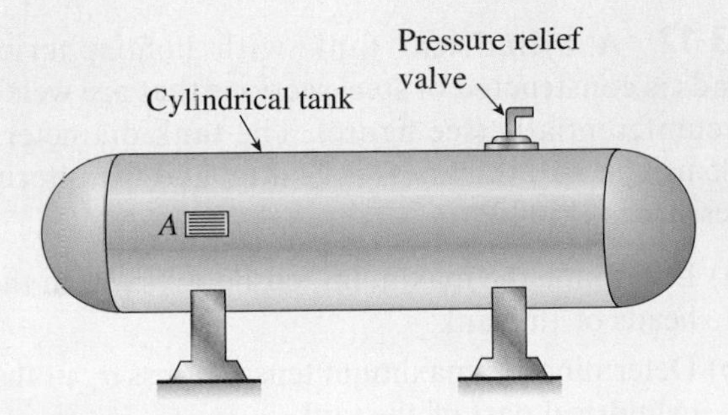

PROBLEM 8.3-8

8.3-9 A cylinder filled with oil is under pressure from a piston, as shown in the figure. The diameter d of the piston is 48 mm and the compressive force F is 16 kN. The maximum allowable shear stress τ_{allow} in the wall of the cylinder is 42 MPa.

What is the minimum permissible thickness t_{min} of the cylinder wall? (See figure.)

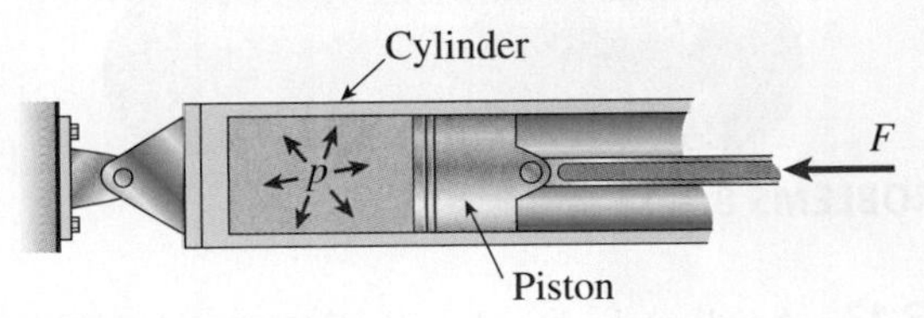

PROBLEMS 8.3-9 and 8.3-10

8.3-10 Solve the preceding problem if $d = 90$ mm, $F = 42$ kN, and $\tau_{allow} = 40$ MPa.

8.3-11 A standpipe in a water-supply system (see figure) is 3.8 m in diameter and 150 mm thick. Two horizontal pipes carry water out of the standpipe; each is 0.6 m in diameter and 25 mm thick. When the system is shut down and water fills the pipes but is not moving, the hoop stress at the bottom of the standpipe is 900 kPa.

(a) What is the height h of the water in the standpipe?

(b) If the bottoms of the pipes are at the same elevation as the bottom of the standpipe, what is the hoop stress in the pipes?

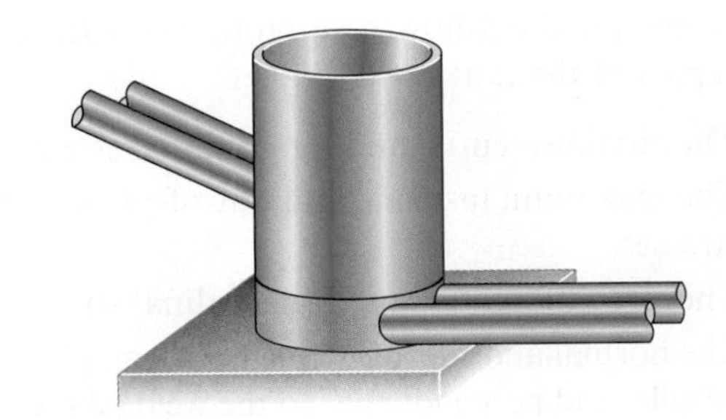

PROBLEM 8.3-11

8.3-12 A cylindrical tank with hemispherical heads is constructed of steel sections that are welded circumferentially (see figure). The tank diameter is 1.25 m, the wall thickness is 22 mm, and the internal pressure is 1750 kPa.

(a) Determine the maximum tensile stress σ_h in the heads of the tank.

(b) Determine the maximum tensile stress σ_c in the cylindrical part of the tank.

(c) Determine the tensile stress σ_w acting perpendicular to the welded joints.

(d) Determine the maximum shear stress τ_h in the heads of the tank.

(e) Determine the maximum shear stress τ_c in the cylindrical part of the tank.

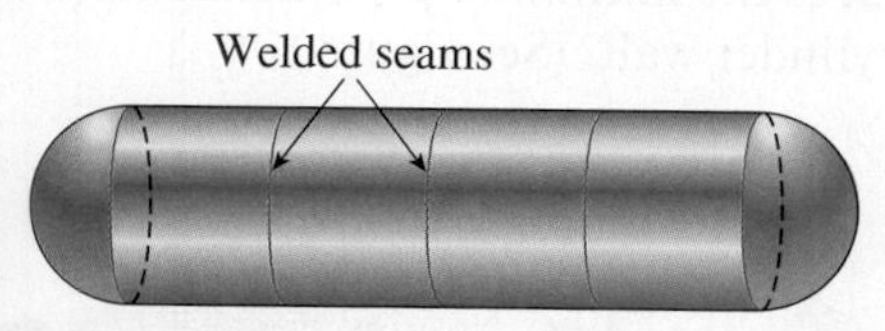

PROBLEMS 8.3-12 and 8.3-13

8.3-13 A cylindrical tank with diameter $d = 300$ mm is subjected to internal gas pressure $p = 2$ MPa. The tank is constructed of steel sections that are welded circumferentially (see figure). The heads of the tank are hemispherical. The allowable tensile and shear stresses are 60 MPa and 24 MPa, respectively. Also, the allowable tensile stress perpendicular to a weld is 40 MPa.

Determine the minimum required thickness t_{min} of (a) the cylindrical part of the tank and (b) the hemispherical heads.

8.3-14 A pressurized steel tank is constructed with a helical weld that makes an angle $\alpha = 55°$ with the longitudinal axis (see figure). The tank has radius $r = 0.6$ m, wall thickness $t = 18$ mm, and internal pressure $p = 2.8$ MPa. Also, the steel has modulus of elasticity $E = 200$ GPa and Poisson's ratio $\nu = 0.30$.

Determine the following quantities for the cylindrical part of the tank.

(a) The circumferential and longitudinal stresses.

(b) The maximum in-plane and out-of-plane shear stresses.

(c) The circumferential and longitudinal strains.

(d) The normal and shear stresses acting on planes parallel and perpendicular to the weld (show these stresses on a properly oriented stress element).

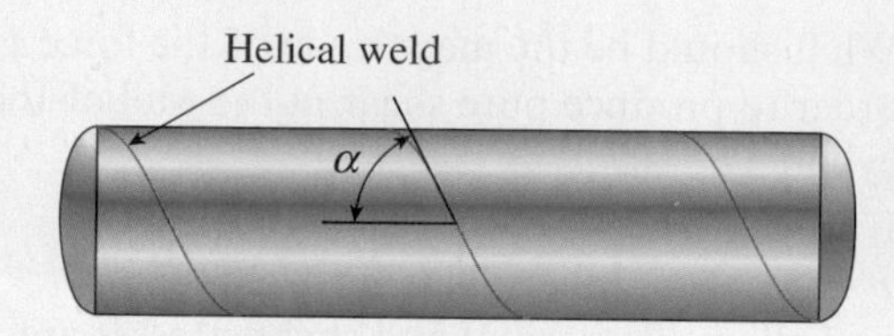

PROBLEMS 8.3-14 and 8.3-15

8.3-15 Solve the preceding problem for a welded tank with $\alpha = 75°$, $r = 450$ mm, $t = 15$ mm, $p = 1.4$ MPa, $E = 200$ GPa, and $\nu = 0.30$.

8.4 Maximum Stresses in Beams

To solve the problems for Section 8.4, consider only the in-plane stresses and disregard the weights of the beams.

Introductory Problems

8.4-1 A wood beam with a cross section 100 mm × 150 mm is simply supported at A and B. The beam has a length of 2.7 m and is subjected to point load $P = 22.25$ kN at mid-span. Calculate the state of stress at point C located 100 mm below the top of the beam and 1.35 m to the right of support A. Neglect the weight of the beam.

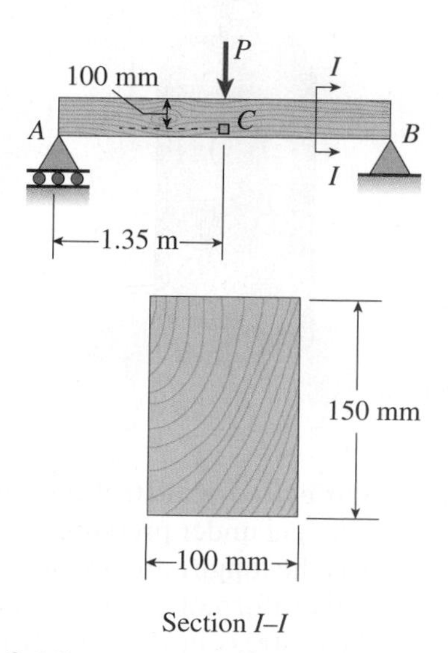

PROBLEM 8.4-1

8.4-2 Solve the preceding problem using the following data: beam cross section is 100 × 150 mm, length is 3 m, and point load is $P = 5$ kN at mid-span. Point C is located 25 mm below the top of the beam and 0.5 m to the right of support A.

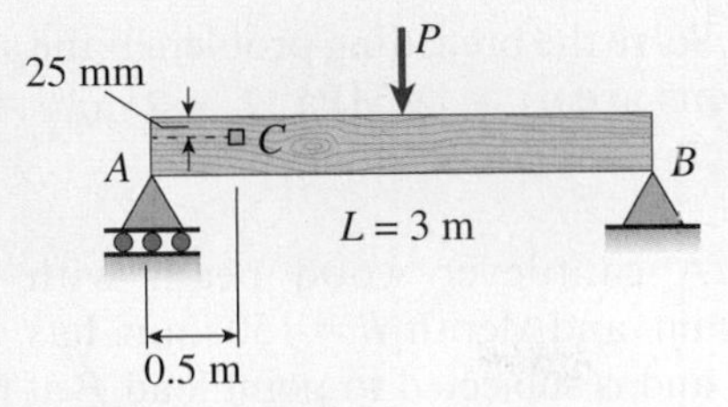

PROBLEM 8.4-2

8.4-3 A simply supported beam is subjected to two point loads, each $P = 2.25$ kN, as shown in the figure. The beam has a cross section of 150 mm × 300 mm. Find the state of plane stress at point C located 200 mm below the top of the beam and 160 mm to the right of support A. Also find the principal stresses and the maximum shear stress at C.

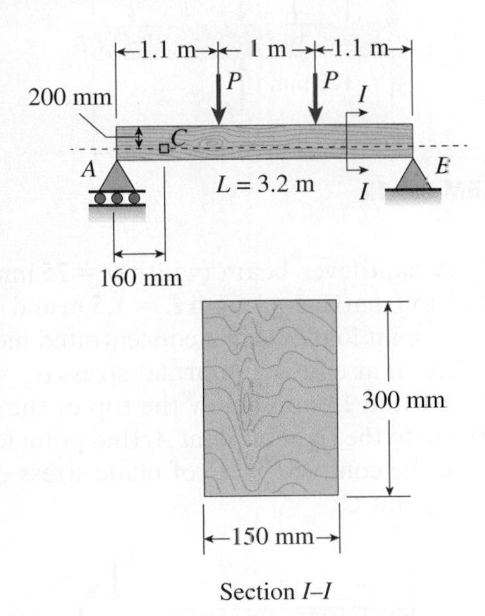

PROBLEM 8.4-3

8.4-4 A cantilever beam with a width $b = 100$ mm and depth $h = 150$ mm has a length $L = 2$ m and is subjected to a point load $P = 500$ N at B. Calculate the state of plane stress at point C located 50 mm below the top of the beam and 0.5 m to the right of point A. Also find the principal stresses and the maximum shear stress at C. Neglect the weight of the beam.

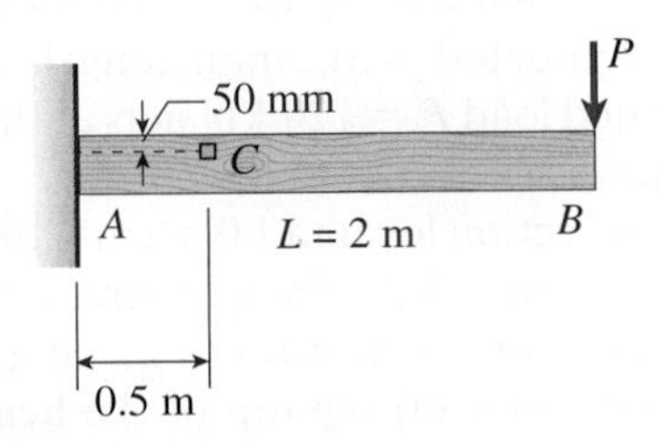

PROBLEM 8.4-4

8.4-5 A beam with a width $b = 150$ mm and depth $h = 200$ mm is simply supported at A and B. The beam has a length $L = 3$ m and is subjected to a linearly varying distributed load with peak intensity $q_0 = 22$ kN/m. Calculate the state of plane stress at point C located 75 mm below the top of the beam and 30 mm to the left of point B. Also find the principal stresses on element C. Neglect the weight of the beam.

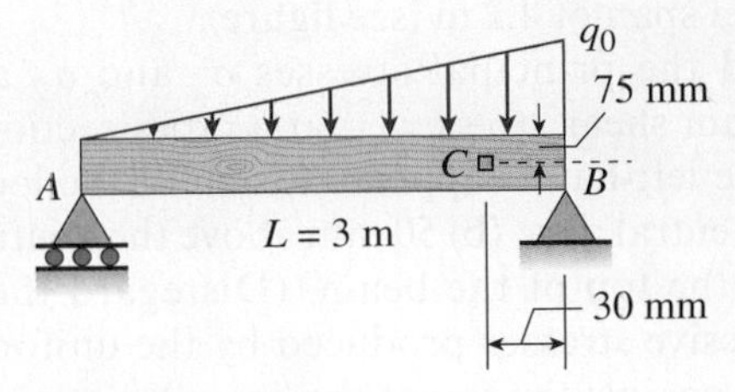

PROBLEM 8.4-5

8.4-6 Beam ABC with an overhang BC is subjected to a linearly varying distributed load on span AB with peak intensity $q_0 = 2500$ N/m and a point load $P = 1250$ N applied at C. The beam has a width $b = 100$ mm and depth $h = 200$ mm. Find the state of plane stress at point D located 150 mm below the top of the beam and 0.2 m to the left of point B. Also find the principal stresses at D. Neglect the weight of the beam.

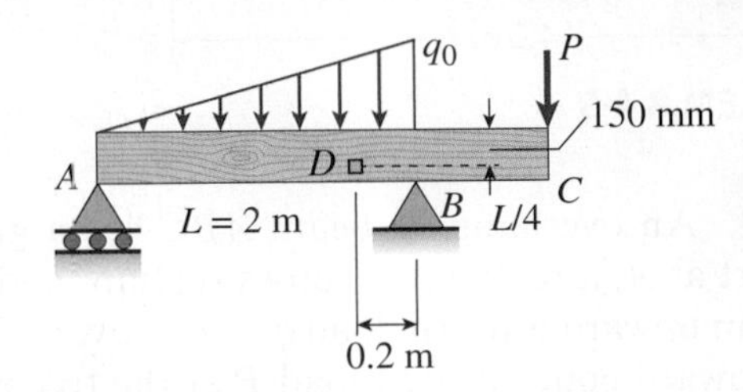

PROBLEM 8.4-6

8.4-7 A cantilever beam ($L = 2$ m) with a rectangular cross section ($b = 90$ mm, $h = 300$ mm) supports an upward load P = 160 kN at its free end.

(a) Find the state of stress (σ_x, σ_y, and τ_{xy} in MPa) on a plane-stress element at $L/2$ that is $d = 200$ mm up from the bottom of the beam. Find the principal normal stresses and maximum shear stress. Show these stresses on sketches of properly oriented elements.

(b) Repeat part (a) if an axial compressive centroidal load $N = 180$ kN is added at B.

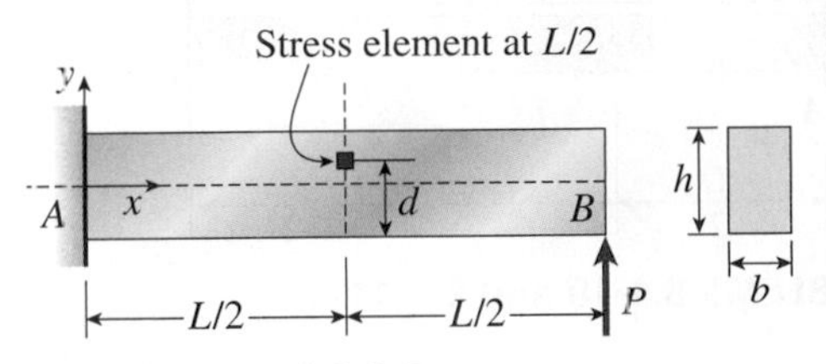

PROBLEMS 8.4-7 and 8.4-8

8.4-8 Solve the preceding problem for the following data: $P = 160$ kN, $N = 200$ kN, $L = 2$ m, $b = 95$ mm, $h = 300$ mm, and $d = 200$ mm.

Representative Problems

8.4-9 A simple beam with a rectangular cross section (width, 90 mm; height, 300 mm) carries a trapezoidally distributed load of 20 kN/m at A and 15 kN/m at B on a span of 4.2 m (see figure).

Find the principal stresses σ_1 and σ_2 and the maximum shear stress τ_{max} at a cross section 0.6 m from the left-hand support at each of the locations: (a) the neutral axis, (b) 50 mm above the neutral axis, and (c) the top of the beam. (Disregard the direct compressive stresses produced by the uniform load bearing against the top of the beam.)

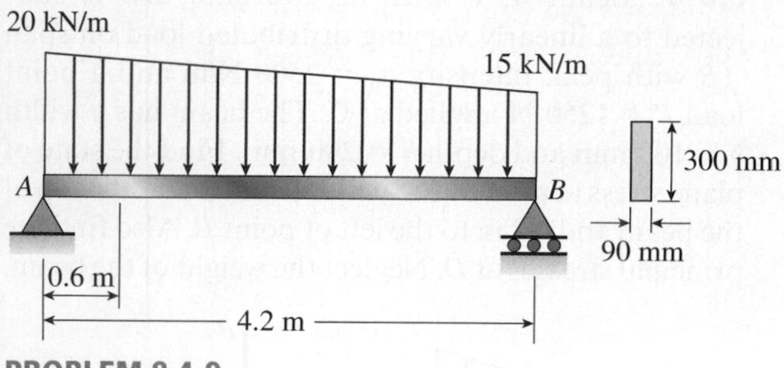

PROBLEM 8.4-9

8.4-10 An overhanging beam ABC has a guided support at A, a rectangular cross section, and supports an upward uniform load $q = P/L$ over AB and a downward concentrated load P at the free end C (see figure). The span length from A to B is L, and the length of the overhang is $L/2$. The cross section has a width of b and a height h. Point D is located midway between the supports at a distance d from the top face of the beam.

Knowing that the maximum tensile stress (principal stress) at point D is $\sigma_1 = 38$ MPa, determine the magnitude of the load P. Data for the beam are $L = 1.75$ m, $b = 50$ mm, $h = 220$ mm, and $d = 55$ mm.

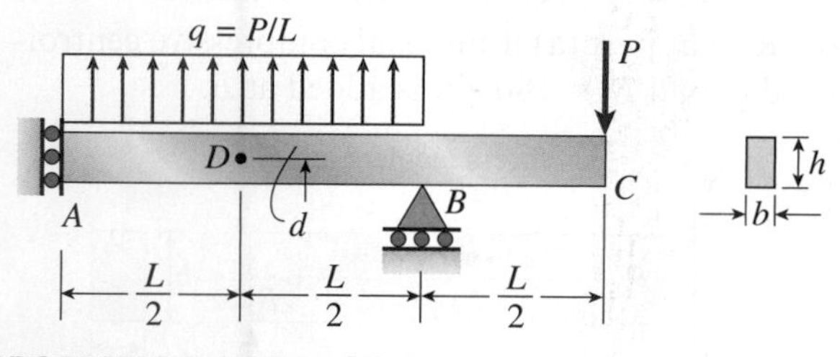

PROBLEMS 8.4-10 and 8.4-11

8.4-11 Solve the preceding problem if the stress and dimensions are $\sigma_1 = 18$ MPa, $L = 2$ m, $b = 60$ mm, $h = 250$ mm, and $d = 65$ mm.

8.4-12 A cantilever wood beam with a width $b = 100$ mm and depth $h = 150$ mm has a length $L = 2$ m and is subjected to point load P at mid-span and uniform load $q = 15$ N/m. (a) If the normal stress $\sigma_x = 0$ at point C, located 120 mm below the top of the beam at the fixed support A, calculate the point load P. Also show the complete state of plane stress on the element at point C. (b) Repeat Part a if $\sigma_x = 220$ kPa. Assume that element C is a sufficient distance from support A so that stress concentration effects are negligible.

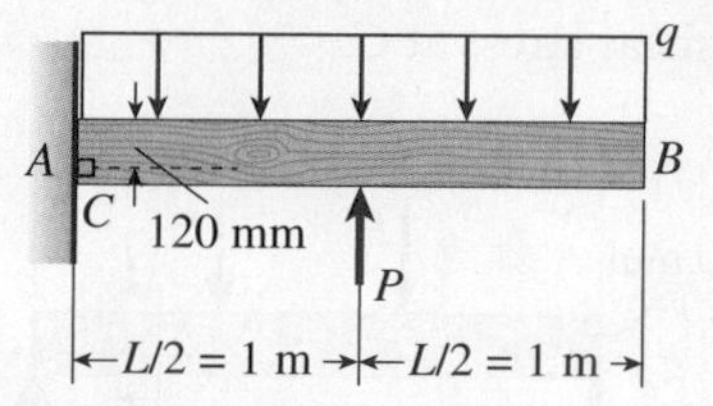

PROBLEM 8.4-12

8.4-13 A cantilever beam (width $b = 75$ mm and depth $h = 150$ mm) has a length $L = 1.5$ m and is subjected to a point load P and a concentrated moment $M = 27$ kN·m at end B. If normal stress $\sigma_x = 0$ at point C, located 13 mm below the top of the beam and 300 mm to the right of point A, find point load P. Also show the complete state of plane stress on the element at point C.

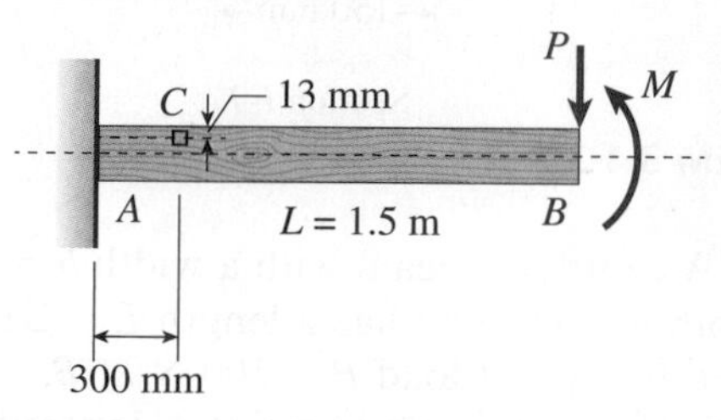

PROBLEM 8.4-13

8.4-14 A beam with a wide-flange cross section (see figure) has the following dimensions: $b = 120$ mm, $t = 10$ mm, $h = 300$ mm, and $h_1 = 260$ mm. The beam is simply supported with span length $L = 3.0$ m. A concentrated load $P = 120$ kN acts at the midpoint of the span.

At a cross section located 1.0 m from the left-hand support, determine the principal stresses σ_1 and σ_2 and the maximum shear stress τ_{max} at each of the following locations: (a) the top of the beam, (b) the top of the web, and (c) the neutral axis.

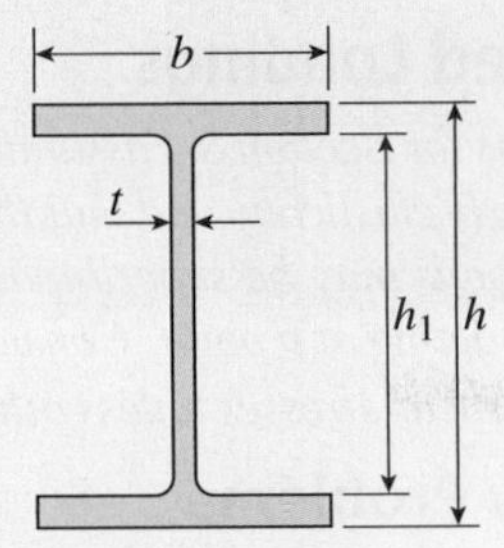

PROBLEMS 8.4-14 and 8.4-15

8.4-15 A beam with a wide-flange cross section (see figure) has the following dimensions: $b = 120$ mm, $t = 10$ mm, $h = 300$ mm, and $h_1 = 260$ mm. The beam is simply supported with span length $L = 3$ m and supports a uniform load $q = 80$ kN/m.

Calculate the principal stresses σ_1 and σ_2 and the maximum shear stress τ_{max} at a cross section located 0.6 m from the left-hand support at each of the following locations: (a) the bottom of the beam, (b) the bottom of the web, and (c) the neutral axis.

8.4-16 An IPN 240 standard beam (see Table F-2, Appendix F) is simply supported with a span length of 2.5 m (see figure). The beam supports a concentrated load of 100 kN at 0.9 m from support B.

At a cross section located 0.7 m from the left-hand support, determine the principal stresses σ_1 and σ_2 and the maximum shear stress τ_{max} at each of the following locations: (a) the top of the beam, (b) the top of the web, and (c) the neutral axis.

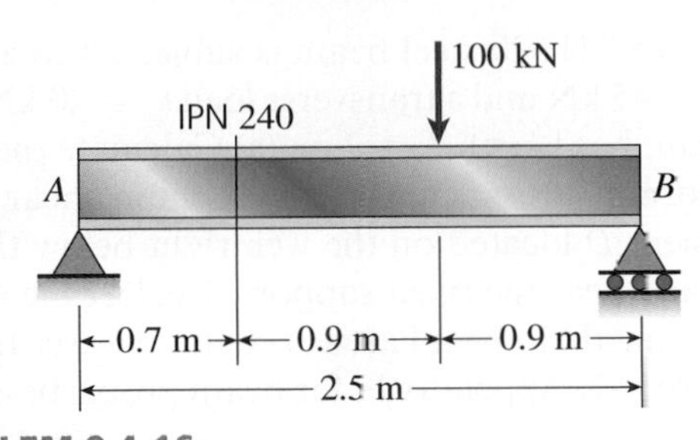

PROBLEM 8.4-16

8.4-17 An IPN 300 steel beam is fixed at A. The beam has length $L = 2$ m and is subjected to a linearly varying distributed load with peak intensity $q_0 = 12$ kN/m. Calculate the state of plane stress at point C located 75 mm below the top of the beam at mid-span. Also find the principal normal stresses and the maximum shear stress at C. Include the weight of the beam. See Table F-2, Appendix F, for beam properties.

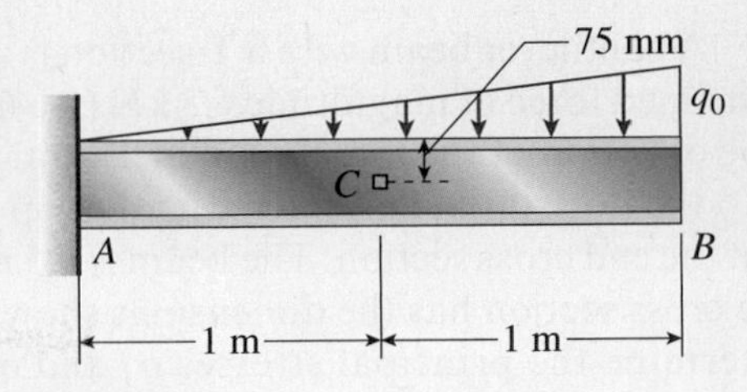

PROBLEM 8.4-17

8.4-18 An IPN 360 steel beam is fixed at A. The beam has a length of 2.5 m and is subjected to a linearly varying distributed load with maximum intensity $q_0 = 500$ N/m on segment AB and a uniformly distributed load of intensity q_0 on segment BC. Calculate the state of plane stress at point D located 220 mm below the top of the beam and 0.3 m to the left of point B. Find the principal normal stresses and the maximum shear stress at D. Include the weight of the beam. See Table F-2, Appendix F, for beam properties.

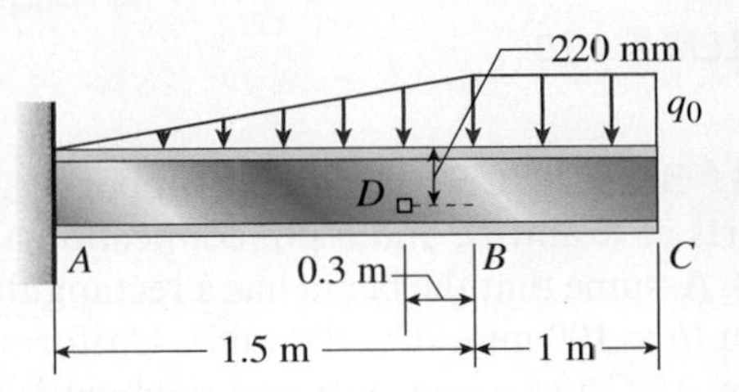

PROBLEM 8.4-18

8.4-19 An IPN 220 standard beam (see Table F-2, Appendix F) is simply supported with a span length of 3 m (see figure). The beam supports two anti-symmetrically placed concentrated loads of 20 kN each.

At a cross section located 0.5 m from the right-hand support, determine the principal stresses σ_1 and σ_2 and the maximum shear stress τ_{max} at each of the following locations: (a) the top of the beam, (b) the top of the web, and (c) the neutral axis.

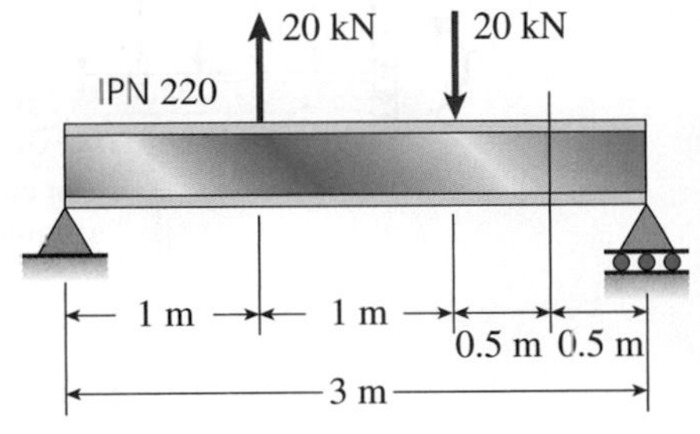

PROBLEM 8.4-19

8.4-20 A cantilever beam with a T-section is loaded by an inclined force of magnitude 6.5 kN (see figure). The line of action of the force is inclined at an angle of 60° to the horizontal and intersects the top of the beam at the end cross section. The beam is 2.5 m long and the cross section has the dimensions shown.

Determine the principal stresses σ_1 and σ_2 and the maximum shear stress τ_{max} at points A and B in the web of the beam near the support.

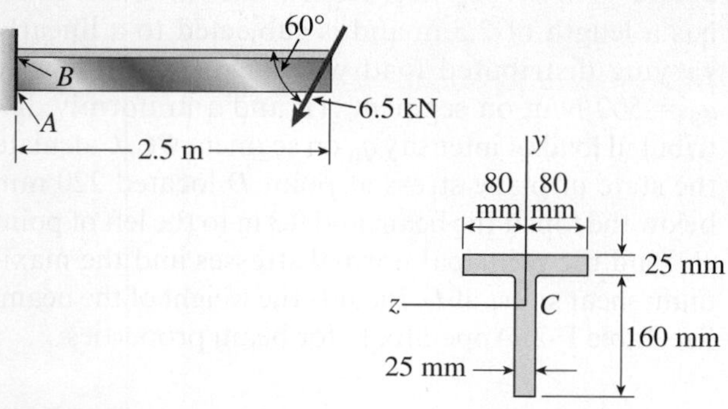

PROBLEM 8.4-20

8.4-21 Beam $ABCD$ has a sliding support at A, roller supports at C and D, and a pin connection at B (see figure). Assume that the beam has a rectangular cross section ($b = 100$ mm, $h = 400$ mm). Uniform load q acts on ABC and a concentrated moment is applied at D. Let load variable $q = 25$ kN/m, and assume that dimension variable $L = 1.25$ m. First, use statics to confirm the reaction moment at A and the reaction forces at C and D, as given in the figure. Then find the *ratio* of the magnitudes of the principal stresses (σ_1/σ_2) just left of support C at a distance $d = 250$ mm up from the bottom.

8.4-22 Solve the preceding problem using the numerical data: $b = 90$ mm, $h = 280$ mm, $d = 210$ mm, $q = 14$ kN/m, and $L = 1.2$ m.

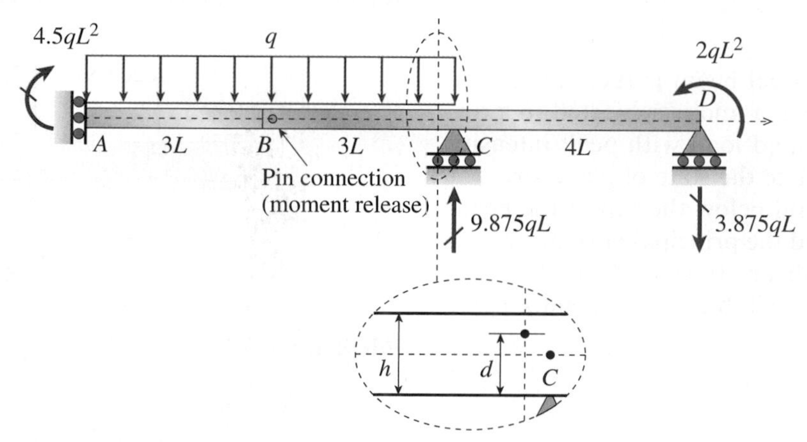

PROBLEMS 8.4-21 and 8.4-22

8.5 Combined Loadings

Solve the problems for Section 8.5 assuming that the structures behave linearly elastically and that the stresses caused by two or more loads may be superimposed to obtain the resultant stresses acting at a point. Consider both in-plane and out-of-plane shear stresses unless otherwise specified.

Introductory Problems

8.5-1 An IPN 300 steel cantilever beam is subjected to an axial load $P = 45$ kN and a transverse load $V = 65$ kN. The beam has length $L = 1.8$ m. (a) Calculate the principal normal stresses and the maximum shear stress for an element located at C near the fixed support. Neglect the weight of the beam. (b) Repeat Part a for point D which is 100 mm above point C (see figure). See Table F-2, Appendix F, for beam properties.

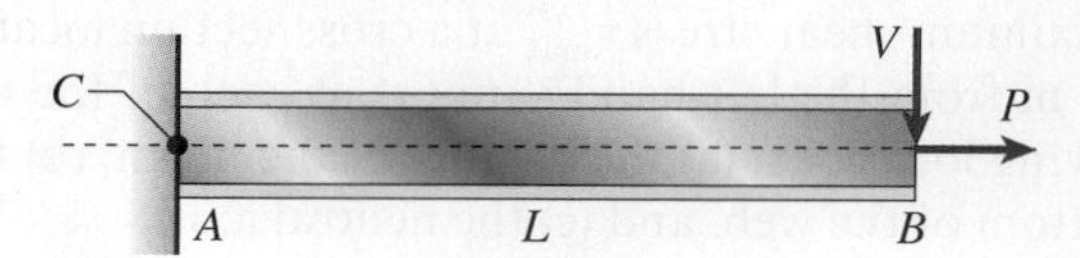

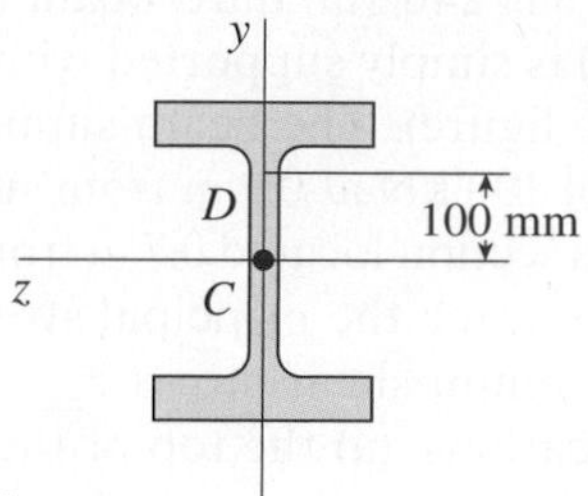

PROBLEM 8.5-1

8.5-2 An IPN 320 steel beam is subjected to a point load $P = 45$ kN and a transverse load $V = 20$ kN at B. The beam has length $L = 2$ m. (a) Calculate the principal normal stresses and the maximum shear stress on element D located on the web right below the top flange and near the fixed support. Neglect the weight of the beam. (b) Repeat Part a at centroid C (see figure). See Table F-2, Appendix F, for beam properties.

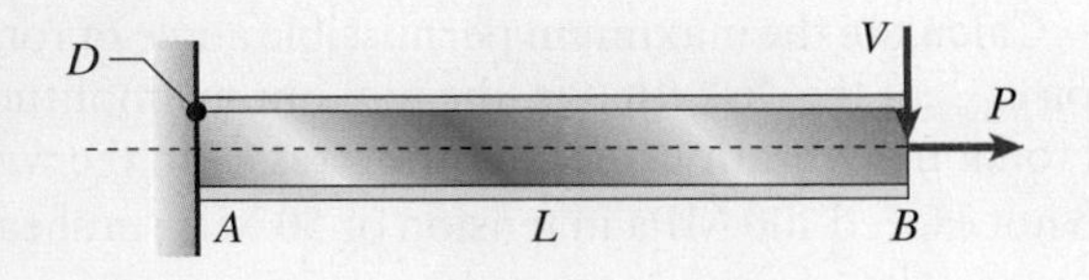

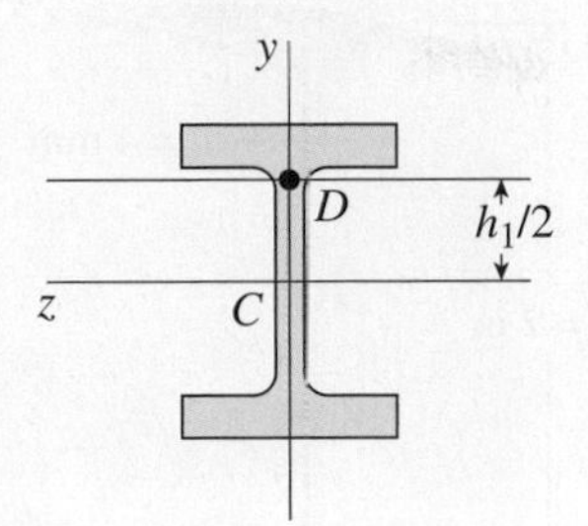

PROBLEM 8.5-2

8.5-3 A solid circular bar is fixed at point A. The bar is subjected to transverse load $V = 300$ N and torque $T = 35$ N·m at point B. The bar has a length $L = 1.5$ m and diameter $d = 75$ mm. Calculate the principal normal stresses and the maximum shear stress at element 1 located on the bottom surface of the bar at fixed end A (see figure).

Assume that element 1 is a sufficient distance from support A so that stress concentration effects are negligible.

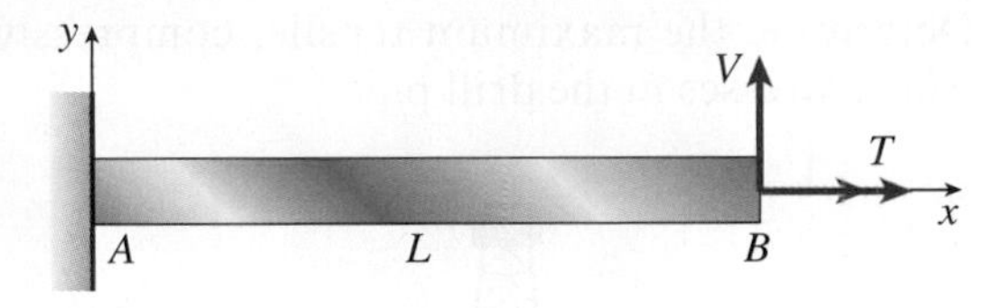

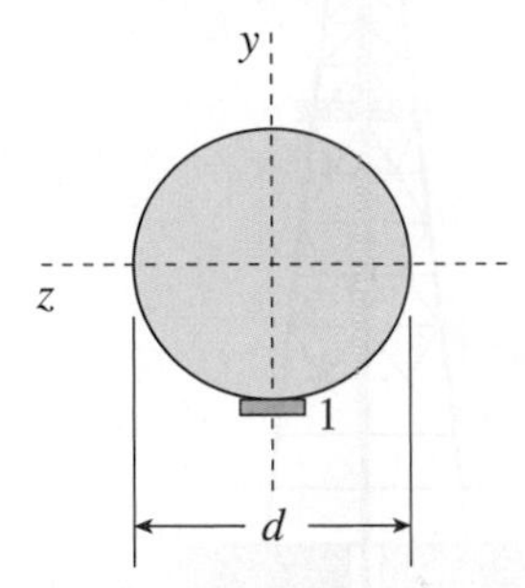

PROBLEM 8.5-3

8.5-4 Solve the preceding problem using transverse load $V = 300$ N and torque $T = 3.5$ N·m applied at point B. The bar has length $L = 1.5$ m and diameter $d = 8$ mm. Calculate the principal stresses and the maximum shear stress for element 2 located at the side of the bar at fixed end A (see figure).

Assume that element 2 is a sufficient distance from support A so that stress concentration effects are negligible.

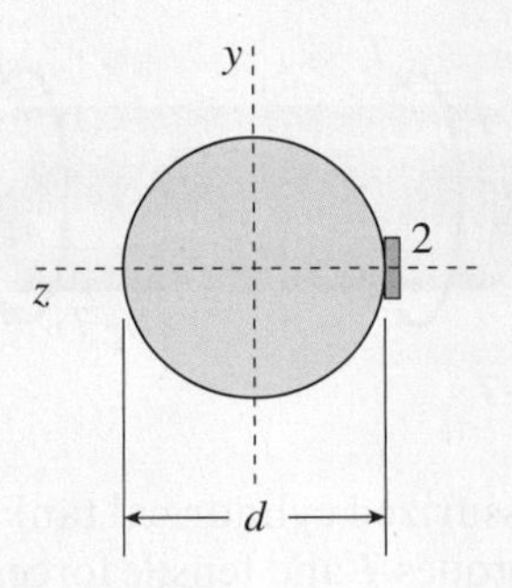

PROBLEM 8.5-4

8.5-5 A cylindrical tank having a diameter $d = 60$ mm is subjected to internal gas pressure $p = 4$ MPa and an external tensile load $T = 4.5$ kN (see figure).

Determine the minimum thickness t of the wall of the tank based upon an allowable shear stress of 20 MPa.

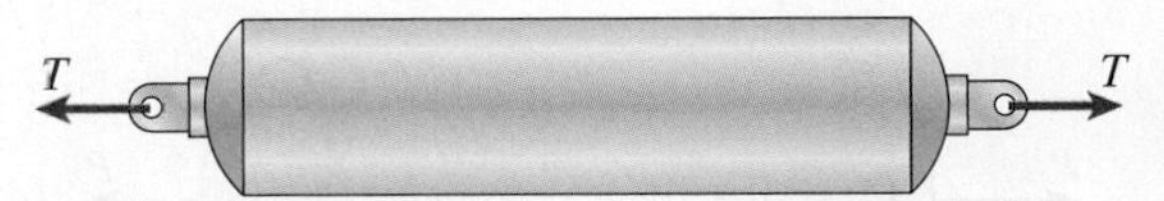

PROBLEM 8.5-5

8.5-6 A cylindrical tank subjected to internal pressure p is simultaneously compressed by an axial force $F = 72$ kN (see figure). The cylinder has diameter $d = 100$ mm and wall thickness $t = 4$ mm.

Calculate the maximum allowable internal pressure p_{max} based upon an allowable shear stress in the wall of the tank of 60 MPa.

PROBLEM 8.5-6

8.5-7 A cylindrical pressure vessel having a radius $r = 300$ mm and wall thickness $t = 15$ mm is subjected to internal pressure $p = 2.5$ MPa. In addition, a torque $T = 120$ kN·m acts at each end of the cylinder (see figure).

(a) Determine the maximum tensile stress σ_{max} and the maximum in-plane shear stress τ_{max} in the wall of the cylinder.

(b) If the allowable in-plane shear stress is 30 MPa, what is the maximum allowable torque T?

(c) If $T = 200$ kN·m and allowable in-plane shear and allowable normal stresses are 30 MPa and 76 MPa, respectively, what is the minimum required wall thickness?

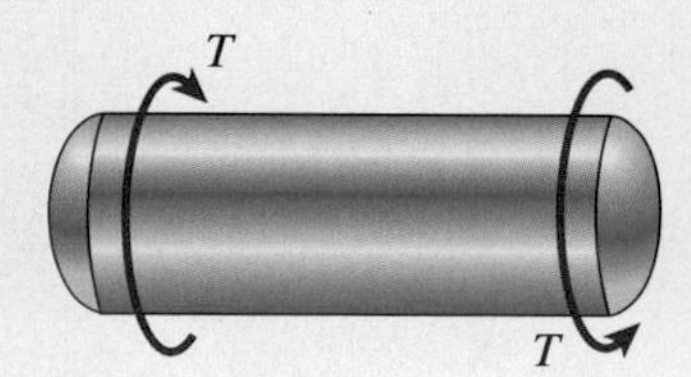

PROBLEM 8.5-7

8.5-8 A pressurized cylindrical tank with flat ends is loaded by torques T and tensile forces P (see figure). The tank has a radius of $r = 125$ mm and wall thickness $t = 6.5$ mm. The internal pressure $p = 7.25$ MPa and the torque $T = 850$ N·m.

(a) What is the maximum permissible value of the forces P if the allowable tensile stress in the wall of the cylinder is 160 MPa?
(b) If forces $P = 400$ kN, what is the maximum acceptable internal pressure in the tank?

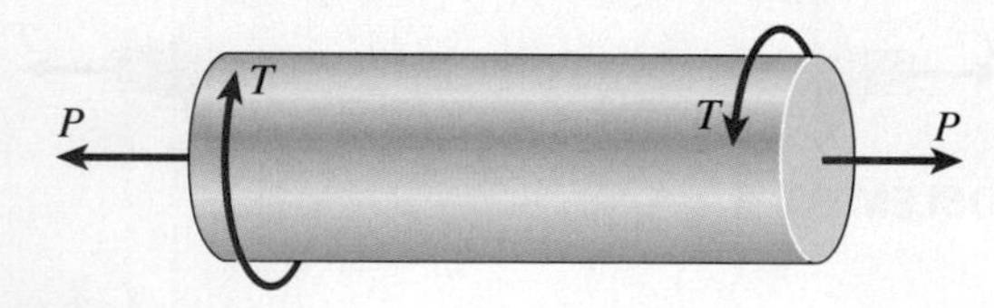

PROBLEM 8.5-8

8.5-9 A cylindrical pressure vessel with flat ends is subjected to a torque T and a bending moment M (see figure). The outer radius is 300 mm and the wall thickness is 25 mm. The loads are $T = 90$ kN·m, $M = 100$ kN·m, and the internal pressure $p = 6.25$ MPa.

Determine the maximum tensile stress σ_t, maximum compressive stress σ_c, and maximum shear stress τ_{max} in the wall of the cylinder.

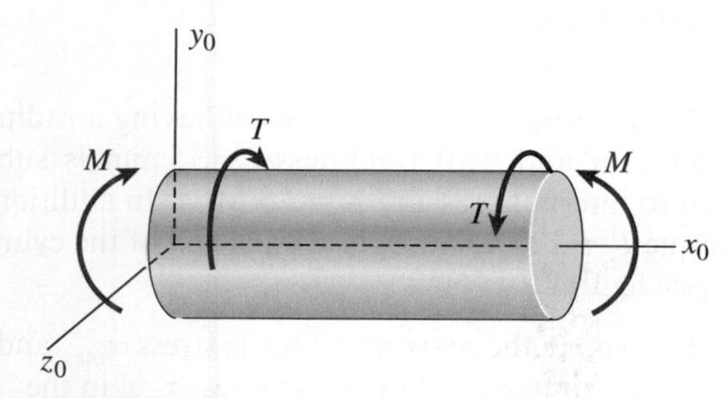

PROBLEM 8.5-9

8.5-10 The torsional pendulum shown in the figure consists of a horizontal circular disk of a mass $M = 60$ kg suspended by a vertical steel wire ($G = 80$ GPa) of a length $L = 2$ m and diameter $d = 4$ mm.

Calculate the maximum permissible angle of rotation ϕ_{max} of the disk (that is, the maximum amplitude of torsional vibrations) so that the stresses in the wire do not exceed 100 MPa in tension or 50 MPa in shear.

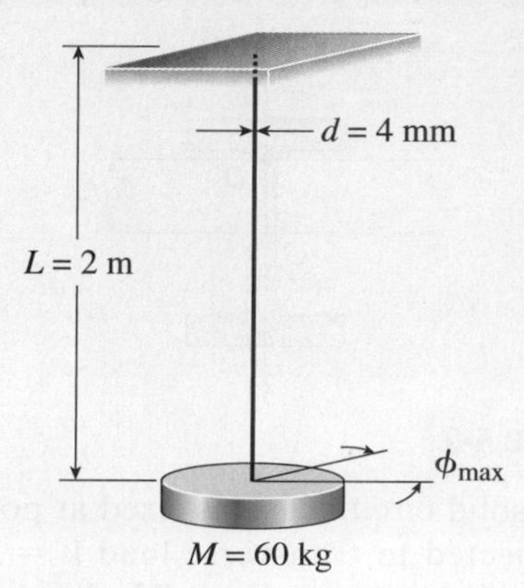

PROBLEM 8.5-10

Representative Problems

8.5-11 The hollow drill pipe for an oil well (see figure) is 150 mm in outer diameter and 15 mm in thickness. Just above the bit, the compressive force in the pipe (due to the weight of the pipe) is 265 kN and the torque (due to drilling) is 19 kN·m.

Determine the maximum tensile, compressive, and shear stresses in the drill pipe.

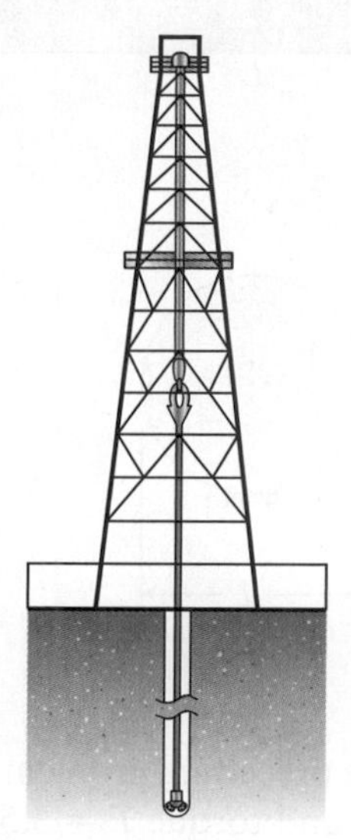

PROBLEM 8.5-11

8.5-12 A segment of a generator shaft is subjected to a torque T and an axial force P, as shown in the figure. The shaft is hollow (outer diameter $d_2 = 300$ mm and inner diameter $d_1 = 250$ mm) and delivers 1800 kW at 4.0 Hz.

If the compressive force $P = 540$ kN, what are the maximum tensile, compressive, and shear stresses in the shaft?

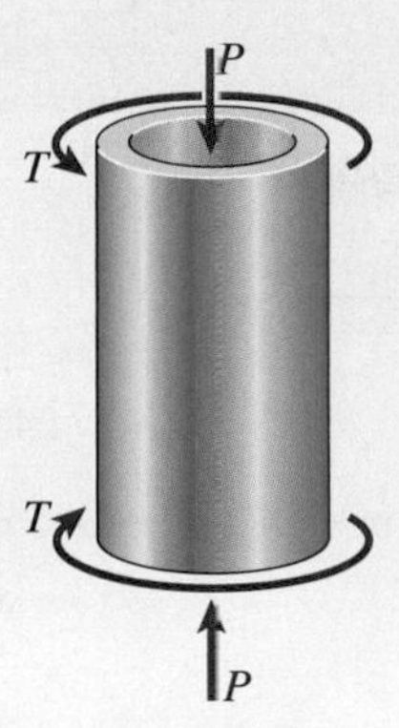

PROBLEMS 8.5-12 and 8.5-13

8.5-13 A segment of a generator shaft with a hollow circular cross section is subjected to a torque $T = 25$ kN·m (see figure). The outer and inner diameters of the shaft are 200 mm and 160 mm, respectively.

What is the maximum permissible compressive load P that can be applied to the shaft if the allowable in-plane shear stress is $\tau_{allow} = 45$ MPa?

8.5-14 A post having a hollow, circular cross section supports a $P = 3.2$ kN load acting at the end of an arm that is $b = 1.5$ m long (see figure). The height of the post is $L = 9$ m, and its section modulus is $S = 2.65 \times 10^5$ mm^3. Assume that the outer radius of the post is $r_2 = 123$ mm, and the inner radius is $r_1 = 117$ mm.

(a) Calculate the maximum tensile stress σ_{max} and maximum in-plane shear stress τ_{max} at point A on the outer surface of the post along the x axis due to the load P. Load P acts at B along line BC.

(b) If the maximum tensile stress and maximum in-plane shear stress at point A are limited to 90 MPa and 38 MPa, respectively, what is the largest permissible value of the load P?

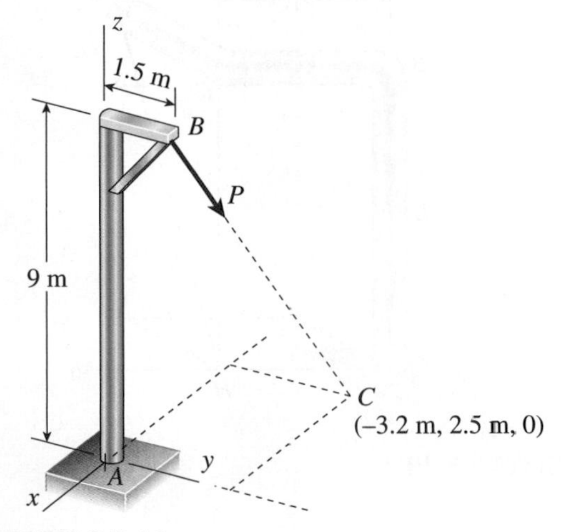

PROBLEM 8.5-14

8.5-15 A sign is supported by a pole of hollow circular cross section, as shown in the figure. The outer and inner diameters of the pole are 250 mm and 200 mm, respectively. The pole is 12 m high and weighs 18 kN. The sign has dimensions 2 m × 1 m and weighs 2.2 kN. Note that its center of gravity is 1.125 m from the axis of the pole. The wind pressure against the sign is 1.5 kPa.

(a) Determine the stresses acting on a stress element at point A, which is on the outer surface of the pole at the "front" of the pole, that is, the part of the pole nearest to the viewer.

(b) Determine the maximum tensile, compressive, and shear stresses at point A.

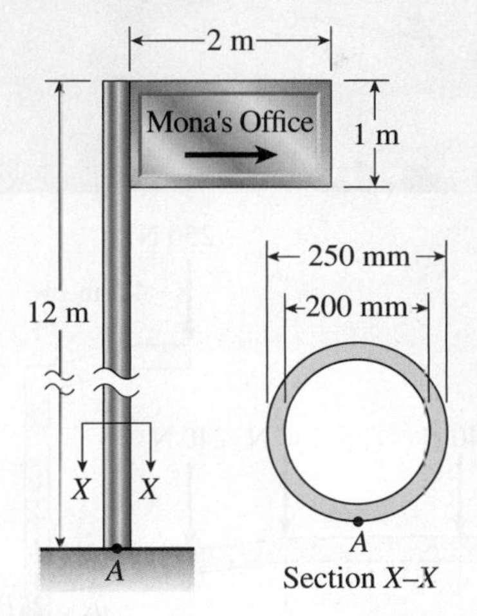

PROBLEM 8.5-15

8.5-16 A sign is supported by a pipe (see figure) having an outer diameter 110 mm and inner diameter 90 mm. The dimensions of the sign are 2.0 m × 1.0 m, and its lower edge is 3.0 m above the base. Note that the center of gravity of the sign is 1.05 m from the axis of the pipe. The wind pressure against the sign is 1.5 kPa.

Determine the maximum in-plane shear stresses due to the wind pressure on the sign at points A, B, and C, located on the outer surface at the base of the pipe.

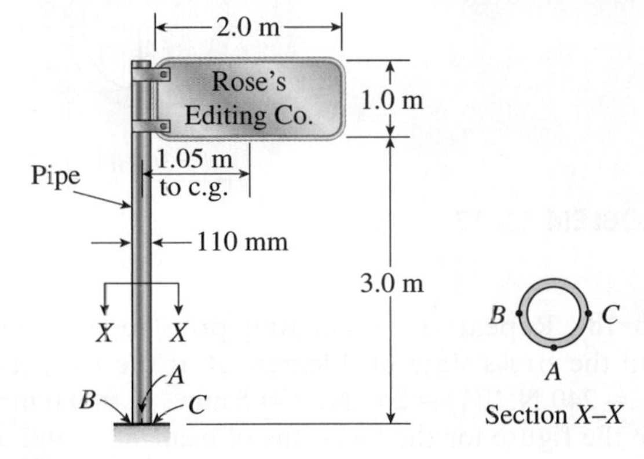

PROBLEM 8.5-16

8.5-17 A traffic light and signal pole is subjected to the weight of each traffic signal $W_S = 240$ N and the weight of the road lamp $W_L = 250$ N. The pole is fixed at the base. Find the principal normal stresses and the maximum shear stress on element B located 5.8 m above the base (see figure). Assume that the weight of the pole and lateral arms is included in the signal and lamp weights.

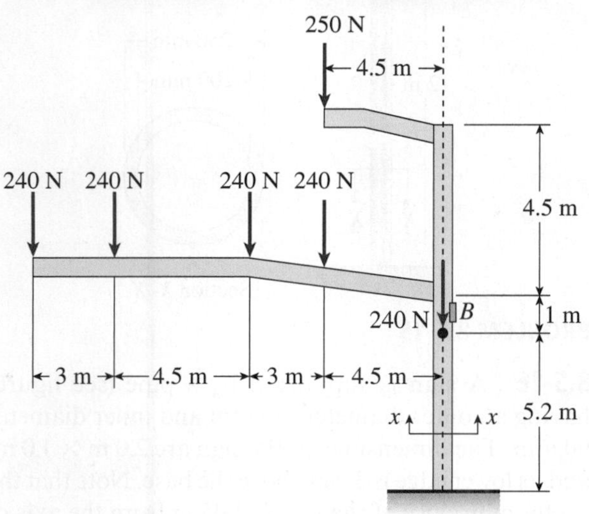

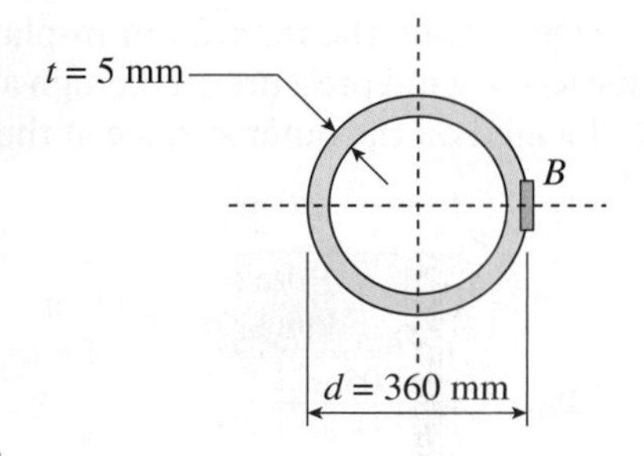

PROBLEM 8.5-17

8.5-18 Repeat the preceding problem but now find the stress state on Element A at the base. Let $W_S = 240$ N, $W_L = 250$ N, $t = 5$ mm, $d = 360$ mm. See the figure for the locations of element A and all loads.

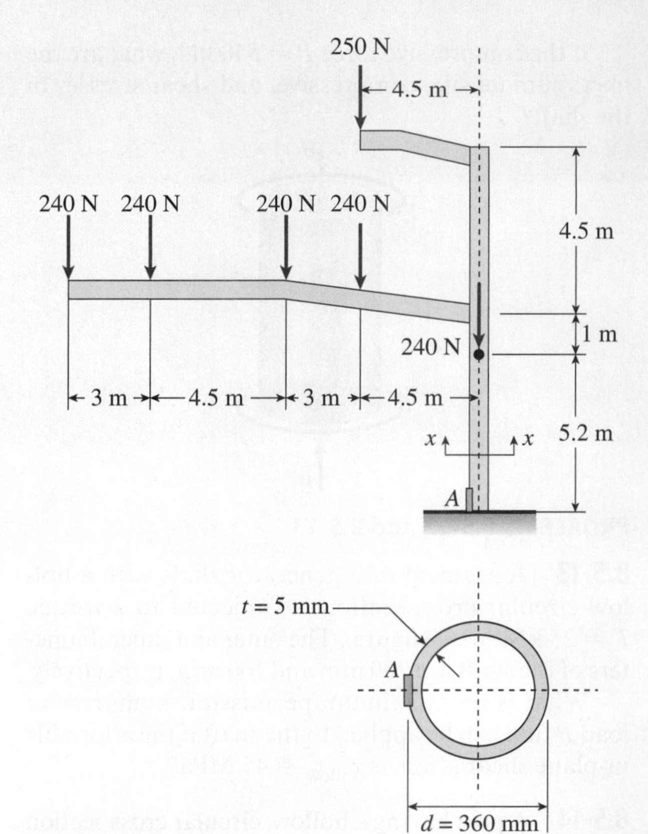

PROBLEM 8.5-18

8.5-19 A bracket $ABCD$ having a hollow circular cross section consists of a vertical arm AB ($L = 1.85$ m), a horizontal arm BC parallel to the x_0 axis, and a horizontal arm CD parallel to the z_0 axis (see figure). The arms BC and CD have lengths $b_1 = 1.1$ m and $b_2 = 0.67$ m, respectively. The outer and inner diameters of the bracket are $d_2 = 190$ mm and $d_1 = 170$ mm. An inclined load $P = 10$ kN acts at point D along line DH. Determine the maximum tensile, compressive, and shear stresses in the vertical arm.

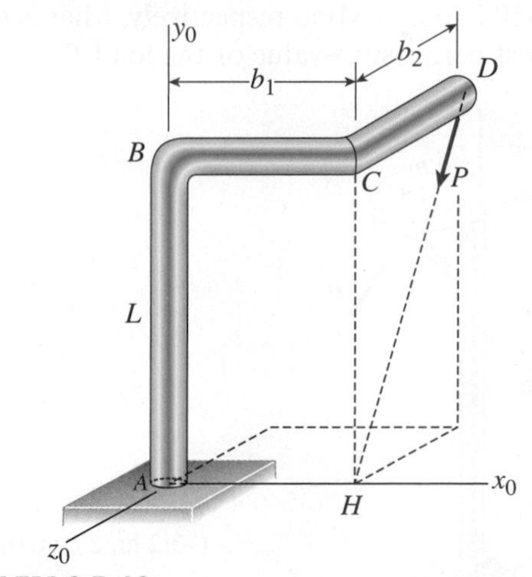

PROBLEM 8.5-19

8.5-20 A gondola on a ski lift is supported by two bent arms, as shown in the figure. Each arm is offset by the distance $b = 180$ mm from the line of action of the weight force W. The allowable stresses in the arms are 100 MPa in tension and 50 MPa in shear.

If the loaded gondola weighs 12 kN, what is the mininum diameter d of the arms?

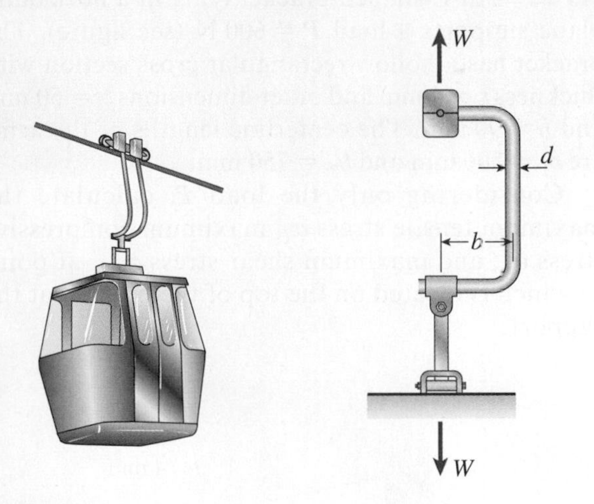

PROBLEM 8.5-20

8.5-21 Determine the maximum tensile, compressive, and shear stresses at points A and B on the bicycle pedal crank shown in the figure.

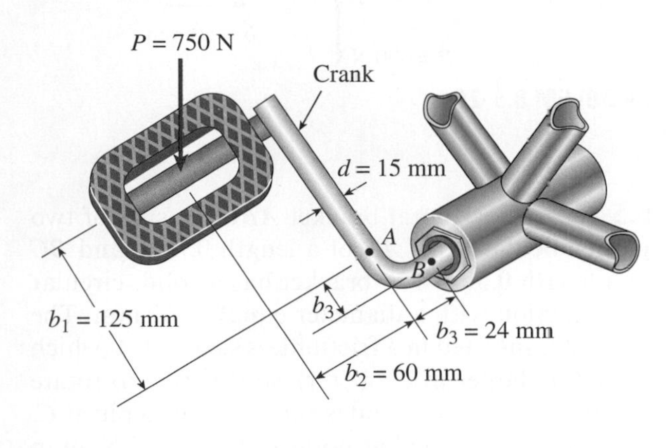

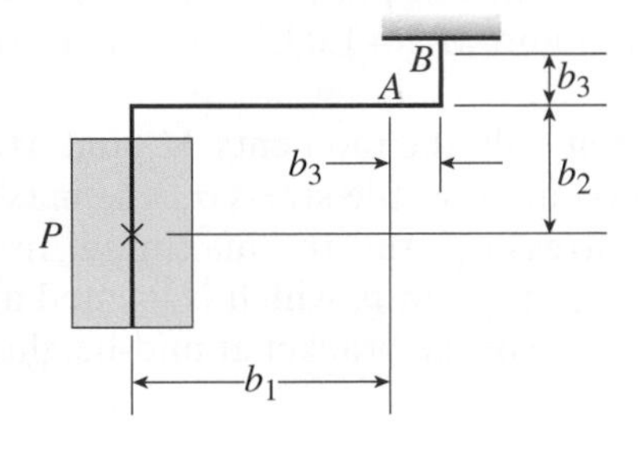

Top view

PROBLEM 8.5-21

The pedal and crank are in a horizontal plane and points A and B are located on the top of the crank. The load $P = 750$ N acts in the vertical direction and the distances (in the horizontal plane) between the line of action of the load and points A and B are $b_1 = 125$ mm, $b_2 = 60$ mm, and $b_3 = 24$ mm. Assume that the crank has a solid circular cross section with diameter $d = 15$ mm.

8.5-22 A double-decker bicycle rack made up of *square* steel tubing is fixed at A (figure a). The weight of a bicycle is represented as a point load applied at B on a plane frame model of the rack (figure b).

(a) Find the state of plane stress on an element C located on the surface at the left side of the vertical tube at the base A. *Include* the weight of the framing system. (Assume weight density $\gamma = 77$ kN/m^3.)

(b) Find the maximum *shear* stresses on an element at C and show them on a sketch of a properly oriented element.

Assume that element C is a sufficient distance from support A so that stress concentration effects are negligible.

(a)

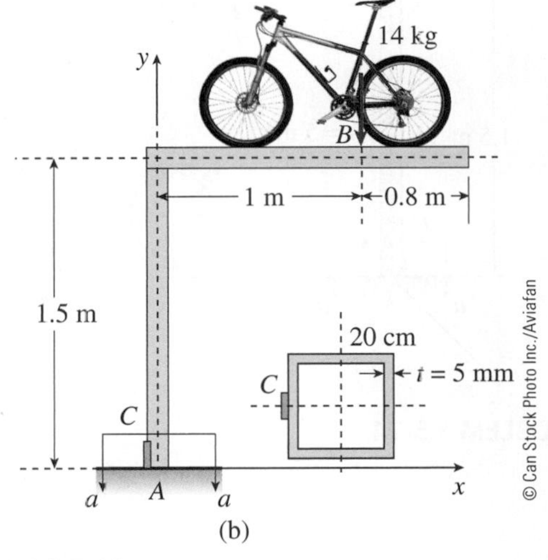

(b)

PROBLEM 8.5-22

8.5-23 A semicircular bar AB lying in a horizontal plane is supported at B (see figure part a). The bar has a centerline radius R and weight q per unit of length (total weight of the bar equals πqR). The cross section of the bar is circular with diameter d.

(a) Obtain formulas for the maximum tensile stress σ_t, maximum compressive stress σ_c, and maximum in-plane shear stress τ_{max} at the top of the bar at the support due to the weight of the bar.

(b) Repeat part (a) if the bar is a quarter-circular segment (see figure part b) but has the same total weight as the semicircular bar.

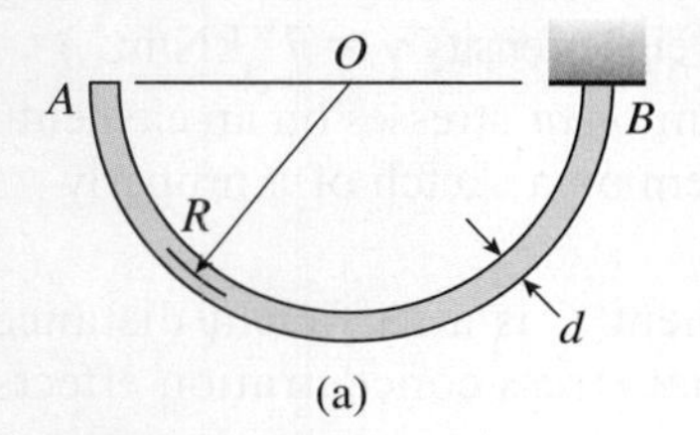

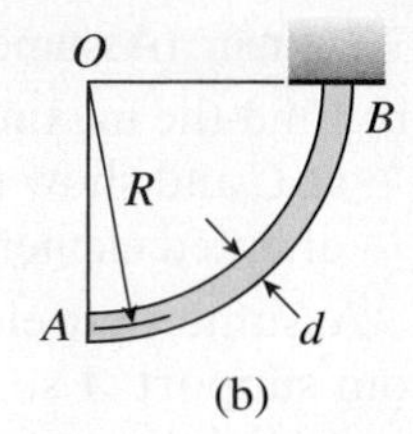

PROBLEM 8.5-23

8.5-24 Repeat Problem 8.5-22 but replace the square tube column with a *circular* tube having a wall thickness $t = 5$ mm and the same cross-sectional area (3900 mm^2) as that of the square tube in figure b in Problem 8.5-22. Also, add force $P_z = 120$ N at B.

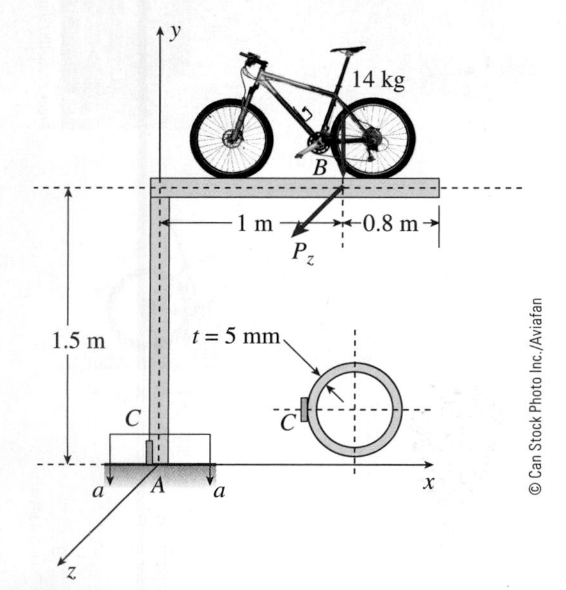

PROBLEM 8.5-24

(a) Find the state of plane stress at C.

(b) Find maximum *normal* stresses and show them on a sketch of a properly oriented element.

(c) Find maximum *shear* stresses and show them on a sketch of a properly oriented element.

8.5-25 An L-shaped bracket lying in a horizontal plane supports a load $P = 600$ N (see figure). The bracket has a hollow rectangular cross section with thickness $t = 4$ mm and outer dimensions $b = 50$ mm and $h = 90$ mm. The centerline lengths of the arms are $b_1 = 500$ mm and $b_2 = 750$ mm.

Considering only the load P, calculate the maximum tensile stress σ_t, maximum compressive stress σ_c, and maximum shear stress τ_{max} at point A, which is located on the top of the bracket at the support.

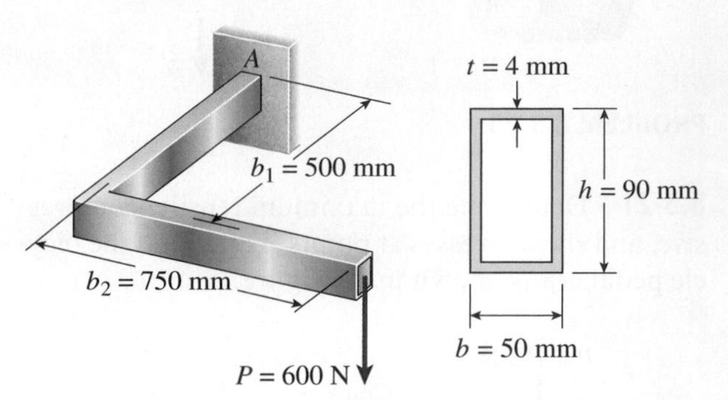

PROBLEM 8.5-25

8.5-26 A horizontal bracket ABC consists of two perpendicular arms AB of a length 0.75 m and BC of a length 0.5 m. The bracket has a solid, circular cross section with a diameter equal to 65 mm. The bracket is inserted in a frictionless sleeve at A (which is slightly larger in diameter), so it is free to rotate about the z_0 axis at A and is supported by a pin at C. Moments are applied at point C $M_1 = 1.5$ kN·m in the x direction and $M_2 = 1.0$ kN·m acts in the $-z$ direction.

Considering only the moments M_1 and M_2, calculate the maximum tensile stress σ_t, the maximum compressive stress σ_c, and the maximum in-plane shear stress τ_{max} at point p, which is located at support A on the side of the bracket at mid-height.

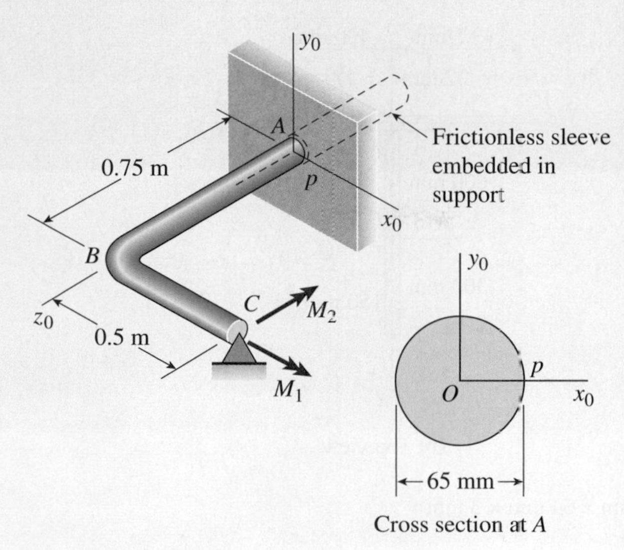

PROBLEM 8.5-26

8.5-27 An arm ABC lying in a horizontal plane and supported at A (see figure) is made of two identical solid steel bars AB and BC welded together at a right angle. Each bar is 0.6 m long.

(a) Knowing that the maximum tensile stress (principal stress) at the top of the bar at support A due solely to the weights of the bars is 7.2 MPa, determine the diameter d of the bars.

(b) If the allowable tensile stress is 10 MPa and each bar has a diameter $d = 50$ mm, what is the maximum downward load P that can be applied at C (in addition to self-weight)?

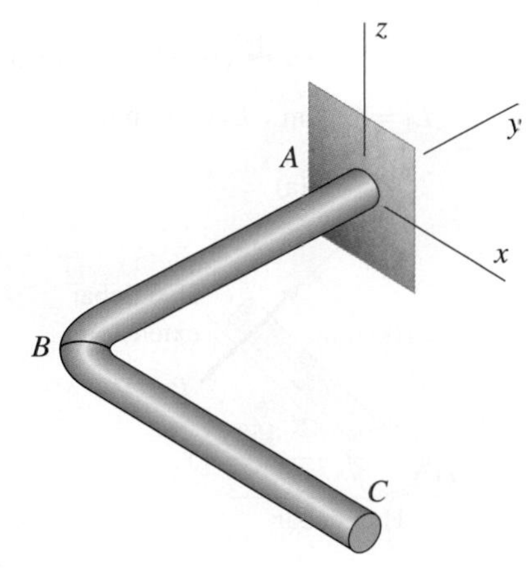

PROBLEM 8.5-27

8.5-28 A crank arm consists of a solid segment of length b_1 and diameter d, a segment of length b_2, and a segment of length b_3, as shown in the figure. Two loads P act as shown: one parallel to $-x$ and another parallel to $-y$. Each load P equals 1.2 kN. The crankshaft dimensions are $b_1 = 75$ mm, $b_2 = 125$ mm, and $b_3 = 35$ mm. The diameter of the upper shaft is $d = 22$ mm.

(a) Determine the maximum tensile, compressive, and shear stresses at point A, which is located on the surface of the shaft at the z axis.

(b) Determine the maximum tensile, compressive, and shear stresses at point B, which is located on the surface of the shaft at the y axis.

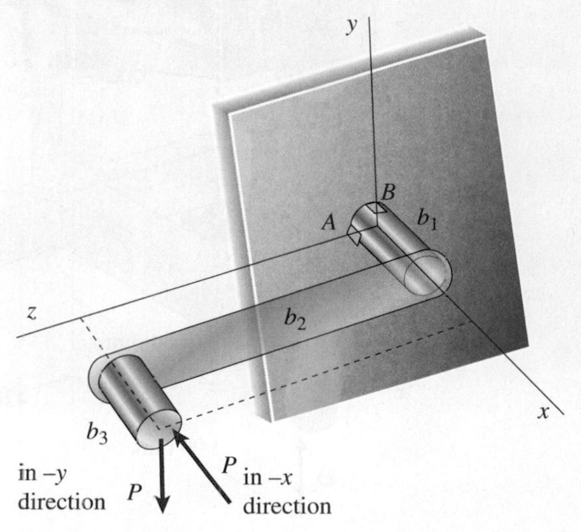

PROBLEM 8.5-28

8.5-29 A moveable steel stand supports an automobile engine weighing $W = 3.4$ kN, as shown in the figure part a. The stand is constructed of 64 mm × 64 mm × 3 mm-thick steel tubing. Once in position, the stand is restrained by pin supports at B and C. Of interest are stresses at point A at the base of the vertical post; point A has coordinates ($x = 32$ mm, $y = 0$, $z = 32$ mm). Neglect the weight of the stand.

(a) Initially, the engine weight acts in the $-z$ direction through point Q, which has coordinates (600 mm, 0, 32 mm). Find the maximum tensile, compressive, and shear stresses at point A.

(b) Repeat part (a) assuming now that, during repair, the engine is rotated about its own longitudinal axis (which is parallel to the x axis) so that W acts through Q' [with coordinates (600 mm, 150 mm, 32 mm)] and force $F_y = 900$ N is applied parallel to the y axis at distance $d = 0.75$ m.

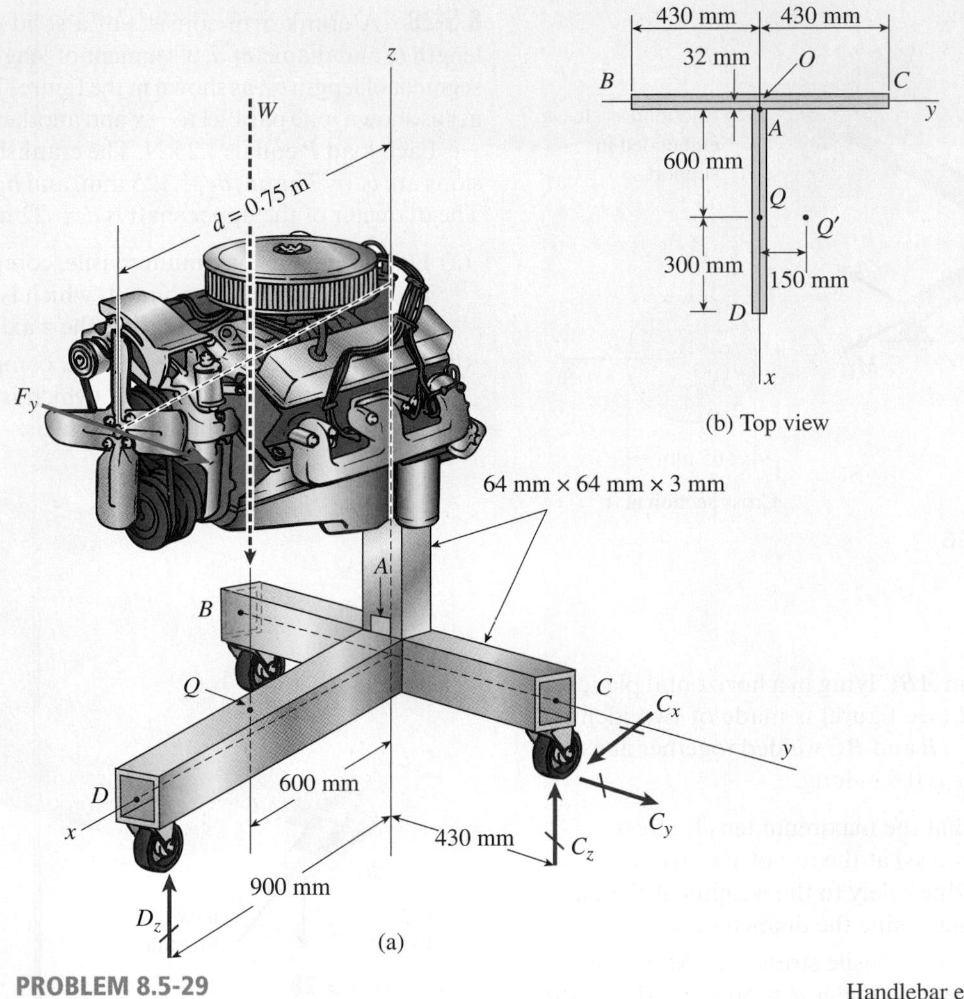

PROBLEM 8.5-29

8.5-30 A mountain bike rider going uphill applies a force $P = 65$ N to each end of the handlebars $ABCD$, made of aluminum alloy 7075-T6, by pulling on the handlebar extenders (DF on right handlebar segment). Consider the right half of the handlebar assembly only (assume the bars are fixed at the fork at A). Segments AB and CD are prismatic with lengths L_1 and L_3 and with outer diameters and thicknesses d_{01}, t_{01} and d_{03}, t_{03}, respectively, as shown. Segment BC of length L_2, however, is tapered, and outer diameter and thickness vary linearly between dimensions at B and C. Consider shear, torsion, and bending effects only for segment AD; assume DF is rigid.

Find the maximum tensile, compressive, and shear stresses adjacent to support A. Show where each maximum stress value occurs.

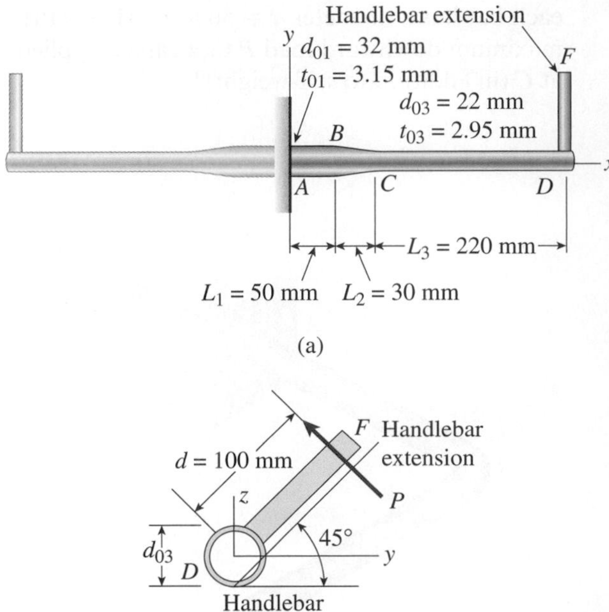

PROBLEM 8.5-30

8.5-31 Determine the maximum tensile, compressive, and shear stresses acting on the cross section of the tube at point A of the hitch bicycle rack shown in the figure.

The rack is made up of 50 mm × 50 mm steel tubing which is 3 mm thick. Assume that the weight of each of four bicycles is distributed evenly between the two support arms so that the rack can be represented as a cantilever beam ($ABCDEF$) in the xy plane. The overall weight of the rack alone is $W = 270$ N directed through C, and the weight of each bicycle is $P = 135$ N.

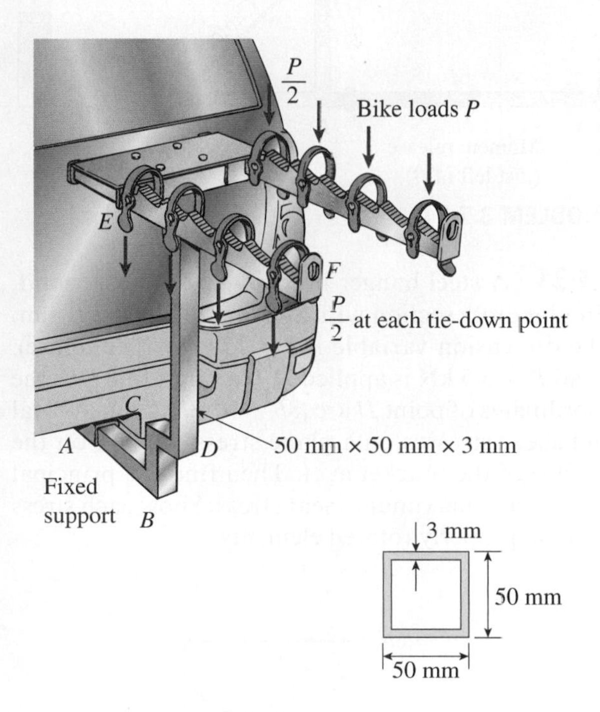

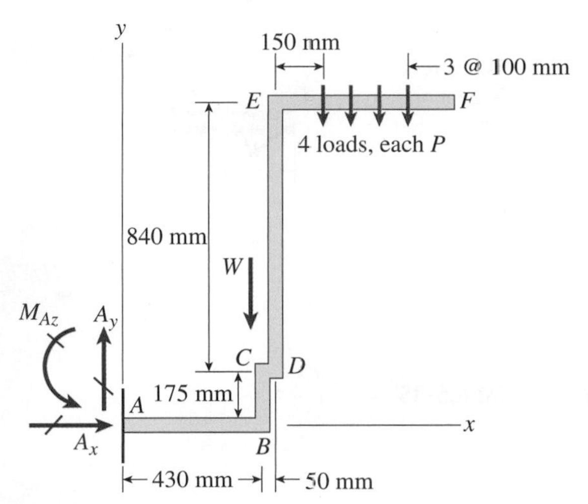

PROBLEM 8.5-31

8.5-32 Consider the mountain bike shown in the figure. To account for impact, crashes, and other loading uncertainties, a *design load* $P = 5000$ N is used to design the seat post. The length of the seat post is $L = 254$ mm.

(a) Find the required diameter of the seat post if it is to be constructed using an *aluminum alloy* with the ultimate stress $\sigma_U = 550$ MPa and a factor of safety of 2.8. Consider only axial and flexural normal stresses in your design.

(b) Repeat part (a) if a *titanium alloy* is used instead. Assume the ultimate stress $\sigma_U = 900$ MPa and a factor of safety of 2.5.

PROBLEM 8.5-32

8.5-33 A plumber's valve wrench is used to replace valves in plumbing fixtures. A simplified model of the wrench (see figure part a) consists of pipe AB (length L, outer diameter d_2, inner diameter d_1), which is fixed at A and has holes of a diameter d_b on either side of the pipe at B. A solid, cylindrical bar CBD (length a, diameter d_b) is inserted into the holes at B and *only one* force $F = 245$ N is applied in the $-Z$ direction at C to loosen the fixture valve at A (see figure part c). Let $G = 81$ GPa, $\nu = 0.30$, $L = 100$ mm, $a = 115$ mm, $d_2 = 32$ mm, $d_1 = 25$ mm, and $d_b = 6$ mm.

Find the state of plane stress on the top of the pipe near A (at coordinates $X = 0, Y = 0, Z = d_2/2$), and show all stresses on a plane stress element (see figure part b). Compute the principal stresses and maximum shear stress, and show them on properly rotated stress elements.

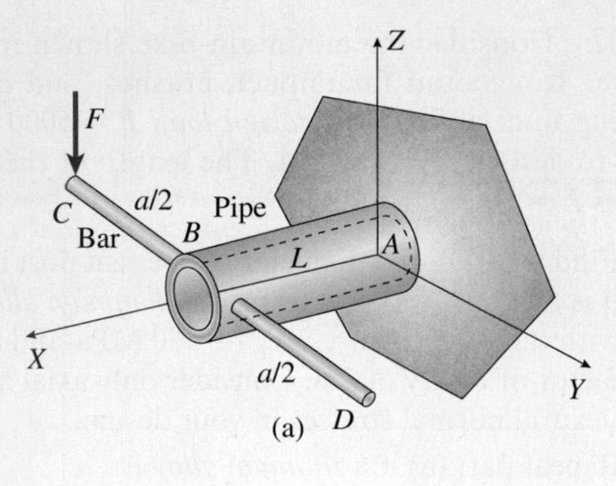

(a)

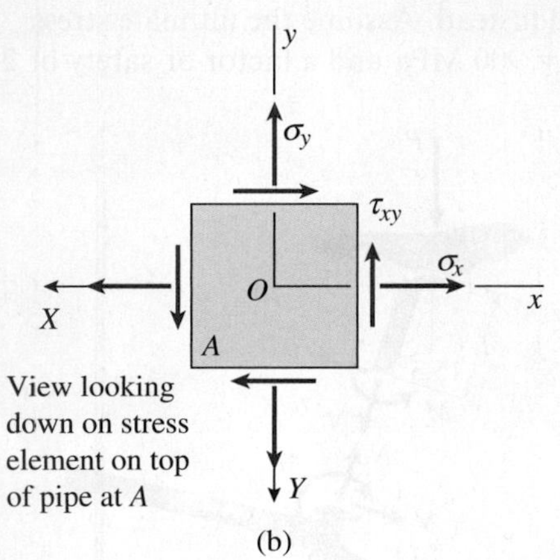

(b)

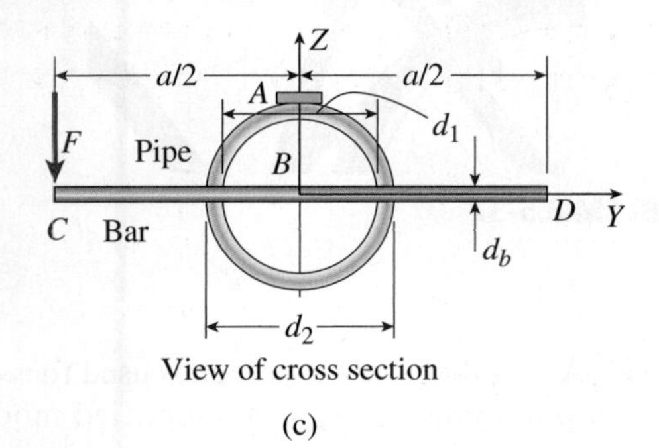

View of cross section

(c)

PROBLEM 8.5-33

8.5-34 A compound beam *ABCD* has a cable with force *P* anchored at *C*. The cable passes over a pulley at *D*, and force *P* acts in the $-x$ direction. There is a moment release just left of *B*. Neglect the self-weight of the beam and cable. Cable force $P = 450$ N and dimension variable $L = 0.25$ m. The beam has a rectangular cross section ($b = 20$ mm, $h = 50$ mm).

(a) Calculate the maximum normal stresses and maximum in-plane shear stress on the bottom surface of the beam at support *A*.

(b) Repeat part (a) for a plane stress element located at mid-height of the beam at *A*.

(c) If the maximum tensile stress and maximum in-plane shear stress at point *A* are limited to 90 MPa and 42 MPa, respectively, what is the largest permissible value of the cable force *P*?

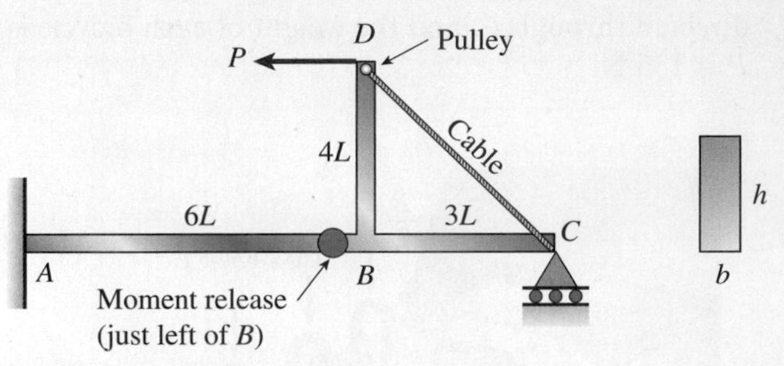

PROBLEM 8.5-34

8.5-35 A steel hanger bracket *ABCD* has a solid, circular cross section with a diameter of $d = 50$ mm. The dimension variable is $b = 150$ mm (see figure). Load $P = 5.5$ kN is applied at *D* along a line *DH*; the coordinates of point *H* are $(8b, -5b, 3b)$. Find normal and shear stresses on a plane stress element on the surface of the bracket at *A*. Then find the principal stresses and maximum shear stress. Show each stress state on properly rotated elements.

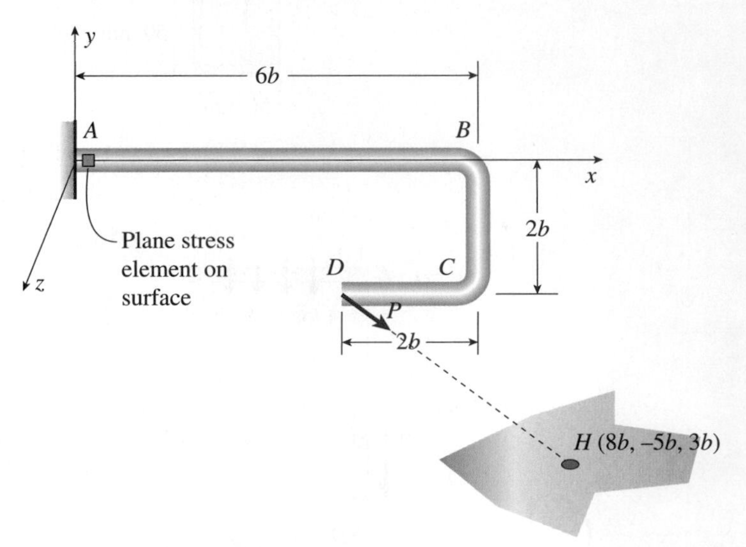

PROBLEM 8.5-35

SOME ADDITIONAL REVIEW PROBLEMS: Chapter 8

R-8.1 A thin-walled spherical tank with a diameter of 1.5 m and wall thickness of 65 mm has an internal pressure of 20 MPa. The maximum shear stress in the wall of the tank is approximately:

(A) 58 MPa
(B) 67 MPa
(C) 115 MPa
(D) 127 MPa

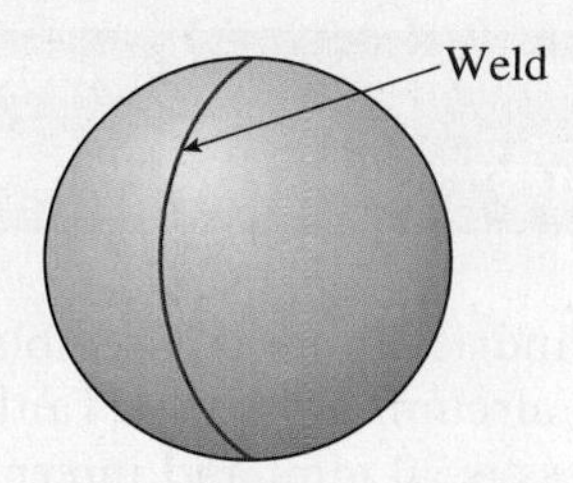

R-8.2 A thin-walled spherical tank has a diameter of 0.75 m and an internal pressure of 20 MPa. The yield stress in tension is 920 MPa, the yield stress in shear is 475 MPa, and the factor of safety is 2.5. The modulus of elasticity is 210 GPa, Poisson's ratio is 0.28, and maximum normal strain is 1220×10^{-6}. The minimum permissible thickness of the tank is approximately:

(A) 8.6 mm
(B) 9.9 mm
(C) 10.5 mm
(D) 11.1 mm

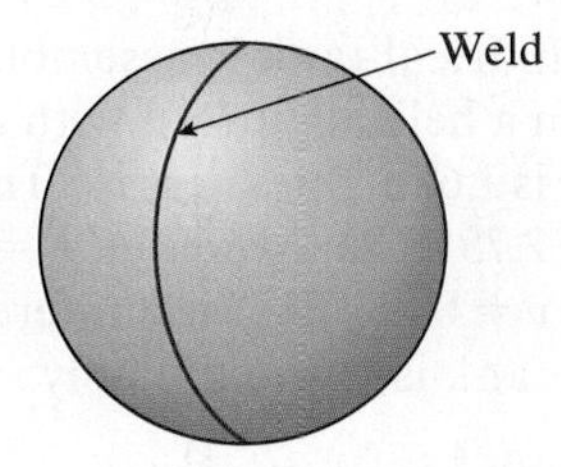

R-8.3 A thin-walled cylindrical tank with a diameter of 200 mm as shown in the above figure has an internal pressure of 11 MPa. The yield stress in tension is 250 MPa, the yield stress in shear is 140 MPa, and the factor of safety is 2.5. The minimum permissible thickness of the tank is approximately:

(A) 8.2 mm
(B) 9.1 mm
(C) 9.8 mm
(D) 11.0 mm

R-8.4 A thin-walled cylindrical tank with a diameter of 2.0 m and wall thickness of 18 mm is open at the top. The height h of water (weight density $= 9.81\ \text{kN/m}^3$) in the tank at which the circumferential stress reaches 10 MPa in the tank wall is approximately:

(A) 14 m
(B) 18 m
(C) 20 m
(D) 24 m

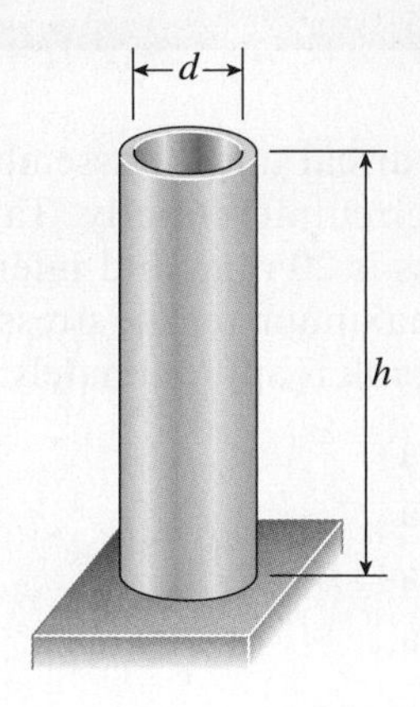

R-8.5 The pressure relief valve is opened on a thin-walled cylindrical tank, with the radius-to-wall thickness ratio of 128, thereby decreasing the longitudinal strain by 150×10^{-6}. Assume $E = 73$ GPa and $\nu = 0.33$. The original internal pressure in the tank was approximately:

(A) 370 kPa
(B) 450 kPa
(C) 500 kPa
(D) 590 kPa

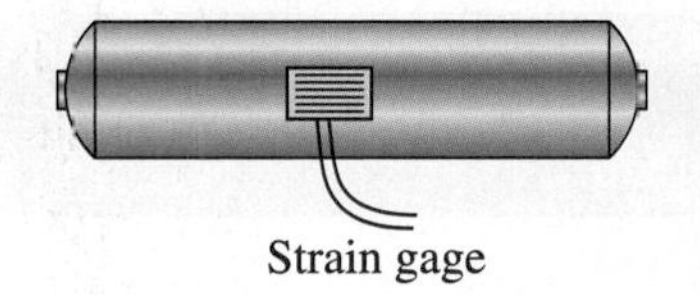

R-8.6 A cylindrical tank is assembled by welding steel sections circumferentially. Tank diameter is 1.5 m, thickness is 20 mm, and internal pressure is 2.0 MPa. The maximum stress in the heads of the tank is approximately:

(A) 38 MPa
(B) 45 MPa
(C) 50 MPa
(D) 59 MPa

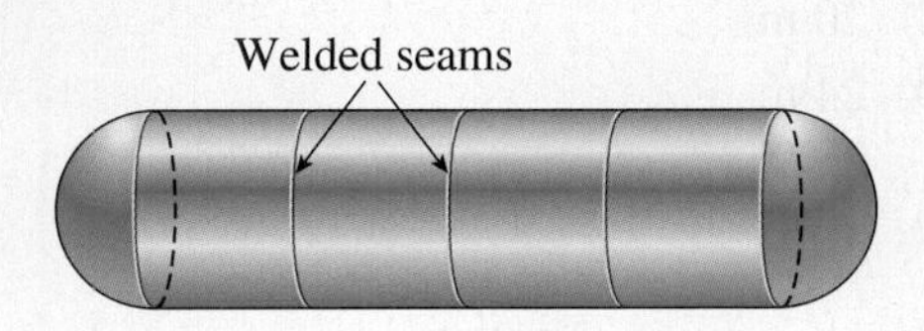

R-8.7 A cylindrical tank is assembled by welding steel sections circumferentially. Tank diameter is 1.5 m, thickness is 20 mm, and internal pressure is 2.0 MPa. The maximum tensile stress in the cylindrical part of the tank is approximately:

(A) 45 MPa
(B) 57 MPa
(C) 62 MPa
(D) 75 MPa

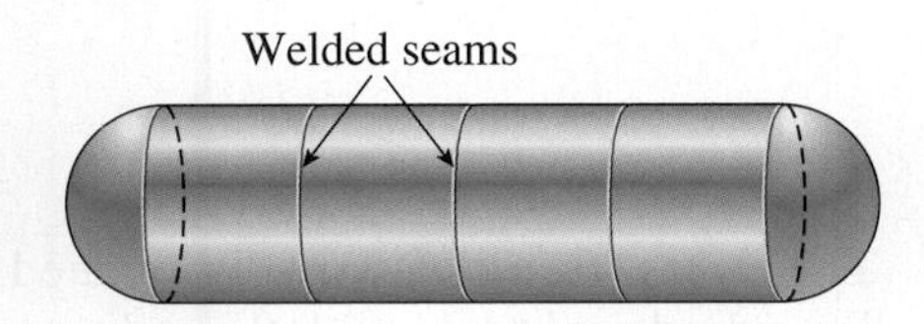

R-8.8 A cylindrical tank is assembled by welding steel sections circumferentially. Tank diameter is 1.5 m, thickness is 20 mm, and internal pressure is 2.0 MPa. The maximum tensile stress perpendicular to the welds is approximately:

(A) 22 MPa
(B) 29 MPa
(C) 33 MPa
(D) 37 MPa

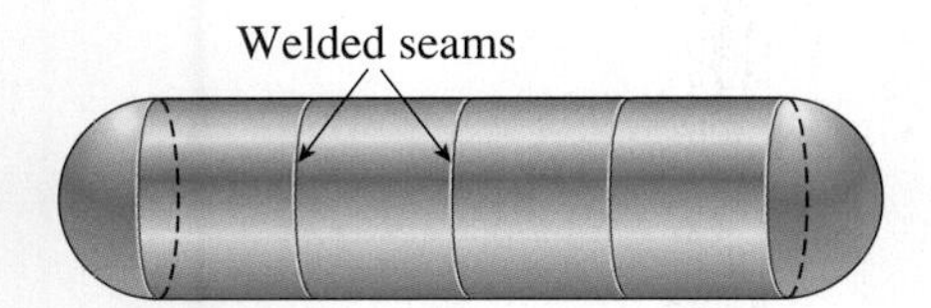

R-8.9 A cylindrical tank is assembled by welding steel sections circumferentially. Tank diameter is 1.5 m, thickness is 20 mm, and internal pressure is 2.0 MPa. The maximum shear stress in the heads is approximately:

(A) 19 MPa
(B) 23 MPa
(C) 33 MPa
(D) 35 MPa

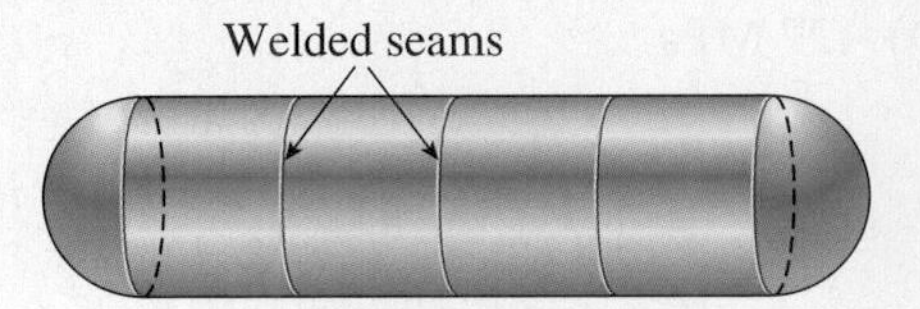

R-8.10 A cylindrical tank is assembled by welding steel sections circumferentially. Tank diameter is 1.5 m, thickness is 20 mm, and internal pressure is 2.0 MPa. The maximum shear stress in the cylindrical part of the tank is approximately:

(A) 17 MPa
(B) 26 MPa
(C) 34 MPa
(D) 38 MPa

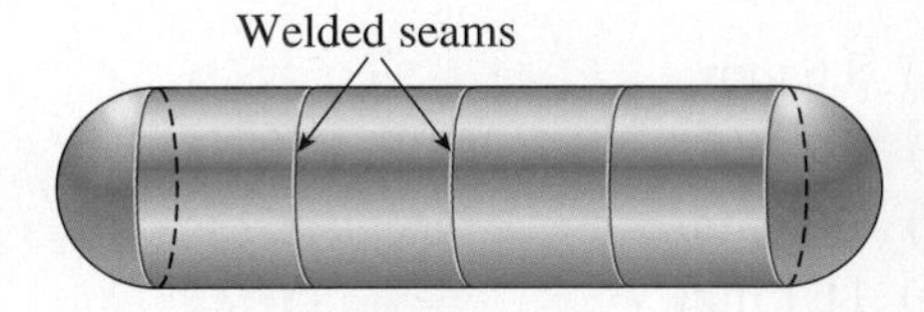

R-8.11 A cylindrical tank is assembled by welding steel sections in a helical pattern with angle $\alpha = 50°$. Tank diameter is 1.6 m, thickness is 20 mm, and internal pressure is 2.75 MPa. Modulus $E = 210$ GPa and Poisson's ratio $\nu = 0.28$. The circumferential strain in the wall of the tank is approximately:

(A) 1.9×10^{-4}
(B) 3.2×10^{-4}
(C) 3.9×10^{-4}
(D) 4.5×10^{-4}

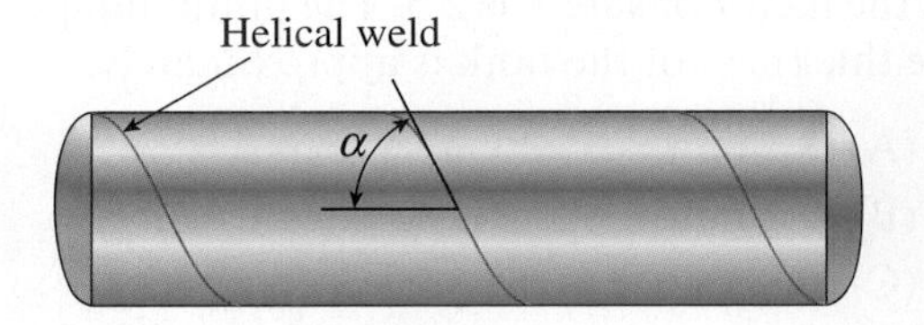

R-8.12 A cylindrical tank is assembled by welding steel sections in a helical pattern with angle $\alpha = 50°$. Tank diameter is 1.6 m, thickness is 20 mm, and internal pressure is 2.75 MPa. Modulus $E = 210$ GPa and Poisson's ratio $\nu = 0.28$. The longitudinal strain in the wall of the tank is approximately:

(A) 1.2×10^{-4}
(B) 2.4×10^{-4}
(C) 3.1×10^{-4}
(D) 4.3×10^{-4}

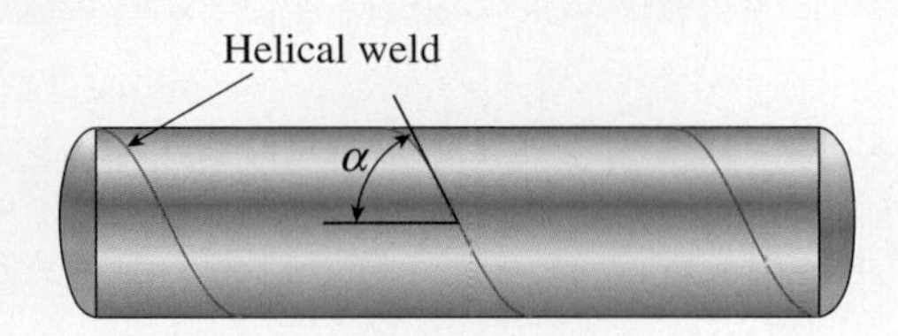

R-8.13 A cylindrical tank is assembled by welding steel sections in a helical pattern with angle $\alpha = 50°$. Tank diameter is 1.6 m, thickness is 20 mm, and internal pressure is 2.75 MPa. Modulus $E = 210$ GPa and Poisson's ratio $\nu = 0.28$. The normal stress acting perpendicular to the weld is approximately:

(A) 39 MPa
(B) 48 MPa
(C) 78 MPa
(D) 84 MPa

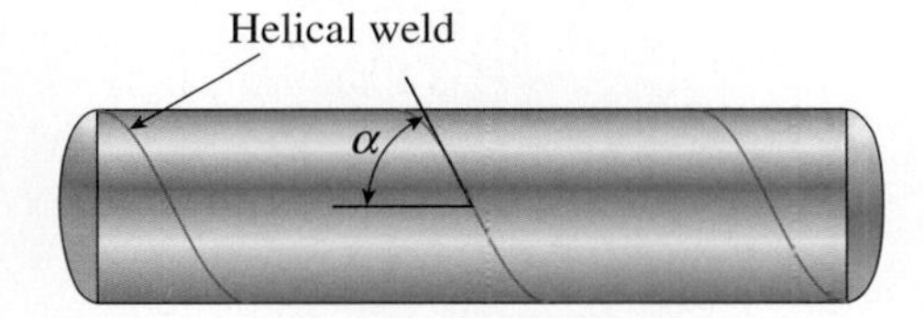

R-8.14 A segment of a drive shaft ($d_2 = 200$ mm, $d_1 = 162$ mm) is subjected to a torque $T = 30$ kN·m. The allowable shear stress in the shaft is 45 MPa. The maximum permissible compressive load P is approximately:

(A) 200 kN
(B) 286 kN
(C) 328 kN
(D) 442 kN

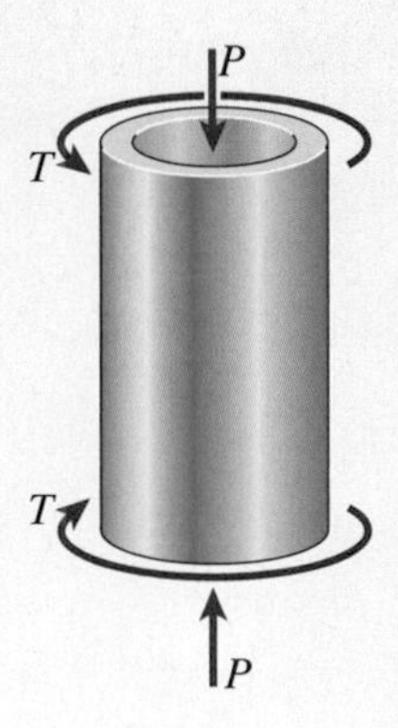

R-8.15 A thin-walled cylindrical tank, under internal pressure p, is compressed by a force $F = 75$ kN. Cylinder diameter is $d = 90$ mm and wall thickness $t = 5.5$ mm. Allowable normal stress is 110 MPa and allowable shear stress is 60 MPa. The maximum allowable internal pressure p_{max} is approximately:

(A) 5 MPa
(B) 10 MPa
(C) 13 MPa
(D) 17 MPa

CHAPTER 9

Deflections of Beams

Deflection of beams is an important consideration in thei· initial design; deflections also must be monitored during construction.

Chapter Objectives

- Study several different methods for computing deflections of beams, including integration of the differential equation of the deflection curve and the method of superposition.
- Study a method for computing beam deflections which is based on the area of the bending-moment diagram.
- Find deflections of nonprismatic beams.
- Apply concepts of work and strain energy to computation of beam deflections using Castigliano's theorem.
- Study beam deflections due to impact effects and differential temperature changes.

Chapter Outline

***Advanced topics**

9.1 Introduction

When a beam with a straight longitudinal axis is loaded by lateral forces, the axis is deformed into a curve, called the **deflection curve** of the beam. In Chapter 5, the curvature of the bent beam was used to find the normal strains and stresses in the beam. However, a method for finding the deflection curve itself was not presented. In this chapter, the equation of the deflection curve and also deflections at specific points along the axis of the beam will be determined.

The calculation of deflections is an important part of structural analysis and design. For example, finding deflections is an essential ingredient in the analysis of statically indeterminate structures (Chapter 10). Deflections are also important in dynamic analyses, as when investigating the vibrations of aircraft or the response of buildings to earthquakes.

Deflections are sometimes calculated in order to verify that they are within tolerable limits. For instance, specifications for the design of buildings usually place upper limits on the deflections. Large deflections in buildings are unsightly (and even unnerving) and can cause cracks in ceilings and walls. In the design of machines and aircraft, specifications may limit deflections in order to prevent undesirable vibrations.

9.2 Differential Equations of the Deflection Curve

Most procedures for finding beam deflections are based on the differential equations of the deflection curve and their associated relationships. Consequently, the discussion here begins by deriving the basic equation for the deflection curve of a beam.

FIGURE 9-1

Deflection curve of a cantilever beam

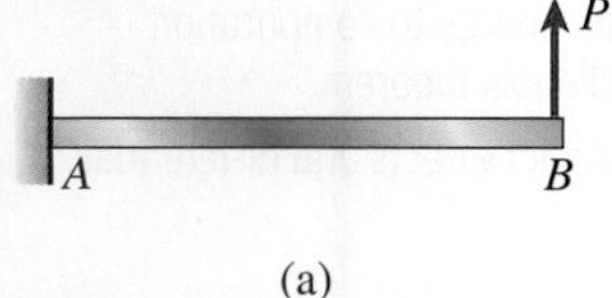

(a)

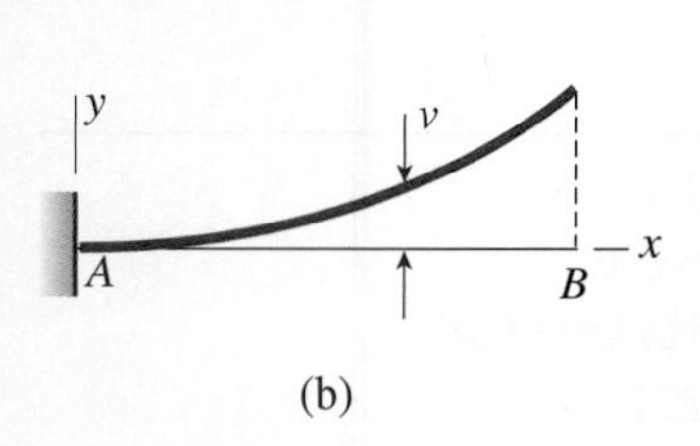

(b)

Consider a cantilever beam with a concentrated load acting upward at the free end (Fig. 9-1a). Under the action of this load, the axis of the beam deforms into a curve, as shown in Fig. 9-1b. The reference axes have their origin at the fixed end of the beam, with the x axis directed to the right and the y axis directed upward. The z axis is directed outward from the figure (toward the viewer).

As in previous discussions of beam bending in Chapter 5, assume that the x-y plane is a plane of symmetry of the beam, and assume that all loads act in this plane (the *plane of bending*).

The **deflection** v is the displacement in the y direction of any point on the axis of the beam (Fig. 9-1b). Because the y axis is positive upward, the deflections are also positive when upward.[1]

To obtain the equation of the deflection curve, express the deflection v as a function of the coordinate x. The deflection v at any point m_1 on the deflection curve is shown in Fig. 9-2a. Point m_1 is located at distance x from the origin (measured along the x axis). A second point m_2, located at distance $x + dx$ from the origin, is also shown. The deflection at this second point is $v + dv$, where dv is the increment in deflection as you move along the curve from m_1 to m_2.

[1]As mentioned in Section 5.1, the traditional symbols for displacements in the x, y, and z directions are u, v, and w, respectively. The advantage of this notation is that it emphasizes the distinction between a *coordinate* and a *displacement*.

When the beam is bent, there is not only a deflection at each point along the axis but also a rotation. The **angle of rotation** θ of the axis of the beam is the angle between the x axis and the tangent to the deflection curve, as shown for point m_1 in the enlarged view of Fig. 9-2b. For the choice of axes (x positive to the right and y positive upward), the angle of rotation is positive when counterclockwise. (Other names for the angle of rotation are *angle of inclination* and *angle of slope*.)

The angle of rotation at point m_2 is $\theta + d\theta$, where $d\theta$ is the increase in angle as you move from point m_1 to point m_2. It follows that if you construct lines normal to the tangents (Figs. 9-2a and b), the angle between these normals is $d\theta$. Also, as discussed earlier in Section 5.3, the point of intersection of these normals is the **center of curvature** O' (Fig. 9-2a) and the distance from O' to the curve is the **radius of curvature** ρ. From Fig. 9-2a, it follows that

$$\rho d\theta = ds \tag{9-1}$$

in which $d\theta$ is in radians and ds is the distance along the deflection curve between points m_1 and m_2. Therefore, the **curvature** κ (equal to the reciprocal of the radius of curvature) is given by the equation

$$\kappa = \frac{1}{\rho} = \frac{d\theta}{ds} \tag{9-2}$$

The **sign convention** for curvature is pictured in Fig. 9-3, which is repeated from Fig. 5-6 of Section 5.3. Note that curvature is positive when the angle of rotation increases as you move along the beam in the positive x direction.

The **slope of the deflection curve** is the first derivative dv/dx of the expression for the deflection v. In geometric terms, the slope is the increment dv in the deflection (as you go from point m_1 to point m_2 in Fig. 9-2) divided by the increment dx in the distance along the x axis. Since dv and dx are infinitesimally small, the slope dv/dx is equal to the tangent of the angle of rotation θ (Fig. 9-2b). Thus,

$$\frac{dv}{dx} = \tan\theta \qquad \theta = \arctan\frac{dv}{dx} \tag{9-3a,b}$$

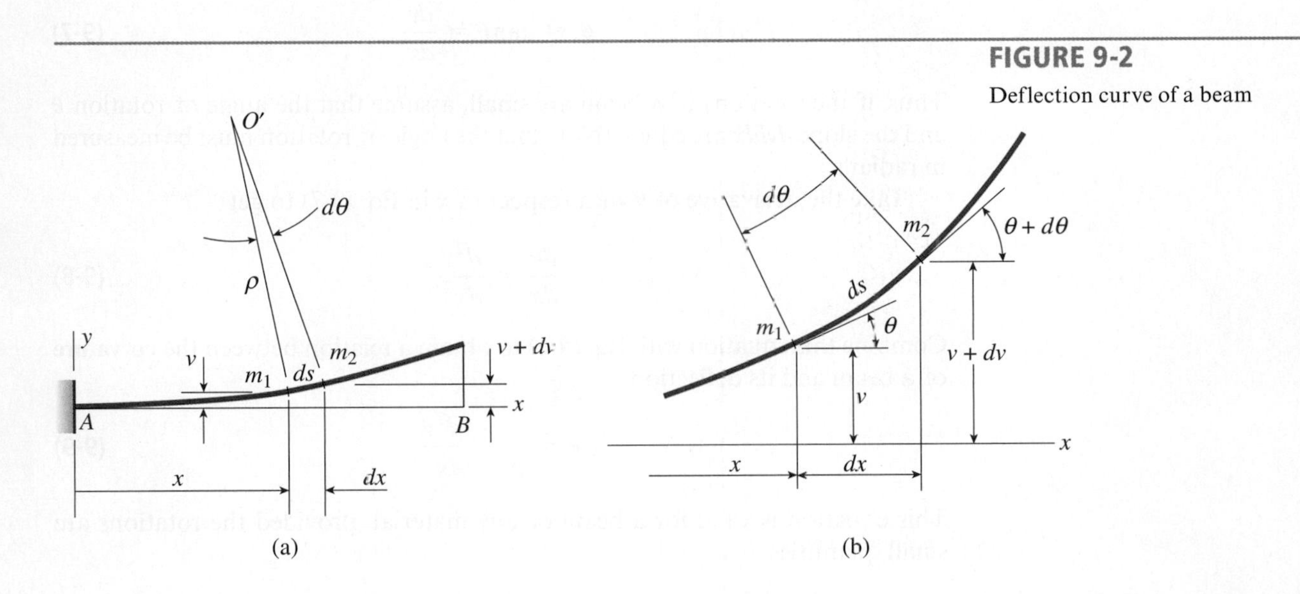

FIGURE 9-2
Deflection curve of a beam

FIGURE 9-3
Sign convention for curvature

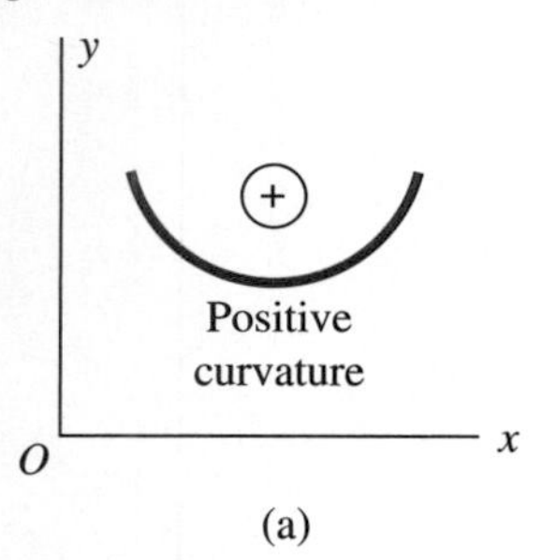

(a)

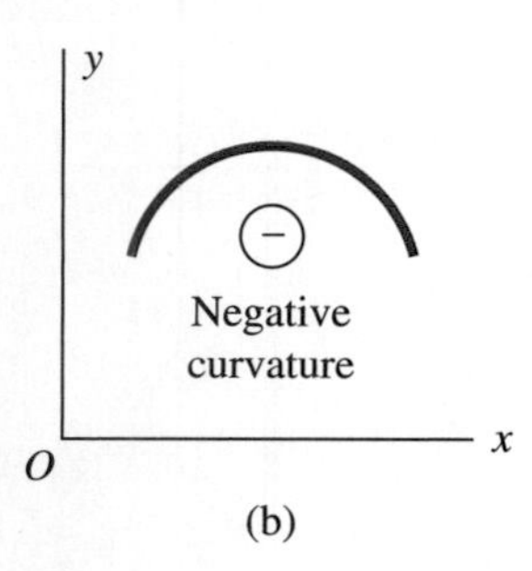

(b)

In a similar manner, the following relationships are obtained:

$$\cos\theta = \frac{dx}{ds} \qquad \sin\theta = \frac{dv}{ds} \tag{9-4a,b}$$

Note that when the x and y axes have the directions shown in Fig. 9-2a, the slope dv/dx is positive when the tangent to the curve slopes upward to the right.

Equations (9-2) through (9-4) are based only upon geometric considerations; therefore, they are valid for beams of any material. Furthermore, there are no restrictions on the magnitudes of the slopes and deflections.

Beams with Small Angles of Rotation

The structures encountered in everyday life, such as buildings, automobiles, aircraft, and ships, undergo relatively small changes in shape while in service. The changes are so small as to be unnoticed by a casual observer. Consequently, the deflection curves of most beams and columns have very small angles of rotation, very small deflections, and very small curvatures. Under these conditions, some mathematical approximations that greatly simplify beam analysis can be made.

Consider, for instance, the deflection curve shown in Fig. 9-2. If the angle of rotation θ is a very small quantity (and hence the deflection curve is nearly horizontal), the distance ds along the deflection curve is practically the same as the increment dx along the x axis. This same conclusion can be obtained directly from Eq. (9-4a). Since $\cos \approx 1$ when the angle θ is small, Eq. (9-4a) gives

$$ds \approx dx \tag{9-5}$$

With this approximation, the curvature becomes [see Eq. (9-2)]

$$\kappa = \frac{1}{\rho} = \frac{d\theta}{dx} \tag{9-6}$$

Also, since $\tan\theta \approx \theta$ when θ is small, the following approximation to Eq. (9-3a) can be made:

$$\theta \approx \tan\theta = \frac{dv}{dx} \tag{9-7}$$

Thus, if the rotations of a beam are small, assume that the angle of rotation θ and the slope dv/dx are equal. (Note that the angle of rotation must be measured in radians.)

Take the derivative of θ with respect to x in Eq. (9-7) to get

$$\frac{d\theta}{dx} = \frac{d^2v}{dx^2} \tag{9-8}$$

Combine this equation with Eq. (9-6) to obtain a relation between the **curvature** of a beam and its deflection:

$$\kappa = \frac{1}{\rho} = \frac{d^2v}{dx^2} \tag{9-9}$$

This equation is valid for a beam of any material, provided the rotations are small quantities.

If the material of a beam is **linearly elastic** and follows Hooke's law, the curvature [from Eq. 5-13, Chapter 5] is

$$\kappa = \frac{1}{\rho} = \frac{M}{EI} \qquad \textbf{(9-10)}$$

in which M is the bending moment and EI is the flexural rigidity of the beam. Equation (9-10) shows that a positive bending moment produces positive curvature and a negative bending moment produces negative curvature, as shown earlier in Fig. 5-10.

Combining Eq. (9-9) with Eq. (9-10) yields the basic **differential equation of the deflection curve** of a beam:

$$\frac{d^2v}{dx^2} = \frac{M}{EI} \qquad \textbf{(9-11)}$$

This equation can be integrated in each particular case to find the deflection v, provided the bending moment M and flexural rigidity EI are known as functions of x.

As a reminder, the **sign conventions** to be used with the preceding equations are repeated here: (1) The x and y axes are positive to the right and upward, respectively; (2) the deflection v is positive upward; (3) the slope dv/dx and angle of rotation θ are positive when counterclockwise with respect to the positive x axis; (4) the curvature κ is positive when the beam is bent concave upward; and (5) the bending moment M is positive when it produces compression in the upper part of the beam.

Additional equations can be obtained from the relations between bending moment M, shear force V, and intensity q of distributed load. In Chapter 4, the following equations between M, V, and q [see Eqs. (4-1) and (4-4)] were derived:

$$\frac{dV}{dx} = -q \qquad \frac{dM}{dx} = V \qquad \textbf{(9-12a,b)}$$

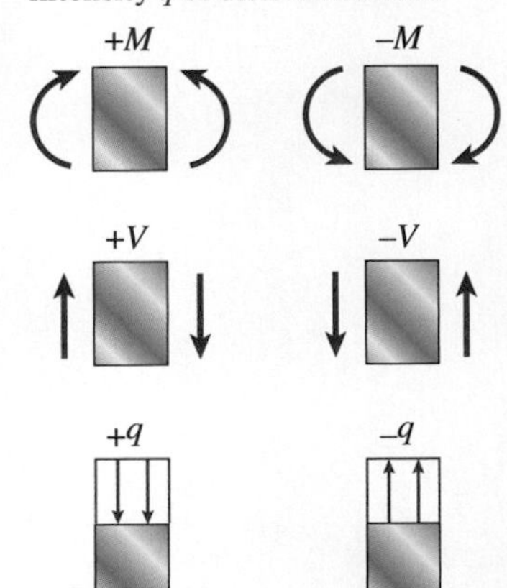

FIGURE 9-4

Sign conventions for bending moment M, shear force V, and intensity q of distributed load

The sign conventions for these quantities are shown in Fig. 9-4. Differentiate Eq. (9-11) with respect to x and then substitute the preceding equations for shear force and load to obtain additional equations. In so doing, consider two cases: nonprismatic beams and prismatic beams.

Nonprismatic Beams

In the case of a nonprismatic beam, the flexural rigidity EI is variable; therefore, Eq. (9-11) is now written in the form

$$EI_x \frac{d^2v}{dx^2} = M \qquad \textbf{(9-13a)}$$

where the subscript x is inserted as a reminder that the flexural rigidity may vary with x. Differentiate both sides of this equation and use Eqs. (9-12a and b) to obtain

$$\frac{d}{dx}\left(EI_x \frac{d^2v}{dx^2}\right) = \frac{dM}{dx} = V \qquad \textbf{(9-13b)}$$

$$\frac{d^2}{dx^2}\left(EI_x \frac{d^2v}{dx^2}\right) = \frac{dV}{dx} = -q \qquad \textbf{(9-13c)}$$

The deflection of a nonprismatic beam can be found by solving (either analytically or numerically) any one of the three preceding differential equations. The choice usually depends upon which equation provides the most efficient solution.

Prismatic Beams

In the case of a prismatic beam (constant EI), the differential equations become

$$EI\frac{d^2v}{dx^2} = M \qquad EI\frac{d^3v}{dx^3} = V \qquad EI\frac{d^4v}{dx^4} = -q \tag{9-14a,b,c}$$

To simplify the writing of these and other equations, **primes** are often used to denote differentiation:

$$v' \equiv \frac{dv}{dx} \qquad v'' \equiv \frac{d^2v}{dx^2} \qquad v''' \equiv \frac{d^3v}{dx^3} \qquad v'''' \equiv \frac{d^4v}{dx^4} \tag{9-15}$$

Using this notation produces the differential equations for a prismatic beam in the forms:

$$EIv'' = M \qquad EIv''' = V \qquad EIv'''' = -q \tag{9-16a,b,c}$$

Refer to these equations as the **bending-moment equation**, the **shear-force equation**, and the **load equation**, respectively.

In the next two sections, the preceding equations are used to find deflections of beams. The general procedure consists of integrating the equations and then evaluating the constants of integration from boundary and other conditions pertaining to the beam.

When deriving the differential equations [Eqs. (9-13), (9-14), and (9-16)], it was assumed that the material followed Hooke's law and that the slopes of the deflection curve were very small. Also any shear deformations were assumed to be negligible; consequently, only the deformations due to pure bending were considered. All of these assumptions are satisfied by most beams in common use.

Exact Expression for Curvature

If the deflection curve of a beam has large slopes, you cannot use the approximations given by Eqs. (9-5) and (9-7). Instead, the exact expressions for curvature and angle of rotation must be used [see Eqs. (9-2) and (9-3b)]. Combine those expressions to get

$$\kappa = \frac{1}{\rho} = \frac{d\theta}{ds} = \frac{d(\arctan v')}{dx}\frac{dx}{ds} \tag{9-17}$$

From Fig. 9-2, note that

$$ds^2 = dx^2 + dv^2 \qquad \text{or} \qquad ds = [dx^2 + dv^2]^{1/2} \tag{9-18a,b}$$

Dividing both sides of Eq. (9-18b) by dx gives

$$\frac{ds}{dx} = \left[1 + \left(\frac{dv}{dx}\right)^2\right]^{1/2} = [1 + (v')^2]^{1/2} \qquad \text{or} \qquad \frac{dx}{ds} = \frac{1}{[1 + (v')^2]^{1/2}} \tag{9-18c,d}$$

Also, differentiation of the arctangent function (see Appendix C) gives

$$\frac{d}{dx}(\arctan v') = \frac{v''}{1 + (v')^2} \tag{9-18e}$$

Substitution of expressions Eqs. (9-18d and e) into the equation for curvature from Eq. (9-17) yields

$$\kappa = \frac{1}{\rho} = \frac{v''}{[1 + (v')^2]^{3/2}} \quad \text{(9-19)}$$

Comparing this equation with Eq. (9-9), note that the assumption of small rotations is equivalent to disregarding $(v')^2$ in comparison to one. Equation (9-19) should be used for the curvature whenever the slopes are large.[2]

9.3 Deflections by Integration of the Bending-Moment Equation

To solve the differential equations of the deflection curve and obtain deflections of beams, the first equation you will use is the bending-moment equation (Eq. 9-16a). Since this equation is of second order, two integrations are required. The first integration produces the slope $v' = dv/dx$, and the second produces the deflection v.

Begin the analysis by writing the equation (or equations) for the bending moments in the beam. Since statically determinate beams are considered first in this chapter, obtain the bending moments from free-body diagrams and equations of equilibrium, using the procedures described in Chapter 4. In some cases, a single bending-moment expression holds for the entire length of the beam, as illustrated in Examples 9-1 and 9-2. In other cases, the bending moment changes abruptly at one or more points along the axis of the beam. Then write separate bending-moment expressions for each region of the beam between points where changes occur, as illustrated in Example 9-3.

Regardless of the number of bending-moment expressions, the general procedure for solving the differential equations is as follows. For each region of the beam, substitute the expression for M into the differential equation and integrate to obtain the slope v'. Each such integration produces one constant of integration. Next, integrate each slope equation to obtain the corresponding deflection v. Again, each integration produces a new constant. Thus, there are two constants of integration for each region of the beam. These constants are evaluated from known conditions pertaining to the slopes and deflections. The conditions fall into three categories: (1) boundary conditions, (2) continuity conditions, and (3) symmetry conditions.

Boundary conditions pertain to the deflections and slopes at the supports of a beam. For example, at a simple support (either a pin or a roller), the deflection is zero (Fig. 9-5), and at a fixed support, both the deflection and the slope are zero (Fig. 9-6). Each such boundary condition supplies one equation that can be used to evaluate the constants of integration.

FIGURE 9-5

Boundary conditions at simple supports

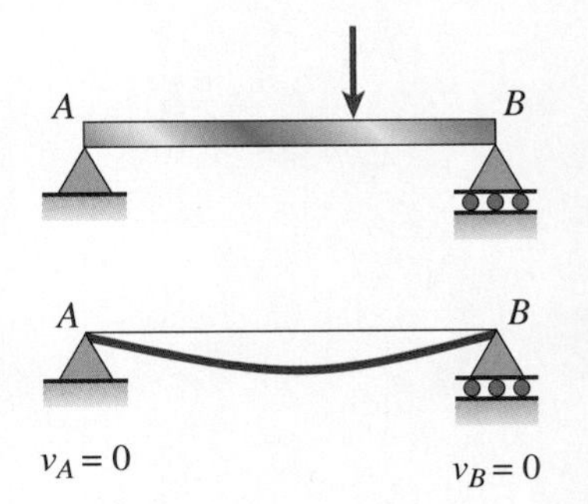

FIGURE 9-6

Boundary conditions at a fixed support

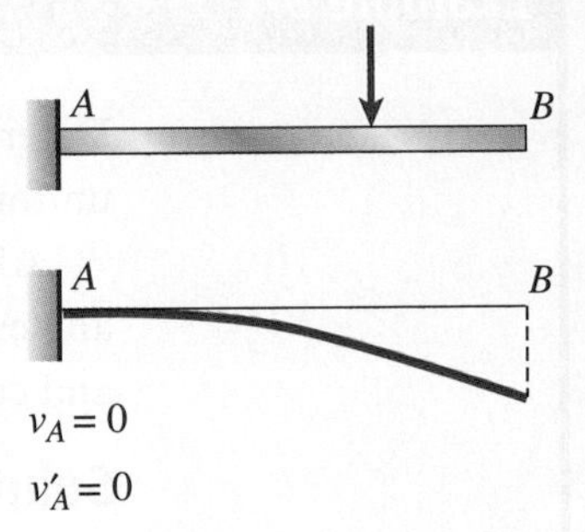

[2]The basic relationship stating that the curvature of a beam is proportional to the bending moment (Eq. 9-10) was first obtained by Jacob Bernoulli, although he obtained an incorrect value for the constant of proportionality. The relationship was used later by Euler, who solved the differential equation of the deflection curve for both large deflections (using Eq. 9-19) and small deflections (using Eq. 9-11).

FIGURE 9-7

Continuity conditions at point C

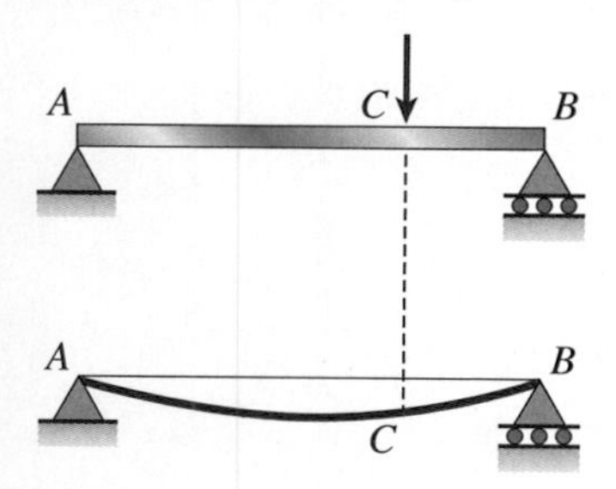

At point C: $(v)_{AC} = (v)_{CB}$
$(v')_{AC} = (v')_{CB}$

Continuity conditions occur at points where the regions of integration meet, such as at point C in the beam of Fig. 9-7. The deflection curve of this beam is physically continuous at point C; therefore, the deflection at point C as determined for the left-hand part of the beam must be equal to the deflection at point C as determined for the right-hand part. Similarly, the slopes found for each part of the beam must be equal at point C. Each of these continuity conditions supplies an equation for evaluating the constants of integration.

Symmetry conditions also may be available. For instance, if a simple beam supports a uniform load throughout its length, the slope of the deflection curve at the midpoint must be zero. This condition supplies an additional equation, as illustrated in Example 9-1.

Each boundary, continuity, and symmetry condition leads to an equation containing one or more of the constants of integration. Since the number of *independent* conditions always matches the number of constants of integration, you can always solve these equations for the constants. (The boundary and continuity conditions alone are always sufficient to determine the constants. Any symmetry conditions provide additional equations, but they are not independent of the other equations. The choice of which conditions to use is a matter of convenience.)

Once the constants are evaluated, they can be substituted back into the expressions for slopes and deflections, thus yielding the final equations of the deflection curve. These equations then can be used to obtain the deflections and angles of rotation at particular points along the axis of the beam.

The preceding method for finding deflections is sometimes called the **method of successive integrations**. The following examples illustrate this method in detail.

Note: Sketches of deflection curves, such as those shown in the following examples and in Figs. 9-5, 9-6, and 9-7, are greatly exaggerated for clarity. However, it always should be kept in mind that the actual deflections are very small quantities.

Example 9-1

Determine the equation of the deflection curve for a simple beam AB supporting a uniform load of intensity q acting throughout the span of the beam (Fig. 9-8a).

Also, determine the maximum deflection δ_{max} at the midpoint of the beam and the angles of rotation θ_A and θ_B at the supports (Fig. 9-8b). *Note:* The beam has length L and constant flexural rigidity EI.

Solution:

Use a four-step problem-solving approach.

1. ***Conceptualize*** [*hypothesize, sketch*]: The beam is statically determinate. Begin by finding reaction R_A. The left-hand free-body diagram in Fig. 9-9 is then used to obtain an expression for internal moment $M(x)$.

   Bending moment in the beam: The bending moment at a cross-section a distance x from the left-hand support is obtained from the free-body diagram

FIGURE 9-8

Example 9-1: Deflections of a simple beam with a uniform load

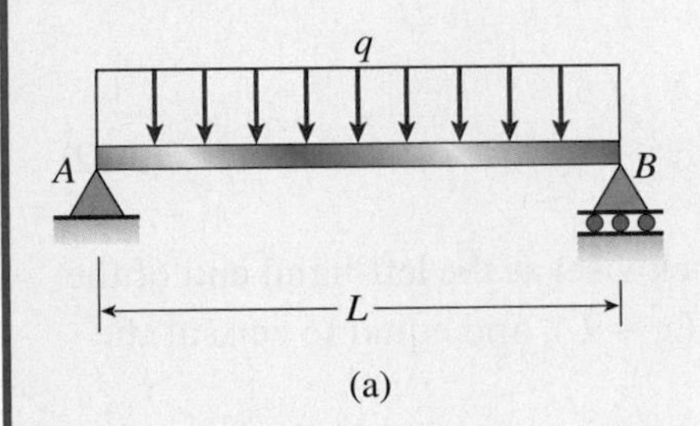

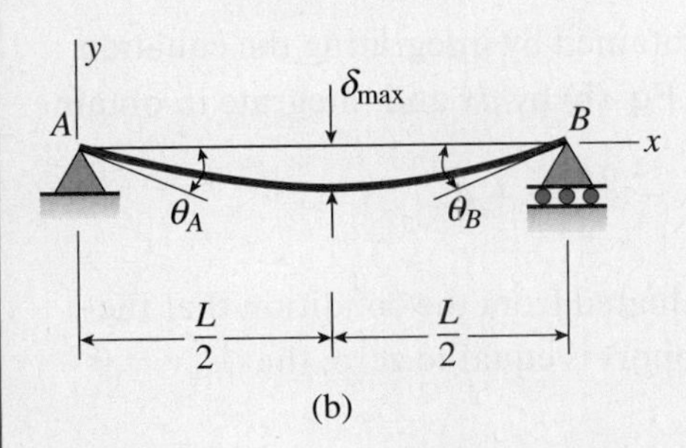

FIGURE 9-9

Example 9-1: Free-body diagram used in determining the bending moment M

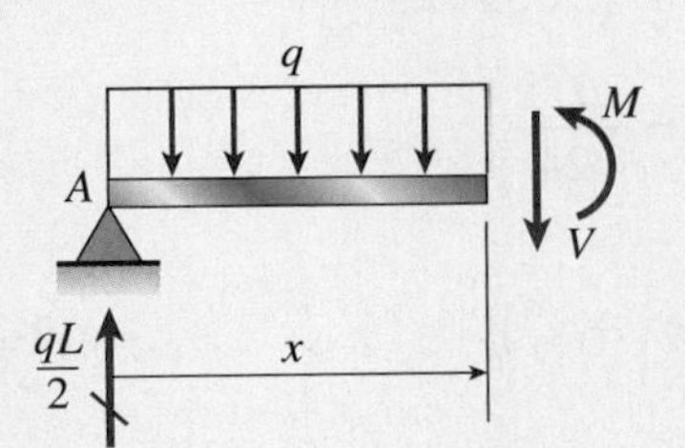

of Fig. 9-9. Since the reaction at the support is $qL/2$, the equation for the bending moment is

$$M = \frac{qL}{2}(x) - qx\left(\frac{x}{2}\right) = \frac{qLx}{2} - \frac{qx^2}{2} \qquad \textbf{(9-20)}$$

2. ***Categorize*** [*simplify, classify*]: Use the expression for internal moment $M(x)$ in the bending-moment equation [Eq. (9-16a)] to find expressions for slopes and deflections for this beam.

Differential equation of the deflection curve: Substitute the expression for the bending moment (Eq. 9-20) into the differential equation (Eq. 9-16a) to obtain

$$EIv'' = \frac{qLx}{2} - \frac{qx^2}{2} \qquad \textbf{(9-21)}$$

3. ***Analyze*** [*evaluate; select relevant equations; carry out mathematical solution*]: This equation can now be integrated to obtain the slope and deflection of the beam.

Slope of the beam: Multiply both sides of the differential equation by dx to get

$$EIv''\,dx = \frac{qLx}{2}dx - \frac{qx^2}{2}dx$$

Integrate each term to obtain

$$EI\int v''dx = \int \frac{qLx}{2}dx - \int \frac{qx^2}{2}dx$$

or

$$EIv' = \frac{qLx^2}{4} - \frac{qx^3}{6} + C_1 \qquad \textbf{(a)}$$

in which C_1 is a constant of integration.

To evaluate the constant C_1, observe from the symmetry of the beam and its load that the slope of the deflection curve at mid-span is equal to zero. Thus, the following symmetry condition applies here:

$$v' = 0 \quad \text{when} \quad x = \frac{L}{2}$$

This condition may be expressed more succinctly as

$$v'\left(\frac{L}{2}\right) = 0$$

Applying this condition to Eq. (a) gives

$$0 = \frac{qL}{4}\left(\frac{L}{2}\right)^2 - \frac{q}{6}\left(\frac{L}{2}\right)^3 + C_1 \quad \text{or} \quad C_1 = -\frac{qL^3}{24}$$

The equation for the slope of the beam [Eq. (a)] then becomes

$$EIv' = \frac{qLx^2}{4} - \frac{qx^3}{6} - \frac{qL^3}{24} \tag{b}$$

or

$$v' = -\frac{q}{24EI}(L^3 - 6Lx^2 + 4x^3) \tag{9-22}$$

As expected, the slope is negative (that is, clockwise) at the left-hand end of the beam ($x = 0$), positive at the right-hand end ($x = L$), and equal to zero at the midpoint ($x = L/2$).

Deflection of the beam: The deflection is obtained by integrating the equation for the slope. Thus, multiply both sides of Eq. (b) by dx and integrate to obtain

$$EIv = \frac{qLx^3}{12} - \frac{qx^3}{24} - \frac{qL^3x}{24} + C_2 \tag{c}$$

The constant of integration C_2 may be evaluated from the condition that the deflection of the beam at the left-hand support is equal to zero; that is, $v = 0$ when $x = 0$, or

$$v(0) = 0$$

Applying this condition to Eq. (c) yields $C_2 = 0$; hence, the equation for the deflection curve is

$$EIv = \frac{qLx^3}{12} - \frac{qx^4}{24} - \frac{qL^3x}{24} \tag{d}$$

or

$$v = -\frac{qx}{24EI}(L^3 - 2Lx^2 + x^3) \tag{9-23}$$

This equation gives the deflection at any point along the axis of the beam. Note that the deflection is zero at both ends of the beam ($x = 0$ and $x = L$) and negative elsewhere (recall that downward deflections are negative).

Maximum deflection: From symmetry, the maximum deflection occurs at the midpoint of the span (Fig. 9-8b). Thus, set x equal to $L/2$ in Eq. (9-23) to obtain

$$v\left(\frac{L}{2}\right) = -\frac{5qL^4}{384EI}$$

in which the negative sign means that the deflection is downward (as expected). Variable δ_{max} represents the magnitude of this deflection and is expressed as

$$\delta_{max} = \left|v\left(\frac{L}{2}\right)\right| = \frac{5qL^4}{384EI} \tag{9-24}$$

Angles of rotation: The maximum angles of rotation occur at the supports of the beam. At the left-hand end of the beam, the angle θ_A, which is a clockwise angle (Fig. 9-8b), is equal to the negative of the slope v'. Thus, substitute $x = 0$ into Eq. (9-22) to find

$$\theta_A = -v'(0) = \frac{qL^3}{24EI} \tag{9-25}$$

In a similar manner, obtain the angle of rotation θ_B at the right-hand end of the beam. Since θ_B is a counterclockwise angle, it is equal to the slope at the end:

$$\theta_B = v'(L) = \frac{qL^3}{24EI} \tag{9-26}$$

4. ***Finalize*** [*conclude; examine answer—Does it make sense? Are units correct? How does it compare to similar problem solutions?*]: Because the beam and loading are symmetric about the midpoint, the angles of rotation at the ends are equal.

This example illustrates the process of setting up and solving the differential equation of the deflection curve. It also illustrates the process of finding slopes and deflections at selected points along the axis of a beam.

Note: Now that formulas for the maximum deflection and maximum angles of rotation have been derived [see Eqs. (9-24), (9-25), and (9-26)], you can evaluate those quantities numerically and observe that the deflections and angles are indeed small, as the theory requires.

Consider a steel beam on simple supports with a span length $L = 2$ m. The cross section is rectangular with width $b = 75$ mm and height $h = 150$ mm. The intensity of uniform load is $q = 100$ kN/m, which is relatively large because it produces a stress in the beam of 178 MPa. (Thus, the deflections and slopes are larger than would normally be expected.)

Substitute into Eq. (9-24), and use $E = 210$ GPa to find that the maximum deflection is $\delta_{max} = 4.7$ mm, which is only 1/500 of the span length. Also, from Eq. (9-25), the maximum angle of rotation is computed as $\theta_A = 0.0075$ radians, or 0.43°, which is a very small angle.

Thus, the assumption that the slopes and deflections are small is validated.

Example 9-2

Determine the equation of the deflection curve for a cantilever beam AB subjected to a uniform load of intensity q (Fig. 9-10a).

Also, determine the angle of rotation θ_B and the deflection δ_B at the free end (Fig. 9-10b). *Note:* The beam has length L and constant flexural rigidity EI.

FIGURE 9-10

Example 9-2: Deflections of a cantilever beam with a uniform load

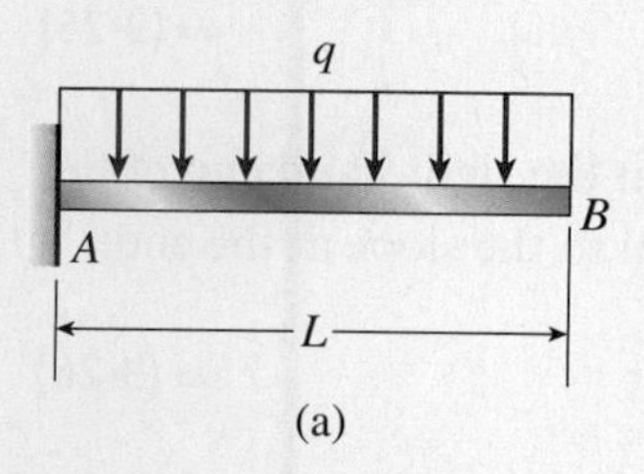

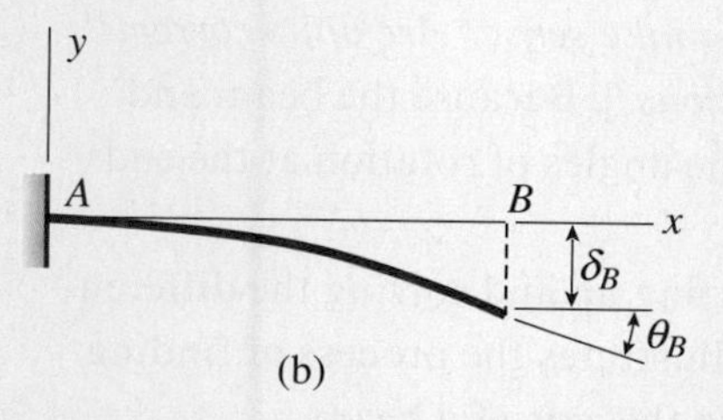

FIGURE 9-11

Example 9-2: Free-body diagram used in determining the bending moment M

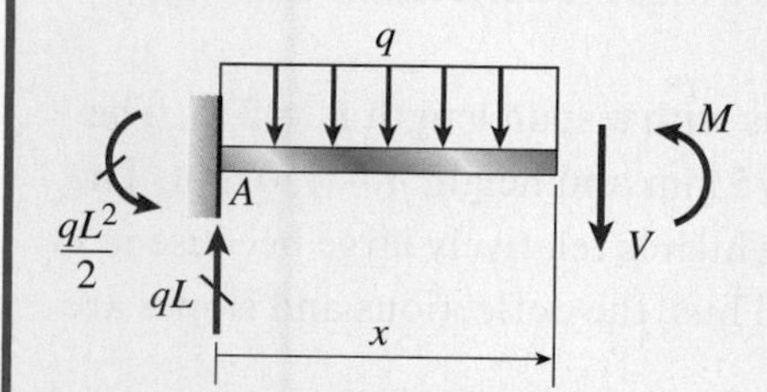

Solution:

Use a four-step problem-solving approach.

1. ***Conceptualize***: The beam is statically determinate. Begin by finding reaction force R_A and reaction moment M_A. The left-hand free-body diagram in Fig. 9-11 is then used to obtain an expression for internal moment $M(x)$.

   Bending moment in the beam: The bending moment at distance x from the fixed support is obtained from the free-body diagram of Fig. 9-11. Note that the vertical reaction at the support is equal to qL and the moment reaction is equal to $qL^2/2$. Consequently, the expression for the bending moment M is

$$M = -\frac{qL^2}{2} + qLx - \frac{qx^2}{2} \qquad \textbf{(9-27)}$$

2. ***Categorize***: Use the expression for internal moment $M(x)$ in the bending moment equation (Eq. 9-14a) to find expressions for slopes and deflections for this beam.

   Differential equation of the deflection curve: When the preceding expression for the bending moment is substituted into the differential equation (Eq. 9-16a), the following equation is obtained:

$$EIv'' = -\frac{qL^2}{2} + qLx - \frac{qx^2}{2} \qquad \textbf{(9-28)}$$

3. ***Analyze***: Now integrate both sides of this equation to obtain the slopes and deflections.

Slope of the beam: The first integration of Eq. (9-28) gives the following equation for the slope:

$$EIv' = -\frac{qL^2x}{2} + \frac{qLx^2}{2} - \frac{qx^3}{6} + C_1 \qquad \textbf{(a)}$$

The constant of integration C_1 can be found from the boundary condition that the slope of the beam is zero at the support; which is expressed as

$$v'(0) = 0$$

When this condition is applied to Eq. (a), the result is $C_1 = 0$. Therefore, Eq. (a) becomes

$$EIv' = -\frac{qL^2x}{2} + \frac{qLx^2}{2} - \frac{qx^3}{6} \qquad \textbf{(b)}$$

and the slope is

$$v' = -\frac{qx}{6EI}(3L^2 - 3Lx + x^2) \qquad \textbf{(9-29)}$$

As expected, the slope obtained from this equation is zero at the support ($x = 0$) and negative (i.e., clockwise) throughout the length of the beam.

Deflection of the beam: Integration of the slope equation [Eq. (b)] yields

$$EIv = -\frac{qL^2x^2}{4} + \frac{qLx^3}{6} - \frac{qx^4}{24} + C_2 \quad \text{(c)}$$

The constant C_2 is found from the boundary condition that the deflection of the beam is zero at the support:

$$v(0) = 0$$

When this condition is applied to Eq. (c), the result is $C_2 = 0$. Therefore, the equation for the deflection v is

$$v = -\frac{qx^2}{24EI}(6L^2 - 4Lx + x^2) \quad \text{(9-30)}$$

As expected, the deflection obtained from this equation is zero at the support ($x = 0$) and negative (that is, downward) elsewhere.

Angle of rotation at the free end of the beam: The clockwise angle of rotation θ_B at end B of the beam (Fig. 9-10b) is equal to the negative of the slope at that point. Thus, use Eq. (9-29) to get

$$\theta_B = -v'(L) = \frac{qL^3}{6EI} \quad \text{(9-31)}$$

This angle is the maximum angle of rotation for the beam.

Deflection at the free end of the beam: Since the deflection δ_B is downward (Fig. 9-10b), it is equal to the negative of the deflection obtained from Eq. (9-30):

$$\delta_B = -v(L) = \frac{qL^4}{8EI} \quad \text{(9-32)}$$

This deflection is the maximum deflection of the beam.

4. ***Finalize***: Equations (9-29) to (9-32) are listed as Case 1 in Table H-1, Appendix H.

Example 9-3

A simple beam AB supports a concentrated load P acting at distances a and b from the left-hand and right-hand supports, respectively (Fig. 9-12a).

Determine the equations of the deflection curve, the angles of rotation θ_A and θ_B at the supports, the maximum deflection δ_{max}, and the deflection δ_C at the midpoint C of the beam (Fig. 9-12b). *Note:* The beam has length L and constant flexural rigidity EI.

FIGURE 9-12

Example 9-3: Deflections of a simple beam with a concentrated load

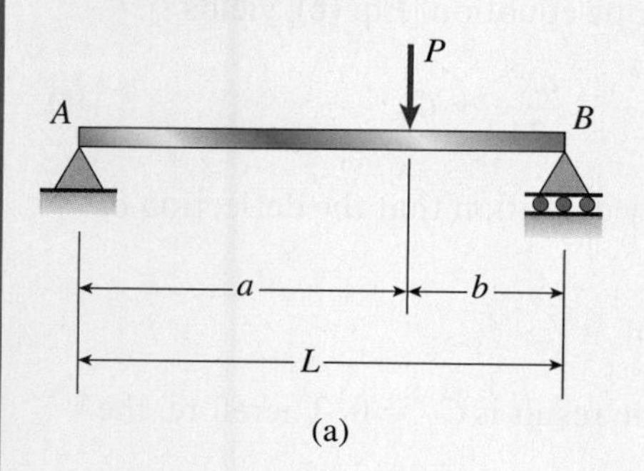

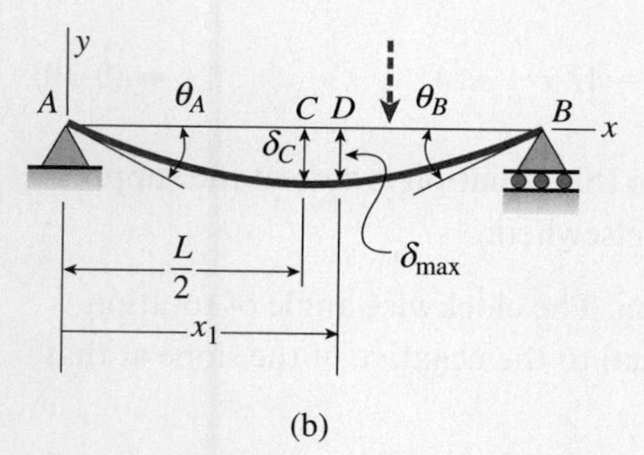

FIGURE 9-13

Example 9-3: Free-body diagrams used in determining the bending moments

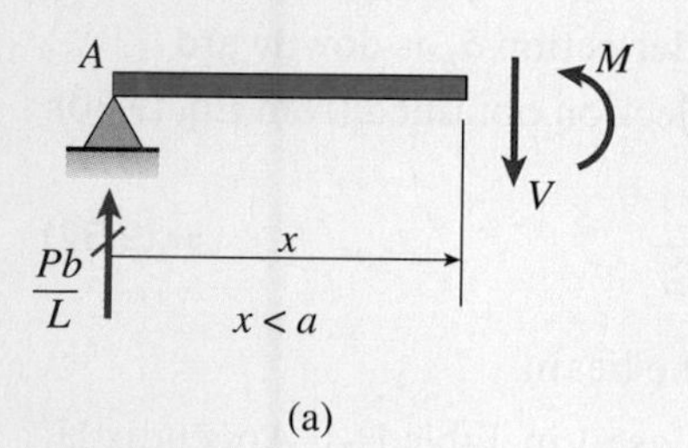

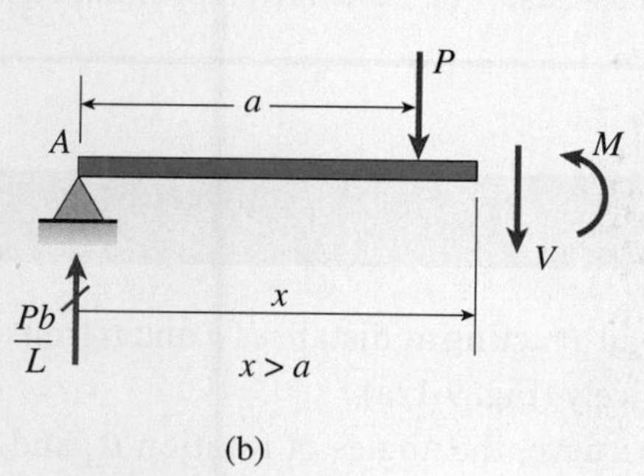

Solution:

Use a four-step problem-solving approach.

1. ***Conceptualize***: Internal shear $V(x)$ and moment $M(x)$ are discontinuous at load point P. Two free-body diagrams are required to obtain moment expressions over the entire length of the beam (see Figs. 9-13a and b).

   Bending moments in the beam: In this example, the bending moments are expressed by two equations—one for each part of the beam. Use the free-body diagrams of Fig. 9-13 to arrive at

$$M = \frac{Pbx}{L} \qquad (0 \le x \le a) \qquad \textbf{(9-33a)}$$

$$M = \frac{Pbx}{L} - P(x-a) \qquad (a \le x \le L) \qquad \textbf{(9-33b)}$$

2. ***Categorize***:

   Differential equations of the deflection curve: The differential equations for the two parts of the beam are obtained by substituting the bending-moment expressions [Eqs. (9-33a and b)] into Eq. (9-16a). The results are

$$EIv'' = \frac{Pbx}{L} \qquad (0 \le x \le a) \qquad \textbf{(9-34a)}$$

$$EIv'' = \frac{Pbx}{L} - P(x-a) \qquad (a \le x \le L) \qquad \textbf{(9-34b)}$$

3. ***Analyze***:

   Slopes and deflections of the beam: The first integrations of the two differential equations yield the following expressions for the slopes:

$$EIv' = \frac{Pbx^2}{2L} + C_1 \qquad (0 \le x \le a) \qquad \textbf{(a)}$$

$$EIv' = \frac{Pbx^2}{2L} - \frac{P(x-a)^2}{2} + C_2 \qquad (a \le x \le L) \qquad \textbf{(b)}$$

   in which C_1 and C_2 are constants of integration. A second pair of integrations gives the deflections:

$$EIv = \frac{Pbx^3}{6L} + C_1x + C_3 \qquad (0 \le x \le a) \qquad \textbf{(c)}$$

$$EIv = \frac{Pbx^3}{6L} - \frac{P(x-a)^3}{6} + C_2x + C_4 \qquad (a \le x \le L) \qquad \textbf{(d)}$$

These equations contain two additional constants of integration, making a total of four constants to be evaluated.

Constants of integration: The four constants of integration can be found from the following four conditions:

i. At $x = a$, the slopes v' for the two parts of the beam are the same.

ii. At $x = a$, the deflections v for the two parts of the beam are the same.

iii. At $x = 0$, the deflection v is zero.

iv. At $x = L$, the deflection v is zero.

The first two conditions are continuity conditions based upon the fact that the axis of the beam is a continuous curve. Conditions (iii) and (iv) are boundary conditions that must be satisfied at the supports.

Condition (i) means that the slopes determined from Eqs. (a) and (b) must be equal when $x = a$; therefore,

$$\frac{Pba^2}{2L} + C_1 = \frac{Pba^2}{2L} + C_2 \quad \text{or} \quad C_1 = C_2$$

Condition (ii) means that the deflections found from Eqs. (c) and (d) must be equal when $x = a$; therefore,

$$\frac{Pba^3}{6L} + C_1 a + C_3 = \frac{Pba^3}{6L} + C_2 a + C_4$$

Inasmuch as $C_1 = C_2$, this equation gives $C_3 = C_4$.

Next, apply condition (iii) to Eq. (c) and obtain $C_3 = 0$; therefore,

$$C_3 = C_4 = 0 \tag{e}$$

Finally, apply condition (iv) to Eq. (d) and obtain

$$\frac{PbL^2}{6} - \frac{Pb^3}{6} + C_2 L = 0$$

Therefore,

$$C_1 = C_2 = -\frac{Pb(L^2 - b^2)}{6L} \tag{f}$$

Equations of the deflection curve: Now substitute the constants of integration [Eqs. (e) and (f)] into the equations for the deflections [Eqs. (c) and (d)] and obtain the deflection equations for the two parts of the beam. The resulting equations, after a slight rearrangement, are

$$v = -\frac{Pbx}{6LEI}(L^2 - b^2 - x^2) \quad (0 \le x \le a) \tag{9-35a}$$

$$v = -\frac{Pbx}{6LEI}(L^2 - b^2 - x^2) - \frac{P(x - a)^3}{6EI} \quad (a \le x \le L) \tag{9-35b}$$

The first of these equations gives the deflection curve for the part of the beam to the left of the load P, and the second gives the deflection curve for the part of the beam to the right of the load.

The slopes for the two parts of the beam can be found either by substituting the values of C_1 and C_2 into Eqs. (a) and (b) or by taking the first derivatives of the deflection equations [Eqs. (9-35a and b)]. The resulting equations are

$$v' = -\frac{Pb}{6LEI}(L^2 - b^2 - 3x^2) \qquad (0 \leq x \leq a) \qquad \textbf{(9-36a)}$$

$$v' = -\frac{Pb}{6LEI}(L^2 - b^2 - 3x^2) - \frac{P(x-a)^2}{2EI} \qquad (a \leq x \leq L) \qquad \textbf{(9-36b)}$$

The deflection and slope at any point along the axis of the beam can be calculated from Eqs. (9-35) and (9-36).

Angles of rotation at the supports: To obtain the angles of rotation θ_A and θ_B at the ends of the beam (Fig. 9-12b), substitute $x = 0$ into Eq. (9-36a) and $x = L$ into Eq. (9-36b):

$$\theta_A = -v'(0) = \frac{Pb(L^2 - b^2)}{6LEI} = \frac{Pab(L+b)}{6LEI} \qquad \textbf{(9-37a)}$$

$$\theta_B = v'(L) = \frac{Pb(2L^2 - 3bL + b^2)}{6LEI} = \frac{Pab(L+a)}{6LEI} \qquad \textbf{(9-37b)}$$

FIGURE (9-12b Repeated)

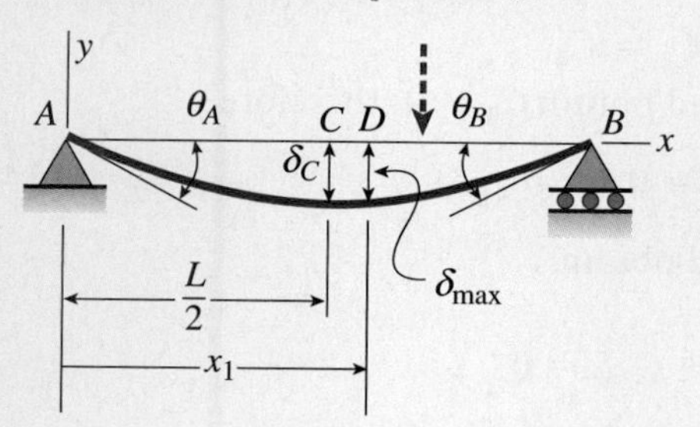

Note that the angle θ_A is clockwise and the angle θ_B is counterclockwise, as shown in Fig. 9-12b.

The angles of rotation are functions of the position of the load and reach their largest values when the load is located near the midpoint of the beam. In the case of the angle of rotation θ_A, the maximum value of the angle is

$$(\theta_A)_{max} = \frac{PL^2\sqrt{3}}{27EI} \qquad \textbf{(9-38)}$$

and occurs when $b = L/\sqrt{3} = 0.577L$ (or $a = 0.423L$). This value of b is obtained by taking the derivative of θ_A with respect to b [using the first of the two expressions for θ_A in Eq. (9-37a)] and then setting it equal to zero.

Maximum deflection of the beam: The maximum deflection δ_{max} occurs at point D (Fig. 9-12b) where the deflection curve has a horizontal tangent. If the load is to the right of the midpoint, that is, if $a > b$, point D is in the part of the beam to the left of the load. Locate this point by equating the slope v' from Eq. (9-36a) to zero and solving for the distance x, which is now denoted as x_1. In this manner, obtain the following formula for x_1:

$$x_1 = \sqrt{\frac{L^2 - b^2}{3}} \qquad (a \geq b) \qquad \textbf{(9-39)}$$

From this equation, note that as the load P moves from the middle of the beam ($b = L/2$) to the right-hand end ($b = 0$), the distance x_1 varies from $L/2$ to $L/\sqrt{3} = 0.577L$. Thus, the maximum deflection occurs at a point very close to the midpoint of the beam, and this point is always between the midpoint of the beam and the load.

The maximum deflection δ_{max} is found by substituting x_1 [from Eq. (9-39)] into the deflection equation [Eq. (9-35a)] and then inserting a minus sign:

$$\delta_{max} = -(v)_{x=x1} = \frac{Pb(L^2 - b^2)^{3/2}}{9\sqrt{3}\, LEI} \qquad (a \geq b) \qquad \text{(9-40)}$$

The minus sign is needed because the maximum deflection is downward (Fig. 9-12b), whereas the deflection v is positive upward.

The maximum deflection of the beam depends on the position of the load P, that is, on the distance b. The maximum value of the maximum deflection (the "max-max" deflection) occurs when $b = L/2$ and the load is at the midpoint of the beam. This maximum deflection is equal to $PL^3/48EI$.

Deflection at the midpoint of the beam: The deflection δ_C at the midpoint C when the load is acting to the right of the midpoint (Fig. 9-12b) is obtained by substituting $x = L/2$ into Eq. (9-35a), as

$$\delta_C = -v\left(\frac{L}{2}\right) = \frac{Pb(3L^2 - 4b^2)}{48EI} \qquad (a \geq b) \qquad \text{(9-41)}$$

4. ***Finalize***: Because the maximum deflection always occurs near the midpoint of the beam, Eq. (9-41) yields a close approximation to the maximum deflection. In the most unfavorable case (when b approaches zero), the difference between the maximum deflection and the deflection at the midpoint is less than 3% of the maximum deflection, as demonstrated in Problem 9.3-9.

Special case (load at the midpoint of the beam): An important special case occurs when the load P acts at the midpoint of the beam ($a = b = L/2$). Then the following results are obtained from Eqs. (9-36a), (9-35a), (9-37), and (9-40), respectively:

$$v' = -\frac{P}{16EI}(L^2 - 4x^2) \qquad \left(0 \leq x \leq \frac{L}{2}\right) \qquad \text{(9-42)}$$

$$v = -\frac{Px}{48EI}(3L^2 - 4x^2) \qquad \left(0 \leq x \leq \frac{L}{2}\right) \qquad \text{(9-43)}$$

$$\theta_A = \theta_B = \frac{PL^2}{16EI} \qquad \text{(9-44)}$$

$$\delta_{max} = \delta_C = \frac{PL^3}{48EI} \qquad \text{(9-45)}$$

Since the deflection curve is symmetric about the midpoint of the beam, the equations for v' and v are given only for the left-hand half of the beam in Eqs. (9-42) and (9-43). If needed, the equations for the right-hand half can be obtained from Eqs. (9-36b) and (9-35b) by substituting $a = b = L/2$.

9.4 Deflections by Integration of the Shear-Force and Load Equations

The equations of the deflection curve in terms of the shear force V and the load q [Eqs. (9-16b and c), respectively] also may be integrated to obtain slopes and deflections. Since the loads are usually known quantities, whereas the bending moments must be determined from free-body diagrams and equations of equilibrium, many analysts prefer to start with the load equation. For this same reason, most computer programs for finding deflections begin with the load equation and then perform numerical integrations to obtain the shear forces, bending moments, slopes, and deflections.

The procedure for solving either the load equation or the shear-force equation is similar to that for solving the bending-moment equation, except that more integrations are required. For instance, with the load equation, four integrations are needed in order to arrive at the deflections. Thus, four constants of integration are introduced for each load equation that is integrated. As before, these constants are found from boundary, continuity, and symmetry conditions. However, these conditions now include conditions on the shear forces and bending moments as well as conditions on the slopes and deflections.

Conditions on the shear forces are equivalent to conditions on the third derivative (because $EIv''' = V$). In a similar manner, conditions on the bending moments are equivalent to conditions on the second derivative (because $EIv'' = M$). When the shear-force and bending-moment conditions are added to those for the slopes and deflections, there are always enough independent conditions to solve for the constants of integration.

The following examples illustrate the techniques of analysis in detail. The first example begins with the load equation and the second begins with the shear-force equation.

Example 9-4

Determine the equation of the deflection curve for a cantilever beam AB supporting a triangularly distributed load of maximum intensity q_0 (Fig. 9-14a).

Also, determine the deflection δ_B and angle of rotation θ_B at the free end (Fig. 9-14b). Use the fourth-order differential equation of the deflection curve (the load equation). *Note:* The beam has length L and constant flexural rigidity EI.

Solution:

Use a four-step problem-solving approach.

1. ***Conceptualize***: The cantilever beam is statically determinate. Use the load equation to find the deflection curve of the beam. An expression for load intensity $q(x)$ can be obtained as shown in Fig. 9-14a.

Differential equation of the deflection curve: The intensity of the distributed load is given by (see Fig. 9-14a):

$$q(x) = \frac{q_0(L - x)}{L} \qquad \textbf{(9-46)}$$

Cantilever portion of roof structure (Courtesy of the National Information Service for Earthquake Engineering EERC, University of California, Berkeley)

FIGURE 9-14

Example 9-4: Deflections of a cantilever beam with a triangular load

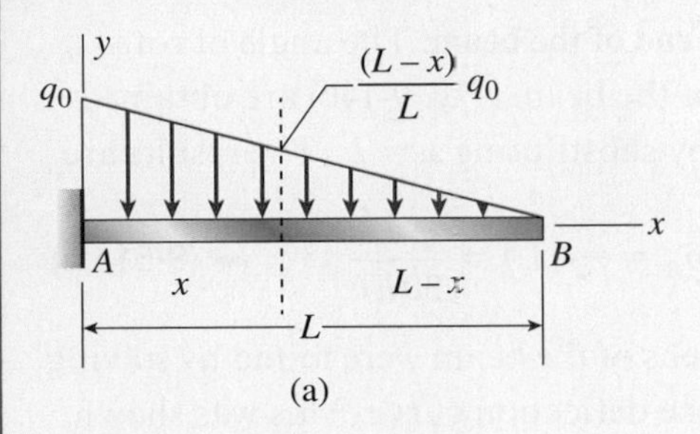

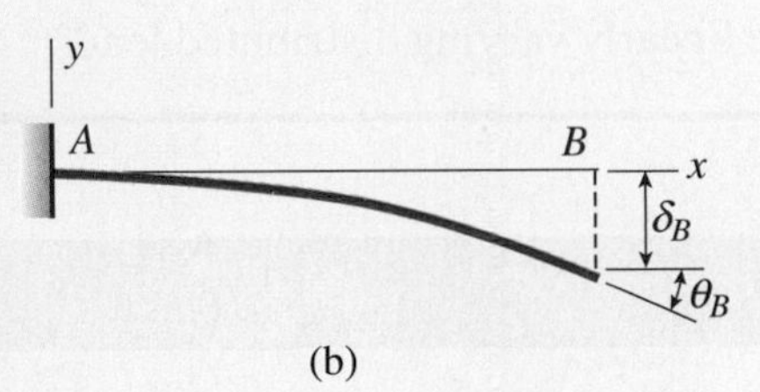

2. ***Categorize***: Consequently, the fourth-order differential equation [Eq. (9-16c)] becomes

$$EIv'''' = -q(x) = -\frac{q_0(L-x)}{L} \qquad \text{(a)}$$

3. ***Analyze:***

Shear force in the beam: The first integration of Eq. (a) gives

$$EIv''' = \frac{q_0}{2L}(L-x)^2 + C_1 \qquad \text{(b)}$$

The right-hand side of this equation represents the shear force V [see Eq. (9-16b)]. Because the shear force is zero at $x = L$, the boundary condition is:

$$v'''(L) = 0$$

Use this condition with Eq. (b), to get $C_1 = 0$. Therefore, Eq. (b) simplifies to

$$EIv''' = \frac{q_0}{2L}(L-x)^2 \qquad \text{(c)}$$

and the shear force in the beam is

$$V = EIv''' = \frac{q_0}{2L}(L-x)^2 \qquad \text{(9-47)}$$

Bending moment in the beam: Integrate a second time to obtain the following equation from Eq. (c):

$$EIv'' = -\frac{q_0}{6L}(L-x)^3 + C_2 \qquad \text{(d)}$$

This equation is equal to the bending moment M [see Eq. (9-16a)]. The bending moment is zero at the free end of the beam, so the boundary condition is

$$v''(L) = 0$$

Apply this condition to Eq. (d) to obtain $C_2 = 0$, so the bending moment is

$$M = EIv'' = -\frac{q_0}{6L}(L-x)^3 \qquad \text{(9-48)}$$

Slope and deflection of the beam: The third and fourth integrations yield

$$EIv' = \frac{q_0}{24L}(L-x)^4 + C_3 \qquad \text{(e)}$$

$$EIv = -\frac{q_0}{120L}(L-x)^5 + C_3x + C_4 \qquad \text{(f)}$$

The boundary conditions at the fixed support, where both the slope and deflection are equal to zero, are

$$v'(0) = 0 \qquad v(0) = 0$$

Apply these conditions to Eqs. (e) and (f), respectively, to find

$$C_3 = -\frac{q_0 L^3}{24} \qquad C_4 = \frac{q_0 L^4}{120}$$

Substitute these expressions for the constants into Eqs. (e) and (f) to obtain the equations for the slope and deflection of the beam:

$$v' = -\frac{q_0 x}{24LEI}(4L^3 - 6L^2x + 4Lx^2 - x^3) \tag{9-49}$$

$$v = -\frac{q_0 x^2}{120LEI}(10L^3 - 10L^2x + 5Lx^2 - x^3) \tag{9-50}$$

Angle of rotation and deflection at the free end of the beam: The angle of rotation θ_B and deflection δ_B at the free end of the beam (Fig. 9-14b) are obtained from Eqs. (9-49) and (9-50), respectively, by substituting $x = L$. The results are

$$\theta_B = -v'(L) = \frac{q_0 L^3}{24EI} \qquad \delta_B = -v(L) = \frac{q_0 L^4}{30EI} \tag{9-51a,b}$$

4. ***Finalize***: The required slopes and deflections of the beam were found by solving the fourth-order differential equation of the deflection curve. This was shown to be an efficient approach for use with the linearly varying distributed load.

Example 9-5

A simple beam AB with an overhang BC supports a concentrated load P at the end of the overhang (Fig. 9-15a). The main span of the beam has length L and the overhang has length $L/2$.

Determine the equations of the deflection curve and the deflection δ_C at the end of the overhang (Fig. 9-15b). Use the third-order differential equation of the deflection curve (the shear-force equation). *Note:* The beam has constant flexural rigidity EI.

FIGURE 9-15

Example 9-5: Deflections of a beam with an overhang

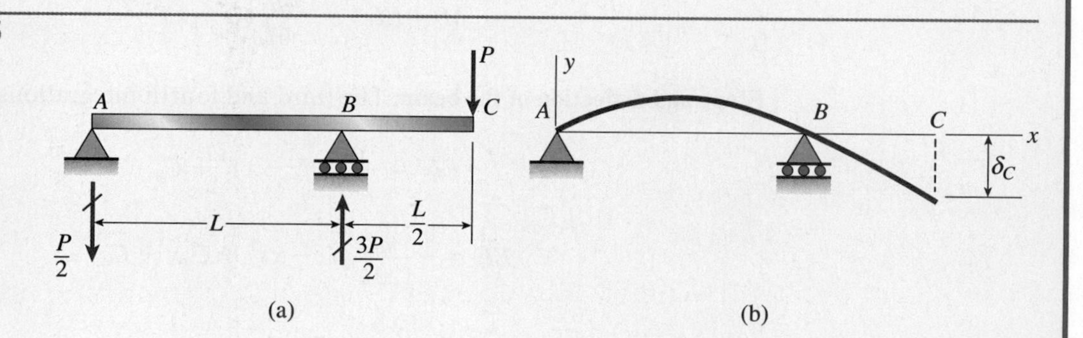

Solution:

Use a four-step problem-solving approach.

1. *Conceptualize*:

Differential equations of the deflection curve: Because reactive forces act at supports A and B, write separate differential equations for parts AB and BC of the beam. Begin by finding the shear forces in each part of the beam.

The downward reaction at support A is equal to $P/2$, and the upward reaction at support B is equal to $3P/2$ (see Fig. 9-15a). It follows that the shear forces in parts AB and BC are

$$V = -\frac{P}{2} \qquad (0 < x < L) \tag{9-52a}$$

$$V = P \qquad \left(L < x < \frac{3L}{2}\right) \tag{9-52b}$$

in which x is measured from end A of the beam (Fig. 9-15b).

2. *Categorize*: The third-order differential equations for the beam now become [see Eq. (9-16b)]:

$$EIv''' = -\frac{P}{2} \qquad (0 < x < L) \tag{a}$$

$$EIv''' = P \qquad \left(L < x < \frac{3L}{2}\right) \tag{b}$$

3. *Analyze*:

Bending moments in the beam: Integration of the preceding two equations yields the bending-moment equations:

$$M = EIv'' = -\frac{Px}{2} + C_1 \qquad (0 \leq x \leq L) \tag{c}$$

$$M = EIv'' = Px + C_2 \qquad \left(L \leq x \leq \frac{3L}{2}\right) \tag{d}$$

Tom Brakefield/Getty Images

Bridge girder with overhang during transport to the construction site

The bending moments at points A and C are zero; hence, the following boundary conditions apply to this beam:

$$v''(0) = 0 \qquad v''\left(\frac{3L}{2}\right) = 0$$

Use these conditions with Eqs. (c) and (d) to get

$$C_1 = 0 \qquad C_2 = -\frac{3PL}{2}$$

Therefore, the bending moments are

$$M = EIv'' = -\frac{Px}{2} \qquad (0 \leq x \leq L) \tag{9-53a}$$

$$M = EIv'' = -\frac{P(3L - 2x)}{2} \qquad \left(L \leq x \leq \frac{3L}{2}\right) \tag{9-53b}$$

These equations can be verified by determining the bending moments from free-body diagrams and equations of equilibrium.

Slopes and deflections of the beam: The next integrations yield the slopes:

$$EIv' = -\frac{Px^2}{4} + C_3 \qquad (0 \leq x \leq L)$$

$$EIv' = -\frac{Px(3L - x)}{2} + C_4 \qquad \left(L \leq x \leq \frac{3L}{2}\right)$$

The only condition on the slopes is the continuity condition at support B. According to this condition, the slope at point B as found for part AB of the beam is equal to the slope at the same point as found for part BC of the beam. Therefore, substitute $x = L$ into each of the two preceding equations for the slopes and obtain

$$-\frac{PL^2}{4} + C_3 = -PL^2 + C_4$$

This equation eliminates one constant of integration because C_4 can be expressed in terms of C_3:

$$C_4 = C_3 + \frac{3PL^2}{4} \qquad \textbf{(e)}$$

The third and last integrations give

$$EIv = -\frac{Px^3}{12} + C_3x + C_5 \qquad (0 \leq x \leq L) \qquad \textbf{(f)}$$

$$EIv = -\frac{Px^2(9L - 2x)}{12} + C_4x + C_6 \qquad \left(L \leq x \leq \frac{3L}{2}\right) \qquad \textbf{(g)}$$

For part AB of the beam (Fig. 9-15a), there are two boundary conditions on the deflections, namely, the deflection is zero at points A and B:

$$v(0) = 0 \quad \text{and} \quad v(L) = 0$$

Apply these conditions to Eq. (f) to obtain

$$C_5 = 0 \quad C_3 = \frac{PL^2}{12} \qquad \textbf{(h,i)}$$

Substitute the preceding expression for C_3 in Eq. (e) to get

$$C_4 = \frac{5PL^2}{6} \qquad \textbf{(j)}$$

For part BC of the beam, the deflection is zero at point B. Therefore, the boundary condition is

$$v(L) = 0$$

Apply this condition to Eq. (g), and also substitute Eq. (j) for C_4 to get

$$C_6 = -\frac{PL^3}{4} \qquad \textbf{(k)}$$

All constants of integration have now been evaluated.

The deflection equations are obtained by substituting the constants of integration (Eqs. h, i, j, and k) into Eqs. (f) and (g). The results are

$$v = \frac{Px}{12EI}(L^2 - x^2) \qquad (0 \le x \le L) \tag{9-54a}$$

$$v = -\frac{P}{12EI}(3L^3 - 10L^2x + 9Lx^2 - 2x^3) \qquad \left(L \le x \le \frac{3L}{2}\right) \tag{9-54b}$$

Note that the deflection is always positive (upward) in part AB of the beam [Eq. (9-54a)] and always negative (downward) in the overhang BC [Eq. (9-54b)].

Deflection at the end of the overhang: Find the deflection δ_C at the end of the overhang (Fig. 9-15b) by substituting $x = 3L/2$ in Eq. (9-54b):

$$\delta_C = -v\left(\frac{3L}{2}\right) = \frac{PL^3}{8EI} \tag{9-55}$$

4. ***Finalize***: The required deflections of the overhanging beam [Eqs. (9-54) and (9-55)] were obtained by solving the third-order differential equation of the deflection curve. The deflection curve is shown in Fig. 9-15b.

FIGURE 9-15 (Repeated)

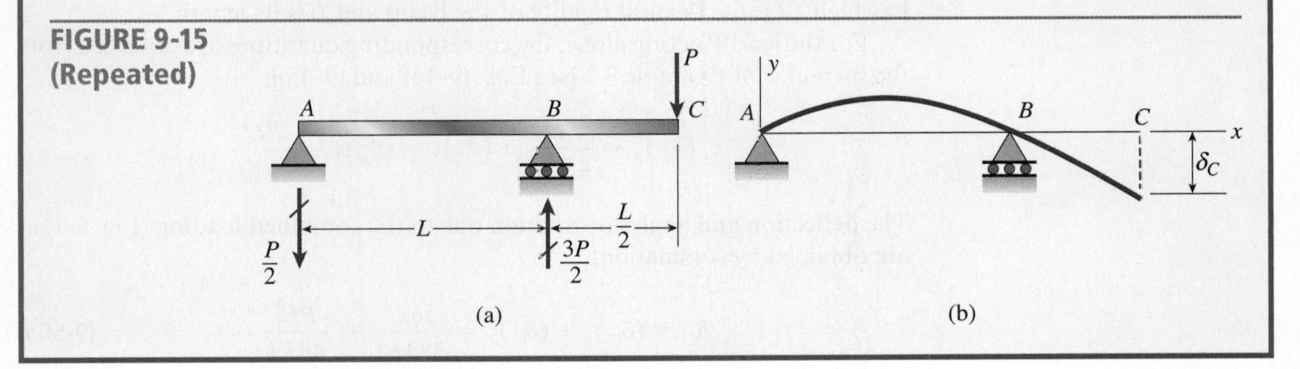

9.5 Method of Superposition

The **method of superposition** is a practical and commonly used technique for obtaining deflections and angles of rotation of beams. The underlying concept is quite simple and may be stated as follows:

Under suitable conditions, the deflection of a beam produced by several different loads acting simultaneously can be found by superposing the deflections produced by the same loads acting separately.

For instance, if v_1 represents the deflection at a particular point on the axis of a beam due to a load q_1, and if v_2 represents the deflection at that same point due to a different load q_2, then the deflection at that point due to loads q_1 and q_2 acting simultaneously is $v_1 + v_2$. (The loads q_1 and q_2 are independent loads and each may act anywhere along the axis of the beam.)

FIGURE 9-16

Simple beam with two loads

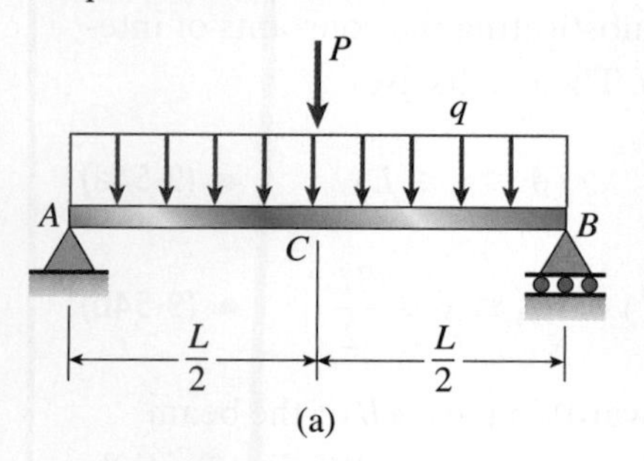

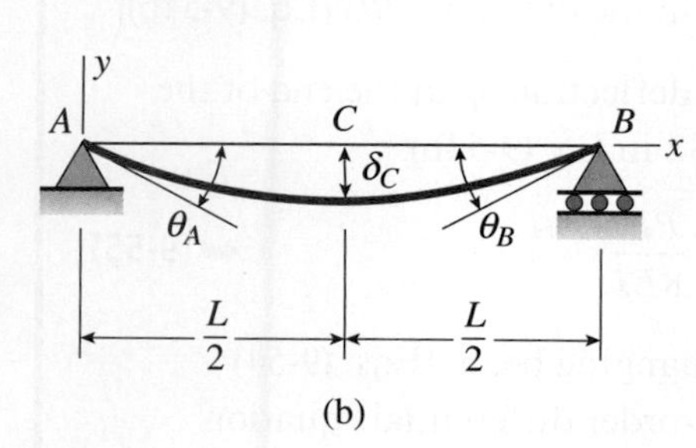

The justification for superposing deflections lies in the nature of the differential equations of the deflection curve [Eqs. (9-16a, b, and c)]. These equations are *linear* differential equations, because all terms containing the deflection v and its derivatives are raised to the first power. Therefore, the solutions of these equations for several loading conditions may be added algebraically, or *superposed*. (The conditions for superposition to be valid are described later in the subsection "Principle of Superposition.")

As an **illustration** of the superposition method, consider the simple beam ACB shown in Fig. 9-16a. This beam supports two loads: (1) a uniform load of intensity q acting throughout the span, and (2) a concentrated load P acting at the midpoint. Suppose you wish to find the deflection δ_C at the midpoint and the angles of rotation θ_A and θ_B at the ends (Fig. 9-16b). Using the method of superposition, first obtain the effects of each load acting separately and then combine the results.

For the uniform load acting alone, the deflection at the midpoint and the angles of rotation are obtained from the formulas of Example 9-1 [see Eqs. (9-24), (9-25), and (9-26)]:

$$(\delta_C)_1 = \frac{5qL^4}{384EI} \qquad (\theta_A)_1 = (\theta_B)_1 = \frac{qL^3}{24EI}$$

in which EI is the flexural rigidity of the beam and L is its length.

For the load P acting alone, the corresponding quantities are obtained from the formulas of Example 9-3 [see Eqs. (9-44) and (9-45)]:

$$(\delta_C)_2 = \frac{PL^3}{48EI} \qquad (\theta_A)_2 = (\theta_B)_2 = \frac{PL^2}{16EI}$$

The deflection and angles of rotation due to the combined loading (Fig. 9-16a) are obtained by summation:

$$\delta_C = (\delta_C)_1 + (\delta_C)_2 = \frac{5qL^4}{384EI} + \frac{PL^3}{48EI} \tag{9-56a}$$

$$\theta_A = \theta_B = (\theta_A)_1 + (\theta_A)_2 = \frac{qL^3}{24EI} + \frac{PL^2}{16EI} \tag{9-56b}$$

The deflections and angles of rotation at other points on the beam axis can be found by this same procedure. However, the method of superposition is not limited to finding deflections and angles of rotation at single points. The method also may be used to obtain general equations for the slopes and deflections of beams subjected to more than one load.

Tables of Beam Deflections

The method of superposition is useful only when formulas for deflections and slopes are readily available. To provide convenient access to such formulas, tables for both cantilever and simple beams are given in Appendix H. Similar tables can be found in engineering handbooks. Use these tables and the method of superposition to find deflections and angles of rotation for many different loading conditions, as illustrated in the examples at the end of this section.

Distributed Loads

Sometimes a distributed load of interest is not included in a table of beam deflections. In such cases, superposition still may be useful. Consider an element of the distributed load as though it were a concentrated load, and then find the required deflection by integrating throughout the region of the beam where the load is applied.

To illustrate this process of integration, consider a simple beam *ACB* with a triangular load acting on the left-hand half (Fig. 9-17a). The goal is to obtain the deflection δ_C at the midpoint *C* and the angle of rotation θ_A at the left-hand support (Fig. 9-17c).

Begin by visualizing an element *qdx* of the distributed load as a concentrated load (Fig. 9-17b). Note that the load acts to the left of the midpoint of the beam. The deflection at the midpoint due to this concentrated load is obtained from Case 5 of Table H-2, Appendix H. The formula given there for the midpoint deflection (for the case in which $a \leq b$) is

$$\frac{Pa}{48EI}(3L^2 - 4a^2)$$

In this example (Fig. 9-17b), substitute *qdx* for *P* and *x* for *a*:

$$\frac{(qdx)(x)}{48EI}(3L^2 - 4x^2) \quad \textbf{(9-57)}$$

This expression gives the deflection at point *C* due to the element *qdx* of the load.

Next, note that the intensity of the uniform load (Figs. 9-17a and b) is

$$q = \frac{2q_0x}{L} \quad \textbf{(9-58)}$$

where q_0 is the maximum intensity of the load. With this substitution for *q*, the formula for the deflection [Eq. (9-57)] becomes

$$\frac{q_0x^2}{24LEI}(3L^2 - 4x^2)dx$$

Now, integrate throughout the region of the load to obtain the deflection δ_C at the midpoint of the beam due to the entire triangular load:

$$\delta_C = \int_0^{L/2} \frac{q_0x^2}{24LEI}(3L^2 - 4x^2)dx \quad \textbf{(9-59)}$$

$$= \frac{q_0}{24LEI}\int_0^{L/2}(3L^2 - 4x^2)x^2dx = \frac{q_0L^4}{240EI}$$

By a similar procedure, you can calculate the angle of rotation θ_A at the left-hand end of the beam (Fig. 9-17c). The expression for this angle due to a concentrated load *P* (see Case 5 of Table H-2) is

$$\frac{Pab(L + b)}{6LEI}$$

Replace *P* with $2q_0xdx / L$, *a* with *x*, and *b* with $L - x$ to obtain

$$\frac{2q_0x^2(L - x)(L + L - x)}{6L^2EI}dx \quad \text{or} \quad \frac{q_0}{3L^2EI}(L - x)(2L - x)x^2dx$$

FIGURE 9-17

Simple beam with a triangular load

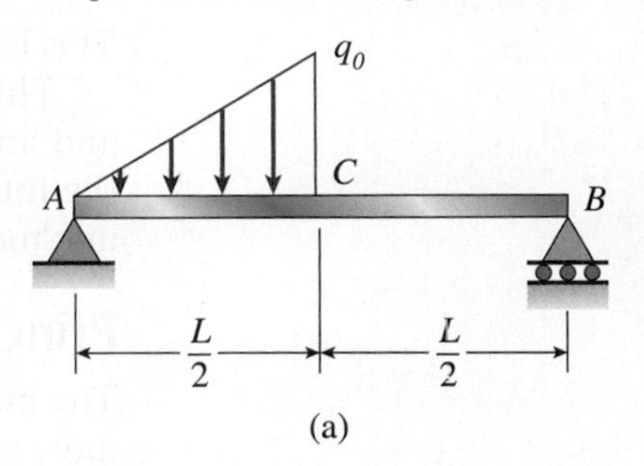

(a)

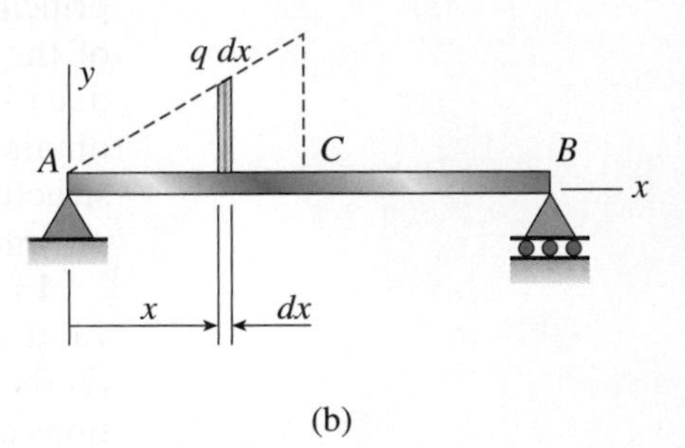

(b)

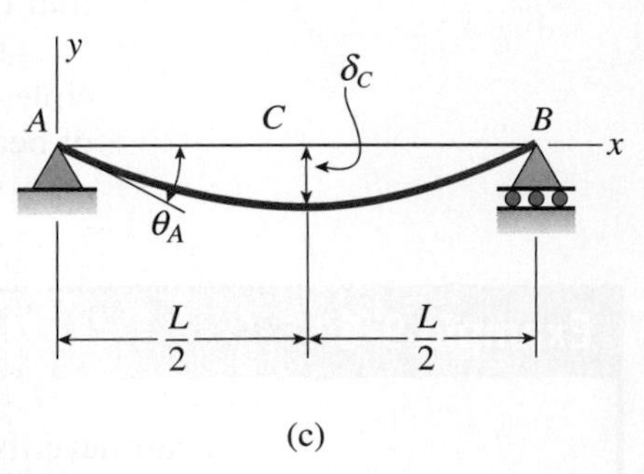

(c)

Finally, integrate throughout the region of the load:

$$\theta_A = \int_0^{L/2} \frac{q_0}{3L^2EI}(L - x)(2L - x)x^2 dx = \frac{41q_0L^3}{2880EI} \quad \textbf{(9-60)}$$

This is the angle of rotation produced by the triangular load.

This illustrates how to use superposition and integration to find deflections and angles of rotation produced by distributed loads of almost any kind. If the integration cannot be performed easily by analytical means, numerical methods can be used.

Principle of Superposition

The method of superposition for finding beam deflections is an example of a more general concept known in mechanics as the **principle of superposition**. This principle is valid whenever the quantity to be determined is a linear function of the applied loads. When that is the case, the desired quantity may be found due to each load acting separately, and then these results may be superposed to obtain the desired quantity due to all loads acting simultaneously. In ordinary structures, the principle is usually valid for stresses, strains, bending moments, and many other quantities besides deflections.

In the particular case of **beam deflections**, the principle of superposition is valid under the following conditions: (1) Hooke's law holds for the material, (2) the deflections and rotations are small, and (3) the presence of the deflections does not alter the actions of the applied loads. These requirements ensure that the differential equations of the deflection curve are linear.

The following examples provide additional illustrations in which the principle of superposition is used to calculate deflections and angles of rotation of beams.

Example 9-6

A cantilever beam AB supports a uniform load of intensity q acting over part of the span and a concentrated load P acting at the free end (Fig. 9-18a).

Determine the deflection δ_B and angle of rotation θ_B at end B of the beam (Fig. 9-18b). *Note:* The beam has length L and constant flexural rigidity EI.

Solution:

Use a four-step problem-solving approach.

1. ***Conceptualize***: Obtain the deflection and angle of rotation at end B of the beam by combining the effects of the loads acting separately.
2. ***Categorize***: If the uniform load acts alone, the deflection and angle of rotation (obtained from Case 2 of Table H-1, Appendix H) are

$$(\delta_B)_1 = \frac{qa^3}{24EI}(4L - a) \qquad (\theta_B)_1 = \frac{qa^3}{6EI}$$

FIGURE 9-18

Example 9-6: Cantilever beam with a uniform load and a concentrated load

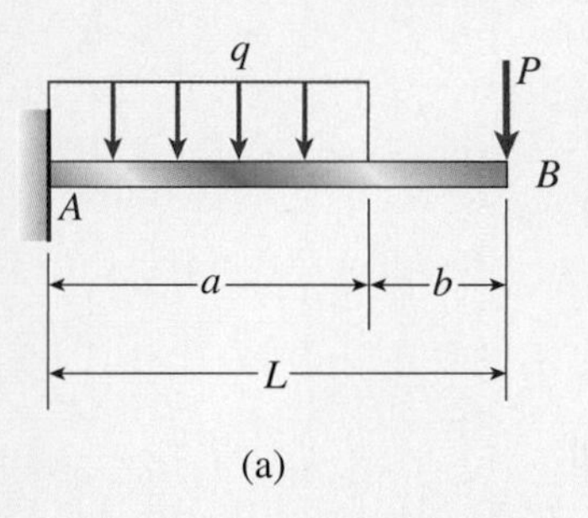

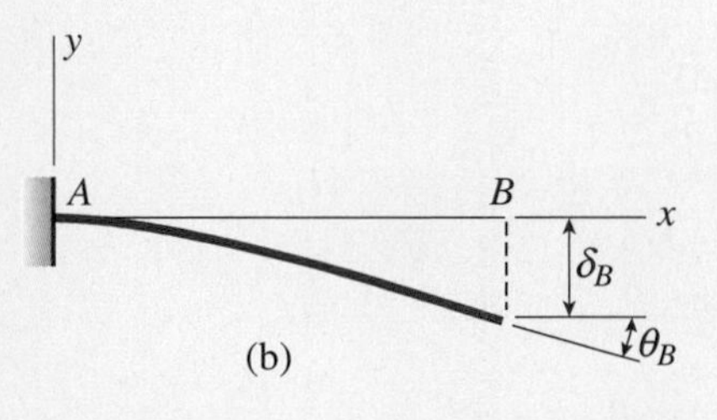

If the load P acts alone, the corresponding quantities (from Case 4, Table H-1) are

$$(\delta_B)_2 = \frac{PL^3}{3EI} \qquad (\theta_B)_2 = \frac{PL^2}{2EI}$$

3. ***Analyze***: Therefore, the deflection and angle of rotation due to the combined loading (Fig. 9-18a) are

$$\delta_B = (\delta_B)_1 + (\delta_B)_2 = \frac{qa^3}{24EI}(4L - a) + \frac{PL^3}{3EI} \qquad \text{(9-61)}$$

$$\theta_B = (\theta_B)_1 + (\theta_B)_2 = \frac{qa^3}{6EI} + \frac{PL^2}{2EI} \qquad \text{(9-62)}$$

4. ***Finalize***: The required quantities were found by using tabulated formulas and the method of superposition. Note that you would have to integrate over two separate regions ($0 \leq x \leq a$, $a \leq x \leq L$) to find the deflection curve shown in Fig. 9-18b.

Example 9-7

A cantilever beam AB with a uniform load of intensity q acting on the right-hand half of the beam is shown in Fig. 9-19a.

Obtain formulas for the deflection δ_B and angle of rotation θ_B at the free end (Fig. 9-19c). *Note:* The beam has length L and constant flexural rigidity EI.

Solution:

Use a four-step problem-solving approach.

1. ***Conceptualize***: In this example, determine the deflection and angle of rotation by treating an element of the uniform load as a concentrated load and then integrating (see Fig. 9-19b).
2. ***Categorize***: The element of the load has a magnitude qdx and is located at distance x from the support. The resulting differential deflection $d\delta_B$ and differential angle of rotation $d\theta_B$ at the free end are found from the corresponding formulas in Case 5 of Table H-1, Appendix H, by replacing P with qdx and a with x; thus,

$$d\delta_B = \frac{(qdx)(x^2)(3L - x)}{6EI} \qquad d\theta_B = \frac{(qdx)(x^2)}{2EI}$$

FIGURE 9-19

Example 9-7: Cantilever beam with a uniform load acting on the right-hand half of the beam

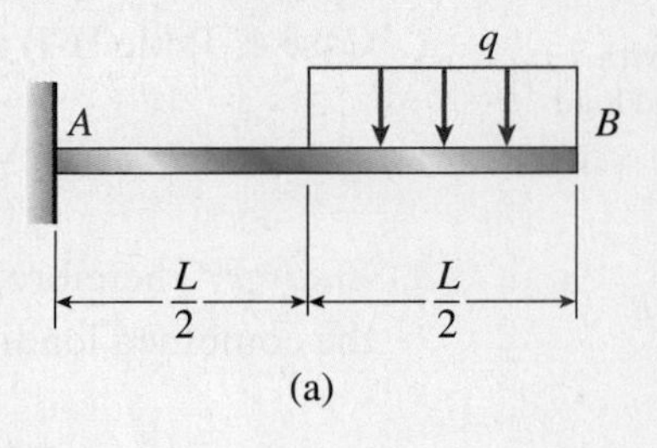

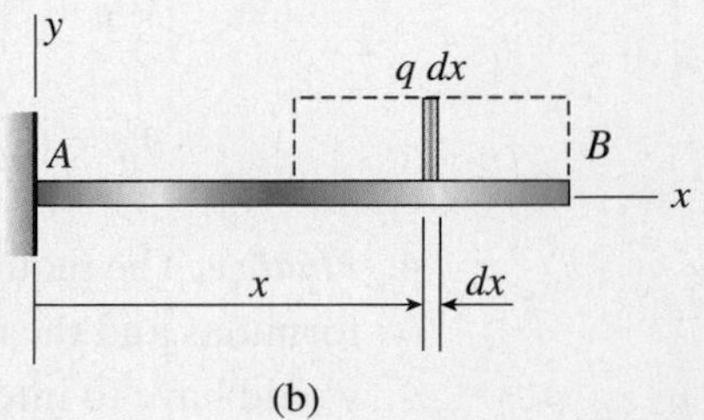

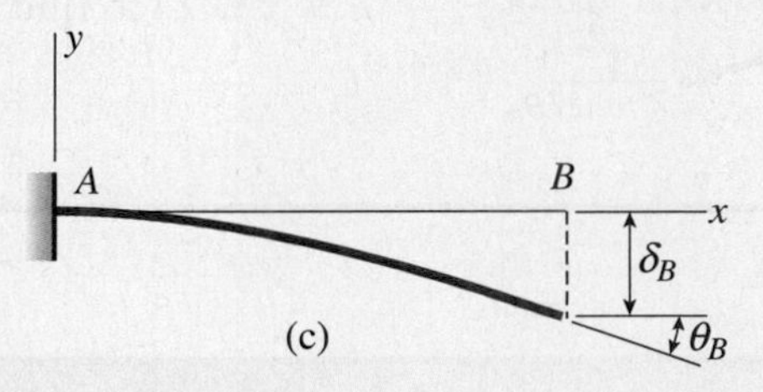

3. ***Analyze***: Integrate over the loaded region to get

$$\delta_B = \int d\delta_B = \frac{q}{6EI} \int_{L/2}^{L} x^2(3L - x)dx = \frac{41qL^4}{384EI} \qquad \text{(9-63)}$$

$$\theta_B = \int d\theta_B = \frac{q}{2EI} \int_{L/2}^{L} x^2 dx = \frac{7qL^3}{48EI} \qquad \text{(9-64)}$$

4. ***Finalize***: These same results can be obtained by using the formulas in Case 3 of Table H-1 and substituting $a = b = L/2$. Also, note that subtraction of Case 2 from Case 1 in Table H-1 (with $a = b = L/2$) leads to the same results.

Example 9-8

A compound beam *ABC* has a roller support at *A*, an internal hinge (that is, moment release) at *B*, and a fixed support at *C* (Fig. 9-20a). Segment *AB* has a length of *a* and segment *BC* has a length of *b*. A concentrated load *P* acts at a distance $2a/3$ from support *A*, and a uniform load of intensity *q* acts between points *B* and *C*.

Determine the deflection δ_B at the hinge and the angle of rotation θ_A at support *A* (Fig. 9-20d). *Note:* The beam has constant flexural rigidity *EI*.

FIGURE 9-20

Example 9-8: Compound beam with a hinge

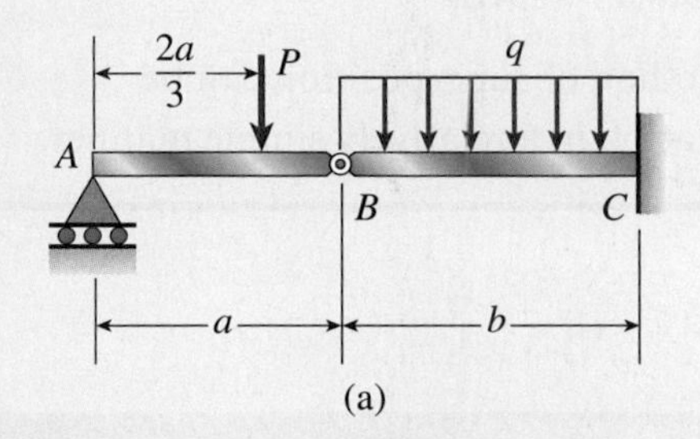

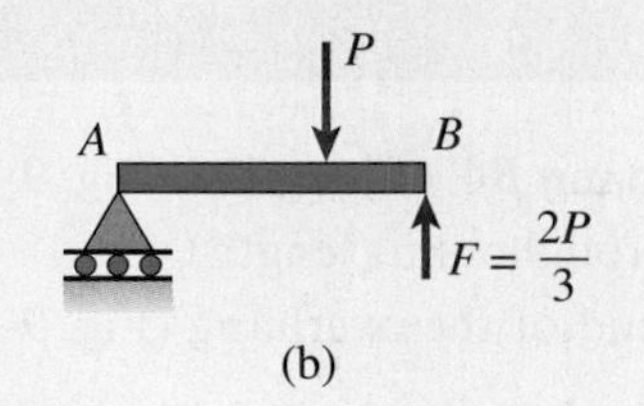

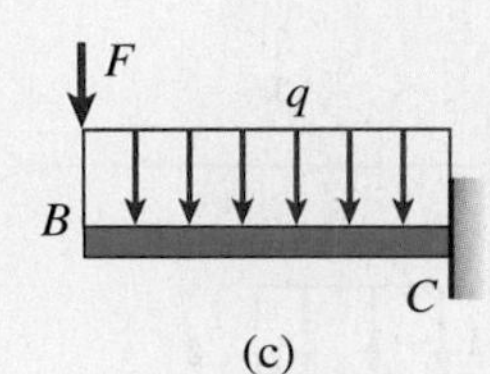

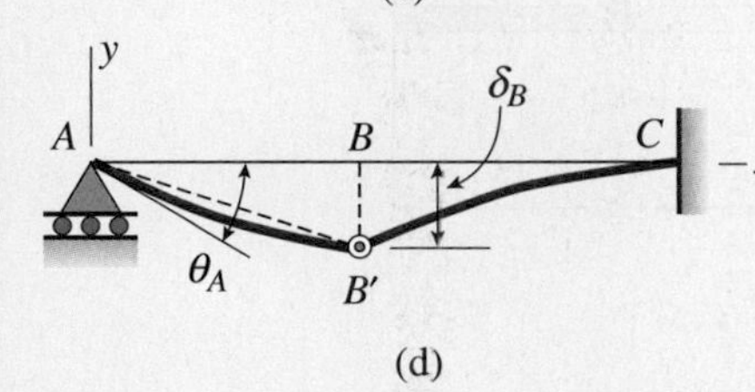

Solution:

Use a four-step problem-solving approach.

1. ***Conceptualize***: For purposes of analysis, consider the compound beam to consist of two individual beams: a simple beam AB of length a and a cantilever beam BC of length b. The two beams are linked together by a pin connection at B.

2. ***Categorize***: If you separate beam AB from the rest of the structure (Fig. 9-20b), there is a vertical force F at end B equal to $2P/3$. This same force acts downward at end B of the cantilever (Fig. 9-20c). Consequently, the cantilever beam BC is subjected to two loads: a uniform load and a concentrated load.

3. ***Analyze***: The deflection at the end of this cantilever (which is the same as the deflection δ_B of the hinge) is readily found from Cases 1 and 4 of Table H-1, Appendix H:

$$\delta_B = \frac{qb^4}{8EI} + \frac{Fb^3}{3EI}$$

or, since $F = 2P/3$,

$$\delta_B = \frac{qb^4}{8EI} + \frac{2Pb^3}{9EI} \qquad \text{(9-65)}$$

The angle of rotation θ_A at support A (Fig. 9-20d) consists of two parts: *an angle BAB'* produced by the downward displacement of the hinge and *an additional angle of rotation* produced by the bending of beam AB (or beam AB') as a simple beam. *The angle BAB'* is

$$(\theta_A)_1 = \frac{\delta_B}{a} = \frac{qb^4}{8aEI} + \frac{2Pb^3}{9aEI}$$

The angle of rotation at the end of a simple beam with a concentrated load is obtained from Case 5 of Table H-2. The formula given there is

$$\frac{Pab(L + b)}{6LEI}$$

in which L is the length of the simple beam, a is the distance from the left-hand support to the load, and b is the distance from the right-hand support to the load. Thus, in the notation of this example (Fig. 9-20a), the angle of rotation is

$$(\theta_A)_2 = \frac{P\left(\frac{2a}{3}\right)\left(\frac{a}{3}\right)\left(a + \frac{a}{3}\right)}{6aEI} = \frac{4Pa^2}{81EI}$$

Combine the two angles to obtain the total angle of rotation at support A:

$$\theta_A = (\theta_A)_1 + (\theta_A)_2 = \frac{qb^4}{8aEI} + \frac{2Pb^3}{9aEI} + \frac{4Pa^2}{81EI} \quad \text{(9-66)}$$

4. ***Finalize***: This example illustrates how the method of superposition can be adapted to handle a seemingly complex situation in a relatively simple manner.

Example 9-9

A simple beam AB of a span length L has an overhang BC of length a (Fig. 9-21a). The beam supports a uniform load of intensity q throughout its length.

Obtain a formula for the deflection δ_C at the end of the overhang (Fig. 9-21c). *Note:* The beam has constant flexural rigidity EI.

FIGURE 9-21

Example 9-9: Simple beam with an overhang

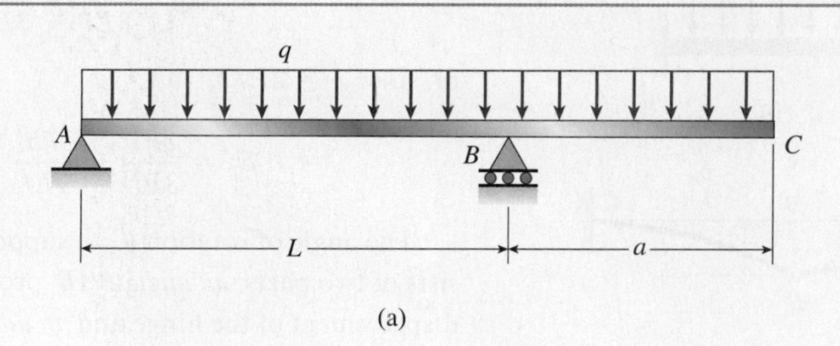

(a)

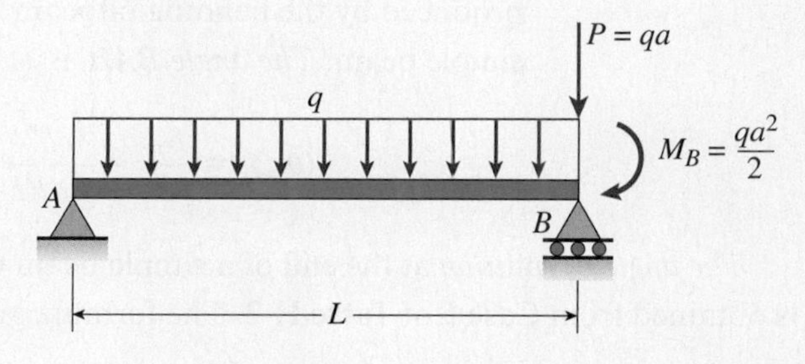

(b)

y
Point of inflection
D
A
B
θ_B
C
x
δ_1
δ_2
δ_C

(c)

Beam with overhang loaded by gravity uniform load (Courtesy of the National Information Service for Earthquake Engineering EERC, University of California, Berkeley)

Solution:

Use a four-step problem-solving approach.

1. *Conceptualize*: Find the deflection of point C by imagining the overhang BC (Fig. 9-21a) to be a cantilever beam subjected to two actions. The first action is the rotation of the support of the cantilever through an angle θ_B, which is the angle of rotation of beam ABC at support B (Fig. 9-21c). (Assume that a clockwise angle θ_B is positive.) This angle of rotation causes a rigid-body rotation of the overhang BC, resulting in a downward displacement δ_1 of point C.

The second action is the bending of BC as a cantilever beam supporting a uniform load. This bending produces an additional downward displacement δ_2 (Fig. 9-21c). The superposition of these two displacements gives the total displacement δ_C at point C.

2. *Categorize*:

Deflection δ_1: Begin by finding the deflection δ_1 caused by the angle of rotation θ_B at point B. To find this angle, observe that part AB of the beam is in the same condition as a simple beam (Fig. 9-21b) subjected to the following loads: a uniform load of intensity q, a couple M_B (equal to $qa^2/2$), and a vertical load P (equal to qa). Only the loads q and M_B produce angles of rotation at end B of this simple beam. These angles are found from Cases 1 and 7 of Table H-2, Appendix H. Thus, the angle θ_B is

$$\theta_B = -\frac{qL^3}{24EI} + \frac{M_B L}{3EI} = -\frac{qL^3}{24EI} + \frac{qa^2 L}{6EI} = \frac{qL(4a^2 - L^2)}{24EI} \quad \textbf{(9-67)}$$

in which a clockwise angle is positive, as shown in Fig. 9-21c.

The downward deflection δ_1 of point C, due solely to the angle of rotation θ_B, is equal to the length of the overhang times the angle (Fig. 9-21c):

$$\delta_1 = a\theta_B = \frac{qaL(4a^2 - L^2)}{24EI} \quad \textbf{(a)}$$

Deflection δ_2: Bending of the overhang BC produces an additional downward deflection δ_2 at point C. This deflection is equal to the deflection of a cantilever beam of length a subjected to a uniform load of intensity q (see Case 1 of Table H-1):

$$\delta_2 = \frac{qa^4}{8EI} \quad \textbf{(b)}$$

3. *Analyze*:

Deflection δ_C: The total downward deflection of point C is the algebraic sum of δ_1 and δ_2:

$$\delta_C = \delta_1 + \delta_2 = \frac{qaL(4a^2 - L^2)}{24EI} + \frac{qa^4}{8EI} = \frac{qa}{24EI}[L(4a^2 - L^2) + 3a^3]$$

or

$$\delta_C = \frac{qa}{24EI}(a + L)(3a^2 + aL - L^2) \qquad \text{(9-68)}$$

4. ***Finalize***: In the preceding equation, the deflection δ_C may be upward or downward, depending upon the relative magnitudes of the lengths L and a. If a is relatively large, the last term in the equation (the three-term expression in parentheses) is positive, and the deflection δ_C is downward. If a is relatively small, the last term is negative, and the deflection is upward. The deflection is zero when the last term is equal to zero:

$$3a^2 + aL - L^2 = 0$$

or

$$a = \frac{L\left(\sqrt{13} - 1\right)}{6} = 0.4343L \qquad \text{(c)}$$

From this result, note that if a is greater than $0.4343L$, the deflection of point C is downward; if a is less than $0.4343L$, the deflection is upward.

Deflection curve: The shape of the deflection curve for the beam in this example is shown in Fig. 9-21c for the case where a is large enough ($a > 0.4343L$) to produce a downward deflection at C and small enough ($a < L$) to ensure that the reaction at A is upward. Under these conditions, the beam has a positive bending moment between support A and a point such as D. The deflection curve in region AD is concave upward (positive curvature). From D to C, the bending moment is negative; therefore, the deflection curve is concave downward (negative curvature).

Point of inflection: At point D, the curvature of the deflection curve is zero because the bending moment is zero. A point such as D, where the curvature and bending moment *change signs*, is called a **point of inflection** (or *point of contraflexure*). The bending moment M and the second derivative d^2v/dx^2 always vanish at an inflection point.

However, a point where M and d^2v/dx^2 equal zero is not necessarily an inflection point because it is possible for those quantities to be zero without changing signs at that point; for example, they could have maximum or minimum values.

9.6 Moment-Area Method

This section presents another method for finding deflections and angles of rotation of beams. Because the method is based upon two theorems related to the *area of the bending-moment diagram*, it is called the **moment-area method**.

The assumptions used in deriving the two theorems are the same as those used in obtaining the differential equations of the deflection curve. Therefore, the moment-area method is valid only for linearly elastic beams with small slopes.

First Moment-Area Theorem

To derive the first theorem, consider a segment AB of the deflection curve of a beam in a region where the curvature is positive (Fig. 9-22). Of course, the deflections and slopes shown in the figure are highly exaggerated for clarity. At point A, the tangent AA' to the deflection curve is at an angle θ_A to the x axis, and at point B, the tangent BB' is at an angle θ_B. These two tangents meet at point C.

The **angle between the tangents**, denoted $\theta_{B/A}$, is equal to the difference between θ_B and θ_A:

$$\theta_{B/A} = \theta_B - \theta_A \tag{9-69}$$

Thus, the angle $\theta_{B/A}$ can be described as the angle to the tangent at B measured relative to, or with respect to, the tangent at A. Note that the angles θ_A and θ_B, which are the angles of rotation of the beam axis at points A and B, respectively, are also equal to the slopes at those points because in reality the slopes and angles are very small quantities.

Next, consider two points m_1 and m_2 on the deflected axis of the beam (Fig. 9-22). These points are a small distance ds apart. The tangents to the deflection curve at these points are shown in the figure as lines m_1p_1 and m_2p_2. The normals to these tangents intersect at the center of curvature (not shown in the figure).

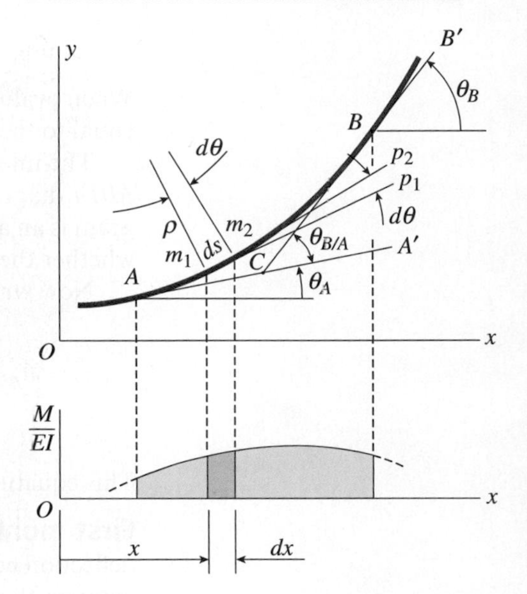

FIGURE 9-22

Derivation of the first moment-area theorem

The angle $d\theta$ between the normals (Fig. 9-22) is given by

$$d\theta = \frac{ds}{\rho} \tag{9-70a}$$

in which ρ is the radius of curvature and $d\theta$ is measured in radians [see Eq. (9-2)]. Because the normals and the tangents (m_1p_1 and m_2p_2) are perpendicular, it follows that the angle between the tangents is also equal to $d\theta$.

For a beam with small angles of rotation, replace ds with dx, as explained in Section 9.2. Thus,

$$d\theta = \frac{dx}{\rho} \tag{9-70b}$$

Also, from Eq. (9-10),

$$\frac{1}{\rho} = \frac{M}{EI} \tag{9-71}$$

therefore,

$$d\theta = \frac{Mdx}{EI} \tag{9-72}$$

in which M is the bending moment and EI is the flexural rigidity of the beam.

The quantity Mdx/EI has a simple geometric interpretation. To see this, refer to Fig. 9-22, which shows the M/EI diagram directly below the beam. At any point along the x axis, the height of this diagram is equal to the bending moment M at that point divided by the flexural rigidity EI at that point. Thus, the M/EI diagram has the same shape as the bending-moment diagram whenever EI is constant. The term Mdx/EI is the area of the shaded strip of width dx within the M/EI diagram. Note that since the curvature of the deflection curve in Fig. 9-22 is positive, the bending moment M and the area of the M/EI diagram are also positive.

Now integrate $d\theta$ from Eq. (9-72) between points A and B of the deflection curve:

$$\int_A^B d\theta = \int_A^B \frac{Mdx}{EI} \tag{9-73}$$

When evaluated, the integral on the left-hand side becomes $\theta_B - \theta_A$, which is equal to the angle $\theta_{B/A}$ between the tangents at B and A from Eq. (9-69).

The integral on the right-hand side of Eq. (9-73) is equal to the area of the M/EI diagram between points A and B. Note that the area of the M/EI diagram is an algebraic quantity and may be positive or negative, depending upon whether the bending moment is positive or negative.

Now write Eq. (9-73) as

$$\theta_{B/A} = \int_A^B \frac{Mdx}{EI}$$
$$= \text{Area of the } M/EI \text{ diagram between points } A \text{ and } B \tag{9-74}$$

This equation may be stated as a theorem.

First moment-area theorem: The angle $\theta_{B/A}$ between the tangents to the deflection curve at two points A and B is equal to the area of the M/EI diagram between those points.

These are the sign conventions used in deriving the preceding theorem.

1. The angles θ_A and θ_B are positive when they are counterclockwise.
2. The angle $\theta_{B/A}$ between the tangents is positive when the angle θ_B is algebraically larger than the angle θ_A. Also, note that point B must be to the right of point A; that is, it must be further along the axis of the beam as you move in the x direction.
3. The bending moment M is positive according to the usual sign convention, that is, M is positive when it produces compression in the upper part of the beam.
4. The area of the M/EI diagram is given a positive or negative sign according to whether the bending moment is positive or negative. If part of the bending-moment diagram is positive and another part is negative, then the corresponding parts of the M/EI diagram are given those same signs.

The preceding sign conventions for θ_A, θ_B, and $\theta_{B/A}$ are often ignored in practice because (as explained later) the directions of the angles of rotation are usually obvious from an inspection of the beam and its loading. For this case, simplify the calculations by ignoring signs and using only absolute values when applying the first moment-area theorem.

Second Moment-Area Theorem

Now consider the second theorem, which is related primarily to deflections rather than to angles of rotation. Consider again the deflection curve between points A and B (Fig. 9-23). Draw the tangent at point A and note that its intersection with a vertical line through point B is at point B_1. The vertical distance between points B and B_1 is denoted $t_{B/A}$ in the figure. This distance is referred to as the **tangential deviation** of B with respect to A. More precisely, the distance $t_{B/A}$ is the vertical deviation of point B on the deflection curve from the tangent at point A. The tangential deviation is positive when point B is above the tangent at A.

To determine the tangential deviation, again select two points m_1 and m_2 a small distance apart on the deflection curve (Fig. 9-23). The angle between the tangents at these two points is $d\theta$, and the segment on line BB_1 between these tangents is dt. Since the angles between the tangents and the x axis are actually very small, the vertical distance dt is equal to $x_1\, d\theta$, where x_1 is the horizontal distance from point B to the small element m_1m_2. Since $d\theta = Mdx / EI$ [Eq. (9-72)], distance dt is

$$dt = x_1 d\theta = x_1 \frac{Mdx}{EI} \tag{9-75}$$

FIGURE 9-23

Derivation of the second moment-area theorem

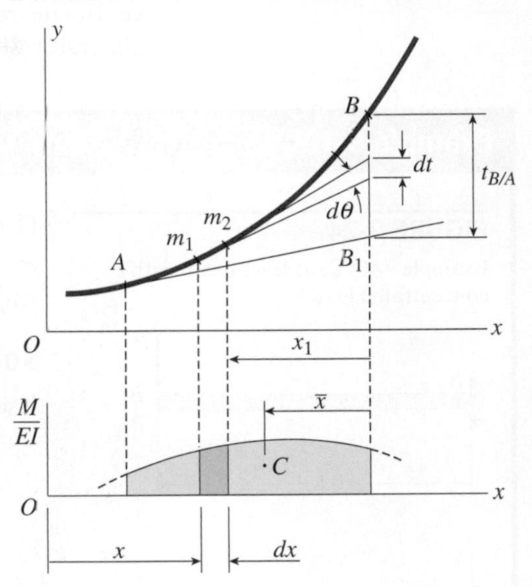

The distance dt represents the contribution made by the bending of element m_1m_2 to the tangential deviation $t_{B/A}$. The expression x_1Mdx / EI may be interpreted geometrically as the first moment of the area of the shaded strip of width dx within the M/EI diagram. This first moment is evaluated with respect to a vertical line through point B.

Integrating Eq. (9-75) between points A and B gives

$$\int_A^B dt = \int_A^B x_1 \frac{Mdx}{EI} \tag{9-76}$$

The integral on the left-hand side is equal to $t_{B/A}$, that is, it is equal to the deviation of point B from the tangent at A. The integral on the right-hand side represents the first moment with respect to point B of the area of the M/EI diagram between A and B. Therefore, write Eq. (9-76) as

$$t_{B/A} = \int_A^B x_1 \frac{Mdx}{EI}$$

$$= \text{First moment of the area of the } M / EI \text{ diagram between points } A \text{ and } B, \text{ evaluated with respect to } B \tag{9-77}$$

This equation represents the second theorem.

Second moment-area theorem: The tangential deviation $t_{B/A}$ of point B from the tangent at point A is equal to the first moment of the area of the M/EI diagram between A and B, evaluated with respect to B.

If the bending moment is positive, then the first moment of the M/EI diagram is also positive, provided point B is to the right of point A. Under these conditions, the tangential deviation $t_{B/A}$ is positive and point B is above the tangent at A (as shown in Fig. 9-23). If, in moving from A to B in the x direction, the area of the M/EI diagram is negative, then the first moment is also negative and the tangential deviation is negative, which means that point B is below the tangent at A.

The first moment of the area of the M/EI diagram can be obtained by taking the product of the area of the diagram and the distance $\bar{x}$ from point B to the centroid C of the area (Fig. 9-23). This procedure is usually more convenient than integrating, because the M/EI diagram usually consists of familiar geometric figures such as rectangles, triangles, and parabolic segments. The areas and centroidal distances of such figures are tabulated in Appendix E.

As a method of analysis, the moment-area method is feasible only for relatively simple kinds of beams. Therefore, it is usually obvious whether the beam deflects upward or downward and whether an angle of rotation is clockwise or counterclockwise. Consequently, it is seldom necessary to follow the formal (and somewhat awkward) sign conventions described previously for the tangential deviation. Instead, determine the directions by inspection and use only absolute values when applying the moment-area theorems.

Example 9-10

FIGURE 9-24

Example 9-10: Cantilever beam with a concentrated load

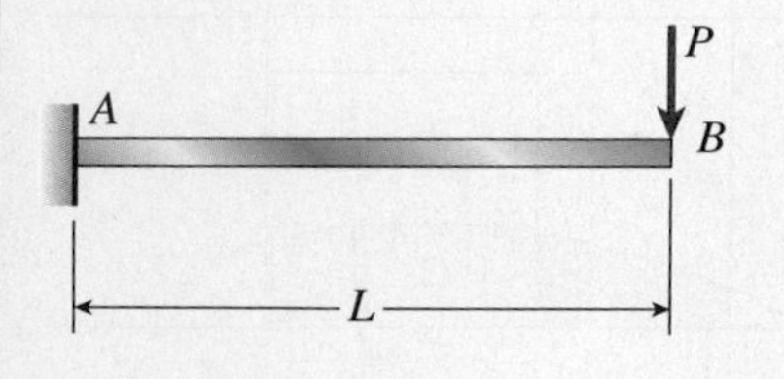

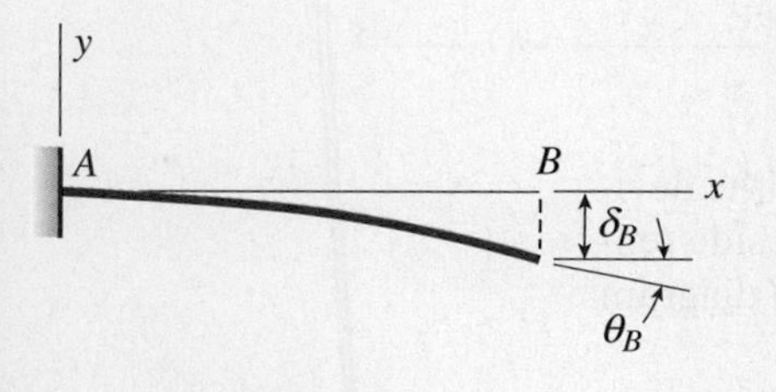

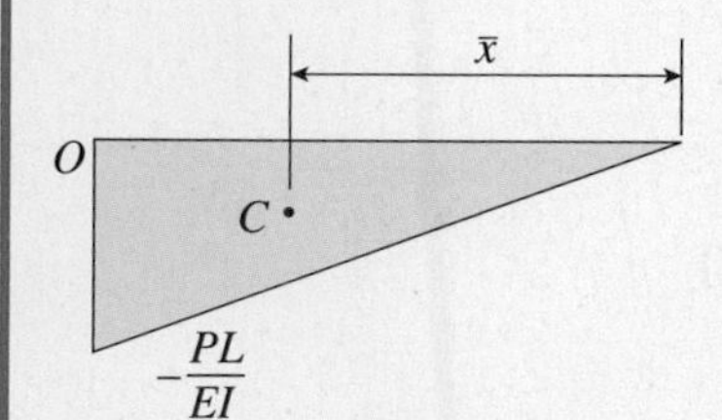

Determine the angle of rotation θ_B and deflection δ_B at the free end B of a cantilever beam AB supporting a concentrated load P Fig. 9-24). *Note*: The beam has a length L and constant flexural rigidity EI.

Solution:

Use a four-step problem-solving approach.

1. *Conceptualize*: By inspection of the beam and its loading, the angle of rotation θ_B is clockwise and the deflection δ_B is downward (Fig. 9-24). Therefore, use absolute values when applying the moment-area theorems.
2. *Categorize*:
   M/EI diagram: The bending-moment diagram is triangular in shape with the moment at the support equal to $-PL$. Since the flexural rigidity EI is constant, the M/EI diagram has the same shape as the bending-moment diagram, as shown in the last part of Fig. 9-24.
3. *Analyze*:
   Angle of rotation: From the first moment-area theorem, the angle $\theta_{B/A}$ between the tangents at points B and A is equal to the area of the M/EI diagram between those points. This area, labeled as A_1, is determined as

$$A_1 = \frac{1}{2}(L)\left(\frac{PL}{EI}\right) = \frac{PL^2}{2EI}$$

This is the absolute value of the area.

The relative angle of rotation between points A and B (from the first theorem) is

$$\theta_{B/A} = \theta_B - \theta_A = A_1 = \frac{PL^2}{2EI}$$

The tangent to the deflection curve at support A is horizontal ($\theta_A = 0$), so

$$\theta_B = \frac{PL^2}{2EI} \qquad \textbf{(9-78)}$$

This result agrees with the formula for θ_B given in Case 4 of Table H-1, Appendix H.

Deflection: The deflection δ_B at the free end is obtained from the second moment-area theorem. In this case, the tangential deviation $t_{B/A}$ of point B

from the tangent at A is equal to the deflection δ_B itself (see Fig. 9-24). The first moment of the area of the M/EI diagram, evaluated with respect to point B, is

$$Q_1 = A_1\bar{x} = \left(\frac{PL^2}{2EI}\right)\left(\frac{2L}{3}\right) = \frac{PL^3}{3EI}$$

Note again that only absolute values are used here.

From the second moment-area theorem, the deflection δ_B is equal to the first moment Q_1. Therefore,

$$\delta_B = \frac{PL^3}{3EI} \qquad \text{(9-79)}$$

4. ***Finalize***: This result also appears in Case 4 of Table H-1.

Example 9-11

FIGURE 9-25

Example 9-11: Cantilever beam supporting a uniform load on the right-hand half of the beam

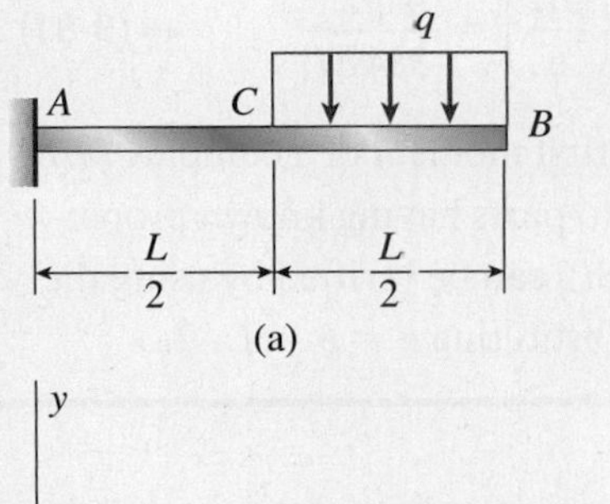

(a)

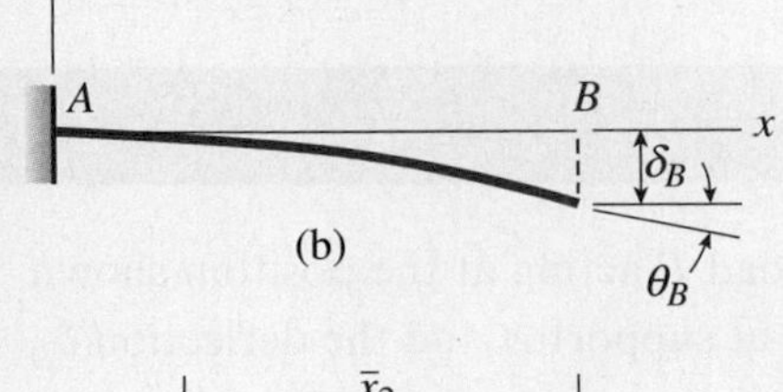

(b)

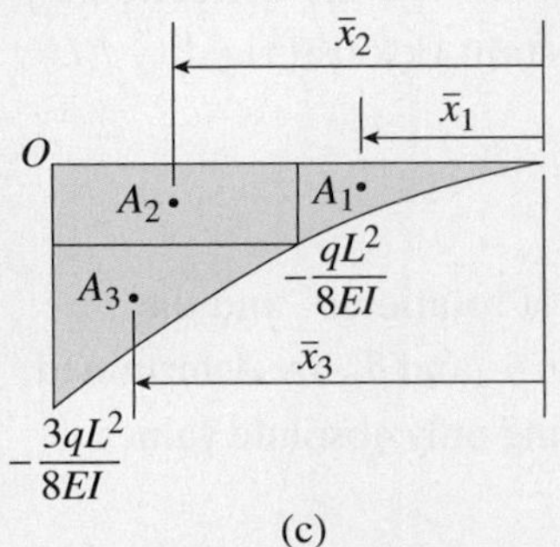

(c)

Find the angle of rotation θ_B and deflection δ_B at the free end B of a cantilever beam ACB supporting a uniform load of intensity q acting over the right-hand half of the beam (Fig. 9-25). *Note*: The beam has a length L and constant flexural rigidity EI.

Solution:

Use a four-step problem-solving approach.

1. ***Conceptualize***: The deflection and angle of rotation at end B of the beam have the directions shown in Fig. 9-25b. These directions are known in advance, so write the moment-area expressions using only absolute values.

2. ***Categorize***:

M/EI diagram: The bending-moment diagram consists of a parabolic curve in the region of the uniform load and a straight line in the left-hand half of the beam. Since EI is constant, the M/EI diagram has the same shape (see Fig. 9-25c). The values of M/EI at points A and C are $-3qL^2/8EI$ and $-qL^2/8EI$, respectively.

3. ***Analyze***:

Angle of rotation: For the purpose of evaluating the area of the M/EI diagram, it is convenient to divide the diagram into three parts: a parabolic spandrel of area A_1, a rectangle of area A_2, and a triangle of area A_3. These areas are

$$A_1 = \frac{1}{3}\left(\frac{L}{2}\right)\left(\frac{qL^2}{8EI}\right) = \frac{qL^3}{48EI} \qquad A_2 = \frac{L}{2}\left(\frac{qL^2}{8EI}\right) = \frac{qL^3}{16EI}$$

$$A_3 = \frac{1}{2}\left(\frac{L}{2}\right)\left(\frac{3qL^2}{8EI} - \frac{qL^2}{8EI}\right) = \frac{qL^3}{16EI}$$

According to the first moment-area theorem, the angle between the tangents at points A and B is equal to the area of the M/EI diagram between those points. Since the angle at A is zero, it follows that the angle of rotation θ_B is equal to the area of the diagram; thus,

$$\theta_B = A_1 + A_2 + A_3 = \frac{7qL^3}{48EI} \tag{9-80}$$

Deflection: The deflection δ_B is the tangential deviation of point B with respect to a tangent at point A (Fig. 9-25b). Therefore, from the second moment-area theorem, δ_B is equal to the first moment of the M/EI diagram and is evaluated with respect to point B:

$$\delta_B = A_1\bar{x}_1 + A_2\bar{x}_2 + A_3\bar{x}_3 \tag{a}$$

in which $\bar{x}_1$, $\bar{x}_2$, and $\bar{x}_3$ are the distances from point B to the centroids of the respective areas. These distances are

$$\bar{x}_1 = \frac{3}{4}\left(\frac{L}{2}\right) = \frac{3L}{8} \quad \bar{x}_2 = \frac{L}{2} + \frac{L}{4} = \frac{3L}{4} \quad \bar{x}_3 = \frac{L}{2} + \frac{2}{3}\left(\frac{L}{2}\right) = \frac{5L}{6}$$

Substitute into Eq. (a) to find

$$\delta_B = \frac{qL^3}{48EI}\left(\frac{3L}{8}\right) + \frac{qL^3}{16EI}\left(\frac{3L}{4}\right) + \frac{qL^3}{16EI}\left(\frac{5L}{6}\right) = \frac{41qL^4}{384EI} \tag{9-81}$$

4. ***Finalize***: This example illustrates how the area and first moment of a complex M/EI diagram can be determined by dividing the area into parts having known properties. The results of this analysis [Eqs. (9-80) and (9-81)] can be verified by using the formulas of Case 3, Table H-1, Appendix H, and substituting $a = b = L/2$.

Example 9-12

A simple beam ADB supports a concentrated load P acting at the position shown in Fig. 9-26. Determine the angle of rotation θ_A at support A and the deflection δ_D under the load P. *Note:* The beam has a length L and constant flexural rigidity EI.

Solution:

Use a four-step problem-solving approach.

1. ***Conceptualize***: The deflection curve, showing the angle of rotation θ_A and the deflection δ_D, is sketched in Fig. 9-26b. The directions of θ_A and δ_D are determined by inspection, so write the moment-area expressions using only absolute values.
2. ***Categorize***:

M/EI diagram: The bending-moment diagram is triangular, with the maximum moment (equal to Pab/L) occurring under the load. Since EI is constant, the M/EI diagram has the same shape as the moment diagram (see Fig. 9-26c).

FIGURE 9-26

Example 9-12: Simple beam with a concentrated load

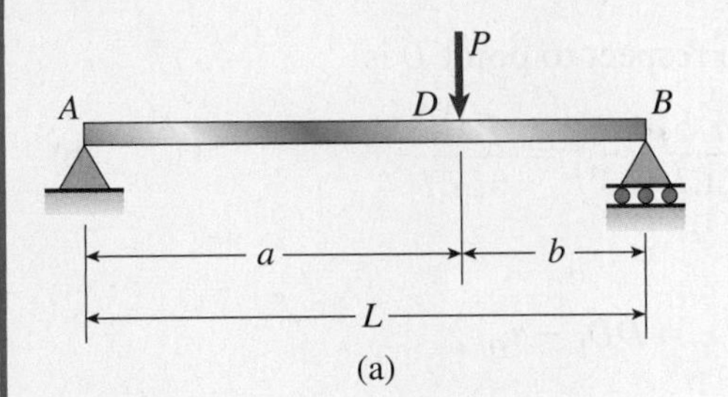

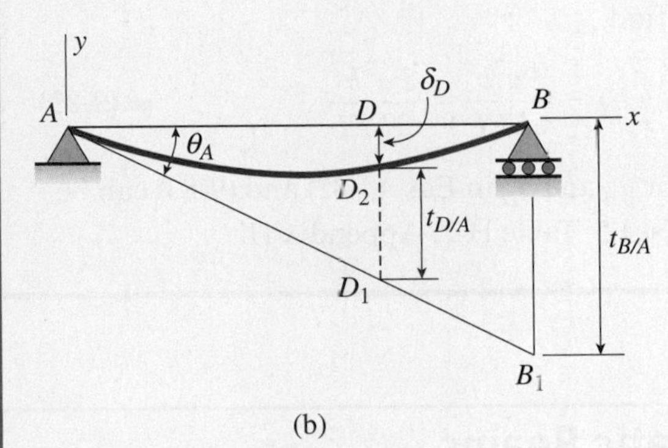

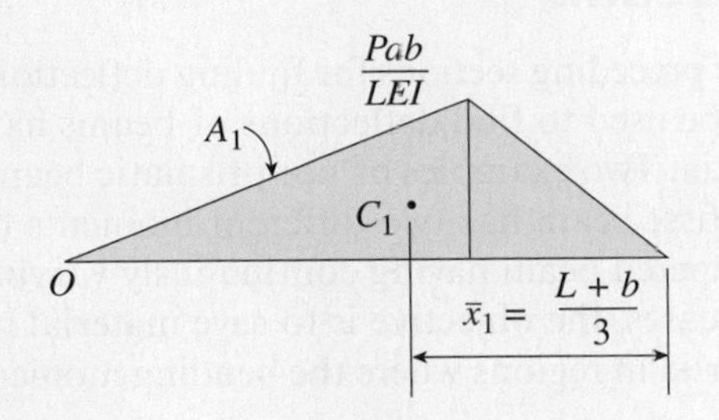

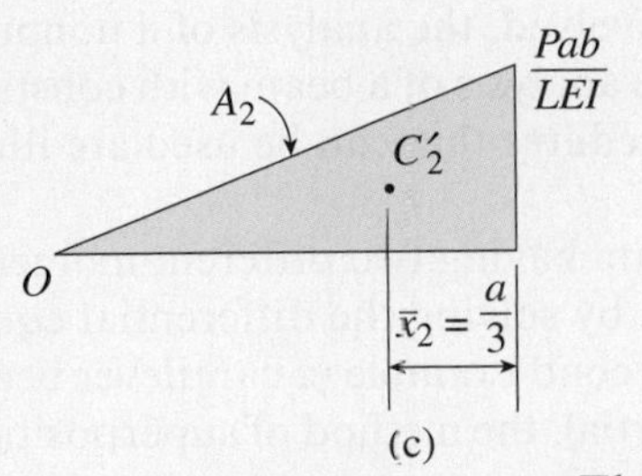

3. *Analyze:*

Angle of rotation at support A: To find this angle, construct the tangent AB_1 at support A. Note that the distance BB_1 is the tangential deviation $t_{B/A}$ of point B from the tangent at A. Calculate this distance by evaluating the first moment of the area of the M/EI diagram with respect to point B and then applying the second moment-area theorem.

The area of the entire M/EI diagram is

$$A_1 = \frac{1}{2}(L)\left(\frac{Pab}{LEI}\right) = \frac{Pab}{2EI}$$

The centroid C_1 of this area is at distance $\bar{x}_1$ from point B (see Fig. 9-26c). This distance, obtained from Case 3 of Appendix E, is

$$\bar{x}_1 = \frac{L+b}{3}$$

Therefore, the tangential deviation is

$$t_{B/A} = A_1\bar{x}_1 = \frac{Pab}{2EI}\left(\frac{L+b}{3}\right) = \frac{Pab}{6EI}(L+b)$$

The angle θ_A is equal to the tangential deviation divided by the length of the beam:

$$\theta_A = \frac{t_{B/A}}{L} = \frac{Pab}{6LEI}(L+b) \qquad \textbf{(9-82)}$$

Thus, the angle of rotation at support A has been found.

Deflection under the load: As shown in Fig. 9-26b, the deflection δ_D under the load P is equal to the distance DD_1 minus the distance D_2D_1. The distance DD_1 is equal to the angle of rotation θ_A times the distance a; thus,

$$DD_1 = a\theta_A = \frac{Pa^2b}{6LEI}(L+b) \qquad \textbf{(a)}$$

The distance D_2D_1 is the tangential deviation $t_{D/A}$ at point D; that is, it is the deviation of point D from the tangent at A. This distance can be found from the second moment-area theorem by taking the first moment of the area of the M/EI diagram between points A and D with respect to D (see Fig. 9-26c). The area of this part of the M/EI diagram is

$$A_2 = \frac{1}{2}(a)\left(\frac{Pab}{LEI}\right) = \frac{Pa^2b}{2LEI}$$

and its centroidal distance from point D is

$$\overline{x}_2 = \frac{a}{3}$$

Thus, the first moment of this area with respect to point D is

$$t_{D/A} = A_2\overline{x}_2 = \left(\frac{Pa^2b}{2LEI}\right)\left(\frac{a}{3}\right) = \frac{Pa^3b}{6LEI} \quad \textbf{(b)}$$

The deflection at point D is

$$\delta_D = DD_1 - D_2D_1 = DD_1 - t_{D/A}$$

Substitute from Eqs. (a) and (b) to find

$$\delta_D = \frac{Pa^2b}{6LEI}(L + b) - \frac{Pa^3b}{6LEI} = \frac{Pa^2b^2}{3LEI} \quad \textbf{(9-83)}$$

4. ***Finalize***: The preceding formulas for θ_A and δ_D in Eqs. (9-82) and (9-83) can be verified by using the formulas of Case 5, Table H-2, Appendix H.

Malcolm Fife/Getty Images

Nonprismatic beams with cutouts in their webs

9.7 Nonprismatic Beams

The methods presented in the preceding sections for finding deflections of prismatic beams also can be used to find deflections of beams having varying moments of inertia. Two examples of nonprismatic beams are shown in Fig. 9-27. The first beam has two different moments of inertia, and the second is a tapered beam having continuously varying moments of inertia. In both cases, the objective is to save material by increasing the moment of inertia in regions where the bending moment is largest.

Although no new concepts are involved, the analysis of a nonprismatic beam is more complex than the analysis of a beam with constant moment of inertia. Some of the procedures that can be used are illustrated in Examples 9-13 and 9-14.

In the first example (a simple beam having two different moments of inertia), the deflections are found by solving the differential equation of the deflection curve. In the second example (a cantilever beam having two different moments of inertia), the method of superposition is used.

These two examples, as well as the problems for this section, involve relatively simple and idealized beams. When more complex beams (such as tapered beams) are encountered, numerical methods of analysis are usually required. (Computer programs for the numerical calculation of beam deflections are readily available.)

FIGURE 9-27

Beams with varying moments of inertia (see also Fig. 5-25)

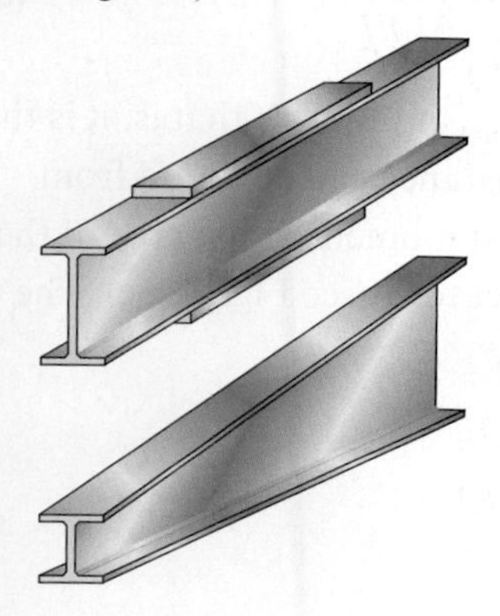

Example 9-13

FIGURE 9-28

Example 9-13: Simple beam with two different moments of inertia

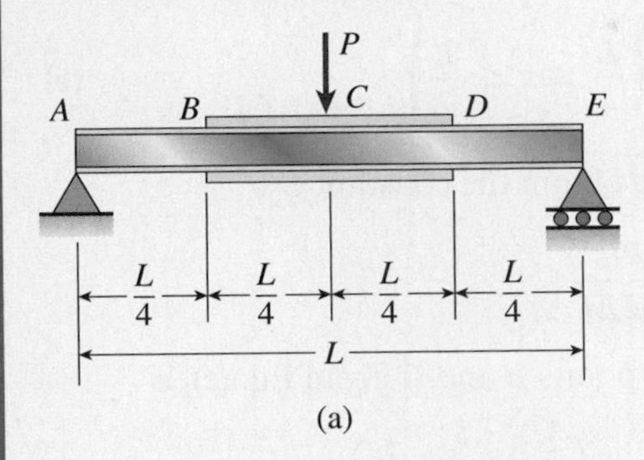

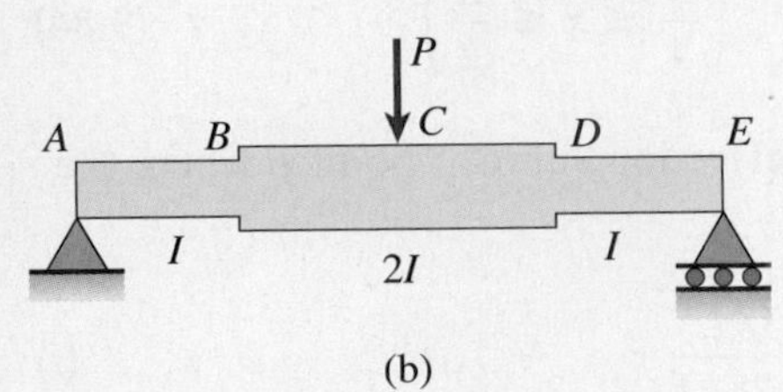

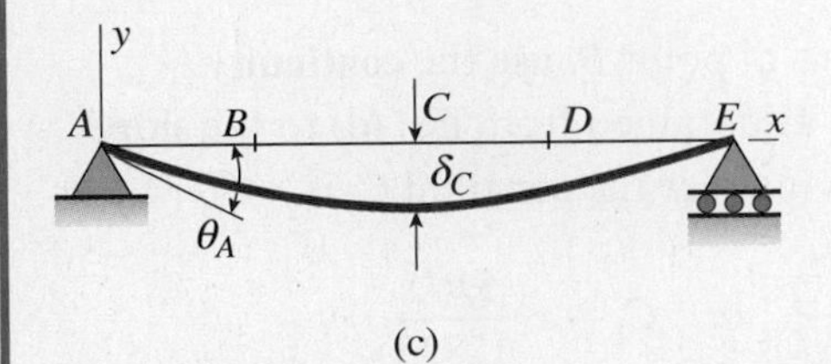

A beam *ABCDE* on simple supports is constructed from a wide-flange beam by welding cover plates over the middle half of the beam (Fig. 9-28a). The effect of the cover plates is to double the moment of inertia (Fig. 9-28b). A concentrated load P acts at the midpoint C of the beam.

Determine the equations of the deflection curve, the angle of rotation θ_A at the left-hand support, and the deflection δ_C at the midpoint (Fig. 9-28c).

Solution:

Use a four-step problem-solving approach. Combine steps for an efficient solution.

1, 2. ***Conceptualize, Categorize:***

Differential equations of the deflection curve: In this example, determine the slopes and deflections of the beam by integrating the bending-moment equation, that is, the second-order differential equation of the deflection curve [Eq. (9-16a)]. Since the reaction at each support is $P/2$, the bending moment throughout the left-hand half of the beam is

$$M = \frac{Px}{2} \qquad \left(0 \le x \le \frac{L}{2}\right) \tag{a}$$

Therefore, the differential equations for the left-hand half of the beam are

$$EIv'' = \frac{Px}{2} \qquad \left(0 \le x \le \frac{L}{4}\right) \tag{b}$$

$$E(2I)v'' = \frac{Px}{2} \qquad \left(\frac{L}{4} \le x \le \frac{L}{2}\right) \tag{c}$$

Each of these equations can be integrated twice to obtain expressions for the slopes and deflections in their respective regions. These integrations produce four constants of integration that can be found from the following four conditions:

- **Boundary condition:** At support A $(x = 0)$, the deflection is zero $(v = 0)$.
- **Symmetry condition:** At point C $(x = L/2)$, the slope is zero $(v' = 0)$.
- **Continuity condition:** At point B $(x = L/4)$, the slope obtained from part *AB* of the beam is equal to the slope obtained from part *BC* of the beam.
- **Continuity condition:** At point B $(x = L/4)$, the deflection obtained from part *AB* of the beam is equal to the deflection obtained from part *BC* of the beam.

3. ***Analyze:***

Slopes of the beam: Integrate each of the differential equations of Eqs. (b) and (c), to obtain the equations for the slopes in the left-hand half of the beam:

$$v' = \frac{Px^2}{4EI} + C_1 \qquad \left(0 \le x \le \frac{L}{4}\right) \tag{d}$$

$$v' = \frac{Px^2}{8EI} + C_2 \qquad \left(\frac{L}{4} \le x \le \frac{L}{2}\right) \tag{e}$$

Apply the **symmetry condition** to Eq. (e) to obtain the constant C_2:

$$C_2 = -\frac{PL^2}{32EI}$$

Therefore, the slope of the beam between points B and C [from Eq. (e)] is

$$v' = -\frac{P}{32EI}(L^2 - 4x^2) \qquad \left(\frac{L}{4} \le x \le \frac{L}{2}\right) \tag{9-84}$$

From this equation, find the slope of the deflection curve at point B where the moment of inertia changes from I to $2I$:

$$v'\left(\frac{L}{4}\right) = -\frac{3PL^2}{128EI} \tag{f}$$

Because the deflection curve is continuous at point B, use the **continuity condition** and equate the slope at point B as obtained from Eq. (d) to the slope at the same point given by Eq. (f). In this manner the constant C_1 is

$$\frac{P}{4EI}\left(\frac{L}{4}\right)^2 + C_1 = -\frac{3PL^2}{128EI} \quad \text{or} \quad C_1 = -\frac{5PL^2}{128EI}$$

Therefore, the slope between points A and B [see Eq. (d)] is

$$v' = -\frac{P}{128EI}(5L^2 - 32x^2) \qquad \left(0 \le x \le \frac{L}{4}\right) \tag{9-85}$$

At support A, where $x = 0$, the angle of rotation (Fig. 9-28c) is

$$\theta_A = -v'(0) = \frac{5PL^2}{128EI} \tag{9-86}$$

Deflections of the beam: Integrate the equations for the slopes [Eqs. (9-85) and (9-84)] to get

$$v = -\frac{P}{128EI}\left(5L^2x - \frac{32x^3}{3}\right) + C_3 \qquad \left(0 \le x \le \frac{L}{4}\right) \tag{g}$$

$$v = -\frac{P}{32EI}\left(L^2x - \frac{4x^3}{3}\right) + C_4 \qquad \left(\frac{L}{4} \le x \le \frac{L}{2}\right) \tag{h}$$

Applying the **boundary condition** at the support to Eq. (g) gives $C_3 = 0$. Therefore, the deflection between points A and B [from Eq. (g)] is

$$v = -\frac{Px}{384EI}(15L^2 - 32x^2) \qquad \left(0 \le x \le \frac{L}{4}\right) \tag{9-87}$$

From this equation, the deflection at point B is

$$v\left(\frac{L}{4}\right) = -\frac{13PL^3}{1536EI} \tag{i}$$

Since the deflection curve is continuous at point B, use the **continuity condition** and equate the deflection at point B as obtained from Eq. (h) to the deflection given by Eq. (i):

$$-\frac{P}{32EI}\left[L^2\left(\frac{L}{4}\right) - \frac{4}{3}\left(\frac{L}{4}\right)^3\right] + C_4 = -\frac{13PL^3}{1536EI}$$

from which

$$C_4 = -\frac{PL^3}{768EI}$$

Therefore, the deflection between points B and C [from Eq. (h)] is

$$v = -\frac{P}{768EI}(L^3 - 24L^2x - 32x^3) \qquad \left(\frac{L}{4} \le x \le \frac{L}{2}\right) \tag{9-88}$$

Thus, the equations of the deflection curve for the left-hand half of the beam are now known. (The deflections in the right-hand half of the beam can be obtained from symmetry.)

Finally, obtain the deflection at the midpoint C by substituting $x = L/2$ into Eq. (9-88):

$$\delta_C = -v\left(\frac{L}{2}\right) = \frac{3PL^3}{256EI} \tag{9-89}$$

4. *Finalize*: All required quantities have now been found, and the analysis of the nonprismatic beam is completed.

   Notes: Using the differential equation for finding deflections is practical only if the number of equations to be solved is limited to one or two and only if the integrations are easily performed, as in this example. In the case of a tapered beam (Fig. 9-27), it may be difficult to solve the differential equation analytically because the moment of inertia is a continuous function of x. In such a case, the differential equation has variable coefficients instead of constant coefficients, and numerical methods of solution are needed.

   When a beam has abrupt changes in cross-sectional dimensions, as in this example, there are stress concentrations at the points where changes occur. However, because the stress concentrations affect only a small region of the beam, they have no noticeable effect on the deflections.

Example 9-14

FIGURE 9-29

Example 9-14: Cantilever beam with two different moments of inertia

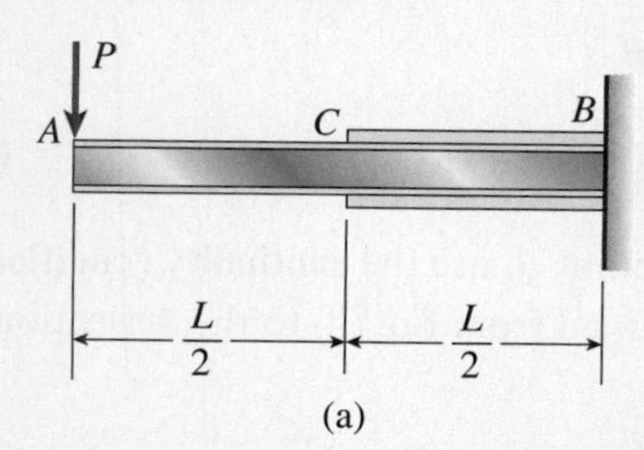

(a)

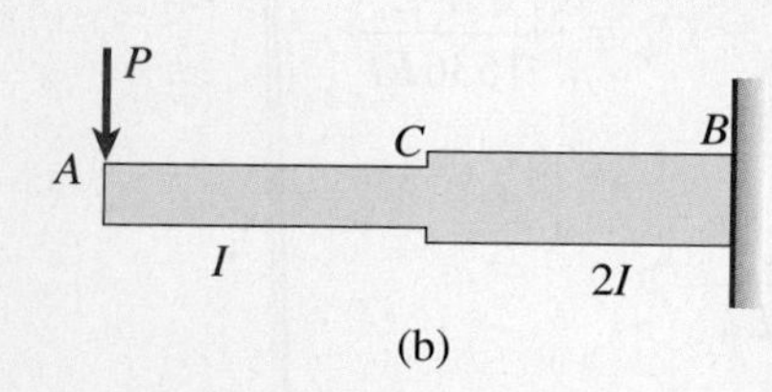

(b)

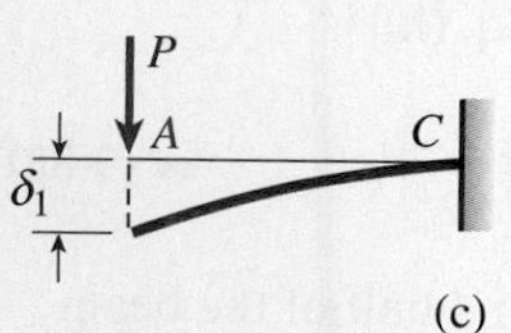

(c)

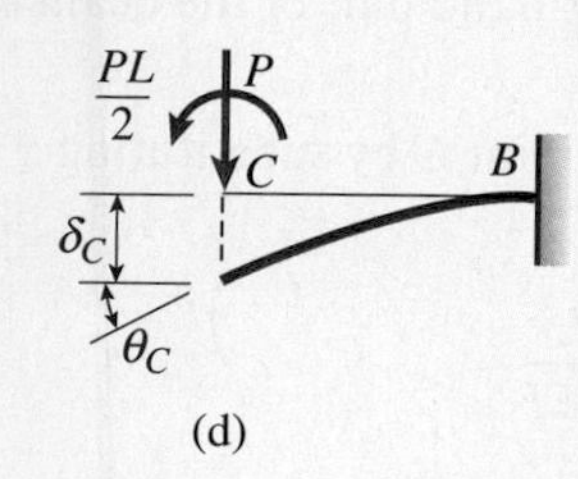

(d)

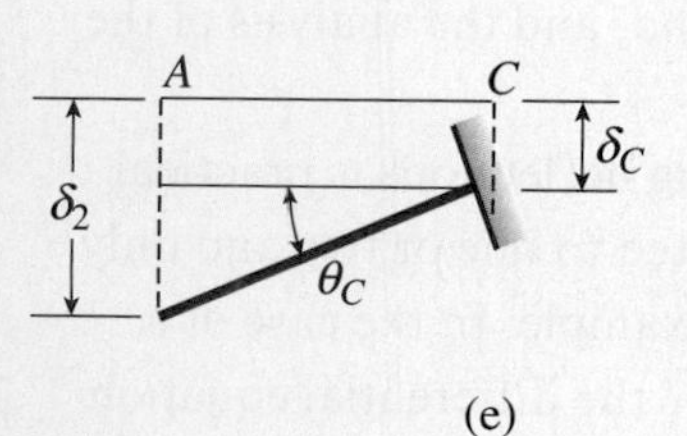

(e)

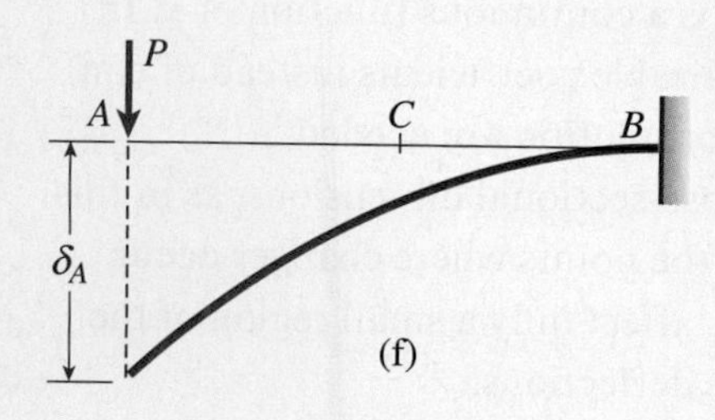

(f)

A cantilever beam ACB having a length L and two different moments of inertia I and $2I$ supports a concentrated load P at the free end A (Figs. 9-29a and b).

Determine the deflection δ_A at the free end.

Solution:

Use a four-step problem-solving approach.

1. ***Conceptualize***: In this example, use the method of superposition to determine the deflection δ_A at the end of the beam. Begin by recognizing that the deflection consists of two parts: the deflection due to bending of part AC of the beam and the deflection due to bending of part CB. Find these deflections separately and then superpose them to obtain the total deflection.
2. ***Categorize***:

Deflection due to bending of part AC of the beam: Imagine that the beam is held rigidly at point C, so that the beam neither deflects nor rotates at that point (Fig. 9-29c). Now calculate the deflection δ_1 of point A in this beam. Since the beam has length $L/2$ and moment of inertia I, its deflection (see Case 4 of Table H-1, Appendix H) is

$$\delta_1 = \frac{P(L/2)^3}{3EI} = \frac{PL^3}{24EI} \qquad \textbf{(a)}$$

Deflection due to bending of part CB of the beam: Part CB of the beam also behaves as a cantilever beam (Fig. 9-29d) and contributes to the deflection of point A. The end of this cantilever is subjected to a concentrated load P and a moment $PL/2$. Therefore, the deflection δ_C and angle of rotation θ_C at the free end (Fig. 9-29d) are (see Cases 4 and 6 of Table H-1):

$$\delta_C = \frac{P(L/2)^3}{3(2EI)} + \frac{(PL/2)(L/2)^2}{2(2EI)} = \frac{5PL^3}{96EI}$$

$$\theta_C = \frac{P(L/2)^2}{2(2EI)} + \frac{(PL/2)(L/2)}{2EI} = \frac{3PL^2}{16EI}$$

This deflection and angle of rotation make an additional contribution δ_2 to the deflection at end A (Fig. 9-29e). Again visualize part AC as a cantilever beam, but now its support (at point C) moves downward by the amount δ_C and rotates counterclockwise through the angle θ_C (Fig. 9-29e). These rigid-body displacements produce a downward displacement at end A equal to

$$\delta_2 = \delta_C + \theta_C\left(\frac{L}{2}\right) = \frac{5PL^3}{96EI} + \frac{3PL^2}{16EI}\left(\frac{L}{2}\right) = \frac{7PL^3}{48EI} \qquad \textbf{(b)}$$

Total deflection: The total deflection δ_A at the free end A of the original cantilever beam (Fig. 9-29f) is equal to the sum of the deflections δ_1 and δ_2:

$$\delta_A = \delta_1 + \delta_2 = \frac{PL^3}{24EI} + \frac{7PL^3}{48EI} = \frac{3PL^3}{16EI} \qquad \text{(9-90)}$$

4. ***Finalize***: This example illustrates one of the many ways that the principle of superposition may be used to find beam deflections.

9.8 Strain Energy of Bending

The general concepts pertaining to strain energy were explained previously in the discussions of bars subjected to axial loads and shafts subjected to torsion (Sections 2.7 and 3.9, respectively). In this section, the same concepts are applied to beams. Once again, use the equations for curvature and deflection derived earlier in this chapter, so this discussion of strain energy applies only to beams that behave in a linearly elastic manner. This requirement means that the material must follow Hooke's law and the deflections and rotations must be small.

Begin with a simple beam AB in pure bending under the action of two couples, each having a moment M (Fig. 9-30a). The deflection curve (Fig. 9-30b) is a nearly flat circular arc of constant curvature $\kappa = M / EI$ (see Eq. 9-10). The angle θ subtended by this arc equals L / ρ, where L is the length of the beam and ρ is the radius of curvature. Therefore,

$$\theta = \frac{L}{\rho} = \kappa L = \frac{ML}{EI} \qquad \text{(9-91)}$$

This linear relationship between the moments M and the angle θ is shown graphically by line OA in Fig. 9-31. As the bending couples gradually increase in magnitude from zero to their maximum values, they perform work W represented by the shaded area below line OA. This work, equal to the strain energy U stored in the beam, is

$$W = U = \frac{M\theta}{2} \qquad \text{(9-92)}$$

This equation is analogous to Eq. (2-35) for the strain energy of an axially loaded bar.

By combining Eqs. (9-91) and (9-92), express the strain energy stored in a beam in pure bending in either of the following forms:

$$U = \frac{M^2L}{2EI} \qquad U = \frac{EI\theta^2}{2L} \qquad \text{(9-93a,b)}$$

The first of these equations expresses the strain energy in terms of the applied moments M, and the second equation expresses it in terms of the angle θ. The equations are similar in form to those for strain energy in an axially loaded bar [Eqs. (2-37a and b)].

FIGURE 9-30

Beam in pure bending by couples of moment M

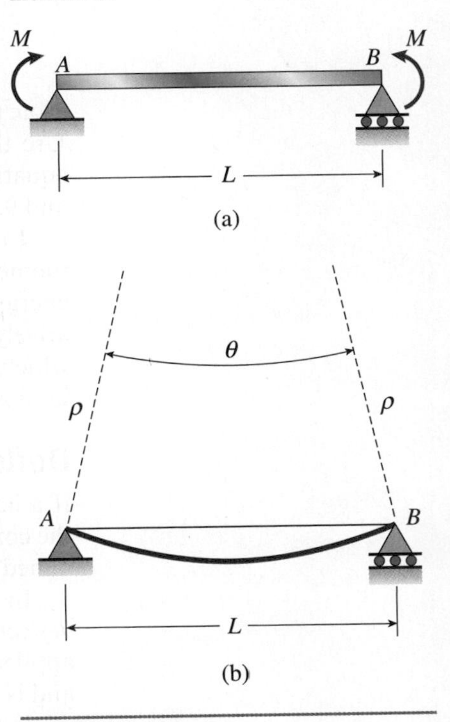

FIGURE 9-31

Diagram showing linear relationship between bending moments M and the angle θ

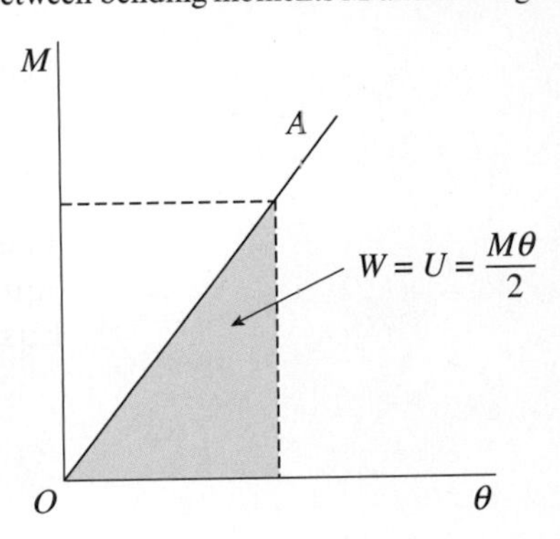

FIGURE 9-32

Side view of an element of a beam subjected to bending moments M

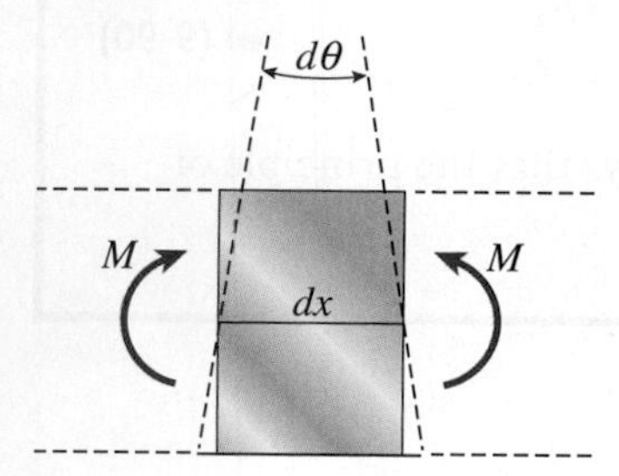

If the bending moment in a beam varies along its length (nonuniform bending), the strain energy is obtained by applying Eqs. (9-93a and b) to an element of the beam (Fig. 9-32) and integrating along the length of the beam. The length of the element itself is dx and the angle $d\theta$ between its side faces can be obtained from Eqs. (9-6) and (9-9) as

$$d\theta = \kappa dx = \frac{d^2v}{dx^2}dx \qquad \textbf{(9-94a)}$$

Therefore, the strain energy dU of the element is given by either of the following equations [see Eqs. (9-93a and b)]:

$$dU = \frac{M^2 dx}{2EI} \quad dU = \frac{EI(d\theta)^2}{2dx} = \frac{EI}{2dx}\left(\frac{d^2v}{dx^2}dx\right)^2 = \frac{EI}{2}\left(\frac{d^2v}{dx^2}\right)^2 dx \qquad \textbf{(9-94b,c)}$$

By integrating the preceding equations throughout the length of a beam, express the strain energy stored in a beam in either of the following forms:

$$U = \int \frac{M^2 dx}{2EI} \quad U = \int \frac{EI}{2}\left(\frac{d^2v}{dx^2}\right)^2 dx \qquad \textbf{(9-95a,b)}$$

Note that M is the bending moment in the beam and may vary as a function of x. Use the first equation when the bending moment is known, and the second equation when the equation of the deflection curve is known. (Examples 9-15 and 9-16 illustrate the use of these equations.)

In the derivation of Eqs. (9-95a and b), only the effects of the bending moments were considered. If shear forces are also present, additional strain energy will be stored in the beam. However, the strain energy of shear is relatively small (in comparison with the strain energy of bending) for beams in which the lengths are much greater than the depths (say, $L/d > 8$). Therefore, in most beams the strain energy of shear may safely be disregarded.

Deflections Caused by a Single Load

If a beam supports a single load, either a concentrated load P or a couple M_0, the corresponding deflection δ or angle of rotation θ, respectively, can be determined from the strain energy of the beam.

In the case of a beam supporting a **concentrated load**, the *corresponding deflection* δ is the deflection of the beam axis at the point where the load is applied. The deflection must be measured along the line of action of the load and is positive in the direction of the load.

In the case of a beam supporting a couple as a load, the *corresponding angle of rotation* θ is the angle of rotation of the beam axis at the point where the couple is applied.

The strain energy of a beam is equal to the work done by the load, where δ and θ correspond to P and M_0, respectively, so the following equations are obtained:

$$U = W = \frac{P\delta}{2} \quad U = W = \frac{M_0\theta}{2} \qquad \textbf{(9-96a,b)}$$

The first equation applies to a beam loaded *only* by a force P, and the second equation applies to a beam loaded *only* by a couple M_0. It follows from Eqs. (9-96a and b) that

$$\delta = \frac{2U}{P} \quad \theta = \frac{2U}{M_0} \qquad \textbf{(9-97a,b)}$$

As explained in Section 2.7, this method for finding deflections and angles of rotation is extremely limited in its application because only one deflection (or one angle) can be found. Furthermore, the only deflection (or angle) that can be found is the one corresponding to the load (or couple). However, the method occasionally is useful and is illustrated in Example 9-16.

Example 9-15

FIGURE 9-33

Example 9-15: Strain energy of a beam

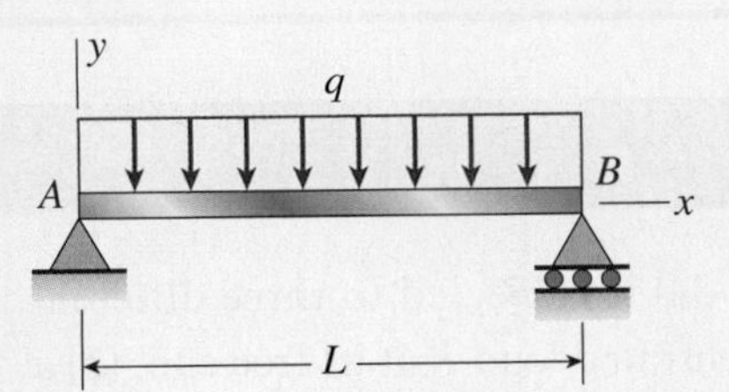

A simple beam AB of a length L supports a uniform load of an intensity q (Fig. 9-33). (a) Evaluate the strain energy of the beam from the bending moment in the beam. (b) Evaluate the strain energy of the beam from the equation of the deflection curve. *Note:* The beam has constant flexural rigidity EI.

Solution:

Use a four-step problem-solving approach. Combine steps as needed for an efficient solution.

Part (a): Strain energy from the bending moment.

1, 2. ***Conceptualize, Categorize*:** The reaction of the beam at support A is $qL/2$, so the expression for the bending moment in the beam is

$$M = \frac{qLx}{2} - \frac{qx^2}{2} = \frac{q}{2}(Lx - x^2) \tag{a}$$

3. ***Analyze*:** The strain energy of the beam [from Eq. (9-95a)] is

$$U = \int_0^L \frac{M^2 dx}{2EI} = \frac{1}{2EI}\int_0^L \left[\frac{q}{2}(Lx - x^2)\right]^2 dx = \frac{q^2}{8EI}\int_0^L (L^2x^2 - 2Lx^2 + x^4)dx \tag{b}$$

which leads to

$$U = \frac{q^2L^5}{240EI} \tag{9-98}$$

Note that the load q appears to the second power, which is consistent with the fact that strain energy is always positive. Furthermore, Eq. (9-98) shows that strain energy is *not* a linear function of the loads, even though the beam itself behaves in a linearly elastic manner.

Part (b): Strain energy from the deflection curve.

1, 2. ***Conceptualize, Categorize*:** The equation of the deflection curve for a simple beam with a uniform load is given in Case 1 of Table H-2, Appendix H, as

$$v = -\frac{qx}{24EI}(L^3 - 2Lx^2 + x^3) \tag{c}$$

3. ***Analyze*:** Taking two derivatives of this equation gives

$$\frac{dv}{dx} = -\frac{q}{24EI}(L^3 - 6Lx^2 + 4x^3) \qquad \frac{d^2v}{dx^2} = \frac{q}{2EI}(Lx - x^2)$$

Substitute the latter expression into the equation for strain energy [Eq. (9-95b)] to obtain

$$U = \int_0^L \frac{EI}{2}\left(\frac{d^2v}{dx^2}\right)^2 dx = \frac{EI}{2}\int_0^L \left[\frac{q}{2EI}(Lx - x^2)\right]^2 dx$$

$$= \frac{q^2}{8EI}\int_0^L (L^2x^2 - 2Lx^3 + x^4)dx \quad \text{(d)}$$

4. ***Finalize***: The final integral in this equation is the same as the final integral in Eq. (b) which leads to the same result as before [Eq. (9-98)].

Example 9-16

FIGURE 9-34

Example 9-16: Strain energy of a beam

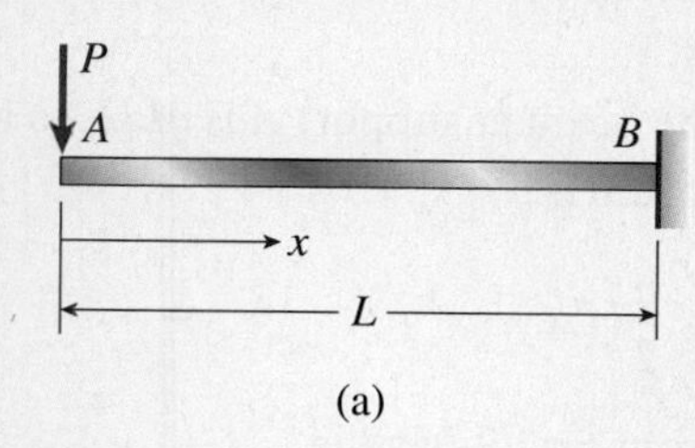

(a)

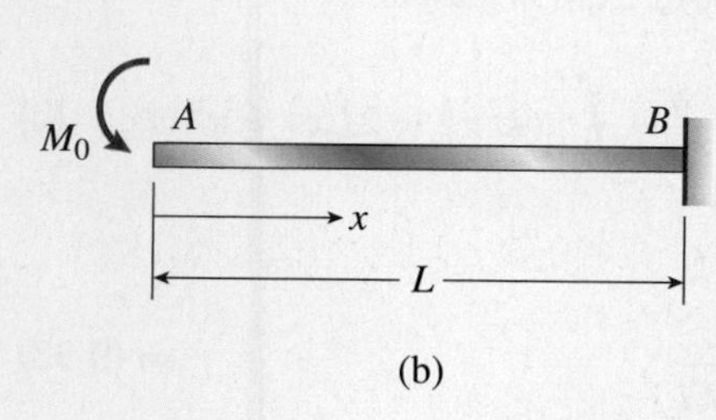

(b)

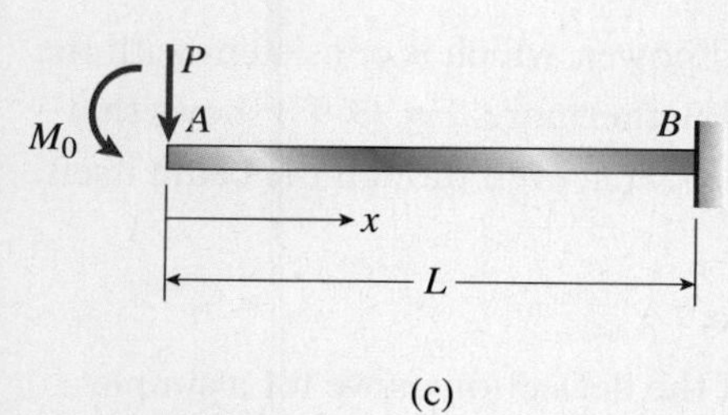

(c)

A cantilever beam AB (Fig. 9-34) is subjected to three different loading conditions: (a) a concentrated load P at its free end, (b) a couple M_0 at its free end, and (c) both loads acting simultaneously.

For each loading condition, determine the strain energy of the beam. Also, determine the vertical deflection δ_A at end A of the beam due to the load P acting alone (Fig. 9-34a), and determine the angle of rotation θ_A at end A due to the moment M_0 acting alone (Fig. 9-34b). *Note:* The beam has constant flexural rigidity EI.

Solution:

Use a four-step problem-solving approach. Combine steps as needed for an efficient solution.

Part (a): Beam with concentrated load P.

1, 2. ***Conceptualize, Categorize***: The bending moment in the beam (Fig. 9-34a) at distance x from the free end is $M = -Px$. Substituting this expression for M into Eq. (9-95a) leads to the expression for the strain energy of the beam:

$$U = \int_0^L \frac{M^2 dx}{2EI} = \int_0^L \frac{(-Px)^2 dx}{2EI} = \frac{P^2L^3}{6EI} \quad \text{(9-99)}$$

3. ***Analyze***: To obtain the vertical deflection δ_A under the load P, equate the work done by the load to the strain energy:

$$W = U \quad \text{or} \quad \frac{P\delta_A}{2} = \frac{P^2L^3}{6EI}$$

from which

$$\delta_A = \frac{PL^3}{3EI}$$

The deflection δ_A is the only deflection that can be found by this procedure, because it is the only deflection that corresponds to the load P.

Part (b): Beam with moment M_0.

1, 2. ***Conceptualize, Categorize***: In this case, the bending moment (Fig. 9-34b) is constant and equal to $-M_0$. Therefore, the strain energy [from Eq. (9-95a)] is

$$U = \int_0^L \frac{M^2 dx}{2EI} = \int_0^L \frac{(-M_0)^2 dx}{2EI} = \frac{M_0^2 L}{2EI} \qquad \text{(9-100)}$$

3. ***Analyze***: The work W done by the couple M_0 during loading of the beam is $M_0\theta_A/2$, where θ_A is the angle of rotation at end A. Therefore,

$$W = U \quad \text{or} \quad \frac{M_0\theta_A}{2} = \frac{M_0^2 L}{2EI}$$

and

$$\theta_A = \frac{M_0 L}{EI}$$

The angle of rotation has the same sense as the moment (counterclockwise in this example).

Part (c): Beam with both loads acting simultaneously.

1, 2. ***Conceptualize, Categorize***: When both loads act on the beam (Fig. 9-34c), the bending moment in the beam is

$$M = -Px - M_0$$

3. ***Analyze***: Therefore, the strain energy is

$$U = \int_0^L \frac{M^2 dx}{2EI} = \frac{1}{2EI}\int_0^L (-Px - M_0)^2 dx$$

$$= \frac{P^2 L^3}{6EI} + \frac{PM_0 L^2}{2EI} + \frac{M_0^2 L}{2EI} \qquad \text{(9-101)}$$

The first term in this result gives the strain energy due to P acting alone [Eq. (9-99)], and the last term gives the strain energy due to M_0 alone [Eq. (9-100)]. However, when both loads act simultaneously, an additional term appears in the expression for strain energy.

4. ***Finalize***: Therefore, *the strain energy in a structure due to two or more loads acting simultaneously cannot be obtained by adding the strain energies due to the loads acting separately.* The reason is that strain energy is a quadratic function of the loads, not a linear function. Therefore, *the principle of superposition does not apply to strain energy.*

Also observe that a deflection for a beam with two or more loads cannot be calculated by equating the work done by the loads to the strain energy. For instance, equating work and energy for the beam of Fig. 9-34c gives

$$W = U \quad \text{or} \quad \frac{P\delta_{A2}}{2} + \frac{M_0\theta_{A2}}{2} = \frac{P^2L^3}{6EI} + \frac{PM_0L^2}{2EI} + \frac{M_0^2 L}{2EI} \qquad \text{(a)}$$

in which δ_{A2} and θ_{A2} represent the deflection and angle of rotation at end A of the beam with two loads acting simultaneously (Fig. 9-34c). Although the work done by the two loads is indeed equal to the strain energy, and Eq. (a) is correct, a solution for either δ_{A2} or θ_{A2} cannot be obtained because there are two unknowns and only one equation.

*9.9 Castigliano's Theorem

FIGURE 9-35

Beam supporting a single load P

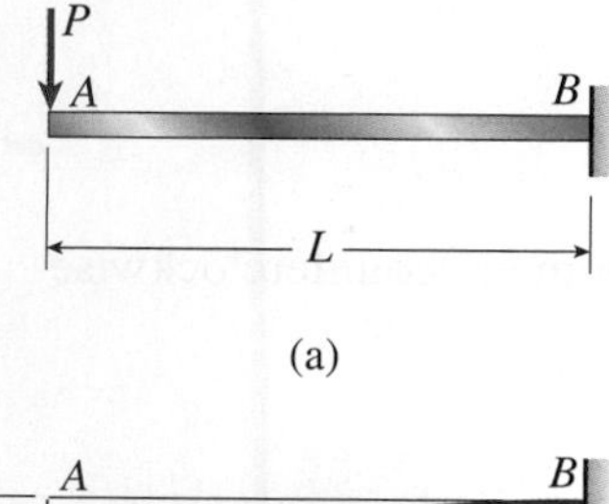

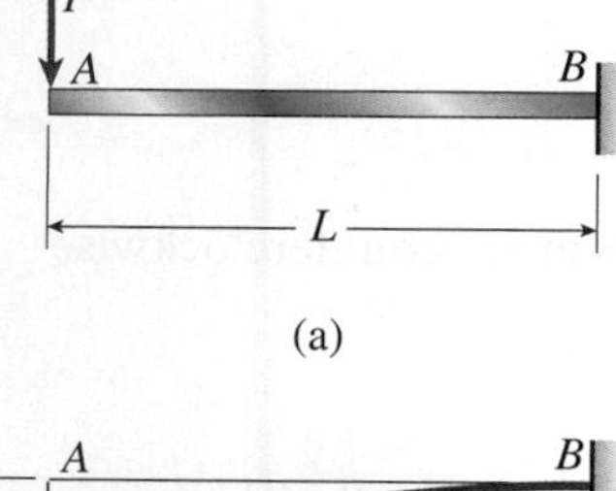

Castigliano's theorem provides a means for finding the deflections of a structure from the strain energy of the structure. To illustrate, consider a cantilever beam with a concentrated load P acting at the free end (Fig. 9-35a). The strain energy of this beam is obtained from Eq. (9-99) of Example 9-16:

$$U = \frac{P^2L^3}{6EI} \tag{9-102a}$$

Now take the derivative of this expression with respect to the load P:

$$\frac{dU}{dP} = \frac{d}{dP}\left(\frac{P^2L^3}{6EI}\right) = \frac{PL^3}{3EI} \tag{9-102b}$$

This result is the deflection δ_A at the free end A of the beam (see Fig. 9-35b). Note especially that the deflection δ_A *corresponds* to the load P itself. (Recall that a deflection corresponding to a concentrated load is the deflection at the point where the concentrated load is applied. Furthermore, the deflection is in the direction of the load.) Thus, Eq. (9-102b) shows that *the derivative of the strain energy with respect to the load is equal to the deflection corresponding to the load.* Castigliano's theorem is a generalized statement of this observation, and it now will be derived in more general terms.

Derivation of Castigliano's Theorem

FIGURE 9-36

Beam supporting n loads

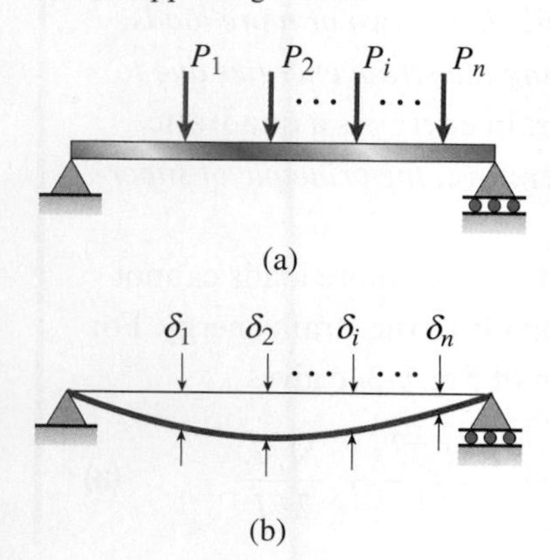

Consider a beam subjected to any number of loads, say n loads $P_1, P_2, ..., P_i, ..., P_n$ (Fig. 9-36a). The deflections of the beam corresponding to the various loads are denoted $\delta_1, \delta_2, ..., \delta_i, ..., \delta_n$, as shown in Fig. 9-36b. As in earlier discussions of deflections and strain energy, assume that the principle of superposition is applicable to the beam and its loads.

Now determine the strain energy of this beam. When the loads are applied to the beam, they gradually increase in magnitude from zero to their maximum values. At the same time, each load moves through its corresponding displacement and does work. The total work W done by the loads is equal to the strain energy U stored in the beam:

$$W = U \tag{9-103}$$

Note that W (and hence U) is a function of the loads $P_1, P_2, ..., P_n$ acting on the beam.

Next, suppose that one of the loads, say the ith load, is increased slightly by the amount dP_i, while the other loads are held constant. This increase in load will cause a small increase dU in the strain energy of the beam. This increase in strain energy may be expressed as the rate of change of U with respect to P_i times the small increase in P_i. Thus, the increase in strain energy is

$$dU = \frac{\partial U}{\partial P_i} dP_i \tag{9-104}$$

where $\partial U / \partial P_i$ is the rate of change of U with respect to P_i. (Since U is a function of *all* the loads, the derivative with respect to any one of the loads is a partial derivative.) The final strain energy of the beam is

$$U + dU = U + \frac{\partial U}{\partial P_i} dP_i \tag{9-105}$$

in which U is the strain energy referred to in Eq. (9-103).

Because the principle of superposition holds for this beam, the total strain energy is independent of the order in which the loads are applied. That is, the final displacements of the beam (and the work done by the loads in reaching those displacements) are the same regardless of the order in which the loads are applied. In arriving at the strain energy given by Eq. (9-105), the n loads $P_1, P_2,\ldots, P_n$ were first applied and then the load dP_i was applied. However, the order of application can be reversed and the load dP_i applied first, followed by the loads P_1, P_2, ..., P_n. The final amount of strain energy is the same in either case.

When the load dP_i is applied first, it produces strain energy equal to one-half the product of the load dP_i and its corresponding displacement $d\delta_i$. Thus, the amount of strain energy due to the load dP_i is

$$\frac{dP_i d\delta_i}{2} \tag{9-106a}$$

When the loads P_1, P_2, ..., P_n are applied, they produce the same displacements as before (δ_1, δ_2, ..., δ_n) and do the same amount of work as before [Eq. (9-103)]. However, during the application of these loads, the force dP_i automatically moves through the displacement δ_i. In so doing, it produces additional work equal to the product of the force and the distance through which it moves. (Note that the work does not have a factor 1/2, because the force dP_i acts at full value throughout this displacement.) Thus, the additional work, equal to the additional strain energy, is

$$dP_i\delta_i \tag{9-106b}$$

Therefore, the final strain energy for the second loading sequence is

$$\frac{dP_i d\delta_i}{2} + U + dP_i\delta_i \tag{9-106c}$$

Equate this expression for the final strain energy to the earlier expression [Eq. (9-105)], which was obtained for the first loading sequence, to get

$$\frac{dP_i d\delta_i}{2} + U + dP_i\delta_i = U + \frac{\delta U}{\delta P_i} dP_i \tag{9-106d}$$

Discard the first term because it contains the product of two differentials and is infinitesimally small compared to the other terms. The following relationship is obtained:

$$\delta_i = \frac{\partial U}{\partial P_i} \tag{9-107}$$

This equation is known as **Castigliano's theorem**.[3]

Although Castigliano's theorem was derived by using a beam as an illustration, any other type of structure (for example, a truss) and any other kinds of loads (for example, loads in the form of couples) could have been used. The important requirements are that the structure be linearly elastic and that the principle of superposition be applicable. Also, note that the strain energy must be expressed as a function of the loads (and not as a function of the displacements), a condition which is implied in the theorem itself, since the partial derivative is taken with respect to a load. With these limitations in mind, Castigliano's theorem is stated in general terms as follows:

The partial derivative of the strain energy of a structure with respect to any load is equal to the displacement corresponding to that load.

The strain energy of a linearly elastic structure is a *quadratic* function of the loads [for instance, see Eq. (9-102a)], and therefore, the partial derivatives and the displacements [Eq. (9-107)] are *linear* functions of the loads (as expected).

When using the terms *load* and *corresponding displacement* in connection with Castigliano's theorem, it is understood that these terms are used in a generalized sense. The load P_i and corresponding displacement δ_i may be a force and a corresponding translation, or a couple and a corresponding rotation, or some other set of corresponding quantities.

Application of Castigliano's Theorem

FIGURE 9-37

Application of Castigliano's theorem to a beam

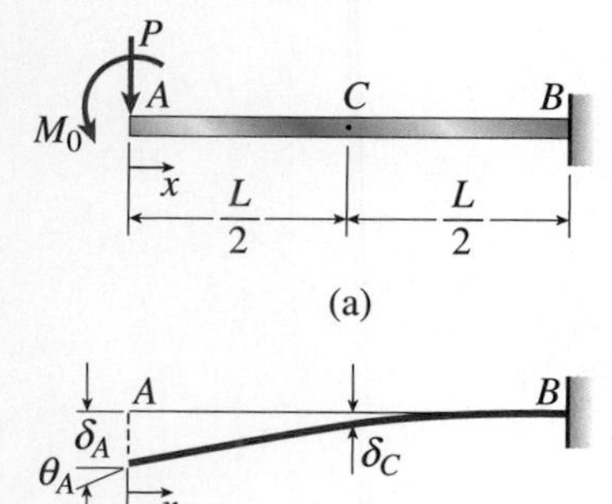

As an application of Castigliano's theorem, consider a cantilever beam AB carrying a concentrated load P and a couple of moment M_0 acting at the free end (Fig. 9-37a). The objective is to determine the vertical deflection δ_A and angle of rotation θ_A at the end of the beam (Fig. 9-37b). Note that δ_A is the deflection corresponding to the load P, and θ_A is the angle of rotation corresponding to the moment M_0.

The first step in the analysis is to determine the strain energy of the beam. For that purpose, write the equation for the bending moment as

$$M = -Px - M_0 \tag{9-108}$$

in which x is the distance from the free end (Fig. 9-37a). The strain energy is found by substituting this expression for M into Eq. (9-95a):

$$U = \int_0^L \frac{M^2 dx}{2EI} = \frac{1}{2EI}\int_0^L (-Px - M_0)^2 dx = \frac{P^2L^3}{6EI} + \frac{PM_0L^2}{2EI} + \frac{M_0^2L}{2EI} \tag{9-109}$$

[3]Castigliano's theorem, one of the most famous theorems in structural analysis, was discovered by Carlos Alberto Pio Castigliano (1847–1884), an Italian engineer (Ref. 9-2). The theorem quoted here [Eq. (9-107)] is actually the second of two theorems presented by Castigliano and is properly called *Castigliano's second theorem*. The first theorem is the reverse of the second theorem in the sense that it gives the loads on a structure in terms of the partial derivatives of the strain energy with respect to the *displacements*.

in which L is the length of the beam and EI is its flexural rigidity. Note that the strain energy is a quadratic function of the loads P and M_0.

To obtain the vertical deflection δ_A at the end of the beam, use Castigliano's theorem [Eq. (9-107)] and take the partial derivative of the strain energy with respect to P:

$$\delta_A = \frac{\partial U}{\partial P} = \frac{PL^3}{3EI} + \frac{M_0 L^2}{2EI} \quad \textbf{(9-110)}$$

This expression for the deflection can be verified by comparing it with the formulas of Cases 4 and 6 of Table H-1, Appendix H.

In a similar manner, find the angle of rotation θ_A at the end of the beam by taking the partial derivative with respect to M_0:

$$\theta_A = \frac{\partial U}{\partial M_0} = \frac{PL^2}{2EI} + \frac{M_0 L}{EI} \quad \textbf{(9-111)}$$

This equation can also be verified by comparing with the formulas of Cases 4 and 6 of Table H-1.

Use of a Fictitious Load

The only displacements that can be found from Castigliano's theorem are those that correspond to loads acting on the structure. To calculate a displacement at a point on a structure where there is no load, apply a fictitious load *corresponding to the desired displacement* to the structure. Then determine the displacement by evaluating the strain energy and taking the partial derivative with respect to the fictitious load. The result is the displacement produced by the actual loads and the fictitious load acting simultaneously. Set the fictitious load equal to zero to obtain the displacement produced only by the actual loads.

To illustrate this concept, find the vertical deflection δ_C at the midpoint C of the cantilever beam shown in Fig. 9-38a. Since the deflection δ_C is downward (Fig. 9-38b), the load corresponding to that deflection is a downward vertical force acting at the same point. Therefore, add a fictitious load Q acting at point C in the downward direction (Fig. 9-39a). Then use Castigliano's theorem to determine the deflection $(\delta_C)_0$ at the midpoint of this beam (Fig. 9-39b). From that deflection, obtain the deflection δ_C in the beam of Fig. 9-38 by setting Q equal to zero.

Begin by finding the bending moments in the beam of Fig. 9-39a:

$$M = -Px - M_0 \qquad \left(0 \le x \le \frac{L}{2}\right) \quad \textbf{(9-112a)}$$

$$M = -Px - M_0 - Q\left(x - \frac{L}{2}\right) \qquad \left(\frac{L}{2} \le x \le L\right) \quad \textbf{(9-112b)}$$

Next, determine the strain energy of the beam by applying Eq. (9-95a) to each half of the beam. For the left-hand half of the beam (from point A to point C), the strain energy is

$$U_{AC} = \int_0^{L/2} \frac{M^2 dx}{2EI} = \frac{1}{2EI}\int_0^{L/2} (-Px - M_0)^2 dx$$

$$= \frac{P^2L^3}{48EI} + \frac{PM_0L^2}{8EI} + \frac{M_0^2 L}{4EI} \quad \textbf{(9-113a)}$$

FIGURE 9-38

Beam supporting loads P and M_0

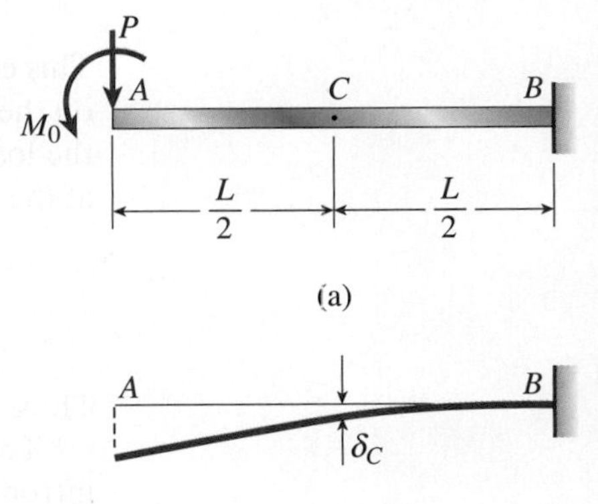

FIGURE 9-39

Beam with a fictitious load Q

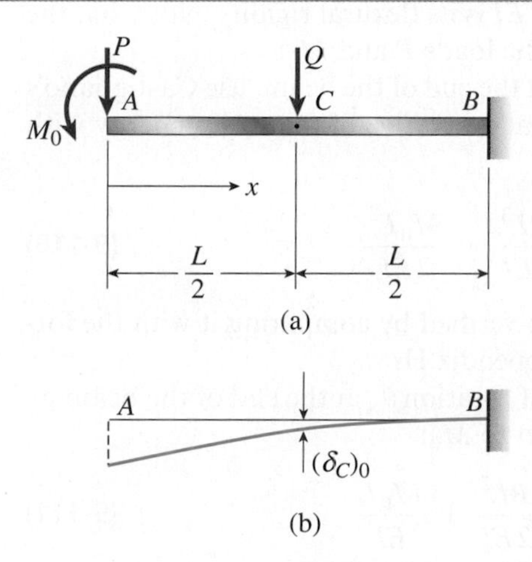

For the right-hand half, the strain energy is

$$U_{CB} = \int_{L/2}^{L} \frac{M^2 dx}{2EI} = \frac{1}{2EI} \int_{L/2}^{L} \left[-Px - M_0 - Q\left(x - \frac{L}{2}\right) \right]^2 dx$$

$$= \frac{7P^2L^3}{48EI} + \frac{3PM_0L^2}{8EI} + \frac{5PQL^3}{48EI} + \frac{M_0^2 L}{4EI} + \frac{M_0 QL^2}{8EI} + \frac{Q^2L^3}{48EI} \quad \textbf{(9-113b)}$$

which requires a very lengthy process of integration. Adding the strain energies for the two parts of the beam gives the strain energy for the entire beam (Fig. 9-39a):

$$U = U_{AC} + U_{CB}$$

$$= \frac{P^2L^3}{6EI} + \frac{PM_0L^2}{2EI} + \frac{5PQL^3}{48EI} + \frac{M_0^2 L}{2EI} + \frac{M_0 QL^2}{8EI} + \frac{Q^2L^3}{48EI} \quad \textbf{(9-114)}$$

The deflection at the midpoint of the beam shown in Fig. 9-39a now can be obtained from Castigliano's theorem:

$$(\delta_C)_0 = \frac{\partial U}{\partial Q} = \frac{5PL^3}{48EI} + \frac{M_0L^2}{8EI} + \frac{QL^3}{24EI} \quad \textbf{(9-115)}$$

This equation gives the deflection at point C produced by all three loads acting on the beam. To obtain the deflection produced by the loads P and M_0 only, set the load Q equal to zero in the preceding equation. The result is the deflection at the midpoint C for the beam with two loads (Fig. 9-38a):

$$\delta_C = \frac{5PL^3}{48EI} + \frac{M_0L^2}{8EI} \quad \textbf{(9-116)}$$

Thus, the deflection in the original beam has been obtained.

This method is sometimes called the *dummy-load method*, because of the introduction of a fictitious, or dummy, load.

Differentiation Under the Integral Sign

As shown in the preceding discussion, the use of Castigliano's theorem for determining beam deflections may lead to lengthy integrations, especially when more than two loads act on the beam. The reason is clear—finding the strain energy requires the integration of the *square* of the bending moment [Eq. (9-95a)]. For instance, if the bending moment expression has three terms, its square may have as many as six terms, each of which must be integrated.

After the integrations are completed and the strain energy has been determined, differentiate the strain energy to obtain the deflections. However, you can bypass the step of finding the strain energy by differentiating *before* integrating. This procedure does not eliminate the integrations, but it does make them much simpler.

To derive this method, begin with the equation for the strain energy [Eq. (9-95a)] and apply Castigliano's theorem [Eq. (9-107)]:

$$\delta_i = \frac{\partial U}{\partial P_i} = \frac{\partial}{\partial P_i} \int \frac{M^2 dx}{2EI} \qquad \textbf{(9-117)}$$

Following the rules of calculus, differentiate the integral by differentiating under the integral sign:

$$\delta_i = \frac{\partial}{\partial P_i} \int \frac{M^2 dx}{2EI} = \int \left(\frac{M}{EI}\right)\left(\frac{\partial M}{\partial P_i}\right) dx \qquad \textbf{(9-118)}$$

This equation is referred to as the **modified Castigliano's theorem**.

When using the modified theorem, integrate the product of the bending moment and its derivative. By contrast, when using the standard Castigliano's theorem [see Eq. (9-117)], it was necessary to integrate the square of the bending moment. Since the derivative is a shorter expression than the moment itself, this new procedure is much simpler. To show this, solve the preceding examples using the modified theorem [Eq. (9-118)].

Begin with the beam shown in Fig. 9-37 and again find the deflection and angle of rotation at the free end. The bending moment and its derivatives [see Eq. (9-108)] are

$$M = -Px - M_0$$

$$\frac{\partial M}{\partial P} = -x \qquad \frac{\partial M}{\partial M_0} = -1$$

From Eq. (9-118), obtain the deflection δ_A and angle of rotation θ_A:

$$\delta_A = \frac{1}{EI}\int_0^L (-Px - M_0)(-x)dx = \frac{PL^3}{3EI} - \frac{M_0 L^2}{2EI} \qquad \textbf{(9-119a)}$$

$$\theta_A = \frac{1}{EI}\int_0^L (-Px - M_0)(-1)dx = \frac{PL^2}{2EI} - \frac{M_0 L}{EI} \qquad \textbf{(9-119b)}$$

These equations agree with the earlier results in Eqs. (9-110) and (9-111). However, the calculations are shorter than those performed earlier, because integration of the square of the bending moment [see Eq. (9-109)] was avoided.

The advantages of differentiating under the integral sign are even more apparent when there are more than two loads acting on the structure, as in the example of Fig. 9-38. In that example, the deflection δ_C at the midpoint C of the beam due to the loads P and M_0 was computed. To do so, a fictitious load Q was added at the midpoint (Fig. 9-39). The deflection $(\delta_C)_0$ at the midpoint of the beam was obtained when all three loads (P, M_0, and Q) were acting. Finally, Q was set equal to zero to obtain the deflection δ_C due to P and M_0 alone. The solution was time-consuming, because the integrations were extremely long. However, using the modified theorem and differentiating first leads to much shorter calculations.

With all three loads acting (Fig. 9-39), the bending moments and their derivatives are [see Eqs. (9-112) and (9-113)]

$$M = -Px - M_0 \qquad \frac{\partial M}{\partial Q} = 0 \qquad \left(0 \le x \le \frac{L}{2}\right)$$

$$M = -Px - M_0 - Q\left(x - \frac{L}{2}\right) \qquad \frac{\partial M}{\partial Q} = -\left(x - \frac{L}{2}\right) \qquad \left(\frac{L}{2} \le x \le L\right)$$

Therefore, the deflection $(\delta_C)_0$ from Eq. (9-118) is

$$(\delta_C)_0 = \frac{1}{EI}\int_0^{L/2} (-Px - M_0)(0)dx$$

$$+ \frac{1}{EI}\int_{L/2}^{L} \left[-Px - M_0 - Q\left(x - \frac{L}{2}\right)\right]\left[-\left(x - \frac{L}{2}\right)\right]dx$$

Since Q is a fictitious load, and since the partial derivatives already have been taken, you can set Q equal to zero before integrating and obtain the deflection δ_C due to the two loads P and M_0 as

$$\delta_C = \frac{1}{EI}\int_{L/2}^{L} [-Px - M_0]\left[-\left(x - \frac{L}{2}\right)\right]dx = \frac{5PL^3}{48EI} + \frac{M_0L^2}{8EI}$$

which agrees with the earlier result in Eq. (9-116). Again, the integrations are greatly simplified by differentiating under the integral sign and using the modified theorem.

The partial derivative that appears under the integral sign in Eq. (9-118) has a simple physical interpretation. It represents the rate of change of the bending moment M with respect to the load P_i, that is, it is equal to the bending moment M produced by a load P_i of unit value. This observation leads to a method of finding deflections known as the *unit-load method*. Castigliano's theorem also leads to a method of structural analysis known as the *flexibility method*. Both the unit-load method and the flexibility method are widely used in structural analysis and are described in textbooks on that subject.

The following examples provide additional illustrations of the use of Castigliano's theorem for finding deflections of beams. However, it should be remembered that the theorem is not limited to finding beam deflections—it applies to any kind of linearly elastic structure for which the principle of superposition is valid.

Example 9-17

A simple beam AB supports a uniform load of intensity $q = 20$ kN/m and a concentrated load $P = 25$ kN (Fig. 9-40). The load P acts at the midpoint C of the beam. The beam has a length $L = 2.5$ m, modulus of elasticity $E = 210$ GPa, and moment of inertia $I = 3120$ cm^4.

Determine the downward deflection δ_C at the midpoint of the beam by the following methods: (a) Obtain the strain energy of the beam and use Castigliano's theorem. (b) Use the modified form of Castigliano's theorem (differentiation under the integral sign).

FIGURE 9-40

Example 9-17: Simple beam with two loads

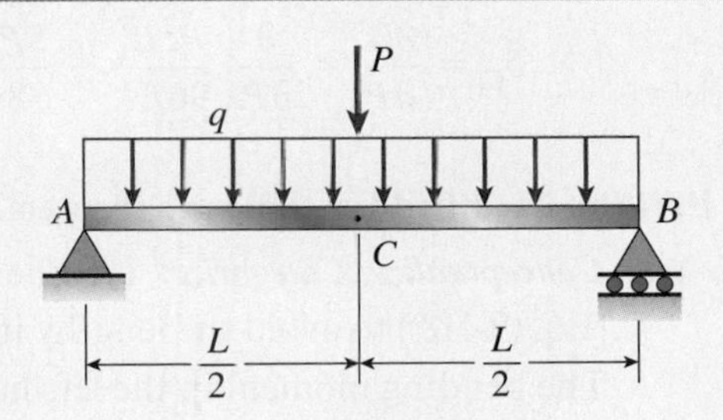

Solution:

Use a four-step problem-solving approach. Combine steps as needed for an efficient solution.

Part (a): Strain energy and Castigliano's theorem.

1, 2. *Conceptualize, Categorize*: Because the beam and its loading are symmetrical about the midpoint, the strain energy for the entire beam is equal to twice the strain energy for the left-hand half of the beam. Therefore, only the left-hand half of the beam must be analyzed.

The reaction at the left-hand support A (Figs. 9-40 and 9-41) is

$$R_A = \frac{P}{2} + \frac{qL}{2}$$

Therefore, the bending moment M is

$$M = R_A x - \frac{qx^2}{2} = \frac{Px}{2} + \frac{qLx}{2} - \frac{qx^2}{2} \quad \text{(a)}$$

in which x is measured from support A.

FIGURE 9-41

Example 9-17: Free-body diagram for determining the bending moment M in the left-hand half of the beam

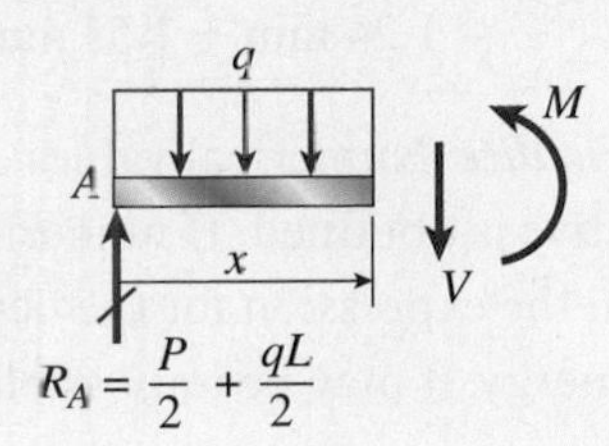

The strain energy of the entire beam [from Eq. (9-95a)] is

$$U = \int \frac{M^2 dx}{2EI} = 2\int_0^{L/2} \frac{1}{2EI}\left(\frac{Px}{2} + \frac{qLx}{2} - \frac{qx^2}{2}\right)^2 dx$$

Square the term in parentheses and perform a lengthy integration to find

$$U = \frac{P^2L^3}{96EI} + \frac{5PqL^4}{384EI} + \frac{q^2L^5}{240EI}$$

3. ***Analyze***: Since the deflection at the midpoint C (Fig. 9-40) corresponds to the load P, find the deflection by using Castigliano's theorem (Eq. 9-107) as

$$\delta_C = \frac{\partial U}{\partial P} = \frac{\partial}{\partial P}\left(\frac{P^2L^3}{96EI} + \frac{5PqL^4}{384EI} + \frac{q^2L^5}{240EI}\right) = \frac{PL^3}{48EI} + \frac{5qL^4}{384EI} \quad \text{(b)}$$

Part (b): Modified Castigliano's theorem.

1, 2. ***Conceptualize, Categorize***: Use the modified form of Castigliano's theorem [Eq. (9-118)] to avoid the lengthy integration for finding the strain energy. The bending moment in the left-hand half of the beam has already been determined [see Eq. (a)], and its partial derivative with respect to the load P is

$$\frac{\partial M}{\partial P} = \frac{x}{2}$$

3. ***Analyze***: Therefore, the modified Castigliano's theorem becomes

$$\delta_C = \int\left(\frac{M}{EI}\right)\left(\frac{\partial M}{\partial P}\right)dx$$

$$= 2\int_0^{L/2} \frac{1}{EI}\left(\frac{Px}{2} + \frac{qLx}{2} - \frac{qx^2}{2}\right)\left(\frac{x}{2}\right)dx = \frac{PL^3}{48EI} + \frac{5qL^4}{384EI} \quad \text{(c)}$$

which agrees with the earlier result [Eq. (b)] but requires a much simpler integration.

Numerical solution: Now using this expression for the deflection at point C, substitute numerical values, as

$$\delta_C = \frac{PL^3}{48EI} + \frac{5qL^4}{383EI}$$

$$= \frac{(25 \text{ kN})(2.5 \text{ m})^3}{48(210 \text{ GPa})(31.2 \times 10^{-6} \text{ m}^4)} + \frac{5(20 \text{ kN/m})(2.5 \text{ m})^4}{383(210 \text{ GPa})(31.2 \times 10^{-6} \text{ m}^4)}$$

$$= 1.24 \text{ mm} + 1.55 \text{ mm} = 2.79 \text{ mm}$$

4. ***Finalize***: Numerical values cannot be substituted until *after* the partial derivative is obtained. If numerical values are substituted prematurely, either in the expression for the bending moment or the expression for the strain energy, it may be impossible to take the derivative.

Example 9-18

A simple beam with an overhang supports a uniform load of an intensity q on span AB and a concentrated load P at end C of the overhang (Fig. 9-42).

Determine (a) the deflection δ_C and (b) angle of rotation θ_C at point C. Use the modified form of Castigliano's theorem.

FIGURE 9-42

Example 9-18: Beam with an overhang

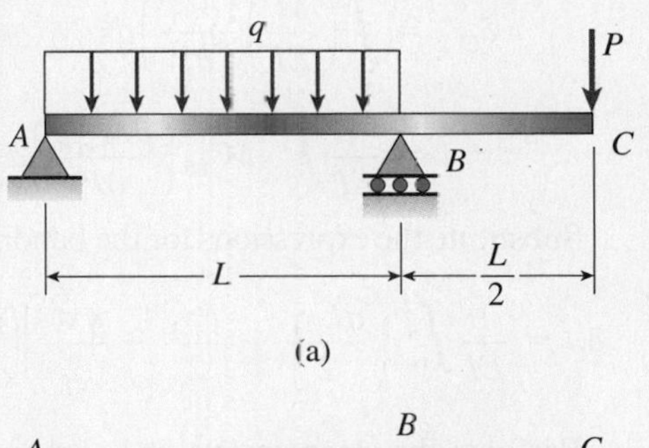

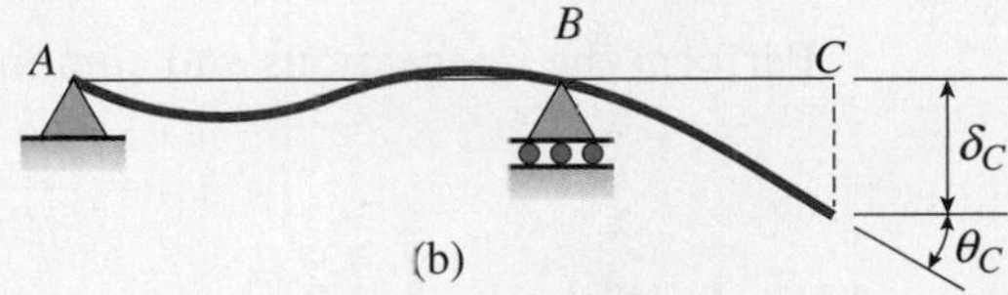

Solution:

Use a four-step problem-solving approach. Combine steps as needed for an efficient solution.

Part (a): Deflection δ_C at the end of overhang.

1, 2. ***Conceptualize, Categorize***: Since the load P corresponds to this deflection (Fig. 9-42), a fictitious load is not needed. Instead, begin immediately to find the bending moments throughout the length of the beam. The reaction at support A is

$$R_A = \frac{qL}{2} - \frac{P}{2}$$

FIGURE 9-43

Reaction at support A and coordinates x_1 and x_2 for the beam of Example 9-18

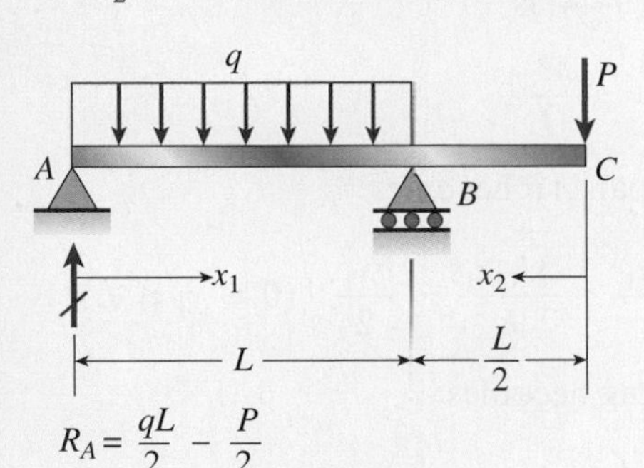

as shown in Fig. 9-43. Therefore, the bending moment in span AB is

$$M_{AB} = R_A x_1 - \frac{qx_1^2}{2} = \frac{qLx_1}{2} - \frac{Px_1}{2} - \frac{qx_1^2}{2} \quad (0 \le x_1 \le L)$$

where x_1 is measured from support A (Fig. 9-43). The bending moment in the overhang is

$$M_{BC} = -Px_2 \quad \left(0 \le x_2 \le \frac{L}{2}\right)$$

where x_2 is measured from point C (Fig. 9-43).

Next, determine the partial derivatives with respect to the load P:

$$\frac{\partial M_{AB}}{\partial P} = -\frac{x_1}{2} \quad (0 \le x_1 \le L)$$

$$\frac{\partial M_{BC}}{\partial P} = -x_2 \quad \left(0 \le x_2 \le \frac{L}{2}\right)$$

3. *Analyze*: Now use the modified form of Castigliano's theorem [Eq. (9-118)] to obtain the deflection at point C:

$$\delta_C = \int \left(\frac{M}{EI}\right)\left(\frac{\partial M}{\partial P}\right) dx$$

$$= \frac{1}{EI}\int_0^L M_{AB}\left(\frac{\partial M_{AB}}{\partial P}\right) dx + \frac{1}{EI}\int_0^{L/2} M_{BC}\left(\frac{\partial M_{BC}}{\partial P}\right) dx$$

Substitute the expressions for the bending moments and partial derivatives to get

$$\delta_C = \frac{1}{EI}\int_0^L \left(\frac{qLx_1}{2} - \frac{Px_1}{2} - \frac{qx_1^2}{2}\right)\left(-\frac{x_1}{2}\right) dx_1 + \frac{1}{EI}\int_0^{L/2} (-Px_2)(-x_2)dx_2$$

Perform the integrations and combining terms to obtain the deflection:

$$\delta_C = \frac{PL^3}{8EI} - \frac{qL^4}{48EI} \qquad \text{(9-120)}$$

4. *Finalize*: Since the load P acts downward, the deflection δ_C is also positive downward. In other words, if the preceding equation produces a positive result, the deflection is downward. If the result is negative, the deflection is upward.

Compare the two terms in Eq. (9-120) to see that the deflection at the end of the overhang is downward when $P > qL/6$ and upward when $P < qL/6$.

Part (a): Angle of rotation θ_C at the end of overhang.

1, 2. *Conceptualize, Categorize*: Since there is no load on the original beam (Fig. 9-42a) corresponding to this angle of rotation (Fig. 9-42b), supply a fictitious load. Place a couple of moment M_C at point C (Fig. 9-44). Note that the couple M_C acts at the point on the beam where the angle of rotation is to be determined. Furthermore, it has the same clockwise direction as the angle of rotation (Fig. 9-42).

Follow the same steps as when determining the deflection at C. First, note that the reaction at support A (Fig. 9-44) is

$$R_A = \frac{qL}{2} - \frac{P}{2} - \frac{M_C}{L}$$

Consequently, the bending moment in span AB becomes

$$M_{AB} = R_A x_1 - \frac{qx_1^2}{2} = \frac{qLx_1}{2} - \frac{Px_1}{2} - \frac{M_C x_1}{L} - \frac{qx_1^2}{2} \quad (0 \le x_1 \le L)$$

Also, the bending moment in the overhang becomes

$$M_{BC} = -Px_2 - M_C \quad \left(0 \le x_2 \le \frac{L}{2}\right)$$

FIGURE 9-44

Fictitious moment M_C acting on the beam of Example 9-18

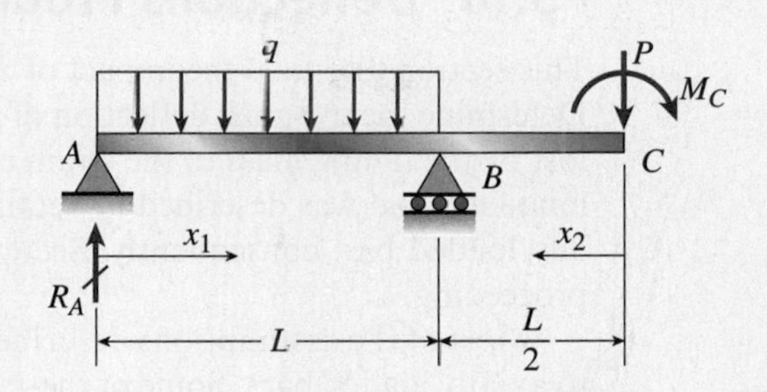

The partial derivatives are taken with respect to the moment M_C, which is the load corresponding to the angle of rotation. Therefore,

$$\frac{\partial M_{AB}}{\partial M_C} = -\frac{x_1}{L} \quad (0 \le x_1 \le L)$$

$$\frac{\partial M_{BC}}{\partial M_C} = -1 \quad \left(0 \le x_2 \le \frac{L}{2}\right)$$

3. ***Analyze***: Now use the modified form of Castigliano's theorem [Eq. (9-118)] to obtain the angle of rotation at point *C*:

$$\theta_C = \int \left(\frac{M}{EI}\right)\left(\frac{\partial M}{\partial M_C}\right) dx$$

$$= \frac{1}{EI}\int_0^L M_{AB}\left(\frac{\partial M_{AB}}{\partial M_C}\right) dx + \frac{1}{EI}\int_0^{L/2} M_{BC}\left(\frac{\partial M_{BC}}{\partial M_C}\right) dx$$

Substitute the expressions for the bending moments and partial derivatives to obtain

$$\theta_C = \frac{1}{EI}\int\left(\frac{qLx_1}{2} - \frac{Px_1}{2} - \frac{M_C x_1}{L} - \frac{qx_1^2}{2}\right)\left(-\frac{x_1}{L}\right) dx_1$$

$$+ \frac{1}{EI}\int_0^{L/2} (-Px_2 - M_C)(-1)dx_2$$

Since M_C is a fictitious load, and since the partial derivatives have already been taken, set M_C equal to zero at this stage of the calculations and simplify the integrations:

$$\theta_C = \frac{1}{EI}\int_0^L \left(\frac{qLx_1}{2} - \frac{Px_1}{2} - \frac{qx_1^2}{2}\right)\left(-\frac{x_1}{L}\right) dx_1 + \frac{1}{EI}\int_0^{L/2} (-Px_2)(-1)dx_2$$

After carrying out the integrations and combining terms, the rotation at *C* is

$$\theta_C = \frac{7PL^2}{24EI} - \frac{qL^3}{24EI} \tag{9-121}$$

4. ***Finalize***: If this equation produces a positive result, the angle of rotation is clockwise. If the result is negative, the angle is counterclockwise.

Comparing the two terms in Eq. (9-121), note that the angle of rotation is clockwise when $P > qL/7$ and counterclockwise when $P < qL/7$.

If numerical data are available, it is now a routine matter to substitute numerical values into Eqs. (9-120) and (9-121) and calculate the deflection and angle of rotation at the end of the overhang.

*9.10 Deflections Produced by Impact

This section discusses the impact of an object falling onto a beam (Fig. 9-45a). Determine the dynamic deflection of the beam by equating the potential energy lost by the falling mass to the strain energy acquired by the beam. This approximate method was described in detail in Section 2.8 for a mass striking an axially loaded bar; consequently, Section 2.8 should be fully understood before proceeding.

Most of the assumptions described in Section 2.8 apply to beams as well as to axially loaded bars. Some of these assumptions are as follows: (1) The falling weight sticks to the beam and moves with it, (2) no energy losses occur, (3) the beam behaves in a linearly elastic manner, (4) the deflected shape of the beam is the same under a dynamic load as under a static load, and (5) the potential energy of the beam due to its change in position is relatively small and may be disregarded. In general, these assumptions are reasonable if the mass of the falling object is very large compared to the mass of the beam. Otherwise, this approximate analysis is not valid and a more advanced analysis is required.

As an example, consider the simple beam AB shown in Fig. 9-45. The beam is struck at its midpoint by a falling body of mass M and weight W. Based upon the preceding idealizations, assume that all of the potential energy lost by the body during its fall is transformed into elastic strain energy that is stored in the beam. Since the distance through which the body falls is $h + \delta_{max}$, where h is the initial height above the beam (Fig. 9-45a) and δ_{max} is the maximum dynamic deflection of the beam (Fig. 9-45b), the potential energy lost is

$$\text{Potential energy} = W(h + \delta_{max}) \tag{9-122}$$

The strain energy acquired by the beam can be determined from the deflection curve by using Eq. (9-95b), which is repeated here:

$$U = \int \frac{EI}{2}\left(\frac{d^2v}{dx^2}\right)^2 dx \tag{9-123}$$

The deflection curve for a simple beam subjected to a concentrated load acting at the midpoint (see Case 4 of Table H-2, Appendix H) is

$$v = -\frac{Px}{48EI}(3L^2 - 4x^2) \quad \left(0 \le x \le \frac{L}{2}\right) \tag{9-124}$$

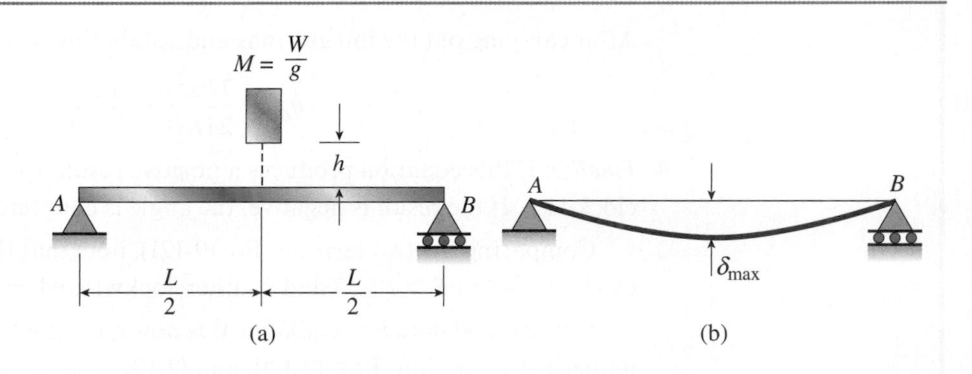

FIGURE 9-45
Deflection of a beam struck by a falling body

Also, the maximum deflection of the beam is

$$\delta_{max} = \frac{PL^3}{48EI} \tag{9-125}$$

Eliminating the load P between Eqs. (9-124) and (9-125) gives the equation of the deflection curve in terms of the maximum deflection:

$$v = -\frac{\delta_{max}x}{L^3}(3L^2 - 4x^2) \qquad \left(0 \le x \le \frac{L}{2}\right) \tag{9-126}$$

Take two derivatives to find

$$\frac{d^2v}{dx^2} = \frac{24\delta_{max}x}{L^3} \tag{9-127}$$

Finally, substitute the second derivative into Eq. (9-123) and obtain the expression for the strain energy of the beam in terms of the maximum deflection:

$$U = 2\int_0^{L/2} \frac{EI}{2}\left(\frac{d^2v}{dx^2}\right)^2 dx = EI\int_0^{L/2}\left(\frac{24\delta_{max}x}{L^3}\right)^2 dx = \frac{24EI\delta_{max}^2}{L^3} \tag{9-128}$$

Equating the potential energy lost by the falling mass in Eq. (9-122) to the strain energy acquired by the beam in Eq. (9-128) gives

$$W(h + \delta_{max}) = \frac{24EI\delta^2_{max}}{L^3} \tag{9-129}$$

This equation is quadratic in δ_{max} and can be solved for its positive root:

$$\delta_{max} = \frac{WL^3}{48EI} + \left[\left(\frac{WL^3}{48EI}\right)^2 + 2h\left(\frac{WL^3}{48EI}\right)\right]^{1/2} \tag{9-130}$$

Note that the maximum dynamic deflection increases if either the weight of the falling object or the height of fall is increased, and it decreases if the stiffness EI/L^3 of the beam is increased.

To simplify the preceding equation, denote the *static deflection* of the beam due to the weight W as δ_{st}:

$$\delta_{st} = \frac{WL^3}{48EI} \tag{9-131}$$

Then Eq. (9-130) for the maximum dynamic deflection becomes

$$\delta_{max} = \delta_{st} + (\delta_{st}^2 + 2h\delta_{st})^{1/2} \tag{9-132}$$

This equation shows that the dynamic deflection is always larger than the static deflection.

If the height h equals zero, which means that the load is applied suddenly but without any free fall, the dynamic deflection is twice the static deflection.

If h is very large compared to the deflection, then the term containing h in Eq. (9-132) predominates, and the equation can be simplified to

$$\delta_{max} = \sqrt{2h\delta_{st}} \tag{9-133}$$

These observations are analogous to those discussed previously in Section 2.8 for impact on a bar in tension or compression.

The deflection δ_{max} calculated from Eq. (9-132) generally represents an upper limit, because no energy losses were considered during impact. Several other factors also tend to reduce the deflection, including localized deformation of the contact surfaces, the tendency of the falling mass to bounce upward, and inertia effects of the mass of the beam. Thus, the phenomenon of impact is complex, and if a more accurate analysis is needed, books and articles devoted specifically to that subject must be consulted.

*9.11 Temperature Effects

Preceding sections of this chapter considered the deflections of beams due to lateral loads. This section considers the deflections caused by **nonuniform temperature changes**. As a preliminary matter, recall that the effects of *uniform* temperature changes already have been described in Section 2.5, where it was shown that a uniform temperature increase causes an unconstrained bar or beam to have its length increased by the amount

$$\delta_T = \alpha(\Delta T)L \tag{9-134}$$

In this equation, α is the coefficient of thermal expansion, ΔT is the uniform increase in temperature, and L is the length of the bar [see Fig. (2-38) and Eq. (2-18) in Chapter 2].

If a beam is supported in such a manner that longitudinal expansion is free to occur, as is the case for all of the statically determinate beams considered in this chapter, then a uniform temperature change will not produce any stresses in the beam. Also, there will be no lateral deflections of such a beam, because there is no tendency for the beam to bend.

The behavior of a beam is different if the temperature is not constant across its height. For example, assume that a simple beam, initially straight and at a uniform temperature T_0, has its temperature changed to T_1 on its upper surface and T_2 on its lower surface, as pictured in Fig. 9-46a. If the variation in temperature is assumed to be linear between the top and bottom of the beam, then the *average temperature* of the beam is

$$T_{aver} = \frac{T_1 + T_2}{2} \tag{9-135}$$

and occurs at mid-height. Any difference between this average temperature and the initial temperature T_0 results in a change in length of the beam, given by Eq. (9-134), as

$$\delta_T = \alpha(T_{aver} - T_0)L = \alpha\left(\frac{T_1 + T_2}{2} - T_0\right)L \tag{9-136}$$

In addition, the temperature differential $T_2 - T_1$ between the bottom and top of the beam produces a *curvature* of the axis of the beam with the accompanying lateral deflections (Fig. 9-46b).

To investigate the deflections due to a temperature differential, consider an element of length dx cut out from the beam (Figs. 9-46a and c). The changes in length of the element at the bottom and top are $\alpha(T_2 - T_0)dx$ and $\alpha(T_1 - T_0)dx$, respectively. If T_2 is greater than T_1, the sides of the element will rotate with respect to each other through an angle $d\theta$, as shown in Fig. 9-46c. The angle $d\theta$ is related to the changes in dimension by the following equation, obtained from the geometry of the figure:

$$hd\theta = \alpha(T_2 - T_0)dx - \alpha(T_1 - T_0)dx$$

from which

$$\frac{d\theta}{dx} = \frac{\alpha(T_2 - T_1)}{h} \tag{9-137}$$

in which h is the height of the beam.

The quantity $d\theta / dx$ represents the curvature of the deflection curve of the beam [see Eq. (9-6)]. Since the curvature is equal to d^2v / dx^2 [Eq. (9-9)], write the following **differential equation of the deflection curve**:

$$\frac{d^2v}{dx^2} = \frac{\alpha(T_2 - T_1)}{h} \tag{9-138}$$

Note that when T_2 is greater than T_1, the curvature is positive and the beam is bent concave upward, as shown in Fig. 9-46b. The quantity $\alpha(T_2 - T_1)/h$ in Eq. (9-138) is the counterpart of the quantity M/EI, which appears in the basic differential equation [Eq. (9-11)].

Solve Eq. (9-138) by the same integration techniques described earlier for the effects of bending moments (see Section 9.3). Integrate the differential

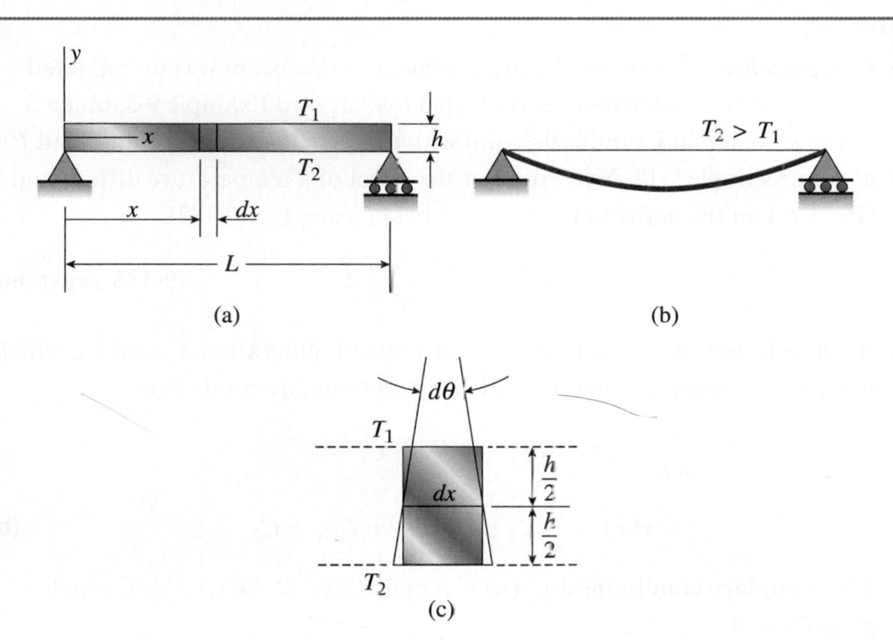

FIGURE 9-46

Temperature effects in a beam

equation to obtain dv/dx and v, and use boundary or other conditions to evaluate the constants of integration. In this manner, obtain the equations for the slopes and deflections of the beam, as illustrated by Probs. 9.11-1 through 9.11-5 at the end of this chapter.

If the beam is able to change in length and deflect freely, there will be no stresses associated with the temperature changes described in this section. However, if the beam is restrained against longitudinal expansion or lateral deflection, or if the temperature changes do not vary linearly from top to bottom of the beam, internal temperature stresses will develop. The determination of such stresses requires the use of more advanced methods of analysis. Statically indeterminate beams subject to temperature effects are discussed in Section 10.5.

Example 9-19

An overhanging beam ABC of height h has a pin support at A and a roller support at B. The beam is heated to a temperature T_1 on the top and T_2 on the bottom (see Fig. 9-47). Determine the equation of the deflection curve of the beam, and the deflection δ_C at end C.

FIGURE 9-47

Example 9-19: Simple beam with overhang and temperature change

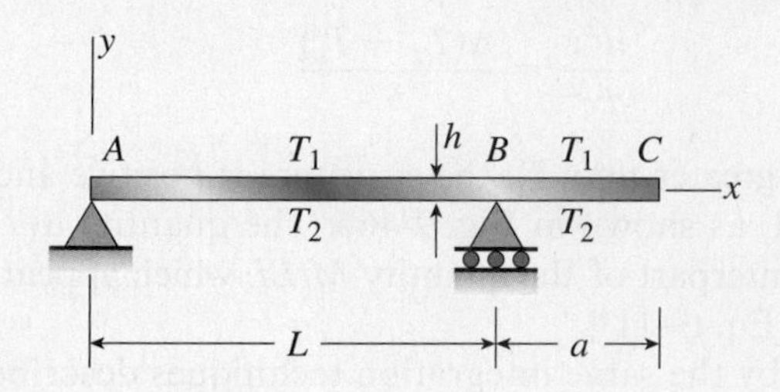

Solution:

Use a four-step problem-solving approach. Combine steps as needed for an efficient solution.

1, 2. ***Conceptualize, Categorize***: The displacement of this beam was investigated at selected points due to a concentrated load at C in Example 9-5, under a uniform load q in Example 9-9, and with uniform load q on AB and load P at C in Example 9-18. Now consider the effect of a temperature differential $(T_2 - T_1)$ on the deflection $v(x)$ of the beam using Eq. (9-137).

$$\frac{d^2v}{dx^2} = \frac{\alpha(T_2 - T_1)}{h} \quad \textbf{(9-138, repeated)}$$

3. ***Analyze***: Integration results in two constants of integration, C_1 and C_2, which must be determined using two independent bounday conditions:

$$\frac{d}{dx}v(x) = \frac{\alpha}{h}(T_2 - T_1)x + C_1 \quad \textbf{(a)}$$

$$v(x) = \frac{\alpha}{h}(T_2 - T_1)\frac{x^2}{2} + C_1x + C_2 \quad \textbf{(b)}$$

The boundary conditions are $v(0) = 0$ and $v(L) = 0$. So $v(0) = 0$, which gives $C_2 = 0$.

Also, $v(L) = 0$, which leads to **(c)**

$$C_1 = \frac{1}{L}\left[\frac{-\alpha L^2}{2h}(T_2 - T_1)\right] = -\left[\frac{L\alpha(T_2 - T_1)}{2h}\right] \tag{d}$$

Substituting C_1 and C_2 into Eq. (b) results in the equation of the elastic curve of the beam due to temperature differential $(T_2 - T_1)$ as

$$v(x) = \frac{\alpha x(T_2 - T_1)(x - L)}{2h} \tag{e}$$

If $x = L + a$ in Eq. (e), an expression for the deflection of the beam at C is

$$\delta_C = v(L + a) = \frac{\alpha(L + a)(T_2 - T_1)(L + a - L)}{2h} = \frac{\alpha(T_2 - T_1)a(L + a)}{2h} \tag{f}$$

4. ***Finalize***: Linear elastic behavior was assumed here and in earlier examples, so (if desired) the *principle of superposition* can be used to find the total deflection at C due to simultaneous application of all loads considered in Examples 9-5, 9-9, and 9-18 and for the temperature differential studied here.

Numerical example: If beam ABC is a steel, wide flange HE 700B [see Table F-1] with a length of $L = 9$ m and with an overhang $a = L/2$, compare the deflection at C due to self-weight (see Example 9-9; let $q = 2.36$ kN/m) to the deflection at C due to temperature differential $(T_2 - T_1) = 3°C$. From Table I-4, the coefficient of thermal expansion for structural steel is $\alpha = 12 \times 10^{-6}/°C$. The modulus for steel is 210 GPa. **(g)**

From Eq. (9-68), the deflection at C due to self-weight is

$$\begin{aligned}\delta_{Cq} &= \frac{qa}{24EI_z}(a + L)(3a^2 + aL - L^2) \\ &= \frac{(2.36 \text{ kN/m})(4.5 \text{ m})}{24(210 \text{ GPa})(256900 \text{ cm}^4)}(4.5 \text{ m} + 9 \text{ m}) \\ &\quad \times [3(4.5 \text{ m})^2 + 4.5 \text{ m}(9 \text{ m}) - (9 \text{ m})^2] \\ &= 0.224 \text{ mm}\end{aligned}$$

where $a = 4.5$ m and $L = 9.0$ m.

The deflection at C due to a temperature differential of only 3° C is from Eq. (f):

$$\begin{aligned}\delta_{CT} &= \frac{\alpha(T_2 - T_1)a(L + a)}{2h} \\ &= \frac{(12 \times 10^{-6})(3)(4.5 \text{ m})(9 \text{ m} + 4.5 \text{ m})}{2(700 \text{ mm})} = 1.562 \text{ mm}\end{aligned} \tag{h}$$

The deflection at C due to a small temperature differential is seven times that due to self-weight.

CHAPTER SUMMARY AND REVIEW

Chapter 9 discussed the linear elastic, small displacement behavior of beams of different types, with different support conditions, and acted upon by a wide variety of loadings, including impact and temperature effects. Methods based on integration of the second-, third-, or fourth-order differential equations of the deflection curve were described. Equations were derived for the deflected shape of the entire beam, and translational and rotational displacements were computed at specific points along a beam. The principle of superposition was used to solve more complicated beams and loadings by combining the simpler standard solutions. A method for calculating displacements of beams based on the area of the moment diagram and an energy-based method for computing beam displacements were also presented. Here are major concepts presented in this chapter.

1. Combining expressions for linear curvature ($\kappa = d^2v / dx^2$) and the moment curvature relation ($\kappa = M / EI$) resulted in the **ordinary differential equation of the deflection curve** for a beam, which is valid only for linear elastic behavior.

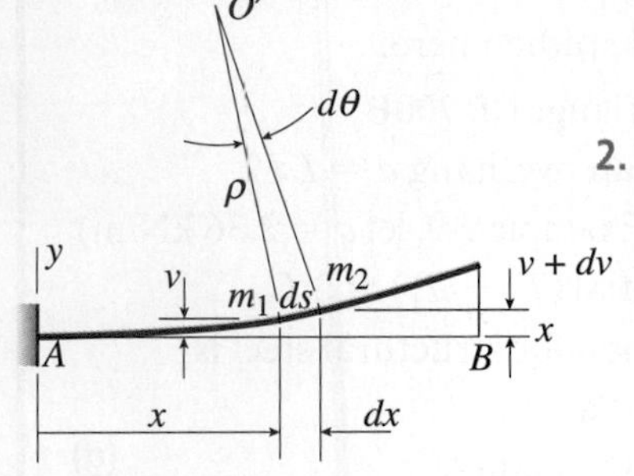

$$EI\frac{d^2v}{dx^2} = M$$

2. The differential equation of the deflection curve can be differentiated once to obtain a third-order equation relating shear force V and first derivative of moment dM/dx or twice to obtain a fourth-order equation relating intensity of distributed load q and first derivative of shear dV/dx.

$$EI\frac{d^3v}{dx^3} = V$$

$$EI\frac{d^4v}{dx^4} = -q$$

The choice of second-, third-, or fourth-order differential equations depends on which is most efficient for a particular beam support case and applied loading.

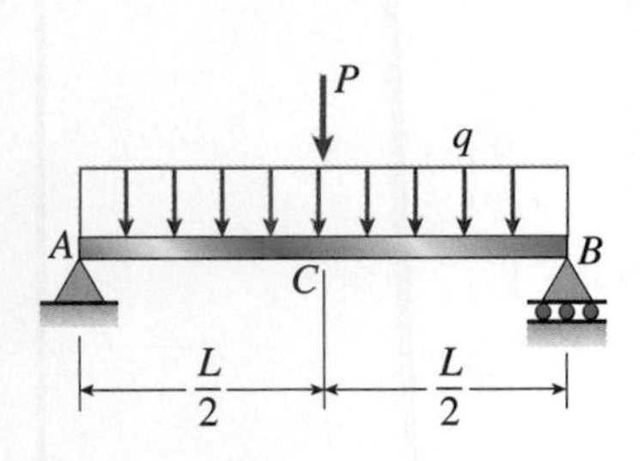

3. Write expressions for either moment (M), shear (V), or load intensity (q) for each separate region of the beam whenever q, V, M, or EI vary. Then apply **boundary**, **continuity**, or **symmetry conditions**, as appropriate, to solve for unknown constants of integration. The beam deflection equation, $v(x)$, can be evaluated at a particular value of x to find the translational displacement at that point; evaluation of dv/dx at that same point provides the slope of the deflection equation.

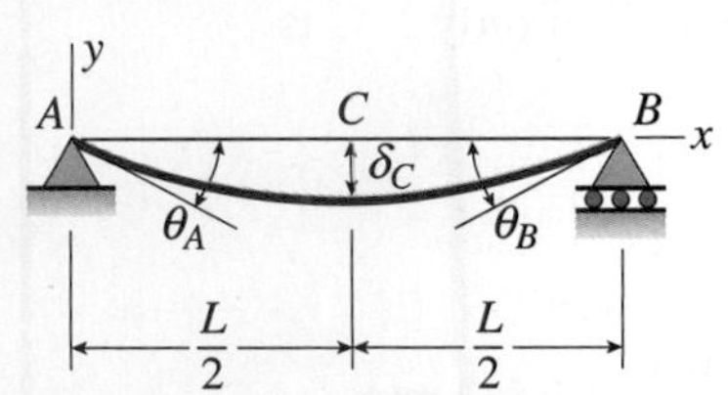

4. The **method of superposition** can be used to solve for displacements and rotations for more complicated beams and loadings. Superposition is only applicable to beams undergoing small displacements and behaving in a linear elastic manner.

5. The **moment-area method** is an alternative approach for finding beam displacements; it is based on two theorems that are related to the area of the bending-moment diagram.

6. Equating the strain energy of bending (U) to the work (W) of a concentrated load or moment and then taking a partial derivative with respect to a particular load (P, M) provides another method for computing beam deflections and rotations; this method is known as **Castigliano's theorem**.

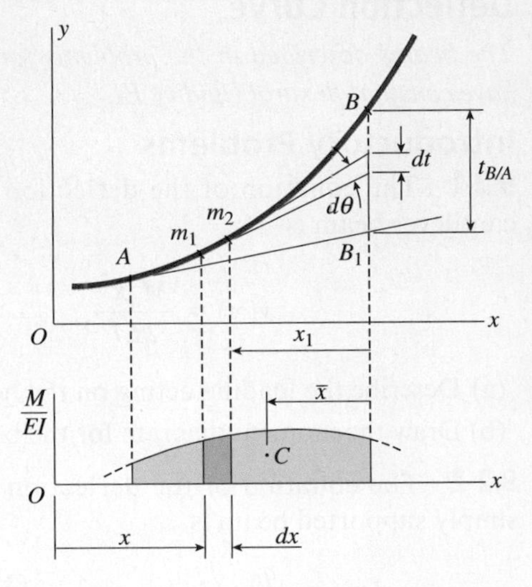

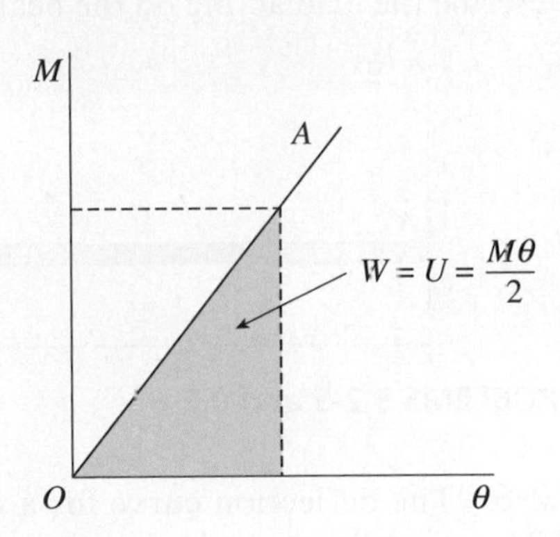

7. By equating the potential energy of a falling mass to strain energy acquired by the beam, **deflections due to impact** can be approximated.

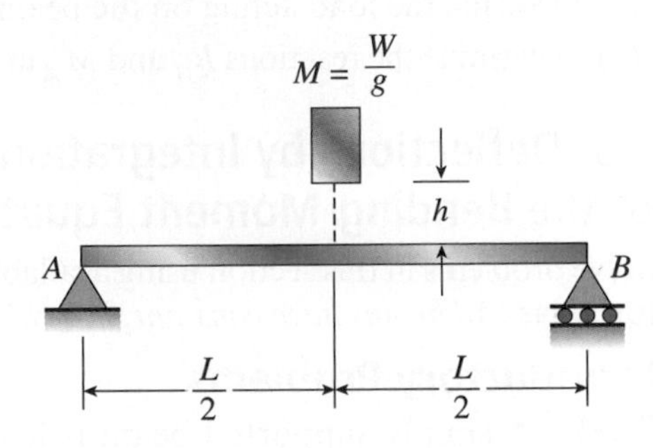

8. Finally, if a beam experiences a **temperature differential**, $T_2 - T_1$, over height h, it produces a curvature of the axis of the beam:

$$\kappa = d\theta / dx = d^2 v / dx^2 = \alpha(T_2 - T_1) / h$$

This equation can be integrated to obtain the equation of the deflection curve using successive integration.

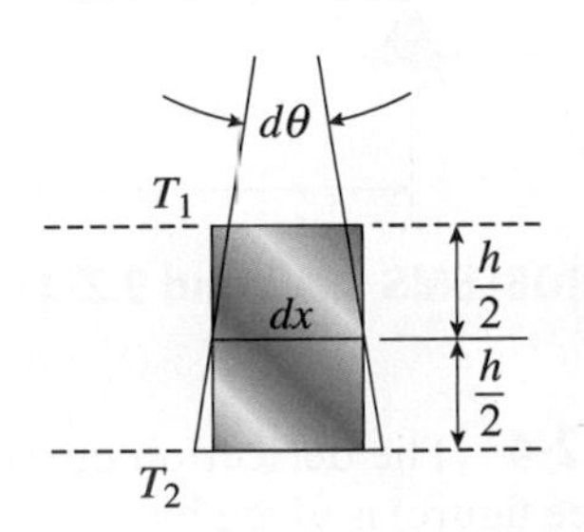

PROBLEMS Chapter 9

9.2 Differential Equations of the Deflection Curve

The beams described in the problems for Section 9.2 have constant flexural rigidity EI.

Introductory Problems

9.2-1 The equation of the deflection curve for a cantilever beam is

$$v(x) = \frac{M_0 x^2}{2EI}$$

(a) Describe the loading acting on the beam.
(b) Draw the moment diagram for the beam.

9.2-2 The equation of the deflection curve for a simply supported beam is

$$v(x) = \frac{q_0}{24EI}(2Lx^3 - x^4 - L^3x)$$

(a) Derive the slope equation of the beam.
(b) Derive the bending-moment equation of the beam.
(c) Derive the shear-force equation of the beam.
(d) Describe the loading acting on the beam.

9.2-3 The deflection curve for a simple beam AB (see figure) is given by

$$v = -\frac{q_0 x}{360LEI}(7L^4 - 10L^2x^2 + 3x^4)$$

Describe the load acting on the beam.

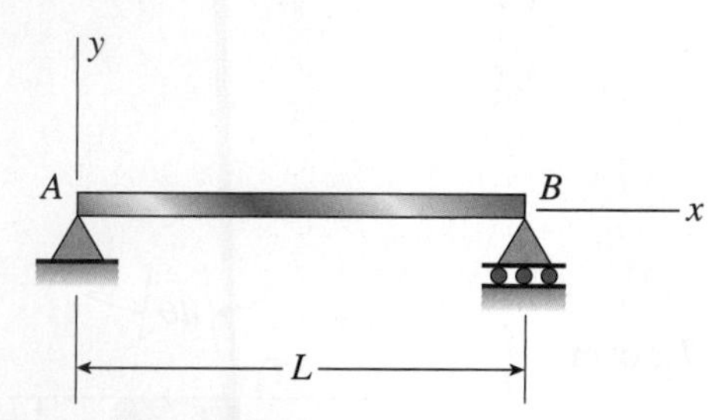

PROBLEMS 9.2-3 and 9.2-4

9.2-4 The deflection curve for a simple beam AB (see figure) is given by

$$v = -\frac{q_0 L^4}{\pi^4 EI}\sin\frac{\pi x}{L}$$

(a) Describe the load acting on the beam.
(b) Determine the reactions R_A and R_B at the supports.
(c) Determine the maximum bending moment M_{max}.

Representative Problems

9.2-5 The deflection curve for a cantilever beam AB (see figure) is given by

$$v = -\frac{q_0 x^2}{120LEI}(10L^3 - 10L^2x + 5Lx^2 - x^3)$$

Describe the load acting on the beam.

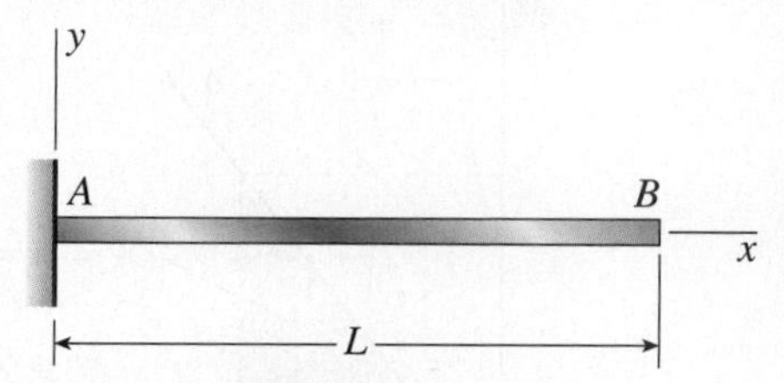

PROBLEMS 9.2-5 and 9.2-6

9.2-6 The deflection curve for a cantilever beam AB (see figure) is given by

$$v = -\frac{q_0 x^2}{360L^2EI}(45L^4 - 40L^3x + 15L^2x^2 - x^4)$$

(a) Describe the load acting on the beam.
(b) Determine the reactions R_A and M_A at the support.

9.3 Deflections by Integration of the Bending-Moment Equation

Solve problems in this section using available deflection formulas. *All beams have constant flexural rigidity EI.*

Introductory Problems

9.3-1 A simply supported beam is loaded with a point load, as shown in the figure. The beam is a steel shape (IPN 300) in strong axis bending. Calculate

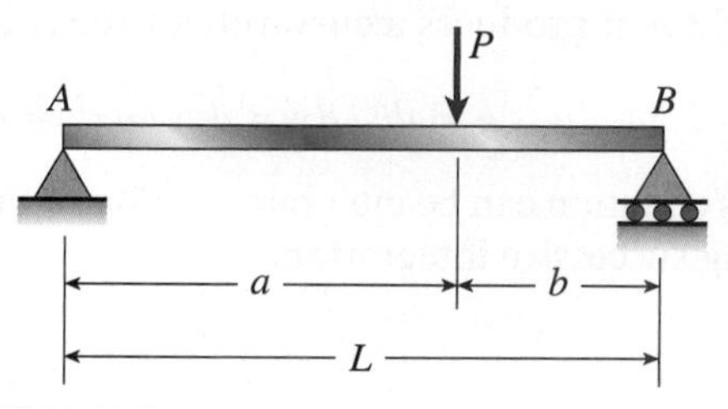

PROBLEM 9.3-1

the maximum deflection of the beam and the rotation at joint A if $L = 3$ m, $a = 2.1$ m, $b = 0.9$ m, and $P = 45$ kN. Neglect the weight of the beam.

9.3-2 A 1-meter-long, simply supported copper beam ($E = 117$ GPa) carries uniformly distributed load q. The maximum deflection is measured as 1.5 mm.

(a) Calculate the magnitude of the distributed load q if the beam has a rectangular cross section (width $b = 20$ mm, height $h = 40$ mm).

(b) If instead the beam has circular cross section and $q = 500$ N/m, calculate the radius r of the cross section.

Neglect the weight of the beam.

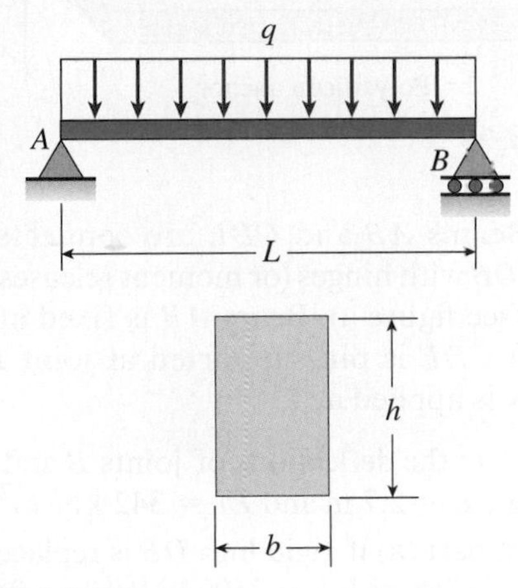

PROBLEM 9.3-2

9.3-3 A wide-flange beam (HE 220B) supports a uniform load on a simple span of length $L = 4.25$ m (see figure).

Calculate the maximum deflection δ_{max} at the midpoint and the angles of rotation θ at the supports if $q = 26$ kN/m and $E = 210$ GPa. (Use the formulas of Example 9-1.)

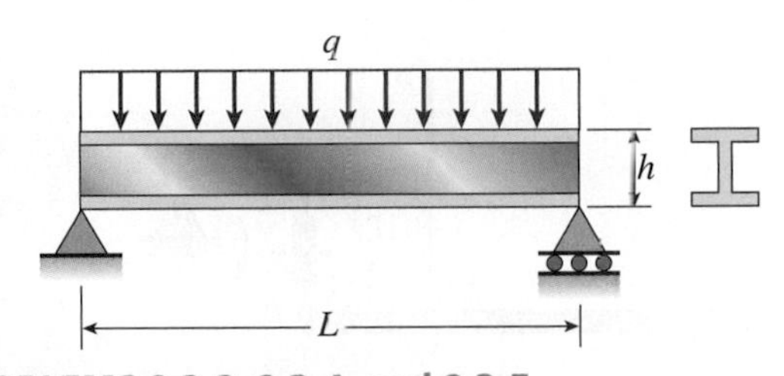

PROBLEMS 9.3-3, 9.3-4, and 9.3-5

9.3-4 A uniformly loaded, steel wide-flange beam with simple supports (see figure) has a downward deflection of 10 mm at the midpoint and angles of rotation equal to 0.01 radians at the ends.

Calculate the height h of the beam if the maximum bending stress is 90 MPa and the modulus of elasticity is 200 GPa. (Use the formulas of Example 9-1.)

9.3-5 What is the span length L of a uniformly loaded, simple beam of wide-flange cross section (see figure) if the maximum bending stress is 84 MPa, the maximum deflection is 2.5 mm, the height of the beam is 300 mm, and the modulus of elasticity is 210 GPa? (Use the formulas of Example 9-1.)

9.3-6 Calculate the maximum deflection δ_{max} of a uniformly loaded simple beam if the span length $L = 2.0$ m, the intensity of the uniform load $q = 2.0$ kN/m, and the maximum bending stress $\sigma = 60$ MPa.

The cross section of the beam is square, and the material is aluminum having modulus of elasticity $E = 70$ GPa. (Use the formulas of Example 9-1.)

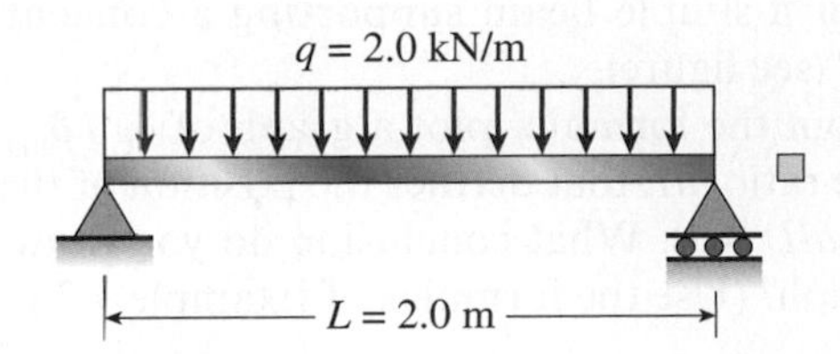

PROBLEM 9.3-6

Representative Problems

9.3-7 A cantilever beam with a uniform load (see figure) has a height h equal to 1/8 of the length L. The beam is a steel wide-flange section with $E = 208$ GPa and an allowable bending stress of 130 MPa in both tension and compression. Calculate the ratio δ/L of the deflection at the free end to the length, assuming that the beam carries the maximum allowable load. (Use the formulas of Example 9-2.)

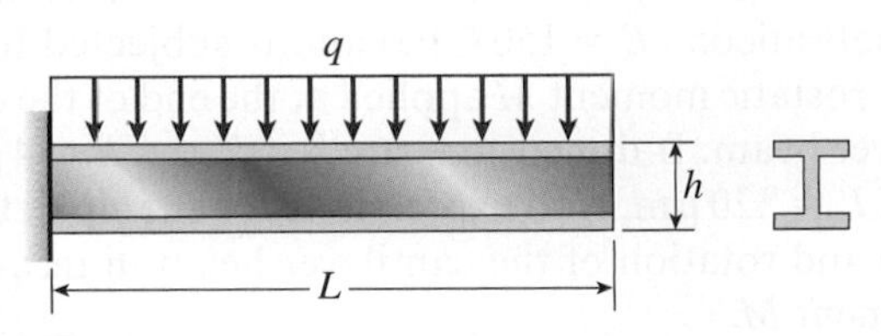

PROBLEM 9.3-7

9.3-8 A gold-alloy microbeam attached to a silicon wafer behaves like a cantilever beam subjected to a uniform load (see figure). The beam has a length $L = 27.5\ \mu m$ and rectangular cross section of a width $b = 4.0\ \mu m$ and thickness $t = 0.88\ \mu m$. The total load on the beam is $17.2\ \mu N$. If the deflection at the end of the beam is $2.46\ \mu m$, what is the modulus of elasticity E_g of the gold alloy? (Use the formulas of Example 9-2.)

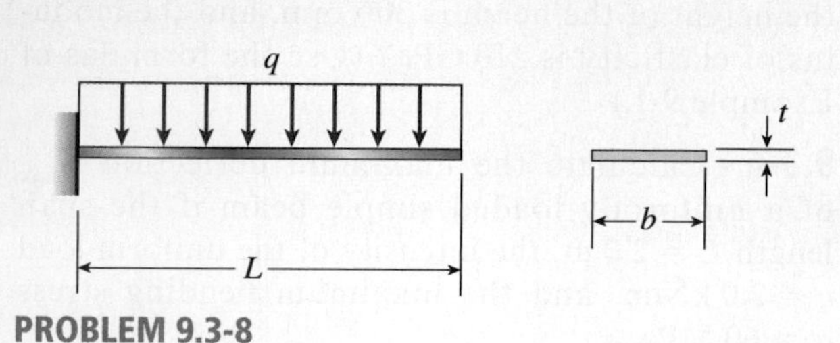

PROBLEM 9.3-8

9.3-9 Obtain a formula for the ratio δ_C / δ_{max} of the deflection at the midpoint to the maximum deflection for a simple beam supporting a concentrated load P (see figure).

From the formula, plot a graph of δ_C / δ_{max} versus the ratio a/L that defines the position of the load ($0.5 < a/L < 1$). What conclusion do you draw from the graph? (Use the formulas of Example 9-3.)

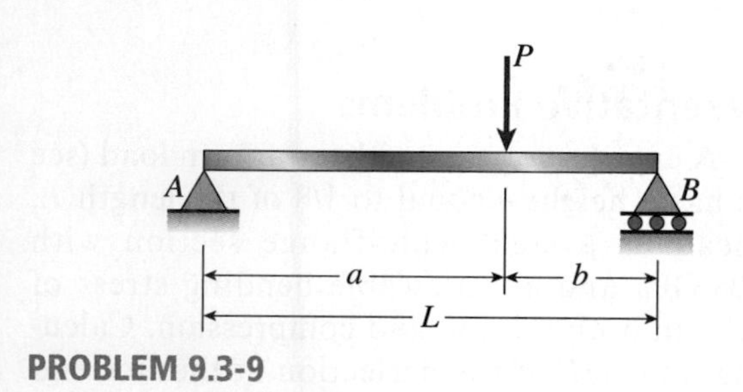

PROBLEM 9.3-9

9.3-10 A cantilever beam model is often used to represent micro-electrical-mechanical systems (MEMS) (see figure). The cantilever beam is made of polysilicon ($E = 150$ GPa) and is subjected to an electrostatic moment M applied at the end of the cantilever beam. If dimensions are $b = 2\ \mu m$, $h = 4\ \mu m$, and $L = 320\ \mu m$, find expressions for the tip deflection and rotation of the cantilever beam in terms of moment M.

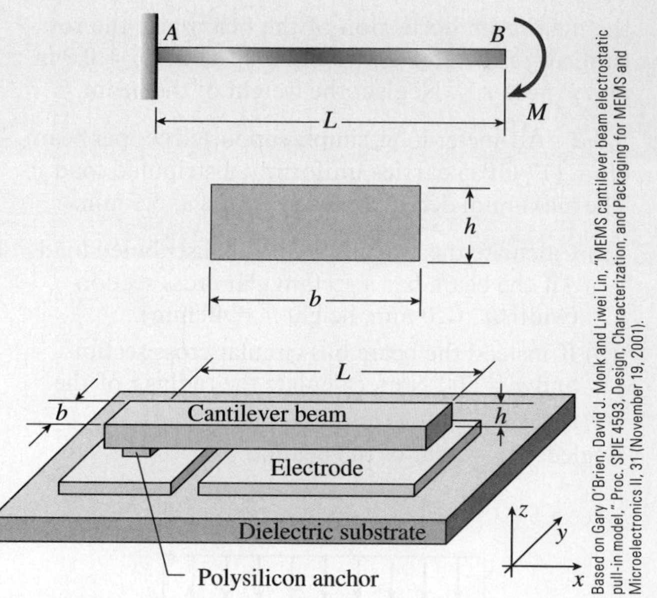

PROBLEM 9.3-10

Based on Gary O'Brien, David J. Monk and Liwei Lin, "MEMS cantilever beam electrostatic pull-in model," Proc. SPIE 4593, Design, Characterization, and Packaging for MEMS and Microelectronics II, 31 (November 19, 2001).

9.3-11 Beams AB and CDE are connected using rigid link DB with hinges (or moment releases) at ends D and B (see figure a). Beam AB is fixed at joint A and beam CDE is pin-supported at joint E. Load $P = 665$ N is applied at C.

(a) Calculate the deflections of joints B and joint C. Assume $L = 2.7$ m and $EI = 342\ kN{\cdot}m^2$.

(b) Repeat part (a) if rigid link DB is replaced by a linear spring with $k = 3500$ kN/m (see figure b).

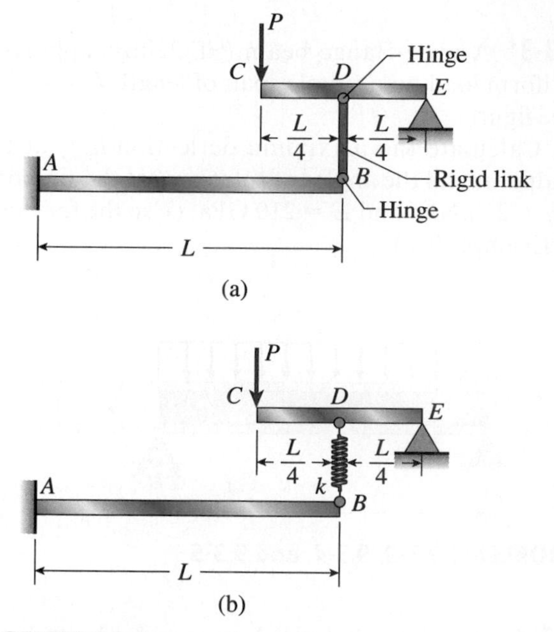

PROBLEM 9.3-11

Solve the problems in this section by integrating the second-order differential equation of the deflection curve (the bending-moment equation). The origin of coordinates is at the left-hand end of each beam, and all beams have constant flexural rigidity EI.

Introductory Problems

9.3-12 Derive the equation of the deflection curve for a cantilever beam AB supporting a load P at the free end (see figure). Also, determine the deflection δ_B and angle of rotation θ_B at the free end. Use the second-order differential equation of the deflection curve.

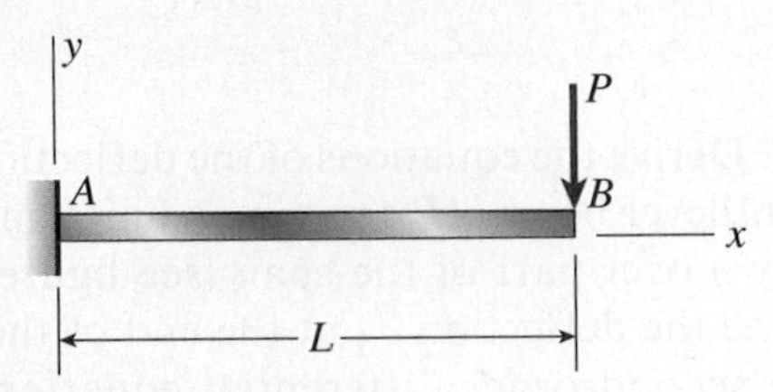

PROBLEM 9.3-12

9.3-13 Derive the equation of the deflection curve for a simple beam AB loaded by a couple M_0 at the left-hand support (see figure). Also, determine the maximum deflection δ_{max}. Use the second-order differential equation of the deflection curve.

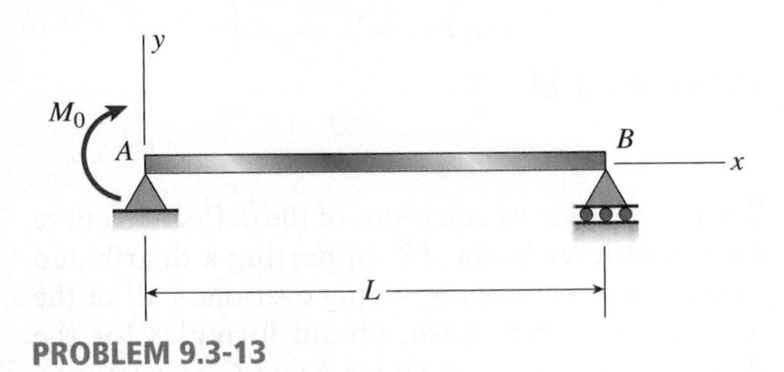

PROBLEM 9.3-13

9.3-14 A cantilever beam AB supporting a triangularly distributed load of maximum intensity q_0 is shown in the figure.

Derive the equation of the deflection curve and then obtain formulas for the deflection δ_B and angle of rotation θ_B at the free end. Use the second-order differential equation of the deflection curve.

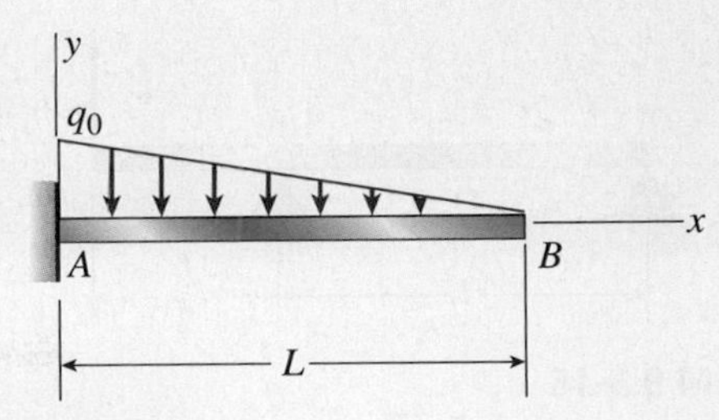

PROBLEM 9.3-14

9.3-15 A cantilever beam has a length $L = 3.6$ m and a rectangular cross section ($b = 400$ mm, $h = 600$ mm). A linearly varying distributed load with peak intensity q_0 acts on the beam.

(a) Find peak intensity q_0 if the deflection at joint B is known to be 4.5 mm. Assume that modulus $E = 205$ GPa.

(b) Find the location and magnitude of the maximum rotation of the beam.

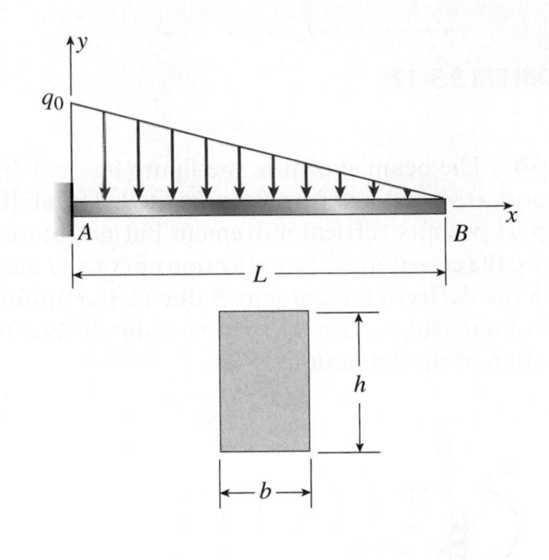

PROBLEM 9.3-15

9.3-16 A simple beam with an overhang is subjected to a point load $P = 6$ kN. If the maximum allowable deflection at point C is 0.5 mm, select the lightest IPN section from Table F-2 that can be used for the beam. Assume that $L = 3$ m and ignore the distributed weight of the beam.

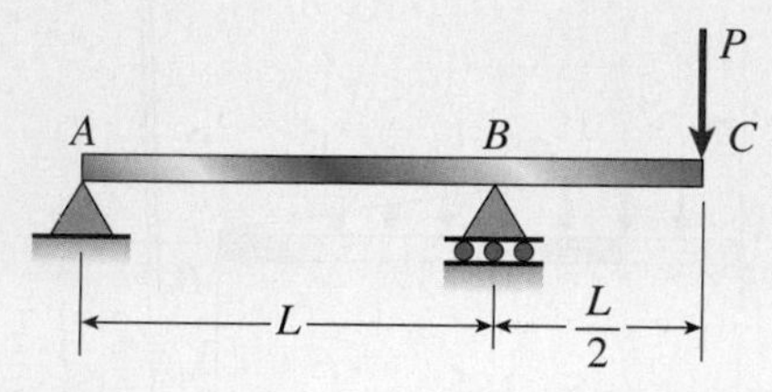

PROBLEM 9.3-16

Representative Problems

9.3-17 A cantilever beam AB is acted upon by a uniformly distributed moment (bending moment, not torque) of intensity m per unit distance along the axis of the beam (see figure).

Derive the equation of the deflection curve and then obtain formulas for the deflection δ_B and angle of rotation θ_B at the free end. Use the second-order differential equation of the deflection curve.

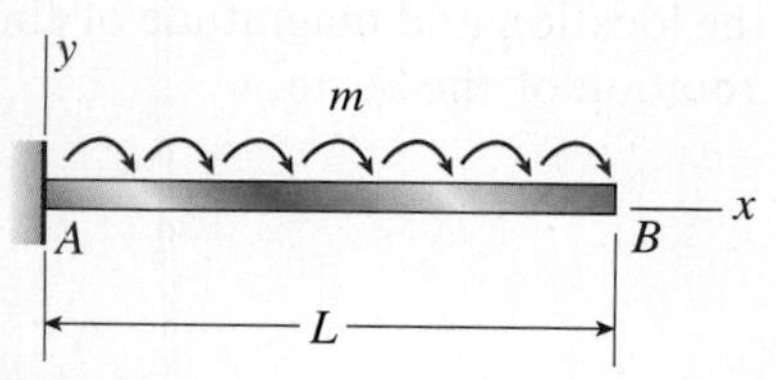

PROBLEM 9.3-17

9.3-18 The beam shown in the figure has a sliding support at A and a spring support at B. The sliding support permits vertical movement but no rotation. Derive the equation of the deflection curve and determine the deflection δ_B at end B due to the uniform load of intensity q. Use the second-order differential equation of the deflection curve.

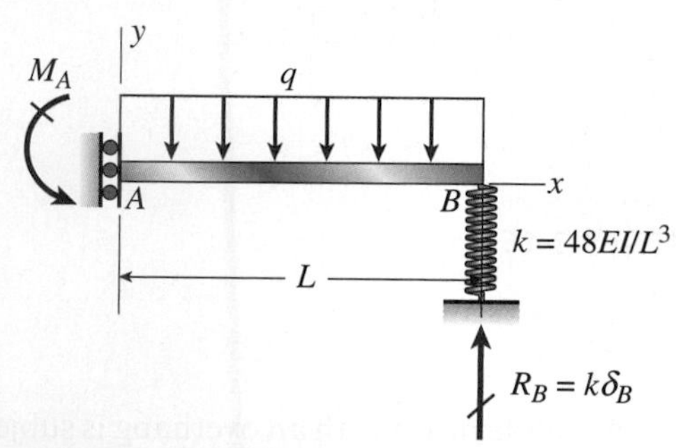

PROBLEM 9.3-18

9.3-19 Derive the equations of the deflection curve for a simple beam AB loaded by a couple M_0 acting at distance a from the left-hand support (see figure). Also, determine the deflection δ_0 at the point where the load is applied. Use the second-order differential equation of the deflection curve.

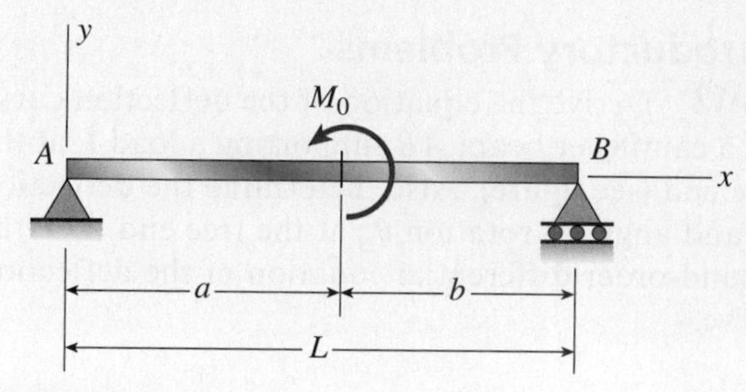

PROBLEM 9.3-19

9.3-20 Derive the equations of the deflection curve for a cantilever beam AB carrying a uniform load of intensity q over part of the span (see figure). Also, determine the deflection δ_B at the end of the beam. Use the second-order differential equation of the deflection curve.

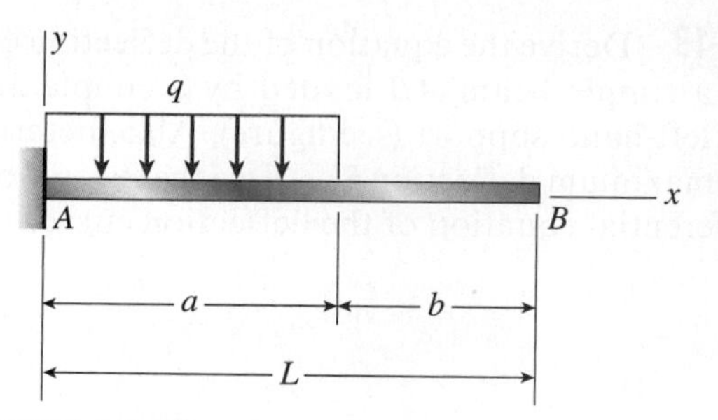

PROBLEM 9.3-20

9.3-21 Derive the equations of the deflection curve for a cantilever beam AB supporting a distributed load of peak intensity q_0 acting over one-half of the length (see figure). Also, obtain formulas for the deflections δ_B and δ_C at points B and C, respectively. Use the second-order differential equation of the deflection curve.

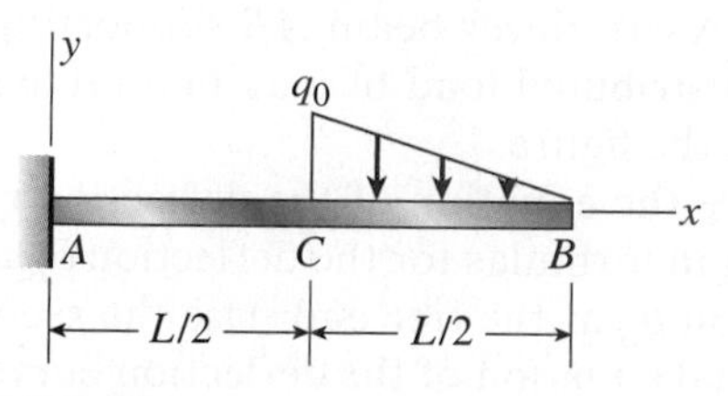

PROBLEM 9.3-21

9.3-22 Derive the equations of the deflection curve for a simple beam AB with a distributed load of peak intensity q_0 acting over the left-hand half of the span (see figure). Also, determine the deflection δ_C at the midpoint of the beam. Use the second-order differential equation of the deflection curve.

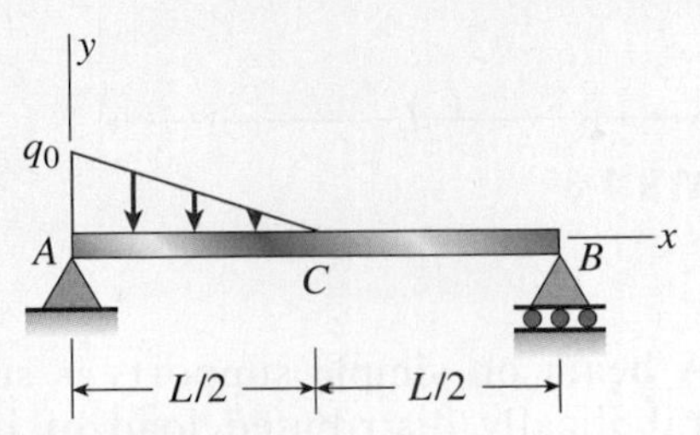

PROBLEM 9.3-22

9.3-23 The beam shown in the figure has a sliding support at A and a roller support at B. The sliding support permits vertical movement but no rotation. Derive the equation of the deflection curve and determine the deflection δ_A at end A and also δ_C at point C due to the uniform load of intensity $q = P/L$ applied over segment CB and load P at $x = L/3$. Use the second-order differential equation of the deflection curve.

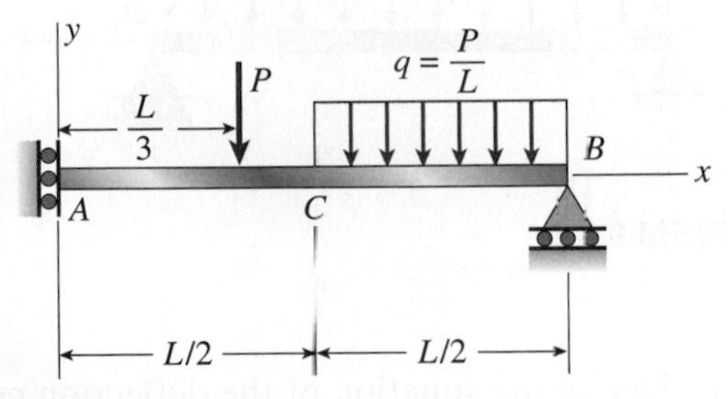

PROBLEM 9.3-23

9.4 Deflections by Integration of the Shear-Force and Load Equations

The beams described in the problems for Section 9.4 have constant flexural rigidity EI. Also, the origin of coordinates is at the left-hand end of each beam.

Introductory Problems

9.4-1 Derive the equation of the deflection curve for a cantilever beam AB when a couple M_0 acts counterclockwise at the free end (see figure). Also, determine the deflection δ_B and slope θ_B at the free end. Use the third-order differential equation of the deflection curve (the shear-force equation).

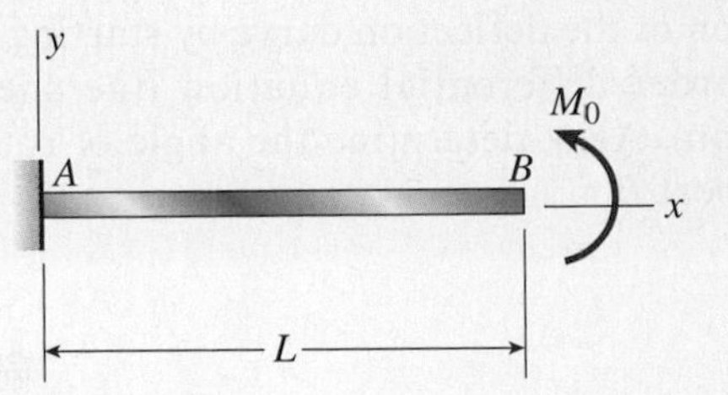

PROBLEM 9.4-1

9.4-2 A simple beam AB is subjected to a distributed load of intensity $q(x) = q_0 \sin \pi x / L$, where q_0 is the maximum intensity of the load (see figure).

Derive the equation of the deflection curve, and then determine the deflection δ_{max} at the midpoint of the beam. Use the fourth-order differential equation of the deflection curve (the load equation).

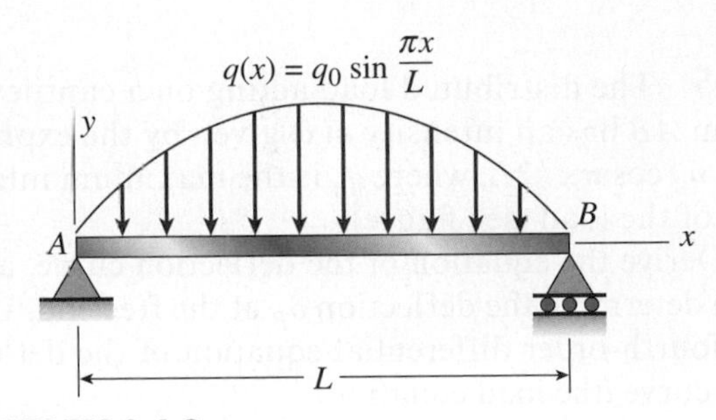

PROBLEM 9.4-2

9.4-3 The simple beam AB shown in the figure has moments $2M_0$ and M_0 acting at the ends.

Derive the equation of the deflection curve, and then determine the maximum deflection δ_{max}. Use the third-order differential equation of the deflection curve (the shear-force equation).

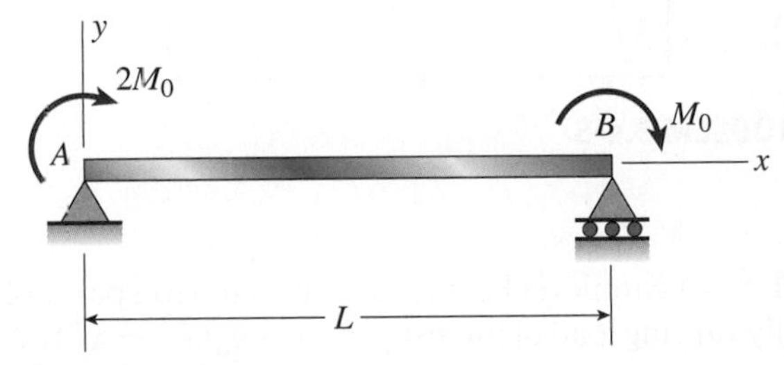

PROBLEM 9.4-3

Representative Problems

9.4-4 A beam with a uniform load has a sliding support at one end and spring support at the other. The spring has a stiffness $k = 48EI/L^3$. Derive the equation of the deflection curve by starting with the third-order differential equation (the shear-force equation). Also, determine the angle of rotation θ_B at support B.

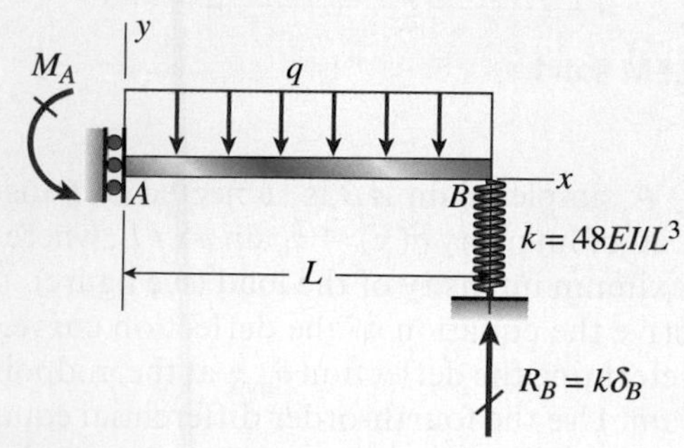

PROBLEM 9.4-4

9.4-5 The distributed load acting on a cantilever beam AB has an intensity $q(x)$ given by the expression $q_0 \cos \pi x / 2L$, where q_0 is the maximum intensity of the load (see figure).

Derive the equation of the deflection curve, and then determine the deflection δ_B at the free end. Use the fourth-order differential equation of the deflection curve (the load equation).

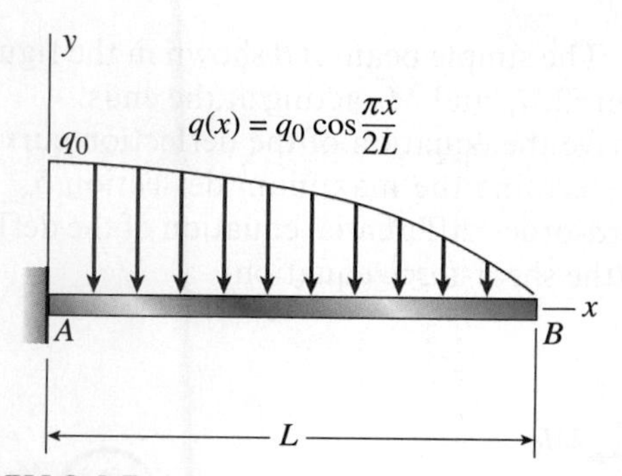

PROBLEM 9.4-5

9.4-6 A cantilever beam AB is subjected to a parabolically varying load of intensity $q(x) = q_0(L^2 - x^2)/L^2$, where q_0 is the maximum intensity of the load (see figure).

Derive the equation of the deflection curve, and then determine the deflection δ_B and angle of rotation θ_B at the free end. Use the fourth-order differential equation of the deflection curve (the load equation).

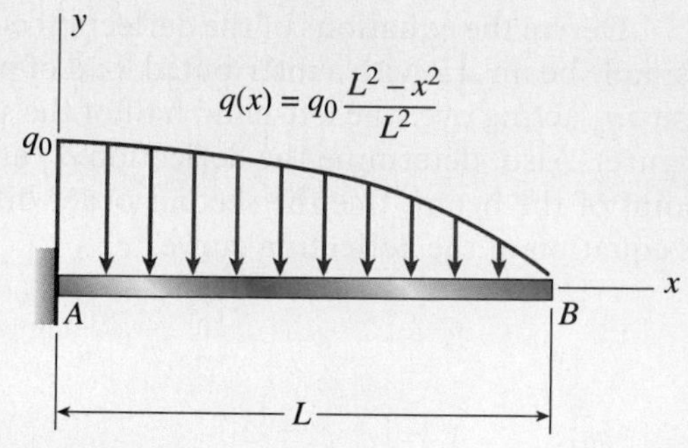

PROBLEM 9.4-6

9.4-7 A beam on simple supports is subjected to a parabolically distributed load of intensity $q(x) = 4q_0x(L - x)/L^2$, where q_0 is the maximum intensity of the load (see figure).

Derive the equation of the deflection curve, and then determine the maximum deflection δ_{max}. Use the fourth-order differential equation of the deflection curve (the load equation).

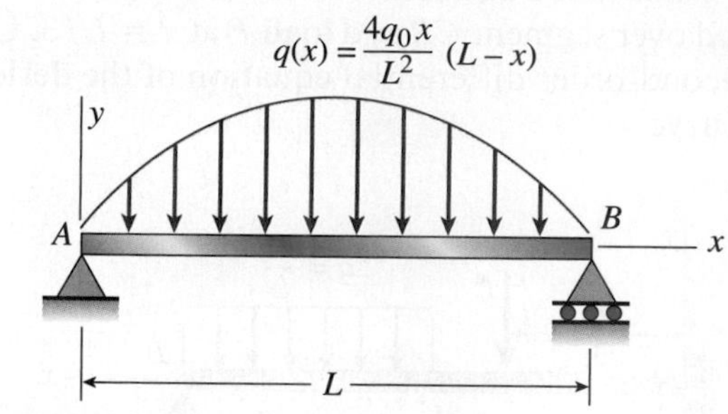

PROBLEM 9.4-7

9.4-8 Derive the equation of the deflection curve for beam AB with sliding support at A and roller at B, carrying a triangularly distributed load of maximum intensity q_0 (see figure). Also, determine the maximum deflection δ_{max} of the beam. Use the fourth-order differential equation of the deflection curve (the load equation).

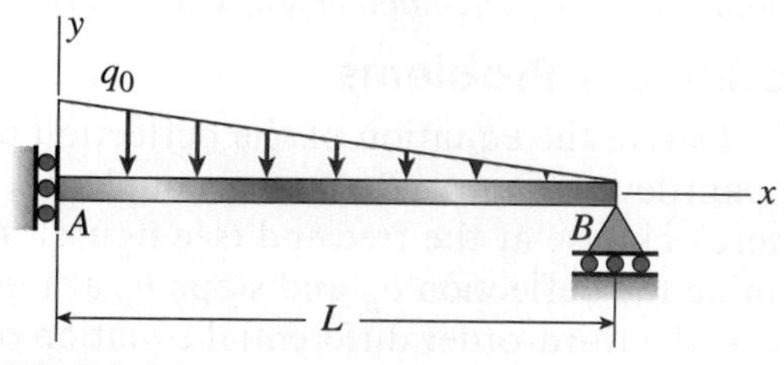

PROBLEM 9.4-8

9.4-9 Derive the equations of the deflection curve for beam ABC with sliding support at A and roller support at B, supporting a uniform load of intensity q acting on the overhang portion of the beam (see figure). Also, determine deflection δ_C and angle of rotation θ_C. Use the fourth-order differential equation of the deflection curve (the load equation).

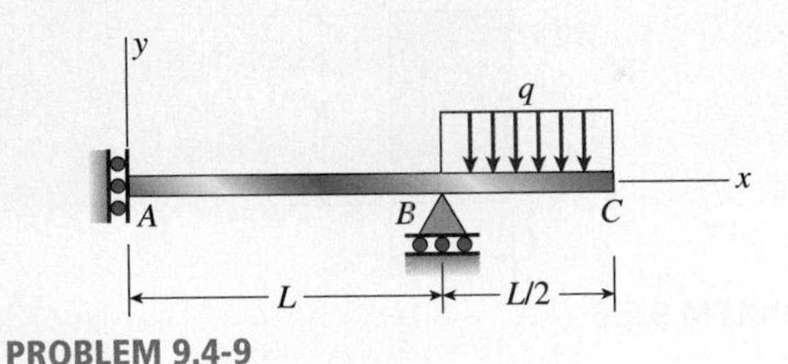

PROBLEM 9.4-9

9.4-10 Derive the equations of the deflection curve for beam AB with sliding support at A and roller support at B, supporting a distributed load of maximum intensity q_0 acting on the right-hand half of the beam (see figure). Also, determine deflection δ_A, angle of rotation θ_B, and deflection δ_C at the midpoint. Use the fourth-order differential equation of the deflection curve (the load equation).

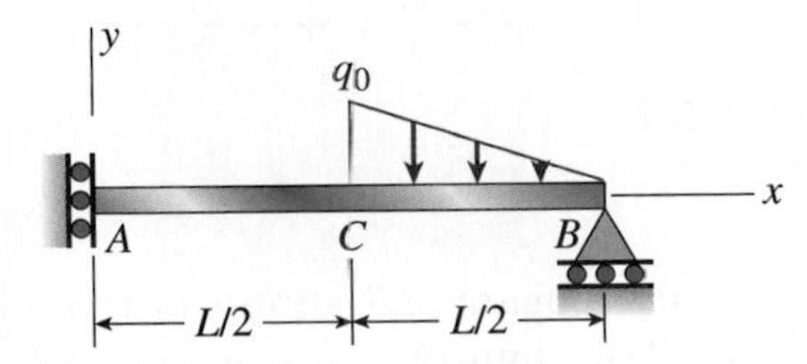

PROBLEM 9.4-10

9.5 Method of Superposition

Solve the problems for Section 9.5 by the method of superposition. All beams have constant flexural rigidity EI.

Introductory Problems

9.5-1 A simply supported beam ($E = 11$ GPa) is loaded by a triangular distributed load from A to C (see figure). The load has a peak intensity $q_0 = 145$ N/m, and the deflection is known to be 0.25 mm at point C. The length of the beam is 3.6 m, and the ratio of the height to the width of the cross section is (h:b) 2:1. Find the height h and width b of the cross section of the beam.

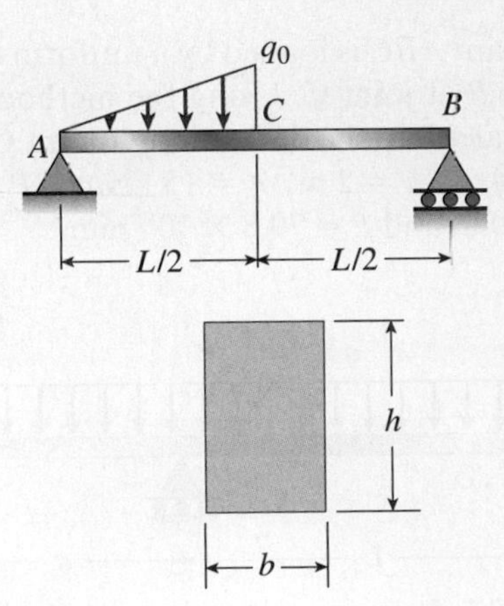

PROBLEM 9.5-1

9.5-2 A simply supported beam ($E = 12$ GPa) carries a uniformly distributed load $q = 125$ N/m, and a point load $P = 200$ N at mid-span. The beam has a rectangular cross section ($b = 75$ mm, $h = 200$ mm) and a length of 3.6 m. Calculate the maximum deflection of the beam.

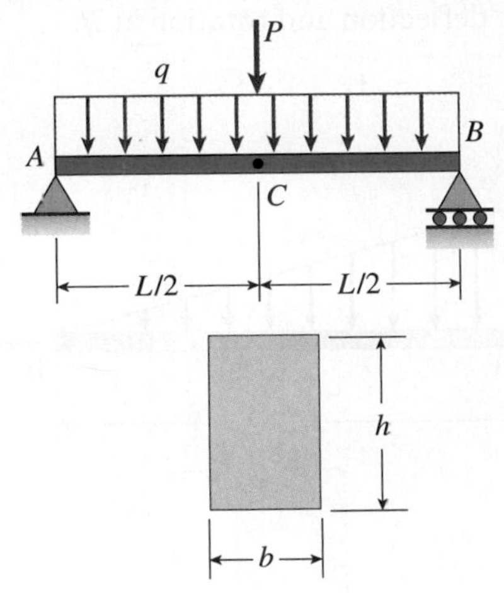

PROBLEM 9.5-2

9.5-3 Copper beam AB has circular cross section with a radius of 6.5 mm and length $L = 0.9$ m. The beam is subjected to a uniformly distributed load $w = 50$ N/m. Calculate the required load P at joint B so that the total deflection at joint B is zero. Assume that $E = 110$ GPa.

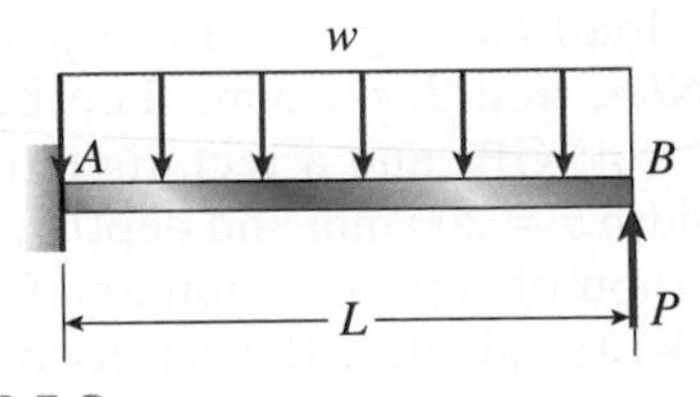

PROBLEM 9.5-3

9.5-4 Beam *ABC* is loaded by a uniform load q and point load P at joint C. Using the method of superposition, calculate the deflection at joint C. Assume that $L = 4$ m, $a = 2$ m, $q = 15$ kN/m, $P = 7.5$ kN, $E = 200$ GPa, and $I = 70.8 \times 10^6$ mm^4.

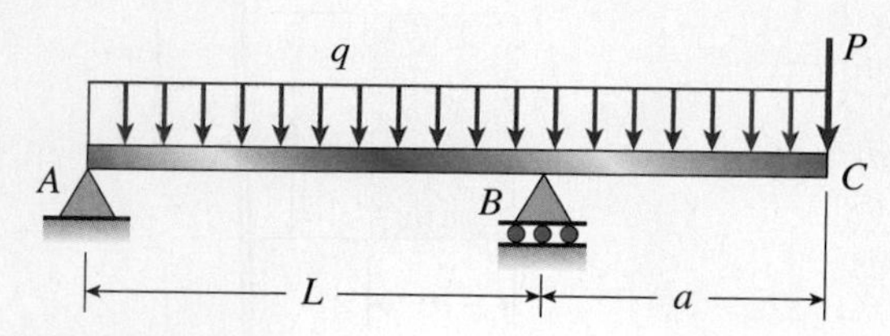

PROBLEM 9.5-4

9.5-5 A cantilever beam of a length $L = 750$ mm has a rectangular cross section ($b = 100$ mm, $h = 200$ mm) and modulus $E = 70$ GPa. The beam is subjected to a linearly varying distributed load with a peak intensity $q_0 = 13$ kN/m. Use the method of superposition and Cases 1 and 9 in Table H-1 to calculate the deflection and rotation at B.

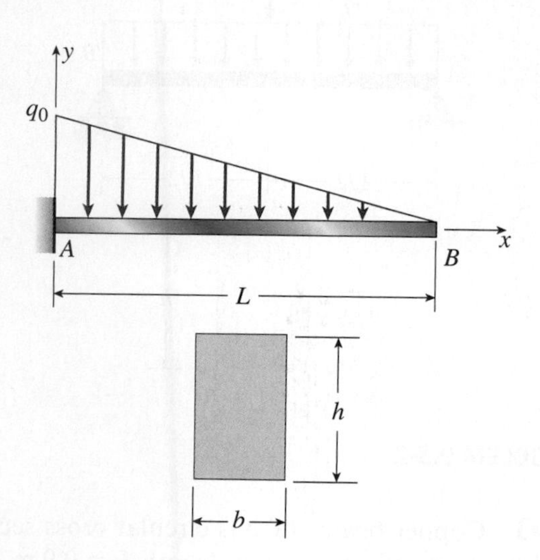

PROBLEM 9.5-5

9.5-6 A cantilever beam carries a trapezoidal distributed load (see figure). Let $w_B = 2.5$ kN/m, $w_A = 5.0$ kN/m, and $L = 2.5$ m. The beam has a modulus $E = 45$ GPa and a rectangular cross section with width $b = 200$ mm and depth $h = 300$ mm. Use the method of superposition and Cases 1 and 8 in Table H-1 to calculate the deflection and rotation at B.

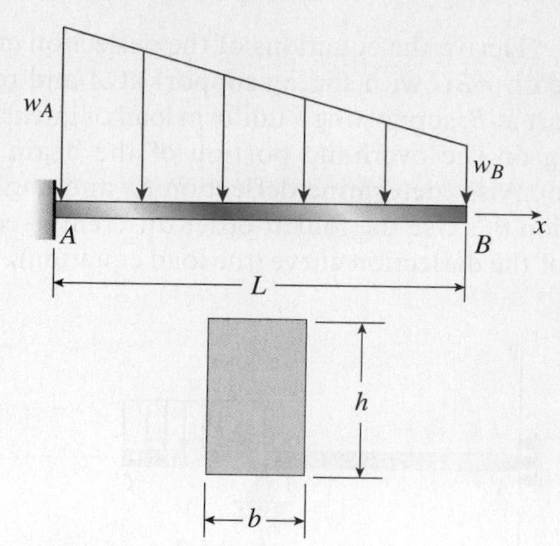

PROBLEM 9.5-6

9.5-7 A cantilever beam AB carries three equally spaced concentrated loads, as shown in the figure. Obtain formulas for the angle of rotation θ_B and deflection δ_B at the free end of the beam.

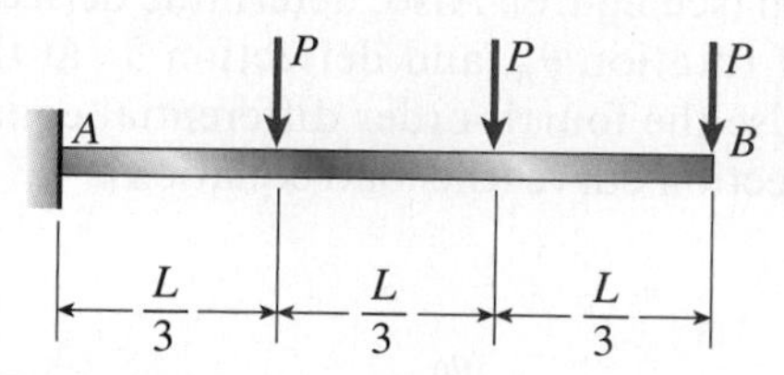

PROBLEM 9.5-7

9.5-8 A simple beam AB supports five equally spaced loads P (see figure).

(a) Determine the deflection δ_1 at the midpoint of the beam.

(b) If the same total load ($5P$) is distributed as a uniform load on the beam, what is the deflection δ_2 at the midpoint?

(c) Calculate the ratio of δ_1 to δ_2.

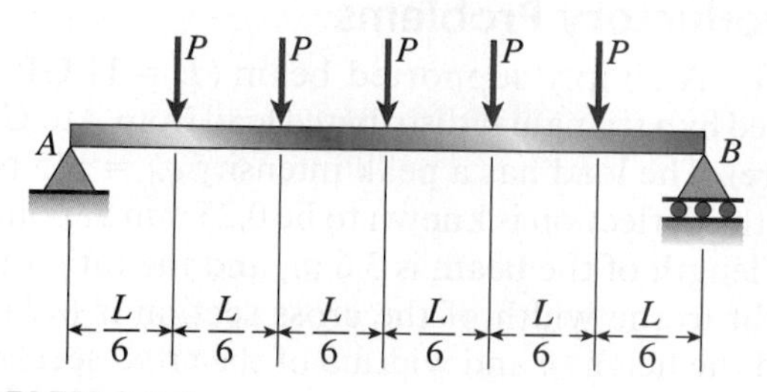

PROBLEM 9.5-8

Representative Problems

9.5-9 The cantilever beam AB shown in the figure has an extension BCD attached to its free end. A force P acts at the end of the extension.

(a) Find the ratio a/L so that the vertical deflection of point B will be zero.

(b) Find the ratio a/L so that the angle of rotation at point B will be zero.

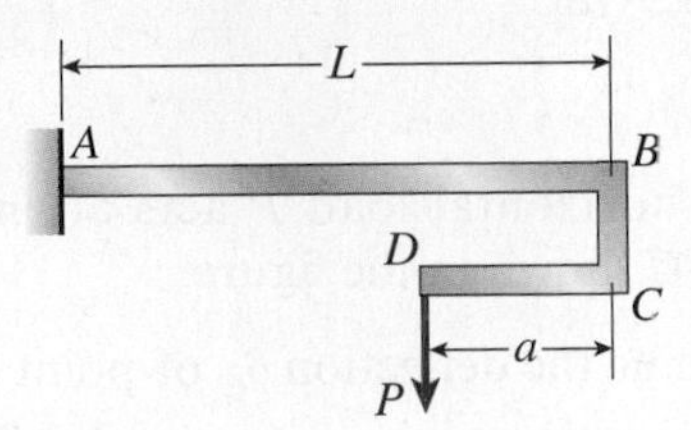

PROBLEM 9.5-9

9.5-10 Beam ACB hangs from two springs, as shown in the figure. The springs have stiffnesses k_1 and k_2 and the beam has flexural rigidity EI.

(a) What is the downward displacement of point C, which is at the midpoint of the beam, when the moment M_0 is applied? Data for the structure are as follows: $M_0 = 10.0$ kN·m, $L = 1.8$ m, $EI = 216$ kN·m^2, $k_1 = 250$ kN/m, and $k_2 = 160$ kN/m.

(b) Repeat part (a), but remove M_0 and apply a uniform load $q = 3.5$ kN/m to the entire beam.

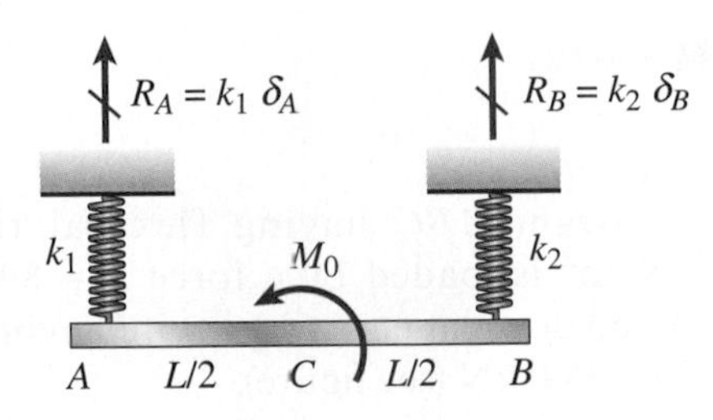

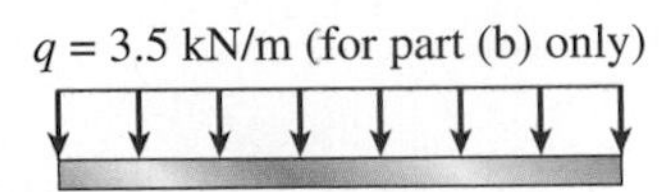

PROBLEM 9.5-10

9.5-11 What must be the equation $y = f(x)$ of the axis of the slightly curved beam AB (see figure) *before* the load is applied in order that the load P, moving along the bar, always stays at the same level?

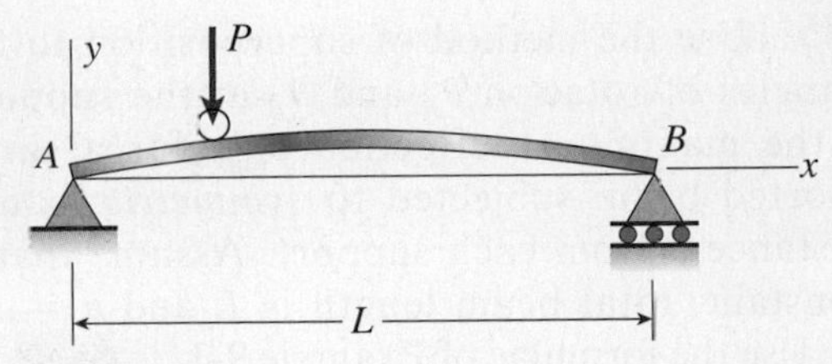

PROBLEM 9.5-11

9.5-12 Determine the angle of rotation θ_B and deflection δ_B at the free end of a cantilever beam AB having a uniform load of intensity q acting over the middle third of its length (see figure).

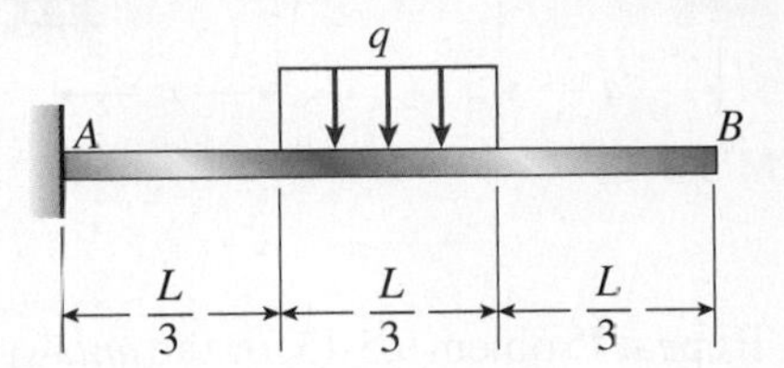

PROBLEM 9.5-12

9.5-13 The cantilever beam ACB shown in the figure has flexural rigidity $EI = 6.1 \times 10^6$ N·m^2. Calculate the downward deflections δ_C and δ_B at points C and B, respectively, due to the simultaneous action of the moment of 4 kN · m applied at point C and the concentrated load of 16 kN applied at the free end B.

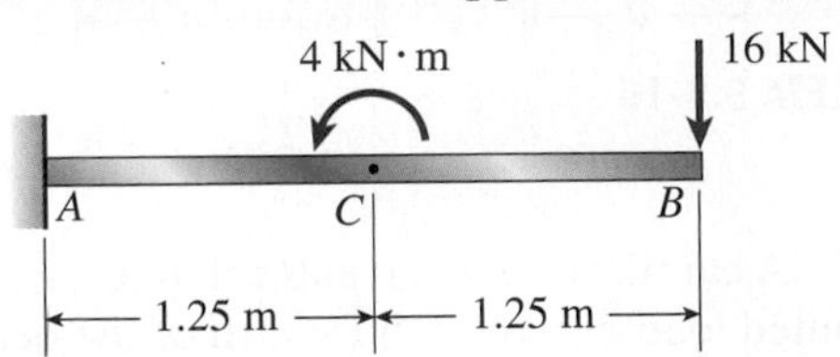

PROBLEM 9.5-13

9.5-14 A cantilever beam is subjected to load P at mid-span and counterclockwise moment M at B (see figure).

(a) Find an *expression* for moment M in terms of the load P so that the reaction moment M_A at A is equal to zero.

(b) Find an *expression* for moment M in terms of the load P so that the deflection is $\delta_B = 0$; *also*, what is rotation θ_B?

(c) Find an *expression* for moment M in terms of the load P so that the rotation $\theta_B = 0$; *also*, what is deflection δ_B?

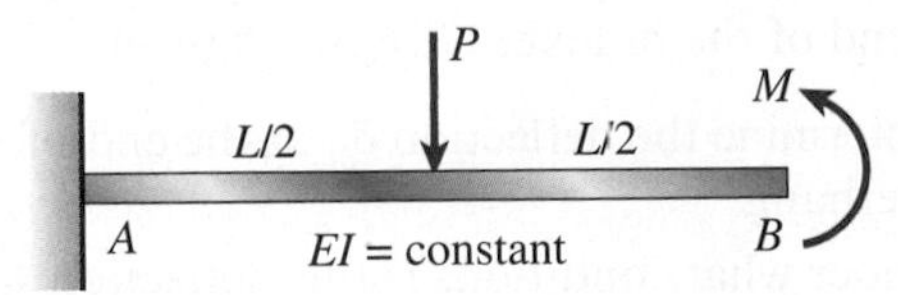

PROBLEM 9.5-14

9.5-15 Use the method of superposition to find the angles of rotation θ_A and θ_B at the supports, and the maximum deflection $\delta_{\max}$ for a simply supported beam subjected to *symmetric* loads P at distance a from each support. Assume that EI is constant, total beam length is L and $a = L/3$. *Hint:* Use the formulas of Example 9-3.

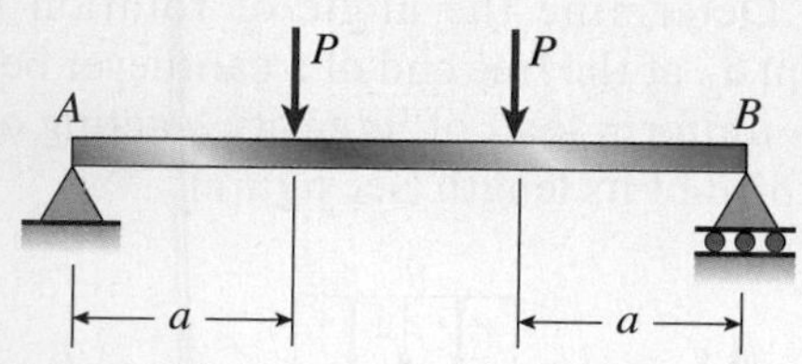

PROBLEM 9.5-15

9.5-16 Repeat Problem 9.5-15 for the *anti-symmetric* loading shown in the figure.

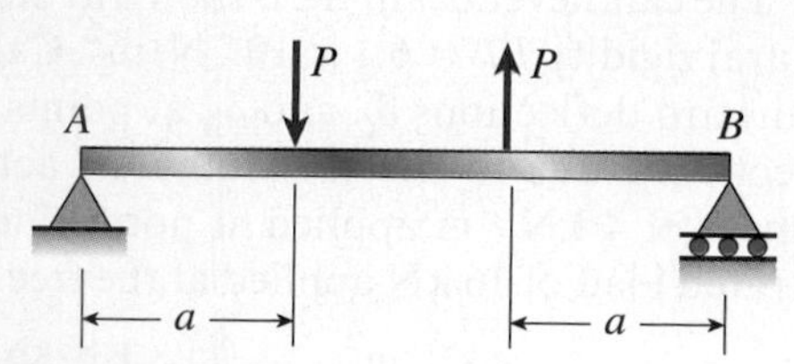

PROBLEM 9.5-16

9.5-17 A cantilever beam is subjected to a quadratic distributed load $q(x)$ over the length of the beam (see figure). Find an *expression* for moment M in terms of the peak distributed load intensity q_0 so that the deflection is $\delta_B = 0$.

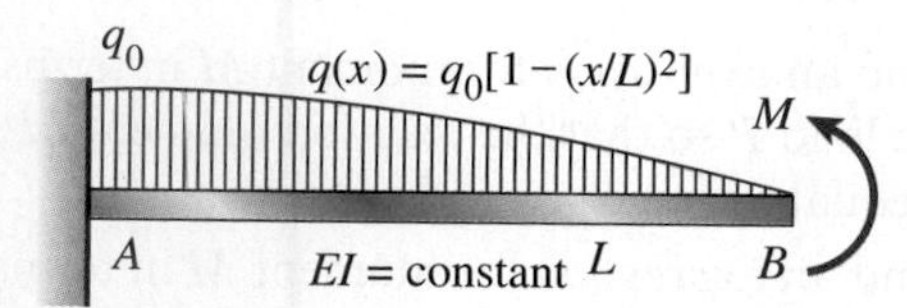

PROBLEM 9.5-17

9.5-18 A beam $ABCD$ consisting of a simple span BD and an overhang AB is loaded by a force P acting at the end of the bracket CEF (see figure).

(a) Determine the deflection δ_A at the end of the overhang.

(b) Under what conditions is this deflection upward? Under what conditions is it downward?

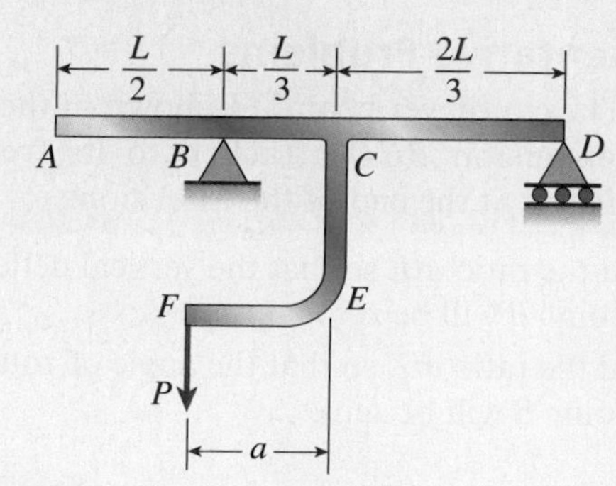

PROBLEM 9.5-18

9.5-19 A horizontal load P acts at end C of the bracket ABC shown in the figure.

(a) Determine the deflection δ_C of point C.

(b) Determine the maximum upward deflection $\delta_{\max}$ of member AB.

Note: Assume that the flexural rigidity EI is constant throughout the frame. Also, disregard the effects of axial deformations and consider only the effects of bending due to the load P.

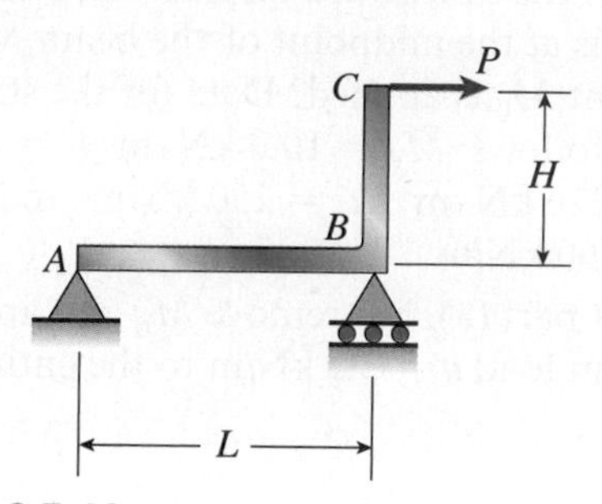

PROBLEM 9.5-19

9.5-20 A beam ABC having flexural rigidity $EI = 75 \text{ kN} \cdot \text{m}^2$ is loaded by a force $P = 800$ N at end C and tied down at end A by a wire having axial rigidity $EA = 900$ kN (see figure).

What is the deflection at point C when the load P is applied?

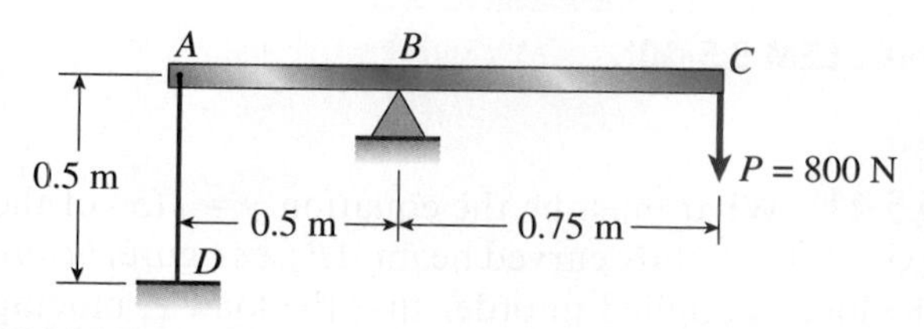

PROBLEM 9.5-20

9.5-21 Determine the angle of rotation θ_B and deflection δ_B at the free end of a cantilever beam AB supporting a parabolic load defined by the equation $q(x) = q_0x^2 / L^2$ (see figure).

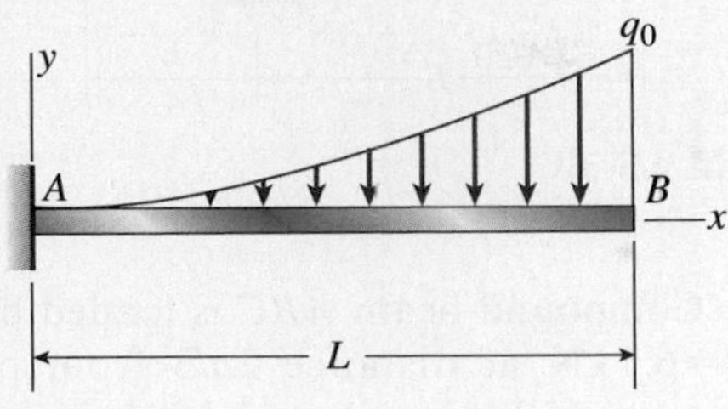

PROBLEM 9.5-21

9.5-22 A simple beam AB supports a uniform load of intensity q acting over the middle region of the span (see figure).

Determine the angle of rotation θ_A at the left-hand support and the deflection δ_{max} at the midpoint.

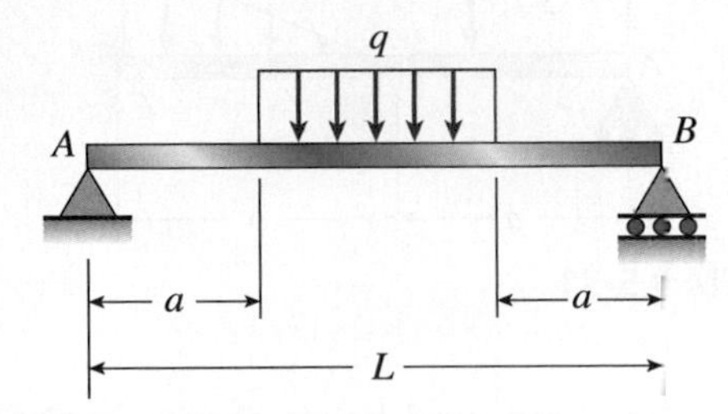

PROBLEM 9.5-22

9.5-23 The overhanging beam $ABCD$ supports two concentrated loads P and Q (see figure).

(a) For what ratio P/Q will the deflection at point B be zero?

(b) For what ratio will the deflection at point D be zero?

(c) If Q is replaced by a uniform load with intensity q (on the overhang), repeat parts (a) and (b), but find ratio $P/(qa)$.

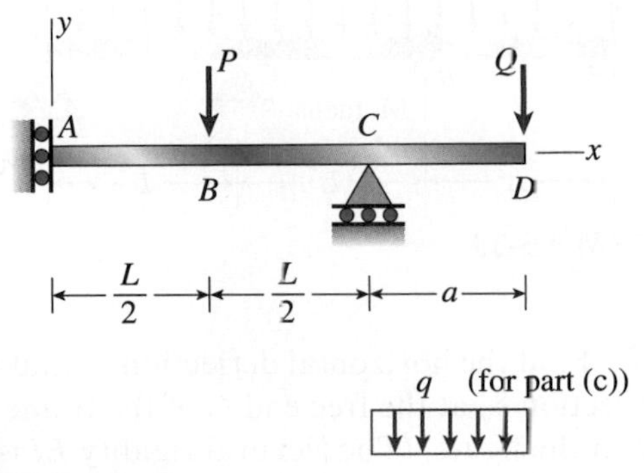

PROBLEM 9.5-23

9.5-24 A thin metal strip of total weight W and length L is placed across the top of a flat table of width $L/3$ as shown in the figure.

What is the clearance δ between the strip and the middle of the table? (The strip of metal has flexural rigidity EI.)

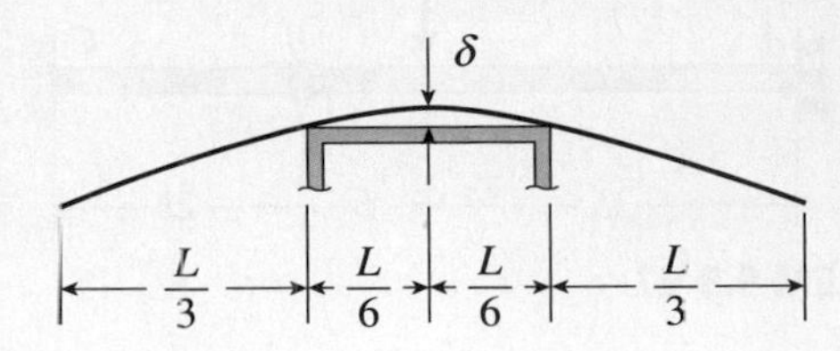

PROBLEM 9.5-24

9.5-25 An overhanging beam ABC with flexural rigidity $EI = 45\ \text{N}\cdot\text{m}^2$ is supported by a sliding support at A and by a spring of stiffness k at point B (see figure). Span AB has a length $L = 0.75$ m and carries a uniform load. The overhang BC has a length $b = 375$ mm. For what stiffness k of the spring will the uniform load produce no deflection at the free end C?

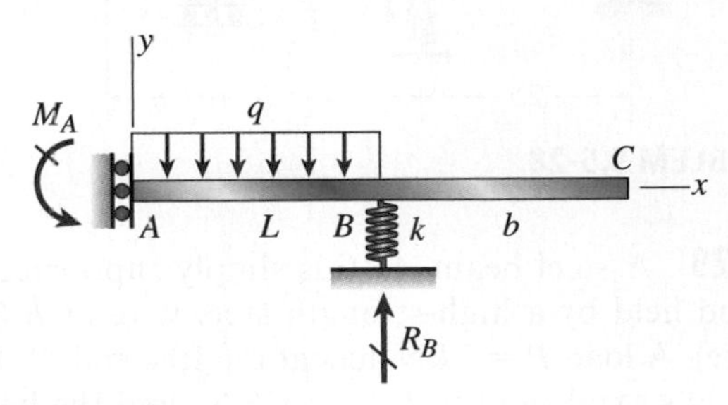

PROBLEM 9.5-25

9.5-26 A beam $ABCD$ rests on simple supports at B and C (see figure). The beam has a slight initial curvature so that end A is 18 mm above the elevation of the supports and end D is 12 mm above. What moments M_1 and M_2, acting at points A and D, respectively, will move points A and D downward to the level of the supports? (The flexural rigidity EI of the beam is $2.5 \times 10^6\ \text{N}\cdot\text{m}^2$ and $L = 2.5\text{m}$).

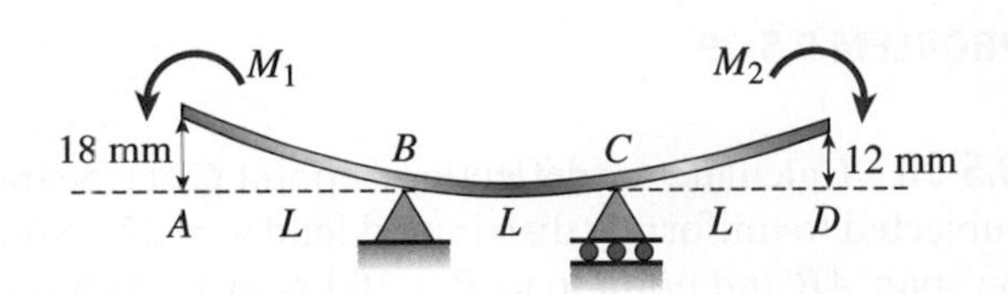

PROBLEM 9.5-26

9.5-27 The compound beam ABC shown in the figure has a sliding support at A and a fixed support at C. The beam consists of two members joined by a pin connection (i.e., moment release) at B. Find the deflection δ under the load P.

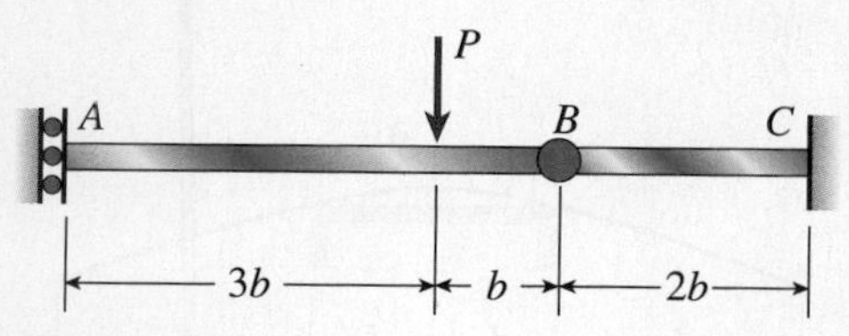

PROBLEM 9.5-27

9.5-28 A compound beam $ABCDE$ (see figure) consists of two parts (ABC and CDE) connected by a hinge (i.e., moment release) at C. The elastic support at B has stiffness $k = EI/b^3$. Determine the deflection δ_E at the free end E due to the load P acting at that point.

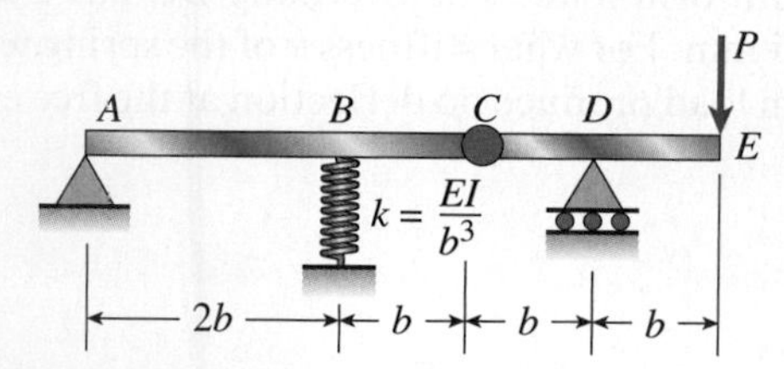

PROBLEM 9.5-28

9.5-29 A steel beam ABC is simply supported at A and held by a high-strength steel wire at B (see figure). A load $P = 1$ kN acts at the free end C. The wire has axial rigidity $EA = 1335$ N, and the beam has flexural rigidity $EI = 86$ kN·m^2.

What is the deflection δ_C of point C due to the load P?

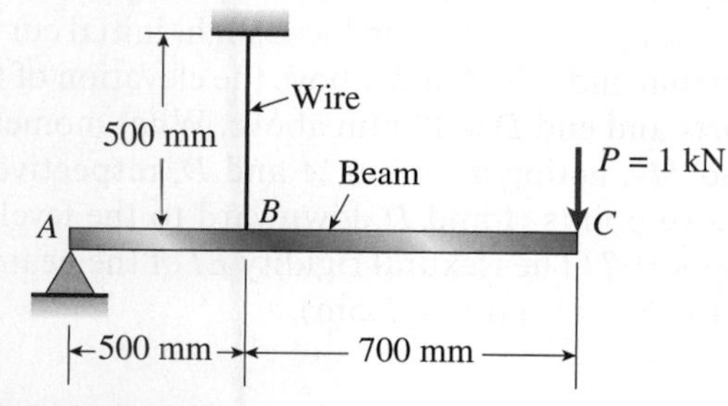

PROBLEM 9.5-29

9.5-30 Calculate the deflection at point C of a beam subjected to uniformly distributed load $w = 275$ N/m on span AB and point load $P = 10$ kN at C. Assume that $L = 5$ m and $EI = 1.50 \times 10^7$ N·m^2.

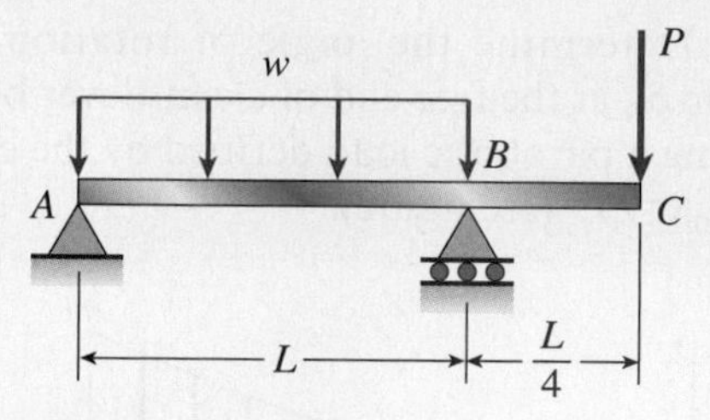

PROBLEM 9.5-30

9.5-31 Compound beam ABC is loaded by point load $P = 6.5$ kN at distance $2a/3$ from point A and a triangularly distributed load on segment BC with peak intensity $q_0 = 7.5$ kN/m. If length $a = 1.5$ m and length $b = 3$ m, find the deflection at B and rotation at A. Assume that $E = 200$ GPa and $I = 2140 \times 10^4$ mm^4.

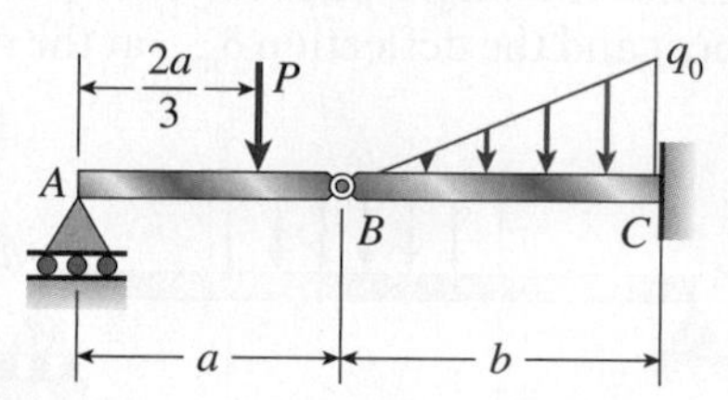

PROBLEM 9.5-31

9.5-32 The compound beam shown in the figure consists of a cantilever beam AB (length L) that is pin-connected to a simple beam BD (length $2L$). After the beam is constructed, a clearance c exists between the beam and a support at C, midway between points B and D. Subsequently, a uniform load is placed along the entire length of the beam.

What intensity q of the load is needed to close the gap at C and bring the beam into contact with the support?

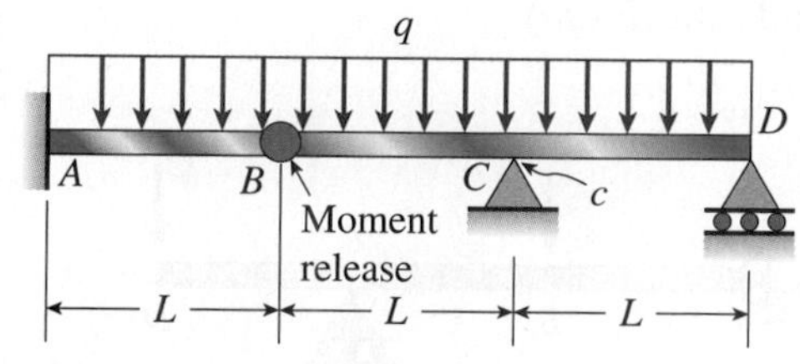

PROBLEM 9.5-32

9.5-33 Find the horizontal deflection δ_h and vertical deflection δ_v at the free end C of the frame ABC shown in the figure. (The flexural rigidity EI is constant throughout the frame.)

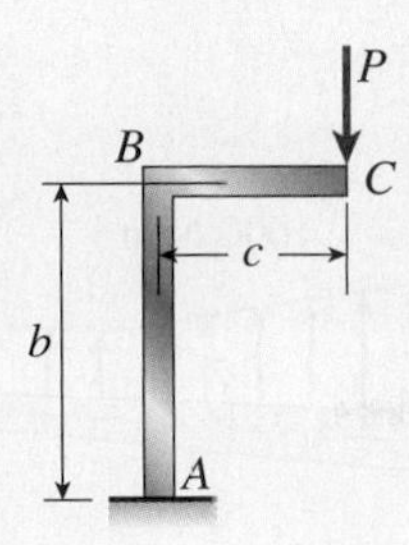

PROBLEM 9.5-33

Note: Disregard the effects of axial deformations and consider only the effects of bending due to the load P.

9.5-34 The frame $ABCD$ shown in the figure is squeezed by two collinear forces P acting at points A and D. What is the decrease δ in the distance between points A and D when the loads P are applied? (The flexural rigidity EI is constant throughout the frame.)

Note: Disregard the effects of axial deformations and consider only the effects of bending due to the loads P.

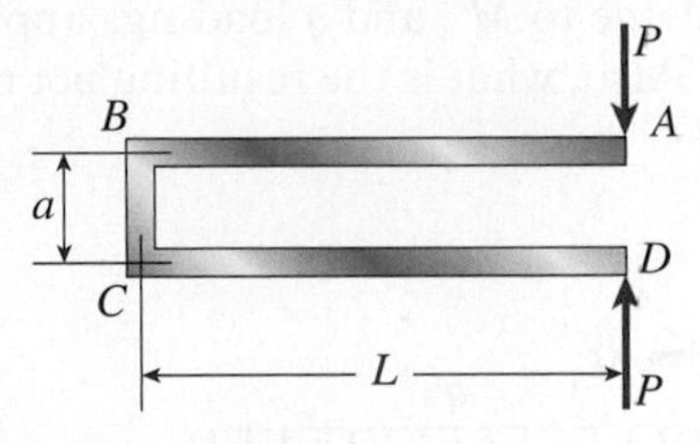

PROBLEM 9.5-34

9.5-35 A framework $ABCD$ is acted on by counterclockwise moment M at A (see figure). Assume that EI is constant.

(a) Find expressions for reactions at supports B and C.
(b) Find expressions for angles of rotation at A, B, C, and D.
(c) Find expressions for horizontal deflections δ_A and δ_D.
(d) If length $L_{AB} = L/2$, find length L_{CD} in terms of L for the absolute value of the ratio $|\delta_A / \delta_D| = 1$.

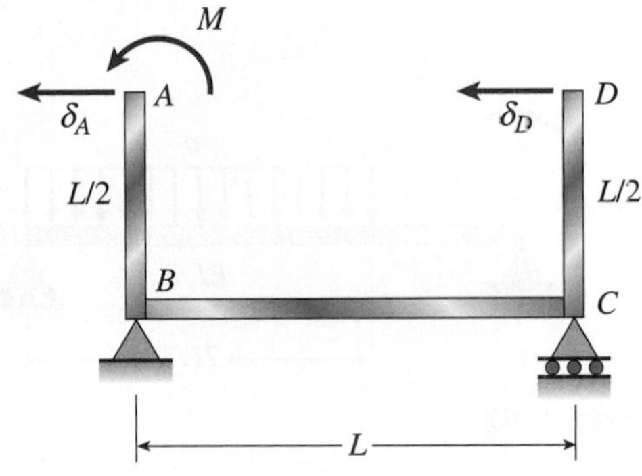

PROBLEM 9.5-35

9.5-36 A framework $ABCD$ is acted on by force P at $2L/3$ from B (see figure). Assume that EI is constant.

(a) Find expressions for reactions at supports B and C.
(b) Find expressions for angles of rotation at A, B, C, and D.
(c) Find expressions for horizontal deflections δ_A and δ_D.
(d) If length $L_{AB} = L/2$, find length L_{CD} in terms of L for the absolute value of the ratio $|\delta_A / \delta_D| = 1$.

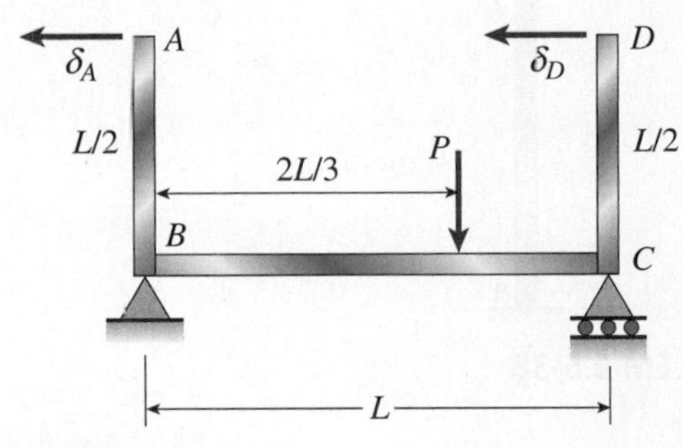

PROBLEM 9.5-36

9.5-37 A beam $ABCDE$ has simple supports at B and D and symmetrical overhangs at each end (see figure). The center span has length L and each overhang has length b. A uniform load of intensity q acts on the beam.

(a) Determine the ratio b/L so that the deflection δ_C at the midpoint of the beam is equal to the deflections δ_A and δ_E at the ends.
(b) For this value of b/L, what is the deflection δ_C at the midpoint?

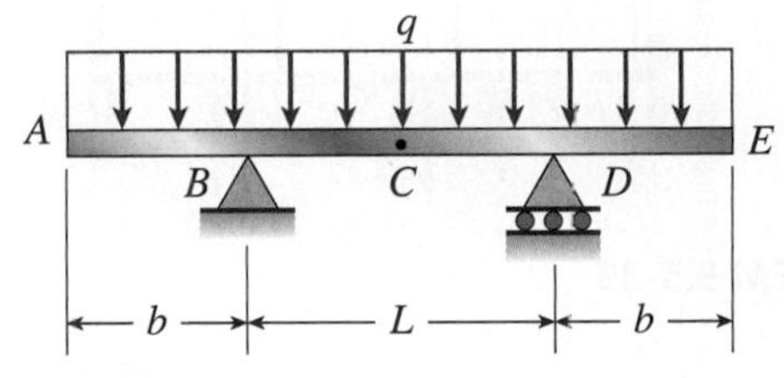

PROBLEM 9.5-37

9.5-38 A frame ABC is loaded at point C by a force P acting at an angle α to the horizontal (see figure). Both members of the frame have the same length and the same flexural rigidity.

Determine the angle α so that the deflection of point C is in the same direction as the load. (Disregard the effects of axial deformations and consider only the effects of bending due to the load P.)

Note: A direction of loading such that the resulting deflection is in the same direction as the load is called a *principal direction*. For a given load on a planar structure, there are two principal directions that are perpendicular to each other.

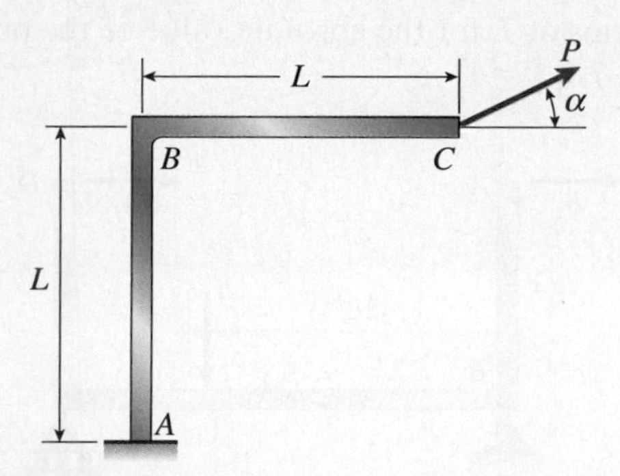

PROBLEM 9.5-38

9.5-39 The wing of a large commercial jet is represented by a simplified prismatic cantilever beam model with uniform load w and concentrated loads P at the two engine locations (see figure). Find expressions for the tip deflection and rotation at D in terms of w, P, L, and EI.

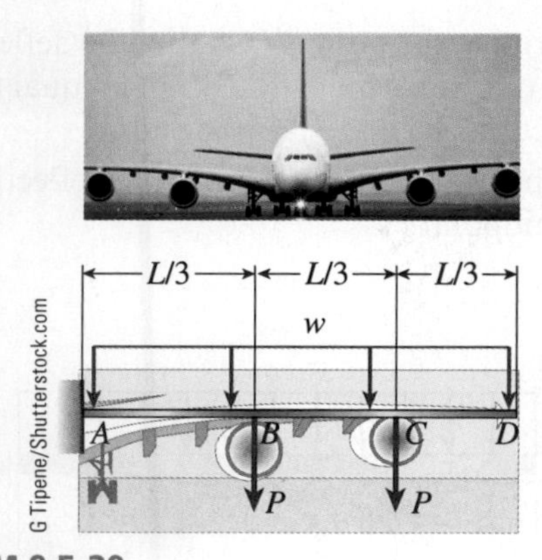

PROBLEM 9.5-39

9.5-40 The wing of a small plane is represented by a simplified prismatic cantilever beam model acted on by the distributed loads shown in the figure. Assume constant $EI = 1200\ \text{kN} \cdot \text{m}^2$. Find the tip deflection and rotation at B.

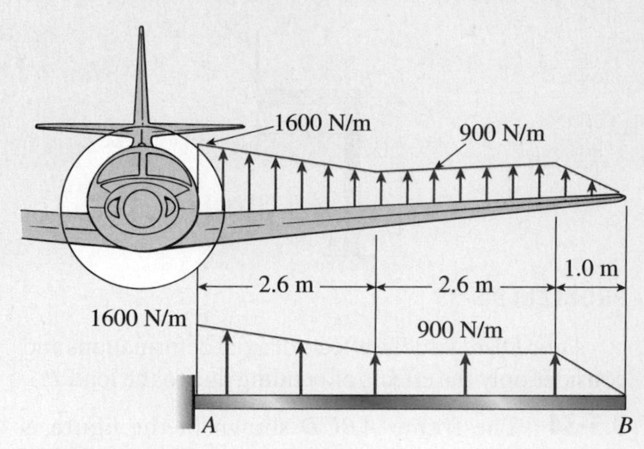

PROBLEM 9.5-40

9.5-41 Find an expression for required moment M_A (in terms of q and L) that will result in rotation $\theta_B = 0$ due to M_A and q loadings applied at the same time. Also, what is the resulting net rotation at support A?

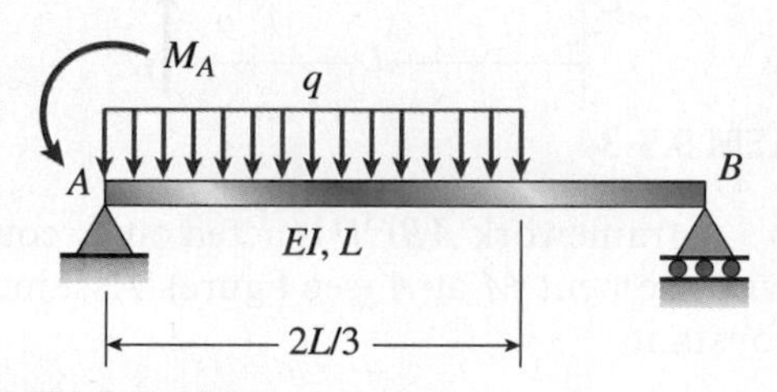

PROBLEM 9.5-41

9.5-42 Find an expression for required moment M_A (in terms of q and L) that will result in rotation $\theta_B = 0$ due to M_A and q loadings applied at the same time. Also, what is the resulting net rotation at support A?

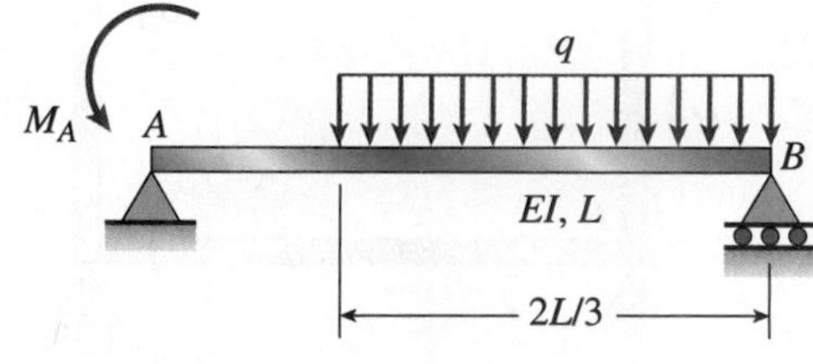

PROBLEM 9.5-42

9.5-43 Find required distance d (in terms of L) so that rotation $\theta_B = 0$ is due to M and q loadings applied at the same time. Also, what is the resulting net rotation θ_A at support A? Moment M is applied at distance d from joint B.

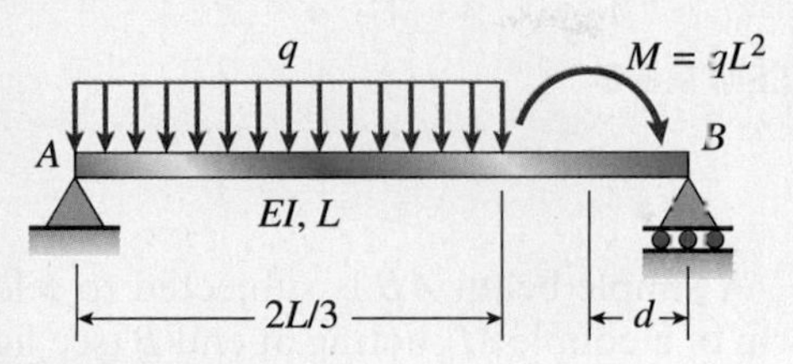

PROBLEM 9.5-43

9.5-44 A cantilever beam has two triangular loads as shown in the figure.

(a) Find an expression for beam deflection δ_C using superposition.

(b) Find the required magnitude of load intensity q_2 in terms of q_1 so that the deflection at C is zero.

(c) Find an expression for the deflection at C if both load intensities, q_1 and q_2, are equal to q_0.

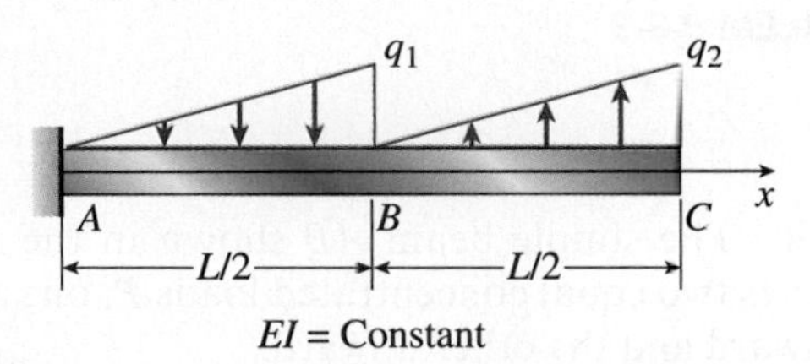

PROBLEM 9.5-44

9.6 Moment-Area Method

Solve the problems for Section 9.6 by the moment-area method. All beams have constant flexural rigidity EI.

Introductory Problems

9.6-1 A cantilever beam AB is subjected to a uniform load of intensity q acting throughout its length (see figure). Determine the angle of rotation θ_B and the deflection δ_B at the free end.

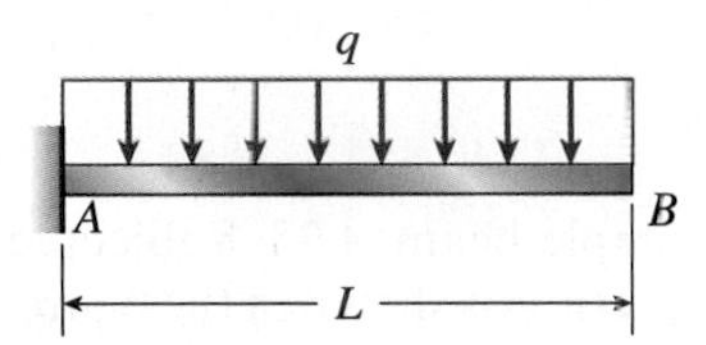

PROBLEM 9.6-1

9.6-2 The load on a cantilever beam AB has a triangular distribution with maximum intensity q_0 (see figure). Determine the angle of rotation θ_B and the deflection δ_B at the free end.

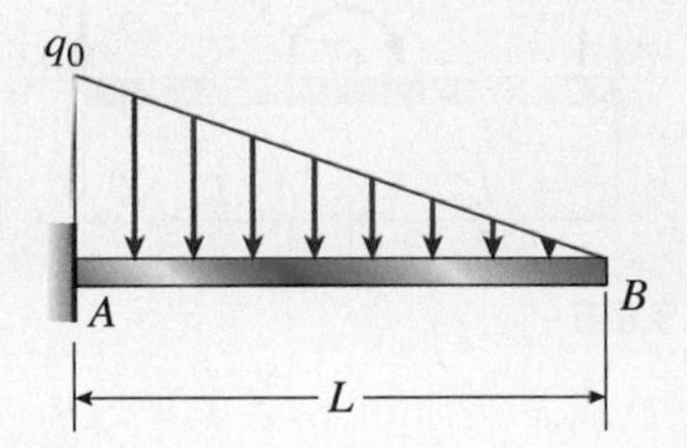

PROBLEM 9.6-2

9.6-3 A cantilever beam AB is subjected to a concentrated load P and a couple M_0 acting at the free end (see figure).

Obtain formulas for the angle of rotation θ_B and the deflection δ_B at end B.

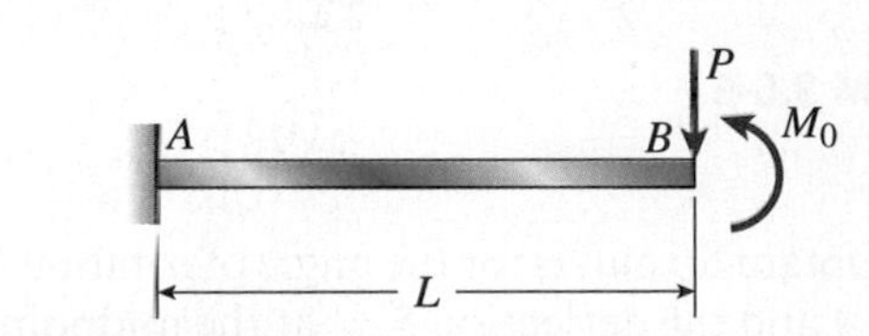

PROBLEM 9.6-3

Representative Problems

9.6-4 Determine the angle of rotation θ_B and the deflection δ_B at the free end of a cantilever beam AB with a uniform load of intensity q acting over the middle third of the length (see figure).

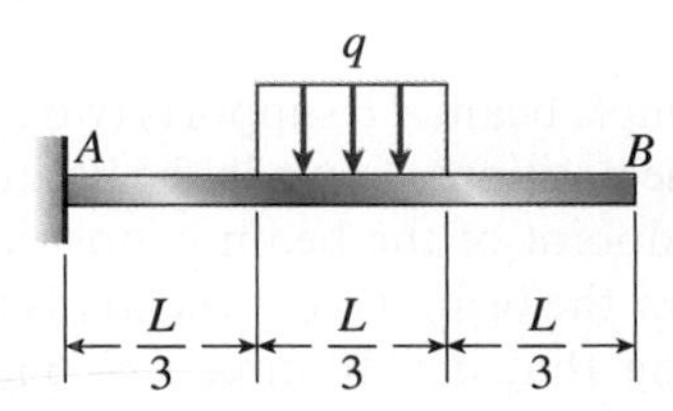

PROBLEM 9.6-4

9.6-5 Calculate the deflections δ_B and δ_C at points B and C, respectively, of the cantilever beam ACB shown in the figure. Assume $M_0 = 4$ kN·m, $P = 16$ kN, $L = 2.4$ m, and $EI = 6.0$ MN·m^2.

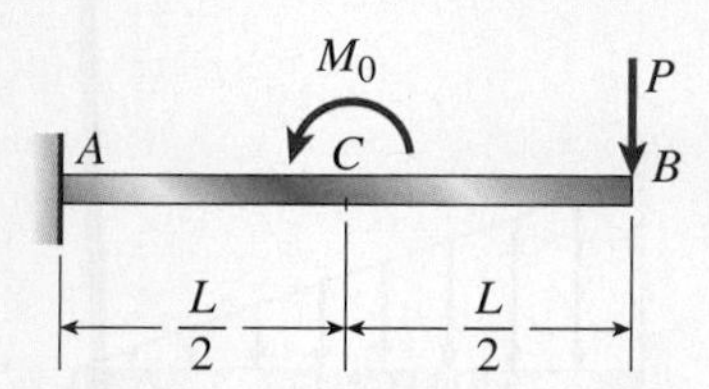

PROBLEM 9.6-5

9.6-6 A cantilever beam ACB supports two concentrated loads P_1 and P_2, as shown in the figure.

Calculate the deflections δ_B and δ_C at points B and C, respectively. Assume $P_1 = 10$ kN, $P_2 = 5$ kN, $L = 2.6$ m, $E = 200$ GPa, and $I = 20.1 \times 10^6$ mm^4.

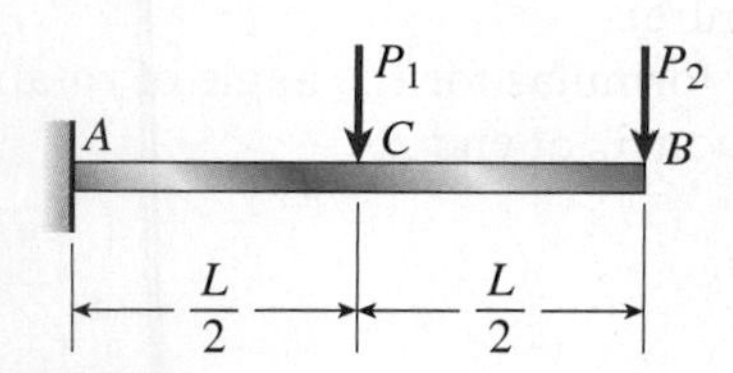

PROBLEM 9.6-6

9.6-7 Obtain formulas for the angle of rotation θ_A at support A and the deflection δ_{max} at the midpoint for a simple beam AB with a uniform load of intensity q (see figure).

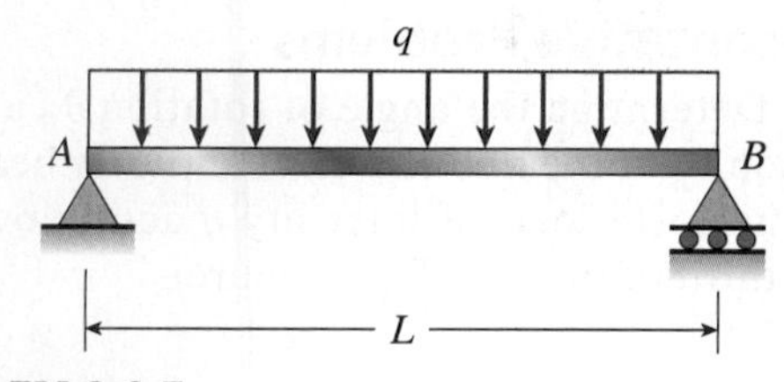

PROBLEM 9.6-7

9.6-8 A simple beam AB supports two concentrated loads P at the positions shown in the figure. A support C at the midpoint of the beam is positioned at distance d below the beam before the loads are applied.

Assuming that $d = 10$ mm, $L = 6$ m, $E = 200$ GPa, and $I = 198 \times 10^6$ mm^4, calculate the magnitude of the loads P so that the beam just touches the support at C.

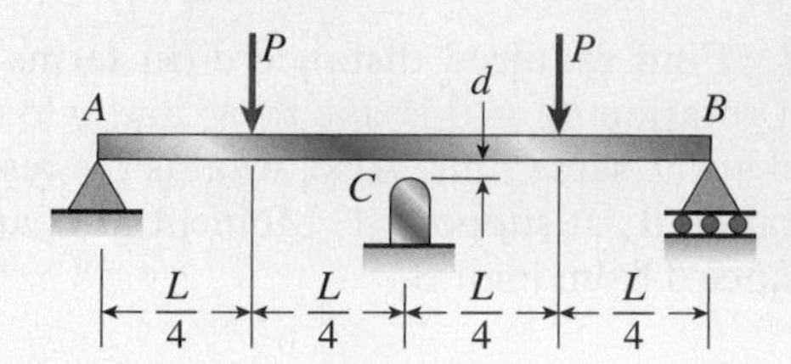

PROBLEM 9.6-8

9.6-9 A simple beam AB is subjected to a load in the form of a couple M_0 acting at end B (see figure). Determine the angles of rotation θ_A and θ_B at the supports and the deflection δ at the midpoint.

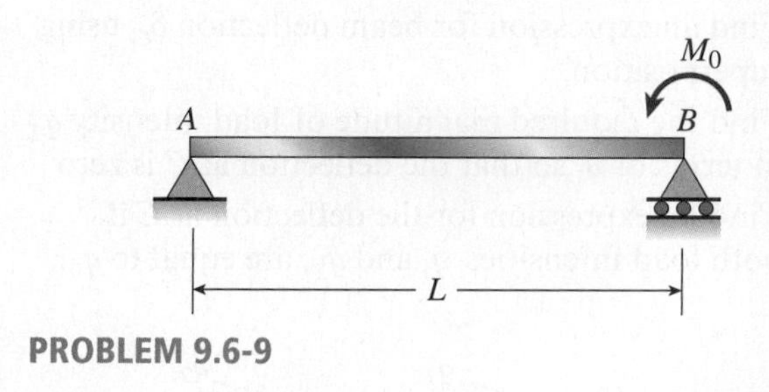

PROBLEM 9.6-9

9.6-10 The simple beam AB shown in the figure supports two equal concentrated loads P: one acting downward and the other upward.

Determine the angle of rotation θ_A at the left-hand end, the deflection δ_1 under the downward load, and the deflection δ_2 at the midpoint of the beam.

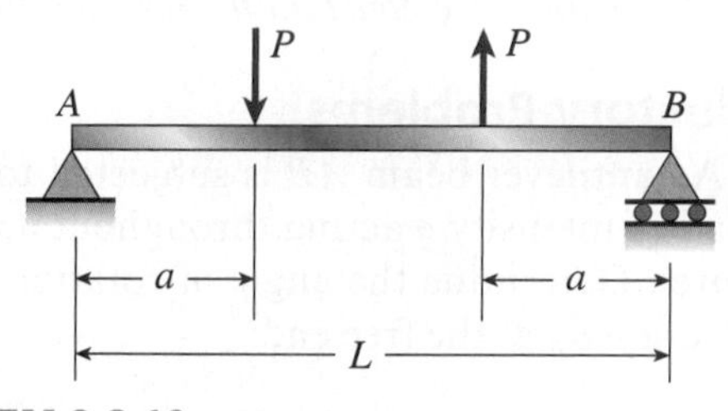

PROBLEM 9.6-10

9.6-11 A simple beam AB is subjected to couples M_0 and $2M_0$ acting as shown in the figure. Determine the angles of rotation θ_A and θ_B at the ends of the beam and the deflection δ at point D where the load M_0 is applied.

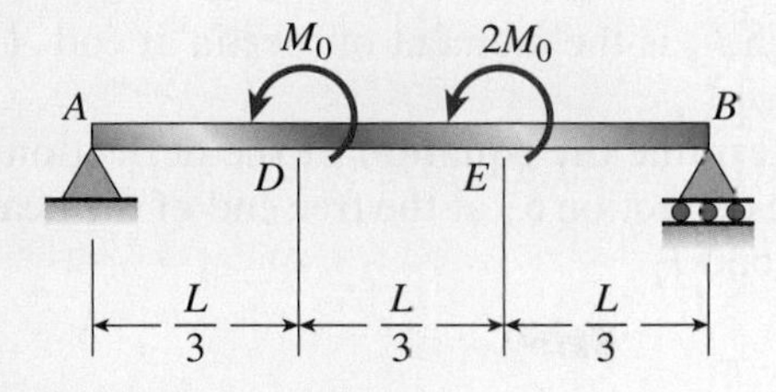

PROBLEM 9.6-11

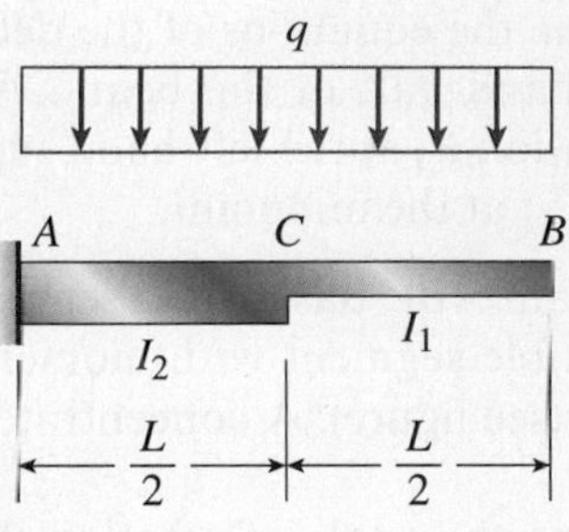

PROBLEM 9.7-2

9.7 Nonprismatic Beams

Introductory Problems

9.7-1 The cantilever beam ACB shown in the figure has moments of inertia I_2 and I_1 in parts AC and CB, respectively.

(a) Using the method of superposition, determine the deflection δ_B at the free end due to the load P.

(b) Determine the ratio r of the deflection δ_B to the deflection δ_1 at the free end of a prismatic cantilever with moment of inertia I_1 carrying the same load.

(c) Plot a graph of the deflection ratio r versus the ratio I_2/I_1 of the moments of inertia. (Let I_2/I_1 vary from 1 to 5.)

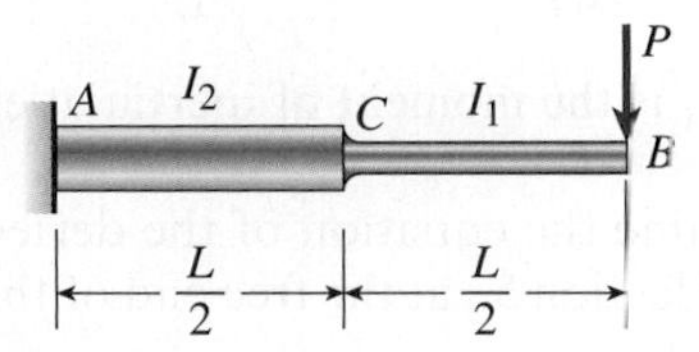

PROBLEM 9.7-1

9.7-2 The cantilever beam ACB shown in the figure supports a uniform load of intensity q throughout its length. The beam has moments of inertia I_2 and I_1 in parts AC and CB, respectively.

(a) Using the method of superposition, determine the deflection δ_B at the free end due to the uniform load.

(b) Determine the ratio r of the deflection δ_B to the deflection δ_1 at the free end of a prismatic cantilever with moment of inertia I_1 carrying the same load.

(c) Plot a graph of the deflection ratio r versus the ratio I_2/I_1 of the moments of inertia. (Let I_2/I_1 vary from 1 to 5.)

9.7-3 Beam ACB hangs from two springs, as shown in the figure. The springs have stiffnesses k_1 and k_2, and the beam has flexural rigidity EI.

(a) What is the downward displacement of point C, which is at the midpoint of the beam, when the moment M_0 is applied? Data for the structure are $M_0 = 3.0$ kN · m, $L = 2.5$ m, $EI = 200$ kN·m², $k_1 = 140$ kN/m, and $k_2 = 110$ kN/m.

(b) Repeat part (a), but remove M_0 and instead apply uniform load q over the entire beam.

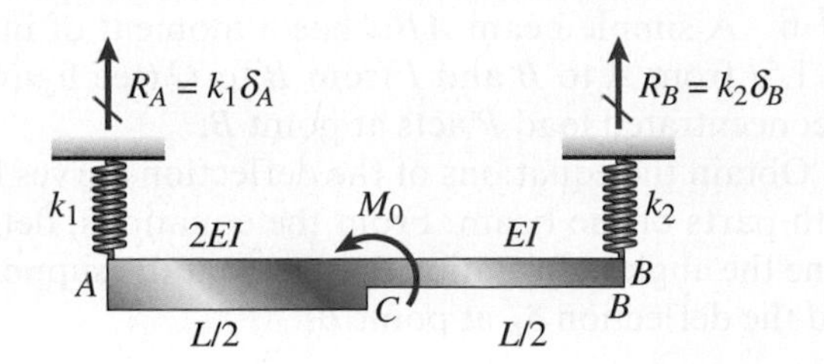

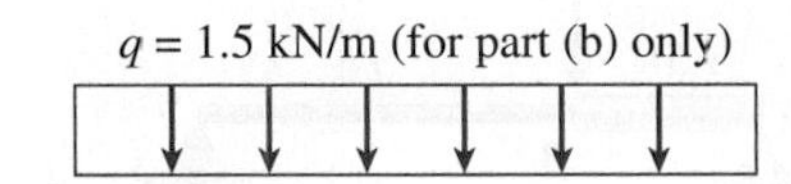

PROBLEM 9.7-3

9.7-4 A simple beam $ABCD$ has moment of inertia I near the supports and moment of inertia $2I$ in the middle region, as shown in the figure. A uniform load of intensity q acts over the entire length of the beam.

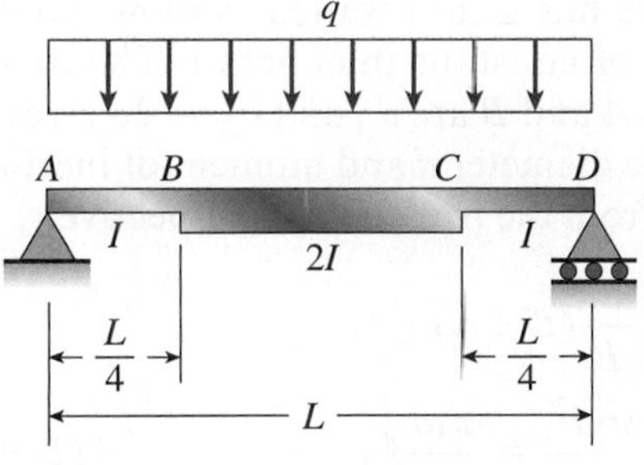

PROBLEM 9.7-4

Determine the equations of the deflection curve for the left-hand half of the beam. Also, find the angle of rotation θ_A at the left-hand support and the deflection δ_{max} at the midpoint.

9.7-5 A beam ABC has a rigid segment from A to B and a flexible segment with moment of inertia I from B to C (see figure). A concentrated load P acts at point B.

Determine the angle of rotation θ_A of the rigid segment, the deflection δ_B at point B, and the maximum deflection δ_{max}.

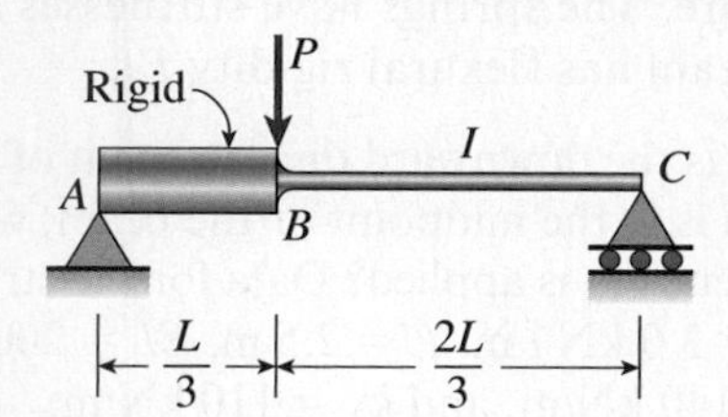

PROBLEM 9.7-5

9.7-6 A simple beam ABC has a moment of inertia $1.5I$ from A to B and I from B to C (see figure). A concentrated load P acts at point B.

Obtain the equations of the deflection curves for both parts of the beam. From the equations, determine the angles of rotation θ_A and θ_C at the supports and the deflection δ_B at point B.

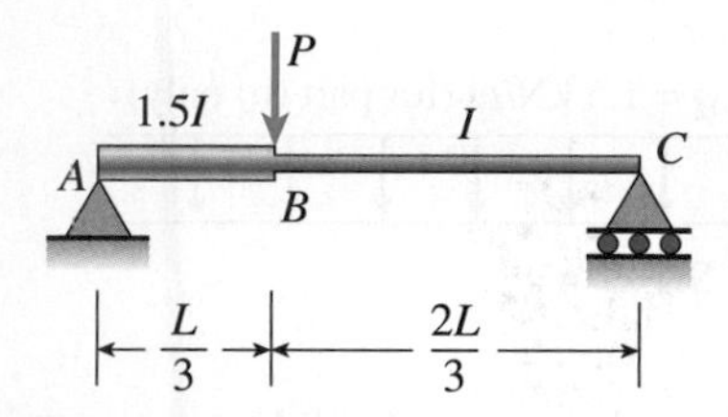

PROBLEM 9.7-6

Representative Problems

9.7-7 The tapered cantilever beam AB shown in the figure has a thin-walled, hollow circular cross sections of constant thickness t. The diameters at the ends A and B are d_A and $d_B = 2d_A$, respectively. Thus, the diameter d and moment of inertia I at distance x from the free end are, respectively,

$$d = \frac{d_A}{L}(L + x)$$

$$I = \frac{\pi t d^3}{8} = \frac{\pi t d_A^3}{8L^3}(L + x)^3 = \frac{I_A}{L^3}(L + x)^3$$

in which I_A is the moment of inertia at end A of the beam.

Determine the equation of the deflection curve and the deflection δ_A at the free end of the beam due to the load P.

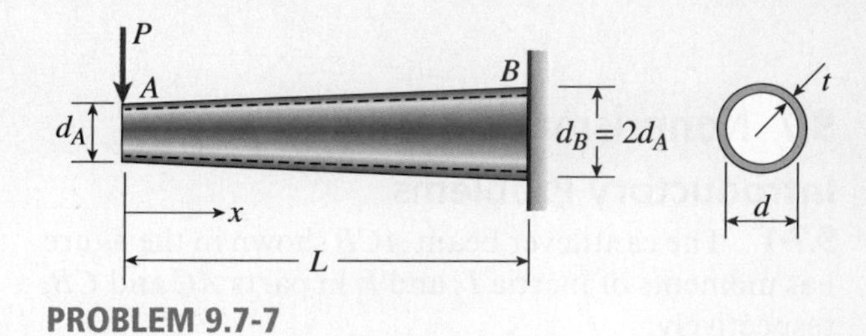

PROBLEM 9.7-7

9.7-8 The tapered cantilever beam AB shown in the figure has a solid circular cross section. The diameters at the ends A and B are d_A and $d_B = 2d_A$, respectively. Thus, the diameter d and moment of inertia I at distance x from the free end are, respectively,

$$d = \frac{d_A}{L}(L + x)$$

$$I = \frac{\pi d^4}{64} = \frac{\pi d_A^4}{64L^4}(L + x)^4 = \frac{I_A}{L^4}(L + x)^4$$

in which I_A is the moment of inertia at end A of the beam.

Determine the equation of the deflection curve and the deflection δ_A at the free end of the beam due to the load P.

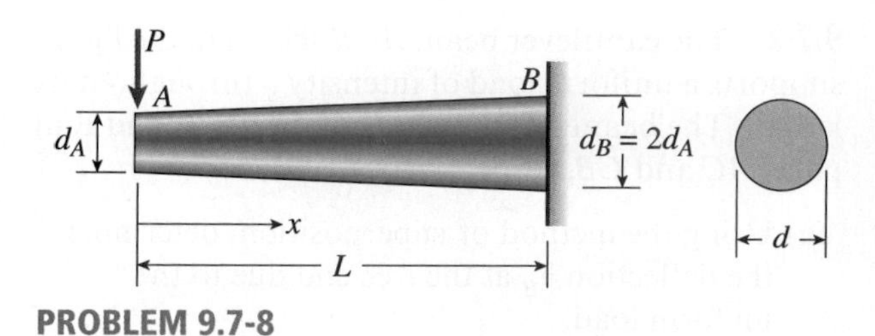

PROBLEM 9.7-8

9.7-9 A tapered cantilever beam AB supports a concentrated load P at the free end (see figure). The cross sections of the beam are rectangular with constant width b, depth d_A at support A, and depth $d_B = 3d_A/2$

at the support. Thus, the depth d and moment of inertia I at distance x from the free end are, respectively,

$$d = \frac{d_A}{2L}(2L + x)$$

$$I = \frac{bd^3}{12} = \frac{bd_A^3}{96L^3}(2L + x)^3 = \frac{I_A}{8L^3}(2L + x)^3$$

in which I_A is the moment of inertia at end A of the beam.

Determine the equation of the deflection curve and the deflection δ_A at the free end of the beam due to the load P.

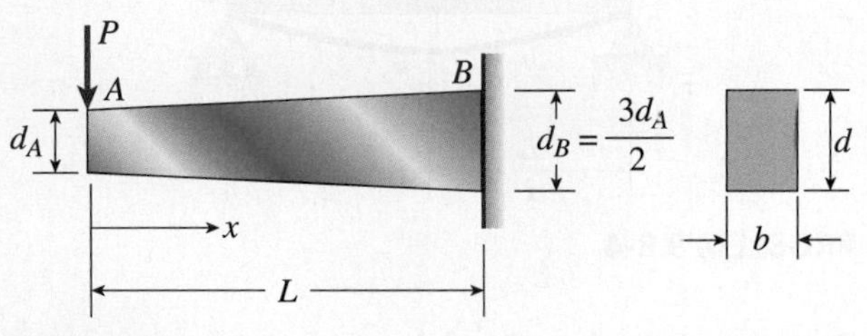

PROBLEM 9.7-9

9.7-10 A tapered cantilever beam AB supports a concentrated load P at the free end (see figure). The cross sections of the beam are rectangular tubes with constant width b, *outer tube* depth d_A at A, and *outer tube* depth $d_B = 3d_A/2$ at support B. The tube thickness is constant, as $t = d_A/20$. I_A is the moment of inertia of the *outer tube* at end A of the beam.

If the moment of inertia of the tube is approximated as $I_a(x)$ as defined, find the *equation* of the deflection curve and the deflection δ_A at the free end of the beam due to the load P.

$$I_a(x) = I_A \times \left(\frac{3}{4} + \frac{10\,x}{27L}\right)^3 \quad I_A = \frac{b\,d^3}{12}$$

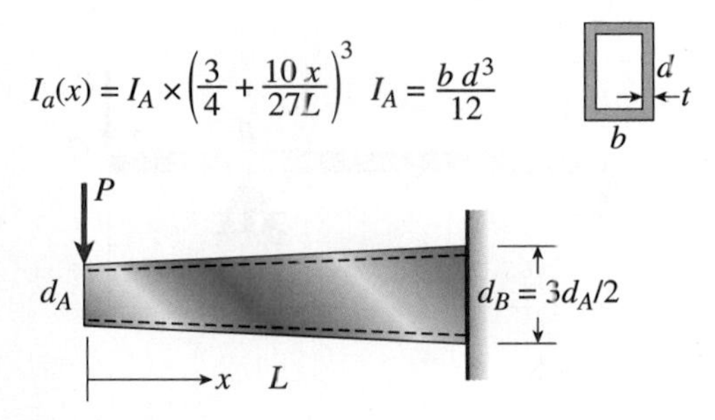

PROBLEM 9.7-10

9.7-11 Repeat Problem 9.7-10, but now use the tapered propped cantilever tube AB with sliding support at B (see figure) that supports a concentrated load P at the sliding end.

Find the equation of the deflection curve and the deflection δ_B at the sliding end of the beam due to the load P.

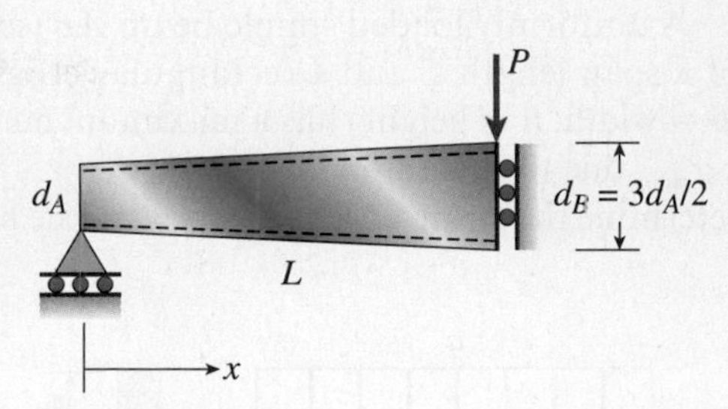

PROBLEM 9.7-11

9.7-12 A simple beam ACB is constructed with square cross sections and a double taper (see figure). The depth of the beam at the supports is d_A and at the midpoint is $d_C = 2d_A$. Each half of the beam has length L. Thus, the depth d and moment of inertia I at distance x from the left-hand end are, respectively,

$$d = \frac{d_A}{L}(L + x)$$

$$I = \frac{d^4}{12} = \frac{d_A^4}{12L^4}(L + x)^4 = \frac{I_A}{L^4}(L + x)^4$$

in which I_A is the moment of inertia at end A of the beam. (These equations are valid for x between 0 and L, that is, for the left-hand half of the beam.)

(a) Obtain equations for the slope and deflection of the left-hand half of the beam due to the uniform load.

(b) From the equations in part (a), obtain formulas for the angle of rotation θ_A at support A and the deflection δ_C at the midpoint.

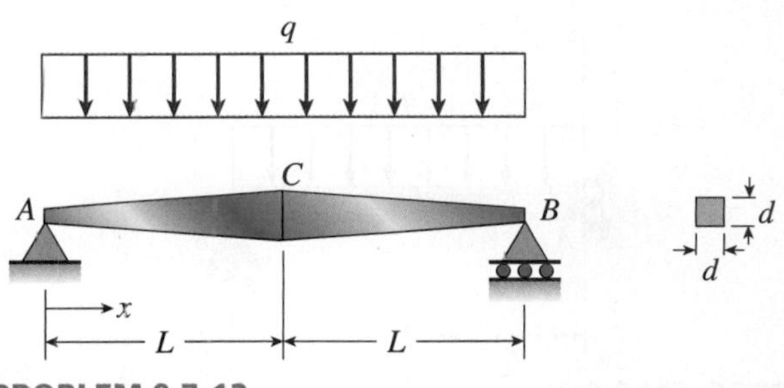

PROBLEM 9.7-12

9.8 Strain Energy of Bending

The beams described in the problems for Section 9.8 have constant flexural rigidity EI.

Introductory Problems

9.8-1 A uniformly loaded simple beam AB (see figure) of a span length L and a rectangular cross section (b = width, h = height) has a maximum bending stress σ_{max} due to the uniform load q.

Determine the strain energy U stored in the beam.

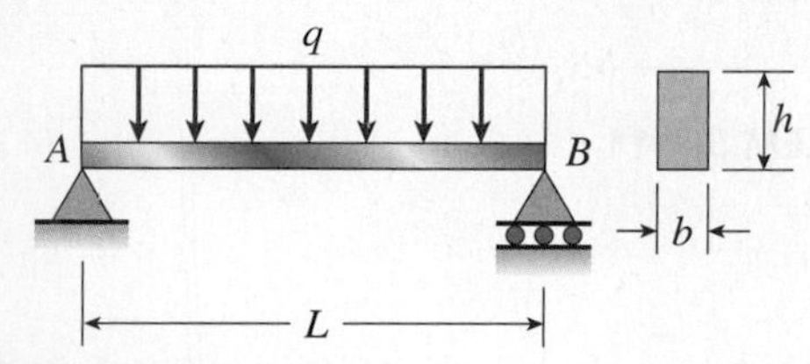

PROBLEM 9.8-1

9.8-2 A simple beam AB of length L supports a concentrated load P at the midpoint (see figure).

(a) Evaluate the strain energy of the beam from the bending moment in the beam.

(b) Evaluate the strain energy of the beam from the equation of the deflection curve.

(c) From the strain energy, determine the deflection δ under the load P.

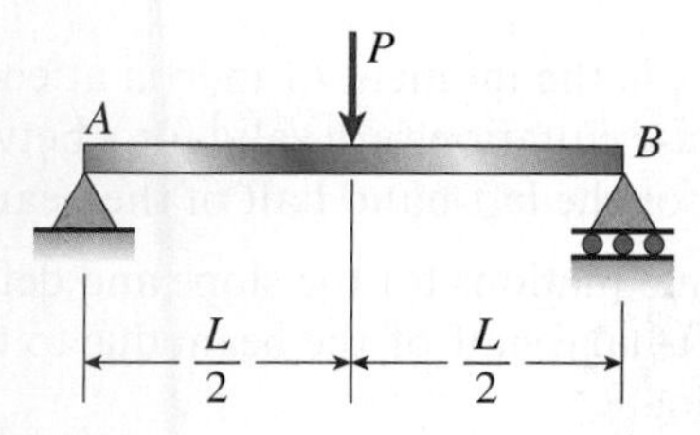

PROBLEM 9.8-2

9.8-3 A propped cantilever beam AB of length L and with a sliding support at A supports a uniform load of intensity q (see figure).

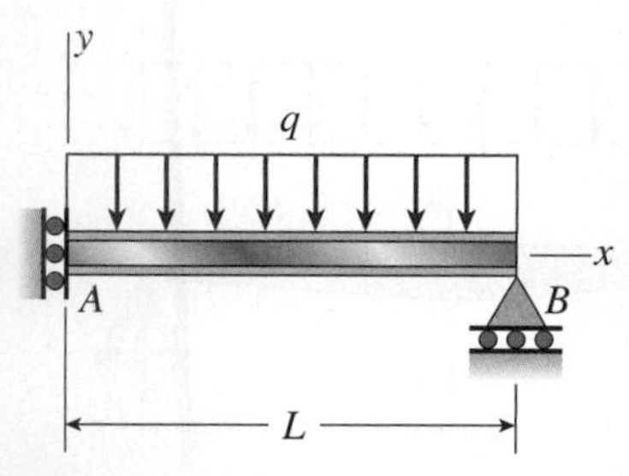

PROBLEM 9.8-3

(a) Evaluate the strain energy of the beam from the bending moment in the beam.

(b) Evaluate the strain energy of the beam from the equation of the deflection curve.

9.8-4 A simple beam AB of length L is subjected to loads that produce a symmetric deflection curve with maximum deflection δ at the midpoint of the span (see figure).

How much strain energy U is stored in the beam if the deflection curve is (a) a parabola and (b) a half wave of a sine curve?

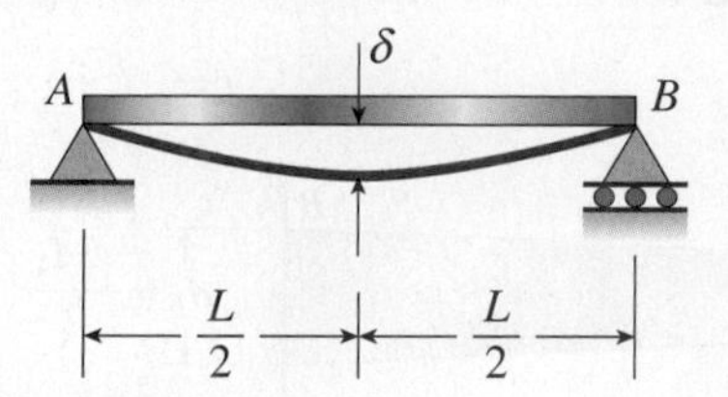

PROBLEM 9.8-4

Representative Problems

9.8-5 A beam ABC with simple supports at A and B and an overhang BC supports a concentrated load P at the free end C (see figure).

(a) Determine the strain energy U stored in the beam due to the load P.

(b) From the strain energy, find the deflection δ_C under the load P.

(c) Calculate the numerical values of U and δ_C if the length L is 2.0 m, the overhang length a is 1 m, the beam is an IPN 200 steel wide-flange section, and the load P produces a maximum stress of 105 MPa in the beam. (Use $E = 210$ GPa.)

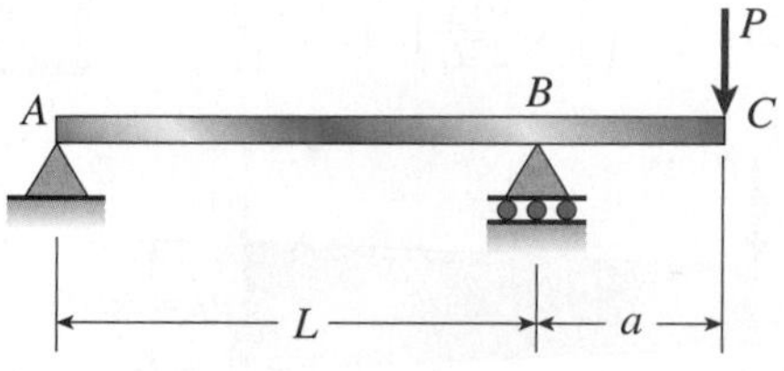

PROBLEM 9.8-5

9.8-6 A simple beam ACB supporting a uniform load q over the first half of the beam and a couple of moment M_0 at end B is shown in the figure.

Determine the strain energy U stored in the beam due to the load q and the couple M_0 acting simultaneously.

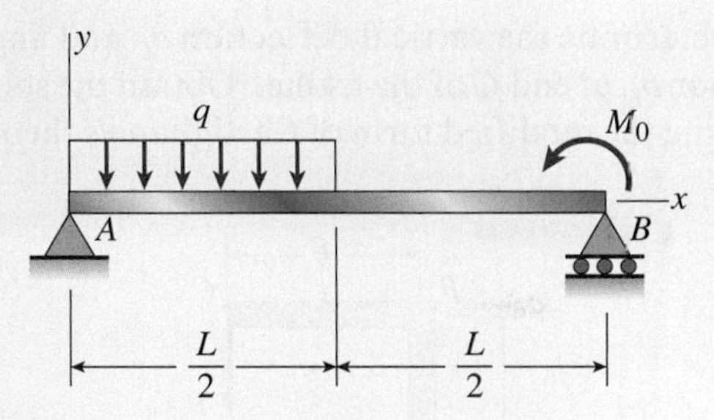

PROBLEM 9.8-6

9.8-7 The frame shown in the figure consists of a beam ACB supported by a strut CD. The beam has length $2L$ and is continuous through joint C. A concentrated load P acts at the free end B.

Determine the vertical deflection δ_B at point B due to the load P.

Note: Let EI denote the flexural rigidity of the beam, and let EA denote the axial rigidity of the strut. Disregard axial and shearing effects in the beam, and disregard any bending effects in the strut.

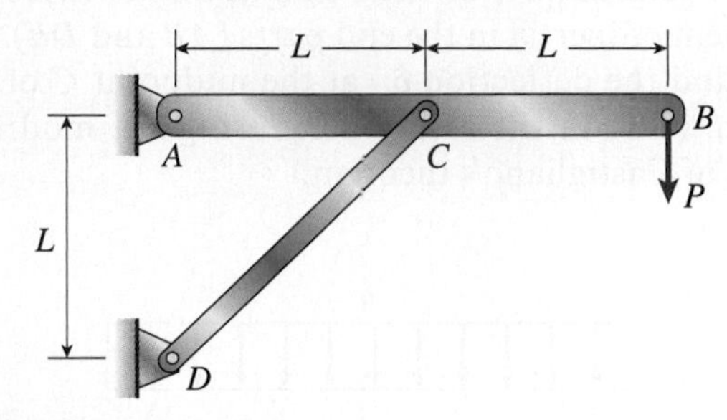

PROBLEM 9.8-7

9.9 Castigliano's Theorem

The beams described in the problems for Section 9.9 have constant flexural rigidity EI.

Introductory Problems

9.9-1 A simple beam AB of length L is loaded at the left-hand end by a couple of moment M_0 (see figure).

Determine the angle of rotation θ_A at support A. (Obtain the solution by determining the strain energy of the beam and then using Castigliano's theorem.)

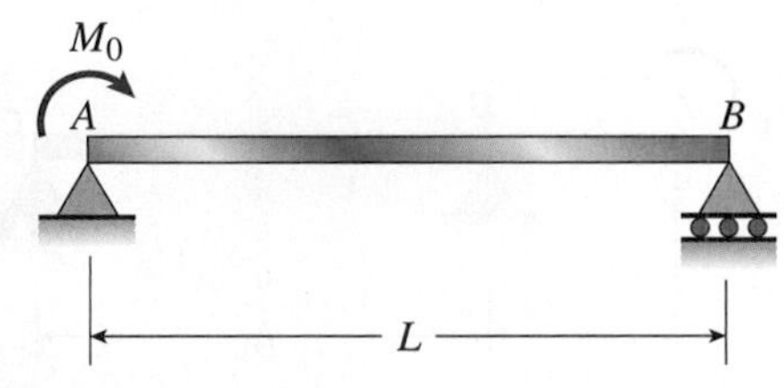

PROBLEM 9.9-1

9.9-2 The simple beam shown in the figure supports a concentrated load P acting at distance a from the left-hand support and distance b from the right-hand support.

Determine the deflection δ_D at point D where the load is applied. (Obtain the solution by determining the strain energy of the beam and then using Castigliano's theorem.)

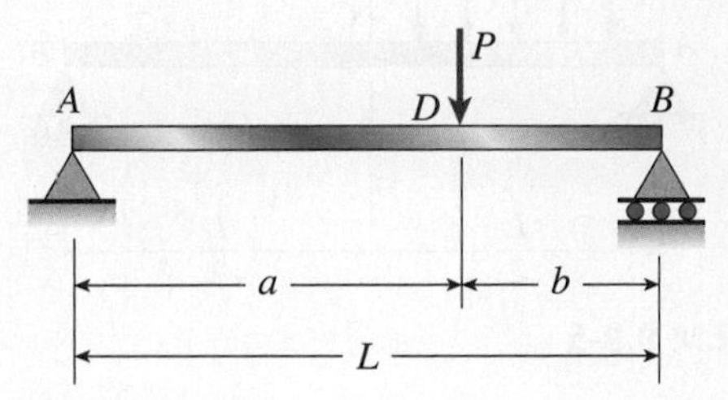

PROBLEM 9.9-2

9.9-3 An overhanging beam ABC supports a concentrated load P at the end of the overhang (see figure). Span AB has length L, and the overhang has length a.

Determine the deflection δ_C at the end of the overhang. (Obtain the solution by determining the strain energy of the beam and then using Castigliano's theorem.)

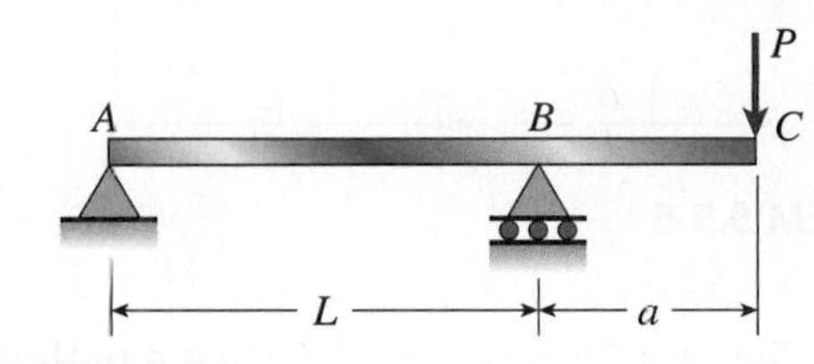

PROBLEM 9.9-3

9.9-4 The cantilever beam shown in the figure supports a triangularly distributed load of maximum intensity q_0.

Determine the deflection δ_B at the free end B. (Obtain the solution by determining the strain energy of the beam and then using Castigliano's theorem.)

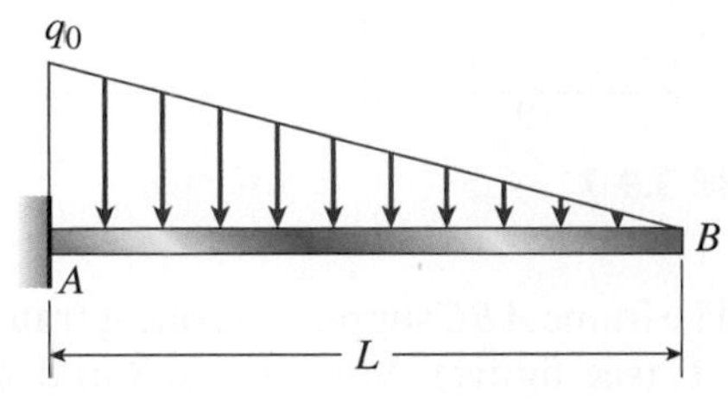

PROBLEM 9.9-4

9.9-5 A simple beam ACB supports a uniform load of intensity q on the left-hand half of the span (see figure).

Determine the angle of rotation θ_B at support B. (Obtain the solution by using the modified form of Castigliano's theorem.)

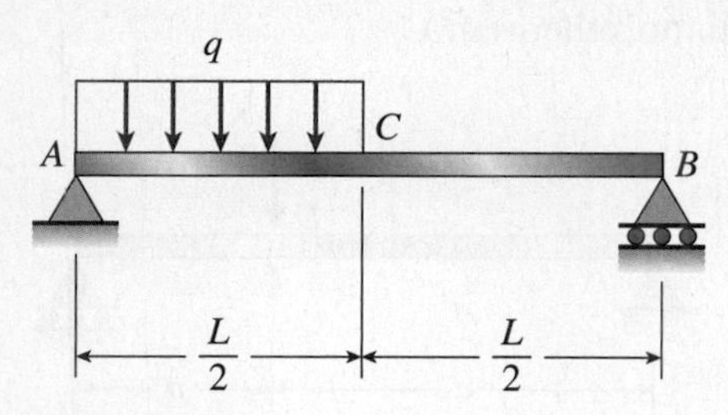

PROBLEM 9.9-5

Representative Problems

9.9-6 A cantilever beam ACB supports two concentrated loads P_1 and P_2, as shown in the figure. Determine the deflections δ_C and δ_B at points C and B, respectively. (Obtain the solution by using the modified form of Castigliano's theorem.)

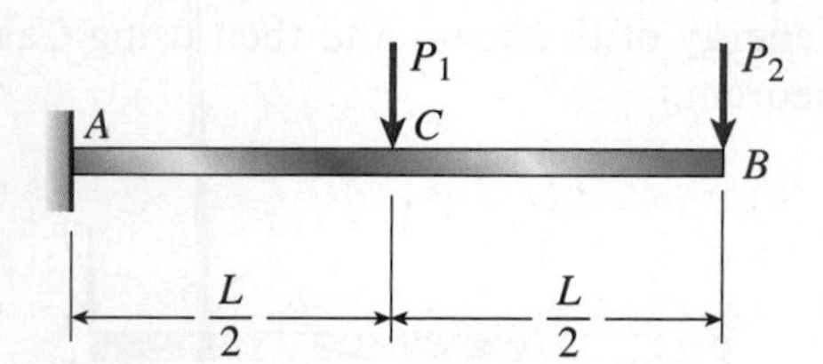

PROBLEM 9.9-6

9.9-7 The cantilever beam ACB shown in the figure is subjected to a uniform load of intensity q acting between points A and C.

Determine the angle of rotation θ_A at the free end A. (Obtain the solution by using the modified form of Castigliano's theorem.)

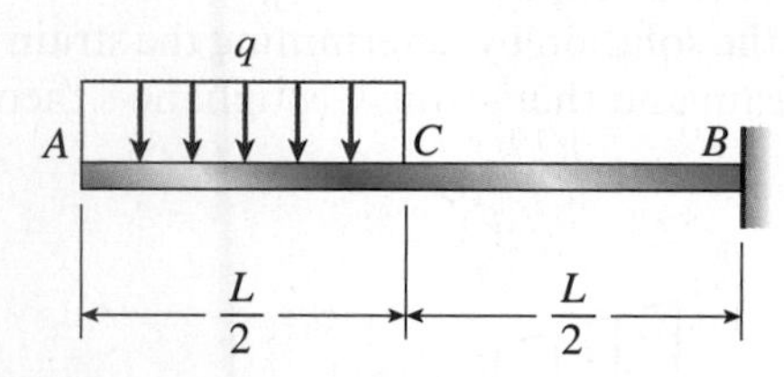

PROBLEM 9.9-7

9.9-8 The frame ABC supports a concentrated load P at point C (see figure). Members AB and BC have lengths h and b, respectively.

Determine the vertical deflection δ_C and angle of rotation θ_C at end C of the frame. (Obtain the solution by using the modified form of Castigliano's theorem.)

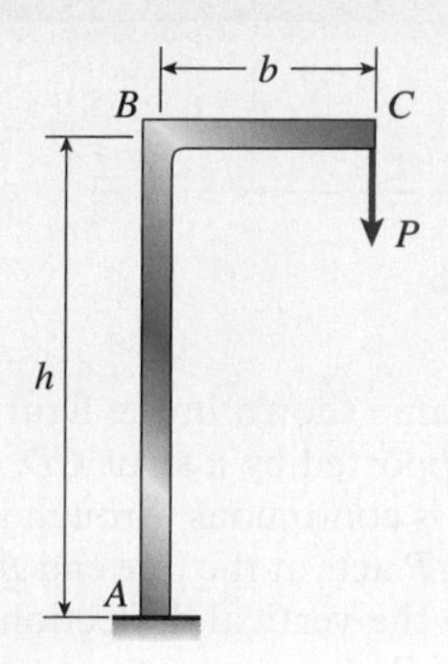

PROBLEM 9.9-8

9.9-9 A simple beam $ABCDE$ supports a uniform load of intensity q (see figure). The moment of inertia in the central part of the beam (BCD) is twice the moment of inertia in the end parts (AB and DE).

Find the deflection δ_C at the midpoint C of the beam. (Obtain the solution by using the modified form of Castigliano's theorem.)

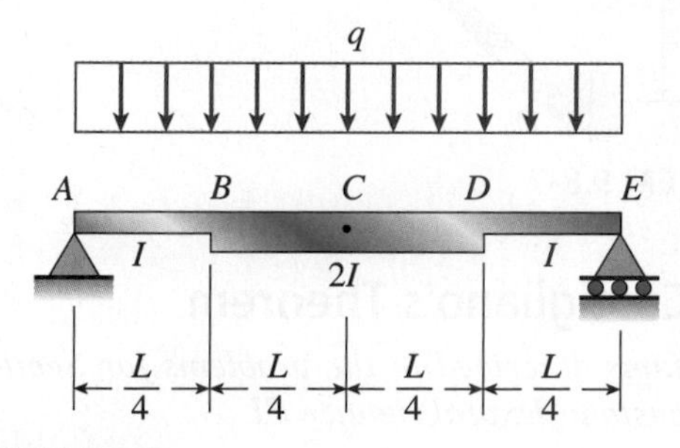

PROBLEM 9.9-9

9.9-10 An overhanging beam ABC is subjected to a couple M_A at the free end (see figure). The lengths of the overhang and the main span are a and L, respectively.

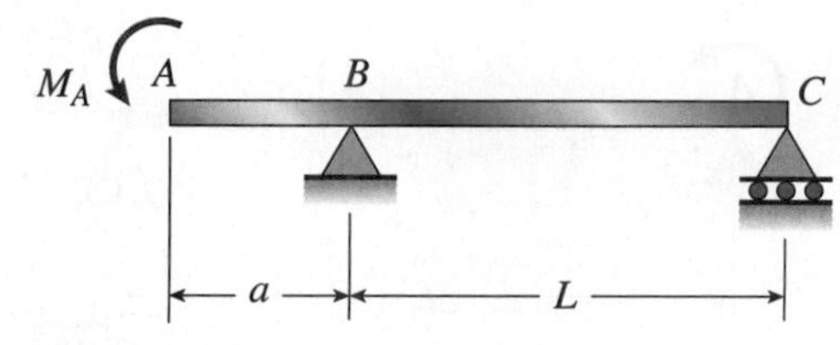

PROBLEM 9.9-10

Determine the angle of rotation θ_A and deflection δ_A at end A. (Obtain the solution by using the modified form of Castigliano's theorem.)

9.9-11 An overhanging beam ABC rests on a simple support at A and a spring support at B (see figure). A concentrated load P acts at the end of the overhang. Span AB has length L, the overhang has length a, and the spring has stiffness k.

Determine the downward displacement δ_C of the end of the overhang. (Obtain the solution by using the modified form of Castigliano's theorem.)

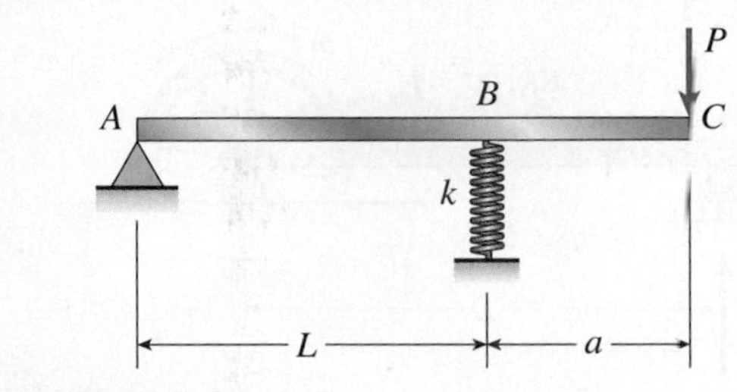

PROBLEM 9.9-11

9.9-12 A symmetric beam $ABCD$ with overhangs at both ends supports a uniform load of intensity q (see figure).

Determine the deflection δ_D at the end of the overhang. (Obtain the solution by using the modified form of Castigliano's theorem.)

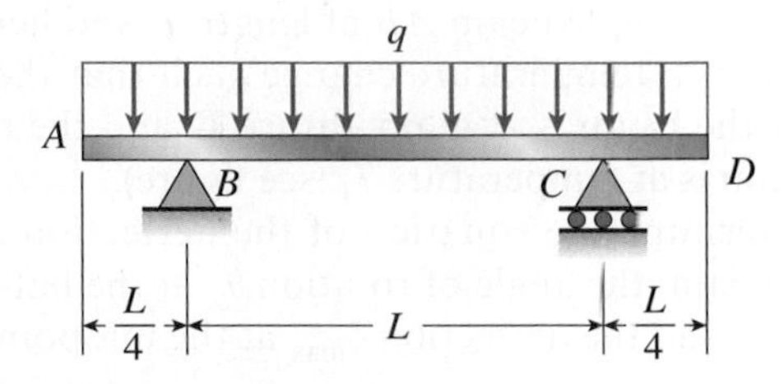

PROBLEM 9.9-12

9.10 Deflections Produced by Impact

The beams described in the problems for Section 9.10 have constant flexural rigidity EI. Disregard the weights of the beams themselves and consider only the effects of the given loads.

Introductory Problems

9.10-1 A heavy object of weight W is dropped onto the midpoint of a simple beam AB from a height h (see figure).

Obtain a formula for the maximum bending stress σ_{max} due to the falling weight in terms of h, σ_{st}, and δ_{st}, where σ_{st} is the maximum bending stress and δ_{st} is the deflection at the midpoint when the weight W acts on the beam as a statically applied load.

Plot a graph of the ratio σ_{max}/σ_{st} (that is, the ratio of the dynamic stress to the static stress) versus the ratio h/δ_{st}. (Let h/δ_{st} vary from 0 to 10.)

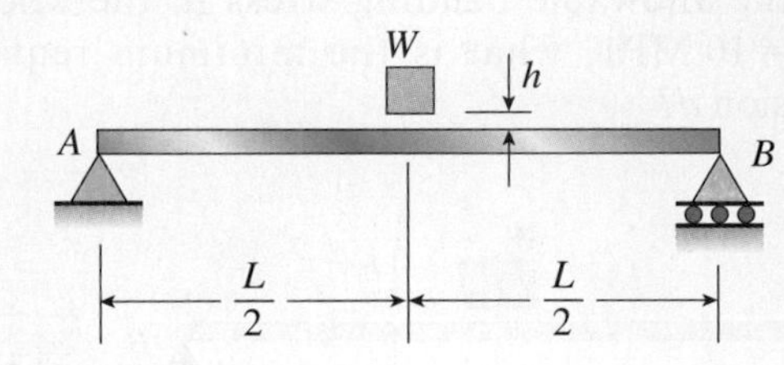

PROBLEM 9.10-1

9.10-2 An object of weight W is dropped onto the midpoint of a simple beam AB from a height h (see figure). The beam has a rectangular cross section of area A.

Assuming that h is very large compared to the deflection of the beam when the weight W is applied statically, obtain a formula for the maximum bending stress σ_{max} in the beam due to the falling weight.

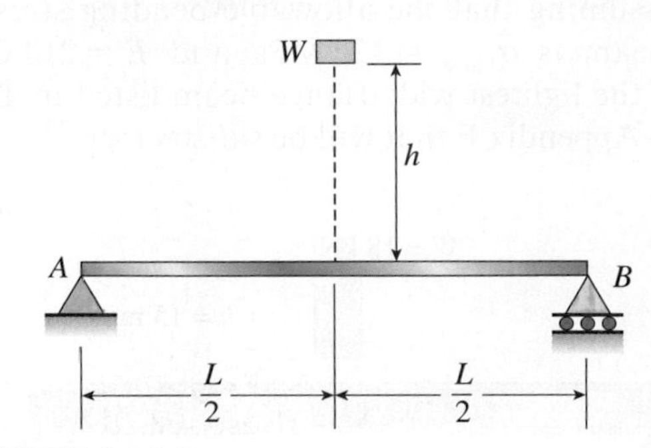

PROBLEM 9.10-2

9.10-3 A cantilever beam AB of length $L = 2$ m is constructed of an IPN 300 wide-flange section (see figure). A weight $W = 7$ kN falls through a height $h = 6$ mm onto the end of the beam.

Calculate the maximum deflection δ_{max} of the end of the beam and the maximum bending stress σ_{max} due to the falling weight. (Assume $E = 210$ GPa.)

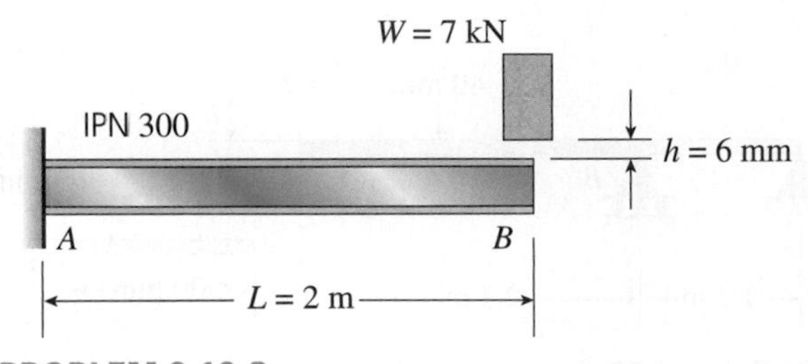

PROBLEM 9.10-3

Representative Problems

9.10-4 A weight $W = 20$ kN falls through a height $h = 1.0$ mm onto the midpoint of a simple beam of length $L = 3$ m (see figure). The beam is made of wood with square cross section (dimension d on each side) and $E = 12$ GPa.

If the allowable bending stress in the wood is $\sigma_{allow} = 10$ MPa, what is the minimum required dimension d?

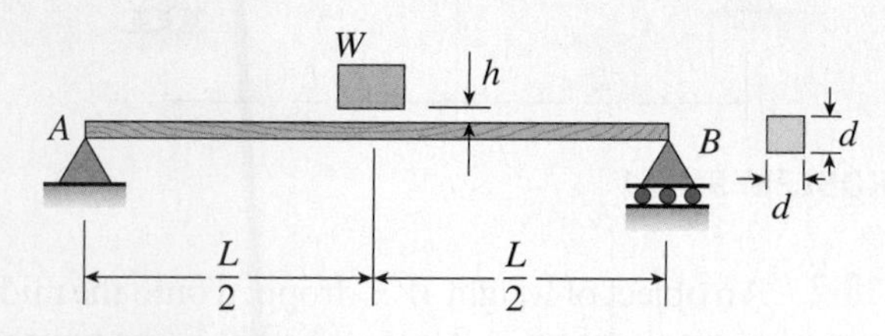

PROBLEM 9.10-4

9.10-5 A weight $W = 18$ kN falls through a height $h = 15$ mm onto the midpoint of a simple beam of length $L = 3$ m (see figure).

Assuming that the allowable bending stress in the beam is $\sigma_{allow} = 125$ MPa and $E = 210$ GPa, select the lightest wide-flange beam listed in Table F-1 in Appendix F that will be satisfactory.

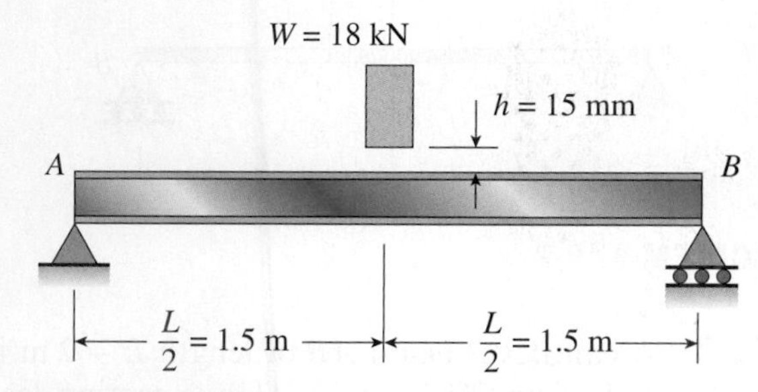

PROBLEM 9.10-5

9.10-6 An overhanging beam ABC with a rectangular cross section has the dimensions shown in the figure. A weight $W = 750$ N drops onto end C of the beam.

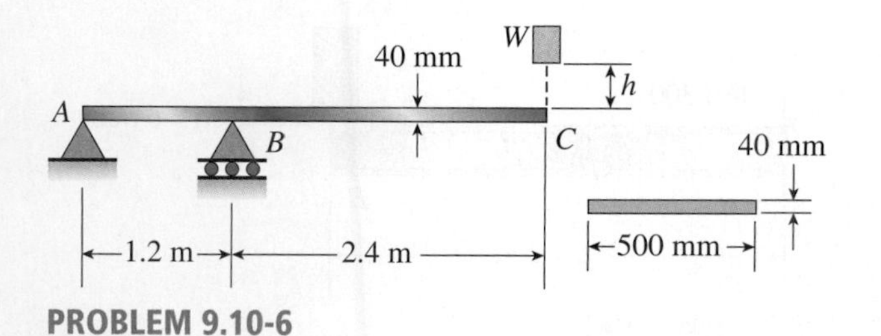

PROBLEM 9.10-6

If the allowable normal stress in bending is 45 MPa, what is the maximum height h from which the weight may be dropped? (Assume $E = 12$ GPa.)

9.10-7 A heavy flywheel rotates at an angular speed ω (radians per second) around an axle (see figure). The axle is rigidly attached to the end of a simply supported beam of flexural rigidity EI and length L (see figure). The flywheel has mass moment of inertia I_m about its axis of rotation.

If the flywheel suddenly freezes to the axle, what will be the reaction R at support A of the beam?

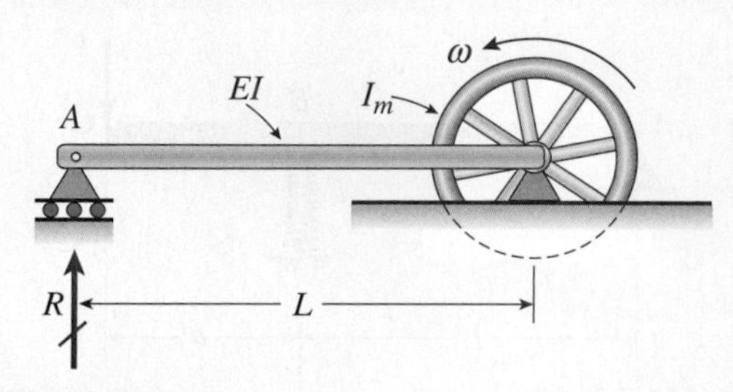

PROBLEM 9.10-7

9.11 Temperature Effects

The beams described in the problems for Section 9.11 have constant flexural rigidity EI. In every problem, the temperature varies linearly between the top and bottom of the beam.

Introductory Problems

9.11-1 A simple beam AB of length L and height h undergoes a temperature change such that the bottom of the beam is at temperature T_2 and the top of the beam is at temperature T_1 (see figure).

Determine the equation of the deflection curve of the beam, the angle of rotation θ_A at the left-hand support, and the deflection δ_{max} at the midpoint.

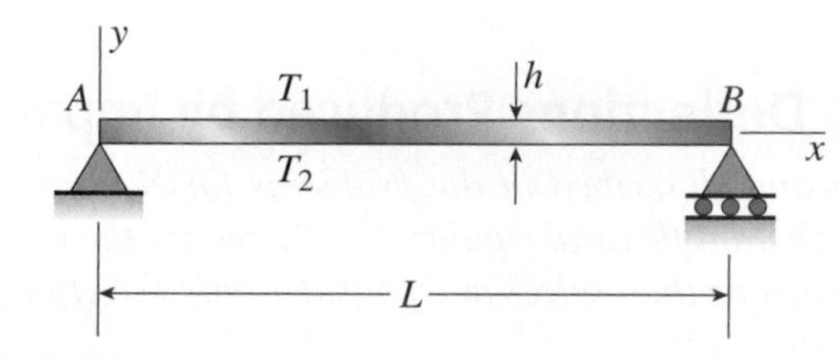

PROBLEM 9.11-1

9.11-2 A cantilever beam AB of length L and height h (see figure) is subjected to a temperature change such that the temperature at the top is T_1 and at the bottom is T_2.

Determine the equation of the deflection curve of the beam, the angle of rotation θ_B at end B, and the deflection δ_B at end B.

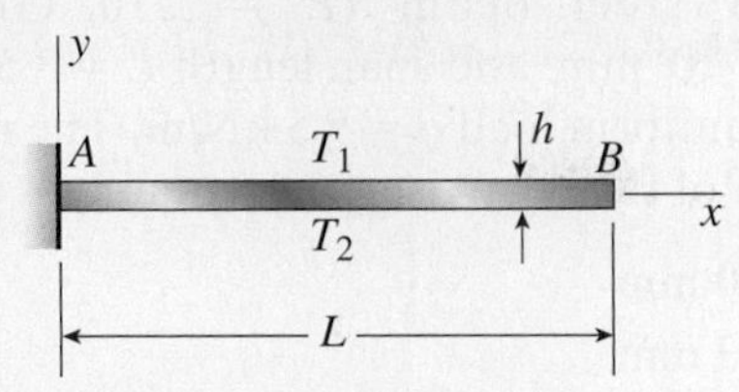

PROBLEM 9.11-2

Representative Problems

9.11-3 An overhanging beam ABC of height h has a sliding support at A and a roller at B. The beam is heated to a temperature T_1 on the top and T_2 on the bottom (see figure).

Determine the equation of the deflection curve of the beam, the angle of rotation θ_C at end C, and the deflection δ_C at end C.

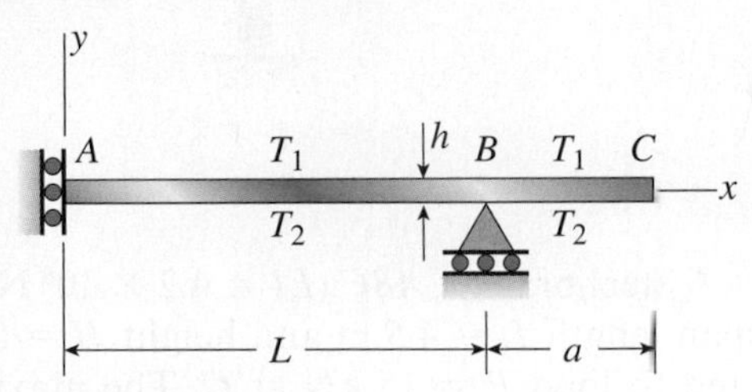

PROBLEM 9.11-3

9.11-4 A simple beam AB of length L and height h (see figure) is heated in such a manner that the temperature difference $T_2 - T_1$ between the bottom and top of the beam is proportional to the distance from support A; that is, assume the temperature difference varies *linearly* along the beam:

$$T_2 - T_1 = T_0 x$$

in which T_0 is a constant having units of temperature (degrees) per unit distance.

(a) Determine the maximum deflection δ_{max} of the beam.

(b) Repeat for a *quadratic* temperature variation along the beam, so $T_2 - T_1 = T_0 x^2$.

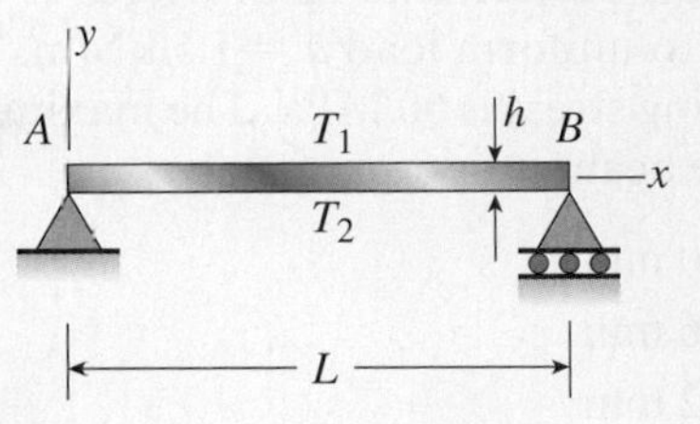

PROBLEM 9.11-4

9.11-5 Beam AB has an elastic support k_R at A, pin support at B, length L, height h (see figure), and is heated in such a manner that the temperature difference $T_2 - T_1$ between the bottom and top of the beam is proportional to the distance from support A. Assume the temperature difference varies *linearly* along the beam:

$$T_2 - T_1 = T_0 x$$

in which T_0 is a constant having units of temperature (degrees) per unit distance. Assume the spring at A is unaffected by the temperature change.

(a) Determine the maximum deflection δ_{max} of the beam.

(b) Repeat for a *quadratic* temperature variation along the beam, so $T_2 - T_1 = T_0 x^2$.

(c) What is δ_{max} for parts (a) and (b) if k_R goes to infinity?

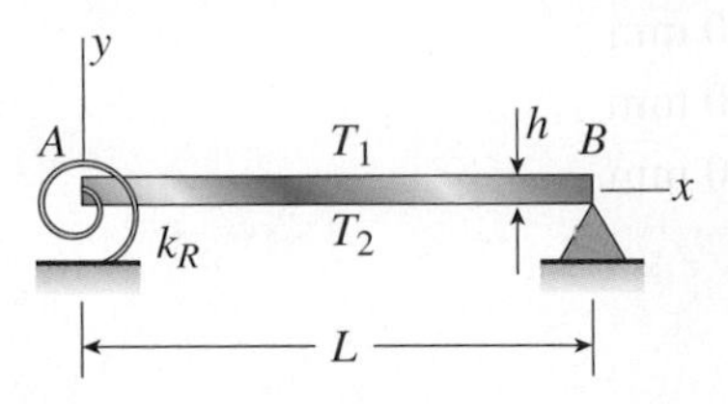

PROBLEM 9.11-5

SOME ADDITIONAL REVIEW PROBLEMS: Chapter 9

R-9.1 An aluminum beam ($E = 72$ GPa) with a square cross section and span length $L = 2.5$ m is subjected to uniform load $q = 1.5$ kN/m. The allowable bending stress is 60 MPa. The maximum deflection of the beam is approximately:

(A) 10 mm
(B) 16 mm
(C) 22 mm
(D) 26 mm

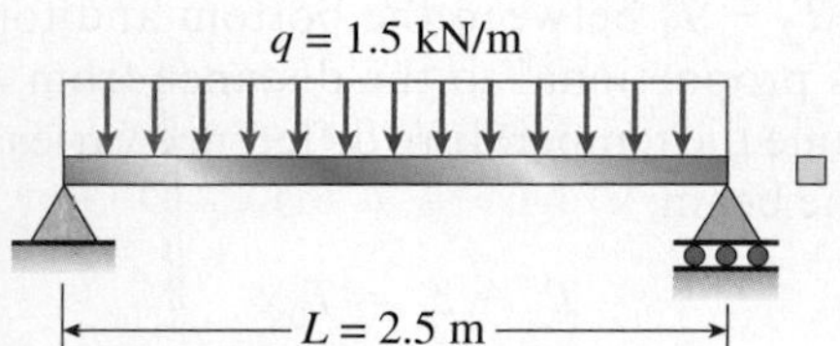

R-9.2 An aluminum cantilever beam ($E = 72$ GPa) with a square cross section and span length $L = 2.5$ m is subjected to uniform load $q = 1.5$ kN/m. The allowable bending stress is 55 MPa. The maximum deflection of the beam is approximately:

(A) 10 mm
(B) 20 mm
(C) 30 mm
(D) 40 mm

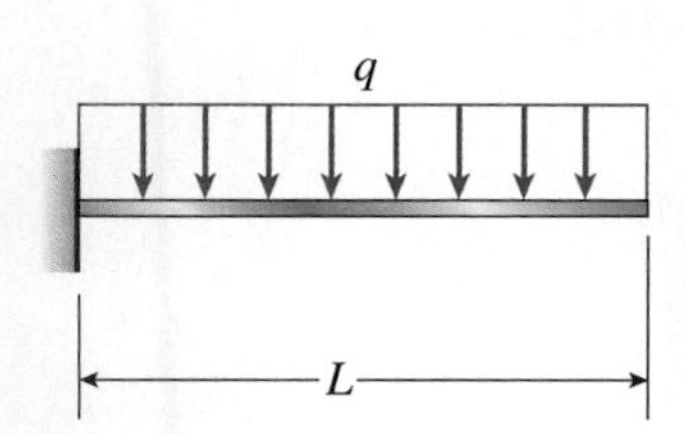

R-9.3 A steel beam ($E = 210$ GPa) with $I = 119 \times 10^6$ mm^4 and span length $L = 3.5$ m is subjected to uniform load $q = 9.5$ kN/m. The maximum deflection of the beam is approximately:

(A) 10 mm
(B) 13 mm
(C) 17 mm
(D) 19 mm

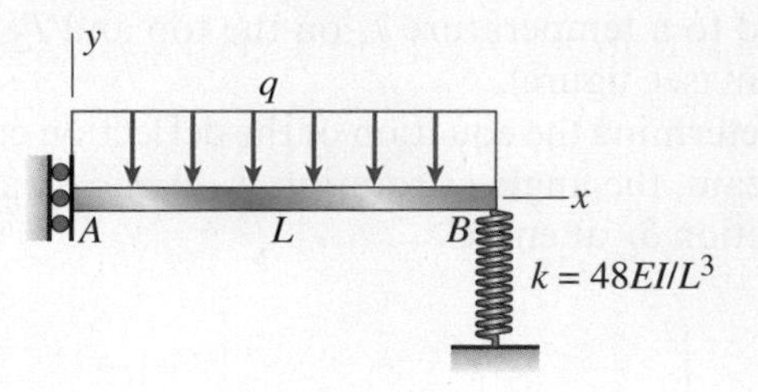

R-9.4 A steel bracket ABC ($EI = 4.2 \times 10^6$ N · m^2) with span length $L = 4.5$ m and height $H = 2$ m is subjected to load $P = 15$ kN at C. The maximum rotation of joint B is approximately:

(A) 0.18°
(B) 0.38°
(C) 0.68°
(D) 0.98°

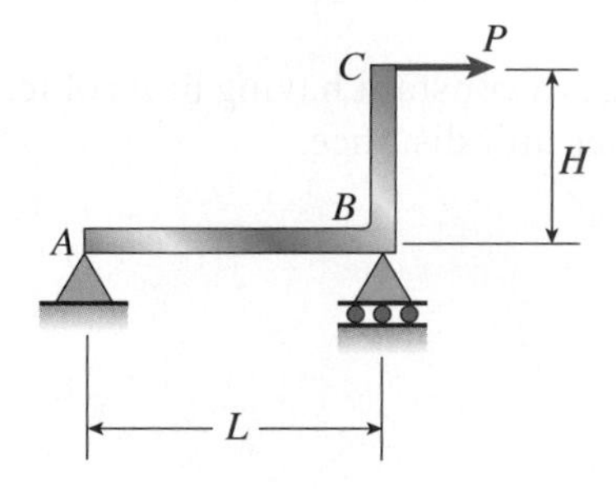

R-9.5 A steel bracket ABC ($EI = 4.2 \times 10^6\ \text{N} \cdot \text{m}^2$) with span length $L = 4.5$ m and height $H = 2$ m is subjected to load $P = 15$ kN at C. The maximum horizontal displacement of joint C is approximately:

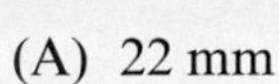

(A) 22 mm
(B) 31 mm
(C) 38 mm
(D) 40 mm

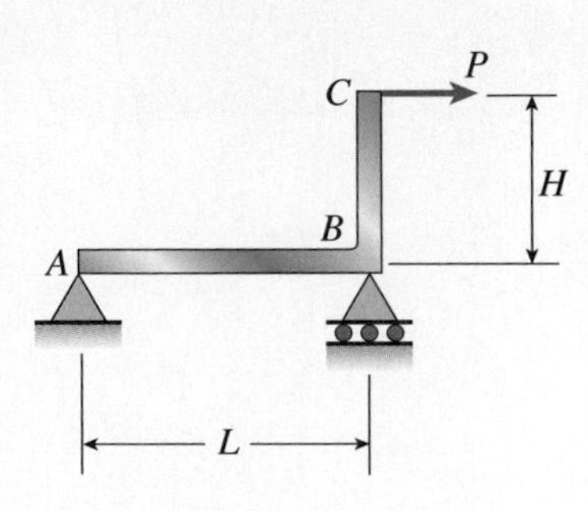

R-9.6 A nonprismatic cantilever beam of one material is subjected to load P at its free end. Moment of inertia $I_2 = 2I_1$. The ratio r of the deflection δ_B to the deflection δ_1 at the free end of a prismatic cantilever with moment of inertia I_1 carrying the same load is approximately:

(A) 0.25
(B) 0.40
(C) 0.56
(D) 0.78

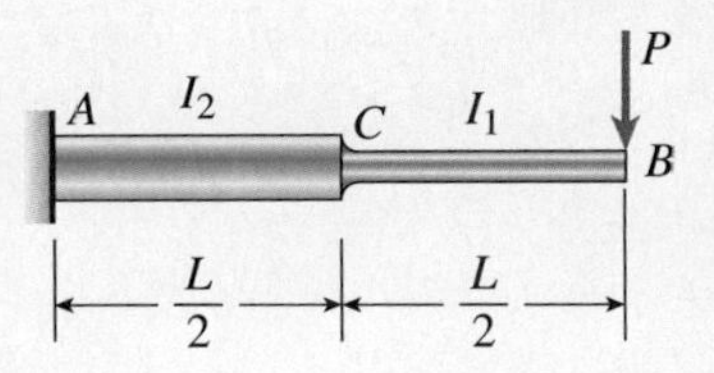

R-9.7 A steel bracket $ABCD$ ($EI = 4.2 \times 10^6\ \text{N} \cdot \text{m}^2$), with span length $L = 4.5$ m and dimension $a = 2$ m, is subjected to load $P = 15$ kN at D. The maximum deflection at B is approximately:

(A) 10 mm
(B) 14 mm
(C) 19 mm
(D) 24 mm

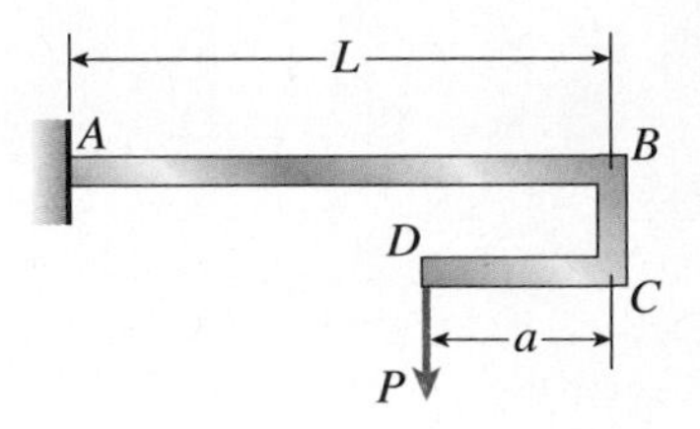

CHAPTER 10

Statically Indeterminate Beams

© Kristiina Paul

Large and complex structures, such as this roof structure for the atrium of a shopping mall, are usually statically indeterminate and require a computer to find forces in members and displacements of joints.

Chapter Objectives

- Investigate the behavior of different types of statically indeterminate beams.
- Determine the degree of indeterminacy of statically indeterminate beams.
- Develop a solution approach based on integration of the equation of the elastic curve.
- Study a superposition approach (known as the force or flexibility method) for solving for redundant forces and moments in statically indeterminate beams.
- Investigate differential temperature changes and longitudinal displacements that induce reaction forces only in statically indeterminate beams.

Chapter Outline

***Advanced topics**

10.1 Introduction

This chapter discusses the analysis of beams in which the number of reactions exceeds the number of independent equations of equilibrium. Since the reactions of such beams cannot be determined by statics alone, the beams are said to be **statically indeterminate**.

The analysis of statically indeterminate beams is quite different from that of statically determinate beams. When a beam is statically determinate, all reactions, shear forces, and bending moments can be obtained from free-body diagrams and equations of equilibrium. Then, knowing the shear forces and bending moments, the stresses and deflections are computed.

However, when a beam is statically indeterminate, the equilibrium equations are not sufficient, and additional equations are needed. The most fundamental method for analyzing a statically indeterminate beam is to solve the differential equations of the deflection curve, as described in Section 10.3. Although this method serves as a good starting point in the analysis, it is practical for only the simplest types of statically indeterminate beams.

Therefore, the focus here is on the method of superposition (Section 10.4), a method that is applicable to a wide variety of structures. In the method of superposition, the equilibrium equations are supplemented with compatibility equations and force-displacement equations. (This same method was described earlier in Section 2.4, where statically indeterminate bars subjected to tension and compression were analyzed.)

In the last part of this chapter, two specialized topics are presented pertaining to statically indeterminate beams, namely, beams with temperature changes (Section 10.5) and longitudinal displacements at the ends of beams (Section 10.6). Throughout this chapter, the beams are assumed to be made of **linearly elastic materials**.

Although only statically indeterminate beams are discussed in this chapter, the fundamental ideas have much wider application. Most structures, including automobile frames, buildings, and aircraft, are statically indeterminate. However, they are much more complex than beams and must be designed using very sophisticated analytical techniques. Many of these techniques rely on the concepts described in this chapter; therefore, this chapter may be viewed as an introduction to the analysis of statically indeterminate structures of all kinds.

10.2 Types of Statically Indeterminate Beams

Statically indeterminate beams are usually identified by the arrangement of their supports. For instance, a beam that is fixed at one end and simply supported at the other (Fig. 10-1a) is called a **propped cantilever beam**. The reactions of the beam shown in the figure consist of horizontal and vertical forces at support A, a moment at support A, and a vertical force at support B. Because there are only three independent equations of equilibrium for this beam, it is not possible to calculate all four of the reactions from equilibrium alone. The number of reactions in *excess* of the number of equilibrium equations is called the **degree of static indeterminacy**. Thus, a propped cantilever beam is statically indeterminate to the first degree.

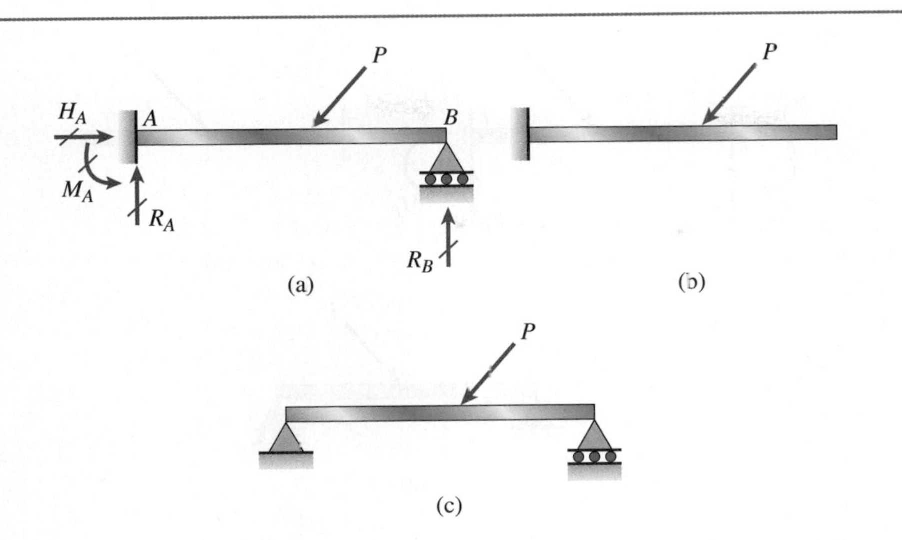

FIGURE 10-1

Propped cantilever beam: (a) beam with load and reactions; (b) released structure when the reaction at end B is selected as the redundant; and (c) released structure when the moment reaction at end A is selected as the redundant

The excess reactions are called **static redundants** and must be selected in each particular case. For example, the reaction R_B of the propped cantilever beam shown in Fig. 10-1a may be selected as the redundant reaction. Since this reaction is in excess of those needed to maintain equilibrium, it can be released from the structure by removing the support at B. When support B is removed, the resulting structure is a cantilever beam (Fig. 10-1b). The structure that remains when the redundants are released is called the **released structure** or the **primary structure**. The released structure must be stable (so that it is capable of carrying loads), and it must be statically determinate (so that all force quantities can be determined by equilibrium alone).

Another possibility for the analysis of the propped cantilever beam of Fig. 10-1a is to select the reactive moment M_A as the redundant. Then, when the moment restraint at support A is removed, the released structure is a simple beam with a pin support at one end and a roller support at the other (Fig. 10-1c).

A special case arises if all loads acting on the beam are vertical (Fig. 10-2). Then the horizontal reaction at support A vanishes, and three reactions remain. However, only two independent equations of equilibrium are now available; therefore, the beam is still statically indeterminate to the first degree. If the reaction R_B is chosen as the redundant, the released structure is a cantilever beam; if the moment M_A is chosen, the released structure is a simple beam.

FIGURE 10-2

Propped cantilever beam with vertical loads only

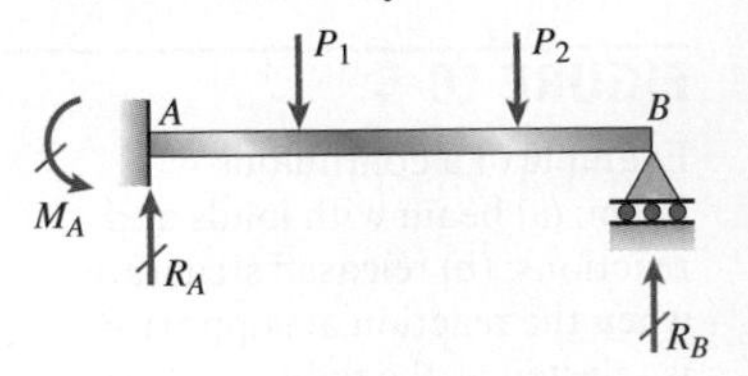

Another type of statically indeterminate beam, known as a **fixed-end beam**, is shown in Fig. 10-3a. This beam has fixed supports at both ends, resulting in a total of six unknown reactions (two forces and a moment at each support). Because there are only three equations of equilibrium, the beam is statically indeterminate to the third degree. (Other names for this type of beam are *clamped beam* and *built-in beam*.)

If the three reactions at end B of the beam are selected as the redundants and if the corresponding restraints are removed, a cantilever beam is left as the released structure (Fig. 10-3b). If the two fixed-end moments and one horizontal reaction are removed instead, the released structure is a simple beam (Fig. 10-3c).

FIGURE 10-3

Fixed-end beam: (a) beam with load and reactions; (b) released structure when the three reactions at end B are selected as the redundants; and (c) released structure when the two moment reactions and the horizontal reaction at end B are selected as the redundants

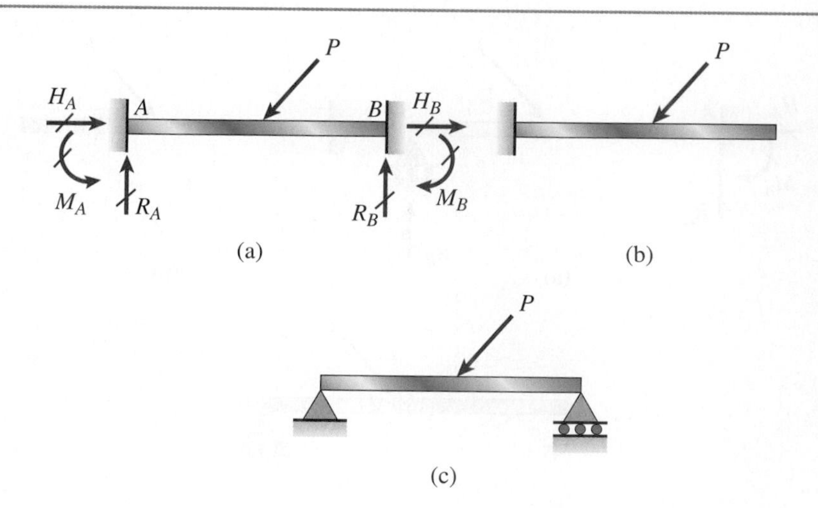

FIGURE 10-4

Fixed-end beam with vertical loads only

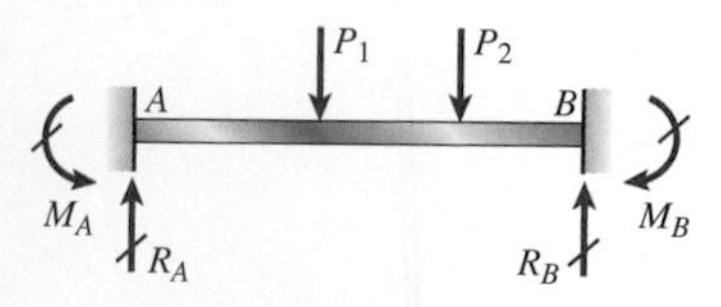

Long-span bridges are often constructed using continuous beams

Again considering the special case of only vertical loads (Fig. 10-4), the fixed-end beam now has only four nonzero reactions (one force and one moment at each support). The number of available equilibrium equations is two, so the beam is statically indeterminate to the second degree. If the two reactions at end B are selected as the redundants, the released structure is a cantilever beam; if the two moment reactions are selected, the released structure is a simple beam.

The beam shown in Fig. 10-5a is an example of a **continuous beam**, so called because it has more than one span and is continuous over an interior support. This particular beam is statically indeterminate to the first degree because there are four reactive forces and only three equations of equilibrium.

If the reaction R_B at the interior support is selected as the redundant, and if the corresponding support is removed from the beam, there remains a released structure in the form of a statically determinate simple beam (Fig. 10-5b). If the reaction R_C is selected as the redundant, the released structure is a simple beam with an overhang (Fig. 10-5c).

The following sections discuss two methods for analyzing statically indeterminate beams. The objective in each case is to determine the redundant

FIGURE 10-5

Example of a continuous beam: (a) beam with loads and reactions; (b) released structure when the reaction at support B is selected as the redundant; and (c) released structure when the reaction at end C is selected as the redundant

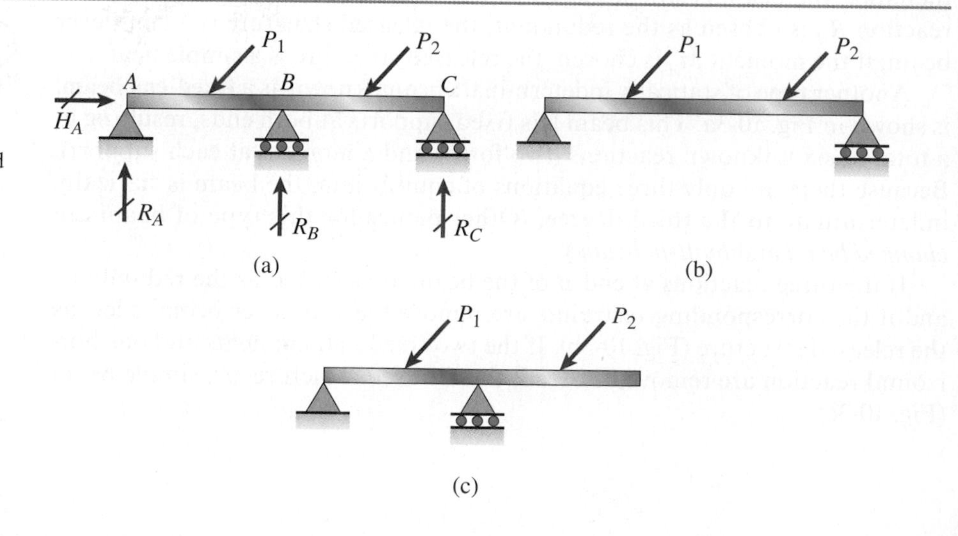

reactions. Once they are known, all remaining reactions (plus the shear forces and bending moments) can be found from equations of equilibrium. In effect, the structure has become statically determinate. Therefore, as the final step in the analysis, the stresses and deflections can be found by the methods described in preceding chapters.

10.3 Analysis by the Differential Equations of the Deflection Curve

Statically indeterminate beams may be analyzed by solving any one of the three differential equations of the deflection curve: (1) the second-order equation in terms of the bending moment [Eq. (9-13a) or Eq. (9-16a) if prismatic], (2) the third-order equation in terms of the shear force [Eq. (9-13b) or Eq. (9-16b)], or (3) the fourth-order equation in terms of the intensity of distributed load [Eq. (9-13c) or Eq. (9-16c)].

The procedure is essentially the same as that for a statically determinate beam (see Sections 9.2, 9.3, and 9.4) and consists of writing the differential equation, integrating to obtain its general solution, and then applying boundary and other conditions to evaluate the unknown quantities. The unknowns consist of the redundant reactions as well as the constants of integration.

The differential equation for a beam may be solved in symbolic terms only when the beam and its loading are relatively simple and uncomplicated. The resulting solutions are in the form of general-purpose formulas. However, in more complex situations, the differential equations must be solved numerically, using computer programs intended for that purpose. In such cases, the results apply only to specific numerical problems.

The following examples illustrate the analysis of statically indeterminate beams by solving the differential equations in symbolic terms.

Example 10-1

FIGURE 10-6

Example 10-1: Propped cantilever beam with a uniform load

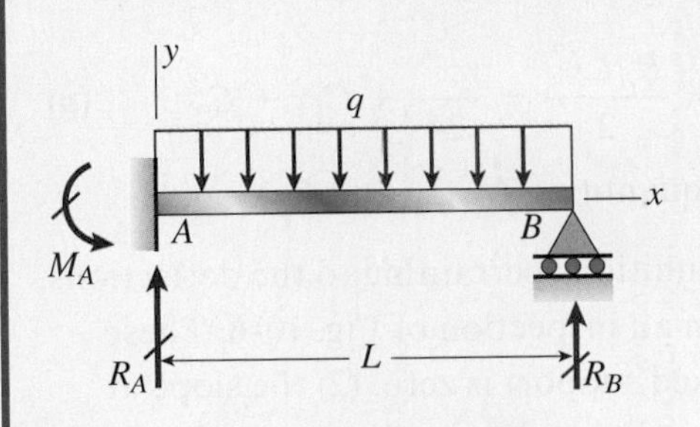

A propped cantilever beam AB of a length L supports a uniform load of intensity q (Fig. 10-6). Analyze this beam by solving the second-order differential equation of the deflection curve (the bending-moment equation). Determine the reactions, shear forces, bending moments, slopes, and deflections of the beam.

Solution:

Use a four-step problem-solving approach.

1. ***Conceptualize*** [*hypothesize, sketch*]: Because the load on this beam acts in the vertical direction (Fig. 10-6), there is no horizontal reaction at the fixed support. Therefore, the beam has three unknown reactions (M_A, R_A, and R_B). Only two equations of equilibrium are available for determining these reactions; therefore, the beam is statically indeterminate to the first degree.

2. ***Categorize*** [*simplify, classify*]: To analyze this beam by solving the bending-moment equation, begin with a general expression for the moment. This expression is in terms of both the load and the selected redundant.

Redundant reaction: Choose the reaction R_B at the simple support as the redundant. Then, by considering the equilibrium of the entire beam, express the other two reactions in terms of R_B:

$$R_A = qL - R_B \qquad M_A = \frac{qL^2}{2} - R_B L \tag{a,b}$$

Bending moment: The bending moment M at distance x from the fixed support is expressed in terms of the reactions as

$$M = R_A x - M_A - \frac{qx^2}{2} \tag{c}$$

This equation is obtained by constructing a free-body diagram of part of the beam and solving an equation of equilibrium.

Substitute into Eq. (c) from Eqs. (a) and (b) to obtain the bending moment in terms of the load and the redundant reaction:

$$M = qLx - R_B x - \frac{qL^2}{2} + R_B L - \frac{qx^2}{2} \tag{d}$$

3. ***Analyze*** [*evaluate; select relevant equations; carry out mathematical solution*]:

Differential equation: The second-order differential equation of the deflection curve [Eq. (9-16a)] now becomes

$$EIv'' = M = qLx - R_B x - \frac{qL^2}{2} + R_B L - \frac{qx^2}{2} \tag{e}$$

After two successive integrations, the following equations are obtained for the slopes and deflections of the beam:

$$EIv' = \frac{qLx^2}{2} - \frac{R_B x^2}{2} - \frac{qL^2 x}{2} + R_B Lx - \frac{qx^3}{6} + C_1 \tag{f}$$

$$EIv = \frac{qLx^3}{6} - \frac{R_B x^3}{6} - \frac{qL^2 x^2}{4} + \frac{R_B L x^2}{2} - \frac{qx^4}{24} + C_1 x + C_2 \tag{g}$$

These equations contain three unknown quantities (C_1, C_2, and R_B).

Boundary conditions: Three boundary conditions pertaining to the deflections and slopes of the beam are apparent from an inspection of Fig. 10-6. These conditions are (1) the deflection at the fixed support is zero, (2) the slope at the fixed support is zero, and (3) the deflection at the simple support is zero. Thus,

$$v(0) = 0 \qquad v'(0) = 0 \qquad v(L) = 0$$

Apply these conditions to the equations for slopes and deflections given in Eqs. (f) and (g) to find that $C_1 = 0, C_2 = 0$, and

$$R_B = \frac{3qL}{8} \quad \text{(10-1)}$$

Thus, the redundant reaction R_B is now known.

Reactions: With the value of the redundant established, the remaining reactions from Eqs. (a) and (b) are

$$R_A = \frac{5qL}{8} \qquad M_A = \frac{qL^2}{8} \quad \text{(10-2a,b)}$$

Knowing these reactions, the shear forces and bending moments in the beam can now be found.

FIGURE 10-7

Shear-force and bending-moment diagrams for the propped cantilever beam of Fig. 10-6

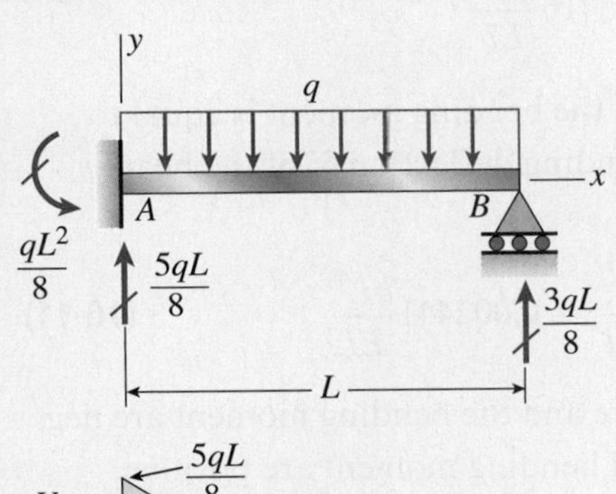

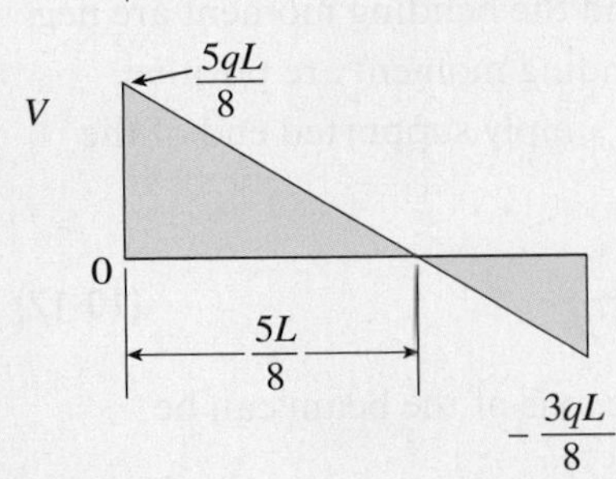

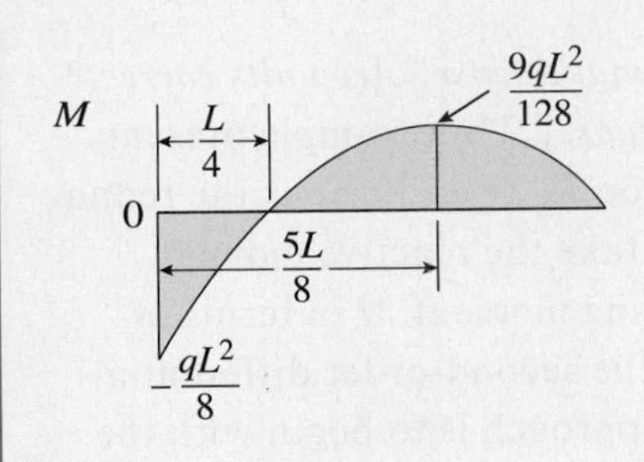

Shear forces and bending moments: These quantities are obtained by the usual techniques involving free-body diagrams and equations of equilibrium. The results are

$$V = R_A - qx = \frac{5qL}{8} - qx \quad \text{(10-3)}$$

$$M = R_A x - M_A - \frac{qx^2}{2} = \frac{5qLx}{8} - \frac{qL^2}{8} - \frac{qx^2}{2} \quad \text{(10-4)}$$

Shear-force and bending-moment diagrams for the beam are drawn with the aid of these equations (see Fig. 10-7).

From the diagrams, note that the maximum shear force occurs at the fixed support and is equal to

$$V_{\text{max}} = \frac{5qL}{8} \quad \text{(10-5)}$$

Also, the maximum positive and negative bending moments are

$$M_{\text{pos}} = \frac{9qL^2}{128} \qquad M_{\text{neg}} = -\frac{qL^2}{8} \quad \text{(10-6a,b)}$$

Finally, note that the bending moment is equal to zero at distance $x = L/4$ from the fixed support.

Slopes and deflections of the beam: Return to Eqs. (f) and (g) for the slopes and deflections; now substitute the values of the constants of integration ($C_1 = 0$ and $C_2 = 0$) as well as the expression for the redundant R_B [Eq. (10-1)] and obtain

$$v' = \frac{qx}{48EI}(-6L^2 + 15Lx - 8x^2) \quad \text{(10-7)}$$

$$v = -\frac{qx^2}{48EI}(3L^2 - 5Lx + 2x^2) \quad \text{(10-8)}$$

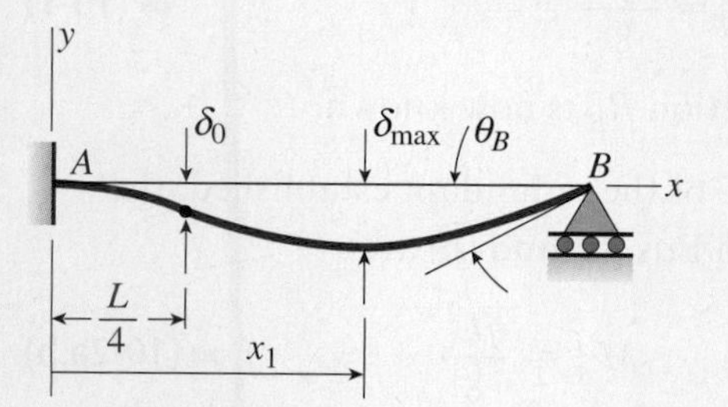

FIGURE 10-8

Deflection curve for the propped cantilever beam of Fig. 10-6

The deflected shape of the beam as obtained from Eq. (10-8) is shown in Fig. 10-8.

To determine the maximum deflection of the beam, set the slope [Eq. (10-7)] equal to zero and solve for the distance x_1 to the point where this deflection occurs:

$$v' = 0 \quad \text{or} \quad -6L^2 + 15Lx - 8x^2 = 0$$

from which

$$x_1 = \frac{15 - \sqrt{33}}{16} L = 0.5785L \tag{10-9}$$

Substitute this value of x into the equation for the deflection [Eq. (10-8)] and also change the sign to get the maximum deflection:

$$\begin{aligned} \delta_{max} &= -v(x_1) = \frac{qL^4}{65{,}536\,EI}(39 + 55\sqrt{33}) \\ &= \frac{qL^4}{184.6\,EI} = 0.005416\frac{qL^4}{EI} \end{aligned} \tag{10-10}$$

The point of inflection is located where the bending moment is equal to zero, that is, where $x = L/4$. The corresponding deflection δ_0 of the beam [from Eq. (10-8)] is

$$\delta_0 = -v(L/4) = \frac{5qL^4}{2048\,EI} = 0.002441\frac{qL^4}{EI} \tag{10-11}$$

Note that when $x < L/4$, both the curvature and the bending moment are negative, and when $x > L/4$, the curvature and bending moment are positive.

To determine the angle of rotation θ_B at the simply supported end of the beam, use Eq. (10-7), as

$$\theta_B = v'(L) = \frac{qL^3}{48EI} \tag{10-12}$$

Slopes and deflections at other points along the axis of the beam can be obtained by similar procedures.

4. ***Finalize*** [*conclude; examine answer—Does it make sense? Are units correct? How does it compare to similar problem solutions?*]: This example presents the analysis of the beam by taking the reaction R_B (Fig. 10-6) as the redundant reaction. An alternative approach is to take the reactive moment M_A as the redundant. Then express the bending moment M in terms of M_A, substitute the resulting expression into the second-order differential equation, and solve as before. Still another approach is to begin with the fourth-order differential equation, as illustrated in the next example.

Example 10-2

FIGURE 10-9

Example 10-2: Fixed-end beam with a concentrated load at the midpoint

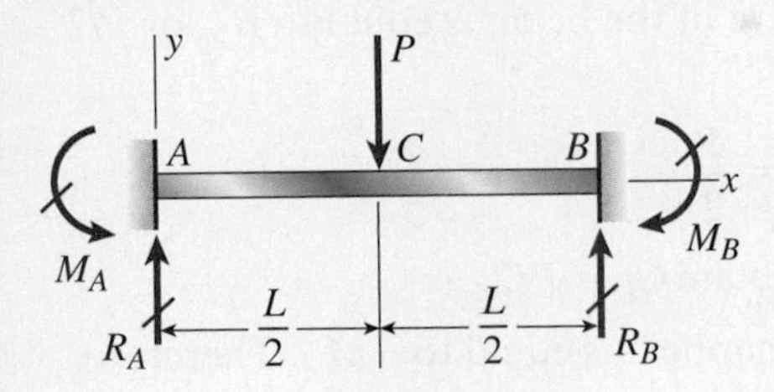

The fixed-end beam *ACB* shown in Fig. 10-9 supports a concentrated load *P* at the midpoint. Analyze this beam by solving the fourth-order differential equation of the deflection curve (the load equation). Determine the reactions, shear forces, bending moments, slopes, and deflections of the beam.

Solution:

Use a four-step problem-solving approach.

1. ***Conceptualize***: Because the load on this beam acts only in the vertical direction, there are no horizontal reactions at the supports. Therefore, the beam has four unknown reactions—two at each support. Since only two equations of equilibrium are available, the beam is statically indeterminate to the second degree.

2. ***Categorize***: The analysis can be simplified by observing from the symmetry of the beam and its loading that the forces and moments at supports *A* and *B* are equal, that is,

$$R_A = R_B \quad \text{and} \quad M_A = M_B$$

The vertical reactions at the supports are equal, so from equilibrium of forces in the vertical direction, each force is equal to $P/2$:

$$R_A = R_B = \frac{P}{2} \tag{10-13}$$

Thus, the only unknown quantities that remain are the moment reactions M_A and M_B. For convenience, select the moment M_A as the redundant quantity.

3. ***Analyze***:

Differential equation: Because there is no load acting on the beam between points *A* and *C*, the fourth-order differential equation [Eq. (9-19c)] for the left-hand half of the beam is

$$EIv'''' = -q = 0 \quad (0 < x < L/2) \tag{a}$$

Successive integrations of this equation yield the following equations, which are valid for the left-hand half of the beam:

$$EIv''' = C_1 \tag{b}$$

$$EIv'' = C_1x + C_2 \tag{c}$$

$$EIv' = \frac{C_1x^2}{2} + C_2x + C_3 \tag{d}$$

$$EIv = \frac{C_1x^3}{6} + \frac{C_2x^2}{2} + C_3x + C_4 \tag{e}$$

These equations contain four unknown constants of integration. There are now five unknowns (C_1, C_2, C_3, C_4, and M_A), so five boundary conditions are needed.

Boundary conditions: The boundary conditions applicable to the left-hand half of the beam are

i. The shear force in the left-hand segment of the beam is equal to R_A, or $P/2$. Therefore, from Eq. (9-16b)

$$EIv''' = V = \frac{P}{2}$$

Combine this equation with Eq. (b) to obtain $C_1 = P/2$.

ii. The bending moment at the left-hand support is equal to $-M_A$. Therefore, from Eq. (9-16a),

$$EIv'' = M = -M_A \quad \text{at} \quad x = 0$$

Combining this equation with Eq. (c) leads to $C_2 = -M_A$.

iii. The slope of the beam at the left-hand support ($x = 0$) is equal to zero. Therefore, Eq. (d) yields $C_3 = 0$.

iv. The slope of the beam at the midpoint ($x = L/2$) is also equal to zero (from symmetry). Therefore, from Eq. (d),

$$M_A = M_B = \frac{PL}{8} \qquad \textbf{(10-14)}$$

Thus, the reactive moments at the ends of the beam have been determined.

v. The deflection of the beam at the left-hand support ($x = 0$) is equal to zero. Therefore, from Eq. (e), $C_4 = 0$.

In summary, the four constants of integration are

$$C_1 = \frac{P}{2} \quad C_2 = -M_A = -\frac{PL}{8} \quad C_3 = 0 \quad C_4 = 0 \qquad \textbf{(f,g,h,i)}$$

Shear forces and bending moments: The shear forces and bending moments are found by substituting the appropriate constants of integration into Eqs. (b) and (c). The results are

$$EIv''' = V = \frac{P}{2} \qquad (0 < x < L/2) \qquad \textbf{(10-15)}$$

$$EIv'' = M = \frac{Px}{2} - \frac{PL}{8} \qquad (0 \leq x \leq L/2) \qquad \textbf{(10-16)}$$

Since the reactions of the beam are known, these expressions also can be obtained directly from free-body diagrams and equations of equilibrium.

The shear-force and bending moment diagrams are shown in Fig. 10-10.

FIGURE 10-10

Shear-force and bending-moment diagrams for the fixed-end beam of Fig. 10-9

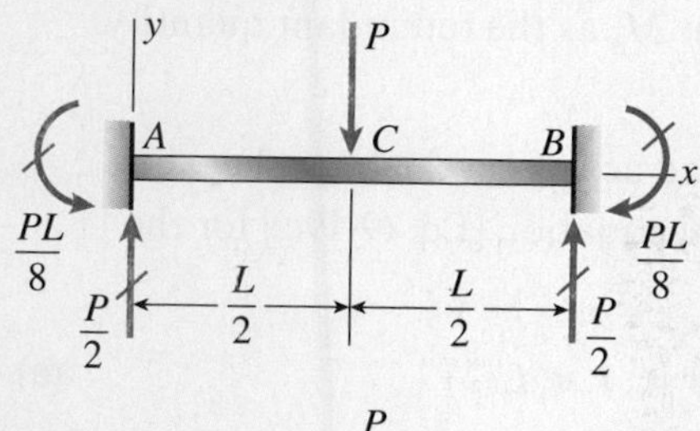

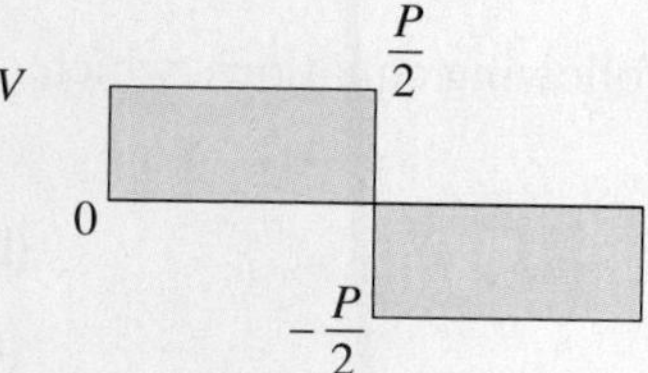

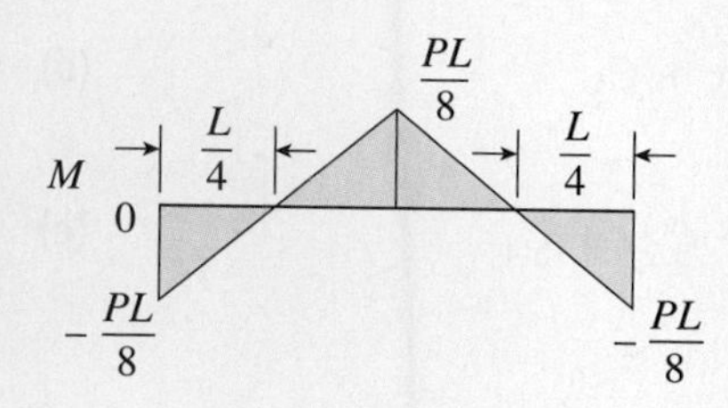

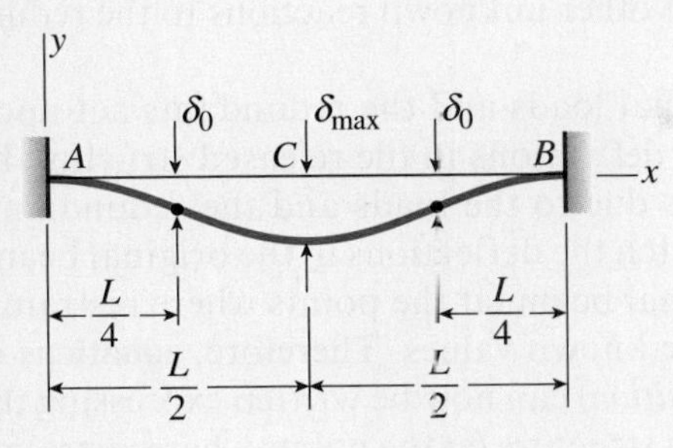

FIGURE 10-11
Deflection curve for the fixed-end beam of Fig. 10-9

Slopes and deflections: The slopes and deflections in the left-hand half of the beam can be found from Eqs. (d) and (e) by substituting the expressions for the constants of integration. The resulting expressions are

$$v' = -\frac{Px}{8EI}(L - 2x) \qquad (0 \le x \le L/2) \tag{10-17}$$

$$v = -\frac{Px^2}{48EI}(3L - 4x) \qquad (0 \le x \le L/2) \tag{10-18}$$

The deflection curve of the beam is shown in Fig. 10-11.

To find the maximum deflection δ_{max}, set x equal to $L/2$ in Eq. (10-18) and change the sign; thus,

$$\delta_{max} = -v(L/2) = \frac{PL^3}{192EI} \tag{10-19}$$

The point of inflection in the left-hand half of the beam occurs where the bending moment M is equal to zero, that is, where $x = L/4$ [see Eq. (10-16)]. The corresponding deflection δ_0 [from Eq. (10-18)] is

$$\delta_0 = -v(L/4) = \frac{PL^3}{384EI} \tag{10-20}$$

which is equal numerically to one-half of the maximum deflection. A second point of inflection occurs in the right-hand half of the beam at distance $L/4$ from end B.

4. ***Finalize***: This example demonstrates that the number of boundary and other conditions is always sufficient to evaluate not only the constants of integration but also the redundant reactions.

Note: Sometimes it is necessary to set up differential equations for more than one region of the beam and use conditions of continuity between regions, as illustrated in Examples 9-3 and 9-5 of Chapter 9 for statically determinate beams. Such analyses are likely to be long and tedious because of the large number of conditions that must be satisfied. However, if deflections and angles of rotation are needed at only one or two specific points, the method of superposition may be useful, as shown in next section.

10.4 Method of Superposition

The method of superposition is of fundamental importance in the analysis of statically indeterminate bars, trusses, beams, frames, and many other kinds of structures. The superposition method already has been used to analyze statically indeterminate structures composed of bars in tension and compression

(Section 2.4) and shafts in torsion (Section 3.8). In this section, the method is applied to beams.

Begin the analysis by noting the degree of static indeterminacy and selecting the redundant reactions. Then, having identified the redundants, write **equations of equilibrium** that relate the other unknown reactions to the redundants and the loads.

Next, assume that both the original loads and the redundants act upon the released structure. Then find the deflections in the released structure by superposing the separate deflections due to the loads and the redundants. The sum of these deflections must match the deflections in the original beam. However, the deflections in the original beam (at the points where restraints were removed) are either zero or have known values. Therefore, **equations of compatibility** (or *equations of superposition*) can now be written expressing the fact that the deflections of the released structure (at the points where restraints were removed) are the same as the deflections in the original beam (at those same points).

The released structure is statically determinate, so determine its deflections by using the techniques described in Chapter 9. The relationships between the loads and the deflections of the released structure are called **force-displacement relations**. When these relations are substituted into the equations of compatibility, you obtain equations in which the redundants are the unknown quantities. Therefore, solving those equations gives the redundant reactions. Then, with the redundants known, all other reactions are found using the equations of equilibrium. Furthermore, the shear forces and bending moments are also obtained from equilibrium.

The preceding steps can be made clearer by considering a particular case, namely, a propped cantilever beam supporting a uniform load (Fig. 10-12a). Two analyses are presented here: the first with the force reaction R_B selected as the redundant and the second with the moment reaction M_A as the redundant. (This same beam was analyzed in Example 10-1 of Section 10.3 by solving the differential equation of the deflection curve.)

FIGURE 10-12

Analysis of a propped cantilever beam by the method of superposition with the reaction R_B selected as the redundant

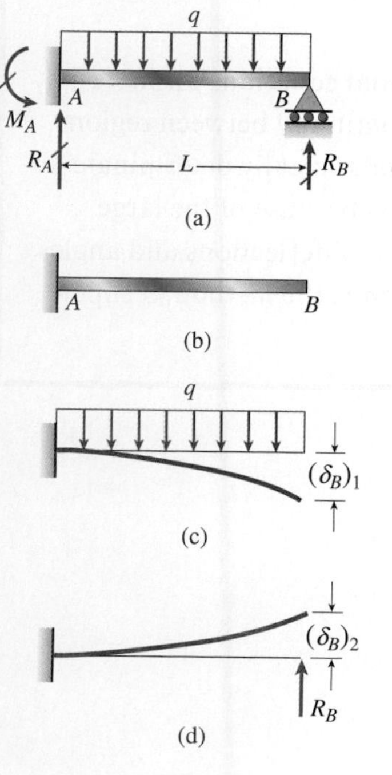

Analysis with R_B as Redundant

For the first case, select the reaction R_B at the simple support (Fig. 10-12a) as the redundant. Then the *equations of equilibrium* that express the other unknown reactions in terms of the redundant are

$$R_A = qL - R_B \quad M_A = \frac{qL^2}{2} - R_B L \tag{10-21a,b}$$

These equations come from equations of equilibrium that apply to the entire beam taken as a free body (Fig. 10-12a).

The next step is to remove the restraint corresponding to the redundant (in this case, remove the support at end B). The *released structure* that remains is a cantilever beam (Fig. 10-12b). The uniform load q and the redundant force R_B are now applied as loads on the released structure (Figs. 10-12c and d).

The deflection at end B of the released structure due solely to the uniform load is denoted $(\delta_B)_1$, and the deflection at the same point due solely to the redundant is denoted $(\delta_B)_2$. The deflection δ_B at point B in the original structure is obtained by superposing these two deflections. The deflection in the original beam is equal to zero, so the *equation of compatibility* is

$$\delta_B = (\delta_B)_1 - (\delta_B)_2 = 0 \tag{10-22}$$

The minus sign appears in this equation because $(\delta_B)_1$ is positive downward whereas $(\delta_B)_2$ is positive upward.

The *force-displacement relations* that give the deflections $(\delta_B)_1$ and $(\delta_B)_2$ in terms of the uniform load q and the redundant R_B, respectively, are found with the aid of Table H-1 in Appendix H (see Cases 1 and 4). Use the formulas given there to get

$$(\delta_B)_1 = \frac{qL^4}{8EI} \qquad (\delta_B)_2 = \frac{R_B L^3}{3EI} \tag{10-23a,b}$$

Substitute these force-displacement relations into the equation of compatibility to find

$$\delta_B = \frac{qL^4}{8EI} - \frac{R_B L^3}{3EI} = 0 \tag{10-23c}$$

which can be solved for the *redundant reaction*:

$$R_B = \frac{3qL}{8} \tag{10-24}$$

Note that this equation gives the redundant in terms of the loads acting on the original beam.

The remaining reactions (R_A and M_A) can be found from the equilibrium equations [Eqs. (10-21a and b)]; the results are

$$R_A = \frac{5qL}{8} \qquad M_A = \frac{qL^2}{8} \tag{10-25a,b}$$

Knowing all reactions, you can now obtain the shear forces and bending moments throughout the beam and plot the corresponding diagrams (see Fig. 10-7 for these diagrams).

The *deflections and slopes* of the original beam also can be detemined using the principle of superposition. The procedure consists of superposing the deflections of the released structure when acted upon by the loads shown in Figs. 10-12c and d. For instance, the equations of the deflection curves for those two loading systems are obtained from Cases 1 and 4, respectively, of Table H-1, Appendix H:

$$v_1 = -\frac{qx^2}{24EI}(6L^2 - 4Lx + x^2)$$

$$v_2 = \frac{R_B x^2}{6EI}(3L - x)$$

Substituting for R_B from Eq. (10-24) and then adding the deflections v_1 and v_2 leads to the following equation for the deflection curve of the original statically indeterminate beam (Fig. 10-12a):

$$v = v_1 + v_2 = -\frac{qx^2}{48EI}(3L^2 - 5Lx + 2x^2)$$

FIGURE 10-13

Analysis of a propped cantilever beam by the method of superposition with the moment reaction M_A selected as the redundant

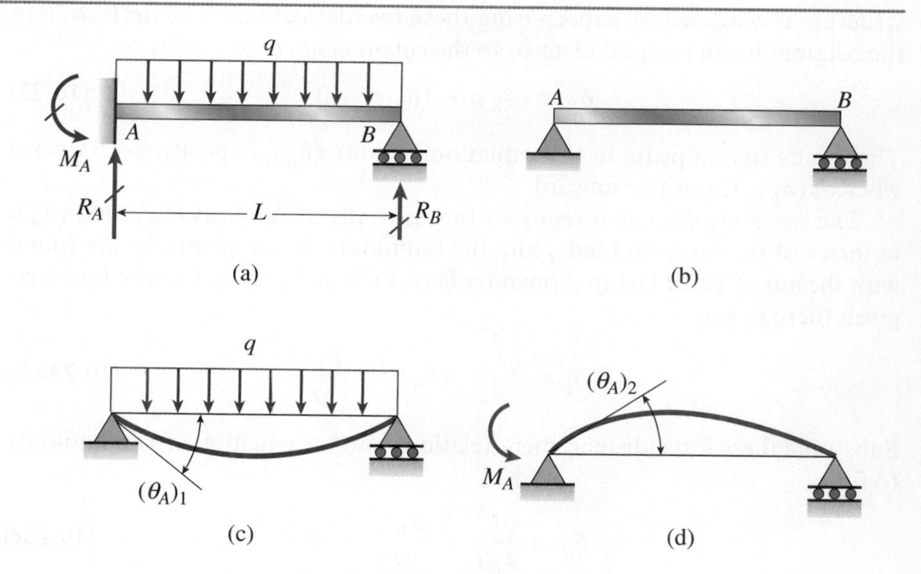

This equation agrees with Eq. (10-8) of Example 10-1. Other deflection quantities can be found in an analogous manner.

Analysis with M_A as Redundant

Now analyze the same propped cantilever beam by selecting the moment reaction M_A as the redundant (Fig. 10-13). In this case, the released structure is a simple beam (Fig. 10-13b). The equations of equilibrium for the reactions R_A and R_B in the original beam are

$$R_A = \frac{qL}{2} + \frac{M_A}{L} \qquad R_B = \frac{qL}{2} - \frac{M_A}{L} \tag{10-26a,b}$$

The equation of compatibility expresses the fact that the angle of rotation θ_A at the fixed end of the original beam is equal to zero. Since this angle is obtained by superposing the angles of rotation $(\theta_A)_1$ and $(\theta_A)_2$ in the released structure (Figs. 10-13c and d), the *compatibility equation* becomes

$$\theta_A = (\theta_A)_1 - (\theta_A)_2 = 0 \tag{10-27a}$$

In this equation, the angle $(\theta_A)_1$ is assumed to be positive when clockwise, and the angle $(\theta_A)_2$ is assumed to be positive when counterclockwise.

The angles of rotation in the released structure are obtained from the formulas given in Table H-2 of Appendix H (see Cases 1 and 7). Thus, the *force-displacement relations* are

$$(\theta_A)_1 = \frac{qL^3}{24EI} \qquad (\theta_A)_2 = \frac{M_A L}{3EI}$$

Substituting into the compatibility equation (Eq. 10-27a) gives

$$\theta_A = \frac{qL^3}{24EI} - \frac{M_A L}{3EI} = 0 \tag{10-27b}$$

Solve this equation for the redundant to get $M_A = qL^2/8$, which agrees with the previous result [Eq. (10-25b)]. Also, the equations of equilibrium

[Eqs. (10-26a and b)] yield the same results as before for the reactions R_A and R_B [see Eqs. (10-25a) and (10-24), respectively].

Now that all reactions have been found, you can determine the shear forces, bending moments, slopes, and deflections by the techniques already described.

General Comments

The method of superposition described in this section is also called the *flexibility method* or the *force method.* The latter name arises from the use of force quantities (forces and moments) as the redundants; the former name is used because the coefficients of the unknown quantities in the compatibility equation [terms such as $L^3/3EI$ in Eq. (10-27a) and $L/3EI$ in Eq. (10-27b)] are *flexibilities* (that is, deflections or angles produced by a unit load).

Since the method of superposition involves the superposition of deflections, it is applicable only to linearly elastic structures. (Recall that this same limitation applies to all topics discussed in this chapter.)

The following examples, and the problems at the end of the chapter, are concerned primarily with finding the reactions, since this is the key step in the solutions.

Example 10-3

A two-span continuous beam ABC supports a uniform load of intensity q, as shown in Fig. 10-14a. Each span of the beam has a length L. Using the method of superposition, determine all reactions for this beam.

FIGURE 10-14

Example 10-3: Two-span continuous beam with a uniform load

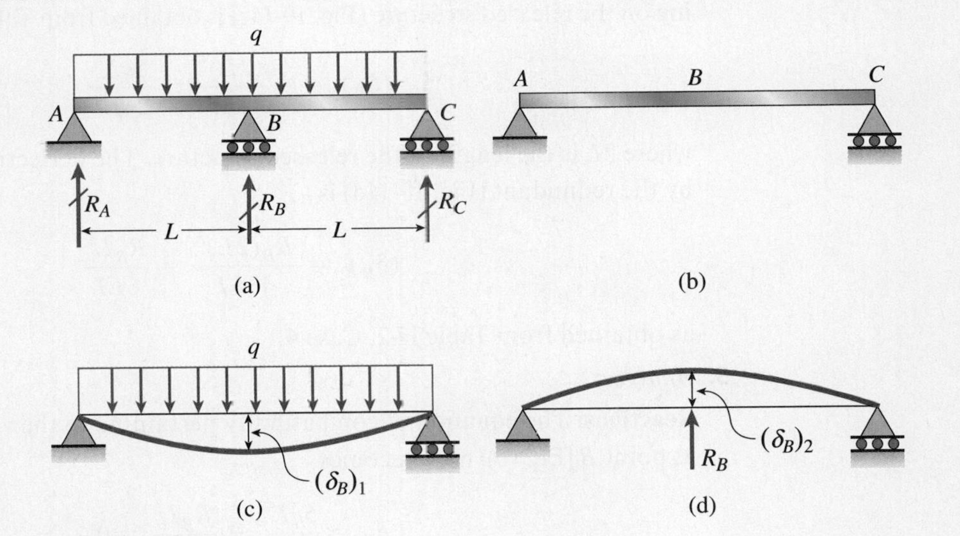

Solution:

Use a four-step problem-solving approach.

1. ***Conceptualize***: This beam has three unknown reactions (R_A, R_B, and R_C). Since there are two equations of equilibrium for the beam as a whole, it is statically

indeterminate to the first degree. For convenience, select the reaction R_B at the middle support as the redundant.

2. *Categorize:*

Equations of equilibrium: Express the reactions R_A and R_C in terms of the redundant R_B by means of two equations of equilibrium. The first equation, which is for equilibrium of moments about point B, shows that R_A and R_C are equal. The second equation, which is for equilibrium in the vertical direction, yields

$$R_A = R_C = qL - \frac{R_B}{2} \tag{a}$$

Equation of compatibility: Because the reaction R_B is selected as the redundant, the released structure is a simple beam with supports at A and C (Fig. 10-14b). The deflections at point B in the released structure due to the uniform load q and the redundant R_B are shown in Figs. 10-14c and d, respectively. Note that the deflections are denoted $(\delta_B)_1$ and $(\delta_B)_2$. The superposition of these deflections must produce the deflection δ_B in the original beam at point B. Since the latter deflection is equal to zero, the equation of compatibility is

$$\delta_B = (\delta_B)_1 - (\delta_B)_2 = 0 \tag{b}$$

in which the deflection $(\delta_B)_1$ is positive downward, and the deflection $(\delta_B)_2$ is positive upward.

Force-displacement relations: The deflection $(\delta_B)_1$ caused by the uniform load acting on the released structure (Fig. 10-14c) is obtained from Table H-2, Case 1 as

$$(\delta_B)_1 = \frac{5q(2L)^4}{384EI} = \frac{5qL^4}{24EI}$$

where $2L$ is the length of the released structure. The deflection $(\delta_B)_2$ produced by the redundant (Fig. 10-14d) is

$$(\delta_B)_2 = \frac{R_B(2L)^3}{48EI} = \frac{R_BL^3}{6EI}$$

as obtained from Table H-2, Case 4.

3. *Analyze:*

Reactions: The equation of compatibility pertaining to the vertical deflection at point B [Eq. (b)] now becomes

$$\delta_B = \frac{5qL^4}{24EI} - \frac{R_BL^3}{6EI} = 0 \tag{c}$$

from which the reaction at the middle support is

$$R_B = \frac{5qL}{4} \tag{10-28}$$

The other reactions are obtained from Eq. (a):

$$R_A = R_C = \frac{3qL}{8} \qquad (10\text{-}29)$$

With the reactions known, the shear forces, bending moments, stresses, and deflections can be found without difficulty.

4. ***Finalize***: The purpose of this example is to demonstrate the method of superposition, so all steps were described in the analysis. However, this particular beam (Fig. 10-14a) can be analyzed by inspection because of the symmetry of the beam and its loading.

   From symmetry, the slope of the beam at the middle support must be zero; therefore, each half of the beam is in the same condition as a propped cantilever beam with a uniform load (see, for instance, Fig. 10-6). Consequently, all of the previous results for a propped cantilever beam with a uniform load [Eqs. (10-1) to (10-12)] can be adapted immediately to the continuous beam of Fig. 10-14a.

Example 10-4

A fixed-end beam AB (Fig. 10-15a) is loaded by a force P acting at an intermediate point D. Find the reactive forces and moments at the ends of the beam using the method of superposition. Also, determine the deflection at point D where the load is applied.

Solution:

Use a four-step problem-solving approach.

1. ***Conceptualize***: This beam has four unknown reactions (a force and a moment at each support) but only two independent equations of equilibrium are available. Therefore, the beam is statically indeterminate to the second degree. In this example, select the reactive moments M_A and M_B as the redundants.

2. ***Categorize***:

   Equations of equilibrium: The two unknown force reactions (R_A and R_B) are expressed in terms of the redundants (M_A and M_B) with the aid of two equations of equilibrium. The first equation is for moments about point B, and the second is for moments about point A. The resulting expressions are

$$R_A = \frac{Pb}{L} + \frac{M_A}{L} - \frac{M_B}{L} \qquad R_B = \frac{Pa}{L} - \frac{M_A}{L} + \frac{M_B}{L} \qquad \text{(a,b)}$$

   Equations of compatibility: When both redundants are released by removing the rotational restraints at the ends of the beam, a simple beam remains as

FIGURE 10-15

Example 10-4: Fixed-end beam with a concentrated load

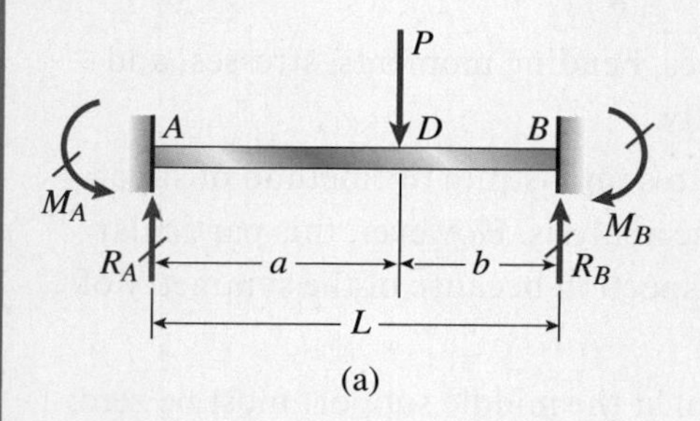

(a)

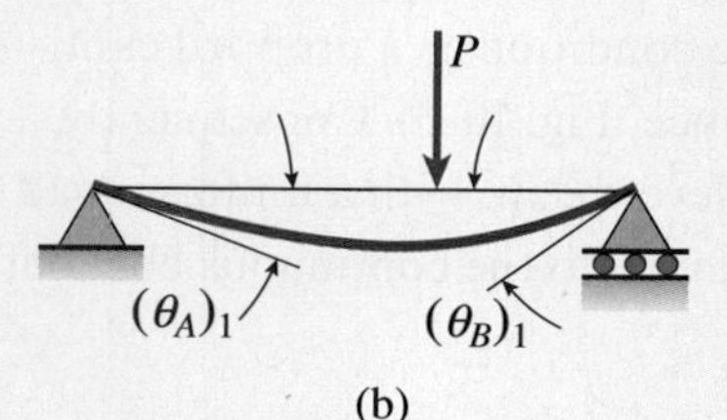

(b)

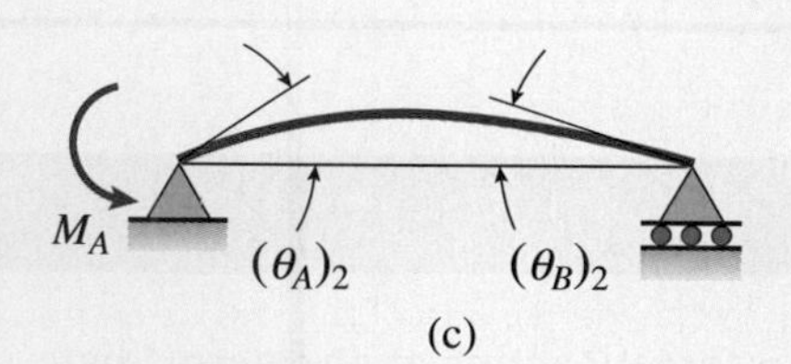

(c)

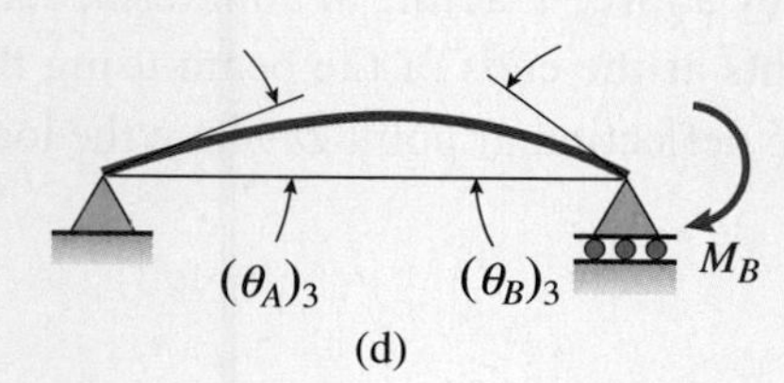

(d)

the released structure (Figs. 10-15b, c, and d). The angles of rotation at the ends of the released structure due to the concentrated load P are denoted $(\theta_A)_1$ and $(\theta_B)_1$, as shown in Fig. 10-15b. In a similar manner, the angles at the ends due to the redundant M_A are denoted $(\theta_A)_2$ and $(\theta_B)_2$, and the angles due to the redundant M_B are denoted $(\theta_A)_3$ and $(\theta_B)_3$.

Since the angles of rotation at the supports of the original beam are equal to zero, the two equations of compatibility are

$$\theta_A = (\theta_A)_1 - (\theta_A)_2 - (\theta_A)_3 = 0 \tag{c}$$

$$\theta_B = (\theta_B)_1 - (\theta_B)_2 - (\theta_B)_3 = 0 \tag{d}$$

in which the signs of the various terms are determined by inspection from the figures.

Force-displacement relations: The angles at the ends of the beam due to the load P (Fig. 10-15b) are obtained from Case 5 of Table H-2:

$$(\theta_A)_1 = \frac{Pab(L+b)}{6LEI} \qquad (\theta_B)_1 = \frac{Pab(L+a)}{6LEI}$$

in which a and b are the distances from the supports to point D where the load is applied.

Also, the angles at the ends due to the redundant moment M_A are (see Case 7 of Table H-2):

$$(\theta_A)_2 = \frac{M_A L}{3EI} \qquad (\theta_B)_2 = \frac{M_A L}{6EI}$$

Similarly, the angles due to the moment M_B are

$$(\theta_A)_3 = \frac{M_B L}{6EI} \qquad (\theta_B)_3 = \frac{M_B L}{3EI}$$

3. *Analyze:*

Reactions: When the preceding expressions for the angles are substituted into the equations of compatibility [Eqs. (c) and (d)], two simultaneous equations containing M_A and M_B as unknowns are obtained:

$$\frac{M_A L}{3EI} + \frac{M_B L}{6EI} = \frac{Pab(L+b)}{6LEI} \tag{e}$$

$$\frac{M_A L}{6EI} + \frac{M_B L}{3EI} = \frac{Pab(L+a)}{6LEI} \tag{f}$$

Solving these equations for the redundants gives

$$M_A = \frac{Pab^2}{L^2} \quad M_B = \frac{Pa^2b}{L^2} \tag{10-30a,b}$$

Substitute these expressions for M_A and M_B into the equations of equilibrium [Eqs. (a) and (b)], to obtain the vertical reactions:

$$R_A = \frac{Pb^2}{L^3}(L + 2a) \qquad R_B = \frac{Pa^2}{L^3}(L + 2b) \tag{10-31a,b}$$

Thus, all reactions for the fixed-end beam have been determined.

The reactions at the supports of a beam with fixed ends are commonly referred to as **fixed-end moments** and **fixed-end forces**. They are widely used in structural analysis, and formulas for these quantities are listed in engineering handbooks.

Deflection at point D: To obtain the deflection at point D in the original fixed-end beam (Fig. 10-15a), again use the principle of superposition. The deflection at point D is equal to the sum of three deflections: (1) the downward deflection $(\delta_D)_1$ at point D in the released structure due to the load P (Fig. 10-15b); (2) the upward deflection $(\delta_D)_2$ at the same point in the released structure due to the redundant M_A (Fig. 10-15c); and (3) the upward deflection $(\delta_D)_3$ at the same point in the released structure due to the redundant M_B (Fig. 10-15d). This superposition of deflections is expressed by

$$\delta_D = (\delta_D)_1 - (\delta_D)_2 - (\delta_D)_3 \tag{g}$$

in which δ_D is the downward deflection in the original beam.

The deflections appearing in Eq. (g) are obtained from the formulas given in Table H-2 of Appendix H (see Cases 5 and 7) by making the appropriate substitutions and algebraic simplifications. The results of these manipulations are

$$(\delta_D)_1 = \frac{Pa^2b^2}{3LEI} \quad (\delta_D)_2 = \frac{M_Aab}{6LEI}(L + b) \quad (\delta_D)_3 = \frac{M_Bab}{6LEI}(L + a)$$

Substitute the expressions for M_A and M_B from Eqs. (10-30a and b) into the last two expressions to get

$$(\delta_D)_2 = \frac{Pa^2b^3}{6L^3EI}(L + b) \quad (\delta_D)_3 = \frac{Pa^3b^2}{6L^3EI}(L + a)$$

Therefore, the deflection at point D in the original beam, obtained by substituting $(\delta_D)_1$, $(\delta_D)_2$, and $(\delta_D)_3$ into Eq. (g) and simplifying, is

$$\delta_D = \frac{Pa^3b^3}{3L^3EI} \tag{10-32}$$

4. ***Finalize***: The method described in this example for finding the deflection δ_D can be used not only to find deflections at individual points, but also to find the equations of the deflection curve.

FIGURE 10-16

Fixed-end beam with a concentrated load acting at the midpoint

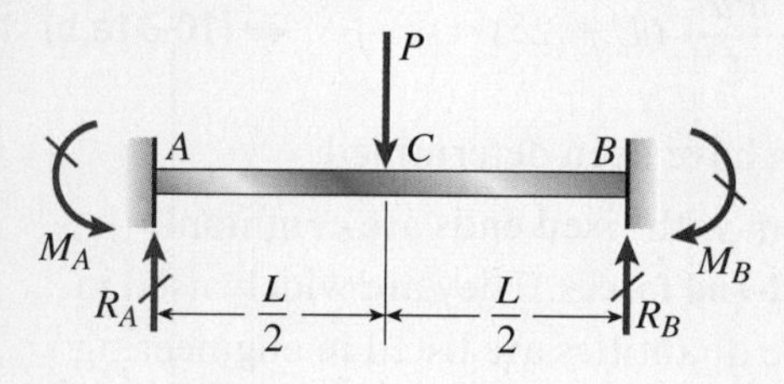

Concentrated load acting at the midpoint of the beam: When the load P acts at the midpoint C (Fig. 10-16), the reactions of the beam [from Eqs. (10-30) and (10-31) with $a = b = L/2$] are

$$M_A = M_B = \frac{PL}{8} \quad R_A = R_B = \frac{P}{2} \tag{10-33a,b}$$

Also, the deflection at the midpoint [from Eq. (10-32)] is

$$\delta_C = \frac{PL^3}{192EI} \tag{10-34}$$

This deflection is only one-fourth of the deflection at the midpoint of a simple beam with the same load, which shows the stiffening effect of clamping the ends of the beam.

The preceding results for the reactions at the ends and the deflection at the middle [Eqs. (10-33) and (10-34)] agree with those found in Example 10-2 by solving the differential equation of the deflection curve [see Eqs. (10-13), (10-14), and (10-19)].

Example 10-5

A fixed-end beam AB supports a uniform load of intensity q acting over part of the span (Fig. 10-17a).

Determine the reactions of this beam (that is, find the fixed-end moments and fixed-end forces).

FIGURE 10-17

Example 10-5: (a) Fixed-end beam with a uniform load over part of the span and (b) reactions produced by an element qdx of the uniform load

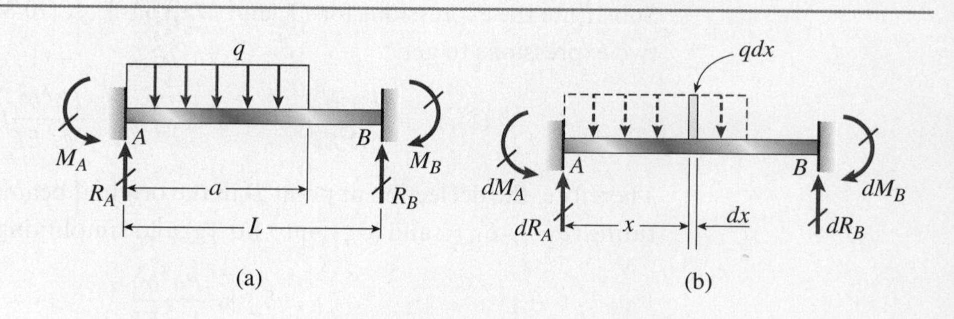

Solution:

Use a four-step problem-solving approach.

1. ***Conceptualize***: Find the reactions of this beam by using the principle of superposition together with the results obtained in Example 10-4. In that example, reactions were found for a fixed-end beam subjected to a concentrated load P acting at distance a from the left-hand end [see Fig. 10-15a and Eqs. (10-30) and (10-31)].

   In order to apply those results to the uniform load of Fig. 10-17a, treat an element of the uniform load as a concentrated load of magnitude qdx acting at distance x from the left-hand end (Fig. 10-17b). Then, using the formulas derived in Example 10-4, find the reactions caused by this element of load. Finally, by integrating over the length a of the uniform load, obtain the reactions due to the entire uniform load.

2. ***Categorize***:

   Fixed-end moments: Begin with the moment reactions using Eqs. (10-30a and b) of Example 10-4. To obtain the moments caused by the element qdx of the uniform load (compare Fig. 10-17b with Fig. 10-15a), replace P with qdx, a with x, and b with $L - x$. Thus, the fixed-end moments due to the element of load (Fig. 10-17b) are

$$dM_A = \frac{qx(L-x)^2\,dx}{L^2} \qquad dM_B = \frac{qx^2(L-x)dx}{L^2}$$

3. ***Analyze***: Integrate over the loaded part of the beam to get the fixed-end moments due to the entire uniform load:

$$M_A = \int dM_A = \frac{q}{L^2}\int_0^a x(L-x)^2\,dx = \frac{qa^2}{12L^2}(6L^2 - 8aL + 3a^2) \quad \textbf{(10-35a)}$$

$$M_B = \int dM_B = \frac{q}{L^2}\int_0^a x^2(L-x)dx = \frac{qa^3}{12L^2}(4L - 3a) \quad \textbf{(10-35b)}$$

   Fixed-end forces: Proceed in a similar manner as for the fixed-end moments, but use Eqs. (10-31a and b) to obtain the following expressions for the fixed-end forces due to the element qdx of load:

$$dR_A = \frac{q(L-x)^2(L+2x)dx}{L^3} \qquad dR_B = \frac{qx^2(3L-2x)dx}{L^3}$$

   Integration gives

$$R_A = \int dR_A = \frac{q}{L^3}\int_0^a (L-x)^2(L+2x)dx = \frac{qa}{2L^3}(2L^3 - 2La^2 + a^3) \quad \textbf{(10-36a)}$$

$$R_B = \int dR_B = \frac{q}{L^3}\int_0^a x^2(3L-2x)dx = \frac{qa^3}{2L^3}(2L - a) \quad \textbf{(10-36b)}$$

   Thus, all reactions (fixed-end moments and fixed-end forces) have been found.

FIGURE 10-18

Fixed-end beam with a uniform load

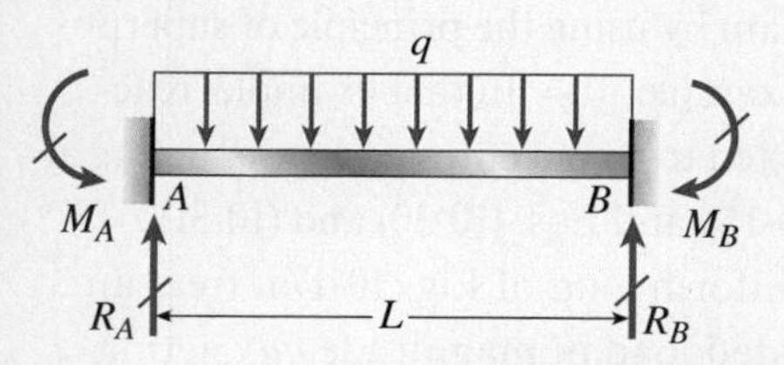

4. ***Finalize:***

Uniform load acting over the entire length of the beam: When the load acts over the entire span (Fig. 10-18), obtain the reactions by substituting $a = L$ into the preceding equations, yielding

$$M_A = M_B = \frac{qL^2}{12} \quad R_A = R_B = \frac{qL}{2} \tag{10-37a,b}$$

The deflection at the midpoint of a uniformly loaded beam is also of interest. The simplest way to obtain this deflection is to use the method of superposition. The first step is to remove the moment restraints at the supports and obtain a released structure in the form of a simple beam. The downward deflection at the midpoint of a simple beam due to a uniform load (from Case 1, Table H-2) is

$$(\delta_C)_1 = \frac{5qL^4}{384EI} \tag{a}$$

and the upward deflection at the midpoint due to the end moments (from Case 10, Table H-2) is

$$(\delta_C)_2 = \frac{M_A L^2}{8EI} = \frac{(qL^2/12)L^2}{8EI} = \frac{qL^4}{96EI} \tag{b}$$

Thus, the final downward deflection of the original fixed-end beam (Fig. 10-18) is

$$\delta_C = (\delta_C)_1 - (\delta_C)_2$$

Substitute for the deflections from Eqs. (a) and (b) to get

$$\delta_C = \frac{qL^4}{384EI} \tag{10-38}$$

This deflection is one-fifth of the deflection at the midpoint of a simple beam with a uniform load [Eq. (a)], again illustrating the stiffening effect of fixity at the ends of the beam.

Example 10-6

A beam ABC (Fig. 10-19a) rests on simple supports at points A and B and is supported by a cable at point C. The beam has total length $2L$ and supports a uniform load of intensity q. Prior to the application of the uniform load, there is no force in the cable nor is there any slack in the cable.

When the uniform load is applied, the beam deflects downward at point C and a tensile force T develops in the cable.

Find the magnitude of this force.

FIGURE 10-19

Example 10-6: Beam ABC with one end supported by a cable

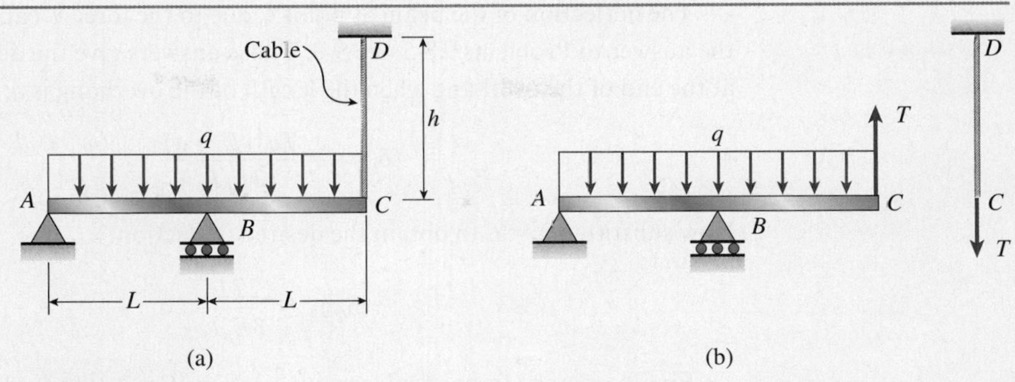

Solution:

Use a four-step problem-solving approach.

1. *Conceptualize*:

Redundant force: The structure $ABCD$, consisting of the beam and cable, has three vertical reactions (at points A, B, and D). However, only two equations of equilibrium are available from a free-body diagram of the entire structure. Therefore, the structure is statically indeterminate to the first degree, and one redundant quantity must be selected for purposes of analysis.

The tensile force T in the cable is a suitable choice for the redundant. Release this force by removing the connection at point C, thereby cutting the structure into two parts (Fig. 10-19b). The released structure consists of the beam ABC and the cable CD as separate elements, with the redundant force T acting upward on the beam and downward on the cable.

2. *Categorize*:

Equation of compatibility: The deflection at point C of beam ABC (Fig. 10-19b) consists of two parts: a downward deflection $(\delta_C)_1$ due to the uniform load and an upward deflection $(\delta_C)_2$ due to the force T. At the same time, the lower end C of cable CD displaces downward by an amount $(\delta_C)_3$, which is equal to the elongation of the cable due to the force T. Therefore, the *equation of compatibility*, which expresses the fact that the downward deflection of end C of the beam is equal to the elongation of the cable, is

$$(\delta_C)_1 - (\delta_C)_2 = (\delta_C)_3 \quad \text{(a)}$$

Having formulated this equation, now turn to the task of evaluating all three displacements.

3. *Analyze*:

Force-displacement relations: The deflection $(\delta_C)_1$ at the end of the overhang (point C in beam ABC) due to the uniform load can be found from the results given in Example 9-9 of Section 9.5 (see Fig. 9-21). Use Eq. (9-68) of that example and substitute $a = L$ to get

$$(\delta_C)_1 = \frac{qL^4}{4E_bI_b} \quad \text{(b)}$$

where E_bI_b is the flexural rigidity of the beam.

The deflection of the beam at point C due to the force T can be taken from the answer to Problems 9.8-5 or 9.9-3. Those answers give the deflection $(\delta_C)_2$ at the end of the overhang when the length of the overhang is a:

$$(\delta_C)_2 = \frac{Ta^2(L + a)}{3E_bI_b}$$

Now substitute $a = L$ to obtain the desired deflection:

$$(\delta_C)_2 = \frac{2TL^3}{3E_bI_b} \quad \text{(c)}$$

Finally, use the force-displacement relation (Eq. 2-3) to find the elongation of the cable

$$(\delta_C)_3 = \frac{Th}{E_cA_c} \quad \text{(d)}$$

where h is the length of the cable and E_cA_c is its axial rigidity.

Force in the cable: Substitute the three displacements [Eqs. (b), (c), and (d)] into the equation of compatibility [Eq. (a)] to get

$$\frac{qL^4}{4E_bI_b} - \frac{2TL^3}{3E_bI_b} = \frac{Th}{E_cA_c}$$

Solve for the force T to find

$$T = \frac{3qL^4E_cA_c}{8L^3E_cA_c + 12hE_bI_b} \quad \text{(10-39)}$$

4. ***Finalize***: With the force T known, all reactions, shear forces, and bending moments can be found by means of free-body diagrams and equations of equilibrium.

This example illustrates how an internal force quantity (instead of an external reaction) can be used as the redundant.

FIGURE 10-19 (Repeated)

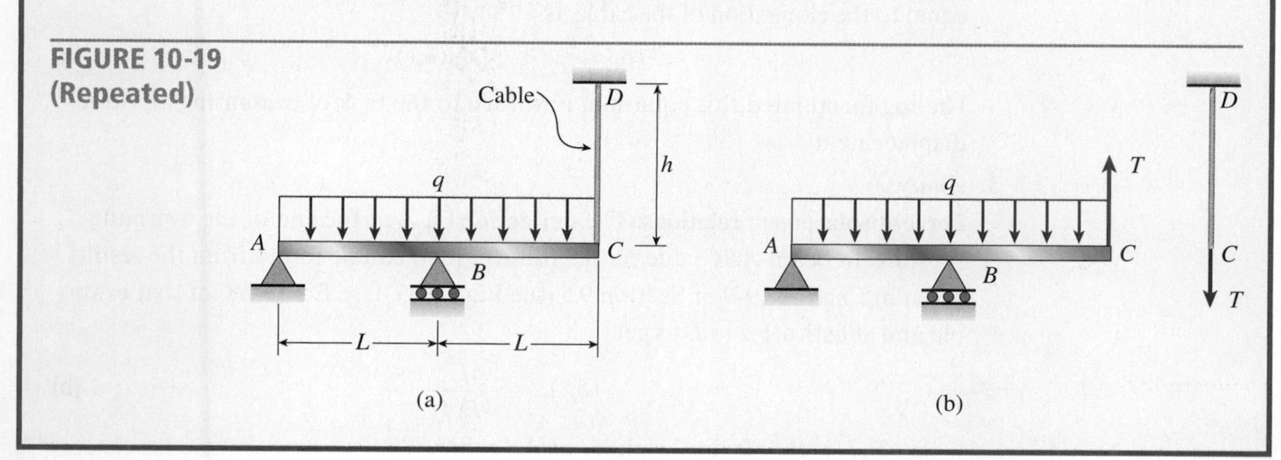

*10.5 Temperature Effects

Temperature changes may produce changes in lengths of bars and lateral deflections of beams, as discussed previously in Sections 2.5 and 9.11. If these length changes and lateral deflections are restrained, thermal stresses will be produced in the material. In Section 2.5, you saw how to find these stresses in statically indeterminate bars. Now consider some of the effects of temperature changes in statically indeterminate beams.

The stresses and deflections produced by temperature changes in a statically indeterminate beam can be analyzed by methods that are similar to those already described for the effects of loads. To begin the discussion, consider the propped cantilever beam AB shown in Fig. 10-20. Assume that the beam was originally at a uniform temperature T_0, but later its temperature is increased to T_1 on the upper surface and T_2 on the lower surface. The variation of temperature over the height h of the beam is assumed to be linear.

FIGURE 10-20

Propped cantilever beam with a temperature differential

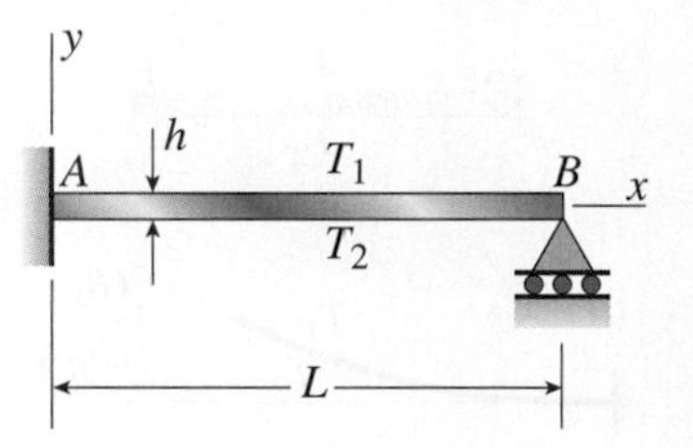

Because the temperature varies linearly, the *average* temperature of the beam is

$$T_{aver} = \frac{T_1 + T_2}{2} \tag{10-40}$$

and occurs at mid-height of the beam. The difference between this average temperature and the initial temperature T_0 results in a tendency for the beam to change in length. If the beam is free to expand longitudinally, its length will increase by an amount δ_T given by Eq. (9-134), which is repeated here:

$$\delta_T = \alpha(T_{aver} - T_0)L = \alpha\left(\frac{T_1 + T_2}{2} - T_0\right)L \tag{10-41}$$

In this equation, α is the coefficient of thermal expansion of the material and L is the length of the beam. If longitudinal expansion is free to occur, no axial stresses will be produced by the temperature changes. However, if longitudinal expansion is restrained, axial stresses will develop, as described in Section 2.5.

Now consider the effects of the temperature differential $T_2 - T_1$, which tends to produce a *curvature* of the beam but no change in length. Curvature due to temperature changes is described in Section 9.11, where the following differential equation of the deflection curve is derived [see Eq. (9-137)]:

$$\frac{d^2v}{dx^2} = \frac{\alpha(T_2 - T_1)}{h} \tag{10-42}$$

This equation applies to a beam that is unrestrained by supports and is free to deflect and rotate. Note that when T_2 is greater than T_1, the curvature is positive, and the beam tends to bend concave upward. Deflections and rotations of simple beams and cantilever beams due to a temperature differential can be determined with the aid of Eq. (10-42), as discussed in Section 9.11. Now use those results when analyzing statically indeterminate beams using the method of superposition.

Method of Superposition

To illustrate the use of superposition, find the reactions of the fixed-end beam of Fig. 10-21a due to the temperature differential. As usual, begin the analysis by selecting the redundant reactions. Although other choices result in more efficient

FIGURE 10-21

(a) Fixed-end beam with a temperature differential, (b) released structure, and (c) deflection curve for the released structure

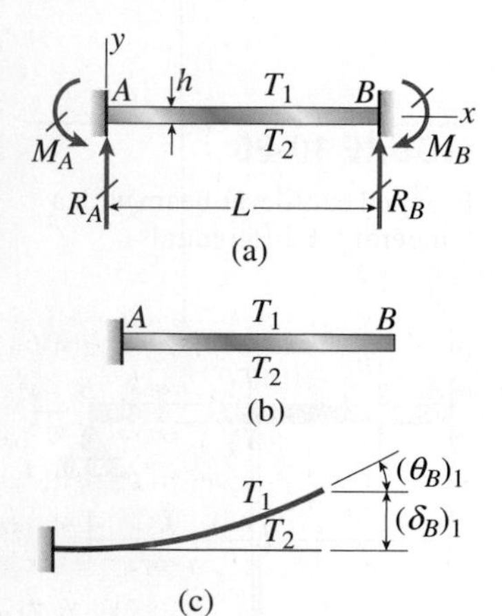

calculations, select the reactive force R_B and reactive moment M_B as the redundants in order to illustrate the general methodology.

Remove the supports corresponding to the redundants to obtain the released structure shown in Fig. 10-21b (a cantilever beam). The deflection and angle of rotation at end B of this cantilever (due to the temperature differential) are (see Fig. 10-21c):

$$(\delta_B)_1 = \frac{\alpha(T_2 - T_1)L^2}{2h} \quad (\theta_B)_1 = \frac{\alpha(T_2 - T_1)L}{h}$$

These equations are obtained from the solution to Problem 9.11-2 in the preceding chapter. Note that when T_2 is greater than T_1, the deflection $(\delta_B)_1$ is upward, and the angle of rotation $(\theta_B)_1$ is counterclockwise.

Next, find the deflections and angles of rotation in the released structure (Fig. 10-21b) due to the redundants R_B and M_B. These quantities are obtained from Cases 4 and 6, respectively, of Table H-1:

$$(\delta_B)_2 = \frac{R_B L^3}{3EI} \quad (\theta_B)_2 = \frac{R_B L^2}{2EI}$$

$$(\delta_B)_3 = -\frac{M_B L^2}{2EI} \quad (\theta_B)_3 = -\frac{M_B L}{EI}$$

In these expressions, upward deflection and counterclockwise rotation are positive (as in Fig. 10-21c).

Now write the equations of compatibility for the deflection and angle of rotation at support B as

$$\delta_B = (\delta_B)_1 + (\delta_B)_2 + (\delta_B)_3 = 0 \quad \textbf{(10-43a)}$$

$$\theta_B = (\theta_B)_1 + (\theta_B)_2 + (\theta_B)_3 = 0 \quad \textbf{(10-43b)}$$

or, upon substituting the appropriate expressions,

$$\frac{\alpha(T_2 - T_1)L^2}{2h} + \frac{R_B L^3}{3EI} - \frac{M_B L^2}{2EI} = 0 \quad \textbf{(10-43c)}$$

$$\frac{\alpha(T_2 - T_1)L}{h} + \frac{R_B L^2}{2EI} - \frac{M_B L}{EI} = 0 \quad \textbf{(10-43d)}$$

These equations can be solved simultaneously for the two redundants:

$$R_B = 0 \quad M_B = \frac{\alpha EI(T_2 - T_1)}{h}$$

The fact that R_B is zero could have been anticipated initially from the symmetry of the fixed-end beam. If this fact had been utilized from the outset, the preceding solution would have been simplified because only one equation of compatibility would have been required.

From symmetry (or from equations of equilibrium), reaction R_B is equal to the reaction R_A and the moment M_A is equal to the moment M_B. Therefore, the reactions for the fixed-end beam shown in Fig. 10-21a are

$$R_A = R_B = 0 \quad M_A = M_B = \frac{\alpha EI(T_2 - T_1)}{h} \quad \textbf{(10-44a,b)}$$

Based on these results, observe that the beam is subjected to a constant bending moment due to the temperature changes.

Differential Equation of the Deflection Curve

The fixed-end beam of Fig. 10-21a also can be analyzed by solving the differential equation of the deflection curve. When a beam is subjected to both a bending moment M and a temperature differential $T_2 - T_1$, the differential equation becomes [see Eqs. (9-11) and (10-42)]:

$$\frac{d^2v}{dx^2} = \frac{M}{EI} + \frac{\alpha(T_2 - T_1)}{h} \quad \text{(10-45a)}$$

or

$$EIv'' = M + \frac{\alpha EI(T_2 - T_1)}{h} \quad \text{(10-45b)}$$

For the fixed-end beam of Fig. 10-21a, the expression for the bending moment in the beam is

$$M = R_A x - M_A \quad \text{(10-46)}$$

where x is measured from support A. Substitute into the differential equation and integrate to obtain the following equation for the slope of the beam:

$$EIv' = \frac{R_A x^2}{2} - M_A x + \frac{\alpha EI(T_2 - T_1)x}{h} + C_1 \quad \text{(10-47)}$$

The two boundary conditions on the slope ($v' = 0$ when $x = 0$ and $x = L$) give $C_1 = 0$ and

$$\frac{R_A L}{2} - M_A = -\frac{\alpha EI(T_2 - T_1)}{h} \quad \text{(10-48)}$$

A second integration gives the deflection of the beam:

$$EIv = \frac{R_A x^3}{6} - \frac{M_A x^2}{2} + \frac{\alpha EI(T_2 - T_1)x^2}{2h} + C_2 \quad \text{(10-49)}$$

The boundary conditions on the deflection ($v = 0$ when $x = 0$ and $x = L$) give $C_2 = 0$ and

$$\frac{R_A L}{3} - M_A = -\frac{\alpha EI(T_2 - T_1)}{h} \quad \text{(10-50)}$$

Solve Eqs. (10-48) and (10-50) simultaneously to find

$$R_A = 0 \quad M_A = \frac{\alpha EI(T_2 - T_1)}{h}$$

The equilibrium of the beam leads to $R_B = 0$ and $M_B = M_A$. Thus, these results agree with those found by the method of superposition [see Eqs. (10-44a and b)].

Note that the preceding solution was carried out without taking advantage of symmetry to illustrate the general approach of the integration method.

Knowing the reactions of the beam, you can now find the shear forces, bending moments, slopes, and deflections. The simplicity of the results may surprise you.

Example 10-7

FIGURE 10-22

Example 10-7: Two-span beam subject to differential temperature change

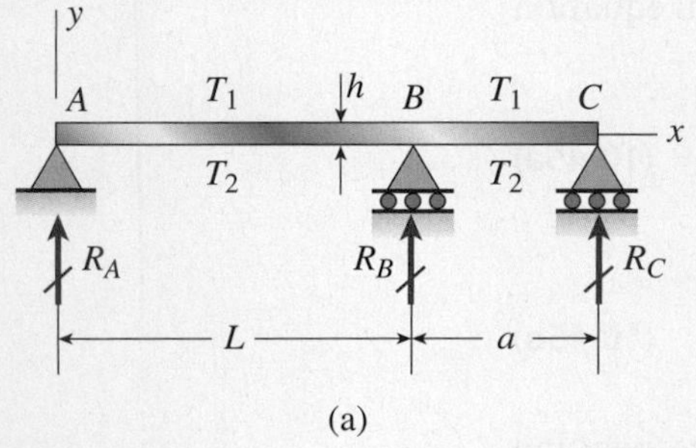

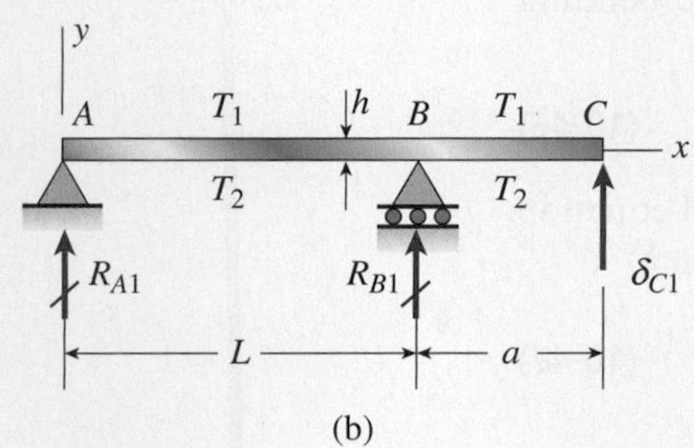

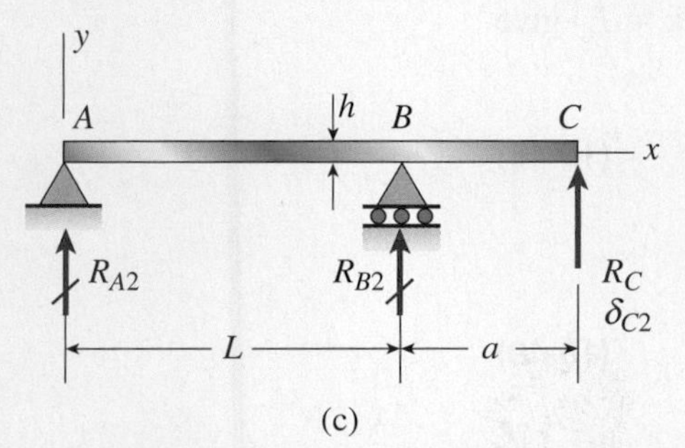

The two-span beam ABC in Fig. 10-22 has a pin support at A, a roller support at B, and *either* a roller (Fig. 10-22) or elastic spring support (Fig. 10-23) (spring constant k) at C. The beam has a height of h and is subjected to a temperature differential with temperature T_1 on its upper surface and T_2 on its lower surface (see Figs. 10-22a and b). Assume that the elastic spring is unaffected by the temperature change.

(a) If support C is a roller support, find all support reactions using the *method of superposition*.

(b) Find all support reactions if the roller at C is replaced by the *elastic spring support*; also find the displacement at C.

Solution:

Use a four-step problem-solving approach.

Part (a): Roller support at C.

1. ***Conceptualize:***

   Roller support at C: This beam (Fig. 10-22a) is statically indeterminate to the first degree (see discussion in Example 10-3 solution). Select reaction R_C as the redundant in order to use the analyses of the released structure (with the support at C removed) presented in Examples 9-5 (concentrated load applied at C) and 9-19 (subject to temperature differential). Use the method of superposition, also known as the *force* or *flexibility* method, to find the solution.

2. ***Categorize:***

   Superposition: The superposition process is shown in the Figs. 10-22b and c in which redundant R_C is removed to produce a *released* (or statically determinate) structure.

   First apply the "actual loads" [here, temperature differential $(T_2 - T_1)$], and then *apply the redundant R_C as a load* to the second released structure.

3. ***Analyze:***

   Equilibrium: Sum forces in the y direction in Fig. 10-22a (using a statics sign convention in which upward forces in the y direction are positive) to find that

$$R_A + R_B = -R_C \quad \textbf{(a)}$$

   Sum moments about B (again using a statics sign convention in which counter-clockwise is positive) to find

$$-R_A L + R_C a = 0$$

   so

$$R_A = \left(\frac{a}{L}\right) R_C \quad \textbf{(b)}$$

which can be subsitituted back into Eq. (a) to give

$$R_B = -R_C - \left(\frac{a}{L}\right)R_C = -R_C\left(1 + \frac{a}{L}\right) \tag{c}$$

(Note that you could also find reactions R_A and R_B using superposition of the reactions shown in Figs. 10-22b and c: $R_A = R_{A1} + R_{A2}$ and $R_B = R_{B1} + R_{B2}$, where R_{A1} and R_{B1} are known to be zero.)

Compatibility: Displacement $\delta_C = 0$ in the actual structure (Fig. 10-22a), so compatibility of displacements requires that

$$\delta_{C1} + \delta_{C2} = \delta_C = 0 \tag{d}$$

where δ_{C1} and δ_{C2} are shown in Figs. 10-22b and c for the released structures subject to temperature differential and applied redundant force R_C, respectively. Initially, δ_{C1} and δ_{C2} are assumed positive (upward) when using a statics sign convention, and a negative result indicates that the reverse is true.

Force-displacement and temperature-displacement relations: Now use the results of Examples 9-5 and 9-19 to find displacements δ_{C1} and δ_{C2}. First, from Eq. (f) in Example 9-19,

$$\delta_{C1} = \frac{\alpha(T_2 - T_1)a(L + a)}{2h} \tag{e}$$

and from Eq. (9-55) (modified to include variable a as the length of member BC and replacing load P with redundant force R_C - see solution to Problems 9.8-5(b) or 9.9-3),

$$\delta_{C2} = \frac{R_C a^2(L + a)}{3EI} \tag{f}$$

Reactions: Now substitute Eqs. (e) and (f) into Eq. (d) and then solve for redundant R_C:

$$\frac{\alpha(T_2 - T_1)}{2h}(a)(L + a) + \frac{R_C a^2(L + a)}{3EI} = 0$$

so

$$R_C = \frac{-3EI\alpha(T_2 - T_1)}{2ah} \tag{g}$$

noting that the negative result means that reaction force R_C is downward [for positive temperature differential $(T_2 - T_1)$]. Now substitute the expression for R_C into Eqs. (b) and (c) to find reactions R_A and R_B as

$$R_A = \left(\frac{a}{L}\right)R_C = \left(\frac{a}{L}\right)\left[\frac{-3EI\alpha(T_2 - T_1)}{2ah}\right] = \frac{-3EI\alpha(T_2 - T_1)}{2Lh} \tag{h}$$

$$R_B = -R_C\left(1 + \frac{a}{L}\right) = \frac{3EI\alpha(T_2 - T_1)}{2ah}\left(1 + \frac{a}{L}\right)$$
$$= \frac{3EI\alpha(T_2 - T_1)(L + a)}{2Lah} \quad \text{(i)}$$

where R_A acts downward and R_B acts upward.

4. *Finalize*:

Numerical example: In Example 9-19, the upward displacement at joint C was computed [see Eq. (h), Example 9-19] assuming that beam ABC is a steel wide flange, HE 700B (see Table F-1), with a length $L = 9$ m, an overhang $a = L/2$, and subject to temperature differential $(T_2 - T_1) = 3°\text{C}$. From Table I-4, the coefficient of thermal expansion for structural steel is $\alpha = 12 \times 10^{-6}/°\text{C}$. The modulus for steel is 210 GPa. Now find numerical values of reactions R_A, R_B, and R_C using Eqs. (g), (h), and (i):

$$R_A = \frac{-3EI\alpha(T_2 - T_1)}{2Lh} = \frac{-3(210\text{ GPa})(256900\text{ cm}^4)(12 \times 10^{-6})(3)}{2(9\text{ m})(700\text{ mm})}$$
$$= -4.62\text{ kN (downward)}$$

$$R_B = \frac{3EI\alpha(T_2 - T_1)(L + a)}{2Lah}$$
$$= \frac{3(210\text{ GPa})(256900\text{ cm}^4)(12 \times 10^{-6})(3)(9\text{ m} + 4.5\text{ m})}{2(9\text{ m})(4.5\text{ m})(700\text{ mm})}$$
$$= 13.87\text{ kN (upward)}$$

$$R_C = \frac{-3EL\alpha(T_2 - T_1)}{2ah} = \frac{-3(210\text{ GPa})(256900\text{ cm}^4)(12 \times 10^{-6})(3)}{2(4.5\text{ m})\ 700\text{ mm}}$$
$$= -9.25\text{ kN (downward)}$$

Note that the reactions sum to zero as required for equilibrium.

Part (b): Elastic spring support at C.

1. *Conceptualize*:

Spring support at C: Once again, select reaction R_C as the redundant. However, R_C is now at the base of the elastic spring support (see Fig. 10-23). When redundant reaction R_C is applied to the second released structure (Fig. 10-23c), it will first compress the spring and then be applied to the beam at C, causing upward deflection.

2. *Categorize*:

Superposition: The superposition solution approach (i.e., force or flexibility method) follows that used previously and is shown in Fig. 10-23.

3. *Analyze*:

Equilibrium: The addition of the spring support at C does not alter the expressions of static equilibrium in Eqs. (a), (b), and (c).

FIGURE 10-23

Example 10-7: Two-span beam with elastic support and subject to differential temperature change

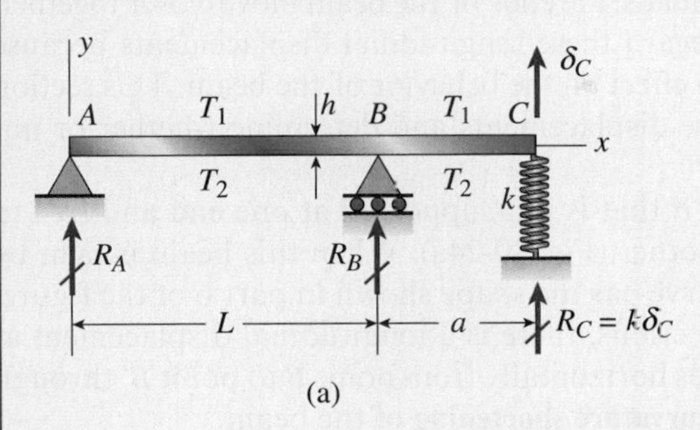

(a)

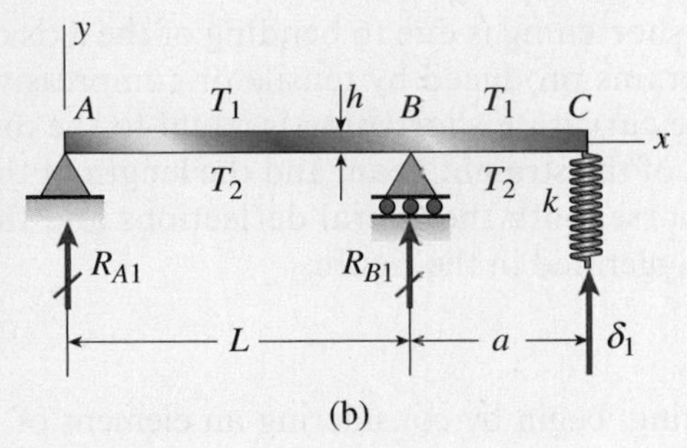

(b)

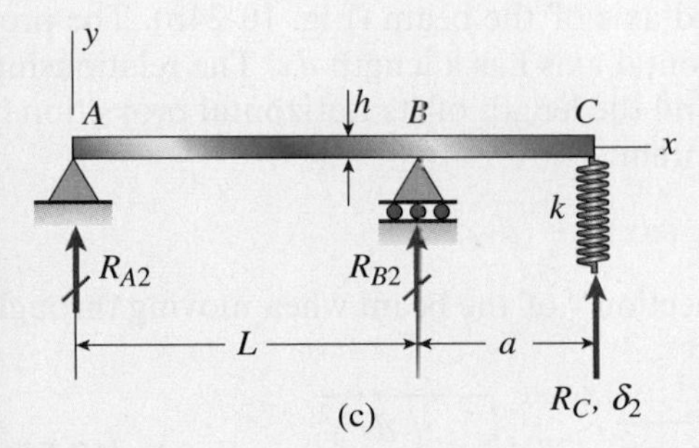

(c)

Compatibility: The compatibility equation is now written for the *base of the spring* (not the top of the spring, where it is attached to the beam at *C*). From Fig. 10-23, compatibility of displacements requires:

$$\delta_1 + \delta_2 = \delta = 0 \tag{j}$$

Force-displacement and temperature-displacement relations: The spring is assumed to be unaffected by the temperature differential, so the top and base of the spring displace the same in Fig. 10-23b, which means that Eq. (e) is still valid and $\delta_1 = \delta_{C1}$. However, the compression of the spring must be included in the expression for δ_2, so

$$\delta_2 = \frac{R_C}{k} + \delta_{C2} = \frac{R_C}{k} + \frac{R_C a^2(L+a)}{3EI} \tag{k}$$

where the expression for δ_{C2} comes from Eq. (f).

Reactions: Now substitute Eqs. (e) and (k) into compatibility with Eq. (j) and solve for redundant R_C:

$$\frac{\alpha(T_2 - T_1)}{2h}(a)(L+a) + \frac{R_C a^2(L+a)}{3EI} + \frac{R_C}{k} = 0$$

so

$$R_C = \frac{-a\alpha(T_2 - T_1)(L+a)}{2h\left[\dfrac{1}{k} + \dfrac{a^2(L+a)}{3EI}\right]} \tag{l}$$

From statics [Eqs. (b) and (c)], reactions at *A* and *B* are

$$R_A = \left(\frac{a}{L}\right)R_C = \frac{-a\alpha(T_2 - T_1)a(L+a)}{2Lh\left[\dfrac{1}{k} + \dfrac{a^2(L+a)}{3EI}\right]} \tag{m}$$

$$R_B = -R_C\left(1 + \frac{a}{L}\right) = \frac{a\alpha(T_2 - T_1)(L+a)^2}{2Lh\left[\dfrac{1}{k} + \dfrac{a^2(L+a)}{3EI}\right]} \tag{n}$$

4. ***Finalize***: Once again, the minus signs for R_A and R_C indicate that they are downward [for positive $(T_2 - T_1)$], while R_B is upward. Finally, if spring constant *k* goes to infinity, the support at *C* is once again a roller support, as in Fig. 10-22, and Eqs. (l), (m), and (n) reduce to Eqs. (g), (h), and (i).

*10.6 Longitudinal Displacements at the Ends of a Beam

When a beam is bent by lateral loads, the ends of the beam move closer together. It is common practice to disregard these longitudinal displacements because usually they have no noticeable effect on the behavior of the beam. This section discusses how to evaluate these displacements and determine whether or not they are important.

Consider a simple beam AB that is pin-supported at one end and free to displace longitudinally at the other (Fig. 10-24a). When this beam is bent by lateral loads, the deflection curve has the shape shown in part b of the figure. In addition to the lateral deflections, there is a longitudinal displacement at end B of the beam. End B moves horizontally from point B to point B' through a small distance λ, called the **curvature shortening** of the beam.

As the name implies, curvature shortening is due to bending of the axis of the beam and is *not* due to axial strains produced by tensile or compressive forces. As shown in Fig. 10-24b, the curvature shortening is equal to the difference between the initial length L of the straight beam and the length of the chord AB' of the bent beam. Of course, both the lateral deflections and the curvature shortening are highly exaggerated in the figure.

Curvature Shortening

To determine the curvature shortening, begin by considering an element of a length ds measured along the curved axis of the beam (Fig. 10-24b). The projection of this element on the horizontal axis has a length dx. The relationship between the length of the element and the length of its horizontal projection is obtained from the Pythagorean theorem:

$$(ds)^2 = (dx)^2 + (dv)^2$$

where dv is the increment in the deflection v of the beam when moving through the distance dx. Thus,

$$ds = \sqrt{(dx)^2 + (dv)^2} = dx\sqrt{1 + \left(\frac{dv}{dx}\right)^2} \tag{10-51a}$$

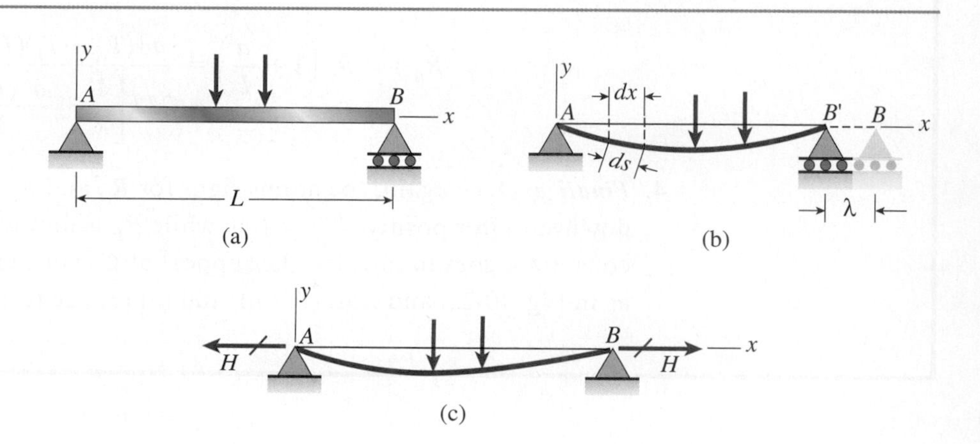

FIGURE 10-24

(a) Simple beam with lateral loads, (b) horizontal displacement λ at the end of the beam, and (c) horizontal reactions H for a beam with immovable supports

The difference between the length of the element and the length of its horizontal projection is

$$ds - dx = dx\sqrt{1 + \left(\frac{dv}{dx}\right)^2} - dx = dx\left[\sqrt{1 + \left(\frac{dv}{dx}\right)^2} - 1\right] \quad \textbf{(10-51b)}$$

Now introduce the following binomial series (see Appendix C):

$$\sqrt{1+t} = 1 + \frac{t}{2} - \frac{t^2}{8} + \frac{t^3}{16} - \cdots \quad \textbf{(10-52)}$$

which converges when t is numerically less than 1. If t is very small compared to 1, disregard the terms involving t^2, t^3, and so on in comparison with the first two terms. It follows that

$$\sqrt{1+t} \approx 1 + \frac{t}{2} \quad \textbf{(10-53)}$$

The term $(dv/dx)^2$ in Eq. (10-51b) is ordinarily very small compared to 1. Therefore, use Eq. (10-53) with $t = (dv/dx)^2$ and rewrite Eq. (10-51b) as

$$ds - dx = dx\left[1 + \frac{1}{2}\left(\frac{dv}{dx}\right)^2 - 1\right] = \frac{1}{2}\left(\frac{dv}{dx}\right)^2 dx \quad \textbf{(10-54)}$$

Integrate the left- and right-hand sides of this expression over the length of the beam to obtain an expression for the difference between the length of the beam and the length of the chord AB' (Fig. 10-24b):

$$L - \overline{AB'} = \int_0^L \frac{1}{2}\left(\frac{dv}{dx}\right)^2 dx$$

Thus, the curvature shortening is

$$\lambda = \frac{1}{2}\int_0^L \left(\frac{dv}{dx}\right)^2 dx \quad \textbf{(10-55)}$$

This equation is valid provided the deflections and slopes are small.

Note that when the equation of the deflection curve is known, substitute into Eq. (10-55) and determine the shortening λ.

Horizontal Reactions

Now suppose that the ends of the beam are prevented from translating longitudinally by immovable supports (Fig. 10-24c). Because the ends cannot move toward each other, a horizontal reaction H will develop at each end. This force will cause the axis of the beam to elongate as bending occurs.

In addition, the force H itself will have an effect upon the bending moments in the beam, because an additional bending moment (equal to H times the deflection) will exist at every cross section. Thus, the deflection curve of the beam depends not only upon the lateral loads but also upon the reaction H, which in turn depends upon the shape of the deflection curve, as shown by Eq. (10-55).

Rather than attempt an exact analysis of this complicated problem, obtain an approximate expression for the force H in order to ascertain its importance.

For that purpose, use any reasonable approximation to the deflection curve. In the case of a pin-ended beam with downward loads (Fig. 10-24c), a good approximation is a parabola having the equation

$$v = -\frac{4\delta x(L - x)}{L^2} \tag{10-56}$$

where δ is the downward deflection at the midpoint of the beam. [Note that Eq. (10-56) also can be obtained using expressions from Case 10, *pure bending of a simply supported beam*, in Table H-2 of Appendix H]. The curvature shortening λ corresponding to this assumed deflected shape is found by substituting the expression for the deflection v into Eq. (10-55) and integrating; the result is

$$\lambda = \frac{8\delta^2}{3L} \tag{10-57}$$

The horizontal force H required to elongate the beam by this amount (see Eqs. 2-1a and 2-4a) is

$$H = \frac{EA\lambda}{L} = \frac{8EA\delta^2}{3L^2} \tag{10-58}$$

in which EA is the axial rigidity of the beam. The corresponding axial tensile stress in the beam is

$$\sigma_t = \frac{H}{A} = \frac{8E\delta^2}{3L^2} \tag{10-59}$$

This equation gives a close estimate of the tensile stress produced by the immovable supports of a simple beam.

General Comments

Now substitute some numerical values to assess the significance of the curvature shortening. The deflection δ at the midpoint of the beam is usually very small compared to the length; for example, the ratio δ/L might be 1/500 or smaller. Use this value and assume that the material is steel (with $E = 200$ GPa) to find from Eq. (10-59) that the tensile stress is only 2.1 MPa. Since the allowable tensile stress in the steel is typically 100 MPa or larger, it becomes clear that the axial stress due to the horizontal force H may be disregarded when compared to the ordinary working stresses in the beam.

Furthermore, in the derivation of Eq. (10-55), the ends of the beam were assume to be held rigidly against horizontal displacements, which is not physically possible. In reality, small longitudinal displacements always occur, thereby reducing the axial stress calculated from Eq. (10-59).[1]

Based on the preceding discussions, the customary practice of disregarding the effects of any longitudinal restraints and assuming that one end of the beam is on a roller support (regardless of the actual construction) is justified. The stiffening effect of longitudinal restraints is significant only when the beam is very long and slender and supports large loads. This behavior is sometimes referred to as "string action," because it is analogous to the action of a cable, or string, supporting a load.

[1]For a more complete analysis of beams with immovable supports, see Ref. 10-1.

CHAPTER SUMMARY AND REVIEW

Chapter 10 discussed the behavior of statically indeterminate beams acted on by concentrated and distributed loads, such as self-weight; thermal effects and longitudinal displacement due to curvature shortening also were considered. Two analysis approaches were presented: (1) **integration** of the equation of the elastic curve and (2) the more general approach based on **superposition**. In the superposition procedure, the **equilibrium** equations from statics were used along with **compatibility** equations and the **force-displacement relations** to generate the additional equations needed to solve the problem. The number of additional equations required is dependent on the **degree of static indeterminacy** of the structure. The superposition approach is limited to structures made of linearly elastic materials. Here are the major concepts presented in this chapter.

1. Several types of statically indeterminate beam structures, such as propped cantilever, fixed-end, and continuous beams were discussed. The **degree of static indeterminacy** was determined for each beam type, and a **released** structure was defined by removing a sufficient number of **redundant** reaction forces.

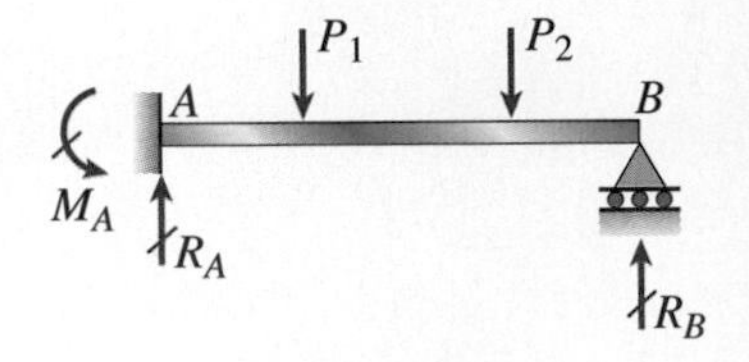

2. The released structure must be statically determinate and **stable** under the action of the applied loadings. It is also possible to insert **internal releases** on axial force, shear, and moment to create a released structure.

3. For simple, statically indeterminate beam structures, the **differential equation of the elastic curve** can be written as a second-, third-, or fourth-order equation in terms of moment, shear force, and distributed load, respectively. Applying boundary and other conditions leads to a solution for the constants of integration and the redundant reactions.

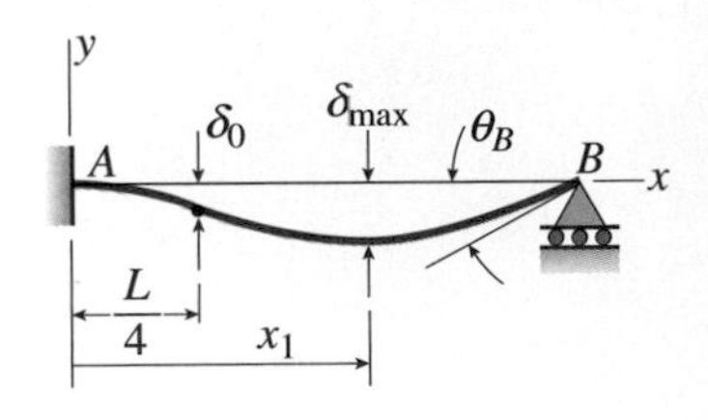

4. A more general solution approach for more complex beam and other structures is the **method of superposition** (also known as the **force** or **flexibility** method). Here, additional equations that describe the **compatibility** of displacements and incorporate the appropriate **force-displacement relations** for beams are used to supplement the **equilibrium** equations.

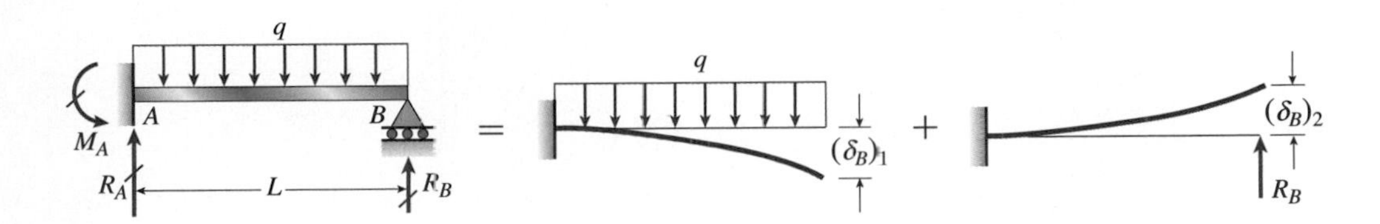

5. In most cases, there are multiple paths to the same solution depending upon the choice of the redundant reactions.

6. Differential **temperature changes** and **longitudinal displacements** induce reaction forces only in statically indeterminate beams; if the beam is statically determinate, joint displacements occur, but no internal forces result from these effects.

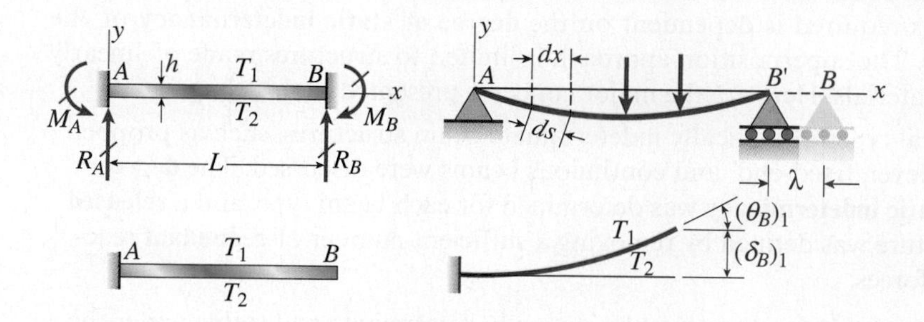

PROBLEMS Chapter 10

10.3 Differential Equations of the Deflection Curve

Solve the problems for Section 10.3 by integrating the differential equations of the deflection curve. All beams have constant flexural rigidity EI. When drawing shear-force and bending-moment diagrams, label all critical ordinates, including maximum and minimum values.

Introductory Problems

10.3-1 A propped cantilever steel beam is constructed from an IPN 300 section. The beam is loaded by its self-weight with intensity q. The length of the beam is 3.5 m. Let $E = 205$ GPa.

(a) Calculate the reactions at joints A and B.

(b) Find the location of zero moment within span AB.

(c) Calculate the maximum deflection of the beam and the rotation at joint B.

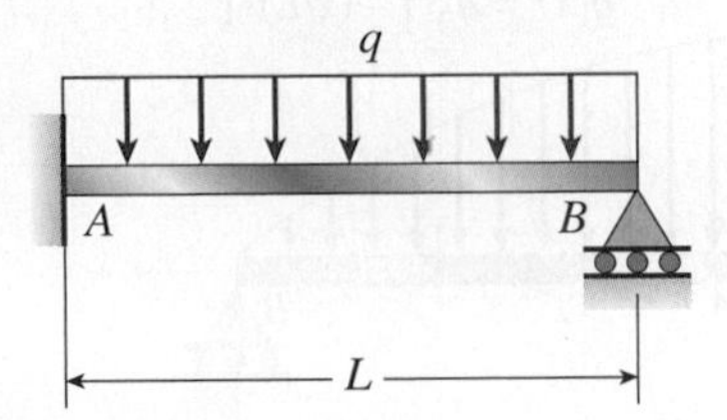

PROBLEM 10.3-1

10.3-2 A fixed-end beam is subjected to a point load at mid-span. The beam has a rectangular cross section (assume that the h/b ratio is 2) and is made of wood ($E = 11$GPa).

(a) Find height h of the cross section if the maximum displacement of the beam is 2 mm.

(b) Calculate the displacement of the beam at the inflection points.

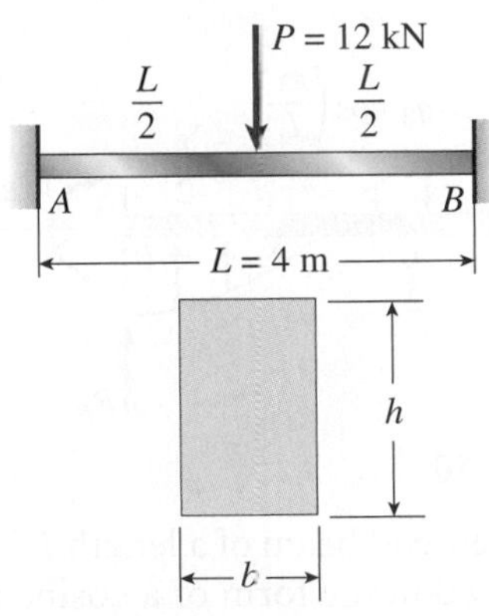

PROBLEM 10.3-2

10.3-3 A propped cantilever beam AB of a length L is loaded by a counterclockwise moment M_0 acting at support B (see figure).

Beginning with the second-order differential equation of the deflection curve (the bending-moment equation), obtain the reactions, shear forces, bending moments, slopes, and deflections of the beam. Construct the shear-force and bending-moment diagrams, labeling all critical ordinates.

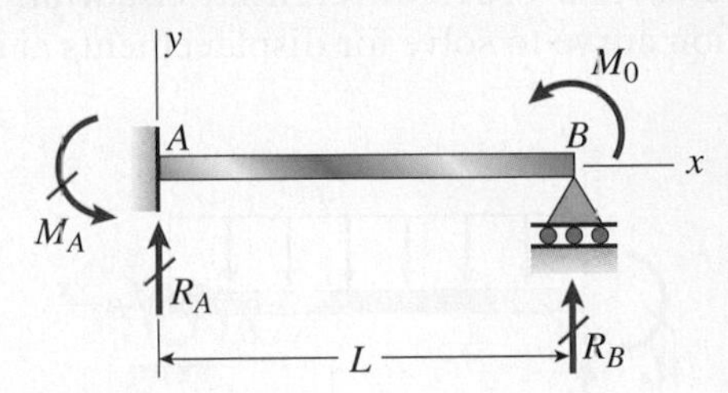

PROBLEM 10.3-3

10.3-4 A fixed-end beam AB of a length L supports a uniform load of intensity q (see figure).

Beginning with the second-order differential equation of the deflection curve (the bending-moment equation), obtain the reactions, shear forces, bending moments, slopes, and deflections of the beam. Construct the shear-force and bending-moment diagrams, labeling all critical ordinates.

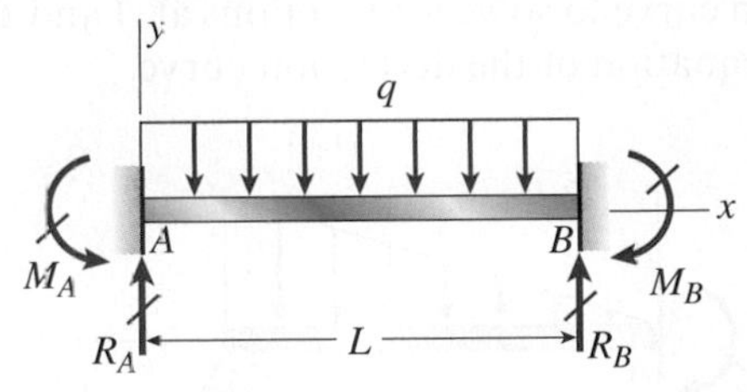

PROBLEM 10.3-4

Representative Problems

10.3-5 A cantilever beam AB of a length L has a fixed support at A and a roller support at B (see figure). The support at B is moved downward through a distance δ_B.

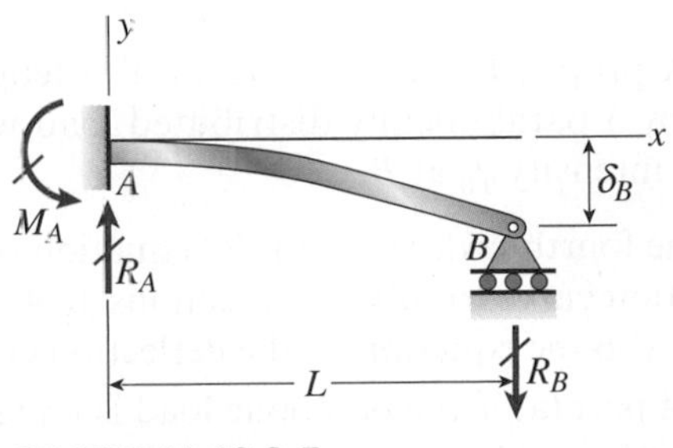

PROBLEM 10.3-5

Using the fourth-order differential equation of the deflection curve (the load equation), determine the reactions of the beam and the equation of the deflection curve. *Note:* Express all results in terms of the imposed displacement δ_B.

10.3-6 A cantilever beam of a length L and loaded by a uniform load of intensity q has a fixed support at A and spring support at B with rotational stiffness k_R. A rotation θ_B at B results in a reaction moment $M_B = k_R \times \theta_B$.

Find rotation θ_B and displacement δ_B at end B. Use the second-order differential equation of the deflection curve to solve for displacements at end B.

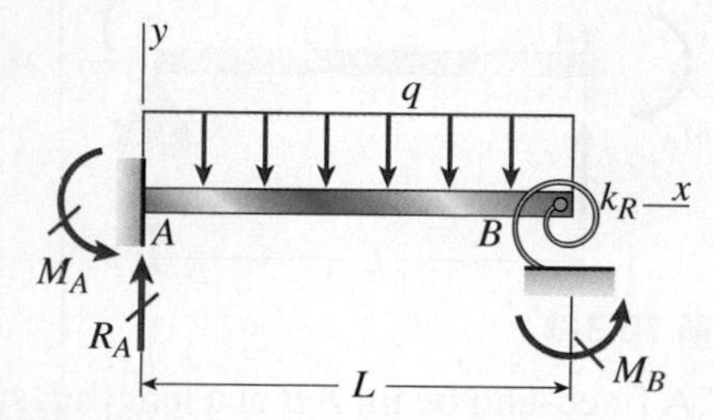

PROBLEM 10.3-6

10.3-7 A cantilever beam has a length L and is loaded by a triangularly distributed load of maximum intensity q_0 at B.

Use the fourth-order differential equation of the deflection curve to solve for reactions at A and B and also the equation of the deflection curve.

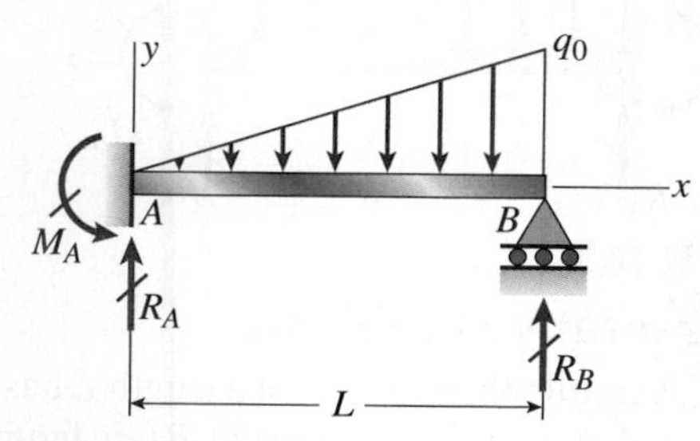

PROBLEM 10.3-7

10.3-8 A propped cantilever beam of a length L is loaded by a parabolically distributed load with a maximum intensity q_0 at B.

(a) Use the fourth-order differential equation of the deflection curve to solve for reactions at A and B and also the equation of the deflection curve.

(b) Repeat part (a) if the parabolic load is replaced by $q_0 \sin(\pi x / 2L)$.

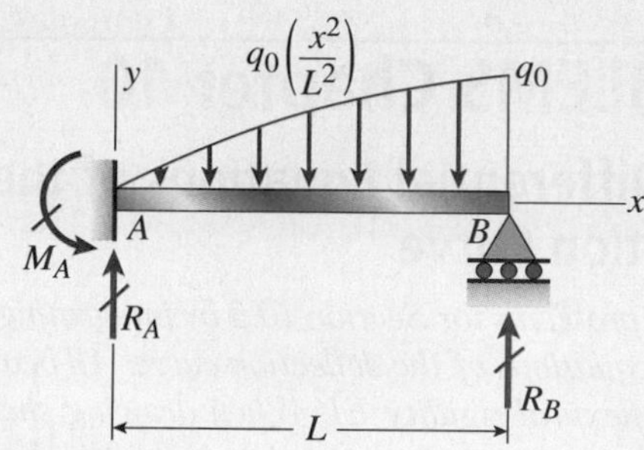

PROBLEM 10.3-8

10.3-9 A propped cantilever beam of a length L is loaded by a parabolically distributed load with a maximum intensity q_0 at A.

(a) Use the fourth-order differential equation of the deflection curve to solve for reactions at A and B and also the equation of the deflection curve.

(b) Repeat part (a) if the parabolic load is replaced by $q_0 \cos(\pi x/2L)$.

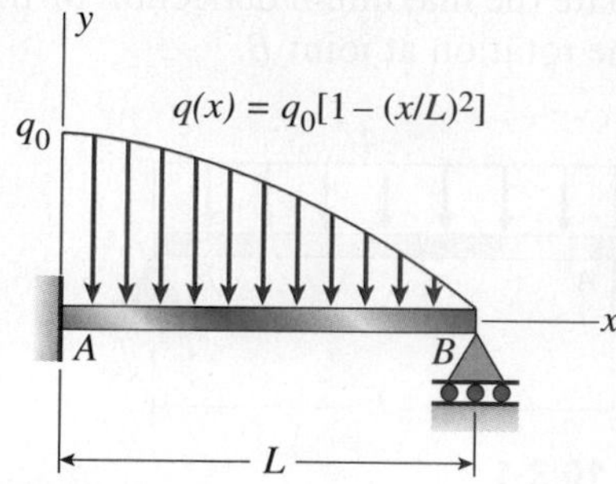

PROBLEM 10.3-9

10.3-10 A fixed-end beam of a length L is loaded by a distributed load in the form of a cosine curve with a maximum intensity q_0 at A.

(a) Use the fourth-order differential equation of the deflection curve to solve for reactions at A and B and also the equation of the deflection curve.

(b) Repeat part (a) using the distributed load $q_0 \sin(\pi x/L)$.

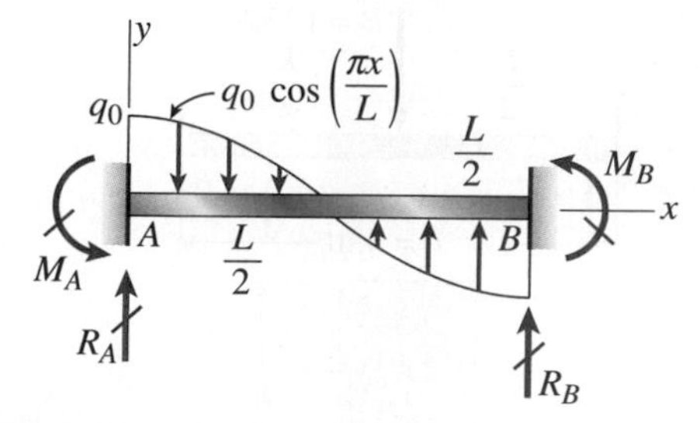

PROBLEM 10.3-10

10.3-11 A fixed-end beam of a length L is loaded by a distributed load in the form of a cosine curve with a maximum intensity q_0 at A.

(a) Use the fourth-order differential equation of the deflection curve to solve for reactions at A and B and also the equation of the deflection curve.

(b) Repeat part (a) if the distributed load is now $q_0(1 - x^2/L^2)$.

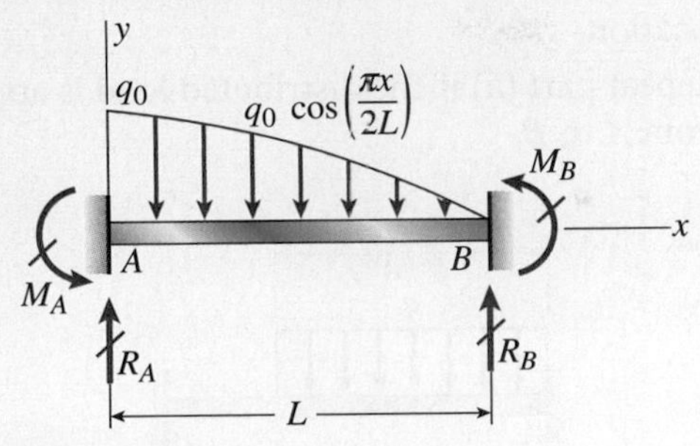

PROBLEM 10.3-11

10.3-12 A fixed-end beam of a length L is loaded by triangularly distributed load of a maximum intensity q_0 at B.

Use the fourth-order differential equation of the deflection curve to solve for reactions at A and B and also the equation of the deflection curve.

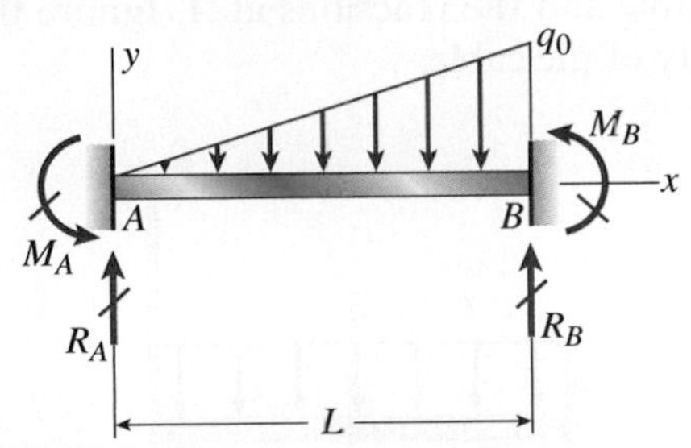

PROBLEM 10.3-12

10.3-13 A counterclockwise moment M_0 acts at the midpoint of a fixed-end beam ACB of length L (see figure).

Beginning with the second-order differential equation of the deflection curve (the bending-moment equation), determine all reactions of the beam and obtain the equation of the deflection curve for the left-hand half of the beam.

Then construct the shear-force and bending-moment diagrams for the entire beam, labeling all critical ordinates. Also, draw the deflection curve for the entire beam.

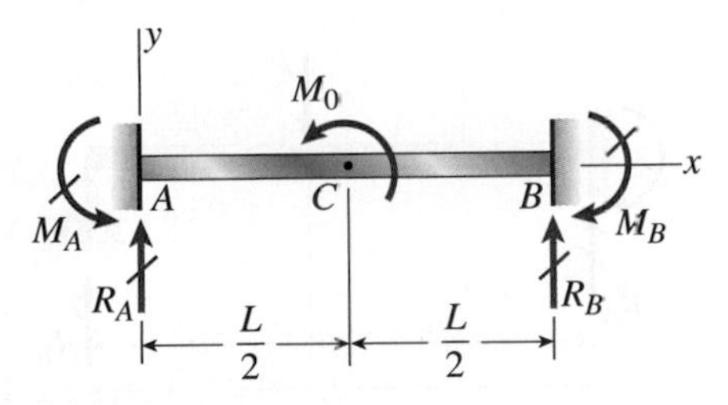

PROBLEM 10.3-13

10.3-14 A propped cantilever beam of a length L is loaded by a concentrated moment M_0 at midpoint C. Use the second-order differential equation of the deflection curve to solve for reactions at A and B. Draw shear-force and bending-moment diagrams for the entire beam. Also find the equations of the deflection curves for both halves of the beam, and draw the deflection curve for the entire beam.

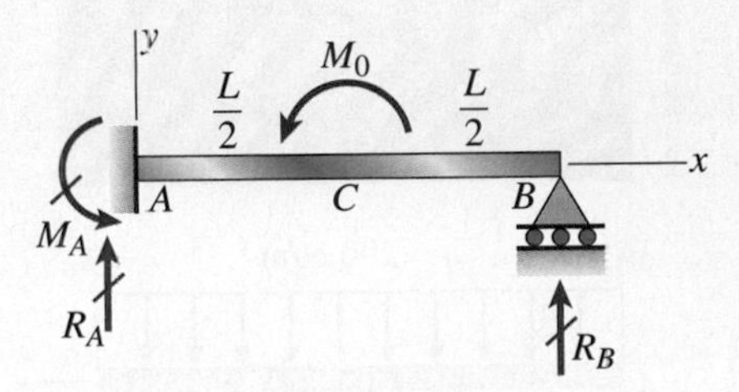

PROBLEM 10.3-14

10.3-15 A propped cantilever beam is subjected to uniform load q. The beam has flexural rigidity $EI = 2973\ \text{kN}\cdot\text{m}^2$ and the length of the beam is 4 m. Find the intensity q of the distributed load if the maximum displacement of the beam is $\delta_{max} = 4$ mm.

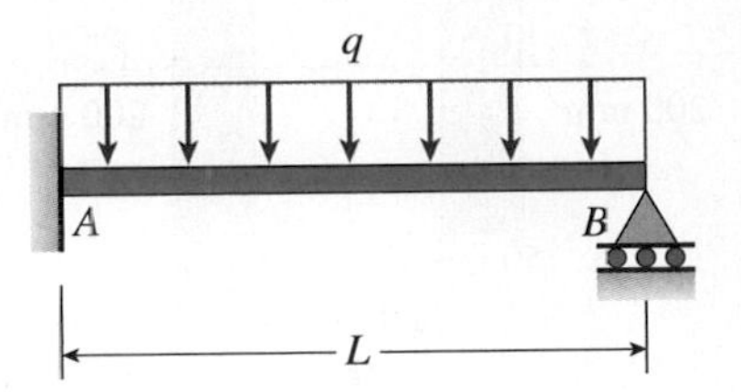

PROBLEMS 10.3-15 and 10.3-16

10.3-16 Repeat Problem 10.3-15 using $L = 3.5$ m, $\delta_{max} = 3$ mm, and $EI = 800\ \text{kN}\cdot\text{m}^2$.

10.4 Method of Superposition

Solve the problems for Section 10.4 by the method of superposition. All beams have constant flexural rigidity EI unless otherwise stated. When drawing shear-force and bending-moment diagrams, label all critical ordinates, including maximum and minimum values.

Introductory Problems

10.4-1 A two-span, continuous wood girder ($E = 12$ GPa) supports a roof patio structure (figure part a). A uniform load of intensity q acts on the girder, and each span is of length 2.4 m. The girder is made up using two 50 mm × 200 mm wood members (see figure part b). Ignore the weight of the beam. Use the nominal dimensions of the beam in your calculations.

(a) Find the reactions at A, B, and C.

(b) Use the method of superposition to calculate the displacement of the beam at the mid-span of segment AB. *Hint*: See Figs. 10-14c and 10-14d in Example 10-3.

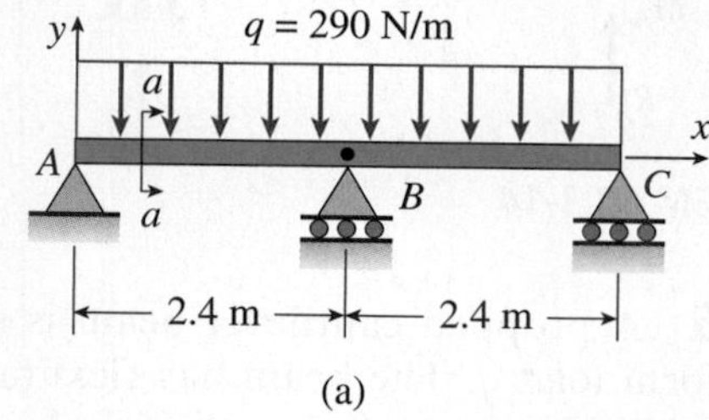

(a)

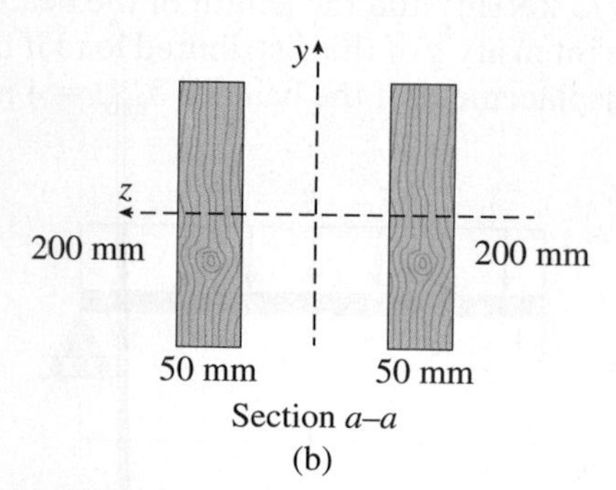

Section a–a

(b)

PROBLEM 10.4-1

10.4-2 A fixed-end beam AB carries point load P acting at point C. The beam has a rectangular cross section ($b = 75$ mm, $h = 150$ mm). Calculate the reactions of the beam and the displacement at point C. Assume that $E = 190$ GPa.

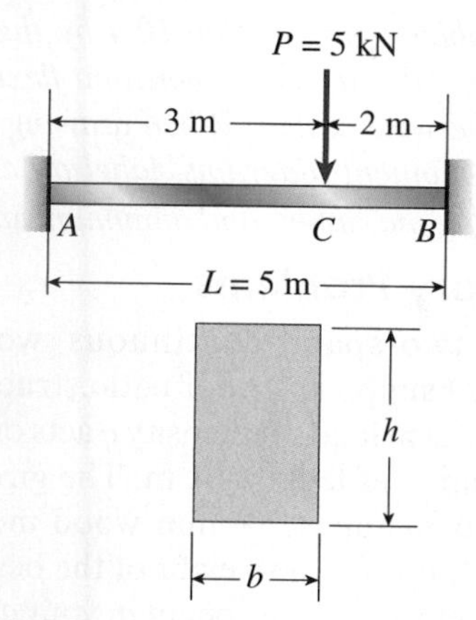

PROBLEM 10.4-2

10.4-3 A fixed-end beam AB supports a uniform load of intensity $q = 1$ kN/m acting over part of the span. Assume that $EI = 125$ kN·m^2.

(a) Calculate the reactions at A and B.

(b) Find the maximum displacement and its location.

(c) Repeat part (a) if the distributed load is applied from A to B.

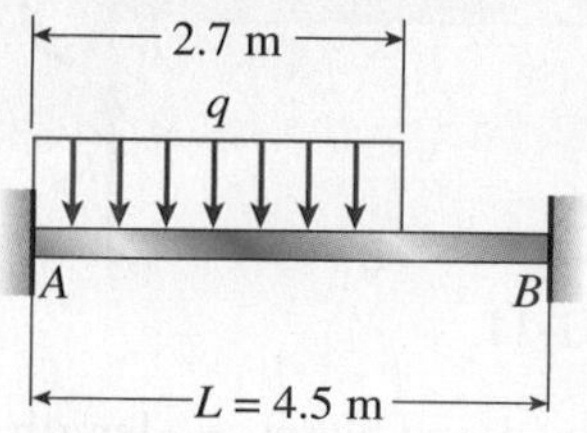

PROBLEM 10.4-3

10.4-4 A cantilever beam is supported at B by cable BC. The beam carries a uniform load $q = 200$ N/m. If the length of the beam is $L = 3$ m, find the force in the cable and the reactions at A. Ignore the axial flexibility of the cable.

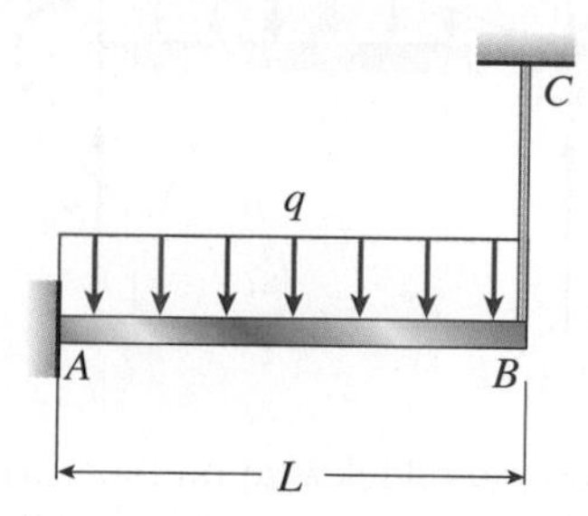

PROBLEM 10.4-4

10.4-5 A propped cantilever beam AB of a length L carries a concentrated load P acting at the position shown in the figure.

Determine the reactions R_A, R_B, and M_A for this beam. Also, draw the shear-force and bending-moment diagrams, labeling all critical ordinates.

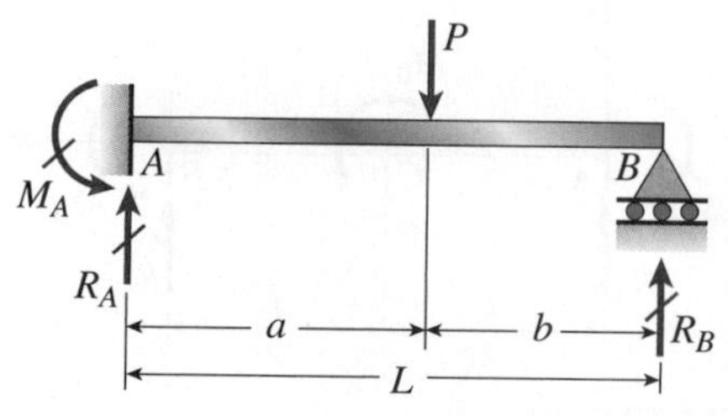

PROBLEM 10.4-5

Representative Problems

10.4-6 A beam with a sliding support at B is loaded by a uniformly distributed load with intensity q. Use the method of superposition to solve for all reactions. Also draw shear-force and bending-moment diagrams, labeling all critical ordinates.

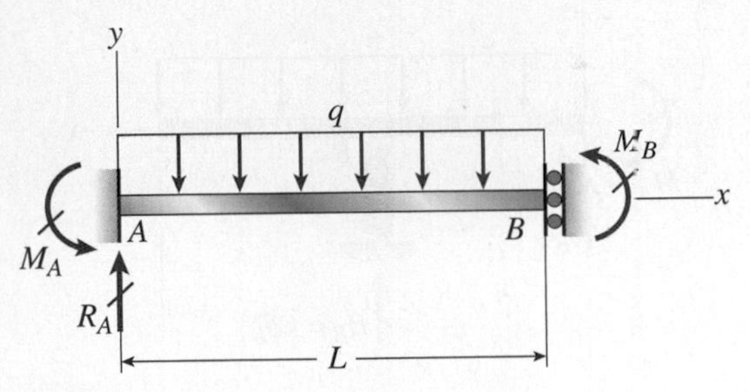

PROBLEM 10.4-6

10.4-7 A propped cantilever beam of a length $2L$ with a support at B is loaded by a uniformly distributed load with intensity q. Use the method of superposition to solve for all reactions. Also draw shear-force and bending-moment diagrams, labeling all critical ordinates.

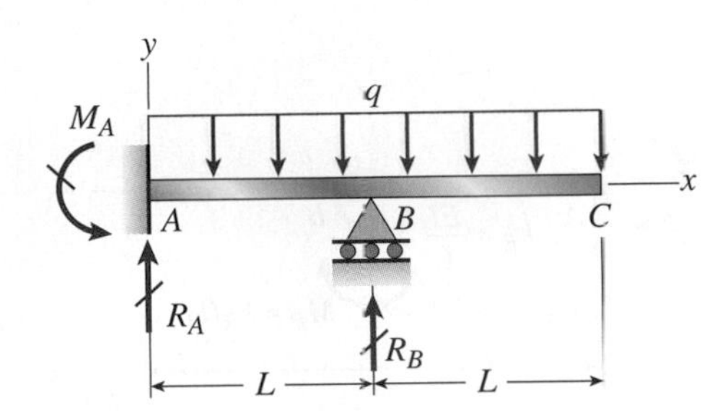

PROBLEM 10.4-7

10.4-8 The continuous frame ABC has a pin support at A, roller supports at B and C, and a rigid corner connection at B (see figure). Members AB and BC each have flexural rigidity EI. A moment M_0 acts counterclockwise at B. *Note:* Disregard axial deformations in member AB and consider only the effects of bending.

(a) Find all reactions of the frame.

(b) Find joint rotations θ at A, B, and C.

(c) Find the required new length of member BC in terms of L, so that θ_B in part (b) is doubled in size.

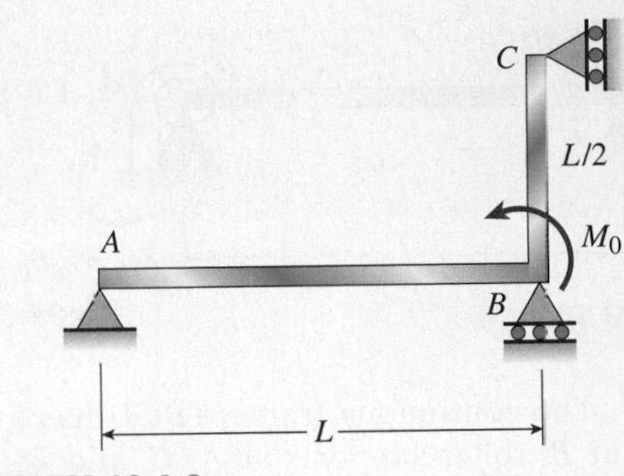

PROBLEM 10.4-8

10.4-9 The continuous frame ABC has a pin support at A, roller supports at B and C, and a rigid corner connection at B (see figure). Members AB and BC each have flexural rigidity EI. A moment M_0 acts counterclockwise at A. *Note:* Disregard axial deformations in member AB and consider only the effects of bending.

(a) Find all reactions of the frame.

(b) Find joint rotations θ at A, B, and C.

(c) Find the required new length of member AB in terms of L, so that θ_A in part (b) is doubled in size.

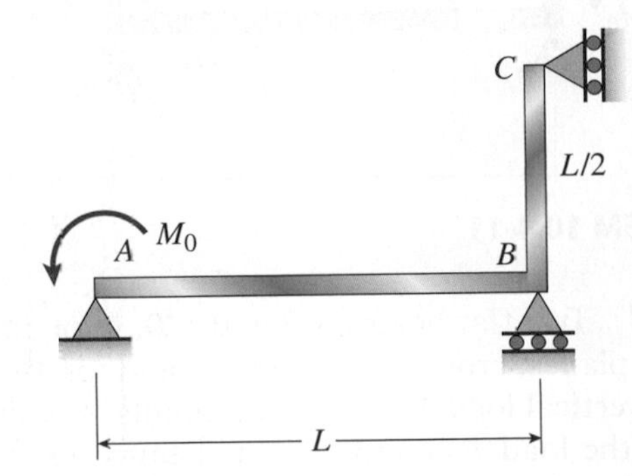

PROBLEM 10.4-9

10.4-10 Beam AB has a pin support at A and a roller support at B. Joint B is also restrained by a linearly elastic rotational spring with stiffness k_R, which provides a resisting moment M_B due to rotation at B. Member AB has flexural rigidity EI. A moment M_0 acts counterclockwise at B.

(a) Use the method of superposition to solve for all reactions.

(b) Find an expression for joint rotation θ_A in terms of spring stiffness k_R. What is θ_A when $k_R \to 0$? What is θ_A when $k_R \to \infty$? What is θ_A when $k_R = 6EI/L$?

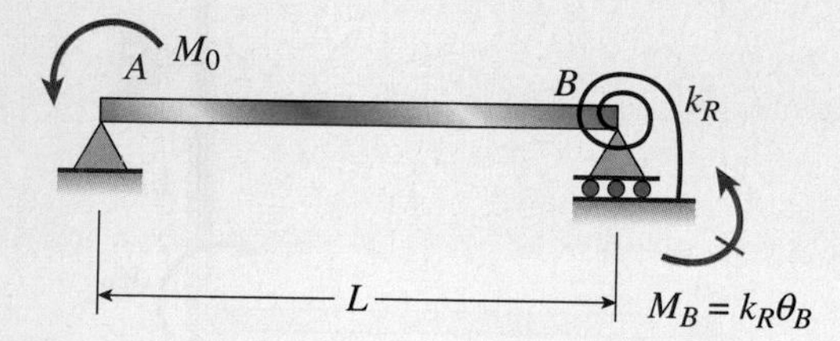

PROBLEM 10.4-10

10.4-11 The continuous frame $ABCD$ has a pin support at B; roller supports at A, C, and D; and rigid corner connections at B and C (see figure). Members AB, BC, and CD each have flexural rigidity EI. Moment M_0 acts counterclockwise at B and clockwise at C. *Note:* Disregard axial deformations in member AB and consider only the effects of bending.

(a) Find all reactions of the frame.
(b) Find joint rotations θ at A, B, C, and D.
(c) Repeat parts (a) and (b) if both moments M_0 are counterclockwise.

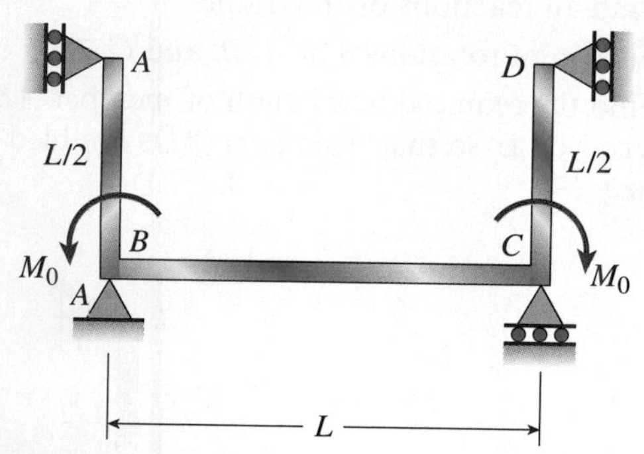

PROBLEM 10.4-11

10.4-12 Two flat beams AB and CD, lying in horizontal planes, cross at right angles and jointly support a vertical load P at their midpoints (see figure). Before the load P is applied, the beams just touch each other. Both beams are made of the same material and have the same widths. Also, the ends of both beams are simply supported. The lengths of beams AB and CD are L_{AB} and L_{CD}, respectively.

What should be the ratio t_{AB}/t_{CD} of the thicknesses of the beams if all four reactions are to be the same?

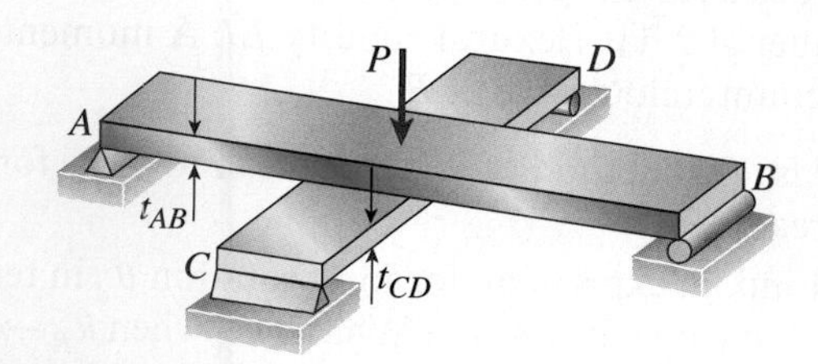

PROBLEM 10.4-12

10.4-13 A propped cantilever beam of a length $2L$ is loaded by a uniformly distributed load with intensity q. The beam is supported at B by a linearly elastic spring with stiffness k. Use the method of superposition to solve for all reactions. Also draw shear-force and bending-moment diagrams, labeling all critical ordinates. Let $k = 6EI/L^3$.

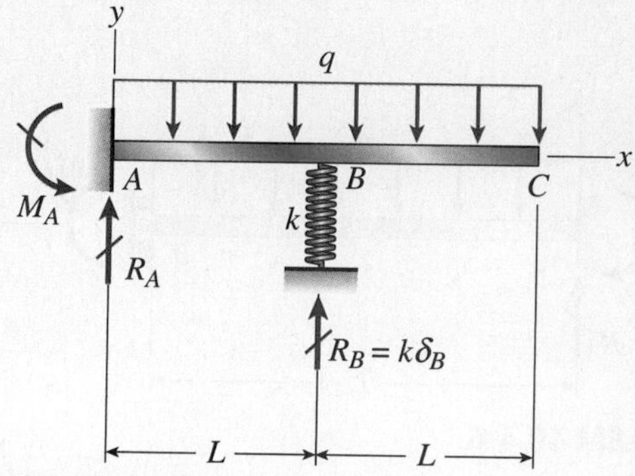

PROBLEM 10.4-13

10.4-14 A propped cantilever beam of a length $2L$ is loaded by a uniformly distributed load with intensity q. The beam is supported at B by a linearly elastic rotational spring with stiffness k_R, which provides a resisting moment M_B due to rotation θ_B. Use the method of superposition to solve for all reactions. Also draw shear-force and bending-moment diagrams, labeling all critical ordinates. Let $k_R = EI/L$.

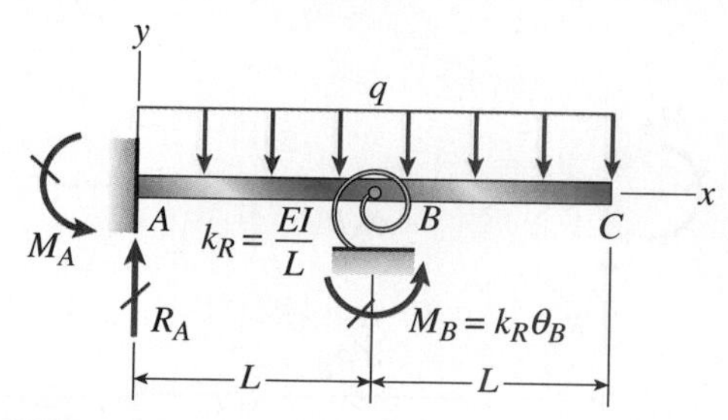

PROBLEM 10.4-14

10.4-15 Determine the fixed-end moments (M_A and M_B) and fixed-end forces (R_A and R_B) for a beam of length L supporting a triangular load of maximum intensity q_0 (see figure). Then draw the shear-force and bending-moment diagrams, labeling all critical ordinates.

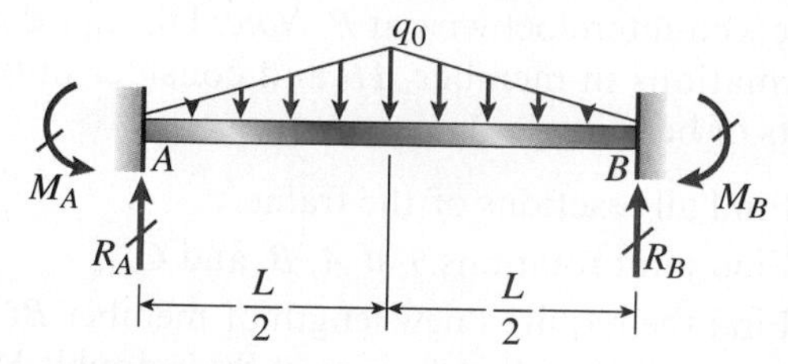

PROBLEM 10.4-15

10.4-16 A continuous beam ABC with two unequal spans, one of length L and one of length $2L$, supports a uniform load of intensity q (see figure).

Determine the reactions R_A, R_B, and R_C for this beam. Also, draw the shear-force and bending-moment diagrams, labeling all critical ordinates.

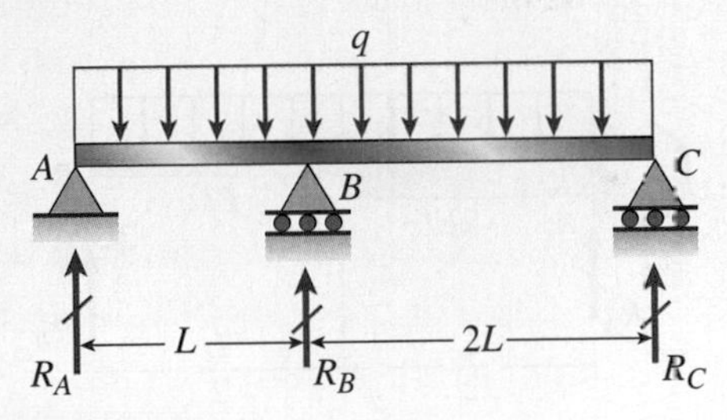

PROBLEM 10.4-16

10.4-17 Beam ABC is fixed at support A and rests (at point B) upon the midpoint of beam DE (see part a of the figure). Thus, beam ABC may be represented as a propped cantilever beam with an overhang BC and a linearly elastic support of stiffness k at point B (see part b of the figure).

The distance from A to B is $L = 3$ m, the distance from B to C is $L/2 = 1.5$ m, and the length of beam DE is $L = 3$ m. Both beams have the same flexural rigidity EI. A concentrated load $P = 8$ kN acts at the free end of beam ABC.

Determine the reactions R_A, R_B, and M_A for beam ABC. Also, draw the shear-force and bending-moment diagrams for beam ABC, labeling all critical ordinates.

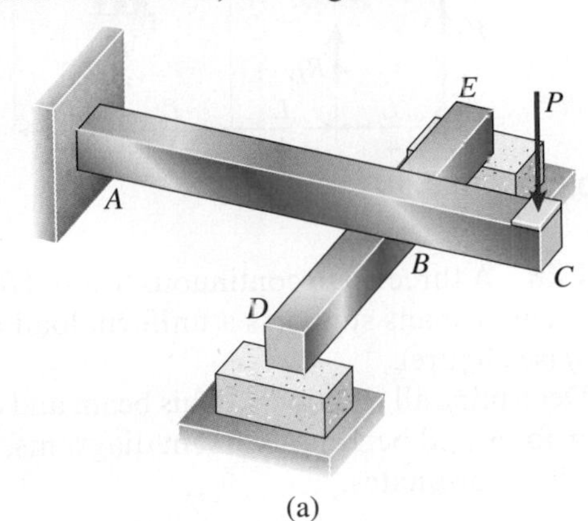

(a)

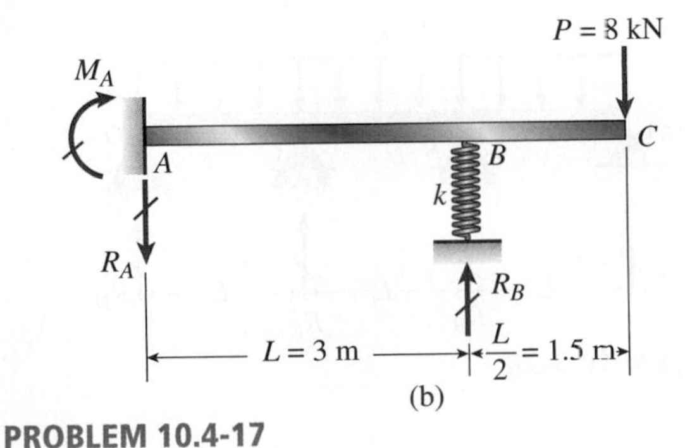

(b)

PROBLEM 10.4-17

10.4-18 A propped cantilever beam has flexural rigidity $EI = 4.5\,\text{MN}\cdot\text{m}^2$. When the loads shown are applied to the beam, it settles at joint B by 5 mm. Find the reaction at joint B.

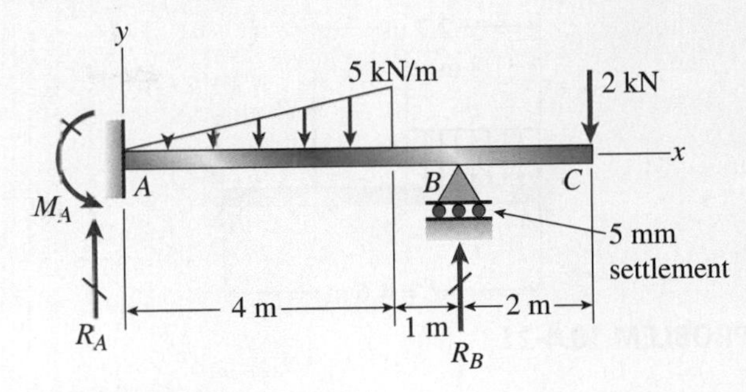

PROBLEM 10.4-18

10.4-19 A triangularly distributed load with a maximum intensity of $q_0 = 145$ N/m acts on propped cantilever beam AB. If the length L of the beam is 3 m, find the reactions at A and B.

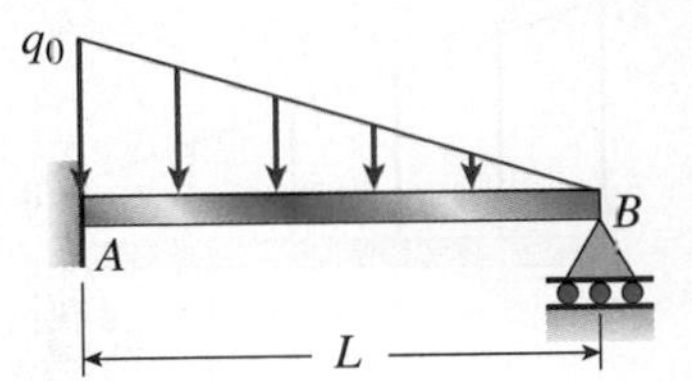

PROBLEM 10.4-19

10.4-20 A fixed-end beam is loaded by a uniform load $q = 15$ kN/m and a point load $P = 30$ kN at mid-span. The beam has a length of 4 m and modulus of elasticity of 205 GPa.

(a) Find reactions at A and B.

(b) Calculate the height of the beam if the displacement at mid-span is known to be 3 mm. Assume that the beam has rectangular cross section with $h/b = 2$.

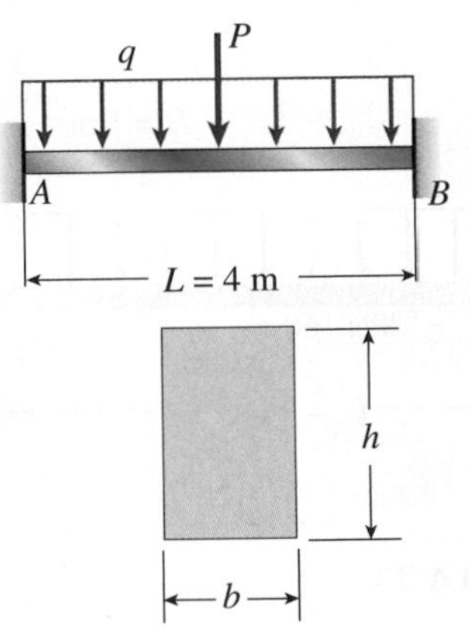

PROBLEM 10.4-20

10.4-21 Uniform load $q = 145$ N/m acts over part of the span of fixed-end beam AB (see figure). Upward load $P = 1100$ N is applied 2.7 m to the right of joint A. Find the reactions at A and B.

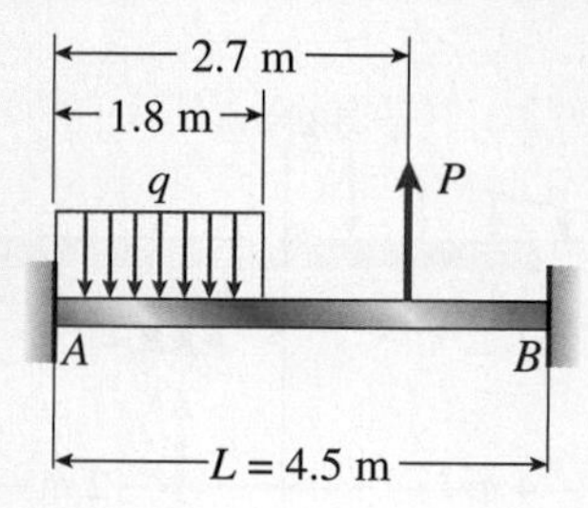

PROBLEM 10.4-21

10.4-22 A propped cantilever beam with a length $L = 4$ m is subjected to a trapezoidal load with intensities $q_0 = 10$ kN/m and $q_1 = 15$ kN/m. Find the reactions at A and B. *Hint:* The loading is the sum of uniform and triangular loads.

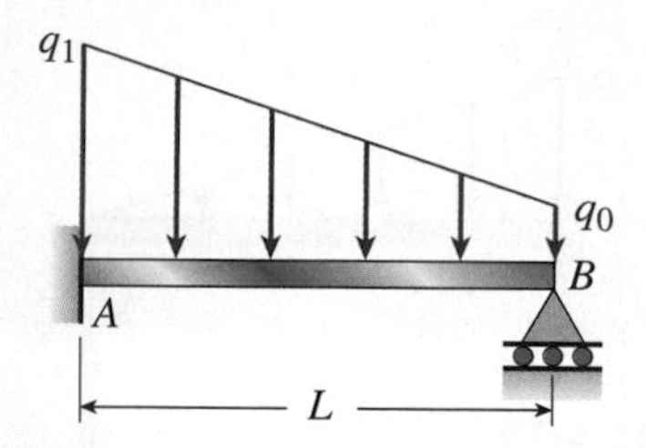

PROBLEM 10.4-22

10.4-23 A cantilever beam is supported by a tie rod at B as shown. Both the tie rod and the beam are steel with $E = 210$ GPa. The tie rod is just taut before the distributed load $q = 3$ kN/m is applied.

(a) Find the tension force in the tie rod.

(b) Draw shear-force and bending-moment diagrams for the beam, labeling all critical ordinates.

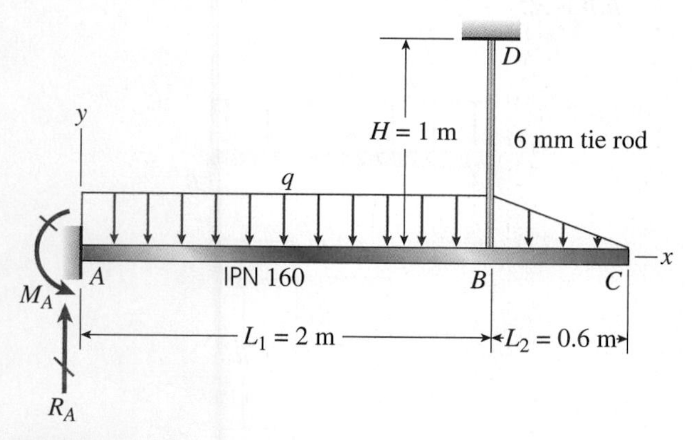

PROBLEM 10.4-23

10.4-24 The figure shows a nonprismatic, propped cantilever beam AB with flexural rigidity $2EI$ from A to C and EI from C to B.

Determine all reactions of the beam due to the uniform load of intensity q. *Hint:* Use the results of Problems 9.7-1 and 9.7-2.

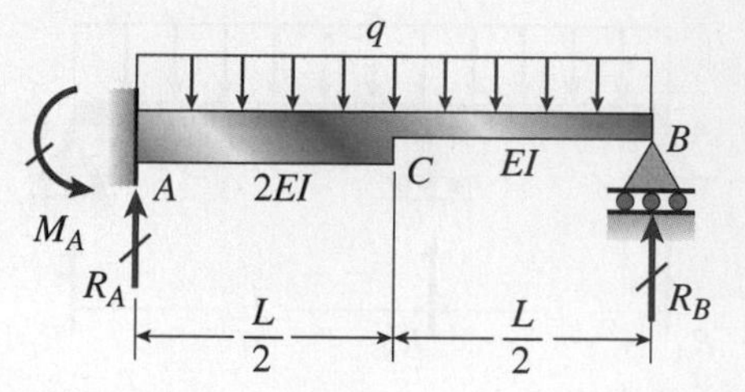

PROBLEM 10.4-24

10.4-25 A beam ABC is fixed at end A and supported by beam DE at point B (see figure). Both beams have the same cross section and are made of the same material.

(a) Determine all reactions due to the load P.

(b) What is the numerically largest bending moment in either beam?

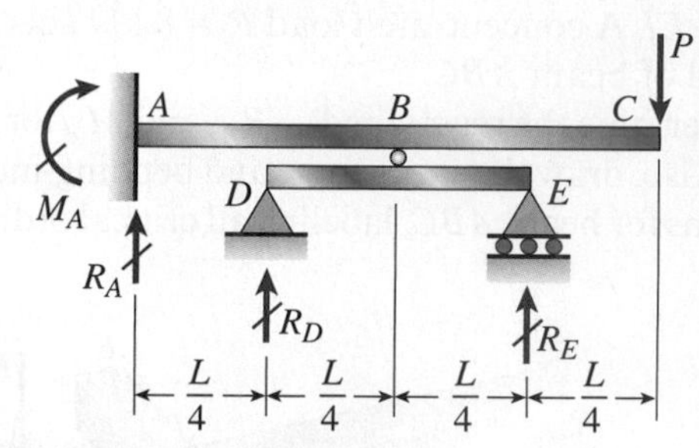

PROBLEM 10.4-25

10.4-26 A three-span continuous beam $ABCD$ with three equal spans supports a uniform load of intensity q (see figure).

Determine all reactions of this beam and draw the shear-force and bending-moment diagrams, labeling all critical ordinates.

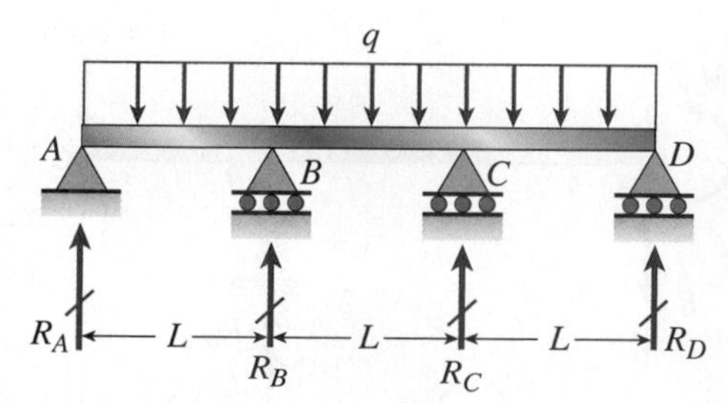

PROBLEM 10.4-26

10.4-27 A beam rests on supports at A and B and is loaded by a distributed load with intensity q as shown. A small gap Δ exists between the unloaded beam and the support at C. Assume that span length $L = 1$ m and flexural rigidity of the beam $EI = 1.2 \times 10^6$ N·m^2. Plot a graph of the bending moment at B as a function of the load intensity q.

Hint: See Example 9-9 for guidance on computing the deflection at C.

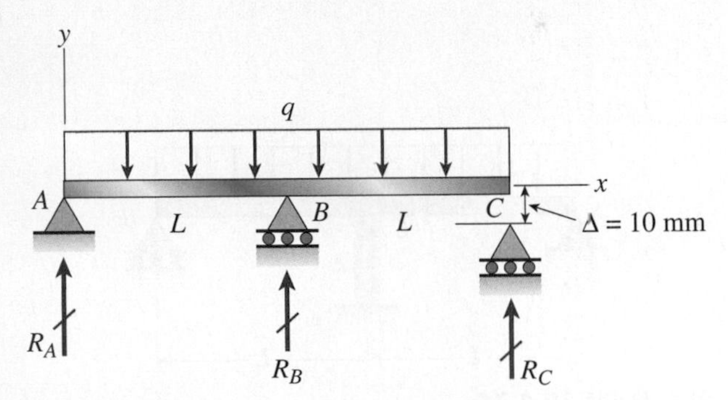

PROBLEM 10.4-27

10.4-28 A propped cantilever beam is subjected to two triangularly distributed loads, each with a peak load intensity equal to q_0 (see figure). Find the expressions for reactions at A and C using superposition. Plot shear and moment diagrams.

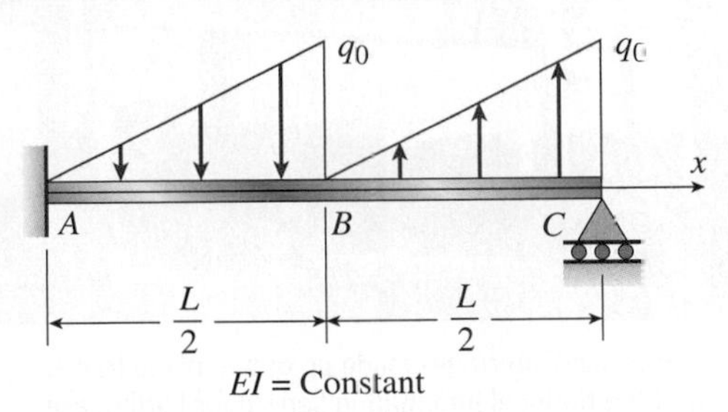

PROBLEM 10.4-28

10.4-29 A propped cantilever beam is loaded by a triangular distributed load from A to C (see figure). The load has a peak intensity $q_0 = 145$ N/m. The length of the beam is 3.6 m. Find support reactions at A and B.

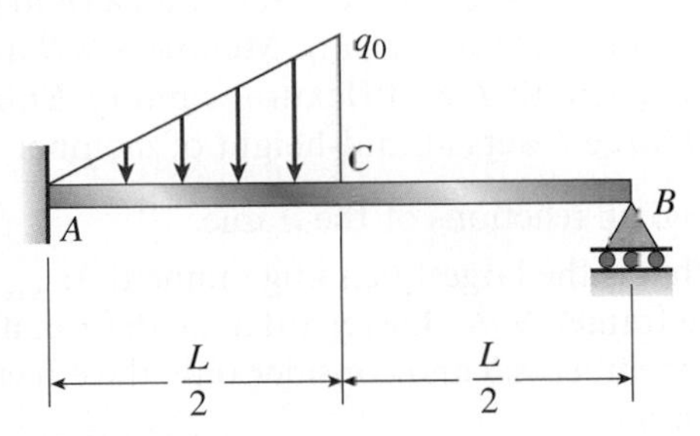

PROBLEM 10.4-29

10.4-30 A fixed-end beam AB of a length L is subjected to a moment M_0 acting at the position shown in the figure.

(a) Determine all reactions for this beam.

(b) Draw shear-force and bending-moment diagrams for the special case in which $a = b = L/2$.

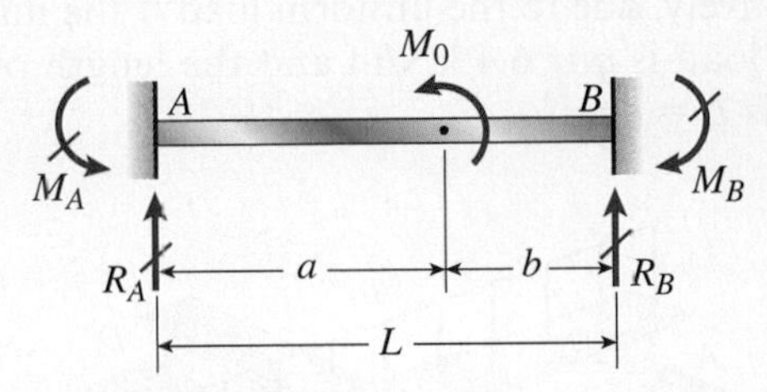

PROBLEM 10.4-30

10.4-31 A temporary wood flume serving as a channel for irrigation water is shown in the figure. The vertical boards forming the sides of the flume are sunk in the ground, which provides a fixed support. The top of the flume is held by tie rods that are tightened so that there is no deflection of the boards at that point. Thus, the vertical boards may be modeled as a beam AB, supported and loaded as shown in the last part of the figure.

Assuming that the thickness t of the boards is 40 mm, the depth d of the water is 1 m, and the height h to the tie rods is 1.3 m, what is the maximum bending stress σ in the boards? *Hint:* The numerically largest bending moment occurs at the fixed support.

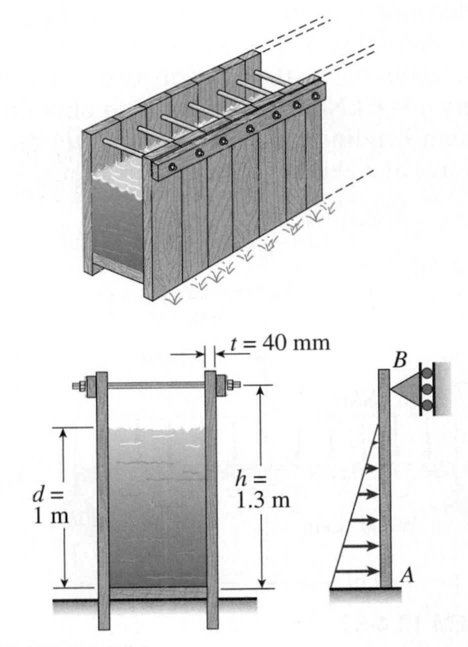

PROBLEM 10.4-31

10.4-32 Two identical, simply supported beams AB and CD are placed so that they cross each other at their midpoints (see figure). Before the uniform load is applied, the beams just touch each other at the crossing point.

Determine the maximum bending moments $(M_{AB})_{\max}$ and $(M_{CD})_{\max}$ in beams AB and CD, respectively, due to the uniform load if the intensity of the load is $q = 6.4$ kN/m and the length of each beam is $L = 4$ m.

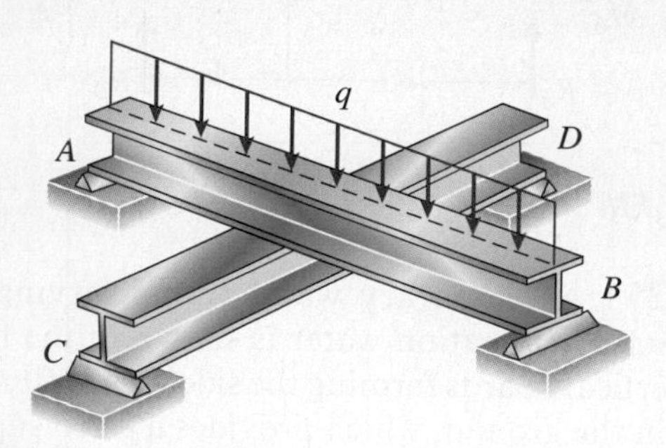

PROBLEM 10.4-32

10.4-33 The cantilever beam AB shown in the figure is an IPN 180 steel I-beam with $E = 200$ GPa. The simple beam DE is a wood beam 100 mm × 300 mm (nominal dimensions) in cross section with $E = 10$ GPa. A steel rod AC of diameter 6 mm, length 3 m, and $E = 200$ GPa serves as a hanger joining the two beams. The hanger fits snugly between the beams before the uniform load is applied to beam DE.

Determine the tensile force F in the hanger and the maximum bending moments M_{AB} and M_{DE} in the two beams due to the uniform load, which has an intensity $q = 6$ kN/m. *Hint:* To aid in obtaining the maximum bending moment in beam DE, draw the shear-force and bending-moment diagrams.

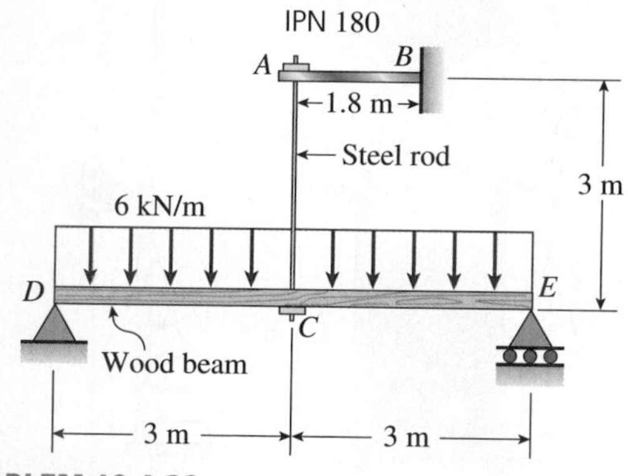

PROBLEM 10.4-33

10.4-34 The beam AB shown in the figure is simply supported at A and B and supported on a spring of stiffness k at its midpoint C. The beam has flexural rigidity EI and length $2L$.

What should be the stiffness k of the spring in order that the maximum bending moment in the beam (due to the uniform load) will have the smallest possible value?

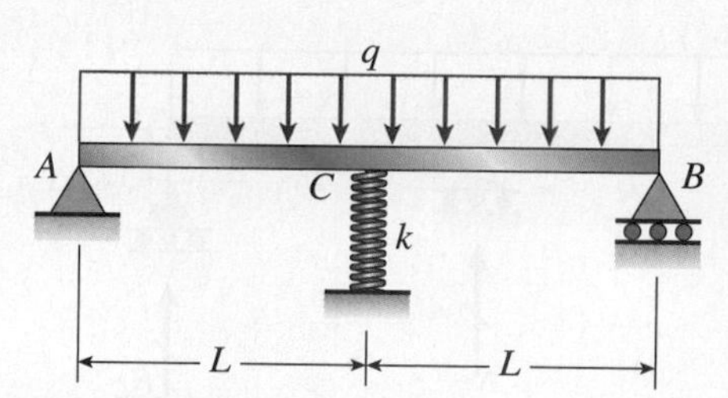

PROBLEM 10.4-34

Ship container handling cranes made up of two plane frames (Courtesy of the National Information Service for Earthquake Engineering EERC, University of California Berkeley)

10.4-35 The continuous frame ABC has a fixed support at A, a roller support at C, and a rigid corner connection at B (see figure). Members AB and BC each have length L and flexural rigidity EI. A horizontal force P acts at mid-height of member AB.

(a) Find all reactions of the frame.

(b) What is the largest bending moment $M_{\max}$ in the frame? *Note:* Disregard axial deformations in member AB and consider only the effects of bending.

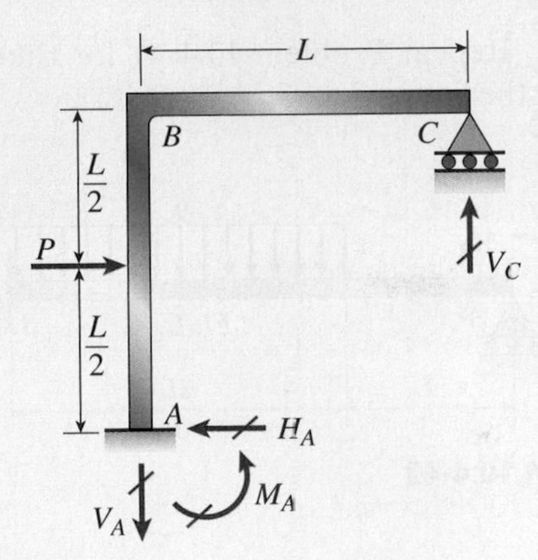

PROBLEM 10.4-35

10.4-36 The continuous frame ABC has a pinned support at A, a sliding support at C, and a rigid corner connection at B (see figure). Members AB and BC each have length L and flexural rigidity EI. A horizontal force P acts at mid-height of member AB.

(a) Find all reactions of the frame.
(b) What is the largest bending moment M_{max} in the frame? *Note:* Disregard axial deformations in members AB and BC and consider only the effects of bending.

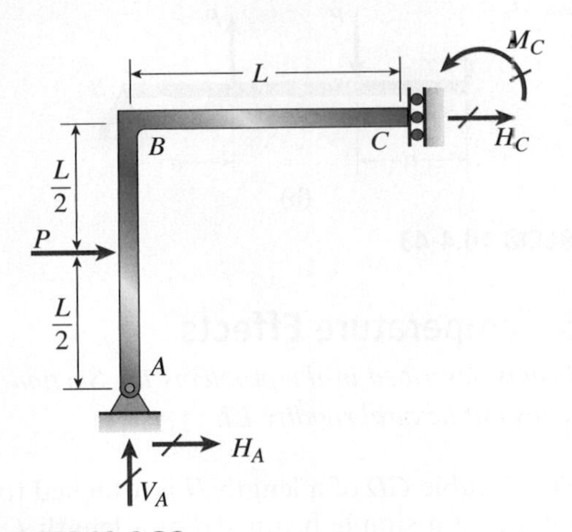

PROBLEM 10.4-36

10.4-37 A wide-flange beam ABC rests on three identical spring supports at points A, B, and C (see figure). The flexural rigidity of the beam is $EI = 20 \times 10^6\,\text{N}\cdot\text{m}^2$, and each spring has stiffness $k = 10$ MN/m. The length of the beam is $L = 5$ m.

If the load P is 25 kN, what are the reactions R_A, R_B, and R_C? Also, draw the shear-force and bending-moment diagrams for the beam, labeling all critical ordinates.

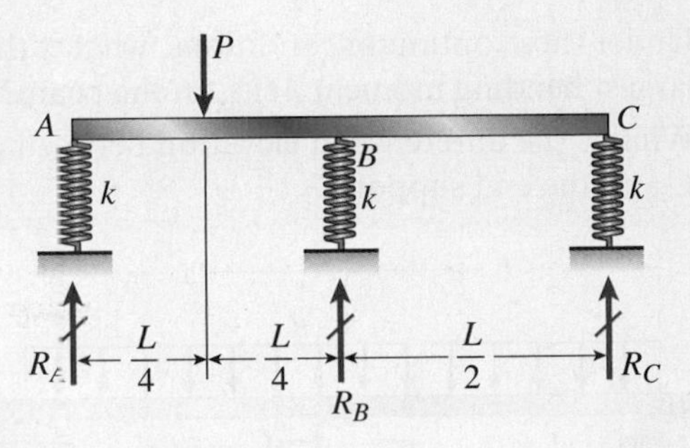

PROBLEM 10.4-37

10.4-38 A fixed-end beam AB of a length L is subjected to a uniform load of intensity q acting over the middle region of the beam (see figure).

(a) Obtain a formula for the fixed-end moments M_A and M_B in terms of the load q, the length L, and the length b of the loaded part of the beam.
(b) Plot a graph of the fixed-end moment M_A versus the length b of the loaded part of the beam. For convenience, plot the graph in the following nondimensional form:

$$\frac{M_A}{qL^2/12} \quad \text{versus} \quad \frac{b}{L}$$

with the ratio b/L varying between its extreme values of 0 and 1.
(c) For the special case in which $a = b = L/3$, draw the shear-force and bending-moment diagrams for the beam, labeling all critical ordinates.

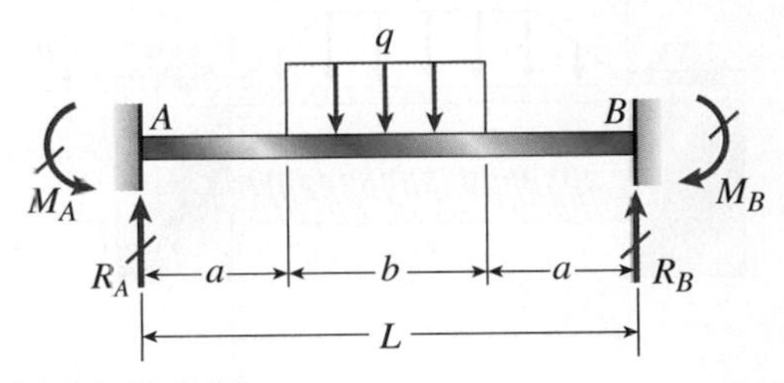

PROBLEM 10.4-38

10.4-39 A beam supporting a uniform load of intensity q throughout its length rests on pistons at points A, C, and B (see figure). The cylinders are filled with oil and are connected by a tube so that the oil pressure on each piston is the same. The pistons at A and B have diameter d_1, and the piston at C has diameter d_2.

(a) Determine the ratio of d_2 to d_1 so that the largest bending moment in the beam is as small as possible.

(b) Under these optimum conditions, what is the largest bending moment M_{max} in the beam?

(c) What is the difference in elevation between point C and the end supports?

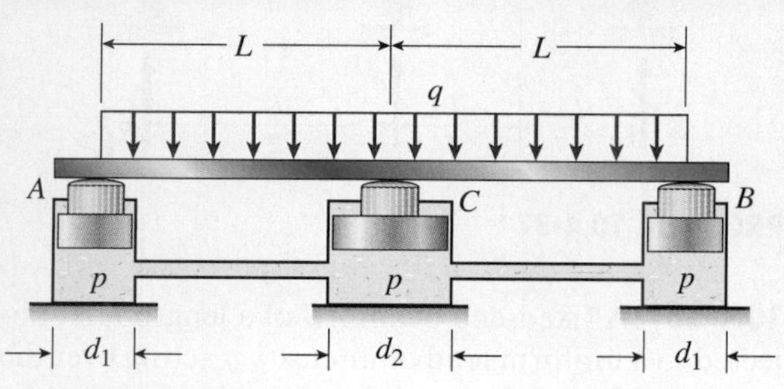

PROBLEM 10.4-39

10.4-40 A thin steel beam AB used in conjunction with an electromagnet in a high-energy physics experiment is securely bolted to rigid supports (see figure). A magnetic field produced by coils C results in a force acting on the beam. The force is trapezoidally distributed with maximum intensity $q_0 = 18$ kN/m. The length of the beam between supports is $L = 200$ mm, and the dimension c of the trapezoidal load is 50 mm. The beam has a rectangular cross section with width $b = 60$ and height $h = 20$ mm.

Determine the maximum bending stress σ_{max} and the maximum deflection δ_{max} for the beam. (Disregard any effects of axial deformations and consider only the effects of bending. Use $E = 200$ GPa.)

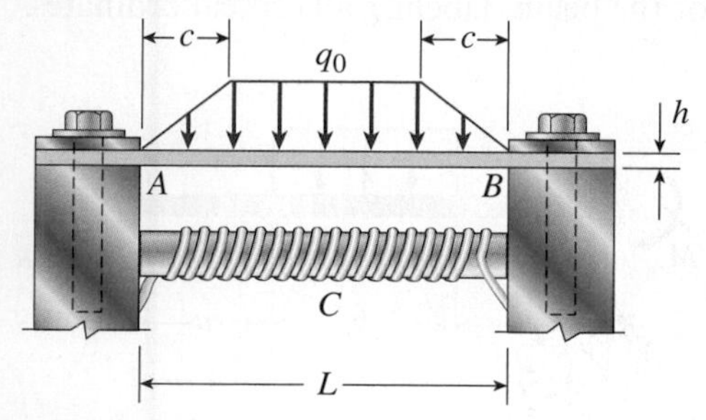

PROBLEM 10.4-40

10.4-41 Find an expression for required moment M_A (in terms of q and L) that will result in rotation $\theta_A = 0$ due to M_A and q loadings applied at the same time.

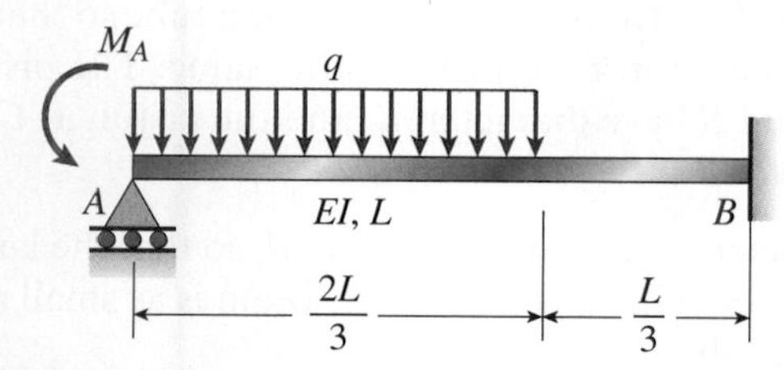

PROBLEM 10.4-41

10.4-42 Repeat Problem 10.4-41 for the loading shown in the figure.

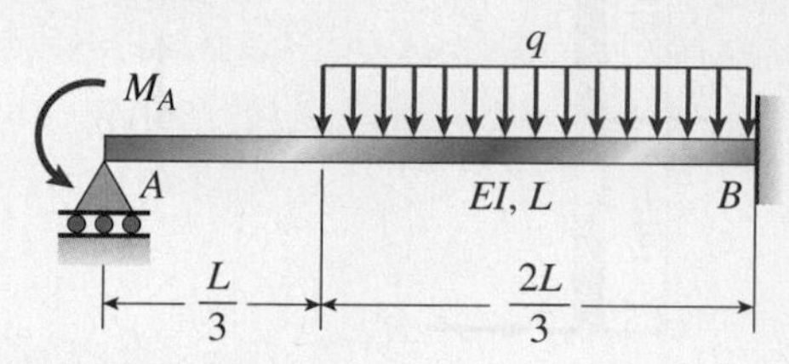

PROBLEM 10.4-42

10.4-43 A propped cantilever beam is loaded by two different load patterns (see figures a and b). Assume that EI is constant and the total beam length is L. Find expressions for reactions at A and B for each beam. Plot shear and moment diagrams. Assume that $a = L/3$.

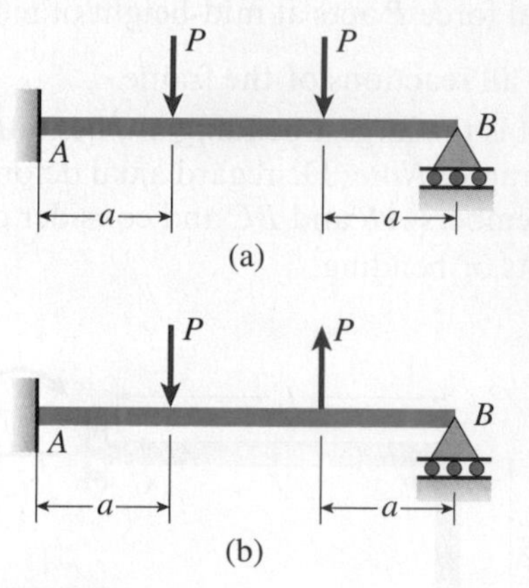

PROBLEM 10.4-43

10.5 Temperature Effects

The beams described in the problems for Section 10.5 have constant flexural rigidity EI.

10.5-1 A cable CD of a length H is attached to the third point of a simple beam AB of a length L (see figure). The moment of inertia of the beam is I, and the effective cross-sectional area of the cable is A. The cable is initially taut but without any initial tension.

(a) Obtain a formula for the tensile force S in the cable when the temperature drops uniformly by ΔT degrees, assuming that the beam and cable are made of the same material (modulus of elasticity E and coefficient of thermal expansion α). Use the method of superposition in the solution.

(b) Repeat part (a), assuming a wood beam and steel cable.

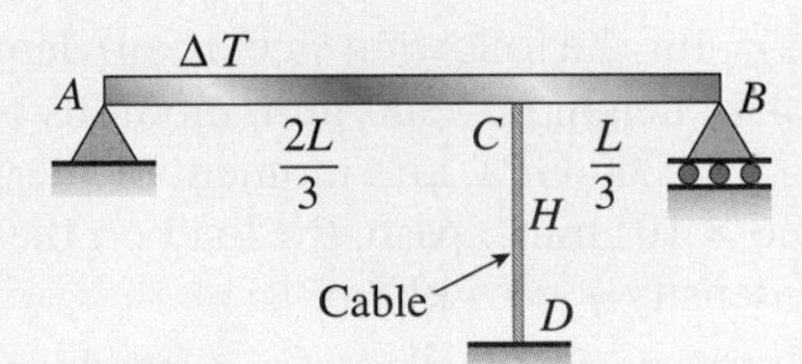

PROBLEM 10.5-1

10.5-2 A propped cantilever beam, fixed at the left-hand end A and simply supported at the right-hand end B, is subjected to a temperature differential with temperature T_1 on its upper surface and T_2 on its lower surface (see figure).

(a) Find all reactions for this beam. Use the method of superposition in the solution. Assume the spring support is unaffected by temperature.
(b) What are the reactions when $k \to \infty$?

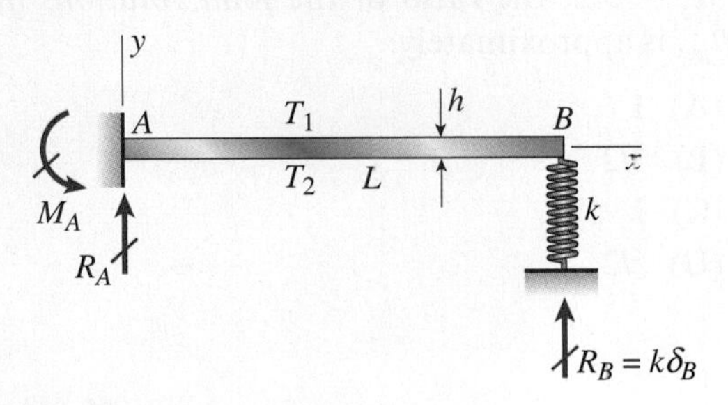

PROBLEMS 10.5-2 and 10.5-3

10.5-3 Solve the preceding problem by integrating the differential equation of the deflection curve.

10.5-4 A two-span beam with spans of lengths L and $L/3$ is subjected to a temperature differential with temperature T_1 on its upper surface and T_2 on its lower surface (see figure).

(a) Determine all reactions for this beam. Use the method of superposition in the solution. Assume the spring support is unaffected by temperature.
(b) What are the reactions when $k \to \infty$?

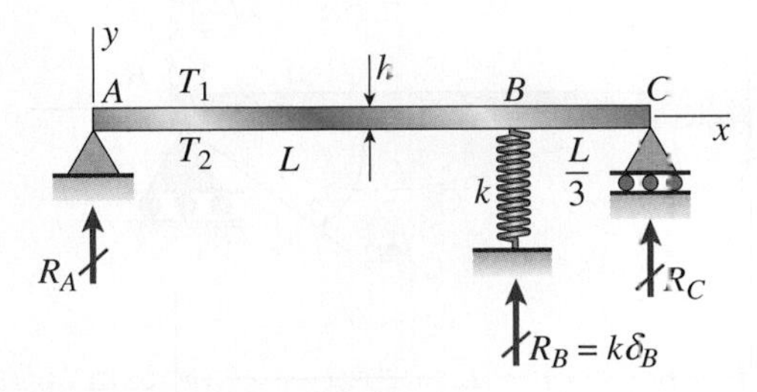

PROBLEMS 10.5-4 and 10.5-5

10.5-5 Solve the preceding problem by integrating the differential equation of the deflection curve.

10.6 Longitudinal Displacements at the Ends of Beams

10.6-1 Assume that the deflected shape of a beam AB with *immovable* pinned supports (see figure) is given by the equation $v = -\delta \sin \pi x/L$, where δ is the deflection at the midpoint of the beam and L is the length. Also, assume that the beam has constant axial rigidity EA.

(a) Obtain formulas for the longitudinal force H at the ends of the beam and the corresponding axial tensile stress σ_t.
(b) For an aluminum-alloy beam with $E = 70$ GPa, calculate the tensile stress σ_t when the ratio of the deflection δ to the length L equals 1/200, 1/400, and 1/600.

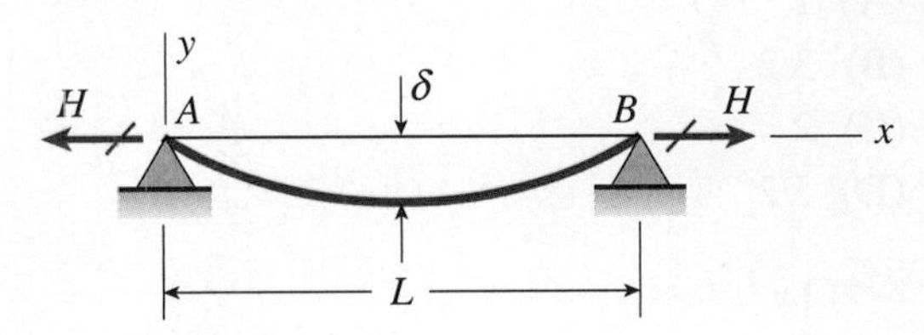

PROBLEM 10.6-1

10.6-2 (a) A simple beam AB with length L and height h supports a uniform load of intensity q (see the figure part a). Obtain a formula for the curvature shortening λ of this beam. Also, obtain a formula for the maximum bending stress σ_b in the beam due to the load q.

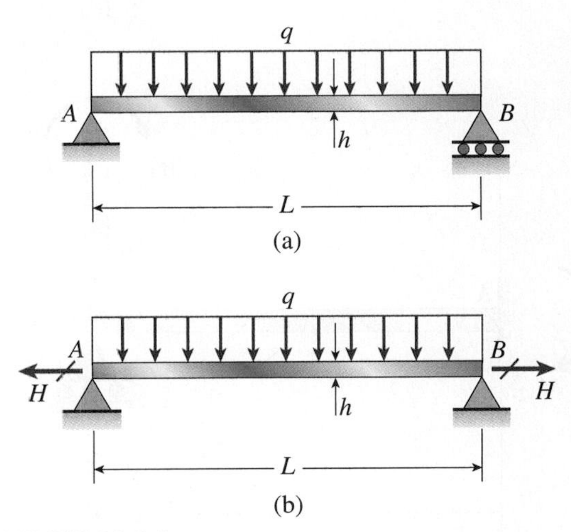

PROBLEM 10.6-2

(b) Now assume that the ends of the beam are pinned so that curvature shortening is prevented and a horizontal force H develops at the supports (see the figure part b). Obtain a formula for the corresponding axial tensile stress σ_t.

(c) Using the formulas obtained in parts (a) and (b), calculate the curvature shortening λ, the maximum bending stress σ_b, and the tensile stress σ_t for the following steel beam: length $L = 3$ m, height $h = 300$ mm, modulus of elasticity $E = 200$ GPa, and moment of inertia $I = 36 \times 10^6$ mm^4. Also, the load on the beam has intensity $q = 25$ kN/m.

Compare the tensile stress σ_t produced by the axial forces with the maximum bending stress σ_b produced by the uniform load.

SOME ADDITIONAL REVIEW PROBLEMS: Chapter 10

R-10.1 Propped cantilever beam AB has moment M_1 applied at joint B. Framework ABC has moment M_2 applied at C. Both structures have constant flexural rigidity EI. If the ratio of the *applied moments* $M_1/M_2 = 3/2$, the ratio of the *reactive moments* M_{A1}/M_{A2} at clamped support A is approximately:

(A) 1
(B) 3/2
(C) 2
(D) 5/2

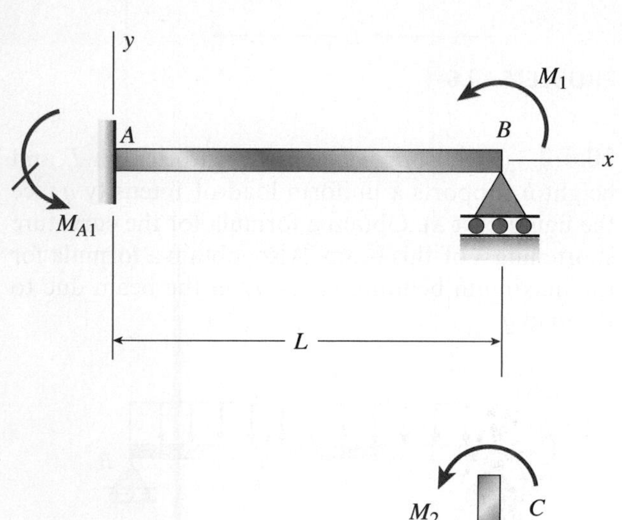

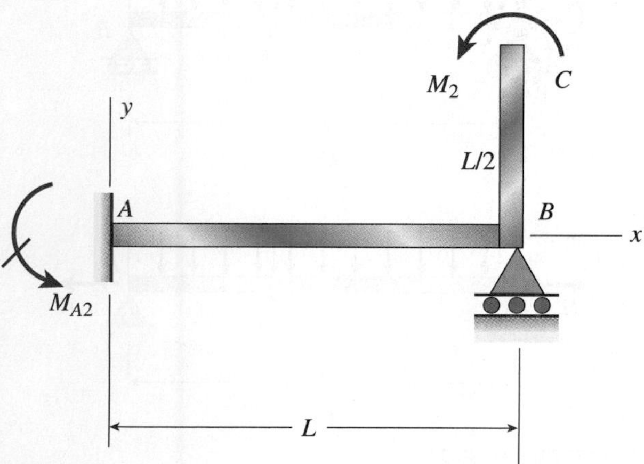

R-10.2 Propped cantilever beam AB has moment M_1 applied at joint B. Framework ABC has moment M_2 applied at C. Both structures have constant flexural rigidity EI. If the ratio of the *applied moments* $M_1/M_2 = 3/2$, the ratio of the *joint rotations at B*, θ_{B1}/θ_{B2}, is approximately:

(A) 1
(B) 3/2
(C) 2
(D) 5/2

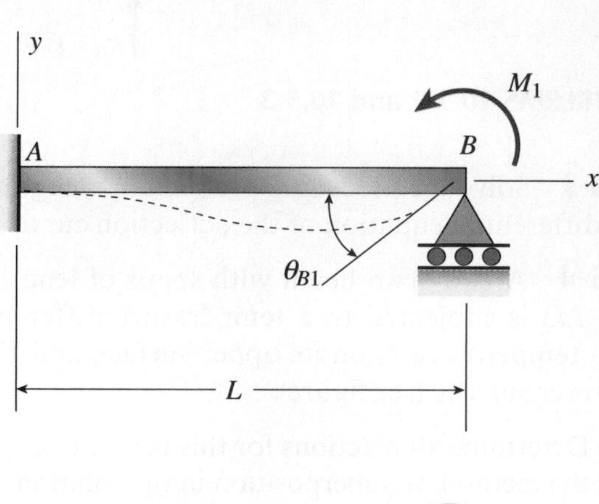

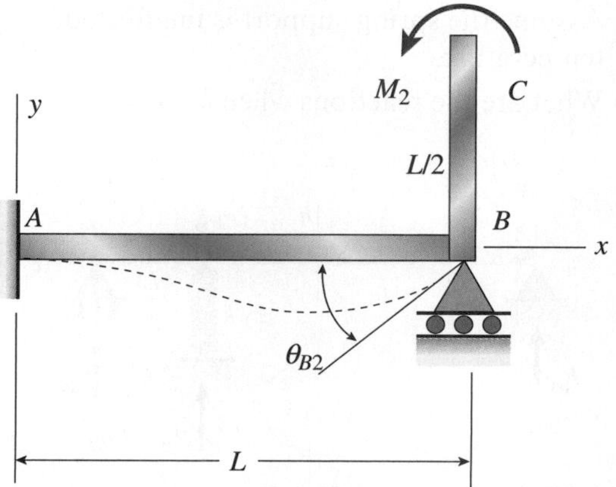

R-10.3 Structure 1 with member BC of length $L/2$ has force P_1 applied at joint C. Structure 2 with member BC of length L has force P_2 applied at C. Both structures have constant flexural rigidity EI. If the ratio of the *applied forces* $P_1/P_2 = 5/2$, the ratio of the *joint B rotations* θ_{B1}/θ_{B2} is approximately:

(A) 1
(B) 5/4
(C) 3/2
(D) 2

R-10.4 Structure 1 with member BC of length $L/2$ has force P_1 applied at joint C. Structure 2 with member BC of length L has force P_2 applied at C. Both structures have constant flexural rigidity EI. The *required* ratio of the *applied forces* P_1/P_2 so that *joint B rotations* θ_{B1} and θ_{B2} are *equal* is approximately:

(A) 1
(B) 5/4
(C) 3/2
(D) 2

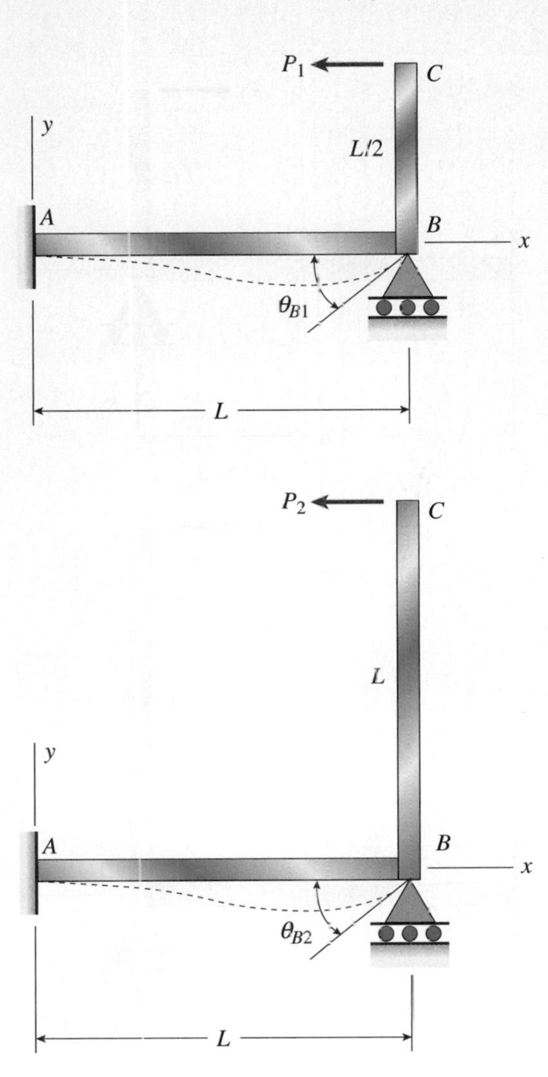

R-10.5 Structure 1 with member BC of length $L/2$ has force P_1 applied at joint C. Structure 2 with member BC of length L has force P_2 applied at C. Both structures have constant flexural rigidity EI. If the ratio of the *applied forces* $P_1/P_2 = 5/2$, the ratio of the *joint B reactions* R_{B1}/R_{B2} is approximately:

(A) 1
(B) 5/4
(C) 3/2
(D) 2

R-10.6 Structure 1 with member BC of length $L/2$ has force P_1 applied at joint C. Structure 2 with member BC of length L has force P_2 applied at C. Both structures have constant flexural rigidity EI. The *required* ratio of the *applied forces* P_1/P_2 so that *joint B reactions* R_{B1} and R_{B2} are *equal* is approximately:

(A) 1
(B) 5/4
(C) 3/2
(D) 2

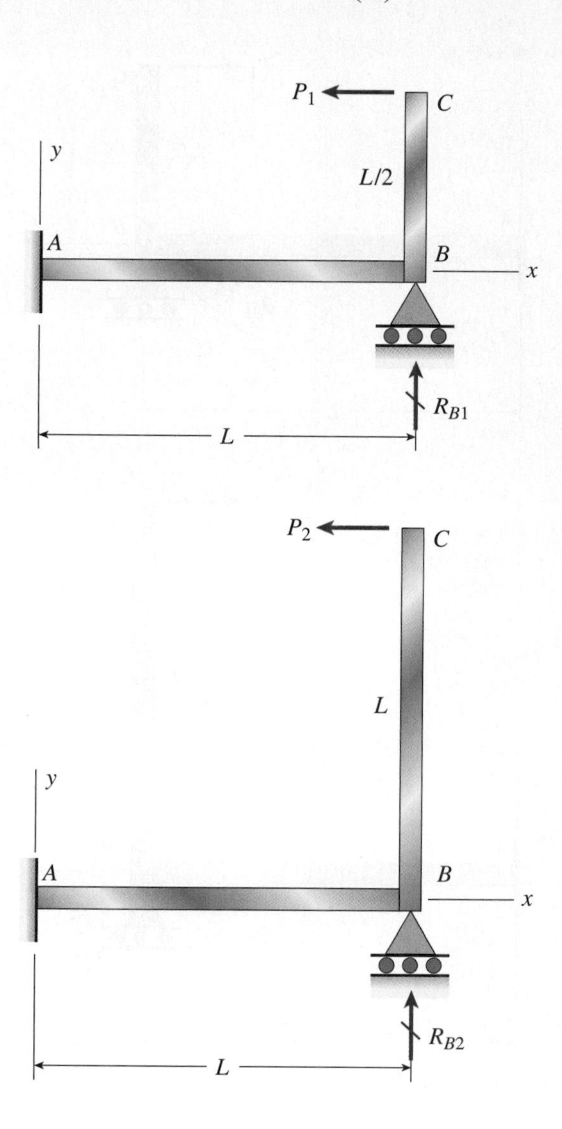

R-10.7 Structure 1 with member BC of length $L/2$ has force P_1 applied at joint C. Structure 2 with member BC of length L has force P_2 applied at C. Both structures have constant flexural rigidity EI. If the ratio of the *applied forces* $P_1/P_2 = 5/2$, the ratio of the *joint C lateral deflections* δ_{C1}/δ_{C2} is approximately:

(A) 1/2
(B) 4/5
(C) 3/2
(D) 2

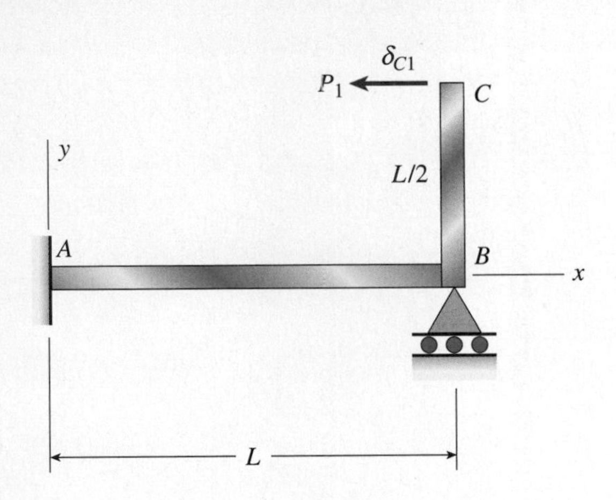

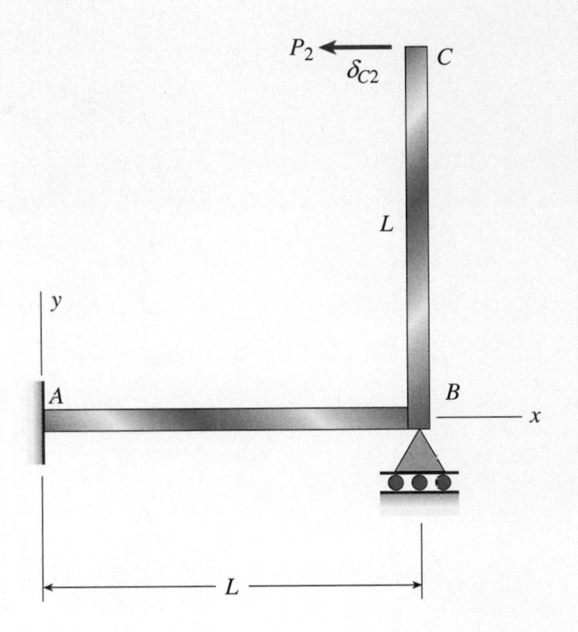

CHAPTER 11

Columns

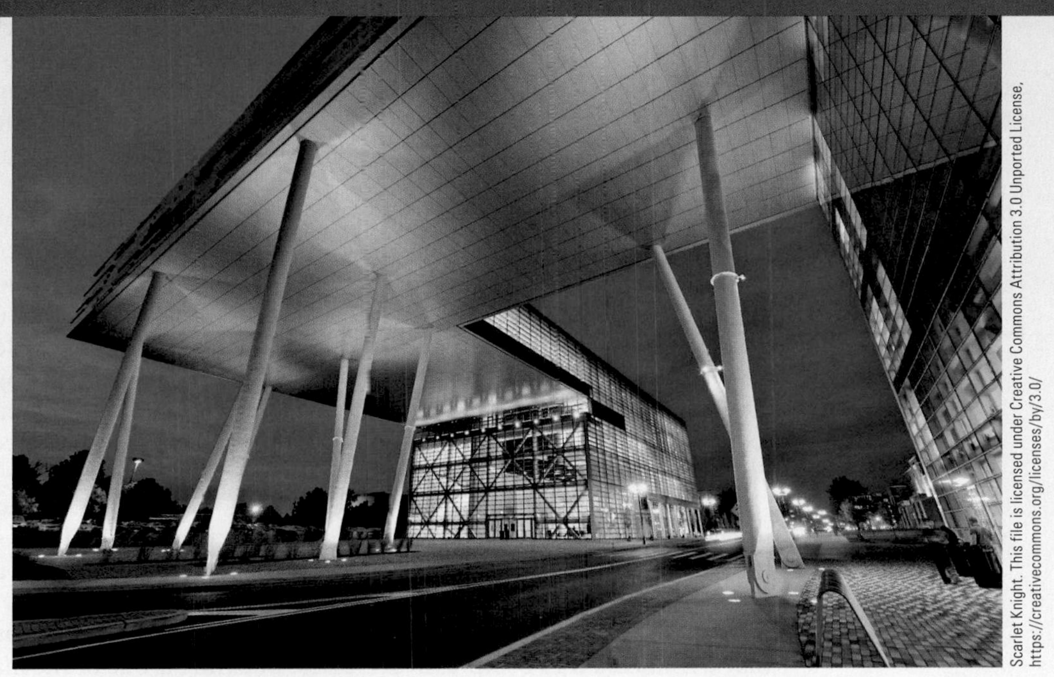

Critical load carrying elements in structures such as columns and other slender compression members are susceptible to buckling failure.

Chapter Objectives

- Study basic concepts of elastic buckling of slender columns.
- Derive formulas for the Euler buckling load of slender columns having a variety of different support conditions.
- Study the buckled mode shapes and compute the effective lengths of columns with different support conditions.
- Study the secant formula, which gives maximum compressive stresses in slender columns acted on by eccentric axial compressive loads.
- Investigate three available theories for inelastic buckling when the material is stressed beyond the proportional limit.

Chapter Outline

FIGURE 11-1

Buckling of a slender column due to an axial compressive load P

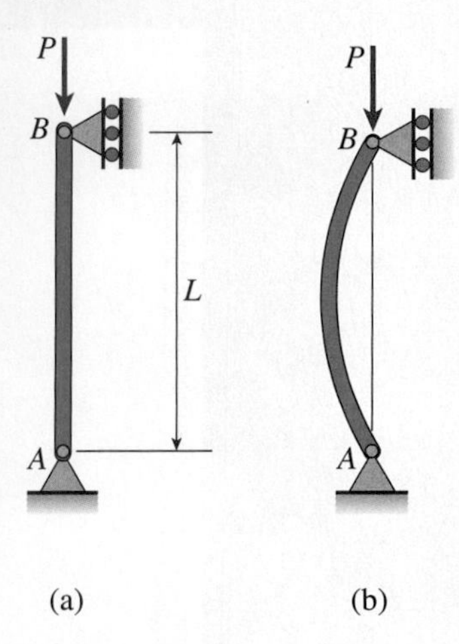

11.1 Introduction

Load-carrying structures may fail in a variety of ways, depending upon the type of structure, the conditions of support, the kinds of loads, and the materials used. For instance, an axle in a vehicle may fracture suddenly from repeated cycles of loading, or a beam may deflect excessively, so that the structure is unable to perform its intended functions. These kinds of failures are prevented by designing structures so that the maximum stresses and maximum displacements remain within tolerable limits. Thus, **strength** and **stiffness** are important factors in design, as discussed throughout the preceding chapters.

Another type of failure is **buckling**, which is the subject matter of this chapter. The focus here is the buckling of **columns**, which are long, slender structural members loaded axially in compression (Fig. 11-1a). If a compression member is relatively slender, it may deflect laterally and fail by bending (Fig. 11-1b) rather than failing by direct compression of the material. You can demonstrate this behavior by compressing a plastic ruler or other slender object. When lateral bending occurs, the column has *buckled.* Under an increasing axial load, the lateral deflections will increase too, and eventually the column will collapse completely.

The phenomenon of buckling is not limited to columns. Buckling can occur in many kinds of structures and can take many forms. When you step on the top of an empty aluminum can, the thin cylindrical walls buckle under your weight and the can collapses. When a large bridge collapsed a few years ago, investigators found that failure was caused by the buckling of a thin steel plate that wrinkled under compressive stresses. Buckling is one of the major causes of failures in structures; therefore, the possibility of buckling should always be considered in design.

11.2 Buckling and Stability

To illustrate the fundamental concepts of buckling and stability, consider the **idealized structure**, or **buckling model**, shown in Fig. 11-2a. This hypothetical structure consists of two rigid bars AB and BC, each of a length $L/2$. They are joined at B by a pin connection and held in a vertical position by a rotational spring having a stiffness β_R.[1]

This idealized structure is analogous to the column of Fig. 11-1a, because both structures have simple supports at the ends and are compressed by an axial load P. However, the elasticity of the idealized structure is "concentrated" in the rotational spring, whereas a real column can bend throughout its length (Fig. 11-1b).

In the idealized structure, the two bars are perfectly aligned and the axial load P has its line of action along the longitudinal axis (Fig. 11-2a). Consequently, the spring is initially unstressed, and the bars are in direct compression.

Now suppose that the structure is disturbed by some external force that causes point B to move a small distance laterally (Fig. 11-2b). The rigid bars

[1]The general relationship for a rotational spring is $M = \beta_R \theta$, where M is the moment acting on the spring, β_R is the rotational stiffness of the spring, and θ is the angle through which the spring rotates. Thus, rotational stiffness has units of moment divided by angle, such as N · m/rad. The analogous relationship for a translational spring is $F = \beta\delta$, where F is the force acting on the spring, β is the translational stiffness of the spring (or spring constant), and δ is the change in length of the spring. Thus, translational stiffness has units of force divided by length, such as N/m.

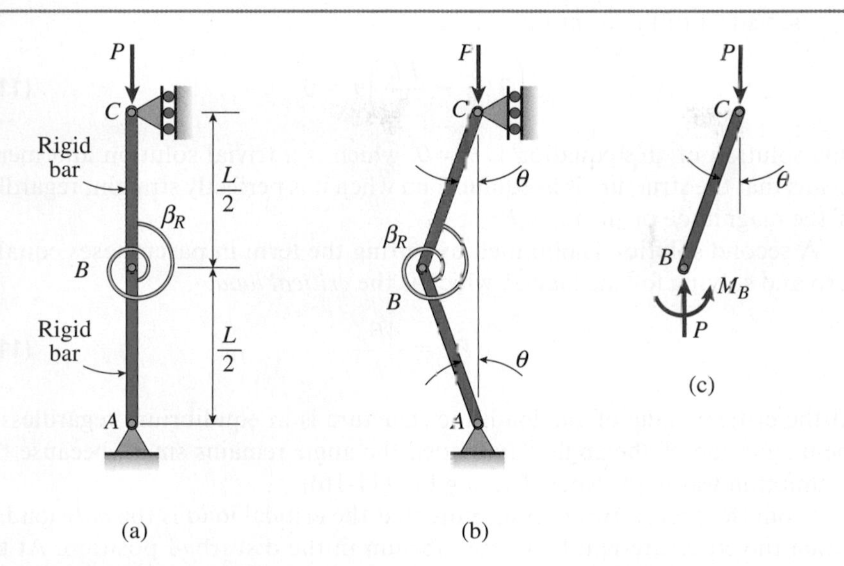

FIGURE 11-2

Buckling of an idealized structure consisting of two rigid bars and a rotational spring

rotate through small angles θ, and a moment develops in the spring. The direction of this moment is such that it tends to return the structure to its original straight position; therefore, it is called a **restoring moment**. At the same time, however, the tendency of the axial compressive force is to increase the lateral displacement. Thus, these two actions have opposite effects—the restoring moment tends to *decrease* the displacement and the axial force tends to *increase* it.

Now consider what happens when the disturbing force is removed. If the axial force P is relatively small, the action of the restoring moment will predominate over the action of the axial force, and the structure will return to its initial straight position. Under these conditions, the structure is said to be **stable**. However, if the axial force P is large, the lateral displacement of point B will increase, and the bars will rotate through larger and larger angles until the structure collapses. Under these conditions, the structure is **unstable** and fails by lateral buckling.

Critical Load

The transition between the stable and unstable conditions occurs at a special value of the axial force known as the **critical load** (denoted by the symbol P_{cr}). The critical load of this buckling model can be determined by considering the structure in the disturbed position (Fig. 11-2b) and investigating its equilibrium.

First, consider the entire structure as a free body and sum moments about support A. This step leads to the conclusion that there is no horizontal reaction at support C. Second, consider bar BC as a free body (Fig. 11-2c) and note that it is subjected to the action of the axial forces P and the moment M_B in the spring. The moment M_B is equal to the rotational stiffness β_R times the angle of rotation 2θ of the spring; thus,

$$M_B = 2\beta_R\theta \tag{11-1a}$$

Since the angle θ is a small quantity, the lateral displacement of point B is $\theta L/2$. Therefore, the equation of equilibrium is obtained by summing moments about point B for bar BC (Fig. 11-2c):

$$M_B - P\left(\frac{\theta L}{2}\right) = 0 \tag{11-1b}$$

or, upon substituting from Eq. (11-1a),

$$\left(2\beta_R - \frac{PL}{2}\right)\theta = 0 \tag{11-2}$$

One solution of this equation is $\theta = 0$, which is a trivial solution and merely means that the structure is in equilibrium when it is perfectly straight, regardless of the magnitude of the force P.

A second solution is obtained by setting the term in parentheses equal to zero and solving for the load P, which is the *critical load*:

$$P_{cr} = \frac{4\beta_R}{L} \tag{11-3}$$

At the critical value of the load, the structure is in equilibrium regardless of the magnitude of the angle θ [provided the angle remains small, because that assumption was made when deriving Eq. (11-1b)].

From the preceding analysis, note that the critical load is the *only* load for which the structure will be in equilibrium in the disturbed position. At this value of the load, the restoring effect of the moment in the spring just matches the buckling effect of the axial load. Therefore, the critical load represents the boundary between stable and unstable conditions.

If the axial load is less than P_{cr}, the effect of the moment in the spring predominates, and the structure returns to the vertical position after a slight disturbance; if the axial load is larger than P_{cr}, the effect of the axial force predominates and the structure buckles:

If $P < P_{cr}$, the structure is *stable*.

If $P > P_{cr}$, the structure is *unstable*.

Equation (11-3) shows that the stability of the structure is increased either by *increasing its stiffness* or by *decreasing its length*. Later in this chapter, these same observations apply when critical loads are determined for various types of columns.

Summary

In summary, the idealized structure (Fig. 11-2a) behaves as follows as the axial load P increases from zero to a large value.

When the axial load is less than the critical load ($0 < P < P_{cr}$), the structure is in equilibrium when it is perfectly straight. Because the equilibrium is **stable**, the structure returns to its initial position after being disturbed. Thus, the structure is in equilibrium *only* when it is perfectly straight ($\theta = 0$).

When the axial load is greater than the critical load ($P > P_{cr}$), the structure is still in equilibrium when $\theta = 0$ (because it is in direct compression and there is no moment in the spring), but the equilibrium is **unstable** and cannot be maintained. The slightest disturbance will cause the structure to buckle.

At the critical load ($P = P_{cr}$), the structure is in equilibrium even when point B is displaced laterally by a small amount. In other words, the structure is in equilibrium for *any* small angle θ, including $\theta = 0$. However, the structure is neither stable nor unstable—it is at the boundary between stability and instability. This condition is referred to as **neutral equilibrium**.

The three equilibrium conditions for the idealized structure are shown in the graph of axial load P versus angle of rotation θ (Fig. 11-3). The two heavy lines—one vertical and one horizontal—represent the equilibrium conditions. Point B, where the equilibrium diagram branches, is called a *bifurcation point*.

The horizontal line for neutral equilibrium extends to the left and right of the vertical axis because the angle θ may be clockwise or counterclockwise. The line extends only a short distance, however, because this analysis is based upon the assumption that θ is a small angle. (This assumption is quite valid, because θ is indeed small when the structure first departs from its vertical position. If buckling continues and θ becomes large, the line labeled "Neutral equilibrium" curves upward, as shown later in Fig. 11-12.)

The three equilibrium conditions represented by the diagram of Fig. 11-3 are analogous to those of a ball placed upon a smooth surface (Fig. 11-4). If the surface is concave upward, like the inside of a dish, the equilibrium is stable, and the ball always returns to the low point when disturbed. If the surface is convex upward, like a dome, the ball theoretically can be in equilibrium on top of the surface, but the equilibrium is unstable and in reality the ball rolls away. If the surface is perfectly flat, the ball is in neutral equilibrium and remains wherever it is placed.

In the next section, the behavior of an ideal elastic column is shown to be analogous to that of the buckling model shown in Fig. 11-2. Furthermore, many other kinds of structural and mechanical systems fit this model.

FIGURE 11-3

Equilibrium diagram for buckling of an idealized structure

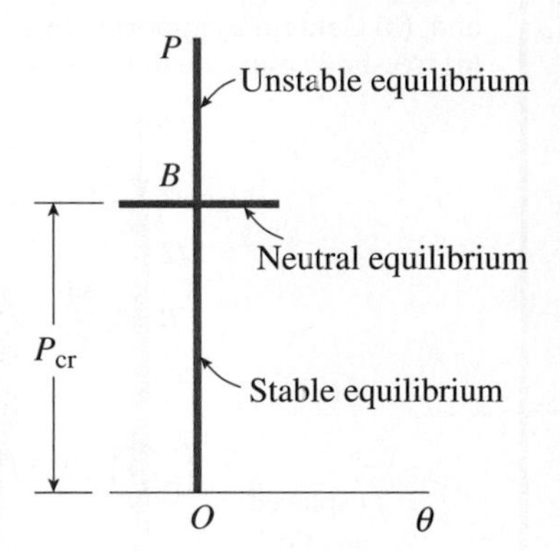

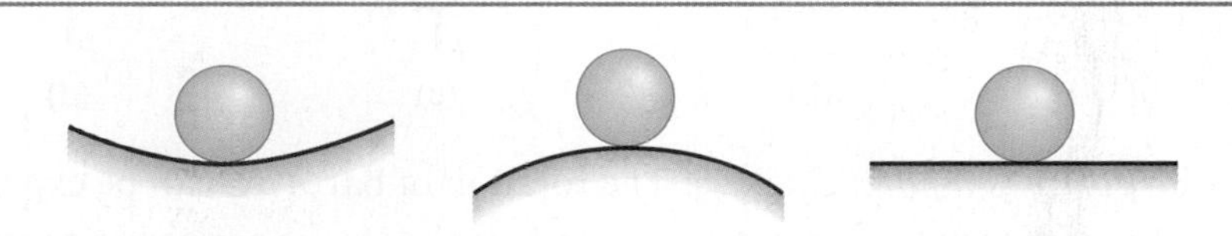

FIGURE 11-4

Ball in stable, unstable, and neutral equilibrium

Example 11-1

Two idealized columns are shown in Fig. 11-5. Both columns are initially straight and vertical with load P applied at joint A.

(a) The first column (Column 1, Fig. 11-5a) consists of a rigid bar ABC with a roller support at B and is connected to bar CD by an elastic connection with rotational stiffness $\beta_R = \beta L^2$.

(b) The second column (Column 2, Fig. 11-5b) is a single rigid bar $ABCD$ that has a roller support at B and a spring support at D.

Column 1 has a sliding support at D that cannot rotate, while Column 2 has a roller support at D with a translational spring with stiffness β. Find an expression for the critical load P_{cr} for each column.

Solution:

Use a four-step problem-solving approach. Combine steps as needed for an efficient solution.

Part (a): Column 1.

1. ***Conceptualize*** *[hypothesize, sketch]*: Begin by considering the equilibrium of Column 1 in a displaced position caused by some external load P applied at A causing *small* lateral displacement Δ_A (Fig. 11-5a). Sum lateral forces for the entire column to find that reaction $H_B = 0$. Next, draw the free-body diagram of bar ABC (Fig. 11-5c). Note that $\Delta_C = \Delta_A$ and that bar ABC rotation is $\Delta_A/(L/2) = 2\Delta_A/L$.

FIGURE 11-5

Example 11-1: Two idealized columns made up of rigid bars: (a) Column 1 with rotational elastic connection at C and, (b) Column 2 supported by a translational elastic support at D, (c) free-body diagram of bar ABC in Column 1, (d) free-body diagram of Column 2

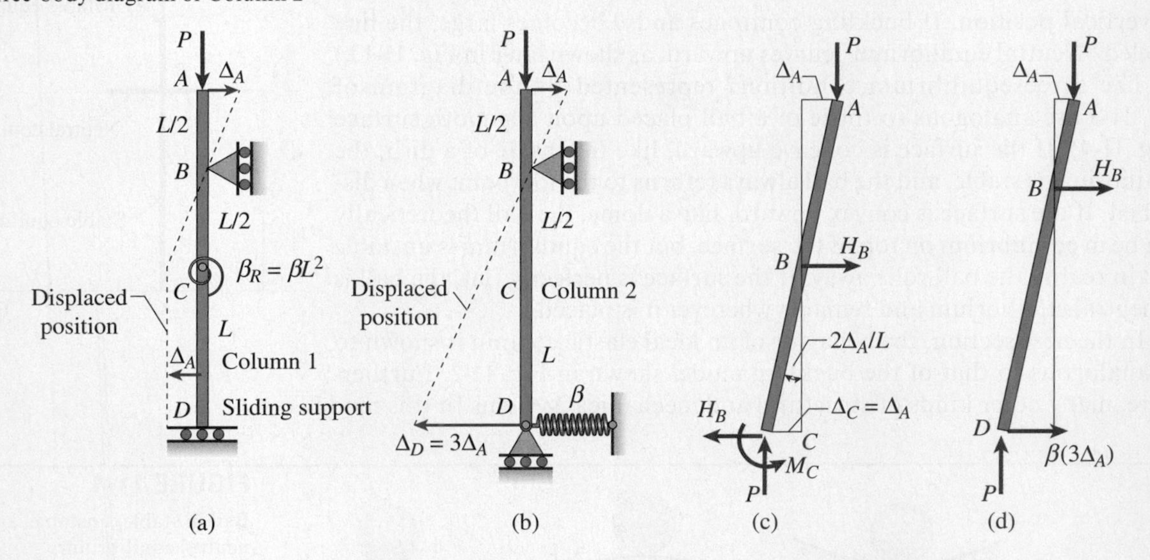

2. ***Categorize*** [*simplify, classify*]: The rotation of bar ABC can be expressed as $2\Delta_A/L$ because the rotation is small, so the elastic connection resisting moment at C is equal to rotational stiffness β_R times the total relative rotation of the spring or $M_C = (\beta_R)(2\Delta_A/L)$.
3. ***Analyze*** [*evaluate; select relevant equations; carry out mathematical solution*]: Sum moments about C in Fig. 11-5c and solve for P to obtain the critical load P_{cr1} for Column 1:

$$P_{cr1} = \frac{M_C}{2\Delta_A} = \frac{\beta_R\left(\dfrac{2\Delta_A}{L}\right)}{2\Delta_A} = \frac{\beta_R}{L} = \beta L \quad \text{(a)}$$

4. ***Finalize*** [*conclude; examine answer—Does it make sense? Are units correct? How does it compare to similar problem solutions?*]: The buckled *mode shape* for Column 1 is the displaced position shown in Fig. 11-5a.

Part (a): Column 2.

1, 2. ***Conceptualize, Categorize***: Investigate the equilibrium of Column 2 in a *displaced* position, once again defined by lateral displacement Δ_A, as shown in Fig. 11-5b. Use a free-body diagram of the entire column (Fig. 11-5d) and sum lateral forces to find that

$$H_B = -(\beta)(3\Delta_A) \quad \text{(b)}$$

3. ***Analyze***: Sum moments about D in Fig. 11-5d to find the critical load:

$$P_{cr2} = \frac{-H_B\left(\dfrac{3L}{2}\right)}{4\Delta_A} = \frac{9}{8}\beta L \quad \text{(c)}$$

4. ***Finalize***: The buckled *mode shape* for Column 2 is the displaced position shown in Fig. 11-5b.

Combined model and analysis.

1. ***Conceptualize***: Now create a more advanced structure model by combining the features of Column 1 and Column 2 into a single column as shown in Fig. 11-5e. This idealized structure has elastic rotational springs at both C and D with rotational stiffnesses β_{R1} and β_{R2}, respectively. The roller support remains at B, and the sliding support at D is now restrained by an elastic translational spring with stiffness β. Small lateral displacements Δ_C and Δ_D are selected as the degrees of freedom that define the possible displaced positions of the column. (Alternatively, rotation angles such as θ_B and θ_D could be used as degrees of freedom to uniquely describe any arbitrary position of the displaced structure.) Hence, the *combined* structure has *two degrees of freedom*; therefore, it has two possible buckled mode shapes and two different critical loads—each of which causes the associated buckling mode. In contrast, Columns 1 and 2 are *single degree-of-freedom* structures because only Δ_A is needed to define the buckled shape of each structure depicted in Figs. 11-5a and b.
2. ***Categorize***: Observe that if rotational spring β_{R2} becomes infinitely stiff and translational spring stiffness $\beta = 0$ in the combined structure (Fig. 11-5e) (while β_{R1} remains finite), the two degree of freedom (2DOF) combined model reduces to the single degree-of-freedom (SDOF) model of Fig. 11-5a. Similarly, if rotational spring stiffness β_{R1} goes to infinity and spring $\beta_{R2} = 0$ in Fig. 11-5e (while translational spring stiffness β remains finite), the model becomes that shown in Fig. 11-5b. It follows *that the solutions for P_{cr} for Columns 1 and 2* [Eqs. (a) *and* (c)] *are simply two special case solutions of the general combined model in* Fig. 11-5e.
3. ***Analyze***: The goal now is to find a general solution for the 2DOF model in Fig. 11-5e and then show that solutions for P_{cr} for Columns 1 and 2 can be obtained from this general solution.

   First, consider the equilibrium of the entire 2DOF model in the displaced position shown in Fig. 11-5f. Sum horizontal forces to find that $H_B = \beta\Delta_D$. Rotation angles θ_B and θ_D in Fig. 11-5f can be expressed in terms of translations Δ_C and Δ_D as

$$\theta_D = \frac{\Delta_D - \Delta_C}{L} \qquad \theta_B = \frac{2}{L}\Delta_C \tag{d}$$

   Sum moments about B in free-body diagram ABC (Fig. 11-5g), noting that the moment at C is equal to rotational spring stiffness β_{R1} times the *relative* rotation $(\theta_D - \theta_B)$ at C, to get

$$2P\Delta_C - \beta\Delta_D\frac{L}{2} + \beta_{R1}(\theta_D - \theta_B) = 0$$

FIGURE 11-5 (*Continued*)

(e) combined model of Columns 1 and 2, (f) free-body diagram of combined model, (g) free-body diagram of bar ABC in combined model

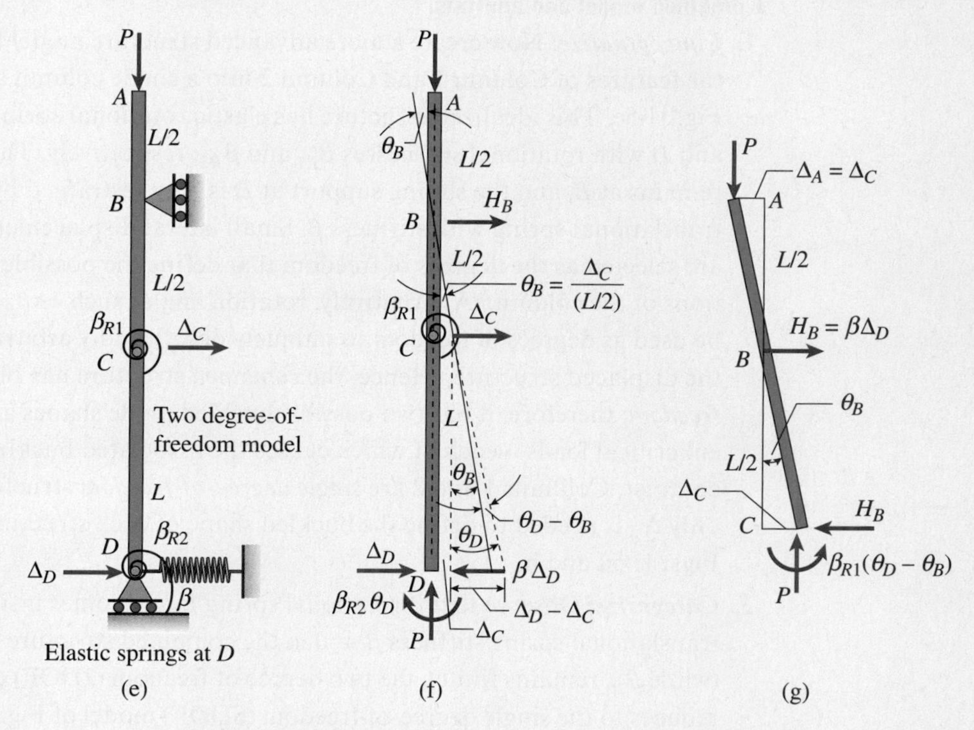

Substituting expressions for θ_D and θ_B from Eqs. (d) produces the following equation in terms of unknown displacements Δ_C and Δ_D:

$$\left(2P - \frac{3\beta_{R1}}{L}\right)\Delta_C + \left(\frac{\beta_{R1}}{L} - \frac{L\beta}{2}\right)\Delta_D = 0 \tag{e}$$

Obtain a second equation that describes the equilibrium of the displaced structure from the free-body diagram of bar CD alone. Summing moments about C for bar CD alone gives

$$P(\Delta_D - \Delta_C) - \beta_{R1}(\theta_D - \theta_B) - \beta_{R2}\theta_D - \beta\Delta_D L = 0$$

which can be rewritten as

$$\left(\frac{3\beta_{R1}}{L} - P + \frac{\beta_{R2}}{L}\right)\Delta_C + \left(P - \frac{\beta_{R1}}{L} - \frac{\beta_{R2}}{L} - L\beta\right)\Delta_D = 0 \tag{f}$$

There are now two algebraic equations [Eqs. (e) and (f)] and two unknowns (Δ_C and Δ_D). Expressing Eqs. (e) and (f) in matrix form gives

$$\begin{pmatrix} 2P - \frac{3\beta_{R1}}{L} & \frac{\beta_{R1}}{L} - \frac{L\beta}{2} \\ \frac{3\beta_{R1}}{L} - P + \frac{\beta_{R2}}{L} & P - \frac{\beta_{R1}}{L} - \frac{\beta_{R2}}{L} - L\beta \end{pmatrix} \begin{pmatrix} \Delta_C \\ \Delta_D \end{pmatrix} = \begin{pmatrix} 0 \\ 0 \end{pmatrix} \tag{g}$$

FIGURE 11-5 (*Continued*)

(h) buckled mode shapes for combined model

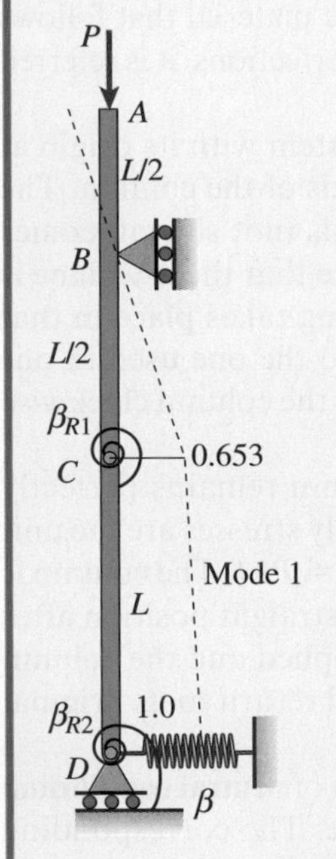

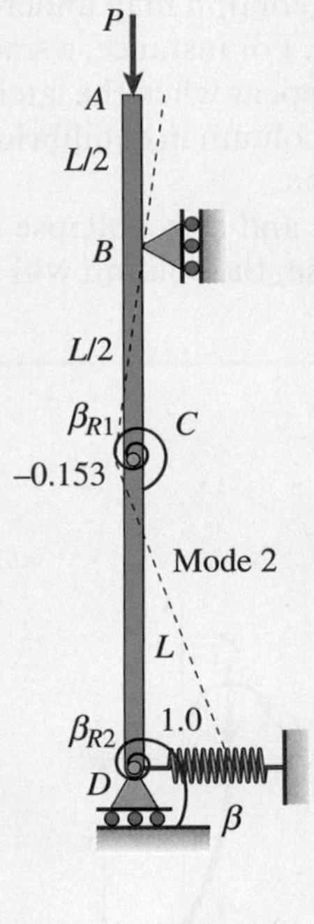

(h)

These homogeneous equations have a nonzero (nontrivial) solution only if the determinant of the coefficient matrix in Eq. (g) is equal to zero. If both elastic connections have the same rotational stiffness, $\beta_{R1} = \beta_{R2} = \beta L^2$, and Eq. (g) becomes

$$\begin{pmatrix} 2P - 3L\beta & \dfrac{L\beta}{2} \\ 4L\beta - P & P - 3L\beta \end{pmatrix} \begin{pmatrix} \Delta_C \\ \Delta_D \end{pmatrix} = \begin{pmatrix} 0 \\ 0 \end{pmatrix} \tag{h}$$

So the determinant of the coefficient matrix in Eq. (h) (known as the characteristic equation) is

$$P^2 - \left(\frac{17}{4}\beta L\right)P + \frac{7}{2}(\beta L)^2 = 0 \tag{i}$$

Solving Eq. (i) using the quadratic formula results in two values of the critical load:

$$P_{cr1} = \beta L\left(\frac{17 - \sqrt{65}}{8}\right) = 1.117\beta L$$

$$P_{cr2} = \beta L\left(\frac{17 + \sqrt{65}}{8}\right) = 3.13\beta L$$

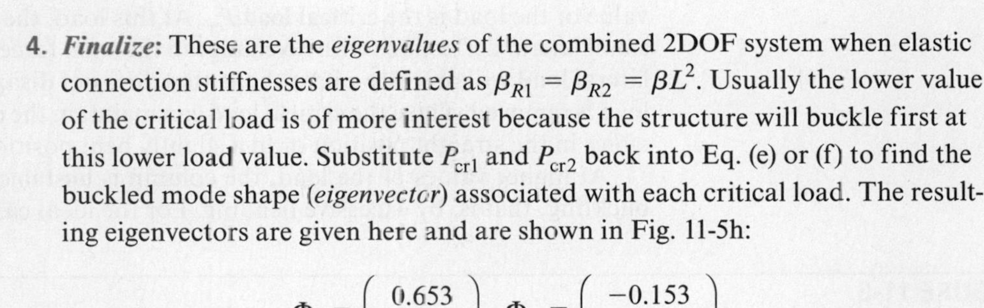

4. ***Finalize***: These are the *eigenvalues* of the combined 2DOF system when elastic connection stiffnesses are defined as $\beta_{R1} = \beta_{R2} = \beta L^2$. Usually the lower value of the critical load is of more interest because the structure will buckle first at this lower load value. Substitute P_{cr1} and P_{cr2} back into Eq. (e) or (f) to find the buckled mode shape (*eigenvector*) associated with each critical load. The resulting eigenvectors are given here and are shown in Fig. 11-5h:

$$\Phi_1 = \begin{pmatrix} 0.653 \\ 1 \end{pmatrix} \quad \Phi_2 = \begin{pmatrix} -0.153 \\ 1 \end{pmatrix}$$

Application of combined model to Columns 1 and 2.
If the rotational spring stiffness β_{R2} goes to infinity and translational spring stiffness β goes to zero while rotational stiffness β_{R1} remains finite, the combined model (Fig. 11-5e) reduces to Column 1, and the critical load obtained from the solution of Eq. (g) is that given in Eq. (a). [Alternatively, equating Δ_C and Δ_D and setting $\beta = 0$ in Eq. (e) confirms P_{cr1} in Eq. (a)]. If, instead, rotational spring stiffness β_{R1} goes to infinity and rotational stiffness β_{R2} goes to zero while translational stiffness β remains finite, the solution of Eq. (g) gives the critical load for Column 2 in Eq. (c).

11.3 Columns with Pinned Ends

The investigation of the stability behavior of columns begins by analyzing a slender column with pinned ends (Fig. 11-6a). The column is loaded by a vertical force P that is applied through the centroid of the end cross section. The column itself is perfectly straight and is made of a linearly elastic material that follows Hooke's law. Since the column is assumed to have no imperfections, it is referred to as an **ideal column**.

For purposes of analysis, construct a coordinate system with its origin at support A and with the x axis along the longitudinal axis of the column. The y axis is directed to the left in the figure, and the z axis (not shown) comes out of the plane of the figure toward the viewer. Assume that the x-y plane is a plane of symmetry of the column and that any bending takes place in that plane (Fig. 11-6b). The coordinate system is identical to the one used in our previous discussions of beams, as can be seen by rotating the column clockwise through an angle of 90°.

When the axial load P has a small value, the column remains perfectly straight and undergoes direct axial compression. The only stresses are the uniform compressive stresses obtained from the equation $\sigma = P/A$. The column is in **stable equilibrium**, which means that it returns to the straight position after a disturbance. For instance, if a small lateral load is applied and the column bends, the deflection will disappear and the column will return to its original position when the lateral load is removed.

As the axial load P is gradually increased, a condition of **neutral equilibrium** is reached in which the column may have a bent shape. The corresponding value of the load is the **critical load** P_{cr}. At this load, the column may undergo small lateral deflections with no change in the axial force. For instance, a small lateral load will produce a bent shape that does not disappear when the lateral load is removed. Thus, the critical load can maintain the column in equilibrium *either* in the straight position or in a slightly bent position.

At higher values of the load, the column is **unstable** and may collapse by buckling, that is, by excessive bending. For the ideal case, the column will be

FIGURE 11-6

Column with pinned ends: (a) ideal column, (b) buckled shape, and (c) axial force P and bending moment M acting at a cross section

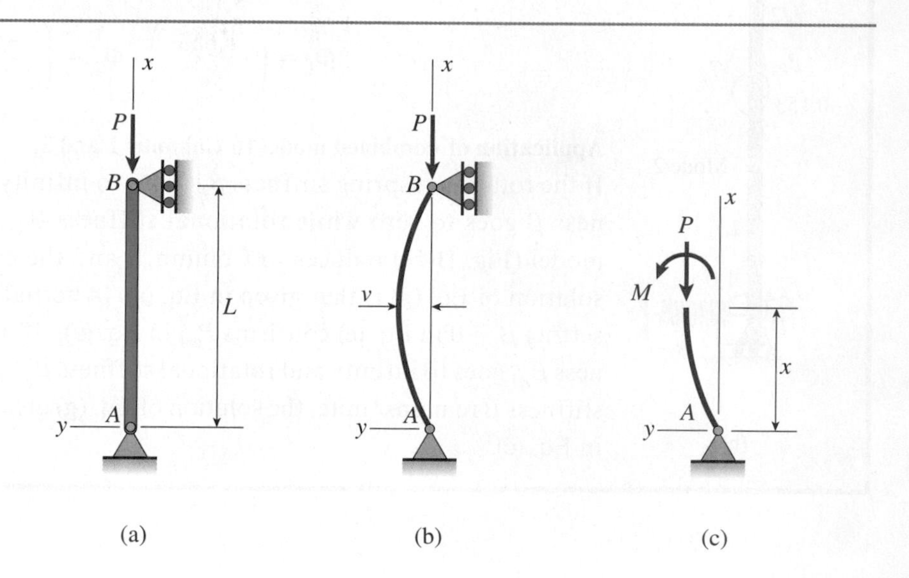

in equilibrium in the straight position even when the axial force P is greater than the critical load. However, since the equilibrium is unstable, the smallest imaginable disturbance will cause the column to deflect sideways. Once that happens, the deflections will immediately increase, and the column will fail by buckling. The behavior is similar to that described in the preceding section for the idealized buckling model (Fig. 11-2).

The behavior of an ideal column compressed by an axial load P (Figs. 11-6a and b) may be summarized as follows.

If $P < P_{cr}$, the column is in stable equilibrium in the straight position.

If $P = P_{cr}$, the column is in neutral equilibrium in either the straight or a slightly bent position.

If $P > P_{cr}$, the column is in unstable equilibrium in the straight position and will buckle under the slightest disturbance.

Of course, a real column does not behave in this idealized manner because imperfections are always present. For instance, the column is not *perfectly* straight, and the load is not *exactly* at the centroid. Nevertheless, this discussion begins by studying ideal columns because they provide insight into the behavior of real columns.

Differential Equation for Column Buckling

To determine the critical loads and corresponding deflected shapes for an ideal pin-ended column (Fig. 11-6a), use one of the differential equations of the deflection curve of a beam [see Eqs. (9-16a, b, and c) in Section 9.2]. These equations are applicable to a buckled column because the column bends as though it were a beam (Fig. 11-6b).

Although both the fourth-order differential equation (the load equation) and the third-order differential equation (the shear-force equation) are suitable for analyzing columns, use the second-order equation (the bending-moment equation) because its general solution is usually the simplest. The **bending-moment equation** (Eq. 9-16a) is

$$EIv'' = M \tag{11-4}$$

in which M is the bending moment at any cross section, v is the lateral deflection in the y direction, and EI is the flexural rigidity for bending in the x-y plane.

The bending moment M at distance x from end A of the buckled column is shown acting in its positive direction in Fig. 11-6c. Note that the bending moment sign convention is the same as that used in earlier chapters, namely, positive bending moment produces positive curvature (see Figs. 9-3 and 9-4).

The axial force P acting at the cross section is also shown in Fig. 11-6c. Since there are no horizontal forces acting at the supports, there are no shear forces in the column. Therefore, equilibrium of moments about point A gives

$$M + Pv = 0 \quad \text{or} \quad M = -Pv \tag{11-5}$$

where v is the deflection at the cross section.

This same expression for the bending moment is obtained if it is assumed that the column buckles to the right instead of to the left (Fig. 11-7a). When the

FIGURE 11-7

Column with pinned ends (alternative direction of buckling)

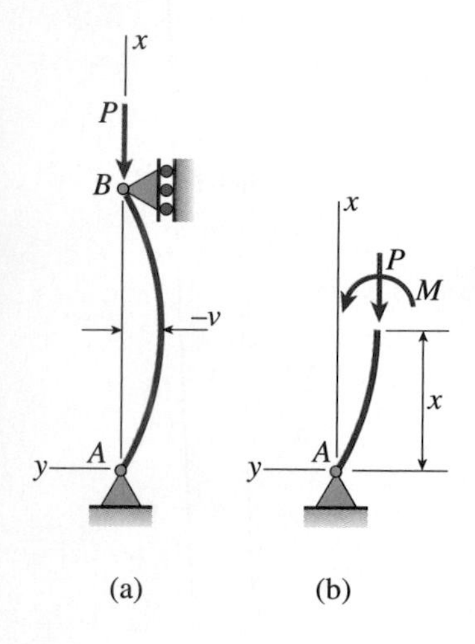

column deflects to the right, the deflection itself is $-v$, but the moment of the axial force about point A also changes sign. Thus, the equilibrium equation for moments about point A (see Fig. 11-7b) is

$$M - P(-v) = 0$$

which gives the same expression for the bending moment M as before.

The **differential equation of the deflection curve** (Eq. 11-4) now becomes

$$EIv'' + Pv = 0 \qquad \textbf{(11-6)}$$

By solving this equation, which is a *homogeneous, linear, differential equation of second order with constant coefficients*, you can determine the magnitude of the critical load and the deflected shape of the buckled column.

Note that the buckling of columns involves solving the same basic differential equation as the one solved in Chapters 9 and 10 when finding beam deflections. However, there is a fundamental difference in the two types of analysis. In the case of beam deflections, the bending moment M appearing in Eq. (11-4) is a function of the loads only—it does not depend upon the deflections of the beam. In the case of buckling, the bending moment is a function of the deflections themselves (Eq. 11-5).

Thus, a new aspect of bending analysis is revealed. In the previous work, the deflected shape of the structure was not considered, and the equations of equilibrium were based upon the geometry of the *undeformed* structure. Now, however, the geometry of the *deformed* structure is taken into account when writing equations of equilibrium.

Solution of the Differential Equation

For convenience in writing the solution of the differential equation of Eq. (11-6), use the notation

$$k^2 = \frac{P}{EI} \quad \text{or} \quad k = \sqrt{\frac{P}{EI}} \qquad \textbf{(11-7a,b)}$$

in which k is always taken as a positive quantity. Note that k has units of the reciprocal of length; therefore, quantities such as kx and kL are nondimensional.

Using this notation, rewrite Eq. (11-6) in the form

$$v'' + k^2 v = 0 \qquad \textbf{(11-8)}$$

From mathematics, you know that the **general solution** of this equation is

$$v = C_1 \sin kx + C_2 \cos kx \qquad \textbf{(11-9)}$$

in which C_1 and C_2 are constants of integration (to be evaluated from the boundary conditions, or end conditions, of the column). Note that the number of arbitrary constants in the solution (two in this case) agrees with the order of the differential equation. Also, note that the solution can be verified by substituting the expression for v [Eq. (11-9)] into the differential equation [Eq. (11-8)] and reducing it to an identity.

To evaluate the **constants of integration** appearing in the solution [Eq. (11-9)], use the boundary conditions at the ends of the column; namely, the deflection is zero when $x = 0$ and $x = L$ (see Fig. 11-6b):

$$v(0) = 0 \quad \text{and} \quad v(L) = 0 \qquad \textbf{(11-10a,b)}$$

The first condition gives $C_2 = 0$ so

$$v = C_1 \sin kx \qquad \textbf{(11-10c)}$$

The second condition gives

$$C_1 \sin kL = 0 \qquad \textbf{(11-10d)}$$

From this equation, either $C_1 = 0$ or $\sin kL = 0$. Both of these possibilities are now considered.

Case 1. If the constant C_1 equals zero, the deflection v is also zero [see Eq. (11-10c)]; therefore, the column remains straight. In addition, note that when C_1 equals zero, Eq. (11-10d) is satisfied for *any* value of the quantity kL. Consequently, the axial load P also may have any value [see Eq. (11-7b)]. This solution of the differential equation (known in mathematics as the *trivial solution*) is represented by the vertical axis of the load-deflection diagram (Fig. 11-8). It gives the behavior of an ideal column that is in equilibrium (either stable or unstable) in the straight position (no deflection) under the action of the compressive load P.

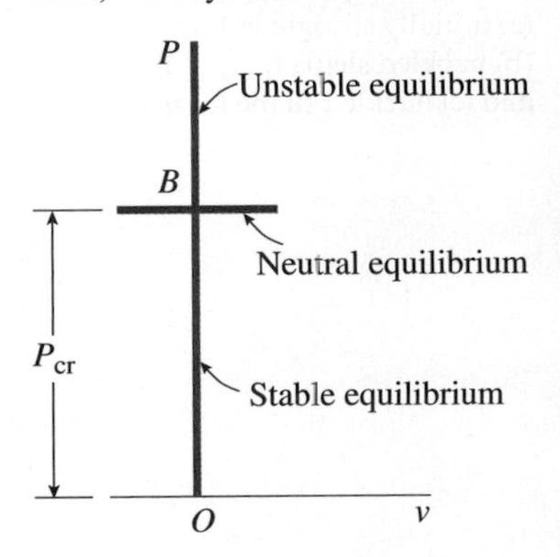

FIGURE 11-8
Load-deflection diagram for an ideal, linearly elastic column

Case 2. The second possibility for satisfying Eq. (11-10d) is given by the equation known as the **buckling equation**:

$$\sin kL = 0 \qquad \textbf{(11-11)}$$

This equation is satisfied when $kL = 0, \pi, 2\pi, \ldots$. However, since $kL = 0$ means that $P = 0$, this solution is not of interest. Therefore, the solutions considered here are

$$kL = n\pi \qquad n = 1, 2, 3, \ldots \qquad \textbf{(11-12)}$$

or [see Eq. (11-7a)]:

$$P = \frac{n^2\pi^2 EI}{L^2} \qquad n = 1, 2, 3, \ldots \qquad \textbf{(11-13)}$$

This formula gives the values of P that satisfy the buckling equation and provide solutions (other than the trivial solution) to the differential equation.

The equation of the **deflection curve** from Eqs. (11-10c) and (11-12) is

$$v = C_1 \sin kx = C_1 \sin \frac{n\pi x}{L} \qquad n = 1, 2, 3, \ldots \qquad \textbf{(11-14)}$$

Only when P has one of the values given by Eq. (11-13) is it theoretically possible for the column to have a bent shape [given by Eq. (11-14)]. For all other values of P, the column is in equilibrium only if it remains straight. Therefore, the values of P given by Eq. (11-13) are the **critical loads** for this column.

Critical Loads

The lowest critical load for a column with pinned ends (Fig. 11-9a) is obtained when $n = 1$:

$$P_{cr} = \frac{\pi^2 EI}{L^2} \qquad \textbf{(11-15)}$$

The corresponding buckled shape (sometimes called a *mode shape* - see Example 11-1) is

$$v = C_1 \sin \frac{\pi x}{L} \qquad \textbf{(11-16)}$$

FIGURE 11-9

Buckled shapes for an ideal column with pinned ends: (a) initially straight column, (b) buckled shape for $n = 1$, and (c) buckled shape for $n = 2$

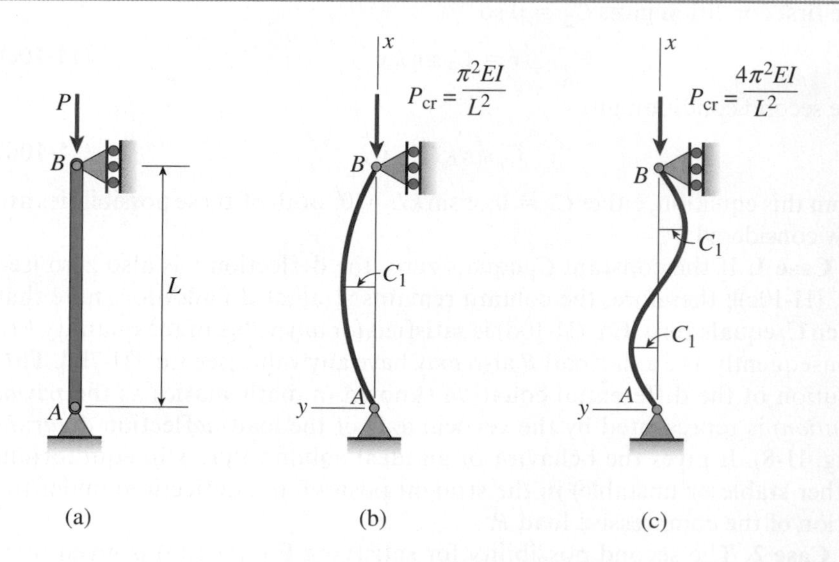

as shown in Fig. 11-9b. The constant C_1 represents the deflection at the midpoint of the column and may have any small value, either positive or negative. Therefore, the part of the load-deflection diagram corresponding to P_{cr} is a horizontal straight line (Fig. 11-8). Thus, the deflection at the critical load is *undefined*, although it must remain small for our equations to be valid. Above the bifurcation point *B*, the equilibrium is unstable, and below point *B* it is stable.

Buckling of a pinned-end column in the first mode is called the **fundamental case** of column buckling.

The type of buckling described in this section is called **Euler buckling**, and the critical load for an ideal elastic column is often called the **Euler load**. The famous mathematician Leonhard Euler (1707–1783), generally recognized as the greatest mathematician of all time, was the first person to investigate the buckling of a slender column and determine its critical load (Euler published his results in 1744); see Ref. 11-1.

Taking higher values of the index n in Eqs. (11-13) and (11-14) leads to an infinite number of critical loads and corresponding mode shapes. The mode shape for $n = 2$ has two half-waves, as pictured in Fig. 11-9c. The corresponding critical load is four times larger than the critical load for the fundamental case. The magnitudes of the critical loads are proportional to the square of n, and the number of half-waves in the buckled shape is equal to n.

Buckled shapes for the **higher modes** are often of no practical interest because the column buckles when the axial load P reaches its lowest critical value. The only way to obtain modes of buckling higher than the first is to provide lateral support of the column at intermediate points, such as at the midpoint of the column shown in Fig. 11-9 (see Example 11-2 at the end of this section).

General Comments

From Eq. (11-15), note that the critical load of a column is proportional to the flexural rigidity EI and inversely proportional to the square of the length. Of particular interest is the fact that the *strength* of the material itself, as represented by a quantity such as the proportional limit or the yield stress, does not appear in the equation for the critical load. Therefore, increasing a strength property does not raise the critical load of a slender column. It only can be raised by increasing the flexural rigidity, reducing the length, or providing additional lateral support.

The *flexural rigidity* can be increased by using a "stiffer" material (that is, a material with larger modulus of elasticity E) or by distributing the material in such a way as to increase the moment of inertia I of the cross section—just as a beam can be made stiffer by increasing the moment of inertia. The moment of inertia is increased by distributing the material farther from the centroid of the cross section. Hence, a hollow tubular member is generally more economical for use as a column than a solid member having the same cross-sectional area.

Reducing the *wall thickness* of a tubular member and increasing its lateral dimensions (while keeping the cross-sectional area constant) also increases the critical load because the moment of inertia is increased. This process has a practical limit, however, because eventually the wall itself will become unstable. When that happens, localized buckling occurs in the form of small corrugations or wrinkles in the walls of the column. Thus, a distinction must be made between *overall buckling* of a column, which is discussed in this chapter, and *local buckling* of its parts. The latter requires more detailed investigations and is beyond the scope of this book.

The preceding analysis (see Fig. 11-9) assumed that the *x-y* plane was a plane of symmetry of the column and that buckling took place in that plane. The latter assumption will be met if the column has lateral supports perpendicular to the plane of the figure, so that the column is constrained to buckle in the *x-y* plane. If the column is supported only at its ends and is free to buckle in *any* direction, the bending will occur about the principal centroidal axis having the smaller moment of inertia.

For instance, consider the rectangular and wide-flange cross sections shown in Fig. 11-10. In each case, the moment of inertia I_1 is greater than the moment of inertia I_2; hence, the column will buckle in the 1–1 plane, and the smaller moment of inertia I_2 should be used in the formula for the critical load. If the cross section is square or circular, all centroidal axes have the same moment of inertia, and buckling may occur in any longitudinal plane.

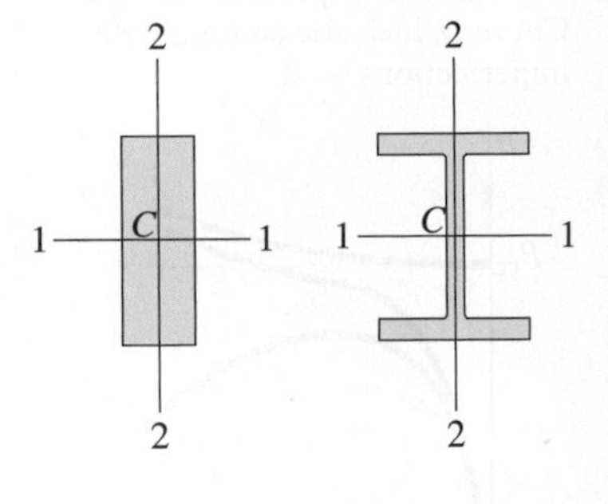

FIGURE 11-10

Cross sections of columns showing principal centroidal axes with $I_1 > I_2$

Critical Stress

After finding the critical load for a column, calculate the corresponding **critical stress** by dividing the load by the cross-sectional area. For the fundamental case of buckling (Fig. 11-9b), the critical stress is

$$\sigma_{cr} = \frac{P_{cr}}{A} = \frac{\pi^2 EI}{AL^2} \tag{11-17}$$

in which I is the moment of inertia for the principal axis about which buckling occurs. This equation can be written in a more useful form by introducing the notation

$$r = \sqrt{\frac{I}{A}} \tag{11-18}$$

in which r is the **radius of gyration** of the cross section in the plane of bending.[2] Then the equation for the critical stress becomes

$$\sigma_{cr} = \frac{\pi^2 E}{(L/r)^2} \tag{11-19}$$

[2]Radius of gyration is described in Section D.3, Appendix D.

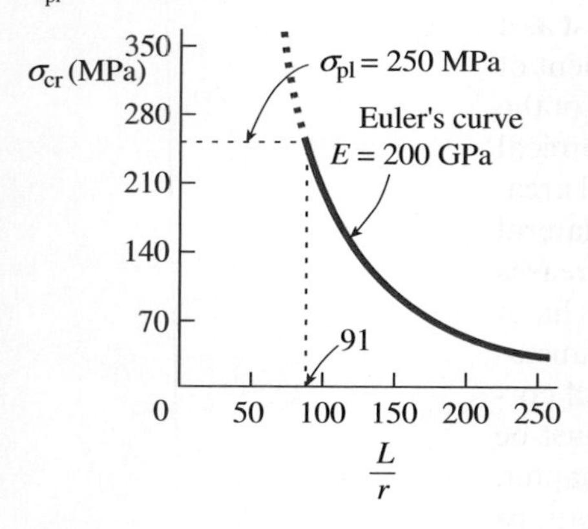

FIGURE 11-11

Graph of Euler's curve from Eq. (11-19) for structural steel with $E = 200$ GPa and $\sigma_{pl} = 250$ MPa

in which L/r is a nondimensional ratio called the **slenderness ratio**:

$$\text{Slenderness ratio} = \frac{L}{r} \qquad \textbf{(11-20)}$$

Note that the slenderness ratio depends only on the dimensions of the column. A column that is long and slender will have a high slenderness ratio and therefore a low critical stress. A column that is short and stubby will have a low slenderness ratio and will buckle at a high stress. Typical values of the slenderness ratio for actual columns are between 30 and 150.

The critical stress is the average compressive stress on the cross section at the instant the load reaches its critical value. You can plot a graph of this stress as a function of the slenderness ratio and obtain a curve known as **Euler's curve** (Fig. 11-11). The curve shown in the figure is plotted for a structural steel with $E = 200$ GPa. The curve is valid only when the critical stress is less than the proportional limit of the steel, because the equations were derived using Hooke's law. Therefore, draw a horizontal line on the graph at the proportional limit of the steel (assumed to be 250 MPa) and terminate Euler's curve at that level of stress.[3]

Effects of Large Deflections, Imperfections, and Inelastic Behavior

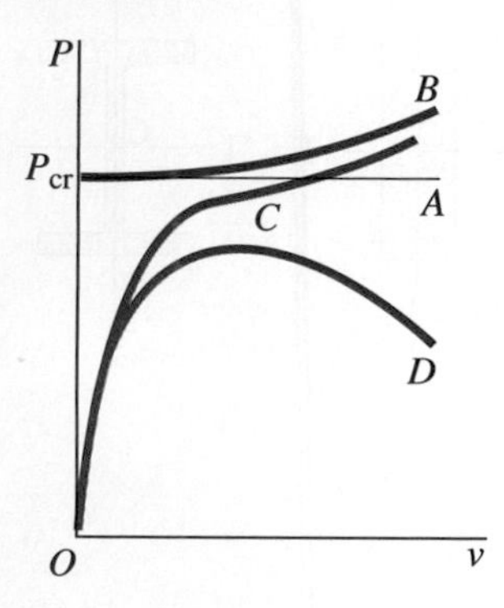

FIGURE 11-12

Load-deflection diagram for columns: Line A, ideal elastic column with small deflections; Curve B, ideal elastic column with large deflections; Curve C, elastic column with imperfections; and Curve D, inelastic column with imperfections

The equations for critical loads were derived for ideal columns, that is, columns for which the loads are precisely applied, the construction is perfect, and the material follows Hooke's law. As a consequence, the magnitudes of the small deflections at buckling were undefined.[4] Thus, when $P = P_{cr}$, the column may have any small deflection, which is a condition represented by the horizontal line labeled A in the load-deflection diagram of Fig. 11-12. (This figure shows only the right-hand half of the diagram, but the two halves are symmetric about the vertical axis.)

The theory for ideal columns is limited to small deflections because the second derivative v'' for the curvature was used. A more exact analysis, based upon the exact expression for curvature [Eq. (9-19) in Section 9.2], shows that there is no indefiniteness in the magnitudes of the deflections at buckling. Instead, for an ideal, linearly elastic column, the load-deflection diagram goes upward in accord with curve B of Fig. 11-12. Thus, after a linearly elastic column begins to buckle, an increasing load is required to cause an increase in the deflections.

Now suppose that the column is not constructed perfectly; for instance, the column might have an imperfection in the form of a small initial curvature, so that the unloaded column is not perfectly straight. Such imperfections produce deflections from the onset of loading, as shown by curve C in Fig. 11-12. For small deflections, curve C approaches line A as an asymptote. However, as the deflections become large, it approaches curve B. The larger the imperfections, the further curve C moves to the right, away from the vertical line. Conversely, if the column is constructed with considerable accuracy, curve C approaches the

[3]Euler's curve is not a common geometric shape. It is sometimes mistakenly called a hyperbola, but hyperbolas are plots of polynomial equations of the second degree in two variables, whereas Euler's curve is a plot of an equation of the third degree in two variables.

[4]In mathematical terminology, you solved a *linear eigenvalue problem* (see Example 11-1). The critical load is an *eigenvalue*, and the corresponding buckled mode shape is an *eigenfunction*.

vertical axis and the horizontal line labeled *A*. By comparing lines *A*, *B*, and *C*, note that for practical purposes the critical load represents the maximum load-carrying capacity of an elastic column, because large deflections are not acceptable in most applications.

Finally, consider what happens when the stresses exceed the proportional limit and the material no longer follows Hooke's law. Of course, the load-deflection diagram is unchanged up to the level of load at which the proportional limit is reached. Then the curve for inelastic behavior (curve *D*) departs from the elastic curve, continues upward, reaches a maximum, and turns downward.

The precise shapes of the curves in Fig. 11-12 depend upon the material properties and column dimensions, but the general nature of the behavior is typified by the curves shown.

Only extremely slender columns remain elastic up to the critical load. Stockier columns behave inelastically and follow a curve such as *D*. Thus, the maximum load that can be supported by an inelastic column may be considerably less than the Euler load for that same column. Furthermore, the descending part of curve *D* represents sudden and catastrophic collapse because it takes smaller and smaller loads to maintain larger and larger deflections. By contrast, the curves for elastic columns are quite stable because they continue upward as the deflections increase; therefore, it takes larger and larger loads to cause an increase in deflection. (Inelastic buckling is described in more detail in Sections 11.7 and 11.8.)

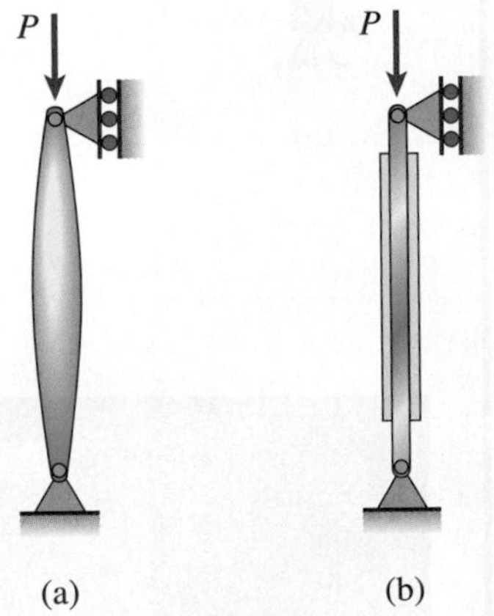

Optimum Shapes of Columns

Compression members usually have the same cross sections throughout their lengths, so only prismatic columns are analyzed in this chapter. However, prismatic columns are not the optimum shape if minimum weight is desired. The critical load of a column consisting of a given amount of material may be increased by varying the shape so that the column has larger cross sections in those regions where the bending moments are larger. Consider, for instance, a column of solid circular cross section with pinned ends. A column shaped as shown in Fig. 11-13a will have a larger critical load than a prismatic column made from the same volume of material. As a means of approximating this optimum shape, prismatic columns are sometimes reinforced over part of their lengths (Fig. 11-13b).

FIGURE 11-13

Nonprismatic columns

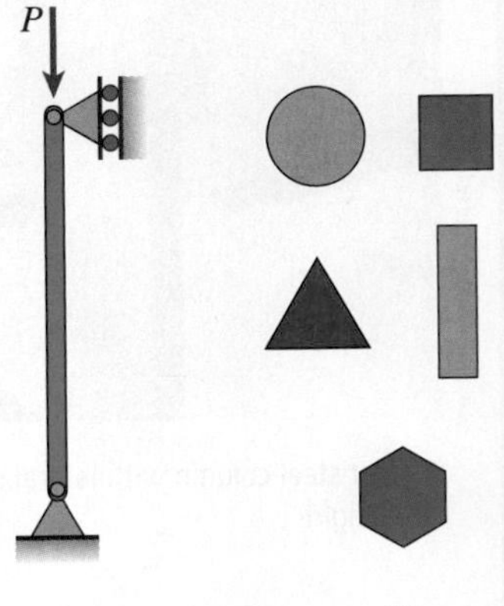

Now consider a prismatic column with pinned ends that is free to buckle in *any* lateral direction (Fig. 11-14a). Also, assume that the column has a solid cross section, such as a circle, square, triangle, rectangle, or hexagon (Fig. 11-14b). An interesting question arises: For a given cross-sectional area, which of these shapes makes the most efficient column? Or, in more precise terms, which cross section gives the largest critical load? Of course, the assumption is that the critical load is calculated from the Euler formula $P_{cr} = \pi^2 EI/L^2$ using the smallest moment of inertia for the cross section.

FIGURE 11-14

Which cross-sectional shape is the optimum shape for a prismatic column?

While a common answer to this question is "the circular shape," you can readily demonstrate that a cross section in the shape of an equilateral triangle gives a 21% higher critical load than does a circular cross section of the same area (see Prob. 11.3-17). The critical load for an equilateral triangle is also higher than the loads obtained for the other shapes; hence, an equilateral triangle is the optimum cross section (based only upon theoretical considerations). For a mathematical analysis of optimum column shapes, including columns with varying cross sections, see Ref. 11-4.

Example 11-2

FIGURE 11-15

Example 11-2: Euler buckling of a slender column

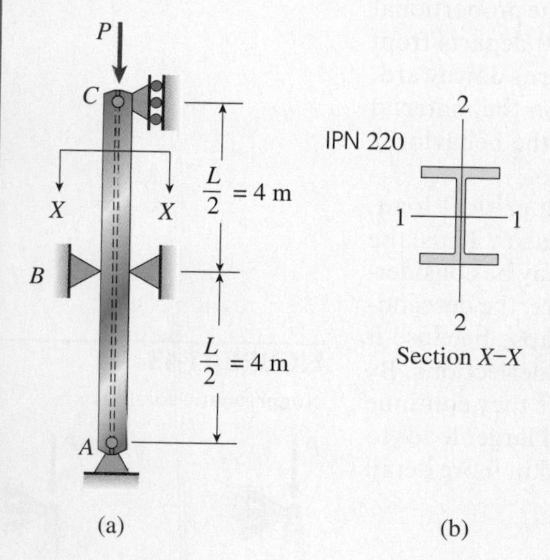

A long, slender column *ABC* is pin-supported at the ends and compressed by an axial load *P* (Fig. 11-15). Lateral support is provided at the midpoint *B* in the plane of the figure. However, lateral support perpendicular to the plane of the figure is provided only at the ends.

The column is constructed of a standard steel shape (IPN 220) having modulus of elasticity $E = 200\,\text{GPa}$ and proportional limit $\sigma_{pl} = 300$ MPa. The total length of the column is $L = 8\,\text{m}$.

Determine the allowable load P_{allow} using a factor of safety $n = 2.5$ with respect to Euler buckling of the column.

Solution:

Use a four-step problem-solving approach.

1. ***Conceptualize***: Because of the manner in which it is supported, this column may buckle in either of the two principal planes of bending. As one possibility, it may buckle in the plane of the figure, in which case the distance between lateral supports is $L/2 = 4\,\text{m}$ and bending occurs about axis 2–2 (see Fig. 11-9c for the mode shape of buckling).

As a second possibility, the column may buckle perpendicular to the plane of the figure with bending about axis 1–1. Because the only lateral support in this direction is at the ends, the distance between lateral supports is $L = 8\,\text{m}$ (see Fig. 11-9b for the mode shape of buckling).

Column properties: From Table F-2 obtain the following moments of inertia and cross-sectional area for an IPN 220 column:

$$I_1 = 3060\ \text{cm}^4 \qquad I_2 = 162\ \text{cm}^4 \qquad A = 39.5\ \text{cm}^2$$

2. ***Categorize***:

Critical loads: If the column buckles in the plane of the figure, the critical load is

$$P_{cr} = \frac{\pi^2 EI_2}{(L/2)^2} = \frac{4\pi^2 EI_2}{L^2}$$

3. ***Analyze***: Substitute numerical values to obtain

$$P_{cr} = \frac{4\pi^2 EI_2}{L^2} = \frac{4\pi^2(200\ \text{GPa})(162\ \text{cm}^4)}{(8\ \text{m})^2} = 200\ \text{kN}$$

Lester Lefkowitz / Getty Images

Slender steel column with lateral support near mid-height

If the column buckles perpendicular to the plane of the figure, the critical load is

$$P_{cr} = \frac{\pi^2 EI_1}{L^2} = \frac{\pi^2(200\text{ GPa})(3060\text{ cm}^4)}{(8\text{ m})^2} = 943.8\text{ kN}$$

Therefore, the critical load for the column (the smaller of the two preceding values) is

$$P_{cr} = 200\text{ kN}$$

and buckling occurs in the plane of the figure.

Critical stresses: Since the calculations for the critical loads are valid only if the material follows Hooke's law, verify that the critical stresses do not exceed the proportional limit of the material. For the larger critical load, the critical stress is

$$\sigma_{cr} = \frac{P_{cr}}{A} = \frac{943.8\text{ kN}}{39.5\text{ cm}^2} = 238.9\text{ MPa}$$

Since this stress is less than the proportional limit ($\sigma_{pl} = 300$ MPa), both critical-load calculations are satisfactory.

4. *Finalize*:

Allowable load: The allowable axial load for the column, based on Euler buckling, is

$$P_{allow} = \frac{P_{cr}}{n} = \frac{200\text{ kN}}{2.5} = 79.9\text{ kN}$$

in which $n = 2.5$ is the desired factor of safety.

11.4 Columns with Other Support Conditions

Buckling of a column with pinned ends (described in the preceding section) is usually considered as the most basic case of buckling. However, in practice, engineers encounter many other end conditions, such as fixed ends, free ends, and elastic supports. The critical loads for columns with various kinds of support conditions can be determined from the differential equation of the deflection curve by following the same procedure that was used when analyzing a pinned-end column.

Digital Vision / Getty Images

Slender concrete columns fixed at the base and free at the top during construction

The procedure is as follows. First, with the column assumed to be in the buckled state, obtain an expression for the bending moment in the column. Second, set up the differential equation of the deflection curve, using the bending-moment equation ($EIv'' = M$). Third, solve the equation and obtain its general solution, which contains two constants of integration plus any other unknown quantities. Fourth, apply boundary conditions pertaining to the deflection v and the slope v' and obtain a set of simultaneous equations. Finally, solve those equations to obtain the critical load and the deflected shape of the buckled column.

This straightforward mathematical procedure is illustrated in the following discussion of three types of columns.

Column Fixed at the Base and Free at the Top

The first case to consider is an ideal column that is fixed at the base, free at the top, and subjected to an axial load P (Fig. 11-16a).[5] The deflected shape of the buckled column is shown in Fig. 11-16b. From this figure, note that the bending moment at distance x from the base is

$$M = P(\delta - v) \tag{11-21}$$

where δ is the deflection at the free end of the column. The **differential equation** of the deflection curve then becomes

$$EIv'' = M = P(\delta - v) \tag{11-22}$$

in which I is the moment of inertia for buckling in the x-y plane.

Using the notation $k^2 = P/EI$ from Eq. (11-7a), rearrange Eq. (11-22) into the form

$$v'' + k^2 v = k^2 \delta \tag{11-23}$$

which is a linear differential equation of second order with constant coefficients. However, it is a more complicated equation than the equation for a column with pinned ends [see Eq. (11-8)] because it has a nonzero term on the right-hand side.

The **general solution** of Eq. (11-23) consists of two parts: (1) the *homogeneous solution*, which is the solution of the homogeneous equation obtained by replacing the right-hand side with zero, and (2) the *particular solution*, which is the solution of Eq. (11-23) that produces the term on the right-hand side.

The homogeneous solution (also called the *complementary solution*) is the same as the solution of Eq. (11-8); hence,

$$v_H = C_1 \sin kx + C_2 \cos kx \tag{11-24a}$$

where C_1 and C_2 are constants of integration. Note that when v_H is substituted into the left-hand side of the differential equation of Eq. (11-23), it produces zero.

FIGURE 11-16

Ideal column fixed at the base and free at the top: (a) initially straight column, (b) buckled shape for $n = 1$, (c) buckled shape for $n = 3$, and (d) buckled shape for $n = 5$

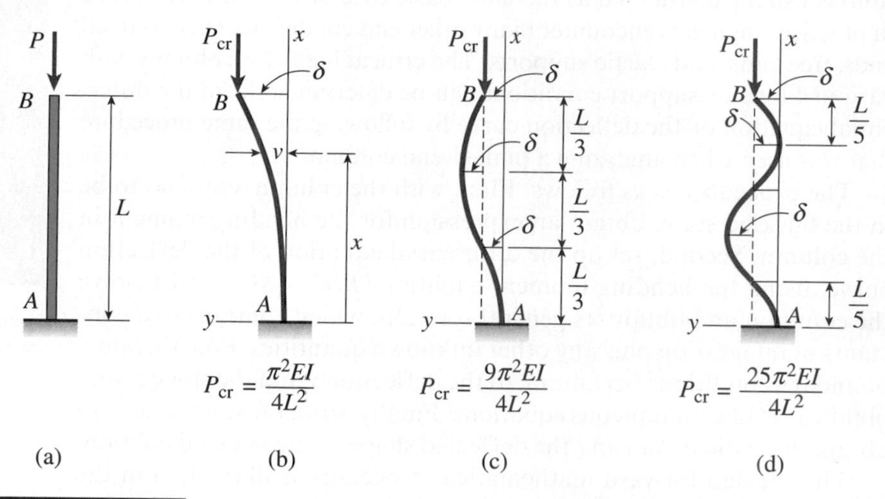

[5]This column is of special interest because it is the one first analyzed by Euler in 1744.

The particular solution of the differential equation is

$$v_P = \delta \tag{11-24b}$$

When v_P is substituted into the left-hand side of the differential equation, it produces the right-hand side, that is, it produces the term $k^2\delta$. Consequently, the *general solution* of the equation, equal to the sum of v_H and v_P, is

$$v = C_1 \sin kx + C_2 \cos kx + \delta \tag{11-25}$$

This equation contains three unknown quantities (C_1, C_2, and δ); therefore, three **boundary conditions** are needed to complete the solution.

At the base of the column, the deflection and slope are each equal to zero. Therefore, the boundary conditions for this case are

$$v(0) = 0 \qquad v'(0) = 0$$

Apply the first condition to Eq. (11-25) to find

$$C_2 = -\delta \tag{11-26}$$

To apply the second condition, first differentiate Eq. (11-25) to obtain the slope:

$$v' = C_1 k \cos kx - C_2 k \sin kx \tag{11-27}$$

Apply the second condition to this equation to find $C_1 = 0$.

Now substitute the expressions for C_1 and C_2 into the general solution of Eq. (11-25) and obtain the **equation of the deflection curve** for the buckled column:

$$v = \delta(1 - \cos kx) \tag{11-28}$$

Note that this equation gives only the *shape* of the deflection curve—the amplitude δ remains undefined. Thus, when the column buckles, the deflection given by Eq. (11-28) may have any arbitrary magnitude, except that it must remain small (because the differential equation is based upon small deflections).

The third boundary condition applies to the upper end of the column, where the deflection v is equal to δ:

$$v(L) = \delta$$

Use this condition with Eq. (11-28) to get

$$\delta \cos kL = 0 \tag{11-29}$$

It follows from this equation that either $\delta = 0$ or $\cos kL = 0$. If $\delta = 0$, there is no deflection of the bar [see Eq. (11-28)] and you have the *trivial solution*—the column remains straight and buckling does not occur. In that case, Eq. (11-29) will be satisfied for any value of the quantity kL, that is, for any value of the load P. This conclusion is represented by the vertical line in the load-deflection diagram of Fig. 11-8.

The other possibility for solving Eq. (11-29) is

$$\cos kL = 0 \tag{11-30}$$

which is the **buckling equation**. In this case, Eq. (11-29) is satisfied regardless of the value of the deflection δ. Thus, as already observed, δ is undefined and may have any small value.

The equation $\cos kL = 0$ is satisfied when

$$kL = \frac{n\pi}{2} \qquad n = 1, 3, 5, \ldots \tag{11-31}$$

Use the expression $k^2 = P/EI$ to obtain the following formula for the **critical loads**:

$$P_{cr} = \frac{n^2\pi^2 EI}{4L^2} \qquad n = 1, 3, 5, ... \qquad \textbf{(11-32)}$$

Also, the **buckled mode shapes** are obtained from Eq. (11-28):

$$v = \delta\left(1 - \cos\frac{n\pi x}{2L}\right) \qquad n = 1, 3, 5, ... \qquad \textbf{(11-33)}$$

The lowest critical load is obtained by substituting $n = 1$ in Eq. (11-32):

$$P_{cr} = \frac{\pi^2 EI}{4L^2} \qquad \textbf{(11-34)}$$

The corresponding buckled shape (from Eq. 11-33) is

$$v = \delta\left(1 - \cos\frac{\pi x}{2L}\right) \qquad \textbf{(11-35)}$$

and is shown in Fig. 11-16b.

Taking higher values of the index n leads to an infinite number of critical loads from Eq. (11-32). The corresponding buckled mode shapes have additional waves in them. For instance, when $n = 3$, the buckled column has the shape shown in Fig. 11-16c, and P_{cr} is nine times larger than for $n = 1$. Similarly, the buckled shape for $n = 5$ has even more waves (Fig. 11-16d), and the critical load is 25 times larger.

Effective Lengths of Columns

The critical loads for columns with various support conditions can be related to the critical load of a pinned-end column through the concept of an **effective length**. To demonstrate this idea, consider the deflected shape of a column fixed at the base and free at the top (Fig. 11-17a). This column buckles in a curve that is one-quarter of a complete sine wave. If the deflection curve is extended (Fig. 11-17b), it becomes one-half of a complete sine wave, which is the deflection curve for a pinned-end column.

FIGURE 11-17

Deflection curves showing the effective length L_e for a column fixed at the base and free at the top

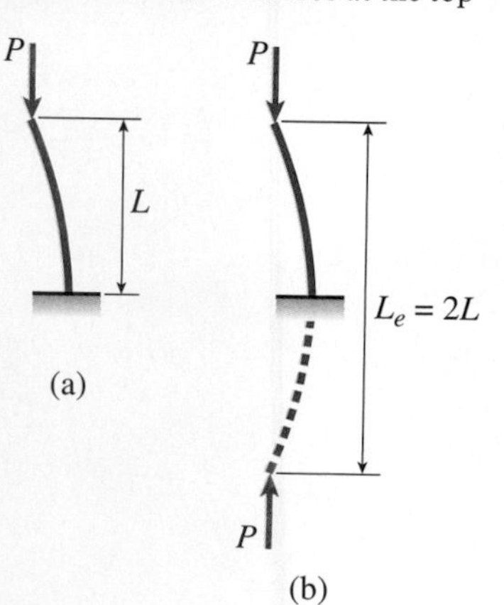

The effective length L_e for any column is the length of the equivalent pinned-end column, that is, it is the length of a pinned-end column having a deflection curve that exactly matches all or part of the deflection curve of the original column.

Another way of expressing this idea is to say that the effective length of a column is the distance between points of inflection (that is, points of zero moment) in its deflection curve, assuming that the curve is extended (if necessary) until points of inflection are reached. Thus, for a fixed-free column (Fig. 11-17), the effective length is

$$L_e = 2L \qquad \textbf{(11-36)}$$

Because the effective length is the length of an equivalent pinned-end column, a general formula for critical loads can be written as

$$P_{cr} = \frac{\pi^2 EI}{L_e^2} \qquad \textbf{(11-37)}$$

If the effective length of a column is known (no matter how complex the end conditions may be), you can substitute into the preceding equation and determine the critical load. For instance, in the case of a fixed-free column, substitute $L_e = 2L$ and obtain Eq. (11-34).

The effective length is often expressed in terms of an **effective-length factor** K:

$$L_e = KL \tag{11-38}$$

where L is the actual length of the column. Thus, the critical load is

$$P_{cr} = \frac{\pi^2 EI}{(KL)^2} \tag{11-39}$$

The factor K equals 2 for a column fixed at the base and free at the top and equals 1 for a pinned-end column. The effective-length factor is often included in design formulas for columns, as illustrated later in Section 11.9.

Column with Both Ends Fixed Against Rotation

Next, consider a column with both ends fixed against rotation (Fig. 11-18a). Note that in this figure the standard symbol for the fixed support is used at the base of the column. However, since the column is free to shorten under an axial load, a new symbol is introduced at the top of the column. This new symbol shows a rigid block that is constrained in such a manner that rotation and horizontal displacement are prevented but vertical movement can occur. (As a convenience when drawing sketches, this more accurate symbol is replaced with the standard symbol for a fixed support—see Fig. 11-18b—with the understanding that the column is free to shorten.)

The buckled shape of the column in the first mode is shown in Fig. 11-18c. Note that the deflection curve is symmetrical (with zero slope at the midpoint) and has zero slope at the ends. Because rotation at the ends is prevented, reactive moments M_0 develop at the supports. These moments, as well as the reactive force at the base, are shown in the figure.

From the previous solutions of the differential equation, the equation of the deflection curve involves sine and cosine functions. Also, the curve is symmetric about the midpoint. Therefore, the curve must have inflection points at distances $L/4$ from the ends. It follows that the middle portion of the deflection curve has the same shape as the deflection curve for a pinned-end column. Thus, the effective length of a column with fixed ends, equal to the distance between inflection points, is

$$L_e = \frac{L}{2} \tag{11-40}$$

Substituting into Eq. (11-37) gives the critical load:

$$P_{cr} = \frac{4\pi^2 EI}{L^2} \tag{11-41}$$

This formula shows that the critical load for a column with fixed ends is four times that for a column with pinned ends. As a check, this result may be verified by solving the differential equation of the deflection curve (see Prob. 11.4-11).

FIGURE 11-18

Buckling of a column with both ends fixed against rotation

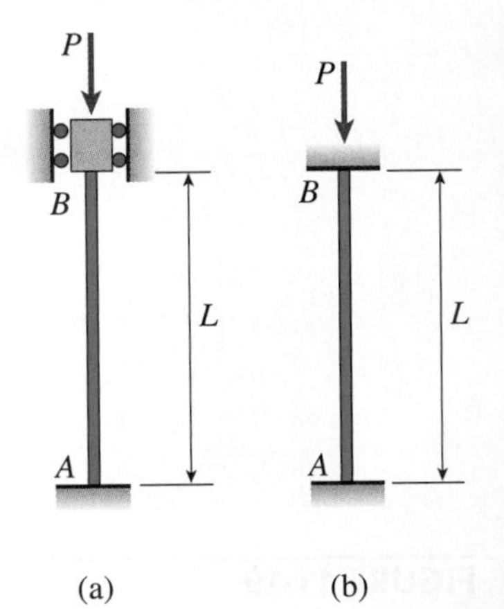

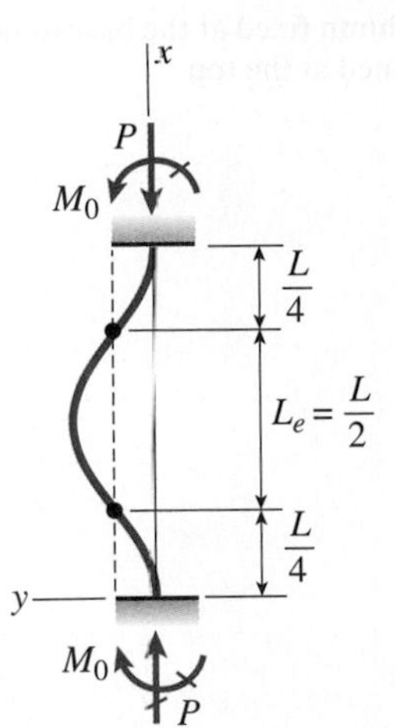

Column Fixed at the Base and Pinned at the Top

The critical load and buckled mode shape for a column that is fixed at the base and pinned at the top (Fig. 11-19a) can be determined by solving the differential equation of the deflection curve. When the column buckles (Fig. 11-19b), a reactive moment M_0 develops at the base because there can be no rotation at that point. Then from the equilibrium of the entire column, there must be horizontal reactions R at each end such that

$$M_0 = RL \tag{11-42}$$

The bending moment in the buckled column, at distance x from the base, is

$$M = M_0 - Pv - Rx = -Pv + R(L - x) \tag{11-43}$$

therefore, the **differential equation** is

$$EIv'' = M = -Pv + R(L - x) \tag{11-44}$$

Again substitute $k^2 = P/EI$ and rearrange to get

$$v'' + k^2 v = \frac{R}{EI}(L - x) \tag{11-45}$$

The **general solution** of this equation is

$$v = C_1 \sin kx + C_2 \cos kx + \frac{R}{P}(L - x) \tag{11-46}$$

in which the first two terms on the right-hand side constitute the homogeneous solution and the last term is the particular solution. This solution can be verified by substitution into the differential equation of Eq. (11-44).

Since the solution contains three unknown quantities (C_1, C_2, and R), the following three **boundary conditions** are required:

$$v(0) = 0 \qquad v'(0) = 0 \qquad v(L) = 0$$

FIGURE 11-19

Column fixed at the base and pinned at the top

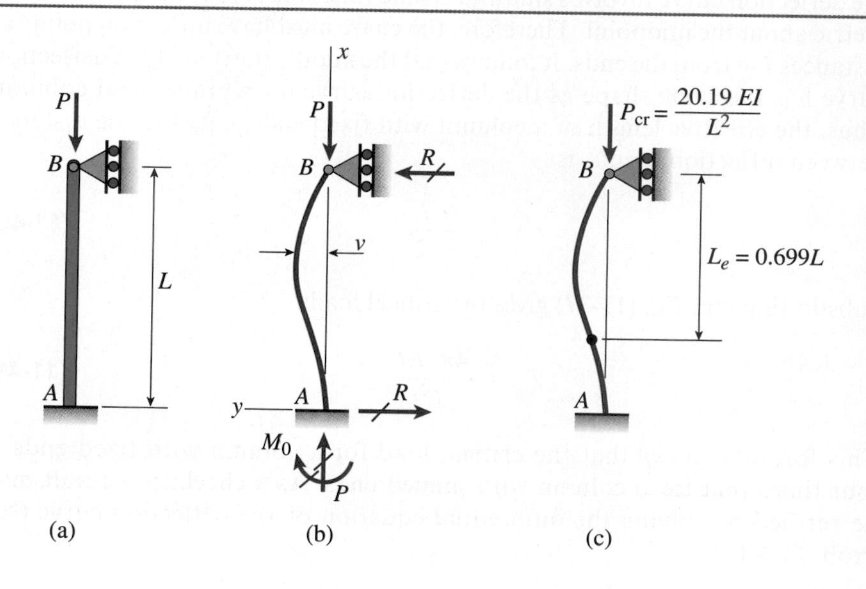

Applying these conditions to Eq. (11-46) yields

$$C_2 + \frac{RL}{P} = 0 \qquad C_1 k - \frac{R}{P} = 0 \qquad \textbf{(11-47a,b,c)}$$
$$C_1 \tan kL + C_2 = 0$$

All three equations are satisfied if $C_1 = C_2 = R = 0$, in which case the trivial solution results and the deflection is zero.

To obtain the solution for buckling, solve Eqs. (11-47a, b, and c) in a more general manner. One method of solution is to eliminate R from the first two equations, which yields

$$C_1 kL + C_2 = 0 \quad \text{or} \quad C_2 = -C_1 kL \qquad \textbf{(11-47d)}$$

Next, substitute this expression for C_2 into Eq. (11-47c) and obtain the **buckling equation**:

$$kL = \tan kL \qquad \textbf{(11-48)}$$

The solution of this equation gives the critical load.

Since the buckling equation is a transcendental equation, it cannot be solved explicitly.[6] Nevertheless, the values of kL that satisfy the equation can be determined numerically by using a computer program for finding roots of equations. The smallest nonzero value of kL that satisfies Eq. (11-48) is

$$kL = 4.4934 \qquad \textbf{(11-49)}$$

The corresponding **critical load** is

$$P_{cr} = \frac{20.19EI}{L^2} = \frac{2.046\pi^2 EI}{L^2} \qquad \textbf{(11-50)}$$

which (as expected) is higher than the critical load for a column with pinned ends and lower than the critical load for a column with fixed ends [see Eqs. (11-15) and (11-41)].

The **effective length** of the column may be obtained by comparing Eqs. (11-50) and (11-37); thus,

$$L_e = 0.699\,L \approx 0.7\,L \qquad \textbf{(11-51)}$$

This length is the distance from the pinned end of the column to the point of inflection in the buckled shape (Fig. 11-19c).

The equation of the **buckled mode shape** is obtained by substituting $C_2 = -C_1 kL$ [Eq. (11-47d)] and $R/P = C_1 k$ [Eq. (11-47b)] into the general solution [Eq. (11-46)]:

$$v = C_1[\sin kx - kL\cos kx + k(L - x)] \qquad \textbf{(11-52)}$$

in which $k = 4.4934/L$. The term in brackets gives the mode shape for the deflection of the buckled column. However, the amplitude of the deflection curve is undefined because C_1 may have any value (within the usual limitation that the deflections must remain small).

[6] In a transcendental equation, the variables are contained within transcendental functions. A transcendental function cannot be expressed by a finite number of algebraic operations; hence, trigonometric, logarithmic, exponential, and other such functions are transcendental.

FIGURE 11-20

Critical loads, effective lengths, and effective-length factors for ideal columns

(a) Pinned-pinned column	(b) Fixed-free column	(c) Fixed-fixed column	(d) Fixed-pinned column
$P_{cr} = \frac{\pi^2 EI}{L^2}$	$P_{cr} = \frac{\pi^2 EI}{4L^2}$	$P_{cr} = \frac{4\pi^2 EI}{L^2}$	$P_{cr} = \frac{2.046\, \pi^2 EI}{L^2}$
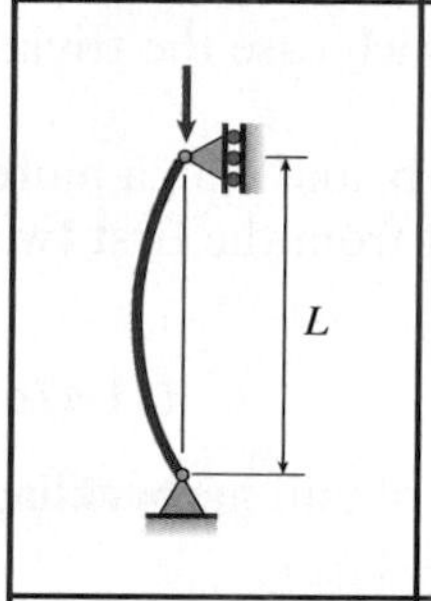 L	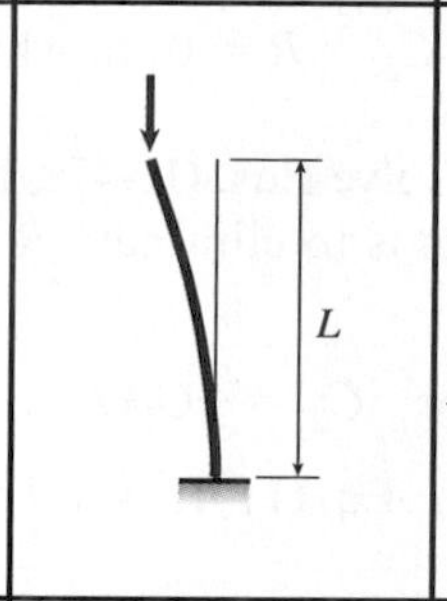 L	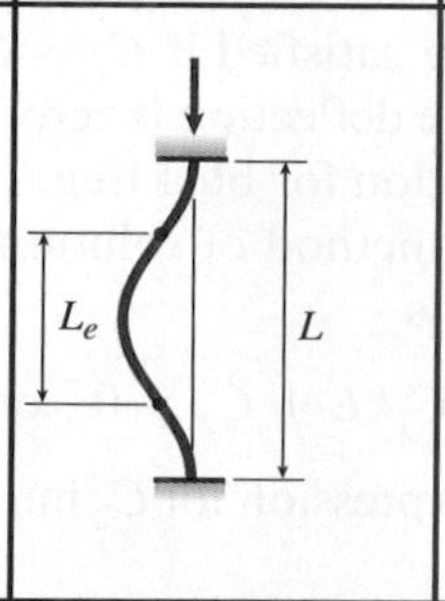 L_e L	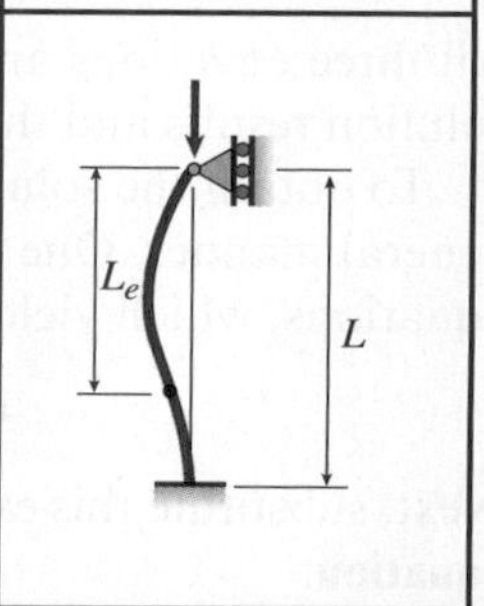 L_e L
$L_e = L$	$L_e = 2L$	$L_e = 0.5L$	$L_e = 0.699L$
$K = 1$	$K = 2$	$K = 0.5$	$K = 0.699$

Limitations

In addition to the requirement of small deflections, the Euler buckling theory used in this section is valid only if the column is perfectly straight before the load is applied, the column and its supports have no imperfections, and the column is made of a linearly elastic material that follows Hooke's law. These limitations were explained previously in Section 11.3.

Summary of Results

The lowest critical loads and corresponding effective lengths for the four columns analyzed here are summarized in Fig. 11-20.

Example 11-3

A viewing platform in a wild-animal park (Fig. 11-21a) is supported by a row of aluminum pipe columns having a length $L = 3.25$ m and an outer diameter $d = 100$ mm The bases of the columns are set in concrete footings, and the tops of the columns are supported laterally by the platform. The columns are being designed to support compressive loads where $P = 100$ kN.

Determine the minimum required thickness t of the columns (Fig. 11-21b) if a factor of safety $n = 3$ is required with respect to Euler buckling. (For the aluminum, use 72 GPa for the modulus of elasticity and use 480 MPa for the proportional limit.)

FIGURE 11-21

Example 11-3: Aluminum pipe column

Solution:

Use a four-step problem-solving approach. Combine steps as needed for an efficient solution.

1, 2. ***Conceptualize, Categorize:***

Critical load: Because of the manner in which the columns are constructed, each column is modeled as a fixed-pinned column (see Fig. 11-20d). Therefore, the critical load is

$$P_{cr} = \frac{2.046\pi^2 EI}{L^2} \tag{a}$$

in which I is the moment of inertia of the tubular cross section:

$$I = \frac{\pi}{64}[d^4 - (d - 2t)^4] \tag{b}$$

Substitute $d = 100$ mm (or 0.1 m) to get

$$I = \frac{\pi}{64}[(0.1 \text{ m})^4 - (0.1 \text{ m} - 2t)^4] \tag{c}$$

in which t is expressed in meters.

3. ***Analyze:***

Required thickness of the columns: Since the load per column is 100 kN and the factor of safety is 3, each column must be designed for the critical load:

$$P_{cr} = nP = 3(100 \text{ kN}) = 300 \text{ kN}$$

Substitute this value for P_{cr} in Eq. (a), and also replace I with its expression from Eq. (c) to obtain

$$300 \text{ kN} = \frac{2.046\,\pi^2(72 \text{ GPa})}{(3.25 \text{ m})^2}\left[\frac{\pi}{64}[(0.1 \text{ m})^4 - (0.1 \text{ m} - 2t)^4]\right]$$

Note that all terms in this equation are expressed in units of newtons and meters.

Solve the previous equation to find that the minimum required thickness of the column to meet the specified conditions is

$$t_{min} = 6.83 \text{ mm}$$

4. ***Finalize:***

Supplementary calculations: Knowing the diameter and thickness of the column, now calculate its moment of inertia, cross-sectional area, and radius of gyration. Using the minimum thickness of 6.83 mm gives

$$I = \frac{\pi}{64}[d^4 - (d - 2t)^4] = 2.18 \times 10^6 \text{mm}^4$$

$$A = \frac{\pi}{4}[d^2 - (d - 2t)^2] = 1999 \text{ mm}^2 \qquad r = \sqrt{\frac{I}{A}} = 33.0 \text{ mm}$$

The slenderness ratio L/r of the column is approximately 98, which is in the customary range for slender columns, and the diameter-to-thickness ratio d/t is approximately 15, which should be adequate to prevent local buckling of the walls of the column.

The critical stress in the column must be less than the proportional limit of the aluminum if the formula for the critical load of Eq. (a) is to be valid. The critical stress is

$$\sigma_{cr} = \frac{P_{cr}}{A} = \frac{300 \text{ kN}}{1999 \text{ mm}^2} = 150 \text{ MPa}$$

which is less than the proportional limit (480 MPa). Therefore, the calculation for the critical load using the Euler buckling theory is satisfactory.

11.5 Columns with Eccentric Axial Loads

Sections 11.3 and 11.4 covered the analysis of ideal columns in which the axial loads acted through the centroids of the cross sections. Under these conditions, the columns remain straight until the critical loads are reached, after which bending may occur.

Now assume that a column is compressed by loads P that are applied with a small eccentricity e measured from the axis of the column (Fig. 11-22a). Each eccentric axial load is equivalent to a centric load P and a couple of moment $M_0 = Pe$ (Fig. 11-22b). This moment exists from the instant the load is first applied; therefore, the column begins to deflect at the onset of loading. The deflection then becomes steadily larger as the load increases.

To analyze the pin-ended column shown in Fig. 11-22, make the same assumptions as in previous sections; namely, the column is initially perfectly straight, the material is linearly elastic, and the x-y plane is a plane of symmetry. The bending moment in the column at distance x from the lower end (Fig. 11-22b) is

$$M = M_0 + P(-v) = Pe - Pv \qquad \textbf{(11-53)}$$

where v is the deflection of the column (positive when in the positive direction of the y axis). Note that the deflections of the column are negative when the eccentricity of the load is positive.

The **differential equation** of the deflection curve is

$$EIv'' = M = Pe - Pv \tag{11-54}$$

or

$$v'' + k^2 v = k^2 e \tag{11-55}$$

in which $k^2 = P/EI$, as before. The general solution of this equation is

$$v = C_1 \sin kx + C_2 \cos kx + e \tag{11-56}$$

in which C_1 and C_2 are constants of integration in the homogeneous solution and e is the particular solution. As usual, verify the solution by substituting it into the differential equation.

The **boundary conditions** for determining the constants C_1 and C_2 are obtained from the deflections at the ends of the column (Fig. 11-22b):

$$v(0) = 0 \qquad v(L) = 0$$

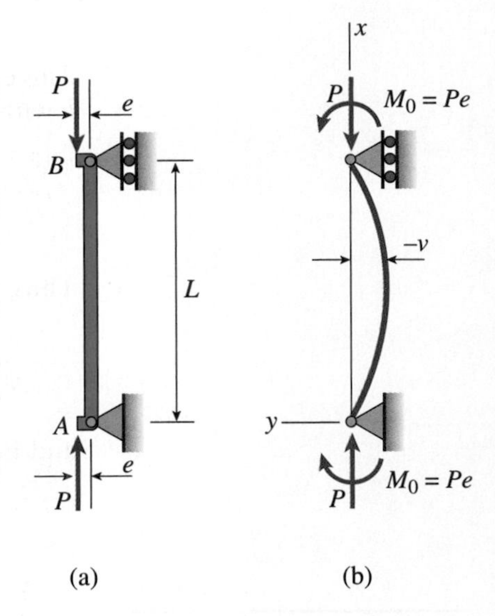

FIGURE 11-22
Column with eccentric axial loads

These conditions yield

$$C_2 = -e \qquad C_1 = -\frac{e(1 - \cos kL)}{\sin kL} = -e \tan\frac{kL}{2}$$

Therefore, the **equation of the deflection curve** is

$$v = -e\left(\tan\frac{kL}{2}\sin kx + \cos kx - 1\right) \tag{11-57}$$

For a column with known loads P and known eccentricity e, use this equation to calculate the deflection at any point along the x axis.

The behavior of a column with an eccentric load is quite different from that of a centrally loaded column, as can be seen by comparing Eq. (11-57) with Eqs. (11-16), (11-33), and (11-52). Equation (11-57) shows that each value of the eccentric load P produces a definite value of the deflection, just as each value of the load on a beam produces a definite deflection. In contrast, the deflection equations for centrally loaded columns give the buckled mode shape (when $P = P_{cr}$) but with the amplitude undefined.

Because the column shown in Fig. 11-22 has pinned ends, its critical load (when centrally loaded) is

$$P_{cr} = \frac{\pi^2 EI}{L^2} \tag{11-58}$$

Use this formula as a reference quantity in some of the equations that follow.

Maximum Deflection

The maximum deflection δ produced by the eccentric loads occurs at the midpoint of the column (Fig. 11-23) and is obtained by setting x equal to $L/2$ in Eq. (11-57):

$$\delta = -v\left(\frac{L}{2}\right) = e\left(\tan\frac{kL}{2}\sin\frac{kL}{2} + \cos\frac{kL}{2} - 1\right)$$

FIGURE 11-23
Maximum deflection δ of a column with eccentric axial loads

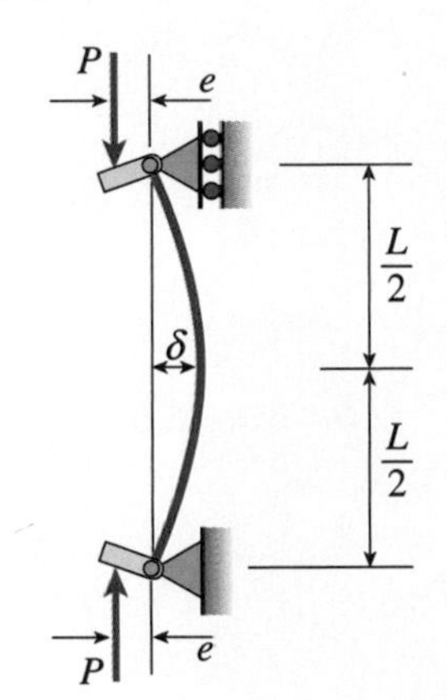

or, after simplifying,

$$\delta = e\left(\sec\frac{kL}{2} - 1\right) \tag{11-59}$$

This equation can be written in a slightly different form by replacing the quantity k with its equivalent in terms of the critical load [see Eq. (11-58)]:

$$k = \sqrt{\frac{P}{EI}} = \sqrt{\frac{P\pi^2}{P_{cr}L^2}} = \frac{\pi}{L}\sqrt{\frac{P}{P_{cr}}} \tag{11-60}$$

Thus, the nondimensional term kL becomes

$$kL = \pi\sqrt{\frac{P}{P_{cr}}} \tag{11-61}$$

and Eq. (11-59) for the **maximum deflection** becomes

$$\delta = e\left[\sec\left(\frac{\pi}{2}\sqrt{\frac{P}{P_{cr}}}\right) - 1\right] \tag{11-62}$$

FIGURE 11-24

Load-deflection diagram for a column with eccentric axial loads [see Fig. 11-23 and Eq. (11-62)]

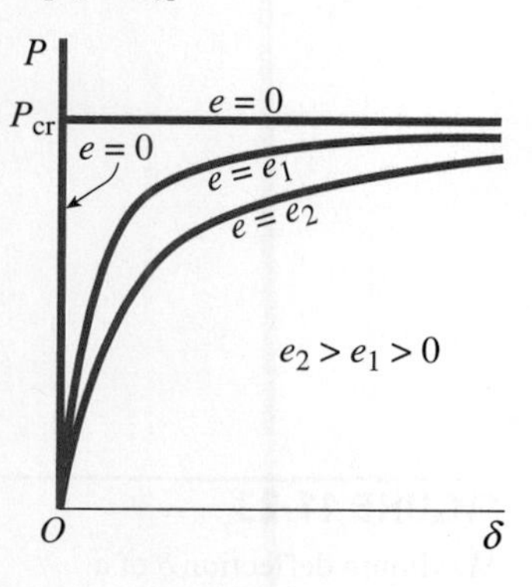

As special cases, note: (1) The deflection δ is zero when the eccentricity e is zero and P is not equal to P_{cr}, (2) the deflection is zero when the axial load P is zero, and (3) the deflection becomes infinitely large as P approaches P_{cr}. These characteristics are shown in the **load-deflection diagram** of Fig. 11-24.

To plot the load-deflection diagram, select a particular value e_1 of the eccentricity and then calculate δ for various values of the load P. The resulting curve is labeled $e = e_1$ in Fig. 11-24. Note that the deflection δ increases as P increases, but the relationship is nonlinear. Therefore, the principle of superposition cannot be used for calculating deflections due to more than one load, even though the material of the column is linearly elastic. As an example, the deflection due to an axial load $2P$ is *not* equal to twice the deflection caused by an axial load P.

Additional curves, such as the curve labeled $e = e_2$, are plotted in a similar manner. Since the deflection δ is linear with e in Eq. (11-62), the curve for $e = e_2$ has the same *shape* as the curve for $e = e_1$, but the abscissas are larger by the ratio e_2/e_1.

As the load P approaches the critical load, the deflection δ increases without limit, and the horizontal line corresponding to $P = P_{cr}$ becomes an asymptote for the curves. In the limit, as e approaches zero, the curves on the diagram approach two straight lines: one vertical and one horizontal (compare with Fig. 11-8). Thus, as expected, an ideal column with a centrally applied load ($e = 0$) is the limiting case of a column with an eccentric load ($e > 0$).

Although the curves plotted in Fig. 11-24 are mathematically correct, keep in mind that the differential equation is valid only for small deflections. Therefore, when the deflections become large, the curves are no longer physically valid and must be modified to take into account the presence of large deflections and (if the proportional limit of the material is exceeded) inelastic bending effects (see Fig. 11-12).

The reason for the nonlinear relationship between loads and deflections, even when the deflections are small and Hooke's law holds, can be understood if you observe once again that the axial loads P are equivalent to centrally applied loads P plus couples Pe acting at the ends of the column (Fig. 11-22b). The couples Pe, if acting alone, would produce bending deflections of the column in the same manner as for a beam. In a beam, the presence of the deflections does not change the action of the loads, and the bending moments are the same whether the deflections exist or not. However, when an axial load is applied to the member, the existence of deflections increases the bending moments (the increases are equal to the product of the axial load and the deflections). When the bending moments increase, the deflections are further increased—hence, the moments increase even more, and so on. Thus, the bending moments in a column depend upon the deflections, which in turn depend upon the bending moments. This type of behavior results in a nonlinear relationship between the axial loads and the deflections.

In general, a straight structural member subjected to both bending loads and axial compressive loads is called a **beam-column**. In the case of a column with eccentric loads (Fig. 11-22), the bending loads are the moments $M_0 = Pe$ and the axial loads are the forces P.

Maximum Bending Moment

The maximum bending moment in an eccentrically loaded column occurs at the midpoint where the deflection is a maximum (Fig. 11-23):

$$M_{max} = P(e + \delta) \tag{11-63}$$

Substitute for δ from Eqs. (11-59) and (11-62) to obtain

$$M_{max} = Pe\sec\frac{kL}{2} = Pe\sec\left(\frac{\pi}{2}\sqrt{\frac{P}{P_{cr}}}\right) \tag{11-64}$$

The manner in which M_{max} varies as a function of the axial load P is shown in Fig. 11-25.

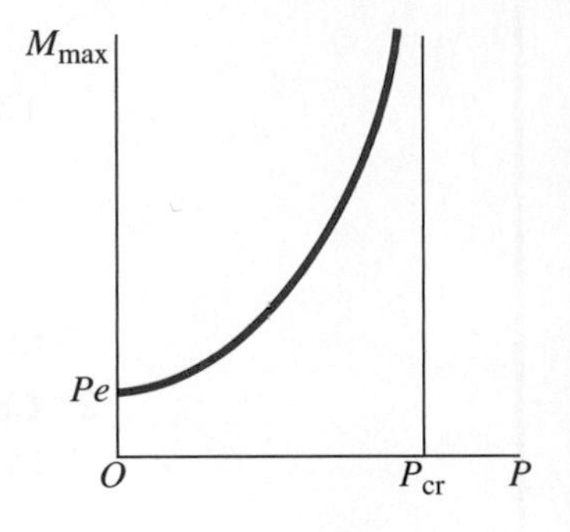

FIGURE 11-25
Maximum bending moment in a column with eccentric axial loads [see Fig. 11-23 and Eq. (11-64)]

When P is small, the maximum moment is equal to Pe, which means that the effect of the deflections is negligible. As P increases, the bending moment grows nonlinearly and theoretically becomes infinitely large as P approaches the critical load. However, as explained before, these equations are valid only when the deflections are small, and they cannot be used when the axial load approaches the critical load. Nevertheless, the preceding equations and accompanying graphs indicate the general behavior of beam-columns.

Other End Conditions

The equations given in this section were derived for a pinned-end column, as shown in Figs. 11-22 and 11-23. If a column is fixed at the base and free at the top (Fig. 11-20b), use Eqs. (11-59) and (11-64) by replacing the actual length L with the equivalent length $2L$ (see Prob. 11.5-11). However, the equations do not apply to a column that is fixed at the base and pinned at the top (Fig. 11-20d). The use of an equivalent length equal to $0.699L$ gives erroneous results; instead, you must return to the differential equation and derive a new set of equations.

In the case of a column with both ends fixed against rotation (Fig. 11-20c), the concept of an eccentric axial load acting at the end of the column has no meaning. Any moment applied at the end of the column is resisted directly by the supports and produces no bending of the column itself.

Example 11-4

FIGURE 11-26

Example 11-4: Brass bar with an eccentric axial load

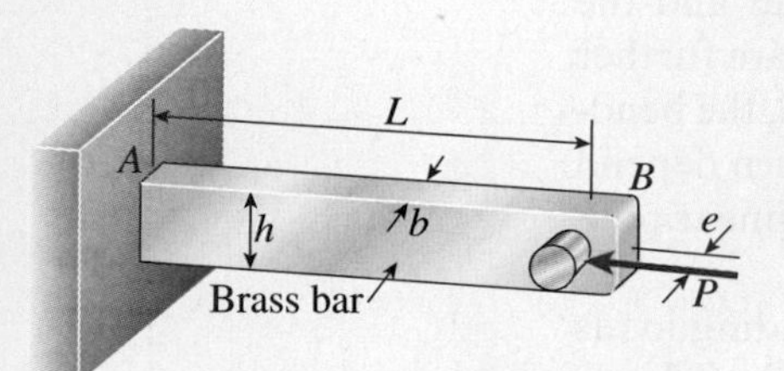

A brass bar AB projecting from the side of a large machine is loaded at end B by a force $P = 7\,\text{kN}$ acting with an eccentricity $e = 11\,\text{mm}$ (Fig. 11-26). The bar has a rectangular cross section with height $h = 30\,\text{mm}$ and width $b = 15\,\text{mm}$.

What is the longest permissible length L_{max} of the bar if the deflection at the end is limited to 3 mm? (For the brass, use $E = 110\,\text{GPa}$.)

Solution:

Use a four-step problem-solving approach.

1. ***Conceptualize:***

Critical load: This bar is a slender column that is fixed at end A and free at end B. Therefore, the critical load (see Fig. 11-20b) is

$$P_{cr} = \frac{\pi^2 EI}{4L^2} \tag{a}$$

The moment of inertia for the axis about which bending occurs is

$$I = \frac{hb^3}{12} = \frac{(30\text{ mm})(15\text{ mm})^3}{12} = 8.44 \times 10^3\text{ mm}^4$$

Therefore, the expression for the critical load becomes

$$P_{cr} = \frac{\pi^2(110\text{ GPa})(8.44 \times 10^3\text{ mm}^4)}{4L^2} = \frac{2.29\text{ kN}\cdot\text{m}^2}{L^2} \tag{b}$$

in which P_{cr} has units of kN and L has units of meters.

2. ***Categorize:***

Deflection: The deflection at the end of the bar is given by Eq. (11-62), which applies to a fixed-free column as well as a pinned-end column:

$$\delta = e\left[\sec\left(\frac{\pi}{2}\sqrt{\frac{P}{P_{cr}}}\right) - 1\right] \tag{c}$$

In this equation, P_{cr} is given by Eq. (a).

3. ***Analyze:***

Length: To find the maximum permissible length of the bar, substitute for δ its limiting value of 3 mm. Also, substitute $e = 11\,\text{mm}$ and $P = 7\,\text{kN}$, and substitute for P_{cr} from Eq. (b). Thus,

$$3\text{ mm} = (11\text{ mm})\left[\sec\left(\frac{\pi}{2}\sqrt{\frac{7\text{ kN}}{2.29/L^2}}\right) - 1\right]$$

The only unknown in this equation is the length L (meters). To solve for L, perform the various arithmetic operations in the equation and then rearrange the terms. The result is

$$0.2727 = \sec(2.746L) - 1$$

Use radians and solve this equation to get $L = 0.243$ m. Thus, the maximum permissible length of the bar is

$$L_{max} = 0.243 \text{ m}$$

4. ***Finalize***: If a longer bar is used, the deflection will exceed the allowable value of 3 mm.

11.6 The Secant Formula for Columns

In the preceding section, the maximum deflection and maximum bending moment were found for a pin-ended column subjected to eccentric axial loads. This section discusses the maximum stresses in the column and a special formula for calculating them.

The maximum stresses in a column with eccentric axial loads occur at the cross section where the deflection and bending moment have their largest values, that is, at the midpoint (Fig. 11-27a). Acting at this cross section are the compressive force P and the bending moment M_{max} (Fig. 11-27b). The stresses due to the force P are equal to P/A, where A is the cross-sectional area of the column, and the stresses due to the bending moment M_{max} are obtained from the flexure formula.

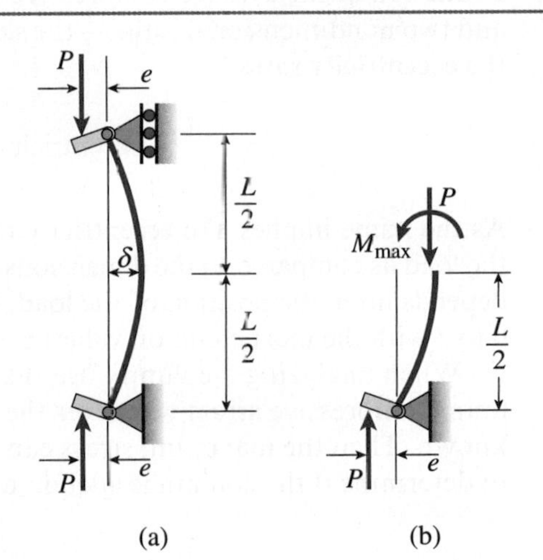

FIGURE 11-27

Column with eccentric axial loads

Thus, the maximum compressive stress, which occurs on the concave side of the column, is

$$\sigma_{max} = \frac{P}{A} + \frac{M_{max}c}{I} \tag{11-65}$$

in which I is the moment of inertia in the plane of bending and c is the distance from the centroidal axis to the extreme point on the concave side of the column. Note that in this equation compressive stresses are considered to be positive, since these are the important stresses in a column.

The bending moment M_{max} is obtained from Eq. (11-64), which is repeated here:

$$M_{max} = Pe\sec\left(\frac{\pi}{2}\sqrt{\frac{P}{P_{cr}}}\right)$$

Since $P_{cr} = \pi^2 EI/L^2$ for a pinned-end column, and since $I = Ar^2$, where r is the radius of gyration in the plane of bending, the preceding equation becomes

$$M_{max} = Pe\sec\left(\frac{L}{2r}\sqrt{\frac{P}{EA}}\right) \tag{11-66}$$

Substitute into Eq. (11-65) to obtain the following formula for the **maximum compressive stress**:

$$\sigma_{max} = \frac{P}{A} + \frac{Pec}{I}\sec\left(\frac{L}{2r}\sqrt{\frac{P}{EA}}\right)$$

or

$$\sigma_{max} = \frac{P}{A}\left[1 + \frac{ec}{r^2}\sec\left(\frac{L}{2r}\sqrt{\frac{P}{EA}}\right)\right] \tag{11-67}$$

Equation (11-67) is commonly known as the **secant formula** for an eccentrically loaded column with pinned ends.

The secant formula gives the maximum compressive stress in the column as a function of the average compressive stress P/A, the modulus of elasticity E, and two nondimensional ratios—the slenderness ratio L/r from Eq. (11-20) and the **eccentricity ratio**:

$$\text{Eccentricity ratio} = \frac{ec}{r^2} \tag{11-68}$$

As the name implies, the eccentricity ratio is a measure of the eccentricity of the load as compared to the dimensions of the cross section. Its numerical value depends upon the position of the load, but typical values are in the range from 0 to 3 with the most common values being less than 1.

When analyzing a column, use the secant formula to calculate the maximum compressive stress whenever the axial load P and its eccentricity e are known. Then the maximum stress can be compared with the allowable stress to determine if the column is adequate to support the load.

The secant formula also can be used in the reverse manner, that is, if the allowable stress is known calculate the corresponding value of the load P. However, because the secant formula is transcendental, it is not practical to derive a formula for the load P. Instead, solve Eq. (11-67) numerically in each individual case.

A **graph of the secant formula** is shown in Fig. 11-28. The abscissa is the slenderness ratio L/r, and the ordinate is the average compressive stress P/A. The graph is plotted for a steel column with modulus of elasticity $E = 200$ GPa and maximum stress $\sigma_{max} = 250$ MPa. Curves are plotted for several values of the eccentricity ratio ec/r^2. These curves are valid only when the maximum stress is less than the proportional limit of the material, because the secant formula was derived using Hooke's law.

A special case arises when the eccentricity of the load disappears ($e = 0$), because then you have an ideal column with a centrally applied load. Under these conditions the maximum load is the critical load ($P_{cr} = \pi^2 EI/L^2$) and the corresponding maximum stress is the critical stress [see Eqs. (11-17) and (11-19)]:

$$\sigma_{cr} = \frac{P_{cr}}{A} = \frac{\pi^2 EI}{AL^2} = \frac{\pi^2 E}{(L/r)^2} \tag{11-69}$$

Since this equation gives the stress P/A in terms of the slenderness ratio L/r, you can plot it on the graph of the secant formula (Fig. 11-28) as **Euler's curve**.

Now assume that the proportional limit of the material is the same as the selected maximum stress, that is, 250 MPa. Construct a horizontal line on the graph at a value of 250 MPa, and terminate Euler's curve at that stress. The horizontal line and Euler's curve represent the limits of the secant-formula curves as the eccentricity e approaches zero.

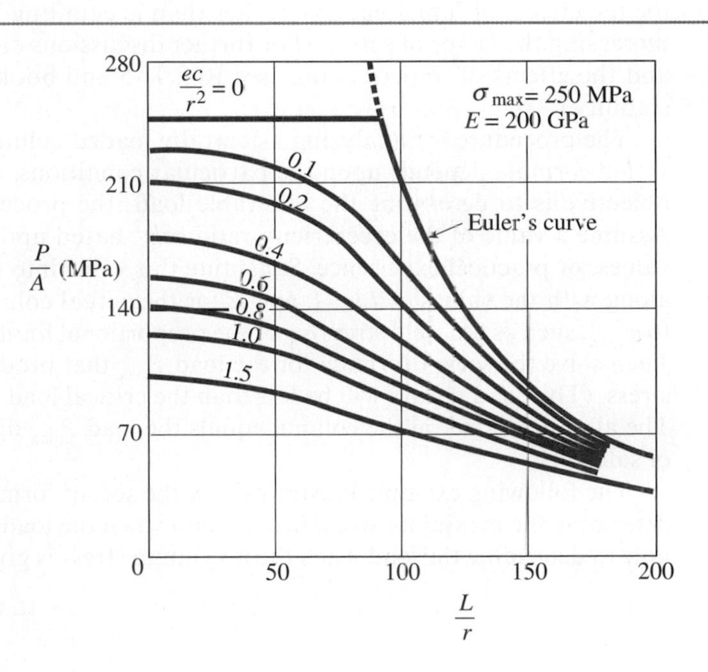

FIGURE 11-28

Graph of the secant formula (Eq. 11-67) for $\sigma_{max} = 250$ MPa and $E = 200$ GPa

Discussion of the Secant Formula

The graph of the secant formula shows that the load-carrying capacity of a column decreases significantly as the slenderness ratio L/r increases, especially in the intermediate region of L/r values. Thus, long slender columns are much less stable than short, stocky columns. The graph also shows that the load-carrying capacity decreases with increasing eccentricity e; furthermore, this effect is relatively greater for short columns than for long ones.

The secant formula was derived for a column with pinned ends, but it also can be used for a column that is fixed at the base and free at the top. All that is required is to replace the length L in the secant formula with the equivalent length $2L$. However, because it is based upon Eq. (11-64), the secant formula is not valid for the other end conditions discussed.

Now consider an actual column that inevitably differs from an ideal column because of imperfections, such as initial curvature of the longitudinal axis, imperfect support conditions, and nonhomogeneity of the material. Furthermore, even when the load is supposed to be centrally applied, there will be unavoidable eccentricities in its direction and point of application. The extent of these imperfections varies from one column to another, so there is considerable scatter in the results of laboratory tests performed with actual columns.

All imperfections have the effect of producing bending in addition to direct compression. Therefore, it is reasonable to assume that the behavior of an imperfect, centrally loaded column is similar to that of an ideal, eccentrically loaded column. In such cases, the secant formula can be used by choosing an approximate value of the eccentricity ratio ec/r^2 to account for the combined effects of the various imperfections. For instance, a commonly used value of the eccentricity ratio for pinned-end columns in structural-steel design is $ec/r^2 = 0.25$. The use of the secant formula in this manner for columns with centrally applied loads provides a rational means of accounting for the effects of imperfections, rather than accounting for them simply by increasing the factor of safety. (For further discussions of the secant formula and the effects of imperfections, see Ref. 11-5 and books on buckling and stability.)

The procedure for analyzing a centrally loaded column by means of the secant formula depends upon the particular conditions. For instance, if the objective is to determine the allowable load, the procedure is as follows. Assume a value of the eccentricity ratio ec/r^2 based upon test results, code values, or practical experience. Substitute this value into the secant formula, along with the values of L/r, A, and E for the actual column. Assign a value to σ_{max}, such as the yield stress σ_Y or the proportional limit σ_{pl} of the material. Then solve the secant formula for the load P_{max} that produces the maximum stress. (This load always will be less than the critical load P_{cr} for the column.) The allowable load on the column equals the load P_{max} divided by the factor of safety n.

The following example illustrates how the secant formula may be used to determine the maximum stress in a column when the load is known, and also how to determine the load when the maximum stress is given.

Example 11-5

FIGURE 11-29

Example 11-5: Column with an eccentrically applied axial load

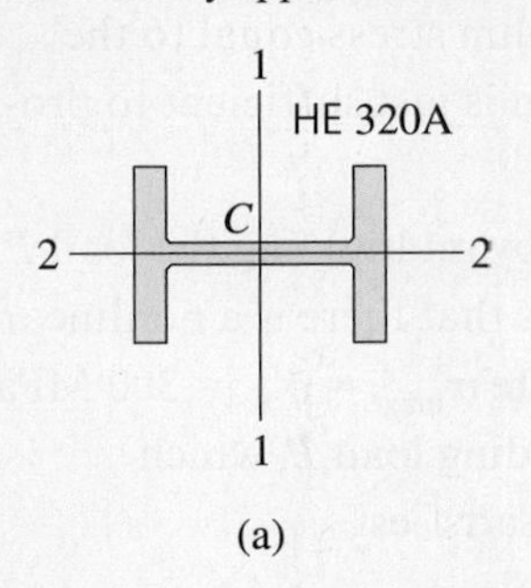

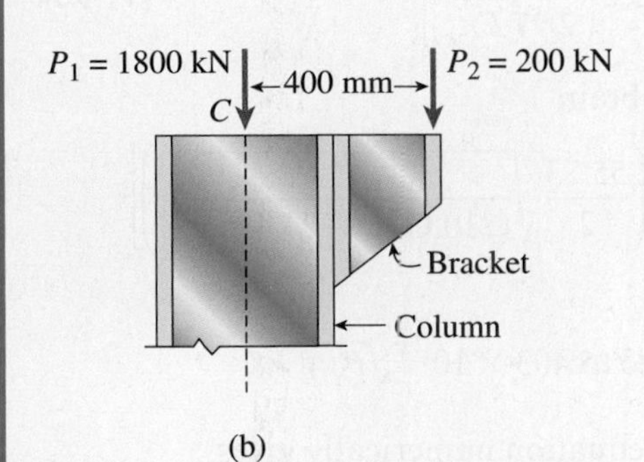

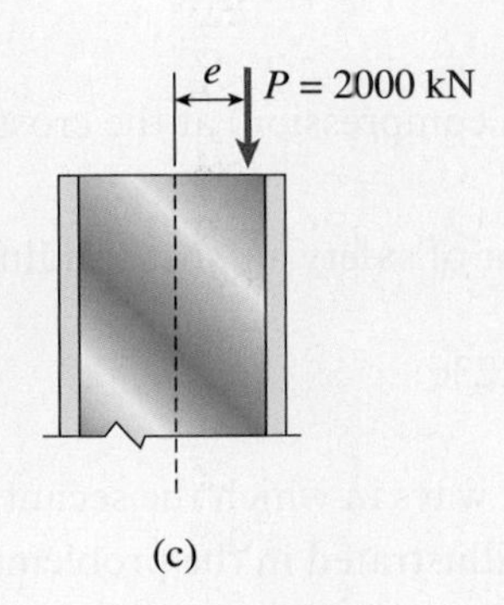

A steel wide-flange column of a HE 320A shape (Fig. 11-29a) is pin-supported at the ends and has a length of 7.5 m. The column supports a centrally applied load $P_1 = 1800$ kN and an eccentrically applied load $P_2 = 200$ kN (Fig. 11-29b). Bending takes place about axis 1–1 of the cross section, and the eccentric load acts on axis 2–2 at a distance of 400 mm from the centroid C.

(a) Using the secant formula, and assuming $E = 210$ GPa, calculate the maximum compressive stress in the column.

(b) If the yield stress for the steel is $\sigma_Y = 300$ MPa, what is the factor of safety with respect to yielding?

Solution:

Use a four-step problem-solving approach.

Part (a): Maximum compressive stress.

1. ***Conceptualize***: The two loads P_1 and P_2 acting as shown in Fig. 11-29b are statically equivalent to a single load $P = 2000$ kN acting with an eccentricity $e = 40$ mm (Fig. 11-29c). Since the column is now loaded by a single force P having an eccentricity e, use the secant formula to find the maximum stress.

   The required properties of the HE 320A wide-flange shape are obtained from Table F-1 in Appendix F:

$$A = 124.4 \text{ cm}^2 \qquad r = 13.58 \text{ cm} \qquad c = \frac{310 \text{ mm}}{2} = 155 \text{ mm}$$

2. ***Categorize***: The required terms in the secant formula of Eq. (11-67) are calculated as

$$\frac{P}{A} = \frac{2000 \text{ kN}}{124.4 \text{ cm}^2} = 160.77 \text{ MPa}$$

$$\frac{ec}{r^2} = \frac{(40 \text{ mm})(155 \text{ mm})}{(13.58 \text{ cm})^2} = 0.336$$

$$\frac{L}{r} = \frac{(7.5 \text{ m})}{13.58 \text{ cm}} = 55.23$$

$$\frac{P}{EA} = \frac{2000 \text{ kN}}{(210 \text{ GPa})(124.4 \text{ cm}^2)} = 765.6 \times 10^{-6}$$

3. ***Analyze***: Substitute these values into the secant formula to get

$$\sigma_{max} = \frac{P}{A}\left[1 + \frac{ec}{r^2}\sec\left(\frac{L}{2r}\sqrt{\frac{P}{EA}}\right)\right]$$

$$= (160.77 \text{ MPa})(1 + 0.466) = 235.6 \text{ MPa}$$

4. ***Finalize***: This compressive stress occurs at mid-height of the column on the concave side (the right-hand side in Fig. 11-29b).

Part (b): Factor of safety with respect to yielding.

1. ***Conceptualize***: To find the factor of safety, determine the value of the load P, acting at the eccentricity e, that will produce a maximum stress equal to the yield stress $\sigma_Y = 300$ MPa. Since this value of the load is just sufficient to produce initial yielding of the material, denote it as P_Y.
2. ***Categorize***: Note that force P_Y cannot be determined by multiplying the load P (equal to 2000 kN) by the ratio σ_Y/σ_{max}. The reason is that there is a nonlinear relationship between load and stress. Instead, substitute $\sigma_{max} = \sigma_Y = 300$ MPa in the secant formula and then solve for the corresponding load P, which becomes P_Y. In other words, find the value of P_Y that satisfies

$$\sigma_Y = \frac{P_Y}{A}\left[1 + \frac{ec}{r^2}\sec\left(\frac{L}{2r}\sqrt{\frac{P_Y}{EA}}\right)\right] \qquad \textbf{(11-70)}$$

3. ***Analyze***: Substitute numerical values to obtain

$$300 \text{ MPa} = \frac{P_Y}{124.4 \text{ cm}^2}\left[1 + 0.336 \sec\left(\frac{55.23}{2}\sqrt{\frac{P_Y}{(210 \text{ GPa})(124.4 \text{ cm}^2)}}\right)\right]$$

or

$$3732 \text{ kN} = P_Y[1 + 0.336 \sec(5.403 \times 10^{-4}\sqrt{P_Y})]$$

in which P_Y has units of kN. Solving this equation numerically gives

$$P_Y = 2473 \text{ kN}$$

This load will produce yielding of the material (in compression) at the cross section of maximum bending moment.

Since the actual load is $P = 2000$ kN, the factor of safety against yielding is

$$n = \frac{P_Y}{P} = \frac{2473 \text{ kN}}{2000 \text{ kN}} = 1.236$$

4. ***Finalize***: This example illustrates two of the many ways in which the secant formula may be used. Other types of analysis are illustrated in the problems at the end of the chapter.

11.7 Elastic and Inelastic Column Behavior

The preceding sections described the behavior of columns when the material is stressed below the proportional limit. First an ideal column subjected to a centrally applied load (Euler buckling) was considered, leading to the concept of a critical load P_{cr}. Then columns with eccentric axial loads were analyzed and the secant formula derived. The results of these analyses were plotted on a diagram of average compressive stress P/A versus the slenderness ratio L/r (see

Fig. 11-28). The behavior of an ideal column is represented in Fig. 11-28 by Euler's curve, and the behavior of columns with eccentric loads is represented by the family of curves having various values of the eccentricity ratio ec/r^2.

Now the discussion is extended to include **inelastic buckling**, that is, the buckling of columns when the proportional limit is exceeded. The behavior will be displayed on the same kind of diagram as before, namely, a diagram of average compressive stress P/A versus slenderness ratio L/r (see Fig. 11-30). Note that Euler's curve is shown on this diagram as curve ECD. This curve is valid only in the region CD where the stress is below the proportional limit σ_{pl} of the material. Therefore, the part of Euler's curve above the proportional limit is shown by a dashed line.

The value of the slenderness ratio above which Euler's curve is valid is obtained by setting the critical stress from Eq. (11-69) equal to the proportional limit σ_{pl} and solving for the slenderness ratio. Thus, letting $(L/r)_c$ represent the **critical slenderness ratio** (Fig. 11-30) gives

$$\left(\frac{L}{r}\right)_c = \sqrt{\frac{\pi^2 E}{\sigma_{pl}}} \tag{11-71}$$

As an example, consider structural steel with $\sigma_{pl} = 250$ MPa and $E = 210$ GPa. Then the critical slenderness ratio $(L/r)_c$ is equal to 90.7. Above this value, an ideal column buckles elastically, and the Euler load is valid. Below this value, the stress in the column exceeds the proportional limit and the column buckles inelastically.

Taking into account the effects of eccentricities in loading or imperfections in construction, while still assuming that the material follows Hooke's law, leads to a curve such as the one labeled "Secant formula" in Fig. 11-30. This curve is plotted for a maximum stress σ_{max} equal to the proportional limit σ_{pl}.

When comparing the secant-formula curve with Euler's curve, keep in mind an important distinction. In the case of Euler's curve, the stress P/A not only is proportional to the applied load P; it also is the actual maximum stress in the column when buckling occurs. Consequently, in moving from C to D along Euler's curve, both the maximum stress P/A (equal to the critical stress) and

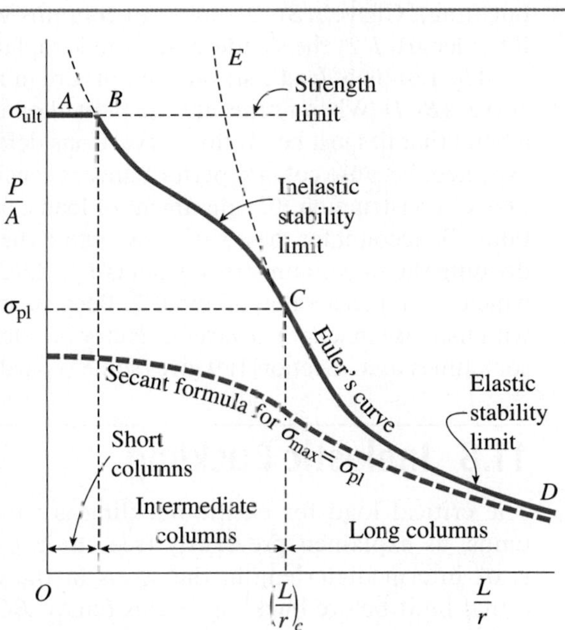

FIGURE 11-30
Diagram of average compressive stress P/A versus slenderness ratio L/r

the axial load P decrease. However, in the case of the secant-formula curve, the *average* stress P/A decreases when moving from left to right along the curve (therefore the axial load P also decreases), but the maximum stress (equal to the proportional limit) remains constant.

Euler's curve shows that **long columns** with large slenderness ratios buckle at low values of the average compressive stress P/A. This condition cannot be improved by using a higher-strength material, because collapse results from instability of the column as a whole and not from failure of the material itself. The stress only can be raised by reducing the slenderness ratio L/r or by using a material with higher modulus of elasticity E.

When a compression member is very **short**, it fails by yielding and crushing of the material, and no buckling or stability considerations are involved. In such a case, define an ultimate compressive stress σ_{ult} as the failure stress for the material. This stress establishes a **strength limit** for the column, which is represented by the horizontal line AB in Fig. 11-30. The strength limit is much higher than the proportional limit, since it represents the ultimate stress in compression.

Between the regions of short and long columns, there is a range of **intermediate slenderness ratios** too small for elastic stability to govern and too large for strength considerations alone to govern. Such an intermediate-length column fails by inelastic buckling, which means that the maximum stresses are above the proportional limit when buckling occurs. Because the proportional limit is exceeded, the slope of the stress-strain curve for the material is less than the modulus of elasticity; hence, the critical load for inelastic buckling is always less than the Euler load (see Section 11.8).

The dividing lines between short, intermediate, and long columns are not precise. Nevertheless, it is useful to make these distinctions because the maximum load-carrying capacity of columns in each category is based upon quite different types of behavior. The maximum load-carrying capacity of a particular column (as a function of its length) is represented by curve $ABCD$ in Fig. 11-30. If the length is very small (region AB), the column fails by direct compression; if the column is longer (region BC), it fails by inelastic buckling; and if it is even longer (region CD), it fails by elastic buckling (that is, Euler buckling). Curve $ABCD$ applies to columns with various support conditions if the length L in the slenderness ratio is replaced by the effective length L_e.

The results of **load tests** on columns are in reasonably good agreement with curve $ABCD$. When test results are plotted on the diagram, they generally form a band that lies just below this curve. Considerable scatter of test results is to be expected, because column performance is sensitive to such matters as the accuracy of construction, the alignment of loads, and the details of support conditions. To account for these variables, obtain the allowable stress for a column by dividing the maximum stress (from curve $ABCD$) by a suitable **factor of safety**, which often has a value of about 2. Because imperfections are apt to increase with increase in length, a variable factor of safety (increasing as L/r increases) is sometimes used. Section 11.9 gives some typical formulas for allowable stresses.

11.8 Inelastic Buckling

The critical load for elastic buckling is valid only for relatively long columns, as explained previously (see curve CD in Fig. 11-30). If a column is of intermediate length, the stress in the column will reach the proportional limit before buckling begins (curve BC in Fig. 11-30). Calculation of critical loads in this intermediate range requires a theory of **inelastic buckling**. Three such theories are described in this section: the tangent-modulus theory, the reduced-modulus theory, and the Shanley theory.

Tangent-Modulus Theory

Consider again an ideal, pinned-end column subjected to an axial force P (Fig. 11-31a). The column is assumed to have a slenderness ratio L/r that is less than the critical slenderness ratio [Eq. (11-71)]; therefore, the axial stress P/A reaches the proportional limit before the critical load is reached.

The compressive **stress-strain diagram** for the material of the column is shown in Fig. 11-32. The proportional limit of the material is indicated as σ_{pl}, and the actual stress σ_A in the column (equal to P/A) is represented by point A (which is above the proportional limit). If the load is increased so that a small increase in stress occurs, the relationship between the increment of stress and the corresponding increment of strain is given by the *slope* of the stress-strain diagram at point A. This slope, which is equal to the slope of the tangent line at A, is called the **tangent modulus** and is denoted by E_t; thus,

$$E_t = \frac{d\sigma}{d\varepsilon} \tag{11-72}$$

Note that the tangent modulus *decreases* as the stress increases beyond the proportional limit. When the stress is below the proportional limit, the tangent modulus is the same as the ordinary elastic modulus E.

According to the **tangent-modulus theory** of inelastic buckling, the column shown in Fig. 11-31a remains straight until the inelastic critical load is reached. At that value of load, the column may undergo a small lateral deflection (Fig. 11-31b). The resulting bending stresses are superimposed upon the axial compressive stresses σ_A. Since the column starts bending from a straight position, the initial bending stresses represent only a small increment of stress. Therefore, the relationship between the bending stresses and the resulting strains is given by the tangent modulus. Since the strains vary linearly across the cross section of the column, the initial bending stresses also vary linearly; therefore, the expressions for curvature are the same as those for linearly elastic bending—except that E_t replaces E:

$$\kappa = \frac{1}{\rho} = \frac{d^2v}{dx^2} = \frac{M}{E_t I} \tag{11-73}$$

[compare with Eqs. (9-9) and (9-11)].

Because the bending moment $M = -Pv$ (see Fig. 11-31b), the differential equation of the deflection curve is

$$E_t I v'' + Pv = 0 \tag{11-74}$$

This equation has the same form as the equation for elastic buckling [Eq. (11-6)] except that E_t appears in place of E. Therefore, the equation can be solved in the same manner as before to obtain the equation for the **tangent-modulus load**:

$$P_t = \frac{\pi^2 E_t I}{L^2} \tag{11-75}$$

FIGURE 11-31
Ideal column of intermediate length that buckles inelastically

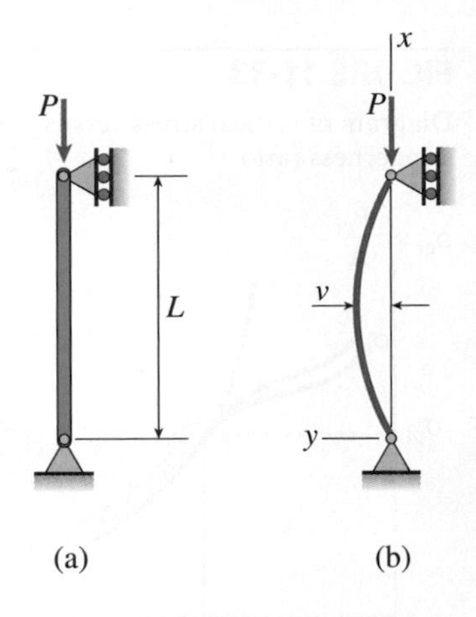

FIGURE 11-32
Compression stress-strain diagram for the material of the column shown in Fig. 11-31

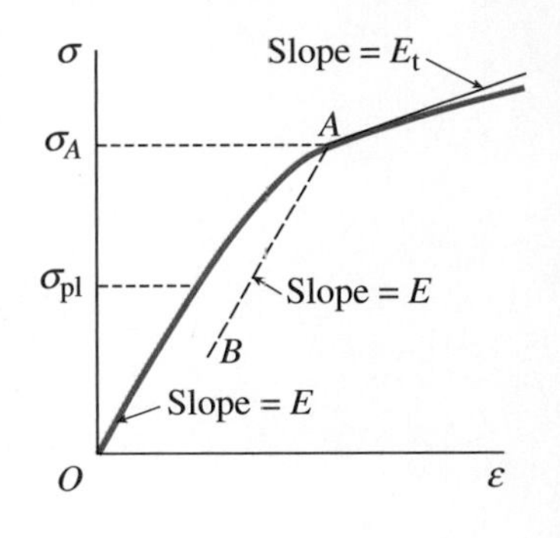

This load represents the critical load for the column according to the tangent-modulus theory. The corresponding critical stress is

$$\sigma_t = \frac{P_t}{A} = \frac{\pi^2 E_t}{(L/r)^2} \tag{11-76}$$

which is similar in form to Eq. (11-69) for the Euler critical stress.

Since the tangent modulus E_t varies with the compressive stress $\sigma = P/A$ (Fig. 11-32), the tangent-modulus load must be obtained by an iterative procedure. Begin by estimating the value of P_t. This trial value, call it P_1, should be slightly larger than $\sigma_{pl}A$, which is the axial load when the stress just reaches the proportional limit. Knowing P_1, next calculate the corresponding axial stress $\sigma_1 = P_1/A$ and determine the tangent modulus E_t from the stress-strain diagram. Then use Eq. (11-75) to obtain a second estimate of P_t. Call this value P_2. If P_2 is very close to P_1, accept the load P_2 as the tangent-modulus load. However, it is more likely that additional cycles of iteration will be required until a load is reached that is in close agreement with the preceding trial load. This value is the tangent-modulus load.

A diagram showing how the critical stress σ_t varies with the slenderness ratio L/r is given in Fig. 11-33 for a typical metal column with pinned ends. Note that the curve is above the proportional limit and below Euler's curve.

The tangent-modulus formulas may be used for columns with various support conditions by using the effective length L_e in place of the actual length L.

FIGURE 11-33

Diagram of critical stress versus slenderness ratio

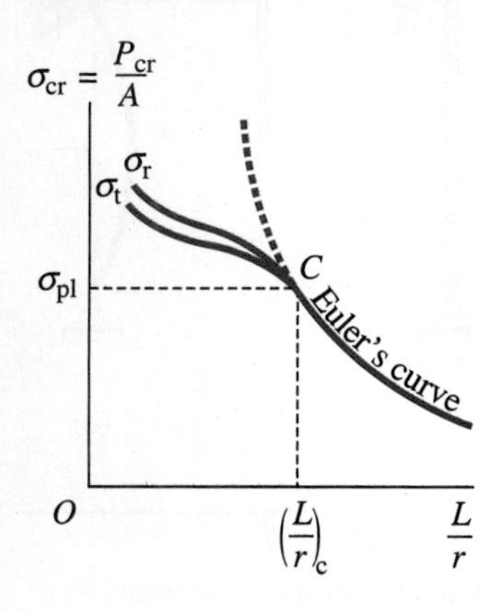

Reduced-Modulus Theory

The tangent-modulus theory is distinguished by its simplicity and ease of use. However, it is conceptually deficient because it does not account for the complete behavior of the column. To explain the difficulty, consider again the column shown in Fig. 11-31a. When this column first departs from the straight position (Fig. 11-31b), bending stresses are added to the existing compressive stresses P/A. These additional stresses are compressive on the concave side of the column and tensile on the convex side. Therefore, the compressive stresses in the column become larger on the concave side and smaller on the other side.

Now imagine that the axial stress P/A is represented by point A on the stress-strain curve (Fig. 11-32). On the concave side of the column (where the compressive stress is increased), the material follows the tangent modulus E_t. However, on the convex side (where the compressive stress is decreased), the material follows the unloading line AB on the stress-strain diagram. This line is parallel to the initial linear part of the diagram, so its slope is equal to the elastic modulus E. Thus, at the onset of bending, the column behaves as if it were made of two different materials: a material of modulus E_t on the concave side and a material of modulus E on the convex side.

A bending analysis of such a column can be made using the bending theories for a beam of two materials (Sections 6.2 and 6.3). The results of such analyses show that the column bends as though the material had a modulus of elasticity between the values of E and E_t. This "effective modulus" is known as the **reduced modulus** E_r, and its value depends not only upon the magnitude of the stress (because E_t depends upon the magnitude of the stress) but also upon the shape of the cross section of the column. Thus, the reduced modulus E_r is more difficult to determine than is the tangent modulus E_t.

In the case of a column having a *rectangular cross section*, the equation for the reduced modulus is

$$E_r = \frac{4EE_t}{(\sqrt{E} + \sqrt{E_t})^2} \quad \textbf{(11-77)}$$

For a *wide-flange beam* with the area of the web disregarded, the reduced modulus for bending about the strong axis is

$$E_r = \frac{2EE_t}{E + E_t} \quad \textbf{(11-78)}$$

The reduced modulus E_r is also called the *double modulus*.

The reduced modulus represents an effective modulus that governs the bending of the column when it first departs from the straight position, so a reduced-modulus theory of inelastic buckling can be formulated. Proceeding in the same manner as for the tangent-modulus theory, begin with an equation for the curvature and then write the differential equation of the deflection curve. These equations are the same as Eqs. (11-73) and (11-74), except that E_r appears instead of E_t. Thus, the equation for the **reduced-modulus load** is

$$P_r = \frac{\pi^2 E_r I}{L^2} \quad \textbf{(11-79)}$$

The corresponding equation for the critical stress is

$$\sigma_r = \frac{\pi^2 E_r}{(L/r)^2} \quad \textbf{(11-80)}$$

To find the reduced-modulus load P_r, again use an iterative procedure, because E_r depends upon E_t. The critical stress according to the reduced-modulus theory is shown in Fig. 11-33. Note that the curve for σ_r is above that for σ_t, because E_r is always greater than E_t.

The reduced-modulus theory is difficult to use in practice because E_r depends upon the shape of the cross section as well as the stress-strain curve and must be evaluated for each particular column. Moreover, this theory also has a conceptual defect. In order for the reduced modulus E_r to apply, the material on the convex side of the column must be undergoing a reduction in stress. However, such a reduction in stress cannot occur until bending actually takes place. Therefore, the axial load P, applied to an ideal straight column, can never actually reach the reduced-modulus load P_r. To reach that load would require that bending already exists, which is a contradiction.

Shanley Theory

From the preceding discussions neither the tangent-modulus theory nor the reduced-modulus theory is entirely rational in explaining the phenomenon of inelastic buckling. Nevertheless, an understanding of both theories is necessary in order to develop a more complete and logically consistent theory. Such a theory was developed by F. R. Shanley in 1946 (see the historical note that follows), and today it is called the *Shanley theory of inelastic buckling*.

The Shanley theory overcomes the difficulties with both the tangent-modulus and reduced-modulus theories by recognizing that it is not possible

FIGURE 11-34

Load-deflection diagram for elastic and inelastic buckling

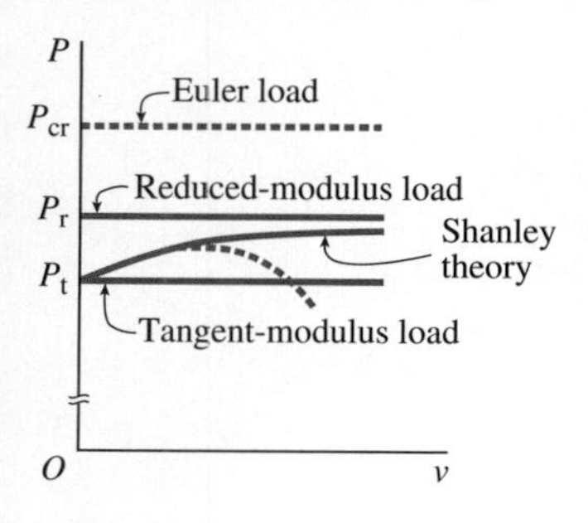

for a column to buckle inelastically in a manner that is analogous to Euler buckling. In Euler buckling, a critical load is reached at which the column is in neutral equilibrium, represented by a horizontal line on the load-deflection diagram (Fig. 11-34). As already explained, neither the tangent-modulus load P_t nor the reduced-modulus load P_r can represent this type of behavior. Both cases lead to a contradiction if the load is associated with a condition of neutral equilibrium.

Instead of neutral equilibrium, wherein a deflected shape suddenly becomes possible with no change in load, think of a column that has an ever-increasing axial load. When the load reaches the tangent-modulus load (which is less than the reduced-modulus load), bending can begin only if the load continues to increase. Under these conditions, bending occurs simultaneously with an increase in load, resulting in a decrease in strain on the convex side of the column. Thus, the effective modulus of the material throughout the cross section becomes greater than E_t, so an increase in load is possible. However, the effective modulus is not as great as E_r, because E_r is based upon full strain reversal on the convex side of the column. In other words, E_r is based upon the amount of strain reversal that exists if the column bends without a change in the axial force, whereas the presence of an increasing axial force means that the reduction in strain is not as great.

Thus, instead of neutral equilibrium, where the relationship between load and deflection is undefined, there is now a definite relationship between each value of the load and the corresponding deflection. This behavior is shown by the curve labeled "Shanley theory" in Fig. 11-34. Note that buckling begins at the tangent-modulus load; then the load increases, but it does not reach the reduced-modulus load until the deflection becomes infinitely large (theoretically). However, other effects become important as the deflection increases, and in reality, the curve eventually goes downward, as shown by the dashed line.

The Shanley concept of inelastic buckling has been verified by numerous investigators and by many tests. However, the maximum load attained by real columns (see the dashed curve trending downward in Fig. 11-34) is only slightly above the tangent-modulus load P_t. In addition, the tangent-modulus load is very simple to calculate. Therefore, for many practical purposes, it is reasonable to adopt the **tangent-modulus load** as the critical load for inelastic buckling of columns.

The preceding discussions of elastic and inelastic buckling are based upon idealized conditions. Although theoretical concepts are important in understanding column behavior, the actual design of columns must take into account additional factors not considered in the theory. For instance, steel columns always contain residual stresses produced by the rolling process. These stresses vary greatly in different parts of the cross section; therefore, the stress level required to produce yielding varies throughout the cross section. For such reasons, a variety of empirical design formulas have been developed for use in designing columns. Some of the commonly used formulas are given in the next section.

Historical Note Over 200 years had elapsed between the first calculation of a buckling load by Euler (in 1744) and the final development of the theory by Shanley (in 1946). Several famous investigators in the field of mechanics contributed to this development, and their work is described in this note.

After Euler's pioneering studies (Ref. 11-1), little progress was made until 1845, when the French engineer A. H. E. Lamarle pointed out that Euler's formula should be used only for slenderness ratios beyond a certain limit and that experimental data should be relied upon for columns with smaller ratios (Ref. 11-6).

Then in 1889, another French engineer, A. G. Considère, published the results of the first comprehensive tests on columns (Ref. 11-7). He pointed out that the stresses on the concave side of the column increased with E_t and the stresses on the convex side decreased with E. Thus, he showed why the Euler formula was not applicable to inelastic buckling, and he stated that the effective modulus was between E and E_t. Although he made no attempt to evaluate the effective modulus, Considère was responsible for beginning the reduced-modulus theory.

In the same year, and quite independently, the German engineer F. Engesser suggested the tangent-modulus theory (Ref. 11-8). He denoted the tangent modulus by the symbol T (equal to $d\sigma/d\varepsilon$) and proposed that T be substituted for E in Euler's formula for the critical load. Later, in March 1895, Engesser again presented the tangent-modulus theory (Ref. 11-9), obviously without knowledge of Considère's work. Today, the tangent-modulus theory is often called the *Engesser theory.*

Three months later, Polish-born F. S. Jasinsky, then a professor in St. Petersburg, pointed out that Engesser's tangent-modulus theory was incorrect, called attention to Considère's work, and presented the reduced-modulus theory (Ref. 11-10). He also stated that the reduced modulus could not be calculated theoretically. In response, and only one month later, Engesser acknowledged the error in the tangent-modulus approach and showed how to calculate the reduced modulus for any cross section (Ref. 11-11). Thus, the reduced-modulus theory is also known as the *Considère-Engesser theory.*

The reduced-modulus theory was also presented by the famous scientist Theodore von Kármán in 1908 and 1910 (Refs. 11-12, 11-13, and 11-14), apparently independently of the earlier investigations. In Ref. 11-13 he derived the formulas for E_r for both rectangular and idealized wide-flange sections (that is, wide-flange sections without a web). He extended the theory to include the effects of eccentricities of the buckling load, and he showed that the maximum load decreases rapidly as the eccentricity increases.

The reduced-modulus theory was the accepted theory of inelastic buckling until 1946, when the American aeronautical-engineering professor F. R. Shanley pointed out the logical paradoxes in both the tangent-modulus and reduced-modulus theories. In a remarkable one-page paper (Ref. 11-15), Shanley not only explained what was wrong with the generally accepted theories, but he also proposed his own theory that resolved the paradoxes. In a second paper, five months later, he gave further analyses to support his earlier theory and gave results from tests on columns (Ref. 11-16). Since that time, many other investigators have confirmed and expanded Shanley's concept.

For excellent discussions of the column-buckling problem, see the comprehensive papers by Hoff (Refs. 11-17 and 11-18), and for a historical account, see the paper by Johnston (Ref. 11-19).

CHAPTER SUMMARY AND REVIEW

Chapter 11 discussed the elastic behavior of axially loaded members known as columns. First, the concepts of **buckling and stability** of these slender compression elements were discussed using the equilibrium of simple column models made up of rigid bars and elastic springs. Then elastic columns with pinned ends acted on by centroidal compressive loads were considered, and the differential equation of the deflection curve was solved to obtain the **buckling load** (P_{cr}) and **buckled mode shape**; linear elastic behavior was assumed. Three additional support cases were investigated, and the buckling load for each case was expressed in terms of the **column's effective length**. Behavior of pinned-end columns with **eccentric axial loads** was discussed, and the **secant formula** that defines the maximum stress in these columns was derived.

Here are the major concepts presented in this chapter.

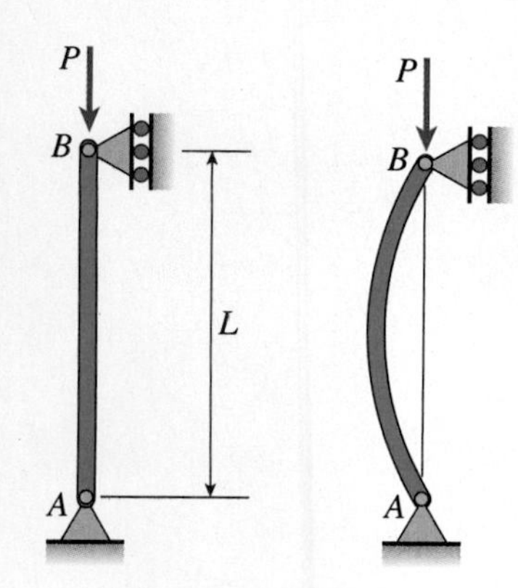

1. Buckling instability of slender columns is an important mode of failure that must be considered in their design (in addition to strength and stiffness).
2. A slender column with pinned ends and length L acted on by a compressive load at the centroid of the cross section and restricted to linear elastic behavior buckles at the **Euler buckling load**

$$P_{cr} = \pi^2 EI / L^2$$

in the fundamental mode; hence, the buckling load depends on the flexural rigidity (EI) and length (L) but not the strength of the material.

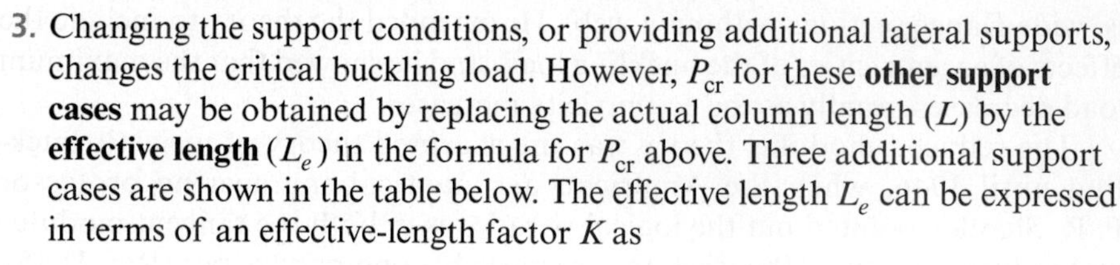

3. Changing the support conditions, or providing additional lateral supports, changes the critical buckling load. However, P_{cr} for these **other support cases** may be obtained by replacing the actual column length (L) by the **effective length** (L_e) in the formula for P_{cr} above. Three additional support cases are shown in the table below. The effective length L_e can be expressed in terms of an effective-length factor K as

$$L_e = KL$$

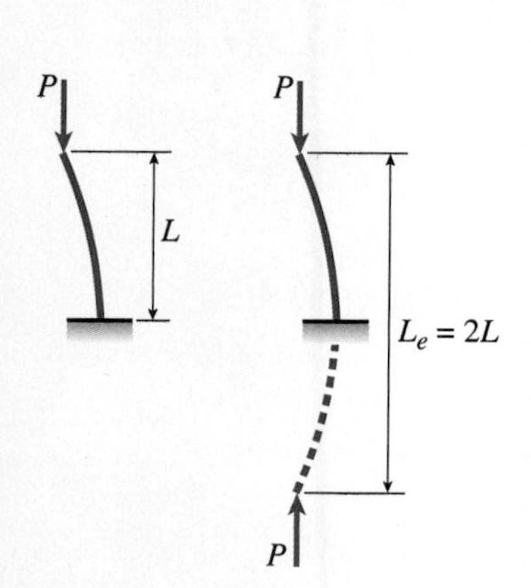

(a) Pinned-pinned column	(b) Fixed-free column	(c) Fixed-fixed column	(d) Fixed-pinned column
$P_{cr} = \frac{\pi^2 EI}{L^2}$	$P_{cr} = \frac{\pi^2 EI}{4L^2}$	$P_{cr} = \frac{4\pi^2 EI}{L^2}$	$P_{cr} = \frac{2.046\ \pi^2 EI}{L^2}$
L	L	L_e L	L_e L
$L_e = L$	$L_e = 2L$	$L_e = 0.5L$	$L_e = 0.699L$
$K = 1$	$K = 2$	$K = 0.5$	$K = 0.699$

where $K = 1$ for a pinned-end column and $K = 2$ for a column fixed at its base. The critical load P_{cr} then is expressed as

$$P_{cr} = \frac{\pi^2 EI}{(KL)^2}$$

Effective-length factor K is often used in column design formulas.

4. Columns with **eccentric axial loads** behave quite differently from those with centroidal loads. The maximum compressive stress in pinned-end columns with load P applied at eccentricity e is defined by the **secant formula**; a graph of this formula (see graph below) shows that column load-carrying capacity decreases with increasing eccentricity. The secant formula gives the maximum compressive stress σ_{max} in an eccentrically loaded, pinned-end column in terms of average compressive stress P/A, modulus of elasticity E, slenderness ratio L/r, and eccentricity ratio ec/r^2 as

$$\sigma_{max} = \frac{P}{A}\left[1 + \frac{ec}{r^2}\sec\left(\frac{L}{2r}\sqrt{\frac{P}{EA}}\right)\right]$$

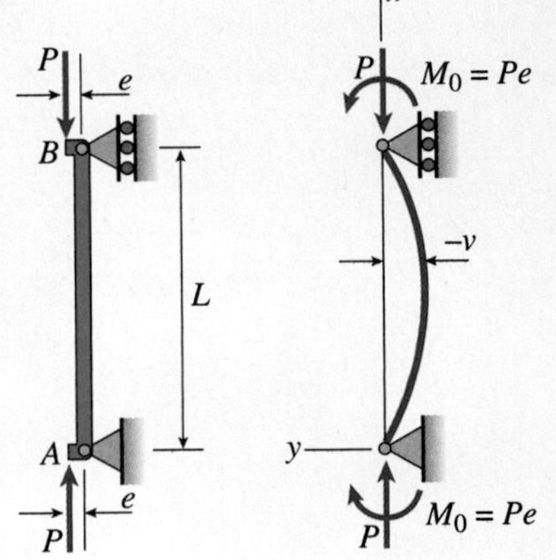

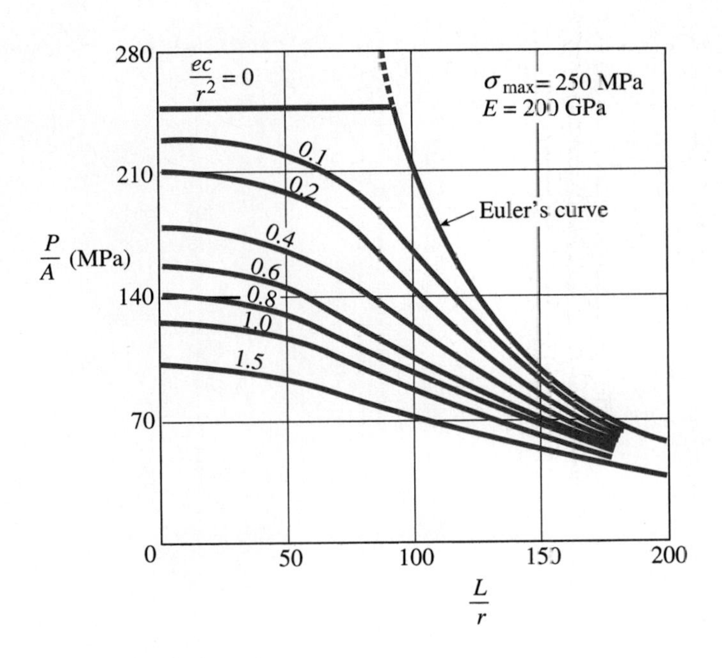

5. **Long columns** (i.e., large slenderness ratios L/r) buckle at low values of compressive stress; **short columns** (i.e., low L/r) fail by yielding and crushing of the material; and **intermediate columns** (with values of L/r which lie between those for long and short columns) fail by inelastic buckling. The critical buckling load for inelastic buckling is always less than the Euler buckling load. The dividing lines between short, intermediate, and long columns are not precisely defined.

6. Three **theories for inelastic buckling of intermediate columns** are the **tangent-modulus theory**, the **reduced-modulus theory**, and the **Shanley theory**. However, empirical formulas are actually used for the design of columns because the theoretical formulas do not account for such things as **residual stresses** in steel columns and other factors.

PROBLEMS Chapter 11

11.2 Buckling and Stability

Introductory Problems

11.2-1 A rigid bar of length L is supported by a linear elastic rotational spring with rotational stiffness β_R at A.

Determine the critical load P_{cr} for the structure.

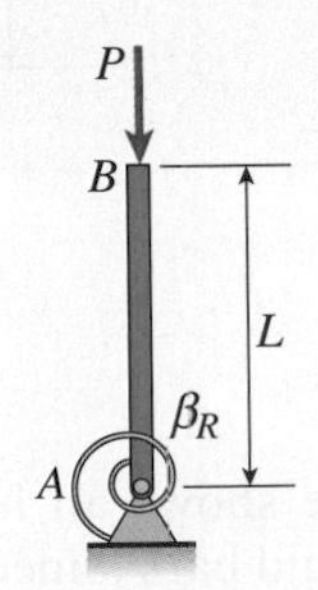

PROBLEM 11.2-1

11.2-2 The figure shows an idealized structure consisting of a rigid bar with pinned connections and linearly elastic springs. Rotational stiffness is denoted β_R, and translational stiffness is denoted β.

(a) Determine the critical load P_{cr} for the structure from the figure part a.

(b) Find P_{cr} if another rotational spring is added at B from the figure part b.

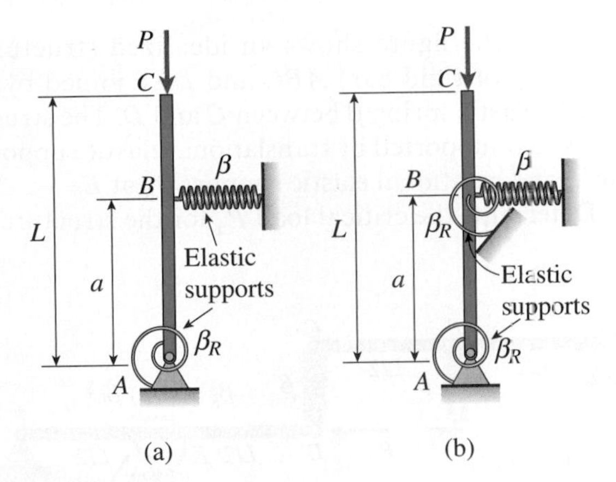

PROBLEM 11.2-2

11.2-3 Two rigid bars are connected with a rotational spring, as shown in the figure. Assume that the elastic rotational spring constant is $\beta_R = 8.5$ kN·m/rad. Calculate the critical load P_{cr} of the system. Assume that $L = 1.8$ m.

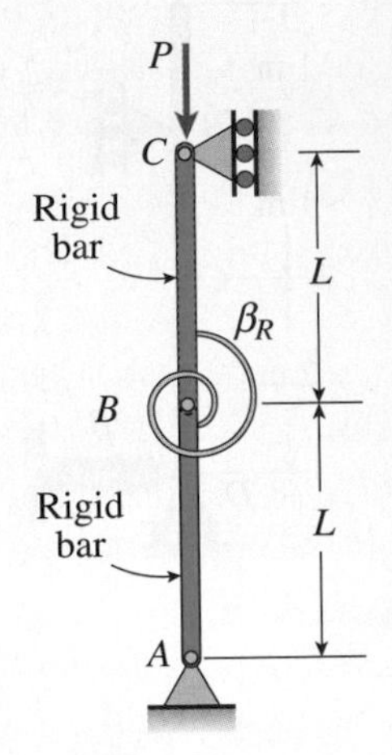

PROBLEMS 11.2-3 and 11.2-4

11.2-4 Repeat Problem 11.2-3 assuming that $\beta_R = 10$ kN·m/rad and $L = 2$ m.

11.2-5 The figure shows an idealized structure consisting of two **rigid bars** with pinned connections and linearly elastic rotational springs. Rotational stiffness is denoted β_R.

Determine the critical load P_{cr} for the structure.

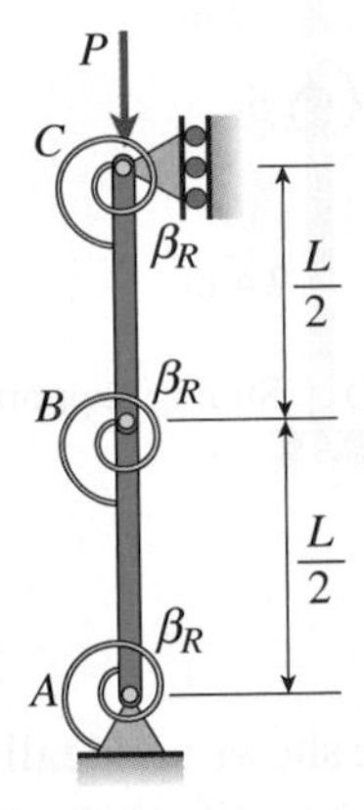

PROBLEM 11.2-5

Representative Problems

11.2-6 An idealized column consists of rigid bar $ABCD$ with a roller support at B and a roller and spring support at D. The spring constant at D is $\beta = 750$ N/m. Find the critical load P_{cr} of the column.

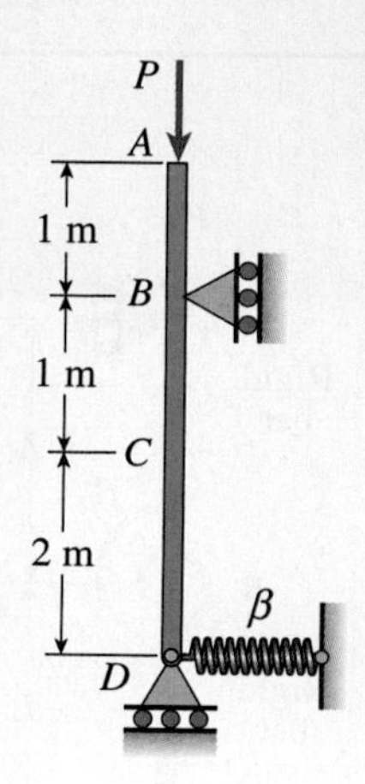

PROBLEM 11.2-6

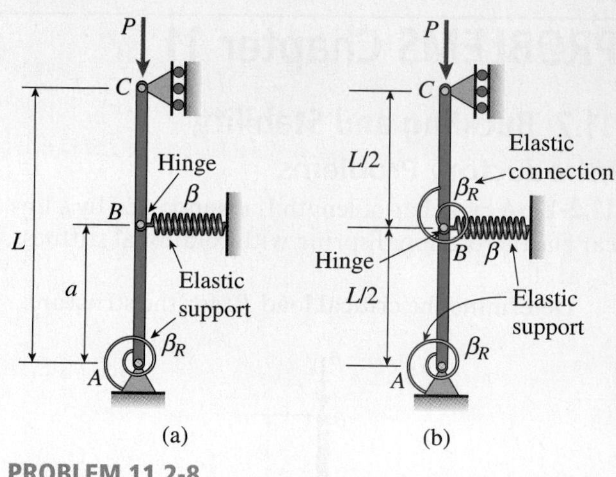

PROBLEM 11.2-8

11.2-7 An idealized column is made up of rigid segments *ABC* and *CD* that are joined by an elastic connection at *C* with rotational stiffness $\beta_R = 11$ kN·m/rad. The column has a roller support at *B* and a sliding support at *D*. Calculate the critical load P_{cr} of the column.

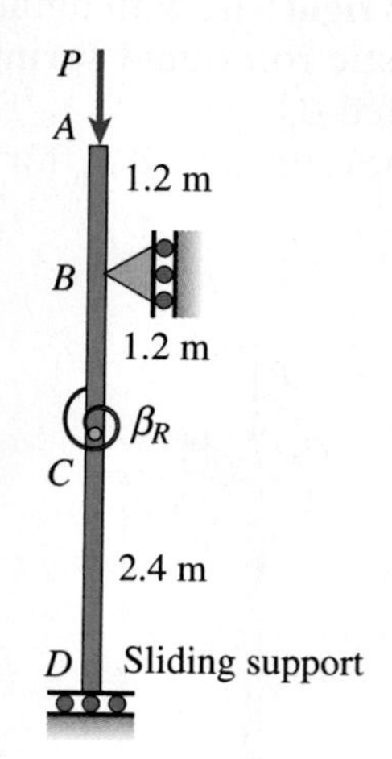

PROBLEM 11.2-7

11.2-8 The figure shows an idealized structure consisting of bars *AB* and *BC* that are connected using a hinge at *B* and linearly elastic springs at *A* and *B*. Rotational stiffness is denoted β_R and translational stiffness is denoted β.

(a) Determine the critical load P_{cr} for the structure from the figure part a.

(b) Find P_{cr} if an elastic connection is now used to connect bar segments *AB* and *BC* from the figure part b.

11.2-9 The figure shows an idealized structure consisting of two rigid bars joined by an elastic connection with rotational stiffness β_R. Determine the critical load P_{cr} for the structure.

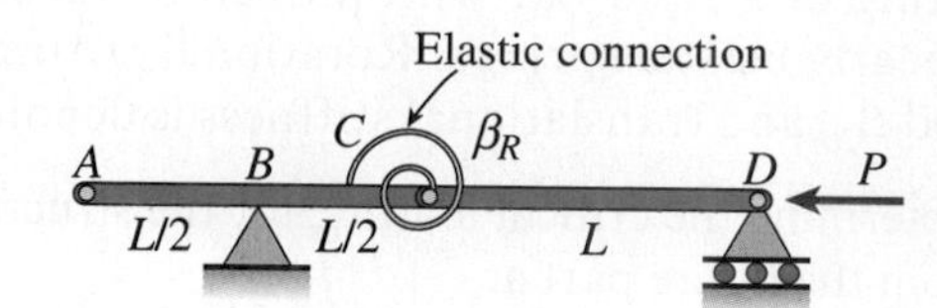

PROBLEM 11.2-9

11.2-10 The figure shows an idealized structure consisting of rigid bars *ABC* and *DEF* joined by a linearly elastic spring β between *C* and *D*. The structure is also supported by translational elastic support β at *B* and rotational elastic support β_R at *E*.

Determine the critical load P_{cr} for the structure.

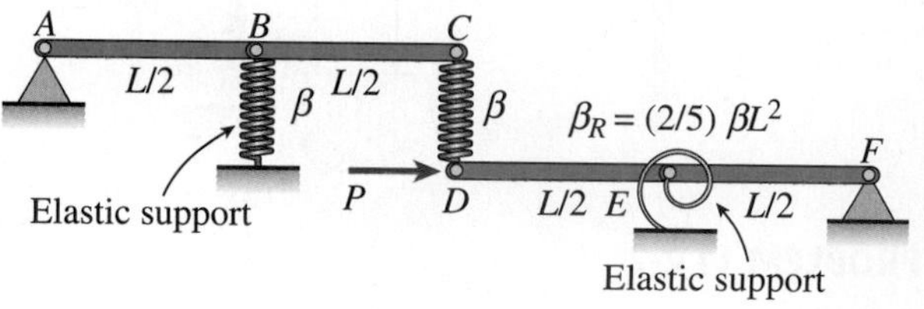

PROBLEM 11.2-10

11.2-11 The figure shows an idealized structure consisting of an L-shaped rigid bar structure supported by linearly elastic springs at *A* and *C*. Rotational stiffness is denoted β_R and translational stiffness is denoted β.

Determine the critical load P_{cr} for the structure.

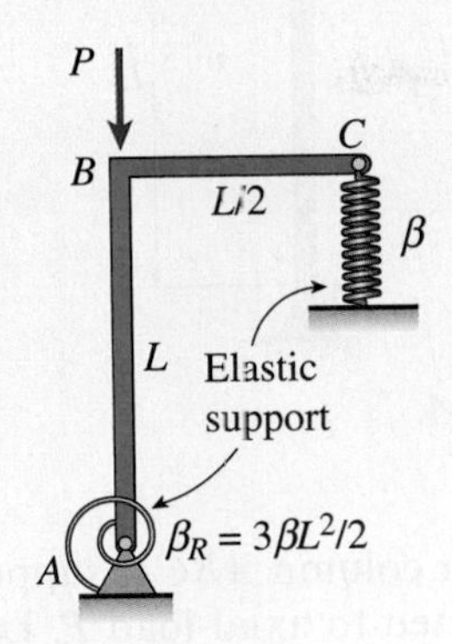

PROBLEM 11.2-11

11.2-12 Rigid column $ABCD$ has an elastic support at B with translational stiffness β. Find an expression for the critical load P_{cr} of the column.

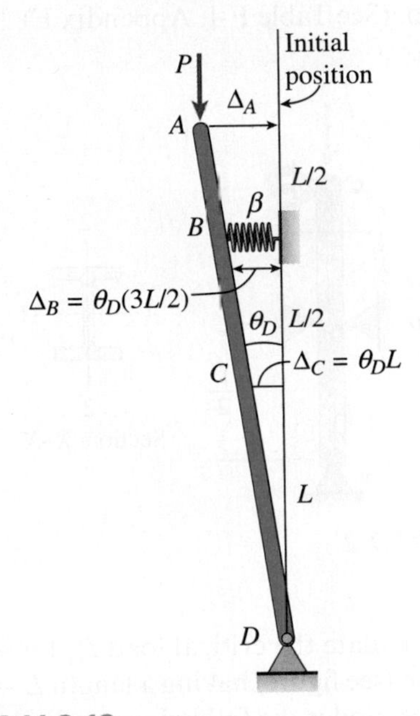

PROBLEM 11.2-12

11.2-13 An idealized column is made up of rigid bars ABC and CD that are joined by a rotational elastic connection at C with stiffness β_R. The column has a roller support at B and a pin support at D. Find an expression for the critical load P_{cr} of the column.

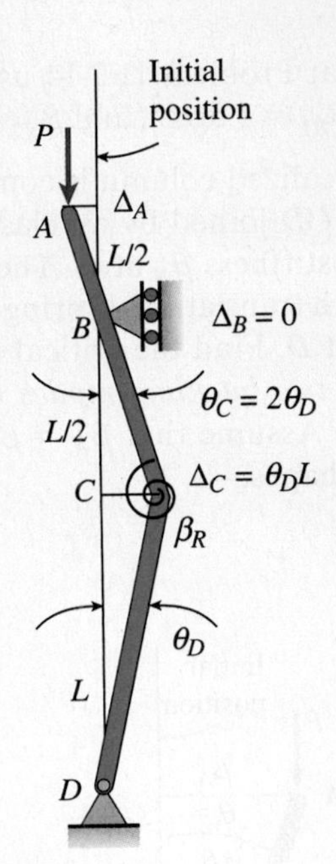

PROBLEM 11.2-13

11.2-14 An idealized column is composed of rigid bars ABC and CD joined by an elastic connection with rotational stiffness β_{R1} at C. There is a roller support at B and an elastic support at D with translational spring stiffness β and rotational stiffness β_{R2}. Find the critical buckling loads for *each of the two buckling modes* of the column. Assume that $L = 3$ m, $\beta = 9$ kN/m, and $\beta_{R1} = \beta_{R2} = \beta L^2$. Sketch the buckled mode shapes.

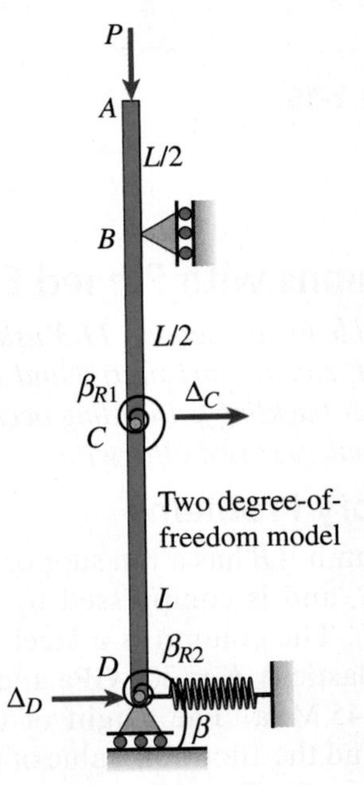

PROBLEM 11.2-14 and 11.2-15

11.2-15 Repeat Problem 11.2-14 using $L = 3.6$ m, $\beta = 45$ kN/m, $\beta_{R1} = 1.5\beta L^2$, and $\beta_{R2} = 2\,\beta_{R1}$.

11.2-16 An idealized column is composed of rigid bars ABC and CD joined by an elastic connection with rotational stiffness β_R at C. There is an elastic support at B with translational spring stiffness β and a pin support at D. Find the critical buckling loads for *each of the two buckling modes* of the column in terms of βL. Assume that $\beta_R = \beta L^2$. Sketch the buckled mode shapes.

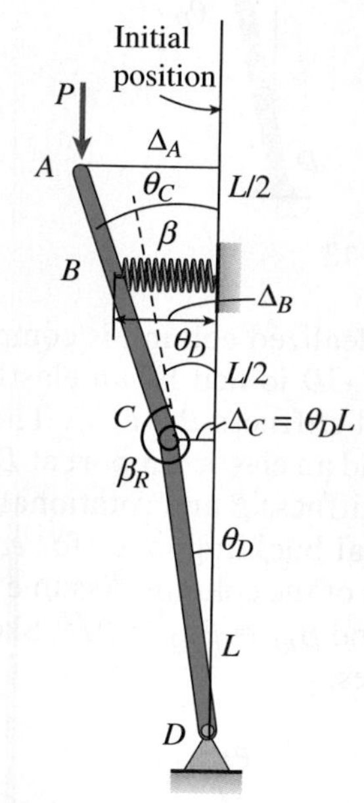

PROBLEM 11.2-16

11.3 Columns with Pinned Ends

Solve the problems for Section 11.3 using the assumptions of ideal, slender, prismatic, and linearly elastic columns (Euler buckling). Buckling occurs in the plane of the figure unless stated otherwise.

Introductory Problems

11.3-1 Column AB has a pin support at A, a roller support at B, and is compressed by an axial load P (see figure). The column is a steel HE 160B with modulus of elasticity $E = 200$ GPa and proportional limit $\sigma_{pl} = 345$ MPa. The height of the column is $L = 3.6$ m. Find the allowable value of load P assuming a factor of safety $n = 2.5$.

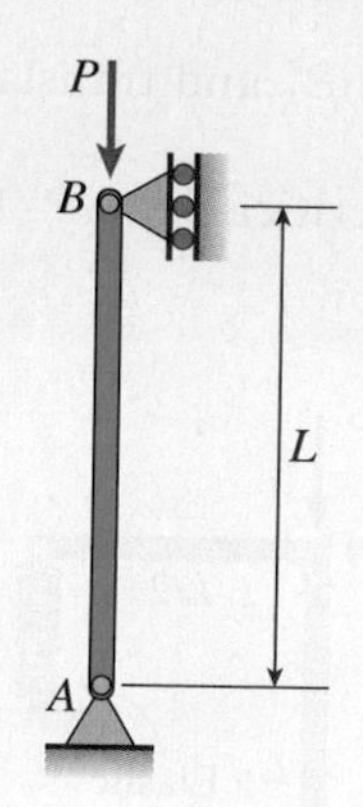

PROBLEM 11.3-1

11.3-2 Slender column ABC is supported at A and C and is subjected to axial load P. Lateral support is provided at mid-height B but only in the plane of the figure; lateral support perpendicular to the plane of the figure is provided only at ends A and C. The column is a steel HE shape with modulus of elasticity $E = 200$ GPa and proportional limit $\sigma_{pl} = 400$ MPa. The total length of the column $L = 9$ m. If the allowable load is 150 kN and the factor of safety is 2.5, determine the lightest HE section that can be used for the column. (See Table F-1, Appendix F).

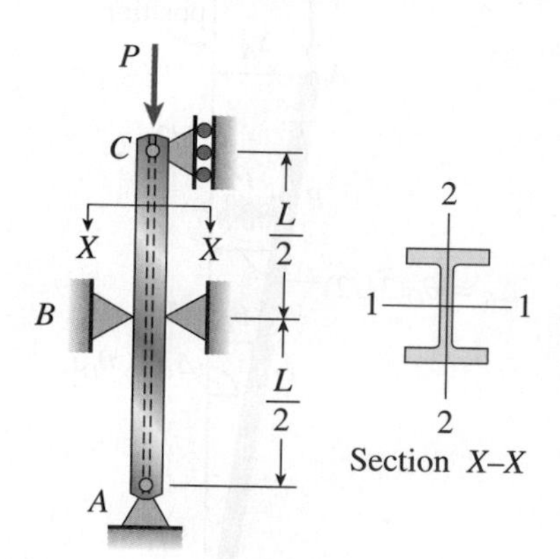

PROBLEM 11.3-2

11.3-3 Calculate the critical load P_{cr} for a HE 140B steel column (see figure) having a length $L = 8$ m and $E = 200$ GPa under the following conditions:

(a) The column buckles by bending about its strong axis (axis 1–1).

(b) The column buckles by bending about its weak axis (axis 2–2).

In both cases, assume that the column has pinned ends.

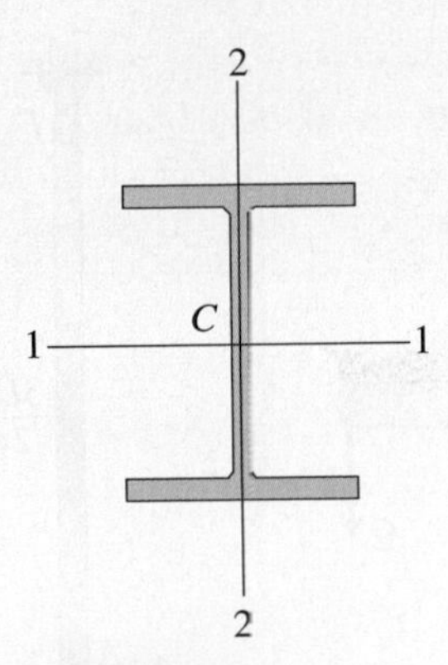

PROBLEMS 11.3-3 through 11.3-5

11.3-4 Solve the preceding problem for an IPN 140 steel column having a length $L = 8$ m. Let $E = 200$ GPa.

11.3-5 Solve Problem 11.3-3 for a HE 140A steel column having a length $L = 8$ m.

Representative Problems

11.3-6 A horizontal beam AB is pin-supported at end A and carries a clockwise moment M at joint B, as shown in the figure. The beam is also supported at C by a pinned-end column of length L; the column is restrained laterally at $0.6L$ from the base at D. Assume the column can only buckle in the plane of the frame. The column is a solid steel bar ($E = 200$ GPa) of square cross section having length $L = 2.4$ m and side dimensions $b = 70$ mm. Let dimensions $d = L/2$. Based upon the critical load of the column, determine the allowable moment M if the factor of safety with respect to buckling is $n = 2.0$.

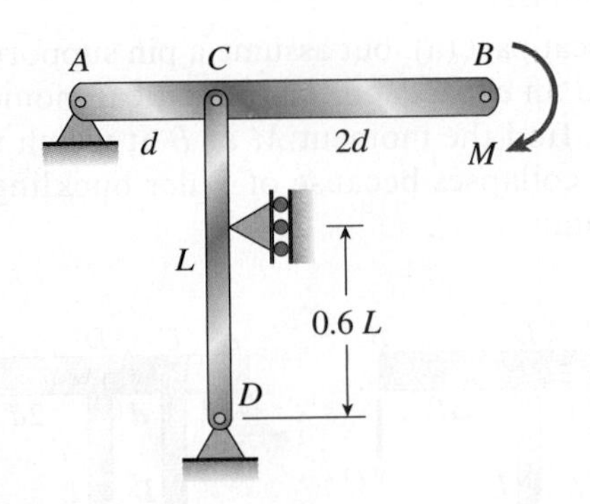

PROBLEM 11.3-6

11.3-7 A column ABC is supported at ends A and C and compressed by an axial load P (figure a). Lateral support is provided at point B but only in the plane of the figure; lateral support perpendicular to the plane of the figure is provided only at A and C. The column is constructed of two channel sections (UPN 140) back to back (see figure b). The modulus of elasticity of the column is $E = 200$ GPa and the proportional limit is 345 MPa. The height of the column is $L = 4.5$ m. Find the allowable value of load P using a factor of safety of 2.5.

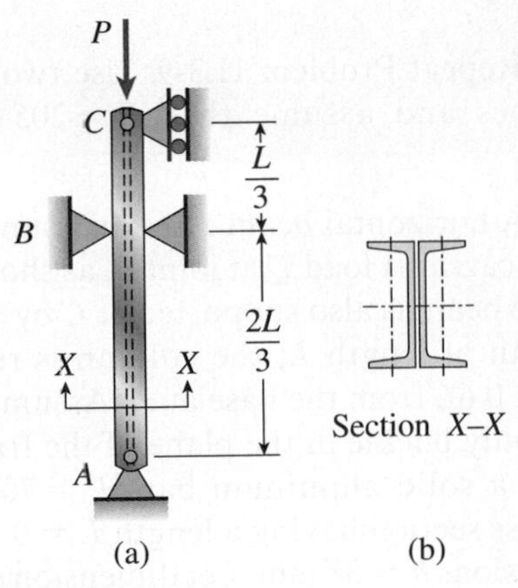

PROBLEM 11.3-7

11.3-8 Find the controlling buckling load (kN) for the steel column shown in the figure. The column is pinned at top and bottom and is made up of two UPN 120 shapes that act together. Assume that $E = 205$ GPa and $L = 6$ m.

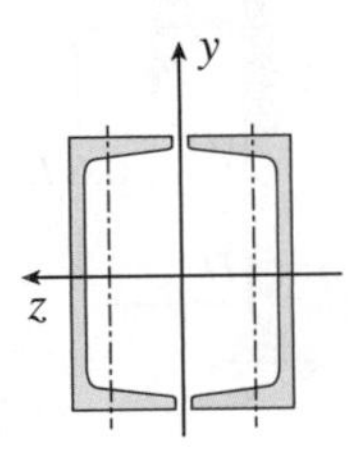

PROBLEM 11.3-8

11.3-9 A column, pinned at top and bottom, is made up of two UPN 160 steel shapes (see figure) that act together.

(a) Find the buckling load (kN) if the gap is zero.

(b) Find required separation distance d (mm) so that the buckling load is the same in y and z directions.

Assume that $E = 205$ GPa and $L = 5.5$ m. Note that distance d is measured between the centroids of the two UPN shapes.

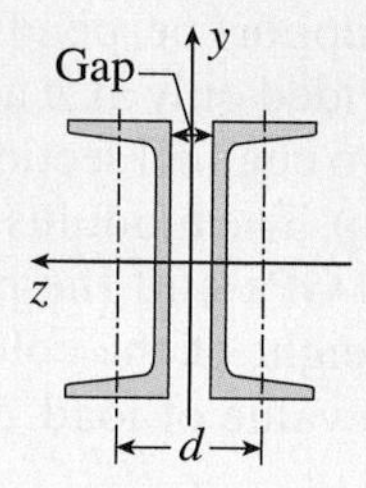

PROBLEM 11.3-9 and 11.3-10

11.3-10 Repeat Problem 11.3-9. Use two UPN 120 steel shapes and assume that $E = 205$ GPa and $L = 6$ m.

11.3-11 A horizontal beam AB is pin-supported at end A and carries a load Q at joint B, as shown in the figure. The beam is also supported at C by a pinned-end column of length L; the column is restrained laterally at $0.6L$ from the base at D. Assume the column can only buckle in the plane of the frame. The column is a solid aluminum bar ($E = 70$ GPa) of square cross section having a length $L = 0.75$ m and side dimensions $b = 38$ mm. Let dimension $d = L/2$. Based upon the critical load of the column, determine the allowable force Q if the factor of safety with respect to buckling is $n = 1.8$.

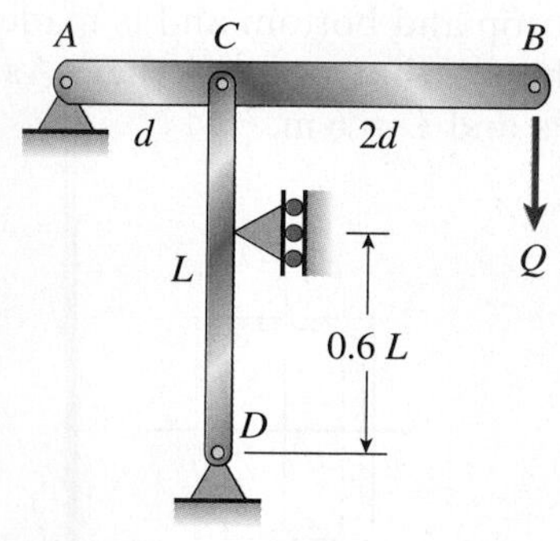

PROBLEM 11.3-11

11.3-12 A horizontal beam AB is supported at end A and carries a load Q at joint B, as shown in the figure part a. The beam is also supported at C by a pinned-end column of length L. The column has flexural rigidity EI.

(a) For the case of a sliding support at A (figure part a), what is the critical load Q_{cr}? (In other words, at what load Q_{cr} does the system collapse because of Euler buckling of the column DC?)

(b) Repeat part (a) if the sliding support at A is replaced by column AF with a length $3L/2$ and flexural rigidity EI (see figure part b).

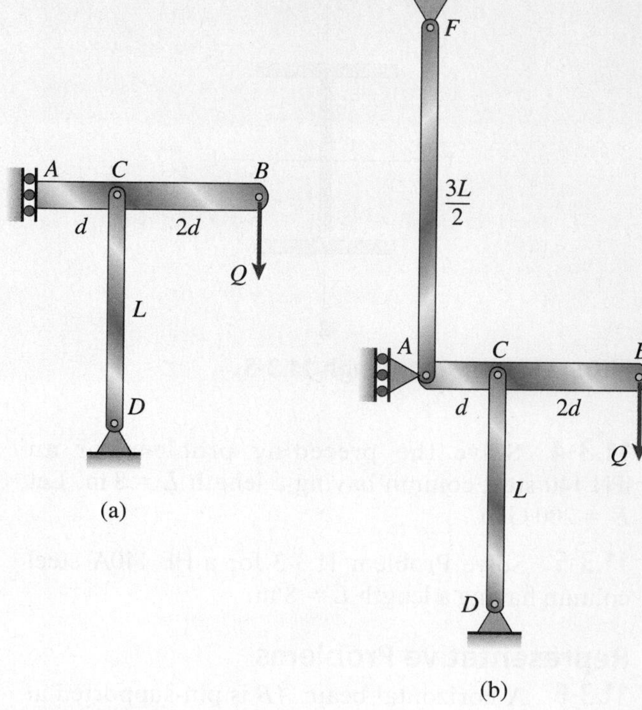

PROBLEM 11.3-12

11.3-13 A horizontal beam AB has a sliding support at end A and carries a load Q at end B, as shown in the figure part a. The beam is supported at C and D by two identical pinned-end columns of length L. Each column has flexural rigidity EI.

(a) Find an expression for the critical load Q_{cr}. (In other words, at what load Q_{cr} does the system collapse because of Euler buckling of the columns?)

(b) Repeat part (a), but assume a pin support at A. Find an expression for the critical moment M_{cr} (i.e., find the moment M at B at which the system collapses because of Euler buckling of the columns).

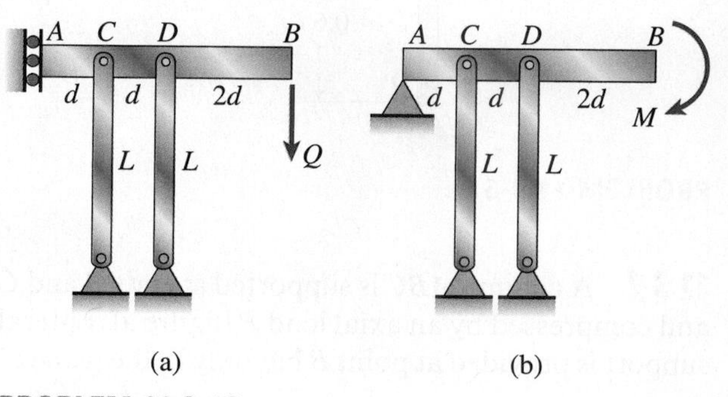

PROBLEM 11.3-13

11.3-14 A slender bar AB with pinned ends and length L is held between immovable supports (see figure).

What increase ΔT in the temperature of the bar will produce buckling at the Euler load?

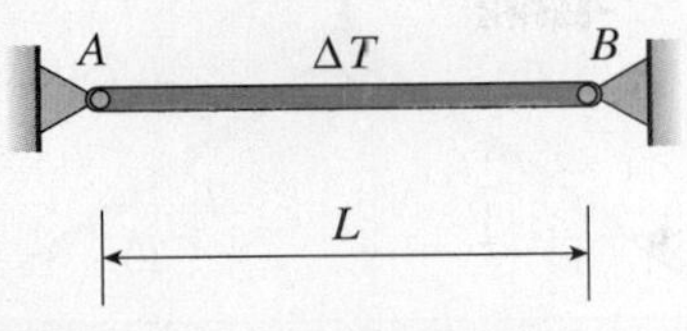

PROBLEM 11.3-14

11.3-15 A rectangular column with cross-sectional dimensions b and h is pin-supported at ends A and C (see figure). At mid-height, the column is restrained in the plane of the figure but is free to deflect perpendicularly to the plane of the figure.

Determine the ratio h/b such that the critical load is the same for buckling in the two principal planes of the column.

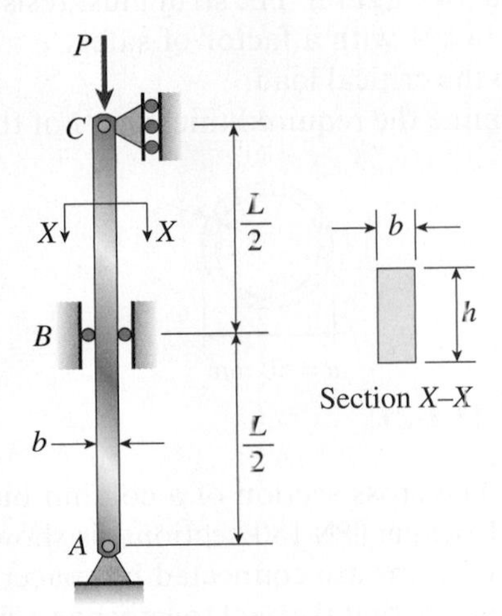

PROBLEM 11.3-15

11.3.16 Three identical, solid circular rods, each of radius r and length L, are placed together to form a compression member (see the cross section shown in the figure).

Assuming pinned-end conditions, determine the critical load P_{cr}.

(a) The rods act independently as individual columns.

(b) The rods are bonded by epoxy throughout their lengths so that they function as a single member.

What is the effect on the critical load when the rods act as a single member?

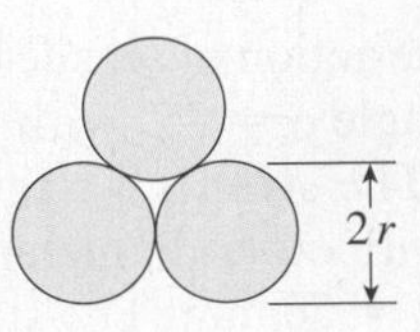

PROBLEM 11.3-16

11.3-17 Three pinned-end columns of the same material have the same length and the same cross-sectional area (see figure). The columns are free to buckle in any direction. The columns have cross sections as: (a) a circle, (b) a square, and (c) an equilateral triangle.

Determine the ratios $P_a : P_b : P_c$ of the critical loads for these columns.

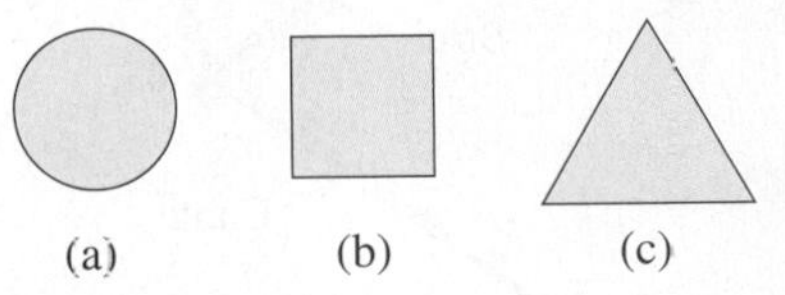

PROBLEM 11.3-17

11.3-18 A long slender column ABC is pinned at ends A and C and compressed by an axial force P (see figure). At the midpoint B, lateral support is provided to prevent deflection in the plane of the figure. The column is a steel wide-flange section (HE 260A) with $E = 200$ GPa. The distance between lateral supports is $L = 5.5$ m.

Calculate the allowable load P using a factor of safety $n = 2.4$, taking into account the possibility of Euler buckling about either principal centroidal axis (i.e., axis 1–1 or axis 2–2).

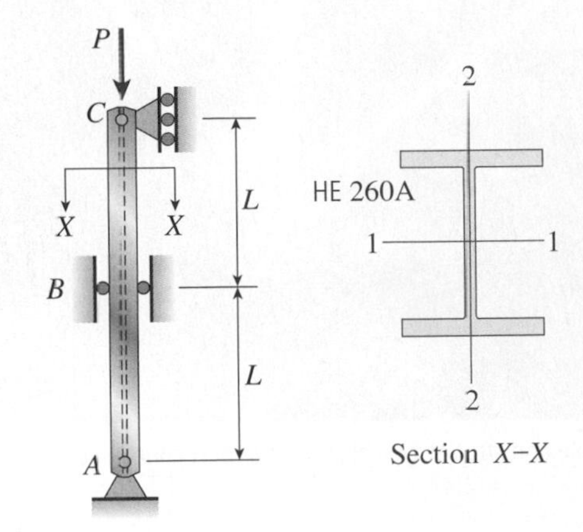

PROBLEM 11.3-18

11.3-19 The roof over a concourse at an airport is supported by the use of pretensioned cables. At a typical joint in the roof structure, a strut AB is compressed by the action of tensile forces F in a cable that makes an angle $\alpha = 75°$ with the strut (see figure and photo). The strut is a circular tube of steel ($E = 200$ GPa) with outer diameter $d_2 = 60$ mm and inner diameter $d_1 = 50$ mm. The strut is 1.75 m long and is assumed to be pin-connected at both ends.

Using a factor of safety $n = 2.5$ with respect to the critical load, determine the allowable force F in the cable.

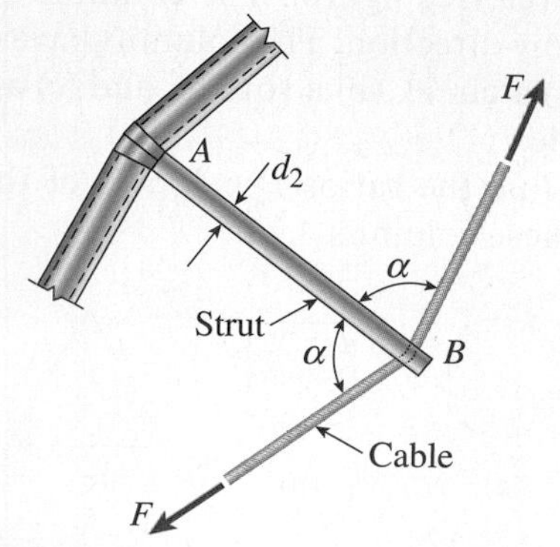

PROBLEM 11.3-19

Cable and strut at typical joint of airport concourse roof

11.3-20 The hoisting arrangement for lifting a large pipe is shown in the figure. The spreader is a steel tubular section with outer diameter 70 mm and inner diameter 57 mm. Its length is 2.6 m, and its modulus of elasticity is 200 GPa.

Based upon a factor of safety of 2.25 with respect to Euler buckling of the spreader, what is the maximum weight of pipe that can be lifted? (Assume pinned conditions at the ends of the spreader.)

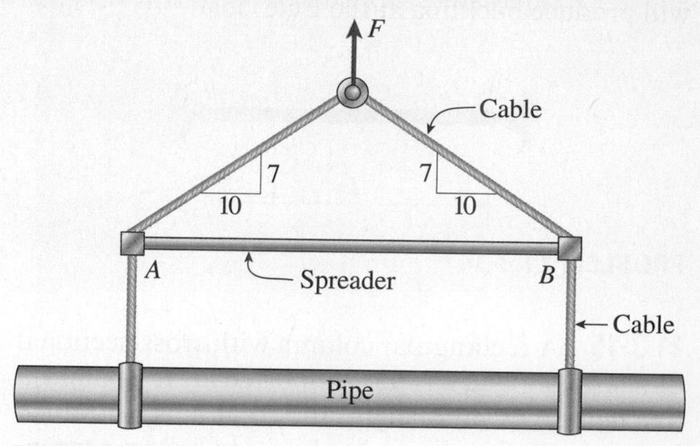

PROBLEM 11.3-20

11.3-21 A pinned-end strut of aluminum ($E = 70$ GPa) with a length $L = 1.8$ m is constructed of circular tubing with an outside diameter $d = 50$ mm (see figure). The strut must resist an axial load $P = 18$ kN with a factor of safety $n = 2.0$ with respect to the critical load.

Determine the required thickness t of the tube.

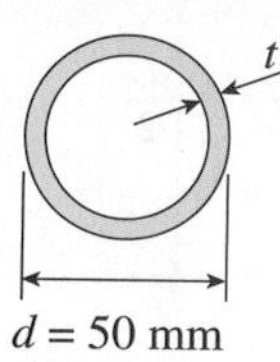

PROBLEM 11.3-21

11.3-22 The cross section of a column built up of two steel I-beams (IPN 180 sections) is shown in the figure. The beams are connected by spacer bars, or *lacing*, to ensure that they act together as a single column. (The lacing is represented by dashed lines in the figure.)

The column is assumed to have pinned ends and may buckle in any direction. Assuming $E = 200$ GPa and $L = 8.5$ m, calculate the critical load P_{cr} for the column.

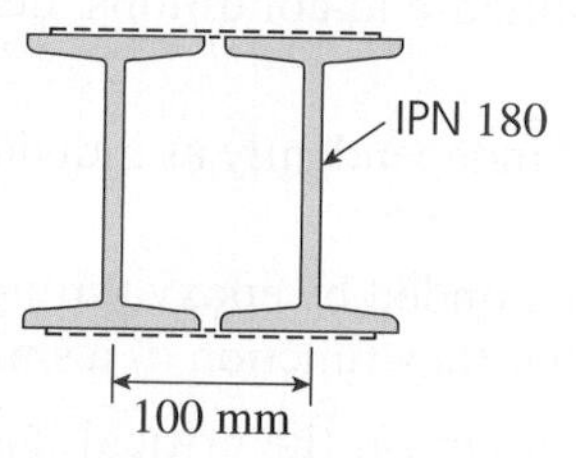

PROBLEM 11.3-22

11.3-23 The truss ABC shown in the figure supports a vertical load W at joint B. Each member is a slender circular steel pipe ($E = 200$ GPa) with an outside diameter of 100 mm and wall thickness 6 mm. The distance between supports is 7 m. Joint B is restrained against displacement perpendicular to the plane of the truss.

Determine the critical value W_{cr} of the load.

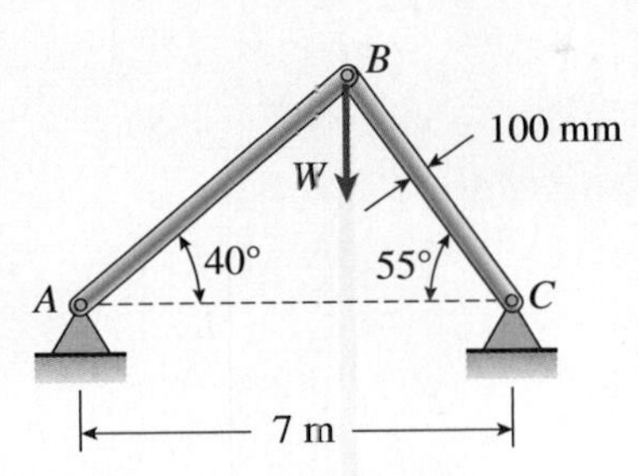

PROBLEM 11.3-23

11.3-24 A truss ABC supports a load W at joint B, as shown in the figure. The length L_1 of member AB is fixed, but the length of strut BC varies as the angle θ is changed. Strut BC has a solid circular cross section. Joint B is restrained against displacement perpendicular to the plane of the truss.

Assuming that collapse occurs by Euler buckling of the strut, determine the angle θ for minimum weight of the strut.

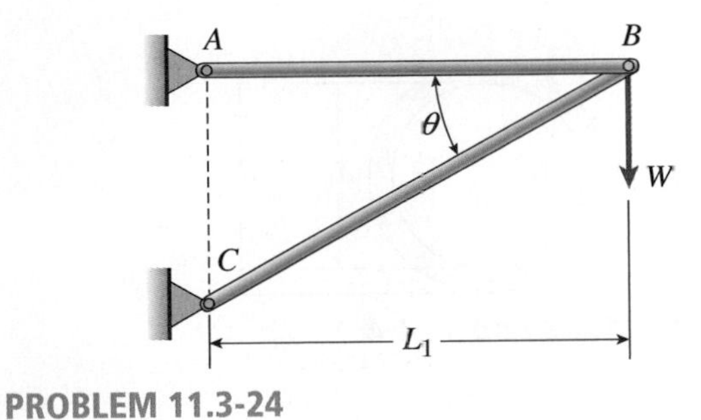

PROBLEM 11.3-24

11.3-25 An IPN 160 steel cantilever beam AB is supported by a steel tie rod at B as shown. The tie rod is just taut when a roller support is added at C at a distance s to the left of B, then the distributed load q is applied to beam segment AC. Assume $E = 200$ GPa and neglect the self-weight of the beam and tie rod. See Table F-2 in Appendix F for the properties of the IPN beam.

(a) What value of uniform load q will, if exceeded, result in buckling of the tie rod if $L_1 = 2$ m, $s = 0.6$ m, $H = 1$ m, and $d = 6$ mm?

(b) What minimum beam moment of inertia I_b is required to prevent buckling of the tie rod if $q = 2$ kN/m, $L_1 = 2$ m, $H = 1$ m, $d = 6$ mm, and $s = 0.6$ m?

(c) For what distance s will the tie rod be just on the verge of buckling if $q = 2$ kN/m, $L_1 = 2$ m, $H = 1$ m, and $d = 6$ mm?

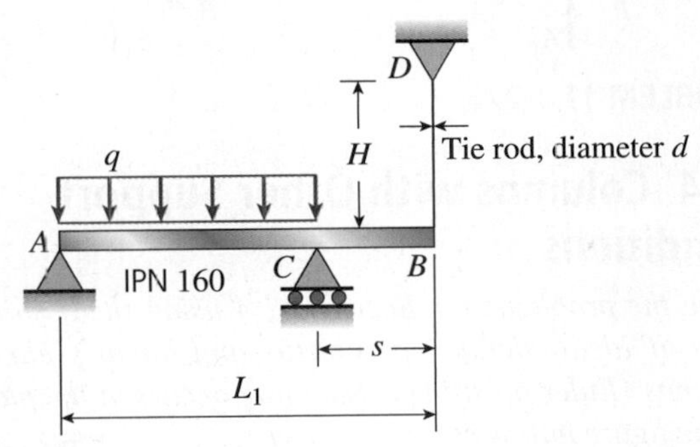

PROBLEM 11.3-25

11.3-26 The plane truss shown in the figure supports vertical loads F at joint D, $2F$ at joint C, and $3F$ at joint B. Each member is a slender circular pipe ($E = 70$ GPa) with an outside diameter of 60 mm and wall thickness of 5 mm. Joint B is restrained against displacement perpendicular to the plane of the truss. Determine the critical value of load variable F (kN) at which member BF fails by Euler buckling.

PROBLEM 11.3-26

11.3-27 A space truss is restrained at joints O, A, B, and C, as shown in the figure. Load P is applied at joint A and load $2P$ acts downward at joint C. Each member is a slender, circular pipe ($E = 73$ GPa) with an outside diameter of 90 mm and wall thickness of 6.5 mm. Length variable $L = 3.5$ m. Determine the critical value of load variable P (kN) at which member OB fails by Euler buckling.

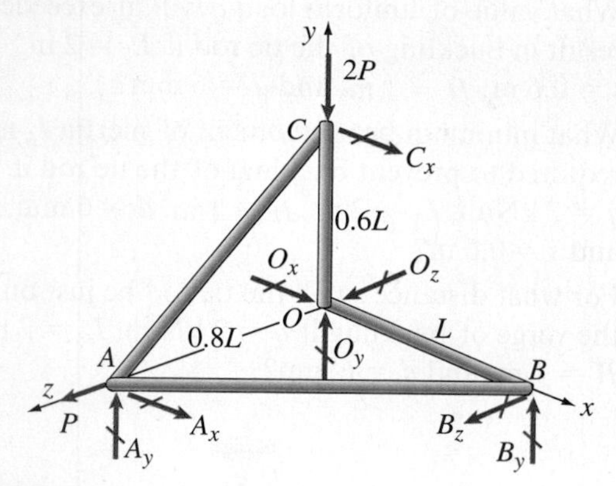

PROBLEM 11.3-27

11.4 Columns with Other Support Conditions

Solve the problems for Section 11.4 using the assumptions of ideal, slender, prismatic, and linearly elastic columns (Euler buckling). Buckling occurs in the plane of the figure unless stated otherwise.

Introductory Problems

11.4-1 A fixed-end column with circular cross section is acted on by compressive axial load P. The 5.5 m-long-column has an outer diameter of 130 mm, a thickness of 13 mm, and is made of aluminum with a modulus of elasticity of 70 GPa. Find the buckling load of the column.

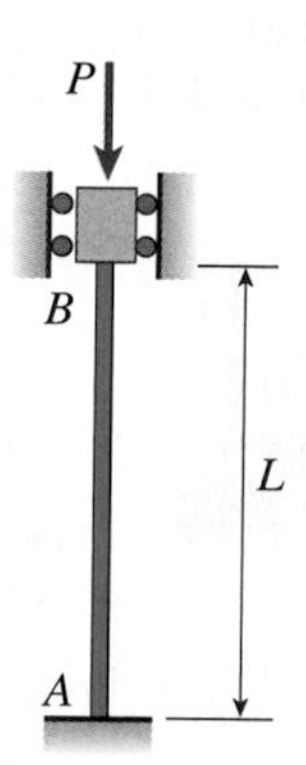

PROBLEM 11.4-1

11.4-2 A cantilever aluminum column has a square tube cross section with an outer dimension of 150 mm. The column has a length $L = 4$ m and is designed to support an axial load of 45 kN. Find the minimum required thickness of the section if the factor of safety $n = 2.5$ with respect to buckling. Assume that the modulus of elasticity is 72 GPa and the proportional limit is 480 MPa.

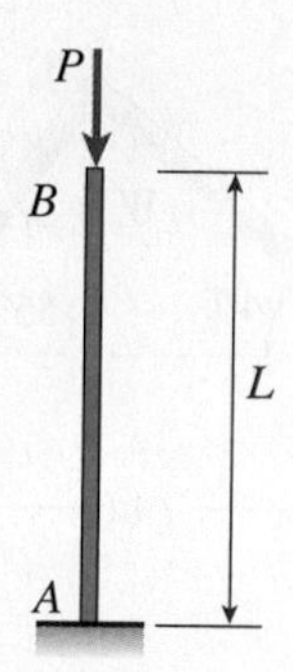

PROBLEM 11.4-2

11.4-3 An aluminum pipe column ($E = 70$ GPa) with a length $L = 3$ m has inside and outside diameters $d_1 = 130$ mm and $d_2 = 150$ mm, respectively (see figure). The column is supported only at the ends and may buckle in any direction.

Calculate the critical load P_{cr} for the following end conditions: (a) pinned-pinned, (b) fixed-free, (c) fixed-pinned, and (d) fixed-fixed.

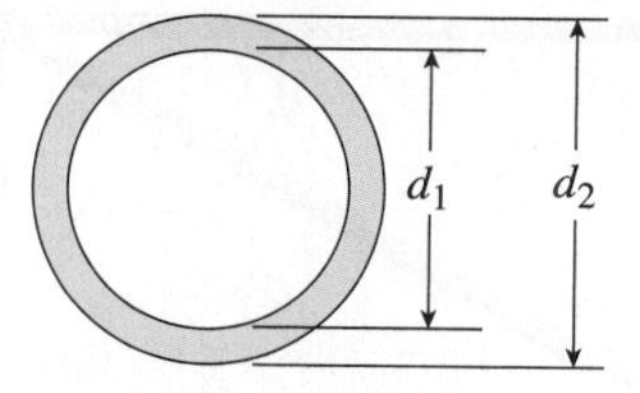

PROBLEMS 11.4-3 and 11.4-4

11.4-4 Solve the preceding problem for a steel pipe column ($E = 210$ GPa) with length $L = 1.2$ m, inner diameter $d_1 = 36$ mm, and outer diameter $d_2 = 40$ mm.

11.4-5 A wide-flange steel column ($E = 200$ GPa) of HE 450A shape (see figure) has a length $L = 9$ m. It is supported only at the ends and may buckle in any direction.

Calculate the allowable load P_{allow} based upon the critical load with a factor of safety $n = 2.5$. Consider the following end conditions: (a) pinned-pinned, (b) fixed-free, (c) fixed-pinned, and (d) fixed-fixed.

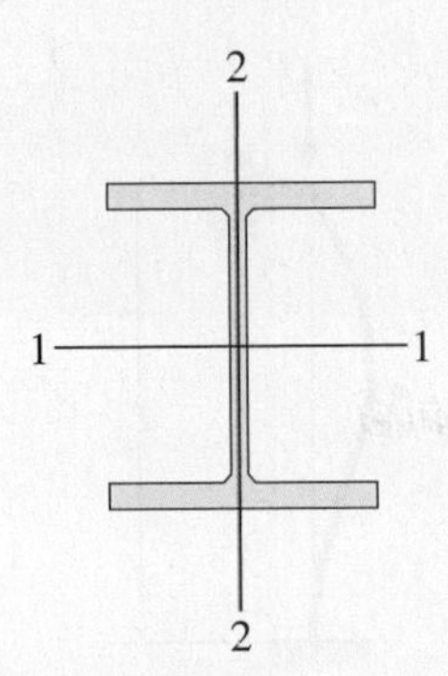

PROBLEMS 11.4-5 and 11.4-6

11.4-6 Solve the preceding problem for a HE 100A shape with length $L = 7.5$ m and $E = 200$ GPa.

Representative Problems

11.4-7 The upper end of an IPN 200 steel column ($E = 200$ GPa) is supported laterally between two pipes (see figure). The pipes are not attached to the column, and friction between the pipes and the column is unreliable. The base of the column provides a fixed support, and the column is 4 m long.

Determine the critical load for the column, considering Euler buckling in the plane of the web and also perpendicular to the plane of the web.

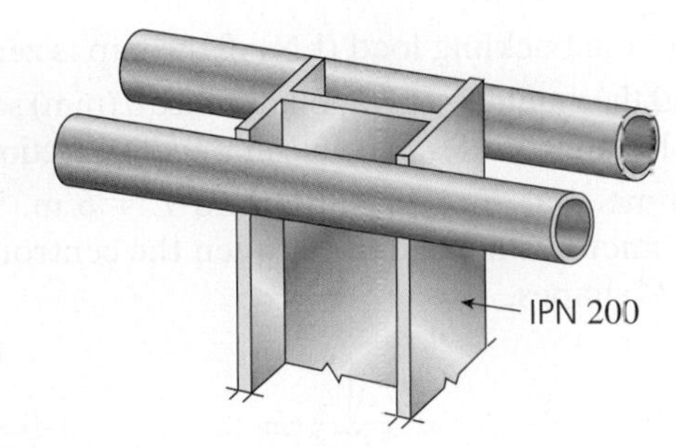

PROBLEM 11.4-7

11.4-8 A vertical post AB is embedded in a concrete foundation and held at the top by two cables (see figure). The post is a hollow steel tube with modulus of elasticity 200 GPa, outer diameter of 40 mm, and thickness of 5 mm. The cables are tightened equally by turnbuckles.

If a factor of safety of 3.0 against Euler buckling in the plane of the figure is desired, what is the maximum allowable tensile force T_{allow} in the cables?

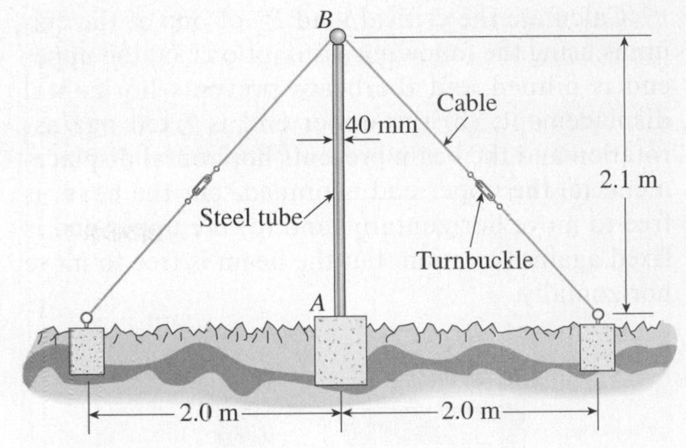

PROBLEM 11.4-8

11.4-9 The horizontal beam ABC shown in the figure is supported by columns BD and CE. The beam is prevented from moving horizontally by the pin support at end A. Each column is pinned at its upper end to the beam, but at the lower ends, support D is a sliding support and support E is pinned. Both columns are solid steel bars ($E = 200$ GPa) of square cross section with width equal to 16 mm. A load Q acts at distance a from column BD.

(a) If the distance $a = 0.5$ m, what is the critical value Q_{cr} of the load?

(b) If the distance a can be varied between 0 and 1.0 m, what is the maximum possible value of Q_{cr}? What is the corresponding value of the distance a?

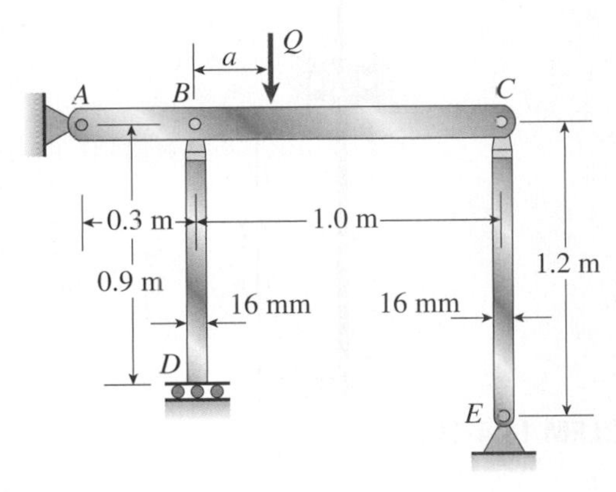

PROBLEM 11.4-9

11.4-10 The roof beams of a warehouse are supported by pipe columns (see figure) having an outer diameter $d_2 = 100$ mm and inner diameter $d_1 = 90$ mm. The columns have a length $L = 4.0$ m, modulus $E = 210$ GPa, and fixed supports at the base.

Calculate the critical load P_{cr} of one of the columns using the following assumptions: (a) the upper end is pinned and the beam prevents horizontal displacement; (b) the upper end is fixed against rotation and the beam prevents horizontal displacement; (c) the upper end is pinned, but the beam is free to move horizontally; and (d) the upper end is fixed against rotation, but the beam is free to move horizontally.

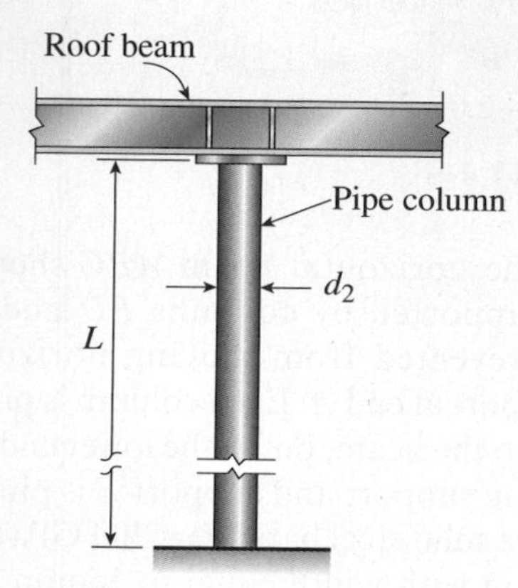

PROBLEM 11.4-10

11.4-11 Determine the critical load P_{cr} and the equation of the buckled shape for an ideal column with ends fixed against rotation (see figure) by solving the differential equation of the deflection curve. (See also Fig. 11-18.)

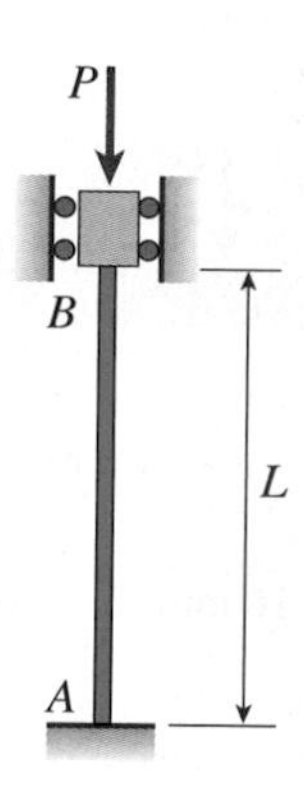

PROBLEM 11.4-11

11.4-12 A fixed-pinned column is a HE 100A steel shape and is designed to carry an axial load of 125 kN. Determine the maximum permissible height L of the column if a factor of safety $n = 2.5$ is required with respect to the buckling of the column. Use $E = 205$ GPa and assume that the proportional limit is 340 MPa. The column may buckle about either axis of the cross section.

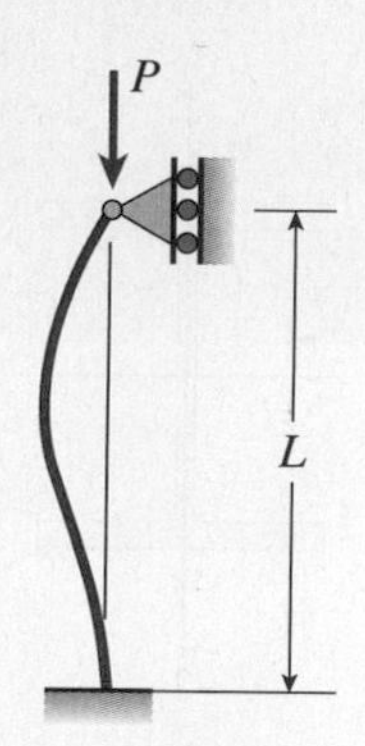

PROBLEM 11.4-12

11.4-13 Find the controlling buckling load (kN) for the steel column shown in the figure. The column is fixed at the base and free at the top and is made up of two UPN 140 shapes that act together. Assume that $E = 205$ GPa and $L = 4.25$ m.

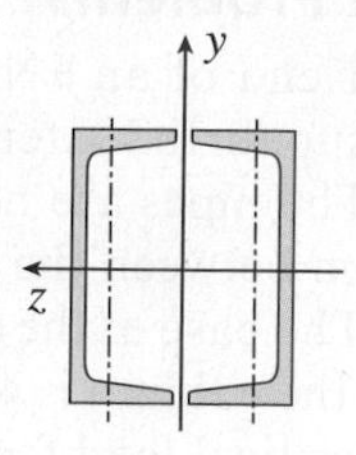

PROBLEM 11.4-13

11.4-14 A column, fixed at the base and free at the top, is made up of two UPN 100 steel shapes (see figure) that act together.

(a) Find the buckling load (kN) if the gap is zero.
(b) Find the required separation distance d (mm) so that the buckling load is the same in y and z directions.

Assume that $E = 205$ GPa and $L = 6$ m. Note that distance d is measured between the centroids of the two C shapes.

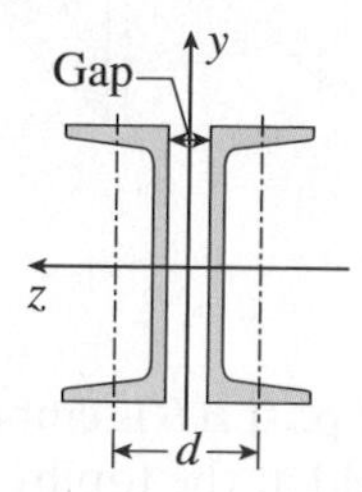

PROBLEM 11.4-14

11.4-15 A rigid L-shaped frame is supported by a steel pipe column AB (see figure) and is subjected to a horizontal load $F = 625$ kN. If the pipe has an outside diameter $d = 100$ mm and a factor of safety of 2.5 is required with respect to Euler buckling, what is the minimum acceptable thickness t of the pipe? Assume that $E = 205$ GPa for AB.

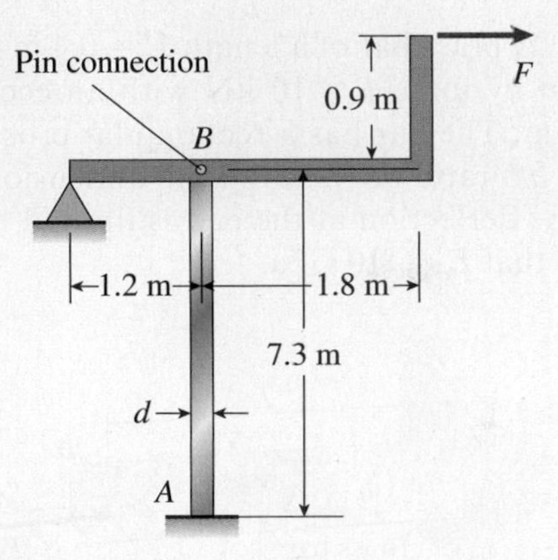

PROBLEM 11.4-15

11.4-16 An aluminum tube AB with a circular cross section has a sliding support at the base and is pinned at the top to a horizontal beam supporting a load $Q = 200$ kN (see figure).

Determine the required thickness t of the tube if its outside diameter d is 200 mm and the desired factor of safety with respect to Euler buckling is $n = 3.0$. (Assume $E = 72$ GPa.)

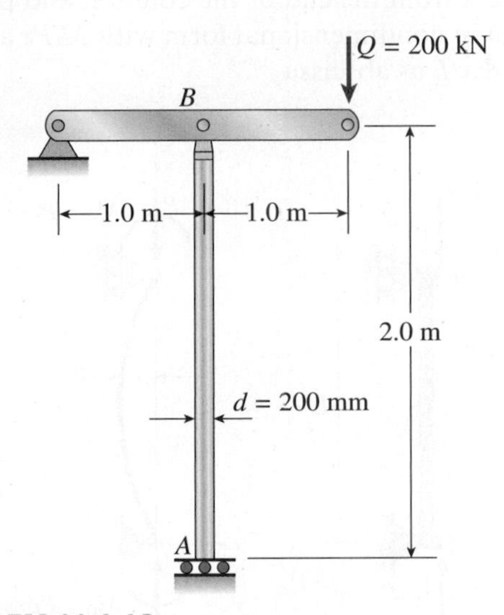

PROBLEM 11.4-16

11.4-17 The frame ABC consists of two members AB and BC that are rigidly connected at joint B, as shown in part a of the figure. The frame has pin supports at A and C. A concentrated load P acts at joint B, thereby placing member AB in direct compression.

To assist in determining the buckling load for member AB, represent it as a pinned-end column, as shown in part b of the figure. At the top of the column, a rotational spring of stiffness β_R represents the restraining action of the horizontal beam BC on the column (note that the horizontal beam provides resistance to rotation of joint B when the column buckles). Also, consider only bending effects in the analysis (i.e., disregard the effects of axial deformations).

(a) By solving the differential equation of the deflection curve, derive the buckling equation for this column:

$$\frac{\beta_R L}{EI}(kL \cot kL - 1) - k^2L^2 = 0$$

in which L is the length of the column and EI is its flexural rigidity.

(b) For the particular case when member BC is identical to member AB, the rotational stiffness β_R equals $3EI/L$ (see Case 7, Table H-2, Appendix H). For this special case, determine the critical load P_{cr}.

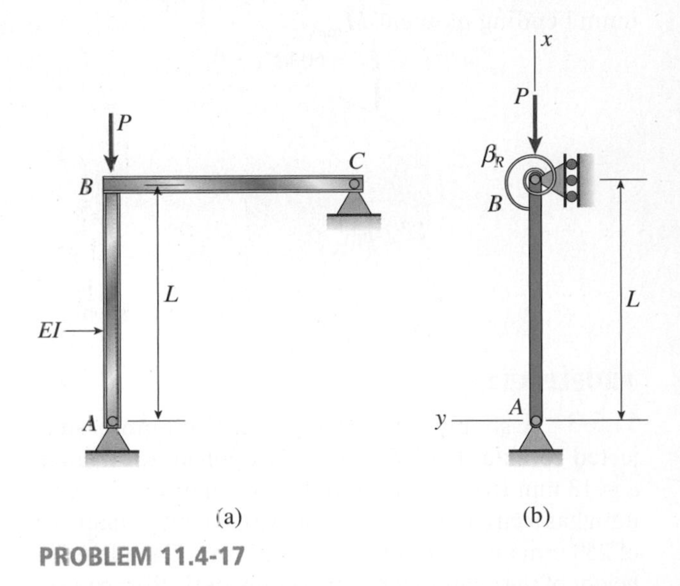

PROBLEM 11.4-17

11.5 Columns with Eccentric Axial Loads

To solve the problems for Section 11.5, assume that bending occurs in the principal plane containing the eccentric axial load.

Introductory Problems

11.5-1 An aluminum bar having a rectangular cross section (50 mm × 25 mm) and length $L = 1.0$ m is compressed by axial loads that have a resultant $P = 12.5$ kN acting at the midpoint of the long side of the cross section (see figure).

Assuming that the modulus of elasticity E is equal to 70 GPa and that the ends of the bar are pinned, calculate the maximum deflection δ and the maximum bending moment M_{max}.

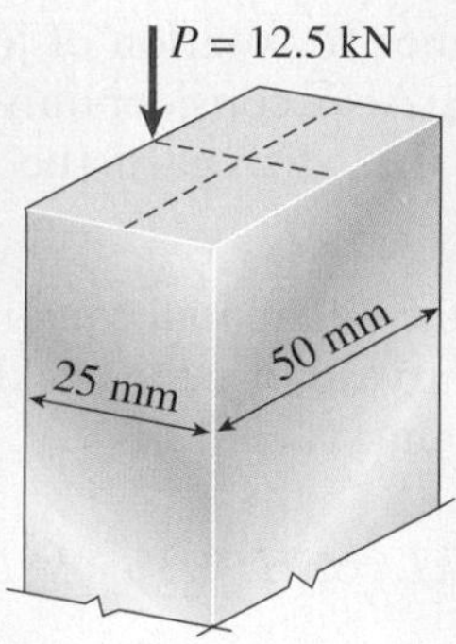

PROBLEM 11.5-1

11.5-2 A steel bar having a square cross section (50 mm × 50 mm) and length $L = 2.0$ m is compressed by axial loads that have a resultant $P = 60$ kN acting at the midpoint of one side of the cross section (see figure).

Assuming that the modulus of elasticity E is equal to 210 GPa and that the ends of the bar are pinned, calculate the maximum deflection δ and the maximum bending moment M_{max}.

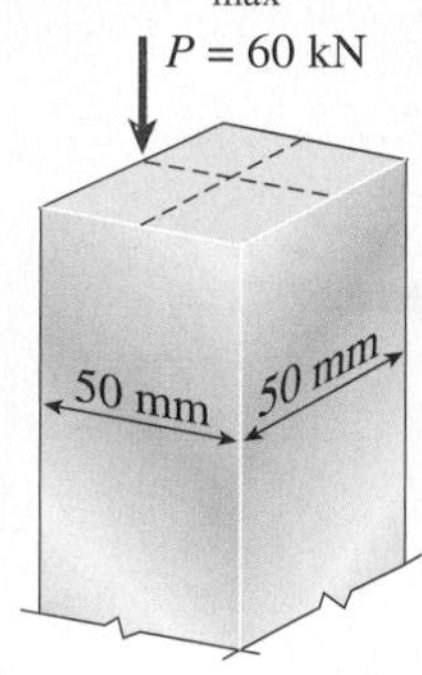

PROBLEM 11.5-2

11.5-3 A simply supported slender column is subjected to axial load $P = 775$ kN applied at distance $e = 13$ mm from joints A and B (see figure). The column has a circular cross section with an outer diameter of 250 mm and wall thickness of 13 mm. Calculate the height of the column if the maximum deflection caused by the axial loads is 2.5 mm. Assume that $E = 200$ GPa.

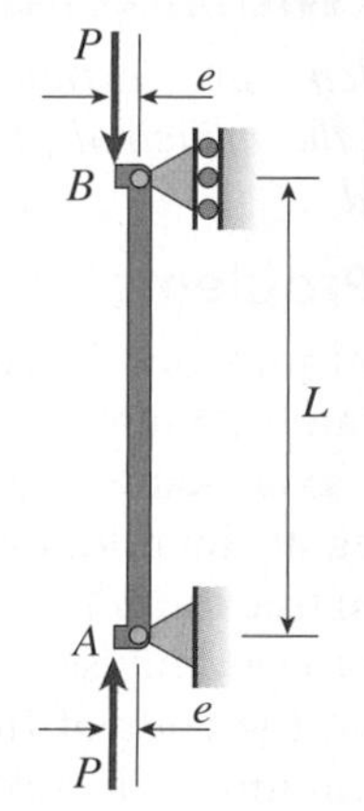

PROBLEM 11.5-3

11.5-4 A brass bar of a length $L = 0.4$ m is loaded at end B by force $P = 10$ kN with an eccentricity $e = 6$ mm. The bar has a rectangular cross section with an h/b ratio of 1.5. Find the dimensions of the bar if the deflection at the end is limited to 4 mm. Assume that $E = 110$ GPa.

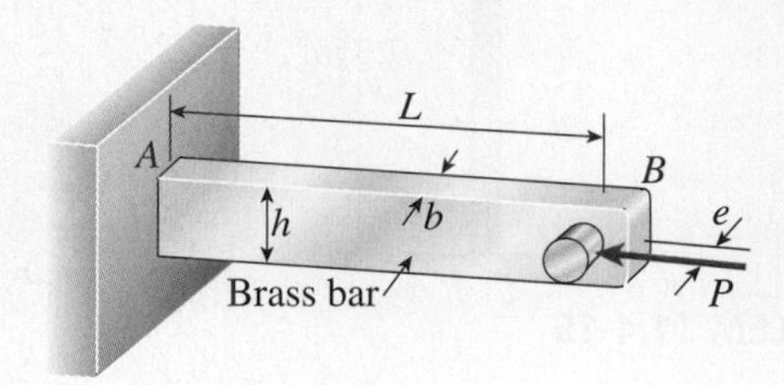

PROBLEM 11.5-4

Representative Problems

11.5-5 Determine the bending moment M in the pinned-end column with eccentric axial loads shown in the figure. Then plot the bending-moment diagram for an axial load $P = 0.3P_{cr}$.

Note: Express the moment as a function of the distance x from the end of the column, and plot the diagram in nondimensional form with M/Pe as ordinate and x/L as abscissa.

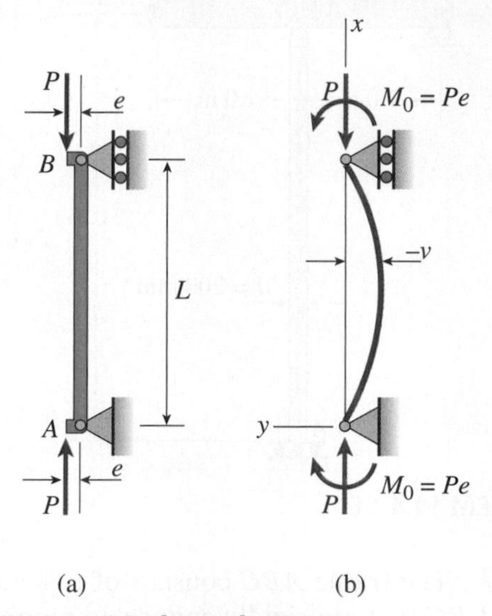

PROBLEMS 11.5-5 through 11.5-7

11.5-6 Plot the load-deflection diagram for a pinned-end column with eccentric axial loads (see figure) if the eccentricity e of the load is 5 mm and the column has a length $L = 3.6$ m, moment of inertia $I = 9.0 \times 10^6$ mm^4, and modulus of elasticity $E = 210$ GPa.

Note: Plot the axial load as ordinate and the deflection at the midpoint as abscissa.

11.5-7 Solve the preceding problem for a column with $e = 5$ mm, $L = 4$ m, $I = 935$ cm^4, and $E = 70$ GPa.

11.5-8 An IPN 200 steel beam is compressed by axial loads that have a resultant P acting at the point shown in the figure. The member has modulus of elasticity $E = 200$ GPa and pinned conditions at the ends. Lateral supports prevent any bending about the weak axis of the cross section.

If the length of the member is 6.2 m and the deflection is limited to 6.5 mm, what is the maximum allowable load P_{allow}?

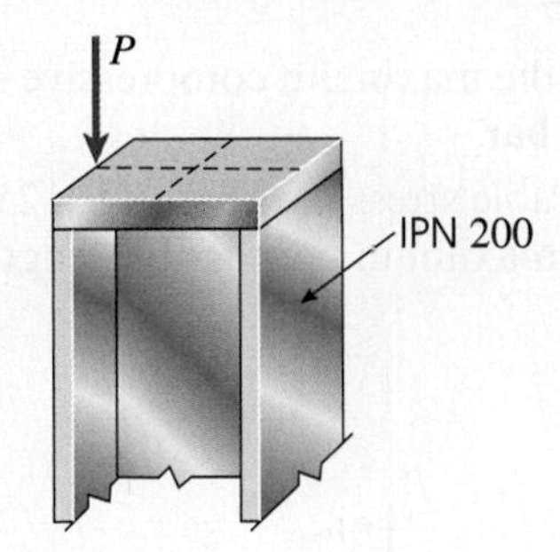

PROBLEM 11.5-8

11.5-9 An IPN 340 steel beam is compressed by axial loads that have a resultant $P = 90$ kN acting at the point shown in the figure. The material is steel with modulus of elasticity $E = 200$ GPa. Assuming pinned-end conditions, determine the maximum permissible length L_{max} if the deflection is not to exceed 1/400th of the length.

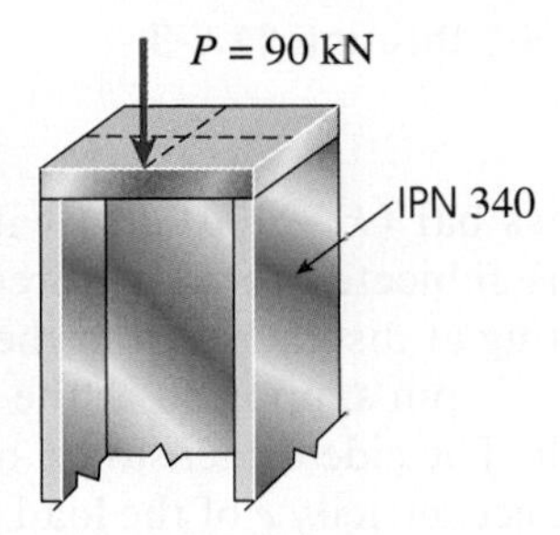

PROBLEMS 11.5-9 and 11.5-10

11.5-10 Solve the preceding problem (HE 160B) if the resultant force P equals 110 kN and $E = 200$ GPa.

11.5-11 The column shown in the figure is fixed at the base and free at the upper end. A compressive load P acts at the top of the column with an eccentricity e from the axis of the column.

Beginning with the differential equation of the deflection curve, derive formulas for the maximum deflection δ of the column and the maximum bending moment M_{max} in the column.

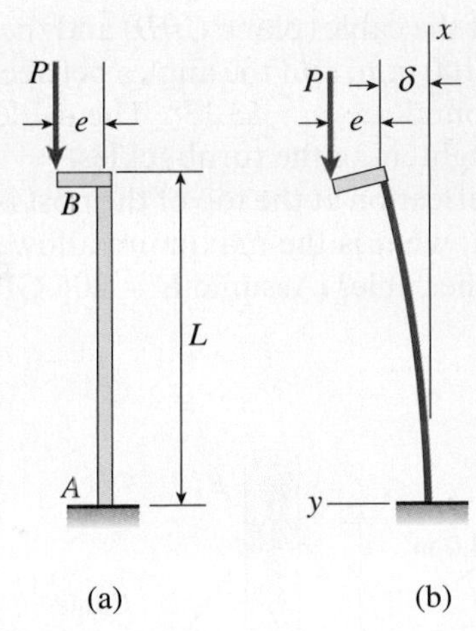

PROBLEM 11.5-11

11.5-12 An aluminum box column with a square cross section is fixed at the base and free at the top (see figure). The outside dimension b of each side is 100 mm and the thickness t of the wall is 8 mm. The resultant of the compressive loads acting on the top of the column is a force $P = 50$ kN acting at the outer edge of the column at the midpoint of one side.

What is the longest permissible length L_{max} of the column if the deflection at the top is not to exceed 30 mm? (Assume $E = 73$ GPa.)

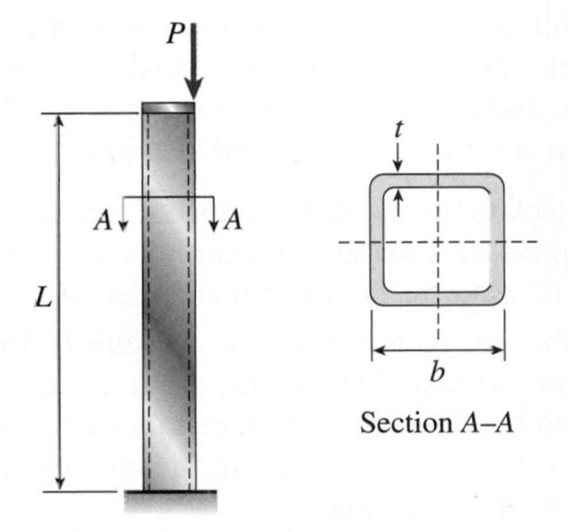

PROBLEMS 11.5-12 and 11.5-13

11.5-13 Solve the preceding problem for a cast iron column with $b = 150$ mm, $t = 12$ mm, $P = 110$ kN, and $E = 90$ GPa. The deflection at the top is limited to 50 mm.

11.5-14 A steel post AB with a hollow circular cross section is fixed at the base and free at the top (see figure). The inner and outer diameters are $d_1 = 96$ mm and $d_2 = 110$ mm, respectively, and the length is $L = 4.0$ m.

A cable CBD passes through a fitting that is welded to the side of the post. The distance between the plane of the cable (plane CBD) and the axis of the post is $e = 100$ mm, and the angles between the cable and the ground are $\alpha = 53.13°$. The cable is pretensioned by tightening the turnbuckles.

If the deflection at the top of the post is limited to $\delta = 20$ mm, what is the maximum allowable tensile force T in the cable? (Assume $E = 205$ GPa.)

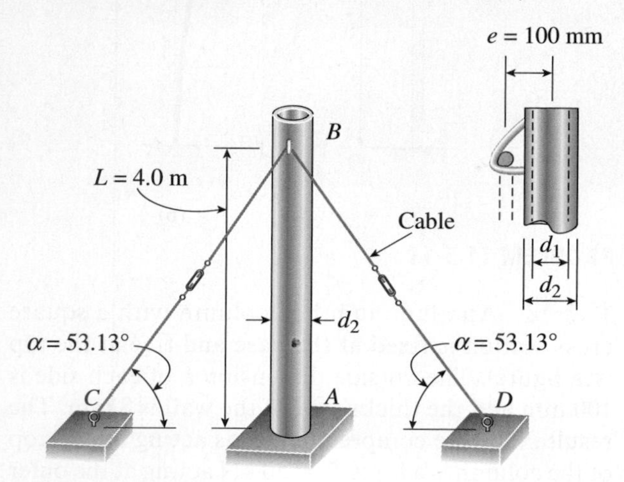

PROBLEM 11.5-14

11.5-15 A frame $ABCD$ is constructed of steel wide-flange members (HE 140A; $E = 200$ GPa) and subjected to triangularly distributed loads of maximum intensity q_0 acting along the vertical members (see figure). The distance between supports is $L = 6$ m and the height of the frame is $h = 1.2$ m. The members are rigidly connected at B and C.

(a) Calculate the intensity of load q_0 required to produce a maximum bending moment of 9 kN·m in the horizontal member BC.

(b) If the load q_0 is reduced to one-half of the value calculated in part (a), what is the maximum bending moment in member BC? What is the ratio of this moment to the moment of 9 kN·m in part (a)?

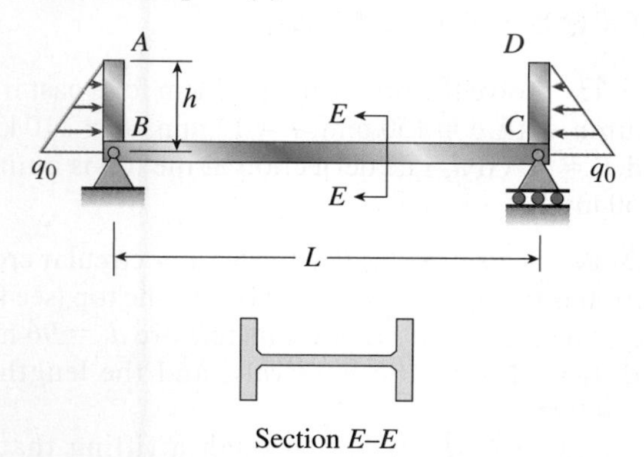

PROBLEM 11.5-15

11.6 The Secant Formula for Columns

To solve the problems for Section 11.6, assume that bending occurs in the principal plane containing the eccentric axial load.

Introductory Problems

11.6-1 A steel bar has a square cross section of width $b = 50$ mm (see figure). The bar has pinned supports at the ends and is 1 m long. The axial forces acting at the end of the bar have a resultant $P = 80$ kN located at distance $e = 20$ mm from the center of the cross section. Also, the modulus of elasticity of the steel is 200 GPa.

(a) Determine the maximum compressive stress σ_{max} in the bar.

(b) If the allowable stress in the steel is 125 MPa, what is the maximum permissible length L_{max} of the bar?

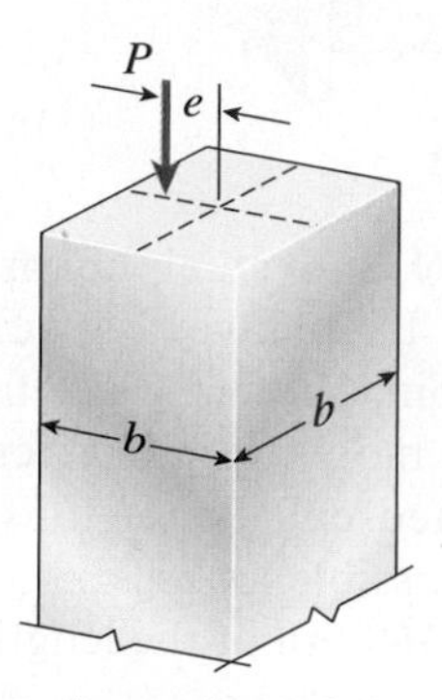

PROBLEMS 11.6-1 through 11.6-3

11.6-2 A brass bar ($E = 100$ GPa) with a square cross section is subjected to axial forces having a resultant P acting at distance e from the center (see figure). The bar is pin supported at the ends and is 0.6 m in length. The side dimension b of the bar is 30 mm and the eccentricity e of the load is 10 mm.

If the allowable stress in the brass is 150 MPa, what is the allowable axial force P_{allow}?

11.6-3 A square aluminum bar with pinned ends carries a load $P = 120$ kN acting at distance $e = 50$ mm from the center (see figure). The bar has a length $L = 1.5$ m and modulus of elasticity $E = 70$ GPa.

If the stress in the bar is not to exceed 42 MPa, what is the minimum permissible width b_{min} of the bar?

11.6-4 A pinned-end column of a length $L = 2.1$ m is constructed of steel pipe ($E = 210$ GPa) having an inside diameter $d_1 = 60$ mm and outside diameter

$d_2 = 68$ mm (see figure). A compressive load $P = 10$ kN acts with eccentricity $e = 30$ mm.

(a) What is the maximum compressive stress σ_{max} in the column?

(b) If the allowable stress in the steel is 50 MPa, what is the maximum permissible length L_{max} of the column?

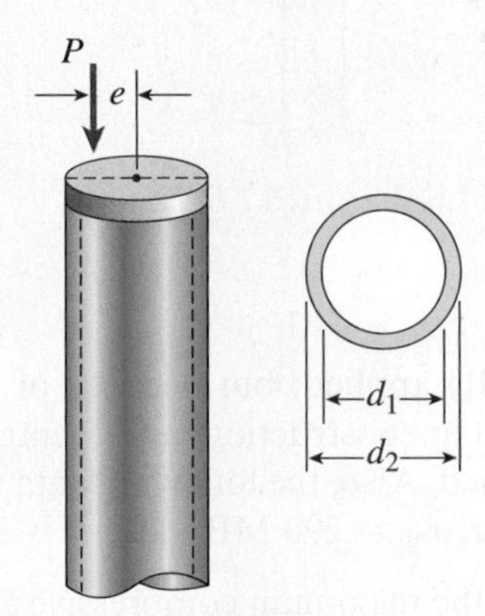

PROBLEMS 11.6-4 through 11.6-6

11.6-5 A pinned-end strut of a length $L = 1.6$ m is constructed of steel pipe ($E = 200$ GPa) having an inside diameter $d_1 = 50$ mm and outside diameter $d_2 = 56$ mm (see figure). A compressive load $P = 10$ kN is applied with eccentricity $e = 25$ mm.

(a) What is the maximum compressive stress σ_{max} in the strut?

(b) What is the allowable load P_{allow} if a factor of safety $n = 2$ with respect to yielding is required? (Assume that the yield stress σ_Y of the steel is 300 MPa.)

11.6-6 A circular aluminum tube with pinned ends supports a load $P = 18$ kN acting at a distance $e = 50$ mm from the center (see figure). The length of the tube is 3.5 m, and its modulus of elasticity is 73 GPa.

If the maximum permissible stress in the tube is 20 MPa, what is the required outer diameter d_2 if the ratio of diameters is to be $d_1 / d_2 = 0.9$?

Representative Problems

11.6-7 A steel HE 240A column is pin-supported at the ends. The column carries an axial load $P = 665$ kN with eccentricity $e = 75$ mm (see figure). Find the length of the column if the maximum stress is restricted to the proportional limit $\sigma_{pl} = 345$ MPa and a factor of safety of 2.0 is assumed. Use modulus of elasticity $E = 200$ GPa.

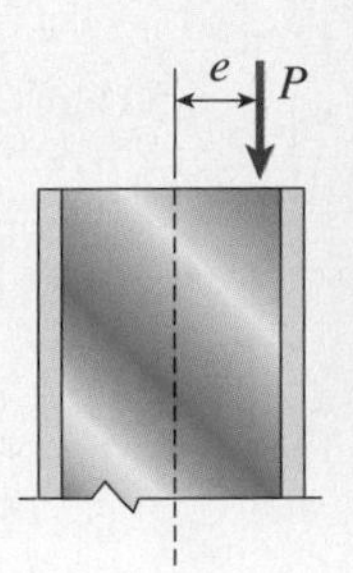

PROBLEM 11.6-7

11.6-8 A steel HE 260A column is pin-supported at the ends and has a length $L = 4$ m. The column supports two eccentrically applied loads $P_1 = 750$ kN and $P_2 = 500$ kN (see figure). Bending occurs about axis 1–1 of the cross section, and the eccentric loads are applied on axis 2–2 at distances of 200 mm from centroid C. Calculate the maximum compressive stress in the column. Assume that modulus elasticity $E = 200$ GPa.

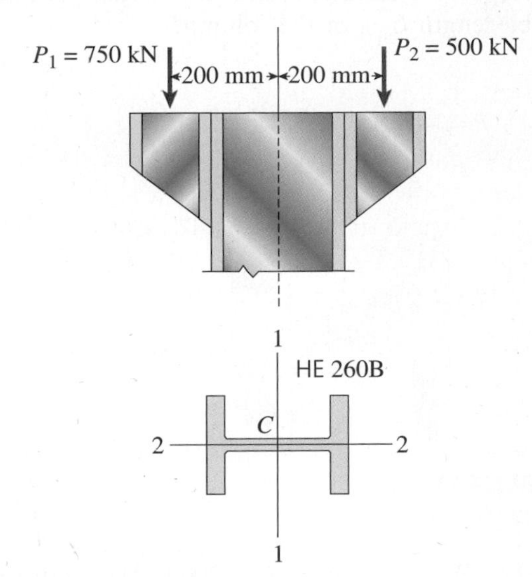

PROBLEM 11.6-8

11.6-9 A steel column ($E = 200$ GPa) with pinned ends is constructed of a HE 260B wide-flange shape (see figure). The column is 7 m long. The resultant of the axial loads acting on the column is a force P acting with an eccentricity $e = 50$ mm.

(a) If $P = 500$ kN, determine the maximum compressive stress σ_{max} in the column.

(b) Determine the allowable load P_{allow} if the yield stress is $\sigma_Y = 300$ MPa and the factor of safety with respect to yielding of the material is $n = 2.5$.

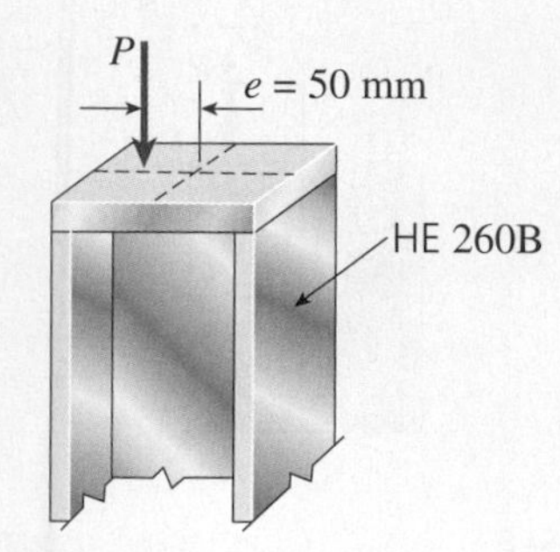

PROBLEM 11.6-9

11.6-10 An IPN 450 steel column is compressed by a force $P = 340$ kN acting with an eccentricity $e = 38$ mm, as shown in the figure. The column has pinned ends and a length L. Also, the steel has a modulus of elasticity $E = 200$ GPa and yield stress $\sigma_Y = 250$ MPa.

(a) If the length $L = 3$ m, what is the maximum compressive stress σ_{max} in the column?
(b) If a factor of safety $n = 2.0$ is required with respect to yielding, what is the longest permissible length L_{max} of the column?

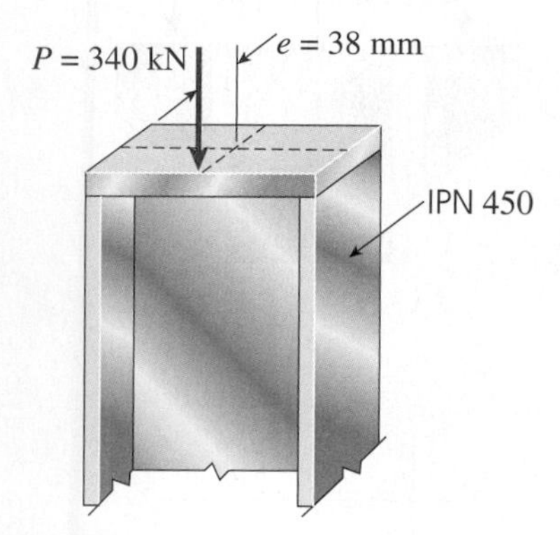

PROBLEM 11.6-10

11.6-11 A steel column ($E = 200$ GPa) that is fixed at the base and free at the top is constructed of a HE 180B wide-flange member (see figure). The column is 3 m long. The force P acting at the top of the column has an eccentricity $e = 32$ mm.

(a) If $P = 180$ kN, what is the maximum compressive stress in the column?
(b) If the yield stress is 240 MPa and the required factor of safety with respect to yielding is 2.1, what is the allowable load P_{allow}?

11.6-12 A HE 240A wide-flange steel column with length $L = 3.8$ m is fixed at the base and free at the top (see figure). The load P acting on the column is intended

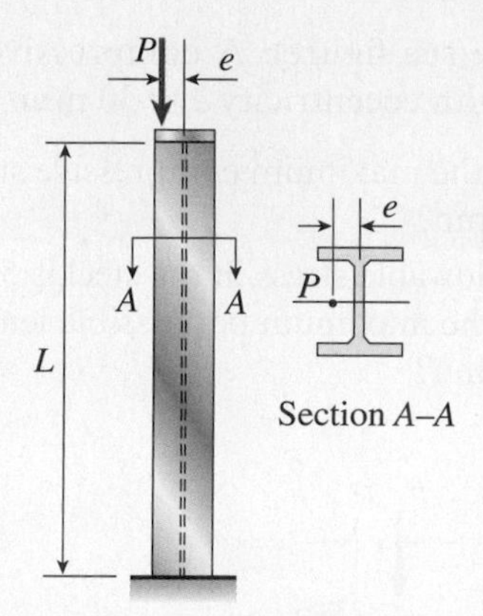

PROBLEMS 11.6-11 and 11.6-12

to be centrally applied, but because of unavoidable discrepancies in construction, an eccentricity ratio of 0.25 is specified. Also, the following data are supplied: $E = 200$ GPa, $\sigma_Y = 290$ MPa, and $P = 310$ kN.

(a) What is the maximum compressive stress σ_{max} in the column?
(b) What is the factor of safety n with respect to yielding of the steel?

11.6-13 A pinned-end column with a length $L = 6$ m is constructed from a HE 320B wide-flange shape (see figure). The column is subjected to a centrally applied load $P_1 = 800$ kN and an eccentrically applied load $P_2 = 350$ kN. The load P_2 acts at a distance $s = 125$ mm from the centroid of the cross section. The properties of the steel are $E = 200$ GPa and $\sigma_Y = 290$ MPa.

(a) Calculate the maximum compressive stress in the column.
(b) Determine the factor of safety with respect to yielding.

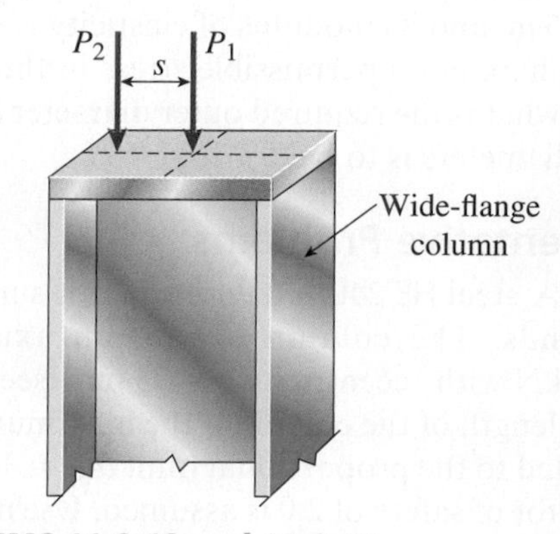

PROBLEMS 11.6-13 and 11.6-14

11.6-14 The wide-flange, pinned-end column shown in the figure carries two loads: a force $P_1 = 450$ kN acting at the centroid and a force $P_2 = 270$ kN acting at a distance $s = 100$ mm from the centroid. The column is a HE 240B shape with $L = 4.2$ m, $E = 200$ GPa, and $\sigma_Y = 290$ MPa.

(a) What is the maximum compressive stress in the column?

(b) If the load P_1 remains at 450 kN, what is the largest permissible value of the load P_2 in order to maintain a factor of safety of 2.0 with respect to yielding?

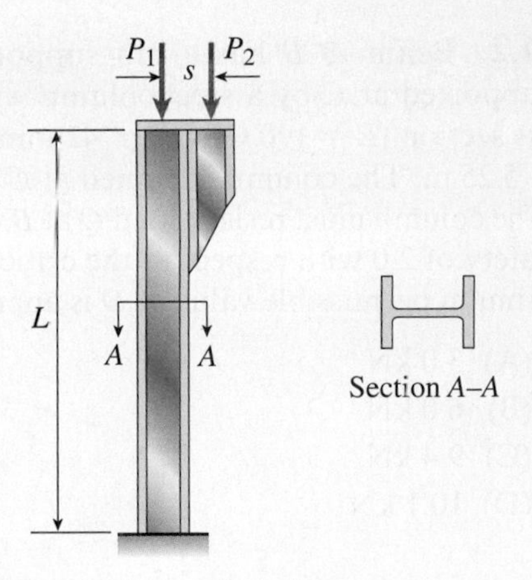

PROBLEMS 11.6-15 and 11.6-16

11.6-15 A HE 320 A wide-flange column of a length $L = 4.5$ m is fixed at the base and free at the top (see figure). The column supports a centrally applied load $P_1 = 530$ kN and a load $P_2 = 180$ kN supported on a bracket. The distance from the centroid of the column to the load P_2 is $s = 300$ mm. Also, the modulus of elasticity is $E = 200$ GPa, and the yield stress is $\sigma_Y = 250$ MPa.

(a) Calculate the maximum compressive stress in the column.

(b) Determine the factor of safety with respect to yielding.

11.6-16 A wide-flange column with a bracket is fixed at the base and free at the top (see figure). The column supports a load $P_1 = 340$ kN acting at the centroid and a load $P_2 = 110$ kN acting on the bracket at a distance $s = 250$ mm from the load P_1. The column is a HE 240A shape with $L = 5$ m, $E = 200$ GPa, and $\sigma_Y = 290$ MPa.

(a) What is the maximum compressive stress in the column?

(b) If the load P_1 remains at 340 kN, what is the largest permissible value of the load P_2 in order to maintain a factor of safety of 1.8 with respect to yielding?

SOME ADDITIONAL REVIEW PROBLEMS: Chapter 11

R-11.1 Beam ACB has a sliding support at A and is supported at C by a pinned-end steel column with square cross section ($E = 200$ GPa, $b = 40$ mm) and height $L = 3.75$ m. The column must resist a load Q at B with a factor of safety of 2.0 with respect to the critical load. The maximum permissible value of Q is approximately:

(A) 10.5 kN
(B) 11.8 kN
(C) 13.2 kN
(D) 15.0 kN

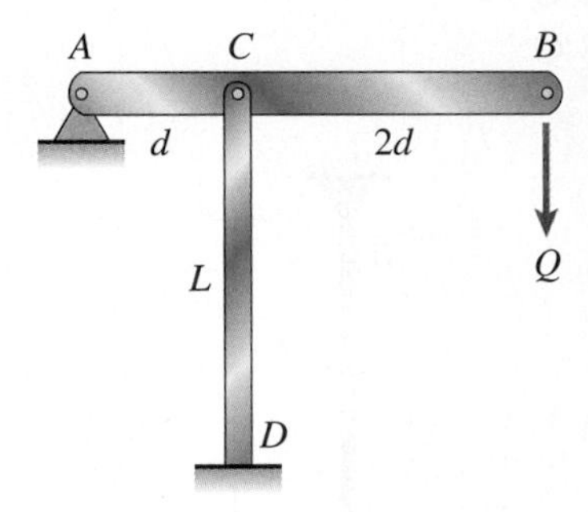

R-11.2 Beam ACB has a pin support at A and is supported at C by a steel column with a square cross section (E = 190 GPa, b = 42 mm) and height L = 5.25 m. The column is pinned at C and fixed at D. The column must resist a load Q at B with a factor of safety of 2.0 with respect to the critical load. The maximum permissible value of Q is approximately:

(A) 3.0 kN
(B) 6.0 kN
(C) 9.4 kN
(D) 10.1 kN

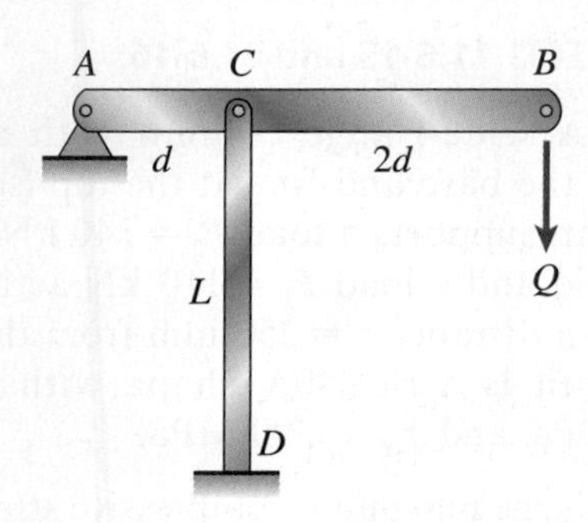

R-11.3 A steel pipe column (E = 190 GPa, $\alpha = 14 \times 10^{-6}/°C$, d_2 = 82 mm, and d_1 = 70 mm) of length L = 4.25 m is subjected to a temperature increase ΔT. The column is pinned at the top and fixed at the bottom. The temperature increase at which the column will buckle is approximately:

(A) 36°C
(B) 42°C
(C) 54°C
(D) 58°C

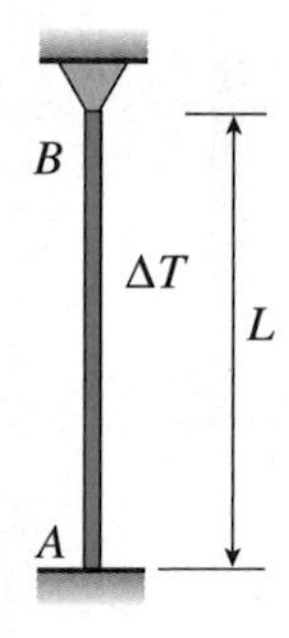

R-11.4 A steel pipe (E = 190 GPa, $\alpha = 14 \times 10^{-6}/°C$, d_2 = 82 mm, and d_1 = 70 mm) of length L = 4.25 m hangs from a rigid surface and is subjected to a temperature increase ΔT = 50 °C. The column is fixed at the top and has a small gap at the bottom. To avoid buckling, the minimum clearance at the bottom should be approximately:

(A) 2.55 mm
(B) 3.24 mm
(C) 4.17 mm
(D) 5.23 mm

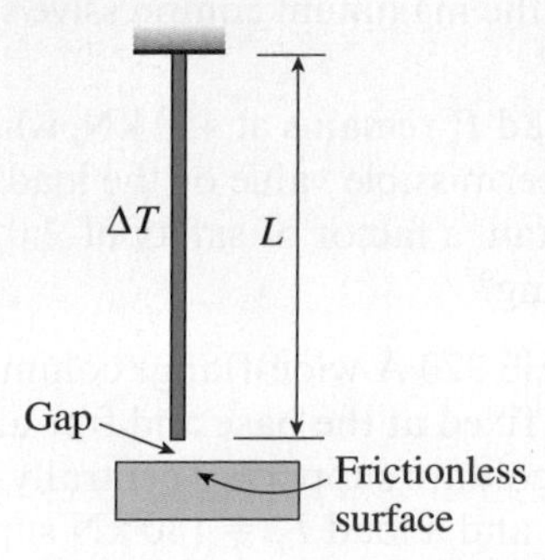

R-11.5 A pinned-end copper strut (E = 110 GPa) with length L = 1.6 m is constructed of circular tubing with outside diameter d = 38 mm. The strut must resist an axial load P = 14 kN with a factor of safety of 2.0 with respect to the critical load. The required thickness t of the tube is approximately:

(A) 2.75 mm
(B) 3.15 mm
(C) 3.89 mm
(D) 4.33 mm

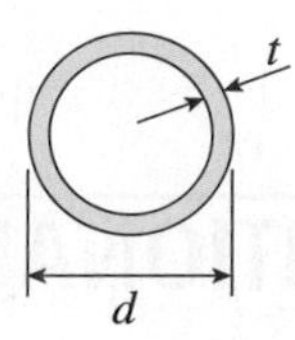

R-11.6 A plane truss composed of two steel pipes (E = 210 GPa, d = 100 mm and wall thickness = 6.5 mm is subjected to vertical load W at joint B. Joints A and C are L = 7 m apart. The critical value of load W for buckling in the plane of the truss is nearly:

(A) 138 kN
(B) 146 kN
(C) 153 kN
(D) 164 kN

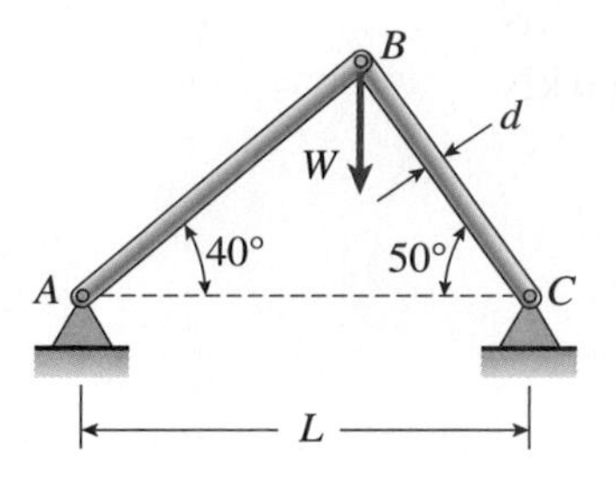

R-11.7 A beam is pin connected to the tops of two identical pipe columns, each of height h, in a frame. The frame is restrained against sidesway at the top of column 1. Only buckling of columns 1 and 2 in the plane of the frame is of interest here. The ratio (a/L) defining the placement of load Q_{cr}, which causes both columns to buckle simultaneously, is approximately:

(A) 0.25
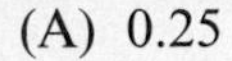
(B) 0.33
(C) 0.67
(D) 0.75

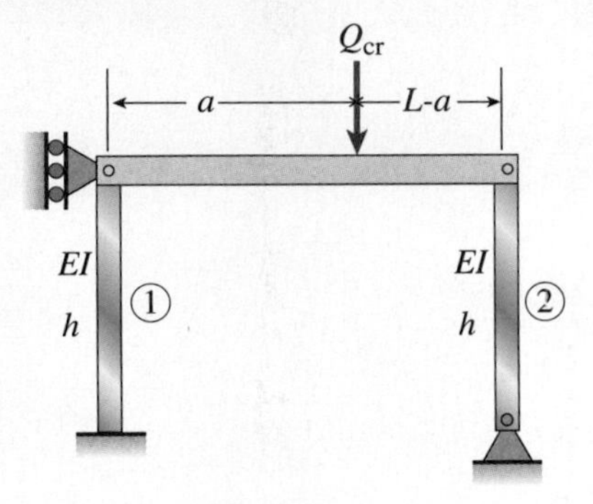

R-11.8 A steel pipe column (E = 210 GPa) with length L = 4.25 m is constructed of circular tubing with outside diameter d_1 = 90 mm and inner diameter d_2 = 64 mm. The pipe column is fixed at the base and pinned at the top and may buckle in any direction. The Euler buckling load of the column is most nearly:

(A) 303 kN
(B) 560 kN
(C) 690 kN
(D) 720 kN

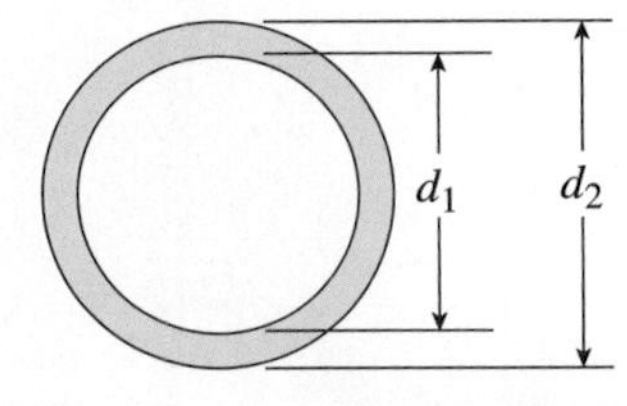

R-11.9 An aluminum tube (E = 72 GPa) AB of circular cross section has a pinned support at the base and is pin-connected at the top to a horizontal beam supporting a load Q = 600 kN. The outside diameter of the tube is 200 mm and the desired factor of safety with respect to Euler buckling is 3.0. The required thickness t of the tube is most nearly:

(A) 8 mm
(B) 10 mm
(C) 12 mm
(D) 14 mm

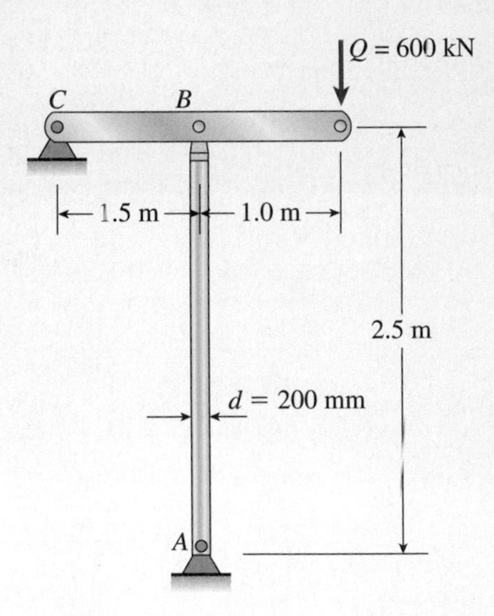

R-11.10 Two pipe columns are required to have the same Euler buckling load P_{cr}. Column 1 has flexural rigidity EI and height L_1; column 2 has flexural rigidity $(4/3)EI$ and height L_2. The ratio L_2/L_1 at which both columns will buckle under the same load is approximately:

(A) 0.55
(B) 0.72
(C) 0.81
(D) 1.10

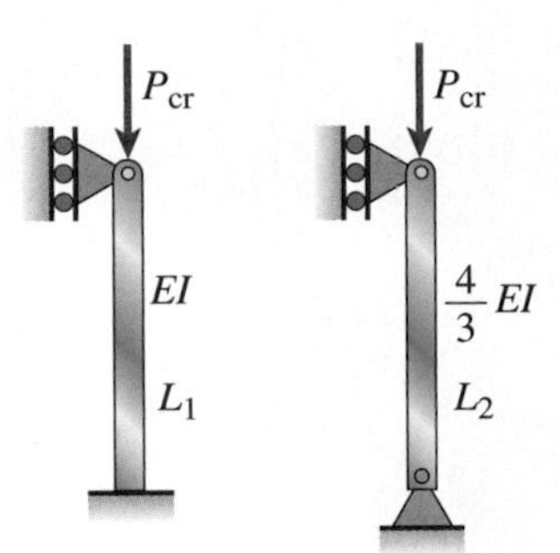

R-11.11 Two pipe columns are required to have the same Euler buckling load P_{cr}. Column 1 has flexural rigidity EI_1 and height L; column 2 has flexural rigidity $(2/3)EI_2$ and height L. The ratio I_2/I_1 at which both columns will buckle under the same load is approximately:

(A) 0.8
(B) 1.0
(C) 2.2
(D) 3.1

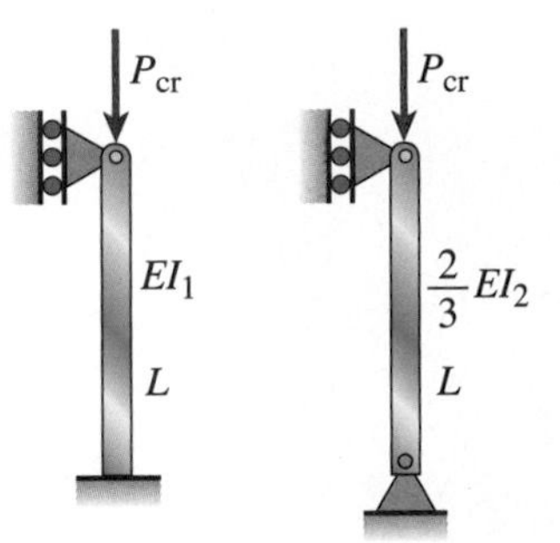

REFERENCES AND HISTORICAL NOTES

1-1 Timoshenko, S. P., *History of Strength of Materials*, Dover Publications, Inc., New York, 1983 (originally published by McGraw-Hill Book Co., Inc., New York, 1953).

S. P. Timoshenko (1878–1972)

Courtesy of Stanford University Archives

Note: Stephen P. Timoshenko (1878–1972) was a famous scientist, engineer, and teacher. Born in Russia, he came to the United States in 1922. He was a researcher with the Westinghouse Research Laboratory, a professor at the University of Michigan, and later a professor at Stanford University, where he retired in 1944.

Timoshenko made many original contributions, both theoretical and experimental, to the field of applied mechanics, and he wrote twelve pioneering textbooks that revolutionized the teaching of mechanics in the United States. These books, which were published in as many as five editions and translated into as many as 35 languages, covered the subjects of statics, dynamics, mechanics of materials, vibrations, structural theory, stability, elasticity, plates, and shells.

1-2 Todhunter, I., and Pearson, K., *A History of the Theory of Elasticity and of the Strength of Materials*, Vols. I and II, Dover Publications, Inc., New York, 1960 (originally published by the Cambridge University Press in 1886 and 1893). *Note*: Isaac Todhunter (1820–1884) and Karl Pearson (1857–1936) were English mathematicians and educators. Pearson was especially noteworthy for his original contributions to statistics.

1-3 Love, A. E. H., *A Treatise on the Mathematical Theory of Elasticity*, 4th Ed., Dover Publications, Inc., New York, 1944 (originally published by the Cambridge University Press in 1927); see "Historical Introduction," pp. 1–31.

Note: Augustus Edward Hough Love (1863–1940) was an outstanding English elastician who taught at Oxford University. His many important investigations included the analysis of seismic surface waves, now called *Love waves* by geophysicists.

1-4 Jacob Bernoulli (1654–1705), also known by the names James, Jacques, and Jakob, was a member of the famous family of mathematicians and scientists of Basel, Switzerland (see Ref. 9-1). He did important work in connection with elastic curves of beams. Bernoulli also developed polar coordinates and became famous for his work in theory of probability, analytic geometry, and other fields.

Jean Victor Poncelet (1788–1867) was a Frenchman who fought in Napoleon's campaign against Russia and was given up for dead on the battlefield. He survived, was taken prisoner, and later returned to France to continue his work in mathematics. His major contributions to mathematics are in geometry; in mechanics he is best known for his work on properties of materials and dynamics. (For the work of Bernoulli and Poncelet in connection with stress-strain diagrams, see Ref. 1-1, p. 88, and Ref. 1-2, Vol. I, pp. 10, 533, and 873.)

1-5 James and James, *Mathematics Dictionary*, Van Nostrand Reinhold, New York (latest edition).

1-6 Robert Hooke (1635–1703) was an English scientist who performed experiments with elastic bodies and developed improvements in timepieces. He also formulated the laws of gravitation independently of Newton, of whom he was a contemporary. Upon the founding of the Royal Society of London in 1662, Hooke was appointed its first curator. (For the origins of Hooke's law, see Ref. 1-1, pp. 17–20, and Ref. 1-2, Vol. I, p. 5.)

1-7 Thomas Young (1773–1829) was an outstanding English scientist who did pioneering work in optics, sound, impact, and other subjects. (For information about his work with materials, see Ref. 1-1, pp. 90–98, and Ref. 1-2, Vol. I, pp. 80–86.)

Thomas Young (1773–1829)

Photos.com/Thinkstock

1-8 Siméon Denis Poisson (1781–1840) was a great French mathematician. He made many contributions to both mathematics and mechanics, and his name is preserved in numerous ways besides Poisson's ratio. For instance, there is Poisson's equation in partial differential equations and the Poisson distribution in theory of probability. (For information about Poisson's theories of material behavior, see Ref. 1-1, pp. 111–114; Ref. 1-2, Vol. I, pp. 208–318; and Ref. 1-3, p. 13.)

S. D. Poisson (1781–1840)

Bettmann/Contributor/Getty Images

2-1 Timoshenko, S. P., and Goodier, J. N., *Theory of Elasticity*, 3rd Ed., McGraw-Hill Book Co., Inc., New York, 1970 (see p. 110). *Note:* James Norman Goodier (1905–1969) was well known for his research contributions to theory of elasticity, stability, wave propagation in solids, and other branches of applied mechanics. Born in England, he studied at Cambridge University and later at the University of Michigan. He was a professor at Cornell University and subsequently at Stanford University, where he headed the program in applied mechanics.

2-2 Leonhard Euler (1707–1783) was a famous Swiss mathematician, perhaps the greatest mathematician of all time. Ref. 11-1 gives information about his life and works. (For his work on statically indeterminate structures, see Ref. 1-1, p. 36, and Ref. 2-3, p. 650.)

2-3 Oravas, G. A., and McLean, L., "Historical development of energetical principles in elastomechanics," *Applied Mechanics Reviews*, Part I, Vol. 19, No. 8, August 1966, pp. 647–658, and Part II, Vol. 19, No. 11, November 1966, pp. 919–933.

2-4 Louis Marie Henri Navier (1785–1836), a famous French mathematician and engineer, was one of the founders of the mathematical theory of elasticity. He contributed to beam, plate, and shell theory, to theory of vibrations, and to the theory of viscous fluids. (See Ref. 1-1, p. 75; Ref. 1-2, Vol. I, p. 146; and Ref. 2-3, p. 652, for his analysis of statically indeterminate structures.)

2-5 Piobert, G., Morin, A.-J., and Didion, I., "Commission des Principes du Tir," *Mémorial de l'Artillerie*, Vol. 5, 1842, pp. 501–552.

Note: This paper describes experiments made by firing artillery projectiles against iron plating. On page 505 appears the description of the markings that are the slip bands. The description is quite brief, and there is no indication that the authors attributed the markings to inherent material characteristics. Guillaume Piobert (1793–1871) was a French general and mathematician who made many studies of ballistics; when this paper was written, he was a captain in the artillery.

2-6 Lüders, W., "Ueber die Äusserung der elasticität an stahlartigen Eisenstäben und Stahlstäben, und über eine beim Biegen solcher Stäbe beobachtete Molecularbewegung," *Dingler's Polytechnisches Journal*, Vol. 155, 1860, pp. 18–22.

Note: This paper clearly describes and illustrates the bands that appear on the surface of a polished steel specimen during yielding. Of course, these bands are only the surface manifestation of three-dimensional zones of deformation; hence, the zones should probably be characterized as "wedges" rather than bands.

2-7 Benoit Paul Emile Clapeyron (1799–1864) was a famous French structural engineer and bridge designer; he taught engineering at the École des Ponts et Chaussées in Paris. It appears that Clapeyron's theorem, which states that the work of the external loads acting on a linearly elastic body is equal to the strain energy, was first published in 1833. (See Ref. 1-1, pp. 118 and 288; Ref. 1-2, Vol. I, p. 578; and Ref. 1-2, Vol. II, p. 418.)

2-8 Poncelet investigated longitudinal vibrations of a bar due to impact loads (see Ref. 1-1, p. 88). See Ref. 1-4 for additional information about his life and works.

2-9 Budynas, R., and Young, W. C., *Roark's Formulas for Stress and Strain*, McGraw-Hill Book Co., Inc., New York, 2002.

2-10 Barré de Saint-Venant (1797–1886) is generally recognized as the most outstanding elastician of all time. Born near Paris, he studied briefly at the École Polytechnique and later graduated from the École des Ponts et Chaussées. His later professional career suffered greatly from his refusal, as a matter of conscience and politics, to join his schoolmates in preparing for the defense of Paris in March 1814, just prior to Napoleon's abdication. As a consequence, his achievements received greater recognition in other countries than they did in France.

Some of his most famous contributions are the formulation of the fundamental equations of elasticity and the development of the exact theories of bending and torsion. He also developed theories for plastic deformations and vibrations. His full name was Adéhmar Jean Claude Barré, Count de Saint-Venant. (See Ref. 1-1, pp. 229–242; Ref. 1-2, Vol. I, pp. 833–872, Vol. II, Part I, pp. 1–286, Vol. II, Part II, pp. 1–51; and Ref. 2-1, pp. 39–40.)

2-11 Zaslavsky, A., "A note on Saint-Venant's principle," *Israel Journal of Technology*, Vol. 20, 1982, pp. 143–144.

2-12 Ramberg, W. A., and Osgood, W. R., "Description of stress-strain curves by three parameters," *National Advisory Committee for Aeronautics*, Technical Note No. 902, July 1943.

3-1 The relationship between torque and angle of twist in a circular bar was correctly established in 1784 by Charles Augustin de Coulomb (1736–1806), a famous French scientist (see Ref. 1-1, pp. 51–53, 82, and 92, and Ref. 1-2, Vol. I, p. 69). Coulomb made contributions in electricity and magnetism, viscosity of fluids, friction, beam bending, retaining walls and arches, torsion and torsional vibrations, and other subjects (see Ref. 1-1, pp. 47–54).

Thomas Young (Ref. 1-7) observed that the applied torque is balanced by the shear stresses on the cross section and that the shear stresses are proportional to the distance from the axis. The French engineer Alphonse J. C. B. Duleau (1789–1832) performed tests on bars in torsion and also developed a theory for circular bars (see Ref. 1-1, p. 82).

C. A. de Coulomb (1736–1806)
Bettmann/Contributor/Getty Images

3-2 Bredt, R., "Kritische Bemerkungen zur Drehungselastizität," *Zeitschrift des Vereines Deutscher Ingenieure*, Vol. 40, 1896, pp. 785–790, and 813–817.

Note: Rudolph Bredt (1842–1900) was a German engineer who studied in Karlsruhe and Zürich. Then he worked for a while in Crewe, England, at a train factory, where he learned about the design and construction of cranes. This experience formed the basis for his later work as a crane manufacturer in Germany. His theory of torsion was developed in connection with the design of box-girder cranes.

5-1 A proof of the theorem that cross sections of a beam in pure bending remain plane can be found in the paper by Fazekas, G. A., "A note on the bending of Euler beams," *Journal of Engineering Education*, Vol. 57, No. 5, January 1967. The validity of the theorem has long been recognized, and it was used by early investigators such as Jacob Bernoulli (Ref. 1-4) and L. M. H. Navier (Ref. 2-4). For a discussion of the work done by Bernoulli and Navier in connection with bending of beams, see Ref. 1-1, pp. 25–27 and 70–75.

5-2 Galilei, Galileo, *Dialogues Concerning Two New Sciences*, translated from the Italian and Latin into English by Henry Crew and Alfonso De Salvio, The Macmillan Company, New York, 1933 (translation first published in 1914).

Note: This book was published in 1638 by Louis Elzevir in Leida, now Leiden, Netherlands. *Two New Sciences* represents the culmination of Galileo's work on dynamics and mechanics of materials. It can truly be said that these two subjects began with Galileo and the publication of this famous book.

Galileo Galilei was born in Pisa in 1564. He made many famous experiments and discoveries, including those on falling bodies and pendulums that initiated the science of dynamics. Galileo was an eloquent lecturer and attracted students from many countries. He pioneered in astronomy and developed a telescope with which he made many astronomical discoveries, including the mountainous character of the moon, Jupiter's satellites, the phases of Venus, and sunspots. Because his scientific views of the solar system were contrary to theology, he was condemned by the church in Rome and spent the last years of his life in seclusion in Florence; during this period he wrote *Two New Sciences.* Galileo died in 1642 and was buried in Florence.

Galileo Galilei (1564–1642)

5-3 The history of beam theory is described in Ref. 1-1, pp. 11–47 and 135–141, and in Ref. 1-2. Edme Mariotte (1620–1684) was a French physicist who made developments in dynamics, hydrostatics, optics, and mechanics. He made tests on beams and developed a theory for calculating load-carrying capacity; his theory was an improvement on Galileo's work, but still not correct. Jacob Bernoulli (1654–1705), who is described in Ref. 1-4, first determined that the curvature is proportional to the bending moment. However, his constant of proportionality was incorrect.

Leonhard Euler (1707–1783) obtained the differential equation of the deflection curve of a beam and used it to solve many problems of both large and small deflections (Euler's life and work are described in Ref. 11-1). The first person to obtain the distribution of stresses in a beam and correctly relate the stresses to the bending moment probably was Antoine Parent (1666–1716), a French physicist and mathematician. Later, a rigorous investigation of strains and stresses in beams was made by Saint-Venant (1797–1886); see Ref. 2-10. Important contributions were also made by Coulomb (Ref. 3-1) and Navier (Ref. 2-4).

5-4 (*1*) EN 1993 Eurocode 3, *Design of steel structures* (see http://eurocodes.jrc.ec.europa.eu/showpage.php?id=332); (*2*) National Structural Steelwork Specification, 5th ed., The British Constructional Steelwork Association Limited, 4 Whitehall Court, Westminster, London SW1A 2ES (www.steelconstruction.org); (*3*) Steel Construction Manual, Fourteenth Edition, American Institute of Steel Construction, 130 East Randolph Street Suite 2000, Chicago, IL 60601-6219. (For other publications and additional information, go to their website: www.aisc.org.) (*4*) For available steel structural shapes in Europe and elsewhere: http://sections.arcelormittal.com/products-services/products-ranges.html

5-5 (*1*) EN 1999 Eurocode 9, *Design of aluminum structures* (see http://eurocodes.jrc.ec.europa.eu/showpage.php?id=332); (*2*) The Aluminum Association, 1400 Crystal Drive, Suite 430, Arlington, VA 22202. (For other publications and additional information, go to their website: www.aluminum.org.) (*3*) For available aluminum structural shapes in Europe: see http://blecha.at/en/engl.htm

5-6 (*1*) EN 1995 Eurocode 5, *Design of timber structures* (see http://eurocodes.jrc.ec.europa.eu/showpage.php?id=332); also see http://www.eurocodes.co.uk/; (*2*) The Timber Research and Development Association (TRADA): http://www.trada.co.uk/; (*3*) *National Design Specification for Wood Construction* (*ASD/LRFD*), published by the American Wood Council, a division of the American Forest and Paper Association, 1101 K Street, NW, Suite 700, Washington, DC 20005. (For other publications and additional information, go to their websites: www.awc.org and www.afandpa.org.) (*4*) See Appendix G for selected shapes available in Europe.

5-7 D. J. Jourawski (1821–1891) was a Russian bridge and railway engineer who developed the now widely used approximate theory for shear stresses in beams (see Ref. 1-1, pp. 141–144, and Ref. 1-2, Vol. II, Part I, pp. 641–642). In 1844, only two years after graduating from the Institute of Engineers of Ways of Communication in St. Petersburg, he was assigned the task of designing and constructing a major bridge on the first railway line from Moscow to St. Petersburg. He noticed that some of the large timber beams split longitudinally in the centers of the cross sections, where he knew the bending stresses were zero. Jourawski drew free-body diagrams and quickly discovered the existence of horizontal shear stresses in the beams. He derived the shear formula and applied his theory to various shapes of beams. Jourawski's paper on shear in beams is cited in Ref. 5-8. His name is sometimes transliterated as Dimitrii Ivanovich Zhuravskii.

5-8 Jourawski, D. J., "Sur la résistance d'un corps prismatique . . .," *Annales des Ponts et Chaussés*, Mémoires et Documents, 3rd Series, Vol. 12, Part 2, 1856, pp. 328–351.

5-9 Zaslavsky, A., "On the limitations of the shearing stress formula," *International Journal of Mechanical Engineering Education*, Vol. 8, No. 1, 1980, pp. 13–19. (See also Ref. 2-1, pp. 358–359.)

5-10 Maki, A. C., and Kuenzi, E. W., "Deflection and stresses of tapered wood beams," Research Paper FPL 34, U. S. Forest Service, Forest Products Laboratory, Madison, Wisconsin, September 1965, 54 pages.

6-1 Timoshenko, S. P., "Use of stress functions to study flexure and torsion of prismatic bars," (in Russian), St. Petersburg, 1913 (reprinted in Vol. 82 of the *Memoirs of the Institute of Ways of Communication*, pp. 1–21).

Note: In this paper, the point in the cross section of a beam through which a concentrated force should act in order to eliminate rotation was found. Thus, this work contains the first determination of a shear center. The particular beam under investigation had a solid semicircular cross section (see Ref. 2-1, pp. 371–373).

7-1 Augustin Louis Cauchy (1789–1857) was one of the greatest mathematicians. Born in Paris, he entered the École Polytechnique at the age of 16, where he studied under Lagrange, Laplace, Fourier, and Poisson. He was quickly recognized for his mathematical prowess, and at age 27 he became a professor at the École and a member of the Academy of Sciences. His major works in pure mathematics were in group theory, number theory, series, integration, differential equations, and analytical functions.

In applied mathematics, Cauchy introduced the modern day concept of stress, developed the equations of theory of elasticity, and introduced the notion of principal stresses and principal strains (see Ref. 1-1, pp. 107–111). An entire chapter is devoted to his work on theory of elasticity in Ref. 1-2 (see Vol. I, pp. 319–376).

7-2 See Ref. 1-1, pp. 229–242. *Note*: Saint-Venant was a pioneer in many aspects of theory of elasticity, and Todhunter and Pearson dedicated their book, *A History of the Theory of Elasticity* (Ref. 1-2), to him. For further information about Saint-Venant, see Ref. 2-10.

7-3 William John Macquorn Rankine (1820–1872) was born in Edinburgh, Scotland, and taught engineering at Glasgow University. He derived the stress transformation equations in 1852 and made many other contributions to theory of elasticity and applied mechanics (see Ref. 1-1, pp. 197–202, and Ref. 1-2, Vol. II, Part I, pp. 86 and 287–322). His engineering subjects included arches, retaining walls, and structural theory.

Rankine also achieved scientific fame for his work with fluids, light, sound, and behavior of crystals, and he is especially well known for his contributions to molecular physics and thermodynamics. His name is preserved by the Rankine cycle in thermodynamics and the Rankine absolute temperature scale.

7-4 The famous German civil engineer Otto Christian Mohr (1835–1918) was both a theoretician and a practical designer. He was a professor at the Stuttgart Polytechnikum and later at the Dresden Polytechnikum. He developed the circle of stress in 1882 (Ref. 7-5 and Ref. 1-1, pp. 283–288).

Mohr made numerous contributions to the theory of structures, including the Williot-Mohr diagram for truss displacements, the moment-area method for beam deflections, and the Maxwell-Mohr method for analyzing statically indeterminate structures. (*Note:* Joseph Victor Williot, 1843–1907, was a French engineer, and James Clerk Maxwell, 1831–1879, was a famous British scientist.)

7-5 Mohr, O., "Über die Darstellung des Spannungszustandes und des Deformationszustandes eines Körperelementes," Zivilingenieur, 1882, p. 113.

8-1 Karl Culmann (1821–1881) was a famous German bridge and railway engineer. In 1849–1850 he spent two years traveling in England and the United States to study bridges, which he later wrote about in Germany. He designed numerous bridge structures in Europe, and in 1855 he became professor of structures at the newly organized Zürich Polytechnicum. Culmann made many developments in graphical methods and wrote the first book on graphic statics, published in Zürich in 1866. Stress trajectories are one of the original topics presented in this book (see Ref. 1-1, pp. 190–197).

9-1 The work of Jacob Bernoulli, Euler, and many others with respect to elastic curves is described in Ref. 1-1, pp. 27 and 30–36, and Ref. 1-2. Another member of the Bernoulli family, Daniel Bernoulli (1700–1782), proposed to Euler that he obtain the differential equation of the deflection curve by minimizing the strain energy, which Euler did. Daniel Bernoulli, a nephew of Jacob Bernoulli, is renowned for his work in hydrodynamics, kinetic theory of gases, beam vibrations, and other subjects. His father, John Bernoulli (1667–1748), a younger brother of Jacob, was an equally famous mathematician and scientist who first formulated the principle of virtual displacements, and solved the problem of the brachystochrone.

Jacob Bernoulli (1654–1705)

Universal History Archive/Contributor/Getty Images

John Bernoulli established the rule for obtaining the limiting value of a fraction when both the numerator and denominator tend to zero. He communicated this last rule to G. F. A. de l'Hôpital (1661–1704), a French nobleman who wrote the first book on calculus (1696) and included this theorem, which consequently became known as *L'Hôpital's rule.*

Daniel's nephew, Jacob Bernoulli (1759–1789), also known as James or Jacques, was a pioneer in the theory of plate bending and plate vibrations.

Much interesting information about the many prominent members of the Bernoulli family, as well as other pioneers in mechanics and mathematics, can be found in books on the history of mathematics.

9-2 Castigliano, A., *Théorie de l'équilibre des systèmes élastiques et ses applications*, A. F. Negro, Turin, 1879, 480 pages.

Note: In this book, Castigliano presented in very complete form many fundamental concepts and principles of structural analysis. Although Castigliano was Italian, he wrote this book in French in order to gain a wider audience for his work. It was translated into both German and English (Refs. 9-3 and 9-4). The English translation was republished in 1966 by Dover Publications and is especially valuable because of the introductory material by Gunhard A. Oravas (Refs. 9-5 and 9-6).

Castigliano's first and second theorems appear on pp. 15–16 of the 1966 edition of his book. He identified them as Part 1 and Part 2 of the "Theorem of the Differential Coefficients of the Internal Work." In mathematical form, they appear in his book as

$$F_p = \frac{dW_i}{dr_p} \quad \text{and} \quad r_p = \frac{dW_i}{dF_p}$$

where W_i is the internal work (or strain energy), F_p represents any one of the external forces, and r_p is the displacement of the point of application of F_p.

Castigliano did not claim complete originality for the first theorem, although he stated in the Preface to his book that his presentation and proof were more general than anything published previously. The second theorem was original with him and was part of his thesis for the civil engineering degree at the Polytechnic Institute of Turin in 1873.

Carlo Alberto Pio Castigliano was born of a poor family in Asti in 1847 and died of pneumonia in 1884, while at the height of his productivity. The story of his life is told by Oravas in the introduction to the 1966 edition, and a bibliography of Castigliano's works and a list of his honors and awards are also given there. His contributions are also documented in Refs. 2-3 and 1-1. He used the name Alberto Castigliano when signing his writings.

9-3 Hauff, E., *Theorie des Gleichgewichtes elastischer Systeme und deren Anwendung*, Carl Gerold's Sohn, Vienna, 1886. (A translation of Castigliano's book, Ref. 9-2.)

9-4 Andrews, E. S., *Elastic Stresses in Structures*, Scott, Greenwood and Son, London, 1919. (A translation of Castigliano's book, Ref. 9-2.)

9-5 Castigliano, C. A. P., *The Theory of Equilibrium of Elastic Systems and Its Applications*, translated by E. S. Andrews, with a new introduction and biographical portrait by G. A. Oravas, Dover Publications, Inc., New York, 1966. (A republication of Ref. 9-4 but with the addition of historical material by Oravas.)

9-6 Oravas, G. A., "Historical Review of Extremum Principles in Elastomechanics," an introductory section (pp. xx–xlvi) of the book, *The Theory of Equilibrium of Elastic Systems and Its Applications*, by C. A. P. Castigliano, translated by E. S. Andrews, Dover Publications, Inc., New York, 1966 (Ref. 9-5).

9-7 Macaulay, W. H., "Note on the deflection of beams," *The Messenger of Mathematics*, vol. XLVIII, May 1918–April 1919, Cambridge, 1919, pp. 129–130.

Note: William Herrick Macaulay, 1853–1936, was an English mathematician and Fellow of King's College, Cambridge. In this paper he defined "by $\{f(x)\}_a$ a function of x which is zero when x is less than a and equal to $f(x)$ when x is equal to or greater than a." Then he showed how to use this function when finding beam deflections. Unfortunately, he did not give any references to the earlier work of Clebsch and Föppl; see Refs. 9-8 through 9-10.

9-8 Clebsch, A., *Theorie der Elasticität fester Körper*, B. G. Teubner, Leipzig, 1862, 424 pages. (Translated into French and annotated by Saint-Venant, *Théorie de l'Élasticité des Corps Solides*, Paris, 1883. Saint-Venant's notes increased Clebsch's book threefold in size.)

Note: The method of finding beam deflections by integrating across points of discontinuity was presented first in this book; see Ref. 1-1, pp. 258–259 and Ref. 9-10. Rudolf Friedrich Alfred Clebsch, 1933–1872, was a German mathematician and scientist. He was a professor of engineering at the Karlsruhe Polytechnicum and later a professor of mathematics at Göttingen University.

9-9 Föppl, A., *Vorlesungen über technische Mechanik*, Vol. III: Festigkeitslehre, B. G. Teubner, Leipzig, 1897.

Note: In this book, Föppl extended Clebsch's method for finding beam deflections. August Föppl, 1854–1924, was a German mathematician and engineer. He was a professor at the University of Leipzig and later at the Polytechnic Institute of Munich.

9-10 Pilkey, W. D., "Clebsch's method for beam deflections," *Journal of Engineering Education*, vol. 54, no. 5, January 1964, pp. 170–174. This paper describes Clebsch's method and gives a very complete historical account, with many references.

10-1 Zaslavsky, A., "Beams on immovable supports," *Publications of the International Association for Bridge and Structural Engineering*, Vol. 25, 1965, pp. 353–362.

11-1 Euler, L., "Methodus inveniendi lineas curvas maximi minimive proprietate gaudentes . . .," Appendix I, "De curvis elasticis," Bousquet, Lausanne and Geneva, 1744. (English translation: Oldfather, W. A., Ellis, C. A., and Brown, D. M., *Isis*, Vol. 20, 1933, pp. 72–160. Also, republished in *Leonhardi Euleri Opera Omnia*, series 1, Vol. 24, 1952.)

Note: Leonhard Euler (1707–1783) made many remarkable contributions to mathematics and mechanics, and he is considered by most mathematicians to be the most productive mathematician of all time. His name, pronounced "oiler," appears repeatedly in present-day textbooks; for instance, in mechanics there are Euler's equations of motion of a rigid body, Euler's angles, Euler's equations of fluid flow, the Euler load in column buckling, and much more; and in mathematics there is the famous Euler constant, as well as Euler's numbers, the Euler identity ($e^{i\theta} = \cos\theta + i\sin\theta$), Euler's formula ($e^{i\pi} + 1 = 0$), Euler's differential equation, Euler's equation of a variational problem, Euler's quadrature formula, the Euler summation formula, Euler's theorem on homogeneous functions, Euler's integrals, and even Euler squares (square arrays of numbers possessing special properties).

Leonhard Euler (1707–1783)

In applied mechanics, Euler was the first to derive the formula for the critical buckling load of an ideal, slender column and the first to solve the problem of the elastica. This work was published in 1744, as cited previously. He dealt with a column that is fixed at the base and free at the top. Later, he extended his work on columns (Ref. 11-2). Euler's numerous books include treatises on celestial mechanics, dynamics, and hydromechanics, and his papers include subjects such as vibrations of beams and plates and statically indeterminate structures.

In the field of mathematics, Euler made outstanding contributions to trigonometry, algebra, number theory, differential and integral calculus, infinite series, analytic geometry, differential equations, calculus of variations, and many other subjects. He was the first to conceive of trigonometric values as the ratios of numbers and the first to present the famous equation $e^{i\theta} = \cos\theta + i\sin\theta$. Within his books on mathematics, all of which were classical references for many generations, are found the first development of the calculus of variations as well as such intriguing items as the proof of Fermat's "last theorem" for $n = 3$ and $n = 4$. Euler also solved the famous problem of the seven bridges of Königsberg, which is a problem of topology, another field he pioneered.

Euler was born near Basel, Switzerland, and attended the University of Basel, where he studied under John Bernoulli (1667–1748). From 1727 to 1741 he lived and worked in St. Petersburg, where he established a great reputation as a mathematician. In 1741 he moved to Berlin upon the invitation of Frederick the Great, King of Prussia. He continued his mathematical research in Berlin until the year 1766, when he returned to St. Petersburg at the request of Catherine II, Empress of Russia.

Euler continued to be prolific until his death in St. Petersburg at the age of 76; during this final period of his life he wrote more than 400 papers. In his entire lifetime, the number of books and papers written by Euler totaled 886; he left many manuscripts at his death and they continued to be published by the Russian Academy of Sciences in St. Petersburg for 47 years afterward. All this in spite of the fact that one of his eyes went blind in 1735 and the other in 1766. The story of Euler's life is told in Ref. 1-1, pp. 28–30, and some of his contributions to mechanics are described in Ref. 1-1, pp. 30–36 (see also Refs. 1-2, 1-3, 2-2, and 5-3).

11-2 Euler, L., "Sur la force des colonnes," *Histoire de L'Académie Royale des Sciences et Belles Lettres*, 1757, published in *Memoires* of the Academie, Vol. 13, Berlin, 1759, pp. 252–282. (See Ref. 11-3 for a translation and discussion of this paper.)

11-3 Van den Broek, J. A., "Euler's classic paper 'On the strength of columns,'" *American Journal of Physics*, Vol. 15, No. 4, July–August 1947, pp. 309–318.

11-4 Keller, J. B., "The shape of the strongest column," *Archive for Rational Mechanics and Analysis*, Vol. 5, No. 4, 1960, pp. 275–285.

11-5 Young, D. H., "Rational design of steel columns," *Transactions of the American Society of Civil Engineers*, Vol. 101, 1936, pp. 422–451. *Note:* Donovan Harold Young (1904–1980) was a well-known engineering educator. He was a professor at the University of Michigan and later at Stanford University. His five textbooks in the field of applied mechanics, written with S. P. Timoshenko, were translated into many languages and used throughout the world.

11-6 Lamarle, A. H. E., "Mémoire sur la flexion du bois," *Annales des Travaux Publiques de Belgique*, Part 1, Vol. 3, 1845, pp. 1–64, and Part 2, Vol. 4, 1846, pp. 1–36. *Note:* Anatole Henri Ernest Lamarle (1806–1875) was an engineer and professor. He was born in Calais, studied in Paris, and became a professor at the University of Ghent, Belgium. For his work on columns, see Ref. 1-1, p. 208.

11-7 Considère, A., "Résistance des pièces comprimées," *Congrès International des Procédés de Construction*, Paris, September 9–14, 1889, proceedings published by Librairie Polytechnique, Paris, Vol. 3, 1891, p. 371. *Note:* Armand Gabriel Considère (1841–1914) was a French engineer.

11-8 Engesser, F., "Ueber die Knickfestigkeit gerader Stäbe," *Zeitschrift für Architektur und Ingenieurwesen*, Vol. 35, No. 4, 1889, pp. 455–462. *Note:* Friedrich Engesser (1848–1931) was a German railway and bridge engineer. Later, he became a professor at the Karlsruhe Polytechnical Institute, where he made important advances in the theory of structures, especially in buckling and energy methods. For his work on columns, see Ref. 1-1, pp. 292 and 297–299.

11-9 Engesser, F., "Knickfragen," *Schweizerische Bauzeitung*, Vol. 25, No. 13, March 30, 1895, pp. 88–90.

11-10 Jasinski, F., "Noch ein Wort zu den 'Knickfragen,'" *Schweizerische Bauzeitung*, Vol. 25, No. 25, June 22, 1895, pp. 172–175. *Note:* Félix S. Jasinski (1856–1899) was born in Warsaw and studied in Russia. He became a professor at the Institute of Engineers of Ways of Communication in St. Petersburg.

11-11 Engesser, F., "Ueber Knickfragen," *Schweizerische Bauzeitung*, Vol. 26, No. 4, July 27, 1895, pp. 24–26.

11-12 von Kármán, T., "Die Knickfestigkeit gerader Stäbe," *Physikalische Zeitschrift*, Vol. 9, No. 4, 1908, pp. 136–140 (this paper also appears in Vol. I of Ref. 11–14).

Note: Theodore von Kármán (1881–1963) was born in Hungary and later worked at the University of Göttingen in the field of aerodynamics. After coming to the United States in 1929, he founded the Jet Propulsion Laboratory and pioneered in aircraft and rocket problems. His research also included inelastic buckling of columns and stability of shells.

11-13 von Kármán, T., "Untersuchungen über Knickfestigkeit," *Mitteilungen über Forschungsarbeiten auf dem Gebiete des Ingenieurwesens, Verein Deutscher Ingenieure*, Berlin, Heft 81, 1910 (this paper also appears in Ref. 11-14).

11-14 *Collected Works of Theodore von Kármán*, Vols. I–IV, Butterworths Scientific Publications, London, 1956.

11-15 Shanley, F. R., "The column paradox," *Journal of the Aeronautical Sciences*, Vol. 13, No. 12, December 1946, p. 678. *Note:* Francis Reynolds Shanley (1904–1968) was a professor of aeronautical engineering at the University of California, Los Angeles.

11-16 Shanley, F. R., "Inelastic column theory," *ibid.*, Vol. 14, No. 5, May 1947, pp. 261–267.

11-17 Hoff, N. J., "Buckling and stability," The Forty-First Wilbur Wright Memorial Lecture, *Journal of the Royal Aeronautical Society*, Vol. 58, January 1954, pp. 3–52.

11-18 Hoff, N. J., "The idealized column," *Ingenieur-Archiv*, Vol. 28, 1959 (Festschrift Richard Grammel), pp. 89–98.

11-19 Johnston, B. G., "Column buckling theory: Historical highlights," *Journal of Structural Engineering*, Structural Division, American Society of Civil Engineers, Vol. 109, No. 9, September 1983, pp. 2086–2096.

APPENDIX A

Systems of Units

A.1 Systems of Units

Measurement systems have been a necessity since people first began to build and barter, and every ancient culture developed some sort of measurement system to serve its needs. Standardization of units took place gradually over the centuries, often through royal edicts. Development of the **British Imperial System** from earlier measurement standards began in the 13th century and was well established by the 18th century. The British system spread to many parts of the world, including the United States, through commerce and colonization. In the United States, the system gradually evolved into the **U.S. Customary System (USCS)** that is in common use today.

The concept of the **metric system** originated in France about 300 years ago and was formalized in the 1790s (at the time of the French Revolution). France mandated the use of the metric system in 1840, and since then, many other countries have done the same. In 1866, the United States Congress legalized the metric system without making it compulsory.

A new system of units was created when the metric system underwent a major revision in the 1950s. Officially adopted in 1960 and named the **International System of Units** (Système International d'Unités), this newer system is commonly referred to as **SI**. Although some SI units are the same as in the old metric system, SI has many new features and simplifications. Thus, SI is an improved metric system.

Length, time, mass, and force are the basic concepts of mechanics for which units of measurement are needed. However, only three of these quantities are independent, since all four of them are related by Newton's second law of motion:

$$F = ma \qquad \textbf{(A-1)}$$

in which F is the force acting on a particle, m is the mass of the particle, and a is its acceleration. Since acceleration has units of length divided by time squared, all four quantities are involved in the second law.

The International System of Units, like the metric system, is based upon length, time, and mass as fundamental quantities. In these systems, force is derived from Newton's second law. Therefore, the unit of force is expressed in terms of the basic units of length, time, and mass, as shown in the next section.

SI is classified as an **absolute system of units** because measurements of the three fundamental quantities are independent of the locations at which the measurements are made; that is, the measurements do not depend upon the effects of gravity. Therefore, the SI units for length, time, and mass may be used anywhere on earth, in space, on the moon, or even on another planet. This is one of the reasons why the metric system has always been preferred for scientific work.

The British Imperial System and the U.S. Customary System are based upon length, time, and force as the fundamental quantities with mass being derived from the second law. Therefore, in these systems, the unit of mass is expressed in terms of the units of length, time, and force. The unit of force is defined as the force required to give a certain standard mass an acceleration equal to the acceleration of gravity, which means that the unit of force varies with location and altitude. For this reason, these systems are called **gravitational systems of units**. Such systems were the first to evolve, probably because weight is such a readily discernible property and because variations in gravitational attraction were not noticeable. It is clear, however, that in the modern technological world an absolute system is preferable.

A.2 SI Units

The International System of Units has seven **base units** from which all other units are derived. The base units of importance in mechanics are the meter (m) for length, second (s) for time, and kilogram (kg) for mass. Other SI base units pertain to temperature, electric current, amount of substance, and luminous intensity.

The **meter** was originally defined as one ten-millionth of the distance from the North Pole to the equator. Later, this distance was converted to a physical standard, and for many years, the standard for the meter was the distance between two marks on a platinum-iridium bar stored at the headquarters of the International Bureau of Weights and Measures (Bureau International des Poids et Mesures) in Sèvres, which is a suburb of Paris, France.

Because of the inaccuracies inherent in the use of a physical bar as a standard, the definition of the meter was changed in 1983 to the length of the path traveled by light in a vacuum during a time interval of 1/299792458 of a second.[1] The advantages of this "natural" standard are that it is not subject to physical damage and is reproducible at laboratories anywhere in the world.

The **second** was originally defined as 1/86400 of a mean solar day (24 hours equals 86,400 seconds). However, since 1967, a highly accurate atomic clock has set the standard, and a second is now defined to be the duration of 9,192,631,770 periods of the radiation corresponding to the transition between the two hyperfine levels of the ground state of the cesium-133 atom. (Most engineers would probably prefer the original definition over the new one, which hasn't noticeably changed the second but is necessary because the earth's rotation rate is gradually slowing down.)

Of the seven base units in SI, the **kilogram** is the only one that is still defined by a physical object. Since the mass of an object only can be determined by comparing it experimentally with the mass of some other object, a physical standard is needed. For this purpose, a one-kilogram cylinder of platinum-iridium, called the International Prototype Kilogram (IPK), is kept by the International Bureau of Weights and Measures at Sèvres. (At the present time, attempts are

[1]Taking the reciprocal of this number gives the speed of light in a vacuum (299,792,458 meters per second).

being made to define the kilogram in terms of a fundamental constant, such as the Avogadro number, thus removing the need for a physical object.)

Other units used in mechanics, called **derived units**, are expressed in terms of the base units of meter, second, and kilogram. For instance, the unit of **force** is the **newton**, which is defined as the force required to impart an acceleration of one meter per second squared to a mass of one kilogram.[2] Use Newton's second law ($F = ma$) to derive the unit of force in terms of base units:

$$1 \text{ newton} = (1 \text{ kilogram})(1 \text{ meter per second squared})$$

Thus, the newton (N) is given in terms of base units by the formula

$$1 \text{ N} = 1 \text{ kg} \cdot \text{m/s}^2 \tag{A-2}$$

To provide a point of reference, note that a small apple weighs approximately one newton.

The unit of **work** and **energy** is the **joule**, which is defined as the work done when the point of application of a force of one newton is displaced a distance of one meter in the direction of the force.[3] Therefore,

$$1 \text{ joule} = (1 \text{ newton})(1 \text{ meter}) = 1 \text{ newton meter}$$

or

$$1 \text{ J} = 1 \text{ N} \cdot \text{m} \tag{A-3}$$

When you raise this book from desktop to eye level, you do about one joule of work, and when you walk up one flight of stairs, you do about 200 joules of work.

The names, symbols, and formulas for SI units of importance in mechanics are listed in Table A-1. Some of the derived units have special names, such as newton, joule, hertz, watt, and pascal. These units are named for notable persons in science and engineering and have symbols (N, J, Hz, W, and Pa) that are capitalized, although the unit names themselves are written in lowercase letters. Other derived units have no special names (for example, the units of acceleration, area, and density) and must be expressed in terms of base units and other derived units.

The relationships between various SI units and some commonly used metric units are given in Table A-2. Metric units such as dyne, erg, gal, and micron are no longer recommended for engineering or scientific use.

The **weight** of an object is the **force of gravity** acting on that object; therefore, weight is measured in newtons. Since the force of gravity depends upon altitude and position on the earth, weight is not an invariant property of a body. Furthermore, the weight of a body as measured by a spring scale is affected not only by the gravitational pull of the earth but also by the centrifugal effects associated with the rotation of the earth.

As a consequence, two kinds of weight, **absolute weight** and **apparent weight** must be recognized. The former is based upon the force of gravity alone, and the latter includes the effects of rotation. Thus, apparent weight is always less than absolute weight (except at the poles). Apparent weight, which is the weight of an object as measured with a spring scale, is the weight customarily used in business and everyday life; absolute weight is used in astroengineering and certain kinds of scientific work. In this book, the term "weight" will always mean "apparent weight."

---

[2]Sir Isaac Newton (1642–1727) was an English mathematician, physicist, and astronomer. He invented calculus and discovered the laws of motion and gravitation.

[3]James Prescott Joule (1818–1889) was an English physicist who developed a method for determining the mechanical equivalent of heat. His last name is pronounced "jool."

Table A-1

Principal units used in mechanics

Quantity	International System (SI)		
	Unit	Symbol	Formula
Acceleration (angular)	radian per second squared		rad/s^2
Acceleration (linear)	meter per second squared		m/s^2
Area	square meter		m^2
Density (mass) (Specific mass)	kilogram per cubic meter		kg/m^3
Density (weight) (Specific weight)	newton per cubic meter		N/m^3
Energy; work	joule	J	$N \cdot m$
Force	newton	N	$kg \cdot m/s^2$
Force per unit length (Intensity of force)	newton per meter		N/m
Frequency	hertz	Hz	s^{-1}
Length	meter	m	(base unit)
Mass	kilogram	kg	(base unit)
Moment of a force; torque	newton meter		$N \cdot m$
Moment of inertia (area)	meter to fourth power		m^4
Moment of inertia (mass)	kilogram meter squared		$kg \cdot m^2$
Power	watt	W	$J/s\ (N \cdot m/s)$
Pressure	pascal	Pa	N/m^2
Section modulus	meter to third power		m^3
Stress	pascal	Pa	N/m^2
Time	second	s	(base unit)
Velocity (angular)	radian per second		rad/s
Velocity (linear)	meter per second		m/s
Volume (liquids)	liter	L	$10^{-3}\,m^3$
Volume (solids)	cubic meter		m^3

Notes: 1 joule (J) = 1 newton meter ($N \cdot m$) = 1 watt second ($W \cdot s$)
1 hertz (Hz) = 1 cycle per second (cps) or 1 revolution per second (rev/s)
1 watt (W) = 1 joule per second (J/s) = 1 newton meter per second ($N \cdot m/s$)
1 pascal (Pa) = 1 newton per meter squared (N/m^2)
1 liter (L) = 0.001 cubic meter (m^3) = 1000 cubic centimeters (cm^3)

The **acceleration of gravity**, denoted by the letter g, is directly proportional to the force of gravity; therefore, it depends upon position. In contrast, **mass** is a measure of the amount of material in a body and does not change with location.

The fundamental relationship between weight, mass, and acceleration of gravity can be obtained from Newton's second law ($F = ma$), which in this case becomes

$$W = mg \tag{A-4}$$

Table A-2 Additional units in common use

SI and Metric Units	
1 gal = 1 centimeter per second squared (cm/s^2) for example, g ≈ 981 gals 1 are (a) = 100 square meters (m^2) 1 hectare (ha) = 10,000 square meters (m^2) 1 erg = 10^{-7} joules (J) 1 kilowatt-hour (kWh) = 3.6 megajoules (MJ) 1 dyne = 10^{-5} newtons (N) 1 kilogram-force (kgf) = 1 kilopond (kp) = 9.80665 newtons (N)	1 centimeter (cm) = 10^{-2} meters (m) 1 cubic centimeter (cm^3) = 1 milliliter (mL) 1 micron = 1 micrometer (μm) = 10^{-6} meters (m) 1 gram (g) = 10^{-3} kilograms (kg) 1 metric ton (t) = 1 megagram (Mg) = 1000 kilograms (kg) 1 watt (W) = 10^7 ergs per second (erg/s) 1 dyne per square centimeter (dyne/cm^2) = 10^{-1} pascals (Pa) 1 bar = 10^5 pascals (Pa) 1 stere = 1 cubic meter (m^3)

In this equation, W is the weight in newtons (N), m is the mass in kilograms (kg), and g is the acceleration of gravity in meters per second squared (m/s^2). Equation (A-4) shows that *a body having a mass of one kilogram has a weight in newtons numerically equal to g*. The values of the weight W and the acceleration g depend upon many factors, including latitude and elevation. However, for scientific calculations, a standard international value of g has been established as

$$g = 9.806650 \text{ m/s}^2 \tag{A-5}$$

This value is intended for use under standard conditions of elevation and latitude (sea level at a latitude of approximately 45°). The recommended value of g for ordinary engineering purposes on or near the surface of the earth is

$$g = 9.81 \text{ m/s}^2 \tag{A-6}$$

Thus, a body having a mass of one kilogram has a weight of 9.81 newtons.

Atmospheric pressure varies considerably with weather conditions, location, altitude, and other factors. Consequently, a standard international value for the pressure at the earth's surface has been defined as

$$1 \text{ standard atmosphere} = 101.325 \text{ kilopascals} \tag{A-7}$$

The following simplified value is recommended for ordinary engineering work:

$$1 \text{ standard atmosphere} = 101 \text{ kPa} \tag{A-8}$$

Of course, the values given in Eqs. (A-7) and (A-8) are intended for use in calculations and do not represent the actual ambient pressure at any given location.

A basic concept in mechanics is **moment** or **torque**, especially the moment of a force and the moment of a couple. Moment is expressed in units of force times length, or newton meters (N·m). Other important concepts in mechanics are **work** and **energy**, both of which are expressed in joules, which is a derived unit that happens to have the same units (newton meters) as the units of moment. However, moment is a distinctly different quantity from work or energy, and the joule should *never* be used for moment or torque.

Frequency is measured in units of **hertz** (Hz), which is a derived unit equal to the reciprocal of seconds (1/s or s^{-1}). The hertz is defined as the frequency of a periodic phenomenon for which the period is one second; thus, it is equivalent to one cycle per second (cps) or one revolution per second (rev/s). It is customarily used for mechanical vibrations, sound waves, and electromagnetic waves, and occasionally it is used for rotational frequency instead of the traditional units of revolution per minute (rpm) and revolution per second (rev/s).[4]

Two other derived units that have special names in SI are the **watt** (W) and the **pascal** (Pa). The watt is the unit of power, which is work per unit of time, and one watt is equal to one joule per second (J/s) or one newton meter per second ($N \cdot m/s$). The pascal is the unit of pressure and stress, or force per unit area, and is equal to one newton per square meter (N/m^2).[5]

The **liter** is not an accepted SI unit, yet it is so commonly used that it cannot be discarded easily. Therefore, SI permits its use under limited conditions for volumetric capacity, dry measure, and liquid measure. Both uppercase L and lowercase l are permitted as symbols for the liter in SI, but in the United States only L is permitted (to avoid confusion with the numeral 1). The only prefixes permitted with the liter are milli and micro.

Loads on structures, whether due to gravity or other actions, are usually expressed in force units, such as newtons, newtons per meter, or pascals (newtons per square meter). Examples of such loads are a concentrated load of 25 kN acting on an axle, a uniformly distributed load of intensity 800 N/m acting on a small beam, and air pressure of intensity 2.1 kPa acting on an airplane wing.

However, there is one circumstance in SI in which it is permissible to express a load in mass units. If the load acting on a structure is produced by gravity acting on a mass, then that load may be expressed in mass units (kilograms, kilograms per meter, or kilograms per square meter). The usual procedure in such cases is to convert the load to force units by multiplying by the acceleration of gravity ($g = 9.81\ m/s^2$).

SI Prefixes

Multiples and submultiples of SI units (both base units and derived units) are created by attaching prefixes to the units (see Table A-3 for a list of prefixes). The use of a prefix avoids unusually large or small numbers. The general rule is that prefixes should be used to keep numbers in the range 0.1 to 1000.

All of the recommended prefixes change the size of the quantity by a multiple or submultiple of three. Similarly, when powers of 10 are used as multipliers, the exponents of 10 should be multiples of three (for example, 40×10^3 N is satisfactory but 400×10^2 N is not). Also, the exponent on a unit with a prefix refers to the entire unit; for instance, the symbol mm^2 means $(mm)^2$ and not $m(m)^2$.

Styles for Writing SI Units

Rules for writing SI units have been established by international agreement, and some of the most pertinent ones are described here. Examples of the rules are shown in parentheses.

1. Units are always written as symbols (kg) in equations and numerical calculations. In text, units are written as words (kilograms) unless numerical values are being reported, in which case either words or symbols may be used (12 kg or 12 kilograms).

[4]Heinrich Rudolf Hertz (1857–1894) was a German physicist who discovered electromagnetic waves and showed that light waves and electromagnetic waves are identical.
[5]James Watt (1736–1819) was a Scottish inventor and engineer who developed a practical steam engine and discovered the composition of water. Watt also originated the term "horsepower." Blaise Pascal (1623–1662) was a French mathematician and philosopher. He founded probability theory, constructed the first calculating machine, and proved experimentally that atmospheric pressure varies with altitude.

Table A-3
SI prefixes

Prefix	Symbol	Multiplication Factor		
tera	T	10^{12}	=	1 000 000 000 000
giga	G	10^{9}	=	1 000 000 000
mega	M	10^{6}	=	1 000 000
kilo	k	10^{3}	=	1 000
hecto	h	10^{2}	=	100
deka	da	10^{1}	=	10
deci	d	10^{-1}	=	0.1
centi	c	10^{-2}	=	0.01
milli	m	10^{-3}	=	0.001
micro	μ	10^{-6}	=	0.000 001
nano	n	10^{-9}	=	0.000 000 001
pico	p	10^{-12}	=	0.000 000 000 001

Note: The use of the prefixes hecto, deka, deci, and centi is not recommended in SI.

2. Multiplication is shown in a compound unit by a raised dot (kN · m). When the unit is written in words, no dot is required (kilonewton meter).
3. Division is shown in a compound unit by a slash (or *solidus*) or by multiplication using a negative exponent (m/s or m · s^{-1}). When the unit is written in words, the slash is always replaced by "per" (meter per second).
4. A space is always used between a number and its units (200 Pa or 200 pascals) with the exception of the degree symbol (either angle or temperature), where no space is used between the number and the symbol (45°, 20°C).
5. Units and their prefixes are always printed in roman type (that is, upright or vertical type) and never in italic type (slanted type), even when the surrounding text is in italic type.
6. When written as words, units are not capitalized (newton) except at the beginning of a sentence or in capitalized material such as a title. When written as a symbol, units are capitalized when they are derived from the name of a person (N). An exception is the symbol for liter, which may be either L or l, but the use of uppercase L is preferred to avoid confusion with the numeral 1. Also, some prefixes are written with capital letters when used in symbols (MPa) but not when used in words (megapascal).
7. When written as words, units are singular or plural as appropriate to the context (1 kilometer, 20 kilometers, 6 seconds). When written as symbols, units are always singular (1 km, 20 km, 6 s). The plural of hertz is hertz; the plurals of other units are formed in the customary manner (newtons, watts).
8. Prefixes are not used in the denominator of a compound unit. An exception is the kilogram (kg),which is a base unit and therefore the letter "k" is not considered as a prefix. For example, write kN/m but not N/mm, and write J/kg but not mJ/g.

Pronunciation of SI Prefixes and Units

A guide to the pronunciation of a few SI names that are sometimes mispronounced is given in Table A-4. For instance, kilometer is pronounced *kill-oh-meter*, not *kil-om-eter*. The only prefix that generates arguments is giga—the official pronunciation is *jig-uh*, but many people say *gig-uh*.

Table A-4

Pronunciation of SI prefixes and units

Prefix	Pronunciation
tera	same as *terra*, as in *terra firma*
giga	pronounced *jig-uh*; with *a* pronounced as in *about* (Alternate pronunciation: *gig-uh*)
mega	same as *mega* in *megaphone*
kilo	pronounced *kill-oh*; rhymes with *pillow*
milli	pronounced *mill-eh*, as in *military*
micro	same as *micro* in *microphone*
nano	pronounced *nan-oh*; rhymes with *man-oh*
pico	pronounced *pea-ko*
	Note: The first syllable of every prefix is accented.

Unit	Pronunciation
joule	pronounced *jool*; rhymes with *cool* and *pool*
kilogram	pronounced *kill-oh-gram*
kilometer	pronounced *kill-oh-meter*
pascal	pronounced *pas-kal*, with the accent on *kal*

A.3 Temperature Units

Temperature is measured in SI by a unit called the kelvin (K), and the corresponding scale is the **Kelvin temperature scale**. The Kelvin scale is an absolute scale, which means that its origin (zero kelvins, or 0 K) is at absolute zero temperature, a theoretical temperature characterized by the complete absence of heat. On the Kelvin scale, water freezes at approximately 273 K and boils at approximately 373 K.

For nonscientific purposes, the **Celsius temperature scale** is normally used. The corresponding unit of temperature is the degree Celsius (°C), which is equal to one kelvin. On this scale, water freezes at approximately zero degrees (0°C) and boils at approximately 100 degrees (100°C) under certain standard conditions. The Celsius scale is also known as the ***Centigrade temperature scale***.

The relationship between Kelvin temperature and Celsius temperature is given by

Temperature in degrees Celsius = temperature in kelvins − 273.15

or

$$T(°C) = T(K) - 273.15 \quad \textbf{(A-9)}$$

where T denotes the temperature. When working with *changes* in temperature, or *temperature intervals*, as is usually the case in mechanics, either unit can be used because the intervals are the same.[6]

[6]Lord Kelvin (1824–1907), William Thomson, was a British physicist who made many scientific discoveries, developed theories of heat, and proposed the absolute scale of temperature. Anders Celsius (1701–1744) was a Swedish scientist and astronomer. In 1742, he developed the temperature scale in which 0 and 100 correspond, respectively, to the freezing and boiling points of water.

Problem Solving

B.1 Types of Problems

Some of the homework problems in this book require symbolic solutions and others require numerical solutions. In the case of **symbolic problems** (also called *analytical*, *algebraic*, or *literal problems*), the data are supplied in the form of symbols for the various quantities, such as P for load, L for length, and E for modulus of elasticity. Such problems are solved in terms of algebraic variables, and the results are expressed as formulas or mathematical expressions. Symbolic problems usually do not involve numerical calculations, except when numerical data are substituted into the final symbolic result in order to obtain a numerical value. However, this final substitution of numerical data should not obscure the fact that the problem was solved in symbolic terms.

In contrast, **numerical problems** are those in which the data are given in the form of numbers (with appropriate units); for example, a load might be given as 12 kN, a length as 3 m, and a dimension as 150 mm. The solution of a numerical problem is carried out by performing calculations from the beginning, and the results, both intermediate and final, are in the form of numbers.

An advantage of a numerical problem is that the magnitudes of all quantities are evident at every stage of the solution, thereby providing an opportunity to observe whether the calculations are producing reasonable results. Also, a numerical solution makes it possible to keep the magnitudes of quantities within prescribed limits. For instance, suppose the stress at a particular point in a beam must not exceed a certain allowable value. If this stress is calculated as an intermediate step in the numerical solution, you can verify immediately whether or not it exceeds the limit.

Symbolic problems have several advantages too. Because the results are algebraic formulas or expressions, you can see immediately how the variables affect the answers. For instance, if a load appears to the first power in the numerator of the final result, you know that doubling the load will double the result. Equally important is the fact that a symbolic solution shows what variables do *not* affect the result. For instance, a certain quantity may cancel out of the solution, a fact that might not even be noticed in a numerical solution. Furthermore, a symbolic solution makes it convenient to check the dimensional homogeneity of all terms in the solution. And most important, a symbolic solution provides a general formula that is applicable to many different problems, each with a different set of numerical data. In contrast, a numerical solution is good for only one set of circumstances, and a completely new solution is required if the data are changed. Of course, symbolic solutions are not feasible when the formulas become too complex to manipulate; when that happens, a numerical solution is required.

In more advanced work in mechanics, problem solving requires the use of **numerical methods**. This term refers to a wide variety of computational methods, including standard mathematical procedures (such as numerical integration and numerical solution of differential equations) and advanced methods of analysis (such as the finite element method). Computer programs for these methods are readily available. More specialized computer programs are also available for performing routine tasks, such as finding deflections of beams and finding principal stresses. However, when studying mechanics of materials, concentrate on the concepts rather than on the use of particular computer programs.

B.2 Steps in Solving Problems

The procedures used in solving problems in this text vary according to the type of problem. One method to consider is the four-step **problem-solving approach** (PSA)* presented in Chapter 1 and illustrated in example problems throughout the text. This approach is explained here using part of Example 1-1 from Chapter 1 to highlight key steps.

Example B-1

FIGURE B-1

(a) Plane truss model of section of side panel in (b) model airplane

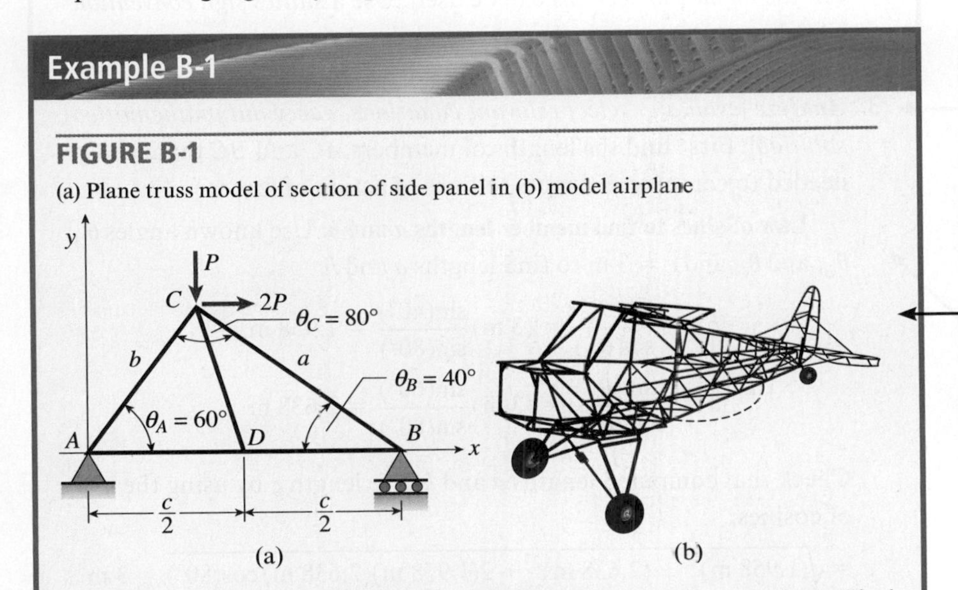

In many cases, the problem involves the analysis of a model of a real physical structure, such as this truss structure representing the fuselage of a model air plane (Fig. B-1b). Begin by sketching the portion of the structure of interest showing relevant supports, dimensions and loadings. This ***Conceptualization*** step in the analysis often leads to a free-body diagram in mechanics problems.

The plane truss shown in Figure B-1 has four joints and five members. Find support reactions at A and B and then use the *methods of joints and sections* to find all member forces. Let $P = 150$ kN and $c = 3$ m.

Solution:

Use the following four-step problem-solving approach.

1. ***Conceptualize*** [*hypothesize, sketch*]: First sketch a free-body diagram of the entire truss model (Figure B-2). Only known applied forces at C and unknown reaction forces at A and B are shown and then are used in an equilibrium analysis to find the reactions.

*The four step Problem-Solving Approach presented here is patterned after that presented by R. Serway and J. Jewett in *Principles of Physics, 5e,* Cengage Learning, 2013.

The next step is to simplify the problem, identify all unknowns, and make necessary assumptions to create a suitable model for analysis. This is the ***Categorize*** step.

Use appropriate mathematical and computational techniques to solve the equations and obtain results, either in the form of mathematical formulas or numerical values. In this example, the equilibrium of the truss is of interest so all member lengths and orientations, coordinates of joints, and so on are needed to solve the problem. One or more methods of analysis are selected and a mathematical solution is carried out in numerical or symbolic form. This is the ***Analysis*** step.

List the major steps in your analysis procedure so that it is easy to review or check at a later time.

2. ***Categorize*** [*simplify, classify*]: Overall equilibrium requires that the force components in x and y directions and the moment about the z axis must sum to zero; this leads to reaction force components A_x, A_y, and B_y. The truss is statically determinate (*unknowns:* $m + r = 5 + 3 = 8$, knowns: $2j = 8$), so all member forces can be obtained using the *method of joints*. If only a few selected member forces are of interest, the *method of sections* can be used. Use a *statics sign convention* when computing external reactions and a *deformation sign convention* when solving for member forces.

FIGURE B-2

Free-body diagram of truss model

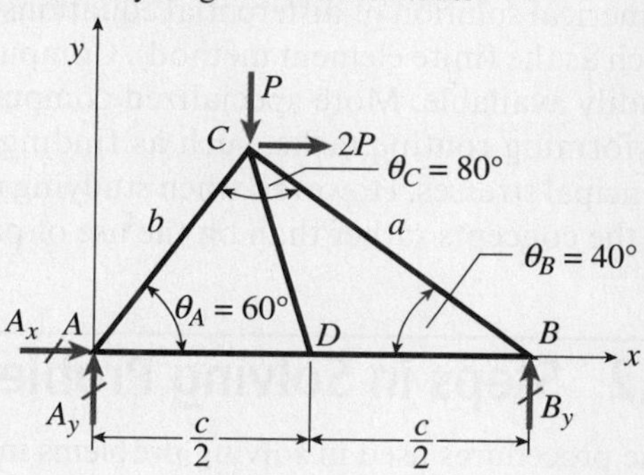

3. ***Analyze*** [*evaluate: select relevant equations, carry out mathematical solution*]: First find the lengths of members AC and BC that are needed to compute distances to lines of action of forces.

Law of sines to find member lengths *a* and *b*: Use known angles θ_A, θ_B, and θ_C and $c = 3$ m to find lengths a and b:

$$b = c\frac{\sin(\theta_B)}{\sin(\theta_C)} = (3\text{ m})\frac{\sin(40°)}{\sin(80°)} = 1.958\text{ m},$$

$$a = c\frac{\sin(\theta_A)}{\sin(\theta_C)} = (3\text{ m})\frac{\sin(60°)}{\sin(80°)} = 2.638\text{ m}$$

Check that computed lengths a and b give length c by using the law of cosines:

$$c = \sqrt{(1.958\text{ m})^2 + (2.638\text{ m})^2 - 2(1.958\text{ m})(2.638\text{ m})\cos(80°)} = 3\text{ m}$$

Support reactions: Using the truss model free-body diagram in Figure B-2, sum forces in x and y directions and moments about joint 4:

$$\Sigma M_A = 0 \quad B_y = \frac{1}{c}[P(b\cos(\theta_A)) + 2P(b\sin(\theta_A)] = 218.5\text{ kN}$$

$$\Sigma F_x = 0 \quad A_x = -2P = -300\text{ kN}$$

$$\Sigma F_y = 0 \quad A_y = P - B_y = -68.5\text{ kN}$$

The support reactions are computed using a *statics sign convention* (forces in the $+x$ direction are positive). However, it is common to use a *deformation sign convention* (tension is positive) for member force calculations.

FIGURE B-3

Free-body diagram of pin at each truss joint

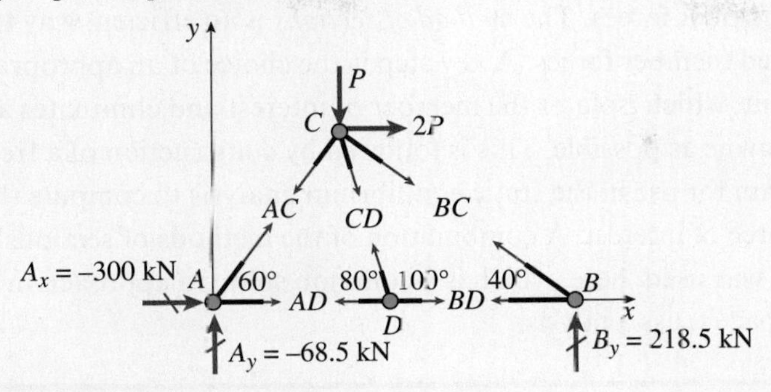

Additional sketches, such as this free body diagram of each joint in the truss (Fig. B-3), are often needed as the analysis proceeds.

Member forces using method of joints: Begin by drawing free-body diagrams of the pin at each joint (Figure B-3). Use a *deformation sign convention* in which each member is assumed to be in tension (so the member force arrows act away from the two joints to which each member is connected). The forces are concurrent at each joint, so use force equilibrium at each location to find the unknown member forces.

In the full Example 1-1 solution in Chapter 1, a second analysis for member forces, the Method of Sections, is described but is not repeated here.

First sum forces in the y direction at joint A to find member force AC, and then sum forces in the x direction to get member force AD:

$$\Sigma F_y = 0 \quad AC = \frac{-1}{\sin(60°)} A_y = 79.1 \text{ kN}$$

$$\Sigma F_x = 0 \quad AD = -A_x - AC\cos(60°) = 260.4 \text{ kN}$$

Summing forces at joint B gives member forces BC and BD as

$$\Sigma F_y = 0 \quad BC = \frac{-1}{\sin(40°)} B_y = -340 \text{ kN}$$

$$\Sigma F_x = 0 \quad BD = -BC\cos(40°) = 260.4 \text{ kN}$$

The minus sign means that member BC is in compression, not in tension as assumed. Finally, observe that CD is a zero-force member because forces in the y direction must sum to zero at joint D.

Selected member forces using method of sections: An alternative approach is to make a section cut all the way through the structure to expose member forces of interest, such as AD, CD, and BC in Figure 1-9. Summing moments about joint B confirms that the force in member CD is zero.

4. ***Finalize*** [*conclude; examine answer—Does it make sense? Are units correct? How does it compare to similar problem solutions?*]: There are $2j = 8$ equilibrium equations for the simple plane truss considered, and using the *method of joints*, these are obtained by applying

In the ***Finalize*** step, the solution is reviewed to make sure that it is presented in a clear fashion so that it can be easily reviewed and checked by others. Are the expressions and numerical values obtained reasonable? Do they agree with your initial expectations? The results must be checked in as many ways as possible because errors can be disastrous and expensive so engineers should never rely on a single solution. The results are interpreted in terms of the physical behavior of the mechanical or structural system to give meaning to the results and draw conclusions about the behavior of the system. Is further analysis required perhaps using other loadings or support conditions?

$\Sigma F_x = 0$ and $\Sigma F_y = 0$ at each joint in succession. A computer solution of these simultaneous equations leads to the three reaction forces and five member forces. The *method of sections* is an efficient way to find selected member forces. A key step is the choice of an appropriate section cut, which isolates the member of interest and eliminates as many unknowns as possible. This is followed by construction of a free-body diagram for use in the static equilibrium analysis to compute the member force of interest. A combination of the methods of sections and joints was used, here, which is a common solution approach in plane and space truss analysis.

B.3 Dimensional Homogeneity

The basic concepts in mechanics are length, time, mass, and force. Each of these physical quantities has a **dimension**, that is, a generalized unit of measurement. For example, consider the concept of length. There are many units of length, such as the meter, kilometer, and micrometer, yet all of these units have something in common—each one represents a distinct length and not some other quantity such as volume or force. Therefore, you can refer to the *dimension of length* without being specific as to the particular unit of measurement. Similar comments can be made for the dimensions of time, mass, and force. These four dimensions are customarily denoted by the symbols L, T, M, and F, respectively.

Every equation, whether in numerical form or symbolic form, must be **dimensionally homogeneous**, that is, the dimensions of all terms in the equation must be the same. To check the dimensional correctness of an equation, disregard numerical magnitudes and write only the dimensions of each quantity in the equation. The resulting equation must have identical dimensions in all terms.

As an example, consider the following equation for the deflection δ at the midpoint of a simple beam with a uniformly distributed load:

$$\delta = \frac{5qL^4}{384EI}$$

The corresponding dimensional equation is obtained by replacing each quantity by its dimensions; thus, the deflection δ is replaced by the dimension L, the intensity of uniform load q is replaced by F/L (force per unit of length), the length L of the beam is replaced by the dimension L, the modulus of elasticity E is replaced by $\mathrm{F/L^2}$ (force per unit of area), and the moment of inertia I is replaced by $\mathrm{L^4}$. Therefore, the dimensional equation is

$$\mathrm{L} = \frac{(\mathrm{F/L})\mathrm{L}^4}{(\mathrm{F/L}^2)\mathrm{L}^4}$$

When simplified, this equation reduces to the dimensional equation L = L, as expected.

Dimensional equations can be written either in generalized terms using the LTMF notation or in terms of the actual units being used in the problem. For instance, when making calculations for the preceding beam deflection using USCS units, write the dimensional equation as

$$\text{mm} = \frac{(\text{N/mm})\text{mm}^4}{(\text{N/mm}^2)\text{mm}^4}$$

which reduces to mm = mm and is dimensionally correct. Frequent checks for dimensional homogeneity (or *consistency of units*) help to eliminate errors when performing derivations and calculations.

B.4 Significant Digits

Engineering calculations are performed by calculators and computers that operate with great precision. For instance, some computers routinely perform calculations with more than 25 digits in every numerical value, and output values with 10 or more digits are available in even the most inexpensive hand-held calculators. Under these conditions, it is important to realize that the accuracy of the results obtained from an engineering analysis is determined not only by the calculations but also by factors such as the accuracy of the given data, the approximations inherent in the analytical models, and the validity of the assumptions used in the theories. In many engineering situations, these considerations mean that the results are valid to only two or three significant digits.

As an example, suppose that a computation yields the result $R = 6287.46$ N for the reaction of a statically indeterminate beam. To state the result in this manner is misleading, because it implies that the reaction is known to the nearest 1/100 of a newton even though its magnitude is over 6000 newtons. Thus, it implies an accuracy of approximately 1/600,000 and a precision of 0.01 N, neither of which is justified. Instead, the accuracy of the calculated reaction depends upon matters such as the following: (1) how accurately the loads, dimensions, and other data used in the analysis are known and (2) the approximations inherent in the theories of beam behavior. Most likely, the reaction R in this example would be known only to the nearest 10 newtons, or perhaps only to the nearest 100 newtons. Consequently, the result of the computation should be stated as either $R = 6290$ N or $R = 6300$ N.

To make clear the accuracy of a given numerical value, it is common practice to use **significant digits**. A significant digit is a digit from 1 to 9 or any zero not used to show the position of the decimal point; for instance, the numbers 417, 8.29, 7.30, and 0.00254 each have three significant digits. However, the number of significant digits in a number such as 29,000 is not apparent. It may have two significant digits, with the three zeros serving only to locate the decimal point, or it may have three, four, or five significant digits if one or more of the zeros is valid. By using powers of ten, the accuracy of a number such as 29,000 can be made clearer. When written as 29×10^3 or 0.029×10^6, the number is understood to have two significant digits; when written as 29.0×10^3 or 0.0290×10^6, it has three significant digits.

When a number is obtained by calculation, its accuracy depends upon the accuracy of the numbers used in performing the calculations. A rule of thumb that serves for *multiplication* and *division* is the following: The number of significant digits in the calculated result is the same as the least number of significant digits in any of the numbers used in the calculation. As an illustration, consider the product of 2339.3 and 35.4. The calculated result is 82,811.220 when recorded to eight digits. However, stating the result in this manner is misleading because it implies much greater accuracy than is warranted by either of the original numbers. Inasmuch as the number 35.4 has only three significant digits, the proper way to write the result is 82.8×10^3.

For calculations involving *addition* or *subtraction* of a column of numbers, the last significant digit in the result is found in the last column of digits that has significant digits in all of the numbers being added or subtracted. To make this notion clearer, consider the following three examples:

	459.637	838.49	856,400
	+7.2	−7	−847,900
Result from calculator:	466.837	831.49	8,500
Write the result as:	466.8	831	8,500

In the first example, the number 459.637 has six significant digits and the number 7.2 has two. When added, the result has four significant digits because all digits in the result to the right of the column containing the 2 are meaningless. In the second example, the number 7 is accurate to one significant digit (that is, it is not an exact number). Therefore, the final result is accurate only as far as the column containing the 7, which means it has three significant digits and is recorded as 831. In the third example, the numbers 856,400 and 847,900 are assumed to be accurate to four significant digits, but the result of the subtraction is accurate to only two significant digits since none of the zeros is significant. In general, subtraction results in reduced accuracy.

These three examples show that numbers obtained by calculation may contain superfluous digits having no physical meaning. Therefore, when reporting such numbers as final results, you should give only those digits that are significant.

In mechanics of materials, the data for problems are usually accurate to about 1%, or perhaps 0.1% in some cases; therefore, the final results should be reported to a comparable accuracy. When greater accuracy is warranted, it will be obvious from the statement of the problem.

Although the use of significant digits provides a handy way to deal with the matter of **numerical accuracy**, it should be recognized that significant digits are not valid indicators of accuracy. To illustrate this fact, consider the numbers 999 and 101. Three significant digits in the number 999 correspond to an accuracy of 1/999, or 0.1%, whereas the same number of significant digits in the number 101 corresponds to an accuracy of only 1/101, or 1.0%. This disparity in accuracy can be reduced by always using one additional significant digit for numbers beginning with the digit 1. Thus, four significant digits in the number 101.1 gives about the same accuracy as three significant digits in the number 999.

In this book, the rule used is that *final* numerical results beginning with the digits 2 through 9 should be recorded to three significant digits and those beginning with the digit 1 should be recorded to four significant digits. However, to preserve numerical accuracy and avoid round-off errors during the calculation process, the results of *intermediate* calculations usually will be recorded with additional digits.

Many of the numbers entering into our calculations are exact, for example, the number π, fractions such as 1/2, and integers such as the number 48 in the formula $PL^3/48EI$ for a beam deflection. Exact numbers are significant to an infinite number of digits and therefore have no role in determining the accuracy of a calculated result.

B.5 Rounding of Numbers

The process of discarding the insignificant digits and keeping only the significant ones is called *rounding*. To illustrate the process, assume that a number is to be rounded to three significant digits. Then the following rules apply:

a. If the fourth digit is less than 5, the first three digits are left unchanged and all succeeding digits are dropped or replaced by zeros. For example, 37.44 rounds to 37.4 and 673,289 rounds to 673,000.

b. If the fourth digit is greater than 5, or if the fourth digit is 5 and is followed by at least one digit other than zero, then the third digit is increased by 1 and all following digits are dropped or replaced by zeros. For example, 26.37 rounds to 26.4 and 3.245002 rounds to 3.25.

c. Finally, if the fourth digit is 5 and all following digits (if any) are zeros, then the third digit is unchanged if it is an even number and increased by 1 if it is an odd number, and the 5 is replaced by a zero. (Trailing and leading zeros are retained only if they are needed to locate the decimal point.) This process is usually described as "rounding to the even digit." Since the occurrence of even and odd digits is more or less random, the use of this rule means that numbers are rounded upward about as often as downward, thereby reducing the chances of accumulating round-off errors.

The rules described in the preceding paragraphs for rounding to three significant digits apply in the same general manner when rounding to any other number of significant digits.

APPENDIX C

Mathematical Formulas

Mathematical Constants

$\pi = 3.14159\ldots \qquad e = 2.71828\ldots \qquad 2\pi \text{ radians} = 360 \text{ degrees}$

$$1 \text{ radian} = \frac{180}{\pi} \text{degrees} = 57.2958°$$

$$1 \text{ degree} = \frac{\pi}{180} \text{radians} = 0.0174533 \text{ rad}$$

Conversions: Multiply degrees by $\frac{\pi}{180}$ to obtain radians

Multiply radians by $\frac{180}{\pi}$ to obtain degrees

Exponents

$$A^n A^m = A^{n+m} \qquad \frac{A^m}{A^n} = A^{m-n} \qquad (A^m)^n = A^{mn} \qquad A^{-m} = \frac{1}{A^m}$$

$$(AB)^n = A^n B^n \qquad \left(\frac{A}{B}\right)^n = \frac{A^n}{B^n} \qquad A^{m/n} = \sqrt[n]{A^m} \qquad A^0 = 1 (A \neq 0)$$

Logarithms

log $\equiv$ common logarithm (logarithm to the base 10)

$10^x = y \qquad \log y = x$

ln $\equiv$ natural logarithm (logarithm to the base e) $\qquad e^x = y \qquad \ln y = x$

$$e^{\ln A} = A \qquad 10^{\log A} = A \qquad \ln e^A = A \qquad \log 10^A = A$$

$$\log AB = \log A + \log B \qquad \log \frac{A}{B} = \log A - \log B \qquad \log \frac{1}{A} = -\log A$$

$$\log A^n = n \log A \qquad \log 1 = \ln 1 = 0 \qquad \log 10 = 1 \qquad \ln e = 1$$

$\ln A = (\ln 10)(\log A) = 2.30259 \log A$

$\log A = (\log e)(\ln A) = 0.434294 \ln A$

Trigonometric Functions

$$\tan x = \frac{\sin x}{\cos x} \qquad \cot x = \frac{\cos x}{\sin x} \qquad \sec x = \frac{1}{\cos x} \qquad \csc x = \frac{1}{\sin x}$$

$$\sin^2 x + \cos^2 x = 1 \qquad \tan^2 x + 1 = \sec^2 x \qquad \cot^2 x + 1 = \csc^2 x$$

$$\sin(-x) = -\sin x \qquad \cos(-x) = \cos x \qquad \tan(-x) = -\tan x$$

$$\sin(x \pm y) = \sin x \cos y \pm \cos x \sin y$$

$$\cos(x \pm y) = \cos x \cos y \mp \sin x \sin y$$

$$\sin 2x = 2 \sin x \cos x \qquad \cos 2x = \cos^2 x - \sin^2 x$$

$$\tan 2x = \frac{2 \tan x}{1 - \tan^2 x}$$

$$\tan x = \frac{1 - \cos 2x}{\sin 2x} = \frac{\sin 2x}{1 + \cos 2x}$$

$$\sin^2 x = \frac{1}{2}(1 - \cos 2x) \qquad \cos^2 x = \frac{1}{2}(1 + \cos 2x)$$

For any triangle with sides a, b, c and opposite angles A, B, C:

Law of sines $\dfrac{a}{\sin A} = \dfrac{b}{\sin B} = \dfrac{c}{\sin C}$

Law of cosines $c^2 = a^2 + b^2 - 2ab \cos C$

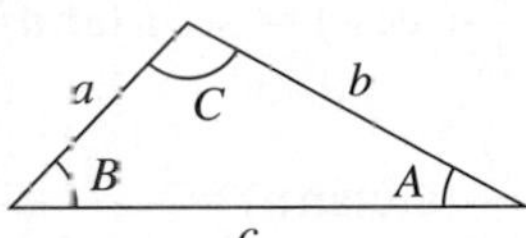

Quadratic Equation and Quadratic Formula

$$ax^2 + bx + c = 0 \qquad x = \frac{-b \pm \sqrt{b^2 - 4ac}}{2a}$$

Infinite Series

$$\frac{1}{1 + x} = 1 - x + x^2 - x^3 + \cdots \qquad (-1 < x < 1)$$

$$\sqrt{1 + x} = 1 + \frac{x}{2} - \frac{x^2}{8} + \frac{x^3}{16} - \cdots \qquad (-1 < x < 1)$$

$$\frac{1}{\sqrt{1 + x}} = 1 - \frac{x}{2} + \frac{3x^2}{8} - \frac{5x^3}{16} + \cdots \qquad (-1 < x < 1)$$

$$e^x = 1 + x + \frac{x^2}{2!} + \frac{x^3}{3!} + \cdots \qquad (-\infty < x < \infty)$$

$$\sin x = x - \frac{x^3}{3!} + \frac{x^5}{5!} - \frac{x^7}{7!} + \cdots \qquad (-\infty < x < \infty)$$

$$\cos x = 1 - \frac{x^2}{2!} + \frac{x^4}{4!} - \frac{x^6}{6!} + \cdots \qquad (-\infty < x < \infty)$$

Note: If x is very small compared to 1, only the first few terms in the series are needed.

Derivatives

$$\frac{d}{dx}(ax) = a \qquad \frac{d}{dx}(x^n) = nx^{n-1} \qquad \frac{d}{dx}(au) = a\frac{du}{dx}$$

$$\frac{d}{dx}(uv) = u\frac{dv}{dx} + v\frac{du}{dx} \qquad \frac{d}{dx}\left(\frac{u}{v}\right) = \frac{v(du/dx) - u(dv/dx)}{v^2}$$

$$\frac{d}{dx}(u^n) = nu^{n-1}\frac{du}{dx} \qquad \frac{dy}{dx} = \frac{dy}{du}\frac{du}{dx} \qquad \frac{du}{dx} = \frac{1}{dx/du}$$

$$\frac{d}{dx}(\sin u) = \cos u\frac{du}{dx} \qquad \frac{d}{dx}(\cos u) = -\sin u\frac{du}{dx}$$

$$\frac{d}{dx}(\tan u) = \sec^2 u\frac{du}{dx} \qquad \frac{d}{dx}(\cot u) = -\csc^2 u\frac{du}{dx}$$

$$\frac{d}{dx}(\sec u) = \sec u \tan u\frac{du}{dx} \qquad \frac{d}{dx}(\csc u) = -\csc u \cot u\frac{du}{dx}$$

$$\frac{d}{dx}(\arctan u) = \frac{1}{1+u^2}\frac{du}{dx} \qquad \frac{d}{dx}(\log u) = \frac{\log e}{u}\frac{du}{dx} \qquad \frac{d}{dx}(\ln u) = \frac{1}{u}\frac{du}{dx}$$

$$\frac{d}{dx}(a^u) = a^u \ln a\frac{du}{dx} \qquad \frac{d}{dx}(e^u) = e^u\frac{du}{dx}$$

Indefinite Integrals

Note: A constant must be added to the result of every integration

$$\int a\,dx = ax \qquad \int u\,dv = uv - \int v\,du \quad \text{(integration by parts)}$$

$$\int x^n dx = \frac{x^{n+1}}{n+1} \quad (n \neq -1) \qquad \int \frac{dx}{x} = \ln|x| \quad (x \neq 0)$$

$$\int \frac{dx}{x^n} = \frac{x^{1-n}}{1-n} \quad (n \neq 1) \qquad \int (a+bx)^n dx = \frac{(a+bx)^{n+1}}{b(n+1)} \quad (n \neq -1)$$

$$\int \frac{dx}{a+bx} = \frac{1}{b}\ln(a+bx) \qquad \int \frac{dx}{(a+bx)^2} = \frac{1}{b(a+bx)}$$

$$\int \frac{dx}{(a+bx)^n} = -\frac{1}{(n-1)(b)(a+bx)^{n-1}} \quad (n \neq 1)$$

$$\int \frac{dx}{a^2 + b^2x^2} = \frac{1}{ab}\tan^{-1}\frac{bx}{a} \qquad (x \text{ in radians}) \quad (a > 0,\ b > 0)$$

$$\int \frac{dx}{a^2 - b^2x^2} = \frac{1}{2ab}\ln\left(\frac{a + bx}{a - bx}\right) \quad (x \text{ in radians}) \quad (a > 0,\ b > 0)$$

$$\int \frac{xdx}{a + bx} = \frac{1}{b^2}[bx - a\ln(a + bx)]$$

$$\int \frac{xdx}{(a + bx)^2} = \frac{1}{b^2}\left[\frac{a}{a + bx} + \ln(a + bx)\right]$$

$$\int \frac{xdx}{(a + bx)^3} = -\frac{a + 2bx}{2b^2(a + bx)^2} \qquad \int \frac{xdx}{(a + bx)^4} = -\frac{a - 3bx}{6b^2(a + bx)^3}$$

$$\int \frac{x^2dx}{a + bx} = \frac{1}{2b^3}[(a + bx)(-3a + bx) + 2a^2\ln(a + bx)]$$

$$\int \frac{x^2dx}{(a + bx)^2} = \frac{1}{b^3}\left[\frac{bx(2a + bx)}{a + bx} - 2a\ln(a + bx)\right]$$

$$\int \frac{x^2dx}{(a + bx)^3} = \frac{1}{b^3}\left[\frac{a(3a + 4bx)}{2(a + bx)^2} + \ln(a + bx)\right]$$

$$\int \frac{x^2dx}{(a + bx)^4} = -\frac{a^2 + 3abx + 3b^2x^2}{3b^3(a + bx)^3}$$

$$\int \sin ax\,dx = -\frac{\cos ax}{a} \qquad \int \cos ax\,dx = \frac{\sin ax}{a}$$

$$\int \tan ax\,dx = \frac{1}{a}\ln(\sec ax) \qquad \int \cot ax\,dx = \frac{1}{a}\ln(\sin ax)$$

$$\int \sec ax\,dx = \frac{1}{a}\ln(\sec ax + \tan ax)$$

$$\int \csc ax\,dx = \frac{1}{a}\ln(\csc ax - \cot ax)$$

$$\int \sin^2 ax\,dx = \frac{x}{2} - \frac{\sin 2ax}{4a}$$

$$\int \cos^2 ax\,dx = \frac{x}{2} + \frac{\sin 2ax}{4a} \qquad (x \text{ in radians})$$

$$\int x\sin ax\,dx = \frac{\sin ax}{a^2} - \frac{x\cos ax}{a} \quad (x \text{ in radians})$$

$$\int x\cos ax\,dx = \frac{\cos ax}{a^2} + \frac{x\sin ax}{a} \quad (x \text{ in radians})$$

$$\int e^{ax}dx = \frac{e^{ax}}{a} \qquad \int xe^{ax}dx = \frac{e^{ax}}{a^2}(ax - 1)$$

$$\int \ln ax\,dx = x(\ln ax - 1)$$

$$\int \frac{dx}{1 + \sin ax} = -\frac{1}{a}\tan\left(\frac{\pi}{4} - \frac{ax}{2}\right)$$

$$\int \sqrt{a + bx}\,dx = \frac{2}{3b}(a + bx)^{3/2}$$

$$\int \sqrt{a^2 + b^2x^2}\,dx = \frac{x}{2}\sqrt{a^2 + b^2x^2} + \frac{a^2}{2b}\ln\left(\frac{bx}{a} + \sqrt{1 + \frac{b^2x^2}{a^2}}\right)$$

$$\int \frac{dx}{\sqrt{a^2 + b^2x^2}} = \frac{1}{b}\ln\left(\frac{bx}{a} + \sqrt{1 + \frac{b^2x^2}{a^2}}\right)$$

$$\int \sqrt{a^2 - b^2x^2}\,dx = \frac{x}{2}\sqrt{a^2 - b^2x^2} + \frac{a^2}{2b}\sin^{-1}\frac{bx}{a}$$

Definite Integrals

$$\int_a^b f(x)dx = -\int_b^a f(x)dx \qquad \int_a^b f(x)dx = \int_a^c f(x)dx + \int_a^b f(x)dx$$

Review of Centroids and Moments of Inertia

Only plane areas are considered. A table of centroids and moments of inertia for a variety of common geometric shapes is given in Appendix E for convenient reference.

Outline

D.1 Centroids of Plane Areas

This appendix is a review of the definitions and formulas pertaining to centroids and moments of inertia of plane areas. The position of the centroid of a plane area is an important geometric property. To obtain formulas for locating centroids, refer to Fig. D-1, which shows a plane area of irregular shape with its centroid at point C. The xy coordinate system is oriented arbitrarily with its origin at any point O. The **area** of the geometric figure is defined by the following integral:

$$A = \int dA \tag{D-1}$$

in which dA is a differential element of area having coordinates x and y (Fig. D-1) and A is the total area of the figure.

The **first moments** of the area with respect to the x and y axes are defined, respectively, as

$$Q_x = \int y dA \qquad Q_y = \int x dA \tag{D-2a,b}$$

FIGURE D-1

Plane area of arbitrary shape with centroid C

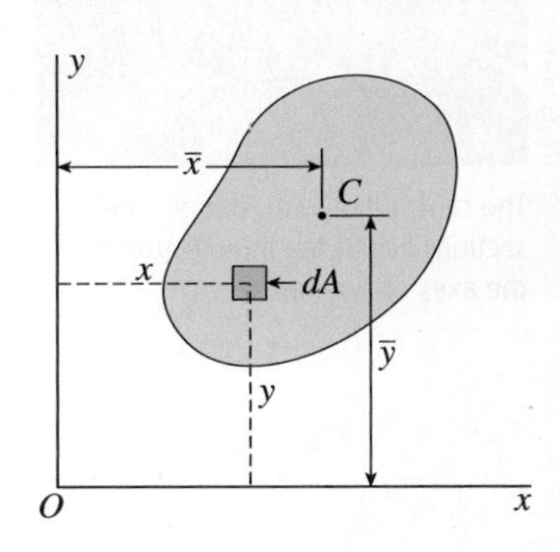

FIGURE D-2
Area with one axis of symmetry

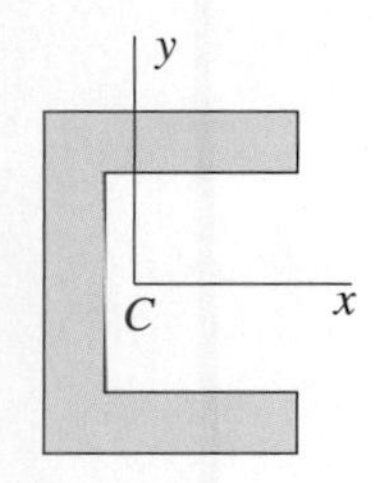

Thus, the first moments represent the sums of the products of the differential areas and their coordinates. First moments may be positive or negative, depending upon the position of the *xy* axes. Also, first moments have units of length raised to the third power; for instance, mm^3.

The coordinates $\bar{x}$ and $\bar{y}$ of the **centroid** *C* (Fig. D-1) are equal to the first moments divided by the area:

$$\bar{x} = \frac{Q_y}{A} = \frac{\int x dA}{\int dA} \qquad \bar{y} = \frac{Q_x}{A} = \frac{\int y dA}{\int dA} \tag{D-3a,b}$$

FIGURE D-3
Area with two axes of symmetry

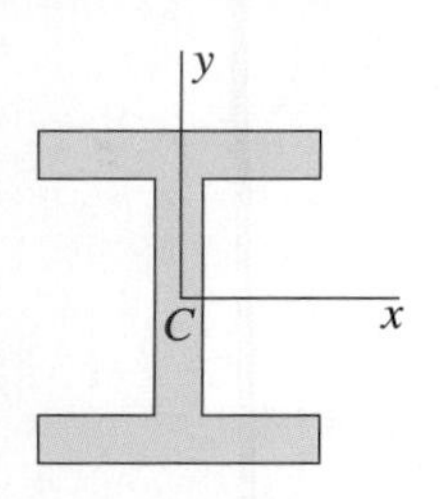

If the boundaries of the area are defined by simple mathematical expressions, evaluate the integrals appearing in Eqs. (D-3a) and (D-3b) in closed form and thereby obtain formulas for $\bar{x}$ and $\bar{y}$. The formulas listed in Appendix E were obtained in this manner. In general, the coordinates $\bar{x}$ and $\bar{y}$ may be positive or negative, depending upon the position of the centroid with respect to the reference axes.

If an area is **symmetric about an axis**, the centroid must lie on that axis because the first moment about an axis of symmetry equals zero. For example, the centroid of the singly symmetric area shown in Fig. D-2 must lie on the *x* axis, which is the axis of symmetry. Therefore, only one coordinate must be calculated in order to locate the centroid *C*.

If an area has **two axes of symmetry**, as illustrated in Fig. D-3, the position of the centroid can be determined by inspection because it lies at the intersection of the axes of symmetry.

FIGURE D-4
Area that is symmetric about a point

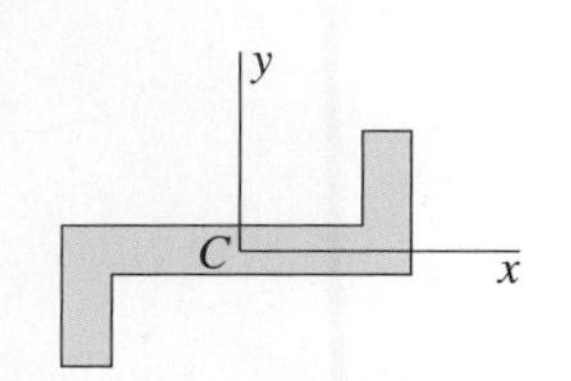

An area of the type shown in Fig. D-4 is **symmetric about a point**. It has no axes of symmetry, but there is a point (called the **center of symmetry**) such that every line drawn through that point contacts the area in a symmetrical manner. The centroid of such an area coincides with the center of symmetry, so the centroid can be located by inspection.

If an area has **irregular boundaries** not defined by simple mathematical expressions, locate the centroid by numerically evaluating the integrals in Eqs. (D-3a and b). The simplest procedure is to divide the geometric figure into small finite elements and replace the integrations with summations. Denote the area of the *i*th element by ΔA_i; then the expressions for the summations are

$$A = \sum_{i=1}^{n} \Delta A_i \quad Q_x = \sum_{i=1}^{n} \bar{y}_i \Delta A_i \quad Q_y = \sum_{i=1}^{n} \bar{x}_i \Delta A_i \tag{D-4a,b,c}$$

(Photo courtesy of Louis Geschwinder)

The centroid of wide-flange steel sections lies at the intersection of the axes of symmetry

in which *n* is the total number of elements, $\bar{y}_i$ is the *y* coordinate of the centroid of the *i*th element, and $\bar{x}_i$ is the *x* coordinate of the centroid of the *i*th element. Replacing the integrals in Eqs. (D-3a and b) by the corresponding summations leads to the following formulas for the coordinates of the centroid:

$$\bar{x} = \frac{Q_y}{A} = \frac{\sum_{i=1}^{n} \bar{x}_i \Delta A_i}{\sum_{i=1}^{n} \Delta A_i} \qquad \bar{y} = \frac{Q_x}{A} = \frac{\sum_{i=1}^{n} \bar{y}_i \Delta A_i}{\sum_{i=1}^{n} \Delta A_i} \tag{D-5a,b}$$

The accuracy of the calculations for $\bar{x}$ and $\bar{y}$ depends upon how closely the selected elements fit the actual area. If they fit exactly, the results are exact. Many computer programs for locating centroids use a numerical scheme similar to the one expressed by Eqs. (D-5a and b).

Example D-1

FIGURE D-5

Example D-1: Centroid of a parabolic semisegment

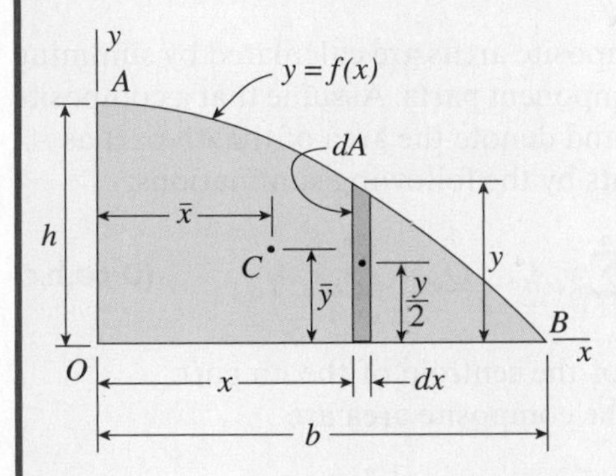

A parabolic semisegment OAB is bounded by the x axis, the y axis, and a parabolic curve having its vertex at A (Fig. D-5). The equation of the curve is

$$y = f(x) = h\left(1 - \frac{x^2}{b^2}\right) \tag{a}$$

in which b is the base and h is the height of the semisegment.

Locate the centroid C of the semisegment.

Solution:

To determine the coordinates $\bar{x}$ and $\bar{y}$ of the centroid C (Fig. D-5), use Eqs. (D-3a and b). Begin by selecting an element of area dA in the form of a thin vertical strip of width dx and height y. The area of this differential element is

$$dA = ydx = h\left(1 - \frac{x^2}{b^2}\right)dx \tag{b}$$

Therefore, the area of the parabolic semisegment is

$$A = \int dA = \int_0^b h\left(1 - \frac{x^2}{b^2}\right)dx = \frac{2bh}{3} \tag{c}$$

Note that this area is two-thirds of the area of the surrounding rectangle.

The first moment of an element of area dA with respect to an axis is obtained by multiplying the area of the element by the distance from its centroid to the axis. Since the x and y coordinates of the centroid of the element shown in Fig. D-5 are x and $y/2$, respectively, the first moments of the element with respect to the x and y axes are

$$Q_x = \int \frac{y}{2}dA = \int_0^b \frac{h^2}{2}\left(1 - \frac{x^2}{b^2}\right)^2 dx = \frac{4bh^2}{15} \tag{d}$$

$$Q_y = \int xdA = \int_0^b hx\left(1 - \frac{x^2}{b^2}\right)dx = \frac{b^2h}{4} \tag{e}$$

in which dA has been substituted from Eq. (b).

Now determine the coordinates of the centroid C:

$$\bar{x} = \frac{Q_y}{A} = \frac{3b}{8} \qquad \bar{y} = \frac{Q_x}{A} = \frac{2h}{5} \tag{f,g}$$

These results agree with the formulas listed in Appendix E, Case 17.

Notes: The centroid C of the parabolic semisegment also may be located by taking the element of area dA as a horizontal strip of height dy and width:

$$x = b\sqrt{1 - \frac{y}{h}} \tag{h}$$

This expression is obtained by solving Eq. (a) for x in terms of y.

FIGURE D-6

Centroid of a composite area consisting of two parts

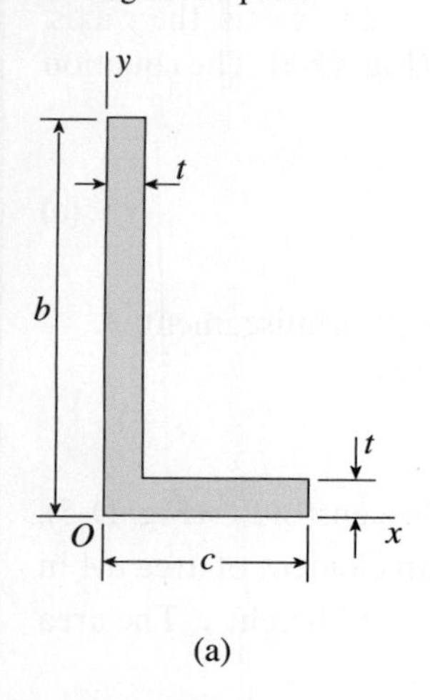

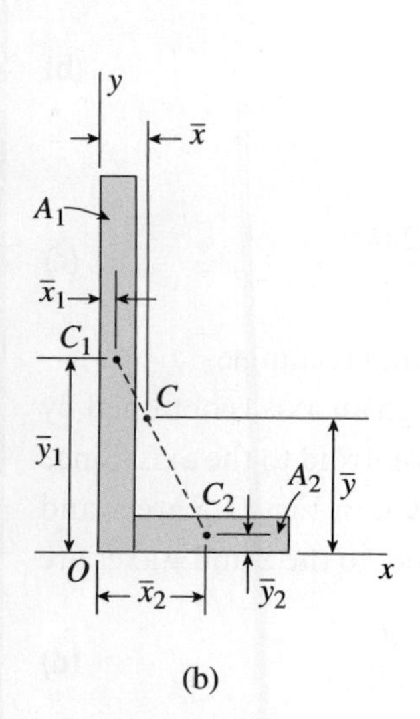

D.2 Centroids of Composite Areas

In engineering work it is often of interest to locate the centroids of areas composed of several parts, each part having a familiar geometric shape, such as a rectangle or a circle. Examples of such **composite areas** are the cross sections of beams and columns, which usually consist of rectangular elements (for instance, see Figs. D-2, D-3, and D-4).

The **areas and first moments** of composite areas are calculated by summing the corresponding properties of the component parts. Assume that a composite area is divided into a total of n parts, and denote the area of the ith part as A_i. Then obtain the area and first moments by the following summations:

$$A = \sum_{i=1}^{n} A_i \qquad Q_x = \sum_{i=1}^{n} \bar{y}_i A_i \qquad Q_y = \sum_{i=1}^{n} \bar{x}_i A_i \tag{D-6a,b,c}$$

in which $\bar{x}_i$ and $\bar{y}_i$ are the coordinates of the centroid of the ith part.

The **coordinates of the centroid** of the composite area are

$$\bar{x} = \frac{Q_y}{A} = \frac{\sum_{i=1}^{n} \bar{x}_i A_i}{\sum_{i=1}^{n} A_i} \qquad \bar{y} = \frac{Q_x}{A} = \frac{\sum_{i=1}^{n} \bar{y}_i A_i}{\sum_{i=1}^{n} A_i} \tag{D-7a,b}$$

Since the composite area is represented exactly by the n parts, the preceding equations give exact results for the coordinates of the centroid.

To illustrate the use of Eqs. (D-7a and b), consider the L-shaped area (or angle section) shown in Fig. D-6a. This area has side dimensions b and c and thickness t. The area can be divided into two rectangles of areas A_1 and A_2 with centroids C_1 and C_2, respectively (Fig. D-6b). The areas and centroidal coordinates of these two parts are

$$A_1 = +bt \qquad \bar{x}_1 = \frac{t}{2} \qquad \bar{y}_1 = \frac{b}{2}$$

$$A_2 = (c - t)t \qquad \bar{x}_2 = \frac{c + t}{2} \qquad \bar{y}_2 = \frac{t}{2}$$

Therefore, the area and first moments of the composite area [from Eqs. (D-6a, b, and c)] are

$$A = A_1 + A_2 = t(b + c - t)$$

$$Q_x = \bar{y}_1 A_1 + \bar{y}_2 A_2 = \frac{t}{2}(b^2 + ct - t^2)$$

$$Q_y = \bar{x}_1 A_1 + \bar{x}_2 A_2 = \frac{t}{2}(bt + c^2 - t^2)$$

Finally, obtain the coordinates $\bar{x}$ and $\bar{y}$ of the centroid C of the composite area (Fig. D-6b) from Eqs. (D-7a and b):

$$\bar{x} = \frac{Q_y}{A} = \frac{bt + c^2 - t^2}{2(b + c - t)} \qquad \bar{y} = \frac{Q_x}{A} = \frac{b^2 + ct - t^2}{2(b + c - t)} \tag{D-8a,b}$$

A similar procedure can be used for more complex areas, as illustrated in Example D-2.

Note 1: When a composite area is divided into only two parts, the centroid C of the entire area lies on the line joining the centroids C_1 and C_2 of the two parts (as shown in Fig. D-6b for the L-shaped area).

Note 2: When using the formulas for composite areas [see Eqs. (D-6) and (D-7)], treat the *absence* of an area using subtraction. This procedure is useful when there are cutouts or holes in the figure.

For instance, consider the area shown in Fig. D-7a. Analyze this figure as a composite area by subtracting the properties of the inner rectangle *efgh* from the corresponding properties of the outer rectangle *abcd*. (From another viewpoint, think of the outer rectangle as a "positive area" and the inner rectangle as a "negative area.")

Similarly, if an area has a hole (Fig. D-7b), subtract the properties of the area of the hole from those of the outer rectangle.

FIGURE D-7

Composite areas with a cutout and a hole

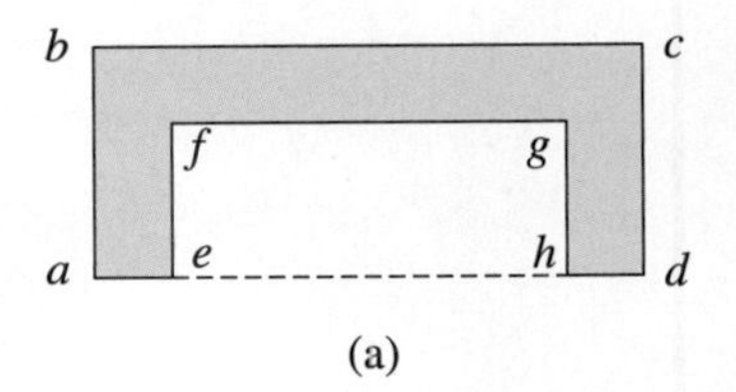

(a)

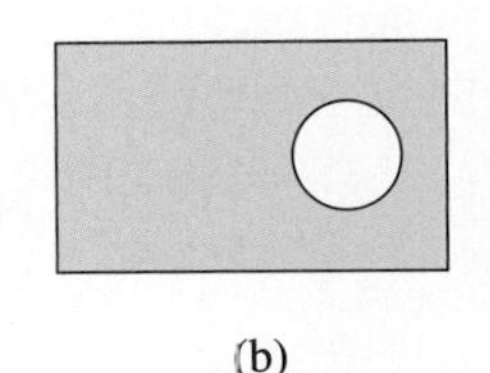
(b)

Example D-2

FIGURE D-8

Example D-2: Centroid of a composite area

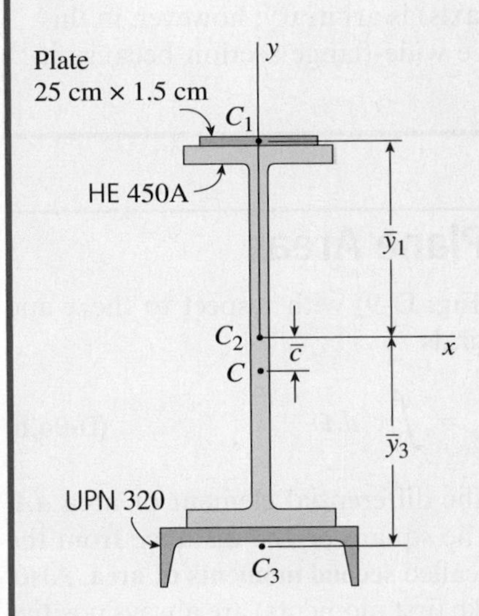

The cross section of a steel beam is constructed of a HE 450A wide-flange section with a 25 cm × 1.5 cm cover plate welded to the top flange and a UPN 320 channel section welded to the bottom flange (Fig. D-8).

Locate the centroid C of the cross-sectional area.

Solution:

Denote the areas of the cover plate, the wide-flange section, and the channel section as areas A_1, A_2, and A_3, respectively. The centroids of these three areas are labeled C_1, C_2, and C_3, respectively, in Fig. D-8. Note that the composite area has an axis of symmetry, so all centroids lie on that axis. The three partial areas are

$$A_1 = (25\,\text{cm})(1.5\,\text{cm}) = 37.5\,\text{cm}^2 \qquad A_2 = 178\,\text{cm}^2 \qquad A_3 = 75.8\,\text{cm}^2$$

in which the areas A_2 and A_3 are obtained from Tables F-1 and F-3 of Appendix F.

Place the origin of the x and y axes at the centroid C_2 of the wide-flange section. Then the distances from the x axis to the centroids of the three areas are as follows:

$$\bar{y}_1 = \frac{440\text{ mm}}{2} + \frac{15\text{ mm}}{2} = 227.5\text{ mm}$$

$$\bar{y}_2 = 0 \quad \bar{y}_3 = \frac{440\text{ mm}}{2} + 26\text{ mm} = 246\text{ mm}$$

in which the pertinent dimensions of the wide-flange and channel sections are obtained from Tables F-1 and F-3.

The area A and first moment Q_x of the entire cross section are obtained from Eqs. (D-6a and b) as

$$A = \sum_{i=1}^{n} A_i = A_1 + A_2 + A_3$$

$$= 37.5\,\text{cm}^2 + 178\,\text{cm}^2 + 75.8\,\text{cm}^2 = 291.3\,\text{cm}^2$$

$$Q_x = \sum_{i=1}^{n} \overline{y}_i A_i = \overline{y}_1 A_1 + \overline{y}_2 A_2 + \overline{y}_3 A_3$$

$$= (22.75\,\text{cm})(37.5\,\text{cm}^2) + 0 - (24.6\,\text{cm})(75.8\,\text{cm}^2) = -1012\,\text{cm}^3$$

Now find the coordinate $\overline{y}$ to the centroid C of the composite area from Eq. (D-7b):

$$\overline{y} = \frac{Q_x}{A} = \frac{-1012\,\text{cm}^3}{291.3\,\text{cm}^2} = -34.726\,\text{mm}$$

Since $\overline{y}$ is positive in the positive direction of the y axis, the minus sign means that the centroid C of the composite area is located below the x axis, as shown in Fig. D-8. Thus, the distance $\overline{c}$ between the x axis and the centroid C is

$$\overline{c} = 34.73\,\text{mm}$$

Note that the position of the reference axis (the x axis) is arbitrary; however, in this example it was placed through the centroid of the wide-flange section because it slightly simplifies the calculations.

FIGURE D-9

Plane area of arbitrary shape

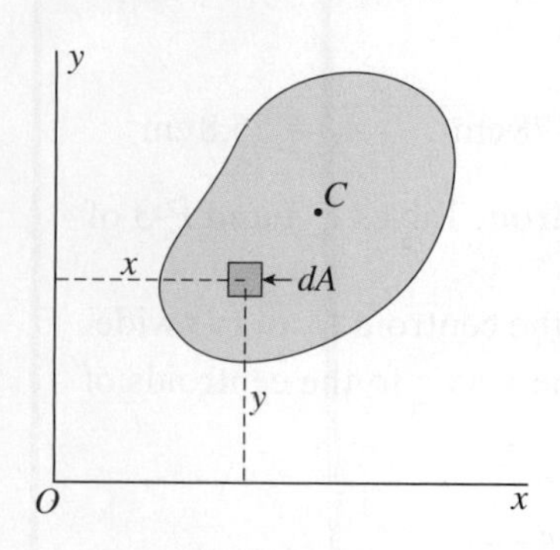

D.3 Moments of Inertia of Plane Areas

The **moments of inertia** of a plane area (Fig. D-9) with respect to the x and y axes, respectively, are defined by the integrals

$$I_x = \int y^2 dA \qquad I_y = \int x^2 dA \qquad \textbf{(D-9a,b)}$$

in which x and y are the coordinates of the differential element of area dA. Because the element dA is multiplied by the square of the distance from the reference axis, moments of inertia are also called **second moments of area**. Also, note that moments of inertia of areas (unlike first moments) are always positive quantities.

To illustrate how moments of inertia are obtained by integration, consider a rectangle having width b and height h (Fig. D-10). The x and y axes have their origin at the centroid C. For convenience, use a differential element of area dA in the form of a thin horizontal strip of width b and height dy (therefore, $dA = bdy$). Since all parts of the elemental strip are the same distance from the x axis, express the moment of inertia I_x with respect to the x axis as

$$I_x = \int y^2 dA = \int_{-h/2}^{h/2} y^2 b dy = \frac{bh^3}{12} \qquad \textbf{(D-10)}$$

In a similar manner, use an element of area in the form of a vertical strip with area $dA = h\,dx$ and obtain the moment of inertia with respect to the y axis:

$$I_y = \int x^2 dA = \int_{-b/2}^{b/2} x^2 h dx = \frac{hb^3}{12} \quad \text{(D-11)}$$

If a different set of axes is selected, the moments of inertia will have different values. For instance, consider axis B–B at the base of the rectangle (Fig. D-10). If this axis is selected as the reference, define y as the coordinate distance from that axis to the element of area dA. Then the calculations for the moment of inertia become

$$I_{BB} = \int y^2 dA = \int_0^h y^2 b dy = \frac{bh^3}{3} \quad \text{(D-12)}$$

Note that the moment of inertia with respect to axis B–B is larger than the moment of inertia with respect to the centroidal x axis. In general, the moment of inertia increases as the reference axis is moved parallel to itself farther from the centroid.

The moment of inertia of a **composite area** with respect to any particular axis is the sum of the moments of inertia of its parts with respect to that same axis. An example is the hollow box section shown in Fig. D-11a, where the x and y axes are axes of symmetry through the centroid C. The moment of inertia I_x with respect to the x axis is equal to the algebraic sum of the moments of inertia of the outer and inner rectangles. (As explained earlier, think of the inner rectangle as a "negative area" and the outer rectangle as a "positive area.") Therefore,

$$I_x = \frac{bh^3}{12} - \frac{b_1 h_1^3}{12} \quad \text{(D-13)}$$

This same formula applies to the channel section shown in Fig. D-11b, where the cutout is considered a "negative area."

For the hollow box section, use a similar technique to obtain the moment of inertia I_y with respect to the vertical axis. However, in the case of the channel section, the determination of the moment of inertia I_y requires the use of the parallel-axis theorem, which is described in the next section (Section D.4).

Formulas for moments of inertia are listed in Appendix E. For shapes not shown, the moments of inertia usually can be obtained by using the listed formulas in conjunction with the parallel-axis theorem. If an area is of such irregular shape that its moments of inertia cannot be obtained in this manner, then numerical methods can be used. The procedure is to divide the area into small elements of area ΔA_i, multiply each such area by the square of its distance from the reference axis, and then sum the products.

Radius of Gyration

A distance known as the **radius of gyration** is defined as the square root of the moment of inertia of the area divided by the area itself; thus,

$$r_x = \sqrt{\frac{I_x}{A}} \qquad r_y = \sqrt{\frac{I_y}{A}} \quad \text{(D-14a,b)}$$

in which r_x and r_y denote the radii of gyration with respect to the x and y axes, respectively. Since a moment of inertia has units of length to the fourth power and area has units of length to the second power, radius of gyration has units of length.

FIGURE D-10

Moments of inertia of a rectangle

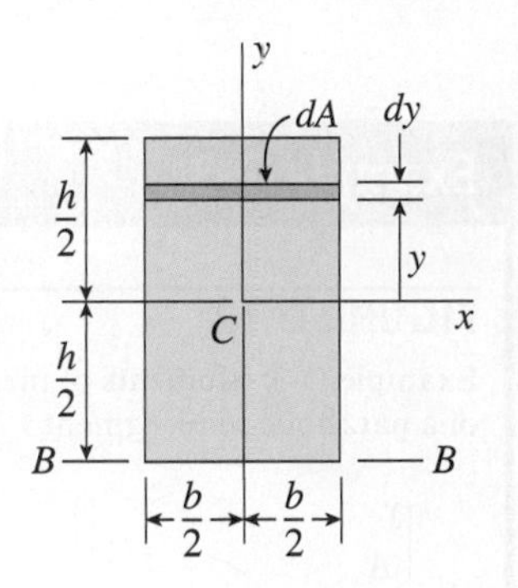

FIGURE D-11

Composite areas

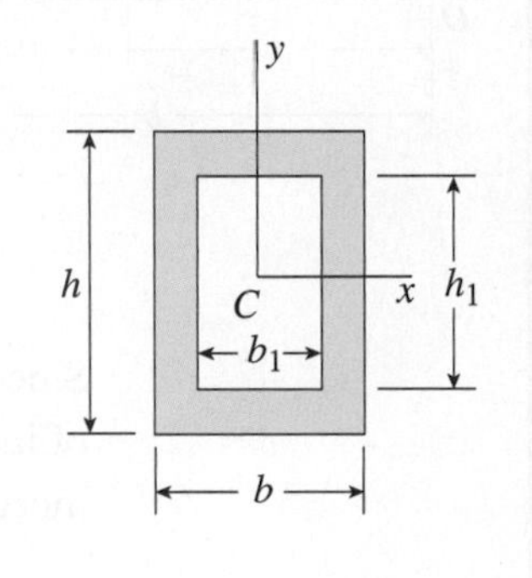

(a)

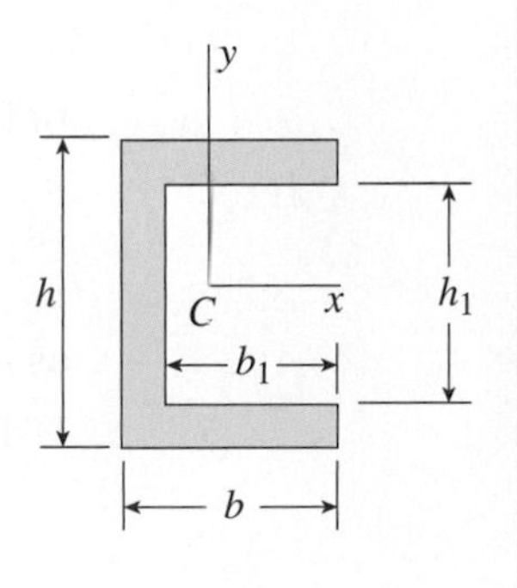

(b)

Although the radius of gyration of an area does not have an obvious physical meaning, consider it to be the distance (from the reference axis) at which the entire area could be concentrated and still have the same moment of inertia as the original area.

Example D-3

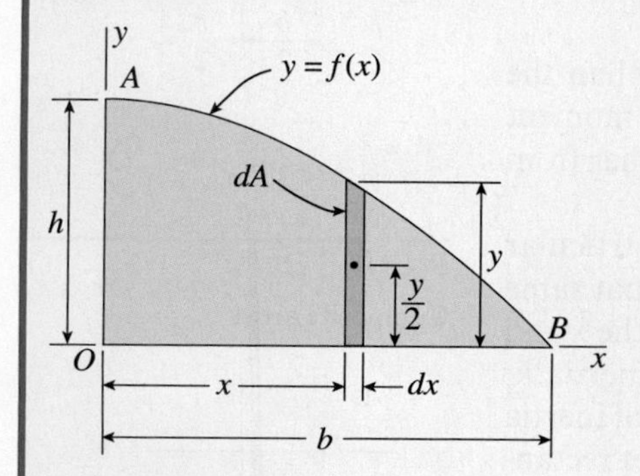

FIGURE D-12

Example D-3: Moments of inertia of a parabolic semisegment

Determine the moments of inertia I_x and I_y for the parabolic semisegment OAB shown in Fig. D-12. The equation of the parabolic boundary is

$$y = f(x) = h\left(1 - \frac{x^2}{b^2}\right) \quad \text{(a)}$$

(This same area was considered previously in Example D-1.)

Solution:

To determine the moments of inertia by integration, use Eqs. (D-9a and b). The differential element of area dA is selected as a vertical strip of width dx and height y, as shown in Fig. D-12. The area of this element is

$$dA = ydx = h\left(1 - \frac{x^2}{b^2}\right)dx \quad \text{(b)}$$

Since every point in this element is at the same distance from the y axis, the moment of inertia of the element with respect to the y axis is $x^2 dA$. Therefore, the moment of inertia of the entire area with respect to the y axis is obtained as

$$I_y = \int x^2 dA = \int_0^b x^2 h\left(1 - \frac{x^2}{b^2}\right)dx = \frac{2hb^3}{15} \quad \text{(c)}$$

To obtain the moment of inertia with respect to the x axis, note that the differential element of area dA has a moment of inertia dI_x with respect to the x axis equal to

$$dI_x = \frac{1}{3}(dx)y^3 = \frac{y^3}{3}dx$$

as obtained from Eq. (D-12). Hence, the moment of inertia of the entire area with respect to the x axis is

$$I_x = \int_0^b \frac{y^3}{3}dx = \int_0^b \frac{h^3}{3}\left(1 - \frac{x^2}{b^2}\right)^3 dx = \frac{16bh^3}{105} \quad \text{(d)}$$

These same results for I_x and I_y can be obtained by using an element in the form of a horizontal strip of area $dA = xdy$ or by using a rectangular element of area $dA = dxdy$ and performing a double integration. Also, note that the preceding formulas for I_x and I_y agree with those given in Case 17 of Appendix E.

D.4 Parallel-Axis Theorem for Moments of Inertia

The **parallel-axis theorem** gives the relationship between the moment of inertia with respect to a centroidal axis and the moment of inertia with respect to any parallel axis.

To derive the theorem, consider an area of arbitrary shape with centroid C (Fig. D-13). Also consider two sets of coordinate axes: (1) the $x_c y_c$ axes with origin at the centroid and (2) a set of parallel xy axes with origin at any point O. The distances between the two sets of parallel axes are denoted d_1 and d_2. Also, identify an element of area dA having coordinates x and y with respect to the centroidal axes.

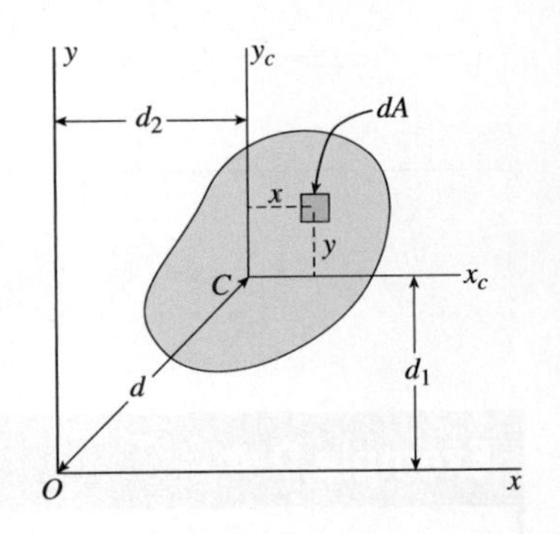

FIGURE D-13

Derivation of parallel-axis theorem

From the definition of the moment of inertia, write the following equation for the moment of inertia I_x with respect to the x axis:

$$I_x = \int (y + d_1)^2 dA = \int y^2 dA + 2d_1 \int y\, dA + d_1^2 \int dA \qquad \textbf{(D-15)}$$

The first integral on the right-hand side is the moment of inertia I_{xc} with respect to the x_c axis. The second integral is the first moment of the area with respect to the x_c axis (this integral equals zero because the x_c axis passes through the centroid). The third integral is the area A itself. Therefore, the preceding equation reduces to

$$I_x = I_{xc} + Ad_1^2 \qquad \textbf{(D-16a)}$$

Proceeding in the same manner for the moment of inertia with respect to the y axis gives

$$I_y = I_{yc} + Ad_2^2 \qquad \textbf{(D-16b)}$$

Equations (D-16a and b) represent the **parallel-axis theorem for moments of inertia**:

The moment of inertia of an area with respect to any axis in its plane is equal to the moment of inertia with respect to a parallel centroidal axis plus the product of the area and the square of the distance between the two axes.

To illustrate the use of the theorem, consider again the rectangle shown in Fig. D-10. Knowing that the moment of inertia about the x axis, which is through the centroid, is equal to $bh^3/12$ [see Eq. (D-10) of Section D.3]), determine the moment of inertia I_{BB} about the base of the rectangle by using the parallel-axis theorem:

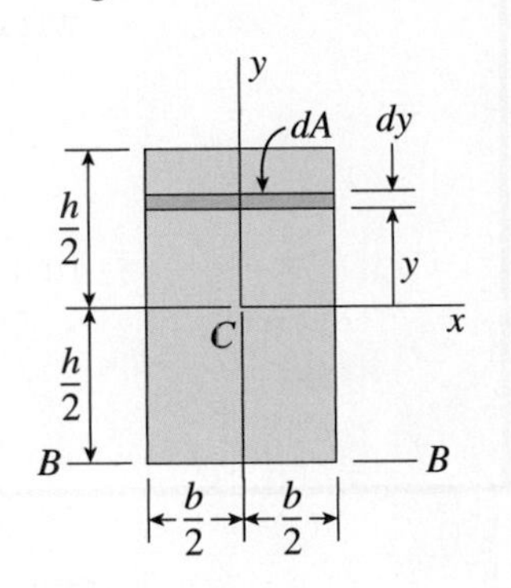

FIGURE D-10 (REPEATED)

Moments of inertia of a rectangle

$$I_{BB} = I_x + Ad^2 = \frac{bh^3}{12} + bh\left(\frac{h}{2}\right)^2 = \frac{bh^3}{3}$$

This result agrees with the moment of inertia obtained previously by integration [see Eq. (D-12) of Section D.3].

From the parallel-axis theorem, note that the moment of inertia increases as the axis is moved parallel to itself farther from the centroid. Therefore, the moment of inertia about a centroidal axis is the least moment of inertia of an area (for a given direction of the axis).

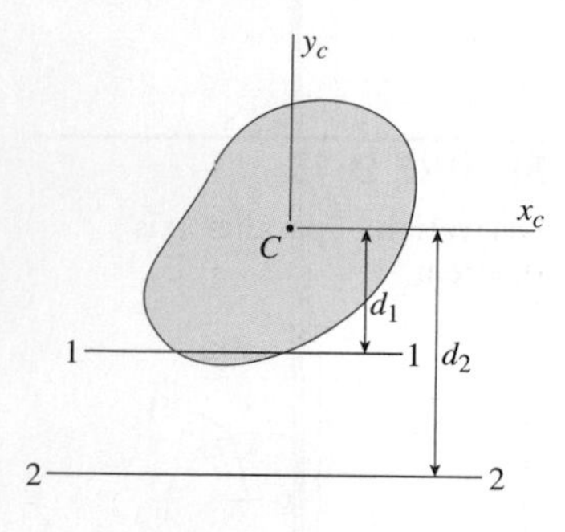

FIGURE D-14

Plane area with two parallel noncentroidal axes (axes 1–1 and 2–2)

When using the parallel-axis theorem, it is essential to remember that one of the two parallel axes *must* be a centroidal axis. If it is necessary to find the moment of inertia I_2 about a noncentroidal axis 2–2 (Fig. D-14) when the moment of inertia I_1 about another noncentroidal (and parallel) axis 1–1 is known, apply the parallel-axis theorem twice. First, find the centroidal moment of inertia I_{xc} from the known moment of inertia I_1:

$$I_{xc} = I_1 - Ad_1^2 \tag{D-17}$$

Then find the moment of inertia I_2 from the centroidal moment of inertia:

$$I_2 = I_{xc} + Ad_2^2 = I_1 + A(d_2^2 - d_1^2) \tag{D-18}$$

This equation shows again that the moment of inertia increases with increasing distance from the centroid of the area.

Example D-4

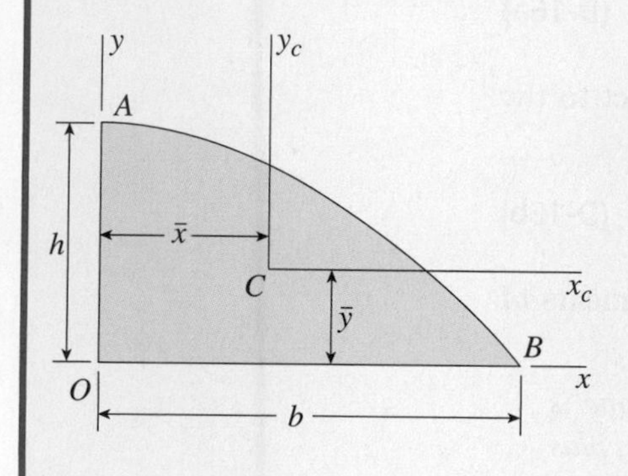

FIGURE D-15

Example D-4: Parallel-axis theorem

The parabolic semisegment OAB shown in Fig. D-15 has base b and height h. Using the parallel-axis theorem, determine the moments of inertia I_{xc} and I_{yc} with respect to the centroidal axes x_c and y_c.

Solution:

Use the parallel-axis theorem (rather than integration) to find the centroidal moments of inertia because the area A, the centroidal coordinates $\bar{x}$ and $\bar{y}$, and the moments of inertia I_x and I_y with respect to the x and y axes are already known. These quantities were obtained earlier in Examples D-1 and D-3 (they also are listed in Case 17 of Appendix E) and are repeated here:

$$A = \frac{2bh}{3} \quad \bar{x} = \frac{3b}{8} \quad \bar{y} = \frac{2h}{5} \quad I_x = \frac{16bh^3}{105} \quad I_y = \frac{2hb^3}{15}$$

To obtain the moment of inertia with respect to the x_c axis, use Eq. (D-17) and write the parallel-axis theorem as

$$I_{xc} = I_x - A\bar{y}^2 = \frac{16bh^3}{105} - \frac{2bh}{3}\left(\frac{2h}{5}\right)^2 = \frac{8bh^3}{175} \tag{D-19a}$$

In a similar manner, obtain the moment of inertia with respect to the y_c axis:

$$I_{yc} = I_y - A\bar{x}^2 = \frac{2hb^3}{15} - \frac{2bh}{3}\left(\frac{3b}{8}\right)^2 = \frac{19hb^3}{480} \tag{D-19b}$$

Example D-5

FIGURE D-16

Example D-5: Moment of inertia of a composite area

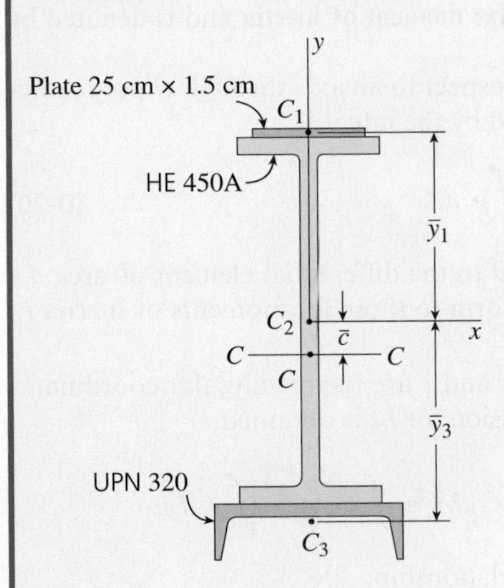

Determine the moment of inertia I_c with respect to the horizontal axis C–C through the centroid C of the beam cross section shown in Fig. D-16. (The position of the centroid C was determined previously in Example D-2 of Section D.2.)

Note: From beam theory (Chapter 5), axis C–C is the neutral axis for bending of this beam; therefore, the moment of inertia I_c must be determined in order to calculate the stresses and deflections of this beam.

Solution:

Find the moment of inertia I_c with respect to axis C–C by applying the parallel-axis theorem to each individual part of the composite area. The area divides naturally into three parts: (1) the cover plate, (2) the wide-flange section, and (3) the channel section. The following areas and centroidal distances were obtained previously in Example D-2:

$$A_1 = 37.5\,\text{cm}^2 \quad A_2 = 178\,\text{cm}^2 \quad A_3 = 75.8\,\text{cm}^2$$

$$\bar{y}_1 = 227.5\,\text{mm} \quad \bar{y}_2 = 0 \quad \bar{y}_3 = 246\,\text{mm} \quad \bar{c} = 34.73\,\text{mm}$$

The moments of inertia of the three parts with respect to horizontal axes through their own centroids C_1, C_2, and C_3 are

$$I_1 = \frac{bh^3}{12} = \frac{1}{12}(25\,\text{cm})(1.5\,\text{cm})^3 = 7.031\,\text{cm}^4$$

$$I_2 = 63720\,\text{cm}^4 \quad I_3 = 597\,\text{cm}^4$$

The moments of inertia I_2 and I_3 are obtained from Tables F-1 and F-3, respectively, of Appendix F.

Now use the parallel-axis theorem to calculate the moments of inertia about axis C–C for each of the three parts of the composite area:

$$(I_c)_1 = I_1 + A_1(\bar{y}_1 + \bar{c})^2 = 7.031\,\text{cm}^4 + (37.5\,\text{cm}^2)(26.22\,\text{cm})^2 = 25790\,\text{cm}^4$$

$$(I_c)_2 = I_2 + A_2\bar{c}^2 = 63720\,\text{cm}^4 + (178\,\text{cm}^2)(34.73\,\text{cm})^2 = 65870\,\text{cm}^4$$

$$(I_c)_3 = I_3 + A_3(\bar{y}_3 - \bar{c})^2 = 597\,\text{cm}^4 + (75.8\,\text{cm}^2)(21.13\,\text{cm})^2 = 34430\,\text{cm}^4$$

The sum of these individual moments of inertia gives the moment of inertia of the entire cross-sectional area about its centroidal axis C–C:

$$I_c = (I_c)_1 + (I_c)_2 + (I_c)_3 = 1.261 \times 10^5\,\text{cm}^4$$

This example shows how to calculate moments of inertia of composite areas by using the parallel-axis theorem.

FIGURE D-17

Plane area of arbitrary shape

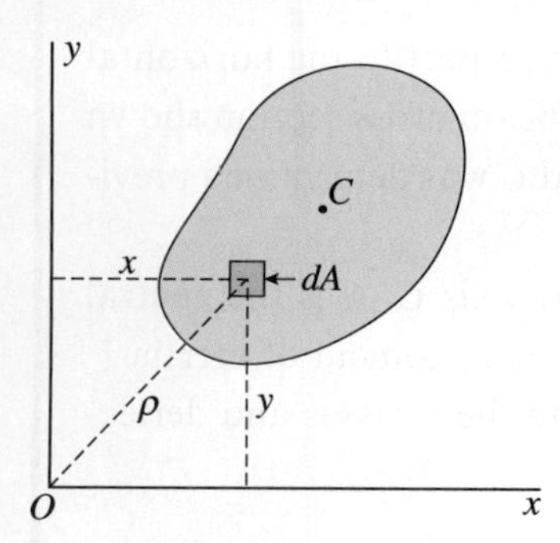

D.5 Polar Moments of Inertia

The moments of inertia discussed in the preceding sections are defined with respect to axes lying in the plane of the area itself, such as the x and y axes in Fig. D-17. Now consider an axis *perpendicular* to the plane of the area and intersecting the plane at the origin O. The moment of inertia with respect to this perpendicular axis is called the **polar moment of inertia** and is denoted by the symbol I_P.

The polar moment of inertia with respect to an axis through O perpendicular to the plane of the figure is defined by the integral

$$I_P = \int \rho^2 dA \tag{D-20}$$

in which ρ is the distance from point O to the differential element of area dA (Fig. D-17). This integral is similar in form to those for moments of inertia I_x and I_y [see Eqs. (D-9a and b)].

Inasmuch as $\rho^2 = x^2 + y^2$, where x and y are the rectangular coordinates of the element dA, the following expression for I_P is obtained:

$$I_P = \int \rho^2 \, dA = \int (x^2 + y^2) dA = \int x^2 dA + \int y^2 dA$$

This leads to the following important relationship:

$$I_P = I_x + I_y \tag{D-21}$$

This equation shows that the polar moment of inertia with respect to an axis perpendicular to the plane of the figure at any point O is equal to the sum of the moments of inertia with respect to *any* two perpendicular axes x and y passing through that same point and lying in the plane of the figure.

For convenience, I_P is usually referred to simply as the polar moment of inertia with respect to point O, without mentioning that the axis is perpendicular to the plane of the figure. Also, to distinguish them from **polar** moments of inertia, I_x and I_y can be referred to as **rectangular** moments of inertia.

Polar moments of inertia with respect to various points in the plane of an area are related by the **parallel-axis theorem for polar moments of inertia**. This theorem can be derived by referring again to Fig. D-13. Denote the polar moments of inertia with respect to the origin O and the centroid C by $(I_P)_O$ and $(I_P)_C$, respectively. Then, using Eq. (D-21), write

$$(I_P)_O = I_x + I_y \qquad (I_P)_C = I_{xc} + I_{yc} \tag{D-22}$$

FIGURE D-13 (REPEATED)

Derivation of parallel-axis theorem

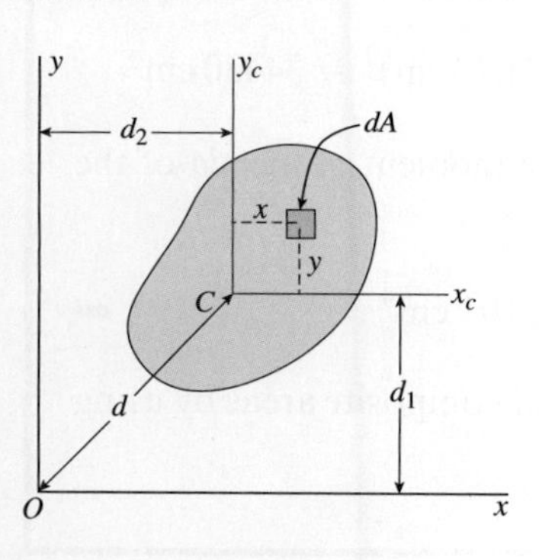

Now refer to the parallel-axis theorems derived in Section D.4 for rectangular moments of inertia [see Eqs.(D-16a and b)]. Adding those two equations gives

$$I_x + I_y = I_{xc} + I_{yc} + A(d_1^2 + d_2^2)$$

Substitute from Eqs. (D-22), noting that $d^2 = d_1^2 + d_2^2$ (Fig. D-13) to obtain

$$(I_P)_O = (I_P)_C + Ad^2 \tag{D-23}$$

This equation represents the **parallel-axis theorem** for polar moments of inertia:

The polar moment of inertia of an area with respect to any point O in its plane is equal to the polar moment of inertia with respect to the centroid C plus the product of the area and the square of the distance between points O and C.

To illustrate the determination of polar moments of inertia and the use of the parallel-axis theorem, consider a circle of radius r (Fig. D-18). Take a differential element of area dA in the form of a thin ring of radius ρ and thickness $d\rho$ (thus, $dA = 2\pi\rho\, d\rho$). Since every point in the element is at the same distance ρ from the center of the circle, the polar moment of inertia of the entire circle with respect to the center is

$$(I_P)_C = \int \rho^2 dA = \int_0^r 2\pi\rho^3 d\rho = \frac{\pi r^4}{2} \quad \textbf{(D-24)}$$

This result is listed in Case 9 of Appendix E.

The polar moment of inertia of the circle with respect to any point B on its circumference (Fig. D-18) can be obtained from the parallel-axis theorem:

$$(I_P)_B = (I_P)_C + Ad^2 = \frac{\pi r^4}{2} + \pi r^2(r^2) = \frac{3\pi r^4}{2} \quad \textbf{(D-25)}$$

As an incidental matter, note that the polar moment of inertia has its smallest value when the reference point is the centroid of the area.

A circle is a special case in which the polar moment of inertia can be determined by integration. However, most of the shapes encountered in engineering work do not lend themselves to this technique. Instead, polar moments of inertia are usually obtained by summing the rectangular moments of inertia for two perpendicular axes [Eq. (D-21)].

FIGURE D-18

Polar moment of inertia of a circle

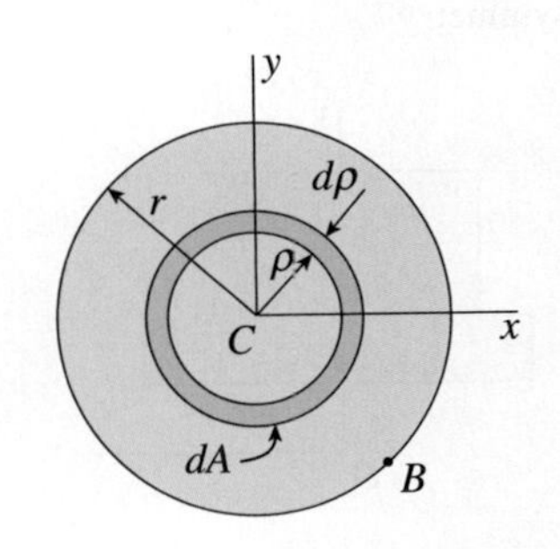

D.6 Products of Inertia

Referring to the area shown in Fig. D-19, define the **product of inertia** with respect to the x and y axes as

$$I_{xy} = \int xy\, dA \quad \textbf{(D-26)}$$

From this definition, note that each differential element of area dA is multiplied by the product of its coordinates. As a consequence, products of inertia may be positive, negative, or zero, depending upon the position of the xy axes with respect to the area.

If the area lies entirely in the first quadrant of the axes (as in Fig. D-19), the product of inertia is positive because every element dA has positive coordinates x and y. If the area lies entirely in the second quadrant, the product of inertia is negative because every element has a positive y coordinate and a negative x coordinate. Similarly, areas entirely within the third and fourth quadrants have positive and negative products of inertia, respectively. When the area is located in more than one quadrant, the sign of the product of inertia depends upon the distribution of the area within the quadrants.

FIGURE D-19

Plane area of arbitrary shape

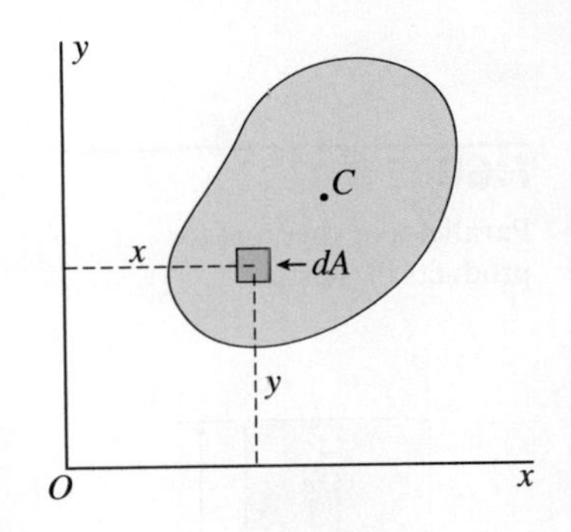

FIGURE D-20

The product of inertia equals zero when one axis is an axis of symmetry

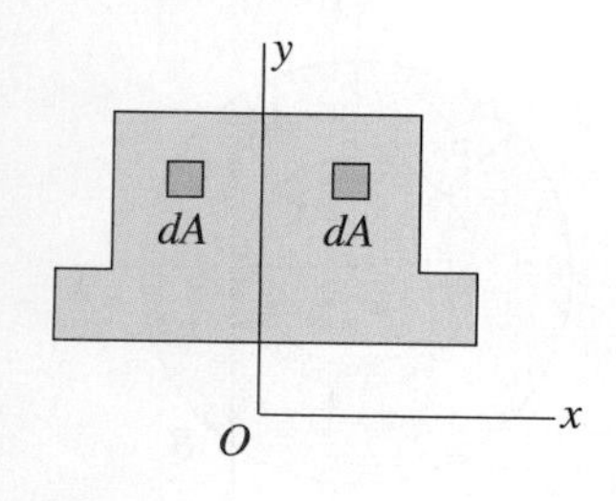

A special case arises when one of the axes is an **axis of symmetry** of the area. For instance, consider the area shown in Fig. D-20, which is symmetric about the y axis. For every element dA having coordinates x and y, there exists an equal and symmetrically located element dA having the same y coordinate but an x coordinate of opposite sign. Therefore, the products $xydA$ cancel each other, and the integral in Eq. (D-26) vanishes. Thus, *the product of inertia of an area is zero with respect to any pair of axes in which at least one axis is an axis of symmetry of the area.*

As examples of the preceding rule, the product of inertia I_{xy} equals zero for the areas shown in Figs. D-10, D-11, D-16, and D-18. In contrast, the product of inertia I_{xy} has a positive nonzero value for the area shown in Fig. D-15. (These observations are valid for products of inertia with respect to the particular xy axes shown in the figures. If the axes are shifted to another position, the product of inertia may change.)

Products of inertia of an area with respect to parallel sets of axes are related by a **parallel-axis theorem** that is analogous to the corresponding theorems for rectangular moments of inertia and polar moments of inertia. To obtain this theorem, consider the area shown in Fig. D-21, which has centroid C and centroidal $x_c y_c$ axes. The product of inertia I_{xy} with respect to any other set of axes, parallel to the $x_c y_c$ axes, is

FIGURE D-21

Plane area of arbitrary shape

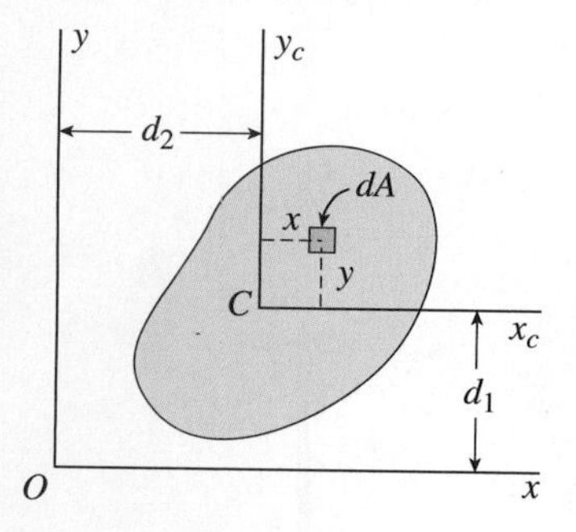

$$I_{xy} = \int (x + d_2)(y + d_1)dA$$

$$= \int xydA + d_1 \int xdA + d_2 \int ydA + d_1 d_2 \int dA$$

in which d_1 and d_2 are the coordinates of the centroid C with respect to the xy axes (thus, d_1 and d_2 may have positive or negative values).

The first integral in the last expression is the product of inertia $I_{xc\,yc}$ with respect to the centroidal axes; the second and third integrals equal zero because they are the first moments of the area with respect to the centroidal axes; and the last integral is the area A. Therefore, the preceding equation reduces to

$$I_{xy} = I_{xcyc} + Ad_1 d_2 \quad \textbf{(D-27)}$$

This equation represents the **parallel-axis theorem for products of inertia**:

The product of inertia of an area with respect to any pair of axes in its plane is equal to the product of inertia with respect to parallel centroidal axes plus the product of the area and the coordinates of the centroid with respect to the pair of axes.

FIGURE D-22

Parallel-axis theorem for products of inertia

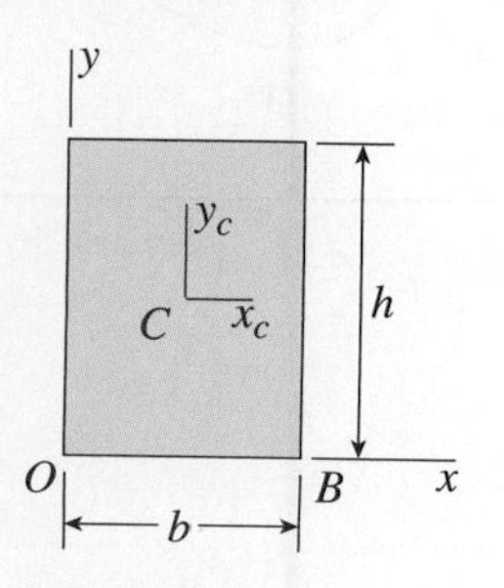

To demonstrate the use of this parallel-axis theorem, determine the product of inertia of a rectangle with respect to x-y axes having their origin at point O at the lower left-hand corner of the rectangle (Fig. D-22). The product of inertia with respect to the centroidal $x_c y_c$ axes is zero because of symmetry. Also, the coordinates of the centroid with respect to the x-y axes are

$$d_1 = \frac{h}{2} \qquad d_2 = \frac{b}{2}$$

Substitute into Eq. (D-27) to obtain

$$I_{xy} = I_{xcyc} + Ad_1 d_2 = 0 + bh\left(\frac{h}{2}\right)\left(\frac{b}{2}\right) = \frac{b^2 h^2}{4} \quad \textbf{(D-28)}$$

This product of inertia is positive because the entire area lies in the first quadrant. If the x-y axes are translated horizontally so that the origin moves to point B at the lower right-hand corner of the rectangle (Fig. D-22), the entire area lies in the second quadrant and the product of inertia becomes $-b^2h^2/4$.

The following example also illustrates the use of the parallel-axis theorem for products of inertia.

Example D-6

FIGURE D-23

Example D-6: Product of inertia of a Z-section

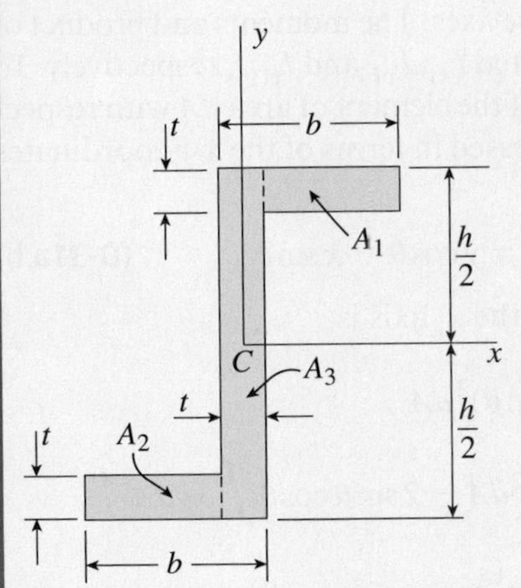

Determine the product of inertia I_{xy} of the Z-section shown in Fig. D-23. The section has width b, height h, and constant thickness t.

Solution:

To obtain the product of inertia with respect to the x-y axes through the centroid, divide the area into three parts and use the parallel-axis theorem. The parts are (1) a rectangle of width $b - t$ and thickness t in the upper flange, (2) a similar rectangle in the lower flange, and (3) a web rectangle with height h and thickness t.

The product of inertia of the web rectangle with respect to the x-y axes is zero (from symmetry). The product of inertia $(I_{xy})_1$ of the upper flange rectangle (with respect to the x-y axes) is determined by using the parallel-axis theorem:

$$(I_{xy})_1 = I_{xcyc} + Ad_1d_2 \tag{a}$$

in which I_{xcyc} is the product of inertia of the rectangle with respect to its own centroid, A is the area of the rectangle, d_1 is the y coordinate of the centroid of the rectangle, and d_2 is the x coordinate of the centroid of the rectangle. Thus,

$$I_{xcyc} = 0 \qquad A = (b - t)(t) \qquad d_1 = \frac{h}{2} - \frac{t}{2} \qquad d_2 = \frac{b}{2}$$

Substitute into Eq. (a) to obtain the product of inertia of the rectangle in the upper flange:

$$(I_{xy})_1 = I_{xcyc} + Ad_1d_2 = 0 + (b - t)(t)\left(\frac{h}{2} - \frac{t}{2}\right)\left(\frac{b}{2}\right) = \frac{bt}{4}(h - t)(b - t)$$

The product of inertia of the rectangle in the lower flange is the same. Therefore, the product of inertia of the entire Z-section is twice $(I_{xy})_1$, or

$$I_{xy} = \frac{bt}{2}(h - t)(b - t) \tag{D-29}$$

Note that this product of inertia is positive because the flanges lie in the first and third quadrants.

FIGURE D-24

Rotation of axes

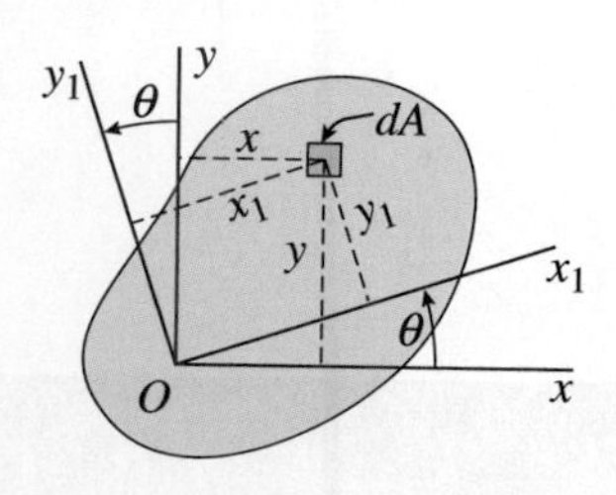

D.7 Rotation of Axes

The moments of inertia of a plane area depend upon the position of the origin and the orientation of the reference axes. For a given origin, the moments and product of inertia vary as the axes are rotated about that origin. The manner in which they vary, and the magnitudes of the maximum and minimum values, are discussed in this and the following section.

Consider the plane area shown in Fig. D-24, and assume that the x-y axes are a pair of arbitrarily located reference axes. The moments and products of inertia with respect to those axes are

$$I_x = \int y^2 dA \quad I_y = \int x^2 dA \quad I_{xy} = \int xy\, dA \qquad \textbf{(D-30a,b,c)}$$

in which x and y are the coordinates of a differential element of area dA.

The x_1-y_1 axes have the same origin as the x-y axes but are rotated through a counterclockwise angle θ with respect to those axes. The moments and product of inertia with respect to the x_1-y_1 axes are denoted I_{x1}, I_{y1}, and I_{x1y1}, respectively. To obtain these quantities, use the coordinates of the element of area dA with respect to the x_1-y_1 axes. These coordinates are expressed in terms of the x-y coordinates and the angle θ by geometry, as

$$x_1 = x\cos\theta + y\sin\theta \qquad y_1 = y\cos\theta - x\sin\theta \qquad \textbf{(D-31a,b)}$$

Then the moment of inertia with respect to the x_1 axis is

$$I_{x1} = \int y_1^2 dA = \int (y\cos\theta - x\sin\theta)^2 dA$$
$$= \cos^2\theta \int y^2 dA + \sin^2\theta \int x^2 dA - 2\sin\theta\cos\theta \int xy dA$$

or, by using Eqs. (D-30a, b, and c),

$$I_{x1} = I_x \cos^2\theta + I_y \sin^2\theta - 2I_{xy}\sin\theta\cos\theta \qquad \textbf{(D-32)}$$

Now introduce the following trigonometric identities (see Appendix C):

$$\cos^2\theta = \frac{1}{2}(1 + \cos 2\theta) \quad \sin^2\theta = \frac{1}{2}(1 - \cos 2\theta)$$
$$2\sin\theta\cos\theta = \sin 2\theta$$

Then Eq. (D-32) becomes

$$I_{x1} = \frac{I_x + I_y}{2} + \frac{I_x - I_y}{2}\cos 2\theta - I_{xy}\sin 2\theta \qquad \textbf{(D-33)}$$

In a similar manner, obtain the product of inertia with respect to the x_1-y_1 axes:

$$I_{x1y1} = \int x_1 y_1 dA = \int (x\cos\theta + y\sin\theta)(y\cos\theta - x\sin\theta)dA$$
$$= (I_x - I_y)\sin\theta\cos\theta + I_{xy}(\cos^2\theta - \sin^2\theta) \qquad \textbf{(D-34)}$$

Again use the trigonometric identities to obtain

$$I_{x1y1} = \frac{I_x - I_y}{2}\sin 2\theta + I_{xy}\cos 2\theta \qquad \textbf{(D-35)}$$

Equations (D-33) and (D-35) give the moment of inertia I_{x1} and the product of inertia I_{x1y1} with respect to the rotated axes in terms of the moments and product of inertia for the original axes. These equations are called the **transformation equations for moments and products of inertia.**

Note that these transformation equations have the same form as the transformation equations for plane stress [Eqs. (7-4a and b) of Section 7.2]. Upon comparing the two sets of equations, note that I_{x1} corresponds to σ_{x1}, I_{x1y1} corresponds to $-\tau_{x1y1}$, I_x corresponds to σ_x, I_y corresponds to σ_y, and I_{xy} corresponds to $-\tau_{xy}$. Therefore, moments and products of inertia can be analyzed using **Mohr's circle** (see Section 7.4).

The moment of inertia I_{y1} may be obtained by the same procedure used for finding I_{x1} and I_{x1y1}. However, a simpler procedure is to replace θ with $\theta + 90°$ in Eq. (D-33). The result is

$$I_{y1} = \frac{I_x + I_y}{2} - \frac{I_x - I_y}{2}\cos 2\theta + I_{xy}\sin 2\theta \tag{D-36}$$

This equation shows how the moment of inertia I_{y1} varies as the axes are rotated about the origin.

A useful equation related to moments of inertia is obtained by taking the sum of I_{x1} and I_{y1} [Eqs. (D-33) and (D-36)]. The result is

$$I_{x1} + I_{y1} = I_x + I_y \tag{D-37}$$

This equation shows that the sum of the moments of inertia with respect to a pair of axes remains constant as the axes are rotated about the origin. This sum is the polar moment of inertia of the area with respect to the origin. Note that Eq. (D-37) is analogous to Eq. (7-6) for stresses and Eq. (7-84) for strains.

D.8 Principal Axes and Principal Moments of Inertia

The transformation equations for moments and products of inertia [Eqs. (D-33), (D-35), and (D-36)] show how the moments and products of inertia vary as the angle of rotation θ varies. Of special interest are the maximum and minimum values of the moment of inertia. These values are known as the **principal moments of inertia**, and the corresponding axes are known as **principal axes.**

Principal Axes

To find the values of the angle θ that make the moment of inertia I_{x1} a maximum or a minimum, take the derivative with respect to θ of the expression on the right-hand side of Eq. (D-33) and set it equal to zero:

$$(I_x - I_y)\sin 2\theta + 2I_{xy}\cos 2\theta = 0 \tag{D-38}$$

Solving for θ from this equation gives

$$\tan 2\theta_p = -\frac{2I_{xy}}{I_x - I_y} \tag{D-39}$$

in which θ_p denotes the angle defining a principal axis. This same result is obtained by taking the derivative of I_{y1} (Eq. D-36).

Equation (D-39) yields two values of the angle $2\theta_p$ in the range from 0 to 360°; these values differ by 180°. The corresponding values of θ_p differ by 90° and define the two perpendicular principal axes. One of these axes corresponds to the maximum moment of inertia, and the other corresponds to the minimum moment of inertia.

Now examine the variation in the product of inertia I_{x1y1} as θ changes [see Eq. (D-35)]. If $\theta = 0$, the result is $I_{x1y1} = I_{xy}$, as expected. If $\theta = 90°$, the result is $I_{x1y1} = -I_{xy}$. Thus, during a 90° rotation, the product of inertia changes sign, which means that for an intermediate orientation of the axes the product of inertia must equal zero. To determine this orientation, set I_{x1y1} [Eq. (D-35)] equal to zero:

$$(I_x - I_y)\sin 2\theta + 2I_{xy}\cos 2\theta = 0$$

This equation is the same as Eq. (D-38), which defines the angle θ_p to the principal axes. This means that *the product of inertia is zero for the principal axes.*

The discussion in Section D.6 shows that the product of inertia of an area with respect to a pair of axes equals zero if at least one of the axes is an axis of symmetry. It follows that if an area has an axis of symmetry, that axis and any axis perpendicular to it constitute a set of principal axes.

The preceding observations may be summarized as follows: (1) principal axes through an origin O are a pair of orthogonal axes for which the moments of inertia are a maximum and a minimum; (2) the orientation of the principal axes is given by the angle θ_p obtained from Eq. (D-39); (3) the product of inertia is zero for principal axes; and (4) an axis of symmetry is always a principal axis.

Principal Points

Now consider a pair of principal axes with origin at a given point O. If there exists a *different* pair of principal axes through that same point, then *every* pair of axes through that point is a set of principal axes. Furthermore, the moment of inertia must be constant as the angle θ is varied.

The preceding conclusions follow from the nature of the transformation equation for I_{x1} [Eq. (D-33)]. Because this equation contains trigonometric functions of the angle 2θ, there is one maximum value and one minimum value of I_{x1} as 2θ varies through a range of 360° (or as θ varies through a range of 180°). If a second maximum exists, then the only possibility is that I_{x1} remains constant, which means that every pair of axes is a set of principal axes and all moments of inertia are the same.

A point located so that every axis through the point is a principal axis, hence the moments of inertia are the same for all axes through the point, is called a **principal point**.

An illustration of this situation is the rectangle of width $2b$ and height b shown in Fig. D-25. The x-y axes, with origin at point O, are principal axes of the rectangle because the y axis is an axis of symmetry. The x'-y' axes, with the same origin, are also principal axes because the product of inertia $I_{x'y'}$ equals zero (because the triangles are symmetrically located with respect to the x' and y' axes). It follows that every pair of axes through O is a set of principal axes and every moment of inertia is the same (and equal to $2b^4/3$). Therefore, point O is a principal point for the rectangle. (A second principal point is located where the y axis intersects the upper side of the rectangle.)

A useful corollary of the concepts described in the preceding four paragraphs applies to axes through the centroid of an area. Consider an area having

FIGURE D-25

Rectangle for which every axis (in the plane of the area) through point O is a principal axis

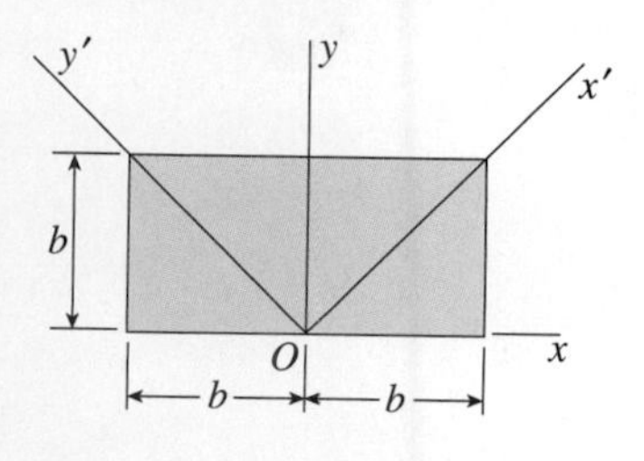

two different pairs of centroidal axes such that at least one axis in each pair is an axis of symmetry. In other words, there exist two different axes of symmetry that are not perpendicular to each other. Then it follows that the centroid is a principal point.

Two examples—a square and an equilateral triangle—are shown in Fig. D-26. In each case, the x-y axes are principal centroidal axes because their origin is at the centroid C, and at least one of the two axes is an axis of symmetry. In addition, a second pair of centroidal axes (the x'-y' axes) has at least one axis of symmetry. It follows that both the x-y and x'-y' axes are principal axes. Therefore, every axis through the centroid C is a principal axis, and every such axis has the same moment of inertia.

If an area has *three different axes of symmetry*, even if two of them are perpendicular, the conditions described in the preceding paragraph are automatically fulfilled. Therefore, if an area has three or more axes of symmetry, the centroid is a principal point, and every axis through the centroid is a principal axis and has the same moment of inertia. These conditions are fulfilled for a circle, for all regular polygons (equilateral triangle, square, regular pentagon, regular hexagon, and so on), and for many other symmetric shapes.

In general, every plane area has two principal points. These points lie equidistant from the centroid on the principal centroidal axis having the larger principal moment of inertia. A special case occurs when the two principal centroidal moments of inertia are equal; then the two principal points merge at the centroid, which becomes the sole principal point.

FIGURE D-26

Examples of areas for which every centroidal axis is a principal axis and the centroid C is a principal point

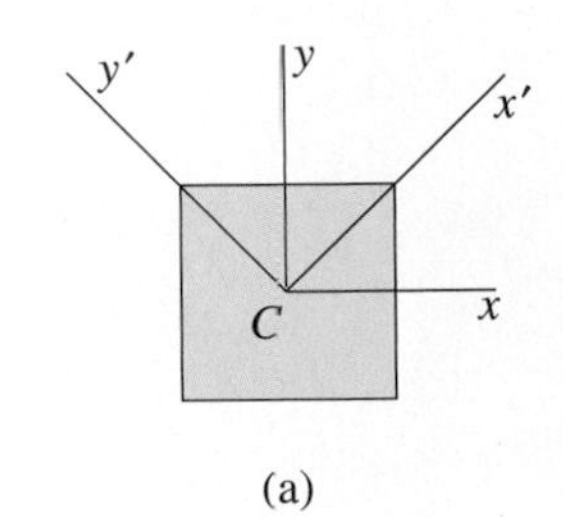

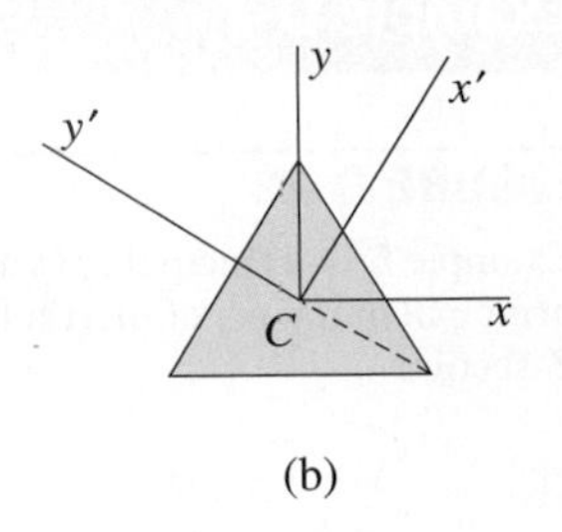

Principal Moments of Inertia

Now determine the principal moments of inertia, assuming that I_x, I_y, and I_{xy} are known. One method is to determine the two values of θ_p (differing by 90°) from Eq. (D-39) and then substitute these values into Eq. (D-33) for I_{x_1}. The resulting two values are the principal moments of inertia, denoted by I_1 and I_2. The advantage of this method is that it is now known which of the two principal angles θ_p corresponds to each principal moment of inertia.

It is also possible to obtain general formulas for the principal moments of inertia. Note from Eq. (D-39) and Fig. D-27 [which is a geometric representation of Eq. (D-39)] that

$$\cos 2\theta_p = \frac{I_x - I_y}{2R} \qquad \sin 2\theta_p = \frac{-I_{xy}}{R} \qquad \textbf{(D-40a,b)}$$

FIGURE D-27

Geometric representation of Eq. (D-39)

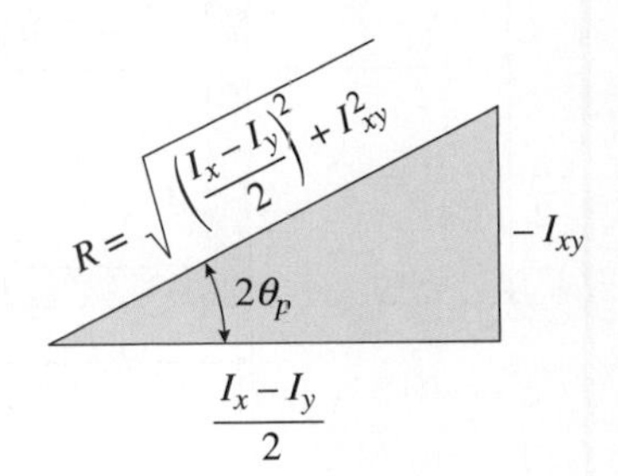

in which

$$R = \sqrt{\left(\frac{I_x - I_y}{2}\right)^2 + I_{xy}^2} \qquad \textbf{(D-41)}$$

is the hypotenuse of the triangle. When evaluating R, always take the positive square root.

Now substitute the expressions for $\cos 2\theta_p$ and $\sin 2\theta_p$ [from Eqs. (D-40a and b)] into Eq. (D-33) for I_{x1} and obtain the algebraically larger of the two principal moments of inertia, denoted by the symbol I_1:

$$I_1 = \frac{I_x + I_y}{2} + \sqrt{\left(\frac{I_x - I_y}{2}\right)^2 + I_{xy}^2} \qquad \textbf{(D-42a)}$$

The smaller principal moment of inertia, denoted as I_2, may be obtained from the equation

$$I_1 + I_2 = I_x + I_y$$

[see Eq. (D-37)]. Substituting the expression for I_1 into this equation and solving for I_2 gives

$$I_2 = \frac{I_x + I_y}{2} - \sqrt{\left(\frac{I_x - I_y}{2}\right)^2 + I_{xy}^2} \quad \textbf{(D-42b)}$$

Equations (D-42a and b) provide a convenient way to calculate the principal moments of inertia.

The following example illustrates the method for locating the principal axes and determining the principal moments of inertia.

Example D-7

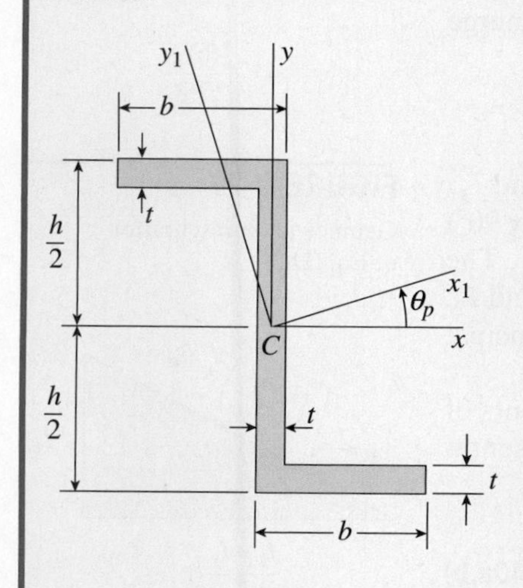

FIGURE D-28

Example D-7: Principal axes and principal moments of inertia for a Z-section

Determine the orientations of the principal centroidal axes and the magnitudes of the principal centroidal moments of inertia for the cross-sectional area of the Z-section shown in Fig. D-28. Use the following numerical data: height h = 200 mm, width b = 90 mm, and constant thickness t = 15 mm.

Solution:

Use the x-y axes (Fig. D-28) as the reference axes through the centroid C. The moments and product of inertia with respect to these axes can be obtained by dividing the area into three rectangles and using the parallel-axis theorems. The results of such calculations are

$$I_x = 29.29 \times 10^6 \text{ mm}^4 \qquad I_y = 5.667 \times 10^6 \text{ mm}^4$$

$$I_{xy} = -9.366 \times 10^6 \text{ mm}^4$$

Substitute these values into the equation for the angle θ_p [Eq. (D-39)] to get

$$\tan 2\theta_p = -\frac{2I_{xy}}{I_x - I_y} = 0.7930 \qquad 2\theta_p = 38.4° \text{ and } 218.4°$$

Thus, the two values of θ_p are

$$\theta_p = 19.2° \text{ and } 109.2°$$

Use these values of θ_p in the transformation equation for I_{x1} [Eq. (D-33)] to find $I_{x1} = 32.6 \times 10^6 \text{ mm}^4$ and $2.4 \times 10^6 \text{ mm}^4$, respectively. These same values are obtained by substituting into Eqs. (D-42a and b). Thus, the principal moments of inertia and the angles to the corresponding principal axes are

$$I_1 = 32.6 \times 10^6 \text{ mm}^4 \quad \theta_{p1} = 19.2°$$

$$I_2 = 2.4 \times 10^6 \text{ mm}^4 \quad \theta_{p2} = 109.2°$$

The principal axes are shown in Fig. D-28 as the x_1y_1 axes.

REVIEW PROBLEMS—Appendix D

D.1 Centroids of Plane Areas

Solve the problems for Section D.1 by integration.

D.1-1 Determine the distances $\bar{x}$ and $\bar{y}$ to the centroid C of a right triangle having base b and altitude h (see Case 6, Appendix E).

D.1-2 Determine the distance $\bar{y}$ to the centroid C of a trapezoid having bases a and b and altitude h (see Case 8, Appendix E).

D.1-3 Determine the distance $\bar{y}$ to the centroid C of a semicircle of radius r (see Case 10, Appendix E).

D.1-4 Determine the distances $\bar{x}$ and $\bar{y}$ to the centroid C of a parabolic spandrel of base b and height h (see Case 18, Appendix E).

D.1-5 Determine the distances $\bar{x}$ and $\bar{y}$ to the centroid C of a semisegment of nth degree having base b and height h (see Case 19, Appendix E).

D.2 Centroids of Composite Areas

Solve the problems for Section D.2 by using the formulas for composite areas.

D.2-1 Determine the distance $\bar{y}$ to the centroid C of a trapezoid having bases a and b and altitude h (see Case 8, Appendix E) by dividing the trapezoid into two triangles.

D.2-2 One quarter of a square of side a is removed (see figure). What are the coordinates $\bar{x}$ and $\bar{y}$ of the centroid C of the remaining area?

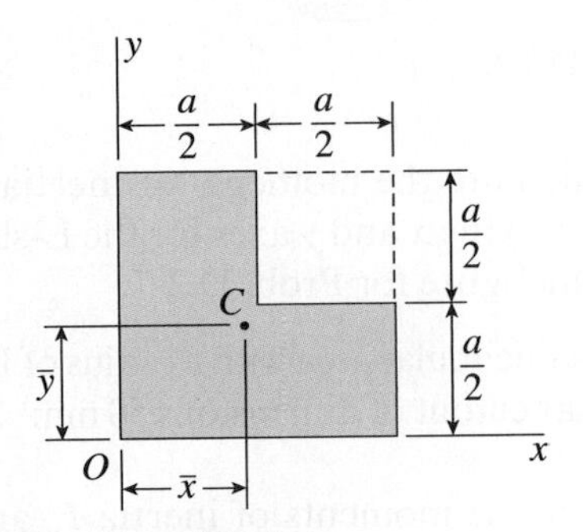

PROBLEMS D.2-2 and D.4-2

D.2-3 Calculate the distance $\bar{y}$ to the centroid C of the channel section shown in the figure if $a = 150$ mm, $b = 25$ mm, and $c = 50$ mm.

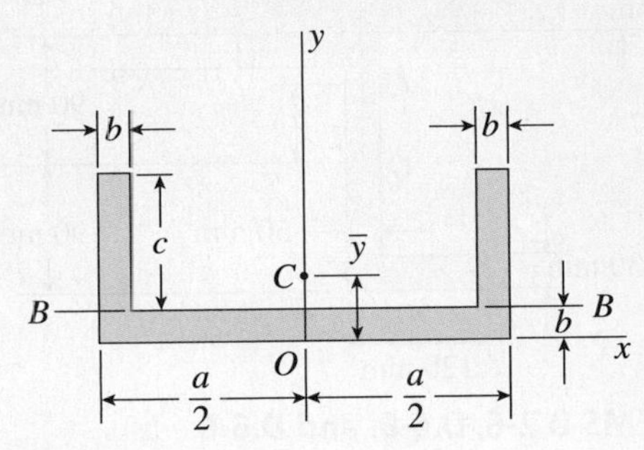

PROBLEMS D.2-3, D.2-4, and D.4-3

D.2-4 What must be the relationship among the dimensions a, b, and c of the channel section shown in the figure in order that the centroid C will lie on line B–B?

D.2-5 The cross section of a beam constructed of a HE 600B wide-flange section with an 200 mm × 20 mm cover plate welded to the top flange is shown in the figure.

Determine the distance $\bar{y}$ from the base of the beam to the centroid C of the cross-sectional area.

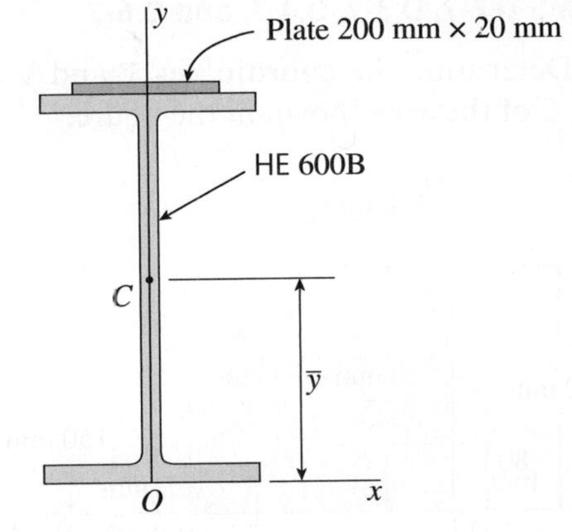

PROBLEMS D.2-5 and D.4-5

D.2-6 Determine the distance $\bar{y}$ to the centroid C of the composite area shown in the figure.

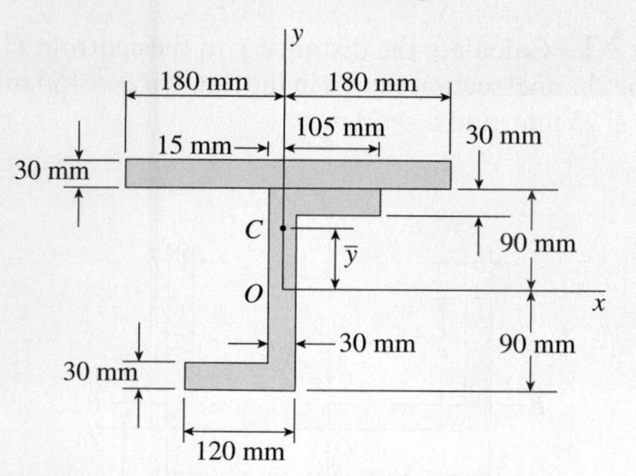

PROBLEMS D.2-6, D.4-6, and D.6-6

D.2-7 Determine the coordinates $\bar{x}$ and $\bar{y}$ of the centroid C of the L-shaped area shown in the figure.

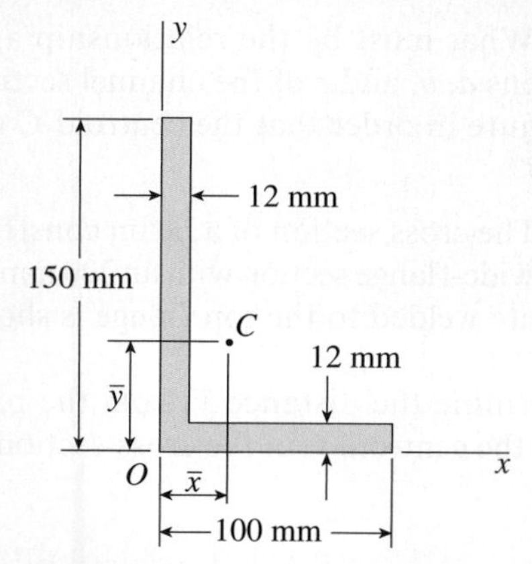

PROBLEMS D.2-7, D.3-7, D.4-7, and D.6-7

D.2-8 Determine the coordinates $\bar{x}$ and $\bar{y}$ of the centroid C of the area shown in the figure.

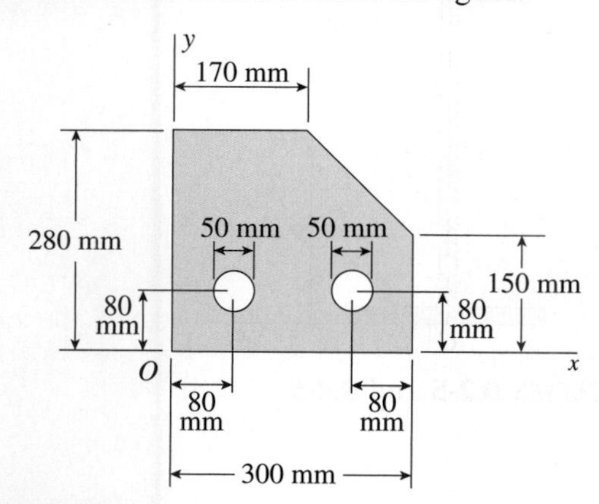

PROBLEM D.2-8

D.3 Moments of Inertia of Plane Areas

Solve problems D.3-1 through D.3-4 by integration.

D.3-1 Determine the moment of inertia I_x of a triangle of base b and altitude h with respect to its base (see Case 4, Appendix E).

D.3-2 Determine the moment of inertia I_{BB} of a trapezoid having bases a and b and altitude h with respect to its base (see Case 8, Appendix E).

D.3-3 Determine the moment of inertia I_x of a parabolic spandrel of base b and height h with respect to its base (see Case 18, Appendix E).

D.3-4 Determine the moment of inertia I_x of a circle of radius r with respect to a diameter (see Case 9, Appendix E).

Solve problems D.3-5 through D.3-9 by considering the area to be a composite area.

D.3-5 Determine the moment of inertia I_{BB} of a rectangle having sides of lengths b and h with respect to a diagonal of the rectangle (see Case 2, Appendix E).

D.3-6 Calculate the moment of inertia I_x for the composite circular area shown in the figure. The origin of the axes is at the center of the concentric circles, and the three diameters are 20, 40, and 60 mm.

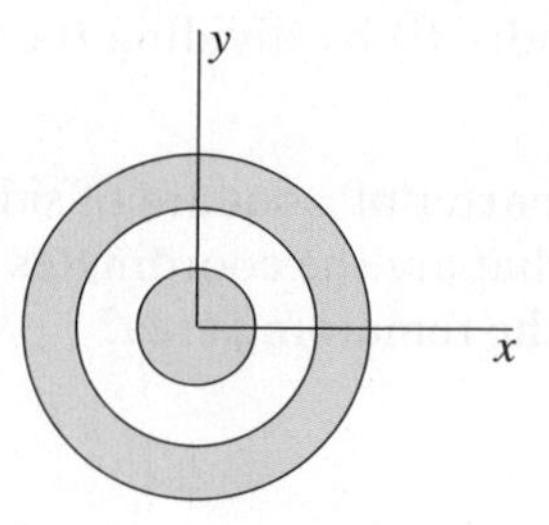

PROBLEM D.3-6

D.3-7 Calculate the moments of inertia I_x and I_y with respect to the x and y axes for the L-shaped area shown in the figure for Prob. D.2-7.

D.3-8 A semicircular area with a radius of 150 mm has a rectangular cutout of dimensions 50 mm × 100 mm (see figure).

Calculate the moments of inertia I_x and I_y with respect to the x and y axes. Also, calculate the corresponding radii of gyration r_x and r_y.

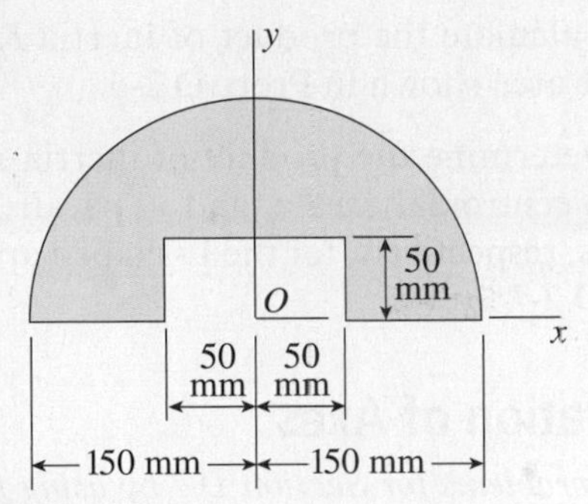

PROBLEM D.3-8

D.3-9 Calculate the moments of inertia I_1 and I_2 of a HE 450A wide-flange section using the cross-sectional dimensions given in Table F-1, Appendix F. (Disregard the cross-sectional areas of the fillets.) Also, calculate the corresponding radii of gyration r_1 and r_2, respectively.

D.4 Parallel-Axis Theorem for Moments of Inertia

D.4-1 Calculate the moment of inertia I_b of a HE 320B wide-flange section with respect to its base. (Use data from Table F-1, Appendix F.)

D.4-2 Determine the moment of inertia I_c with respect to an axis through the centroid C and parallel to the x axis for the geometric figure described in Prob. D.2-2.

D.4-3 For the channel section described in Prob. D.2-3, calculate the moment of inertia I_{xc} with respect to an axis through the centroid C and parallel to the x axis.

D.4-4 The moment of inertia with respect to axis 1–1 of the scalene triangle shown in the figure is 90×10^3 mm^4. Calculate its moment of inertia I_2 with respect to axis 2–2.

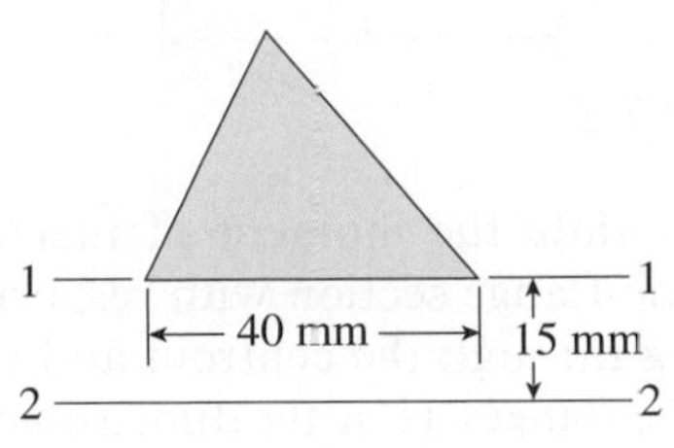

PROBLEM D.4-4

D.4-5 For the beam cross section described in Prob. D.2-5, calculate the centroidal moments of inertia I_{xc} and I_{yc} with respect to axes through the centroid C such that the x_c axis is parallel to the x axis and the y_c axis coincides with the y axis.

D.4-6 Calculate the moment of inertia I_{xc} with respect to an axis through the centroid C and parallel to the x axis for the composite area shown in the figure for Prob. D.2-6.

D.4-7 Calculate the centroidal moments of inertia I_{xc} and I_{yc} with respect to axes through the centroid C and parallel to the x and y axes, respectively, for the L-shaped area shown in the figure for Prob. D.2-7.

D.4-8 The wide-flange beam section shown in the figure has a total height of 250 mm and a constant thickness of 15 mm.

Determine the flange width b if it is required that the centroidal moments of inertia I_x and I_y be in the ratio 3 to 1, respectively.

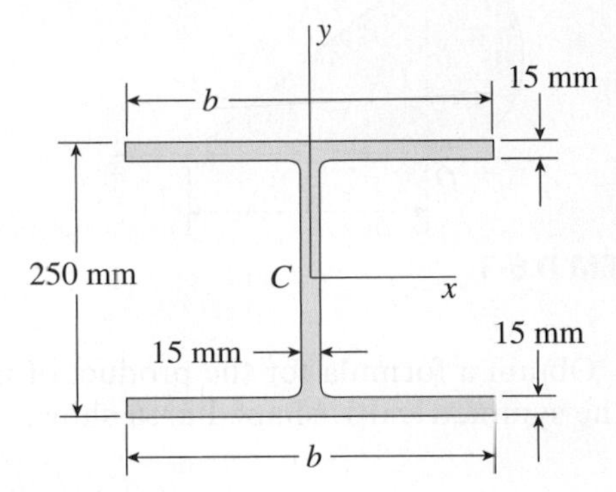

PROBLEM D.4-8

D.5 Polar Moments of Inertia

D.5-1 Determine the polar moment of inertia I_P of an isosceles triangle of base b and altitude h with respect to its apex (see Case 5, Appendix E).

D.5-2 Determine the polar moment of inertia $(I_P)_C$ with respect to the centroid C for a circular sector (see Case 13, Appendix E).

D.5-3 Determine the polar moment of inertia I_P for a HE 220B wide-flange section with respect to one of its outermost corners.

D.5-4 Obtain a formula for the polar moment of inertia I_P with respect to the midpoint of the hypotenuse for a right triangle of base b and height h (see Case 6, Appendix E).

D.5-5 Determine the polar moment of inertia $(I_p)_C$ with respect to the centroid C for a quarter-circular spandrel (see Case 12, Appendix E).

D.6 Products of Inertia

D.6-1 Using integration, determine the product of inertia I_{xy} for the parabolic semisegment shown in Fig. D-5 (see also Case 17 in Appendix E).

D.6-2 Using integration, determine the product of inertia I_{xy} for the quarter-circular spandrel shown in Case 12, Appendix E.

D.6-3 Find the relationship between the radius r and the distance b for the composite area shown in the figure in order that the product of inertia I_{xy} will be zero.

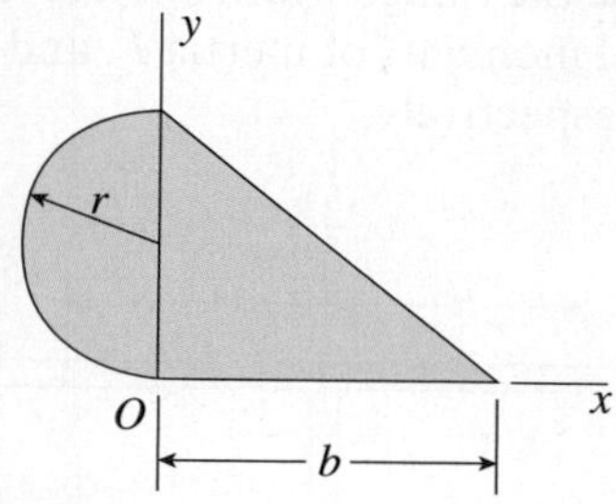

PROBLEM D.6-3

D.6-4 Obtain a formula for the product of inertia I_{xy} of the symmetrical L-shaped area shown in the figure.

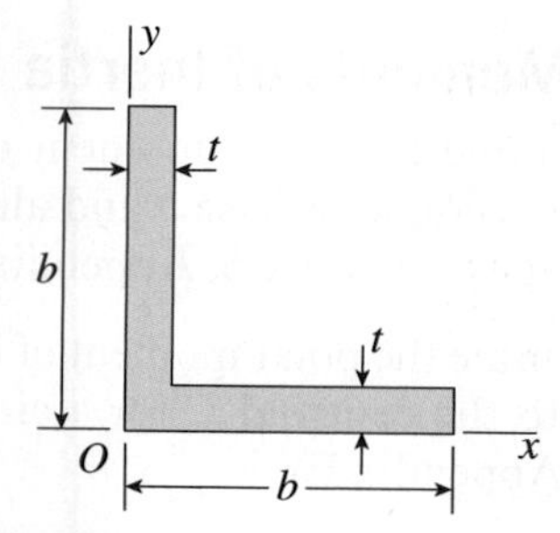

PROBLEM D.6-4

D.6-5 Calculate the product of inertia I_{12} with respect to the centroidal axes 1–1 and 2–2 for an L150 × 150 × 15 mm angle section (see Table F-4, Appendix F). (Disregard the cross-sectional areas of the fillet and rounded corners.)

D.6-6 Calculate the product of inertia I_{xy} for the composite area shown in Prob. D.2-6.

D.6-7 Determine the product of inertia I_{xcyc} with respect to centroidal axes x_c and y_c parallel to the x and y axes, respectively, for the L-shaped area shown in Prob. D.2-7.

D.7 Rotation of Axes

Solve the problems for Section D.7 by using the transformation equations for moments and products of inertia.

D.7-1 Determine the moments of inertia I_{x1} and I_{y1} and the product of inertia I_{x1y1} for a square with sides b, as shown in the figure. (Note that the x_1y_1 axes are centroidal axes rotated through an angle θ with respect to the x-y axes.)

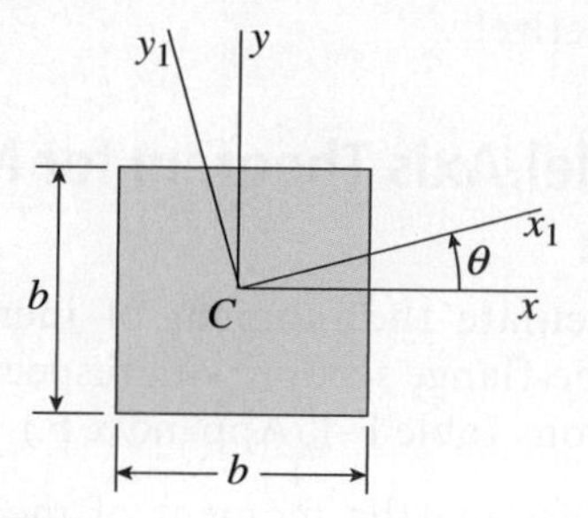

PROBLEM D.7-1

D.7-2 Determine the moments and product of inertia with respect to the x_1y_1 axes for the rectangle shown in the figure. (Note that the x_1 axis is a diagonal of the rectangle.)

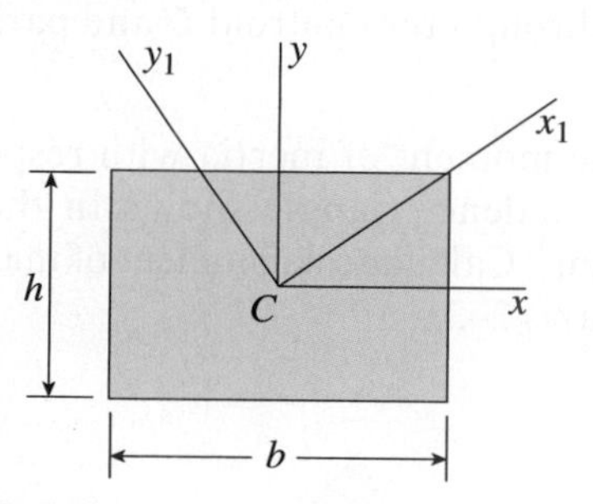

PROBLEM D.7-2

D.7-3 Calculate the moment of inertia I_d for a HE 320A wide-flange section with respect to a diagonal passing through the centroid and two outside corners of the flanges. (Use the dimensions and properties given in Table F-1.)

D.7-4 Calculate the moments of inertia I_{x1} and I_{y1} and the product of inertia I_{x1y1} with respect to the x_1y_1 axes for the L-shaped area shown in the figure if $a = 150$ mm, $b = 100$ mm, $t = 15$ mm, and $\theta = 30°$.

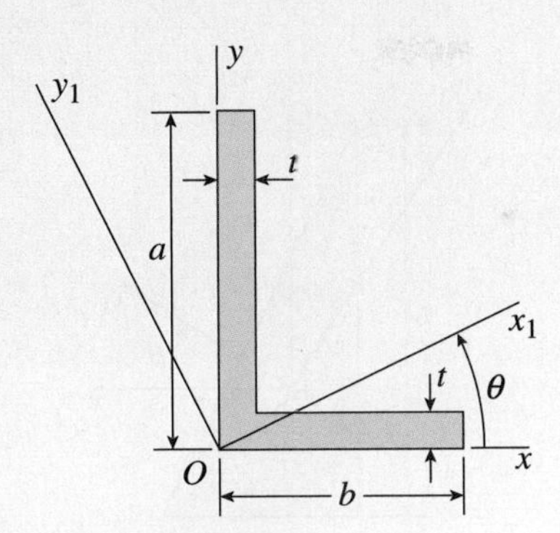

PROBLEMS D.7-4 and D.8-4

D.7-5 Calculate the moments of inertia I_{x1} and I_{y1} and the product of inertia I_{x1y1} with respect to the x_1y_1 axes for the Z-section shown in the figure if $b = 75$ mm, $h = 100$ mm, $t = 12$ mm, and $\theta = 60°$.

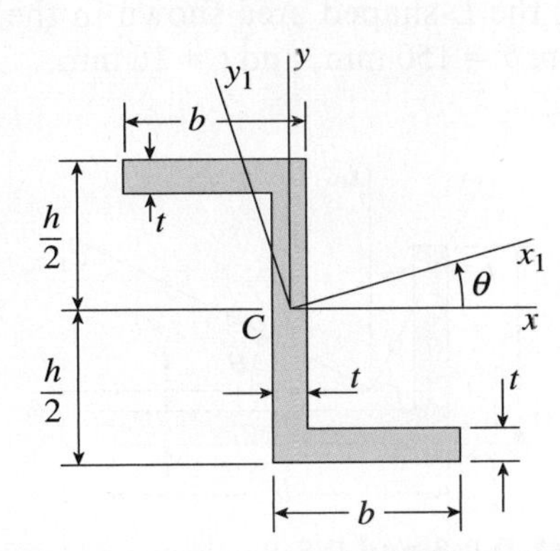

PROBLEMS D.7-5, D.7-6, D.8-5, and D.8-6

D.7-6 Solve the preceding problem if $b = 80$ mm, $h = 120$ mm, $t = 12$ mm, and $\theta = 30°$.

D.8 Principal Axes and Principal Moments of Inertia

D.8-1 An ellipse with major axis of length $2a$ and minor axis of length $2b$ is shown in the figure.

(a) Determine the distance c from the centroid C of the ellipse to the principal points P on the minor axis (y axis).

(b) For what ratio a/b do the principal points lie on the circumference of the ellipse?

(c) For what ratios do they lie inside the ellipse?

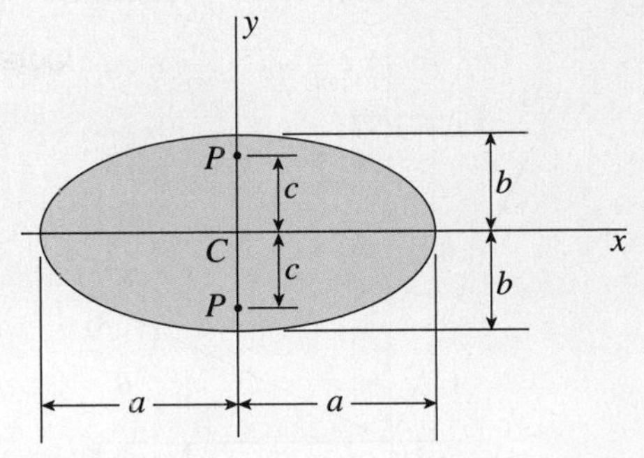

PROBLEM D.8-1

D.8-2 Demonstrate that the two points P_1 and P_2, located as shown in the figure, are the principal points of the isosceles right triangle.

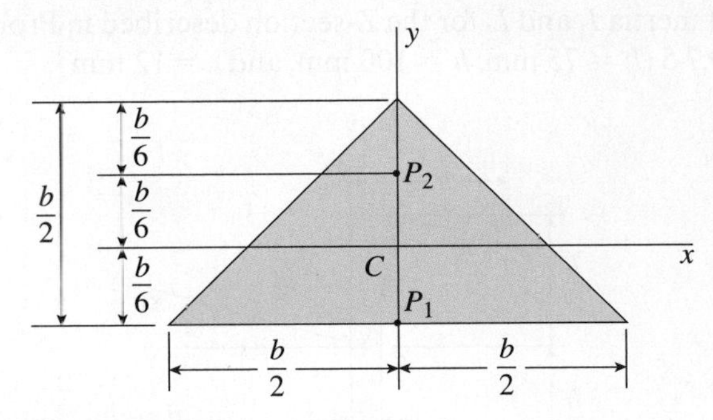

PROBLEM D.8-2

D.8-3 Determine the angles θ_{p1} and θ_{p2} defining the orientations of the principal axes through the origin O for the right triangle shown in the figure if $b = 150$ mm and $h = 200$ mm. Also, calculate the corresponding principal moments of inertia I_1 and I_2.

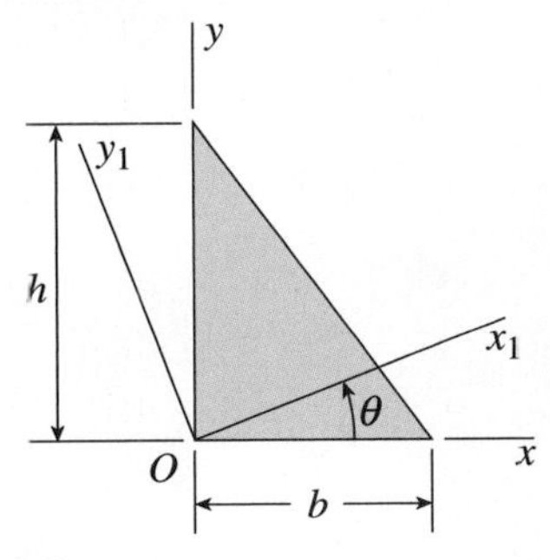

PROBLEM D.8-3

D.8-4 Determine the angles θ_{p1} and θ_{p2} defining the orientations of the principal axes through the origin O and the corresponding principal moments of inertia I_1 and I_2 for the L-shaped area described in Prob. D.7-4 ($a = 150$ mm, $b = 100$ mm, and $t = 15$ mm).

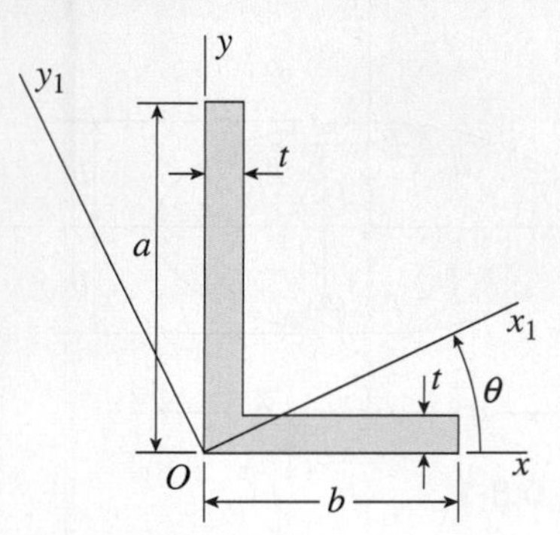

PROBLEMS D.7-4 and D.8-4

D.8-5 Determine the angles θ_{p1} and θ_{p2} defining the orientations of the principal axes through the centroid C and the corresponding principal centroidal moments of inertia I_1 and I_2 for the Z-section described in Prob. D.7-5 ($b = 75$ mm, $h = 100$ mm, and $t = 12$ mm).

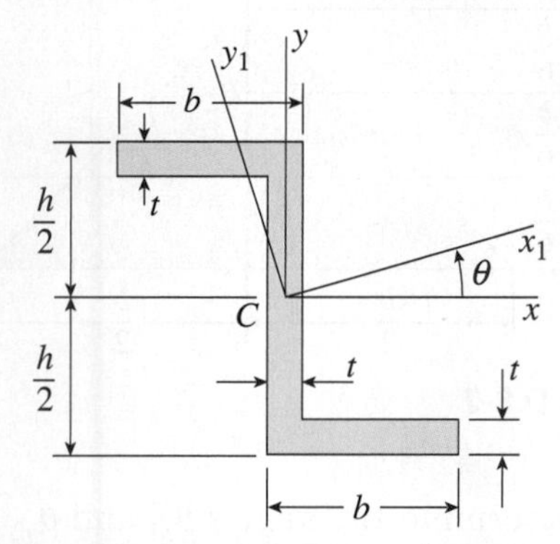

PROBLEMS D.7-5, D.7-6, D.8-5, and D.8-6

D.8-6 Solve the preceding problem for the Z-section described in Prob. D.7-6 ($b = 80$ mm, $h = 120$ mm, and $t = 12$ mm).

D.8-7 Determine the angles θ_{p1} and θ_{p2} defining the orientations of the principal axes through the centroid C for the right triangle shown in the figure if $h = 2b$. Also, determine the corresponding principal centroidal moments of inertia I_1 and I_2.

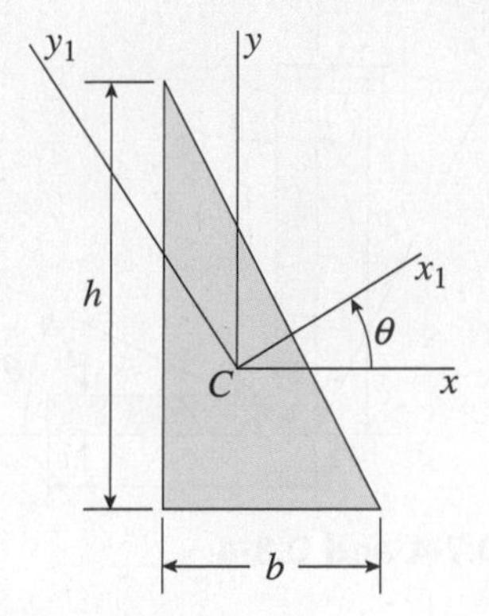

PROBLEM D.8-7

D.8-8 Determine the angles θ_{p1} and θ_{p2} defining the orientations of the principal centroidal axes and the corresponding principal moments of inertia I_1 and I_2 for the L-shaped area shown in the figure if $a = 80$ mm, $b = 150$ mm, and $t = 16$ mm.

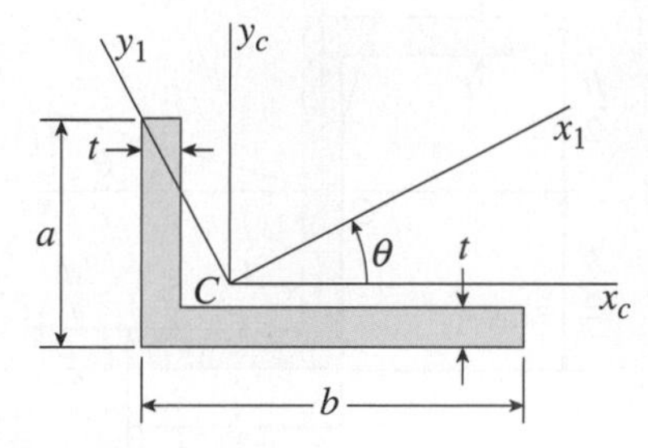

PROBLEMS D.8-8 and D.8-9

D.8-9 Solve the preceding problem if $a = 75$ mm, $b = 150$ mm, and $t = 12$ mm.

Properties of Plane Areas

Notation:

A = area
$\bar{x}, \bar{y}$ = distances to centroid C
I_x, I_y = moments of inertia with respect to the x and y axes, respectively
I_{xy} = product of inertia with respect to the x and y axes
$I_P = I_x + I_y$ = polar moment of inertia with respect to the origin of the x and y axes
I_{BB} = moment of inertia with respect to axis B–B

1

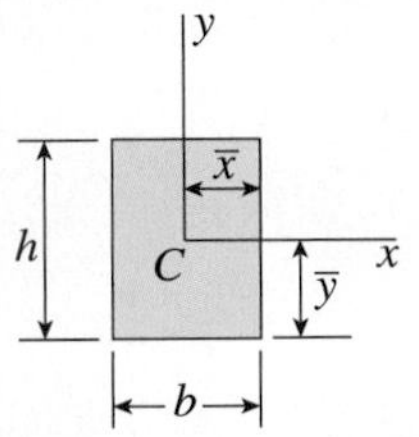

Rectangle (Origin of axes at centroid)

$$A = bh \quad \bar{x} = \frac{b}{2} \quad \bar{y} = \frac{h}{2}$$

$$I_x = \frac{bh^3}{12} \quad I_y = \frac{hb^3}{12} \quad I_{xy} = 0 \quad I_P = \frac{bh}{12}(h^2 + b^2)$$

2

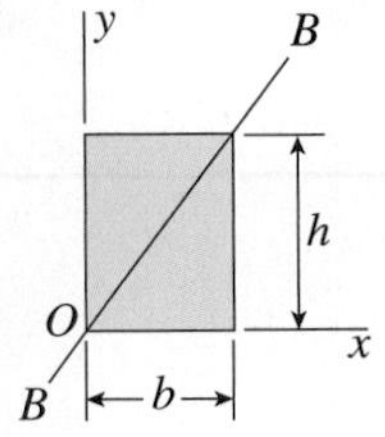

Rectangle (Origin of axes at corner)

$$I_x = \frac{bh^3}{3} \quad I_y = \frac{hb^3}{3} \quad I_{xy} = \frac{b^2h^2}{4} \quad I_P = \frac{bh}{3}(h^2 + b^2)$$

$$I_{BB} = \frac{b^3h^3}{6(b^2 + h^2)}$$

(see Problem D.3-5)

3

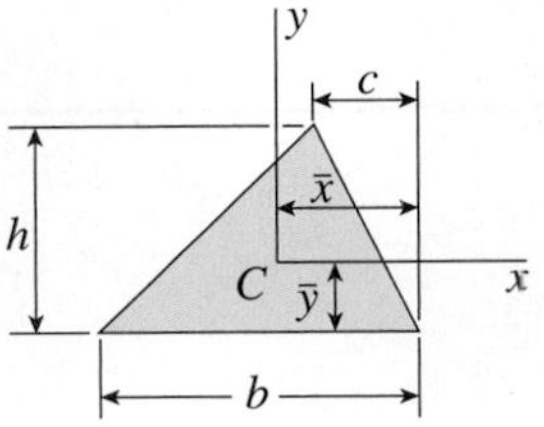

Triangle (Origin of axes at centroid)

$$A = \frac{bh}{2} \quad \bar{x} = \frac{b + c}{3} \quad \bar{y} = \frac{h}{3}$$

$$I_x = \frac{bh^3}{36} \quad I_y = \frac{bh}{36}(b^2 - bc + c^2)$$

$$I_{xy} = \frac{bh^2}{72}(b - 2c) \quad I_P = \frac{bh}{36}(h^2 + b^2 - bc + c^2)$$

4

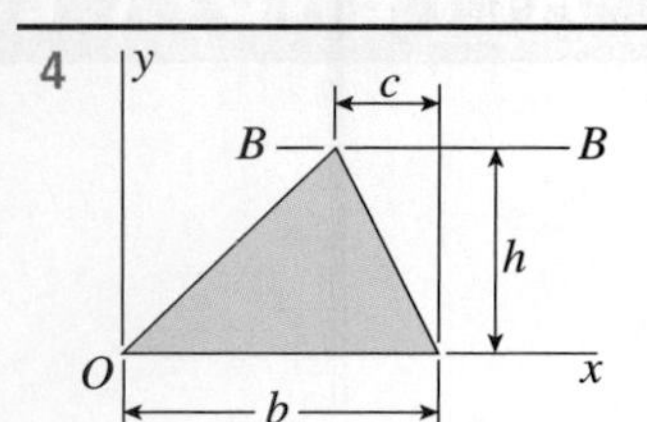

Triangle (Origin of axes at vertex)

$$I_x = \frac{bh^3}{12} \quad I_y = \frac{bh}{12}(3b^2 - 3bc + c^2)$$

$$I_{xy} = \frac{bh^2}{24}(3b - 2c) \quad I_{BB} = \frac{bh^3}{4}$$

(see Problem D.3-1)

5

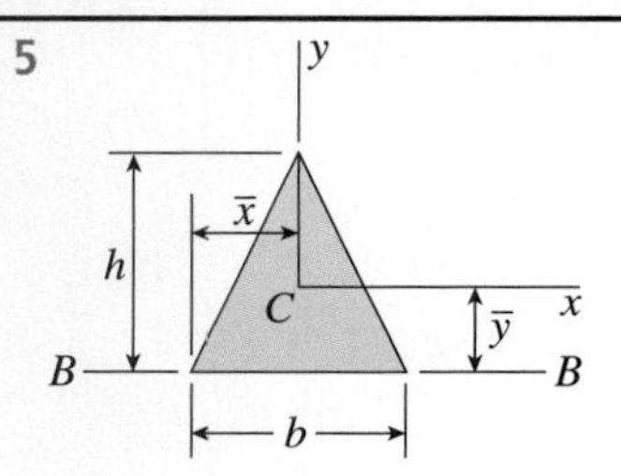

Isosceles triangle (Origin of axes at centroid)

$$A = \frac{bh}{2} \quad \bar{x} = \frac{b}{2} \quad \bar{y} = \frac{h}{3}$$

$$I_x = \frac{bh^3}{36} \quad I_y = \frac{hb^3}{48} \quad I_{xy} = 0$$

$$I_P = \frac{bh}{144}(4h^2 + 3b^2) \quad I_{BB} = \frac{bh^3}{12}$$

(*Note:* For an equilateral triangle, $h = \sqrt{3}\,b/2$.)
(see Problem D.5-1)

6

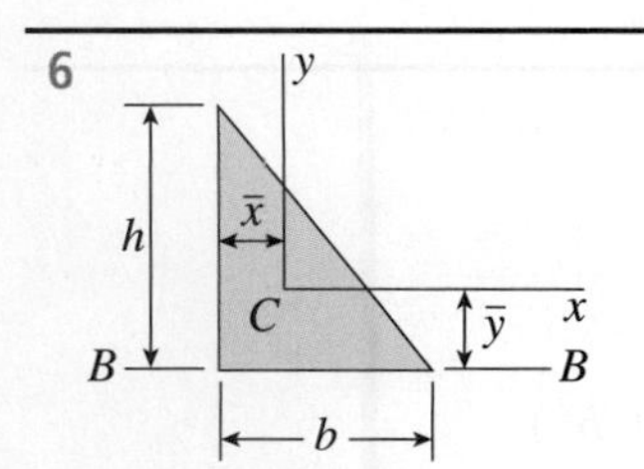

Right triangle (Origin of axes at centroid)

$$A = \frac{bh}{2} \quad \bar{x} = \frac{b}{3} \quad \bar{y} = \frac{h}{3}$$

$$I_x = \frac{bh^3}{36} \quad I_y = \frac{hb^3}{36} \quad I_{xy} = -\frac{b^2h^2}{72}$$

$$I_P = \frac{bh}{36}(h^2 + b^2) \quad I_{BB} = \frac{bh^3}{12}$$

(see Problems D.1-1, D.5-4)

7

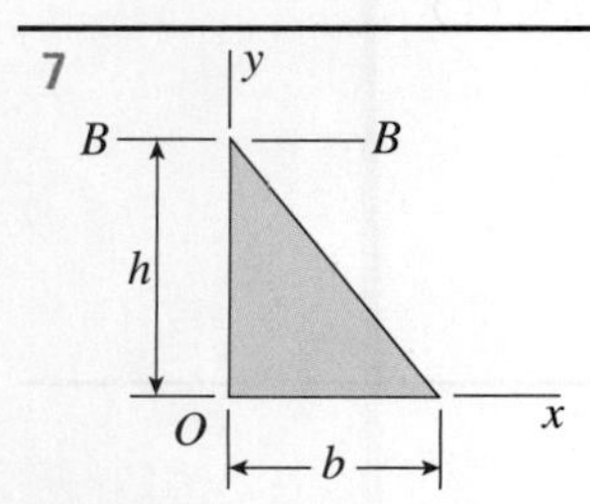

Right triangle (Origin of axes at vertex)

$$I_x = \frac{bh^3}{12} \quad I_y = \frac{hb^3}{12} \quad I_{xy} = \frac{b^2h^2}{24}$$

$$I_P = \frac{bh}{12}(h^2 + b^2) \quad I_{BB} = \frac{bh^3}{4}$$

8

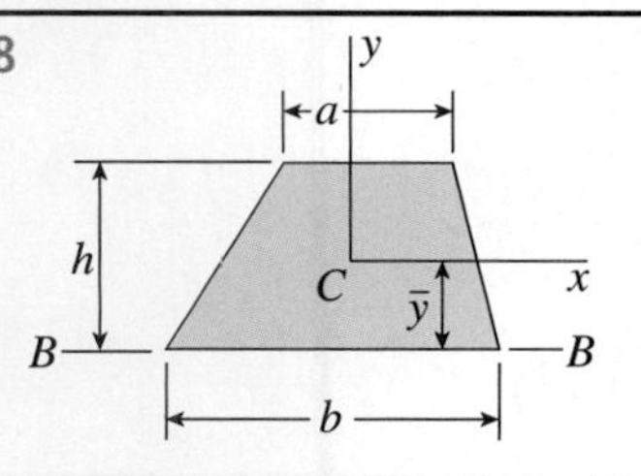

Trapezoid (Origin of axes at centroid)

$$A = \frac{h(a + b)}{2} \quad \bar{y} = \frac{h(2a + b)}{3(a + b)}$$

$$I_x = \frac{h^3(a^2 + 4ab + b^2)}{36(a + b)} \quad I_{BB} = \frac{h^3(3a + b)}{12}$$

(see Problems D.1-2, D.3-2)

9

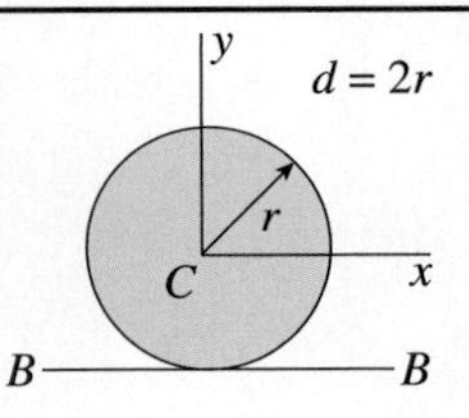

Circle (Origin of axes at center)

$$A = \pi r^2 = \frac{\pi d^2}{4} \quad I_x = I_y = \frac{\pi r^4}{4} = \frac{\pi d^4}{64}$$

$$I_{xy} = 0 \quad I_P = \frac{\pi r^4}{2} = \frac{\pi d^4}{32} \quad I_{BB} = \frac{5\pi r^4}{4} = \frac{5\pi d^4}{64}$$

(see Problem D.3-4)

10

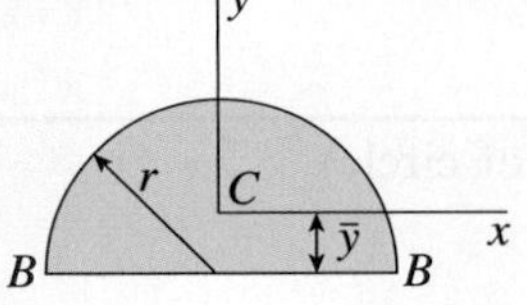

Semicircle (Origin of axes at centroid)

$$A = \frac{\pi r^2}{2} \quad \bar{y} = \frac{4r}{3\pi}$$

$$I_x = \frac{(9\pi^2 - 64)r^4}{72\pi} \approx 0.1098r^4 \quad I_y = \frac{\pi r^4}{8} \quad I_{xy} = 0 \quad I_{BB} = \frac{\pi r^4}{8}$$

(see Problem D.1-3)

11

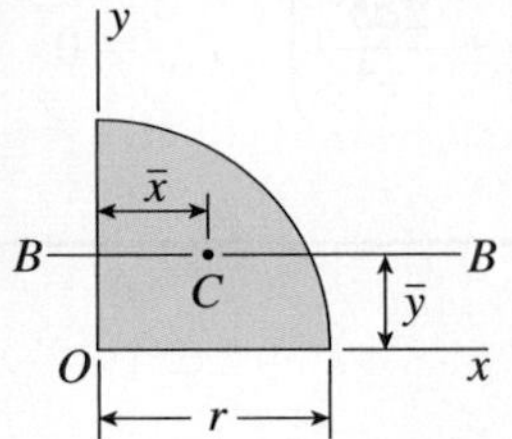

Quarter circle (Origin of axes at center of circle)

$$A = \frac{\pi r^2}{4} \quad \bar{x} = \bar{y} = \frac{4r}{3\pi}$$

$$I_x = I_y = \frac{\pi r^4}{16} \quad I_{xy} = \frac{r^4}{8} \quad I_{BB} = \frac{(9\pi^2 - 64)r^4}{144\pi} \approx 0.05488r^4$$

12

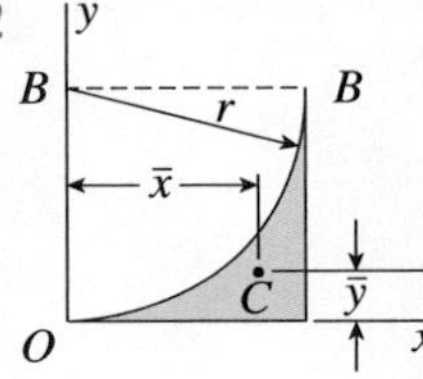

Quarter-circular spandrel (Origin of axes at point of tangency)

$$A = (1 - \frac{\pi}{4})r^2 \quad \bar{x} = \frac{2r}{3(4 - \pi)} \approx 0.7766r \quad \bar{y} = \frac{(10 - 3\pi)r}{3(4 - \pi)} \approx 0.2234r$$

$$I_x = \left(1 - \frac{5\pi}{16}\right)r^4 \approx 0.01825r^4 \quad I_y = I_{BB} = \left(\frac{1}{3} - \frac{\pi}{16}\right)r^4 \approx 0.1370r^4$$

(see Problems D.5-5, D.6-2)

13

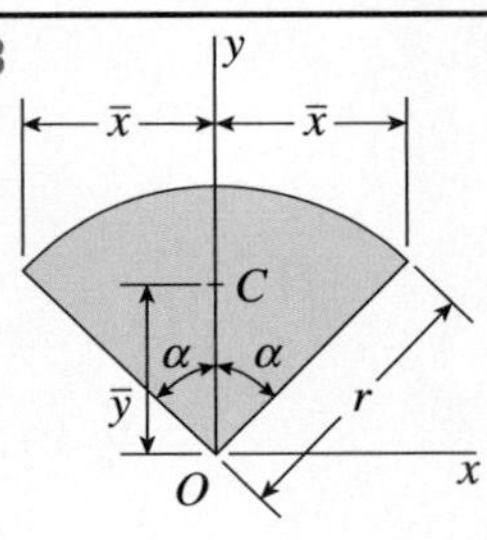

Circular sector (Origin of axes at center of circle)

α = angle in radians $\quad (\alpha \leq \pi/2)$

$$A = \alpha r^2 \quad \bar{x} = r\sin\alpha \quad \bar{y} = \frac{2r\sin\alpha}{3\alpha}$$

$$I_x = \frac{r^4}{4}(\alpha + \sin\alpha\cos\alpha) \quad I_y = \frac{r^4}{4}(\alpha - \sin\alpha\cos\alpha) \quad I_{xy} = 0 \quad I_P = \frac{\alpha r^4}{2}$$

(see Problem D.5-2)

14

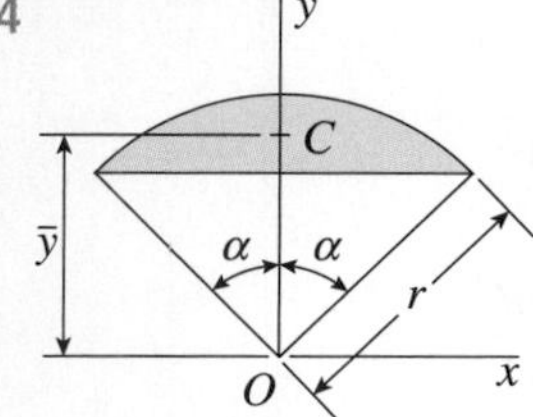

Circular segment (Origin of axes at center of circle)

α = angle in radians $\quad (\alpha \leq \pi/2)$

$$A = r^2(\alpha - \sin\alpha\cos\alpha) \qquad \bar{y} = \frac{2r}{3}\left(\frac{\sin^3\alpha}{\alpha - \sin\alpha\cos\alpha}\right)$$

$$I_x = \frac{r^4}{4}(\alpha - \sin\alpha\cos\alpha + 2\sin^3\alpha\cos\alpha) \qquad I_{xy} = 0$$

$$I_y = \frac{r^4}{12}(3\alpha - 3\sin\alpha\cos\alpha - 2\sin^3\alpha\cos\alpha)$$

15

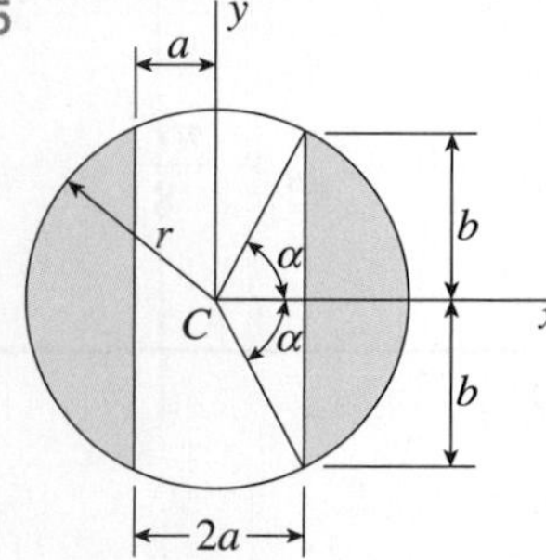

Circle with core removed (Origin of axes at center of circle)

α = angle in radians $\quad (\alpha \leq \pi/2)$

$$\alpha = \arccos\frac{a}{r} \qquad b = \sqrt{r^2 - a^2} \qquad A = 2r^2\left(\alpha - \frac{ab}{r^2}\right)$$

$$I_x = \frac{r^4}{6}\left(3\alpha - \frac{3ab}{r^2} - \frac{2ab^3}{r^4}\right) \qquad I_y = \frac{r^4}{2}\left(\alpha - \frac{ab}{r^2} + \frac{2ab^3}{r^4}\right) \qquad I_{xy} = 0$$

16

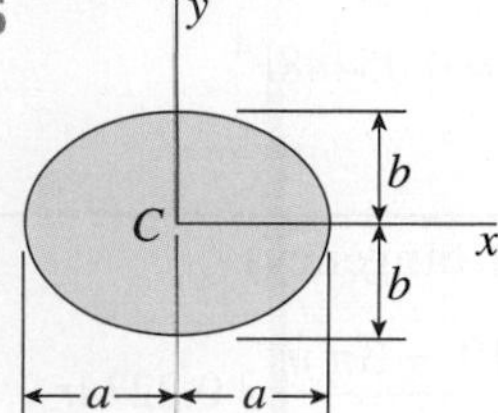

Ellipse (Origin of axes at centroid)

$$A = \pi ab \qquad I_x = \frac{\pi ab^3}{4} \qquad I_y = \frac{\pi ba^3}{4}$$

$$I_{xy} = 0 \qquad I_P = \frac{\pi ab}{4}(b^2 + a^2)$$

Circumference $\approx \pi[1.5(a + b) - \sqrt{ab}] \quad (a/3 \leq b \leq a)$

$\approx 4.17b^2/a + 4a \quad (0 \leq b \leq a/3)$

17

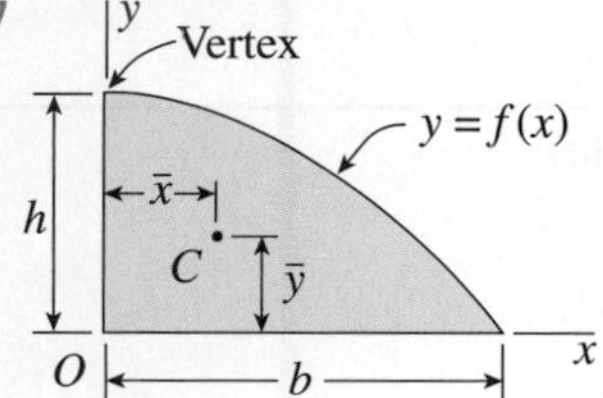

Parabolic semisegment (Origin of axes at corner)

$$y = f(x) = h\left(1 - \frac{x^2}{b^2}\right)$$

$$A = \frac{2bh}{3} \qquad \bar{x} = \frac{3b}{8} \qquad \bar{y} = \frac{2h}{5}$$

$$I_x = \frac{16bh^3}{105} \qquad I_y = \frac{2hb^3}{15} \qquad I_{xy} = \frac{b^2h^2}{12}$$

(see Problem D.6-1)

18

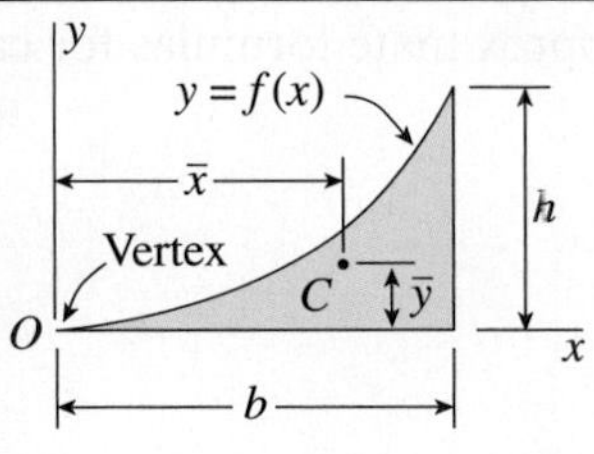

Parabolic spandrel (Origin of axes at vertex)

$$y = f(x) = \frac{hx^2}{b^2}$$

$$A = \frac{bh}{3} \quad \bar{x} = \frac{3b}{4} \quad \bar{y} = \frac{3h}{10}$$

$$I_x = \frac{bh^3}{21} \quad I_y = \frac{hb^3}{5} \quad I_{xy} = \frac{b^2h^2}{12}$$

(see Problems D.1-4, D.3-3)

19

Semisegment of *n*th degree (Origin of axes at corner)

$$y = f(x) = h\left(1 - \frac{x^n}{b^n}\right) \quad (n > 0)$$

$$A = bh\left(\frac{n}{n+1}\right) \quad \bar{x} = \frac{b(n+1)}{2(n+2)} \quad \bar{y} = \frac{hn}{2n+1}$$

$$I_x = \frac{2bh^3n^3}{(n+1)(2n+1)(3n+1)} \quad I_y = \frac{hb^3n}{3(n+3)} \quad I_{xy} = \frac{b^2h^2n^2}{4(n+1)(n+2)}$$

(see Problem D.1-5)

20

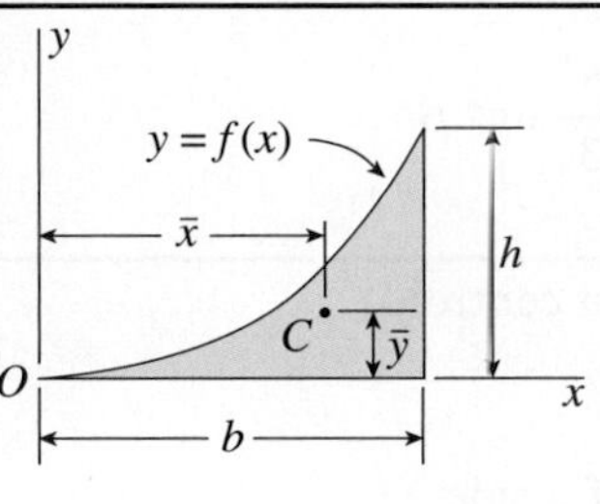

Spandrel of *n*th degree (Origin of axes at point of tangency)

$$y = f(x) = \frac{hx^n}{b^n} \quad (n > 0)$$

$$A = \frac{bh}{n+1} \quad \bar{x} = \frac{b(n+1)}{n+2} \quad \bar{y} = \frac{h(n+1)}{2(2n+1)}$$

$$I_x = \frac{bh^3}{3(3n+1)} \quad I_y = \frac{hb^3}{n+3} \quad I_{xy} = \frac{b^2h^2}{4(n+1)}$$

21

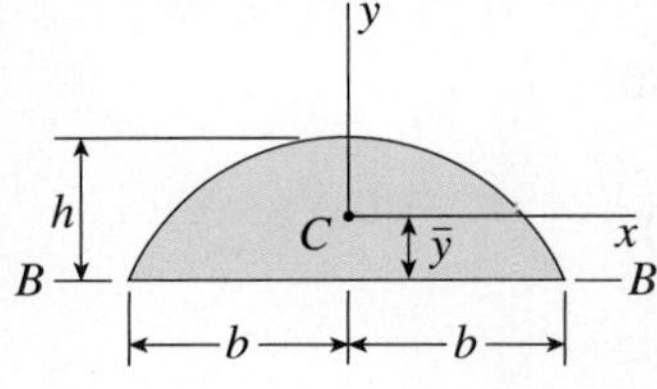

Sine wave (Origin of axes at centroid)

$$A = \frac{4bh}{\pi} \quad \bar{y} = \frac{\pi h}{8}$$

$$I_x = \left(\frac{8}{9\pi} - \frac{\pi}{16}\right)bh^3 \approx 0.08659bh^3 \quad I_y = \left(\frac{4}{\pi} - \frac{32}{\pi^3}\right)hb^3 \approx 0.2412hb^3$$

$$I_{xy} = 0 \quad I_{BB} = \frac{8bh^3}{9\pi}$$

22

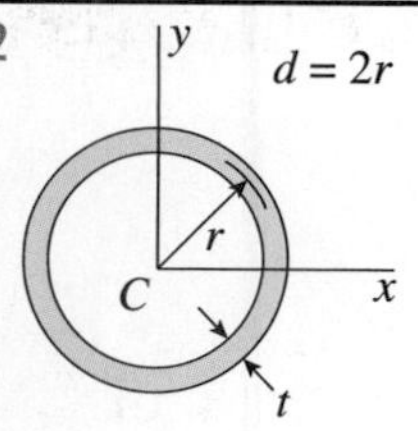

Thin circular ring (Origin of axes at center) Approximate formulas for case when t is small

$$A = 2\pi rt = \pi dt \quad I_x = I_y = \pi r^3 t = \frac{\pi d^3 t}{8}$$

$$I_{xy} = 0 \quad I_P = 2\pi r^3 t = \frac{\pi d^3 t}{4}$$

23

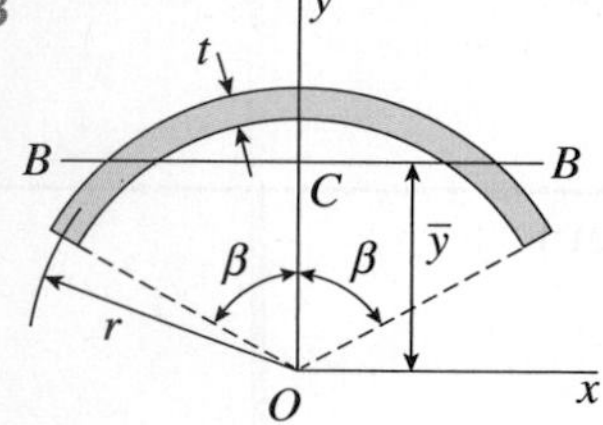

Thin circular arc (Origin of axes at center of circle) Approximate formulas for case when t is small

β = angle in radians (*Note*: For a semicircular arc, $\beta = \pi/2$.)

$$A = 2\beta rt \quad \bar{y} = \frac{r \sin\beta}{\beta}$$

$$I_x = r^3 t(\beta + \sin\beta\cos\beta) \quad I_y = r^3 t(\beta - \sin\beta\cos\beta)$$

$$I_{xy} = 0 \quad I_{BB} = r^3 t\left(\frac{2\beta + \sin 2\beta}{2} - \frac{1 - \cos 2\beta}{\beta}\right)$$

24

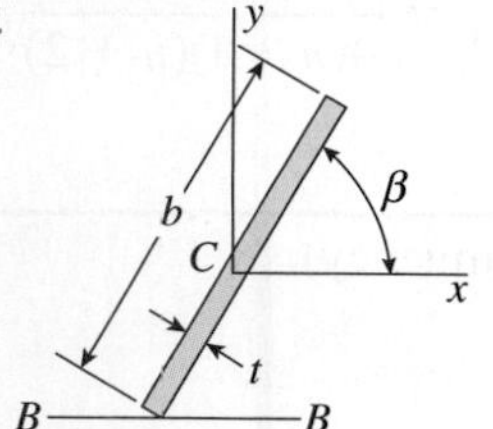

Thin rectangle (Origin of axes at centroid) Approximate formulas for case when t is small

$$A = bt$$

$$I_x = \frac{tb^3}{12}\sin^2\beta \quad I_y = \frac{tb^3}{12}\cos^2\beta \quad I_{BB} = \frac{tb^3}{3}\sin^2\beta$$

25

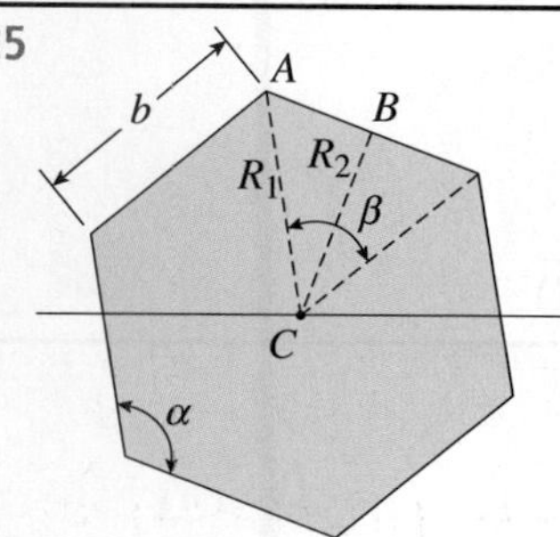

Regular polygon with n sides (Origin of axes at centroid)

C = centroid (at center of polygon)

n = number of sides ($n \geq 3$) $\quad b$ = length of a side

β = central angle for a side $\quad \alpha$ = interior angle (or vertex angle)

$$\beta = \frac{360°}{n} \quad \alpha = \left(\frac{n-2}{n}\right)180° \quad \alpha + \beta = 180°$$

R_1 = radius of circumscribed circle (line CA)

R_2 = radius of inscribed circle (line CB)

$$R_1 = \frac{b}{2}\csc\frac{\beta}{2} \quad R_2 = \frac{b}{2}\cot\frac{\beta}{2} \quad A = \frac{nb^2}{4}\cot\frac{\beta}{2}$$

I_c = moment of inertia about any axis through C (the centroid C is a principal point and every axis through C is a principal axis)

$$I_c = \frac{nb^4}{192}\left(\cot\frac{\beta}{2}\right)\left(3\cot^2\frac{\beta}{2} + 1\right) \quad I_P = 2I_c$$

Properties of Structural-Steel Shapes

In the following tables, the properties of a few structural-steel shapes are presented as an aid to the reader in solving problems in the text. These tables were compiled from the extensive tables of properties for steel shapes commonly used in Europe (Ref. 5-4).

Notation:

I = moment of inertia

S = section modulus

$r = \sqrt{I/A}$ = radius of gyration

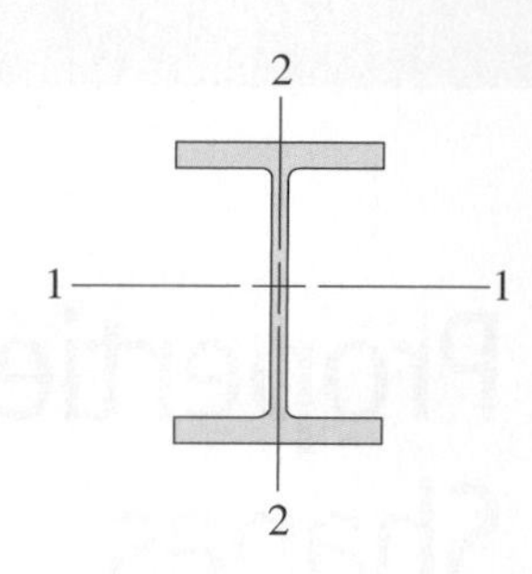

Table F-1

Properties of European Wide-Flange Beams

Designation	Mass per meter	Area of section	Depth of section	Width of section	Thickness		Strong axis 1-1			Weak axis 2-2		
	G	A	h	b	t_w	t_f	I_1	S_1	r_1	I_2	S_2	r_2
	kg/m	cm²	mm	mm	mm	mm	cm⁴	cm³	cm	cm⁴	cm³	cm
HE 1000 B	314	400	1000	300	19	36	644700	12890	40.15	16280	1085	6.38
HE 900 B	291	371.3	900	300	18.5	35	494100	10980	36.48	15820	1054	6.53
HE 700 B	241	306.4	700	300	17	32	256900	7340	28.96	14440	962.7	6.87
HE 650 B	225	286.3	650	300	16	31	210600	6480	27.12	13980	932.3	6.99
HE 600 B	212	270	600	300	15.5	30	171000	5701	25.17	13530	902	7.08
HE 550 B	199	254.1	550	300	15	29	136700	4971	23.2	13080	871.8	7.17
HE 600 A	178	226.5	590	300	13	25	141200	4787	24.97	11270	751.4	7.05
HE 450 B	171	218	450	300	14	26	79890	3551	19.14	11720	781.4	7.33
HE 550 A	166	211.8	540	300	12.5	24	111900	4146	22.99	10820	721.3	7.15
HE 360 B	142	180.6	360	300	12.5	22.5	43190	2400	15.46	10140	676.1	7.49
HE 450 A	140	178	440	300	11.5	21	63720	2896	18.92	9465	631	7.29
HE 340 B	134	170.9	340	300	12	21.5	36660	2156	14.65	9690	646	7.53
HE 320 B	127	161.3	320	300	11.5	20.5	30820	1926	13.82	9239	615.9	7.57
HE 360 A	112	142.8	350	300	10	17.5	33090	1891	15.22	7887	525.8	7.43
HE 340 A	105	133.5	330	300	9.5	16.5	27690	1678	14.4	7436	495.7	7.46
HE 320 A	97.6	124.4	310	300	9	15.5	22930	1479	13.58	6985	465.7	7.49
HE 260 B	93	118.4	260	260	10	17.5	14920	1148	11.22	5135	395	6.58
HE 240 B	83.2	106	240	240	10	17	11260	938.3	10.31	3923	326.9	6.08
HE 280 A	76.4	97.26	270	280	8	13	13670	1013	11.86	4763	340.2	7
HE 220 B	71.5	91.04	220	220	9.5	16	8091	735.5	9.43	2843	258.5	5.59
HE 260 A	68.2	86.82	250	260	7.5	12.5	10450	836.4	10.97	3668	282.1	6.5
HE 240 A	60.3	76.84	230	240	7.5	12	7763	675.1	10.05	2769	230.7	6
HE 180 B	51.2	65.25	180	180	8.5	14	3831	425.7	7.66	1363	151.4	4.57
HE 160 B	42.6	54.25	160	160	8	13	2492	311.5	6.78	889.2	111.2	4.05
HE 140 B	33.7	42.96	140	140	7	12	1509	215.6	5.93	549.7	78.52	3.58
HE 120 B	26.7	34.01	120	120	6.5	11	864.4	144.1	5.04	317.5	52.92	3.06
HE 140 A	24.7	31.42	133	140	5.5	8.5	1033	155.4	5.73	389.3	55.62	3.52
HE 100 B	20.4	26.04	100	100	6	10	449.5	89.91	4.16	167.3	33.45	2.53
HE 100 A	16.7	21.24	96	100	5	8	349.2	72.76	4.06	133.8	26.76	2.51

Note: Axes 1-1 and 2-2 are principal centroidal axes.

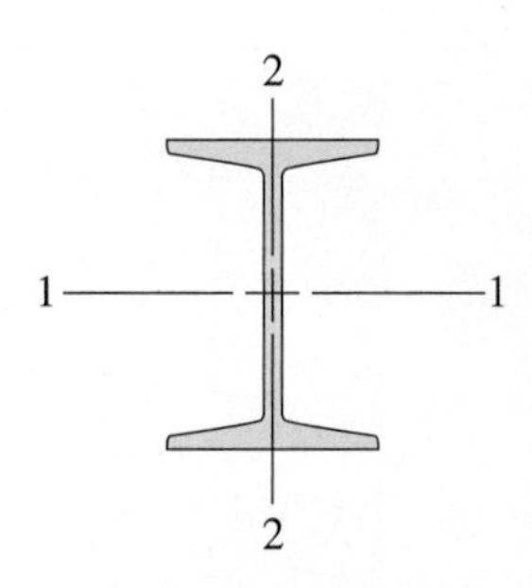

Table F-2

Properties of European Standard Beams

Designation	Mass per meter	Area of section	Depth of section	Width of section	Thickness		Strong axis 1-1			Weak axis 2-2		
	G	A	h	b	t_w	t_f	I_1	S_1	r_1	I_2	S_2	r_2
	kg/m	cm^2	mm	mm	mm	mm	cm^4	cm^3	cm	cm^4	cm^3	cm
IPN 550	166	212	550	200	19	30	99180	3610	21.6	3490	349	4.02
IPN 500	141	179	500	185	18	27	68740	2750	19.6	2480	268	3.72
IPN 450	115	147	450	170	16.2	24.3	45850	2040	17.7	1730	203	3.43
IPN 400	92.4	118	400	155	14.4	21.6	29210	1460	15.7	1160	149	3.13
IPN 380	84	107	380	149	13.7	20.5	24010	1260	15	975	131	3.02
IPN 360	76.1	97	360	143	13	19.5	19610	1090	14.2	818	114	2.9
IPN 340	68	86.7	340	137	12.2	18.3	15700	923	13.5	674	98.4	2.8
IPN 320	61	77.7	320	131	11.5	17.3	12510	782	12.7	555	84.7	2.67
IPN 300	54.2	69	300	125	10.8	16.2	9800	653	11.9	451	72.2	2.56
IPN 280	47.9	61	280	119	10.1	15.2	7590	542	11.1	364	61.2	2.45
IPN 260	41.9	53.3	260	113	9.4	14.1	5740	442	10.4	288	51	2.32
IPN 240	36.2	46.1	240	106	8.7	13.1	4250	354	9.59	221	41.7	2.2
IPN 220	31.1	39.5	220	98	8.1	12.2	3060	278	8.8	162	33.1	2.02
IPN 200	26.2	33.4	200	90	7.5	11.3	2140	214	8	117	26	1.87
IPN 180	21.9	27.9	180	82	6.9	10.4	1450	161	7.2	81.3	19.8	1.71
IPN 160	17.9	22.8	160	74	6.3	9.5	935	117	6.4	54.7	14.8	1.55
IPN 140	14.3	18.3	140	66	5.7	8.6	573	81.9	5.61	35.2	10.7	1.4
IPN 120	11.1	14.2	120	58	5.1	7.7	328	54.7	4.81	21.5	7.41	1.23
IPN 100	8.34	10.6	100	50	4.5	6.8	171	34.2	4.01	12.2	4.88	1.07
IPN 80	5.94	7.58	80	42	3.9	5.9	77.8	19.5	3.2	6.29	3	0.91

Note: Axes 1-1 and 2-2 are principal centroidal axes.

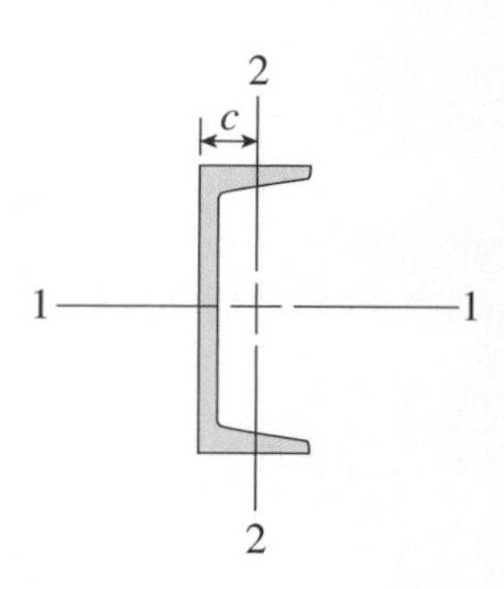

Table F-3

Properties of European Standard Channels

Designation	Mass per meter	Area of section	Depth of section	Width of section	Thickness		Strong axis 1-1			Weak axis 2-2			
	G	A	h	b	t_w	t_f	I_1	S_1	r_1	I_2	S_2	r_2	c
	kg/m	cm^2	mm	mm	mm	mm	cm^4	cm^3	cm	cm^4	cm^3	cm	cm
UPN 400	71.8	91.5	400	110	14	18	20350	1020	14.9	846	102	3.04	2.65
UPN 380	63.1	80.4	380	102	13.5	16	15760	829	14	615	78.7	2.77	2.38
UPN 350	60.6	77.3	350	100	14	16	12840	734	12.9	570	75	2.72	2.4
UPN 320	59.5	75.8	320	100	14	17.5	10870	679	12.1	597	80.6	2.81	2.6
UPN 300	46.2	58.8	300	100	10	16	8030	535	11.7	495	67.8	2.9	2.7
UPN 280	41.8	53.3	280	95	10	15	6280	448	10.9	399	57.2	2.74	2.53
UPN 260	37.9	48.3	260	90	10	14	4820	371	9.99	317	47.7	2.56	2.36
UPN 240	33.2	42.3	240	85	9.5	13	3600	300	9.22	248	39.6	2.42	2.23
UPN 220	29.4	37.4	220	80	9	12.5	2690	245	8.48	197	33.6	2.3	2.14
UPN 200	25.3	32.2	200	75	8.5	11.5	1910	191	7.7	148	27	2.14	2.01
UPN 180	22	28	180	70	8	11	1350	150	6.95	114	22.4	2.02	1.92
UPN 160	18.8	24	160	65	7.5	10.5	925	116	6.21	85.3	18.3	1.89	1.84
UPN 140	16	20.4	140	60	7	10	605	86.4	5.45	62.7	14.8	1.75	1.75
UPN 120	13.4	17	120	55	7	9	364	60.7	4.62	43.2	11.1	1.59	1.6
UPN 100	10.6	13.5	100	50	6	8.5	206	41.2	3.91	29.3	8.49	1.47	1.55
UPN 80	8.64	11	80	45	6	8	106	26.5	3.1	19.4	6.36	1.33	1.45

Notes:
1. Axes 1-1 and 2-2 are principal centroidal axes.
2. The distance c is measured from the centroid to the back of the web.
3. For axis 2-2, the tabulated value of S is the smaller of the two section moduli for this axis.

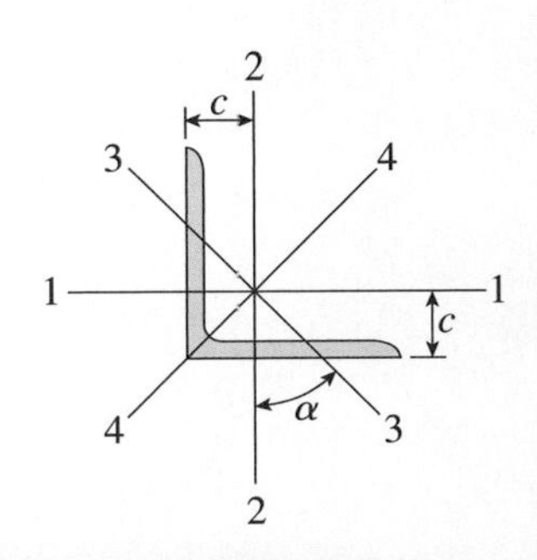

Table F-4

Properties of Angle Sections with Equal Legs (L Shapes)—USCS Units (Abridged List)

Designation	Thickness	Mass per meter	Area of section	Axis 1-1 and Axis 2-2				Axis 3-3	
		G	A	I	S	r	c	I_{min}	r_{min}
	mm	kg/m	cm²	cm⁴	cm³	cm	cm	cm⁴	cm
L 200 × 200 × 26	26	76.6	97.59	3560	252.7	6.04	5.91	1476	3.89
L 200 × 200 × 22	22	65.6	83.51	3094	217.3	6.09	5.76	1273	3.9
L 200 × 200 × 19	19	57.1	72.74	2726	189.9	6.12	5.64	1117	3.92
L 180 × 180 × 20	20	53.7	68.35	2043	159.4	5.47	5.18	841.3	3.51
L 180 × 180 × 19	19	51.1	65.14	1955	152.1	5.48	5.14	803.8	3.51
L 200 × 200 × 16	16	48.5	61.79	2341	161.7	6.16	5.52	957.1	3.94
L 180 × 180 × 17	17	46	58.66	1775	137.2	5.5	5.06	727.8	3.52
L 180 × 180 × 15	15	40.9	52.1	1589	122	5.52	4.98	650.5	3.53
L 160 × 160 × 17	17	40.7	51.82	1225	107.2	4.86	4.57	504.1	3.12
L 160 × 160 × 15	15	36.2	46.06	1099	95.47	4.88	4.49	450.8	3.13
L 180 × 180 × 13	13	35.7	45.46	1396	106.5	5.54	4.9	571.6	3.55
L 150 × 150 × 15	15	33.8	43.02	898.1	83.52	4.57	4.25	368.9	2.93
L 150 × 150 × 14	14	31.6	40.31	845.4	78.33	4.58	4.21	346.8	2.93
L 150 × 150 × 12	12	27.3	34.83	736.9	67.75	4.6	4.12	302	2.94
L 120 × 120 × 15	15	26.6	33.93	444.9	52.43	3.62	3.51	184.1	2.33
L 120 × 120 × 13	13	23.3	29.69	394	46.01	3.64	3.44	162.2	2.34
L 150 × 150 × 10	10	23	29.27	624	56.91	4.62	4.03	256	2.96
L 140 × 140 × 10	10	21.4	27.24	504.4	49.43	4.3	3.79	206.8	2.76
L 120 × 120 × 11	11	19.9	25.37	340.6	39.41	3.66	3.36	139.7	2.35
L 100 × 100 × 12	12	17.8	22.71	206.7	29.12	3.02	2.9	85.42	1.94
L 110 × 110 × 10	10	16.6	21.18	238	29.99	3.35	3.06	97.72	2.15
L 100 × 100 × 10	10	15	19.15	176.7	24.62	3.04	2.82	72.64	1.95
L 90 × 90 × 9	9	12.2	15.52	115.8	17.93	2.73	2.54	47.63	1.75
L 90 × 90 × 8	8	10.9	13.89	104.4	16.05	2.74	2.5	42.87	1.76
L 90 × 90 × 7	7	9.6	12.24	92.5	14.13	2.75	2.45	38.02	1.76

Notes:

1. Axes 1–1 and 2–2 are centroidal axes parallel to the legs.
2. The distance c is measured from the centroid to the back of the legs.
3. For axes 1–1 and 2–2, the tabulated value of S is the smaller of the two section moduli for those axes.
4. Axes 3–3 and 4–4 are principal centroidal axes.
5. The moment of inertia for axis 3–3, which is the smaller of the two principal moments of inertia, can be found from the equation $I_{33} = Ar^2_{\text{min}}$.
6. The moment of inertia for axis 4–4, which is the larger of the two principal moments of inertia, can be found from the equation $I_{44} + I_{33} = I_{11} + I_{22}$.

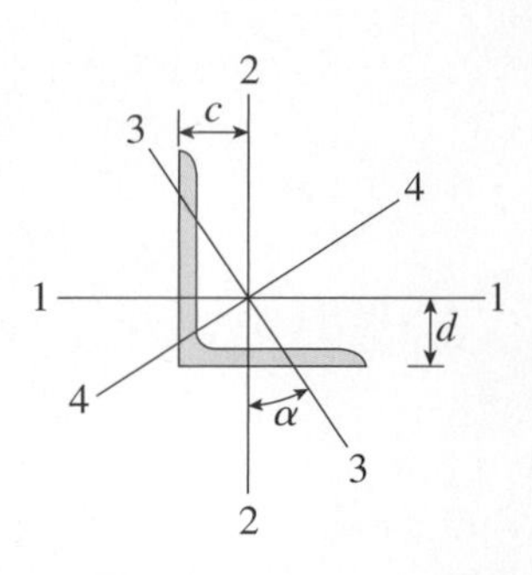

Table F-5

Properties of Angle Sections with Unequal Legs (L Shapes)—USCS Units (Abridged List)

Designation	Thickness	Mass per meter	Area of section	Axis 1-1				Axis 2-2				Axis 3-3		Angle α
		G	A	I	S	r	d	I	S	r	c	I_{min}	r_{min}	tan α
	mm	kg/m	cm2	cm4	cm3	cm	cm	cm4	cm3	cm	cm	cm4	cm	
L 200 × 100 × 14	14	31.6	40.28	1654	128.4	6.41	7.12	282.2	36.08	2.65	2.18	181.7	2.12	0.261
L 150 × 100 × 14	14	26.1	33.22	743.5	74.12	4.73	4.97	264.2	35.21	2.82	2.5	153	2.15	0.434
L 200 × 100 × 12	12	25.1	34.8	1440	111	6.43	7.03	247.2	31.28	2.67	2.1	158.5	2.13	0.263
L 200 × 100 × 10	10	23	29.24	1219	93.24	6.46	6.93	210.3	26.33	2.68	2.01	134.5	2.14	0.265
L 150 × 100 × 12	12	22.6	28.74	649.6	64.23	4.75	4.89	231.9	30.58	2.84	2.42	133.5	2.16	0.436
L 160 × 80 × 12	12	21.6	27.54	719.5	69.98	5.11	5.72	122	19.59	2.1	1.77	78.77	1.69	0.260
L 150 × 90 × 11	11	19.9	25.34	580.7	58.3	4.79	5.04	158.7	22.91	2.5	2.08	95.71	1.94	0.360
L 150 × 100 × 10	10	19	24.18	551.7	54.08	4.78	4.8	197.8	25.8	2.86	2.34	113.5	2.17	0.439
L 150 × 90 × 10	10	18.2	23.15	533.1	53.29	4.8	5	146.1	20.98	2.51	2.04	87.93	1.95	0.361
L 160 × 80 × 10	10	18.2	23.18	611.3	58.94	5.14	5.63	104.4	16.55	2.12	1.69	67.01	1.7	0.262
L 120 × 80 × 12	12	17.8	22.69	322.8	40.37	3.77	4	114.3	19.14	2.24	2.03	66.46	1.71	0.432
L 120 × 80 × 10	10	15	19.13	275.5	34.1	3.8	3.92	98.11	16.21	2.26	1.95	56.6	1.72	0.435
L 130 × 65 × 10	10	14.6	18.63	320.5	38.39	4.15	4.65	54.2	10.73	1.71	1.45	35.02	1.37	0.259
L 120 × 80 × 8	8	12.2	15.49	225.7	27.63	3.82	3.83	80.76	13.17	2.28	1.87	46.39	1.73	0.438
L 130 × 65 × 8	8	11.8	15.09	262.5	31.1	4.17	4.56	44.77	8.72	1.72	1.37	28.72	1.38	0.262

Notes:

1. Axes 1–1 and 2–2 are centroidal axes parallel to the legs.
2. The distances *c* and *d* are measured from the centroid to the backs of the legs.
3. For axes 1–1 and 2–2, the tabulated value of *S* is the smaller of the two section moduli for those axes.
4. Axes 3–3 and 4–4 are principal centroidal axes.
5. The moment of inertia for axis 3–3, which is the smaller of the two principal moments of inertia, can be found from the equation $I_{33} = Ar_{min}^2$.
6. The moment of inertia for axis 4–4, which is the larger of the two principal moments of inertia, can be found from the equation $I_{44} + I_{33} = I_{11} + I_{22}$.

Properties of Structural Timber

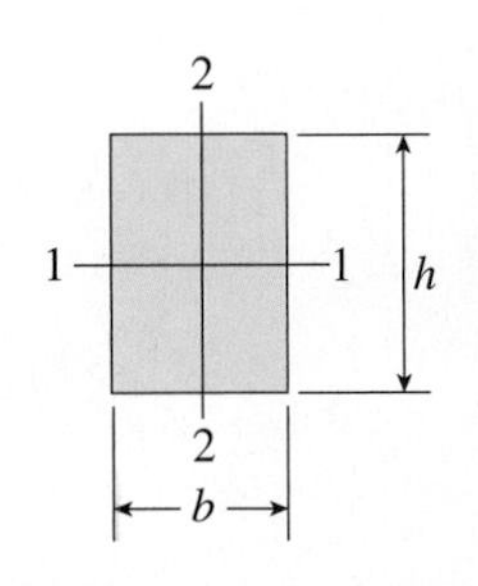

Properties of Sawn Solid Timber Sizes (for the most readily available stress graded European sizes, dry)

Nominal dimensions $b \times h$	Net dimensions $b \times h$	Area $A = bh$	Axis 1–1: Moment of inertia $I_1 = \frac{bh^3}{12}$	Axis 1–1: Section modulus $S_1 = \frac{bh^2}{6}$	Axis 2–2: Moment of inertia $I_2 = \frac{hb^3}{12}$	Axis 2–2: Section modulus $S_2 = \frac{hb^2}{6}$	Weight per linear meter (based on 560 kg/m³)
mm	mm	10^3mm^2	10^6mm^4	10^6mm^3	10^6mm^4	10^6mm^3	N
38 × 75	35 × 72	2.52	1.09	0.0302	0.257	0.0147	13.83
38 × 100	35 × 97	3.4	2.66	0.0549	0.347	0.0198	18.64
38 × 125	35 × 122	4.27	5.3	0.0868	0.436	0.0249	23.45
50 × 75	47 × 72	3.38	1.46	0.0406	0.623	0.0265	18.64
50 × 100	47 × 97	4.56	3.57	0.0737	0.839	0.0357	25.02
50 × 125	47 × 122	5.73	7.11	0.117	1.06	0.0449	31.49
50 × 150	47 × 147	6.91	12.4	0.169	1.27	0.0541	37.96
50 × 200	47 × 195	9.17	29	0.298	1.69	0.0718	50.33
50 × 250	47 × 245	11.5	57.6	0.47	2.12	0.0902	63.27
75 × 100	72 × 97	6.98	5.48	0.113	3.02	0.0838	38.36
75 × 150	72 × 147	10.6	19.1	0.259	4.57	0.127	58.17
75 × 200	72 × 147	14	44.5	0.456	6.07	0.168	77.11
75 × 250	75 × 245	17.6	88.2	0.72	7.62	0.212	96.92
100 × 100	97 × 97	9.41	7.38	0.152	7.38	0.152	51.7
100 × 150	97 × 147	14.3	25.7	0.349	11.2	0.231	78.38
100 × 200	97 × 195	18.9	59.9	0.615	14.8	0.306	103.89
100 × 250	97 × 295	23.8	119	0.97	18.6	0.384	130.57
100 × 300	97 × 295	28.6	208	1.41	22.4	0.463	157.16
150 × 150	147 × 195	21.6	38.9	0.529	38.9	0.529	118.7
150 × 200	147 × 195	28.7	90.8	0.932	51.6	0.702	157.45
150 × 300	147 × 295	43.4	314	2.13	78.1	1.06	238.19
200 × 200	195 × 195	38	120	1.24	120	1.24	208.85
200 × 300	195 × 295	57.5	417	2.83	182	1.87	315.98
300 × 300	295 × 295	87	631	4.28	631	4.28	478.04

Note: Axes 1–1 and 2–2 are principal centroidal axes.

APPENDIX H

Deflections and Slopes of Beams

Table H-1

Deflections and Slopes of Cantilever Beams

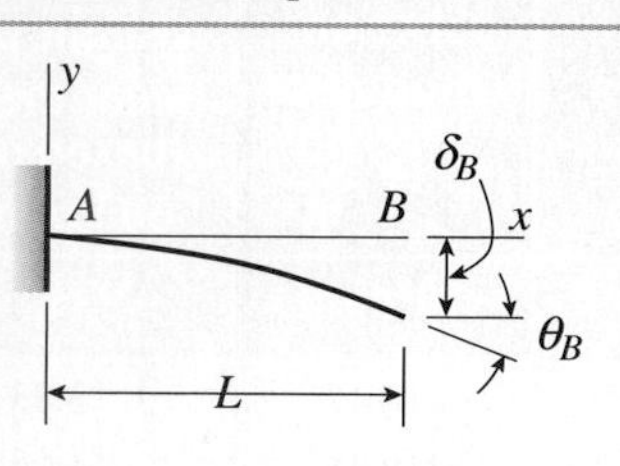

Notation:

v = deflection in the y direction (positive upward)

$v' = dv/dx$ = slope of the deflection curve

$\delta_B = -v(L)$ = deflection at end B of the beam (positive downward)

$\theta_B = -v'(L)$ = angle of rotation at end B of the beam (positive clockwise)

EI = constant

1

q

$$v = -\frac{qx^2}{24EI}(6L^2 - 4Lx + x^2) \qquad v' = -\frac{qx}{6EI}(3L^2 - 3Lx + x^2)$$

$$\delta_B = \frac{qL^4}{8EI} \qquad \theta_B = \frac{qL^3}{6EI}$$

2

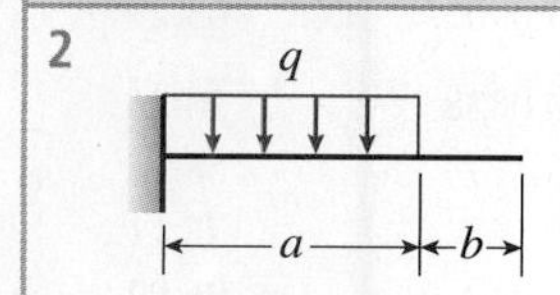

$$v = -\frac{qx^2}{24EI}(6a^2 - 4ax + x^2) \qquad (0 \le x \le a)$$

$$v' = -\frac{qx}{6EI}(3a^2 - 3ax + x^2) \qquad (0 \le x \le a)$$

$$v = -\frac{qa^3}{24EI}(4x - a) \qquad v' = -\frac{qa^3}{6EI} \qquad (a \le x \le L)$$

$$\text{At } x = a: v = -\frac{qa^4}{8EI} \qquad v' = -\frac{qa^3}{6EI}$$

$$\delta_B = \frac{qa^3}{24EI}(4L - a) \qquad \theta_B = \frac{qa^3}{6EI}$$

Table H-1 (Continued)

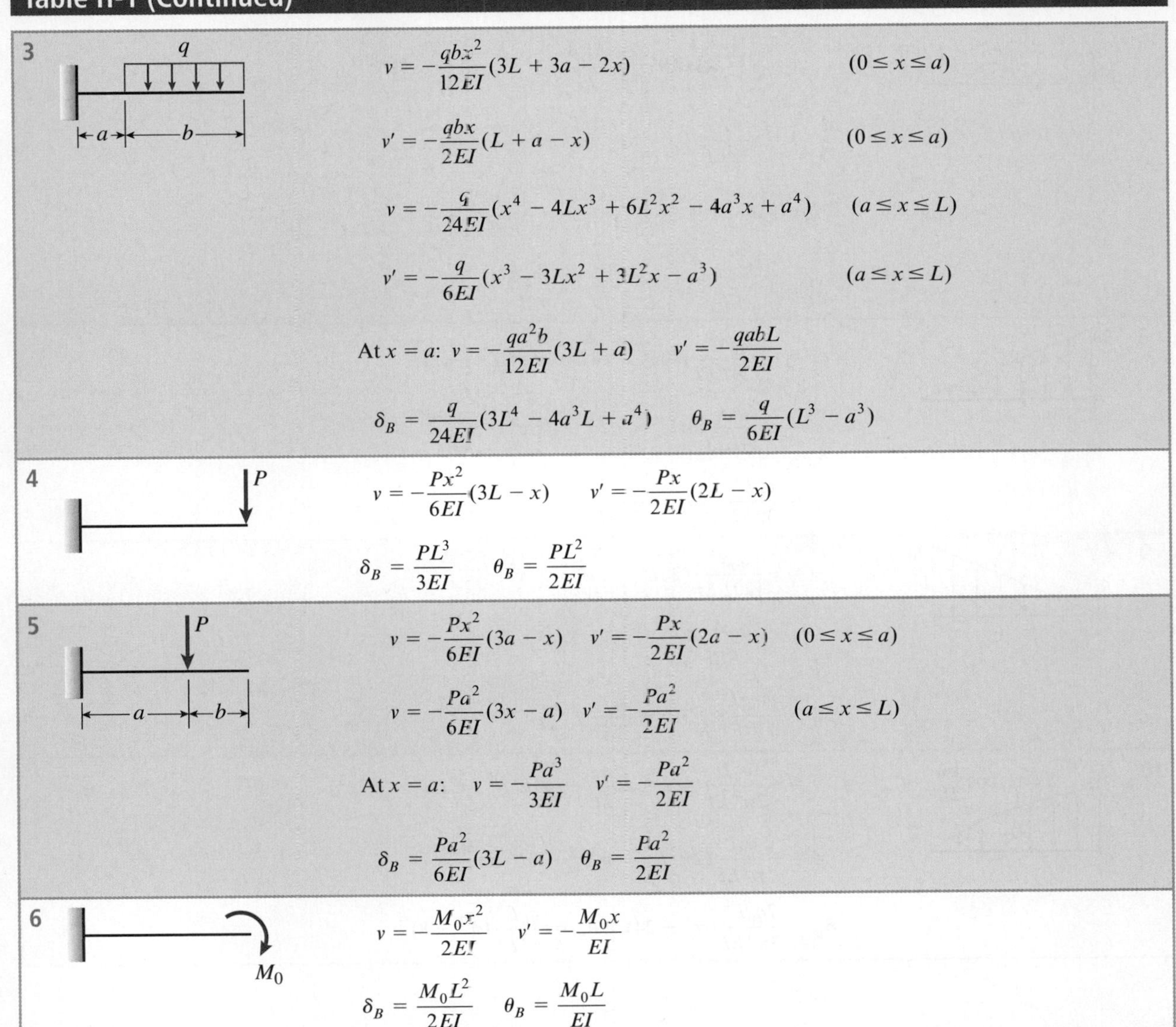

3

$$v = -\frac{qbx^2}{12EI}(3L + 3a - 2x) \qquad (0 \leq x \leq a)$$

$$v' = -\frac{qbx}{2EI}(L + a - x) \qquad (0 \leq x \leq a)$$

$$v = -\frac{q}{24EI}(x^4 - 4Lx^3 + 6L^2x^2 - 4a^3x + a^4) \qquad (a \leq x \leq L)$$

$$v' = -\frac{q}{6EI}(x^3 - 3Lx^2 + 3L^2x - a^3) \qquad (a \leq x \leq L)$$

At $x = a$: $v = -\frac{qa^2b}{12EI}(3L + a) \qquad v' = -\frac{qabL}{2EI}$

$$\delta_B = \frac{q}{24EI}(3L^4 - 4a^3L + a^4) \qquad \theta_B = \frac{q}{6EI}(L^3 - a^3)$$

4

$$v = -\frac{Px^2}{6EI}(3L - x) \qquad v' = -\frac{Px}{2EI}(2L - x)$$

$$\delta_B = \frac{PL^3}{3EI} \qquad \theta_B = \frac{PL^2}{2EI}$$

5

$$v = -\frac{Px^2}{6EI}(3a - x) \qquad v' = -\frac{Px}{2EI}(2a - x) \qquad (0 \leq x \leq a)$$

$$v = -\frac{Pa^2}{6EI}(3x - a) \qquad v' = -\frac{Pa^2}{2EI} \qquad (a \leq x \leq L)$$

At $x = a$: $v = -\frac{Pa^3}{3EI} \qquad v' = -\frac{Pa^2}{2EI}$

$$\delta_B = \frac{Pa^2}{6EI}(3L - a) \qquad \theta_B = \frac{Pa^2}{2EI}$$

6

$$v = -\frac{M_0x^2}{2EI} \qquad v' = -\frac{M_0x}{EI}$$

$$\delta_B = \frac{M_0L^2}{2EI} \qquad \theta_B = \frac{M_0L}{EI}$$

(Continued)

Table H-1 (Continued)

7

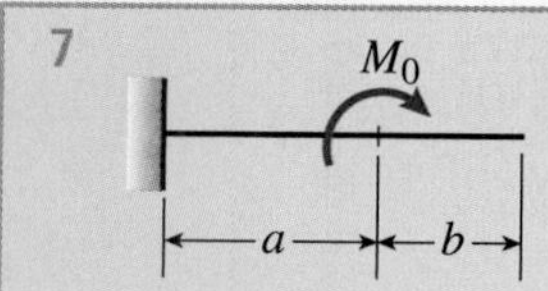

$$v = -\frac{M_0x^2}{2EI} \qquad v' = -\frac{M_0x}{EI} \qquad (0 \le x \le a)$$

$$v = -\frac{M_0a}{2EI}(2x - a) \quad v' = -\frac{M_0a}{EI} \quad (a \le x \le L)$$

$$\text{At } x = a\text{:} \quad v = -\frac{M_0a^2}{2EI} \quad v' = -\frac{M_0a}{EI}$$

$$\delta_B = \frac{M_0a}{2EI}(2L - a) \quad \theta_B = \frac{M_0a}{EI}$$

8

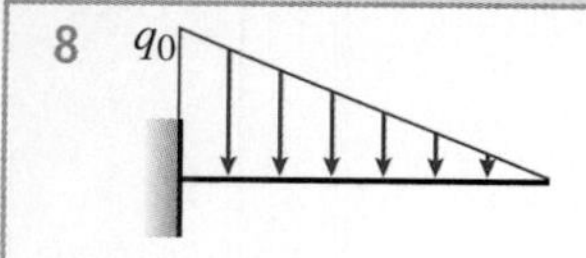

$$v = -\frac{q_0x^2}{120LEI}(10L^3 - 10L^2x + 5Lx^2 - x^3)$$

$$v' = -\frac{q_0x}{24LEI}(4L^3 - 6L^2x + 4Lx^2 - x^3)$$

$$\delta_B = \frac{q_0L^4}{30EI} \quad \theta_B = \frac{q_0L^3}{24EI}$$

9

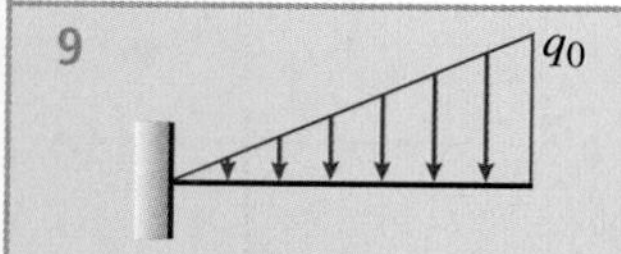

$$v = -\frac{q_0x^2}{120LEI}(20L^3 - 10L^2x + x^3)$$

$$v' = -\frac{q_0x}{24LEI}(8L^3 - 6L^2x + x^3)$$

$$\delta_B = \frac{11q_0L^4}{120EI} \quad \theta_B = \frac{q_0L^3}{8EI}$$

10

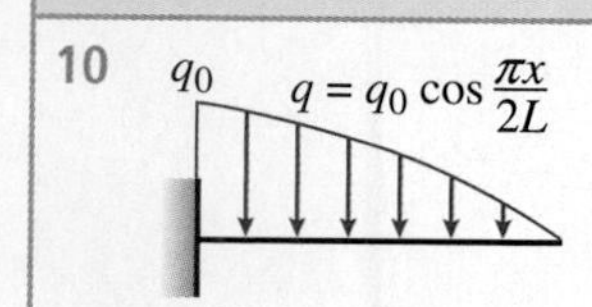

$$v = -\frac{q_0L}{3\pi^4EI}\left(48L^3 \cos\frac{\pi x}{2L} - 48L^3 + 3\pi^3Lx^2 - \pi^3x^3\right)$$

$$v' = -\frac{q_0L}{\pi^3EI}\left(2\pi^2Lx - \pi^2x^2 - 8L^2 \sin\frac{\pi x}{2L}\right)$$

$$\delta_B = \frac{2q_0L^4}{3\pi^4EI}(\pi^3 - 24) \quad \theta_B = \frac{q_0L^3}{\pi^3EI}(\pi^2 - 8)$$

Table H-2

Deflections and Slopes of Simple Beams

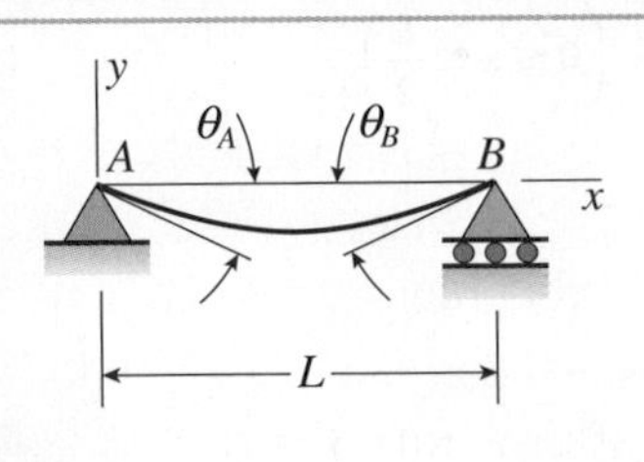

Notation:

v = deflection in the y direction (positive upward)

$v' = dv/dx$ = slope of the deflection curve

$\delta_C = -v(L/2)$ = deflection at midpoint C of the beam (positive downward)

x_1 = distance from support A to point of maximum deflection

$\delta_{max} = -v_{max}$ = maximum deflection (positive downward)

$\theta_A = -v'(0)$ = angle of rotation at left-hand end of the beam (positive clockwise)

$\theta_B = v'(L)$ = angle of rotation at right-hand end of the beam (positive counterclockwise)

EI = constant

1

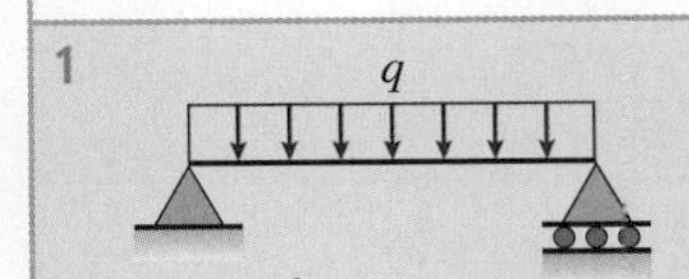

$$v = -\frac{qx}{24EI}(L^3 - 2Lx^2 + x^3)$$

$$v' = -\frac{q}{24EI}(L^3 - 6Lx^2 + 4x^3)$$

$$\delta_C = \delta_{max} = \frac{5qL^4}{384EI} \qquad \theta_A = \theta_B = \frac{qL^3}{24EI}$$

2

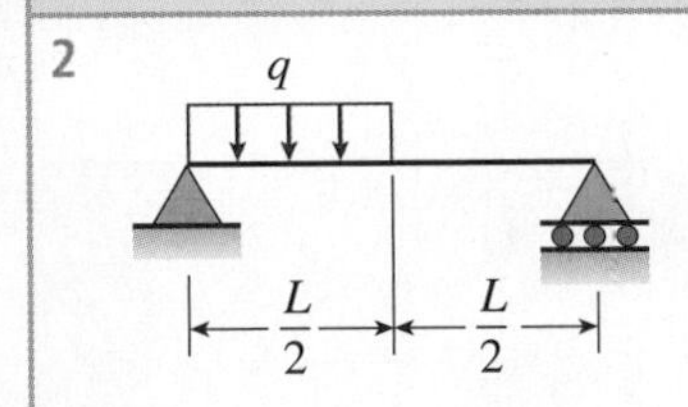

$$v = -\frac{qx}{384EI}(9L^3 - 24Lx^2 + 16x^3) \qquad \left(0 \le x \le \frac{L}{2}\right)$$

$$v' = -\frac{q}{384EI}(9L^3 - 72Lx^2 + 64x^3) \qquad \left(0 \le x \le \frac{L}{2}\right)$$

$$v = -\frac{qL}{384EI}(8x^3 - 24Lx^2 + 17L^2x - L^3) \qquad \left(\frac{L}{2} \le x \le L\right)$$

$$v' = -\frac{qL}{384EI}(24x^2 - 48Lx + 17L^2) \qquad \left(\frac{L}{2} \le x \le L\right)$$

$$\delta_C = \frac{5qL^4}{768EI} \qquad \theta_A = \frac{3qL^3}{128EI} \qquad \theta_B = \frac{7qL^3}{384EI}$$

3

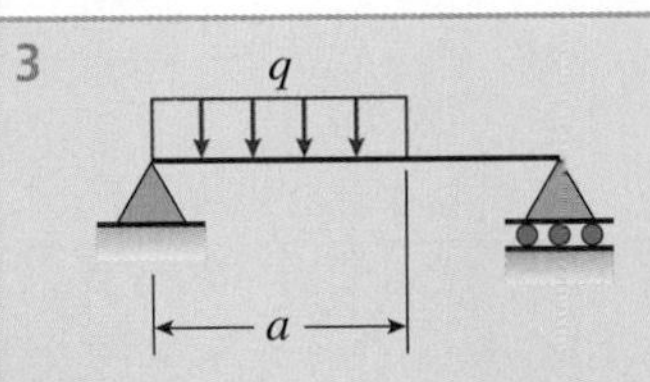

$$v = -\frac{qx}{24LEI}(a^4 - 4a^3L + 4a^2L^2 + 2a^2x^2 - 4aLx^2 + Lx^3) \qquad (0 \le x \le a)$$

$$v' = -\frac{q}{24LEI}(a^4 - 4a^3L + 4a^2L^2 + 6a^2x^2 - 12aLx^2 - 4Lx^3) \qquad (0 \le x \le a)$$

$$v = -\frac{qa^2}{24LEI}(-a^2L + 4L^2x + a^2x - 6Lx^2 + 2x^3) \qquad (a \le x \le L)$$

$$v' = -\frac{qa^2}{24LEI}(4L^2 + a^2 - 12Lx + 6x^2) \qquad (a \le x \le L)$$

$$\theta_A = \frac{qa^2}{24LEI}(2L - a)^2 \qquad \theta_B = \frac{qa^2}{24LEI}(2L^2 - a^2)$$

(Continued)

Table H-2 (Continued)

4

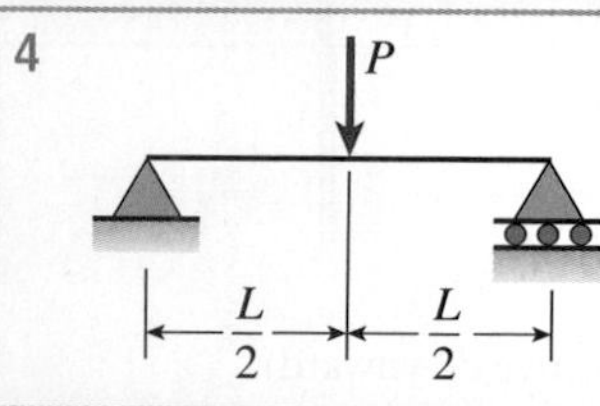

$$v = -\frac{Px}{48EI}(3L^2 - 4x^2) \quad v' = -\frac{P}{16EI}(L^2 - 4x^2) \quad \left(0 \le x \le \frac{L}{2}\right)$$

$$\delta_C = \delta_{max} = \frac{PL^3}{48EI} \quad \theta_A = \theta_B = \frac{PL^2}{16EI}$$

5

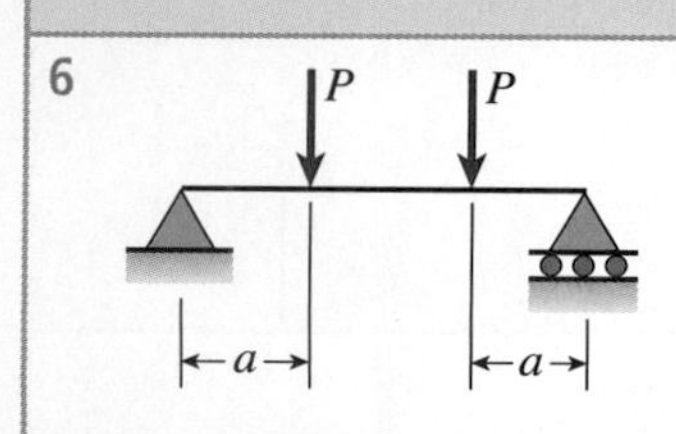

$$v = -\frac{Pbx}{6LEI}(L^2 - b^2 - x^2) \quad v' = -\frac{Pb}{6LEI}(L^2 - b^2 - 3x^2) \quad (0 \le x \le a)$$

$$\theta_A = \frac{Pab(L + b)}{6LEI} \quad \theta_B = \frac{Pab(L + a)}{6LEI}$$

If $a \ge b$, $\delta_C = \dfrac{Pb(3L^2 - 4b^2)}{48EI}$ If $a \le b$, $\delta_C = \dfrac{Pa(3L^2 - 4a^2)}{48EI}$

If $a \ge b$, $x_1 = \sqrt{\dfrac{L^2 - b^2}{3}}$ and $\delta_{max} = \dfrac{Pb(L^2 - b^2)^{3/2}}{9\sqrt{3}LEI}$

6

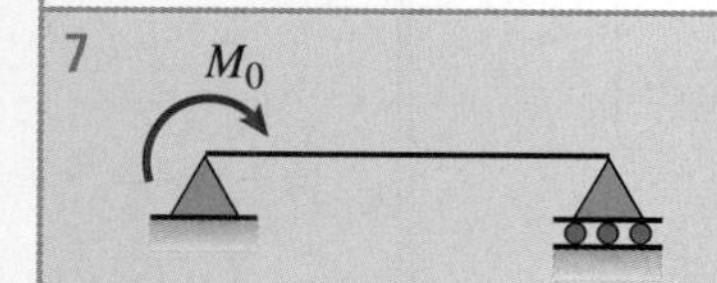

$$v = -\frac{Px}{6EI}(3aL - 3a^2 - x^2) \quad v' = -\frac{P}{2EI}(aL - a^2 - x^2) \quad (0 \le x \le a)$$

$$v = -\frac{Pa}{6EI}(3Lx - 3x^2 - a^2) \quad v' = -\frac{Pa}{2EI}(L - 2x) \quad (a \le x \le L - a)$$

$$\delta_C = \delta_{max} = \frac{Pa}{24EI}(3L^2 - 4a^2) \quad \theta_A = \theta_B = \frac{Pa(L - a)}{2EI}$$

7

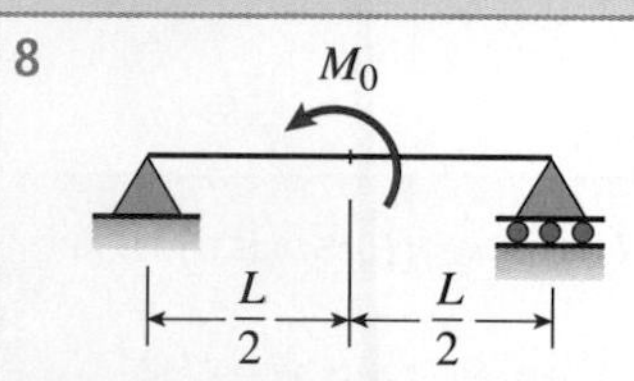

$$v = -\frac{M_0 x}{6LEI}(2L^2 - 3Lx + x^2) \quad v' = -\frac{M_0}{6LEI}(2L^2 - 6Lx + 3x^2)$$

$$\delta_C = \frac{M_0 L^2}{16EI} \quad \theta_A = \frac{M_0 L}{3EI} \quad \theta_B = \frac{M_0 L}{6EI}$$

$$x_1 = L\left(1 - \frac{\sqrt{3}}{3}\right) \text{ and } \delta_{max} = \frac{M_0 L^2}{9\sqrt{3}EI}$$

8

$$v = -\frac{M_0 x}{24LEI}(L^2 - 4x^2) \quad v' = -\frac{M_0}{24LEI}(L^2 - 12x^2) \quad \left(0 \le x \le \frac{L}{2}\right)$$

$$\delta_C = 0 \quad \theta_A = \frac{M_0 L}{24EI} \quad \theta_B = -\frac{M_0 L}{24EI}$$

9

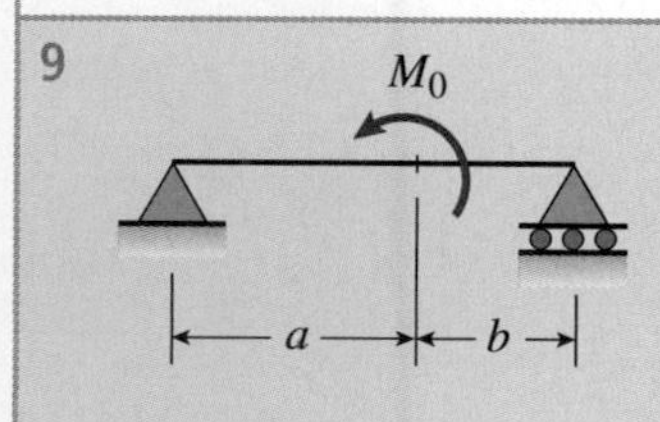

$$v = -\frac{M_0 x}{6LEI}(6aL - 3a^2 - 2L^2 - x^2) \quad (0 \le x \le a)$$

$$v' = -\frac{M_0}{6LEI}(6aL - 3a^2 - 2L^2 - 3x^2) \quad (0 \le x \le a)$$

At $x = a$: $v = -\dfrac{M_0 ab}{3LEI}(2a - L)$ $v' = -\dfrac{M_0}{3LEI}(3aL - 3a^2 - L^2)$

$$\theta_A = \frac{M_0}{6LEI}(6aL - 3a^2 - 2L^2) \quad \theta_B = \frac{M_0}{6LEI}(3a^2 - L^2)$$

Table H-2 (Continued)

Case	Beam	Deflections and slopes
10	M_0 M_0	$v = -\dfrac{M_0 x}{2EI}(L - x)$ $\quad v' = -\dfrac{M_0}{2EI}(L - 2x)$ $\delta_C = \delta_{max} = \dfrac{M_0 L^2}{8EI}$ $\quad \theta_A = \theta_B = \dfrac{M_0 L}{2EI}$
11	q_0	$v = -\dfrac{q_0 x}{360LEI}(7L^4 - 10L^2x^2 + 3x^4)$ $v' = -\dfrac{q_0}{360LEI}(7L^4 - 30L^2x^2 + 15x^4)$ $\delta_C = \dfrac{5q_0L^4}{768EI}$ $\quad \theta_A = \dfrac{7q_0L^3}{360EI}$ $\quad \theta_B = \dfrac{q_0L^3}{45EI}$ $x_1 = 0.5193L$ $\quad \delta_{max} = 0.00652\dfrac{q_0L^4}{EI}$
12	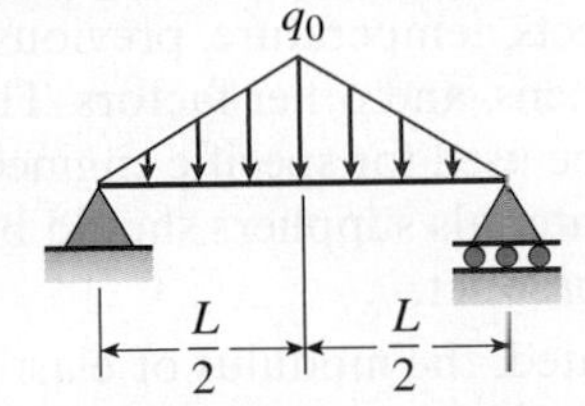	$v = -\dfrac{q_0 x}{960LEI}(5L^2 - 4x^2)^2$ $\quad \left(0 \le x \le \dfrac{L}{2}\right)$ $v' = -\dfrac{q_0}{192LEI}(5L^2 - 4x^2)(L^2 - 4x^2)$ $\quad \left(0 \le x \le \dfrac{L}{2}\right)$ $\delta_C = \delta_{max} = \dfrac{q_0L^4}{120EI}$ $\quad \theta_A = \theta_B = \dfrac{5q_0L^3}{192EI}$
13	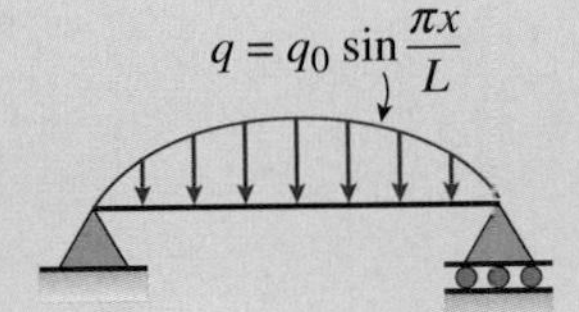	$v = -\dfrac{q_0L^4}{\pi^4 EI}\sin\dfrac{\pi x}{L}$ $\quad v' = -\dfrac{q_0L^3}{\pi^3 EI}\cos\dfrac{\pi x}{L}$ $\delta_C = \delta_{max} = \dfrac{q_0L^4}{\pi^4 EI}$ $\quad \theta_A = \theta_B = \dfrac{q_0L^3}{\pi^3 EI}$

APPENDIX I

Properties of Materials

Notes:

1. Properties of materials vary greatly depending upon manufacturing processes, chemical composition, internal defects, temperature, previous loading history, age, dimensions of test specimens, and other factors. The tabulated values are typical but should never be used for specific engineering or design purposes. Manufacturers and materials suppliers should be consulted for information about a particular product.
2. Except when compression or bending is indicated, the modulus of elasticity E, yield stress σ_Y, and ultimate stress σ_U are for materials in tension.

Table I-1

Weights and Mass Densities

Material	Weight Density γ	Mass Density ρ
	kN/m^3	kg/m^3
Aluminum alloys	26–28	2600–2800
2014-T6, 7075-T6	28	2800
6061-T6	26	2700
Brass	82–85	8400–8600
Bronze	80–86	8200–8800
Cast iron	68–72	7000–7400
Concrete		
Plain	23	2300
Reinforced	24	2400
Lightweight	11–18	1100–1800
Copper	87	8900
Glass	24–28	2400–2800
Magnesium alloys	17–18	1760–1830
Monel (67% Ni, 30% Cu)	87	8800
Nickel	87	8800
Plastics		
Nylon	8.6–11	880–1100
Polyethylene	9.4–14	960–1400
Rock		
Granite, marble, quartz	26–28	2600–2900
Limestone, sandstone	20–28	2000–2900
Rubber	9–13	960–1300
Sand, soil, gravel	12–21	1200–2200
Steel	77.0	7850
Titanium	44	4500
Tungsten	190	1900
Water, fresh	9.81	1000
sea	10.0	1020
Wood (air dry)		
Douglas fir	4.7–5.5	480–560
Oak	6.3–7.1	640–720
Southern pine	5.5–6.3	560–640

Table I-2

Moduli of Elasticity and Poisson's Ratios

Material	Modulus of Elasticity E	Shear Modulus of Elasticity G	Poisson's Ratio ν
	GPa	GPa	
Aluminum alloys	70–79	26–30	0.33
2014-T6	73	28	0.33
6061-T6	70	26	0.33
7075-T6	72	27	0.33
Brass	96–110	36–41	0.34
Bronze	96–120	36–44	0.34
Cast iron	83–170	32–69	0.2–0.3
Concrete (compression)	17–31		0.1–0.2
Copper and copper alloys	110–120	40–47	0.33–0.36
Glass	48–83	19–35	0.17–0.27
Magnesium alloys	41–45	15–17	0.35
Monel (67% Ni, 30% Cu)	170	66	0.32
Nickel	210	80	0.31
Plastics			
Nylon	2.1–3.4		0.4
Polyethylene	0.7–1.4		0.4
Rock (compression)			
Granite, marble, quartz	40–100		0.2–0.3
Limestone, sandstone	20–70		0.2–0.3
Rubber	0.0007–0.004	0.0002–0.001	0.45–0.50
Steel	190–210	75–80	0.27–0.30
Titanium alloys	100–120	39–44	0.33
Tungsten	340–380	140–160	0.2
Wood (bending)			
Douglas fir	11–13		
Oak	11–12		
Southern pine	11–14		

Table I-3

Mechanical Properties

Material	Yield Stress σ_Y	Ultimate Stress σ_U	Percent Elongation (25 mm gage length)
	MPa	MPa	
Aluminum alloys	35–500	100–550	1–45
2014-T6	410	480	13
6061-T6	270	310	17
7075-T6	480	550	11
Brass	70–550	200–620	4–60
Bronze	82–690	200–830	5–60
Cast iron (tension)	120–290	69–480	0–1
Cast iron (compression)		340–1400	
Concrete (compression)		10–70	
Copper and copper alloys	55–760	230–830	4–50
Glass		30–1000	0
Plate glass		70	
Glass fibers		7000–20,000	
Magnesium alloys	80–280	140–340	2–20
Monel (67% Ni, 30% Cu)	170–1100	450–1200	2–50
Nickel	100–620	310–760	2–50
Plastics			
Nylon		40–80	20–100
Polyethylene		7–28	15–300
Rock (compression)			
Granite, marble, quartz		50–280	
Limestone, sandstone		20–200	
Rubber	1–7	7–20	100–800
Steel			
High-strength	340–1000	550–1200	5–25
Machine	340–700	550–860	5–25
Spring	400–1600	700–1900	3–15
Stainless	280–700	400–1000	5–40
Tool	520	900	8
Steel, structural	200–700	340–830	10–40
ASTM-A36	250	400	30
ASTM-A572	340	500	20
ASTM-A514	700	830	15

(*Continued*)

Table I-3 (Continued)

Mechanical Properties

Material	Yield Stress σ_Y	Ultimate Stress σ_U	Percent Elongation (25 mm gage length)
	MPa	MPa	
Steel wire	280–1000	550–1400	5–40
Titanium alloys	760–1000	900–1200	10
Tungsten		1400–4000	0–4
Wood (bending)			
Douglas fir	30–50	50–80	
Oak	40–60	50–100	
Southern pine	40–60	50–100	
Wood (compression parallel to grain)			
Douglas fir	30–50	40–70	
Oak	30–40	30–50	
Southern pine	30–50	40–70	

Table I-4

Coefficients of Thermal Expansion

Material	Coefficient of Thermal Expansion α
	$10^{-6}/°C$
Aluminum alloys	23
Brass	19.1–21.2
Bronze	18–21
Cast iron	9.9–12
Concrete	7–14
Copper and copper alloys	16.6–17.6
Glass	5–11
Magnesium alloys	26.1–28.8
Monel (67% Ni, 30% Cu)	14
Nickel	13

Material	Coefficient of Thermal Expansion α
	$10^{-6}/°C$
Plastics	
Nylon	70–140
Polyethylene	140–290
Rock	5–9
Rubber	130–200
Steel	10–18
High-strength	14
Stainless	17
Structural	12
Titanium alloys	8.1–11
Tungsten	4.3

CHAPTER 1

1.3-1 $A_y = 130.4\text{ N}, B_x = 135\text{ N}, B_y = 421.5\text{ N}, N = 0, V = 37.5\text{ N}, M = -198.7\text{ N}\cdot\text{m}$

1.3-2 $A_y = 198.3\text{ N}, B_x = -120\text{ N}, B_y = -38.3\text{ N}, N = 0, V = -41.7\text{ N}, M = 510\text{ N}\cdot\text{m}$

1.3-3 (a) $A_y = 22.7\text{ N}, B_y = -22.7\text{ N}, C_x = 220\text{ N}, C_y = 0$; (b) $N = 220\text{ N}, V = 22.7\text{ N}, M = 102\text{ N}\cdot\text{m}$

1.3-4 (a) $M_A = 0, C_y = 236\text{ N}, D_y = -75.6\text{ N}$; (b) $N = 0, V = -70\text{ N}, M = -36.7\text{ N}\cdot\text{m}$; (c) $M_A = 0, C_y = 236\text{ N}, D_y = -75.6\text{ N}; N = 0, V = -70\text{ N}, M = -36.7\text{ N}\cdot\text{m}$

1.3-5 (a) $A_x = 57.7\text{ N}, A_y = -66.7\text{ N}, C_y = 464.3\text{ N}, D_x = 50.3\text{ N}, D_y = -87.2\text{ N}$; (b) $\text{Resultant}_B = 88.2\text{ N}$; (c) $A_x = 191.7\text{ N}, A_y = 165.5\text{ N}, M_A = 696.5\text{ N}\cdot\text{m}, D_x = -83.7\text{ N}, D_y = 145\text{ N}, \text{Resultant}_B = 253\text{ N}$

1.3-6 (a) $R_{3x} = 40\text{ N}, R_{3y} = -25\text{ N}, R_{5x} = 20\text{ N}$; (b) $F_{11} = 0, F_{13} = 28.3\text{ N}$

1.3-7 (a) $A_x = 0, A_y = 4.5\text{ kN}, E_y = 22.5\text{ kN}$; (b) $F_{FE} = 8.13\text{ kN}$

1.3-8 (a) $F_x = 0, F_y = 12.0\text{ kN}, D_y = 6.0\text{ kN}$; (b) $F_{FE} = 0$

1.3-9 $A_x = 0, A_y = -86.1\text{ kN}, B_y = 86.1\text{ kN}, AC = 99.4\text{ kN}, AD = -49.7\text{ kN}, BC = -134\text{ kN}, BD = 102.6\text{ kN}, DC = -94.2\text{ kN}$

1.3-10 $A_x = 0, A_y = 8.67\text{ kN}, B_y = 71.3\text{ kN}, AD = -10.01\text{ kN}, AB = 5.01\text{ kN}, DB = 141.3\text{ kN}, DC = 42.2\text{ kN}, CB = -181.3\text{ kN}$

1.3-11 $A_x = 0, A_y = -86.1\text{ kN}, B_y = 86.1\text{ kN}, AC = 99.4\text{ kN}, BD = 102.6\text{ kN}$

1.3-12 $A_x = 0, A_y = 8.67\text{ kN}, B_y = 71.3\text{ kN}, AB = 5.01\text{ kN}, DC = 42.2\text{ kN}$

1.3-13 (a) $B_x = -0.8P, B_z = 2.0P, O_z = -1.25P$; (b) $F_{AC} = 0.960P$

1.3-14 (a) $A_x = -1.25P, B_y = 0, B_z = -P$; (b) $F_{AB} = 1.601P$

1.3-15 (a) $A_y = 4.67P, A_z = -4.0P$; (b) $F_{AB} = -8.33P$

1.3-16 (a) $A_z = 0, B_x = -3.75\text{ kN}$; (b) $F_{AB} = 6.73\text{ kN}$

1.3-17 (a) $T_A = 1270\text{ N}\cdot\text{m}$; (b) $T(L_1/2) = -1270\text{ N}\cdot\text{m}, T(L_1 + L_2/2) = 1130\text{ N}\cdot\text{m}$

1.3-18 (a) $T_A = -1225\text{ N}\cdot\text{m}$; (b) $T(L_1/2) = 62.5\text{ N}\cdot\text{m}, T(L_1 + L_2/2) = -T_2 = -1100\text{ N}\cdot\text{m}$

1.3-19 (a) $A_x = -2405\text{ N}, A_y = -247\text{ N}, M_A = 5932\text{ N}\cdot\text{m}, C_y = 247\text{ N}$; (b) $N = 247\text{ N}, V = 2229\text{ N}, M = -3586\text{ N}\cdot\text{m}$

1.3-20 (a) $A_x = 280\text{ N}, A_y = 8.89\text{ N}, M_A = -1120\text{ N}\cdot\text{m}, D_y = 151.1\text{ N}$; (b) $\text{Resultant}_B = 280\text{ N}$

1.3-21 $A_y = -173.1\text{ N}, D_x = -236.9\text{ N}, D_y = 724.9\text{ N}, N = 0, V = -266.1\text{ N}, M = -33.6\text{ N}\cdot\text{m}$

1.3-22 $A_x = -80\text{ N}, A_y = 11.67\text{ N}, D_y = 308\text{ N}, N = -308\text{ N}, V = -80\text{ N}, M = -53.3\text{ N}\cdot\text{m}$

1.3-23 (a) $A_x = 285\text{ N}, A_y = 1330\text{ N}, C_x = -285\text{ N}, C_y = 570\text{ N}$; (b) $N = -222\text{ N}, V = -190\text{ N}, M = 95\text{ N}\cdot\text{m}$

1.3-24 (a) $A_x = 10.98\text{ kN}, A_y = 29.0\text{ kN}, E_x = -8.05\text{ kN}, E_y = -22\text{ kN}$; (b) $\text{Resultant}_C = 23.4\text{ kN}$

1.3-25 (a) $A_y = -5625$ N, $E_x = 0$, $E_y = 7875$ N; (b) $N = 7875$ N, $V = 2250$ N, $M = 776$ N·m

1.3-26 (a) $A_x = 320$ N, $A_y = -240$ N, $C_y = 192$ N, $E_y = -192$ N; (b) $N = -312$ N, $V = -57.9$ N, $M = 289$ N·m; (c) $\text{Resultant}_C = 400$ N

1.3-27 (a) $A_x = -125$ N, $A_y = 217$ N, $B_x = -281$ N; (b) $T = 310$ N

1.3-28 (a) $A_x = -10$ kN, $A_y = -2.17$ kN, $C_y = 9.83$ kN, $E_y = 1.333$ kN; (b) $\text{Resultant}_D = 12.68$ kN

1.3-29 (a) $O_x = -213$ N, $O_y = 180$ N, $O_z = 56.9$ N, $M_{Ox} = 37.1$ N·m, $M_{Oy} = 75.9$ N·m, $M_{Oz} = -37.8$ N·m; (b) $N = -180$ N, $V = 220$ N, $T = -75.9$ N·m, $M = 53$ N·m

1.3-30 (a) $A_y = -120$ N, $A_z = -60$ N, $M_{Ax} = -70$ N·m, $M_{Ay} = -142.5$ N·m, $M_{Az} = -180$ N·m, $D_x = -60$ N, $D_y = 120$ N, $D_z = 30$ N; (b) $N = 120$ N, $V = 41.3$ N, $T = 142.5$ N·m, $M = 180.7$ N·m

1.3-31 (a) $A_x = 26$ N, $A_y = 208$ N, $A_z = -10.39$ N, $M_{Az} = 21.9$ N·m; (b) $T_{DC} = 17.13$ N, $T_{EC} = 30.6$ N

1.3-32 $C_x = 120$ N, $C_y = -160$ N, $C_z = 506$ N, $D_z = 466$ N, $H_y = 320$ N, $H_z = 499$ N

1.3-33 $A_y = 253$ N, $B_x = 195.8$ N (to the left), $B_y = 504$ N, $C_x = 129.2$ N, $C_y = 26.2$ N

1.3-34 (a) $H_B = -104.6$ N, $V_B = 516$ N, $V_F = 336$ N; (b) $N = -646$ N, $V = 176.8$ N, $M = 44.9$ kN·m

1.4-1 (a) $\sigma_{AB} = 9.95$ MPa; (b) $P_2 = 6$ kN; (c) $t_{BC} = 12.62$ mm

1.4-2 (a) $\sigma_1 = 23.2$ MPa, $\sigma_2 = 10.61$ MPa; (b) $\varepsilon_1 = 6.8 \times 10^{-3}$, $\varepsilon_2 = 3.13 \times 10^{-3}$

1.4-3 $\sigma_{max} = 3.9$ MPa

1.4-4 (a) $\sigma = 130.2$ MPa; (b) $\varepsilon = 4.652 \times 10^{-4}$

1.4-5 (a) $R_B = 400$ N (cantilever), 848 N (V-brakes); $\sigma_C = 1.0$ MPa (cantilever), 2.12 MPa (V-brakes); (b) $\sigma_{cable} = 185.7$ MPa (both)

1.4-6 (a) $\varepsilon_s = 3.101 \times 10^{-4}$; (b) $\delta = 0.1526$ mm; (c) $P_{max} = 89.5$ kN

1.4-7 (a) $\sigma_C = 16.28$ MPa; (b) $x_C = 489$ mm, $y_C = 489$ mm

1.4-8 (a) $\sigma_t = 132.7$ MPa; (b) $\alpha_{max} = 34.4°$

1.4-9 (a) $\sigma_1 = 245$ MPa, $\sigma_2 = 206$ MPa; (b) $d_{1\,new} = 0.818$ mm; (c) $\sigma_1 = 120.5$ MPa, $\sigma_2 = 119.2$ MPa, $\sigma_3 = 141.6$ MPa

1.4-10 $\sigma_C = 5.21$ MPa

1.4-11 (a) $T = 1.298$ kN, $\sigma = 118$ MPa; (b) $\varepsilon_{cable} = 8.4 \times 10^{-4}$

1.4-12 (a) $T = 819$ N, $\sigma = 74.5$ MPa; (b) $\varepsilon_{cable} = 4.923 \times 10^{-4}$

1.4-13 (a) $T = \begin{pmatrix} 25951 \\ 20662 \\ 31616 \end{pmatrix}$ N; (b) $\sigma = \begin{pmatrix} 337 \\ 268 \\ 411 \end{pmatrix}$ MPa; (c) $T = \begin{pmatrix} 18890 \\ 28532 \\ 14755 \\ 18890 \end{pmatrix}$ N, $\sigma = \begin{pmatrix} 245 \\ 371 \\ 192 \\ 245 \end{pmatrix}$ MPa

1.4-14 (a) $\sigma_x = \gamma\omega^2(L^2 - x^2)/2g$; (b) $\sigma_{max} = \gamma\omega^2 L^2/2g$

1.4-15 (a) $T_{AB} = 6757$ N, $T_{BC} = 6358$ N, $T_{CD} = 6859$ N; (b) $\sigma_{AB} = 87.8$ MPa, $\sigma_{BC} = 82.6$ MPa, $\sigma_{CD} = 89.1$ MPa

1.4-16 (a) $T_{AQ} = T_{BQ} = 50.5$ kN; (b) $\sigma = 166$ MPa

1.4-17 (a) $T_A = 5.56$ kN, $T_B = 4.53$ kN; (b) $\sigma_A = 18.54$ MPa, $\sigma_B = 15.09$ MPa

1.5-1 (a) $L_{max} = 3377$ m; (b) $L_{max} = 3881$ m

1.5-2 (a) $L_{max} = 7143$ m; (b) $L_{max} = 8209$ m

1.5-3 % elongation = 9.0, 26.4, 38.3; % reduction = 8.8, 37.6, 74.5; Brittle, ductile, ductile

1.5-4 11.9×10^3 m; 12.7×10^3 m; 6.1×10^3 m; 6.5×10^3 m; 23.9×10^3 m

1.5-5 $\sigma = 345$ MPa

1.5-6 $\sigma_{pl} \approx 47$ MPa, Slope ≈ 2.4 GPa, $\sigma_Y \approx 53$ MPa; Brittle

1.5-7 $\sigma_{pl} \approx 486$ MPa, Slope ≈ 224 GPa, $\sigma_Y \approx 520$ MPa, $\sigma_U \approx 852$ MPa; Elongation = 6%, Reduction = 32%

1.6-1 (a) 2.375 mm longer (b) $\sigma = 278.5$ MPa, no permanent set

1.6-2 4.0 mm longer

1.6-3 5.5 mm longer

1.6-4 (a) $\delta_{pset} = 4.28$ mm; (b) $\sigma_B = 65.6$ MPa

1.6-5 (a) $\delta_{pset} = 48.6$ mm; (b) $\sigma_B = 220$ MPa

1.6-6 (a) $\delta = 11.23$ mm; (b) no permanent set

1.6-7 (b) 32 mm; (c) 30 mm; (d) 328 MPa

1.7-1 $P_{max} = 654$ kN

1.7-2 $P = 27.4$ kN (tension)

1.7-3 $P = -38.5$ kN

1.7-4 (a) $L_p = 405.25$ mm; (b) $\dfrac{V_i}{V_f} = 1.003$

1.7-5 $P = 612.5$ kN, $gap = 5.03$ mm

1.7-6 (a) $P = 74.1$ kN; (b) $\delta = \varepsilon L = 0.469$ mm shortening; $\dfrac{A_f - A}{A} = +0.052\%$, $\Delta V_1 = V_{1f} - \text{Vol}_1 = -207\text{ mm}^3$; (c) $d_3 = 65.4$ mm

1.7-7 $\Delta d = -4.17 \times 10^{-3}$ mm, $P = 10.45$ kN

1.7-8 (a) $E = 104$ GPa; (b) $v = 0.34$

1.7-9 (a) $\Delta d_{BCinner} = 0.022$ mm; (b) $v_{brass} = 0.34$; (c) $\Delta t_{AB} = 6.90 \times 10^{-3}$ mm, $\Delta d_{ABinner} = 4.02 \times 10^{-3}$ mm

1.7-10 (a) $\Delta L_1 = 12.66$ mm; $\Delta L_2 = 5.06$ mm; $\Delta L_3 = 3.8$ mm; (b) $\Delta\text{Vol}_1 = 21{,}548\text{ mm}^3$; $\Delta\text{Vol}_2 = 21{,}601\text{ mm}^3$; $\Delta\text{Vol}_3 = 21{,}610\text{ mm}^3$

1.8-1 $\sigma_b = 46.9$ MPa, $\tau_{ave} = 70.9$ MPa

1.8-2 $\sigma_b = 139.86$ MPa; $P_{ult} = 144.45$ kN

1.8-3 (a) $\tau = 89.1$ MPa; (b) $\sigma_{bf} = 140$ MPa, $\sigma_{bg} = 184.2$ MPa

1.8-4 (a) $B_x = -252.8$ N, $A_x = -B_x$, $A_y = 1150.1$ N; (b) $A_{resultant} = 1178$ N; (c) $\tau = 5.86$ MPa, $\sigma_{rail} = 18.4$ MPa

1.8-5 (a) $\tau_{max} = 22.9$ MPa; (b) $\sigma_{bmax} = 6.75$ MPa

1.8-6 $T_1 = 13.176$ kN, $T_2 = 10.772$ kN, $\tau_{1ave} = 25.888$ MPa, $\tau_{2ave} = 21.166$ MPa, $\sigma_{b1} = 9.15$ MPa, $\sigma_{b2} = 7.48$ MPa

1.8-7 (a) Resultant = 4882 N; (b) $\sigma_b = 33.3$ MPa; (c) $\tau_{nut} = 21.6$ MPa, $\tau_{pl} = 4.28$ MPa

1.8-8 $G = 2.5$ MPa

1.8-9 (a) $\gamma_{aver} = 0.004$; (b) $V = 384$ kN

1.8-10 $\tau_{pl} = 348$ MPa, $\sigma_u = 154.7$ MPa, $\sigma_l = 348$ MPa

1.8-11 $d = 0.213$ mm, $\gamma = 0.0164$

1.8-12 $\tau_{ave} = 0.299$ MPa

1.8-13 (a) $\sigma_b = 482$ MPa, $\sigma_{brg} = 265$ MPa, $\tau_f = 45.7$ MPa; (b) $\sigma_b = 419$ MPa, $\sigma_{brg} = 231$ MPa, $\tau_f = 39.7$ MPa

1.8-14 (a) $\gamma_{aver} = 0.50$; (b) $\delta = 4.92$ mm

1.8-15 (a) $A_x = 0$, $A_y = 765$ N, $M_A = 520$ N·m; (b) $B_x = 1.349$ kN, $B_y = 0.72$ kN, $B_{res} = 1.53$ kN, $C_x = -B_x$; (c) $\tau_B = 27.1$ MPa, $\tau_C = 13.4$ MPa; (d) $\sigma_{bB} = 42.5$ MPa, $\sigma_{bC} = 28.1$ MPa

1.8-16 $\tau_{aver} = 42.9$ MPa

1.8-17 (a) $\tau = \dfrac{P}{2\pi rh}$; (b) $\delta = \dfrac{P}{2\pi hG}\ln\dfrac{b}{d}$

1.8-18 (a) $\tau_1 = 2.95$ MPa, $\tau_4 = 0$; (b) $\sigma_{b1} = 1.985$ MPa, $\sigma_{b4} = 0$; (c) $\sigma_{b4} = 41$ MPa; (d) $\tau = 10.62$ MPa; (e) $\sigma_3 = 75.1$ MPa

1.8-19 (a) $O_x = 55.7$ N, $O_y = 5.69$ N, $O_{res} = 56.0$ N; (b) $\tau_O = 3.96$ MPa, $\sigma_{bO} = 6.22$ MPa; (c) $\tau = 2.83$ MPa

1.8-20 (a) $F_x = 153.9$ N, $\sigma = 3.06$ MPa; (b) $\tau_{ave} = 1.96$ MPa; (c) $\sigma_b = 1.924$ MPa

1.8-21 (a) $P = 1736$ N; (b) $C_x = 1647$ N, $C_y = -1041$ N, $C_{res} = 1948$ N; (c) $\tau = 137.8$ MPa, $\sigma_{bC} = 36.1$ MPa

1.8-22 (a) $R_x = -145.4$ kN, $R_y = -229$ kN, $R_z = -51.5$ kN, $M_x = -23.2$ kN·m, $M_y = 0$, $M_z = 65.4$ kN·m; (b) $\tau_{ave} = 42.6$ MPa

1.8-23 (a) $R_O = \begin{Bmatrix} 0 \\ 937.5 \\ -321.7 \end{Bmatrix}$ N, $M_O = \begin{Bmatrix} 116.7 \\ -134.6 \\ -438.4 \end{Bmatrix}$ N·m;
(b) $\tau = 10$ MPa

1.8-24 For a bicycle with $L/R = 1.8$: (a) $T = 1440$ N; (b) $\tau_{aver} = 147$ MPa

1.9-1 $P_{allow} = 2.53$ kN

1.9-2 $T_{max} = 33.4$ kN·m

1.9-3 $P_{allow} = 2.67$ kN

1.9-4 (a) $P_{allow} = 8.74$ kN; (b) $P_{allow} = 8.69$ kN; (c) $P_{allow} = 21.2$ kN, $P_{allow} = 8.69$ kN (shear controls)

1.9-5 $P = 1.126$ MN

1.9-6 $d = 56.8$ mm

1.9-7 air: $n = 247$, water: $n = 284$

1.9-8 (a) $F = 1.171$ kN; (b) Shear: $F_a = 2.86$ kN

1.9-9 $W_{max} = 21.6$ kN

1.9-10 (a) $F_A = \sqrt{2}T$, $F_B = 2T$, $F_C = T$; (b) Shear at A: $W_{max} = 66.5$ kN

1.9-11 $P_a = 49.1$ kN

1.9-12 $C_{ult} = 5739$ N: $P_{max} = 445$ N

1.9-13 $W_{max} = 1.382$ kN

1.9-14 Shear in rivets in CG & CD controls: $P_{allow} = 45.8$ kN

1.9-15 (a) $P_a = \sigma_a(0.587d^2)$; (b) $P_a = 98.7$ kN

1.9-16 $P_{allow} = 96.5$ kN

1.9-17 $p_{max} = 557$ Pa

1.9-18 (a) $P_{allow} = \sigma_c(\pi d^2/4)\sqrt{1-(R/L)^2}$; (b) $P_{allow} = 9.77$ kN

1.10-1 (a) $d_{min} = 99.5$ mm; (b) $d_{min} = 106.2$ mm

1.10-2 (a) $d_{min} = 164.6$ mm; (b) $d_{min} = 170.9$ mm

1.10-3 (a) $d_{min} = 17.84$ mm; (b) $d_{min} = 18.22$ mm

1.10-4 $d_{min} = 63.3$ mm

1.10-5 $d_{pin} = 26$ mm

1.10-6 $A_{cable} = 75$ mm^2, $d_O = 7.71$ mm, $d_B = d_D = 8.64$ mm

1.10-7 (a) $A_{cable} = 42.9$ mm^2; (b) $d_O = 10.3$ mm, $d_A = d_B = 8.9$ mm, $d_D = 12.6$ mm

1.10-8 (b) $A_{min} = 435$ mm^2

1.10-9 $d_{min} = 9.50$ mm

1.10-10 $d_{min} = 5.96$ mm

1.10-11 $n = 11.8$, or 12 bolts

1.10-12 $(d_2)_{min} = 131$ mm

1.10-13 $A_c = 1043$ mm^2

1.10-14 (a) $t_{min} = 18.8$ mm, use $t = 20$ mm; (b) $D_{min} = 297$ mm

1.10-15 (a) $\sigma_{DF} = 65.2$ MPa $< \sigma_{allow}$, $\sigma_{bF} = 2.72$ MPa $< \sigma_{ba}$; (b) new $\sigma_{BC} = 158.9$ MPa, so increase rod BC to 6 mm diameter; required diameter of washer at B to 34 mm

1.10-16 (a) $d_m = 24.7$ mm; (b) $P_{max} = 49.4$ kN

1.10-17 (a) $P_{max} = 11.56$ kN; (b) $d_A = d_B = 12.1$ mm, $d_D = 16.9$ mm

1.10-18 $d_A = d_B = 8.32$ mm, $d_D = 11.60$ mm

1.10-19 $\theta = \arccos 1/\sqrt{3} = 54.7°$

R-1.1 C

R-1.2 C

R-1.3 D

R-1.4 A

R-1.5 B

R-1.6 A

R-1.7 A

R-1.8 D

R-1.9 A

R-1.10 C

R-1.11 D

R-1.12 D

R-1.13 D

R-1.14 A

R-1.15 B

R-1.16 C

CHAPTER 2

2.2-1 $\delta_A = 21.43$ mm

2.2-2 $\delta_{rod} = 0.1471$ mm $\delta_B = 0.1912$ mm

2.2-3 (a) $\delta = \frac{6W}{5k}$; (b) $\delta = \frac{4W}{5k}$

2.2-4 (a) $\delta = 12.5$ mm; (b) $n = 5.8$

2.2-5 (a) $\frac{\delta_a}{\delta_s} = \frac{E_s}{E_a} = 2.711$; (b) $\frac{d_a}{d_s} = \sqrt{\frac{E_s}{E_a}} = 1.646$;
(c) $\frac{L_a}{L_s} = 1.5\frac{E_a}{E_s} = 0.553$;
(d) $E_1 = \frac{E_s}{1.7} = 121$ GPa (cast iron or copper alloy) (see App. I)

2.2-6 $h = 13.4$ mm

2.2-7 $\Delta_{DC} = 7.05 \times 10^{-2}$ mm, $\delta_B = 0.264$ mm

2.2-8 $\Delta_A = 0.542$ mm

2.2-9 $h = L - \pi \rho_{max} d^2/4k$

2.2-10 (a) $x = 102.6$ mm; (b) $x = 205$ mm; (c) $P_{max} = 12.51$ N; (d) $\theta_{init} = 1.325°$; (e) $P = 20.4$ N

2.2-11 $W = 11.76$ N

2.2-12 $n = 8.9$

2.2-13 (a) $\delta_4 = \frac{26P}{3k}$;
(b) $\delta_4 = \frac{104P}{45k}$, ratio $= \frac{15}{4} = 3.75$

2.2-14 (a) $\delta_B = 1.827$ mm; (b) $P_{max} = 390$ kN; (c) $\delta_{Bx} = 6.71$ mm, $P_{max} = 106.1$ kN

2.2-15 $P_{max} = 186$ N

2.2-16 (a) $x = 134.7$ mm; (b) $k_1 = 0.204$ N/mm; (c) $b = 74.1$ mm; (d) $k_3 = 0.638$ N/mm

2.2-17 (a) $t_{c,min} = 0.580$ mm; (b) $\delta_r = 0.912$ mm; (c) $h_{min} = 1.412$ mm

2.2-18 $\delta_A = 0.200$ mm, $\delta_D = 0.880$ mm

2.2-19 (a) $\delta_D = \frac{P}{16}(28f_2 - 9f_1)$; (b) $\frac{L_1}{L_2} = \frac{27}{16}$;
(c) $\frac{d_1}{d_2} = 1.225$; (d) $x = \frac{365L}{236}$

2.2-20 (a) $\theta = 35.1°$, $\delta = 44.6$ mm, $R_A = 25$ N, $R_C = 25$ N; (b) $\theta = 43.3°$, $\delta = 8.19$ mm, $R_A = 31.5$ N, $R_C = 18.5$ N, $M_A = 1.882$ N · m

2.2-21 (a) $\theta = 52.7°$, $\delta = 19.54$ mm, $R_A = 50$ N, $R_C = 50$ N;
(b) $\theta = 54.4°$, $\delta = 4.89$ mm, $R_A = 60$ N, $R_C = 40$ N, $M_A = 3.5$ N·m

2.3-1 (a) $\delta = 0.838$ mm; (b) $d_B = 29.4$ mm

2.3-2 (a) $\delta = 0.675$ mm; (b) $P_{max} = 267$ kN

2.3-3 (a) $\delta = 0.296$ mm (elongation);
(b) So new value of P_3 is 7530 N, an increase of 1750 N; (c) $A_{AB} = 491$ mm^2

2.3-4 $\delta_B = 2.29 \times 10^{-4}$ mm, $\delta_D = 4.67 \times 10^{-4}$ mm, $\delta_E = 6.85 \times 10^{-4}$ mm

2.3-5 $\delta_B = -3.18 \times 10^{-4}$ mm,
$\delta_C = -5.31 \times 10^{-4}$ mm,
$\delta_D = -1.38 \times 10^{-3}$ mm

2.3-6 $\delta_B = 4.22 \times 10^{-4}$ mm, $\delta_D = 7.84 \times 10^{-4}$ mm, $\delta_E = 10.51 \times 10^{-4}$ mm

2.3-7 $\delta_B = 1.85 \times 10^{-5}$ mm, $\delta_C = 4.68 \times 10^{-5}$ mm, $\delta_D = -6.1 \times 10^{-4}$ mm

2.3-8 (a) $\delta = \frac{7PL}{6Ebt}$; (b) $\delta = 0.5$ mm;
(c) $L_{slot} = 244$ mm

2.3-9 (a) $\delta = \frac{7PL}{6Ebt}$; (b) $\delta = 0.53$ mm;
(c) $L_{slot} = 299$ mm

2.3-10 (a) $\delta_{AC} = 3.72$ mm; (b) $P_0 = 44.2$ kN

2.3-11 (a) $\delta_a = 3.1$ mm; (b) $\delta_b = 2.41$ mm;
(c) $\frac{\delta_c}{\delta_a} = 1.0$, $\frac{\delta_c}{\delta_b} = 1.284$

2.3-12 (a) $d_{max} = 23.9$ mm; (b) $b = 4.16$ mm; (c) $x = 183.3$ mm

2.3-13 (a) $\delta = \frac{PL}{2EA}$; (b) $\sigma(y) = \frac{P}{A}\left(\frac{y}{L}\right)$;
(c) $\delta = \frac{PL}{EA}\left(\frac{2}{3}\right)$, $\sigma(y) = \frac{P}{A}\left[\frac{y}{L}\left(2 - \frac{y}{L}\right)\right]$

2.3-14 (a) $\delta_{2\text{-}4} = 0.024$ mm; (b) $P_{max} = 8.15$ kN; (c) $L_2 = 9.16$ mm

2.3-15 (a) $R_1 = -3P/2$; (b) $N_1 = 3P/2$ (tension), $N_2 = P/2$ (tension); (c) $x = L/3$; (d) $\delta_2 = 2PL/3EA$; (e) $\beta = 1/11$

2.3-16 (a) $\delta_C = W(L^2 - h^2)/2EAL$; (b) $\delta_B = WL/2EA$; (c) $\beta = 3$;

(d) $\delta = \dfrac{WL}{2EA} = 359$ mm (in sea water);

$\delta = \dfrac{WL}{2EA} = 412$ mm (in air)

2.3-17 (b) $\delta = 0.304$ mm

2.3-18 $\delta_B = 0.937$ mm, $\delta_C = 2.95$ mm

2.3-19 $\delta_B = 2.12$ mm, $\delta_C = 6.35$ mm

2.3-20 $P_1 = 415$ kN to left, $\delta_B = -1.389$ mm

2.3-21 $L = 2.83$ m

2.3-22 $\delta = 4.14$ mm

2.3-23 $\delta = 2WL/\pi d^2 E$

2.3-24 $\delta = 2PH/3Eb^2$

2.3-25 (b) $\delta = 3.55$ m

2.3-26 (a) $\delta = 2.18$ mm; (b) $\delta = 6.74$ mm

2.3-27 $\delta_C = 0.653$ mm downward

2.3-28 $\delta_B = -4\dfrac{PL}{EA}, \delta_F = -\dfrac{11}{3}\dfrac{PL}{EA}$

2.3-29 $\delta_B = -\dfrac{3}{4}\dfrac{PL}{EA}, \delta_D = -\dfrac{5}{4}\dfrac{PL}{EA}$

2.3-30 $\delta_B = \delta_D = -\dfrac{1}{2}\dfrac{PL}{EA}$

2.4-1 (a) $P = 9.24$ kN; (b) $P_{allow} = 7.07$ kN

2.4-2 (a) $P = 104$ kN; (b) $P_{max} = 116$ kN

2.4-3 $A_y = -6.67$ kN, $B_y = -3.33$ kN (both downward), $\delta_C = 6.54 \times 10^{-3}$ mm (upward)

2.4-4 $A = 70.3\ \text{mm}^2$

2.4-5 $P_{allow} = 3090$ kN

2.4-6 (a) $P_B/P = 3/11$; (b) $\sigma_B/\sigma_A = 1/2$; (c) Ratio = 1

2.4-7 (a) If $x \le L/2$, $R_A = (-3PL)/(2(x + 3L))$, $R_B = -P(2x + 3L)/(2(x + 3L))$. If $x \ge L/2$, $R_A = (-P(x + L))/(x + 3L)$, $R_B = (-2PL)/(x + 3L)$; (b) If $x \le L/2$, $\delta = PL(2x + 3L)/[(x + 3L)E\pi d^2]$. If $x \ge L/2$, $\delta = 8PL(x + L)/[3(x + 3L)E\pi d^2]$; (c) $x = 3L/10$ or $x = 2L/3$; (d) $R_B = \rho g \pi d^2 L/8$, $R_A = 3\rho g \pi d^2 L/32$

2.4-8 $R_A = -80$ kN, $R_C = -120$ kN, $\delta_B = 0.466$ mm

2.4-9 $R_A = -69.52$ kN, $R_C = -20.48$ kN, $\delta_B = 0.474$ mm

2.4-10 (a) $\delta = 1.91$ mm; (b) $\delta = 1.36$ mm; (c) $\delta = 2.74$ mm

2.4-11 (a) 41.2%; (b) $\sigma_M = 238$ MPa, $\sigma_O = 383$ MPa

2.4-12 (a) $R_A = 10.5$ kN (to the left), $R_D = 2.0$ kN (to the right); (b) $F_{BC} = 15.0$ kN (compression)

2.4-13 (a) $R_A = 2P/3$, $R_E = -5P/3$;

(b) $\delta_B = -\dfrac{LP}{6EA}, \delta_C = \dfrac{LP}{6EA}, \delta_D = \dfrac{5LP}{6EA}$;

(c) $\delta_{max} = \dfrac{5LP}{6EA}$ (to the right), $\delta_A = \delta_E = 0$;

(d) $P_{max} = 53$ kN

2.4-14 (a) $P = 13.73$ kN, $R_1 = 9.07$ kN, $R_2 = 4.66$ kN, $\sigma_2 = 7$ MPa; (b) $\delta_{cap} = 190.9$ mm, Axial Force Diagram: $N(x) = -R_2$ if $x \le L_2$, $N(x) = R_1$ if $x > L_2$; Axial Displacement Diagram:

$$\delta(x) = \left[\frac{-R_2}{EA_2}(x)\right] \text{if } x \le L_2,$$

$$\delta(x) = \left[\frac{-R_2 L_2}{EA_2} + \frac{R_1}{EA_1}(x - L_2)\right] \text{if } x > L_2;$$

(c) $q = 1.552$ kN/m

2.4-15 (b) $\sigma_a = 10.6$ MPa (compression), $\sigma_s = 60.6$ MPa (tension)

2.4-16 (a) $P_{allow} = 1504$ N; (b) $P_{allow} = 820$ N; (c) $P_{allow} = 703$ N

2.4-17 (a) $P_1 = PE_1/(E_1 + E_2)$;
(b) $e = b(E_2 - E_1)/[2(E_2 + E_1)]$;
(c) $\sigma_1/\sigma_2 = E_1/E_2$

2.4-18 (a) $A_x = -41.2\,\text{kN}$, $A_y = -71.4\,\text{kN}$,
$B_x = -329\,\text{kN}$, $B_y = 256\,\text{kN}$;
(b) $P_{max} = 233\,\text{kN}$

2.4-19 $d_2 = 9.28$ mm, $L_2 = 1.10$ m

2.4-20 $P_{max} = 1800$ N

2.4-21 (a) $\sigma_C = 50.0$ MPa, $\sigma_D = 60.0$ MPa;
(b) $\delta_B = 0.320$ mm

2.4-22 $M_A = \dfrac{2}{3}PL$, $A_x = 0$, $D_x = 0$, $D_y = \dfrac{2}{3}P$,
$R_F = \dfrac{4}{3}P$

2.4-23 $\sigma_s = 58.9$ MPa, $\sigma_b = 28.0$ MPa,
$\sigma_c = 33.6$ MPa

2.4-24 $A_x = 0$, $A_y = -6P$, $D_x = 0$, $D_y = \dfrac{18}{5}P$,
$R_F = \dfrac{27}{5}P$

2.5-1 $\sigma = 100.8$ MPa

2.5-2 $T = 40.3°\text{C}$

2.5-3 $\Delta T = 90°\text{C}$

2.5-4 (a) $\Delta T = 24°\text{C}$, $\sigma_{rod} = 57.6$ MPa;
(b) Clevis: $\sigma_{bc} = 42.4$ MPa, Washer:
$\sigma_{bw} = 74.1$ MPa; (c) $d_b = 10.68$ mm

2.5-5 (a) $\sigma_c = E\alpha(\Delta T_B)/4$;
(b) $\sigma_c = E\alpha(\Delta T_B)/[4(EA/kL + 1)]$

2.5-6 $\sigma_T = -0.74$ MPa, $\delta_B = 1.609$ mm

2.5-7 $\sigma_T = -24$ MPa

2.5-8 (a) $N = 51.8$ kN, max. $\sigma_c = 26.4$ MPa,
$\delta_C = -0.314$ mm;
(b) $N = 31.2$ kN, max. $\sigma_c = 15.91$ MPa,
$\delta_C = -0.546$ mm

2.5-9 $R_A = -R_C = 5.2$ kN, $\delta_B = 5.32 \times 10^{-3}$ mm
to right

2.5-10 $R_A = -R_C = 117.4$ kN, $\delta_B = 0$

2.5-11 $\delta = 5$ mm

2.5-12 ends $\sigma_T = -22.7$ MPa,
middle $\sigma_T = -27.0$ MPa

2.5-13 $\tau = 67.7$ MPa

2.5-14 $\Delta T = 34°\text{C}$

2.5-15 (a) $T_A = 1760$ N, $T_B = 880$ N;
(b) $T_A = 2008$ N, $T_B = 383$ N;
(c) $\Delta T = 177°\text{C}$

2.5-16 $P_{allow} = 39.5$ kN

2.5-17 (a) $\sigma = -6.62$ MPa; (b) $F_k = 12.99$ kN (C);
(c) $\sigma = -17.33$ MPa

2.5-18 (a) $\sigma = 98$ MPa; (b) T = 35°C

2.5-19 $F_{tube} = -1036$ kN, $F_{each-cable} = 517.9$ kN,
$\delta_{tube} = -2.45$ mm

2.5-20 $s = PL/6EA$

2.5-21 (a) $P_1 = 1027$ kN, $R_A = -249$ kN,
$R_B = 249$ kN; (b) $P_2 = 656$ kN,
$R_A = -249$ kN, $R_B = 249$ kN; (c) For P_1,
$\tau_{max} = 93.8$ MPa, for P_2, $\tau_{max} = 133.3$ MPa;
(d) $\Delta T = 35°\text{C}$, $R_A = 0$, $R_B = 0$;
(e) $R_A = -249$ kN, $R_B = 249$ kN

2.5-22 (a) $R_A = [-s + \alpha\Delta T(L_1 + L_2)]/[(L_1/EA_1) + (L_2/EA_2) + (L/k_3)]$, $R_D = -R_A$;
(b) $\delta_B = \alpha\Delta T(L_1) - R_A(L_1/EA_1)$,
$\delta_C = \alpha\Delta T(L_1 + L_2) - R_A[(L_1/EA_1) + (L_2/EA_2)]$

2.5-23 $T_B = 2541$ N, $T_C = 4623$ N

2.5-24 $P_{allow} = 1.8$ MN

2.5-25 (a) $\sigma_p = -1.231$ MPa, $\sigma_r = 17.53$ MPa;
(b) $\sigma_b = 11.63$ MPa, $\tau_c = 1.328$ MPa

2.5-26 $\sigma_p = 25.0$ MPa

2.5-27 $\sigma_p = 15.0$ MPa

2.5-28 (a) $P_B = 25.4$ kN, $P_s = -P_B$;
(b) $s_{reqd} = 25.7$ mm; (c) $\delta_{final} = 0.35$ mm

2.5-29 (a) $F_k = -727$ N; (b) $F_t = 727$ N;
(c) $L_f = 305.2$ mm; (d) $\Delta T = 76.9°\text{C}$

2.5-30 $\sigma_a = 500$ MPa (tension), $\sigma_c = 10$ MPa
(compression)

2.5-31 (a) $F_k = 727$ N; (b) $F_t = -727$ N;
(c) $L_f = 304.8$ mm; (d) $\Delta T = -76.8°\text{C}$

2.6-1 $P_{max} = 312$ kN

2.6-2 $d_{min} = 6.81\,mm$

2.6-3 $P_{max} = 104\ kN$

2.6-4 (a) $\Delta T_{max} = -46°C$; (b) $\Delta T = +9.93°C$

2.6-5 (a) $\tau_{max} = 84.7\,MPa$; (b) $\Delta T_{max} = -17.38°C$; (c) $\Delta T = +42.7°C$

2.6-6 (a) $\sigma_x = 84\,MPa$; (b) $\tau_{max} = 42\,MPa$; (c) On rotated x face: $\sigma_{x1} = 42\,MPa$, $\tau_{x1y1} = 42\,MPa$; On rotated y face: $\sigma_{y1} = 42\,MPa$; (d) On rotated x face: $\sigma_{x1} = 71.7\,MPa$, $\tau_{x1y1} = -29.7\,MPa$; On rotated y face: $\sigma_{y1} = 12.3\,MPa$

2.6-7 (a) $\sigma_{max} = 16.8\ MPa$; (b) $\tau_{max} = 8.4\ MPa$

2.6-8 (a) Element A: $\sigma_x = 105\,MPa$ (compression), Element B: $\tau_{max} = 52.5\,MPa$; (b) $\theta = 33.1°$

2.6-9 $\sigma_\theta = -5.81\,MPa$, $\sigma_{\theta + \pi/2} = -2.85\,MPa$, $\tau_\theta = 4.07\ MPa$

2.6-10 $\sigma_\theta = -30\,MPa$, $\sigma_{\theta + \pi/2} = -30\,MPa$, $\tau_\theta = 30\,MPa$

2.6-11 (a) $\tau_{max\,AC} = \dfrac{\sigma_{AC}}{2} = 13.1\ MPa$, $\tau_{max\,AB} = \dfrac{\sigma_{AB}}{2} = 52.3\ MPa$, $\tau_{max\,BC} = \dfrac{\sigma_{BC}}{2} = -66.3\ MPa$; (b) $P_{max} = 159.3\ kN$

2.6-12 (a) (1) $\sigma_x = -945\,kPa$; (2) $\sigma_\theta = -807\,kPa$, $\tau_\theta = 334\,kPa$; (3) $\sigma_\theta = -472\,kPa$, $\tau_\theta = 472\,kPa$, $\sigma_{max} = -945\,kPa$, $\tau_{max} = -472\,kPa$; (b) $\sigma_{max} = -378\,kPa$, $\tau_{max} = -189\,kPa$

2.6-13 (a) $\tau_{pq} = 4.85\ MPa$; (b) $\sigma_{pq} = -8.7\ MPa$, $\sigma(pq + \pi/2) = -2.7\ MPa$; (c) $P_{max} = 127.7\ kN$

2.6-14 (a) $\Delta T_{max} = 31.3°C$; (b) $\sigma_{pq} = -21.0\,MPa$ (compression), $\tau_{pq} = 30\ MPa$ (CCW); (c) $\beta = 0.62$

2.6-15 $N_{AC} = 34.6\ kN$; $d_{min} = 32.4\ mm$

2.6-16 (a) $\sigma_\theta = 0.57\,MPa$, $\tau_\theta = -1.58\,MPa$; (b) $\alpha = 33.3°$; (c) $\alpha = 26.6°$

2.6-17 (a) $\theta = 30°$, $\tau_0 = -34.6\ MPa$; (b) $\sigma_{max} = 80\ MPa$, $\tau_{max} = 40\ MPa$

2.6-18 $\sigma_{\theta 1} = 54.9\,MPa$, $\sigma_{\theta 2} = 18.3\,MPa$, $\tau_\theta = -31.7\,MPa$

2.6-19 $\sigma_{max} = 64.4\ MPa$, $\tau_{max} = 34.7\ MPa$

2.6-20 (a) $\theta = 30.96°$; (b) $P_{max} = 1.53\,kN$

2.6-21 (a) $\tau_\theta = 2.15\ MPa$, $\theta = 22°$; (b) $\sigma_{x1} = -5.3\ MPa$, $\sigma_{y1} = -0.869\ MPa$; (c) $k_{max} = 3962\ kN/m$; (d) $L_{max} = 0.755\ m$; (e) $\Delta T_{max} = 53.7°C$

2.6-22 $P_{max} = 5.63\,kN$

2.7-1 (a) $U = 23P^2L/12EA$; (b) $U = 14.02\ N \cdot m$

2.7-2 (a) $U = 5P^2L/4\pi Ed^2$; (b) $U = 1.036\,J$

2.7-3 $U = 788\ J$

2.7-4 (c) $U = P^2L/2EA + PQL/2EA + Q^2L/4EA$

2.7-5 Aluminum: 826 kPa, 31 m

2.7-6 (a) $U = P^2L/EA$; (b) $\delta_B = 2PL/EA$

2.7-7 (a) $U_1 = 0.00422\ J$; (b) $U_2 = 0.305\ J$; (c) $U_3 = 0.264\ J$

2.7-8 (a) $U = 5k\delta^2$; (b) $\delta = W/10k$; (c) $F_1 = 3W/10$, $F_2 = 3W/20$, $F_3 = W/10$

2.7-9 (a) $U = \dfrac{P^2L}{2Et(b_2 - b_1)} \ln\dfrac{b_2}{b_1}$; (b) $\delta = \dfrac{PL}{Et(b_2 - b_1)} \ln\dfrac{b_2}{b_1}$

2.7-10 (a) $P_1 = 270\,kN$; (b) $\delta = 1.321\,mm$; (c) $U = 243\,J$

2.7-11 (a) $x = 2s$, $P = 2(k_1 + k_2)s$; (b) $U_1 = (2k_1 + k_2)s^2$

2.7-12 (a) $U = 6.55\,J$; (b) $\delta_C = 168.8\,mm$

2.8-1 (a) $\delta_{max} = 0.869\,mm$; (b) $\sigma_{max} = 152.1\ MPa$; (c) Impact factor = 117

2.8-2 (a) $\delta_{max} = 6.33\,mm$; (b) $\sigma_{max} = 359\,MPa$; (c) Impact factor = 160

2.8-3 (a) $\delta_{max} = 0.762\ mm$; (b) $\sigma_{max} = 177.7\ MPa$; (c) Impact factor = 133

2.8-4 (a) $\delta_{max} = 215\ mm$; (b) Impact factor = 3.9

2.8-5 (a) $\delta_{max} = 270\ mm$; (b) Impact factor = 4.2

2.8-6 $v = 13.1\,m/s$

2.8-7 $h_{max} = 0.27\ m$

2.8-8 $L_{min} = 9.25$ m

2.8-9 $L_{min} = 4.59$ m

2.8-10 $v_{max} = 5.40$ m/s

2.8-11 $\delta_{max} = 200$ mm

2.8-12 $L = 25.5$ m

2.8-13 (a) Impact factor $= 1 + (1 + 2EA/W)^{1/2}$; (b) 10

2.8-14 $\sigma_{max} = 33.3$ MPa

2.10-1 (a) $\sigma_{max} \approx 45$ and 50 MPa; (b) $\sigma_{max} \approx 76$ and 61 MPa

2.10-2 (a) $\sigma_{max} \approx 26$ MPa and 29 MPa; (b) $\sigma_{max} \approx 25$ MPa and 22 MPa

2.10-3 $P_{max} = \sigma_t bt/3$

2.10-4 $\sigma_{max} \approx 46$ MPa

2.10-5 $\sigma_{max} \approx 41.9$ MPa

2.10-6 (a) No, it makes it weaker: $P_1 = 25.1$ kN, $P_2 \approx 14.4$ kN; (b) $d_0 \approx 15.1$ mm

2.10-7 $d_{max} \approx 13$ mm

2.11-2 (a) $\delta_C = 1.67$ mm; (b) $\delta_C = 5.13$ mm; (c) $\delta_C = 11.88$ mm

2.11-3 (b) $P = 79.9$ kN

2.11-4 For $P = 30$ kN: $\delta = 6.2$ mm; for $P = 40$ kN: $\delta = 12.0$ mm

2.11-5 For $P = 106$ kN: $\delta = 4.5$ mm; for $P = 180$ kN: $\delta = 17.5$ mm

2.11-6 For $P = 3.2$ kN: $\delta_B = 4.85$ mm; for $P = 4.8$ kN: $\delta_B = 17.3$ mm

2.12-1 $P_Y = P_P = 2\sigma_Y A \sin\theta$

2.12-2 $P_P = 201$ kN

2.12-3 (a) $P_P = 5\sigma_Y A$

2.12-4 $P_P = 2\sigma_Y A(1 + \sin\alpha)$

2.12-5 $P_P = 220$ kN

2.12-6 $P_P = 82.5$ kN

2.12-7 (a) $P_P = 102$ kN; (b) no change

2.12-8 (a) $P_Y = \sigma_Y A$, $\delta_Y = 3\sigma_Y L/2E$; (b) $P_P = 4\sigma_Y A/3$, $\delta_P = 3\sigma_Y L/E$

2.12-9 (a) $P_Y = \sigma_Y A$, $\delta_Y = \sigma_Y L/E$; (b) $P_P = 5\sigma_Y A/4$, $\delta_P = 2\sigma_Y L/E$

2.12-10 (a) $W_Y = 28.8$ kN, $\delta_Y = 125$ mm; (b) $W_P = 48$ kN, $\delta_P = 225$ mm

2.12-11 (a) $P_Y = 300$ kN, $\delta_Y = 0.475$ mm; (b) $P_P = 456$ kN, $\delta_P = 0.735$ mm

R-2.1 D

R-2.2 B

R-2.3 A

R-2.4 A

R-2.5 D

R-2.6 A

R-2.7 B

R-2.8 C

R-2.9 A

R-2.10 D

R-2.11 D

R-2.12 C

R-2.13 A

R-2.14 C

R-2.15 D

R-2.16 C

CHAPTER 3

3.2-1 $\gamma_{min} = 4.2 \times 10^{-3}$ rad., $\gamma_{med} = 4.6 \times 10^{-3}$ rad.

3.2-2 (a) $L_{min} = 162.9$ mm; (b) $d_{max} = 68.8$ mm

3.2-3 (a) $d_{max} = 10.54$ mm; (b) $L_{min} = 545$ mm

3.2-4 (a) $\gamma_1 = 393 \times 10^{-6}$ rad; (b) $r_{2,\,max} = 50.9$ mm

3.2-5 (a) $\gamma_1 = 1.967 \times 10^{-4}$ rad; (b) $r_{2,\,max} = 65.1$ mm

3.2-6 (a) $\gamma_1 = 267 \times 10^{-6}$ rad; (b) $r_{2,\,min} = 183.3$ mm

3.3-1 $\phi = 4.85 \times 10^{-3}$ rad, $\tau_{max} = 8.2$ MPa, $\tau_A = 6.5$ MPa

3.3-2 $d_{min} = 35.7$ mm

3.3-3 $\phi = 5.59 \times 10^{-3}$ rad, $\tau_{max} = 9.4$ MPa, $\tau_A = 7.5$ MPa

3.3-4 $d_{min} = 64.7$ mm

3.3-5 (a) $\tau_{max} = 60.4$ MPa; (b) $d_{min} = 15.87$ mm

3.3-6 (a) $\tau_{max} = 23.8$ MPa; (b) $T_{max} = 0.402$ N·m; (c) $\theta = 9.12°$/m

3.3-7 (a) $\tau_{max} = 133$ MPa; (b) $\phi = 3.65°$

3.3-8 (a) $k_T = 2059$ N·m; (b) $\tau_{max} = 27.9$ MPa, $\gamma_{max} = 997 \times 10^{-6}$ radians; (c) $\dfrac{k_{Thollow}}{k_{Tsolid}} = 0.938$, $\dfrac{\tau_{max\,H}}{\tau_{max\,S}} = 1.067$; (d) $d_2 = 32.5$ mm

3.3-9 (a) $L_{min} = 838$ mm; (b) $L_{min} = 982$ mm

3.3-10 $T_{max} = 6.03$ N·m, $\phi = 2.20°$

3.3-11 (a) $\tau_{max} = 46$ MPa; $\gamma_{max} = 2.094 \times 10^{-3}$, $G = 22$ GPa; (b) $T_{max} = 548$ N·m

3.3-12 (a) $T_{max} = 9164$ N·m; (b) $T_{max} = 7765$ N·m;

3.3-13 $\tau_{max} = 54.0$ MPa

3.3-14 (a) $d_{min} = 63.3$ mm; (b) $d_{min} = 66$ mm (4.2% increase in diameter)

3.3-15 (a) $\tau_2 = 25.7$ MPa; (b) $\tau_1 = 18.4$ MPa; (c) $\theta = 0.21°$/m

3.3-16 (a) $\tau_2 = 30.1$ MPa; (b) $\tau_1 = 20.1$ MPa; (c) $\theta = 0.306°$/m

3.3-17 (a) $d_{min} = 100$ mm; (b) $k_T = 648 \dfrac{\text{kN·m}}{\text{rad}}$; (c) $d_{min} = 88.9$ mm

3.3-18 (a) $d_{min} = 64.4$ mm; (b) $k_T = 134.9$ kN·m/rad; (c) $d_{min} = 50$ mm

3.3-19 (a) $T_{1,max} = 424$ N·m; (b) $T_{1,max} = 398$ N·m; (c) Torque: 6.25%, Weight: 25%

3.3-20 (a) $\phi = 5.19°$; (b) $d = 88.4$ mm; (c) Ratio = 0.524

3.3-21 (a) $r_2 = 35.2$ mm; (b) $P_{max} = 6486$ N

3.4-1 (a) $\tau_{max} = \tau_{BC} = 50.3$ MPa, $\phi_C = 0.138°$; (b) $d_{BC} = 50.1$ mm, $\phi_C = -0.161°$

3.4-2 (a) $\tau_{bar} = 79.6$ MPa, $\tau_{tube} = 32.3$ MPa; (b) $\phi_A = 9.43°$

3.4-3 (a) $\tau_{max} = \tau_{BC} = 66$ MPa, $\phi_D = 2.44°$; (b) $d_{AB} = 76.5$ mm, $d_{BC} = 60$ mm, $d_{CD} = 39.5$ mm, $\phi_D = 2.6°$

3.4-4 $T_{allow} = 459$ N·m

3.4-5 $d_1 = 20.7$ mm

3.4-6 (a) $d = 77.5$ mm; (b) $d = 71.5$ mm

3.4-7 $d_{min} = 32.5$ mm, $\phi_{BD} = -0.459°$

3.4-8 $\tau_{max\,AB} = -367$ MPa, $\tau_{max\,BC} = -466$ MPa, $\phi_{max} = -5.63°$

3.4-9 (a) $d = 44.4$ mm; (b) $d = 51.5$ mm

3.4-10 $d_B/d_A = 1.45$

3.4-11 Minimum $d_A = 63.7$ mm

3.4-12 Minimum $d_B = 48.6$ mm

3.4-13 (a) $R_1 = -3T/2$; (b) $T_1 = 1.5T$, $T_2 = 0.5T$; (c) $x = 7L/17$; (d) $\phi_2 = (12/17)(TL/GI_p)$

3.4-14 $\phi = 3TL/2\pi Gtd_A^3$

3.4-15 (a) $\phi = 2.48°$; (b) $\phi = 1.962°$

3.4-16 (a) $R_1 = \dfrac{-T}{2}$; (b) $\phi_3 = \dfrac{19}{8}\dfrac{TL}{\pi Gtd^3}$

3.4-17
$$\phi_D = \frac{4Fd}{\pi G}\left\{\frac{L_1}{t_{01}d_{01}^{\,3}} + \int_0^{L_2} \frac{L_2^4}{(d_{01}L_2 - d_{01}x + d_{03}x)^3(t_{01}L_2 - t_{01}x + t_{03}x)}\,dx + \frac{L_3}{t_{03}d_{03}^{\,3}}\right\},$$
$\phi_D = 0.133°$

3.4-18 (a) $\tau_{max} = 16tL/\pi d^3$; (b) $\phi = 16tL^2/\pi Gd^4$

3.4-19 (a) $\tau_{max} = 8t_A L/\pi d^3$; (b) $\phi = 16t_A L^2/3\pi Gd^4$

3.4-20 (a) $L_{max} = 4.42$ m; (b) $\phi = 170°$

3.4-21 (a) $R_A = \dfrac{T_0}{6}$;

(b) $T_{AB}(x) = \left(\dfrac{T_0}{6} - \dfrac{x^2}{L^2}T_0\right)$, $0 \le x \le \dfrac{L}{2}$,

$T_{BC}(x) = -\left[\left(\dfrac{x-L}{L}\right)^2 \dfrac{T_0}{3}\right]$, $\dfrac{L}{2} \le x \le L$;

(c) $\phi_c = \dfrac{T_0 L}{144 GI_p}$;

(d) $\tau_{max} = \dfrac{8}{3\pi}\dfrac{T_0}{d_{AB}^{\,3}}$

3.4-22 (a) $T_{max} = 875\ N \cdot m$; (b) $\tau_{max} = 25.3\ MPa$

3.5-1 $\tau_{max} = 41.05\ MPa$, $\gamma_{max} = 1.0 \times 10^{-3}$ rad, $\sigma_{max} = 41.05\ MPa$, $\sigma_{min} = -41.05\ MPa$, $\varepsilon_{max} = 5 \times 10^{-4}$

3.5-2 $T_{max} = 2.71\ kN \cdot m$

3.5-3 (a) $\sigma_{max} = 48$ MPa; (b) $T = 8836\ N \cdot m$

3.5-4 (a) $\varepsilon_{max} = 320 \times 10^{-6}$; (b) $\sigma_{max} = 51.2$ MPa; (c) $T = 20.0\ kN \cdot m$

3.5-5 (a) $d_1 = 60.0$ mm; (b) $\phi = 2.30°$; (c) $\gamma_{max} = 1670 \times 10^{-6}$ rad

3.5-6 $G = 30.0$ GPa

3.5-7 $T = 234\ N \cdot m$

3.5-8 (a) $d_{min} = 37.7$ mm; (b) $T_{max} = 431\ N \cdot m$

3.5-9 (a) $d_1 = 14.39$ mm; (b) $d_{1max} = 16.25$ mm

3.5-10 (a) $d_2 = 79.3$ mm; (b) $d_2 = 80.5$ mm

3.5-11 (a) $\tau_{max} = 36.7$ MPa; (b) $\gamma_{max} = 453 \times 10^{-6}$ rad

3.5-12 (a) $\tau_{max} = 23.9$ MPa; (b) $\gamma_{max} = 884 \times 10^{-6}$ rad

3.5-13 (a) $T_{1allow} = 1.928\ kN \cdot m$, $T_{2allow} = 1.536\ kN \cdot m$; (b) $L_{mid} = 597$ mm; (c) $d_{3new} = 65.9$ mm; (d) $T_{max1} = 1.881\ kN \cdot m$, $T_{max2} = 1.498\ kN \cdot m$, $\phi_{max1} = 1.492°$, $\phi_{max2} = 1.29°$

3.7-1 (a) $\tau_{max} = 36.5$ MPa; (b) $d_{min} = 81.9$ mm

3.7-2 (a) $\tau_{max} = 50.0$ MPa; (b) $d_{min} = 32.3$ mm

3.7-3 $n = 820$ rpm

3.7-4 $\tau_{max} = 63.3$ MPa, $\phi_{AC} = 0.0535$ rad

3.7-5 (a) $H = 20.2$ MW; (b) Shear stress is halved.

3.7-6 (a) $\tau_{max} = 16.8$ MPa; (b) $P_{max} = 267$ kW

3.7-7 $d_{min} = 90.3$ mm

3.7-8 $d_{min} = 110$ mm

3.7-9 Minimum $d_1 = 1.221d$

3.7-10 $P_{max} = 91.0$ kW

3.7-11 $d = 69.1$ mm

3.7-12 $d = 53.4$ mm

3.8-1 (a) $\phi_{max} = 3T_0L/5GI_P$; (b) $\phi_{max} = \dfrac{9LT_0}{25GI_p}$

3.8-2 (a) $x = L/4$; (b) $\phi_{max} = T_0L/8GI_P$

3.8-3 $\phi_{max} = 2b\tau_{allow}/Gd$

3.8-4 $P_{allow} = 2717$ N

3.8-5 (a) $T_{0,max} = 419\ N \cdot m$; (b) $T_{0,max} = 436\ N \cdot m$

3.8-6 (a) $T_{0,max} = 150\ N \cdot m$; (b) $T_{0,max} = 140\ N \cdot m$

3.8-7 (a) $a/L = d_A/(d_A + d_B)$; (b) $a/L = d_A^4/(d_A^4 + d_B^4)$

3.8-8 $T_A = T_B = 225\ N \cdot m$, $\phi_{L/2} = -4.34°$

3.8-9 $T_A = -265.4\ kN \cdot m$, $T_C = 85.4\ kN \cdot m$, $\tau_{AB\,max} = 92.3$ MPa, $\tau_{BC\,max} = -77.1$ MPa

3.8-10 (a) $T_A = \dfrac{Lt_0}{6}$, $T_B = \dfrac{Lt_0}{3}$; (b) $\phi_{max} = \phi\left(\dfrac{L}{\sqrt{3}}\right) = -\dfrac{\sqrt{3}L^2t_0}{27GI_P}$

3.8-11 (a) $x = 767$ mm; (b) $\phi_{max} = -1.031°$ (at $x = 767$ mm)

3.8-12 (a) $\tau_1 = 32.7$ MPa, $\tau_2 = 49.0$ MPa; (b) $\phi = 1.030°$; (c) $k_T = 22.3\ kN \cdot m/rad$

3.8-13 (a) $\tau_1 = 25.4$ MPa, $\tau_2 = 38.1$ MPa; (b) $\phi = 0.48°$; (c) $k_T = 238\ kN \cdot m/rad$

3.8-14 (a) $T_{max} = 1.521\ kN \cdot m$; (b) $d_2 = 56.9$ mm

3.8-15 (a) $T_{max} = 1.041\ kN \cdot m$; (b) $d_2 = 53.3$ mm

3.8-16 (a) $T_{1,allow} = 7.14\ kN \cdot m$; (b) $T_{2,allow} = 6.35\ kN \cdot m$; (c) $T_{3,allow} = 7.41\ kN \cdot m$; (d) $T_{max} = 6.35\ kN \cdot m$

3.8-17 (a) $T_A = 1720\ N \cdot m$, $T_B = 2780\ N \cdot m$; (b) $T_A = 983\ N \cdot m$, $T_B = 3517\ N \cdot m$

3.8-18 (a) $R_1 = -0.77T$, $R_2 = -0.23T$; (b) $T_{max} = 2.79\ kN \cdot m$; (c) $\phi_{max} = 7.51°$; (d) $T_{max} = 2.48\ kN \cdot m$ (shear in flange plate bolts controls); (e) $R_2 = \dfrac{\beta}{f_{T1} + f_{T2}}$, $R_1 = -R_2$, with $f_{T1} = \dfrac{L_1}{G_1I_{p1}}$, $f_{T2} = \dfrac{L_2}{G_2I_{p2}}$; (f) $\beta_{max} = 29.1°$

3.9-1 (a) $U = 41.9$ J; (b) $\phi = 1.29°$

3.9-2 (a) $U = 5.36$ J; (b) $\phi = 1.53°$

3.9-3 $U = 15.2$ J

3.9-4 $U = 1.84 \text{ J}$

3.9-5 $U = 81 \text{ kN}\cdot\text{m}$

3.9-6 (c) $U_3 = T^2L/2GI_P + TtL^2/2GI_P + t^2L^3/6GI_P$

3.9-7 $U = 19T_0^2L/32GI_P$

3.9-8 $\phi = T_0L_AL_B/[G(L_BI_{PA} + L_AI_{PB})]$

3.9-9 $U = t_0^2L^3/40GI_P$

3.9-10 (a) $U = \dfrac{T^2L(d_A + d_B)}{\pi Gtd_A^2d_B^2}$; (b) $\phi = \dfrac{2TL(d_A + d_B)}{\pi Gtd_A^2d_B^2}$

3.9-11 $U = \dfrac{\beta^2GI_{PA}I_{PB}}{2L(I_{PA} + I_{PB})}$

3.9-12 $\phi = \dfrac{2n}{15d^2}\sqrt{\dfrac{2\pi I_mL}{G}}$; $\tau_{max} = \dfrac{n}{15d}\sqrt{\dfrac{2\pi GI_m}{L}}$

3.11-1 (a) $\tau_{approx} = 42.7$ MPa; (b) $\tau_{exact} = 47$ MPa

3.11-2 $t_{min} = \pi d/64$

3.11-3 (a) $\tau = 15.0$ MPa; (b) $\phi = 0.578°$

3.11-4 (a) $\tau = 9.17$ MPa; (b) $\phi = 0.140°$

3.11-5 $\tau_{max} = 21.54$ MPa, $\phi = 2.96 \times 10^{-3}$ rad

3.11-6 $U_1/U_2 = 2$

3.11-7 $\tau = 46.7$ MPa, $\theta = 0.543°/\text{m}$

3.11-8 $\tau = 35.0$ MPa, $\phi = 0.570°$

3.11-9 $\tau = T\sqrt{3}/9b^2t$, $\theta = 2T/9Gb^3t$

3.11-10 (a) $\phi_1/\phi_2 = 1 + 1/4\beta^2$

3.11-11 $t_{min} = 4.57$ mm

3.11-12 (a) $t = 6.66$ mm; (b) $t = 7.02$ mm

3.11-13 $\tau = 2T(1 + \beta)^2/tL_m^2\beta$

3.12-1 $T_{max} \approx 1770 \text{ N}\cdot\text{m}$

3.12-2 $R_{min} \approx 4.0$ mm

3.12-3 For $D_1 = 18$ mm; $\tau_{max} \approx 121.5$ MPa

3.12-4 $D_2 \approx 115$ mm; lower limit

3.12-5 $D_1 \approx 33.5$ mm

R-3.1 D

R-3.2 A

R-3.3 C

R-3.4 A

R-3.5 D

R-3.6 D

R-3.7 B

R-3.8 B

R-3.9 C

R-3.10 B

R-3.11 D

R-3.12 B

R-3.13 B

R-3.14 D

R-3.15 B

CHAPTER 4

4.3-1 $V = 1.46$ kN, $M = 3.72 \text{ kN}\cdot\text{m}$

4.3-2 $V = -0.938$ kN, $M = 5.06 \text{ kN}\cdot\text{m}$

4.3-3 (a) $V_{mid} = 0$, $M_{mid} = 0$; (b) $V_{mid} = 0$, $M_{mid} = 0$

4.3-4 $V = 3$ kN, $M = -6 \text{ kN}\cdot\text{m}$

4.3-5 (a) $V = -786$ N, $M = 21.8 \text{ kN}\cdot\text{m}$; (b) $q = 5.37$ kN/m (upward)

4.3-6 (a) $V = -1.0$ kN, $M = -7 \text{ kN}\cdot\text{m}$; (b) $P_2 = 4$ kN; (c) $P_1 = -8$ kN (acts to right)

4.3-7 (a) $b/L = 1/2$; (b) $b/L = 1$

4.3-8 $M = 108 \text{ N}\cdot\text{m}$

4.3-9 $N = P\sin\theta$, $V = P\cos\theta$, $M = Pr\sin\theta$

4.3-10 $V = -3172$ N, $M = 5553 \text{ N}\cdot\text{m}$

4.3-11 (a) $P = 37.5$ kN; (b) $P = 4.17$ kN

4.3-12 $V = -4.17$ kN, $M = 75 \text{ kN}\cdot\text{m}$

4.3-13 (a) $V_B = 24$ kN, $M_B = 12 \text{ kN}\cdot\text{m}$; (b) $V_m = 0$, $M_m = 30 \text{ kN}\cdot\text{m}$

4.3-14 (a) $V_C = -15.5\,\text{kN}, M_C = -44.8\,\text{kN}\cdot\text{m}$
(b) $M_A = -81\,\text{kN}\cdot\text{m}, V_C = -25\,\text{kN}$, $M_C = -30.5\,\text{kN}\cdot\text{m}$

4.3-15 $V_{\text{mid}} = \dfrac{2}{45}q_0L, M_{\text{mid}} = \dfrac{13}{180}q_0L^2$

4.3-16 (a) $V_{23} = \dfrac{-5}{36}q_0L, M_{23} = \dfrac{23}{324}q_0L^2$
(b) $V_{\text{mid}} = \dfrac{5}{144}q_0L, M_{\text{mid}} = \dfrac{23}{288}q_0L^2$

4.3-17 (a) $V_{23} = \dfrac{-1}{36}q_0L, M_{23} = \dfrac{-5}{324}q_0L^2$
(b) $V_{23} = \dfrac{-5}{72}q_0L, M_{23} = \dfrac{35}{648}q_0L^2$

4.3-18 $V\left(\dfrac{L}{2}\right) = \dfrac{\sqrt{2}-4}{6}q_0L, M\left(\dfrac{L}{2}\right) = \dfrac{\sqrt{2}+2}{30}q_0L^2$

4.3-19 $V\left(\dfrac{2L}{3}\right) = \dfrac{-19}{81}q_0L, M\left(\dfrac{2L}{3}\right) = \dfrac{43}{972}q_0L^2$

4.3-20 $V_{\text{mid}} = \dfrac{-q_0L}{6}, M_{\text{mid}} = 0$

4.3-21 $V_{\text{mid}} = \dfrac{P}{2}(\sqrt{2}-1), M_{\text{mid}} = \dfrac{-PL}{4}$

4.3-22 (a) $V_{\text{mid}} = \dfrac{41}{216}q_0L, M_{\text{mid}} = \dfrac{59}{432}q_0L^2$
(b) $V_{\text{mid}} = \dfrac{-4}{27}q_0L, M_{\text{mid}} = \dfrac{2}{27}q_0L^2$

4.3-23 $V\left(\dfrac{L}{3}\right) = \dfrac{-P}{\sqrt{5}}, M\left(\dfrac{L}{3}\right) = \dfrac{2\sqrt{5}}{15}PL$

4.3-24 $V\left(\dfrac{L}{3}\right) = \dfrac{11P}{36}, M\left(\dfrac{L}{3}\right) = \dfrac{-43}{81}PL$

4.3-25 $N = \dfrac{-P}{2}, V = 0, M = 0$

4.3-26 (a) $N = 21.6\,\text{kN}$ (compression), $V = 7.2\,\text{kN}$, $M = 50.4\,\text{kN}\cdot\text{m}$; (b) $N = 21.6\,\text{kN}$ (compression), $V = -5.4\,\text{kN}$, $M = 0$ (at moment release)

4.3-27 $V_{\text{max}} = 91wL^2\alpha/30g, M_{\text{max}} = 229wL^3\alpha/75g$

4.5-1 (a) $V_{\text{max}} = P, M_{\text{max}} = Pa$;
(b) $A_y = -B_y = P\left(2\dfrac{a}{L} - 1\right)$

4.5-2 (a) $V_{\text{max}} = \dfrac{M_0}{L}, M_{\text{max}} = M_0\dfrac{a}{L}$;
(b) $V_{\text{max}} = 2\dfrac{M_0}{L}, M_{\text{max}} = 2M_0\dfrac{a}{L}$

4.5-3 $V_{\text{max}} = qL/2, M_{\text{max}} = -3qL^2/8$

4.5-4 $V_{\text{max}} = P, M_{\text{max}} = PL/4$

4.5-5 (a) $q_1 = 3q$; (b) $q_0 = 4.5q$

4.5-6 $V_{\text{max}} = -2P/3, M_{\text{max}} = PL/9$

4.5-7 $V_{\text{max}} = 2M_1/L, M_{\text{max}} = 7M_1/3$

4.5-8 (a) $V_{\text{max}} = \dfrac{P}{2}$ (on AB),
$M_{\text{max}} = R_C\left(\dfrac{3L}{4}\right) = \dfrac{3LP}{8}$ (just right of B);
(b) $N_{\text{max}} = P$ (tension on AB), $V_{\text{max}} = \dfrac{P}{5}$,
$M_{\text{max}} = \dfrac{-P}{5}\left(\dfrac{3L}{4}\right) = -\dfrac{3LP}{20}$ (just right of B)

4.5-9 $N_{\text{max}} = P, V_{\text{max}} = P, M_{\text{max}} = -\dfrac{2}{3}PL$

4.5-10 (a) $V_{\text{max}} = P, M_{\text{max}} = -Pa$; (b) $M = 3Pa$ (CCW); $V_{\text{max}} = 2P, M_{\text{max}} = 2Pa$

4.5-11 (a) $V_{\text{max}} = \dfrac{1}{2}qL, M_{\text{max}} = \dfrac{5}{72}qL^2$;
(b) $V_{\text{max}} = \dfrac{7}{20}qL, M_{\text{max}} = \dfrac{5}{54}qL^2$

4.5-12 (a) $V_{\text{max}} = -q_0L/2, M_{\text{max}} = -q_0L^2/6$;
(b) $V_{\text{max}} = -\dfrac{2Lq_0}{3}, M_{\text{max}} = -\dfrac{4L^2q_0}{15}$ (at B)

4.5-13 $V_{\text{max}} = -758.3\,\text{N}, M_{\text{max}} = 470.7\,\text{N}\cdot\text{m}$

4.5-14 (a) $V_{\text{max}} = 1200\,\text{N}, M_{\text{max}} = 960\,\text{N}\cdot\text{m}$;
(b) $V_{\text{max}} = 675\,\text{N}, M_{\text{max}} = 640\,\text{N}\cdot\text{m}$

4.5-15 $V_{\text{max}} = 4\,\text{kN}, M_{\text{max}} = -13\,\text{kN}\cdot\text{m}$

4.5-16 $V_{\text{max}} = 4.5\,\text{kN}, M_{\text{max}} = -12.67\,\text{kN}\cdot\text{m}$

4.5-17 $V_{\text{max}} = -8549\,\text{N}, M_{\text{max}} = -3240\,\text{N}\cdot\text{m}$

4.5-18 $V_{\text{max}} = 13.15\,\text{kN}, M_{\text{max}} = 11.21\,\text{kN}\cdot\text{m}$

4.5-19 (a) $V_{\text{max}} = \dfrac{4}{5}P, M_{\text{max}} = -\dfrac{17}{20}PL$
(b) $V_{\text{max}} = -\dfrac{7}{5}P, M_{\text{max}} = \dfrac{1}{2}PL$

4.5-20 (a) $N_{\text{max}} = \dfrac{-3}{5}P, V_{\text{max}} = \dfrac{-12}{5}P$,
$M_{\text{max}} = PL$; (b) $N_{\text{max}} = \dfrac{-3}{5}P, V_{\text{max}} = \dfrac{-4}{5}P$,
$M_{\text{max}} = \dfrac{-4}{5}PL$; (c) $N_{\text{max}} = \dfrac{-3}{5}P$,
$V_{\text{max}} = \dfrac{4}{5}P, M_{\text{max}} = \dfrac{-6}{5}PL$

4.5-21 $V_{max} = 8.0$ kN, $M_{max} = -8.0$ kN·m

4.5-22 $V_{max} = -7$ kN, $M_{max} = -12$ kN·m

4.5-23 Two cases have the same maximum moment: (PL).

4.5-24 $V_{max} = 31.5$ kN, $M_{max} = -73.5$ kN·m, $M_{mid} = -72.6$ kN·m

4.5-25 (a) $V_{max} = 6.4$ kN, $M_{max} = 11.5$ kN·m; (b) $V_{max} = 6.4$ kN, $M_{max} = -11.5$ kN·m

4.5-26 $M_{Az} = -PL$ (clockwise), $A_x = 0$, $A_y = 0$, $C_y = \frac{1}{12}P$ (upward), $D_y = \frac{1}{6}P$ (upward), $V_{max} = P/12$, $M_{max} = PL$

4.5-27 (a) $V_{max} = -2.88$ kN, $M_{max} = 2.28$ kN·m; (b) $a = 1.72$ m, $V_{max} = -2.54$ kN, $M_{max} = 2.5$ kN·m; (c) $a = 1.41$ m, $M_{max} = 2.53$ kN·m

4.5-28 (a) $V_{max} = -2.4$ kN, $M_{max} = 1.266$ kN·m (b) $M_0 = 0.523$ kN·m clockwise

4.5-29 (a) $a = 0.586L$, $V_{max} = 0.293qL$, $M_{max} = 0.0214qL^2$; (b) $a = 0.404L$, $V_{max} = 0.161qL$, $M_{max} = 0.00882qL^2$

4.5-30 $V_{max} = -2.25$ kN, $M_{max} = 3.5$ kN·m

4.5-31 (a) $V_{max} = -\frac{1}{2}q_0L$, $M_{max} = \frac{1}{3}q_0L^2$ (b) $V_{max} = -\frac{2}{3}q_0L$, $M_{max} = \frac{2}{5}q_0L^2$

4.5-32 $A_y = 6.25$ kN, $B_y = 8.75$ kN, $M_{max} = 8.75$ kN·m at $x = 3$ m

4.5-33 $A_y = 2.67$ kN at $x = 1.2$ m, $B_y = 1.774$ kN at $x = 6.0$ m, $M_{max} = 1.9$ kN·m at $x = 3.9$ m

4.5-34 $V_{max} = -w_0L/3$, $M_{max} = -w_0L^2/12$

4.5-35 $M_A = -\frac{w_0}{30}L^2$ (clockwise), $A_x = -3w_0L/10$ (leftward), $A_y = -3w_0L/20$ (downward), $C_y = w_0L/12$ (upward), $D_y = w_0L/6$ (upward), $V_{max} = w_0L/4$, $M_{max} = -w_0L^2/24$ at B

4.5-36 (a) $x = 0$ and $V_{max} = 56.8$ kN; (b) $x_m = 7.1$ m and $M_{max} = 202$ kN·m

4.5-37 (a) $A_x = 233$ N (right), $A_y = 936$ N (upward), $B_x = -233$ N (left), $N_{max} = -959$ N, $V_{max} = -219$ N, $M_{max} = 373$ N·m; (b) $A_x = 0$, $A_y = 298$ N, $B_x = 0$, $B_y = 638$ N, $N_{max} = 600$ N, $V_{max} = -219$ N, $M_{max} = 373$ N·m

4.5-38 (a) $N_{max} = \frac{1}{2}q_0L$, $V_{max} = \frac{19}{18}q_0L$, $M_{max} = q_0L^2$; (b) $N_{max} = \frac{-1}{2}q_0L$, $V_{max} = \frac{1}{2}q_0L$, $M_{max} = q_0L^2$

4.5-39 (a) $N_{AB} = 0$, $V_{AB} = 0$, $M_{AB\,max} = \frac{1}{6}q_0L^2$

$N_{BC} = 0$, $V_{BC\,max} = \frac{-1}{2}q_0L$,

$M_{BC\,max} = \frac{1}{6}q_0L^2$;

(b) $N_{AB} = 0$, $V_{AB\,max} = \frac{4}{3}q_0L$,

$M_{AB\,max} = \frac{-9}{10}q_0L^2$,

$N_{BC} = 0$, $V_{BC\,max} = \frac{-1}{2}q_0L$,

$M_{BC\,max} = \frac{1}{6}q_0L^2$

4.5-40 $M_A = 0$, $A_x = 0$, $A_y = -18.41$ kN, (downward), $M_D = 0$, $D_x = -63.0$ kN, (leftward), $D_y = 62.1$ kN (upward), $N_{max} = -62.1$ kN, $V_{max} = 63.0$ kN, $M_{max} = 756$ kN·m

R-4.1 D

R-4.2 C

R-4.3 D

R-4.4 A

R-4.5 A

R-4.6 C

R-4.7 B

CHAPTER 5

5.4-1 (a) $\varepsilon_{max} = 8.88 \times 10^{-4}$; (b) $R_{min} = 243$ mm; (c) $d_{max} = 5.93$ mm

5.4-2 (a) $L_{min} = 5.24$ m; (b) $d_{max} = 4.38$ mm

5.4-3 (a) $\varepsilon_{max} = 5.89 \times 10^{-3}$; (b) $d_{max} = 124.3$ mm; (c) $L_{min} = 15.45$ m

5.4-4 (a) $\rho = 85$ m, $\kappa = 0.0118\frac{1}{\text{m}}$, $\delta = 23.5$ mm; (b) $h_{max} = 136$ mm; (c) $\delta = 75.3$ mm

5.4-5 (a) $\varepsilon = 8.4 \times 10^{-4}$; (b) $t_{max} = 6.7$ mm; (c) $\delta = 18.27$ mm; (d) $L_{max} = 936$ mm

5.4-6 (a) $\varepsilon = 4.57 \times 10^{-4}$; (b) $L_{max} = 2$ m

5.4-7 $\delta = 11.0$ mm, $\varepsilon_x = 8.6 \times 10^{-4}$

5.4-8 $\rho = 600$ m, $\kappa = 1.667 \times 10^{-3}\frac{1}{\text{m}}$, $\delta_B = 7.5$ mm

5.5-1 (a) $\sigma_{max} = 361$ MPa; (b) 33.3%; (c) $L_{new} = 3.07$ m

5.5-2 (a) $\sigma_{max} = 250$ MPa; (b) -19.98%; (c) $+25\%$

5.5-3 (a) $\sigma_{max} = 186.2$ MPa; (b) $+10\%$; (c) $+55.8\%$

5.5-4 (a) $\sigma_{max} = 8.63$ MPa; (b) $\sigma_{max} = 6.49$ MPa

5.5-5 $\sigma_{max} = -2.4$ MPa

5.5-6 $q_{0,max} = 351\frac{\text{kN}}{\text{m}}$

5.5-7 $\sigma_{max} = 122.3$ MPa

5.5-8 $\sigma_{max} = 203$ MPa

5.5-9 $\sigma_{max} = 19.1$ MPa

5.5-10 $\sigma_{max} = 101$ MPa

5.5-11 $\sigma_{max} = 70.6$ MPa

5.5-12 $\sigma_{max} = 7.0$ MPa

5.5-13 (a) $\sigma_{max} = 5264$ kPa; (b) $s = 0.58579\,L$, $\sigma_{min} = 1884$ kPa; (c) $s = 0$ or L, $\sigma_{max} = 10.98$ MPa

5.5-14 $\sigma_{max} = 2.10$ MPa

5.5-15 (a) $\sigma_t = 30.93\,M/d^3$; (b) $\sigma_t = 360\,M/(73bh^2)$; (c) $\sigma_t = 85.24\,M/d^3$

5.5-16 $\sigma_{max} = 10.965\,M/d^3$

5.5-17 (a) $\sigma_{max} = 147.4$ MPa; (b) $L = 6.1$ m; (c) $d = 2.54$ m

5.5-18 (a) $\sigma_t = 35.4$ MPa, $\sigma_c = 61$ MPa; (b) $d_{max} = \frac{L}{2}$, $\sigma_t = 37.1$ MPa, $\sigma_c = 64.1$ MPa

5.5-19 (a) $\sigma_t = 105.8$ MPa, $\sigma_c = 28.8$ MPa; (b) $P_{max} = 893$ N; (c) 1.34 m

5.5-20 (a) $\sigma_c = 1.456$ MPa, $\sigma_t = 1.514$ MPa; (b) $\sigma_c = 1.666$ MPa $(+14\%)$, $\sigma_t = 1.381$ MPa (-9%); (c) $\sigma_c = 0.728$ MPa (-50%), $\sigma_t = 0.757$ MPa (-50%)

5.5-21 $P_{max} = 2.5$ kN

5.5-22 $\sigma_{max} = 3\rho L^2 a_0/t$

5.5-23 (a) $\sigma_t = 101.3$ MPa, $\sigma_c = 180$ MPa; (b) $a = 4.24$ m

5.5-24 (a) $\sigma = 25.1$ MPa, 17.8 MPa, -23.5 MPa

5.5-25 (a) $\sigma_t = 132.8$ MPa, $\sigma_c = 99.6$ MPa; (b) $h = 82.8$ mm; (c) $q = 1.867$ kN/m, $P = 3.9$ kN

5.5-26 (a) $c_1 = 91.7$ mm, $c_2 = 108.3$ mm, $I_z = 7.969 \times 10^7$ mm^4; (b) $\sigma_t = 4659$ kPa (top of beam at C), $\sigma_c = 5506$ kPa (bottom of beam at C)

5.5-27 $d = 1.0$ m, $\sigma_{max} = 1.55$ MPa, $d = 2$ m, $\sigma_{max} = 7.52$ MPa

5.5-28 (a) $\sigma_t = 0.335$ MPa, $\sigma_c = 0.288$ MPa (b) $P_{max} = 21.9$ kN

5.5-29 (a) $F_{res} = 441$ N; (b) $\sigma_{max} = 257$ MPa (compression at base); (c) $\sigma_{max} = 231$ MPa (tension at base)

5.5-30 $P_{max} = 675\ N$

5.6-1 100×300

5.6-2 IPN 240

5.6-3 $d_{min} = 100$ mm

5.6-4 (a) $d_{min} = 12.62$ mm; (b) $P_{max} = 39.8$ N

5.6-5 (a) UPN 260; (b) IPN 180; (c) HE 240A

5.6-6 (a) HE 180B; (b) HE 260B

5.6-7 (a) IPN 280; (b) $P_{max} = 17.24$ kN

5.6-8 (a) $b_{min} = 161.6$ mm; (b) $b_{min} = 141.2$ mm, $\text{area}_{(b)}/\text{area}_{(a)} = 1.145$

5.6-9 (a) 50×250; (b) $w_{max} = 8.36$ kN/m^2

5.6-10 (a) $s_{max} = 429$ mm; (b) $h_{min} = 214$ mm

5.6-11 (a) $q_{0,allow} = 13.17$ kN/m; (b) $q_{0,allow} = 6.9$ kN/m

5.6-12 $h_{min} = 30.6$ mm

5.6-13 (a) $S_{reqd} = 249$ cm^3; (b) IPN 220

5.6-14 (a) $d_{min} = 37.6$ mm; (b) $d_{min} = 42.4$ mm, $\text{area}_{(b)}/\text{area}_{(a)} = 0.557$

5.6-15 (a) 100×250; (b) $q_{max} = 332$ N/m

5.6-16 $b = 152$ mm, $h = 202$ mm

5.6-17 $b = 259$ mm

5.6-18 $t = 13.61$ mm

5.6-19 $W_1 : W_2 : W_3 : W_4 = 1 : 1.260 : 1.408 : 0.794$

5.6-20 (a) $q_{max} = 6.61$ kN/m; (b) $q_{max} = 9.37$ kN/m

5.6-21 6.03%

5.6-22 (a) $b_{min} = 11.91$ mm; (b) $b_{min} = 11.92$ mm

5.6-23 (a) $s_{max} = 1.732$ m; (b) $d = 292$ mm

5.6-24 IPN 360

5.6-25 (a) $\beta = 1/9$; (b) 5.35%

5.6-26 Increase when $d/h > 0.6861$; decrease when $d/h < 0.6861$

5.7-1 (a) $x = L/4$, $\sigma_{max} = 4PL/9h_A^3$, $\sigma_{max}/\sigma_B = 2$; (b) $x = 0.209\,L$, $\sigma_{max} = 0.394\,PL/h_A^3$, $\sigma_{max}/\sigma_B = 3.54$

5.7-2 (a) $x = 4$ m, $\sigma_{max} = 37.7$ MPa, $\sigma_{max}/\sigma_B = 9/8$; (b) $x = 2$ m, $\sigma_{max} = 25.2$ MPa, $\sigma_{max}/\sigma_m = 4/3$

5.7-3 (a) $x = 182$ mm, $\sigma_{max} = 8.94$ MPa, $\sigma_{max}/\sigma_B = 1.047$; (b) $x = 109$ mm, $\sigma_{max} = 8.85$ MPa, $\sigma_{max}/\sigma_m = 1.220$

5.7-4 (a) $\sigma_A = 210$ MPa; (b) $\sigma_B = 221$ MPa; (c) $x = 0.625$ m; (d) $\sigma_{max} = 231$ MPa; (e) $\sigma_{max} = 214$ MPa

5.7-5 (a) $1 \le d_B/d_A \le 1.5$; (b) $\sigma_{max} = \sigma_B = 32PL/\pi d_B^3$

5.7-6 $h_x = h_B x/L$

5.7-7 $b_x = 2b_B x/L$

5.7-8 $h_x = h_B\sqrt{x/L}$

5.8-2 (a) $\tau_{max} = 731$ kPa, $\sigma_{max} = 4.75$ MPa; (b) $\tau_{max} = 1462$ kPa, $\sigma_{max} = 19.01$ MPa

5.8-3 $\sigma_x = -1.5$ MPa, $\tau_{xy} = -0.136$ MPa

5.8-4 $q_{max} = 24\dfrac{\text{kN}}{m}$

5.8-5 (a) $M_{max} = 41.7$ kN·m; (b) $M_{max} = 3.24$ kN·m

5.8-6 $\tau_{max} = 500$ kPa

5.8-7 $\tau_{max} = 22.5$ MPa

5.8-8 (a) $L_0 = h(\sigma_{allow}/\tau_{allow})$; (b) $L_0 = (h/2)(\sigma_{allow}/\tau_{allow})$

5.8-9 (a) $P_{max} = 8.2$ kN; (b) $P_{max} = 8.77$ kN

5.8-10 (a) $M_{max} = 72.2$ N·m; (b) $M_{max} = 36.0$ N·m

5.8-11 (a) 150 mm × 300 mm beam; (b) 200 mm × 300 mm beam

5.8-12 (a) $P = 38.0$ kN; (b) $P = 35.6$ kN

5.8-13 (a) $w_1 = 9.46$ kN/m^2; (b) $w_2 = 19.88$ kN/m^2; (c) $w_{allow} = 9.46$ kN/m^2

5.8-14 (a) $b = 89.3$ mm; (b) $b = 87.8$ mm

5.9-1 $d_{min} = 158$ mm

5.9-2 (a) $W = 28.6$ kN; (b) $W = 38.7$ kN

5.9-3 $\tau_{max} = 7.6$ MPa

5.9-4 $d_0 = 47.6$ mm

5.9-5 (a) $d = 328$ mm; (b) $d = 76.4$ mm

5.9-6 (a) $q_{0,max} = 55.7$ kN/m; (b) $L_{max} = 2.51$ m

5.10-1 (a) $\tau_{max} = 41.9$ MPa; (b) $\tau_{min} = 31.2$ MPa; (c) $\tau_{aver} = 40.1$ MPa; (d) $V_{web} = 124.3$ kN

5.10-2 (a) $\tau_{max} = 28.43$ MPa; (b) $\tau_{min} = 21.86$ MPa; (c) $\tau_{aver} = 27.41$ MPa; (d) $V_{web} = 119.7$ kN

5.10-3 (a) $\tau_{max} = 38.56$ MPa; (b) $\tau_{min} = 34.51$ MPa; (c) $\tau_{aver} = 41.98$ MPa; (d) $V_{web} = 39.9$ kN

5.10-4 (a) $\tau_{max} = 32.28$ MPa; (b) $\tau_{min} = 21.45$ MPa; (c) $\tau_{aver} = 29.24$ MPa; (d) $V_{web} = 196.1$ kN

5.10-5 (a) $\tau_{max} = 19.01$ MPa; (b) $\tau_{min} = 16.21$ MPa; (c) $\tau_{aver} = 19.66$ MPa; (d) $V_{web} = 82.73$ kN

5.10-6 (a) $\tau_{max} = 28.40$ MPa; (b) $\tau_{min} = 19.35$ MPa; (c) $\tau_{aver} = 25.97$ MPa; (d) $V_{web} = 58.63$ kN

5.10-7 $q_{max} = 131.5$ kN/m

5.10-8 (a) $q_{max} = 184.7$ kN/m; (b) $q_{max} = 247$ kN/m

5.10-9 IPN 220

5.10-10 $V = 273$ kN

5.10-11 $\tau_{max} = 10.17$ MPa, $\tau_{min} = 7.38$ MPa

5.10-12 $\tau_{max} = 19.7$ MPa

5.10-13 $\tau_{max} = 13.87$ MPa

5.11-1 $V_{max} = 3067$ N

5.11-2 $V_{max} = 1.924$ MN

5.11-3 $F = 323$ kN/m

5.11-4 $V_{max} = 10.7$ kN

5.11-5 (a) $s_{max} = 67.0$ mm; (b) $s_{max} = 53.8$ mm

5.11-6 (a) $s_A = 78.3$ mm; (b) $s_B = 97.9$ mm

5.11-7 (a) $s_{max} = 67.4$ mm; (b) $s_{max} = 44.9$ mm

5.11-8 $s_{max} = 92.3$ mm

5.11-9 $V_{max} = 103.0$ kN

5.11-10 $s_{max} = 61.5$ mm

5.11-11 (a) Case (2); (b) Case (3); (c) Case (1); (d) Case (1)

5.11-12 $s_{max} = 165.2$ mm

5.12-1 $\sigma_A = 92.6$ MPa, $\sigma_B = -171.5$ MPa

5.12-2 $d = 89.4$ mm

5.12-3 $\sigma_t = 94.6$ MPa, $\sigma_c = -96.7$ MPa

5.12-4 $\sigma_t = 5770$ kPa, $\sigma_c = -6668$ kPa

5.12-5 $t_{min} = 12.38$ mm

5.12-6 $\sigma_t = 11.83$ MPa, $\sigma_c = -12.33$ MPa

5.12-7 $\sigma_t = 2.42$ MPa, $\sigma_c = -2.50$ MPa

5.12-8 $T_{max} = 108.6$ kN

5.12-9 $\alpha = \arctan[(d_2^2 + d_1^2)/(4hd_2)]$

5.12-10 (a) $d_{min} = 8.46$ cm; (b) $d_{min} = 8.91$ cm

5.12-11 $H_{max} = 9.77$ m

5.12-12 $W = 33.3$ kN

5.12-13 (a) $\sigma_t = 826$ kPa, $\sigma_c = -918$ kPa; (b) $d_{max} = 0.750$ m

5.12-14 (a) $b = \pi \times d/6$; (b) $b = \pi \times d/3$; (c) Rectangular post

5.12-15 (a) $\sigma_t = 36.7$ MPa, $\sigma_c = -30.9$ MPa; (b) Both stresses increase in magnitude.

5.12-16 (a) $\sigma_t = 8P/b^2$, $\sigma_c = -4P/b^2$ (b) $\sigma_t = 9.11P/b^2$, $\sigma_c = -6.36P/b^2$

5.12-17 (a) $\sigma_t = 7.23$ MPa, $\sigma_c = -36.2$ MPa; (b) $y_0 = -100.1$ mm; (c) $\sigma_t = 3.77$ MPa, $\sigma_c = -19.04$ MPa, $y_0 = -139.2$ mm

5.12-18 (a) $\sigma_t = 3.27$ MPa, $\sigma_c = -24.2$ MPa; (b) $y_0 = -76.2$ mm; (c) $\sigma_t = 1.587$ MPa, $\sigma_c = -20.3$ MPa, $y_0 = -100.8$ mm

5.12-19 (a) $\sigma_t = 112.1$ MPa; (b) $\sigma_t = 21.4$ MPa

5.12-20 (a) $y_0 = -31.2$ mm; (b) $P = 115.3$ kN; (c) $y_0 = 891$ mm, $P = 296$ kN

5.13-1 (a) $d = 6, 12, 18, 24$ mm, $\sigma_{max} = 94.1, 96.4, 103.1, 143.5$ MPa; (b) $R = 1.25, 2.5, 3.75, 5.0$ mm, $\sigma_{max} \approx 500, 433, 367, 333$ MPa

5.13-2 (a) $d = 16$ mm, $\sigma_{max} = 81$ MPa (b) $R = 4$ mm, $\sigma_{max} \approx 200$ MPa

5.13-3 $b_{min} \approx 6.39$ mm

5.13-4 $b_{min} \approx 33$ mm

5.13-5 (a) $R_{min} \approx 11.5$ mm; (b) $d_{max} = 104.1$ mm

R-5.1 A

R-5.2 C

R-5.3 A

R-5.4 D

R-5.5 A

R-5.6 B

R-5.7 B

R-5.8 C

R-5.9 C

R-5.10 A

R-5.11 B

R-5.12 B

CHAPTER 6

6.2-1 $\sigma_w = 5$ MPa, $\sigma_s = 85$ MPa

6.2-2 $\sigma_w = 4.66$ MPa, $\sigma_s = 91.5$ MPa

6.2-3 $\sigma_{\text{face}} = \pm 21.17$ MPa, $\sigma_{\text{core}} = \pm 5.29$ MPa

6.2-4 (a) $M_{\max} = 58.7$ kN·m; (b) $M_{\max} = 90.9$ kN·m; (c) $t = 7.08$ mm

6.2-5 (a) $M_{\max} = 20.8$ kN·m; (b) $M_{\max} = 11.7$ kN·m

6.2-6 (a) $M_{\text{allowTi}} = \dfrac{\sigma_{\text{Ti}}(E_{\text{Ti}} I_{\text{Ti}} + E_{\text{Cu}} I_{\text{Cu}})}{E_{\text{Ti}}\left(\dfrac{d_2}{2}\right)}$,

$M_{\text{allowCu}} = \sigma_{\text{Cu}} \dfrac{(E_{\text{Ti}} I_{\text{Ti}} + E_{\text{Cu}} I_{\text{Cu}})}{E_{\text{Cu}}\left(\dfrac{d_1}{2}\right)}$;

(b) $M_{\max \text{Ti}} = 4989$ N·m; (c) $d_1 = 36.4$ mm

6.2-7 (a) $\sigma_w = 4.99$ MPa, $\sigma_s = 109.7$ MPa; (b) $q_{\max} = 3.57$ kN/m; (c) $M_{0,\max} = 2.44$ kN·m

6.2-8 (a) $M_{\text{allow}} = 768$ N·m; (b) $\sigma_{\text{sa}} = 47.9$ MPa, $M_{\max} = 1051$ N·m

6.2-9 (a) $\sigma_{\text{face}} = 26.9$ MPa, $\sigma_{\text{core}} = 0$; (b) $\sigma_{\text{face}} = 26.7$ MPa, $\sigma_{\text{core}} = 0$

6.2-10 (a) $\sigma_{\text{face}} = 14.1$ MPa, $\sigma_{\text{core}} = 0.214$ MPa; (b) $\sigma_{\text{face}} = 14.9$ MPa, $\sigma_{\text{core}} = 0$

6.2-11 $\sigma_a = 3753$ kPa, $\sigma_c = 4296$ kPa

6.2-12 (a) $\sigma_w = 5.1$ MPa (compression), $\sigma_s = 37.6$ MPa (tension); (b) $t_s = 3.09$ mm

6.2-13 (a) $\sigma_{\text{plywood}} = 7.29$ MPa, $\sigma_{\text{pine}} = 6.45$ MPa (b) $q_{\max} = 1.43$ kN/m

6.2-14 $q_{0,\max} = 15.53$ kN/m

6.2-15 $t_s = 22.7$ mm; (b) $t_s = 2.32$ mm; (c) $t_s = 9.33$ mm

6.2-16 ratio = 0.723

6.2-17 top of beam: $\sigma_w = -5.4$ MPa, $\sigma_s = -97.9$ MPa
bottom of beam: $\sigma_w = 6.3$ MPa
bottom of steel: $\sigma_s = 86$ MPa

6.3-1 same as 6.2-17

6.3-2 top of core: $\sigma_p = -0.253$ MPa
top of beam: $\sigma_s = -66$ MPa

6.3-3 (a) $M_{\max} = 103$ kN·m; (b) $M_{\max} = 27.2$ kN·m

6.3-4 $t_{\min} = 15.0$ mm

6.3-5 (a) $q_{\text{allow}} = 11.98$ kN/m; (b) $\sigma_{\text{wood}} = 1.10$ MPa, $\sigma_{\text{steel}} = 55.8$ MPa

6.3-6 (a) $\sigma_B = 60.3$ MPa, $\sigma_w = 7.09$ MPa; (b) $t_B = 25.1$ mm, $M_{\max} = 80$ kN·m

6.3-7 $\sigma_a = 23.0$ MPa, $\sigma_p = 0.891$ MPa

6.3-8 $\sigma_a = 12.14$ MPa, $\sigma_p = 0.47$ MPa

6.3-9 (a) $q_{\text{allow}} = 4.16$ kN/m; (b) $q_{\text{allow}} = 4.39$ kN/m

6.3-10 (a) $\sigma_s = 93.5$ MPa; (b) $h_s = 5.08$ mm, $h_a = 114.92$ mm

6.3-11 $M_{\max} = 10.4$ kN·m

6.3-12 $S_A = 50.6\ \text{mm}^3$, Metal A

6.3-13 $\sigma_s = 77.8$ MPa (tension), $\sigma_c = 6.86$ MPa (compression).

6.3-14 (a) $\sigma_c = 8.51$ MPa, $\sigma_s = 118.3$ MPa; (b) $M_{\max} = M_c = 172.9$ kN·m; (c) $A_s = 2254\ \text{mm}^2$, $M_{\text{allow}} = 167.8$ kN·m

6.3-15 (a) $\sigma_c = 4.24$ MPa, $\sigma_s = 85.9$ MPa; (b) $M_{\text{allow}} = M_s = 349$ kN·m

6.3-16 (a) $M_{\max} = M_s = 10.59$ kN·m; (b) $A_s = 1262\ \text{mm}^2$, $M_{\text{allow}} = 15.79$ kN·m

6.3-17 top of beam: $\sigma_w = -4.5$ MPa, $\sigma_s = -82.2$ MPa,
bottom of wood: $\sigma_w = 2.5$ MPa,
bottom of steel: $\sigma_s = 87$ MPa

6.3-18 $M_{\text{allow}} = M_w = 16.39$ kN·m

6.4-1 $\tan\beta = h/b$, so NA lies along other diagonal

6.4-2 $\beta = 51.8°$, $\sigma_{\max} = 17.5$ MPa

6.4-3 $\beta = 53.1°$, $\sigma_{\max} = 440$ MPa

6.4-4 $\beta = 84.5°, \sigma_A = -\sigma_E = 141$ MPa, $\sigma_B = -\sigma_D = -89$ MPa

6.4-5 $\beta = 47.7°, \sigma_A = -\sigma_E = 160$ MPa, $\sigma_B = -\sigma_D = -12$ MPa

6.4-6 $\beta = -79.3°, \sigma_{max} = 8.87$ MPa

6.4-7 $\beta = -75.9°, \sigma_{max} = 12.7$ MPa

6.4-8 $P_{max} = 22.4$ kN

6.4-9 $q_{max} = 6.7$ kN/m

6.4-10 $\beta = -83.2°, \sigma_{max} = 62.3$ MPa

6.4-11 $\beta = 84.6°, \sigma_{max} = 31.0$ MPa

6.4-12 $\beta = 62.6°, \sigma_{max} = 21.6$ MPa

6.4-13 (a) $\sigma_A = 103.94 \sin\alpha + 11.48 \cos\alpha$ (MPa); (b) $\tan\beta = 21.729 \tan\alpha$

6.4-14 $\beta = 70.3°, \sigma_{max} = 11.9$ MPa

6.4-15 (a) $\beta = -76.6°, \sigma_{max} = 48.5$ MPa; (b) $\beta = -80.9°, \sigma_{max} = 48.8$ MPa

6.4-16 $\sigma_{max} = 14.6$ MPa

6.5-1 $\beta = 74.5°, \sigma_t = 37.6$ MPa, $\sigma_c = -63.7$ MPa

6.5-2 $\beta = 78.0°, \sigma_t = 7.2$ MPa, $\sigma_c = -13.2$ MPa

6.5-3 $\beta = 75.6°, \sigma_t = 30.4$ MPa, $\sigma_c = -37.7$ MPa

6.5-4 $\beta = 75.5°, \sigma_t = 22.7$ MPa, $\sigma_c = -27.9$ MPa

6.5-5 (a) $\beta = -30.6°, \sigma_t = 43.5$ MPa, $\sigma_c = -43.6$ MPa; (b) $\beta = -41.9°$, $\sigma_t = 52.5$ MPa, $\sigma_c = -42.6$ MPa

6.5-6 $\beta = 78.1°, \sigma_t = 40.7$ MPa, $\sigma_c = -40.7$ MPa

6.5-7 $\beta = 73.0°, \sigma_t = 10.3$ MPa, $\sigma_c = -9.5$ MPa

6.5-8 $\beta = 2.93°, \sigma_t = 6.56$ MPa, $\sigma_c = -6.54$ MPa

6.5-9 For $\theta = 0$: $\sigma_t = -\sigma_c = 2.546M/r^3$; for $\theta = 45°$: $\sigma_t = 4.535M/r^3, \sigma_c = -3.955M/r^3$; for $\theta = 90°$: $\sigma_t = 3.867M/r^3, \sigma_c = -5.244M/r^3$

6.5-10 $\beta = -78.9°, \sigma_t = 131.1$ MPa, $\sigma_t = -148.5$ MPa

6.5-11 $\beta = -11.7°, \sigma_t = 118.8$ MPa, $\sigma_c = -103.2$ MPa

6.5-12 $\beta = -48.6°, \sigma_t = 16.85$ MPa, $\sigma_c = -15.79$ MPa

6.5-13 $\sigma_A = -87.7$ MPa, $\sigma_B = 65.1$ MPa

6.8-1 (a) $\tau_{max} = 29.9$ MPa; (b) $\tau_B = 2.7$ MPa

6.8-2 (a) $\tau_{max} = 21.0$ MPa; (b) $\tau_B = 3.4$ MPa

6.8-3 (a) $\tau_{max} = 21.7$ MPa; (b) $\tau_{max} = 21.1$ MPa

6.8-4 (a) $\tau_{max} = 16.6$ MPa; (b) $\tau_{max} = 16.3$ MPa

6.9-1 $e = 31.0$ mm

6.9-2 $e = 19.12$ mm

6.9-6 (b) $e = \dfrac{63\pi r}{24\pi + 38} = 1.745r$

6.9-8 (a) $e = \dfrac{b}{2}\left(\dfrac{2h + 3b}{h + 3b}\right)$; (b) $e = \dfrac{b}{2}\left(\dfrac{43h + 48b}{23h + 48b}\right)$

6.10-1 $f = 2(2b_1 + b_2)/(3b_1 + b_2)$

6.10-2 (a) $f = 16r_2(r_2^3 - r_1^3)/3\pi(r_2^4 - r_1^4)$; (b) $f = 4/\pi$

6.10-3 $q = 125$ kN/m

6.10-4 (a) 56.7%; (b) $M = 12.3$ kN·m

6.10-5 $f = 1.10$

6.10-6 $f = 1.15$

6.10-7 $Z = 188.4$ cm^3, $f = 1.17$

6.10-8 $Z = 1230$ cm^3, $f = 1.07$

6.10-9 $M_Y = 688$ kN·m, $M_P = 814$ kN·m, $f = 1.18$

6.10-10 $M_Y = 365$ kN·m, $M_P = 431$ kN·m, $f = 1.18$

6.10-11 $M_Y = 154.3$ kN·m, $M_P = 205$ kN·m, $f = 1.33$

6.10-12 $M_Y = 672$ kN·m, $M_P = 878$ kN·m, $f = 1.31$

6.10-13 $M_Y = 345$ kN·m, $M_P = 453$ kN·m, $f = 1.31$

6.10-14 $M_Y = 122$ kN·m, $M_P = 147$ kN·m, $f = 1.20$

6.10-15 (a) $M = 378$ kN·m; (b) 23.3%

6.10-16 (a) $M = 524$ kN·m; (b) 36%

6.10-17 (a) $M = 209$ kN·m; (b) 7.6%

6.10-18 $Z = 136 \times 10^3$ mm^3, $f = 1.79$

6.10-19 $M_P = 117.3$ kN·m

6.10-20 $M_P = 295$ kN·m

R-6.1 B

R-6.2 C

R-6.3 B

R-6.4 B

R-6.5 D

CHAPTER 7

7.2-1 For $\theta = 52°$: $\sigma_{x1} = 30.2$ MPa, $\sigma_{y1} = 27.8$ MPa, $\tau_{x1y1} = -21.9$ MPa

7.2-2 For $\theta = 40°$: $\sigma_{x1} = 117.2$ MPa, $\sigma_{y1} = 62.8$ MPa, $\tau_{x1y1} = -10.43$ MPa

7.2-3 $\sigma_{x1} = 8.2$ MPa, $\sigma_{y1} = 2.3$ MPa, $\tau_{x1y1} = 0.2$ MPa

7.2-4 $\sigma_{x1} = 185$ MPa, $\sigma_{y1} = 35$ MPa, $\tau_{x1y1} = -10$ MPa

7.2-5 For $\theta = 32°$: $\sigma_{x1} = -23.8$ MPa, $\sigma_{y1} = -51.2$ MPa, $\tau_{x1y1} = -85.1$ MPa

7.2-6 For $\theta = 52°$: $\sigma_{x1} = -136.6$ MPa, $\sigma_{y1} = 16.6$ MPa, $\tau_{x1y1} = -84$ MPa

7.2-7 For $\theta = 48°$: $\sigma_{x1} = -9.84$ MPa, $\sigma_{y1} = -46.2$ MPa, $\tau_{x1y1} = 10.16$ MPa

7.2-8 For $\theta = -40°$: $\sigma_{x1} = -5.5$ MPa, $\sigma_{y1} = -27$ MPa, $\tau_{x1y1} = -28.1$ MPa

7.2-9 For $\theta = 36°$: $\sigma_{x1} = -94.1$ MPa, $\sigma_{y1} = -22.9$ MPa, $\tau_{x1y1} = 32.1$ MPa

7.2-10 For $\theta = -40°$: $\sigma_{x1} = -66.5$ MPa, $\sigma_{y1} = -6.52$ MPa, $\tau_{x1y1} = -14.52$ MPa

7.2-11 Normal stress on seam, 1370 kPa tension. Shear stress, 1158 kPa (clockwise)

7.2-12 Normal stress on seam, 1440 kPa tension. Shear stress, 1030 kPa (clockwise)

7.2-13 $\sigma_{x1} = 41.25$ MPa, $\sigma_{y1} = -54.25$ MPa, $\tau_{x1y1} = 69.26$ MPa

7.2-14 $\sigma_{x1} = -14.44$ MPa, $\sigma_{y1} = 1.44$ MPa, $\tau_{x1y1} = 83.75$ MPa

7.2-15 $\sigma_w = -912$ kPa, $\tau_w = 2647$ kPa

7.2-16 $\sigma_w = 10.0$ MPa, $\tau_w = -5.0$ MPa

7.2-17 $\sigma_{x1} = 5$ MPa, $\sigma_{y1} = 5$ MPa, $\tau_{x1y1} = 5$ MPa

7.2-18 $\sigma_{x1} = 9$ MPa, $\sigma_{y1} = 3$ MPa, $\tau_{x1y1} = 5.2$ MPa

7.2-19 $\theta = 51.8°$, $\sigma_{y1} = 3.7$ MPa, $\tau_{x1y1} = -7.63$ MPa

7.2-20 $\theta = 36.6°$, $\sigma_{y1} = -9$ MPa, $\tau_{x1y1} = -14.83$ MPa

7.2-21 For $\theta = -36°$: $\sigma_{x1} = -81.1$ MPa, $\sigma_{y1} = -31.9$ MPa, $\tau_{x1y1} = -27.2$ MPa

7.2-22 For $\theta = -50°$: $\sigma_{x1} = +51.4$ MPa, $\sigma_{y1} = -14.4$ MPa, $\tau_{x1y1} = -31.3$ MPa

7.2-23 $\sigma_y = 25.3$ MPa, $\tau_{xy} = 10$ MPa

7.2-24 $\sigma_y = -77.7$ MPa, $\tau_{xy} = -27.5$ MPa

7.2-25 $\sigma_b = -32.0$ MPa, $\tau_b = 18.79$ MPa, $\theta_1 = 47.9°$

7.2-26 $\sigma_y = -211$ kPa

7.2-27 $\sigma_{x1} = -1.4$ MPa, $\sigma_{y1} = 2.4$ MPa, $\tau_{x1y1} = 0$

7.3-1 $\sigma_1 = 7$ MPa, $\sigma_2 = 1$ MPa, $\tau_{max} = 3$ MPa

7.3-2 $\sigma_1 = 8.04$ MPa, $\sigma_2 = 1.463$ MPa, $\tau_{max} = 3.29$ MPa

7.3-3 $\sigma_1 = 40.8$ MPa, $\sigma_2 = 7.24$ MPa, $\theta_{p1} = 8.68°$

7.3-4 $\sigma_1 = 119.2$ MPa, $\sigma_2 = 60.8$ MPa, $\theta_{p1} = 29.52°$

7.3-5 $\sigma_1 = -43.7$ MPa, $\sigma_2 = -8.31$ MPa, $\theta_{p1} = -23.6°$

7.3-6 $\sigma_1 = 53.6$ MPa, $\theta_{p1} = -14.2°$

7.3-7 $\sigma_1 = 39.9$ MPa, $\sigma_2 = 124.1$ MPa, $\tau_{max} = -42.1$ MPa, $\theta_{p1} = -14.2°$

7.3-8 $\tau_{max} = 24.2$ MPa, $\sigma_{x1} = -14.25$ MPa, $\sigma_{y1} = -14.25$ MPa, $\theta_{s1} = 60.53°$

7.3-9 $\tau_{max} = 47.3$ MPa, $\theta_{s1} = 61.7°$

7.3-10 $\tau_{max} = 26.7$ MPa, $\theta_{s1} = 19.08°$

7.3-11 $(\tau_{max})_{z1} = 1$ MPa, $(\tau_{max})_{x1} = -2$ MPa, $(\tau_{max})_{y1} = -1$ MPa

7.3-12 $\sigma_1 = 20$ kPa, $\sigma_2 = -20$ kPa, $\tau_{max} = 20$ kPa

7.3-13 (a) $\sigma_1 = 1$ MPa, $\theta_{p1} = -18.43°$; (b) $\tau_{max} = 5$ MPa, $\theta_{s1} = -63.4°$

7.3-14 $\sigma_1 = 0$, $\sigma_2 = -5$ MPa, $\tau_{max} = 2.5$ MPa

7.3-15 $\sigma_1 = 0$, $\sigma_2 = -7.5$ MPa, $\tau_{max} = 3.75$ MPa

7.3-16 (a) $\sigma_1 = 25$ MPa, $\sigma_2 = -130$ MPa; (b) $\tau_{max} = 77.5$ MPa, $\sigma_{ave} = -52.5$ MPa

7.3-17 (a) $\sigma_1 = 18.45$ MPa, $\sigma_2 = 4.55$ MPa; (b) $\tau_{max} = 6.95$ MPa, $\sigma_{ave} = 11.5$ MPa

7.3-18 (a) $\sigma_1 = 2262$ kPa, $\theta_{p1} = -13.70°$ (b) $\tau_{max} = 1000$ kPa, $\theta_{s1} = -58.7°$

7.3-19 (a) $\sigma_1 = 101.8$ MPa, $\theta_{p1} = 7.85°$ (b) $\tau_{max} = 48$ MPa, $\theta_{s1} = -37.2°$

7.3-20 (a) $\sigma_1 = 29.2$ MPa, $\theta_{p1} = -17.98°$ (b) $\tau_{max} = 66.4$ MPa, $\theta_{s1} = -63.0°$

7.3-21 (a) $\sigma_1 = -8.72$ MPa, $\theta_{p1} = 24.7°$ (b) $\tau_{max} = 40.8$ MPa, $\theta_{s1} = -20.3°$

7.3-22 (a) $\sigma_1 = 76.3$ MPa, $\theta_{p1} = 107.5°$ (b) $\tau_{max} = 101.3$ MPa, $\theta_{s1} = -62.5°$

7.3-23 $20.6 \text{ MPa} \le \sigma_y \le 65.4 \text{ MPa}$

7.3-24 $18.5 \text{ MPa} \le \sigma_y \le 85.5 \text{ MPa}$

7.3-25 (a) $\sigma_y = 28.6$ MPa; (b) $\theta_{p1} = -40.24°$, $\sigma_1 = 44$ MPa, $\theta_{p2} = 49.76°$, $\sigma_2 = 17.64$ MPa

7.3-26 (a) $\sigma_y = 23.3$ MPa; (b) $\theta_{p1} = 65.6°$, $\sigma_1 = 41$ MPa, $\theta_{p2} = -24.4°$, $\sigma_2 = -62.7$ MPa

7.3-27 $\sigma_1 = 3.6$ kPa, $\sigma_2 = -638$ kPa, $\tau_{max} = 321$ kPa

7.3-28 $\sigma_1 = 12$ MPa, $\sigma_2 = 0$, $\tau_{max} = 6$ MPa

7.4-1 (a) $\sigma_{x1} = 75$ MPa, $\sigma_{y1} = 23$ MPa, $\tau_{x1y1} = -41.6$ MPa; (b) $\tau_{max} = 49$ MPa, $\sigma_{ave} = 49$ MPa

7.4-2 (a) $\sigma_{x1} = 40.1$ MPa, $\sigma_{y1} = 16.91$ MPa, $\tau_{x1y1} = 26$ MPa; (b) $\tau_{max} = 28.5$ MPa, $\sigma_{ave} = 28.5$ MPa

7.4-3 (a) $\sigma_{x1} = -37.6$ MPa, $\sigma_{y1} = -9.4$ MPa, $\tau_{x1y1} = 18.8$ MPa; (b) $\tau_{max} = -23.5$ MPa, $\sigma_{aver} = -23.5$ MPa

7.4-4 For $\theta = 25°$: (a) $\sigma_{x1} = -36.0$ MPa, $\tau_{x1y1} = 25.7$ MPa; (b) $\tau_{max} = 33.5$ MPa, $\theta_{s1} = 45.0°$

7.4-5 For $\theta = 55°$: (a) $\sigma_{x1} = 6.09$ MPa, $\tau_{x1y1} = -25.8$ MPa; (b) $\tau_{max} = 27.5$ MPa, $\theta_{s1} = -45.0°$, $\sigma_{aver} = 15.50$ MPa

7.4-6 For $\theta = 21.80°$: (a) $\sigma_{x1} = -17.1$ MPa, $\tau_{x1y1} = 29.7$ MPa; (b) $\tau_{max} = 43.0$ MPa, $\theta_{x1} = 45.0°$

7.4-7 For $\theta = 52°$: (a) $\sigma_{x1} = 18.44$ MPa, $\tau_{x1y1} = -4.60$ MPa; (b) $\sigma_1 = 19.00$ MPa, $\theta_{p1} = 45.0°$

7.4-8 $\sigma_{x1} = 1.75$ MPa, $\sigma_{y1} = 13.25$ MPa, $\tau_{x1y1} = 4.82$ MPa

7.4-9 (a) $\sigma_{x1} = -3.3$ MPa, $\sigma_{y1} = -6.7$ MPa, $\tau_{x1y1} = 4.7$ MPa; (b) $\sigma_1 = 0$, $\sigma_2 = -10$ MPa, $\tau_{max} = 5$ MPa

7.4-10 (a) $\sigma_{x1} = -60.8$ MPa, $\sigma_{y1} = 128.8$ MPa, $\tau_{x1y1} = -46.7$ MPa; (b) $\sigma_1 = 139.6$ MPa, $\sigma_2 = -71.6$ MPa, $\tau_{max} = 105.6$ MPa

7.4-11 For $\theta = 36.87°$: (a) $\sigma_{x1} = 25.0$ MPa, $\tau_{x1y1} = 7.28$ MPa; (b) $\sigma_1 = 26.00$ MPa, $\theta_{p1} = 45.0°$

7.4-12 For $\theta = 40°$: $\sigma_{x1} = 27.5$ MPa, $\tau_{x1y1} = -5.36$ MPa

7.4-13 For $\theta = -51°$: $\sigma_{x1} = 82.7$ MPa, $\tau_{x1y1} = -24.6$ MPa

7.4-14 For $\theta = -33°$: $\sigma_{x1} = -61.7$ MPa, $\tau_{x1y1} = 51.7$ MPa, $\sigma_{y1} = -171.3$ MPa

7.4-15 For $\theta = 14°$: $\sigma_{x1} = -10.42$ MPa, $\tau_{x1y1} = 3.85$ MPa, $\sigma_{y1} = -6.58$ MPa

7.4-16 For $\theta = 35°$: $\sigma_{x1} = 46.4$ MPa, $\tau_{xlyl} = -9.81$ MPa

7.4-17 For $\theta = 65°$: $\sigma_{x1} = -12.71$ MPa, $\tau_{x1y1} = 27.3$ MPa

7.4-18 (a) $\sigma_1 = 10{,}865$ kPa, $\theta_{p1} = 115.2°$; (b) $\tau_{max} = 4865$ kPa, $\theta_{s1} = 70.2°$

7.4-19 (a) $\sigma_1 = 17.72$ MPa, $\theta_{p1} = 31.4°$; (b) $\tau_{max} = 22.47$ MPa, $\theta_{s1} = -13.57°$

7.4-20 (a) $\sigma_1 = 18.2$ MPa, $\theta_{p1} = 123.3°$; (b) $\tau_{max} = 15.4$ MPa, $\theta_{s1} = 78.3°$

7.4-21 (a) $\sigma_1 = -47.67$ MPa, $\theta_{p1} = -32.9°$; (b) $\tau_{max} = 54.8$ MPa, $\theta_{s1} = -77.9°$

7.4-22 (a) $\sigma_1 = 40.0$ MPa, $\theta_{p1} = 68.8°$; (b) $\tau_{max} = 40.0$ MPa, $\theta_{s1} = 23.8°$

7.4-23 (a) $\sigma_1 = 4.4$ MPa, $\theta_{p1} = -26.8°$; (b) $\tau_{max} = 23.6$ MPa, $\theta_{s1} = -71.8°$

7.4-24 (a) $\sigma_1 = 3.43$ MPa, $\theta_{p1} = -19.68°$; (b) $\tau_{max} = 15.13$ MPa, $\theta_{s1} = -64.7°$

7.4-25 (a) $\sigma_1 = 51.6$ MPa, $\theta_{p1} = 9.9°$; (b) $\tau_{max} = 26.6$ MPa, $\theta_{s1} = -35.1°$

7.5-1 $\sigma_x = 175.2$ MPa, $\sigma_y = 135.3$ MPa, $\Delta t = -7.2 \times 10^{-3}$ mm

7.5-2 $\sigma_x = 102.6$ MPa, $\sigma_y = -11.21$ MPa, $\Delta t = -1.646 \times 10^{-3}$ mm

7.5-3 $e = -3.17 \times 10^{-4}$

7.5-4 (a) $\varepsilon_x = 7.25 \times 10^{-5}$, $\varepsilon_y = -9 \times 10^{-5}$, $\varepsilon_z = 7.5 \times 10^{-6}$, $\gamma_{xy} = 6.5 \times 10^{-5}$; (b) $u = 1.2$ kPa

7.5-5 (a) $\varepsilon_z = -v(\varepsilon_x + \varepsilon_y)/(1 - v)$; (b) $e = (1 - 2v)(\varepsilon_x + \varepsilon_y)/(1 - v)$

7.5-6 $v = 0.24$, $E = 112.1$ GPa

7.5-7 $v = 0.3$, $E = 204.1$ GPa

7.5-8 (a) $\gamma_{max} = 5.85 \times 10^{-4}$; (b) $\Delta t = -1.32 \times 10^{-3}$ mm; (c) $\Delta V = 387\ \text{mm}^3$

7.5-9 (a) $\gamma_{max} = 1.921 \times 10^{-3}$; (b) $\Delta t = -3.55 \times 10^{-3}$ mm (decrease); (c) $\Delta V = 1372\ \text{mm}^3$ (increase)

7.5-10 (a) $\Delta V_b = -49.2\ \text{mm}^3$, $U_b = 3.52$ J; (b) $\Delta V_a = -71.5\ \text{mm}^3$, $U_a = 4.82$ J

7.5-11 $\Delta V = -602\ \text{mm}^3$, $U = 6.2452$ J

7.5-12 (a) $\Delta V = 2766\ \text{mm}^3$, $U = 56$ J; (b) $t_{max} = 36.1$ mm; (c) $b_{min} = 640$ mm

7.5-13 (a) $\Delta V = 623\ \text{mm}^3$, $U = 64.5$ J; (b) $t_{max} = 17.02$ mm; (c) $b_{min} = 262$ mm

7.5-14 (a) $\Delta ac = \varepsilon_x d = 0.1296$ mm (increase); (b) $\Delta bc = \varepsilon_y d = -0.074$ mm (decrease); (c) $\Delta t = \varepsilon_z t = -2.86 \times 10^{-3}$ mm (decrease); (d) $\Delta V = eV_0 = 430\ \text{mm}^3$; (e) $U = uV_0 = 71.2\ \text{N}\cdot\text{m}$; (f) $t_{max} = 22.0$ mm; (g) $\sigma_{x\max} = 63.9$ MPa

7.5-15 (a) $\varepsilon_x = 0.182$, $\varepsilon_y = -0.318$, $\varepsilon_z = 0.182$; (b) $e = 0.045$

7.5-16 $\nu = 0.477$, $E = 1.006$ MPa

7.6-1 $\varepsilon_x = -3.98 \times 10^{-4}$, $\varepsilon_y = 3.24 \times 10^{-4}$, $\varepsilon_z = 2.1 \times 10^{-4}$, $e = 1.36 \times 10^{-4}$

7.6-2 $\varepsilon_x = -3.33 \times 10^{-4}$, $\varepsilon_y = 5.79 \times 10^{-4}$, $\varepsilon_z = -2.95 \times 10^{-4}$, $u = 14.09$ kPa

7.6-3 (a) $\tau_{max} = 60$ MPa; (b) $\Delta a = 0.1955$ mm, $\Delta b = -0.0944$ mm, $\Delta c = -0.034$ mm; (c) $\Delta V = 287\ \text{mm}^3$; (d) $U = 109.9$ J; (e) $\sigma_{x\max} = 88.5$ MPa; (f) $\sigma_{x\max} = 82$ MPa

7.6-4 (a) $\tau_{max} = \dfrac{\sigma_1 - \sigma_3}{2} = 8.5$ MPa; (b) $\Delta a = a\varepsilon_x = -0.0525$ mm, $\Delta b = \varepsilon_y b = -9.67 \times 10^{-3}$ mm, $\Delta c = \varepsilon_z c = -9.67 \times 10^{-3}$ mm; (c) $\Delta V = eV_0 = -2.052 \times 10^3\ \text{mm}^3$; (d) $U = uV_0 = 56.2\ \text{N}\cdot\text{m}$; (e) $\sigma_{x\max} = -50$ MPa; (f) $\sigma_{x\max} = -65.1$ MPa

7.6-5 (a) $\sigma_x = -28.8$ MPa, $\sigma_y = -14.4$ MPa, $\sigma_z = -14.4$ MPa; (b) $\tau_{max} = 7.2$ MPa; (c) $\Delta V = -300\ \text{mm}^3$; (d) $U = 3.78$ J; (e) $\sigma_{x\max} = -26.5$ MPa; (f) $\varepsilon_{x\max} = -242(10^{-6})$

7.6-6 (a) $\sigma_x = -82.6$ MPa, $\sigma_y = -54.7$ MPa, $\sigma_z = -54.7$ MPa; (b) $\tau_{max} = \dfrac{\sigma_1 - \sigma_3}{2} = 13.92$ MPa; (c) $\Delta V = eV_0 = -846\ \text{mm}^3$; (d) $U = uV_0 = 29.9\ \text{N}\cdot\text{m}$; (e) $\sigma_{x\max} = -73$ MPa; (f) $\varepsilon_{x\max} = -741(10^{-6})$

7.6-7 (a) $K_{Al} = 74.1$ GPa; (b) $E = 42$ GPa, $v = 0.35$

7.6-8 (a) $K = 4.95$ GPa; (b) $E = 1.297$ GPa, $v = 0.40$

7.6-9 (a) $p = vF/[A(1 - v)]$; (b) $\delta = FL(1 + v)(1 - 2v)/EA(1 - v)]$

7.6-10 (a) $p = vp_0$; (b) $e = -p_0(1 + v)(1 - 2v)/E$; (c) $u = p_0^2(1 - v^2)/2E$

7.6-11 (a) $p = -3.1$ kPa; (b) $\Delta V = -0.2175\ \text{cm}^3$

7.6-12 $P = -81.5$ kN

7.6-13 (a) $\Delta d = 0.036$ mm, $\Delta V = 2863\ \text{mm}^3$, $U = 34.4$ J; (b) $h = 1620$ m

7.6-14 (a) $p = 700$ MPa; (b) $K = 175$ GPa; (c) $U = 2470$ J

7.6-15 $\varepsilon_0 = 2.77 \times 10^{-4}$, $e = 8.3 \times 10^{-4}$, $u = 0.0344$ MPa

7.7-1 (a) $\varepsilon_{x1} = -5.53 \times 10^{-4}$, $\varepsilon_{y1} = 1.053 \times 10^{-3}$, $\gamma_{x1y1} = 1.915 \times 10^{-3}$; (b) $\varepsilon_1 = 1.5 \times 10^{-3}$ $\varepsilon_2 = -1.0 \times 10^{-3}$, $\gamma_{max} = 2.5 \times 10^{-3}$

7.7-2 (a) $\varepsilon_1 = 2.0 \times 10^{-3}$, $\varepsilon_2 = 1.5 \times 10^{-3}$; (b) $\gamma_{max} = 5.0 \times 10^{-4}$

7.7-3 (a) $\varepsilon_{x1} = 1.299 \times 10^{-4}$, $\varepsilon_{y1} = -1.299 \times 10^{-4}$, $\gamma_{x1y1} = 1.5 \times 10^{-4}$; (b) $\varepsilon_1 = 1.5 \times 10^{-4}$, $\varepsilon_2 = -1.5 \times 10^{-4}$, $\gamma_{max} = 3 \times 10^{-4}$

7.7-4 (a) $\varepsilon_{x1} = 1.398 \times 10^{-3}$, $\varepsilon_{y1} = -9.76 \times 10^{-5}$, $\gamma_{x1y1} = -8.63 \times 10^{-4}$; (b) $\varepsilon_1 = 1.513 \times 10^{-3}$, $\varepsilon_2 = -2.13 \times 10^{-4}$, $\gamma_{max} = 1.726 \times 10^{-3}$

7.7-5 (a) $\Delta d = 0.0476$ mm; (b) $\Delta\phi = -\alpha = 1.419 \times 10^{-4}$ (decrease, radians); (c) $\Delta\psi = -\alpha = 1.419 \times 10^{-4}$ (increase, radians)

7.7-6 (a) $\Delta d = \varepsilon_{x1} L_d = 0.062$ mm; (b) $\Delta\phi = -\alpha = 1.89 \times 10^{-4}$ (decrease, radians); (c) $\Delta\psi = -\alpha = 1.89 \times 10^{-4}$ (increase, radians)

7.7-7 (a) $\Delta d = 0.1146$ mm (increase); (b) $\Delta\phi = 150 \times 10^{-6}$ rad (decrease); (c) $\gamma = -314 \times 10^{-6}$ rad (angle *ced* increases)

7.7-8 (a) $\Delta d = 0.168$ mm (increase); (b) $\Delta\phi = 317 \times 10^{-6}$ rad (decrease); (c) $\gamma = -634 \times 10^{-6}$ rad (angle *ced* increases)

7.7-9 $\varepsilon_{x1} = 3.97 \times 10^{-4}$, $\varepsilon_{y1} = 3.03 \times 10^{-4}$, $\gamma_{x1y1} = 1.829 \times 10^{-4}$

7.7-10 $\varepsilon_{x1} = 9.53 \times 10^{-5}$, $\varepsilon_{y1} = -1.353 \times 10^{-4}$, $\gamma_{x1y1} = -3.86 \times 10^{-4}$

7.7-11 $\varepsilon_1 = 554 \times 10^{-6}$, $\theta_{p1} = -22.9°$, $\gamma_{max} = 488 \times 10^{-6}$

7.7-12 $\varepsilon_1 = 172 \times 10^{-6}$, $\theta_{p1} = 163.9°$, $\gamma_{max} = 674 \times 10^{-6}$

7.7-13 For $\theta = 75°$: (a) $\varepsilon_{x1} = 202 \times 10^{-6}$, $\gamma_{x1y1} = -569 \times 10^{-6}$; (b) $\varepsilon_1 = 568 \times 10^{-6}$, $\theta_{p1} = 22.8°$; (c) $\gamma_{max} = 587 \times 10^{-6}$

7.7-14 For $\theta = 45°$: (a) $\varepsilon_{x1} = -385 \times 10^{-6}$, $\gamma_{x1y1} = 690 \times 10^{-6}$; (b) $\varepsilon_1 = -254 \times 10^{-6}$, $\theta_{p1} = 65.7°$; (c) $\gamma_{max} = 1041 \times 10^{-6}$

7.7-15 $\tau_{max\,xy} = 29.5$ MPa, $\gamma_{xy\,max} = 7.19 \times 10^{-4}$, $\gamma_{xz\,max} = 9.01 \times 10^{-4}$, $\gamma_{yz\,max} = 1.827 \times 10^{-4}$

7.7-16 $\tau_{max\,xy} = \dfrac{\sigma_x - \sigma_y}{2} = 33.7$ MPa,

$$\gamma_{xy\,max} = 2\sqrt{\left(\frac{\varepsilon_x - \varepsilon_y}{2}\right)^2 + \left(\frac{\gamma_{xy}}{2}\right)^2},$$
$$= 1.244 \times 10^{-3},$$
$$\gamma_{xz\,max} = 2\sqrt{\left(\frac{\varepsilon_x - \varepsilon_z}{2}\right)^2 + \gamma_{xz}^2} = 1.459 \times 10^{-3},$$
$$\gamma_{yz\,max} = 2\sqrt{\left(\frac{\varepsilon_y - \varepsilon_z}{2}\right)^2 + \gamma_{yz}^2} = 2.15 \times 10^{-4}$$

7.7-17 For $\theta = 30°$: (a) $\varepsilon_{x1} = -7.61 \times 10^{-4}$, $\gamma_{x1y1} = 8.62 \times 10^{-4}$; (b) $\varepsilon_1 = 4.27 \times 10^{-4}$, $\theta_{p1} = 100.1°$; (c) $\gamma_{max} = 1.345 \times 10^{-3}$

7.7-18 For $\theta = 50°$: (a) $\varepsilon_{x1} = -1469 \times 10^{-6}$, $\gamma_{x1y1} = -717 \times 10^{-6}$; (b) $\varepsilon_1 = -732 \times 10^{-6}$, $\theta_{p1} = 166.0°$; (c) $\gamma_{max} = 911 \times 10^{-6}$

7.7-19 $\varepsilon_1 = 551 \times 10^{-6}$, $\theta_{p1} = 12.5°$, $\gamma_{max} = 662 \times 10^{-6}$

7.7-20 $\varepsilon_1 = 332 \times 10^{-6}$, $\theta_{p1} = 12.0°$, $\gamma_{max} = 515 \times 10^{-6}$

7.7-21 (a) $P = 23.6$ kN, $T = -114.9$ N·m; (b) $\gamma_{max} = 2.84 \times 10^{-4}$, $\tau_{max} = 23.1$ MPa

7.7-22 $P = 121.4$ kN, $\alpha = 56.7°$

7.7-23 $P = 44.1$ kN, $\alpha = 75.2°$

7.7-24 $\varepsilon_x = \varepsilon_a$, $\varepsilon_y = (2\varepsilon_b + 2\varepsilon_c - \varepsilon_a)/3$, $\gamma_{xy} = 2(\varepsilon_b - \varepsilon_c)/\sqrt{3}$

7.7-25 For $\theta_{p1} = 30°$: $\varepsilon_1 = 1550 \times 10^{-6}$, $\varepsilon_2 = -250 \times 10^{-6}$, $\sigma_1 = 68.3$ MPa, $\sigma_2 = 13.67$ MPa

7.7-26 $\sigma_x = 91.6$ MPa

7.7-27 $\varepsilon_{x1} = 3.97 \times 10^{-4}$, $\varepsilon_{y1} = 3.03 \times 10^{-4}$, $\gamma_{x1y1} = 1.829 \times 10^{-4}$

7.7-28 $\varepsilon_{x1} = 9.53 \times 10^{-5}$, $\varepsilon_{y1} = -1.353 \times 10^{-4}$, $\gamma_{x1y1} = -3.86 \times 10^{-4}$

7.7-29 $\varepsilon_1 = 554 \times 10^{-6}$, $\theta_{p1} = 157.1°$, $\gamma_{max} = 488 \times 10^{-6}$

7.7-30 $\varepsilon_1 = 172 \times 10^{-6}$, $\theta_{p1} = 163.9°$, $\gamma_{max} = 674 \times 10^{-6}$

7.7-31 For $\theta = 75°$: (a) $\varepsilon_{x1} = 202 \times 10^{-6}$, $\gamma_{x1y1} = -569 \times 10^{-6}$; (b) $\varepsilon_1 = 568 \times 10^{-6}$, $\theta_{p1} = 22.8°$; (c) $\gamma_{max} = 587 \times 10^{-6}$

7.7-32 For $\theta = 45°$: (a) $\varepsilon_{x1} = -385 \times 10^{-6}$, $\gamma_{x1y1} = 690 \times 10^{-6}$; (b) $\varepsilon_1 = -254 \times 10^{-6}$, $\theta_{p1} = 65.7°$; (c) $\gamma_{max} = 1041 \times 10^{-6}$

R-7.1 C

R-7.2 C

R-7.3 D

R-7.4 A

R-7.5 B

R-7.6 A

R-7.7 C

R-7.8 D

CHAPTER 8

8.2-1 $p_\sigma = 16.82$ kPa, $p_\tau = 9.6$ kPa

8.2-2 $p_a = 1.748$ kPa

8.2-3 (a) Use $t = 112$ mm; (b) $p_{max} = 3.14$ MPa

8.2-4 (a) Use $t = 98$ mm; (b) $p_{max} = 3.34$ MPa

8.2-5 (a) $F = 15.81$ kN, $\sigma = 3.14$ MPa; (b) $d_b = 13.23$ mm; (c) $r = 334$ mm

8.2-6 (a) $\sigma_{max} = 3.12$ MPa, $\varepsilon_{max} = 0.438$; (b) $t_{reqd} = 1.29$ mm

8.2-7 (a) $\sigma_{max} = 4.17$ MPa, $\varepsilon_{max} = 0.655$; (b) $p_{max} = 129.8$ kPa

8.2-8 (a) $p_{max} = 3.51$ MPa; (b) $p_{max} = 2.93$ MPa

8.2-9 (a) $f = 5100$ kN/m; (b) $\tau_{max} = 51$ MPa; (c) $\varepsilon_{max} = 344 \times 10^{-6}$

8.2-10 (a) $f = 5.5$ MN/m; (b) $\tau_{max} = 57.3$ MPa; (c) $\varepsilon_{max} = 3.87 \times 10^{-4}$

8.2-11 (a) $t_{min} = 11.67$ mm; (b) $p = 16.33$ MPa

8.2-12 (a) $t_{min} = 7.17$ mm; (b) $p = 19.25$ MPa

8.2-13 $D_0 = 26.6$ m

8.3-1 $\sigma_1 = 152.6$ MPa, $\sigma_2 = 76.3$ MPa, $\tau_{max\,In} = 38.2$ MPa, $\tau_{max\,Out} = 76.3$ MPa

8.3-2 $\sigma_1 = 76.2$ MPa, $\sigma_2 = 38.1$ MPa, $\tau_{max\,In} = 19.05$ MPa, $\tau_{max\,Out} = 38.1$ MPa

8.3-3 (a) $t_{min} = 6.43$ mm; (b) $p_{max} = 11.2$ MPa

8.3-4 (a) $h = 22.2$ m; (b) zero

8.3-5 $n = 2.38$

8.3-6 (a) $F = 3\pi pr^2$; (b) $t_{reqd} = 10.91$ mm

8.3-7 (a) $p = 350$ kPa; (b) $\varepsilon_r = 8.35 \times 10^{-4}$

8.3-8 (a) $\varepsilon_{max} = 6.67 \times 10^{-5}$; (b) $\varepsilon_r = 2.83 \times 10^{-4}$

8.3-9 $t_{min} = 2.53$ mm

8.3-10 $t_{min} = 3.71$ mm

8.3-11 (a) $h = 7.24$ m; (b) $\sigma_1 = 817$ kPa

8.3-12 (a) $\sigma_h = 24.9$ MPa; (b) $\sigma_c = 49.7$ MPa; (c) $\sigma_w = 24.9$ MPa; (d) $\tau_h = 12.43$ MPa; (e) $\tau_c = 24.9$ MPa

8.3-13 (a) $t_{min} = 6.25$ mm; (b) $t_{min} = 3.12$ mm

8.3-14 (a) $\sigma_1 = 93.3$ MPa, $\sigma_2 = 46.7$ MPa; (b) $\tau_1 = 23.2$ MPa, $\tau_2 = 46.7$ MPa; (c) $\varepsilon_1 = 3.97 \times 10^{-4}$, $\varepsilon_2 = 9.33 \times 10^{-5}$; (d) $\theta = 35°$, $\sigma_{x_1} = 62.0$ MPa, $\sigma_{y1} = 78.0$ MPa, $\tau_{x1y1} = 21.9$ MPa;

8.3-15 (a) $\sigma_1 = 42$ MPa, $\sigma_2 = 21$ MPa; (b) $\tau_1 = 10.5$ MPa, $\tau_2 = 21$ MPa; (c) $\varepsilon_1 = 178.5 \times 10^{-6}$, $\varepsilon_2 = 42 \times 10^{-6}$; (d) $\theta = 15°$, $\sigma_{x1} = 22.4$ MPa, $\sigma_{y1} = 40.6$ MPa, $\tau_{x1y1} = 5.25$ MPa

8.4-1 $\sigma_x = 13.35$ MPa, $\sigma_y = 0$, $\tau_{xy} = -0.989$ MPa

8.4-2 $\sigma_x = -2.22$ MPa, $\sigma_y = 0$, $\tau_{xy} = -0.139$ MPa

8.4-3 $\sigma_x = 53.3$ kPa, $\sigma_y = 0$, $\tau_{xy} = -66.7$ kPa, $\sigma_1 = 98.5$ kPa, $\sigma_2 = -45.1$ kPa, $\tau_{max} = 71.8$ kPa

8.4-4 $\sigma_x = 667$ kPa, $\sigma_y = 0$, $\tau_{xy} = -44.4$ kPa, $\sigma_1 = 670$ kPa, $\sigma_2 = -2.95$ kPa, $\tau_{max} = 336$ kPa

8.4-5 $\sigma_x = -162.5$ kPa, $\sigma_y = 0$, $\tau_{xy} = 1$ MPa, $\sigma_1 = 923$ kPa, $\sigma_2 = -1085$ kPa, $\tau_{max} = 1004$ kPa

8.4-6 $\sigma_x = -208$ kPa, $\sigma_y = 0$, $\tau_{xy} = 84.6$ kPa, $\sigma_1 = 30.1$ kPa, $\sigma_2 = -238$ kPa, $\tau_{max} = 134.1$ kPa

8.4-7 (a) $\sigma_x = -39.5$ MPa, $\sigma_y = 0$, $\tau_{xy} = 7.9$ MPa, $\sigma_1 = 1.522$ MPa, $\sigma_2 = -41$ MPa, $\tau_{max} = 21.3$ MPa; (b) $\sigma_x = -46.2$ MPa, $\sigma_y = 0$, $\tau_{xy} = 7.9$ MPa, $\sigma_1 = 1.315$ MPa, $\sigma_2 = -47.5$ MPa, $\tau_{max} = 24.4$ MPa

8.4-8 (a) $\sigma_x = -37.4$ MPa, $\sigma_y = 0$, $\tau_{xy} = 7.49$ MPa, $\sigma_1 = 1.442$ MPa, $\sigma_2 = -38.9$ MPa, $\tau_{max} = 20.2$ MPa; (b) $\sigma_x = -44.4$ MPa, $\sigma_y = 0$, $\tau_{xy} = 7.49$ MPa, $\sigma_1 = 1.227$ MPa, $\sigma_2 = -45.7$ MPa, $\tau_{max} = 23.4$ MPa

8.4-9 (a) $\sigma_1 = 1484$ kPa, $\sigma_2 = -1484$ kPa, $\tau_{max} = 1484$ kPa; (b) $\sigma_1 = 337$ kPa, $\sigma_2 = -5163$ kPa, $\tau_{max} = 2750$ kPa; (c) $\sigma_1 = 0$, $\sigma_2 = -14.48$ MPa, $\tau_{max} = 7238$ kPa

8.4-10 $P = 20$ kN

8.4-11 $P = 13.38$ kN

8.4-12 (a) $P = 30$ N, $\sigma_x = \sigma_y = \tau_{xy} = 0$; (b) $P = 167.5$ N, $\sigma_x = 220$ kPa, $\sigma_y = 0$, $\tau_{xy} = 8.8$ kPa

8.4-13 $P = 22.5$ kN, $\sigma_x = \sigma_y = 0$, $\tau_{xy} = -950$ kPa

8.4-14 (b) $\sigma_1 = 4.5$ MPa, $\sigma_2 = -76.1$ MPa, $\tau_{max} = 40.3$ MPa

8.4-15 (b) $\sigma_1 = 75.3$ MPa, $\sigma_2 = -6.5$ MPa, $\tau_{max} = 40.9$ MPa

8.4-16 (b) $\sigma_1 = 3.52$ MPa, $\sigma_2 = -66.9$ MPa, $\tau_{max} = 35.2$ MPa

8.4-17 $\sigma_x = 4.03$ MPa, $\sigma_y = 0$, $\tau_{xy} = -3.19$ MPa, $\sigma_1 = 5.78$ MPa, $\sigma_2 = -1.754$ MPa, $\tau_{max} = 3.77$ MPa

8.4-18 $\sigma_x = -215$ kPa, $\sigma_y = 0$, $\tau_{xy} = -398$ kPa, $\sigma_1 = 305$ kPa, $\sigma_2 = -519$ kPa, $\tau_{max} = 412$ kPa

8.4-19 (b) $\sigma_1 = 0.961$ MPa, $\sigma_2 = -11.62$ MPa, $\tau_{max} = 6.29$ MPa

8.4-20 (b) $\sigma_1 = 17.86$ MPa, $\sigma_2 = -0.145$ MPa, $\tau_{max} = 9.0$ MPa

8.4-21 $\dfrac{\sigma_1}{\sigma_2} = -58$

8.4-22 $\dfrac{\sigma_1}{\sigma_2} = -663$

8.5-1 (a) $\sigma_1 = 27.1$ MPa, $\sigma_2 = -20.6$ MPa, $\tau_{max} = 23.8$ MPa; (b) $\sigma_1 = 129.1$ MPa, $\sigma_2 = -3.18$ MPa, $\tau_{max} = 66.1$ MPa

8.5-2 (a) $\sigma_1 = 51.9$ MPa, $\sigma_2 = -0.438$ MPa, $\tau_{max} = 26.2$ MPa; (b) $\sigma_1 = 9.9$ MPa, $\sigma_2 = -4.1$ MPa, $\tau_{max} = 7.0$ MPa

8.5-3 $\sigma_1 = 10.9$ MPa, $\sigma_2 = -16.41$ kPa, $\tau_{max} = 5.5$ MPa

8.5-4 $\sigma_1 = 42.8$ MPa, $\sigma_2 = -42.8$ MPa, $\tau_{max} = 42.8$ MPa

8.5-5 $t_{min} = 3$ mm

8.5-6 $p_{max} = 9.60$ MPa

8.5-7 (a) $\sigma_{max} = \sigma_1 = 57.2$ MPa, $\tau_{max} = 19.72$ MPa; (b) $T_{max} = 215$ kN·m; (c) $t_{min} = 14.11$ mm

8.5-8 (a) $P_{max} = 461$ kN; (b) $p_{max} = 8.32$ MPa

8.5-9 $\sigma_t = 74.2$ MPa: No compressive stresses. $\tau_{max} = 37.1$ MPa

8.5-10 $\phi_{max} = 0.552$ rad $= 31.6°$

8.5-11 $\sigma_t = 32.0$ MPa, $\sigma_c = -73.7$ MPa, $\tau_{max} = 52.8$ MPa

8.5-12 $\sigma_t = 16.43$ MPa, $\sigma_c = -41.4$ MPa, $\tau_{max} = 28.9$ MPa

8.5-13 $P = 815\ \text{kN}$

8.5-14 (a) $\sigma_{max} = \sigma_1 = 35.8\ \text{MPa}$, $\tau_{max} = 18.05\ \text{MPa}$; (b) $P_{max} = 6.73\ \text{kN}$

8.5-15 (a) $\sigma_x = 0, \sigma_y = 37\ \text{MPa}, \tau_{xy} = 1.863\ \text{MPa}$; (b) $\sigma_1 = 37\ \text{MPa}, \sigma_2 = -0.094\ \text{MPa}$, $\tau_{max} = 18.57\ \text{MPa}$

8.5-16 $\tau_A = 76.0\ \text{MPa}, \tau_B = 19.94\ \text{MPa}$, $\tau_C = 23.7\ \text{MPa}$

8.5-17 $\sigma_1 = 21.3\ \text{MPa}, \sigma_2 = 0, \tau_{max} = 10.6\ \text{MPa}$

8.5-18 $\sigma_1 = 0, \sigma_2 = -21.7\ \text{MPa}, \tau_{max} = 10.87\ \text{MPa}$

8.5-19 $\sigma_1 = 6.96\ \text{MPa}$ (max. tensile stress at base of pole), $\sigma_2 = -46.2\ \text{MPa}$ (max. compressive stress at base of pole), $\tau_{max} = 24\ \text{MPa}$ (max. shear stress at base of pole)

8.5-20 $d_{min} = 48.4\ \text{mm}$

8.5-21 $\sigma_t = 298\ \text{MPa}$ at A, $\sigma_c = -15.45\ \text{MPa}$ at A, $\tau_{max} = 156.9\ \text{MPa}$ at A

8.5-22 (a) $\sigma_x = 0, \sigma_y = 2.23\ \text{MPa}, \tau_{xy} = 0$; (b) $\tau_{max} = 1.116\ \text{MPa}$

8.5-23 (a) $\sigma_t = 29.15\frac{qR^2}{d^3}, \sigma_c = -8.78\frac{qR^2}{d^3}$, $\tau_{max} = 18.97\frac{qR^2}{d^3}$; (b) $\sigma_t = 14.04\frac{qR^2}{d^3}$, $\sigma_c = -2.41\frac{qR^2}{d^3}, \tau_{max} = 8.22\frac{qR^2}{d^3}$

8.5-24 (a) $\sigma_x = 0, \sigma_y = 2.34\ \text{MPa}$, $\tau_{xy} = -0.1912\ \text{MPa}$; (b) $\sigma_1 = 2.35\ \text{MPa}$, $\sigma_2 = -0.0155\ \text{MPa}$; (c) $\tau_{max} = 1.185\ \text{MPa}$

8.5-25 $\sigma_t = 21.6\ \text{MPa}, \sigma_c = -9.4\ \text{MPa}$, $\tau_{max} = 15.5\ \text{MPa}$

8.5-26 Pure shear: $\tau_{max} = 0.804\ \text{MPa}$

8.5-27 (a) $d_{min} = 47.4\ \text{mm}$; (b) $P_{max} = 55.9\ \text{N}$

8.5-28 (a) $\sigma_1 = 29.3\ \text{MPa}, \sigma_2 = -175.9\ \text{MPa}$, $\tau_{max} = 102.6\ \text{MPa}$; (b) $\sigma_1 = 156.1\ \text{MPa}$, $\sigma_2 = -33\ \text{MPa}, \tau_{max} = 94.5\ \text{MPa}$

8.5-29 (a) $\sigma_1 = 0, \sigma_2 = -148.1\ \text{MPa}$, $\tau_{max} = 74.0\ \text{MPa}$; (b) $\sigma_1 = 7.02\ \text{MPa}$, $\sigma_2 = -155.1\ \text{MPa}, \tau_{max} = 81.1\ \text{MPa}$

8.5-30 Maximum: $\sigma_t = 18.35\ \text{MPa}$, $\sigma_C = -18.35\ \text{MPa}, \tau_{max} = 9.42\ \text{MPa}$

8.5-31 Top of beam: $\sigma_1 = 64.4\ \text{MPa}, \sigma_2 = 0$, $\tau_{max} = 32.2\ \text{MPa}$

8.5-32 (a) $d_{Al} = 26.3\ \text{mm}$; (b) $d_{Ti} = 21.4\ \text{mm}$

8.5-33 $\sigma_y = 0, \sigma_x = \dfrac{(FL)\frac{d_2}{2}}{\frac{I_p}{2}} = 12.14\ \text{MPa}$,

$\tau_{xy} = \dfrac{-T\frac{d_2}{2}}{I_p} = -3.49\ \text{MPa}, \sigma_1 = 13.07\ \text{MPa}$,

$\sigma_2 = -0.932\ \text{MPa}, \tau_{max} = 7\ \text{MPa}$

8.5-34 (a) $\sigma_1 = 0, \sigma_2 = \sigma_x = -108.4\ \text{MPa}$, $\tau_{max} = \frac{\sigma_x}{2} = -54.2\ \text{MPa}$; (b) $\sigma_1 = 0.703\ \text{MPa}, \sigma_2 = -1.153\ \text{MPa}$, $\tau_{max} = 0.928\ \text{MPa}$; (c) $P_{max} = 348\ \text{N}$

8.5-35 $\sigma_x = -136.4\ \text{MPa}, \sigma_y = 0, \tau_{xy} = 32.7\ \text{MPa}$, $\sigma_1 = 7.42\ \text{MPa}, \sigma_2 = -143.8\ \text{MPa}$, $\tau_{max} = 75.6\ \text{MPa}$

R-8.1 A

R-8.2 C

R-8.3 D

R-8.4 B

R-8.5 C

R-8.6 A

R-8.7 D

R-8.8 D

R-8.9 A

R-8.10 D

R-8.11 D

R-8.12 A

R-8.13 C

R-8.14 B

R-8.15 C

CHAPTER 9

9.2-1 moment M_0 applied at $x = L$

9.2-2 uniform load

9.2-3 $q = q_0 x/L$; Triangular load, acting downward

9.2-4 (a) $q = q_0 \sin \pi x/L$, Sinusoidal load; (b) $R_A = R_B = q_0 L/\pi$; (c) $M_{max} = q_0 L^2/\pi^2$

9.2-5 $q = q_0(1 - x/L)$; Triangular load, acting downward

9.2-6 (a) $q = q_0(L^2 - x^2)/L^2$; Parabolic load, acting downward; (b) $R_A = 2q_0 L/3$, $M_A = -q_0 L^2/4$

9.3-1 $\delta_{max} = 1.04$ mm, $\theta_A = 9.4 \times 10^{-4}$ rad

9.3-2 (a) $q = 1.438 \frac{kN}{m}$; (b) $r = 14.74$ mm

9.3-3 $\delta_{max} = 6.5$ mm, $\theta = 0.28°$

9.3-4 $h = 96$ mm

9.3-5 $L = 3.0$ m

9.3-6 $\delta_{max} = 15.4$ mm

9.3-7 $\delta/L = 1/320$

9.3-8 $E_g = 80.0$ GPa

9.3-9 Let $\beta = a/L$: $\frac{\delta_C}{\delta_{max}}$

$$\frac{\delta_C}{\delta_{max}} = \frac{(3\sqrt{3})(-1 + 8\beta - 4\beta^2)}{16(2\beta - \beta^2)^{3/2}}$$

The deflection at the midpoint is close to the maximum deflection. The maximum difference is only 2.6%.

9.3-10 $\frac{\delta_B}{M} = 3.2 \times 10^4 \left[\frac{1}{N}\right]$, $\frac{\theta_B}{M} = 200 \left[\frac{1}{N \cdot \mu m}\right]$

9.3-11 (a) $\delta_B = 25.5$ mm, $\delta_C = 51.4$ mm; (b) $\delta_B = 25.5$ mm, $\delta_C = 52.2$ mm

9.3-15 (a) $q_0 = 1186$ kN/m

(b) $\theta_B = 1.563 \times 10^{-3}$ rad

9.3-16 IPN 380

9.3-17 $v(x) = -mx^2(3L - x)/6EI$, $\delta_B = mL^3/3EI$, $\theta_B = mL^2/2EI$

9.3-18 $v(x) = -\frac{q}{48EI}(2x^4 - 12x^2L^2 + 11L^4)$,

$$\delta_B = \frac{qL^4}{48EI}$$

9.3-19 See Table H-2, Case 9.

9.3-20 See Table H-1, Case 2.

9.3-21 $v(x) = \frac{q_0 L}{24EI}(x^3 - 2Lx^2)$ for $0 \le x \le \frac{L}{2}$,

$v(x) = \frac{-q_0}{960LEI}(-160L^2x^3 + 160L^3x^2 + 80Lx^4 - 16x^5 - 25L^4x + 3L^5)$

for $\frac{L}{2} \le x \le L$, $\delta_B = \frac{7}{160}\frac{q_0 L^4}{EI}$,

$$\delta_C = \frac{1}{64}\frac{q_0 L^4}{EI}$$

9.3-22 $v(x) = \frac{q_0 x}{5760LEI}(200x^2L^2 - 240x^3L + 96x^4 - 53L^4)$ for $0 \le x \le \frac{L}{2}$,

$v(x) = \frac{-q_0 L}{5760EI}(40x^3 - 120Lx^2 + 83L^2x - 3L^3)$ for $\frac{L}{2} \le x \le L$,

$$\delta_C = \frac{3q_0 L^4}{1280EI}$$

9.3-23 $v(x) = -\frac{PL}{10{,}368EI}(-4104x^2 + 3565L^2)$

for $0 \le x \le \frac{L}{3}$, $v(x) = -\frac{P}{1152EI}(-648Lx^2 + 192x^3 + 64L^2x + 389L^3)$

for $\frac{L}{3} \le x \le \frac{L}{2}$, $v(x) = -\frac{P}{144EIL}(-72L^2x^2 + 12Lx^3 + 6x^4 + 5L^3x + 49L^4)$

for $\frac{L}{2} \le x \le L$, $\delta_A = \frac{3565PL^3}{10{,}368EI}$,

$$\delta_C = \frac{3109PL^3}{10{,}368EI}$$

9.4-3 $v = -M_0 x(L - x)^2/2LEI$, $\delta_{max} = 2M_0 L^2/27EI$ (downward)

9.4-4 $v(x) = -\frac{q}{48EI}(2x^4 - 12x^2L^2 + 11L^4)$, $\theta_B = -\frac{qL^3}{3EI}$

9.4-5 See Table H-1, Case 10.

9.4-6 $v = -q_0x^2(45L^4 - 40L^3x + 15L^2x^2 - x^4)/360L^2EI, \delta_B = 19q_0L^4/360EI$, $\theta_B = q_0L^3/15EI$

9.4-7 $v = -q_0x(3L^5 - 5L^3x^2 + 3Lx^4 - x^5)/90L^2EI, \delta_{max} = 61q_0L^4/5760EI$

9.4-8 $v(x) = \frac{q_0}{120EIL}(x^5 - 5Lx^4 + 20L^3x^2 - 16L^5)$, $\delta_{max} = \frac{2q_0L^4}{15EI}$

9.4-9 $v(x) = -\frac{qL^2}{16EI}(x^2 - L^2)$ for $0 \le x \le L$, $v(x) = -\frac{q}{48EI}(-20L^3x + 27L^2x^2 - 12Lx^3 + 2x^4 + 3L^4)$ for $L \le x \le \frac{3L}{2}$, $\delta_C = \frac{9qL^4}{128EI}, \theta_C = \frac{7qL^3}{48EI}$

9.4-10 $v(x) = -\frac{q_0L^2}{480EI}(-20x^2 + 19L^2)$ for $0 \le x \le \frac{L}{2}$, $v(x) = -\frac{q_0}{960EIL}(80Lx^4 - 16x^5 - 120L^2x^3 + 40L^3x^2 - 25L^4x + 41L^5)$ for $\frac{L}{2} \le x \le L, \delta_A = \frac{19q_0L^4}{480EI}$, $\theta_B = -\frac{13q_0L^3}{192EI}, \delta_C = \frac{7q_0L^4}{240EI}$

9.5-1 $b = 86.25$ mm, $h = 172.5$ mm

9.5-2 $\delta_{max} = 0.78$ mm

9.5-3 $P = 16.88$ N

9.5-4 $\delta_C = 5.9$ mm

9.5-5 $\delta_B = 0.03$ mm, $\theta_B = 4.9 \times 10^{-5}$ rad

9.5-6 $\delta_B = 0.764$ mm, $\theta_B = 4.02 \times 10^{-4}$ rad

9.5-7 $\theta_B = 7PL^2/9EI, \delta_B = 5PL^3/9EI$

9.5-8 (a) $\delta_1 = 11PL^3/144EI$; (b) $\delta_2 = 25PL^3/384EI$; (c) $\delta_1/\delta_2 = 88/75 = 1.173$

9.5-9 (a) $a/L = 2/3$; (b) $a/L = 1/2$

9.5-10 (a) $\delta_C = 6.25$ mm (upward); (b) $\delta_C = 18.36$ mm (downward)

9.5-11 $y = Px^2(L - x)^2/3LEI$

9.5-12 $\theta_B = 7qL^3/162EI, \delta_B = 23qL^4/648EI$

9.5-13 $\delta_C = 3.76$ mm, $\delta_B = 12.12$ mm

9.5-14 (a) $M = PL/2$; (b) $M = 5PL/24, \theta_B = PL^2/12EI$; (c) $M = PL/8, \delta_B = -PL^3/24EI$

9.5-15 $\theta_A = \frac{PL^2}{9EI}CW, \theta_B = \theta_A$ but CCW $\delta_{max} = \frac{23}{648}\frac{PL^3}{EI}$ at $x = \frac{L}{2}$

9.5-16 $\theta_A = \theta_B = \frac{PL^2}{81EI}CW, \delta_{max} = \frac{2\sqrt{6}}{2187}\frac{PL^3}{EI}$

9.5-17 $M = (19/180)q_0L^2$

9.5-18 (a) $\delta_A = PL^2(10L - 9a)/324EI$ (positive upward); (b) Upward when $a/L < 10/9$, downward when $a/L > 10/9$

9.5-19 (a) $\delta_C = PH^2(L + H)/3EI$; (b) $\delta_{max} = PHL^2/9\sqrt{3}EI$

9.5-20 $\delta_C = 3.5$ mm

9.5-21 $\theta_B = q_0L^3/10EI, \delta_B = 13q_0L^4/180\,EI$

9.5-22 $\theta_A = q(L^3 - 6La^2 + 4a^3)/24EI$, $\delta_{max} = q(5L^4 - 24L^2a^2 + 16a^4)/384EI$

9.5-23 (a) $P/Q = 9a/4L$; (b) $P/Q = 8a(3L + a)/9L^2$; (c) $P/qa = 9a/8L$ for $\delta_B = 0$, $P/qa = a(4L + a)/3L^2$ for $\delta_D = 0$

9.5-24 $\delta = 19WL^3/31,104EI$

9.5-25 $k = 640$ N/m

9.5-26 $M_1 = 7800$ N·m, $M_2 = 4200$ N·m

9.5-27 $\delta = \frac{6Pb^3}{EI}$

9.5-28 $\delta_E = \frac{47Pb^3}{12EI}$

9.5-29 $\delta_C = 5.07$ mm

9.5-30 $\delta_C = 2.05$ mm downward

9.5-31 $\delta_B = 13.84$ mm, $\theta_A = 0.538°$

9.5-32 $q = 16cEI/7L^4$

9.5-33 $\delta_h = Pcb^2/2EI, \delta_v = Pc^2(c + 3b)/3EI$

9.5-34 $\delta = PL^2(2L + 3a)/3EI$

9.5-35 (a) $H_B = 0, V_B = \dfrac{M}{L}, V_C = -V_B$;
(b) $\theta_A = \dfrac{5ML}{6EI}, \theta_B = \dfrac{ML}{3EI}, \theta_C = \dfrac{-ML}{6EI}$,
$\theta_D = \theta_C$; (c) $\delta_A = (7/24)ML^2/EI$ (to the left), $\delta_D = (1/12)ML^2/EI$ (to the right);
(d) $L_{CD} = \dfrac{\sqrt{14}}{2}L = 1.871L$

9.5-36 (a) $H_B = 0, V_B = \dfrac{P}{3}, V_C = \dfrac{2P}{3}$;
(b) $\theta_A = \left(\dfrac{-4}{81}\right)\dfrac{ML}{EI}, \theta_B = \theta_A$,
$\theta_C = \left(\dfrac{5}{81}\right)\dfrac{ML}{EI}, \theta_D = \theta_C$;
(c) $\delta_A = -\theta_B\left(\dfrac{L}{2}\right) = \dfrac{2L^2M}{81EI}$ (to the right),
$\delta_D = \theta_C\left(\dfrac{L}{2}\right) = \dfrac{5L^2M}{162EI}$ (to the left);
(d) $L_{CD} = \dfrac{2\sqrt{5}L}{5} = 0.894L$

9.5-37 (a) $b/L = 0.403$;
(b) $\delta_C = 0.00287qL^4/EI$

9.5-38 $\alpha = 22.5°, 112.5°, -67.5°$, or $-157.5°$

9.5-39 $\delta_D = \dfrac{2}{9}\dfrac{PL^3}{EI} + \dfrac{wL^4}{8EI}, \theta_D = \dfrac{5}{18}\dfrac{PL^2}{EI} + \dfrac{wL^3}{6EI}$

9.5-40 $\delta_B = 113.6$ mm, $\theta_B = 0.0238$ rad

9.5-41 $M_A = \dfrac{14}{81}qL^2, \theta_A = \dfrac{2}{81}\dfrac{qL^3}{EI}$ CCW

9.5-42 $M_A = \dfrac{16}{81}qL^2, \theta_A = \dfrac{qL^3}{27EI}$ CCW

9.5-43 $d = 0.375\,L, \theta_A = 0.0635\dfrac{qL^3}{EI}$ CCW

9.5-44 (a) $\delta_C = \dfrac{121}{1920}\dfrac{q_2L^4}{EI} - \dfrac{13}{960}\dfrac{q_1L^4}{EI}$ upward;
(b) $q_2 = \dfrac{26}{121}q_1$; (c) $\delta_C = \dfrac{19}{384}\dfrac{q_CL^4}{EI}$ upward

9.6-4 $\theta_B = 7qL^3/162EI, \delta_B = 23qL^4/648EI$

9.6-5 $\delta_B = 10.85$ mm, $\delta_C = 3.36$ mm

9.6-6 $\delta_B = 11.8$ mm, $\delta_C = 4.10$ mm

9.6-8 $P = 64$ kN

9.6-9 $\theta_A = M_0L/6EI, \theta_B = M_0L/3EI$,
$\delta = M_0L^2/16EI$

9.6-10 $\theta_A = Pa(L - a)(L - 2a)/6LEI$,
$\delta_1 = Pa^2(L - 2a)^2/6LEI, \delta_2 = 0$

9.6-11 $\theta_A = M_0L/6EI, \theta_B = 0, \delta = M_0L^2/27EI$ (downward)

9.7-1 (a) $\delta_B = PL^3(1 + 7I_1/I_2)/24EI_1$;
(b) $r = (1 + 7I_1/I_2)/8$

9.7-2 (a) $\delta_B = qL^4(1 + 15I_1/I_2)/128EI_1$;
(b) $r = (1 + 15I_1/I_2)/16$

9.7-3 (a) $\delta_C = -2.14$ mm (upward);
(b) $\delta_C = 18.08$ mm (downward)

9.7-4 $v = -qx(21L^3 - 64Lx^2 + 32x^3)/768EI$ for $0 \le x \le L/4$,
$v = -q(13L^4 + 256L^3x - 512Lx^3 + 256x^4)/12{,}288EI$ for $L/4 \le x \le L/2$,
$\theta_A = 7qL^3/256EI, \delta_{max} = 31qL^4/4096EI$

9.7-5 $\theta_A = 8PL^2/243EI, \delta_B = 8PL^3/729EI$,
$\delta_{max} = 0.01363PL^3/EI$

9.7-6 $v = -2Px(19L^2 - 27x^2)/729EI$ for $0 \le x \le L/3$,
$v = P(13L^3 - 175L^2x + 243Lx^2 - 81x^3)/1458EI$ for $L/3 \le x \le L$,
$\theta_A = 38PL^2/729EI, \theta_C = 34PL^2/729EI$,
$\delta_B = 32PL^3/2187EI$

9.7-7 $v = \dfrac{PL^3}{EI_A}\left[\dfrac{L}{2(L+x)} - \dfrac{3x}{8L} + \dfrac{1}{8} + \ln\left(\dfrac{L+x}{2L}\right)\right]$,
$\delta_A = \dfrac{PL^3}{8EI_A}(8\ln 2 - 5)$

9.7-8 $v = \dfrac{PL^3}{24EI_A}\left[7 - \dfrac{4L(2L + 3x)}{(L + x)^2} - \dfrac{2x}{L}\right]$,
$\delta_A = \dfrac{PL^3}{24EI_A}$

9.7-9 $v = \dfrac{8PL^3}{EI_A}\left[\dfrac{L}{2L+x} - \dfrac{2x}{9L} - \dfrac{1}{9} + \ln\left(\dfrac{2L+x}{3L}\right)\right]$,
$\delta_A = \dfrac{8PL^3}{EI_A}\left(\ln\dfrac{3}{2} - \dfrac{7}{18}\right)$

9.7-10 $v(x) = \frac{19{,}683PL^3}{2000EI_A}\left(\frac{81L}{81L + 40x} + 2\ln\left(\frac{81}{121} + \frac{40x}{121L}\right) - \frac{6440x}{14{,}641L} - \frac{3361}{14{,}641}\right)$,
$\delta_A = \frac{19{,}683PL^3}{7{,}320{,}500EI_A}\left(-2820 + 14{,}641\ln\left(\frac{11}{9}\right)\right) = 0.317\frac{PL^3}{EI_A}$

9.7-11 $v(x) = -\frac{19{,}683PL^3}{2000EI_A}\left(\frac{81L}{81L + 40x} + 2\ln\left(\frac{81}{121} + \frac{40x}{81L}\right) - \frac{6440x}{14{,}641L} - \frac{3361}{14641}\right)$,
$\delta_B = \frac{19{,}683PL^3}{7{,}320{,}500EI_A}\left(-2820 + 14{,}641\ln\left(\frac{11}{9}\right)\right)$

9.7-12 (a) $v' = -\frac{qL^3}{16EI_A}\left[1 - \frac{8Lx^2}{(L + x)^3}\right]$ for $0 \le x \le L$,
$v = -\frac{qL^4}{2EI_A}\left[\frac{(9L^2 + 14Lx + x^2)x}{8L(L + x)^2} - \ln\left(1 + \frac{x}{L}\right)\right]$ for $0 \le x \le L$;
(b) $\theta_A = \frac{qL^3}{16EI_A}$, $\delta_C = \frac{qL^4(3 - 4\ln 2)}{8EI_A}$

9.8-1 $U = 4bhL\sigma_{max}^2/45E$

9.8-2 (a) and; (b) $U = P^2L^3/96EI$; (c) $\delta = PL^3/48EI$

9.8-3 (a) and; (b) $U = \frac{q^2L^5}{15EI}$

9.8-4 (a) $U = 32EI\delta^2/L^3$; (b) $U = \pi^4EI\delta^2/4L^3$

9.8-5 (a) $U = P^2a^2(L + a)/6EI$; (b) $\delta_C = Pa^2(L + a)/3EI$; (c) $U = 56.2$ J, $\delta_C = 5.0$ mm

9.8-6 $U = \frac{L}{15{,}360EI}(17L^4q^2 + 280qL^2M_0 + 2560M_0^2)$

9.8-7 $\delta_B = 2PL^3/3EI + 8\sqrt{2}PL/EA$

9.9-2 $\delta_D = Pa^2b^2/3LEI$

9.9-3 $\delta_C = Pa^2(L + a)/3EI$

9.9-6 $\delta_C = L^3(2P_1 + 5P_2)/48EI$, $\delta_B = L^3(5P_1 + 16P_2)/48EI$

9.9-7 $\theta_A = 7qL^3/48EI$

9.9-8 $\delta_C = Pb^2(b + 3h)/3EI$, $\theta_C = Pb(b + 2h)/2EI$

9.9-9 $\delta_C = 31qL^4/4096EI$

9.9-10 $\theta_A = M_A(L + 3a)/3EI$, $\delta_A = M_Aa(2L + 3a)/6EI$

9.9-11 $\delta_C = Pa^2(L + a)/3EI + P(L + a)^2/kL^2$

9.9-12 $\delta_D = 37qL^4/6144EI$ (upward)

9.10-1 $\sigma_{max} = \sigma_{st}[1 + (1 + 2h/\delta_{st})^{1/2}]$

9.10-2 $\sigma_{max} = \sqrt{18WEh/AL}$

9.10-3 $\delta_{max} = 4.33$ mm, $\sigma_{max} = 102.3$ MPa

9.10-4 d = 281 mm

9.10-5 HE 320A

9.10-6 $h = 360$ mm

9.10-7 $R = \sqrt{3EII_m\omega^2/L^3}$

9.11-1 $v = -\alpha(T_2 - T_1)(x)(L - x)/2h$ (pos. upward), $\theta_A = \alpha L(T_2 - T_1)/2h$ (clockwise), $\delta_{max} = \alpha L^2(T_2 - T_1)/8h$ (downward)

9.11-2 $v = -\alpha(T_2 - T_1)(x^2)/2h$ (upward), $\theta_B = \alpha L(T_2 - T_1)/h$ (counterclockwise), $\delta_B = \alpha L^2(T_2 - T_1)/2h$ (upward)

9.11-3 $v(x) = \frac{\alpha(T_2 - T_1)(x^2 - L^2)}{2h}$, $\theta_C = \frac{\alpha(T_2 - T_1)(L + a)}{h}$ (counterclockwise), $\delta_C = \frac{\alpha(T_2 - T_1)(2La + a^2)}{2h}$ (upward)

9.11-4 (a) $\delta_{max} = \frac{\alpha T_0L^3}{9\sqrt{3}h}$ (downward); (b) $\delta_{max} = \frac{\alpha T_0L^4(2\sqrt{2} - 1)}{48h}$ (downward)

9.11-5 (a) $\delta_{max} = \dfrac{\alpha T_0 L^3}{6h}$ (downward);

(b) $\delta_{max} = \dfrac{\alpha T_0 L^4}{12h}$ (downward);

(c) $\delta_{max} = \dfrac{\alpha T_0 L^3}{6h}$ (downward),

$\delta_{max} = \dfrac{\alpha T_0 L^4}{12h}$ (downward)

R-9.1 C

R-9.2 C

R-9.3 B

R-9.4 C

R-9.5 B

R-9.6 C

R-9.7 D

CHAPTER 10

10.3-1 (a) $R_A = 1163$ N, $R_B = 698$ N, $M_A = 814$ N·m; (b) $x_{zero} = 0.875$ m; (c) $\delta_{max} = 0.022$ mm, $\theta_B = 2.36 \times 10^{-5}$ rad

10.3-2 (a) $h = 257$ mm; (b) $\delta_0 = 1$ mm

10.3-3 $R_A = -R_B = 3M_0/2L$, $M_A = M_0/2$, $v = -M_0x^2(L-x)/4LEI$

10.3-4 $R_A = R_B = qL/2$, $M_A = M_B = qL^2/12$, $v = -qx^2(L-x)^2/24EI$

10.3-5 $R_A = R_B = 3EI\delta_B/L^3$, $M_A = 3EI\delta_B/L^2$, $v = -\delta_B x^2(3L-x)/2L^3$

10.3-6 $\theta_B = \dfrac{qL^3}{6(k_RL - EI)}$,

$$\delta_B = \left(\frac{1}{EI}\right)\left(-\frac{1}{8}qL^4 + \frac{k_RqL^5}{12(k_RL - EI)}\right)$$

10.3-7 $R_A = V(0) = \dfrac{9}{40}q_0L$,

$R_B = -V(L) = \dfrac{11}{40}q_0L$, $M_A = \dfrac{7}{120}q_0L^2$

10.3-8 (a) $R_A = V(0) = \dfrac{7}{60}q_0L$,

$R_B = -V(L) = \dfrac{13}{60}q_0L$, $M_A = \dfrac{1}{30}q_0L^2$,

$$v = \frac{q_0}{360L^2EI}(-x^6 + 7L^3x^3 - 6q_0L^4x^2);$$

(b) $R_A = V(0) = 0.31q_0L$

$$= \left(\frac{2}{\pi} - 6\frac{\pi^2 - 4\pi + 8}{\pi^4}\right)q_0L,$$

$R_B = -V(L) = 0.327q_0L$

$$= \left(6\frac{\pi^2 - 4\pi + 8}{\pi^4}\right)q_0L,$$

$$M_A = -2q_0L^2\frac{\pi^2 - 12\pi + 24}{\pi^4},$$

$$v = \frac{1}{EI} \times \left[-q_0\left(\frac{2L}{\pi}\right)^4 \sin\left(\frac{\pi x}{2L}\right) - 6q_0L\frac{\pi^2 - 4\pi + 8}{\pi^4}\frac{x^3}{6} + 2q_0L^2\frac{\pi^2 - 12\pi + 24}{\pi^4}\frac{x^2}{2} + q_0\left(\frac{2L}{\pi}\right)^3 x\right]$$

10.3-9 (a) $R_A = \dfrac{61Lq_0}{120}$,

$M_A = \dfrac{11L^2q_0}{120}$,

$R_B = \dfrac{19Lq_0}{120}$,

$$v(x) = -\frac{q_0x^2(33L^4 - 61L^3x + 30L^2x^2 - 2x^4)}{720EIL^2};$$

(b) $R_A = \dfrac{48Lq_0}{\pi^4}$,

$$R_B = \int_0^L q(x)dx - R_A = \frac{2Lq_0}{\pi} - \frac{48Lq_0}{\pi^4},$$

$$M_A = \int_0^L q(x)xdx - R_BL = \frac{2L^2q_0(\pi - 2)}{\pi^2} - L\left(\frac{2Lq_0}{\pi} - \frac{48Lq_0}{\pi^4}\right),$$

$$v(x) = \frac{16L^4q_0 - 24L^2q_0x^2 + 8Lq_0x^3 - 16L^4q_0\cos\left(\frac{\pi x}{2L}\right)}{\pi^4EI}$$

10.3-10 (a) $R_A = V(0) = \dfrac{24}{\pi^4}q_0L$,

$R_B = -V(L) = -\dfrac{24}{\pi^4}q_0L$,

$M_A = \left(\frac{12}{\pi^4} - \frac{1}{\pi^2}\right) q_0 L^2$ (counterclockwise),

$M_B = \left(\frac{12}{\pi^4} - \frac{1}{\pi^2}\right) q_0 L^2$ (counterclockwise),

$v = \frac{1}{\pi^4 EI}[- q_0 L^4 cos\left(\frac{\pi x}{L}\right) + 4q_0 L x^3 - 6q_0 L^2 x^2 + q_0 L^4]$;

(b) $R_A = R_B = q_0 L/\pi$, $M_A = M_B = 2q_0 L^2/\pi^3$, $v = -q_0 L^2(L^2 \sin \pi x/L + \pi x^2 - \pi L x)/\pi^4 EI$

10.3-11 (a) $R_A = V(0) = \frac{48(4 - \pi)}{\pi^4} q_0 L$,

$R_B = -V(L) = \left(\frac{2}{\pi} - \frac{48(4 - \pi)}{\pi^4}\right) q_0 L$,

$M_A = -q_0 \left(\frac{2L}{\pi}\right)^2 + \frac{16(6 - \pi)}{\pi^4} q_0 L^2$,

$M_B = -\frac{32(\pi - 3)}{\pi^4} q_0 L^2$,

$v = \frac{1}{\pi^4 EI}[-16 q_0 L^4 \cos\left(\frac{\pi x}{2L}\right) + 8(4 - \pi) q_0 L x^3 - 8(6 - \pi) q_0 L^2 x^2 + 16 q_0 L^4]$;

(b) $R_A = V(0) = \frac{13}{30} q_0 L$,

$R_B = -V(L) = \frac{7}{30} q_0 L$,

$M_A = \frac{1}{15} q_0 L^2$ (counterclockwise),

$M_B = -\frac{1}{20} q_0 L^2$ (counterclockwise),

$v = \frac{q_0}{360 L^2 EI}[x^6 - 15L^2 x^4 + 26 L^3 x^3 - 12 L^4 x^2]$

10.3-12 $R_A = V(0) = \frac{3}{20} q_0 L$,

$R_B = -V(L) = \frac{7}{20} q_0 L$,

$M_A = \frac{1}{30} q_0 L^2$,

$v = \frac{1}{120 LEI}(-q_0 x^5 + 3 q_0 L^2 x^3 - 2 q_0 L^3 x^2)$

10.3-13 $R_A = -R_B = 3M_0/2L$, $M_A = -M_B = M_0/4$, $v = -M_0 x^2 (L - 2x)/8LEI$ for $0 \le x \le L/2$

10.3-14 $R_B = -\frac{9}{8}\frac{M_0}{L}$, $R_A = \frac{9}{8}\frac{M_0}{L}$, $M_A = \frac{1}{8} M_0$,

$v = \frac{1}{EI}\left(\frac{9M_0}{48L} x^3 - \frac{M_0}{16} x^2\right)\left(0 \le x \le \frac{L}{2}\right)$,

$v = \frac{1}{EI} \times \left(\frac{9M_0}{48L} x^3 - \frac{9M_0}{16} x^2 + \frac{M_0 L}{2} x - \frac{M_0 L^2}{8}\right) \left(\frac{L}{2} \le x \le L\right)$

10.3-15 $q = 8.6$ kN/m

10.3-16 $q = 2.95 \frac{\text{kN}}{\text{m}}$

10.4-1 (a) $R_A = R_C = 261$ N, $R_B = 870$ N; (b) $\delta_{AB} = -0.06$ mm

10.4-2 $R_A = 1.76$ kN, $M_A = 2.4$ kN·m, $R_B = 3.24$ kN, $M_B = 3.6$ kN·m, $\delta_C = 0.0674$ mm

10.4-3 (a) $R_A = 2.02$ kN, $M_A = 1.385$ kN·m, $R_B = 0.68$ kN, $M_B = 0.802$ kN·m; (b) $\delta_{max} = 5.98$ mm at $x = 0.466L = 2.1$ m; (c) $R_A = R_B = 2.25$ kN, $M_A = -M_B = 1.688$ kN·m

10.4-4 $R_A = 375$ N, $M_A = 225$ N·m, $R_B = 225$ N

10.4-5 $R_A = Pb(3L^2 - b^2)/2L^3$, $R_B = Pa^2(3L - a)/2L^3$, $M_A = Pab(L + b)/2L^2$

10.4-6 $R_A = qL$, $M_A = \frac{qL^2}{3}$, $M_B = \frac{qL^2}{6}$

10.4-7 $R_A = -\frac{1}{8} qL$, $R_B = \frac{17}{8} qL$, $M_A = -\frac{1}{8} qL^2$

10.4-8 (a) $R_A = M_0/3L$, $H_A = -4M_0/3L$, $R_B = -R_A$, $R_C = -H_A$; (b) $\theta_A = -M_0 L/18EI$, $\theta_B = M_0 L/9EI$, $\theta_C = \theta_A$; (c) $L_{BC} = 2L$

10.4-9 (a) $R_A = \frac{4M_0}{3L}$, $H_A = \frac{2M_0}{3L}$, $R_B = -\frac{4M_0}{3L}$, $R_C = -\frac{2M_0}{3L}$; (b) $\theta_A = \frac{5}{18}\frac{M_0 L}{EI}$, $\theta_B = \frac{-M_0 L}{18EI}$, $\theta_C = \frac{M_0 L}{36EI}$; (c) $L_{AB} = 2.088L$

10.4-10 (a) $R_A = \dfrac{M_0}{L} + \dfrac{M_0 k_R}{2(3EI + Lk_R)}$,

$R_B = -R_A, M_B = \dfrac{LM_0 k_R}{6EI + 2Lk_R}(CCW)$;

(b) $\theta_A = \dfrac{LM_0}{4EI} + \dfrac{LM_0}{4(3EI + Lk_R)}$

For k_R goes to zero:

$\theta_A = \dfrac{LM_0}{4EI} + \dfrac{LM_0}{4(3EI)} = \dfrac{LM_0}{3EI}$

For k_R goes to infinity: $\theta_A = \dfrac{M_0 L}{4EI}$

For k_R goes to $6EI/L$:

$\theta_A = \dfrac{LM_0}{4EI} + \dfrac{LM_0}{4\left[3EI + L\left(\dfrac{6EI}{L}\right)\right]} = \dfrac{5LM_0}{18EI}$

10.4-11 (a) $H_A = \dfrac{3}{2}\dfrac{M_0}{L}$,

$H_B = 0, V_B = 0, V_C = 0, H_D = -H_A$;

(b) $\theta_A = \dfrac{-M_0 L}{16EI}, \theta_D = -\theta_A$,

$\theta_B = \dfrac{M_0 L}{8EI}, \theta_C = -\theta_B$;

(c) $H_A = \dfrac{M_0}{L}, H_B = -2\dfrac{M_0}{L}, V_B = \dfrac{M_0}{L}$,

$V_C = \dfrac{-M_0}{L}, H_D = H_A, \theta_A = \dfrac{-M_0 L}{24EI}$,

$\theta_D = \theta_A, \theta_B = \dfrac{M_0 L}{12EI}, \theta_C = \theta_B$

10.4-12 $t_{AB}/t_{CD} = L_{AB}/L_{CD}$

10.4-13 $R_A = \dfrac{7}{12}qL, R_B = \dfrac{17}{12}qL, M_A = \dfrac{7}{12}qL^2$

10.4-14 $R_A = 2qL, M_B = \dfrac{7}{12}qL^2$

10.4-15 $R_A = R_B = q_0L/4, M_A = M_B = 5q_0L^2/96$

10.4-16 $R_A = qL/8, R_B = 33qL/16, R_C = 13qL/16$

10.4-17 $R_A = 11/17\ P = 5.18$ kN (downward), $R_B = R_A + P = 13.18$ kN (upward), $M_A = 5/34\ PL = 3.53$ kN·m (clockwise)

10.4-18 $R_B = 6.44$ kN

10.4-19 $R_A = 174$ N, $R_B = 43.5$ N, $M_A = 87$ N·m

10.4-20 (a) $R_A = R_B = 45$ kN, $M_A = -M_B = 35$ kN·m; (b) $h = 167.1$ mm

10.4-21 $R_A = -159.6$ N, $R_B = -679$ N, $M_A = -347$ N·m, $M_B = -669$ N·m

10.4-22 $R_A = 33$ kN, $R_B = 17$ kN, $M_A = 25.3$ kN·m

10.4-23 (a) The tension force in the tie rod = $R_D = 2.12$ kN; (b) $R_A = 4.78$ kN, $M_A = 3.74$ kN·m

10.4-24 $R_A = 31qL/48, R_B = 17qL/48, M_A = 7qL^2/48$

10.4-25 (a) $R_A = -23P/17, R_D = R_E = 20P/17$, $M_A = 3PL/17$; (b) $M_{max} = PL/2$

10.4-26 $R_A = R_D = 2qL/5, R_B = R_C = 11qL/10$

10.4-27 $M_B(q) = (-1/2)(q)$ for $q \le 48$ kN/m, $M_B(q) = (-q/8) - 18$ for $q \ge 48$ kN/m

10.4-28 $R_A = \dfrac{19}{128}q_0L, R_C = -\dfrac{19}{128}q_0L$,

$M_A = \dfrac{3}{128}q_0L^2, V_{max} = \dfrac{19}{128}q_0L$,

$M_{max} = 0.0147q_0L^2$

10.4-29 $R_A = 109.3$ N, $R_B = 21.2$ N, $M_A = 80.3$ N·m

10.4-30 $R_A = -R_B = 6M_0ab/L^3$, $M_A = M_0b(3a - L)/L^2$, $M_B = -M_0a(3b - L)/L^2$

10.4-31 $\sigma = 3.14$ MPa

10.4-32 $(M_{AB})_{max} = 121qL^2/2048 = 6.05$ kN·m; $(M_{CD})_{max} = 5qL^2/64 = 8.0$ kN·m

10.4-33 $F = 14.47$ kN, $M_{AB} = 26.0$ kN·m, $M_{DE} = 9.66$ kN·m

10.4-34 $k = 48EI(6 + 5\sqrt{2})/7L^3 = 89.63EI/L^3$

10.4-35 (a) $V_A = V_C = 3P/32, H_A = P$, $M_A = 13PL/32$; (b) $M_{max} = 13PL/32$

10.4-36 $H_A = \dfrac{-29}{64}P, H_C = \dfrac{-35}{64}P, M_{max} = \dfrac{29}{128}PL$

10.4-37 $R_A = 12.53$ kN, $R_B = 12.45$ kN, $R_C = 0.026$ kN

10.4-38 (a) $M_A = M_B = qb(3L^2 - b^2)/24L$; (b) $b/L = 1.0, M_A = qL^2/12$; (c) For $a = b = L/3, (M_{max})_{pos} = 19qL^2/648$

10.4-39 (a) $d_2/d_1 = \sqrt[4]{8} = 1.682$; (b) $M_{max} = qL^2(3 - 2\sqrt{2})/2 = 0.08579qL^2$; (c) Point C is below points A and B by the amount $0.01307qL^4/EI$

10.4-40 $M_{max} = 19q_0L^2/256, \sigma_{max} = 13.4\,\text{MPa}$, $\delta_{max} = 19q_0L^4/7680EI = 0.00891\,\text{mm}$

10.4-41 $M_A = \dfrac{2}{27}qL^2$

10.4-42 $M_A = \dfrac{4}{81}qL^2$

10.4-43 (a) $V_{max} = \dfrac{4}{3}P, M_{max} = -\dfrac{PL}{3}$; (b) $V_{max} = \dfrac{-17}{27}P, M_{max} = -\dfrac{10}{81}PL$

10.5-1 $S = \dfrac{243E_SE_WIAH\alpha(\Delta T)}{4AL^3E_S + 243IHE_W}$

10.5-2 (a) $R_A = -\dfrac{\alpha(T_2 - T_1)L^2}{2h}\left(\dfrac{3EIk}{3EI + L^3k}\right)$, $R_B = \dfrac{\alpha(T_2 - T_1)L^2}{2h}\left(\dfrac{3EIk}{3EI + L^3k}\right)$, $M_A = R_BL = \dfrac{\alpha(T_2 - T_1)L^3}{2h}\left(\dfrac{3EIk}{3EI + L^3k}\right)$; (b) $R_A = -R_B = -\dfrac{3EI\alpha(T_2 - T_1)}{2hL}$ (upward), $R_B = \dfrac{3EI\alpha(T_2 - T_1)}{2hL}$ (downward), $M_A = R_BL = \dfrac{3EI\alpha(T_2 - T_1)}{2h}$ (counterclockwise)

10.5-3 $R_A = -R_B = -\dfrac{\alpha(T_2 - T_1)L^2}{2h}\left(\dfrac{3EIk}{3EI + L^3k}\right)$ (upward), $R_B = \dfrac{\alpha(T_2 - T_1)L^2}{2h}\left(\dfrac{3EIk}{3EI + L^3k}\right)$ (downward), $M_A = R_BL = \dfrac{\alpha(T_2 - T_1)L^3}{2h}\left(\dfrac{3EIk}{3EI + L^3k}\right)$ (counterclockwise)

10.5-4 (a) $R_B = -\dfrac{\alpha(T_1 - T_2)L^2}{h}\left(\dfrac{6EIk}{36EI + L^3k}\right)$ (downward), $R_A = -\dfrac{1}{4}R_B = \dfrac{\alpha(T_1 - T_2)L^3}{2h}\left(\dfrac{3EIk}{36EI + L^3k}\right)$ (upward), $R_C = -\dfrac{3}{4}R_B = \dfrac{\alpha(T_1 - T_2)L^2}{2h}\left(\dfrac{9EIk}{36EI + L^3k}\right)$ (upward); (b) $R_B = -\dfrac{6EI\alpha(T_1 - T_2)}{Lh}$ (downward), $R_A = \dfrac{3EI\alpha(T_1 - T_2)}{2Lh}$ (upward), $R_C = \dfrac{9EI\alpha(T_1 - T_2)}{2Lh}$ (upward)

10.5-5 $R_B = -\dfrac{\alpha(T_1 - T_2)L^2}{h}\left(\dfrac{6EIk}{36EI + L^3k}\right)$ (downward), $R_A = -\dfrac{1}{4}R_B = \dfrac{\alpha(T_1 - T_2)L^2}{2h}\left(\dfrac{3EIk}{36EI + L^3k}\right)$ (upward), $R_C = -\dfrac{3}{4}R_B = \dfrac{\alpha(T_1 - T_2)L^2}{2h}\left(\dfrac{9EIk}{36EI + L^3k}\right)$ (upward)

10.6-1 (a) $H = \pi^2EA\delta^2/4L^2$, $\sigma_t = \pi^2E\delta^2/4L^2$; (b) $\sigma_t = 4.32, 1.08$, and 0.48 MPa

10.6-2 (a) $\lambda = 17q^2L^7/40, 320E^2I^2$, $\sigma_b = qhL^2/16I$; (b) $\sigma_t = 17q^2L^6/40, 320EI^2$; (c) $\lambda = 0.01112\,\text{mm}, \sigma_b = 117.2\,\text{MPa}$, $\sigma_t = 0.741\,\text{MPa}$

R-10.1 B

R-10.2 B

R-10.3 B

R-10.4 D

R-10.5 B

R-10.6 D

R-10.7 A

CHAPTER 11

11.2-1 $P_{cr} = \beta_R / L$

11.2-2 (a) $P_{cr} = \dfrac{\beta a^2 + \beta_R}{L}$; (b) $P_{cr} = \dfrac{\beta a^2 + 2\beta_R}{L}$

11.2-3 $P_{cr} = 9.44$ kN

11.2-4 $P_{cr} = 10$ kN

11.2-5 $P_{cr} = 6\beta_R / L$

11.2-6 $P_{cr} = 1.688$ kN

11.2-7 $P_{cr} = 4.58$ kN

11.2-8 (a) $P_{cr} = \dfrac{(L-a)(\beta a^2 + \beta_R)}{aL}$;

(b) $P_{cr} = \dfrac{\beta L^2 + 20\beta_R}{4L}$

11.2-9 $P_{cr} = \dfrac{3\beta_R}{L}$

11.2-10 $P_{cr} = \dfrac{3}{5}\beta L$

11.2-11 $P_{cr} = \dfrac{7}{4}\beta L$

11.2-12 $P_{cr} = \dfrac{9}{8}\beta L$

11.2-13 $P_{cr} = \dfrac{9}{5}\dfrac{\beta_R}{L}$

11.2-14 $P_{cr1} = 30.2$ kN, $P_{cr2} = 84.6$ kN

11.2-15 $P_{cr1} = 243$ kN, $P_{cr2} = 932$ kN

11.2-16 $P_{cr1} = \beta L$, $P_{cr2} = \dfrac{9}{4}\beta L$

11.3-1 $P_{allow} = 542$ kN

11.3-2 HE 160B

11.3-3 (a) $P_{cr} = 465$ kN; (b) $P_{cr} = 169.5$ kN

11.3-4 (a) $P_{cr} = 176.7$ kN; (b) $P_{cr} = 10.86$ kN

11.3-5 (a) $P_{cr} = 319$ kN; (b) $P_{cr} = 120.1$ kN

11.3-6 $M_{allow} = 1143$ kN·m

11.3-7 $P_{allow} = 220$ kN

11.3-8 $P_{crY} = 339$ kN

11.3-9 (a) $P_{cr} = 223$ kN; (b) $d = 118.3$ mm

11.3-10 (a) $P_{cr} = 97.5$ kN; (b) $d = 86.9$ mm

11.3-11 $Q_{allow} = 109.8$ kN

11.3-12 (a) $Q_{cr} = \dfrac{\pi^2 EI}{L^2}$; (b) $Q_{cr} = \dfrac{2\pi^2 EI}{9L^2}$

11.3-13 (a) $Q_{cr} = \dfrac{2\pi^2 EI}{L^2}$; (b) $M_{cr} = \dfrac{3d\pi^2 EI}{L^2}$

11.3-14 $\Delta T = \pi^2 I / \alpha A L^2$

11.3-15 $h/b = 2$

11.3-16 (a) $P_{cr} = 3\pi^3 Er^4/4L^2$;
(b) $P_{cr} = 11\pi^3 Er^4/4L^2$

11.3-17 $P_1 : P_2 : P_3 = 1.000 : 1.047 : 1.209$

11.3-18 $P_{allow} = 710$ kN

11.3-19 $F_{allow} = 164$ kN

11.3-20 $W_{max} = 124$ kN

11.3-21 $t_{min} = 4.53$ mm

11.3-22 $P_{cr} = 426$ kN

11.3-23 $W_{cr} = 203$ kN

11.3-24 $\theta = \arctan 0.5 = 26.57°$

11.3-25 (a) $q_{max} = 1.045$ kN/m; (b) $I_{b,min} = 2411\ \text{cm}^4$;
(c) $s = 70$ mm, 869 mm

11.3-26 $P_{cr} = 3.56$ kN

11.3-27 $P_{cr} = 70.3$ kN

11.4-1 $P_{cr} = 756$ kN

11.4-2 $t_{min} = 4.98$ mm

11.4-3 $P_{cr} = 831$ kN, 208 kN, 1700 kN, 3330 kN

11.4-4 $P_{cr} = 62.2$ kN, 15.6 kN, 127 kN, 249 kN

11.4-5 $P_{allow} = 923$ kN, 231 kN, 1888 kN, 3690 kN

11.4-6 $P_{allow} = 18.78$ kN, 4.70 kN, 38.4 kN, 75.1 kN

11.4-7 $P_{cr} = 295$ kN

11.4-8 $T_{allow} = 18.1$ kN

11.4-9 (a) $Q_{cr} = 13.41$ kN;
(b) $Q_{cr} = 35.8$ kN, $a = 0$

11.4-10 $P_{cr} = 447$ kN, 875 kN, 54.7 kN, 219 kN

11.4-11 $P_{cr} = 4\pi^2 EI/L^2$, $v = \delta(1 - \cos 2\pi x/L)/2$

11.4-12 $L_{max} = 4.2$ m

11.4-13 $P_{max} = 242$ kN

11.4-14 (a) $P_{cr} = 17.4$ kN, (b) $d = 72.4$ mm

11.4-15 $t_{min} = 7.8$ mm

11.4-16 $t_{min} = 10.0$ mm

11.4-17 (b) $P_{cr} = 13.89EI/L^2$

11.5-1 $\delta = 5.98$ mm, $M_{max} = 231$ N·m

11.5-2 $\delta = 8.87$ mm, $M_{max} = 2.03$ kN·m

11.5-3 $L = 4.8$ m

11.5-4 $b = 11.07$ mm, $h = 16.61$ mm

11.5-5 For $P = 0.3P_{cr}$: $M/Pe = 1.162(\sin 1.721\,x/L) + \cos 1.721x/L$

11.5-6 $P = 583.33\{\arccos[5/(5 + \delta)]\}^2$, in which P = kN and δ = mm; $P = 884$ kN when $\delta = 10$ mm

11.5-7 $P = 163.61\{\arccos[5/(5 + \delta)]\}^2$, in which P = kN and δ = mm; $P = 248$ kN when $\delta = 10$ mm

11.5-8 $P_{allow} = 54.9$ kN

11.5-9 $L_{max} = 3.91$ m

11.5-10 $L_{max} = 3.69$ m

11.5-11 $\delta = e(\sec kL - 1)$, $M_{max} = Pe \sec kL$

11.5-12 $L_{max} = 2.21$ m

11.5-13 $L_{max} = 3.86$ m

11.5-14 $T_{max} = 8.29$ kN

11.5-15 (a) $q_0 = 3.33$ kN/m; (b) $M_{max} = 4.23$ kN·m, ratio = 0.47

11.6-1 (a) $\sigma_{max} = 116.8$ MPa; (b) $L_{max} = 1.367$ m

11.6-2 $P_{allow} = 37.2$ kN

11.6-3 $b_{min} = 106.3$ mm

11.6-4 (a) $\sigma_{max} = 38.8$ MPa; (b) $L_{max} = 5.03$ m

11.6-5 (a) $\sigma_{max} = 63.7$ MPa; (b) $P_{allow} = 18.98$ kN

11.6-6 $d_2 = 131$ mm

11.6-7 $L = 5.2$ m

11.6-8 $\sigma_{max} = 212$ MPa

11.6-9 (a) $\sigma_{max} = 66.5$ MPa; (b) $P_{allow} = 779$ kN

11.6-10 (a) $\sigma_{max} = 94.3$ MPa; (b) $L_{max} = 4.05$ m

11.6-11 (a) $\sigma_{max} = 68.6$ MPa; (b) $P_{allow} = 268$ kN

11.6-12 (a) $\sigma_{max} = 56.6$ MPa; (b) $n = 2.54$

11.6-13 (a) $\sigma_{max} = 96.1$ MPa; (b) $n = 3.06$

11.6-14 (a) $\sigma_{max} = 98.9$ MPa; (b) $P_2 = 459$ kN

11.6-15 (a) $\sigma_{max} = 100.2$ MPa; (b) $n = 2.24$

11.6-16 (a) $\sigma_{max} = 120.4$ MPa; (b) $P_2 = 113.5$ kN

R-11.1 D

R-11.2 B

R-11.3 D

R-11.4 A

R-11.5 D

R-11.6 A

R-11.7 B

R-11.8 B

R-11.9 B

R-11.10 C

R-11.11 D

APPENDIX D

D.2-2 $\bar{x} = \bar{y} = 5a/12$

D.2-3 $\bar{y} = 27.5$ mm

D.2-4 $2c^2 = ab$

D.2-5 $\bar{y} = 340$ mm

D.2-6 $\bar{y} = 52.5$ mm

D.2-7 $\bar{x} = 24.5$ mm, $\bar{y} = 49.5$ mm

D.2-8 $\bar{x} = 137$ mm, $\bar{y} = 132$ mm

D.3-6 $I_x = 518 \times 10^3$ mm^4

D.3-7 $I_x = 13.55 \times 10^6$ mm^4, $I_y = 4.08 \times 10^6$ mm^4

D.3-8 $I_x = I_y = 194.6 \times 10^6$ mm^4, $r_x = r_y = 80.1$ mm

D.3-9 $I_1 = 61{,}390$ cm^4, $I_2 = 9455$ cm^4, $r_1 = 18.90$ cm, $r_2 = 7.42$ cm

D.4-1 $I_b = 72{,}113 \text{ cm}^4$

D.4-2 $I_c = 11a^4/192$

D.4-3 $I_{xc} = 2.82 \times 10^6 \text{ mm}^4$

D.4-4 $I_2 = 405 \times 10^3 \text{ mm}^4$

D.4-5 $I_{xc} = 204{,}493 \text{ cm}^4, I_{yc} = 14{,}863 \text{ cm}^4$

D.4-6 $I_{xc} = 106 \times 10^6 \text{ mm}^4$

D.4-7 $I_{xc} = 656 \text{ cm}^4, I_{yc} = 237 \text{ cm}^4$

D.4-8 $b = 250 \text{ mm}$

D.5-1 $I_P = \text{bh}(b^2 + 12h^2)/48$

D.5-2 $(I_P)_C = r^4(9\alpha^2 - 8\sin^2\alpha)/18\alpha$

D.5-3 $I_P = 32{,}966 \text{ cm}^4$

D.5-4 $I_P = \text{bh}(b^2 + h^2)/24$

D.5-5 $(I_P)_C = r^4(176 - 84\pi + 9\pi^2)/[72(4 - \pi)]$

D.6-2 $I_{xy} = r^4/24$

D.6-3 $b = 2r$

D.6-4 $I_{xy} = t^2(2b^2 - t^2)/4$

D.6-5 $I_{12} = -540 \text{ cm}^4$

D.6-6 $I_{xy} = 24.3 \times 10^6 \text{ mm}^4$

D.6-7 $I_{xcyc} = -230 \text{ cm}^4$

D.7-1 $I_{x1} = I_{y1} = b^4/12, I_{x1y1} = 0$

D.7-2 $I_{x1} = \dfrac{b^3h^3}{6(b^2 + h^2)}, I_{y1} = \dfrac{bh(b^4 + h^4)}{12(b^2 + h^2)},$

$I_{x1y1} = \dfrac{b^2h^2(h^2 - b^2)}{12(b^2 + h^2)}$

D.7-3 $I_d = 14{,}696 \text{ cm}^4$

D.7-4 $I_{x1} = 12.44 \times 10^6 \text{ mm}^4,$
$I_{y1} = 9.68 \times 10^6 \text{ mm}^4,$
$I_{x1y1} = 6.03 \times 10^6 \text{ mm}^4$

D.7-5 $I_{x1} = 480.2 \text{ cm}^4, I_{y1} = 145.8 \text{ cm}^4,$
$I_{x1y1} = 181.2 \text{ cm}^4$

D.7-6 $I_{x1} = 8.75 \times 10^6 \text{ mm}^4,$
$I_{y1} = 1.02 \times 10^6 \text{ mm}^4,$
$I_{x1y1} = -0.356 \times 10^6 \text{ mm}^4$

D.8-1 (a) $c = \sqrt{a^2 - b^2}/2$; (b) $a/b = \sqrt{5}$;
(c) $1 \le a/b < \sqrt{5}$

D.8-2 Shows that two different sets of principal axes exist at each point.

D.8-3 $\theta_{p1} = -29.9°, \theta_{p2} = 60.1°,$
$I_1 = 121.5 \times 10^6 \text{ mm}^4, I_2 = 34.7 \times 10^6 \text{ mm}^4$

D.8-4 $\theta_{p1} = -8.54°, \theta_{p2} = 81.46°,$
$I_1 = 17.24 \times 10^6 \text{ mm}^4,$
$I_2 = 4.88 \times 10^6 \text{ mm}^4$

D.8-5 $\theta_{p1} = 37.7°, \theta_{p2} = 127.7°,$
$I_1 = 5.87 \times 10^6 \text{ mm}^4, I_2 = 0.714 \times 10^6 \text{ mm}^4$

D.8-6 $\theta_{p1} = 32.63°, \theta_{p2} = 122.63°,$
$I_1 = 8.76 \times 10^6 \text{ mm}^4, I_2 = 1.00 \times 10^6 \text{ mm}^4$

D.8-7 $\theta_{p1} = 16.85°, \theta_{p2} = 106.85°, I_1 = 0.2390b^4,$
$I_2 = 0.0387b^4$

D.8-8 $\theta_{p1} = 74.08°, \theta_{p2} = -15.92°,$
$I_1 = 8.29 \times 10^6 \text{ mm}^4, I_2 = 1.00 \times 10^6 \text{ mm}^4$

D.8-9 $\theta_{p1} = 75.3°, \theta_{p2} = -14.7°,$
$I_1 = 6.28 \times 10^6 \text{ mm}^4, I_2 = 0.66 \times 10^6 \text{ mm}^4$

INDEX

D

E

M

N

O

P

R

S